608

Use these cards to...

- order extra copies
- share your comments or suggestions with the editor

...or call toll-free
1-800-877-GALE

Gale Research Inc.

S0-BOF-049

Please send me ____copy(s) of **Inventing & Patenting Sourcebook.**
To save 5% on future editions, I've checked the "standing order" space.

Copies			Enter as Standing Order (5% discount)
_____	**Inventing & Patenting Sourcebook**	**$75.00**	_____

Institution ____________________

Address ____________________

City, State & Zip ____________________

Attention ____________________
Name Title/Department

Phone (   ) ____________________

Gale Research Inc.

Please send me ____copy(s) of **Inventing & Patenting Sourcebook.**
To save 5% on future editions, I've checked the "standing order" space.

Copies			Enter as Standing Order (5% discount)
_____	**Inventing & Patenting Sourcebook**	**$75.00**	_____

Institution ____________________

Address ____________________

City, State & Zip ____________________

Attention ____________________
Name Title/Department

Phone (   ) ____________________

COMMENT CARD

Please use this postage paid card to make suggestions regarding the content, arrangement, indexing, or other features of **Inventing & Patenting Sourcebook.**

Name/Title ____________________

Institution ____________________

Address ____________________

City, State, & Zip ____________________

Phone (   ) ____________________

No postage necessary if mailed in the United States

BUSINESS REPLY MAIL
FIRST CLASS PERMIT NO. 17022 DETROIT, MI 48226
POSTAGE WILL BE PAID BY ADDRESSEE

Order Department
Gale Research Inc.
P.O. Box 441914
Detroit, MI 48244-9980

INVENTING AND PATENTING SOURCEBOOK

ISSN 1044-4742

INVENTING AND PATENTING SOURCEBOOK

How to Sell and Protect Your Ideas

FIRST EDITION

RICHARD C. LEVY

Robert J. Huffman,
Editor

Gale Research Inc.

DETROIT • NEW YORK • FORT LAUDERDALE • LONDON

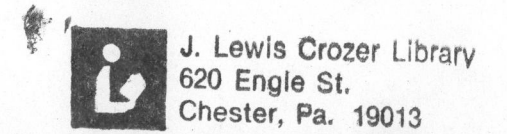

Richard C. Levy

Editor: Robert J. Huffman

Senior Editor: Martin Connors

Aided by: Donna Batten, Shawn Brennan, Kathleen J. Edgar, Karen Hill, Amy Lucas, Edward F. Nakfoor, Annette Novallo, Brian Partin, Terrance W. Peck, John Salerno, Carol A. Schwartz, Scott Stange, Gwen Turecki

Supervisor of Systems and Programming: Theresa Rocklin
Programmer: Michael A. Hagen

Production Manager: Mary Beth Trimper
External Production Assistant: Anthony J. Scolaro

Art Director: Arthur Chartow
Graphic Designers: Cynthia D. Baldwin and Bernadette M. Gornie
Keyliner: C.J. Jonik

Production Supervisor: Laura Bryant
Internal Production Associate: Louise Gagne
Internal Production Assistants: Shelly Andrews and Sharana Wier

835 Penobscot Bldg.
Detroit, MI 48226-4094

ISBN: 0-8103-4871-3
ISSN: 1044-4742

Printed in the United States of America

To my wife, Sheryl, who brings to every one of our creative adventures a great intelligence, imagination, and spirit.

To my daughter, Bettie, who sets the pace, and has taught me more about the creative process than she may ever realize.

Best-selling author and inventor Richard C. Levy has developed and licensed more than 50 innovative concepts in the last decade. He is the author of ***Secrets of Selling Inventions*** and the creator of the popular game **Adver*teasing***, which can be found on the shelves of department and toy stores nationwide.

"A rare and multi-talented individual . . ."—*Inventor's Digest*

CONTENTS

Acknowledgments	ix
Introduction	xi
Dark Constitution Afficial Inner	
Part One: Inventor Miscellanea	
Top 100 Corporations Receiving U.S. Patents in 1988	XVII
Significant Inventors and Inventions in America's First Two Hundred Years	
Forms and Documents (includes Table of Figures)	xxıx
Part Two: How to Protect and License Your Invention, by Richard C. Levy	1
How to Protect Your Ideas (Patents)	5
Trademarks and Copyrights	47
How to License Your Invention	
How to Select Which Company to Approach	
How to Present a Concept	73
TION to 1 100011 a Consoptimination	
Part Three: Directory of Inventing and Patenting Information	
National and Regional Inventor Associations	81
Invention Consultants and Research Firms	97
University Innovation Research Centers	119
Regional Small Business and State Innovation Programs	151
Business Incubators	183
Venture Capitalists	
Federal Funding Sources for the Inventor	229
Publications for the Inventor	255
Youth Innovation Programs and Publications: Project XL	285
Online Resources for the Inventor	299
Invention Trade Shows	309
Patent Depository and Special Libraries	317
Registered Patent Attorneys and Agents	335
1 그리는 BT 이번 2명 열명 역사 문제 기계 전에 가장하는 경기 보고 있다.	
Part Four: Appendixes and Master Index	
Appendix A: U.S. Patent Classifications	623
Appendix B: Patent and Trademark Office Telephone Directory	
Master Index	887

Acknowledgments

A book such as this depends heavily upon the cooperation and assistance of many people. During the course of my research, I contacted an extensive number of independent inventors, federal and state government officials, educators, patent attorneys, corporate executives, trade association directors, librarians, and members of inventor organizations. Their warm reception, hospitality, and understanding made my assignment one that I'll fondly remember.

While many of those who have helped me are mentioned at appropriate places in the book, I do want to express special thanks to certain people.

Oscar Mastin, Public Affairs Officer, Patent and Trademark Office, a good friend for many years, was an invaluable source of assistance. He provided much of the material for the patenting section of the book, guided me through the PTO's bureaucracy to get answers to seemingly endless queries, and often went above and beyond the call of duty to see that I received everything I needed to meet my deadlines. Don Corrigan, Operations Group Leader, National Institute of Standards and Technology, Office of Energy-Related Inventions, who became a good friend through this book, took countless telephone calls from me and provided an outstanding number of leads, background materials, and suggestions. Both men are outstanding, savvy, unselfish professionals who love their work and care deeply for the fate of America's inventors.

I owe a debt of appreciation to many people at the Patent and Trademark Office, including: Donald J. Quigg, Commissioner of Patent and Trademarks; Ed Kazenske, Executive Assistant to the Commissioner; William O. Craig, Director, Office of Public Affairs; Lou Massel, Editor, *Manual of Patent Examining Procedure;* Carole A. Shores, Director of Patent Depository Libraries; Jane Myers, Director, Office of Documentation Information; Martha Crocket, Technical Information Specialist, Patent Depository Libraries; Don Kelly, Group Director, Group 320; Jeff Nase, Supervisory Petition Examiner; J. Michael Thesz, Special Program Examiner; Jan G. Casner, Special Projects Coordinator, Project XL; Elizabeth Weimar, Patent Examiner, Group 180; Anne Faris, Patent Examiner, Group 260; and Kuang Y. Lin, Senior Primary Examiner.

Also of great assistance were: Ray Barns, Invention Coordinator, Office of Invention and Innovation, Department of Energy; Brenda Mack, Press Officer, Federal Trade Commission; Keith Golden, Freedom of Information Officer, Federal Trade Commission; Marianne K. Clarke, Senior Policy Analyst, Center for Policy Research and Analysis, National Governors' Association; Jeff Norris and Joyce Hamaty, Public Affairs Specialists, National Science Foundation; Richard Sparks, Program Manager, Defense Technical Information Center; Gil Young, (Acting) Director, Minnesota Department of Trade and Economic Development; G. Thomas Cator, Executive Director, Association of Small Business Development Centers; Herbert C. Wamsley, Executive Director, Intellectual Property Owners; John T. Farady, President, Affiliated Inventors Foundation, Inc.; Jan Kosko, Public Affairs Officer, National Institute of Standards and Technology; Robert Faris, Esq., Nixon & Vanderhye; Howard Doesher, Esq., Cushman, Darby & Cushman; and Ginny Panholzer, Intellectual Property Owners.

Special thanks to Michael Ross, a good friend and confidant, and to Warren Greenberg.

At Gale Research, I wish to thank Elizabeth A. Geiser for taking the phone call that started this project; Bob Elster and John Schmittroth, for their belief in my proposal and for being its champions; and Bob Huffman, an exceptional editor and fellow inventor for whom I have the greatest admiration.

Richard C. Levy

Introduction

Today's inventors often find themselves faced with a dilemma: while they usually have a surplus of creative ideas, they often don't have a good understanding of how to move their concepts into the marketplace. The labyrinthine journey from concept to fully marketed product can be a frustrating, time-consuming effort. The inventor is faced with a wide variety of lawyers, invention marketing services, research centers, venture capitalists, government agencies, and other organizations involved with protecting and marketing inventions, many of which charge hefty fees for their services. Simple errors in filling out government forms or negligence in returning the forms may lead to penalties, delays, or an outright refusal by the government to consider the invention. Consequently, inventors might well spend considerable time and money trying to turn their ideas into reality.

Inventing and Patenting Sourcebook (IPS) provides inventors, innovators, and marketers of new products and inventions with a comprehensive and practical "how-to" guide to developing, patenting, licensing, and marketing their ideas and concepts. Inventors are led step-by-step through the complexities of the inventing and patenting process, guided by the expert advice provided by successful inventor and lecturer Richard C. Levy. Covered are such topics as independent patent and trademark search, laws protecting inventors from phony invention marketing companies, selecting companies to approach for licensing agreements, and procuring federal research and development funds. *IPS* also directs inventors to thousands of agencies, programs, professionals, and other organizations and resources that can provide invaluable assistance and information. Assisted by Mr. Levy's contacts and expertise in the field, *IPS* is a unique, convenient compilation featuring diverse and often hard to access government documents, insider lists, and other information resources.

Content and Arrangement of *IPS*

IPS is divided into four main parts: Inventor Miscellanea, How to Protect and License Your Invention, the Directory of Inventing and Patenting Information, and the Appendixes and Master Index. A detailed description of each is provided below.

PART ONE: Inventor Miscellanea

Inventor Miscellanea contains two lists, "Top 100 Corporations Receiving U.S. Patents in 1988" and "Significant Inventors and Inventions in U.S. History," and the collection of copyright-free "Forms and Documents."

The two lists provide an historical perspective by contrasting the fame of inventors from the past with the more anonymous corporate patent record holders of today.

The forms and documents section contains 35 reproducible items that inventors can use to file for patents and copyrights, saving the expense of an attorney in these processes. Included are necessary official government forms as well as a selection of custom-designed forms and examples of preliminary contractual agreements that allow inventors to approach lawyers, contractors, fellow inventors, and potential licensees without fear of their ideas being stolen prior to any formal protection. For example, the Agreement to Hold Secret and Confidential (Fig. 32) should be used anytime an inventor reveals a new product for evaluation, prototype design, technical drawing, or licensing agreement. In addition to these forms, this section provides schedules, flow charts, and illustrations that will be invaluable to any inventor. For example, a sample U.S. patent is included as Fig. 1. It illustrates the form a patent takes, including examples of perspective drawings, declaration, and specification and claims. Fig. 2 shows the Patent and Trademark Office (PTO) Fee Schedule. And Fig. 18 illustrates the channels of appeal for inventors whose patent application has been rejected by the PTO. A "Table of Figures" precedes the actual documents in this section.

PART TWO: How to Protect and License Your Invention

In this informal 77-page essay, Mr. Levy provides a lively first-hand account of the invention protection and licensing process. He speaks equally to the seasoned independent inventor or the novice who needs to know how to get started in the invention business. Frequent references are made throughout this essay to specific chapters and figures in other parts of *IPS,* alerting the user to related directory information, forms, and documents. Sample letters and agreement forms are highlighted in the essay by italic type.

The essay is broken into five sections:

▶**How to Protect Your Ideas (Patents).** A discussion of the patenting process, including the latest information on facsimile transmissions to the PTO, examples of how to write patent claims and specifications, standards for patent drawings, a list of bonded draftsmen (the only draftsmen who can change original patent drawings on file at the PTO), and much more.

▶**Trademarks and Copyrights.** Includes information on the recent changes in law that make it much easier to register a trademark, specifications for trademark drawings, and the direct HOTLINE for questions on how to fill out copyright forms.

▶**How to License Your Invention.** Covers legislation protecting independent inventors from fraudulent invention marketing services, how to identify legitimate consulting firms, and marketing via inventor expositions.

▶**How to Select Which Company to Approach.** Discusses how to get the insider information on both public and private companies.

▶**How to Present a Concept.** Offers proven methods on how to prepare proposals, how to make a prototype that works, and how to submit ideas to manufacturers.

PART THREE: Directory of Inventing and Patenting Information

The Directory comprises 13 chapters listing various indispensable information sources for the inventor. Each section begins with an introduction by Mr. Levy that explains how best to make use of the government contacts, inventor groups, and other organizations included.

Individual chapters in the Directory are arranged alphabetically by the name of the organization or service, unless it was determined that a geographic arrangement is more complimentary to the material at hand. In an attempt to simplify the organizational complexities present in chapters that deal primarily with government agencies and universities ("Federal Funding Sources for the Inventor," "University Innovation Research Centers," and "Regional Small Business and State Innovation Programs"), entries are arranged by major department, agency, or university. Subunits are arranged under their parent organization, and repetitive contact information is presented only for the parent entry. For example, if the Connecticut State Department of Economic Development has several programs that are listed under its jurisdiction and they can all be contacted through the main office, the *IPS* entries for these organizations will appear as shown below:

★76★ Connecticut State Department of Economic Development
Corner Loan Program
210 Washington St.
Hartford, CT 06106
(203) 566-3322

Description: Offers financing for working capital to manufacturers that have been doing business in the region for at least one year.

★77★ Naugatuck Valley Revolving Loan Program
Description: Offers financing for fixed assets and for working capital.

★78★ Sales Contacts Centers
(203) 566-4051
Description: Helps businesses generate new sales by matching manufacturers with purchasers.

If even a slight variation exists in the address of a subordinate organization, the entire address will be given for that subunit.

The following brief description of each Directory chapter provides specific information about the content and arrangement of each. (For a more detailed discussion, see the introductory material beginning on the first page of each chapter.)

National and Regional Inventor Associations. An alphabetic listing of 160 associations.

Invention Consultants and Research Firms. An alphabetic listing of 240 new product development firms, invention marketing organizations, and small business consultants.

University Innovation Research Centers. An alphabetic listing of 390 university research parks, technology transfer programs, invention evaluation centers, and patent and trademark offices.

Regional Small Business and State Innovation Programs. Describes 500 state technology offices, research granting agencies, seed capital lending organizations, small business clearinghouses, and regional U.S. Small Business Administration offices. Entries are arranged alphabetically by state.

Business Incubators. Profiles 125 organizations established to encourage entrepreneurship and provide low-cost office and production space to innovative organizations. Entries in this section are arranged alphabetically by state.

Venture Capitalists. Provides information on 600 firms—organized alphabetically by state—that provide loans to small businesses and innovative companies.

Federal Funding Sources for the Inventor. Supplies information on 120 federal agencies that participate in the Small Business Innovation Research Program and/or other granting and concept evaluation programs available to inventors. Entries are organized by cabinet-level department.

Publications for the Inventor. Provides an alphabetical listing of 300 periodicals of interest to inventors, entrepreneurs, and small business owners.

Youth Innovation Programs and Publications: Project XL. This alphabetically arranged chapter identifies 150 programs operated by school districts in coordination with the U.S. Department of Commerce's Project XL, a newly initiated program to provide youth with development in creative thinking, problem-solving, and invention skills. The chapter is divided into two parts, Programs and Publications.

Online Resources for the Inventor. Offers information on 75 international databases that deal with patents, copyrights, and trademarks. Entries are arranged alphabetically.

Invention Trade Shows. This chapter describes 40 trade shows and exhibitions of interest to inventors, marketers, and product development companies. Entries are arranged alphabetically.

Patent Depository and Special Libraries. Lists 173 patent depository libraries (PDLs) and special libraries with collections that deal with patents, trademarks, and copyrights. Entries are arranged alphabetically.

Registered Patent Attorneys and Agents. Lists names, addresses, and telephone numbers of more than 12,000 patent attorneys and agents registered to practice before the U.S. Department of Commerce's Patent and Trademark Office. Arranged by state, then alphabetically by city and name of agent.

PART FOUR: Appendixes and Master Index

As a convenience to the user, *IPS* includes two government documents which are often hard to access:

Appendix A, U.S. Patent Classifications, contains the entire index to the government patent classification system. Familiarity with the classification system will give the inventor a head start when using the resources of a patent depository library. This section facilitates telephone inquiries to librarians, lawyers, or consultants, who will need the patent classifications in order to conduct business with the inventor.

Appendix B, Patent and Trademark Office Telephone Directory, provides the official directory of the more than 1,500 professionals who work at the PTO. With this tool, users can call patent examiners directly to clarify their actions. Information numbers offer access to the latest in the changes in forms and fees. This can save a tremendous amount of time and money.

The **Master Index** is an alphabetical listing of all relevant organizations, programs, publications, conferences, and other entities that appear in Part 3 of *IPS,* the Directory of Inventing and Patenting Information. It also includes references to organization acronyms when these are commonly known, and citations to proper names when they are part of organization names. References to important keywords within organization names provide additional subject and geographic access to the organizations listed. Publication names are presented in italic type.

All citations in the Master Index refer to entry numbers, not page numbers. Citations that include a star (★) next to the entry number refer to important programs listed or described within the descriptive text of the main entries indicated by the number given.

To illustrate the Master Index, the following example is provided:

AIM-Associated Inventors of Minnesota 8876
Associated Inventors of Minnesota 8876
Entrepreneurial News 3
Inventors of Minnesota, Associated 8876
Inventors Workshop .... ★256
Jonas Barr Patent Group 221
Minnesota, Associated Inventors of 8876
Patent Group, Jonas Barr 221

The Inventor Miscellanea, How to Protect and License Your Inventions, and Appendixes sections of *IPS* are not referenced in the Master Index due to the nature of the material presented in these sections. For example, one of the Appendixes, "U.S. Patent Classifications," is itself a 236-page index to the U.S. Patent Classification system. In addition, the Directory section covering "Registered Patent Attorneys and Agents" is also not indexed since it presents contact information for the 12,000 individuals listed in a geographic index-type arrangement.

Compilation of this Sourcebook

The database resources of Gale Research were used to supply information for various Chapters in Part Three, Directory of Inventing and Patenting Information; however, most of these chapters have been augmented by original research and selected lists compiled outside of Gale. Other chapters have been compiled from original research, including "Federal Funding Sources for the Inventor," "Youth Innovation Programs and Publications: Project XL," and "Regional Small Business and State Innovation Programs." Additionally, many government documents that are difficult to locate have been reproduced here, including *Index to the U.S. Patent Classification* and the patent and trademark section of the *U.S. Department of Commerce Telephone Directory.*

Suggestions and Comments Welcome

The editors encourage users to bring new or unlisted information concerning the inventing, patenting, and licensing process to their attention. Comments on or suggestions for improving *IPS* are also welcome. Please contact:

Editor
Inventing and Patenting Sourcebook
Gale Research Inc.
835 Penobscot Bldg.
Detroit, MI 48226-4094
Toll-free: 800-347-GALE
FAX: (313) 961-6241
Telex: 8102217087

Robert J. Huffman

PART ONE
Inventor Miscellanea

TOP 100 CORPORATIONS RECEIVING U.S. PATENTS IN 1988

NOTE: Intellectual Property Owners, Inc. compiled this list from data provided by the U.S. Patent and Trademark Office. The numbers reflect patents issued during calendar year 1988 for which an assignment of title was recorded in the Office by the date on which the patent was issued. Patents issued in the names of subsidiaries, other related companies, or divisions have not been combined with patents issued in the name of the parent company. The list does not reflect recent mergers, acquisitions, and name changes.

Rank Co. Name & Patents

1. Hitachi Ltd. 907
2. Toshiba Corp. 750
3. Cannon K.K. 723
4. General Electric Co. 690
5. Fuji Photo Film Co., Ltd. 589
6. U.S. Phillips Corp. 581
7. Siemens A.G. 562
8. IBM Corp. 549
9. Mitsubishi Denki K.K. 543
10. Bayer A.G. 442
11. Westinghouse Electric Corp. 435
12. Eastman Kodak Co. 433
13. Dow Chemical Co. 421
14. General Motors Corp. 383
15. E.I. du Pont de Nemours & Co. 375
16. AT&T Co. 372
17. Honda Motor Co., Ltd. 364
18. NEC Corp. 353
19. Motorola, Inc. 341
20. Toyota Jidosha K.K. 301
21. Sony Corp. 300
22. BASF A.G. 296
23. Ciba-Geigy Corp.
    3M Co. 279
25. Matsushita Electric Industrial Co., Ltd. 277
26. Xerox Corp. 258
27. Mobil Oil Corp. 248
28. Fujitsu Limited 245
29. Allied-Signal, Inc. 244
30. Nissan Motor Company, Ltd. 238
31. Sharp Corp. 237
32. Hoechst A.G. 226
33. Robert Bosch GmbH 223
34. Hughes Aircraft Co. 216
35. Shell Oil Co. 206
36. AMP, Inc. 201
37. Alps Electric Co., Ltd. 198
38. Texas Instrument, Inc. 195
39. United Technologies Corp. 189
40. Olympus Optical Co., Ltd. 180
41. Ricoh Co., Ltd. 178
42. Minolta Camera Co., Ltd. 171
    Tektronix, Inc. 171
44. Boeing Co. 169
45. Amoco Corp. 164
46. Imperial Chemical Industries PLC 161
47. Hewlett-Packard Co. 153
    Phillips Petroleum Co. 153
49. Mazda Motor Co. 148
50. Honeywell, Inc. 147
51. GTE Products Corp. 142
52. Union Carbide Corp. 141

Rank Co. Name & Patents

53. Aisin Seiki K.K. 140
54. Ford Motor Co. 139
55. Rockwell International Corp. 134
56. RCA Corp. 129
57. Thomson-CSF 128
58. Merck & Co., Inc. 125
59. Nippondenso Co., Ltd. 124
60. Konishiroku Photo Industry Co., Ltd. 122
    Sumitomo Chemical Co., Ltd. 122
62. PPG Industries, Inc. 118
63. Warner-Lambert Co., Inc. 113
64. Exxon Research & Engineering Co. 106
65. Eaton Corp. 103
66. Henkel KGAA 102
67. American Cyanamid Co. 101
68. Northern Telecom Ltd. 98
69. Unisys Corp. 97
70. Texaco, Inc. 96
71. Daimler-Benz A.G. 95
    Pioneer Electronic Corp. 95
73. Sumitomo Electric Industries, Ltd. 94
74. Chevron Research Co. 92
    Hoechst Celanese Corp. 92
    Procter & Gamble Co. 92
77. Commissariat A L'Energie Atomique 90
78. Victor Company of Japan, Ltd. 88
79. Deere & Co. 85
80. Dow Corning Corp. 82
    Goodyear Tire & Rubber Co. 82
82. Advanced Micro Devices, Inc. 81
    Brother Kogyo K.K. 81
    Sundstrand Corp. 81
85. Sanyo Electric Co., Ltd. 80
    Zenith Electronics Corp. 80
87. NCR Corp. 79
    RCA Licensing Corp. 79
89. Agency of Indus. Science & Tech. 78
    Atlantic Richfield Co. 78
    Fanuc Ltd. 78
92. Monsanto Co. 76
93. B.F. Goodrich Co. 75
    Kimberly Clark Corp. 75
95. Alfred Teves GmbH 74
    Aluminum Company of America 74
    Corning Glass Works 74
    E.R. Squibb & Sons, Inc. 74
    Nippon Kogaku K.K. 74
100. Bridgestone Corp. 73
    Fuji Jokogyo K.K. 73
    TDK Corp. 73

SIGNIFICANT INVENTORS & INVENTIONS IN AMERICA'S FIRST TWO HUNDRED YEARS

The U.S. patent system has played an important role in the development of our nation. As President and patentee Abraham Lincoln said, it gives "the fuel of interest to the fire of genius." President McKinley told Americans that the country must turn to invention "as one of the most powerful aids...in the enlargement and advance of science, industry, and commerce."

"The patent system is a cornerstone of America's free enterprise economy," wrote C. Marshall Dann, Commissioner of Patents and Trademarks, at the time of our Bicentennial celebration. "The individuals who have been named as inventors in U.S. patents have made great contributions to the nation's technological strength."

The Patent and Trademark Office compiled the following lists of patents to draw attention to at least some of the more significant patents during our first two hundred years. These lists are by no means comprehensive, but interesting and inspiring reading nevertheless.

Many of the inventors listed in this section are better known for their social, political, or other endeavors, but have also been granted U.S. patents. Others on the list, whose names are not commonly known, merit recognition as the first members of a minority and other groups to obtain patents. Still others have been selected by the National Council of Patent Law Associations as especially significant and have been inducted into the National Inventors Hall of Fame.

Some Important Patents cites a sampling of important patents that have had an impact on our nation. Some of the patentees are well-known; others have received little publicity, but their contributions have been substantial.

Design Patents features a few design patents whose subjects you will instantly recognize.

Plant Patents highlights a few plant patents for flora you are sure to know.

SIGNIFICANT INVENTORS AND INVENTIONS IN U.S. HISTORY

There are 9,957 patents that were issued prior to July 13, 1836, at which time patent numbering began. Inventors are listed in alphabetic order by last name. The title of the invention is surrounded by quotation marks.

Bardeen, John Walter H. Brattain. Transistor; "Three-Electrode Circuit Element Utilizing Semiconductive Materials" (1950). Patent Number: 2,524,035. Members of National Inventors Hall of Fame.

Bell, Alexander Graham. Telephone; "Improvement in Telegraphy" (1876). Patent Number: 174,465. Member of National Inventors Hall of Fame.

Blair, Henry. "Corn Planter" (October 14, 1834). First black patentee.

Boone, Sarah. "Ironing-Board" (1892). Patent Number: 473,653. One of the first black female inventors.

Brattain, Walter H.; John Bardeen. Transistor; "Three-Electrode Circuit Element Utilizing Semiconductive Materials" (1950). Patent Number: 2,524,035. Members of National Inventors Hall of Fame.

Carver, George Washington. Cosmetics; "Cosmetic and Process of Producing the Same" (1925). Patent Number: 1,522,176. Black educator.

Clemens, Samuel L. (Mark Twain). Suspenders; "Improvement in Adjustable and Detachable Straps for Garments" (1871). Patent Number: 121,992. Author, humorist, and steamboat pilot.

Coolidge, William D. Tungsten lamp filament; "Tungsten and Method of Making Same for Use as Filaments of Incandescent Electric Lamps and for Other Purposes" (1913). Patent Number: 1,082,933. "X-Ray Tube" (1933). Patent Number: 1,917,099. "X-Ray Tube" (1934). Patent Number: 1,946,312. "Cathode Ray Tube" (1935). Patent Number: 2,010,712. Member of National Inventors Hall of Fame.

Diesel, Rudolf. Diesel engine; "Internal-Combustion Engine" (1898). Patent Number: 608,845. Member of National Inventors Hall of Fame.

Edison, Thomas A. Quadruplex telegraph; "Improvement in Printing-Telegraphs" (1873). Patent Number: 140,488. Phonograph; "Improvement in Phonograph or Speaking Machines" (1878). Patent Number: 200,521. Incandescent lamp; "Electric Lamp" (1880). Patent Number: 223,898. "Speaking Telegraph" (1892). Patent Number: 474,230. Motion picture projector; "Apparatus for Exhibiting Photographs of Moving Objects" (1893). Patent Number: 493,426. "Kinetographic Camera" (1897). Patent Number: 589,168. Member of National Inventors Hall of Fame; 1,093 patents were granted to him.

Fermi, Enrico; Leo Szilard. "Neutronic Reactor" (1955). Patent Number: 2,708,656. Member of National Inventors Hall of Fame.

Goodyear, Charles. Vulcanized rubber; "Improvement in India-Rubber Fabrics" (1844). Patent Number: 3,633. Member of National Inventors Hall of Fame.

Goodwin, Hannibal. Celluloid photographic film; "Photographic Pellicle and Process of Producing Same" (1898). Patent Number: 610,861. Clergyman patentee.

Hall, Charles M. Process for the "Manufacture of Aluminum" (1889). Patent Number: 400,665. Member of National Inventors Hall of Fame.

Hollerith, Herman. Punch card accounting; "Art of Compiling Statistics" (1889). Patent Number: 395,782. Former Patent Office employee.

Hopkins, Samuel. Process of making potash and pearl ash; "Improvement, not known before such Discovery, in the making of Pot ash and Pearl ash by a new Apparatus and Process" (July 31, 1780). First U.S. patent.

Kies, Mary. "Weaving Straw With Silk or Thread" (May 5, 1809). First woman patentee.

Lincoln, Abraham. A device for "Bouying Vessels Over Shoals" (1849). Patent Number: 6,469. U.S. President patentee.

Marconi, Guglielmo. Wireless telegraphy; "Transmitting Electrical Signals" (1897). Patent Number: 586,193. Member of National Inventors Hall of Fame.

McCormick, Cyrus H. Reaper; "Improvement in

Machines for Reaping Small Grain" (June 21, 1834). Member of National Inventors Hall of Fame.

Morse, Samuel F.B. Telegraph signs; "Improving in the Mode of Communicating Information by Signals by the Application of Electro-Magnetism" (1840). Patent Number: 1,647. Member of National Inventors Hall of Fame.

Mullikin, Samuel. Four machine patents; "Machine for Threshing Grain and Corn"; "Machine for Breaking and Swingling Hemp"; "Machine for Cutting and Polishing Marble"; and "Machine for Raising a Nap on Cloths" (March 11, (1791). First patentee to obtain more than one patent.

Ruggles, John. Locomotive steam engine; "Locomotive Steam-Engine for Rail and Other Roads" (1836). Patent Number: 1. First chairman of the Senate Committee on Patents, whose report culminated in the 1836 law which is the "basis ofthe patent system as it functions today."

Russell, Lillian. "Dresser-Trunk" (1912). Patent Number: 1,014,853. Movie star patentee.

Schawlow, Arthur L; Charles H. Townes. Laser (optical maser); "Masers and Maser Communications System" (1960). Patent Number: 2,929,922.

Shockley, William. Transistor; "Circuit Element Utilizing Semi-conductive Material" (1951). Patent Number: 2,569,347. Member of National Inventors Hall of Fame.

Stuart, James E.B. "Method of Attaching Sabers to Belts" (1859). Patent Number: 25,684. Well-known Confederate General.

Szilard, Leo; Enrico Fermi. "Neutronic Reactor" (1955). Patent Number: 2,708,656.

Tesla, Nikola. Induction type of electric motor; "Electrical Transmission of Power" (1888). Patent Number: 382,280. "Pyromagneto-Electric Generator" (1890). Patent Number: 428,057. Member of National Inventors Hall of Fame.

Townes, Charles H. Maser; "Production of Electromagnetic Energy" (1959). Patent Number: 2,879,439. Laser (optical maser) with Arthur L. Schawlow; "Masers and Maser Communications System" (1960). Patent Number: 2,929,922. Member of National Inventors Hall of Fame.

Whitney, Eli. Cotton gin; "Description of New Invented Cotton Gin, or Machine for Separating Cotton From Its Seed" (March 14, 1794). Member of National Inventors Hall of Fame.

Winslow, Samuel. Salt extracting method (1641). First patent granted in North America, by Massachusetts Bay Colony.

Wright, Orville and Wilbur. Airplane with motor; "Flying-Machine" (1906). Patent Number: 821,393. Members of National Inventors Hall of Fame.

SIGNIFICANT PATENTS

There are 9,957 patents that were issued prior to July 13, 1836, at which time patent numbering began. Inventions are listed in chronological order. The title of the invention is surrounded by quotation marks.

NAIL-MAKING MACHINE; "a machine for making nails." Patented by Samuel Briggs, Sr. and Samuel Briggs, Jr. on August 2, 1791.

CARDING MACHINE; "an improvement in manufacturing cards." Patented by Amos Whittemore on June 5, 1797.

CAST IRON PLOW; "an improvement in the art of plow making." Patented by Charles Newbold on June 26, 1797.

HIGH-PRESSURE STEAM ENGINE. Patented by Oliver Evans on February 14, 1804.

STEAMBOAT, EXPERIMENTAL. Patented by Robert Fulton on February 11, 1809.

STEAMBOAT, PRACTICAL. Patented by Robert Fulton on February 9, 1811.

WASHING MACHINE. Patented by Chester Stone on February 17, 1827.

FOUNTAIN PEN. Patented by D. Hyde on May 20, 1830.

MOWING MACHINE; "improvement in the machine for cutting grass and grain." Patented by W. Manning on May 3, 1831.

STEAM BOILER; "improvement in the construction of steam boilers or Cooper's Steam Boiler." Patented by Peter Cooper on October 13, 1831.

REVOLVER; "improvement in fire arms." Patented by Samuel Colt on February 25, 1836.

MATCH; "manufacture of friction—Matches." Patented by Alonzo D. Phillips in 1836. Patent Number: 68.

SOWING MACHINE; "machine for sowing plaster, ashes, seed, and other separable substances." Patented by Julius Hatch in 1838. Patent Number: Re. 1. First reissued patent.

WATER WHEEL; "spiral—bucket water—wheel." Patented by Lorenzo Dow Adkins in 1839. Patent Number: 1,154.

COLLAPSIBLE TUBE (as for toothpaste); "improvement in the construction of vessels or apparatus for preserving paint, etc." Patented by John Rand in 1841. Patent Number: 2,252.

CYLINDER LOCK; "door—Lock." Patented by Linus Yale in 1844. Patent Number: 3,630.

SEWING MACHINE; "improvement in sewing machines." Patented by Elias Howe, Jr. in 1846. Patent Number: 4,750.

RUBBER TIRE; "improvement in carriage wheels, etc." Patented by Robert W. Thomson in 1847. Patent Number: 5,104.

ROTARY PRINTING PRESS; "improvement in printing-presses." Patented by Richard M. Hoe in 1847. Patent Number: 5,188.

VALVE GEAR, STEAM ENGINE; "cut-off and working the valves of steam-engines." Patented by George H. Corliss in 1849. Patent Number: 6,162.

SAFETY PIN; "dress—pin." Patented by Walter Hunt in 1849. Patent Number: 6,281.

ELECTRO-MAGNETIC ENGINE; "improvement in electro-magnetic engines." Patented by Jacob Neff in 1851. Patent Number: 7,889.

ICE-MAKING MACHINE; "improved process for the artificial production of ice." Patented by John Gorrie in 1851. Patent Number: 8,080.

RAILROAD-CAR, TRUCK BRAKE; "railroad-car truck and brake." Patented by E.G. Otis in 1852. Patent Number: 8,973.

"IMPROVEMENT IN FIRE-ARMS." Patented by Horace Smith and Daniel B. Wesson in 1854. Patent Number: 10,535.

DECOMPOSING FAT; "improvement in process for purifying fatty bodies." Patented by Richard A. Tilghman in 1854. Patent Number: 11,766.

"IMPROVEMENT IN SEWING-MACHINES." Patented by Isaac M. Singer in 1855. Patent Number: 13,661.

STEEL PROCESS; "improvement in the manufacture of iron and steel." Patented by Henry Bessemer in 1856. Patent Number: 16,082.

MASON JAR; "improvement in screw-neck bottles." Patented by John L. Mason in 1858. Patent Number: 22,186.

RAILROAD AIR BRAKE; "improvement in railroad-brakes." Patented by Nehemiah Hodge in 1860. Patent Number: 28,670.

BREECH LOADING GUN. Patented by Van Houten in 1861.

MACHINE GUN; "improvement in revolving battery-guns." Patented by Richard J. Gatling in 1862. Patent Number: 36,836.

WEB PRINTING PRESS; "printing-machine." Patented by William Bullock in 1863. Patent Number: 38,200.

TYPEWRITER; "improvement in typewriting machine." Patented by C. Latham Sholes, Carlos Glidden, and Samuel W. Soule in 1868. Patent Number: 79,265.

TAPE MEAUSRE; "improvement in tape measures." Patented by Alvin J. Fellows in 1868. Patent Number: 79,965,

AIR BRAKE; "improvement in steam-power-brake devices." Patented by George Westinghouse, Jr. in 1869. Patent Number: 88,929.

VACUUM CLEANER; "improved sweeping-machine." Patented by Ives W. McGaffey in 1869. Patent Number: 91,145.

GUN CARRIAGE; "improvement in gun carriages." Patented by James B. Eads. Patent Number: 93,691.

"IMPROVEMENT IN POWER-LOOM FOR WEAVING INGRAIN CARPET." Patented by William and John W. Murkland in 1869. Patent Number: 97,106.

"IMPROVEMENT IN TREATING AND MOLDING PYROXYLINE" (resulted in the development of celluloid). Patented by John W. Hyatt, Jr. and Isaiah S. Hyatt in 1870. Patent Number: 105,338.

GOODYEAR WELT; "improvement in machines for sewing boots and shoes." Patented by Charles Goodyear, Jr. in 1871. Patent Number: 111,197.

VASELINE rgt; "improvement in products from petroleum." Patented by Robert A. Chesebrough in 1872. Patent Number: 127,568.

ELECTRIC TRAIN SIGNALING APPARATUS; "improvement for electric-signaling apparatus for railroads." Patented by William Robinson in 1872. Patent Number: 130,661.

PASTEURIZATION; "improvement in brewing beer and ale." Patented by Louis Pasteur in 1873. Patent Number: 135,245.

AUTOMATIC CAR COUPLINGS; "improvement in car-couplings." Patented by Eli H. Janney in 1873. Patent Number: 138,405.

BARBED WIRE; "improvement in wire fences." Patented by Joseph F. Glidden in 1874. Patent Number: 157,124.

"IMPROVEMENT IN CARPET-SWEEPERS." Patented by Melville R. Bissell in 1876. Patent Number: 182,346.

INTERNAL COMBUSTION ENGINE; "improvement in gas-motor engines." Patented by Nicolaus A. Otto in 1877. Patent Number: 194,047.

REINFORCED CONCRETE; "improvement in composition floors, roofs, pavements, etc." Patented by Thaddeus Hyatt in 1878. Patent Number: 206,112.

"IMPROVEMENT IN VEGETABLE-ASSORTERS." Patented by John H. Heinz in 1879. Patent Number: 212,000.

CARBON ARC LAMP; "electric lamp." Patented by Charles F. Brush in 1879. Patent Number: 219,208.

"MICROPHONE." Patented by Emile Berliner in 1880. Patent Number: 224,573.

"AIR-ENGINE." Patented by John Ericsson in 1880. Patent Number: 226,052.

CREAMER; "centrifugal creamer." Patented by Edwin J. Houston and Elihu Thomson in 1881. Patent Number: 239,659.

"BUTTON-HOLE SEWING-MACHINE." Patented by John Reece in 1881. Patent Number: 240,546.

"LIFE-RAFT." Patented by Frederick S. Allen in 1881. Patent Number: 240,634.

PLAYER PIANO; "mechanical musical instrument." Patented by John McTammany, Jr. in 1881. Patent Number: 242,786.

"SPEAKING-TELEPHONE." Patented by Francis Blake in 1881. Patent Number: 250,126.

"STETHOSCOPE." Patented by William F. Ford in 1882. Patent Number: 257,487.

"ELECTRIC FLAT-IRON." Patented by Henry W. Seeley in 1882. Patent Number: 259,054.

"CASH REGISTER AND INDICATOR." Patented by James Ritty and John Birch in 1883. Patent Number: 271,363.

TRANSPARENT PHOTOGRAPHIC PAPER STRIP FILM; "photographic film." Patented by George Eastman in 1884. Patent Number: 306,954.

ELECTRICAL WELDING; "apparatus for electric welding." Patented by Elihu Thomson in 1886. Patent Number: 347,140.

RECORD, DISC; "gramophone." Patented by Emile Berliner in 1887. Patent Number: 372,786.

ELECTRIC METER; "meter for alternating electric currents." Patented by Oliver B. Shallenberger in 1888. Patent Number: 388,003.

ROLL FILM CAMERA (KODAK™); "camera." Patented by George Eastman in 1888. Patent Number: 388,850.

BALLPOINT PEN; "pen." Patented by John J. Loud in 1888. Patent Number: 392,046.

LINOTYPE™; "machine for producing linotypes, type-matrices, etc." Patented by Ottmar Mergenthaler in 1890. Patent Number: 436,532.

"PHOTOGRAPHIC FILM." Patented by George Eastman in 1890. Patent Number: 441,831.

"AUTOMATIC TELEPHONE-EXCHANGE." Patented by Almon B. Strowger in 1891. Patent Number: 447,918.

AUTOMATIC TELEPHONE; "automactic telephone or other electrical exchange." Patented by Almon B. Stowger in 1892. Patent Number: 486,909.

HALF-TONE PRINTING; "photogravure-printing plate." Patented by Frederick E. Ives in 1893. Patent Number: 495,341.

ELECTRIC TROLLEY CAR; "travelling contact for electric railways." Patented by Charles J. Van Depoele in 1893. Patent Number: 495,443.

ZIPPER™; "clasp locker or unlocker for shoes." Patented by Whitcomb L. Judson in 1893. Patent Number: 504,038.

GASOLINE AUTOMOBILE; "road-vehicle." Patented by Charles E. Duryea in 1895. 540,648.

ROAD CARRIAGE; "road-engine." Patented by George B. Selden in 1895. Patent Number: 549,160.

"ROTARY-DISC PLOW." Patented by Clement A. Hardy in 1896. Patent Number: 556,972.

MONOTYPE™; "machine for making justified lines of type. Patented by Tolbert Lanston in 1896. Patent Number: 557,994.

PRODUCTION OF CARBORUNDUM™; "electrical furnace." Patented by Edward Goodrich Acheson in 1896. Patent Number: 560,291.

EVEN KEEL SUBMARINE; "submarine vessel." Patented by Simon Lake in 1897. Patent Number: 581,213.

"CARBURETOR." Patented by Henry Ford in 1898. Patent Number: 610,040.

"NAVIGABLE BALLOON." Patented by Ferdinand Graf Zeppelin in 1899. Patent Number: 621,195.

MOTOR-DRIVEN VACUUM CLEANER; "pneumatic carpet-renovator." Patented by John S. Thurman in 1899. Patent Number: 634,042.

GAS TURBINE; "apparatus for generating mechanical power." Patented by Charles G. Curtis in 1899. Patent Number: 635,919.

ASPIRIN; "acetyl salicylic acid." Patented by Felix Hoffmann in 1900. Patent Number: 644,077.

STANLEY STEAMER; "motor-vehicle." Patented by Francis E. and Freelan O. Stanley in 1900. Patent Number: 657,711.

MAGNETIC TAPE RECORDER; "method of recording and reproducing sounds or signals." Patented by Valdemar Poulsen in 1900. Patent Number: 661,619.

HIGH SPEED STEEL TOOLS; "metal cutting tool and method of making same." Patented by Frederick W. Taylor and Maunsel White in 1901. Patent Number: 668,269.

"MOTOR CARRIAGE." Patented by Henry Ford in 1901. Patent Number: 686,046.

"SUBMARINE BOAT OR VESSEL." Patented by John P. Holland in 1902. Patent Number: 694,154.

AUTOMATIC STEREOTYPE-PRINTING; "automatic stereotype-printing—plate casting and finishing apparatus." Patented by Henry A. Wood in 1903. Patent Number: 721,117.

STARTING MOTOR FOR AUTO; "means for operating motor-vehicles." Patented by Clyde J. Coleman in 1903. Patent Number: 745,157.

AUTOMOBILE TIRE MAKING MACHINE; "machine for making outer casings for double-tube tires." Patented by Frank A. Seiberling and William C. Stevens in 1904. Patent Number: 762,561.

GLASS SHAPING BOTTLE MACHINE; "glass shaping machine." Patented by Michael J. Owens in 1904. Patent Number: 766,768.

SAFETY RAZOR; "razor" (title for both patents). Patented by King C. Gillette in 1904. Patent Numbers: 775,134 and 775,135.

TWO-ELEMENT VACUUM TUBE; "instrument for converting alternating electric currents into continuous currents." Patented by John Ambrose Fleming in 1905. Patent Number: 803,684.

RADIO TUBE DETECTOR; "oscillation-responsive device." Patented by Lee De Forest in 1906. Patent Number: 836,070.

RADIO AMPLIFIER TUBE; "device for amplifying feeble electrical currents." Patented by Lee De Forest in 1907. Patent Number: 841,387.

THREE-ELEMENT VACUUM TUBE (AUDION); "space-telegraphy." Patented by Lee De Forest in 1908. Patent Number: 879,532.

BAKELITE™; "condensation product and method of making same." Patented by Leo H. Baekeland in 1909. Patent Number: 942,809.

ELECTRICAL INSULATORS; "insulating material." Patented by James C. Dow in 1910. Patent Number: 952,513.

SLEEVE VALVE ENGINE; "internal-combustion engine." Patented by Charles Y. Knight in 1910. Patent Number: 968,166.

SYNTHETIC AMMONIA; "production of ammonia." Patented by Fritz Habor and Robert Le Rossignol in 1910. Patent Number: 971,501.

MERCURY VAPOR LAMP; "apparatus for the electrical production of light." Patented by Peter Cooper Hewitt in 1912. Patent Number: 1,030,178.

CRACKING PROCESS; "manufacture of gasoline." Patented by William M. Burton in 1913. Patent Number: 1,049,667.

ROCKET ENGINE; "rocket apparatus." Patented by Robert H. Goddard in 1914. Patent Number: 1,103,503.

"WIRELESS RECEIVING SYSTEM." Patented by Edwin H. Armstrong in 1914. Patent Number: 1,113,149.

"BRASSIERE." Patented by Mary Phelps Jacob (Caresse Crosby) in 1914. Patent Number: **1,115,674.**

HYDRO AIRPLANE; "heavier-than-air flying-machine." Patented by Glenn H. Curtiss in 1915. Patent Number: 1,156,215.

SELECTIVE RADIO TUNING SYSTEM; "selective tuning system." Patented by Ernst F.W. Alexanderson in 1916. Patent Number: 1,173,079.

INCANDESCENT GAS LAMP; "incandescent electric lamp." Patented by Irving Langmuir in 1916. Patent Number: 1,180,159.

"RADIOTELEPHONY." Patented by Carl R. Englund in 1917. Patent Number: 1,245,446.

GYROCOMPASS; "gyroscopic compass." Patented by Elmer A. Sperry in 1918. Patent Number: 1,279,471.

"DIVER'S SUIT" (permitting escape). Patented by Harry Houdini in 1921. Patent Number: 1,370,376.

HIGH VACUUM RADIO TUBE; "electrical discharge apparatus and process of preparing and using the same." Patented by Irving Langmuir in 1925. Patent Number: 1,558,436.

LEAD ETHYL GASOLINE; "method and means for using motor fuels." Patented by Thomas Midgley, Jr. in 1926. Patent Number: 1,573,846.

ELECTRIC RAZOR; "shaving implement." Patented by Jacob Schick in 1928. Patent Number: 1,721,530.

PACKAGED FROZEN FOODS; "method of preparing food products." Patented by Clarence Birdseye in 1930. Patent Number: 1,773,079.

"TELEVISION SYSTEM." Patented by Philo T. Farnsworth in 1930. Patent Number: 1,773,980. "TELEVISION RECEIVING SYSTEM." Patented by Philo T. Farnsworth in 1930. Patent Number: 1,773,981.

REFRIGERATION APPARATUS; "refrigeration." Patented by Albert Einstein and Leo Szilard in 1930. Patent Number: 1,781,541.

MICROFILMING CAMERA; "photographing apparatus." Patented by George Lewis McCarthy in 1931. Patent Number: 1,806,763.

REFRIGERANTS, LOW-BOILING FLUORINE COMPOUND (FREON™); "heat transfer." Patented by Thomas Midgley, Jr., Albert R. Henne, and Robert R. McNary in 1931. Patent Number: 1,833,847.

"POLARIZING REFRACTING BODIES." Patented by Edwin H. Land and Joseph S. Friedman in 1933. Patent Number: 1,918,848.

TWO-PATH FM RADIO; "radio signalling system." Patented by Edwin H. Armstrong in 1933. Patent Number: 1,941,066.

CYCLOTRON; "method and apparatus for the acceleration of ions." Patented by Ernest O. Lawrence in 1934. Patent Number: 1,948,384.

MONOPOLY™; "board game apparatus." Patented by Charles B. Darrow in 1935. Patent Number: 2,026,082.

NYLON; "linear condensation polymers," and "Fiber and Method of Producing It." Patented by Wallace H. Carothers in 1937. Patent Numbers: 2,071,250 and 2,071,251.

CATHODE RAY TUBE, TELEVISION; "cathode ray tube." Patented by Vladimir K. Zworykin in 1938. Patent Number: 2,139,296.

XEROGRAPHY; "electron photography." Patented by Chester F. Carlson in 1940. Patent Number: 2,221,776.

D.D.T. (dichlorodiphenyltrichloroethane); "devitalizing composition of matter." Patented by Paul Muller in 1943. Patent Number: 2,329,074.

WIRE RECORDER, METHOD AND MEANS OF MAGNETIC RECORDING. Patented by Marvin Camras in 1944: "method and means of magnetic recording" Patent number: 2,351,004.

SULFONAMIDE; "diazine compounds." Patented by James M. Sprague in 1946. Patent Number: 2,407,966.

ANTIBIOTIC; "process and culture media for producing new penicillins." Patented by Otto K. Behrens, Joseph W. Corse, Reuben G. Jones, and Quentin F. Soper in 1949. Patent Number: 2,479,295.

DUCTILE CAST IRON; "cast ferrous alloy." Patented by Keith Dwight Millis, Albert Paul Gagnebin, and Norman Baden Pilling in 1949. Patent Number: 2,485,760.

DIURETIC; "heterocyclic sulfonamides and **methods of preparation thereof." Patented by James** W. Clapp and Richard O. Robin in 1951. Patent Number: 2,554,816.

CORTICOSTEROID; "oxygenation of steroids by mucorales fungi." Patented by Herbert C. Murray and Durey H. Peterson in 1952. Patent Number: 2,602,769.

TRANQUILIZER; "anti-excitatory compositions." Patented by Joseph Seifter, Anthony L. Monaco, and Franklin Judson Hoover in 1957. Patent Number: 2,799,619.

"SATELLITE STRUCTURE." Patented by Robert C. Baumann in 1958. Patent Number: 2,835,548.

VIDEOTAPE RECORDER; "broad band magnetic tape system and method." Patented by Charles P. Ginsburg, Shelby F. Henderson, Jr., Ray M. Dolby, and Charles E. Anderson in 1960. Patent Number: 2,956,114.

INTEGRATED CIRCUIT; "semiconductor device-and-lead structure." Patented by Robert N. Noyce in 1961. Patent Number: 2,981,877.

GOEDESIC DOME; "suspension building." Patented by Richard Buckminster Fuller in 1964. Patent Number: 3,139,957.

"APPARATUS FOR COOLING AND SOLAR HEATING A HOUSE." Patented by Harry E. Thomason in 1967. Patent Number: 3,295,591.

"COMBINATION SMOKE AND HEAT DETECTOR ALARM." Patented by Sidney Jacoby in 1976. Patent Number: 3,938,115.

DESIGN PATENTS

TYPE FACE; "printing type." Patented by George Bruce in 1842. Patent Number: Des. 1.

STATUE OF LIBERTY; "design for a statue." Patented by Auguste Bartholdi in 1879. Patent Number: Des. 11,023.

CONGRESSIONAL MEDAL OF HONOR; "design for a badge." Patented by George L. Gillespie in 1904. Patent Number: Des. 37,236.

LADY'S STOCKING; "design for a lady's stocking."

Patented by William G. Bley in 1948. Patent Number: Des. 151,732.

PLANT PATENTS

"CLIMBING OR TRAILING ROSE." Patented by Henry F. Bosenberg in 1931. Patent Number: Plant Pat. 1.

"GRAPEVINE." Patented by Chester A. Sanderson in 1948. Patent Number: Plant Pat. 782.

"STRAWBERRY PLANT." Patented by Frank J. Keplinger in 1953. Patent Number: Plant Pat. 1,183.

"APPLE TREE." Patented by Ralph Banta in 1971. Patent Number: Plant Pat. 3,045.

"WALNUT TREE." Patented by Louis Rodhouse in 1972. Patent Number: Plant Pat. 3,159.

"GRAPEFRUIT TREE." Patented by Richard A. Hensz in 1972. Patent Number: Plant Pat. 3,222.

Forms and Documents

TABLE OF FIGURES

Patents

United States Patent .... Fig. 1
PTO Fee Schedule .... Fig. 2
Patent Application Transmittal Letter .... Fig. 3
Notice of Informal Application .... Fig. 4
Notice of Incomplete Application .... Fig. 5
Notice to File Missing Parts of Application — No Filing Date .... Fig. 6
Notice to File Missing Parts of Application — Filing Date Granted .... Fig. 7
Declaration for Patent Application .... Fig. 8
Power of Attorney or Authorization of Agent, Not Accompanying Application .... Fig. 9A
Revocation of Power of Attorney or Authorization of Agent .... Fig. 9B
Declaration Claiming Small Entity Status — Nonprofit Organization .... Fig. 10
Declaration by a Non-Inventor Supporting a Claim by Another for Small Entity Status .... Fig. 11
Declaration Claiming Small Entity Status — Small Business Concern .... Fig. 12
Declaration Claiming Small Entity Status — Independent Inventor .... Fig. 13
Information Disclosure Citation .... Fig. 14
Notice of References Cited .... Fig. 15
Examiner's Action .... Fig. 16
Amendment Transmittal Letter .... Fig. 17
Channels of Ex Parte Review .... Fig. 18
Assignment of a Patent .... Fig. 19A
Assignment of Patent Application .... Fig. 19B
Design Patent — Shoe .... Fig. 20
Patent Worksheet .... Fig. 21

Trademarks

Trademark Examination Activities .... Fig. 22
International Schedule of Classes of Goods and Services .... Fig. 23

Copyrights

Copyright Form TX .... Fig. 24
Copyright Form VA .... Fig. 25
Copyright Form RE .... Fig. 26
Copyright Form CA .... Fig. 27

Miscellaneous

Request for Evaluation of an Energy-Related Invention .... Fig. 28
Defense Small Business Innovation Research Program Cost Proposal .... Fig. 29
Small Business Innovation Research Program Proposal Cover Sheet .... Fig. 30
Small Business Innovation Research Program Proposal Summary .... Fig. 31
Agreement to Hold Secret and Confidential .... Fig. 32
Idea Submission Agreement I .... Fig. 33
Idea Submission Agreement II .... Fig. 34
NIST Agreement of Nondisclosure .... Fig. 35

United States Patent [19]

DeLay, Jr.

[11] **Patent Number: 4,557,395**

[45] **Date of Patent: Dec. 10, 1985**

[54] **PORTABLE CONTAINER WITH INTERLOCKING FUNNEL**

[75] Inventor: **Victor A. DeLay, Jr.**, Largo, Fla.

[73] Assignee: **E-Z Out Container Corp.**, Clearwater, Fla.

[21] Appl. No.: **717,439**

[22] Filed: **Mar. 28, 1985**

[51] **Int. Cl.**[4] **B65D 3/04**

[52] **U.S. Cl.** .. .. **220/86 R**; 220/85 F; 220/1 C; 141/98

[58] **Field of Search** .... 220/86 R, 85 F, 1 C; 141/98, 331; 220/360

[56] **References Cited**

U.S. PATENT DOCUMENTS

1,554,589	9/1925	Long	220/1 C X
3,410,438	11/1968	Bartz	220/1 C
4,010,863	3/1977	Ebel	220/1 C
4,149,575	4/1979	Fisher	220/85 F X
4,162,020	7/1979	Kirkland	220/1 C X
4,296,838	10/1981	Cohen	220/1 C X
4,301,841	11/1981	Sandow	220/1 C X

Primary Examiner—Steven M. Pollard
Attorney, Agent, or Firm—Stanley M. Miller

[57] **ABSTRACT**

A portable, vented container for dirty oil, of the type having a small fill spout and having increased utility when used in conjunction with a funnel. A vent closure member and a funnel securing latch are integral with the funnel so that when the funnel is inverted and positioned in surmounting relation to the container, the vent closure member closes the vent and the securing latch is engaged by a fill spout cap which engagement secures the funnel against movement and hence maintains the vent closure as well. Removal of the fill spout cap releases the funnel, and positioning the funnel into its operative position relative to an automotive oil drain plug separates the vent closure portion of the funnel from the vent. An elongate extension member having a flexible medial portion is further provided.

20 Claims, 12 Drawing Figures

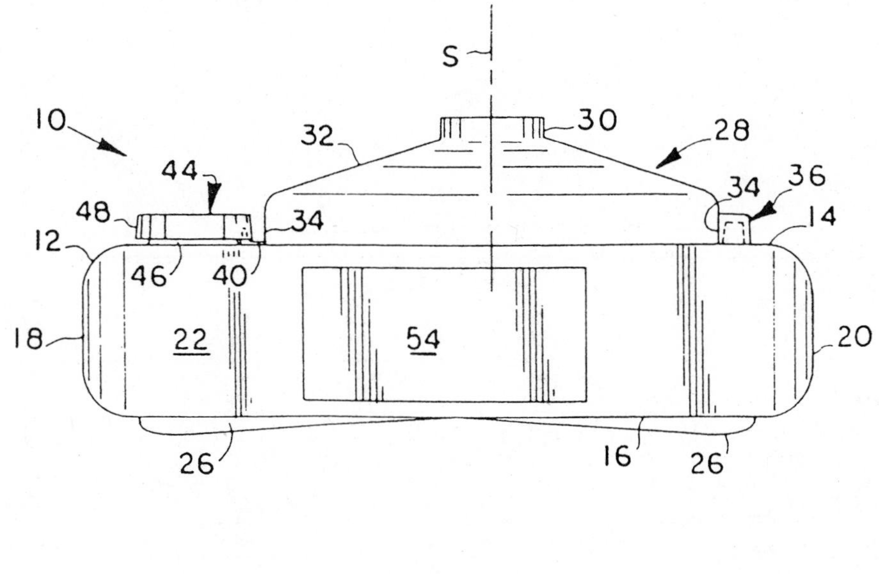

Fig. 1

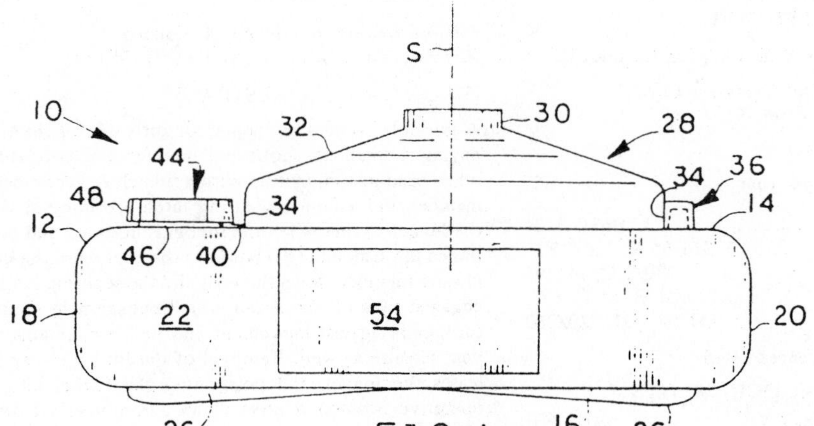

FIG_1

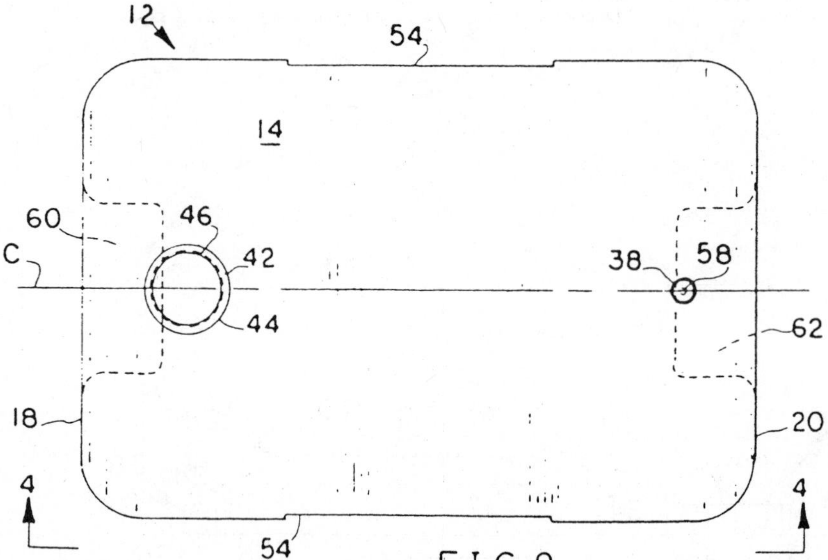

FIG_2

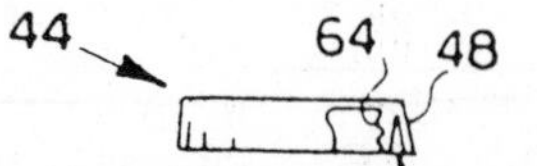

FIG_3

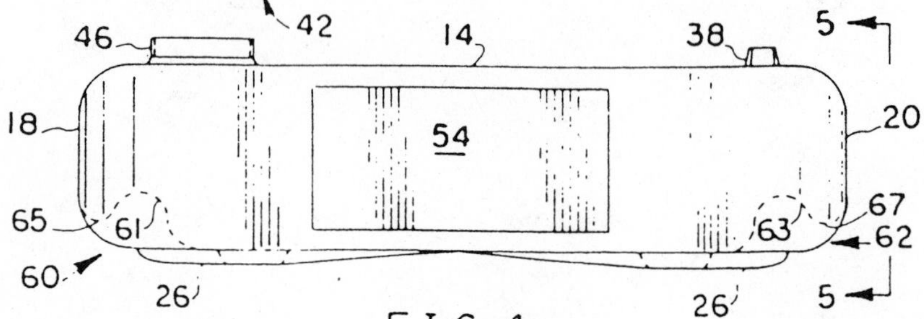

FIG_4

Fig. 1 (continued)

FIG_5

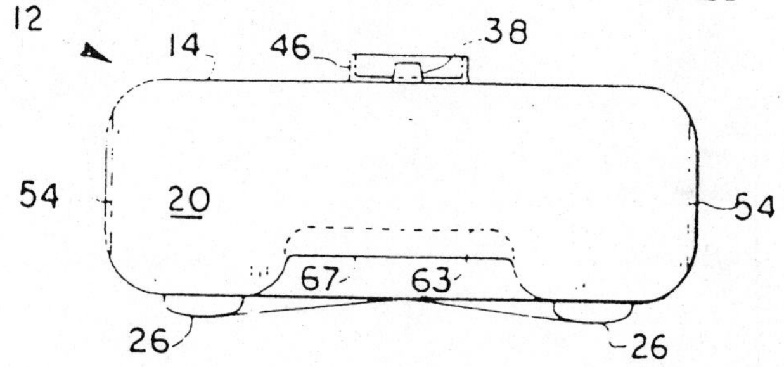

FIG_6

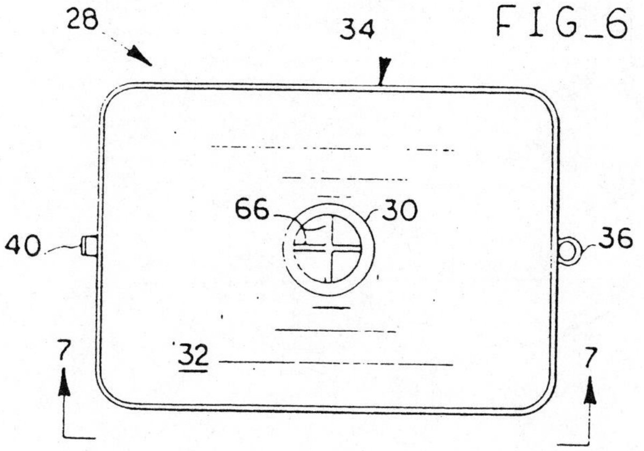

FIG_7

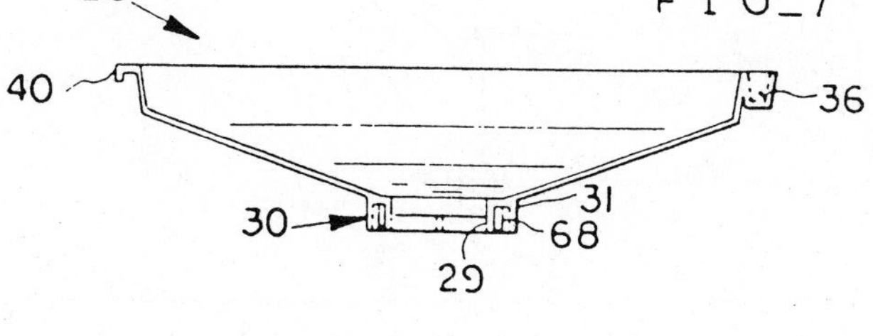

FIG_8

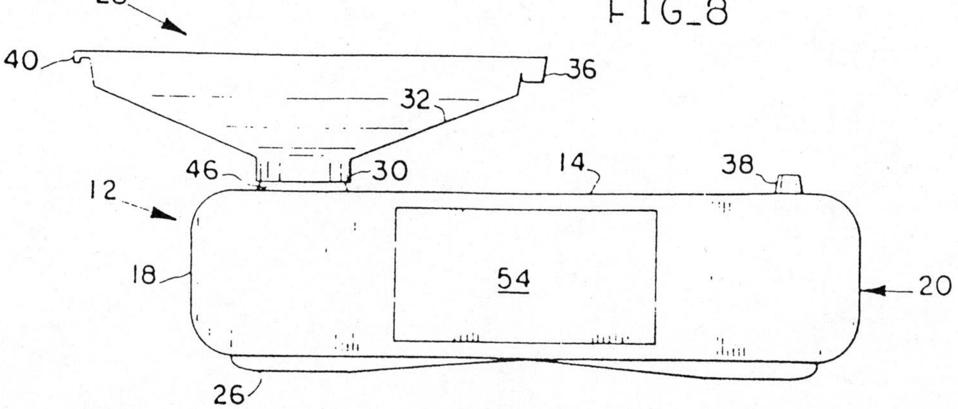

Fig. 1 (continued)

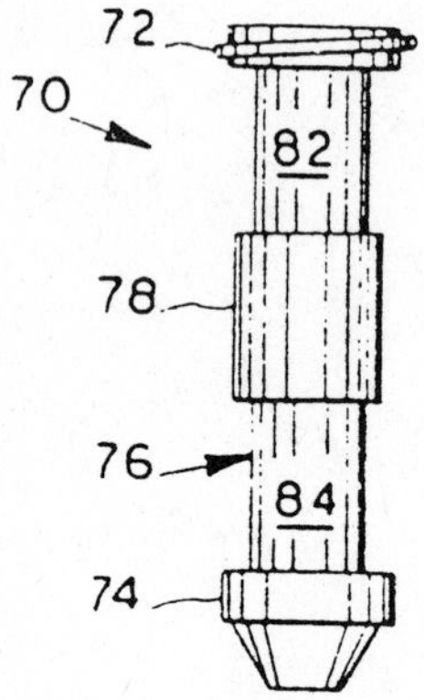

FIG_9

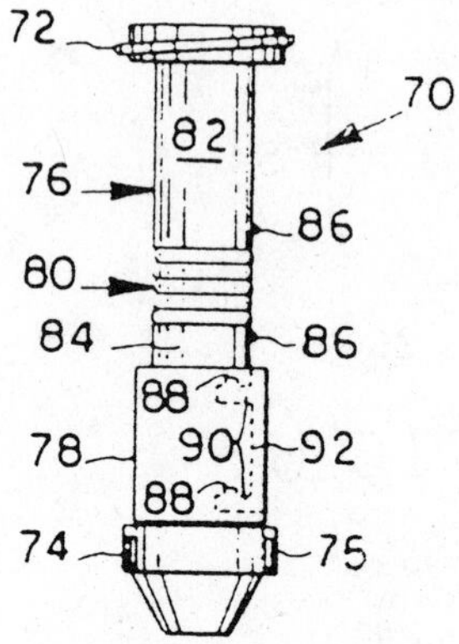

FIG_10

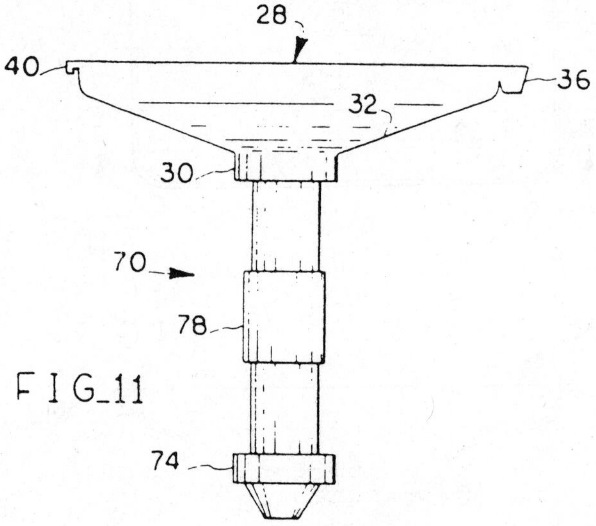

FIG_11

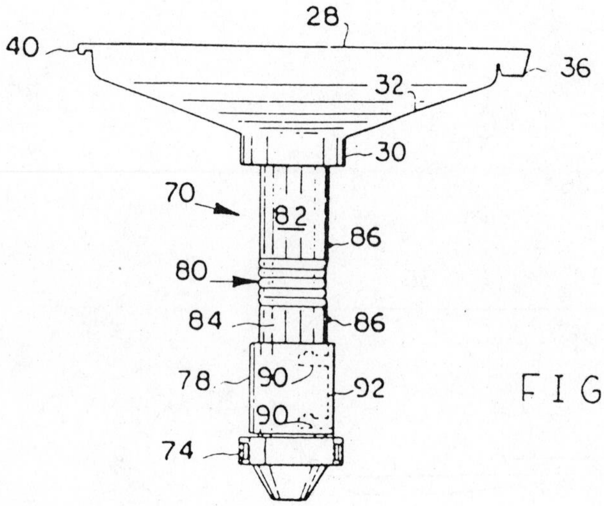

FIG_12

Fig. 1 (continued)

PORTABLE CONTAINER WITH INTERLOCKING FUNNEL

BACKGROUND OF THE INVENTION

1. Field of the Invention

This invention relates generally to containers having small fill spouts, and more particularly this invention relates to a vented container the vent of which is closed when the funnel is stored in latching engagement with the container body.

2. Description of the Prior Art

A thorough description of the prior art in the the field to which this invention pertains may be found in my co-pending application having a filing date of Sept. 14, 1983, Ser. No. 06/531,948. Moreover, the most pertinent prior art is believed to be the container for dirty oil disclosed in said application.

Other patents of interest are: U.S. Pat. Nos. 4,403,692 to Pollacco (1983); 822,854 to Cosgrave (1906); 2,576,154 to Trautvetter (1951); 4,098,393 to Meyers (1978); 4,217,940 to Wheeler and others (1980); and 4,301,841 to Sandow (1981).

Of the known containers, only the container provided by the present inventor and disclosed in the above-identified patent application contains a means whereby the funnel of the container can be conveniently stored when not in use.

Containers having small fill spouts are normally vented to allow the air inside the container to escape as liquid fluids are charged thereinto. Typically, the vent is provided in the form of an upstanding coupling which is provided with a closure member in the form of a cap which may or may not be attached to the coupling itself. Where the cap is attached to the coupling, its loss is safeguarded against but still the user of the container must remember to open and close the vent as needed. Vent caps that are not attached to their couplings are usually lost.

There is a need, therefore, for a vent cap that is safeguarded against loss, and which also opens and closes the vent as needed without requiring the user thereof to remember to open and close such vent.

Another common problem with small-mouthed containers is that the funnels which must be used therewith are often lost. Pollacco solves this problem by permanently securing his funnel to his container. This storage expedient is unsatisfactory because it is important to maintain funnels of the type used to fill automotive crankcases in a substantially clean condition as the introduction of dirt into a crankcase can damage engine parts.

Therefore, there is a need for a funnel storage apparatus capable of storing a funnel in an inverted position when it is not in use. The storage apparatus that is needed would also safeguard against the loss of the funnel.

The art has heretofore developed elongate funnel extension members of the type disclosed by Cosgrave, Trautvetter, and the present inventor, but the same are inflexible and thus inadequate and lacking in utility in certain specific environments.

SUMMARY OF THE INVENTION

The longstanding but heretofore unfulfilled need for a portable container for dirty oil having the desireable features of a self-opening and self-closing vent, a funnel that is storable in an inverted position and which is also secured against loss, is now fulfilled by the invention disclosed hereinafter and summarized as follows.

The container is of parallelepiped form and has finger-receiving recesses formed in its opposite ends, on the underside thereof, which recesses are grasped by an individual when transporting the container.

The top of the container includes a large, imperforate medial portion against which the rim of the funnel is seated when the funnel is in its storage position.

A fill spout of small diameter projects upwardly from the top of the container, and is disposed near the periphery of the container so the medial portion of the container can receive the stored funnel, as aforesaid.

A sleeve member which defines a vent opening projects upwardly from the top of the container as well, but is disposed in longitudinally spaced relation to the fill spout so that it is near the periphery of the container opposite from the fill spout.

The longitudinal axis of symmetry of the container bisects the finger-receiving recesses or handles, the fill spout, the vent-defining sleeve, and the funnel when the latter is in its stored position. In this manner, the container is stable when transported.

The funnel has an integral vent closure member that projects outwardly from the rim of the funnel, in radial relation to the funnel's axis of symmetry. A latch member used to secure the stored funnel against movement is also formed integral to the funnel, extends radially with respect to said axis from the rim thereof, and is positioned in opposition to the vent closure member.

The funnel's size and the amount of space between the fill spout and the vent opening are selected so that when the funnel is inverted and placed in the center of the medial portion of the top wall of the container, and properly rotated about its axis of symmetry, the vent closure member will align with and seal the vent opening and the latch which is opposed to the vent closure member will be positioned in close proximity to the fill spout.

A novel fill spout closure member in the form of a double-walled cap, when brought into screw threaded engagement with the fill spout, will seal the spout and simultaneously overlie the funnel latch to secure the funnel against displacement.

The novel cap's first wall is internally threaded and thus adapted for screw threaded engagement with the externally threaded fill spout. It outer wall defines an annular recess having an open bottom, which recess surrounds the first wall and which recess receives the funnel latch therewithin. The annular configuration of the recess eliminates any need for aligning the cap with respect to the latch.

In this manner, the act of inverting the funnel and placing it in its storage position on the top wall of the funnel will close the vent if the proper alignment is made. Once the vent has been closed, no further alignment is required as the sealing of the fill spout by the novel cap will also secure the funnel as desired.

Thus, when the funnel is deployed into its operative configuration, the user of the invention need only remove the fill spout cap, as such will release the funnel from its stored position. The act of placing the funnel's spout into the container's fill spout then serves to open the vent.

A funnel extension member having a flexible medial portion is also disclosed hereinafter. A slideably mounted rigid sleeve member serves to delete the flexibility function of the extension member when desired when such sleeve member is positioned in registration with the flexible portion of the member. However, the flexibility of the member is restored upon slidingly displacement of the sleeve away from the flexible medial portion.

An important object of this invention, therefore, is to provide a container for dirty oil that includes a funnel as an attachment to the container so that the funnel is not easily misplaced.

Another object is to provide an attachment means that protects the sloping inside walls of the funnel contamination when the funnel is stored.

Another object of this invention is to provide a means whereby the vent of a container can be automatically opened and closed at the time the container's funnel is placed into its operative position and its storage position, respectively.

Other objects will become apparent as this description proceeds.

The invention accordingly comprises the features of construction, combination of elements and arrangement of parts that will be exemplified in the construction hereinafter set forth, and the scope of the invention will be indicated in the claims.

BRIEF DESCRIPTION OF THE DRAWINGS

For a fuller understanding of the nature and objects of the invention, reference should be made to the following detailed description, taken in connection with the accompanying drawings, in which:

FIG. 1 is a side elevational view of the container with the funnel stored in its inverted position thereatop;

FIG. 2 is a top plan view of the container body member;

FIG. 3 is a partially cut away side elevational view of the novel fill spout closure means;

FIG. 4 is a side elevational view taken along line 4—4 of FIG. 2;

FIG. 5 is an end view taken along line 5—5 of FIG. 4;

FIG. 6 is a top plan view of the novel funnel member;

FIG. 7 is a side elevational view of the funnel member taken along line 7—7 of FIG. 6;

FIG. 8 is a side elevational view, like that of FIG. 4, which shows the funnel member engaging the fill spout of the container body;

FIG. 9 is a side elevational view of the novel funnel downspout extension member with the rigid sleeve in its locked position;

FIG. 10 is a side elevational view of the funnel downspout extension member with the rigid sleeve in its unlocked position;

FIG. 11 is a side elevational view showing the extension member operatively coupled to the funnel member with the sleeve in its locked position; and

FIG. 12 is a side elevational view showing the extension member operatively coupled to the funnel member with the sleeve in its unlocked position.

Similar reference numerals refer to similar parts throughout the several views of the drawings.

DETAILED DESCRIPTION OF THE PREFERRED EMBODIMENT

Referring now to FIG. 1, it will there be seen that an illustrative embodiment of the invention is designated by the reference numeral 10 as a whole. The container body 12 has a parallelepiped construction when seen in perspective. Visible in FIG. 1 are the container's top wall 14, bottom wall 16, its left and right end walls 18, 20, a side wall 22, and support members collectively designated 26.

The novel funnel is indicated generally by the numeral 28. Funnel 28 includes downspout 30, sloping or converging walls 32, and an annular rim 34.

A vent closure member 36 is integrally formed with the rim 34 and extends therefrom as shown. The closure member 36 overlies a vent shroud 38 which is shown in phantom lines in FIG. 1.

A latch 40 is also integrally formed with the funnel rim 34 and is on the opposite side thereof relative to the vent closure member 36. The latch 40 has an "L" shape as shown. The horizontal leg of the latch abuts the top wall 14 of the container 12 and extends radially with respect to the axis of symmetry S of the funnel 28. It terminates in an upstanding leg (shown in phantom lines in FIG. 1) that extends into a cavity 42, which cavity 42 is an annular recess as shown in FIG. 2.

Referring again to FIG. 1, fill spout cap 44 is internally threaded to mate with the external threads of the fill spout 46. The annular latch-receiving recess 42 is formed by the provision of annular wall 48 that surrounds the spout 46, said annular wall depending to the periphery of the top wall of cap 44. The diameter of the top wall of cap 44 is greater than the diameter of the fill spout 46 by an amount substantially equal to the width of the latch-receiving recess 42.

The placement of the upstanding portion of latch 40 in the annular cavity 42 maintains the funnel 28 in its inverted, stored position until the cap 44 is removed.

The space designated 54 in FIG. 1 is a display space and accommodates a label which may have imprinted thereon the trademark of the device and other information.

Returning now to FIG. 2, it will there be seen that the longitudinal axis of symmetry of the device 10 is indicated by the centerline C. It bisects the vent 58 which is formed in the top wall 14 of the container 10 and which is surrounded by vent shroud 38, the fill spout 46, and the longitudinally spaced handles 60, 62 of the invention. The width of the handles 60, 62 is sufficient to accommodate four fingers of a human hand. Both of the label-accommodating recesses 54, 54 mentioned in connection with the description of FIG. 1 are shown in FIG. 2 as well.

The vent closure member 36 slideably and snugly engages the outer walls of the shroud 38, thereby closing the vent opening 58, when funnel 28 is in the inverted storaage position, as aforesaid.

FIG. 3 shows the internal threads 64 on the cap 44 and the annular wall 48 that depends to the periphery of the cap top wall to define the annular cavity 42 into which the upstanding portion of latch 40 extends.

The externally threaded fill spout 46 is shown in FIG. 4, which FIG. shows the container 12 with funnel 28 and cap 44 separated therefrom.

The handles 60, 62 include concave surfaces 61, 63, respectively, and convex surfaces 65, 67, the former of which are abutted by fingertips when the container is carried and the latter of which provide a comfortable rounded weight bearing surface.

An end view of the container 12 is provided in FIG. 5.

A top view of the novel funnel 28 appears in FIG. 6. A strainer 66 formed by a pair of cross bars is formed where the downwardly sloping walls 32 of the funnel 28 merge with the funnel's downspout. The generally rectangular planform of the funnel 28 conforms to the planform of the container body 12 as shown in FIG. 2, but the corresponding dimensions of the funnel are smaller.

The downspout 30 of funnel 28 is internally threaded as indicated by the reference numeral 68 appearing in FIG. 7, and is thus adapted for screw threaded engagement with the externally threaded fill spout 46. Accordingly, the downspout 30 of the funnel 28 is coupled to fill spout 46 when it is desired to charge the container with dirty oil. This operative positioning of the funnel 28 and fill spout 46 is depicted in FIG. 8. A comparison of FIGS. 1 and 8 indicates that the removal of cap 44 from spout 46 releases latch 40 so that funnel 28 can be separated from its engagement with top wall 14 of container 12, restored to its upright configuration, and coupled with the spot 46. The separation of the funnel 28 and the container body top wall 14 also separates the vent closure member 36 from vent shroud 38, which separation exposes vent 58 (FIG. 2) to ambient. The internal threads 68 of downspout 30 are formed in outer wall 31 thereof. An inner wall 29 is spaced radially inwardly of outer wall 31, and is concentric therewith. Accordingly, dirty oil contacts inner wall 29 only.

The truncate downspout 30 of funnel 28 is provided because some vehicle are built close to the ground. However, other vehicles are built higher from the ground and the use of a downspout extension member becomes advisable.

An improved downspout extension member is shown in FIGS. 9-12, and is designated 70 as a whole. It includes an externally threaded adapter 72 which is coupled to the internally threaded downspout 30 of funnel 28 when in use, as shown in FIGS. 11 and 12. Another adapter 74 at the lower end of the extension member 70 is internally threaded as at 75 (FIG. 10) to mate with the external threads of the fill spout 46. An elongate medial portion 76 interconnects the upper and lower adapters 72 and 74.

A slideably mounted rigid sleeve member 78 is shown mid-length of the medial portion 76 in FIG. 9. When the sleeve member 78 is locked into this position by means disclosed hereinafter, the novel extension member 70 can be used in the same manner as conventional downspout extension members, which use is depicted in FIG. 11.

However, when the sleeve 78 is unlocked and slideably displaced to its lowermost position, which position is depicted in FIG. 10, such displacement frees a flexible member 80 from confinement so that it is free to bend. More specifically, upper portion 82 of the downspout extension member medial portion 76 and lower portion 84 thereof may be displaced from their axial alignment with each other, i.e., their respective axes of longitudinal symmetry may be made oblique to one another. As shown in FIG. 12, when the flexible member 80 is free, funnel 28 can be moved in any direction relative to lower coupling 74, or vice versa.

FIGS. 10 and 12 both show the means employed to lock and unlock sleeve 78 as desired. A pair of vertically spaced beads, collectively designated 86, are formed on upper and lower portions 82, 84 of the extension member medial portion 76. A pair of vertically spaced bead-receiving cavities, collectively designated 88, are formed internally of sleeve member 78, so that the sleeve 78 is locked into overlying relation to the flexible member 80 when beads 86 are disposed therein.

To unlock the sleeve 78, the user of the inventive apparatus grasps sleeve 78 and slides it upwardly by a distance equal to the depth of the bead-receiving cavities 88. Each bead 86 will then be positioned in channels 90 which are also formed internally of sleeve 78. The user of the device then rotates the sleeve 78 until the beads 86 have traveled the length of the arcuate channels 90, which length could be a quarter of an inch, for example. This rotation of sleeve 78 will bring the beads 86 into registration with a vertically extending channel 92 so that the sleeve 78 can be moved to the position shown in FIGS. 10 and 12.

It will thus be seen that the objects set forth above, and those made apparent from the foregoing description, are effectively attained and since certain changes may be made in the above construction without departing from the scope of the invention, it is intended that all matters contained in the foregoing description or shown in the accompanying drawings shall be interpreted as illustrative and not in a limiting sense.

It is also to be understood that the following claims are intended to cover all of the generic and specific features of the invention herein described, and all statements of the scope of the invention which, as a matter of language, might be said to fall therebetween.

Now that the invention has been described,

What is claimed is:

1. A container of the type having a small fill spout and having increased utility when used in conjunction with a funnel, comprising:
 a container body member of generally parallelepiped configuration,
 a fill spout formed in a top wall of said container body member and projecting upwardly therefrom,
 a vent means in the form of an aperture formed in said top wall,
 a funnel member having a rim, converging sidewalls, and a downspout,
 said fill spout and funnel downspout adapted for releasable engagement with one another,
 a vent closure member secured to said funnel rim and projecting outwardly therefrom,
 said vent closure member closing said vent when brought into registration therewith.

2. The container of claim 1, further comprising,
 a fill spout closure means in the form of a cap member,
 a latch member secured to and projecting outwardly from said funnel rim,
 said cap member adapted to releasably engage said latch member when said funnel member is inverted and disposed atop said container top wall and when said cap member is releasably engaged to said fill spout.

3. The container of claim 2, wherein said vent closure member and said latch member are secured to said rim in opposed relation to each other.

4. The container of claim 3, further comprising,
 a sleeve-shaped shroud member disposed in surrounding relation to said aperture and projecting upwardly from said container top wall,
 said vent closure member adapted to engage said shroud member when said funnel is inverted and said vent closure member is brought into releasable engagement with said shroud member.

5. The container of claim 4, further comprising,
 a first handle means formed in said container body member at a first end thereof,

a second handle means formed in said container body member at a second end thereof which is longitudinally spaced from said first end,

each of said first and second handle means defined by a concavity formed in the bottom wall of said container body member and by a convexity contiguous thereto and continuous therewith, said convexity merging with an end wall of said container body member.

6. The container of claim 5, wherein the depth of the concavity forming a handle means is greater than the height of the convexity contiguous thereto.

7. The container of claim 5, wherein said first and second handle means are disposed transverse to and are bisected by the longitudinal axis of symmetry of said container body member.

8. The container of claim 3, wherein said cap member has a top wall having a diameter greater than the outer diameter of said fill spout, wherein an annular wall depends to the periphery of said cap top wall, wherein an annular cavity is defined between said fill spout and said depending wall, and wherein said latch member is specifically configured to enter into said annular cavity when brought into registration therewith.

9. The container of claim 8, wherein said latch member has a generally L-shaped configuration.

10. The container of claim 3, wherein said fill spout, said vent and said funnel member, latch member and vent closer member are collectively aligned with the longitudinal axis of symmetry of said container body member when said funnel member is inverted, when said vent closure member is disposed in engaging relation to said vent, and when said latch member is disposed in engaging relation to said fill spout cap.

11. The container of claim 1, wherein a strainer means is positioned within said funnel member at the juncture of said converging sidewalls and said downspout.

12. The container of claim 1, wherein said funnel member has a generally rectangular configuration when seen in plan view, and wherein said latch member and vent closure member are disposed mid-length of the opposite truncate sidewalls of said funnel member.

13. The container of claim 1, wherein said fill spout is externally threaded and wherein said funnel member downspout is internally threaded.

14. The container of claim 1, further comprising,

an elongate funnel downspout extension member having a first end adapted to releasably engage said funnel downspout and a second end adapted to releasably engage said fill spout,

and said downspout extension member having a flexible medial portion.

15. The container of claim 14, further comprising,

a rigid sleeve-shaped locking member, having a length greater than the length of said flexible medial portion and having an inside diameter slightly greater than the outside diameter of said downspout extension member, disposed in ensleeving relation to said flexible medial portion and restricting said downspout extension member from flexing at said medial portion.

16. The container of claim 15, further comprising,

means for selectively locking and unlocking said sleeve member into and out of its restricting engagement with said medial portion, respectively.

17. The container of claim 16, wherein said means for selectively locking and unlocking said sleeve member includes a pair of vertically spaced bead members formed on said downspout extension member, one of which is positioned above said flexible medial portion and one of which is positioned below said flexible medial portion, and wherein said sleeve member has a pair of cooperatively spaced bead-receiving cavities formed therein, which cavities are interconnected by a vertical slot and which cavities are formed at the end of associated channels orthogonal to said vertical slot.

18. The container of claim 13, wherein said funnel member downspout further comprises a cylindrical outer wall within which said internal threads are formed, and a cylindrical inner wall spaced radially inwardly of said outer wall so that dirty oil contacts only said inner wall when the container is used.

19. The container of claim 18, wherein said downspout inner wall is concentric with said downspout outer wall.

20. The container of claim 19, wherein the spacing between said downspout outer and inner walls is sufficient to receive therebetween said externally threaded fill spout.

* * * * * *

Fig. 1 (continued)

PTO FEE SCHEDULE

FEE CODE	DESCRIPTION	FEE
	PATENT FEES	
	Group 1 - Patent Filing Fees	
101	BASIC FILING FEE - UTILITY	370.00
102	INDEPENDENT CLAIMS IN EXCESS OF THREE	36.00
103	CLAIMS IN EXCESS OF TWENTY	12.00
104	MULTIPLE DEPENDENT CLAIM	120.00
105	SURCHARGE - LATE FILING FEE OR OATH/DECL.	120.00
106	DESIGN FILING FEE	150.00
107	PLANT FILING FEE	250.00
108	REISSUE FILING FEE	370.00
109	REISSUE INDEPENDENT CLAIMS OVER PATENT	36.00
110	REISSUE CLAIMS IN EXCESS OF TWENTY & PATENT	12.00
139	NON-ENGLISH SPECIFICATION	30.00
	Group 2 - Small Entity Patent Filing Fees	
201	BASIC FILING FEE - UTILITY	185.00
202	INDEPENDENT CLAIMS IN EXCESS OF THREE	18.00
203	CLAIMS IN EXCESS OF TWENTY	6.00
204	MULTIPLE DEPENDENT CLAIM	60.00
205	SURCHARGE - LATE FILING FEE OR OATH	60.00
206	DESIGN FILING FEE	75.00
207	PLANT FILING FEE	125.00
208	REISSUE FILING FEE	185.00
209	REISSUE INDEPENDENT CLAIMS OVER PATENT	18.00
210	REISSUE CLAIMS IN EXCESS OF TWENTY & PATENT	6.00
	Group 3 - Patent Extension Fees	
115	EXTENSION - ONE MONTH	62.00
116	EXTENSION - TWO MONTHS	180.00
117	EXTENSION - THREE MONTHS	430.00
118	EXTENSION - FOUR MONTHS	680.00
	Group 4 - Small Entity Patent Extension Fees	
215	EXTENSION - ONE MONTH	31.00
216	EXTENSION - TWO MONTHS	90.00
217	EXTENSION - THREE MONTHS	215.00
218	EXTENSION - FOUR MONTHS	340.00
	Group 5 - Patent Appeals/Interference Fees	
119	NOTICE OF APPEAL	140.00
120	FILING A BRIEF	140.00
121	REQUEST FOR ORAL HEARING	120.00
	Group 6 - Small Entity Patent Appeals/Interference Fees	
219	NOTICE OF APPEAL	70.00
220	FILING A BRIEF	70.00
221	REQUEST FOR ORAL HEARING	60.00
	Group 7 - Patent Petition Fee	

Fig. 2

PETITIONS TO THE COMMISSIONER:

122	-NOT ALL INVENTORS; NOT THE INVENTOR	120.00
123	-CORRECTION OF INVENTORSHIP IN APPL.	120.00
124	-NOT PROVIDED FOR QUESTIONS	120.00
125	-SUSPEND RULES	120.00
160	-EXPEDITED LICENSE	120.00
161	-CHANGE SCOPE OF LICENSE	120.00
162	-RETROACTIVE LICENSE	120.00
163	-REFUSING MAINTENANCE FEE	120.00
164	-REINSTATEMENT OF EXPIRED PATENT	120.00
165	-INTERFERENCE	120.00
166	-RECONSIDER INTERFERENCE PET. DECISION	120.00
167	-LATE FILING OF INTERFERENCE SETTLEMENT	120.00
168	-REFUSAL TO PUBLISH SIR	120.00
127	-ACCESS TO ASSIGNMENT RECORD	120.00
128	-ACCESS TO APPLICATION	120.00
129	-LATE PRIORITY PAPERS	120.00
131	-SUSPEND ACTION	120.00
132	-DIVISIONAL REISSUES	120.00
133	-ACCESS TO INTERFERENCE AGREEMENT	120.00
134	-AMENDMENT AFTER ISSUE FEE PAID	120.00
135	-WITHDRAWAL FROM ISSUE	120.00
136	-DEFER ISSUE	120.00
137	-ISSUE TO LATE RECORDED ASSIGNEE	120.00
130	PET. TO COMM. TO MAKE APPL. SPECIAL	80.00
138	PET. TO COMM. - PUBLIC USE PROCEEDING	1200.00
140	PET. - REVIVE ABAND. APPL. - UNAVOIDABLE	62.00
141	PET. - REVIVE ABAND. APPL. - UNINTENTIONAL	620.00
146	PET. - CORRECTION OF INVENTORSHIP IN PATENT	120.00

Group 8 - Small Entity Patent Petition Fees

240	PET. - REVIVE ABAND. APPL. - UNAVOIDABLE	31.00
241	PET. - REVIVE ABAND. APPL. - UNINTENTIONAL	310.00

Group 9 - Patent Issue Fees

142	UTILITY ISSUE FEE	620.00
143	DESIGN ISSUE FEE	220.00
144	PLANT ISSUE FEE	310.00
148	STATUTORY DISCLAIMER	62.00

Group 10 - Small Entity Patent Issue Fees

242	UTILITY ISSUE FEE	310.00
243	DESIGN ISSUE FEE	110.00
244	PLANT ISSUE FEE	155.00
248	STATUTORY DISCLAIMER	31.00

Group 11 - Patent Post-Allowance Fees

112	SIR - PRIOR TO EXAMINER'S ACTION	400.00
113	SIR - AFTER EXAMINER'S ACTION	800.00
145	CERTIFICATE OF CORRECTION	60.00
147	RE-EXAMINATION	2000.00
111	EXTENSION OF THE TERM OF PATENT	600.00

Group 12 - Patent Maintenance Fees - Applications Filed December 12, 1980 - August 26, 1982

Fig. 2 (continued)

170	DUE AT 3.5 YEARS	245.00
171	DUE AT 7.5 YEARS	495.00
172	DUE AT 11.5 YEARS	740.00
176	SURCHARGE - LATE PAYMENT WITHIN SIX MONTHS	120.00

Group 13 - Patent Maintenance Fees - Applications Filed On Or After August 26, 1982

173	DUE AT 3.5 YEARS	490.00
174	DUE AT 7.5 YEARS	990.00
175	DUE AT 11.5 YEARS	1480.00
177	SURCHARGE - LATE PAYMENT WITHIN SIX MONTHS	120.00
178	SURCHARGE AFTER EXPIRATION	550.00

Group 14 - Small Entity Patent Maintenance Fees - Applications Filed On Or After August 27, 1982

273	DUE AT 3.5 YEARS	245.00
274	DUE AT 7.5 YEARS	495.00
275	DUE AT 11.5 YEARS	740.00
277	SURCHARGE - LATE PAYMENT WITHIN SIX MONTHS	60.00

Group 15 - Patent Service Fees

501	COPY OF PATENT	1.50
503	COPY OF PLANT PATENT	10.00
506	COPY OF OFFICE RECORDS, (each thirty pages)	10.00
500	COPY OF UTILITY PATENT IN COLOR	20.00
535	PATENT COPY - EXPEDITED SERVICE	3.00
536	PATENT COPY EXPEDITED SERVICE VIA EOS	25.00
504	COPY OF APPLICATION AS FILED, CERTIFIED	10.00
505	COPY OF FILE WRAPPER, CERTIFIED	170.00
533	COPY OF PATENT ASSIGNMENT, CERTIFIED	5.00
537	CERTIFIED COPY OF PATENT APPLICATION EXPEDITED	20.00
508	CERTIFYING OFFICE RECORDS	3.00
509	SEARCH OF RECORDS	15.00
513	PATENT DEPOSITORY LIBRARY	50.00
514	LIST OF PATENTS IN SUBCLASS	2.00
528	UNCERTIFIED STATEMENT	5.00
532	COPY OF NON-U.S. DOCUMENT	10.00
510	COMPARING COPIES PER DOCUMENT	10.00
534	DUPLICATE OR CORRECTED FILING RECEIPT	15.00
516	FILING A DISCLOSURE DOCUMENT	6.00
522	BOX RENTAL	50.00
526	INTERNATIONAL TYPE SEARCH REPORT	30.00
517	SEARCHING, INVENTOR RECORDS, TEN YEARS	10.00
524	COPISHARE CARD PER PAGE	.15
518	RECORDING PATENT ASSIGNMENT	8.00
520	PUBLICATION IN OFFICIAL GAZETTE	20.00
521	DUPLICATE USER PASS	10.00
523	LOCKER RENTALS	.25
525	UNSPECIFIED OTHER SERVICES	AT COST
529	RETAINING ABANDONED APPLICATION	120.00
530	HANDLING FEE - OMITTED SPEC./DRAWING	15.00
531	HANDLING FEE FOR WITHDRAWAL OF SIR	120.00

Group 16 - Patent Enrollment Service Fees

609	ADMISSION TO EXAMINATION	270.00

Fig. 2 (continued)

610	REGISTRATION TO PRACTICE	90.00
611	REINSTATEMENT TO PRACTICE	10.00
612	COPY OF CERTIFICATE OF GOOD STANDING	10.00
613	CERTIFICATE OF GOOD STANDING - FRAMING	100.00
615	REVIEW OF DECISION OF DIRECTOR, OED	100.00
616	REGRADING OF EXAMINATION	100.00

Group 17 - Patent PCT Fees - International Stage

150	TRANSMITTAL FEE	170.00
151	PCT SEARCH FEE - NO U.S. APPLICATION	550.00
153	PCT SEARCH - PRIOR U.S. APPLICATION	380.00
152	SUPPLEMENTAL SEARCH PER ADDITIONAL INVENTION	150.00
190	PRELIMINARY EXAM. FEE - ISA WAS U.S.	400.00
191	PRELIMINARY EXAM. FEE - ISA NOT U.S.	600.00
192	ADDITIONAL INVENTION - ISA WAS U.S.	130.00
193	ADDITIONAL INVENTION - ISA NOT U.S.	200.00

Group 18 - Patent PCT Fees - National Stage

956	IPEA - U.S.	330.00
958	INTERNATIONAL SEARCHING AUTHORITY - U.S.	370.00
960	PTO NOT ISA OR IPEA	500.00
962	CLAIMS MEET ART. 33(1)-(4) - IPEA - U.S.	50.00
964	CLAIMS - EXTRA INDEPENDENT (over three)	36.00
966	CLAIMS - EXTRA TOTAL (over twenty)	12.00
968	CLAIMS - MULTIPLE DEPENDENT	120.00
154	SURCHARGE - LATE FILING FEE OR OATH/DEC.	120.00
156	ENGLISH TRANSL. - AFTER TWENTY MONTHS	30.00

Group 19 - Small Entity Patent PCT Fees - National Stage

957	IPEA - U.S.	165.00
959	INTERNATIONAL SEARCHING AUTHORITY - U.S.	185.00
961	PTO NOT ISA OR IPEA	250.00
963	CLAIMS MEET ART. 33(1)-(4) - ipea - U.S.	25.00
965	CLAIMS - EXTRA INDEPENDENT (over three)	18.00
967	CLAIMS - EXTRA TOTAL (over twenty)	6.00
969	CLAIMS - MULTIPLE DEPENDENT	60.00
254	SURCHARGE - LATE FILING FEE OR OATH/DEC.	60.00

Group 20 - Patent PCT Fees to WIPO

800	BASIC FEE (first thirty pages)	485.00
801	BASIC SUPPLEMENTAL FEE (for each page over thirty)	10.00
803	HANDLING FEE	150.00
	DESIGNATION FEE PER COUNTRY	120.00

Group 21 - Patent PCT Fees to EPO

802	INTERNATIONAL SEARCH	1160.00

TRADEMARK FEES

Group 22 - Trademark Filing Fees

301	APPLICATION FOR REGISTRATION, PER CLASS	175.00

Group 23 - Trademark Post Registration Fees

Fig. 2 (continued)

302	APPLICATION FOR RENEWAL, PER CLASS	300.00
303	SPECIAL HANDLING FOR LATE RENEWAL	100.00
304	PUBLICATION OF MARK UNDER 12c, PER CLASS	100.00
309	FILING 8 AFFIDAVIT, PER CLASS	100.00
310	FILING 15 AFFIDAVIT, PER CLASS	100.00
311	FILING COMBINED 8 & 15 AFFIDAVIT, PER CLASS	200.00

Group 24 - Trademark Amended Registration Fees

305	ISSUING NEW CERTIFICATE OF REGISTRATION	100.00
306	CERT. OF CORRECTION, REGISTRANT'S ERROR	100.00
307	FILING DISCLAIMER TO REGISTRATION	100.00
308	FILING AMENDMENT TO REGISTRATION	100.00

Group 25 - Trademark Petition Fee

312	PETITION TO THE COMMISSIONER, PER CLASS	100.00

Group 26 - Trademark Trial and Appeal Board Fees

313	PETITION FOR CANCELLATION, PER CLASS	200.00
314	NOTICE OF OPPOSITION, PER CLASS	200.00
315	EX PARTE APPEAL, PER CLASS	100.00

Group 27 - Trademark Service Fees

401	PRINTED COPY OF EACH REGISTERED MARK	1.50
403	CERTIFY TM RECORDS, PER CERTIFICATE	3.50
404	PHOTOCOPIES OF TM RECORDS, PER PAGE	.30
405	RECORDING ASSIGNMENT, PER MARK, PER DOCUMENT	8.00
407	ABSTRACTS OF TITLE, PER REGISTRATION	12.00
408	COPY OF REG. MARK WITH TITLE OR STATUS	6.50
410	MAKE CERTIFICATION SPECIAL	25.00
409	UNSPECIFIED OTHER SERVICES	AT COST
424	COPISHARE CARD, PER PAGE	.15

Group 28 - [Reserved]

GENERAL FEES

Group 29 - Finance Service Fees

607	ESTABLISH DEPOSIT ACCOUNT	10.00
608	SERVICE CHARGE FOR BELOW MIN. BALANCE	20.00
617	PROCESSING RETURNED CHECKS	50.00

Group 30 - Computer Service Fees

618	COMPUTER RECORDS	AT COST

Fig. 2 (continued)

OMB No. 0651-0011 (12/31/86)

PATENT APPLICATION TRANSMITTAL LETTER	ATTORNEY'S DOCKET NO.

TO THE COMMISSIONER OF PATENTS AND TRADEMARKS:

Transmitted herewith for filing is the patent application of ______________________________

____

for ______________________________

____

Enclosed are:

☐ __________ sheets of drawing.

☐ an assignment of the invention to ______________________________

____

☐ a certified copy of a ______________________________ application.

☐ associate power of attorney.

☐ verified statement to establish small entity status under 37 CFR 1.9 and 1.27. ______

CLAIMS AS FILED

FOR	NO. FILED	NO. EXTRA		SMALL ENTITY RATE	SMALL ENTITY FEE		OTHER THAN A SMALL ENTITY RATE	OTHER THAN A SMALL ENTITY FEE
			OR					
BASIC FEE					$	OR		$
TOTAL CLAIMS	-20 -	*		x $6=	$	OR	x $12=	$
INDEP CLAIMS	-3 -	*		x $17=	$	OR	x $34=	$
MULTIPLE DEPENDENT CLAIM PRESENT				+ $55=	$	OR	+ $110=	$
				TOTAL	$	OR	TOTAL	$

* If the difference in col. 1 is less than zero, enter "0" in col. 2

☐ Please charge my Deposit Account No. __________ in the amount of $ __________

☐ A duplicate copy of this sheet is enclosed.

☐ A check in the amount of $ ________ to cover the filing fee is enclosed.

☐ The Commissioner is hereby authorized to charge payment of the following fees associated with this communication or credit any overpayment to Deposit Account No. __________. A Duplicate copy of this sheet is enclosed.

☐ Any additional filing fees required under 37 CFR 1.16.

☐ Any patent application processing fees under 37 CFR 1.17

☐ The Commissioner is hereby authorized to charge payment of the following fees during the pendency of this application or credit any overpayment to Deposit Account No. __________. A duplicate copy of this sheet is enclosed.

☐ Any filing fees under 37 CFR 1.16 for presentation of extra claims.

☐ Any patent application processing fees under 37 CFR 1.17.

☐ The issue fee set in 37 CFR 1.18 at or before mailing of the Notice of Allowance, pursuant to 37 CFR 1.311(b).

__________ date

______________________________ signature

Form PTO-FB-A510 (10-85)
(also form PTO-1082)

Patent and Trademark Office - U.S. DEPARTMENT of COMMERCE

Fig. 3

UNITED STATES DEPARTMENT OF COMMERCE
Patent and Trademark Office
Address: COMMISSIONER OF PATENTS AND TRADEMARKS
Washington, D.C. 20231

SERIAL NUMBER	FILING DATE	FIRST NAMED APPLICANT	ATTY. DOCKET NO.

DATE MAILED:

NOTICE OF INFORMAL APPLICATION

(Attachment to Office Action)

This application does not conform with the rules governing applications for the reason(s) checked below. The period within which to correct these requirements and avoid abandonment is set in the accompanying Office action.

A. A new oath or declaration, identifying this application by the serial number and filing date is required. The oath or declaration does not comply with 37 CFR 1.63 in that it:

1. ☐ was not executed in accordance with either 37 CFR 1.66 or 1.68.
2. ☐ does not identify the city and state or foreign country of residence of each inventor.
3. ☐ does not identify the citizenship of each inventor.
4. ☐ does not state whether the inventor is a sole or joint inventor.
5. ☐ does not state that the person making the oath or declaration:
   a. ☐ has reviewed and understands the contents of the specification, including the claims, as amended by any amendment specifically referred to in the oath or declaration.
   b. ☐ believes the named inventor or inventors to be the original and first inventor or inventors of the subject matter which is claimed and for which a patent is sought.
   c. ☐ acknowledges the duty to disclose information which is material to the examination of the application in accordance with 37 CFR 1.56(a).
6. ☐ does not identify the foreign application for patent or inventor's certificate on which priority is claimed pursuant to 37 CFR 1.55, and any foreign application having a filing date before that of the application on which priority is claimed, by specifying the application serial number, country, day, month, and year of its filing.
7. ☐ does not state that the person making the oath or declaration acknowledges the duty to disclose material information as defined in 37 CFR 1.56(a) which occurred between the filing date of the prior application and filing date of the continuation-in-part application which discloses and claims subject matter in addition to that disclosed in the prior application (37 CFR 1.63(d)).
8. ☐ does not include the date of execution.
9. ☐ does not use permanent ink, or its equivalent in quality, as required under 37 CFR 1.52(a) for the: ☐ signature ☐ oath/declaration.
10. ☐ contains non-initialed alterations (See 37 CFR 1.52(c) and 1.56).
11. ☐ does not contain the clause regarding "willful false statements..." as required by 37 CFR 1.68.
12. ☐ Other:

B. Applicant is required to provide:

1. ☐ A statement signed by applicant giving his or her complete name. A full name must include at least one given name without abbreviation as required by 37 CFR 1.41(a).
2. ☐ Proof of authority of the legal representative under 37 CFR 1.44.
3. ☐ An abstract in compliance with 37 CFR 1.72(b).
4. ☐ A statement signed by applicant giving his or her complete post office address (37 CFR 1.33(a)).
5. ☐ A copy of the specification written, typed, or printed in permanent ink, or its equivalent in quality as required by 37 CFR 1.52(a).
6. ☐ Other:

FORM PTO-152 (REV. 4-87)

Fig. 4

UNITED STATES DEPARTMENT OF COMMERCE
Patent and Trademark Office
Address: COMMISSIONER OF PATENTS AND TRADEMARKS
Washington, D.C. 20231

SERIAL NUMBER	FILING DATE	FIRST NAMED APPLICANT	ATTY. DOCKET NO.

DATE MAILED:

Notice of Incomplete Application

A filing date has NOT been assigned to the above identified application papers for the reason(s) shown below.

1. ☐ The specification (description and claims):
   a. ☐ is missing
   b. ☐ has pages________missing.
   c. ☐ does not include a written description of the invention.
   d. ☐ does not include at least one claim in compliance with 35 U.S.C. 112.

A complete specification in compliance with 35 U.S.C. 112 is required.

2. ☐ A drawing of Figure(s) ________ described in the specification is required in compliance with 35 U.S.C. 111.
3. ☐ A drawing of applicant's invention is required since it is necessary for the understanding of the subject matter of the invention in compliance with 35 U.S.C. 113.
4. ☐ The inventor's name(s) is missing. The full names of all inventors are required in compliance with 37 CFR 1.41.
5. ☐ Other items missing but not required for a filing date:

All of the above-noted omissions, unless otherwise indicated, must be submitted within TWO MONTHS of the date of this notice or the application will be returned or otherwise disposed of. Any fee which has been submitted will be refunded less a $50.00 handling fee. See 37 CFR 1.53(c).

The filing date will be the date of receipt of all the items required above, unless otherwise indicated. Any assertions that the items required above were submitted, or are not necessary for a filing date, must be by way of a petition directed to the attention of the Office of the Assistant Commissioner for Patents accompanied by the $140.00 petition fee (37 CFR 1.17(h)). If the petition alleges that no defect exists, a request for refund of the petition fee may be included in the petition.

Direct the response to, and questions about, this notice to the undersigned, Attention: Application Branch, and include the above Serial Number and Receipt Date.

Enclosed:

☐ "General Information Concerning Patents". See page ________.
☐ Copy of a patent to assist applicant in making corrections.
☐ "Notice to File Missing Parts of Application", Form PTO-1532.
☐ Other: ____________________

____
For: Manager, Application Branch
(703) 557-______

FORM PTO-1123 (REV. 4-87)

Fig. 5

UNITED STATES DEPARTMENT OF COMMERCE
Patent and Trademark Office
Address: COMMISSIONER OF PATENTS AND TRADEMARKS
Washington, D.C. 20231

SERIAL NUMBER	FILING DATE	FIRST NAMED APPLICANT	ATTY DOCKET NO

DATE MAILED:

NOTICE TO FILE MISSING PARTS OF APPLICATION— NO FILING DATE

(Attachment to Form PTO-1123)

In order to avoid payment by applicant of the surcharge required if items 1 and 3-6 are filed after the filing date the following items are also brought to applicant's attention at this time.

If all missing parts of this form and on the "Notice of Incomplete Application" are filed together, the total amount owed by applicant as a ☐ large entity ☐ small entity (verified statement filed) is $ ________.

1. ☐ The statutory basic filing fee is: ☐ missing ☐ insufficient. Applicant as a ☐ large entity ☐ small entity must submit $_____ to complete the basic filing fee and MUST ALSO SUBMIT THE SURCHARGE, IF REQUIRED, AS INDICATED BELOW.

2. ☐ Additional claim fees of $_______ as a ☐ large entity, ☐ small entity, including any required multiple dependent claim fee, are required. Applicant must submit the additional claim fees or cancel the additional claims for which fees are due. NO SURCHARGE IS REQUIRED FOR THIS ITEM.

3. ☐ The oath or declaration:
   ☐ is missing.
   ☐ does not cover items required on the "Notice of Incomplete Application".
   An oath or declaration in compliance with 37 CFR 1.63, referring to the above Serial Number and Receipt Date is required. A SURCHARGE, IF REQUIRED, MUST ALSO BE SUBMITTED AS INDICATED BELOW.

4. ☐ The oath or declaration does not identify the application to which it applies. An oath or declaration in compliance with 37 CFR 1.63, identifying the application by the above Serial Number and Receipt Date is required. A SURCHARGE, IF REQUIRED, MUST ALSO BE SUBMITTED AS INDICATED BELOW.

5. ☐ The signature to the oath or declaration is: ☐ missing; ☐ a reproduction; ☐ by a person other than the inventor or a person qualified under 37 CFR 1.42, 1.43, or 1.47. A properly signed oath or declaration in compliance with 37 CFR 1.63, referring to the above Serial Number and Recipt Date is required. A SURCHARGE, MUST ALSO BE SUBMITTED AS INDICATED BELOW.

6. ☐ The signature of the following joint inventor(s) is missing from the oath or declaration: ________________ Applicant(s) should provide, if possible, an oath or declaration signed by the omitted inventor(s), identifying this application by the above Serial Number and Receipt Date. A SURCHARGE, IF REQUIRED, MUST ALSO BE SUBMITTED AS INDICATED BELOW.

7. ☐ **A $20.00 processing fee is required for returned checks.** (37 CFR 1.21(m)).

8. ☐ **Other:**

Required items 1-7 above SHOULD be filed, if possible, with any items required on the "Notice of Incomplete Application" enclosed with this form. If concurrent filing of all required items is not possible, items 1-7 above must be filed no later than two months from the filing date of this application. The filing date will be the date of receipt of the items required on the "Notice of Incomplete Application." If items 1 and 3-6 above are submitted after the filing date. THE PAYMENT OF A SURCHARGE OF $110.00 for large entities, or $55.00 for small entities who have filed a verified statement claiming such status, is required. (37 CFR 1.16(e)).

Applicant must file all the required items 1-7 indicated above within two months from any filing date granted to avoid abandonment. Extensions of time may be obtained by filing a petition accompanied by the extension fee under the provisions of 37 CFR 1.136(a).

Direct the response to, and any questions about, this notice to the undersigned. Attention: Application Branch.

A copy of this notice MUST be returned with response.

____
For: Manager, Application Branch
(703) 557-3254

For Office Use Only	
☐ 102	☐ 202
☐ 103	☐ 203
☐ 104	☐ 204
☐ 105	☐ 205

FORM PTO 1532 (REV. 7-87)

OFFICE COPY

Fig. 6

UNITED STATES DEPARTMENT OF COMMERCE
Patent and Trademark Office
Address: COMMISSIONER OF PATENTS AND TRADEMARKS
Washington, D.C. 20231

SERIAL NUMBER	FILING DATE	FIRST NAMED APPLICANT	ATTY DOCKET NO

DATE MAILED:

NOTICE TO FILE MISSING PARTS OF APPLICATION— FILING DATE GRANTED

A filing date has been granted to this application. However, the following parts are missing.

If all missing parts are filed within the period set below, the total amount owed by applicant as a ☐ large entity, ☐ small entity (verified statement filed), is $ ______.

1. ☐ The statutory basic filing fee is: ☐ missing. ☐ insufficient. Applicant as a ☐ large entity, ☐ small entity, must submit $ ______ to complete the basic filing fee and MUST ALSO SUBMIT THE SURCHARGE AS INDICATED BELOW.
2. ☐ Additional claim fees of $ ______ as a ☐ large entity, ☐ small entity, including any required multiple dependent claim fee, are required. Applicant must submit the additional claim fees or cancel the additional claims for which fees are due. NO SURCHARGE IS REQUIRED FOR THIS ITEM.
3. ☐ The oath or declaration:
   ☐ is missing.
   ☐ does not cover items omitted at the time of execution.

   An oath or declaration in compliance with 37 CFR 1.63, identifying the application by the above Serial Number and Filing Date is required. A SURCHARGE MUST ALSO BE SUBMITTED AS INDICATED BELOW.
4. ☐ The oath or declaration does not identify the application to which it applies. An oath or declaration in compliance with 37 CFR 1.63 identifying the application by the above Serial Number and Filing Date is required. A SURCHARGE MUST ALSO BE SUBMITTED AS INDICATED BELOW.
5. ☐ The signature to the oath or declaration is: ☐ missing; ☐ a reproduction; ☐ by a person other than the inventor or a person qualified under 37 CFR 1.42, 1.43, or 1.47. A properly signed oath or declaration in compliance with 37 CFR 1.63, identifying the application by the above Serial Number and Filing Date is required. A SURCHARGE MUST ALSO BE SUBMITTED AS INDICATED BELOW.
6. ☐ The signature of the following joint inventor(s) is missing from the oath or declaration: ______________. Applicant(s) should provide, if possible an oath or declaration signed by the omitted inventor(s), identifying this application by the above Serial Number and Filing Date. A SURCHARGE MUST ALSO BE SUBMITTED AS INDICATED BELOW.
7. ☐ The application was filed in a language other than English. Applicant must file a verified English translation of the application and a fee of $26.00 under 37 CFR 1.17(k), unless this fee has already been paid NO SURCHARGE UNDER 37 CFR 1.16(e) IS REQUIRED FOR THIS ITEM.
8. ☐ A $20.00 processing fee is required for returned checks. (37 CFR 1.21(m)).
9. ☐ Your filing receipt was mailed in error because check was returned.
10. ☐ Other:

A Serial Number and Filing Date have been assigned to this application. However, to avoid abandonment under 37 CFR 1.53(d), the missing parts and fees identified above in items 1 and 3-6 must be timely provided ALONG WITH THE PAYMENT OF A SURCHARGE OF $110.00 for large entities or $55.00 for small entities who have filed a verified statement claiming such status. The surcharge is set forth in 37 CFR 1.16(e). Applicant is given ONE MONTH FROM THE DATE OF THIS LETTER, OR TWO MONTHS FROM THE FILING DATE of this application, WHICHEVER IS LATER, within which to file all missing parts and pay any fees. Extensions of time may be obtained by filing a petition accompanied by the extension fee under the provisions of 37 CFR 1.136(a).

Direct the response to, and any questions about, this notice to the undersigned, Attention: Application Branch.

A copy of this notice MUST be returned with response.

____
For: Manager, Application Branch
(703) 557-3254

For Office Use Only	
☐ 102	☐ 202
☐ 103	☐ 203
☐ 104	☐ 204
☐ 105	☐ 205

FORM PTO 1533 (REV. 7-85)

OFFICE COPY

Fig. 7

DECLARATION FOR PATENT APPLICATION Docket No. ____________

As a below named inventor, I hereby declare that:

My residence, post office address and citizenship are as stated below next to my name.

I believe I am the original, first and sole inventor (if only one name is listed below) or an original, first and joint inventor (if plural names are listed below) of the subject matter which is claimed and for which a patent is sought on the invention entitled ______________________________, the specification of which

(check one) ☐ is attached hereto.
☐ was filed on ______________________________ as
Application Serial No. ______________________________
and was amended on ______________________________ (if applicable).

I hereby state that I have reviewed and understand the contents of the above identified specification, including the claims, as amended by any amendment referred to above.

I acknowledge the duty to disclose information which is material to the examination of this application in accordance with Title 37, Code of Federal Regulations, §1.56(a).

I hereby claim foreign priority benefits under Title 35, United States Code, §119 of any foreign application(s) for patent or inventor's certificate listed below and have also identified below any foreign application for patent or inventor's certificate having a filing date before that of the application on which priority is claimed:

Prior Foreign Application(s)			Priority Claimed	
(Number)	(Country)	(Day/Month/Year Filed)	Yes	No
(Number)	(Country)	(Day/Month/Year Filed)	Yes	No
(Number)	(Country)	(Day/Month/Year Filed)	Yes	No

I hereby claim the benefit under Title 35, United States Code, §120 of any United States application(s) listed below and, insofar as the subject matter of each of the claims of this application is not disclosed in the prior United States application in the manner provided by the first paragraph of Title 35, United States Code, §112, I acknowledge the duty to disclose material information as defined in Title 37, Code of Federal Regulations, §1.56(a) which occurred between the filing date of the prior application and the national or PCT international filing date of this application:

(Application Serial No.)	(Filing Date)	(Status—patented, pending, abandoned)
(Application Serial No.)	(Filing Date)	(Status—patented, pending, abandoned)

I hereby appoint the following attorney(s) and/or agent(s) to prosecute this application and to transact all business in the Patent and Trademark Office connected therewith:

______________________________.

Address all telephone calls to ______________________________ at telephone no. ______________.
Address all correspondence to ______________________________

______________________________.

I hereby declare that all statements made herein of my own knowledge are true and that all statements made on information and belief are believed to be true; and further that these statements were made with the knowledge that willful false statements and the like so made are punishable by fine or imprisonment, or both, under Section 1001 of Title 18 of the United States Code and that such willful false statements may jeopardize the validity of the application or any patent issued thereon.

Full name of sole or first inventor ______________________________
Inventor's signature ______________________________ Date ______________
Residence ______________________________ Citizenship ______________
Post Office Address ______________________________

Full name of second joint inventor, if any ______________________________
Second Inventor's signature ______________________________ Date ______________
Residence ______________________________ Citizenship ______________
Post Office Address ______________________________

(Supply similar information and signature for third and subsequent joint inventors.)

Form PTO-FB-A110 (8-83)

Fig. 8

POWER OF ATTORNEY OR AUTHORIZATION OF AGENT, NOT ACCOMPANYING APPLICATION

TO THE COMMISSIONER OF PATENTS AND TRADEMARKS:

The undersigned having, on or about the .... day of ...., 19 ...., made application for letters patent for an improvement in ...., Serial Number.................................., hereby appoints .... of................................., State of ...., Registration No..................................., his attorney (or agent), to prosecute said application, and to transact all business in the Patent and Trademark Office connected therewith.

....
(Signature)

REVOCATION OF POWER OF ATTORNEY OR AUTHORIZATION OF AGENT

TO THE COMMISSIONER OF PATENTS AND TRADEMARKS:

The undersigned having, on or about the .... day of ...., 19 ...., appointed...................................., of...................................., State of ...., his attorney (or agent) to prosecute an application for letters patent which application was filed on or about the.. day of..., 19 ..., for an improvement in ...., Serial Number............................, hereby revokes the power of attorney (or authorization of agent) then given.

....
(Signature)

Fig. 9 (A,B)

Applicant or Patentee:______________________________ Attorney's
Serial or Patent No.:______________________________ Docket No.:______
Filed or Issued:____
For:____

VERIFIED STATEMENT (DECLARATION) CLAIMING SMALL ENTITY STATUS
(37 CFR 1.9(f) & 1.27(c)) - NONPROFIT ORGANIZATION

I hereby declare that I am an official empowered to act on behalf of the nonprofit organization identified below:

NAME OF ORGANIZATION ______________________

ADDRESS OF ORGANIZATION ______________________

____

TYPE OF ORGANIZATION

[ ] UNIVERSITY OR OTHER INSTITUTION OF HIGHER EDUCATION

[ ] TAX EXEMPT UNDER INTERNAL REVENUE SERVICE CODE (26 U.S.C. 501(a) and 501(c)(3)

[ ] NONPROFIT SCIENTIFIC OR EDUCATIONAL UNDER STATUTE OF STATE OF THE UNITED STATES OF AMERICA

(NAME OF STATE ______________________)

(CITATION OF STATUTE ______________________)

[ ] WOULD QUALIFY AS TAX EXEMPT UNDER INTERNAL REVENUE SERVICE CODE (26 U.S.C. 501(a) and 501(c) IF LOCATED IN THE UNITED STATES OF AMERICA

[ ] WOULD QUALIFY AS NONPROFIT SCIENTIFIC OR EDUCATIONAL UNDER STATUTE OF STATE OF THE UNITED STATES OF AMERICA IF LOCATED IN THE UNITED STATES OF AMERICA

(NAME OF STATE ______________________)

(CITATION OF STATUTE ______________________)

I hereby declare that the nonprofit organization identified above qualifies as a small business concern as defined in 37 CFR 1.9(e) for purposes of paying reduced fees under section 41(a) and (b) of Title 35, United States Code with regard to the invention entitled ______________________ by inventor(s)______ described in

[ ] the specification filed herewith

[ ] application serial no.___________, filed _______

[ ] patent no.______________________ , issued ______

I hereby declare that rights under contract or law have conveyed to and remain with the nonprofit organization with regard to the above identified invention.

If the rights held by nonprofit organization are not exclusive, each individual, concern or organization having rights to the invention is listed below* and no rights to the invention are held by any person, other than the inventor, who would not qualify as a small business concern under 37 CFR 1.9(d) or by any concern which would not qualify as a small business concern under 37 CFR 1.9(d) or a nonprofit organization under 37 CFR 1.9(e).

*NOTE: Separate verified statements are required from each named person, concern or organization having rights to the invention averring to their status as small entities. (37 CFR 1.27)

NAME ______________________
ADDRESS ______________________
[ ] INDIVIDUAL [ ] SMALL BUSINESS CONCERN [ ] NONPROFIT ORGANIZATION

NAME ______________________
ADDRESS ______________________
[ ] INDIVIDUAL [ ] SMALL BUSINESS CONCERN [ ] NONPROFIT ORGANIZATION

I acknowledge the duty To file, in this application or patent, notification of any change in status resulting in loss of entitlement to small entity status prior to paying, or at the time of paying, the earliest of the issue fee or any maintenance fee due after the date an which status as a small entity is no longer appropriate. (37 CFR 1.28(b))

I hereby declare that all statements made herein of my own knowledge are true and that all statements made on information and belief are believed to be true; and further that these statements were wade with the knowledge that willful false statements and the like so made are punishable by fine or imprisonment, or both, under section 1001 of Title 18 of the United States Code, and that such willful false statements may jeopardize the validity of the application, any patent issuing thereon, or any patent to which this verified statement is directed.

NAME OF PERSON SIGNING ______________________
TITLE IN ORGANIZATION ______________
ADDRESS OF PERSON SIGNING ______________________
____

SIGNATURE ______________ DATE ______________

VERIFIED STATEMENT (DECLARATION) BY A NON-INVENTOR
SUPPORTING A CLAIM BY ANOTHER FOR SMALL ENTITY STATUS

I hereby declare that I am making this verified statement to support a claim by ________________ for small entity status for purposes of paying reduced fees under section 41(a) and (b) of Title 35, United States Code, with regard to the invention entitled ________________________________ by inventor(s) __________________

[ ] the specification filed herewith
[ ] application serial number ___________, filed ________________
[ ] patent number _____________________, issued ____________ .

I hereby declare that I would qualify as an independent inventor as defined in 37 CFR 1.9(c) for purposes of paying fees under section 41(a) and (b) of Title 35, United States Code, if I had made the above identified invention.

I have not assigned, granted, conveyed or licensed and am under no obligation under contract or law to assign, grant, convey or license, any rights to the invention to any person who could not be classified as an independent inventor under 37 CFR 1.9(c) if that person had made the invention, or to any concern which would not qualify as a small business concern under 37 CFR 1.9(d) or a nonprofit organization under 37 CFR 1.9(a).

Each person, concern or organization to which I have assigned, granted, conveyed, or licensed or am under an obligation under contract or law to assign, grant, convey, or license any rights in the invention is listed below:

[ ] No such person, concern, or organization
[ ] Persons, concerns or organizations listed below*

* Note: Separate verified statements are required from each named person, concern or organization having rights to the invention averring to their status as small entities. (37 CFR 1.27)

NAME ________________________
ADDRESS ________________________
[ ] INDIVIDUAL [ ] SMALL BUSINESS CONCERN [ ] NONPROFIT ORGANIZATION

NAME ________________________
ADDRESS ________________________
[ ] INDIVIDUAL [ ] SMALL BUSINESS CONCERN [ ] NONPROFIT ORGANIZATION

NAME ________________________
ADDRESS ________________________
[ ] INDIVIDUAL [ ] SMALL BUSINESS CONCERN [ ] NONPROFIT ORGANIZATION

I acknowledge the duty To file, in this application or patent, notification of any change in status resulting in loss of entitlement to small entity status prior to paying, or at the time of paying, the earliest of the issue fee or any maintenance fee due after the date an which status as a small entity is no longer appropriate. (37 CFR 1.28(b))

I hereby declare that all statements mad
e herein of my own knowledge are true and that all statements made on information and belief are believed to be true; and further that these statements were wade with the knowledge that willful false statements and the like so made are punishable by fine or imprisonment, or both, under section 1001 of Title 18 of the United States Code, and that such willful false statements may jeopardize the validity of the application, any patent issuing thereon, or any patent to which this verified statement is directed.

NAME OF PERSON SIGNING ________________________
ADDRESS OF PERSON SIGNING ________________
SIGNATURE ____________________ DATE ______________

Fig. 11

Applicant or Patentee:______________________________ Attorney's
Serial or Patent No.:______________________________ Docket No.:_____ Filed or Is-
sued:____
For:____

VERIFIED STATEMENT (DECLARATION) CLAIMING SMALL ENTITY STATUS
(37 CFR 1.9(f) & 1.27(c)) - SMALL BUSINESS CONCERN

I hereby declare that I am

[ ] the owner of the small business concern identified below:
[ ] an official of the small business concern empowered to act on behalf of the concern identified below:
NAME OF CONCERN __________________________
ADDRESS OF CONCERN __________________________
____

I hereby declare that the above identified small business concern qualifies as a small business concern as defined in 13 CFR 121.12, and reproduced in 37 CFR 1.9(d), for purposes of paying reduced fees under section 41(a) and (b) of Title 35, United States Code, in that the number of employees of the concern, including those of its affiliates, does not exceed 500 persons. For purposes of this statement. (l) the number of employees of the business concern is the average over the previous fiscal year of the concern of the persons employed on a full-time, part-time or temporary basis during each of the pay periods of the fiscal year, and (2) concerns are affiliates of each other when either, directly or indirectly, one concern controls or has the power to control the other, or a third party or parties controls or has the power to control both.

I hereby declare that rights under contract or law have been conveyed to and remain with the small business concern identified above with regard to the invention, entitled __________________________
______________________________________by inventor(s)

described in

[ ] the specification filed herewith
[ ] application serial no.__________, filed ______
[ ] patent no._____________________, issued _____

If the rights held by the above identified small business concern are not exclusive, each individual, concern or organization having rights to the invention is listed below and no rights to the invention are held by any person, other than the inventor, who would not qualify as a small business concern under 37 CFR 1.9(d) or by any concern which would not qualify as a small business concern under 37 CFR 1.9(d) or a nonprofit organization under 37 CFR 1.9(e). NOTE: Separate verified statements are required from each named person, concern or organization having rights to the invention averring to their status as small entities. (37 CFR 1.27)

NAME __________________________
ADDRESS __________________________
[ ] INDIVIDUAL [ ] SMALL BUSINESS CONCERN [ ] NONPROFIT ORGANIZATION

NAME __________________________
ADDRESS __________________________
[ ] INDIVIDUAL [ ] SMALL BUSINESS CONCERN [ ] NONPROFIT ORGANIZATION

I acknowledge the duty To file, in this application or patent, notification of any change in status resulting in loss of entitlement to small entity status prior to paying, or at the time of paying, the earliest of the issue fee or any maintenance fee due after the date an which status as a small entity is no longer appropriate. (37 CFR 1.28(b))

I hereby declare that all statements made herein of my own knowledge are true and that all statements made on information and belief are believed to be true; and further that these statements were wade with the knowledge that willful false statements and the like so made are punishable by fine or imprisonment, or both, under section 1001 of Title 18 of the United States Code, and that such willful false statements may jeopardize the validity of the application, any patent issuing thereon, or any patent to which this verified statement is directed.

NAME OF PERSON SIGNING ______________________
TITLE OF PERSON OTHER THAN OWNER ____________
ADDRESS OF PERSON SIGNING __________________________
____

SIGNATURE ________________ DATE ____________________

Fig. 12

Applicant or Patentee:______________________________ Attorney's
Serial or Patent No.:______________________________ Docket No.:______________________
Filed or Issued:______________________________
For:______________________________

VERIFIED STATEMENT (DECLARATION) CLAIMING SMALL ENTITY STATUS
(37 CFR 1.9(f) & 1.27(c)) - INDEPENDENT INVENTOR

As a below named inventor,I hereby declare that I qualify as an independent inventor as defined in 37 CFR 1.9(c) for purposes of paying reduced fees under section 41(a) and (b) of Title 35, United States Code, to the Patent and Trademark Office with regard to the invention entitled ______________________
described in

[ ] the specification filed herewith
[ ] application serial number ___________, filed _____________
[ ] patent number ______________________, issued _____________ .

I have not assigned, granted, conveyed or licensed and am under no obligation under contract or law to assign, grant, convey or license, any rights to the invention to any person who could not be classified as an independent inventor under 37 CFR 1.9(c) if that person made the invention, or to any concern which would not qualify as a small business concern under 37 CFR 1.9(d) or a nonprofit organization under 37 CFR 1.9(a).

Each person, concern or organization to which I have assigned , granted, conveyed, or licensed or am under an obligation under contract or law to assign, grant, convey, or license any rights in the invention is listed below:

[ ] No such person, concern, or organization
[ ] Persons, concerns or organizations listed below*

* Note: Separate verified statements are required from each named person, concern or organization having rights to the invention averring to their status as small entities. (37 CFR 1.27)

NAME ________________________
ADDRESS ________________________
[ ] INDIVIDUAL [ ] SMALL BUSINESS CONCERN [ ] NONPROFIT ORGANIZATION

NAME ________________________
ADDRESS ________________________
[ ] INDIVIDUAL [ ] SMALL BUSINESS CONCERN [ ] NONPROFIT ORGANIZATION

NAME ________________________
ADDRESS ________________________
[ ] INDIVIDUAL [ ] SMALL BUSINESS CONCERN [ ] NONPROFIT ORGANIZATION

I acknowledge the duty To file, in this application or patent, notification of any change in status resulting in loss of entitlement to small entity status prior to paying, or at the time of paying, the earliest of the issue fee or any maintenance fee due after the date an which status as a small entity is no longer appropriate. (37 CFR 1.28(b))

I hereby declare that all statements made herein of my own knowledge are true and that all statements made on information and belief are believed to be true; and further that these statements were wade with the knowledge that willful false statements and the like so made are punishable by fine or imprisonment, or both, under section 1001 of Title 18 of the United States Code, and that such willful false statements may jeopardize the validity of the application, any patent issuing thereon, or any patent to which this verified statement is directed.

NAME OF INVENTOR	NAME OF INVENTOR	NAME OF INVENTOR
Signature of inventor	Signature of inventor	Signature of inventor
Date	Date	Date

Fig. 13

INFORMATION DISCLOSURE CITATION *(Use several sheets if necessary)*	ATTY. DOCKET NO.	SERIAL NO.
	APPLICANT	
	FILING DATE	GROUP

U.S. PATENT DOCUMENTS

*EXAMINER INITIAL		DOCUMENT NUMBER	DATE	NAME	CLASS	SUBCLASS	FILING DATE IF APPROPRIATE

FOREIGN PATENT DOCUMENTS

		DOCUMENT NUMBER	DATE	COUNTRY	CLASS	SUBCLASS	TRANSLATION	
							YES	NO

OTHER DOCUMENTS *(Including Author, Title, Date, Pertinent Pages, Etc.)*

EXAMINER	DATE CONSIDERED

*EXAMINER: Initial if reference considered, whether or not citation is in conformance with MPEP 609; Draw line through citation if not in conformance and not considered. Include copy of this form with next communication to applicant.

Form PTO-FB-A820
(also form PTO-1449)

Patent and Trademark Office - U.S. DEPARTMENT of COMMERCE

Fig. 14

FORM PTO-892 (REV 3-78) U.S. DEPARTMENT OF COMMERCE PATENT AND TRADEMARK OFFICE	SERIAL NO.	GROUP ART UNIT	ATTACHMENT TO PAPER NUMBER	
NOTICE OF REFERENCES CITED	APPLICANT(S)			

U.S. PATENT DOCUMENTS

*		DOCUMENT NO.	DATE	NAME	CLASS	SUB-CLASS	FILING DATE IF APPROPRIATE
	A						
	B						
	C						
	D						
	E						
	F						
	G						
	H						
	I						
	J						
	K						

FOREIGN PATENT DOCUMENTS

*		DOCUMENT NO.	DATE	COUNTRY	NAME	CLASS	SUB-CLASS	PERTINENT SHTS. DWG	PERTINENT PP. SPEC.
	L								
	M								
	N								
	O								
	P								
	Q								

OTHER REFERENCES (Including Author, Title, Date, Pertinent Pages, Etc.)

R	
S	
T	
U	

EXAMINER	DATE	

* A copy of this reference is not being furnished with this office action.
(See Manual of Patent Examining Procedure, section 707.05 (a).)

Fig. 15

UNITED STATES DEPARTMENT OF COMMERCE
Patent and Trademark Office
Address: COMMISSIONER OF PATENTS AND TRADEMARKS
Washington, D.C. 20231

☐ This application has been examined ☐ Responsive to communication filed on ________ ☐ This action is made final.

A shortened statutory period for response to this action is set to expire________ month(s), ________ days from the date of this letter.
Failure to respond within the period for response will cause the application to become abandoned. 35 U.S.C. 133

Part I THE FOLLOWING ATTACHMENT(S) ARE PART OF THIS ACTION:

1. ☐ Notice of References Cited by Examiner, PTO-892.
2. ☐ Notice re Patent Drawing, PTO-948.
3. ☐ Notice of Art Cited by Applicant, PTO-1449.
4. ☐ Notice of informal Patent Application, Form PTO-152.
5. ☐ Information on How to Effect Drawing Changes, PTO-1474.
6. ☐ ________

Part II SUMMARY OF ACTION

1. ☐ Claims ________ are pending in the application.

   Of the above, claims ________ are withdrawn from consideration.

2. ☐ Claims ________ have been cancelled.
3. ☐ Claims ________ are allowed.
4. ☐ Claims ________ are rejected.
5. ☐ Claims ________ are objected to.
6. ☐ Claims ________ are subject to restriction or election requirement.
7. ☐ This application has been filed with informal drawings under 37 C.F.R. 1.85 which are acceptable for examination purposes.
8. ☐ Formal drawings are required in response to this Office action.
9. ☐ The corrected or substitute drawings have been received on ________. Under 37 C.F.R. 1.84 these drawings are ☐ acceptable. ☐ not acceptable (see explanation or Notice re Patent Drawing, PTO-948).
10. ☐ The proposed additional or substitute sheet(s) of drawings, filed on ________ has (have) been ☐ approved by the examiner. ☐ disapproved by the examiner (see explanation).
11. ☐ The proposed drawing correction, filed on ________, has been ☐ approved. ☐ disapproved (see explanation).
12. ☐ Acknowledgment is made of the claim for priority under U.S.C. 119. The certified copy has ☐ been received ☐ not been received ☐ been filed in parent application, serial no. ________ ; filed on ________ .
13. ☐ Since this application appears to be in condition for allowance except for formal matters, prosecution as to the merits is closed in accordance with the practice under Ex parte Quayle, 1935 C.D. 11; 453 O.G. 213.
14. ☐ Other

EXAMINER'S ACTION

PTOL-326 (Rev. 6-88)

Fig. 16

AMENDMENT TRANSMITTAL LETTER			ATTORNEY'S DOCKET NO.
SERIAL NO.	FILING DATE	EXAMINER	GROUP ART UNIT
INVENTION			

TO THE COMMISSIONER OF PATENTS AND TRADEMARKS:

Transmitted herewith is an amendment in the above-identified application.

Small entity status of this application under 37 CFR 1.27 has been established by a verified statement previously submitted.

A verified statement to establish small entity status under 37 CFR 1.9 and 1.27 is enclosed.

No additional fee is required.

The fee has been calculated as shown below:

	(1) CLAIMS REMAINING AFTER AMENDMENT		(2) HIGHEST NO. PREVIOUSLY PAID FOR	(3) PRESENT EXTRA	SMALL ENTITY RATE	SMALL ENTITY ADDIT FEE		OTHER THAN A SMALL ENTITY RATE	OTHER THAN A SMALL ENTITY ADDIT FEE
TOTAL	*	MINUS	**	=	x $6=	$	OR	x $12=	$
INDEP	*	MINUS	***	=	x$17=	$		x $34=	$
FIRST PRESENTATION OF MULTIPLE DEP CLAIM					+$55=	$		+$110=	$
					TOTAL ADDIT. FEE	$	OR	TOTAL	$

* If the entry in Col. 1 is less than the entry in Col. 2, write "0" in Col. 3.

** If the "Highest No. Previously Paid For" IN THIS SPACE is less than 20, enter "20".

*** If the "Highest No. Previously Paid For" IN THIS SPACE is less than 3, enter "3".

The "Highest No. Previously Paid For" (Total or Indep.) is the highest number found in the appropriate box in Col. 1.

Please charge my Deposit Account No. ____________ in the amount of $ ________ . A duplicate copy of this sheet is enclosed.

A check in the amount of $________ to cover the filing fee is enclosed.

The Commissioner is hereby authorized to charge payment of the following fees associated with this communication or credit any overpayment to Deposit Account No. ______________ . A Duplicate copy of this sheet is enclosed.

Any additional filing fees required under 37 CFR 1.16.

Any patent application processing fees under 37 CFR 1.17

________________________ (date)

________________________ (signature)

Form PTO-FB-A520 (10-85)
(also form PTO-1083)

Patent and Trademark Office - U.S. DEPARTMENT of COMMERCE

Fig. 17

Channels of Ex Parte Review

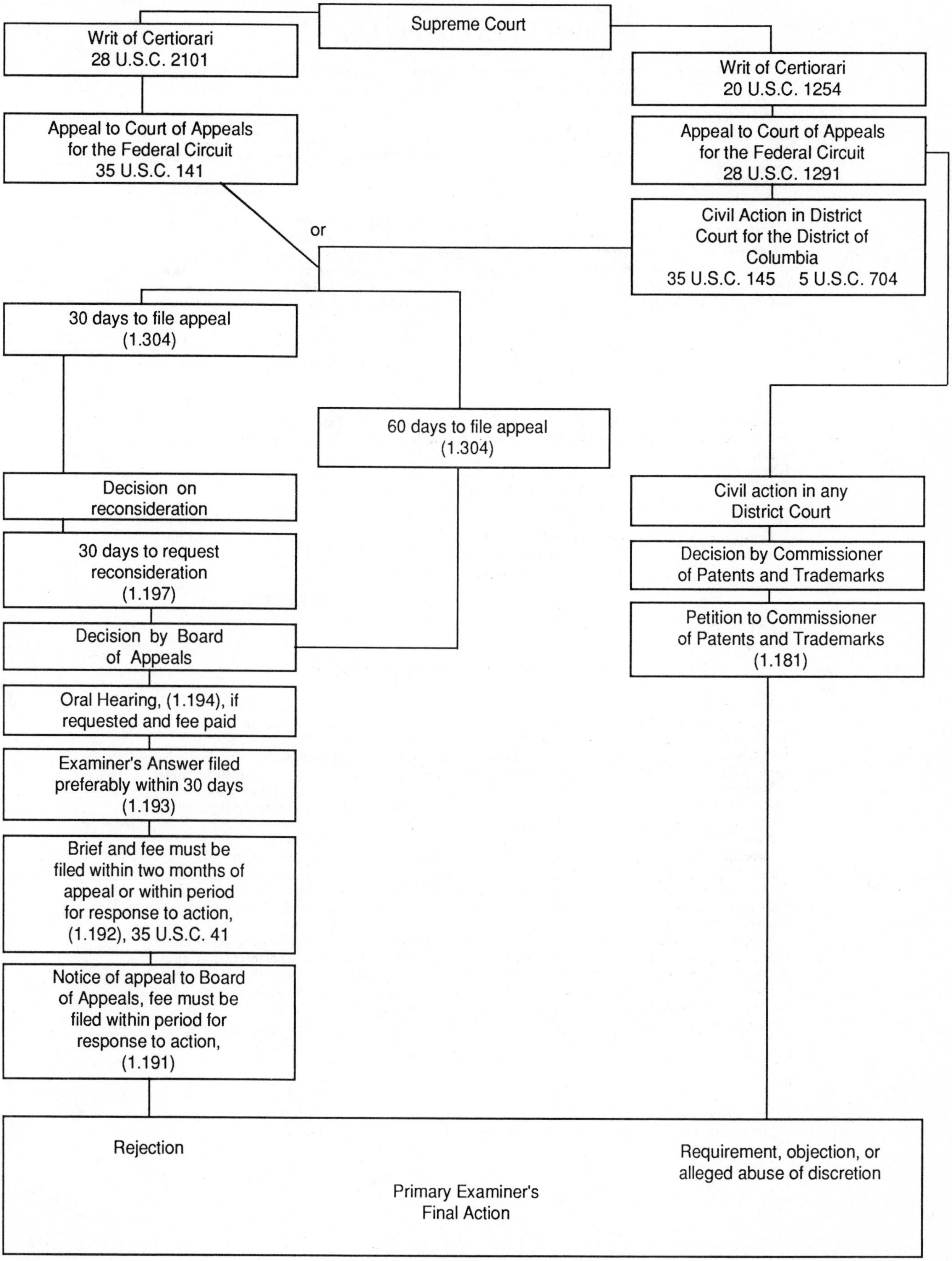

Fig. 18

ASSIGNMENT OF PATENT

(No special form is prescribed for assignments, which may contain various provisions depending upon the agreement of the parties. The following two formsare specimens of assignments which have been used in some cases.)

WHEREAS, I,.., of......................................., did obtain Letters Patent of the United States for an improvement in ....
No............................., dated............................; and whereas, I am now the sole owner of said patent; and,

WHEREAS,..., of..., whose post-office address is ...., City of.., and State of ...., is desirous of acquiring the entire interest in the same;
NOW, THEREFORE, in consideration of the sum of..................... dollars ($......................), the receipt of which is hereby acknowledged, and other good and valuable considerations, I, ..., by these presents do sell, assign, and transfer unto the said .... the entire right, title, and interest in and to the said Letters Patent aforesaid; the same to be held and enjoyed by the said ...., for his own used and hehoof, and for his legal representatives and assigns, to the full end of the term for which said Letters Patent are granted, as fully and entirely as the same would have been held by me had this assignment and sale not been made.

EXecuted, this .... day of ...., 19 ...., at ....
STATE...
County of ....
ss:

Before me personally appeared said .... and acknowledged the foregoing instruments to be his free act and deed this.......................... day of.........................,

....
(NOTARY PUBLIC)

[SEAL]

ASSIGNMENT OF APPLICATION

Whereas, I,..., of.. have invented certain new anduseful Improvements in ...., for which an application for United States Letters Patent was filed on ....,
Serial No................., [if the application has been prepared but not yet filed, state "for which an application for United States Letters Patent was executed on ....," instead] and

Wheareas,..., of ...., whose post-office address is ...., is desirous of acquiring the entire right, title and interest in the same;

Now, THEREFORE, in consideration of the sum of..................... dollars ($......................), the receipt whereof is hereby acknowledged, and other good and valuable consideration, I, the said ...., by these presents do sell, assign nd transfer unto said ...., the full and exclusive right to the said invention in the United States and the entire right, title, and interest in and to any and all Letters Patent which may be granted, therefore in the United States.

I hereby authorize and request the Commissioner of Patents and Trademarks to issue said Letters Patent to said ...., as the assignee of the entire right, title, and interest in and to the same, for his sole use and behoof; and for the use and behoof of his legal representatives, to the full end of the term for which said Letters Patent may be granted, as fully and entirely as the same would have been held by me had this assignment and sale not be made.
Excuted this .... day of.........................., 19 ...., at ....
State of ....
County of ....

ss:

Before me personally appeared said .... and acknowledged the foregoing instrument to be his free act an deed this......................... day of ...., 19.....

....
(Notary Public)

[SEAL]

Fig. 19 (A,B)

United States Patent

Des. 237,427
Patented Nov. 4, 1975

237,427

SHOE

William H. Thornberry, Newtown, Conn., assignor to Uniroyal, Inc.

Filed July 26, 1974, Ser. No. 492,307

Term of patent 14 years

Int. Cl. D2—*04*

U.S. Cl. D2—310

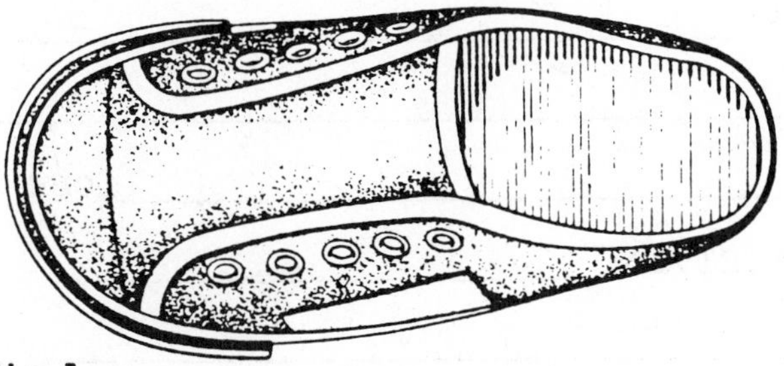

Fig.1.

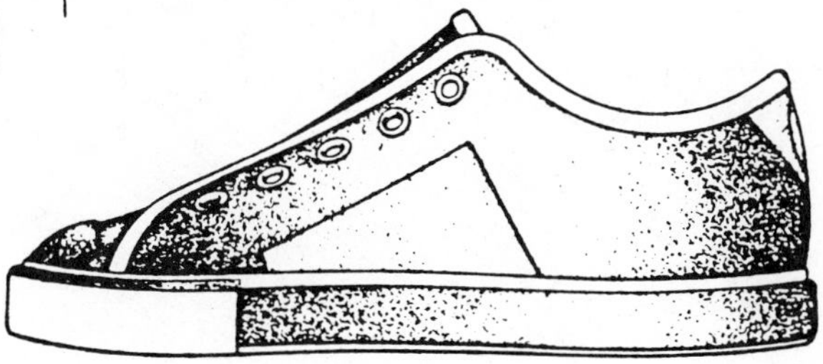

Fig.2.

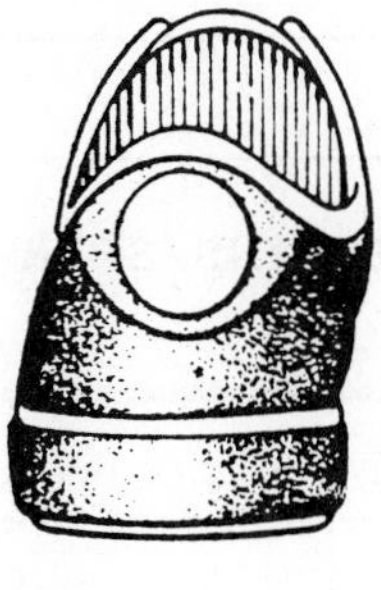

Fig.4.

Fig.3.

FIG. 1 is a plan view of a shoe embodying my new design;

FIG. 2 is a side elevational view of the FIG. 1 article;

FIG. 3 is a side elevational view of the FIG. 1 article; and

FIG. 4 is an end elevational view of the FIG. 1 article.

I claim:

The ornamental design for a shoe, substantially as shown and described.

References Cited

UNITED STATES PATENTS

D. 118,131	12/1939	Pick	D2—313
D. 173,699	12/1954	Hosker	D2—310
D. 226,461	3/1973	Nelson	D2—309.

LOIS S. LANIER, Primary Examiner

Fig. 20

Patent Worksheet

TRADE MARK (WORKING): ____________________

TAG LINE: ____________________

DESCRIPTION: ____________________

____

____

____

PATENT NOTES: ____________________

____

____________________ LODE # __________

WHO/DATE CONCEIVED: ____________________

WITNESSED: ____________________

SKETCH/PHOTO:

NOTES: ____________________

____

____

LICENSES/TIE-INS/SPIN-OFFS/ACCESSORIES/LINE CONCEPTS: ___

____

____

____

POTENTIAL MANUFACTURERS: ____________________

____

____

SEEN BY/DATE: __________ __________

__________ __________ __________

__________ __________ __________

Fig. 21

TRADEMARK EXAMINATION ACTIVITIES*

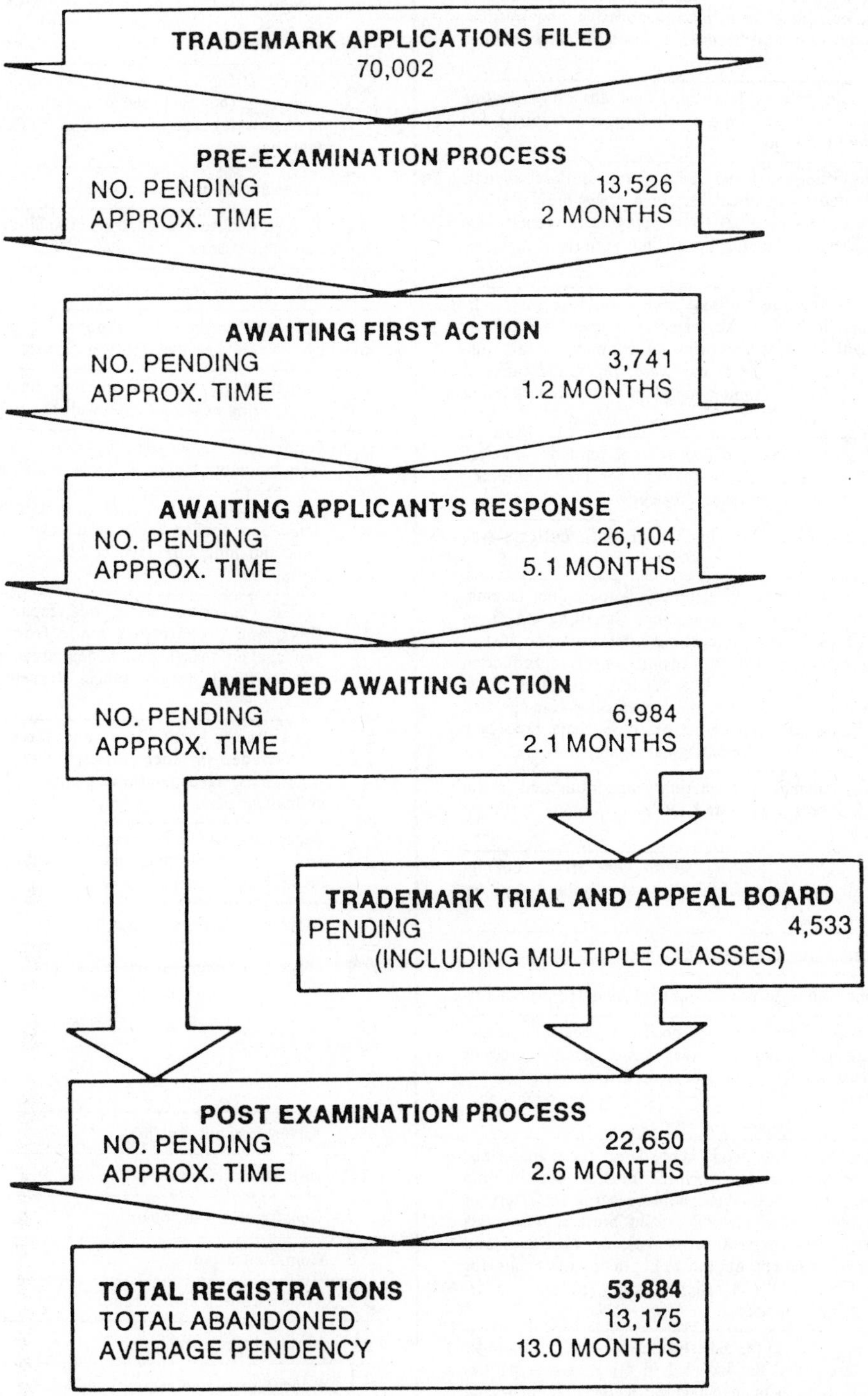

*This figure represents a simplified work-flow diagram with statistics on cases at various stages of processing at the end of FY 1987.

Fig. 22

International schedule of classes of goods and services

Goods

1	Chemicals used in industry, science, photography, as well as in agriculture, horticulture, and forestry; unprocessed artificial resins; unprocessed plastics; manures; fire extinguishing compositions; tempering and soldering preparations; chemical substances for preserving foodstuffs; tanning substances; adhesives used in industry.
2	Paints, varnishes, lacquers; preservatives against rust and against deterioration of wood; colourants; mordants; raw natural resins; metals in foil and powder form for painters, decorators, printers and artists.
3	Bleaching preparations and other substances for laundry use; cleaning, polishing, scouring and abrasive preparations; soaps; perfumery, essential oils, cosmetics, hair lotions; dentifrices.
4	Industrial oils and greases; lubricants; dust absorbing, wetting and binding compositions; fuels (including motor spirit) and illuminants; candles, wicks.
5	Pharmaceutical, veterinary, and sanitary preparations; dietetic substances adapted for medical use, food for babies; plasters, materials for dressings material for stopping teeth, dental wax, disinfectants; preparations for destroying vermin; fungicides, herbicides.
6	Common metals and their alloys; metal building materials; transportable buildings of metal; materials of metal for railway tracks; non-electric cables and wires of common metal; ironmongery, small items of metal hardware; pipes and tubes of metal; safes; goods of common metal not included in other classes; ores
7	Machines and machine tools; motors (except for land vehicles); machine coupling and belting (except for land vehicles); agricultural implements; incubators for eggs.
8	Hand tools and implements (hand operated); cutlery; side arms; razors.
9	Scientific, nautical, surveying, electric, photographic, cinematographic, optical, weighing, measuring, signalling, checking (supervision), life-saving and teaching apparatus and instruments; apparatus for recording transmission or reproduction of sound or images; magnetic data carriers, recording discs; automatic vending machines and mechanisms for coin-operated apparatus; cash registers, calculating machines, data processing equipment and computers; fire-extinguishing apparatus.
10	Surgical, medical, dental, and veterinary apparatus and instruments, artificial limbs, eyes and teeth; orthopedic articles; suture materials.
11	Apparatus for lighting, heating, steam generating, cooking, refrigerating, drying, ventilating, water supply, and sanitary purposes.
12	Vehicles; apparatus for locomotion by land, air or water.
13	Firearms; ammunition and projectiles; explosives; fireworks.
14	Precious metals and their alloys and goods in precious metals or coated therewith, not included in other classes; jewelry, precious stones; horological and other chronometric instruments.
15	Musical instruments.
16	Paper and cardboard and goods made from these materials, not included in other classes; printed matter; bookbinding material; photographs; stationery; adhesives for stationery or household purposes; artists' materials; paint brushes; typewriters and office requisites (except furniture); instructional and teaching material (except apparatus); plastic materials for packaging (not included on other classes); playing cards; printers' type; printing blocks.
17	Rubber, gutta-percha, gum, asbestos, mica and goods made from these materials and not included in other classes; plastics in extruded form for use in manufacture; packing, stopping and insulating materials; flexible pipes, not of metal.
18	Leather and imitations of leather, and goods made from these materials and not included in other classes; animal skins, hides; trunks and travelling bags; umbrellas, parasols and walking sticks; whips, harness and saddlery.
19	Building materials (non-metallic); non-metallic rigid pipes for building; asphalt, pitch and bitumen; non-metallic transportable buildings; monuments, not of metal.
20	Furniture, mirrors, picture frames; goods (not included in other classes) of wood, cork, reed, cane, wicker, horn, bone, ivory, whalebone, shell, amber, mother-of-pearl, meerschaum and substitutes for all these materials, or of plastics.
21	Household or kitchen utensils and containers (not of precious metal or coated therewith); combs and sponges; brushes (except paint brushes); brush-making materials; articles for cleaning purposes; steel wool; unworked or semi-worked glass (except glass used in building); glassware, porcelain and earthenware, not included in other classes.
22	Ropes, string, nets, tents, awnings, tarpaulins, sails, sacks; and bags (not included other classes); padding and stuffing materials (except of rubber or plastics); raw fibrous textile materials.
23	Yarns and threads, for textile use.
24	Textile and textile goods, not included in other classes; bed and table covers.
25	Clothing, footwear, headgear.
26	Lace and embroidery, ribbons and braid; buttons, hooks and eyes, pins and needles; artificial flowers.
27	Carpets, rugs, mats and matting; linoleum and other materials for covering existing floors; wall hangings (non-textile).
28	Games and playthings; gymnastic and sporting articles not included in other classes; decorations for Christmas trees.
29	Meats, fish, poultry and game; meat extracts; preserved, dried and cooked fruits and vegetables; jellies, jams; eggs, milk and milk products; edible oils and fats; salad dressings; preserves.
30	Coffee, tea, cocoa, sugar, rice, tapioca, sago, artificial coffee; flour, and preparations made from cereals, bread, pastry and confectionery, ices; honey, treacle; yeast, baking-powder; salt, mustard, vinegar, sauces, (except salad dressings) spices; ice.
31	Agricultural, horticultural and forestry products and grains not included in other classes; living animals; fresh fruits and vegetables; seeds, natural plants and flowers; foodstuffs for animals, malt.
32	Beers; mineral and aerated waters and other non-alcoholic drinks; fruit drinks and fruit juices; syrups and other preparations for making beverages.
33	Alcholic beverages (except beers).
34	Tobacco; smokers' articles; matches.

Services

35	Advertising and business.
36	Insurance and financial.
37	Construction and repair.
38	Communication.
39	Transportation and storage.
40	Material treatment.
41	Education and entertainment.
42	Miscellaneous.

Fig. 23

Filling Out Application Form TX

Detach and read these instructions before completing this form. Make sure all applicable spaces have been filled in before you return this form.

BASIC INFORMATION

When to Use This Form: Use Form TX for registration of published or unpublished non-dramatic literary works, excluding periodicals or serial issues. This class includes a wide variety of works: fiction, non-fiction, poetry, textbooks, reference works, directories, catalogs, advertising copy, compilations of information, and computer programs. For periodicals and serials, use Form SE.

Deposit to Accompany Application: An application for copyright registration must be accompanied by a deposit consisting of copies or phonorecords representing the entire work for which registration is to be made. The following are the general deposit requirements as set forth in the statute:

Unpublished Work: Deposit one complete copy (or phonorecord).

Published Work: Deposit two complete copies (or phonorecords) of the best edition.

Work First Published Outside the United States: Deposit one complete copy (or phonorecord) of the first foreign edition.

Contribution to a Collective Work: Deposit one complete copy (or phonorecord) of the best edition of the collective work.

The Copyright Notice: For published works, the law provides that a copyright notice in a specified form "shall be placed on all publicly distributed copies from which the work can be visually perceived." Use of the copyright notice is the responsibility of the copyright owner and does not require advance permission from the Copyright Office. The required form of the notice for copies generally consists of three elements: (1) the symbol "©", or the word "Copyright," or the abbreviation "Copr."; (2) the year of first publication; and (3) the name of the owner of copyright. For example: "© 1981 Constance Porter." The notice is to be affixed to the copies "in such manner and location as to give reasonable notice of the claim of copyright."

For further information about copyright registration, notice, or special questions relating to copyright problems, write:

Information and Publications Section, LM-455
Copyright Office
Library of Congress
Washington, D.C. 20559

PRIVACY ACT ADVISORY STATEMENT Required by the Privacy Act of 1974 (Public Law 93-579)

AUTHORITY FOR REQUESTING THIS INFORMATION:
- Title 17, U.S.C., Secs. 409 and 410

FURNISHING THE REQUESTED INFORMATION IS:
- Voluntary

BUT IF THE INFORMATION IS NOT FURNISHED:
- It may be necessary to delay or refuse registration
- You may not be entitled to certain relief, remedies, and benefits provided in chapters 4 and 5 of title 17, U.S.C.

PRINCIPAL USES OF REQUESTED INFORMATION:
- Establishment and maintenance of a public record
- Examination of the application for compliance with legal requirements

OTHER ROUTINE USES:
- Public inspection and copying
- Preparation of public indexes
- Preparation of public catalogs of copyright registrations
- Preparation of search reports upon request

NOTE:
- No other advisory statement will be given you in connection with this application
- Please keep this statement and refer to it if we communicate with you regarding this application

LINE-BY-LINE INSTRUCTIONS

1 SPACE 1: Title

Title of This Work: Every work submitted for copyright registration must be given a title to identify that particular work. If the copies or phonorecords of the work bear a title (or an identifying phrase that could serve as a title), transcribe that wording *completely* and *exactly* on the application. Indexing of the registration and future identification of the work will depend on the information you give here.

Previous or Alternative Titles: Complete this space if there are any additional titles for the work under which someone searching for the registration might be likely to look, or under which a document pertaining to the work might be recorded.

Publication as a Contribution: If the work being registered is a contribution to a periodical, serial, or collection, give the title of the contribution in the "Title of this Work" space. Then, in the line headed "Publication as a Contribution," give information about the collective work in which the contribution appeared.

2 SPACE 2: Author(s)

General Instructions: After reading these instructions, decide who are the "authors" of this work for copyright purposes. Then, unless the work is a "collective work," give the requested information about every "author" who contributed any appreciable amount of copyrightable matter to this version of the work. If you need further space, request additional Continuation sheets. In the case of a collective work, such as an anthology, collection of essays, or encyclopedia, give information about the author of the collective work as a whole.

Name of Author: The fullest form of the author's name should be given. Unless the work was "made for hire," the individual who actually created the work is its "author." In the case of a work made for hire, the statute provides that "the employer or other person for whom the work was prepared is considered the author."

What is a "Work Made for Hire"? A "work made for hire" is defined as: (1) "a work prepared by an employee within the scope of his or her employment"; or (2) "a work specially ordered or commissioned for use as a contribution to a collective work, as a part of a motion picture or other audiovisual work, as a translation, as a supplementary work, as a compilation, as an instructional text, as a test, as answer material for a test, or as an atlas, if the parties expressly agree in a written instrument signed by them that the work shall be considered a work made for hire." If you have checked "Yes" to indicate that the work was "made for hire," you must give the full legal name of the employer (or other person for whom the work was prepared). You may also include the name of the employee along with the name of the employer (for example: "Elster Publishing Co., employer for hire of John Ferguson").

"Anonymous" or "Pseudonymous" Work: An author's contribution to a work is "anonymous" if that author is not identified on the copies or phonorecords of the work. An author's contribution to a work is "pseudonymous" if that author is identified on the copies or phonorecords under a fictitious name. If the work is "anonymous" you may: (1) leave the line blank; or (2) state " anonymous" on the line; or (3) reveal the author's identity. If the work is "pseudonymous" you may : (1) leave the line blank; or (2) give the pseudonym and identify it as such (for example: "Huntley Haverstock, pseudonym"); or (3) reveal the author's name, making clear which is the real name and which is the pseudonym (for example: "Judith Barton, whose pseudonym is Madeline Elster"). However, the citizenship or domicile of the author **must** be given in all cases.

Dates of Birth and Death: If the author is dead, the statute requires that the year of death be included in the application unless the work is anonymous or pseudonymous. The author's birth date is optional, but is useful as a form of identification. Leave this space blank if the author's contribution was a "work made for hire."

Author's Nationality or Domicile: Give the country of which the author is a citizen, or the country in which the author is domiciled. Nationality or domicile **must** be given in all cases.

Nature of Authorship: After the words "Nature of Authorship" give a brief general statement of the nature of this particular author's contribution to the work. Examples: "Entire text"; "Coauthor of entire text"; "Chapters 11-14"; "Editorial revisions"; "Compilation and English translation"; "New text."

Fig. 24

3 SPACE 3: Creation and Publication

General Instructions: Do not confuse "creation" with "publication." Every application for copyright registration must state "the year in which creation of the work was completed." Give the date and nation of first publication only if the work has been published.

Creation: Under the statute, a work is "created" when it is fixed in a copy or phonorecord for the first time. Where a work has been prepared over a period of time, the part of the work existing in fixed form on a particular date constitutes the created work on that date. The date you give here should be the year in which the author completed the particular version for which registration is now being sought, even if other versions exist or if further changes or additions are planned.

Publication: The statute defines "publication" as "the distribution of copies or phonorecords of a work to the public by sale or other transfer of ownership, or by rental, lease, or lending"; a work is also "published" if there has been an "offering to distribute copies or phonorecords to a group of persons for purposes of further distribution, public performance, or public display." Give the full date (month, day, year) when, and the country where, publication first occurred. If first publication took place simultaneously in the United States and other countries, it is sufficient to state "U.S.A."

4 SPACE 4: Claimant(s)

Name(s) and Address(es) of Copyright Claimant(s): Give the name(s) and address(es) of the copyright claimant(s) in this work even if the claimant is the same as the author. Copyright in a work belongs initially to the author of the work (including, in the case of a work made for hire, the employer or other person for whom the work was prepared). The copyright claimant is either the author of the work or a person or organization to whom the copyright initially belonging to the author has been transferred.

Transfer: The statute provides that, if the copyright claimant is not the author, the application for registration must contain "a brief statement of how the claimant obtained ownership of the copyright." If any copyright claimant named in space 4 is not an author named in space 2, give a brief, general statement summarizing the means by which that claimant obtained ownership of the copyright. Examples: "By written contract"; "Transfer of all rights by author"; "Assignment"; "By will." Do not attach transfer documents or other attachments or riders.

5 SPACE 5: Previous Registration

General Instructions: The questions in space 5 are intended to find out whether an earlier registration has been made for this work and, if so, whether there is any basis for a new registration. As a general rule, only one basic copyright registration can be made for the same version of a particular work.

Same Version: If this version is substantially the same as the work covered by a previous registration, a second registration is not generally possible unless: (1) the work has been registered in unpublished form and a second registration is now being sought to cover this first published edition; or (2) someone other than the author is identified as copyright claimant in the earlier registration, and the author is now seeking registration in his or her own name. If either of these two exceptions apply, check the appropriate box and give the earlier registration number and date. Otherwise, do not submit Form TX; instead, write the Copyright Office for information about supplementary registration or recordation of transfers of copyright ownership.

Changed Version: If the work has been changed, and you are now seeking registration to cover the additions or revisions, check the last box in space 5, give the earlier registration number and date, and complete both parts of space 6 in accordance with the instructions below.

Previous Registration Number and Date: If more than one previous registration has been made for the work, give the number and date of the latest registration.

6 SPACE 6: Derivative Work or Compilation

General Instructions: Complete space 6 if this work is a "changed version," "compilation," or "derivative work," and if it incorporates one or more earlier works that have already been published or registered for copyright, or that have fallen into the public domain. A "compilation" is defined as "a work formed by the collection and assembling of preexisting materials or of data that are selected, coordinated, or arranged in such a way that the resulting work as a whole constitutes an original work of authorship." A "derivative work" is "a work based on one or more preexisting works." Examples of derivative works include translations, fictionalizations, abridgments, condensations, or "any other form in which a work may be recast, transformed, or adapted." Derivative works also include works "consisting of editorial revisions, annotations, or other modifications" if these changes, as a whole, represent an original work of authorship.

Preexisting Material (space 6a): For derivative works, complete this space **and** space 6b. In space 6a identify the preexisting work that has been recast, transformed, or adapted. An example of preexisting material might be: "Russian version of Goncharov's 'Oblomov'." Do not complete space 6a for compilations.

Material Added to This Work (space 6b): Give a brief, general statement of the new material covered by the copyright claim for which registration is sought. **Derivative work** examples include: "Foreword, editing, critical annotations"; "Translation"; "Chapters 11-17." If the work is a **compilation**, describe both the compilation itself and the material that has been compiled. Example: "Compilation of certain 1917 Speeches by Woodrow Wilson." A work may be both a derivative work and compilation, in which case a sample statement might be: "Compilation and additional new material."

7 SPACE 7: Manufacturing Provisions

Due to the expiration of the Manufacturing Clause of the copyright law on June 30, 1986, this space has been deleted.

8 SPACE 8: Reproduction for Use of Blind or Physically Handicapped Individuals

General Instructions: One of the major programs of the Library of Congress is to provide Braille editions and special recordings of works for the exclusive use of the blind and physically handicapped. In an effort to simplify and speed up the copyright licensing procedures that are a necessary part of this program, section 710 of the copyright statute provides for the establishment of a voluntary licensing system to be tied in with copyright registration. Copyright Office regulations provide that you may grant a license for such reproduction and distribution solely for the use of persons who are certified by competent authority as unable to read normal printed material as a result of physical limitations. The license is entirely voluntary, nonexclusive, and may be terminated upon 90 days notice.

How to Grant the License: If you wish to grant it, check one of the three boxes in space 8. Your check in one of these boxes, together with your signature in space 10, will mean that the Library of Congress can proceed to reproduce and distribute under the license without further paperwork. For further information, write for Circular R63.

9,10,11 SPACE 9, 10, 11: Fee, Correspondence, Certification, Return Address

Deposit Account: If you maintain a Deposit Account in the Copyright Office, identify it in space 9. Otherwise leave the space blank and send the fee of $10 with your application and deposit.

Correspondence (space 9): This space should contain the name, address, area code, and telephone number of the person to be consulted if correspondence about this application becomes necessary.

Certification (space 10): The application can not be accepted unless it bears the date and the **handwritten signature** of the author or other copyright claimant, or of the owner of exclusive right(s), or of the duly authorized agent of author, claimant, or owner of exclusive right(s).

Address for Return of Certificate (space 11): The address box must be completed legibly since the certificate will be returned in a window envelope.

Fig. 24 (continued)

FORM TX
UNITED STATES COPYRIGHT OFFICE

REGISTRATION NUMBER

TX TXU

EFFECTIVE DATE OF REGISTRATION

Month Day Year

DO NOT WRITE ABOVE THIS LINE. IF YOU NEED MORE SPACE, USE A SEPARATE CONTINUATION SHEET.

1

TITLE OF THIS WORK ▼

PREVIOUS OR ALTERNATIVE TITLES ▼

PUBLICATION AS A CONTRIBUTION If this work was published as a contribution to a periodical, serial, or collection, give information about the collective work in which the contribution appeared. **Title of Collective Work ▼**

If published in a periodical or serial give: **Volume ▼** **Number ▼** **Issue Date ▼** **On Pages ▼**

2

a **NAME OF AUTHOR ▼** **DATES OF BIRTH AND DEATH** Year Born ▼ Year Died ▼

Was this contribution to the work a "work made for hire"? ☐ Yes ☐ No

AUTHOR'S NATIONALITY OR DOMICILE Name of Country OR { Citizen of ▶ ______ Domiciled in ▶ ______

WAS THIS AUTHOR'S CONTRIBUTION TO THE WORK Anonymous? ☐ Yes ☐ No Pseudonymous? ☐ Yes ☐ No If the answer to either of these questions is "Yes," see detailed instructions.

NATURE OF AUTHORSHIP Briefly describe nature of the material created by this author in which copyright is claimed. ▼

b **NAME OF AUTHOR ▼** **DATES OF BIRTH AND DEATH** Year Born ▼ Year Died ▼

Was this contribution to the work a "work made for hire"? ☐ Yes ☐ No

AUTHOR'S NATIONALITY OR DOMICILE Name of country OR { Citizen of ▶ ______ Domiciled in ▶ ______

WAS THIS AUTHOR'S CONTRIBUTION TO THE WORK Anonymous? ☐ Yes ☐ No Pseudonymous? ☐ Yes ☐ No If the answer to either of these questions is "Yes," see detailed instructions.

NATURE OF AUTHORSHIP Briefly describe nature of the material created by this author in which copyright is claimed. ▼

c **NAME OF AUTHOR ▼** **DATES OF BIRTH AND DEATH** Year Born ▼ Year Died ▼

Was this contribution to the work a "work made for hire"? ☐ Yes ☐ No

AUTHOR'S NATIONALITY OR DOMICILE Name of Country OR { Citizen of ▶ ______ Domiciled in ▶ ______

WAS THIS AUTHOR'S CONTRIBUTION TO THE WORK Anonymous? ☐ Yes ☐ No Pseudonymous? ☐ Yes ☐ No If the answer to either of these questions is "Yes," see detailed instructions.

NATURE OF AUTHORSHIP Briefly describe nature of the material created by this author in which copyright is claimed. ▼

NOTE

Under the law, the "author" of a "work made for hire" is generally the employer, not the employee (see instructions). For any part of this work that was "made for hire" check "Yes" in the space provided, give the employer (or other person for whom the work was prepared) as "Author" of that part, and leave the space for dates of birth and death blank.

3

YEAR IN WHICH CREATION OF THIS WORK WAS COMPLETED **This information must be given in all cases.** ______ ◀ Year

DATE AND NATION OF FIRST PUBLICATION OF THIS PARTICULAR WORK **Complete this information ONLY if this work has been published.** Month ▶ ______ Day ▶ ______ Year ▶ ______ ◀ Nation

4

COPYRIGHT CLAIMANT(S) Name and address must be given even if the claimant is the same as the author given in space 2.▼

See instructions before completing this space.

TRANSFER If the claimant(s) named here in space 4 are different from the author(s) named in space 2, give a brief statement of how the claimant(s) obtained ownership of the copyright.▼

APPLICATION RECEIVED

ONE DEPOSIT RECEIVED

TWO DEPOSITS RECEIVED

REMITTANCE NUMBER AND DATE

DO NOT WRITE HERE OFFICE USE ONLY

MORE ON BACK ▶ • Complete all applicable spaces (numbers 5-11) on the reverse side of this page. • See detailed instructions. • Sign the form at line 10.

DO NOT WRITE HERE

Page 1 of ______ pages

Fig. 24 (continued)

EXAMINED BY

FORM TX

CHECKED BY

☐ CORRESPONDENCE Yes

☐ DEPOSIT ACCOUNT FUNDS USED

FOR COPYRIGHT OFFICE USE ONLY

DO NOT WRITE ABOVE THIS LINE. IF YOU NEED MORE SPACE, USE A SEPARATE CONTINUATION SHEET.

5

PREVIOUS REGISTRATION Has registration for this work, or for an earlier version of this work, already been made in the Copyright Office?
☐ **Yes** ☐ **No** If your answer is "Yes," why is another registration being sought? (Check appropriate box) ▼

☐ This is the first published edition of a work previously registered in unpublished form.

☐ This is the first application submitted by this author as copyright claimant.

☐ This is a changed version of the work, as shown by space 6 on this application.

If your answer is "Yes," give: **Previous Registration Number ▼** **Year of Registration ▼**

6

DERIVATIVE WORK OR COMPILATION Complete both space 6a & 6b for a derivative work; complete only 6b for a compilation.

a. Preexisting Material Identify any preexisting work or works that this work is based on or incorporates. ▼

b. Material Added to This Work Give a brief, general statement of the material that has been added to this work and in which copyright is claimed. ▼

See instructions before completing this space.

7

—space deleted—

8

REPRODUCTION FOR USE OF BLIND OR PHYSICALLY HANDICAPPED INDIVIDUALS A signature on this form at space 10, and a check in one of the boxes here in space 8, constitutes a non-exclusive grant of permission to the Library of Congress to reproduce and distribute solely for the blind and physically handicapped and under the conditions and limitations prescribed by the regulations of the Copyright Office: (1) copies of the work identified in space 1 of this application in Braille (or similar tactile symbols); or (2) phonorecords embodying a fixation of a reading of that work; or (3) both.

a ☐ Copies and Phonorecords **b** ☐ Copies Only **c** ☐ Phonorecords Only

See instructions.

9

DEPOSIT ACCOUNT If the registration fee is to be charged to a Deposit Account established in the Copyright Office, give name and number of Account.
Name ▼ **Account Number ▼**

CORRESPONDENCE Give name and address to which correspondence about this application should be sent. Name/Address/Apt/City/State/Zip ▼

Area Code & Telephone Number ▶

Be sure to give your daytime phone ◀ number.

10

CERTIFICATION* I, the undersigned, hereby certify that I am the
Check one ▶
☐ author
☐ other copyright claimant
☐ owner of exclusive right(s)
☐ authorized agent of ______
Name of author or other copyright claimant, or owner of exclusive right(s) ▲

of the work identified in this application and that the statements made by me in this application are correct to the best of my knowledge.

Typed or printed name and date ▼ If this is a published work, this date must be the same as or later than the date of publication given in space 3.

date ▶

☞ **Handwritten signature (X) ▼**

11

MAIL CERTIFICATE TO

Certificate will be mailed in window envelope

Name ▼

Number/Street/Apartment Number ▼

City/State/ZIP ▼

YOU MUST:
- Complete all necessary spaces
- Sign your application in space 10

SEND ALL 3 ELEMENTS IN THE SAME PACKAGE:
1. Application form
2. Non-refundable $10 filing fee in check or money order payable to *Register of Copyrights*
3. Deposit material

MAIL TO:
Register of Copyrights
Library of Congress
Washington, D.C. 20559

* 17 U.S.C. § 506(e): Any person who knowingly makes a false representation of a material fact in the application for copyright registration provided for by section 409, or in any written statement filed in connection with the application, shall be fined not more than $2,500.

U.S. GOVERNMENT PRINTING OFFICE 1988—241-428 80,009

February 1989—65,000

Fig. 24 (continued)

Filling Out Application Form VA

Detach and read these instructions before completing this form. Make sure all applicable spaces have been filled in before you return this form.

BASIC INFORMATION

When to Use This Form: Use Form VA for copyright registration of published or unpublished works of the visual arts. This category consists of "pictorial, graphic, or sculptural works," including two-dimensional and three-dimensional works of fine, graphic, and applied art, photographs, prints and art reproductions, maps, globes, charts, technical drawings, diagrams, and models.

What Does Copyright Protect? Copyright in a work of the visual arts protects those pictorial, graphic, or sculptural works, either alone or in combination, represent an "original work of authorship." The statute declares: "In no case does copyright protection for an original work of authorship extend to any idea, procedure, process, system, method of operation, concept, principle, or discovery, regardless of the form in which it is described, explained, illustrated, or embodied in such work."

Works of Artistic Craftsmanship and Designs: "Works of artistic craftsmanship" are registrable on Form VA, but the statute makes clear that protection extends to "their form" and not to "their mechanical or utilitarian aspects." The "design of a useful article" is considered copyrightable "only if, and only to the extent that, such design incorporates pictorial, graphic, or sculptural features that can be identified separately from, and are capable of existing independently of, the utilitarian aspects of the article."

Labels and Advertisements: Works prepared for use in connection with the sale or advertisement of goods and services are registrable if they contain "original work of authorship." Use Form VA if the copyrightable material in the work you are registering is mainly pictorial or graphic; use Form TX if it consists mainly of text. **NOTE:** Words and short phrases such as names, titles, and slogans cannot be protected by copyright, and the same is true of standard symbols, emblems, and other commonly used graphic designs that are in the public domain. When used commercially, material of that sort can sometimes be protected under state laws of unfair competition or under the Federal trademark laws. For information about trademark registration, write to the Commissioner of Patents and Trademarks, Washington, D.C. 20231.

Deposit to Accompany Application: An application for copyright registration must be accompanied by a deposit consisting of copies representing the entire work for which registration is to be made.

Unpublished Work: Deposit one complete copy.

Published Work: Deposit two complete copies of the best edition.

Work First Published Outside the United States: Deposit one complete copy of the first foreign edition.

Contribution to a Collective Work: Deposit one complete copy of the best edition of the collective work.

The Copyright Notice: For published works, the law provides that a copyright notice in a specified form "shall be placed on all publicly distributed copies from which the work can be visually perceived." Use of the copyright notice is the responsibility of the copyright owner and does not require advance permission from the Copyright Office. The required form of the notice for copies generally consists of three elements: (1) the symbol "©", or the word "Copyright," or the abbreviation "Copr."; (2) the year of first publication; and (3) the name of the owner of copyright. For example: "© 1981 Constance Porter." The notice is to be affixed to the copies "in such manner and location as to give reasonable notice of the claim of copyright."

For further information about copyright registration, notice, or special questions relating to copyright problems, write:
Information and Publications Section, LM-455
Copyright Office, Library of Congress, Washington, D.C. 20559

PRIVACY ACT ADVISORY STATEMENT Required by the Privacy Act of 1974 (P.L. 93-579)
The authority for requesting this information is title 17, U.S.C., secs. 409 and 410. Furnishing the requested information is voluntary. But if the information is not furnished, it may be necessary to delay or refuse registration and you may not be entitled to certain relief, remedies, and benefits provided in chapters 4 and 5 of title 17, U.S.C.
The principal uses of the requested information are the establishment and maintenance of a public record and the examination of the application for compliance with legal requirements.
Other routine uses include public inspection and copying, preparation of public indexes, preparation of public catalogs of copyright registrations, and preparation of search reports upon request.
NOTE: No other advisory statement will be given in connection with this application. Please keep this statement and refer to it if we communicate with you regarding this application.

LINE-BY-LINE INSTRUCTIONS

1 SPACE 1: Title

Title of This Work: Every work submitted for copyright registration must be given a title to identify that particular work. If the copies of the work bear a title (or an identifying phrase that could serve as a title), transcribe that wording *completely* and *exactly* on the application. Indexing of the registration and future identification of the work will depend on the information you give here.

Previous or Alternative Titles: Complete this space if there are any additional titles for the work under which someone searching for the registration might be likely to look, or under which a document pertaining to the work might be recorded.

Publication as a Contribution: If the work being registered is a contribution to a perodical, serial, or collection, give the title of the contribution in the "Title of This Work"space. Then, in the line headed "Publication as a Contribution," give information about the collective work in which the contribution appeared.

Nature of This Work: Briefly describe the general nature or character of the pictorial, graphic, or sculptural work being registered for copyright. Examples: "Oil Painting"; "Charcoal Drawing"; "Etching"; "Sculpture"; "Map"; "Photograph"; "Scale Model"; "Lithographic Print"; "Jewelry Design"; "Fabric Design."

2 SPACE 2: Author(s)

General Instructions: After reading these instructions, decide who are the "authors" of this work for copyright purposes. Then, unless the work is a "collective work," give the requested information about every "author" who contributed any appreciable amount of copyrightable matter to this version of the work. If you need further space, request additional Continuation Sheets. In the case of a collective work, such as a catalog of paintings or collection of cartoons by various authors, give information about the author of the collective work as a whole.

Name of Author: The fullest form of the author's name should be given. Unless the work was "made for hire," the individual who actually created the work is its "author." In the case of a work made for hire, the statute provides that "the employer or other person for whom the work was prepared is considered the author."

What is a "Work Made for Hire"? A "work made for hire" is defined as: (1) "a work prepared by an employee within the scope of his or her employment"; or (2) "a work specially ordered or commissioned for use as a contribution to a collective work, as a part of a motion picture or other audiovisual work, as a translation, as a supplementary work, as a compilation, as an instructional text, as a test, as answer material for a test, or as an atlas, if the parties expressly agree in a written instrument signed by them that the work shall be considered a work made for hire." If you have checked "Yes" to indicate that the work was "made for hire," you must give the full legal name of the employer (or other person for whom the work was prepared). You may also include the name of the employee along with the name of the employer (for example: "Elster Publishing Co., employer for hire of John Ferguson").

"Anonymous" or "Pseudonymous" Work: An author's contribution to a work is "anonymous" if that author is not identified on the copies or phonorecords of the work. An author's contribution to a work is "pseudonymous" if that author is identified on the copies or phonorecords under a fictitious name. If the work is "anonymous" you may: (1) leave the line blank; or (2) state "anonymous" on the line; or (3) reveal the author's identity. If the work is "pseudonymous" you may: (1) leave the line blank; or (2) give the pseudonym and identify it as such (for example: "Huntley Haverstock, pseudonym"); or (3) reveal the author's name, making clear which is the real name and which is the pseudonym (for example: "Henry Leek, whose pseudonym is Priam Farrel"). However, the citizenship or domicile of the author **must** be given in all cases.

Dates of Birth and Death: If the author is dead, the statute requires that the year of death be included in the application unless the work is anonymous or pseudonymous. The author's birth date is optional, but is useful as a form of identification. Leave this space blank if the author's contribution was a "work made for hire."

Fig. 25

Author's Nationality or Domicile: Give the country of which the author is a citizen, or the country in which the author is domiciled. Nationality or domicile **must** be given in all cases.

Nature of Authorship: Give a brief general statement of the nature of this particular author's contribution to the work. Examples: "Painting"; "Photograph"; "Silk Screen Reproduction"; "Co-author of Cartographic Material"; "Technical Drawing"; "Text and Artwork."

3 SPACE 3: Creation and Publication

General Instructions: Do not confuse "creation" with "publication." Every application for copyright registration must state "the year in which creation of the work was completed." Give the date and nation of first publication only if the work has been published.

Creation: Under the statute, a work is "created" when it is fixed in a copy or phonorecord for the first time. Where a work has been prepared over a period of time, the part of the work existing in fixed form on a particular date constitutes the created work on that date. The date you give here should be the year in which the author completed the particular version for which registration is now being sought, even if other versions exist or if further changes or additions are planned.

Publication: The statute defines "publication" as "the distribution of copies or phonorecords of a work to the public by sale or other transfer of ownership, or by rental, lease, or lending"; a work is also "published" if there has been an "offering to distribute copies or phonorecords to a group of persons for purposes of further distribution, public performance, or public display." Give the full date (month, day, year) when, and the country where, publication first occurred. If first publication took place simultaneously in the United States and other countries, it is sufficient to state "U.S.A."

4 SPACE 4: Claimant(s)

Name(s) and Address(es) of Copyright Claimant(s): Give the name(s) and address(es) of the copyright claimant(s) in this work even if the claimant is the same as the author. Copyright in a work belongs initially to the author of the work (including, in the case of a work made for hire, the employer or other person for whom the work was prepared). The copyright claimant is either the author of the work or a person or organization to whom the copyright initially belonging to the author has been transferred.

Transfer: The statute provides that, if the copyright claimant is not the author, the application for registration must contain "a brief statement of how the claimant obtained ownership of the copyright." If any copyright claimant named in space 4 is not an author named in space 2, give a brief, general statement summarizing the means by which that claimant obtained ownership of the copyright. Examples: "By written contract"; "Transfer of all rights by author"; "Assignment"; "By will." Do not attach transfer documents or other attachments or riders.

5 SPACE 5: Previous Registration

General Instructions: The questions in space 5 are intended to find out whether an earlier registration has been made for this work and, if so, whether there is any basis for a new registration. As a rule, only one basic copyright registration can be made for the same version of a particular work.

Same Version: If this version is substantially the same as the work covered by a previous registration, a second registration is not generally possible unless: (1) the work has been registered in unpublished form and a second registration is now being sought to cover this first published edition; or (2) someone other than the author is identified as copyright claimant in the earlier registration, and the author is now seeking registration in his or her own name. If either of these two exceptions apply, check the appropriate box and give the earlier registration number and date. Otherwise, do not submit Form VA; instead, write the Copyright Office for information about supplementary registration or recordation of transfers of copyright ownership.

Changed Version: If the work has been changed, and you are now seeking registration to cover the additions or revisions, check the last box in space 5, give the earlier registration number and date, and complete both parts of space 6 in accordance with the instructions below.

Previous Registration Number and Date: If more than one previous registration has been made for the work, give the number and date of the latest registration.

6 SPACE 6: Derivative Work or Compilation

General Instructions: Complete space 6 if this work is a "changed version," "compilation," or "derivative work," and if it incorporates one or more earlier works that have already been published or registered for copyright, or that have fallen into the public domain. A "compilation" is defined as "a work formed by the collection and assembling of preexisting materials or of data that are selected, coordinated, or arranged in such a way that the resulting work as a whole constitutes an original work of authorship." A "derivative work" is "a work based on one or more preexisting works." Examples of derivative works include reproductions of works of art, sculptures based on drawings, lithographs based on paintings, maps based on previously published sources, or "any other form in which a work may be recast, transformed, or adapted." Derivative works also include works "consisting of editorial revisions, annotations, or other modifications" if these changes, as a whole, represent an original work of authorship.

Preexisting Material (space 6a): Complete this space **and** space 6b for derivative works. In this space identify the preexisting work that has been recast, transformed, or adapted. Examples of preexisting material might be "Grunewald Altarpiece"; or "19th century quilt design." Do not complete this space for compilations.

Material Added to This Work (space 6b): Give a brief, general statement of the **additional** new material covered by the copyright claim for which registration is sought. In the case of a derivative work, identify this new material. Examples: "Adaptation of design and additional artistic work"; "Reproduction of painting by photolithography"; "Additional cartographic material"; "Compilation of photographs." If the work is a compilation, give a brief, general statement describing both the material that has been compiled **and** the compilation itself. Example: "Compilation of 19th Century Political Cartoons."

7,8,9 SPACE 7, 8, 9: Fee, Correspondence, Certification, Return Address

Deposit Account: If you maintain a Deposit Account in the Copyright Office, identify it in space 7. Otherwise leave the space blank and send the fee of $10 with your application and deposit.

Correspondence (space 7): This space should contain the name, address, area code, and telephone number of the person to be consulted if correspondence about this application becomes necessary.

Certification (space 8): The application cannot be accepted unless it bears the date and the **handwritten signature** of the author or other copyright claimant, or of the owner of exclusive right(s), or of the duly authorized agent of the author, claimant, or owner of exclusive right(s).

Address for Return of Certificate (space 9): The address box must be completed legibly since the certificate will be returned in a window envelope.

MORE INFORMATION

Form of Deposit for Works of the Visual Arts

Exceptions to General Deposit Requirements: As explained on the reverse side of this page, the statutory deposit requirements (generally one copy for unpublished works and two copies for published works) will vary for particular kinds of works of the visual arts. The copyright law authorizes the Register of Copyrights to issue regulations specifying "the administrative classes into which works are to be placed for purposes of deposit and registration, and the nature of the copies or phonorecords to be deposited in the various classes specified." For particular classes, the regulations may require or permit "the deposit of identifying material instead of copies or phonorecords," or "the deposit of only one copy or phonorecord where two would normally be required."

What Should You Deposit? The detailed requirements with respect to the kind of deposit to accompany an application on Form VA are contained in the Copyright Office Regulations. The following does not cover all of the deposit requirements, but is intended to give you some general guidance.

For an Unpublished Work, the material deposited should represent the entire copyrightable content of the work for which registration is being sought.

For a Published Work, the material deposited should generally consist of two complete copies of the best edition. Exceptions: (1) For certain types of works, one complete copy may be deposited instead of two. These include greeting cards, postcards, stationery, labels, advertisements, scientific drawings, and globes; (2) For most three-dimensional sculptural works, and for certain two-dimensional works, the Copyright Office Regulations require deposit of identifying material (photographs or drawings in a specified form) rather than copies; and (3) Under certain circumstances, for works published in five copies or less or in limited, numbered editions, the deposit may consist of one copy or of identifying reproductions.

Fig. 25 (continued)

FORM VA

UNITED STATES COPYRIGHT OFFICE

REGISTRATION NUMBER

VA VAU

EFFECTIVE DATE OF REGISTRATION

Month Day Year

DO NOT WRITE ABOVE THIS LINE. IF YOU NEED MORE SPACE, USE A SEPARATE CONTINUATION SHEET.

1

TITLE OF THIS WORK ▼

NATURE OF THIS WORK ▼ See instructions

PREVIOUS OR ALTERNATIVE TITLES ▼

PUBLICATION AS A CONTRIBUTION If this work was published as a contribution to a periodical, serial, or collection, give information about the collective work in which the contribution appeared. **Title of Collective Work ▼**

If published in a periodical or serial give: **Volume ▼** **Number ▼** **Issue Date ▼** **On Pages ▼**

2

NOTE

Under the law, the "author" of a "work made for hire" is generally the employer, not the employee (see instructions). For any part of this work that was "made for hire" check "Yes" in the space provided, give the employer (or other person for whom the work was prepared) as "Author" of that part, and leave the space for dates of birth and death blank.

a **NAME OF AUTHOR ▼**

DATES OF BIRTH AND DEATH
Year Born ▼ Year Died ▼

Was this contribution to the work a "work made for hire"?
☐ Yes
☐ No

AUTHOR'S NATIONALITY OR DOMICILE
Name of Country
OR { Citizen of ▶ ______
Domiciled in ▶ ______

WAS THIS AUTHOR'S CONTRIBUTION TO THE WORK
Anonymous? ☐ Yes ☐ No
Pseudonymous? ☐ Yes ☐ No
If the answer to either of these questions is "Yes," see detailed instructions.

NATURE OF AUTHORSHIP Briefly describe nature of the material created by this author in which copyright is claimed. ▼

b **NAME OF AUTHOR ▼**

DATES OF BIRTH AND DEATH
Year Born ▼ Year Died ▼

Was this contribution to the work a "work made for hire"?
☐ Yes
☐ No

AUTHOR'S NATIONALITY OR DOMICILE
Name of country
OR { Citizen of ▶ ______
Domiciled in ▶ ______

WAS THIS AUTHOR'S CONTRIBUTION TO THE WORK
Anonymous? ☐ Yes ☐ No
Pseudonymous? ☐ Yes ☐ No
If the answer to either of these questions is "Yes," see detailed instructions.

NATURE OF AUTHORSHIP Briefly describe nature of the material created by this author in which copyright is claimed. ▼

c **NAME OF AUTHOR ▼**

DATES OF BIRTH AND DEATH
Year Born ▼ Year Died ▼

Was this contribution to the work a "work made for hire"?
☐ Yes
☐ No

AUTHOR'S NATIONALITY OR DOMICILE
Name of Country
OR { Citizen of ▶ ______
Domiciled in ▶ ______

WAS THIS AUTHOR'S CONTRIBUTION TO THE WORK
Anonymous? ☐ Yes ☐ No
Pseudonymous? ☐ Yes ☐ No
If the answer to either of these questions is "Yes," see detailed instructions.

NATURE OF AUTHORSHIP Briefly describe nature of the material created by this author in which copyright is claimed. ▼

3

YEAR IN WHICH CREATION OF THIS WORK WAS COMPLETED **This information must be given in all cases.**
______ ◀ Year

DATE AND NATION OF FIRST PUBLICATION OF THIS PARTICULAR WORK
Complete this information ONLY if this work has been published. Month ▶ ______ Day ▶ ______ Year ▶ ______
______ ◀ Nation

4

COPYRIGHT CLAIMANT(S) Name and address must be given even if the claimant is the same as the author given in space 2.▼

See instructions before completing this space.

TRANSFER If the claimant(s) named here in space 4 are different from the author(s) named in space 2, give a brief statement of how the claimant(s) obtained ownership of the copyright.▼

DO NOT WRITE HERE
OFFICE USE ONLY

APPLICATION RECEIVED

ONE DEPOSIT RECEIVED

TWO DEPOSITS RECEIVED

REMITTANCE NUMBER AND DATE

MORE ON BACK ▶
- Complete all applicable spaces (numbers 5-9) on the reverse side of this page.
- See detailed instructions.
- Sign the form at line 8.

DO NOT WRITE HERE

Page 1 of ______ pages

Fig. 25 (continued)

EXAMINED BY

FORM VA

CHECKED BY

☐ CORRESPONDENCE Yes

☐ DEPOSIT ACCOUNT FUNDS USED

FOR COPYRIGHT OFFICE USE ONLY

DO NOT WRITE ABOVE THIS LINE. IF YOU NEED MORE SPACE, USE A SEPARATE CONTINUATION SHEET.

5

PREVIOUS REGISTRATION Has registration for this work, or for an earlier version of this work, already been made in the Copyright Office?

☐ **Yes** ☐ **No** If your answer is "Yes," why is another registration being sought? (Check appropriate box) ▼

☐ This is the first published edition of a work previously registered in unpublished form.

☐ This is the first application submitted by this author as copyright claimant.

☐ This is a changed version of the work, as shown by space 6 on this application.

If your answer is "Yes," give: **Previous Registration Number** ▼ **Year of Registration** ▼

6

DERIVATIVE WORK OR COMPILATION Complete both space 6a & 6b for a derivative work; complete only 6b for a compilation.

a. **Preexisting Material** Identify any preexisting work or works that this work is based on or incorporates. ▼

See instructions before completing this space.

b. **Material Added to This Work** Give a brief, general statement of the material that has been added to this work and in which copyright is claimed. ▼

7

DEPOSIT ACCOUNT If the registration fee is to be charged to a Deposit Account established in the Copyright Office, give name and number of Account.

Name ▼ **Account Number** ▼

CORRESPONDENCE Give name and address to which correspondence about this application should be sent. Name/Address/Apt/City/State/Zip ▼

Area Code & Telephone Number ▶

Be sure to give your daytime phone ◀ number.

8

CERTIFICATION* I, the undersigned, hereby certify that I am the

Check only one ▼

☐ author

☐ other copyright claimant

☐ owner of exclusive right(s)

☐ authorized agent of ______

Name of author or other copyright claimant, or owner of exclusive right(s) ▲

of the work identified in this application and that the statements made by me in this application are correct to the best of my knowledge.

Typed or printed name and date ▼ If this is a published work, this date must be the same as or later than the date of publication given in space 3.

date ▶

Handwritten signature (X) ▼

9

MAIL CERTIFICATE TO

Certificate will be mailed in window envelope

Name ▼

Number/Street/Apartment Number ▼

City/State/ZIP ▼

Have you:

- Completed all necessary spaces?
- Signed your application in space 8?
- Enclosed check or money order for $10 payable to *Register of Copyrights?*
- Enclosed your deposit material with the application and fee?

MAIL TO: Register of Copyrights, Library of Congress, Washington, D.C. 20559.

* 17 U.S.C. § 506(e): Any person who knowingly makes a false representation of a material fact in the application for copyright registration provided for by section 409, or in any written statement filed in connection with the application, shall be fined not more than $2,500.

U.S. GOVERNMENT PRINTING OFFICE: 1988—202-133 80,008

Fig. 25 (continued)

August 1988—60,000

FORM RE

UNITED STATES COPYRIGHT OFFICE
LIBRARY OF CONGRESS
WASHINGTON, D.C. 20559

APPLICATION FOR *Renewal Registration*

HOW TO REGISTER A RENEWAL CLAIM:

- ***First:*** Study the information on this page and make sure you know the answers to two questions:

  (1) What are the renewal time limits in your case?

  (2) Who can claim the renewal?

- ***Second:*** Turn this page over and read through the specific instructions for filling out Form RE. Make sure, before starting to complete the form, that the copyright is now eligible for renewal, that you are authorized to file a renewal claim, and that you have all of the information about the copyright you will need.

- ***Third:*** Complete all applicable spaces on Form RE, following the line-by-line instructions on the back of this page. Use typewriter, or print the information in dark ink.

- ***Fourth:*** Detach this sheet and send your completed Form RE to: Register of Copyrights, Library of Congress, Washington, D.C. 20559. Unless you have a Deposit Account in the Copyright Office, your application must be accompanied by a check or money order for $6, payable to: *Register of Copyrights*. Do not send copies, phonorecords, or supporting documents with your renewal application.

WHAT IS RENEWAL OF COPYRIGHT? For works originally copyrighted between January 1, 1950 and December 31, 1977, the statute now in effect provides for a first term of copyright protection lasting for 28 years, with the possibility of renewal for a second term of 47 years. If a valid renewal registration is made for a work, its total copyright term is 75 years (a first term of 28 years, plus a renewal term of 47 years). Example: For a work copyrighted in 1960, the first term will expire in 1988, but if renewed at the proper time the copyright will last through the end of 2035.

SOME BASIC POINTS ABOUT RENEWAL:

(1) There are strict time limits and deadlines for renewing a copyright.

(2) Only certain persons who fall into specific categories named in the law can claim renewal.

(3) The new copyright law does away with renewal requirements for works first copyrighted after 1977. However, copyrights that were already in their first copyright term on January 1, 1978 (that is, works originally copyrighted between January 1, 1950 and December 31, 1977) **still have to be renewed** in order to be protected for a second term.

TIME LIMITS FOR RENEWAL REGISTRATION: The new copyright statute provides that, in order to renew a copyright, the renewal application and fee must be received in the Copyright Office "within one year prior to the expiration of the copyright." It also provides that all terms of copyright will run through the end of the year in which they would otherwise expire. Since all copyright terms will expire on December 31st of their last year, all periods for renewal registration will run from December 31st of the 27th year of the copyright, and will end on December 31st of the following year.

To determine the time limits for renewal in your case:

(1) First, find out the date of original copyright for the work. (In the case of works originally registered in unpublished form, the date of copyright is the date of registration; for published works, copyright begins on the date of first publication.)

(2) Then add 28 years to the year the work was originally copyrighted.

Your answer will be the calendar year during which the copyright will be eligible for renewal, and December 31st of that year will be the renewal deadline. Example: a work originally copyrighted on April 19, 1957, will be eligible for renewal between December 31, 1984, and December 31, 1985.

WHO MAY CLAIM RENEWAL: Renewal copyright may be claimed only by those persons specified in the law. Except in the case of four specific types of works, the law gives the right to claim renewal to the individual author of the work, regardless of who owned the copyright during the original term. If the author is dead, the statute gives the right to claim renewal to certain of the author's beneficiaries (widow and children, executors, or next of kin, depending on the circumstances). The present owner (proprietor) of the copyright is entitled to claim renewal only in four specified cases, as explained in more detail on the reverse of this page.

CAUTION: Renewal registration is possible only if an acceptable application and fee are **received** in the Copyright Office during the renewal period and before the renewal deadline. If an acceptable application and fee are not received before the renewal deadline, the work falls into the public domain and the copyright cannot be renewed. The Copyright Office has no discretion to extend the renewal time limits.

PRIVACY ACT ADVISORY STATEMENT
Required by the Privacy Act of 1974 (Public Law 93-579)

AUTHORITY FOR REQUESTING THIS INFORMATION:
- Title 17, U.S.C., Sec. 304

FURNISHING THE REQUESTED INFORMATION IS:
- Voluntary

BUT IF THE INFORMATION IS NOT FURNISHED:
- It may be necessary to delay or refuse renewal registration
- If renewal registration is not made, the copyright will expire at the end of its 28th year

PRINCIPAL USES OF REQUESTED INFORMATION:
- Establishment and maintenance of a public record
- Examination of the application for compliance with legal requirements

OTHER ROUTINE USES:
- Public inspection and copying
- Preparation of public indexes
- Preparation of public catalogs of copyright registrations
- Preparation of search reports upon request

NOTE:
- No other advisory statement will be given you in connection with this application
- Please retain this statement and refer to it if we communicate with you regarding this application

Fig. 26

INSTRUCTIONS FOR COMPLETING FORM RE

SPACE 1: RENEWAL CLAIM(S)

• ***General Instructions:*** In order for this application to result in a valid renewal, space 1 must identify one or more of the persons who are entitled to renew the copyright under the statute. Give the full name and address of each claimant, with a statement of the basis of each claim, using the wording given in these instructions.

• ***Persons Entitled to Renew:***

A. The following persons may claim renewal in all types of works except those enumerated in Paragraph B, below:

1. The author, if living. State the claim as: *the author.*

2. The widow, widower, and/or children of the author, if the author is not living. State the claim as: *the widow (widower) of the author* . . . . (Name of author) and/or *the child (children) of the deceased author* . . . . (Name of author)

3. The author's executor(s), if the author left a will and if there is no surviving widow, widower, or child. State the claim as: *the executor(s) of the author* . . . . (Name of author)

4. The next of kin of the author, if the author left no will and if there is no surviving widow, widower, or child. State the claim as: *the next of kin of the deceased author* . . . . *there being no will.* (Name of author)

B. In the case of the following four types of works, the proprietor (owner of the copyright at the time of renewal registration) may claim renewal:

1. Posthumous work (a work as to which no copyright assignment or other contract for exploitation has occurred during the author's lifetime). State the claim as: *proprietor of copyright in a posthumous work.*

2. Periodical, cyclopedic, or other composite work. State the claim as: *proprietor of copyright in a composite work.*

3. "Work copyrighted by a corporate body otherwise than as assignee or licensee of the individual author." State the claim as: *proprietor of copyright in a work copyrighted by a corporate body otherwise than as assignee or licensee of the individual author.* (This type of claim is considered appropriate in relatively few cases.)

4. Work copyrighted by an employer for whom such work was made for hire. State the claim as: *proprietor of copyright in a work made for hire.*

SPACE 2: WORK RENEWED

• ***General Instructions:*** This space is to identify the particular work being renewed. The information given here should agree with that appearing in the certificate of original registration.

• ***Title:*** Give the full title of the work, together with any subtitles or descriptive wording included with the title in the original registration. In the case of a musical composition, give the specific instrumentation of the work.

• ***Renewable Matter:*** Copyright in a new version of a previous work (such as an arrangement, translation, dramatization, compilation, or work republished with new matter) covers only the additions, changes, or other new material appearing for the first time in that version. If this work was a new version, state in general the new matter upon which copyright was claimed.

• ***Contribution to Periodical, Serial, or other Composite Work:*** Separate renewal registration is possible for a work published as a contribution to a periodical, serial, or other composite work, whether the contribution was copyrighted independently or as part of the larger work in which it appeared. Each contribution published in a separate issue ordinarily requires a separate renewal registration. However, the new law provides an alternative, permitting groups of periodical contributions by the same individual author to be combined under a single renewal application and fee in certain cases.

If this renewal application covers a single contribution, give all of the requested information in space 2. If you are seeking to renew a group of contributions, include a reference such as "See space 5" in space 2 and give the requested information about all of the contributions in space 5.

SPACE 3: AUTHOR(S)

• ***General Instructions:*** The copyright secured in a new version of a work is independent of any copyright protection in material published earlier. The only "authors" of a new version are those who contributed copyrightable matter to it. Thus, for renewal purposes, the person who wrote the original version on which the new work is based cannot be regarded as an "author" of the new version, unless that person also contributed to the new matter.

• ***Authors of Renewable Matter:*** Give the full names of all authors who contributed copyrightable matter to this particular version of the work.

SPACE 4: FACTS OF ORIGINAL REGISTRATION

• ***General Instructions:*** Each item in space 4 should agree with the information appearing in the original registration for the work. If the work being renewed is a single contribution to a periodical or composite work that was not separately registered, give information about the particular issue in which the contribution appeared. You may leave this space blank if you are completing space 5.

• ***Original Registration Number:*** Give the full registration number, which is a series of numerical digits, preceded by one or more letters. The registration number appears in the upper right hand corner of the certificate of registration.

• ***Original Copyright Claimant:*** Give the name in which ownership of the copyright was claimed in the original registration.

• ***Date of Publication or Registration:*** Give only one date. If the original registration gave a publication date, it should be transcribed here; otherwise the registration was for an unpublished work, and the date of registration should be given.

SPACE 5: GROUP RENEWALS

• ***General Instructions:*** A single renewal registration can be made for a group of works if **all** of the following statutory conditions are met: (1) all of the works were written by the same author, who is named in space 3 and who is or was an individual (not an employer for hire); (2) all of the works were first published as contributions to periodicals (including newspapers) and were copyrighted on their first publication; (3) the renewal claimant or claimants, and the basis of claim or claims, as stated in space 1, is the same for all of the works; (4) the renewal application and fee are "received not more than 28 or less than 27 years after the 31st day of December of the calendar year in which all of the works were first published"; and (5) the renewal application identifies each work separately, including the periodical containing it and the date of first publication.

Time Limits for Group Renewals: To be renewed as a group, all of the contributions must have been first published during the same calendar year. For example, suppose six contributions by the same author were published on April 1, 1960, July 1, 1960, November 1, 1960, February 1, 1961, July 1, 1961, and March 1, 1962. The three 1960 copyrights can be combined and renewed at any time during 1988, and the two 1961 copyrights can be renewed as a group during 1989, but the 1962 copyright must be renewed by itself, in 1990.

Identification of Each Work: Give all of the requested information for each contribution. The registration number should be that for the contribution itself if it was separately registered, and the registration number for the periodical issue if it was not.

SPACES 6, 7 AND 8: FEE, MAILING INSTRUCTIONS, AND CERTIFICATION

• ***Deposit Account and Mailing Instructions (Space 6):*** If you maintain a Deposit Account in the Copyright Office, identify it in space 6. Otherwise, you will need to send the renewal registration fee of $6 with your form. The space headed "Correspondence" should contain the name and address of the person to be consulted if correspondence about the form becomes necessary.

• ***Certification (Space 7):*** The renewal application is not acceptable unless it bears the handwritten signature of the renewal claimant or the duly authorized agent of the renewal claimant.

• ***Address for Return of Certificate (Space 8):*** The address box must be completed legibly, since the certificate will be returned in a window envelope.

Fig. 26 (continued)

FORM RE

UNITED STATES COPYRIGHT OFFICE

REGISTRATION NUMBER

EFFECTIVE DATE OF RENEWAL REGISTRATION

(Month) (Day) (Year)

DO NOT WRITE ABOVE THIS LINE. FOR COPYRIGHT OFFICE USE ONLY

1 Renewal Claimant(s)

RENEWAL CLAIMANT(S), ADDRESS(ES), AND STATEMENT OF CLAIM: (See Instructions)

1
Name ....
Address ....
Claiming as .... (Use appropriate statement from instructions)

2
Name ....
Address ....
Claiming as .... (Use appropriate statement from instructions)

3
Name ....
Address ....
Claiming as .... (Use appropriate statement from instructions)

2 Work Renewed

TITLE OF WORK IN WHICH RENEWAL IS CLAIMED:

RENEWABLE MATTER:

CONTRIBUTION TO PERIODICAL OR COMPOSITE WORK:

Title of periodical or composite work: ....

If a periodical or other serial, give: Vol. .... No. .... Issue Date ....

3 Author(s)

AUTHOR(S) OF RENEWABLE MATTER:

4 Facts of Original Registration

ORIGINAL REGISTRATION NUMBER: ....

ORIGINAL COPYRIGHT CLAIMANT:

ORIGINAL DATE OF COPYRIGHT:

- If the original registration for this work was made in published form, give:
  DATE OF PUBLICATION: .... (Month) (Day) (Year)

OR

- If the original registration for this work was made in unpublished form, give:
  DATE OF REGISTRATION: .... (Month) (Day) (Year)

Fig. 26 (continued)

	EXAMINED BY: .... CHECKED BY:	RENEWAL APPLICATION RECEIVED:	FOR COPYRIGHT OFFICE USE ONLY
	CORRESPONDENCE ☐ Yes	REMITTANCE NUMBER AND DATE:	
	DEPOSIT ACCOUNT FUNDS USED: ☐		

DO NOT WRITE ABOVE THIS LINE. FOR COPYRIGHT OFFICE USE ONLY

5 Renewal for Group of Works

RENEWAL FOR GROUP OF WORKS BY SAME AUTHOR: To make a single registration for a group of works by the same individual author published as contributions to periodicals (see instructions), give full information about each contribution. If more space is needed, request continuation sheet (Form RE/CON).

1
Title of Contribution: ....
Title of Periodical: .... Vol. .... No. .... Issue Date ....
Date of Publication: .... Registration Number: ....
(Month) (Day) (Year)

2
Title of Contribution: ....
Title of Periodical: .... Vol. .... No. .... Issue Date ....
Date of Publication: .... Registration Number: ....
(Month) (Day) (Year)

3
Title of Contribution: ....
Title of Periodical: .... Vol. .... No. .... Issue Date ....
Date of Publication: .... Registration Number: ....
(Month) (Day) (Year)

4
Title of Contribution: ....
Title of Periodical: .... Vol. .... No. .... Issue Date ....
Date of Publication: .... Registration Number: ....
(Month) (Day) (Year)

5
Title of Contribution: ....
Title of Periodical: .... Vol. .... No. .... Issue Date ....
Date of Publication: .... Registration Number: ....
(Month) (Day) (Year)

6
Title of Contribution: ....
Title of Periodical: .... Vol. .... No. .... Issue Date ....
Date of Publication: .... Registration Number: ....
(Month) (Day) (Year)

7
Title of Contribution: ....
Title of Periodical: .... Vol. .... No. .... Issue Date ....
Date of Publication: .... Registration Number: ....
(Month) (Day) (Year)

6 Fee and Correspondence

DEPOSIT ACCOUNT: (If the registration fee is to be charged to a Deposit Account established in the Copyright Office, give name and number of Account.)

Name: ....
Account Number: ....

CORRESPONDENCE: (Give name and address to which correspondence about this application should be sent.)

Name: ....
Address: .... (Apt.)
.... (City) (State) (ZIP)

7 Certification (Application must be signed)

CERTIFICATION: I, the undersigned, hereby certify that I am the: (Check one)
☐ renewal claimant ☐ duly authorized agent of: ....
(Name of renewal claimant)

of the work identified in this application, and that the statements made by me in this application are correct to the best of my knowledge.

Handwritten signature: (X) ....
Typed or printed name: ....
Date: ....

8 Address for Return of Certificate

MAIL CERTIFICATE TO

.... (Name)
.... (Number, Street and Apartment Number)
.... (City) (State) (ZIP code)

(Certificate will be mailed in window envelope)

U.S. GOVERNMENT PRINTING OFFICE: 1987—181-531 40,019 March 1987—30,000

Fig. 26 (continued)

USE THIS FORM WHEN:

- An earlier registration has been made in the Copyright Office; and
- Some of the facts given in that registration are incorrect or incomplete; and
- You want to place the correct or complete facts on record.

FORM CA

UNITED STATES COPYRIGHT OFFICE
LIBRARY OF CONGRESS
WASHINGTON, D.C. 20559

Application for Supplementary Copyright Registration

To Correct or Amplify Information Given in the Copyright Office Record of an Earlier Registration

What is "Supplementary Copyright Registration"? Supplementary registration is a special type of copyright registration provided for in section 408(d) of the copyright law.

Purpose of Supplementary Registration. As a rule, only one basic copyright registration can be made for the same work. To take care of cases where information in the basic registration turns out to be incorrect or incomplete, the law provides for "the filing of an application for supplementary registration, to correct an error in a copyright registration or to amplify the information given in a registration."

Earlier Registration Necessary. Supplementary registration can be made only if a basic copyright registration for the same work has already been completed.

Who May File. Once basic registration has been made for a work, any author or other copyright claimant, or owner of any exclusive right in the work, who wishes to correct or amplify the information given in the basic registration, may submit Form CA.

Please Note:

- Do not use Form CA to correct errors in statements on the copies or phonorecords of the work in question, or to reflect changes in the content of the work. If the work has been changed substantially, you should consider making an entirely new registration for the revised version to cover the additions or revisions.
- Do not use Form CA as a substitute for renewal registration. For works originally copyrighted between January 1, 1950 and December 31, 1977, registration of a renewal claim within strict time limits is necessary to extend the first 28-year copyright term to the full term of 75 years. This cannot be done by filing Form CA.
- Do not use Form CA as a substitute for recording a transfer of copyright or other document pertaining to rights under a copyright. Recording a document under section 205 of the statute gives all persons constructive notice of the facts stated in the document and may have other important consequences in cases of infringement or conflicting transfers. Supplementary registration does not have that legal effect.

How to Apply for Supplementary Registration:

First: Study the information on this page to make sure that filing an application on Form CA is the best procedure to follow in your case.

Second: Turn this page over and read through the specific instructions for filling out Form CA. Make sure, before starting to complete the form, that you have all of the detailed information about the basic registration you will need.

Third: Complete all applicable spaces on this form, following the line-by-line instructions on the back of this page. Use typewriter, or print the information in dark ink.

Fourth: Detach this sheet and send your completed Form CA to: Register of Copyrights, Library of Congress, Washington, D.C. 20559. Unless you have a Deposit Account in the Copyright Office, your application must be accompanied by a non-refundable filing fee in the form of a check or money order for $10 payable to: *Register of Copyrights.* Do not send copies, phonorecords, or supporting documents with your application, since they cannot be made part of the record of a supplementary registration.

What Happens When a Supplementary Registration is Made? When a supplementary registration is completed, the Copyright Office will assign it a new registration number in the appropriate registration category, and issue a certificate of supplementary registration under that number. The basic registration will not be expunged or cancelled, and the two registrations will both stand in the Copyright Office records. The supplementary registration will have the effect of calling the public's attention to a possible error or omission in the basic registration, and of placing the correct facts or the additional information on official record. Moreover, if the person on whose behalf Form CA is submitted is the same as the person identified as copyright claimant in the basic registration, the Copyright Office will place a note referring to the supplementary registration in its records of the basic registration.

Fig. 27

PLEASE READ DETAILED INSTRUCTIONS ON REVERSE

Please read the following line-by-line instructions carefully and refer to them while completing Form CA.

INSTRUCTIONS

For Completing FORM CA (Supplementary Registration)

PART A: BASIC INSTRUCTIONS

• ***General Instructions:*** The information in this part identifies the basic registration to be corrected or amplified. Each item must agree exactly with the information as it already appears in the basic registration (even if the purpose of filing Form CA is to change one of these items).

• ***Title of Work:*** Give the title as it appears in the basic registration, including previous or alternative titles if they appear.

• ***Registration Number:*** This is a series of numerical digits, preceded by one or more letters. The registration number appears in the upper right hand corner of the certificate of registration.

• ***Registration Date:*** Give the year when the basic registration was completed.

• ***Name(s) of Author(s) and Name(s) of Copyright Claimant(s):*** Give all of the names as they appear in the basic registration.

PART B: CORRECTION

• ***General Instructions:*** Complete this part **only** if information in the basic registration was incorrect at the time that basic registration was made. Leave this part blank and complete Part C, instead, if your purpose is to add, update, or clarify information rather than to rectify an actual error.

• ***Location and Nature of Incorrect Information:*** Give the line number and the heading or description of the space in the basic registration where the error occurs (for example: "Line number 3 . . . Citizenship of author").

• ***Incorrect Information as it Appears in Basic Registration:*** Transcribe the erroneous statement exactly as it appears in the basic registration.

• ***Corrected Information:*** Give the statement as it should have appeared.

• ***Explanation of Correction (Optional):*** If you wish, you may add an explanation of the error or its correction.

PART C: AMPLIFICATION

• ***General Instructions:*** Complete this part if you want to provide any of the following: (1) additional information that could have been given but was omitted at the time of basic registration; (2) changes in facts, such as changes of title or address of claimant, that have occurred since the basic registration; or (3) explanations clarifying information in the basic registration.

• ***Location and Nature of Information to be Amplified:*** Give the line number and the heading or description of the space in the basic registration where the information to be amplified appears.

• ***Amplified Information:*** Give a statement of the added, updated, or explanatory information as clearly and succinctly as possible.

• ***Explanation of Amplification (Optional):*** If you wish, you may add an explanation of the amplification.

PARTS D, E, F, G: CONTINUATION, FEE, MAILING INSTRUCTIONS AND CERTIFICATION

• ***Continuation (Part D):*** Use this space if you do not have enough room in Parts B or C.

• ***Deposit Account and Mailing Instructions (Part E):*** If you maintain a Deposit Account in the Copyright Office, identify it in Part E. Otherwise, you will need to send the non-refundable filing fee of $10 with your form. The space headed "Correspondence" should contain the name and address of the person to be consulted if correspondence about the form becomes necessary.

• ***Certification (Part F):*** The application is not acceptable unless it bears the handwritten signature of the author, or other copyright claimant, or of the owner of exclusive right(s), or of the duly authorized agent of such author, claimant, or owner.

• ***Address for Return of Certificate (Part G):*** The address box must be completed legibly, since the certificate will be returned in a window envelope.

PRIVACY ACT ADVISORY STATEMENT

Required by the Privacy Act of 1974 (Public Law 93-579)

AUTHORITY FOR REQUESTING THIS INFORMATION:
- Title 17, U.S.C., Sec. 408 (d)

FURNISHING THE REQUESTED INFORMATION IS:
- Voluntary

BUT IF THE INFORMATION IS NOT PROVIDED:
- It may be necessary to delay or refuse supplementary registration

PRINCIPAL USES OF REQUESTED INFORMATION:
- Establishment and maintenance of a public record
- Evaluation for compliance with legal requirements

OTHER ROUTINE USES:
- Public inspection and copying
- Preparation of public indexes
- Preparation of public catalogs of copyright registrations
- Preparation of search reports upon request

NOTE:
- No other Advisory Statement will be given you in connection with the application
- Please retain this statement and refer to it if we communicate with you regarding this application

Fig. 27 (continued)

FORM CA
UNITED STATES COPYRIGHT OFFICE

REGISTRATION NUMBER								
TX	TXU	PA	PAU	VA	VAU	SR	SRU	RE

Effective Date of Supplementary Registration

....
MONTH DAY YEAR

DO NOT WRITE ABOVE THIS LINE. FOR COPYRIGHT OFFICE USE ONLY

(A) Basic Instructions

TITLE OF WORK:

REGISTRATION NUMBER OF BASIC REGISTRATION:

YEAR OF BASIC REGISTRATION:

NAME(S) OF AUTHOR(S):

NAME(S) OF COPYRIGHT CLAIMANT(S):

(B) Correction

LOCATION AND NATURE OF INCORRECT INFORMATION IN BASIC REGISTRATION:

Line Number .... Line Heading or Description ....

INCORRECT INFORMATION AS IT APPEARS IN BASIC REGISTRATION:

CORRECTED INFORMATION:

EXPLANATION OF CORRECTION: (Optional)

(C) Amplification

LOCATION AND NATURE OF INFORMATION IN BASIC REGISTRATION TO BE AMPLIFIED:

Line Number .... Line Heading or Description ....

AMPLIFIED INFORMATION:

EXPLANATION OF AMPLIFIED INFORMATION: (Optional)

Fig. 27 (continued)

	EXAMINED BY: .... CHECKED BY:	FORM CA RECEIVED:	FOR COPYRIGHT OFFICE USE ONLY
	CORRESPONDENCE: ☐ YES	REMITTANCE NUMBER AND DATE:	
	REFERENCE TO THIS REGISTRATION ADDED TO BASIC REGISTRATION: ☐ YES ☐ NO	DEPOSIT ACCOUNT FUNDS USED: ☐	

DO NOT WRITE ABOVE THIS LINE. FOR COPYRIGHT OFFICE USE ONLY

CONTINUATION OF: (Check which) ☐ PART B OR ☐ PART C

D — Continuation

DEPOSIT ACCOUNT: If the registration fee is to be charged to a Deposit Account established in the Copyright Office, give name and number of Account.

Name .... Account Number ....

CORRESPONDENCE: Give name and address to which correspondence should be sent:

Name .... Apt. No. ....

Address ....
(Number and Street) (City) (State) (ZIP Code)

E — Deposit Account and Mailing Instructions

CERTIFICATION * I, the undersigned, hereby certify that I am the: (Check one)

☐ author ☐ other copyright claimant ☐ owner of exclusive right(s) ☐ authorized agent of: ....
(Name of author or other copyright claimant, or owner of exclusive right(s))

of the work identified in this application and that the statements made by me in this application are correct to the best of my knowledge.

Handwritten signature: (X) ....

Typed or printed name ....

Date: ....

* 17 USC §506(e): FALSE REPRESENTATION – Any person who knowingly makes a false representation of a material fact in the application for copyright registration provided for by section 409, or in any written statement filed in connection with the application, shall be fined not more than $2,500.

F — Certification (Application must be signed)

....
(Name)

....
(Number, Street and Apartment Number)

....
(City) (State) (ZIP code)

MAIL CERTIFICATE TO

(Certificate will be mailed in window envelope)

G — Address for Return of Certificate

U.S. GOVERNMENT PRINTING OFFICE: 1988—202-133 60,014

Fig. 27 (continued)

April 1988—15,000

U.S. Department of Commerce
National Bureau of Standards

OFFICE OF ENERGY-RELATED INVENTIONS

REQUEST FOR EVALUATION OF AN ENERGY-RELATED INVENTION

Instructions for Submission of Invention Disclosures and Substantiating Material for Evaluation.
After reading this page and the following page, complete page 3 of both OERI and Submitter copies. Check appropriate box on page 4 and sign, date and complete the Memorandum of Understanding. Retain the Submitter's Copy for your records. Detach the OERI copy (pages 3 and 4) and send with your invention disclosure to:

Office of Energy-Related Inventions
National Bureau of Standards
Gaithersburg, Maryland 20899

A written disclosure of your invention, *in the English language,* must be attached to the OERI copy of this form. This disclosure should include an outline, a complete description of the invention and information to substantiate any claims for performance. Drawings or patents, where appropriate, should be included.

The quality of the evaluation will depend upon the quality of your submission. It should include or cover the following:

(1) **Purpose** of the invention. Discuss, if appropriate, where it can be used to best advantage: By industry? By individuals? By the Government? Emphasize the energy conservation or energy production potential.
(2) The **existing method(s)**, if any, of performing the function of the invention. Disadvantages of the existing method(s).
(3) The **new method**, using your invention. Details of the operation of the invention, identifying specific features which are new. If the invention is conceptual in nature, discuss typical applications.
(4) **Construction** of the invention, showing changes, deletions, improvement over the old method(s).
(5) **Data and calculations.** If tests have been conducted, detail the test conditions, controls, and results. Energy savings or efficiency estimates should be documented by calculations and data if available. Theoretical analyses should include the pertinent equations, definitions of terminology, and references.
(6) **Status of development:** Include information on stage of research, development, preproduction or production. Discuss proprietary nature, circumstances of public disclosure, instances of disclosure to government agencies, etc.
(7) **Difficulties** encountered or to be expected in exploiting your invention. Reasons why it has not been patented, manufactured, used, or accepted. What needs to be done to bring the invention closer to use?

Program Description and Statement of Policy

The Federal Nonnuclear Energy Research and Development Act of 1974 (Pub. L. 93-577) recognized the importance of encouraging invention and innovation in a national energy program. Section 14 of the Act directs the National Bureau of Standards (NBS) to give particular attention to the evaluation of promising energy-related inventions, particularly those from individual inventors and small companies. The Office of Energy-Related Inventions (OERI) was established at NBS to carry out the provisions of Section 14. Its duties include conducting analyses of submitted inventions to determine their technical and commercial feasibility for saving or producing energy, and bringing noteworthy concepts to the attention of the Department of Energy (DOE).

Fig. 28

The principal objective of the OERI effort is to assist DOE in identifying inventions that are ready to be moved into the private sector but may require business management assistance, or inventions that require further research and development (R&D), prototype fabrication, or laboratory tests in order to bring them to the point where they can compete with other DOE projects for program R&D funds. The evaluation of inventions submitted will, therefore, be performed principally as a service to DOE. Thus, the outcome of an evaluation will be either a recommendation for action by DOE in connection with the invention, or notification to the inventor that his invention is not being so recommended. It should be noted that a recommendation by OERI is no guarantee that DOE will provide assistance in developing a given invention.

A decision not to recommend action by DOE does not necessarily mean that the invention is considered scientifically unsound or without practical value. Also, a favorable evaluation by OERI should not be construed as being a ruling as to the patentability of any feature of an invention. The inventor should apply for a patent whenever such action is thought to be appropriate. OERI will provide no assistance in filing or prosecuting patent applications. Inventors interested in patent protection should discuss the matter with a registered patent attorney or agent.

To safeguard such proprietary rights as may exist in a submission, OERI will restrict access to invention disclosures to those persons having a need for purposes of administration or evaluation. Accordingly, in accepting invention disclosures for evaluation, an explicit statement is required (see page 4) that the information does or does not come within one of the exemptions of the Freedom of Information Act. If, for example, the disclosure contains information that is (a) a trade secret or (b) commercial or financial information that is privileged or confidential, such information falls within the exemption that is set out in 5 U.S.C. 552(b) (4). Thus, if the disclosure is protectable, the following or a similar statement should be applied to the title page or first page of the disclosure: "The disclosure contains information which is (a) a trade secret or (b) commercial or financial information that is privileged or confidential."

The Privacy Act of 1974 (Public Law 93-579), 5 U.S.C. 552a, requires that you be provided with certain information in connection with this form. You should know that:

a. The authority for collecting this data is the Federal Nonnuclear Energy Research and Development Act of 1974 (Public Law 93-577).
b. The furnishing of the information is entirely voluntary on your part.
c. The principal purpose for which the data will be used is to conduct an evaluation of your invention to determine its technical validity and potential for saving or producing energy.
d. The routine uses which may be made of the information submitted in this form are as follows:

   1) Disclosure to those employees of the Office of Energy-Related Inventions or other Federal agencies having need for the information, either to perform evaluations or administer the evaluation program.
   2) Disclosure to a contractor of the National Bureau of Standards having need for the information in the performance of a contract to perform evaluations of inventions and having agreed to hold the information in confidence.
   3) Disclosure to a Member of Congress submitting a request involving your invention, when you have requested his assistance.
   4) Disclosure to any persons with your written authorization.

Fig. 28 (continued)

OMB Number 0652-0020
Approval Expires Feb. 28, 1987

Submitter's Copy *(Keep this copy and return original OERI copy with disclosure)*

NBS-1019
(Rev. 9-84)

U.S. DEPARTMENT OF COMMERCE
National Bureau of Standards

Energy-Related Invention Evaluation Request

Follow attached instructions and submit the OERI copy of this form and other descriptive material of invention to:
Office of Energy-Related Inventions
National Bureau of Standards
Gaithersburg, Maryland 20899
(Please print or type all information)

Name and address of Inventor

Telephone No.

Name and address of Owner, if different from above

Telephone No.

This box is for office use only

Date ______ ER No. ______

Classification ______

Tech. Category ______

Analyst ______ Date ______

How did you learn of this program?

Request is being submitted by *(check which)*:

☐ inventor ☐ owner ☐ other

Name and address of Submitter, if not inventor or owner

Other *(identify)*

Size of company involved *(write: No. of employees; $ gross last year; N/N if none)*: ______
No. of employees $ gross last year

The name or title of this invention is:

Status of invention development *(check to indicate both the steps completed and the current status; highest number checked will indicate current status)*

0. ☐ Concept Definition 1. ☐ Concept Development 2. ☐ Lab. Test 3. ☐ Eng'g Design 4. ☐ Working Model
5. ☐ Prototype Development 6. ☐ Prototype Test 7. ☐ Production Eng'g. 8. ☐ Limited Prod./Mktg. 9. ☐ Production and Mktg.

Patent Status *(check one)*

0. ☐ Not patentable 1. ☐ Not applied for 2. ☐ Disclosure Document Program
3. ☐ Patent applied for 4. ☐ Patent granted (Patent Nos: ______

Check the item below that most nearly describes why you are requesting evaluation:

☐ 1. I wish the U.S. Government to provide funds to support development of the invention or new concept. Support is *first* needed for *(write in)*:
______
______

☐ 2. Development is complete. I need assistance to bring my invention or product into full utilization. Assistance is needed in: *(check whichever applies)* ☐ General Marketing ☐ Selling to the Government ☐ Business Management ☐ Other ______

☐ 3. I only desire an opinion that the disclosure describes a technically valid invention. This information is for:
☐ use in obtaining private development support ☐ other *(specify in disclosure)*

☐ 4. The Small Business Administration suggested I request evaluation from NBS in connection with a loan application.

☐ 5. Other *(specify)* ______
______

☐ Yes ☐ No Has the invention been described to other agencies of the Government? *(If yes, discuss in disclosure.)*
☐ Yes ☐ No Has the invention been disclosed to any private companies, patent attorneys, etc.? *(If yes, identify in disclosure.)*

Fig. 28 (continued)

OMB Number 0652-0020
Approval Expires Feb. 28, 1987

OERI Copy *(Place carbon paper behind this sheet to make SUBMITTER's Copy of page 3)*

NBS-1019
(Rev. 9-84)

U.S. DEPARTMENT OF COMMERCE
National Bureau of Standards

Energy-Related Invention Evaluation Request

Follow attached instructions and submit the OERI copy of this form and other descriptive material of invention to:

Office of Energy-Related Inventions
National Bureau of Standards
Gaithersburg, Maryland 20899
(Please print or type all information)

Name and address of Inventor

Telephone No.

Name and address of Owner, if different from above

Telephone No.

This box is for office use only

Date ____________ ER No. ____________

Classification ____________

Tech. Category ____________

Analyst ____________ Date ____________

How did you learn of this program?

Request is being submitted by *(check which)*:

☐ inventor ☐ owner ☐ other

Name and address of Submitter, if not inventor or owner

Other *(identify)*

Size of company involved *(write: No. of employees; $ gross last year; N/N if none)*: ____________
No. of employees $ gross last year

The name or title of this invention is:

Status of invention development *(check to indicate both the steps completed and the current status; highest number checked will indicate current status)*

0. ☐ Concept Definition 1. ☐ Concept Development 2. ☐ Lab. Test 3. ☐ Eng'g Design 4. ☐ Working Model
5. ☐ Prototype Development 6. ☐ Prototype Test 7. ☐ Production Eng'g. 8. ☐ Limited Prod./Mktg. 9. ☐ Production and Mktg.

Patent Status *(check one)*

0. ☐ Not patentable 1. ☐ Not applied for 2. ☐ Disclosure Document Program
3. ☐ Patent applied for 4. ☐ Patent granted (Patent Nos: ____________

Check the item below that most nearly describes why you are requesting evaluation:

☐ 1. I wish the U.S. Government to provide funds to support development of the invention or new concept. Support is *first* needed for *(write in)*:
____
____

☐ 2. Development is complete. I need assistance to bring my invention or product into full utilization. Assistance is needed in: *(check whichever applies)* ☐ General Marketing ☐ Selling to the Government ☐ Business Management ☐ Other ____________

☐ 3. I **only** desire an opinion that the disclosure describes a technically valid invention. This information is for:
☐ use in obtaining private development support ☐ other *(specify in disclosure)*

☐ 4. The Small Business Administration suggested I request evaluation from NBS in connection with a loan application.

☐ 5. Other *(specify)* ____________
____

☐ Yes ☐ No Has the invention been described to other agencies of the Government? *(If yes, discuss in disclosure.)*
☐ Yes ☐ No Has the invention been disclosed to any private companies, patent attorneys, etc.? *(If yes, identify in disclosure.)*

Fig. 28 (continued)

Place carbon behind this sheet to make Submitter's Copy of page 4.

MEMORANDUM OF UNDERSTANDING

I have read the Program Description and Statement of Policy on pages 1 and 2 of this form. As the owner, or with the authority from the owner who is listed on Page 3, I have attached (or previously submitted) a disclosure of the identified invention for the purpose of evaluation by the National Bureau of Standards (NBS) pursuant to Section 14 of Public Law 93-577.

I understand that to help protect property rights in an unpatented invention an appropriate statement or notation should be applied to the title page or first page of the invention description, and that if the description is so marked, the Government will consider all information that is in fact (a) trade secret or (b) commercial or financial information that is privileged or confidential, as coming within the exemption set out in 5 U.S.C. 552(b) (4). Accordingly, I have checked directly below, the box which is applicable to this invention.

Yes No

☐ ☐ An appropriate statement has been applied to the information I have submitted.

☐ ☐ Please apply an appropriate statement to all material I have submitted describing the invention to which this request pertains. (Example: This material contains commercial or financial information which is confidential.)

☐ No statement is required because the information submitted is not confidential.

I also understand that NBS will evaluate the invention described in the invention disclosure on the following conditions:

(a) The Government will, in the evaluation process, restrict access to the description to those persons, within or without the Government, who have a need for purposes of administration or evaluation and will restrict their use of this information to such purposes.

(b) The information submitted will not be returned and may be retained as a Government record.

(c) The Government may make additional copies of the material submitted if required to facilitate the review process.

(d) The acceptance of the information for evaluation does not, in itself, imply a promise to pay, a recognition of novelty or originality, or a contractual relationship such as would render the Government liable to pay for use of the information submitted.

(e) The provisions of this Memorandum of Understanding shall also apply to additions to the disclosure made by me incidental to the evaluation of the invention.

Date	Signature
Status (Owner, Business or Company Representative, Patent Attorney, Interested Party, etc.)	Printed or Typed Name

Fig. 28 (continued)

MEMORANDUM OF UNDERSTANDING

I have read the Program Description and Statement of Policy on pages 1 and 2 of this form. As the owner, or with the authority from the owner who is listed on Page 3, I have attached (or previously submitted) a disclosure of the identified invention for the purpose of evaluation by the National Bureau of Standards (NBS) pursuant to Section 14 of Public Law 93-577.

I understand that to help protect property rights in an unpatented invention an appropriate statement or notation should be applied to the title page or first page of the invention description, and that if the description is so marked, the Government will consider all information that is in fact (a) trade secret or (b) commercial or financial information that is privileged or confidential, as coming within the exemption set out in 5 U.S.C. 552(b) (4). Accordingly, I have checked directly below, the box which is applicable to this invention.

Yes No

☐ ☐ An appropriate statement has been applied to the information I have submitted.

☐ ☐ Please apply an appropriate statement to all material I have submitted describing the invention to which this request pertains. (Example: This material contains commercial or financial information which is confidential.)

☐ No statement is required because the information submitted is not confidential.

I also understand that NBS will evaluate the invention described in the invention disclosure on the following conditions:

(a) The Government will, in the evaluation process, restrict access to the description to those persons, within or without the Government, who have a need for purposes of administration or evaluation and will restrict their use of this information to such purposes.

(b) The information submitted will not be returned and may be retained as a Government record.

(c) The Government may make additional copies of the material submitted if required to facilitate the review process.

(d) The acceptance of the information for evaluation does not, in itself, imply a promise to pay, a recognition of novelty or originality, or a contractual relationship such as would render the Government liable to pay for use of the information submitted.

(e) The provisions of this Memorandum of Understanding shall also apply to additions to the disclosure made by me incidental to the evaluation of the invention.

Date	Signature
Status (Owner, Business or Company Representative, Patent Attorney, Interested Party, etc.)	Printed or Typed Name

Fig. 28 (continued)

U.S. DEPARTMENT OF DEFENSE

DEFENSE SMALL BUSINESS INNOVATION RESEARCH (SBIR) PROGRAM PHASE I—FY 19__ COST PROPOSAL

Background:

The following items, as appropriate, should be included in proposals responsive to the DOD Solicitation Brochure.

Cost Breakdown Items (in this order, as appropriate):

1. Name of offeror
2. Home office address
3. Location where work will be performed
4. Title of proposed effort
5. Topic number and topic title from DOD Solicitation Brochure
6. Total Dollar amount of the proposal (dollars)
7. Direct material costs
   a. Purchased parts (dollars)
   b. Subcontracted items (dollars)
   c. Other
      (1) Raw material (dollars)
      (2) Your standard commercial items (dollars)
      (3) Interdivisional transfers (at other than cost) (dollars)
   d. Total direct material (dollars)
8. Material overhead (rate ______%) × total direct material = dollars
9. Direct labor (specify)
   a. Type of labor, estimated hours, rate per hour and dollar cost for each type.
   b. Total estimated direct labor (dollars)
10. Labor overhead
    a. Identify overhead rate, the hour base and dollar cost.
    b. Total estimated labor overhead (dollars)
11. Special testing (include field work at Government installations)
    a. Provide dollar cost for each item of special testing
    b. Estimated total special testing (dollars)
12. Special equipment
    a. If direct charge, specify each item and cost of each
    b. Estimated total special equipment (dollars)
13. Travel (if direct charge)
    a. Transportation (detailed breakdown and dollars)
    b. Per Diem or subsistence (details and dollars)
    c. Estimated total travel (dollars)
14. Consultants
    a. Identify each, with purpose, and dollar rates
    b. Total estimated consultants costs (dollars)
15. Other direct costs (specify)
    a. Total estimated direct cost and overhead (dollars)
16. General and administrative expense
    a. Percentage rate applied
    b. Total estimated cost of G&A expense (dollars)
17. Royalties (specify)
    a. Estimated cost (dollars)
18. Fee or profit (dollars)
19. Total estimate cost and fee or profit (dollars)
20. The cost breakdown portion of a proposal must be signed by a responsible official, and the person signing must have typed name and title and date of signature must be indicated.
21. On the following items offeror must provide a yes or no answer to each question.
    a. Has any executive agency of the United States Government performed any review of your accounts or records in connection with any other government prime contract or subcontract within the past twelve months? If yes, provide the name and address of the reviewing office, name of the individual and telephone/extension.
    b. Will you require the use of any government property in the performance of this proposal? If yes, identify.
    c. Do you require government contract financing to perform this proposed contract? If yes, then specify type as advanced payments or progress payments.
22. Type of contract proposed, either cost-plus-fixed-free or firm-fixed price.

Fig. 29

U.S. DEPARTMENT OF DEFENSE

SMALL BUSINESS INNOVATION RESEARCH (SBIR) PROGRAM
PHASE 1—FY 19__
PROPOSAL COVER SHEET

Topic Number: ______ ☐ Army ☐ Navy ☐ Air Force ☐ DARPA ☐ DNA ☐ SDIO

Proposal Title: ______________________________

____

Submitted By: Firm ______________________________

Address ______________________________

City ______________ State ______ Zip Code ______

Submitted To: (Activity identified with the topic) ______________________________

____

Address ______________________________

City ______________ State ______ Zip Code ______

Small Business Certification:

The above firm certifies it is a small business firm and meets the definition stated in the Small Business Act 15 U.S.C. 631 and in the Definition Section of the Program Solicitation.

The above firm certifies that it qualifies as a minority or disadvantaged small business as defined in the Definition Section of the Program Announcement. Yes ____ No ____.

The above firm certifies that it qualifies as a woman-owned small business firm : Yes ____ No ____.

This proposal has been submitted to other US Government agency/agencies: or DOD components, or the same DOD component. If SBIR proposal, list Topic Number.

Yes ________ ; Name(s) ______________________________

No ________

Disclosure permission statement as follows:

All data on Appendix A is releasable information. All data on Appendix B, for an awarded contract, is also releasable.

Will you permit the Government to disclose the information on Appendix B, if your proposal does not result in an award, to any party that may be interested in contacting you for further information or possible investment? Yes ____No ____.

Number of employees including all affiliates (average for preceding 12 months): ________

Proposed Cost (Phase I): ________ Proposed Duration: ____months (not to exceed six months).

Project Manager/Principal Investigator	Corporate Official (Business)
Name ______________	Name ______________
Title ______________	Title ______________
Signature ______________	Signature ______________
Date ______________	Date ______________
Telephone ______________	Telephone ______________

For any purpose other than to evaluate the proposal, this data except Appendix A and B shall not be disclosed outside the Government and shall not be duplicated, used, or disclosed in whole or in part, provided that if a contract is awarded to this proposer as a result of or in connection with the submission of this data, the Government shall have the right to duplicate, use, or disclose the data to the extent provided in the funding agreement. This restriction does not limit the Government's right to use information contained in the data if it is obtained from another source without restriction. The data subject to this restriction is contained in page(s) ________ of this proposal. Failure to fill in all appropriate spaces may cause your proposal to be disqualified.

Nothing on this page is classified or proprietary information/data

Fig. 30

U.S. DEPARTMENT OF DEFENSE

SMALL BUSINESS INNOVATION RESEARCH (SBIR) PROGRAM
PHASE 1—FY 19__
PROJECT SUMMARY

Topic No. ________ Military Department/Agency ________.

Name and Address of Proposing Small Business Firm

Name and Title of Principal Investigator

Proposal Title

Technical Abstract (Limit your abstract to 200 words with no classified or proprietary information/data.)

Anticipated Benefits/Potential Commercial Applications of the Research or Development

List a maximum of 8 Key Words that describe the Project.

Nothing on this page is classified or proprietary information/data

Fig. 31

AGREEMENT TO HOLD SECRET AND CONFIDENTIAL

The below described invention, idea or concept (hereinafter referred to as INVENTION is being submitted to ______________________________ of ________________(hereinafter referred to as COMPANY) by __________ __________of_______________ ________on_____,_____, 19_ (hereinafter referred to as INVENTOR) who is the inventor of record. The undersigned, in consideration of examining said INVENTION, with a purpose to opening negotiations to obtain a license to manufacture and sell said INVENTION, hereby agrees on behalf of himself/herself and said COMPANY that he/she represents, that:

1) He/she (during or after the termination of employment with said COMPANY) and said COMPANY, will keep said INVENTION and any information pertaining to it, in confidence.

2) He/she will not disclose said INVENTION or data related thereto to anyone save for employees of said COMPANY, sufficient information about said INVENTION to enable said COMPANY to continue with negotiations for said license, and that anyone in said COMPANY to whom said INVENTION is revealed, shall be informed of the confidential nature of the disclosure and shall agree to hold confidential the information, and be bound by the terms hereof, to the same extent as if they had signed this Agreement.

3) Neither he/she nor said COMPANY shall use any of the information provided to produce said INVENTION until agreement is reached with INVENTOR.

4) He/she has the authority to make this Agreement on behalf of said COMPANY.

It is understood, nevertheless, that the undersigned and said COMPANY shall not be prevented by the Agreement from selling any product heretofore sold by said COMPANY, or any product in the development or planning stage, as of the date first above written, or any product disclosed in any heretofore issued U.S. Letters Patent or otherwise known to the general public.

The terms of the preceding section releasing, under certain conditions, the obligation to hold the disclosure in confidence does not however, constitute a waiver of any patent, copyright or other rights which said Inventor or any licensee thereof may have against the undersigned or said COMPANY.

Fig. 32

IDEA SUBMISSION AGREEMENT I

While ________________ wishes to take every opportunity to improve its products and add profitable ones to its line, it has found certain precautions necessary in accepting disclosures from persons not in its employ. For an idea to be considered, this form must be completed in full, signed and returned with any disclosure of an idea or invention.

(Date)________________, 19_____

To

I am submitting to you, for your evaluation and permanent record, copies of certain ideas, suggestions or other materials having to do with

The information I am submitting to you consists of the following:

_____ *1) Description*

_____ *2) Drawing or sketches*

_____ *3) Samples*

_____ *4) Copy of a patent application(s)*

_____ *5) Other*

Please check the appropriate blank(s).

In doing so, I agree to the conditions printed on the back of this sheet and further agree that such conditions shall apply to any additional disclosures made incidental to the original material submitted.

(Signature) ____________________________________

(Name Printed) ____________________________________

(Address) ____________________________________

CONDITIONS OF SUBMISSION

1) All submissions or disclosures of ideas are voluntary on the part of the submitter. No confidential relationship is established by submission or implied

Fig. 33

from receipt or consideration of the submitted material.

2) Patented ideas and ideas covered by pending applications for patent are considered only with the understanding that the submitter agrees to rely for his/her protection solely on such rights as he/she may have under the patent laws of the United States.

3) Ideas which have not been covered by a patent or a pending application for patent are considered only with the understanding that the use to be made of such ideas and the compensation, if any, to be paid for them are matters resting solely in the discretion of the Company.

4) If the subject matter offered the Company is a proposed trademark, advertising slogan, or merchandising plan, susceptible to trademark or copyright protection, the Company will examine it only under the terms set forth in this Agreement. The submitter shall rely for his/her protection solely on such rights as he/she may have under the Copyright and Trademark Laws of the United States.

5) The foregoing conditions may not be modified or waived.

Fig. 33 (continued)

IDEA SUBMISSION AGREEMENT II

AGREEMENT made and entered into this_____day of_____, 19_____, by and between ____
("Disclosure's" full name and address)
and ____________________ , a corporation organized under the laws of the State of_____, with offices located at ____________, ________________.

WHEREAS Discloser is a developer of, or has licensing rights to, concepts for ____, and

WHEREAS, Discloser represents that he/she has developed a certain concept, device or other proprietary subject matter more specifically described at the end of this Agreement and on the attachment hereto (hereinafter referred to as the "Item"), and

WHEREAS, _______________desires to evaluate the commercial utility of the item, and

WHEREAS, in order to make this evaluation possible, it will be necessary for Discloser to disclose confidential information concerning the Item to _____________________.

NOW, THEREFORE, in consideration of the mutual promises hereinafter contained, and for other good and valuable consideration, the parties agree as follows:

1) Discloser shall make full disclosure with respect to the Item to employees of ________________ or one of its affiliates (collectively, ______________) and shall submit to ______________ all relevant data in connection therewith. The disclosure by Discloser to __________________ is solely to enable _____ to evaluate the Item in order to determine its commercial utility. ____________ is under no obligation to market or produce the Item, unless and until a formal written agreement is entered into, and the obligations of______________shall be only those which are set forth in any such agreement.

2) Discloser hereby represents to _________________that the Item is his own individual creation and wholly and solely the property of Discloser and that Discloser has not assigned, sold, licensed, mortgaged, pledged, or otherwise transferred or encumbered the Item or entered into any agreement to do any of the foregoing with respect to the Item. The execution and performance of this Agreement by Discloser does not violate any contract, agreement or other restriction to which Discloser is a party or by which it is bound or any rights of any third party.

3) The disclosure of the Item and all information incidental thereto is confidential and shall be received by_____ _______in confidence. ______________shall not disclose such confidential information to others and shall take reasonable steps to prevent such disclosure. _______________agrees to use the same degree of care in protecting and safeguarding the confidentiality of the concepts and information disclosed hereunder as it uses for its own information of like

Fig. 34

importance. _______________ shall not be liable for inadvertent disclosure or use of the Item by persons who are or have been in its employ, unless _____ fails to exercise the degree of care set forth above.

4) It is understood that________________s willingness to evaluate the Item is not to be construed as an admission of the Item's novelty, priority, or originality. Discloser understands that _________________may have rights to the Item or particular elements thereof, due to prior access to information similar to the Item or elements thereof, including, by way of illustration and not limitation, prior patents, prior publication, prior submissions to _________________by others, prior development by______________s personnel or representatives, prior use by________________________, prior knowledge, or prior sale. Accordingly, consideration of the Item by_____ shall not deprive _________of its existing rights, if any, with respect to the Item or any element thereof.

5) Without limiting the generality of the provisions of paragraph 4 hereof, the obligations of_____________ hereunder are not applicable to such information which:

a) prior to disclosure by Discloser, was already know to ________________ as evidenced by records kept in the ordinary course of business of________________or by proof of actual use by __________

b) was known to the public or generally available to the public prior to the date of disclosure.

c) becomes known to the public or is generally available to the public subsequent to the date of said disclosure through no act of________________ contrary to the obligations imposed by this Agreement.

d) is disclosed by Discloser to an unrelated third party without restriction.

e) is approved for public release by Discloser.

f) is rightfully received from a third party without similar restriction and without breach of this Agreement.

g) is independently developed by ______________without breach of this Agreement.

h) is required to be disclosed by judicial or government action.

i) is disclosed in a judicial or governmental proceeding subject to a protective order.

______________ shall be free of any obligations restricting disclosure and use of the information provided by Discloser hereunder, subject to Discloser's patent rights, if any of the provisions of a) through i) of this paragraph 5 are applicable to the information disclosed.

6) Upon submission of the Item to__________ , _____________ shall consider the Item and as promptly as practicable advise Discloser of_____________s interest or lack of interest therein, all subject to the terms, conditions and provisions of this Agreement.

Fig. 34 (continued)

7)______________shall not be obligated to take any action with regard to the Item other than pursuant to paragraphs 3 and 6 hereof.

8)_____________will, upon request, return any letters, drawings, descriptions, specifications or other materials submitted to it in connection with the Item.

9) The provisions of this Agreement shall apply to any additional or supplemental information pertaining to the Item provided by Discloser to_______________.

10) This writing reflects the entire agreement between the parties concerning the Item, and no modification, amendment, waiver or cancellation of this Agreement or any provision hereof shall have any validity or effect whatsoever unless in writing and signed by both parties hereto. Without limiting the generality of the foregoing, no agreement relating to the purchase or use of the Item by ___________or any of its affiliates, or relating to the terms of or consideration of such purchase or use, or relating to any compensation to, or reimbursement or any expenses of, Discloser, shall be binding upon either party hereto unless in writing and signed by both parties hereto.

11) This Agreement shall be governed by, construed and enforced in accordance with the internal laws of the State of _____, without reference to principles or conflict of laws.

12) This Agreement shall be binding upon, and inure to the benefit of, the Discloser and________________(and the affiliates of________) and their respective heirs, executors, administrators, successors, and assigns.

IN WITNESS WHEREOF, the parties have signed this Agreement on the respective dates hereinafter written.

The Item is generally described as follows:

SEE ATTACHED

____

(company)

*By:*_____________________

____

(Discloser)

*Date:*____________________

Fig. 34 (continued)

NIST Agreement of Nondisclosure

I agree to handle Invention Disclosures received by me from the Office of Energy-Related Inventions pursuant to Section 14 of the Federal Nonnuclear Energy Research and Development Act of 1974.

I further agree that I shall hold in confidence for NIST any such Invention Disclosures provided to me by the Office of Energy-Related Inventions, and shall not disclose any Invention Disclosure or any portion thereof to anyone without the written authorization of the Contracting Officer.

My obligations under the Agreement of Nondisclosure shall not extend to any information or technical data

a) which is now available or which later becomes available to the general public, other than by any breach of this agreement;

b) which is obtained from any source other than the Officer of Energy-Related Inventions by proper means and without notice of any obligation to hold such information or technical data in confidence; or

c) which is developed without the use of any Invention Disclosure provided to me by the Office of Energy-Related Inventions.

I further agree not to make, have made, or permit to be made, any copies of any Invention Disclosure or portions thereof, except with the written permission of Mr. George P. Lewett, Chief, Office of Energy-Related Inventions. Upon completion of the task called for in the letter, I shall return the Invention Disclosure and any copies thereof to Mr. Lewett.

If, upon examination of an Invention Disclosure, I feel that I have any financial interest or any relation with a third party which might be deemed likely to affect the integrity and impartiality of the performance of the task specified in the letter, I shall provide Mr. Lewett with a complete written disclosure of such interest or relationship prior to undertaking the task and shall not proceed with the task without the written authorization of Mr. Lewett.

If any invention or discovery is conceived or first actually reduced to practice by me in the course of or under this task, I shall promptly furnish Mr. Lewett with complete information thereon; and NIST shall have the sole power to determine whether or not and where a patent application shall be filed, and to determine the disposition of the title and rights in and to any invention or discovery and any patent application or patent that may result. The judgment of NIST on these matters shall be accepted as final and I agree to execute all documents and do all things necessary or proper to carry out the judgment of NIST.

____

(signature)

____

(typed name)

____

(date)

Fig. 35

PART TWO

How to Protect and License Your Invention

CHAPTER 1

HOW TO PROTECT AND LICENSE YOUR INVENTION

by Richard C. Levy

"Heavier-than-air flying machines are impossible," the celebrated British mathematician and physicist William Kelvin assured everyone back in 1895.

"...man can never tap the power of the atom," said Nobel Laureate and physicist Robert Andrews Millikan, credited with being first to isolate the electron and measure its charge.

"Everything that can be invented has been invented," offered another man of vision, Charles H. Duell, Director of the U.S. Patent Office in 1899.

TRW reminded us of these profound observations in a 1985 *Wall Street Journal* ad which was tagged, "There's no future in believing something can't be done. The future is in making it happen."

What the above gentlemen did not know, quite obviously, is what inventors from fully equipped labs to basement workshops across America are proving everyday: there is no future in the word impossible. Results first. Theory second.

We are privileged! We live in a land of opportunity. Nowhere in the world do people have more freedom and encouragement to innovate, to be different and individual than in America.

We are always looking to challenge the previous and reach new levels of interest and involvement by doing things in novel ways. Our history is replete with examples of Yankee ingenuity, independent and courageous individuals who succeeded by doing things differently, dreamers who believed in themselves and their ideas.

"It takes a special kind of independence to invent something. You put yourself and your ideas on the line. And maybe people will say that you're crazy or that you're impractical," said President George Bush when, as vice president, he spoke at the National Museum of American History to commemorate the 150th anniversary of the Patent Act of 1836.

"But for over more than two centuries, millions of Americans have ignored the ridicule," he continued. "They've worked on ideas. From those ideas, they've started businesses. And many of those businesses have grown and are today our great industrial companies — companies like Xerox, Ford Motor Company, American

Telephone and Telegraph and Apple Computer. Think of what America would be like if the skeptics had silenced the inventors."

We have been on the leading edge of uncertainty, experimentation, and exploration since the first Pilgrims set out for the New World and came ashore at Plymouth, Massachusetts in 1620.

In 1850, British subject James Nasmyth, best known for inventing the steam hammer, said about inventive Americans, "There is not a working boy of average ability in the New England States...who has not an idea of some mechanical invention or improvement...by which he hopes to better his position, or rise to fortune."

Alexis de Tocqueville, the French writer who visited America in 1831, wrote, "They [Americans] all have a lively faith in the perfectibility of man, and judge that the diffusion of knowledge must necessarily be advantageous, and the consequences of ignorance fatal. They [Americans] consider society as a body in a state of improvement, humanity as a changing scene, in which nothing is, or ought to be, permanent; and what appears to them today to be good, may be superseded by something better tomorrow."

Our heritage is rich in examples of American inventors, tinkerers, daydreamers, and gadgeteers, working from basements and garages, who dared to be different and refused to trade incentive for security. Many of their names have become celebrated: George Westinghouse created the first air brake; W.H. Carrier gave us air conditioning; Benjamin Franklin put us on bicycles; Edwin H. Land developed the 60-second camera; Charles Goodyear first vulcanized rubber; George Pullman designed double-deck sleeping cars on trains; Robert Goddard started the race to space; King S. Gillette made the first safety razor; and William S. Burroughs patented the first mechanical adding machine.

Other Americans are not so well known, but their inventions are: W.H. Carothers created nylon; Whitcomb L. Judson patented the zipper; Luther Childs Crowell made the square bottom paper bag; Walter Hunt stuck us with the safety pin; James Ritty rang up tremendous sales with the first cash register; Mary Anderson improved motoring safety with the first windshield wipers; and Garrett A. Morgan changed our driving habits and city streets with the traffic signals and is also credited with a pioneering gas mask.

Not all inventions are so conventional. Two virtually unknown American inventors, Philip Leder of Chestnut Hill, Massachusetts and Timothy A. Stewart of San Francisco, California made history on April 12, 1988 when U.S. Patent No. 4,736,866, entitled "Transgenic Non-Human Mammals" was issued. The Harvard University researchers were awarded the first patent covering an animal. Their technique introduces activated cancer genes into early-stage embryos of mice. The resulting mice are born with activated cancer genes in all their cells. These mice are extremely sensitive to cancer-causing chemicals developing tumors quickly if exposed even to small amounts. The resulting value to medical and scientific research may be immense.

As diverse a group of individuals as this appears to be, these people share many things in common, things that you too will require to fulfill your aspirations and see your inventions patented, licensed, manufactured, and marketed.

On an intellectual level, they knew how to detect and follow what Emerson called that "gleam of light that flashes across the mind from within." They did not dismiss their own thoughts without notice. They abided by their own spontaneous impression. Successful inventors permit nothing to affect the integrity of their minds.

Donald J. Quigg, Assistant Secretary of Commerce and Commissioner of Patents and Trademarks, says of American inventors, "They are more than just thinkers and problem solvers. They are our economic future."

These people shared the challenges every inventor faces in how to best protect and exploit a new product. Whether you have invented a wheel mounting device, a method for manufacturing semiconductors, an adjustable hoe attachment for a rake, fluidic rotational speed sensors, a U-joint mount, or a better mouse trap, the same basic steps apply.

This is not a theoretical book. I've personally developed or co-developed and licensed more than fifty original products and concepts to companies ranging from Fortune 500 diversified manufacturers to small independent businesses. This book will describe the pragmatic "how-to" as well as serve as a guide to the enormous amount of information and services available to the independent inventor.

In addition to the formal information on how to go about protecting your ideas through U.S. government devices such as patents, copyrights, and trademarks, the *Inventing and Patenting Sourcebook* is crammed full of suggestions, modus operandi, and "insider" tips and pointers on invention marketing, prototyping, corporate licensing, and funding research and development. This is information every independent inventor requires, but only a fortunate few possess. The objective is to guide the inventor down the frequently murky road leading from concept to market fulfillment.

Educational Background is Irrelevant

My university degree is in television and film with a minor in English. I have never had a course in marketing. I am a totally self-taught inventor, designer, and marketeer. I know, therefore, better than most, what the independent inventor requires in terms of information. I also know the limited resources and time restraints the independent inventor faces.

I have done my best to make this book as comprehensive as possible, designing it to deliver empirical information on what to do and how to do it. It is not just another directory that tells you the where but not the how.

The Golden Rules

To be successful in the exciting business of product development it takes much more than a good idea and a strong patent. Understanding and practicing the following six points is critical to the inventor who attempts to beat what often appear to be insurmountable odds.

1) **Don't take yourself too seriously**. Don't take your idea too seriously either. The world will probably survive without your idea. Industry will probably survive without your idea. You might need it to survive, but no one else will.

2) **You can't do it all yourself**. Remember the words of John Donne, "No man is an island, entire of itself; every man is a piece of the continent, a part of the main."

The success I have experienced is the result of unselfish, highly talented, and creative associates willing to face the frustrations, rejections, and seemingly open-ended time frames that are inherent in any product development and marketing exercise. I have also been lucky to have met and worked with very creative, understanding, and courageous corporate executives willing to believe in me and gamble on our concepts.

It is the cross-pollination and subsequent synergism that results in success, success in which all parties share. For if any link in this often complex and serpentine chain breaks, an entire project could flag.

3) **Keep egos under control**. Creative and inventive people, according to profile, hate to be rejected or criticized for any reason. They are usually highly critical of others. And they are extremely defensive where their creations are concerned.

I have always found that my concepts are enhanced by the right touch. Working together or in competition, other people contribute time and time again to making an idea more useful or marketable.

Share an idea and get a better one! Products are actually a series of improvements. Inventors are people who see how to organize what exists in a novel method. They are creative people with vision. They see the ordinary in a new light. Gutenberg combined the wine press and the coin punch to give the world movable type and the printing press. Wilbur and Orville Wright would be amazed by the U.S. Space Shuttle, but it is only a modern extension of their celebrated glider that took flight near Kill Devil Hills, North Carolina in 1911.

Unchecked egocentricity can be the source for major failure in the development and licensing of new concepts. Arrogance has no place in the process.

4) **Learn to take rejection**. Rejection can be positive if it is turned into constructive growth. Don't let it shake you or your confidence.

I have rarely licensed a product to the first manufacturer who sees it. And for every product I've licensed there are at least five that did not make it.

5) **Don't just do it for the financial rewards**. You should be motivated by the gamesmanship. It may sound trite, but people who do things just for the money usually come up shortchanged.

6) **Be patient**. It's going to take time. It is a misconception that anything is made overnight other than baked goods and newspapers. Even though the market is big, the competition is ferocious. Murphy's first corollary is: **Nothing is as easy as it looks**. Murphy's second corollary is: **Everything takes longer than you think**.

The Beatles worked for ten years in the clubs of Liverpool before getting a break. Michael Jackson had made zillions of dollars by age twenty two, but let us not forget he began performing professionally at age five. Closer to home, it took me three years of pitching the idea for this book before Gale Research decided to publish it; ideas generally don't diminish with time, but grow and find their proper environment.

Longfellow said it best in the work, "The Ladder of St. Augustine." "The heights by great men reached and kept were not attained by sudden flight. But they, while their companions slept, were toiling upward in the night."

Information is power, but only when properly understood and utilized. This volume is packed with power and the instructions for its most effective use. If you carefully apply this information to an innovative and novel concept, putting your own English on the lessons implied, you'll have that added edge that could mean success.

In closing, let me quote again from President Bush. "Let us rededicate ourselves to ensuring that America is always a land in which, for those who dare to create new technologies and new businesses, the air is clear and the sky is open and the energies of man are free, where the enthusiasm for invention and the spirit of enterprise is always part of the American spirit, where men and women who want to build and dream can always find a home...Today, Tomorrow, and for all time to come."

HOW TO PROTECT YOUR IDEAS

It is smart to initiate every form of appropriate protection available under the law for an invention. First, this sends a signal to those to whom you disclose a concept, whether they be potential licensees or investors, that you are serious, committed, and willing to go that extra mile. Secondly, it tends to keep honest people honest, much in the same way locks do on doors. A dishonest person or company bent on misappropriating a piece of intellectual property will do it regardless of the protective steps you may take. Locks have never stood in the way of a good second-story man. In most cases, strong protection is a required prerequisite to any licensing agreement. You'll find that companies favor licensing protected products over nonprotected products as extra insurance against competition.

The U.S. Government supplies several methods for protecting ideas: **patents**, **copyrights**, and **trademarks**. As each of these can be important to the independent inventor, in this essay, I will attempt to cut through the dim "bureaucratese" and explain each in the simplest and most widely understandable of terms. I will address the questions most frequently asked and then some.

This chapter is not intended to be a comprehensive text or total legal guide on patent, copyright, and trademark law. The source material for this writing weighed just over fifteen pounds! My purpose is to enlighten you, make you comfortable with the material and processes, share some personal experiences, and, in the end, equip you better to handle matters, make decisions, save some money, and otherwise defend yourself and your innovative concepts in what can be a very rough-and-tumble marketplace. Protecting your invention can be a tortuous journey—the steps outlined here should remove some of the tougher obstacles.

People occasionally confuse patents, copyrights, and trademarks. Although a resemblance exists in the rights granted of these three kinds of intellectual property, they are different and serve different purposes. Let's begin with patents.

PATENTS

What is a Patent?

A patent for an invention is a grant of property right by the U.S. Government to the inventor (or his heirs or assigns), acting through the Patent and Trademark Office (PTO).

The right conferred by the patent grant is, in the language of the statute and of the grant itself, "the right to exclude others from making, using, or selling" the invention. What is granted is not the right to make, use, or sell, but the right to exclude others from making, using, or selling the invention.

There are three types of patents:

1) Utility patent—granted for a new, useful, and nonobvious process, machine, manufactured article, composition, or an improvement in any of the above;

2) Design patent—available for the invention of a new, original, and ornamental design for an article of manufacture. A design patent only protects the appearance of an article and not its structure or utilitarian features;

3) Plant patent—provided to anyone who has invented or discovered and asexually reproduced any distinct and new variety of plant, including cultivated spores, mutants, hybrids, and newly found seedlings, other than a tuber-propagated plant or a plant found in an uncultivated state.

What Can Be Patented?

The patent law specifies the general field of subject matter that can be patented.

In the language of the statute, a person who "invents or discovers any new and useful process, machine, manufacture, or composition of matter or any new and useful improvements thereof, may obtain a [utility] patent," subject to the conditions and requirements of the law. The word **"process"** means a process or method, and new processes, primarily industrial or technical processes, that may be patented. The term **"machine"** used in the statute needs no explanation. The term **"manufacture"** refers to articles which are made, and includes all manufactured articles. The term **"composition of matter"** relates to chemical compositions and may include mixtures of ingredients as well as new chemical compounds. These classes of subject matter taken together include practically everything which is made by man and the process for making them.

In a 1980 Supreme Court decision the Court indicated that the PTO could not refuse a patent simply because it covered living subject matter. The PTO subsequently announced on April 17, 1987, that it would consider patents on animals after an agency appellate board ruled that oysters are patentable subject matter. This patent, however, did not issue. But it wasn't long until one did.

On April 12, 1988, patent history was made when the PTO issued the first patent covering an animal, specifically genetically engineered mice for cancer research.

You might also find it of interest that increased numbers of patents on computer software are issued each week. If you design such programs, don't fail to consider your work for possible patentability. You can go for it simultaneously with copyright protection.

What Cannot Be Patented?

Many things are not open to patent protection.

The laws of nature, physical phenomena, and abstract ideas are not patentable subject matter.

Human beings cannot be patented because, according to the PTO, the "grant of a limited, but exclusive property right in a human being is prohibited by the Constitution."

A new mineral discovered in the earth or a new plant found in the wild is not patentable subject matter. Likewise, Einstein could not patent his celebrated E=mc2; nor could Newton have patented the law of gravity. Such discoveries are manifestations of nature, free to all men and reserved exclusively to none.

The Atomic Energy Act of 1954 excludes the patenting of inventions useful solely in the utilization of special nuclear material or atomic energy for atomic weapons.

The patent law specifies that the subject matter must be "useful." The term "useful" in this connection refers to the condition that the subject matter has a useful purpose and that also includes operativeness, i.e., a machine which will not operate to perform the intended purpose would not be called useful, and therefore would not be granted patent protection.

Interpretations of the statute by the courts have defined the limits of the field of subject matter which can be patented. Thus it has been held that methods of doing business and printed matter cannot be patented.

In the case of mixtures of ingredients, such as medicines, a patent cannot be granted unless the mixture is more than the effect of its components. It is of interest to note that so-called "patent medicines" are generally not patented; the phrase "patent medicine" in this connection does not mean that the medicine has been awarded a patent.

A patent cannot be obtained on a mere idea or suggestion. The patent is granted for the new machine, manufacture, etc., and not for the idea or suggestion for a new machine.

Do Patents Matter?

Yes! Some companies in certain industries may regard patents as unimportant, but for the independent inventor experience dictates that it is wise to proceed under the assumption that acquiring a patent is important. In point of fact, few companies will license an idea unless it has been or can be protected by a patent. "If there is no patent, what are we licensing?" is a standard query when an inventor has not made application for patent protection.

According to *Business Week,* in 1988, 6,059 intellectual property suits were filed in the U.S., 60 percent more than in 1980. Whether it's Walt Disney Company protecting Snow White or Xerox guarding the proprietary copier, almost every week a property rights case is brought to court. Chief Judge Howard T. Markey of the U.S. Court of Appeals for Washington, D.C. Federal Circuit, a patent's court of last resort, is upholding patents 80 percent of the time.

Typically, the stronger the patent protection, the better the contract you will be able to negotiate. This is because a meaningful patent should keep any competition from manufacturing, using or selling your idea for at least a limited time. If nothing else, your manufacturer will have a head start, and manufacturing is an extremely competitive business. The winners usually have an edge, and a patent provides one.

In 1985, Polaroid's patents for its instant cameras pushed Eastman Kodak out of the market and pending damage trials could cost Kodak upwards of $10 billion.

Japan's Sumitomo Electric Industries was closed down by Corning Glass Works in 1987 when Corning won a patent infringement suit over its design of optical fibers.

As Donald J. Quigg, Commissioner of Patents and Trademarks, impresses upon new graduates of the Patent Academy, "Quality patent grants ensure an opportunity for rewards for creativity and security for investment, and they present an incomparable resource for new technological information."

Patent Laws

The U.S. Constitution gives Congress the power to enact laws relating to patents, in Article I, section 8, which reads, "Congress shall have power...to promote the progress of science and useful arts, by securing for limited times to authors and inventors the exclusive right to their respective writings and discoveries." Under this power Congress has from time to time enacted various laws relating to patents. The first patent law was enacted in 1790. The law now in effect is a general revision which was enacted July 19, 1952, and which went into effect January 1, 1953. It is codified in Title 35, United States Code.

The patent law specifies the subject matter for which a patent may be obtained and the conditions of patentability. The law establishes the Patent and Trademark Office (PTO) for administering the law relating to the granting of patents, and contains various other provisions relating to patents.

The Patent and Trademark Office

"The Patent Office is a curious record of the fertility of the mind of man when left to its own resources..." wrote an English lady who visited American in 1828.

The Patent and Trademark Office (PTO) administers the patent laws as they relate to the granting of patents for inventions, and performs other duties vis-a-vis patents. It examines applications for patents to determine if the applicants are entitled to patents under the law, and grants patents when they are so entitled; it publishes issued patents; records assignments of patents; maintains a search room for the use of the public to examine issued patents; and supplies copies of records and other papers. **The PTO has no jurisdiction over questions of infringement and the enforcement of patents, nor over matters relating to the promotion or utilization of patents or inventions.**

Today the PTO has 3,100 employees. More than 125,000 patent applications are received annually. Over five million pieces of mail arrive each year at the PTO.

It is no surprise, then, that the PTO is understaffed and the examiners overloaded with work in certain areas. For example, in the biotechnology field, the shortage of examiners who understand genetic engineering is severe. The PTO has been criticized for taking up to four years to decide biotechnology cases.

More than 4.5 million patents have been awarded since the Office was established in 1802. In Fiscal Year 1988, 83,584 patents issued.

Office Hours: The PTO's working hours are 8:30 a.m. to 5:00 p.m., Monday through Friday, excluding federal holidays in the District of Columbia.

Term of a U.S. Patent

The term of a **utility patent** is 17 years from the date of issue, subject to the payment of maintenance fees. The right conferred by a utility patent grant extends throughout the U.S.A. and its territories and possessions.

The term of a **design patent** is 14 years from the date of issue, and is not subject to the payment of maintenance fees. The right conferred by a design patent grant extends throughout the U.S. and its territories.

The term of a **plant patent** is 17 years and is not subject to maintenance fees.

See Fig. 2 for a full menu of current fees and charges.

Who May Apply for a Patent?

According to the law, only the inventor may apply for a patent, with certain exceptions. If you are a co-inventor, you may make a joint application. Financial contributors are not considered joint inventors, and cannot be joined in the application as an inventor. If you make an innocent mistake of erroneously omitting or naming an inventor, the application can be corrected. If a partisan who is not the inventor should apply for a patent, the patent, if awarded, would be invalid. Any person who falsely applies as the inventor is subject to criminal penalties.

If the inventor is dead, the application may be made by legal representatives, i.e., the administrator or executor of the estate. If the inventor is not mentally competent, the application for patent may be made by a guardian. If an inventor refuses to apply for a patent or cannot be found, a joint inventor or a person having a proprietary interest in the invention may apply on behalf of the missing inventor.

Officers and employees of the PTO are prohibited by law from applying for a patent or acquiring, directly or indirectly, except by inheritance or bequest, any patent or any right or interest in any patent.

Can Foreigners Apply For U.S. Patents?

Yes. The patent laws of the United States do not discriminate with respect to the citizenship of the inventor. Any inventor, regardless of citizenship, may apply for a patent on the same basis as a U.S. citizen. In fact, the number of foreigners obtaining U.S. patents continues to increase.

In 1987, foreign inventors were issued 48 percent of the U.S. utility patents, compared with 37 percent in 1976. In 1987, to no one's surprise, Japan placed five companies among the top 12 corporations receiving U.S. utility patents.

See 100 Corporations Receiving U.S. Patents for more information.

"The world is waking up as far as the value of intellectual property is concerned, and ours is the richest of markets," commented PTO Commissioner Donald J. Quigg, referring to the growth of foreign patenting in the United States.

Conditions for Obtaining a Patent

In order for an invention to be patentable it must be new as defined in the patent law, which provides that an invention cannot be patented if:

> "a) The invention was known or used by others in this country, or patented or described in a printed publication in this or a foreign country, before the invention thereof by the applicant for a patent, or
>
> "b) The invention was patented or described in printed publication in this or a foreign country or in a public use or on sale in this country more than one year prior to the application for patent in the United States..."

For these reasons care must be taken to **not** permit any public disclosure of a patentable idea unless you are prepared to make the application before one year goes by from the date it was first disclosed.

Exhibiting a new invention at an invention exposition or new product fair, writing a magazine piece about a new idea, or permitting someone else to report in print on your idea will make it ineligible for patent protection if a patent application is not filed within one year from the time of your invention's first public exposure.

Inventors should be very careful, therefore, about disclosing their inventions prior to making patent application. Many inventors who do not have the money to patent their products think that by exposing them at expos and fairs, and in the mass media, investors and/or licensees might see them, make an offer, and cover the patent with advance monies. Maybe so. But if you take this dangerous route before making a patent application, be aware that the clock starts ticking down when you go public. On top of this, you have exposed an unprotected idea to the world, a dangerous and questionable move under any circumstances. Knock-off artists frequent such events.

While I support such public displays as a method to promote and publicize inventions, they are simply not recommended as a place for unprotected products that you might hope one day to patent.

Even if your idea is not exactly shown by the prior art, i.e., patents on file, and involves one or more differences over the most nearly similar thing already known, a patent may still be refused if the differences would be obvious. The subject matter you seek to patent must be sufficiently different from what has been used or described before so that it may be said to be not obvious to a person having ordinary skill in the area of technology related to the invention. For example, the substitution of one material for another, or changes in size, are ordinarily not patentable.

Patent Application Time

The time to process a patent application from filing to issue or abandonment is called the "patent pendency time." The average patent pendency time decreased in fiscal year 1988 to 19.9 months for utility, reissue, and plant patents. The previous year averaged 20.8 months. The average patent pendency time for design patents was 30.7 months in fiscal year 1988.

The process may go more rapidly if you qualify for a Special Status Patent.

What are Special Status Patents?

The PTO allows special cases to qualify for accelerated processing. No special forms are required. A letter of explanation need only be attached to your application explaining why you feel acceleration handling should be allowed.

Certain pre-set categories automatically permit special status:

1) **Age (65).** If you are 65 years of age or older, your application may be made special by sending along a copy of your birth certificate. No fee is required with such a petition.

2) **Illness (Terminal).** If the state of health of the applicant is such that he or she might not be available to assist in the normal prosecution of the application, it can be made special by sending along a physician's letter attesting to that state of health. No fee is required with such a petition.

3) **Infringer (Actual).** Subject to a requirement for a further showing as may be necessitated by the facts of your particular case, an application may be made special because of actual infringement (but not for prospective infringement) upon payment of a fee and the filing of a petition alleging facts under oath or declaration to show, or indicating why it is not possible to show (1) that there is an infringing

device or product actually on the market or method in use; (2) when the device, product, or method alleged to infringe was first discovered to exist; supplemented by an affidavit or declaration from your attorney or agent to show; (3) that a rigid comparison of the alleged infringing device, product, or method with the claims of the application has been made; (4) that, in the applicant's opinion, some of the claims are unquestionably infringed; (5) that the applicant has made or caused to be made a careful and thorough search of the prior art or has good knowledge of the salient prior art; and, (6) that the applicant believes all of the claims in their application are allowable.

Models or specimens of the infringing product or your product should not be submitted unless requested.

4) **Manufacture (Ready $$).** If you have a manufacturer ready to go, this may also qualify your application for special handling. I have requested and received this on numerous occasions. You pay a fee and allege under oath or declaration: (a) that the licensee has the capital available (stating the approximate amount) and the facilities (stating briefly the nature thereof) to manufacture your invention in quantity or that sufficient capital and facilities will be made available if a patent is granted; (b) that the prospective licensee will not manufacture, or will not increase present manufacture, unless certain that the patent will be granted; and, (c) that you and/or your licensee stand obligated to produce the product immediately if the patent issues which protects the investment of capital and facilities.

Furthermore, you or your attorney must file an affidavit or declaration to show: (1) that you have searched or caused to be searched the prior art; and, (2) that you believe all the claims in your application to be allowable.

5) **Energy.** No fee is required to make special an application for an invention that will materially contribute to: (1) the discovery or development of energy resources; or, (2) the more efficient use and conservation of energy resources. Examples of inventions in category 1 would be developments in fossil fuels (natural gas, coal, and petroleum), nuclear energy, and solar energy. Category 2 would include inventions relating to the reduction of energy consumption in combustion systems, industrial equipment, household appliances, etc.

6) **Superconductors.** In accordance with the President's proposal directing the PTO to accelerate the processing of patent applications and adjudication of disputes involving superconductivity technologies when requested by the applicant to do so, the PTO will comply accordingly. No fee is required.

Examples of such inventions would include those directed to the superconductive materials themselves as well as to their manufacture and application.

7) **Environment.** If the invention will materially enhance the quality of the environment by contributing to the restoration or maintenance of the basic life-sustaining natural elements—air, water, and soil—you can request to make it special. No fee is required for such a petition.

8) **DNA.** In recent years revolutionary genetic research has been conducted involving recombinant deoxyribonucleic acid (recombinant DNA). Recombinant DNA appears to have extraordinary potential benefit for mankind. It has been suggested, for example, that research in this field may lead to ways of controlling or treating cancer and hereditary defects.

A petition to make special in this category must contain an affidavit or declaration in writing to the effect that your invention is directly related to safety of research in the field of recombinant DNA. A fee of must be paid.

General Information and Correspondence

The Patent and Trademark Office is a huge bureaucracy. Without a map and compass, so to speak, it can be frustrating or near impossible for a beginner (especially one living outside the National Capital area), to find out who handles what and how to contact any particular person directly. **To help you quickly reach the most appropriate person for your needs, see Appendix B, Patent and Trademark Office Telephone Directory.**

If you wish to mail a letter to someone at the Patent and Trademark Office, address your letter to the addressee care of: Patent and Trademark Office, Washington, D.C. 20231.

Since the PTO does not have resources for picking up any mail, including Express Mail, Post Office-to-Post Office Express Mail will not reach the Patent and Trademark Office. Express mail sent to a specific street address and office will arrive.

Special PTO mail box numbers should be used to allow forwarding of particular types of mails to the appropriate areas as quickly as possible. Mail will be forwarded directly to the appropriate area without being opened. If a document other than the specified type identified for each box is addressed to that box, expect delays in correct delivery.

Special Post Office Boxes

Commissioner of Patents and Trademarks, Washington, D.C. 20231

Box 3—Mail for the Office of Personnel from NFC.

Box 4—Mail for the Assistant Commissioner for External Affairs, or the Office of Legislation and International Inquiries.

Box 5—Non-fee mail related to trademarks.

Box 6—Mail for the Office of Procurement.

Box 7—Reissue applications for patents involved in litigation.

Box 8—All papers for Office of Solicitor except for letters relating to pending litigation.

Box 9—Coupon orders for U.S. patents and trademarks.

Box 10—Orders for certified copies of patents and trademark applications.

Box 11—Electronic Ordering Services (EOS).

Box AF—Expedited procedure for processing amendments and other responses after final rejection.

Box FWC—Requests for File Wrapper Continuation Applications.

Box Issue Fee—Issue Fee Transmittals and associated fees and corrected drawings.

Box M Fee—Letters related to interferences and applications and patents involved in interference.

Box Non Fee—Non-fee amendments to patent applications.

Amendment

Box Pat. Ext.—Applications for patent term extensions.

Box PCT—Mail related to applications filed under the Patent Cooperation Treaty.

Box reexam—Mail related to re-examination.

Patent

Application—New patent application and associated papers and fees.

Trademark

Application—New trademark application and associated papers and fees.

Separate letters (but not necessarily in separate envelopes) should be written in relation to each distinct subject of inquiry, such as assignments, payments, orders for printed copies of patents, orders for copies of records, and requests for other services. None of these should be included with letters responding to Office actions in applications.

If your letter concerns a patent application, include the serial number, filing date, and Group Art Unit number. When a letter concerns a patent, include the name of the patentee, the title of the invention, the patent number, and date of issue.

If ordering a copy of an assignment, provide the book and page or reel and frame record, as well as the name of the inventor; otherwise, the PTO will assess an additional charge to cover the time consumed in making the search for the assignment.

The PTO will not send or show you applications for patents. They are not open to the public, and no information concerning patent applications is released except on written authority of the applicant, assignee, or designated attorney, or when necessary to the conduct of the PTO's business. You can write for and receive, however, records of any decisions, the records of assignments other than those relating to assignment of patent applications; books; or other records and papers in the PTO that are open to the public.

The PTO will not respond to inquiries concerning the novelty and patentability of an invention in advance of the filing of an application; give advice as to possible infringement of a patent; advise of the propriety of filing an application; respond to inquiries as to whether or to whom any alleged invention has been patented; or act as an expounder of the patent law or as a counselor for individuals, except in deciding questions arising before it in regularly filed cases.

Keep copies of everything you send the PTO. Don't forget to put your name and return address on all papers and the envelope.

In case of emergency the PTO has established a contingency plan for filing any paper or paying any fee in the Office in the event of an emergency caused by any major interruption in the country's mail service.

Upon the determination by the Commissioner of Patents and Trademarks that such an emergency exists, the Commissioner will cause to be printed a notice of the plan in the *Wall Street Journal* and make it available by telephone at (703) 557-3158. Also, certain publications, patent bar groups, and other organizations closely associated with the patent system will be notified. Termination of the emergency program will be similarly announced. Where the postal emergency is not nationwide, the Commissioner will designate the areas of the country in which the procedures outlined will be in effect.

The plan calls for the U.S. Department of Commerce District Offices to be designated as emergency receiving stations for filing papers and paying fees in the PTO.

Disclosure Document Program

If you are not ready, or don't care to apply for a patent yet but want to officially evidence and register the conception date of an invention, the PTO offers a disclosure document program. For a fee of $6.00, the Office will preserve your idea on file for a period of two years. This inexpensive recognition will strengthen your case if any conflict arises as to the date of your conception, but is not meant as a replacement of an inventor's notebook or actual patent.

The requirements are simple. Send the PTO a paper disclosing the invention. Although there are no stipulations as to content, and claims are not required, the benefits afforded by the Disclosure Document Program will depend directly upon the adequacy of your disclosure. Therefore, it is strongly recommended that the document contain a clear and complete explanation of the manner and process of making and using the invention in sufficient detail to enable a person having ordinary knowledge in the field of the invention to make and use the invention. When the nature of the invention permits, a drawing or sketch should be included. The use or utility of the invention should be described, especially in chemical inventions.

This disclosure is limited to written matter or drawings on paper other than thin, flexible material, such as linen or plastic drafting material, having dimensions or being folded to dimensions not to exceed 8-1/2 x 13 inches (21.6 by 33 cm.). Photographs are also acceptable. Number each page, and make sure that the text and drawings are of such quality as to permit reproduction.

In addition to the $6.00 fee, the Disclosure Document must be accompanied by a stamped, self-addressed envelope (SASE) and a separate paper in duplicate, signed by you as the inventor. These papers will be stamped by the PTO upon receipt, and the duplicate request will be returned in the SASE together with a notice indicating that the Disclosure Document may be relied upon only as evidence that a patent application should be diligently filed if patent protection is desired.

Your request may take the following form:

> *"The undersigned, being the inventor of the disclosed invention, requests that the enclosed papers be accepted under the Disclosure Document Program, and that they be preserved for a period of two years."*

Warning: The two-year retention period should not be considered to be a "grace period" during which you can wait to file a patent application without possible loss of benefits. It must be recognized that in establishing priority of invention an affidavit or testimony referring to the Disclosure Document must usually also establish diligence in completing the invention or in filing the patent application since the filing of the Disclosure Document.

Also be reminded that any public use or sale in the U.S. or publication of the invention anywhere in the world more than one year prior to the filing of your patent application on the invention disclosed will prohibit the granting of a patent on it.

The Disclosure Document is not a patent application, and the date of its receipt in the Patent and Trademark Office will not become the effective filing date of any patent application subsequently filed. It will be retained for two years and then be destroyed unless referred to in a separate letter in a related application within two years.

The program does not diminish the value of the conventional witnessed and notarized records as evidence of conception of your invention, but it should provide as more credible form of evidence than that provided by the popular practice of mailing a disclosure to oneself or another person by registered mail.

IT ALL BEGINS WITH A SEARCH

Why Make a Patent Search?

Before making an application for patent protection, it is advisable to see whether or not any prior art exists, i.e. if the same concept has been patented by someone else. This is done through what is called a **patent search**.

A thorough search is recommended for numerous reasons. First, making a search involves less expense than trying to obtain a patent and having it rejected on the basis of prior art. If the search reveals that the invention cannot be protected as engineered, the cost of preparing and filing an application, as well as significant time and energy will be saved.

Further, even if none of the earlier patents show all the details of the invention, they may point out important features or better ways of doing the job. If this is the case, you may not want to get patent protection on an invention that could encounter commercial difficulty.

In the event that nothing is found in the search that would prevent or delay application, the information gathered during the search will prove helpful, acquainting you with the details of patents related to the invention.

Who Conducts a Patent Search?

There are three ways to approach the search. You could have a patent attorney do it for you; engage the services of a professional patent search firm or individual; or do it yourself.

Law Firm Initiated Search

Nothing is wrong with having a law firm handle the search. Just understand that lawyers do not usually personally do patent searches, but typically hire a search firm or individual and mark-up the fees they are charged. Law firms may step on search fees 40% or more depending upon the firm and city.

If you decide to retain a lawyer for this task, remember to insist upon an estimate of the costs in advance. The fee will be based upon how far back and encompassing you want the search.

See Chapter 14, Registered Patent Attorneys and Agents for information on the hiring of a patent lawyer or agent.

Direct Hire Professional Search

If you want to save yourself the lawyer's mark-up, consider hiring a patent searcher. Typically, they are listed in the yellow pages under "Patent Searchers." Be careful to contact a firm or person who specializes in this work and not an invention marketing organization listed under a toll-free number. Many such firms use the search as a hook to sucker unsuspecting inventors into their rapacious grasp. Get all the facts before committing yourself. **Always ask for an estimate of charges in advance.**

Presently in Washington, D.C., searchers charge between $30.00 and $60.00 per hour. This range depends a great deal upon whether it is an independent or large firm. A simple search can run between six to seven hours for a mechanical patent. Chemical and electrical patents average eight to ten hours. Fixed price deals for $180 are available, providing three hours and six references.

There will be a fee for making copies as well. Some searchers mark this up while some go at cost. I've received estimates from 40 cents per page to $3.00 per patent no matter how many pages.

If you live in a city with one of the PTO's 61 Patent Depository Libraries (PDLs), a librarian can usually assist you in finding a list of those who make a living searching patents. Librarians are not encouraged to personally recommend any patent searcher or search organization.

See Chapter 13, Patent Depository and Special Libraries for more information.

The better the lawyer and/or searcher understands the invention, the better the search. Highlight the novel features of your invention. This explanation can be made through drawings or sketches, models, written description, or a discussion, or a combination of these. Take the time; it'll be worth it.

Correspondence concerning the search should be kept in a safe place, you may need it to prove dates and other facts about the invention later.

Do-it-Yourself Search

You may opt to do the patent search. Several methods are available.

1) The PTO runs a **Patent Public Search Room** located in Arlington, Virginia. Here, every U.S. patent granted since 1836 (over four million) may be searched and examined. Patents are arranged according to a classification system of more than 400 classes and 115,000 subclasses. **See Appendix A, U.S. Patent Classifications, for a complete list.**

The Patent Public Search Room is really something to behold. You can touch and feel the original documents, including everything from Abraham Lincoln's 1849 patent (No. 6,469) for a device to buoy vessels over shoals to Auguste Bartholdi's design patent on a statue entitled, Liberty Enlightening The World (a/k/a The Statue of Liberty).

For more information on important historical patents, see Significant Inventions and Inventors in U.S. History.

Many inventors like to make at least one pilgrimage to the Crystal Plaza facility. It is located less than five minutes from National Airport by taxi and Metro Rail (Blue and Yellow Lines - Crystal City Station). Many fine hotels are within walking distance.

Upon your arrival the PTO will issue you, at no cost, a non-transferable User Pass for the day. It is wise to double check the hours of operation by calling (703) 557-2276. Depending where in the facility you want to search, the Patent Public Search Room is typically open from 8:00 a.m. to 8:00 p.m.

2) Nearby the Patent Public Search Room is the **The Scientific Library of the Patent and Trademark Office.** The Scientific Library makes publicly available over 120,000 volumes of scientific and technical books in various languages, about 90,000 bound volumes of periodicals devoted to science and technology, the official journals of 77 foreign patent organizations, and over 12 million foreign patents. The hours are from 8:45 a.m. to 4:45 p.m.

3) If you cannot make the trip to Washington, D.C., you may inspect copies of the patents at a **Patent Depository Library.** The Patent Depository Libraries, a nationwide network of prestigious academic, research, and public libraries, continues to be one of the PTO's most effective mechanisms for publicly disseminating patent information. PDLs receive current issues of U.S. Patents and maintain collections of earlier issued patents. The scope of these collections varies from library to library, ranging from patents of only recent years to all or most of the patents issued since 1790. Due to the variations in the scope of patent collections among the PDLs and their hours of service to the public, you should call first to find out when it is open, to avoid any possible inconvenience.

Online Patent Searches

Available at all PDLs is the **Classification and Search Support Information System (CASSIS)**, an online computer database.

CASSIS brings you all the data available at the Patent and Trademark Search Room. And it can streamline the manual search procedure by providing electronic access to basic patent search tools. CASSIS can help you define a "field of search" and identify the patents in that field. It provides classifications of a patent; supplies patents in a classification; displays patent titles and/or company names; structures classification titles; finds key words in classification titles and patent abstracts; and much more.

In an effort to reduce cost of providing CASSIS online, the PTO has begun providing a CD-ROM (Compact Disk Read-Only Memory) to over 1,500 libraries. The disks contain full abstracts of all the patents issued in the most recent twelve months, the full text of the U.S Patent Classification Manual, and the Assignee File. Issued every two months, the disks are available to the public at low cost.

Contact local PDLs listed in Chapter 13, Patent Depository and Special Libraries, for Access to the CASSIS system.

Other Databases

Databases are available that savvy searchers use to augment their work. Access through host systems such as ORBIT and DIALOG, these files include, for example, Derwent's World Patent Index covering 33 countries and technology back to 1963. The information provided by Derwent is in the form of abstracts and bibliographic background. Some databases just supply bibliographic data. If you decide to use one, find out about scope of its information.

Currently in the option period of a five-year contract with the PTO, Derwent will be will be the only database contractor permitted to operate out of the PTO's Public Search Room for the next two years. Derwent's direct dial telephone number is (703) 486-8155.

Computer-aided searches are not by any means the final word. The best they can do is generate abstracts of patents. Someone must ultimately retrieve copies of the full patents and study them.

For more information on databases, see Chapter 11, Online Resources for the Inventor.

Patent Search Steps

No matter where you decide to conduct a search, certain steps must be taken in any patent search. Here is a brief guide to manual searches for U.S. patents.

Step 1: If you know the PATENT NUMBER, go to the *Official Gazette* to read a summary of the patent. This publication is available at any of the above-mentioned patent search facilities and in many public library reference rooms.

Step 2: If you know the PATENTEE or ASSIGNEE, look at the *Patent/Assignee Index* to locate the PATENT NUMBER. This is available at any of the patent search facilities. In Crystal City, it is on microfiche and in card catalogues.

Step 3: If you know the SUBJECT, start with: *Index to the U.S. Patent Classification.* **The most recent *Index* is included in this book as Appendix A, U.S. Patent Classifications.** This will help you prepare for a search and familiarize yourself with the PTO's indexing system By doing some homework before you begin searching, i.e., familiarizing yourself with the *Index* and locating the Class and Subclass numbers for terms that pertain to your invention, you'll save time.

Step 4: Once you have jotted down the class(es) and subclass(es) out of the *Index,* refer to the *Manual of Classification* and check this information vis-a-vis the hierarchy to see if they are close to what you need. The *Manual of Classification* is available at all patent search facilities.

Step 5: Using the Class/Subclass numbers you have found, look at the *U.S. Patent Classification Subclass and Numeric Listing* and copy the patent numbers of patents assigned to the selected Class/Subclass. If you are at the Crystal City facility, take the Class/Subclass numbers into the stacks of patents and begin "pulling shoes". To pull shoes is to physically remove patent groupings from the open shelves.

Step 6: Then, using the *Official Gazette* again, look at the patent summaries of those patents. At the Crystal City facility, you will not have to go back to this publication since the actual patents are there.

Step 7: Upon locating the relevant patents, examine the complete patent in person or on microfilm—depending upon where you conduct the search.

If you do not locate the patent you are seeking, try again using another subject class and subclass.

More Help—If you Have the Time

If you want help after referring to the *Index to the U.S. Patent Classification,* you may write for free assistance to the Commissioner of Patents and Trademarks.

Sample Letter

Date:

Commissioner of Patents and Trademarks
Att: Patent Search Division
Washington, D.C. 20231

Please let me know what subject area or class(es) and subclass(es) cover my idea. The enclosed sketches on the back of this sheet, with each part labeled, show my intended invention.

My idea has certain features of structure, mode of operation, and intended uses which I have defined below.

I understand that there is no charge for this information.

Thank you for your help.

Sincerely,

(1) Features of structure, or how it is constructed.

(2) Mode of operation, or how it works.

(3) Intended uses, or purpose of idea.

(4) Rough sketches of idea, viewed from all sides, with labels to identify each part. [If necessary use extra sheets of paper, either plain or lined. These sketches may be made in pencil and need not be drawn to scale.]

How to Order Copies of Searched Patents

You may order copies of original patents or cross-referenced patents contained in subclasses comprising the field of search from the Patent and Trademark Office. Mail your requests to: Patent and Trademark Office, Box 9, Washington, D.C. 20231.

Payment may be made by check, coupons, or money. Expect a wait of up to four weeks when ordering copies of patents from the PTO. Postage is free.

For the convenience of attorneys, agents, and the general public in paying any fees due, deposit accounts may be established in the PTO with a minimum deposit of $50.00. For information on this service, call (703) 557-3227.

What to Do with Your Search Results?

Study the results of the patent search. You're out of luck if anyone of the patented inventions is exactly like yours; your invention may even be infringing on another invention. On the other hand, one or more patents may describe inventions that are intended for the same purpose as yours, but are different in various ways.

Look these over and decide whether it is worthwhile to proceed.

If the features that make your invention different from the prior art provide important advantages, you should discuss the situation with your attorney to determine whether there is a fair chance of obtaining a patent covering these features.

I have found from experience that a good patent attorney can often get some claim(s) to issue, albeit not always strong ones. A patent for patent's sake is usually possible. Whether it will be worth the paper it is printed on is yet another matter. Do not take this decision lightly because the patent process is not inexpensive. The average utility patent will cost about $2,500.

HOW TO APPLY FOR A PATENT

Your application must be directed to the Commissioner of Patents and Trademarks and include:

1) A written document which comprises a specification (description and claims), and an oath or declaration. Two forms, are used in association with patent applications: Power of Attorney or Authorization of Agent, Not Accompanying Application and Revocation of Power of Attorney or Authorization of Patent Agent. **These forms may be found as part of Fig. 9.**

Power of Attorney... If you file your application with an attorney or agent, no power of attorney form is required (it will be written into the oath or declaration). If however, you decide to appoint an attorney or agent *after* your application, you may use the form provided.

Revocation of Power of Attorney... If, after your application, you decide to change attorney or agent use this form.

2) A drawing in those cases in which a drawing is required; and

3) The correct filing fee.

The specification and oath or declaration must be legibly written or printed in permanent ink on one side of the paper only. The PTO prefers typewriting on legal size paper, 8 to 8 1/2 by 10 1/2 to 13 inches, 1 1/2 or double spaced, with margins of one inch on the left-hand side and at the top. If the papers filed are not correctly, legibly, or clearly written, the Office may require typewritten or printed papers.

The application for patent will not be sent for examination until all of the required parts, complying with the rules of presentation, are received. If the papers and parts are incomplete or so defective they cannot be accepted as a complete application for examination, you will be notified about the found deficiencies and given a time period in which to remedy them.

To help make sure everything is together, a copy of a **Patent Application Transmittal Letter is shown as Fig. 3.** This will serve as your cover document.

The four formal responses used by the Application Division to notify inventors of defects in their applications are reproduced as **Fig. 4, Notice of Informal Application; Fig. 5, Notice of Incomplete Application; Fig. 6, Notice to File Missing Parts of an Application—No Filing Date; and Fig. 7, Notice to File Missing Parts of an Application—Filing Date Granted.** Become familiar with them, for sooner or later you are sure to receive one if you do patent work.

Upon receiving one of these notices, a surcharge may be required. If you do not respond within the prescribed time period, your patent application will be returned or otherwise discarded. The filing fee may be refunded if the application is refused acceptance as incomplete; however, a handling fee will likely be charged.

File all the application documents together; otherwise, each part must be signed and accompanied by a letter accurately and clearly connecting it with the other parts of your application. This can be a nightmare, so send everything in together from the start and save yourself headaches.

Every application received by the Office is numbered in serial order, and you are informed of the serial number and filing date of the application by a filing receipt. **This document does not mean that a patent has been awarded, only that your application is at the PTO.**

Keep copies of everything you submit. If you ask the PTO for copies of what you've submitted, there will be a charge for the service.

Applications for patents are not open to the public, and no information concerning them is released except on written authority of the applicant, assignee, or designated attorney, or when necessary to the conduct of the PTO's business.

Oath or Declaration

By law the inventor must file an oath or declaration. The inventor must make an oath or declare that he/she is believed to be the original and first inventor of the subject matter of the application. Additionally, the inventor must make various other allegations required by law and by PTO rules.

If you opt for an oath, it must be sworn before a notary public or other officer authorized to administer oaths.

I favor using the declaration in lieu of an oath, making it part of the original application. A declaration does not require notarization.

A sample blank application, which includes the declaration, is shown as Fig. 8. The Office will not supply blank forms.

Filing Fees

Caution: Fees change from time to time. People not in the PTO information loop are not alerted to price hikes. It is not exactly front-page news. There are penalties for making incorrect payments, so double check the required amount before sending in a check. The PTO Fees list, printed in the Forms and Documents section of this book, is current per publication, but still should be confirmed. The latest fee information is available by calling (703) 557-5168.

FACSIMILE TRANSMISSIONS TO THE PTO

Effective November 1, 1988, certain papers to be filed in national patent applications and re-examination proceedings for consideration by the Office of the Assistant Commissioner for Patents, the Office of the Deputy Assistant Commissioner for Patents, and the Patent Examining Groups (Patent Examining Corps) may be submitted to the Patent and Trademark Office (PTO) by facsimile transmission.

The provision of 37 CFR 1.33(a), requiring signatures on amendments and other papers filed in applications, is hereby waived to the extent that a facsimile signature is acceptable. The paper that is used as the original for the facsimile transmission must have an original signature, and should be retained by applicant or the representative as evidence of the content of the facsimile transmission. No special format, addressing information or written ratification is required for facsimile transmissions. However, the paper size must be 8 1/2 inches by 14 inches or smaller to be accepted.

A facsimile center has been established in the Patent Examining Corps to receive and process submissions. The filing date accorded the submission will be the date the complete transmission is received by the PTO facsimile unit unless that date is a Saturday, Sunday or Federal holiday within the District of Columbia, in which case the official date of receipt will be the next business day.

Each transmission session must be limited to papers to be filed in a single national patent application or re-examination proceeding. It is recommended that the Serial Number of the application or Control Number of the re-examination be entered as part of the sender's identification, if possible. It is also recommended that the sending facsimile machine generate a report confirming transmission for each transmission session. The transmitting activity report should be retained along with the paper used as the original.

The papers, including authorizations to charge deposit accounts, which may be submitted using this procedure, are limited to those which may be filed in national patent applications and re-examination proceedings and which are to be considered by the PTO organizations named above. Examples of such papers are amendments, responses to restriction requirements, requests for reconsideration before the examiner, petitions, terminal disclaimers, powers of attorney, notices of appeal, and appeal briefs.

New or continuing patent applications of any type, assignments, issue fee payments, maintenance fee payments, declarations or oaths under 37 CFR 1.63 or 1.67, and formal drawings are excluded, as are all papers relating to international patent applications. Papers to be filed in applications that are subject to a secrecy order under 37 CFR 5.1-5.8, and directly related to the secrecy order content of the application, are also excluded. Informal communications between applicant and the examiner, such as proposed claims for interview purposes, are permissible and are encouraged. Informal communications from applicants will not be made of record in the application or re-examination and must be clearly identified as informal such as by including the word "DRAFT" on each paper. To facilitate informal communications from examiners, applicants are encouraged to supply their facsimile phone numbers on communications to the Office.

The facsimile submissions may include a certificate for each paper stating the date of transmission. A copy of the facsimile submission with a certificate attached thereto will be evidence of transmission of the paper should the original be misplaced. The person signing the certificate should have a reasonable basis to expect that the paper would be facsimile transmitted on the date indicated. An example of a preferred certificate is:

Certification of Facsimile Transmission

I hereby certify that this paper is being facsimile transmitted to the Patent and Trademark Office on the date shown below.

____

Type or print name of person signing certification

________________________________ __________

Signature *Date*

When possible, the certification should appear on a portion of the paper being transmitted. If the certification is presented on a separate paper, it must identify the application to which it relates, and the type of paper being transmitted, e.g., amendment, notice of appeal, etc.

In the event that the facsimile submission is misplaced or lost in the PTO, the submission will be considered filed as of the date of the transmission, if the party who transmitted the paper:

1) Informs the PTO of the previous facsimile transmission promptly after becoming aware that the submission has been misplaced or lost;

2) Supplies another copy of the previously transmitted submission with the Certification of the Transmission; and

3) Supplies a copy of the sending unit's report confirming transmission of the submission. In the event that a copy of the report is not available, the party who transmitted the paper may file a declaration under 37 CFR 1.68 which attests on a personal knowledge basis or to the satisfaction of the Commissioner to the previous timely transmission.

If all criteria above cannot be met, the PTO will require applicant to submit a verified showing of facts. Such a showing must show to the satisfaction of the Commissioner the date the PTO received the submission.

The facsimile center will have five facsimile units and will be staffed during the business hours of 8:30 a.m. and 5:00 p.m., Monday through Friday, excluding holidays. Although the units may normally be accessed at all times, including non-business hours, there may be times when reception is not possible due to equipment failure or maintenance requirements. Accordingly, applicants are cautioned not to rely on the availability of this service at the end of response periods.

The telephone number for accessing the facsimile machines is (703) 557-9564. In the event that the transmission cannot be accepted at the telephone number above, a backup number has been established at (703) 557-9567. The facsimile center staff can be reached at telephone number (703) 557-4277 during normal business hours.

Small Entity Status Reduces Fees by Half

On August 27, 1982, Public Law 97-247 provided that effective October 1, 1982, funds would be made available to the PTO to reduce by 50 percent the payment of fees by independent inventors, small business concerns, and nonprofit organizations.

The reduced fees include those for patent application, extension of time, revival, appeal, patent issues, statutory disclaimer, and maintenance on patents based on applications filed on or after August 27, 1982.

Fees that are **not** reduced include petition and processing, (other than revival), document supply, certificate of correction, request for re-examination, international application fees, and certain maintenance fees.

What is an Independent Inventor?

The PTO considers an inventor as independent if the inventor (1) has not assigned, granted, conveyed, or licensed, and (2) is under no obligation under contract or law to assign, grant, convey, or license any rights in the invention to any person who could not likewise be classified as an independent inventor if that person had made the invention, or to any concern which would not qualify as a small concern or a nonprofit organization.

What is a Small Business Concern?

The PTO defines a small business as one whose number of employees, including those of its affiliates, does not exceed 500 persons. The definition also requires a small business, for this purpose, to be one which has not assigned, granted, conveyed, or licensed, and is under no obligation under contract or law to assign, grant, convey, or license, any rights in the invention to any person who could not be classified as an independent inventor if that person had made the invention, or to any concern which would not qualify as a small business concern or a nonprofit organization.

What is a Nonprofit Organization?

To be recognized as a nonprofit organization it must be so accredited by a nationally recognized accrediting agency or association or of the type described in Section 501 (c) (3) of the IRS Code of 1954 (26 U.S.C. 501 (c) (3)) and which is exempt from taxation under 26 U.S.C. 501 (a).

Four PTO forms are available for claiming small entity status. See Figs. 10, 11, 12, and 13.

Specification (Description and Claims - Utility Patents)

The specification is a written description of the invention and of the manner and process of making and using it, and is required to be in such full, clear, concise, and exact language as to enable any person skilled in the technological area to which the invention pertains, or with which it is most nearly connected, to make and use it. For the record, even after all these years and a familiarity with patent descriptions.

The specification must set forth the precise invention for which a patent is solicited, in such a manner as to distinguish it from other inventions and from what is old. It must describe completely a specific embodiment of the process, machine, manufacture, composition of matter, or improvement invented, and must explain the method of operation or principle whenever applicable. The best way you see to carry out the invention also must be described.

In the case of an improvement, your specification must particularly point out the part(s) of the process, machine, manufacture, or composition of matter to which the improvement relates, and the description should be confined to the specific improvement and to such parts as necessarily co-operate with it or as may be required to complete understanding or description of it.

Title and Abstract

The title of your invention, which should be as short and specific as possible, should be a heading on the first page of the specification, if it does not otherwise appear at the beginning of the application.

A brief abstract of the technical disclosure in the specification must be set forth in a separate page immediately following the claims in a separate paragraph under the heading "Abstract of the Disclosure." The purpose of the abstract is to enable the PTO and the public generally to determine quickly from a cursory inspection the nature and gist of the technical disclosure. It is not used for interpreting the scope of the claims. I have included the following as an example of an Abstract:

> An electronic toy doll including electronic circuitry for selectively generating a number of simulation sounds typically associated with a mystic or science-fantasy character. In respective operational modes, the sound of wind, the sound of breathing, an eerie pseudo-random sequence of musical notes and sounds representing the operation of a weapon are selectively generated. In a further operational mode, a random one of a predetermined number of responses are provided upon generation of an actuation signal. The random response may be considered to be an answer to inquiries. Preferred circuitry for generating the simulation sounds is described, accessories, adapted for removable interconnection with the circuitry are also described.

Summary of the invention

A short summary of your invention indicating its nature and substance, which may include a statement of the object of the invention, should precede the detailed description. Your summary should, when set forth, be commensurate with the invention you claim and any object recited should be that of the invention as claimed. I have included the following as an example of a Summary:

> The present invention provides a relatively inexpensive, rugged electronic doll. The doll selectively simulates the sounds typically associated with a mystic or science-fantasy character. The sounds of wind, breathing, weapons operation, and an eerie tune of pseudo-random musical notes are selectively produced. Additionally, the future is "foretold" through generation of a random one of a predetermined number of gong sounds generated in response to an actuation signal. The operational mode is controlled by a central function select logic circuit in cooperation with remote switches. The remote switches may be located in the doll body, or may be disposed in removably interconnectable remote accessories.

Reference to Drawings

If you have drawings, the PTO requires a brief description of several views of the drawing and the detailed description of the invention shall refer to the different views by specifying the numbers of the figures and to the different parts by use of reference letters or numerals.

A sample United States Patent, which includes drawings and references, is shown as Fig. 1.

Claims

I would recommend that you hire a lawyer to prepare and prosecute a utility patent **(see Chapter 14, Registered Patent Attorneys and Agents).** The preparation of a patent application is a highly complex business and not to be taken lightly.

If you are bent, however, on writing your own description and claims, it is best to study many patents for style and content. Then locate an updated copy of the PTO's *Statutory Requirements of Claims* at the nearest Patent Depository Library. You can photocopy them from the *Manual of Patent Examining Procedure* which is available at each PDL.

For an example of Claims, see below; also Fig. 1, United States Patent.

In the meantime, here are some points to get you going:

> a) The specification section of the patent must conclude with a claim particularly pointing out and distinctly claiming the subject matter which the application regards as your invention or discovery.
>
> b) The claims are brief descriptions of the subject matter of your invention, eliminating unnecessary details and reciting all essential features necessary to distinguish the invention from what is old. The claims are the operative part of the patent. Novelty and patentability are judged by the claims, and, when a patent is granted, questions of infringement are judged by the courts on the basis of the claims.

c) When more than one claim is presented, they may be placed in dependent form in which a claim may refer back to and further restrict one or more preceding claims.

d) A claim in multiple dependent form shall contain a reference, in the alternative only, to more than one claim previously set forth and then specify a further limitation of the subject matter claimed. A multiple dependent claim shall not serve as a basis for any other multiple dependent claim. A multiple dependent claim shall be construed to incorporate by reference all the limitations of the particular claim in relation to which it is being considered.

e) The claims or claims must conform to the invention as set forth in the remainder of your specification and the terms and phrases used in the claims must find clear support or antecedent basis in the description so that the meaning of the terms in the claims may be ascertainable by reference to the description.

A sample of Claims is shown, in part, below:

1) An electronic doll comprising:

a body;

function select means, disposed on said body, for selectively generating respective mode control signals;

electronic signal generator means, disposed in said body and responsive to said mode control signals, including:

means for generating a wind simulation signal representative of the sounds of wind;

means for generating a breathing simulation signal representative of the sounds of breathing;

means for generating a weapons simulation signal representative of weapons fire;

means for generating pseudo-random musical notes; and

means for generating one of a predetermined number of responses to an actuation signal, said one response being determined in a random manner;

transducer means for generating audible output signals representative of electrical input signals applied thereto; and

means, responsive to said mode control signals, for selectively applying said respective electrical simulation signals as input signals to said transducer means.

2) The electronic doll of claim 1 wherein said means for generating a wind simulation signal comprises:

pseudo-random signal generator means, responsive to clock signals applied thereto for producing a pseudo-random signal, said pseudo-random signal having a digital value changing in a pseudo-random manner at a rate in accordance with said clock signal;

oscillator means for generating said clock signal at a frequency in accordance with a frequency control signal applied thereto; and

irregular signal generator means, responsive to said pseudo-random signal, for generating an irregular signal which alternatively rises or falls in amplitude in an irregular manner, said irregular signal being applied to said oscillator means as said frequency control signal.

3) The electronic doll of claim 2 wherein said pseudo-random signal generator means comprises:...[etc.].

Arrangement of Application Elements

The elements of the application should appear in the following order:

1) Title of the invention; or an introductory portion stating the inventor(s) name(s), citizenship, and residence of the applicant, and the title of the invention may be used.

2) Cross-references to related applications, if any.

3) Brief summary of the invention.

4) Brief description of the several views of the drawing, if there are drawings.

5) Detailed description.

6) Claim(s).

7) Abstract of the disclosure.

You are also encouraged to file an **Information Disclosure Citation (Fig. 14)** at the time of filing your patent application or within three months after the filing date of the application or two months after a filing receipt is received. If filed separately, be sure to tag it with the Group Art Unit to which the action is assigned as indicated on the filing receipt. This form will enable you to provide the PTO with a uniform listing of citations.

While the filing of information disclosure statements is voluntary, the form must be accompanied by an explanation of relevance of each listed item, a copy of each listed patent or publication of other item of information and a translation of the pertinent portions of foreign documents (only if an existing translation is available to you).

Examiners will consider all citations submitted and place their initials adjacent the citations in the boxes provided on the form.

Drawings

When you apply for a patent, it is required by law in most cases that you furnish a drawing of the invention together with your application. This includes practically all inventions except compositions of matter or processes, but a drawing may also be useful in the case of many processes.

The drawing must show every feature of the invention specified in your claims and is required by the PTO rules to be in a particular form. The Office specifies the size of the sheet on which the drawing is made, the type of paper, the margins, and other details relating to the making of the drawing. The reason for specifying the standards in detail is that the drawings are printed and published in a uniform style when the patent issues, and the drawings must also be such that they can be readily understood by persons using the patent descriptions.

No names or other identification will be permitted within the "sight" of the drawing, and applicants are expected to use the space above and between the hole locations to identify each sheet of drawings. This identification may consist of the attorney's name and docket number or the inventor's name and case number and may include the sheet number and the total number of sheets filed (for example, "sheet 2 of 4").

Drawings in colors other than black and white are not acceptable unless the drawing requirements are waived. Only the Deputy Assistant Commissioner for Patents can make the decision.

Color drawings are permitted for plant patents where color is a distinctive feature.

See Fig. 1, United States Patent for an example of format used in patent drawings.

HOW TO GET PATENT DRAWINGS DONE

You have several options. You can do-it-yourself. To this end standards for drawings are included below to guide your efforts and keep you to the letter of the required standards. I have never attempted to do the drawings for a utility patent, but I have been successful at doing a few uncomplicated design patent drawings.

You could ask that your patent attorney arrange for the drawings. This is all right, but understand that just as in the search business, the lawyer or law firm will be adding a premium of typically no less than 40% to the fees charged by the draftsman. If you go with the law firm, get a guarantee that the drawings will pass PTO muster. If they are rejected for any technical reason whatsoever, you do not want to be double billed, i.e. pay for the draftsman's mistakes.

On more than one occasion I have had to have extra charges taken off my bills for the reworking of unacceptable drawings.

The surest and least expensive way, if you are not a draftsman, is to take bids and hire your own draftsman. Find the fair market price by calling around. Look in the telephone directory or ask at a regional patent library for candidates. Get the same guarantee for PTO acceptance from the draftsman if you contract the work yourself.

Standards for Patent Drawings

a) **Paper and ink.** Drawings must be made upon paper which is flexible, strong, white, smooth, nonshiny, and durable. Two-ply or three-ply bristol board is preferred. The surface of the paper should be calendered and of a quality which will permit erasure and correction with India ink. India ink, or its equivalent in quality, is preferred for pen drawings to secure perfectly black solid lines. The use of white pigment to cover lines is not normally acceptable.

b) **Size of sheet and margins.** The size of the sheets on which drawings are made may either be exactly 8 1/2 by 14 inches (21.6 by 35.6 cm.) or exactly 21.0 by 29.7 cm. (DIN size A4). All drawing sheets in a particular application must be the same size. One of the shorter sides of the sheet is regarded as its top.

1) On 8 1/2 by 14 inch drawing sheets, the drawing must include a top margin of 2 inches (5.1 cm.) and bottom and side margins of 1/4 inch (6.4 mm.) from the edges, thereby leaving a "sight" precisely 8 by 11 3/4 inches (20.3 by 29.8 cm.). Margin border lines are not permitted. All work must be included within the "sight." The sheets may be provided with two 1/4 inch (6.4 mm.) diameter holes having their centerlines spaced 11/16 inch (17.5 mm.) below the top edge and 2 3/4 inches (7.0 cm.) apart, said holes being equally spaced from the respective side edges.

2) On 21.0 by 29.7 cm. drawing sheets, the drawing must include a top margin of at least 2.5 cm., a left side margin of 2.5 cm., a right side margin of 1.5 cm., and a bottom margin of 1.0 cm. Margin border lines are not permitted. All work must be contained within a sight size not to exceed 17 by 26.2 cm.

c) **Character of lines.** All drawings must be made with drafting instruments or by a process which will give them satisfactory reproduction characteristics. Every line and letter must be durable, black, sufficiently dense and dark, uniformly thick and well defined; the weight of all lines and letters must be heavy enough to permit adequate reproduction. This direction applies to all lines however fine, to shading, and to lines representing cut surfaces in sectional views. All lines must be clean, sharp, and solid. Fine or crowded lines should be avoided. Solid black should not be used for sectional or surface shading. Freehand work should be avoided wherever it is possible to do so.

d) **Hatching and shading.** 1) Hatching should be made by oblique parallel lines spaced sufficiently apart to enable the lines to be distinguished without difficulty. 2) Heavy lines on the shade side of objects should preferably be used except where they tend to thicken the work and obscure reference characters. The light should come from the upper left-hand corner at a 45-degree angle. Surface delineations should preferably be shown by proper-shading, which should be open.

e) **Scale.** The scale to which a drawing is made ought to be large enough to show the mechanism without crowding when the drawing is reduced in size to two-thirds in reproduction, and views of portions of the mechanism on a larger scale should be used when necessary to show details clearly; two or more sheets should be used if one does not give sufficient room to accomplish this end, but the number of sheets should not be more than is necessary.

f) **Reference characters.** The different views should be consecutively numbered figures. Reference numerals (and letters, but numerals are preferred) must be plain, legible, and carefully formed, and not be encircled. They should, if possible, measure at least one-eighth of an inch (3.2 mm.) in height so that they may bear reduction to one twenty-fourth of an inch (1.1 mm.); and they may be slightly larger when there is sufficient room. They should not be so placed in the close and complex parts of the drawing as to interfere with a thorough comprehension of the same, and therefore should rarely cross or mingle with the lines. When necessarily grouped around a certain part, they should be placed at a little distance, at the closest point where there is available space, and connected by lines with the parts to which they refer. They should not be placed upon hatched or shaded surfaces but when necessary, a blank space may be left in the hatching or shading where the character occurs so that it shall appear perfectly distinct and separate from the work. The same part of an invention appearing in more than one view of the drawing must always be designated by the same character, and the same character must never be used to designate different parts. Reference signs not mentioned in the description shall not appear in the drawing and vice versa.

g) **Symbols, legends.** Graphic drawing symbols and other labeled representations may be used for conventional elements when appropriate, subject to approval by the Patent and Trademark Office. The elements for which such symbols and labeled representations are used must be adequately identified in the specification. While descriptive matter on drawings is not permitted, suitable legends may be used, or may be required, in proper cases, as in diagrammatic views and flow sheets or to show materials or where labeled representations are employed to illustrate conventional elements. Arrows may be required, in proper cases, to show direction of movement. The lettering should be as large as, or larger than, the reference characters.

h) **[Reserved]**

i) **Views.** The drawing must contain as many figures as may be necessary to show the invention; the figures should be consecutively numbered if possible in the order in which they appear. The figure may be plain, elevation, section, or perspective views, and detail views of portions of elements, on a larger scale if necessary, may also be used. Exploded views, with the separated parts of the same figure embraced by a bracket, to show the relationship or order of assembly of various parts are permissible. When necessary, a view of a large machine or device in its entirety may be broken and extended over several sheets if there is no loss in facility of understanding the view. Where figures on two or more sheets form in effect a single complete figure, the figures on the several sheets should be so arranged that the complete figure can be understood by laying the drawing sheets adjacent to one another. The arrangement should be such that no part of any of the figures appearing on the various sheets are concealed and that the complete figure can be understood even though spaces will occur in the complete figure because of the margins on the drawing sheets. The plane upon which a sectional view is taken should be indicated on the general view by a broken line, the ends of which should be designated by numerals corresponding to the figure number of the sectional view and have arrows applied to indicate the direction in which the view is taken. A moved position may be shown by a broken line superimposed upon a suitable figure if this can be done without crowding, otherwise a separate figure must be used for this purpose. Modified forms of construction can only be shown in separate figures. Views should not be connected by projection lines nor should centerlines be used.

j) **Arrangement of views.** All views on the same sheet should stand in the same direction and, if possible, stand so that they can be read with the sheet held in an upright position. If views longer than the width of the sheet are necessary for the clearest illustration of the invention, the sheet may be turned on its side so that the top of the sheet with the appropriate top margin is on the right-hand side. One figure must not be placed upon another or within the outline of another.

k) **Figure for Official Gazette.** The drawing should, as far as possible, be so planned that one of the views will be suitable for publication in the *Official Gazette* as the illustration of the invention.

l) **Extraneous matter.** Identifying indicia (such as the attorney's docket number, inventor's name, number of sheets, etc.) not to exceed 2 3/4 inches (7.0 cm.) in width may be placed in a centered location between the side edges within three-fourths inch (19.1 mm.) of the top edge. Authorized security markings may be placed on the drawings provided they be outside the illustrations and are removed when the material is declassified. Other extraneous matter will not be permitted upon the face of a drawing.

m) **Transmission of drawings.** Do **not** send original patent drawings with your application. Under new regulations in effect since January, 1989, a clean photocopy is all that needs to be submitted so long as the lines are uniformly thick, black, and solid. Should the PTO request a change in the drawing, you may submit a revised copy of the original.

Drawings transmitted to the PTO should be sent flat, protected by a sheet of heavy binder's board, or may be rolled in a suitable mailing tube; but must never be folded. If received creased or mutilated, new drawings will be required.

How to Effect Drawing Changes

If your patent was filed before January, 1, 1989, it was required that you submit an original copy of the patent drawing to the PTO. If you find yourself in need of changing such a drawing you have two options. The first is to file new drawings with the changes incorporated therein. The art unit number, serial number, and number of drawing sheets should be written on the reverse side of the drawings. You may delay filing the new drawings until receipt of the "Notice of Allowability." If delayed, the new drawings must be filed within the three month shortened statutory period set for response in the before-mentioned form. You may request an extension. Drawings should be filed as a separate paper, with a transmittal letter addressed to the PTO Official Draftsman or examiner.

If you do not wish to go through the expense of having drawings totally reworked, contact one of the bonded patent draftsmen listed below. Bonded draftsmen are the only draftsmen authorized by the PTO to touch-up original patent drawings filed at the PTO.

Under new regulations, the authorization of bonded patent draftsmen will expire as of January 1, 1991; thereafter, no one will be permitted to change original patent drawings. After this date, new drawings will have to be submitted should there be a need for changes or corrections in original patent drawings filed in the PTO.

In any case, no changes will be permitted, other than correction of informalities, unless the examiner has approved the proposed changes.

Should the bonded system be dismantled? Robert MacCollum, a bonded draftsman from Silver Spring, Maryland, feels the inventor should have access through a bonded system, "so I don't think that it's such a hot idea." On the other hand, he feels that not having it will do away with a lot of confusion and running around for he and his colleagues. "Some draftsmen dropped their bond because it was such a nuisance to have to go to the PTO in person and pick up these drawings. Also, with the bonded system, it is tough to estimate your work. It might take three trips to find and get drawings. This must be added into the cost of you work. Some out-of-towners will never send you that extra money."

The following is the complete roster of PTO bonded draftsmen. Publication of this list does not constitute a recommendation or endorsement by me or the PTO.

Bonded Draftsmen

John A. Ballard
2001 Jefferson Davis Hwy.
Suite 705, Crystal Plz. 1
Arlington, VA 22202
(703) 685-7228

Anthony L. Costantino
17300 Lafayette Dr.
Olney, MD 20832
(301) 924-3491

Fleit, Jacobson, Cohn
& Price
Jennifer Bldg.
400 7th St., N.W.
Washington, D.C. 20004
(202) 638-6666

Ellsworth G. Jackson
101 Rittenhouse St., N.E.
Washington, D.C. 20011
(202) 726-0908

Litman Law Offices, Ltd.
P.O. Box 15035
Crystal City Station
Arlington, VA 22215-1728
(301) 920-6004

Robert MacCollum
Patent Drafting Services
13108 Engelwood Dr.
Silver Spring, MD 20904
(301) 622-3940

Mason, Fenwick &
Larence
1730 Rhode Island Ave.,
N.W.
Washington, D.C. 20036
(202) 295-2010

Thomas E. Melvin
TEM Patent Drafting
P.O. Box 15809
Arlington, VA 22215
(703) 349-8518

Mil-R Productions
3110 Mt. Vernon Ave.
Suite 100
Alexandria, VA 22305
(703) 548-3879

J. A. Mortenson
Patent Drafting
P.O. Box 518
Northeast, MD 21901
(301) 287-8669

Gerald M. Murphy
P.O. Box 2098
Eads Street Station
Arlington, VA 22202
(703) 881-1928

Suzanne Nahmias
D S N Patent Drafting
P.O. Box 2431
Gaithersburg, MD 20879
(301) 869-0756

Edward J. Oliver
Oliver Patent
Drafting Service
1205 Darlington St.
Forestville, MD 20747
(301) 336-0351

Patent Reproduction Co.
26 N St., SE.
Washington, D.C. 20003
(202)488-7096

Quality Patent Printing
P.O. Box 2404
556 S. 22nd St.
Arlington, VA 22202
(703) 892-6212

Quinn Patent
Drawing Service
1416 Duke St.
Alexandria, VA 22314
(703) 521-5940

Philip Sweet T.A. and for
Oblon, Fisher, Spivak
1755 S. Jefferson Davis
Crystal Square Five,
Suite 400
Arlington, VA 22202
(703) 521-5940

Applications Filed Without Drawings

Not all applications require drawings. It has been a long approved procedure, for example, to accept a process case (i.e., a case having only process or method claims) which is filed without drawings.

Other situations where drawings are usually not considered essential for a filing date are:

1) **Coated articles or products.** If the invention resides only in coating or impregnating a conventional sheet, e.g., paper or cloth, or an article of known and conventional character with particular composition.

2) **Articles made from a particular material or composition.** If the invention consists in making an article of a particular material or composition, unless significant details of structure or arrangement are involved in the claims.

3) **Laminated structures.** If the invention involves only laminations of sheets (and coatings) of specified material, unless significant details of structure or arrangement (other than the mere order of layers) are involved in the claims.

4) **Articles, apparatus, or systems where sole distinguishing feature is presence of a particular material.** If the invention resides solely in the use of particular material in an otherwise old article, apparatus, or system recited broadly in its claim.

Photographs are not normally considered to be proper drawings. Photographs are acceptable for a filing date and are generally considered to be informal drawings. Photographs are only acceptable where they come within special categories. Photolithographs are **never** acceptable.

The PTO is willing to accept black and white photographs or photomicrographs (not photolithographs or other reproductions of photographs made using screens) printed on sensitized paper in lieu of India ink drawings, to illustrate inventions which are incapable of being accurately or adequately depicted by India ink drawings restricted to the following categories: crystalline structures, metallurgical microstructures, textile fabrics, grain structures, and ornamental effects.

Models are generally **not** required as part of an application or patent. They have not been for many, many years. And save for cases involving perpetual motion, or a composition of matter, you will probably never have to submit one.

If your invention is, however, a microbiological invention, a deposit of the microorganism is required.

WHAT HAPPENS TO YOUR APPLICATION AT THE PTO?

If your application passes initial muster, it will be assigned to the appropriate examining group, and then to an examiner. Applications are handled in the order received.

The application examination inspects for compliance with the legal requirements and includes a search through U.S. patents, prior foreign patent documents which are available in the PTO, and available literature to ensure that the invention is new. A decision is reached by the examiner in light of the study and the result of the search.

First Office Action

You or your attorney will be notified of the examiner's decision by what the PTO refers to as an "action." An action is actually a letter that gives the reasons for any adverse action or any objection or requirement. Noted will be any appropriate references or information that you'll find useful in making the decision to continue the prosecution of the application or to drop it.

If the invention is not considered patentable subject matter, the claims will be rejected. If the examiner finds that the invention is not new, the claims will be rejected; but the claims may also be rejected if they differ somewhat from what is found to be obvious. It is not uncommon for some or all of the claims to be rejected on the examiner's first action; very few applications sail through as first submitted.

A sample Examiner's Action is shown as Fig. 16.

Your First Response

Let's say the examiner gives you the thumbs down on all or some of your claims. Your next move, if you wish to continue prosecuting the patent, is to respond, specifically pointing out the supposed errors in the examiner's action. Patent examiners have a lot on their plates and their units are typically understaffed for the amount of work they handle. For example, the PTO reported in November of 1988 that 6,500 biotechnology patents were awaiting a decision or about 70 per Government examiner. In this area alone the PTO predicts that the number of applications will grow by 12 percent a year through the early 1990s.

Examiners must process a specific number of patents to be considered productive by their superiors for periodic job performance ratings. The bottom line is that as careful as they try to be, they make mistakes that can be reversed with careful argument.

Your response should address every ground of objection and/or rejection. Show where the examiner is wrong. The mere allegation that the examiner has erred is not enough.

Your response will cause the examiner to reconsider, and you'll be notified if the claims are rejected, or objections or requirements made, in the same manner as after the first examination. This second Office action usually will be made final.

Feel free to call your examiner up on the telephone to discuss your case. I have always found them to be most hospitable and helpful. His or her telephone number will appear at the end of the Office action or you can look it up in **Appendix B, Patent and Trademark Office Telephone Directory.**

Depending upon how serious the matter, you or your attorney might wish to make an appointment to personally visit the examiner. Don't just drop in unannounced. It is to your benefit that the examiner has the time to prepare for your visit and get up to speed on the case. Remember that personal interviews do not remove the necessity for response to Office actions within the required time, and the action of the Office is based solely on the written record.

If you feel that you can handle the matter alone, try it and save the attorney's fees. Patent examiners will meet with inventors. I've always found it to be a rewarding experience.

Final Rejection

On the second or later consideration, the rejection of claims may be made final. Your response is then limited to appeal and further amendment is restricted. You may petition the Commissioner in the case of objections or requirements not involved in the rejection of any claim. Response to a final rejection must include cancellation of, or appeal from the rejection of each claim so rejected and, if any claim stands allowed, compliance with any requirement or objection as to its form.

In determining such final rejection, your examiner will repeat or state all grounds of rejection then considered applicable to your claims as stated in the application.

The odds? As in the case of the examination by the Office, patents are granted in the case of about two out of every three applications filed.

Making Amendments to Your Application

The preceding section referred to amendments to an application. Following are some details concerning amendments:

1) The applicant may amend before or after the first examination and action as specified in the rules, or when and as specifically required by the examiner.

2) After final rejection or action, amendments may be made canceling claims or complying with any requirement of form which has been made but the admission of any such amendment or its refusal, and any proceedings relative thereto, shall not operate to relieve the application from its condition as subject to appeal or to save it from abandonment.

3) If amendments touching the merits of the application are presented after final rejection, or after appeal has been taken, or when such amendment might not otherwise be proper, they may be admitted upon a showing of good and sufficient reasons why they are necessary and were not earlier presented.

4) No amendment can be made as a matter of right in appealed cases. After decision on appeal, amendments can only be made as provided in the rules.

5) The specifications, claims, and drawing must be amended and revised when required, to correct inaccuracies of description and definition of unnecessary words, and to secure correspondence between the claims, the description, and the drawing.

All amendments of the drawings or specifications, and all additions thereto, must conform to at least one of them as it was at the time of the filing of the application. Matter not found in either, involving a departure from or an addition to the original disclosure, cannot be added to the application even though supported by a supplemental oath or declaration, and can be shown or claimed only in a separate application.

The claims may be amended by canceling particular claims, by presenting new claims, or by amending the language of particular claims (such amended claims being in effect new claims). In presenting new or amended claims, the applicant must point out how they avoid any reference or ground rejection of record which may be pertinent.

Erasures, additions, insertions, or alterations of the papers and records must not be made by the applicant. Amendments are made by filing a paper, directing or requesting that specified changes or additions be made. The exact word or words to be stricken out or inserted in the application must be specified and the precise point indicated where the deletion or insertion is to be made.

Amendments are "entered" by the Office through the making of proposed deletions by drawing a line in red ink through the word or words canceled and by making the proposed substitutions or insertions in red ink, small insertions being written in at the designated place and larger insertions being indicated by reference.

> No change in the drawing may be made except by permission of the Office. Permissible changes in the construction shown in any drawing may be made only by the bonded draftsmen. A sketch in permanent ink showing proposed changes, to become part of the record, must be filed for approval by the Office before the corrections are made. The paper requesting amendments to the drawing should be separate from other papers.
>
> If the number or nature of the amendments render it difficult to consider the case, or to arrange the papers for printing or copying, the examiner may require the entire specification or claims, or any part thereof, to be rewritten.
>
> The original numbering of the claims must be preserved throughout the prosecution. When claims are canceled, the remaining claims must not be renumbered. When claims are added by amendment or substituted for canceled claims, they must be numbered by the applicant consecutively beginning with the number next following the highest numbered claim previously presented. When the application is ready for allowance, the examiner, if necessary, will renumber the claims consecutively in the order in which they appear or in such order as may have been requested by applicant.

To better organize your request for amendment, a copy of the PTO's Amendment Transmittal Letter format appears as Fig. 17.

Time for Response and Abandonment

The maximum period given for response is six months, but the Commissioner has the right to shorten the period to no less than thirty days. The typical response time allowed to an Office action is three months. If you want a longer time, you usually have to pay some extra money for an extension. The amount of the fee depends upon the response time desired. If you miss any target date, your application will be abandoned by the PTO and made no longer pending. However, if you can show whereby your failure to prosecute was unavoidable or unintentional, the application can be revived by the Commissioner.

The revival requires a petition to the Commissioner and a fee for petition, which should be filed without delay. The proper response must also accompany the petition if it has not yet been filed.

Appeal to the Board of Patent Appeals and Interferences and to the Courts

If the examiner circles his or her wagons and begins to stonewall, there is a higher court. Rejections that have been made final may be appealed to the Board of Patent Appeals and Interferences. This august body consists of the Commissioner of Patents and Trademarks, the Deputy Commissioner, the Assistant Commissioners, and the examiners-in-chief, but typically each appeal is heard by only three members. An appeal fee is required and you must file a brief in support of your position. You can even get an oral hearing if you pay enough.

An alternative to appeal in situations where you wish consideration of different claims or further evidence is to file a new continuation application. This requires a filing fee and the claims and evidence for which consideration is desired.

If the Board goes against you, there is yet a higher court, the Court of Appeals for the Federal Circuit. Or you might file a civil action against the Commissioner in the U.S. District Court for the District of Columbia. He won't take it personally. It goes with the territory. The Court of Appeals for the Federal Circuit will review the record made in the Office and may affirm or reverse the Office's action. In a civil action, you may present testimony in the court, and the court will make a decision.

Channels of Ex Parte Review are illustrated as Fig. 18.

Interferences

Parallel development is a phenomenon that should not be discounted. On numerous occasions a company executive has said to me, "I've seen that concept twice in the last month," or something to this effect. At times, two or more applications may be filed by different inventors claiming substantially the same patentable invention. A patent can only be granted to one of them, and a proceeding known as an "interference" is instituted by the Office to determine who is the first inventor and entitled to the patent. About one percent of all applications filed become engaged in an interference proceeding.

Interference proceedings may also be instituted between an application and a patent already issued, if the patent has not been issued for more than one year prior to the filing of the conflicting application, and if that the conflicting application is not barred from being patentable for some other reason.

The priority question is determined by a board of three examiners-in-chief on the evidence submitted. From the decision of the Board of Patent Appeals and Interferences, the losing party may appeal to the Court of Appeals for the Federal Circuit or file a civil action against the winning party in the appropriate U.S. district court.

The terms "conception of the invention" and "reduction to practice" are encountered in connection with priority questions. "Conception of the invention" refers to the completion of the devising of the means for accomplishing the result. "Reduction to practice" refers to the actual construction of the invention in physical form. In the case of a machine it includes the actual building of the machine. In the case of an article or composition it includes the actual carrying out of the steps in the process; actual operation, demonstration, or testing for the intended use is usually required. The filing of a regular application for patent completely disclosing the invention is treated as equivalent to reduction to practice. The inventor who proves to be the first to conceive the invention and the first to reduce it to practice will be held to be the prior inventor, but more complicated situations cannot be stated this simply.

Here is a case when it is important to be able to have evidence that proves when you first had an idea and when the prototype was made. This is why you should keep careful and accurate records throughout the development of an idea. The Disclosure Document Program was established by the PTO for this purpose.

Fig. 21, Patent Worksheet, was designed to help inventors keep track of when ideas originated; for information on the Disclosure Document Program, see page 14.

ALLOWANCE AND ISSUE OF PATENT

If your utility patent is found to be allowable, a notice of allowance will be sent to you or your attorney. Within three months from the date of the notice you must pay an issue fee.

What Rights Does a Patent Give You?

It's a pretty exciting moment when you get your first patent. It comes bound inside a beautiful oyster white folder that has the United States Constitution screened in blue as its background. The large official gold seal of the Patent and Trademark Office is embossed thereon with two red ribbons furcate as a tail.

Between the covers of this folder is your patent, a grant that gives you the inventor(s) "the right to exclude others from making, using or selling the invention throughout the United States" and its territories and possessions for a period of 17 years (14 for a design patent) subject to the payment of maintenance fees as provided by law. The patent does not give you the right to make, use, or sell the invention. Any person is ordinarily free to make, use, or sell anything he or she pleases, and a grant from Uncle Sam is not required.

If you receive a patent for a new soda pop and the marketing of said beverage is prohibited by law, the patent will not help you. Nor may you market said soda pop if by doing so you infringe the prior rights of others.

Since the essence of the right granted by a patent is the right to exclude others from commercial exploitation of the invention, the patentee is the only one who may make, use, or sell the invention. Others may not do so without your authorization. You may assign your rights in the invention to another person or company.

After a patent has expired, anyone may make, use, or sell the invention without permission of the patentee, provided the matter covered by other unexpired patents is not used. The terms may not be extended save for by a special act of Congress.

Maintenance Fees

All utility patents which issue from applications filed on or after December 12, 1980 are subject to the payment of maintenance fees which must be paid to keep the patent in force. These fees are due at 3.5, 7.5, and 11.5 years from the date the patent is granted and can be paid without a surcharge during the six-month period preceding each due date, e.g. 3 years to 3 years and six months, etc. The amounts of the maintenance fees are subject to change every three years.

The PTO does not mail notices to patent owners advising them that a maintenance fee is due.

If you have a patent attorney tracking your business, he or she will let you know when the money is due. An attorney gets paid every time your business moves across his or her desk. But if you are doing it by yourself and you miss a payment, it may result in the expiration of the patent. A six month grace period is provided when the maintenance fee may be paid with a surcharge.

Patents relating to some pharmaceutical inventions may be extended by the Commissioner for up to five years to compensate for marketing delays due to federal FDA pre-marketing regulatory procedures. Patents relating to all other types of inventions can only be extended by congressional legislation.

Can Two People Own A Patent?

Yes. Two or more people may jointly own patents as either inventors, investors, or licensees. Most of my patents are joint ownerships. Anyone who shares in the ownership of a patent, no matter how small a part they might own, has the right to make, use, or sell it for his or her own profit, unless prohibited from doing so by prior agreement. It is accordingly dangerous to assign part interest in a patent of yours without having a definite agreement hammered out vis-a-vis respective rights and obligations to each other.

Can You Sell Your Patent?

Yes. The patent law provides for the transfer or sale of a patent, or of an application for patent, by a contract. When assigned the patent, the assignee becomes the owner of the patent and has rights identical to those of the original patentee.

Assignment of Patents Applications

Should you wish to assign your patent or patent application to a third party (manufacturer, investor, university, employer, etc.), this is possible by filing the appropriate form, **Assignment of Patent, or Assignment of Patent Application, Fig. 19.**

You can sell all or part interest in a patent. If you prefer, you could even sell it by geographic region.

I consider patents as valuable properties—personal assets. Never assume that because you have been unsuccessful in selling a patent it has no value. You might sell it eventually or find someone infringing it, thus turning it to positive account.

Infringement of Patents

Infringement of a patent consists in the unauthorized making, using, or selling of the patented invention within the territory of the U.S. during the term of the patent. If your patent is infringed, it is your right to seek relief in the appropriate federal court.

When I see an apparent infringement of a patent of ours, as has occurred occasionally over the years, the first thing I do is call the company and set a meeting. I am not litigious. Things can often be worked out between parties. Thus far, I have always been able to do this. Court battles over patents can be long and expensive affairs. And, if you want to continue working in your particular field, it is wise to not make too many corporate enemies.

I have a close friend who beat a large manufacturer in court to the tune of a million dollars. After splitting the award with his partner and attorney, and paying taxes and costs, he came away with about $250,000. It sounded good at first. But he soon realized that every door in his industry was closed to him. None of the major companies wanted to see his concepts. They still don't. It was a high price to pay. The right idea with the right company can be worth millions.

Several years ago I saw an infringement of a patent we hold. One call to the company's president, and a quick fax of our patent, brought immediate relief in the form of a royalty on all items made to date and in the future. It was an early Christmas. Not only that, but I was invited to submit ideas for licensing consideration.

If your friendly approach is turned away, and you are sure of your position, then the next step is to get a lawyer and decide if a Temporary Restraining Order (TRO) is appropriate. A TRO is an injunction to prevent the continuation of the infringement. You may also ask the court for a award of damages because of the infringement. In such an infringement suit, the defendant may raise the question of the validity of the patent, which is then decided by the court. The defendant may also aver that what is being done does not constitute infringement.

Infringement is determined primarily by the language of the claims of the patent and, if what the defendant is making does not fall within the language of any of the claims of the patent, there is no infringement.

The PTO has no jurisdiction over questions relating to infringement of patents. In examining applications for patent, no determination is made as to whether the patent-seeking invention infringes any prior patent.

What Does "Patent Pending" Signify?

These words are put on a product by a manufacturer to inform the public (and competition) that an application for patent on that item is on file at the PTO. It means stay away. The law imposes a fine on those who use these words falsely to deceive the public.

DESIGN PATENTS

If you want an inexpensive patent that will give you little actual protection because they are easy to end-run, but still meet the requirement that a patent issue, maybe a design patent is for you. It appears to be popular with lots of folks. Over 6,000 new designs were patented last year.

Inventors get design patents on just about anything, e.g. baby bibs, sweatbands, tissue box holders, dishes, ammo boxes, game boards, vending machines, telephones, pencils and pens, and even internal combustion engines.

Rubbermaid has one for a cereal container. Totes protects its umbrella handle designs. The Parker Pen Company takes them out for writing instruments. And the Ford Motor Company has them on car parts such as automobile quarter panels.

For little expense and effort a design patent is another way to stake out a claim. It permits you to legally post a no trespass sign in the form of **Patent Pending**, **Patent Applied For**, or **Patent** and the number of the patent.

The range of ornamental appearances that have been patented during the 147-year history of design patents is most impressive. Over 270,000 designs have received design patent protection since the first one was granted to George Bruce for "Printing Type" on November 9, 1842.

If you've invented any new, original, and ornamental design for an article of manufacture, a design patent may be appropriate. A design patent protects only the appearance of an article, and not its structure or utilitarian features. The proceedings relating to granting of design patents are the same as those relating to other patents with few differences.

In design cases as in "mechanical" cases, novelty and unobviousness are necessary prerequisites to the grant of a patent. In the case of designs, the inventive novelty resides in the shape or configuration or ornamentation as determining the appearance or visual aspect of the object or article of manufacture in contradistinction to the structure of a machine, article of manufacture, or the constitution of a composition of matter. Simply put, it is the appearance presented by the object that creates an impression upon the mind of the observer.

A utility patent and a design patent may be based upon the same object matter; however, there must be a clearly patentable distinction between them.

Are Patent Attorneys Required for Design Patents?

This is, of course, a personal decision. I never engage a patent attorney's services for design patent applications. Unlike the complicated business of utility patents, the design patent application process is very easy and uncomplicated. I have found that paying a lawyer to do my design patents is like tossing money out the window because the application form is simple, esoteric language is not required to draft the claims, and searches are actually so pleasant an experience that I will often do them myself.

There is not much to read in a design patent search. You're just looking at a lot of line drawings, many of which are fascinating.

To give you an idea how they are presented, refer to Fig. 20, Design Patent —Shoe.

It cost me $500 to learn that lawyers were not necessary in such matters. At the time I had been using a fine law firm for utility patents, and I naturally released our first design patent to them. I knew no better.

My lawyer said that he would "write up the specifications." I was asked to provide his draftsman with a prototype of the design. It was a tricycle with a mainframe shaped like a toothpaste tube. It became Procter & Gamble's Crest Fluorider. When the design patent arrived, I saw the simplicity of the specifications (see Design Patent Specification below). Design patents are obvious cash cows for patent attorneys. I have removed the product name so that you will be able to use its format for your design patents by inserting a title. This is a sample only and, unlike other Figures in this book, should not be returned to the Patent Office. The petition, specification, and claim must be typed on legal-sized paper.

Design Patent Title

The title is of great importance in a design patent application. It serves to identify the article in which the design is embodied and which is shown in the drawing, by a name generally used by the public. The title should be to a specific definite article. Thus a stove would be called a "Stove" and not "Heating Device." The same title is used in the preamble to the specification, in the description of the drawing, and in the claim.

To allow latitude of construction it is permissible to add to the title - "or similar article." The title must be in the singular.

Design Patent Specification

Be it known that I (we), _________ and __________ have invented a new, original, and ornamental design for a___________, over which the following is a specification. Reference is made to the accompanying drawing, which forms a part hereof, wherein:

FIGURE 1 is a perspective view of a _______ in accordance with the present design;

FIGURE 2 is a rear elevational view of said __________________;

FIGURE 3 is a side elevational view of said _____________, the side of the ___________ not shown being a mirror image of the side shown in this view;

FIGURE 4 is a front elevational view of said _______.

WHAT I (we) CLAIM IS:

The ornamental design for a _____________ or similar article, as shown and described.

(Note: attached to this application are the draftsman drawings)

Design Patent Drawings

Unless you are capable of doing this work, hire a competent, experienced patent draftsman to make your drawings. The requirements for drawings are strictly enforced. Professional draftsmen will stand behind their work and guarantee revisions if requested by the Office due to inconsistencies in the drawings.

The claim of the design patent determines its classification, i.e. the appropriate class and subclass into which the design patent will be placed. This classification is designated as the "original classification" of the patent. Copies of a patent may be placed in other subclasses for the convenience of the examiner as an aid in searching during the examination process. Additional copies are designated "cross reference."

The PTO changes the Design Patent Classification System as required to provide an appropriate area for each patented design. Areas which show great activity are expanded while other areas, used infrequently as activity fades, are compressed into other subclasses.

A copy of the operative Design Patent Classification System may be found in Appendix A, U.S. Patent Classifications.

PLANT PATENTS

The law provides for the granting of a patent to anyone who has invented or discovered and asexually reproduced any distinct and new variety of plant, including cultivated sports, mutants, hybrids, and newly found seedlings, other than a tuber-propagated plant or a plant found in an uncultivated state.

Asexually propagated plants are those that are reproduced by means other than from seeds, such as by the rooting of cuttings, by layering, budding, grafting, and inarching.

With reference to tuber-propagated plants, for which a plant patent cannot be obtained, the term "tuber" is used in its narrow horticultural sense as meaning a short, thickened portion of an underground branch. The only plants covered by the term "tuber-propagated" are the Irish potato and the Jerusalem artichoke.

Plant Variety Protection Act

The Plant Variety Protection Act (Public Law 91-557), approved December 24, 1970, provides for a system of protection for sexually reproduced varieties, for which protection was not previously provided, under the administration of a Plant Variety Protection Office within the Department of Agriculture. Requests for information regarding the protection of sexually reproduced varieties should be addressed to Commissioner, Plant Variety Protection Office, Consumer and Marketing Service, Grain Division, 6525 Bellcrest Road, Hyattsville, Maryland 20782.

Elements of a Plant Application

An application for a plant patent consists of the same parts as other applications and must be filed in duplicate, but only one need be signed and executed; the second copy may be a legible carbon copy of the original. Two copies of color drawings must be submitted.

The reason for submitting two copies is that one, the photocopy, must go to the Department of Agriculture for an advisory report on plant variety. The original is retained at the PTO.

Applications for a plant patent which fail to include two copies of the specification and two copies of the drawing when in color, will be accepted for filing only. The Application Division will notify you immediately if something is missing from the filing. You'll be given one month to rectify the situation. Failure to do so will result in loss of the filing date and the fee paid.

Plant Patent Specification and Claim

The specification should include a complete, detailed description of the plant and its characteristics that distinguish the plant over related known varieties and its antecedents. The specification must be expressed in botanical terms (in the general form followed in standard botanical text books or publications dealing with the varieties of the kind of plant involved), rather than the non-botanical characterizations commonly found in nursery or seed catalogs. The specification should also include the origin or parentage of the plant variety sought to be patented and must particularly point out where and in what manner the variety of plant has been asexually reproduced.

When color is a distinctive feature of the plant, it should be positively identified in the specification by reference to a designated color as given by a recognized color dictionary, for example, cherry red blooms. Where the plant variety originated as a newly found seedling, the specification must fully describe the conditions (cultivation, environment, etc.) under which the seedling was found growing to establish that it was not found in an uncultivated state.

A plant patent is granted on the entire plant. It, therefore, follows that only one claim is necessary and only one is permitted.

Plant Patent Oath or Declaration

Your oath or declaration as inventor, in addition to the statements required for other applications, must include the statement that you asexually reproduced the new plant variety. Where the plant is a newly found plant the oath or declaration must also state that it was found in a cultivated area.

Plant Patent Drawings

Plant patent drawings are not mechanical drawings and should be artistically and competently executed. The drawing must disclose all the distinctive characteristics of the plant capable of visual representation. When color is a distinguishing characteristic of the new variety, the drawing must be in color. Two duplicate copies of color drawings must be submitted. Color drawings may be made either in permanent water color or oil, or in lieu thereof may be photographs made by color photography or properly colored on sensitized paper. The paper in any case must correspond in size, weight, and quality to the paper required for other drawings. Mounted photographs are acceptable.

All color drawings should be mounted so as to provide a two-inch margin at the top for Office markings when the patent is printed.

Specimens

Specimens of the plant variety, its flower, or fruit, should not be submitted unless specifically called for by the examiner. If the PTO wants to inspect a plant that cannot be physically submitted, it will send an examiner to the growing site.

Fees and Correspondence

All inquiries relating to plant patents and pending plant patent applications should be directed to the Patent and Trademark Office and not to the Department of Agriculture.

For more information on fees, see Fig. 2, PTO Fee Schedule.

TREATIES AND FOREIGN PATENTS

The rights granted by a U.S. patent extend only throughout the territory of the U.S. and have no effect in a foreign country. Therefore, to receive patent protection in other countries you'll have to make separate application(s) in each of the other countries or in regional patent offices. Almost every country has its own patent law.

The laws in many countries differ from our own. In most foreign countries, publication of the invention before the date of the application will bar the right to a patent. Most foreign countries require maintenance fees and that the patented invention be manufactured in that country within a certain period, usually about three years. If no manufacturing occurs within that period, the patent may be subject to the grant of compulsory licenses to any person who may apply for a license.

The Paris Convention

There is a treaty relating to patents which is followed by 93 countries, including the United States, and is known as the Paris Convention for the Protection of Industrial Property. It provides that each country guarantees to the citizens of the other countries the same rights in patent and trademark matters that it gives to its own citizens. The Paris Convention is administered by the World Intellectual Property Organization (WIPO) in Geneva, Switzerland.

The treaty also provides for the right of priority in the case of patents, design patents, and trademarks. This right means that, on the basis of a regular first application filed in one of the member countries, the applicant may, within a certain period of time, apply for protection in all the other member countries. These later applications will than be regarded as if they had been filed on the same day as the first application. Thus, these later applications will have priority over applications for the same invention which may have been filed during the same period by other persons. Moreover, these later applications, being based on the first application, will not be invalidated by any acts accomplished in the interval, such as, for example, publication or exploitation of the invention, sale of copies of the design, or use of the trademark.

The time frame allowed for subsequent applications in other member countries is 12 months in the case of utility patents and six months in the case of design patents and trademarks.

Patent Cooperation Treaty

Negotiated at a diplomatic conference in Washington, D.C., in June of 1970, the Patent Cooperation Treaty (PCT) came into force on January 24, 1978, and is presently adhered to by the 39 countries listed alphabetically below.

LIST OF PATENT COOPERATION TREATY MEMBER STATES

1) Austria
2) Australia
3) Barbados

4) Belgium
5) Brazil
6) Bulgaria
7) Cameroon
8) Central Africa Republic
9) Chad
10) Congo
11) Democratic People's Republic of Korea (North Korea)
12) Denmark
13) Finland
14) France
15) Gabon
16) Germany, Federal Republic of
17) Hungry
18) Italy
19) Japan
20) Liechtenstein
21) Luxembourg
22) Madagascar
23) Malawi
24) Mali
25) Mauritania
26) Monaco
27) Netherlands
28) Norway
29) Republic of Korea (South Korea)
30) Romania
31) Senegal
32) Soviet Union
33) Sri Lanka
34) Sudan
35) Sweden
36) Switzerland
37) Togo
38) United Kingdom
39) United States of America

The PCT facilitates the filing of applications for patent on the same invention in member countries by providing, among other benefits, for centralized filing procedures and a standardized application format.

Under U.S. law it is necessary, in the case of inventions made in that country, to obtain a license from the Commissioner of Patents and Trademarks before applying for a patent in a foreign country. Such a license is required if the foreign application is to be filed before an application is filed in the United States or before the expiration of six months from the filing of an application in the U.S.

If the invention has been ordered to be kept secret, the consent to the filing abroad must be obtained from the Commissioner of Patents and Trademarks during the period the secret is in effect.

There is a basic fee for the first thirty pages, a basic supplemental fee for each page over thirty, a designation fee per member country or region, etc.

PCT fees are listed in Fig. 2, PTO Fee Schedule.

Foreign Applications for U.S. Patents

Any person of any nationality may make application for a U.S. patent so long as that person is the inventor of record. The inventor must sign the same oath and declaration (with certain exceptions).

No U.S. patent can be obtained if the invention was patented abroad more than one year before filing in the U.S. Six months are allowed in the case of a design patent.

An application for a patent filed in the U.S. by any person who has previously regularly filed an application for a patent for the same invention in a foreign country (which affords similar privileges to citizens of the U.S.) shall have the same force and effect for the purpose of overcoming intervening acts of others. The requirement is that it be filed in the U.S. on the date on which the application for a patent (for the same invention) was first filed in the foreign country, provided that the application in the U.S. is filed within 12 months (six months in the case of a design patent) from the earliest date on which any such foreign application was filed. A copy of the foreign application certified by the patent office of the country in which it was filed is required to secure this right of priority.

If any application for patent has been filed in any foreign country prior to application in the U.S., the applicant must, in the oath or declaration accompanying the application, state the country in which the earliest such application has been filed, giving the date of filing the application. All applications filed more than a year before the filing in the U.S. must also be recited in the oath or declaration.

An oath or declaration must be made with respect to each and every application. When the applicant is in a foreign country, the oath or affirmation may be before any diplomatic or consular officer of the U.S. It may also be made before any officer having an official seal and authorized to administer oaths in the foreign country, whose authority shall be proved by a certificate of a diplomatic or consular officer of the U.S. In all cases, the oath is to be attested by the proper official seal of the officer before whom the oath is made.

When the oath is taken before an officer in the country foreign to the U.S., all the application papers (except the drawings) must be attached together and a ribbon passed one or more times through all the sheets. The ends of the ribbons are to be brought together under the seal before the latter is affixed and impressed, or each sheet must be impressed with the official seal of the officer before whom the oath was taken.

If the application is filed by the legal representative (executive, administrator, etc.) of a deceased inventor, the legal representative must make the oath or declaration.

When a declaration is used, the ribboning procedure is not necessary, nor is it necessary to appear before an official in connection with the making of a declaration.

A foreign applicant may be represented by any patent attorney or agent who is registered to practice before the U.S. Patent and Trademark Office.

TRADEMARKS AND COPYRIGHTS

Coke. Greyhound. Scrabble. Crest. NBC's peacock. Mr. Goodwrench. The Campbell Kids. Morris the Cat. Paramount's mountain and stars. The MGM "pussy cat". They are all well known trademarks and are very important to their owners as tools to sell products and services.

Trademarks can be very important to inventors as a tool for helping to sell an invention or new concept. A good trademark creates fast product identification and can help tell the product's story. I always spend a great deal of time creating an appropriate trademark for a product that requires one as part of its release package. For example, my suggestion to Procter & Gamble that it license our patented tricycle for Crest to use as a premium would not have had the same sex appeal without the trademark, Crest Fluorider.

When told the name of our proposed original ride-on, the product's brand manager immediately responded, "Wow! That sounds exciting, a Crest Fluorider. When can we see a prototype?"

TRADEMARK OFFICE

Trademarks are handled by the Trademark Office of the Patent and Trademark Office, under the directorship of an Assistant Commissioner for Trademarks. It employs a total of 110 Examining Attorneys.

In 1987, this office handled over 70,000 trademark applications. It disposed of about 67,000 applications (54,000 registered and 13,000 abandoned). **Trademark Examination Activities, Fig. 25, will give you some idea of how applications are routed and the time period needed to get them on the official trademark register.**

The federal government has issued over one million trademarks since the passage of the Trademark Act of 1905. Today, about 620,000 are in active status.

Correspondence & Information

All correspondence should be addressed to: The Commissioner of Patents and Trademarks, Washington, D.C. 20231, unless you have the name of a particular official.

Call (703) 557-INFO with any questions about filing requirements.

For names of officials and their direct dial phone numbers, refer to Appendix B, Patent and Trademark Office Telephone Directory.

Definition of Trademarks

Trademarks. A "Trademark," as defined in section 45 of the 1946 Trademark Act (Lanham Act) "includes any word, name, symbol, or device, or any combination thereof adopted and used by a manufacturer or merchant to identify his goods and distinguish them from those manufactured or sold by others."

Service Marks. A mark used in the sale or advertising of services to identify the services of one person and distinguish them from the services of others. Titles, character names, and other distinctive features of radio or television programs may be registered as service marks notwithstanding that they, or the programs, may advertise the goods of the sponsor.

Certification Marks. A mark used upon or in connection with the products or services of one or more persons other than the owner of the mark to certify regional or other origin, material, mode of manufacture, quality, accuracy, or other characteristics of such goods or services, or that the work or labor on the goods or services was performed by members of a union or other organization.

Collective Marks. Trademark or service mark used by the members of a cooperative, an association, or other collective group or organization. Marks used to indicate membership in a union, an association, or other organization may be registered as Collective Membership Marks.

Trade and Commercial Names. Marks differ from trade and commercial names that are used by manufacturers, industrialists, merchants, agriculturists, and others to identify their businesses, vocations, or occupations, or other names or titles lawfully adopted by persons, firms, associations, companies, unions, and other organizations. The latter are not subject to registration unless actually used as trademarks.

Function of Trademarks

The primary function of a trademark is to indicate origin. However, trademarks also serve to guarantee the quality of the goods bearing the mark and, through advertising, serve to create and maintain a demand for the product. Rights in a trademark are acquired **only** by use and the use must ordinarily continue if the rights you acquire are to be preserved. Registration of a trademark in the Patent and Trademark Office does not in itself create or establish any exclusive rights, but it is recognition by the government of your right to use the mark in commerce to distinguish your goods from those of others.

Use in Commerce

In public hearing held June 29, 1989 by the Department of Commerce discussed proposed major changes that would make it easier for companies to apply for registration of trademarks. The proposed changes, if accepted, will be effective as of November 16, 1989. They include a dual system whereby trademark owners may apply for registration on the basis of a bona fide **intent** to use the mark commercially. This is a significant change from the policy that in the past allowed owners to register their marks only if the mark has already been used to identify goods or services moving in interstate commerce.

Specifically, the rule changes would allow an application based on "intent-to-use." Providing the mark meets the criteria for registration, it will be published for opposition in the *Official Gazette.* If there is no successful opposition, the application will receive a Notice of Allowance. The applicant will then have six months to file a statement at the PTO verifying that the mark has been used, together with the prescribed fee and specimens showing the use of the mark. Upon examination and acceptance of the statement of use, the mark will be registered. If the applicant does not file a statement, the application will be abandoned. Provisions are made for extending the time for filing a statement of use from six-months up to three years from the mailing of the Notice of Allowance.

In addition, under the proposed new rules, the term of a trademark registration will be reduced from 20 to 10 years. At the end of the 10-year period, trademark owners would need to apply for renewal.

Marks Not Subject to Registration

A trademark cannot be registered if it:

a) Consists of or comprises immoral, deceptive, or scandalous matter or matter that may disparage or falsely suggest a connection with persons, living or dead, institutions, beliefs, or national symbols, or bring them into contempt or disrepute.

b) Consists of or comprises the flag or coat of arms or other insignia of the United States, or of any state or municipality, or of any foreign nation, or any simulation thereof.

c) Consists of or comprises a name, portrait, or signature identifying a particular living individual except by his written consent, or the name, signature, or portrait of a deceased President of the United States during the life of his widow, if any, except by the written consent of the widow.

d) Consists of or comprises a mark which so resembles a mark registered in the Patent and Trademark Office or a mark or trade name previously used in the United States by another and not abandoned, as to be likely when applied to the goods of another person, to cause confusion, or to cause mistake, or to deceive.

Registrable Marks

Principal Register: The trademark, if otherwise eligible, may be registered on the Principal Register unless it consists of a mark which, 1) when applied to the goods/service of the applicant is merely descriptive or deceptively mis-descriptive of them, except as indications of regional origin, or 2) is primarily merely a surname.

Such marks, however, may be registered on the Principal Register, provided they have become distinctive as applied to the applicant's goods in commerce. The Commissioner may accept as prima facie evidence that the mark has become distinctive as applied to applicant's goods/services in commerce, proof of substantially exclusive and continuous use thereof as a mark by the applicant in commerce for the five years next preceding the date of filing of the application for registration.

Supplemental Register: All marks capable of distinguishing your goods and not registrable on the Principal Register, which have been in lawful use in commerce for the year preceding your filing for registration, may be registered on the Supplemental Register. A mark on this register may consist of any trademark, symbol, label, package, configuration of goods, name, word, slogan, phrase, surname, geographical name, numeral, or device, or any combination of the foregoing.

TRADEMARK SEARCH

Just as in the case of patent application, it is advisable for you to make a search of registered marks before filing an application.

There are three ways available. Have your patent attorney do it; engage the services of a professional trademark search organization; or do it yourself.

Law Firm Trademark Search

A patent attorney will gladly handle a trademark search. The attorney will not do it personally, but instead will rely on the services of a professional trademark search firm. Law firms may step on search fees 40 percent or more depending upon what the market will bear.

If you decide to hire a lawyer for this job, receive an estimate of costs up front. Lawyers most often use computer-aided search methods, which can take a few minutes. There may be a minimum charge. To avoid surprises, find out in advance how much the lawyer will charge for copies.

Lawyers and professional searchers often charge more for logo searches. In my opinion, both take an equal amount of effort. Yet you'll find trademarks at $75.00 and logos $150.00 from the same firm.

Professional Trademark Search

If you want to save yourself the lawyer's mark-up, consider hiring a trademark searcher yourself. They are listed in the yellow pages under "Trademark Searchers" or "Patent Searchers." Patent searchers will often handle trademark searches as well.

Get an estimate. An average price is $75.00 per hour with a one hour minimum. It should not take anywhere near one hour to do a single mark. Significant portions of the trademark operation are already fully automated. For over five years, the public has been able to access the Office's electronic files concerning registrations and applications through the Trademark Reporting and Application Monitoring system (TRAM). Via this computer-aided search, it can take a few minutes to access bibliographic, status, and location data. Access to TRAM often obviates the need to consult the actual file.

Add to this price about $3.00 per copy of each trademark copied.

Do-it-Yourself Trademark Search

You do the search utilizing two methods:

1) The Trademark Office has a **Public Search Room**, located in Crystal Plaza 2 - 2C08, Arlington, Virginia. Once the layout is learned (which takes about five minutes), you can breeze through a manual search in 20 minutes.

2) If a visit to Arlington is inconvenient, see if there is a Patent Depository Library (PDL) is nearby. It will offer numerous trademark reference books, e .g. Gale's *Tradenames Dictionary*, Compu-Mark's *Directory of U.S. Trademarks,* etc., as

well as an online computer search capability (See TRAM reference above).

Depending upon how extensive the search, costs can run anywhere from $10.00 and up for computer time. Often appointments must be made to access the computer services.

For a complete list of PDL's, See Chapter 13, Patent Depository and Special Libraries.

T-Search

T-Search is the PTO's online trademark search system to research word marks and design elements of trademarks. The PTO currently offers an experimental public use program on this system.

How Long is a Trademark in Force?

The term of a trademark is potentially infinite.

How to Register for a Trademark

A Trademark application package consists of four items:

1) A written application;
2) A drawing of the mark;
3) Five specimens or facsimiles of the mark;
4) The required filing fee.

Trademark Application

Due to the recent changes in the trademark law, the PTO was unable to provide the updated forms for this publication. Trademark forms are available from the PTO, (703) 557-4636 or Department of Commerce Offices.

Trademark Classification

In the upper right hand corner of the application form, put down the classification number that refers to your product. An application in which a single fee is submitted must be limited to the goods or to the services comprised in a single class.

The International Schedule of Goods and Services is reproduced as Fig. 23.

Trademark Drawings

Your drawing must be a fairly exact representation of the mark as actually used in connection with the invention. You do not have to draw a service mark if it is not capable of representation by a drawing. If your application is for registration of a word, letter or numeral, or any combination thereof, not depicted in special form, the drawing may be simply the mark typed in upper case letters on paper, otherwise complying with the requirements.

Here are the exact specifications required by the PTO for trademark drawings:

Paper and Ink. The drawing must be made upon pure white durable paper, the surface of which must be calendered and smooth. A good grade of bond paper is suitable. India ink alone must be used for pen drawings to secure perfectly black solid lines. The use of white pigment to cover lines is not acceptable.

Size of Sheet and Margins. The size of the sheet on which a drawing is made must be 8.5 inches wide and 11 inches long. One of the shorter sides of the sheet should be regarded as its top. When the figure is longer than the width of the sheet, the sheet should be turned on its side with the top at the right. The size of the mark must be such as to leave a margin of at least one inch on the sides and bottom of the paper and at least one inch between it and the heading.

Heading. Across the top of the drawing, beginning one inch from the top edge and not exceeding one-fourth of the sheet, there must be placed a heading, listing in separate lines, applicant's name, applicant's post office address, date of first use, date of first use in commerce, and the goods or services recited in the application (or typical items of goods or services if a number are recited in the application). This heading may be typewritten.

Character of Lines. All drawings, except as otherwise provided, must be made with the pen or by a process which will give them satisfactory reproduction characteristics. Every line and letter, names included, must be black. This direction applies to all lines, however fine, and to shading. All lines must be clean, sharp, and solid, and they must not be too fine or crowded. Surface shading, when used, should be open. A photolithographic reproduction or printer's proof copy may be used if otherwise suitable. Photocopies are not acceptable.

Extraneous Matter. Extraneous matter must not appear upon the face of the drawing.

Linings for Color. Where color is a feature of the mark, the color or colors may be designated in the drawing by means of conventional linings as shown in the following chart:

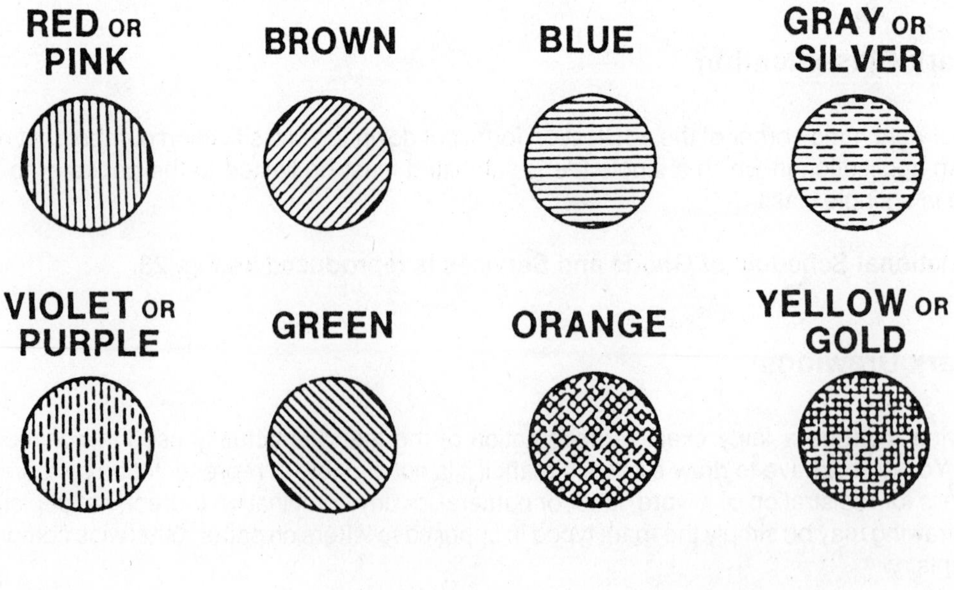

Transmission of Drawings. Other than typed drawings, you should send your artwork to the PTO flat and well protected by a sheet of heavy binder's board, rolled and posted in a suitable mailing tube.

Informal Drawings. A drawing that does not conform with these requirements may be accepted for the purpose of examination only, but it must be ultimately corrected or a new one furnished, as required, before the application will be allowed or the mark can be published.

Trademark Specimens/Facsimiles

Trademark Specimens. A trademark may be placed in any manner on the goods, or their containers, tags, labels, or displays. The five specimens submitted must be duplicates of the actual labels, tags, containers, or displays or portions thereof, when made of suitable material and capable of being arranged flat and of a size no larger than 8 1/2 x 13 inches. Three-dimensional or bulky material submitted as specimens will not be accepted, and may delay your filing date.

Trademark Facsimiles. If you must submit facsimiles of your trademark, send in five copies of a suitable photograph or other acceptable reproduction, not larger than 8 1/2 x 13 inches and clearly and legibly showing the mark and all matter used in connection with it.

Service Mark Specimens or Facsimiles. In the case of a service mark, specimens or facsimiles of the mark as used in the sale or advertising of the services must be furnished unless impossible from the nature of the mark or the manner in which it is used, in which event some other representation must be submitted.

If your service mark is not used in a printed or written form, only three recordings will be required.

Trademark Examination Procedure

Applications are docketed and examined in the order of receipt. If the examiner finds any reason that the proposed mark will not pass, you'll be advised of the objections.

You are given six months to respond. If you fail to respond, the Application for trademark will be abandoned.

In fiscal year 1988, the wait for a trademark to be awarded was an average of 13.3 months.

For more information, see Fig. 22, Trademark Examination Activities. For a schedule of fees, see Fig. 2, PTO Fee Schedule. You may also check the current fees by calling (703)-557-INFO.

COPYRIGHTS

What do Hal David's song "Do You Know The Way To San Jose?," Cadaco's hit game "Adver*teasing," Baby Talk* magazine, "The Graduate," and this book have in common? If you said they are all protected by copyright, you are correct.

Copyrights are very different than patents and trademarks. A patent primarily prevents inventions, discoveries, or advancements of useful processes from being manufactured, used, or marketed. A trademark is a word, name, or symbol to indicate origin, and in so doing distinguish the products and services of one company from those of another. Copyrights protect the form of expression rather than the subject matter of the writing. They protect the original works of authors and other creative people against copying and unauthorized public performance.

The Copyright Office

Copyrights are not handled by the Patent and Trademark Office. For this discussion we move across the Potomac River from the PTO's Crystal City, Virginia headquarters, up Independence Avenue, and onto Capitol Hill to the **Library of Congress** which is primarily responsible for administering copyright law.

What is a Copyright?

Copyright is a form of protection provided by the laws of the United States (Title 17, U.S. Code) to the authors of "original works of authorship" including literary, dramatic, musical, artistic, and certain other intellectual works. The U.S. Constitution authorized Congress to establish the copyright legislation. The first federal copyright act was passed in 1790. The most recent change in copyright law took effect on March 1, 1989, which states under the Berne Convention treaty that the created works of U.S. citizens are now afforded copyright protection in 79 foreign countries.

What Can Be Copyrighted?

This protection is available to both published and unpublished works. I slap copyright notices on everything I create. Copyrights can be as important to an inventor as to an author, hence the reason I have included it in this book. Instructions and other written instruments such as background papers, concept papers, drawings, photographs, and the like that relate to inventions are all protected under U.S. copyright laws.

How to Secure a Copyright

The way in which copyright protection is secured under the present law is frequently misunderstood. In years past it was required that you fill out special forms and send them to the Library of Congress together with a check and a number of copies of the original work. Today, no publication or registration or other action in the Copyright Office is required to secure copyright under the new law.

Under present law, copyright is secured "automatically" when the work is created, and the work is "created" when it is fixed in a copy or phonographically recorded for the first time. In general "copies" are material objects from which a work can be read or visually perceived either directly or with the aid of a machine or device, such as books, manuscripts, sheet music, film, videotape, or microfilm.

However, it is still prudent to make a formal application with the Library of Congress. This is to establish a "public record" of your claim; receive a certificate of registration (required if you ever have to go into a court of law over infringement); and if you receive a copyright within five years of publication, it will be considered to be prima facie evidence in a court of law.

If you would like to get specific information on the process, call a Copyright Public Information Specialist at (202) 287-8700 weekdays between the hours of 8:30 a.m. and 5:00 p.m. These folks are a wealth of information. You can also write for information to Information and Publications Section, LM-455, Copyright Office, Library of Congress, Washington, D.C. 20559.

Notice of Copyright

Before you publicly show or distribute your work, notice of copyright is required. The use of the copyright notice is your responsibility and does not need any special advance permission from, or registration with, the Copyright Office.

The notice for visually perceptible copies should contain all the following three elements:

1) **The Symbol** © (the letter C in a circle), or the word "Copyright," or the abbreviation "Copr."

2) **The year of first publication** of said work. In the case of complications or derivative works incorporating previously published material, the year of first publication of the compilation or derivative work is enough. The year may be omitted where a pictorial, graphic, or sculptural work, with accompanying text (if any) is reproduced in or on greeting cards, postcards, stationary, jewelry, dolls, toys, or any useful articles.

3) **The name of the owner of copyright** in the work, or an abbreviation by which the name can be recognized, or a generally known alternative of the owner.

Example: © 1989 Richard C. Levy

You should affix the notice in such a way as to give it "reasonable notice of the claim of copyright."

How Long Copyright Protection Endures

A work that was created on or after January 1, 1978, is automatically protected from the moment of its creation, and is usually given a term enduring for the author's life, plus an extra fifty years after the author's death. In the case of a "joint work prepared by two or more authors who did not work for hire," the term lasts for fifty years after the last surviving author's death. For works made for hire, and for anonymous and pseudonymous works (unless the author's identity is revealed in Copyright Office records), the duration of copyright will be seventy-five years from publication or one hundred years from creation, whichever is shorter.

Works that were created before the present law came into effect, but had neither been published nor registered for copyright before January 1, 1978, have been automatically brought under the statute and are now given Federal copyright protection. The duration of copyright in these works will generally be computed in the same way as for works created on or after January 1, 1978 (the life-plus-50 or 75/100-year terms apply to them as well). However, all works in this category are guaranteed at least twenty-five years of statutory protection.

Under the law in effect before 1978, copyright was secured either on the date a work was published, or on the date of registration if the work was registered in unpublished form. In either case, the copyright endured for a first term of twenty-eight years from the date secured. During the last (28th) year of the first term, the copyright was eligible for renewal. The new copyright law has extended the renewal term from twenty-eight to forty-seven years for copyrights that were subsisting on January 1, 1978, making these works eligible for a total term of protection of seventy-five years. However, the copyright must be timely renewed to receive the forty-seven year period of added protection.

Deposit: Under the new law, there is a requirement that you deposit two copies of the published Work in the Copyright Office for use by the Library of Congress. If the work is unpublished, one copy is required. This should be done within three months of publication of any notice of copyright. A failure to do so may mean a fine. Even so, such an omission would not affect your copyright protection.

What Copyrights Do Not Protect

Ideas cannot be copyrighted. The same is true of the name or title given to a product and the method or methods for doing something.

Copyright protects only the particular manner in which you express yourself in a literary, artistic, or musical form. Copyright protection does not extend to ideas, systems, devices, or trademark material involved in the development, merchandising, or usage of a product.

Application Forms

The forms you will likely require include Form TX, Form VA, Form CA, and Form RE.

Form TX: For published and unpublished non-dramatic literary works. This comprises the broadest category, covering everything from novels to computer programs, game instructions, and invention proposals.

Form VA: For published and unpublished works of the visual arts. This would be for artwork you may have developed as an adjunct to your invention, charts, technical drawings, diagrams, models, and works of artistic craftsmanship.

Form CA: For application for supplementary Copyright Registration. Use when an earlier registration has been made in the Copyright Office; and some of the facts given in that registration are incorrect or incomplete; and you want to place the correct or complete facts on record.

Form RE: Renewal registration. Use when you wish to renew a copyright.

These forms are reproduced as figures 24, 25, 26, and 27.

The cost is only $10.00 per registration. This fee has been the same since 1978, and, unlike PTO fees, an act of Congress is needed to increase it. **If you desire original forms to register copyrights, call the forms HOTLINE at (202) 707-9100 and leave your request as a recorded message.** The Copyright Office has established this new service to speed up service. Requests made on the HOTLINE are handled promptly. The recorders are connected 24-hours per day, seven days a week.

HOW TO LICENSE YOUR INVENTION

Competition drives American industry and creates a fertile market for new inventions. No scarcity exists of progressive manufacturers willing to consider appropriate, new, and innovative products from outside, independent inventors. Many companies in every field strive to produce better and more competitive products on a continuing basis, and often rely upon the independent inventor.

What Victor Hugo wrote in 1852 has never rung truer than today: "Greater than the tread of mighty armies is an idea whose time has come." The difficulty is bridging the supply and demand gap between the creators of new ideas and those who possess the capabilities and facilities to manufacture and market them successfully.

America constantly demands new products. We are, after all, a "throw away" society. We purchase something, use it for a while, and upgrade to the next generation of the product as soon as it's affordable. Generally, we do not take good care of things, nor do we spend a lot of money fixing them when they break.

The biggest problem inventors—experienced and inexperienced—have is how to license their innovations. While many people have the ability to dream and innovate, few of them also have the ability to sell.

There are two ways to get your inventions manufactured and to market. One is to license them to a firm that specializes in manufacturing and selling products. The other method is to raise venture capital and become the manufacturer. Both have pluses and minuses. Neither is risk free. There are no guarantees either way, just different opportunities. Both are in the best and proudest traditions of American entrepreneurism.

I personally subscribe to the school of thought that says it is best to license inventions to companies run by executives who know more than I do about how to cost-effectively develop, manufacture, package, market, and promote products to specifically targeted users. I have no desire to go through the learning curve to acquire an expertise in all that's needed to do what it is a manufacturer does. And if you are thinking of doing this, remember Murphy. Murphy's first corollary is: Nothing is as easy as it looks. Murphy's second corollary is: Everything takes longer than you think.

My aim is to create and maintain an atmosphere in which I can be most creative and productive, a situation free from supervisors, meetings, suits, power lunches, time-tested methods, office politics, and assorted "administrivia." You don't do this by raising venture capital and starting a manufacturing company. You do this by accepting the risks inherent in independent product development and licensing your creative yield.

You have three methods to address the business of invention licensing. The first is the do-it-yourself approach. I prefer this method. Over the years I have found that putting together my own team and making my own deals works best. I prefer to take my chances, not someone else's. It is more fun, I have better control, and the rewards are far greater.

Another way is to take your inventions to an inventor's exposition and run it up the flag pole for everyone to salute.

And finally, you can engage the services of an agent or broker. Good agents are few and far between. The best agents become so involved that they become virtual co-developers. I speak from personal experience, based on occasionally representing products other than my own. Nothing less will do. Jon W. Bayless, a partner in Sevin Rosen Bayless Venture Fund, says "A third party who is not committing his future to the business has no part to play in the process. Third party involvement is frequently a reason for a project to lose the consideration it may deserve."

Let's look at the latter first because as soon as the patent issues, you are likely to hear from one species of agent, whether you want to or not, the infamous invention marketing service.

INVENTION MARKETING SERVICES

How Do They Get Your Name?

Don't be surprised if you first learn of your patent award from an "invention marketing" company. When your patent issues, a notice of it will be automatically carried in the *Official Gazette* of the United States Patent and Trademark Office. This publication, issued weekly since 1872, comes out each Tuesday. It publicly records the following information about each new patent:

1) the name, city, and state of residence of the applicant, with the Post Office address in the case of unassigned patents;
2) the same data for the assignee, if any;
3) the filing date;
4) the serial number of the application;
5) the patent number;
6) the title of invention;
7) the number of claims;
8) the U.S. classification by class and subclass;
9) a selected figure of the drawing, if any, except in the case of a plant patent;
10) a claim or claims;
11) international classification;
12) U.S. patent application data, if any; and,
13) foreign priority application data, if any.

In the case of a reissue patent they publish the additional data of the number and date of the original patent and original application.

Invention marketing companies subscribe to the *Official Gazette* as an inexpensive way to obtain a qualified mailing list of inventors recently granted patent protection. They typically approach the independent inventors rather than their assignee corporations.

The "Pros" and the "Cons"

Caveat emptor! Learning to recognize the good invention marketing companies from the bad ones is not easy. The invention marketing business is rife with nonperforming, paracreative slugs who prey on and take advantage of unsuspecting, inexperienced, frustrated, and hungry independent inventors. In my experience, most invention marketing organizations that advertise via direct mail, classified ads, radio, or television spots are less than effective. Often, they carry a very appropriate appellation "front money frauds."

Alan A. Tratner, president of Inventors Workshop International, says about such parasites, "We view the unsavory practices of many of these firms and individuals as a 'cancer' in the inventing community that needs to be eradicated immediately. Too often inventors lose large amounts of money and are derailed by the unfulfilled promises and come-ons of these companies."

Florida inventor Frederick L. Jones, a member of the Palm Beach Society of American Inventors, agrees about invention marketing companies, "They prey on people. They'll take all the money they can."

Frank Smirne, president of F. Smirne Plastic Co., Allentown, Pennsylvania, suggests that inventors get advice from other inventors before dealing with invention marketeers.

Lawrence Smith of the Patent and Trademark Office cautions, "Beware of these people." He suggests that inventors only permit marketeers to have their patented products on a contingency, non-exclusive basis.

"They [invention marketeers] scare the living daylights out of me," says James Kubiatowicz, director of product development/toys at Spearhead Industries. "They're leeches. They prey on the novices." Then pulling open his desk drawer, he adds, "I've got thirty-two post cards from these guys and I am saving the stamps. One of these days I'll steam them off and put them into a retirement fund. Anybody who is breathing and has a dollar can get strokes from these companies. They work on a person's ego."

It is almost a guarantee that if you engage the services of an invention marketing company, the only ones who'll get rich will be the invention marketeers on your payroll.

Watch out for them. You'll find their ads running on all night radio talk shows, at 3:00 a.m. sponsoring Million Dollar Movie reruns, and in the classified advertising sections of newspapers and magazines that appeal to do-it-yourselfers.

The direct mail pieces they send might be in the form of a postal card. They usually read something like this:

> ***IMPORTANT NOTICE***
>
> *To: Owner of U.S. Patent*
>
> *We have located six companies that produce, market, or sell products in a field to which your invention might apply. You may wish to contact these manufacturers if you are interested in one or more of the following:*
>
> *A. Licensing your patent.*
>
> *B. Finding a company to produce your invention.*
>
> *C. Securing marketing, distribution, and/or sales help.*
>
> *D. Hiring design or technical support.*

Then comes the kicker. You are asked to remit a fee of $75.00 plus $1.00 for each company named.

Other direct mail offers come in #10 envelopes overflowing with all kinds of slick fliers and official-looking confidential disclosure agreements with diploma-like borders. Rarely will you be able to make out the signature of the "authorized agent" who signed off on the standard "Dear Inventor" cover letter. If they feature any products which have been successfully licensed to industry, it is doubtful you will recognize any of them.

They use all kinds of subtle techniques to compensate for the fact that they have no successful products to show off. For example, I recall seeing one promotional flier that depicted a beautiful, modern building with the caption "4th Floor, Chicago, Illinois." The company wanted the reader to assume that this was the base of its corporate headquarters. However, no reference was made to the picture or office building anywhere in the flier's copy.

The more sophisticated invention marketing companies wrap themselves in Old Glory and apple pie. Their names incorporate words like American, Federal, and National and they have Washington, D.C. postal addresses.

How to Identify the Good Guys

One of the acid tests I use to separate the honest and serious professional invention marketing companies from those involved in unfair business practices is whether or not they want any advance payments from the inventor for anything whatsoever.

"The firm we dealt with appears to be interested in marketing our idea for us," a woman from Virginia wrote me, "but they are quoting us a price of from $3,000-$9,000 for this service. Also, they want a slice of any profits (up to 60 percent)."

The most reputable invention marketing services will not require front money from the inventor. You invested your ingenuity, time, and money in creating the product. The marketer must invest ingenuity, time, and money in causing the product to be licensed. What is more fair than that? The moment you pay for marketing services, the carrot is removed. There is no risk. With nothing to lose, and a gain already in the bank, the incentive for your marketing partner is reduced.

Charles F. Mullen, an inventor from Houston, Texas, works part time as sales manager for national accounts at an advertising agency, and says he spends the rest of his time inventing and helping others market their products. "We work straight percentage. We don't charge them anything up front," he says. "That's the trouble with the industry. Most of the people get them up front." Mr. Mullen claims to have licensed fourteen inventions all in different fields.

A reputable and active marketing company will be able to demonstrate a track record of successful products and satisfied inventor clients. References should be available. Ask for a list of clients and their telephone numbers. The products they have marketed should be available somewhere other than their office waiting rooms.

Read the small print. I saw one flier that showed numerous products leaving the impression that they had been licensed by the invention marketing group. Then I saw the disclaimer. "These products are not intended to represent success for inventors who have worked with our firm." They should be able to substantiate licensing agreements claimed and royalty earnings. Do they make the majority of their earnings from royalties or from inventor service fees? If it is not from royalties, you're in the wrong place.

Always check the name of the invention marketing company and/or individual marketer with the local Better Business Bureau, state agency for consumer fraud, and your Attorney General's office. See if any complaints have ever been registered against the company or its officers and representatives.

Ask fellow inventors. Word about bad apples spreads like wild fire. Many inventor organizations maintain files on the worst invention marketing organizations and frequently write them up in their newsletters.

See Chapters 2, National and Regional Inventor Organizations and 9, Publications for the Inventor, for more information on publications.

It is also wise to check invention companies out with your State Attorney General, and even the Federal Trade Commission.

FEDERAL TRADE COMMISSION

The Federal Trade Commission (FTC) has received so many complaints about invention marketing companies that it has a complaint category for them.

If you wish to register a complaint against an invention marketing company, write to: **Stanley Ciurczak, Chief, Complaint Division, Room 692, Federal Trade Commission, Washington, D.C. 20580.**

Based upon a case decided in 1978, the FTC has determined that the following practices used in the advertising and marketing of idea or invention promotion or development services are unfair or deceptive trade practices and are unlawful under Section 5 (a) (1) of the FTC Act:

1) For a seller of idea or invention promotion or development services to misrepresent, directly or indirectly, that potential purchasers will be provided with evaluations or appraisals of the patentability, merit, or marketability of ideas or inventions.

2) To represent, directly or indirectly, that the seller of idea or invention promotion or development services, or its officers, agents, representatives, or employees are registered patent attorneys or patent agents, or are qualified to practice before the U.S. Patent and Trademark Office, unless such is a fact.

3) To misrepresent, directly or indirectly, the scope, nature, or quality of the services performed to develop or refine ideas or inventions.

4) To misrepresent, directly or indirectly, the scope, nature, or quality of the services performed to introduce or promote ideas or inventions to industry.

5) To represent, directly or indirectly, that a seller of idea or invention development service has special access to manufacturers or has been retained to locate new product ideas, unless such is a fact.

6) To misrepresent, directly or indirectly, that a person, partnership, corporation, government agency, or other entity endorses or uses the services of a seller or provider of services.

7) For sellers of idea or invention promotion or development services to fail to disclose, when price information is provided to potential purchasers, all significant fees or charges that may be incurred by purchasers in connection with such services.

8) To misrepresent, directly or indirectly, the background, qualifications, experience, or expertise of a seller or provider of services.

9) For a seller of idea or invention promotion or development services to induce through misleading or deceptive representations the purchase of services that have little or no inherent value, or to offer to provide services that grossly exceed the value of the services actually provided. It is also an unfair or deceptive act or practice to retain money from the sale of such services.

If you want to find out whether the FTC has any ongoing investigation into a particular invention marketing company, use a **Freedom of Information Act (FOIA)** request. The FTC will not divulge any particulars about an active investigation, legal or otherwise, or even about complaints received; however, if a consent order has been issued against the invention marketing company, you'll find out.

You may also read between the lines. If under an FOIA request you ask for copies of any and all FOIA requests for information about a specific company(s), you just may get in return copies of requests sent to the FTC by law firms representing said company(s). The law firm would not be making a request about its client unless something was amiss.

For example, I once made an FOIA request to find out whether anyone else was interested in a specific invention marketing firm. The FTC sent me copies of requests, although denied, from a television news department, and a law firm. The fact that the FTC had denied their requests told me that said invention marketing firm was being looked at for some reason. **Send your Freedom of Information Requests to: Keith Golden, FOIA Officer, Federal Trade Commission, 6th Street and Pennsylvania Avenue, N.W., Washington, D.C. 20580. A general telephone number for the FTC is (202) 326-2222.**

PROTECTIVE LEGISLATION

The invention marketing business went unregulated for years. Now, however, some states such as Minnesota, Virginia, Washington, and Tennessee have enacted protective legislation on behalf of resident inventors.

Let's take a quick look at some of the issues that the states address. This will give you a good idea of what to watch for and ask prospective invention marketing companies. If you want a full text of the legislation—which I recommend you acquire—contact the state legislature and it will be sent to you free of charge.

If your state does not have protective legislation for inventors yet, get the word to the appropriate elected officials. Every state should have some sort of regulation on the its books that sets fair standards for invention marketing companies.

Let's take a look at the laws of Minnesota and Virginia. I shall highlight some important points they make.

Minnesota (Invention Services 325A.02)

1) Notwithstanding any contractual provision to the contrary, inventors have the unconditional right to cancel a contract for invention development services for any reason at any time before midnight of the third business day following the date the inventor gets a fully executed copy.

2) A contract for invention development services shall be set in no less than 10-point type.

3) An invention developer who is not a lawyer may not give you legal advice with

respect to patents, copyrights, or trademarks.

4) The invention marketer must tell you (1) the total number of customers who have contracted with him up to the last thirty days; (2) the number of customers who have received, by virtue of the invention marketer's performance, an amount of money in excess of the amount of money paid by such customers to the invention marketer pursuant to a contract for invention development services.

5) The contract shall state the expected date of completion of invention marketing services.

6) Every invention marketer rendering invention development services must maintain a bond issued by a surety admitted to do business in the state, and equal to either ten percent of the marketer's gross income from the invention development business during the preceding fiscal year, or $25,000, whichever is larger.

Virginia, Chapter 18, 59.1-209

1) No invention developer may acquire **any interest**, partial or whole, in the title to the inventor's invention or patent rights, unless the invention developer contracts to manufacture the invention and acquires such interest for this purpose at or about the time the contract for manufacture is executed.

2) The developer must tell you if they intend to spend more for their services than the cash fee you will have to pay.

3) The Attorney General has the mandate to enforce the provisions of this chapter, and recover civil penalties.

MARKETING VIA INVENTOR EXPOS

While some inventor expositions are worth while and well-meaning, such as the annual National Inventors Day Exhibit sponsored by the Patent and Trademark Office or regional shows sponsored by inventor organizations, many slick operators disguise their motives by organizing inventor expositions and fairs. Know thy promoters and their motives.

Some wolves in sheep's clothing invite inventors suffering from "sellitus" to display prototypes, even drawings or photographs of their inventions. They charge for booth exhibition space and advertising in publications that are released in conjunction with the event. The general public is charged an admission fee to see the inventions. Then to top it all off, some hit exhibitors with broker or commission fees should an invention be licensed through its exposure at the show. Another scheme is to offer the inventor a large cash buy out should a product sell, and the promoter walks off with the royalty points.

When all is said and done, my experience confirms that few if any meaningful contacts ever come out of these shows from the inventor's standpoint, while the promoters make lots of money and get publicity to boot.

IMPORTANT! Should you be contemplating such a show, keep one very important thing in mind. If you intend to apply for patent protection on an invention, its display may foul your chances. Exhibiting a new invention in public for sale or otherwise will make it ineligible for patent protection if a patent application is not filed within one year from the time of the invention's first public exposure.

See Chapter 12, Invention Tradeshows for more information on inventor expositions.

DO-IT-YOURSELF MARKETING

This book is a guide to locating information and doing it yourself. All the information required in this regard is provided in the chapters that follow. But remember this rule: never get so wrapped up in the marketing of the invention that you forget to market yourself.

First Impressions

Inventors are always selling two things: the concept and themselves. Personal credibility is often more important than any single creative concept.

It is critical that a corporate executive buy the inventor as much as the invention. You may be capable of dreaming up numerous innovative products for a company to consider, and will want to be invited back again and again. Without respect from corporate executives, your products will never be taken seriously. And you cannot put a dollar value on the ability to make an encore.

PTO Can Help You Get the Word Out

The Patent and Trademark Office cannot help develop or market your invention, but it will publish, at the request of the patent owner, a notice in its *Official Gazette* that the patent is available for licensing or sale. The fee for this is a very reasonable $7.00.

Organizations Offer a Helping Hand

Inventor and business organizations, as well as state and university assistance programs can be of great value to the inventor preparing presentation packages for prospective licensees. **For more information, see Chapters 2, National and Regional Inventor Organizations and 4, University Innovation Research Centers.**

Marketing is a Science

Marketing inventions is a science; no less a discipline than electronics, mechanical engineering, or any other field of study. It can be learned. Of course, as with any talent, some natural abilities may make it easier for some, but nothing is out-of-reach if you take the time.

HOW TO SELECT WHICH COMPANY TO APPROACH

Remember the old saw that says if you have a better mousetrap, the world will beat a path to your door? Maybe so back in 1889 when Ralph Waldo Emerson suggested this, but today the reverse is true for inventors without track records. Amateur inventors spend a lot of tiem beating paths to the doors of companies looking to license better mourse traps.

Selecting which companies to approach as possible licensees for your invention is a matter that requires a great deal of thought. It is not something to be taken lightly. You don't want to make the same mistake Rick Blaine (Humphrey Bogart) made in the movie Casablanca. "I came to Casablanca for the waters," he said. When told that he was in the middle of a desert, he adds, "I was misinformed."

You must be very well informed, indeed. Your first approach to a company requires the kind of detailed analysis, imagination, and forethought associated with championship chess. In fact, your approach resembles the game of chess insomuch as it consists of a planned attack and defense, and has as its object, the king's surrender.

Just as much effort, and often more, is required to deal with a small company as with a large one. Don't be afraid to approach the major players. Remember, eighty percent of the business in any particular industry is typically done by twenty percent of the companies. You'll want a company capable of engineering and then following through on a success. Also the larger the company, the better and faster it pays.

Put a lot of time into your decision on which manufacturer to approach. Don't make the error of insufficient options. Some rejection is to be expected, so you'll want as many targets as possible.

Study corporate product lines. Do store checks. Get new product catalogues. Companies do what they do, and bringing them something out of their discipline is usually a lost effort. You would not, for example, go to Black & Decker with a new type of record player. Black & Decker manufactures innovative tools and labor saving devices. Round pegs don't fit square holes.

Corporations exist to make money. Executives, especially those in lucrative profit-sharing plans and incentive programs, want their corporations to be successful, but at what risk? I have found that most senior executives will listen to any scheme that rings of potential profit. That profit can come in the form of a new product or a labor-saving device. But it is rare that an executive will rock the proverbial boat for untested, unfamiliar products, especially those that fall outside the company's expertise or channels of distribution.

Unless you are dealing with an executive in charge of new business opportunities, trying to get a company to purchase a product that is inappropriate to its line-up or retail outlet channels is not only a waste of everyone's time, but it does nothing for your reputation. And that reputation is more precious then your invention.

After a while in this business you learn that many large firms are guided by numbers as much as products. Lawyers and accountants tend to become CEOs before R&D executives do. These bean counters see products in terms of SKUs (stock keeping units) only. They like to keep the pipeline filled with line extensions of already proven and profitable products. Their attitude is that a major breakthrough in medicine, for example, would be nice, but let's keep the mouthwashes and toothpastes coming. These types would rather take a popular pudding and put it on a stick than gamble on creating a new novelty food. They are so busy listening to statistics that they forget companies can create them.

One company licensed a product of ours, and was going to produce it until a consumer study showed that its popularity would upset existing business. The company opted for the status quo and dropped the item. "Why should we erode our market share in an industry we almost totally control?," reasoned the company president. "If we bring your product out, the consumers will obviously love it, but we'll just point to opportunities our competition does not realize exist."

On the other end of the spectrum, a senior research and development executive from a $1 billion plus company told me once, "It's my responsibility to waste $2 million per year on long shots. I would not be doing my job if I didn't." Unfortunately, there are too few corporate executives with this entrepreneurial attitude.

The first move is yours. It is not an easy one. Going to the wrong company with the wrong product can cost valuable time, do nothing to enhance your contacts, and even bring grief. You want your idea in the right hands because, as advertising legend Bill Burnbach said, "An idea can turn to dust or magic depending on the talent that rubs against it."

NIH SYNDROME

NIH, as used here, is not the acronym for the government's National Institutes of Health. NIH is corporate jargon for "Not Invented Here," a syndrome from which many companies suffer. It means that such companies have in-house research and development staffs and do not entertain outside submissions at all, or do, but only to see what is being done independent of themselves, not to license.

It is hard to tell from the outside which companies fall into the NIH category. Those you think would, do not, and those who you would bet don't, do. This learning curve is part of every selection exercise.

There are two sides to the NIH issue, and corporate policy can change with different administrations. Many executives feel that no insulated group of salaried product development people, no matter how brilliant, can come up with winning products day in and day out. There are companies that cannot afford to pay engineers and designers to sit around and blue-sky ideas all day long. These kinds of companies are always worth approaching.

Other firms are against outside licensing of patented ideas because they would rather see the millions of dollars paid in royalties kept inside for its own research and development activities. They are not typically structured to interact with independent inventors.

I like to tell executives who believe they can do it all in-house, that it is the spirit of the independent inventor that built America. To completely shut the independent inventor out is to severely limit one's opportunities and horizons. Then I remind them of these stories:

Kodak, America's largest manufacturer of photographic products, should have developed the instant camera. It didn't. The 60-second camera was invented and produced by Edwin Land, a maverick inventor. And when Kodak ultimately decided to imitate Land's invention, it was stopped in its tracks by the courts.

IBM, a name synonymous with computer innovation, completely missed the hand-held calculator market. It should not have. The Japanese captured the lucrative market and never gave an inch.

The Swiss, the once undisputed world leaders in watchmaking, got their chimes rung when the Japanese came out with cheap, plastic timepieces utilizing microchips, capturing billions of consumer dollars in the process. Only in recent years, with the plastic Swatch, have the Swiss begun to rebound.

And the U.S. television networks, with worldwide news-gathering operations, let CNN get started because they thought no one would pay to see news 24 hours a day. CNN has become such a success that the nets are now all over the cable market.

Breaking the Code

Many companies that do work with outside developers do not encourage "unknowns" and return inqu i ries with a letter like this:

> *"Our advertising, research, marketing, and new product planning staffs are primarily responsible for creativity and development. Corporate policy precludes us from either encouraging or accepting unsolicited ideas from persons outside the Company. While an idea may seem feasible to the submitter, there are usually a number of factors that would make it impractical for us to implement. Moreover, many of the unsolicited ideas that we receive from both nonprofessional and professional sources have previously been submitted in one form or another."*

Reading through the lines this letter is not as negative as it appears to be at a glance. The first clue, that the company does not do everything internally, is that its internal staffs are not "exclusively responsible," but rather "primarily responsible" for creativity and development. This means that outside people do back-up their company's research and development.

The next good news comes when the letter states that the company cannot encourage or accept "unsolicited ideas from persons outside the Company." I read this to signify that the company probably solicits ideas from a trusted base of outside creative sources.

Companies that send back letters similar to this are worth a second look.

PUBLIC COMPANIES vs. PRIVATE COMPANIES

I have no preference for either one. My decision is based upon what I am able to find out about a company and whether it is best for my product. I maintain, at all costs, the mental frame of mind that I am evaluating the company rather than an inferior position of being considered by the company.

Public Companies

The best place I know to obtain deep and detailed information on a publicly traded company is at the Securities and Exchange Commission (SEC), Washington, D.C. This independent, bipartisan, quasi-judicial federal agency was created July 2, 1934 by act of Congress. It requires a public disclosure of financial and other data about companies whose securities are offered for public sale. There are 11,000 companies registered.

All companies whose securities are registered on a national securities exchange and, in general, companies whose assets exceed $3,000,000 with a class of equity securities held by 500 or more investors must register their securities under the 1934 Act.

In the national capital area, the SEC operates a public reference room at 450 5th Street, N.W., Washington, DC 20549. It is open from 9:00 a.m. to 5:00 p.m. daily. The telephone number is: (202) 272-7450. Consumer Telecommunications for the Deaf-TTY-Voice (202) 272-7065.

This specially staffed and equipped facility provides, for your inspection, all of the publicly available records of the Commission.

These include corporate registration statements, periodic company reports, annual reports to shareholders, tender offers and acquisition reports, and much more.

SEC reference libraries located in New York City and Chicago are open to the public. They are located at 26 Federal Plaza (212) 264-1685 and 219 South Dearborn Street (312) 353-7433, respectively.

Requests by Mail

If you find it inconvenient to visit one of the Public Reference Rooms, the Commission will, upon written request, send copies of any document or information. Send a written request stating the documents or information needed, and indicate a willingness to pay the copying and shipping charges. Also include a daytime telephone number. Address all correspondence to: **Securities and Exchange Commission, Public Reference Branch, Stop 1-2, 450 Fifth Street, N.W., Washington, D.C. 20549.**

Bechtel Information Services also provides prompt and low-cost research and copying services. It is located at 15740 Shady Grove Road, Gaithersburg, MD 20877-1454. In Maryland, the telephone number is (301) 258-4300, Outside Maryland, phone toll free: 1-800-231-DATA.

Corporate annual reports are on file with the SEC, or can be obtained directly from the company. There is no charge for annual reports that are ordered from the company. The SEC copies would require copying. Contact the executive in charge of Investor Relations or the Senior Vice President and Chief Financial Officer at the particular company that you are researching.

Another way to obtain annual reports and SEC filings is to tell a stockbroker that you are interested in purchasing stock in a particular company and to acquire the company's annual report.

Somewhere within an annual report it will say something like this: A copy of the company's annual report on Form 10-K, as filed with the SEC, will be furnished without charge upon written request to the Office of the Corporate Secretary. The 10-K is research pay dirt!

FORM 10-K

I find this annual report to be the most useful of all SEC filings. In summary, it will tell you the registrant's state of business. This form is filed within 90 days after the end of the company's fiscal year. The SEC retains 10-Ks for ten years.

Part I of the 10-K reveals, among other things:

a) **When the company was organized and incorporated.** You will want to know how long the company has been in business to gauge its experience. What you would expect from an established company may vary from what you would tolerate at a start-up firm.

b) **What the company produces, percentages of sales any one item may be; seasonal/nonseasonal, etc.** It is critical to have a complete picture of company's product lines, their strengths and markets, any seasonality or other restrictions to the appropriateness of your item, and if and how your product could be positioned.

c) **How the company markets, for instance via independent sales representatives or its own regional staff offices.** It is important to know how a company gets something onto the market and where sales staff loyalty is, i.e. company employees typically have more loyalty than independent sales reps who handle more than one company's line.

d) **Whether or not it pays royalties and how much per year.** You can often see how much work the company does with outside developers and whether it licenses anything at all. An example of such wording is this from one corporate 10-K: "We review several thousand ideas from professionals outside the Company each year."

I recall another 10-K that read, "The Company is actively planning to expand its business base as a licenser of its products." Statements such as these show that doors are open!

e) **What amount of money the company spends to advertise and promote its products.** If your product will require heavy promotion, and the company does not promote its lines, you may be at the wrong place. It is counterproductive to take promotional products to companies that don't advertise.

f) **Details on design and development.** You should know before approaching a company whether an internal design and development group exists and how strong it is. I found one 10-K in which a company stated, "Management believes that expansion of its R&D department will reduce expenses associated with the use of independent designers and engineers and enable the Company to exert greater control over the design and quality of its products." It could not be more obvious that outside inventors were not wanted.

g) **Significant background on production capabilities.** Often it is valuable to know in advance what the company's in-house production capabilities are, and what its out-sourcing experiences are in your field of invention. It's no use taking a technology to a company that does not have the experience to produce it.

h) **Terms of long-term leases.** It can be important to know whether a company owns or rents its facilities as a measurement of its strength and capabilities. An inventory of real estate can also give you an excellent overview of warehouses, plants, offices, etc.

i) **If the company is involved in any legal proceedings, law suits, for instance.** You may not want to go with a manufacturer that is being sued right and left. Maybe it has just risen from a bankruptcy and is still not strong financially. All of this kind of information is an excellent indicator of corporate health.

j) **The security ownership of certain beneficial owners and management.** This is vital to understanding the pecking order and power structure. Here is where you'll see who owns how much stock (including family members), and what percentage of the company this represents. The ages and years with the company are also shown.

l) **Competition.** This section will give you a frank assessment of the company's competition and its ability to compete. One 10-K I read once admitted, "The Company competes with many larger, better capitalized companies in design and development..." It is unlawful to paint a rosy picture when it doesn't exist. The 10-K is one of the few places you can get an accurate picture. Would you want to license a product to a company that states, for example, "...most of the Company's competitors have financial resources, manufacturing capability, volume and marketing expertise which the Company does not have." This tells me to check out the competition!

m) **Exhibits.** On occasion, a company will attach to its 10-K exhibits such as employee stock option plans, licensing agreements, executive employment contracts, leases, letters of credit, etc. Information from any one of such documents could be important in a future negotiation. If a company denies your request for a certain contractual term, would it not be nice to be able to point out that there is corporate precedent for your receiving said stipulation?

I once had read the employment agreement between the executive with whom I was negotiating and his company. This helped me to estimate his worth to the company and what he could and could not make happen on my behalf. I even knew the type of car he was leased.

FORM 10-Q

The Form 10-Q is a report filed quarterly by most registered companies. It includes unaudited financial statements and provides a continuing view of the company's financial position during the year.

The 10-K and 10-Q are the two filings I find most valuable. However, the SEC has many other reports available. The best way to view a full inventory of records on file is to contact the nearest SEC regional office and pick up or have sent to you one of the Commission's booklets.

By combining the information available in 10-Ks with your personal observations, you will be much better prepared to do business.

PRIVATE COMPANIES

It is not as easy to find out detailed information on privately held companies. No regulations require that they fill out the kinds of revealing reports public companies must. Nevertheless, it is important to gather as much background information as possible.

Here are some questions I get answered before approaching a private company. The answers come to me from a combination of sources ranging from state incorporation records to interviews with competition, suppliers, retailers (as appropriate), and the owners themselves.

1) **Is the company a corporation, partnership, or sole proprietorship?** This can have legal ramifications from the standpoint of liabilities the licensee

assumes. A lawyer can advise you on the the pluses and minuses of each situation.

2) **When was the company organized or incorporated? If a corporation, in which state is it registered?** When a company was organized will give you some idea as to its experience. The more years in business, the more tracks in the sand are left. The state it is registered to do business in will tell you where you may have to go to sue it.

3) **Who are the company's owners, partners, or officers?** Always know with whom you are going into business. In the end companies are people, not just faceless institutions.

4) **What are the company's bank and credit references?** How a company pays its bills is important for obvious reasons, and its capital base is worth assessing.

5) **Is the manufacturer the source for raw material? Does it do the fabrication?** Such information will help estimate a company's capabilities for bringing your invention to the marketplace.

6) **How many plants does the company own (lease), and what is the total square footage? Does it warehouse?** This kind of information will help complete the corporate picture.

7) **What products are currently being manufactured or distributed?** You don't want to waste time pitching companies that do not manufacture your type of invention. Maybe a company you thought to be a manufacturer is really only a distributor.

8) **How does the company distribute?** Find out about the direct sales force. Make inquiries concerning outside sales representatives and number of jobbers. Does the company use mail order, house-to-house, mass marketing, or some other form of distribution? This information will quickly reveal how a company delivers its product and whether its system is appropriate to your product. With a mass market item, it would be foolish to approach a firm that markets door-to-door, regardless of its success.

9) **What kind of marketing and advertising support can you expect your product to receive?** Specify whether or not your product will receive national, regional, or local promotion. Will announcements be carried on television, radio, or in print? Does the company provide point of purchase promotion? It makes no sense to license a product requiring promotion to a firm that does not promote its lines.

IF YOU NEED HELP IN LOCATING A COMPANY

The Thomas Register

One of the best sources for product and corporate profiles is the *Thomas Register.* Available in most public library reference rooms, "Thomcat", as it is known, contains information on more than 120,000 U.S. companies in alphabetical order, including addresses and phone numbers, asset ratings, company executives, the location of sales offices, distributors, plants, and service and engineering offices. If you know a brand name, you can locate it in the *Thomcat Brand Names Index*.

If you wish to purchase your own set of books, the *Thomas Register* costs $175.00 for the 18 volumes (1989). Like an encyclopedia, it requires considerable shelf space. To order call: 1-800-222-7900.

S&P's Register

Another excellent source is *Standard & Poor's Register of Corporations, Directors and Executives,* available at public libraries. Consisting of three volumes, it carries data on more than 45,000 corporations including their zip codes; telephone numbers; names, titles, and functions of approximately 400,000 officers, directors, and executives. A separate volume selects 70,000 key executives for special biographical sketches. The last volume contains a classified industrial index.

The *S&P Register of Corporations, Directors and Executives* (1988-89) costs $450.00. To order call: 1-800-221-5277.

After you have read all the literature and investigated the company inside and out, you must ask yourself, can the company deliver and will you be comfortable working with its people. The abbreviation "Inc." after a company's name is not significant. Nice offices, a few secretaries, a fax and copying machine do not a successful licensee make. Just as with your inventions, steak must be under the sizzle. And the way you are treated should be in good taste.

HOW TO PRESENT A CONCEPT

Avoid Cold Calls

Presentations are best when carefully choreographed and staged. Nothing is usually gained ambushing executives outside of their offices. As Agesilaus II, King of Sparta, said, "It is circumstance and proper timing that give action its character and make it either good or bad."

For every rule ever told about when to sell, another rule proves it wrong. The best rule on when to sell ideas is whenever possible. Timing is, of course, everything. I operate under the principle that when you have something hot, burn it. When it gets cold, sell it for ice.

I licensed our first electronic toy, StarBird, to Milton Bradley after Thanksgiving, and the manufacturer premiered it only weeks later at American International Toy Fair. Milton Bradley put its resources on the line to make it happen in spite of the toy's complicated structure.

I developed the game "Adver*teasing,*" which was licensed in June to Cadaco, and manufactured and in stores by mid-September of the same year. By early December it had racked up sales of over 250,000 units.

While these examples are exceptions to the rule, had I waited until the "best" time to make the pitch, opportunity might have been delayed or the product may never have been licensed. The marketplace is temperamental and erratic.

Another rule to consider is this: the faster a product is made available, the faster it begins to generate income.

Curtain Up. Light The Lights.

Once you have been extended an invitation to display your concept to a manufacturer, it's major show time. And if you thought inventing was tough, you haven't experienced hardship. The moment you walk into a company's conference room with your invention at the ready, you pass into an eerie twilight zone. You are the hero in a Nintendo video game and the corporate executives comprise an array of varied personalities. Some pray for you to succeed. Others are gremlins out to gobble you up, shoot you down, and otherwise obliterate your ideas faster than Octorok can toss stones at Link.

Thomas Alva Edison once observed, "Society is never prepared to receive any inventions. Every new thing is resisted, and it takes years for the inventor to get people to listen to him before it can be introduced."

"Creation is a stone thrown uphill against the downward rush of habit," another inventor said.

World history is rich in stories about people's resistance to new ideas and change. Some seem unbelievable in retrospect. I find it helpful and comforting to recall some of these stories before I begin a presentation.

Impressionist art that sells for millions of dollars today was met with onslaughts of condemnation and cavil when first introduced. The innovative Stravinsky was once labeled "cynically hell-bent to destroy music as an art form." Today he is regarded as a genius.

When railroads were established, farmers protested that the "iron horse" would scare their cattle to death and stop hens from laying eggs. The British Association for the Advancement of Science insisted the automobile would fail because a human driver "has not the advantage of the intelligence of the horse in shaping his path."

Back in the 1930s, Charles B. Darrow invented a game called Monopoly, it was rejected by six or seven companies, including Parker Brothers. Parker Brothers eventually saw the light and published Monopoly, which annually sells more than three million units. Darrow became the first millionaire game inventor.

Inflexibility is Inherent in the System

Remember that in order to operate, companies must have rules and controls. Loose cannons do not last long in corporate environments. Any organization must understandably have somewhat of an established routine to survive. Therefore, executives, especially in larger companies, often fall into predictable and ordered routines. Your job is to break this routine, make people believe in you, and interest the company in buying your concept.

PREPARING YOUR PROPOSAL

I always back-up my verbal presentations with written proposals. The inventor cannot always go out with every prototype, and as the saying goes, if it ain't on the page, it ain't on the stage.

Do not depend on the company to copy your proposals. Bring enough originals with you for the meeting and then some. You cannot afford to have someone inadvertently lose a critical page.

All of my proposals begin with a concept summary. This is nothing elaborate, just a solid paragraph that paints a picture of the product and its overall concept and objective.

This is followed by a technical section that addresses costing and manufacture. You are well advised to have a rough estimate of what the invention will cost to manufacture. Don't expect the company to know.

I always try to provide the following data with every item (as appropriate):

1) A sheet listing all components with respective prices from various sources. Pricing from three different sources is a good bet. When possible, a mix of

domestic and offshore numbers is best. Do not forget to include the volume the quotes are based upon, plus vendor contacts.

2) Note the type of material(s) desirable, e.g. polyethylene, wood, board, etc. Provide substitutions and options for consideration.

3) When you calculate the item's cost, do not forget to consider the price of assembly (if any). The quoting vendors will be helpful here.

Every written proposal should also contain:

1) **Detailed operating instructions:** Take nothing for granted. The worse thing that can happen is a client's inability to use your item after you have departed. Illustrate with pictures if required. No item is too easy.

2) **A marketing plan:** Highlight your item's advantages over existing product(s) and what makes it unique. Define its appeal and target audience. Suggest follow-ups, including second generations and line extensions of your product (as appropriate). Manufacturers like products that have a future, especially if they will be required to spend lots of start-up dollars in the development and launch phases.

3) **Trademarks:** Offer possible trademarks. If a trademark search has been done, include the results of your search or status of any applications. The right mark can go a long way in securing a sale.

4) **Patents:** Include an update on any patent searches, PTO actions, etc. If a patent has issued, attach a copy of it. The more detailed and comprehensive your work is here, the quicker and easier the corporate evaluation of your product should be.

5) **Ad Campaign/Copy:** Suggest advertising direction, artwork, and slogans whenever appropriate. This type of work makes presentations even more persuasive and polished.

6) **Test results:** Report the results of any formal testing or focus group sessions. Include photographs and videotapes whenever appropriate.

7) **Inventor's Background/Capabilities:** If you are unknown to the company, provide personal background and describe yourself and your capabilities to support further development and manufacture of the invention under consideration. After all, who understands an invention better than its inventor? Your experience may save the manufacturer a great deal of money.

The aim here is to show the manufacturer that your front line is strong and the bench is deep. The more confident the company feels about you the better.

Watch Your Language

You cannot be too specific in proposals. Spell out everything. Take nothing for granted. And do not assume that words and terms have the same meaning to every reader, especially when you're making presentations to potential foreign licensees.

Although much is known about how new products are developed, there is no consensus on the meaning of key terms and definitions.

"With the increased emphasis in developing new commercial products and processes to help the U.S. become more competitive internationally, the need for a common language in the innovation process is becoming even more important," says George Lewett, chief of the National Bureau of Energy-Related Inventions at National Institute of Standards and Technology (NIST).

It is of interest to note that the NIST, the Department of Energy, and the National Society of Professional Engineers are collaborating to develop a consensus language for use in describing the innovation process. The project is expected to take about one year.

So far, seven organizations have nominated members to a task group. They are: American Institute of Chemical Engineers, American Society for Engineering Education, American Society of Mechanical Engineers, Commercial Development Association, Industrial Research Institute, Institute of Industrial Engineers, and NSPE.

I hope to incorporate their glossary in a future edition of this book.

MANUFACTURERS DON'T LICENSE IDEAS

It is imperative that you develop a prototype of an invention prior to disclosing it to potential licensees. A prototype is defined as an original model on which something is patterned. If you do not have the time, money, or commitment to build a prototype, the odds of licensing the idea are nil.

The most effective kind of prototype is what is called a "looks like-works like" version. There are no short cuts. Manufacturers don't license "ideas". They react to physical matter. Don't count on people being able to "imagine" what your product will look like or how it will operate. Even if they could, busy executives do not usually have the time or interest to engage in such typically futile exercises.

Executives love to touch and feel prototypes. Kick the tires, so to speak. Knowing this, do your best to have prototypes that most resemble and operate like a production model. Go that extra mile to ensure the prototypes are solid and have perceived value. Prototypes must be solid because often they take quite a beating at the hands of potential licensees. Don't be surprised to get them back broken. It happens even at the best of companies. This comes with the territory.

Don't take any of this lightly. You must be as sophisticated and slick in your presentation to a potential licensee as it will have to be in its pitch to the trade and/or the consumer. While getting a product known is relatively easy, marketing a need is something else. And that's your ultimate goal.

Making Prototypes

If you cannot make the prototype, plenty of places will provide assistance. In some cities prototype makers can be found in the telephone directory. Universities and engineering schools often have workshops where connections can be made to get something done. Local inventor groups are a wonderful source of information as well. **See Chapters 2, National and Regional Inventor Associations and 4, University Innovation Research Centers for more information.**

Many invention marketing companies offer prototype making services. Be careful. Know what you are getting into when contracting to have a product prototyped. I would not engage anyone to do any prototype work without inspecting their shop and checking references.

It is prudent to require that any prototype maker sign a **Confidential Disclosure Agreement** before you show or discuss your product. A sample agreement is shown as Fig. 32.

Multiple Submissions

If you have more than one prototype, and the situation is appropriate, you may wish to consider making submissions to more than one manufacturer at the same time. I have no set rule about this and take it case by case, guided by experience.

If a company asks to hold off further presentations until it has an opportunity to review the item more in depth, try to set guidelines. In all fairness to everyone, some products require a reasonable number of days to be properly considered. However, if you feel the company is asking for an unreasonable period of time, seek some earnest money to hold the product out of circulation. The amount of time and money is negotiable. Also insist that the product not be shown to anyone outside the company.

To keep track of who has seen what and when, a Patent Worksheet appears as Fig. 21.

Mutual Dependency

You need the company or you would not be there. Show yourself as being independently creative, while at the same time taking the "we-approach" and not the "I-approach."

In order for your product to sustain itself through the review and development process, it will need a champion. Typically this standard-bearer will come from among those attending your first meeting. Get others involved. Turn "your idea" into "our idea."

If You Sign My Paper, I'll Sign Yours

Many companies ask that an agreement be signed for the submission of outside ideas. This may surface when you first approach the company or happen on the day you appear to make the formal presentation. I have never had a problem with such requests. I always know with whom I am dealing and feel confident in the relationship. If I did not, I wouldn't be there in the first place.

A suspicious attitude may seriously inhibit your progress. Put your time and energies into creating concepts versus over protecting them.

Idea Submission by Outsiders

Outside submission agreements take many forms. **Figs. 33, Idea Submission Agreement I and Fig. 34, Idea Submission Agreement II are commonly used formats.** Most companies use variations of these forms. Use these forms to familiarize yourself with the types of document, and as examples, in case you want to use one to protect yourself before looking at another individual's ideas.

Agreement to Hold Secret and Confidnetial

It is appropriate, in some instances for the inventor to have the company sign an agreement to hold a product secret and confidential. **A sample agreement is shown as Fig. 32.**

A Fish Story

The art of product presentation is not unlike the sport of fly fishing. It takes time and patience. In casting, the lure is presented and then pulled back. The lure's movement, aided by twitches, pauses, and jerks of the rod by the angler, entices the fish to strike. In both cases the object is to hook a big one!

PART THREE
Directory of Inventing and Patenting Information

CHAPTER 2

NATIONAL AND REGIONAL INVENTOR ASSOCIATIONS

Inventors are not known for their business acumen or marketing expertise. Getting from the workshop drawing board to the corporate conference room can be a long, tough, and lonely haul. Operating on the frontiers of an emerging idea is difficult enough. Spearheading the development, licensing, and manufacture of an invention for commercial utilization can leave even the most experienced players nonplused.

A wide range of support organizations available to independent inventors are covered in this chapter. I encourage you to be selective in your association with any organization and make sure that those charging fees provide the appropriate kinds of services and data.

Inventors can receive guidance and support from myriad sources, patent lawyers, government agencies, universities, reference books, business and technical resource centers, and so forth. But often the most sincere information, albeit not always the most useful, comes from colleague inventors, peers who understand and also pursue the "quest for fire."

Inventors are no different than any group of people with a common interest. They love to get together and explore professional issues and share empirical experiences--success stories, and heartbreaks, "insider" information, personal contacts, methods, techniques, and dreams. They gain strength from each other in a symbiotic, cross-circuiting of spirits and concepts.

Inventor groups appear to take root anywhere inventors practice their trade. Seeking to build relationships among themselves, independent inventors have formed organizations throughout the country that provide professional and social forums. These assemblies offer members product development support and guidance and resource information. A common objective is to stimulate self-fulfillment, creativity, and problem-solving.

Some inventor groups are more sophisticated and active than others, some more organized, but most have something beneficial to offer.

There are two basic kinds of inventor organizations, local and national.

Local Clubs

Local clubs are real grass roots. They typically encompass only a limited geographic area, for example

Albuquerque Invention Club or the Inventor's Council of Hawaii. Often they go under names that are not so obvious. For example, one Cincinnati, Ohio inventors group calls itself The Salmon Club. They say this name was chosen because "inventors always swim upstream."

The local clubs can be great fun and informative, not to mention wonderful places to strike up friendships. They range in size from several members to several hundred members. Rarely do they have formal offices. Meetings are held at hotels, restaurants, or members' homes.

The Salmon Club is large enough to run occasional seminars for its members and guests. One of its recent seminars was a two-day event with speakers presenting oral and written information on topics encompassing invention evaluation, licensing, financing, basic patent law, incubators, and marketing.

Such clubs don't spend their money on slick newsletters and staff salaries. They are what you would expect from smallish, local organizations. Volunteers do their best with what they have at hand.

Local clubs do not promise to deliver anything more than you would expect. They are usually not the best place to pick up fast-breaking information on intellectual property legislation. But what I particularly like about these kinds of clubs is their lack of pretension, and the sense of warmth, hospitality, community, and comradeship they engender.

Frederick L. Jones, a member of the Palm Beach Society of American Inventors, says his club has approximately 150 members. Asked how many of his colleagues make a success at inventing, he responed, "Practically none. We're all dreamers. Inventors are all dreamers. If we weren't we wouldn't be inventors."

Typical of most independent inventors in his club, Fred Jones holds nine patents and has yet to license one. "I've been too busy inventing," he says.

Chuck Mullen, who bills himself as an inventor and a consultant for new idea development, is chairman of the board of advisors of the Houston Inventors Association, which has reported around 220 active members. How many have been successful at licensing their inventions? "Not many," he says, "Maybe five or six. Most join the Association to learn."

Typically run by inventors for inventors, the local clubs are not the businesses of businesspersons, as is often the case with national organizations.

Alexander T. Marinaccio, president of Inventors Clubs of America, says the average inventor's club lasts only two years. "Everybody joins, wants to get rich and then they all drop out."

National Inventor Organizations

National inventor organizations are often businesses owned and operated by business people, many strictly for profit. They run the gamut from reputable, high-powered lobbying organizations based in the nation's capital, to "nonprofit" societies that make money selling magazine subscriptions, official-looking certificates, book clubs, group insurance policies, discounted car rentals, and kindred fare.

When you join a national organization your expectations and dues are much higher than at the local level. And so is the organization's overhead required to support high-paid, ever-traveling executives and toll-free telephones numbers. In return for the steeper membership dues, you'll want the kind of information unavailable via local clubs, for example, access to major corporate executives, state-of-the-art marketing and patenting, trademark and copyright advice, quality product evaluations, and so forth. Very few national organizations can deliver all of this under one roof.

Ask questions before you join. Compare membership offers. Even the best can disappoint when seen in the light of unrealistic expectations. Don't be mislead by slick brochures, certificates, directories, travel and insurance discounts, nonprofit status, press release services, and book clubs. The bottom line is what can the organization do for you.

Ask for member and business references. You want to know which companies rely upon the organization for input. Talk to corporate R&D executives who express faith in the organization. Money can shotgun press releases and newsletters out to a range of diversified manufacturers; it does not guarantee results.

Over and above dues, find out whether the organization expects a piece of your invention in return for its assistance. Make sure it is just not an invention marketing company in disguise. **Query the salary of full-time organization executives and relevant budget information. As a potential member, this kind of information should be made available to you.**

Selected Organization Profiles

AFFILIATED INVENTORS FOUNDATION, INC. (AIF)
2132 East Bijou Street
Colorado Springs, CO 80909-5950
Tel.: (719) 635-1234
Executive Director: John T. Farady

The Foundation was established in 1975 as a national organization with honorary members in 41 states. Membership is by invitation only, and the majority of its members are holders of U.S. patents. It was founded by John Farady and the late Tom Wille, an inventor, electrical engineer, and an avid student of Nickola Tesla.

1987 was a very busy year for AIF. It claims to have provided more than 2,500 independent inventors with free preliminary appraisals for 2,625 inventions. Of these, 759 were rejected in this first stage. According to Foundation executive director, John Farady , about 30 percent of all inventions they see have "fatal" flaws and are thus rejected.

Since it opened for business fifteen years ago, the Foundation has evaluated a reported 18,000 plus inventions.

"Our educational goal is to provide independent inventors with sufficient information about each phase of invention development to help them make better decisions in their own best interests," says executive director Farady .

"We do not require any advance cash fees," he continues. "And our contract negotiation services are on a no-fee, straight commission basis should we be successful."

The Foundation states that "unlike the front-end-cash-fee companies, it is not the aim of the Foundation to offer marketing services to so many inventors that it cannot focus on anything properly. This organization keeps the membership controlled in such a way that it can be most effective."

The Foundation provides the following services to member inventors: free educational materials; free consultations; free preliminary appraisals; invention evaluations; low-cost patent and trademark services; funding opportunities; marketing assistance services; commission-based negotiation services (5 to 15 percent range); and an invention publicity program, for example, features member inventions in the bimonthly newsletter, *Invention News*, which is mailed to thousands of manufacturers.

I have always liked AIF and what co-founder John Farady is trying to do for his membership. The Foundation is one national organization that has lots to offer independent inventors and its publications, especially *Inventors' Digest*, are most informative.

Write for an information package. Check out the Foundation's references with any of the following organizations; Colorado Springs Chamber of Commerce, 100 Chase Stone Center, Colorado Springs, CO 80902; Dun & Bradstreet (any local office) Report No. DUNS-06-062-9748; and Central Bank of Colorado Springs, 2308 East Pikes Peak Avenue, Colorado Springs, CO 80909.

INTELLECTUAL PROPERTY OWNERS, INC.
1255 Twenty-Third Street, N.W., Suite 850
Washington, D.C. 20037
Telephone: (202) 466-2396
Telecopier: (202) 833-3636
Telex: 248959
Executive Director: Herbert C. Wamsley

Intellectual Property Owners (IPO) is a nonprofit association representing people who own patents, trademarks, and copyrights. It was founded in 1972 by a group of individuals who were concerned about the lack of understanding of intellectual property rights in the United States. Members include about 100 large and medium sized corporations and a growing number of small businesses, universities, patent attorneys and independent inventors.

IPO's corporate members include the likes of Monsanto, Standard Oil, Westinghouse Electric, United Technologies, Procter & Gamble, IBM, Ciba-Geigy, Upjohn, AT&T, Amoco, and Union Carbide.

IPO conducts an active government and public relations program in Washington, D.C. One of its best known programs is the Inventor of the Year Award, given each spring to an inventor whose invention was either patented or first made commercially available during the previous year.

In my opinion, IPO is outstanding and well-respected, and the flagship among such organizations. It offers its members the most current information on patent, trademark, and copyright issues, as well as a strong voice on Capitol Hill and at the PTO.

For example, during 1988, IPO mailed about twenty editions of its *Washington Briefs,* usually within a few hours after significant intellectual property events occurred. More detailed information is published in its printed newsletter, *IPO News.* Much of this information was available from no other source.

Herb Wamsley, IPO executive director, says that he is interested in expanding his organization's membership base to include more individual members such as independent inventors, investors, and attorneys.

Chapter Arrangement

This chapter is arranged alphabetically by association name. All association names in this chapter are referenced in the Master Index. The Master Index may also include additional associations that are mentioned in the text of other chapters.

★1★ Affiliated Inventors Foundation, Inc. (AIF)

2132 Bijou St.
Colorado Springs, CO 80909 (719) 635-1234
John T. Farady, Executive Director

Founded: 1975. **Description:** Offers encouragement and assistance to independent inventors. Supplies members with sufficient information on each phase of their invention. Provides low-cost or free patent attorney and consultant services. **Members:** 400. **Publications:** (1) Invention News (newsletter), bimonthly; (2) Inventors' Digest (newsletter), bimonthly; also publishes informational material. **Meetings:** Annual - always October, Colorado Springs, CO. For a further discussion of this groups activities, see the introduction at the beginning of this chapter.

★2★ Albuquerque Invention Club

Box 30062
Albuquerque, NM 87190 (505) 266-3541
Dr. Albert Goodman, Contact

★3★ American Association of Entrepreneurial Dentists (AAED)

420 Magazine St.
Tupelo, MS 38801 (601) 842-1036
Dr. Charles E. Moore, Exec.Dir.

Founded: 1983. **Description:** Dentists and other dental professionals involved in research, industry, manufacturing, marketing, publication, and other entrepreneurial activities. Objectives are to: promote dental research and product development and assist entrepreneurial dentists; offer expertise in technical, professional, and scientific skills; exchange information among members on how to improve dentistry, creativity, and free enterprise; encourage high ethical standards and professional conduct in dental service, products, and marketing. Informs the public, dentists, educators, and manufacturing companies of new and beneficial techniques, products, and services; coordinates the review of specifications for dental materials and products by regulatory agencies; and encourages the dental industry to become more actively involved in public dental education. Offers consultation and placement services; evaluates and presents new ideas and products to manufacturing companies at convention trade expositions. Provides lists of foreign dental dealers and buyers, export technique information and shipping procedures, hospitality accommodations for members, and private entrepreneurial activities. Encourages creativity; sponsors competitions; bestows awards. Maintains biographical archives, and hall of fame of plaques awarded for recognition of dentist of the year, of the decade, and of the century. **Publications:** Entrepreneurial News, periodic; also publishes pamphlet. **Meetings:** Four dental meetings per year (with exhibits).

★4★ American Association of Inventors (AAI)

6562 E. Curtis Rd.
Bridgeport, MI 48722 (517) 799-8208
Dennis Ray Martin, Pres.

Founded: 1862. **Description:** Private inventors and other interested individuals. Provides assistance with development of ideas, patent applications, marketing, and production. Offers workshops on product development; maintains 485 volume library and speakers' bureau. Compiles statistics; bestows awards. **Members:** 6512. **Publications:** (1) The Success System, bimonthly; (2) Threw the Key Ring, bimonthly. **Meetings:** Annual conference.

★5★ American Copyright Council (ACC)

1600 I St., N.W.
Washington, DC 20006 (202) 293-1966
Fritz E. Attaway, V.Pres & Sec.

Founded: 1984. **Description:** Coalition of computer, film, law, magazine, music, publishing, recording, and television organizations created to unite the copyright community. Seeks to initiate a concerted effort to preserve copyright principles and to educate the public about the value of copyrights and the harm caused by copyright infringement. Maintains speakers' bureau; compiles statistics. Publishes the Copyright Industries in the United States. **Members:** 25.

★6★ American Copyright Society (ACS)

345 W. 58th St.
New York, NY 10019 (212) 582-5705
Gerard Delachapelle, Mng.Dir.

Founded: 1952. **Description:** Authors, composers, and music publishers. Seeks to protect and enhance the corporate interests of members. Arranges for the collection of performance and mechanical royalties. Conducts periodic conferences, meetings, and research on copyrights.

★7★ American Entrepreneurs Association (AEA)

2392 Morse Ave.
Irvine, CA 92714 (714) 261-2393
Wellington Ewen, Pres.

Founded: 1973. **Description:** Persons interested in business opportunities and in starting profitable businesses. Conducts in-depth research on all types of small businesses. **Members:** 150,000. **Publications:** Entrepreneur Magazine, monthly; also publishes research reports.

★8★ American Society of Inventors (ASI)

P.O. Box 58426
Philadelphia, PA 19102-8426 (215) 546-6601
Henry H. Skillman, Treas.

Founded: 1953. **Description:** Engineers, scientists, businessmen, and others who are interested in a cooperative effort to serve both the short- and long-term needs of the inventor and society. Works with government and industry to improve the environment for the inventor. Goals include: encouraging invention and innovation; helping the independent inventor become self-sufficient; establishing an invention market and a communication system for inventors and businessmen to solve problems; seeking the optimization of employee disclosure agreements for use by employers; providing a voice for the independent inventor in reference to patent law changes. Acts as clearinghouse and pilot group for the formation of additional chapters; chapters develop special service projects, hold regular meetings featuring speakers, and provide a meeting place for other inventors. Sponsors periodic seminars. **Members:** 100. **Publications:** ASI Newsletter, periodic. **Meetings:** Periodic National Fall Inventor's Conference - usually November, Philadelphia, PA.

★9★ Arkansas Inventors Congress, Inc.
AIDC/Energy Division
One State Capital Mall
Little Rock, AR 72201 (501) 371-1370
Morris Jenkins, Contact

★10★ Association for Science, Technology and Innovation (ASTI)
P.O. Box 1242
Arlington, VA 22210
Ben Sands, Pres.

Founded: 1978. **Description:** Aim is to establish dialogue among different disciplines such as engineering, medicine, education, and the physical, social, and biological sciences, which share the common problem of effective management of innovation. Objectives are: to share ideas, knowledge, and experience among diverse communities; to expand and organize knowledge of the factors that effect productivity of science and technology efforts; to promote development, demonstration, and application of policies, standards, and techniques for improving management of innovation. Although the association operates primarily in the Washington, DC area, it has members throughout the U.S., Europe, Africa, and Central America. Conducts seminars; holds monthly management luncheon, roundtable, and symposium. Sponsor competitions. **Members:** 175. **Publications:** Newsletter, bimonthly.

★11★ Association of Collegiate Entrepreneurs (ACE)
Center for Entrepreneurship
Box 147
Wichita State Univ.
Wichita, KS 67208 (316) 689-3000
Douglas Mellinger, Dir.

Founded: 1983. **Description:** Student entrepreneurs, community entrepreneurs, and sponsoring firms. Seeks to enhance opportunities for student entrepreneurial pursuits by providing opportunities for students to network with faculty and entrepreneurs. Encourages venture capitalists, consultants, and other entrepreneurs to become involved as mentors in assisting students with entrepreneurial ventures. Works with Center for Entrepreneurial Management (see separate entry). Maintains speakers' bureau; compiles statistics; bestows awards honoring the top 100 young entrepreneurs. **Members:** 2000. **Publications:** Newsletter, monthly. **Meetings:** Annual conference (with exhibits) - usually March.

★12★ Association of Small Business Development Centers (ASBDC)
1050 17th St., N.W.
Suite 810
Washington, DC 20036 (202) 887-5599
Tom Cator, Executive Director

Founded: 1977. **Description:** State centers providing advice for those planning to establish a small business. **Publications:** ASBDC News.

★13★ Association of Venture Founders (AVF)
805 Third Ave., 26th Fl.
New York, NY 10022 (212) 319-9220
Anni J. Lipper, Exec.Dir.

Founded: 1979. **Description:** Successful entrepreneurs who seek to enhance the wealth, knowledge, and business success of members. Provides educational networking for the continuing education of members. Holds three seminars per year. **Members:** 150. **Publications:** (1) Venture Magazine, monthly; (2) Who's Who in AVF (directory), annual.

★14★ California Engineering Foundation
913 K St. Mall
Ste. A
Sacramento, CA 95814 (916) 440-5411
Robert J. Kuntz, Contact

★15★ California Inventors Council
P.O. Box 2732
Castro Valley, CA 94546 (415) 652-3138
Lawrence Udell, Contact

Contact: Barrett Johnson, P.O. Box 2036, Sunnyvale, CA 94087 (408) 732-4314.

★16★ Center for Entrepreneurial Management (CEM)
180 Varick St., Penthouse Suite
New York, NY 10014 (212) 633-0060
Joseph R. Mancuso, Pres.

Founded: 1978. **Description:** Serves as a management resource for entrepreneurial managers and their professional advisors. Selects and makes available published materials on developing business plans, organizing an entrepreneurial team, attracting venture capital, and obtaining patents, trademarks, and copyrights. Develops, collects, and disseminates current information on business trends, new laws and regulations, and tax guidance. Conducts intensive-study courses and seminars. Has identified stages of the entrepreneurial process and, through essays and audiocassettes, addresses problems pertinent to each stage. Maintains library of small business and venture capital information. **Members:** 3000. **Publications:** The Entrepreneurial Manager (newsletter), monthly.

★17★ Central Florida Inventors Club
6402 Gamble Dr.
Orlando, FL 32808 (305) 299-8598
Steve Chandler, Contact

★18★ Central Florida Inventors Council
4855 Big Oaks Lane
Orlando, FL 32806 (305) 859-4855
Dr. David Flinchbaugh, Contact

Contact: Alternative address: P.O. Box 13416, Orlando, FL 32859.

★19★ Columbus Inventors Association
2480 East Ave.
Columbus, OH 43202 (614) 267-9033
Tim Nyros, Contact

★20★ Confederacy of Mississippi Inventors
4759 Nailor Rd.
Vicksburg, MS 39180 (601) 636-6561
Rudy Paine, Contact

★21★ Copyright Clearance Center (CCC)
27 Congress St.
Salem, MA 01970 (508) 744-3350
Eamon T. Fennessy, Pres.

Founded: 1977. **Description:** Photocopy users (corporations, academic and research libraries, information brokers, government agencies, and others who systematically utilize or distribute photocopy material) and publishers, authors, or other owners of copyrights. Established in response to 1978 copyright law which requires that permission of copyright owners be obtained by anyone doing systematic photocopying or photocopying not permitted under the fair use provision of the law. Provides owners with a centralized agency through which permission to use registered materials may be granted and fees may be collected. Operates Annual Authorizations Service, a photocopy license service for U.S. corporations. **Members:** 4000. **Publications:** (1) Photocopy Authorizations Report (newsletter), quarterly; (2) Publishers' Photocopy Fee Catalog (with quarterly supplements), semiannual; (3) Annual Review. **Telecommunication Access:** Fax, (617)741-2318.

★22★ Copyright Society of the U.S.A. (CSUSA)

New York Univ. School of Law
40 Washington Sq., S.
New York, NY 10012 (212) 998-6194
Elizabeth Botta, Sec.

Founded: 1953. **Description:** Lawyers and laymen; libraries, universities, publishers, and firms interested in the protection and study of rights in music, literature, art, motion pictures, and other forms of intellectual property. Promotes research in the field of copyright; encourages study of economic and technological aspects of copyright by those who deal with problems of communication, book publishing, motion picture production, and television and radio broadcasting. Seeks better understanding among students and scholars of copyright in foreign countries, to lay a foundation for development of international copyright. Cosponsors (with New York University School of Law) the Walter J. Derenberg Copyright and Trademark Library, which includes foreign periodicals dealing with literary and artistic property and related fields. Sponsors symposia and lectures on copyright. Encourages study of copyright in U.S. law schools. **Members:** 1000. **Publications:** Journal of the Copyright Society, quarterly. **Meetings:** Annual conference - always June. Also holds mid-winter meeting - always February.

★23★ Educators' Ad Hoc Committee on Copyright Law (EAHCCL)

c/o August W. Steinhilber
Natl. School Boards Assn.
1680 Duke St.
Alexandria, VA 22314 (703) 838-6710
August W. Steinhilber, Gen. Counsel

Founded: 1963. **Description:** Coalition of organizations representing educators, librarians, and scholars dedicated to protecting the rights of educators and scholars to free access to copyrighted materials. Charges that legitimate photocopying of published works is being discouraged by exaggerated copyright warnings which go far beyond the restrictions in the federal copyright law. This law permits "the fair use of a copyrighted work" and defines "fair use" to include reproduction "for purposes such as criticism, comment, news reporting, teaching, scholarship or research." Opposes broad, strict copyright notices because they could "readily cause a scholar, teacher, or librarian to refrain from exercising his rights in making or obtaining a copy of material," resulting in a serious restraint on legitimate professional activity. Asserts that the field of education has a limited right to record television programs, use copyrighted material in computer-assisted instruction, and photocopy printed materials for use in the classroom. Proposes that the Copyright Office refuse to register copyrights for works in which the published statement exceeds the terms of the law.

★24★ The Entrepreneurship Institute (TEI)

3592 Corporate Dr., Suite 100
Columbus, OH 43229 (614) 895-1153
Mr. Jan W. Zupnick, Exec. Officer

Founded: 1976. **Description:** Provides encouragement and assistance to entrepreneurs who are creating and developing new companies. Unites financial, legal, and community resources to help foster the success of new companies. Promotes sharing of information and interaction between members. Offers consulting on accounting, marketing, banking, and legal issues. Operates Community Entrepreneurial Development Projects which are designed to improve communication between businesses, develop one-to-one business relationships between entrepreneurs and local resources, provide networking, and stimulate the start-up and growth of companies. **Publications:** Brochure; also distributes list of available publications. **Meetings:** Periodic forum, with lectures and workshops. Local groups hold periodic meetings. **Telecommunication Access:** 24-hour hot line.

★25★ Foundation for Innovation in Medicine (FIM)

411 North Ave., E.
Cranford, NJ 07016 (201) 272-2967
Stephen L. DeFelice, M.D., Chm.

Description: Seeks to regenerate interest in medical discovery and innovation, which the foundation believes flourished in the U.S. in the 1940s and 1950s, but has since declined despite "vastly increased public and private expenditures in research and development." Intends to monitor the state of innovation by conducting seminars and conferences. Encourages clinical research on natural substances and substances with little commercial value. Publishes books; offers educational videocassettes. **Meetings:** Semiannual; also holds annual symposium. **Telecommunication Access:** FAX (201) 272-4583; telex, 475-4248.

★26★ Gulf Coast Breeder

109 E. Scenic Dr.
Pass Christian, MS 39571 (601) 452-2007
Frank J. Wilem, Jr., President

★27★ High Technology Entrepreneurs Council

P.O. Box 72791
Las Vegas, NV 89170-2791 (702) 736-3794
George S. Sanders, Contact

★28★ Houston Inventors Association

600 W. Gray
Houston, TX 77019 (713) 520-1443
Greg Micek, Contact

★29★ Inno-Media

230 Tenth Ave., S.
Minneapolis, MN 55415 (612) 342-4311
Sam Koutavas, Director

Description: Acts as a resource center for inventors, innovators, and industry.

★30★ Innovation Development Institute

45 Beach Fluff Ave., Suite 300
Swampscott, MA 01907 (617) 595-2920

★31★ Innovation Group

Innovative Products Group
2325 Ulmerton Rd., Suite 16
Clearwater, FL 33520
George Feldcamp, Vice President

★32★ Innovation Institute

Box 429
larimore, ND 58215 (701) 343-2237
Dr. Jerry Udell, Contact

★33★ Innovation Invention Network
13 Benjamin Rd.
Worcester, MA 01602 (617) 799-4951
Andrew Stidsen, Contact

★34★ Innovators International (II)
P.O. Box 4636
Rolling Bay, WA 98061 (206) 842-7833
Paul von Minden, Acting Director

Founded: 1980. **Description:** Purposes are to: formalize innovation as a profession by encouraging individuals and organizations to develop services and products through creative, innovative thinking and establishing guidelines for professional innovation; promote innovation as a means for global communication. Projects include: Business Builders, helping underemployed and unemployed individuals to create their own work; Ideacraft, providing educational programs and developing activities and tools to help people to be creative and innovative; Innovator Builder Entrepreneur, encouraging educators, counselors, and administrators to promote and provide courses and program curricula associated with innovation; Earth Plan, motivating people to be creative and innovative in their efforts to meet social responsibilities and develop alternative methods of promoting world peace and stability; Product Town, involving the development of a model environment that could eventually be established in space and utilize technologically advanced, tested, and perfected products specially created for such an environment; Seton-Language Based Services, establishing links between human and computer language and developing computer software that performs what are called intelligent tasks such as the checking of spelling and proofreading. Publishes A Time To Build. **Meetings:** Periodic.

★35★ Innovex Inc.
Ste. 1125 LB37
4144 N. Central Expressway
Dallas, TX 75204 (214) 265-1540
Floyd Murphy, President

★36★ Intellectual Property Owners, Inc.
1255 Twenty-Third St., N.W.
Ste. 850
Washington, DC 20037 (202) 466-2396
Herbert Wamsley, Executive Director

Description: Founded: 1972. People who own patents, trademarks, and copyrights. **Publications:** IPO News and Washington Briefs. Offers Inventor of the Year Award. For a further discussion of this group's activities, see the introduction at the beginning of this chapter.

★37★ Intermountain Society of Inventors
1395 E. Greenfield Ave.
Salt Lake City, UT 84121 (801) 278-5679
M. Charles Faux, Contact

★38★ Intermountain Society of Inventors and Designers
P.O. Box 1056
Tooele, UT 84074 (801) 882-4943
Edward C. DeVore, Contact

Contact: Alternate address: P.O. Box 1514, Salt Lake City, UT 84110.

★39★ International Association of Professional Inventors
Route 10
4412 Greenhill Way
Anderson, IN 46011 (317) 644-2104
Jack L. Banther, Contact

Contact: Tom Nix, 818 Westminster, Kokomo, IN 46901 (317) 459-3553.

★40★ International Copyright Information Center (INCINC)
c/o Assn. of Amer. Publishers
2005 Massachusetts Ave., N.W.
Washington, DC 20036 (202) 232-3335
Carol A. Risher, Dir.

Founded: 1970. **Description:** Formed as a result of a meeting of the Joint Study Group comprising members of United Nations Educational, Scientific and Cultural Organization (UNESCO) and Bureaux Internationaux Reunis pour la Protection de la Propriete Intellectuelle (International Bureau for the Protection of Intellectual Property) (BIRPI) in the fall of 1969, which passed a proposal calling for the establishment of national information centers on copyright clearance in major publishing countries throughout the world. INCINC serves as the U.S. center where it assists publishers in developing countries in their efforts to contact U.S. publishers regarding the licensing of translation and English-language reprint rights to U.S. books.

★41★ International Licensing Industry and Merchandisers' Association (ILIMA)
350 Fifth Ave., Suite 6210
New York, NY 10118 (212) 244-1944
Murray Altchuler, Exec.Dir.

Founded: 1984. **Description:** Companies and individuals engaged in the marketing of licensed properties, both as agents and as property owners; manufacturers and retailers in the licensing business. Objectives are: to establish a standard reflecting a professional and ethical management approach to the marketing of licensed properties; to become the leading source of information in the industry; to communicate this information to members and others in the industry through publishing, public speaking, seminars, and an open line; to represent the industry in trade and consumer media and in relationships with the government, retailers, manufacturers, other trade associations, and the public. Compiles statistics; bestows awards; maintains hall of fame and placement service. **Members:** 235. **Publications:** (1) Licensing Directory, annual; (2) LIMA Light (newsletter), quarterly. **Meetings:** Annual trade show, with exhibit and seminar.

★42★ International Patent and Trademark Association (IPTA)
33 W. Monroe
Chicago, IL 60603 (312) 641-1500
Jeremiah D. McAuliffe, Pres.

Founded: 1930. **Description:** Lawyers who have professional qualifications and interest in the international protection of patents, designs, trademarks, copyrights, and other intellectual property rights. IPTA is the American group of the International Association for the Protection of Industrial Property. Monitors international developments that may affect industrial property and related rights. Studies, discusses, and reports on proposed national and foreign legislation treaties and conventions that are likely to affect national and international intellectual property interests. Bestows Sydney Diamond Award. **Members:** 700. **Publications:** (1) Annuaire, 3-4/year; (2) Membership Directory, annual; also publishes minutes of committee and executive meetings. **Meetings:** Annual meeting; also holds triennial international congress - 1989 June, the Netherlands.

★43★ Invention Development Society
8502A S.W. 8th St.
Oklahoma City, OK 73128 (405) 376-2362
William L. Enter, Sr., President

★44★ Invention Marketing Institute (IMI)
345 W. Cypress St.
Glendale, CA 91204 (818) 246-6540
Ted DeBoer, Exec.Dir.

Founded: 1964. **Description:** Inventors (8,175) and manufacturers (700). Purpose is to help inventors get their products into the marketplace and to help manufacturers find new products to make and market. Brings together inventors and manufacturers for mutual benefit. Maintains speakers' bureau, 1500 volume library, small museum, and hall of fame. **Members:** 8875. **Telecommunication Access:** Electronic bulletin board, (818)246-6546.

★45★ Inventor Associates of Georgia
637 Linwood Ave. N.W.
Atlanta, GA 30306 (404) 656-5361
Tom J. Sutor, Contact

★46★ Inventors & Entrepreneurs Association of Austin
6714 Spicewood Springs Rd.
Austin, TX 78759 (512) 452-9043
Veronique Berardino, Contact

★47★ Inventors & Entrepreneurs Society of Indiana, Inc.
Box 2224
Hammond, IN 46323 (219) 989-2354
Prof. Daniel J. Yovich, Contact

★48★ Inventors & Technology Transfer Corporation
P.O. Box 2185
Minneapolis, MN 55402 (612) 339-1863
William Braddock, Contact

Telecommunication Access: Alternate telephone (612) 379-7387.

★49★ Inventors Assistance League
345 West Cypress
Glendale, CA 91204 (818) 246-6540
Arthur Ryan, Contact

Contact: Ted DeBoer.

★50★ Inventors Association of America (IAA)
P.O. Box 1531
Rancho Cucamonga, CA 91730 (714) 980-6446
L. Troy Hall, Pres.

Founded: 1985. **Description:** Private inventors. Seeks to assist members in patenting, producing, and marketing their ideas. Maintains a Board of Evaluators to judge the quality and potential of members' creations; aims to thwart efforts by large corporations to improve or manufacture members' inventions and thereby obtain patents. Fulfills commissions to manufacture inventions through parent company, the Corporation of Inventive Minded People. Sponsors educational program in conjunction with schools on inventions and the inventive process. Conducts research and educational seminars. Maintains library. Also sponsors compeitions; bestows awards. Although currently active only in the Los Angeles, CA, area, plans to expand nationally. **Members:** 191. **Publications:** Inventor's Gazette, monthly; makes available audio- and videocassette tapes. **Meetings:** Annual; also holds quarterly meeting.

★51★ Inventors Association of Connecticut
9 Sylvan Rd. South
Westport, CT 06880 (203) 226-9155
Murray Schiffman, President

★52★ Inventors Association of Georgia
241 Freyer Dr. N.E.
Marietta, GA 30060 (404) 427-8024
Hal Stribling, Contact

★53★ Inventors Association of Indiana (IAI)
612 Ironwood Dr.
Plainfield, IN 46168 (317) 745-5597
Randall N. Redelman, President

Founded: 1979. **Description:** Provides technical, business, and educational information to members. **Members:** 51. **Publications:** Newsletter, monthly

★54★ Inventors Association of Metro Detroit
19813 E. Nine Mile Rd.
St. Clair Shores, MI 48080 (303) 772-7888
Peter P. Ruppe, Jr., Contact

★55★ Inventors Association of New England
P.O. Box 335
Lexington, MA 02173 (617) 862-5008
Dr. Donald Job, Contact

★56★ Inventors Association of New England, Connecticut Chapter
9 Sylvan Rd. South
Westport, CT 06880 (203) 226-9155
Murray Schiffman, Contact

★57★ Inventors Association of St. Louis
P.O. Box 16544
St. Louis, MO 63105 (314) 534-2677
Roberta Toole, Director

Contact: Alternate address: P.O. Box 30062, St. Loius, MO 63105.

★58★ Inventors Association of Washington
P.O. Box 1725
Bellevue, WA 98009 (206) 455-5520
David V. Sires, Contact

★59★ The Inventors Club
Rte. 11, Box 379
Pensacola, FL 32514 (904) 433-5619
Bill Bowman, Contact

★60★ Inventors Club of America (ICA)
P.O. Box 450261
Atlanta, GA 30345 (404) 938-5089
Alexander T. Marinaccio, Pres.

Founded: 1935. **Description:** Clubs of inventors, scientists, manufacturers, and others involved in problem-solving and inventing. Stimulates inventiveness and helps inventors in all phases of their work, including patenting, development, manufacturing, marketing, and advertising. Seeks to prevent abuses to the individual inventor, such as theft of ideas. Works to create new industry and lower the tax rate. Conducts research program and competitions. Sponsors children's services, seminars, and professional training; presents awards. Resources include hall of fame, biographical archives, 2000 volume library, and museum. Operates charitable program and speakers' bureau. **Members:** 6000. **Publications:** Inventors News, monthly; also publishes How to Shape an Idea Into an Invention and How to Protect an Idea Before Patent, and produces educational books. **Meetings:** annual conference (with exhibits) - always second Friday in November. 1989 Atlanta, GA; 1990 Orlando, FL.

★61★ Inventors Club of Greater Cincinnati (ICGC)
c/o William M. Selenke
18 Gambier Circle
Cincinnati, OH 45218 (513) 825-1222
William M. Selenke, Pres.

Founded: 1983. **Description:** Seeks to further education and professional contact among members. **Members:** 40. **Publications:** Newsletter, 9/year. **Telecommunication Access:** Alternate telephone (513) 922-9462.

★62★ Inventors Club of Minnesota
Inventors & Technology Transfer Corporation
P.O. Box 14-235
St. Paul, MN 55114

★63★ Inventors Connection of Greater Cleveland
1496 Lakewood Ave.
Lakewood, OH 44107 (216) 226-9681
Calvin Wight, Contact

★64★ Inventors' Council
53 W. Jackson, Suite 1041
Chicago, IL 60604 (312) 939-3329
Don Moyer, President

★65★ Inventors Council of Dayton
140 E. Monument Ave.
Dayton, OH 45402 (513) 439-4497
Leonard E. Smith, Contact

Contact: Ron Versic, P.O. Box 77, Dayton, OH 45409.

★66★ Inventors Council of Greater Lorain County
246 Harvard Ave.
Elyria, OH 44035
Dana N. Clark, Sr., Contact

★67★ Inventors Council of Hawaii
P.O. Box 27844
Honolulu, HI 96827 (808) 595-4296
George K.C. Lee, President

★68★ Inventors Council of Illinois (ICI)
c/o Donald F. Mayer
53 Jackson Blvd.
Chicago, IL 60604 (312) 922-5616
Donald F. Mayer, Exec. Officer

★69★ Inventors Council of Michigan (INCOM)
c/o University of Michigan
Industrial Development Division
Institute of Science and Technology
2200 Bonisteel Blvd.
Ann Arbor, MI 48109 (313) 764-5260
J. Downs Herold, Chairman

Founded: 1983. **Description:** Inventors, entrepreneurs, and business, education, and government leaders. Seeks to: establish support network for inventors; aid members with marketing strategies; facilitate the exchange of ideas between business and industry. **Members:** 927. **Publications:** INCOM Newsletter, monthly. **Meetings:** monthly - always fourth Wednesday of the month.

★70★ Inventors Council of Ohio—Columbus
3539 Lacon Rd.
Hilliard, OH 45459
Charles R. Morrison, Contact

★71★ Inventors Education Network
P.O. Box 14775
Minneapolis, MN 55414 (612) 379-7387
Marge Braddock, Contact

★72★ Inventors League
403 Longfield Rd.
Philadelphia, PA 19118

★73★ Inventors Network of Columbus
Business Technology Center
1445 Summit St.
Columbus, OH 43201 (614) 291-7900
Kris Akins, Contact

★74★ Inventors of California
15 Rheem Blvd.
Moraga, CA 94556 (415) 376-7541
Norman C. Parrish, Contact

Contact: Alternate address: P.O. Box 158, Rheem Valley, CA 94570.

★75★ Inventors Workshop International Education Foundation (Tarzana) (IWIEF)
P.O. Box 251
Tarzana, CA 91356 (818) 344-3375
Alan Arthur Tratner, Exec.Dir.

Founded: 1971. **Description:** Amateur and professional inventors. To provide for inventors and creative persons instruction, assistance, and guidance in such areas as: the steps needed to get patent protection for inventions, including how to make patent searches; how to offer inventions for sale; how to get inventions and products designed, production engineered, and manufactured; how to choose the experts who may be required for these services; how inventors may perform as many of these vital actions as capabilities and resources provide. Conducts research, seminars, and semiannual programs on invention promotion and "Reduction to Practice." Maintains exhibit at Los Angles County Museum of Science and Industry in California. Works in cooperation with Inventors Licensing and Marketing Agency which was formed to help inventor/members sell or license their products. Operates library of 500 volumes; maintains speakers' bureau; sponsors competitions and bestows awards. **Members:** 15,000. **Publications:** Lightbulb (magazine), bimonthly; also publishes Complete Guide to Making Money With Your Inventions; Little Inventions That Made Big Money; Inventor's Guidebook; Inventor's Journal; Science of Creating Ideas for Industry; and Simplicity: The Key to Success. **Meetings:** Annual Invention Convention (with exhibits). **Contact:** Maggie Weisberg.

★76★ Inventors Workshop International (Newbury Park)
3537 Old Conejo Rd.
Ste. 120
Newbury Park, CA 91320-2157 (805) 499-1626
Melvin L. Fuller, Contact

★77★ Inventrepreneurs' Forum (IF)
Five Riverside Dr.
New York, NY 10023 (212) 874-7362
David A. Lee, Co-Founder

Founded: 1986. **Description:** Individual inventors and entrepreneurs. To promote the success of emerging proprietary technologies. Serves as a referral service, pairing inventors with entrepreneurs who provide financial backing for new ideas and inventions.

★78★ Kearney Inventors Association
c/o Kearney Development Council
2001 Ave. A, Box 607
Kearney, NE 68847 (308) 237-3101
Steve Buttress, President

★79★ Licensing Executives Society (LES)
71 East Ave., Suite S
Norwalk, CT 06851 (203) 852-7168
James E. Menge, Exec.Dir.

Founded: 1966. **Description:** U.S. and foreign businessmen, scientists, engineers, and lawyers having direct responsibility for the transfer of technology. Maintains placement service. Sponsors quarterly technology transfer seminar. **Members:** 5200. **Publications:** (1) LES Nouvelles (journal), quarterly; (2) Membership Directory, annual; (3) International Technology Transfer Directory, biennial; has also published Law and Business of Licensing. **Meetings:** Annual - always October. 1989 Maui, HI; 1990 Homestead, WV; 1991 San Diego, CA; 1992 Boca Raton, FL.

★80★ Lincoln Inventors Association
P.O. Box 94666
Lincoln, NE 68509 (402) 471-3782
Steve Williams, Contact

★81★ Los Angeles Copyright Society (LACS)
c/o Donald L. Zachary
3000 W. Alameda Ave.
Burbank, CA 91523 (818) 840-3508
Donald L. Zachary, Exec. Officer

Founded: 1952. **Description:** Sponsors lectures and discussions on all phases of entertainment law, both domestic and international. **Members:** 300. **Meetings:** Monthly (September through May).

★82★ Michigan Biotechnology Institute
P.O. Box 27609
Lansing, MI 48909 (517) 355-2277
Dr. Jack H. Pincus, Director

Description: Promotes the commercialization of biotechnology in the state. Provides in-house research and provides technology transfer.

★83★ Michigan Energy and Resource Research Association
328 Executive Plaza
1200 6th St.
Detroit, MI 48226 (313) 964-5030
Todd Anuskiewics, Director

Description: Serves as a statewide partnership between industry, university, and government, promoting energy-resource-technology research and working to bring public and private research and development grants and contracts into the state. Provides information about the Small Business Innovation Research (SBIR) program.

★84★ Michigan Patent Law Association (MPLA)
c/o James A. Kushman
Brook and Kushman
2000 Town Center, Suite 200
Southfield, MI 48075 (313) 358-4400
James A. Kushman, Pres.

Founded: 1913. **Description:** Patent attorneys and agents organized for the promotion of high ethical standards, the exchange of information concerning the industry, and fellowship among members. Sponsors annual judges banquet. **Members:** 265. **Publications:** Newsletter, monthly

★85★ Midwest Inventors Group (MIG)
P.O. Box 518
Chippewa Falls, WI 54729 (715) 723-5061
Steve Henry, Exec. Officer

★86★ Minnesota Entrepreneurs Club
511 11th Ave. S.
Minneapolis, MN 55415

★87★ Minnesota Inventors Congress
129 S.E. Bedford St.
Minneapolis, MN 55414 (612) 379-7387
Marge Braddock, Contact

★88★ Minnesota Inventors Congress, Inventors Resource Center
P.O. Box 71
230 Second Street West
Redwood Falls, MN 56283-0071 (507) 637-2344
Penny Becker, Coordinator

Telecommunication Access: Toll-free in Minnesota (800) INVENT-1.

★89★ Minnesota Project Innovation, Inc.
Hazeltine Gates Office Bldg.
1107 Hazeltine Blvd.
Chaska, MN 55318 (612) 448-8826
James Swiderski, Executive Director

Telecommunication Access: Toll-free in Minnesota (800) 247-0864.

★90★ Mississippi Inventors Workshop
4729 Kings Hwy.
Jackson, MS 39206 (601) 366-3661
Karl Rabe, President

★91★ Mississippi Research and Development Center
3825 Ridgewood Rd.
Jackson, MS 39211-6453 (601) 982-6425
R.W. Parkin, Marketing Consultant

★92★ Mississippi Society of Scientists
508 Cindy Lane
Pearl, MS 39208

★93★ Mississippi Society of Scientists and Inventors
Box 2244
Jackson, MS 39205

★94★ National Association of Black Women Entrepreneurs (NABWE)
P.O. Box 1375
Detroit, MI 48231 (313) 341-7400
Marilyn French-Hubbard, Founder

Founded: 1979. **Description:** Black women who own and operate their own businesses; black women interested in starting businesses; organizations and companies desiring mailing lists. Acts as a national support system for black businesswomen in the U.S. and focuses on the unique problems they face. Objective is to enhance business, professional, and technical development of both present and future black businesswomen. Maintains speakers' bureau and national networking program. Offers symposia, workshops, and forums aimed at increasing the business awareness of black women. Shares resources, lobbies, provides placement service, and bestows annual Black Woman Entrepreneur of the Year Award. **Members:** 3000. **Publications:** (1) Making Success Happen Newsletter, bimonthly; (2) Membership Directory, annual. **Meetings:** Annual conference (with exhibits).

★95★ National Association of Minority Entrepreneurs (Balch Springs) (NAME)
3300 Shepherd Ln., #E
Balch Springs, TX 75180 (214) 286-9705
Leon Johnson, Pres.

Founded: 1985. **Description:** Minority professionals. Seeks to: develop a minority professional organization and form alliances with other professional groups; improve and increase the minority entrepreneurial community. Works to educate members in areas such as management, marketing, finance, and networking. Serves as a referral service and offers users reduced rates for the services of members. Holds workshops and seminars. Plans to: act as a clearinghouse for information; establish a credit union and an entrepreneurial institute; encourage and negotiate joint ventures between majority and minority firms; offer moral and financial support to minority businesses. Also plans to provide a telex service and hold annual convention. **Members:** 150. **Publications:** (1) Focus, bimonthly; (2) NAME Directory, annual.

★96★ National Association of Minority Entrepreneurs (New York) (NAME)
271 W. 125th St., Rm. 302
New York, NY 10027 (212) 316-3706
William J. Hampton, Exec.Dir.

Founded: 1972. **Description:** Minority individuals working for economic development in minority communities. **Members:** 500.

★97★ National Association of Plant Patent Owners (NAPPO)
1250 I St., N.W., Suite 500
Washington, DC 20005 (202) 789-2900
Robert F. Lederer, Exec.V.Pres.

Founded: 1940. **Description:** Owners of patents on newly propagated trees, shrubs, fruits, and other plants. Seeks to keep members informed of plant patents issued, provisions of patent laws, changes in practice, and new legislation. **Members:** 60. **Publications:** Membership Roster, annual; (2) Bulletin, periodic. **Meetings:** Annual - always July.

★98★ National Business Incubation Association (NBIA)
153 S. Hanover St.
Carlisle, PA 17013 (717) 249-4508
Carlos Morales, Exec.Dir.

Founded: 1985. **Description:** Incubator owners and operators; real estate developers; venture capital investors; economic development professionals. (Incubators are facilities where office space, consultants, receptionists, and other related services are shared.) Helps keep newly formed businesses operating, in part, by lowering business expenses; provides information to members. Educates businesses and investors on incubator benefits; offers specialized training in incubator formation. Conducts research and referral services. Maintains speakers' bureau; bestows awards. **Members:** 400. **Publications:** (1) NBIA Review, quarterly; (2) National Directory of Incubators, quarterly; (3) Newsletter, quarterly; also distributes books and sponsors publications on incubators. **Meetings:** Annual conference (with exhibits).

★99★ National Congress of Inventors Organizations (NCIO)
P.O. Box 6158
Rheem Valley, CA 94570 (415) 376-7541
Norman C. Parrish, Pres.

Founded: 1981. **Description:** Inventors' groups. Coordinates information relating to inventor education and programs such as wanted and available inventions and credible organizations offering development and marketing assistance. Conducts National Innovation Workshops in cooperation with the National Bureau of Standards. Sponsors competitions; operates children's services; maintains speakers' bureau and library of books and tapes relating to invention, innovation, marketing, and idea development and protection. **Members:** 32. **Publications:** (1) NCIO Newsletter, monthly; (2) NCIO Membership Directory, annual; (3) NCIO Proceedings, annual; (4) NCIO Bulletin, periodic. **Meetings:** Bimonthly conference.

★100★ National Council for Industrial Innovation (NCII)

105 Charles St., Suite 530
Boston, MA 02114 (617) 367-0072
Harry G. Pars, Ph.D., Chm.

Founded: 1970. **Description:** Encourages the abilities of individuals and small innovative enterprises to create new products, thus stimulating the economy. **Meetings:** Annual.

★101★ National Council of Patent Law Associations (NCPLA)

c/o Office of Public Affairs
U.S. Patent and Trade Office
Crystal Plaza 2, Rm. 1A05
2021 Jefferson Davis Hwy.
Arlington, VA 22202 (703) 557-3341

Founded: 1934. **Description:** To inform member associations of matters of interest in patent, trademark, and copyright fields and to facilitate exchange of information among member associations. Annually inducts inventors into National Inventors Hall of Fame. Maintains numerous committees. Cosponsors, with the Patent and Trademark Office, the National Inventors Hall of Fame Foundation. Bestows awards; maintains speakers' bureau; compiles statistics. **Members:** 53. **Publications:** (1) Chairman's Letter, monthly; (2) Legislative Letter, monthly; (3) Newsletter, quarterly. **Meetings:** Quarterly.

★102★ National Inventors Cooperative Association

P.O. Box 6585
Denver, CO 80206 (303) 756-0034
Morton J. Levand, Contact

★103★ National Inventors Foundation (NIF)

345 W. Cypress St.
Glendale, CA 91204 (818) 246-6540
Ted DeBoer, Exec.Dir.

Founded: 1963. **Description:** Independent inventors united to educate individuals with regard to the protection and promotion of inventions and new products. Instructs potential inventors on how to protect their inventions through the use of methods developed by the foundation and patent laws. Has assisted individuals throughout the U.S. and in 44 other countries. Maintains speakers' bureau, hall of fame, and museum. **Members:** 8089. **Telecommunication Access:** 24-hour on-line bulletin board with information for potential inventors.

★104★ National Patent Council (NPC)

Crystal Plaza One
2001 Jefferson Davis Hwy., Suite 301
Arlington, VA 22202 (703) 521-1669
Eric P. Schellin, Exec.V.Pres.

Founded: 1945. **Description:** Manufacturers, patent attorneys, inventors, and others interested in patents. **Publications:** Patent Trends, monthly. Presently inactive.

★105★ National Society of Inventors

539 Laurel Place
South Orange, NJ 07079 (201) 596-3322
Lawrence Jay Schmerzler, President

Contact: Frank Sowa, P.O. Box 434, Cranford, NJ 07016 (201) 276-0213. **Telecommunication Access:** Alternate phone (201) 763-6197.

★106★ National Venture Capital Association (NVCA)

1655 N. Fort Myer Dr., Suite 700
Arlington, VA 22209 (202) 528-4370
Daniel T. Kingsley, Exec.Dir.

Founded: 1973. **Description:** Venture capital organizations, corporate financiers, and individual venture capitalists who are responsible for investing private capital in young companies on a professional basis. Organized to foster a broader understanding of the importance of venture capital to the vitality of the U.S. economy and to stimulate the free flow of capital to young companies. Seeks to improve communications among venture capitalists throughout the country and to improve the general level of knowledge of the venturing process in government, in the universities, and in the business community. Maintains speakers' bureau. **Members:** 223. **Publications:** Annual Membership Directory. **Meetings:** Annual - always Washington, DC.

★107★ Nevada Innovation & Technology Council

c/o U.S. Dept. of Commerce - ITA
1755 E. Plumb Ln., Rm. 152
Reno, NV 89502-3680 (702) 784-5203
J.J. Jeremy, Contact

★108★ New York Society of Professional Inventors

State University of New York at Farmingdale
Lupton Hall
Farmingdale, NY 11735 (516) 420-2397
J.E. Manuel, Contact

Contact: Phillip Knapp, 116 Stuart Ave., Amityville, NY 11701 (516) 598-0036 or (516) 331-3606.

★109★ Ohio Inventors Association (OIA)

2480 East Ave.
Columbus, OH 43202 (614) 267-9033
Tim Nyros, Exec. Officer

★110★ Oklahoma Inventors Congress

P.O. Box 75635
Oklahoma City, OK 73147
Ken Addison, Jr., Contact

Contact: Albert Janco, P.O. Box 54625, Oklahoma City, OK 73152 (405) 848-1991.

★111★ Omaha Inventors Club

c/o U.S. Small Business Administration
11145 Mill Valley Rd.
Omaha, NE 68145 (402) 221-3604
Bob Simon, Contact

★112★ Palm Beach Society of American Inventors
P.O. Box 26
Palm Beach, FL 33480 (305) 736-6594
Robert E. White, Contact

Contact: Kiki Shapero (305) 655-0536.

★113★ Pan Hellenic Society Inventors of Greece in U.S.A. (PHSIG)
c/o Dr. Kimon M. Louvaris
2053 Narwood Ave.
South Merrick, NY 11566 (516) 223-5958
Dr. Kimon M. Louvaris, Exec.Dir.

Founded: 1969. **Description:** Inventors of Greek-American descent whose inventions are patented in both the U.S. and Greece; individuals with patents pending in the U.S., South America, or in Athens, Greece. To assist the Greek-American individual with the patenting, protection, and marketing of his/her invention; to promote the marketing of inventions which contribute to health. Operates biographical archives and museum. Compiles statistics. Conducts triennial seminars in Athens and research programs. **Members:** 150. **Publications:** Newsletter, monthly; also publishes Inventors Greek-English Guide and technical books. **Meetings:** Annual.

★114★ Patent and Trademark Office Society (PTOS)
P.O. Box 2089
Arlington, VA 22202 (703) 557-6511
Richard T. Stouffer, Pres.

Founded: 1917. **Description:** Professional patent examiners in the U.S. Patent and Trademark Office (850); patent and trademark attorneys outside the patent office and former examiners (450). Purposes are educational, social, and legislative. Legislative program concerns federal patent and trademark legislation. Social program is for members and their families and patent and trademark bench and bar members. Educational programs are designed for the public and for continuing professional education. Supplies judges for annual International Science Fair Competition and local fairs. Bestows annual Rossman Award for the best article in the Journal. **Members:** 1300. **Publications:** (1) Journal, monthly; (2) Unofficial Gazette, monthly; also publishes monographs. **Meetings:** Annual.

★115★ Patent Office Professional Association (POPA)
P.O. Box 2745
Arlington, VA 22202 (703) 557-2975
Randall P. Myers, Sec.

Founded: 1962. **Description:** Professional, nontrademark, and nonmanagement employees of the U.S. Patent and Trademark Office. To establish better working conditions and professionalism in the U.S. Patent and Trademark Office. Maintains library of labor reports and legal texts. **Members:** 1500. **Publications:** Newsletter, monthly. **Meetings:** Annual - always first Thursday in December, Arlington, VA.

★116★ Rocky Mountain Inventors Congress
P.O. Box 4365
Boulder, CO 80204-0365 (303) 231-7724
Ken Richardson, Contact

★117★ Silicon Valley Entrepreneurs Club, Inc.
TECHMART Bldg., Ste. 360
5201 Great America Parkway
Santa Clara, CA 95054 (408) 562-6040
Robert Hanson, President

Contact: Dianne Craig, Executive Director.

★118★ Society for Inventors and Entrepreneurs
306 Georgetown Dr.
Casselberry, FL 32707 (305) 859-4855
Dr. Frank Dumont, Contact

★119★ Society for the Encouragement of Research and Invention (SERI)
P.O. Box 412
100 Summit Ave.
Summit, NJ 07901 (201) 273-1088
Dr. J. A. M. LeDuc, Exec.Dir.

Founded: 1976. **Description:** Independent organization devoted exclusively to the advancement of research and invention. Principal function is to recognize persons of diverse activities, of all ages, at different times in their lives, who have demonstrated significant achievements. Encourages researchers and scientists of all disciplines and fields and honors those who have distinguished themselves by their activities and contributions furthering the evolution and growth of research and invention. Also rewards authors of scientific research, inventions, and technical realizations. Cooperates in the development of the national economy by providing the atmosphere and means to facilitate achievements in research and invention. Fosters international understanding, relationships, and interchange of ideas. Has established an Invention Center for the advancement of research, invention, and innovation. Compiles statistics; bestows annual awards. **Publications:** Newsletter, bimonthly. **Meetings:** Annual; also holds annual general meeting - always October.

★120★ Society of American Inventors (SAI)
505 E. Jackson St., Suite 204
Tampa, FL 33602 (813) 221-2343
Donald R. Hinst, Exec.Dir.

Founded: 1987. **Description:** Inventors and inventors' groups, patent attorneys, and patent agents. Promotes, encourages, and assists American inventors. Works with government agencies, local inventors' groups, and individuals to gather and disseminate information about new inventions to members, corporations, and the public. Encourages government and industry to support American creativity and innovation. Co-sponsors Project XL, a national education program for public and private schools, which fosters creativity and innovation in problem solving. Maintains 600 volume library. Bestows Inventor of the Month Award and cash grants. **Members:** 1600. **Publications:** (1) Inventor USA, bimonthly; (2) Newsletter, bimonthly; also publishes Financial Resource Guide; makes available attorney and business prep kits. **Meetings:** Anual. **Telecommunication Access:** Toll-free number, (800)443-7892.

★121★ Society of Minnesota Inventors
20231 Basalt St., N.W.
Anoka, MN 55303 (612) 753-2766
Paul Paris, Contact

Contact: Helen Saatzer, P.O, Box 355, St. Cloud, MN 56302 (612) 253-2537.

★122★ Society of University Patent Administrators (SUPA)
c/o Spencer L. Blaylock
315 Beardshear Hall
Iowa State Univ.
Ames, IA 50511 (515) 294-4740
Spencer L. Blaylock, Pres.

Founded: 1975. **Description:** Patent administrators from institutions of higher learning and affiliated hospitals; attorneys; foundations. Assists administrators in the licensing of technology and reporting of inventions. Recommends more effective technology transfer procedures. Sponsors national workshops and regional information sessions. **Members:** 520. **Publications:** (1) Newsletter, quarterly; (2) SUPA Journal, annual; (3) SUPA Membership Directory, annual. **Meetings:** Semiannual conference.

★123★ South Dakota Inventors Congress

Watertown Area Chamber of Commerce
P.O. Box 1113
Watertown, SD 57201 (605) 886-5814
Barry Wilfahrt, Executive Vice President

★124★ Spirit of the Future Creative Institute (SFCI)

3308 1/2 Mission St., Suite 300
San Francisco, CA 94110
Gary Marchi, Founder & Dir.

Founded: 1976. **Description:** "Idea creators, pioneers, innovators, inventors, and futurists"; research and development groups, educators, consultants, and researchers. Acts as a "creative innovation center" whose purpose is to explore, discover, create, and utilize new ideas and concepts, technologies, systems, processes, and applications of existing technologies. Seeks to facilitate the "process of positive potential" and assist "emerging growth industries." Primary areas of interest include: future science and space technology; economics and the free enterprise system; the U.S. Constitution; mental development, such as applied logic, creativity, and learning improvement systems; natural health and fitness; conservation; historical "re-discoveries" and mysterious phenomena. Provides advisory services on future planning, evaluations, forecasts, and applied research. Produces Future Consumer radio and television documentary series. Holds seminars and workshops; sponsors internship program. Maintains speakers' bureau. Operates library of 425 volumes, 100 cassette tapes, biographical archives, and clippings. Compiles statistics. **Members:** 20.

★125★ Tampa Bay Inventor's Council

805 W. 118th Ave.
Tampa, FL 33612 (813) 933-9124
F. MacNeill MacKay, Contact

Contact: Alternate address: P.O. Box 2254, Largo, FL 34294-2254.

★126★ Television Licensing Center (TLC)

5547 N. Ravenswood Ave.
Chicago, IL 60640 (312) 878-2600

Founded: 1980. **Description:** Colleges and universities, school districts, public and private schools, and public libraries. Purpose is to provide members with a legally-protected means of videotaping commercial and public television programs and of copying and retaining those videotapes for educational purposes. Offers members free written copyright information pertaining to their specific situations. Acts as a liaison between consumers and copyright holders by licensing members to videotape and retain programs, then billing members and making payments to copyright holders. Compiles statistics. **Members:** 3100. **Publications:** TLC Guide (educational newsletter), 7/year; also publishes an educational catalog. **Telecommunication Access:** Toll-free number, (800) 323-4222.

★127★ Tennessee Inventors Association

P.O. Box 11225
Knoxville, TN 37939-1225 (615) 534-0105
Martin J. Skinner, Contact

Telecommunication Access: Alternate phone (615) 376-6894.

★128★ Texas Inventors Association

4000 Rock Creek Dr., #100
Dallas, TX 75204 (214) 256-1540
Tom E. Workman, President

Telecommunication Access: Alternate telephone (214) 528-8050.

★129★ Trademark Society (TMS)

P.O. Box 2631, Eads Station
Arlington, VA 22202 (703) 557-3277
Jerry Price, Pres.

Description: Labor union of trademark attorneys in the U.S. Department of Commerce. **Members:** 100.

★130★ Trilon Discovery Center

3334 Cemetery Rd.
Hilliard, OH 43026 (614) 771-2312
Edgar D. Young, Contact

★131★ United States Trademark Association (USTA)

Six E. 45th St.
New York, NY 10017 (212) 986-5880
Robin A. Rolfe, Exec.Dir.

Founded: 1878. **Description:** Trademark owners; associate members are lawyers, law firms, advertising agencies, designers, market researchers, and public relations practitioners. Seeks to: protect the interests of the public in the use of trademarks and trade names; promote the interests of members and of trademark owners generally in the use of their trademarks and trade names; disseminate information concerning the use, registration, and protection of trademarks in the United States, its territories, and in foreign countries. Maintains informal placement service, speakers' bureau, and library of 1500 volumes. Presents essay award. **Members:** 1750. **Publications:** (1) Bulletins, weekly; (2) Trademark Reporter, bimonthly; (3) Executive Newsletter, quarterly; (4) Trademark Stylesheets, semiannual; (5) Roster of Members, annual; also publishes books about trademarks. **Meetings:** Annual (with exhibits) - always April and May.

★132★ U.S. Patent Model Foundation (USPMF)

1331 Pennsylvania Ave., N.W., Suite 903
Washington, DC 20004 (202) 737-1836
Nancy Metz, Exec.Dir.

Founded: 1984. **Description:** Foundation established to stimulate American inventiveness and productivity. Seeks to enhance public awareness of American inventions, both past and present. Works to recover many patent models of the 19th century, most of which were sold at public auction by the U.S. Patent Office in 1925; plans to donate these models to the Smithsonian Institution. (Patent models are working replicas of devices for which patents are sought.) Sponsors Invent America Program, an educational competition for elementary school students; provides training seminars for teachers involved in the program; bestows awards and grants to the winning students, teachers, and schools. Plans to develop a resource center to provide educators and potential inventors with information on the patent process;develops curriculum materials. Compiles statistics. **Publications:** Newsletter, quarterly; also publishes Invent America! Creative Resource Guide; plans to publish a quarterly newsletter for the IAP. **Meetings:** annual conference (with exhibits).

★133★ Venture Capital Network, Inc.

P.O. Box 882
Durham, NH 03824-0882 (603) 862-3556
Helen Goodman, Director

Description: Introduces entrepreneurs to individual venture investors and venture capital firms. Maintains a confidential database.

★134★ Women Entrepreneurs (WE)

1275 Market St., Suite 1300
San Francisco, CA 94103 (415) 929-0129
Sharon Cannon, Pres.

Founded: 1974. **Description:** Offers the woman business owner support, recognition, and access to vital information and resources. Has participated in government studies and in the 1980 White House Conference on Small Business. Conducts monthly programs featuring speakers and technical assistance educational seminars and workshops; sponsors Advice Forum, providing business and problem-solving information; bestows Appreciation Awards. **Members:** 150. **Publications:** (1) Prospectus (newsletter), monthly; (2) Membership Roster, annual.

★135★ Worcester Area Inventors USA

132 Sterling St.
Worcester, MA 01583 (617) 835-6435
Barbara N. Wyatt, Contact

★136★ Young Entrepreneurs Organization (YEO)

Campus Box 147
Wichita State Univ.
Wichita, KS 67208 (316) 689-3000
Douglas K. Melligan, Dir.

Founded: 1985. **Description:** Entrepreneurs under the age of 30. Facilitates communication among members; sponsors YEO 100 listing, which notes the top 100 young entrepreneurs worldwide based on gross revenues. Cooperates with the Association of Collegiate Entrepreneurs (see separate entry). Maintains speakers' bureau; bestows awards; compiles statistics. **Members:** 750. **Publications:** ACE-YEO Newsletter, monthly. **Meetings:** Annual conference. **Telecommunication Access:** Electronic news network.

CHAPTER 3

INVENTION CONSULTANTS AND RESEARCH FIRMS

I was introduced to the product development business when I was hired as a marketing consultant by an entrepreneur who had an idea for a child's computer. As you can imagine, therefore, I believe in using consultants whenever their expertise can contribute to the progress of a project.

What is consulting? I like the definition offered by England's Institute of Management Consultants.

"The service provided by an independent and qualified person or persons in identifying and investigating problems concerned with policy, organization, procedures and methods; recommending appropriate action and helping to implement these recommendations."

There are as many types of consultants as there are problems to solve. These experts can bring new techniques and approaches to bear on an inventor's work. This contribution can range from helping to bridge a technological gap to the special knowledge and talent required to successfully license or market a particular innovation.

In the case of the child's first computer project, an electronics wizard originated the basic product with help from several consultants. A former Texas Instruments engineer, for example, consulted regarding the best microprocessor to use. My services were engaged to provide information on how to best position the item within the marketplace.

Advisors can provide impartial points-of-view by seeing challenges in a fresh light. They operate outside existing frameworks and free from existing beliefs, politics, problems, and procedures inherent in many organizations or situations.

Most consultants operate on the basis of an hourly rate plus expenses. However, inventors, by the nature of their work, are often able to make equity deals whereby in return for their advice, consultants are given participation in any profits said invention might generate. Inventors should think long and hard before doing something like this because it is often less expensive to risk the cash and hold all the points possible in-house.

One recent example of an inventor who took in consulting partners to make his project materialize is Seattle waiter, Robert Angel, inventor of the popular game "Pictionary." Angel knew that he had a terrific concept, but needed some graphic support. He asked artist Gary Everson to help him design the game board in return for points in the venture. Everson agreed. The then 26-year old inventor also needed assistance in making business decisions, so he gave points, as well, to an accountant named Terry Langston.

"Pictionary" went on to be the best-selling game of 1987, grossing more than $52 million at retail. In 1988 the product's sales soared to an astounding $120 million. To date more than 14 million units of the quick draw game have been sold. The three partners have become wealthy, receiving royalties on every game sold.

Do not think that consultants and research organizations have all the answers. They do not. Consulting is very hard work and not everything can be solved as quickly as one would like. Do not look for miracle solutions. Shop around. Get references on any consultant or research organization you are considering. Don't be impressed by a consultant's or organization's professional association alone, (for example, if they are part of a university). Their success rate in fields related to yours is what matters. What matters is know-who as much as know-how. How much can they do with a single phone call? Results is what you want, not just paper reports.

Chapter Arrangement

This chapter is arranged alphabetically by consulting organization name. All organization names listed in this chapter are referenced in the Master Index. The Master Index may also include references to similar organizations (state government agencies or associations, for example) that occur in other chapters.

★137★ A. Olivetti Associates, Inc.

P.O. Box 515
Saddle River, NJ 07458 (201) 934-9063
Alfred C. Olivetti, President

Founded: 1982. **Description:** Offers assistance in the standardization of products and submittal to testing labs, such as UL, CSA, AGA, CGA, ETL, FM, DAR, and MEA. Serves U.S. firms seeking entry into the European market and European firms seeking markets in the United States.

★138★ Access Management Corporation

7 Woodlawn Green, Suite 212
Charlotte, NC 28217 (704) 525-7030
George A. Suter, President

Founded: 1984. **Description:** International marketing specialist in international technology transfer, joint ventures, licensing, international marketing consulting, productivity engineering and crisis management.

★139★ Albert L. Emmons

580 Jackson Ave.
Westwood, NJ 07675 (201) 666-0225
Albert L. Emmons

Founded: 1980. **Description:** General business consultant whose expertise includes strategic business planning, financial counsel, handling of mergers and acquisitions, and marketing systems and services. Clients include investors, entrepreneurs, venture capital firms, investment and commercial banks, manufacturers, distributors, retailers, and small business.

★140★ American European Consulting Company, Inc.

P.O. Box 20147
Houston, TX 77225 (713) 666-8797
Arthur W. Krueger, Principal

Founded: 1980. **Description:** International business consultants offering specialized market entry and market development services including market research, assistance with joint ventures, technology transfer, and licensing. Emphasis is on working with European companies entering or reassessing their approach to the United States market. Industries served include energy, oil/gas, process industries, construction, engineering and design - offshore and onshore, international projects, industrial ceramics, rubber/plastic, electronics/electrical, communications (including satellite) health care, medical and hospitality. **Telecommunication Access:** Alternate telephone (713) 652-2008.

★141★ American Indian Economic Development Fund

Box 5474
New York, NY 10163
David McCord, Principal

Description: A non-profit organization involved in developing business activities on Indian reservations, and in developing off- reservation businesses owned by American Indians. Also provides consulting services to Indian groups on trademark licensing, and works with manufacturers interested in using Indian-related trademarks under license.

★142★ AML Information Services

Box 405
Corte Madera, CA 94925 (415) 927-0340
Anne Morgens, Owner

Founded: 1982. **Description:** Patent monitoring and ordering in the areas of drugs, vaccines, biochemistry, genetic engineering, molecular biology, microbiology, antibody technology, clinical diagnostics, biomedical devices, and diagnostic imaging. **Publications:** Patent Alert (monthly newsletter).

★143★ Anderson-Bache Financial Group

10555 N.W. Freeway
Suite 240
Houston, TX 77092 (713) 956-8950
Dan Fries, President

Founded: 1981. **Description:** Corporate financing consultants specializing in the financial needs of small to medium-sized companies from the start-up phase through expansion or mergers and acquisitions. Financial advice offered includes assisting with business loans, venture capital, equipment leasing and commercial real estate. Additional services include loan/investment analysis, proposal packaging assistance, and computerized funding searches. Active with a variety of industries.

★144★ API International Business Development Group

8976 N. Santa Monica Blvd.
Milwaukee, WI 53217 (414) 352-7205
George Madden, Principal

Founded: 1977. **Description:** Offers counsel on international/domestic business development. Works with industrialists who wish to become successfully established in major industrial areas overseas by means of consortia, joint working relationships, joint ventures, cross licensing, technology exchange, co-marketing arrangements, or other methods of technology transfer. Activity limited to working with industrial/manufacturing firms interested in corporate development by means of geographical diversification.

★145★ Applied Information Resources, Inc.

P.O. Box 10021
1759 Weston Dr.
Clearwater, FL 34617 (813) 446-3769
Ted Nohren, President

Founded: 1987. **Description:** Information retrieval consultants accessing over 1800 international databases to perform market/industry analysis, competitive analysis and product life cycle evaluations. Specializes in strategic market planning, tracking technology for licensing or acquisitions and/or joint ventures. Has established network with information brokers.

★146★ ARA Corporation
2844 Cascadia Ave.
Seattle, WA 98144 (206) 723-4600
James A. Curry, President

Founded: 1972. **Description:** Constituted to encourage the transfer of technology and the application of research achievement, the corporation maintains extensive licensing contacts regionally and nationally as well as continuing contact with sources of available new products/processes. Applications are primarily related to personal health, nutrition, agriculture, fishery products, small business growth through licensing in, market expansion through licensing out, and the introduction of new products for regional manufacturing growth.

★147★ Armbruster Associates Chemical Consulting Services
43 Stockton Rd.
Summit, NJ 07901 (201) 277-1614
David C. Armbruster, President

Founded: 1982. **Description:** Chemical consultants serving clients worldwide in the chemical, electronics, petroleum, pharmaceutical, plastics, and polymer industries, offering business, marketing, planning and technical consulting services and studies. Activities cover: marketing, technology, research and development, new business opportunities, new markets/products/ventures, product line expansion and introduction, competitive analysis, marketing research, feasibility, commercial development, corporate profiles, patent analysis, seminars/training, technical writing, custom manufacturing, expert witness, licensing/technology transfer and technology acquisition/assessment/forecasting. Specialization covers the following products: coatings, inks, paints, adhesives and sealants, plastics, organic chemicals/intermediates, heterocycles, monomers, polymers, radiation curing/photopolymers, emulsion/water soluble polymers, specialty chemicals, acrylics, epoxies, urethanes, photoresists and water treatment chemicals. **Publications:** D. Armbruster, Hydrophilic Polymers: New Markets, Technologies and Opportunities 1985-1991, Jeffrey R. Ellis, Inc. (1986).

★148★ Arthur Fakes
Eight S. Michigan Ave., Suite 900
Chicago, IL 60603 (312) 372-8050
Arthur Fakes, Principal

Founded: 1978. **Description:** Specialist in the relationship of data processing to the practice of law in the following areas: transactions involving computers, software, and databases; copyright, trademark, and trade secret law; print and electronic publishing transactions; and technology protection and transfer. **Publications:** "Choosing Your Computer Law Counsel," Information Management Review (Summer 1988); "The Illinois Software License Enforcement Act," Computer Law Reporter (1986); serves on the editorial advisory group of the Software Law Journal. **Meetings:** Copyright Law for Software Development, Publishers, and Distributors; The Importance of Trademark to Small Businesses; Computer Software Law; International Transactions in High-Tech Products and Services.

★149★ Asset Timing Corporation
500 Stephenson Hwy.
Suite 405
Troy, MI 48083 (313) 589-1111
Donald F. Chamberlin, President

Founded: 1967. **Description:** Concentrates its consulting activities on small companies that need assistance in their financial plan and capital formation. Active in the selection, monitoring, and disposition of investments.

★150★ Associated Production Music
6255 Sunset Blvd.
Suite 724
Los Angeles, CA 90028 (213) 461-3211
Phil Spieller, Principal

Founded: 1983. **Description:** The firm represents production music libraries and supplies music and scoring services to the film, video and multi-image production industries. Also represents the publishing of over 30,000 titles and licenses this music to the theatrical and non-theatrical markets.

★151★ Author Consultation Services
P.O. Box 288
Tallahassee, FL 32302 (904) 681-0019
William S. Loiry, President

Founded: 1982. **Description:** Comprehensive nationwide consulting to self-publishers, especially to those writing and publishing their life histories. Includes help in writing, editing, book design, typesetting, layout, cover design, printing coordination, marketing and distribution design.

★152★ Avery Business Development Services
80 Bacon St.
Waltham, MA 02154 (617) 894-6722
Henry Avery, President

Founded: 1981. **Description:** Offers general business consulting that covers mergers/acquisitions, locating capital venture funds, licensing, business strategy planning, interim or project management, and technology transfer.

★153★ Baldhard G. Falk & Associates
P.O. Box 315
Belvedere, CA 94920 (415) 435-3070
Baldhard G. Falk, Principal

Founded: 1959. **Description:** Offers business consulting services in international corporate planning, international licensing, and international negotiations. Recent experience with the electronics industry and manufacturers of scientific instruments.

★154★ Barrington BioIndustries, Inc.
64 Warwick Rd.
Great Neck, NY 11023 (516) 487-3110
Lawrence P. Kunstadt, President

Founded: 1985. **Description:** Active in all phases of the commercial development of technologies in the life sciences. Focuses on human and veterinary medicine, agriculture and industrial processes in both conventional areas and cutting-edge biotechnology. Works with individual inventors, universities and small and medium-sized companies to profitably commercialize their technologies. Also works with major domestic and foreign corporations to find new technologies and acquisitions. Involved in the areas of commercial development and financing, intellectual property rights, technology transfer and acquisition. Assists in joint ventures between American technology firms and foreign companies. **Meetings:** High Tech Entrepreneurship; Innovation in Medical Devices; Biotechnology and Diagnostic Product Packaging.

★155★ Belden Associates
2900 Turtle Creek Plaza
Suite 200
Dallas, TX 75219 (214) 522-8630
Ralph Bubis, Chairman

Founded: 1940. **Description:** Offers product, opinion, advertising, communications, market and sales, and legal research services. Tests and develops products. Conducts new product acceptance studies, prototype research, package and label tests, brand name tests, brand share-of-market surveys, consumer characteristics analysis, consumer behavior surveys, readership studies, audience and circulation analysis, media selection studies, readability tests, political polls, image studies, public relations surveys, shareholder relations surveys, employee attitude surveys, community relations surveys, company name identification studies, health care surveys, trademark confusion studies, packaging studies, and product design research. **Telecommunication Access:** FAX (214) 522-0926.

★156★ Bio-Metric Systems, Inc.

9932 West 74th Street
Eden Prairie, MN 55344 (612) 941-0080
Patrick Guire, Director, R&D

Founded: 1979. **Description:** Biotechnology and sensor systems for detection of biological materials, toxins, drugs, and other biologically active components in the environment. Provides patent technology in biomembrane chemical sensors, biocompatibility control technology, detection/diagnostic kits, and proprietary data and reagents. Technical strengths include bio-organics, bioanalysis, protein and immunochemistry, immunoassay technology, microbiology, enzymology, cell biology, and immunology.

★157★ BioLabs, Inc.

15 Sheffield Ct.
Lincolnshire, IL 60069 (312) 945-2767
Clyde R. Goodheart, President

Founded: 1985. **Description:** Assists with start-up of new companies in the biotechnology/biopharmaceutical/medical area. Performs services in planning, writing business plan, raising capital, and serving as CEO during initial phases. Primarily active with pharmaceutical companies.

★158★ BIONICS

Box 1553
Owosso, MI 48867
Ben Campbell, President

Founded: 1988. **Description:** Advises scientists, designers, investors, and stockbrokers who are interested in the bionics industry. Introduces scientists, designers, and inventors to financers, investors, and security-brokers. Helps raise venture capital and private placements financing. Publishes special reports and proposals. **Publications:** Bionics newsletter.

★159★ Blalack-Loop, Incorporated

696 E. Colorado Blvd.
Suite 220
Pasadena, CA 91101 (818) 449-3411
Charles M. Blalack, Chairman/Treasurer

Founded: 1969. **Description:** Investment advisors who provide investment advisory services and manage portfolios for individuals and institutions. Also involved in venture capital.

★160★ Bob West Associates, Incorporated

P.O. Box 2001
Stamford, CT 06906 (203) 322-4466
Bob West, President

Founded: 1976. **Description:** Provides scientific, regulatory and management consulting services for a wide variety of product lines and industries including pharmaceutical, industrial and agricultural chemicals, household chemical specialties, toiletries and cosmetics, food additives, diagnostics and devices. Areas of expertise include product integrity, scientific affairs, management, medical communications, environmental science, pharmacology and toxicology, clinical studies, FDA liaison, and product licensing.

★161★ Bobbe Siegel, Rights Representative

41 W. 83rd St.
New York, NY 10024 (212) 877-4985
Bobbe Siegel, Principal

Founded: 1975. **Description:** Handles foreign rights for publishers, literary agents and represents foreign agencies for United States rights. Also consults with publishers on problems relating to foreign rights.

★162★ Boice Dunham Group

437 Madison Ave.
New York, NY 10022 (212) 752-5550
Craig K. Boice, Principal

Founded: 1982. **Description:** Specialized business development service offering identification, assessment and positioning of new business opportunities; acquisition of new corporate resources (e.g., joint ventures, investors, technology); and business plan implementation for new ventures based on innovative technologies and growing markets.

★163★ Bradley Agri-Consulting Inc.

P.O. Box 1668
Palatine, IL 60078 (312) 632-0050
Bernard L. Bradley, President

Founded: 1980. **Description:** Complete agri-business and agrimarketing consultancy, serving both domestic and international clients. Provides marketing and market surveys, product development and introduction, competitive intelligence, regulatory and environmental concerns, acquisitions and licensing, confidential contacts, export/import concerns, foreign negotiations and representation, training, information systems, applications and data transfer and lead management.

★164★ Bruce W. McGee and Associates

7826 Eastern Ave., N.W.
Suite 306
Washington, DC 20002 (202) 726-7272
Bruce W. McGee, Principal

Founded: 1985. **Description:** Business consultants experienced in office automation, small business management, invention and patent counseling, technology commercialization, loan packaging, and business plan development. **Meetings:** Marketing Technological Products to Industry; How to Evaluate Your Technical Idea; Patenting Your Own Invention.

★165★ Business Development Associates, Inc.

46 Main St.
New Canaan, CT 06840 (203) 972-3812
Gordon J. Bockner, President

Founded: 1981. **Description:** International and domestic technology transfer consultants who provide assistance with international licensing, international business and trade, marketing programs, new product introduction, and mergers and acquisitions. Serve the following industries: food and beverage, plastics, and manufacturing.

★166★ Business Development Associates

2865 Taylor St.
Eugene, OR 97405 (503) 345-8683
Stephen Robinson, President

Founded: 1979. **Description:** Management consultant advising businesses on matters of strategic and long-range planning, budgeting and organizational development. Also offers counsel on mergers and acquisitions and venture capital.

★167★ Business Matters
P.O. Box 287
Fairbanks, AK 99707 (907) 452-5650
Kathi Minsky

Description: General business consultant experienced in the development of policies and procedures and loan packages.

★168★ Business Planning Consultants
S. 3510 Ridgeview Dr.
Spokane, WA 99206 (509) 926-2483
Michael R. Cady, Principal

Founded: 1981. **Description:** Prepares business plans and feasibility studies for shopping centers, manufacturing, and mining. Also prepares financing proposals for seed and venture capital and bank loans. Plans community economic development programs: recruiting new business, business retention programs, financing programs (seed capital and revolving loan funds), and central business district revitalization. Involved in international trade and investment activities in Japan and Korea. Also offers start-up real estate development management and planning services.

★169★ Business Sciences Corporation
Arizona Bar Center, Suite 101
363 N. First Ave.
Phoenix, AZ 85003 (602) 971-0828
Richard P. Hubbard, Principal

Founded: 1979. **Description:** Business consultants experienced in business planning, financial planning, strategy development and execution, bank loan request packages, venture capital packages, entrepreneurial consultation, management information systems, and accounting systems. Serves high-technology start-ups and developing companies, manufacturing organizations, banks, and professional societies, but excludes any retail concerns.

★170★ Byron Boothe & Associates, Inc.
234 N. Rock Island
P.O. Box 47128
Wichita, KS 67201 (316) 263-1166
Byron W. Boothe, President

Founded: 1978. **Description:** An investment firm offering guidance in employee benefit packages, estate planning, insurance and venture capital.

★171★ BZ/Rights & Permissions, Inc.
145 W. 86th St.
New York, NY 10024 (212) 580-0615
Barbara Zimmerman, President

Description: Offers a rights clearance service. Will obtain rights on behalf of clients to use anything that's copyrighted (i.e., music, film-TV footage, photographs, graphic and fine arts, literary works and celebrity photos and testimonials). Serves ad agencies, producers of film and video tape (both corporate and commercial), record companies, television and cable companies, and publishers.

★172★ C. Scanlon & Company
3545 Long Beach Blvd.
Long Beach, CA 90807 (213) 492-1552
Carlyle Scanlon, President

Founded: 1968. **Description:** Provides consulting services to manufacturers, producers and service companies in the United States and abroad that are seeking to enter or expand overseas markets. Assists companies in the sale of products, technology or services by establishment abroad of effective distribution channels, manufacturing licenses, joint ventures or subsidiary operations. Specializes in ancillary services such as export licensing and administration under the jurisdiction of the Department of Commerce, Department of State and the Nuclear Regulatory Commission.

★173★ Calmac Manufacturing Corporation
150 S. Van Brunt St.
Englewood, NJ 07631 (201) 569-0420
Calvin D. MacCracken, President

Founded: 1947. **Description:** Offers counsel on product development in the energy field, principally related to heating and cooling.

★174★ CAMS, Ltd.
3000 Dundee Rd., Suite 303
Northbrook, IL 60062 (312) 272-0888
Carl J. Kosnar, President

Founded: 1973. **Description:** Develops business expansion plans and strategies for services, retail trade and manufacturing industries. These expansion plans and strategies include franchising, licensing, joint ventures, limited partnerships, public and private stock issuances and other expansion vehicles.

★175★ Capitol Services
926 J St., Suite 919
Sacramento, CA 95814 (916) 443-0657
David E. Kalb, President

Founded: 1982. **Description:** Offers consulting and assistance to out-of-state businesses wishing to relocate or expand in California, and to existing businesses within the State. Specific activities include helping building and construction trade firms secure a California contractors license, research and assistance in obtaining state and local government permits and licenses, assisting lawyers with public document search and retrieval, public policy analysis, and state government contract assistance. Also acts as business liaison to state agencies.

★176★ Carlton Group
P.O. Box 29015
Richmond, VA 23229 (804) 289-5734
Robert C. McGee, Principal

Founded: 1985. **Description:** Financial services consultants specializing in management, business plan development, cash and inventory management, and acquisitions and mergers analysis. All services provided to small business and entrepreneurs. **Meetings:** How to Finance Your Business; Financing Alternatives for Small Business.

★177★ CEDCO
180 N. Michigan Ave.
Suite 333
Chicago, IL 60601 (312) 984-5950
Webster Daniels

Founded: 1965. **Description:** Provides management advisory services to new and existing commercial enterprises that are for-profit oriented. Specific areas of service are business planning, management, finance, marketing, construction and economic feasibility studies.

★178★ Centennial Engineering & Research, Inc.
237 North Main
Suite 1
Sheridan, WY 82801 (307) 672-6439
Thomas L. Barker, President

Founded: 1976. **Description:** Information collection, systems design, product development, patent evaluation, and economic analysis of electronic and mechanical systems, remote water systems, control systems, photo cell control equipment, and microprocessors. Also offers water, soils, concrete, and asphalt testing.

★179★ Cerberus Group
P.O. Box 470
Frenchtown, NJ 08825 (201) 996-2648
Charles C. Varga, Jr., Chairman

Founded: 1980. **Description:** Provides research on information processing and technology in selected areas within the computer industry, applied to: strategic and long range business planning; acquisitions, mergers and divestitures; market planning, analysis and development; financial and investment analysis; and venture capital financing. The focus of research activities is on the establishment of a comprehensive reference base of firms and includes structured profiles describing company operations.

★180★ CERES
417 Wakara Way
Salt Lake City, UT 84108 (801) 582-0144
Mark Walton, General Manager

Founded: 1973. **Description:** Commercialization of technology via services and technology licenses in plant biotechnology, including molecular biology, cell and tissue biology, microbiology, phytochemistry, plant breeding and genetics, and applied ecology. **Telecommunication Access:** Telex 820832 (NPISLC).

★181★ Challenger Industries, Ltd.
1552 Palisade Avenue
Fort Lee, NJ 07024 (201) 592-9090
P.C. Torigian, Contact

Description: Firm providing marketing research and development, including product development, quality control monitoring, product marketing consulting, patent development, and development of injectable medicaments.

★182★ Chandler & Associates Ltd.
4220 S. Pratt St.
Omaha, NE 68111 (402) 453-4560
Virgil R. Chandler, President

Founded: 1980. **Description:** Consultant on developing and implementation of small and minority businesses. Serves as resource firm for search and discovery of possible financial sources and offers assistance and guidance in preparation of business plans. Develops public and community relations programs for entrepreneurs. **Publications:** Editor and publisher of The Bottom Line newsletter. **Telecommunication Access:** Alternate telephone (402) 451-3335.

★183★ Charles Scott Pugh, Jr.
9607 Tilehurst Ct.
Richmond, VA 23237 (804) 748-2364
Charles Scott Pugh, Jr., Principal

Founded: 1979. **Description:** Offers consulting services for small businesses in new venture analysis and organization, profitability and payback studies, project cost control, estimating and scheduling, business valuation analysis and appraisals, business plans, capital formation, and small business operating services.

★184★ Charles Shano
3255 E. Camelback Rd.
Phoenix, AZ 85018 (602) 956-6206
Charles Shano, Principal

Founded: 1979. **Description:** Specializes in electronic control products for automotive, consumer and industrial applications. Designs, develops and can manufacture, if required. Patent application support is available. Patentable developments can be assigned or licensed.

★185★ Charles W. O'Conor and Associates Wright Wyman, Inc.
211 Congress St.
Boston, MA 02110 (617) 482-8330
Franklin Wyman, Jr., Chairman/Treasurer

Founded: 1965. **Description:** Specialists in programs of corporate growth and diversification in the United States and Canada. Principal activities include the planning and implementation of corporate acquisition and/or merger programs, forward planning audits, private placements and arranging licensing and joint ventures agreements.

★186★ CHI Research/Computer Horizons, Inc.
10 White Horse Pike
Haddon Heights, NJ 08035 (609) 779-0911
Dr. Francis Narin, President

Founded: 1968. **Description:** Quantitative measurement of research and development activity, including assessment of papers, patents, citations, domains, subfields, laboratories, programs, companies, institutions, agencies, economic sectors, and countries. Specializes in patent and literature citation analysis. **Telecommunication Access:** Telex 279630 (CHI UR); FAX (609) 546-9633.

★187★ Chris Enterprises
13220 Wye Oak Dr.
Gaithersburg, MD 20878 (301) 948-7517
Stephen J. Chris, President

Founded: 1970. **Description:** Aeronautical and mechanical engineering consultants offering services in research and development, design and fabrication, and accident and safety investigation/analysis. Also work with patent attorneys.

★188★ Clark Engineering
916 West 25 Street
Norfolk, VA 23517 (804) 625-1140
Stephen E. Clark, President

Founded: 1979. **Description:** Energy, specifically low temperature thermal energy conversion; environmental sciences, including pollution control device development and testing; engineering, economic, and marketing studies; computer simulation and analysis; patent evaluation; and prototype performance testing. Research results published in project reports.

★189★ Cleveland Area Development Corporation
690 Huntington Bldg.
Cleveland, OH 44115 (216) 621-3300
William R. Plato

Founded: 1975. **Description:** Provides services in business development to firms expanding or relocating in the Greater Cleveland area, business finance and loan services, export trade assistance, minority business assistance, research on local market, and economic issues.

★190★ Combined Agencies Corporation
4900 Massachusetts Ave., N.W.
Washington, DC 20016 (202) 966-4500
Joseph Hasek, President/Treasurer

Founded: 1948. **Description:** Offers consulting on international economic projects and serves as liaison to United States and overseas governments and international agencies in economic matters. Also acts as export manager for United States manufacturers and as American Purchasing Office for overseas industries and negotiates licensing contracts.

★191★ Commercial Associates Incorporated
1001 Connecticut Ave.
Washington, DC 20036 (202) 331-1363
V. S. Choslowsky, Managing Director

Founded: 1955. **Description:** Consultants on international commerce and sales development, including patent licensing and exploitation, and representatives maintaining liaison with Congressional and governmental offices, as well as trade, financial and professional organizations in Washington, D.C.

★192★ Communication Certification Laboratory
1940 West Alexander Street
Salt Lake City, UT 84119 (801) 972-6146
Thomas C. Jackson, President

Founded: 1971. **Description:** Offers certification, evaluation, quality assurance, and technical documentation services. Prepares quality assurance procedure manuals for federal and state regulatory commissions. Conducts product evaluations and on-site inspections of manufacturing procedures and evaluates designs for compliance with copyright law standards. Assists in filing applications with the Federal Communications Commission and provides type acceptance and certification testing to meet regulations. Also performs testing and filing for certification of telecommunications products to meet Canadian and Japanese government regulations. Research results published in reports to the Federal Communications Commission and clients. **Publications:** Quarterly newsletter. **Telecommunication Access:** Telex 709966 (CCLSLCUD); FAX (801) 972-8432.

★193★ Compass Materials Group
P.O. Box 5
Tolland, CT 06084 (203) 871-2500
Leon Fenocketti

Founded: 1985. **Description:** Offers materials science engineering consulting and contract services for plastic/chemical, laser/fiber optic technologies as well as EMI/RFI and ESD or conductive composites. Serves as applications consultants for concept development through product introduction; formulation, compounding, fabrication expertise; process/product design and engineering; testing specification and interpretation; and troubleshooting and turnarounds. Industries served include aerospace, automotive, chemical, communications electronics, fiber optics, plastics and transportation.

★194★ Computer Age Associates Inc.
P.O. Box 9712
Santa Fe, NM 87504 (505) 982-8792
Artthur V. Rubino, President

Founded: 1978. **Description:** Performs specialized venture capital placements and financial consulting to the computer industry.

★195★ Comquest, Inc.
512 West Lancaster Avenue
Wayne, PA 19087 (215) 688-3288
Rosemary H. Driehaus, President

Founded: 1982. **Description:** Business research and information, including computerized literature searches and analysis, marketing and opinion reports, patent and trademark information, and executive and expert identification. Research results published in reports to clients. **Publications:** Update (monthly newsletter). **Telecommunication Access:** Telex 902320; Electronic mail 62801611 (EASYLINK) or 1245992 (MCI).

★196★ Connecticut Venture Management Corporation
184 Atlantic St.
Stamford, CT 06901 (203) 324-7075
F. Eugene Davis, Principal

Founded: 1985. **Description:** Offers venture management, financial management, and financing for regional venture companies and venture operations of established companies. Markets currently include early stage ventures; industrial corporations; financial institutions; venture capitalists and private investors.

★197★ Consortium House, Ltd.
922 Stratford Ct.
P.O. Box 728
Del Mar, CA 92014 (619) 481-5268
Eugene G. Schwartz, President

Founded: 1978. **Description:** Offers consulting services that cover business planning, new product development, venture investment, project management, and operating systems troubleshooting for book, magazine, multi-media and electronic publishers. Experienced in personal computer applications to business.

★198★ Consultant Affiliates, Inc.
10,500 N. Port Washington Rd.
Mequon, WI 53092 (414) 241-8100
Hubert O. Ranger, President

Founded: 1973. **Description:** Offers consulting in corporate development, long range planning, acquisition and divestment, new products, and market analysis—especially for technical manufacturing industries. Also counsels would-be consultants on how to enter the business and on the use of partners for specific projects in consulting.

★199★ Consulting Industries International
Worldway Center, No. 91435
Los Angeles, CA 90009 (213) 670-1177
Robert S. Feller, Sr., CEO

Founded: 1959. **Description:** Offers management consulting toward the goals of profit, productivity, quality, and morale improvements; new products research and development and marketing; office and factory automation; and research and development of concepts-to-systems of virtually anything from concept-to-completion. **Publications:** Fortunes From Ideas; Protecting & Selling Ideas & Inventions; Improving Profits, Productivity, Quality, Morale & Results in any Organization.

★200★ Contact - Europe
18 E. 68th St.
New York, NY 10021 (212) 288-9999
Herbert H. Hauser, Managing Partner

Founded: 1970. **Description:** Establishes links and maintains liaison between the USA and Europe in the field of high-technology, especially the electronics industry. Specializes in searches for new products or licenses, technology transfer, and export marketing. Performs acquisitions and investment evaluation studies for the benefit of industry and Wall Street.

★201★ Copyright Information Services
P.O. Box 1460-A
Friday Harbor, WA 98250 (217) 356-7590
Jerome K. Miller, President

Founded: 1983. **Description:** Consulting responsibilities for schools, colleges, churches, and libraries consist of helping these systems establish reasonable copyright policies and procedures so they can fully employ their rights under the copyright law, without infringing on the rights of others. Also helps them to establish appropriate procedures to obtain permission to reproduce, transmit, or otherwise use copyrighted works. The firm does not negotiate copyright permissions nor trace hard-to-find copyright owners. **Publications:** Firm has published Copyright Policy Development: A Resource Book for Educators.

★202★ Coramar Inc.
142 Selborne Way
Moraga, CA 94556 (415) 376-5033
Coral L. DePriester, President

Founded: 1982. **Description:** Offers consultation on research, development and technology marketing through licensing, joint ventures or direct sale. Broad experience in oil and gas producing technology, computer program licensing, and digital-electro-mechanical devices.

★203★ Corporate Management, Inc.
1400 Fifth Ave., Suite 301
San Diego, CA 92101 (619) 234-4733
Kenneth C. Henry, CEO

Founded: 1984. **Description:** Business consultants experienced in venture management, turnaround and crisis management, bankruptcy planning, chief financial officer responsibilities, and strategic planning.

★204★ CR Chambers & Associates
3 Franklin Dr.
Voorhees, NJ 08043 (609) 770-0833
Cecil Chambers, Jr., President

Founded: 1982. **Description:** Provides sources of financing for small business development, start-ups, franchise acquisition, real estate development, working capital, accounts receivable, machinery leasing and commercial loans from the private and public sectors. Also offers management and financial consulting to business.

★205★ Cresheim Company, Inc.
803 East Willow Grove Avenue
P.O. Box 27785
Philadelphia, PA 19118 (215) 836-1400
James E. Barrett, Managing Director

Founded: 1968. **Description:** Business negotiating and strategy planning, product and program development and management, organization assessment, opinion research, market and economic research, venture management, industrial marketing, telemarketing, and forecasting, specializing in the areas of construction, distribution, insurance, and transportation. Research results published in reports to clients and through the National Technical Information Service of the U.S. Department of Commerce. **Telecommunication Access:** Telex 703814; FAX (215) 836-1403.

★206★ Cross Gates Consultants
P.O. Box 822
East Brunswick, NJ 08816 (201) 238-3381
Elliott L. Weinberg, Principal

Founded: 1977. **Description:** Consultants to the chemical, pharmaceutical, and plastics industries offering counsel on business development via commercial development, mergers and acquisitions, specialized scientific, technical and engineering services, and confidential assignments (domestic and foreign). Also serves as expert witness in liability and patent litigations.

★207★ Crystal Resources, Inc.
64 E. Broadway, Suite 210
Tempe, AZ 85282 (602) 968-8858
Burke Files, Principal

Description: Offers small business and venture consulting.

★208★ Cuisine Crafts, Inc.
P.O. Box 1141
Ridgewood, NJ 07451 (201) 670-7371
Steven Kingsley, Owner

Founded: 1981. **Description:** Food processing and biochemical engineering, including product design and development, laboratory experimentation, information research, editorial services, patent application assistance, and other evaluation and consulting services. Research results published in trade journals. **Telecommunication Access:** Telex 984939.

★209★ Darden Research Corporation
1534 N. Decatur Rd., N.E.
Atlanta, GA 30307 (404) 377-9294
Claibourne H. Darden, Jr., President

Founded: 1968. **Description:** Public opinion and marketing research consultants offering the following expertise: test marketing of new products and services, advertising effectiveness studies, image (perceived strengths and weaknesses) studies, customer satisfaction studies, product and service positioning studies, public opinion polls, merger studies, trademark violation studies, and delivery of service studies.

★210★ DAVCO Solar Corporation
P.O. Box 335
Coldwater, MI 49036 (517) 238-2155
C.E. Davis, President

Description: Offers space solar energy research and development as well as solar energy consulting and management services. Also arranges licensing and royalty agreements and performs energy and public utilities economic impact studies.

★211★ Derwent, Inc.
6845 Elm Street
Suite 500
McLean, VA 22101 (703) 790-0400
Holly Chong, Online Training Executive

Founded: 1981. **Description:** Analyzes, classifies, indexes, abstracts, and codes patent documents from 29 patent issuing authorities. The resulting information is available in printed material, microform, or online. **Publications:** On-Line News; Patent News; literature services include Ringdoc, pharmaceutical abstract service; Chemical Reations Documentation Service; Biotechnology Abstracts; and Vetdoc and Pestdoc, providing abstracts in the fields of veterinary products and pest control.

★212★ Domestic & International Technology
115 W. Ave.
Jenkintown, PA 19046 (215) 885-7670
Richard Stollman, President

Founded: 1975. **Description:** Consulting specialties include international licensing of technology and products, international marketing of industrial products, financing of exports and imports, new product development, and manufacturing engineering.

★213★ Dr. Dvorkovitz and Associates
Box 1748
Ormond Beach, FL 32075 (904) 677-7033
Dr. Vladimir Dvorkovitz, President

Founded: 1961. **Description:** Provides new product and process development. Investigates technological developments with objective of obtaining licenses for clients. Provides data, prototypes, samples, assistance in arranging meetings, and negotiation consultation. **Telecommunication Access:** Telex 4940321; FAX (904) 677-7113.

★214★ E.J. Kuuttila & Associates
10934 Dryden Ave.
Cupertino, CA 95014 (408) 725-1100
Elizabeth J. Kuuttila, President

Founded: 1982. **Description:** Offers services in technology analysis and transfer, marketing and licensing for the high-technology industries.

★215★ E. Janet Berry
274 Madison Ave.
New York, NY 10016 (212) 679-0580
E.J. Berry, Principal

Founded: 1965. **Description:** Offers services in the following areas: patents and trademarks; licensing, technology management; copyrights; industry - academia relations; professional associations; research management and objectives; and chemical (science) education.

★216★ Energy-Environmental Research and Development Company
4638 Adkins Drive
Corpus Christi, TX 47811 (512) 857-2158
Frank D. Kodras, P.E., Chief Operating Officer

Founded: 1980. **Description:** Offers product planning, invention searches, and marketing in the areas of chemical processes, cryogenics, nuclear power, engineering, gold and silver ore leaching, management information systems, and patents. Services include market planning, product development, survey taking, patent trend analysis, technology transfer, and development of environmental control systems. Research focuses on coal bacterial desulfurization, flue gas cleaning systems, chemical and natural gas engineering, pollution control, injection additives, coal/water/oil boiler fuels, hazardous wastes, liquid fertilizers, and byproduct utilization and recycling. Research results published in technical society journals and company reports and by the Environmental Protection Agency and the U.S. Government Printing Office in confidential research reports. **Publications:** Innovations & Inventions Newsletter (quarterly). **Telecommunication Access:** Alternate telephone (512) 592-6820.

★217★ EOS Technologies, Inc.
606 Wilshire Boulevard
Suite 700
Santa Monica, CA 90401 (213) 458-1791
Bryan Gabbard, Contact

Description: Provides research, analysis, and consulting on systems engineering and concept definitions as applied to national defense, energy resources, and commercial technology transfer.

★218★ Europacific Inc.
Box 1108
Cupertino, CA 95015 (408) 446-9699
Walter Rhiner, President

Founded: 1978. **Description:** Offers management services, particularly to high technology industries. Services include: crisis/turnaround management, international technology transfer, venture capital assistance, and corporate strategy development. Technology specialties include: semiconductor equipment, medical imaging, fiber optics, and basic physics and engineering.

★219★ Fry Consultants, Inc.
1 Park Place, Suite 450
1900 Emery Steet, N.W.
Atlanta, GA 30318 (404) 352-2293
Dr. Kenneth Bernhardt, Research Director

Founded: 1942. **Description:** Conducts merger and acquisition studies, industrial marketing research, consumer research, long-range planning, competitive situation analysis, new venture research, strategic analysis of markets and industries, analysis of the buying decision process, and technology transfer studies. Research results published in client reports. **Telecommunication Access:** Telex 4611041 (COMTEL ATL); FAX (404) 352-2299.

★220★ Gaylord Research Institute
28 Newcomb Dr.
New Providence, NJ 07974 (201) 665-1255
Norman G. Gaylord, President

Founded: 1961. **Description:** Provides technical consultation on- and off-premises in polymer science and technology, and product and process development. Prepares literature analyses (technical articles and patents) in polymer field. Also prepares patent applications and assists attorneys in patent prosecution through U.S. and foreign patent offices. Carries out reverse engineering on polymer products using own and subcontracted laboratory facilities and instrumentations. Also assists with expert witness in polymer-related ligation. **Publications:** N. Gaylord has authored 205 technical articles in polymer field including 85 U.S. patents and more than 200 foreign patents, 5 books in polymer field, and is the founding editor of Encyclopedia of Polymer Science and Technology.

★221★ Gellman Research Associates
115 West Avenue
Suite 201
Jenkintown, PA 19046 (215) 884-7500
Aaron J. Gellman, President

Founded: 1972. **Description:** Data collection and analysis, information/literature research, economic impact studies, cost allocation, benefit-cost analysis, new product feasibility, technology transfer, and microeconomic analysis of technology, transportation, and public policy. Research results published in technical journals. **Telecommunication Access:** Telex 834653; FAX (215) 884-1385.

★222★ George W.K. King Associates Consulting Engineers
1050 Eagle Rd.
Newtown, PA 18940 (215) 968-4483
G.W.K. King, Principal

Founded: 1951. **Description:** Offers long-term product planning, new product procurement, licensing in-out, joint ventures, international business development, product assistance in a wide range of commercial, industrial and technical industries. **Publications:** Firm publishes Alert newsletter to all clients listing hundreds of product licensing and joint venture opportunities in 54 countries. Maintains a new product licensing and technology transfer database of companies throughout the world. **Meetings:** International Licensing.

★223★ Gorham International Incorporated
P.O. Box 8
Gorham, ME 04038 (207) 892-2216
Philip E. MacLean, Chairman

Founded: 1956. **Description:** Consultants in technical business development in the paper, energy, chemical and related industries. Activities include acquisitions, corporate development planning, feasibility studies, product and process development, market surveys, R & D management and planning, licensing (foreign and domestic), plant site searches, technical audits, United States representation (for overseas firms), new product development and introduction and sales audits and surveys. **Publications:** Reports concerning paper production and ink jet printing are available from the firm.

★224★ GRADTECH
415 South 19th Street
Philadelphia, PA 19146 (215) 893-7497
Robert F. Lovelace, Executive Director

Founded: 1985. **Description:** Identification, research and development, and commercialization of inventions and other new technologies that address the health care concerns of society. Services include market reserch, product evaluation and feasibility studies, intellectual property protection, prototype research and development, and licensing and marketing negotiation. Acts as a general partner whose purpose is to manage and provide seed funding for product development projects.

★225★ H.B. Hindin Associates, Inc.
P.O. Box 2035
Grass Valley, CA 95945 (916) 265-8424
H.B. Hindin, President

Description: Offers new product evaluation, legal services including consultation and expert witness, patent evaluation and investigation, market research and analysis, product development, market surveys and economic viability.

★226★ H.J. Barrington Nevitt
Two Clarendon Ave., Apartment 207
Toronto, ON, Canada M4V 1H9 (416) 925-9628
H.J. Barrington Nevitt, Principal

Founded: 1947. **Description:** Consultant specializing in communication and innovation. Offers private counsel and seminars on material, mental, and social effects of communication technologies; international design, marketing and management problems; and on economic development, innovation and communication strategy. Active with government, university and private business organizations, large and small, to aid them in reorganizing for the electronic information revolution now in progress.

★227★ Harrold Organization
P.O. Box 741
1508 Terrace Dr.
Newton, KS 67114 (316) 283-8176
H. Eugene Knutson, President

Founded: 1981. **Description:** Provides engineering and marketing studies both North American and international. Also handles domestic and foreign license agreements. **Publications:** H.E. Knutson, "How to Export in Tough Times," Kansas Business News (March 1986); "Let's Get Innovative in the Import-Export Business," Wichita Business Journal (May 1986).

★228★ Harry Prebluda
4101 Pine Tree Dr., Suite 803
Miami Beach, FL 33140 (305) 531-6927
Harry Prebluda, Principal

Founded: 1961. **Description:** Marketing consultant active with patent inventions/developments in nutrition, and with the market development of industrial and fermentation wastes. Also offers literature retrieval services in the life sciences, with expertise in enzymes, fermentation processes, vitamins, and food additives.

★229★ Henry R. Friedberg & Associates
P.O. Box 45220
Cleveland, OH 44145 (216) 333-2843
Henry R. Friedberg, President

Founded: 1978. **Description:** Consultants to industry specializing in metal finishing including organic and inorganic coatings, automated paint lines and systems, as well as electroplating systems. Areas of expertise include waste treatment and disposal systems, EPA compliance, hazardous waste management, Hazard Communication Act training, analysis and testing. Also provides expert witness and patent litigation counsel. Also conducts overseas cargo inspections.

★230★ Herbert E. Mecke Associates, Inc.
25 Cross Hill
Hartsdale, NY 10530 (914) 997-2591
Herbert E. Mecke, President

Founded: 1971. **Description:** Provides general management, marketing, and licensing consulting for industrial and financial corporations. Emphasis on emerging technology, investment planning and marketing strategy appraisals. Clients served include textile, fiber, chemical, printing, metals and machinery industries.

★231★ Hereld Organization
2 Park Pl., Suite 15K
Hartford, CT 06106 (203) 951-0888
P.C. Hereld, Principal

Founded: 1971. **Description:** Consulting to management in the food, drug, biotechnology, and chemical industries in the following areas: technology licensing, economic studies, market research, acquisitions, mergers, divestitures, joint ventures, international trade, and commercial development. **Publications:** Reports on natural food colors are available from the firm.

★232★ Hi-Tech Venture Consultants, Inc.
8 W. 40th St., 10th Fl.
New York, NY 10018 (212) 819-9500
Robert A. Friedenberg, Principal

Founded: 1983. **Description:** Consultants to entrepreneurs, venture capital sources, corporations and the legal profession providing a synthesis of business and technical consulting services. For entrepreneurs services include: strategic positioning, business plan production, and capital seeking; for capital sources: technology evaluation, venture screening, and opportunity locating; for corporations: product development, market strategy, and corporate planning; and for the legal profession: due diligence review, litigation support, and company valuation. **Publications:** Principals have authored three publications on a proprietary technique called the Z-Plan(sm): "Strategic Planning for Artificial Intelligence Companies: A Guide for Entrepreneurs," AI Magazine (August 1986); "Synthesizing Technology and Strategic Planning," Intrapreneurial Excellence (October 1986); "Coordinating Technical and Strategic Planning," Presidents (December 1986).

★233★ High Technology Associates

P.O. Box 42808
No. 234
Houston, TX 77042 (713) 963-9300
Marvin L. Baker, President

Founded: 1984. **Description:** Consultants on United States business development for high-technology European firms. Activities include acquisitions, establishing joint ventures, technology licensing, market development, and manufacturing investments.

★234★ Hillard-Lyons Patent Management, Inc.

545 S. Third St.
Louisville, KY 40202 (502) 588-8638
Robert W. Fletcher, Principal

Founded: 1976. **Description:** Patent assistance consultants offering assessment of the legal strength of patents and related know-how which support research and development limited partnerships and new venture capital investments. Also offers risk analysis and rating for casualty underwriters who are issuing patent infringement insurance. **Meetings:** Enhancing Options in Patent Dispute Resolutions.

★235★ Hoeft Associates

P.O. Box 2323
Arlington, VA 22202 (703) 920-2216
Pamela A. Hoeft, President

Founded: 1978. **Description:** Data collection and analysis, information and literature research, market research, product liability research, and patent, trademark, and copyright services in the areas of chemistry, electrical and electronics engineering, and physics.

★236★ Holland W. Phillips & Associates, Ltd.

P.O. Box 570
Naperville, IL 60566 (815) 886-0956
Holland W. Phillips, President

Founded: 1976. **Description:** Engaged in consulting to American and foreign companies who wish to expand their export opportunities. Also consults on the international level with market studies, countertrade arrangement, licensing and joint ventures, and international business seminars.

★237★ Hub Financial Network Entrepreneurial Services and Seminars

Box 2223
Palos Verdes, CA 90274 (213) 831-5625
Dick Royce, President

Founded: 1980. **Description:** Franchising consultants offering services in network marketing, real estate development, joint ventures, venture capital, and equity sharing. Also conducts seminars on entrepreneurial services and multi-level consultancies.

★238★ Hyland/Associates

5613 DTC Pkwy., Suite 700
Englewood, CO 80111 (303) 773-3292
Warren Hyland, President

Founded: 1975. **Description:** General business consultants whose services emphasize profit improvement, turnaround venture management, and technology start-ups.

★239★ ICS Group International (US)

P.O. Box 4082
Irvine, CA 92716 (714) 552-8494
A.J. Siposs, President

Founded: 1946. **Description:** Offers international and United States consulting and implementary services for United States companies seeking international business developments (exports, imports, offshore operations, etc.), and for foreign clients interested in entering or expanding in North American markets, through direct exports, acquisitions, joint ventures, or subsidiary operations. Active assistance in the funding of international projects, and in the short-term financing of export and import transactions. Corporate management programs on international business topics. Also offers assistance in new ventures and diversification projects, and international business arbitration, and serves as expert witnesses in business and management.

★240★ Industrial Marketing Arm

369 Pine St., Eighth Fl.
San Francisco, CA 94104 (415) 391-6240
D. Charles White, Chairman

Founded: 1978. **Description:** New product consultants specializing in helping identify and create opportunities for introducing uniquely positioned goods and services bought by executives. Extensive experience in conducting creative marketing research on offerings to business, with emphasis on probing focus groups and senior level concept/problem definition interviews with key commercial decision-makers. Skilled at developing marketing plans designed to capitalize on new directions in rapidly changing marketplaces, based on innovative competitive strategies and targeting of prime customer prospects. Activity includes serving as staff marketing arm for clients from abroad seeking industry information and action programs for entering the United States with new ventures or acquisitions. Categories include chemicals, computers, fuel/energy, financial services, extractives, agriculture, trade/transportation, beverages/food ingredients and plant and office equipment.

★241★ Industrial Marketing Projects

P.O. Box 6665
Grand Rapids, MI 49516 (616) 942-8820
Rafael O. Diaz, Principal

Founded: 1978. **Description:** Offers a variety of services designed to assist industrial firms in the development and expansion of domestic and international markets for their products and technologies. Provides this assistance to domestic as well as foreign firms and has carried out client assignments in over 25 countries and the United States. Services offered include market research studies, strategic planning, technology transfer negotiations, acquisitions and divestitures. Industries served include electronics, electromechanical products, chemicals and advanced materials and the office furniture industry.

★242★ Influentia

Suite 201
3954 Peachtree Rd., N.E.
Atlanta, GA 30319 (404) 266-1977
Dain Schult, President

Founded: 1987. **Description:** A turnkey operation to assist inventors, creators or anyone with a new idea for services or products with research, product viability analysis, product design, new product introduction, funding sources, business plan development, advertising and product promotion.

★243★ Info/Consult

Box 204
Bala Cynwyd, PA 19004 (215) 667-0266
Gabrielle S. Revesz, Contact

Founded: 1980. **Description:** Reviews new developments in the fields of chemistry, biomedicine, pharmaceuticals, engineering, patents, and technical sciences. Research results published in confidential reports to clients. **Telecommunication Access:** Telex 509166 (INFO CONSULT); Electronic mail 62770201 (Easylink) or REVESZ (DIALMAIL).

★244★ Information Brokers of Colorado

2888 Bluff St., Suite 152
Boulder, CO 80301 (303) 449-8896
Cassandra Geneson, Principal

Founded: 1979. **Description:** A professional research and information service for business and industry offering a complete line of information-gathering that includes computer database searching, corporate and competitor intelligence, market and industry, investigation and document, and patent fulfillment.

★245★ Information Consulting, Incorporated

2584 Coventry Rd.
P.O. Box 21865
Columbus, OH 43221 (614) 486-7755
E.N. Mimnaugh, President

Founded: 1975. **Description:** Engages in on-line and manual searching in support of chemical research and patent searching; also undertakes information retrieval systems for the chemical and allied industries, as well as library start-ups and maintanance. Market research and activities supporting acquisitions analysis are also done, but no laboratory work.

★246★ Information Guild

Box 254
Lexington, MA 02173 (617) 862-9278
Alice Sizer Warner, Principal

Founded: 1980. **Description:** Offers consulting on fee-based information services with emphasis on money matters planning, including pricing, selling, and collecting. Serves libraries, academic institutions, nonprofit and for-profit organizations as well as entrepreneurships. **Publications:** A. Warner, Mind Your Own Business: A Guide for the Information Entrepreneur, Neal-Schuman (1987); also writes column "What Do You Want to Know," for Information Brokers, published by Burwell Enterprises.

★247★ Innovators & Associates, Inc.

4917 St. Charles Ave.
New Orleans, LA 70115 (504) 581-2526
C. Emmett Pugh, President

Founded: 1972. **Description:** Consultants in the marketing and promotion of inventions and innovations. Experienced in problems of licensing and franchising.

★248★ Integrated Circuit Engineering Corporation

15022 N. 75 Street
Scottsdale, AZ 85260 (602) 998-9780
Glen Madland, Chairman

Founded: 1964. **Description:** Data collection and analysis, market and opinion research, technological forecasting, testing and analysis, and failure and construction analysis in the field of microelectronics. Provides technical consulting to users, manufacturers, and suppliers to the semiconductor industry; legal consulting on patents; and expert witness testimony. Long-range project involves the development of a failure analysis laboratory in Bordeaux, France. **Publications:** Status Report (annually); Icecap (monthly); Report on the Integrated Circuit Industry (annually). **Telecommunication Access:** Telex 165755 (ICE SCOT).

★249★ Intelligent Systems Technology, Inc.

700 North Fairfax Street #600
Arlington, VA 22314-2026 (703) 739-0885
Andrew P. Varrieuv, Contact

Founded: 1985. Firm providing high-technology invention, including intelligent processors and memory elements and systems.

★250★ Interaction Business Services

Austin Centre Tower
701 Brazos, Suite 500
Austin, TX 78701 (512) 320-9067
Ray R. Hufford, President

Founded: 1985. **Description:** Offers assistance in the following types of business development transactions: marketing of products/services/ technology through direct marketing programs, representative agreements, agency formations, distributor systems, special incentive programs, joint marketing, countertrade, technology licensing, know how sharing, barters/offsets, trade shows, and trade delegations; modifications of operations or ownership via mergers, acquisitions, divestitures, partnerships, consortiums, joint ventures, strategic alliances, new startups, research pooling, outsourcing, free trade zones, and syndications. Services available to all corporations, particularly small to mid-size companies intent on applying innovative programs to improve competitive position.

★251★ Intercon Research Associates, Ltd.

6865 Lincoln Ave.
Lincolnwood, IL 60646 (312) 982-1100
John V. Donovan, President

Founded: 1963. **Description:** A professional service firm assisting Japanese, European, North American and Middle East companies to find new products, new technologies, company acquisitions, and perform information research on an international basis. Special expertise in international licensing, joint ventures, and cross border acquisitions for diversification or market expansion.

★252★ Intercontinental Enterprises Ltd.

256 S. Robertson, Suite 3194
Beverly Hills, CA 90211 (213) 276-6525
Gordon S. Riess, President

Founded: 1976. **Description:** Consultants in international trade and technology transfer. Specializes in finding and supervising foreign distributors, licensees, and joint-venture partners; identifying markets; preparing market studies; and implementing sales programs. Assists small, high- technology firms in setting up international operations. Clients include both United States and European corporations. Specializes in medical devices, high-tech engineering and testing equipment, computer peripherals, and pharmaceuticals. Acts as the client company's "International Department" at home, and its "Local Office" abroad. Assists in negotiating technology transfers, mergers, and acquisitions.

★253★ Interdevelopment, Inc.
2001 S. Jefferson Davis Hwy.
Suite 307
Arlington, VA 22202 (703) 521-1020
Frank Faris, President

Founded: 1967. **Description:** Offers management consulting services, particularly in connection with international business, in areas of strategic planning, market research and development, new products and technology, joint ventures and acquisitions. Company activities are directed to the private industrial sector in North America, Western Europe and Latin America. **Publications:** F.E. Faris and M. Luddemann, State, Local, and Cooperative Programs to Promote Innovation and Attract High Technology Investment in the United States: A Survey and Assessment of Effectiveness in Six States (March 1987); "International Marketing," AMA Handbook (1988).

★254★ Intermarketing Group Management Consultants
29 Holt Rd.
Amherst, NH 03031 (603) 672-0499
Linda L. Gerson, President

Founded: 1985. **Description:** A full service international consulting organization specializing in licensing representation, marketing/sales management, international trade and related matters. Objective is to assist corporations, manufacturers and professional companies in the development, execution and management of strategic plans and programs to achieve profitable business growth and market penetration of their products, services, licensed properties or patented inventions on a worldwide basis. Serves all consumer related products industries.

★255★ International Book Marketing Ltd.
210 Fifth Ave.
New York, NY 10010 (212) 683-3411
Allan Lang, President

Founded: 1979. **Description:** Marketing consultants specializing in book publishing, distribution, subsidiary rights and sale of foreign books in United States market. Also active in the licensing of book character rights.

★256★ International Business Consulting Group
Lincoln Bldg.
60 E. 42nd St.
New York, NY 10165 (212) 286-0707
Chat Gurtu, Principal

Founded: 1985. **Description:** Services provided by the business consulting group include counseling on general investment matters; assistance in all aspects of establishing a corporate base in the United States, including finding the appropriate U.S. bankers and brokers; and formation of U.S. entities. Offers consultation on regulatory matters applicable to foreigners, U.S. taxation and foreign tax matters; also assists with matters of repatriation and investment of funds outside the United States, including formation of foreign corporations and trusts. Additional expertise in marketing and sales support as well as invoicing and collection management. Employs attorneys, CPAs and engineers. **Publications:** Principals have authored Doing Business in the U.S., a 109-page booklet; International Transfer of Technology: Regulatory Aspects (1986).

★257★ International Business Resources, Inc.
900 Fort St., Suite 1777
P.O. Box 6475
Honolulu, HI 96818 (808) 521-8204
Dennis T. Oshiro, President

Founded: 1982. **Description:** Provides practical international management expertise to companies seeking capital from Japan or expansion in Asia. Areas of experience include export development, overseas start-ups, technology transfer, project finance, joint ventures and acquisitions. Works with chief executive officer or as member of corporate project team to minimize the risks of doing business in Asia. International business know-how is imparted by consultant to client firm during the consulting process as objectives and results are being achieved. Emphasis is on actual hands-on implementation rather than written reports. Strong in problem-solving and troubleshooting in crisis situations. Locates Japanese joint venture partners for real estate developers in the United States. Numerous contacts within companies and banks throughout Asia.

★258★ International Creative
10929 S. St., Suite 116B
Cerritos, CA 90701 (213) 402-1356
William Uglow, President

Founded: 1985. **Description:** Offers services in product development, premium development, licensing, brand name and identity development, and marketing. Specializes in Far East importation in toy, gift, housewares, and premium fields.

★259★ International Defense Consultant Services, Inc.
7315 Hooking Rd.
McLean, VA 22101 (703) 827-9032
Leo D. Carl, CEO

Founded: 1974. **Description:** Consultants in defense marketing and sales, manufacturing license and technical assistance agreements, patent assignments and licenses, economic compensation (offset) agreements, export licensing problems, international representation, and co-production arrangements.

★260★ International Science and Technology Institute, Inc.
1129 20th Street, N.W.
Suite 800
Washington, DC 20036 (202) 785-0831
B.K. Wesley Copeland, Chairman & C.E.O.

Founded: 1977. **Description:** Technology transfer, especially energy applications for developing countries. Provides information on technologies available, appraises technological needs, conducts feasibility studies, and offers conference management services. Research results published in professional journals, books, magazines, and project reports. **Telecommunication Access:** Telex 272785 (ISTI UR).

★261★ Invention Submission Corp.
903 Liberty Ave.
Pittsburgh, PA 15222 (412) 288-1300

Founded: 1984. **Description:** Provides services to inventors who seek to present new ideas, inventions, products or technologies to industry. Packages ideas in a professional format and submits ideas to manufacturers for review. Also provides supplementary marketing services for new products. Company sells its services through a network of regional sales offices in over thirty cities in the United States, Canada, and England.

★262★ J.C. Croft & Associates International Consultants
3905 N.E. 21st Ave.
P.O. Box 11507
Fort Lauderdale, FL 33339 (305) 564-9620
J.C. Croft, Jr., President

Founded: 1962. **Description:** International technology and marketing consultants. Services include new product and new market development, technology transfer, sales, marketing, corporate management consulting, capital project development, and securing financial resources. Offers revitalization programs for technical and industrial turnaround to increase profits and market position. Particular expertise in export/import licensing, representation, market development and indigenous manufacturing requirements. Affiliated with Techmark Corporation, a Far East Trade Corporation located in Hong Kong.

★263★ J.P. Hughes & Company

26 Warrington Dr.
Lake Bluff, IL 60044 (312) 234-0867
J. Peter Hughes, President

Founded: 1978. **Description:** Serves manufacturers in international business with emphasis on establishing overseas subsidiaries joint venture, licensing and export development.

★264★ Jack J. Bulloff

399 Ridge Hill Rd.
Schenectady, NY 12303 (518) 355-0977
Jack J. Bulloff, Principal

Founded: 1968. **Description:** Consultant involved in research, applied research, technical development and technological innovation in chemical and graphic arts industries; hazardous chemicals and wastes management; expert witness in the areas of occupational and environmental safety and health; development of information for legislation, regulation, regulatory compliance, and criminal and civil court proceedings or administrative hearings; and technical reporting and writing.

★265★ Japan/Pacific Associates

467 Hamilton Ave., Suite 2
Palo Alto, CA 94301 (415) 322-8441
Yoriko Kishimoto

Founded: 1982. **Description:** Provides an integrated analysis of the wide range of problems posed in United States/Pacific Basin business relationships. Activities are as follows: (1) Private business consulting on specific business dealings - marketing strategies/information, studies of competitive strategies, evaluation and auditing services, and negotiation and mediation. (2) Research projects on business management and international policy topics. Specializes in serving high-technology, entrepreneurial corporations. Biotechnology, telecommunications, and venture capital are areas of focus. **Publications:** J. Kotlan and Y. Kishimoto, The Third Century: America in the Era of Asian Renaissance, Crown Publishers (1988); also publishes a monthly newsletter Biotechnology in Japan News Service.

★266★ Jeffrey D. Marshall

21 Central St.
Topsfield, MA 01983 (617) 887-9788
Jeffrey D. Marshall, Principal

Founded: 1981. **Description:** Offers assistance to small businesses with federal procurement problems, financing of new products and ideas and protecting those ideas or products. Firm specializes in planning for new ventures including business plans, capital sources for development, and evaluation of patentability. **Publications:** J. Marshall, Manual on SBIR Proposal Preparation, Marshall Associates (1986); Lowering Patent Costs, Marshall Associates (1986). Principal serves as an agent of the U.S. Patent and Trademark Office; publisher of Patent News, newsletter on changes in intellectual property protection.

★267★ JoAnn Johnson

6 Thomas Rd.
Westport, CT 06880 (203) 226-9725
JoAnn Johnson, President

Founded: 1981. **Description:** Literary agent offering services in rights representation, and general consulting regarding the publishing business.

★268★ Johnson Engineering Company

17 N. Drexel Ave.
Havertown, PA 19083 (215) 446-4540
Herbert G. Johnson, Principal

Founded: 1943. **Description:** Product and process development consultants. Services include patent assistance, pilot and experimental operations, market research and sales promotion, and siting and plant layout.

★269★ Josef K. Murek Company

4454 Casitas St.
San Diego, CA 92107 (619) 224-6810
Josef K. Murek, President

Founded: 1969. **Description:** Provides technical and financial consulting services for ventures in North America and Europe. These services include international marketing and research; long-range planning and strategic analysis of markets and technological trends; aid in the selection of qualified representation in form of manufacturing licensees, distributors, franchisees, partner in joint ventures; and counsel on the transfer of technology.

★270★ Joseph W. Prane

213 Church Rd.
Elkins Park, PA 19117 (215) 635-2008
Joseph W. Prane, Principal

Founded: 1969. **Description:** Provides advisory services to the chemical and polymer industries specializing in coatings, sealants, adhesives and plastics. Services offered include technical, economic, marketing and business consultation in the areas of product development, marketing research, commercial development, diversification studies, distribution analysis, market expansion, and technology and licensing arrangements. Other activities include: worldwide acquisition and merger arrangements; new technology seminars; expert witness in litigations. **Publications:** Over 100, including column in Polymer News and Adhesives Age.

★271★ Kappa Associates

462 Chestnut Hill Ave.
Brookline, MA 02146 (617) 277-8134
Kenneth D. Krantz, President

Founded: 1985. **Description:** Firm provides consulting and managerial services to clients in areas related to medical research and development, including pharmaceutical clinical research, product license evaluation, research personnel assessment and training, strategic planning, and product development programs. Also retained as expert counsel regarding drug development and applicable FDA regulations, drug and biological product technology, adverse drug reactions, and drug interactions.

★272★ Keller Research Services

1343 Mill Pond Way
Palmyra, PA 17078 (717) 838-4168
George H. Keller, Principal

Founded: 1983. **Description:** Offers expertise in ELISA immunoassay technology; anitbody preparation and evaluation, antigen and antibody immobilization, enzyme-antibody conjugate selection and preparation, colorimetric substrate selection and preparation; non-colorimetric detection methods such as fluorescent and luminescent; microwell, microscope slide, membrane, bead and dipstick formats. Also experienced with DNA probe technology, probe preparation and non-isotopic labeling, incorporation of various hybridization formats, and non-isotopic probe detection. Can provide evaluation of diagnostic technology including test kits, reagents, processes, patents; preparation of literature searches and comprehensive reports. **Publications:** Sixteen publications in the areas of DNA probes, nucleic acid hybridization, gene expression, growth factors, histones and hormonal control.

★273★ Ken-Quest Limited
425 S. Chester Ave.
Park Ridge, IL 60068 (312) 692-2509
Dolores Torres Kenney, President

Founded: 1974. **Description:** A minority-owned consulting firm offering technical support services for computer systems and networks bringing together hardware, software, data capture, and in-house personnel. Also serves as financial broker for systems and network development programs and joint venture brokers for licensing, patents, and copyrights program relating to computers or telecommunications technology.

★274★ Kendall Square Associates
Twin City Office Plaza
P.O. Box 277
Cambridge, MA 02141 (617) 776-8830
Jose Gerson Bloch, Principal

Founded: 1975. **Description:** Offers general management assistance to small and minority owned companies, new business and venture capital consulting, and project and program management.

★275★ Kevin C. McGuire
P.O. Box 684
Stevensville, MT 59870 (406) 777-3833
Kevin C. McGuire, Principal

Description: Consultant in new business ventures, venture capital, new products, and investment capital.

★276★ Korea Strategy Associates, Inc.
University Pl., Suite 200
124 Mount Auburn St.
Cambridge, MA 02138 (617) 576-5773
Karl Moskowitz, President

Founded: 1984. **Description:** Assists clients to establish and develop successful businesses in Korea dnd provides focused industry, market, and country analysis. Principal practices are new business venture development, market and industry research, cross-cultural management systems, and organizational development. **Publications:** K. Moskowitz, "Protecting Intellectual Property in Korea," Asian Wall Street Journal (November 28, 1986).

★277★ Korey International Ltd.
55 Tanbark Crescent
Don Mills, ON, Canada M3B 1N7 (416) 445-4968
George Korey, President

Founded: 1960. **Description:** Offers counsel to industry and governments on economic planning, strategic planning and management, industrial development, organization, long range planning, marketing and general management, as well as international marketing strategy, new products and new technology transfer, foreign licensing, joint ventures and all aspects of international business.

★278★ Lawrence M. Liggett
1856 Piedras Cir.
Danville, CA 94526 (415) 820-9304
Lawrence M. Liggett, Principal

Founded: 1982. **Description:** Business management consultant specializing in technology licensing, research and development management, and joint ventures. Experienced with the following technologies: photovoltaics, optical memories, and vacuum processing equipment for semiconductor manufacturers.

★279★ Leo Beiser Inc.
151-77 28th Ave.
Flushing, NY 11354 (718) 353-7298
Leo Beiser, President

Founded: 1976. **Description:** Consultation and research company specializing in electro-optical information handling including use of lasers, CRTs and CCDs. Expertise in laser scanning and recording for input and output of image and data systems. Serves commercial, industrial and governmental communities in image capture, printing and display for business graphics, publishing, printing, video and data storage and retrieval. Services provided are analysis of technical requirements as well as design, experimentation and guidance for satisfying requirements. Additional assistance offered in patent analyses and situations requiring expert witness. **Publications:** L. Beiser, "Imaging with Laser Scanners," Optics News (November 1986); Holographic Scanning, John Wiley (1988).

★280★ Lewis B. Weisfeld
Suite 703 - Benson E.
Township Line & Old York Roads
Jenkintown, PA 19046 (215) 886-2315
Lewis B. Weisfeld, Principal

Founded: 1979. **Description:** Offers consulting services on research, development, technology and commercial aspects of specialty chemicals, especially plastics additives and thermal stabilizers; market/technological surveys; governmental regulatory liaison and compliance. Services include assistance on chemical and technological problems arising in connection with patent matters; and specific special tactical and strategic project assignments. **Publications:** Regulatory Update: Government Regulations Related to Plastics Industry (monthly).

★281★ M. Kleiman and Associates
P.O. Box 59014
Chicago, IL 60659 (312) 679-4000
Morton Kleiman, Principal

Founded: 1954. **Description:** Offers research and development services in agricultural chemicals, pesticides, organic and pharmaceutical chemicals. Engages in special and/or ongoing research projects in own or in client's laboratories. Also offers service in chemical patent matters and expert testimony in patent suits.

★282★ M.W. Probol & Associates
P.O. Box 854
Battle Creek, MI 49016 (616) 964-3774
Manfred W. Probol, Principal

Founded: 1985. **Description:** International engineering systems consultants involved in licensing of technology, engineering systems and procedures, microfilming programs, and metrication. Works toward maximum interchange ability of parts and product built in multi-plant, multi- country locations. Involved in distribution and control of technology worldwide. Offers vendor part number control programs to eliminate duplicate stocking of parts at manufacturing and dealer locations.

★283★ Maehara & Associates

700 S. Flower St.
Suite 2200
Los Angeles, CA 90017 (213) 629-8055
Toshio Maehara, Principal

Founded: 1980. **Description:** Offers consulting services in international joint ventures, financial planning, and venture capital arrangements.

★284★ Mankind Research Unlimited, Inc.

1315 Apple Avenue
Silver Spring, MD 20910 (301) 587-8686
Dr. Carl Schleicher, President

Founded: 1972. **Description:** Human systems design, alternate energy sources, research and development for innovative technology, and bionic, biocybernetics, biomedical, and high speed learning systems. Research results published in project reports.

★285★ Margiloff & Associates

26 Windmill Rd.
Armonk, NY 10504 (914) 273-3949
Irwin B. Margiloff, Principal

Founded: 1982. **Description:** Business consultants offering assistance in commercial development, new products and new processes. Services include analysis of research and development activities, procurement assistance, market research, licensing, energy studies, and economics. Projects involve development, manufacturing economics, business planning, operations, and expert witness services. Clients are in the food, chemical, fermentation, energy, financial and legal services and general manufacturing fields.

★286★ Mark J. Egyed

Morningside Dr.
Ossining, NY 10562 (914) 762-4907
Mark J. Egyed, Principal

Founded: 1986. **Description:** Offers assistance in research and development efforts including prototype fabrication, design assistance, construction and operation of sophisticated or specialized tools or systems. Also provides employee training in industrial electrical safety.

★287★ Materials Research, Inc.

790 East 700 South
Centerville, UT 84014 (801) 298-4000
Dr. R. Natesh, President

Founded: 1973. **Description:** Product and process development and testing in the materials industry. Provides characterization of metals and materials, analysis of ultrafine particles, analysis of air and waterborne pollutants, inspection of microelectric devices, failure analysis, corrosion studies, investigation of product contamination, examination of welds and diffusion bonding problems, technical feasibility studies, forensic investigations and accident reconstruction, patent infringement and product liability studies, geothermal kinetic studies, and legal testimony. Also studies problems related to caking, porosity, and discoloration; production, fabrication, and tooling; and lubrication, viscosity, and adhesion. Long-range projects include analysis of defect structure in silicon and chemical reduction of nuclear steam generator crevice deposits. **Publications:** Analysis of Defect Structures in Silicon (series of 20); Chemical Reduction of Nuclear Steam Generator Crevice Deposites (series of 2); and Boron Nitride Coating for High Performance Electric Heaters.

★288★ Matryxx Corporation

P.O. Box 1201
Piscataway, NJ 08854 (201) 752-8235
Darryl L. North, Contact

Description: Technical services for the chemical processing industry with specialization in scrubbers/towers, aerators and oxidation pond units, hoods and ducts, oil separators, plate settlers, and plating solution evaporator systems. Specific services include data collection and analysis, site reviews, process surveys, hazardous waste permit review assistance, patent services, and chemical process equipment design reviews. Also recommends, sizes, and designs equipment for process or pollution control, especially fiberglass and thermoplastics. **Publications:** Newsbreak (newsletter).

★289★ MG Consulting

1016 Lemon St.
Menlo Park, CA 94025 (415) 326-1611
Morton Grosser, Principal

Founded: 1967. **Description:** Technology assessment consultant specializing in management assistance for new ventures and venture capitalists. Particular expertise in transfer of innovation techniques. Provides explanation and evaluation of technical concepts to non-technical clients, including large audiences. Major industries served are: aerospace, banking and finance, data processing, electronics, education, energy, entertainment, health care, manufacturing, petroleum, transportation, and the legal profession. **Publications:** M. Grosser, "Nerve Repair at the Axon Level: A Merger of Microsurgery and Microelectronics," in Artificial Organs, VCH Publishers (1987); author of six books, numerous papers, articles in Atlantic, Harper's, Industrial Design, Natural History, The New Yorker, and Technology Review. **Meetings:** Venture Capital Conference; Sessions on Innovation and Technology.

★290★ Michael Gigliotti and Associates Incorporated

42 Rogers St.
Box 591
Gloucester, MA 01931 (617) 283-7795
M. F. X. Gigliotti, President

Founded: 1977. **Description:** Offers counsel, conducts studies, implements projects and presents educational seminars on: product development, product introduction, technical appraisals, acquisitions and divestiture, plant location surveys, designs and estimates, process engineering, operations analysis, technical personnel—organization, planning, objectives setting, counseling and development, and machinery design and development. Clients in manufacturing, financial, government and educational organizations. Active with both small and large-scale organizations, in Europe (United Kingdom, France and Germany) and in the United States. Clients are plastics or packaging or machinery oriented. **Telecommunication Access:** Alternate telephone (617) 281-3310.

★291★ Minerva Consulting Group, Inc.

Box 5474
New York, NY 10163 (212) 972-1020
Adam Starchild, President

Founded: 1978. **Description:** Arranges licensing structures; creates and expands product lines; researches legal and capital structures; analyzes regulatory environment related to manufacture, import, and distribution of goods; and monitors investment portfolios. Research activities concentrate on international business, tax, new venture development, oceanography and fisheries, and securities and real estate. Research results published in professional journals.

★292★ Mobion International
P.O. Box 572
Wayzata, MN 55391 (612) 476-0975
Winthrop A. Eastman, President

Founded: 1979. **Description:** Offers technology transfer consulting services with emphasis on locating and negotiating purchase, sale or licensing of technology for companies in the United States and Spain. Recent emphasis on animal health, equine, and livestock related products.

★293★ Morris J. Root Associates, Limited
900 Green Bay Rd.
Highland Park, IL 60035 (312) 432-8614
Morris J. Root, President

Founded: 1978. **Description:** Provides consulting services in the areas of cosmetics - aerosol products; patents - product liability; government regulatory agencies research and development - seminars; and chemical engineering.

★294★ Murray Associates, Incorporated
73 Thunder Mountain Rd.
P.O. Box 1406
Greenwich, CT 06836 (203) 869-8183
Charles K. Murray, President

Founded: 1970. **Description:** Offers business and marketing consulting in the following areas: (1) technology marketing - domestic and foreign know- how marketing through licensing, etc.; (2) new product introduction - preparation of marketing plan based on thorough research of market conditions; assistance in planning business strategy including plant location; (3) technology sourcing - procuring technology for corporate diversification; and (4) related services - searches for distributors in the United States and overseas; mergers, divestitures and acquisitions - including financial analysis. The company has associates which carry out field work throughout the industrialized world in assignments requiring overseas product-technology marketing/sourcing.

★295★ Murray Jelling
21 Spring Hill Rd.
Roslyn Heights, NY 11577 (516) 621-0060
Murray Jelling, Principal

Founded: 1957. **Description:** Technical consultant active in the chemical field, with emphasis on: new products and processes; industrial, trade and consumer products. Technical specialization is in organic chemistry, wetting agents, fatty oils and derivatives, polymers, bituminous additives, germicides, fungicides, laundry and drycleaning products, and other chemical specialities. For inventors, consultant offers evaluation of patentability of new products and processes; assistance in patent matters; licensing in chemical and other fields; and expert court testimony.

★296★ Neurological Research & Development
120 N. Abington Rd.
Clarks Summit, PA 18411 (800) 327-6759
Jerome M. Thier, President

Founded: 1972. **Description:** A medical products research firm and a manufacturer of products marketed to hospitals, nursing homes and the home health care market. Consults in the areas of general management, marketing, finance and production. Have extensive entrepreneurial experience in the development and growth of start-up and expanding companies. The fields covered are medical, food, chemical, textile and banking.

★297★ Nicoletti Productions, Inc.
P.O. Box 2818
Newport Beach, CA 92663 (714) 494-0181
J. Nicoletti, President

Founded: 1976. **Description:** Specialists in the entertainment, music and sound industries. Offers consulting to music publishing, record companies, artists, producers and songwriters. Also serves any users of music or sound.

★298★ Nilson's Inventors League
2911 16 Street, No. 210
San Francisco, CA 94103 (415) 552-5518
Nilson V. Ortiz, Inventor

Description: Systems design and development and market and opinion research in the fields of electrical motion, magnetism, and perpetualtronics. Research projects include feedback systems, permanent magnets and electron negarive energy, and electric vehicles.

★299★ Noone Associates
813 Columbus Dr.
Teaneck, NJ 07666 (201) 833-2461
Thomas M. Noone, President

Founded: 1984. **Description:** Technology scout searches out business opportunities in the chemical and allied industries. Services include: market research, technology assessment, commercial and product development, licensing, strategic planning, new ventures, and profit improvement programs. All services conducted for the North American market. **Publications:** T.M. Noone, "Managing Innovation," Inside Research and Development, Part I (October 14, 1987), Part II (November 11, 1987). **Meetings:** Licensing as a Business Strategy for Small Companies; Commercializing University Technology; The Business Side of Patents; Evaluating Patents.

★300★ O.A. Battista Research Institute
3863 S.W. Loop 820
Suite 100
Fort Worth, TX 76133 (817) 292-4272
O.A. Battista, President

Founded: 1974. **Description:** Consultants to research management with emphasis on research for profit.

★301★ O A Laboratories and Research, Inc.
410 W. 10th St.
Indianapolis, IN 46202 (317) 639-2626
Thomas M. Crawford, President

Founded: 1986. **Description:** Provides consulting services regarding production material problems, workplace atmosphere, product liability, chemical product development and patent application, interrogatories and depositions in environmental impairment cases and product liability cases where chemical or technical knowledge is required.

★302★ OMEC International, Inc.
727 Fifteenth Street N.W.
Washington, DC 20005 (212) 639-8900
Donna Zubris, Editorial Director

Founded: 1982. **Description:** Patent retrieval and information searches in the areas of biotechnology, chemical technology, inventions, and government regulations. **Publications:** BioINVENTION (monthly). **Telecommunication Access:** FAX (212) 639-8993.

★303★ Packaged Facts

274 Madison Avenue
New York, NY 10016 (212) 532-5533
David A. Weiss, President

Founded: 1962. **Description:** Trademark research, syndicated market studies, legal research, back-dated clipping, and advertisement tracking of a particular product, company, or industry. Provides consumer market studies in such areas as the food and beverage industry, personal care products, household products, and advertising specialties. Specific projects include prospective client searches, aquisition/merger research, and historical research for presentations and promotional materials. Research results published in reports. **Publications:** Consumer Market Studies.

★304★ Pan-Atlantic Consultants

1818 N. St., N.W., Suite 600
Washington, DC 20036 (202) 223-4001
Hugh J. Gownley, President

Founded: 1986. **Description:** Consulting firm dedicated to assisting medium and small European companies that wish to engage in, or expand, production and/or marketing activities in the United States. Special emphasis is placed on working with French companies. At the same time, PAC has an interest in identifying opportunities for American firms that wish to transfer their products and technologies to Europe.

★305★ Patentrol Corporation

6660 Greenbriar Drive
Cleveland, OH 44130 (216) 842-0828
Dr. R.S. Pauliukonis, P.E., President

Founded: 1971. **Description:** Provides research, development, and consulting services in the areas of manufacturing techniques, production cost reduction, product improvement, and performance processes. Also studies patents and licensing, international markets, fluid power components, and energy conservation controls and systems related to power, heat, hydraulics, pneumatics, underwater technology, cryogenic science and technology, heat transfer, and thermodynamics.

★306★ Pixel Instruments Corp.

2336-D Walsh Ave.
Santa Clara, CA 95051 (408) 986-9090
Lorelei Cooper, President

Founded: 1981. **Description:** Offers complete analog and digital design services from initial engineering studies through pilot manufacturing runs for video circuits and instruments.

★307★ Princeton Polymer Laboratories, Inc.

501 Plainsboro Road
Plainsboro, NJ 08536 (609) 799-2060
Dr. Donald E. Hudgin, President

Founded: 1969. **Description:** Research and development on new products and processes, market and technological studies, economic evaluation, and consulting in the area of polymers. Research and development capabilities include monomer and polymer synthesis and modification, compounding, and polymer blending, grafting, characterization, testing, and end use evaluation. Studies high-temperature, liquid crystal polymers, ethylene-carbon monoxide copolymers, photodegradable polymers, and polyethyene waxes. Also provides patent confirmation, research corroboration, and economic feasibility studies in the areas of plastics, adhesives, coatings, films, thermoplastics, fibers, foams, additives, and elastomers. **Telecommunication Access:** FAX (609) 655-0613.

★308★ Rain Hill Group, Inc.

90 Broad Street
New York, NY 10004 (212) 483-9162
Nancy E. Haar, Contact

Founded: 1977. **Description:** Identifies technology and/or market-based opportunities to augment diversification and growth strategies. Services include screening and review licensing, joint venture research, marketing and distribution studies, minority equity investment research, and acquisition opportunity forecasting in business and technology. **Telecommunication Access:** Telex 960883 (RAINHILL NYK); FAX (212) 514-6217.

★309★ RCS Associates

1603 Danbury Dr.
Claremont, CA 91711 (714) 624-1801
James Constant, President

Founded: 1968. **Description:** Business consultants specializing in development, protection, and litigation of high-technology products and patents. Expertise in both business and technical aspects of research and development including licensing (foreign and domestic), venture capital, and partnerships. Active with aerospace and electronics industry.

★310★ Reference Desk

Box 12122
La Jolla, CA 92037 (619) 459-1513
Ellen Slotoroff Zyroff, Ph.D., Chief Operating Officer

Founded: 1983. **Description:** Provides computerized database searching, library research, consulting on information sources, bibliography compilation, article photocopy and document retrieval, company profiles, product and industry surveys, patent searches, and current awareness updates in the areas of business and industry, engineering, psychology, medicine, education, law, humanities, history, ancient studies, classics, linguistics, and current events.

★311★ Rescorp International

3792 Dust Commander Dr.
Fairfield, OH 45011 (513) 867-0718
Richard S. Cremisio, President

Founded: 1969. **Description:** A metallurgical and materials consulting firm offering a broad range of technical management and manufacturing know-how. The total facilities of the corporation provide a solid basis for such functions as: design and planning of new metalworking facilities; design and construction of prototype equipment and pilot plants; generation of process and quality control procedures including manuals and internal specifications for plant operations; preparation of job descriptions, assistance with staffing or "in-plant" training of personnel; short term project management and augmentation for existing R & D efforts; technical marketing of new or existing products; merger and acquisition studies; patent advisory service and assistance with patent exploitation; and metallurgical investigations, failure analyses, accident investigation studies, and expert witness services.

★312★ Researched Products

7443-C Somerset Bay
Indianapolis, IN 46240 (317) 255-1855
Stuart I. Felton, Principal

Founded: 1969. **Description:** Performs research projects in technical feasibility, marketing, economic investments, and venture capital.

★313★ Richard E. Wolf & Associates
P.O. Box 968
Arlington, VA 22216 (703) 276-0270
Richard E. Wolf, Principal

Founded: 1985. **Description:** Information broker/consultant providing a service located in the Washington, D.C. metropolitan area. Document delivery and patent searches have been provided to engineers working in a variety of areas. Literature is mainly obtained from major government suppliers in this area. Clients located in Pennsylvania, Switzerland and Peru. Client and employee interests will determine future areas of activity.

★314★ Richard W. Hynes & Associates
495 W. Edgerton Dr.
Sky Harbor
San Bernardino, CA 92405 (714) 886-8261
Richard W. Hynes

Founded: 1974. **Description:** Sales and marketing consultants providing planning and implementation for consumer products companies. Also offers small business development and start-ups, new product ideas, venture capital, and business organizations and communications.

★315★ Robert B. Leisy
14408 E. Whittier Blvd.
Suite B-5
Whittier, CA 90605 (213) 698-4862
Robert B. Leisy, Principal

Founded: 1972. **Description:** Consultant in market research and venture capital, offering counsel in these areas: industrial, business-to-business, market research, through an affiliated organization, Industrial Market Research Company. Related venture capital consulting is done for venture capital investment firms, venture financed portfolio companies, and in organizing venture capital investment management training seminars for a major venture capital trade association.

★316★ Robert Talmage
163 Bowery Rd.
New Canaan, CT 06840 (203) 966-2082
Robert Talmage

Description: Offers advice and counsel on powder metallurgy (P/M) materials, processing and applications; research and development of new P/M applications; and invention licensing of new P/M applications, materials, and processes.

★317★ Royalco International Inc.
749 N. Leonard St.
Montebello, CA 90640 (213) 888-2006
Mike S. Misawa, President/CEO

Founded: 1976. **Description:** Offers business consulting research and development for Mexico, Korea, Taiwan and Japan. Consults on import and export business, on establishing new companies and joint venture with foreign companies and foreign countries, and on applying for Japanese patent, trademark design patent. Also consults on problems of unfair competition, monopoly laws and related fields.

★318★ Saul Soloway
180 Broadview Ave.
New Rochelle, NY 10804 (914) 636-6970
Saul Soloway, Principal

Founded: 1946. **Description:** Provides consulting services on chemical research and development in the areas of synthesis of organic compounds including formulation of consumer products - cosmetic, food, automotive; polymers and resins; fats and oils. Services provided also include literature surveys, market analyses, and patent applications.

★319★ Scientific Conversion, Inc.
2800 Third St.
San Francisco, CA 94107 (415) 821-6464
Jon D. Paul, Principal

Founded: 1973. **Description:** Electronic research, development, and design consultants. Specialty is power supplies, lamp ballasts, circuit design and analog circuitry. Capabilities include: magnetic design, high voltage, and power supply design. Complete testing laboratory, prototype and printed circuit design facility. Serves the audio, lighting, telephone and electronics industry.

★320★ Scott-Brown Engineers
575 Ethan Ln.
Carson City, NV 89704 (702) 323-5677
Peter Scott-Brown, Principal

Founded: 1970. **Description:** Consultants in the design of low-pollution power plants for air, sea and land transportation. Services include new product design, development and patent counseling. Also provides analytical services in thermodynamics and structural dynamics, including collision analysis.

★321★ SEARCH Corporation
655 Mine Ridge Road
Great Falls, VA 22066 (703) 759-3560
Jean Tibbetts, President

Founded: 1973. **Description:** Specializes in finding available new products, processes, and technology for industry. Conducts technology, market, information, patent, acquisition, and photograph searches. Monitors activity in biotechnology, medical devices, electronics, and any other field of technology.

★322★ SEARCHLINE
4914 Columbia Avenue
P.O. Box F
Lisle, IL 60532-0188 (312) 964-0127
Ronald J. Scheidelman, Ph.D., Director

Founded: 1968. **Description:** Information and literature research, market research, technological forecasting, and product development in the areas of science, engineering, biomedicine and law, especially regarding new products, product liability, patents, and health. Current project focuses on the origin, distribution, fate, and chemical speciation of copper in the environment. **Telecommunication Access:** Telex 4990991 (SEARCH).

★323★ Spier Corp.
50 Park Ave.
New York, NY 10016 (212) 679-4180
I. Martin Spier, President

Founded: 1946. **Description:** Design, development and supervision of manufacture of plastic products including design of molds, specifications of materials and other technical aspects of production. Available to financial institutions and manufacturing companies to evaluate existing facilities and new ventures both from a financial as well as production standpoint. Also offers expert witness services in patent infringement and product liability. **Meetings:** Designing Plastics Products.

★324★ Techni Research Associates, Inc.
Willow Grove Plaza
York & Davisville Roads
Willow Grove, PA 19090 (215) 657-1753
Louis F. Schiffman, President

Founded: 1968. **Description:** Offers technology management services that include but are not limited to, product development, technology transfer, licensing, and technology assessment. **Publications:** World Technology/Patent Licensing Gazette (bimonthly periodical focusing on licensing and technology transfer); Guide to Available Technologies (annual).

★325★ Technology Research Corporation

Springfield Professional Park
8328-A Traford Lane
Springfield, VA 22152 (703) 451-8830
V. Daniel Hunt, President

Founded: 1979. **Description:** Collects and analyzes data, performs experimental review and synthesis, conducts information and literature research, and provides systems design and technological forecasting in such fields as defense systems, advanced manufacturing technology, applied expert systems, computer-aided design and manufacturing, computer-integrated manufacturing, applied arctificial intelligence, robotics, machine vision, and expert systems. Related services include technical and management consulting, requirements analysis, program planning and evaluation, technology assessment, risk assessment, and technolgy transfer. Long-range projects focus on superconductivity technology assessment. **Publications:** Mechatronics: Japan's Newest Threat.

★326★ Technology Search International, Inc.

500 E. Higgins Rd.
Elk Grove Village, IL 60007 (312) 593-2111
John W. Morehead, President

Founded: 1981. **Description:** New product search consultants retained by industrial firms and development agencies in North America, Europe and Japan. Also search for technology-based business opportunities compatible with client- firm's capabilities and interests, which can be made available to client-firm through licensing, joint venture, distribution, contract manufacturing, acquisition and technology transfer arrangements. Also offer training and executive search services. Typical clients represent most major industries. **Publications:** J.W. Morehead, Finding and Licensing New Products and Technology From the U.S.A. (January 1988), available from firm.

★327★ Thomson & Thomson

500 Victory Road
North Quincy, MA 02171 (617) 479-1600
Peter B. Johnson, President

Founded: 1920. **Description:** Conducts trademark, company name, design, and ancillary research. **Publications:** Trademark Alert (periodical listing of new trade names). **Telecommunication Access:** Toll-free 800-692-8833; Toll-free in Canada 800-338-1867; Telex 6971430; FAX (617) 786-8273.

★328★ Trans Energy Corporation

4040 E. McDowell Rd.
Phoenix, AZ 85008 (602) 267-7250
Roy E. McAlister, President

Founded: 1969. **Description:** Consultants in mechanical, chemical, and industrial engineering who specialize in energy and materials engineering and new product development. Also provide financial planning, business plan development, and financing assistance, materials and product testing, and new product feasibility studies.

★329★ Transcience Associates, Inc.

1112 Hinman Avenue
Evanston, IL 60202 (312) 475-1125
Harold Kantner, President

Founded: 1979. **Description:** Computer and biomedical sciences; control systems; hydraulic device development, including photovoltaic-controlled agricultural systems; electrochemistry and electronics; mathematical sciences; physics and solar energy; and technology transfer. Offers development, promotion, brokerage, and licensing support services. Research results published in project reports.

★330★ TransTech Services USA

P.O. Box 21003
Alexandria, VA 22320 (703) 548-8543
William B. Miller, President

Founded: 1979. **Description:** Provides guidance and counsel to small and medium-sized enterprises interested in entering international marketing and international technology transfer activities, with particular reference to licensing and joint venturing, as well as to export. Services include assistance in organizing for technology transfer, in locating the best overseas partners, and in negotiating, concluding, and implementing technology transfer agreement. Services are also available to foreign firms seeking to license or joint venture in the United States.

★331★ 21ST Century Research

8200 Blvd. E.
North Bergen, NJ 07047 (201) 868-0881
Bohdan O. Szuprowicz, Principal

Founded: 1971. **Description:** Offers business intelligence pertaining to high technology investment and markets in automation, aerospace, electronic warfare, expert systems, computers, communications, machine tools, VLSI semiconductors, and technology transfer. Special knowledge of Soviet, East European, Chinese, South African, and Arab markets. Clients drawn from aerospace, finance, banking, data processing, and electronics manufacturing industries.

★332★ Vencon Management, Incorporated

301 W. 53rd St.
New York, NY 10019 (212) 581-8787
Irvin Barash, President

Founded: 1973. **Description:** Venture and management consultants to corporations and entrepreneurs. Specializes in the areas of mergers and acquisitions evaluation and negotiation, and the preparation of marketing and business plans. Assists small or new businesses in expansion plans and financing. Also involved in new enterprise planning with industry and communities for rural economic analysis and for fostering minority business development.

★333★ Venture Development Corporation

1 Apple Hill
Natick, MA 01760 (617) 653-9000
Lewis Solomon, President

Founded: 1971. **Description:** Provides customized market research, market planning, strategic planning, and acquisition and new venture research in the areas of computers, communication systems, instrumentation, office equipment, electronic components, and consumer electronics. Offers the VENTURESEARCH quick response service, including database access and consultation regarding specific market questions. **Publications:** Industry Reports (identifying key technology trends, competitive strategies, and areas of risk and opportunity); The Uniterruptible Power Supply Industry: 1987-1991; The Intelligent Copier/ Printer Industry; Venturecasts (annual compilation). **Telecommunication Access:** Telex 709190 (VENTURE).

★334★ Venture Economics
16 Laurel Ave.
P.O. Box 348
Wellesley Hills, MA 02181 (617) 431-8100
Stanley E. Pratt, Chairman

Founded: 1961. **Description:** Venture capital and business development specialists providing customized research for industrial corporations. Services include: identification of high-potential companies for investment, alliance or acquisition; assistance in establishing venture capital or strategic alliance programs; and assistance with specific acquisition searches. Services for institutional investors in venture capital include: basic education and due diligence evaluation of venture capital as an investment; development of venture capital investment strategy; and identification of investment opportunities, portfolio monitoring and analysis. **Publications:** Principals contribute to many publications including: Venture Capital Journal; UK Venture Capital Journal; Canadian Venture Capital; Pratt's Guide to Venture Capital Sources; Trends in Venture Capital; The Venture Capital Industry: Opportunities and Considerations for Investors.

★335★ Venture Management Associates
400 W. Cummings Park, Suite 2400
Woburn, MA 01801 (617) 933-1550
George Keramas, President

Founded: 1987. **Description:** Assists entrepreneurs in planning, starting, financing, staffing and managing a business. Although primary emphasis is on technology- based companies, the firm is involved in a wide variety of fields. In addition to consulting services, provides financial brokerage services as well as a specialized service whereby the firm matches working partners/investors with appropriate opportunities.

★336★ W. John Foxwell
2557 Lake Charnwood Blvd.
Troy, MI 48098 (313) 879-0173
W. John Foxwell, Principal

Founded: 1975. **Description:** Mechanical and agricultural consulting engineer specializing in the review of design and development of machinery and components (engines, drive lines, hydraulics, hitches, and adaptation of these to electronic control). Areas covered include: agriculture, construction, implements, industrial designs, off-road vehicles, prime movers, and mechanical handling.

★337★ Walter C. Beard, Inc.
P.O. Dr.
Middlebury, CT 06762 (203) 758-2194
Walter C. Beard, President

Description: Offers counsel on aerosols-pressurized packaging, valves, components, and systems; on research and marketing including new production introduction; on product development; and on patent evaluation, development, and studies. Also prepares testimony and expert witness in areas of aerosols, patents, and certain other technical fields.

★338★ Walter C. Beckwith Associates
205 Pennsylvania Avenue, S.E.
Washington, DC 20003 (202) 544-6960
W.C. Beckwith, III, Chief Operating Officer

Description: Product development, technology transfer, patent investigation and representation, and marketing support related to government research and development.

★339★ Weintritt Testing Laboratories
305 Guidry Road
P.O. Box 30162
Lafayette, LA 70503 (318) 981-1560

Description: Provides identification and analysis of drilling muds, polymers, contaminants, water and wastewater, cores, scales, and minerals. Other capabilities include failure analysis of oilfield equipment, corrosion tests, and testing and evaluation of production chemicals. Specializes in torque and drag reduction, shale control drilling fluids, environmental analysis, patent assistance, market and product development, and evaluation of lubricants.

★340★ William B. Hardy
260 Metape S.
Bound Brook, NJ 08805 (201) 356-7093
William B. Hardy, Principal

Founded: 1981. **Description:** Technical consultant specializing in process research, aromatic intermediates, and solutions to related patent problems. Serves as expert witness. Advises clients on all aspects (research and development, product introduction, market development) of plastics additives (light stabilizers, antioxidants, flame retardants, etc.) for polyolefins, PVC and other polymers. Also provides advice on synthesis and use of compounds absorbing near IR radiation.

★341★ Wolff Consultants
One Buck Rd.
P.O. Box 1003
Hanover, NH 03755 (603) 643-6015
Nikolaus E. Wolff, Principal

Founded: 1976. **Description:** Technical consultant to management working to further both national and international business development in the high technology sector. Provides advisory, informational, tutorial, and research services to assist management in the decision process for business planning. Specific activities include: technology assessment involving electronics, materials science, chemistry, and physics; technology information services, patent support, R & D management services such as planning, risk assessment, and staffing; and overhead value analysis, general management problems; bilingual (German/English) technical/business negotiations. Clients in North America and Europe.

★342★ Writers Publishing Service Co.
1512 Western Ave.
Seattle, WA 98101 (206) 284-9954
William R. Griffin, Principal

Founded: 1976. **Description:** Offers manuscript evaluation and consultation services, including the following: gives authors an overview of the publishing process from manuscript to sold book, then evaluates the project from editorial, production and marketing points of view, letting the author know what is involved and required. Also covers copyrights, Library of Congress registration and ISBN-Books-in-Print listings. Provides rough estimates of the costs involved for each step of the process as well as how various decisions may affect the marketability of the project.

★343★ Ychem International Corporation
9108 Goldamber Garth
Columbia, MD 21045 (301) 596-6339
Niteen A. Vaidya, President

Founded: 1986. **Description:** Research and development activities emphasize technology transfer, custom synthesis, speciality chemicals, idea generating literature/patent search, and patent evaluation. Serves chemical, energy, and pharmaceutical industries.

CHAPTER 4

UNIVERSITY INNOVATION RESEARCH CENTERS

Many universities have formal programs specifically designed to assist inventors with ideas or patents. Nearly every state has at least one college or university that can provide research and development facilities to technically oriented companies or individuals.

State governments are heavily involved with university research through cooperative programs and support of research parks and technology transfer programs. Known at some universities as "Advanced Technology Centers" or "Centers of Excellence," these programs are designed to increase cooperation between academic institutions and state-based industries. They assist in the creation of new firms through the development of new technology, attracting new business to the state, and increasing competition.

Specifically, technology transfer programs facilitate the transmission of new technologies from the laboratory to the private sector. Such technologies can become the impetus for the creation of new businesses, new products, or revitalization of industries.

Research parks are planned groupings of technology companies, often near universities, that encourage university/private partnerships. They draw industry to a particular location and provide incubator services.

Other kinds of services that are available at many university research centers include innovation evaluation; invention testing; counseling on the start-up of new high technology businesses; assistance in identifying potential research funding from the federal government under programs such as the Small Business Innovation Research (SBIR) Grants and the Energy Related Inventions Program (ERIP); counseling regarding patents, trademarks, and copyrights; and assistance in identifying sources of government contracts for goods or services needed. Many centers also have excellent libraries of technical information and computer literature searching capabilities.

Whenever I require specialized engineering skills on a project, especially to build complicated prototypes, one of the first places I check is the talent pool available at the local university research center. The costs are reasonable and there is no better place to get fresh, young minds working on a resolution to a particular problem.

Universities do have specialities, so it is best to check in advance before going out to the campus. For example, some universities concentrate on biomedical and life sciences research, others handle aerospace engineering, electronic systems, mechanics, physical sciences, fabrication technologies, and so forth.

Idea evaluation is the first major step after a concrete, detailed idea has been developed. This is a critical phase since every following phase requires the investment of more time and money. University research centers are an excellent place to have an invention evaluated to determine its overall technical and commercial feasibility.

The university research centers listed below may be useful in these evaluation activities. They have people to help determine whether the new invention is a marked improvement over its competition; whether it is likely to be commercially viable; what the probable demand for it will be; who will produce it; and how it will be distributed. These specialists can help an inventor arrive at the decision to go ahead to the commercialization stage, to redesign the invention, or to kill the project altogether.

The failure rate of technological innovators is estimated by one State of Michigan study to be one quarter that of all entrepreneurs during the first five years of business. One way to decrease the chances of failure is to ensure that all the resources are available when required. A good university research center program can help towards this end.

Listed below are principal university research institutes, technology transfer centers, and research parks whose purpose is to promote innovation, invention, and product development. For additional information on the related services, see Chapter 6, Business Incubators.

Chapter Arrangement

This chapter is arranged alphabetically by the research center name. If the research center is an integral part of a university, it will occur under the university name. If additional centers are listed under a university, they will be displayed directly below, set off by indentation. Both the research center name and the university name are listed in the Master Index. The Master Index may also list additional research centers mentioned in other chapters.

★344★ Advanced Science & Technology Institute

University of Oregon
319 Hendricks Hall
Eugene, OR 97403 (503) 686-3189
Dr. Robert S. McQuate, Executive Director

Description: Bridges university research activities and resources with the corporate community. Encourages cooperative research projects with industry through its Industrial Associates Program Serves as a broker for patenting and licensing arrangements between industry and Oregon State University and University of Oregon. **Publications:** Newsletter; also a research directory of Oregon State University and University of Oregon researchers. Sponsors technical seminars, workshops, and conferences.

★345★ Alaska Small Business Development Center

Anchorage Community College
430 W. Seventh Ave., Suite 115
Anchorage, AK 99501 (907) 274-7232
Janet Nye, Director

Telecommunication Access: FAX (907) 271-4545.

★346★ Ann Arbor Technology Park

Richard Wood & Company, Inc.
230 Huron View Boulevard
Ann Arbor, MI 48103 (313) 769-6030
David R. Tyler, Vice President

Description: 820-acre site providing tenants access to research, educational, and computing resources at the University.

★347★ Applied Information Technologies Research Center

1880 Mackenzie Drive
Suite 111
Columbus, OH 43220 (614) 442-1955
George Minot, President

Description: Information environment, information representation, and user interfaces and search and retrieval technologies. Performs contract reseach and development for member and nonmember companies and assists entrepreneurial ventures and start-up companies in the area of information product and/or service development.

★348★ Argonne National Laboratory Technology Transfer Center

9700 South Cass Avenue
Argonne, IL 60439 (312) 972-4929
Dr. Brian Frost, Director

Description: Facilitates the exchange of Argonne research resources and inventions with industry on the state and national level and develops industrywide partnerships for the Laboratory. Collaborates with Argonne/University of Chicago Development Corporation to oversee patenting and licensing activities.

★349★ Arizona Small Business Development Center

108 North 40th St.
Phoenix, AZ 85343 (602) 392-5224
William Dewey, Acting Director

★350★ Arizona State University Engineering Excellence Program

Center of Research
College of Engineering and Applied Science
Tempe, AZ 85287 (602) 965-1725
Charles Bachus, Director

Description: Partnership of local industries, universities, and state government striving to promote economic development by attracting new businesses to the state. Provides general advice and lobbying in the state legislature and promotes economic development.

★351★ Arizona State University Research Park

2049 East ASU Circle
Arizona State University
Tempe, AZ 85284 (602) 752-1000
Michael S. Ammann, Executive Director

Description: Seeks to link technological results of university and individual research with private industry. The Park provides tenants with leased land for construction of research and development facilities, laboratories, offices, pilot plants, facilities for production or assembly of prototype products, and University and government research facilities.

★352★ Arkansas Small Business Development Center

University of Arkansas at Little Rock
Research and Public Service Library, Fifth Floor
33rd St. and University Ave.
Little Rock, AR 72204 (501) 371-5381

★353★ Association of University Related Research Parks

4500 South Lakeshore Drive
Suite 336
Tempe, AZ 85282-7055 (602) 752-2002
Chris Boettcher, Executive Director

Description: Serves as forum for the exchange of information on planning, construction, marketing, and managing of university-related research parks, particularly information on university-industry relations, innovation, and technology transfer to the private sector. Monitors legislative and regulatory actions affecting the development and operation of research parks. Acts as a clearinghouse for career opportunities. **Publications:** The Research Park Forum (quarterly newsletter).

★354★ Auburn University
Auburn Technical Assistance Center
111 Drake Center
Auburn University, AL 36849-5350 (205) 826-4659
Henry Burdg, Contact

Description: Provides management assistance and applied research to private sector organizations.

★355★ Babson College
Board of Research
Babson Park, MA 02157 (617) 235-1200
Prof. Edward Handler, Director

Description: Business administration of a professional character related to management and nonmanagement subjects taught in the College, including experimentation and studies on identification of personal characteristics of entrepreneurs, influence of technical executives in decision making, multi-product inventory analysis, development of Know Nothing Party in Massachusetts, and curriculum development. Includes design of a basic accounting computerized game, development of a freshman seminar, and support for design of a major in society and technology at the College.

★356★ Ball State University
Center for Entrepreneurial Resources
Carmichael Hall, 2nd Floor
Muncie, IN 47306 (317) 285-1588
Dr. B.J. Bischoff Whittaker, Acting Director

Description: Entrepreneurship and development of existing businesses, including corporate training, executive development, intrapreneurship, adult literacy, technology transfer, employee honesty testing, strategic planning, human resource development, needs assessment, creativity, program evaluation, consultant selection, computer software, new product development, and sales training. **Publications:** Practical Guides for Professions (annually).

★357★ SBIR Proposal Assistance
Office of Research
1825 Riverside Avenue
Muncie, IN 47306 (317) 285-1600

Description: Offers assistance on submissions to the Federal Small Business Innovation Research (SBIR) program.

★358★ Battelle Memorial Institute
505 King Avenue
Columbus, OH 43201 (614) 424-6424
Dr. Douglas E. Olesen, President and Chief Executive

Description: Advanced materials, biological and chemical sciences, electronic and defense systems, and engineering and manufacturing technology and information systems. Activities include research and development, commercialization of technology, and management services. Conducts marine research at three coastal locations: Florida Marine Research Facility (Daytona Beach), Northwest Marine Research Laboratory (Sequim, Washington), and the Ocean Sciences and Technology Laboratory (Duxbury, Massachusetts). Performs additional activities at the following facilities: Center for High-Speed Commercial Flight, Aviation Safety Reporting System Project Office, Advanced Materials Center for the Commercial Development of Space, Copper Data Center, Metals and Ceramics and Information Center, Magnesium Research Center, National Center for Biomedical Infrared Spectroscopy, Health and Population Center, Human Factors and Organizational Effectiveness Research Center, Technology and Society Research Center, Office of Low-Level Waste Technology, Office of Waste Technology Development, Office of Transportation Systems and Planning, Battelle Pressure Vessel and Piping Test Facility, Battelle Center for Materials Fabrication, Battelle Industrial Technology Center, Battelle Toxicology and Bioscience Center, Battelle Applied Economics Center. **Publications:** Battelle Today (four times per year); Annual Report; Published Papers and Articles (annually).

★359★ Baylor University
Center for Private Enterprise
Hankamer School of Business
Waco, TX 76798 (817) 755-3766
Dr. Calvin A. Kent, Director

Description: Entrepreneurship and private enterprise, including studies on characteristics of entrepreneurs, student and teacher attitudes toward the private enterprise system, business taxation, and women entrepreneurs.

★360★ Biomedical Research Zone
740 Court Street
Memphis, TN 38105 (901) 526-1165
Daniel S. Beasley, Ph.D., Acting Director

Description: Serves as a development center for companies involved with the health care industry and as an international headquarters for pharmaceutical, biomedical, dental, and veterinary firms and professional health care societies. **Publications:** Newsletter (quarterly).

★361★ Boise State University
Idaho Business and Economic Development Center
1901 University Drive
Boise, ID 83725 (208) 385-1511
Ronald R. Hall, Director

Description: Based in the College of Business. Manages the outreach functions of the College. Research Unit provides high technology support and research capabilities to private companies and functions as local affiliate to the National Aeronautics and Space Administration Industrial Applications Center (Boise), which promotes technology transfer. Operates the Idaho Small Business Development Center and Idaho Economic Development Center, which provide business assistance.

★362★ Bradley University
Technology Commercialization Center
Lovelace Technology Center
Peoria, IL 61625 (309) 677-2263
Dr. William M. Hammond, Director

Description: Multidisciplinary organization assisting in the development and commercialization of new products and in the transfer of new technologies and manufacturing processes. Activities include product and material testing, prototype development, microelectronic design layout, very-large-scale-intergrated circuit design, medical instrumentation, accelerated corrosive testing, printed circuit board layout, sports medicine instrumentation, optical electronics investigations, and microprocessory control system design.

★363★ British Columbia Research Corporation
3650 Wesbrook Mall
Vancouver, BC, Canada V6S 2L2 (604) 224-4331
Dr. T.E. Howard, President

Description: Conducts cost-accountable research, development, and other technical work, including industrial development, operations management, educational planning, and industrial assistance through technology transfer. Research services to industry include product development, process development, and systems and operational analysis. **Publications:** Annual Report.

★364★ California University of Pennsylvania Mon Valley Renaissance
Box 62
California, PA 15419 (412) 938-5938
Richard H. Webb, Executive Director

Description: Multiprogram consortium with a focus on applied research and economic development services to business and industry.

★365★ Canada Centre for Mineral and Energy Technology, Office of Technology Transfer
555 Booth Street
Ottawa, ON, Canada K1A 0G1 (613) 995-4267
J. Kuryllowicz, Director

Description: Addresses technology transfer issues. Develops guidelines to aid at all stages of research and development, including planning, bench-scale, pilot plant, and demonstration phases, obtaining financial assistance, and commercial applications. Also assists with patenting procedures and handles intellectual property, including licensing and legal matters arising from research and development contracts.

★366★ Canadian Industrial Innovation Centre/ Waterloo
156 Columbia Street West
Waterloo, ON, Canada N2L 3L3 (519) 885-5870
Gordon F. Cummer, Chief Operating Officer

Description: Processes involved in invention, innovation, and entrepreneurship. Projects include an evaluation and implementation model for new product innovations, development of research and development strategies in high technology companies, innovation in small and medium sized firms (lessons from Japan), and the role of University of Waterloo as a technology growth pole in a regional development strategy. **Publications:** Innovation Showcase (quarterly); The Canadian Inventors Newsletter (quarterly); Eureka!.

★367★ Carnegie Mellon University Biotechnology Center
4400 Fifth Avenue
Mellon Institute
Pittsburgh, PA 15213-2683 (412) 268-3188
Dr. Edwin Minkley, Director

Description: Transfers molecular biology technology into viable commercial processes. Activities emphasize large-sale production of recombinantly engineeered proteins, from cost-efficient expression systems through in-house commercial production.

★368★ Center for Education and Research with Industry
213 Northcott Hall
Huntington, WV 25701 (304) 696-3367
William A. Edwards, Director

Description: Facilitates joint ventures between the academia, business, and government by linking Marshall University campus resources and faculty expertise with technology transfer activities around the state. Administers joint research programs between academia and industry.

★369★ Center for Entrepreneurial Development
120 South Whitfield Street
Pittsburgh, PA 15206 (412) 621-0700
Prof. Dwight M. Baumann, Executive Director

Description: Entrepreneurial activities, including small business and industrial projects as a vehicle for research and experimentation in teaching, advancement of technology and management sources, innovation and entrepreneurship, and transfer of university-based technology in these fields. Aids potential businessmen in overcoming barriers in moving projects from their conception to realization within the business community, assists professionals in starting their own businesses, and provides technical advice and introductions to lending institutions, potential clients, and suppliers.

★370★ Center for Entrepreneurial Studies and Development, Inc.
West Virginia University
College of Engineering
P.O. Box 6101
Morgantown, WV 26506 (304) 293-4607
Dr. Jack Byrd, Executive Director

Description: Business operations improvement, employee relations, management, and systems. Operations improvement studies focus on materials handling systems, cost reduction, work standards development, training manual development, facilities utilization and planning, work methods, inventory control systems, production automation, computer applications, quality control, and facility relocation models. Employee relations studies focus on job incentive systems, wage payments, job enrichment, employee motivation programs, job evaluation, labor/ management relations, and employee staffing. Management studies focus on development programs, organization development, management incentives, succession planning, policy manual development, and small business organizations. Systems studies focus on financial planning, computer systems, sales forecasting, strategic planning, office systems improvements, competition, insurance policies, and business plan development.

★371★ Center for Innovation & Business Development
Box 8103
University Station
Grand Forks, ND 58202 (701) 777-3132
Bruce Gjovig, Director

Description: Works with applied research personnel at the University of North Dakota and other institutions to facilitate the commercialization of new technologies by providing research support services to entrepreneurs, inventors, and small manufacturers in the areas of invention evaluation, technology transfer, SBIR applications, technical development, licensing, and business development, including market feasibility studies and marketing and business plans. Acts as the University of North Dakota National Aeronautics and Space Administration Industrial Applications Center. **Publications:** Annual Report; Entrepreneur Kit; Business Plan for Start-ups; Marketing Plan for Start-ups.

★372★ Center for Innovative Technology
Hallmark Building
Suite 201
13873 Park Center Road
Herndon, VA 22071 (703) 689-3000
Dr. Edward M. Davis, President

Description: Facilitates the transfer of technology for commonwealth universities to industry in biotechnology, computer-aided engineering, information technology, and materials science engineering.

★373★ Center for the New West
Suite 1700 South Tower
600 Seventeenth Street
Denver, CO 80202 (303) 623-9298
Philip M. Burgess, President

Description: Seeks to improve the quality and usefulness of information on the new economy, to promote the economic growth and diversification of the western United States, and to improve the competitiveness of western enterprise in a changing global economy. Focuses on national and global trends and their impact on the economic vitality of the western regions. Emphasizes enterprise development, including the formation, retention, expansion, and revitalization of traditional industries, especially agriculture, mining, and forestry; capital formation, including the formation, availability, and cooperative use of public and private capital; technology and innovation development, diffusion, and use; expanded international commerce; human capital, including demographic trends and their impact on economic growth and productivity; and area and regional development, including infrastructure requirements and innovative institutional arrangements. **Publications:** Profile of the West (annual compendium of 200 demographic, economic, and social indicators on the western states).

★374★ Central Florida Research Park
12424 Research Parkway
Suite 100
Orlando, FL 32826 (407) 282-3944
Joe Wallace, Director of Marketing

Description: 1,027-acre site zoned for commercial and light manufacturing adjacent to the University of Central Florida. Established to create an environment which promotes and fosters relationships between industry and the University.

★375★ Centre for Advanced Technology
P.O. Box 483
601 South Howes Street
Fort Collins, CO 80522 (303) 482-2916
Robert B. Hutchinson, III, Contact

Description: 235-acre site that facilitates the exchange of Colorado State University research resources with Park tenants.

★376★ Chicago Technology Park
2201 West Campbell Park Drive
Chicago, IL 60612 (312) 829-7252
Nina M. Klarich, President

Description: Seeks to coordinate industry, university, and government partnerships to stimulate the formation of science-based companies and economic development in the Chicago area. Provides access to university and hospital resources, offers assistance in the creation of new venture companies, and provides space in an incubator building.

★377★ Clemson University
Emerging Technology Development and Marketing Center
338 University Square
P.O. Box 5703
Clemson, SC 29634-5703 (803) 656-4237

Description: Seeks to enhance economic development in the state. Provides technical and marketing assistance to new technology-oriented product manufacturing business firms; seeks to stimulate the transfer of technology and new business concepts to existing companies; enters research and development partnerships with existing businesses that intend to manufacture in the state; and provides assistance to University faculty, staff, and students with products, patents, and ideas that may be commercialized and manufactured.

★378★ College of DuPage
Technology Commercialization Center
Business and Professional Institute
22nd and Lambert Road
Glen Ellyn, IL 60137-6599 (312) 858-6870
Nancy Pfahl, Manager

Description: Links high technology businesses to university and other resources to assist in the production and commercialization of new ideas and products and to enhance the transfer of technologies from University laboratories into the marketplace.

★379★ College of St. Thomas
Entrepreneurial Enterprise Center
1107 Hazeltine Boulevard
Chaska, MN 55318 (612) 448-8800

Description: Incubator facility.

★380★ Colorado School of Mines
Table Mountain Research Center
5930 McIntyre Street
Golden, CO 80403 (303) 279-2581
Gary E. Butts, Director

Description: Provides laboratory, pilot-plant, and office space to engineers, scientists, and researchers in the fields of geology, mining, and ore processing for scientific and technological discoveries, processes, and inventions.

★381★ Colorado Small Business Development Center
1391 N. Spear Blvd.
Suite 600
Denver, CO 80204 (303) 620-2422
Richard Wilson, Director

★382★ Columbia University
Center for Law and Economic Studies
435 West 116 Street, Box E-2
School of Law
New York, NY 10027 (212) 854-3739
Prof. Jeffrey Gordon, Codirector

Description: Fundamental economic and legal problems of the modern industrial society, including studies on takeovers and the market for corporate control, new directions in law and economics, international taxation, competition in international business, the impact of the modern corporation, and the relationship between administrative law and political economy. Also studies contracts, torts, nuisance, takeovers and the market for corporate control, regulation, antitrust, intellectual property, and the economics of the legal profession.

★383★ Center for Studies in Innovation and Entrepreneurship
315 Uris Hall
New York, NY 10027 (212) 280-2830
Michael Tushman, Director

Description: Designs for innovation, entrepreneurship, and technology and organizations.

Columbia University *(Cont.)*

★384★ Office of Science and Technology Development
411 Low Memorial Library
New York, NY 10027 (212) 280-8444
Jack M. Granowitc, Director

Description: University technology transfer office specializing in biotechnology.

★385★ Competitive Enterprise Institute
611 Pennsylvania Avenue S.E.
Washington, DC 20003 (202) 547-1010
Fred L. Smith, Jr., President

Description: Domestic economic policy issues, including tax reform, deregulation of industry, deficit reduction, privatization, antitrust, free trade, free-market environmentalism, intellectual property, transportation deregulation, and risk and insurance. **Publications:** Washington Antitrust Report (quarterly newsletter); CEI Update (monthly newsletter on Institute events).

★386★ Connecticut Small Business Development Center
University of Connecticut
School of Business Administration
368 Fairfield Rd., Room 442
Box U-41
Storrs, CT 06226 (203) 486-4135
John O'Connor, Director

★387★ Cornell Research Foundation Office of Patents and Licensing
East Hill Plaza
Ithaca, NY 14850 (607) 255-7367
Walter Haeussler, Director

Description: University-affiliated technology transfer office specializing in agriculture and engineering.

★388★ Cornell University Cornell Business and Technology Park
102 Langmuir Lab
Ithaca, NY 14853 (607) 255-5341
Richard E. deVito, Marketing Manager

Description: The Park serves as a conduit between Cornell University and business, especially in electronics and biotechnology, and is a home for the independent development of technologies resulting from efforts by Cornell researchers. Leases space for business incubator activities and venture start-up companies.

★389★ Corporation for Enterprise Development
1725 K Street, N.W.
Suite 1401
Washington, DC 20006 (202) 293-7963
Robert E. Friedman, President

Description: Research, development, and dissemination of entrepreneurial policy initiatives at the local, state, and federal levels. Specific studies include economic climate of the states, transfer payment investment, pension funds, job creation, seed capital assessment, international exchange, flexible manufacturing networks, and state capital market analysis. Emphasizes economic empowerment of economically disadvataged populations. **Publications:** The Entrepreneurial Economy (monthly); Making the Grade: The Development Report Card for the States; State Enterprise Development Implementation Packets (paper series); State Strategy Memoranda (paper series); Investing In ... (technical report series); Business Climate Tax Index.

★390★ Cummings Research Park
Chamber of Commerce
P.O. Box 408
Huntsville, AL 35804-0408 (205) 535-2008
Jim Reichardt, Vice President

Description: 3,600-acre site linking University research and educational resources with Park tenants, particularly in the areas of aerospace, missile research and development, aplied optics, artificial intelligence, software development, and data communications.

★391★ Dandini Research Park
DRI Research Foundation
P.O. Box 60220
Reno, NV 89506 (702) 673-7315
Dale F. Schulke, Contact

Description: 470-acre site that links the research and development activities of Park tenants with the Desert Research Institute's technological equipment, personnel, laboratories, and training programs. Instrument design, environmental testing, and research and development services may be done in cooperation with DRI staff.

★392★ Delaware Small Business Development Center
University of Delaware
Purnell Hall, Suite 005
Newark, DE 19716 (302) 451-2747
Helene Butler, Director

★393★ Discovery Foundation
220-3700 Gilmore Way
Burnaby, BC, Canada V5G 4M1 (604) 430-3533
F.C. Hodges, General Manager

Description: Spearheads joint efforts by industry, government, and higher education in the province. Activities are carried out through: Discovery Parks Incorporated, which manages the industrial research parks adjacent to University of British Columbia, Simon Fraser University, British Columbia Institute of Technology, and University of Victoria; Discovery Enterprises Inc., which provides seed money to high-technology enterprises, products, and processes through the commercial stage of development; and Discovery Innovation Office, which offers counseling and referral services to innovators. **Publications:** BC-R&D (bimonthly); Hi-TECH (monthly); BC Technology Directory.

★394★ Discovery Parks Incorporated
220-3700 Gilmore Way
Burnaby, BC, Canada V5G 4M1 (604) 430-3533
F. Hodges, Vice President

Description: Manages 55-acre park at University of British Columbia, 75-acre park at Simon Fraser University, 80-acre park at British Columbia Institute of Technology, and 10-acre park at University of Victoria that link university research resources with technological and research companies in the parks. Offers office and laboratory space to encourage exchange between University and tenant researchers. Serves as a network of investment and business contacts for new companies. Provides serviced land for construction of corporate owned facilities.

★395★ East Carolina University Center for Applied Technology
Greenville, NC 27834 (919) 757-6708

Description: Provides business planning, information, management, and technical assitance. Acts as the coordinating agency between the business community and the University.

★396★ Eastern Michigan University
Center for Entrepreneurship
121 Pearl Street
Ypsilanti, MI 48197 (313) 487-0225
Dr. Patricia B. Weber, Director

Description: Applied research toward the development of entrepreneurship and growth management. Focuses on the vital transistion from start-up to sustained long-term growth and stability. A major project involves following the progress of 150 companies during their first year. **Publications:** Working Paper Series.

★397★ Edmonton Research Park
203 Advanced Technology Centre
9650-20 Avenue
Edmonton, AB, Canada T6N 1G1 (403) 462-2121
Glenn A. Mitchell, General Manager

Description: Serves as a link between University of Alberta research resources and tenants involved in basic, applied, and developmental research, product development, light production, advanced technology activities, and related support services.

★398★ First Coast Technology Park
4567 St. John's Bluff Road South
Jacksonville, FL 32216 (904) 646-2500
William A. Ingram, Executive Director

Description: Facilitates cooperative research and development activities between Park tenants and the University of North Florida community.

★399★ Florida Atlantic Research Park
Office of Academic Affairs
P.O. Box 3091
Boca Raton, FL 33431-0991 (407) 393-3066
Dr. J.S. Tennant, Associate Vice President

Description: Serves as a bridge between the research interests of the tenant companies and research activities of the University of South Florida community. The Park features an innovation center to aid in transferring technology and houses University functions such as the Small Business Development Center and National Aeronautics and Space Administration/Southern Technology Application Center.

★400★ Florida Small Business Development Center
University of West Florida
College of Business
Building 38
Pensacola, FL 32514-5750 (904) 474-3016
John Jefferies, Acting Director

Telecommunication Access: FAX (904) 474-2030.

★401★ Florida State University
Florida Economic Development Center
335 College of Business
Tallahassee, FL 32306-1007 (904) 644-1044
Roy Thompson, Director

Description: Business planning, venture capital, insurance, purchasing, workman's compensation, community development, and management. Conducts target industry studies, area business analysis studies, and downtown business surveys. **Publications:** Special Reports/Studies; Small Business News and Views (weekly column); Florida Venture Capital Handbook; Special Sources of Credit; Business Planning Guide; Florida Entrepreneurial Network (quarterly leaflets announcing all small business workshops in Florida).

★402★ Florida Entrepreneurial Network
Florida Economic Development Center
College of Business
Room 335
Tallahassee, FL 32306-1007
Roy Thompson, Contact

Description: Provides information sharing, networking, and business development. **Publications:** North Florida Entrepreneurial Network and South Florida Entrepreneurial Network.

★403★ Fox Valley Technical College
Technical Research Innovation Park
1825 North Bluemound Drive
P.O. Box 2277
Appleton, WI 54913-2277 (414) 735-5600
Stanley Spanbauer, President

Description: Seeks to encourage the growth of business and industry in the region through the organization and establishment of regional industry and association headquarters. Encourages marketing ventures, supports inventors and entrepreneurs, fosters research and development for the paper industries, and promotes new technology industries as tenants in the Park. Plans include establishing the D.J. Burdini Technological Innovation Center of the College, which will include a technical library, economic development center, communications network area, product development service center, flexography laboratory, high technology demonstration laboratories, and facilities for conferences and classes.

★404★ George Mason University
Center for the Productive Use of Technology
3401 North Fairfax Drive
Suite 322
Arlington, VA 22201-4498 (703) 841-2675
David S. Bushnell, Director

Description: Technology transfer, work group dynamics and motivation, production measurement, human factors and sociotechnical systems design, and management development. Assists public and private sector organizations in planning, implementing, and evaluating alternative strategies for productivity improvement. **Publications:** Monographs.

★405★ George Mason Entrepreneur Center
4400 University Drive
Fairfax, VA 22030 (703) 323-2568

Description: Assists businesses through financial contacts and provides access to the Commonwealth Technology Information Service, a catalog of the state's technology and research resources.

★406★ Institute for Advanced Study in the Integrative Sciences
Thompson Hall, Room 219
4400 University Drive
Fairfax, VA 22030 (703) 425-3998
Dr. John N. Warfield, Director

Description: Acts to stimulate on-campus research activity by encouraging cooperation between the University and high-technology industry. Principal investigators specialize in systems research, communications research, systems design, and computer science.

★407★ George Washington University
Center for International Science and Technology Policy
Gelman Library, Suite 714
2130 H Street, Northwest
Washington, DC 20052 (202) 676-7292
Robert W. Rycroft, Director

Description: Interdisciplinary research and policy analysis. Program includes such disciplines as public administration, physics, political science, sociology, economics, international affairs, urban planning, and environmental resources for application to science and technology policy, international science policy, technology transfer, research and development policy, risk analysis and management, regulatory process, institutional analysis, public perception assessment, telecommunications policy, space policy, environmental quality, disaster recovery, and emerging technologies.

★408★ Georgia Institute of Technology
Advanced Technology Development Center
430 Tenth Street, N.W.
Suite N-116
Atlanta, GA 30318 (404) 894-3575
Dr. Richard Meyer, Director

Description: Serves as a conduit to the University system's research programs, faculty, and facilities for developing technology businesses, particularly in the areas of aerospace vehicles and equipment, biotechnology products, telecommunications equipment, computers and peripheral devices, computer software, electronic equipment, medical devices, instrumentation and test equipment, pharmaceuticals, new materials, and robotics.

★409★ Economic Development Laboratory
Georgia Tech Research Institute
Atlanta, GA 30332 (404) 894-3841
Dr. David S. Clifton, Jr., Director

Description: Agricultural technology, industrial energy conservation, hazardous waste management, safety engineering, industrial hygiene, asbestos abatement, market planning and research, target industry analysis, cost-benefit analysis, energy modelling, technology transfer, industrial training, productivity improvement, small business assistance, analytical chemistry, and indoor air pollution. Supports 12 regional offices throughout Georgia to assist local industries. **Publications:** Engineering Reviews (occasionally); Technical Briefs (occasionally). Also publishes five quarterly newsletters: The Industrial Advisor; Industrial Energy Conserver; Environmental Spectrum; Poultry and Egg Computing; and Poultry Engineering Progress.

★410★ Georgia Productivity Center
Atlanta, GA 30332 (404) 894-6101

Description: Provides assistance, education, and applied research in productivity in support of its technology transfer programs.

★411★ Patent Assistance Program
Department of Microfilms
Atlanta, GA 30332-0999 (404) 894-4508
Jean Kirkland, Head

★412★ Technology Policy and Assessment Center
Office of Interdisciplinary Programs
Atlanta, GA 30332 (404) 894-2375
Dr. Frederick A. Rossini, Director

Description: Policy and societal aspects of science and technology, both domestic and international. Studies include technology and impact assessment, technological innovation and diffusion of innovations, cost benefit analysis, socioeconomic development, research and development policy and management, energy and environmental policy, and the interdisciplinary research process.

★413★ Georgia Small Business Development Center
University of Georgia
Chicopee Complex
1180 E. Broad St.
Athens, GA 30602 (404) 542-5760
James McGovern, Director

★414★ Georgia State University
International Center for Entrepreneurship
College of Business Administration
University Plaza
Atlanta, GA 30303 (404) 651-3782
Dr. Francis W. Rushing, Director

Description: Conducts and facilitates studies in entrepreneurship. Serves as a resource center.

★415★ Gulf South Research Institute, Technology Management Division
P.O. Box 14787
Baton Rouge, LA 70898 (504) 766-3300
James H. Clinton, President

Description: Performs technology documentation, financial planning, technology protection coordination, preliminary commercial assessment, technology development, market analysis and planning, preproduction development, and technology transfer.

★416★ Harvard Medical School
Office for Technology, Licensing, and Industry-Sponsored Research
221 Longwood Avenue, Room 202
Boston, MA 02115 (617) 732-0920

Description: University technology transfer office.

★417★ Harvard University
Office for Patents, Copyrights, and Licensing
1350 Massachusetts Avenue
Holyoke Center, Room 499
Cambridge, MA 02138 (617) 495-3067
Joyce Brinton, Director

Description: University technology transfer office specializing in applied sciences, biotechnology, chemistry, material science, software, courseware, biomedical technology, and research products.

★418★ Hawaii Ocean Science and Technology Park (HOST)
220 South King Street, Suite 840
Honolulu, HI 96813 (808) 548-8996
William M. Bass, Executive Director

Description: Facilitates cooperative research and development activities between Park tenants and University of Hawaii community, particularly in the areas of marine microbiology, oceanography, and alternative energy production and other forms of ocean-related high technology. **Publications:** Hawaii High Tech Journal (quarterly).

★419★ Idaho Small Business Development Center
Boise State University
College of Business
1910 University Dr.
Boise, ID 83725 (208) 385-1640

★420★ Idaho State University Research Park
Box 8044
Pocatello, ID 83209-0009 (208) 236-2430
T. Les Purce, Director

Description: Facilitates the exchange of University research community with Park tenants, particularly in the areas of health professions, life sciences, pharmacy, engineering, electronics, and business.

★421★ Illinois Institute of Technology Manufacturing Productivity Center
IIT Research Institute
10 West 35th Street
Chicago, IL 60616 (312) 567-4800

Description: Provides information on new manufacturing techniques and maintains a list of manufacturing technology centers.

★422★ Technology Commercialization Center
Stuart School of Business Building
Room 229B
Chicago, IL 60616 (312) 567-5115
Stephen J. Fraenkel, Director

Description: Provides assistance and support to small businesses, entrepreneurs, and inventors through technology transfer, including technical and business feasibility studies.

★423★ Illinois Small Business Development Center
Illinois State Department of Commerce and Community Affairs
620 E. Adams St., Fifth Floor
Springfield, IL 62701 (217) 785-6267

★424★ Illinois State University Technology Commercialization Center
Room 215
Media Services Building
Normal, IL 61761 (309) 438-7127
Jerry W. Abner, Director

Description: General technology commercialization. Provides innovation evaluation service for new product and business proposals, including complete research and development assistance for those proposals which meet the economic development objectives of the I-TEC Illinois program. **Publications:** Annual Report; Quarterly Report (both distributed to I-TEC).

★425★ Indiana Institute of Technology McMillen Productivity and Design Center
1600 East Washington Boulevard
Fort Wayne, IN 46803 (219) 422-5561

Description: Offers technical assistance in industrial production hardware and software and on-campus leasing.

★426★ Indiana University Entrepreneur in Residence
School of Business, Room 640D
Bloomington, IN 47405 (812) 335-9200

Description: Offers business planning, management, and networking through contacts with venture capitalists, bankers, and investors.

★427★ Indiana University-Purdue University Indianapolis Commercial/Industrial Liaison
355 North Lansing
Indianapolis, IN 46202 (317) 264-8285

Description: Helps businesses seeking assistance in identifying faculty expertise. Also helps commercial firms develop university innovation.

★428★ Economic Development Administration University Center
611 North Capitol Avenue
Indianapolis, IN 46204 (317) 262-5083
Lucinda Pile, Contact

Description: Provides technical assistance and technology transfer services with affiliation to National Aeronautics and Space Administration/Indianapolis Center for Advanced Research.

★429★ Tech Net
EDA University Center at IUPUI
611 North Capitol Avenue
Indianapolis, IN 46204 (317) 262-5003

Description: State-wide referral service for scientific, engineering, and technology needs of business.

★430★ Industrial Technology Institute
P.O. Box 1485
2901 Hubbard Road
Ann Arbor, MI 48106 (313) 769-4000
George Kuper, President

Description: Design and application of advanced manufacturing technologies and processes and the human issues new technologies spawn. Specific areas include computer-aided design and engineering, computer-aided manufacturing (flexible machining, assembly, inspection, and materials handling), manufacturing information systems, artificial intelligence, computer communication networks, computer-based information and technology transfer products and services, and economic, organizational, and social impacts. **Publications:** Gateway: The MAP/TOP Reporter (bimonthly newsletter); Modern Michigan (quarterly manufacturing journal).

★431★ Innovation and Productivity Strategies Research Program

606 Hill Hall
360 Dr. Martin Luther King, Jr. Boulevard
Rutgers University
Newark, NJ 07102 (201) 648-5837
Robert Wharton, Codirector

Description: Innovation and productivity improvement strategies in U.S. and international industrial establishments. Studies include productivity strategies for invention and management of new product commercialization from an international perspective, strategic partnerships between large firms and small hi-tech firms, entrepreneurship strategies of small high-tech industries, intrapreneurship strategies for corporations, employee attitudes and behaviors, quality management techniques, and joint ventures and technology transfer strategies. Collaborates with universities in the U.S. and overseas to carry out international projects.

★432★ Innovation Park

1673 West Dirac Drive
Tallahassee, FL 32304 (904) 575-6381
1977 Mike Lea, Contact

Description: Fosters research partnerships between Florida A&M University, Florida State University, and industry tenants.

★433★ Innovation Place

15 Innovation Boulevard
Saskatoon, SK, Canada S7N 2X8 (306) 933-6258
Doug Tastad, Manager

Description: 120-acre research and development park providing office, industrial, and research space for lease to tenants interested in accessing research resources at the University of Saskatchewan.

★434★ Institute for New Enterprise Development

Box 360
Cambridge, MA 02238 (617) 491-0203
Dr. Stewart E. Perry, President

Description: Assesses specialized resources for community economic development; analyzes problems and potential problems of minorities, younger workers, women, and the elderly in relation to local economic revitalization; analyzes the differences in entrepreneurial patterns related to the ethnicity of a community; develops community based ventures, especially in the field of energy; and evaluates federal, state, and municipal policies for economic development.

★435★ Institute for Research on Public Policy

Faculte des sciences de l'administration
Laval University
Cite universitaire
Ste-Foy, PQ, Canada G1K 7P4 (418) 656-7960
Yvon Gasse, Ph.D., Director

Description: Canadian public policy, including small business policy, and entrepreneurship.

★436★ Institute for Technology Development

700 North State Street
Suite 500
Jackson, MS 39202 (601) 982-6545
Dr. David Murphree, President

Description: Identifies the state's technological needs in the areas of acoustics, advanced living systems, hazardous and industrial waste, microelectronics, and space remote sensing and coordinates the commercialization of research and development from the state's university, government and small business laboratories.

★437★ Institute of Advanced Manufacturing Sciences

1111 Edison Drive
Cincinnati, OH 45216 (513) 948-2000
Charles F. Carter, Jr., Executive Director

Description: Improving manufacturing quality, productivity, and innovation. Seeks to minimize variation in manufacturing operations; apply computer-aided design, automated process planning, artificial intelligence, nontraditional manufacturing processes, diagnostic aids, and employee training to increase effectiveness of manufacturing resources; and reduce manufacturing response time. **Publications:** Monographs; Research Reports.

★438★ Inventors Workshop International Education Foundation (Camarillo)

HQ, Inventor Center USA
3201 Corte Malpaso, #304-A
Camarillo, CA 93010 (805) 484-9786
Alan Arthur Tratner, President

Description: New product development and market research focusing on technology-oriented inventions. Also studies patent classifications. **Publications:** Invent! (magazine).

★439★ Iowa Small Business Development Center

Iowa State University
College of Business Administration
Chamberlynn Bldg.
137 Lynn Ave.
Ames, IA 50010 (515) 292-6351
Ronald A. Manning, Director

Telecommunication Access: FAX (515) 292-2009.

★440★ Iowa State University Center for Industrial Research and Service

205 Engineering Annex
Ames, IA 50011 (515) 294-3420
David H. Swanson, Director

Description: Problem areas of business, manufacturing, technology transfer, productivity, new product design, manufacturing processes, marketing, and related topics. Acts as a problem-handling facility and a clearinghouse for efforts to help Iowa's industry grow through studies highlighting not only production and management problems but also markets, marketing, and profit potential of possible new developments.

★441★ Iowa State University Research Park

125 Beardshear Hall
Ames, IA 50011 (515) 294-5121
Leonard C. Goldman, President

Description: 195-acre site on the University's South Campus facilitating interaction between corporate research laboratories and the University research community.

★442★ Jacksonville State University
Center for Economic Development & Business Research
College of Commerce and Business Administration
114 Merrill Hall
Jacksonville, AL 36265 (205) 231-5324
Pat W. Shaddix

Description: Conducts industrial strategy studies and provides management assistance to business community in northeast Alabama.

★443★ Johns Hopkins University
Applied Physics Laboratory
John Hopkins Road
Laurel, MD 20707 (301) 953-5000

Description: University technology transfer office specializing in biomedicine.

★444★ Bayview Research Campus
Dome Corporation
550 North Broadway
Baltimore, MD 21205 (301) 955-7724
David Hash, Director, Property Development

Description: Biomedical research park established to provide government and industry with space for offices and laboratory facilities.

★445★ Kansas Small Business Development Center
Wichita State University
College of Business Administration
021 Clinton Hall
Campus Box 48
Wichita, KS 67208 (316) 689-3193
Tom Hull, State Director

Telecommunication Access: FAX (316) 689-3770.

★446★ Kansas State University
Engineering Extension
Ward Hall 133
Manhattan, KS 66506 (913) 532-6026

Description: Promotes technology transfer in Kansas.

★447★ Kansas State University Foundation
1408 Denison
Manhattan, KS 66502 (913) 532-6266
Arthur Loub, President

Description: Provides financial assistance to research and educational activities of the University. Owns and operates the TechniPark, a research and development site.

★448★ Kentucky Small Business Development Center
University of Kentucky
18 Porter Bldg.
Lexington, KY 40506-0205 (606) 257-1751
Jerry Owen, Director

★449★ Knowledge Transfer Institute
1308 4th Street S.W.
Washington, DC 20024 (202) 554-9434
Dr. Ronald G. Havelock, Director

Description: Knowledge dissemination, transfer, and use, including scientific knowledge and technology transfer and use in education, medicine, and other areas. Conducts studies of networks involving schools and universities, the role of external innovations agents in dissemination and use of educational innovations, transfer of new cancer technologies into routine patient care, the effects of the mandating of legislation on use of new knowledge, and utilization of new technologies by managers.

★450★ Laval University
Administrative Sciences Research Laboratory
Cite Universitaire
Ste. Foy, PQ, Canada G1K 7P4 (418) 656-3675
Dr. Maurice Landry, Director

Description: Manpower planning, small business management, program evaluation, consumer behavior, management of technology, decision support systems, office automation, management of innovation, and insurance and risk management. Funds research projects, stimulates research, forms research groups within the faculty, and assists in editing texts of research papers.

★451★ Industrial Organization Research Group
Faculte des sciences de l'administration
Cite universitaire
Ste-Foy, PQ, Canada G1K 7P4 (418) 656-7973
The-Hiep Nguyen, Ph.D., Contact

Description: Industrial organization, including theory and methodology, technological innovations, technology transfer, investment strategies, and economic policy.

★452★ Small Business and Entrepreneurship Research Group
Faculte des sciences de l'administration
Cite universitaire
Ste-Foy, PQ, Canada G1K 7P4 (418) 656-7960
Yvon Gasse, Ph.D., Contact

Description: Small and medium-sized businesses and entrepreneurship, including growth strategies and technological innovations to promote development.

★453★ Lehigh University
Center for Innovation Management Studies (CIMS)
Johnson Hall 36
Bethlehem, PA 18015 (215) 758-4819
Alden S. Bean, Director

Description: Management of technological innovation. Seeks to understand the reasons for the success or failure of technological innovation emphasizing the role of management in improving industrial innovation. Specific projects include comparison of management of external versus internal technological ventures, analysis of the response of the stock market to corporate financial decisions concerning research and development, examination of the productivity and creativity of research and development teams, success and failure of high-tech innovations, termination decisions in monitoring research and development projects, technological cycles and industrial innovation, leadership and productivity in innovation activities, cooperative ventures in large and small firms, boundary management in innovation groups, and creation of a database on technological entrepreneurship.

Lehigh University *(Cont.)*

★454★ Office of Research and Sponsored Programs
Bethlehem, PA 18015 (215) 758-3020
Dr. Richard B. Streeter, Director

Description: Administers and coordinates, as administrative agency of the University, sponsored and cooperative research supported by government agencies, industry, and technical associations, including studies in physical, natural, social, and engineering sciences, and the humanities. Assists faculty and students in unsponsored research and scholarly efforts.

★455★ Small Business Development Center
301 Broadway
Bethlehem, PA 18015 (215) 758-3980
Dr. John W. Bonge, Director

Description: Problems faced by small businesses, the impact of the general economy on the formation and operation of small business, and characteristics on entrepreneurs. **Publications:** Lehigh University Small Business Reporter (semiannually); Financing Guide for Northampton, Lehigh and Berks County; Market Planning Guide; Export Planning Guide; Lehigh Valley Business Support Services; Financing Your Business.

★456★ Lorain County Community College Advanced Technologies Center
1005 North Abbe Road
Elyria, OH 44035 (216) 366-6618
Eugene Voda, Director

Description: Serves as an application center in the transfer of technological information to the design, manufacture, and marketing of systems in robotics, flexible manufacturing, microelectronics, computer system maintenance and repair, computer-aided design, injection molding simulation, and computer numerical control products.

★457★ Louisiana State University Office of Technology Transfer
60 University Lakeshore Drive
Baton Rouge, LA 70803 (504) 388-6941
Ted Kohn, Director

Description: Identifies, protects, and transfers technology that originates from University research activities. Activities include patent work, finding state or national licensees, and starting new entrepreneurial projects. Projects include work in such areas as computer chips, enzymes, instrumentation, genetic engineering, drugs, and aquaculture. **Publications:** Newsletter (occasionally).

★458★ Maine Small Business Development Center
University of Southern Maine
246 Deering Ave.
Portland, ME 04102 (207) 780-4420
Robert H. Hird, Director

Telecommunication Access: FAX (207) 780-4417.

★459★ Maine Technology Park
Stillwater Ave. at I-95
Orono, ME 04469 (207) 942-6380

Description: Associated with the University of Maine. Offers technical consultants, computer and laboratory access, and financial assistance to businesses.

★460★ Management Advisory Institute
University of Alberta
222 Faculty of Business Building
Edmonton, AB, Canada T6G 2R6 (403) 432-2225
Dr. Charles A. Lee, Executive Director

Description: Technological innovation, new ventures and entrepreneurship, managing technological change, and student/business interface.

★461★ Marquette University Center for the Study of Entrepreneurship
College of Business Administration
Milwaukee, WI 53233 (414) 224-5578
William J. Gleeson, Director

Description: New business formation, including studies on nature of the entrepreneur, his background and training, opportunities for independent venture, social, economic, and legal climates conducive to new firm formation, business failures, and incidence of new firm formation by region, industry type, and growth of industry type.

★462★ Maryland Small Business Development Center
217 E. Redwood St., Room 1006
Baltimore, MD 21202
Elliott Rittenhouse, Director

★463★ Massachusetts Biotechnology Research Park
373 Plantation Street
Worcester, MA 01605 (508) 755-2230
Raymond L. Quinlan, Executive Director

Description: Research and development space designed for biotechnology. Fosters the exchange between the research communities of the affiliated institutions and park tenants.

★464★ Massachusetts Institute of Technology Innovation Center
Room W59-201
Cambridge, MA 02139 (617) 253-5180
Prof. David G. Jansson, Director

Description: Invention and entrepreneurship, including involvement in actual processes of development and commercialization of technological innovations. Attempts to formalize methodology of teaching invention developed at the Center so that this method can be shared with other academic institutions. Investigates methods by which the Institute and industry can cooperate to their mutual benefit. Provides facilities through its product development laboratory for participation of students at the Institute resulting in development of a heating, ventilating, and air conditioning control system, an electronic tennis game for use with home television, a racing bicycle frameset, a small molecule detector for biomedical use, a process for detecting impurities in precious metals, a three-dimensional computer display, and a noncontact ground metal surface measurement system.

★465★ MIT Research Program on the Management of Technology
50 Memorial Drive
E52-535
Cambridge, MA 02139 (617) 253-4934
Prof. Edward B. Roberts, Chairman

Description: Managerial research on technology-based innovation, with emphasis on industry and government. Studies focus on organizations in the United States, Europe, Asia, and Latin America.

Massachusetts Institute of Technology *(Cont.)*

★466★ Technology Licensing Office
28 Carleton Street, Room E32-300
Cambridge, MA 02139 (617) 253-6966
John Preston, Director

Description: University technology transfer office specializing in biotechnology, biomedical science, ceramics, chemistry, computers, electro-optics, integrated circuits, polymers, metallurgy, optics, semiconductors, sensors, signal processing, and software.

★467★ University Park
77 Massachusetts Avenue
Cambridge, MA 02139 (617) 253-5278
Walter Milne, Contact

Description: Fosters interaction between the Institute's research community and park tenants.

★468★ Massachusetts Small Business Development Center
University of Massachusetts
205 School of Management
Amherst, MA 01003 (413) 549-4930
John Ciccarelli, Director

★469★ Maui Research and Technology Park
Maui Economic Development Board, Inc.
P.O. Box 187
Kahului Maui, HI 97632 (808) 871-6802
Donald G. Malcolm, President

Description: 300-acre research park fostering research activities between the academic sector and Park tenants, particularly in the areas of optic systems, electronics design and assembly, information systems and telecommunications, biotechnology, and alternate energy. **Publications:** Newsletter; Proceedings of Symposium.

★470★ Medical College of Wisconsin
MCW Research Foundation
8701 Watertown Plank Road
Milwaukee, WI 53226 (414) 257-8219
D.H. Westermann, Executive Vice President

Description: Creates intellectual property from ideas and processes of College researchers for license of products and services to the marketplace primarily in the areas of biophysics, biomedical engineering, biochemistry, microbiology, pharmacology, medical instrumentation, medical imaging, magnetics, computer software design, and genetic engineering, and immunology.

★471★ Memorial University of Newfoundland
P.J. Gardiner Institute for Small Business Studies
Faculty of Business
St. John's, NF, Canada A1B 3X5 (709) 737-8855
Garfield Pynn, Director

Description: Small businesses and entrepreneurship in Newfoundland and Labrador. Conducts feasibility studies, break-even analyses, market analyses and market planning, and marketing research studies. Specific projects include an international comparison of the response of small business to offshore development, a survey of 2,800 small businesses in Western Newfoundland, and a survey of 1,250 business alumni and 800 engineering alumni on entrepreneurial activity.

★472★ MetroTech
333 Jay Street
Brooklyn, NY 11201 (718) 260-3665
Dr. Seymour Scher, President

Description: 10-block, 16-acre urban research park linking Polytechnic University research resources with information technology industries.

★473★ Miami Valley Research Institute
1850 Kettering Tower
Dayton, OH 45423 (513) 228-7987
John F. Torley, President

Description: Facilitates the transfer of basic and applied scientific technological research from Miami Valley Research Park tenants and member institutions to production, manufacture, and marketing of materials and services. Solicits public and private grants for research personnel and facilities of member institutions and recruits tenants for the Park.

★474★ Miami Valley Research Park (MVRI)
P.O. Box 20026
Dayton, OH 45420 (513) 224-5930
Peter H. Forster, Chairman

Description: Located in close proximity to area universities, Park tenants are offered availability to the research capabilities and expertise of MVRI institutions, especially in the areas of computer and information science, materials, biomedical and human factors engineering, biomedicine, environmental systems, earth resources and energy utilization, applied mathematics, and aeronautical, astronautical, and allied sciences.

★475★ Michigan Small Business Development Center
Wayne State University
2727 Second Ave.
Detroit, MI 48201 (313) 577-4848
Norman J. Schlafmann, Director

Telecommunication Access: FAX (313) 963-7606.

★476★ Michigan State University
Center for the Redevelopment of Industrialized States
403 Olds Hall
College of Social Science
East Lansing, MI 48824-1047 (517) 353-3255
Dr. Jack H. Knott, Director

Description: Economic and social changes affecting the state of Michigan and ways to diversify and improve the state's business climate. Addresses issues such as technological innovation and strategic human resources management, modeling municipal expenditure patterns in Michigan, determinants of state industrial policy expenditures, implementing new technologies, organizational training practices and the facilitation of technological change, models of plant and investment decisions, and implications of changes in world automobile production for local communities. **Publications:** Newsletter; Reprints.

★477★ Michigan Technological University
Bureau of Industrial Development
Houghton, MI 49931 (906) 487-2470
Richard E. Tieder, Director

Description: Small business, business analysis and operations research, natural resource economics, and technology transfer; provides a broad base of knowledge for developing the resources, industries, markets, and communities of Michigan's Upper Peninsula through research, service, and academic activities.

★478★ Microelectronics Center of North Carolina
P.O. Box 12889
Research Triangle Park, NC 27709 (919) 248-1800

Description: A nonprofit institution that works in cooperation with the Research Triangle Institute, North Carolina State University, North Carolina Agricultural and Technical, Duke University, and the University of North Carolina (Chapel Hill and Charlotte campuses). Helps the state's universities to educate more students in fields relating to microelectronics by providing complete research and training facilities.

★479★ Midwest Technology Development Institute
Suite 815 BTC
245 East 6th Street
St. Paul, MN 55101-1940 (612) 297-6300
Wm. C. Morris, President/Chairman

Description: Seeks to enhance competitiveness and economic growth in the Midwest through the establishment of cooperative research and development ventures involving industry, universities, and government. Institute ventures include the Rural Enterprise Partnership, which conducts case studies of midwestern farmers successfully using new tillage techniques and crops and livestock management systems; and the Advanced Ceramics and Composites Partnership, a coalition to promote start-up programs in materials. Also seeks to facilitate the equitable transfer of technology among industry, universities, and domestic and foreign governments. **Publications:** Technology Advance.

★480★ Minnesota Small Business Development Center
College of St. Thomas
1107 Haxeltine Blvd., Suite 245
Chaska, MN 55318 (612) 448-8810
Jerry Cartwright, Director

Telecommunication Access: FAX (612) 448-8832.

★481★ Minnesota Technology Corridor
Midland Square Building
331 Second Avenue South
Suite 700
Minneapolis, MN 55401 (612) 348-7140
Judy Cedar, Project Coordinator

Description: 128-acre site established to foster technology transfer, research and development, and prototype manufacturing activities between the University of Minnesota and related businesses.

★482★ Mississippi Research and Technology Park
P.O. Box 2740
Starkville, MS 39759 (601) 324-3219
Dr. W.M. Bost, President

Description: Operates laboratories, incubator and multitenant facilities, professional and business services, and office space to foster high technology research collaboration between Mississippi State University and industry tenants.

★483★ Mississippi Small Business Development Center
University of Mississippi
School of Business Administration
3825 Ridgewood Rd.
Jackson, MS 39211 (601) 982-6760
Dr. Robert D. Smith, Director

Telecommunication Access: FAX (601) 982-6761.

★484★ Missouri Small Business Development Center
St. Louis University
3674 Lindell Blvd., Room 007
St. Louis, MO 63108 (314) 534-7204
Bob Brockhaus, Acting Director

★485★ Montana State University Entrepreneurship Center
412 Reid Hall
Bozeman, MT 59717 (406) 994-4423

Description: Provides business training, research, and technical assistance to small businesses.

★486★ Morgantown Industrial/Research Park
1000 DuPont Road
Building 510
Morgantown, WV 26505-9654 (304) 292-9453
John R. Snider, Executive Vice President

Description: 600-acre site linking West Virginia University research resources with industrial research and development activities. The 200-acre Industrial Park provides pilot plants for industrial research tenants to test findings for entry into commercial markets and further generate research in the 400-acre Research Park.

★487★ National Aeronautics and Space Administration/Indianapolis Center for Advanced Research
611 North Capitol Avenue
Indianapolis, IN 46204 (317) 262-5000
Dr. Thomas D. Franklin, Jr., President

Description: Diagnostic and therapeutic applications of ultrasound, urban technology, technology transfer, computer engineering, software development for engineering applications, medical instrumentation, automated manufacturing, and advanced electronics. **Publications:** Newsletter (quarterly).

★488★ National Centre for Management Research and Development
School of Business Administration
University of Western Ontario
London, ON, Canada N6A 3K7 (519) 661-3233
Dr. David Leighton, Director

Description: Entrepreneurship, productivity, and international business, including studies on generating profits from new technology, doing business in the U.S., corporate governance, accountability in the public sector, women in management, and entrepreneurship within a corporate environment. Seeks to improve communication between managers and researchers. **Publications:** Working Papers. Sponsors research workshops.

★489★ National Research Council of Canada (NRC)
Building M58
Montreal Road
Ottawa, ON, Canada K1A 0R6 (613) 993-9101
Dr. Larkin Kerwin, President

Description: Conducts basic and directed research programs aimed at improving Canada's economic competitiveness. Provides Canadian industry with services, facilities, and technology transfer programs and collaborative research opportunities. Maintains Industrial Research Assistance Program, a technology network. **Publications:** Bulletins (quarterly); Annual Report; NRC Directory of Research Activities; Fact Sheets.

★490★ Nebraska Small Business Development Center
University of Nebraska at Omaha
Omaha, NE 68182-0248 (402) 554-2521
Robert Bernier, Director

★491★ NERAC, Inc.
One Technology Drive
Tolland, CT 06084 (203) 872-7000
Dr. Daniel U. Wilde, President

Description: Examines results of government and other research and development activities and applies these results to industrial and nonindustrial sectors of the economy. Also studies ramifications of transfer and utilization of scientific and technological developments. Maintains close ties with worldwide information sources, including specialized data centers, libraries, governmental agencies, universities, and research institutes.

★492★ Nevada Small Business Development Center
University of Nevada—Reno
College of Business Administration
Room 411
Reno, NV 89557-0100 (702) 784-1717
Samuel Males, Director

★493★ New Hampshire Small Business Development Center
University of New Hampshire
400 Commercial St., Room 311
University Center
Manchester, NH 03103 (603) 625-4522
James E. Bean, Director

Telecommunication Access: Toll-free telephone 800-322-0390; FAX (603) 624-6658.

★494★ New Jersey Institute of Technology Center for Information Age Technology
Newark, NJ 07102 (201) 596-3035
William R. Kennedy, Executive Director

Description: Serves as a bridge between the Institute and industry to transfer new and existing computer and information technologies from university research laboratories to state and local governments and to small- and medium-sized businesses.

★495★ New Jersey Small Business Development Center
Rutgers University
Ackeron Hall, Third Floor
180 University St.
Newark, NJ 07102 (201) 648-5950
Janet Holloway, Director

Telecommunication Access: FAX (201) 648-5889.

★496★ New Mexico Research Park
New Mexico Institute of Mining and Technology
New Mexico Tech
Socorro, NM 87801 (505) 835-5600
Laurence Lattman, President

Description: Facilitates the exchange of research resources between the Institute's research community and Park tenants.

★497★ New Mexico State University Arrowhead Research Park
Box 30001-Dept. 3RED
Las Cruces, NM 88003 (505) 646-2022
Dr. Averett S. Tombes, Vice Pres./Research & Econ. De

Description: Fosters opportunities for exchange between Park tenants and the University community. Multidisciplinary activities include heavy metal recovery from waste streams, flower production, hearing disorder studies, computer telemetry data, fiber optic imaging, and water purification.

★498★ New York State Small Business Development Center
State University of New York (SUNY)
State University Plaza, Room S523
Albany, NY 12246 (518) 442-5398
James L. King, Director

Description: This small business development center services both upstate and downstate New York and is considered two separate development centers. **Telecommunication Access:** FAX (518) 465-4992.

★499★ New York University Center for Entrepreneurial Studies (CES)
Leonard N. Stern School of Business
90 Trinity Place, Room 421
New York, NY 10006 (212) 285-6150
Prof. William D. Guth, Director

Description: Conducts and funds research on the factors that promote entrepreneurship and lead to the creation of new wealth and business revenues and on business venturing within established firms. Topics include the major pitfalls and obstacles to start-ups, securing of venture capital, psychology and sociology of entrepreneurship, management of new ventures, innovation and innovative problem-solving, organizing for corporate venturing, control and reward systems, and cross-cultural environments that stimulate entrepreneurship. Also operates a program focusing on entrepreneurship in nonprofit corporations. **Publications:** CES Reports (semiannually); Journal of Business Venturing (copublished with Wharton, quarterly); INE Reports (for nonprofit corporations).

★500★ Initiatives for Not-for-Profit Entrepreneurship (INE)
Graduate School of Business Administration
90 Trinity Place
New York, NY 10006 (212) 285-6548
Laura Landy, Director

Description: Works with those who are seeking entrepreneurial solutions to the problem of revenue generation and assists and supports them in their efforts. Promotes approaches that enhance success and reduce the risk of venturing, including the correlation between planning, investment, management, size, and venture success in not-for-profit corporations and competition between not-for-profit corporations. **Publications:** INE Reports (semiannual newsletter); Research Reports.

★501★ North Carolina Small Business Development Center
University of North Carolina
Research Triangle Park Area
820 Clay St.
Raleigh, NC 27605 (919) 733-4643
Scott R. Daugherty, Director

Telecommunication Access: FAX (919) 733-0659.

★502★ North Dakota Small Business Development Center
Liberty Memorial Bldg.
State Capitol
Bismark, ND 58505 (701) 224-2810
Terry J. Stallman, Director

Telecommunication Access: FAX (701) 223-3081.

★503★ Northeast Louisiana University Louisiana Small Business Development Center
College of Business Administration
Monroe, LA 71209 (318) 342-2464
Dr. John Baker, Director

Description: Offers business planning, information, management, and technical assistance to the small business community in Louisiana.

★504★ Northeastern Texas Small Business Development Center
Dallas County Community College
302 N. Market Street, Suite 300
Dallas, TX 75202-1806 (214) 746-0555
Norbert Dettmann, Director

Telecommunication Access: FAX (214) 746-2475.

★505★ Northern Advanced Technologies Corporation (NATCO)
Van Housen Hall
P.O. Box 72
State University College at Potsdam
Potsdam, NY 13676 (315) 265-2194
Steven C. Hychkano, Executive Director

Description: Provides support for technology transfer and for research and development in the areas of nondestructive testing, software development, computer applications, materials processing, and high technology. Clients are offered building space, general assistance, and seed money for a variety of projects to commercialize new technology.

★506★ Northern Illinois University Technology Commercialization Center
Dekalb, IL 60115-2874 (815) 753-1238
Dr. Larry Sill, Director

Description: Assists inventors, entrepreneurs, and small businesses with product research and development, technical and commercial assessments, patent applications, and licensing, particularly in the areas of basic sciences, engineering, computer science, and business.

★507★ Northern Kentucky University Foundation Research/Technology Park
Highland Heights, KY 41076 (606) 572-5126
Paul A. Gibson, Contact

Description: Facilitates the exchange of research resources between the University and park tenants.

★508★ Northwestern Texas Small Business Development Center
Texas Tech University
1313 Avenue L
P.O. Box 5948
Lubbock, TX 79417 (806) 744-0743

★509★ Northwestern University Technology Innovation Center
906 University Place
Evanston, IL 60201 (312) 491-3750
Dr. Jack L. Bishop, Director

Description: Matches University resources with the needs of state business by developing small business innovation research programs, linking businesses to share technologies, developing Japan-Illinois technology cooperatives, commercializing University technology, and providing business planning activities for entrepreneurs.

★510★ Northwestern University/Evanston Research Park
1710 Orrington Avenue
Evanston, IL 60201 (312) 475-7170
Ron Kysiak, Executive Director

Description: Encourages the exchange of research activities between the University and Park tenants and transfer technological advances to basic industry. Current tenants include the Basic Industry Research Laboratory, which is funded by the U.S. Department of Energy and owned and staffed by Northwestern University; the Lab focuses on manufacturing and applied materials research, including ceramics, coatings, friction, harsh environments, and energy-efficiency. The Park also provides a small-business incubator system to support newly-developing, high-technology companies, and well as technical assistance, and a seed capital fund of one million dollars.

★511★ Oakland Technology Park
Schostak Brothers & Co., Inc.
First Center Office Plaza
26913 Northwestern Highway
Southfield, MI 48034 (313) 262-1000
Philip J. Houdek, Vice President

Description: Facilitates technology transfer between Oakland University, Oakland Community College, and Park tenants, enhancing the research strength of the affiliated institutions. Facilities are oriented toward robotics, engineering, automation, computer technology, and advanced manufacturing applications.

★512★ Ohio Small Business Development Center
Ohio State Department of Development
30 E. Broad Street, 23rd Floor
Columbus, OH 43266-0101 (614) 466-5111
Jack Brown, Director

★513★ Ohio State University National Center for Research in Vocational Education
1960 Kenny Road
Columbus, OH 43210 (614) 486-3655
Dr. Ray D. Ryan, Executive Director

Description: Research and development on vocational education, including specialization in adult occupational training, career education, economic development and productivity, entrepreneurship, basic skills, handicaps, and collaboration with business, industry, and education. **Publications:** Centergram (monthly research newsletter); Vocational Educator (quarterly); Facts $ Findings (quarterly).

★514★ Ohio State University Research Park
104 Research Center
1314 Kinnear Road
Columbus, OH 43212 (614) 292-9250

Description: 200-acre park offering research-oriented companies a site for their administrative, research, and development facilities to foster exchange between the University and industry. Seeks to enhance the University's teaching and research capabilities through stimulating the exchange of ideas and sharing of resources between University and tenant researchers. Coordinates University and tenant resources in developing commercial applications for new discoveries and technologies.

★515★ Ohio University
Innovation Center and Research Park
One President Street
Athens, OH 45701-2979 (614) 593-1818
Dr. Wilfred R. Konneker, Director

Description: Created to foster entrepreneurial activities and to provide technical and business assistance to new and expanding companies. Facilitates consulting, product testing, and technical assistance between tenants and the University.

★516★ Oklahoma Small Business Development Center
Southeastern Oklahoma State University
P.O. Box 4229, Station A
Durant, OK 74701 (405) 924-0277
Herb Manning, Acting Director

Telecommunication Access: Toll-free telephone 800-522-6154.

★517★ Oklahoma State University
Technology Transfer Center
OSU District Office
P.O. Box 1378
Ada, OK 74820 (405) 332-4100

Description: Provides evaluations of new technology.

★518★ Oregon Graduate Center Science Park
1600 N.W. Compton Drive
Suite 300
Beaverton, OR 97006 (503) 690-1025
Bert Gredvig, Executive Vice President/COO

Description: Science Park serves as a site for interaction between tenants and Center faculty, students, and facilities, including joint scientific research ventures and internship opportunities for Center students.

★519★ Oregon Small Business Development Center
Lane Community College
1059 Willamette St.
Eugene, OR 97401 (503) 726-2250
Edward Cutler, Director

★520★ Oregon State University
Office of Vice President for Research, Graduate Studies, and International Programs
Corvallis, OR 97331 (503) 754-3437
Dr. George H. Keller, Vice President

Description: Coordinates research activities of the University, including individual projects in various academic schools and special research organizations. Also administers the Technology Transfer Program and international activities of the University.

★521★ Oregon Productivity and Technology Center
School of Industrial and General Engineering
Corvallis, OR 97331 (503) 754-3249
Dr. James Riggs, Director

Description: Assists firms in the state with improvied productivity methods.

★522★ Pan American University
Center for Entrepreneurship and Economic Development (CEED)
School of Business Building, Room 124
1201 West University Drive
Edinburg, TX 78539-2999 (512) 381-3361
Dr. J. Michael Patrick, Director

Description: South Texas business assistance and economic research, focusing on business plans, economic area profiles, economic impact studies, economic development planning, market feasibility studies, international trade, urban and rural commercial revitalization, industrial park development and feasibility, and land use surveys.

★523★ Pennsylvania Small Business Development Center
University of Pennsylvania
Vance Hall, 4th Floor
Philadelphia, PA 19104-6374 (215) 898-1219
Gregory Higgins, Jr., Director

Telecommunication Access: FAX (215) 898-2400.

★524★ Pennsylvania State University
Technology Transfer Office
101 George Building
306 West College Avenue
University Park, PA 16801 (814) 865-6277
Dr. Kenneth J. Yost, Liaison

Description: University technology transfer office specializing in electronics materials, structural ceramics, ploymers, and manufacturing.

★525★ Pittsburg State University
Center for Technology Transfer
School of Technology and Applied Science
Pittsburg, KS 66762 (316) 237-7000
Dr. Victor Sullivan, Dean

Description: Development, introduction, and transfer of technology to Kansas regional industries, particularly to wood and plastics industries. Emphasizes design, testing, and development of products and processing methods, including the applications of computer-aided design, computer numerical control, and robotics to manufacturing.

Pittsburg State University *(Cont.)*

★526★ Institute for Economic Development

Pittsburg, KS 66762 (316) 231-7000

Description: Provides one-stop managerial, financial, and technical assistance to new and expanding businesses in southeast Kansas. Services include financial packaging, business plan development, small business counseling, training and skill development, and research on business operations and markets.

★527★ Princeton University Princeton Forrestal Center

105 College Road East
Princeton, NJ 08540 (609) 452-7720
David H. Knights, Jr., Director of Marketing

Description: The Center is designed as a planned multiuse development area creating an interdependent mix of academia and business enterprise in the Princeton area.

★528★ Progress Center: University of Florida Research and Technology Park

One Progress Boulevard, Box 25
Alachua, FL 32615 (904) 462-4040
Rick Finholt, Director/Research Park Develop

Description: 200-acre research and technology park open to both public and private research and manufacturing organizations. High-technology development is emphasized, including the areas of electronics, biotechnology, advanced materials, pharmacology, and agriculture. Provides a link between University researchers and industry and is designed to transfer new technologies from the laboratory to the marketplace. Mainframe computer resources available.

★529★ PTC Research Foundation

Franklin Pierce Law Center
2 White Street
Concord, NH 03301 (603) 228-1541
Robert Shaw, Director

Description: Patents, trademarks, copyrights, invention, and legal and practical systems for dealing with industrial and intellectual property, both in the U.S. and worldwide. **Publications:** IDEA (a journal of law and technology); Monographs; Project Reports.

★530★ Puerto Rico Small Business Development Center

University of Puerto Rico
College Station, Bldg. B
P.O. Box 5253
Mayaguez, PR 00709 (809) 834-3790
Jose M. Romaguera, Director

★531★ Purdue Industrial Research Park

Purdue Research Foundation
Frederick L. Hovde Hall of Administration
Purdue University
West Lafayette, IN 47907 (317) 494-8642
Winfield F. Hentschel, V.P., Purdue Research Foundati

Description: Provides facilities and University research and technical support services to industrial tenants.

★532★ Purdue University Technical Information Service

West Lafayette, IN 47907 (317) 494-9876

Description: Provides business information through online searches on business and industry in the state.

★533★ R&D Village

Montgomery County
Office of Economic Development
101 Monroe Street, Suite 1500
Rockville, MD 20850 (301) 217-2345
Dyan Lingle, Director

Description: 1,200-acre site linking biomedical research and development activities between government, park tenants, and academia. Houses the Center for Advanced Research in Biotechnology, a joint research venture of the National Institute of Standards and Technology, University of Maryland, and Montgomery county government; also houses a Johns Hopkins University facility focusing on advanced study programs in computer science, electrical engineering, and technical management. The Village encompasses the Shady Grove Life Sciences Center, Shady Grove Executive Center, and The Washingtonian, a mixed use development, and branch facilities of University of Maryland and Johns Hopkins University. **Publications:** Economic Focus Newsletter.

★534★ Rensselaer Polytechnic Institute George M. Low Center for Industrial Innovation

Troy, NY 12180 (518) 276-6023
Dr. Christopher W. LeMaistre, Director

Description: Consists of five component centers—Center for Interactive Computer Graphics, Center for Manufacturing Productivity and Technology Transfer, Center for Integrated Electronics, Decision Sciences and Engineering Systems Department, and Automation and Robotics Department. Coordinates research related to high technology business and industrial productivity and quality, including solid geometric modeling, computer integrated manufacturing, vision systems, and computer-aided design of very-large-scale-integrated circuits.

★535★ Rensselaer Technology Park

100 Jordan Road
Troy, NY 12180 (518) 283-7102
Michael Wacholder, Director

Description: Serves as a conduit for joint research activities, consultancies, refresher studies, associate programs, and human interactions between Park tenants and the Institute.

★536★ Research Corporation Technologies

6840 East Broadway Boulevard
Tucson, AZ 85710-2815 (602) 296-6400
Dr. John P. Schaefer, President

Description: Evaluation, patent, development, and commericialization of inventions from colleges, universities, medical research organizations, and other nonprofit laboratories. Provides incentives for invention disclosure, funds applied research and new start-up and joint ventures, and develops commercialization strategies.

★537★ Research Institute for the Management of Technology (RIMTech)
215 North Marengo
Third Floor
Pasadena, CA 91101 (818) 584-9139
Dr. Steven M. Panzer, President

Description: Management issues, technology policy, and the international competitiveness of technology industries. Develops programs facilitating the commercialization of technology from national research facilities, particularly the National Aeronautics and Space Administration (NASA) and its Jet Propulsion Laboratory (JPL), operated by the California Institute of Technology. Projects focus on the commercialization of environmental technology.

★538★ Research Triangle Park
2 Hanes Drive
P.O. Box 12255
Research Triangle Park, NC 27709 (919) 549-8181
James O. Roberson, President

Description: Facilitates interaction between industrial and governmental research and development organizations with the research communities of North Carolina State University, University of North Carolina at Chapel Hill, and Duke University.

★539★ Rhode Island Small Business Development Center
Bryant College
450 Douglas Pike, Route 7
Smithfield, RI 02917 (401) 232-6111
Doug Jobling, Director

★540★ Rio Grande Research Corridor
Pinion Building, Suite 358
1220 St. Francis Drive
Sante Fe, NM 87501 (505) 827-5964
Ponziano Ferarracio, Senior Program Officer

Description: Links University research resources with government laboratories and private research facilities, particularly in the areas of non-invasive diagnosis, materials, explosives technology, plant genetic engineering, computer research applications, and commercial product development.

★541★ RiverBend
2501 Gravel Drive
Fort Worth, TX 76118-6999 (817) 284-5555
David Newell, Contact

Description: Fosters interaction between the University of Texas at Austin research community and industry, particularly in the areas of robotics and automation.

★542★ Riverfront Research Park
University Planning Office
1295 Franklin Boulevard
Eugene, OR 97403 (503) 686-5566
Diane K. Wiley, Representative

Description: 67-acre site that provides opportunities for research interaction between university and Park tenants engaged in such activities as industrial research and development, biotechnology, materials science, environmental technology, computer software development, and business, educational, and governmental research and consulting services. Seeks to assist in the diversification of the economic base of the state.

★543★ Rochester Institute of Technology Research and Business Park
One Lomb Memorial Drive
Rochester, NY 14623-3435 (716) 475-5316
Eric Hardy, Director

Description: 1,300-acre site linking business and industry tenants with research resources of the Institute, particularly in the areas of optical and imaging sciences, photolithography, printing systems, laser, and microchip production. Operates research, office, and industrial facilities.

★544★ Rose-Hulman Institute of Technology Innovators Forum
5500 Wabash Avenue
Terre Haute, IN 47803 (812) 877-1511
Russell Holcomb, Contact

Description: Offers inventors technical screening and access to faculty expertise.

★545★ Rutgers University Technical Assistance Program
180 University Avenue
Newark, NJ 07102 (201) 648-5891
Patricia Johnson, Contact

Description: Provides managerial and technical assistance to small business community.

★546★ Sangamon State University Entrepreneurship and Enterprise Development Center
Springfield, IL 62708 (217) 786-6571
Richard J. Judd, Ph.D., Director

Description: Performs the following activities for private and nonprofit organizations: marketing research (including customer/client surveys), economic analyses, and organizational analyses and development. **Publications:** Economic Business Review (quarterly). Offers training to private and nonprofit organizations.

★547★ Science and Technology Research Center
P.O. Box 12235
Research Triangle Park, NC 27709-2235 (919) 549-0671
J. Graves Vann, Director

Description: Provides information, technical assistance, referrals, workshops, and seminars on subjects of concern to busines and industry. Provides access to ongoing research in federal laboratories through affiliation with NASA.

★548★ Science Park
Five Science Park
New Haven, CT 06511 (203) 786-5000
William W. Ginsberg, President and C.E.O.

Description: 80-acre technology and light industrial site providing scientific facilities, services, and assistance for tenants to interface with university research resources. Houses the New Enterprise Center, an incubator facility.

★549★ South Carolina Research Authority
P.O. Box 12025
Columbia, SC 29211 (803) 799-4070
Robert E. Henderson, Ph.D., Director

Description: Promotes cooperative research and development activities between the University of South Carolina at Columbia and Clemson

University and park tenants. The Authority serves as the contractor and manager of the American Manufacturing Research Consortium, comprised of four companies and research institutions involved in computer-integrated manufacturing. The Consortium develops factory plans that employ technologies to reduce overhead costs in manufacturing and provide low cost products, including factory design and construction based on RAMP (rapid acquisition of manufactured parts) technologies for small machined parts and printed wire assemblies for the U.S. Navy.

★550★ South Carolina Small Business Development Center

University of South Carolina
College of Business Administration
Columbia, SC 29208 (803) 777-4907
W.F. Littlejohn, Director

Telecommunication Access: FAX (803) 252-2435.

★551★ South Dakota Small Business Development Center

University of South Dakota
School of Business
414 E. Clark
Vermillion, SD 57069 (605) 677-5272
Donald Greenfield, Director

Telecommunication Access: FAX (605) 677-5073.

★552★ Southeastern Oklahoma State University Central Industrial Applications Center

Station A, Box 2584
Durant, OK 74701-2584 (405) 924-6822
Dr. Dickie Deel, Director

Description: Data searches are provided for technology transfer to state businesses.

★553★ Southern Illinois University at Carbondale Technology Commercialization Center

Washington Square C
Carbondale, IL 62901-6706 (618) 536-7551
Martha Cropper, Director

Description: Facilitates the production, transfer, and commercialization of new technology in agriculture, forestry, engineering, materials, mining, and biomedicine by investigating technical feasibility, commercial development, marketing, and cash flow. **Publications:** Connections (ten times per year).

★554★ Southern Illinois University at Edwardsville Technology Commercialization Center

Box 1108
Edwardsville, IL 62026-1108 (618) 692-2166
James W. Macer, Director

Description: Provides assistance to inventors, entrepreneurs, and businesses in the commercialization of ideas and products, particularly in the areas of robotics, CAD/CAM, optical coatings, laser communications, automatic inspection, inferfacing microcomputers to instrumentation, computer simulation, analysis of variance, statistical quality control, statistical process control, plant operations, materials planning, linear and nonlinear programming, radioactive biochemicals, human resources, and marketing.

★555★ Southern Willamette Research Corridor

72 West Broadway
Eugene, OR 97401 (503) 687-5033
Bill Barrons, Chairman

Description: 40-mile research, development, and specialized manufacturing site facilitating cooperative ventures between participating colleges and universities, industry, and local government.

★556★ Southwest State University Science & Technology Resource Center

Marshall, MN 56258 (800) 642-0684
James Babcock, Contact

★557★ Stanford University Office of Technology Licensing

350 Cambridge Avenue, Suite 250
Palo Alto, CA 94306 (415) 723-0651
Niels Reimers, Director

Description: University technology transfer office specializing in biotechnology, electronics, and lasers.

★558★ Stanford Research Park

105 Encina Hall
Stanford, CA 94305-6080 (415) 725-6886
Zera Murphy, Managing Director

Description: 655-acre site linking research resources of the University with Park tenants, particularly in the areas of electronics, space, publishing, pharmaceutics, and chemistry. Activities between park tenants and the University community include cooperative research ventures, instruction, and consulting.

★559★ State University of New York at Binghamton Office of Sponsored Program Development

Administration Building, Room 242
Binghamton, NY 13901 (607) 777-6136
Stephen A. Gilje, Associate Vice Provost for Res

Description: Coordinates preparation and submission of grant and contract proposals for sponsored programs. Assists faculty and administrators in developing preliminary and formal proposals locating potential sponsors, processing and transmitting applications, and negotiating budgets. Negotiates grant and contract awards and assists faculty in technology transfer. **Publications:** Graduate Studies Research News (monthly).

★560★ State University of New York at Oswego Center for Innovative Technology Transfer

209 Park Hall
Oswego, NY 13126 (315) 341-2128
Harry Hawkins, Director

★561★ State University of New York at Plattsburgh Economic Development and Technical Assistance Center

Plattsburgh, NY 12901 (518) 564-2214
Stephen Hyde, Contact

Description: Provides business planning and information, including management, marketing, financial packaging, and feasibility analysis services.

★562★ Sunset Research Park
P.O. Box 809
Corvallis, OR 97339 (503) 929-2477
B. Bond Starker, General Manager

Description: Facilitates the exchange of research resources and expertise between Park tenants and faculty at Oregon State University; special emphasis in areas of biotechnology, instrumentation, superconductivity, food science, and natural resources.

★563★ Syracuse University Institute for Energy Research
103 College Place
Syracuse, NY 13244-4010 (315) 423-3353
Dr. Walter Meyer, Director

Description: Application of advanced industrial processes, forecasting of technological trends, planning and evaluation of research and technology transfer programs, technology assessment of direct coal liquifaction, low-level radioactive waste disposal, development of corrosion-resistant thin films to reduce radiation build-up in nuclear reactors, development of pattern recognition artificial intelligence-based electric and gas load forecasting systems, nondestructive analysis for hydrogen in steel and other materials, and energy-related computer-based information systems.

★564★ Technology and Information Policy Program (TIPP)
103 College Place
Syracuse, NY 13244-1120 (315) 443-1890
Barry Bozeman, Director

Description: Nexus of technology, information, and public policy, particularly computer-based technological forecasting and assessment, management of scientific and technical information flows in organizations, technological innovation studies, and the relationship of management information systems, information technology, and organization design.

★565★ Tampa Bay Area Research & Development Authority
University of South Florida
Administration 275
Tampa, FL 33620 (813) 974-2890
John Hennessey, Executive Director

Description: Research and development park leasing land to industries needing facilities for research and related scientific manufacturing in medicine, engineering, and natural sciences. Tenants share University of South Florida research facilities and services, use faculty as consultants, and utilize graduate students as part-time work force. Oversees development of university-related research parks in the Tampa Bay area, in particular University Center Research and Development Park.

★566★ Tennessee Center for Research and Development
Tennessee Technology Foundation
P.O. Box 23184
Knoxville, TN 37933-1184 (615) 675-9505
David A. Patterson, President

Description: Promotes economic development in Tennessee through two components: a research division to conduct basic and applied studies for industrial development and to perform technology transfer activities, and a for-profit subsidiary to provide management support services and temporary laboratory facilities to new companies. Serves as a resource center for public information on the effects of science and technology on the environment.

★567★ Tennessee Small Business Development Center
Memphis State University
320 Dudley St.
Memphis, TN 38104 (901) 678-2500
Dr. Leonard Rosser, Director

★568★ Tennessee Technology Corridor
10915 Hardin Valley Road
P.O. Box 23184
Knoxville, TN 37933-1184 (615) 694-6772
Dr. David A. Patterson, President

Description: Research park linking the research resources of the University of Tennessee, Oak Ridge National Laboratory , and Tennessee Valley Authority with tenants, particularly in the areas of advanced materials, biotechnology, information sciences, waste technology, measurements and control, and electrotechnology. Fosters the transfer of technology from research and development resources in the Corridor and in Tennessee.

★569★ Texas A&M University Technology Business Development
Texas Engineering Experiment Station
Suite 310
Wisenbaker Engineering Research Center
College Station, TX 77843-3369 (409) 845-0538
Helen Baca Dorsey, Director

Description: Statewide mechanism to commercialize innovative technologies developed in Tecas. Facilitates the start-up of a limited number of commercial operations each year involving university research in engineering and licenses university-developed technology. Provides assistance to the University, Texas entrepreneurs, and small companies in technical research and development, marketing and pricing studies, technical evaluations, and production and financial planning of unique ideas. **Publications:** Diversification Report (quarterly newsletter).

★570★ Texas A&M University Research Park
College Station, TX 77843 (409) 845-7275
Dr. Mark L. Money, Vice Chancellor

Description: Located on 434 acres west of main campus, the Park serves to assist private utilization of University resources, to promote closer ties between industry engaged in research and the University, to improve the quality and productivity of University research activities, and to accelerate the dissemination of new knowledge and the transfer of new technologies.

★571★ Texas Engineering Experiment Station (TEES)
College Station, TX 77843 (409) 845-5510
Dr. Herbert H. Richardson, Director

Description: Engineering and related sciences. Major research thrusts include biotechnical engineering, electrooptics and telecommunications, artificial intelligence and expert systems, electrochemical engineering, engineering toxicology, and manufacturing systems. Assists technology transfer, commercialization, and economic development of University research through the Technology Business Development Division. **Publications:** Windows (four per year); Research Report; Annual Report; Engineering Issues.

★572★ Texas Research and Technology Foundation
8207 Callaghan Road, Suite 345
San Antonio, TX 78230 (512) 342-6063
John F. D'Aprix, President and CEO

Description: Supports basic and applied science and new advanced technology enterprises in the San Antonio area by providing research and development facilities, equipment, endowment, and business and scientific expertise. Manages the 1,500-acre Texas Research Park providing industry with access to the resources of several educational and research organizations. Activities include creation of research centers conducting studies in areas such as connective tissue, human performance, neuroscience, battlefield casualty management, toxicology, and clinical trials. Also conducts an inventory and analysis of scientific and technical resources in San Antonio.

★573★ Texas Research Park
Texas Research and Technology Foundation
8207 Callaghan Road, Suite 345
San Antonio, TX 78230 (512) 342-6063
Chris Harness, Vice President

Description: Supports technology development efforts of private industry by providing facilities for basic and applied research and access to the resources of local educational and research organizations. 1,300 acres of the Park are available to private firms through sale or long-term lease; 200 acres are reserved for nonprofit research and development. The Institute of Biotechnology, supporting basic and applied research in the biosciences, the Institute of Applied Sciences, providing technology transfer to industry and government, and the Invention and Investment Institute, offering business and technical services in support of advanced technology business ventures, are collaborative programs based at the Park.

★574★ Texas Small Business Development Center (Houston)
University of Houston
401 Louisiand, 8th Floor
Houston, TX 77002 (713) 223-1141
Jon P. Goodman, Director

Telecommunication Access: FAX (713) 223-5507.

★575★ Texas Small Business Development Center (Lubbock)
Texas Tech University
2005 Broadway
Lubbock, TX 79401 (806) 744-5343
J.E. (Ted) Cadou, Director

Telecommunication Access: FAX (806) 744-0792.

★576★ Texas Small Business Development Center (San Antonio)
University of Texas at San Antonio
Center for Economic Development
Hemisphere Plaza Bldg., #448
San Antonio, TX 78285 (512) 224-0791
Henry Travieso, Director

★577★ Troy State University Center for Business and Economic Services
Sorrell College of Business
Troy State University, AL 36082-0001 (205) 566-3000
Joseph W. Creek

Description: Supports applied research in business and government throughout Alabama. Topics include small business management, government funds management, income tax, salesmanship, venture capital, and Small Business Administration workshops.

★578★ Tulane University U.S.-Japan Biomedical Research Laboratories
Herbert Research Center
Belle Chase, LA 70037 (504) 394-7199

Description: A biomedical research center being developed as an international research park.

★579★ University Center Research and Development Park
c/o Vantage Companies
7650 West Courtney Campbell Causeway
Suite 1100
Tampa, FL 33607-1432 (813) 882-0601
Daniel Woodward, Project Manager

Description: Park provides an interface between research and development tenants and University resources, particularly in the areas of medical technology and engineering sciences. Tenants access University of South Florida research facilities and services, faculty consulting, and graduate student assistance in research and development and prototype assembly activities.

★580★ University of Alabama in Huntsville Alabama High Technology Assistance Center
101 Morton Hall
Huntsville, AL 35899 (205) 895-6409
M. Carl Ziemke, Director

Description: Offers technology transfer, research, analysis, counseling, education, and information services to small businesses in Alabama, focusing on technical and high technology assistance. Projects include assistance with product prototypes to determine patentability and producibility.

★581★ University of Alaska Alaska Economic Development Center (AEDC)
School of Business and Public Administration
1108 F St.
Juneau, AK 99801 (907) 789-4402
Dr. Henry Kohl, Science Advisor

Description: Provides technical assistance state and local governments, community and civic organizations, nonprofit corportations, and Alaska Native Claims Settlement Act corporations involved in economic development activities.

★582★ University of Alberta Office of Research Services
Room 1-3 University Hall
Edmonton, AB, Canada T6G 2J9 (403) 432-5360
Robert E. Armit, Director

Description: Provides assistance in obtaining grants and contracts. Facilitates the transfer of technology from the University to industry. Administers University patent policy and markets research capabilities of the University to industry and government. Operates University of Alberta PRIME (Principal Researcher's Interests and Major Expertise) database, open to the public.

★583★ University of Arkansas
Arkansas Center for Technology Transfer
Engineering Experiment Station
Fayetteville, AR 72701 (501) 575-3747
William H. Rader, Director

Description: Facilitates research and technology transfer in the areas of robotics and automation, interactive technology, productivity and industrial efficiency, and business incubation and analysis. Specific projects include bar-code reading of trucks at weigh stations, scales to weigh-in-motion, tactile sensing, and advanced brakes, clutches, and transmissions.

★584★ Entrepreneurial Service Center
Bureau of Business and Economic Research
College of Business Administration
Fayetteville, AR 72701 (510) 575-4151
Dr. Phillip Taylor, Director

Description: Operates as a consulting service to Arkansas entrepreneurs. Provides business plans, capital location, problem-solving for new businesses, marketing plans, seminars, and other services tailored to the needs of individual businesses.

★585★ University of Arkansas at Little Rock
Technology Center
100 S. Main, Suite 401
Little Rock, AR 72201 (501) 371-5381
Paul McGinnis, Director

★586★ University of Calgary
Office of Technology Transfer
2500 University Drive, N.W.
Calgary, AB, Canada T2N 1N4 (403) 220-7220
Dr. E.L. Jessop, Acting Director

Description: Offers contract research and tranfers new technologies, inventions, and products conceived by University and hospital researchers to industry. Coordinates University and hospital inventions, products, patents, expertise, facilities, and technologies with research and development needs of industry. Adminsters license and joint venture programs.

★587★ University of California, Berkeley
Patent, Trademark and Copyright Office
1250 Shattuck Avenue
Berkeley, CA 94720 (415) 642-5000
Roger Ditzel, Director

Description: University technology transfer office specializing in biotechnology.

★588★ University of California, Los Angeles
Microelectronics Innovation and Computer Research Opportunities (MICRO)
7514 Boelter Hall
Los Angeles, CA 90024 (213) 206-6814
Prof. C.R. Viswanathan, MICRO Executive Committee

Description: Conducts research in microelectronics in collaboration with industry. Seeks to increase interaction with university researchers to enhance the transfer of technology.

★589★ University of Colorado
East Campus Research Park Project
Boulder, CO 80309 (303) 492-7523

★590★ University of Colorado Foundation Inc.
Office of Patents and Licensing
Box 1140
Boulder, CO 80306 (303) 492-8134
John P. Holloway, Patent Officer

Description: University technology transfer office specializing in optoelectronics, molecular biology, pharmacy, chemistry, and medical technology.

★591★ University of Delaware Research Foundation
210 Hullihen Hall
University of Delaware
Newark, DE 19716 (302) 451-2136
Howard E. Simmons, Jr., President

Description: Supports University research by awarding grants to faculty members, also supports invention and patent activities for faculty members. Facilitates technology transfer between the University and industry.

★592★ University of Georgia Research Foundation, Inc.
Boyd Graduate Studies Research Center
Suite 609
Athens, GA 30602 (404) 542-5969
Dr. Joe L. Key, Executive Vice President

Description: Life sciences, agriculture, veterinary medicine, pharmacy, chemistry, and forestry resources. Subcontracts research projects to the University, handles all University technology transfer, and is instrumental in computer hardware acquisition.

★593★ University of Hawaii
Pacific Business Center Program
College of Business Administration
2404 Maile Way
Honolulu, HI 96822 (808) 948-6286

Description: Provides direct counseling and referral services as well as access to the faculty and physical resources of the University to small business owners.

★594★ University of Illinois
Bureau of Economic and Business Research
428 Commerce West
1206 South Sixth Street
Champaign, IL 61820 (217) 333-2330
Marvin Frankel, Acting Director

Description: Economics and business, including studies in business expectations, forecasting and planning, public utilities, innovation, entrepreneurship, consumer behavior, poverty problems, small business operations and problems, investment and growth, productivity, and research methodology. **Publications:** Project Reports; Monographs; Bulletins; Illinois Business Review (six times a year); Illinois Economic Outlook (annually); Illinois Statistical Abstract; Quarterly Review of Economics and Business; Research Projects and Publications (annual report of the faculty of the College of Commerce and Business Administration).

★595★ Business Development Service
109 Coble Hall
801 South Wright Street
Champaign, IL 61820 (217) 333-8357
Arthur H. Perkins, Director

Description: Business development aimed at University entrepreneurs. Produces and commercializes new ideas and products and promotes technology transfer from laboratories into the marketplace.

★596★ ILLITECH
Office of Federal and Corporate Relations
363 Administration Building
506 South Wright Street
Urbana, IL 61801 (217) 333-8634
Ms. Sunny Christensen, Director

Description: Produces and commercializes new ideas and products and promotes technology transfer from laboratories to the marketplace. Stimulates the development and commercialization of University technologies through a partnership with the State of Illinois, the University, and Illinois business and industry.

★597★ University of Illinois at Chicago
Technology Commercialization Program
P.O. Box 4348
M/C 345
Chicago, IL 60680 (312) 996-9131
Dr. L.F. Barry Barrington, Director

Description: Produces and commercializes new ideas and products and promotes technology transfer in the fields of biotechnology, robotics, mechanical devices, engineering software, and data processing hardware components. Arranges assistance for qualified clients' requirements for testing, prototyping, demonstration, market surveys, and business planning. **Publications:** Workshops on Business Plans; Workshops on Technology Transfer (semiannually).

★598★ University of Illinois at Urbana-Champaign
Office of the Vice Chancellor for Research
601 E. John Street
Swanlund Building, 4th Floor
Champaign, IL 61820 (217) 333-7862
Dillon Mapother, Assoc. Vice Chancellor

Description: University technology transfer office specializing in engineering and computing.

★599★ University of Kansas
Space Technology Center
Raymond Nichols Hall
Lawrence, KS 66045 (913) 864-4775
Prof. B.G. Barr, Director

Description: Supports development of new knowledge, concepts, and technology for surveying earth resources and evaluating environmental quality, including remote sensing, technology transfer, flight research, economic and business research, microprocessor control, computer integrated manufacturing, mineral resource surveys, geochronology, and energy research. Staff also works with industry and government in applying newly developed technology. The Center facilitates the activities of the Applied Energy Research Program, Industrial Innovation Program, Augmented Telerobotic Laboratory, Energy Research/Development Center, Flight Research Laboratory, Isotope Geochemistry Laboratory, Kansas Applied Remote Sensing Program, Kansas Biological Survey, Radiation Physics Laboratory, Radar Systems and Remote Sensing Laboratory, and Telecommunications and Information Sciences Laboratory.

★600★ University of Kansas Center for Research, Inc. (CRINC)
2291 Irving Hill Drive
Lawrence, KS 66045 (913) 864-3441
Dr. Carl E. Locke, Ph.D., Director

Description: Remote sensing technology (development, analysis, and applications), energy, environmental quality, aircraft performance improvement, stress analysis, communications systems, technology transfer, microprocessing, transportation, biomedical engineering, isotope geochemistry, and radiation physics. **Publications:** Annual Report; Focus Newsletter (semiannually).

★601★ University of Kentucky
National Aeronautics and Space Administration/ University of Kentucky Technology Applications Program
109 Kinkead Hall
Lexington, KY 40506-0057 (606) 257-6322
William R. Strong, Acting Director

Description: Provides access to scientific and technical information for use in problem-solving by government and industry. Promotes secondary use of NASA technology and stimulates technology transfer through its information services and technical assistance. Program is also a member of the Federal Laboratory Consortium Technology Transfer Network.

★602★ University of Louisville
Center for Entrepreneurship and Technology
School of Business
Louisvill, KY 40292 (502) 588-7854
Lou Dickie, Director

Description: Provides business counseling and management consulting services to technology-based businesses, including entrepreneurs and inventors.

★603★ University of Manitoba
Office of Industrial Research (OIR)
230 Engineering Building
Winnipeg, MB, Canada R3T 2N2 (204) 474-9463
Prof. R.E. Chant, Director

Description: Industrial problems, including studies related to product development, agriculture, energy, energy conservation, laser holographic techniques, product evaluation and testing, wind power development, home economics, policy and planning analysis, management, and small enterprises. Also provides collaboration on innovative research and development projects.

★604★ University of Maryland
Engineering Research Center (ERC)
College of Engineering
Wind Tunnel Building
Room 2104
College Park, MD 20742 (301) 455-7941
Dr. David Barbe, Executive Director

Description: Engineering and scientific disciplines. Activities include a Technology Extension Service (TES), which offers technical assistance to the Maryland business community; a Technology Advancement Program (TAP), an incubator for start-up companies engaging in the development of technically oriented products and services; a Technology Initiatives Program (TIP) to support technological capabilities within the University; and the Maryland Industrial Partnerships (MIPS) program, a matching fund for industry sponsored research. **Publications:** ERC Update (quarterly newsletter).

★605★ University of Maryland Science and Technology Center
7505 Greenway Center Drive
Suite 203
Greenbelt, MD 20770 (301) 982-9400
Dr. Jon K. Hutchison, Project Director

Description: 466-acre research and development park focusing in the areas of math, physics, computer sciences, electrical engineering, and mechanical engineering. Researchers are provided access on a contract basis to University of Maryland faculty and equipment. The Supercomputing Research Center is housed at the Center.

**★606★ University of Miami
Innovation and Entrepreneurship Institute**
P.O. Box 249117
Coral Gables, FL 33124 (305) 284-4692
Jeanie L. McGuire, Director

Description: Entrepreneurship and innovation in Florida, including studies on high-technology ventures and black and Cuban-American entrepreneurs. Promotes interaction of entrepreneurs and capital and service providers. **Publications:** Research Report Series; Friends of the Forum Directory.

**★607★ University of Michigan
Technology Transfer Center, Special Projects Division**
College of Engineering
2200 Bonisteel Blvd.
Ann Arbor, MI 48109 (313) 763-9000
Larry Crockett, Director

Description: University technology transfer office specializing in automated engineering.

**★608★ University of Minnesota
Center for the Development of Technological Leadership**
107 Lind Hall
Institute of Technology
Minneapolis, MN 55455 (612) 624-2006
W.T. Sackett, Acting Director

Description: Technology transfer and technology leadership, including technology transfer methods and pilot experiments.

★609★ Office of Patents and Licensing
1919 University Avenue
St. Paul, MN 55104 (612) 624-0550
John Thuente, Director

Description: University technology transfer office specializing in health sciences and engineering.

★610★ Office of Research and Technology Transfer Administration
1919 University Avenue
St. Paul, MN 55104 (612) 624-1648
A.R. Potami, Assistant Vice President

Description: Serves as the research support unit for University of Minnesota faculty members by administering non-programmatic aspects of all research, training, and public service projects funded by external sources. Reviews and processes all proposals and awards for research, training, and public service projects and is responsible for financial management, cash receipt, and financial reporting of project funds. Works with faculty to stimulate the disclosure of patentable discoveries resulting from University research. Assists faculty in locating potential sources of support for research, training, and public service programs.

★611★ University Research Consortium
Minneapolis Business and Technology Center
511 11th Avenue, S.
Minneapolis, MN 55415 (612) 341-0422
Dr. Ellen Fitzgerald, Director

Description: Provides consulting and technical services in business planning and information management.

**★612★ University of Missouri
Missouri Research Park**
215 University Hall
Columbia, MO 65211 (314) 882-3397
Dr. Duane Stucky, Executive Director

Description: A 240-acre research park designed to be a link between academia and industry and being developed as a center for research and development in such fields as agriculture, chemicals, medicine, computers, engineering, and food processing. Also planned site for an international, policy-level think tank in the field of agribusiness.

**★613★ University of Missouri—Rolla
Center for Technology Transfer and Economic Development**
One Ngogami Terrace
Rolla, MO 65401 (314) 341-4559
H. Dean Keith, Director

Description: Technological innovation.

**★614★ University of Moncton
Center of Research in Administrative Sciences (CRAS)**
Moncton, NB, Canada E1A 3E9 (506) 858-4555
Dr. Jean Cadieux, Director

Description: Regional development, small businesses, capital venture, and impact of new technologies on the Atlantic region of Canada. **Publications:** Cahiers du CRSA (annually).

**★615★ University of Nebraska, Lincoln
Nebraska Technical Assistance Center**
W191 Nebraska Hall
Lincoln, NE 68588-0635 (402) 472-5600
Herbert Hoover, Director

Description: Provides information about new technologies, nonlegal assistance in patent searches and the patent process, and educational activities.

**★616★ University of New Hampshire
Industrial Research and Consulting Center**
Thompson Hall
Durham, NH 03824 (603) 862-3750
Dr. Donald C. Sundberg, Director

Description: Promotes research and development relationships between the private sector and the University. Organizes problem solving teams and makes available University instrumentation and computer facilities and research and development laboratories. Assists business and industry with product development, process development, long-range research and planning, modeling, software development, technical troubleshooting, feasibility studies, development of laboratory testing procedures, market analysis, risk analysis, and planning and educational programs. Develops patents and licenses technology and other intellectual property of the University.

★617★ University of New Mexico
Technology Application Center
Albuquerque, NM 87131 (505) 277-3622
Dr. Stanley A. Morain, Director

Description: Retrieves, processes, and analyzes satellite and aerial data for earth resources and develops geographic information systems (GIS). Image processing and GIS activities include mineral exploration, cover type mapping, habitat mapping and modeling, and surveys of archeological locations. Photo search and retrieval services include satellite images, aerial photos, maps, digital data, LANDSAT data, and photos from Gemini, Apollo, Apollo-Soyuz, Skylab, and Space Shuttle missions. Designated as University of New Mexico National Aeronautics and Space Administration Industrial Applications Center to assist in commercializing the national program for space especially in the areas of remote sensing and image dissemination. Offers national and international visiting scientist programs providing customized technical assistance and training in remote sensing and image processing. **Publications:** Remote Sensing of Natural Resources: A Quarterly Literature Review.

★618★ Technology Innovation Center
Albuquerque, NM 87131 (505) 277-5934
Gary Smith, Program Director

★619★ University of North Carolina at Charlotte
Engineering Research and Industrial Development Office
Charlotte, NC 28223 (704) 547-4150
Dr. J.M. Roblin, Contact

Description: Seeks to assist business in the region to compete more effectively by providing industrial development facilities, equipment, applied research, and consulting resources for solving practical industrial problems.

★620★ University of Oklahoma
Swearingen Research Park
1700 Lexington Avenue
Norman, OK 73069 (405) 325-7233
George Hargett, Airpark Administrator

Description: Provides on a 900-acre tract facilities and services for research administered by the University's Office of Research Administration, as well as U.S. and state government, industry, and University of Oklahoma laboratories located in the Park.

★621★ University of Pennsylvania
Office of Corporate Programs and Technology
133 South 36th Street, Suite 300
Philadelphia, PA 19104 (215) 898-7293
George C. Farnbach, Director

Description: University technology transfer office specializing in biomedical technology.

★622★ Office of Sponsored Programs
1335 South 36th Street
Philadelphia, PA 19104-3246
Anthony Merritt, Director

Description: Administers extramurally sponsored research for all departments and research units of the University and handles processing of research applications. Responsible for licensing of inventions and other technology transfer activities.

★623★ Sol C. Snider Entrepreneurial Center
Vance Hall, Suite 400
Philadelphia, PA 19104 (215) 898-4856
Dr. Ian C. MacMillan, Director

Description: Entrepreneurship in private sector, public sector, and nonprofit organizations, including venture capital, corporate venturing, development of internal corporate entrepreneurship, technology transfer, and university spin-offs.

★624★ University of Pittsburgh
Center for Applied Science and Technology
100 Loeffler Building
Pittsburgh, PA 15260 (412) 624-2024
Richard K. Olson, Director

Description: Fosters collaborative research efforts between the University and industry for the transfer of technology.

★625★ National Aeronautics and Space Administration Industrial Applications Center
823 William Pitt Union
Pittsburgh, PA 15260 (412) 648-7000
Paul A. McWilliams, Ph.D., Executive Director

Description: Works with business and industrial clients to transfer technology developed by the National Aeronautics and Space Administration (NASA) and other government agencies. Activities center on new product identification and testing, engineering analyses, corporate resource redeployment initiatives, literature searching and document procurement, database system design and electronic publishing, and science curriculum enhancements for grades K-12. The Center is hard-wired to the NASA Recon database and it maintains direct link with researchers in NASA field centers, federal research laboratories, and academic centers. **Publications:** Economic Books: Current Selections (quarterly); United States Political Science Documents (annually).

★626★ Technology Management Studies Institute
1048 BEH
Pittsburgh, PA 15228 (412) 624-9836
Dr. William E. Souder, Director

Description: Management of high technology projects and new product innovation processes, including research and development management, organization design, group dynamics, and strategic management. **Publications:** Annual Report.

★627★ University of Puerto Rico
Center for Business Research
College of Business Administration
Department of Management & Business
Rio Piedras, PR 00931 (809) 764-0000
Prof. Alvin J. Martinez, Director

Description: Commerce and economics, including studies on consumer analysis, marketing, budget analysis, transfer of managerial technology, economic and business development, marketing, budget analysis, and problems of private enterprise. Compiles statistical data on personnel, employment and unemployment, health and nutritional problems management training and proficiency, entrepreneurship, international commerce, economic problems, finance, and accounting in Puerto Rico. **Publications:** Annual Report; Revista de Ciencias Comerciales (semiannually).

★628★ University of Quebec at Hull
Organizational Efficiency Research Group
283 Tache Boulevard
P.O. Box 1250
Hull, PQ, Canada J8X 3X7 (819) 595-2318
Prof. Alain Albert, Director

Description: Office automation and management and high-technology and venture capital.

★629★ University of Rhode Island
Business Assistance Programs
Kingston, RI 02881 (407) 792-2337
Robert Comerford, Director

Description: Provides assistance through Rhode Island Innovation Center, which evaluates potential of inventions and advises inventors on commercialization strategies.

★630★ University of Scranton
Technology Center
Scranton, PA 18510-2192 (717) 961-4050
Jerome P. DeSanto, Executive Director

Description: Works with business and industry to find practical solutions in such areas as computer-aided design, applied research, chemistry and biotechnology, software engineering, computer networking, telecommunications, and business planning and management to aid the economic development of northeastern Pennsylvania. **Publications:** Newsletter.

★631★ University of South Florida
Small Business Development Center
College of Business Administration
Tampa, FL 33620 (813) 974-4274
Bill Manck, Director

Description: Small business operations, entrepreneurship, and success and failure factors for small business development and management, including developing business plans, marketing strategy, and loan packages. **Publications:** Speaking of Small Business (monthly newsletter).

★632★ University of Southern California
Office of Patent and Copyright Administration
376 South Hope Street, Suite 200
Los Angeles, CA 90007 (213) 743-4926
L. Kenneth Rosenthal, Assistant Director

Description: University technology transfer office specializing in medicine and chemistry.

★633★ Urban University Center, Western Research Applications Center
3716 South Hope St.
Los Angeles, CA 90007-4344 (213) 743-2371

Description: Supports the Small Business Innovation Research program through seminars, workshops, conferences, and one-to-one counseling activities.

★634★ University of Southern Maine
Center for Business and Economic Research
118 Bedford Street
Portland, ME 04103 (207) 780-4187
Richard J. Clarey, Dean of SBEM

Description: Business expansion in Maine, including management and small business development. **Publications:** Maine Business Indicators (quarterly); Reports.

★635★ The New Enterprise Institute
Center for Research and Advanced Study
246 Deering Avenue
Portland, ME 04102 (207) 780-4420
Dr. Richard J. Clarey, Director

Description: Creates, tests, and implements economic development mechanisms. Provides technical assistance, feasibility studies, applied research, and productivity enhancement analysis to all economic sectors of the state. Also acts as a connection between users of entrepreneurial and technology-related information.

★636★ Office of Sponsored Research
96 Falmouth Street
Portland, ME 04103 (207) 780-4411
Dr. Robert J. Goettel, Director

Description: Economic development, the formation and management of business enterprises, health and human services, education, marine resources, state and local governance, science and technology, and organized camping. The Center is organized into five cooperating institutes: Marine Law Institute, Human Services Development Institute, The New Enterprise Institute, Health Policy Unit, and National Child Welfare Resource Center.

★637★ University of Tennessee Research Corporation
415 Communications Building
University of Tennessee
Knoxville, TN 37996 (615) 974-1882
Eugene Joyce, President

Description: Supports research at various campuses of the University by investigating the patentability and marketability of faculty ideas, aiding faculty and staff in the procuring of patents and copyrights, and assisting in commercializing inventions and creative works.

★638★ University of Texas
Center for Technology Development and Transfer
Ernest Cockrell Jr., Hall 2:516
College of Engineering
Austin, TX 78712-1080 (512) 471-3700
Dr. Stephen A. Szygenda, Director

Description: Links university researchers with industrial needs and facilitates the commercialization of academic research, university-generated technology, scientific information, and other intellectual property. Activities range from serving as a clearinghouse for technology to product development and the formation of businesses. Also serves as the hub for a statewide technology development and transfer network.

★639★ University of Texas at Arlington
Technology Enterprise Development Center
UTA Station Box 19029
Arlington, TX 76019 (817) 273-2559
Dr. John Troutman, Director

★640★ University of Texas at Austin
IC2 Institute

2815 San Gabriel
Austin, TX 78705 (512) 478-4081
Dr. George Kozmetsky, Director

Description: Management of technology, creative and innovative management, measurement of the state of society, dynamic business development and entrepreneurship, new methods of economic analysis, and the evaluation of attitudes, concerns, and opinions on key issues. **Publications:** Newsletter (quarterly).

★641★ University of Texas Health Science Center at Houston
Institute for Technology Development and Assessment

P.O. Box 20334
6901 Bertner
Houston, TX 77225 (713) 792-4609
R.W. Butcher, Ph.D., Director

Description: Established to evaluate, develop, and promote cost-effective new technologies relating to health care, preventative medicine, and biomedical research. Acts as an administrative umbrella for multidisciplinary research at the Health Science Center.

★642★ University of Utah
University of Utah Research Park

505 Wakara Way
Salt Lake City, UT 84108 (801) 581-8133
Charles A. Evans, Director

Description: 300-acre tract designed to facilitate scientific research projects between government agencies, industrial organizations, and the University. Facilities include general and technical libraries, scientific and research equipment, and computer center. The Park is the site of 22 buildings, including facilities for the Utah Innovation Center, Inc.

★643★ University of Washington
Office of Technology Transfer

201 Administration Building, AG/10
Seattle, WA 98195 (206) 543-5900
Donald Baldwin, Director

Description: University technology transfer office specializing in biotechnology, bioengineering, engineering, and medical applications.

★644★ University of Waterloo
Research Services & Development

Needles Hall
Waterloo, ON, Canada N2L 3G1 (519) 885-1211
Mr. A.H. Headlam, Director

Description: Responsible for administration of research grants and contracts for faculty research at the University. Processes applications for research support and oversees research projects, especially performance of contract research by Waterloo Research Institute. Facilitates technology transfer and commerialization of research results.

★645★ University of Wisconsin—Madison
University-Industry Research Program

1215 WARF Building
610 Walnut Street
Madison, WI 53705 (608) 263-2840

Description: Identifies scientific resources of faculty, information, and facilities at the University for application to needs of industry, commerce, and government. Facilitates technology transfer activities with industry and provides advisory services to faculty on patents and consortia development. **Publications:** Touchstone (quarterly); Faculty Research Directory (annually).

★646★ The Wisconsin Alumni Research Foundation

Box 7365
Madison, WI 53707 (608) 263-2500
Thomas Hinkes, Director of Licensing

Description: University technology transfer office specializing in pharmacy and life sciences.

★647★ University of Wisconsin—Milwaukee
Office of Industrial Research and Technology Transfer

P.O. Box 340
Milwaukee, WI 53201 (414) 229-5000
Irving D. Ross, Jr., Director

Description: Serves as a catalyst in developing university/industry research programs and facilitates the transfer of technology between the University and industry. Provides assistance on patents, copyrights, and proprietary agreements to faculty.

★648★ University of Wisconsin—Stout
Center for Innovation and Development

Menomonie, WI 54751 (715) 232-1252
Dr. John F. Entorf, Director

Description: Manufacturing productivity, electronics, and process evaluation. Evaluates inventions and new products, develops prototype models of innovative products, and evaluates materials for fabrication of new products.

★649★ University of Wisconsin—Whitewater
Wisconsin Innovation Service Center

402 McCutchan Hall
Whitewater, WI 53190-1790 (414) 472-1365
Debra Knox-Malewicki, Program Manager

★650★ University Research Park (Charlotte)

Suite 1980, Two First Union Plaza
Charlotte, NC 28282 (704) 375-6220
Seddon Goode, Jr., President

Description: 2,700-acre site providing companies located in the park with research and educational interaction with University of North Carolina at Charlotte.

★651★ University Research Park (Madison)

946 WARF Building
610 Walnut Street
Madison, WI 53705 (608) 262-3677
Wayne McGown, Director

Description: The 180-acre Park facilitates technology transfer between the research produced on the University of Wisconsin—Madison campus and applied research of industry and provides a long-term endowment income to the University. Leases land to private companies and agencies, particularly those involved with microelectronics, forestry economics, genetic engineering, microbiology, computer research and software development, environmental research, food research, medical instrumentation, robotics, energy conservation, biotechnology, remote sensing, space sciences, manufacturing systems, materials science, polymers, toxicology, and hazardous waste. Utilizes University resources and encourages entrepreneurial innovation among faculty.

★652★ University Technology Incorporated
Research Building #1
11000 Cedar Avenue
Cleveland, OH 44106 (216) 368-5514
William Carlson, President

Description: Established to strengthen regional business and industry, stimulate research, and provide a channel for the commercialization of Case Western University campus technologies. Identifies, evaluates, and implements development strategies for technologies, including intellectual property and protection; patent strategy; defining business and marketing opportunities; designing business structures; and forming business, commercial, and financial relationships. Licenses proprietary rights to companies prepared to invest in further on-campus research and development of technology, including established businesses and newly formed ventures.

★653★ Utah Small Business Development Center
University of Utah
660 South 200 St.
Suite 418
Salt Lake City, UT 84111 (801) 581-7905
Kumen B. Davis, Director

★654★ Utah State University
Utah State University Research and Technology Park
1780 North Research Park Way
Suite 104
North Logan, UT 84321 (801) 750-6924
Wayne Watkins, Director

Description: Fosters the interaction between the University research community and Park tenants. Administers a technology transfer program and provides incubator services to start-up companies.

★655★ Vermont Small Business Development Center
University of Vermont
Extension Service
Small Business Development
Morrill Hall
Burlington, VT 05405 (802) 656-4479
Norris A. Elliott, Director

★656★ Virgin Islands Small Business Development Center
College of the Virgin Islands
Grand Hotel Bldg., Annex B
P.O. Box 1087
St. Thomas, VI 00801 (809) 776-3206
Solomon S. Kabuka, Jr., Director

★657★ Virginia Polytechnic Institute and State University
Virginia Tech Corporate Research Park
Development and University Relations
315 Burruss Hall
Blacksburg, VA 24061 (703) 961-7676
Charles M. Forbes, Vice-President

Description: 120 acres adjacent to the main campus and university airport providing building sites for lease to companies interested in developing or expanding a research relationship with the University. Houses an innovation center to provide facilities for start-up companies requiring support of University programs.

★658★ Washington, D.C., Small Business Development Center
Howard University
Sixth and Fairmount St., N.W., Room 128
Washington, DC 20059 (202) 636-5150
Nancy Flake, Director

★659★ Washington Small Business Development Center
Washington State University
441 Todd Hall
Pullman, WA 99164-4740 (509) 335-1576
Lyle Anderson, Director

Telecommunication Access: FAX (509) 335-3421.

★660★ Washington State University
Innovation Assessment Center
Small Business Development Center
Pullman, WA 99164 (206) 464-5450
Jim Van Orsow, Director

★661★ Research and Technology Park
N.E. 1615 East Gate Boulevard
Pullman, WA 99164 (509) 335-5526
John R. Schade, Director

Description: Encourages research interaction between Washington State University and Universiy of Idaho and industry. Seeks to promote economic development in the 3,000 square miles of Idaho's Panhandle Region and southeastern Washington. The Park leases land to industries engaged in research, development, and light manufacturing. Industrial tenants share University research facilities and services, use faculty as consultants, and utilize graduate students as a part-time work force. Major research areas include agriculture, forestry, mines, veterinary medicine, plant and animal biotechnology and engineering. **Publications:** The Research Connection; Quarterly Report.

★662★ Washington University
Industrial Contracts and Licensing
Campus Box 8013
St. Louis, MO 63130 (314) 362-5866
H.S. Leahy, Director

Description: University technology transfer office specializing in biotechnology.

★663★ Waterloo Centre for Process Development
University of Waterloo
University Avenue
Waterloo, ON, Canada N2L 3G1 (519) 885-1211
E.B. Cross, Executive Director

Description: Industrial research including computerized grain dryer controller, economical flue gas desuphurization scrubber, and a sophisticated software program for monitoring and control using an industrial AT (7531, 7532) computer. Markets processes and technologies developed at Canadian universities to the industrial and commercial sectors; this transfer of technology is facilitated through licensing agreements and through research and development contracts. Products developed at the Centre include a bioconversion process to produce proteinaceous animal feed product, methane fuel gas, and fuel grade alcohol from waste agricultural or forest residues; microwave drying process utilizing microwaves in drying commercial, industrial, and agricultural products; gas scrubber allowing for the removal of submicron particulate matter from gas streams; ash monitoring system for the continuous measurement of ash build-up within large utility boilers; and gravity clarifier that improves separation efficiency in large-scale rectangular of circular clarifiers used for industrial or sewage treatment.

★664★ Wayne State University Technology Transfer Center
5050 Cass Avenue, Room 242
Detroit, MI 48202 (313) 577-2788
Nate R. Borofsky, Director

Description: Assists manufacturers, entrepreneurs, and inventors in gaining access to technical resources at the University.

★665★ West Virginia Small Business Development Center
Governor's Office of Community and Industrial Development
115 Virginia St., E.
Charleston, WV 25305 (304) 348-2960
Eloise Jack, Director

★666★ Western Illinois University Center for Business and Economic Research
College of Business
Macomb, IL 61455 (309) 298-1594
Dr. Richard E. Hattwick, Director

Description: Business and economics, with emphasis on labor markets, occupational education, regional economic analysis, and entrepreneurial case studies. **Publications:** Monographs; Working Papers; Journal of Behavioral Economics; Illinois Business Leader (newspaper); American Business Leader (newspaper).

★667★ Technology Commercialization Center
Room 212 Seal
Macomb, IL 61455 (309) 298-2211
Daniel Voorhis, Director

Description: Produces and commercializes new ideas and products. Promotes the transfer of technology from the laboratory to the marketplace. Provides assistance to develop technology-based products and businesses through technical and commercial assessments and design assistance.

★668★ Western Michigan University Institute for Technological Studies
Kalamazoo, MI 49008 (616) 387-4022
Robert M. Wygant, Director

Description: Provides administrative support for internal and external research of the College. Also links industry with University resources by providing technical assessments, information, referrals, testing, and instruction.

★669★ Wichita State University Center for Entrepreneurship
008 Clinton Hall—Campus Box 147
Wichita, KS 67208 (316) 689-3000
Prof. Fran Jabara, Director

Description: Entrepreneurship, particularly the effect entrepreneurial education has on business start-ups and success rates, and profile studies of entrepreneurs. **Publications:** Center Report (twice yearly).

★670★ Center for Productivity Enhancement
Engineering, Room 203
Wichita, KS 67208-1595 (316) 689-3525
Dr. Richard Graham, Contact

Description: Technology transfer in the areas of computer-integrated manufacturing and advanced composite materials.

★671★ Wisconsin for Research, Inc.
210 North Bassett Street
Madison, WI 53703 (608) 258-7070
Reed Coleman, President

Description: Seeks to encourage economic development by promoting the transfer of University of Wisconsin research and technology to businesses in the state. Operates the Madison Business Incubator to generate economic development in Dane County by providing office space, support services, and business assistance. Administers a seed capital fund.

★672★ Wisconsin Small Business Development Center
University of Wisconsin
432 N. Lake St., Room 423, Ext. Bldg.
Madison, WI 53706 (608) 263-7793
Bill Pinkovits, Director

★673★ Worcester Polytechnic Institute Management of Advanced Automation Technology Center
Department of Management
Worcester, MA 01609 (617) 793-5000
Arthur Gerstenfeld, Director

Description: Management of technological innovations in the areas of computer-aided manufacturing, decision support systems, office automation, group technology, manufacturing communication networks, automation assembly process, application of bar coding, user friendly shop floor control, and manufacturing strategy. Conducts applied research for the specific technological needs of industrial clients. **Publications:** Newsletter (quarterly).

★674★ Wyoming Small Business Development Center
Casper College
130 North Ash, Suite A
Casper, WY 82601 (307) 235-4825
M.C. "Mac" Bryant, Director

★675★ Yale University Office of Cooperative Research
252 JWG, Box 6666
New Haven, CT 06511 (203) 432-3003
Dr. R.K. Bickerton, Director

Description: University technology transfer office specializing in biotechnolgy, computer science, and engineering.

CHAPTER 5

REGIONAL SMALL BUSINESS AND STATE INNOVATION PROGRAMS

This chapter contains a wide variety of state and regional programs that promote technological innovation. Many of these programs fall under the jurisdiction of the Small Business Innovation Development Act (1982), which created the Small Business Innovation Research (SBIR) program. The U.S. Small Business Administration (SBA) monitors and oversees these programs, which range from limited managerial and technical assistance centers to comprehensive investment programs. Additionally, the SBA supports a network of local offices that provide training and assistance to small businesses. This chapter lists many state-supported programs as well as over 100 regional offices of the SBA. For more information on the program and the federal government's participation, see Chapter 7, Federal Funding Sources for the Inventor.

According to the Minnesota Department of Trade and Economic Development's Office of Science and Technology 1988 report *State Technology Programs in the United States*, over $550 million was allocated for state science and technology initiatives for 1988 fiscal year. The largest amount of funding (41.2 percent) was allocated to technology or research centers to promote research and development, however, states provide (or support) a wide variety of programs, including:

*Technology Offices
*Research Grants
*Research Parks
*Incubators
*Technology Transfer Programs
*Technical/Managerial Assistance Programs
*Seed/Venture Capital Programs
*Technical Training Programs
*Clearinghouses
*Equity/Royalty Investment Programs

Because of the overlap between university research, incubators, venture capital groups, state-supported programs, and U.S. government regional small business development centers, this chapter concentrates only on state government agencies and regional office of the U.S. Small Business Administration. Check the Master Index of this book for any state or regional organization not listed in this chapter. It may appear in a more appropriate section (Business Incubators, University Innovation Research Centers, or Venture Capitalists, for example) depending upon its sponsoring organization.

For more information on state programs, contact the Association of Small Business Development Centers,

1050 17th St., N.W., Washington, D.C. 20036; (202) 223-8607.

Information on regional offices of the U.S. Small Business Administration may be obtained by contacting Raymond Marchakitus, U.S. Small Business Administration, SBIR, 1441 L St., N.W., Premier Bldg. #414, Washington, DC 20416; 1-800-368-5855.

Chapter Arrangement

This chapter is arranged geographically by state, then alphabetically by organization name. The subordinate organizations and programs of large divisions are displayed directly below the main entry, set off by indentation. The contact information for the secondary organization is not repeated if it is identical to that of the parent organization. All organizations in this chapter are referenced in the Master Index.

Alabama

★676★ Alabama Small Business Development Consortium
1717 11th Ave., S., Suite 419
Medical Towers Bldg.
Birmingham, AL 35294 (205) 934-7260
Dr. Jeff D. Gibbs, Director

Description: Nonprofit consortium provides managerial and technical assistance and training through twelve small business development centers, the Alabama International Trade Center, Alabama High Technology Assistance Center, Alabama Small Business Procurement System, and six affiliated centers.

★677★ Alabama State Department of Economic and Community Affairs
Planning and Economic Development
3465 Norman Bridge Rd.
P.O. Box 2939
Montgomery, AL 36105-0939 (205) 261-3572
Dr. Don Hines, Director

Description: Assists small and developing businesses through information dissemination for federal grants. Also provides consulting services.

★678★ Science Technology and Energy Division
3465 Norman Bridge Rd.
P.O. Box 2939
Montgomery, AL 36105-0939 (205) 284-8952
Fred Braswell, Director

Description: Provides technical assistance to business and insustry in identifying energy management opportunities and cost avoidance procedures. Also provides limited financial assistance.

★679★ Alabama State Department of Environmental Management
1751 Federal Dr.
Montgomery, AL 36130 (205) 271-7700

Description: Consolidates the major permitting agencies involving water, sewer, air, and solid and hazardous wastes.

★680★ Alabama State Department of Finance
Division of Purchasing
State Capitol Bldg., Room 207
11 S. Union St.
Montgomery, AL 36130 (205) 261-3128

★681★ Alabama State Development Office
Industrial Finance Division
State Capitol
Montgomery, AL 36130 (205) 263-0048

Description: Provides direct funding, packages other funding, and offers information on additional industrial development funding. Maintains the Southern Development Council—the state's U.S. Small Business Administration 503 corporation—which manages an economic development revolving loan fund and assists in arranging the necessary private financing to meet the funding needs of an industrial location. Assists in obtaining other sources of industrial development financing, including Industrial Development Bond Financing, Urban Development Action Grants, State Economic Development Loans, U.S. Small Business Administration 503 Loans, Economic Development Administration Loans and Grants, Appalachian Regional Commission Grants, State Industrial Development Authority Building Loans, and the State Industrial Site Preparation Grant Program. **Publications:** Alabama Industrial Development.

★682★ Alabama State House of Representatives
Small Business Committee
The State House
Montgomery, AL 36130 (205) 261-7688

★683★ Oxmoor West Industrial Park
c/o Metropolitan Development Board
2027 First Ave., N.
600 Commerce Center
Birmingham, AL 35203 (205) 328-3047

Description: Tenants of industrial park are involved in light industry.

★684★ U.S. Small Business Administration
Birmingham District Office
2121 Eighth Ave., N., Suite 200
Birmingham, AL 35203-2398 (205) 731-1338

Alaska

★685★ Alaska Industrial Development Authority
1577 C St., Suite 304
Anchorage, AK 99501 (907) 274-1651

Description: Provides financing for businesses through the use of tax-exempt bonds and the purchase of federally guaranteed loans.

★686★ Alaska State Department of Administration
Purchasing Office
State Office Bldg.
Pouch C
Juneau, AK 99811 (907) 465-2253

★687★ Alaska State Department of Commerce and Economic Development
P.O. Box D
Juneau, AK 99811 (907) 465-2500

Description: Maintains a Business Development Division and an Office of Enterprise. **Telecommunication Access:** Alternate telephone (907) 465-2017.

★688★ U.S. Small Business Administration
Anchorage District Office
701 C St., Room 1068
Anchorage, AK 99513 (907) 271-4022

Arizona

★689★ Arizona Enterprise Development Corporation
Arizona State Department of Commerce
Development Finance Program
State Capitol Tower
1700 W. Washington, Fifth Floor
Phoenix, AZ 85007 (602) 255-1782

Description: A private, nonprofit corporation that offers loans to small businesses through the U.S. Small Business Administration's 504 Program. Grants such loans for the purchase of land, buildings, machinery, and equipment, for construction, for renovation/leasehold improvements, and for related professional fees. Loans are available to for-profit businesses that generated an average net profit below $2 million for the preceding two years and that have a net worth below $6 million. (Businesses in Phoenix and Tucson do not qualify but have similar programs.) **Telecommunication Access:** Toll-free/Additional Phone Number: (602)255-5705.

★690★ Arizona State Department of Administration
Purchasing Office
1688 W. Adams
Phoenix, AZ 85007 (602) 255-5511

★691★ Arizona State Department of Commerce
State Capitol Tower
1700 W. Washington, Fourth Floor
Phoenix, AZ 85007 (602) 255-5371

Description: Offers business counseling and provides information about establishing a business, licensing, taxation, funding, labor and environmental regulations, and foreign trade zones. Participates in the National Small Business Revitalization Program. Administers three federal financing programs for Arizona's small businesses. Also refers loan applicants to other sources such as the loan guarantee programs of the U.S. Small Business Administration, the Economic Development Administration, and the Farmers Home Administration. Publications: Guide to Establishing a Business in Arizona.

★692★ Revolving Loans Program
State Capitol Tower
1700 W. Washington
Phoenix, AZ 85007 (602) 255-5705

Description: A federally funded program that provides loans for economic development projects of small for-profit and nonprofit businesses in Arizona, except Maricopa and Pima counties. Offers below-market interest rates—with negotiable terms—for site/facility acquisition and improvements, construction, machinery and equipment, building rehabilitation, leasehold improvements, and sometimes working capital.

★693★ Urban Development Action Grants
(602) 255-5705

Description: Grants subordinated loans to new and expanding businesses for site improvements, industrial and commercial rehabilitation, machinery and equipment, and like expenses—except working capital, debt refinancing, and consolidation.

★694★ U.S. Small Business Administration
Phoenix District Office
2005 N. Central Ave., Fifth Floor
Phoenix, AZ 85004 (602) 261-3732

★695★ Tucson District Office
300 W. Congress St., Room 3V
Tucson, AZ 85701 (602) 629-6715

Arkansas

★696★ Arkansas Capital Corporation/Arkansas Capital Development Corporation
Governor's Office
State Capitol
Little Rock, AR 72201

Description: Utilizes both state and private funds dedicated to the growth and stimulation of economic activity within the state. The Arkansas Capital Corporation is authorized to borrow $20 million from the interest earned on the state's average daily balances; the ACC works in concert with financial institutions and economic/industrial development agencies. The Arkansas Capital Development Corporation is a for-profit, risk-taking partner of the ACC. It extends financial assistance in the forms of loans, loans with options to purchase equity interest, and the direct acquisition of equity interest. The ACDC serves as a supplement to the ACC and other financial devices within the public and private sectors. The ACDC has the ability to engage in venture capital-type projects.

★697★ Arkansas Development Finance Authority
P.O. Box 8023
Little Rock, AR 72203

Description: Provides Arkansas businesses with long-term, fixed-rate financing. Issues both tax-exempt and taxable revenue bonds for housing, industrial and agricultural processing enterprises, and public facilities. Administers the Bond Guaranty Program.

★698★ Arkansas Industrial Development Commission
One Capitol State Mall
Little Rock, AR 72201 (501) 371-1121

Description: Serves as an information and referral source for small business concerns.

★699★ Arkansas Science and Technology Authority
200 Main St., Suite 450
Little Rock, AR 72201 (501) 371-3554
Dr. John W. Ahlen, Executive Director

Description: Administers an investment fund to provide seed capital in the form of loans, royalty agreements, and the purchase of limited amounts of stock for new and developing technology-based companies. **Contact:** Alan Gumbel, SBIR Assistance.

★700★ Arkansas State Department of Finance and Administration
Purchasing Office
P.O. Box 3278
Little Rock, AR 72203 (501) 371-2336

★701★ U.S. Small Business Administration
Little Rock District Office
320 W. Capitol Ave., Suite 601
Little Rock, AR 72201 (501) 378-5871

California

★702★ California Industrial Development Financing Advisory Commission
P.O. Box 942809
Sacramento, CA 94209-0001 (916) 445-9597

Description: Provides through the issuance of revenue bonds an alternative method of financing for acquiring, constructing, or rehabilitating industrial and energy development facilities.

★703★ California State and Consumer Services Agency
Department of Consumer Affairs
1020 N St., Room 516
Sacramento, CA 95814 (916) 445-4465

Description: Administers the state's licensing program. Also acts as a consumer protection agency.

★704★ Office of Procurement
1823 14th St.
Sacramento, CA 95814 (916) 445-6942

★705★ California State Assembly Select Committee on Small Business
State Capitol
Sacramento, CA 95814

★706★ California State Department of Commerce
Office of Small Business
1121 L St., Suite 600
Sacramento, CA 95814 (916) 445-6545

Description: Provides management and technical assistance to small business professionals. Helps entrepreneurs develop business skills and assists them in problem solving. Directs entrepreneurs to specific financing resources, including sources of seed money and venture capital.

★707★ California State Senate Select Committee on Small Business Enterprise
State Capitol
Sacramento, CA 95814

★708★ U.S. Small Business Administration
Fresno District Office
2202 Monterey St., Suite 108
Fresno, CA 93721 (209) 487-5189

★709★ Los Angeles District Office
350 S. Figueroa St., Sixth Floor
Los Angeles, CA 90071 (213) 894-2956

★710★ Sacramento District Office
660 J St., Suite 215
Sacramento, CA 95814 (916) 551-1445

★711★ San Diego District Office
880 Front St.
San Diego, CA 92188 (619) 557-7269

★712★ San Francisco District Office
211 Main St., Fourth Floor
San Francisco, CA 94105 (415) 974-0642

★713★ U.S. Small Business Administration (Region Nine)
450 Golden Gate Ave.
Box 36044
San Francisco, CA 94102 (415) 556-7487

Colorado

★714★ Colorado Housing Finance Authority
500 E. Eighth Ave.
Denver, CO 80203 (303) 861-8962

Description: Sponsors the Quality Investment Capital (QIC) Program to provide fixed-rate financing for small business loans guaranteed by the U.S. Small Business Administration. Also administers the ACCESS Program—a small business financing tool for fixed assets. Has created the Flex Fund Program to provide a secondary market for the purchase of up to 75 percent participation in loans originated under locally administered economic development loan programs. Also offers export credit insurance.

★715★ Colorado State Department of Administration
Division of Purchasing
1525 Sherman St.
Denver, CO 80203 (303) 866-3261

★716★ Colorado State Department of Economic Development
Small Business Office
1625 Broadway, Suite 1710
Denver, CO 80202 (303) 892-3840

Description: Operates an information referral service for small business professionals. Maintains data bases. Sponsors a hotline for small business-related inquiries. Toll-free/Additional Phone Number: 800-323-7798 (hotline).

★717★ Colorado State Department of Regulatory Agencies
1525 Sherman St.
Denver, CO 80203 (303) 866-3304

Description: Serves as the state's occupational and professional licensing bureau.

★718★ Colorado State Division of Commerce and Development
1313 Sherman St., Room 523
Denver, CO 80203 (303) 833-2205

Description: Provides capital formation assistance.

★719★ CU Business Advancement Centers
1690 38th St., #101
Boulder, CO 80301 (303) 444-5723

Description: Offers business consulting, technology transfer, information retrieval, and information on selling products and services to the U.S. Government. Centers are located in Boulder, Grand Junction, Durango, Colorado Springs, Trinidad, and Burlington.

★720★ Pueblo Memorial Airport Industrial Park
Pueblo, CO 81001 (303) 544-2000

Description: Houses approximately 15 industrial companies.

★721★ U.S. Small Business Administration
Denver District Office
721 19th St., Room 407
Denver, CO 80202-2599 (303) 844-2607

★722★ U.S. Small Business Administration (Region Eight)
999 18th St., Suite 701
Denver, CO 80202 (303) 294-7033

Connecticut

★723★ Connecticut Business Development Corporation

Connecticut State Department of Economic Development
210 Washington St.
Hartford, CT 06106 (203) 566-4051

Description: For businesses whose net worths are less than $6 million and whose after-tax profits of the last two years are less than $2 million. Arranges federally guaranteed debenture financing for fixed-asset projects on a second-mortgage basis. (Second mortgages may not exceed $500,000 or 40 percent of the cost of the project.)

★724★ Connecticut Development Authority

217 Washington St.
Hartford, CT 06106 (203) 522-3730

Description: Helps companies undertake new capital expansions. Offers assistance to manufacturing, research and development, distribution, warehouse, and office facilities and for pollution control and energy conservation projects. Provides financing to cover the cost of land, the purchase and installation of machinery and equipment, and the construction, purchase, improvement, or expansion of buildings. Sponsors the Umbrella Bond/Direct Loan Program for projects up to $850,000, the Self-Sustaining Industrial Revenue Bond Program for projects up to $10 million, and the Mortgage Insurance Program for projects up to $10 million.

★725★ Connecticut Product Development Corporation

93 Oak St.
Hartford, CT 06106 (203) 566-2920

Description: Offers financing for new product development. Provides up to 60 percent of the eligible development costs of specific projects—from initial concept through fabrication of the production-ready prototype. Also offers low-cost loans of up to $200,000. Does not request a share in company ownership or management; requires only a limited royalty on the sales of sponsored products.

★726★ Connecticut State Department of Administrative Services
Bureau of Purchases

460 Silver St.
Middletown, CT 06457 (203) 638-3267

★727★ Connecticut State Department of Consumer Protection
Division of Licensing and Administration

State Office Bldg.
165 Capitol Ave.
Hartford, CT 06106 (203) 566-7177

Description: Issues occupational and professional licenses.

★728★ Connecticut State Department of Economic Development
Corner Loan Program

210 Washington St.
Hartford, CT 06106 (203) 566-3322

Description: Offers financing for working capital or fixed assets to manufacturers and wholesale distributors whose gross sales for their most recently completed fiscal years were less than $10 million and that have been doing business in the region for at least one year. Also offers financing to manufacturers and wholesale distributors who have been doing business in the region for at least two months (but less than one year) and whose gross revenues did not exceed an average of $600,000 per calendar month during that period.

★729★ Naugatuck Valley Revolving Loan Fund

Description: Offers financing for fixed assets such as land or building purchases, construction, renovation, rehabilitation, and the purchase and installation of machinery and equipment; for start-up capital; and for working capital. Funds are available to manufacturers and to wholesale distributors (no gross sales limit).

★730★ Sales Contact Centers

(203) 566-4051

Description: Helps businesses generate new sales by matching manufacturers with the purchasing executives of large corporations. Also assists corporations with locating their own suppliers.

★731★ Set-Aside Program

(203) 566-4051

Description: A procurement program for small businesses and for minority- and women-owned firms. Requires that small businesses provide 15 percent of all goods and services purchased by the state and that 25 percent of this amount be supplied by women- and minority-owned businesses. Eligible companies must have been doing business in Connecticut for at least one year, with annual gross revenues not exceeding $1.5 million.

★732★ Small Business Services

(203) 566-4051

Description: Offers managerial, job training, and employment assistance to small businesses. Sponsors low-cost financing programs for small businesses. Administers a procurement program for small and minority- and women-owned businesses. Conducts workshops and seminars.

★733★ Small Manufacturers Loan Program

Description: For companies whose gross revenues do not exceed $5 million. Offers sole source direct loans for fixed-asset financing. Makes available working capital direct loans on a matching basis.

★734★ Urban Enterprise Zone Program

Description: Offers investment incentives for manufacturers, commercial businesses, retailers, and residential property owners undertaking new capital investments in six designated cities. Special financing programs include low-cost working capital, venture capital, and small business loans. Incentives include a seven-year graduated deferral of any increase in property taxes attributable to new improvements. Added incentives for manufacturers and research-and-development facilities include 80 percent property tax abatements, 50 percent state corporate business tax reductions, and $1000 grants for each new job created. Urban enterprise zones are located in Bridgeport, Hartford, New Britain, New Haven, New London, and Norwalk.

★735★ Urban Jobs Program

Description: Assists firms undertaking new capital investments in key urban areas.

★736★ U.S. Small Business Administration
Hartford District Office

330 Main St., Second Floor
Hartford, CT 06106 (203) 240-4670

Delaware

★737★ Delaware Development Office

99 Kings Hwy.
P.O. Box 1401
Dover, DE 19903 (302) 736-4271
Dr. Dale E. Wolf, Director

Description: Administers the U.S. Small Business Administration's 503 loan program in the state. **Contact:** Donna Murray, Business Development Specialist (302) 736-4271.

★738★ Delaware Economic Development Authority
Delaware State Development Office
99 Kings Hwy.
P.O. Box 1401
Dover, DE 19903 (302) 736-4271

Description: Administers an industrial revenue bond program. Provides long-term, low-interest financing, generally at 75 percent of prime. Grants bonds to finance industrial, commercial, agricultural, and pollution control fixed assets. Limits financing to $10 million per project—except for pollution control projects, which have no maximum.

★739★ Delaware State Chamber of Commerce
One Commerce Center, Suite 200
Wilmington, DE 19801 (302) 655-7221

Description: Sponsors a series of special programs on small business financing.

★740★ Delaware State Department of Administrative Services
Division of Professional Regulations
Townsend Bldg.
P.O. Box 1401
Dover, DE 19903 (302) 736-4522

Description: Serves as the state's professional licensing agency.

★741★ Division of Purchasing
Governor Bacon Health Center
P.O. Box 299
Delaware City, DE 19706 (302) 834-4500

★742★ Delaware State Development Office
99 Kings Hwy.
P.O. Box 1401
Dover, DE 19903 (302) 736-4271

Description: Encourages business development by maintaining contacts with existing industries to stimulate growth and expansion and by providing economic information to new industrial prospects. Offers several special financing programs to assist businesses that will add to Delaware's employment base. Helps small businesses obtain financing. Provides assistance to firms experiencing regulatory problems with governmental agencies. Recommends improvements in state programs that affect small business. Collects, analyzes, and distributes statistical data on the state's economy and business climate. Coordinates employment and training programs to meet the needs of business and industry. Publications: Small Business Start-up Guide; Selling to the State of Delaware: A Guide to Procurement Opportunities.

★743★ Delaware State Division of Revenue (Wilmington)
Carvel State Office Bldg.
820 N. French St.
P.O. Box 2340
Wilmington, DE 19801 (302) 571-3369

Description: Administers Delaware's business licensing program. Responsible for taxation. Offers gross receipts tax information. Supplies businesses with necessary licensing and tax forms.

★744★ New Castle County Economic Development Corporation
12th and Market Sts.
Wilmington, DE 19801 (302) 656-5050

Description: Offers industrial revenue bonds and financing through the U.S. Small Business Administration's 503 Program.

★745★ Sussex County Department of Economic Development
P.O. Box 589
Georgetown, DE 19947 (302) 856-7701

Description: Issues industrial revenue bonds.

★746★ U.S. Small Business Administration
Wilmington District Office
844 King St., Room 5207
Wilmington, DE 19801 (302) 573-6295

★747★ Wilmington Department of Commerce
City of Wilmington
City/County Bldg., Ninth Floor
800 French St.
Wilmington, DE 19801 (302) 571-4610

Description: Maintains an industrial revenue bond and other financing programs.

★748★ Wilmington Economic Development Corporation
City of Wilmington
Two E. Seventh St.
Wilmington, DE 19801 (302) 571-9088

Description: Provides financing through the U.S. Small Business Administration's 503 Program and through other government mechanisms.

District of Columbia

★749★ District of Columbia Department of Administrative Services
Office of Material Management Administration
613 G St., N.W., Room 1009
Washington, DC 20002 (202) 727-0171

Description: Responsible for the District of Columbia's procurement activities.

★750★ District of Columbia Department of Consumer and Regulatory Affairs
614 H St., N.W.
Washington, DC 20001 (202) 727-7100

Description: Issues business and professional licenses, building permits, and certificates of occupancy. Handles corporate registrations. Maintains a one-stop Business and Permit Center for information and processing. (The Business and Permit Center is comprised of the Building and Land Regulation Administration and the Business Regulation Administration.)

★751★ District of Columbia Office of Business and Economic Development
Revolving Loan Fund
1111 E St., N.W.
Washington, DC 20004 (202) 727-6600

Description: Offers direct loans to be used in conjunction with private funds. Provides relatively short-term, gap financing. Also provides direct, below-market, long-term loans for property acquisition and capital equipment.

★752★ U.S. Small Business Administration
Washington, D.C., District Office
1111 18th St., N.W., Sixth Floor
Washington, DC 20036 (202) 634-4950

Florida

★753★ Florida State Department of Commerce
Bureau of Business Assistance
107 W. Gaines St., Room G26
Tallahassee, FL 32301 (904) 487-1314
Murry Hagerman, Development Representative
Description: Assists and counsels small businesses. **Telecommunication Access:** Alternate telephone (904) 487-1314.

★754★ Entrepreneurship Program
107 West Gains St.
Tallahassee, FL 32399-2000 (904) 488-9357
James Hosler, Director
Description: Encourages the creation and development of local organizations that support entrepreneurship.

★755★ Florida State Department of General Services
Division of Purchasing
Larson Bldg.
Tallahassee, FL 32399-0950 (904) 488-1194

★756★ Florida State Department of Professional Regulation
130 N. Monroe St.
Tallahassee, FL 32301 (904) 487-2252

★757★ Product Innovation Center
The Progress Center
One Progress Blvd., Box 7
Alachua, FL 32615 (904) 462-3942
Pamela H. Riddle, Director
Description: Provides small business inventors and entrepreneurs with technical and market feasibility assessments.

★758★ U.S. Small Business Administration
Coral Gables District Office
1320 S. Dixie Hwy., S. 501
Coral Gables, FL 33136 (305) 536-5533

★759★ Jacksonville District Office
400 W. Bay St., Room 261
Jacksonville, FL 32202 (904) 791-3782

★760★ Tampa District Office
700 Twiggs St., Room 607
Tampa, FL 33602 (813) 228-2594

★761★ West Palm Beach District Office
3500 45th St., Suite 6
West Palm Beach, FL 33407 (305) 689-2223

Georgia

★762★ Business Council of Georgia
1280 S. CNN Center
Atlanta, GA 30303-2705 (404) 223-2264
Description: A nonprofit organization that serves as the voice of Georgia's businesses. Sponsors programs in economic development, governmental affairs, human resources, and education. Works to create jobs for the citizens of Georgia. Publications: 1) Business Counselor (monthly); 2) Employer's Desk Manual; 3) How to Write an Employee's Handbook; 4) Membership Directory and Business Guide.

★763★ City of Atlanta
Business Licenses
98 Mitchell St., S.W.
Atlanta, GA 30303 (404) 658-6323

★764★ Clayton County Permits and Licenses
Clayton County Courthouse
Jonesboro, GA 30236 (404) 478-9911

★765★ Cobb County Business Licenses
1772 County Farm Rd.
Marietta, GA 30060 (404) 426-3611

★766★ DeKalb County Business Licenses
1100 Plaza Dr.
Lawrenceville, GA 30245 (404) 962-1405

★767★ Fulton County Business Licenses
Administration Bldg., Room 310
165 Central Ave., S.W.
Atlanta, GA 30303 (404) 572-3226

★768★ Georgia Secretary of State
Examining Boards Division
166 Pryor St., S.W.
Atlanta, GA 30303 (404) 656-3900

★769★ Georgia State Department of Administrative Services
Division of Purchasing
W. Tower, Suite 1308
200 Piedmont Ave., S.E.
Atlanta, GA 30334 (404) 656-3240

★770★ Georgia State Department of Agriculture
Consumer Protection Division
19 Martin Luther King, Jr., Dr.
Atlanta, GA 30334 (404) 656-3627
Description: Administers the permit program for businesses engaged in the processing, handling, storage, or distribution of food products.

★771★ Georgia State Department of Community Affairs
Small Business Revitalization Program
40 Marietta St., N.W., Eighth Floor
Atlanta, GA 30303 (404) 656-6200
Description: Offers assistance in securing long-term financing.

★772★ Georgia State House of Representatives
Industry Committee
State Capitol, Room 226
Atlanta, GA 30334 (404) 656-5115
Description: Handles small business legislation.

★773★ Gwinnett County Business Licenses
1100 Plaza Dr.
Lawrenceville, GA 30245 (404) 962-1405

★774★ U.S. Small Business Administration
Atlanta District Office
1720 Peachtree Rd., N.W., Sixth Floor
Atlanta, GA 30309 (404) 347-4749

★775★ Statesboro District Office
52 N. Main St., Room 225
Statesboro, GA 30458 (912) 489-8719

★776★ U.S. Small Business Administration (Region Four)
1375 Peachtree St., N.E., Fifth Floor
Atlanta, GA 30367 (404) 347-4999

Guam

★777★ U.S. Small Business Administration
Agana District Office
Pacific News Bldg., Room 508
238 Archbishop F. C. Flores St.
Agana, GU 96910 (671) 472-7277

Hawaii

★778★ Chamber of Commerce of Hawaii
Small Business Center
735 Bishop St., Suite 225
Honolulu, HI 96813 (808) 531-4111
Description: Serves as a resource center offering information, referrals, and consulting services. Provides business financial management and/or loan packaging assistance to meet business needs. Conducts special entrepreneur training programs.

★779★ Hawaii State Department of Accounting and General Services
Division of Purchasing and Supply
Kalanimoku Bldg.
P.O. Box 119
Honolulu, HI 96810 (808) 548-4057

★780★ Hawaii State Department of Commerce and Consumer Affairs
Professional and Vocational Licensing Division
1010 Richards St.
Honolulu, HI 96813
Description: Issues professional and occupational licenses.

★781★ Hawaii State Department of Planning and Economic Development
Small Business Information Service
P.O. Box 2359
Honolulu, HI 96814 (808) 548-8741
Carl Swanholm, Science & Technology Officer
Description: Offers information and referrals on permits and licensing, market data, business plan writing, consulting, government procurement, alternative financing, and business classes. Maintain the Small Business Information Service, a one-stop clearinghouse of small business resources. Publications: 1) Hawaii's Business Regulations; 2) Starting a Business in Hawaii.

★782★ Hawaii State Department of Taxation
425 Queen St.
Honolulu, HI 96813 (808) 548-4242
Description: Issues general excise licenses.

★783★ U.S. Small Business Administration
Honolulu District Office
300 Ala Moana, Room 2213
Honolulu, HI 96850 (808) 541-2990

Idaho

★784★ The Idaho Company
P.O. Box 6812
Boise, ID 83707 (208) 334-6308
Description: Offers assistance in obtaining venture capital.

★785★ Idaho First National Bank
Economic Action Council
Communications Department #1-5011
P.O. Box 8247
Boise, ID 83733 (208) 383-7265
Description: Administers a program for economic action fund grants.

★786★ Idaho State Board of Occupational Licenses
2417 Bank Dr., Room 312
Boise, ID 83705 (208) 334-3233

★787★ Idaho State Department of Administration
Division of Purchasing
650 W. State St.
Boise, ID 83720 (208) 334-2465

★788★ Idaho State Department of Commerce
Statehouse, Room 108
Boise, ID 83720 (208) 334-2470
Description: Offers assistance with all types of business inquiries. Provides financial information about community development block grants, urban development action grants, industrial revenue bonds, and Idaho Travel Council grants. Also offers information about regulations, permits, and licenses; state procurement programs; travel and tourism; and international trade. Maintains a census data center that provides economic and demographic data.

★789★ Idaho State Department of Revenue and Taxation
700 W. State St.
Boise, ID 83720 (208) 334-3560
Description: Operates a program for investment and employment tax credits.

★790★ U.S. Small Business Administration
Boise District Office
1020 Main St., Second Floor
Boise, ID 83702 (208) 334-1696

Illinois

★791★ Build Illinois Small Business Development Program
Equity Investment Fund
Illinois State Department of Commerce and Community Affairs
620 E. Adams St.
Springfield, IL 62701 (217) 782-7500
Description: Designed to stimulate the development of technology-based companies. Provides equity financing to companies having significant potential for job creation. Offers up to one-third of anticipated project costs, with a maximum of $250,000. Program funding may be used for the purchase of real estate, machinery, and equipment, for working capital, for research-and-development costs, and for organizational fees.

★792★ Illinois Development Finance Authority
Direct Loan Program
Two N. LaSalle St., Suite 980
Chicago, IL 60602 (312) 793-5586

Description: Offers supplemental financing to small and medium-sized businesses undertaking fixed-asset projects that will increase employment opportunities. Covers no more than 30 percent of a project.

★793★ Illinois Secretary of State
Division of Purchasing
213 Capitol Bldg.
Springfield, IL 62756 (217) 782-1560

★794★ Illinois State Department of Commerce and Community Affairs
Business Innovation Fund
620 E. Adams St.
Springfield, IL 62701 (217) 782-7500

Description: Provides funding for projects that may not otherwise attract traditional lenders or venture capitalists. Offers financial aid to businesses using a university's assistance to advance technology, to create new products, or to improve manufacturing processes. Businesses may qualify for up to $100,000 in financial assistance for projects matched with private resources. A royalty agreement reimburses the state when the product is developed and sold in the marketplace.

★795★ Small Business Advocacy

Description: Responsible for the representation, coordination, legislative, and ombudsman services to small businesses and related organizations throughout the state. Refers small business owners to technical, management, and special financial assistance programs. The department also sponsors a program for entrepreneurial and employee education.

★796★ Small Business Assistance Bureau

Description: Provides information about all state government forms and applications for new and existing businesses. Maintains a one-stop permit center that offers a business start-up kit featuring all state permit forms and license applications. Toll-free/Additional Phone Number: 800-252-2923 (Small Business Hotline).

★797★ Small Business Development Program

Description: Provides—along with a participating financial institution—direct financing to small businesses for expansion and subsequent job creation or retention. Offers long-term, fixed-rate, low-interest loans.

★798★ Small Business Energy Management/Loan Program

Description: Offers small businesses management assistance and financial aid in lowering utility costs. Sponsors the Energy Conservation Loan Program to provide direct financing to small businesses at a below-market interest rate in cooperation with private sector lenders.

★799★ Small Business Financing Program

Description: Helps finance economic development/job creation projects by combining dollars from the U.S. Small Business Administration, private capital, and public block grants. Other funding categories for community improvement activities include a Competitive Economic Development Program, a Public Facilities and Housing Program, and Set-Aside Funds.

★800★ Small Business Fixed-Rate Financing Fund

Description: Purpose is to provide affordable start-up and expansion financing to small businesses and to create jobs for low- and moderate-income people. Offers long-term, fixed-rate financing to new or expanding companies for fixed assets and working capital.

★801★ Small Business Micro Loan Program

Description: Offers direct financing to small businesses at below-market interest rates in cooperation with private sector lenders. Funds may be used for acquisition of land or buildings; construction, renovation, or leasehold improvements; purchase of machinery or equipment; or inventory and working capital.

★802★ Illinois State Department of Commerce and Community Development
Procurement Assistance Program
620 E. Adams St.
Springfield, IL 62701 (217) 782-7500

Description: Helps businesses participate in government markets. Sponsors procurement assistance centers that help firms seeking state and federal government contracts.

★803★ Illinois State Department of Registration and Education
Division of Licensing and Testing
320 W. Washington St.
Springfield, IL 62786 (217) 785-0800

★804★ Illinois Venture Capital Fund
Illinois Development Finance Authority
Two N. LaSalle St., Suite 980
Chicago, IL 60602 (312) 793-5586

Description: Funded by the state of Illinois and the Frontenac Venture Company. Provides seed capital for enterprises seeking to develop and test new products, processes, technologies, and inventions.

★805★ U.S. Small Business Administration
Chicago District Office
219 S. Dearborn St., Room 437
Chicago, IL 60604-1779 (312) 353-4528

★806★ Springfield District Office
Four N. Old State Capitol Plaza
Springfield, IL 62701 (217) 492-4416

★807★ U.S. Small Business Administration (Region Five)
230 S. Dearborn St., Room 510
Chicago, IL 60604-1593 (312) 353-0359

Indiana

★808★ Corporation for Innovation Development
One N. Capitol, Suite 520
Indianapolis, IN 46204 (317) 635-7325
Mr. Marion C. Dietrich, President & C.E.O.

Description: Encourages capital investment, with the primary objective of providing additional employment through the expansion of business and industry—particularly new and existing small business enterprises. Invests in small business investment companies as well as directly in new and existing businesses. Includes common or preferred stock equity investments and various forms of debt, with warrants or other convertible features.

★809★ Indiana Corporation for Science and Technology
One N. Capitol, Ninth Floor
Indianapolis, IN 46204 (317) 635-3058

Description: Administers funds set aside for research and development of new products.

★810★ Indiana Institute for New Business Ventures
One N. Capitol, Suite 420
Indianapolis, IN 46204 (317) 634-8418
Robert Cummings, President

Description: Encourages and supports the creation and development of emerging enterprises that are expected to provide growth opportunities for Indiana's economy. Introduces businesses to appropriate financing sources. Offers management, educational, technical, and financial resources to assist in starting and operating small enterprises. Provides referral services and small business counsellors. Sponsors conferences and educational programs.

★811★ Indiana Seed Capital Network
Indiana Institute for New Business Ventures
One N. Capitol, Suite 420
Indianapolis, IN 46204 (317) 634-8418
David C. Clegg, Director

Description: A computerized matching service for investors and entrepreneurs. Identifies investment opportunities in entrepreneurial ventures as well as informal investors. Acts as a referral service to entrepreneurs and investors.

★812★ Indiana State Department of Administration Division of Procurement
507 State Office Bldg.
Indianapolis, IN 46204 (317) 232-3032

★813★ Indiana State Department of Commerce
One N. Capitol
Indianapolis, IN 46204 (317) 232-8800

Description: Administers financial assistance programs for small businesses.

★814★ Indiana State Professional Licensing Agency
State Office Bldg., Suite 1021
Indianapolis, IN 46204 (317) 232-3997

★815★ U.S. Small Business Administration Indianapolis District Office
575 N. Pennsylvania St., Room 578
Indianapolis, IN 46204-1584 (317) 269-7272

Iowa

★816★ Iowa Business Development Credit Corporation
901 Insurance Exchange Bldg.
Des Moines, IA 50309 (515) 282-2164

Description: Stimulates economic development through loans to new or established firms in conjunction with banks, insurance companies, savings and loan associations, and other financial institutions. Loan proceeds may be used to purchase land, to purchase or construct buildings, machinery, equipment, or inventory, or as working capital. A portion may be used to retire debt. Makes available loans up to $500,000—plus the amount of bank participation. (Most loans range from $100,000 to $200,000.) Nonprofit enterprises, lending or financial institutions, and those firms able to acquire funds at reasonable rates from other sources are ineligible for assistance. Also offers counseling in financial, marketing, and managerial areas.

★817★ Iowa Business Growth Company
901 Insurance Exchange Bldg.
Des Moines, IA 50309 (515) 282-2164

Description: Maintains the U.S. Small Business Administration's 503/504 loan programs for businesses in every Iowa community. Helps stimulate growth and expansion of small businesses by providing long-term, fixed-asset financing at a fixed rate of interest slightly below the market rate. Applicants must demonstrate that—as a result of the loan—they will hire additional employees, fill a needed service in the community, or help upgrade a deprived area. May finance up to 45 percent of the cost of plant construction, conversion, or expansion, including the acquisition of land, existing buildings, leasehold improvements, and equipment (if the equipment has a useful life of at least ten years). May lend up to $500,000 for 15, 20, or 25 years on a second-mortage basis. The small business must provide a minimum of 10 percent of the project costs. The remaining project funds are provided by a conventional first-mortgage lender for at least ten years at market interest rates.

★818★ Iowa State Department of Commerce Division of Professional Licensing and Regulation
1918 S.E. Hulsizer Ave.
Ankeny, IA 50021 (515) 281-5596

★819★ Iowa State Department of Economic Development Call One Program
200 E. Grand Ave.
Des Moines, IA 50309 (515) 281-8310

Description: Answers questions about licenses, permits, and other requirements to operate a business in Iowa. Channels specific requests to proper resources. Toll-free/Additional Phone Number: 800-532-1216 (in Iowa); (515)281-8324.

★820★ Community Development Block Grant Program
(515) 281-3538

Description: Uses federal dollars to finance community improvements such as public works projects, housing rehabilitation, and job-generating expansions in all counties and cities, except the nine largest Iowa cities (which receive funds directly from the federal government). Funds are awarded on a competitive basis.

★821★ Community Economic Betterment Account
(515) 281-3415

Description: Established to invest proceeds of the Iowa Lottery in local economic development projects. Cities, counties, or merged area schools are eligible to apply on behalf of local enterprises that are expanding, modernizing, or locating. Key criteria for approval of applications are the number of jobs created, the cost per job for the funds involved in the project, and significant community involvement and interest. Funds are used to buy-down principal or interest on business loans to acquire land or buildings for construction or reconstruction, to purchase equipment, to prepare industrial sites, and to assist in other business financings.

★822★ Export Finance Program
(515) 281-6295

Description: Provides financial assistance through grant fund monies to qualified exporters of Iowa-manufactured or Iowa-processed products. A company applying for financial assistance on an export transaction may utilize the interest buy-down for both pre-export expenses and post-export term financing.

★823★ Iowa Procurement Outreach Center
c/o Kirkwood Community College
6301 Kirkwood Blvd., S.W.
Cedar Rapids, IA 52406 (319) 398-5665

Description: A service center for Iowa small businesses interested in selling goods and services to the federal government. Uses existing resources and programs to identify interested businesses and assists them in bidding for and receiving government contracts. Toll-free/Additional Phone Number: 800-458-4465 (in Iowa); (319)398-5666.

★824★ Iowa Product Development Corporation
200 E. Grand Ave.
Des Moines, IA 50309 (515) 281-5459
Glen Burmeister, Deputy Director

Description: Provides risk capital to feasible ventures that have the potential to create new jobs or to develop new products for

Iowa State Department of Economic Development *(Cont.)*

manufacture in Iowa. The return on investments received by the IPDC is reinvested in new ventures.

★825★ Iowa Small Business Advisory Council
(515) 281-7252

Description: Works to determine the problems and priorities of Iowa's business owners. Makes recommendations on legislative matters affecting businesses. Serves as an advocate for small businesses in the state. Toll-free/Additional Phone Number: 800-532-1216 (in Iowa).

★826★ Iowa Small Business Loan Program
(515) 281-4058

Description: Assists in the development and expansion of small businesses in Iowa through the sale of bonds and notes that are exempt from federal income tax and through the use of the proceeds to provide limited types of financing for new or existing small businesses. The loans may be used for purchasing land, construction, building improvements, or equipment. Funds cannot be used for working capital inventory or operating purposes. The maximum loan is $10 million. Rates vary with the level of risk.

★827★ Iowa Small Business Vendor Application Program

Description: An aid to small businesses competing for state government contracts. Application forms are available to vendors wanting their names placed on lists for use by state purchasing agencies soliciting bids. Toll-free/Additional Phone Number: 800-532-1216 (in Iowa).

★828★ Self-Employment Loan Program
(515) 281-3600

Description: Assists low-income entrepreneurs by providing low-interest loans for new or expanding small businesses. The loans shall not exceed $5000 or be issued at a rate to exceed 5 percent simple interest per annum.

★829★ Targeted Small Business Loan Guarantee Program
(515) 281-4058

Description: Provides guarantees to lenders for loans made to certified targeted small businesses in the state of Iowa. Provides up to a 75 percent loan guarantee for private lenders making loans to targeted small businesses.

★830★ Iowa State Department of General Services Division of Purchasing and Materials Management
Hoover State Office Bldg.
Des Moines, IA 50319 (515) 281-3089

★831★ Iowa State House Committee on Small Business and Commerce
State Capitol
Des Moines, IA 50319 (515) 281-3221

★832★ U.S. Small Business Administration Cedar Rapids District Office
373 Collins Rd., N.E., Room 100
Cedar Rapids, IA 52402-3118 (319) 399-2571

★833★ Des Moines District Office
210 Walnut St., Room 749
Des Moines, IA 50309 (515) 284-4567

Kansas

★834★ Avenue Area, Inc.
New Brotherhood Bldg.
Seventh and Minnesota
Kansas City, KS 66101 (913) 371-0065

Description: Provides financial packaging services for U.S. Small Business Administration and other financing programs to eligible small businesses in Wyandotte County.

★835★ Big Lakes Certified Development Company
104 S. Fourth
Manhattan, KS 66502-6110 (913) 776-0417

Description: Provides financial packaging services for U.S. Small Business Administration and other financing programs to eligible small businesses located in Clay, Geary, Pottawatomie, Marshall, and Riley counties.

★836★ Citywide Development Corporation of Kansas City, Kansas, Inc.
Kansas State Department of Economic Development
Municipal Office Bldg.
One Civic Plaza
Kansas City, KS 66101 (913) 573-5730

Description: Provides financial packaging services for U.S. Small Business Administration and other financing programs to eligible small businesses in Kansas City, Kansas.

★837★ Four Rivers Development, Inc.
108 E. Main
P.O. Box 365
Beloit, KS 67420 (913) 738-2210

Description: Provides financial packaging services for U.S. Small Business Administration and other financing programs to eligible small businesses in Cloud, Dickinson, Ellsworth, Jewell, Lincoln, Mitchell, Ottawa, Republic, Saline, and Washington counties.

★838★ Great Plains Development, Inc.
1111 Kansas Plaza
Garden City, KS 67846 (316) 275-9176

Description: Provides financial packaging services for U.S. Small Business Administration and other financing programs to eligible small businesses in Barber, Barton, Clark, Cloud, Comanche, Edwards, Finney, Ford, Grant, Gray, Greeley, Hamilton, Haskell, Hodgeman, Kearney, Kiowa, Lane, Meade, Morton, Ness, Pawnee, Pratt, Rush, Scott, Seward, Stafford, Stanton, Stevens, and Wichita counties. Maintains an office at 407 S. Main, P.O. Box 8776, Pratt, KS 67124. Toll-free/Additional Phone Number: (316)672-9421 (Pratt office).

★839★ Kansas State Department of Administration Small Business Procurement Program
Landon State Office Bldg., Room 102
Topeka, KS 66612 (913) 296-2376

Description: The Kansas Small Business Procurement Act of 1978 establishes a set-aside program for small businesses. It is the policy of the state to ensure that a fair proportion of the purchase of and contracts for property and services for the state be placed with qualified small business contractors. Program includes supplies, materials, equipment, maintenance, contractual services, repair services, and construction.

★840★ Kansas State Department of Commerce Economic Development Bond Program
400 W. Eighth St., Fifth Floor
Topeka, KS 66603-3957 (913) 296-5298

Description: Kansas cities and counties are authorized to issue economic development bonds, generally referred to as industrial revenue bonds, to facilitate the expansion of existing businesses and the location of new businesses in the state. Economic development bonds may be used to purchase, construct, or equip buildings; to acquire sites; and to enlarge or remodel buildings for agricultural, commercial, hospital, industrial, and manufacturing facilities. The interest on bonds issued for manufacturing purposes are exempt from federal and state income taxes.

★841★ Kansas Enterprise Zone Act
(913) 296-3485

Description: Allows qualified businesses and industries located in an approved enterprise zone to take enhanced job expansion and investment income tax credits and Kansas sales tax exemptions on specified capital improvements.

★842★ One-Stop Permitting

Description: Serves as a central point of contact for general information needed to establish or operate a business in Kansas. Provides necessary state applications and forms required by agencies that license, regulate, or tax businesses. Answers questions about starting or expanding a business in Kansas.

★843★ Small Cities Community Development Block Grant Program
(913) 296-3004

Description: Approximately $12.5 million will be awarded on a competitive basis to Kansas communities with populations less than 50,000 for projects that stimulate or support local economic activity, principally for persons of low or moderate incomes. These community revitalization funds may be used for public facilities, housing rehabilitation, and economic development projects.

★844★ Tax Increment Financing Program
(913) 296-3485

Description: Authorizes cities to redevelop blighted central business district areas through private investment aided by the issuance of city bonds. The bonds are retired by increased property tax revenues raised as a result of higher assessed values of the redeveloped property. Tax increment financing provisions also may be utilized in designated Kansas enterprise zones.

★845★ Kansas State Procurement Technical Assistance Program
Wichita State University
Kansas Small Business Development Centers
021 Clinton Hall
Campus Box 148
Wichita, KS 67208-1595 (316) 689-3193

Description: Assists Kansas businesses in getting their fair share of government contracts. Offered through the Kansas State Small Business Development Center system, which provides one-on-one counseling assistance and procurement training seminars. (Counseling assistance is extended to any Kansas business free of charge.)

★846★ Kansas Technology Enterprise Corporation
Kansas State Department of Commerce
400 W. Eighth St., Fifth Floor
Topeka, KS 66603-3957 (913) 296-5272
Kevin Carr, Director

Description: A nonprofit corporation designed to foster innovation, with programs ranging from research-and-development grants to equity investments. Offers research matching grants, which fund 40 percent of the cost of industry research-and-development projects that lead to job creation in Kansas. Also maintains a seed capital fund to provide equity financing for high-tech product development. Matching funds are provided for the Small Business Innovation Research Program. Also conducts an annual high-tech expo.

★847★ League of Kansas Municipalities
112 W. Seventh St.
Topeka, KS 66603 (913) 354-9565

Description: Administers tax-exempt bond and tax increment financing programs.

★848★ Lenexa Development Company, Inc.
8700 Monrovia, Suite 300
P.O. Box 14244
Lenexa, KS 66215 (913) 888-1826

Description: Provides financial packaging services for U.S. Small Business Administration and other financing programs to eligible small businesses in the city of Lenexa.

★849★ McPherson County Small Business Development Association
101 S. Main
P.O. Box 1032
McPherson, KS 67460 (316) 241-2100

Description: Provides financial packaging services for U.S. Small Business Administration and other financing programs to eligible small businesses in McPherson County.

★850★ Mid-America, Inc.
1715 Corning
P.O. Box 708
Parsons, KS 67357 (316) 421-6350

Description: Provides financial packaging services for U.S. Small Business Administration and other financing programs to eligible small businesses in Allen, Anderson, Bourbon, Cherokee, Crawford, Labette, Montgomery, Neosho, Wilson, and Woodson counties.

★851★ Neosho Basin Development Company
Emporia State University
1200 Commercial
Box 46
Emporia, KS 66801 (316) 342-7041

Description: Provides financial packaging services for U.S. Small Business Administration and other financing programs to eligible small businesses in Chase, Coffey, Lyon, Morris, Osage, and Wabaunsee counties.

★852★ Pioneer Country Development, Inc.
317 N. Pomeroy Ave.
P.O. Box 248
Hill City, KS 67642 (913) 674-3488

Description: Provides financial packaging services for U.S. Small Business Administration and other financing programs to eligible small businesses in Cheyenne, Decatur, Ellis, Gove, Graham, Logan, Norton, Osborne, Phillips, Rawlins, Rooks, Russell, Sheridan, Sherman, Smith, Thomas, Trego, and Wallace counties.

★853★ South Central Kansas Economic Development District
River Park Place, Suite 565
727 N. Waco
Wichita, KS 67203 (316) 262-5246

Description: Provides financial packaging services for U.S. Small Business Administration and other financing programs to eligible small businesses in Butler, Chautauqua, Cowley, Elk, Greenwood, Harper, Harvey, Kingman, Marion, Reno, Rice, Sedgwick, and Sumner counties.

★854★ Topeka/Shawnee County Development, Inc.
820 S.E. Quincy, Suite 501
Topeka, KS 66612 (913) 234-0072

Description: Provides financial packaging services for U.S. Small Business Administration and other financing programs to eligible small businesses in Shawnee County.

★855★ U.S. Small Business Administration Wichita District Office
110 E. Waterman St.
Wichita, KS 67202 (316) 269-6571

★856★ Wakarusa Valley Development, Inc.
1321 Wakarusa Dr., Suite 2102
P.O. Box 1732
Lawrence, KS 66046 (913) 841-7120

Description: Provides financial packaging services for U.S. Small Business Administration and other financing programs to eligible small businesses in Douglas County.

★857★ Wichita Area Development, Inc.
350 W. Douglas
Wichita, KS 67202 (316) 265-7771

Description: Provides financial packaging services for U.S. Small Business Administration and other financing programs to eligible small businesses in the city of Wichita.

Kentucky

★858★ Kentucky Development Finance Authority
Kentucky State Commerce Cabinet
2400 Capital Plaza Tower
Frankfort, KY 40601 (502) 564-4886

Description: Encourages economic development, business expansion, and job creation by providing financial support to manufacturing, agribusiness, and tourism projects. Issues industrial revenue bonds for eligible projects and second-mortgage loans to private firms in participation with other lenders. Houses the Commonwealth Small Business Development Corporation. Maintains the Kentucky Community Development Block Grant Program. **Telecommunication Access:** Alternate telephone (502) 564-4554.

★859★ Kentucky State Commerce Cabinet
Business Information Celaringhouse
Capital Plaza Tower, 22nd Floor
Frankfort, KY 40601 (502) 564-4252

Description: Provides new and existing businesses with a centralized information source on business regulation. Assists businesses in securing necessary licenses, permits, and other endorsements, including an ombudsman service with regulatory agencies. Acts as a referral service for government financial and management assistance programs. Serves as a regulatory reform advocate for business. Toll-free/Additional Phone Number: 800-626-2250.

★860★ Small Business Division
Capital Plaza Tower, 22nd Floor
Frankfort, KY 40601
Patsy Wallace, Acting Director

Description: Encourages small business development in the state. Assists existing small businesses in Kentucky through referrals and information. Operates a computerized procurement service. Functions as a small business development center in conjunction with other such facilities across the state. Toll-free/Additional Phone Number: 800-626-2250.

★861★ Kentucky State Finance and Administration Cabinet
Division of Purchases
Capitol Annex, Room 348
Frankfort, KY 40601 (502) 564-4510

Description: Kentucky's purchasing law provides an opportunity for state agencies to make a special effort to do business with small companies and with minority- and female-owned firms through the set-aside of certain purchases for bidding by small businesses or small minority firms.

★862★ Occupations and Professions Division
301 Capitol Annex
Frankfort, KY 40601 (502) 564-3296

★863★ Kentucky State Transportation Cabinet
Division of Purchases
State Office Bldg.
Frankfort, KY 40601 (502) 564-4630

★864★ U.S. Small Business Administration
Louisville District Office
600 Federal Place, Room 188
Louisville, KY 40202 (502) 582-5971

Louisiana

★865★ Louisiana Small Business Equity Corporation
4521 Jamestown Ave., Suite 9
Baton Rouge, LA 70808 (504) 925-4112

Description: Administers financial assistance programs for small businesses. **Telecommunication Access:** Alternate telephone (504) 342-9213.

★866★ Louisiana State Department of Commerce and Industry
Small Business Development Corporation
P.O. Box 94185
Baton Rouge, LA 70804 (504) 342-5361
Mr. Nadia Goodman, Acting Executive Director

Description: Maintains a variety of programs to assist owners and operators of small businesses.

★867★ Louisiana State Department of Health and Human Resources
Licensing and Regulation Office
P.O. Box 3776
Baton Rouge, LA 70821 (504) 342-6711

★868★ Louisiana State Office of the Governor
State Purchasing Office
P.O. Box 94095
Baton Rouge, LA 70804-9095 (504) 922-0074

★869★ U.S. Small Business Administration
New Orleans District Office
1661 Canal St., Suite 2000
New Orleans, LA 70112 (504) 589-6685

★870★ Shreveport District Office
500 Fannin St., Room 8A08
Shreveport, LA 71101 (318) 226-5196

Maine

★871★ Finance Authority of Maine
83 Western Ave.
P.O. Box 949
Augusta, ME 04330 (207) 623-3263

Description: An economic development agency that administers all state-operated business financing programs, including a mortgage insurance program, small business loan guarantees, veterans' loans, a natural resources entrants program, and several industrial development bond programs.

★872★ Maine State Department of Administration
Bureau of Purchasing
State House Station #74
Augusta, ME 04333 (207) 289-3521

★873★ Maine State Department of Professional and Financial Regulation
Division of Licensing and Enforcement
State House Station #35
Augusta, ME 04333 (207) 289-3671

★874★ Maine State Development Office
Financial Programs
State House Station #59
Augusta, ME 04333 (207) 289-2656

Description: Administers a variety of financing programs, including job and investment tax credits, revolving loan funds, a development fund, and tax increment financing.

★875★ Maine Growth Program

Description: Offers financial packaging to healthy small businesses. Funds may be used for building new plants, expanding existing facilities, or purchasing new equipment.

★876★ U.S. Small Business Administration
Augusta District Office
40 Western Ave., Room 512
Augusta, ME 04330 (207) 622-8378

Maryland

★877★ Maryland Small Business Development Financing Authority
World Trade Center
401 E. Pratt St.
Baltimore, MD 21202 (301) 659-4270

Description: Assists socially and economically disadvantaged business persons in the development of their businesses. Acts as a funds precipitator and a risk taker. Minority businesses are the primary recipients of assistance, although nonminorities may seek surety bond guarantees. Provides working capital and funds for equipment acquisition through direct loans and loan guarantees for the completion of projects that are under federal, state, and local government and public utility contracts. Offers long-term loan guarantees to lending financial institutions for property improvements and for working capital and equipment financing (up to a ten-year period and 80 percent of the loan request). Provides up to a 4 percent interest rate subsidy on long-term loan guarantees. Insures up to a maximum of 90 percent loss on a $1 million face contract for both minority and nonminority applicants seeking surety bonding. Provides equity participation franchise financing up to a maximum of $100,000 per franchisee or 45 percent of costs.

★878★ Maryland State Department of Economic and Community Development
Maryland Business Assistance Center
45 Calvert St.
Annapolis, MD 21401 (301) 974-2945

Description: Provides expertise in industrial and community development, financial management, and technical assistance to the private sector and to local governments. Offers information on financing programs, facility location, state-funded employee training, procurement, licensing and permits, and starting a business. Toll-free/Additional Phone Number: 800-OK-GREEN (hotline).

★879★ Maryland State Department of General Services
Purchasing Bureau
301 W. Preston St.
Baltimore, MD 21201 (301) 225-4620

Description: Anne Arundel, Howard, Montgomery, and Prince George's counties also maintain procurement offices.

★880★ Maryland State Department of Licensing and Regulations
Division of Occupational and Professional Licensing
501 St. Paul Pl.
Baltimore, MD 21202 (301) 659-6209

★881★ Maryland State License Bureau
County Courthouse
Upper Marlboro, MD 20772

Description: Responsible for business licenses.

★882★ Maryland State Treasury Department
State License Bureau
301 W. Preston St., Fourth Floor
Baltimore, MD 21201 (301) 225-1550

★883★ U.S. Small Business Administration
Baltimore District Office
Ten N. Calvert St.
Baltimore, MD 21202 (301) 962-2054

Massachusetts

★884★ Massachusetts Business Development Corporation
One Boston Place
Boston, MA 02108 (617) 723-7515

Description: Provides loans for small and medium-sized businesses that cannot obtain all of their financial requirements from conventional sources. Makes loans for working capital, leveraged buy outs, second mortgages, government guaranteed loans, U.S. Small Business Administration 503 loans, and long-term loans for new equipment or energy conversion. Lending terms are flexible.

★885★ Massachusetts Capital Resource Company
545 Boylston St.
Boston, MA 02116 (617) 536-3900

Description: Supports the maintenance and growth of Massachusetts businesses by investing in traditional and technology-based industries, high-risk start-up companies, expanding businesses, management buy outs, and turnaround situations. Structures its investments to fit the specific needs of its portfolio companies and to increase employment opportunities within Massachusetts.

★886★ Massachusetts Community Development Finance Corporation
131 State St., Suite 600
Boston, MA 02109 (617) 742-0366

Description: Provides flexible financing for working capital needs and real estate development projects when there is some clear public benefit. Makes investments in conjunction with community development corporations, which are organized to promote economic development in targeted areas of Massachusetts. Financing is available to businesses that are able to provide good employment opportunities, that are unable to meet their capital needs in the traditional markets, and that have the sponsorship of an eligible community development corporation. Includes venture capital investment, community development investment, and small loan guarantee programs.

★887★ The Massachusetts Government Land Bank
Six Beacon St., Suite 900
Boston, MA 02108 (617) 727-8257

Description: A quasi-public state agency functioning as a financier and public developer. Offers below-market mortgage financing to qualifying public-purpose development projects that lack sufficient public/private investment. Can acquire, rehabilitate, prepare, construct, demolish, and dispose of eligible properties. Has access to $40 million in Massachusetts general obligation bonds, half of which are industrial development bonds. Project investments generally range from $200,000 to $3 million.

★888★ Massachusetts Industrial Finance Agency
125 Pearl St.
Boston, MA 02110 (617) 451-2477

Description: Promotes employment growth through incentives that stimulate business investment in Massachusetts. Incentives for industrial companies include industrial revenue bonds, loan guarantees, and pollution control bonds. Financing is available for commercial real estate if these projects are located in locally identified commercial area revitalization districts.

★889★ Massachusetts Product Development Corporation
Industrial Services Program
12 Marshall St.
Boston, MA 02108 (617) 727-8158

Description: Promotes new product development in Massachusetts firms. Offers priority in funding decisions to firms that will experience job loss because of foreign competition, declining markets, or shifts in industry technologies. Investments may be used for product prototypes, marketing efforts, distribution, or new production processes. Firms must detail the number and types of jobs created or maintained in order to receive capital.

★890★ Massachusetts State Department of Commerce
Massachusetts Suppliers and Manufacturers Matching Service
100 Cambridge St.
Boston, MA 02202 (617) 727-3206

Description: Serves as a liaison between manufacturers and suppliers in Massachusetts to meet their contractual purchasing needs of parts, raw materials, or components. Maintains the Massachusetts Matching Service, a computerized listing of suppliers.

★891★ Office of Financial Development
(617) 727-2932

Description: Assists businesses and individuals attempting to utilize the many federal, state, and local finance programs established to help businesses expand. Sponsors the Massachusetts Venture Capital Fair. Publications: 1) Massachusetts Financial Resources Directory; 2) Massachusetts Venture Capital Directory.

★892★ Small Business Assistance Division
(617) 727-4005

Description: Provides technical assistance and programs to support small business development. Serves as an advocate within state government for small business concerns. Provides information on starting a business, developing business plans, doing business with the state, and identifying sources of financing, management, marketing, or other business assistance. Advises firms interested in exporting. Administers and monitors the Small Business Purchasing Program.

★893★ Massachusetts State Department of Revenue
Corporations Bureau
215 First St., Third Floor
Cambridge, MA 02142 (617) 727-4264

Description: Administers a variety of financing programs to assists businesses, including loss carry over for new corporations, local tax exemption on tangible property, expensing research and development and other federal tax incentives, and dividend deductions.

★894★ Taxpayer Assistance Bureau, Corporation Section
100 Cambridge St.
Boston, MA 02204 (617) 727-4271

Description: Administers financial programs such as investment tax credits, sales tax exemptions for machinery, capital gains deductions, and tax incentives for donations of scientific equipment.

★895★ Massachusetts State Executive Office of Administration and Finance
Division of Purchasing Agents
One Ashburton Place
Boston, MA 02108 (617) 727-2882

★896★ Massachusetts State Executive Office of Consumer Affairs and Business Regulation
Division of Registration
100 Cambridge St.
Boston, MA 02202 (617) 727-3076

★897★ Thrift Institution Fund for Economic Development
One Ashburton Place
Boston, MA 02108 (617) 727-7755

Description: A $100 million lending pool to be invested over a ten-year period for a variety of economic development and job-generating purposes, including small business development.

★898★ Massachusetts State Office of Facilities Management
Division of Capital Planning and Operations
One Ashburton Place, 15th Floor
Boston, MA 02108 (617) 727-4028

Description: Offers financial assistance and tax reductions for pollution control facilities and alternative energy sources.

★899★ Massachusetts Technology Development Corporation
84 State St., Suite 500
Boston, MA 02109 (617) 423-9293

Description: Provides capital to new and expanding high-technology companies that have the capacity to generate significant employment growth in Massachusetts and that offer other public benefits.

★900★ Massachusetts Technology Park Corporation
Westborough Executive Park
P.O. Box 663
Westborough, MA 01581 (617) 870-0312
Ms. Christine Sheroff, Public Relations

Description: Will establish educational centers such as the Massachusetts Microelectronics Center. Centers will contain design, fabrication, and testing facilities and equipment for postsecondary academic and practical training programs required to satisfy the education and employment needs of the state's businesses and industries.

★901★ U.S. Small Business Administration
Boston District Office
Ten Causeway St., Room 265
Boston, MA 02222-1093 (617) 565-5561

★902★ U.S. Small Business Administration (Region One)
60 Batterymarch St., Tenth Floor
Boston, MA 02110 (617) 223-3204

Michigan

★903★ Michigan Investment Fund
Michigan State Department of Treasury
Venture Capital Division
P.O. Box 15128
Lansing, MI 48901 (517) 373-4330

Description: Invests in businesses with strong management that show a substantially above-average potential for growth, profitability, and equity appreciation. Will consider both high-technology and existing technology businesses, but the company must have a proprietary competitive edge. Will also consider joint investment with other financial institutions. Real estate investments and requests involving working capital alone will not be considered. A typical investment is $1 million or more. Additional requirements include an annual return on the investment of at least 35 percent and a business location or planned location with at least half of the assets and personnel in Michigan.

★904★ Michigan State Business Ombudsman
P.O. Box 30107
Lansing, MI 48909 (517) 373-6241

Description: Serves as a one-stop center for business permits. Utilizes a computerized process to determine all permits and licenses required for a specific business. Applications for licenses and permits also are available through this office, as well as a business start-up kit and an employer packet, which include all necessary forms, posters, and applicable information on Michigan laws concerning employer rights and responsibilities. Toll-free/Additional Phone Number: 800-232-2727.

★905★ Michigan State Department of Commerce
Procurement Assistance
Law Bldg., Fourth Floor
P.O. Box 30004
Lansing, MI 48909 (517) 373-9626

Description: Helps businesses tap into government and corporate contracts by training companies to use standard procurement procedures.

★906★ Technology Transfer Network
P.O. Box 30225
Lansing, MI 48909 (517) 335-2139
Sharon Woollard, Director

Description: Seeks to link Michigan businesses with the technical innovations. Provides technical assistance, faculty counsulting, research and development, joint research, specialized equipment and facilities, and seminars and workshops through programs at affiliated universities, which include University of Michigan, Michigan State University, Michigan Technological University, Wayne State University, and Western Michigan University.

★907★ Michigan State Department of Licensing and Regulation
P.O. Box 30018
Lansing, MI 48909 (517) 373-1870

★908★ Michigan State Department of Management and Budget
Division of Purchasing
P.O. Box 30026
Lansing, MI 48909 (517) 373-0300

★909★ Michigan State Department of Transportation
Small Business Liaison
425 W. Ottawa
P.O. Box 30050
Lansing, MI 48909 (517) 373-2090

Description: Enforces set-aside procurement programs. Publishes contract notices. Accepts bids for surface transportation projects.

★910★ Michigan Strategic Fund
Seed Capital Companies
Law Bldg.
P.O. Box 30234
Lansing, MI 48909 (517) 373-6378
Dr. James Kenworthy, Manager R&D Technology

Description: Focuses on addressing the financing gap that exists in the early stages of business development to finance pre-start-up activities such as developing a working prototype, preparing a business plan, initial market analysis, and assembling a management team. Also administers the Capital Access Program, which provides financing based on a risk pooling concept for small and medium-sized businesses. Toll-free/Additional Phone Number: (517)373-0349 (Capital Access Program).

★911★ U.S. Small Business Administration
Detroit District Office
477 Michigan Ave., Room 515
Detroit, MI 48226 (313) 226-6075

★912★ Marquette District Office
300 S. Front St.
Marquette, MI 49885 (906) 225-1108

Minnesota

★913★ Indian Affairs Council
Indian Business Loan Program
1819 Bemidji Ave.
Bemidji, MN 56601 (218) 755-3825

Description: Provides Minnesota-based Indians with loans to establish or expand a business in the state. Loan funds cannot be used to repay or consolidate existing debt. Also provides management and technical assistance resources.

★914★ Minnesota Science and Technologies Office
900 American Center Bldg.
150 East Kellogg Blvd.
St. Paul, MN 55101 (612) 297-1554

Description: Provides business access to NASA technologies. Publishes a directory of state high-technology companies and software companies and a guide to business assistance for high-tech companies.

★915★ Minnesota State Department of Administration
Division of Procurement
112 Administration Bldg.
50 Sherburne Ave.
St. Paul, MN 55155 (612) 296-3871

Description: Publishes Selling Your Product to the State of Minnesota. Toll-free/Additional Phone Number: (612)296-6949.

★916★ Minnesota State Department of Energy and Economic Development
Community Development Corporation Program
900 American Center Bldg.
150 E. Kellogg Blvd.
St. Paul, MN 55101 (612) 297-1304

Description: An annual grant cycle whereby community development corporations make applications to the authority for funds for the administrative costs of their operations and for assistance to ventures within the defined community area through direct loan participation.

★917★ Energy Development Loan Program
(612) 297-1940

Description: Finances loans through the sale of tax-exempt or taxable industrial development bonds, providing a lower interest rate to the borrower than is available commercially. Loans may be for up to

Minnesota State Department of Energy and Economic Development *(Cont.)*

90 percent of project costs, the actual percentage determined by credit and project analysis.

★918★ Energy Loan Insurance Program
(612) 297-1940

Description: Operates in partnership with private financial institutions. Applicants and participating financial institutions develop a loan package together, based on the lending institution's credit analysis and determination of appropriate financing. The program then insures up to 90 percent of the loan. (The actual percentage insured is determined by the credit analysis and loan package.) The maximum insurable amount is $2.5 million. Interest rates are negotiated between the borrower and the lender, but may not exceed three percentage points above the prime rate.

★919★ Minnesota Fund
(612) 297-3547

Description: Provides below-market-interest-rate, fixed-asset financing for new and existing manufacturers and other industrial enterprises. Eligible firms are for-profit manufacturing or industrial corporations, partnerships, or sole proprietorships; are independently owned and operated; and are not dominant in their fields of business. Loan funds may be used to purchase land to construct a new manufacturing or industrial facility; for construction of the new facility; to purchase an existing facility; to expand an existing facility; and to purchase machinery and equipment. Loan funds may not be used for debt refinancing, interim financing, or working capital.

★920★ Minnesota State Small Business Assistance Office
(612) 296-3871

Description: Provides information and assistance to businesses in areas of start-up, operation, and expansion. Maintains the Bureau of Business Licenses, which offers information on the number and kind of licenses required for a business venture, the agencies that issue them, and the affirmative burdens imposed on the applicant. Also operates the Bureau of Small Business, which serves as a focal point within state government for information and resources available to small businesses. Sponsors workshops and seminars on small business issues throughout the year. Publications: Guide to Starting a Business in Minnesota; State of Minnesota Directory of Licenses and Permits. Toll-free/Additional Phone Number: 800-652-9747 (in Minnesota).

★921★ Office of Project Management
(612) 296-5021

Description: Responsible for assisting businesses with financial packages utilizing a variety of resources. Assists in developing business plans and in obtaining financing needed for businesses to start-up or expand.

★922★ Small Business Development Loan Program
(612) 297-3547

Description: Targeted to existing manufacturing and industrial firms that want to expand in Minnesota. Funds for business loans are raised through the sale of revenue bonds issued by the authority, the interest on which is generally exempt from state and federal taxes. These funds generally take the form of long-term, fixed-rate loans for land, buildings, and capital equipment. By pooling bonds and inducing the investment of private capital in connection with the business loans, the program facilitates a partnership between the public and private sectors. Funds may be used for interim or long-term financing for certain capital expenditures, including land and/or building acquisition costs; site preparation; construction costs; engineering costs; underwriting or placement fees; and other fees and costs.

★923★ Opportunities Minnesota Incorporated
Minnesota State Department of Energy and Economic Development
900 American Center Bldg.
150 E. Kellogg Bldg.
St. Paul, MN 55101 (612) 296-0582

Description: Provides subordinated financing through the issuance of debentures for businesses that are purchasing buildings or capital assets with useful lives greater than 15 years. Offers financing for fixed assets, including land acquisition, building construction, leasehold improvements, renovation and modernization, and machinery and equipment.

★924★ U.S. Small Business Administration Minneapolis District Office
100 N. Sixth St., Suite 610
Minneapolis, MN 55403 (612) 349-3530

Mississippi

★925★ Mississippi State Department of Economic Development Finance Division
P.O. Box 849
Jackson, MS 39205 (601) 359-3449

Description: Administers a variety of financing programs, including enterprise zones, incentives, and industrial revenue bonds. Assists in securing loans through federal, state, and private sources.

★926★ Mississippi State General Services Office Division of Purchasing
1501 Walter Sillers Bldg.
Jackson, MS 39201 (601) 354-7107

★927★ Mississippi State Research and Development Center
3825 Ridgewood Rd.
Jackson, MS 39211 (601) 982-6606
Dr. James Meredith, Director

Description: Provides management and technical assistance to business professionals.

★928★ U.S. Small Business Administration Gulfport District Office
One Hancock Plaza, Suite 1001
Gulfport, MS 39501-7758 (601) 863-4449

★929★ Jackson District Office
100 W. Capitol St., Suite 322
Jackson, MS 39269 (601) 965-4378

Missouri

★930★ Missouri State Department of Economic Development High Technology Program
Economic Development Programs
P.O. Box 118
Jefferson City, MO 65102 (314) 751-4241
John S. Johnson, Director

Description: Markets the state to outside business firms and entrepreneurs and lends assistance to state businesses and entrepreneurs regarding the high-technology process, including federal and state research programs, research and development parks, innovation centers, venture capital, patents, licenses, and trademarks.

★931★ Missouri Corporation for Science and Technology
P.O. Box 118
Jefferson City, MO 65102 (314) 751-3906
John Johnson, Director

Description: Establishes innovation centers as well as managerial and financial assistance to new firms.

★932★ Small and Existing Business Development
P.O. Box 118
Jefferson City, MO 65102 (314) 751-4982

Description: Offers business skill training and technical assistance to small business professionals.

★933★ Missouri State Department of Professional Registration
P.O. Box 1335
Jefferson City, MO 65102 (314) 751-2334

★934★ Missouri State Office of Administration Division of Purchasing
P.O. Box 809
Jefferson City, MO 65102 (314) 751-3273

★935★ MO-KAN Development, Inc.
1302 Faraon
St. Joseph, MO 64501 (816) 233-8485

Description: Provides financial packaging services for U.S. Small Business Administration and other financing programs to eligible small businesses in Atchison, Brown, Doniphan, Jackson, Jefferson, and Nemaha counties.

★936★ St. Louis Technology Center
5050 Oakland
St. Louis, MO 63110 (314) 534-2600

★937★ U.S. Small Business Administration Kansas City District Office
1103 Grand Ave., Sixth Floor
Kansas City, MO 64106 (816) 374-3416

★938★ St. Louis District Office
815 Olive St., Room 242
St. Louis, MO 63101 (314) 425-6600

★939★ U.S. Small Business Administration (Region Seven)
911 Walnut St., 13th Floor
Kansas City, MO 64106 (816) 374-5288

Montana

★940★ Montana State Department of Administration Purchasing Bureau
Sam W. Mitchell Bldg., Suite 165
Helena, MT 59620 (406) 444-2575

★941★ Montana State Department of Commerce Business Advocacy and Licensing Assistance
1424 Ninth Ave., Room 1
Helena, MT 59620 (406) 444-3494

Description: Maintains a Small Business Advocate that serves as a contact for individuals starting businesses and acts as an ombudsman for businesses with questions or complaints regarding state agencies.

★942★ Business Development Specialist

Description: Provides information and technical assistance to small businesses—particularly manufacturers—on such topics as employee training opportunities, U.S. government contract leads, and sources of loan or grant funds. Maintains a small business advocate. Also offers a wide range of services to prospective, new, and existing businesses. Assists in site selection and development. Provides product promotion and marketing assistance. Helps small businesses wishing to enter foreign markets. Operates the Census and Economic Information Center, which supplies population and economic data for research, planning, and decision making.

★943★ Montana Economic Development Board
Lee Metcalf Bldg.
1520 E. Sixth Ave., Room 050
Helena, MT 59620-0505 (406) 444-2090

Description: Makes investments in qualifying Montana businesses. Offers businesses long-term, fixed-rate financing at competitive interest rates for a variety of needs. Preference is given to small and medium-sized businesses, locally owned enterprises, employee-owned enterprises, and businesses that provide jobs for Montanans, that improve the environment, or that promote Montana's agricultural products. The Department of Commerce also offers development finance technical assistance to businesses. Finance specialists assist in analysis and planning, in preparing loan and bond applications, and with other financing programs available to businesses.

★944★ Montana Science and Technology Alliance
46 North Last Chance Gulch
Suite 2B
Helena, MT 59620 (406) 449-2778
Frank Culver, Director

Description: Established to promote public/private sector partnership. Assists in financing the establishment of technology-intensive businesses in the state. Components include seed capital investment, applied research, technical assistance and technology transfer, and research capability development.

★945★ Montana State Department of Natural Resources Renewable Energy and Conservation Program
Grant & Loan Section
1520 East Sixth Ave.
Helena, MT 59620 (406) 444-6774
Greg Mills, Director

Description: Promotes research, development, and demonstration of energy conservation. Offers grants and contracts with individuals and organizations.

★946★ U.S. Small Business Administration Helena District Office
301 S. Park Ave., Room 528
Helena, MT 59626 (406) 449-5381

Nebraska

★947★ Nebraska Investment Finance Authority
Gold's Galleria, Suite 304
1033 O St.
Lincoln, NE 68508 (402) 477-4406

Description: Provides lower-cost financing for manufacturing facilities, certain farm properties, and health care and residential development. Has established a Small Industrial Development Bond Program for small businesses as well as an Industrial Development Revenue Bond Program.

★948★ Nebraska Secretary of State
State Capitol, Suite 2300
P.O. Box 94608
Lincoln, NE 68509-1608 (402) 471-2554

Description: Registers corporations, trademarks, and service marks. Also responsible for business and occupational licenses.

★949★ Nebraska State Administrative Service Department
State Purchasing Section
301 Centennial Mall, S.
Lincoln, NE 68509 (402) 471-2401

★950★ Nebraska State Department of Economic Development
Research and Development Authority
301 Centennial Mall, S.
P.O. Box 94666
Lincoln, NE 68509 (402) 471-2593

Description: Provides seed capital for commercially viable ideas resulting from basic research-and-development phases. Does not make grants but invests in the idea by taking an equity position in the venture.

★951★ Small Business Division
(402) 471-4668

Description: Advocates and supports the creation, expansion, and retention of businesses. Provides export sales assistance through planning, workshops and seminars, trade missions and trade shows, a network of honorary commercial attaches, and a trade leads advisory service. Assists in the design and implementation of industrial training programs. Publications: 1) Buy Nebraska—matches buyers and sellers of Nebraska products and services; 2) Directory of Nebraska Manufacturers (every two years); 3) Industry Census Findings Report—summarizes data collected about industry in Nebraska; 4) Nebraska International Trade Directory—lists Nebraska firms interested in exporting; 5) Nebraska Statistical Handbook—a fact book for business planning and analysis.

★952★ Nebraska State Energy Office
Nebraska Energy Fund
State Capitol
P.O. Box 95085
Lincoln, NE 68509 (402) 471-2867

Description: Makes revolving loans available to Nebraskans for energy efficiency improvements.

★953★ Nebraska State Ethanol Authority Development Board
301 Centennial Mall, S.
P.O. Box 95108
Lincoln, NE 68509 (402) 471-2941

Description: Funds projects for the expanded use of Nebraska's agricultural products; for efficient, less polluting energy sources; and for the development of protein that will be stored more efficiently. Financing is provided by way of grants, loans, or loan guarantees.

★954★ U.S. Small Business Administration
Omaha District Office
11145 Mill Valley Rd.
Omaha, NE 68154 (402) 221-4691

Nevada

★955★ Nevada State Commission on Economic Development
Capitol Complex
Carson City, NV 89710 (702) 885-4325

Description: Assists businesses in various areas. Operates programs that promote community development activities. Encourages commercial energy conservation. Aids businesses seeking government contracts. Toll-free/Additional Phone Number: 800-992-0900, extension 4420.

★956★ Nevada State Department of General Services
Division of Purchasing
209 E. Musser St.
Carson City, NV 89710 (702) 885-4070

★957★ Nevada State Legislative Counsel Bureau
Research Division
Legislative Bldg.
401 S. Carson St.
Carson City, NV 89710 (702) 885-5637

★958★ Nevada State Office of Community Services
Procurement Outreach Program
Capitol Complex
1100 E. William, Suite 117
Carson City, NV 89710 (702) 885-4420

Description: Assists businesses interested in bidding for government contracts. Maintains an office in Las Vegas at Temple Plaza, 3017 W. Charleston Blvd., Suite 20, Las Vegas, NV 89102. Toll-free/Additional Phone Number: 800-992-0900, extension 4420; (702)468-6174 (Las Vegas office).

★959★ Small Business Revitalization Program
Capitol Complex
1100 E. William St., Suite 117
Carson City, NV 89710 (702) 885-4602

Description: Assists small businesses by reviewing proposed expansion projects, by recommending the most practical financing for those projects, and by structuring loan proposals that will meet the needs of various funding sources.

★960★ U.S. Small Business Administration
Las Vegas District Office
301 E. Stewart
Las Vegas, NV 89125 (702) 388-6611

★961★ Reno District Office
50 S. Virginia St., Room 238
Reno, NV 89505 (702) 784-5268

New Hampshire

★962★ New Hampshire State Department of Administration Services
Division of Plant and Property Management
State House Annex
Concord, NH 03301 (603) 271-2700

★963★ New Hampshire State Department of Resources and Economic Development
Industrial Development Office
105 Loudon Rd.
P.O. Box 856
Concord, NH 03301 (603) 271-2591
Paul Gilderson, Director

Description: Sponsors programs to encourage economic development. Offers market, site location, and import/export information.

★964★ U.S. Small Business Administration
Concord District Office
55 Pleasant St., Room 210
Concord, NH 03301 (603) 225-4400

New Jersey

★965★ Corporation for Business Assistance in New Jersey
200 S. Warren St., Suite 600
Trenton, NJ 08608 (609) 633-7737
Description: Provides businesses with long-term, fixed-asset financing for the acquisition of land and buildings, machinery and equipment, construction, renovation, and restoration.

★966★ New Jersey Commission on Science and Technology
122 West State Street, CN832
Trenton, NJ 08625 (609) 984-1671
Edward Cohen, Director
Description: Awards "bridge" grants of up to $40,000 to small companies that have received seed money from the Federal Small Business Innovation Research (SBIR) program. Grants are awarded on a competitive basis in the following areas: biotechnology, hazardous and toxic waste management, materials science, food technology, fisheries and aquaculture, and plastics recycling.

★967★ New Jersey Economic Development Authority
Capitol Place One
200 S. Warren St.
CN 990
Trenton, NJ 08625 (609) 292-1800
Description: Offers low-interest loans for capital costs (industrial development bonds), loan guarantees, and direct loans. Maintains an Urban Center Small Loans Program to assist existing retail and commercial businesses located in commercial districts of designated municipalities. Provides long-term, fixed-asset financing through the U.S. Small Business Administration's 504 Program. Administers programs for low-interest, fixed-asset loans (local development financing funds) and for recycling loans. Maintains an Urban Industrial Parks Program. Offers technical assistance to manufacturers impacted by imports.

★968★ New Jersey State Department of Commerce and Economic Development
New Jersey Set-Aside Program
One W. State St.
CN 823
Trenton, NJ 08625 (609) 984-4442
Description: Maintains a list of vendors wishing to participate in the state's set-aside program for small, minority-owned, and women-owned firms.

★969★ Office of Small Business Assistance
CN 823
Trenton, NJ 08625-0823
Description: Advises and encourages small, women-owned, and minority-owned businesses in matters related to establishing and operating a small business in the state. General managerial, financial, and economic development assistance are offered to companies with 500 or fewer employees. Certain projects and programs are available to businesses with 100 or fewer employees and to minority- and women-owned firms based in New Jersey.

★970★ New Jersey State Department of the Treasury
Purchase Bureau
CN 230
Trenton, NJ 08625
Description: Maintains a bidders list for state procurement contracts.

★971★ U.S. Small Business Administration
Camden District Office
2600 Mt. Ephrain
Camden, NJ 08104 (609) 757-5183

★972★ Newark District Office
60 Park Place, Fourth Floor
Newark, NJ 07102 (201) 645-3580

New Mexico

★973★ New Mexico Research and Development Institute
Pinon Bldg., Suite 358
1220 S. St. Francis Dr.
Santa Fe, NM 87501 (505) 827-5886
Description: Offers a seed capital program to finance the research-and-development stages of projects by start-up and small companies in New Mexico. Projects should accelerate the commercialization of advanced technology-based products, processes, or services.

★974★ New Mexico State Department of Economic Development and Tourism
Joseph M. Montoya Bldg.
1100 St. Francis Dr.
Santa Fe, NM 87503 (505) 827-0272
Description: Provides businesses with information on markets, site location, hiring and training incentives, international trade, and other general business assistance. **Contact:** Patrick Rodriguez, Senior Program Officer, High Technology Program; Marci Talnack, SBIR Assistance.

★975★ New Mexico State Department of Regulation and Licensing
Bataan Memorial Bldg.
Santa Fe, NM 87503 (518) 827-6318

★976★ New Mexico State Purchasing Division
Joseph M. Montoya Bldg.
1100 St. Francis Dr., Room 2016
Santa Fe, NM 87503 (505) 827-0472
Description: Maintains a special minority and small business development procurement program. Toll-free/Additional Phone Number: (505)827-0425 (Minority and Small Business Development Program).

★977★ U.S. Small Business Administration
Albuquerque District Office
5000 Marble Ave., N.E., Room 320
Patio Plaza Bldg.
Albuquerque, NM 87100 (505) 262-6171

New York

★978★ New York Small Business Innovation Research Promotion Program
99 Washington Ave.
Suite 1730
Albany, NY 12210 (518) 474-4349
Tab Wilkins, Director
Description: Encourages technology companies to participate in the SBIR program.

★979★ New York State Department of Commerce
Division for Small Business
230 Park Ave.
New York, NY 10169 (212) 309-0400
Description: Responsible for developing programs, providing services, and undertaking other initiatives that are responsive to the special needs

of the state's small businesses. Offers small businesses an array of programs and services aimed at helping them prosper and grow. Maintains a Small Business Advocacy Program, Procurement Assistance Unit, a Small Business Advisory Board, and an Interagency Small Business Task Force. Also sponsors a Business Services Ombudsman Program to assist businesses in resolving red-tape difficulties that they may encounter in their interactions with all levels of government. Also offers one-on-one counseling services to both start-up and existing small businesses. **Telecommunication Access:** Alternate telephone (212) 309-0460.

★980★ New York State Department of General Services
Division of Purchasing
Corning Tower
Empire State Plaza
Albany, NY 12242 (518) 474-3695

★981★ New York State Job Development Authority
605 Third Ave.
New York, NY 10158 (212) 818-1700
Description: Offers financing programs.

★982★ New York State Office of Permits and Regulatory Assistance
Governor Alfred E. Smith Bldg., 17th Floor
P.O. Box 7027
Albany, NY 12225 (518) 474-8275
Description: Maintains information on all state-required permits and licenses. Identifies necessary permits for particular businesses. Provides a Permit Assistance Kit, including application forms, instructions, and other materials. Arranges pre-application conferences. Has a Master Application Procedure for complex projects. Toll-free/Additional Phone Number: 800-342-3464 (permit assistance); (518)473-0620 (regulations review); (518)474-7321 (paperwork reduction).

★983★ New York State Science and Technology Foundation
Centers for Advanced Technology Program
99 Washington Ave.
Suite 1730
Albany, NY 12210 (518) 474-4347
Vernon Ozarow, Director
Description: Provides a network of technological centers within the state's research institutes. Designed to meet the needs for increased investment in applied research and development, professional and technical expertise, and links with business and university sectors.

★984★ Corporation for Innovation Development
99 Washington Ave., Room 924
Albany, NY 12210 (518) 474-4349
Mr. H. Graham Jones, Executive Director
Description: Administers financing programs for businesses.

★985★ Industrial Innovation Extension Service
99 Washington Ave.
Suite 1730
Albany, NY 12210 (518) 474-4349
Tab Wilkins, Director
Description: Program objectives include dissemination and educational service based on current research and development strengths in the area of manufacturing technology, assisting business retention and expansion, and improving competitiveness and market share of New York state industries through increased knowledge of new technologies and other innovations.

★986★ Regional Technology Development Corporation
99 Washington Ave.
Suite 1730
Albany, NY 12210 (518) 474-4349
Mark Tebbano, Director
Description: Assists emerging science and technology-oriented businesses, disseminate information, and assist in the development of small business incubator facilities.

★987★ U.S. Small Business Administration
Albany District Office
445 Broadway, Room 242
Albany, NY 12207 (518) 472-6300

★988★ Buffalo District Office
111 W. Huron St., Room 1311
Buffalo, NY 14202 (716) 846-4301

★989★ Elmira District Office
333 E. Water St., Fourth Floor
Elmira, NY 14901 (607) 734-6610

★990★ Melville District Office
35 Pinelawn Rd., Room 102E
Melville, NY 11747 (516) 454-0764

★991★ New York City District Office
26 Federal Plaza, Room 3100
New York, NY 10278 (212) 264-1318

★992★ Rochester District Office
100 State St., Room 601
Rochester, NY 14614 (716) 263-6700

★993★ Syracuse District Office
100 S. Clinton St., Room 1071
Syracuse, NY 13260 (315) 423-5371

★994★ U.S. Small Business Administration (Region Two)
26 Federal Plaza, Room 29-118
New York, NY 10278 (212) 264-7772

North Carolina

★995★ North Carolina Biotechnology Center
P.O. Box 13547
79 /alexander Drive
5401 Building
Research Triangle Park, NC 27709-3547 (919) 541-9366
W. Steven Burke, Director of Public Affiars
Description: Acts as a catalyst to stimulate further development of the state's biotechnology research community. Facilitates collaboration between universities and industry. Advises the governor and university presidents of developments of long-term plans regarding biotechnology. **Contact:** Barry D. Teater, Communications Specialist.

★996★ North Carolina State Department of Administration
Purchase and Contract Division
116 W. Jones St.
Raleigh, NC 27603-8002 (919) 733-3581
Description: Serves as the state's purchasing office.

★997★ North Carolina State Department of Commerce Business Information Referral Center
Dobbs Bldg., Room 2238
430 N. Salisbury St.
Raleigh, NC 27611 (919) 733-9013

Description: Uses a computer-searchable data base of agencies that regulate, license, certify, issue permits, or provide assistance to North Carolina businesses. Offers information on small business assistance; training programs; market, labor, and facilities data; and technical problems.

★998★ Industrial Financing Section
430 N. Salisbury St.
Raleigh, NC 27611 (919) 733-5297

Description: Issues industrial revenue bonds, a form of long-term, low-interest financing.

★999★ Small Business Development Division
Dobbs Bldg., Room 2019
430 N. Salisbury St.
Raleigh, NC 27611 (919) 733-6254

Description: Offers information and assistance to those planning to start a business or to established business professionals. Assists with financing, taxes, regulations, marketing, education, and training. Sponsors workshops and conferences, including the Industrial Buyer/Supplier Exchange. Maintains a business clearinghouse to aid businesses and investors.

★1000★ North Carolina State House Committee on Small Business
1006 Legislative Bldg.
Raleigh, NC 27611 (919) 733-5861

★1001★ North Carolina State Senate Committee on Small Business
2016 Legislative Bldg.
Raleigh, NC 27611 (919) 733-5882

★1002★ North Carolina Technological Development Authority Incubator Facilities Program
Dobbs Bldg., Room 4216
430 N. Salisbury St.
Raleigh, NC 27611 (919) 733-7022
Julie Tenney, Executive Director

Description: Provides one-time grants of not more than $200,000, which must be matched in cash or real estate, to nonprofit organizations for the establishment of facilities that provide low-rent space, shared support services, and basic equipment to resident small businesses. Offers the small business community technical, management, and entrepreneurial advice. Includes facilities in Haywood County, Ahoskie, Goldsboro, and McDowell County.

★1003★ Innovation Research Fund
Dobbs Bldg., Room 4216
430 N. Salisbury St.
Raleigh, NC 27611

Description: Grants royalty financing of not more than $50,000 to North Carolina small businesses for research leading to the improvement or to the development of new products, processes, or services. Enables small businesses to secure technical and management assistance. Conducts research leading to the growth or start-up of a firm, thereby creating new jobs.

★1004★ U.S. Small Business Administration Charlotte District Office
222 S. Church St., Room 300
Charlotte, NC 28202 (704) 371-6563

North Dakota

★1005★ North Dakota State Department of Management and Budget Division of Purchasing
Capitol Bldg.
Bismarck, ND 58505 (701) 224-2680

★1006★ North Dakota State Economic Development Commission Entrepreneurial Assistance
Liberty Memorial Bldg.
Bismarck, ND 58505 (701) 224-2810

Description: Identifies agencies and individuals that assist inventors and entrepreneurs with technical, legal, financial, and marketing issues that must be resolved before launching a new venture. Provides training for inventors and entrepreneurs. Helps North Dakota companies develop international markets for their products. Maintains a marketing committee for home-based manufactured goods. Undertakes special projects.

★1007★ Federal Procurement

Description: Assists in-state businesses in obtaining contracts with the federal government for new markets for goods and services.

★1008★ Financial Assistance

Description: Assists locally referred businesses looking to expand but having difficulty assembling financial packages. Works to improve the financial referral service for North Dakota businesses and industries.

★1009★ U.S. Small Business Administration Fargo District Office
657 Second Ave., N., Room 218
Fargo, ND 58102 (701) 237-5771

Ohio

★1010★ Ohio State Department of Administrative Services Minority Business Enterprise Program
State Office Tower
30 E. Broad St., 18th Floor
Columbus, OH 43266-0408 (614) 466-8380

Description: Ensures minority participation in the state contract bidding process. At least 5 percent of the state's construction contracts shall be set aside for competitive bidding by certified minority contractors. At least 7 to 10 percent of state construction contracts shall be awarded to minority subcontractors and materialmen. At least 15 percent of all state purchases of equipment, materials, supplies, and contracts of insurance or services shall be set aside for competitive bidding by minority suppliers.

★1011★ Office of State Purchasing
(614) 466-5090

★1012★ Ohio State Department of Development Economic Development Financing Division
P.O. Box 1001
Columbus, OH 43266-0101 (614) 466-5420

Description: Provides direct loans and loan guarantees to businesses for new fixed-asset financing for land, buildings, and equipment. Also issues industrial revenue bonds to provide businesses with 100 percent of the financing for eligible fixed assets. Interest rates are below current market rates. Terms are up to 30 years, often with an option to buy the project financed at maturity.

Ohio State Department of Development *(Cont.)*

★1013★ Minority Development Financing Commission
(614) 462-7709

Description: Helps finance minority business expansion through direct loans and through construction bonds. Direct loans are granted to finance up to 40 percent (or $200,000, whichever is less) of an eligible project for an eligible company, including the purchase and/or improvement of fixed assets such as land, buildings, equipment, and machinery. Issues up to $1 million in construction bonds to a minority business.

★1014★ Small and Developing Business Division
(614) 466-4945

Description: Offers a variety of programs to small business professionals, including a One-Stop Business Permit Center and a Women's Business Resource Program. The Department of Development also maintains Small Business Enterprise Centers that offer advice and assistance to small businesses. The department's other services include business relocation, site selection, and international trade programs. Publications: International Business Opportunities (monthly)—a newsletter. Toll-free/Additional Phone Number: 800-282-1085 (business and international trade development); (614)466-5700 (Small Business Enterprise Centers).

★1015★ U.S. Small Business Administration
Cincinnati District Office
550 Main St., Room 5028
Cincinnati, OH 45202 (513) 684-2814

★1016★ Cleveland District Office
1240 E. Ninth St., Room 317
Cleveland, OH 44199 (216) 522-4182

★1017★ Columbus District Office
85 Marconi Blvd., Room 512
Columbus, OH 43215 (614) 469-6860

★1018★ The Withrow Plan of Linked Deposits
c/o Mary Ellen Withrow
Treasurer of the State of Ohio
State Office Tower, Ninth Floor
30 E. Broad St.
Columbus, OH 43215 (800) 228-1102

Description: Allows the treasurer to channel a portion of the state's investment portfolio into reduced-rate investments. These reduced-rate investments are then linked to reduced-rate loans. To qualify, small businesses must document that the proceeds of the loans will help to create or retain jobs.

Oklahoma

★1019★ Oklahoma Industrial Finance Authority
c/o Oklahoma State Department of Commerce
6601 Broadway Extension
Oklahoma City, OK 73116-8214 (405) 843-9770

Description: Has a borrowing capacity of $90 million. Empowered to loan up to 66 2/3 percent of the cost of land and buildings on a secured first mortgage and up to 33 1/3 percent on a second mortgage. Maximum loan amount is $1 million for the state's share of a single project.

★1020★ Oklahoma State Department of Commerce
Business Development Division
6601 Broadway Extension
Oklahoma City, OK 73116-8214 (405) 843-9770

Description: Identifies, counsels, and assists locating new and expanding industries. Helps with site evaluation. Coordinates programs with waterway authorities. Maintains a film office geared toward promoting Oklahoma as a location for film and video productions.

★1021★ Division of Community Affairs and Development

Description: Develops and operates programs designed to stimulate economic development through assistance to cities and local businesses. Also administers programs to improve economic and social environments of communities. Programs include community services block grants and community development block grants.

★1022★ Information and Research Division

Description: Compiles and collects economic data about Oklahoma. Publications: Business Resources Sourcebook.

★1023★ Procurement Section

Description: Maintains a computerized inventory system that allows for the tracking of property and supplies. The department also maintains 21 state bid assistance centers established to assist firms with federal procurement.

★1024★ Special Services Division

Description: Provides financial package analysis to Oklahoma firms. Assists local communities with state and federal loan or grant contract work. Administers the Industrial Development Bond Fund allocation program. Assists with finance program development, particularly when bond issues are used as part of a package.

★1025★ Oklahoma State Department of Public Affairs
Division of Purchasing
104 State Capitol Bldg.
Oklahoma City, OK 73105 (405) 521-2115

★1026★ U.S. Small Business Administration
Oklahoma City District Office
200 N.W. Fifth St., Suite 670
Oklahoma City, OK 73102 (405) 231-4301

Oregon

★1027★ Oregon Resource and Technology Development Corporation
Oregon State Economic Development Department
595 Cottage St., N.E.
Salem, OR 97310 (503) 373-1200

Description: Funded by lottery revenues. An independent public corporation established to provide financing and other services to existing and start-up businesses, particularly small businesses. **Contact:** John Beaulieu, President, One Lincoln Center, Suite 430, 10300 Southwest Greenburg Road, Portland, OR 97233 (503) 246-4844.

★1028★ Oregon State Department of Commerce
Corporation Division
158 12th St., N.E.
Salem, OR 97310 (503) 378-2290

Description: Provides forms and registrations required by the state for starting a business. Maintains a business registry information center and hotline. Also handles state business and occupation licenses and permits. Toll-free/Additional Phone Number: (503)378-4166 (business registry information center).

★1029★ Oregon State Department of Economic Development
Regulation Assistance Service
595 Cottage St., N.E.
Salem, OR 97310 (503) 373-1234

Description: Refers business owners to appropriate state agencies for licenses, permits, and registrations. Maintains a one-stop permit information center. Toll-free/Additional Phone Number: (503)373-1999 (one-stop permit information center).

★1030★ Oregon State Department of General Services
Division of Purchasing

1225 Ferry St., S.E.
Salem, OR 97310 (503) 378-4643

★1031★ Oregon State Economic Development Department
Business Development Division

595 Cottage St., N.E.
Salem, OR 97310 (503) 373-1200

Description: Offers services to businesses in such areas as business recruitment, industrial property, licenses and permits, business information, financing, and international trade. Publications: 1) Directory of Oregon Manufacturers; 2) Exporters Handbook; 3) The Oregon Advantage: Services for Oregon Business Firms; 4) Oregon International Trade Directory; 5) Starting a Business in Oregon.

★1032★ Financial Services Division

Description: Administers the Industrial Development Revenue Bond Program to promote and develop industrial activities, tourism, and international trade in Oregon. Also maintains the Umbrella Revenue Bond Program, which provides the benefits of industrial development revenue bonds to growing small and medium-sized Oregon businesses. Commercial banks provide 10 to 35 percent of the financing for each individual project. Small projects ($100,000)—and large projects ($1 million)—may be funded. **Telecommunication Access:** Alternate telephone (503) 373-1215.

★1033★ Oregon Business Development Fund

(503) 373-1240

Description: Provides loans for both long-term, fixed-assest financing and working capital. Loans may be made only where there is a demonstrated creation of new jobs or retention of existing jobs. Program emphasizes businesses with fewer than 50 employees or rural and lagging areas.

★1034★ U.S. Small Business Administration
Portland District Office

1220 S.W. Third Ave., Room 676
Portland, OR 97204-2882 (503) 423-5221

Pennsylvania

★1035★ New-Penn-Del Regional Minority Purchasing Council, Inc.

5070 Parkside Ave., Suite 1400
Philadelphia, PA 19131 (215) 578-0964

Description: Links major corporate buyers with minority business owners who can provide the needed products or services.

★1036★ Pennsylvania State Department
Bureau of Professional and Occupational Affairs

Transportation and Safety Bldg., Room 618
Harrisburg, PA 17120 (717) 787-8503

★1037★ Pennsylvania State Department of Commerce
Ben Franklin Partnership

c/o Small Business Action Center
P.O. Box 8100
Harrisburg, PA 17105 (717) 783-5700

Description: Includes the Small Business Incubator Loan Program, which funds the development of incubator facilities that provide new manufacturing or product development companies with the space and business development services they can use to start-up and survive the early years of business growth. Four privately-managed Seed "Venture" Capital Funds, established through a BFP challenge grant program, provide equity financing to new businesses during their earliest stages of growth. Also awards Research "Seed" Grants of up to $35,000 to small businesses seeking to develop or introduce advanced technology into the marketplace.

★1038★ Business Infrastructure Development Program

467 Forum Bldg.
Harrisburg, PA 17120 (717) 787-7120

Description: Provides communities with funds for infrastructure improvements that will encourage private firms to locate or expand in Pennsylvania. Locally sponsored manufacturing, industrial, agricultural, and research-and-development enterprises may apply. Eligible projects include energy facilities, fire and safety facilities, sewer systems, waste disposal facilities, water supply systems, drainage systems, and transportation facilities. Job creation requirements vary depending on the amount of the grant or loan.

★1039★ Office of Minority Business Enterprise

491 Forum Bldg.
Harrisburg, PA 17120 (717) 783-1301

Description: The primary liaison for minority-owned companies interested in contracting with state government agencies. Helps strengthen Pennsylvania's minority business community by providing state contracting information to minority companies. Serves as a liaison between minority-owned firms and public sector purchasers.

★1040★ Pennsylvania Capital Loan Fund

405 Forum Bldg.
Harrisburg, PA 17120 (717) 783-1768

Description: Receives funds from both federal and state appropriations. Loans are targeted to small manufacturing, industrial, and export service businesses with the goal of creating at least one new job for each $15,000 loaned; one job for each $10,000 to apparel manufacturers, which includes shoe and garment manufacturers. Loans are meant to encourage private sector investment in small businesses by subordinating its collateral position. Funds may be used for land and building acquisition and renovation, machinery and equipment, and working capital. Funds may not be used to replace private sector money readily available at reasonable rates; refinance existing debt or pay for previously incurred obligations; or speculate.

★1041★ Pennsylvania Energy Development Authority

462 Forum Bldg.
Harrisburg, PA 17120 (717) 787-6554

Description: Promotes development and efficient use of Pennsylvania's energy resources. Provides financial assistance for qualifying energy projects. F unds are appropriated by the General Assembly and deposited in the authority's Energy Development Fund. The amount and type of funds available are tied to the stage of development and include grants for research and feasibility studies; repayable grants (venture capital) for pilot and demonstration projects; and loans for commercial-stage projects. In addition, revenue bond financing may be possible for commercially viable, revenue producing energy projects, depending upon conformance with the Federal Internal Revenue Codes and Internal Revenue Service regulations on tax-exempt bonds and state availability.

★1042★ Pennsylvania Industrial Development Authority

405 Forum Bldg.
Harrisburg, PA 17120 (717) 787-6245

Description: Loans are available to businesses engaged in manufacturing or classified as industrial enterprises. An industrial enterprise includes the areas of research and development, food processing and related agribusiness, computer or clerical operation centers, buildings being used for national and regional headquarters, and certain warehouse and terminal facilities. Excluded are commercial, mercantile, or retail enterprises. Funds may be used for land and building acquisition, new construction, expansion, or renovation in conjunction with an acquistion. Loan applications are

Pennsylvania State Department of Commerce *(Cont.)*
first approved by a local industrial development corporation and then reviewed by the authority's board.

★1043★ Revenue Bond and Mortgage Program
405 Forum Bldg.
Harrisburg, PA 17120 (717) 783-1108
Description: Financing for projects approved through this program is secured from private sector sources such as banks, insurance companies, individuals, and through public bond issues. These funds are borrowed through a local industrial development authority and can be used for manufacturing projects, and specialized projects such as solid waste disposal. Businesses or investors can use the funds to acquire land, buildings, machinery, and equipment.

★1044★ Small Business Action Center
P.O. Box 8100
Harrisburg, PA 17105
Description: Answers general business questions. Publications: 1) Resource Directory for Small Business; 2) Small Business Planning; 3) Starting a Small Business in Pennsylvania.

★1045★ Pennsylvania State Department of General Services
Vendor Information and Support Division
P.O. Box 1365
Harrisburg, PA 17105 (717) 783-2903

★1046★ Pennsylvania State Department of Transportation
Disadvantaged and Women Business Enterprises
Transportation and Safety Bldg., Room 109
Harrisburg, PA 17120 (800) 468-4201
Description: Operates a special procurement program for firms owned by women, minorities, or disadvantaged persons. Toll-free/Additional Phone Number: 800-845-3375 (support services).

★1047★ U.S. Small Business Administration
Harrisburg District Office
100 Chestnut St., Suite 309
Harrisburg, PA 17101 (717) 782-3840

★1048★ Philadelphia District Office
231 St. Asaphs Rd., Suite 400 East
Philadelphia, PA 19004 (215) 596-5801

★1049★ Pittsburgh District Office
960 Penn Ave., Fifth Floor
Pittsburgh, PA 15222 (412) 644-4306

★1050★ Wilkes-Barre District Office
20 N. Pennsylvania Ave., Room 2327
Wilkes-Barre, PA 18701 (717) 826-6497

★1051★ U.S. Small Business Administration (Region Three)
231 St. Asaphs Rd., Suite 640 West
Philadelphia, PA 19004 (215) 596-5962

Puerto Rico

★1052★ Puerto Rico Department of Commerce
P.O. Box S-4275
San Juan, PR 00905 (809) 724-0542

★1053★ Puerto Rico Economic Development Administration
P.O. Box 2350
San Juan, PR 00936 (809) 765-1303

★1054★ Puerto Rico General Services Administration
Division of Purchasing, Services, and Supply
P.O. Box 412
San Juan, PR 00905 (809) 722-0580

★1055★ Puerto Rico Licensing Administration
Minillas Governmental Center
North Bldg.
Santurce, PR 00940 (809) 726-3120

★1056★ U.S. Small Business Administration
Hato Rey District Office
Federico Degatan Federal Bldg., Room 691
Carlos Chardon Ave.
Hato Rey, PR 00918 (809) 753-4003

Rhode Island

★1057★ Ocean State Business Development Authority
c/o Rhode Island State Department of Economic Development
Seven Jackson Walkway
Providence, RI 02903-3189 (401) 277-2601
Description: An independent, private, nonprofit corporation established and supported by businesses and government to encourage economic growth in Rhode Island. The authority can provide 90 percent financing on loan requests from $200,000 up to $1,000,000 at interest rates keyed to long term treasury rates established by the U.S. Department of the Treasury. Loan proceeds may be used to purchase land, renovate and construct buildings, and acquire machinery and equipment. Financing is not available for nonprofit organizations, print media such as newspapers and magazines, lending institutions, gambling facilities, recreation facilities that are not open to the public, and real estate development and speculation.

★1058★ Rhode Island Partnership for Science and Technology
Seven Jackson Walkway
Providence, RI 02903-3189 (401) 277-2601
Bruce Lang, Executive Director
Description: A nonprofit corporation. Provides state matching grants for applied research that creates a linkage between a private business venture and a Rhode Island nonprofit research facility.

★1059★ Rhode Island State Department of Administration
Purchases Office
301 Promenade St.
Providence, RI 02908 (401) 277-2321

★1060★ Rhode Island State Department of Economic Development
Rhode Island Financial Assistance Programs
Seven Jackson Walkway
Providence, RI 02903-3189 (401) 277-2601
Description: Provides new or expanding businesses with financing for applied research, real estate, machinery and equipment, and working capital through a variety of programs. The Taxable Industrial Development Revenue Bond Program provides competitive interest rates as well as tax advantages. Tax Exempt Industrial Development Revenue Bonds, available for manufacturing projects up to $10,000,000, offer lower interest rates and other advantages. The Mortgage Insurance

Program reduces the capital necessary for new manufacturing facilities, renovation of manufacturing facilities, and the purchase of new machinery and equipment in financing projects up to $5,00,000. Insured Industrial Development Bond Financing reduces the interest rate for smaller firms that otherwise would not be eligible for Industrial Revenue Bond financing. Revolving Loan Fund Program provides eligible Small Business Fixed Asset Loans from $25,000 to a maximum of $150,000 and Working Capital Loans to a maximum of $30,000. Rhode Island Business Investment Fund provides small businesses with loans from $30,000 to a maximum of $500,000. Applied Research Grants—Rhode Island Partnership for Science and Technology provides matching state grants for joint ventures between a Rhode Island nonprofit research facility and a business venture. Administers U.S. Small Business Administration 504 Loans.

★1061★ Rhode Island State Department of Labor
Division of Professional Regulation
220 Elmwood Ave.
Providence, RI 02907 (401) 457-1860

★1062★ U.S. Small Business Administration
Providence District Office
380 Westminster Mall, Fifth Floor
Providence, RI 02903 (401) 528-4586

South Carolina

★1063★ South Carolina State Budget and Control Board
Procurement Office
P.O. Box 1244
Columbia, SC 29211 (803) 734-2320

★1064★ South Carolina State Development Board
P.O. Box 927
Columbia, SC 29202 (803) 734-1400
Ron Young, Manager, Entrepreneur Dev.

Description: Offers financing programs and business assistance services. **Telecommunication Access:** Alternate telephone (803) 737-0400.

★1065★ Division of Business Assistance and Development
1301 Gervais St.
P.O. Box 927
Columbia, SC 29202

★1066★ State of South Carolina
Office of Small and Minority Business Assistance
Edgar A. Brown Bldg., Room 305
1205 Pendleton St.
Columbia, SC 29201 (803) 734-0562

Description: Offers procurement assistance, management training, and technical assistance to small and minority firms.

★1067★ U.S. Small Business Administration
Columbia District Office
1835 Assembly St., Third Floor
Columbia, SC 29202 (803) 765-5339

South Dakota

★1068★ South Dakota State Department of Administration
Purchasing and Printing Office
500 E. Capitol
Pierre, SD 57501 (605) 773-3405

★1069★ South Dakota State Department of Commerce and Regulation
Professional and Occupational License Division
910 Sioux
Pierre, SD 57501 (605) 773-3177

★1070★ South Dakota State Governor's Office of Economic Development
Capitol Lake Plaza
711 Wells
Pierre, SD 57501 (605) 773-5032
Roland Dolly, Enterprise Initiation

Description: Offers a variety of financial resources to small businesses.

★1071★ U.S. Small Business Administration
Sioux Falls District Office
101 S. Main Ave., Suite 101
Sioux Falls, SD 57102-0577 (605) 336-2980

Tennessee

★1072★ Tennessee State Department of Commerce and Insurance
Regulatory Boards
1808 West End Bldg.
Nashville, TN 37219 (615) 741-3449

★1073★ Tennessee State Department of Economic and Community Development
Office of Small Business
Rachel Jackson Bldg., Seventh Floor
320 Sixth Ave., N.
Nashville, TN 37219 (615) 741-3282

Telecommunication Access: Toll-free telephone 800-872-7201.

★1074★ Tennessee State Department of General Services
Division of Purchasing
C2-211 Central Services Bldg.
Nashville, TN 37219-5203 (615) 741-6875

★1075★ Tennessee Technology Foundation
P.O. Box 32184
Knoxville, TN 37933
Dr. David A. Patterson, Director

Description: Promotes high-technology economic development in the state. Provides site selection assistance; research access to Oak Ridge National Laboratory, Tennessee Valley Authority, and University of Tennessee; and supports new product commercialization and support of entrepreneurs, including access to venture capital sources.

★1076★ U.S. Small Business Administration
Nashville District Office
404 James Robertson Pkwy., Suite 1012
Nashville, TN 37219 (615) 736-5881

Texas

★1077★ Advanced Robotics Research Institute
2501 Gravel Dr.
Fort Worth, TX 76118-6999 (817) 284-5555

Description: Provides high-technology information on robotics and automation. Institute will be surrounded by related technological research and manufacturing businesses.

★1078★ Texas State Department of Labor and Standards
Division of Licensing and Enforcement
Capitol Station
P.O. Box 12157
Austin, TX 78711 (512) 463-5520

★1079★ Texas State Economic Development Commission
P.O. Box 12728
Austin, TX 78711 (512) 472-5059

★1080★ Texas State Purchasing and General Services Commission
Division of Purchasing
Capitol Station
P.O. Box 13047
Austin, TX 78711-3047 (512) 463-3443

★1081★ U.S. Small Business Administration
Austin District Office
300 E. Eighth St., Room 520
Austin, TX 78701 (512) 482-5288

★1082★ Corpus Christi District Office
400 Mann St., Suite 403
Corpus Christi, TX 78401 (512) 888-3331

★1083★ Dallas District Office
1100 Commerce St., Room 3C36
Dallas, TX 75242 (214) 767-0608

★1084★ El Paso District Office
10737 Gateway West, Suite 320
El Paso, TX 79935 (915) 541-7676

★1085★ Fort Worth District Office
819 Taylor St.
Fort Worth, TX 76102 (817) 334-3777

★1086★ Harlingen District Office
222 E. Van Buren St., Room 500
Harlingen, TX 78550 (512) 427-8533

★1087★ Houston District Office
2525 Murworth, Room 112
Houston, TX 77054 (713) 660-4407

★1088★ Lubbock District Office
1611 Tenth St., Suite 200
Lubbock, TX 79401 (806) 743-7462

★1089★ Marshall District Office
505 E. Travis, Room 103
Marshall, TX 75670 (214) 935-5257

★1090★ San Antonio District Office
727 E. Durango St., Room A-513
San Antonio, TX 78206 (512) 229-6105

★1091★ U.S. Small Business Administration (Region Six)
8625 King George Dr., Bldg. C
Dallas, TX 75235-3391 (214) 767-7643

Utah

★1092★ U.S. Small Business Administration
Salt Lake City District Office
125 S. State St., Room 2237
Salt Lake City, UT 84138-1195 (801) 524-5804

★1093★ Utah Innovation Center
419 Wakara Way
Suite 206, Research Park
Salt Lake City, UT 84108 (801) 584-2500
Dr. Gerald L. Davey, Director

Description: Encourages technical innovation and entrepreneurship. Assists creation of companies, providing legal, administrative, accounting, technical and clerical support.

★1094★ Utah State Department of Administrative Services
Division of Purchasing
2100 State Office Bldg.
Salt Lake City, UT 84114 (801) 533-4620

★1095★ Utah State Department of Business Regulation
Division of Occupational and Professional Licensing
P.O. Box 45802
Salt Lake City, UT 84145-0801 (801) 530-6620

★1096★ Utah State Department of Community and Economic Development
Division of Economic and Industrial Development
6150 State Office Bldg.
Salt Lake City, UT 84114 (801) 533-5325

★1097★ Utah Technology Finance Corporation
419 Wakara Way
Salt Lake City, UT 84108 (801) 583-8832

Vermont

★1098★ U.S. Small Business Administration
Montpelier District Office
87 State St., Room 205
Montpelier, VT 05602 (802) 828-4474

★1099★ Vermont Industrial Development Authority
58 E. State St.
Montpelier, VT 05602 (802) 223-7226

Description: Promotes economic growth and increases employment through a variety of financing programs. Makes low-interest loans available to businesses for the purchase or construction of land, buildings, machinery, and equipment for use in an industrial facility. Issues tax-exempt, low-interest bonds to provide funds for the acquisition of land, buildings, and/or machinery and equipment for use in a manufacturing facility. Provides loans to nonprofit local development corporations for the purchase of land for industrial parks, industrial park planning and development, and the construction or improvement of speculative buildings or small business incubator facilities. Aids businesses by guaranteeing loans of commercial lending institutions.

Offers small family farmers low-interest loans in amounts up to $50,000. Manages the Vermont 503 Corporation, a U.S. Small Business Administration 504 Certified Development Company.

★1100★ Vermont Secretary of State
Division of Licensing and Registration
26 Terrace St.
Montpelier, VT 05602 (802) 828-2458

★1101★ Vermont State Administration Agency
Division of Purchasing and Public Records
133 State St.
Montpelier, VT 05602 (802) 828-2211

★1102★ Vermont State Agency of Development and Community Affairs
Department of Economic Development
Pavilion Office Bldg.
109 State St., Fourth Floor
Montpelier, VT 05602 (802) 828-3221
Graeme Freeman, Development Spcialist

Description: Provides advice, information, and referrals to help prospective or operating businesses develop business plans and create workable financing packages. Creates a network of entrepreneurs and business service providers through entrepreneurship forums and incubator facilities. Mobilizes a network of public agencies to help troubled businesses and their employees. Assists businesses in obtaining state permits and other approvals. Maintains and shares information about new changes in technology affecting business. Disseminates information about Vermont companies and their products and services. Helps Vermont companies sell their products by researching markets, developing merchandising ideas, and informing them about export practices and policies. Helps locate sites and financing for companies who want to find facilities in Vermont. Assists in training workers in the specific skills needed by new and expanding companies. Organizes international trade missions and trade show presentations.

★1103★ Vermont State Department of Economic Development
Entrepreneurship Program
State St.
Montpelier, VT 05602 (802) 828-3221
Curt Carter, Director

Description: Helps develop business incubators.

Virgin Islands

★1104★ U.S. Small Business Administration
St. Thomas District Office
Veterans Dr., Room 210
St. Thomas, VI 00801 (809) 774-8530

★1105★ Virgin Islands Small Business Development Agency
P.O. Box 2058
St. Thomas, VI 00801 (809) 774-8784

Virginia

★1106★ U.S. Small Business Administration
Richmond District Office
400 N. Eighth St., Room 3015
Richmond, VA 23240 (804) 771-2741

★1107★ Virginia State Code Commission
Register of Regulations
P.O. Box 3-AG
Richmond, VA 23208 (804) 786-3591

Description: Publishes Virginia Register of Regulations (26/year)—includes four indexes.

★1108★ Virginia State Corporation Commission
P.O. Box 1197
Richmond, VA 23209 (804) 786-3733

Description: Administers filing procedures for articles of incorporation for profit and nonprofit organizations.

★1109★ Virginia State Department of Commerce
3600 W. Broad St.
Richmond, VA 23230 (804) 257-8500

Description: Regulates some professions, occupations, and businesses in nonhealth areas.

★1110★ Virginia State Department of Economic Development
Office of Small Business and Financial Services
1000 Washington Bldg.
Richmond, VA 23219 (804) 786-3791

Description: Provides expertise in financial management and technical assistance.

★1111★ Virginia State Office of Administration
Division of Purchases and Supplies
805 E. Broad St.
P.O. Box 1199
Richmond, VA 23219 (804) 786-3846

★1112★ Virginia State Small Business Financing Authority
Umbrella Industrial Development Bond Program
1000 Washington Bldg.
Richmond, VA 23219 (804) 786-3791

Description: Provides long-term financing of fixed assets through a mechanism generally available only to larger businesses.

Washington

★1113★ U.S. Small Business Administration
Seattle District Office
915 Second Ave., Room 1792
Seattle, WA 98174 (206) 442-5534

★1114★ Spokane District Office
W920 Riverside Ave., Room 651
Spokane, WA 99210 (509) 456-3781

★1115★ U.S. Small Business Administration (Region Ten)
2615 Fourth Ave., Room 440
Seattle, WA 98121 (206) 442-5676

★1116★ Washington State Department of General Administration
Division of Purchasing
General Administration Bldg., Suite 216
Olympia, WA 98504 (206) 753-6461

★1117★ Washington State Department of Licensing Business and Professions Division
12th and Franklin
Olympia, WA 98504 (206) 753-4202

★1118★ Washington State Department of Trade and Economic Development
101 General Administration Bldg.
Olympia, WA 98504 (206) 753-3065
Description: Offers a variety of financing programs to businesses.

West Virginia

★1119★ U.S. Small Business Administration Charleston District Office
550 Eagan St., Room 309
Charleston, WV 25301 (304) 347-5220

★1120★ Clarksburg District Office
168 W. Main St., Fifth Floor
Clarksburg, WV 26301 (304) 623-5631

★1121★ West Virginia Industrial Trade Jobs Development Corporation
State Capitol
M-146
Charleston, WV 25305 (304) 348-0400
Description: Supplements other financial incentive programs to help create jobs. The State Board of Investment makes loans available to the corporation for up to $10 million per project at negotiable rates and terms.

★1122★ West Virginia State Department of Finance and Administration Division of Purchasing
State Capitol
Building 1, Room E-119
Charleston, WV 25305 (304) 348-2309

★1123★ West Virginia State Economic Development Authority
State Capitol
Building 6, Room 525
Charleston, WV 25305 (304) 348-3650
Description: Provides low-interest loans from a revolving fund for land acquisition, building construction, and equipment purchases; loans are geared toward manufacturing firms with an emphasis on new job creation. Issues low-interest industrial revenue development bonds that are exempt from income tax at the federal and state levels; bonds are available to a wide variety of projects with the potential of reducing fees and interest rates.

★1124★ West Virginia State Office of Community and Industrial Development
State Capitol
M-146
Charleston, WV 25305 (304) 348-0400
Description: Offers low-interest loans and other financial arrangements to businesses. Administers such programs as the West Virginia Certified Development Corporation, which provides long-term fixed-rate loans to small and medium-sized firms, and the Treasurer's Economic Development Deposit Incentive, which offers low-cost financing to businesses that operate exclusively in West Virginia and that employ less than 200 or whose gross receipts total less than $4 million (financing must create or preserve jobs). Maintains contacts with several area and regional banks; helps businesses seeking assistance through federal financing programs.

★1125★ West Virginia State Small Business Office
State Capitol Complex
Charleston, WV 25305 (304) 348-2960
Description: Provides guidance to existing and prospective businesses. Offers technical assistance, loan packaging, and procurement services.

Wisconsin

★1126★ Innovation Network Foundation
P.O. Box 71
Madison, WI 53701 (608) 256-8348
Diane Curtz, Director
Description: Nonprofit group seeking to join marketing and sales ideas to those willing to finance them. Also oversees the development of an entrepreneurial resource library newsletter covering training programs, and telephone contacts generally not available to the entrepreneur.

★1127★ Northwest Wisconsin Business Development Corporation Northwest Wisconsin Business Development Fund
302 Walnut St.
Spooner, WI 54801 (715) 635-2197
Description: Promotes private sector investment in long-lived assets. Creates jobs by addressing capital gaps in the market for long-term debt. Serves businesses primarily in timber and wood, manufacturing, and tourism industries in northwestern Wisconsin.

★1128★ State of Wisconsin Investment Board Private Placements
121 E. Wilson St.
Madison, WI 53702 (608) 266-2381
Description: Meets fiduciary responsibilities to the trusts under the care of the board. Minimum loan is $3 million for Wisconsin-based firms and $10 million for firms based elsewhere.

★1129★ U.S. Small Business Administration Eau Claire District Office
500 S. Barstow St., Room 37
Eau Claire, WI 54701 (715) 834-9012

★1130★ Madison District Office
212 E. Washington Ave., Room 213
Madison, WI 53703 (608) 264-5268

★1131★ Milwaukee District Office
310 W. Wisconsin Ave., Room 400
Milwaukee, WI 53203 (414) 291-3941

★1132★ Wisconsin Housing and Economic Development Authority Linked-Deposit Loan Program
One S. Pinckney St., #500
P.O. Box 1728
Madison, WI 53701 (608) 266-0976
Description: Offers low-interest loans to improve access to capital for small businesses owned or controlled by women or minorities. Toll-free/Additional Phone Number: (608)266-7884.

★1133★ Wisconsin State Department of Administration Procurement Bureau
101 S. Webster
P.O. Box 7864
Madison, WI 53707 (608) 266-0974

★1134★ Wisconsin State Department of Development Financing Programs
123 W. Washington Ave.
P.O. Box 7970
Madison, WI 53707 (608) 266-1018

Description: Offers a variety of financial resources, including tax increment financing, industrial revunue bonds, employee ownership assistance loans, and a technology development fund. Administers the Wisconsin Development Fund-Economic Development Program.

★1135★ Small Business Ombudsman
(608) 266-0562

Description: Identifies the problems of individuals or groups of business people. Gathers facts concerning regulations and policies. Rosolves problems between businesses and state agencies or officials. Recommends changes in procedures, regulations, or laws that are determined to be unfair or discriminatory to the stability of small businesses in the state. Toll-free/Additional Phone Number: 800-HELPBUS (Business Hotline).

★1136★ Wisconsin State Department of Regulation and Licensing
P.O. Box 8935
Madison, WI 53708-8935 (608) 266-8609

★1137★ Wisconsin State Housing and Economic Development Authority
One S. Pinckney St.
Box 1728
Madison, WI 53701 (608) 266-0191

Description: Provides below-market, fixed-rate loans to manufacturers and first-time farmers for fixed assets. Toll-free/Additional Phone Number: (608)266-0976.

★1138★ Wisconsin State Rural Development Loan Fund
Route 2
Box 8
Turtle Lake, WI 54889 (715) 986-4171

Description: Provides loans to create jobs in rural areas of northwestern Wisconsin.

Wyoming

★1139★ U.S. Small Business Administration Casper District Office
100 E. B St., Room 4001
Casper, WY 82602-2839 (307) 261-5761

★1140★ Wyoming Community Development Authority
139 W. Second, Suite E
Casper, WY 82602 (307) 265-0603

Description: Offers financing for new and expanding businesses through direct loans, leases, guarantees to banks, and interest rate subsidies.

★1141★ Wyoming Office of the State Treasurer Small Business Assistance Act
Capitol Bldg.
Cheyenne, WY 82002 (307) 777-7408

Description: Provides businesses with fixed-asset financing and a five-year interest rate subsidy. Allows the state treasurer to purchase a guarnateed portion of U.S. Small Business Administration and Farmers Home Administration loans.

★1142★ State Link Deposit Plan

Description: Provides businesses with a five-year, fixed-rate interest subsidy. State treasurer contracts for deposits with Wyoming financial institutions at a rate up to 3 percent below market rates.

★1143★ Wyoming State Department of Administration and Fiscal Control Division of Purchasing and Property Control
Emerson Bldg.
2001 Capitol Ave.
Cheyenne, WY 82002 (307) 777-7253

★1144★ Wyoming State Economic Development and Stabilization Board
Herschler Bldg., Third Floor East
Cheyenne, WY 82002 (307) 777-7286

Description: Administers the Economic Development Loan Program, which provides Wyoming businesses with loans and loan guarantees at flexible rates. Board's Business/Financing staff administers the Wyoming Economic Development Block Grant Program. This program's funds help communities attract or expand local industry. Provides low-interest loans for low- and moderate-income job creation and retention.

CHAPTER 6

BUSINESS INCUBATORS

Business incubators provide start-up companies with below-market rates for office and lab space. They are designed to encourage entrepreneurship and minimize obstacles to new business formation and growth, particularly for high-technology firms. In addition, these facilities offer shared support for clerical, reception, and computer services.

Business incubators may also provide programs that assist in the development of business plans and marketing strategies, advise firms on personnel, accounting, and legal matters, and identify sources of financing. Professionals may also evaluate product lines and manufacturing processes, assist in the use of state-of-the-art design and manufacturing tools, and identify special expertise at universities and other research centers. Many incubators are located on or near universities or research/technology parks. For additional listings, see the chapter entitled University Innovation Research Centers.

Chapter Arrangement

This chapter is arranged geographically by state, then alphabetically by incubator name. All organization names in this chapter are referenced in the Master Index. The Master Index may also include additional incubator names that are mentioned in other chapters.

Arkansas

★1145★ East Arkansas Business Incubator System, Inc
5501 Krueger Dr.
Jonesboro, AR 72401 (501) 935-8365

California

★1146★ Lancaster Economic Development Corporation
104 E. Avenue K4, Suite A
Lancaster, CA 93537 (805) 945-2741

★1147★ McDowell Industrial Business Center
921 Transport Way
Petaluma, CA 94952 (707) 762-6341

★1148★ Southern California Innovation Center
225 Yale Ave., Suite H
Claremont, CA 91711 (714) 624-7161

★1149★ Victor Valley College Small Business Incubator Project
18422 Bear Valley Rd.
Victorville, CA 92392-9699 (619) 245-4271

Colorado

★1150★ Control Data Business and Technology Center (Pueblo)
Business and Technology Center
301 N. Main
Pueblo, CO 81002 (303) 546-1133

Connecticut

★1151★ Bridgeport Innovation Center
955 Connecticut Ave.
Bridgeport, CT 06607 (203) 336-8864

★1152★ Science Park Development Corporation
Five Science Park
New Haven, CT 06511 (203) 786-5000

★1153★ Waterbury Industrial Commons Project
1875 Thomaston Ave.
Waterbury, CT 06704 (203) 574-7704

Idaho

★1154★ Business Center for Innovation and Development (Hayden Lake)
11100 Airport Dr.
Hayden Lake, ID 83835

★1155★ Idaho Innovation Center
457 Broadway Ave.
Idaho Falls, ID 83402

Illinois

★1156★ Bradley Industrial Incubator Program
c/o Area Jobs Development Association
P.O. Box 845
Kankakee, IL 60901 (815) 933-2537

★1157★ Business Center for New Technology (Rockford)
1300 Rock St.
P.O. Box 1200
Rockford, IL 61101-1200 (815) 968-6833

★1158★ Control Data Business and Technology Center (Champaign)
Business and Technology Center
701 Devonshire Dr.
Champaign, IL 61820 (217) 398-5759

★1159★ The Decatur Industrial Incubator
2121 U.S. Route 51, S.
Decatur, IL 62521 (217) 423-2832

★1160★ Des Plaines River Valley Enterprise Zone Incubator Project
912 E. Washington St.
Joliet, IL 60433 (815) 726-0028

★1161★ Electronic Decisions, Inc.
University Microelectronics Center
1706 E. Washington St.
Urbana, IL 61801 (217) 367-2600

★1162★ Fulton-Carroll Center for Industry
Industrial Council of Northwest Chicago
2023 W. Carroll Ave.
Chicago, IL 60612 (312) 421-3941

★1163★ Galesburg Business and Technology Center
Monmouth Blvd.
Galesburg, IL 61401 (309) 343-1194

★1164★ Macomb Business and Technology Center
Western Illinois University Campus
Seal Hall
P.O. Box 6070
Macomb, IL 61455 (309) 298-2212

★1165★ Maple City Business and Technical Center
620 S. Main St.
Monmouth, IL 61462 (309) 734-8544

★1166★ Shetland Properties of Illinois
1801 S. Lumber
Chicago, IL 60616 (312) 738-3121

★1167★ Sterling Small Business and Technology Center
1741 Industrial Dr.
Sterling, IL 61081 (815) 625-5255

Indiana

★1168★ Control Data Business and Technology Center (South Bend)
Business and Technology Center
300 N. Michigan St.
South Bend, IN 46601 (219) 282-4340

★1169★ Indiana Enterprise Center
2523 Merivale St.
Fort Wayne, IN 46805 (219) 426-5700

Iowa

★1170★ University of Iowa Technology Innovation Center
University of Iowa
109 TIC
Oakdale Campus
Iowa City, IA 52242 (319) 335-4063

Louisiana

★1171★ Northeast Louisiana Incubator Center
State Route 594
Swartz School Rd.
Monroe, LA 71203 (318) 343-2262

Maryland

★1172★ Baltimore Medical Incubator
Baltimore Economic Development Corporation
36 S. Charles St., Suite 24
Baltimore, MD 21201 (301) 837-9305

★1173★ Control Data Business and Technology Center (Baltimore)
Business and Technology Center
2901 Druid Park Dr.
Baltimore, MD 21215 (301) 367-1600

★1174★ Technology Advancement Program
University of Maryland
Engineering Research Center, Bldg. 335
College Park, MD 20742 (301) 454-8827

Massachusetts

★1175★ China Trade Center
Chinese Economic Development Council
31 Beach St., Second Floor
Boston, MA 02111 (617) 482-0111

★1176★ J.B. Blood Building
Lynn Office of Economic Development
One Market St., Suite 4
Lynn, MA 01901 (617) 581-9399

★1177★ Schaeffer Business Center
130 Centre St.
Danvers, MA 01923 (617) 777-4605

Michigan

★1178★ Business Service Center (Albion)
Albion Economic Development Corporation
1104 Industrial Ave.
Albion, MI 49224 (517) 629-3926

★1179★ Center for Business Development
c/o Greater Niles Economic Development Foundation, Inc.
P.O. Box 585
Niles, MI 49120 (616) 683-1833

★1180★ Chamber Innovation Center
912 N. Main
Ann Arbor, MI 48104 (313) 662-0550

★1181★ Delta Market Street Incubator
470 Market St., S.W.
Grand Rapids, MI 49503
Incubator is operated by Delta Properties, 1300 Four Mile Rd., N.W., Grand Rapids, MI 49503; (616)451-2561.

★1182★ Delta Monroe Street Incubator
820 Monroe St., N.W.
Grand Rapids, MI 49503
Incubator is operated by Delta Properties, 1300 Four Mile Rd., N.W., Grand Rapids, MI 49503; (616)451-2561.

★1183★ Flint Industrial Village for Enterprise
2717 N. Saginaw St.
Flint, MI 48505 (313) 235-5555

★1184★ Grand Rapids Terminal Ward
446 Granville Ave., S.W.
Grand Rapids, MI 49503
Incubator is operated by Delta Properties, 1300 Four Mile Rd., N.W., Grand Rapids, MI 49503; (616)451-2561.

★1185★ Manufacturing Resource and Productivity Center (Big Rapids)
Ferris State College
Big Rapids, MI 49306 (616) 796-3100

★1186★ Metropolitan Center for High Technology
2727 Second Ave.
Detroit, MI 48201 (313) 963-0616

★1187★ The Venture Center
Downriver Community Conference
15100 Northline
Southgate, MI 48195 (313) 283-1289

Minnesota

★1188★ Control Data Business and Technology Center (Bemidji)
P.O. Box 602
Bemidji, MN 56601 (218) 751-6480

★1189★ Control Data Business and Technology Center (St. Paul)
Business and Technology Center
245 E. Sixth St.
St. Paul, MN 55101 (612) 292-2693

★1190★ Minneapolis Business and Technology Center Limited Partnership
Business and Technology Center
511 11th Ave., S.
Minneapolis, MN 55415 (612) 375-8066

★1191★ Northstar Community Development Corporation
325 Lake Ave., S., Suite 504
Duluth, MN 55802 (218) 727-6690
Currently incubator service is inactive. Future plans for the service are pending.

★1192★ St. Paul Small Business Incubator
Department of Planning and Economic Development
2325 Endicott
St. Paul, MN 55102 (612) 228-3301

★1193★ University Technology Center (Minneapolis)
1313 Fifth St., S.E.
Minneapolis, MN 55414 (612) 623-7774

Mississippi

★1194★ New Building Workspace for Women
121 S. Harvey
Greenville, MS 38701 (601) 335-3523

Missouri

★1195★ Center for Business Innovation, Inc.
4747 Troost
Kansas City, MO 64110 (816) 561-8567

★1196★ Missouri Incutech Foundation
Route 4
Box 519
Rolla, MO 65401 (314) 364-8570

Nebraska

★1197★ Control Data Business and Technology Center (Omaha)
2505 N. 24th St.
Omaha, NE 68110 (402) 346-8262

New Jersey

★1198★ Princeton Capital Corporation
P.O. Box 384
Princeton, NJ 08540 (609) 924-7614

New Mexico

★1199★ Los Alamos Economic Development Corporation
P.O. Box 715
Los Alamos, NM 87544 (505) 662-0001

★1200★ New Mexico Business Innovation Center, Inc.
3825 Academy Parkway South, N.E.
Albuquerque, NM 87109 (505) 345-8668

★1201★ Ventures in Progress
134 Rio Rancho Dr.
Rio Rancho, NM 87124 (505) 892-2161

New York

★1202★ Broome County Industrial Development Agency
P.O. Box 1026
Binghamton, NY 13902 (607) 772-8212

★1203★ Business Incubator (Brooklyn)
Local Development Corporation of East New York
116 Williams Ave.
Brooklyn, NY 11207 (718) 385-6700

★1204★ Cornell Industry Research Park
Cornell University Industry Research Park
Brown Rd., Bldg. 1
Ithaca, NY 14850 (607) 256-7315

★1205★ Greater Syracuse Business Incubator Center
Greater Syracuse Chamber of Commerce
100 E. Onandaga
Syracuse, NY 13202 (315) 470-1343

★1206★ Incubator Industries Building (Buffalo)
Buffalo Urban Renewal Agency
920 City Hall
Buffalo, NY 14202 (716) 855-5056

★1207★ 190 Willow Avenue Industrial Incubator
c/o South Bronx Development Organization
190 Willow Ave.
Bronx, NY 10454 (212) 402-1300

★1208★ Syracuse Incubator
Cental New York Regional High Technology Development Council
c/o Knowledge Systems and Research, Inc.
500 S. Salina St., Room 826
Syracuse, NY 13202 (315) 470-1350

★1209★ Troy Incubator Program
Rensselaer Polytechnic Institute (RPI)
J Bldg.
Troy, NY 12181 (518) 276-6658

★1210★ Western New York Technology Development Center
2211 Main St.
Buffalo, NY 14214 (716) 831-3472

North Carolina

★1211★ Ahoskie Incubator Facility
c/o Tomorrow, Inc.
ECSU Box 962
Elizabeth City, NC 27909

★1212★ Haywood County Incubator
Smokey Mountain Development Corporation
100 Industrial Park
Waynesville, NC 28786 (704) 452-1967

Ohio

★1213★ Akron Industrial Incubator
One Cascade Plaza
Akron, OH 44308 (216) 253-7918

★1214★ Athens Innovation Center
Ohio University
One President St.
Athens, OH 45701 (614) 593-1818

★1215★ Barberton Incubator
576 W. Park Ave.
Barberton, OH 44203 (216) 753-6611

★1216★ Business and Technology Center (Columbus)
1445 Summit St.
Columbus, OH 43201 (614) 294-0206

★1217★ Control Data Business and Technology Center (Toledo)
Business and Technology Center
1946 N. 13th St.
Toledo, OH 43624 (419) 255-6700

★1218★ River East
615 Front St.
Toledo, OH 43605 (419) 698-2310

★1219★ Springfield Incubator
76 E. High St.
Springfield, OH 45502 (513) 324-7744

Oklahoma

★1220★ Atoka Industrial Incubator
Kiamichi Area Vo-Tech School
Atoka Campus
P.O. Box 220
Atoka, OK 74525 (405) 889-7321

★1221★ Durant Industrial Incubator
Rural Enterprises Corporation
Ten Waldron Dr.
Durant, OK 74701 (405) 924-5094

★1222★ Hugo Industrial Incubator
Kiamichi Area Vo-Tech School
Hugo Campus
107 S. 15th St.
Hugo, OK 74743 (405) 326-6491

★1223★ McAlester Industrial Incubator
Kiamichi Area Vo-Tech School
McAlester Campus
Box 308
McAlester, OK 74502 (918) 426-0940

Oregon

★1224★ Cascade Business Center Corporation
Portland Community College
573 N. Killingsworth
Portland, OR 97217 (503) 244-6111

Pennsylvania

★1225★ Altoona Business Incubator
Sixth Ave. and 45th St.
Altoona, PA 16602 (814) 949-2030

★1226★ Ben Franklin Advanced Technology Center
115 Research Dr.
Bethlehem, PA 18015 (215) 861-0584

★1227★ Control Data Business and Technology Center (West Philadelphia)
5070 Parkside Ave.
Philadelphia, PA 19131 (215) 879-8500

★1228★ East Liberty Incubator
Center for Entreprenuerial Development, Inc.
120 S. Whitfield St.
Pittsburgh, PA 15213 (412) 361-5000

★1229★ Executive Office Network, Inc. (Malvern)
Five Great Valley Pkwy.
Malvern, PA 19335 (215) 648-3900

★1230★ Greenville Incubator
12 N. Diamond St.
Greenview, PA 16125 (412) 588-1161

★1231★ Hunting Park West
Southwest Germantown Community Development Corporation
5002 Wayne Ave.
Philadelphia, PA 19144 (215) 843-2000

★1232★ Lansdale Business Center
650 N. Cannon Ave.
Lansdale, PA 19446 (215) 855-6700

★1233★ Liberty Street Market Place
204 Liberty St.
P.O. Box 547
Warren, PA 16365 (814) 726-2400

★1234★ Matternville Business and Technology Center
Road #1
Box 354
Port Matilda, PA 16870 (814) 234-1829

★1235★ Meadville Industrial Incubator
Meadville Area Industrial Commission
628 Arch St.
Meadville, PA 16335 (814) 724-2975

★1236★ Model Works Industrial Commons
Girard Area Industrial Development Corporation
227 Hathoway St.
Girard, PA 16417 (814) 774-9339

★1237★ North East Tier Advanced Technology Center
Ben Franklin Advanced Technology Center #125
Lehigh University
125 Goodman Dr.
Bethlehem, PA 18015 (215) 758-5200

★1238★ Paoli Technology Enterprise Center
19 E. Central Ave.
Paoli, PA 19301 (215) 251-0505

★1239★ Ridgeway Manufacturing Incubator
North Central Pennsylvania Regional Planning and Development Commission
P.O. Box 488
Ridgeway, PA 15853 (814) 772-6901

★1240★ River Bridge Industrial Center
One S. Olive
Media, PA 19063 (215) 872-4469

★1241★ Southwest Pennsylvania Business Development Center
12300 Perry Hwy.
P.O. Box 216
Wexford, PA 15090 (412) 931-8444

★1242★ Technology Centers International (Montgomeryville)
1060 Route 309
Montgomeryville, PA 18936 (215) 646-7800

★1243★ University City Science Center
3624 Market St.
Philadelphia, PA 19104 (215) 387-2255

★1244★ University Technology Development Center I (Pittsburgh)
4516 Henry St.
Pittsburgh, PA 15213

★1245★ University Technology Development Center II (Pittsburgh)
3400 Forbes Ave.
Pittsburgh, PA 15213

South Carolina

★1246★ Control Data Business and Technology Center (Charleston)
Business and Technology Center
701 E. Bay St.
Charleston, SC 29403 (803) 722-1219

★1247★ Florence Incubator
City of Florence
City/County Complex
Drawer FF
Florence, SC 29501 (803) 665-3141

★1248★ North Augusta Incubator
City of North Augusta
P.O. Box 6400
North Augusta, SC 29841 (803) 278-0816

★1249★ Rock Hill Incubator
Rock Hill Economic Development Corporation
P.O. Box 11706
Rock Hill, SC 29731 (803) 328-6171

★1250★ Spartanburg Incubator
City of Spartanburg
P.O. Box 1749
Spartanburg, SC 29304 (803) 596-2072

Tennessee

★1251★ Knox County Business Incubator
City County Bldg.
Knoxville, TN 37901 (615) 521-2275

★1252★ Tennessee Innovation Center
P.O. Box 607
Oakridge, TN 37830 (615) 576-3375

Texas

★1253★ Control Data Business and Technology Center (San Antonio)
301 S. Frio
San Antonio, TX 78207 (512) 270-4500

Vermont

★1254★ East Creek Center
Star Enterprises, Inc.
38 1/2 Center St.
Rutland, VT 05701 (802) 775-8011

★1255★ Fellows Complex
Precision Valley Development Corporation
100 River St.
Springfield, VT 05156 (802) 885-2138

★1256★ North Bennington Business Incubator
Bennington County Industrial Corporation (BCIC)
P.O. Box 357
North Bennington, VT 05257 (802) 442-8975

Washington

★1257★ The Center (Everett)
917 134th St., S.W., Suite 100
Everett, WA 98204 (206) 743-9669

★1258★ Spokane Business and Technology Center
Spokane International Airport Business Park
3707 S. Godfrey Blvd.
Spokane, WA 99204 (509) 458-6340

Wisconsin

★1259★ The Advocap Small Business Center
19 W. First St.
Fond du Lac, WI 54925 (414) 922-7760

★1260★ Madison Business Incubator
c/o Wisconsin for Research, Inc.
210 N. Bassett St.
Madison, WI 53703 (608) 258-7070

CHAPTER 7

VENTURE CAPITALISTS

Venture capitalists and inventors have much in common. High rollers and risk takers in the finest American tradition, they live on the edge, pushing their respective envelopes in the hopes of generating technological advancement coupled with great financial reward. It is this blend of business acumen and inventive thinking that helps keep the U.S. a nation of suppliers rather than buyers of new technologies.

Banks are creditors. They're interested in the immediate future, yet most heavily influenced by the past. Loan officers examine the product and market position of the company for assurance that the invention can return a steady flow of sales and generate enough cash to repay the loan and interest.

Venture capital companies are owners. They gamble on the future. Venture capitalists can provide the money and management skills that permit inventors to bring their inventions to fruition and market. Inventors, in turn, give the venture capitalists an opportunity to get what they seek, typically a three to five times return on their investment within five to seven years.

Their investment unprotected in the event of failure, most venture capital firms set rigorous policies for venture proposal size, maturity of the seeking company, and requirements and evaluation procedures to reduce risks.

Projects requiring under $250,000 are of limited interest because of the high cost of investigation and administration; however, some venture firms will consider smaller proposals, if the investment is intriguing enough. Most venture capitalists live in the $250,000 to $1.5 million atmosphere.

Experienced in putting together "marriages" between good ideas and good money, these wheeling, deal-makers have relationships with a wide variety of potential investors such as corporations, insurance companies, union pension funds, university endowments, and wealthy individuals. Yet, unlike passive investors and traditional lenders, venture capitalists take a hands-on proactive role in managing the companies in which they invest. In the end, investors invest in the venture capitalist as much as in any invention they may represent.

While the inventor works on product development, the venture capitalist gets the money to keep the pump primed, prepare long-term corporate business programs, marketing plans, and personnel recruitment. Most inventors are not skilled at these activities and welcome the assistance.

The inventor's motivation most often involves more than money, i.e., things like ego and pride. The venture capitalist is in it typically for money alone, the invention being a vehicle to this end. For the relationship to work the parties must be highly compatible. The chemistry between inventor and venture capitalist is, therefore, a major factor in how well their business will develop. Inventors should look at venture capitalists as working partners and not just investors. The relationship begins with the ritual of "contact" which leads to "courtship" and then to "investigation."

According to the Small Business Administration (SBA), the typical venture capital company receives over 1,000 proposals a year. Probably 90 percent will be rejected quickly because they don't fit the established geographical, technical, or market area policies of the firm, or because they have been poorly presented.

The remaining 10 percent are investigated carefully. The venture capitalist may spend between $2,000 and $3,000 per company to hire consultants that will make preliminary investigations. Per the SBA, they result in maybe 10 to 15 proposals of interest. Then, second investigations, more thorough and more expensive than the first, reduce the number of companies under active consideration to only three or four. Eventually the firm invests in one or two of these.

The entire process can take anywhere from three to six months. Before any attempts are made to raise money, the venture capitalist and inventor had best become very familiar with each other. Annulments are rare. Divorces can be messy.

The best way to check out the capabilities of venture capitalists is to speak with the management of other companies the venture capitalist has set-up. And don't just look into successful enterprises, but find people who have experienced total flops to see how the venture capitalist handles both success and failure. Any seasoned venture capitalist will have had a share of good and bad deals.

An inventor would be well advised to engage the services of an attorney and or CPA before signing on with any venture capitalist. The money you spend will be money well spent if it gets you a good deal or saves you from a bad one.

Chapter Arrangement

This chapter is arranged geographically by state, then alphabetically by venture capitalist name. All organization names in this chapter are referenced in the Master Index. The Master Index may also include additional venture capitalists that are mentioned in other chapters.

Alabama

★1261★ Alabama Capital Corporation
16 Midtown Park, E.
Mobile, AL 36606 (205) 476-0700
David C. Delaney, President
Description: Minority enterprise small business investment company. Investment Policy: No industry preference.

★1262★ First SBIC of Alabama
16 Midtown Park, E.
Mobile, AL 36606 (205) 476-0700
David Delaney, President
Description: Small business investment company. Investment Policy: No industry preference.

★1263★ Hickory Venture Capital Corporation
699 Gallatin, Suite A2
Huntsville, AL 35801 (205) 539-1931
J. Thomas Noojin, President
Description: Small business investment company. Investment Policy: No industry preference.

★1264★ Tuskegee Capital Corporation
4853 Richardson Rd.
Montgomery, AL 36108 (205) 281-8059
A. G. Bartholomew, President
Description: Minority enterprise small business investment company. Investment Policy: Diversified.

Alaska

★1265★ Calista Business Investment Corporation
516 Denali St.
Anchorage, AK 99501 (907) 279-5516
Johnny T. Hawk, President
Description: Minority enterprise small business investment company. Investment Policy: No industry preference.

Arizona

★1266★ Navajo Small Business Development Corporation
P.O. Drawer L
Fort Defiance, AZ 86504
Description: Certified development company. Investment Policy: Inquire.

★1267★ Norwest Venture Capital Management, Inc. (Scottsdale)
8777 E. Via de Ventura, Suite 335
Scottsdale, AZ 85258 (602) 483-8940
Robert Zicarelly, Chairman of the Board
Description: Small business investment company. Investment Policy: Prefers to invest in high-technology enterprises. Remarks: An affiliated office of the Norwest Ventur Capital Management headquarters in Minneapolis.

★1268★ Rocky Mountain Equity Corporation
4530 Central Ave.
Phoenix, AZ 85012 (602) 274-7534
Anthony J. Nicoli, President
Description: Small business investment company. Investment Policy: No industry preference.

★1269★ Valley National Investors, Inc.
P.O. Box 71
Phoenix, AZ 85001 (602) 261-1577
J. M. Holliman, III, Managing Director
Description: Small business investment company. Investment Policy: No industry preference.

Arkansas

★1270★ Capital Management Services, Inc.
1910 N. Grant St., Suite 200
Little Rock, AR 72207 (501) 664-8613
David L. Hale, President
Description: Minority enterprise small business investment company. Investment Policy: No industry preference.

★1271★ Kar-Mal Venture Capital, Inc.
2821 Kavanaugh Blvd.
Little Rock, AR 72205 (501) 661-0010
Amelia Karam, President
Description: Minority enterprise small business investment company. Investment Policy: No industry preference.

★1272★ Power Ventures, Inc.
829 Highway 270, N.
Malvern, AR 72104 (501) 332-3695
Dorsey D. Glover, President
Description: Minority enterprise small business investment company. Investment Policy: No industry preference.

★1273★ Small Business Investment Capital, Inc.
P.O. Box 3627
Little Rock, AR 72203 (501) 455-3590
Jerry Davis, President
Description: Small business investment company. Investment Policy: Retail grocery industry preferred.

California

★1274★ ABC Capital Corporation
610 E. Live Oak Ave.
Arcadia, CA 91006-5740 (818) 445-2789
Anne B. Cheng, President

Description: Minority enterprise small business investment company. Investment Policy: No industry preference.

★1275★ Accel Partners (San Francisco)
One Embarcadero Center
San Francisco, CA 94111 (415) 989-5656

Description: Private venture capital firm. Investment Policy: Prefers to invest in telecommunications, software, or biotechnology/medical products industries. **Contact:** Dixon R. Doll; Arthur C. Patterson; James R. Swartz, General Partners.

★1276★ Adler and Company (Sunnyvale)
1245 Oakmead Pkwy., Suite 103
Sunnyvale, CA 94086 (408) 720-8700

Description: Private venture capital supplier. Investment Policy: Provides all stages of financing. Remarks: A branch office of Adler and Company, New York, New York.

★1277★ Allied Business Investors, Inc.
428 S. Atlantic Blvd., Suite 201
Monterey Park, CA 91754 (818) 289-0186
Jack Hong, President

Description: Minority enterprise small business investment company. Investment Policy: No industry preference.

★1278★ Ally Finance Corporation
9100 Wilshire Blvd., Suite 408
Beverly Hills, CA 90212 (213) 550-8100
Percy P. Lin, President

Description: Minority enterprise small business investment company. Investment Policy: No industry preference.

★1279★ AMF Financial, Inc.
9910-D Mira Mesa Blvd.
San Diego, CA 92131 (619) 695-0233
William Temple, President

Description: Small business investment company. Investment Policy: No industry preference.

★1280★ Asian American Capital Corporation
1251 W. Tennyson Rd., Suite 4
Hayward, CA 94544 (415) 887-6888
Jennie Chien, Manager

Description: Minority enterprise small business investment company. Investment Policy: No industry preference.

★1281★ Asset Management Company
2275 E. Bayshore, Suite 150
Palo Alto, CA 94303 (415) 494-7400

Description: Venture capital firm. Investment Policy: High-technology industries preferred. **Contact:** John Shoch or Craig Taylor, Partners.

★1282★ Associates Venture Capital Corporation
425 California St., Suite 2203
San Francisco, CA 94104 (415) 956-1444
Walter P. Stryker, President

Description: Investment Policy: No industry preference.

★1283★ Bankamerica Ventures, Inc.
555 California St.
San Francisco, CA 94104 (415) 953-3001
Patrick Topolski, President

Description: Small business investment company. Investment Policy: No industry preference.

★1284★ Bay Partners
10600 N. De Anza Blvd.
Cupertino, CA 95014 (415) 725-2444

Description: Venture capital supplier. Investment Policy: Provides start-up financing primarily to West Coast technology companies that have highly qualified management teams. Initial investments range from $100,000 to $800,000; where large investments are required, will act as lead investor to bring in additional qualified venture investors. **Contact:** John Freidenrich, W. Charles Hazel, and John Bosch, Partners.

★1285★ Bay Venture Group
One Embarcadero Center, Suite 3303
San Francisco, CA 94111 (415) 989-7680
William R. Chandler, General Partner

Description: Small business investment company. Investment Policy: High-technology industries preferred; geographic region limited to the San Francisco Bay area.

★1286★ Biovest Partners
12520 High Bluff Dr., Suite 250
San Diego, CA 92130 (619) 481-8100
Timothy J. Wollaeger, General Partner

Description: Venture capital partnership. Investment Policy: Prefers medical and biotechnology industries.

★1287★ BNP Venture Capital Corporation
3000 Sand Hill Rd.
Building One, Suite 125
Menlo Park, CA 94025 (415) 854-1084
Edgerton Scott, II, President

Description: Small business investment company. Investment Policy: No industry preference.

★1288★ Brentwood Associates
11661 San Vicente Blvd., Suite 707
Los Angeles, CA 90049 (213) 826-6581

Description: Venture capital supplier. Investment Policy: Provides start-up and expansion financing to technology-based enterprises specializing in computing and data processing, electronics, communications, materials, energy, industrial automation, and bioengineering and medical equipment. Investments generally range from $1 to $3 million. **Contact:** B. Kipling Hagopian or Frederick J. Warren.

★1289★ Burr, Egan, Deleage, and Company (San Francisco)
Three Embarcadero Center, 25th Floor
San Francisco, CA 94111 (415) 362-4022
Jean Deleage

Description: Private venture capital supplier. Investment Policy: Invests start-up, expansion, and acquisitions capital nationwide. Principal concerns are strength of the management team; large, rapidly expanding markets; and unique products for services. Past investments have been made in the fields of electronics, health, and communications. Investments range from $750,000 to $5 million. Remarks: Maintains an office in Boston, Massachusetts.

★1290★ Business Equity and Development Corporation
1411 W. Olympic Blvd., Suite 200
Los Angeles, CA 90015 (213) 385-0351
Leon M. N. Garcia, President

Description: Minority enterprise small business investment company. Investment Policy: No industry preference.

★1291★ Cable and Howse Ventures (Palo Alto)
435 Tasso St., Suite 115
Palo Alto, CA 94301 (415) 322-8400

Description: Venture capital supplier. Investment Policy: Provides start-up and early-stage financing to enterprises in the western United States, although a national perspective is maintained. Interest lies in proprietary or patentable technology. Investments range from $500,000 to $2 million. Remarks: A branch office of Cable and Howse Ventures, Bellevue, Washington.

★1292★ California Capital Investors
11812 San Vicente Blvd.
Los Angeles, CA 90049 (213) 820-7222
Arthur H. Bernstein, Managing General Partner

Description: Small business investment company. Investment Policy: No industry preference.

★1293★ California Partners Draper Association CGP
c/o Alan Brudos
3000 Sand Hill Rd.
Building Four, Suite 235
Menlo Park, CA 94025 (415) 854-1712
Tim Draper, President

Description: Small business investment company. Investment Policy: No industry preference.

★1294★ Canaan Venture Partners (Menlo Park)
3000 Sand Hill Rd.
Building One, Suite 205
Menlo Park, CA 94025 (415) 854-8092
Eric A. Young, General Partner

Description: Venture capital supplier. Investment Policy: Primary concern is strong entrepreneurial management. There are no geographic or industry-specific constraints. Remarks: A branch offic of Canaan Venture Partners in Rowayton, Connecticut.

★1295★ CFB Venture Capital Corporation (Los Angeles)
350 California St.
Los Angeles, CA 94120-7104 (619) 230-3304
Pieter Westerbeek, III, Chief Financial Officer

Description: Small business investment company. Investment Policy: No industry preference. Remarks: Maintains an office in San Francisco, California.

★1296★ CFB Venture Capital Corporation (San Francisco)
350 California St., 12th Floor
San Francisco, CA 94104 (415) 445-0594

Description: Small business investment company. Investment Policy: No industry preference. Remarks: A branch office of CFB Venture Capital Corporation, San Diego, California.

★1297★ Charterway Investment Corporation
222 S. Hill St., Suite 800
Los Angeles, CA 90012 (213) 687-8539
Harold H. M. Chuang, President

Description: Minority enterprise small business investment company. Investment Policy: No industry preference.

★1298★ Churchill International
444 Market St., Suite 2501
San Francisco, CA 94111 (415) 328-4401

Description: International venture capital supplier and investment management company. Investment Policy: Provides start-up, growth and development, and bridge and buy out capital to small and medium-sized high-technology companies in the United States, Europe, and Pacific Basin.

★1299★ Cin Industrial Investments Ltd.
170 Middlefield Rd., Suite 150
Menlo Park, CA 94025 (415) 325-6830

Description: Venture capital fund. Investment Policy: No industry preference. **Contact:** JanetEffland, President/Corporate General Partner.

★1300★ Citicorp Venture Capital Ltd. (Palo Alto)
2200 Geng Rd., Suite 203
Palo Alto, CA 94303 (415) 424-8000

Description: Small business investment company. Investment Policy: Invests in information processing, telecommunications, health care, and other industries. Provides financing to companies in all stages of development. Remarks: A branch office of Citicorp Venture Capital Ltd., New York, New York.

★1301★ Comdisco Venture Lease, Inc. (San Francisco)
101 California St., 38th Floor
San Francisco, CA 94111 (415) 421-1800

Description: Venture capital subsidiary of operating firm. Investment Policy: Involved in all stages of financing except seed. Has no industry preference, but avoids real estate and oil and gas. Investments range from $500,000 to $5 million. **Contact:** William Tenneson or Terrence Fowler.

★1302★ Concord Partners (Palo Alto)
435 Tasso St., Suite 305
Palo Alto, CA 94301 (415) 327-2600

Description: Venture capital supplier. Investment Policy: Diversified in terms of stage of development, industry classification, and geographic location. Areas of special interest include computers, telecommunications, health care and medical products, energy, and leveraged buy outs. Preferred investments range from $2 to $3 million, with a $1 million minimum and $10 million maximum.

★1303★ Continental Investors, Inc.
8781 Seaspray Dr.
Huntington Beach, CA 92646 (714) 964-5207
Lac Thantrong, President

Description: Minority enterprise small business investment company. Investment Policy: No industry preference.

★1304★ Crosspoint Investment Corporation
1951 Landings Dr.
Mountain View, CA 94043 (415) 964-3545
Max Simpson, President

Description: Small business investment company. Investment Policy: No industry preference. Remarks: A division of Crosspoint Venture Partners, Mountain View, California.

★1305★ Crosspoint Venture Partners
1951 Landings Dr.
Mountain View, CA 94043 (415) 964-0100

Description: Venture capital partnership. Investment Policy: Seeks to invest start-up capital in unique products, services, and/or market opportunities located in the Western United States. Primary interests lie in communication devices and systems; biotechnology; medical products, instruments, and equipment; computer software and productivity tools; computers and computer peripherals; industrial automation and controls; semiconductor devices and equipment; instrumentation; and related service and distribution businesses. Investments range from $50,000 to $3 million. Remarks: Maintains an incubator facility in Palo Alto, California, to support new ventures. Also maintains an office in Newport Beach, California. **Contact:** John Mumford; Roger Barry; Bill Cargile; Bob Hoff; Fred Dotzler, General Partners.

★1306★ Developers Equity Capital Corporation
1880 Century Park, E., Suite 311
Los Angeles, CA 90067 (213) 277-0330
Larry Sade, Chairman of the Board

Description: Small business investment company. Investment Policy: Real estate preferred.

★1307★ Dime Investment Corporation
3731 Wilshire Blvd., Suite 535
Los Angeles, CA 90010 (213) 739-1847
Chun Y. Lee, President

Description: Minority enterprise small business investment company. Investment Policy: No industry preference.

★1308★ Dougery, Jones & Wilder (Mountain View)
2003 Landings Dr.
Mountain View, CA 94043 (415) 968-4820

Description: Venture capital supplier. Investment Policy: Prefers to invest in small to medium-sized companies headquartered or primarily operating in the West and Southwest. The typical company sought is in the computers/communications or medical/biotechnology industries; is privately owned; has an experienced management team; serves growth markets; and can achieve at least $30 million in sales. Initial investment ranges from $250,000 to $2.5 million; may be expanded by including other venture capital firms. Remarks: Maintains an office in Dallas, Texas. **Contact:** John R. Dougery; David A. Jones; Henry L.B. Wilder; A. Lawson Howard; Gerald R. Schoonhoven.

★1309★ El Dorado Technology Partners
Two N. Lake Ave., Suite 480
Pasadena, CA 91101 (818) 193-1936
Brent T. Rider, General Partner

Description: Private venture capital firm. Investment Policy: Prefers to invest in communications, electronics, industrial products and services, and medical/health care industries.

★1310★ El Dorado Ventures
Two N. Lake Ave., Suite 480
Pasadena, CA 91101 (818) 793-1936
Brent T. Rider, General Partner

Description: Private venture capital firm. Investment Policy: Prefers to invest in communications, electronics, industrial products and services, and medical/health care industries. Remarks: Affiliated with El Dorado Technology Partners.

★1311★ Enterprise Partners
5000 Birch St., Suite 6200
Newport Beach, CA 92660 (714) 833-3650

Description: Venture capital fund. Investment Policy: Prefers electronics, medical technology, and health care ventures in southern California. **Contact:** Charles D. Martin or James H. Berglund, General Partners.

★1312★ Equitable Capital Corporation
855 Sansome St., Second Floor
San Francisco, CA 94111 (415) 434-4114
James Lee, Vice-President

Description: Minority enterprise small business investment company. Investment Policy: No industry preference.

★1313★ First American Capital Funding, Inc.
38 Corporate Park
Irvine, CA 92714 (714) 660-9288
Lou Trankiem, President

Description: Minority enterprise small business investment company. Investment Policy: No industry preference.

★1314★ First Century Partners (San Francisco)
350 California St.
San Francisco, CA 94104 (415) 955-1612
C. Sage Givens

Description: Private venture capital firm. Investment Policy: High-technology, medical, and specialty retail industries are preferred; has no interest in real estate or entertainment investments. Minimum investment is $1 million.

★1315★ First Interstate Capital, Inc.
5000 Birch St., Suite 10100
Newport Beach, CA 92660 (714) 253-4360
Ronald J. Hall, Managing Director

Description: Small business investment company. Investment Policy: No industry preference. Provides capital for small and medium-sized companies through participation in private placements of subordinated debt, preferred, and common stock. Offers growth-acquisition and later-stage venture capital.

★1316★ First SBIC of California (Costa Mesa)
650 Town Center Dr., 17th Floor
Costa Mesa, CA 92626 (714) 556-1964
Brian Jones, Senior Vice-President

Description: Small business investment company and venture capital company. Investment Policy: No industry preference. Remarks: Maintains offices in Palo Alto and Pasadena, California; Boston, Massachusetts; Washington, Pennsylvania; and London.

★1317★ First SBIC of California (Palo Alto)
Five Palo Alto Sq., Suite 1038
Palo Alto, CA 94306 (415) 424-8011

Description: Small business investment company. Investment Policy: No industry preference. Remarks: A branch office of First Small business investment company of California, Costa Mesa, California.

★1318★ First SBIC of California (Pasadena)
155 N. Lake Ave., Suite 1010
Pasadena, CA 91109 (818) 304-3451

Description: Small business investment company. Investment Policy: No industry preference. Remarks: A branch office of First Small business investment company of California, Costa Mesa, California.

★1319★ G C and H Partners
One Maritime Plaza, 20th Floor
San Francisco, CA 94111 (415) 981-5252
James C. Gaither, General Partner

Description: Small business investment company. Investment Policy: No industry preference.

★1320★ Glenwood Management
3000 Sand Hill Rd.
Building Four, Suite 230
Menlo Park, CA 94025 (415) 854-8070
Dag Tellefsen

Description: Venture capital supplier. Investment Policy: Provides early-stage financing to companies with proprietary technology and exceptional management teams. Areas of interest include telecommunications; computer hardware, software, and peripherals; applications of genetic engineering; medical instruments; and semiconductor equipment and devices. Initial investments range from $300,000 to $500,000, with subsequent rounds totalling $800,000 to $1.2 million. Geographic area limited to the Western United States.

★1321★ Hambrecht & Quist Venture Partners
235 Montgomery St., Fifth Floor
San Francisco, CA 94104 (415) 576-3300
Richard Gorman, Vice-President

Description: Venture capital firm. Investment Policy: Prefers start-up investments in high-technology and biotechnology industries. Investments range from $500,000 to $5 million.

★1322★ Hamco Capital Corporation
235 Montgomery St.
San Francisco, CA 94104 (415) 986-5500
William R. Hambrecht, President

Description: Small business investment company. Investment Policy: High-technology industries preferred.

★1323★ Harvest Ventures, Inc. (Cupertino)
Building SW3, Suite 365
10080 N. Wolfe Rd.
Cupertino, CA 95014 (408) 996-3200

Description: Private venture capital supplier. Investment Policy: Prefers to invest in high-technology, growth-oriented companies with proprietary technology, large market potential, and strong management teams. Specific areas of interest include computers, computer peripherals, semiconductors, telecommunications, factory automation, military electronics, medical products, and health-care services. Provides seed capital of up to $250,000; investments in special high-growth and leveraged buy out situations range from $500,000 to $2 million. Will serve as lead investor and arrange financing in excess of $2 million in association with other groups. Remarks: A branch office of Harvest Ventures, Inc., New York, New York.

★1324★ Helio Capital, Inc.
5900 S. Eastern Ave., Suite 136
Commerce, CA 90040 (213) 721-8053
Frank Remski, General Manager

Description: Minority enterprise small business investment company. Investment Policy: No industry preference.

★1325★ Hillman Ventures, Inc.
2200 Sand Hill Rd., Suite 240
Menlo Park, CA 94025 (415) 854-4653
Philip S. Paul, Chairman

Description: Venture capital firm. Investment Policy: Medical, biotechnology, and computer-related electronics industries preferred.

★1326★ I.K. Capital Loans Ltd.
9460 Wilshire, Suite 608
Beverly Hills, CA 90212 (213) 276-8547
Iraj Kermanshahchi, President

Description: Small business investment company. Investment Policy: No industry preference.

★1327★ Imperial Ventures, Inc.
P.O. Box 92991
Los Angeles, CA 90009 (213) 417-5830

Description: Small business investment company. Investment Policy: No industry preference. **Contact:** H.Wayne Snavely, President; John H. Upshen, Senior Vice-President.

★1328★ Institutional Venture Partners III
3000 Sand Hill Rd.
Building Two, Suite 290
Menlo Park, CA 94025 (415) 854-0132
Reid W. Dennis, General Partner

Description: Venture capital fund. Investment Policy: Will provide $750,000 to $2 million in early-stage financing.

★1329★ Interscope Investments, Inc.
10900 Wilshire Blvd., Suite 1400
Los Angeles, CA 90024 (213) 208-8525
Murray Hill, Chief Financial Officer

Description: Venture capital firm. Investment Policy: No industry preference.

★1330★ Interven II
333 S. Grand Ave.
Los Angeles, CA 90071 (213) 622-1922
Jonatahn E. Funk, General Partner

Description: Venture capital fund. Investment Policy: Semiconductors, factory automation, telecommunications, health care, and medical technologies preferred. Will invest $1.5 to $2 million in seed and first-round technology companies in Oregon, Washington, Idaho, and southern California. Will also handle buy outs of companies with annual sales of $20 to $50 million.

★1331★ Interwest Partners
3000 Sand Hill Rd.
Building Three, Suite 255
Menlo Park, CA 94025 (415) 854-8585
W. Scott Hedrick, General Partner

Description: Venture capital fund. Investment Policy: Both high-tech and low- or non-technology companies are considered. No oil, gas, real estate, or construction projects.

★1332★ Investments Orange Nassau, Inc.
Westerly Place
1500 Quail St., Suite 540
Newport Beach, CA 92660 (714) 752-7811

Description: Small business investment company. Investment Policy: Provides all stages of financing to technology-based and oil and gas service industries; no geographic preference. Experience of management team and high-growth market potential are emphasized. Investments range from $500,000 to $5 million. Remarks: A branch office of Orange Nassau Capital Corporation, Boston, Massachusetts.

★1333★ Ivanhoe Venture Capital Ltd.
737 Pearl St., Suite 201
LaJolla, CA 92037 (619) 454-8882
Alan Toffler, General Partner

Description: Small business investment company. Investment Policy: No industry preference.

★1334★ Jupiter Partners
600 Montgomery St., 35th Floor
San Francisco, CA 94111 (415) 421-9990
John M. Bryan, President

Description: Small business investment company. Investment Policy: No industry preference.

★1335★ Kleiner, Perkins, Caufield, and Byers
Four Embarcadero Center, Suite 3520
San Francisco, CA 94111 (415) 421-3110

Description: Investment Policy: Provides seed, start-up, second- and third-round, and bridge financing to companies on the West Coast. Past investments have been made in the following fields: computers and computer peripherals, software, office equipment, medical products and instruments, microbiology, genetic engineering, telecommunications, instrumentation, semiconductors, lasers and optics, and unique consumer products and services. Areas avoided for investment include real estate, motion pictures, solar energy, hotels and motels, restaurants, resort areas, oil and gas, construction, and metallurgy. Investments range from $1 to $5 million. **Contact:** Thomas J. Perkins; Frank J. Caufield; Brook H. Byers, General Partners.

★1336★ Lailai Capital Corporation
1545 Wilshire Blvd., Suite 510
Los Angeles, CA 90017 (213) 484-5085
Kenneth Chen, President and General Manager

Description: Minority enterprise small business investment company. Investment Policy: No industry preference.

★1337★ Magna Pacific Investments
977 N. Broadway, Suite 301
Los Angeles, CA 90012 (213) 680-2505
David Wong, President

Description: Minority enterprise small business investment company. Investment Policy: No industry preference.

★1338★ Marwit Capital Corporation
180 Newport Center Dr., Suite 200
Newport Beach, CA 92660 (714) 640-6234
Martin W. Witte, President

Description: Small business investment company. Investment Policy: No industry preference.

★1339★ Matrix Partners (Menlo Park)
2500 Sand Hill Rd., Suite 113
Menlo Park, CA 94025 (415) 854-3131

Description: Private venture capital partnership. Investment Policy: Investments range from $500,000 to $1 million. Remarks: A branch office of Matrix Partners in Boston, Massachusetts.

★1340★ Mayfield Fund
2200 Sand Hill Rd., Suite 200
Menlo Park, CA 94025 (415) 854-5560

Description: Venture capital partnership. Investment Policy: Prefers technically oriented companies that have potential to achieve revenues of at least $75 to $100 million in 5 years. Principal interest is experience of management team. Initial investments range from $50,000 to $2 million; for subsequent investments, capital commitment may double or triple the initial investment. **Contact:** Thomas J. Davis, Jr.; A. Grant Heidrich III; F. Gibson Myers, Jr.; Norman A. Fogelsong; Glenn M. Mueller, General Partners.

★1341★ Menlo Ventures
3000 Sand Hill Rd.
Building Four, Suite 100
Menlo Park, CA 94025 (415) 854-8540

Description: Venture capital supplier. Investment Policy: Provides start-up and expansion financing to companies with experienced management teams, distinctive product lines, and large growing markets. Primary interest is in technology-oriented, service, consumer products, and distribution companies. Investments range from $500,000 to $3 million; also provides capital for leveraged buy outs. **Contact:** Douglas C. Carlisle; Ken E. Joy; Richard P. Magnuson; H. Dubose Montgomery, General Partners.

★1342★ Merrill, Pickard, Anderson, and Eyre (MPAE)
Two Palo Alto Sq., Suite 425
Palo Alto, CA 94306 (415) 856-8880

Description: Private venture capital partnership. Investment Policy: Provides start-up and early-stage financing to companies with experienced management teams, with the ability to grow to $50 to $100 million in annual revenues within 4 to 6 years and with distinctive product lines. High-technology industries are preferred; little interest in financial, real estate, or consulting companies. Investments range from $750,000 to $2.5 million. **Contact:** Steven L. Merrill; W. Jeffers Pickard; James C. Anderson; Chris A. Eyre; Stephen E. Coit, General Partners.

★1343★ Metropolitan Venture Company, Inc.
5757 Wilshire Blvd., Suite 670
Los Angeles, CA 90036 (213) 938-3488
Rudolph J. Lowy, Chairman of the Board

Description: Small business investment company. Investment Policy: No industry preference.

★1344★ MIP Equity Advisors, Inc. (Menlo Park)
3000 Sand Hill Rd.
Building Four, Suite 280
Menlo Park, CA 94025 (415) 854-2653
Hans Severiens, President

Description: International venture capital supplier. Investment Policy: Seeks participants from a variety of business activities that wish to expand into European markets by establishing a base in the Netherlands. High-technology ventures are preferred, but not the the exclusion of other industry groups. Minimum investment is $1 million. Remarks: Company headquarters are located in The Hague, the Netherlands. Another U.S. office is located in Boston, Massachusetts.

★1345★ Montgomery Securities
600 Montgomery St.
San Francisco, CA 94111 (415) 627-2454
May Leong

Description: Private venture capital and investment banking firm. Investment Policy: Diversified, but will not invest in real estate or energy-related industries. Involved in both start-up and later-stage financing.

★1346★ Myriad Capital, Inc.
328 S. Atlantic Blvd., Suite 200
Monterey, CA 91754 (818) 570-4548
Kuo Hung Chen, President

Description: Minority enterprise small business investment company. Investment Policy: No industry preference.

★1347★ New Enterprise Associates (Menlo Park)
3000 Sand Hill Rd.
Building Four, Suite 235
Menlo Park, CA 94025 (415) 854-2215
C. Woodrow Rae, General Partner

Description: Venture capital supplier. Investment Policy: Concentrates on technology-based industries that have the potential for product innovation, rapid growth, and high profit margins. Investments range from $250,000 to $2 million. Past investments have been made in the following industries: computer software, medical and life sciences, computers and peripherals, communications, semiconductors, specialty retailing, energy and alternative energy, defense electronics, materials, and specialty chemicals. Management must demonstrate intimate knowledge of its marketplace and have a well-defined strategy for achieving strong market penetration. Remarks: An affiliated office of New Enterprises Associates, Baltimore, Maryland.

★1348★ New Enterprise Associates (San Francisco)
Russ Bldg., Suite 1025
235 Montgomery St.
San Francisco, CA 94104 (415) 956-1579

Description: Venture capital supplier. Investment Policy: Concentrates in technology-based industries that have the potential for product innovation, rapid growth, and high profit margins. Investments range from $250,000 to $2 million. Past investments have been made in the following industries: computer software, medical and life sciences, computers and computer peripherals, communications, semiconductors, specialty training, energy and alternative energy, defense electronics, materials, and specialty chemicals. Management must demonstrate intimate knowledge of its marketplace and have a well-defined strategy for achieving strong market penetration. **Contact:** C. Richard Kramlich or Thomas C. McConnell.

★1349★ New Enterprise Associates (Westlake Village)
200 N. Westlake Blvd., Suite 215
Westlake Village, CA 91362 (805) 373-8537
James Cole, Partner

Description: Venture capital partnership. Investment Policy: Early-stage investors, with initial investments ranging from $500,000 to $2 million. Remarks: An affiliated office of the New Enterprise Associates headquarters in Baltimore.

★1350★ New Kukje Investment Company
958 S. Vermont Ave., Suite C
Los Angeles, CA 90006 (213) 389-8679
George Su Chey, President

Description: Minority enterprise small business investment company. Investment Policy: No industry preference.

★1351★ New West Partners (Newport Beach)
4600 Campus Dr., Suite 103
Newport Beach, CA 92660 (714) 756-8940
Timothy P. Haidinger, General Partner

Description: Small business investment company. Investment Policy: No industry preference. Remarks: Maintains an office in San Diego, California.

★1352★ New West Partners (San Diego)
4350 Executive Dr., Suite 206
San Diego, CA 92121 (619) 457-0722

Description: Small business investment company. Investment Policy: No industry preference. Remarks: A branch office of New West Partners, Newport Beach, California.

★1353★ Northwest Business Investment Corporation
30 Knollcrest Rd.
Burlingame, CA 94010 (415) 626-8598
R. N. Arrington, Secretary

Description: Minority enterprise small business investment company. Investment Policy: Diversified, minimum investment is $1 million.

★1354★ Oak Investment Partners (Menlo Park)
3000 Sand Hill Rd.
Building Three, Suite 240
Menlo Parks, CA 94025 (415) 854-8825

Description: Private venture capital firm. Investment Policy: Computer, biotechnology, and communications industries preferred.

★1355★ Opportunity Capital Corporation
39650 Liberty St., Suite 425
Fremont, CA 94538 (415) 421-5935
J. Peter Thompson, President

Description: Minority enterprise small business investment company. Investment Policy: No industry preference.

★1356★ Oxford Partners (Santa Monica)
233 Wilshire Blvd., Suite 830
Santa Monica, CA 90401 (213) 458-3135

Description: Independent venture capital partnership. Investment Policy: Prefers to invest in high-technology industries. Initial investments range from $500,000 to $1.5 million; will invest up to $3 million over several later rounds of financing. Remarks: A branch office of Oxford Partners, Stamford, Connecticut.

★1357★ Pacific Capital Fund, Inc.
675 Mariners' Island Blvd., Suite 103
San Mateo, CA 94404 (415) 574-4747
Eduardo B. CuUnjieng, President

Description: Minority enterprise small business investment company. Investment Policy: No industry preference.

★1358★ PBC Venture Capital, Inc.
P.O. Box 6008
Bakersfield, CA 93386 (805) 395-3555
Henry L. Wheeler, Manager

Description: Small business investment company. Investment Policy: No industry preference. Area limited to California.

★1359★ PCF Venture Capital Corporation
675 Mariners' Island Blvd., Suite 103
San Mateo, CA 94404 (415) 574-4747
Eduardo B. CuUnjieng, President

Description: Small business investment company. Investment Policy: No industry preference.

★1360★ Positive Enterprises, Inc.
399 Arguello St.
San Francisco, CA 94118 (415) 386-6600
Kwok Szeto, President

Description: Minority enterprise small business investment company. Investment Policy: No industry preference.

★1361★ R&D Funding Corporation
3945 Freedom Circle, Suite 800
Santa Clara, CA 95054 (408) 980-0990
Richard E. Moser, President

Description: Venture capital firm. Investment Policy: Invests in high-growth businesses. Direct investment in research and development.

★1362★ Ritter Partners
150 Isabella Ave.
Atherton, CA 94025 (415) 854-1555
William C. Edwards, President

Description: Small business investment company. Investment Policy: No industry preference; retail, hotel, or real estate ventures are not considered.

★1363★ Robertson, Colman & Stephens
One Embarcadero, Suite 3100
San Francisco, CA 94111 (415) 781-9700
Susan Vican, Associate

Description: Investment banking firm. Investment Policy: Considers investments in any attractive merging-growth area, including product and service companies. Key preferences include health care, hazardous waste services and technology, biotechnology, software, and information services. Maximum investment is $5 million.

★1364★ Round Table Capital Corporation
655 Montgomery St.
San Francisco, CA 94111 (415) 392-7500
Richard Dumke, President

Description: Small business investment company. Investment Policy: No industry preference.

★1365★ RSC Financial Corporation
223 E. Thousand Oaks Blvd., Suite 310
Thousand Oaks, CA 91360 (805) 646-2925
Frederick K. Bae, President

Description: Minority enterprise small business investment company. Investment Policy: No industry preference.

★1366★ San Joaquin Capital Corporation
P.O. Box 2538
Bakersfield, CA 93303 (805) 323-7581
Chester Troudy, President

Description: Small business investment company. Investment Policy: No industry preference.

★1367★ San Jose SBIC
100 Park Center, Suite 427
San Jose, CA 95113 (408) 293-7708
Robert T. Murphy, President

Description: Small business investment company. Investment Policy: No industry preference.

★1368★ Seaport Ventures, Inc.
525 B St., Suite 630
San Diego, CA 92101 (619) 232-4609
Michael Stolper, President

Description: Small business investment company Investment Policy: No industry preference.

★1369★ Security Pacific Capital Corporation
650 Town Center Dr., 17th Floor
Costa Mesa, CA 92626 (714) 556-1964
Brian Jones, Senior Vice-President

Description: Small business investment company. Investment Policy: No industry preference. Remarks: Maintains offices in Boston, Palo Alto, Pasadena, Pennsylvania, and London.

★1370★ Sequoia Capital
3000 Sand Hill Rd.
Building 4-280
Menlo Park, CA 94025 (415) 854-3927

Description: Private venture capital partnership with $300 million under management. Investment Policy: Provides financing for all stages of development of well-managed companies with exceptional growth prospects in fast-growth industries. Past investments have been made in computers and peripherals, communications, health care, biotechnology, and medical instruments and devices. Investments range from $350,000 for early stage companies to $4 million for late stage accelerates. **Contact:** Pierre Lamond; Walter Baumgartner; Nancy Olson; Doug DeVivo; Gordon Russell; Don Valentine; Michael Moritz.

★1371★ Sierra Ventures
3000 Sand Hill Rd., Building 1, Suite 280
Menlo Park, CA 94025 (415) 854-1000
Vince Topkin, General Partner

Description: Venture capital partnership. Investment Policy: Diversified, with interests in early-stage financing. Prior investment experience in technology-related service businesses, environmental technology, telecommunications, semiconductors, and computer companies.

★1372★ Sprout Group (Menlo Park)
3000 Sand Hill Rd.
Building 1, Suite 285
Menlo Park, CA 94025 (415) 854-1550
Keith Geeflin, Partner

Description: Venture capital partnership. Investment Policy: No industry preference. Remarks: An affiliated office of the Sprout Group headquarters in New York City.

★1373★ Sutter Hill Ventures
Two Palo Alto Sq., Suite 700
Palo Alto, CA 94306-0910 (415) 493-5600

Description: Venture capital supplier. Investment Policy: Provides seed, start-up, and second-stage financing to companies that manufacture products with some degree of technology and whose markets are potentially worldwide. Investments range from $100,000 to $2 million; larger equity needs may be combined with other leading venture capital firms.

★1374★ TA Associates (Palo Alto)
435 Tasso St.
Palo Alto, CA 94301 (415) 328-1210
Jack Bunce, Associate

Description: Private venture capital firm. Investment Policy: Prefers technology companies and leveraged buy outs. Provides from $1 to $20 million in investments.

★1375★ Technology Funding, Inc.
2000 Alameda de las Pulgas, Suite 250
San Mateo, CA 94403 (415) 345-2200
M. David Titus

Description: Private venture capital supplier. Investment Policy: Provides primarily late first-stage and early second-stage equity financing. Also offers secured debt with equity participation to venture capital-backed companies. Investments range from $500,000 to $1 million.

★1376★ Technology Venture Investors
3000 Sand Hill Rd.
Building Four, Suite 210
Menlo Park, CA 94025 (415) 854-7472
Mark Wilson, Administrative Partner

Description: Private venture capital partnership. Investment Policy: Primary interest is in technology companies, with minimum investment of $1 million.

★1377★ 3i Ventures (Newport Beach)
450 Newport Center Dr.
Newport Beach, CA 92660 (714) 720-1421

Description: Venture capital supplier. Investment Policy: Provides start-up and early-stage financing to companies in high-growth fields such as microelectronics, computers, telecommunications, biotechnology, health sciences, and industrial automation. Investments generally range from $500,000 to $2 million. Remarks: A branch office of 3i Ventures, Boston, Massachusetts.

★1378★ Trinity Ventures Ltd.
20813 Stevens Creek Blvd., Suite 101
Cupertino, CA 95014 (408) 446-9690

Description: Venture capital fund. Investment Policy: Will provided up to $3 million in seed capital to third-round financing for West Coast companies. Emphasis on high technology, with flexibility to enter all areas at all stages of development. Will not consider real estate, oil and gas, or construction ventures. **Contact:** Gerald S. Casilli, Noel J. Fenton, David Nierenbert, General Partners.

★1379★ Twenty-first Century Venture Partners
Oryx Capital Corporation
170 Lombard St.
San Francisco, CA 94111 (415) 397-5999
Thomas J. Sherwood, Director

Description: Venture capital partnership. Investment Policy: Prefers later-stage financing for companies on the verge of explosive growth.

★1380★ U.S. Venture Partners
2180 Sand Hill Rd, Suite 300
Menlo Park, CA 94025 (415) 854-9080

Description: Venture capital partnership. Investment Policy: Prefers specialty retail, consumer products, technology, and biomedical industry. **Contact:** Nancy Glasser and Steven Krausz, General Partners.

★1381★ USVP-Schlein Marketing Fund
2180 Sand Hill Rd.
Menlo Park, CA 94025 (415) 854-9080

Description: Venture capital fund. Investment Policy: Specialty retailing/consumer products companies preferred. **Contact:** Philip S. Schlein or Nancy E. Glasser.

★1382★ Venrock Associates (Palo Alto)
Two Palo Alto Sq., Suite 528
Palo Alto, CA 94306 (415) 493-5577

Description: Private venture capital supplier. Remarks: Maintains an office in New York City.

★1383★ Vista Capital Corporation
5080 Shoreman Pl., Suite 202
San Diego, CA 92122 (619) 453-0780
Frederick J. Howden, Jr., Chairman

Description: Small business investment company. Investment Policy: No industry preference.

★1384★ The Vista Group (Newport Beach)
610 Newport Center Dr., Suite 400
Newport Beach, CA 92660 (714) 720-1416

Description: Venture capital supplier. Investment Policy: Provides start-up and second-stage financing to technology-related businesses that seek to become major participants in high-growth markets of at least $100 million in annual sales. Areas of investment interest include information systems, communications, computer peripherals, medical products and services, instrumentation, and genetic engineering. Investments range from $500,000 to $3 million. Remarks: Maintains a branch office in New Canaan, Connecticut.

★1385★ Walden Capital Partners
750 Battery St., Seventh Floor
San Francisco, CA 94111 (415) 391-7225
Arthur S. Berliner, President

Description: Small business investment company. Investment Policy: No industry preference.

★1386★ Westamco Investment Company
8929 Wilshire Blvd., Suite 400
Beverly Hills, CA 90211 (213) 652-8288
Leonard G. Muskin, President

Description: Small business investment company. Investment Policy: No industry preference.

★1387★ Wilshire Capital, Inc.
3932 Wilshire Blvd., Suite 303
Los Angeles, CA 90010 (213) 388-1314

Description: Minority enterprise small business investment company. Investment Policy: No industry preference.

★1388★ Wood River Capital (Menlo Park)
300 Sand Hill Rd.
Building 1, Suite 280
Menlo Park, CA 94025 (415) 854-1000
Vince Topkin, General Partner

Description: Small business investment company. Investment Policy: Diversified, with interests in early-stage financing. Prior investment experience in technology-related service businesses, environmental technology, telecommunications, semiconductors, and computer companies.

★1389★ Xerox Venture Capital (Los Angeles)
2029 Century Park, E., Suite 740
Los Angeles, CA 90067 (213) 278-7940
Kathleen Rubenkoknig, Secretary

Description: Venture capital subsidiary of operating firm. Investment Policy: High-technology industries preferred.

Colorado

★1390★ Centennial Business Development Fund Ltd.
1999 Broadway, Suite 2100
Denver, CO 80202 (303) 298-9066
David Bullwinkle, General Partner

Description: Venture capital fund. Investment Policy: Prefers to invest in later-stage companies. Remarks: Managed by Larimer and Company, Denver, Colorado.

★1391★ Centennial Fund II Ltd.
1999 Broadway, Suite 2100
Denver, CO 80202 (303) 298-9066
Mark Dubovoy, General Partner

Description: Venture capital fund. Investment Policy: Prefers to invest in early-stage companies in the Rocky Mountain regin or in the telecommunications industry nationwide. Remarks: Managed by Lorimer and Company, Denver, Colorado.

★1392★ Centennial Fund III LP
1999 Broadway, Suite 2100
Denver, CO 80202 (303) 298-9066
G. Jackson Tankersley, Jr., General partner

Description: Venture capital fund. Investment Policy: Prefers to invest in early-stage companies in the Rocky Mountain region. Remarks: Managed by Larimer and Company, Denver, Colorado.

★1393★ Centennial Fund Ltd.
1999 Broadway, Suite 2100
Denver, CO 80202 (303) 298-9066
Mark Dubovoy, General Partner

Description: Venture capital fund. Investment Policy: Currently not investing in new companies, but still involved in follow-up investments. Remarks: Managed by Larimer and Company, Denver, Colorado.

★1394★ Colorado Growth Capital, Inc.
1600 Broadway, Suite 2125
Denver, CO 80202 (303) 831-0205
Nicholas H. C. Davis, President

Description: Small business investment company. Investment Policy: No industry preference.

★1395★ Intermountain Ventures Ltd.
1100 Tenth St.
Greeley, CO 80632 (303) 356-1200
Norman M. Dean, President

Description: Small business investment company. Investment Policy: No industry preference.

★1396★ William Blair and Company (Denver)
1225 17th St., Suite 2440
Denver, CO 80202 (303) 825-1600

Description: Investment banker and venture capital supplier. Investment Policy: Provides all stages of financing to growth companies; also deals in leveraged acquisitions. Remarks: A branch office of William Blair and Company, Chicago, Illinois.

Connecticut

★1397★ Abacus Ventures
411 W. Putnam Ave.
Greenwich, CT 06830 (203) 629-1100

Description: Venture capital fund. Investment Policy: Will provide $500,000 to $2 million in start-up and first-round financings. **Contact:** Yung Wong; Charles T. Lee; Patrick F. Whelan; Thomas J. Crotty.

★1398★ Canaan Venture Partners (Rowayton)
105 Rowayton Ave.
Rowayton, CT 06853 (203) 855-0400
Harry T. Rein, Managing General Partner

Description: Venture capital supplier. Investment Policy: Primary concern is strong entrepreneurial management. There are no geographic or industry-specific constraints. Remarks: Maintains an office in Menlo Park, California.

★1399★ Capital Impact Corporation
961 Main St.
Bridgeport, CT 06601 (203) 384-5670
William D. Starbuck, President

Description: Small business investment company. Investment Policy: No industry preference.

★1400★ Capital Resource Company of Connecticut, LP
699 Bloomfield Ave.
Bloomfield, CT 06002 (203) 243-1114
I. Martin Fierberg, Partner

Description: Small business investment company. Investment Policy: No industry preference.

★1401★ Fairfield Venture Partners
1275 Summer St.
Stamford, CT 06905 (203) 358-0255
T. Berman

Description: Venture capital firm. Investment Policy: Diversified. Prefers early-stage financing. Minimum investment is $500,000.

★1402★ First Connecticut SBIC
177 State St.
Bridgeport, CT 06604 (203) 366-4726
David Engelson, President

Description: Small business investment company. Investment Policy: Real estate industry preferred.

★1403★ James B. Kobak and Company
774 Hollow Tree Ridge Rd.
Darien, CT 06820 (203) 655-8764
James B. Kobak, Owner

Description: Venture capital supplier and consultant. Investment Policy: Provides assistance to new ventures in the communications field through conceptualization, planning, organization, raising money, and control of actual operations. Special interest is in magazine publishing.

★1404★ Marcon Capital Corporation
49 Riverside Ave.
Westport, CT 06880 (203) 226-6893
Martin A. Cohen, President

Description: Small business investment company. Investment Policy: Media preferred.

★1405★ Marketcorp Venture Associates
285 Riverside Ave.
Westport, CT 06880 (203) 222-1000
Buck Griswold, President

Description: Venture capital firm. Investment Policy: Prefers to invest in consumer-market businesses, including the packaged goods, specialty retailing, communications, and consumer electronics industries. Investments range from $500,000 to 2.5 million.

★1406★ Northeastern Capital Corporation
209 Church St.
New Haven, CT 06510 (203) 865-4500
Mosha Reiss, President

Description: Small business investment company. Investment Policy: No industry preference.

★1407★ Oxford Partners (Stamford)
Soundview Plaza
1266 Main St.
Stamford, CT 06902 (203) 964-0592
Kenneth W. Rind, General Partner

Description: Independent venture capital partnership. Investment Policy: Prefers to invest in high-technology industries. Initial investments range from $500,000 to $1.5 million; up to $3 million over several later rounds of financing. Remarks: Maintains an office in Santa Monica, California.

★1408★ Regional Financial Enterprises (New Canaan)
36 Grove St., Third Floor
New Canaan, CT 06840 (203) 966-2800
Robert M. Williams, Managing Partner

Description: Small business investment company. Investment Policy: Prefers to invest in high-technology companies or nontechnology companies with significant growth prospects, located anywhere in the United States. Investments range from $1 to $5 million. Remarks: Maintains an office in Ann Arbor, Michigan.

★1409★ Saugatuck Capital Company
595 Summer St.
Stamford, CT 06901 (203) 348-6669

Description: Private investment partnership. Investment Policy: Seeks to invest in building products, transportation, health care products and services, energy services and products, process control instrumentation, industrial and automation equipment, test and measurement instrumentation, communications, fasteners, filtration equipment and filters, and valves and pumps. Prefers leveraged buy out situations, but will consider start-up financing. Investments range from $1 to $5 million. **Contact:** Norman W. Johnson; Steve Crittfield; John S. Crowley; Frank J. Hawley, Jr.; Alexander H. Dunbar, General Manager.

★1410★ The SBIC of Connecticut, Inc.
1115 Main St.
Bridgeport, CT 06604 (203) 367-3282
Kenneth F. Zarrilli, President

Description: Small business investment company. Investment Policy: No industry preference.

★1411★ The Vista Group (New Canaan)
36 Grove St.
New Canaan, CT 06840 (203) 972-3400
Gerald B. Bay, Managing Partner

Description: Venture capital supplier. Investment Policy: Provides start-up and second-stage financing to technology-related businesses that seek to become major participants in high-growth markets of at least $100 million in annual sales. Areas of investment interest include information systems, communications, computer peripherals, medical products and services, retailing, agrigenetics, biotechnology, low technology, no technology, instrumentation, and genetic engineering. Investments range from $500,000 to $1.5 million. Remarks: A branch office of The Vista Group in Newport Beach, California.

★1412★ Xerox Venture Capital (Stamford)
Headquarters
800 Long Ridge Rd.
Stamford, CT 06904 (203) 329-8700
Lawrence R. Robinson, Principal

Description: Venture capital subsidiary of operating company. Investment Policy: Prefers to invest in document processing industries.

District of Columbia

★1413★ Allied Investment Corporation (Washington, DC)
1666 K. St., Suite 901
Washington, DC 20006 (202) 331-1112
David Gladstone, President

Description: Small business investment company; Minority enterprise small business investment company. Investment Policy: No industry preference. Remarks: Maintains an office in Fort Lauderdale, Florida.

★1414★ Allied Venture Partnership
1666 K St., Suite 901
Washington, DC 20006 (202) 331-1112
David Gladstone, President and General Partner

Description: Venture capital fund. Investment Policy: Low-technology businesses, such as radio stations or retail companies, preferred. Will provide $1 million in a second- or third-round financing in the form of convertible debentures.

★1415★ American Security Capital Corporation, Inc.
730 15th St., N.W., Suite A-2/304
Washington, DC 20013 (202) 624-4843
Brian K. Mercer, Vice-President and Manager

Description: Small business investment company. Investment Policy: No industry preference.

★1416★ Broadcast Capital, Inc.
1771 N St., N.W., Fourth Floor
Washington, DC 20036 (202) 429-5393
John E. Oxendine, President

Description: Minority enterprise small business investment company. Investment Policy: Communications media industry preferred.

★1417★ Consumers United Capital Corporation
2100 M St., N.W.
Washington, DC 20037 (202) 872-5262
Esther M. Carr-Davis, President

Description: Minority enterprise small business investment company. Investment Policy: No industry preference.

★1418★ DC Bancorp Venture Capital Company
1801 K St., N.W.
Washington, DC 20006 (202) 955-6970
Allen A. Weissburg, President

Description: Small business investment company. Investment Policy: No industry preference; preferred geographic area includes District of Columbia, Maryland, and Virginia.

★1419★ Fulcrum Venture Capital Corporation
1030 15th St., N.W., Suite 203
Washington, DC 20005 (202) 785-4253
Renate Todd, Vice-President

Description: Minority enterprise small business investment company. Investment Policy: No industry preference.

★1420★ Minority Broadcast Investment Corporation
1820 Jefferson Place, N.W.
Washington, DC 20036 (202) 293-1166
Walter Threadgill, President

Description: Minority enterprise small business investment company. Investment Policy: Communications industry preferred.

★1421★ Syncom Capital Corporation
1030 15th St., N.W., Suite 203
Washington, DC 20005 (202) 293-9428
Herbert P. Wilkins, President

Description: Minority enterprise small business investment company. Investment Policy: Telecommunications media industry only.

Florida

★1422★ Allied Investment Corporation (Fort Lauderdale)
111 E. Las Olas Blvd.
Fort Lauderdale, FL 33301 (305) 763-8484

Description: Small business investment company. Investment Policy: No industry preference. Remarks: A branch office of Allied Investment Corporation, Washington, District of Columbia.

★1423★ Caribank Capital Corporation
255 E. Dania Beach Blvd.
Dania, FL 33004 (305) 925-2211
Michael E. Chaney, President

Description: Small business investment company. Investment Policy: No industry preference.

★1424★ First North Florida SBIC
1400 Gadsden St.
P.O. Box 1021
Quincy, FL 32351 (904) 875-2600
J. B. Higdon, President

Description: Small business investment company. Investment Policy: Grocery industry preferred.

★1425★ Gold Coast Capital Corporation
3550 Biscayne Blvd., Room 601
Miami, FL 33137 (305) 576-2012
William I. Gold, President

Description: Small business investment company. Investment Policy: No industry preference.

★1426★ Horn and Hardart Capital Corporation
701 S.E. Sixth Ave.
Del Ray Beach, FL 33483 (305) 265-0751
Morton Farber, Manager

Description: Minority enterprise small business investment company. Investment Policy: No industry preference.

★1427★ Ideal Financial Corporation
780 N.W. 42nd Ave., Suite 501
Miami, FL 33126 (305) 442-4665
Ectore E. Reynaldo, General Manager

Description: Minority enterprise small business investment company. Investment Policy: No industry preference.

★1428★ J and D Capital Corporation
12747 Biscayne Blvd.
North Miami, FL 33181 (305) 893-0303
Jack Carmel, President

Description: Small business investment company. Investment Policy: No industry preference.

★1429★ Market Capital Corporation
1102 N. 28th St.
P.O. Box 22667
Tampa, FL 33622 (813) 247-1357
E. E. Eads, President

Description: Small business investment company. Investment Policy: Grocery industry preferred.

★1430★ Safeco Capital, Inc.
835 S.W. 37th Ave.
Miami, FL 33135 (305) 443-7953
Rene J. Leonard, President

Description: Minority enterprise small business investment company. Investment Policy: No industry preference.

★1431★ SBAC of Panama City, Florida
2612 W. 15th St.
Panama City, FL 32401 (904) 785-9577
Charles Smith, President

Description: Small business investment company. Investment Policy: Lodging and amusement industries preferred.

★1432★ South Atlantic Capital Corporation
614 W. Bay St., Suite 200
Tampa, FL 33606 (813) 253-2500
Donald W. Burton, President

Description: Venture capital supplier. Investment Policy: Provides long-term working capital for privately owned, rapidly growing companies located in the Southeast.

★1433★ Southeast Venture Capital Limited 1
One Southeast Financial Center
Miami, FL 33131 (305) 375-6470
Clement L. Hofmann, Chairman

Description: Small business investment company. Investment Policy: No industry preference, but prefers to invest in firms located in the southwest.

★1434★ Universal Financial Services, Inc.
3550 Biscayne Blvd., Suite 702
Miami Beach, FL 33137 (305) 573-1496
Norman Zipkin, President

Description: Minority enterprise small business investment company. Investment Policy: No industry preference.

★1435★ Venture Group, Inc.
5433 Buffalo Ave.
Jacksonville, FL 32208 (904) 353-7313
Ellis W. Hitzing, President

Description: Minority enterprise small business investment company. Investment Policy: Automotive industry preferred.

★1436★ Verde Capital Corporation
255 Alhambra Circle, Suite 720
Coral Gables, FL 33134 (305) 444-8938
Jose Dearing, President

Description: Minority enterprise small business investment company. Investment Policy: No industry preference.

★1437★ Western Financial Capital Corporation
1380 Miami Cardens Dr., Suite 225
North Miami Beach, FL 33179 (305) 949-5900
Dr. F. M. Rosemore, President

Description: Small business investment company. Investment Policy: Medical industry preferred.

Georgia

★1438★ Investor's Equity, Inc.
2629 First National Bank Tower
Atlanta, GA 30383 (404) 523-3999
I. Walter Fisher, President

Description: Small business investment company. Investment Policy: No industry preference.

★1439★ Noro-Moseley Partners
4200 Northside Pkwy., Building 9
Atlanta, GA 30327 (404) 233-1966
Charles D. Moseley, Jr., General Partner

Description: Venture capital partnership. Investment Policy: Prefers to invest in private, diversified small and medium-size growth companies located in the southeastern United States.

★1440★ North Riverside Capital Corporation
50 Technology Park/Atlanta
Norcross, GA 30092 (404) 446-5556
Tom Barry, President

Description: Small business investment company. Investment Policy: No industry preference.

Hawaii

★1441★ Bancorp Hawaii SBIC, Inc.
111 S. King St., Suite 1060
Honolulu, HI 96813 (808) 537-8557
Thomas T. Triggs, Vice-President and Manager

Description: Small business investment company. Investment Policy: No industry preference.

★1442★ Pacific Venture Capital Ltd.
222 S. Vineyard St., #PH-1
Honolulu, HI 96813 (808) 521-6502
Dexter J. Taniguchi, President

Description: Minority enterprise small business investment company. Investment Policy: Inquire.

Illinois

★1443★ Allstate Venture Capital
Allstate Plaza, E2
Northbrook, IL 60062 (312) 402-5681
Robert L. Lestina, Director

Description: Venture capital supplier. Investment Policy: Investments are not limited to particular industries or geographical locations. Interest is in unique products or services that address large potential markets and offer great economic benefits; strength of management team is also important. Investments range from $500,000 to $5 million.

★1444★ Alpha Capital Venture Partners LP
Three First National Plaza, 14th Floor
Chicago, IL 60602 (312) 372-1556
Andrew H. Kalnow, General Partner

Description: Small business investment company providing venture capital. Investment Policy: No industry preference; however, no real estate, oil and gas, or start-up ventures are considered. All investments are structured to provide equity participation in the business; no straight loans are considered. Minimum investment is $200,000; preferred size of investment is $500,000 to $1 million. Company must be located in the Midwest.

★1445★ Ameritech Development Corporation
10 S. Wacker Dr., 24th Floor
Chicago, IL 60606 (312) 446-5556

Description: Venture capital supplier. Investment Policy: Seeks ideas in high-technology information management that relate to Ameritech's existing and expanding business in telecommunications technologies. Remarks: Ameritech venture teams are assigned to particular projects and are responsible for planning and day-to-day management.

★1446★ Amoco Venture Capital Company
200 E. Randolph Dr.
Chicago, IL 60601 (312) 856-6523
Gordon E. Stone, President

Description: Minority enterprise small business investment company. Investment Policy: Technical industires or those that relate to Amoco Corporation needs.

★1447★ Business Ventures, Inc.
20 N. Wacker Dr., Suite 550
Chicago, IL 60606 (312) 346-1580
Milton Lefton, President

Description: Small business investment company. Investment Policy: No industry preference; considers only ventures in the Chicago area.

★1448★ The Capital Strategy Group, Inc.
20 N. Wacker Dr.
Chicago, IL 60606 (312) 444-1170
Eric Von Bauer, President

Description: Investment banker and venture capital supplier. Investment Policy: Provides financing to start-up and early-stage companies, located in the Midwest, in the manufacturing and service industries. Remarks: Also develops and improves business plans and product strategies, assists with acquisitions, and conducts valuations of closely held companies.

★1449★ Caterpillar Venture Capital, Inc.
100 N.E. Adams St.
Peoria, IL 61629-4390 (309) 675-5503
Robert Powers, President

Description: Venture capital subsidiary of operating firm. Investment Policy: Prefers to invest in industrial electronics, advanced materials, and environmental-related industries.

★1450★ Chicago Community Ventures, Inc.
104 S. Michigan Ave., Suites 215-218
Chicago, IL 60603 (312) 726-6084
Phyllis George, President

Description: Minority enterprise small business investment company. Investment Policy: No industry preference.

★1451★ Comdisco Venture Lease, Inc. (Rosemont)
6400 Shafer Ct.
Rosemont, IL 60018 (312) 698-3000
James Labe

Description: Venture capital subsidiary of operating firm. Investment Policy: Involved in all stages of financing except seed. Has no industry preference but avoids real estate and oil and gas industries. Investments range from $500,000 to $5 million.

★1452★ Continental Illinois Venture Corporation (CIVC)
231 S. LaSalle St.
Chicago, IL 60697 (312) 828-8021

Description: Small business investment company. Investment Policy: Provides start-up and early-stage financing to growth-oriented companies with capable management teams, proprietary products, and expanding markets. Remarks: CIVC is a wholly owned subsidiary of Continental Illinois Equity Corporation (CIEC), which is a wholly owned subsidiary of Continental Illionis Corporation. CIEC invests in more mature companies, with a special interest in leveraged buy outs. **Contact:** John L. Hines or William Putze.

★1453★ First Capital Corporation of Chicago
Three First National Plaza
Chicago, IL 60670 (312) 732-5414
John A. Canning, Jr., President

Description: Small business investment company. Investment Policy: No industry preference.

★1454★ First Chicago Venture Capital (Chicago)
Three First National Plaza, Suite 1330
Chicago, IL 60670-0501 (312) 732-5400
John A. Canning, Jr., President

Description: Venture capital supplier. Investment Policy: Invests a minimum of $1 million in early-stage situations to a maximum of $25 million in mature growth or buy out situations. Emphasis is placed on a strong management team and unique market opportunity. Remarks: Maintains an office in Boston, Massachusetts.

★1455★ Frontenac Capital Corporation
208 S. LaSalle St., Room 1900
Chicago, IL 60604 (312) 368-0047
David A. R. Dullum, President

Description: Small business investment company. Investment Policy: No industry preference.

★1456★ Golder, Thoma and Cressey
120 S. LaSalle St., Suite 630
Chicago, IL 60603 (312) 853-3322

Description: Private equity investors. Investment Policy: Provides financing for start-up and leveraged buy out situations. No geographic or industry limitations, with the exception of real estate. Past investments have been made in the health care field and in information services. Investments range from $2 to $10 million. **Contact:** Stanley C. Golder; Carl D. Thoma; Bryan C. Cressey, Founding Partners.

★1457★ IEG Venture Management, Inc.
401 N. Michigan Ave., Suite 2020
Chicago, IL 60611 (312) 644-0890
Marian Zamlynski, Operations Manager

Description: Venture capital supplier. Investment Policy: Provides start-up financing primarily to companies located in the Midwest that focus on productivity-enhancing technologies in the medical, manufacturing, electronic, telecommunications, agricultural, service, chemical and mineral, and metal products industries.

★1458★ Mesirow Capital Partners SBIC Ltd.
135 S. LaSalle St.
Chicago, IL 60603 (312) 443-5773
James C Tyree

Description: Small business investment company. Investment Policy: No industry preference.

★1459★ Neighborhood Fund, Inc.
1950 E. 71st St.
Chicago, IL 60649 (312) 684-8074
James Fletcher, President

Description: Minority enterprise small business investment company. Investment Policy: No industry preference.

★1460★ Peterson Finance and Investment Company
3300 W. Peterson Ave., Suite A
Chicago, IL 60659 (312) 583-6300
James S. Rhee, President

Description: Minority enterprise small business investment company. Investment Policy: No industry preference.

★1461★ Seidman Jackson Fisher and Company
233 N. Michigan Ave., Suite 1812
Chicago, IL 60601 (312) 856-1812
David S. Seidman, Managing Partner

Description: Private venture capital supplier. Investment Policy: Provides early-stage and growth-equity financing to companies with proprietary or patented products or services that deal with large and rapidly growing industrial markets; limited interest in consumer markets. Leveraged buy outs and turn-around of mature companies are considered under certain circumstances. Investments range from $200,000 to $2 million.

★1462★ Tower Ventures, Inc.
Sears Tower, BSC 43-50
Chicago, IL 60684 (312) 875-0571
Robert T. Smith, President

Description: Minority enterprise small business investment company. Investment Policy: No industry preference.

★1463★ Walnut Capital Corporation
208 S. LaSalle, Suite 1043
Chicago, IL 60604 (312) 346-2033
David L. Bogetz, Vice-President

Description: Small business investment company. Investment Policy: No industry preference.

★1464★ William Blair and Company (Chicago)
135 S. LaSalle St.
Chicago, IL 60603 (312) 236-1600

Description: Investment banker and venture capital supplier. Investment Policy: Provides all stages of financing to growth companies; also deals in leveraged acquisitions. Remarks: William Blair Venture Partners was formed to meet the needs of new clients. William Blair and Company maintains offices in Denver, Colorado, and in London, England. **Contact:** Samuel B. Guren, General Partner, William Blair Venture Partners.

Indiana

★1465★ Circle Ventures, Inc.
2502 Roosevelt Ave.
Indianapolis, IN 46218 (317) 633-7303
Ms. Murray Welch, Investment Manager

Description: Small business investment company. Investment Policy: No industry preference.

★1466★ Equity Resource Company, Inc.
202 S. Michigan St.
South Bend, IN 46601 (219) 237-5255
Michael J. Hammes, Vice-President

Description: Small business investment company. Investment Policy: No industry preference.

★1467★ First Source Capital Corporation
P.O. Box 1602
South Bend, IN 46634 (219) 236-2180
Eugene L. Cavanaugh, Jr., Vice-President

Description: Small business investment company. Investment Policy: No industry preference.

★1468★ Heritage Venture Group, Inc.
2400 One Indiana Sq.
Indianapolis, IN 46204 (317) 635-5696
Arthur A. Angotti, President

Description: Venture capital fund. Investment Policy: Prefers communications industries, especially broadcasting.

★1469★ Mount Vernon Venture Capital Company
P.O. Box 40177
Indianapolis, IN 46240 (317) 253-3606
Thomas J. Grande, General Manager

Description: Small business investment company. Investment Policy: No industry preference.

★1470★ White River Capital Corporation
P.O. Box 929
Columbus, IN 47202 (812) 376-1759
Bradley J. Kime, Vice-President

Description: Small business investment company. Investment Policy: No industry preference.

Iowa

★1471★ InvestAmerica Venture Group, Inc. (Cedar Rapids)
800 American Bldg.
Cedar Rapids, IA 52401 (319) 363-8249

Description: Venture capital fund management company. Investment Policy: No industry preference. Remarks: Manager of Iowa Venture Capital Fund LP, and MorAmerica Capital Corporation. Maintains branch offices in Milwaukee, Wisconsin, and Kansas City, Missouri. **Contact:** Donald E. Flynn, President; David R. Schroder, Executive Vice-President; Robert A. Comey, Vice-President.

Kansas

★1472★ Kansas Venture Capital, Inc.
1030 Bank IV Tower
One Townsite Plaza
Topeka, KS 66603 (913) 235-3437
Rex Wiggins, President

Description: Small business investment company. Investment Policy: No industry preference.

Kentucky

★1473★ Equal Opportunity Finance, Inc.
420 Hurstbourne Lane, Suite 201
Louisville, KY 40222 (502) 423-1943
David Sattich, Vice-President and Manager

Description: Minority enterprise small business investment company. Investment Policy: No industry preference; geographic areas limited to Indiana, Kentucky, Ohio, and West Virginia.

★1474★ Financial Opportunities, Inc.
6060 Dutchman's Ln.
P.O. Box 35680
Louisville, KY 40232 (502) 451-3800
Joe P. Peden, President

Description: Small business investment company. Investment Policy: No industry preference.

★1475★ Mountain Ventures, Inc.
P.O. Box 628
London, KY 40741 (606) 864-5175
Lloyd R. Moncrief, President

Description: Small business investment company. Investment Policy: No industry preference.

Louisiana

★1476★ Capital Equity Corporation
1885 Wooddale Blvd.
Baton Rouge, LA 70806 (504) 924-9209
Arthur J. Mitchell, General Manager

Description: Small business investment company. Investment Policy: No industry preference.

★1477★ Capital for Terrebonne, Inc.
27 Austin Dr.
Houma, LA 70360 (504) 868-3933
Hartwell A. Lewis, President

Description: Small business investment company. Investment Policy: No industry preference.

★1478★ Dixie Business Investment Company
401-1/2 Lake St.
P.O. Box 588
Lake Providence, LA 71254 (318) 559-1558
L. W. Baker, President

Description: Small business investment company. Investment Policy: No industry preference.

★1479★ First Southern Capital Corporation
P.O. Box 14418
Baton Rouge, LA 70898-4418 (504) 769-3004
Charest Thibaut, President

Description: Small business investment company. Investment Policy: No industry preference.

★1480★ Louisiana Equity Capital Corporation
451 Florida St.
Baton Rouge, LA 70821 (504) 389-4421
Melvin L. Rambin, President

Description: Small business investment company. Investment Policy: No industry preference.

★1481★ SCDF Investment Corporation
1006 Surrey St.
P.O. Box 3885
Lafayette, LA 70502 (318) 232-3769
Marvin Beaulieu, President

Description: Minority enterprise small business investment company. Investment Policy: No industry preference.

★1482★ Walnut Street Capital Company
231 Carondelet St., Suite 702
New Orleans, LA 70130 (504) 821-4952
William D. Humphries, Managing General Partner

Description: Small business investment company. Investment Policy: Inquire.

Maine

★1483★ Maine Capital Corporation
70 Center St.
Portland, ME 04101 (207) 772-1001
David M. Coit, President

Description: Small business investment company. Investment Policy: No industry preference.

Maryland

★1484★ Albright Venture Capital, Inc.
8005 Rappahannock Ave.
Jessup, MD 20794 (301) 799-7935
William A. Albright, President

Description: Minority enterprise small business investment company. Investment Policy: No industry preference.

★1485★ Broventure Capital Management
16 W. Madison St.
Baltimore, MD 21201 (301) 727-4520
William M. Gust, Principal

Description: Venture capital partnership. Investment Policy: Provides start-up capital to early-stage companies, expansion capital to companies experiencing rapid growth, and capital for acquisitions. Initial investments range from $400,000 to $750,000.

★1486★ First Maryland Capital, Inc.
107 W. Jefferson St.
Rockville, MD 20850 (301) 251-6630
Joseph A. Kenary, President

Description: Small business investment company. Investment Policy: No industry preference.

★1487★ Greater Washington Investors, Inc.
5454 Wisconsin Ave.
Chevy Chase, MD 20815 (301) 656-0626
Don A. Christensen, President

Description: Small business investment company. Investment Policy: Provides financing to small developing companies primarily in computer-related industries but also in the medical and communications areas.

★1488★ New Enterprise Associates (Baltimore)
1119 St. Paul St.
Baltimore, MD 21202 (301) 244-0115
Frank A. Bonsal, Jr., General Partner

Description: Venture capital supplier. Investment Policy: Concentrates in technology-based industries that have the potential for product innovation, rapid growth, and high profit margins. Investments range from $250,000 to $2 million. Past investments have been made in the following industries: computer software, medical and life sciences, computers and peripherals, communications, semiconductors, specialty retailing, energy and alternative energy, defense electronics, materials, and specialty chemicals. Management must demonstrate intimate knowledge of its marketplace and have a well- defined strategy for achieving strong market penetration. Remarks: New Enterprise Associates operates nationwide, with principal offices in San Francisco and Baltimore with affiliated offices in Menlo Park and Westlake Village, California and in New York, New York.

★1489★ Security Financial and Investment Corporation
7720 Wisconsin Ave., Suite 207
Bethesda, MD 20814 (301) 951-4288
Han Y. Cho, President

Description: Minority enterprise small business investment company. Investment Policy: No industry preference.

★1490★ T. Rowe Price
100 E. Pratt St.
Baltimore, MD 21202 (301) 547-2000
George J. Collins, President

Description: Venture capital supplier. Investment Policy: Offers specialized investment services to meet the needs of companies in various stages of growth.

Massachusetts

★1491★ Advanced Technology Ventures
Ten Post Office Sq.
Boston, MA 02109 (617) 423-4050
Dr. Albert E. Paladino, Managing Partner

Description: Private venture capital firm. Investment Policy: Prefers early-stage financing in high-technology industries.

★1492★ Advent Atlantic Capital Company LP
45 Milk St.
Boston, MA 02109 (617) 338-0800
David D. Croll, Managing Partner

Description: Small business investment company. Investment Policy: Communications industry preferred.

★1493★ Advent Industrial Capital Company LP
45 Milk St.
Boston, MA 02109 (617) 338-0800
David D. Croll, Managing Partner

Description: Small business investment company. Investment Policy: Communications industry preferred.

★1494★ Advent International Corporation
45 Milk St.
Boston, MA 02109 (617) 574-8400
Clinton B. Harris, Vice-President

Description: Venture capital firm. Investment Policy: Specializes in working with companies that need assistance in accessing international markets and major corporations.

★1495★ Advent IV Capital Company
45 Milk St.
Boston, MA 02109 (617) 338-0800
David D. Croll, Managing Partner

Description: Small business investment company. Investment Policy: Communications industry preferred.

★1496★ Advent V Capital Company
45 Milk St.
Boston, MA 02109 (617) 338-0800
David D. Croll, Managing Partner

Description: Small business investment company. Investment Policy: Communications industry preferred.

★1497★ Aeneas Venture Corporation
600 Atlantic Ave.
Boston, MA 02210 (617) 523-4400
Scott Sperling, Managing Partner

Description: Venture capital firm. Investment Policy: Diversified. Minimum investment is $1 million.

★1498★ American Research and Development
45 Milk St.
Boston, MA 02109 (617) 423-7500
Charles J. Coulter, Managing General Partner

Description: Independent private venture capital partnership. Investment Policy: All stages of financing; no minimum or maximum investment.

★1499★ Ampersand Ventures
265 Franklin St., Suite 1501
Boston, MA 02110 (617) 439-8300
Richard A. Charpie, Managing General Partner

Description: Venture capital supplier. Investment Policy: Provides start-up and early-stage financing to technology-based companies. Investments range from $500,000 to $1 million.

★1500★ Analog Devices Enterprises
One Technology Way
P.O. Box 9106
Norwood, MA 02062-9106 (617) 329-4700
Robert A. Boole, Director of Venture Analysis

Description: Venture capital division of Analog Devices, Inc., a supplier of integrated analog circuits. Investment Policy: Prefers to invest in industries involved in analog devices.

★1501★ Atlantic Energy Capital Corporation
260 Franklin, 15th Floor
Boston, MA 02110 (617) 439-6160
Joost S. Tjaden, President

Description: Small business investment company. Investment Policy: No industry preference.

★1502★ Atlas II Capital Corporation
260 Franklin St., Suite 1500
Boston, MA 02110 (617) 451-6220
Joost E. Tjaden, President and Chief Executive

Description: Small business investment company. Investment Policy: Provides all stages of financing to technology-based and oil and gas service industries; no geographic preference. Experience of management team and high-growth market potential are emphasized. Investments range from $200,000 to $400,000.

★1503★ Bain Capital Fund
Two Copley Place
Boston, MA 02116 (617) 572-3000

Description: Private venture capital firm. Investment Policy: No industry preference but avoids investing in high-tech industries. Minimum investment is $500,000.

★1504★ BancBoston Ventures, Inc.
100 Federal St.
Boston, MA 02110 (617) 434-2442
Paul F. Hogan, President

Description: Small business investment company. Investment Policy: Provides start-up and first- and second-round financing to companies in the communications, computer hardware and software, electronic components and instrumentation, industrial products, and health care industries. Investments range from $500,000 to $2 million.

★1505★ Bever Capital Corporation
260 Franklin St., Suite 1500
Boston, MA 02110 (617) 451-6220
Joost E. Tjaden, President and Chief Executive

Description: Small business investment company. Investment Policy: Provides all stages of financing to technology-based and oil and gas service industries; no geographic preference. Experience of management team and high-growth market potential are emphasized. Investments range from $200,000 to $400,000.

★1506★ Boston Capital Ventures LP
Old City Hall
45 School St.
Boston, MA 02108 (617) 227-6550

Description: Venture capital firm. Investment Policy: Prefers health care and high-technology industries. **Contact:** Donald Steiner; A. Dana Callow, Jr.; H.J. vaugh-der-Goltz, General Partners.

★1507★ Burr, Egan, Deleage, and Company (Boston)
One Post Office Sq., Suite 3800
Boston, MA 02109 (617) 482-8020

Description: Private venture capital supplier. Investment Policy: Invests start-up, expansion, and acquisitions capital nationwide. Principal concerns are strength of the management team; large, rapidly expanding markets; and unique products or services. Past investments have been made in the fields of electronics, health, and communications. Investments range from $750,000 to $5 million. Remarks: A branch office of Burr, Egan, Deleage, and Company, San Francisco, California.

★1508★ Business Achievement Corporation
1280 Centre St.
Newton Centre, MA 02159 (617) 965-0550
Michael L. Katzeff, President

Description: Small business investment company. Investment Policy: No industry preference.

★1509★ Charles River Ventures
67 Battery March St., Suite 600
Boston, MA 02110 (617) 439-0477
Barbara A. Piette, Associate

Description: Venture capital partnership. Investment Policy: Diversified, but no real estate. Minimum investment is $750,000.

★1510★ Chestnut Capital International II LP
45 Milk St.
Boston, MA 02109 (617) 338-0800
David D. Croll, Managing Partner

Description: Small business investment company. Investment Policy: Communications industry preferred.

★1511★ Claflin Capital Management, Inc.
185 Devonshire St., Room 700
Boston, MA 02110 (617) 426-6505
Thomas M. Claflin, II, President

Description: Private venture capital firm investing its own capital. Investment Policy: No industry preference but prefers early stage companies.

★1512★ Eastech Management Company
One Liberty Sq., Ninth Floor
Boston, MA 02109 (617) 338-0200
Michael H. Shanahan, Associate

Description: Private venture capital supplier. Investment Policy: Provides start-up and first- and second-stage financing to companies in the following industries: communications, computer-related electronic components and instrumentation, and industrial products and equipment. Will not consider real estate, agriculture, forestry, fishing, finance and insurance, transportation, oil and gas, publishing, entertainment, natural resources, or retail. Investments range from $400,000 to $900,000, with a minimum of $250,000. Eastech prefers that portfolio companies be located in New England or on the East Coast, within two hours of the office.

★1513★ First Chicago Venture Capital (Boston)
One Financial Center
Boston, MA 02110-2621 (617) 542-9185
Kevin McCafferty

Description: Venture capital supplier. Investment Policy: Invests a minimum of $1 million in later-stage situations to a maximum of $25 million in mature growth or buy out situations. Emphasis is placed on a strong management team and unique market opportunity. Remarks: A branch office of First Chicago Venture Capital, Chicago, Illinois.

★1514★ First United SBIC, Inc.
135 Will Dr.
Canton, MA 02021 (617) 828-6150
Alfred W. Ferrera, Vice-President

Description: Small business investment company. Investment Policy: Retail grocery stores.

★1515★ Fleet Venture Partners II
1740 Massachusetts Ave.
Boxborough, MA 01719 (617) 367-6700
Jim Saulfield, General Partner

Description: Venture capital partnership. Investment Policy: No industry preference. Remarks: A branch office of Fleet Venture Resources, Inc., Providence, Rhode Island.

★1516★ Greylock Management Corporation (Boston)
One Federal St.
Boston, MA 02110 (617) 423-5525

Description: Private venture capital partnership. Investment Policy: Minimum investment of $250,000; preferred investment size of over $1 million. Will function either as deal originator or investor in deals created by others. Remarks: Maintains an office in Palo Alto, California. **Contact:** William W. Helman and William S. Kaiser, General Partners.

★1517★ Hambro International Venture Fund (Boston)
160 State St.
Boston, MA 02109 (617) 523-7767

Description: Private venture firm. Investment Policy: Seeks to invest in mature companies as well as in high-technology areas, from start-ups to leveraged buy outs. Investments range from $500,000 to $3 million, with initial investments ranging from $800,000 to $1 million. Remarks: A branch office of Hambro International Venture Fund, New York, New York. **Contact:** Richard A. D'Amore, General Partner; Peter Santeusanio, Associate.

★1518★ John Hancock Venture Capital Management, Inc.
One Financial Center, 39th Floor
Boston, MA 02111 (617) 350-4002
Edward W. Kane, Managing Director

Description: Venture capital supplier. Investment Policy: Primary interest is in high technology-based emerging growth companies with experienced management teams. Investments range from $500,000 to $1.5 million, with a targeted minimum of $1 million.

★1519★ Matrix Partners (Boston)
One Post Office Sq.
Boston, MA 02109 (617) 482-7735
Paul J. Ferri, General Partner

Description: Private venture capital partnership. Investment Policy: Investments range from $500,000 to $1 million. Remarks: Maintains an office in Menlo Park, California.

★1520★ McGowan, Leckinger, Berg
Ten Forbes Rd.
Braintree, MA 02184 (617) 849-0020

Description: Venture capital supplier. Investment Policy: Provides early-stage financing to retail ventures and related businesses whose retail concepts follow the demographic trends of the 1980s. **Contact:** James A. McGowan; Robert T. Leckinger, Principals.

★1521★ MIP Equity Advisors, Inc. (Boston)
283 Franklin St., Suite 200
Boston, MA 02110 (617) 338-3099
Barney Ussher, Vice-President

Description: International venture capital supplier. Investment Policy: Seeks participants from a variety of business activities that wish to expand into European markets by establishing a base in the Netherlands. High-technology ventures are preferred, but not to the exclusion of other industry groups. Minimum investment is $1 million. Remarks: Company headquarters are located in The Hague, the Netherlands. Another U.S. office is located in Menlo Park, California.

★1522★ Morgan, Holland Ventures Corporation
One Liberty Sq.
Boston, MA 02109 (617) 423-1765

Description: Venture capital partnership. Investment Policy: Provides start-up, early-stage, and expansion financing to companies that are pioneering applications of proven technology; also will consider nontechnology-based companies with strong management teams and with plans for expansion. Investments range from $500,000 to $1 million, with a $6 million maximum. **Contact:** James F. Morgan; Daniel J. Holland; Jay Delahanty; Joseph T. McCullen, Jr.; Edwin M. Kania, Jr., Partners.

★1523★ New England Capital Corporation
One Washington Mall, Seventh Floor
Boston, MA 02108 (617) 722-6400
Z. David Patterson, Executive Vice-President

Description: Small business investment company. Investment Policy: Provides early-stage, second-stage, third-stage, and expansion financing to manufacturing, communications, assembly, high-technology, and sophisticated service businesses with experienced management teams and high profit markets. Leveraged buy out situations also are considered. Investments range from $200,000 to $2 million; will co-invest with other institutional investors if more capital is needed.

★1524★ New England MESBIC, Inc.
530 Turnpike St.
North Andover, MA 01845 (617) 688-4326
Etang Chen, President

Description: Minority enterprise small business investment company. Investment Policy: No industry preference.

★1525★ Northeast Small Business Investment Corporation
16 Cumberland St.
Boston, MA 02115 (617) 267-3983
Joseph Mindick, Treasurer

Description: Small business investment company. Investment Policy: No industry preference.

★1526★ Orange Nassau Capital Corporation
260 Franklin St., Suite 1501
Boston, MA 02110 (617) 451-6220
Joost E. Tjaden, President and Chief Executive

Description: Small business investment company. Investment Policy: Provides all stages of financing to technology-based and oil and gas service industries; no geographic preference. Experience of management team and high-growth market potential are emphasized. Investments range from $200,000 to $400,000. Remarks: Maintains offices in Dallas, Texas, and Newport Beach, California. Orange Nassau is able to establish Asian and European manufacturing facilities and local distribution channels for its portfolio companies through an international network. International affiliates are located in The Hague, the Netherlands, in Paris, France, and in Singapore.

★1527★ Palmer Service Corporation
300 Unicorn Park Dr.
Woburn, MA 01801 (617) 933-5445

Description: Venture capital partnership. Investment Policy: Provides financing to new enterprises and existing growth companies. Investments range from $100,000 to $1 million. **Contact:** William H. Congleton; John A. Shane; Stephen J. Ricci; Karen S. Camp; William T. Fitzgerald, General Partners.

★1528★ Sprout Group (Boston)
One Center Plaza
Boston, MA 02108 (617) 570-8720

Description: Venture capital affiliate of Donaldson, Lufkin, and Jenrette. Investment Policy: Provides early-stage financing to companies specializing in retailing, financial services, and high-technology products. Will also invest in selected later-stage, emerging/growing companies and in management buy outs of mature companies. Investments range from $1 to $2 million, with a minimum of $500,000. Remarks: A branch office of Sprout Group, headquartered in New York, New York.

★1529★ Stevens Capital Corporation
P.O. Box 871
Fall River, MA 02722 (617) 679-0044
Edward Capuano, President

Description: Small business investment company. Investment Policy: No industry preference.

★1530★ Summit Ventures
One Boston Place, Suite 3420
Boston, MA 02108 (617) 742-5500
Harrison B. Miller, Associate

Description: Venture capital firm. Investment Policy: Prefers to invest in emerging, profitable, growth companies in the electronic technology, environmental services, and health care industries. Investments range from $1 to $4 million.

★1531★ TA Associates (Boston)
45 Milk St.
Boston, MA 02109 (617) 338-0800
Brian J. Conway, Associate

Description: Private venture capital partnership. Investment Policy: Technology companies, media communications companies, and leveraged buy outs preferred. Will provide from $1 to $20 million in investments.

★1532★ 3i Capital
99 High St., Suite 1530
Boston, MA 02110 (617) 542-8560
William N. Holm, Senior Vice-President (Boston)

Description: Venture capital supplier. Investment Policy: Provides capital for growth, acquisition, share repurchase or leveraged buy out to companies in communications, computers, consumer, distribution, electronics, industrial products, manufacturing equipment, medical/health, and retail industries. Investments range from $500,000 to $5 million. Remarks: Maintains an office in Newport Beach, California.

★1533★ 3i Ventures (Boston)
99 High St., Suite 1530
Boston, MA 02110 (617) 542-8560

Description: Venture capital supplier. Investment Policy: Provides start-up and early-stage financing to companies in high-growth industries such as biotechnology, computers, electronics, health care, industrial automation, and telecommunications. Investments generally range from $500,000 to $2 million. Remarks: Maintains offices in Newport Beach, California and London, England.

★1534★ Transportation Capital Corporation (Boston)
230 Newbury St., No 21
Boston, MA 02116 (617) 536-0344
Jon Hirsch, Assistant Vice-President

Description: Minority enterprise small business investment company. Investment Policy: No industry preference. Specializes in medallion loans. Remarks: A branch office of Transportation Capital Corporation, New York, New York. Toll-free/Additional Phone Number: (617)333-0858.

★1535★ Vadus Capital Corporation
260 Franklin, Suite 1500
Boston, MA 02110 (617) 451-6220
Joost E. Tjaden, President and Chief Executive

Description: Small business investment company. Investment Policy: Provides all stages of financing to technology-based and oil and gas service industries; no geographic preference. Experience of management team and high-growth market potential are emphasized. Investments range from $200,000 to $400,000.

★1536★ Venture Capital Fund of New England II
160 Federal St., 23rd Floor
Boston, MA 02110 (617) 439-4646

Description: Venture capital fund. Investment Policy: Prefers New England high-technology companies that have a commercial prototype or initial product sales. Will provide up to $500,000 in first-round financing. **Contact:** Richard A. Farrell; Harry J. Healer, Managing General Partners.

★1537★ Venture Founders Corporation
One Cranberry Hill
Lexington, MA 02173 (617) 863-0900
Edward Getchell, General Partner

Description: Venture capital fund. Investment Policy: CAD/CAM and robotics, materials components, information technology, biotechnology, and medical industries preferred. Will provide $250,000 to $1 million in start-up and first-round financing. Preferred initial investment size is between $100,000 and $750,000.

★1538★ Vimac Corporation
12 Arlington St.
Boston, MA 02116 (617) 267-2785
Max J. Steinmann, President

Description: Venture capital supplier. Investment Policy: Provides start-up and early-stage financing to businesses in the computer industry, as well as in other suitable high-technology industries.

Michigan

★1539★ Dearborn Capital Corporation
P.O. Box 1729
Dearborn, MI 48121 (313) 337-8577
Michael LaManes, President

Description: Minority enterprise small business investment company. Investment Policy: Loans to minority-owned, operated, and controlled suppliers to Ford Motor Company, Dearborn Capital Corporation's patent.

★1540★ Demery Seed Capital Fund
P.O. Box 550
Franklin, MI 48025 (313) 433-1722
Thomas Demery, President

Description: Seed capital fund. Investment Policy: Diversified, but interested in food processing. Invests in start-up companies in Michigan.

★1541★ Federated Capital Corporation
30955 Northwestern Hwy.
Farmington Hills, MI 48018 (313) 737-1300
Louis P. Ferris, Jr., President

Description: Small business investment company. Investment Policy: No industry preference.

★1542★ MBW Management, Inc.
4251 Plymouth Rd.
P.O. Box 986
Ann Arbor, MI 48106 (313) 747-9401
Richard Goff, Vice-President

Description: Manages an Small business investment company and two venture capital funds. Investment Policy: Prefers high-tech industries and leveraged buy outs for manufacturing companies. No geographic limitations.

★1543★ Metro-Detroit Investment Company
30777 Northwestern Hwy., Suite 300
Farmington Hills, MI 48018 (313) 851-6300
William J. Fowler, President

Description: Minority enterprise small business investment company. Investment Policy: Food store industry preferred.

★1544★ Motor Enterprises, Inc.
3044 W. Grand Blvd.
Detroit, MI 48202 (313) 556-4273
James Kobus, Manager

Description: Minority enterprise small business investment company. Investment Policy: Prefers manufacturing.

★1545★ Mutual Investment Company, Inc.
21415 Civic Center Dr.
Mark Plaza Bldg., Suite 217
Southfield, MI 48076 (313) 357-2020
Tim Taylor

Description: Investment Policy: Inquire.

★1546★ Regional Financial Enterprises (Ann Arbor)
325 E. Eisenhower Pkwy., Suite 103
Ann Arbor, MI 48108 (313) 769-0941

Description: Manages one Small business investment company and two private venture capital firms. Investment Policy: Prefers to invest in high-technology companies or nontechnology companies with significant growth prospects located anywhere in the United States. Investments range from $2 to $3 million. Remarks: A branch office of Regional Financing Enterprises, New Canaan, Connecticut.

Minnesota

★1547★ Capital Dimensions Ventures Fund, Inc.
Two Apple Tree Sq., Suite 244
Minneapolis, MN 55425-1637 (612) 854-3506
T. F. Hunt, Jr., Vice-President

Description: Minority enterprise small business investment company. Investment Policy: No industry preference.

★1548★ Cherry Tree Ventures
1400 Northland Plaza
3800 W. 80th St.
Minneapolis, MN 55431 (612) 893-9012

Description: Venture capital supplier. Investment Policy: Provides start-up and early-stage financing. Fields of interest include communications, medical devices, health care services, applications and systems software, and microprocessor-based systems for the office and factory. There are no minimum or maximum investment limitations. **Contact:** Anton J. Christianson; Gordon F. Stofer, General Partners.

★1549★ Consumer Growth Capital, Inc.
8200 Humboldt Ave., S.
Minneapolis, MN 55431 (612) 888-9561
Bruce A. Thomson, President

Description: Small business investment company. Investment Policy: No industry preference.

★1550★ DGC Capital Company
603 Alworth Bldg.
Duluth, MN 55802 (218) 727-8921
Robert F. Poirier, President

Description: Small business investment company. Investment Policy: No industry preference.

★1551★ FBS Venture Capital Company
First Bank Place East
120 S. Sixth St.
Minneapolis, MN 55480 (612) 370-4764

Description: Small business investment company. Investment Policy: Generally invests in high-technology companies, although they are not necessarily preferred.

★1552★ First Midwest Ventures, Inc.
914 Plymouth Bldg.
12 S. Sixth St.
Minneapolis, MN 55402 (612) 339-9391
Alan K. Ruvelson, President

Description: Venture capital consultants.

★1553★ ITASCA Growth Fund, Inc.
One N.W. Third St.
Grand Rapids, MN 55744 (218) 326-0754

Description: Small business investment company. Investment Policy: No industry preference.

★1554★ North Star Ventures, Inc.
100 S. Fifth St., Suite 2200
Minneapolis, MN 55402 (612) 333-1133
Terrence W. Glarner, President

Description: Small business investment company. Investment Policy: Invests start-up and early-stage capital in all industry segments excluding real estate, with a preference toward companies in high-technology, electronics, and/or medical industries. Investments range from $400,000 to $1 million, with a preferred limit of $1 million.

★1555★ Northland Capital Corporation
613 Missabe Bldg.
Duluth, MN 55802 (218) 722-0545
George G. Barnum, Jr., President

Description: Small business investment company. Investment Policy: No industry preference.

★1556★ Norwest Venture Capital Management, Inc. (Minneapolis)
2800 Piper Jaffray Tower
222 S. Ninth St.
Minneapolis, MN 55402-3388 (612) 372-8770
Robert F. Zicarelli, Chairman and Chief Executive

Description: Venture capital supplier. Investment Policy: Seeks industries that offer long-term growth potential, including information processing, microelectronics, computer software, medical products, health-care delivery, biotechnology, telecommunications, and industrial automation. Investments range from $750,000 to $2 million, with the ability to invest more than $4 million. Remarks: Norwest Venture Capital is the colletive name for the three investment companies of Norwest Venture Capital Management: Norwest Growth Fund, Norwest Equity Capital, and Norwest Venture Partners. Norwest Venture Capital also maintains an office in Portland, Oregon, and Scottsdale, Arizona.

★1557★ Pathfinder Venture Capital Funds
7300 Metro Blvd., Suite 585
Minneapolis, MN 55435 (612) 835-1121
Andrew J. Greenshields, General Partner

Description: Venture capital supplier. Investment Policy: Provides start-up and early-stage financing to emerging companies in the medical, computer, pharmaceuticals, and data communications industries. Emphasis is on companies with proprietary technology or market positions and with substantial potential for revenue growth. Midwest location is preferred. Investments range from $75,000 to $1.5 million, with a $2.5 million maximum.

★1558★ Threshold Ventures, Inc.
430 Oak Grove St., Suite 200
Minneapolis, MN 55403 (612) 874-7199
John L. Shannon, President

Description: Small business investment company. Investment Policy: No industry preference.

Mississippi

★1559★ Sun-Delta Capital Access Center, Inc.
819 Main St.
Greenville, MS 38701 (601) 335-5291
Howard Boutte, Jr., Vice-President

Description: Minority enterprise small business investment company. Investment Policy: No industry preference.

★1560★ Vicksburg SBIC
302 First National Bank Bldg.
Vicksburg, MS 39180 (601) 636-4762
David L. May, President

Description: Small business investment company. Investment Policy: No industry preference.

Missouri

★1561★ Bankers Capital Corporation
3100 Gillham Rd.
Kansas City, MO 64109 (816) 531-1600
Raymond E. Glasnapp, President

Description: Small business investment company. Investment Policy: No industry preference.

★1562★ Capital for Business, Inc. (Kansas City)
1000 Walnut, 18th Floor
Kansas City, MO 64106 (816) 234-2357
Bart S. Bergman, Executive Vice-President

Description: Small business investment company. Investment Policy: No industry preference. Remarks: A branch office of the Capital for Business headquarters located in St. Louis.

★1563★ Capital for Business, Inc. (St. Louis)
11 S. Meramec, Suite 800
St. Louis, MO 63105 (314) 854-7427

Description: Small business investment company. Investment Policy: Focuses primarliy on expanding manufacturing and on leveraged buy outs. Investments range from $300,000 to $1 million. Remarks: Maintains a branch office in Kansas City, Missouri.

★1564★ Intercapco, Inc.
7800 Bonhomme Ave.
Clayton, MO 63105 (314) 863-0600
Mark Lincoln, Chairman

Description: Small business investment company. Investment Policy: No industry preference.

★1565★ Intercapco West, Inc.
7800 Bonhomme Ave.
St. Louis, MO 63105 (314) 863-0600
Thomas E. Phelps, President

Description: Small business investment company. Investment Policy: No industry preference.

★1566★ InvestAmerica Venture Group, Inc. (Kansas City)
911 Main St., Suite 2724A
Commerce Tower Bldg.
Kansas City, MO 64105 (816) 842-0114

Description: Small business investment company. Investment Policy: No industry preference. Remarks: A branch office of InvestAmerica Venture Group, Inc., Cedar Rapids, Iowa.

★1567★ MBI Venture Capital Investors, Inc.
850 Main St.
Kansas City, MO 64105 (816) 471-1700
Anthony Sommers, President

Description: Small business investment company. Investment Policy: No industry preference.

★1568★ United Missouri Capital Corporation
1010 Grand Ave.
Kansas City, MO 64106 (816) 556-7333
Joe Kessinger, Manager

Description: Small business investment company. Investment Policy: No industry preference.

Nebraska

★1569★ Community Equity Corporation of Nebraska
6421 Ames Ave.
Omaha, NE 68104 (402) 455-7722
William C. Moore, President

Description: Minority enterprise small business investment company. Investment Policy: No industry preference.

★1570★ United Financial Resources Corporation
P.O. Box 1131
Omaha, NE 68101 (402) 734-1250
Terrence W. Olsen, President

Description: Small business investment company. Investment Policy: Grocery store industry preferred.

Nevada

★1571★ Enterprise Finance Capital Development Corporation
One E. First St., Suite 1102
Reno, NV 89501 (702) 329-7797
Robert S. Russell, Chairman

Description: Small business investment company. Investment Policy: No industry preference.

New Hampshire

★1572★ Granite State Capital, Inc.
Seven Islington Rd.
P.O. Box 6564
Portsmouth, NH 03801 (603) 436-5044
Richard S. Carey, Treasurer

Description: Small business investment company. Investment Policy: No industry preference.

★1573★ VenCap
1155 Elm St.
Manchester, NH 03101 (603) 644-6100
Richard J. Ash, President

Description: Small business investment company. Investment Policy: No industry preference.

New Jersey

★1574★ Accel Partners (Princeton)
One Palmer Sq.
Princeton, NJ 08542 (609) 683-4500

Description: Venture capital firm. Investment Policy: Telecommunications industry, software, and biotechnology/medical products preferred. Will provide $100,000 to $5 million in first-round and later-stage financing. Remarks: Maintains an office in San Francisco, California. **Contact:** Dixon R. Doll; Arthur C. Patterson; James R. Swartz, General Partners.

★1575★ Bradford Associates
22 Chambers St., Fourth Floor
Princeton, NJ 08540 (609) 921-3880
Winston J. Churchill, General Partner

Description: Venture capital firm. Investment Policy: No industry preference.

★1576★ Bridge Capital Advisors, Inc.
Glen Point Center West
Teaneck, NJ 07666 (201) 836-3900

Description: Venture capital firm. Investment Policy: Prefers later-stage financing. Prefers small growth companies or companies with $5 to $100 million in annual revenues. **Contact:** Donald P. Remey, Managing Director; Hoyt J. Goodrich, Geoffrey Wadsworth; Jay Barton Goodwin, General Partners.

★1577★ Bridge Capital Investors
Glen Point Center West
Teaneck, NJ 07666 (201) 836-3900

Description: Venture capital partnership. Investment Policy: Investments range from $3 to $5 million. Remarks: A subsidiary of Bridge Capital Advisors, Inc. **Contact:** Donald P. Remey; Hoyt J. Goodrich; William I. Spencer, General Partners.

★1578★ DSV Partners
221 Nassau St.
Princeton, NJ 08542 (609) 924-6420

Description: Private venture capital supplier. Investment Policy: Provides seed, research and development, start-up, first-stage, and second-stage financing to companies specializing in communications, computers, electronic components and instrumentation, energy/natural resources, genetic engineering, industrial products and equipment, and medical-related products and services. Will not consider real estate. Preferred investment is $750,000 or more, with a minimum of $250,000. **Contact:** James R. Bergman; John K. Clarke; Morton Collins; Robert S. Hillas, Partners.

★1579★ Edelson Technology Partners
Park 80, W., Plaza 2
Saddle Brook, NJ 07662 (201) 843-4474
Harry Edelson, General Partner

Description: Venture capital partnership. Investment Policy: Prefers high-tech industries.

★1580★ Edison Venture Fund
90 Nassau St.
Princeton, NJ 08540 (609) 683-1900

Description: Private venture capital firm. Investment Policy: No industry preference. **Contact:** John H. Martinton; James F. Mrazek, General Partners.

★1581★ Eslo Capital Corporation
212 Wright St.
Newark, NJ 07114 (201) 242-4488
Leo Katz, President

Description: Small business investment company. Investment Policy: No industry preference.

★1582★ First Princeton Capital Corporation
227 Hamburg Turnpike
Pompton Lakes, NJ 07442 (201) 831-0330
Michael Lytell, Chairman

Description: Small business investment company. Investment Policy: No industry preference.

★1583★ Formosa Capital Corporation
605 King George Post Rd.
Fords, NJ 08863 (201) 738-4710
Philp Chen, President

Description: Minority enterprise small business investment company. Investment Policy: No industry preference.

★1584★ InnoVen Group
Park 80, W., Plaza 1
Saddle Brook, NJ 07662 (201) 845-4900
Gerald A. Lodge, Chief Executive Officer

Description: Venture capital firm. Investment Policy: Prefers to invest in high-tech industries. Also prefers second- and third-round financing.

★1585★ Johnston Associates, Inc.
181 Cherry Valley Rd.
Princeton, NJ 08540 (609) 924-3131
Robert F. Johnston, Managing Director

Description: Venture capital supplier. Investment Policy: Seeks high-technology, medical, and biologically oriented concepts in order to initiate and guide the establishment of a company by providing seed and start-up financing. Remarks: Will work with a scientist or inventor to recruit a management team.

★1586★ Monmouth Capital Corporation
125 Wyckoff Rd.
P.O. Box 335
Eatontown, NJ 07724 (201) 542-4927
Ralph B. Patterson, Executive Vice-President

Description: Small business investment company. Investment Policy: No industry preference.

★1587★ Rutgers Minority Investment Company
92 New St.
Newark, NJ 07102 (201) 648-5627
Oscar Figueroa, President

Description: Minority enterprise small business investment company. Investment Policy: No industry preference.

★1588★ Unicorn Ventures II LP
Six Commerce Dr.
Cranford, NJ 07016 (201) 276-7880
Frank P. Diassi, General Partner

Description: Small business investment company. Investment Policy: No industry preference.

★1589★ Unicorn Ventures Ltd.
Six Commerce Dr.
Cranford, NJ 07016 (201) 276-7880
Frank P. Diassi, General Partner

Description: Small business investment company. Investment Policy: No industry preference.

New Mexico

★1590★ Ads Capital Corporation
524 Camino del Monte Sol
Santa Fe, NM 87501 (505) 983-1769
A. David Silver

Description: Venture capital supplier. Investment Policy: Prefers to invest in manufacturing or distribution companies and health science industries. Average investment is $750,000.

★1591★ Albuquerque SBIC
501 Tijeras Ave., N.W.
P.O. Box 487
Albuquerque, NM 87103 (505) 247-0145
Albert T. Ussery, President

Description: Small business investment company. Investment Policy: No industry preference.

★1592★ Associated Southwest Investors, Inc.
2400 Louisiana, N.E., Buidling 4, Suite 225
Albuquerque, NM 87110 (505) 881-0066
John R. Rice, President

Description: Minority enterprise small business investment company. Investment Policy: No industry preference.

★1593★ Equity Capital Corporation
119 E. Marcy, Suite 101
Santa Fe, NM 87501 (505) 988-4273
Jerry A. Henson, President

Description: Small business investment company. Investment Policy: No industry preference.

★1594★ Fluid Capital Corporation
3707 Ewbank, N.E.
Albuquerque, NM 87111 (505) 292-4747
George T. Slaughter, President

Description: Small business investment company. Investment Policy: No industry preference.

★1595★ Fluid Financial Corporation
3707 Ewbank, N.E.
Albuquerque, NM 87111 (505) 292-4747
George T. Slaughter, President

Description: Minority enterprise small business investment company. Investment Policy: No industry preference.

★1596★ Industrial Development Corporation of Lea County
P.O. Box 1376
Hobbs, NM 88240 (505) 327-2039
Harold Lampe, Executive Director

Description: Certified development company. Investment Policy: Inquire.

★1597★ Meadow Research Venture Capital Group
1650 University Blvd., N.E., #500
Albuquerque, NM 87102 (505) 243-7600

Description: Venture capital firm. Investment Policy: Inquire.

★1598★ Mora San Miguel Guadalupe Development Corporation
131 Bridge St.
Las Vegas, NM 87701
Elroy Aragon

Description: Certified development company. Investment Policy: Inquire.

★1599★ National Central New Mexico Economic Development District
Development Authority of New Mexico, Inc.
P.O. Box 5115
Sante Fe, NM 87501
Leo Murphy

Description: Certified development company. Investment Policy: Inquire.

★1600★ New Mexico Business Development Corporation
6001 Marble, N.E., #6
Albuquerque, NM 87110 (505) 268-1316

Description: Venture capital firm. Investment Policy: Inquire.

★1601★ SBA 503 Development Company
Roswell Chamber of Commerce
131 W. Second St.
Roswell, NM 88201

Description: Certified development company. Investment Policy: Inquire.

★1602★ Southwest Capital Investments, Inc.
The Southwest Building
3500 E. Comanche Rd., N.E.
Albuquerque, NM 87107 (505) 884-7161
Martin J. Roe, President

Description: Small business investment company. Investment Policy: No industry preference.

★1603★ United Mercantile Capital Corporation
P.O. Box 37487
Albuquerque, NM 87176 (505) 883-8201
Joe Justice, General Manager

Description: Small business investment company. Investment Policy: Manufacturing and distribution preferred.

New York

★1604★ Adler and Company (New York City)
375 Park Ave., Suite 3303
New York, NY 10152 (212) 759-2800
Frederick R. Adler, General Partner

Description: Private venture capital supplier. Investment Policy: Provides all stages of financing. Remarks: Maintains an office in Sunnyvale, California.

★1605★ Alan Patricof Associates, Inc.
545 Madison Ave.
New York, NY 10022 (212) 753-6300
Alan Patricof, Chairman

Description: Venture capital firm. Investment Policy: No industry preference, but avoids real estate. Interested in all stages of financing. Also handles turnarounds and leveraged buy outs.

★1606★ AMEV Capital Corporation
One World Trade Center, Suite 5001
New York, NY 10048-0024 (212) 775-9100
Martin S. Orland, President

Description: Venture capital supplier. Investment Policy: Diversified with respect to industry, stage of development, and geographic location. Preferred minimum investment is $1 million; able to lead or participate in syndications of up to $10 million or more.

★1607★ AMEV Venture Management, Inc.
One World Trade Center, 50th Floor
New York, NY 10048 (212) 323-9780
Martin S. Orland, President

Description: Manages two limited partnerships and one Small business investment company. Investment Policy: Prefers leveraged buy outs and later-stage financing. Investments range from $1 to $3 million.

★1608★ ASEA—Harvest Partners II
767 Third Ave.
New York, NY 10017 (212) 838-7776
Harvey Wertheim, General Partner

Description: Small business investment company. Investment Policy: No industry preference.

★1609★ Avdon Capital Corporation
805 Avenue L
Brooklyn, NY 11230 (718) 692-0950
A. M. Donner, President

Description: Minority enterprise small business investment company. Investment Policy: No industry preference.

★1610★ Beneficial Capital Corporation
645 Fifth Ave.
New York, NY 10022 (212) 688-7307
John Hoey, President

Description: Small business investment company. Investment Policy: No industry preference.

★1611★ Bessemer Venture Partners
630 Fifth Ave.
New York, NY 10111 (212) 708-9303
Robert H. Buscher, Partner

Description: Venture capital partnership. Investment Policy: No industry preference.

★1612★ Bohlen Capital Corporation
767 Third Ave.
New York, NY 10017 (212) 838-7776
Harvey J. Wertheim, President

Description: Small business investment company. Investment Policy: No industry preference.

★1613★ BT Capital Corporation
280 Park Ave.
New York, NY 10017 (212) 850-1916
James G. Hellmuth, Deputy Chairman

Description: Small business investment company. Investment Policy: No industry preference.

★1614★ The Business Loan Center
79 Madison Ave.
New York, NY 10016 (212) 696-4334
Henry Fierst

Description: Small business investment company.

★1615★ Capital Investors and Management Corporation
Three Pell St., Suite 3
New York, NY 10013 (212) 964-2480
Rose Chao, Manager

Description: Minority enterprise small business investment company. Investment Policy: No industry preference.

★1616★ Central New York SBIC
351 S. Warren St.
Syracuse, NY 13202 (315) 478-5026
Albert Meertheiner, President

Description: Small business investment company. Investment Policy: Vending machine industry preferred.

★1617★ Chase Manhattan Capital Corporation
One Chase Manhattan Plaza, 13th Floor
New York, NY 10081 (212) 552-6275
Gustav H. Koven, President

Description: Small business investment company. Investment Policy: No industry preference.

★1618★ Chemical Venture Capital Associates LP
277 Park Ave., 11th Floor
New York, NY 10172 (212) 310-4962
Steven J. Gilbert, Managing General Partner

Description: Small business investment company. Investment Policy: No industry preference.

★1619★ Citicorp Venture Capital Ltd. (New York City)
Citicorp Center
153 E. 53rd St., 28th Floor
New York, NY 10043 (212) 559-1127
Peter G. Gerry, President

Description: Small business investment company. Investment Policy: Invests in the fields of information processing and telecommunications, transportation and energy, and health care, providing financing to companies in all stages of development. Also provides capital for leveraged buy out situations. Remarks: Maintains offices in Palo Alto, California, and Dallas, Texas.

★1620★ Clinton Capital Corporation
79 Madison Ave.
New York, NY 10016 (212) 696-4688
Mark Scharfman, President

Description: Small business investment company. Investment Policy: No industry preference.

★1621★ CMNY Capital Company, Inc.
77 Water St.
New York, NY 10005 (212) 437-7078
Robert Davidoff, Vice-President

Description: Small business investment company. Investment Policy: No industry preference.

★1622★ Columbia Capital Corporation
79 Madison Ave.
New York, NY 10016 (212) 696-4334
Mark Scharfman, President

Description: Minority enterprise small business investment company.

★1623★ Concord Partners (New York City)
535 Madison Ave.
New York, NY 10022 (212) 906-7000
John P. Birkelund, President

Description: Venture capital partnership. Investment Policy: Diversified in terms of stage of development, industry classification, and geographic location. Areas of special interest include computers, telecommunications, health care and medical products, energy, and leveraged buy outs. Preferred investments range from $2 to $3 million, with a $1 million minimum and $10 million maximum. Remarks: Maintains an office in Palo Alto, California.

★1624★ Croyden Capital Corporation
45 Rockefeller Plaza, Suite 2168
New York, NY 10111 (212) 974-0184
Lawrence D. Gorfinkle, President

Description: Small business investment company. Investment Policy: No industry preference.

★1625★ CVC Capital Corporation
131 E. 62nd St.
New York, NY 10021 (212) 319-7210
Joerg G. Klebe, President

Description: Minority enterprise small business investment company. Investment Policy: Radio and television industries preferred.

★1626★ CW Group, Inc.
1041 Third Ave.
New York, NY 10021 (212) 308-5266

Description: Venture capital supplier. Investment Policy: Interest is in the health-care field, including diagnostic and therapeutic products, services, and biotechnology. Invests in companies at developing and early stages.

★1627★ Edwards Capital Company
215 Lexington Ave., Suite 805
New York, NY 10016 (212) 686-2568
Edward H. Teitlebaum, Managing Partner

Description: Small business investment company. Investment Policy: Transportation industry preferred.

★1628★ Elk Associates Funding Corporation
600 Third Ave., 38th Floor
New York, NY 10016 (212) 972-8550
Gary C. Granoff, President

Description: Minority enterprise small business investment company. Investment Policy: Transportation industry preferred.

★1629★ Elron Technologies, Inc.
1211 Avenue of the Americas, Suite 2975
New York, NY 10036 (212) 819-1644

Description: Venture capital supplier. Investment Policy: Provides incubation and start-up financing to high-technology companies. Remarks: Elron Technologies is a subsidiary of Elron Electronic Industries Ltd., a diversified high-technology holding company based in Israel.

★1630★ Equico Capital Corporation
1290 Avenue of the Americas, Suite 3400
New York, NY 10019 (212) 397-8660
Duane Hill, President

Description: Minority enterprise small business investment company. Investment Policy: No industry preference.

★1631★ Euclid Partners Corporation
50 Rockefeller Plaza, Suite 1022
New York, NY 10020 (212) 489-1770
Fred Wilson, Associate

Description: Venture capital firm. Investment Policy: Prefers early-stage financing in health care and information processing industries.

★1632★ European Development Corporation LP
767 Third Ave., Seventh Floor
New York, NY 10017 (212) 838-7706
Harvey J. Wertheim, President

Description: Small business investment company. Investment Policy: No industry preference.

★1633★ Everlast Capital Corporation
350 Fifth Ave., Suite 2805
New York, NY 10118 (212) 695-3910
Frank J. Segreto, Vice-President

Description: Minority enterprise small business investment company. Investment Policy: No industry preference.

★1634★ Exim Capital Corporation
290 Madison Ave.
New York, NY 10017 (212) 683-3375
Victor K. Chun, President

Description: Minority enterprise small business investment company. Investment Policy: No industry preference.

★1635★ Fair Capital Corporation
Three Pell St., Second Floor
New York, NY 10013 (212) 964-2480
Robert Yet Sen Chen, President

Description: Minority enterprise small business investment company. Investment Policy: No industry preference.

★1636★ Fairfield Equity Corporation
200 E. 42nd St.
New York, NY 10018 (212) 867-0150
Matthew A. Berdon, President

Description: Small business investment company. Investment Policy: No industry preference.

★1637★ Ferranti High Technology, Inc.
515 Madison Ave.
New York, NY 10022 (212) 688-9828
Michael R. Simon, Vice-President and Director

Description: Small business investment company. Investment Policy: No industry preference.

★1638★ First Boston Corporation
Park Avenue Plaza
New York, NY 10055 (212) 909-2000
Peter T. Buchanan, President and Chief Executive

Description: Investment banker. Investment Policy: Provides financing to the oil and gas pipeline, hydroelectric, medical technology, consumer products, electronics, aerospace, and telecommunications industries. Supplies capital for leveraged buy outs. Remarks: Maintains offices in Atlanta, Boston, Chicago, Cleveland, Dallas, Denver, Geneva, Houston, London, Los Angeles, Melbourne, Philadelphia, San Francisco, San Juan, Sydney, and Tokyo.

★1639★ First Century Partnership (New York City)
1345 Avenue of the Americas
New York, NY 10105 (212) 698-6382
Michael J. Myers

Description: Private venture capital firm. Investment Policy: Diversified; minimum investment is $1.5 million.

★1640★ First New York SBIC
20 Squadron Blvd., Suite 480
New York, NY 10956 (914) 638-1550
Israel Mindick, General Partner

Description: Small business investment company. Investment Policy: No industry preference.

★1641★ Fortieth Street Venture Partners (New York City)
226 W. 37th St., Suite 1700
New York, NY 10018 (212) 967-1800
Joseph Mizrachi, General Partner

Description: Venture capital firm. Investment Policy: No industry preference.

★1642★ Foster Management Company
437 Madison Ave.
New York, NY 10022 (212) 753-4810
John H. Foster, Chairman and President

Description: Private venture capital supplier. Investment Policy: Not restricted to specific industries or geographic locations; diversified with investments in the health-care, transportation, broadcasting, communications, energy, and home furnishings industries. Investments range from $2 to $15 million.

★1643★ Franklin Corporation
767 Fifth Ave.
New York, NY 10153 (212) 486-2323
Norman Stroebel, President

Description: Small business investment company. Investment Policy: No industry preference; no start-ups.

★1644★ Fredericks Michael and Company
One World Trade Center
New York, NY 10048 (212) 466-6200
David Fredericks, Managing Director

Description: Private venture capital supplier. Investment Policy: Provides start-up and early-stage financing to companies located primarily on the East Coast. Also supplies capital for buy outs and acquisitions.

★1645★ Fresh Start Venture Capital Corporation
313 W. 53th St., Third Floor
New York, NY 10019 (212) 265-2249
Zindel Zelmanovich, President

Description: Minority enterprise small business investment company. Investment Policy: No industry preference.

★1646★ Fundex Capital Corporation
525 Northern Blvd.
Great Neck, NY 11021 (516) 466-8551
Howard Sommer, President

Description: Small business investment company. Investment Policy: No industry preference.

★1647★ GHW Capital Corporation
489 Fifth Ave.
New York, NY 10017 (212) 687-1708
Mr. Nesta Stephens, Vice-President

Description: Small business investment company. Investment Policy: No industry preference.

★1648★ Hambro International Venture Fund (New York)
17 E. 71 St.
New York, NY 10021 (212) 288-7778
Anders K. Brag

Description: Venture capital supplier. Investment Policy: Seeks to invest in mature companies as well as in high-technology areas from start-ups to leveraged buy outs. Investments range from $100,000 to $3 million, with initial investments ranging from $500,000 to $800,000. Remarks: The Hambro Bank Group of London, England, organized and sponsored the Hambro International Venture Fund for the purpose of making venture capital investments internationally—with primary emphasis on companies located in the United States. Hambro International Venture Fund also maintains an office in Boston, Massachusetts.

★1649★ Hanover Capital Corporation
150 E. 58th St., Suite 2710
New York, NY 10155 (212) 980-9670
Robert Wilson

Description: Small business investment company. Investment Policy: No industry preference.

★1650★ Harvest Ventures, Inc. (New York City)
767 Third Ave.
New York, NY 10017 (212) 838-7776

Description: Private venture capital supplier. Investment Policy: Prefers to invest in high-technology, growth-oriented companies with proprietary technology, large market potential, and strong management teams. Specific areas of interest include computers, computer peripherals, semiconductors, telecommunications, factory automation, military electronics, medical products, and health-care services. Provides seed capital of up to $250,000; investments in special high-growth and leveraged buy out situations range from $500,000 to $2 million. Will serve as lead investor and arrange financing in excess of $2 million in association with other groups. Remarks: Maintains an office in Cupertino, California. Harvey J. Wertheim; Harvey P. Mallement, Managing Directors.

★1651★ Holding Capital Management Corporation
685 Fifth Ave., 14th Floor
New York, NY 10022 (212) 486-6670
Sash A. Spencer, President

Description: Small business investment company. Investment Policy: No industry preference.

★1652★ Ibero American Investors Corporation
38 Scio St.
Rochester, NY 14604 (716) 262-3440
Emilio Serrano, President

Description: Minority enterprise small business investment company. Investment Policy: No industry preference.

★1653★ Intercontinental Capital Funding Corporation
60 E. 42nd St., Suite 740
New York, NY 10165 (212) 286-9642
James S. Yu, President

Description: Minority enterprise small business investment company. Investment Policy: No industry preference.

★1654★ Intergroup Venture Capital Corporation
230 Park Ave.
New York, NY 10169 (212) 661-5428
Ben Hauben, President

Description: Small business investment company. Investment Policy: No industry preference.

★1655★ Interstate Capital Company, Inc.
380 Lexington Ave.
New York, NY 10017 (212) 972-3445
David Scharf, President

Description: Small business investment company. Investment Policy: No industry preference.

★1656★ Irving Capital Corporation
1290 Avenue of the Americas, 29th Floor
New York, NY 10104 (212) 408-4800
J. Andrew McWethy, President

Description: Small business investment company; bank holding company fund. Investment Policy: No industry preference.

★1657★ ITC Capital Corporation
1290 Avenue of the Americas, 29th Floor
New York, NY 10104 (212) 408-4800
J. Andrew McWethy, President

Description: Small business investment company; bank holding company fund. Investment Policy: No industry preference.

★1658★ Japanese American Capital Corporation
19 Rector St.
New York, NY 10006 (212) 344-4588
Stephen C. Huang, President

Description: Minority enterprise small business investment company. Investment Policy: No industry preference.

★1659★ Josephberg, Grosz and Company, Inc.
344 E. 49th St.
New York, NY 10017 (212) 935-1050

Description: Venture capital firm. Investment Policy: No industry preference. **Contact:** Richard Josephberg, Chairman; Ivan Grosz, President.

★1660★ Kwiat Capital Corporation
576 Fifth Ave.
New York, NY 10036 (212) 391-2461
Sheldon F. Kwiat, President

Description: Small business investment company. Investment Policy: No industry preference.

★1661★ Lambda Funds
Drexel Burnham Lambert, Inc.
55 Broad St., 15th Floor
New York, NY 10004 (212) 480-3965
Alexa Mahnken, Associate

Description: Venture capital partnership. Investment Policy: Prefers to invest in manufacturers, service companies, and leveraged buy outs; does not favor real estate, fashion or style businesses, mining, entertainment, airlines, financial intermediaries, research and development partnerships, and start-ups. Investments range from $500,000 to $1 million. Remarks: The Lambda Funds are three limited partnerships managed by an affiliate of Drexel Burnham Lambert, Inc., an investment banking firm.

★1662★ Lawrence, Tyrrell, Ortale, and Smith
515 Madison Ave.
New York, NY 10022 (212) 826-9080
Brian T. Horey

Description: Venture capital firm. Investment Policy: No industry preference.

★1663★ M and T Capital Corporation
One M and T Plaza
Buffalo, NY 14240 (716) 842-5881
William Randon, President

Description: Small business investment company. Investment Policy: No industry preference.

★1664★ Manufacturers Hanover Venture Capital Corporation
270 Park Ave.
New York, NY 10017 (212) 286-3220

Description: Venture capital and leveraged buy out firm. Investment Policy: Invests in leveraged buy outs and growth equity.

★1665★ Medallion Funding Corporation
205 E. 42nd St., Suite 2020
New York, NY 10017 (212) 682-3300
Alvin Murstein, President

Description: Minority enterprise small business investment company. Investment Policy: Transportation industry preferred.

★1666★ Minority Equity Capital Company, Inc.
275 Madison Ave.
New York, NY 10016 (212) 686-9710
Donald F. Greene, President

Description: Minority enterprise small business investment company. Investment Policy: No industry preference.

★1667★ Monsey Capital Corporation
125 Route 59
Monsey, NY 19052 (914) 425-2229
Shamuel Myski, President

Description: Minority enterprise small business investment company. Investment Policy: No industry preference.

★1668★ Multi-Purpose Capital Corporation
31 S. Broadway
Yonkers, NY 10701 (914) 963-2733
Eli B. Fine, President

Description: Small business investment company. Investment Policy: No industry preference.

★1669★ Nazem and Company
600 Madison Ave.
New York, NY 10022 (212) 644-6433
Frederick F. Nazem, Managing General Partner

Description: Venture capital fund. Investment Policy: Electronics and medical industries preferred. Will provide seed, first-round, and second-round financing.

★1670★ Nelson Capital Corporation
585 Stewart Ave.
Garden City, NY 11530 (516) 222-2555
Irwin Nelson, President

Description: Small business investment company. Investment Policy: No industry preference.

★1671★ New Enterprise Associates (New York City)
119 E. 55th St.
New York, NY 10022 (212) 371-8210

Description: Venture capital supplier. Investment Policy: Concentrates on technology-based industries that have the potential for product innovation, rapid growth, and high profit margins. Investments range from $250,000 to $2 million. Past investments have been made in the following industries: computer software, medical and life sciences, computers and peripherals, communications, semiconductors, specialty retailing, energy and alternative energy, defense electronics, and materials and specialty chemicals. Management must demonstrate intimate knowledge of its marketplace and have a well- defined strategy for achieving strong market penetration. Remarks: A branch office of New Enterprise Associates, Baltimore, Maryland.

★1672★ Noro Capital Ltd.
767 Third Ave., Seventh Floor
New York, NY 10017 (212) 838-9720
Harvey J. Wertheim, President

Description: Small business investment company. Investment Policy: No industry preference.

★1673★ Norstar Venture Capital
One Norstar Plaza
Albany, NY 12207-2796 (518) 447-4043
Raymond A. Lancaster, President

Description: Venture capital supplier. Investment Policy: No industry preference. Typical investment is between $500,000 and $1 million. Remarks: A wholly owned subsidiary of Norstar Bancorp.

★1674★ North American Funding Corporation
177 Canal St.
New York, NY 10013 (212) 226-0080
Franklin F. Y. Wong, President

Description: Minority enterprise small business investment company. Investment Policy: No industry preference.

★1675★ North Street Capital Corporation
250 North St., RA-6S
White Plains, NY 10625 (914) 335-7901
Ralph McNeal, President

Description: Minority enterprise small business investment company. Investment Policy: No industry preference. Toll-free/Additional Phone Number: (914)335-7903.

★1676★ NYBDC Capital Corporation
41 State St.
Albany, NY 12207 (518) 463-2268
Robert Lazar, President
Description: Small business investment company. Investment Policy: No industry preference.

★1677★ Pan Pac Capital Corporation
121 E. Industry Ct.
Deer Park, NY 11729 (516) 586-7653
Dr. In Ping Jack Lee, President
Description: Minority enterprise small business investment company. Investment Policy: No industry preference.

★1678★ Pierre Funding Corporation
270 Madison Ave., Room 1608
New York, NY 10016 (212) 689-9361
Elias Debbas, President
Description: Minority enterprise small business investment company. Investment Policy: No industry preference.

★1679★ Pioneer Associates
113 E. 55th St.
New York, NY 10022 (212) 980-9080
James G. Niven, President
Description: Venture capital partnership. Investment Policy: No industry preference.

★1680★ Pioneer Capital Corporation
113 E. 55th St.
New York, NY 10022 (212) 980-9096
James G. Niven, President
Description: Minority enterprise small business investment company. Investment Policy: No industry preference.

★1681★ Prospect Group, Inc.
645 Madison Ave.
New York, NY 10022 (212) 758-8500
Thomas A. Barron, President
Description: Venture capital supplier. Investment Policy: Investments focus on computer communications and software products, fiber optics, genetic engineering, biotechnology, health care management, and solar energy. Remarks: The Prospect Group's assets include holdings in Sierra Ventures, a technology-oriented venture capital limited partnership, and in Wood River Capital Corporation, an Small business investment company located in New York, with an office in Menlo Park, California.

★1682★ Prudential Venture Capital
717 Fifth Ave., Suite 1600
New York, NY 10022 (212) 753-0901
Description: Venture capital fund. Investment Policy: Specialty retailing, medical and health services, communications, and technology companies preferred. Will provide $3 to $7 million in equity financing for later-stage growth companies. **Contact:** Robert Knox, President; William Field, Chairman.

★1683★ Questech Capital Corporation
600 Madison Ave., 21st Floor
New York, NY 10022 (212) 758-8522
John E. Koonce, III, President
Description: Small business investment company. Investment Policy: No industry preference.

★1684★ R and R Financial Corporation
1451 Broadway
New York, NY 10066 (212) 790-1400
Imre Rosenthal, President
Description: Small business investment company. Investment Policy: No industry preference.

★1685★ Rand SBIC, Inc.
1300 Rand Bldg.
Buffalo, NY 14203 (716) 853-0802
Donald Ross, President
Description: Small business investment company. Investment Policy: No industry preference.

★1686★ Realty Growth Capital Corporation
331 Madison Ave., Fourth Floor East
New York, NY 10017 (212) 661-8380
Lawrence Benenson, President
Description: Small business investment company. Investment Policy: Real estate industry preferred.

★1687★ Revere AE Capital Fund, Inc.
745 Fifth Ave., 19th Floor
New York, NY 10151 (212) 888-6800
Clinton A. Reynolds, President
Description: Closed-end mutual fund. Investment Policy: Concentrates on mezzanine financing. Interested in depth security with equity participation. Investment range is $1 to $2 million.

★1688★ Rothschild Ventures, Inc.
One Rockefeller Plaza
New York, NY 10020 (212) 757-6000
Jess L. Bessler, President
Description: Private venture capital firm. Investment Policy: Prefers seed and all later-stage financing.

★1689★ S and S Venture Associates Ltd.
Seven Penn Plaza, Suite 320
New York, NY 10001 (212) 736-4530
Donald Smith, President
Description: Small business investment company. Investment Policy: No industry preference.

★1690★ 767 Limited Partnership
767 Third Ave.
New York, NY 10017 (212) 838-7776
Description: Small business investment company. Investment Policy: No industry preference. **Contact:** H. Wertheim; H. Mallement, General Partners.

★1691★ Situation Ventures Corporation
502 Flushing Ave.
Brooklyn, NY 11205 (718) 438-4909
Description: Investment Policy: Inquire.

★1692★ Sprout Group (New York City)
140 Broadway
New York, NY 10005 (212) 504-3600
Richard E. Kroon, Managing Partner
Description: Venture capital affiliate of Donaldson, Lufkin, and Jenrette. Investment Policy: Provides early-stage financing to companies specializing in retailing, financial services, and high-technology products. Will also invest in selected later-stage emerging growth companies and management buy outs of mature companies. Investments range from $1 to $2 million, with a minimum of $500,000. Remarks: Maintains offices in Menlo Park, California, and Boston, Massachusetts.

★1693★ TA Associates (New York City)
919 Third Ave.
New York, NY 10022 (212) 230-9870
Jacqueline C. Morby
Description: Private venture capital firm. Investment Policy: Prefers technology companies, media communications companies, and leveraged buy outs. Will provide from $1 to $20 million in investments.

★1694★ Taroco Capital Corporation
19 Rector St., 35th Floor
New York, NY 10006 (212) 344-6690
David R. C. Chang, President

Description: Minority enterprise small business investment company. Investment Policy: Chinese-Americans preferred.

★1695★ TCW Special Placements Fund I
200 Park Ave.
New York, NY 10166 (212) 972-1440

Description: Venture capital fund. Investment Policy: Companies with sales of $25 to $100 million preferred. Will provide up to $20 million in later-stage financing for recapitalizations, restructuring management buy outs, and general corporate purposes. **Contact:** Thomas L. Cassidy; Frank J. Pados, Senior Partner.

★1696★ Telesciences Capital Corporation
26 Broadway, Suite 1391
New York, NY 10004 (212) 425-0320
Mike A. Petrozzo

Description: Small business investment company. Investment Policy: No industry preference.

★1697★ Tessler and Cloherty, Inc.
155 Main St.
Cold Spring, NY 10516 (914) 265-4244

Description: Small business investment company. Investment Policy: No industry preference. **Contact:** Patricia Cloherty; Dan Tessler, General Partners.

★1698★ TLC Funding Corporation
141 S. Central Ave.
Hartsdale, NY 10530 (914) 683-1144
Philip G. Kass, President

Description: Small business investment company. Investment Policy: No industry preference.

★1699★ Transportation Capital Corporation (New York City)
60 E. 42nd St., Suite 3115
New York, NY 10165 (212) 697-4885
Melvin L. Hirsch, President

Description: Minority enterprise small business investment company. Investment Policy: No industry preference. Remarks: Maintains an office in Boston, Massachusetts.

★1700★ Triad Capital Corporation of New York
960 Southern Blvd.
Bronx, NY 10459 (212) 589-6541
Lorenzo J. Barrera, President

Description: Minority enterprise small business investment company. Investment Policy: No industry preference.

★1701★ Vega Capital Corporation
720 White Plains Rd.
Scarsdale, NY 10583 (914) 472-8550
Victor Harz, President

Description: Small business investment company. Investment Policy: No industry preference.

★1702★ Venrock Associates (New York City)
30 Rockefeller Plaza, Room 5508
New York, NY 10112 (212) 247-3700

Description: Private venture capital supplier. Remarks: Maintains an office in Palo Alto, California.

★1703★ Venture Opportunities Corporation
110 E. 59th St., 29th Floor
New York, NY 10022 (212) 832-3737
A. Fred March, President

Description: Minority enterprise small business investment company. Investment Policy: No industry preference.

★1704★ Venture SBIC, Inc.
249-12 Jericho Tnpk.
Floral Park, NY 11001 (516) 352-0068
Arnold Feldman, President

Description: Small business investment company. Investment Policy: No industry preference.

★1705★ Warburg Pincus Ventures, Inc.
466 Lexington Ave.
New York, NY 10017 (212) 878-0600
Christopher Brody, Managing Director

Description: Private venture capital firm. Investment Policy: No industry preference.

★1706★ Watchung Capital Corporation
431 Fifth Ave., Fifth Floor
New York, NY 10016 (212) 889-3466
S. T. Jeng

Description: Investment Policy: Inquire.

★1707★ Welsh, Carson, Anderson, and Stowe
200 Liberty, Suite 3601
New York, NY 10281 (212) 945-2000
Patrick Welsh, General Partner

Description: Venture capital partnership. Investment Policy: High-technology industries preferred. Also interested in leveraged buy outs. Minimum investment is $5 million.

★1708★ WFG-Harvest Partners Ltd.
767 Third Ave.
New York, NY 10017 (212) 838-7776
Harvey J. Wertheim, General Partner

Description: Small business investment company. Investment Policy: No industry preference.

★1709★ Winfield Capital Corporation
237 Mamaroneck Ave.
White Plains, NY 10605 (914) 949-2600
Stanley M. Pechman, President

Description: Small business investment company. Investment Policy: No industry preference.

★1710★ Wood River Capital Corporation (New York City)
645 Madison Ave., 22nd Floor
New York, NY 11542 (212) 758-8500
Thomas A. Barron, President

Description: Small business investment company. Investment Policy: No industry preference.

★1711★ Yang Capital Corporation
41-40 Kissena Blvd.
Flushing, NY 11355 (718) 849-1037
Ms. Maysing Yang, President

Description: Minority enterprise small business investment company. Investment Policy: No industry preference.

★1712★ Yusa Capital Corporation
622 Broadway, Second Floor
New York, NY 10012 (212) 420-1350
Christopher Yeung, Chairman of the Board

Description: Minority enterprise small business investment company. Investment Policy: No industry preference.

North Carolina

★1713★ Delta Capital, Inc.
227 N. Tryon St., Suite 201
Charlotte, NC 28202 (704) 372-1410

Description: Small business investment company. Investment Policy: Real estate industry preferred. **Contact:** Alex B. Wilkins, Jr., President; Rodney C. Pitts.

★1714★ Falcon Capital Corporation
400 W. Fifth St.
Greenville, NC 27834 (919) 752-5918
P. S. Prasad, President

Description: Small business investment company. Investment Policy: No industry preference.

★1715★ Heritage Capital Corporation
2095 Two Union Center
Charlotte, NC 28282 (704) 334-2867
William R. Starnes, President

Description: Small business investment company. Investment Policy: Diversified industries.

★1716★ Kitty Hawk Capital Ltd.
1640 Independence Center
Charlotte, NC 28246 (704) 333-3777

Description: Small business investment company. Investment Policy: No industry preference. **Contact:** Walter H. Wilkinson, President; Chris Hegele, Vice-President.

★1717★ NCNB SBIC Corporation
One NCNB Plaza (T05-2)
Charlotte, NC 28255 (704) 374-5583
Troy S. McCrory, Jr., President

Description: Small business investment company. Investment Policy: No industry preference.

★1718★ NCNB Venture Company LP
One NCNB Plaza (T05-2)
Charlotte, NC 28255 (704) 374-5723
S. Epes Robinson, General Partner

Description: Small business investment company. Investment Policy: No industry preference.

★1719★ Southgate Venture Partners (Charlotte)
227 N. Tryon St., Suite 201
Charlotte, NC 28202 (704) 372-1410
Alex Wilkins

Description: Private venture capital firm. Investment Policy: Diversified.

Ohio

★1720★ A. T. Capital Corporation
900 Euclid Ave., T-18
Cleveland, OH 44101 (216) 687-4970

Description: Small business investment company. Investment Policy: No industry preference. **Contact:** Robert C. Salipante, President; Lisa M. Simecek, Vice-President and Manager.

★1721★ Banc One Capital Corporation
100 E. Broad St.
Columbus, OH 43271-0251 (614) 248-5832
James E. Kolls, Vice-President

Description: Small business investment company. Investment Policy: No industry preference.

★1722★ Capital Funds Corporation
800 Superior Ave.
Cleveland, OH 44114 (216) 344-5775

Description: Small business investment company. Investment Policy: Diversified. **Contact:** Carl G. Nelson, Chief Investment Officer; David B. Chilcote, Vice-President.

★1723★ Cardinal Development Capital Fund I
40 S. Third St.
Columbus, OH 43215 (614) 464-5550

Description: Private venture capital firm. Investment Policy: Provides expansion capital to manufacturing and service firms, particularly in Ohio and other parts of the Midwest. Avoids investments in real estate and resource recovery systems. Minimum investment is $250,000, with a preferred investment of $500,000 or more.

★1724★ Center City MESBIC, Inc.
40 S. Main St., Suite 762
Dayton, OH 45402 (513) 461-6164
Michael A. Robinson, President

Description: Minority enterprise small business investment company. Investment Policy: Diversified industries.

★1725★ Clarion Capital Corporation
35555 Curtis Blvd.
East Lake, OH 44094 (216) 953-0555
Morton A. Cohen, President

Description: Small business investment company. Investment Policy: Specialty chemicals, instrumentation, and health care.

★1726★ First Ohio Capital Corporation
P.O. Box 2061
Toledo, OH 43603 (419) 259-7141
David J. McMacken, General Manager

Description: Small business investment company. Investment Policy: No industry preference.

★1727★ Fortieth Street Venture Partners (Cincinnati)
3712 Carew Tower
Cincinnati, OH 45202 (513) 579-0101
Joseph Mizrachi, General Partner

Description: Venture capital firm. Investment Policy: No industry preference.

★1728★ Gries Investment Corporation
Statler Office Tower, Suite 1500
Cleveland, OH 44115 (216) 861-1146
Robert D. Gries, President

Description: Small business investment company. Investment Policy: No industry preference.

★1729★ Lubrizol Enterprises, Inc.
29400 Lakeland Blvd.
Wickliffe, OH 44092 (216) 943-4200
Bruce H. Grasser, Senior Vice-President

Description: Venture capital supplier. Investment Policy: Provides seed capital and later-stage expansion financing to emerging companies in the biological, chemical, and material sciences whose technology is applicabel to and related to the production and marketing of specialty and fine chemicals. Investments range from $250,000 to $15 million.

★1730★ Miami Valley Capital, Inc.
Talbott Tower, Suite 315
131 N. Ludlow St.
Dayton, OH 45402 (513) 222-7222
Everett F. Telljohann

Description: Small business investment company. Investment Policy: No industry preference.

★1731★ Morgenthaler Ventures
700 National City Bank Bldg.
Cleveland, OH 44114 (216) 621-3070
David T. Morgenthaler, Managing Partner

Description: Private venture capital supplier. Investment Policy: Provides start-up and later-stage financing to all types of business in North America; prefers not to invest in real estate and mining. Investments range from $500,000 to $3 million.

★1732★ National City Capital Corporation
629 Euclid Ave., Mezzanine
Cleveland, OH 44114 (216) 575-2491
John B. Naylor, President

Description: Small business investment company. Investment Policy: No industry preference.

★1733★ Primus Capital Fund
One Cleveland Center, Suite 2140
1375 E. Ninth St.
Cleveland, OH 44114 (216) 621-2185

Description: Venture capital partnership. Investment Policy: Provides seed, early-stage, and expansion financing to companies located in Ohio and the Midwest, preferring to invest in businesses compatible with the manufacturing, service, and technology base in the Midwest. Industries of interest include industrial automation, medical technologies, and consumer and industrial products with technical content. Investments generally range from $1 to $5 million. Equity capital for leverage management buy outs also is available. **Contact:** Loyal Wilson; James Bartlett, Managing Partners.

★1734★ River Capital Corporation (Cleveland)
796 Huntington Bldg.
Cleveland, OH 44114 (216) 781-3655

Description: Small business investment company. Investment Policy: No industry preference. Remarks: A branch office of River Capital Corporation, Alexandria, Virginia.

★1735★ Rubber City Capital Corporation
1144 E. Market St.
Akron, OH 44316 (216) 796-9167
Jesse T. Williams, Sr., President

Description: Minority enterprise small business investment company. Investment Policy: No industry preference.

★1736★ SeaGate Venture Management, Inc.
245 Summit St., Suite 1403
Toledo, OH 43603 (419) 259-8526
Charles A. Brown, Vice-President

Description: Small business investment company. Investment Policy: No industry preference.

★1737★ Seed One
Park Place
Ten W. Statesboro St.
Hudson, OH 44236 (216) 650-2338
Burton D. Morgan

Description: Private venture capital firm. Investment Policy: No industry preference. Equity financing only.

★1738★ Technology Group Ltd.
Scientific Advances, Inc.
601 W. Fifth Ave.
Columbus, OH 43201 (614) 294-5541
Charles G. James, President

Description: Venture capital partnership managed by Scientific Advances, Inc. Investment Policy: Seeks small to medium-sized companies with innovative, proven technologies. Minimum investment is $250,000. Remarks: Scientific Advances is a subsidiary of Battelle Memorial Institute, a large, independent research organization that assists in the technical evaluation of investment proposals.

★1739★ Tomlinson Capital Corporation
13700 Broadway
Garfield Heights, OH 44125 (216) 587-3400
John A. Chernak, President

Description: Small business investment company. Investment Policy: Miniature supermarket industry preferred.

Oklahoma

★1740★ Alliance Business Investment Company (Tulsa)
17 E. Second St.
One Williams Center, Suite 2000
Tulsa, OK 74172 (918) 584-3581
Barry Davis, President

Description: Small business investment company. Investment Policy: No industry preference. Remarks: Maintains an office in Houston, Texas.

★1741★ Rubottom, Dudash and Associates, Inc.
2700 E. 51st St., Suite 340
Tulsa, OK 74105 (918) 742-3031
Donald J. Rubottom, President

Description: Management and investment consultants. Investment Policy: Emphasis on retail, wholesale, and light fabrication. Remarks: Affiliated with Southwest Venture Capital which is currently inactive.

★1742★ Signal Capital Corporaton
One Leadership Sq., Suite 400
Oklahoma City, OK 73102 (405) 235-4440
John Lewis

Description: Small business investment company. Investment Policy: No industry preference.

★1743★ Western Venture Capital Corporation
P.O. Box 702680
Tulsa, OK 74170 (918) 749-7000
William Baker, President

Description: Small business investment company. Investment Policy: Seeks growth companies, turnarounds, and leveraged buy outs in the following areas: oil and gas technology; other natural resources; agribusiness; medical technology and biotechnology; health care; electronics and computer technology; communications and information; leisure and entertainment; and specialty retailing. Investments range from $200,000 to $2 million.

Oregon

★1744★ Cable and Howse Ventures (Portland)
101 S.W. Main, Suite 1800
Portland, OR 97204 (503) 248-9646

Description: Venture capital supplier. Investment Policy: Provides start-up and early-stage financing to enterprises in the western United States, although a national perspective is maintained. Interest lies in proprietary or patentable technology. Investments range from $500,000 to $2 million. Remarks: A branch office of Cable and Howse Ventures, Bellevue, Washington.

★1745★ InterVen II LP
227 S.W. Pine St., Suite 200
Portland, OR 97204 (503) 223-4334

Description: Small business investment company. Investment Policy: No industry preference. Remarks: Affiliated with InterVen II LP, in Los Angeles.

★1746★ Northern Pacific Capital Corporation
P.O. Box 1658
Portland, OR 97207 (503) 241-1255
Joseph P. Tennant, President

Description: Small business investment company. Investment Policy: No industry preference.

★1747★ Norwest Venture Capital Management, Inc. (Portland)
1300 S.W. Fifth Ave., Suite 3018
Portland, OR 97201-5683 (503) 223-6622

Description: Investment Policy: Seeks industries that offer long-term growth potential, including information processing, microelectronics, computer software, medical products, health care delivery, biotechnology, telecommunications, and industrial automation. Investments range from $750,000 to $2 million, with the ability to invest more than $4 million. Remarks: A branch office of Norwest Venture Capital Management, Inc., headquartered in Minneapolis.

★1748★ Trendwest Capital Corporation
803 Main St., Suite 404
Klamath Falls, OR 97601 (503) 882-8059
Mark E. Nicol, President

Description: Small business investment company. Investment Policy: No industry preference.

Pennsylvania

★1749★ Alliance Enterprise Corporation
1801 Market St., Third Floor
Philadelphia, PA 19103 (215) 977-3925
W. B. Priestley, President

Description: Minority enterprise small business investment company. Investment Policy: Broadcasting and manufacturing preferred.

★1750★ Capital Corporation of America
225 S. 15th St., Suite 920
Philadelphia, PA 19102 (215) 732-1666
Martin M. Newman, President

Description: Small business investment company. Investment Policy: No industry preference.

★1751★ Core States Enterprise Fund
One Penn Center, Suite 1360.
1617 J. F. Kennedy Bldg.
Philadelphia, PA 19103 (215) 568-4677

Description: Venture capital supplier. Investment Policy: No industry preference. Remarks: A subsidiary of the Philadelphia National Bank. **Contact:** Paul A. Mitchell, President; Michael F. Donoghue, Vice-President.

★1752★ Enterprise Venture Capital Corporation of Pennsylvania
227 Franklin St., Suite 215
Johnstown, PA 15901 (814) 535-7597
Donald W. Cowie, Vice-President

Description: Small business investment company. Investment Policy: No industry preference.

★1753★ Erie SBIC
32 W. Eighth St., Suite 615
Erie, PA 16501 (814) 453-7964
George R. Heaton, President

Description: Small business investment company. Investment Policy: No industry preference.

★1754★ First SBIC of California (Washington)
P.O. Box 512
Washington, PA 15301 (412) 223-0707
Daniel A. Dye

Description: Small business investment company. Investment Policy: No industry preference. Remarks: A branch office of First Small business investment company of California, Costa Mesa, California.

★1755★ First Valley Capital Corporation
640 Hamilton Mall
Allentown, PA 18101 (215) 776-6760
Matthew W. Thomas, President

Description: Small business investment company. Investment Policy: No industry preference.

★1756★ Greater Philadelphia Venture Capital Corporation, Inc.
225 S. 15th St., Suite 920
Philadelphia, PA 19102 (215) 732-3415
Martin M. Newman, General Manager

Description: Minority enterprise small business investment company. Investment Policy: No industry preference.

★1757★ Keystone Venture Capital Management Company
211 S. Broad St.
Philadelphia, PA 19107 (215) 985-5519
G. Kenneth Macrae, General Partner

Description: Private venture capital partnership. Investment Policy: Provides equity-based expansion financing to companies in the computer, communications, automated equipment, medical, and other industries. Also provides financing for small leveraged buy outs.

★1758★ Meridian Capital Corporation
Suite 222, Blue Bell West
650 Skippack Pike
Blue Bell, PA 19422 (215) 278-8907
Joseph E. Laky, President

Description: Small business investment company. Investment Policy: No industry preference.

★1759★ Peter J. Schmitt SBIC, Inc.
385 Shenango Ave.
Sharon, PA 16146 (412) 981-1500

Description: Small business investment company. Investment Policy: Retail grocery industry preferred.

★1760★ Philadelphia Ventures
1760 Market St.
Philadelphia, PA 19103 (215) 751-9444

Description: Venture capital partnership. Investment Policy: Provides start-up and early-stage financing to companies offering products or services based on technology or other proprietary capabilities. Industries of particular interest are information processing equipment and services, medical products and services, data communications, and industrial automation. Initial investments range from $500,000 to $2 million. Also supplies capital for buy out situations.

★1761★ PNC Venture Capital Group
Pittsburgh National Bank Bldg.
Fifth Ave. and Wood, 19th Floor
Pittsburgh, PA 15222 (412) 762-8892
David Hillman, Executive Vice-President

Description: PNC Venture Capital Group comprises two separate entities; PNC Capital Corporation, an Small business investment company, and PNC Venture Corporation, a venture capital supplier. Investment Policy: Prefers to invest in later-stage and leveraged buy out situations. Interest is in regionally based companies, but will consider investing nationwide. Industry preference is diversified.

★1762★ Salween Financial Services, Inc.
228 N. Pottstown Pike
Exton, PA 19341 (215) 524-1880
Dr. Ramarao Naidu, President

Description: Minority enterprise small business investment company. Investment Policy: No industry preference.

★1763★ Trivest Venture Fund
P.O. Box 136
Ligonier, PA 15658 (412) 471-0151
Thomas W. Courtney, General Partner

Description: Venture capital firm. Investment Policy: No industry preference.

Puerto Rico

★1764★ North America Investment Corporation
P.O. Box 1831
Hato Rey, PR 00919 (809) 754-6177
Santigo Ruz Betacourt, President

Description: Minority enterprise small business investment company. Investment Policy: No industry preference.

Rhode Island

★1765★ Domestic Capital Corporation
815 Reservoir Ave.
Cranston, RI 02910 (401) 946-3310
Nathaniel B. Baker, President

Description: Small business investment company. Investment Policy: No industry preference.

★1766★ Fleet Venture Partners
111 Westminster St.
Providence, RI 02903 (401) 278-6770
Robert M. Van Degna, General Partner

Description: Venture capital fund. Investment Policy: Managed leveraged buy outs. Remarks: Maintains offices in Boston, Massachusetts; and New York, New York.

★1767★ Moneta Capital Corporation
285 Governor St.
Providence, RI 02906 (401) 861-4600
Arnold Kilberg, President

Description: Small business investment company. Investment Policy: No industry preference.

★1768★ Narragansett Venture Corporation
Fleet Center, Ninth Floor
50 Kennedy Plaza
Providence, RI 02903 (401) 751-1000
Robert D. Manchester, Managing Director

Description: Private venture capital supplier. Investment Policy: No industry preference.

★1769★ Old Stone Capital Corporation
One Old Stone Sq.
Providence, RI 02903 (401) 278-2559

Description: Small business investment company. Investment Policy: Real estate industry preferred.

South Carolina

★1770★ Charleston Capital Corporation
111 Church St.
P.O. Box 328
Charleston, SC 29402 (803) 723-6464
Henry Yaschik, President

Description: Small business investment company. Investment Policy: No industry preference.

★1771★ Floco Investment Company, Inc.
P.O. Box 919
Lake City, SC 29560 (803) 389-2731
William H. Johnson, Jr., President

Description: Small business investment company. Investment Policy: Invests only in the retail food industry.

★1772★ Lowcountry Investment Corporation
4444 Daley St.
P.O. Box 10447
Charleston, SC 29411 (803) 554-9880
Joseph T. Newton, Jr., President

Description: Small business investment company. Investment Policy: Grocery industry preferred.

★1773★ Reedy River Ventures, Inc.
P.O. Box 17526
Greenville, SC 29606 (803) 297-9198
John M. Sterling, President

Description: Small business investment company. Investment Policy: No industry preference.

Tennessee

★1774★ Chickasaw Capital Corporation
67 Madison Ave.
P.O. Box 387
Memphis, TN 38147 (901) 523-6404
Tom Moore, President

Description: Minority enterprise small business investment company. Investment Policy: No industry preference.

★1775★ Financial Resources, Inc.
2800 Sterick Bldg.
Memphis, TN 38103 (901) 527-9411
Milton Picard, Chairman of the Board

Description: Small business investment company. Investment Policy: No industry preference.

★1776★ International Paper Capital Formation, Inc.
6400 Poplar Ave., 10-21
Memphis, TN 38197 (901) 763-5951
Richard M. Ludwig, President

Description: Minority enterprise small business investment company. Investment Policy: No industry preference.

★1777★ Lawrence Venture Associates
3401 W. End Ave., Suite 680
Nashville, TN 37203 (615) 383-0982
Tom Gallagher

Description: Private venture capital firm. Investment Policy: Prefers to invest in health care industries.

★1778★ Massey Burch Investment Group
310 25th Ave., N., Suite 103
Nashville, TN 37203 (615) 329-9448
Bob Fisher, Associate

Description: Venture capital firm. Investment Policy: No industry preference. Investments range from $1 to $3 million.

★1779★ Tennessee Venture Capital Corporation
P.O. Box 2567
Nashville, TN 37219 (615) 244-6935
Wendell P. Knox, President

Description: Minority enterprise small business investment company. Investment Policy: No industry preference.

★1780★ Valley Capital Corporation
Krystal Bldg., Suite 806
100 W. Martin Luther King Blvd.
Chattanooga, TN 37402 (615) 265-1557
Lamar J. Partridge, President

Description: Minority enterprise small business investment company. Investment Policy: No industry preference.

★1781★ West Tennessee Venture Capital Corporation
P.O. Box 300
Memphis, TN 38103 (901) 527-6091
Robert E. Fenner, Manager

Description: Minority enterprise small business investment company. Investment Policy: No industry preference.

Texas

★1782★ Acorn Ventures, Inc.
2401 Fountainview, Suite 950
Houston, TX 77057 (713) 977-7421

Description: Investment Policy: No industry preference. Remarks: Maintains an office in Austin, Texas. **Contact:** Stuart Schube; Walter Cunningham.

★1783★ Alliance Business Investment Company (Houston)
910 Louisiana
3990 One Shell Plaza
Houston, TX 77002 (713) 224-8224

Description: Small business investment company. Investment Policy: No industry preference. Remarks: A branch office of Alliance Business Investment Company in Tulsa, Oklahoma.

★1784★ Allied Bancshares Capital Corporation
P.O. Box 3326
Houston, TX 77253 (713) 224-6611
D. Kent Anderson, Chairman of the Board

Description: Small business investment company. Investment Policy: No industry preference.

★1785★ Americap Corporation
7575 San Felipe Ave., No. 160
Houston, TX 77063 (713) 780-8084
James L. Hurn, President

Description: Small business investment company. Investment Policy: No industry preference.

★1786★ Austin Ventures LP
1300 Norwood Tower
114 W. Seventh St.
Austin, TX 78701 (512) 479-0055
Kenneth P. DeAngelis, General Partner

Description: Austin Ventures administers investments through two funds: Austin Ventures LP, and Rust Ventures LP. Investment Policy: Prefers to invest in start-up/emerging growth companies located in the Southwest and in special situations such as buy outs, acquisitions, and mature companies. No geographic limitations are placed on later-stage investments. Past investments have been made in broadcasting, cable television, data communications, factory automation, lodging, telecommunications, and general manufacturing. Investments range from $1 to $2 million for start-up financing and from $2 to $5 million for special situation companies.

★1787★ Brittany Capital Company
1525 Elm St.
2424 LTV Tower
Dallas, TX 75201 (214) 954-1515
Robert E. Clements, General Partner

Description: Small business investment company. Investment Policy: No industry preference.

★1788★ Capital Marketing Corporation
100 Nat Gibbs Dr.
P.O. Box 1000
Keller, TX 76248 (817) 656-7309
Ray Ballard, General Manager

Description: Small business investment company. Investment Policy: Retail grocery industry preferred.

★1789★ Capital Southwest Corporation
12900 Preston Rd., Suite 700
Dallas, TX 75230 (214) 233-8242
William R. Thomas, President

Description: Small business investment company. Investment Policy: No industry preference.

★1790★ Charter Venture Group, Inc.
2600 Citadel Plaza Dr., Suite 600
Houston, TX 77008 (713) 863-0704
Winston C. Davis, President

Description: Small business investment company. Investment Policy: No industry preference.

★1791★ Chen's Financial Group, Inc.
1616 W. Loop, S., Suite 200
Houston, TX 77027 (713) 850-0879
Samuel S. C. Chen, President

Description: Minority enterprise small business investment company. Investment Policy: No industry preference.

★1792★ Citicorp Venture Capital Ltd. (Dallas)
717 N. Harwood, Suite 2920-LB 87
Dallas, TX 75221 (214) 880-9670
Thomas F. McWilliams

Description: Small business investment company. Investment Policy: No industry preference. Remarks: A branch office of Citicorp Venture Capital Ltd. in New York, New York.

★1793★ Criterion Investments
1000 Louisiana St., Suite 6200
Houston, TX 77002 (713) 751-2400
David O. Wick, Jr., President

Description: Venture capital fund. Investment Policy: Criterion Investments raises venture capital, seeking companies headquartered in the Sunbelt region.

★1794★ Dougery, Jones and Wilder (Dallas)
Two Lincoln Centre, Suite 1100
5420 LBJ Fwy.
Dallas, TX 75240 (214) 960-0077

Description: Venture capital supplier. Investment Policy: Prefers to invest in small to medium-sized companies headquartered or primarily operating in the West and Southwest. The typical company sought is in the computers/communications or medical/biotechnological industries; is privately owned; has an experienced management team; serves growth markets; and can achieve at least $30 million in sales. Initial investment ranges from $250,000 to $2.5 million and can be expanded by including other venture capital firms. Remarks: An affiliated office of Dougery, Jones & Wilder in Mountain View, California. **Contact:** John R. Dougery; David A. Jones; Henry L.B. Wilder; Gerald R. Shoonhoven; A. Lawson Howard.

★1795★ Energy Assets, Inc.
4900 Republic Bank Center
Houston, TX 77002 (713) 236-9999
Laurence E. Simmons, Executive Vice-President

Description: Small business investment company. Investment Policy: Energy industry preferred.

★1796★ Energy Capital Corporation
953 Esperson Bldg.
Houston, TX 77002 (713) 236-0006
Herbert F. Poyner, President

Description: Small business investment company. Investment Policy: Specializes in oil and gas energy industries.

★1797★ Enterprise Capital Corporation
4543 Post Oak Place, #130
Houston, TX 77027 (713) 621-9444
Fred Zeidman, President

Description: Small business investment company. Investment Policy: No industry preference.

★1798★ Evergreen Capital Company, Inc.
8502 Tybor Dr., Suite 201
Houston, TX 77074 (713) 778-9770
Shen-Lim Lin, President

Description: Minority enterprise small business investment company. Investment Policy: Inquire.

★1799★ FCA Investment Company
3000 Post Oak, Suite 1790
Houston, TX 77056 (713) 965-0047
Robert S. Baker, Jr., Chairman

Description: Small business investment company. Investment Policy: No industry preference; no start-ups.

★1800★ Grocers Small Business Investment Corporation
3131 E. Holcombe Blvd.
Houston, TX 77221 (713) 747-7913
Milton E. Levit, President

Description: Small business investment company. Investment Policy: Retail grocery industry preferred.

★1801★ Idanta Partners
201 Main St., Suite 3200
Fort Worth, TX 76102 (817) 338-2020

Description: Venture capital partnership. Investment Policy: Provides start-up and second-stage financing; will also invest in special situations such as leveraged buy outs. Minimum investment is $250,000. David J. Dunn, Managing Partner; Dev Purkayastha, General Partner.

★1802★ Livingston Capital Ltd.
P.O. Box 2507
Houston, TX 77252 (713) 872-3213
J. Livingston Kosberg, General Partner

Description: Small business investment company. Investment Policy: No industry preference.

★1803★ Mapleleaf Capital Corporation
55 Waugh, Suite 710
Houston, TX 77007 (713) 880-4494
Edward Fink, Managing General Partner

Description: Small business investment company. Investment Policy: No industry preference.

★1804★ May Financial Corporation
3302 Southland Center
Dallas, TX 75201 (214) 922-0070
Jack McGuire, Head of Corporate Financing

Description: Brokerage firm working with a venture capital firm. Investment Policy: Prefers food, oil and gas, and electronics industries.

★1805★ MESBIC Financial Corporation of Dallas
12655 N. Central Exwy., Suite 814
Dallas, TX 75243 (214) 991-1597
Ira D. Harrison, Vice-President

Description: Minority enterprise small business investment company. Investment Policy: No industry preference.

★1806★ MESBIC Financial Corporation of Houston
811 Rusk, Suite 201
Houston, TX 77002 (713) 228-8321
Lynn H. Miller, President

Description: Minority enterprise small business investment company. Investment Policy: No industry preference.

★1807★ MESBIC of San Antonio, Inc.
2300 W. Commerce, Suite 300
San Antonio, TX 78207 (512) 224-0909
Domingo Bueno, President

Description: Minority enterprise small business investment company. Investment Policy: No industry preference.

★1808★ Minority Enterprise Funding, Inc.
17300 El Camino Real, Suite 107B
Houston, TX 77058 (713) 488-4919
Frederick C. Chang, President

Description: Minority enterprise small business investment company. Investment Policy: No industry preference.

★1809★ Mventure Corp
P.O. Box 662090
Dallas, TX 75266-2090 (214) 939-3131
J. Wayne Gaylord, President

Description: Small business investment company. Investment Policy: Diversified; no start-ups.

★1810★ Omego Capital Corporation
P.O. Box 2173
Beaumont, TX 77704 (409) 832-0221
Theodric E. Moor, Jr., President

Description: Small business investment company. Investment Policy: No industry preference.

★1811★ Red River Ventures, Inc.
777 E. 15th St.
Plano, TX 75074 (214) 422-4999
J. R. Heard

Description: Investment Policy: Inquire.

★1812★ Republic Venture Group, Inc.
P.O. Box 655961
Dallas, TX 75265-5964 (214) 922-5078

Description: Small business investment company. Investment Policy: Prefers to provide second- and third-stage financing (rather than initial investments) to companies in the Southwest, although there is no geographic limitation within the United States. Investments range from $500,000 to $2.5 million. **Contact:** Robert H. Wellborn, President; Brad E. Wiese, Investment Officer.

★1813★ Retzloff Capital Corporation
P.O. Box 41250
Houston, TX 77240-1250 (713) 466-4690
Steven F. Retzloff, President

Description: Small business investment company. Investment Policy: No industry preference.

★1814★ San Antonio Venture Group, Inc.
2300 W. Commerce St., Suite 300
San Antonio, TX 78207 (512) 223-3633
Domingo Bueno, President

Description: Small business investment company. Investment Policy: No industry preference.

★1815★ SBI Capital Corporation
P.O. Box 771668
Houston, TX 77215 (713) 975-1188
William E. Wright, President

Description: Small business investment company. Investment Policy: No industry preference; Texas businesses only.

★1816★ South Texas SBIC
One O'Connor Plaza
Victoria, TX 77902 (512) 573-5151
Kenneth Vickers, President

Description: Small business investment company. Investment Policy: No industry preference.

★1817★ Southern Orient Capital Corporation
2419 Fannin, Suite 200
Houston, TX 77002 (713) 225-3369
Min H. Liang, Chairman of the Board

Description: Minority enterprise small business investment company. Investment Policy: No industry preference.

★1818★ Southwest Enterprise Associates
Two Lincoln Centre, Suite 1266
5420 LBJ Fwy.
Dallas, TX 75240 (214) 991-1620
C. Vincent Prothro

Description: Venture capital supplier. Investment Policy: Concentrates on technology-based industries that have the potential for product innovation, rapid growth, and high profit margins. Investments range from $250,000 to $1.5 million. Past investments have been made in the following industries: computer software, medical and life sciences, computers and peripherals, communications, semiconductors, and defense electronics. Management must demonstrate intimate knowledge of its marketplace and have a well-defined strategy for achieving strong market penetration.

★1819★ Southwest Venture Partnerships
300 Convent, Suite 1400
San Antonio, TX 78205 (512) 227-1010

Description: Venture capital partnership. Investment Policy: Invests in maturing companies located primarily in the Southwest. Invests in$1 million range.

★1820★ Southwestern Venture Capital of Texas, Inc.
1336 E. Court St.
P.O. Box 1719
Seguin, TX 78155 (512) 379-0380
James A. Bettersworth, President

Description: Investment Policy: No industry preference.

★1821★ SRB Partners Fund Ltd.
Sevin Rosen Management Company
Two Galleria Tower, Suite 1670
Dallas, TX 75240 (214) 960-1744
Jon W. Bayless, General Partner

Description: Venture capital firm. Investment Policy: Prefers early-stage financing in high-technology, medical, and biotechnology industries.

★1822★ Sunwestern Capital Corporation
Three Forest Plaza
12221 Merit Dr., Suite 1300
Dallas, TX 75251 (214) 239-5650
Thomas W. Wright, President

Description: Small business investment company. Investment Policy: No industry preference.

★1823★ Tenneco Ventures, Inc.
1010 Milam, 29th Floor, Suite 2919
P.O. Box 2511
Houston, TX 77001 (713) 757-8776
Richard L. Wambold, President

Description: Venture capital supplier. Investment Policy: Provides financing to small, early-stage growth companies. Areas of interest include energy-related technologies, factory automation, biotechnology, and health care services. Prefers to invest in Texas-based companies, but will consider investments elsewhere within the United States. Investments range from $250,000 to $1 million; will commit additional funds over several rounds of financing and will work with other investors to provide larger financing. Remarks: Tenneco Ventures is a subsidiary of Tenneco, Inc., which is headquartered in Houston, Texas.

★1824★ Texas Commerce Investment Company
P.O. Box 2558
Houston, TX 77252 (713) 236-5332
Fred Lummis, Vice-President

Description: Small business investment company. Investment Policy: No industry preference.

★1825★ United Oriental Capital Corporation
908 Town and Country Blvd., Suite 310
Houston, TX 77024 (713) 461-3909
Don J. Wang, President

Description: Minority enterprise small business investment company. Investment Policy: No industry preference.

★1826★ Wesbanc Ventures Ltd.
2401 Fountainview, Suite 950
Houston, TX 77057 (713) 977-7421
Stuart Schube, General Partner

Description: Small business investment company. Investment Policy: No industry preference.

Virginia

★1827★ Atlantic Venture Partners (Alexandria)
801 N. Fairfax St.
Alexandria, VA 22314 (703) 548-6026
Wallace L. Bennett, President

Description: Private venture capital partnership. Investment Policy: Provides all stages of financing primarily to high-technology businesses. Initial investments generally range from $250,000 to $750,000. Under certain circumstances, an investment of less than $100,000 or more than $1 million will be undertaken. May provide financing of $1 to $10 million (or more) in participation with other venture capital firms. Remarks: Maintains an office in Richmond, Virginia.

★1828★ Atlantic Venture Partners (Richmond)
P.O. Box 1493
Richmond, VA 23212 (804) 644-5496
Robert H. Pratt, Chairman

Description: Private venture capital partnership. Investment Policy: Provides all stages of financing primarily to high-technology businesses. Initial investments generally range from $250,000 to $750,000. Under certain circumstances, an investment of less than $100,000 or more than $1 million will be undertaken. May provide financing of $1 to $10 million (or more) in participation with other venture capital firms. Remarks: A branch office of Atlantic Venture Partners, Alexandria, Virginia.

★1829★ Basic Investment Corporation
6723 Whittier Ave.
McLean, VA 22101 (703) 356-4300
Frank F. Luwis, President

Description: Minority enterprise small business investment company. Investment Policy: No industry preference.

★1830★ Crestar Capital
Nine S. 12th St., Third Floor
Richmond, VA 23219 (804) 643-7358
A. Hugh Ewing, Managing General Partner

Description: Small business investment company. Investment Policy: No industry preference.

★1831★ James River Capital Associates
P.O. Box 1776
Richmond, VA 23214 (804) 643-7323
A. Hugh Ewing, Managing Partner

Description: Small business investment company. Investment Policy: No industry preference.

★1832★ Metropolitan Capital Corporation
2550 Huntington Ave.
Alexandria, VA 22303 (703) 960-4698
S. W. Austin, Vice-President

Description: Small business investment company. Investment Policy: Equity or loans with equity features. No retail or real estate.

★1833★ River Capital Corporation (Alexandria)
1033 N. Fairfax, Suite 306
Alexandria, VA 22314 (703) 739-2100
Carl L. Schmitz, Vice-President

Description: Small business investment company. Investment Policy: No industry preference. Remarks: Maintains an office in Cleveland, Ohio.

★1834★ Sovran Funding Corporation
Sovran Center, Sixth Floor
One Commercial Place
Norfolk, VA 23510 (804) 441-4041
David A. King, Jr., President

Description: Small business investment company. Investment Policy: No industry preference.

★1835★ Tidewater Industrial Capital Corporation
United Virginia Bank Bldg., Suite 1424
Norfolk, VA 23510 (804) 622-1501
Armand Caplan, President

Description: Small business investment company. Investment Policy: No industry preference.

★1836★ Tidewater Small Business Investment Corporation (TBBIC)
1214 First Virginia Bank Tower
101 St. Paul's Blvd.
Norfolk, VA 23510 (804) 627-2315
Diane Newell, Manager

Description: Small business investment company. Investment Policy: Prefers manufacturing, distribution, and service industries.

Washington

★1837★ Cable and Howse Ventures (Bellevue)
777 108th Ave., Suite 4300
Bellevue, WA 98004 (206) 646-3030

Description: Venture capital investor. Investment Policy: Provides start-up and early-stage financing to enterprises in the western United States, although a national perspective is maintained. Interest lies in proprietary or patentable technology. Investments range from $50,000 to $2 million. Remarks: Cable and Howse Ventures manages five limited partnerships: CH Partners, CH Partners II, CH Partners III, CH Partners IV, and CH Transition Fund. Other offices are maintained in Portland, Oregon, and Palo Alto, California. **Contact:** Thomas J. Cable; Elwood D. Howse, Jr., Michael A. Ellison; L. Barton Alexander; Gregory H. Turnbull; Wayne C. Wagner, General Partners.

★1838★ Capital Resource Corporation
1001 Logan Bldg.
Seattle, WA 98101 (206) 623-6550
T. Evans Wyckoff, President

Description: Small business investment company. Investment Policy: No industry preference.

★1839★ Palms and Company, Inc.
6702 139 Ave., N.E.
Redmond, WA 98052 (206) 883-3580
Peter J. Palms, President

Description: Private venture capital supplier and investment banker. Investment Policy: Provides all stages of financing to companies located anywhere in the United States and Canada. Investments range from $100,000 to $10 million.

★1840★ Washington Trust Equity Corporation
Washington Trust Financial Center
P.O. Box 2127
Spokane, WA 99210 (509) 455-3821
John M. Snead, President

Description: Small business investment company. Investment Policy: No industry preference.

Wisconsin

★1841★ Bando McGlocklin Capital Corporaton
13555 Bishops Ct., Suite 205
Brookfield, WI 53005 (414) 784-9010
George Schonath, Chief Executive Officer

Description: Small business investment company. Investment Policy: Fixed assets-based lender.

★1842★ Capital Investments, Inc.
Commerce Bldg., Suite 540
Milwaukee, WI 53203 (414) 273-6560
Robert L. Banner, Vice-president

Description: Small business investment company. Investment Policy: No industry preference.

★1843★ Future Value Ventures, Inc.
622 N. Water St., Suite 500
Milwaukee, WI 53202 (414) 278-0377
William P. Beckett, President

Description: Small business investment company. Investment Policy: Provides financing to companies owned by socially and economically disadvantaged persons. Managers should have proven management ability. Prefers businesses with potential to create jobs. Flexible regarding industry preference.

★1844★ InvestAmerica Venture Group, Inc. (Milwaukee)
600 E. Mason St.
Milwaukee, WI 53202 (414) 276-3839

Description: Small business investment company. Investment Policy: Prefers second-stage and leveraged buy out investments of $250,000 to $500,000. Will not consider real estate or retail sales. Remarks: A branch office of InvestAmerica Venture Group, Inc., Cedar Rapids, Iowa.

★1845★ Lubar and Company, Inc.
3060 First Wisconsin Center
Milwaukee, WI 53202 (414) 291-9000
Sheldon B. Lubar, President

Description: Private investment and management firm. Investment Policy: Interest is in energy-related products and services, manufacturers of industrial products, and unique or niche businesses. Remarks: Manages the Wisconsin Venture Capital Fund, Inc., which provides financing to Wisconsin-based businesses. Typical investments range from $250,000 to $750,000.

★1846★ M & I Ventures Corporation
770 N. Water St.
Milwaukee, WI 53202 (414) 765-7910
John T. Byrnes, President

Description: Small business investment company. Investment Policy: Manufacturing, distribution, electronics, and technology-related industries. Approximately $50 million is available for investments.

★1847★ Marine Venture Capital, Inc.
111 E. Wisconsin Ave.
Milwaukee, WI 53202 (414) 765-2274
H. Wayne Foreman, President

Description: Small business investment company. Investment Policy: No industry preference. Remarks: Marine Venture Capital is a subsidiary of a commercial bank.

★1848★ Super Market Investors, Inc.
P.O. Box 473
Milwaukee, WI 53201 (414) 547-7999
David H. Maass, President

Description: Small business investment company. Investment Policy: Retail grocery industry preferred.

★1849★ Twin Ports Capital Company
1230 Poplar Ave.
Superior, WI 54880 (715) 392-8131
Paul Leonidas, President

Description: Small business investment company. Investment Policy: Grocery store industry preferred.

★1850★ Venture Investors of Wisconsin, Inc.
100 State St.
Madison, WI 53703 (608) 256-8185

Description: Venture capital firm. Investment Policy: No industry preference. **Contact:** Roger H. Ganser, President; John Neis, Vice-President.

★1851★ Wisconsin Community Capital, Inc.
14 W. Mifflin St., Suite 314
Madison, WI 53703 (608) 256-3441
Paul J. Eble, President

Description: Small business investment company. Investment Policy: No industry preference.

Wyoming

★1852★ Capital Corporation of Wyoming, Inc.
145 S. Durbin St.
P.O. Box 3599
Casper, WY 82602 (307) 234-5351
Scott Weaver, Manager

Description: Small business investment company. Investment Policy: No industry preference.

CHAPTER 8

FEDERAL FUNDING SOURCES FOR THE INVENTOR

The Federal government funds almost one-half of all U.S. research and development. More than one billion dollars each year goes to small companies for R&D through direct awards. But pursuing it is not for everyone. Tapping the pot of green at the end of Uncle Sam's red, white, and blue rainbow takes a major commitment of both time and money. And unless you are willing to tolerate a cumbersome and inordinate amount of paper work, red tape, and costly bureaucratic excesses and delays, this exercise may not be for you.

On the other hand, there appears to be something for everyone. Washington's interest ranges from innovations in astronomy, earth, environmental, and marine sciences to chemistry, physics, engineering, and materials sciences. It awards money for breakthroughs in robotics, genetics, plastics, adhesives, fusion, optics, and artificial intelligence, to name just a few categories.

And you do not have to be a Fortune 500 diversified manufacturer to qualify for such financing. Under the Small Business Innovation Development Act, Public Law 97-219, each agency with an extramural R&D budget in excess of $100 million must establish a **Small Business Innovation Research (SBIR) Program.** R&D firms from 1 to 500 employees are encouraged to submit proposals. The following agencies are currently participating in the SBIR program.

Department of Agriculture
Department of Commerce
Department of Defense
Department of Education
Department of Energy
Department of Health and Human Services
Department of Transportation
Environmental Protection Agency
National Aeronautics and Space Administration
National Science Foundation
Nuclear Regulatory Commission

SMALL BUSINESS ADMINISTRATION

Each agency makes its own awards using contracts, grants, or co-operative agreements. But the Small Business Administration (SBA) is charged with the formulation and issues policy direction for government-wide SBIR programs. For further information contact:

Raymond Marchakitus
Small Business Administration
SBIR
1441 L Street, N.W.
Premier Building #414
Washington, DC 20416

There are ten SBA Regional Offices that can also provide you with information on the SBIR program and what's happening at any time throughout the participating agencies and departments. **For a complete listing of local SBA offices, see Chapter 5, Regional Small Business and State Innovation Programs.**

SMALL BUSINESS ANSWER DESK: SBA operates a toll free number that is also available to you. Dial 1-800-368-5855

DEPARTMENTS OF COMMERCE AND ENERGY

NIST Expands Invention Evaluation and Funding

The U.S. Department of Commerce's National Institute of Standards and Technology (NIST) (formerly National Bureau of Standards) has been recently reorganized. Its new name reflects a broadened role and area of responsibility. In addition to its traditional functions of providing the measurements, calibrations, data, and quality assurance support which are vital to U.S. commerce and industry, NIST has a new purpose: "to assist industry in the development of technology and procedures needed to improve quality, to modernize manufacturing processes, to ensure product reliability, manufacturability, functionality, and cost-effective-ness, and to facilitate the more rapid commercialization ...of products based on new scientific discoveries."

The Department of Commerce is in the process of centralizing all new technology work. In December of 1988, Commerce Secretary C. William Verity announced the appointment of Ernest Ambler as Under Secretary for Technology (Acting) to head Commerce's new Technology Administration.

"Today's global market realities demand not only the creation of new technologies - at which America continues to excel - but rapid and continuous transfer of these technologies to new products," Verity said. "The undersecretary for technology will serve as a strategic catalyst to promote the use of science and technology by industry and entrepreneurs and be both a listening post and a new voice for business in Washington on technology."

NIST has been directed to develop four new major programs that will be of great interest to independent inventors. Its mandate is to enhance the country's technological competitiveness through the fast and effective transfer of new technologies to U.S. industries.

NIST will accomplish this through:

1) Regional Centers for the Transfer of Manufacturing Technology;

2) Assistance to state technology programs;

3) The Advanced Technology Program; and

4) Clearinghouse for State Technology Programs.

This reorganization will be underway throughout 1989. Funding will be worked out by the 101st Congress.

Although much of the information on these new programs is still sketchy, it has been included to give readers a jump on events that will be unfolding throughout 1989-90. If you are interested in the status of any particular program, contact the designated U.S. government officials.

As of now the Regional Center program has received partial funding. The others will be implemented as resources become available.

What are Regional Centers for the Transfer of Manufacturing Technology?

These regional centers are intended to provide direct support to small- and medium-sized manufacturing firms in automating and modernizing their facilities. If you are an inventor who has his/her own manufacturing facility, NIST could help you develop technically and financially sound plans for modernizing production runs.

NIST is creating these centers in partnership with nonprofit organizations established by state and local governments, universities, or companies. NIST is authorized to provide up to 50 percent (limited to $3 million annually) of the operating funds for these centers for their first three years.

The first three organizations that will establish regional manufacturing technology centers are: The Cleveland Advanced Manufacturing Program (CAMP) in Cleveland, Ohio; Rensselaer Polytechnic Institute in Troy, New York; and the University of South Carolina in Columbia, South Carolina.

For updated information on the Regional Centers, contact Dr. John Lyons, Director of National Engineering Laboratory, NIST, Gaithersburg, Maryland 20899. Telephone (301) 975-2300.

What Assistance is Intended for State Technology Programs?

NIST is taking an active role in promoting the transfer of federal technology through the many existing sate and local technology extension services. Such state and local programs usually highlight business advice rather than inventions. The NIST program will help to coordinate the state and local extension services with Federal technology transfer programs.

For updated information on the advanced technology program and extension services, contact: Dr. Donald Johnson, Director, National Measurement Laboratory, NIST, Gaithersburg, Maryland 20899.

What is Going on in Nonenergy-Related Invention Development?

The most exciting part of this activity for the independent inventor will be the technical evaluation of promising inventions that are **not** energy related—a program modeled after NIST's successful 12 year old program for energy-related inventions managed by NIST for the Department of Energy. Please see below for information on the Office of Energy-Related Inventions and its evaluation and funding program.

Let me again say that this part of the program still has no funding. But it will and for updates, contact: George P. Lewett, Chief, Office of Energy-Related Inventions, NIST, Gaithersburg, Maryland 20899 (301) 975-5500.

What is the Advanced Technology Program?

NIST will provide small amounts of money to leverage private investment in specific projects to develop new products and processes. Candidate projects would include, for example, promising inventions evaluated under the program mentioned above.

What is the Clearinghouse for State Technology Programs?

This is a shared assignment with the U.S. Department of Commerce. And it could be helpful to inventors seeking to get information on state technology policies.

ENERGY-RELATED INVENTIONS PROGRAM (ERIP)

"We don't have the NIH (Not Invented Here) Syndrome rattling around here," says Don Corrigan, Operations Group Leader at the National Institute of Standards and Technology (formerly National Bureau of Standards) Office of Energy-Related Inventions (OERI).

The National Institute of Standards and Technology (NIST) and the Department of Energy (DOE), under provisions of the Federal Non-nuclear Energy Research and Development Act of 1974 (P.L. 93-577), have combined to offer a marvelous opportunity to inventors of energy-related concepts, devices, products, materials, or industrial processes. Called the Energy-Related Inventions Program (ERIP), it is designed as a process to discover and assist the development of worthy inventions which might otherwise never be commercialized.

From an NIST facility in Gaithersburg, Maryland, just outside Washington, D.C., off I-270, the program provides a chance for independent inventors and small businesses with promising energy-related inventions to obtain Federal assistance in the development and commercialization of their inventions.

Since its inception OERI has received more than 25,000 evaluation requests, accepted about half of these for evaluation, and recommended 440 for U.S. Government support. Approximately $21.5 million in grants or contracts have been awarded to date on some 290 recommended inventions; the remainder are either in process (94) or have received other than financial assistance.

OERI is under the experienced leadership of George P. Lewett. "We currently review about 200 inventions a month for evaluation and we recommend, fairly steadily, 3-4 inventions per month," he said. And the system works.

Ask Harrison Robert Woolworth, inventor of a mechanism to pre-heat scrap metal with waste heat. He submitted his invention to OERI in 1975 and in 1977 was awarded $170,000. by DOE.

Ask Phillip Zacuto and Daniel Ben-Shmuel, co-inventors of a heat extractor. They submitted their invention to OERI in 1977 and by 1978 had landed a DOE award in the amount of $125,000.

Ask Maurice W. Lee, inventor of a new way of cooking hamburgers using the dielectric resistance characteristic of meat. He submitted his invention to OERI in 1984 and in 1987 received an award of $75,000 from DOE.

Ask F.J. Perhats and James V. Enright, inventors of a device which uses an automobile engine's heat to heat the car's interior after the engine has been turned off. They received $71,034 through the Department of Energy.

Ask Albert B. Csonka, inventor of a new kind of micro-carburetor which is claimed to be fuel-saving and pollution-reducing. He was awarded $193,500 to build a working prototype fit to a late model, standard 350 cubic inch V-8 engine.

Ask Donald E. Wise, inventor of a convertible flat/drop trailer. He received a grant of $63,069 to build and test his concept. Then, according to DOE, he successfully licensed his technology to Trail King Company of Nebraska.

Ask Karakian Bedrosian, inventor of a method of preserving fruits and vegetables without refrigeration. It only took him one year to win a DOE award of $97,300. Marketed under the trade name "TomAHtoes, DOE reports that" 751,000 25-pound boxes were shipped in 1987 to the tune of $35 million in retail sales.

Inventions Covered by OERI

Invention disclosures evaluated by OERI are categorized as follows:

Fossil Fuel Production
Direct Solar
Other Natural Sources
Combustion Engines and Components
Transportation Systems, Vehicles, and Components
Building, Structures, and Components
Industrial Processes
Miscellaneous

You Keep the Rights. Uncle Sam Keeps Your Secret

It is the Government's policy that inventors retain patent rights to their inventions, and that these rights are not compromised by the OERI evaluation.

OERI maintains the confidentiality of invention disclosures submitted by limiting access to them. All Federal Government personnel involved in the program sign a statement advising them of penalties for disclosure or misuse of information under 18 U.S.C. 1905. Non-government consultants also sign an agreement of non-disclosure which includes agreement not to undertake invention reviews that may subject the consultant to a conflict of interest.

OERI Submission Process

Submissions are handled by the Office of Energy-Related Inventions, U.S. Department of Commerce, Gaithersburg, Maryland 20899 (301) 975-5500.

The disclosure of an invention should include information required by NIST Form 1019 **(See Fig. 28, Evaluation of an Energy-Related Invention)**, but the format may vary widely depending on a number of factors. Here are some suggestions that you should consider in preparing the description of the invention to be submitted with the Evaluation Request Form 1019.

Make a Complete Disclosure. The principal requirement in submitting a request is a thorough and complete invention disclosure which describes the invention in detail. It is most important to submit ALL information which is available even if the method of presentation and organization is not professional in nature. Test data and information on how tests were conducted are particularly critical, when applicable, since **no testing will be done by NIST as part of its evaluation.**

Emphasize the Energy Relation. The program is interested in all energy-related inventions including both those that involve energy conservation and those that involve alternative sources of energy. This means everything from new methods of recovery to drill bits. Your disclosure should emphasize and document, to the extent possible, the amount of energy saved or made available through an alternate source.

It is only after the invention reaches the commercialization stage that its ultimate contribution to the solution of our energy problem can be realized. It is not necessary to calculate energy savings exactly, but the potential should be indicated.

Time to Process Your Request. Do not expect an immediate response to your request for evaluation. The evaluation process is time-consuming and there are a large number of submittals to consider. Remember that submission of an invention to NIST for evaluation is no guarantee that it will be recommended to the Department of Energy (DOE), and a recommendation is no guarantee that your invention will be accepted for a DOE grant award.

Describe Your Competition. Make an effort to find out if there are any similar products on the market. Detail the known competition and document why your innovation is superior technically or from an energy or economic standpoint.

Give the Status of Your Invention. Address the question of what needs to be done to bring your invention closer to fruition. Spell out what you would expect from the U.S. Government in the event of a favorable evaluation.

Highlight the Innovation. As you can imagine, NIST has seen all kinds of concepts involving common devices, such as windmills, wave machines, furnaces, carburetors, internal combustion engines, and space heaters. And like so many ideas, many of those submitted are neither new nor innovative. Be sure, therefore, to point out and highlight novel principles or innovations that make yours different, particularly if it falls into a common device class.

Be Factual, Realistic. Your proposal will be evaluated by savvy technical and business-oriented professionals. Prepare your disclosure with that in mind and do not make outrageous claims that cannot be justified or substantiated by data or information in your disclosure.

The Evaluation Process

The OERI evaluation process, under the direction of Howard Robb, is the formal procedure that will determine, in an objective way, whether the invention you have submitted is appropriate for recommendation to the Department of Energy (DOE) for support. OERI considers an invention appropriate for DOE consideration if it is technically feasible, will offer a favorable impact on the energy situation, and holds potential for commercial success. The focus is on your technology; your credentials or capabilities are not a factor in OERI decision making.

Each disclosure is logged in and tracked through a computerized system that records the number of days spent in each stage of the five-step evaluation process and the number of days required to obtain the opinions of government consultants.

1) **Disclosure Acknowledgment.** OERI will immediately acknowledge by letter the receipt of each disclosure received. Every disclosure must be accompanied by a signed evaluation request, NIST Form 1019. **(Fig. 28).**

2) **Disclosure Review & Analysis.** OERI will review the invention disclosures to determine whether they are complete, readable, technically sufficient, and within the scope of its program. Inventors are notified by letter whether their disclosure is accepted for first-stage evaluation or is rejected. This is, in effect, a preliminary screening.

3) **First-Stage Evaluation.** In this initial screening process, OERI will determine whether an invention is promising enough to warrant an in-depth analysis. OERI evaluators typically seek brief expert opinions (invention reviews) independently from two consultants before deciding whether to place the invention into second-stage evaluation or to reject it.

The consultants are asked to comply with the following rules:

a) Do not transmit information on the invention disclosure to anyone who has not signed an NIST Agreement of Nondisclosure **(Fig.35).**

b) Do not copy the disclosure.

c) Do not write on or otherwise mark the disclosure.

d) Return the complete disclosure with the invention report to OERI by certified mail or other secure transmittal method.

OERI evaluators and their consultants are expected to be liberal and to highlight the positive aspects of an invention rather than seek reasons for rejection. The inventor is notified by letter of said decision. The actual evaluation paper is not made available, but a summary of its finding is communicated to the inventor.

4) **Second-Stage Evaluation.** OERI will determine whether to recommend an invention to DOE for support. OERI evaluators obtain a deep analysis of the invention from a consultant before deciding whether to recommend the invention to DOE or to reject it. The consultant, to the extent possible, will try to provide information and references to pertinent literature that may be of assistance to the OERI evaluator and the inventor. The inventor is notified by letter of this decision too, in either case, and is sent a copy of the second-stage consultant's analysis.

5) **Recommendation to DOE.** OERI will prepare a transmittal document which includes all pertinent material submitted by the inventor, the second-stage consultant's in-depth analysis, any other pertinent material obtained or developed by OERI during the evaluation, and an OERI evaluator's report outlining the reason for recommending the invention and the appropriate next step to DOE. Inventors are provided a copy of this document and the name of an OERI co-ordinator whom they may contact.

At each stage in the evaluation process, you may present additional information, which is fully considered before further decisions are made. If your invention is rejected at any stage, you may submit additional information addressing the problems raised by OERI and request re-evaluation. This "Open Appeal" feature is considered an intrinsic part of the evaluation process.

"We'll reconsider them from now until Doomsday," says Don Corrigan of OERI. "Fifteen to twenty percent of the inventions we recommend to DOE have been turned down at some point in the evaluation process."

The evaluation process is primarily a selection process which does not necessarily include analysis of every aspect of an invention. **OERI will provide you with any useful information it obtains or develops during the evaluation process of your innovation.** The brief first-stage invention reviews made by consultants are not released, but information contained in them may be included in OERI's correspondence with you.

Department of Energy

Upon receipt of a recommendation from OERI, an Invention Co-ordinator from DOE's **Inventions and Innovation Programs Division** analyzes it and opens negotiations on the type and amount of support (if any) to be provided. A letter is written to the inventor.The inventor's response to this letter is treated as a preliminary proposal by DOE.

The OERI recommendation, the inventor's preliminary proposal, and other information available to the Invention Co-ordinator (such as information concerning other DOE projects in related fields) are considered. Also taken into account by the official are the inventor's business/management capabilities and the resources available to the inventor. The Invention Co-ordinator then decides whether federal support is warranted and, if so, what type of support should be provided.

DOE support under this program may consist of a grant award, a contract award, testing service at a DOE facility, guidance in obtaining private venture capital, or other types of assistance.

In practice, each case is handled on an individual basis. There is no standard operational procedure. Each case is unique.

Inventors, especially those in the early stages of development, often require an amount of money that exceeds DOE's authority and guidelines. Grants have ranged from $20,000 to $200,000, averaging about $70,000, with most granted to support technical research, scientific testing, or business planning. DOE cannot support marketing efforts, although it can help inventors obtain market information. As the program has evolved, the DOE staff, under the direction of Anthony J. "Jack" Vitullo has sought to fine-tune financial support to the needs and abilities of individual inventors and their technologies, and to expand the range of non-financial services.

Among the many benefits the Program provides to independent inventors is **credibility**. The NIST evaluation has served many inventors as a kind of "Good Housekeeping Seal of Approval" that they converted into a bankable asset. It was reported that one inventor parlayed the NIST evaluation into an endorsement by, and financial support from, a state agency, then went on to build a business that sold $32 million worth of his product. Another testified that he "couldn't get a venture capitalist to talk to him" until he had the NIST evaluation and DOE support as credentials.

When asked about the benefits of this NIST/DOE combined program, many inventors cited the gain in "credibility" as important as the money awards.

Inquires may be directed care of U.S. Department of Energy, CE-12, Washington, D.C. 20585 (202) 586-1478.

DEPARTMENT OF DEFENSE

In my opinion, Department of Defense (DoD) presents one of the best opportunities for an individual or small firm to land federal R&D funds and/or sell innovative technologies. The Pentagon has an active and wide ranging **Small Business Innovation Research Program** (SBIR) which makes special efforts to identify creative individuals, small businesses, and small disadvantaged businesses with R&D capabilities and/or better ideas. DoD not only has an interest in soliciting business, but it also entertains outside queries and submissions.

Individuals and firms with strong R&D capabilities in science and engineering and other disciplines are encouraged to participate. DoD's SBIR Program goals include stimulating technological innovation in the private sector, strengthening the role of small business in meeting DoD research and development needs, fostering and encouraging participation by minority and disadvantaged persons in technological innovation, and increasing the commercial application of DoD-supported research and development results.

How can you get current DoD R&D information?

There are several ways to approach DoD under the SBIR Program.

a) You can go directly to the different departments to present your concepts and/or to receive information on what they need in terms of outside assistance. The names and addresses of same are provided in this Chapter.

b) Each year the Department of Defense, Office of Small Business Innovation Research, conducts seminars around the country at which it presents "wish lists" and explains the SBIR program. To get general information on the where and when of the seminars contact:

Bob Wrenn, SBIR Coordinator
OSD/SADBU
U.S. Department of Defense
The Pentagon - Room 2A340
Washington, D.C. 20301-3061
(202) 697-1481

"The amazing thing about these SBIR meetings is that I sit there and meet this endless stream of awesomely bright people," says Richard Sparks, program manager for the **Defense Technical Information Center** (DTIC).

To show how active the DoD program is, consider the Fiscal Year '88 numbers as an example. Out of the 8,625 proposals received, 1,011 were identified as of September 1st as offering the greatest potential in their field for meeting the research and development needs of DoD, and were subjected to further evaluations and negotiations leading to contract awards.

There is no charge to register for the SBIR Program and receive its DoD program solicitations. The one for FY-1989 weighed just under two pounds! The best ways to get your name on the list to receive the next program solicitation is via the nearest Small Business Administration (SBA) office. They are listed in your telephone directory under U. S. Government. Or you may wish to contact:

Defense Technical Information Center,
Att: DTIC-SBIR, Alexandria, VA 22304-6145
(202) 274-6902.

c) The Defense Technical Information Center (DTIC) is a great place to learn what the Pentagon is looking for in terms of technologies and gadgets. "We exist to help the independent inventors and small businesses cut through the red tape and bureaucracy," says a most helpful and hospitable Richard Sparks, program manager for DTIC and the pointman on the front line with outside queries. He boasts the only toll free number in the Pentagon (800-368-5211). And he answers his own phone! There are no flacks to block your access.

DTIC is the central storehouse of scientific and technical information resulting from and describing R&D projects that are funded by the Department of Defense. It has on file 1.8 million technical reports.

DTIC users include university libraries, special libraries, DoD agencies, other U.S. Government agencies ranging from the U.S. Coast Guard to the CIA, commercial companies like Union Carbide, General Electric, IBM, independent inventors, and students.

During the FY '88 DTIC provided 2,629 small business requesters approximately 30,000 technical information packages and 10,310 technical reports on the 787 DoD solicitation topics. Additionally, DTIC distributed 11,589 copies of the solicitation documents to small R&D businesses.

URI Program

DTIC information is also available via University Research Initiative (URI), a basic DoD research program. Eighty-six universities throughout the U.S.A. currently participate in the program which provides free access to DTIC products and services.

Terminal Access to DTIC

You may gain DTIC access via the Defense RDT&E Online System (DROLS) by using one of a wide variety of available terminals. DROLS can communicate with any terminal (CRT or typewriter) which employs the standard AASCII asynchronous protocol. Terminal communications speeds are 300, 1200, or 2400 baud (30,120, or 240 characters per second) in even parity. Access may be also gained by using TYMNET commercial data communications network.

This service is limited to unclassified access only. Users requiring classified access will be required to use specialized UNIVAC terminals with dedicated telephone lines.

* There is a charge of $30.00 per connect hour or proportionate share.

* Subscribers to this service must have a deposit account with the National Technical Information Service.

* Users will **not** be charged for time they input technical data into the DTIC databases.

If you are interested in this service, contact:

Defense Technical Information Center
ATTN.: Online Support Office (DTIC-BLD)
Bldg. 5, Cameron Station
Alexandria, VA 22304-6145
(202) 274-7709
AUTOVON 284-7709

What Do DTIC Services Cost?

Most of DTIC services are free to registered users. The only charges are for paper copy documents, which cost $5.00 for 1-100 pages and $.07 for each additional page over 100, and microfiche copies, which are $.95 per demand document and $.35 per document supplied under the Automatic Document Distribution.

How Can You Become A Registered DTIC User?

You can receive an information packet that describes DTIC's services and products and instructions on registration if you contact:

Defense Technical Information Center
ATTN.: Registration and Services Section (DTIC-FDRB)
Bldg. 5, Cameron Station
Alexandria, VA 22304-6145
(202) 274-6871
AUTOVON 284-6871

Mr. Sparks says that each year near the end of August, military departments begin to publish SBIR Requests for Proposals (RFPs) in the "Commerce Business Daily" and mail out formal Program Solicitations. Then beginning on/about October 1st for the next thirteen weeks, he and his **Defense Technical Information Centers** are "open for business" (see list below) to assist people wishing to compete for research and development grants or present concepts.

While "officially" he only has the budget to handle outside queries during this time period, Mr. Sparks is a very warm, knowledgeable, and helpful individual who will never turn away an inventor in need of guidance. "You never know what you'll find," he says. "One that fascinated me most was a three mile-long magnetic gun for putting things into orbit at low cost without expending fuel. These guys were a couple of wiz kids!"

You may call or visit at the following locations.

Defense Technical Information Center
ATTN.: DTIC-SBIR
Building 5, Cameron Station
Alexandria, VA 22304-6145
(800) 368-5211 (Toll Free)
(202) 274-6902 (Commercial for VA, AK, HI)

DTIC Boston On-Line Service Facility
DTIC-BOS
Building 1103, Hanscom AFB
Bedford, MA 01731-5000
(617) 377-2413

DTIC Albuquerque Regional Office
AFWL/SUL Bldg. 419
Kirtland AFB, NM 87117-6008
(505) 846-6797

DTIC Los Angeles On-Line Service Facility
Defense Contract Administration Services Region
222 North Sepulveda Blvd.
El Segundo, CA 90245-4320
(213) 355-4170

GUIDE TO UNSOLICITED PROPOSALS TO DOD

You do not have to follow a particular format for the submission of proposals. However, a proposal should, at a minimum, cover the points set forth below in the order indicated. Elaborate proposals or presentations are neither necessary nor desirable. Two copies of your proposal should be submitted.

The DoD follows established competitive procurement procedures for awarding contracts. In other cases it may issue statements of interest or similar notices to announce various program opportunities in such publications as the *Commerce Business Daily.*

DoD also awards contracts based upon unsolicited proposals.

Competitive Proposals

Normally, competitive procurements will be initiated by the issuance of a formal request for proposal (RFP). The RFP will contain instructions telling you how to prepare the proposal.

Unsolicited Proposals

A formal unsolicited proposal should be in the form of a detailed document signed by the inventor. This document forms the basis for further technical evaluations and for possible contract negotiations. Make sure each proposal includes the following elements as appropriate:

1) Cover Sheet
2) Abstract
3) Narrative
4) Cost Proposal (should be a document that is separate from original proposal)

Cover Sheet. Each proposal must have a cover sheet providing the information set forth in the sample form entitled **Small Business Innovation Research (SBIR) Program Proposal Cover Sheet shown in Fig. 30.**

Abstract. Each proposal should start with a narrative in which you describe the relevance of your invention to the DoD mission. Your personal qualifications should also stated. **A sample Small Business Innovation Research (SBIR) Program Proposal Summary is offered for your use in Fig. 31.**

The Department of Defense is primarily interested in technical competence. The people who'll evaluate your proposals want to know that you understand what the project involves and that you can perform the work required. Once your technical competence has been established, DoD will consider costs.

Be sure to specifically address these points:

1) **Purpose and Objective.** State briefly the primary purpose, general objective, and expected results of the proposal, such as:

a) State the problem or problems your invention will contribute to solving, and the anticipated contribution of any research to that solution.

b) Enumerate the specific objective of any additional research and specify the questions that your research will attempt to answer. This will be important if you hope for DoD moneys to finance further research.

c) State the expected consequences of successful completion of your research, including potential economic and other benefits.

d) Discuss the existing interest by potential beneficiaries or users.

2) **Previous or Ongoing Related Work.** Document any knowledge you may have of other related activities with appropriate references to the literature or currently ongoing R&D. In particular, show how your proposed project relates to these activities and how your invention/research will extend the level of knowledge in the field.

3. **Statement of Work.** If your proposal involves additional R&D, give full and complete technical details of the procedures that will be followed throughout the scope of your work. Outline it phase by phase. Give time frames and objectives.

4. **Organization, Facilities and Qualifications.** Fully describe who you are, if individual, or what your company does.

6. **Patents.** If you believe you are presenting a patentable idea, I suggest that you file, as a protection to yourself and to the government, necessary patent applications with the Patent and Trademark Office, or otherwise identify your intention to do this. Mark your proposal as containing proprietary data.

Cost Estimate

Give a detailed estimate of all costs. To help you formalize this, I am including a **Defense Small Business Innovation Research (SBIR) Program Cost Proposal (Fig. 29).** Take from it those parts that are most appropriate to your project and situation.

Proprietary Data

If you do not want your invention or idea disclosed to the public or used by the government for any other purpose than proposal evaluation, clearly label it as containing proprietary data. Don't just do it on the cover sheet, but rather on every page that holds such information.

Note that this restriction does not limit the government's right to use or disclose any data contained in your proposal if it is obtainable from another source (or from yourself previously) without restriction.

Notification of DoD Action

It may take up to six months for you to get an answer! Be patient. Your proposal will be acknowledged upon its receipt.

Security

Always submit unclassified material when possible. When this is not possible, make sure you properly label documents as being classified. Ask the Pentagon for a copy of its Industrial Security Manual for Safeguarding Classified Information (DoD 5220.22-M). Paragraph 11 of this publication will tell you how to correctly classify your submissions.

Around the Federal Triangle

I have conscientiously read through all federal research and development activity descriptions and selected for this chapter only those agencies and departments that are potential funding sources to independent inventors. I have omitted those agencies and departments whose interest is more in reports and studies than in patentable inventions and innovative concepts. Many federal R&D budgets are applied towards scientific research programs and policy research and analysis studies and the like and, thus, inappropriate.

Because locating and obtaining federal R&D funding requires a significant commitment of time and resources, the advice of NASA Headquarters Small Business Specialist Mark Kilkenny should be remembered. He recommends that you concentrate on the important few and not waste time and paperwork on the unimportant many.

IMPORTANT!! You should critically evaluate yourself before you submit a proposal for a government contract. And you should be equally critical in evaluating any idea you submit. The federal government is interested in only the most experienced people with pertinent novel ideas. Don't waste its time or your own on half-baked ideas or proposals for contracts that you are not qualified to carry out.

If you feel your invention would be of interest to a particular agency or that an agency appears to be funding research and development in your field of expertise, get in touch.

If the system starts to frustrate you either in the finding of appropriate government officials or the the way in which you are being handled by the government officials you find, don't hesitate a moment to enlist the assistance of your elected representatives on Capitol Hill. You'll be amazed at how quickly a call from a U.S. Senator's or Representative's office can make things happen.

Chapter Arrangement

This chapter is arranged alphabetically by department, followed by an alphabetic listing of programs. Organizations and programs this chapter are referenced in the Master Index. U.S. Small Business Administration offices are listed in Chapter 5, Regional Small Business & State Innovation Programs.

National Aeronautics and Space Administration

★1853★ Ames Research Center (ARC)

Moffett Field, CA 94035 (415) 694-4044
Lawrence Milov, External Relations Officer

Description: Research and development programs in aeronautics, life sciences, space science, space technology, and flight research. Other efforts are aimed at human factors in space, earth resources study, thermal protection systems for atmosphere entry vehicles, computational physics and chemistry, and artificial intelligence and autonomous system. **Publications:** NASA Technical Reports..

★1854★ Dryden Flight Research Facility

P.O. Box 273
Edwards Air Force Base, CA 93523 (415) 694-5800
Ms. Carolyn Anderson, Small Business Specialist

Description: Manned flight research, including systems, configuration, and technology developments for high performance aircraft. **Publications:** X-Press Magazine (biweekly).

★1855★ George C. Marshall Space Flight Center

Huntsville, AL 35813 (205) 453-2675
Conrad Walker, Small Business Specialist

Description: Serves as a primary center for the design and development of space transportation systems, elements of the Space Station, scientific applications payloads, and other systems for present and future space exploration.

★1856★ Goddard Space Flight Center (GSFC)

Procurement Analysis Branch
Code 202.3
Greenbelt, MD 20771 (301) 286-5417
Franz W. Hoffman, Small Business Specialist

Description: Research interests include orbital spacecraft development, tracking and data acquisition systems, space physics and astronomy payloads, upper atmospheric research, weather and climate, earth dynamics and resources, information systems, sounding rocket and payload development, planetary science, and sensors and experiments in environmental monitoring and ocean dynamics. Maintains an automated source system comprised of approximately 4,000 firms. The Center encourages businesses to submit a Bidders Mailing List Application (SF129) to be added to this system.

★1857★ Headquarters Contracts and Grants Division

600 Independence Ave.
Washington, DC 20546 (202) 453-2090
Stuart J. Evans, Assistant Administrator

Description: Responsible for planning, negotiating, awarding, and administering contracts based on procurement requirements, including system engineering services, reliability studies, basic and applied research and development, mobile lecture-demonstration units, exhibits, motion picture services, management analysis surveys and automatic data processing equipment, and software services. The Division also has agency-wide responsibility for negotiating and executing NASA contracts with foreign governments and commercial organizations. **Contact:** Mark E. Kilkenny, Procurement System Branch Chief (202) 453-1840; Eugene D. Rosen, Director, Small and Disadvantaged Business Utilization (202) 453-2088; James T. Rose, Assistant Administrator, Commercial Programs Office (202) 453-1123.

★1858★ Jet Propulsion Laboratory (JPL)

California Institute of Technology
4800 Oak Grove Drive
Pasadena, CA 91109 (818) 354-5722
Margo Kuhn, Small Business Specialist

Description: Has the primary responsibility of exploration of the solar system with unmanned spacecraft. Research and development activities include aeronautics and aerospace, communications, computer science and mathematics, earth and space sciences, electronics, and physics. **Publications:** (1) Annual Report, Highlights; (2) Closeup. **Contact:** Mr. T. May, Minority Business Specialist (818) 354-2121.

★1859★ John F. Kennedy Space Center

Kennedy Space Center, FL 32899 (305) 867-7353
Norman Perry, Small Business Specialist

Description: Responsible for assembly, checkout, servicing, launch, recovery, and operational support of space transportation system elements. Principle research activity is the design and development of launch and landing facilities. Technology transfer programs are administered by the Center's Advanced Projects and Technology Office. **Publications:** NASA Tech Briefs.

★1860★ Langley Research Center

Hampton, VA 23365 (804) 865-3751
Norma Patrick, Small Business Specialist

Description: Research and development in aeronautics and space technology, including acoustics and noise reduction, aerodynamics, aerospace vehicle structures and materials, aerothermodynamics, avionics technology, environmental quality monitoring technology, sensor and data acquisition technology, long-haul aircraft, military support, and advanced space vehicle configuration. Activities also include technology transfer and providing development support to government agencies, industry, and other NASA centers.

★1861★ Lewis Research Center

21000 Brookpark Road
Cleveland, OH 44135 (216) 433-4000
Mr. J. Liwosz, Jr., Small Business Officer

Description: Responsible for managing and design and development of the power generation, storage, and distribution system for the U.S. space station. Areas of research include air breathing and space propulsion systems; turbomachinery thermodynamics and aerodynamics; fuels and combustion; aero and space propulsion systems; power transmission; tribology; internal engine computational fluid dynamics; high temperature engine instrumentation, space communications; and space and terrestrial energy processes, systems technology, and applications.

National Aeronautics and Space Administration *(Cont.)*

★1862★ Lyndon B. Johnson Space Center (JSC)
Houston, TX 77058 (713) 483-4511
Mr. R.L. Duppstadt, Small Business Specialist

Description: Primary mission is the development of spacecraft for manned space flight programs and the conduct of manned flight operations; selection and training of astronauts; providing management in systems engineering and integration and business and operations management for the Space Shuttle Program; conducting investigations of earth resources technology; and conducting space life sciences research.

★1863★ National Space Technology Laboratories (NSTL)
NSTL, MS 39529 (601) 688-1636
David Anderson, Small Business Specialist

Description: Primary installation for static test firing of large rocket engines and propulsion systems. Beside NASA, resident agencies include NOAA, and departments of Army, Navy, and Interior; U.S. Coast Guard; as well as groups from various universities. These groups are scientific base for technology exchange. **Contact:** Mr. P.B. Higdon, Contracting Officer, Earth Resources Laboratory (601) 688-1641.

★1864★ Office of Space Science and Applications (OSSA)
OSSA Steering Committee
Code EP-3
Washington, DC 20546 (202) 453-1409
Leonard A. Fisk, Associate Administrator

Description: Operationally responsible for scientific research into the nature and origin of the universe and applying space systems and techniques to solve problems of Earth.

★1865★ Small Business Innovation Research Program
Washington, DC 20546 (202) 453-2848
Harry Johnson, Manager

Description: Areas of soliciting proposals are based on the needs of NASA's programs and missions as described by the roles of each NASA Center.

National Institutes of Health

★1866★ National Heart, Lung, and Blood Institute (NHLBI)
Contracts Operations Branch
Westwood Bldg., Room 650
Bethesda, MD 20892 (301) 496-7666
Robert Carlsen, Chief

Description: Supports basic and clinical research, development, and other activities related to the prevention, diagnosis, and treatment of cardiovascular, lung, and blood diseases. The Institute plans and directs research in the development, trial, and evaluaiton of drugs and devices relating to the prevention such diseases.

National Science Foundation

★1867★ Division of Policy Research and Analysis
1800 G St., N.W., Room 1233
Washington, DC 20550 (202) 357-9689
Dr. Peter House, Division Director

Description: Funds policy research, including business firms specializing in science and innovation policy and technology and resource policy.

★1868★ Industry/University Cooperative Research Centers
Room 1121
1800 G St., N.W.
Washington, DC 20550 (202) 357-7307
Dr. Morris Ojalvo, Program Manager

Description: Fosters university-industry interaction at university centers by supporting a limited number of research projects. **Contact:** Alex Schwarzkopf, Manager.

★1869★ Office of Small Business Research and Development
1800 G St., N.W., Room 1250
Washington, DC 20550 (202) 357-9666
Dr. Donald Senich, Director

Description: Provides information and serves as a referral point for small businesses interested in the Foundation's research or procurement opportunities, or how to submit proposals. Also compiles and publishes information on research grants and contracts awarded by NSF to small business. **Contact:** Linda L. Boutchyard, Program Specialist (202) 357-7464.

★1870★ Small Business Innovation Research Program
1800 G St., N.W., Room 1250
Washington, DC 20550 (202) 357-7527
Roland Tibbetts, Program Manager

Description: Acts as the principal NSF research opportunity organization for small businesses. **Contact:** Mr. Ritchie Coryell, Program Manager (202) 357-7527.

U.S. Department of Agriculture

★1871★ Agricultural Research Service
Bldg. 005, BARC-W
Beltsville, MD 28705 (301) 344-3084
Dr. Gary R. Evans, Deputy Admin.

Description: Research interests include physical, biological, and chemical engineering; clothing, housing, and household economics; industrial and food products and processing methods for agricultural commodities; and soil and water, crop, animal husbandry, entomology, agricultural engineering, and energy. Published solicitations are not made by the office, but proposals are received at any time.

★1872★ Office of Grants and Program Systems (OGPS)
CSRS - USDA
14th & Independence Ave., S.W.
Room 324-A Administration
Washington, DC 20250 (202) 475-5720
Dr. William D. Carlson

Description: Maintains Competitive Research Grants Office, which publishes its solicitations in the Federal Register. Grants are awarded on a competitive basis. Proposals are requested in the areas of biological nitrogen fixation, biological stress on plants, and human nutrition.

★1873★ Small Business Innovation Research Program
901 D St., S.W.
Room 323-J
Washington, DC 20251 (202) 475-7002
Dr. Charles Cleland

U.S. Department of Commerce

★1874★ National Institute of Standards and Technology (NIST)
Building 301, Room B128
Gaithersburg, MD 20855 (301) 975-6353
Keith Chandler, Small Business Specialist

Description: NIST is organized into three technical laboratories as follows: (1) National Engineering Laboratory (NEL) furnishes technology and technical services to public and private sectors to address national needs and to solve problems in the public interest. NEL conducts research in engineering and applied science; builds and maintains competence in the necessary disciplines required to carry out this research; and develops engineering data and measurement capabilities. (2) National Measurement Laboratory (NML) provides the national system of physical, chemical, and materials measurement; coordinates the system with measurement systems of other nations, and furnishes services leading to uniform measurement; and conducts research leading to methods of measurement, standards, and data on the properties of materials. (3) National Computer and Telecommunication Laboratory (NCTL), which develops computer standards, conducts research, and provides scientific and technical services in the selection, acquisition, application, and use of computer technology; manages a government-wide program for standards development and use, including management of federal participation in voluntary standardization.

★1875★ National Institute of Standards and Technology, Office of Energy Related Inventions (OERI)
Gaithersburg, MD 20899 (301) 975-5500
George L. Lewett

Description: Invention disclosures evaluated by OERI are categorized as follows: fossil fuel production; direct solar; other natural resources; combustion engines and components; transportation systems, vehicles, and components; building, structures, and components; industrial processes, and miscellaneous. (For further information, see introduction to this chapter.)

★1876★ National Oceanic and Atmospheric Administration (NOAA)
NBOC-1, Room 205
11420 Rockville Pike
Rockville, MD 20852 (301) 443-8584
Randy Linderholm, Small Business Specialist

Description: Research and development interests in environmental monitoring and prediction include automation of meteorological observations, analysis, and communication; remote sensing; oceanic monitoring; stratospheric measurement; satellite technology and development of ground equipment; weather modification; and modeling of atmospheric and oceanic processes, including climate variation. **Contact:** Edward Tiernan, SBIR Program Manager Chief, NOAA/NESDIS, Orta FB4, Room 3316, Suitland, MD 20233 (202) 763-2418.

U.S. Department of Defense

★1877★ Defense Nuclear Agency (DNA)
Office of Contracts
Washington, DC 20305 (202) 325-7658
Mrs. Patricia Brooks, Director, Business Utilization

Description: Seeks research and development from small business firms with strong capabilities in the nuclear weapons effects, including simulation, instrumentation, directed energy, nuclear hardening and survivabilty, security of nuclear weapons, and operational planning. **Contact:** Small Business Innovation Research Program, Attn: OAAM/SBIR, Jim Gerding (202) 325-7018.

U.S. Department of Energy

★1878★ Assistant Secretary for Conservation and Renewable Energy
1000 Independence Ave., S.W.
Washington, DC 20585 (202) 586-9275
Dr. Robert L. SanMartin

Description: Programs in conservation and renewable energy support research that attracts limited or no venuture capital because the risks are too high or the payoff too long-range or unpredictable.

★1879★ Assistant Secretary for Defense Programs
1000 Independence Ave., S.W.
Washington, DC 20585 (202) 252-1870
Annie Smith

Description: Directs the DOE's programs for nuclear weapons research, development, testing, production, and surveillance; manages programs for the production of the special materials used by the weapons program within the Department; and manages the defense nuclear waste and by-products program.

★1880★ Assistant Secretary for Nuclear Energy
1000 Independence Ave., S.W.
Washington, DC 20585 (202) 353-4380
A.S. Lyman

Description: Responsible for programs and projects for nuclear power generation and fuel technology; evaluation of alternative reactor fuel cycle concepts; development of naval nuclear propulsion plants and cores; and nuclear waste technology and remedial action programs. Unsolicited proposals for projects in nuclear energy should be sent to the Unsolicited Proposal Coordinator, Reports and Analysis Branch, Procurement and Assistance Management Directorate, Department of Energy, Washington, DC 20585.

★1881★ Morgantown Energy Technology Center (METC)
3610 Collins Ferry Rd.
P.O. Box 880
Morgantown, WV 26505-0880 (304) 291-4764
Augustine Petrolo, Director

Description: Fossil fuel research and development laboratory responsible for projects in unconventional gas recovery, fluidized bed combustion, gas stream cleanup, fuel cells, heat engines, component development for coal conversion and utilization devices, surface coal gasification, instrumentation and control, oilshale technology, tar sands, underground goal gasification, and arctic and offshore technologies. **Telecommunication Access:** Alternate telephone (304) 599-4511.

U.S. Department of Energy *(Cont.)*

★1882★ Office of Energy Research (OER)
100 Independence Ave.
Washington, DC 20585 (301) 586-5430
Robert Hunter, Jr., Director-Designate

Description: Manages research programs in the basic energy sciences, high energy physics, and fusion energy; administers DOE programs supporting university researchers; funds research in mathematical and computational sciences critical to the use and development of supercomputers; and administers a financial support program for research and development not funded elsewhere in the Department. The Office also manages the Small Business Innovation Research Program for the Department. **Contact:** Jean Marrow (301) 353-5544.

★1883★ Pittsburgh Energy Technology Center (PETC)
P.O. Box 10940
Pittsburgh, PA 15236 (412) 675-6400
Dr. Sun W. Chun, Director

Description: Primary mission is to promote the use of coal and its derived synthetic fuels in an environmentally sound manner. The Center is also responsible for handling all unsolicited research proposals pertaining to any part of the DOE's fossil energy program. **Publications:** (1) PETC Quarterly Technical Progress Report; (2) D.P. Databeat; and (3) Energizer (all quarterly). **Telecommunication Access:** Alternate telephone (412) 892-6128.

★1884★ Small Business Innovation Research Program (SBIR)
Mail Stop ER-16, GTN
Washington, DC 20545 (301) 353-5544
Jean Marrow, Contact

Description: Stimulates technological innovation; uses small business to meet federal research and development needs; fosters and encourages participation by minority and disadvantaged persons in technologcial innovation; and increases private secor commercializiaiton of innovations derived from federal research and development. **Telecommunication Access:** Alternate telephone (301) 353-5867.

U.S. Department of Health and Human Services

★1885★ National Cancer Institute (NCI)
Westwood Bldg., Room 10A10
Bethesda, MD 20205 (301) 496-5583
Dr. Vincent T. DeVita, Jr., Director

Description: Plans, directs, conducts, and coordinates a national research program on the detectin, diagnosis, cause, prevention, treatment, and palliation of cancers. Conducts and directs research performed in its own laboratories and through contracts; supports construction of facilities necessary for research on cancer; supports demonstration projects on cancer control; and collaborates with cancer research with industrial concerns. **Contact:** Louis P. Greenberg, Acting Chief, Research and Contracts Branch.

★1886★ National Eye Institute
Research Contract Branch
Division of Contracts and Grants
Bldg. 31, Room 1B44
9000 Rockville Pike
Bethesda, MD 20892 (301) 496-4487
Dave Snight

Description: Supports research on the cause, natural history, prevention, diagnosis, and treatment of disorders of the eye and the visual system, especially glaucoma, retinal disease, corneal diseases, cataract, and sensory-motor disorders. Support is provided for basic studies, clinical trials, and the development of animal models for vision disorders. **Publications:** Planning Report (every five years). **Meetings:** Grand Rounds (weekly) and Neuro-Ophthalmology Seminar (monthly).

★1887★ National Institute of Allergy and Infectious Diseases
Westwood Bldg., Room 707
5333 Westbard Ave.
Bethesda, MD 20205 (301) 496-7116
Lew Pollack, Contract Management Branch

Description: Supports research on causes, diagnosis, treatment, and prevention of infections, allergic, and other immunologically mediated diseases. Contract programs are concerned with bacterial and viral vaccines, antiviral substances, transplantation, and viral and allergen reagents.

★1888★ National Institute of Arthritis and Musculoskeletal and Skin Diseases (NIAMS)
NIH Bldg. 31, Room 9A35
9000 Rockville Pike
Bethesda, MD 20892 (310) 496-7111
Patrick Sullivan

Description: Conducts and supports basic and clinical research on arthritis, osteoarthritis, lupus, muscle disease, psoriasis, acne, ichthyosis, and vitiligo. Offers investigator-initiated grants, reseasrch center grants, individual and institutional research training research awards, career development awards, and contracts to public and private research institutions and organizations. **Contact:** Lawrence E. Shulman, Director (301) 496-4353.

★1889★ National Institute of Dental Research (NIDR)
Contract Management Section
Westwood Bldg., Room 521
Bethesda, MD 20892 (301) 496-7311
Ms. Marion Blevins

Description: Supports investigations into the causes, means of prevention, diagnosis, and treatment of oral diseases, including dental caries, periodontal diseases, lesions of soft and hard tissue, and oral-facial abnormalities. Also supports biomaterials research, improvements in anesthesia and analgesia, basic epidemiological research, and clinical trials.

★1890★ National Institute of General Medical Sciences (NIGMS)
Research Contract Branch
Division of Contracts and Grants
Bldg. 31, Room 1B44
9000 Rockville Pike
Bethesda, MD 20892 (301) 496-4487
David Snight

Description: Supports research in the sciences basic to medicine, behavioral sciences, and clinical disciplines. The Institute fosters multidisciplinary approaches to research and employs the full range of support mechanisms, including research project grants, program-project grants, research center grants, career development awards, awards to new investigators, and institutional and individual fellowships.

★1891★ National Institute of Neurological and Communicative Disorders and Stroke (NINCDS)
Contract Management Branch
Federal Bldg., Room 901
7550 Wisconsin Ave.
Bethesda, MD 20892 (301) 496-9203
Lawrence Fitzgerald

Description: Conducts, coordinates, and supports research concerned with the cause, development, diagnosis, therapy, and prevention of disorders and diseases of the central nervous system and the communicative and sensory systems. Research includes neurological and mental development, infectious diseases, multiple sclerosis, epilepsy, head injury and stroke, biomedical engineering and instrumentation, and neural prostheses.

★1892★ Office of Human Development Services (HDS)
Room 324B-HHS North Bldg.
200 Independence Ave., S.W.
Washington, DC 20201 (202) 245-1787
Cynthia Haile Selassie, Small Business Specialist

Description: Oversees programs that contribute to the social and economic well-being of a number of vulnerable populations, including the elderly, children, youth and families, the developmentally disabled, and Native Americans.

U.S. Department of the Air Force

★1893★ Air Force Systems Command
Room 206, Bldg. 1535
Andrews Air Force Base
Washington, DC 20334 (301) 981-6107
Dave McNabb, Executive for Small Business

Description: Command includes project divisions that acquire aircraft, missiles, space systems, electronic systems, and conventional weapons; test centers; laboratories that conduct research in the mathematical, physical, engineering, and environmental sciences; and divisions that are involved in medical research and technology, identification of foreign technology, and contract management.

★1894★ Air Force Systems Command, Aeronautical Systems Division (Eglin Air Force Base)
Eglin Air Force Base, FL 32542-5000 (904) 882-2843
Ralph K. Frangioni, Jr.

Description: Armament systems research and development, testing, and procurement, including total program management responsibility for Air Force non-nuclear munitions, including guided bombs, mines, fuzes, flares, bomb racks, missiles, aerial targets, and munition handling and transportation equipment.

★1895★ Air Force Systems Command, Aeronautical Systems Division (Wright-Patterson Air Force Base)
Wright-Patterson Air Force Base
Dayton, OH 45433 (513) 255-5322
Tom Dickman, Small Business Specialist

Description: Development and acquisition of aeronautical systems, including aircraft engines, aircraft wheels and brakes, airborne communication systems, aircraft bombing and navigation systems, and aircraft instruments, and aeronautical reconnaissance systems, and mobile land-based tactical information processing and interpretaton facilities. **Contact:** Jim Beach, Small and Disadvantaged Business Utilization Specialist.

★1896★ Air Force Systems Command, Space Division
Los Angeles Air Force Station
P.O. Box 92960
Worldway Postal Center
Los Angeles, CA 90009 (213) 643-2855
Chuck Willeme, Deputy for Small Business

Description: Acquisition and evaluation of space systems and equipment; management of research, tracking, telemetry, and recovery; feasibility studies; acquisition, production, quality assurance, and installation of assigned space and missile systems; and quality assurance and installation of boosters, and aerospace ground equipment to support launch control and recovery.

★1897★ Arnold Engineering Development Center (AEDC)
Arnold Air Force Base, TN 37389-5000 (615) 454-4407
William Lamb, Small Business Specialist

Description: Conducts development, certification, and qualification testing on aircraft, missile, and space systems. Also conducts research and technology programs to develop advanced testing techniques and instrumentation and to support the design of new test facilities. **Publications:** AEDC Test Highlights.

★1898★ Ballistic Missile Office
Attn: BMO/BC
Norton Air Force Base
San Bernardino, CA 92409 (714) 382-2304
Mr. Terry Carey, Chief, Small Business Office

Description: Responsible for intercontinental ballistic missile development. Office integrates activities of contractors, retaining overall engineering responsibility for a particular system.

★1899★ Contract Management Division
Kirtland Air Force Base, NM 87117-5000 (505) 844-6644

Description: Acquisition contract management agency for the Air Force. Administers contracts and provides an interface with contractors.

★1900★ Defense Advanced Research Projects Agency (DARPA)
1400 Wilson Boulevard
Arlington, VA 22209 (202) 694-1440
Dr. Robert C. Duncan, Director

Description: Pursues imaginative and innovative research ideas and concepts offering significant military utility. Role in basic research is to develop selected new ideas from conception to hardware prototype for transfer to development agencies. Programs are conducted through contracts with industrial, university, and nonprofit organizations, focussing on improved strategic, conventional, rapid-development, and sea-power forces; and on scientific investigation into technologies for the future. Unsolicited proposals will be accepted if they are related to any of the current technical programs assigned to DARPA. DARPA invites small businesses to submit proposals under its SBIR program, which has as its objectives stimulating technological innovation in the private sector, strengthening the needs, fostering and encouraging participation by minority and disadvantaged persons in technological innovation, and increasing commerical application of Department of Defense-supported research and development. **Contact:** Bud Durand, SBIR Progam Manager (202) 694-1626.

U.S. Department of the Air Force *(Cont.)*

★1901★ Directorate of Research and Development Procurement

Aeronautical Systems Division
Wright-Patterson Air Force Base
Dayton, OH 45433 (513) 255-3825
George R. Laudenslayer, Small Business Specialist

Description: Procurement of exploratory and advanced research and development in the areas of air breathing, electric, and advanced propulsion; fuels and lubricants; and power gerneration, including molecular electronics, bionics, lasers, vehicle environment, photo materials, position and motion sensing devices, navigation, communications, flight dynamics, materials science, and life support. **Contact:** Ms. Dorothy Muhlhauser, Small and Disadvantaged Business Utilization Specialist.

★1902★ Eastern Space and Missile Center (ESMC)

Patrick Air Force Base, FL 32935-5000 (305) 494-2207
Don Hoskins, Small Business Specialist

Description: Launches the Titan III space booster and represents Department of Defense interests in Space Shuttle operations. Research, development, and procurement program is confined primarily to range test instrumentation, including radar, telemetry, electro-optics, impact locations, data reduction, range/mission safety and control, weather timing and fireing, and frequency control and analysis. **Telecommunication Access:** Alternate telephone (305) 494-2208.

★1903★ Electronic Systems Division (ESD)

Hanscom Air Force Base, MA 01730-5000 (617) 861-4441
Al Hart

Description: Development, acquisition, and delivery of electronic systems and equipment for the command, control, and communications function of aerospace foreces. **Telecommunication Access:** Alternate telephone (617) 861-4973/4.

★1904★ Flight Test Center (AFFTC)

Edwards Air Force Base, CA 93523 (805) 277-2619
James A. Beucherie

Description: Flight testing of aerospace vehicles. Specific activities include: (1) conduct and support of aircraft systems tests; (2) flight evaluation and recovery of aerospace research vehicles and development testing of aerodynamic decelerators; (3) management and operation of USAF Test Pilot School; (4) management and operation of the Utah Test and Training Range and the Edwards Flight Test Range; and (5) support for and participation in agency, foreign, and contractor test and evaluation programs. **Telecommunication Access:** Alternate telephone (617) 861-4974.

★1905★ Office of Scientific Research (AFOSR)

Building 410
Bolling Air Force Base
Washington, DC 20332 (202) 767-4943
Louise Harrison, Small Business Specialist

Description: Considers basic research proposals from industrial and small business concerns and nonprofit organizations. Accepts unsolicited proposals. Any scientific investigator may make a preliminary inquiry to obtain advice on the degree of interst in a project, or may submit a specific research proposal. Proposals should indicate the field of the investigation and objectives sought, describing previous work and related grants and contract held, if any; in addition, it should outline the approach planned for the research and should include estimates of the time and cost requirements.

★1906★ Operational Test and Evaluation Center (AFOTEC)

Office of Small and Disadvantaged Business Utilization
Air force Contract Management Division
Kirtland Air Force Base, NM 87117-5023 (505) 844-3819
Ms. Rosina Aragon, Chief

Description: Independent agency responsible for testing and evaluation of systems being developed for Air Force and joint service use. Center's focus is on the operation effectiveness and suitability of the Air Force's future weapons and supporting equipment, as well as identifying deficiencies requiring corrective action. **Contact:** Nancy Lindquist, SADBU Specialist.

★1907★ Rome Air Development Center (RADC)

Griffis Air Force Base, NY 13441 (315) 330-4020
Peter Nicotera, Small Business Specialist

Description: Provides a technology base for projects that pertain to command, control, communications, and intelligence research and development, including radar, airborne and space-based reconnaissance and sensing systems, vast arrays of communications systems, and computers. Has also been designated as a lead laboratory for research and development of the technology of photonics.

★1908★ Space Technology Center

Kirtland Air Force Base, NM 87117-6008

Description: Manages three Air Force laboratories, integrating space technology efforts in order to explore military space capabilities and the needs of future space systems. Specific areas of interest include directed energy research, nuclear weapons effects, survivability issues, rocket propulsion, and the earth and space environment.

★1909★ Strategic Defense Initiative Organization (SDIO)

SDIO/OSD
Washington, DC 20301-7100 (202) 693-1527
Carl Nelson, SBIR Coordinator

Description: Supports programs involved with kinetic energy weapons, and surveillance satellites, and directed energy weapons research. **Contact:** Submission of unsolicited proposals should be directed to Col. Kluter, SDIO/OSD, Washington, DC 20301-7100 (202) 653-0034.

★1910★ Weapons Laboaratory (AFWL)

Kirtland Air Force Base, NM 87117-6008 (505) 844-9426

Description: Responsible for reseasrch efforts in the areas of survivable systems and directed energy weapons, including particle beam technology, advanced weapons technology, high power microwave technology, and space nuclear power.

U.S. Department of the Army

★1911★ Aberdeen Proving Ground Installation Support Activity

Attn: STEAP-SB
Aberdeen Proving Ground, MD 21005-5001 (301) 278-3878
Thomas Rodgers, Small Business Advisor

Description: Research and development, production, and post-production testing of components and complete items of weapons, systems, ammunition, and combat and support vehicles. Also tests items of individual equipment in use throughout the Army.

★1912★ Armament Research, Development, and Engineering Center (ARDC)
Attn: AMSMC-SBD (D)
Dover, NJ 07801 (201) 328-4104
Ed Smith, Small Business Advisor

Description: Responsible for management of research, development, life-cycle engineering, and initial acquisition of weapon systems and support equipment.

★1913★ Aviation Systems Command (AVSCOM)
Attn: AMSAV-V
Building 102E, 4300 Goodfellow Blvd.
St. Louis, MO 63120-1798 (314) 263-2200
Charles Robinson, Small Business Advisor

Description: Aviation design, research, development, maintenance, engineering, stock and supply control, and technical assistance to users of all Army aviation and aerial delivery equipment. Evaluates prototype hardware for fueling and defueling equipment for use in combat areas and in solving fuel contamintation problems.

★1914★ Ballistic Missile Defense Systems Command
P.O. Box 1500
Huntsville, AL 35807 (205) 895-3410
Virginia B. Wright

Description: Research and development in the fields of radar, interceptors, optics, discrimination, and data processing applicable to ballistic missile defense, including analysis of new and novel applications of science and engineering seeking revolutionary approaches to ballistic missile defense.

★1915★ Ballistic Research Laboratory (BRL)
ARRADCOM Ballistic Research Laboratories
Building E4455
Aberdeen Proving Ground, MD 21005-5066 (301) 671-2309
Harold Siler, Small Business Specialist

Description: Provides basic and applied research in mathematics, physics, chemistry, biophysics, and the engineering sciences related to the solution of problems in ballistics and vulnerability technology.

★1916★ Belvior Research, Development, and Engineering Center
Attn: STRBE-V
Fort Belvoir, VA 22060-5606 (703) 664-5134
Susan Irwin, Small Business Advisor

Description: Research, development, engineering, and initial production buys in the areas of mobility/countermobility, survivability, energy, and logistics. Research is carried out by the Combat Engineering Directorate, Logistics Support Directorate, and the Materials, Fuels, and Lubricants Directorate. Related support services are provided by the Advanced Systems Concepts Directorate, Information Management Directorate, Product Assurance and Engineering Directorate, and Resource Management Directorate. **Telecommunication Access:** Alternate telephone (703) 644-2482.

★1917★ Benet Weapons Laboratory
Attn: SMCWV-SB
Watervliet, NY 12180-5000 (518) 266-5765
A.V. Paparian, Small Business Advisor

Description: Research, development, engineering, and design of mortars, recoilless rifles, and cannons for tanks and towed and self-propelled artillery. Laboratory is a division of U.S. Army Close Combat Armament Center (Picatinny Arsenal, New Jersey 07806-5000), which publishes patent disclosures for inventions.

★1918★ Chemical Research, Development, and Engineering Center
Building E4455
Aberdeen Proving Ground, MD 21010-5423 (301) 671-2309
Harold Siler

Description: Research and development in the fields of chemical, smoke, and flame weapons. Activities also include studies in pollution abatement and environmental control technology.

★1919★ Cold Regions Research and Engineering Laboratory
P.O. Box 282
Hanover, NH 03755 (603) 646-4324
Raymond F. May, Jr., Small Business Specialist

Description: Conducts research pertaining to characterists and events unique to cold regions, especially winter conditions, including design of facilities, structures and equipment, and refining methods for building, traveling, living, and working in cold environments.

★1920★ Communications-Electronics Command (CECOM)
Attn: DRSEL-SB AMSEL-SB
Fort Monmouth, NJ 07703 (201) 532-4511
John Meschler

Description: Covers full spectrum of services to the U.S. soldier in the field of communications and electronics. The first steps in converting concepts into new military materiel are taken in the CECOM Research and Development Center and its laboratories. These laboratories are responsible for research and development of communications and electronics equipment and systems.

★1921★ Construction Engineering Research Laboratory (CERL)
Procurement/Supply
P.O. Box 4005
Champaign, IL 61820-1305 (217) 352-6511

Description: Mission is to provide research and development to support Army programs in facility construction, operation, and maintenance. Efforts focus on vertical construction as applied to buildings and structures rather than heavy construction. Laboratory has four main divisions: Engineering and Materials Division, Energy Systems Division, Environmental Division, and Facilities Systems Division. Center emphasizes communication and interchange of information with academic, engineering, and construction activities within the Department of Defense, other governmental agencies, and the private sector. **Publications:** CERL Reports (quarterly).

★1922★ Dugway Proving Ground (DPG)
Attn: STEDP-PR
Dugway, UT 84022-5202 (801) 522-2102
Clyde Harris, Small Business Advisor

Description: Conducts field and laboratory tests to evaluate chemical and radiological weapons and defense systems and materiel. Also conducts biological defense research.

U.S. Department of the Army *(Cont.)*

★1923★ Engineer Topographic Laboratories (USAETL)

Fort Belvoir, VA 22060-5546 (202) 355-2659
Mrs. M.L. Williams, Small Business Specialist

Description: Research and development in the topographic sciences, including mapping, charting, terrain analysis, geodesy, remote sensing, point positioning, surveying, and land navigation. Provides scientific and technical advisory services to meet environmental design criteria requirements of military materiel developers and in support of geographic intelligence and land environmental resources inventory requirements. Composed of the following laboratories and operational centers: Geographic Sciences Laboratory, Research Institute, Space Programs Laboratory, Terrain Analysis Center, and Topographic Developments Laboratory.

★1924★ Engineer Waterways Experiment Station (WES)

P.O. Box 631
Vicksburg, MS 39180 (601) 636-3111
Mr. A.J. Breithaupt, Utilization Specialist

Description: Principle research, testing, and development facility of the Corps of Engineers. Operats six laboratories, including Coastal Engineering Research Center, Environmental Laboratory, Geotechnical Laboratory, Hydraulics Laboratory, Information Technology Laboratory, and Structures Laboratory, which conduct research in soil mechanics, concrete, engineering geology, pavements, weapons effects, protective structures, water quality, and dredge materials. **Publications:** List of Publications of the Waterways Experiment Station. **Meetings:** Annual Aquatic Plant Control Research and Operations Review (November), attendance is open to all.

★1925★ Jefferson Proving Ground (JPG)

STEJP-LD-P
Madison, IN 47250-5100 (812) 273-7226
Ms. Mary N. Gassert, Small Business Advisor

Description: Processing, assembling, and acceptance testing of ammunition and ammunition components. Receives, stores, maintains and issues assigned industrial stocks, including calibrated components.

★1926★ Laboratory Command (LABCOM)

Attn: DRDEL-SB
2800 Powder Mill Road
Adelphi, MD 20783-1145 (202) 394-1076
Arthur Wolters, Small Business Advisor

Description: Research, development, engineering, and initial procurement of assigned items in the areas of electronic signal intelligence, electric warfare, atmospheric sciences, target acquisition and combat surveillance, electronic fuzing, radars, sensors, night vision, radar frequency and optical devices, nuclear weapons effects, instrumentation and simulation, and fluidics. Has direct control and management of seven Army laboratories, including Atmospheric Sciences Laboratory, Ballistic Research Laboratory, Electronics Technology and Devices Laboratory, Human Engineering Laboratory, Materials Technology Laboratory, and Army Vulnerability Assessment Laboratory.

★1927★ Materials Technology Laboratory

Attn: DRXMR-AP AMXMR-K
Watertown, MA 02172 (617) 923-5005
Capt. John Rouse, Small Business Advisor Dir.

Description: Manages Army's research and development programs relating to materials technology, solid mechanics, lightweight technology, and manufacturing testing technology. Is a center for excellence in the area of corrosion protection. **Meetings:** Sponsors Sagamore Army Materials Research Conference (annually in August) for invited participants and Military Handbook-17 Coordinating Group Meeting (semiannually, in May/November).

★1928★ Materiel Command (AMC)

Attn: AMCSB
5001 Eisenhower Avenue
Alexandria, VA 22333-0001 (202) 274-8185
Kurt Wussow, Disadvantaged Utilization

Description: Responsible for the life cycle of U.S. Army hardware, including research, development, procurement, production, supply, and maintenance. Provides supervisory, planning, and budetary direction to installations where contracts are executed and administered. **Contact:** John Flakenham, Doris Agnew, Janet Tull. **Telecommunication Access:** Alternate phone number (202) 274-8186.

★1929★ Medical Research and Development Command

Attn: SGRD-SADBU
Fort Detick
Frederick, MD 21701-5012 (301) 663-2744
Mrs. Audrey L. Wolfe, Assistant Director, SADBU

Description: Research involves assessment, prevention, diagnosis, and treatment of infectious diseases that would hamper military operations; disease vector surveillance; combat casualty care; health hazards of military materiel; factors limiting soldier effectiveness; prevention of oral disease; and dental materials. Operates research and development laboratories, including Aeromedical Research Laboratory, Biomedical Research and Development Laboratory, Institute of Dental Research, Institute of Surgical Research, Medical Research Institute of Chemical Defense, Medical Research Institute of Infectious Diseases, Research Institute of Environmental Medicine, and Walter Reed Institute of Research. Also supports medical research through contract research awards. Announcements of specific proposals solicited are published in Commerce Business Daily.

★1930★ Missile Command (MICOM)

Redstone Arsenal
Attn: DRSMI-B
Huntsville, AL 35809 (205) 876-2561
Dr. Joe Plano, Small Bus. Advisor Cmdr.

Description: Responsible for design and development; product, production, and maintenance engineering; and new equipment design of training devices in the areas of rockets, guided missiles, targets, air defense, fire control equipment, test equipment, missile launching and ground support equipment, and metrology and calibration equipment. **Telecommunication Access:** Alternate telephone (205) 876-5441.

★1931★ Natick Research, Development, and Engineering Center

Attn: STRNC-25B
Natick, MA 01760-5008 (617) 633-4995
Ms. Victoria Tangherlini, Small Business Advisor

Description: Research, development, and engineering in advanced systems concepts, aero-mechanical engineering, food engineering, individual protection, and science and advanced technology.

★1932★ Research and Development Directorate

Attn: DEAN-RDZ
20 Massachusetts Ave., N.W.
Washington, DC 20314-1000 (202) 272-0725
George Wishchmann, Dir., Small Bus. Utilization

Description: Serves as scientific advisor to the Chief of Engineers for Research and Development; directs the activities of all Corps laboratories; exercises responsibility for Army technology-based programs in environmental quality; and maintains responsibility for the planning and budgeting of the Corps research and development program and the management of all resources, including military research, development, test, and evaluaiton; civil works research and development appropriations; and mission support funding.

★1933★ Small Business Innovation Research Program (SBIR)

Headquarters Laboratory
Code AMSLC-TP-TI
2800 Powder Mill Road
Adelphi, MD 20783 (301) 394-4602
Ruth Morarre, SBIR Coordinator

Description: Will provide funding for nearly 500 topics in fiscal year 1989.

★1934★ Tank-Automotive Command (TACOM)

Attn: AMSTA-CB
Warren, MI 48297-5000 (313) 574-5388
F.D. Folette, Small Business Officer

Description: Responsible for research, design, development, engineering, test management, modification, product assurance, integrated logistics support, acquisition, and deployment of wheeled and tracked vehicles and associated automotive equipment. Research and development are the functional responsibility of TACOM's Tank-Automotive Research, Development, and Engineering Center. **Contact:** Tom Clynes, Small Business Office (313) 574-5406.

★1935★ Test and Evaluation Command (TECOM)

Attn: AMSTE-PR
Aberdeen Proving Ground, MD 21005-5055 (301) 278-5184
Edward Snodgrass, Small Business Advisor

Description: Directs research activities, proving grounds, installations, boards, and facilities required to test equipment, weapons, and materiel systems intended for use by U.S. Army.

★1936★ Water Resources Support Center

Humphreys Engineering Center
Kingman Bldg.
Telegraph and Leaf Rds.
Fort Belvoir, VA 22060-5580 (202) 355-2153
John Carpenter

Description: Research and development in the topographic sciences; implementation of emergency operations to provide the capabilities to transfer data between civil organizations; and remote sensing, water control data systems, and telecommunications research planning. Administers the Waterborne Commerce Statistics Center, which collects and reports on traffic and tonnage.

★1937★ White Sands Missile Range (WSMR)

Attn: STEWS-PR
White Sands Missile Range, NM 88002-5031 (915) 678-1401
Luis E. Sosa, Small Business Advisor

Description: Conducts testing and evaluation of Army missiles, rockets, warheads, and special weapons. Major directorates include National Range Operations, Army Materiel Test and Evaluation, Instrumentation Directorate, and Directed Energy Directorate. Major tenants are Naval Ordnance Missile Test Station, Air Fore Range Operations Office, U.S. Army Vulnerability Assessment Laboratory, LABCOM, U.S. Army Atmospheric Sciences Laboratory, and U.S. Army TRADOC Analysis Command.

★1938★ Yuma Proving Ground (YPG)

Attn: STEYP-PC
Yuma Proving Ground, AZ 85364-1530 (602) 328-2825
Mr. Loren Cady, Small Business Advisor

Description: Conducts research and development, production, and post-production testing of components and complete items of weapons, systems, ammunition, and combat and support vehicles in desert environments.

U.S. Department of Interior

★1939★ Bureau of Mines

Room 1024
2401 E St., N.W.
Washington, DC 20241 (202) 634-1303
David S. Brown, Acting Director

Description: Conducts research and collects, interprets, and analyzes information involving mineral reserves and the production, consumption, and recycling of mineral materials. **Publications:** (1) Guide for the Submission of Unsolicited Research and Development; (2) Reports of Investigations Information Circulars; (3) Minerals Yearbook; and (4) Mineral Facts and Problems. **Contact:** Chief Mining Engineer (202) 634-1303; Division of Procurement (202) 634-4704.

★1940★ U.S. Geological Survey, Geologic Division

National Center
12201 Sunrise Valley Dr.
Reston, VA 22092 (703) 648-6600
Benjamin A. Morgan, Chief Geologist

Description: Conducts programs to assess energy and mineral resources, identify and predict geologic hazards, and investigate the effects of climate.

U.S. Department of the Navy

★1941★ David Taylor Research Center (DTRC)

Code 003
Bethesda, MD 20084-5000 (202) 227-1220
David Tychnan, Small Business Specialist

Description: Navy's principal research, development, test, and evaluation center for naval vehicles. Research information may be obtained from the National Technical Information Service, 5285 Port Royal Road, Springfield, VA 22161.

★1942★ Marine Corps Headquarters

U.S. Marine Corps (LS)
Washington, DC 20380 (202) 694-1939
Mrs. Sheila D'Agostino, Small Business Representative

Description: Electronic equipment, specialized vehicles, and equipment peculiar to the Marine Corps.

★1943★ Naval Air Development Center (NADC)

Code 094
Street Road
Warminster, PA 18974-5000 (215) 441-2456
Ms. Janet Koch (Code 094), Small Business Specialist

Description: Navy's principal center for research, development, testing, and evaluation of naval aircraft systems, including simulation, personal safety, aircraft stuctures, aircraft hydraulics and fluids, magnetics, air to air strike, antisubmarine warfare systems, computers, and software. Organized into the following departments: Antisubmarine Warfare,

U.S. Department of the Navy *(Cont.)*

Systems Department, Battle Force Systems Department, Tactical Air Systems Department, Air Vehicles and Crew Systems Technology Department, Communication and Navigation Technology Department, Mission Avionics Technology Department, and Systems and Software Technology Department.

★1944★ Naval Air Engineering Center

Code 00Q, Bldg. 129
Lakehurst, NJ 08733-5028 (201) 323-2064
Constance Hardy, Small Business Specialist

Description: Research, development, test, system evaluation, and engineering in support of aircraft and shipboard interface systems, including launch and recovery systems, support equipment, and visual landing aids.

★1945★ Naval Air Propulsion Center (NAPC)

Box 7176
Trenton, NJ 08628-0176 (609) 896-5653
Robert Reale, Technical Representative

Description: Technical and engineering support for air breathing propulsion systems, including their accessories and components, fuels, and lubricants.

★1946★ Naval Air Systems Command (NAVAIR)

Jefferson Plaza, Room 478
Washington, DC 20361-0001 (202) 692-0935
Sarah Cross, Small Business Representative

Description: Design, development, testing, and evaluation of aircraft, airborne weapon systems, avionics, related photographic and meteorological equipment, ranges, and targets. Principal research and engineering components include the Naval Air Engineering Center, Naval Air Propulsion Center, Naval Air Test Center, Naval Training Systems Center, Naval Weapons Evaluation Facility, and Pacific Missile Test Center. **Contact:** Barbara Williams, Small Business Representative (202)692-0936; Mr. J. A. Johnson, Technical Representative, (AIR 303M), Room 424, Washington, DC 20361.

★1947★ Naval Avionics Center

6000 East 21st St.
Indianapolis, IN 46218 (317) 353-7009
James F. Wilson, Small Business Specialist

Description: Research, development, pilot, and limited manufacturing and depot maintenance on avionics and related equipment.

★1948★ Naval Civil Engineering Laboratory (NCEL)

Naval Construction Battalion Center (Code 6641)
Port Hueneme, CA 93043 (805) 982-5992
Mary Lorenzana, Small Business Representative

Description: Principal Navy research, development, test, and evaluation center for shore and seafloor facilities.

★1949★ Naval Coastal Systems Center (NCSC)

Panama City, FL 32407 (904) 234-4347
Johnny L. Peace, Small Business Specialist

Description: Research, development, test, and evaluation center for mine and undersea countermeasures, special warfare, amphibious warfare, diving, and other naval mission that take place primarily in the coastal regions. Hosts several tenant activities, including the Navy Experimental Diving Unit and the Naval Diving and Salvage Training Center. **Contact:** Mr. W.R. Donaldson (904) 234-4862.

★1950★ Naval Explosive Ordnance Disposal Technology Center

Attn: Code 604
Naval Ordnance Station
Indian Head, MD 20640-5070 (301) 743-4530
Edward W. Rice, Technical Representative

Description: Research and development of specialized equipment, tool, techniques, and procedures required to support operational explosive ordnance disposal units in the location, neutralization, and disposal of surface and underwater explosive ordnance. Also provides support toward the demilitarization of chemical weapons. **Telecommunication Access:** Alternate telephone (301) 743-4841.

★1951★ Naval Medical Research and Development Command (NMRDC)

National Capital Region
Building 54
Bethesda, MD 20814-5044 (202) 295-0548
Ms. Sandy Shepard, Small Business Representative

Description: Research, development, test, and evaluation programs in submarine and diving medicine, aviation medicine, fleet occupational health, human performance, combat casualty care, infectious disease, oral and dental health, and electromagnetic radiation. Major components include Naval Aerospace Medical Research Laboratory, Naval Dental Research Institute, Naval Health Research Center, Naval Medical Research Institute, and Naval Submarine Medical Research Laboratory.

★1952★ Naval Ocean Systems Center (NOSC)

271 Catalina Boulevard
San Diego, CA 92152-5000 (619) 225-2707
Forrest L. Hodges, Small Business Representative

Description: Research, development, test, and evaluation for command, control, and communications; ocean surveillance, surface- and air-launched weapons systems; and submarine arctic warfare. Also involved in development of technologies that support these activities, including ocean engineering, environmental sciences, marine bioscience, electronics, computer sciences, and atmospheric sciences. **Publications:** (1) NOSC Technical Manuals; (2) Technical Documents; and (3) Technical Reports.

★1953★ Naval Oceanographic Office (NAVOCEANO)

NSTL Station
NSTL, MS 39522-5001 (601) 688-4166
Don Hutchison

Description: Collects, analyzes, and displays oceanographic data in support of Naval operations and establishment. Activities include oceanographic, hydrographic, magnetic, gravity, navigational, and acoustic surveys in support of mapping, charting, and geodesy. **Publications:** (1) NAVOCEANO's Reference Publications; (2) Special Publications; and (3) Technical Reports.

★1954★ Naval Ordnance Missile Test Station (NOMTS)

Building N-103
White Sands Missile Range, NM 88002 (915) 678-6115
Mr. H.L. Hendon, Technical Representative

Description: Supports research, development, test, and evaluation programs in flight testing and guided missiles, rocket gun, and directed energy programs.

★1955★ Naval Ordnance Station
Code 1141D
Naval Ordnance Station
Indian Head, MD 20640-5070 (301) 743-4404
Ernest K. Tunney, Small Business Specialist

Description: Research, development, test, and evaluation of ammunition, pyrotechnics, and solid propellant used in missiles, rockets, and guns.

★1956★ Naval Research Laboratory (NRL)
Code 2490
4555 Overlook Ave., S.W.
Washington, DC 20375 (202) 767-2914
Burt Copson, Small Business Representative

Description: Multidisciplinary scientific research and technological development directed toward materials, equipment, techniques, systems, and related operational procedures for the Navy. **Publications:** (1) NRL Review; (2) Fact Book (annually).

★1957★ Naval Sea Systems Command (NAVSEA)
SEA 00311
Washington, DC 20362-5101 (202) 692-7713
Mr. S. Tatigian, Small Business Representative

Description: Provides material support to the Navy and Marine Corps for ships, submarines, and other sea platforms; shipboard combat systems and components; other surface and undersea warfare and weapons systems; and ordnance expendables. **Contact:** George W. Gatling, Jr., SBIR Coordinator, Crystal Plaza 6, Room 850, Crystal City, VA 20362.

★1958★ Naval Surface Warfare Center (NSWC)
Attn: Code S095
10901 New Hampshire Avenue
Silver Spring, MD 20910-5000 (703) 663-8391
Hugh H. Snider, Jr.

Description: Principal research, development, test, and evaluation center for surface ship weapons systems, ordnance, mines, and strategic systems support. Current areas of interest include low observables technology, applications of artificial intelligence to naval systems, mission and weapons analysis in support of the Navy's use of space systems, and development of technology for advanced autonomous weapons. **Publications:** (1) NSWC Briefs; (2) On the Surface (both weekly).

★1959★ Naval Training Systems Center
Code N-005
Orlando, FL 32813-7100 (305) 646-5515
Dr. John H. Rhodes, Utilization Specialist

Description: Responsible for procurement of training aids and equipment for the Army, Navy, Marine Corps, Air Force, and other government activities, including research and development in simulation, training psychology, human factors and human engineering, and design and engineering of training equipment. **Contact:** Wiley V. Dykes, Technical Representitive, Code N-731 (305) 646-4629/5464.

★1960★ Naval Underwater Systems Center (NUSC)
Newport, RI 02840 (401) 841-2675
Robert Bowerman, Technical Representative

Description: Research, development, test, and evaluation for submarine warfare and weapons systems, including sonar, electromagnetics combat control, weapon systems, launchers, undersea ranges, and combat systems analysis. **Contact:** Marvin Berger, Technical Representative (New London Laboratory), New London, CT 06320 (203) 440-4811.

★1961★ Naval Weapons Station
Yorktown, VA 23691 (804) 887-4644
LCDR J. Kinlaw, Small Business Representative

Description: Development of explosives processing and explosives loading methods for Navy weapons. **Telecommunication Access:** Alternate telephone (804) 887-4744.

★1962★ Naval Weapons Support Center
Code C-2
Crane, IN 47522 (812) 854-1542
Mrs. Reva Swango, Small Business Representative

Description: Design, engineering, and inservice engineering, evaluation, and analysis programs required in providng support for ships and craft components, shipboard weapons systems, and assigned expendable and non-expendable ordnance items. **Contact:** James F. Short. Jr., Technical Representative, Code 505, (812) 854-1625/1626.

★1963★ Naval Weapons Systems Center (NWC)
Code 005
China Lake, CA 93555-6001 (619) 939-2712
Lois Herrington, Small Business Specialist

Description: Research, development, test, and evaluation of air warfare and missile systems, including missile propulsion, warheads, fuzes, avionics and fire control, and missile guidance. Also acts as national range facility for parachute test and evaluation. **Publications:** (1) Naval Weapons Center Technical Publications; (2) Current Technical Events (irregular newsletter). **Contact:** George F. Linsteadt, Technical Representative, Code O1T3 (619) 939-2305.

★1964★ Office of Naval Research (ONR)
Code 1111MA
800 North Quincy Street
Arlington, VA 22217-5000 (202) 696-4313
Dr. Neal Glassman, Program Manager

Description: Mission is to encourage and promote naval research. Offers broad programs to encourage and assist young scientists, including (1) ONR Naval Science Awards Program, which recognizes achievements of high school science students; (2) ONR Young Investigators Program, which is intended to attract the best young academc researchers at U.S. universities; (3) ONR Graduate Fellowship Program which supports 135 students at about 80 universities in nine fields of science and engineering; (4) ONR University Research Initiatives Block Research Program, which offers interaction resesarch between university and naval laboratory personnel; (5) ONR Summer Faculty Research Program, which allows university researchers to spend the summer working at a naval research activity; and (6) ONR Instrumentation Program, which allows for the purchase of major, high-cost university research equipment. **Contact:** Joseph C. Ely (202) 696-6525.

★1965★ Pacific Missile Test Center (PMTC)
Code 6009
Point Mugu, CA 93042 (805) 989-8432
Eric D. Duncan, Small Business Specialist

Description: Primary test and evaluation facility for air-launched weapons and airborne electronic warfare systems.

U.S. Department of the Navy *(Cont.)*

★1966★ Space and Naval Warfare Systems Command (SPAWAR)
National Center Building 1
Room 1E58
Washington, DC 20363-5100 (202) 692-6091
Mrs. Betty Geesey, Asst. Dir. for Small Business

Description: Research, development, test, and evaluation for command, control, and communications; underseas and space surveillance; electronic test equipment; and electronic materials, components, and devices. **Contact:** Robert H. Branner, Head of Operational ASW Systems Branch (SPAWAR 661), (202) 433-4729.

U.S. Department of Transportation

★1967★ Coast Guard Office of Engineering and Development
G-FCP-S
2100 2nd St., N.W.
Washington, DC 20593 (202) 267-1844
Rear Adm. Kenneth G. Wilman, Chief

Description: Expects to spend $19-million in 1989 to support reseach and maintain and improve search and rescue systems, environmental protection, marine safety, and aids to navigation. Majority of the research and development tasks are conducted by private business on contract.

★1968★ Federal Aviation Administration
Small Business Coordinator
800 Independence Ave., S.W.
Washington, DC 20591 (202) 426-8230

Description: Expects to spend $165-million in 1989 to conduct engineering and development programs in air traffic control, advanced computer applications, navigation, weather, aviation medicine, aircraft safety, and enviornment. For procurement information on the Federal Aviation Administration Technical Center contact: Procurement Officer, ANA-51B, FAA Technical Center, Atlantic City International Airport, NJ 08405 (609) 484-4000.

★1969★ Federal Highway Administration (FHWA)
Office of Contracts and Procurement
400 7th St., S.W.
Washington, DC 20509 (202) 366-0650
Robert E. Farris, Administrator

Description: Research on civil, mechanical, geotechnical, chemical, hydraulic, electrical, environmental, and human factors engineering, including direct and contract research and development relating to traffic operations, new construction techniques, and social and environmental aspects of highways and programs. **Contact:** George H. Duffy, Disadvantaged Business Enterprise Division (202) 366-1586.

★1970★ Federal Railroad Administration
Office of Procurement
400 7th St., S.W., Room 8206
Washington, DC 20590 (202) 366-0563
Joseph Kerner, Contact

Description: Will obligate $9-million to continue emphasis on safety of train operations, including testing and evaluation, computer modeling, systems engineering, and safety hazard analysis.

★1971★ National Highway Traffic Safety Adminstration
Office of Contracts and Procurement
400 7th St., S.W.
Washington, DC 20590 (202) 366-0607
Thomas Stafford, Director

Description: Plans to spend $30-million for motor vehicle research, traffic safety research and demonstration, and other statistical and analytical studies in 1989. Enters into contracts with private industry, educational institutions, nonprofit organizations, and state and local governments for defects investigations, crashworthyness programs, alcohol traffic safety programs, systems operations, emergency medical services, safety manpower development, driver/vehicle interaction, experimental vehicles, occupant packaging testing, biomechanics, passive restraint tests, computer support, management studies, and data acquisition.

★1972★ Procurement Operations Division
Office of the Secretary of Transportation
Washington, DC 20590 (202) 366-4952

Description: Anticipates to spend $7-million on broad-based policy research on domestic and international transportation issues of importance to the nation. Procurement Operations Division contracts for studies covering social, economic, environmental, safety and policy-oriented transportation research. Such studies include data collection, modeling, economic, and financial studies and projects to support national transportation policy development and evaluation.

★1973★ Transportation Systems Center (TSC)
DOT SBIR Program Office, DTS-23
Kendall Square
Cambridge, MA 02142 (617) 494-2051
Dr. George Kovatch

Description: Industrial-funded research and analysis organization that applies technical skills to national transportation and logistics problems. Expertise includes radionavigation, human factors, telecommunications, structural analysis, railroad inspection technology, industry analysis, emergency and readiness planning, air traffic automation, explosives detection, security and surveillance, and information systems development. Manages the Department of Transportation, Small Business Innovation Research Program. **Publications:** (1) National Transportation Statistics; (2) Transportation Safety Information (annual); and (3) Project Directory.

CHAPTER 9

PUBLICATIONS FOR THE INVENTOR

Publications can provide a wide variety of information and opinion to inventors. I cannot put too great an emphasis on their value and the time you should dedicate to researching and reading them.

In the United States, an estimated 12,300 magazines are listed as commercial publications. If you add on smaller publications and newsletters, the total jumps to no less than 25,000.

I love prospecting through publications. Often when I need creative stimulation, I will go to a nearby library and spend hours leafing through a variety of publications. If nothing else, this exercise tends to focus my direction.

It is a sure bet that even the most general circulation magazines will run a few stories each year on inventions and creativity, and maybe even something on your particular field of interest. The way I keep up with what is being covered is through the *Reader's Guide to Periodical Literature* (The H.W. Wilson Co.) at my local library. The *Reader's Guide* keeps track of hundreds of publications day-to-day.

There are some publications I read on a regular basis, however. I mention them in hopes that it might help you organize your reading list.

Newspapers and News Magazines

The Wall Street Journal. The *Journal* provides the most current, fast-breaking information on industry. It was through a piece in this newspaper that I learned what was going on at Proctor & Gamble's Crest brand, information that led to my selling P&G what was to become a $12-million premium program (Crest Fluorider).

The Wall Street Journal, Forbes, Fortune, Business Week. These publications are the best way to track the ins and outs of senior executives and the ups and downs of corporate earnings, trends, and competition. *Time* and *Newsweek* are also must reads.

Commerce Business Daily. In order to alert potential sources to emerging government R&D interests, agencies must, by law, publish advance notice of R&D opportunities in the *Commerce Business Daily (CBD).*

> Published Monday through Saturday (except on federal holidays), copies of *CBD* may be obtained at the Department of Commerce field offices and at most major urban and university libraries. If you want your own personnel copies, write or call the Government Printing Office (GPO). Subscriptions to *CBD* may be obtained from The Superintendent of Documents, Government Printing Office, Washington, D.C. 20402; (202) 275-3054. The GPO takes Visa and Mastercard.

Trade Journals and Newsletters

Although the news is not as fresh as you'll find in a daily newspaper, industry trade magazines and newsletters are excellent sources for in-depth information. And they carefully track and report a very broad range of executive assignments, not just the upper-most echelon.

It was through a toy trade magazine, for example, that I obtained the name of the Milton Bradley senior vice president of research and development who licensed our first toy, STARBIRD.

Advertising Age and *Ad Week*. There are no better sources than these weekly publications for the latest information on Fortune 500 new introductions, and who's spending how much to promote and introduce what product.

Here are some controlled circulation publications that I find of great use, overall interest, and inspiration. Contact the publishers to see if you qualify for a free subscription. There are hundreds of such magazines available to inventors and product developers in almost every field.

These magazines make their money through the sale of advertising, not the sale of subscriptions. One of the best things about them is the abundance of trade advertisements.

Tech Briefs. The National Aeronautics and Space Administration (NASA) sponsors the publication of *Tech Briefs,* a monthly high-gloss magazine dedicated to the transfer of technology from the space program to industry. Its technical section covers new product ideas, electronic components and circuits, materials, machinery, computer programs, fabrication technology and life sciences.

> For information about how to receive *Tech Briefs,* contact Associated Business Publications, 41 East 42nd Street, Suite 921, New York, N.Y. 10017-5391; (212) 490-3999. FAX (212) 986-7864.

NASA Spinoffs. If you have applied NASA technology to your products and processes, you can receive free publicity in *NASA Spinoffs,* an annual publication designed to tell consumers how NASA technologies are being applied by industry.

> To find out if you qualify, contact Linda Watts at (301) 621-0241.

Product Design & Development. Chilton's *PD&D* is billed as a product news magazine for design engineers. Its dedicated coverage ranges from printers/plotters to powered metals and motion control devices. If you have an interest in valves and switches, there is no better source.

> For information about how to receive *PD&D,* contact Chilton, Chilton Way, Radnor, PA 19089; (215) 964-4000. FAX (215) 964-4100.

Designfax. *Designfax* is dedicated to product design engineering. It has a fine section on new materials, for example, long-fiber composites, epoxy potting compounds, ceramic coatings, and optical release films. It has high advertisement content and provides reader service cards for more information.

> For information about how to receive *Designfax,* contact Huebcore Communications, Inc., 6521 Davis Industrial Parkway, Solon, OH 44139; (216) 248-1125.

Electronics. This is a monthly magazine that reports on a wide range of subjects from corporate reorganizations and news analysis to new product and technology introductions. It includes executive profiles, reviews, and news of upcoming meetings.

For information about how to receive *Electronics,* contact VNU Business Publications, Inc., 10 Holland Dr., Hasbrouck Heights, N.J. 07604; (201) 393-6000. FAX (201) 393-6388.

Electronic Engineering Times. Billing itself as the "Industry Newspaper for Engineers and Technical Management," *Electronic Engineering Times* is one of the better free publications in its field. It serves up the latest news on technologies, business, and new products. Its reporters cover major trade shows.

For information about how to receive *Electronic Engineering Times,* contact CMP Publications, 600 Community Drive, Manhasset, NY 11030; (516) 562-5882.

These kinds of esoteric business trades are found in many major libraries that offer comprehensive business reference rooms. University technical libraries also usually have a wide range of these kinds of magazines.

Inventors may also find general science magazines such as *Popular Mechanics, Popular Science, Discover, Omni,* and so forth of interest.

In addition to the publications listed in this chapter, I have included a large section on publications for the young inventor in the chapter entitled Project XL.

As valuable as these publications are, it is important to remember that there is no substitute for personal contacts. It's "know how" combined with "know who" that makes the difference. I recommend a phone call to authors and editors as the best way to get information on R&D dollar opportunities out of many publications.

Chapter Arrangement

This chapter is arranged alphabetically by publication title. All titles in this chapter are referenced in the Master Index. The Master Index may also include additional publication titles that are mentioned in the text of other chapters.

★1974★ *Academic Research and Public Service Centers in California: A Guide*

California Institute of Public Affairs
c/o Claremont College
Box 10
Claremont, CA 91711 (714) 624-5212
Lizanne Fleming, Editor

Description: Research institutes and public service centers at California colleges and universities. **Price:** $18.50, plus $1.50 shipping.

★1975★ *Accent On Living*

P.O. Box 700
Bloomington, IL 61702-0700 (309) 378-2961
Betty Garee, Editor

Description: Magazine for people with physical disabilities, their families, and the professional and lay persons working with disabled people. Features motivational articles that emphasize success stories of handicapped people and contains new inventions and ideas for making daily living easier. **Frequency:** Quarterly. **Subscription:** $6.00.

★1976★ *Advanced Manufacturing Technology*

Technical Insights, Inc.
32 N. Dean St.
Englewood, NJ 07631 (201) 568-4744
Charles Joslin, Editor

Description: Designed to inform readers about "the cutting edge" of manufacturing technology, including products being currently developed. Deals with both traditional and electronics manufacturing technology, covering all phases of discrete products manufacturing: automation, robotics, laser, assembly, welding, soldering, finishing, material treatment, and new techniques for machining. Recurring features include news of research, book reviews, a calendar of events, and a monthly insert titled Plant Data Flow about the use of computers in manufacturing plants. **Frequency:** Semimonthly. **Price:** $357/yr., U.S. and Canada; $405 elsewhere. **Send Orders To:** P.O. Box 1304, Fort Lee, NJ 07024.

★1977★ *Advertising Age*

Crain Communications, Inc.
220 E. 42nd St.
New York, NY 10017 (212) 210-0100
Fred Danzig, Editor

Description: Newspaper covering news, trends, and analysis in advertising agency-media-advertiser relationships. **Frequency:** Weekly. **Subscription:** $64.00.

★1978★ *Adweek/East*

A/S/M Communications, Inc.
49 E. 21st St.
New York, NY 10010 (212) 529-5500
Geoffrey Precourt, Editor

Description: Advertising news magazine. **Frequency:** Weekly. **Subscription:** $50.00.

★1979★ *AFHRL Newsletter*

Air Force Human Resources Laboratory
AFHRL/TSRE
Brooks AFB, TX 78235-5601 (512) 536-3879
Dr. Ruth M. Buescher, Editor

Description: Presents research and development techniques and findings related to manpower and personnel, education and training, simulation and aircrew training devices, and logistics and human factors that were developed by the AFHRL. Promotes the exchange of ideas between researchers and Air Force and other defense personnel. Recurring features include news of research and notices of publications available. **Frequency:** Quarterly. **Price:** Free.

★1980★ *Agency Sales Magazine*

Manufacturers' Agents National Association
23016 Mill Creek Rd.
P.O. Box 3467
Laguna Hills, CA 92654 (714) 859-4040
Bert Holtje, Editor

Description: Magazine for manufacturers' agents. Includes trend identifying market data, classified ads, industry trend show calendar, and new product information. **Frequency:** Monthly. **Subscription:** $37.50 U.S. $50.00 for Canada and other foreign subscriptions.

★1981★ *Air Market News*

General Publications, Inc.
P.O. Box 2368
Columbia, MD 21045 (301) 381-9295
Jennifer Prill, Editor

Description: Magazine on new products fot the aviation industry. **Frequency:** 6x/yr. **Subscription:** $20.00.

★1982★ *American Association of Small Research Companies—Members Directory*

American Association of Small Research Companies
1200 Lincoln Avenue, Suite 5
Prospect Park, PA 19076 (215) 522-1500
Joanne Martin, Editor

Description: About 400 small research and development companies covering most scientific disciplines. **Price:** $17.00.

★1983★ *American Bulletin of International Technology Transfer*
International Advancement, Inc.
Box 75537
Los Angeles, CA 90075

★1984★ *American Industry*
Publications for Industry
21 Russell Woods Rd.
Great Neck, NJ 11021 (516) 487-0990
Jack S. Panes, Editor

Description: Tabloid featuring new products of interest to managers of manufacturing plants with 100 or more employees. **Frequency:** Monthly. **Subscription:** $25.00.

★1985★ *American Inventor*
10310 Menhart Lane
Cupertino, CA 95014

★1986★ *American Investors Market Finance Directory*
American Investors Market
Box 122
Watertown, MA 02172
G. B. Alperin, Editor

Description: Small Business Investment Companies, (SBICs), financial companies, and institutions lending money for business investment purposes such as venture capital, equity financing, mortgages, real estate development, and equipment and machinery capital purchases.

★1987★ *Annual Report of the Financial Institutions Bureau*
Financial Institutions Bureau
Michigan Department of Commerce
Box 30224
Lansing, MI 48909 (517) 373-8674
R. La Coursier, Economic Specialist

Description: Banks, savings and loan associations, credit unions, and finance companies throughout Michigan. **Price:** Free.

★1988★ *Appliance New Product Digest*
Dana Chase Publications, Inc.
1110 Jorie Blvd.
CS 9019
Oak Brook, IL 60522-9019 (312) 990-3484
James Stevens, Editor

Description: New product magazine (tabloid) for production, engineering, and purchasing functions for companies producing consumer, commercial, and business appliances. **Frequency:** Quarterly. **Subscription:** $10.00; $20.00 foreign.

★1989★ *ASTI Newsletter*
Association for Science, Technology, and Innovation
P.O. Box 1242
Arlington, VA 22210 (703) 241-2850
Sally Rood, Editor

Description: Publishes news of ASTI, which promotes the development, demonstration, and application of policies, standards, and techniques for improving management of innovation, and news of other organizations with similar goals. Recurring features include book reviews. **Frequency:** Bimonthly. **Price:** Included in membership.

★1990★ *The Authority Report*
Arkansas Science and Technology Authority
100 Main St., Suite 450
Little Rock, AR 72201 (501) 371-3554
Kay Speed Kelly, Editor

Description: Concentrates on issues related to the development of Arkansas' scientific and technological resources. Contains notices of publications available and related conferences as well as news of research, including research and grants funded by the Authority. Recurring features include interviews, reports of meetings, and news of educational opportunities. **Frequency:** Quarterly.

★1991★ *Behind Small Business*
Dona M. Risdall
P.O. Box 37147
Minneapolis, MN 55431 (612) 881-5364
Dona M. Risdall, Editor

Description: Offers investment advice, marketing tips, public relations assistance, success stories, money-making ideas, and other advice for the start-up and operation of a small business. Gives small business owners the opportunity for free publicity on a national scale through press releases. Recurring features include editorials, news of research, book reviews, a calendar of events, and columns titled Small Business Publicity Handbook, Peace of Mind Investing, and Your Image. **Frequency:** Bimonthly. **Price:** $14/yr., U.S. and Canada; $18 elsewhere.

★1992★ *Bio Engineering News*
P.O. Box 1210
Port Angeles, WA 98362 (206) 928-3176
Thomas G. Mysiewicz, Editor

Description: Biotechnical newsletter. **Frequency:** Weekly. **Subscription:** $595.00 (includes free subscription to World Biolicensing Report); $395.00 (public libraries); $195.00 (5-10 subscriptions).

★1993★ *Biochemical & Biophysical Research Communications*
Academic Press, Inc.
1250 Sixth Ave.
San Diego, CA 92101 (619) 699-6825
John N. Abelson, Editor

Description: International scientific journal reporting on experimental results in the field of modern biology. Articles emphasize the innovative aspects of research reports. **Frequency:** Semimonthly. **Subscription:** $712.00 (U.S. and Canada); $879.00 (other countries).

★1994★ *Biographical Dictionary of Scientists: Engineers and Inventors*
Peter Bedrick Books
2112 Broadway, Room 318
New York, NY 10023 (212) 496-0751
David Abbott, Editor

Description: About 200 engineers and inventors. **Price:** $28.00. **Send Orders To:** Peter Bedrick Books, 125 E. 23 Street, New York, NY 10010 (212-777-1187).

★1995★ *Bioprocessing Technology*
Technical Insights, Inc.
32 N. Dean St.
Englewood, NJ 07631 (201) 568-4744
Carol J. Mickel, Editor

Description: Reports developments in the biological production of chemicals and energy. Lists new patent introductions. Recurring features include news of research, book reviews, notices of publications available, and a calendar of events. **Frequency:** Monthly. **Price:** $320/yr., U.S. and Canada; $356 elsewhere. **Send Orders To:** P.O. Box 1304, Fort Lee, NJ 07024.

★1996★ *BioScience*

American Institute of Biological Sciences
730 11th St., N.W.
Washington, DC 20001-4584 (202) 628-1500
Julie Ann Miller, Editor

Description: Review journal for biologists, containing articles, book reviews, new product reviews, features, and announcements. **Frequency:** Monthly. **Subscription:** $42.00; $93.00 foreign; $8.75 single issue.

★1997★ *Biotechnology Law Report*

Mary Ann Liebert, Inc.
Bourse Bldg., Suite 900
21 S. Fifth St.
Philadelphia, PA 19106 (215) 592-9440
Gerry Elman, Editor

Description: Covers legal developments affecting the fields of biotechnology and genetic engineering. Discusses patent law, product liability, biomedical law, contract and licensing law, and international law. Describes pertinent legislation, regulatory actions, personnel and company changes, litigation resolution, and international developments. Publishes complete texts of significant court decisions, regulations, and legislation. Recurring features include book reviews, news of research, and a calendar of events. **Frequency:** Bimonthly. Indexed: Annually. **Price:** $295/yr., U.S. and Canada; $325 elsewhere. **Send Orders To:** 157 E. 86th St., New York, NY 10028; (212) 289-2300.

★1998★ *BNA's Patent, Trademark & Copyright Journal*

Bureau of National Affairs, Inc.
1231 25th St., N.W.
Washington, DC 20037 (202) 452-4500
Jeffrey Samuels, Editor

Description: Monitors developments in the intellectual property field, including patents, trademarks, and copyrights. Covers proposed and enacted legislation, litigation, Patent and Trademark Office decisions, Copyright Office practices, activities of professional associations, government contracting, and international developments. **Frequency:** Weekly. **Price:** $648/yr.

★1999★ *Boardroom Reports—Breakthrough*

Boardroom Reports, Inc.
330 W. 42nd St.
New York, NY 10036 (212) 239-9000

Description: Disseminates information on emerging technology in lay terms. Reports on the fields of electronics, biotechnology, medicine, business, and computers; identifies the breakthrough and often supplies data on its background, application, and developers. Recurring features include periodic "special report" supplements focusing on specific fields. **Frequency:** Semimonthly. **Price:** $69/yr., U.S. **Send Orders To:** RGK, 21 Bleeker St., Millburn, NJ 07041; (201) 379-4642.

★2000★ *Brand Names: Who Owns What*

Facts on File, Inc.
460 Park Avenue South
New York, NY 10016 (212) 683-2244

Description: Over 750 firms and their 15,000 brand names. **Price:** $65.00.

★2001★ *Business & Acquisition Newsletter*

Newsletters International, Inc.
2600 S. Gessner
Houston, TX 77063-3297 (713) 783-0100

Description: Contains highly confidential information about companies that want to buy or sell companies, divisions, subsidiaries, product lines, and patents. Includes information on sources of capital to finance purchases of such operations and suggestions on how to structure, negotiate, and complete such deals. **Frequency:** Monthly. **Price:** $250/yr., U.S.; $350 elsewhere.

★2002★ *Business Capital Sources*

IWS, Inc.
24 Canterbury Road
Rockville Centre, NY 11570 (516) 766-5850
Tyler G. Hicks, Editor

Description: About 1,500 banks, insurance and mortgage companies, commercial finance, leasing, and venture capital firms that lend money for business investment. **Price:** $15.00.

★2003★ *Business Ideas & Shortcuts*

ICS Group
P.O. Box 4082
Irvine, CA 92716 (714) 552-8494
Peter Joseph, Editor

Description: Contains special reports, practical ideas, tips, and guidelines on current business opportunities and shortcuts to profits. Covers specific topics, such as how to be a manufacturer without investing; how to get free national advertising; how to protect your business; and how to tap overlooked sources of financing. Recurring features include columns titled Business Shortcut of the Month and Unique Ideas for Entrepreneurs. **Frequency:** Monthly. **Price:** $59/yr.

★2004★ *Business Ideas Newsletter*

1051 Bloomfield Ave., 2-A
Clifton, NJ 07012 (201) 778-6677
Doris Leff, Editor

Description: Newsletter containing information on new products, promotion, and advertising of interest to businessmen, advertising managers, and marketing professionals. **Frequency:** 10x/yr. **Subscription:** $40.00.

★2005★ *Business Organizations, Agencies, and Publications Directory*

Gale Research Inc.
Book Tower
Detroit, MI 48226 (313) 961-2242
Donald P. Boyden, Editor

Description: More than 22,000 organizations and publications of all kinds which are helpful in business, including trade, business, commercial, and labor associations; government agencies; commodity and stock exchanges; United States and foreign diplomatic offices; regional planning and development agencies; convention, fair, and trade organizations; franchise companies; banks and savings and loans; hotel/motel systems; publishers; newspapers; information centers; computer information services; research centers; graduate schools of business; special libraries, periodicals, directories, indexes, etc. **Price:** $298.00 per set.

★2006★ *Business Review*
National Office of Smaller Business Services
Price Waterhouse
1801 K St., N.W.
Washington, DC 20006 (202) 822-8579
Mary Marks, Editor

Description: Features items of interest to small businesses, including tax considerations for privately held companies, executive tax planning, business financing, employee benefits, and federal tax legislation. **Frequency:** 6-8/yr. **Price:** Free.

★2007★ *Business Week—R&D Scoreboard Issue*
McGraw-Hill, Inc.
1221 Avenue of the Americas
New York, NY 10020 (212) 512-2000

Description: Price: $2.00.

★2008★ *Canadian Research*
Sentry Communications
245 Fairview Mall Dr., Suite 500
Willowdale, ON, Canada M2J 4T1 (416) 490-0220
Tom Gale, Editor

Description: Magazine covering research and development in the life and physical sciences in Canada and around the world. **Frequency:** Monthly. **Subscription:** $30.00.

★2009★ *Catalog of Government Inventions Available for Licensing to U.S. Businesses*
Center for the Utilization of Federal Technology
National Technical Information Service
Department of Commerce
5285 Port Royal Road
Springfield, VA 22161 (703) 487-4838
Edward J. Lehman, Editor

Description: About 1,800 federal inventions, developed during the previous year, that are available for licensing. **Price:** $36.00, plus $3.00 shipping.

★2010★ *Category Reports*
International Product Alert
Marketing Intelligence Service
33 Academy St.
New York, NY 14512 (716) 374-6326
Donna Maschiano, Editor

Description: New product reporting service (newsletter) in the following categories: foods, beverages, snacks, health and beauty aids, and household, pets, and miscellaneous. **Frequency:** Monthly. **Subscription:** $1,000.00; overseas postage extra.

★2011★ *CDA Reporter*
Commercial Development Association
1133 15th St., N.W.
Washington, DC 20005 (202) 429-9440
Trecey St. Pierre, Editor

Description: Covers the activities of the Association, whose members are engaged in "identifying, evaluating, and establishing profitable new businesses." Provides limited coverage of related industry events. Recurring features include news of members, a calendar of events, and the column titled President's Message. **Frequency:** Quarterly. **Price:** Included in membership.

★2012★ *Chemical Marketing Reporter*
Schnell Publishing Co., Inc.
100 Church St.
New York, NY 10007-2694 (212) 732-9820
Harry Van, Editor

Description: International tabloid newspaper for the chemical process industries. Includes new technology. **Frequency:** Weekly. **Subscription:** $65.00.

★2013★ *China Exchange Newsletter*
Committee on Scholarly Communication with the People's Republic of China
National Academy of Sciences
2101 Constitution Ave., N.W.
Washington, DC 20418 (202) 334-2718
Kyna Rubin, Editor

Description: Covers recent developments in science, technology, and scholarship in the People's Republic of China. Discusses exchange activities with the Committee and worldwide exchange activities of China, often listing participants. Includes an extensive bibliography of recent publications on China. **Frequency:** Quarterly. **Price:** Free.

★2014★ *Circuit News*
JIT Publishing, Inc.
P.O. Box 3709
San Bernardino, CA 92404 (714) 887-8083
Robert J. Blanset, Editor

Description: News and current events in electronics and production, covering engineering, design, production, new products. **Frequency:** Monthly. **Subscription:** $50.00. $100.00 foreign.

★2015★ *The Communications Industries Report*
International Communications Industries Association
3150 Spring St.
Fairfax, VA 22031-2399 (703) 273-7200
R. Jane Gould, Editor

Description: Trade newsletter for audio, audio-visual, and microcomputer dealers, agents, manufacturers, producers, and distributors. Focus is on current issues and new products. **Frequency:** Monthly. **Subscription:** $15.00.

★2016★ *Computer Systems News*
CMP Publications, Inc.
600 Community Dr.
Manhasset, NY 11030 (516) 365-4600
Mike Azzara, Editor

Description: Computer trade newspaper covering all aspects of the industry: news, trends, new products, prople, business events, jobs, mergers/acquisitions, hardware/software developments, and product/technology updates. **Frequency:** Weekly. **Subscription:** $15.00.

★2017★ *Contacto*
National Eye Research Foundation
910 Skokie Blvd.
Northbrook, IL 60062 (312) 564-4652
Catherine Stahl, Editor

Description: Magazine containing general clinical information on eye and health care. New product and services column highlights companies with innovative items. **Frequency:** Quarterly. **Subscription:** $125.00.

★2018★ *Copyright Law Reports*
Commerce Clearing House, Inc.
4025 W. Peterson Ave.
Chicago, IL 60646 (312) 583-8500
Allen E. Schechter, Editor

Description: Copyright law publication. **Frequency:** Monthly. **Subscription:** $395.00.

★2019★ *Copyright Management*
Communications Publishing Group, Inc.
1505 Commonwealth Ave., No. 32
Boston, MA 02135
Jerry Cohen, Editor

Description: Discusses legal, tax, management, and business aspects of the copyright industries: publishing, entertainment, records and tapes, and software. Recurring features include editorials, news of research, letters to the editor, book reviews, and a calendar of events. **Frequency:** Monthly. Indexed: Annually. **Price:** $257/yr., U.S. and Canada; $307 elsewhere.

★2020★ *Corporate Finance Sourcebook*
National Register Publishing Company, Inc.
Macmillan, Inc.
3004 Glenview Road
Wilmette, IL 60091 (312) 441-2344
Cathy Patruno, Publication Manager

Description: Securities research analysts; major private lenders; merger and acquisition specialists; investment banking personnel; commercial banks; United States-based foreign banks; commercial finance firms; leasing companies; foreign investment bankers in the United States; pension managers; banks that offer master trusts; cash managers; business insurance brokers; business real estate specialists; lists about 2,500 firms. **Price:** $212.00.

★2021★ *Corporate Venturing News*
Venture Economics, Inc.
16 Laurel Ave.
P.O. Box 81348
Wellesley Hills, MA 02181 (617) 431-8100
Lindsay R. Jones, Editor

Description: Concerned with corporate development strategies, particularly strategic alliances and business relationships such as joint ventures, research and development contracts, licensing agreements, and supplier/purchaser relationships. Lists recent deals by industry, profiles companies, and provides articles on strategic partnering. Includes coverage of internal venturing activities. Recurring features include interviews, news of research, and columns titled Deals in the News, Deal Log, and For Your Information. **Frequency:** 18/yr. Indexed: Semiannually. **Price:** $245/yr. for institutions, U.S. and Canada; $270 for institutions elsewhere.

★2022★ *COSSA Washington Update*
Consortium of Social Science Associations
1625 I St., N.W., Suite 911
Washington, DC 20006 (202) 887-6166
David Jenness, Editor

Description: Reports on developments in Congress and in federal agencies which affect funding for social and behavioral science research. Discusses current issues of federal science policy and the policies and practices of federal research agencies. Recurring features include a column titled Sources of Research Support, which profiles a particular federal agency's research program. **Frequency:** Biweekly. **Price:** $40/yr. for individuals, $90 for institutions, U.S. and Canada; $50 for individuals elsewhere.

★2023★ *Coup*
Box 1553
Owosso, MI 48867

Description: Venture capital newsletter. **Frequency:** Quarterly. **Subscription:** $100.00. $25.00 single issue.

★2024★ *CPIA Bulletin*
Chemical Propulsion Information Agency
Johns Hopkins Rd.
Laurel, MD 20707 (301) 935-5000
Sharon Poppe, Editor

Description: Updates and summarizes the progress of research-development-test-and-evaluation programs on chemical guns and rocket propulsion; CPIA products and services; rockets, missiles, and guns; ramjets, boosters, and sustainers; and space vehicles. Recurring features include notices of the Joint Army-Navy-NASA-Air Force (JANNAF) Interagency Propulsion Committee and columns titled People in Propulsion, Readers Write, and Recent CPIA Publications and Literature Searches. **Frequency:** Bimonthly. **Price:** Free.

★2025★ *CRS Venture Directory—Florida*
CRS Publishing Division
7270 N.W. 12th Street, Suite 760
Miami, FL 33126 (305) 591-9475
David M. Sowers, Managing Director

Description: Organizations and individuals seeking venture capital funding in Florida; companies and government agencies offering services and information related to venture capital; financial institutions offering venture capital funding nationally. **Price:** $49.95.

★2026★ *C3I News*
Washington Defense Reports, Inc.
7043 Wimsatt Rd.
Springfield, VA 22151 (703) 941-6600
Clay Wick, Editor

Description: Focuses on the area of command, control, and communications intelligence (C3I). Provides information on government contracts, research and development programs, effectiveness of new systems, and related political and legislative issues. Recurring features include news of research and reports of meetings. **Frequency:** Monthly. **Price:** $120/yr., U.S. and Canada; $129 elsewhere. **Send Orders To:** P.O. Box 34312, Bethesda, MD 20817.

★2027★ *Current Controversy*
Institute for Scientific Information
3501 Market St.
Philadelphia, PA 19104 (215) 386-0100
Thomas G. DiRenzo, Editor

Description: "Designed to provide concise, balanced overviews of current and emerging issues in science and technology." Consists of abstracts with citations of articles and research in controversial areas of science and technology. **Frequency:** 12/yr. **Price:** $50/yr., U.S.; $55, Canada, Mexico, and Europe; $60 elsewhere.

★2028★ *Current Energy Patents*
P.O. Box 62
Oak Ridge, TN 37831 (615) 576-1170
Lila Smith, Editor

Description: Worldwide coverage of energy patents. **Frequency:** Monthly. **Subscription:** $85.00.

★2029★ *The Defense Monitor*
Center for Defense Information
1500 Massachusetts Ave., N.W.
Washington, DC 20005 (202) 484-9490

Description: Concerned with U.S. military issues such as nuclear and conventional weapons; research, development, and procurement; armed force levels; foreign commitments; arms control and SALT; annual military budget; and economic and political implications. Provides analysis and conclusions for a single subject in each issue, and recommends desirable changes and areas for further study. **Frequency:** 10/yr. Cumulative index available for May 1972 to February 1988. **Price:** $1/yr.

★2030★ *Defense R&D*
Pasha Publications, Inc.
1401 Wilson Blvd., Suite 900
Arlington, VA 22209 (703) 528-1244
Harry Baisden, Editor

Description: Tracks trends in defense research and development by scrutinizing Department of Defense programs and budgets. Highlights particularly promising technologies and dramatic technological innovations in each issue. Recurring features include a calendar of events and a column titled R&D Briefs. **Frequency:** Biweekly. **Price:** $167/yr., U.S. and Canada; $182 elsewhere.

★2031★ *Dental Lab Products*
7400 Skokie Blvd.
Skokie, IL 60077-3339 (312) 674-0110
Jeanne K. Matson, Editor

Description: Tabloid serving dental laboratory owners and managers. Includes new product introduction, new literature, technical training seminars and videotapes, major laboratory conferences, and technique features. **Frequency:** 6x/yr. **Subscription:** $18.00.

★2032★ *Design Engineering*
McLean Hunter Ltd.
777 Bay St.
Toronto, ON M5W 1A7 (416) 596-5819
Steve Purwitsky, Editor

Description: Magazine on product design engineering. **Frequency:** Monthly. **Subscription:** $26.00.

★2033★ *Designfax*
International Thompson Industrial Press, Inc.
6521 Davis Industrial Pkwy.
Solon, OH 44139 (216) 248-1125
David Curry, Editor

Description: Magazine covering design engineering. **Frequency:** Monthly. **Subscription:** $20.00.

★2034★ *Development Directory*
Editorial Pkg.
108 Neck Road
Madison, CT 06443 (203) 245-9513

Description: Organizations, foundations, academic institutions, government agencies, professional associations, and firms pursuing research and development interests worldwide. **Price:** $60.00.

★2035★ *Directory of American Research and Technology*
Jaques Cattell Press
R. R. Bowker Company
245 W. 17th Street
New York, NY 10011 (212) 337-7050
Beverley McDonough, Senior Editor

Description: 11,270 publicly and privately owned industrial research facilities. **Price:** $185.00 (ISSN 0073-7623).

★2036★ *Directory of Financing Sources for Mergers, Buyouts & Acquisitions*
Venture Economics in Massachusetts
16 Laurel Avenue
Wellesley Hills, MA 02181 (616) 431-8100
Keith Stephenson, Editor

Description: Over 250 sources of acquisition financing, including banks, asset-based lenders, small business investment companies, insurance companies, and venture capital firms. **Price:** $85.00, plus $3.00 shipping.

★2037★ *Directory of Intellectual Property Lawyers and Patent Agents*
Clark Boardman Company Ltd.
435 Hudson Street
New York, NY 10014 (212) 929-7500
Lynn M. LoPucki, Editor

Description: More than 13,000 patent agents, lawyers, and law firms specializing in intellectual property law, including acquisitions and divestitures, biotechnology, consumer products, and tax aspects. **Price:** $145.00, cloth; $125.00, paper.

★2038★ *Directory of Operating Small Business Investment Companies*
Small Business Administration
1441 L Street, N. W., Room 808
Washington, DC 20416 (202) 653-6672
John R. Wilmeth, Editor

Description: About 570 operating small business investment companies holding regular licenses and licenses under the section of the Small Business Investment Act covering minority enterprise SBICs. **Price:** Free.

★2039★ *Directory of Public High Technology Corporations*
American Investor, Inc.
311 Bainbridge Street
Philadelphia, PA 19147 (215) 925-2761
Ronald P. Smolin, Editor

Description: 2,000 high-technology publicly held corporations in all aspects of computer technology, electronics, aerospace, telecommunications, medical devices and services, biotechnology, artificial intelligence, pharmaceuticals, optics and electro-optics, lasers, chemicals, materials, environmental control, robotics, scientific instruments, and technical services. **Price:** Base edition, $195.00 (1987 edition) (ISSN 0738-7369).

★2040★ *Directory of Scientific Resources in Georgia*
Economic Development Laboratory
Georgia Tech Research Institute
Georgia Institute of Technology
Atlanta, GA 30332 (404) 894-3863

Description: Over 660 research facilities in Georgia, including industrial laboratories, consulting engineering firms, government laboratories, and colleges and universities. **Price:** Free to Georgia residents; $10.00 to others.

★2041★ *Directory of Venture Capital Clubs*
International Venture Capital Institute, Inc.
Baxter Associates, Inc.
Box 1333
Stamford, CT 06904 (203) 323-3143

Description: Approximately 95 venture capital clubs comprised of entrepreneurs, inventors, and small businessmen. **Price:** $7.50 per issue; $35.00 per year.

★2042★ *East/West Technology Digest*
Welt Publishing Company
1413 K St., N.W., Suite 800
Washington, DC 20005 (202) 371-0555
Daniel Bliss, Editor

Description: Supplies information on new technology and license offers, with addresses for obtaining further information. Recurring features include news of research. **Frequency:** Monthly. **Price:** $68/yr.

★2043★ *Edison Entrepreneur*
Thomas Edison Program
Ohio State Department of Development
65 E. State St.
P.O. Box 1001
Columbus, OH 43266-0330

Description: Covers a variety of subjects dealing with research and development throughout the state of Ohio. Recurring features include interviews and news of research. **Frequency:** Quarterly. **Price:** Free.

★2044★ *EE: Electonic/Electrical Product News*
Sutton Publishing Co., Inc.
707 Westchester Ave.
White Plains, NY 10604 (914) 949-8500
Ed Walter, Editor

Description: New product tabloid. **Frequency:** Monthly.

★2045★ *Electric Light & Power*
PennWell Publishing Co.
1250 S. Grove Ave., Suite 302
Barrington, IL 60010 (312) 382-2450
Robert A. Lincicome, Editor

Description: Tabloid providing news of electrical utility industry developments and activities and coverage of new products and technology. **Frequency:** Monthly. **Subscription:** $38.00. $4.00 plus postage, single issue.

★2046★ *Electronic Engineering Times*
CMP Publications INc.
600 Community Dr.
Manhasset, NY 11030 (516) 356-4600
Steve Weitzner, Editor

Description: Tabloid newspaper reporting on electronic news, developments, and products. **Frequency:** Weekly. **Subscription:** Free (controlled).

★2047★ *Electronics*
VNU Business Publications, Inc.
Ten Holland Dr.
Hasbrouck Heights, NJ 07604 (201) 393-6000
J. Robert Lineback, Editor

Description: Magazine providing managers of electronic O.E.M. companies with reports on new technology for engineers and business trends for managers in the growth markets of computers, communication systems, industrial equipment, military components, medical equipment, and testing and measurement instrumentation. **Frequency:** Monthly. **Subscription:** $60.00. $6.00 per issue.

★2048★ *Electronics and Technology Today*
Moorshead Publications Ltd.
Toronto, ON, Canada M3B 3M8 (416) 445-5600
Halvor W. Moorshead, Editor

Description: "Canada's magazine for high-tech discovery." **Frequency:** Monthly. **Subscription:** $22.95.

★2049★ *Electronics of America*
P.O. Box 848
Holly Hill, FL 32017 (904) 788-8617
G.E. Hopper, Editor

Description: Lists new electronic products, technical developments, techniques, and research programs initiated by American manufacturers and institutions. Includes technical description, price, and name and address of manufacturer. Recurring features include occasional book reviews and special reports on specific aspects of the electronics industry. **Frequency:** Weekly. **Price:** $205/yr.

★2050★ *Emerging High Tech Ventures*
Technical Insights, Inc.
32 N. Dean Street
Englewood, NJ 07631 (201) 568-4744

Description: About 40 companies involved in high technology research and development for applications in genetics, light robotics, acoustics, computers, artificial intelligence, and medical fields. **Price:** $890.00. **Send Orders To:** Technical Insights, Inc. Box 1304, Fort Lee, NJ 07024.

★2051★ *Encyclopedia of Associations: National Organizations of the U.S.*
Gale Research Inc.
Book Tower
Detroit, MI 48226 (313) 961-2242
Karin E. Koek, Editor

Description: 22,000 nonprofit United States membership organizations of national scope divided into 18 classifications: trade, business, and commercial; agricultural organizations and commodity exchanges; legal, governmental, public administration, and military; scientific, engineering, and technical; educational; cultural; social welfare; health and medical; public affairs; fraternal, foreign interest, nationality, and ethnic; religious organizations; veterans, hereditary, and patriotic; hobby and avocational; athletic and sports; labor unions, associations, and federations;

chambers of commerce and trade and tourism; Greek and non-Greek letter societies, associations and federations, fan clubs. **Price:** Volume 1, "National Organizations of the United States," $240.00; volume 2, "Geographic and Executive Index," $220.00; "New Association and Projects," (supplement), $220.00; interedition updating service, $195.00.

★2052★ *Engineering Journal*
Delta Communications, Inc.
400 N. Michigan Ave., Suite 1216
Chicago, IL 60611 (312) 670-5424

Description: Magazine featuring articles from its readers on new developments or techniques in steel design, research, the design and/or construction of new projects, steel fabrication methods, or new products or techniques of significance to the uses of steel in building and bridge construction. **Frequency:** Quarterly. **Subscription:** $11.00. $3.50 single issue.

★2053★ *Entrepreneur Magazine*
Entrepreneur, Inc.
2392 Morse Ave.
Irvine, CA 92714 (714) 261-2325
Rieva Lesonsky, Editor

Description: Small business magazine. **Frequency:** Monthly. **Subscription:** $19.97. $2.95 single issue.

★2054★ *Entrepreneurial Manager's Newsletter*
Center for Entrepreneurial Management, Inc.
180 Varick St., Penthouse
New York, NY 10014 (212) 633-0060
Joseph Mancuso, Editor

Description: "Designed to provide accurate and authoritative information relative to subjects of concern to entrepreneurial managers." Covers management, taxes, finance, marketing, information sources, and educational programs. Recurring features include news of seminars, book reviews, news of research and survey results, and columns titled Entrepreneurs's Hall of Fame, Mind Your Own Business, Resources, Accounting, Personal, Miscellaneous, and The Business Exchange. **Frequency:** Monthly. **Price:** Included in membership; $71/yr. for nonmembers.

★2055★ *Entrepreneurship: Theory and Practice*
The John F. Buagh Center for Entrepreneurship
Baylor University Box 8011
Waco, TX 76798-8011 (817) 755-2265
Dr. D. Ray Bagby, Editor

Description: Academic journal on small business management, entrepreneurship, and family-owned businesses. Formerly known as American Journal of Small Business. **Frequency:** 4x/yr. **Subscription:** $35.00; $18.00 individuals; add $15.00 outside North America.

★2056★ *Explosives & Pyrotechnics*
Applied Physics Laboratory
Franklin Research Center
20th and Race Sts.
Philadelphia, PA 19103 (215) 448-1555
R.H. Thompson, Editor

Description: Emphasizes educational and technology transfer aspects of basic and applied science in the field of explosives and pyrotechnics. Covers new developments and applications, training and safety techniques, and related meetings and seminars. Recurring features include reviews of new books and technical reports, and listings of U.S. manufacturers and their products. **Frequency:** Monthly. **Price:** $40/yr.

★2057★ *Export/Exportador*
Johnson International Publishing Corp.
386 Park Ave., S.
New York, NY 10016 (212) 689-0120
Robert Weingarten, Editor

Description: New product and merchandising magazine (printed in English and Spanish). **Frequency:** Bimonthly.

★2058★ *Extended Care Product News*
Health Management Publications
649 S. Henderson Rd.
King of Prussia, PA 19406 (215) 337-4466
Laurie Gustafson, Editor

Description: Health care magazine (tabloid) focusing on new products and news in the field of ostomy, incontinence, nutrition, skin care, and wound care. Geared toward industry purchasing directors. **Frequency:** Quaterly. **Subscription:** Controlled.

★2059★ *F&S/Japanese Information Technology Industry Letter*
Frost & Sullivan, Inc.
106 Fulton St.
New York, NY 10038 (212) 233-1080
Hiro Shibuya, Editor

Description: Examines developments in the Japanese information technology industry. Reports on research and development, start-up companies, and export relations. Includes "analysis, observations, and conclusions on trends and their impact on North American and European competitors and markets." Recurring features include news of research and editorials. **Frequency:** Monthly. **Price:** $340/yr.

★2060★ *F&S/Japanese Venture Capital & OTC Opportunities*
Frost & Sullivan, Inc.
106 Fulton St.
New York, NY 10038 (212) 233-1080

Description: Written by "Japanese analysts in Japan for American and European executives and investors about the opportunities and developments in the new Venture Capital and Over-the-Counter Companies" in Japan. Covers government policy on such investments, trends in the industry, and growth of small companies in Japan. **Frequency:** Monthly. **Price:** $340/yr.

★2061★ *Federal Research in Progress* (FEDRIP)
Office of Product Management
National Technical Information Service
Department of Commerce
5285 Port Royal Road
Springfield, VA 22161 (703) 487-4929
Linda J. LaGarde, Product Manager

★2062★ *Federal Research Report*
Business Publishers, Inc.
951 Pershing Dr.
Silver Spring, MD 20910 (301) 587-6300
Leonard A. Eiserer, Ph.D., Editor

Description: Provides information on research and development funds available from federal agencies and bureaus or associations that provide support money for research and development. Lists items in categories including environment/energy, transportation, medicine/health, education, and social sciences. **Frequency:** Weekly. **Price:** $136/yr.

★2063★ *Flame*
Horatio Alger Association of Distinguished Americans
One Rockefeller Plaza, Suite 1609
New York, NY 10020 (212) 581-6433

Description: Promotes the private enterprise system to the youth of America and shows them that "opportunity still knocks in America for anyone willing to work." Contains features and news that recognize modern-day individuals whose own initiative and efforts led to significant career success. **Frequency:** Quarterly.

★2064★ *The FLC News*
Federal Laboratory Consortium for Technology Transfer
1945 N. Fine Ave., Suite 109
Fresno, CA 93727 (209) 251-3830
D.M. DelaBarre, Editor

Description: Summarizes current projects involving domestic technology transfer among FLC-member laboratories. Describes technological innovations developed in member laboratories and reports on related funding and grant awards. Recurring features include news of research and educational opportunities, reports of meetings, notices of publications available, and a calendar of events. **Frequency:** Bimonthly. **Price:** Free.

★2065★ *For Your Eyes Only*
Tiger Publications
P.O. Box 8759
Amarillo, TX 79114-8759 (806) 655-2009
Stephen V. Cole, Editor

Description: Digests the specialty military press. Carries reports from published and unpublished sources on military events, developments, arms sales, technology, and research programs. Recurring features include book reviews and statistics. **Frequency:** Biweekly. **Price:** $45/yr., U.S. and Canada; $60 elsewhere.

★2066★ *Foreign Technology: An Abstract Newsletter*
National Technical Information Service
U.S. Department of Commerce
5285 Port Royal Rd.
Springfield, VA 22161 (703) 487-4630
Albert Eggerton, Editor

Description: Carries abstracts of reports on the field of foreign technology. Covers biomedical technology; civil, construction, structural, and building engineering; communications; computer, electro, and optical technology; energy, manufacturing, and industrial engineering; and physical and materials sciences. Recurring features include notices of publications available and a form for ordering reports from NTIS. **Frequency:** Weekly. Indexed: Annually. **Price:** $125/yr., U.S., Canada, and Mexico; $175 elsewhere.

★2067★ *Foreign Trade Fairs New Products Newsletter*
Printing Consultants, Publishers
Box 636, Federal Sq.
Newark, NJ 07101 (201) 686-2382

Description: Provides descriptions of and manufacturer's addresses for new foreign products. **Frequency:** Monthly.

★2068★ *French Advances in Science & Technology*
Science and Technology Office
Embassy of France
4101 Reservoir Rd., N.W.
Washington, DC 20007-2176 (202) 944-6246
Jane Alexander, Editor

Description: Explores recent French basic and applied research and development in the fields of aerospace and transportation; telecommunications and computers; health, genetic engineering, and environmental protection; astronomy and nuclear science; and materials and manufacturing. Recurring features include editorials by French decision-makers in government, industry and education, in-depth special reports, news in brief, and sections titled News From France and International Cooperation. **Frequency:** Quarterly. **Price:** Free.

★2069★ *Get Rich News*
Get Rich News Inc.
P.O. Box 126
Lake Worth, FL 33460 (407) 586-0978
Brian Hogan, Editor

Description: Magazine (tabloid) containing mass market business news for lower to middle income entrepreneurs. **Frequency:** Monthly. **Subscription:** $9.95. $1.25 single issue.

★2070★ *Gorman's New Product News*
Gorman Publishing Company
8750 W. Bryn Mawr Ave.
Chicago, IL 60631 (312) 693-3200
Martin J. Friedman, Editor

Description: Reports on new consumer product introductions in the U.S. and abroad that will be sold in food and drug stores. Gives a brief description of each product and lists the manufacturer. **Frequency:** Monthly. **Price:** $245/yr.

★2071★ *Government Data Systems*
Media Horizons, Inc.
50 W. 23rd St.
New York, NY 10010 (212) 645-1000
Mark Baven, Editor

Description: Magazine examing computer and systems solutions within government and focusing on computer applications, new products, and product reviews. **Frequency:** 8x/yr. **Subscription:** $25.00. $3.00 per issue.

★2072★ *Government Inventions for Licensing: An Abstract Newsletter*
National Technical Information Service
U.S. Department of Commerce
5285 Port Royal Rd.
Springfield, VA 22161 (703) 487-4630
Albert Eggerton, Editor

Description: Abstracts reports on mechanical devices and equipment and other government inventions in chemistry, nuclear technology, biology and medicine, metallurgy, and electrotechnology, as well as optics and lasers, patent applications, and miscellaneous instruments. Recurring features include a form for ordering reports from NTIS. **Frequency:** Weekly. Indexed: Annually. **Price:** $225/yr., U.S., Canada, and Mexico; $330 elsewhere.

★2073★ *Government Research Directory*
Gale Research Inc.
Book Tower
Detroit, MI 48226 (313) 961-2242
Kay Gill, Editor

Description: About 4,000 research and development facilities operated by or partly or fully funded by the United States government, including research centers, bureaus, and institutes; research and development installations; testing and experiment stations; and major research-supporting service units; units are active in all areas of physical, social, and life sciences, technology, etc. **Price:** Base edition, $375.00; supplement, $210.00.

★2074★ *GRID*
Gas Research Institute
8600 W. Bryn Mawr
Chicago, IL 60631 (312) 399-8100
R. Busby, Editor

Description: Reports on gas energy research and development sponsored by the Institute. Carries announcements of technical reports available. **Frequency:** Quarterly. **Price:** Free.

★2075★ *Handbook of Business Finance and Capital Sources*
AMACOM, Book Division
American Management Association
135 W. 50th Street
New York, NY 10020 (212) 586-8100
Dileep Rao, Editor

Description: Price: $95.00.

★2076★ *Healthcare Technology & Business Opportunities*
Biomedical Business International, Inc.
17722 Irvine Blvd., Suite 3
Tustin, CA 92680 (714) 838-8350
Michael Gibb, Editor

Description: Features patents, business and joint venture opportunities and technology, products for development or sale, as well as editorial material relevant to R & D management. Also contains international opportunities from primary, hard-to-find sources, and carries complete contact information for reader convenience. Recurring features include columns titled New Technology, Business Opportunities, Patents, and Information Resources. **Frequency:** Monthly. **Price:** $325/yr.

★2077★ *Hi-Tech Alert*
Communication Research Associates, Inc.
10606 Mantz Rd.
Silver Spring, MD 20903 (301) 445-3230
Michael R. Naver, Editor

Description: Provides "fresh, timely, usable news of hi-tech developments in plain English." Monitors the areas of electronic mail, office automation, desktop publishing, computer-aided research, online information services, and personal computer applications. Recurring features include notices of publications available and news of educational opportunities. **Frequency:** Monthly. **Price:** $98/yr. for individuals, $108 for institutions, U.S. and Canada; $108 for individuals, $118 for institutions elsewhere.

★2078★ *High-Tech Materials Alert*
Technical Insights, Inc.
32 N. Dean St.
Englewood, NJ 07631 (201) 568-4744
Alan Brown, Editor

Description: Alerts research directors and business executives to advances in materials research, development, testing, manufacture, and application. Emphasizes technology transfer; covers a wide range of materials, including metals, glasses, ceramics, plastics, and electronics materials. Recurring features include news of research, listings of new patents, and a calendar of events. **Frequency:** Monthly. **Price:** $325/yr., U.S. and Canada; $361 elsewhere. **Send Orders To:** P.O. Box 1304, Fort Lee, NJ 07024.

★2079★ *High Tech Tomorrow*
High Tech Information, Inc.
330 W. 42nd St.
New York, NY 10036
Laurie Meisler, Editor

Description: Provides an overview of developments in the field of high technology for potential investors. Covers the areas of computer-aided technologies, mainframes/electronics, biotechnology, and computer-aided drafting and design. Rates and makes recommendations for buying technology stocks. **Frequency:** Monthly. **Price:** $95/yr., U.S. and Canada; $120 elsewhere.

★2080★ *Hispanic Business Magazine*
360 S. Hope Ave., Suite 300C
P.O. Box 30794
Santa Barbara, CA 93130-0794 (805) 682-5843
Joel Russell, Editor

Description: Business magazine catering to Hispanic professionals, executives and entrepreneurs. **Frequency:** Monthly. **Subscription:** $12.00.

★2081★ *Home Business News*
12221 Beaver Pike
Jackson, OH 45640 (614) 988-2331
Ed Simpson, Editor

Description: "The voice of America's homebased business owner." Magazine for small business entrepreneurs operating from their homes. **Frequency:** 6x/yr. **Subscription:** $18.00. $3.00 single issue.

★2082★ *HomeBased Entrepreneur Newsletter*
JEB Publications
5520 S. Cornell
Chicago, IL 60637 (312) 324-5802
Joanne Esters-Brown, Editor

Description: Furnishes guidance and information for homebased business owners. Features items on marketing, advertising, financing, taxes, recordkeeping, insurance, starting a business, and other issues of interest to entrepreneurs. Recurring features include book reviews and columns titled Business Basics Q & A, Entrepreneur's Corner, Business Planning, Ideas, and News and Notes. **Frequency:** Monthly. **Price:** $18/yr., U.S. and Canada; $24 elsewhere. **Send Orders To:** P.O. Box 19036, Chicago, IL 60619.

★2083★ *I.C. Asia*
Dataquest, Inc.
Dun & Bradstreet Corp.
1290 Ridder Park Dr.
San Jose, CA 95131-2398 (408) 971-0910
Patricia S. Cox, Editor

Description: Provides interpretation and analysis of information generally available to the public on Japanese and Asian high technology industries. Covers major industry events, the companies themselves, the products manufactured, materials and equipment, and new technology. Presents indicators of the integrated circuit industry in Asia. Recurring features include editorials and news of research. **Frequency:** Semimonthly. **Price:** $430/yr.

★2084★ *ICSB Bulletin*
International Council for Small Business
Entrepreneur Program, BRI 6
Graduate School of Business Administration
University of Southern California
Los Angeles, CA 90089-1421 (213) 743-2098
Alan L. Carsrud, Editor

Description: Deals with management assistance projects and programs for entrepreneurial ventures and small business. Covers innovations in entrepreneurship and small business management development and focuses on material of international impact. Recurring features include editorials, news of research, government policy, a calendar of events, and columns titled President's Message and News From Abroad. **Frequency:** Quarterly. **Price:** Free.

★2085★ *IMPACT Compressors*
IMPACT Publications
P.O. Box 93
Somerset, MI 49281 (517) 688-9654
E.L. Farrah, Editor

Description: Reviews new compressor products, recent patents, industry software, and literature in the field. Recurring features include notices of publications available, news of research, a calendar of events, reports of meetings, book reviews, and news of educational opportunities. **Frequency:** 10/yr. Indexed: Annually. **Price:** $55/yr., U.S.; $65, Canada; $75 elsewhere.

★2086★ *IMPACT Pumps*
IMPACT Publications
P.O. Box 93
Somerset, MI 49281 (517) 688-9654
E.L. Farrah, Editor

Description: Provides data and major claim of newly issued U.S. pump patents. Supplies designers with ideas to improve existing designs or develop ones for specific needs. **Frequency:** 10/yr. **Price:** $75/yr.

★2087★ *IMPACT Valves*
IMPACT Publications
P.O. Box 93
Somerset, MI 49281 (517) 688-9654
E. L. Farrah, Editor

Description: Reviews data and the major claim of newly issued U.S. valve patents. Provides designers with new ideas to improve existing designs, and develop new applications for specific needs **Frequency:** 10/yr. **Price:** $125/yr.

★2088★ *In Business*
The JG Press
18 South Seventh St.
Emmaus, PA 18049 (215) 967-4135
Jerome Goldstein, Editor/Publisher

Description: Small business management magazine. **Frequency:** 6x/yr. **Subscription:** $21.00.

★2089★ *Incubator Times*
Office of Private Sector Initiatives
U.S. Small Business Administration
1441 L St., N.W., Rm. 720A
Washington, DC 20416 (202) 653-7880
Samantha Silva, Editor

Description: Relates information on incubator projects and initiatives for entrepreneurs and small businesses around the country. Discusses new and existing legislative programs, strategies for incubator operators, and economic development success stories. Recurring features include a calendar of events and columns titled Innovators in the Field, Legislative Update, A Helping Hand, A Closer Look, and Ask the Network. **Frequency:** 4/yr. **Price:** Free.

★2090★ *Industrial Equipment News*
11 Penn Plaza, Suite 1005
New York, NY 10001 (212) 868-5661
Mark F. Devlin, Editor

Description: Magazine containing new product information for manufacturing industries. **Frequency:** Monthly. **Subscription:** $35.00.

★2091★ *Industrial Product Bulletin*
P.O. Box 1952
Dover, NJ 07801-0952 (201) 361-9060
Anita LaFord, Editor

Description: Magazine reporting new product information on manufacturing equipment, maintenance supplies, and high technology innovations. **Frequency:** 12x/yr. **Subscription:** $45.00.

★2092★ *Industrial Product Ideas*
Sentry Communications
245 Fairfiew Mall Dr., Suite 500
Willowdale, ON M2J 4T1 (416) 490-0220
Michael Shelley, Editor

Description: New product tabloid. **Frequency:** 8x/yr.

★2093★ *Industrial Research and Development Magazine*
Technical Publishing
1301 S. Groove Street
Barrington, IL 60010 (312) 381-1840

★2094★ *Ingenuity*
1630 16th Lane
Lake Worth, FL 33463

Description: Serves as a "nationwide communication network to and among inventors." Explores the inventing process as well as the researching, developing, financing, marketing, licensing, and distributing of new inventions. **Frequency:** Quarterly. **Price:** $10/yr., U.S.; $22 elsewhere.

★2095★ *InKnowVation Newsletter*
Innovation Development Institute
45 Beach Bluff Ave., Suite 300
Swampscott, MA 01907 (617) 595-2920

Description: Focuses on SBIR (Small Business Innovative Research) program opportunities for small businesses entering early-stage, high-risk research and development ventures. Discusses how SBIR programs can be used to obtain initial funding or to test promising ideas.

★2096★ *Innovation News*
Aremco Products, Inc.
P.O. Box 429
Ossining, NY 10562-0429 (914) 762-0685

Description: Features technical research, updates on process equipment, and photographs of ceramic materials. Covers high temperature design, including process information and case histories.

★2097★ *Inside R&D*
Technical Insights, Inc.
32 N. Dean St.
Englewood, NJ 07631 (201) 568-4744
Richard Consolas, Editor

Description: Describes research and development breakthroughs in industry, government, and academic labs, with an emphasis on technology transfer. Emphasizes how the results of the research covered can be applied practically by industry. Recurring features include news of research and columns titled Managing Innovation, Technology Transfer, and Future Tech. **Frequency:** Weekly. **Price:** $457/yr., U.S. and Canada; $529 elsewhere. **Send Orders To:** P.O. Box 1304, Fort Lee, NJ 07024.

★2098★ *Instruments & Computers-Applications in the Laboratory*
154 E. Boston Post Rd.
Mamaroneck, NY 10543-2826 (914) 698-6655
M. Lovetta Francis, Editor

Description: Journal featuring test reports on new laboratory software and systems, original papers on programming for lab research and experimentation, and new products related to lab software, instruments, and computers. **Frequency:** 12x/yr. **Subscription:** $75.00.

★2099★ *Integrated Manufacturing Technology*
American Society of Mechanical Engineers
345 E. 47th St.
New York, NY 10017
Donald Porteous, Advertising Manager

Description: Journal of engineering research and development for designers, researchers, developers, managers, and users of integrated manufacturing technology and related management and economic issues. **Frequency:** Quarterly.

★2100★ *Intelligence*
Edward Rosenfeld
P.O. Box 20008
New York, NY 10025 (212) 222-1123
Edward Rosenfeld, Editor

Description: Covers technologies that affect the future of computing and offers viewpoints. Concentrates on business, research and government activities in artificial intelligence, neural networks, parallel processing, pattern recognition, expert systems, natural language interfaces, voice and speech technologies, art and graphics, optical storage, and industrial policy. Recurring features include editorials and news of research. **Frequency:** Monthly. **Price:** $295/yr., U.S. and Canada. **Telecommunication Access:** Alternate telephone (800)638-7257.

★2101★ *International High Technology Report*
Amersham Associates
1209 G St., N.E.
Washington, DC 20002 (202) 396-2568
David S. Harvey, Editor

Description: Provides information on trends, key developments, and research and development plans in foreign high technology. Includes stock market information from around the world, a Japanese high technology news roundup, and a listing of significant new research and development patents granted to Japanese firms by the U.S. Recurring features include news of research and a calendar of events. **Frequency:** Monthly. **Price:** $197/yr., U.S. and Canada; $222 elsewhere.

★2102★ *International Invention Register*
Catalyst
Box 547
Fallbrook, CA 92028
Dudley Rosborough, Editor

Description: Price: $18.00 per year.

★2103★ *The International Lawyer*
International Law & Practice Section, American Bar Assn.
750 N. Lake Shore Dr.
Chicago, IL 60611 (312) 988-5000
Marla Hillery, Editor

Description: Journal featuring practical papers on legal issues. Emphasizes international trade, licensing, direct investment, finance, taxation, litigation and dispute resolution. **Frequency:** Quarterly. **Subscription:** $23.00; $28.00 outside the U.S.

★2104★ *International New Product Newsletter*
Transcommunications International, Inc.
Six St. James Ave.
Boston, MA 02116-3819 (617) 426-6647
Pamela H. Michaelson, Editor

Description: Provides "advance news of new products and processes, primarily from sources outside the U.S." Emphasizes new products which can cut costs and improve efficiency. Recurring features include the column Special Licensing Opportunities which lists new products and processes that are available for manufacture under license, or are for sale or import. **Frequency:** Monthly. **Price:** $130/yr., U.S.; $175 elsewhere.

★2105★ *International New Products Newsletter*
U.S. International Marketing Co., Inc.
17057 Bellflower Blvd., Suite 205
Bellflower, CA 90706 (213) 925-2918
R. Mervyn Heaton, Editor

Description: Newsletter focusing on new products. **Frequency:** 6x/yr. **Subscription:** $48.00.

★2106★ *International Product Alert*
Marketing Intelligence Service, Ltd.
33 Academy St.
Naples, NY 14512 (716) 374-6326
Cherie Barton, Editor

Description: Provides concise reports on new products in 18 countries outside of the U.S. and Canada. Covers foods, beverages, nonprescription drugs, cosmetics, toiletries, pet products, and miscellaneous household items. Also lists products that are extensions of existing lines and lists packaging changes. Recurring features include occasional copies of advertising. **Frequency:** Semimonthly. **Price:** $600/yr.

★2107★ *International Research Centers Directory*

Gale Research, Inc.
Book Tower
Detroit, MI 48226 (313) 961-2242
Darren L. Smith, Editor

Description: 6,000 research and development facilities outside the United States operated by governments, universities, or private companies, and concerned with all areas of physical, social, and life sciences, technology, business, and military science. **Price:** $360.00.

★2108★ *International Technology Review*

Amber Research Group, Inc.
10907 Hunt Club Rd.
Reston, VA 22090 (703) 471-9113
Maggie Waters, Editor

Description: Covers international issues pertaining to agriculture, money, and communication as they advance technologically. **Frequency:** Monthly. **Price:** $125/yr.

★2109★ *International Trade Names Dictionary: Company Index*

Gale Research, Inc.
Book Tower
Detroit, MI 48226 (313) 961-2242
Donna Wood, Editor

Description: 10,000 companies that manufacture, distribute, import, or otherwise market consumer-oriented products in countries other than the United States. **Price:** $210.00.

★2110★ *International Venture Capital Institute—Directory of Venture Capital Clubs*

International Venture Capital Institute
Box 1333
Stamford, CT 06904 (203) 323-3143
Carroll A. Greathouse, President

Description: Over 100 venture capital clubs; international coverage. **Price:** $9.95 per issue; $14.95 per year.

★2111★ *International Wealth Success*

Tyler G. Hicks
24 Canterbury Rd.
Rockville Centre, NY 11570 (516) 766-5850
Tyler G. Hicks, Editor

Description: Covers methods of making money in a successful business, including sources of business capital, real estate income methods, mail order, import/export, franchising, and licensing of products. Recurring features include news of export/import opportunities and a column on capital sources. **Frequency:** Monthly. **Price:** $24/yr.

★2112★ *Invent!*

Mindsight Publishing
3201 Corte Malpaso, Suite 304
Camarillo, CA 93010 (805) 388-3097
David Alan Foster, Editor

Description: International magazine for inventors, innovators, designers, engineers, and entrepreneurs. **Frequency:** Bimonthly. **Subscription:** $35.00. $6.00 per single issue.

★2113★ *Inventors' Digest*

Affiliated Inventors Foundation, Inc.
2132 E. Bijou St.
Colorado Springs, CO 80909-5950 (719) 635-1234
Joanne H. Mordus, Editor

Description: Magazine for inventors and others interested in the invention process, development and marketing. **Frequency:** 6x/yr. **Subscription:** $15.00; $32.00 overseas.

★2114★ *Inventors News*

Inventors Clubs of America, Inc.
P.O. Box 450261
Atlanta, GA 30345 (404) 938-5089
Alexander T. Marinaccio, Editor

Description: Designed to keep inventors abreast with the latest information on patents and trademarks. Carries news of the Club and information on exhibits and inventions. Recurring features include editorials, news of research, letters to the editor, news of members, a calendar of events, and a column titled Patents for Sale. **Frequency:** Monthly. **Price:** $50/yr., U.S. and Canada; $60 elsewhere.

★2115★ *Investing Licensing & Trading Conditions Abroad*

Business International Corp.
215 Park Ave., S., 15th Fl.
New York, NY 10003 (212) 460-0600
Robert Harris, Editor

Description: International business magazine. **Frequency:** Monthly. **Subscription:** $1,275.00.

★2116★ *IPS Industrial Products & Services*

Clifford Elliot Publishing
Royal Life Building
277 Lakeshore Rd., E. Suite 209
Oakville, ON L6J 6J3 (416) 842-2884
Carol Radford, Editor

Description: Magazine covering new products and services with emphasis on the Canadian market. **Frequency:** 6x/yr. **Subscription:** $15.00; $30.00 U.S.

★2117★ *IVCI Venture Capital Digest*

International Venture Capital Institute
Box 1333
Stamford, CT 06904 (203) 323-3143

Description: Price: $40.00.

★2118★ *Jewelers' Circular/Keystone Brand Name and Trademark Guide*
Chilton Company
Chilton Way
Radnor, PA 19089 (215) 964-4480
Helena Matlack, Editor

Description: Owners of 11,000 brand names, trademarks, and symbols used by 4,700 manufacturers of jewelry store products. **Price:** $14.95, plus $1.50 shipping; payment must accompany order.

★2119★ *Jobless Newsletter*
Al Baker
14902 Running Deer Trail
Austin, TX 78734 (512) 266-2338
Al Baker, Editor

Description: Presents suggestions for how to make a living without a job. Provides ideas for businesses that can be started with little or no capital. Focuses on one or two businesses in each issue. Discusses the relationship between one's attitude and the success of one's businesses. **Frequency:** Monthly. Indexed: Annually. **Price:** $24/yr.

★2120★ *Journal of Business Venturing*
Elsevier Science Publishing Company, Inc.
Journal Information Center
52 Vanderbilt Ave.
New York, NY 10017 (212) 370-5520
Ian C. MacMillan, Editor

Description: Journal presenting empirically based research on entrepreneurship, either as independent start-ups or within existing corporations. **Frequency:** Quarterly. **Subscription:** $52.00; $96.00 institutions; add $16.00 outside the U.S.

★2121★ *Journal of Small Business Management*
Bureau of Business Research
West Virginia University
College of Business and Economics
P.O. Box 6025
Morgantown, WV 26506-6025 (304) 293-5837
J.H. Thompson, Editor

Description: Magazine dedicated to the development of entrepreneurship and small business management through education, research, and the free exchange of ideas. **Frequency:** Quarterly. **Subscription:** $25.00; $30.00 institutions. $7.50 per issue.

★2122★ *Journal of the Copyright Society of the U.S.A.*
Copyright Society of the U.S.A.
c/o New York University Law School
40 Washington Sq. S.
New York, NY 10012 (212) 998-6194

Description: Journal focusing on copyright law. **Frequency:** Quarterly. **Subscription:** $125.00.

★2123★ *Laboratory Times*
Sentry Communications
245 Fairview Mall Dr., Suite 500
Willowdale, ON M2J 4T1 (416) 490-0220
Doug Dingeldein, Editor

Description: Professional tabloid presenting new product reviews for the laboratory market. **Frequency:** 6x/yr. **Subscription:** Controlled circulation to qualified laboratory personnel.

★2124★ *Lasers/Electro-Optics Patents Newsletter*
Communications Publishing Group, Inc.
209 W. Central St.
Natick, MA 01760 (617) 651-9904
Jeffrey L. Swartz, Editor

Description: Identifies newly published U.S. and international patent documents in laser and electro-optics. Provides insights into research and development results and industry marketing strategies. **Frequency:** Monthly. **Price:** $297/yr.

★2125★ *Les Nouvelles*
Licensing Executives Society International
71 E. Ave., Suite S
Norwalk, CT 06851 (216) 771-2600
Jack Stuart Ott, Editor

Description: Concerned with licensing and related subjects. Covers technology, patents, trade marks, and licensing "know-how" world-wide. **Frequency:** Quarterly. **Price:** Included in membership. **Telecommunication Access:** Alternate telephone (203)852-7168.

★2126★ *Licensing Book*
Adventure Publishing Group, Inc.
264 W. 40th St., 7th Fl.
New York, NY 10018 (212) 575-4510
Janet Kilcommons, Editor

Description: Tabloid for retailers covering licensed merchandising industry, including character, entertainment, design and fashion, and sports licensing. **Frequency:** Monthly. **Subscription:** $36.00.

★2127★ *Licensing Law and Business Report*
Clark Boardman Company, Ltd.
435 Hudson St.
New York, NY 10014 (212) 929-7500

Description: Focuses on a specific area within the licensing field in each issue. Provides analysis of court cases and recent developments affecting the design of licensing agreements worldwide, including such topics as antitrust law, tax considerations, and technology management consulting. Recurring features include an annual table of cases. **Frequency:** 6/yr. **Price:** $125/yr.

★2128★ *The Licensing Letter*
New Market Enterprises
P.O. Box 1665
Scottsdale, AZ 85252 (602) 948-1527
Arnold R. Bolka, Editor

Description: Concerned with all aspects of licensed merchandising, "the business of associating someone's name, likeness or creation with someone else's product or service, for a consideration." Recurring features include statistics, news of research, a calendar of events, mechanics, properties, and lists of licensors and licensees. **Frequency:** Monthly. **Price:** $115/yr., U.S. and Canada; $145 elsewhere.

★2129★ *Licensing Today*
International Thomson Retail Press
345 Park Ave., S.
New York, NY 10010 (212) 686-7744
James K. Willcox, Editor

Description: Newsletter covering the licensing industry, with an emphasis on toy licensing. Supplement to Toy & Hobby World. **Frequency:** 6x/yr. **Subscription:** $60.00. Included in subscription to Toy & Hobby World.

★2130★ *Lookout*
Marketing Intelligence Service, Ltd.
33 Academy St.
Naples, NY 14512 (716) 374-6326
Tom Vierhile, Editor

Description: Carries photographs and detailed descriptions of the most innovative products, package design, line extensions, and marketing background in consumer goods categories. Includes food and beverages, nonprescription drugs, cosmetics and toiletries, and miscellaneous household items. Also copies advertising support, including layouts and storyboards. Recurring features include product information: name of manufacturer, ingredients, nutritional information, directions for use, background data, and marketing strategies. **Frequency:** Semimonthly. Indexed: Annually. **Price:** $600/yr.

★2131★ *MacRae's Industrial Directory Connecticut*
MacRae's Blue Book, Inc.
Business Research Publications, Inc.
817 Broadway
New York, NY 10003 (212) 673-4700
Barry Lee, Editor

Description: Price: $85.00, plus $4.50 shipping (ISSN 0740-2937).

★2132★ *MacRae's Industrial Directory Maine/New Hampshire/Vermont*
MacRae's Blue Book, Inc.
Business Research Publications, Inc.
817 Broadway
New York, NY 10003 (212) 673-4700
Barry Lee, Editor

Description: Price: $95.00, plus $4.50 shipping.

★2133★ *MacRae's Industrial Directory Maryland/D.C./Delaware*
MacRae's Blue Book, Inc.
Business Research Publications, Inc.
817 Broadway
New York, NY 10003 (212) 673-4700
Barry Lee, Editor

Description: Price: $95.00 (ISSN 0740-2929).

★2134★ *MacRae's Industrial Directory Massachusetts/Rhode Island*
MacRae's Blue Book, Inc.
Business Research Publications, Inc.
817 Broadway
New York, NY 10003 (212) 673-4700
Barry Lee, Editor

Description: Price: $110.00, plus $4.50 shipping (ISSN 0732-1112).

★2135★ *MacRae's Industrial Directory Minnesota*
MacRae's Blue Book, Inc.
Business Research Publications, Inc.
817 Broadway
New York, NY 10003
Harry P. Dedyo, Publisher

★2136★ *MacRae's Industrial Directory New Jersey*
MacRae's Blue Book, Inc.
Business Research Publications, Inc.
817 Broadway
New York, NY 10003 (212) 673-4700
Barry Lee, Editor

Description: About 10,100 firms with six or more employees in New Jersey. **Price:** $135.00, cloth; $110.00, paper; plus $4.50 shipping (ISSN 0739-8492).

★2137★ *MacRae's Industrial Directory New York State*
MacRae's Blue Book, Inc.
Business Research Publications, Inc.
817 Broadway
New York, NY 10003 (212) 673-4700
Barry Lee, Editor

Description: Price: $135.00, cloth; $110.00, paper; plus $4.50 shipping (ISSN 0740-2953).

★2138★ *MacRae's Industrial Directory North Carolina/South Carolina/Virginia*
MacRae's Blue Book, Inc.
Business Research Publications, Inc.
817 Broadway
New York, NY 10003 (212) 673-4700
Barry Lee, Editor

Description: Price: $110.00, plus $4.50 shipping.

★2139★ *MacRae's Industrial Directory Oregon*
MacRae's Blue Book, Inc.
Business Research Publications, Inc.
817 Broadway
New York, NY 10003 (212) 673-4700
Barry Lee, Editor

★2140★ *MacRae's Industrial Directory Pennsylvania*
MacRae's Blue Book, Inc.
Business Research Publications, Inc.
817 Broadway
New York, NY 10003 (212) 673-4700
Barry Lee, Editor

Description: Price: $135.00, cloth; $110.00, paper; plus $4.50 shipping (ISSN 0740-4298).

★2141★ *Manufacturing Technology: An Abstract Newsletter*
National Technical Information Service (NTIS)
U.S. Department of Commerce
5285 Port Royal Rd.
Springfield, VA 22161 (703) 487-4630

Description: Reports on Computer Aided Design, Computer Aided Manufacturing, technology transfer, and other matters related to manufacturing technology. Also provided information on subjects such as planning, marketing and economics, and research program administration. **Frequency:** Weekly. Indexed: Annually. **Price:** $125/yr., U.S., Canada, and Mexico; $175 elsewhere.

★2142★ *Marketeer*
1602 E. Glen Ave.
Peoria, IL 61614 (309) 688-8106
V.B. Cook, Editor

Description: Magazine focusing on new product merchandising for industries. **Frequency:** Monthly. **Subscription:** $5.00.

★2143★ *Martindale-Hubbell Law Directory*
Martindale-Hubbell, Inc.
Box 1001
Summit, NJ 07901 (201) 464-6800

Description: Lawyers and law firms in the United States and its possessions, Canada, and abroad; includes a biographical section by firm, and a separate list of patent lawyers and attorneys in government service. **Price:** $195.00.

★2144★ *McGraw-Hill's Biotechnology PatentWatch*
McGraw-Hill, Inc.
1221 Ave. of the Americas, 36th Fl.
New York, NY 10020 (212) 512-6090
David N. Leff, Editor

Description: Provides analytical summaries of issued U.S. patents and published European and Japanese patent disclosures in the following biotechnology areas: industrial; pharmaceutical; recombinant-DNA; assays and probes; and vaccines. Organizes summarized information under the headings Product, Process, and Benefit; includes data on application number, date filed, priority, and inventors' names. **Frequency:** Semimonthly. Cross-indexed in McGraw-Hill's Biotechnology Newswatch (see separate listing). **Price:** $397/yr.

★2145★ *MCIC Current Awareness Bulletin*
Metals and Ceramics Information Center
Battelle Columbus Division
505 King Ave.
Columbus, OH 43201-2693 (614) 424-5000
Harold Hucek, Editor

Description: Features articles and reference information on metals and ceramics of interest to the Department of Defense. Carries abstracts and critiques of reports primarily related to Defense Department funded materials development and research. Recurring features include new developments update, a state-of-the-art summary, and a calendar of meetings and symposia. **Frequency:** Monthly. **Price:** Free.

★2146★ *Metalworking Production and Purchasing*
Action Communications Inc.
135 Spy Ct.
Markham, ON L3R 5H6 (416) 477-3222
Maurice Holtham, Editor

Description: Metalworking industry journal (tabloid) emphasizing new products. **Frequency:** 6x/yr. **Subscription:** $30.00.

★2147★ *Military Fiber Optic News*
Phillips Publishing, Inc.
7811 Montrose Rd.
Potomac, MD 20854 (301) 340-2100
C. David Chaffee, Editor

Description: Covers specialized fiber optic applications in the defense industry and in the federal government. Tracks developments relating to military fiber optics projects, including the Strategic Defense Initiative (SDI). Also deals with networks and standards and includes updates of more than 100 projects currently underway in the military that include fiber optics. **Frequency:** Biweekly. **Price:** $247/yr., U.S. and Canada; $280 elsewhere. **Telecommunication Access:** Alternate telephone (800)558-8851.

★2148★ *Military Research Letter*
Callahan Publications
P.O. Box 3751
Washington, DC 20007 (703) 356-1925
Vincent F. Callahan, Editor

Description: Provides information on contracting opportunities for military research, development, testing, and evaluation. Includes news of installations, programs, legislation, and new developments. **Frequency:** Semimonthly. **Price:** $150/yr., U.S. and Canada.

★2149★ *Military Robotics*
L&B, Ltd.
19 Rock Creek Church Rd., N.W.
Washington, DC 20011-6005 (202) 723-5031
Joseph A. Lovece, Editor

Description: Contains "timely and accurate information in the area of government and defense applications of robotics." Covers remotely-piloted aircraft, unmanned submarines, teleoperated combat vehicles, cruise missiles, unmanned spacecraft, and teleoperated and autonomous weapons. Lists solicitations and contract awards. **Frequency:** Biweekly. **Price:** $310/yr., U.S. and Canada; $335 elsewhere.

★2150★ *Minorities and Women in Business*
Venture X, Inc.
1701 Link Rd.
Winston-Salem, NC 27103 (919) 722-3927
John D. Enoch, Editor

Description: Magazine networks with major corporations and small businesses owned and operated by minority and female entrepreneurs. **Frequency:** Bimonthly. **Subscription:** $12.00 yearly; $29.00 for three years.

★2151★ *Minority Business Entrepreneur*
924 N. Market St.
Inglewood, CA 90302 (213) 673-9398
Jeanie M. Barnett, Editor

Description: Business magazine aimed primarily at Black and Hispanic readership. **Frequency:** Bimonthly. **Subscription:** $12.00.

★2152★ *NASA Tech Briefs*
Associated Business Publications Co.
41 E. 42nd St.
New York, NY 10017 (212) 490-3999
Joseph Pramberger, Editor

Description: Publication transferring technology to American industry and government in the fields of electronics, computers, physical sciences, materials, mechanics, machinery, fabrication technology, math and information sciences, and the life sciences. **Frequency:** Monthly, except July/Aug. and Nov./Dec. **Subscription:** Free.

★2153★ *National Association of Investment Companies—Membership Directory*
National Association of Investment Companies
915 15th Street, N. W., Suite 700
Washington, DC 20005 (202) 347-8600
Benita M. Gore, Publications Director

Description: About 150 venture capital firms for minority small businesses; licensed by the Small Business Administration. **Price:** $3.39, postpaid.

★2154★ *National Venture Capital Association—Membership Directory*
National Venture Capital Association
1655 N. Fort Myer Drive, Suite 700
Arlington, VA 22209 (703) 528-4370
Molly M. Myers, Editor

Description: Nearly 225 venture capital firms, including subsidiaries of banks and insurance companies. **Price:** Free; send self-addressed, business envelope, stamped with $1.50 postage.

★2155★ *NBIA Review*
National Business Incubation Association
153 S. Hanover St.
Carlisle, PA 17013-3437 (703) 765-0927

Description: Serves as an information exchange and network for individuals interested in business incubation, a concept based on shared resources and services among entrepreneurs and small businesses. Carries Association reports, interviews with executives and entrepreneurs, and notices of business opportunities. Recurring features include a calendar of events and columns titled Notable & Quotable, For Your Reading, and In My Opinion. **Frequency:** 6/yr. **Price:** Included in membership.

★2156★ *NCPLA Newsletter*
National Council of Patent Law Associations
1819 H St., N.W., Suite 1100
Washington, DC 20006 (202) 659-2811
Charles P. Baker, Editor

Description: Provides legal information concerning the patent, trademark, and copyright fields and news of the member state and local associations. **Frequency:** Quarterly. **Price:** $50/yr.

★2157★ *NCST Quarterly Briefing*
National Coalition for Science and Technology
2000 P St., N.W., Suite 305
Washington, DC 20036 (202) 833-2322
Deborarh A. Cohn, Editor

Description: Concentrates on the political activities of the National Coalition for Science & Technology regarding items such as animal rights, technology transfer, technology innovation, and other areas in which science policy affects society. Recurring features include book reviews, a calendar of events, and columns titled The Chairman's Column, The Director's Column, and Nest Action Alert. **Frequency:** Quarterly. **Price:** Included in membership; $30/yr. for nonmembers.

★2158★ *New From Europe*
Prestwick Publications, Inc.
390 N. Federal Hwy., No. 401
Deerfield Beach, FL 33441 (305) 427-2924
Roy H. Roecker, Editor

Description: Contains market forecasts, trends, and descriptions of new products and technologies from Europe. Descriptions include the developer's name and address and an explanation of why the new product is superior to existing products or processes. Provides an overview of the European economy, its research and new product emphasis, and governmental actions that will affect future market activity. Recurring features include news of research. **Frequency:** Monthly. **Price:** $275/yr.

★2159★ *New From Japan*
Prestwick Publications, Inc.
390 N. Federal Hwy., No. 401
Deerfield Beach, FL 33441 (305) 427-2924
Roy H. Roecker, Editor

Description: Describes new Japanese products and technologies and explains why they are superior to existing products or processes. Covers consumer products, energy conserving processes and products, manufacturing methods, and electronic products. Recurring features include news of research. **Frequency:** Monthly. **Price:** $275/yr.

★2160★ *New From U.S.*
Prestwick Publications, Inc.
390 N. Federal Hwy., No. 401
Deerfield Beach, FL 33441 (305) 427-2924
Roy H. Roecker, Editor

Description: Describes new products and technologies researched and developed in the U.S. Examines a single product and its use and applications in depth in each issue. **Frequency:** Monthly. **Price:** $275/yr.

★2161★ *New Mexico R&D Forum*
New Mexico Research and Development Institute
Research and Development Communications Office
University of New Mexico
457 Washington, S.E., Suite M
Albuquerque, NM 87108 (505) 277-3661
Richard W. Cole, Editor

Description: Concerned with developments relating to technology in New Mexico. Reports on new projects being funded by the Institute, profiles technology firms and laboratories in New Mexico, and contains news briefs on pertinent legislation. Recurring features include information on workshops and useful publications, news of research, book reviews, and a calendar of events. **Frequency:** Monthly. **Price:** Free.

★2162★ *New Product Development*
Point Publishing Company, Inc.
P.O. Box 1309
Point Pleasant, NJ 08742 (201) 295-8258
Jim Betts, Editor

Description: Concentrates on issues relating to new product research and development, idea generation, marketing, distribution, design, and other aspects of product development within national and international companies. Recurring features include interviews, news of research, reports of meetings, book reviews, and a calendar of events. **Frequency:** Monthly. **Price:** $70/yr., U.S. and Canada; $85 elsewhere.

★2163★ *New Product Monthly Reports*
Berliner Research Center, Inc.
Berliner Research Building
Danbury, CT 06810 (203) 744-2333

★2164★ *New Products and Processes*
Newsweek International
P.O. Box 424
Livingtston, NJ 07039

★2165★ *New Products Bulletin*
Tiffany Products, Inc.
239 Main St.
West Orange, NJ 07052 (201) 731-9111

★2166★ *New Products News*
8576 Mesa Drive
Sandy, UT 84070 (801) 561-3259

★2167★ *New Technology Week*
King Communications Group, Inc.
627 National Press Bldg.
Washington, DC 20045 (202) 638-4260
Richard McCormack, Editor

Description: Carries news on evolving technologies, especially those in defense-related fields. Follows legislation and government agency action affecting defense and high-tech industries. Lists recipients of foundation and research grants in the U.S. Recurring features include a calendar of events and news of employment opportunities. **Frequency:** Weekly. **Price:** $295/yr.

★2168★ *Newsletter for Independent Businessowners*
Earl D. Brodie
465 California St.
San Francisco, CA 94104 (415) 986-4834
Earl D. Brodie, Editor

Description: Offers specific recommendations for dealing with the wide range of problems facing independent business owners. Discusses topics in the areas of finance, manufacturing, production, wholesaling and retailing, services, and personnel, as well as the overall economic scene. **Frequency:** Semimonthly. **Price:** $85/yr.

★2169★ *NTIAC Newsletter*
Nondestructive Testing Information Analysis Center
Defense Information Analysis Center
P.O. Box 28510
San Antonio, TX 78284 (512) 684-5111
G.A. Matzkanin, Editor

Description: Features articles on methods, applications, and happenings in the field of nondestructive testing of defense systems. Carries items on new equipment and techniques, listings of contract awards and negotiations, literature surveys, and conference reports. Recurring features include news of research, calls for papers, and a calendar of events. **Frequency:** Quarterly. **Price:** Free.

★2170★ *Official Gazette of the United States Patent and Trademark Office: Patents*
Patent and Trademark Office
Department of Commerce
Washington, DC 20231 (703) 557-3158

Description: Price: $18.00 per issue; $375.00 per year (S/N 003-004-80001-1; ISSN 0098-1133). **Send Orders To:** Government Printing Office, Washington, DC 20402.*

★2171★ *Official Gazette of the United States Patent and Trademark Office: Trademarks*
Department of Commerce
Patent and Trademark Office
Washington, DC 20231 (703) 557-3158

Description: Price: $7.00 per issue; $246.00 per year (S/N 003-004-80002-0). **Send Orders To:** Government Printing Office, Washington, DC 20402.

★2172★ *Oil, Gas & Petrochem Equipment*
PennWell Publishing Company
1412 S. Sheridan
Tulsa, OK 74112 (918) 835-3161
J.B. Avants, Editor

Description: Tabloid of new products and services for the petroleum industry in the fields of drilling, refining, production, petrochemical manufacturing, natrual gas processing, pipeline, enhanced oil recovery, maintenance, safety, and instrumentation. **Frequency:** Monthly.

★2173★ *OTTO News*
Ohio Technology Transfer Organization
Ohio State University
1712 Neil Ave.
Columbus, OH 43210 (614) 422-5485
Barbara J. Ayres, Editor

Description: Provides information about members and the Organization, which "serves the needs of Ohio's business and industry by brokering information from Ohio's two-year colleges and universities to Ohio's businesses, federal laboratories, and other sources. **Frequency:** Quarterly. **Price:** Free.

★2174★ *Partners*
Partners of the Americas
1424 K St., N.W., No. 700
Washington, DC 20005 (202) 332-7332
Cynthia Kenny, Editor

Description: Carries news of the association, a non-profit organization which sponsors technical assistance projects and exchanges between the U.S. and Latin America. Reports on projects in agriculture, public health, education, and development. Recurring features include relevant clippings from local newspapers and notices of workshops and conferences. **Frequency:** 5-6/yr.

★2175★ *Patent Attorneys*
American Business Directories, Inc.
American Business Lists, Inc.
5707 S. 86th Circle
Omaha, NE 68127 (402) 331-7169

Description: Price: $90.00, payment with order. Significant discounts offered for standing orders.

★2176★ *The Patent Trader*
Tucker Communications, Inc.
P.O. Box 1000
Cross River, NY 10518 (914) 763-3700
Carll Tucker, Editor

Description: Community newspaper. **Frequency:** Weekly. **Subscription:** $45.00.

★2177★ *PharmIndex*
Skyline Publishers, Inc.
P.O. Box 1029
Portland, OR 97207 (503) 228-6568
Frank D. Portash, Editor

Description: Compiles information on new, changed, and forthcoming pharmaceutical products, including description, adverse reactions, warnings, cautions, pharmacology, and related products. Divides information into areas such as Hormones, Ear, Nose and Throat, Geriatric Therapy, and Immunological Agents. Recurring features include reviews of continuing education programs and information on new and changed products, package sizes, drug prices, discontinued items, and investigational drugs. **Frequency:** Monthly. Indexed: Monthly; cumulated annually. **Price:** $93/yr.

★2178★ *Photocopy Authorization Report*
Copyright Clearance Center
27 Congress St.
Salem, MA 01970 (617) 744-3350

Description: Publicizes Center services and activities and reports general news of the copyright community. Recurring features include news of research and columns titled Coverage Update, Items of Interest, and Readers Ask. **Frequency:** Quarterly. **Price:** Included in membership; $10/yr. for nonmembers, U.S.; $12 for nonmembers elsewhere.

★2179★ *Playthings*
51 Madison Ave.
New York, NY 10010 (212) 689-4411
Frank Reysen, Editor

Description: Magazine focusing on toys, games, hobbycraft, and licensing. **Frequency:** Monthly. **Subscription:** $20.00.

★2180★ *Pratt's Guide to Venture Capital Sources*
Venture Economics, Inc.
16 Laurel Avenue
Wellesley Hills, MA 02181 (617) 431-8100

Description: Over 700 venture capital firms, principally in the United States; small business investment corporations (SBICs); corporate venture groups; and selected consultants and "deal men." **Price:** $125.00. **Send Orders To:** Bernan Associates - UNIPUB, 4611-F Assembly Drive, Lanham, MD 20706 (800-233-0506).

★2181★ *Product Alert*
Marketing Intelligence Service, Ltd.
33 Academy St.
Naples, NY 14512 (716) 374-6326
Diane Seager, Editor

Description: Reports on new consumer goods launched in American retailing, including foods and beverages, non-prescription drugs, cosmetics and toiletries, and miscellaneous household items. Lists products that are an extension of an existing product line, package changes, and marketing plans. Recurring features include pictures as well as descriptions of the products. **Frequency:** Weekly. Indexed: Annually. **Price:** $600/yr.

★2182★ *Product Design and Development*
Chilton Company
Chilton Way
Radnor, PA 19089 (215) 964-4354
Robert Bierwirth, Editor

Description: Magazine on desing and development of durable goods. **Frequency:** Monthly.

★2183★ *Product Engineering*
Morgan-Grampian Publications
2 Park Avenue
New York, NY 10016 (212) 573-8133

★2184★ *Project Summaries*
Division of Science Resources Studies
National Science Foundation
1800 G Street, N. W.
Washington, DC 20550 (202) 655-4000
Millicent Gough, Editor

Description: About 70 projects in information collection and analysis sponsored by the National Science Foundation in the fiscal year. **Price:** Free.

★2185★ *Prospectus*
Women Entrepreneurs
1275 Market St., No. 1300
San Francisco, CA 94103-1424 (415) 929-0129

Description: Offers women business owners support, recognition, and access to vital information and resources. Monitors legislative developments affecting business and covers programs, workshops, and technical assistance educational seminars conducted by the organization. Recurring features include news of members. **Frequency:** Monthly. **Price:** Included in membership.

★2186★ *Public Information Contact Directory*
American Association for the Advancement of Science
Office of Communications
1333 H. Street, N.W.
Washington, DC 20005 (202) 326-6400
Carol L. Rogers, Head of Communications

Description: Public information contacts at more than 400 colleges and universities, foundations, government agencies and laboratories, museums, nonprofit and industrial research institutions, and scientific and related organizations in the United States, Canada, and Puerto Rico. **Price:** $10.00; payment must accompany order.

★2187★ *Pump News*
IMPACT Publications
P.O. Box 93
Somerset, MI 80517 (517) 688-9654
E. L. Farrah, Editor

Description: Reviews new pump products, software, and current literature in the field. Recurring features include news of research, a calendar of events, reports of meetings, news of educational opportunities, book reviews, notices of publications available, notices of upcoming seminars, and technical reports. **Frequency:** 10/yr. Indexed: Annually in December. **Price:** $30/yr., U.S.; $40, Canada; $50 elsewhere.

★2188★ *Quarterly Counselor*
Vidas & Arrett, P.A.
2925 Multifoods Tower
33 S. 6th St.
Minneapolis, MN 55402 (612) 339-8801
Oliver F. Arrett, Editor

Description: Focuses on intellectual property law. Supplies general information and helpful guidelines concerning patent, trademark, copyright, and trade secret law. **Frequency:** Quarterly. **Price:** Free.

★2189★ *R&D Management Digest*
Lomond Publications, Inc.
P.O. Box 88
Mt. Airy, MD 21771 (301) 875-5475
Dr. Lowell H. Hattery, Editor

Description: Summarizes new literature and science and technology policies and events. Provides news of developments in specialized, practical problems of laboratory management. Recurring features include book reviews and news of research. **Frequency:** Monthly. **Price:** $75/yr., U.S., Canada, and Mexico; $88.50 elsewhere. **Telecommunication Access:** Alternate telephone (800)443-6299.

★2190★ *Research and Development*
Cahners Publishing
1350 E. Touhy Ave.
P.O. Box 5080
Des Plaines, IL 60017-5080 (312) 635-8800
Robert R. Jones, Editor

Description: Magazine serving research scientists, engineers, and technical managers. Reports significant advances, problems, and trends that affect the performance, funding, and administration of research. **Frequency:** Monthly. **Subscription:** $45.00.

★2191★ *Research & Development Directory*
Government Data Publications
1661 McDonald Avenue
Brooklyn, NY 11230 (718) 627-0819
Siegfried Lobel, Editor

Description: Firms which received research and development contracts from the federal government during preceding fiscal year. **Price:** $15.00.

★2192★ *Research & Development Telephone Directory*
Cahners Publishing Company
275 Washington Street
Newton, MA 02158 (617) 964-3030

Description: About 4,000 manufacturers, distributors, and suppliers of products and equipment to industrial research facilities. **Price:** $15.00 (ISSN 0160-4074).

★2193★ *Research & Invention*
Research Corp.
6840 E. Broadway Blvd.
Tucson, AZ 85710-2815 (602) 296-6400
W. Stevenson Bacon, Editor

Description: Examines academic scientific and technological research and invention. Provides information on patenting and licensing inventions and on foundation developments, programs, and personnel. Recurring features include columns titled Patent Highlights, Patent Pitfalls, and Grants Update. **Frequency:** Quarterly. **Price:** Free. ISSN 0276-0401.

★2194★ *Research & Technology Management*
The Industrial Research Institute, Inc.
100 Park Ave., Suite 3600
New York, NY 10017 (212) 683-7626
Michael F. Wolff, Editor

Description: Magazine for research and development managers. **Frequency:** 6x/yr. **Subscription:** $33.00; $55.00 institutions, libraries and companies. $10.00 per issue.

★2195★ *Research Centers Directory*
Gale Research Inc.
Book Tower
Detroit, MI 48226 (313) 961-2242
Karen Ann Hill, Editor

Description: About 10,000 university-related and other nonprofit research organizations which are established on a permanent basis and carry on continuing research programs in all areas of study; includes research institutes, laboratories, experiment stations, computing centers, and other facilities and activities; coverage includes Canada. **Price:** Base edition, $380.00; supplement service, $240.00.

★2196★ *Research Horizons*
Research Communications Office
Georgia Institute of Technology
223 Centennial Research Bldg.
Atlanta, GA 30332 (404) 894-6987
Mark Hodges, Editor

Description: Reports highlights of engineering research conducted by Georgia Tech Research Institute and related Georgia Tech academic departments. A recent issue included articles on a remote life detection device, a study of elite women runners, a new synthetic pigment of ultramarine blue, and research and alcohol and drug abuse in the workplace. **Frequency:** Quarterly.

★2197★ *Research Money*
Evert Communications, Ltd.
982 Wellington St.
Ottawa, ON, Canada K1Y 2X8 (613) 728-4621
Vincent Wright, Editor

Description: Supplies "reports and analyses of the forces driving science and technology investment in Canada," with special emphasis on government policies, granting programs, and other incentives for industry and universities. Tracks major expenditures on research and development and highlights areas where research monies are available. Recurring features include interviews, news of research, reports of meetings, news of educational opportunities, and a calendar of events. **Frequency:** 20/yr. **Price:** $175/yr.

★2198★ *Research Services Directory*
Gale Research Inc.
Book Tower
Detroit, MI 48226 (313) 961-2242
Robert J. Huffman, Editor

Description: Over 3,000 laboratories, consultants, firms, data collection and analysis centers, individuals, and facilities in the private sector which conduct research in all areas of business, government, humanities, social science, and science and technology. **Price:** $290.00.

★2199★ *Research-Technology Management*
Sheridan Press
Fame Ave.
Hanover, PA 17331 (717) 632-3535
Michael Wolff, Editor

Description: Journal about management of research and development companies. **Frequency:** 6x/yr. **Subscription:** $55.00.

★2200★ *The Review of Scientific Instruments*
American Institute of Physics
335 45th Street
New York, NY 10017 (212) 661-9404

Description: Magazine focussing on instruments and methods. **Frequency:** Monthly. **Subscription:** $455.00.

★2201★ *Rights Alert*
Knowledge Industry Publications, Inc.
701 Westchester Ave.
White Plains, NY 10604 (914) 328-9157
Janet Bailey, Editor

Description: Concerned with the protection of proprietary rights "affecting all major information and entertainment markets, including print, data, software, video, film and music." Analyzes current suits, court decisions, and legislative actions. **Frequency:** Monthly. **Price:** $175/yr., U.S. and Canada; $185 elsewhere.

★2202★ *Robotics Patents Newsletter*
Communications Publishing Group, Inc.
309 W. Central St., Suite 226
Natick, MA 01760 (617) 651-9904
Jeffrey L. Swartz, Editor

Description: Focuses on recently published U.S. and international patent documents in the field of robotics. Supplies evaluations of research and development results; predicts industry marketing strategies. **Frequency:** Monthly. **Price:** $357/yr.

★2203★ *Rocky Mountain High Technology Directory*
Leading Edge Communications, Inc.
2620 South Parker Road, Suite 185
Aurora, CO 80014 (303) 752-2400
Charles Koelsch, Managing Editor

Description: About 2,000 manufacturers and research and development firms in Arizon, Colorado, New Mexico, Montana, Nevada, Utah, and Wyoming engaged in high technology activities, including work with aerospace equipment and systems, biotechnology devices and materials, communications, computers, electronics, genetics, instruments, material handling systems, medical diagnostics, medical electronics, microelectronics, office automation, pharmaceuticals, robotics, video equipment, and other categories. Also lists about 60 venture capital firms, law, accounting, and merger and acquisition firms. **Price:** $129.00, plus $4.00 shipping.

★2204★ *Roster of Attorneys and Agents Registered to Practice before the United States Patent and Trademark Office*
Patent and Trademark Office
Department of Commerce
Washington, DC 20231 (703) 557-3341

Description: **Price:** Free (S/N 003-004-00609-0). **Send Orders To:** Government Printing Office, Washington, DC 20402.

★2205★ *SBIC Directory and Handbook of Small Business Finance*
International Wealth Success, Inc.
24 Canterbury Road
Rockville Centre, NY 11570 (516) 766-5850
Tyler G. Hicks, Editor in Chief

Description: Over 400 small business investment companies (SBIC's) which lend money for periods from 5 to 20 years to small businesses. **Price:** $15.00, payment with order.

★2206★ *Science Trends*
Trends Publishing, Inc.
National Press Bldg.
Washington, DC 20045 (202) 393-0031
Arthur Kranish, Editor

Description: Reports on developments in general science, in education and throughout society. Covers research and development, current trends, information on scientific and technical publications, and the high-technology outlook. Recurring features include news of research, book reviews, items on publications available, calls for papers, and notices of conferences, seminars, and symposia. **Frequency:** Weekly; monthly in July and August. **Price:** $480/yr., U.S. and Canada.

★2207★ *Scientific and Technical Organizations and Agencies Directory*
Gale Research Inc.
Book Tower
Detroit, MI 48226 (313) 961-2242
Margaret Labash Young, Editor

Description: Nearly 15,000 national and international organizations and agencies concerned with the physical sciences, engineering, and technology, including associations, computer information services, consulting firms, educational institutions, federal government agencies, general grant and assistance programs, libraries and information centers, patent sources and services, research and development centers, science-technology centers, standards organizations, state academies of science, and state government agencies in the fields of aeronautics and space sciences, chemistry, computer science specialties, geography, geology, machinery, mathematics, metallurgy, meteorology, mineralogy, nuclear science, petroleum, and gas, physics, plastics, transportation, water resources, and other areas. **Price:** $185.00.

★2208★ *Semiconductors/ICs Patents Newsletter*
Communications Publishing Group, Inc.
209 W. Central St., Suite 226
Natick, MA 01760 (617) 651-9904
Steven Weissman, Editor

Description: Newsletter updating U.S. and international patent activity in the semiconductor field. **Frequency:** Monthly. **Subscription:** $357.00; $407.00 foreign.

★2209★ *Silicon Mountain Report*
Silicon Mountain Associates, Inc.
c/o Jon Fitzgerald
9250 E. Costilla Ave., Suite 600
Englewood, CO 80112 (303) 499-0215
Alys Novak, Editor

Description: Focuses on high technology companies engaged in biotechnology, communications, data processing, electronics, medical instrumentation, optics, robotics, software, and storage in the geographic region between Colorado Springs and Fort Collins, Colorado. Announces new products, contract awards, key personnel changes, and financial results. Profiles two local high technology companies in each issue. Recurring features include news briefs, editorials, and a calendar of events. **Frequency:** Monthly. **Price:** $195/yr., U.S. and Canada; $219 elsewhere.

★2210★ *The Silver Prescription*

Silver Prescription Press, Inc.
524 Camino del Monte Sol
Santa Fe, NM 87501 (505) 983-1769
A. David Silver, Editor

Description: Designed to "instruct the reader to become a more competent entrepreneur," with the intent of developing more well-trained entrepreneurs in the work force who are confident enough to "solve society's big problems, ones that affect all of us on a personal level." Describes how companies are tackling illiteracy, AIDS, cancer, cardiovascular disease, crime, drug and alcohol abuse, and other problems that are counterproductive to society. Presents a business plan for solving one such problem in each issue. **Frequency:** Monthly. Indexed: Annually. **Price:** $80/yr.

★2211★ *Small Business Guide to Federal R&D Funding Opportunities*

Office of Small Business Research and Development
National Science Foundation
1800 G Street
Washington, DC 20550 (202) 357-7464

Description: Federal agencies and their major components with significant research and development programs. **Price:** Free (S/N 038-000-00522-7). **Send Orders To:** Government Printing Office, Washington DC 20402.

★2212★ *Small Business: The Magazine for Canadian Entrepreneurs*

McLean Hunter Ltd.
777 Bay St.
McLean Hunter Bldg., 4th Fl., Suite 412
Toronto, ON, Canada M5W 1A7 (416) 596-5914
Randall Litchfield, Editor

Description: Magazine addressing the needs and concerns of Canada's entrepreneurs. **Frequency:** 10x/yr. **Subscription:** $19.95; $12.50 students. $2.50 single copy. $3.50 special June issue.

★2213★ *Software Protection*

Law & Technology Press
P.O. Box 3280
Manhattan Beach, CA 90266 (213) 372-1678

Description: Addresses current issues in the development and maintenance of computer software and database security for attorneys, government agencies, and software company executives. Recurring features include letters to the editor, book reviews, case summaries, and a calendar of events. **Frequency:** Monthly. **Price:** $147/yr., U.S. and Canada; $172 elsewhere.

★2214★ *Space R&D Alert*

Aerospace Communications
350 Cabrini Blvd.
New York, NY 10040 (212) 927-8919
Jeffrey K. Manber, Editor

Description: Devoted to the transfer of technology from research centers to the commercial space market. Covers patents, product developments, conferences, and reports in the materials processing, satellite communications, and space technology industries. Recurring features include news of research and a calendar of events. **Price:** $195/yr.

★2215★ *Spacenews Capsules*

QW Communications Company
P.O. Box 1272
Boston, MA 02254-1272 (617) 899-3868
Michael A. O'Bryant, Editor

Description: Contains short news pieces on the space industry, covering the space shuttle, space stations, space vehicles, remote sensing and satellites, launchers and rockets, space defense and the Strategic Defense Initiative (SDI), commercialization of space, and planetary exploration. Also contains information on people in the industry, a publications review, a list of information sources, and related abstracts. **Frequency:** Monthly. **Price:** $15/yr., U.S.; $18 elsewhere.

★2216★ *Stack Gas Control Patents*

IMPACT Publications
P.O. Box 1972
Estes Park, CO 80517 (303) 586-5636
Arthur L. Anderson, Editor

Description: Reviews newly issued U.S. patents related to air pollution control, particularly in the areas of sulfur oxide control, electrostatic precipitators, filters, hydrogen sulfide control, nitrous oxide control, and scrubber systems. **Frequency:** Monthly. **Price:** $75/yr.

★2217★ *State & Regional Directory*

Pennsylvania Chamber of Business and Industry
222 N. Third Street
Harrisburg, PA 17101 (717) 255-3252
Susan E. Smith, Director of Chamber Services

Description: About 1,500 Pennsylvania organizations, including civic, cultural, educational, health and welfare, professional, research, taxpayer, trade associations, and venture capital sources. **Price:** $25.00, postpaid (ISSN 0098-5368).

★2218★ *Status Report of the Energy-Related Inventions Program*

Office of Energy-Related Inventions
National Bureau of Standards
Department of Commerce
Gaithersburg, MD 20899 (301) 975-5500
A. J. Vitullo, Manager, Inventions Program

Description: Inventors of items recommended for possible Department of Energy support. **Price:** Free; limited supply.*

★2219★ *SUNY Research '88*

Research Foundation
State University of New York (SUNY)
State University Plaza
Albany, NY 12246-0001 (518) 434-7180
Nancy Sweeney, Editor

Description: Carries scientific and educational research news of the State University of New York. Recurring features include articles on campus research and columns titled Research Reflections, Perspective on Research, Campus Commentary, Technology Transfer Trends (SUNY patents and inventions), the SUNY Press, Research Newsline, and Research Foundation Update. **Frequency:** Bimonthly during the academic year. **Price:** Free.

★2220★ *Superconductivity Research, Development and Commercialization Report*
Business Publishers, Inc.
951 Pershing Dr.
Silver Spring, MD 20910-4464 (301) 587-6300
Allan L. Franl, Editor

Description: Summarizes developments within the superconductivity industry, focusing on high-temperature superconductors. Reports on governmental and legislative actions, technological advancements, industry trends, and financial issues concerning the superconductivity industry. **Frequency:** Weekly. **Price:** $464.72/yr., U.S., Canada, and Mexico; $484 elsewhere.

★2221★ *Supergrowth Technology USA*
21st Century Research
8200 Blvd. E.
North Bergen, NJ 07047 (201) 868-0881
Maria R. Hendrie, Editor

Description: Ranks the top 100 fastest growing high technology billion dollar markets, including data on current and previous month's rank, rapid growth market segment, and estimated annual compound growth in years ahead. Identifies the most promising new ventures financed by leading venture capital firms and monitors initial public offerings of venture-backed firms and institutional acquisition rates. **Frequency:** Monthly. **Price:** $750/yr., U.S. and Canada; $795 elsewhere.

★2222★ *Tech Notes*
National Technical Information Service
U.S. Department of Commerce
5285 Port Royal Rd.
Springfield, VA 22161 (703) 487-4805
Edward J. Lehmann, Editor

Description: Presents fact sheets on recently developed federal government technology, selected as having potential commercial or practical application, in the following fields: agriculture and food, computers, electrotechnology, energy, engineering, environmental science and technology, manufacturing, machinery and tools, materials, medicine and biology, natural resources technology and engineering, physical sciences, and transportation. Recurring features include news of research. **Frequency:** Monthly. **Price:** $147/yr., U.S., Canada, and Mexico; $294 elsewhere. **Telecommunication Access:** Alternate telephone (703)487-4630.

★2223★ *Technology Forecasts and Technology Surveys*
PWG Publications
205 S. Beverly Dr., Suite 208
Beverly Hills, CA 90212 (213) 273-3486
Irwin Stambler, Editor

Description: Covers new developments in advanced technology and predicts future trends in areas such as sales volumes, consumer demand, new technological advances, and developments in the methodology for forecasting future trends. Concerned with a range of technologies, including electronics, computers, medical technology, chemicals, pulp and paper, food, and materials. **Frequency:** Monthly. **Price:** $115/yr., U.S. and Canada; $129 elsewhere.

★2224★ *Technology Management News*
Communications Publishing Group, Inc.
1505 Commonwealth Ave., No. 32
Boston, MA 02135
Jeffrey L. Swartz, Editor

Description: Concerned with inventions of significance and the patenting and marketing processes. Discusses developments in law, taxes, patenting and licensing. Recurring features include announcements of noteworthy publications, a calendar of events, and a column titled Subscribers' Forum. **Frequency:** Semimonthly. Indexed: Annually. **Price:** $257/yr., U.S.; $307 elsewhere.

★2225★ *Technology Mart*
Thomas Publishing Company
One Penn Plaza
50 West 34th Street
New York, NY 10001

★2226★ *Technology Newsletter*
Honeywell, Inc.
Honeywell Plaza
Minneapolis, MN 55408

Description: Addresses upcoming technologies pertinent to computer engineering.

★2227★ *Technology NY*
Anderson Research & Communications
Advanced Technology Support Center
Rensselaer Technology Park
Troy, NY 12180 (518) 283-1749
Olga K. Anderson, Editor

Description: Provides a comprehensive analysis on all aspects of state-wide technology developments in companies and universities. Focuses on specific products, projects, and corporate ventures, including news briefs on capital availability and news concerning research and development issues. Recurring features include news of relevant legislative and regulatory activity, people and company news, reports on regional economic development, special reports, and a calendar of events. **Frequency:** Monthly. Indexed: Annually. **Price:** $60/yr. individuals, $87 institutions, $30 students, U.S.; $75 for individuals elsewhere.

★2228★ *Technology Stock Monitor*
HMR Publishing Company
P.O. Box 471
Barrington, IL 60010 (312) 382-7857
Alexander P. Paris, Editor

Description: Tracks "more than 1,000 companies competing in the technology boom," providing "all the information you need to uncover tomorrow's winners." Evaluates selected companies, reports on mergers and upcoming public offerings, and announces new product breakthroughs. Recurring features include a model portfolio, a high tech monitor list (earnings reports), and commentary on the technology sector. **Frequency:** Monthly. **Price:** $125/yr., U.S. and Canada.

★2229★ *Technology Transfer Society—Newsletter*
Technology Transfer Society
611 N. Capitol Ave.
Indianapolis, IN 46204
Dr. F. Timothy Janis, Editor

Description: Serves the communication and information needs of technology transfer professionals and policy makers. Focuses on issues and methodologies critical to the effective transfer of technology. **Frequency:** 12/yr. **Price:** Included in membership; $35/yr. for nonmembers.

★2230★ *Technology Update*
Predicasts, Inc.
11001 Cedar Ave.
Cleveland, OH 44106 (216) 795-3000
Cynthia Lenox, Editor

Description: Compiles "technology news abstracted from more than 1000 industry and trade journals, government reports, research studies and other documents." Covers areas such as technical management, agriculture, chemistry, health and medicine, energy, engineering, transportation, communications, environment, lifestyle and leisure, and education. **Frequency:** Weekly. **Price:** $200/yr., U.S.; $225 elsewhere. **Telecommunication Access:** Alternate telephone (800)321-6388.

★2231★ *Test Engineering and Management*
The Mattingley Publishing Company
61 Monmouth Rd.
Oakhurst, NJ 07755

★2232★ *TOWERS Club, U.S.A.—Newsletter*
TOWERS Club, U.S.A.
P.O. Box 2038
Vancouver, WA 98668-2038 (206) 574-3084
Jerry Buchanan, Editor

Description: Intended for freelance writers, publishers, entrepreneurs, and those engaged in marketing their own creative efforts. Provides an exchange of news, quotes, and clippings. Recurring features include columns titled News, Tips, and Sources, Readin' Jerry's Mail, and M/O Mini Clinic. **Frequency:** Monthly, except August and December. **Price:** $60/yr.

★2233★ *Trade Name Directory*
Carpet and Rug Institute, Inc.
310 Holiday Drive
Dalton, GA 30720 (404) 278-3176

Description: Manufacturers of more than 10,000 carpet and rug styles. **Price:** $28.00.

★2234★ *Trade Names Dictionary: Company Index*
Gale Research Inc.
Book Tower
Detroit, MI 48226 (313) 961-2242
Donna Wood, Editor

Description: Over 45,000 companies that manufacture, distribute, import, or otherwise market consumer-oriented products. **Price:** $310.00.

★2235★ *Trademark Design Register*
Trademark Register
National Press Building
Washington, DC 20045 (202) 662-1233

Description: Owners of over 15,000 registered logos, symbols, and design trademarks. **Price:** $274.00; payment must accompany order.

★2236★ *Trademark Directory*
National Paint & Coatings Association
1500 Rhode Island Avenue, N. W.
Washington, DC 20005 (202) 462-6272
Bruce Hamil, Director

Description: Over 60,000 trademarks representing about 6,000 paint and coatings companies. **Price:** Base edition, $20.00; supplement, $5.00.

★2237★ *Trademark Reporter*
The U.S. Trademark Assn.
6 E. 45th St.
New York, NY 10017 (212) 986-5880

Description: Legal journal focusing on trademarks. **Frequency:** 6x/yr. **Subscription:** $80.00 membership only.

★2238★ *TRADEMARKSCAN—FEDERAL*
Thomson & Thomson
International Thomson Information, Inc.
500 Victory Road
North Quincy, MA 02171 (617) 479-1600

Description: More than 780,000 pending and active federal trademark registrations on file in the United States Patent and Trademark Office (USPTO).

★2239★ *TRADEMARKSCAN—STATE*
Thomson & Thomson
International Thomson Information, Inc.
500 Victory Road
North Quincy, MA 02171 (617) 479-1600

★2240★ *20/20*
Jobson Publishing Co.
352 Park Ave., S.
New York, NY 10010 (212) 685-4848
Pat McMillan, Editor

Description: New products, marketing, and merchandinsing tabloid directed to retailers of eyewear, including opticians, optometrists, managers of optical chains, and dispensing ophthalmologists. **Frequency:** 12x/yr. **Subscription:** $50.00.

★2241★ *Two's News*
Technocracy, Inc.
Section 2, Regional Division 11833
435 E. Market St.
Long Beach, CA 90805 (213) 428-4915
John L. Berge, Editor

Description: Reports on the activities of Section 11833-2 of Technocracy, Inc., as well as on trends and events in the field of technological change. **Frequency:** 8/yr. **Price:** $6/yr.

★2242★ *United States Patents Quarterly*
Bureau of National Affairs, Inc.
1231 25th St., N.W.
Washington, DC 20037 (202) 452-4200
Cynthia J. Bolbach, Editor

Description: Reports all important decisions dealing with patents, trademarks, copyrights, unfair competition, trade secrets, and computer chip protection. **Frequency:** Weekly. Indexed: Monthly; cumulated annually. **Price:** $776/yr.

★2243★ *U.S. Executive Report*
18 Blooms Corners Rd.
Warwick, NY 10990 (914) 986-7755
Francesca Lupton, Editor

Description: Magazine containing articles authored by leading U.S. CEO's on new technologies, financial trends, business management, and marketing innovations for their counterparts in Europe and the Pacific rim. **Frequency:** Every six weeks (except Dec.). **Subscription:** $200.00.

★2244★ *USSR Technology Update*
Delphic Associates
c/o Mary Heslin
7700 Leesburg Pike, No. 250
Falls Church, VA 22043 (703) 556-0278
Mary Heslin, Editor

Description: Provides current information on Soviet activities in trade and technology. Presents articles on topics "ranging from fifth generation computer research to industrial automation." Compiles information from Soviet scientific and technical journals in areas including lasers, energy technology, fiber optics, low temperature physics, and computer technology. Also lists U.S. patents granted to Soviet and East European countries. **Frequency:** Biweekly. Indexed: Annually. **Price:** $375/yr. for individuals, $200 for institutions, U.S. and Canada; $400 for individuals, $200 for institutions elsewhere.

★2245★ *Valve News*
IMPACT Publications
P.O. Box 93
Somerset, MI 49281 (517) 688-9654
E. L. Farrah, Editor

Description: Examines new valve products, software, and technical developments. Includes literature reviews and manufacturers' addresses. Recurring features include news of research, a calendar of events, reports of meetings, news of educational opportunities, book reviews, notices of publications available, notices of upcoming seminars, and technical reports. **Frequency:** 10/yr. Indexed: Annually in December. **Price:** $30/yr., U.S.; $40, Canada; $50 elsewhere.

★2246★ *Venture*
521 Fifth Ave.
New York, NY 10175 (212) 682-7373
Jim Jubak, Editor

Description: Magazine about business owners and investors, containing profiles of entrepreneurs, companies, and examinations of industries and areas of opportunity. **Frequency:** Monthly. **Subscription:** $18.00. $3.00 per issue.

★2247★ *Venture Capital*
Venture Capital Sons of America
509 Madison Ave., Suite 812
New York, NY 10022 (212) 838-5577
B. Henry Campbell, Editor

Description: Reports on all aspects of corporate financing, including venture capital, mergers and acquisitions, leveraged buyouts, private placements, spin-offs and divestitures, SOPs, and IPOs. Recurring features include editorials, news of research, book reviews, and columns titled Interview and Funding Sources. **Frequency:** Monthly. **Price:** $125/ yr.; $10/single copy.

★2248★ *Venture Capital Journal*
Venture Economics, Inc.
75 Second Ave., Suite 700
Needham, MA 02194-2813 (617) 449-2100
Jane Koloski Morris, Editor

Description: Magizine on new business development and venture capital investment. **Frequency:** Monthly. **Subscription:** $595.00.

★2249★ *Venture Capital Resource Directory*
Office of Urban Assistance
Illinois Department of Commerce and Community Affairs
620 E. Adams Street
Springfield, IL 62701 (217) 782-7500
Dennis R. Whetstone, Editor

Description: Over 65 venture capital firms, clubs, and networks in Illinois; branch offices of the Department of Commerce and Community Affairs. **Price:** Free.

★2250★ *Venture Capital Sources for Book Publishers*
Ad-Lib Publications
51 N. Fifth Street
Fairfield, IA 52556 (515) 472-6617

Description: About 20 venture capital firms that have expressed an interest in financing publishers and have responded to the publisher's survey; about 30 venture capital firms that did not respond to the survey. **Price:** $25.00.

★2251★ *Venture Capital: Where to Find It*
National Association of Small Business Investment Companies
1156 15th Street, N. W., Suite 1101
Washington, DC 20005 (202) 833-8230
Eileen E. Denne, Editor

Description: About 400 member firms licensed as small business investment companies (SBICs) under the Small Business Investment Act of 1958; associate and sustaining members who are non-SBIC investors in small businesses or suppliers of services are included. **Price:** $2.00, payment with order; send self-addressed, business-size envelope.

★2252★ *Venture/Product News*
Technology Information Operation
Genium Publishing Corp.
1145 Catalyn St.
Schenectady, NY 12303 (518) 377-8857
Robert A. Roy, Editor

Description: Carries "concise descriptions of products and processes from . research and development firms, universities, industries, and government sources that are available for license/acquisition. Provides information on technology transfer, acquisition, business opportunities, and new products." Recurring features include licensing tips, meeting announcements, and descriptions of firms with unique product-related skills. **Frequency:** Monthly. **Price:** $250/yr., U.S. and Canada; $280 elsewhere.

★2253★ *Washington D.C. Area R&D Firms Directory*
WJB Company
7704 Massena Road
Bethesda, MD 20817 (301) 320-5076
Walter J. Bank, Editor

Description: Over 400 firms located in the Washington, D. C., area engaged in research and development in all fields. **Price:** $45.00.

★2254★ *Western Association of Venture Capitalists—Directory of Members*
Western Association of Venture Capitalists
3000 Sand Hill Road, Building 2, Suite 215
Menlo Park, CA 94025 (415) 854-1322

Description: About 120 venture capital firms; coverage limited to the western United States. **Price:** $25.00.

★2255★ *Who's Who in Technology*

Research Publications
12 Lunar Drive
Woodbridge, CT 06525 (203) 397-2600

Description: 36,500 engineers, scientists, inventors, and researchers; volume 1, electronics and computer science; volume 2, mechanical engineering and materials science; volume 3, chemistry and plastics; volume 4, civil engineering, energy, and earth science; volume 5, physics and optics; volume 6, biotechnology. **Price:** Seven volume set, $545.00; biographical volumes, $95.00 each; index volume, $150.00.

★2256★ *Who's Who in Venture Capital*

John Wiley & Sons, Inc.
605 Third Avenue
New York, NY 10158 (212) 850-6331

Description: Companies employing about 650 individuals involved in investment and venture capital. **Price:** $29.95.

★2257★ *The Woman Entrepreneur Tax Letter*

Richard S. Greenwood
P.O. Box 43204
Detroit, MI 48243
Richard S. Greenwood, Editor

Frequency: Monthly. **Price:** $8/yr.

★2258★ *Women & Co.*

Bantam/Doubleday/Dell Publishing Group, Inc.
666 Fifth Ave., 21st Fl.
New York, NY 10103 (212) 554-9614

Description: Provides financial, legal, and tax strategies for women entrepreneurs. Also profiles successful firms owned by women. **Price:** $29/yr.

★2259★ *World Biolicensing & Patent Report TM*

Deborah J. Mysiewicz Publisher, Inc.
P.O. Box 1210
Port Angeles, WA 98362 (206) 928-3176
Thomas Mysiewicz, Editor

Description: Newsletter reporting on licensing opportunities available in biotechnology. Lists U.S. and European patent applications with summaries of important applications. **Frequency:** 10x/yr. **Subscription:** Free with subscription to BioEngineering News.

★2260★ *World Electronic Developments*

Prestwick Publications, Inc.
390 N. Federal Hwy., No. 401
Deerfield Beach, FL 33441 (305) 427-2924

Description: Contains forecasts, trends, and descriptions of new developments in electronics in western Europe, Japan, and the United States. Covers automation and robotics, communications and information processing, computer processing and computer-aided design, and circuits and electronics. Recurring features include statistics and news of research. **Frequency:** Monthly. **Price:** $195/yr., U.S. and Canada; $225 elsewhere.

★2261★ *World Patent Information*

Pergamon Journals, Inc.
Maxwell House
Fairview Park
Elmsford, NY 10523 (914) 592-7700
V.S. Dodd, Editor

Description: Journal serving as worldwide forum for the exchange of information among professionals in the patent information and documentation field. **Frequency:** Quarterly. **Subscription:** $115.00.

★2262★ *World Technology/Patent Licensing Gazette*

Techni Research Associates, Inc.
Willow Grove Plaza
Willow Grove, PA 19090 (215) 657-1753
Louis F. Schiffman, Editorial Director

Description: Leading firms, private and government research laboratories, universities, inventors, consultants, and others that have new products, new process developments, and new technologies available for license or acquisition; also lists related seminars and meetings. **Price:** $120.00 per year.

★2263★ *World Weapons Review*

Forecast Associates, Inc.
22 Commerce Rd.
Newtown, CT 06470 (203) 426-0800

Description: Identifies applications and problems relating to all types of weapons and weapons systems, from small arms to Intercontinental Ballistic Missiles (ICBMs), on an international scale. Discusses the impact of new weapons development; funding; reported arms sales, transfers, and assistance deals worldwide and their impact on balance of power; weapons retrofit and modernization programs; and the outlook for procurement of new weapons or weapons systems. **Frequency:** 24/yr. **Price:** $385/yr., U.S. and Canada; $460 elsewhere.

CHAPTER 10

YOUTH INNOVATION PROGRAMS AND PUBLICATIONS: PROJECT XL

Project XL is an outreach program of the Patent and Trademark Office (PTO) and an integral part of the U.S. Department of Commerce's Private Sector Initiative Program. Designed to encourage the development of inventive thinking and problem-solving skills among America's youth, the Project's principal focus is the promotion of educational programs that teach critical and creative thinking, and on fostering national proliferation of such programs. The overall objective of Project XL, which was initiated by Donald J. Quigg, Assistant Secretary, Commissioner of Patents and Trademarks, is to ensure the nation's position as a world technological leader in the next century—to guarantee that Americans will have the innovative skills to meet the challenges of an increasingly competitive world.

"I believe that the schools of this great nation are filled with Edisons, Wrights, Marconis, Whitneys, and Bells, along with other potential thinkers who can change the world," says Commissioner Quigg. "The very least we can do is to help them realize their potential—to nurture those young people who will inherit and build the future."

A secondary benefit from Project XL is that young people and their parents and teachers will gain an increased awareness of new technology's importance to advancing society and strengthening the domestic economy. Project XL aims to instill an increased appreciation of the contributions inventors make to our way of life and recapture the spirit of those golden years at the turn of the century when inventors were heralded as true American heroes.

Project XL is comprised of the following components:

*National coordination of efforts to teach inventive thinking and problem-solving skills at every level of public and private education throughout the country.

*Presentation of national and regional conferences to promote the teaching of critical and creative thinking skills and the inventive process.

*Establishment of an Education Roundtable, an open forum and national discussion network, drawing upon the talents and resources of public and private sector leaders to develop and promote programs in this area.

*Development of broad-based speakers' bureau on the topics of invention, problem-solving, creativity, thinking skills, and related topics.

*Dissemination of an informational guide called the *Inventive Thinking Project,* designed to channel students in grades K-12 into the inventive thinking process through the creation of their own unique inventions or innovations.

*Creation of an educator's resource guide to include programs, materials, literature, organizations, and other sources that promote thinking across all disciplines.

*Curriculum development for special teaching materials designed to stress problem-solving, the value of creative thinking, and the importance of American inventors.

*Identification of government programs and resources that focus on the development of future problem-solvers in all fields.

*Establishment of an Inventive Thinking Center, a collection of literature, videotapes, and other curriculum materials.

To this end, the following directory is comprised of programs of Project XL and includes a section on publications for the teaching of creativity, critical thinking, problem-solving, and invention.

Chapter Arrangement

This chapter is organized in two sections: Programs and Publications. Each section is arranged alphabetically by program or publication. All programs and publications in this chapter are referenced in the Master Index. The Master Index may also include additional publications mentioned in other chapters.

Programs

★2264★ Connecticut Invention Convention
Connecticut Educators' Network for the Talented and Gifted
85 Fern St.
Hartford, CT 06105 (203) 677-1791
Michelle Munson, Chairperson

Description: Competition for Connecticut students who have created original inventions. Designed to encourage and share creative productivity among today's youth.

★2265★ Creativity and Innovation
New York State Education Department
Technology Education
1 Commerce Plaza, Room 1619
Albany, NY 12234 (518) 474-3954
John Fabozzi, Bureau Chief

Description: Study course designed to enhance 9-12 grade students' creative thinking, problem-solving skills, and production of innovative products and/or processes.

★2266★ Design and Research Exhibition (DARE)
New Jersey Department of Education
Center for Technology Education
Armstrong Hall
Trenton State College CN4700
Trenton, NJ 08650-4700 (609) 771-2068
Dr. Patricia Hutchinson, Project Director

Description: Designed to enhance problem-solving skills and technological literacy among public, private, and vocational high school students through a noncompetitive design and invention program. Outstanding designs are exhibited in an annual state exhibition.

★2267★ Imagination Celebration/Invention Convention
Imagination Celebration Patent Office
Cultural Education Center, Room 9B38
Empire State Plaza
Albany, NY 12230 (518) 473-0823
Dr. Vivianne Anderson, Coordinator

Description: Awards certificates to students in grades 6-9 in New York who submit descriptions of their inventions. Top inventors receive an Imagination Celebration patent and their inventions are displayed at the State Museum.

★2268★ INVENT AMERICA!
United States Patent Model Foundation
510 King St., Suite 420
Alexandria, VA 22314 (703) 684-1836
Kevin O'Brien, Vice President

Description: Encourages creativity and productivity by developing the problem-solving and analytical skills of K-8 grade students. Includes state, regional, and national invention competition with grants and awards for students, teachers, and schools. Distributes educational material to elementary schools and sponsors a national, annual conference to encourage classroom instruction on invention and creativity.

★2269★ The Invention Convention (Morristown)
Silver Burdett and Ginn
250 James St.
Morristown, NJ 07960 (201) 285-7740
Andrew Socha, Product Manager

Description: Encourages students to apply basic science skills in a creative and productive manner through classroom, school, or district-wide events. Sponsors international convention where student inventions are exhibited, judged, and awards are presented. Publishes guidelines for teachers and procedures for students.

★2270★ Invention Convention (Plattsburgh)
Imagination Celebration/Council on the Arts for Clinton County
P.O. Box 451
64 Margaret St.
Plattsburgh, NY 12901 (518) 563-5222
JoAnn Perry, Contact

Description: Exhibits and judges inventions of Clinton County K-12 grade students at the annual Imagination Celebration Community Showcase Festival.

★2271★ Invention Convention (Richardson)
Richardson Independent School District
400 S. Greenville Ave.
Richardson, TX 75080 (214) 470-5202
Dr. Leonard Molotsky, Contact

Description: Designed to stimulate K-12 grade students' imaginations and promote problem-solving and creative thinking skills through invention conception, development, and marketing. Sponsors school competitions; winners compete at the district Invention Convention.

★2272★ Inventors Association of New England Youth Education Program
Youth Education Committee
P.O. Box 335
Lexington, MA 02173 (617) 244-4679
Samuel C. Smith, Chairman

Description: Offers technical assistance to elementary and middle schools whose curricula include invention education. Compiles information on school invention programs and contests.

★2273★ Inventors Association of St. Louis Youth Programs
P.O. Box 16544
St. Louis, MO 63105 (314) 534-2677
Roberta Toole, Executive Director

Description: Sponsors programs on innovation and invention for children in grades 2-9, gifted children, and adults. Is developing a program to use inventions and innovation to study science.

★2274★ InVenture
Inventors Council of Dayton, Ohio
c/o Ronald T. Dodge Company
P.O. Box 9488
Dayton, OH 45409 (513) 439-4497
Dr. Ronald J. Versic, Chairman

Description: Classroom program designed to promote innovation by encouraging middle and high school students to create a product, process, or art form and take the first step to protect the invention. Publishes teachers manual, guide for students, and a videotape of lessons.

★2275★ Midland Public Schools Invention Program
Northeast Intermediate School
1305 E. Sugnet
Midland, MI 48640 (517) 835-7128
Jody Pagel, Contact

Description: Uses inventions as a way of teaching critical thinking and problem solving and encourages students to design products to solve every day problems. Designed for grades 7-9; also taught to special education students at all levels and gifted students in elementary schools.

★2276★ Mini-Invention Innovation Team Contest
New Jersey Department of Education
Division of Vocational Education
225 W. State St.
Trenton, NJ 08625 (609) 292-5720
Sylvia M. Kaplan, Director

Description: Sponsors competitions to inspire children in grades K-9 to think creatively. Competition winners within school districts progress to the regional competition and top contestants enter statewide finals. Conducts teacher in-service workshops and provides technical assistance to students.

★2277★ Minnesota Student Inventors Congress
Minnesota Inventors Congress
P.O. Box 71
Redwood Falls, MN 56283-0071 (507) 637-2344
Penny Becker, Coordinator

Description: Statewide competition and exhibition for winners of student (grades K-12) invention fairs.

★2278★ San Diego Invention Program
San Diego Unified School District
Education Center, Room 2005
4100 Normal St.
San Diego, CA 92103-2682 (619) 293-8552
Jo Anne Schaper, Contact

Description: Encourages strong involvement from K-12 students in the national INVENT AMERICA! contest.

★2279★ Toledo Public Schools Invention Convention
Toledo Public Schools
Manhattan Blvd. & Elm St.
Toledo, OH 43608 (419) 729-8315
Robert Frisch, Contact

Description: Designed to promote cognitive and creative thinking skills in 3-6 grade students through the development and exhibition of inventions, display boards, and tributes to inventors.

★2280★ Tualatin Invention Program
Tualatin Elementary School
19945 S.W. Boones Ferry Rd.
Tualatin, OR 97062 (503) 684-2359
Evelyn Andrews, Contact

Description: Trains teachers to work with students in grades K-6 to develop inventions that are judged at the Invention Convention.

★2281★ Weekly Reader National Invention Contest
Weekly Reader
245 Long Hill Rd.
Middletown, CT 06457 (203) 638-2638
Dr. Irwin Siegelman, Editorial Director

Description: Presents awards to K-8 students nationwide for inventions or innovations that are judged on originality, usefulness in addressing real needs, workability, and clarity of presentation. Awards a grand prize for elementary school students and for middle school students and bestows awards for specific grade levels.

★2282★ Western New York Invention Program
Buffalo Public Schools
419 City Hall
Buffalo, NY 14202 (716) 842-3693
Marge Korzelius, Contact

Description: Draws innovation and invention entries from pre-kindergarten through grade 12 students and displays the top invention from each school or grade at the Buffalo Science Museum. Sponsors a teacher in-service workshop and publishes instructional materials.

★2283★ Young Inventors Program
New York City Teacher Centers Consortium
48 E. 21st St.
New York, NY 10010 (212) 475-3737
Myrna Cooper, Director

Description: Trains K-6 grade teachers in New York City to foster inventive and creative thinking in heterogeneous settings. Encourages students to learn how to investigate real problems using scientific processes of inquiry, understand technology, and develop new products and designs as a result of these experiences. Sponsors competitions and bestows awards at each grade level.

Publications

★2284★ *ABCs of Books and Thinking Skills K-8*
Book Lures, Inc.
P.O. Box 9450
O'Fallon, MO 63366 (314) 272-4242

Description: Applications of thinking skills for K-8 grade students.
Author/Editor: Nancy Polette.

★2285★ *ABCs of Reading, Thinking and Literacy 7-12*
Book Lures, Inc.
P.O. Box 9450
O'Fallon, MO 63366 (314) 272-4242

Description: Step-by-step directions for multi-level activities covering 50 separate thinking skills for grades 7-12. **Author/Editor:** Nancy Polette and Gloria Levine.

★2286★ *Acting, Creating and Thinking* (ACT)
Sundance Publishers & Distributors, Inc.
Newton Rd.
Littleton, MA 01460 (508) 486-9201

Description: Each ACT packet contains 10 activity sheets used to reinforce oral language and critical-creative thinking skills with elementary and middle school students. ACT is designed for use with Take Part books which are children's stories in play form. **Author/Editor:** Dr. Olive Stafford Niles and Audrey A. Friedman.

★2287★ *Analytical Reading and Reasoning*
Innovative Sciences, Inc.
300 Broad St.
Stamford, CT 06901 (203) 359-1311

Description: Text book designed to enhance reading comprehension of high school students by improving the thinking and reasoning process that support higher-order comprehension. **Author/Editor:** Dr. Arthur Whimbey.

★2288★ *Analyze*
Kolbe Concepts, Inc.
P.O. Box 15050
Phoeniz, AZ 85060 (602) 840-9770

Description: Activity book containing problem solving situations that stimulate the use of analysis.

★2289★ *Apple Shines*
Good Apple, Inc.
1204 Buchanan St.
P.O. Box 299
Carthrage, IL 62321 (217) 357-3981

Description: "Think-and-then-write" activities to stimulate elementary and middle school students to use original ideas as a source. Each activity includes an introduction, instruction, and follow-up. **Author/Editor:** Bob Eberle.

★2290★ *Art and Perception - The Flexible Line*
Foxtail Press
P.O. Box 2996
La Habra, CA 90632 (214) 552-3922

Description: Information and activity-based book focusing on line. **Author/Editor:** Betty Lewis and Marge Tezak.

★2291★ *The Art of Perceiving Problems*
Foxtail Press
P.O. Box 2996
La Habra, CA 90632 (214) 522-3922

Description: Booklet to help intermediate students become aware of problem situations, understand mental blocks, and learn techniques of dealing with problems in unique and creative ways. **Author/Editor:** Eileen Babcock and Marilyn Brown.

★2292★ *The Art of Resolving Problems*
Foxtail Press
P.O. Box 2996
La Habra, CA 90632 (214) 522-3922

Description: Booklet presenting middle school teachers and students with sequential methods for creative problem solving. **Author/Editor:** Marilyn Brown and Eileen Babcock.

★2293★ *Basic Thinking Skills*
Society for Visual Education
1345 W. Diversey Parkway
Chicago, IL 60614 (312) 525-1500

Description: Four sound filmstrips designed to teach middle school students to think: Finding the Main Idea, Coming to Conclusions, Developing Ideas, and Deciding on the Facts. Includes teacher's guide.

★2294★ *Be an Inventor*
Harcourt Brace Jovanovich
111 Fifth Ave.
New York, NY 10003 (212) 614-3000

Description: Shows children how to turn their ideas into inventions by leading them step-by-step through the inventive process. Designed for teachers, students, and parents. **Author/Editor:** Barbara Taylor.

★2295★ *The Book for Women Who Invent or Want To*
Women Inventors Project
P.O. Box 689
Waterloo, ON, Canada N2J 4B8

Description: Discusses the creative process from the idea to product stage, including the patent process and marketing. Covers how to create a network of women inventors. **Author/Editor:** Elizabeth Wallace.

★2296★ *Brain Muscle Builders - Games to Increase Your Natural Intelligence*
Trillium Press, Inc.
Box 921, Madison Square Station
New York, NY 10159 (914) 783-2999

Description: Book encouraging game playing as a means of improving thinking skills. **Author/Editor:** Marco Meirovitz and Paul I. Jacobs.

★2297★ *Brain Scratchers*
Kolbe Concepts, Inc.
P.O. Box 15050
Phoenix, AZ 85060 (602) 840-9770

Description: Instructions for designing crossword, hidden puzzle, and maze games.

★2298★ *A Catalog of Programs for Teaching Thinking*
Research For Better Schools
444 N. Third St.
Philadelphia, PA 19123 (215) 574-9300

Description: Catalog for teachers providing concise summaries of some major commercial published programs that teach thinking. **Author/Editor:** Janice Kruse and Barbara Z. Presseisen.

★2299★ *Catalog of Publications and Services*
Center for Creative Learning
P.O. Box 619
Honeoye, NY 14471

Description: Catalog listing publications related to thinking skills and creativity as well as programs and workshops offered at the Center.

★2300★ *Challenge*
Dale Seymour Publications
P.O. Box 10888
Palo Alto, CA 94303 (415) 324-2800

Description: Magazine for students in grades 5-8. **Author/Editor:** Carole Greenes, George Immerzeel, Linda Schulman, and Rita Spungin.

★2301★ *Challenge Boxes*
Dale Seymour Publications
P.O. Box 10888
Palo Alto, CA 94303 (415) 324-2800

Description: Book of 50 activities, each with a creative thinking theme, that can be used with middle school students as a learning center or enrichment challenge. **Author/Editor:** Katherine Valentino.

★2302★ *The Challenge of the Unknown*
W.W. Norton Company, Inc.
500 Fifth Ave.
New York, NY 10110 (212) 354-5500

Description: Videotape with teacher's manual for a seven-part film series on mathematics and problem solving for grades 5-9. **Author/Editor:** Hillary C. Maddus.

★2303★ *Challenges for Children*
NL Associates, Inc.
P.O. Box 1199
Highstown, NJ 08250

Description: Contains activities to stimulate creative thinking and challenge elementary school children to solve problems.

★2304★ *Chrysalis - Nurturing Creative and Independent Thought in Children*
Zephr Press
430 S. Essex Lane
Tucson, AZ 85711

Description: Designed to provide middle and high school teachers with a framework for educational experiences for developing creativity, self-reliance, and an independent approach to learning. Seven units cover thinking and feeling, health, scientific phenomena, esthetics, fine art, world problems, exploring the future, and a seqenced plan for independent learning. **Author/Editor:** Micki McKisson.

★2305★ *Circles of Creativity*
Point Publishing Company
P.O. Box 1309
Point Pleasant Beach, NJ 08742 (201) 295-8258

Description: Brainstorming aid consisting of 276 idea stimulator words and phrases located on rotating disks. The disks are used to prompt ideas through the use of random stimuli. **Author/Editor:** Arthur B. VanGundy and James Betts.

★2306★ *Citizenship Decision-Making*
Addison-Wesley Publishing Company
South St.
Reading, MA 01867 (617) 944-3700

Description: Set of 25 social studies supplementary lessons for elementary and junior high school teachers that focus on building students' skills with the tasks of making, judging, and influencing decisions. **Author/Editor:** Roger La Raus and Richard C. Remy.

★2307★ *Cognetics*
The Talent Network
Educational Information and Resource Center
700 Hollydell Ct.
Sewell, NJ 08080 (609) 582-7000

Description: Annual school program to stimulate creativity and creative problem solving. Subscription includes newsletters and other supplemental material and three manuals providing a background in creativity and creative problem solving, six problems to be solved, and a calendar, forms, and procedures for the current program year. **Author/Editor:** Theodore J. Gourley and Judith Burr.

★2308★ *The Cort Thinking Program*
Pergamon Press, Inc.
Maxwell House, Fairview Park
Elmsford, NY 10523 (914) 592-7700

Description: Program of six units with each unit containing 10 lessons that teach separate thinking skills. Includes teacher's handbook and 10 student workcards for each lesson in the first five units; unit six includes a teacher's handbook and student textbook. **Price:** $185.00. **Author/Editor:** Edward de Bono.

★2309★ *Creative Encounters with Creative People*
Good Apple, Inc.
1204 Buchanan St.
P.O. Box 299
Carthrage, IL 62321 (217) 357-3981

Description: Contains analysis of creative personalities such as Henry Ford, Walt Disney, and Joni Eareckson. Each unit includes a biographical sketch, creative encounters, and work sheets. **Author/Editor:** Janice Gudeman.

★2310★ *Creative Kids*
GCT Inc.
P.O. Box 6448
Mobile, AL 36660 (205) 478-4700

Description: Magazine containing stories, poetry, music, artwork, plays, experiments, and other features contributed by children aged five to eighteen. The work represents children's ideas, questions, fears, concerns, and pleasures. **Author/Editor:** Fay L. Gold.

★2311★ *Creative Problem Solving: A Guide for Trainers and Management*
Quorum Books
88 Post Rd. West
Westport, CT 06881 (203) 226-3571

Description: Presents a six-step problem solving model designed to deal with problems that require creative solutions. Includes guidelines for teachers. **Author/Editor:** Dr. Arthur B. VanGundy.

★2312★ *Creative Problem Solving Techniques*

Trillium Press
Box 921, Madison Square Station
New York, NY 10159 (914) 783-2999

Description: Handbook of creative problem solving procedures for middle and high school students. **Author/Editor:** Julie L. Ellis.

★2313★ *Creative Writing in Action*

Good Apple, Inc.
1204 Buchanan St.
P.O. Box 299
Carthrage, IL 62321 (217) 357-3981

Description: Book designed to enhance writing and speaking skills. Student activities focus on characterizations, dialogue development, setting, and plot. **Author/Editor:** Elizabeth Marten and Nina Crosby.

★2314★ *Creativity 1,2,3*

Trillium Press
Box 921, Madison Square Station
New York, NY 10159 (914) 738-2999

Description: Book providing ideas on how to make creativity an integral part of the elementary classroom. Explains ideas for incorporating creativity and encourages teachers to increase their own creativity. **Author/Editor:** Susan Ellis Baum and Martha Cray Andrews.

★2315★ *Critical Thinking and Thinking Skills: State of the Art Definitions and Practice in Public Schools*

Research for Better Schools
444 N. Third St.
Philadelphia, PA 19123 (215) 574-9300

Description: Book providing an overview of the history of critical thinking from 1938 to the 1980s. Contains a reference list and appendix listing thinking skill meetings and conferences in the United States. **Author/Editor:** Barbara Z. Pressisen.

★2316★ *Critical Thinking: How to Evaluate Information and Draw Conclusions*

The Center for Humanities, Inc.
Communications Park, Box 3000
Mt. Kisco, NY 10549-9989 (914) 666-4100

Description: Videotape uses scenarios with teenagers to explain concept evaluation and establish critical thinking skills. Includes teacher's manual. **Price:** $197.00.

★2317★ *Daughters of Invention: An Invention Workshop for Girls*

Women Inventors Project
P.O. Box 689
Waterloo, ON, Canada N2J 4B8

Description: Handbook suggests guidelines for planning invention workshops for girls. **Author/Editor:** Rachelle Beauchamp and Lisa Avedon.

★2318★ *Delta Science Modules*

Delta Education
P.O. Box M
Nashua, NH 03061-6012 (603) 889-8899

Description: Activity-based science series consisting of science topic modules for grades K-6. Modules cover concepts in life, earth, and phyical science and are accompanied by a classroom kit of materials for activities.

★2319★ *Detecting and Deducing - Preparing for Logical Thinking*

Foxtail Press
P.O. Box 2996
La Habra, CA 90632 (214) 522-3922

Description: Booklet to assist intermediate school students with logic, become aware of assumptions, use facts to reach valid conclusions, avoid generalizations, consider evidence, and recognize and use the syllogistic pattern for logic. **Author/Editor:** Eileen Babcock, Marilyn Brown, and Betty Lewis.

★2320★ *The Discovery*

Foxtail Press
P.O. Box 2996
La Habra, CA 90632 (214) 522-3922

Description: Describes a creative problem solving process for teachers and students integrating all areas of study. **Author/Editor:** Ken Mittan.

★2321★ *Do-It-Yourself Critical and Creative Thinking*

Kolbe Concepts, Inc.
P.O. Box 15050
Phoeniz, AZ 85060 (602) 840-9770

Description: Explains thinking skills and provides confidence-building activities.

★2322★ *Education and Learning to Think*

National Academy Press
2102 Constitution Ave., N.W.
Washington, DC 20418 (202) 334-3318

Description: Addresses the question of what American educators can do to teach higher order skills more effectively. **Author/Editor:** Lauren B. Resnick.

★2323★ *Effective Questions to Strengthen Thinking*

Foxtail Press
P.O. Box 2996
La Habra, CA 90632 (214) 522-3922

Description: Booklet addresses logical, critical, and creative thinking through questioning. Questions are for all grade levels; most material is for middle and high school students. **Author/Editor:** Marilyn Brown.

★2324★ *Elementary Science Study* (ESS)

Delta Education
P.O. Box M
Nashua, NH 03061-6012 (603) 889-8899

Description: Designed to teach children to think critically about the scientific world through hands-on experience. Incorporates different instructional strategies and styles; includes activity kits for scientific investigation with materials for eight students.

★2325★ *Evaluate*

Kolbe Concepts, Inc.
P.O. Box 15050
Phoeniz, AZ 85060 (602) 840-9770

Description: More than 50 activities designed to develop levels of logical thinking and evaluation with elementary and middle school children.

★2326★ *Expanding Creative Imagination*
NL Associates, Inc.
P.O. Box 1199
Highstown, NJ 08250

Description: Student workbook containing over 100 exercises and activities to develop major aspects of creative thinking. Teacher's guide contains information and directions to accompany student book; activities require no teacher preparation.

★2327★ *Exploring the Lives of Gifted People in the Arts*
Good Apple, Inc.
1204 Buchanan St.
P.O. Box 299
Carthrage, IL 62321 (217) 357-3981

Description: Profiles contemporary heroes such as Bart Connor, Dr. Norman Vincent Peale, and Peter Strauss. Encourages students to become involved in creative problem solving and decision making processes and to expand career awareness. **Author/Editor:** Kathy Balsamo.

★2328★ *Exploring the Lives of Gifted People in the Sciences*
Good Apple, Inc.
1204 Buchanan St.
P.O. Box 1204
Carthrage, IL 62321 (217) 357-3981

Description: Six units containing an interview with contemporary heroes including Dr. Sally Ride, Barbara Jordan, and Dr. Isaac Asimov, followed by over 40 activities. **Author/Editor:** Kathy Balsamo.

★2329★ *Eye Cue Puzzles Sets*
Dale Seymour Publications
P.O. Box 10888
Palo Alto, CA 94303 (415) 324-2800

Description: Set of 40 challenges designed to promote visual thinking skills.

★2330★ *Fact Fantasy and Folklore*
Good Apple, Inc.
1204 Buchanan St.
P.O. Box 299
Carthrage, IL 62321 (217) 357-3981

Description: Stimulates critical thinking using 11 well-known fairy tales through role play, creative activities, and open-ended questions. Includes background clues for the teacher and lesson plans. **Author/Editor:** Greta B. Lipson and Baxter Morrison.

★2331★ *Focus on Thinking*
Focus on Thinking Foundation
Box 430
Ivermere, BC, Canada V0A 1K0

Description: Focus on Thinking I - An Introduction to Thinking Skills contains interviews with many of Canada's top thinking skills consultants. Focus on Thinking II is a television program of interviews of top consultants from Canada and the United States.

★2332★ *Future Options - Unlimited*
Foxtail Press
P.O. Box 2996
La Habra, CA 90632 (214) 522-3922

Description: Booklet on students' future needs as adults. Includes exercises requiring research and recordkeeping, vocabulary needed for jobs and careers of the future, background materials, news items, and study units. **Author/Editor:** Dr. Eldon Meyler and Donald David Zielinski.

★2333★ *Game Puzzles for the Joy of Thinking*
Kadon Enterprises, Inc.
1227 Lorene Dr., Suite 16
Pasadena, MD 21122 (301) 437-2163

Description: Catalog of original games and puzzles based on mathematical principles and geometric sets. **Author/Editor:** Kathy Jones.

★2334★ *The Gifted Child Today*
GCT Inc.
P.O. Box 6448
Mobile, AL 36660 (205) 478-4700

Description: Magazine designed to meet the needs of parents and teachers in working with gifted, creative, and talented children. **Author/Editor:** Marvin J. Gold.

★2335★ *Gifted Children Monthly*
Gifted and Talented Publications, Inc.
213 Hollydell Dr.
Sewell, NJ 08080 (609) 582-0277

Description: Provides columns on a variety of topics. **Subscription:** $24.00.

★2336★ *The Great Bridge Lowering*
Trillium Press
Box 921, Madison Square Station
New York, NY 10159 (914) 783-2999

Description: Guide to creative thinking for elementary and middle school students. Encourages development of the ability to generate ideas, break mind sets, create new ideas, and elaborate on ideas. **Author/Editor:** Sandra Warren.

★2337★ *How Can We Teach Intelligence?*
Research For Better Schools
444 N. Third St.
Philadelphia, PA 19123 (215) 574-9300

Description: Report on three programs that train aspects of intelligence. Includes references and bibliography. **Author/Editor:** Robert J. Sternberg.

★2338★ *How do You Figure That? Ways of Problem Solving*
The Center for Humanities, Inc.
Communications Park, Box 3000
Mt. Kisco, NY 10549-9989 (914) 666-4100

Description: Set of four filmstrips/cassettes covering a problem to be systematically solved using a five-step procedure by students in grades 3-6. Includes teacher's guides. **Price:** $139.00.

★2339★ *How to Invent*
IFI/Plenum Data Corporation
302 Swann Ave.
Alexandria, VA 22301 (703) 683-1085

Description: Introduces basic principles, methods, and tools of the inventing process. Includes drawings from famous patents, information on protecting inventions, and a history of inventions in the United States. **Author/Editor:** B. Edward Shelesinger, Jr.

★2340★ *The Idea Generator*
Experience in Software, Inc.
2039 Shattuck Ave., Suite 401
Berkeley, CA 94704 (415) 644-0694

Description: Offers middle and high school students a step-by-step approach to solving problems. **Author/Editor:** Gerald I. Nierenberg.

★2341★ *Imagine That*
Rainbow Planet
5110 Comwell Dr.
Gig Harbor, WA 98335

Description: Album of children's songs that paint a picture of children's fantasies. **Author/Editor:** Jim Valley.

★2342★ *Improving the Quality of Student Thinking*
Association for Supervision and Curriculum Development
Order Processing Department
125 N. West St.
Alexandria, VA 22314-2798 (703) 549-9110

Description: Videotape created for staff development. Contains explanations of methods for improving the quality of student thinking.

★2343★ *Invention Convention Procedural Manual*
Richardson Independent School District
400 S. Greenville Ave.
Richardson, TX 75081 (214) 238-8111

Description: Covers administrative aspects of organizing an invention convention at the classroom, building, or district level. A companion book to Inventive Thinking: A Teacher-Student Handbook.

★2344★ *Inventioneering*
Good Apple, Inc.
1204 Buchanan St.
P.O. Box 299
Carthrage, IL 62321 (217) 357-3981

Description: Provides invention instructions, content, process and product objectives, and photographs of student inventions. **Author/Editor:** Bob Stanish and Carol Singletary.

★2345★ *The Inventive Child*
Encyclopedia Britannica Educational Corporation
425 N. Michigan Ave.
Chicago, IL 60611 (312) 347-7400

Description: Series of 20 film programs providing a basis for creative thinking and problem solving activities in science, social studies, communication arts, industrial arts, home economics, art, music, and interdisciplinary humanities programs. **Author/Editor:** Grace N. Lacey.

★2346★ *The Inventive Imagination to Illumination*
Foxtail Press
P.O. Box 2996
La Habra, CA 90632 (214) 522-3922

Description: Activity booklet of exercises to enhance inventive thinking. Based on the work of Guilford, Torrance, and Parnes. **Author/Editor:** Marilyn Brown.

★2347★ *The Inventive Innovation to Ingenuity*
Foxtail Press
P.O. Box 2996
La Habra, CA 90632 (214) 522-3922

Description: Booklet containing background information, teacher directions, and student activities on invention, imagination, resourceful thinking, and the invention process. **Author/Editor:** Marilyn Brown.

★2348★ *Inventive Thinking: A Teacher-Student Handbook*
Richardson Independent School District
400 S. Greenville Ave.
Richardson, TX 75081 (214) 238-8111

Description: Handbook for teachers and students provides activities on creativity and a four step method for teaching the inventive process; includes bibliography.

★2349★ *The Inventor's Guide*
Haley Publications
Box 335
Lexington, MA 02173

Description: Guide to the inventing process, documenting the process, inventing on and off the job, assessing market and financial worth, protecting an invention, building a business, and producing a product. **Author/Editor:** Dr. Donald D. Job.

★2350★ *The Journal of Creative Behavior*
Creative Education Foundation, Inc.
437 Franklin St.
Buffalo, NY 14202 (716) 884-2744

Description: Journal containing articles which tend to be theoretical on the fields of creativity and problem solving. **Frequency:** Quarterly.

★2351★ *Just Think Program Series and Stretch Think Program Series*
Thomas Geal Publications, Inc.
P.O. Box 370540
Montara, CA 94037

Description: Curricula to teach thinking from the pre-school through grade 8. Each full year thinking program includes lessons, instructions, worksheets, and a matrix of curricular content objectives. **Author/Editor:** Sydney Billig Tyler.

★2352★ *Keep Them Thinking*
Illinois Renewal Institute, Inc.
200 E. Wood St., Suite 250
Palatine, IL 60067 (312) 991-6300

Description: Three books present lesson designs to teach, rehearse, practice, and transfer explicit thinking skills: Level I: Primary (K-4), Level II: Intermediate (5-8), and Level III: Advanced (9-12). **Author/Editor:** Jim Bellanca, Kay Opeka, and Robin Fogarty.

★2353★ *The Lateral Thinking Machine*
Gemini Group
RD #2, Box 117
Bedford, NY 10506

Description: IBM personal computer-based program for problem solving. **Author/Editor:** Ken Finn.

★2354★ *Lessons in Logic: Unravelling Common Complexities*
Kolbe Concepts, Inc.
P.O. Box 15050
Phoenix, AZ 85060 (602) 840-9770

Description: Course teaching creative problem solving methodologies. Provides information needed for specific problem solving approaches and includes practice activities.

★2355★ *Life After School*
Simon & Schuster
Julian Mesmer Division
1230 Avenue of the Americas
New York, NY 10020 (212) 698-7000

Description: Self-help career development book for 11th and 12th graders which enables students to identify and determine their creative skills and possible creative careers.

★2356★ *Literature is for Thinking* (LIFT)
Sundance Publishers & Distributors, Inc.
Newton Rd.
Littleton, MA 01460 (508) 486-9201

Description: Series of supplementary materials to be used with literary books to stimulate high school students' development of critical and creative thinking and language skills. Includes rational for the program, teacher's guide, and 14-18 reproducable activity sheets for students' use. **Author/Editor:** Katherine Paterson.

★2357★ *Logic Number Problems*
Dale Seymour Publications
P.O. Box 10888
Palo Alto, CA 94303 (415) 324-2800

Description: Collection of 50 number puzzles designed to provide experience in problem solving and thinking skills for middle and high school mathematics students. **Author/Editor:** Wade H. Sherard III.

★2358★ *Looking Glass Logic: Problems and Solutions*
The Perfection Form Company
1000 N. Second Ave.
Logan, IA 51546 (712) 644-2831

Description: Workbook uses illustrations and quotations from Lewis Carroll's classics to demostrate the principles of logic. **Author/Editor:** Kristin Kalsem.

★2359★ *Managing Group Creativity*
American Management Associations
135 W. 50th St.
New York, NY 10020 (212) 903-8089

Description: Provides teachers with information on group dynamics, effectiveness, and problem solving. Includes references to research literature and questionnaires for evaluating problem-solving techniques. **Author/Editor:** Dr. Arthur B. VanGundy.

★2360★ *Mastering Reading Through Reading*
Innovative Sciences, Inc.
300 Broad St.
Stamford, CT 06901 (203) 359-1311

Description: Secondary level text book geared to improving the thinking and reasoning processes that support higher-order comprehension. Includes teacher's guide. **Author/Editor:** Dr. Arthur Whimbey.

★2361★ *Mathematics Pentathlon*
Pentathlon Institute
P.O. Box 20590
Indianapolis, IN 46220 (317) 782-1553

★2362★ *Max Think*
Max Think, Inc.
230 Crocker Ave.
Piedmont, CA 94610

Description: IBM-based computer software for processing ideas. Includes instruction manual. **Author/Editor:** Neil Larson.

★2363★ *Mental Menus*
Illinois Renewal Institute, Inc.
200 E. Wood St., Suite 250
Palatine, IL 60067 (312) 991-6300

Description: Lesson plans for 24 critical and creative thinking skills including inferencing, contrasting, determining bias, analyzing assumptions, and hypothesizing. **Price:** $14.95. **Author/Editor:** Rubin Fogarty and James Bellanca.

★2364★ *The Million Dollar Idea*
New Product Development
P.O. Box 1309
Point Pleasant, NJ 08742 (201) 295-8258

Description: Paperback book on new product ideas and developing them. **Author/Editor:** Jim Betts.

★2365★ *Mind Games: Puzzles and Logic*
The Perfection Form Company
1000 N. Second Ave.
Logan, IA 51546 (712) 644-2831

Description: Explains three types of logic problems and instructs students how to break problems into smaller units, examine each part, and identify relationships between parts. **Author/Editor:** Kristin Kalsem.

★2366★ *Mind Joggers*
NL Associates, Inc.
P.O. Box 1199
Highstown, NJ 08250

Description: Contains activities for elementary and middle school students covering thinking and reasoning, math, language and writing, and listening and remembering. **Author/Editor:** Susan S. Petreshene.

★2367★ *Mind Movers - Creative Homework Assignments*
Addison-Wesley Publishing Company
South St.
Reading, MA 01867 (617) 944-3700

Description: Weekly homework assignments, most of which contian higher level thinking skills, for students in grades 3-6. **Author/Editor:** Diane Hart and Margaret Rechif.

★2368★ *New Frontiers/Nuevas Fronteras*
Pergamon Press, Inc.
Maxwell House, Fairview Park
Elmsford, NY 10523 (914) 592-7700

Description: Oral, early learning congitive curriculum program with a strong emphasis on mathematics and science. Designed to encourage young Spanish-speaking children to learn English and develop thinking skills and problem solving strategies. **Author/Editor:** Barbara Coffingan Cox, Janet McCaulay, and Manual Ramierz III.

★2369★ *A New Way to Use Your Bean - Developing Thinking Skills in Children*
Trillium Press
Box 921, Madison Square Station
New York, NY 10159 (914) 783-2999

Description: Book containing 30 cooking activities that introduce critical thinking, creative thinking, logic, and problem solving. Includes descriptions of thinking processes utilized, materials and equipment needed, precooking activities, cooking suggestions, and suggestions for additional activities. **Author/Editor:** Darlene Freeman.

★2370★ *Oceanography*
Engine-Uity, Ltd.
P.O. Box 9610
Phoeniz, AZ 85068 (602) 997-7144

Description: Writing activities to help children learn how to think creatively and to use critical thinking skills.

★2371★ *Odyssey of the Mind: Problems to Develop Creativity*
Creative Competitions, Inc.
P.O. Box 27
Glassboro, NJ 08028 (609) 881-1603

Description: Book of 57 long-term and spontaneous problems used in Odyssey of the Mind competitions. **Price:** $12.50. **Author/Editor:** C. Samuel Micklus.

★2372★ *OM-AHA! Problems to Develop Creative Thinking Skills*
Creative Competitions, Inc.
P.O. Box 27
Glassboro, NJ 08028 (609) 881-1603

Description: Book of problems designed to allow students to use their imaginations. Includes nine long-term problems, more than 40 spontaneous problems, and nine Odyssey of the Mind World Finals warm-up problems. **Price:** $15.50. **Author/Editor:** C. Samuel Micklus.

★2373★ *Patent Pending*
Social Studies School Service
P.O. Box 802
Culver City, CA 90232-0802 (213) 839-2436

Description: Film providing an overview of the U.S. patent system and highlights inventions in use today.

★2374★ *Perception, Inc.*
Illinois Renewal Institute, Inc.
200 E. Wood St., Suite 250
Palatine, IL 60067 (312) 991-6300

Description: Software program designed to provide early elementary students with practice in perception, attributing, and classification. Includes student activity book and teacher's guide.

★2375★ *Planning for Thinking*
Illinois Renewal Institute, Inc.
200 W. Wood St., Suite 250
Palatine, IL 60067 (312) 991-6300

Description: Book to guide instructional leaders in designing and implementing an effective thinking curriculum for their districts. **Author/Editor:** James Bellanca and Robin Fogarty.

★2376★ *Problem Solving and Comprehension*
Lawrence Erlbaum Associates, Inc.
365 Broadway
Hillsdale, NJ 07642 (201) 666-4110

Description: Addresses how to increase secondary students' problem solving and test taking skills and learning comprehension. **Author/Editor:** Arthur Whimbey and Jack Lockhead.

★2377★ *Problem Solving in Science*
Curriculum Associates, Inc.
5 Esquire Rd.
North Billerica, MA 01862 (508) 667-8000

Description: Activity-based program that builds reading comprehension and creative problem solving in the content area of science. **Author/Editor:** Louis James Taris and James Robert Taris.

★2378★ *Problems! Problems! Problems! Discussions and Activities Designed to Enhance Creativity*
Creative Competitions, Inc.
P.O. Box 27
Glassboro, NJ 08028 (609) 881-1603

Description: Features a discussion on creativity and over 60 creative long-term and spontaneous problems. **Price:** $10.95. **Author/Editor:** C. Samuel Micklus and T. Gourley.

★2379★ *Product Improvement Checklist*
Point Publishing Company
P.O. Box 1309
Point Pleasant Beach, NJ 08742 (201) 295-8258

Description: Contains 576 stimulator words and phrases to be used as a brainstorming aid to prompt ideas by provoking associations between unrelated stimuli and aspects of a problem. **Price:** $10.00. **Author/Editor:** Arthur B. VanGundy.

★2380★ *Project: Problem Solving*
Kolbe Concepts, Inc.
P.O. Box 15050
Phoenix, AZ 85060 (602) 840-9770

Description: Kit for do-it-yourself patterns and constructions for bending light, designing mazes, creating film strip viewers, discovering variances in air rockets, and other activities.

★2381★ *Project Success Enrichment*
Creative Child Concepts
Station 111, P.O. Box 61100
Seattle, WA 98121

Description: Nationally-validated (NDN) language arts and art curriculum appropriate for grades 2-8 that uses cooperative learning strategies to address critical and creative thinking processes, self-management, and social skills. **Author/Editor:** Carolyn G. Bronson, Sally Maryatt, and Karlene George.

★2382★ *Prolific Thinkers Guide*
Dale Seymour Publications
P.O. Box 10888
Palo Alto, CA 94303 (415) 324-2800

Description: Guide to teaching critical and creative thinking using a hands-on approach. **Author/Editor:** Gary A. Carnow and Constance Gibson.

★2383★ *Quizzles - Logic Problem Puzzles*
Dale Seymour Publications
P.O. Box 10888
Palo Alto, CA 94303 (415) 324-2800

Description: Booklet of 38 problems designed as student worksheets that require inventive thinking. Also available is More Quizzles - Logic Problem Puzzles containing 48 puzzles of increasing difficulty. **Author/Editor:** Wayne Williams.

★2384★ *Risk Taking*
Kolbe Concepts, Inc.
P.O. Box 15050
Phoenix, AZ 85060 (602) 840-9770

Description: Activities designed to help elementary and middle school students practice decision making skills.

★2385★ *Science Curriculum Improvement Study*
Delta Education
P.O. Box M
Nashua, NH 03061-6012 (603) 889-8899

Description: Elementary science series using hands-on activities to develop thinking skills. Through each topic, students proceed in discover, concept, introduction, and concept application activities.

★2386★ *Secrets and Surprises*
Good Apple, Inc.
1204 Buchanan St.
P.O. Box 299
Carthrage, IL 62321 (217) 357-3981

Description: Contains 183 activities focusing on creative thinking to encourage language development for elementary and middle school children. **Author/Editor:** Joe Wayman and Lorraine Plum.

★2387★ *Simulations*
Knolbe Concepts, Inc.
P.O. Box 15050
Phoeniz, AZ 85060

Description: Presents 20 realistic situations that involve exercising judgement, wighing alternatives, and considering consequences.

★2388★ *Solve - Action Problem Solving*
Curriculum Associates, Inc.
5 Esquire Rd.
North Billerica, MA 01862 (508) 667-8000

Description: Five step program for students in grades 4-9 that integrates strategies with a problem solving procedure. Problems are presented through short stories and are illustrated by action photos. Student books and teacher's guides are available for: Book I - Whole Numbers; Book II - Fractions; Book III - Decimals and Percents. **Price:** $3.95 for student books; $5.95 for teacher's guides. **Author/Editor:** Brian E. Enright.

★2389★ *Sound Ideas*
Foxtail Press
P.O. Box 2996
La Habra, CA 90632 (214) 522-3922

Description: Activity booklets with teacher directions and background information on critical thinking: Critical Thinking Grades 1-2, Critical Thinking Grades 4-5, Logical Thinking Grades 7-8, and Creative Thinking Grades 7-8.

★2390★ *Springboards to Creative Thinking*
NL Associates, Inc.
P.O. Box 1199
Highstown, NJ 08250

Description: Contains 101 activites to stimulate creative thinking in students in grades 3-6. **Author/Editor:** Patricia Tyler Muncy.

★2391★ *Start Them Thinking*
Illinois Renewal Institute, Inc.
200 E. Wood St., Suite 250
Palatine, IL 60067 (312) 991-6300

Description: Handbook of 37 classroom strategies designed to promote thinking and stimulate active participation in the K-4 classroom. **Author/Editor:** Robin Fogarty and Kay Opeka.

★2392★ *Stories to Strech Minds*
NL Associates, Inc.
P.O. Box 1199
Highstown, NJ 08250

Description: Presents short story situations designed to be read aloud. Listeners are to ask yes or no questions to decide how or why situations occured. Available in four volumes. **Author/Editor:** Nathan Levy.

★2393★ *Strategic Reasoning*
Innovative Sciences, Inc.
300 Broad St.
Stamford, CT 06901 (203) 359-1311

Description: Systematic approach to improving the thinking process. **Author/Editor:** Dr. Arthur Whimbey.

★2394★ *Teachers Teaching Thinking*
Illinois Renewal Institute, Inc.
200 E. Wood St., Suite 250
Palatine, IL 60067 (312) 991-6300

Description: Videotape series for teachers showing thinking skills being taught in a classroom within the context of a subject. Each of four tapes addresses a different age group and subject: Structured Interaction with Thinking, Teaching Explicit Skills of Thinking, Setting the Climate for Thinking, and Metacognitive Processing About Thinking. **Price:** $79.00 each or $295.00 for set of 4. **Author/Editor:** Robin Fogarty and Jim Bellanca.

★2395★ *Teaching and Learning Mathematical Problem Solving: Multiple Research Perspectives*
Lawrence Erlbaum Associates, Inc.
365 Broadway
Hillsdale, NJ 07642 (201) 666-4110

Description: Compilation of 24 papers presented at San Diego State University in 1983 on learning mathematical problem solving. **Author/Editor:** Edward A. Silver.

★2396★ *Techniques of Structured Problem Solving*
Van Nostrand Reinhold Co., Inc.
115 Fifth Ave.
New York, NY 10003 (212) 254-3232

Description: Booklet containing descriptions and evaluations of over 100 techniques for analyzing and redefining problems, and generating, evaluating, selecting, and implementing ideas. Includes bibliography and index. **Author/Editor:** Dr. Arthur B. VanGundy.

★2397★ *Technology, Innovation & Entrepreneurship for Students Magazine* (TIES Magazine)
Drexel University
College of Design Arts
Philadelphia, PA 19104 (215) 895-2386

Description: Drawn upon the information resources of industry, business, government, and education to assist teachers interested in helping students increase technological literacy and capability. **Frequency:** 6 per year. **Subscription:** Free to teachers and administrators; $15 for others.

★2398★ *Think and Reason*
Weekly Reader
245 Long Hill Rd.
Middletown, CT 06457 (203) 638-2400

Description: Paperback activity books designed to teach students thinking skills through problem-oriented activities and puzzles. **Author/Editor:** Donald Barnes and Arlene Burgdorf.

★2399★ *A Thinker's Log*
Illinois Renewal Institute, Inc.
200 E. Wood St., Suite 250
Palatine, IL 60067 (312) 991-6300

Description: Booklet designed to provide high school students with a recording device for reflections on concepts encountered in current studies and class interactions. **Author/Editor:** Robin Fogarty.

★2400★ *The Thinker's Toolbox*
Dale Seymour Publications
P.O. Box 10888
Palo Alto, CA 94303 (415) 324-2800

Description: Book introduces ways to teach and strengthen divergent thinking skills. Presents 16 tools used in problem solving and contains more than 50 problems that allow for a choice of tools to create alternative solutions. **Author/Editor:** Pamela and David Thornbury.

★2401★ *The Thinking Log*
Illinois Renewal Institute, Inc.
200 E. Wood St., Suite 250
Palatine, IL 60067 (312) 991-6300

Description: A 24-page workbook for elementary school students to be used to respond to ideas presented in class. **Price:** $52.50 for 30. **Author/Editor:** Robin Fogarty.

★2402★ *Thinking Posters: Keys to Critical Thinking*
Sundance Publishers & Distributors, Inc.
Newton Rd.
Littleton, MA 01460 (508) 468-9201

Description: Series of lessons designed to improve students' (grades 5-8) critical thinking by teaching them to distinguish good from bad critical thinking. **Author/Editor:** D.N. Perkins.

★2403★ *Thinking Skills: Meanings, Models, and Materials*
Research For Better Schools
444 N. Third St.
Philadelphia, PA 19123 (215) 574-9300

Description: Brief synopsis, models, and discussions of available thinking programs. **Author/Editor:** Barbara Z. Presseisen.

★2404★ *Thinking Skills Set*
Kolbe Concepts, Inc.
P.O. Box 15050
Phoenix, AZ 85060 (602) 840-9770

Description: Presents examples of critical thinking and eight creative behaviors, with practical applications, to enhance student understanding of the thinking process.

★2405★ *Thinking to Write - A Work Journal Program*
Curriculum Development Associates, Inc.
1211 Connecticut Ave., N.W., Suite 414
Washington, DC 20036 (202) 293-1760

Description: Journals intended to provide students with an opportunity to assume control of their thinking and to reflect on their thinking process and problem solving abilities. Includes a teacher's edition. **Author/Editor:** Frances R. Link and Shannon Almquist.

★2406★ *Thinking Visually*
Dale Seymour Publications
P.O. Box 10888
Palo Alto, CA 94303 (415) 324-2800

Description: Strategy manual that describes approaches to problem solving through visual thinking. **Author/Editor:** Robert McKim.

★2407★ *The Thoughtwave Curriculum - Applying Creative Thinking to Problem Solving*
Foxtail Press
P.O. Box 2996
La Habra, CA 90632 (214) 522-3922

Description: Model for primary-level teachers interested in using a computer with students to develop higher-order thinking skills.

★2408★ *The Unconventional Invention Book*
Good Apple, Inc.
1204 Buchanan St.
P.O. Box 299
Carthrage, IL 62321 (217) 357-3981

Description: Collection of reproducible activities to stimulate inventive thinking and creating inventions. **Author/Editor:** Bob Standish.

★2409★ *Wake Up Your Creative Genius*
William Kaufmann, Inc.
95 First St.
Los Altos, CA 94022 (415) 948-5810

Description: Sourcebook of ideas providing techniques and practical examples that can be applied in realistic settings. Topics cover ways to increase and nurture creativity and methods for having ideas accepted and protected. **Author/Editor:** Kurt Hanks and Jay Parry.

★2410★ *Warm-up to Creativity*
Good Apple, Inc.
1204 Buchanan St.
P.O. Box 299
Carthrage, IL 62321 (217) 357-3981

Description: Games drawn from language arts and social studies skills to provide a conscious and deliberate crossover to divergent ways of thinking. **Author/Editor:** Bob Eberle.

★2411★ *What to Do?*
Trillium Press
Box 921, Madison Square Station
New York, NY 10159 (914) 783-2999

Description: Book of problem solving ideas with reproducible worksheets to assist students with fact collecting, defining the problem, and arriving at a solution. **Author/Editor:** Kathryn T. Hegeman.

★2412★ *What's Next*
Kolbe Concepts, Inc.
P.O. Box 15050
Phoeniz, AZ 85060 (602) 840-9770

Description: Set of 38 drawings and a cassette tape designed to encourage young children to develop observation skills by identifying discrepant events and predicting outcomes.

★2413★ *Wise Owl*
Sundance Publishers & Distributors, Inc.
Newton Rd.
Littleton, MA 01460 (508) 486-9201

Description: Educational package for teachers of grades 1-3. Designed to develop critical and creative thinking skills. Includes classic children's paperback book, reproducible activity cards with philosopical discussion questions, and teacher's guide for directing activities. **Author/Editor:** Lenore Carlisle and Gareth B. Matthews.

Chapter 11

ONLINE RESOURCES FOR THE INVENTOR

BIZ. ORBIT. CompuServe. DIALOG. BYTE. NEXIS. These words look and sound like ITT cable addresses, but represent just six of the major players in an industry encompassing more than 4,000 online databases. Such an abundance of information is available electronically, that choosing the appropriate database can be a complicated business.

The Gale publication *Computer-Readable Databases: A Directory and Data Sourcebook* reports on 4,500 databases representing a 57 percent increase over the years 1985 through 1988.

Years ago, electronic databases were used and understood only by specially trained librarians. Not any more. Today online databases are available to anyone, and no prior knowledge of computers is required. The learning curve is short. A few online databases can be learned with less than 20 minutes of practice.

Online databases are not the end-all by any means. The personal touch is often required to iron out the finer points and make sure that some original material has not been omitted. The data are, after all, input by human beings. But, online databases provide more information, faster and easier and more cost-effectively than manual searches.

The inventor is required to stay informed about myriad fields of research and development, many outside one's specialty. Before the advent of online databases this was a tough, time-consuming task. But with the present network of electronic tracking systems, the inventor is capable of monitoring fast-breaking events as they occur. All one requires is a PC connected to a phone line, and entry to numerous online databases. And, of course, you can now maintain select databases on your PC utilizing CD-ROM technology.

"Eighty to ninety percent of all the scientists that have ever lived are alive now," historian D.J. Price pointed out in his 1961 book *Science Since Babylon.* Thus, it is a good guess to venture that 80 to 90 percent of man's written communication has been put to paper in our lifetime.

Many of the publications cited in the chapter entitled Publications for the Inventor, and many more, are available via online databases. And your access to online databases may be as close as your own computer terminal or as far away as your nearest public library. For more information on databases for use by inventors, see the section entitled Online Patent Searches in Chapter 1, How to Protect and License Your Invention.

Chapter Arrangement

This chapter is arranged alphabetically by database name. All names in this chapter are referenced in the Master Index.

★2414★ AgPat
American Chemical Society (ACS)
Chemical Abstracts Service (CAS)
2540 Olentangy River Rd.
P.O. Box 3012
Columbus, OH 43210 (614) 447-3600

Description: Provides detailed abstracts for patent documents relating to agriculture and pest control and indexed in the printed Chemical Abstracts and its online counterpart, CA File. **Type of Database:** Bibliographic. **Record Items:** Title of document; inventor; patent assignee and location; patent application country, number, date, and pagination; CODEN; patent information; designated states; patent application information; patent priority application information; language of patent; graphic image (structure diagram which appears with abstract in printed Chemical Abstracts); abstract, including overview, independent claims, additional claims, uses and advantages, an illustrative example, and specifics including testing procedures, toxicity and process conditions, and other information; International Patent Classification (main and secondary); Chemical Abstracts section code and title; Chemical Abstracts section cross reference and code; index entries (substance and qualifiers); CAS Registry number.

★2415★ APIPAT
American Petroleum Institute (API)
Central Abstracting and Indexing Service
156 William St.
New York, NY 10038 (212) 587-9660

Description: Provides worldwide coverage of patents related to petroleum refining, the petro-chemical industry, and synthetic fuels. **Type of Database:** Bibliographic. **Record Items:** Author(s); corporate author(s); title of item; date; patent information. **Contact:** E.H. Brenner, Manager, Central Abstracting and Indexing Service.

★2416★ BioPatents
BIOSIS
BIOSIS Connection
2100 Arch St.
Philadelphia, PA 19103-1399 (215) 587-4800

Description: Contains references to recently granted U.S. patents in biotechnology, biomedicine, agriculture, and food technology. **Type of Database:** Bibliographic. **Record Items:** Patent number, patent title, U.S. Patent Classification number, inventor, inventor address, assignee, journal name, date granted, publication year. **Contact:** A.W. Elias, Director, Marketing and Distribution Division, or Gina S. Waserstein, Marketing Section Chief, BIOSIS.

★2417★ BNA's Patent, Trademark & Copyright Journal (Online)
The Bureau of National Affairs, Inc. (BNA)
BNA ONLINE
1231 25th St., N.W.
Washington, DC 20037 (202) 452-4132

Description: Interprets and analyzes developments in intellectual property issues. **Type of Database:** Full-text. **Contact:** BNA ONLINE Help Desk.

★2418★ BRANDY
Toyo Information Systems Co., Ltd.
Shinbashi-Sanwa-Toyo Bldg.
1-11-7, Shinbashi, Minato-ku
Tokyo 105, Japan

Description: Contains trademarks and patents registered in Japan. **Type of Database:** Full-text. **Contact:** Mr. K. Miyazaki, Database Service Section, Marketing and Sales Department, Toyo Information Systems Co., Ltd.

★2419★ BREV
Belgium Ministere des Affaires Economiques
Office de la Propriete Industrielle (OPRI)
24-26, rue J.A. De Mot
B-1040 Brussels, Belgium

Description: Contains references to all sectors of patentable activities in Belgium. **Type of Database:** Bibliographic. **Record Items:** International Patent Classification codes and English translation; descriptive text.

★2420★ BYTE
McGraw-Hill, Inc.
1221 Avenue of the Americas
48th Floor
New York, NY 10020 (212) 512-2911

Description: Provides information of interest to computer programmers, including new products, product reviews and comparisons, and highlights of technological developments in the computer industry. **Type of Database:** Full-text. **Record Items:** Title; journal name; publication date and year; page number; journal code; ISSN; section heading; byline; text; special feature (accompanying captions in graphs, tables, illustrations, and photographs). **Contact:** Andrea D. Broadbent, Manager, Marketing Services, McGraw-Hill, Inc.

★2421★ Canadian Patent Reporter (CPR)
Canada Law Book Inc.
240 Edward St.
Aurora, ON, Canada L4G 3S9 (416) 773-6300

Description: Provides case law data concerning copyright, patent, trademark, design, and intellectual property decisions. **Type of Database:** Full-text. **Contact:** Catherine Campbell, Manager, CAN/LAW Projects, Canada Law Book Inc.

★2422★ Chinese Patent Abstracts in English Data Base
International Patent Documentation Center (INPADOC)
Mollwaldplatz 4
A-1041 Vienna, Austria

Description: Contains bibliographic information and English-language abstracts of all patents published in the People's Republic of China since the opening of the Chinese Patent Office on April 1, 1985. **Type of Database:** Bibliographic. **Record Items:** The initial publication for a given invention in China (called a patent basic), plus any subsequent Chinese patent documents concerning the same invention; application country, date, or number; authors (inventors); International Patent Classification (IPC) Code; number of patents; patent assignee; patent country code; publication date or year; patent country code and number. **Contact:** Dipl.-Kfm Norbert Fux, Director, Sales Department, International Patent Documentation Center.

★2423★ CIBERPAT

Registro de la Propiedad Industrial (RPI)
Departamento de Informacion Tecnologica
Calle Panama, 1
28036 Madrid, Spain

Description: Provides bibliographic information and abstracts on patents and models registered in Spain. **Type of Database:** Bibliographic. **Record Items:** International classification, applicant, description of invention, significant date, and foreign country where the invention is registered. **Contact:** Dr. Josefina Aljaro Martinez, Chief, Documentation Research and Dissemination Service.

★2424★ CLAIMS/CITATION

IFI/Plenum Data Corporation
302 Swann Ave.
Alexandria, VA 22301 (703) 683-1085

Description: Provides more than 5,000,000 patent references cited during the patent examination process against each United States patent, and references to the patents in which it has subsequently been cited. **Type of Database:** Bibliographic. **Contact:** Harry M. Allcock, Vice President, IFI/Plenum Data Corporation.

★2425★ CLAIMS/CLASS

IFI/Plenum Data Corporation
302 Swann Ave.
Alexandria, VA 22301 (703) 683-1085

Description: Provides a classification code and title dictionary for all classes and subclasses of the U.S. Patent Classification System. **Type of Database:** Full-text. **Record Items:** Uniterm code/text, USC code, general or compound term code, CDB fragment code/text, molecular formula. **Contact:** Harry M. Allcock, Vice President, IFI/Plenum Data Corporation.

★2426★ CLAIMS/Comprehensive Data Base

IFI/Plenum Data Corporation
302 Swann Ave.
Alexandria, VA 22301 (703) 683-1085

Description: Provides bibliographic and major claim information on chemical patents issued by the United State Patent and Trademark Office since 1950; general, electrical, and mechanical patents since 1963; and design patents since 1980; abstracts are included since 1971. **Type of Database:** Bibliographic. **Record Items:** Abstract, claim text, CAS Registry Numbers, application country/date/number, controlled term, document type, family member country/date/number, field availability, file segment, fragment code, IPC code, inventors, patent assignee/country/number, publication date, role indicator, title, uniterm code, UPC code. **Contact:** Harry M. Allcock, Vice President, IFI/Plenum Data Corporation.

★2427★ CLAIMS/Reassignment & Reexamination

IFI/Plenum Data Corporation
302 Swann Ave.
Alexandria, VA 22301 (703) 683-1085

Description: Provides information on patents whose ownership has been reassigned from the original assignee to another company or individual since 1975, and patents reexamined since 1981 by the U.S. Patent and Trademark Office at the request of a second party who has raised substantial new questions regarding the patentability of the patent's claims. **Type of Database:** Bibliographic. **Record Items:** Reexamination request number/date, requestor and requestor location, reexamination certificate date/number/text; reassignment date/type, new patent assignee; expiration date. **Contact:** Harry M. Allcock, Vice President, IFI/Plenum Data Corporation.

★2428★ CLAIMS/UNITERM

IFI/Plenum Data Corporation
302 Swann Ave.
Alexandria, VA 22301 (703) 683-1085

Description: Provides bibliographic and major claim information on chemical patents issued by the United States Patent and Trademark Office since 1950; general, electrical, and mechanical patents since 1963; and design patents since 1980. **Type of Database:** Bibliographic. **Record Items:** Title, abstract, claim text, country code, application country/date/number, inventor name/country, class code, CAS Registry Number, document type, International Patent Classification (IPC) code, patent assignee, patent assignee country/code, issue date, patent number, publication year. **Contact:** Harry M. Allcock, Vice President, IFI/Plenum Data Corporation.

★2429★ CLAIMS/U.S. Patent Abstracts

IFI/Plenum Data Corporation
302 Swann Ave.
Alexandria, VA 22301 (703) 683-1085

Description: Provides bibliographic and major claim information on chemical patents issued by the United States Patent and Trademark Office since 1950; general, electrical, and mechanical patents since 1963; design patents since l980; abstracts are included since 1971. **Type of Database:** Bibliographic. **Record Items:** Title, abstract, claim text, country code, application country/date/number, inventor name/country, class code, CA reference number, document type, International Patent Classification (IPC) code, patent assignee, patent assignee country/code, issue date, patent number, publication year, CAS Registry Number. **Contact:** Harry M. Allcock, Vice President, IFI/Plenum Data Corporation.

★2430★ CLINPAT

Registro de la Propiedad Industrial (RPI)
Departamento de Informacion Tecnologica
Calle Panama, 1
28036 Madrid, Spain

Description: Contains the full text of the International Patent Classification. **Type of Database:** Full-text. **Contact:** Dr. Josefina Aljaro Martinez, Chief, Documentation Research and Dissemination Service.

★2431★ Compu-Mark Rechtsstandlexicon

Telesystemes
Questel
83-85, blvd. Vincent Auriol
F-75013 Paris, France (014) 582-6464

Description: Contains all trademarks currently in force in West Germany. **Type of Database:** Bibliographic. **Record Items:** Trademark name, registration number, international class, legal status. **Contact:** In the United States: Questel, Inc., 5201 Leesburg Pike, Suite 603, Falls Church, VA 22041.

★2432★ Compu-Mark U.K. On-Line

Compu-Mark (UK) Ltd.
93 Chancery Lane
London WC2A 1DT, England

Description: Covers more than 250,000 registrations, applications, and pending applications filed with the British Patent Office. **Type of Database:** Full-text. **Record Items:** Trademark; registration of application number; Part B marks; class(es) of goods and/or services;

status: pending unpublished applications, active published applications/ registrations, inactive marks (abandoned, cancelled, or expired); WHO, INN's, or ISO pesticide names; year and page of publication in Trade Marks Journal; owner; codes indicating changes to any aspect of the trademark's status. **Contact:** David I. Sheppard, Manager, Conpu-Mark (UK) Ltd.

★2433★ Compu-Mark U.S. On-Line

Compu-Mark U.S.
1333 F St., N.W.
Washington, DC 20004 (202) 737-7900

Description: Covers all trademarks contained in the U.S. federal and state registers. **Type of Database:** Full-text. **Record Items:** Trademark; owner; registration of application numbers, with code indicating federal or state origins; class(es) of goods or services by U.S. and international classification; status: pending applications, active registrations, inactive (abandoned, cancelled, or expired), state registration, affidavit(s); publication date in official gazette; texts to provide information on changes of ownership, assignment, other. **Contact:** Alison Grove, Customer Relations, Compu-Mark U.S.

★2434★ Computerized Administration of Patent Documents Reclassified According to the IPS (CAPRI)

International Patent Documentation Center (INPADOC)
Mollwaldplatz 4
A-1041 Vienna, Austria

Description: Contains references to worldwide patent documentation issued before 1973 which have been or are being reclassified according to the International Patent Classification. **Type of Database:** Bibliographic. **Record Items:** Country of publication; type of document; document number; IPC symbols; edition(s) in which the given IPC symbol is valid (IPC 1, 2, 3.). **Contact:** Dipl.-Kfm. Norbert Fux, Director, Sales Department, International Patent Documentation Center.

★2435★ COMPUTERPAT

Pergamon ORBIT InfoLine, Inc.
8000 Westpark Dr.
McLean, VA 22102 (703) 442-0900

Description: Contains data for all U.S. digital data processing patent documents as classified by the U.S. Patent and Trademark Office in subclasses 364/200 and 364/900, beginning the year the first patent for this technology was issued. **Type of Database:** Bibliographic. **Record Items:** Author(s); author's address; corporate author(s); title of item; date; cited references by source item: bibliographic description; country codes for priority data and cited references; abstract; patent information; patent claim. **Contact:** Michael Jones, Manager, Marketing, Pergamon ORBIT InfoLine, Inc.

★2436★ Deutsche Patent Datenbank (PATDPA)

Deutsches Patentamt
Zweibrueckenstr. 12
D-8000 Munich 2, Federal Republic of Germany

Description: Contains bibliographic information, abstracts, and graphics from patent documents published by the Deutsches Patentamt. **Type of Database:** Bibliographic. **Record Items:** Title of invention; inventor; system number; application country, date, kind, number, and type; document type; entry date and week; family member country, publication date, kind, number, and publication type; International Patent Classification; language; patent country; publication date. **Contact:** A. Dollt, Library Information, Deutsches Patentamt.

★2437★ DYNIS

Control Data Canada, Ltd.
Information Services Group
130 Albert St., Suite 1105
Ottawa, ON, Canada K1P 5G4 (613) 598-0200

Description: Contains information on more than 300,000 registered and pending trademark applications in Canada. **Type of Database:** Full-text. **Contact:** Michael Vincent, Marketing Representative, Control Data Canada, Ltd.

★2438★ ECLATX

Institut National de la Propriete Industrielle (INPI)
26 bis, rue de Leningrad
F-75800 Paris Cedex 8, France (014) 293-2120

Description: Contains the complete text of the 4th edition of the International Patent Classification. Includes all codes and terms that are included in the 86,700 groups and subgroups of the classification scheme. **Type of Database:** Full-text. **Contact:** Catherine Pagis, Marketing Manager, Institut National de la Propriete Industrielle.

★2439★ EDOC

Institut National de la Propriete Industrielle (INPI)
26 bis, rue de Leningrad
F-75800 Paris Cedex 8, France (014) 293-2120

Description: Provides cross-referenced numbers to patents issued by different countries for the same invention. **Type of Database:** Bibliographic. **Contact:** Catherine Pagis, Marketing Manager, Institut National de la Propriete Industrielle.

★2440★ EPAT

Institut National de la Propriete Industrielle (INPI)
26 bis, rue de Leningrad
F-75800 Paris Cedex 8, France (014) 293-2120

Description: Lists patents applied for and published in the European Patent Office's printed European Patent Bulletin **Type of Database:** Bibliographic. **Record Items:** Depositor; title; technical section; country of origin; place of publication; date of filing; date of publication; publication and registration numbers; applicant name; inventor name; representative priority rights; title; technical sections according to the International Patent Classification (IPC); opposition; designated states; and original titles in French, English, and German. **Contact:** Catherine Pagis, Marketing Manager, Institut National de la Propriete Industrielle.

★2441★ Federal Applied Technology Database (FATD)

U.S. National Technical Information Service (NTIS)
Center for the Utilization of Federal Technology (CUFT)
5285 Port Royal Rd., Room 8R
Springfield, VA 22161 (703) 487-4838

Description: Contains the following: descriptions of federal laboratory resources available to technology-oriented professionals, including facilities and equipment for sharing, expertise, and special services; technology fact sheets covering expertise and technologies selected as having better than average potential; and U.S. government-owned inventions available for licensing by U.S. businesses. Also covers technologies that companies can use to develop new products and processes. **Type of Database:** Bibliographic. **Record Items:** Title, author, information type, year, journal announcement, report number, availability, subject classification codes, descriptors, abstract. **Contact:** Edward J. Lehmann, Center for the Utilization of Federal Technology.

★2442★ Financial Times Business Reports
Financial Times Business Information (FTBI)
Financial Times Electronic Publishing
126 Jermyn St.
London SW1Y 4UJ, England

Description: Covers specialist financial, business, technology, and media markets. **Type of Database:** Full-text; bibliographic. **Record Items:** Trade names, company and organizational names, personal names and affiliation, country, topic, article type, section codes, length of text. **Contact:** Sarah Pebody, Marketing Manager, Financial Times Electronic Publishing.

★2443★ FPAT
Institut National de la Propriete Industrielle (INPI)
26 bis, rue de Leningrad
F-75800 Paris Cedex 8, France (014) 293-2120

Description: Lists French patents applied for and published in the printed Bulletin Officiel de la Propriete Industrielle. **Type of Database:** Bibliographic. **Record Items:** Depositor; title; technical section; country of origin; place of publication; date of filing; date of publication; publication and registration numbers; applicant name; inventor name; representative priority rights; title; technical sections according to the International Patent Classification (IPC). **Contact:** Catherine Pagis, Marketing Manager, Institut National de la Propriete Industrielle.

★2444★ Friday Memo
Information Industry Association (IIA)
555 New Jersey Ave., N.W., Suite 800
Washington, DC 20001 (202) 639-8262

Description: Reports on developments in the information industry and within the Information Industry Association and its divisions and service councils. Covers fair use, copyright, computer software, database publishing, micropublishing, privacy, videotext, teletext, and artificial intelligence. **Type of Database:** Full-text.

★2445★ Government-Industry Data Exchange Program (GIDEP)
U.S. Navy
Naval Fleet Analysis Center
GIDEP Operations Center
Corona, CA 91720 (714) 736-4677

Description: Facilitates the exchange of technical data between government and private industry on parts, components, materials, and processes. **Type of Database:** Numeric; full-text. **Contact:** Edwin T. Richards, Program Manager for Reliability, Government-Industry Data Exchange Program.

★2446★ High Tech International
High Technology Verlag GmbH
Leopoldstr. 70
D-8000 Munich 40, Federal Republic of Germany

Description: Provides information on high technology applications as they develop in West Germany. **Type of Database:** Full-text.

★2447★ INPADOC Data Base (IDB)
International Patent Documentation Center (INPADOC)
Mollwaldplatz 4
A-1041 Vienna, Austria

Description: Contains references to worldwide patent documentation for 55 countries which accounts for 96 percent of the world's currently published patent documents. **Type of Database:** Bibliographic. **Record Items:** Country of publication; type of document; document publication date; country of priority; number of application (serves as the basis of priority); priority date; International Patent Classification (IPC) symbol (if present). For certain countries, the following are also provided: inventor name; name of owner; applicant name; invention title; national classification symbol; and other legally related domestic application. **Contact:** Dipl.-Kfm. Norbert Fux, Director, Sales Department, International Patent Documentation Center.

★2448★ INPAMAR
Registro de la Propiedad Industrial (RPI)
Departamento de Informacion Tecnologica
Calle Panama, 1
28036 Madrid, Spain

Description: Contains current references, national marks, trade names, business signs, and international marks registered in Spain. **Type of Database:** Full-text. **Contact:** Dr. Josefina Aljaro Martinez, Chief, Documentation Research and Dissemination Service.

★2449★ INPI-MARQUES
Institut National de la Propriete Industrielle (INPI)
26 bis, rue de Leningrad
F-75800 Paris Cedex 8, France (014) 293-2120

Description: Covers all French trademarks currently in use. **Type of Database:** Full-text. **Record Items:** Trademark name; designated products; classification code; agent; applicant; filing number and date; publication number; renewal information; and design and color. **Contact:** Catherine Pagis, Marketing Manager, Institut National de la Propriete Industrielle.

★2450★ Japan High Tech Review
Kyodo News International, Inc. (KNI)
50 Rockefeller Plaza, Suite 832
New York, NY 10020 (212) 586-0152

Description: Contains news and analyses of Japan's high technology industries. **Type of Database:** Full-text. **Contact:** Pamela Stein, Director, Online Services, Kyodo News International, Inc.

★2451★ Japio
Japan Patent Information Organization
Bansui Bldg.
5-16, Toranomon, 1-Chome, Minato-ku
Tokyo 105, Japan

Description: Contains abstracts and drawings of Japanese patent and utility model documents and covers design and trademark information as well. **Type of Database:** Bibliographic; graphic. **Contact:** Yasushi Furukawa, Manager, International Affairs Section, Japan Patent Information Organization.

★2452★ JURINPI
Institut National de la Propriete Industrielle (INPI)
Bureau de Documentation Juridique et Technique
26 bis, rue de Leningrad
F-75800 Paris Cedex 8, France

Description: Contains references and abstracts of French legal decisions concerning patents issued since 1823 and trademarks registered since 1904. **Type of Database:** Bibliographic. **Record Items:** Jurisdiction, date of decision, parties involved, patent number and title, abstract, precedent cases, and bibliographic reference.

★2453★ LATIPAT
Registro de la Propiedad Industrial (RPI)
Departamento de Informacion Tecnologica
Calle Panama, 1
28036 Madrid, Spain

Description: Provides bibliographic data of granted Latin American inventions; currently covers Argentina, Mexico, and Colombia. **Type of Database:** Bibliographic. **Contact:** Dr. Josefina Aljaro Martinez, Chief, Documentation Research and Dissemination Service.

★2454★ LEXIS Federal Patent, Trademark, & Copyright Library (PATCOP)
Mead Data Central, Inc. (MDC)
LEXIS
9393 Springboro Pike
P.O. Box 933
Dayton, OH 45401 (513) 865-6800

Description: Provides the complete text of court decisions and publications issues relating to patent, trademark, and copyright laws. **Type of Database:** Full-text. **Contact:** LEXIS Customer Service.

★2455★ LEXPAT
Mead Data Central, Inc. (MDC)
9393 Springboro Pike
P.O. Box 933
Dayton, OH 45401 (513) 865-6800

Description: Provides the complete text of utility patents issued by the U.S. Patent and Trademark Office (PTO) since January 1975, and plant and design patents since December 1976. **Type of Database:** Full-text.

★2456★ LitAlert
Research Publications, Inc.
Rapid Patent Service
P.O. Box 2527, Eads Station
Arlington, VA 22202 (703) 920-5050

Description: Provides information on unpublished, unresolved, and current U.S. patent and trademark litigation. **Type of Database:** Bibliographic. **Record Items:** Patent or trademark number, patent or trademark title, classification title, patent publication date, patent or trademark document type, inventor, patent assignee, other patent or trademark numbers, trademark status date, trademark, court location, defendant, plaintiff, docket number, filing date, action date, description of the action, note, and names of law firms or attorneys. **Contact:** Eleanor Roberts, Rapid Patent Service.

★2457★ Master Search TM
Tri Star Publishing
475 Virginia Dr.
Fort Washington, PA 19034 (215) 641-6000

Description: Contains the complete text of some 600,000 active U.S. trademark applications and registrations. **Type of Database:** Full-text. **Record Items:** Trademark text and image; status; status date; registration number and date; serial number; date filed and published; international class; U.S. class; use; foreign registration information; owner address information; disclaimers, translations, color lining, and related statements; and other items. **Contact:** Richard E. Sharp, Executive Vice President, Tri Star Publishing.

★2458★ McGraw-Hill Publications Online
McGraw-Hill, Inc.
1221 Avenue of the Americas
48th Floor
New York, NY 10020 (212) 512-2911

Description: Contains the full text of 30 industry-specific magazines and newsletters which serve as a source of background information on companies and industry, the economy and international markets, labor and management, government, and technology. **Type of Database:** Full-text. **Record Items:** Title; journal name; publication date and year; page number; journal code; ISSN; section heading; byline; text; special feature (accompanying captions in graphs, tables, illustrations, and photographs). **Contact:** Andrea D. Broadbent, Manager, McGraw-Hill, Inc. Marketing Services.

★2459★ NUANS
Control Data Canada, Ltd.
Information Services Group
130 Albert St., Suite 1105
Ottawa, ON, Canada K1P 5G4 (613) 598-0200

Description: Contains corporate names and trademarks in Canada currently in use. **Type of Database:** Full-text. **Record Items:** Name, evaluation of corporate name based on phonetics, letter content, root word, coined words, synonyms, distinctive versus descriptive terms, line of business, geographical proximity. **Contact:** Mr. Jean Millette, NUANS Administrator, Control Data Canada, Ltd.

★2460★ ORBPAT
Pergamon ORBIT InfoLine, Inc.
ORBIT Search Service
8000 Westpark Dr.
McLean, VA 22102 (703) 442-0900

Description: Assists user in crossfile searching of patents databases available through the ORBIT Search Service. **Type of Database:** Bibliographic.

★2461★ PAPERCHEM
Institute of Paper Chemistry (IPC)
Division of Information Services
P.O. Box 1039
Appleton, WI 54912 (414) 734-9251

Description: Covers worldwide literature dealing with pulp and paper technology, including patents, equipment and products, and processes. **Type of Database:** Bibliographic. **Record Items:** Author; editor; corporate author; article title; journal title; publication date; volume and issue numbers; page numbers; publisher; abstract; language; price. **Contact:** Marianne Fiscus, Information Retrieval Specialist.

★2462★ PATDATA
BRS Information Technologies
1200 Route 7
Latham, NY 12110 (518) 783-1161

Description: Contains citations and abstracts of U.S. patents issued since 1975. **Type of Database:** Bibliographic. **Record Items:** Author(s); author's address; title of item; date; abstract; patent information; patent assignee, application data, reissued patent, foreign priority, U.S. Classification Code, Cross-Ref-U.S. Classification Code, International Patent Classification Code, U.S. & foreign patents cited.

★2463★ PharmPat

American Chemical Society (ACS)
Chemical Abstracts Service (CAS)
2540 Olentangy River Rd.
P.O. Box 3012
Columbus, OH 43210 (614) 447-3600

Description: Provides detailed abstracts for patent documents relating to drugs and other agents for the treatment or diagnosis of disease and indexed in the printed Chemical Abstracts and its online counterpart, CA File. **Type of Database:** Bibliographic. **Record Items:** Title of document; inventor; patent assignee and location; patent application country, number, date, and pagination; CODEN; patent information; designated states; patent application information; patent priority application information; language of patent; graphic image (structure diagram which appears with abstract in printed Chemical Abstracts); abstract, including overview, independent claims, additional claims, uses and advantages, an illustrative example, and specifics including testing procedures, process conditions, and other information; International Patent Classification (main and secondary); Chemical Abstracts section code and title; Chemical Abstracts section cross reference and code; index entries (substance and qualifiers); CAS Registry Number; formulation components and use.

★2464★ PTD-BASEN

Patentdirektoratet
Nyropsgade 45
DK-1602 Copenhagen K, Denmark

Description: Contains patents, trademarks, and patterns recognized in Denmark as well as new patents applications. **Type of Database:** Full-text. **Record Items:** Application, submitting and date of issue, classification code, applicant's name and address, inventor(s), appellation.

★2465★ PTS New Product Announcements/Plus (NPA/Plus)

Predicasts
11001 Cedar Ave.
Cleveland, OH 44106 (216) 795-3000

Description: Contains the complete text of news releases that announce new product introductions and modifications, new and applied technologies, and facilities expansions as issued by the developing company or its marketing agent. **Type of Database:** Full-text. **Record Items:** Dateline, release date, word count, company name, telephone number, address, additional access numbers, contact name and phone, date, title, text, company name, product name and code, use name and code, tradename, geographic name and code, special features. **Contact:** Customer Service Department, Predicasts.

★2466★ PTS PROMT

Predicasts
11001 Cedar Ave.
Cleveland, OH 44106 (216) 795-3000

Description: Provides citations and abstracts of journal articles, newspaper articles, and other sources of worldwide market and technology information relating to more than 120,000 companies and organizations. **Type of Database:** Bibliographic. **Record Items:** Article title; source journal title; publication date; page numbers; abstract; country code and name; product code and name; event code and description; company name; D-U-N-S number; ticker symbol; CUSIP number. **Contact:** Customer Service Department, Predicasts.

★2467★ RAPRA Tradenames (RAPTN)

Rapra Technology Ltd.
Information Centre
Shawbury
Shrewsbury, Shrops. SY4 4NR, England (093) 925-0383

Description: Contains abstracts and indexes of worldwide literature on tradenames of interest to companies producing rubber and plastics materials and products, as well as to suppliers to the industry and users of its products. **Type of Database:** Bibliographic. **Record Items:** Trade name, company name, product category, product description, graphic design, and source citation. **Contact:** Paul Cantrill, Controller, Information Services, or Judy Stubbington, Group Leader, Information Marketing Unit, Rapra Technology Ltd.

★2468★ SITADEX

Registro de la Propiedad Industrial (RPI)
Departamento de Informacion Tecnologica
Calle Panama, 1
28036 Madrid, Spain

Description: Includes information on the legal status of all property rights registered in Spain. **Type of Database:** Full-text. **Contact:** Dr. Josefina Aljaro Martinez, Chief, Documentation Research and Dissemination Service.

★2469★ TECHNO-SEARCH

Nihon Data Base Development Company
No. 1 Moritoku Bldg. 3 F
7-7-27, Nishi-shinjuku, Shinjuku-ku
Tokyo 360, Japan

Description: Contains the titles and abstracts of articles appearing in five major Japanese industrial and engineering newspapers. **Type of Database:** Bibliographic. **Record Items:** Source newspaper code; publication date; subject(s); keywords(s); technical field code; organization name; organizational and regional codes; length of article; title; abstract.

★2470★ Technology Assessment and Forecast Reports Data Base

U.S. Patent and Trademark Office
Technology Assessment and Forecast Program (TAF)
Washington, DC 20231 (703) 557-0400

Description: Contains statistical and other data relating to patents granted by the U.S. Patent and Trademark Office (PTO). **Type of Database:** Statistical. **Record Items:** Patent number; assignee for specific corporation or government; inventor's residence as state or country; date of patent application; independent inventors (street address, city, state and ZIP Code); abstract. **Contact:** Jane S. Myers, Manager, Technology Assessment and Forecast Program.

★2471★ Technology Information Exchange-Innovation Network (TIE-IN)

Ohio State Department of Development
Division of Technological Innovation
30 E. Broad St.
P.O. Box 1001
Columbus, OH 43266-0101 (614) 466-2115

Description: An inventory of research and development activity in Ohio. **Type of Database:** Bibliographic; directory. **Record Items:** Venture Opportunities: type of funding desired, amount of funding, and technology of interest. Ohio Patents: patent title, date assigned, company name, and investor's name. Corporate Research and Development: research category, location, company name, type of technology, and training interests. Faculty Research Interests: institution name, location, department, type of technology, and training interest. Sponsored

University Grants: title, keywords, investigator name, awarding agency. Technical Publications of Ohio Authors: author, title, institutions, or corporate source.

★2472★ Technology Transfer Databank (TECTRA)
California State University
School of Business and Public Administration
6000 J St.
Sacramento, CA 95819 (916) 929-8454

Description: Contains descriptions of cases exemplifying the successful transfer of technology from government-supported research laboratories to the public and/or private sectors. **Type of Database:** Directory. **Record Items:** Specific technology; name, address and telephone number of the person who generated the technology and that of the person using it; keyword and/or classification, such as year, laboratory, type of technology, and type of user. **Contact:** Dr. James A. Jolly, Director, Technology Transfer Databank.

★2473★ Technology Transfer Directory of People
California State University
School of Business and Public Administration
6000 J St.
Sacramento, CA 95819 (916) 929-8454

Description: Lists personnel involved in the transfer of technology from government-supported research laboratories to the public and/or private sectors. **Type of Database:** Directory. **Contact:** Dr. James A. Jolly, Director.

★2474★ Thomas New Industrial Products Database
Thomas Publishing Company
Buckwalter Rd.
Phoenixville, PA 19460 (215) 935-7875

Description: Provides technical information on new industrial products and systems introduced by American and foreign manufacturers and sellers. **Type of Database:** Full-text. **Record Items:** Product name, synonymous names for the product, product features, attributes, performance specifications, press release publication date, trade names, model numbers, manufacturer name and address. **Contact:** Charles Dremann, Managing Editor, Thomas New Industrial Products Database.

★2475★ Thomas Register Online
Thomas Publishing Company
One Penn Plaza
New York, NY 10119 (212) 290-7291

Description: Provides information on products made in the United States, the companies that make them, and where they are made. **Type of Database:** Directory. **Record Items:** Company name; address; telephone number and other communications information; assets; number of employees; parent company name and officers; products; trade names. **Contact:** Scott Safran, Associate Editor, Thomas Publishing Company.

★2476★ TMA Trademark Report
Tobacco Merchants Association of the United States (TMA)
231 Clarkville Rd.
P.O. Box 8019
Princeton, NJ 08543-8019 (609) 275-4900

Description: Contains a listing of tobacco-related trademark activity as reported in the U.S. Patent Office Official Gazette. **Type of Database:** Full-text. **Contact:** Lyn Guaciaro, Manager, Marketing, Tobacco Merchants Association of the United States.

★2477★ TMINT
Institut National de la Propriete Industrielle (INPI)
26 bis, rue de Leningrad
F-75800 Paris Cedex 8, France (014) 293-2120

Description: Covers all 270,000 international trademarks in force, filed, and renewed with the World Intellectual Property Organization under the Madrid Agreement. **Type of Database:** Full-text. **Record Items:** Trademark name; designated products; classification code; registered owner; registration number, date, and duration of the mark; country of origin of registration; countries for which protection is claimed or refused; colors claimed and classification of figurative elements of device marks. **Contact:** Catherine Pagis, Marketing Manager, Institut National de la Propriete Industrielle.

★2478★ Trade Marks (TMRK)
Canada Systems Group (CSG)
Federal Systems Division
Electronic Publishing Division
Product Sales Directorate
Ottawa, ON, Canada K2C 3V4 (613) 727-5445

Description: Provides information on more than 300,000 registered and pending marks in Canada. **Type of Database:** Directory. **Record Items:** Registration number and date, application number and date, priority date, registered owner, agent or representative, trademark name, disclaimer, products or services for which the trademark is registered, basis of claim, associated marks, and footnotes. Formatted fields include: wordmark, design, certification mark, distinguishing guise marks, date the trademark was first used, and filing, priority, and registration dates. **Contact:** David Macdonald, Marketing Manager, Electronic Publishing Division.

★2479★ Trademarkscan-Federal
Thomson & Thomson
One Monarch Dr.
North Quincy, MA 02171-2126 (617) 479-1600

Description: Provides information on all active registered and pending trademarks on file in the U.S. Patent and Trademark Office and all inactive trademarks since October 1983. **Type of Database:** Full-text. **Record Items:** Trademark text and enhancements; serial number; U.S. class number; status date; date of first use of mark; goods/services description; mark type; owner name; registration number; series code; status text and code; permuted trademark text and enhancements. **Contact:** Anthea Gotto, Manager, Online Marketing, Thomson & Thomson.

★2480★ Trademarkscan-State
Thomson & Thomson
One Monarch Dr.
North Quincy, MA 02171-2126 (617) 479-1600

Description: Provides information on trademarks registered with the Secretaries of State of all 50 U.S. states and in Puerto Rico. **Type of Database:** Full-text. **Record Items:** Trademark text and enhancements; rotated trademark; design type; state of registration; U.S. class number; international class; goods/service description; mark type; registration number; status; date of registration; date of renewal; date of cancellation; date of first use; owner name. Corporate name records are not included. Trade names, assumed names, and fictitious names are also not generally included, but may be identified for some states. **Contact:** Anthea Gotto, Manager, Online Marketing, Thomson & Thomson.

★2481★ TRANSIN
Transinove International
INPI
26 bis, rue de Leningrad
F-75800 Paris Cedex 8, France (014) 294-5250

Description: Indexes offers and requests for patented technologies, new products and inventions, and innovative ideas in need of development from the private and public sectors worldwide. **Type of Database:** Bibliographic. **Record Items:** International Patent Classification (IPC) code; type of opportunity code; development stage; descriptive title; technology level and descriptors; description; technical skill descriptors; agency or company source; contact person name and address; publication reference number; file input date. **Contact:** Olivier Arondel, Assistant Manager, Transinove International.

★2482★ UK Trade Marks (UKTM)
The Patent Office (London)
State House
66-71 High Holborn
London WC1R 4TP, England

Description: Contains information on nearly 500,000 British trademarks that are active, pending, or lapsed since January 1976. **Type of Database:** Directory. **Record Items:** Name and address of owner, class of goods, goods specification.

★2483★ U.S. Patent Classification System
U.S. Patent and Trademark Office
Office of Documentation Planning and Support
Patent Documentation Organizations
Washington, DC 20231 (703) 557-0400

Description: A classified file of patent documents designed to assist patent examiners and others in the field in assessing the novelty, intrinsic merit, and utility of inventions deposited at the U.S. Patent and Trademark Office. **Contact:** Edward J. Earls, Director, Office of Documentation Planning and Support.

★2484★ U.S. Patents Files
Derwent, Inc.
6845 Elm St., Suite 500
McLean, VA 22101 (703) 790-0400

Description: Contains patent information for inventions patented in the United States. **Type of Database:** Full-text. **Record Items:** Title of item; date; total number; abstract; patent information; inventor and assignee name, patent number, application number, attorney name, examiner name. **Contact:** Jeffrey L. Forman, Vice President, Marketing, Derwent, Inc.

★2485★ USCLASS
Derwent, Inc.
6845 Elm St., Suite 500
McLean, VA 22101 (703) 790-0400

Description: Contains classification information for nearly 5 million patents issued by the U S Patent and Trademark Office (PTO). **Type of Database:** Bibliographic. **Contact:** Jeffrey L. Forman, Vice President, Marketing, Derwent, Inc.

★2486★ WESTLAW Copyright, Patent and Trademark Library
West Publishing Company
50 W. Kellogg Blvd.
P.O. Box 64526
St. Paul, MN 55164-0526 (612) 228-2500

Description: Contains the complete text of federal court decisions, statutes and regulations, specialized files, and texts and periodicals dealing with copyright, patent, and trademark law. **Type of Database:** Full-text. **Contact:** Gary Schmidt, Assistant Manager Marketing, or Ron Anderson, Research Attorney, West Publishing Company.

★2487★ World Bank of Technology
Dr. Dvorkovitz and Associates (DDA)
P.O. Box 1748
Ormond Beach, FL 32074 (904) 677-7033

Description: Covers licensable technology around the world. **Type of Database:** Directory. **Record Items:** Title, description, subject category, individual. **Contact:** Nancy Driver, Data Processing Manager, Dr. Dvorkovitz and Associates.

★2488★ World Patents Index (WPI)
Derwent Publications Ltd.
Rochdale House
128 Theobalds Rd.
London WC1X 8RP, England

Description: Supplies titles and other details of general, mechanical, electrical, and chemical patents, covering the patent literature of leading industrial countries and using IPC codes for areas such as human necessities, performing operations, transporting, chemistry, textiles, building, construction, mechanics, lighting, heating, instruments, nuclear science, and electricity. **Type of Database:** Bibliographic. **Record Items:** Author(s); corporate author(s); title of item; date; publisher; patent number; abstract; patent information; equivalents which are patent numbers in other countries and priorities (filing date and serial number for first filing).

★2489★ Zeitschrift fur Urheber und Medienrecht (ZUM)
NOMOS Datapool
Waldseestr. 3-5
Postfach 610
D-7570 Baden-Baden, Federal Republic of Germany

Description: Reports developments in the areas of copyright, communications, video technology, and satellite television. **Type of Database:** Full-text.

CHAPTER 12

INVENTION TRADE SHOWS

Every industry takes part in trade fairs, including the butchers, the bakers, and the candlestick makers. According to the Trade Show Bureau, in 1988, 9,000 trade shows took place in the United States, with 3,289 of them in excess of 10,000 square feet or 100 booths.

The proliferation of trade shows in America has created a trade show industry, complete with its own newsletters, magazines, and professional organizations.

National, regional, and local events promote the sale of almost anything you can imagine. There are trade shows for everything from hardware, consumer electronics, apparel, and aircraft to nuclear medicine, dental equipment, toys, comic books, and musical instruments. If it is manufactured and sold, you can be sure there is a trade show somewhere, sometime.

The Patent and Trademark Office (PTO) even co-sponsors an annual trade show.The **National Inventors Expo** is held in February at the PTO's search room in Crystal City, Virginia. It features exhibits by independent inventors of their patented inventions, and large and small businesses of interest to the inventor. For current information, call (703) 557-3341. Admission is free and the event is open to the public.

In fact, most trade shows are admission free "to the trade." All you usually need is a business card to enter the exhibition area.

While I do not recommend trade shows as the best place to present or license patented inventions, they are a "must" for getting the beat on any particular market and its dynamics. It's all there for you to see. Competitors line up side by side for the all important buyers to compare the manufacturers' products and pricing.

If you go to the shows, remember that companies have paid many thousands of dollars to participate. Their primary reason to do a trade fair is to ring up sales. They are not there to license concepts.

The sales force does not exist to review new concepts. It is responsible to sell. It is both fruitless and dangerous to impose on and expose inventions to sales people. They are, however, excellent sources of information and normally delighted to chat about their products, the state-of-the-market, and industry particulars. Information on the market can be gathered in a very convenient and time-effective manner.

There are exceptions, of course. In some industries, R&D executives attend trade shows to get a feel for the competition as well as host "invited" outside inventors with product to show. Presentations are typically conducted in hotel suites away from the exhibition site. It is best to call the corporate headquarters in advance of the trade show and check policy.

I attend the shows to scout for new product introductions, pick-up information handouts and samples, and make personal contacts. No better or more cost-effective way exists to acquire product literature than at a trade show. Manufacturers publish fliers and information kits just for trade show distribution. And most come with price lists!

I take empty flight bags with me in which to transport home everything collected. Sturdy ones. I never rely on the paper or plastic bags some of the companies supply. The material is too valuable (and heavy!). Thanks to trade shows over the past decade, I have today a most comprehensive reference library.

Trade shows are also an excellent place to meet and network with executives to whom you would otherwise not have access. They rarely take their "bodyguards" to trade shows; it is too expensive and, after all, they also go to meet new people. They even make it easier by wearing name tags!

I have made super contacts in convention hotel elevators, lobby queues, and taxi-shares to and from the exhibition centers.

The best kinds of shows at which to meet senior executives are the national or international ones. The smaller regional or local trade shows are typically staffed by sales people alone. Nevertheless, such shows provide a less hectic atmosphere and many of the same resource materials.

Many exhibitors host receptions and special events for buyers. If you get lucky, you may be invited to attend.

Where and When

There are several ways to find out about where and when trade shows for any particular industry will be taking place. Gale Research annually publishes the *Trade Shows and Professional Exhibits Directory,* a comprehensive listing of nearly 5,000 shows worldwide. Check your local library for a copy.

Other methods:

1) Ask a manufacturer or distributor in your field of invention. The sales and marketing people will have such information realily available.

2) Contact the trade association that covers your field of invention. About 3,600 trade associations operate on the national level in America. A great start is through Gale's latest *Encyclopedia of Associations,* available at most public libraries. The 1990 edition features detailed entries describing over 22,000 active associations, organizations, clubs and other nonprofit membership groups in virtually every field of human endeavor.

Conferences and Meetings

There are perhaps more conferences and meetings going on than trade shows. It doesn't take much to have either. All you technically require is a number of experts sitting around a table discussing a field of interest.

The biggest difference between trade shows and conferences and meetings is that you almost always pay to attend conferences. This is because the primary reason for a conference is to hear experts speak, pick their brains, share your ideas, and network.

Conferences are excellent places to get to know the people behind the products. Socializing is encouraged and the atmosphere is calmer than at trade shows. There is no pressure to sell or to buy. The object is to brainstorm and trade in ideas. Participants can increase their "know how" and "know who" at the same time.

Many trade fairs have conferences or seminars as part of the program. And many conferences offer simultaneous resource fairs.

Where and When

There are several ways to find out about where and when conferences for any particular industry will be taking place.

1) Ask a manufacturer or distributor in your field of invention. Many larger manufacturers have training departments which can provide helpful information.

2) Contact the trade association that covers your field of invention. There are about 3,600 trade associations operating on the national level in America. Again, a great start is through Gale's latest *Encyclopedia of Associations* available at most public libraries. If the particular association holds a conference, it will usually be listed in its entry.

3) Ask department heads and professors at a nearby university where your field of interest is taught. Universities, especially those teaching engineering and kindred technical fields, will have a current schedule of conferences on hand.

National Innovation Workshops

National Institute of Standards and Technology's OERI supports six National Innovation Workshops each year in different regions of the country to bring inventors in contact with local innovation sources. Evolving over the years, this program provides a network for innovation and is recognized as successful throughout the inventor community.

The National Innovation Workshop series was initiated in the Spring of 1980. Its format is a standardized two-day seminar. There are two addresses each day (keynote and luncheon) by nationally known speakers; a wide variety of free, how-to and technical publications available from federal, state and local government agencies and other organizations; and daily periods of 8 to 10 concurrent workshops, totaling 48 to 60 individual sessions.

There are seven cosponsors at a national level: The National Congress of Inventor Organizations; American Intellectual Property Law Association; Licensing Executives Society; Association of Small Business Development Centers; National Society of Professional Engineers; U.S. Department of Commerce; and the U.S. Department of Energy.

Workshop topics include, patenting and protection; estimating the worth of an invention; licensing; marketing; new business start-up; the business plan; R&D and venture financing; the DOE-NIST Energy-Related Inventions Program; and SBIR Programs. The cost of attending is about $85.

Call OERI at (301) 975-5500 or write OERI, Building 202, Room 209, Gaithersburg, MD 20899 for information on future locations.

Chapter Arrangement

This chapter is arranged alphabetically by trade show name. All trade show names in this chapter are referenced in the Master Index.

★2490★ American International Toy Fair
Toy Manufacturers of America
200 Fifth Ave.
New York, NY 10010 (212) 675-1141
Charles Riotto

Principal Exhibits: Toys, games, puzzles, hobby and craft items, dolls and accessories, electronic games, model kits, wheel goods, holiday decorations, gift wrappings, greeting cards, juvenile furniture, sporting goods, backyard play equipment, costumes, playsuits, and back-to-school lines. **Exhibition Frequency:** Annual. **Dates and Locations:** 1990 Feb. 16-19; New York, NY.

★2491★ Baby Expo
ExpoGroup, Inc.
220 Felspar Ridge
Peachtree City, GA 30269 (404) 631-3976
Larry Shinglton, Contact

Principal Exhibits: Infant-related products. **Dates and Locations:** 1989 Sept. 8-10; Atlanta, GA.

★2492★ Barclays Techmart-New Technology Transfer Exhibition
National Exhibition Centre Ltd.
Exhibition and Events Division
National Exhibition Centre
Birmingham B40 1NT, England

Principal Exhibits: High-, medium-, and low-technology equipment. **Exhibition Frequency:** Annual.

★2493★ British Toy and Hobby Fair
Earl's Court
80 Camberwell Rd.
London SE5 OEG, England

★2494★ Chicago Model Hobby Show
2400 E. Devon
Des Plaines, IL 60018 (312) 299-3131
Kerry Connolly, Contact

Principal Exhibits: Radio-controlled cars, planes, boats, and trains. **Dates and Locations:** 1989 Nov. 2-5; Chicago, IL • 1990 March 8-11; Los Angeles, CA • 1990 May 2-4; Philadelphia, PA.

★2495★ Conference on Innovation and Creativity
Patent and Trademark Office
Office of Finance
2101 Crystal Plaza, Suite 270
Arlington, VA 22202 (703) 557-3341

Principal Exhibits: Trade.

★2496★ Consumer Electronics Show
1722 Eye St., N.W.
Washington, DC 20006 (202) 457-8700
Dennis Corcoran, Vice President

Principal Exhibits: Audiovisual equipment. **Dates and Locations:** 1990 Jan. 6-9; Las Vegas, NV • 1990 June 2-5; Chicago, IL.

★2497★ Great American Family Expo
Trade Show Marketing
555 Old Country Road
Suite 201
San Carlos, CA 94070 (415) 594-0452
Frank Cort, President

Principal Exhibits: Aimed at the professional couple with children under 5 years old. **Dates and Locations:** 1989 Oct. 7-8; Oakland, CA • 1990 April; Sacramento, CA • 1990 June; San Jose, CA • 1990 Aug.; Los Angeles, CA.

★2498★ Hobby Industry Convention and Trade Show
Hobby Industry Association of America
319 E. 54th St.
Elmwood Park, NJ 07407 (201) 794-1133
Susan Danker

Principal Exhibits: Hobby related materials. **Exhibition Frequency:** Annual. **Dates and Locations:** 1990 Jan. 25-28; Dallas, TX • 1991 Jan. 20-23; San Francisco, CA • 1992 Jan. 19-22; San Diego, CA.

★2499★ International Craft Exposition/Chicago Craft and Creative Industries Show and Convention
Officer Management Co.
1100-H Brandywine Blvd.
P.O. Box 2188
Zanesville, OH 43702 (614) 452-4541
Walter Offinger, President

Principal Exhibits: Craft, art, miniature, floral, and needlework supplies. **Exhibition Frequency:** Annual.

★2500★ International Exhibition of Inventions and New Techniques of Geneva
Promex SA
8, rue du 31 December
CH-1027 Geneva, Switzerland
Mr. Jean-Luc Vincent, Contact

Principal Exhibits: Inventions and new products. **Exhibition Frequency:** Annual. **Dates and Locations:** 1990 Mar. 3-4; Geneva, Switzerland • 1991 Apr. 12-21; Geneva, Switzerland.

★2501★ The International Licensing and Merchandising Conference and Expo/Licensing

Expocon Management Associates, Inc.
3695 Post Rd.
Southport, CT 06490 (203) 259-5734
Susan Reuter

Principal Exhibits: Trade.

★2502★ Invention Convention (Los Angeles)

International Convention Services Corporation
6753 Hollywood Blvd., Suite 212
Los Angles, CA 90028 (213) 460-4408
Jean Paul LeClaire, Exhibits Manager

Principal Exhibits: Product and innovation opportunities. Brings together inventors and technologies with executives and professionals. Offers seminars, panel discussions, workshops, and awards **Exhibition Frequency:** Three times a year. **Dates and Locations:** 1990 Apr.; San Diego, CA • 1990 June; San Francisco, CA • 1990 Sept.; Pasadena, CA. **Telecommunication Access:** FAX (213) 469-1892; Toll-free (outside California) 800-458-5624.

★2503★ Inventors Expo (Camarillo)

Inventors Workshop
3201 Corte Malpaso, Suite 304
Camarillo, CA 93010 (805) 484-9786

Principal Exhibits: New inventions and products. **Exhibition Frequency:** Annual.

★2504★ Inventors Expo (Tarzana)

Education Foundation
P.O. Box 251
Tarzana, CA 91356 (818) 344-3375
Alan A. Tratner, Exhibits Manager

Principal Exhibits: New inventions and products. **Exhibition Frequency:** Annual.

★2505★ INVEX-International Exhibition of Inventions and Novel Features

BVV-Trade Fairs and Exhibitons
Vystaviste 1
602 00 Brno 1, Czechoslovakia
Dr. Ing. Bretislav Fabian, Exhibits Manager

Principal Exhibits: Inventions and improvements in electrical and water engineering, agriculture, metallurgy, machine tools, optics, office equipment, scientific instruments, packaging equipment, transport engineering, building industry, and pharmaceuticals. **Exhibition Frequency:** Biennial. **Dates and Locations:** 1990 Oct. 24-30; Brno, Czechoslovakia.

★2506★ Juvenile Products Manufacturers Association

Juvenile Products Manufacturers Association
66 E. Main St.
Moorestown, NJ 08057 (609) 234-9155
William L. MacMillian, III

Exhibition Frequency: Annual. **Dates and Locations:** 1990 Oct. 4-7; Dallas, TX.

★2507★ LIZ '89 Licensing Fair for Design Ideas, Publication, and Production Rights

Wittlesbacherstr. 10
Postf. 4129
D-6200 Wiesbaden, Federal Republic of Germany

★2508★ MICEL-International Market for Creativity and Licensing

MIDEM Organisation
176 ave. Victor-Hugo
F-75116 Paris, France
Elizabeth Peyraus, Exhibits Manager

Principal Exhibits: Licensing services. **Exhibition Frequency:** Annual. **Contact:** Perard Associates, Inc., 38 W. 32nd St., Suite 1512, New York, NY 10001; (212) 967-7600.

★2509★ Mid-Year Variety Merchandise Show

Thalheim Expositions, Inc.
42 Bayview Ave.
P.O. Box 4200
Manhasset, NY 11030 (516) 627-4000
David Thalheim, Contact

Principal Exhibits: General merchandise. **Exhibition Frequency:** Annual.

★2510★ Miniatures Industrial Association of America

Offinger Management Co.
1100-H Brandywine Blvd.
P.O. Box 2188
Zanesville, OH 43702 (614) 452-4541
Susan Danker, Contact

Exhibition Frequency: Annual.

★2511★ National Back-to-School Merchandise Show

Thalheim Expositions, Inc.
42 Bayview Ave.
P.O. Box 4200
Manhasset, NY 11030 (516) 627-4000
David T. Thalheim, President

Principal Exhibits: Business and home office equipment, supplies, and services. **Exhibition Frequency:** Annual. **Dates and Locations:** 1990 Feb. 11-13; New York, NY • 1991 Feb. 10-12; New York, NY.

★2512★ National Inventors Expo

Patent and Trademark Office
2011 Crystal Dr., Suite 208B
Arlington, VA 22202 (703) 557-3341
William Craig, Contact

Principal Exhibits: Inventions. **Exhibition Frequency:** Annual.

★2513★ National Merchandise Show

Thalheim Expositions, Inc.
4200 Bayview Rd.
P.O. Box 4200
Manhasset, NY 11030 (516) 627-4000
David T. Thalheim, Manager

Principal Exhibits: Giftware, housewares, toys, stationery, leather goods, health and beauty aids, novelties, hardware, electronics, apparel, and related goods. **Exhibition Frequency:** Annual. **Dates and**

Locations: 1990 Sept. 8-11; New York, NY • 1991 Sept. 21-24; New York, NY.

★2514★ National Premium Incentive Show
Hall-Erickson, Inc.
150 Burlington Ave.
Clarendon Hills, IL 60514 (312) 850-7779
Peter H. Erickson, Exhibits Manager

Principal Exhibits: Electronics, housewares, sporting goods, food, and giftware. **Exhibition Frequency:** Annual. **Dates and Locations:** 1990 Oct. 9-11; Chicago, IL • 1991 Oct. 15-17; Chicago, IL.

★2515★ National Stationery Show
George Little Management, Inc.
Two Park Ave., Suite 1100
New York, NY 10016
Mary Ann Burke, Exhibit Manager

Principal Exhibits: Greeting cards, social stationery, and realted products such as calenders, desk accessories, small leather goods, and photo frames. **Exhibition Frequency:** Annual.

★2516★ NECHIA Trade Show/North East Craft and Hobby
North East Craft and Hobby Industry Association
P.O. Box 85
Fair Hills, PA 19030 (215) 943-6378
Robert Stover, Contact

Exhibition Frequency: Annual.

★2517★ New York International Gift Fair
George Little Management, Inc.
Two Park Ave.
Suite 1100
New York, NY 10016 (212) 686-6070
Susan Corwin, Exhibit Manager

Principal Exhibits: General giftware, tabletop and housewares, decorative accessories, personal accessories, contemporary design products, and traditional and contemporary gifts. **Exhibition Frequency:** Semiannual.

★2518★ Patent Information Fair
Japan Industrial Journal
1-28-5 Kanda-Jimbocho, Chiyoda-ku
Tokyo 101, Japan

Principal Exhibits: Automated office equipment; office computers; word processors; microfilmed information on patents; printers; online terminals; and database services for patent administration. **Exhibition Frequency:** Biennal.

★2519★ Premium Incentive Show
Thalheim Expositions, Inc.
42 Bayview Ave.
P.O. Box 4200
Manhasset, NY 11030 (516) 627-4000
Jay Thalheim, Chairman of the Board

Principal Exhibits: Premium incentive products, ideas, and services. **Exhibition Frequency:** Annual. **Dates and Locations:** 1990 May 7-10; New York, NY • 1991 May 6-9; New York, NY.

★2520★ SITOY-Seoul International Toy Fair
Korea Trade Promotion Corporation
C.P.O. Box 3109
Seoul, Republic of Korea

★2521★ SME - Society of Manufacturing Engineers
Society of Manufacturing Engineers
One SME Dr.
P.O. Box 930
Dearborn, MI 48121 (313) 271-0023
Nancy Berg, Exhibits Manager

Principal Exhibits: High-tech manufacturing processes and products. **Exhibition Frequency:** Annual.

★2522★ Southeast Craft and Hobby Show
Southeast Craft and Hobby Association
6175 Barfield Rd., Suite 220
Atlanta, GA 30328 (404) 252-2454

Principal Exhibits: Crafts, hobbies, needlework, and publications. **Exhibition Frequency:** Annual.

★2523★ Southwestern Craft and Hobby Christmas in July Show
Offinger Management
1100-H Brandywine Blvd.
P.O. Box 2188
Zanesville, OH 43702 (614) 452-4541
Walter Offinger, President

Exhibition Frequency: Annual.

★2524★ Taipei International Toy Fair
China External Trade Development Council
CETRA Tower, 4-7th Floor
International Trade Bldg., Taipei World
333 Keelung Rd., Section 1
Taipei 10548, Taiwan

Principal Exhibits: Toys.

★2525★ TECHEX Americas
Dvorkovitz and Associates
P.O. Box 1748
Ormond Beach, FL 32075 (904) 677-7033
Anne E. Klenner, Exhibits Manager

Principal Exhibits: New products and technology available for license from industry and universities. **Exhibition Frequency:** Annual.

★2526★ Techno Tokyo - International Licensing and Joint Venture Expo
Japan Industrial Journal
Promotion Division
1-28-5, Kanda-Jimbocho, Chiyoda-ku
Tokyo 101, Japan
Akihiro Hirose, Exhibits Manager

Principal Exhibits: New technologies and technology transfer services.

★2527★ Transworld Housewares and Variety Exhibit
TransWorld Exhibits, Inc.
1850 Oak St.
Northfield, IL 60093 (312) 446-8434

Principal Exhibits: General merchandise.

CHAPTER 13

PATENT DEPOSITORY AND SPECIAL LIBRARIES

Every inventor should do at least one hands-on patent search to fully understand and appreciate the process. Obviously, not everyone can visit the Patent and Trademark Office's (PTO's) Public Search Room in Arlington, Virginia. But, a system of Patent Depository Libraries (PDLs) has been established throughout the country, where collections of patents may be examined and researchers have access to a computerized patent finding aid(s), for example, CASSIS, which provides direct, online access to PTO data. Additionally, there are many special collections in corporate and university libraries that may contain information on specific aspects of patents. Engineering and law libraries are typical examples that have been included in this chapter.

Many of the following libraries receive current issues of U.S. patents and maintain collections of earlier issued patents. The scope of these collections varies from library to library, ranging from patents of only recent years to all or most of the patents issued since 1790.

The patent collections in the PDLs are open to the general public and I have always found the librarians most willing to take the time to help newcomers gain effective access to the information contained in patents. In addition to the patents, PDLs usually have all the publications of the U.S. Patent Classification System, for example, *The Manual of Classification, Index to the U.S. Patent Classification, Classifications Definitions, Official Gazette of the United States Patent and Trademark Office,* and so forth.

Facilities for making paper copies from either microfilm in reader-printers or from the bound volumes in paper-to-paper copies are generally made available for a fee.

Even though I live near enough to the Alexandria, Virginia, Public Search Room, often I opt to do my work at the University of Maryland's PDL within its Engineering and Physical Sciences Library. Not much larger than a small meeting room, it is crowded with books and offers only a couple of chairs. However, not only can I do my search work, but the bonus is that there is a very extensive collection of technical publications outside the PDL that I enjoy browsing through for ideas and technologies. Such "extras" are not available at the main government facility.

Because of the variations in the scope of patent collections among the PDLs and special libraries and in their hours of service to the public, anyone contemplating use of the patents at a particular library is well-advised to contact the library, in advance, about its collection and hours, so as to avoid possible inconveniences.

Chapter Arrangement

This chapter is arranged alphabetically by library name. All names in this chapter are referenced in the Master Index.

★2528★ AKZO Chemicals Inc., Research Library
8401 W. 47th St.
McCook, IL 60525 (312) 442-7100
Robyn Petry, Head Librarian

Subjects: Organic chemistry, fatty acid chemistry. **Holdings:** 3500 books; 3800 bound periodical volumes; 15,000 patents; 10 file boxes of company reprints; 7500 internal reports; chemical patents on microfilm. **Online Access:** DIALOG Information Services, Pergamon ORBIT InfoLine, Inc., National Pesticide Information Retrieval System (NPIRS), NLM, Chemical Information Systems, Inc. (CIS), Occupational Health Services, Inc.

★2529★ Allied Corporation, Buffalo Research Center Library
20 Peabody St.
Buffalo, NY 14210 (716) 827-6229
Janice Hood, Library Service Manager

Subjects: Organic chemicals, organofluoro chemicals, inorganic chemicals, polymers. **Holdings:** 10,000 books; 12,000 bound periodical volumes; 12,000 internal reports; 2 million U.S. chemical patents; 500 dissertations; 200 reels of microfilm. **Online Access:** DIALOG Information Services, STN International; Datatrieve (internal database).

★2530★ Allied-Signal Engineered Materials Research Center, Technical Information Center
50 E. Algonquin Rd.
Des Plaines, IL 60017-5016 (312) 391-3109
Suzanne M. Gaumond, Manager

Subjects: Petroleum refining processes and technology, petrochemical processes, chemical engineering, air and water conservation, special purpose chemicals, catalysis, patents, trademarks, copyrights. **Special Collections:** American Petroleum Institute project publications; Technical Oil Mission reports (microfilm); official patent publications of Australia, Brazil, Canada, France, Germany, Great Britain, India, South Africa, and United States (150,000 patents). **Holdings:** 20,000 books; 12,250 bound periodical volumes; 25 VF drawers of clippings; 50 VF drawers and 2500 microfiche of government documents; 450,000 U.S. patents. **Online Access:** STN International, DIALOG Information Services, Pergamon ORBIT InfoLine, Inc., BRS Information Technologies, I.P. Sharp Associates Limited, LEXIS, NEXIS, Chemical Information Systems, Inc. (CIS), Data Resources (DRI); Research Reports (internal database).

★2531★ Allied Signal, Inc., Syracuse Research Laboratory, Library
Box 500
Solvay, NY 13209-0006 (315) 487-4151
Linda Griffo, Librarian

Subjects: Applied technology, inorganic chemistry, chemical engineering. **Holdings:** 6000 books; 6000 bound periodical volumes; 19 VF drawers of U.S. patents; 15 VF drawers of foreign patents; 21 VF drawers of pamphlets, specifications, house organs. **Online Access:** DIALOG Information Services, CAS ONLINE. Performs searches on fee basis.

★2532★ Anchorage Municipal Libraries, Z.J. Loussac Public Library
3600 Denali St.
Anchorage, AK 99503-6093 (907) 261-2907

Special Collections: U.S. patent depository library.

★2533★ Arizona State University, Daniel E. Noble Science and Engineering Library
Tempe, AZ 85287 (602) 965-7607
Vladimir T. Borovansky, Head

Subjects: Engineering; physical, life, health sciences; mathematics; agriculture; geography. **Special Collections:** Solar energy (50,000 archival materials including DOE reports, pamphlets); patent depository library (complete collection; in microform). **Holdings:** 225,000 books; 105,000 bound periodical volumes; 150,000 maps; 600,000 microforms. **Online Access:** DIALOG Information Services, Pergamon ORBIT InfoLine, Inc., BRS Information Technologies, NASA/RECON, Integrated Technical Information System (ITIS), U.S. Patent Classification System, OCLC, STN International, WILSONLINE. Performs searches on fee basis. Contact Person: Joyce Plaza, Coord., Comp.Ref.Serv., 965-7608.

★2534★ Arkansas State Library
One Capitol Mall
Little Rock, AR 72201-1081 (501) 682-2053

Special Collections: U.S. patent depository library.

★2535★ Arnold, White & Durkee, Library
Box 4433
Houston, TX 77210 (713) 787-1400
Genel F. Moran, Librarian

Subjects: Law - patent, trademark, copyright, antitrust, franchise. **Special Collections:** Official Gazette of Patents; Official Gazette of Trademarks. **Holdings:** 7900 books; 360 bound periodical volumes. **Online Access:** DIALOG Information Services, LEXIS, WESTLAW, Pergamon ORBIT InfoLine, Inc., Dun & Bradstreet Corporation, TRW Business Profiles, Information America, VU/TEXT Information Services, DataTimes; ABA/net (electronic mail service).

★2536★ Ashland Chemical Company, Library and Information Services
Box 2219
Columbus, OH 43216 (614) 889-3281
Priscilla Ratliff, Supervisor

Subjects: Chemistry - organic, polymer, catalysis and surface, analytical; chemical industry. **Special Collections:** Chemical Abstracts, 1907 to present (complete). **Holdings:** 8500 books; 12,000 bound periodical volumes; 33,500 U.S. patents; 20,700 foreign patents; 750 U.S. and foreign annual reports; U.S. chemical patents, 1960 to present, on microfilm. **Online Access:** DIALOG Information Services, Pergamon ORBIT InfoLine, Inc., NLM, Chemical Information Systems, Inc. (CIS), OCLC, STN International, Telesystemes Questel, VU/TEXT Information Services, LEXIS, NEXIS.

★2537★ Asija Associates, Library
7 Woonsocket Ave.
Shelton, CT 06484 (203) 736-9934

Subjects: Patent law and inventions in the areas of energy, environment, computers, communications. **Holdings:** 2000 books; 25 reports.

★2538★ Association of University Related Research Parks, Reference Library
4500 S. Lakeshore Dr., Suite 336
Tempe, AZ 85282 (602) 752-2002
Chris Boettcher, Executive Director

Subjects: Research parks, technology transfer, planning issues, high technology, university/industrial relations, real estate. **Special Collections:** United Kingdom Science Parks Collection. **Holdings:** 45 books; 200 bound periodical volumes; 450 periodical articles and reports; 200 vertical files. **Online Access:** Internal database.

★2539★ Auburn University Libraries, Science and Technology Department
Auburn University, AL 36849-5606 (205) 826-4500

Special Collections: U.S. patent depository library.

★2540★ B.F. Goodrich Chemical Company, Avon Lake Technical Center, Information Center
Box 122
Avon Lake, OH 44012 (216) 933-0524
Peter W. Bowler, Supervisor, Information Center

Subjects: Polymerization technology, plastics applications, rubber applications, specialty polymers. **Holdings:** 10,000 volumes; 100,000 company technical reports on microfilm; U.S. and foreign patents, 1964 to present, with translations. **Online Access:** Pergamon ORBIT InfoLine, Inc., OCLC, DIALOG Information Services, STN International, WILSONLINE; internal databases.

★2541★ B.F. Goodrich Company, Research and Development Center, Brecksville Information Center
9921 Brecksville Rd.
Brecksville, OH 44141 (216) 447-5299
Lillian DeVault, Supervisor

Subjects: Plastics, polymer chemistry, rubber. **Holdings:** 10,000 books; 15,000 bound periodical volumes; 500 pamphlets and reports; 100,000 internal reports; 1800 reels of microfilm of journals; U.S. patents, 1952 to present, on microfilm; Chemical Abstracts, 1907 to present, hardcopy and microfilm. **Online Access:** DIALOG Information Services, Pergamon ORBIT InfoLine, Inc., STN International, Dow Jones News/Retrieval, BRS Information Technologies, INKA; internal database.

★2542★ Baker & Botts, Law Library
3000 One Shell Plaza
Houston, TX 77002 (713) 229-1412
Robert K. Downie, Librarian

Subjects: Law - tax, intellectual property, public utilities, labor, bankruptcy, corporate, energy, real estate, oil and gas, probate, international, trial, banking, environmental, admiralty, antitrust. **Holdings:** 72,000 books. **Online Access:** DIALOG Information Services, WESTLAW, LEXIS, Dow Jones News/Retrieval, Pergamon ORBIT InfoLine, Inc., VU/TEXT Information Services, DataTimes.

★2543★ Baker & Mc Kenzie, Library
2800 Prudential Plaza
Chicago, IL 60601 (312) 861-2915
Frank Lukes, Law Librarian

Subjects: U.S. taxation, corporations, foreign trade and investment, Illinois law, foreign taxation, foreign industrial property laws, securities, products liability, medical malpractice. **Special Collections:** Complete tax library in original languages on Germany, Switzerland, Netherlands, France, Italy, British Commonwealth of Nations; substantial holdings on Latin American, European, Asian, and Middle Eastern countries of interest to American corporations. **Holdings:** 42,000 volumes. **Online Access:** LEXIS, DIALOG Information Services; internal database.

★2544★ Baxter Healthcare Corporation, Information Resource Center
Route 120 and Wilson Rd., RLT-10
Box 490
Round Lake, IL 60073 (312) 546-6311
Elizabeth Tan, Section Manager

Subjects: Plastics, biomedical engineering, engineering, medicine. **Holdings:** 1800 books; 6500 unbound periodical volumes; 900 government and industry specifications and standards; 200 reports; 35,000 U.S. and foreign patents.

★2545★ Birmingham Public and Jefferson County Free Library, Government Documents Department
2100 Park Place
Birmingham, AL 35203 (205) 226-3680
Rebecca Scarborough, Head

Subjects: Census, patents, statistics, federal legislation, laws and regulations. **Special Collections:** ASI microfiche collection (complete, 1974 to present); U.S. Patent Depository (mechanical, electrical, and chemical patents, 1967 to present; design patents, 1923 to present); CIS full microfiche collection (1970 to present; pre-1970 reports, prints, serial set); SRI microfiche collection (January 1981 to present). **Holdings:** 250,000 documents; 2500 shelves of federal documents; 5800 reels of microfilm; 16 cabinets of microfiche. **Online Access:** DIALOG Information Services, U.S. Patent Classification System.

★2546★ Boston Public Library, Government Documents, Microtext, Newspapers
Copley Square
Box 286
Boston, MA 02117 (617) 536-5400
V. Lloyd Jameson, Division Coordinator

Subjects: The Government Documents Section serves as a depository for United Nations, General Agreement on Tariffs and Trade (GATT), Colombo Plan, Danube Commission, European Community, U.S. Government Printing Office (1859 to present; regional, 1971 to present); U.S. Arms Control and Disarmament Agency, U.S. Employment and Training Administration, U.S. Air Quality Control Commission, U.S. Geological Survey, U.S. Patent Office. **Online Access:** DIALOG Information Services.

★2547★ Boston Public Library, Science Reference Department
Copley Square
Box 286
Boston, MA 02117 (617) 536-5400
Marilyn T. McLean, Curator

Subjects: Science and technology; patents - U.S., British, German, foreign. **Special Collections:** Patent Depository Library. **Holdings:** Figures not available. **Online Access:** BRS Information Technologies, DIALOG Information Services, Pergamon ORBIT InfoLine, Inc., CAS ONLINE, VU/TEXT Information Services, U.S. Patent Classification System, MEDLARS.

★2548★ Brooklyn Public Library, Science and Industry Division
Grand Army Plaza
Brooklyn, NY 11238 (718) 780-7745
Walter Wolff, Div.Chf.

Subjects: Anthropology, automobiles, biology, birds, chemistry, cookery, engineering, geology, mathematics, product technology, radio, standards, television, health, nursing, patents and trademarks. **Holdings:** 160,000 books; 4700 bound periodical volumes; 40 vertical file drawers of pamphlets.

★2549★ Broward County Main Library, Government Documents Department
100 S. Andrews Ave.
Fort Lauderdale, FL 33301 (305) 357-7444

Special Collections: U.S. patent depository library.

★2550★ Brown and Williamson Tobacco Corporation, Research Library
1600 W. Hill St.
Louisville, KY 40210 (502) 568-7683
Carol S. Lincoln, Research Librarian

Subjects: Chemistry, tobacco, agriculture, chemical engineering, physics, statistics. **Holdings:** 4000 books; 5000 bound periodical volumes; 30,000 patents; 6000 company reports; 15 vertical file drawers of pamphlets and miscellanea. **Online Access:** DIALOG Information Services, Pergamon ORBIT InfoLine, Inc., NLM, DARC Pluridata System (DPDS), CAS ONLINE; internal database.

★2551★ Brown Maroney Rose Barber & Dye, Law Library
1300 One Republic Plaza
333 Guadalupe St.
Austin, TX 78701 (512) 472-5456
Sandra J. Bieri, Librarian

Subjects: Law - environmental, business, utilities, intellectual, public; litigation. **Holdings:** 15,000 books; 400 bound periodical volumes; 15 drawers of microfiche; 300 audio cassettes; 50 video cassettes. **Online Access:** DIALOG Information Services, WESTLAW, LEXIS, NEXIS, MEDIS, Information America, VU/TEXT Information Services; internal databases; ABA/net (electronic mail service).

★2552★ Brownstein, Zeidman & Schomer, Law Library
1401 New York Ave., N.W., Suite 900
Washington, DC 20005 (202) 879-5700
Richard F. Cousins, Librarian

Subjects: Law - housing, franchising, patent and trademark, securities, real estate. **Holdings:** 12,000 books. **Online Access:** LEXIS.

★2553★ Buckman Laboratories International, Technical Information Center
1256 N. McLean Blvd.
Memphis, TN 38108 (901) 278-0330
W. Ellen McDonell, Manager

Subjects: Chemistry, microbiology, corrosion, pulp and paper, water treatment, agriculture. **Special Collections:** National Technical Information Service (NTIS) Collection in Environmental Sciences. **Holdings:** 8000 books; 300 bound periodical volumes; U.S. and foreign patents; microfilm; reports. **Online Access:** DIALOG Information Services, STN International, Pergamon ORBIT InfoLine, Inc., MEDLARS, Toxicology Data Network (TOXNET), National Pesticide Information Retrieval System (NPIRS), Dow Jones News/Retrieval, Dun & Bradstreet Corporation.

★2554★ Buffalo & Erie County Public Library, Science and Technology Department
Lafayette Square
Buffalo, NY 14203 (716) 846-7101
Sharon L. Edward, Head

Subjects: Engineering, mathematics, life sciences, earth sciences, agriculture, astronomy, military science, naval science, computer science. Special Collections: Patent depository library. **Holdings:** 214,000 books; 47,000 bound periodical volumes. **Online Access:** DIALOG Information Services, U.S. Patent Classification System, Chemical Information Systems, Inc. (CIS). Performs searches on fee basis.

★2555★ C-I-L Inc., Research Centre Library
2101 Hadwen Rd.
Sheridan Park
Mississauga, ON, Canada L5K 2L3 (416) 823-7160
Joan L. Leishman, Senior Resident Librarian

Subjects: Chemistry, chemical engineering, agriculture, biotechnology, pulping and bleaching. **Holdings:** 3000 books; 4000 bound periodical volumes; 4 vertical file drawers of government reports; 25,000 Canadian, U.S., foreign patents; 700 internal reports. **Online Access:** Pergamon ORBIT InfoLine, Inc., DIALOG Information Services, CAN/OLE, Canada Systems Group (CSG).

★2556★ California State Library, Government Publications Section
Library-Courts Bldg.
P.O. Box 942837
Sacramento, CA 94237-0001 (916) 322-4572

Special Collections: U.S. patent depository library.

★2557★ Canadian Industrial Innovation Centre/ Waterloo, Resource Centre
156 Columbia St., W.
Waterloo, ON, Canada N2L 3L3 (519) 885-5870
Carol Stewart, Librarian

Subjects: Technological innovation, invention, entrepreneurship, beginning businesses, patents, licensing. **Holdings:** 600 books and bound periodical volumes. **Online Access:** DIALOG Information Services, iNET 2000; Spires (University of Waterloo internal database); COSY (electronic mail service). Performs searches on fee basis. Contact Person: Donna Hewitt.

★2558★ Carnegie Library of Pittsburgh, Science and Technology Department
4400 Forbes Ave.
Pittsburgh, PA 15213 (412) 622-3138
Margery Peffer, Acting Head

Subjects: Chemistry, physics, engineering, technology, geology, health, metallurgy, botany, biology, zoology. **Special Collections:** U.S. and British patent depository library; American National Standards (ANSI); British Standards (BSI); U.S. military and federal specifications; Brutcher translations; ISI translations; AEC translations; AEC reports; ERDA reports. **Holdings:** 376,600 books; 420,000 bound periodical volumes; 870,000 reels of microfilm and microfiche; 22,000 historical trade catalogs; complete sets of U.S. and British patents. **Online Access:** DIALOG Information Services, OCLC, Pergamon ORBIT InfoLine, Inc., U.S. Patent Classification System, STN International. Performs searches on fee basis. Contact Person: David Murdock, Senior Librarian.

★2559★ Catalytica, Inc., Information Center
430 Ferguson Dr., Bldg. 3
Mountain View, CA 94043 (415) 960-3000
Cliff Mills, Manager

Subjects: Catalysis, chemical engineering, biotechnology, catalysts. **Special Collections:** U.S. and foreign patents (catalysis and chemical engineering; 6000). **Holdings:** 3000 books; 100 bound periodical volumes. **Online Access:** DIALOG Information Services, CAS ONLINE, STN International, INKA, Syracuse Research Corporation, Pergamon ORBIT InfoLine, Inc.

★2560★ CBS Inc., Law Library
51 W. 52nd St., 36th Fl.
New York, NY 10019 (212) 975-4260
Donna M. Severino, Law Librarian

Subjects: Communications, copyright, cable television, entertainment law, broadcasting. **Holdings:** 4700 books; 12 bound periodical volumes. **Online Access:** LEXIS, NEXIS, DIALOG Information Services.

★2561★ Celanese Chemical Company, Inc., Technical Center Library
1901 Clarkwood Rd.
Box 9077
Corpus Christi, TX 78469 (512) 241-2343
Betty Goodridge, Coordinator

Subjects: Organic aliphatic chemistry and derived technologies, physical chemistry of combustion, production of petroleum chemicals via direct oxidation. **Holdings:** 17,000 books; 9000 bound periodical volumes; 225,000 U.S. and foreign patents; internal and government technical reports. **Online Access:** Online systems.

★2562★ Centre de Recherche Industrielle du Quebec, Industrie Information
8475, rue Christophe Colomb
P.O. Box 2000
Montreal, PQ, Canada H2P 2X1 (514) 383-1550
Claude Lafrance, Director

Subjects: Electronics, robotics. **Holdings:** 3000 books; patents. **Online Access:** DIALOG Information Services, Pergamon ORBIT InfoLine, Inc., BRS Information Technologies, IST-Informatheque Inc., CAN/OLE; Banque des Sources d'Information (internal database); Envoy 100 (electronic mail service). Performs searches on fee basis. Contact Person: Helene Vachon, Librarian.

★2563★ Chevron Corporation, Corporate Law Department, Library
Box 7141
San Francisco, CA 94120-7141 (415) 894-1714
Julia T. Duboczy, Librarian

Subjects: Patents, trademarks, law, licensing. **Holdings:** 1155 books; 1286 bound periodical volumes; documents; microfilm. **Online Access:** LEXIS; internal database.

★2564★ Chevron Research Company, Technical Information Center
576 Standard Ave.
Box 1627
Richmond, CA 94802-0627 (415) 620-2105
M.S. Wawrzonek, Manager

Subjects: Chemistry, petroleum refining, petrochemicals, engineering, fuels, lubricants. **Holdings:** 12,500 books; 8500 bound periodical volumes; preprints of meeting papers of the Society of Automotive Engineers and American Society of Chemical Engineers; 2 million U.S. and foreign patents; 60,000 pamphlets; 20 file drawers of trade literature; 700 vertical file drawers of company reports and correspondence; 2 million pages of documents on microfilm. **Online Access:** DIALOG Information Services, Pergamon ORBIT InfoLine, Inc., RLIN, BRS Information Technologies, CAS ONLINE; OnTyme Electronic Message Network Service, DIALMAIL (electronic mail services).

★2565★ Chicago Public Library Central Library, Business/Science/Technology Division
425 N. Michigan Ave.
Chicago, IL 60611
David R. Rouse, Division Chief

Subjects: Small business, marketing, technology, corporate reports, investments, management, personnel, patents, physical and biological sciences, medicine, health, computer science, careers, environmental information, gardening, cookbooks. **Special Collections:** U.S. and British Patents complete (38,265 volumes and 5581 reels of U.S. patents); gazettes complete; domestic and foreign automobile manuals; radio, TV, electrical, computer schematics; industrial, corporate, product directories; career information; science fair projects; standards and specifications, including International Organization for Standardization (ISO; on microfilm). **Holdings:** 22,000 books; 39,539 bound periodical volumes; Securities and Exchange Commission (SEC) reports; 8 vertical file drawers of federal specifications and standards; 20 vertical file drawers of American National Standards Institute standards; 50 vertical file drawers of pamphlets and corporate annual reports. **Online Access:** DIALOG Information Services, Pergamon ORBIT InfoLine, Inc., BRS Information Technologies, Dow Jones News/Retrieval, Mead Data Central, OCLC, VU/TEXT Information Services, Info Globe, LEGI-SLATE, DataTimes.

★2566★ Ciba Corning Diagnostics Corporation, Steinberg Information Center
63 North St.
Medfield, MA 02052 (617) 359-7722

Subjects: Clinical medicine, market research, engineering, biotechnology, business. **Special Collections:** Market research collection (medical instruments and diagnostics markets; 3500 items); competitor files. **Holdings:** 3000 books; 2200 bound periodical volumes; 24,000 patents; 850 internal research reports. **Online Access:** DIALOG Information Services, MEDLARS, Pergamon ORBIT InfoLine, Inc., BRS Information Technologies, Dow Jones News/Retrieval, INVESTEXT, VU/TEXT Information Services, LabNet; internal database.

★2567★ Cleveland Public Library, Documents Collection
325 Superior Ave.
Cleveland, OH 44114-1271 (216) 623-2870
Siegfried Weinhold, Department Head

Subjects: United States Government publications. **Holdings:** Government Publication Office (GPO) depository, 1886 to present; patent depository, 1790 to present (microfilm); U.S. Census, decennial and nondecennial, 1790 to present (microform); U.S. NASA depository, 1968 to present (microfiche); U.S. Bureau of Mines publications, 1910 to present (microfiche); National Technical Information Service (NTIS) subscription to Selected Research in Microfiche (SRIM), science and technology, 1975 to present (microfiche); Atomic Energy Commission (AEC) and U.S. Department of Energy (DOE) Depository, 1940s to present (microform and hardcopy); U.S. Congressional Committee Hearings, 23rd Congress, 1833 to present (microfiche); American Statistics Index/Abstracts, 1974 to present (microfiche). **Online Access:** OCLC, DIALOG Information Services, BRS Information Technologies, OhioPI (Ohio Public Information Utility), U.S. Patent Classification System, Hannah Legislative Service, Pergamon ORBIT InfoLine, Inc.

★2568★ The Coca-Cola Company, Law Library
P.O. Drawer 1734
Atlanta, GA 30301 (404) 676-2096
Glenn Cooper, Manager

Subjects: Antitrust and trade regulations; food and drug laws; corporation law; Securities and Exchange Commission; labor and employee benefits; patents, trademarks, and copyright; international law. **Special Collections:** Roy D. Stubbs Collection (40 volumes). **Holdings:** 15,000 books; 800 bound periodical volumes; Federal Register, 1970 to present, on microfiche; Code of Federal Regulations, 1977 to present, on microfiche; FTC Decisions on microfiche. **Online Access:** LEXIS, NEXIS, DIALOG Information Services, Dow Jones News/Retrieval, Easylink.

★2569★ Cominco Ltd., Central Technical Library
Technical Research Centre
P.O. Box 2000
Trail, BC, Canada V1R 4S4 (604) 364-4408
S.R. (Stan) Greenwood, Information Specialist

Subjects: Chemistry, extractive metallurgy, mining and milling, engineering. **Holdings:** 6000 books; 12,000 bound periodical volumes; 2000 pamphlets; 15,000 technical patents; 2000 U.S. Bureau of Mines reports and circulars; Chemical Abstracts, 1906 to present; U.S. Patent Office Gazette, 1897 to present; Canadian Patent Office Record, 1907 to present. **Online Access:** Online systems, DIALOG Information Services, QL Systems. Performs searches on fee basis.

★2570★ Computer Horizons, Inc., Library
10 White Horse Pike
Haddon Heights, NJ 08035 (609) 546-0600

Subjects: Bibliometrics, science and technology indicators, science, social sciences. **Special Collections:** Bibliometrics collection (75 books; approximately 750 papers); U.S. patents and patent citations, 1971 to present (600,000). **Holdings:** 1000 books; 3000 papers in subject areas; 250 reels of magnetic computer tapes. **Online Access:** Internal database.

★2571★ Conoco, Inc., Research and Development Department, Technical Information Services
Ponca City, OK 74603 (405) 767-2334
Patsy S. Hoskins, Supervisor

Subjects: Petroleum, chemicals. **Holdings:** 11,500 books; 29,000 bound periodical volumes; 3000 pamphlets; 270,500 patents; 3000 reports in vertical files; API Abstracts, Petroleum Abstracts (University of Tulsa), Chemical Abstracts, U.S. Patents, 1952 to present, and U.S. Patent Gazette, 1930 to present, on microfilm; 23,000 Proprietary Research Reports (online). **Online Access:** Pergamon ORBIT InfoLine, Inc., DIALOG Information Services, NLM, STN International; internal database.

★2572★ Consumer and Corporate Affairs Canada, Patent Library
Place du Portage
Ottawa, ON, Canada K1A 0C9 (613) 997-2964
John Marosi, Head

Subjects: Patents of the world, patent acts and rules, copyrights. **Holdings:** 60,000 books; 5 million copies of patents of various countries; 15,000 gazettes and journals. **Online Access:** DIALOG Information Services.

★2573★ Dallas Public Library, J. Erik Jonsson Central Library, Government Publications Division
1515 Young St.
Dallas, TX 75201 (214) 670-1468
Marie R. Hartman, Manager

Subjects: Official publications - United States, Texas, international, United Nations and affiliates; maps; atlases. **Special Collections:** U.S. Patents depository, 1790 to present; U.S. Geological Survey depository; Texas documents depository, NASA depository. **Holdings:** 10,000 books; 6000 bound periodical volumes; 750,000 U.S. Government publications; 10,000 Texas state publications; 10,000 international government documents; 3000 reels of microfilm; 25,000 maps; 150,000 microfiche; geological publications from most of the 50 states. **Online Access:** DIALOG Information Services, U.S. Patent Classification System, BRS Information Technologies, Pergamon ORBIT InfoLine, Inc.. Performs searches on fee basis. Contact Person: Johanna Johnson.

★2574★ Denver Public Library, Business, Science & Government Publications Department
1357 Broadway
Denver, CO 80203 (303) 571-2122
Michael Espinosa, Department Manager

Subjects: Business, technology, pure science. **Special Collections:** Automobile repair manuals; U.S. Patents depository (20 books; Official Gazettes; all patents, 1935 to present). **Holdings:** 200,000 volumes; regional depository for U.S. Government documents. **Online Access:** DIALOG Information Services, Pergamon ORBIT InfoLine, Inc., BRS Information Technologies, U.S. Patent Classification System. Performs searches on fee basis.

★2575★ Detroit Public Library, Technology and Science Department
5201 Woodward Ave.
Detroit, MI 48202 (313) 833-1450
George Unterburger, Chief

Subjects: Metals and metal technology; engineering - automotive, mechanical, civil, electronic, nuclear; biological sciences; space sciences. **Holdings:** 266,000 books; 90,000 bound periodical volumes; U.S. patent collection; 250,000 government reports; 50 vertical file drawers of pamphlets and trade catalogs; 170,000 microcards and microfiche. **Online Access:** NEXIS, U.S. Patent Office Classification System.

★2576★ Dickinson School of Law, Sheely-Lee Law Library
150 S. College St.
Carlisle, PA 17013 (717) 243-4611
James R. Fox, Law Librarian

Subjects: Law. **Special Collections:** Intellectual property law; Jewish, Israeli, Italian law; law and medicine. **Holdings:** 210,000 volumes; 20 vertical file drawers of records and briefs of the U.S. Supreme Court in microform; 365 shelves of records and briefs of Pennsylvania Supreme, Superior, Commonwealth Courts. **Online Access:** LEXIS, WESTLAW, OCLC.

★2577★ Dravo Engineers Inc., Library
1 Oliver Plaza
Pittsburgh, PA 15222 (412) 566-5070
Judy Hoover, Manager

Subjects: Engineering, business. **Holdings:** 15,000 books; 900 bound periodical volumes; specifications and standards from 200 associations; U.S. patents, 1966 to present, on microfilm. **Online Access:** DIALOG Information Services. Performs searches on fee basis.

★2578★ Du Pont Canada, Inc., Patents Library
Streetsville Postal Sta., Box 2200
Mississauga, ON, Canada L5M 2H3 (416) 821-5504
Francine Port, Librarian

Subjects: Patents, trademarks, law. **Holdings:** 3000 books; 150,000 patents. **Online Access:** DIALOG Information Services, Pergamon ORBIT InfoLine, Inc., Canadian Systems Group (CSG).

★2579★ Engelhard Corporation, Technical Information Center
Menlo Park
Edison, NJ 08818 (201) 321-5271
Roger L. Meyer, Manager

Subjects: All aspects of noble metals-platinum group metals; gold and silver; catalysis; air pollution control; Kaolin; clays; ore dressing; paper making and converting; pigments. **Holdings:** 25,000 books; 5000 bound periodical volumes; 1000 other cataloged items; 20,000 special reports; 50,000 foreign patents; complete U.S. patents, 1974 to present. **Online Access:** Online systems.

★2580★ Estee Lauder Inc., Information Center
125 Pine Lawn Rd.
Melville, NY 11747 (516) 531-1174
Dr. Dorothy A. Kramer, Manager

Subjects: Cosmetics, dermatology, biochemistry, chemistry, computer science, engineering. **Special Collections:** U.S. and foreign patents (1000). **Holdings:** 1500 books; 1500 bound periodical volumes; supplier literature. **Online Access:** DIALOG Information Services, BRS Information Technologies, Pergamon ORBIT InfoLine, Inc., Chemical Information Systems, Inc. (CIS), STN International, CAS ONLINE, Telesystemes Questel, Data-Star. Performs searches free of charge.

★2581★ Eveready Battery Company, Inc., Technical Information Center
25225 Detroit Rd.
Westlake, OH 44145 (216) 835-7631
Claire Marie Langkau, Manager

Subjects: Batteries, electrochemistry. **Holdings:** 8000 books; 10,000 documents; 30,000 patents. **Online Access:** DIALOG Information Services, Pergamon ORBIT InfoLine, Inc., BRS Information Technologies, Dow Jones News/Retrieval, STN International; internal databases.

★2582★ Firestone Tire and Rubber Company, Central Research Library
1200 Firestone Pkwy.
Akron, OH 44317 (216) 379-7430
S. Koo, Librarian

Subjects: Rubber, plastics, textiles, chemical and mechanical engineering, polymer chemistry. **Holdings:** 7500 books; 11,000 bound periodical volumes; 54 vertical file drawers of patents; 12 vertical file drawers of pamphlets; 2000 government documents; 3750 reels of microfilm; 3000 microfiche; 10,000 internal research reports. **Online Access:** DIALOG Information Services, Pergamon ORBIT InfoLine, Inc.

★2583★ Fish and Neave, Library
875 Third Ave.
New York, NY 10022 (212) 715-0672
Janet M. Stark, Librarian

Subjects: Law - patent, trademark, unfair competition. **Holdings:** 15,000 books; 34 bound periodical volumes. **Online Access:** LEXIS, DIALOG Information Services, WESTLAW.

★2584★ Free Library of Philadelphia, Government Publications Department
Logan Square
Philadelphia, PA 19103 (215) 686-5330

Special Collections: U.S. patent depository library.

★2585★ Frost & Jacobs, Library
2500 Central Trust Center
201 E. 5th St.
Cincinnati, OH 45202 (513) 651-6810
Yvonne M. Davis, Librarian

Subjects: Law. **Special Collections:** Patents. **Holdings:** 25,000 books; 600 bound periodical volumes; 1200 microfiche. **Online Access:** LEXIS, NEXIS, DIALOG Information Services, PHINet FedTax Database, Hanna Legislative Service.

★2586★ GAF Corporation, Technical Information Services
1361 Alps Rd.
Wayne, NJ 07470 (201) 628-3321
Ira Naznitsky, Manager

Subjects: Chemistry, engineering, pulp and paper, petroleum, polymers, surfactants, brighteners, chemical intermediates, textile assistants, organic chemistry. **Holdings:** 9500 books; 10,000 bound periodical volumes; 500 unbound journal volumes; 500,000 U.S. patents; 3000 foreign patents; 950 reels of microfilm; 63 vertical file drawers of pamphlets. **Online Access:** Pergamon ORBIT InfoLine, Inc., NLM, BRS Information Technologies, DIALOG Information Services, CAS ONLINE.

★2587★ General Electric Company, Technology Center, Library
Box 68
Washington, WV 26181 (304) 863-7335
Jo Ellen Butcher, Library Coordinator

Subjects: Polymers, organic chemistry, plastics, rubber, management, adhesives, chemical engineering, petrochemicals. **Holdings:** 25,000 volumes; 32 vertical file drawers of pamphlets; 24 vertical file drawers of U.S. patents; 13 drawers of U.S. patents on microfiche; 12 vertical file drawers of foreign patents; 10 drawers of government reports; audio cassette collection; AD and PB reports on microfiche. **Online Access:** DIALOG Information Services, Pergamon ORBIT InfoLine, Inc., NLM, Dow Jones News/Retrieval, CAS ONLINE, The Source Information Network, NLM, WILSONLINE, Chemical Information Services, Inc. (CIS), CIS, NTIS, Occupational Health Services, Inc., Superindex Inc., SEC Online, Corrosion Data Base; internal database.

★2588★ Georgia Institute of Technology, Price Gilbert Memorial Library
225 North Ave.
Atlanta, GA 30332-0900 (404) 894-4501
Miriam A. Drake, Director

Subjects: Engineering - aerospace, ceramic, chemical, civil, electrical, industrial, systems, mechanical, textile; city planning; architecture; biology; chemistry; science and mechanics; geophysical science; information and computer sciences; mathematics; physics; psychology. **Special Collections:** Maps (147,000); patents (4.8 million); technical reports (1.9 million); government documents (680,000). **Holdings:** 2.2 million volumes; 19,000 photographs; 52,000 pamphlets; 2000 films; 2.6 million microforms. **Online Access:** DIALOG Information Services, Pergamon ORBIT InfoLine, Inc., BRS Information Technologies, Wharton Econometric; BITNET, ALANET (electronic mail services). Performs searches on fee basis. Contact Person: Kathy Tomajko, 894-4511.

★2589★ Greyhound Corporation, Patent Library
Dial Technical Center
15101 N. Scottsdale Rd.
Scottsdale, AZ 85254 (602) 998-6365
Shirley C. Blazer, Patent Coordinator

Subjects: Packaging, shelf-stable foods, soaps and detergents, personal care products, household products. **Special Collections:** U.S. Patent Office Gazette, 1950 to present. **Holdings:** 60,000 U.S. patents; 40,000 foreign patents. **Online Access:** DIALOG Information Services, Pergamon ORBIT InfoLine, Inc., VU/TEXT Information Services, WESTLAW, NEXIS, CAS ONLINE.

★2590★ Henkel Corporation, Process Chemicals Library
350 Mt. Kemble Ave., CN 1931
Morristown, NJ 07960-1931 (201) 267-1000
Amy J. Meskin, Librarian

Subjects: Chemistry - organic, textile, paper, polymer, leather; business. **Holdings:** 11,000 volumes; 32 vertical file drawers of U.S. patents; U.S. Chemicals patents, 1959 to present, on microfilm; 28 vertical file drawers of technical data sheets. **Online Access:** DIALOG Information Services, STN International.

★2591★ Hoechst Celanese Corporation, R.L. Mitchell Technical Information Center
86 Morris Ave.
Summit, NJ 07901 (201) 522-7500
Maddy Urken, Information Specialist

Subjects: Polymers; plastics; fibers; chemicals; coatings; biochemistry; chemistry - organic, physical, analytical, inorganic. **Holdings:** 20,000 books; 20,000 bound and microform periodical volumes; 35,000 internal reports in hardcopy and microform; 1.5 million patents in microform. **Online Access:** DIALOG Information Services, Pergamon ORBIT InfoLine, Inc., CAS ONLINE, BRS Information Technologies, Dun & Bradstreet Corporation, Telesystemes Questel, NLM; internal database.

★2592★ Howard University, University Libraries
P.O. Box 708
Washington, DC 20059 (202) 636-5060

Special Collections: U.S. patent depository library.

★2593★ Howrey & Simon, Library
1730 Pennsylvania Ave., N.W.
Washington, DC 20006-4793 (202) 783-0800
Marie Coleman Kaddell, Director

Subjects: Law - antitrust, patent and trademark, government contracts, international trade. **Holdings:** 30,000 books. **Online Access:** DIALOG Information Services, LEXIS, NewsNet, Inc., VU/TEXT Information Services, DataTimes; internal database.

★2594★ ICI Americas Inc., Atlas Library
Concord Pike & New Murphy Rd.
Wilmington, DE 19897 (302) 575-8235
Frieda S. Mecray, Librarian

Subjects: Biomedicine, chemistry, chemical technology, business economics, pharmaceuticals. **Holdings:** 15,000 volumes; 10 vertical file drawers of government documents; 60 linear feet of patents; 7 vertical file drawers of specifications; 25 vertical file drawers of trade catalogs; 35 vertical file drawers; U.S. patents, 1964 to present, on microfilm; 108 titles on microfilm. **Online Access:** DIALOG Information Services, BRS Information Technologies, NEXIS, Datext Corporate Database, OCLC.

★2595★ IFI/Plenum Data Company, Library
302 Swann Ave.
Alexandria, VA 22301 (703) 683-1085
Harry M. Allcock, Vice President

Subjects: Chemistry, mechanics, electronics. **Special Collections:** U.S. Chemical Patent Index, 1950 to present; U.S. Mechanical and Electrical Patent Bibliographic Data Collection, 1963. **Holdings:** Figures not available. **Online Access:** Online systems.

★2596★ Illinois State Library
Centennial Bldg.
Springfield, IL 62706 (217) 782-2994
Bridget L. Lamont, Director

Subjects: U.S. and Illinois state government, business. **Special Collections:** State documents (461,110); federal documents (1.6 million); U.S. patent depository. **Holdings:** 4.5 million items. **Online Access:** BRS Information Technologies, DIALOG Information Services, LEXIS, CAS ONLINE, OCLC, U.S. Patent Classification System, TechCentral; CLSI (internal database); ALANET (electronic mail service).

★2597★ Indianapolis-Marion County Public Library, Business, Science and Technology Division
Box 211
Indianapolis, IN 46206 (317) 269-1741
Mark Leggett, Head

Subjects: Science, engineering, space science, agriculture, electronics, computer science, building, health, cookery, television, accounting, advertising, economics, business, insurance, investment management. **Special Collections:** Wright Marble Collection of rare cookbooks; Arthur Stumpf Collection of old menus; U.S. patent depository. **Holdings:** 65,000 titles; 90 VF drawers. **Online Access:** Access to DIALOG Information Services, NEXIS, U.S. Patent Classification System, Statistical Information System (STATIS).

★2598★ Institute of Textile Technology, Textile Information Services, Roger Milliken Textile Library
Rte. 250 W.
Box 391
Charlottesville, VA 22902 (804) 296-5511
Terry L. Beckwith, Manager

Subjects: Textile technology, dyeing, polymers, apparel manufacture. **Holdings:** 13,000 books; 15,000 bound periodical volumes; 4500 cataloged translations; 2500 technical reports; 66,000 patents; 25 shelves of reprints; 5 shelves of trade literature; 36 VF drawers; 900 microforms. **Online Access:** Pergamon ORBIT InfoLine, Inc., DIALOG Information Services; internal database.

★2599★ Inventors Clubs of America, Library
Box 450261
Atlanta, GA 30345 (404) 938-5089
Alexander T. Marinaccio, President

Subjects: Inventions - patenting, development, manufacturing, marketing, advertising. **Holdings:** 2000 volumes.

★2600★ James River Corporation, Neenah Technical Information Center
1915 Marathon Ave.
Box 899
Neenah, WI 54956 (414) 729-8169
Cheryl Lamb, Information Scientist

Subjects: Pulp, paper, and paperboard chemistry and technology; plastics; packaging technology; food technology; chemical engineering.

Special Collections: All U.S. patents, 1978 to present (microfilm). **Holdings:** 6000 books; 2500 bound periodical volumes; 12,000 technical documents; 175 vertical file drawers. **Online Access:** DIALOG Information Services, Pergamon ORBIT InfoLine, Inc., OCLC, STN International, WILSONLINE, BRS Information Technologies.

★2601★ John Marshall Law School, Library
315 S. Plymouth Court
Chicago, IL 60604 (312) 427-2737
Randall T. Peterson, Director

Subjects: Law, taxation, intellectual property. **Holdings:** 217,045 volumes; 75,515 volumes in microform; NRS ultrafiche (1st series); Congressional Information Service (CIS) publications, 1970 to present; Congressional Record, 1873 to present; Code of Federal Regulations (CFR) microfiche edition, 1938 to present; IHS microfiche of legislative histories (10 subject areas); CCH Tax Library; U.S. Supreme Court Records and Briefs, 1930 to present; Federal Register, 1936 to present. **Online Access:** LEXIS, WESTLAW, DIALOG Information Services.

★2602★ Kaye, Scholer, Fierman, Hays & Handler, Law Library
425 Park Ave.
New York, NY 10022 (212) 836-8312
Gerald Goodhartz, Librarian

Subjects: Law. **Special Collections:** Law - antitrust, tax, copyright, trademark, corporate, real estate, banking, bankruptcy, labor. **Holdings:** 58,500 books; 1000 bound periodical volumes; 5319 microfiche; 501 ultrafiche; 1600 reels of microfilm; 100 vertical file drawers. **Online Access:** LEXIS, NEXIS, Pergamon ORBIT InfoLine, Inc., DIALOG Information Services, Info Globe, WESTLAW, Dow Jones News/Retrieval, Information America.

★2603★ Kraft, Inc., Technology Center Library
801 Waukegan Rd.
Glenview, IL 60025 (312) 998-3707
Helen Pettway, Librarian

Subjects: Food technology, dairy science, nutrition, microbiology, packaging. **Holdings:** 9550 books; 1900 bound periodical volumes; 15,300 patents; 4000 reels of microfilm; 15 boxes of bulletins; 21,393 research reports. **Online Access:** DIALOG Information Services, Pergamon ORBIT InfoLine, Inc., BRS Information Technologies, CAS ONLINE, ChicagoFile, Leatherhead Food Research Association, LEXIS, NEXIS, VU/TEXT Information Services, WILSONLINE.

★2604★ Levi Strauss & Company, Corporate Law Library
1155 Battery St.
San Francisco, CA 94106 (415) 544-7676
Yvonne B. Marty, Manager

Subjects: Antitrust law, trademarks, copyrights. **Holdings:** 3500 books; 2000 bound periodical volumes; 25 cassettes; 45 loose-leaf services; worldwide listing of trademarks and copyrights. **Online Access:** DIALOG Information Services, WESTLAW.

★2605★ Library of Congress, Copyright Public Information Office
James Madison Memorial Bldg., LM-401
101 Independence Ave., S.E.
Washington, DC 20559 (202) 479-0700
Victor Marton, Head

Subjects: Copyright. **Holdings:** Copyright Card Catalog (40 million cards).

★2606★ Linda Hall Library
5109 Cherry St.
Kansas City, MO 64110 (816) 363-4600

Special Collections: U.S. patent depository library.

★2607★ Los Angeles Public Library, Science & Technology Department
630 W. Fifth St.
Los Angeles, CA 90071-2097 (213) 612-0503
Billie M. Connor, Department Manager

Subjects: Physical and biological sciences, consumer health, medicine and drugs, alternative medicine, earth and natural sciences, applied technology, ecology, astronomy, cookery, computer sciences, natural history, climatology, oceanography. **Special Collections:** U.S. patents depository; specifications and standards. **Holdings:** 332,766 volumes; 80,819 documents. **Online Access:** DIALOG Information Services, LEXIS, NEXIS, U.S. Patent Classification System, Bookline, LEXPAT; internal database; OnTyme Electronic Message Network Service (electronic mail service). Performs searches on fee basis.

★2608★ Louisiana State University, Business Administration/Government Documents Department
Troy H. Middleton Library
Baton Rouge, LA 70803 (504) 388-2570
Myrtle S. Bolner, Head

Subjects: Business administration, government, energy, nuclear sciences, international agencies, agriculture. **Special Collections:** U.S. and United Nations documents depository; NRC Collection; U.S. Patent Depository Library. **Holdings:** 450,000 volumes; 500,000 microforms; 10K reports. **Online Access:** DIALOG Information Services, Pergamon ORBIT InfoLine, Inc., BRS Information Technologies, U.S. Patent Classification System; DIALMAIL (electronic mail service). Performs searches on fee basis. Contact Person: W. David Gay, Assistant Librarian.

★2609★ Louisiana State University, Chemistry Library
301 Williams Hall
Baton Rouge, LA 70803 (504) 388-2530
Silvia D. Espinosa, Head

Subjects: Chemistry, biochemistry, chemical engineering. **Holdings:** 44,942 volumes; U.S. chemical patents, 1955-1975, on microcards and microfiche. **Online Access:** DIALOG Information Services, BRS Information Technologies, Pergamon ORBIT InfoLine, Inc., Chemical Information Systems, Inc. (CIS), STN International, CAS ONLINE, MEDLINE. Performs searches on fee basis.

★2610★ Louisville Free Public Library, Reference and Adult Services
301 York St.
Louisville, KY 40203-2257 (502) 561-8614

Special Collections: U.S. patent depository library.

★2611★ M and T Chemicals, Inc., Technical & Business Information Center
Rahway Rd. & Randolph Ave.
Box 1104
Rahway, NJ 07065 (201) 499-2437
Louis P. Torre, Director

Subjects: Chemistry, ceramics, plastics, organometallic chemistry, glass coatings, tin chemistry, electroplating, electronic chemicals. **Special Collections:** Organotin Literature. **Holdings:** 12,000 books; 11,500 bound periodical volumes; 7000 technical reports and pamphlets; 145,000 patents; 16,000 photocopies; 150 theses; microfilm. **Online Access:** DIALOG Information Services, STN International, Pergamon ORBIT InfoLine, Inc., Telesystemes Questel, Mead Data Central, Dow Jones News/Retrieval; internal database.

★2612★ Manitoba Research Council, Industrial Technology Centre Library

1329 Niakwa Rd.
Winnipeg, MB, Canada R2J 3T4 (204) 945-6000
George Montgomery, Head

Subjects: Engineering - mechanical, electronics, industrial, computer-aided, chemical. **Holdings:** 3000 books; 800 unbound periodicals; 5000 government reports; 3000 technical society papers on microfiche; 100,000 Canadian patent abstracts. **Online Access:** DIALOG Information Services, CAN/OLE, Pergamon ORBIT InfoLine, Inc., Telesystemes Questel; internal database; Envoy 100, CompuServe, Inc. (electronic mail services). Performs searches on fee basis.

★2613★ Mc Carthy and Mc Carthy, Library

Toronto Dominion Bank Tower
Toronto-Dominion Center
P.O. Box 48
Toronto, ON, Canada M5K 1E6 (416) 362-1812
Mary Percival, Librarian

Subjects: Law - computer/telecommunications, patent, trademark. **Holdings:** 17,000 volumes. **Online Access:** QL Systems, Info Globe, Canada Systems Group (CSG), DIALOG Information Services, WESTLAW, Dow Jones News/Retrieval, LEXIS, NEXIS, CAN/LAW, The Financial Post Information Service, VU/TEXT Information Services.

★2614★ Mc Gean-Rohco, Inc., Research Library

2910 Harvard Ave.
Box 09087
Cleveland, OH 44109 (216) 441-4900
Jeanne R. Winters, Librarian

Subjects: Metallurgy, organic and inorganic chemistry, electroplating. **Special Collections:** Plating information. **Holdings:** 1000 books; 1600 bound periodical volumes; 10,000 patents; 4 vertical file drawers of pamphlets; 4 vertical file drawers of catalogs; 3 vertical file drawers of articles. **Online Access:** DIALOG Information Services.

★2615★ Medical University of South Carolina Library

171 Ashley Ave.
Charleston, SC 29425 (803) 792-2371

Special Collections: U.S. patent depository library.

★2616★ Memphis and Shelby County Public Library and Information Center, Business/Science Department

1850 Peabody Ave.
Memphis, TN 38104 (901) 725-8876

Special Collections: U.S. patent depository library.

★2617★ Metropolitan Toronto Reference Library, Science & Technology Department

789 Yonge St.
Toronto, ON, Canada M4W 2G8 (416) 393-7086
Jean Forde, Manager

Subjects: Physical, biological, and medical sciences; engineering sciences; natural history; horticulture; technology and food technology; cookery; sports and recreation. **Special Collections:** Geological Survey of Canada publications; Atomic Energy of Canada publications; workshop manuals for motor vehicles and household appliances. **Holdings:** 125,250 books; 5982 volumes of Canadian, American, and British patent abstracts; 12,000 volumes of standards; 4250 volumes of Canadian and American radio, television, and personal computer schematics; 7956 vertical file folders; 4019 maps; 5000 automotive shop manuals; 1250 cookbooks. **Online Access:** DIALOG Information Services, Info Globe, Pergamon ORBIT InfoLine, Inc., MEDLARS, WESTLAW, VU/TEXT Information Services, CAN/OLE, WILSONLINE, Canada Systems Group (CSG), Dunserve II, med Data, Internatio nal Development Research (IDRC), QL Systems, UTLAS; Envoy 100 (electronic mail service). Performs searches on fee basis. Contact Person: Helen Baltais, 393-7005.

★2618★ Miami-Dade Public Library, Business, Science and Technology Department

101 W. Flagler St.
Miami, FL 33130-1504 (305) 375-2664

Subjects: Business, economics, international trade, investments, layperson's medical reference, pure and applied sciences. **Special Collections:** Annual reports of companies and Florida corporations; U.S. patent depository; manufacturing, specialized, and foreign business directories (600); microfiche collection. **Holdings:** 42,000 volumes; 80 vertical file drawers of pamphlets. **Online Access:** DIALOG Information Services, BRS Information Technologies, U.S. Patent Classification System, LOGIN.

★2619★ Miller Brewing Company, Scientific and Technical Information Facility

3939 W. Highland Blvd.
Milwaukee, WI 53201 (414) 931-3640
Joanne L. Schwarz, Head Librarian

Subjects: Brewing, chemistry, microbiology, chemical engineering, genetics, enzymology. **Special Collections:** Brewing science. **Holdings:** 500 books; 1500 bound periodical volumes; 10,000 patents; 80 reels of microfilm; 2000 microfiche; 200 research reports. **Online Access:** Pergamon ORBIT InfoLine, Inc., DIALOG Information Services, RLIN, BRS Information Technologies, MEDLINE; internal database.

★2620★ Milwaukee Public Library, Science & Business Division

814 W. Wisconsin Ave.
Milwaukee, WI 53233 (414) 278-3043
Theodore Cebula, Coordinator

Subjects: Agriculture, business, census, chemistry, economics, engineering, industrial labor, natural and physical sciences, physics, statistics. **Special Collections:** U.S. and British patent depository; industrial and government standards. **Holdings:** Figures not available. **Online Access:** OCLC, U.S. Patent Classification System.

★2621★ Minneapolis Public Library & Information Center, Technology and Science Department

300 Nicollet Mall
Minneapolis, MN 55401 (612) 372-6570
Edythe Abrahamson, Head

Subjects: Natural and applied sciences. **Special Collections:** Environmental Conservation Library (ECOL; 17,500 volumes); Minnesota Regional Copper-Nickel Project documents; U.S. Nuclear Regulatory Commission Public Documents Room; patent depository library, 1790 to present; SAMS Photofact Service, Volume 1 to present. **Holdings:** 162,000 books; 192,683 microforms; 3000 environmental impact statements; 2800 auto repair manuals; 425 computer programs. **Online Access:** DIALOG Information Services, Pergamon ORBIT InfoLine, Inc., BRS Information Technologies, U.S. Patent Classification System; internal database. Performs searches free of charge.

★2622★ Mobil Research & Development Corporation, Dallas Research Laboratory, Library

Box 819047
Dallas, TX 75381 (214) 851-8140
Janet Wolford, Manager

Subjects: Petroleum exploration and production; basic sciences. **Holdings:** 40,000 books; 40,000 bound periodical volumes; 9200 maps; U.S. patents on microfilm; U.S. Bureau of Mines documents on microfilm. **Online Access:** DIALOG Information Services, BRS Information Technologies, OCLC, Pergamon ORBIT InfoLine, Inc.

★2623★ Monsanto Company, Patent Department Library

Mail Code G3ND
800 N. Lindbergh Blvd.
St. Louis, MO 63141 (314) 694-3065

Subjects: Chemical patents, patent and trademark law, licensing, trade secrets. **Special Collections:** U.S. Chemical Patents, 1952 to present (580,000). **Holdings:** 700 books; 125 bound periodical volumes; patents on microfilm; patent abstracts; U.S. Official Gazette on microfilm. **Online Access:** DIALOG Information Services, Pergamon ORBIT InfoLine, Inc., Telesystemes Questel, LEXIS, NEXIS; internal databases.

★2624★ Montana College of Mineral Science and Technology, Library

W. Park St.
Butte, MT 59701 (406) 496-4281
Joanne V. Lerud, Director

Subjects: Geology, mining, mineral processing, geochemistry, geophysics, petroleum, environmental engineering, occupational safety, mineral economics, technology and society, industrial hygiene. **Special Collections:** Mining and geology of Montana; international and state geological documents; U.S. patents depository, 1981 to present; federal document depository (378,500 documents, bound and on microfilm). **Holdings:** 40,000 books; 35,400 bound periodical volumes. **Online Access:** DIALOG Information Services, Pergamon ORBIT InfoLine, Inc., BRS Information Technologies, STN International; FAPRS (internal database). Performs searches on fee basis.

★2625★ Morgan & Finnegan, Library

345 Park Ave.
New York, NY 10154 (212) 758-4800
Lucy Curci, Librarian

Subjects: Law - patent, trademark, copyright. **Holdings:** 10,000 volumes. **Online Access:** LEXIS, WESTLAW, Pergamon ORBIT InfoLine, Inc., VU/TEXT Information Services, Dow Jones News/Retrieval, DIALOG Information Services.

★2626★ Motor Vehicle Manufacturers Association, Patent Research Library

320 New Center Bldg.
Detroit, MI 48202 (313) 872-4311
James A. Wren, Manager

Subjects: Automotive patents, technology, automotive history. **Special Collections:** Patents specifically related to motor vehicle technology (one million); foreign and U.S. automotive sales brochures, pre-1900 to present. **Holdings:** 13,000 bound periodical volumes; 1000 textbooks; manufacturers' brochures and instruction books. **Online Access:** Motor Vehicle Manufacturers Association Trademark Data Base (internal database).

★2627★ National Association of Watch and Clock Collectors, Inc., Watch & Clock Museum Library

514 Poplar St.
Box 33
Columbia, PA 17512 (717) 684-8261
Donald J. Summar, Librarian

Subjects: Horology. **Special Collections:** Hamilton Watch Company business records (73 bound volumes; 10 cubic feet of documents; 352 reels of microfilm). **Holdings:** 3250 books; 561 bound periodical volumes; 21,000 patents; 423 reels of microfilm; 23 vertical file drawers; archive boxes of manuscripts, catalogs, ephemera. **Online Access:** Internal database. Performs searches on fee basis.

★2628★ National Video Clearinghouse, Inc., NVC Library and Information Service

100 Lafayette Dr.
Syosset, NY 11791 (516) 364-3686
Marie Siegel, Director

Subjects: Video - film programs, equipment, products; advertising and marketing; book trade; publishing; broadcasting. **Special Collections:** National Video Clearinghouse publications; collected material relating to Public Domain; Motion Picture Association of America ratings; copyright; movie industry; educational films; video industry. **Holdings:** 200 books; 120 bound periodical volumes; 20 bound newsletters. **Online Access:** Internal databases.

★2629★ New York Public Library, Annex Section, Patents Collection

521 W. 43rd St.
New York, NY 10036-4396 (212) 714-8529
Richard L. Hill, First Assistant

Subjects: Complete U.S. and British patents, extensive holdings from France, Germany, Belgium, Denmark, and Sweden, abstracts from other nations, complete files of U.S. indexes. **Holdings:** 100,000 volumes; 5000 reels of microfilm; 4800 microfiche. **Online Access:** U.S. Patent Classification System.

★2630★ New York State Library, Sciences/Health Sciences/Technology Reference Services

Cultural Education Center
Empire State Plaza
Albany, NY 12230 (518) 474-7040
Christine A. Bain, Associate Librarian

Subjects: Science, medicine and allied health sciences, technology. **Special Collections:** National Technical Information Service (NTIS) reports (microfiche); all U.S. patents (depository library); Department of Energy reports (microfiche); U.S. Government Printing Office publications (depository and nondepository); Rand reports; standards and specifications. **Holdings:** 75,000 titles in the health sciences collection; 332,000 titles in the science and technology collections; microform collection. **Online Access:** OCLC, U.S. Patent Classification System, BRS Information Technologies, DIALOG Information Services, NLM, Mead Data Central, LEXIS, NEXIS, STN International, WILSONLINE. Performs searches on fee basis.

★2631★ Newark Public Library, Sciences Division

5 Washington St.
Box 630
Newark, NJ 07101-0630 (201) 733-7782
Lawrence C. Schwartz, Supervising Librarian

Subjects: Social studies, education, science, consumer affairs, applied technology. **Special Collections:** U.S. patent depository library; American National Standards Institute (ANSI) standards (24 vertical file drawers); U.S. Government Documents Regional Depository. **Holdings:**

100,000 books; 50,000 bound periodical volumes. **Online Access:** DIALOG Information Services, BRS Information Technologies, NEXIS, OCLC, Pergamon ORBIT InfoLine, Inc., VU/TEXT Information Services, WILSONLINE, U.S. Patent Classification System. Performs searches on fee basis.

★2632★ NL Industries, Inc., Spencer Kellogg Products, Research Center Library
4201 Genesee St.
Box 210
Buffalo, NY 14225 (716) 852-5850

Subjects: Coatings technology; resins - epoxy, alkyd, urethane; vegetable oils. **Holdings:** 2500 books; 1500 bound periodical volumes; 40 vertical file drawers of technical literature; chemical patents, 1967-1981, on microfilm; 6 vertical file drawers of foreign and domestic patents.

★2633★ NL Industries, Inc., Technology Systems, Technical Information Center
3000 North Belt
Box 60070
Houston, TX 77205 (713) 987-4544
Leah A. Bartlett-Cahill, Supervisor

Subjects: Oil and gas drilling, engineering, chemistry, petroleum, materials science, physical sciences. **Special Collections:** Society of Petroleum Engineers papers; National Technical Institute Service (NTIS) reports; U.S. Patents, 1972 to present (complete). **Holdings:** 10,000 books; 10,000 bound periodical volumes; 2000 unbound reports; 15,000 microfiche; 3500 microfilm cartridges; vendor catalogs; industry standards. **Online Access:** DIALOG Information Services, Pergamon ORBIT InfoLine, Inc.; internal database.

★2634★ North Carolina State University, D.H. Hill Library, Documents Department
Box 7111
Raleigh, NC 27695-7111 (919) 737-3280
Jean Porter, Head

Subjects: Agriculture, science and technology, health, aerospace. **Special Collections:** U.S. patent depository library; National Technical Information Service (NTIS) reports (625,988 microfiche); Department of Energy (DOE) reports (620,313 microfiche); NASA reports (118,988 microfiche and hardcopy reports); North Carolina maps (topographic, geographic, outline, road; 13,365). **Holdings:** 730,000 U.S. Government publications. **Online Access:** U.S. Patent Classification System, DIALOG Information Services. Performs searches on fee basis.

★2635★ Nova Scotia Research Foundation Corporation, Library
100 Fenwick St.
P.O. Box 790
Dartmouth, NS, Canada B2Y 3Z7 (902) 424-8670
Helen I. Hendry, Librarian

Subjects: Technology, marine biology, chemistry, geosciences, electronics, production management, small business, ocean engineering. **Special Collections:** Canada Patent Office Record (current 17 years). **Holdings:** 5000 books; 800 bound periodical volumes; 1500 unbound periodicals; 2000 government documents, reports, pamphlets.

★2636★ Ohio State University Libraries, Information Services Department
1858 Neil Ave. Mall
Columbus, OH 43210 (614) 292-6286

Special Collections: U.S. patent depository library.

★2637★ Oklahoma State University, Documents Department
University Library
Stillwater, OK 74078 (405) 624-6546
Vicki W. Phillips, Head

Special Collections: U.S. Government Regional Depository; NASA depository; U.S. patent depository; Oklahoma documents depository. **Holdings:** 1.3 million items; 157,237 accessioned volumes; 1.3 million microforms.

★2638★ Oregon State Library
State Library Bldg.
Summer and Court Sts.
Salem, OR 97310 (503) 378-4274
Wesley A. Doak, State Librarian

Subjects: Oregon history and government, business, librarianship, social sciences, humanities, science and technology. **Special Collections:** U.S. patent depository library. **Holdings:** 282,592 books; 1.2 million government documents and publications; 152,000 microforms; 400 video cassettes; clippings; pamphlets. **Online Access:** DIALOG Information Services, LEXIS, ISIS, LEGISNET, BRS Information Technologies, NLM, EROS Data Center, Oregon Legislative Information System; ALANET, TYMNET, OCLC, Oregon Public Access Catalog (OPAC) (electronic mail services).

★2639★ Patent, Trademark and Copyright Research Foundation, Library
Franklin Pierce Law Center
2 White St.
Concord, NH 03301 (603) 228-1541

Subjects: Patents, trademarks, copyright. **Holdings:** 1000 items.

★2640★ Pennie & Edmonds, Law Library
1155 Ave. of the Americas
New York, NY 10036 (212) 790-0909
Mary Gilligan, Librarian

Subjects: Law - patent, copyright, trademark; biotechnology; chemistry; electronics. **Special Collections:** Foreign Collection (intellectual property). **Holdings:** 20,000 books; 100 bound periodical volumes; videotapes. **Online Access:** DIALOG Information Services, WESTLAW, LEXIS, Pergamon ORBIT InfoLine, Inc., VU/TEXT Information Services, Telesystemes Questel.

★2641★ Pennsylvania State University Libraries, C207 Pattee Library, Documents Section
C207 Pattee Library
University Park, PA 16802 (814) 865-4861

Special Collections: U.S. patent depository library.

★2642★ Petro-Canada Products Inc., Technical Library
2489 N. Sheridan Way
Mississauga, ON, Canada L5K 1A8 (416) 822-6770
Roy E. Metcalfe, Supervisor

Subjects: Petroleum chemistry and technology. **Special Collections:** Society of Automotive Engineers Transactions, 1965 to present; Canadian Patents, 1977-1985; U.S. Chemical Patents, 1952 to present. **Holdings:** 9000 books; 1600 bound periodical volumes. **Online Access:** CAN/OLE, DIALOG Information Services, Info Globe, STN International, Pergamon ORBIT InfoLine, Inc., BRS Information Technologies.

★2643★ Petrolite Corporation, Information Center
369 Marshall Ave.
St. Louis, MO 63119 (314) 968-6008
Pauline C. Beinbrech, Manager

Subjects: Chemistry - organic, petroleum, corrosion; water treatment; wax. **Holdings:** 7200 books; 8500 bound periodical volumes; 94,000 patents; 33,000 microfiche; 10 vertical file drawers of trade literature. **Online Access:** DIALOG Information Services, Pergamon ORBIT InfoLine, Inc., Chemical Information Systems, Inc. (CIS), STN International, Oil & Gas Journal Energy Database, NLM, CHEMEST, OCLC; internal database.

★2644★ Phillips Petroleum Company R&D Library
102 PLB
Bartlesville, OK 74004 (918) 661-3433
Annabeth Robin, Library Supervisor

Subjects: Chemistry, petroleum science and technology, polymer science and technology, biotechnology, geosciences, plastics, physics. **Holdings:** 30,000 books; 25,000 bound periodical volumes; 2 million U.S. and foreign patents; 5500 microfilm cartridges; 20,000 U.S. Government reports on microfiche. **Online Access:** Pergamon ORBIT InfoLine, Inc., DIALOG Information Services, NEXIS, OCLC; internal database.

★2645★ Pillsbury Company, Technical Information Center
311 Second St., S.E.
Minneapolis, MN 55414 (612) 330-4750
James B. Tchobanoff, Manager

Subjects: Food science and technology, cereal chemistry, microbiology, mathematics, statistics, agriculture, plant science. **Holdings:** 6200 books; 8500 bound periodical volumes; 40,000 patents; 35,000 internal reports. **Online Access:** DIALOG Information Services, MEDLINE, CAS ONLINE, Pergamon ORBIT InfoLine, Inc., OCLC; internal database; OnTyme Electronic Message Network Service, DIALMAIL (electronic mail services).

★2646★ Platt Saco Lowell Corporation, Engineering Library
Drawer 2327
Greenville, SC 29602 (803) 859-3211
Alice K. Dill, Patent Technician

Subjects: Engineering, textile machinery and manufacture, patent and trademark law. **Special Collections:** U.S. and British patents on textile machinery. **Holdings:** 3000 books; 1500 bound periodical volumes; 300,000 patent copies; 1500 microfiche cards of abstracts and patents; 2000 paper copies of abstracts and patents.

★2647★ Procter & Gamble Company, Cellulose & Specialties Division Technical Information Services
949 Tillman Ave.
Memphis, TN 38108 (901) 320-8311

Subjects: Chemistry - cellulose, physical, organic, analytical; polymer sciences; colloid science; textiles. **Holdings:** 5000 books; 5000 bound periodical volumes; 10,000 technical reports; 8 vertical file drawers of government documents; 10,000 U.S. and foreign patents; 500 reels of microfilm. **Online Access:** Pergamon ORBIT InfoLine, Inc., NLM, DIALOG Information Services.

★2648★ Providence Public Library, Knight Memorial Branch
275 Elmwood Ave.
Providence, RI 02907 (401) 521-8726
Shirley Long, Head

Special Collections: U.S. patent depository library. Holdings: indexes to U.S. patents, 1790 to present; official gazettes, 1872 to present, on microfilm; pamphlets. **Online Access:** DIALOG Information Services, OCLC. Performs searches on fee basis.

★2649★ Public Library of Cincinnati and Hamilton County, Science and Technology Department
800 Vine St.
Cincinnati, OH 45202 (513) 369-6936
Rosemary Dahmann, Head

Subjects: Pure and applied science, especially chemistry. **Special Collections:** U.S. Depository Library, including U.S. patents, 1790 to present, and Official Gazette, 1872 to present; U.S. military and federal, ASTM, and ANSI standards and specifications; trade directories; Rand Corporation documents. **Holdings:** 197,000 books; 62,300 bound periodical volumes; 54,000 pamphlets; 261,000 microforms. **Online Access:** DIALOG Information Services, WILSONLINE, BRS Information Technologies, NEXIS, U.S. Patent Classification System. Performs U.S. Patent searches free of charge; other searches on fee basis.

★2650★ Quaker Chemical Corporation, Information Resources Center
Conshohocken, PA 19428-0873 (215) 828-4250
Ellen B. Morrow, Manager

Subjects: Chemical technology for the metals and paper specialty fields. **Holdings:** 5500 volumes; 5500 pamphlets; 18 vertical file drawers of vendor literature; 85 vertical file drawers of documents and miscellanea; chemical patents, 1966 to present, on microfilm; 425 reels of microfilm of journals; 365 reels of microfilm of documents; 4 vertical file drawers of government reports on microfiche; 45 audio cassettes. **Online Access:** DIALOG Information Services, Pergamon ORBIT InfoLine, Inc., U.S. Patents Files; internal databases.

★2651★ Queens Borough Public Library, Science & Technology Division
89-11 Merrick Blvd.
Jamaica, NY 11432 (718) 990-0760
John D. Brady, Jr., Head

Subjects: Mathematics, engineering, accounting, chemistry, biological sciences, advertising, nursing, physics, business administration, aeronautics, patents. **Special Collections:** Telephone directories for all major U.S. and foreign cities; automobile and household repair manuals; Sams Photofacts and Computerfacts. **Holdings:** 217,373 books; 29,612 bound periodical volumes; 12,374 reels of microfilm of back issue periodicals; 50 vertical file drawers. **Online Access:** ALANET (electronic mail service).

★2652★ R.J. Reynolds Tobacco Company, R&D Scientific Information Services Library
Bowman Gray Technical Ctr., 611-12, 205C
Winston-Salem, NC 27102 (919) 741-4360
Randy D. Ralph, Contact

Subjects: Tobacco, chemistry, biochemistry, agriculture, chemical engineering. **Holdings:** 25,528 books; 25,346 bound periodical volumes; 1405 unbound periodicals; 6500 internal reports; 1.5 million patents on microfilm.

★2653★ Rice University, Division of Government Publications & Special Resources
Fondren Library
Box 1892
Houston, TX 77251-1892 (713) 527-8101
Barbara Kile, Director

Special Collections: U.S. patent depository library. **Holdings:** 230,000 government documents; 500,000 technical reports; 1.5 million microforms; U.S. patents: utility, 1960 to present; design, 1842 to present; reissues, 1838 to present; plant, 1978 to present. **Online Access:** U.S. Patent Classification System, LEGI-SLATE. Performs searches free of charge.

★2654★ Rochester Public Library, Science and Technology Division
115 South Ave.
Rochester, NY 14604 (716) 428-7327
Jeffrey Levine, Head

Subjects: Physical and natural sciences, applied science and technology, health sciences, environmental sciences, agriculture, home economics. **Special Collections:** Trade catalogs of national firms; automobile shop manuals; Sam's Photofacts Service; Official Gazette of U.S. Patent Office, 1846 to present. **Holdings:** 54,000 books; 15 VF drawers of pamphlets; 130 slide sets; 100 phonograph records; 70 cassettes.

★2655★ St. Louis Public Library
1301 Olive St.
St. Louis, MO 63103 (314) 241-2288

Special Collections: U.S. patent depository library.

★2656★ San Diego Public Library, Science & Industry Section
820 E St.
San Diego, CA 92101 (619) 236-5813
Joanne Anderson, Supervisor

Subjects: Business, industry, science, cookery, automobile repair. **Special Collections:** U.S. patent depository library; space and aeronautics historical collection; depository for U.S., California, and San Diego city and county government publications (over 1.3 million); American National Standards Institute (ANSI) and American Society for Testing and Materials (ASTM) standards on microfiche; U.S. Utility Patents, 1955 to present; U.S. Design Patents, 1951 to present. **Holdings:** 61,600 books; 23,000 maps; 265,000 microforms; Sams Photofacts (complete collection); Atomic Energy Commission (AEC) and NASA depository collections. **Online Access:** DIALOG Information Services, U.S. Patent Classification System. Performs searches free of charge.

★2657★ Science Park Library
5 Science Park
New Haven, CT 06511 (203) 786-5447

Special Collections: U.S. patent depository library.

★2658★ Sunnyvale Patent Information Clearinghouse
1500 Partridge Ave., Bldg. 7
Sunnyvale, CA 94087 (408) 730-7290
Beverley J. Simmons, Director

Subjects: U.S. patents, 1836 to present; patent, trademark, and copyright registration information. **Holdings:** 7800 volumes; 4 million patents; Federal Trademark Register; Report of the Commissioner of Patents, 1790-1835; Official Gazette, 1836 to present; list of patentees, 1870 to present; English language abstracts of Japanese patents (physical field, 1980 to present; electrical field, 1979 to present); European Patent Bulletin, 1979 to present; PCT Gazette, 1981 to present. **Online Access:** DIALOG Information Services, U.S. Patent Classification System. Performs searches on fee basis. Contact Person: Mary-Jo Di Muccio, Administrative Librarian.

★2659★ Technology Transfer Society, Library
611 N. Capitol Ave.
Indianapolis, IN 46204 (317) 262-5022
Barbara Ostermeier, Office Manager

Subjects: Technology transfer. **Special Collections:** Proceedings of the Technology Transfer Society. **Holdings:** 200 volumes.

★2660★ Texas A&M University, Evans Library—Documents Division
College Station, TX 77843-5000 (409) 845-2551

Special Collections: U.S. patent depository library.

★2661★ Thomas J. Lipton, Inc., Library/Information Services
800 Sylvan Ave.
Englewood Cliffs, NJ 07632 (201) 894-7568
Gloria S. Bernstein, Manager

Subjects: Tea, food technology. **Holdings:** 9000 books; 3500 bound periodical volumes; U.S. chemical patents, 1970 to present, on microfilm. **Online Access:** DIALOG Information Services, NEXIS, STN International, MEDLINE; Database of Tea Technology (internal database).

★2662★ Thomson, Rogers, Barristers & Solicitors, Library
390 Bay St., Suite 3100
Toronto, ON, Canada M5H 1W2 (416) 868-3100
Dianne D. Sydij, Librarian

Subjects: Law - commercial, motion picture, entertainment, copyright, insurance, aviation, taxation, real estate, municipal. **Holdings:** 10,000 volumes.

★2663★ 3M, Patent and Technical Communications Services
3M Center, 201-2C-12
St. Paul, MN 55144 (612) 733-7670
Victoria K. Veach, Manager

Subjects: U.S. and foreign patents. **Special Collections:** Complete U.S. Patent collection, 1963 to present (microfilm). **Holdings:** 2700 bound periodical volumes; foreign patents on aperture cards; 110,000 3M reports on microfiche.

★2664★ Toledo-Lucas County Public Library, Science and Technology Department
325 Michigan St.
Toledo, OH 43624 (419) 255-7055
Mary B. Hubbard, Department Manager

Subjects: Physical and natural sciences, applied science and technology. **Special Collections:** U.S. patent depository library. **Holdings:** 87,099 books; depository for federal documents; 11,118 bound patent specifications, 1871-1965; patent specifications, 1966 to present, on microfilm; 13 drawers of microforms; pamphlets; clippings. **Online Access:** DIALOG Information Services, BRS Information

Technologies, Dow Jones News/Retrieval, CAS ONLINE, U.S. Patent Classification System; TLM (internal database).

★2665★ Townley & Updike, Law Library

405 Lexington Ave.
New York, NY 10174 (212) 682-4567
John S. Kostecky, Librarian

Subjects: Law - labor, product liability, antitrust, securities, patent, trademark and copyright. **Holdings:** 20,000 volumes. **Online Access:** LEXIS, WESTLAW, DIALOG Information Services, Dow Jones News/Retrieval, ABA/net, NEXIS.

★2666★ Union Carbide Corporation, Law Department Library

Section N2, 39 Old Ridgebury Rd.
Danbury, CT 06817 (203) 794-6396
Carolyn A. Mariani, Manager

Subjects: Law - antitrust, tax, patent, trademark, labor, corporation. **Holdings:** 30,000 volumes; 3 vertical file drawers; 7 titles in microform; Federal Register, 1970 to present, in microform. **Online Access:** LEXIS, WESTLAW, DIALOG Information Services.

★2667★ UNISYS Corporation, Law Library

Box 500
Blue Bell, PA 19424 (215) 542-4789
Marsha A. Frederick, Chief

Subjects: Law - labor, contract, patent, trademark, copyright. **Holdings:** 11,000 volumes.

★2668★ United Catalysts, Inc., Technical Library

Box 32370
Louisville, KY 40232 (502) 634-7200
Betty B. Simms, Technical Librarian

Subjects: Catalysis, chemistry, physics, engineering, mathematics, clays, management. **Special Collections:** Catalysis. **Holdings:** 4000 books; 1300 bound periodical volumes; 18,000 patents; 18 vertical file drawers of indexed technical reports; microfilm. **Online Access:** DIALOG Information Services, Pergamon ORBIT InfoLine, Inc.

★2669★ United States Department of Commerce, Office of Productivity, Technology and Innovation, Commerce Productivity Center

14th St. & Constitution Ave., N.W., Rm. 7413
Washington, DC 20230 (202) 377-0940
Carol Ann Meares, Manager

Subjects: Technology and innovation, productivity, quality of working life, economics, management, labor relations, public administration. **Holdings:** 5580 volumes; 2000 microfiche; 2000 clippings.

★2670★ United States Patent & Trademark Office, Scientific Library

Crystal Plaza Bldg. 3
2021 Jefferson Davis Hwy.
Arlington, VA 22202 (703) 557-2955
Henry Rosicky, Program Manager

Subjects: Technology, applied science. **Special Collections:** Foreign patents (12 million in numerical arrangement). **Holdings:** 250,000 books; 87,800 bound periodical volumes; 58,790 titles on microfiche; 430 titles on microfilm; U.S. Government documents depository (selective). **Online Access:** DIALOG Information Services, OCLC, Pergamon ORBIT InfoLine, Inc., BRS Information Technologies, MEDLINE, Integrated Technical Information System (ITIS), DTIC.

★2671★ United States Trademark Association, Law Library

6 E. 45th St.
New York, NY 10017 (212) 986-5880
Charlotte Jones, Librarian

Subjects: Trademarks. **Holdings:** 1800 books; 73 bound periodical volumes.

★2672★ University of California, Berkeley, Chemistry Library

100 Hildebrand Hall
Berkeley, CA 94720 (415) 642-3753
Alison Howard, Acting Head

Subjects: Chemistry - inorganic, organic, physical; chemical kinetics, thermodynamics, and engineering; electrochemistry; transport and mass transfer; polymer chemistry. **Special Collections:** Russian monographs and serials obtained on exchange; U.S. chemical patents (20,000). **Holdings:** 47,078 volumes; 31,300 microforms; 322 pamphlets. **Online Access:** DIALOG Information Services, BRS Information Technologies, CAS ONLINE.

★2673★ University of California, Los Angeles, Chemistry Library

4238 Young Hall
Los Angeles, CA 90024-1569 (213) 825-3342
Marion C. Peters, Head Librarian

Subjects: Chemistry - organic, inorganic, analytical, physical; biochemistry. **Special Collections:** Morgan Memorial Collection (history of chemistry). **Holdings:** 59,591 volumes; 1275 reels of microfilm; 37,720 microcards; U.S. chemical patents, 1952 to present, in microform. **Online Access:** DIALOG Information Services, BRS Information Technologies, CAS ONLINE, Institute for Scientific Information (ISI); ORION, MELVYL (internal databases); BITNET (electronic mail service). Performs searches on fee basis.

★2674★ University of Delaware Library, Reference Department

Newark, DE 19717-5267 (302) 451-2965

Special Collections: U.S. patent depository library.

★2675★ University of Idaho Library

Moscow, ID 83843 (208) 885-6235

Special Collections: U.S. patent depository library.

★2676★ University of Maryland, College Park Libraries, Engineering & Physical Sciences Library

College Park, MD 20742 (301) 454-3037
Herbert N. Foerstel, Head

Subjects: Aeronautics and astronautics; astronomy; computer science; engineering - chemical, civil, electrical, industrial, mechanical; geology; materials science; mathematics; oceanography; physics; transportation. **Special Collections:** Patent Depository Library (complete patent backfile, 1789 to present); R. von Mises Collection (1100 titles; 217 boxes of reprints); Max Born Collection (theoretical mathematics and physics; 650 titles; 6 boxes of reprints). **Holdings:** 125,000 books; 120,000 bound periodical volumes; 125,000 hardcopy reports; 1.1 million reports on microfiche; 90,460 reports on microcard; 5100 reels of microfilm. **Online**

Access: DIALOG Information Services, BRS Information Technologies, DTIC, U.S. Patent Classification System, OCLC.

★2677★ University of Massachusetts Physical Sciences Library
Graduate Research Center
Amherst, MA 01003 (413) 545-1370

Special Collections: U.S. patent depository library.

★2678★ University of Michigan, Engineering Transportation Library
312 UGL
Ann Arbor, MI 48109-1185 (313) 764-7494

Special Collections: U.S. patent depository library.

★2679★ University of Nebraska, Lincoln, Engineering Library
Nebraska Hall, 2nd Fl. W.
Lincoln, NE 68588-0516 (402) 472-3411
Alan V. Gould, Assistant Professor

Subjects: Engineering. **Special Collections:** Government Printing Office depository; patent depository. **Holdings:** 36,849 books; 47,194 bound periodical volumes; 238,476 microfiche; 4033 reels of microfilm. **Online Access:** DIALOG Information Services. Performs searches on fee basis.

★2680★ University of Nevada, Reno, Government Publications Department
Library
Reno, NV 89557-0044 (702) 784-6579
Duncan M. Aldrich, Acting Head

Subjects: Government - Nevada, federal; United Nations. **Special Collections:** U.S. patent depository library. **Holdings:** 1.05 million documents; 1.5 million microforms. **Online Access:** DIALOG Information Services, BRS Information Technologies, RLIN, U.S. Patent Classification System, LIBS 100 System; Nevada Documents Online (internal database). Performs searches on fee basis.

★2681★ University of New Hampshire, University Library, Patent Collection
Durham, NH 03824 (603) 862-1777

Special Collections: U.S. patent depository library.

★2682★ University of New Mexico, Centennial Science and Engineering Library
100 S. Andrews Ave.
Albuquerque, NM 87131 (505) 277-5441

Special Collections: U.S. patent depository library.

★2683★ University of Southern California, Law Library
University Park - MC 0072
Los Angeles, CA 90089-0072 (213) 743-6487
Albert Brecht, Director

Subjects: Anglo-American law, legal history, legal literature, taxation, law and social sciences. **Special Collections:** Legislative history of Internal Revenue Acts; depository of Copyright Office publications. **Holdings:** 212,897 books; 305,265 microforms; 230 shelves of documents; 718 audiotapes and video cassettes. **Online Access:** LEXIS, WESTLAW, DIALOG Information Services, NEXIS, DataQuick Information Network, ORION, Legaltrac, RLIN; RLG, ABA/net (electronic mail services).

★2684★ University of Texas at Austin, McKinney Engineering Library
Rm. 1.3 ECJ
Austin, TX 78713 (512) 471-1610

Special Collections: U.S. patent depository library.

★2685★ University of Utah, Documents Division
Marriott Library
Salt Lake City, UT 84112 (801) 581-8394
Julianne P. Hinz, Head

Subjects: Energy research and development, business and economics, statistics, geological and earth sciences, legislative documents, presidential materials, patents. **Special Collections:** U.S. patent depository library. Energy research and development reports, 1950 to present; Congressional committee prints on microfiche (15,100 prints); Congressional committee hearings on microfiche (29,400 hearings); American Statistics Index Nondepository Collection on microfiche (complete set); United Nations Depository Collection; Federal Documents Depository Collection. **Holdings:** 350,000 volumes. **Online Access:** DIALOG Information Services, Pergamon ORBIT InfoLine, Inc., BRS Information Technologies.

★2686★ University of Washington, Engineering Library and Information Services
Engineering Library Bldg., FH-15
Seattle, WA 98195 (206) 543-0740
Harold N. Wiren, Head

Subjects: Applied mathematics, applied physics, computer science, energy, engineering, environment, social management of technology, theoretical and applied mechanics. Special Collections: U.S. patent depository library. **Holdings:** 116,724 volumes; 45,935 paper copy technical reports; 1 million technical reports in microform; patent specifications, 1966 to present, on microfilm. **Online Access:** DIALOG Information Services, STN International, Pergamon ORBIT InfoLine, Inc., BRS Information Technologies, OCLC.

★2687★ University of Wisconsin—Madison, Kurt F. Wendt Library
215 North Randall Ave.
Madison, WI 53706 (608) 262-6845

Special Collections: U.S. patent depository library.

★2688★ Upjohn Company, Corporate Patents and Trademarks, Library
Unit 1920-32-1
301 Henrietta St.
Kalamazoo, MI 49001 (616) 385-7012
Sandra Williams, Contact

Subjects: U.S. patent, trademark, and copyright law; licensing and unfair competition. **Special Collections:** U.S. chemical patents; U.S. Official Gazette (microform). **Holdings:** 2000 books; 1500 bound periodical volumes. **Online Access:** LEXIS, DIALOG Information Services.

★2689★ U.S. Court of Appeals for the Federal Circuit, National Courts' Library
717 Madison Pl., N.W., Rm. 218
Washington, DC 20439 (202) 633-5871
Patricia M. McDermott, Librarian

Subjects: Law, taxation, government contracts, patents and trademarks, customs, international trade. **Holdings:** 36,263 books. **Online Access:** WESTLAW, LEXIS, NEXIS, LEGI-SLATE, VERALEX.

★2690★ U.S. Department of Agriculture, Southern Regional Research Center
1100 Robert E. Lee Blvd.
Box 19687
New Orleans, LA 70179 (504) 589-7072
Dorothy B. Skau, Librarian

Subjects: Chemistry, textiles, food processing, plant sciences, aquaculture, mechanical and chemical engineering, microscopy, electron microscopy, vegetable fats and oils, microbiology, statistics. **Special Collections:** Trade literature; U.S. and foreign patents in laboratory's fields of interest. **Holdings:** 35,000 volumes; 59 vertical file drawers of pamphlets, foreign patents, trade literature, reprints, translations and manuscripts; 68 shelves of U.S. patents. **Online Access:** DIALOG Information Services, OCLC.

★2691★ U.S. Department of Energy, Morgantown Energy Technology Center Library
Box 880
Morgantown, WV 26505 (304) 291-4184
S. Elaine Pasini, Information Specialist

Subjects: Coal and fossil fuel, petroleum, chemistry, chemical engineering, geology, coal gasification. **Special Collections:** U.S. Office of Coal Research reports (100); U.S. Dept. of Energy publications; U.S. Bureau of Mines publications (complete). **Holdings:** 10,000 books; 7000 bound periodical volumes; 1500 reports; 20 vertical file drawers of patents. **Online Access:** OCLC, Pergamon ORBIT InfoLine, Inc., DIALOG Information Services, BRS Information Technologies.

★2692★ U.S. Department of Justice, Civil Branch Library
10th & Pennsylvania Ave., N.W., Rm. 3344
Washington, DC 20530 (202) 633-3523
Roger N. Kerr, Librarian

Subjects: Law, customs, bankruptcy, government contracts, commercial law, admiralty, aviation, patents, trademarks, copyright. **Special Collections:** Legislative histories. **Holdings:** 45,000 volumes. **Online Access:** DIALOG Information Services, Pergamon ORBIT InfoLine, Inc., LEXIS, NEXIS, JURIS, LEGI-SLATE, Dow Jones News/Retrieval, WESTLAW, MEDLINE, DataTimes, WILSONLINE, Washington Alert Service, DATALIB, BRS Information Technologies, OCLC; Email (electronic mail service).

★2693★ U.S. International Trade Administration, U.S. and Foreign Commercial Service, Charleston District Office Library
Federal Office Bldg.
500 Quarrier St.
Charleston, WV 25301 (304) 347-5123
Roger L. Fortner, Director

Subjects: Exporting, patents, copyright. **Holdings:** 4400 volumes.

★2694★ U.S. National Park Service, Edison National Historic Site, Archives
Main St. and Lakeside Ave.
West Orange, NJ 07052 (201) 736-0550
Mary Bowling, Archives

Subjects: Invention, science, electricity, botanic research, chemistry, geology. **Holdings:** 10,000 volumes; 3.5 million pages of Edison's personal and laboratory correspondence and documents; business records of Edison Industries and Thomas Alva Edison, Inc.; 3000 notebooks kept by Edison and his workers; 60,000 photographic images.

★2695★ USX Corporation, USS Division, Information Resource Center
4000 Tech Center Dr.
MS 88
Monroeville, PA 15146 (412) 825-2344
Angela R. Pollis, Staff Supervisor

Subjects: Metallurgy, materials science, steel manufacture and finishing, chemistry and physics, coal and coke technology, physical chemistry, business. **Holdings:** 25,000 books; 1500 dissertations; 30,000 translations; 10,000 government and university reports; U.S. patents and chemical abstracts on microfilm. **Online Access:** DIALOG Information Services, NEXIS, Dow Jones News/Retrieval, Dun & Bradstreet Corporation, Inforonics, Inc.; internal database. Performs searches on fee basis.

★2696★ Vanderbilt University, Jean and Alexander Heard Library, Science Library
419 21st Ave., S.
Nashville, TN 37240 (615) 322-2775
Timothy F. Richards, Director

Subjects: Biology, chemistry, engineering, geology, mathematics, physics, astronomy. **Special Collections:** Foreign and State Geological Survey Collections; patent depository library. **Holdings:** 225,000 volumes; 385 boxes of microcards of Landmarks of Science I and II; 96 drawers of microfiche of AEC Reports through February 1971; 8 drawers of microfiche of NASA Reports, 1976 to present; 708 reels of microfilm of U.S. Patent Official Gazette, 1872 to present; 39 reels of microfilm of U.S. Patent Official Gazette Trademarks, 1971 to present; 20 reels of microfilm of U.S. Annual Report of the Commissioner of Patents, 1790-1871; 3190 reels of microfilm of U.S. patents, January 1965 to present; 24 drawers of microfiche of miscellaneous materials. **Online Access:** DIALOG Information Services, OCLC, BRS Information Technologies, CAS ONLINE, U.S. Patent Classific ation System. Performs searches on fee basis.

★2697★ Virginia Commonwealth University, University Library Services, Documents and Interlibrary Loan
Box 2033 - 901 Park Ave.
Richmond, VA 23284-2033 (804) 367-1104

Special Collections: U.S. patent depository library.

★2698★ Wichita Public Library, Business and Technology Division
223 S. Main
Wichita, KS 67202 (316) 262-0611
Brian Beattie, Head

Subjects: Aeronautics, petroleum, geology, economics, mathematics, taxes, finances, firearms, automobiles, business management. **Holdings:** 90,128 volumes; Patent Gazette, 1872 to present, on microfilm; 22,839 pamphlets; 2737 auto repair manuals; 1385 telephone directories. **Online Access:** DIALOG Information Services.

★2699★ Wilfrid Laurier University, Research Centre for Management of New Technology, Library
University Ave.
Waterloo, ON, Canada N2L 3C5 (519) 884-1970

Subjects: Cost effectiveness of implementation and acquisition of new technology, including computerization, robotics, and flexible manufacturing systems. **Holdings:** Figures not available.

★2700★ Witco Corporation, Technical Information Center
100 Bauer Dr.
Oakland, NJ 07436 (201) 337-5812
Jo Therese Smith, Manager

Subjects: Chemistry - organic, analytic, petroleum; chemical engineering. **Special Collections:** Petroleum technology; surfactant technology. **Holdings:** 14,000 books; 13,500 bound periodical volumes; 15 vertical file drawers of patents, pamphlets, reprints; 10 drawers of patents and journals on microfiche; Chemical Abstracts, 1980 to present. **Online Access:** DIALOG Information Services, STN International, CAS ONLINE.

★2701★ Yonkers Public Library, Information Services, Technical & Business Division
7 Main St.
Yonkers, NY 10701 (914) 337-1500
Frances C. Roberts, Acting Head

Subjects: Business and finance, transportation, engineering, automobiles, building trades, plumbing and heating, mathematics, physics, chemistry. **Special Collections:** Annual reports of most corporations; telephone directories of U.S. cities; state, business, and biographical directories; local history. **Holdings:** 19,500 books; 107 bound periodical volumes; Official Patent Gazette, 1925 to present; 7000 government depository publications; 17,000 pamphlets. **Online Access:** DIALOG Information Services; WESNEWS (internal database). Performs searches on fee basis.

CHAPTER 14

REGISTERED PATENT ATTORNEYS AND AGENTS

Do you need a patent attorney?

The answer is probably yes. It is perfectly legal to prepare your own patent application. You can conduct your own proceedings in the Patent and Trademark Office (PTO). But, unless you are familiar with such matters and have studied them in detail, you could experience considerable difficulty and extreme frustration.

The preparation of an application for a **utility** patent and conducting proceedings at the PTO requires a knowledge of patent law and PTO practice, and a knowledge of the scientific or technical matters involved with the particular invention. In my essay at the beginning of this book, entitled How to Protect and License Your Invention, I provide a comprehensive, step-by-step explanation of the entire patenting process. It will help you understand the various steps your lawyer will take, and identify those steps you might feel comfortable handling yourself.

While a patent may be obtained in many cases by persons not skilled in such esoteric work, there would be no assurance that the patent awarded would adequately protect the particular invention. I, therefore, highly recommend that a qualified patent attorney be retained for your **utility** patent work. Most inventors employ the services of registered patent attorneys or patent agents.

The law gives the PTO the power to make rules and regulations governing conduct and the recognition of patent attorneys and agents to practice before the PTO. Persons who are not recognized by the PTO for this practice are not permitted by law to represent inventors in their patent actions.

The PTO has the power to disbar, or suspend from practicing before it, persons guilty of gross misconduct, but this can only be done after a full hearing with the presentation of clear and convincing evidence concerning the misconduct.

The PTO will receive and, in appropriate cases, act upon complaints against attorneys and agents. If you wish to register a complaint, contact Cam Weiffenbach, Director, Office of Enrollment, P.O. Box OED, U.S. Patent and Trademark Office, Washington, D.C. 20231 (703) 557-2012.

How Much do Legal Services Cost?

It is not inexpensive to retain a patent attorney. The amount of time a patent attorney will have to put into any particular matter will depend a great deal upon the complexity of the invention. Below are some guidelines for costs, arranged according to the ordered sequence of steps a lawyer will take.

Patent Search. The first thing a patent lawyer will rightfully suggest is that you authorize a patent search to see what, if any, prior art exists. The cost of a patent search will depend upon the scope of your patent. Rarely does a lawyer do the search. Lawyers normally engage the services of a professional patent searcher or patent search organization. To whatever the lawyer is charged by the patent searcher, a premium will be added. Some inventors save this added cost by either independently doing the patent search or by hiring a patent searcher directly. The results are then handed over to an attorney for an opinion.

A good manual search can take up to seven hours. Searchers work at hourly rates. In Washington, D.C., fees range from $30.00 to $60.00 per hour. Generally it is less expensive to employ an independent searcher than a large firm. Some firms require minimums of up to $230. Independents often do not insist upon minimums.

Chemical and electrical patent searches typically cost more than mechanical patent searches because, according to the searchers, they take more time (eight to ten hours, average).

Some searchers will augment their work through the use of computer databases. See the introductory essay heading Online Patent Searches and the chapter entitled Online Resources for the Inventor for details on how to conduct a patent search yourself.

Add to this a photocopying fee. Some independent searchers charge their cost or 40 cents per page; larger firms get $3.00 per copy no matter how few or how many pages.

If you are quoted too low a figure from a lawyer, it will probably not be much of a search. Searches done fast may not worth much. Done correctly, a search takes some time.

I have found some search firms that have $150.00 specials. Typically, you get six pieces of prior art, copies included, for the package price. As in most things in life, you get what you pay for. If the searcher stops at three and there are six pieces of prior art, you may have lost $150. To be useful, a search must be comprehensive.

Patent Drawings. Lawyers do not personally do patent drawings either. They employ the services of a draftsman skilled in such matters. Sometimes the draftsmen are in-house staffers. Smaller practitioners use free-lancers, whose fees they step on as much as 40 percent. In Washington, D.C., the average price charged by a bonded draftsman is $35.00 per hour. It takes an average of three hours to do one sheet. Patents comprise numerous sheets of drawings—depending upon the complexity of the invention. You can expect to pay around $100.00 per sheet.

Patent Application. To prepare a patent application, attorneys charge anywhere from $2,500 to $4,500, depending upon the complexity of the invention. The more complex the patent, the more you can expect to pay. On the other hand, simple inventions can be done for as little as $1,000.

It is a buyers market. And just like anything else, it usually pays to shop around. Patent attorneys are able to give pretty close estimates of expected charges once they see the scope of an invention and its claims.

When you have an acceptable estimate, get your attorney to agree in writing to that price and cap it off. If you do not do this, you may find yourself caught in what I call **fee creep.** Make a package deal. A price cap gives the lawyer no incentive to draw out the case.

Get a handle on the photocopying charges in advance. Patent work generates a great deal of paper and it can be a gold mine for law firms. I found one large firm charging 50-cents per photocopied page and sending three copies of each document, one for each of my partners. I quickly put a stop to this abuse, requesting one copy of each document. The little things add up.

The fees charged by patent attorneys and agents for their professional services are not subject to regulation by the PTO. Solid evidence of overcharging may afford a basis for PTO action, but the Office rarely intervenes in disputes concerning fees.

Legal Forms. There are five forms which lawyers charge to prepare. Copies of these are found in the Forms and Documents Section of this book.

1. Declaration for a Patent Application (Fig. 8).
2. Power of Attorney or Authorization of Agent, Not Accompanying Application (Fig. 9).
3. Revocation of Power of Attorney or Authorization of Agent (Fig. 9).
4. Assignment of Patent (Fig. 19).
5. Assignment of Patent Application (Fig. 19).

What to Look for in a Patent Attorney

You should look for an attorney who specializes in your field of invention. Just as you would not hire a dermatologist to do the job of a cardiologist, even though both are licensed physicians, you would not want to hire a patent attorney with a background in mechanical engineering to do an electronic patent.

Larger firms offer quite an array of patent specialists from which to choose. But the smaller firms and many independent practitioners will often take anything that comes along. Do not be timid about requesting the technical qualifications of any attorney you are considering.

If your attorney cannot read your schematics, engineering drawings, or similar technical specifications, chances are your patent will not be nearly as complete and strong as it could be. Furthermore, you will be paying a premium to educate the attorney.

Bigger firms are not always better. At very large patent firms you will be small potatoes compared to lucrative corporate retainers. In such cases, your account may be assigned to a "spear carrier," who may be very good, but will seldom get the time to concentrate on your work. The primary responsibility of junior associates is to carry the workload for senior partners, the so-called "rainmakers," who bring in the accounts that pay lunch, health club, limo, and kindred perks.

Big patent law firms make their real money defending the patents of corporate clients in court, not from two and three thousand dollar application jobs on behalf of independents. Experienced patent attorneys know that the chances of an entrepreneur inventor being worth much more than what is earned on the patent application are slim. The majority of independents are dead ends financially. Big corporate clients can be "cash cows."

I am most comfortable with the independent specialist, even if this approach means a different attorney every time I change disciplines.

Do not rely on any patent attorney for prototyping advice, manufacturing processes, or insights into the day-to-day complexities of marketing.

Attorneys and Agents Registered to Practice before U.S. Patent and Trademark Office

I cannot over stress that the preparation and prosecution of an application that will adequately protect your invention is an undertaking that requires knowledge of patent law and PTO practices. It also requires a knowledge of technical aspects of the invention. It is for this reason that I highly recommend that you hire only a **registered** patent attorney or agent to do your patent work.

To be admitted to this select register, a person must comply with the regulations prescribed by the PTO, which require showing that the person is of good moral character and repute and that he or she has the legal and scientific credentials necessary to render a valuable service. Certain of these qualifications must be demonstrated by the passing of an examination. Those admitted to the examination must have a college degree in engineering or physical science or the equivalent of such a degree.

The PTO registers both attorneys at law and persons who are not members of the bar. The former persons are now referred to as "patent attorneys," and the latter persons are referred to as "patent agents." Insofar as the work of preparing an application for patent and conducting the prosecution in the PTO is concerned, patent agents are typically as well qualified as patent attorneys, although patent agents **cannot** conduct patent litigation in the courts or perform various services which the local jurisdiction considers as practicing law. For example, a patent agent would not be allowed to draw up a contract relating to a patent, such as an assignment or a license, if the agent resides in a state that considers such contracts as practicing law.

Some individuals and organizations that are not registered advertise their services in the fields of patent searching and invention marketing and development. Such individuals and organizations **cannot** represent inventors before the PTO. Caveat emptor! They are not subject to PTO discipline, and the Office cannot assist you in dealing with them.

While calling to acquire information on organizations that I felt might be helpful to inventors, I came across a nifty scam used by some lawyers to attract patent, trademark, and copyright clients. What appeared in telephone directories and source material to be professional councils specific to patents, trademarks, and copyrights were actually law offices. In one case the phone numbers for three different councils led to the same law firm. The person listed as president of the three respective organizations was a senior partner in the law firm.

The secretary who answered the phone pitched me on the fees charged for various services. This is a highbrow bait and switch that should be avoided.

The PTO cannot recommend any particular attorney or agent, or aid in the selection of an attorney or agent, as by stating in response to inquiry that a named patent attorney, agent, or firm is "reliable" or "capable."

This chapter lists the most current names and addresses of 12,623 individuals authorized to represent inventors before the PTO.

In the listings, a number sign (#) appears beside the name of each registrant who is a patent agent. An asterisk (*) appears in the listings besides the name of each registrant who is an officer or employee of the U.S. government. Registrants who are officers or employees of the U.S. government cannot, otherwise than in the discharge of their official duties, represent inventors before the PTO.

NATIONAL COUNCIL OF PATENT LAW ASSOCIATIONS

Patent attorneys who are active in the National Council of Patent Law Associations (NCPLA) should be particularly up-to-date on PTO matters. NCPLA is a 35-year-old organization consisting of some 40 local and regional patent law associations.

In addition to promoting exchange of information and lobbying, NCPLA keeps its members current on legislative and executive actions which might affect the nation's intellectual property system.

For more information on NCPLA, contact your nearest member association, or write to:
NCPLA, Crystal Plaza 3, Room 1DO1
2021 Jefferson Davis Highway
Arlington, VA 22202
Tel: (703) 557-3341.

Getting In Touch

Here are the current NCPLA officers, addresses and telephone numbers.

CALIFORNIA

Los Angeles Patent Law Association

Don W. Martens
Knobbe, Martens, Olson & Bear
620 Newport Center Drive, Suite 1600
Newport Beach, CA 92660-8016
(714) 760-0404

William J. Robinson
Poms, Smith, Lande & Rose
2121 Avenue of the Stars, Suite 1400
Los Angeles, CA 90067-5010

(213) 277-8141

Orange County Patent Law Association

Howard J. Klein
Klein & Szekeres
4199 Campus Drive, Suite 700
Irvine, CA 92715
(714) 854-5502

Bruce B. Brunda
Stetina & Brunda
24221 Calle De La Louisa, Suite 401
Laguna Hills, CA 92653
(714) 855-1246

Peninsula Patent Law Association of California

Kate Murashige
Cotti, Murashige, Irell & Manella
545 Middlefield Road, Suite 200
Menlo Park, CA 94025
(415) 327-7250

Henry K. Woodward
Flehr, Hohbach, Test, Albritton & Herbert
Four Embarcadero Center, Suite 3400
San Francisco, CA 94111
(415) 326-0747

San Francisco Patent and Trademark Association

Harry A. Pacini
Stauffer Chemical Co.
1200 South 47th Street, Box 4023
Richmond, CA 94804-0023
(415) 231-1202

Neil A. Smith
Limbach, Limbach and Sutton
1200 Ferry Building
San Francisco, CA 94111
(415) 433-4150

CAROLINA

Carolina Patent, Trademark and Copyright Law Association

John B. Hardaway, III
Bailey & Hardaway
125 Broadus Avenue
Greenville, SC 29601
(803) 233-1338

Howard A. MacCord, Jr.
Burlington Industries, Inc.
P.O. Box 21207
Greensboro, NC 27420
(919) 379-4517

COLORADO

Colorado Bar Association, Patent, Trademark, and Copyright Section

Kenneth L. Richardson
Solar Energy Research Institute
1617 Cole Blvd.
Golden, CO 80401
(303) 231-7724

CONNECTICUT

Albert W. Hilburger, Perman and Green
425 Post Road
Fairfield, CT 06430
(203) 259-1800

Edward R. Hyde
261 Danbury Road
P.O. Box 494
Wilton, CT 06897
(203) 762-5444

Connecticut Patent Law Association

Howard S. Reiter
Corporate Patent Counsel
Colt Industries, Inc.
Charter Oak Boulevard
West Hartford, CT 06110-0651
(203) 236-0651

F. Eugene Davis IV
184 Atlantic Street
Stamford, CT 06905
(203) 324-9662

DISTRICT OF COLUMBIA

Bar Association of D.C., Patent, Trademark, and Copyright Section

Archie W. Umphlett
Phillips Petroleum Co.
1825 K Street, N.W.
Washington, DC 20006
(202) 785-1252

Robert G. Weilacher
Beveridge, DeGrandi & Weilacher
Federal Bar Building West
1819 H Street, N.W.
Washingtion, DC 20006
(202) 659-2811

D.C. Bar

Howard D. Doescher
Cushman, Darby & Cushman
1615 L Street, N.W.
Washington, DC 20036
(202) 861-3000

Herbert C. Wamsley
2725 Fort Scott Drive
Arlington, VA 22202
(202) 466-2396

FLORIDA

South Florida Patent Law Association

Jack E. Dominik
Domink & Saccocio
6175 N.W. 153rd Street, Suite 225
Miami Lakes, FL 33014
(305) 556-9889

Henry W. Collins
Patent Cousel
Cordis Corporation
P.O. Box 025700
Miami, FL 33102-5700
(305) 551-2707

GEORGIA

State Bar of Georgia, Patent, Trademark, and Copyright Section

Todd Devau
Hurt, Richardson, Garner, Todd & Cadenhead
1400 Peachtree Place Tower
999 Peachtree Street, N.E.
Atlanta, GA 30309-3999
(404) 870-3999

ILLINOIS

Patent Law Association of Chicago

Ronald B. Coolley
Arnold, White & Durkee
800 Quaker Tower
321 North Clark Street
Chicago, IL 60610
(312) 744-0090

John Chrystal
Ladas & Parry
104 S. Michigan Avenue
Chicago, IL 60603
(312) 236-9021

INDIANA

Indiana State Bar Association, Patent, Trademark, and Copyright Section

Steven T. Belsheim
322 Main Street, Suite 100
Clarksville, TN 37040
(615) 647-8337

Gilbert E. Alberding
Ball Corporation
Legal Department
345 S. High Street
Muncie, IN 47302
(317) 747-6422

IOWA

Iowa Patent Law Association

H. Robert Henderson
Henderson & Sturm
1213 Midland Financial Building
Des Moines, IA 50309
(515) 288-9589

MARYLAND

Maryland Patent Law Association

Frank E. Robbins
Robbins & Laramie
1919 Pennsylvania Ave., N.W.
Washington, DC 20006
(202) 887-5050

Jim Haight
Mackler and Associates
P.O. Box 2187
Gaithersburg, MD 20879
(202) 842-1690

MASSACHUSETTS

Boston Patent Law Association

John M. Skenyon
Fish & Richardson
One Financial Center
Boston, MA 02111
(617) 542-5070

Jacob N. Erlich
424 Trapelo Road, Bldg. 104
Waltham, MA 02154
(617) 377-4072

MICHIGAN

Michigan Patent Law Association

Robert F. Hess
Federal-Mogul Corporation
P.O. Box 1966
Detroit, MI 48235
(313) 354-9926

George A. Grove
General Motors Corp., Patent Section
P.O. Box 33114
Detroit, MI 48232
(313) 974-1322

Saginaw Valley Patent Law Association

James Bittell
Dow Corning Corporation
Patent Department - C01232
P.O. Box 994
Midland, MI 48686-0994
(517) 496-5882

Robert Spector
Dow Corning Corporation
Patent Department - C01232
P.O. Box 994

Midland, MI 48686-0994
(517) 496-5523

State Bar of Michigan, Patent, Trademark, and Copyright Section

Stephen A. Grace
Dow Chemical
Corporate Center, Bldg. 1776
Midland, MI 48674
(517) 636-3052

Robert A. Armitage
The Upjohn Company
301 Henrietta Street
Kalamazoo, MI 49001
(616) 385-7345

MINNESOTA

Minnesota Intellectual Property Law Association

Terryl K. Qualey
Office of Patent Counsel - 3M
P.O. Box 33427
St. Paul, MN 55133-3427
(612) 733-1940

Richard E. Brink
3M Company
P.O. Box 33427
St. Paul, MN 55133
(612) 733-1517

MISSOURI

Bar Association of Metropolitan St. Louis, Patent, Trademark, and Copyright Section

Veo Peoples, Jr.
1221 Locust Street, Suite 100
St. Louis, MO 63103
(314) 231-9775

Edward H. Renner
Cohn, Powell, & Hind, P.C.
7700 Clayton Road, Suite 103
St. Louis, MO 63117
(314) 645-2442

NEW JERSEY

New Jersey Patent Law Association

Raymond M. Speer
Merck & Co., Inc.
P.O. Box 2000
Rahway, NJ 07065-0907
(201) 574-4481

New Jersey State Bar Assn., Patent, Trademark, and Copyright Section

Arthur J. Plantamura
10 Butterworth Drive
Morristown, NJ 07960
(201) 455-3781

Teresa Cheng
Merck & Co., Inc.
P.O. Box 2000
Rahway, NJ 07068-0907
(201) 574-4982

NEW YORK

Central New York Patent Law Association

John S. Gasper
Patent Operations (Dept. N50)
IBM Corporation
1701 North Street - Bldg. 251-2
Endicott, NY 13760
(607) 755-3342

Eastern New York Patent Law Association

John W. Harbour
General Electric Company
Silicone Prod. Business Division
Waterford, NY 12188
(518) 266-2471

William S. Teoli
General Electric Company
Corporate Research & Development
P.O. Box 8
Bldg. K1, Room 3A68
Schenectady, NY 12302
(518) 387-5872

New York Patent, Trademark & Copyright Law Association

David H.T. Kane
Kane, Dalsimer, Sulivan, Kurucz, Levy, Eisele & Richard
711 Third Avenue, 20th Floor
New York, NY 10017
(212) 687-6000

Douglas W. Wyatt
Wyatt, Gerber, Shoup, Scobey & Badie
261 Madison Avenue
New York, NY 10016
(212) 687-0911

Niagara Frontier Patent Law Association

William G. Gosz
Occidental Chemical Corporation
P.O. Box 189
Niagara Falls, NY 14302
(716) 773-8459

Thomas Gordon
Law Offices of Eric Schellin
2001 Jefferson Davis Highway
Suite 301, Crystal Plaza 1
Arlington, VA 22202
(703) 521-1666

Rochester Patent Law Association

Stephen B. Salai
850 Crossroads Office Building
Rochester, NY 14614
(716) 325-5553

Paul F. Morgan
Xerox Corporation
Xerox Square - 020
Rochester, NY 14644
(716) 423-3015

OHIO

Cincinnati Patent Law Association

Leonard Williamson
The Procter & Gamble Company
Patent Division
11520 Reed Hartman Highway
Cincinnati, OH 45241
(513) 530-3387

Eric W. Guttag
The Procter & Gamble Company
P.O. Box 39175
Cincinnati, OH 45247
(513) 659-2736

Cleveland Patent Law Association

James V. Tura
Pearne, Gordon, McCoy & Granger
1200 Leader
Cleveland, OH 44114
(216) 579-1700

Edwin W. (Ned) Oldham
Oldham, Oldham, Webber Co., L.P.A.
Twin Oaks Estate
1225 West Market Street
Akron, OH 44313
(216) 864-5550

Columbus Patent Law Association

Robert B. Watkins
Kremblas, Foster, Millard & Watkins
2941 Kenny Rd.
Columbus, OH 43221
(614) 457-5700

Richard C. Stevens
1400 First National Plaza
Dayton, OH 45402
(513) 223-2050

Dayton Patent Law Association

William Weigl
Hobart Corporation
World Headquarters
Troy, OH 45374
(513) 332-2111

Joseph J. Nauman
Biebel, French & Nauman
2500 Kettering Tower
Dayton, OH 45423
(513) 461-4543

Ohio State Bar Association, Patent, Trademark, and Copyright Section

Frank Foster
Kremblas, Foster, Millard & Watkins
50 West Broad Street
Columbus, OH 43215
(614) 464-2700

Ralph Jocke
Parker Hannifin Corporation
17325 Euclid Avenue
Cleveland, OH 44112
(216) 531-3000

Toledo Patent Law Section

Oliver E. Todd, Jr.
Champion Spark Plug Company
P.O. Box 910
Toledo, OH 43661
(419) 535-2364

David D. Murray
William Brinks Olds Hofer Gilson & Lione
930 National Bank Building
Toledo, OH 43604
(419) 244-6578

OKLAHOMA

Oklahoma Bar Association, Patent, Trademark, and Copyright Section

Gary Peterson
Dunlap, Codding & Peterson
9400 North Broadway, Suite 420
Oklahoma City, OK 73114
(405) 478-5344

Allen W. Richmond
Phillips Petroleum Company
208 PLB
Bartlesville, OK 74006
(918) 661-0561

OREGON

Oregon Patent Law Association

William O. Geny
Chernoff Vilhauser McCung & Stenzel
600 Benjamin Franklin Plaza
1 SW Columbia
Portland, OR 97258
(503) 227-5631

Francine Gray
Kolisch, Hartwell & Dickenson
200 Pacific Bldg.
520 SW Yamhill St.
Portland, OR 97204
(503) 224-6655

PENNSYLVANIA

Allegheny County Bar Association Intellectual Property Section

Michael D. Fox
Berkman, Ruslander, Pohl, Lieber & Engle
One Oxford Centre
Pittsburgh, PA 15219
(412) 261-6161

John W. Jordan, IV
Grigsby, Gaca & Davies
One Gateway Center, 10th Floor
Pittsburgh, PA 15222
(412) 281-0737

Philadelphia Patent Law Association

Eugene G. Seems
FMC Corporation
2000 Market Street
Philadelphia, PA 19103
(215) 299-6971

Paul F. Prestia
500 North Gulph Road
Valley Forge, PA 19482
(215) 265-6666

Intellectual Property Law Association of Pittsburgh

David S. Urey
U.S. Steel Corporation
600 Grant Street
Pittsburgh, PA 15230
(412) 422-2873

Patrick J. Viccaro
Allegheny Ludlum Steel Corp.
1000 Six PPG Place
Pittsburgh, PA 15272
(412) 394-2839

TEXAS

Austin Intellectual Property Law Association

Dudley R. Dobie, Jr.
Fulbright and Jaworski
600 Congress, #2400
Austin, TX 78701
(512) 474-5201

Andrea P. Bryant
IBM Corporation
11400 Burnet Road
Austin, TX 78758
(512) 838-1003

Dallas-Ft. Worth Patent Law Association

Charles Gunter, Jr.
Felsman, Bradley, Gunter & Kelly
2850 Continental Plaza
777 Main Street
Ft. Worth, TX 76102
(817) 332-8143

Robert A. Felsman
Felsman, Bradley, Gunter & Kelly
2850 Continental Plaza
777 Main Street
Ft. Worth, TX 76102
(817) 332-8143

Houston Intellectual Property Law Association

Kenneth E. Kuffner
Arnold, White & Durkee
P.O. Box 4433
Houston, TX 77210
(713) 789-7600

Daivd A. Rose
Butler A. Binion
1600 Allied Bank Plaza
Houston, TX 77002
(713) 237-3640

State Bar of Texas, Intellectual Property Law Section

Robert A. Felsman
Felsman, Bradley, Gunter & Kelly
2850 Continental Plaza
777 Main Street
Fort Worth, TX 76102
(817) 332-8143

William L. LaFuze
Vinson & Elkins
3300 First City Tower
Houston, TX 77002-6760
(712) 236-2595

UTAH

State Bar of Utah, Patent, Trademark, and Copyright Section

John Christiansen
Van Cott, Bagley, Cornwall & McCarthy
50 South Main Street, Suite 1600
Salt Lake City, UT 84144
(801) 532-3333

Allen R. Jensen
Workemen, Nydegger & Jensen
American Plaza II, Third Floor
57 West 200 South
Salt Lake City, UT 84101
(801) 522-9800

VIRGINIA

Virginia State Bar, Patent, Trademark, and Copyright Section

Antohony J. Zelano
Millen and White
1911 Jefferson Davis Highway
Arlington, VA 22202
(703) 892-2200

James H. Laughlin, Jr.
Benoit, Smith & Laughlin
2001 Jefferson Davis Highway
Arlington, VA 22202
(703) 521-1677

WASHINGTON

Washington State Patent Law Association

James P. Hamley
The Boeing Company
P.O. Box 3707, Mail Stop 7E-25
Seattle, WA 98124-2207
(206) 251-0262

WISCONSIN

Wisconsin Intellectual Property Law Association

Ramon A. Klitzke, Professor of Law
Marquette University Law School
1103 West Wisconsin Avenue
Milwaukee, WI 53233
(414) 224-7094

Howard W. Bremer
Wisconsin Alumni Research Foundation
P.O. Box 7365
Madision, WI 53707
(608) 263-2831

Chapter Arrangement

This chapter is arranged geographically by state, then alphabetically by city and name of agent or attorney. The cross-hatch (#) after the name indicates that the person is a patent agent; the asterisk (*) after the name indicates that the person is employed by the U.S. Government and not available to accept clients. Names in this chapter are not listed in the Master Index.

Alabama

Kenneth L. Cleveland
Cleveland & Cleveland
2326 Highland Ave. S.
Birmingham, AL 35205 (205) 252-8473

William B. Hairston, III
Engel, Hairston & Johanson
4th, Fl, 109 N. 20th St.
Birmingham, AL 35203 (205) 328-4600

Donald H. Jones
1425 21st St., S., Suite 200
Birmingham, AL 35205 (205) 933-2525

James D. Long
P.O. Box 590052
Six Office Park Circle, Suite 100
Birmingham, AL 35259 (205) 871-1443

Thad G. Long
Bradley, Arant, Rose & White
1400 Park Place Tower
Birmingham, AL 35203 (205) 252-4500

Charles F. Martin, Jr.
928 S. Forrest Dr.
Birmingham, AL 35209 (205) 879-7164

Wm. Randall May
4513 Valleydale Rd., Suite 1
Birmingham, AL 35242 (205) 991-6367

Roy Leon Mims
3712 Wioodvale Rd.
Birmingham, AL 35223 (205) 967-5271

Theodore T. Robin, Jr.
4524 Pine Mtn. Rd.
Birmingham, AL 35213 (205) 870-7268

Rebecca B. Schoumacher#
114 Three Sons Dr.
Birmingham, AL 35226

Woodford R. Thompson, Jr.
Jennings, Carter, Thompson & Veal
1150 Bank For Savings Bldg.
Birmingham, AL 35203 (205) 324-1524

Robert J. Veal
Jennings, Carter, Thompson & Veal
1150 Bank For Savings Bldg.
Birmingham, AL 35203 (205) 324-1524

George L. Williamson
108 Black Oak Way
Daphne, AL 36526 (205) 661-1888

Arthur B. Beindorff#
2812 Burning Tree Mountain Rd.
Decatur, AL 35603 (205) 350-1256

Charles H. Hittson#
1901 Erskine Dr.
Florence, AL 35630 (205) 766-6955

James Everette Staudt#
406 Audubon Street
Hartselle, AL 35640 (205) 773-6292

L. Frederick Hilbers
P.O. Box 19393
Homewood, AL 35219 (205) 871-1939

Leonard Flank
1502 Elmwood Dr., S.E.
Huntsville, AL 35801 (205) 539-9704

Harold W. Hilton#
7233 Statton Drive
Huntsville, AL 35802 (205) 882-2431

Herbert H. Murray#
8905 Strong Drive
Huntsville, AL 35802 (205) 881-4952

Charles A. Phillips
1100 Jordan Lane, Suite K
Huntsville, AL 35816 (205) 536-8261

George J. Porter
P.O. Box 4123
10011 Allison Drive
Huntsville, AL 35815 (205) 883-9212

Wayland H. Riggins
8802 Louis Drive
Huntsville, AL 35802 (205) 881-1428

Gary L. Rigney
115 Manning Drive, Suite B-202
Huntsville, AL 35801 (205) 536-3264

Robert Clinton Sims, Sr.*#
U.S. Army Missile Comm.
A M S M I - L P
Huntsville, AL 35898 (205) 876-5106

Jack Wendel Voigt, Sr.#
2601 Vista Drive
Huntsville, AL 35803 (205) 881-1594

Leon Daniel Wofford, Jr.*
N A S A
Geo. C. Marshall, Space Flight Ctr.
Huntsville, AL 35812 (205) 453-0020

Joseph H. Beumer*
N A S A
Off. Of Chief Counl.
Marshall Space Flight Ctr., AL 35812
(205) 544-0013

William J. Sheehan, III*
N A S A
Attn. C C 01 Patent Coun.
Marshall Space Flight Ctr., AL 35812
(205) 544-0021

James A. Beneburg
Dravo Natural Resources Company
P.O. Box 1685, 61 St. Joseph Street
Mobile, AL 36633 (205) 438-3531

James A. Berneburg
Dravo Natural Resources Company
P.O. Box 1685, 61 St. Joseph Street
Mobile, AL 36633 (205) 438-3531

Clifford Claborne Carter*
P.O. Box 2646
Mobile, AL 36652 (205) 690-2872

Gregory M. Friedlander
Marr & Friedlander
955 Downtowner Blvd.
Mobile, AL 36609 (205) 460-0303

L. Daniel Morris, Jr.
Blount Inc.
P.O. Box 949
4520 Executive Park Drive
Montgomery, AL 36192 (205) 272-8020

Robert A. Petrusek*#
Tennessee Valley Authority
Off. Of General Coun.
Natl. Fertilizer Dev. Center
Muscle Shoals, AL 35660 (205) 386-2363

Harold C. Hogencamp#
Rt. 4 - Box 406
Phenix City, AL 36867 (205) 297-7580

Donald W. Phillion, Sr.
203 Carlisle Way
Rainbow City, AL 35901 (205) 442-1676

Robert L. Broad, Jr.*
U.S. Army Missile Comm.
Attn: A M S M I - G C - Ip
Redstone Arsenal, AL 35981 (205) 876-1121

Freddie M. Bush*#
U.S. Army Missile Comm.
Attn: A M S M I-G
Redstone Arsenal, AL 35898 (205) 876-5107

James T. Deaton*#
U.S. Army Missile Comm.
Attn: A M S M I - G C
Redstone Arsenal, AL 35898 (205) 876-5106

J. Keith Fowler*#
U.S. Army Missile Command
Attn: Amsmi-Gc-Ip
Redstone Arsenal, AL 35898 (205) 876-1121

John Calder Garvin, Jr.*
U.S. Army Missile Comm.
Attn: Amsmi-Gc-Ip
Redstone Arsenal, AL 35898 (205) 876-1121

Jack M. Glandon*
U.S. Army Missile Command
Amsmi-Gc-Ip
Redstone Arsenal, AL 35898 (205) 876-1121

Isaac Pugh Espy
Gray Espy & Nettles
P.O. Box 2786
Tuscaloosa, AL 35403 (205) 758-5591

Alaska

Lloyd V. Anderson, Jr.
Birch, Horton, Bittner, Pestinger & Anderson
1127 W. 7th Ave.
Anchorage, AK 99501 (907) 276-1550

Daniel Anthony Gerety
Delaney, Wiles, Hayes, Reitman & Brubaker, Inc.
1007 W. 3rd Ave.
Anchorage, AK 99501 (907) 279-3581

Kenneth P. Jacobus
Hughes, Thorsness, Gantz, Powell & Brundin
509 W. 3rd Ave.
Anchorage, AK 99501 (907) 274-7522

Michael J. Tavella#
6900 Rovena Street
Anchorage, AK 99502 (907) 349-2495

Terrance A. Turner
Owens & Turner, P.C.
1500 West 33rd Avenue, Suite 200
Archorage, AK 99503 (907) 276-3963

Richard L. Blackmer#
P.O. Box 80286
College, AK 99708

Arizona

Lockwood D. Burton
4424 W. Keating Circle
Glendale, AZ 85308 (602) 843-4825

Karen M. Casto
5208 W. Surrey Ave.
Glendale, AZ 85304 (602) 843-3852

Joseph S. Failla#
4201 W. Angela Dr.
Glendale, AZ 85308 (602) 978-8750

Arthur S. Stewart
6540 W. Butler Dr.
#85 Court 3 Glencroft
Glendale, AZ 85302 (602) 939-0569

George Aichele
21300 S. Heather Ridge Cir.
Green Valley, AZ 85614 (602) 648-8241

John C. L. Cowen
1093 S. Paseo Del Prado
Green Valley, AZ 85614 (602) 625-0565

John M. Johnson#
110 W. Paseo Tesoro
Green Valley, AZ 85614 (602) 625-3190

Karl E. Sager#
160 Calle Del Chancero
NBU 2716
Green Valley, AZ 85614 (602) 625-3829

Richard W. Gurtler
1304 W. Mtn. View Dr.
Mesa, AZ 85201 (602) 969-2504

James H. Gray#
210 E. Chateau Circle
Payson, AZ 85541 (602) 474-5015

Joe E. Barbee
Motorola, Inc.
4250 E. Camelback Road, Suite 300K
Phoenix, AZ 85018 (602) 952-4704

Frank Timothy Barber
Grehound Corp.
Grehound Tower
Phoenix, AZ 85077 (602) 248-5676

Michael D. Bingham
Motorola, Inc.
Pat. Dept. Suite 300K
4250 E. Camelback Rd.
Phoenix, AZ 85018 (602) 952-4703

Frank J. Bogacz
Motorola, Inc.
4250 E. Camelback Road, Suite 300-K
Phoenix, AZ 85018 (602) 952-4701

Kenneth R. Bowers, Jr.
2346 E. Orangewood Ave.
Phoenix, AZ 85020 (602) 870-9818

William C. Cahill
Cahill, Sutton & Thomas
1400 Valley Bank Center
Phoenix, AZ 85073 (602) 258-8008

Charles E. Cates
Cates & Phillips
2700 N. Central Ave., Suite 1210
Phoenix, AZ 85004 (602) 248-0982

Vincent F. Chiappetta
Meyer, Hendricks, Victor, Osborn & Maledon
2700 North Third Street, Suite 4000
Phoenix, AZ 85004 (602) 263-8700

Lowell E. Clark#
5901 E. Calle Del Sud
Phoenix, AZ 85018 (602) 945-5818

John H. Colter
7107 N. 13th Pl.
Phoenix, AZ 85020 (602) 997-8720

Thomas William DeMond
2936 W. Larkspur
Phoenix, AZ 85029 (602) 866-9724

James F. Duffy
13430 N. 2nd St.
Phoenix, AZ 85022 (602) 942-8615

Don J. Flickinger#
1700 North 7th Street, Suite three
Phoenix, AZ 85006 (602) 271-0092

William John Foley
Cahill, Sutton & Thomas
1400 Valley Bank Ctr.
Phoenix, AZ 85073 (602) 258-8008

Donald P. Gabrielson
3335 E. Garfield
Phoenix, AZ 85008 (602) 275-8351

Edward R. J. Glady, Jr.
Fennemore, Craig, Von Ammon, Udall & Powers
Two North Central Avenue, Suite 2200
Phoenix, AZ 85004 (602) 251-2527

Marvin A. Glazer
Cahill, Sutton & Thomas
1400 Valley Bank Ctr.
201 N. Central
Phoenix, AZ 85073 (602) 258-8000

Robert M. Handy
Motorola, Inc.
Pat. Dept. - Suite 300K
4350 E. Camelback Rd.
Phoenix, AZ 85018 (602) 952-4704

Richard Grant Harrer
Cates & Phillips
2700 North Central Ave., Suite 1210
Phoenix, AZ 85004 (602) 248-0982

Herbert E. Haynes, Jr.#
Karsten Manufacturing Corp.
2201 W. Desert Cove
Phoenix, AZ 85029 (602) 277-1300

Louise S. Heim#
1518 W. Hazelwood Apt. Four
Phoenix, AZ 85015 (602) 266-9068

Gregory G. Hendricks
Gte Service Corp.
2500 W. Utopia Rd.
Phoenix, AZ 85027 (602) 581-4136

Robert A. Hirschfeld
4723 N. 44th Street
Phoenix, AZ 85018 (602) 840-0342

Charles R. Hoffman
Cahill, Sutton & Thomas
1400 Valley Bank Ctr.
Phoenix, AZ 85073 (602) 258-8000

William P. Hovell
Robbins & Green, P.A.
3000 N. Central, Suite 1800
Phoenix, AZ 85012 (602) 248-7999

Bernard L. Howard
Greyhound Corp.
Greyhound Tower
Phoenix, AZ 85077 (602) 248-5776

Vincent B. Ingrassia
Motorola, Inc.
4350 E. Camelback Road, Suite 300K
Phoenix, AZ 85018 (602) 952-4703

Dale E. Jepsen
Motorola, Inc.
4250 E. Camelback Road, Suite 300K
Phoenix, AZ 85018 (602) 952-4702

Charles W. Jirauch
Streich, Lang, Weeks & Cardon
P.O. Box 471
Phoenix, AZ 85001 (602) 229-5200

Maurice J. Jones, Jr.
Motorola Inc.
Pat. Dept. - Suite 300K
4250 E. Camelback Road
Phoenix, AZ 85018 (602) 952-4701

William E. Koch
Motorola, Inc.
4350 E. Camelback Rd., Suite 200-F
Phoenix, AZ 85018 (602) 994-6334

William J. Kubida
Lisa & Kubida, P.C.
2700 N. Central Avenue, Suite 1225
Phoenix, AZ 85004 (602) 285-4455

Elliott Kurzman
5115 E. Windsor Avenue
Phoenix, AZ 85008 (602) 952-9035

Charles R. Lewis
Motorola, Inc.
Pat. Dept. - Suite 300K
4350 E. Camelback
Phoenix, AZ 85018 (602) 994-6300

Warren F. B. Lindsley, Sr.
Camel Square, Suite 200E
4350 E. Camelback Rd.
Phoenix, AZ 85018 (602) 840-7310

Robert S. Linne
The Garrett Corporation
Pat. Dept., Bldg. 301-1R A
111 S. 34th Street
Phoenix, AZ 85010 (602) 231-3333

Donald Julius Lisa
2700 N. Central Avenue, Suite 1400
Phoenix, AZ 85004 (602) 285-4455

Raymond F. Maldoon#
4631 E. Walatowa St.
Phoenix, AZ 85044 (602) 893-3682

Jordan M. Meschkow
1700 N. Seventh Street, Suite One
Phoenix, AZ 85006 (602) 256-6996

John K. Mickevicius
Tandem Computers, Inc.
3300 N. Central, Suite 700
Phoenix, AZ 85012 (602) 264-2206

Terry L. Miller
Garrett Corporation
111 S. 34th Street
Phoenix, AZ 85010 (602) 267-3887

Anthony Miologos#
GTE Communication Systems Corp.
2500 W. Utopia Rd.
Phoenix, AZ 85027 (602) 581-4314

Foorman L. Mueller
4350 E. Camelback Road, Suite 250F
Phoenix, AZ 85018 (602) 994-6335

Victor Myer
Cates & Roediger
3800 N. Central Avenue, Suite 920
Phoenix, AZ 85012 (602) 248-0982

Gregory J. Nelson
Nelson & Roediger
2623 North Seventh Street
Phoenix, AZ 85006 (602) 263-8782

Walter W. Nielsen
Motorola, Inc.
Pat. Dept. - 200F
4350 E. Camelback Rd.
Phoenix, AZ 85018 (602) 994-6300

Tod R. Nissle#
Drummond & Nissle
4041 N. Central Avenue
Phoenix, AZ 85012 (602) 263-0920

Eugene Arthur Parsons
Motorola, Inc.
4350 E. Camelback Rd., Suite 200F
Phoenix, AZ 85018 (602) 994-6301

James Harold Phillips
Cates & Phillips
2700 North Central Avenue, Suite 1210
Phoenix, AZ 85004 (602) 248-0982

Edward C. Rapp
Maricopa County Superior Court Juvenile Div.
3125 W. Durango
Phoenix, AZ 85009 (602) 269-4404

Janelle F. Raupp
Jennings, Strouss & Salmon
111 West Monroe
Phoenix, AZ 85003 (602) 262-5810

Joseph H. Roediger
Nelson & Roediger
2623 North Seventh Street
Phoenix, AZ 85006 (602) 263-8782

Arthur A. Sapelli
Honeywell
P.O. Box 8000, M/s B55
Phoenix, AZ 85029 (602) 862-6542

M. David Shapiro
Kaplan, Jacobowitz, Hendricks & Bosse, P.A.
3003 North Central Avenue, Suite 1500
Phoenix, AZ 85012 (602) 264-3134

H. Gordon Shields
7830 N. 23rd Ave.
Phoenix, AZ 85021 (602) 995-0490

Martin Lee Stoneman
525 East Cherry Lynn Road
Phoenix, AZ 85012

Samuel J. Sutton, Jr.
Cahill, Sutton & Thomas
1400 Valley Bank Center
Phoenix, AZ 85073 (602) 258-8000

Joel E. Thompson
13 West Jefferson
Phoenix, AZ 85003 (602) 258-8451

Carl R. VonHellens
Cahill, Sutton & Thomas
1400 Valley Bank Ctr.
201 N. Central
Phoenix, AZ 85073 (602) 258-8000

Raymond J. Warren
Motorola Inc.
4250 E. Camelback Rd., Suite 300K
Phoenix, AZ 85018 (602) 952-4701

Thomas G. Watkins, III
Cahill, Sutton & Thomas
1400 Valley Bank Ctr.
201 N. Central Ave.
Phoenix, AZ 85073 (602) 258-8000

Leonard Weiss
4204 N. Brown Ave.
Phoenix, AZ 85251 (601) 875-6069

Paul F. Wille
Motorola Inc.
4250 E. Camelback Rd., Suite 300K
Phoenix, AZ 85018 (602) 952-4700

Harry A. Wolin#
Motorola, Inc.
4250 E. Camelback Rd., Suite 300K
Phoenix, AZ 85018 (602) 952-4702

John Joseph Roethel
715 Pauley Dr.
Prescott, AZ 86301 (602) 778-3232

J. Stanley Edwards
4301 Winfield Scott Plaza
Scottsdale, AZ 85251 (602) 941-8863

Lannas S. Henderson, Jr
6801 E. Camelback Rd., Apt. P-103
Scottsdale, AZ 85251 (602) 990-0684

Edward W. Hughes
6451 E. Cholla St.
Scottsdale, AZ 85254 (602) 948-3356

John S. Lieb
4251 N. Brown Ave., Suite A-1
Scottsdale, AZ 85251 (602) 941-5342

Richard R. Mybeck
4251 N. Brown Ave., Suite A-1
Scottsdale, AZ 85251 (602) 941-5342

Henry T. Olsen
6514 N. 85th Pl.
Scottsdale, AZ 85253 (602) 948-9938

Herschel Croft Omohundro, Sr.
5631 E. Windsor Ave.
Scottsdale, AZ 85257 (602) 945-2460

Lavalle D. Ptak
4301 Winfield Scott Plaza
Scottsdale, AZ 85251 (602) 994-1003

David G. Rosenbaum
4204 N. Brown Ave.
Scottsdale, AZ 85251 (602) 994-8888

Harry M. Weiss
4204 N. Brown Ave.
Scottsdale, AZ 85251 (602) 994-8888

Noel G. Artman
15830 Nicklaus Ln.
Sun City, AZ 85351 (602) 977-2431

J. Edwin Coates
9414 Cedar Hill Circle
Sun City, AZ 85351 (602) 974-0284

Harry P. Eichin#
10023 Lancaster Dr.
Sun City, AZ 85351 (602) 977-8019

Albert B. Griggs
9114 Long Hills Dr.
Sun City, AZ 85351 (602) 974-2430

James A. Hauer
10211 W. Edgewood Dr.
Sun City, AZ 85351 (602) 933-4942

Mary B. Moshier
16807 103rd Ave.
Sun City, AZ 85351

George C. Nebesar
10327 Prairie Hills Circle
Sun City, AZ 85351

John W. Overman
10909 Palmeras Dr.
Sun City, AZ 85351 (602) 974-5697

Hilmond O. Vogel
10320 Desert Rock Dr.
Sun City, AZ 85351 (602) 977-8288

Frank Cristiano, Jr.#
17430 Conquistador Dr.
Sun City West, AZ 85375 (602) 584-5274

Norman C. Fulmer
20443 135th Ave.
Sun City West, AZ 85375 (602) 584-7470

Roe D. McBurnett, Jr.
12970 Ballad Dr.
Sun City West, AZ 85375 (602) 584-2585

Frederick Burton Sellers#
12637 Rampart Dr.
Sun City West, AZ 85375 (602) 975-1760

Ronald M. Halvorsen
10310 E. Silvertree Court
Sun Lakes, AZ 85248 (602) 895-7389

Nedwin Berger
2645 E. Southern, Apt. A-229
Tempe, AZ 85282 (602) 831-3449

Charles P. Padgett, Jr.
1634 E. Del Rio Dr.
Tempe,, AZ 85282 (602) 839-6913

H. Walter Clum#
3940 N. Romero Rd., No. 11
Tucson, AZ 85705 (602) 293-1394

Victor Flores
3721 W. Goret Rd.
Tucson, AZ 85705 (602) 743-7990

Milton C. Hansen
P.O. Box 30011
Tucson, AZ 85751 (602) 299-2819

John H. Holcombe
IBM Corp.
Intellectual Property Law Dept., 90A/301
Tucson, AZ 85744 (602) 629-4102

James M. McClanahan
Broadway Executive Plaza North
7473 E. Broadway
Tucson, AZ 85710 (602) 881-0060

James V. McDonald#
2345 N. Craycroft Rd., Apt. 312
Tucson, AZ 85712 (602) 326-4218

S. Debra Miller
2028 E. Mabel St.
Tucson, AZ 85719 (602) 325-2402

Mark E. Ogram
780 S. Freeman Rd.
Tucson, AZ 85748 (602) 298-1210

James A. Pershon
IBM Corp.
Dept. 90A/301, Patent Operations
Tucson, AZ 85744 (602) 629-4101

Robert James Sanders, Jr.
Research Corp.
6840 E. Broadway Blvd.
Tucson, AZ 85710 (602) 296-6400

Manny W. Schecter
IBM Corp.
GDP Division
Intellectual Property Law, 90A/301
Tucson, AZ 85744 (602) 629-4104

Herbert F. Somermeyer
8421 E. Fernhill Dr.
Tucson, AZ 85715

George M. Stadler#
Research Corp.
6840 E. Broadway Blvd.
Tucson, AZ 85710 (602) 296-6400

Jerome M. Teplitz
Research Corp.
6840 E. Broadway Blvd.
Tucson, AZ 85710 (602) 296-6400

James M. Thomson#
IBM General Products Div.
Dept. 90A/301
Tucson, AZ 85744 (602) 629-4106

David A. Wiersma#
Research Corp.
6840 E. Broadway Blvd.
Tucson, AZ 85710 (602) 296-6400

Herbert L. Martin
P.O. Box 3506
W. Sedona, AZ 86340 (602) 282-3688

Arkansas

Boyd D. Cox
26 E. Center St.
Fayetteville, AR 72701 (501) 521-2052

Robert Raymond Keegan
112 W. Center St.
Suite 615, First Pl.
Fayetteville, AR 72701 (502) 521-4412

R. Donald Pitts#
410 Skyline Dr.
Harrison, AR 72601 (501) 741-3914

Stephen D. Carver
Pleasant Valley Corp. Ctr.
2024 Arkansas Valley Dr., Suite 800
Little Rock, AR 72212 (501) 224-1500

Hermann Ivester
Ivester, Henry, Skinner & Camp
212 Center St., Suite 900
Little Rock, AR 72201 (501) 376-7788

Joseph A. Strode
Bridges, Young, Matthews, Holmes & Drake
315 E. 8th Ave.
Pine Bluff, AR 71611 (501) 534-5532

Clayton O. Obenland#
1300 Birch Dr.
Rogers, AR 72756 (501) 631-1792

California

Jack Calvin Munro#
5210 Lewis Rd., Unit 10
Agoura, CA 91301 (818) 991-1687

Taylor M. Belt
1825 Shoreline Dr., Apt. 305
Alameda, CA 94501 (415) 523-1284

Yuan Chao#
1918 Kofman Parkway
Alameda, CA 94501 (415) 865-1668

Karen B. Dow
Triton Biosciences, Inc.
1501 Harbor Bay Parkway
Alameda, CA 94501 (415) 769-5360

Valentin D. Fikovsky#
University Of California Patent Office
1320 Harbor Bay Parkway
Alameda, CA 94501 (415) 642-5000

Albert A. Jecminek
Triton Biosciences Inc.
1501 Harbor Bay Pkwy.
Alameda, CA 94501 (415) 769-5203

Francis H. Lewis, Jr.
Lewis & Lewis
Bank Of America Bldg.
2411 Santa Clara Ave., Suite 17
Alameda, CA 94501 (415) 865-1600

Sherri E. Vinyard
Triton Biosciences, Inc.
1501 Harbor Bay Pkwy.
Alameda, CA 94501 (415) 769-4786

Margaret A. Connor*
U.S. Dept. Of Agriculture
Western Regional Res. Center
800 Buchanan St.
Albany, CA 94710 (415) 486-3208

Matthias L. Tam
248 E. Main St., Apt. 201
Alhambra, CA 91801 (818) 289-9616

Harold Burg
P.O. Box 18776
Anaheim, CA 92807 (714) 637-8691

Wilfred G. Caldwell
Rockwell International Corp.
3370 Miraloma Ave.
Anaheim, CA 92803 (714) 762-4517

Thomas M. Deforest
Fujitsu Bus. Communications Of America, Inc.
3190 Miraloma Ave.
Anaheim, CA 92806 (714) 630-7721

Allen A. Dicke, Jr.#
224 Mall Way
Anaheim, CA 92804 (714) 527-3766

H. Frederick Hamann
Rockwell International Corp.
3370 Miraloma Ave., Ha 52
Anaheim, CA 92803 (714) 762-1663

Grant L. Hubbard
300 S. Harbor Blvd., Suite 805
Anaheim, CA 92805 (714) 491-9076

James F. Kirk
Rockwell International
Electronic Operations
Pat. Dept. Ad06
3370 Miraloma Ave., P.O. Box 3105
Anaheim, CA 92803 (714) 762-6120

George A. Montanye
Rockwell International Corp.
3370 Miraloma Ave.
Anaheim, CA 92803 (714) 762-6300

Jonathan B. Orlick
Rockwell International Corp.
Mail Code A D 06
3370 Miraloma Ave.
Anaheim, CA 92803 (714) 762-5662

James R. Thornton#
U.S. Borax Research Corp.
412 Crescent Way
Anaheim, CA 92803 (714) 774-2670

Steven D. Goldby
180 Stockbridge Ave.
Atherton, CA 94025 (415) 364-1791

Harold I. Johnson
172 Austin Ave.
Atherton, CA 94025 (415) 366-6942

Chester Martin McCloskey#
Norac Co., Inc.
P.O. Box F
Azusa, CA 91702 (213) 334-2908

Sidney Magnes#
16212 Bellflower Blvd.
Bellflower, CA 90706 (213) 920-1493

Ellis A. Pangborn#
309 Daken Brook Dr.
P.O. Box 266
Ben Lomond, CA 95005 (408) 336-5485

Peter J. Szabo
810 Oxford Way
Benicia, CA 94510 (707) 746-0396

David J. Aston
Cutter Laboratories
4th & Parker St.
Berkeley, CA 94701 (415) 420-5357

Bertram Bradley
Miles Laboratories, Inc.
4th & Parker St.
Berkeley, CA 94710 (415) 420-4326

Roy L. Brown
2834 Garber St.
Berkeley, CA 94705 (408) 365-4888

Roger G. Ditzel#
University Of Calif.
Pat. Tdmk. & Copyright
2150 Shattuck Ave., Suite 1000A
Berkeley, CA 94720 (415) 642-5000

James A. Giblin
Miles Inc.
4th & Parker Sts.
Berkeley, CA 94710 (415) 420-5511

George C. Gorman
952 The Alameda
Berkeley, CA 94707 (415) 524-9520

Marcus Lothrop
Lothrop & West
726 Euclid Ave.
Berkeley, CA 94708 (415) 986-5833

Edward H. Maker
University of California
2200 University Ave.
Berkeley, CA 94720 (415) 642-5000

Keiichi Nishimura
461 Grizzly Peak Blvd.
Berkeley, CA 94708 (415) 524-6196

Alvah Levern Snow
1806 San Antonio Ave.
Berkeley, CA 94707 (415) 525-6647

Louis J. Strom
University Of California
Pat., Tdmk & Copyright
2490 Channing Way
Berkeley, CA 94720 (415) 642-5000

William Takacs
886 Arlington Ave.
Berkeley, CA 94707 (415) 524-0257

Stanley M. Teigland
630 Gravatt Dr.
Berkeley, CA 94705 (415) 841-5716

Allen B. Wagner
Univ. Of Calif.
Off. Of Gen. Coun.
2199 Addison St.
Berkeley, CA 94720 (415) 642-2822

Steve W. Ackerman
Cooper, Epstein & Hurewitz
9465 Wilshire Blvd., 8th Fl.
Beverly Hills, CA 90212 (213) 205-8331

I. Morley Drucker
Drucker & Sommers
9465 Wilshire Blvd., Suite 328
Beverly Hills, CA 90212 (213) 278-6852

Joseph R. Evanns
9465 Wilshire Blvd., Suite 428
Beverly Hills, CA 90212 (213) 272-8671

Charles L. Hartman
Drucker & Sommers
9465 Wilshire Blvd., Suite 328
Beverly Hills, CA 90212 (213) 278-6852

Harvey Sander Hertz
9777 Wilshire Blvd., Suite 500
Beverly Hills, CA 90212 (213) 278-9673

Albert M. Herzig
Herzig & Walsh, Inc.
9465 Wilshire Blvd., Suite 428
Beverly Hills, CA 90212 (213) 272-8671

Philip Hoffman
9454 Wilshire Blvd., Suite 900
Beverly Hills, CA 90212 (213) 655-4164

Booker T. Hogan
9595 Wilshire Blvd., Suite 900
Beverly Hills, CA 90212 (213) 278-0966

Robert H. Lentz
Litton Industries, Inc.
360 N. Crescent Dr.
Beverly Hills, CA 90210 (213) 859-5153

Michael F. McEntee
9595 Wilshire Blvd., Suite 700
Beverly Hills, CA 90212 (213) 273-3342

Michael A. Painter
Cooper, Epstein & Hurewitz
9465 Wilshire Blvd., Suite 800
Beverly Hills, CA 90212 (213) 278-1111

Seymour A. Scholnick
Turner & Scholnick
9100 Wishire Blvd., Suite 800
Beverly Hills, CA 90212 (213) 273-1870

Howard N. Sommers
Drucker & Sommers
9465 Wilshire Blvd., Suite 328
Beverly Hills, CA 90212 (213) 278-6852

Paul D. Supnik
9601 Wilshire Blvd., Suite 700
Beverly Hills, CA 90210 (213) 274-8281

Walter Richard Thiel
Litton Industries, Inc.
Litton Plaza S., Room 2030
360 N. Crescent Dr.
Beverly Hills, CA 90210 (213) 859-5451

Thomas A. Turner
9100 Wilshire Blvd., Suite 800
Beverly Hills, CA 90212 (213) 273-1870

Edward C. Walsh
Evanns & Walsh
119 N. San Vicente Blvd., Suite 206
Beverly Hills, CA 90211 (213) 273-0938

Abraham Wasserman
440 S. Doheny Dr.
Beverly Hills, CA 90211 (213) 273-5522

June M. Bostich#
Union Oil Company Of Calif.
376 S. Valencia Ave.
P.O. Box 76
Brea, CA 92621 (714) 528-7201

Timothy H. Briggs
P.O. Box 401
Brea, CA 92622 (714) 733-9868

Daniel R. Farrell
Unocal Corp.
376 S. Valencia Ave.
P.O. Box 76
Brea, CA 92621 (714) 528-7201

Yale S. Finkle
Unocal Corp.
376 S. Valencia Ave.
Science & Tech. Div.
Brea, CA 92621 (714) 528-7201

Gerald L. Floyd
Union Oil Co. Of Calif.
P.O. Box 76
Brea, CA 92621 (714) 528-7201

Richard G. Jackson
Unocal Corp.
Science & Tech. Div., Pat. Dept.
376 S. Valencia Ave.
Brea, CA 92621 (714) 528-7201

Michael H. Laird
Union Oil Co. Of Calif.
P.O. Box 76
Brea, CA 92621 (714) 528-7201

Howard R. Lambert
Unocal Science & Tech. Div.
Unocal Corporation
376 S. Valencia Ave.
P.O. Box 76
Brea, CA 92621 (714) 528-7201

Arthur E. Oaks
Unocal Science & Technology Div.
376 S. Valencia Ave.
P.O. Box 76
Brea, CA 92621 (714) 528-7201

Dean Sandford
Union Oil Co. Of Calif.
Research Center
P.O. Box 76
Brea, CA 92621 (714) 528-7201

Gregory F. Wirzbicki
Union Oil Co. Of Calif.
P.O. Box 76
Brea, CA 92621 (714) 528-7201

Stacey R. Condon Sias#
430 Valley Dr.
Brisbane, CA 94005 (415) 468-6300

Louis L. Dachs
Lockheed-California Co.
P.O. Box 551
Burbank, CA 91520 (818) 847-5291

Harry P. Levin#
Electro Energy Corp.
120 Elm Ct.
Burbank, CA 91502 (213) 845-2666

Frederic Paul Smith
Lockheed Corp.
P.O. Box 551
Burbank, CA 91520 (213) 847-5291

Kenneth B. Salomon
1652 Balboa Way
Burlingame, CA 94010 (415) 692-9729

Milton F. Custer#
4885 N. Point
Byron, CA 94514 (415) 634-5322

Billy G. Corber
Lockheed Corp.
4500 Park Granada Blvd.
Calabasas, CA 91399 (818) 712-2390

David O'Reilly
23603 Park Sorrento, Suite 103
Calabasas, CA 91302 (213) 883-3600

John Michael Koch
23108 Village 23
Camarillo, CA 93010 (805) 482-6401

Fay I. Konzem#
American Patent Institute
2707 N. Los Pinos Cir.
Camarillo, CA 93010 (805) 987-1880

Van Wesley Smart
26123 Atherton Dr.
Camel, CA 93921 (408) 624-5598

John J. Casparro
23326 Sandalwood St.
Canoga Park, CA 91307 (818) 992-4067

Robert P. Egermeier
22354 Malden St.
Canoga Park, CA 91304 (818) 348-1670

David C. Faulkner
Rockwell International Corp.
Rocketdyne Patent Dept. F B18
6633 Canoga Ave.
Canoga Park, CA 91304 (818) 700-4616

Harry Bruce Field
Rockwell Internatl. Corp.
6633 Canoga Ave.
Canoga Park, CA 91303 (818) 700-4616

Lawrence N. Ginsberg
Rockwell International Corp.
6633 Canoga Ave.
Canoga Park, CA 91304 (818) 700-3629

Gilbert Kivenson#
22030 Wyandotte St.
Canoga Park, CA 91303 (213) 883-5707

Henry Kolin
Rockwell International Corp.
6633 Canoga Ave.
Canoga Park, CA 91304 (213) 700-3629

Philip Schneider
Rockwell International
6633 Canoga Ave.
Canoga Park, CA 91304 (213) 700-4616

Thomas R. Waite
8609 De Sota Ave., Suite 147
Canoga Park, CA 91304 (213) 341-4474

John Stelmah
P.O. Box 5108
30157 Little Harbor Dr.
Canyon Lake, CA 92380 (714) 679-9597

Richard E. Cummins
870 Park Ave., Apt. 116
Capitola, CA 95010 (408) 462-4325

Duane C. Bowen
2551 State Street
Carlsbad, CA 92008 (714) 729-8446

John J. Murphey
Pacific Center One
701 Palomar Airport Rd., Suite 230
Carlsbad, CA 92009 (619) 431-0091

Charles G. Miller
20014 S. Camba Ave.
Carson, CA 90746 (213) 604-1494

Alfred Fafarman#
Applied Physics Consultants
P.O. Box 2994
Castro Valley, CA 94546 (415) 530-2326

James R. Naughten#
17485 Almond Rd.
Castro Valley, CA 94546 (415) 881-4458

Steven P. Brown#
Optical Disc Corp.
17517-H Fabrica Way
Cerritos, CA 90701 (714) 522-2370

Gilbert P. Hyatt#
P.O. Box 3357
Cerritos, CA 90703 (714) 995-1087

Thomas J. Clough
Ensei Inc.
20717 Prairie St.
Chatsworth, CA 91311 (818) 700-7501

Herbert Eckerling
20336 Coraline Circle
Chatsworth, CA 91311 (213) 998-9037

Casey Heeg
Patley Corp.
20415 Nordhoff St.
Chatsworth, CA 91311 (818) 407-3999

Don A. Hollingsworth#
10511 Keokuk Ave.
Chatsworth, CA 91311 (818) 998-3465

John B. Miller, Jr.
Pertron Controls Corp.
20630 Plummer St.
Chatsworth, CA 91311 (818) 998-4444

Stephen J. Church*
Naval Weapons Center
Off. Of Pat. Coun.
Code 006
China Lake, CA 93555 (619) 939-3733

Woodie D. English, III#*
Office Of Naval Research
Off. Of Patent Coun.
Naval Weapons Center
China Lake, CA 93555 (619) 939-3733

William Thomas Skeer*
Off. Of Naval Res.
N W C - Code 012
China Lake, CA 93555 (714) 939-3733

Patrick J. Schlesinger
Rohr Inds., Inc.
Foot of H St.
Chula Vista, CA 92012 (619) 691-2555

Tai S. Cho#
Marshall & Cho
987 W. Foothhill Blvd., Suite H
Claremont, CA 91711 (714) 625-5321

Doris Drucker#
636 Wellesley Dr.
Claremont, CA 91711 (714) 626-3172

Harold S. Gault
2549 N. Mountain Ave.
Claremont, CA 91711 (714) 626-5500

Harry Loberman
876 Occidental Dr.
Claremont, CA 91711 (714) 624-4198

Charles O. Marshall, Jr.
Marshall & Cho
987 W. Foothill Blvd., Suite H
Claremont, CA 91711 (714) 625-5321

Bernard V. Ousley
140 Foothill Blvd., W.
Claremont, CA 91711 (714) 621-7949

Robert Nathan Schlesinger#
RNS Research Institute
P.O. Box 1117
Claremont, CA 91711 (714) 625-4260

Thomas J. Murphy
Box 432
Coloma, CA 95613 (916) 626-4162

Edward J. Holzrichter#
Whittaker Coatings Research Center
1231 S. Lincoln Ave.
P.O. Box 825
Colton, CA 92324 (714) 825-6292

James W. Lucas#
1401 Bonnie Doone Terrace
Corona Del Mar, CA 92625 (714) 644-9500

Sean P. Fitzgerald
Discovision Associates
2183 Fairview Rd., Suite 211
P.O. Box 6600
Costa Meas, CA 92628 (714) 957-3000

Roy A. Ekstrand
125 E. Baker St., Suite 240
Costa Mesa, CA 92626 (714) 662-7733

David L. Fehrman#
Spensley, Horn, Jubas & Lubitz
650 Town Center Dr., Suite 1930
Costa Mesa, CA 92626 (714) 557-2047

Gideon Gimlan
Spensley, Horn, Jubas & Lubitz
650 Town Center Dr., Suite 1930
Costa Mesa, CA 92626 (714) 557-2047

David A. Hall
Spensley, Horn, Jubas & Lubitz
650 Town Center Dr., Suite 1930
Costa Mesa, CA 92626

David L. Henty
Spensley, Horn, Jubas & Lubitz
650 Town Center Dr., Suite 1930
Costa Mesa, CA 92626 (714) 557-2047

Ronald C. Hudgens
Discovision Associates
2183 Fairview Rd., Suite 211
Costa Mesa, CA 92627 (714) 957-3000

William J. Kearns#
3350 California St.
Costa Mesa, CA 92626 (714) 641-9435

Stephen G. Mican
Discovision Associates
2183 Fairview Rd., Suite 211
Costa Mesa, CA 92627 (714) 957-3000

Samuel Beckner Stone
Lyon & Lyon
3200 Park Center Dr., Suite 1170
Costa Mesa, CA 92626 (714) 751-6606

Robert M. Taylor, Jr.
Lyon & Lyon
3200 Park Center Dr., Suite 1170
Costa Mesa, CA 92626 (714) 751-6606

Roger C. Turner
702 Shalimar Dr., Suite A
Costa Mesa, CA 92627 (714) 548-5649

Robert A. Westerlund, Jr.
Discovision Associates
Ibm-Mca J.V.
2183 Fairview Rd., Suite 211
Costa Mesa, CA 92627 (714) 957-3000

Leonard R. Cool#
9646 Wagner Rd.
Greeley Hill
Coulterville, CA 95311 (209) 878-3271

James R. Eckel#
5104 Copperfield Lane
Culver City, CA 90230 (213) 839-0108

Richard K. Ehrlich
4901 S. Overland Ave.
Culver City, CA 90230 (213) 559-7415

John H. Kusmiss#
12200 Allin St.
Culver City, CA 90230 (213) 313-0926

Hal Jay Bohner
Measurex Corp.
One Results Way
Cupertino, CA 95014 (408) 255-1500

Anthony T. Cascio
Tandem Computers Inc.
19191 Vallco Parkway
Cupertino, CA 95014 (408) 725-7369

Robert T. Martin#
Apple Computer, Inc.
20525 Mariani Ave.
M/s 28-B
Cupertino, CA 95014 (408) 973-4700

Theodore S. Park
1147 Stafford Dr.
Cupertino, CA 95014 (408) 253-9724

Stephen J. Phillips
Fairchild Semiconductor Corp.
10400 Ridgeview Ct.
Cupertino, CA 95014 (408) 864-6012

Donald Penprase
11355 Orrs Ct.
Cypress, CA 90630 (714) 894-3759

John Helms Warden
6666 Vinalhaven Court
Cypress, CA 90630 (714) 893-8443

Stephen L. Hurst
142 Wyandotte Ave.
Daly City, CA 94014 (415) 992-7473

Nita J. Almquist
24271 Philemon
Dana Point, CA 92629 (619) 455-5015

Robert Edward Havranek
130 Virginia Ct.
Danville, CA 94526 (415) 820-8082

Pauline M Naillon#
17 Diamond Dr.
Danville, CA 94526 (415) 837-5848

Michael D. Nelson
417 Front St.
Danville, CA 94526 (415) 837-8019

Carey B. Huscroft
4849 El Cemonte Ave., Apt. 113
Davis, CA 95616 (916) 758-2608

Lowell R. Wedemeyer
3002 Catalina Dr.
Davis, CA 95616 (916) 757-2200

Gerald T. Richards
P.O. Box 728
2210 Caballo Ranchero Ct.
Diablo, CA 94528 (415) 837-1887

William S. Bernheim
Whitaker & Bernheim
255 N. Lincoln St.
Dixon, CA 95620 (916) 678-4447

Joseph Furrow#
Rockwell Internatl. Inc.
12214 Lakewood Blvd.
Downey, CA 90241 (213) 922-1309

Ronald L. Juniper
9025 Florence Ave., Suite A
Downey, CA 90240 (213) 861-0796

Dominick Nardelli
11015 Paramont Blvd.
Downey, CA 90241 (213) 869-1078

Kenneth T. Theodore
9012 Suva St.
Downey, CA 90240 (213) 420-7017

Harold C. Weston
Rockwell Internatl. Corp.
12214 Lakewood Blvd.
Downey, CA 90241 (213) 922-1636

Sidney Sternick
6500 Village Parkway, Suite 106
Dublin, CA 94568 (415) 829-9270

Ervin Frederic Johnston
1904 Ventana Way
El Cajon, CA 92020 (619) 448-2228

Alan Lloyd Newman
Friedman Homes Inc.
4455 Rowland Ave.
El Monte, CA 91731 (213) 579-1450

William B. Wong
South Coast Air Quality Management District
Legal Division
9150 Flair Dr.
El Monte, CA 91731 (818) 572-6200

Robert E. Cunha
Xerox Corp.
701 S. Aviation Blvd.
E S A E - 335
El Segundo, CA 90245 (213) 333-7292

Robert J. Fasnacht
Aerospace Corp.
2350 E. El Segundo Blvd.
El Segundo, CA 90245 (213) 648-6608

Howard A. Kenyon#
Rockwell International Corp.
827 N. Douglas St.
El Segundo, CA 90245 (213) 647-6846

Mary E. Lachman#
Hughes Aircraft Co.
Pats & Licensing
Bldg. C2, Mail Station A126
P.O. Box 1042
El Segundo, CA 90245 (213) 414-6096

Earnest F. Oberheim#
Hughes Aircraft Co.
Pats & Licensing
Bldg. C2, Mail Station A126
P.O. Box 1042
El Segundo, CA 90245 (213) 414-6091

Edward Polosky
Rockwell International Corp.
2230 Imperial Hwy., E.
El Segundo, CA 90245 (213) 647-5315

Sheldon F. Raizes
Xerox Corp.
701 S. Aviation Blvd.
El Segundo, CA 90245 (213) 576-9802

Wayne P. Sobon#
Hughes Aircraft
P.O. Box 902
E54/f22E
Rosecrans Ave.
El Segundo, CA 90294 (213) 616-0710

Franklyn Charles Weiss
Xerox Corp.
701 S. Aviation Blvd.
El Segundo, CA 90245 (213) 536-7935

Leonard Zalman
Xerox Corp.
Pat. Dept.
701 S. Aviation Blvd.
El Segundo, CA 90245 (213) 536-7245

Gregory J. Giotta#
Cetus Corp.
1400 53rd St.
Emeryville, CA 94608 (415) 420-3152

Albert Price Halluin
Cetus Corp.
1400 Fifty-Third St.
Emeryville, CA 94608 (415) 420-3414

Paul R. Martin
4 Captain Dr., Apt. 312
Emeryville, CA 94608 (415) 655-8133

Jeffrey K. Weaver#
1067 48th St.
Emeryville, CA 94608

Lewis Anten
17530 Ventura Blvd., Suite 201
Encino, CA 91316 (213) 501-3535

Albert O. Cota#
5460 White Oak Ave., Suite A-331
Encino, CA 91316 (818) 905-0848

Donald Diamond
16133 Ventura Blvd., 7th Fl.
Encino, CA 91436 (213) 986-2221

Robert Louis Finkel
16055 Ventura Blvd., Suite 915
Encino, CA 91436 (213) 986-5000

Thomas Gunzler#
16929 Escalon Dr.
Encino, CA 91436 (818) 988-1040

Thomas D. Linton, Jr.
Atrium Bldg., Suite 600
16530 Ventura Blvd.
Encino, CA 91436 (213) 990-7282

Matthew P. Lynch
16633 Ventura Blvd.
Encino, CA 91436 (213) 990-7282

Seymour Rosenberg
15915 Ventura Blvd., Suite 201
Encino, CA 91436 (213) 905-6888

Bette M. Light#
8334 Woodborough
Fair Oaks, CA 95628 (916) 966-0363

James M. Ritchey
4420 New York Ave.
Fair Oaks, CA 95628 (916) 966-8631

Leslie K. Loehr#
530 Georgine Rd.
Fallbrook, CA 92028 (619) 728-9569

Kenneth A. Cox
704 Somerset Lane
Forest City, CA 94404 (415) 574-0179

T. R. Zegree#
421 Beach Park Blvd.
Foster City, CA 94404 (415) 574-2639

Harland L. Burge, Jr.
Putman, Strid & Burge
17330 Brookhurst, Suite 350
Fountain Valley, CA 92708 (714) 842-4484

Walter F. Krstulja
16393 Rosewood St.
Fountain Valley, CA 92708 (714) 839-6488

Thomas L. Peterson
ITT Corporation
Patent Dept.
10550 Talbert Ave.
Fountain Valley, CA 92708 (714) 964-8444

Vergil L. Gerard
1341 E. San Jose Ave., Suite B
Fresno, CA 93710 (209) 225-5872

George G. Grigel
Worrel & Worrel
1171 Fulton Mall, Suite 1202
Fresno, CA 93721 (209) 486-4526

Mark D. Miller
Kimble, Mac Michael & Upton
4201 W. Shaw Ave., Suite 100
Fresno, CA 93722 (209) 275-5200

Edward J. Pavsek, Jr.
4974 N. Fresno St., Suite 378
Fresno, CA 93726

Victor Sepulveda
5070 W. Sixth St., Suite 144
Fresno, CA 93710 (209) 255-5561

Richard Milton Worrel
Worrel & Worrel
1171 Fulton Mall, Suite 1202
Fresno, CA 93721 (209) 486-4526

Rodney K. Worrel
Worrel & Worrel
1171 Fulton Mall, Suite 1202
Fresno, CA 93721 (209) 486-4526

Julia E. Abers
Beckman Instruments, Inc.
2500 Harbor Blvd.
Fullerton, CA 92634 (714) 773-6968

Dale E. Bennett#
801 N. Mountain View Pl.
Fullerton, CA 92631 (714) 871-8577

Norman E. Carte#
924 White Water Dr.
Fullerton, CA 92633 (714) 992-2354

William C. Daubenspeck*
Hughes Aircraft Co.
P.O. Box 3310
Bldg. 618, M.S. E425
Fullerton, CA 92634 (714) 732-8097

Arnold Grant
Beckman Instruments, Inc.
2500 Harbor Blvd.
Fullerton, CA 92634 (714) 773-6922

Gary T. Hampson
Beckman Instruments, Inc.
2500 Harbor Blvd.
Fullerton, CA 92634 (714) 773-6922

Paul R. Harder
Beckman Instruments, Inc.
2500 Harbor Blvd.
Fullerton, CA 92634 (714) 773-6909

Richard C. Hartman
419 Thunderbird Ct.
Fullerton, CA 92635 (714) 870-0585

Ferd L. Mehlhoff
Beckman instruments Inc.
2500 Harbor Blvd.
Fullerton, CA 92634 (714) 773-6970

Robert T. Spaulding
550 Elinor Dr.
Fullerton, CA 92635 (714) 879-9313

Robert Jay Steinmeyer
609 Lemon Hill Terrace
Fullerton, CA 92632 (714) 526-5229

Cleveland R. Williams
1506 N. Sycamore
Fullerton, CA 92631 (714) 773-0748

Martin G,. Reiffin
9262 Royal Palm Blvd.
Garden Grove, CA 92641 (714) 530-0493

Thomas L. Venezia
9301 Shannon Ave.
Garden Grove, CA 92641 (714) 638-8011

Randall G. Wick
Perkin-Elmer Corp.
7421 Orangewood Ave.
Garden Grove, CA 92641 (714) 895-1667

Howard L. Johnson
16010 Crenshaw Blvd.
Gardena, CA 90249 (213) 323-6396

Chris Papageorge
14625 S. Vermont, Apt. 5
Gardena, CA 90247

Vincent C. Tyrrell#
American Honda Motors Co., Inc.
100 W. Alondra Blvd.
Gardena, CA 90247 (213) 604-2663

Michael Aguilar
3508 Angelus Ave.
Glendale, CA 91208 (213) 249-9898

James E. Brunton
225 W. Broadway, Suite 500
Glendale, CA 91204 (213) 956-7154

Ted DeBoer#
National Inventors Foundation Inc.
345 W. Cypress St.
Glendale, CA 91204 (213) 246-6540

Robert B. Langford#
644 Haverkamp Dr.
Glendale, CA 91206 (213) 240-3172

Theodore Hawley Lassagne
1627 Sheridan Rd.
Glendale, CA 91206 (818) 246-1256

Clarence J. Morrissey
P.O. Box 11036
Glendale, CA 91206 (818) 354-6834

John Emery Wagner
Wagner & Middlebrook
3541 Ocean View Blvd.
Glendale, CA 91208 (818) 957-3340

Alan H. Thompson
630 N. Wildwood Ave.
Glendora, CA 91740 (818) 335-3045

William C. Schubert
Santa Barbara Research Center
75 Coromar Dr.
Goleta, CA 93117 (805) 968-3511

Cecilia L. Yu
16617 Echo Hill Way
Hacienda Heights, CA 91745 (818) 968-2189

Melvin Arthur Klein
Mattel, Inc.
5150 Rosecrans
Hawthorne, CA 90250 (213) 978-5150

Michael B. Lachuk
Northrop Corp.
One Northrop Ave.
Hawthorne, CA 90250 (213) 332-1226

John Ernest Peele, Jr.
Northrop Corp.
Corp. Pat. Dept. 110/31
One Northrop Ave.
Hawthorne, CA 90250 (213) 970-2318

Daniel F. Sullivan
Mattel, Inc.
Pat. Dept.
M.S. 01-111-A03
5150 Rosecrans Ave.
Hawthorne, CA 90250 (213) 978-6106

Linval B. Castle
22693 Hesperian Blvd., Suite 270
Hayward, CA 94541 (415) 887-1346

Michael L. Sherrard
4175 China Ct.
Hayward, CA 94541 (415) 538-3950

Jerome B. Rockwood#
1400 W. Florida Ave., Apt. 64
Hemet, CA 92343 (714) 929-4536

Sanford S. Wadler
Bio-Rad Laboratories
1000 Alfred Nobel Dr.
Hercules, CA 94547 (415) 724-3167

Harry R. Lubcke#
2443 Creston Way
Hollywood, CA 90068 (213) 469-3266

Barry A. Bisson
17111 Beach Blvd., Suite 207
Huntington Beach, CA 92646 (714) 848-0479

Elbert D. Craft#
17311 Almelo Lane
Huntington Beach, CA 92649 (714) 846-5317

Walter C. Glowski
McDonnell Douglas Corp.
5301 Bolsa Ave.
Huntington Bch, CA 92647 (714) 896-3713

Charles S. Gumpel
9601 Onset Circle
Huntington Bch, CA 92646 (714) 964-3361

Merrill G. Hinton, Jr.
16161 Ballantine Lane
Huntington Beach, CA 92647 (714) 846-4887

James G. O Neill
325-21st St.
Huntington Beach, CA 92648 (714) 960-3436

Edward D. O'Brian
16152 Beach Blvd., Suite 145
Huntington Beach, CA 92647 (714) 841-1592

Thomas A. Schenach#
6531 Meath Circle
Huntington Beach, CA 92647 (714) 892-8886

Paul H. Ware
8910 2nd Ave.
Ingelwood, CA 90305 (213) 354-3318

Leslie Badin, Jr.
4882 Basswood Lane
Irvine, CA 92715 (714) 786-1113

Allan R. Fowler
18662 Mac Arthur Blvd., Suite 460
Irvine, CA 92715 (714) 833-8311

Gilbert H. Friedman
Three Creekwood
Irvine, CA 92714 (714) 551-2308

Albin h. Gess
Price, Gess & Ubell
2100 S.E. Main St., Suite 250
Irvine, CA 92714 (714) 261-8433

Walter A. Hackler
2302 Martin St., Suite 320
Irvine, CA 92715 (714) 851-5010

Howard J. Klein
Klein & Szekeres
4199 Campus Dr., Suite 700
Irvine, CA 92715 (714) 854-5502

Newton H. Lee, Jr.#
Gradco Systems, Inc.
Seven Morgan
Irvine, CA 92718 (714) 770-1223

John H. Lynn
Kendrick, Netter & Bennett
Two Park Plaza, Suite 800
Irvine, CA 92714 (714) 474-2010

Richard L. Myers
American Hospital Supply Corp.
2132 Michelson Dr.
Irvine, CA 92715 (714) 975-1800

Thomas J. Plante
11 Solana
Irvine, CA 92715 (714) 752-2323

Sandra S. Schultz
Baxter Travenol Labs., Inc.
2132 Michelson Dr.
Irvine, CA 92715 (714) 476-5323

John R. Shewmaker
Shiley Inc.
17600 Gillette Ave.
Irvine, CA 92714 (714) 250-8222

Gabor L. Szekeres
Klein & Szekeres
4199 Campus Dr., Suite 700
Irvine, CA 92715 (714) 854-5502

Robert G. Upton#
Smith Tool
P.O. Box C-19511
17871 Von Karman Ave.
Irvine, CA 92713 (714) 660-5344

Corwin R. Horton
The Park
25 Mann Dr., Suite 2001
Kentfield, CA 94904 (415) 453-5443

William T. O'Neil#
2142 La Canada Crest Dr., No. 3
La Canada, CA 91011 (818) 248-3252

Louis J. Bachand, Jr.
P.O. Box 12330
La Crescenta, CA 91214 (818) 352-8841

Selwyn S. Berg
7730 A Hershel Ave.
La Jolla, CA 92037 (619) 459-2374

William M. Dooley
Aerojet-General Corp.
10300 N. Torrey Pines Rd.
La Jolla, CA 92037 (714) 455-8589

Natalie Jensen#
Calbiochem-Behring Corp.
10933 N. Torrey Pines Rd.
La Jolla, CA 92037 (714) 453-7331

Karol J. Mysels#
8327 La Jolla Scenic Dr.
La Jolla, CA 92037 (714) 453-6988

Ernest Arthur Polin
5810 Caminito Cardelina
La Jolla, CA 92037 (619) 459-5334

Richard J. Reilly
1556 Virginia Way
La Jolla, CA 92037 (614) 456-0184

Charles Chalmers Logan, II
7373 University Ave., Suite 214
La Mesa, CA 92041 (619) 463-7344

Richard M. Stanley
8764 Glenira Ave.
La Mesa, CA 92041 (619) 464-0197

Derrick M. Reid
5600 Orangethrope Ave., Suite 1305
La Palma, CA 90623 (714) 521-1585

Arthur James Wagner
285 Camino A. Barranco
La Selva Beach, CA 95076 (408) 684-0893

Sidney Levy#
4433 Dawn Ave.
La Verne, CA 91750 (714) 596-7641

Robert J. Henry
3812-C Happy Valley Rd.
Lafayette, CA 94549 (415) 283-5146

Donald J. Mc Rae#
1050 Via Roble
Lafayette, CA 94549 (415) 284-4760

George W. Wasson
3123 Indian Way
Lafayette, CA 94549 (415) 283-4420

Herb Boswell#
La Paz Office Plaza
25283 Cabot Rd., Suite 209
Laguna Hills, CA 92653 (714) 380-4890

Bruce B. Brunda
Stetina & Brunda
24221 Calle De La Louisa, Suite 401
Laguna Hills, CA 92653 (714) 855-1246

Robert D. Buyan
24121 A Hollyoak
Laguna Hills, CA 92656 (714) 855-1246

Grover A. Frater
23041 Avenida De La Corlata, Suite 310
Laguna Hills, CA 92653 (714) 855-6040

Loyal M. Hanson
Weissenberger & Peterson
24012 Calle De La Plata, Suite 470
Laguna Hills, CA 92653 (714) 380-4046

Gordon Lloyd Peterson
Weissenberger & Peterson
24012 Calle De La Plata, Suite 470
Laguna Hills, CA 92653 (714) 380-4046

Kit M. Stetina
Hubbard & Stetina
24221 Calle De La Louisa, Suite 401
Laguna Hills, CA 92653 (714) 855-1246

Harry G. Weissenberger
Weissenberger & Peterson
24012 Calle De La Plata, Suite 470
Laguna Hills, CA 92653 (714) 380-4046

John L. Hummer
P.O. Box 6482
Laguna Niguel, CA 92677

Martin Andries Voet
Voet Pharmaceutical Intl.
30101 Town Center, Suite 104
Laguna Niguel, CA 92677 (714) 831-2431

David F. O'Brien
2102 E. Tern Bay Ln.
Lakewood, CA 90712 (213) 616-3336

Don A. Fischer
390 Ledroit
Languna Beach, CA 92651 (714) 494-7937

Robert J. Baran
29131 Ridgeview Dr.
Languna Niguel, CA 92677 (714) 831-5247

Max Elwin Shirk
P.O. Box 550
Lebec, CA 93243 (805) 248-6190

Lafayette E. Carnahan#*
U.S. Dept. Of Energy
Off. Of Pat. Coun.
P.O. Box 808, L-376
Livermore, CA 94550 (415) 422-1430

Clifton E. Clouise#*
U.S. Dept. Of Energy
Off. Of Pat. Coun.
P.O. Box 808, L-376
Livermore, CA 94550 (415) 422-1429

Martin I. Finston#
Lawrence Livermore National Laboratory
Classification Office, L-302
P.O. Box 808
Livermore, CA 94550 (415) 423-3055

Roger Sherwin Gaither
U.S. Dept. Of Energy
Off. Of Pat. Coun.
P.O. Box 808, L-376
Livermore, CA 94550 (415) 422-4367

Michael B. K. Lee
LLNL
P.O. Box 808, L-668
East Ave.
Livermore, CA 94550 (415) 423-8051

Shyamala Rajender
Lawrence Livermore National Lab.
L-376, Box 808
Livermore, CA 94550 (415) 422-4367

Frederick A. Robertson
704 Wimbldeon Lne
Livermore, CA 94550 (415) 447-6787

Gary C. Roth
Lawrence Livermore Lab.
P.O. Box 808
Livermore, CA 94550 (415) 422-7819

Henry P. Sartorio#
Univ. Of Calif.
Lawrence Livermore Lab.
7000 East Ave.
Livermore, CA 94550 (415) 422-7816

Palmer Martin Simpson, Jr.
Lawrence Livermore National Lab.
P.O. Box 808, L-668
Livermore, CA 94550 (415) 423-2778

John H. G. Wallace#
598 Escondido Circle
Livermore, CA 94550 (415) 443-4565

Berthold J. Weis#
Lawrence Livermore Lab.
Univ. Of Calif.
P.O. Box 808
Livermore, CA 94550 (415) 422-7274

Basil B. Travis
1209 W. Tokay St., Apt. 9
P.O. Box 287
Lodi, CA 95241 (209) 333-8379

I. Michael Bak-Boychuk
400 Oceangate Plaza, Suite 325
Long Beach, CA 90802 (213) 432-8419

Donald L. Carlson
6301 Bixby Hill Rd.
Long Beach, CA 90815 (213) 431-0644

Gregory A. Cone
McDonnell Douglas Corp.
C1-H009 (122-23)
3855 Lakewood Blvd.
Long Beach, CA 90846 (213) 593-6812

George Walter Finch
McDonnell Douglas Corp.
M S 78-81
3855 Lakewood Blvd.
Long Beach, CA 90846 (213) 593-6834

Curtis L. Harrington
McDonnell Douglas
C1-H009 78-81
3855 Lakewood Blvd.
Long Beach, CA 90846 (213) 593-6812

Benjamin Hudson
McDonnell Douglas Corp.
C1-H009 (78-81)
3855 Lakewood Blvd.
Long Beach, CA 90846 (213) 593-7579

L. Lee Humphries
7821 Tibana St.
Long Beach, CA 90808 (213) 596-6962

Paul T. Loef
McDonnell Douglas Corp.
C1-H009 (122-23)
3855 Lakewood Blvd.
Long Beach, CA 90846 (213) 593-6812

George R. Loftis
P.O. Box 1602
Long Beach, CA 90801

Joseph F. McLellan
Fulwider, Patton, Rieber, Lee & Utecht
11 Golden Shore, Suite 510
Long Beach, CA 90802 (213) 432-0453

James Thomas McMillan
McDonnell Douglas Finance Corp.
100 Oceangate, Suite 900
Long Beach, CA 90802 (213) 593-1722

Richard J. Otto, Jr.
144 Santa Ana Ave.
Long Beach, CA 90803 (213) 434-4614

James W. Paul
Fulwider, Patton, Rieber, Lee & Utecht
11 Golden Shore, Suite 510
Long Beach, CA 90802 (213) 432-0453

Jerry R. Potts
110 W. Ocean Blvd., Suite C
Long Beach, CA 92408 (213) 491-4647

Donald L. Royer
McDonnell Douglas Corp.
C1-H009 (122-23)
3855 Lakewood Blvd.
Long Beach, CA 90846 (213) 593-6834

John P. Scholl
McDonnell Douglas Corp.
3855 Lakewood Blvd.
M S 122-23
Long Beach, CA 90846 (213) 593-6834

Vern D. Schooley
Fulwider, Patton, Rieber, Lee & Utecht
P.O. Box 22615
11 Golden Shore, Suite 510
Long Beach, CA 90802 (213) 432-0453

James M. Skorich
McDonnell Douglas Corp.
Pat. Dept., Mail Station 122-23
3855 Lakewood Blvd.
Long Beach, CA 90846 (213) 593-6812

Charles H. Thomas, Jr.
Cislo, O'Reilly & Thomas
4201 long Beach Blvd., Suite 405
Long Beach, CA 90807 (213) 595-8422

Francis A. Utecht
Fulwider, Patton, Rieber, Lee & Utecht
11 Golden Shore, Suite 510
Long Beach, CA 90802 (213) 432-0453

Georges A. Maxwell
11362 Wallingsford Road
Los Alamitos, CA 90720 (213) 431-6255

Victor R. Beckman
175 S. Antonio Rd., Suite 216
Los Altos, CA 94022 (415) 949-3103

Mervin Halstead
1131 Hillslope Pl.
Los Altos, CA 94022 (415) 948-8324

Leon F. Herbert
610 Twelve Acres Dr.
Los Altos, CA 94022 (415) 948-8653

Albert C. Smith
P.O. Box 782
Los Altos, CA 94022 (415) 857-2441

H. Donald Volk
374 Benvenue Ave.
Los Altos, CA 94022 (408) 742-0691

Edward Y. Wong
P.O. Box 1161
Los Altos, CA 94022 (415) 857-3873

Arthur J. Deex
25396 La Loma Dr.
Los Altos Hills, CA 94022 (415) 949-1830

Glenn H. Lenzen, Jr.
Cambridge Systems Group
24275 Elise
Los Altos Hills, CA 94022 (408) 249-1048

John C. Oberlin
26140 Robb Rd.
Los Altos Hills, CA 94022 (415) 948-8157

David B. Abel#
Garrett Corp.
9851 Sepulveda Blvd.
Los Angeles, CA 90009 (213) 417-6552

Colin P. Abrahams
3600 Wilshire Blvd., Suite 1520
Los Angeles, CA 90010 (213) 385-4281

Carolyn R. Adler
Romney, Golant, Martin, Seldon & Ashen
10920 Wilshire Blvd., Suite 1000
Los Angeles, CA 90024 (213) 208-1100

Matthew L. Ajeman#
3010 Wilshire Blvd., Apt. #486
Los Angeles, CA 90010

Naveed Alam
453 S. Spring, Suite 1017
Los Angeles, CA 90013 (213) 489-3131

Alan K. Aldous#
3346 Canfield Ave., Apt. 203
Los Angeles, CA 90034 (213) 204-4186

Leonard A. Alkov
Hughes Aircraft Co.
Corp. Pats. & Licensing
P.O. Box 45066
7200 Hughes Terrace Bldg.
Los Angeles, CA 90045 (213) 568-6081

Gary M. Anderson
Lyon & Lyon
611 W. Sixth St., 34th Fl.
Los Angeles, CA 90017 (213) 489-1600

Roy L. Anderson
Lyon & Lyon
611 W. Sixth St., 34th Fl.
Los Angeles, CA 90017 (213) 489-1600

William L. Androlia
Koda & Androlia
1880 Century Park, E.
Los Angeles, CA 90067 (213) 277-1391

Gene Wesley Arant
Arant, Kleinberg & Lerner
2049 Century Park, E.
Los Angeles, CA 90067 (213) 557-1511

Erik M. Arnhem
4113 Beverly Blvd.
Los Angeles, CA 90004 (213) 660-5067

Robert M. Ashen
Romney, Golant, Martin, Seldon & Ashen
10920 Wilshire Blvd., Suite 1000
Los Angeles, CA 90024 (213) 208-1100

Craig B. Bailey
Fulwider, Patton, Rieber, Lee & Utecht
3435 Wilshire Blvd., Suite 2400
Los Angeles, CA 90010 (213) 380-6800

Michael Barclay
Spensley, Horn, Jubas, & Lubitz
1880 Century Park, E., 5th Fl.
Los Angeles, CA 90067 (213) 553-5050

Richard A. Bardin
Fulwider, Patton, Rieber, Lee & Utecht
3435 Wilshire Blvd., Suite 2400
Los Angeles, CA 90010 (213) 380-6800

Vernon D. Beehler
Beehler, Pavitt, Siegemund, Jagger, Martella & Dawes
3435 Wilshire Blvd., Suite 1100
Los Angeles, CA 90010 (213) 385-7087

Vincent J. Belusko
Spensley, Horn, Jubas, & Lubitz
1880 Century Park, E., 5th Fl.
Los Angeles, CA 90067 (213) 553-5050

John M. Benassi
Lyon & Lyon
611 W. 6th St., 34th Fl.
Los Angeles, CA 91745 (213) 489-1600

William J. Benman, Jr.
10850 Wilshire Blvd., Suite 800
Los Angeles, CA 90024 (213) 475-3112

Luc P. Benoit
Benoit Law Corp.
2551 Colorado Blvd.
Los Angeles, CA 90041 (213) 255-0000

Richard P. Berg
Ladas & Parry
3600 Wilshire Blvd., Suite 1520
Los Angeles, CA 90010 (213) 385-4281

Stephen L. Berger#
10564 Eastborne Ave.
Los Angeles, CA 90024 (213) 475-9913

Robert Berliner
Nilsson, Robbins, Dalgarn, Berliner, Carson & Wurst
201 N. Figueroa St. 5th Fl.
Los Angeles, CA 90012 (213) 977-1001

Rod S. Berman
Spensley, Horn, Jubas, & Lubitz
1880 Century Park, E., 5th Fl.
Los Angeles, CA 90067 (213) 553-5050

Bruce L. Birchard
Birchard & Birchard
6363 Wilshire Blvd., Suite 406
Los Angeles, CA 90048 (213) 653-2758

Herbert A. Birenbaum
Atlantic Richfield Co.
Legal Dept.
515 S. Flower St.
Los Angeles, CA 90071 (213) 486-1569

Henry Martyn Bissell
6820 La Tijera Blvd., Suite 106
Los Angeles, CA 90045 (213) 776-3122

Roger W. Blakely, Jr.
Blakely, Sokoloff, Taylor & Zafman
12400 Wilshire Blvd., 7th Fl.
Los Angeles, CA 90025 (213) 207-3800

Jeffrey J. Blatt
Blakely, Sokoloff, Taylor & Zafman
12400 Wilshire Blvd., 7th Fl.
Los Angeles, CA 90025 (213) 207-3800

Coe A. Bloomberg
Lyon & Lyon
611 W. Sixth St., 34th Fl.
Los Angeles, CA 90017 (212) 489-1600

Breton A. Bocchieri
Blakely, Sokoloff, Taylor & Zafman
12400 Wilshire Blvd., 7th Fl.
Los Angeles, CA 90025 (213) 207-3800

Raymond A. Bogucki
Bogucki, Scherlacher, Mok & Roth
3345 Wilshire Blvd.
Los Angeles, CA 90010 (213) 386-7701

Richard J. Botos
Lyon & Lyon
611 W. Sixth St., 34th Fl.
Los Angeles, CA 90017 (213) 489-1600

Louis J. Bovasso
Poms, Smith, Lande, & Rose
2121 Ave. of the Stars, Suite 1400
Los Angeles, CA 90067 (213) 277-8141

Ralph M. Braunstein
Reagin & King
12400 Wilshire Blvd., Suite 500
Los Angeles, CA 90025 (213) 820-5864

E. Lawrence Brevik
B Two M Industries, Inc.
12923 S. Spring St.
Los Angeles, CA 90061 (213) 770-1871

Patrick Francis Bright
Kendrick, Netter & Bennett
612 S. Flower St., Suite 600
Los Angeles, CA 90017 (213) 626-7792

James C. Brooks
Lyon & Lyon
611 W. Sixth St., 34th Fl.
Los Angeles, CA 90017 (213) 489-1600

Charles D. Brown
Hughes Aircraft Co.
Corp. Pats. & Licensing
Bldg. C1, Mail Station A126
P.O. Box 45066
Los Angeles, CA 90045 (213) 568-6084

Marc E. Brown
Fleishman & Damon
1901 Ave. of the Stars, Suite 931
Los Angeles, CA 90067 (213) 277-3338

Clarence L. Browning
Whann & Connors
315 W. Ninth St., Suite 920
Los Angeles, CA 90015 (213) 622-7163

James R. Brueggemann
Pretty, Schroeder, Brueggemann & Clark
444 S. Flower St., Suite 2000
Los Angeles, CA 90071 (213) 489-4442

Norman, E Brunell
4249 San Rafael Ave.
Los Angeles, CA 90042 (213) 258-3000

William J. Burke
Aerospace Corp.
P.O. Box 92957
Mail Stop M1/040
Los Angeles, CA 90009 (213) 336-6708

Cathryn A. Campbell
Irell & Manella
1800 Ave. of the Stars, Suite 900
Los Angeles, CA 90067 (213) 277-1010

M. Michael Carpenter
Poms, Smith, Lande & Rose
2121 Ave. of the Stars, Suite 1400
Los Angeles, CA 90067 (213) 277-8141

M. John Carson, III
Nilsson, Robbins, Dalgarn, Berliner, Carson & Wurst
201 N. Figueroa St., 5th Fl.
Los Angeles, CA 90012 (213) 977-1001

David Charness
612 N. Sepulveda Blvd.
Los Angeles, CA 90049 (213) 472-8012

John Stirling Christopher
Fulwider, Patton, Rieber, Lee & Utecht
3435 Wilshire Blvd., Suite 2400
Los Angeles, CA 90010 (213) 380-6800

Eric T. S. Chung
Chung & Stein
10100 Santa Monica Blvd., Suite 750
Los Angeles, CA 90067 (213) 556-2104

Gary Alan Clark
Pretty, Schroeder, Brueggemann & Clark
444 S. Flower St., 20th Fl.
Los Angeles, CA 90071 (213) 489-4442

Vincent W. Cleary*
U.S. Air Force
SD/JA
Box 92960
L.A. Air Force Station
Los Angeles, CA 90009 (213) 643-0916

Adam Cochran
Nilsson, Robbins, Dalgarn, Berliner, Carson & Wurst
201 N. Figueroa St., 5th Fl.
Los Angeles, CA 90012 (213) 977-1001

Marc S. Colen
Jones, Day, Reavis & Pogue
355 S. Grand Ave., Suite 3000
Los Angeles, CA 90071 (213) 625-3939

David W. Collins
Romney, Golant, Martin, Seldon & Ashen
10920 Wilshire Blvd., Suite 1000
Los Angeles, CA 90024 (213) 208-1100

John J. Connors
Whann & Connors
315 W. Ninth St., Suite 920
Los Angeles, CA 90015 (213) 622-7163

Mary S. Consalvi
Lyon & Lyon
611 W. Sixth St., 34th Fl.
Los Angeles, CA 90017 (213) 489-1600

Lewis M. Dalgarn
Nilsson, Robbins, Dalgarn, Berliner, Carson & Wurst
201 N. Figueroa St., 5th Fl.
Los Angeles, CA 90012 (213) 977-1001

Christopher Darrow
Poms, Smith, Lande & Rose
2121 Ave. of the Stars, Suite 1400
Los Angeles, CA 90067 (213) 277-8141

Daniel L. Dawes#
Beehler, Pavitt, Siegemunnd, Jagger, Martella & Dawes
3435 Wilshire Blvd., Suite 1100
Los Angeles, CA 92649 (213) 385-7087

George K. DeBrucky
Nilsson, Robbins, Dalgarn, Berliner, Carson & Wurst
201 N. Figueroa St., 5th Fl.
Los Angeles, CA 90012 (213) 977-1001

Wanda K. Denson-Low
Hughes Aircraft Co.
7200 Hughes Terrace
C1, A126
Los Angeles, CA 90045 (213) 568-6972

Robert W. Dickerson, Jr.
Lyon & Lyon
611 W. Sixth St., 34th Fl.
Los Angeles, CA 90017 (213) 489-1600

Arthur V. Doble
12301 Wilshire Blvd., Suite 512
Los Angeles, CA 90025 (213) 826-5505

Bernard Philip Drachlis
5012 Eagle Rock Blvd.
Los Angeles, CA 90041 (213) 255-1918

Bradford J. Duft
Lyon & Lyon
611 West Sixth St., 34th Fl.
Los Angeles, CA 90017 (213) 489-1600

Walter W. Duft
Lyon & Lyon
611 W. Sixth St., 34th Fl.
Los Angeles, CA 90017 (213) 489-1600

Vijayalakshmi D. Duraiswamy
Hughes Aircraft Co. Pats. & Licensing
7200 Hughes Terrace
Bldg. C1, M. Station A126
Los Angeles, CA 90045 (213) 568-6076

Robert D. Eastham
Lyon & Lyon
611 W. Sixth St., 34th Fl.
Los Angeles, CA 90017 (213) 489-1600

Michael S. Elkind
Nilsson, Robbins, Dalgarn, Berliner, Carson & Wurst
210 N. Figueroa St., 5th Fl.
Los Angeles, CA 90012 (213) 977-1001

Natan Epstein
Beehler, Pavitt, Siegemund, Jagger, Martella & Dawes
3435 Wilshire Blvd., Suite 1100
Los Angeles, CA 90010 (213) 385-7087

Saul Epstein
1880 Century Park, E., Suite 500
Los Angeles, CA 90067 (213) 553-2223

John F. Feldsted#
Rogers & Wells
201 N. Figueroa St., 16th Fl.
Los Angeles, CA 90012 (213) 580-1236

A. M. Fernandez
Freilich, Hornbaker, Rosen & Fernandez
10960 Wilshire Blvd., Suite 1434
Los Angeles, CA 90024 (213) 477-0578

Andra M. Finkel
10920 Wilshire Blvd., Suite 1000
Los Angeles, CA 90024 (213) 208-4900

Don Berry Finkelstein
700 S. Flower St., Suite 1218
Los Angeles, CA 90017 (213) 622-9502

Fred Flam
Flam & Flam
2049 Century Park, E., Suite 475
Los Angeles, CA 90067 (213) 277-7771

Arthur Freilich
Freilich, Hornbaker, Rosen & Fernandez, P.C.
10960 Wilshire Blvd., Suite 1434
Los Angeles, CA 90024 (213) 477-0578

Frank Frisenda, Jr.
Frisenda, Morris & Nicholson
11755 Wilshire Blvd., 10th Fl.
Los Angeles, CA 90025 (213) 478-4540

Bernard R. Gans
Poms, Smith, Lande, & Rose
2121 Ave. of the Stars, Suite 1400
Los Angeles, CA 90067 (213) 277-8141

Paul Lawrence Gardner
Spensley, Horn, Jubas & Lubitz
1880 Century Park, E., 5th Fl.
Los Angeles, CA 90067 (213) 553-5050

James W. Geriak
Lyon & Lyon
611 W. Sixth St., 34th Fl.
Los Angeles, CA 90017 (213) 489-1600

Rabindra N. Ghose
8167 Mulholland Terrace
Los Angeles, CA 90046 (818) 880-4533

Larry F. Gitlin
Smyth, Pavitt, Siegemund & Martella
4262 Wilshire Blve., Suite 320
Los Angeles, CA 90010 (213) 938-6251

William West Glenny
2121 Ave. of the Stars, Suite 1400
Los Angeles, CA 90067 (213) 557-0384

Joseph H. Golant
Ashen, Golant, Martin & Selden
10920 Wilshire Blvd., Suite 1000
Los Angeles, CA 90024 (213) 208-1100

Leonard Golove
Warner Bros. music
9200 Sunset Blvd., Suite 222
Los Angeles, CA 90069 (213) 273-3323

Lawrence W. Granatelli
Spensley, Horn, Jubas & Lubitz
1880 Century Park, E., 5th Fl.
Los Angeles, CA 90067 (213) 553-5050

John C. Grant#
Grant Investment Co.
3932 Wilshire Blvd., Suite 203
Los Angeles, CA 90010 (213) 385-0992

Robert A. Green
Nilsson, Robbins, Dalgarn, Berliner, Carson & Wurst
201 N. Figueroa St., 5th Fl.
Los Angeles, CA 90012 (213) 977-1001

William P. Green
201 N. Figueroa St., 5th Fl.
Los Angeles, CA 90012 (213) 977-1001

Stephen D. Gross
Blakely, Sokoloff, Taylor & Zafman
12400 Wilshire Blvd., 7th Fl.
Los Angeles, CA 90025 (213) 207-3800

David S. Guttman
Spensley, Horn, Jubas & Lubitz
1880 Century Park, E., 5th Fl.
Los Angeles, CA 90067 (213) 553-5050

Michael M. Hachigian
4250 Wilshire Blvd., 2nd Fl.
Los Angeles, CA 90010 (213) 933-5743

Joseph M. Hageman
Herzig & Yanny
1900 Ave. of the Stars, Suite 1520
Los Angeles, CA 90067 (213) 551-2966

John J. Hall
1631 Beverly Blvd.
Los Angeles, CA 90026 (213) 250-1145

John D. Harriman, II
Blakely, Sokoloff, Taylor & Zafman
12400 Wilshire Blvd., 7th Fl.
Los Angeles, CA 90025 (213) 207-3800

Michael D. Harris
Poms, Smith, Lande & Rose
2121 Ave. of the Stars, Suite 1400
Los Angeles, CA 90067 (213) 277-8141

Charles S. Haughey
Hughes Aircraft Co.
Pats. & Licensing
Bldg. C1, Mail Station A126
P.O. Box 45066
Los Angeles, CA 90045 (213) 568-7025

Robert A. Hays
Hughes Aircraft Co.
Pats. & Licensing
Bldg. C1, Mail Station A126
P.O. Box 45066
Los Angeles, CA 90045 (213) 568-6090

Gary A. Hecker
Blakely, Sokolof, Taylor & Zafman
12400 Wilshire Blvd., 7th Fl.
Los Angeles, CA 90025 (213) 207-3800

Paul J. Hedlund
6535 Wilshire Blvd., Suite 800
Los Angeles, CA 90048 (213) 658-6411

Steven D. Hemminger#
Lyon & Lyon
611 W. Sixth St., 34th Fl.
Los Angeles, CA 90017 (213) 489-1600

James A. Henricks
Pretty, Schroeder, Brueggemann & Clark
444 S. Flower St., Suite 2000
Los Angeles, CA 90071 (213) 489-4442

Henry L. Herold
1412 Butler Ave., Apt. 24
Los Angeles, CA 90025 (213) 390-8673

Kenjiro Hidaka#
2040 Pelham Ave.
Los Angeles, CA 90025 (213) 474-2668

Howard L. Hoffenberg
1409 Midvale Ave., Apt. 320
Los Angeles, CA 90024 (213) 478-3328

Jon E. Hokanson
Lyon & Lyon
611 W. Sixth St., 34th Fl.
Los Angeles, CA 90017 (213) 489-1600

Martin R. Horn
Spensley, Horn, Jubas & Lubitz
1880 Century Park, E., 5th Fl.
Los Angeles, CA 90067 (213) 553-5050

Robert David Hornbaker
Freilich, Hornbaker, Rosen & Fernandez
10960 Wilshire Blvd., Suite 1434
Los Angeles, CA 90024 (213) 477-4039

Harlan P. Huebner
900 Wilshire Blvd., Suite 100
Los Angeles, CA 90017 (213) 626-7766

Eric S. Hyman
Blakely, Sokoloff, Taylor & Zafman
12400 Wilshire Blvd., 7th Fl.
Los Angeles, CA 90025 (213) 207-3800

Keiichiro Imai
3264 Granville Ave.
Los Angeles, CA 90066 (213) 398-8444

Bryan T. Inoue#
Nilsson, Robbins, Dalgarn, Berliner, Carson & Wurst
707 Wilshire Blvd., 47th Fl.
Los Angeles, CA 90017 (213) 620-0600

Richard Thomas Ito
Fulwider, Patton, Rieber, Lee & Utecht
3435 Wilshire Blvd., Suite 2400
Los Angeles, CA 90010 (213) 380-6800

Harry Irwin Jacobs
TRW Electronics
10880 Wilshire Blvd., Suite 510
Los Angeles, CA 90024 (213) 475-6777

Bruce A. Jagger
Beehler, Pavitt, Siegemund, Jagger, Martella & Dawes
3435 Wilshire Blvd., Suite 1100
Los Angeles, CA 90010 (213) 385-7087

Allan W. Jansen
Lyon & Lyon
611 W. Sixth St., Suite 3400
Los Angeles, CA 90017 (213) 489-1600

Edward F. Jaros
Poms, Smith, Lande & Rose
2121 Avenue Of The Stars, Suite 1400
Los Angeles, CA 90067 (213) 277-8141

Matthew F. Jodziewicz
Arant, Kleinberg & Lerner
2049 Century Park E., Suite 1080
Los Angeles, CA 90067 (213) 557-1511

Walter E. Johansen, III
11661 San Vicente Blvd.
Los Angeles, CA 90049 (213) 826-4902

Philip D. Junkins#
11500 Olympic Blvd., Suite 425
Los Angeles, CA 90064 (213) 479-1550

Anthony W. Karambelas
Hughes Aircraft Co.
Pats. & Licensing
Bldg. C1, Mail Station A126
P.O. Box 45066
Los Angeles, CA 90045 (213) 568-7233

Stephen C. Kaufman
Hughes Aircraft Company
Pats. & Licensing
Bldg. C1, Mail Station A126
P.O. Box 45066
Los Angeles, CA 90045 (213) 568-6093

Richard W. Keefe
Ladas & Parry
3600 Wilshire Blvd.
Los Angeles, CA 90010 (213) 385-4281

Elwood S. Kendrick
612 S. Flower St., Suite 1200
Los Angeles, CA 90017 (213) 626-7792

Warren L. Kern
Harris, Kern, Wallen & Tinsley
Quinby Bldg., Top Floor
650 S. Grand Ave.
Los Angeles, CA 90017 (216) 626-5251

William W. Kidd
Blakely, Sokoloff, Taylor & Zafman
12400 Wilshire Blvd., 7th Fl.
Los Angeles, CA 90025 (213) 207-3800

Wellesley R. Kime
8745 Appian Way
Los Angeles, CA 90046 (213) 656-4218

Stephen Lowell King
Reagin & King
5959 W. Century Blvd., Suite 1118
Los Angeles, CA 90045 (213) 645-2824

Marvin H. Kleinberg
Arant, Kleinberg & Lerner
2049 Century Park E., Suite 1080
Los Angeles, CA 90067 (213) 557-1511

H. Henry Koda#
Koda & Androlia
1880 Century Park E. Suite 519
Los Angeles, CA 90067 (213) 277-1391

Michael Albert Kondzella
Nilsson, Robbins, Dalgarn, Berliner, Carson & Wurst
201 N. Figueroa St., 5th Fl.
Los Angeles, CA 90012 (213) 977-1001

William K. Konrad
Spensley, Horn, Jubas & Lubitz
1880 Century Pk. E.
Los Angeles, CA 90067 (213) 553-5050

Gerald A. Koris
International Rectifier Corp.
9220 Sunset Blvd.
Los Angeles, CA 90069 (213) 278-3100

Gilbert Gerald Kovelman
Fulwider, Patton, Rieber, Lee & Utecht
2400 Equitable Plaza
3435 Wilshire Blvd.
Los Angeles, CA 90010 (213) 380-6800

Bernard Kriegel
953 N. Highland Ave.
Los Angeles, CA 90038 (213) 461-8141

Robert D. Kummel
Carnation Co.
Carnation Bldg.
5045 Wilshire Blvd.
Los Angeles, CA 90036 (213) 932-6223

Carl Kustin, Jr.
Fulwider, Patton, Rieber, Lee & Utecht
3435 Wilshire Blvd., Suite 2400
Los Angeles, CA 90010 (213) 380-6800

John F. Land
Spensley, Horn, Jubas & Lubitz
1880 Century Park E., Suite 500
Los Angeles, CA 90067 (213) 553-5050

Gary E. Lande
Poms, Smith, Lande & Rose
2121 Avenue Of The Stars, Suite 1400
Los Angeles, CA 90067 (213) 277-8141

Victor G. Laslo
Hughes Aircraft Co.
Pats. & Licensing
Bldg. C1, Mail Station A126
P.O. Box 45066
Los Angeles, CA 90045 (213) 568-7233

Don C. Lawrence
Poms, Smith, Lande & Rose
2121 Ave. of the Stars, Suite 1400
Los Angeles, CA 90067 (213) 277-8141

Marshall A. Lerner
Arant, Kleinberg & Lerner
2049 Century Pk. E., Suite 1080
Los Angeles, CA 90067 (213) 557-1511

Irving J. Levin
P.O. Box 49166
Los Angeles, CA 90049 (213) 550-8060

Donald G. Lewis#
10960 Wilshire Blvd., Suite 1434
Los Angeles, CA 90024 (213) 477-4030

Lawrence V. Link, Jr.
Hughes Aircraft Co.
Pats. & Licensing
Bldg. C1, Mail Station A126
P.O. Box 45066
Los Angeles, CA 90045 (213) 568-7233

Peter I. Lippman
Romney, Golant, Martin, Disner & Ashen
10920 Wilshire Blvd., Suite 1000
Los Angeles, CA 90024 (213) 208-1100

Don F. Livornese#
Spensley, Horn, Jubas & Lubitz
1880 Century Park, E., Suite 500
Los Angeles, CA 90067 (213) 553-5050

Stuart Lubitz
Spensley, Horn, Jubas & Lubitz
1880 Century Pk. E., Suite 500
Los Angeles, CA 90067 (213) 553-5050

Edward Joseph Lynch
Fulwider, Patton, Rieber, Lee & Utecht
2400 Equitable Plaza
3435 Wilshire Blvd.
Los Angeles, CA 90010 (213) 380-6800

Richard E. Lyon, Jr.
Lyon & Lyon
611 W. Sixth St., 34th Fl.
Los Angeles, CA 90017 (213) 489-1600

Robert Douglas Lyon
Lyon & Lyon
611 W. Sixth St., 34th Fl.
Los Angeles, CA 90017 (213) 489-1600

Robert Edward Lyon
Lyon & Lyon
611 W. Sixth St., 34th Fl.
Los Angeles, CA 90017 (213) 489-1600

Michael J. MacDermott
Harris, Kern, Wallen & Tinsley
650 S. Grand Ave., 14th Fl.
Los Angeles, CA 90017 (213) 626-5251

Kurt A. MacLean
Poms, Smith, Lande & Rose
2121 Avenue Of The Stars, Suite 1400
Los Angeles, CA 90067 (213) 277-8141

Thomas H. Majcher
Nilsson, Robbins, Dalgarn, Berliner, Carson & Wurst
707 Wilshire Blvd., Suite 4750
Los Angeles, CA 90017 (213) 620-0600

David N. Makous
Fulwider, Patton, Rieber, Lee & Utecht
2400 Equitable Plaza
3435 Wilshire Blvd.
Los Angeles, CA 90010 (213) 380-6800

Sara L. Mandel
Lyon & Lyon
611 W. Sixth St., 34th Fl.
Los Angeles, CA 90017 (213) 489-1600

Roy J. Mankovitz
Reagin & King
5959 W. Century Blvd., Suite 1118
Los Angeles, CA 90045 (213) 645-2824

Mario A. Martella
Beehler, Pavitt, Siegemund, Jagger & Martella
3435 Wilshire Blvd., Suite 1100
Los Angeles, CA 90010 (213) 385-7087

Kenneth Watson Mateer
Teledyne, Inc.
1901 Ave. Of The Stars, Suite 1800
Los Angeles, CA 90067 (213) 277-3311

John T. Matlago
3435 Wilshire Blvd., Suite 1100
Los Angeles, CA 90010 (213) 380-4606

William H. Maxwell, Jr.#
12301 Wilshire Blvd.
Los Angeles, CA 90025 (213) 820-3911

J. Donald McCarthy
Lyon & Lyon
611 W. Sixth St., 34th Fl.
Los Angeles, CA 90017 (213) 489-1600

John D. McConaghy
Lyon & Lyon
611 W. Sixth St., 34th Fl.
Los Angeles, CA 90017 (213) 489-1600

James W. McFarland
Garrett Corp.
9851 Sepulveda Blvd.
P.O. Box 92248
Los Angeles, CA 90009 (213) 417-6551

Daniel Joseph Meaney, Jr.
Morganstern & Meaney
1900 Ave. Of The Stars, 18th Fl.
Los Angeles, CA 90067 (213) 201-4300

Paul H. Meier
Lyon & Lyon
611 W. Sixth St., 34th Fl.
Los Angeles, CA 90017 (213) 489-1600

Mark J. Meltzer
Hughes Aircraft Co.
Pats. & Licensing
Bldg. C1, Mail Station A126
P.O. Box 45066
Los Angeles, CA 90045 (213) 568-7233

Thomas Joe Mielke
Spensley, Horn, Jubas & Lubitz
1880 Century Park, E., 5th Fl.
Los Angeles, CA 90067 (213) 553-5050

Albert J. Miller
Garrett Corp.
9851 Sepulveda Blvd.
Los Angeles, CA 90009 (213) 417-6550

Scott R. Miller
Lyon & Lyon
611 W. Sixth St., 34th Fl.
Los Angeles, CA 90017 (213) 489-1600

Wendell Smith Miller#
Intertechnical Associates
1341 Comstock Ave.
Los Angeles, CA 90024 (213) 274-1203

Steven M. Mitchell
Hughes Aircraft Co.
Pats. & Licensing
Bldg. C1, Mail Station A126
P.O. Box 45066
Los Angeles, CA 90045 (213) 568-6091

Louis Mok
Bogucki, Scherlacher, Mok & Roth
3345 Wishire Blvd., Suite 704
Los Angeles, CA 90010 (213) 386-7701

Richard Morganstern
Morganstern & Meaney
1900 Ave. Of The Stars, 18th Fl.
Los Angeles, CA 90067 (213) 201-4300

Frederick E. Mueller
Quinby Bldg., 14th Fl.
650 S. Grand Ave.
Los Angeles, CA 90017 (213) 626-6061

Joseph E. Mueth
Wills, Green & Mueth
700 S. Flower St., Suite 1120
Los Angeles, CA 90017 (213) 688-7407

John H. Muetterties
Garrett Corp.
9851 Sepulveda Blvd.
P.O. Box 92248
Los Angeles, CA 90009 (213) 417-6552

Robert W. Mulcahy
Twentieth Century-Fox Film Corp.
10201 W. Pico Blvd.
Los Angeles, CA 90035 (213) 203-2945

Frederick F. Mumm#
Walter, Finestone, Richter & Kane
10920 Wilshire Blvd., Suite 1400
Los Angeles, CA 90024 (213) 824-0800

David B. Murphy
Lyon & Lyon
611 W. Sixth St., 34th Fl.
Los Angeles, CA 90017 (213) 489-1600

John S. Nagy
Fulwider, Patton, Rieber, Lee & Utecht
2400 Equitable Plaza
3435 Wilshire Blvd.
Los Angeles, CA 90010 (213) 380-3600

Roger Nash#
5123 Cavanagh Rd.
Los Angeles, CA 90032 (213) 227-6418

George J. Netter
Kendrick, Netter & Bennett
612 S. Flower St.
Los Angeles, CA 90017 (213) 626-7792

Andrew J. Nilles
Lyon & Lyon
611 W. Sixth St., 34th Fl.
Los Angeles, CA 90017 (213) 489-1600

Byard G. Nilsson
Nilsson, Robbins, Dalgarn, Berliner, Carson & Wurst
201 N. Figueroa St., 5th Fl.
Los Angeles, CA 90012 (213) 977-1001

Donald Edward Nist
Whittaker Corp.
10880 Wilshire Blvd.
Los Angeles, CA 90024 (213) 475-9411

Cyrus S. Nownejad
10100 Santa Monica Blvd., Suite 2500
Los Angeles, CA 90067 (213) 552-9005

Kenneth H. Ohriner#
Lyon & Lyon
611 W. Sixth St., 34th Fl.
Los Angeles, CA 90017 (213) 489-1600

David J. Oldenkamp
Poms, Smith, Lande & Rose
2121 Ave. Of The Stars, Suite 1400
Los Angeles, CA 90067 (213) 277-8141

Douglas E. Olson
Lyon & Lyon
611 W. Sixth St., 34th Fl.
Los Angeles, CA 90017 (213) 489-1600

Jeffrey M. Olson
Lyon & Lyon
611 W. Sixth St., 34th Fl.
Los Angeles, CA 90017 (213) 489-1600

Mildred Oncken
455 S. Berendo St., Apt. 301
Los Angeles, CA 90020 (213) 385-0975

David G. Parkhurst
Fulwider, Patton, Rieber, Lee & Utecht
3435 Wilshire Blvd., Suite 2400
Los Angeles, CA 90010 (213) 380-6800

William N. Patrick
Occidental Petroleum Corp.
10889 Wilshire Blvd.
Los Angeles, CA 90024 (213) 879-1700

Warren L. Patton
Fulwider, Patton, Rieber, Lee & Utecht
3435 Wilshire Blvd., Suite 2400
Los Angeles, CA 90010 (213) 380-6800

William H. Pavitt, Jr.
Beehler, Pavitt, Siegemund, Jagger & Martella
1100 Equitable Plaza
3435 Wilshire Blvd.
Los Angeles, CA 90010 (213) 385-7087

Bradley D. Pedersen
Gibson, Dunn & Crutcher
2029 Century Park E., Suite 4000
Los Angeles, CA 90067 (213) 557-8094

William Poms
Poms, Smith, Lande & Rose
2121 Ave. Of The Stars, Suite 1400
Los Angeles, CA 90067 (213) 277-8141

Laurence H. Pretty
Pretty, Schroeder, Bureggemann & Clark
444 S. Flower St., Suite 2000
Los Angeles, CA 90071 (213) 489-4442

John A. Rafter, Jr.
Lyon & Lyon
611 W. Sixth St., 34th Fl.
Los Angeles, CA 90017 (213) 489-1600

Robert Clifton Rasche
2437 Federal Ave.
Los Angeles, CA 90064

Ronald W. Reagin
Reagin & King
5959 W. Century, Suite 1118
Los Angeles, CA 90045 (213) 645-2824

Jerrold B. Reilly
Lyon & Lyon
611 W. Sixth St., 34th Fl.
Los Angeles, CA 90017 (213) 489-1600

Thomas C. Reynolds
Spensley, Horn, Jubas & Lubitz
1880 Century Park E., Fifth Fl.
Los Angeles, CA 90067 (213) 553-5050

Janine P. Rickman
Lyon & Lyon
611 W. Sixth St., 34th Fl.
Los Angeles, CA 90017 (213) 489-1600

William K. Rieber
Fulwider, Patton, Rieber, Lee & Utecht
2400 Equitable Plaza
3435 Wilshire Blvd.
Los Angeles, CA 90010 (213) 380-6800

David B. Ritchie
Lyon & Lyon
611 W. Sixth St., 34th Fl.
Los Angeles, CA 90017 (213) 489-1600

Billy A. Robbins
Nilsson, Robbins, Dalgarn, Berliner, Carson & Wurst
201 N. Figueroa St., 5th Fl.
Los Angeles, CA 90012 (213) 977-1001

William J. Robinson
Poms, Smith, Lande & Rose
2121 Ave. Of The Stars, Suite 1400
Los Angeles, CA 90067 (213) 277-8141

Alan C. Rose
Poms, Smith, Lande & Rose, P.C.
2121 Ave. Of The Stars, Suite 1400
Los Angeles, CA 90067 (213) 277-8141

Leon David Rosen
Freilich, Hornbaker, Rosen & Fernandez
10960 Wilshire Blvd., Suite 1434
Los Angeles, CA 90024 (213) 477-0578

Charles Rosenberg
Poms, Smith, Lande & Rose
2121 Ave. Of The Stars, Suite 1400
Los Angeles, CA 90067 (213) 277-8141

L. Kenneth Rosenthal#
University Of Southern California
3716 S. Hope St., Suite 200
Los Angeles, CA 90007 (213) 743-4926

Gregory L. Roth
Bogucki, Scherlacher, Mok & Roth
3345 Wilshire Blvd., Suite 704
Los Angeles, CA 90010 (213) 386-7701

W. Norman Roth
Roth & Goldman
Pacific Mutual Bldg.
523 W. Sixth St., Suite 840
Los Angeles, CA 90014 (213) 688-1143

Thomas I. Rozsa
Tilles & Webb
6320 Commodore Sloat Dr.
Los Angeles, CA 90048 (213) 931-1800

Julius L. Rubinstein
Ahmanson Center East, 7th Fl.
3701 Wilshire Blvd.
Los Angeles, CA 90010 (213) 380-2208

Thomas A. Runk
Hughes Aircraft Co.
Pats. & Licensing
Bldg. C1, Mail Station A126
P.O. Box 45066
Los Angeles, CA 90045 (213) 568-7233

Michael W. Sales
Hughes Aircraft Co.
Pats. & Licensing
Bldg. C1, Mail Station A126
P.O. Box 45066
Los Angeles, CA 90045 (213) 568-7028

Carol K. Samek
Pettit & Martin
355 S. Grand Ave., Suite 3300
Los Angeles, CA 90071 (213) 626-1717

John A. Sarjeant
Hughes Aircraft Co.
Pats. & Licensing
Bldg. C1, Mail Station A126
P.O. Box 45066
Los Angeles, CA 90045 (213) 568-7233

Joseph A. Sawyer, Jr.
Hughes Aircraft Co.
Pats. & Licensing
Bldg. C1, Mail Station A126
P.O. Box 45066
Los Angeles, CA 90045 (213) 568-6093

Robert Jay Schaap
Herzig, Schaap & Yanny
6820 La Tijera Blvd., Suite 107
Los Angeles, CA 90045 (213) 645-6460

James C. Scheller, Jr.
Blakely, Sokoloff, Taylor & Zafman
12400 Wilshire Blvd., 7th Fl.
Los Angeles, CA 90025 (213) 207-3800

John Paul Scherlacher
Bogucki, Scherlacher, Mok & Roth
3345 Wilshire Blvd., Suite 704
Los Angeles, CA 90010 (213) 386-7701

Michael C. Schiffer
Nilsson, Robbins, Dalgarn, Berliner, Carson & West
201 N. Figueroa St., 5th Fl.
Los Angeles, CA 90012 (213) 977-1001

Robert A. Schroeder
Pretty, Schroeder, Brueggemann & Clark
444 S. Flower St., Suite 2000
Los Angeles, CA 90071 (213) 489-4442

Charles H. Schwartz
Roston & Schwartz
5900 Wilshire Blvd., Suite 1430
Los Angeles, CA 90036 (213) 938-3657

Robert A. Seldon
Brunell & Seldon
10920 Wilshire Blvd., Suite 1000
Los Angeles, CA 90024 (213) 208-0080

Todd B. Serota
Poms, Smith, Lande & Rose
2121 Ave. Of The Stars, Suite 1400
Los Angeles, CA 90067 (213) 277-8141

James H. Shalek
Lyon & Lyon
611 W. Sixth St., 34th Fl.
Los Angeles, CA 90017 (213) 489-1600

Edmond F. Shanahan
725 S. Norton Ave.
Los Angeles, CA 90005 (213) 385-1331

James J. Short
Lyon & Lyon
611 W. Sixth St., 34th Fl.
Los Angeles, CA 90017 (213) 489-1600

John F. Sicotte
8186 Gould Ave.
Los Angeles, CA 90046 (213) 654-8576

Ira M. Siegel
Blakely, Sokoloff, Taylor & Zafman
12400 Wilshire Blvd., 7th Fl.
Los Angeles, CA 90025 (213) 207-3800

Ralf H. Siegemund
Beehler, Pavitt, Siegemund, Jagger & Martella
3435 Wilshire Blvd.
1100 Equitable Plaza
Los Angeles, CA 90010 (213) 385-7087

Howard A. Silber
Spensley, Horn, Jubas & Lubitz
1880 Century Park E., Suite 500
Los Angeles, CA 90067 (213) 553-5050

Marvin O. Sleven#
13257 Chalon Rd.
Los Angeles, CA 90049 (213) 476-1063

Thomas M. Small
Fulwider, Patton, Rieber, Lee & Utecht
2400 Equitable Plaza
3435 Wilshire Blvd.
Los Angeles, CA 90010 (213) 380-6800

Guy Porter Smith
Poms, Smith, Lande & Rose
2121 Ave. Of The Stars, Suite 1400
Los Angeles, CA 90067 (213) 277-8141

Roland N. Smoot
Lyon & Lyon
611 W. Sixth St., 34th Fl.
Los Angeles, CA 90017 (213) 489-1600

George Frederick Smyth
Beehler, Pavitt, Siegemund, Jagger & Martella
3435 Wilshire Blvd.
Los Angeles, CA 90010 (213) 938-6251

Stanley William Sokoloff
Blakley, Sokoloff, Taylor & Zafman
12400 Wilshire Blvd., 7th Fl.
Los Angeles, CA 90025 (213) 207-3800

Conrad R. Solum, Jr.
Lyon & Lyon
611 W. Sixth St., 34th Fl.
Los Angeles, CA 90017 (213) 489-1600

W. Robert Spensley
Spensley, Horn, Jubas & Lubitz
1880 Century Pk. E., Suite 500
Los Angeles, CA 90067 (213) 553-5050

John P. Spitals#
Spensley, Horn, Jubas & Lubitz
1880 Century Pk. E., Fifth Fl.
Los Angeles, CA 90067 (213) 553-5050

William C. Steffin
Lyon & Lyon
611 W. Sixth St., 34th Fl.
Los Angeles, CA 90017 (213) 489-1600

Lewis B. Sternfels
Hughes Aircraft Co.
Pats. & Licensing
Bldg. C1, Mail Station A126
P.O. Box 45066
Los Angeles, CA 90045 (213) 568-7233

Craig S. Summers
Pretty, Schroeder, Bruggemann & Clark
444 S. Flower St., Suite 2000
Los Angeles, CA 90071 (213) 489-4442

Joseph E. Szabo
Hughes Aircraft Co.
Pats. & Licensing
Bldg. C1, Mail Station A126
P.O. Box 45066
Los Angeles, CA 90045 (213) 568-7233

Edwin H. Taylor
Blakley, Sokoloff, Taylor & Zafman
12400 Wilshire Blvd., 7th Fl.
Los Angeles, CA 90025 (213) 207-3800

Ronald L. Taylor
Hughes Aircraft Co.
Pats. & Licensing
Bldg. C1, Mail Station A126
P.O. Box 45066
Los Angeles, CA 90045 (213) 568-7233

Robert Thompson
Hughes Aircraft Co.
Pats. & Licensing
Bldg. C1, Mail Station A126
P.O. Box 45066
Los Angeles, CA 90045 (213) 568-7027

William Edward Thomson, Jr.
Lyon & Lyon
611 W. Sixth St., 34th Fl.
Los Angeles, CA 90017 (213) 489-1600

Robert R. Thornton
515 S. Figueroa St., Suite 900
Los Angeles, CA 90071 (213) 689-1400

Walton Eugene Tinsley
Harris, Kern, Wallen & Tinsley
650 S. Grand Ave.
Los Angeles, CA 90017 (213) 626-5251

Michael L. Wachtell
Rosen, Wachtell & Gilbert
1888 Century Pk. E., Suite 2400
Los Angeles, CA 90067 (213) 553-2900

Louis P. Walsh#
Tmsi Arabia Ltd.
3250 Wilshire Blvd.
Los Angeles, CA 90010 (213) 381-1338

Paul A. Weilein
1010 S. Flower St., Suite 301
Los Angeles, CA 90015 (213) 748-2277

Les J. Weinstein
Blecher, Collins & Weinstein
611 W. Sixth St., 28th Fl.
Los Angeles, CA 90020 (213) 622-4222

David Weiss
2251 Colorado Blvd.
Los Angeles, CA 90041 (213) 254-5020

Robert Charles Weiss
Lyon & Lyon
611 W. Sixth St., 34th Fl.
Los Angeles, CA 90017 (213) 489-1600

R. Welton Whann
Whann & McManigal
315 W. Ninth St., Suite 920
Los Angeles, CA 90015 (213) 622-7163

John T. Wiedemann
Pretty, Schroeder, Brueggemann & Clark
444 S. Flower St., Suite 2000
Los Angeles, CA 90071 (213) 489-4442

Wayne E. Willenberg
Spensley, Horn, Jubas & Lubitz
1880 Century Pk. E., Suite 500
Los Angeles, CA 90067 (213) 553-5050

Charles E. Wills
725 S. Figueroa St., 34th Fl.
Los Angeles, CA 90017 (213) 689-5123

Roger R. Wise
Spensley, Horn, Jubas & Lubitz
1880 Century Pk. E., Suite 500
Los Angeles, CA 90067 (213) 827-9058

Gregory B. Wood
Nilsson, Robbins, Dalgarn, Berliner, Carson & Wurst
201 N. Figueroa St., 5th Fl.
Los Angeles, CA 90012

Harold E. Wurst
Nilsson, Robbins, Dalgarn, Berliner, Carson & Wurst
201 N. Figueroa St., 5th Fl.
Los Angeles, CA 90012 (213) 977-1001

Paul Richter Wylie, Jr.
Dart Inds. Inc.
8480 Beverly Blvd.
Los Angeles, CA 90048 (213) 651-2622

Joseph A. Yanny
Herzig & Yanny
1900 Ave. Of The Stars, Suite 1520
Los Angeles, CA 90067 (213) 551-2966

Norman Zafman
Blakley, Sokoloff, Taylor & Zafman
12400 Wilshire Blvd., 7th Fl.
Los Angeles, CA 90025 (213) 207-3800

Richard H. Zaitlen
Spensley, Horn, Jubas & Lubitz
1880 Century Pk. E., Suite 500
Los Angeles, CA 90067 (213) 553-5050

Harris Zeitzew
Meyers, Bianchi & McConnell
12301 Wilshire Blvd., Suite 206
Los Angeles, CA 90025 (213) 820-1500

Eugene Carl Ziehm#
Carnation Co.
Legal Dept.
5045 Wilshire Blvd.
Los Angeles, CA 90036 (213) 932-6000

William W. Burns
15720 Winchester Blvd.
Los Gatos, CA 95030 (408) 395-5111

Gerald L. Moore
17510 Farley Rd.
Los Gatos, CA 95030 (408) 395-6161

George Edgar Roush
16250 Jacaranda Way
Los Gatos, CA 95030 (408) 246-1405

Arthur R. Sorkin#
12340 Indiana Trail Rd.
Los Gatos, CA 95030 (408) 867-5830

Thomas A. Fournier
6805 Zumirez Dr.
Malibu, CA 90265 (213) 275-5366

James E. McTaggart#
21470 Rambla Vista
Malibu, CA 90265 (213) 456-8854

John D. Raiford
Hughes Research Labs.
M S R L 85
3011 Malibu Canyon Rd.
Malibu, CA 90265 (213) 456-6411

Gordon A. Shifrin#
7145 Fernhill Dr.
Malibu, CA 90265 (213) 457-2317

Rodolfo Aquirre
P.O. Box 746
Manhattan Beach, CA 90266 (213) 643-8500

Newell C. Rodewald#
P.O. Box 652
Manhattan Beach, CA 90266 (213) 372-0454

Joseph Shulsinger
Suite 117, Box J
Manhattan Beach, CA 90266 (213) 647-0544

Robert M. Unruh
2100 Blanche Rd.
Manhattan Beach, CA 90266 (213) 545-5226

Frank Wattles
P.O. Box 3514
Manhattan Beach, CA 90266 (213) 372-0454

Robert H. Fraser
3812 Via Dolce
Marina Del Rey, CA 91292 (213) 821-1242

Gerald J. Woloson
4151 Via Marina
Marina Del Rey, CA 90292 (213) 827-4173

William H. Benz
Ciotti & Murashige
545 Middlefield Rd., Suite 200
Menlo Park, CA 94025 (415) 327-7250

Robert P. Blackburn
Ciotti & Murashige
545 Middlefield Rd., Suite 200
Menlo Park, CA 94025 (415) 327-7250

Herbert G. Burkard
Raychem Corp.
300 Constitution Dr.
Menlo Park, CA 94025 (415) 361-3338

John Y. Chen#
SRI International
333 Ravenswood Ave.
Menlo Park, CA 94025 (415) 859-2446

Thomas Edward Ciotti
Ciotti & Murashige
545 Middlefield Rd., Suite 200
Menlo Park, CA 94025 (415) 327-7250

Peter J. Dehlinger
Ciotti & Murashige
545 Middlefield Rd., Suite 200
Menlo Park, CA 94025 (415) 327-7250

Troy G. Dillahunty
Raychem Corporation
300 Constitution Dr.
Menlo Park, CA 94025 (415) 361-5106

Janet E. Farrant#
Membrane Technology & Res. Inc.
1360 Willow Rd.
Menlo Park, CA 94025 (415) 328-2228

Urban Hart Faubion
SRI International
333 Ravenswood Ave.
Menlo Park, CA 94025 (415) 859-4550

Ronald C. Fish
Ciotti & Murashige
545 Middlefield Rd., Suite 200
Menlo Park, CA 94025 (415) 327-7250

Grant D. Green
Ciotti & Murashinge
545 Middlefield Rd., Suite 200
Menlo Park, CA 94025 (415) 327-7250

William Hinman Hooper
Nine Vasilakos Ct.
Menlo Park, CA 94025 (415) 321-9553

Dennis E. Kovach
Raychem Corp.
300 Constitution Dr.
Menlo Park, CA 94025 (415) 361-4153

Harvey Gunther Lowhurst
P.O. Box 7187
Menlo Park, CA 94025 (415) 854-2874

Albert Macovski#
2505 Alpine Rd.
Menlo Park, CA 94025 (415) 497-2708

Stephen L. Malaska#
385 Waverley St.
Menlo Park, CA 94025 (415) 327-9049

Gladys H. Monroy
Ciotti & Murashige
545 Middlefield Rd.
Menlo Park, CA 94025 (415) 327-7250

Kate H. Murashige
Ciotti & Murashige
545 Middlefield Rd.
Menlo Park, CA 94025 (415) 327-7250

Lisabeth Feix Murphy
Ciotti & Murashige
545 Middlefield Rd., Suite 200
Menlo Park, CA 94025 (415) 327-7250

William Thomas Nye
Commtech International Management Corp.
545 Middlefield Rd., Suite 180
Menlo Park, CA 94025 (415) 328-0190

Dianne E. Reed
Ciotti & Murashige
545 Middlefield Rd., Suite 200
Menlo Park, CA 94025 (415) 327-7250

Norman E. Reitz
P.O. Box 2630
Menlo Park, CA 94026

Edith A. Rice
Raychem Corp.
300 Constitution Dr.
Menlo Park, CA 94025 (415) 361-3331

Timothy H.P. Richardson#
Raychem Corp.
300 Constitution Dr.
Menlo Park, CA 94025 (415) 361-3069

Richard M. Ladden
1404 Acadia Ave.
Milpitas, CA 95035 (408) 262-5993

Thomas H. Williams
Greyhawk Systems, Inc.
1601 Centre Pointe Dr.
Milpitas, CA 95035 (408) 945-1776

Fred N. Schwend
456 Ivy Glen Dr.
Mira Loma, CA 91752 (714) 685-0327

Ronald B. Blanchard#
24481 Dardania
Mission Viejo, CA 92691 (714) 770-5161

Clifford B. Boehmer
25266 Pacifica
Mission Viejo, CA 92691 (714) 896-1914

Bradley Lionel Jacobs#
26816 La Sirra Dr.
Mission Viejo, CA 92691 (714) 582-1554

Frank R. Lafontaine
Shell Dev. Co.
Pats & Licensing
P.O. Box 4248
Modesto, CA 95352 (209) 545-8176

Darrell Gene Brekke*
NASA-Ames Research Center
Mail Stop 200-11
Moffett Field, CA 94035 (415) 694-5104

F. David La Riviere
Schroeder, Davis & Orliss Inc.
215 W. Franklin St., 4th Fl.
P.O. Box 3080
Monterey, CA 93942 (408) 649-1122

Marvin Jabin
Jabin & Jabin
701 S. Atlantic Blvd.
Monterey Park, CA 91754 (213) 570-1117

John R. Murtha
1253 Larch Ave.
Moraga, CA 94556 (415) 376-1498

Daniel C. Mc Kown
355 Fiarview Ave.
Morro Bay, CA 93442 (805) 541-5148

Wayne O. Hadland
721 June Hollow Rd.
P.O. Box 179
Moss Beach, CA 94038 (415) 728-7932

Keith E. Archer
836 Bay St.
Mountain View, CA 94041 (408) 756-9530

J. L. Bohan
P.O. Box 4720
Mountain View, CA 94040 (408) 272-2688

John H. Grate#
Catalytica, Inc.
430 Ferguson Dr., Bldg. 3
Mountain View, CA 94043 (415) 960-3000

Saundra S. Hand#
750 Stierlin Rd., No. 99
Mountain View, CA 94043

Paul L. Hickman
P.O. Box 391837
Mountainview, CA 94039 (415) 961-1950

John F. Lawler
P.O. Box 638
Mountain View, CA 94042 (415) 948-1214

Kenneth J. Nussbacher
Daisy Systems Corp.
P.O. Box 7006
Mountain View, CA 94039 (415) 960-0123

Karen S. Perkins
1166 Nilda Ave.
Mountain View, CA 94040 (415) 961-1166

Michael G. Petit#
California Biotechnology Inc.
2450 Bayshore Frontage Rd.
Mountain View, CA 94043 (415) 966-1550

Reynold J. Wong#
330 Sierra Vista Way, Apt. 26
Mountain View, CA 94043 (415) 968-7606

Roy W. Latham#
6812 Jarvis Ave.
Newark, CA 94560 (415) 791-2518

Lowell Anderson
Knobbe, Martens, Olson & Bear
620 Newport Center Dr., 16th Fl.
Newport Beach, CA 92660 (714) 760-0404

James Barth Bear
Knobbe, Martens, Olson, & Bear
620 Newport Center Dr., 16th Fl.
Newport Beach, CA 92660 (714) 760-0404

George Frazier Bethel
Beehler, Pavitt, Siegemund, Jagger, Martella & Bethel
610 Newport Center Dr., Suite 1420
Newport Beach, CA 92660 (714) 640-0900

Patience K. Bethel#
Beehler, Pavitt, Siegemund, Jagger, Martella & Bethel
610 Newport Center Dr., Suite 1420
Newport Beach, CA 92660 (714) 640-0900

William B. Bunker
Knobbe, Martens, Olson & Bear
620 Newport Center Dr., 16th Fl.
Newport Beach, CA 92660 (714) 760-0404

William P. Christie
Christie, Parker & Hale
P.O. Box 1730
One Newport Pl., Suite 1000
Newport Beach, CA 92658 (714) 476-0757

Marcia A. Devon
Knobbe, Martens, Olson & Bear
620 Newport Center Dr., 16th Fl.
Newport Beach, CA 92660 (714) 760-0404

William H. Drummond
4041 Mac Arthur Blvd., Suite 170
Newport Beach, CA 92660 (714) 851-1981

Dennis H. Epperson
Knobbe, Martens, Olson & Bear
620 Newport Center Dr., 16th Fl.
Newport Beach, CA 92660 (714) 760-0404

Morland C. Fischer
Fischer, Tachner & Strauss
1301 Dove St., Suite 270
Newport Beach, CA 92660 (714) 752-8525

Kenneth W. Float
Hughes Aircraft Co.
500 Superior Ave.
Newport Beach, CA 92658 (714) 759-7335

Hans O. Hagna#
127 Agate
Balboa Island
Newport Beach, CA 92662 (714) 675-7022

James E. Hawes
Beehler & Pavitt
610 Newport Center Dr., Suite 1420
Newport Beach, CA 92660 (714) 644-7740

Martin J. Hirsch
Knobbe, Martens, Olson & Bear
620 Newport Center Dr., 16th Fl.
Newport Beach, CA 92660 (714) 760-0404

Scott Hunter
1300 Bristol St., N., Suite 180
Newport Beach, CA 92660 (714) 833-9922

Ned A. Israelsen
Knobbe, Martens, Olson & Bear
620 Newport Center Dr., 16th Fl.
Newport Beach, CA 92660 (714) 760-0404

Louis J. Knobbe
Knobbe, Martens, Olson, & Bear
620 Newport Center Dr., 16th Fl.
Newport Beach, CA 92660 (714) 760-0404

John C. Lambertsen
Knobbe, Martens, Olson & Bear
620 Newport Center Dr., 16th Fl.
Newport Beach, CA 92660 (714) 760-0404

William Gregory Lane
Christie, Parker & Hale
P.O. Box 1730
Newport Beach, CA 92658 (714) 476-0757

James Francis Lesniak
Knobbe, Martens, Olson & Bear
620 Newport Center Dr., 16th Fl.
Newport Beach, CA 92660 (714) 760-0404

Thomas P. Mahoney
Mahoney & Schick
4000 Mac Arthur Blvd., Suite 6200
Newport Beach, CA 92660 (714) 851-8081

Don W. Martens
Knobbe, Martens, Olson & Bear
620 Newport Center Dr., 16th Fl.
Newport Beach, CA 92660 (714) 760-0404

Robert B. C. Newcomb#
CLA-VAL Co.
Box 1325
Newport Beach, CA 92663 (714) 548-2201

William H. Nieman
Knobbe, Martens, Olson & Bear
620 Newport Center Dr., 16th Fl.
Newport Beach, CA 92660 (714) 760-0404

Harvey Charles Nienow
1300 Dove St., Suite 200
Newport Beach, CA 92660 (714) 851-8585

Darrell L. Olson
Knobbe, Martens, Olson & Bear
620 Newport Center Dr., 16th Fl.
Newport Beach, CA 92660 (714) 760-0404

Gordon H. Olson
Knobbe, Martens, Olson & Bear
620 Newport Center Dr., 16th Fl.
Newport Beach, CA 92660 (714) 760-0404

Joseph W. Price, Jr.
Price, Gess & Ubell
4740 Von Karman, Suite 100
Newport Beach, CA 92660 (717) 955-1170

Manuel Quioque
Roberts & Quiogue
660 Newport Center Dr., Suite 1400
Newport Beach, CA 92660 (714) 640-6200

Joseph R. Re#
Knobbe, Martens, Olson & Bear
620 Newport Center Dr.
Newport Beach, CA 92660 (714) 760-0404

Larry K. Roberts
Robert & Quiogue
660 Newport Center Dr., Suite 1400
Newport Beach, CA 92660 (714) 640-6200

William C. Rooklidge
Knobbe, Martens, Olson & Bear
620 Newport Center Dr., 16th Fl.
Newport Beach, CA 92660 (716) 760-0404

Arthur S. Rose
Knobbe, Martens, Olson & Bear
620 Newport Center Dr., 16th Fl.
Newport Beach, CA 92660 (714) 760-0404

Edward A. Schlatter
Knobbe, Martens, Olson & Bear
620 Newport Center Dr., 16th Fl.
Newport Beach, CA 92660 (714) 760-0404

Jerry T. Sewell
Knobbe, Martens, Olson & Bear
620 Newport Center Dr., 16th Fl.
Newport Beach, CA 92660 (714) 760-0404

John B. Sganga, Jr.
Knobbe, Martens, Olson & Bear
620 Newport Center Dr., 16th Fl.
Newport Beach, CA 92660 (714) 760-0404

Michael A. Shimokaji
Price, Gess & Ubell, P.C.
4740 Von Karman, Suite 100
Newport Beach, CA 92660 (714) 955-1170

Andrew H. Simpson#
Knobbe, Martens, Olson & Bear
620 Newport Center Dr., 16th Fl.
Newport Beach, CA 92660 (714) 760-0404

Leonard Tachner
3990 Westerly Pl., Suite 295
Newport Beach, CA 92660 (714) 752-8525

Franklin D. Ubell#
Price, Gess & Ubell
4740 Von Karman
Newport Beach, CA 92660 (714) 955-1170

Robert Michael Vargo
Smith International, Inc.
P.O. Box 1860
Newport Beach, CA 92660 (714) 476-6201

William P. Waters
5140 Campus Dr., Suite 100
Newport Beach, CA 92660

Frank L. Zugelter
10221 Riverside Dr., Suite 207
North Hollywood, CA 91602 (213) 769-3411

William H. Fleeson#
11019 Nestle Ave.
Northridge, CA 91324 (213) 363-1465

David A. Kemper#
10913 Reseda Blvd.
Northridge, CA 91326 (213) 885-1101

Joseph Kriensky
10214 Topeka Dr.
Northridge, CA 91324 (213) 368-1326

Alexander Linger
10420-C Zelzah Ave.
Northridge, CA 91326 (213) 360-0035

Francis R. Reilly
19000 Merridy St.
Northridge, CA 91324 (213) 886-3065

Perry E. Turner
P.O. Drawer E
Northridge, CA 91328 (818) 360-6485

Bruce E. Francone*
U.S. Dept. Of Air Force
HQBMO/JA
Norton, A.F.B., CA 92409 (714) 382-6433

Donald J. De Geller
2725 Topaz Dr.
Novato, CA 94947 (415) 892-2147

Jack M. Whitney
2 Irene Dr.
Novato, CA 94947 (415) 897-6388

Donald L. Beeson
Beilock, Collins & Beeson
One Kaiser Plaza, Suite 1350
Oakland, CA 94612 (415) 832-8700

Theodore J. Bielen, Jr.
Bielen & Peterson
100 Webster St., Suite 101
Oakland, CA 94607 (415) 893-1515

H. Michael Brucker
166 Santa Clara Ave.
Oakland, CA 94601 (415) 658-2500

Robert Boyd Chickering
Warren, Chickering & Grunewald
166 Santa Clara Ave.
Oakland, CA 94610 (415) 658-2500

Howard S. Cohen#
1330 Broadway, Suite 1150
Oakland, CA 94612 (415) 465-0828

James Robert Cypher
405 - 14th St., Suite 1607
Oakland, CA 94612 (415) 832-4111

Harold M. Dixon*
U.S. Dept. Of Energy
Off. Of pat. Coun.
1333 Broadway
Oakland, CA 94612 (415) 273-6428

Glen R. Grunewald
Grunewald & Lampe
166 Santa Clara Ave.
Oakland, CA 94610 (415) 658-2500

James M. Hanley*
U.S. Dept. Of Energy
1333 Broadway
Oakland, CA 94612 (415) 273-6240

Frank M. Hansen
Kaiser Aluminum & Chem. Corp.
Kaiser Center, Rm. 533
300 Lakeside Dr.
Oakland, CA 94643 (415) 271-3638

Henry G. Hardy
6500 Chabot Rd.
Oakland, CA 94618 (415) 658-2244

Joel J. Hayashida
Clorox Co.
P.O. Box 24305
Oakland, CA 94623 (415) 271-7847

William H.F. Howard
921 Rose Ave.
Oakland, CA 94610 (415) 654-8636

Henry H. Johnson#
494 58th St.
Oakland, CA 94609 (415) 655-4464

Paul Maria Klein, Jr.
719 54th St.
Oakland, CA 94609

Thomas Raymond Lampe
Grunewald & Lampe
166 Santa Clara Ave.
Oakland, CA 94610 (415) 658-2500

Jane R. Mc Laughlin#
Cetus Corp.
511 Florence Ave.
Oakland, CA 94618 (415) 658-6250

Malcolm Caven Mc Quarrie
Kaiser Aluminum & Chem. Corp.
300 Lakeside Dr., Rm. 842
Oakland, CA 94643 (415) 271-3380

Ernest H. McCoy
Bruce & Mc Coy
One Kaiser Plaza, Suite 2385
Oakland, CA 94612 (415) 836-2400

Clinton H. Neagley
Advanced Genetic Sciences, Inc.
6701 San Pablo Ave.
Oakland, CA 94608 (415) 547-2395

Richard E. Peterson
Bielen & Peterson
100 Webster St., Suite 101
Oakland, CA 94607 (415) 893-1515

John Stephen Rhoades
Kaiser Aluminum & Chemical Corp.
300 Lakeside Dr.
Oakland, CA 94643 (415) 271-3478

Kathleen A. Skinner
One Kaiser Plaza, Suite 2385
Oakland, CA 94612 (415) 636-2400

Robert R. Tipton
2101 Webster St. Tower, Suite 1700
Oakland, CA 94612 (415) 465-9330

Stephen M. Westbrook
Clorox Co.
1221 Broadway
Oakland, CA 94612 (415) 271-7296

Harris Zimmerman
1330 Broadway, Suite 1150
Oakland, CA 94612 (415) 465-0828

Arthur C. Baker
3615-3 Vista Bella
Oceanside, CA 92056 (619) 757-5566

Claron N. White
3571 Papaya Way
Oceanside, CA 92054 (619) 439-4020

Lewis E. Massie#
Abaris Corp.
2218 13th St.
Olivenhain, CA 92024 (714) 436-0061

William G. Anderson
Sunkist Growers Inc.
Patent Law Dept.
760 E. Sunkist St.
Ontario, CA 91761 (714) 983-9811

Richard F. Carr
Gausewitz, Carr & Rothenberg
One City Blvd. W., Suite 830
Orange, CA 92668 (714) 634-4003

William L. Chapin
2410 W. Palm Ave.
Orange, CA 92668 (714) 978-6130

Elgin C. Edwards
Gausewitz, Carr, Rothberg & Edwards
Bank Of Amer. Tower
One City Blvd. W., Suite 830
Orange, CA 92668

Richard L. Gausewitz
Gausewitz, Carr & Rothberg
One City Blvd., W., Suite 830
Orange, CA 92668 (714) 634-4003

Willie Krawitz#
3001 Chapel Hill Rd.
Orange, CA 92667 (714) 974-1190

Garald J. Orman
51 Town & Country Business Plaza
1111 Town & Country Rd.
Orange, CA 92668 (714) 972-8855

Allan Rothenberg
Gausewitz, Carr, Rothenberg & Edwards
1 City Blvd. W., Suite 830
Orange, CA 92668 (714) 547-6735

G. Donald Weber, Jr.
505 City Parkway W., Suite 1000
Orange, CA 92668 (714) 634-4540

Steven W. Wilcox#
9520 Beacon Ave.
Orangevale, CA 95662 (916) 988-2179

Merwyn G. Brosler#
1 Meadow Ct.
Orinda, CA 94563 (415) 254-5593

David Lowell Hagmann
464 Camino Sobrante
Orinda, CA 94563 (415) 254-3146

Robert N. Richards
1130 Lookout Dr.
Oxnard, CA 93033 (805) 985-0985

Gary L. Jordan
P.O. Box 51,790
Pacific Grove, CA 93950 (408) 647-4148

Frank A. Campbell
P.O. Box 25
Pacific Palisades, CA 90272 (213) 333-5552

Thomas Long Flattery
439 Via De La Paz
Pacific Palisades, CA 90272 (213) 454-3768

Roy H. Davies
679 Parkview Court
Pacifica, CA 94044 (415) 355-1344

I. Louis Wolk
1540 Sierra Way
Palm Springs, CA 92262 (619) 325-3659

Serge Abend
Xerox Corp.
Pat. Dept.
3333 Coyote Hill Rd.
Palo Alto, CA 94304 (415) 494-4262

Kenneth R. Allen
Townsend & Townsend
Five Palo Alto Square, Suite 900
Palo Alto, CA 94306 (415) 493-2590

Robert A. Barr
3787 Redwood Circle
Palo Alto, CA 94306

Carole F. Barrett
Syntex
3401 Hillview Ave.
P.O. Box 10850
Palo Alto, CA 94304 (415) 852-1309

Patrick J. Barrett
Hewlett-Packard Co.
Legal Dept., M/s 20B0
3000 Hanover St.
Palo Alto, CA 94304 (415) 857-3489

James T. Beran
Xerox Corp.
3333 Coyote Hill Rd.
Palo Alto, CA 94304 (415) 494-4253

Edward H. Berkowitz
Varian Associates
611 Hansen Way
Palo Alto, CA 94303 (415) 424-5403

Gerald Alan Blaufarb
Syntex Inc.
3401 Hillview Ave.
P.O. Box 10850
Palo Alto, CA 94304 (415) 855-6176

Roger S. Borovoy
Brown & Bain
600, Suite 100
Palo Alto, CA 94306 (415) 856-9411

Charles Lee Botsford
P.O. Box 1116
Palo Alto, CA 94302 (415) 277-1561

Julian Caplan
Flehr, Hohbach, Test, Albritton & Herbert
200 Page Mill Rd., Suite 200
Palo Alto, CA 94306 (415) 324-8888

W. Douglas Carothers, Jr.
Xerox Corp.
Pat. Dept.
3333 Coyote Hill Rd.
Palo Alto, CA 94304 (415) 494-4264

David H. Carroll
Fairchild Semiconductor Corp.
4001 Miranda Ave.
Palo Alto, CA 94304 (415) 858-4590

Steven F. Caserza
Leydig, Voit & Mayer
350 Cambridge Ave., Suite 200
Palo Alto, CA 94306 (415) 324-8999

Jean C. Chognard
P.O. Box 406
Palo Alto, CA 94302 (415) 857-2541

Y. Ping Chow#
Syntex Inc.
3401 Hillview Ave.
P.O. Box 10850
Palo Alto, CA 94304 (415) 852-1356

Pauline A. Clarke#
Syntex Inc.
3401 Hillview Ave.
P.O. Box 10850
Palo Alto, CA 94303 (415) 852-1355

Stanley Z. Cole
Varian Associates, Inc.
611 Hansen Way
Palo Alto, CA 94303 (415) 424-5408

John A. Dhuey
Syntex Inc.
3401 Hillview Ave.
P.O. Box 10850
Palo Alto, CA 94303 (415) 855-6118

Hana H. Dolezalova
Zoecon Corp.
975 California Ave.
Palo Alto, CA 94304 (415) 354-3593

Terrence Enroth Dooher
Varian Associates
611 Hansen Way
Palo Alto, CA 94303 (414) 424-5167

Gerald M. Fisher
Varian Associates
611 Hansen Way
Palo Alto, CA 94301 (415) 424-5407

Stephen P. Fox
Hewlett-Packard Co.
3000 Hanover St., M S 20B0
Palo Alto, CA 94304 (415) 847-3510

John A. Frazzini
Hewlett-Packard Co.
3000 Hanover St., 20B0
Palo Alto, CA 94304 (415) 857-2177

Derek P. Freyberg
Syntex Inc.
3401 Hillview Ave.
P.O. Box 10850
Palo Alto, CA 94303 (415) 855-6166

William E. Green
550 Hamilton Ave., Suite 301
Palo Alto, CA 94301 (415) 321-9992

Edward B. Gregg
Flehr, Hohbach, Test, Albritton & Herbert
200 Page Mill Rd., Suite 200
Palo Alto, CA 94306 (415) 324-8888

Marshall C. Gregory
2211 Park Blvd.
Palo Alto, CA 94306 (415) 321-5030

Roland I. Griffin
Hewlett-Packard Co.
Mail Stop 20B0
3000 hanover St.
Palo Alto, CA 94304 (415) 857-2805

Paul C. Haughey
Townsend & Townsend
379 Lytton Ave.
Palo Alto, CA 94301 (415) 326-2400

Paul M. Hentzel
441 Nevada Ave.
Palo Alto, CA 94301 (415) 326-8254

James M. Heslin
Townsend & Townsend
379 Lytton Ave.
Palo Alto, CA 94301 (415) 326-9800

Willis Edward Higgins
Flehr, Hohbach, Test, Albritton & Herbert
200 Page Mill Rd., Suite 200
Palo Alto, CA 94306 (415) 324-8888

James P. Hillman
Electric Power Res. Inst., Inc.
3412 Hillview Ave.
Palo Alto, CA 94303 (415) 855-2636

Joseph I. Hirsch
Syntex Inc.
3401 Hillview Ave.
P.O. Box 10850
Palo Alto, CA 94304 (415) 855-6186

James M. Kanagy
Syntex Inc.
3401 Hillview Ave.
P.O. Box 10850
Palo Alto, CA 94303 (415) 855-5986

Alan M. Krubiner
Syntex Inc.
3401 Hillview Ave.
P.O. Box 10850
Palo Alto, CA 94304 (415) 855-6133

Katharine Ku#
1130 Waverley St.
Palo Alto, CA 94301 (415) 327-2290

Douglas A. Kundrat
Hewlett-Packard Co.
P.O. Box 10301
Palo Alto, CA 94303 (415) 857-5143

Jacqueline S. Larson
Townsend & Townsend
379 Lytton Ave.
Palo Alto, CA 94301 (415) 326-2400

Theodore J. Leitereg*
Syntex Inc.
3401 Hillview Ave.
P.O. Box 10850
Palo Alto, CA 94304 (415) 855-5050

Brian Lewis#
Syntex Inc.
3401 Hillview Ave.
P.O. Box 10850
Palo Alto, CA 94303 (415) 852-3097

David A. Lowin
Syntex Inc.
3401 Hillview Ave.
P.O. Box 10850
Palo Alto, CA 94304 (415) 855-6167

William H. Mac Allister, Jr.
Hewlett-Packard Co.
Corporate Legal Dept.
3000 Hanover St.
Palo Alto, CA 94303 (415) 857-1501

Stephen C. Macevicz
DNAX Research Institute
901 California Ave.
Palo Alto, CA 94304 (415) 852-9196

Edward Lewis Mandell
Alza Corp.
950 Page Mill Rd.
Palo Alto, CA 94304 (415) 494-5223

John B. Mc Gowan
Hewlett-Packard Co.
Federal Systems Operations
1501 Page Mill Rd., Bldg. 6A
Palo Alto, CA 94303 (415) 857-7417

Thomas J. Mc Naughton
Syntex Inc.
3401 Hillview Ave.
P.O. Box 10850
Palo Alto, CA 94303 (415) 855-1560

Michelle M. Mc Spadden#
Syntex Inc.
3401 Hillview Ave.
P.O. Box 10850
Palo Alto, CA 94303 (415) 855-6184

Thomas V. Michaelis
104 Seale Ave.
Palo Alto, CA 94301 (415) 326-6108

Annette M. Moore#
Syntex Inc.
3401 Hillview Ave.
P.O. Box 10850
Palo Alto, CA 94304 (415) 855-5564

Tom M. Moran
Syntex Inc.
3401 Hillview Ave.
P.O. Box 10850
Palo Alto, CA 94304 (415) 855-6137

Richard L. Neeley
Leydig, Voit & Mayer, Ltd.
350 Cambridge Ave., Suite 200
Palo Alto, CA 94306 (213) 324-8999

Richard B. Nelson#
Varian Associates
Pat. Dept.
611 Hansen Way
Palo Alto, CA 94303 (415) 493-4000

Linda J. Nyari
Syntex Usa Inc.
3401 Hillview Ave.
P.O. Box 10850
Palo Alto, CA 94303

Thomas Henry Olson
415 Cambridge Ave.
Palo Alto, CA 94306 (415) 321-7464

Bradley A. Perkins
Hewlet-Packard Co.
3000 Hanover St.
Mail Stop 20B0
Palo Alto, CA 94304 (415) 857-4377

Edward J. Radlo
Ford Aerospace & Comm. Corp.
Mail Station A09
3939 Fabian Way
Palo Alto, CA 94303 (415) 852-5379

Bloor Redding, Jr.
Hewlett-Packard Co.
P.O. Box 10301
Palo Alto, CA 94303 (206) 254-8110

David Stewart Romney
Romney, Golant, Martin, Disner & Ashen
2501 Park Blvd.
Palo Alto, CA 94306 (415) 327-2353

Norman P. Rousseau
Syntex Inc.
3401 Hillview Ave.
P.O. Box 10850
Palo Alto, CA 94304 (415) 855-6446

Bertram I. Rowland
Leydig, Voit & Mayer
350 Cambridge Ave., Suite 200
Palo Alto, CA 94306 (415) 324-8999

Paul L. Sabatine
455 Marlowe
Palo Alto, CA 94301 (415) 325-3504

Eberhard G. H. Schmoller
Consolidated Freightways, Inc.
3350 W. Bayshore Rd.
P.O. Box 10110
Palo Alto, CA 94303 (415) 855-9100

David Schnapf
Varian Associates, Inc.
611 Hansen Way
Palo Alto, CA 94303 (415) 424-5406

Peter J. Sgarbossa
Varian Associates, Inc.
611 Hansen Way
Palo Alto, CA 94303 (415) 424-5406

James A. Sheridan
Fler, Hohbach, Test, Albritton & Herbert
260 Sheridan Ave.
Palo Alto, CA 94306 (415) 326-0747

Barry P. Smith
Xerox Corp.
Pat. Dept.
3333 Coyote Hill Rd.
Palo Alto, CA 94304 (415) 494-4268

William M. Smith
DNAX Research Institute
901 California Ave.
Palo Alto, CA 94304 (415) 852-9196

Robert K. Stoddard#
560 Oxford Ave., Apt. 8
Palo Alto, CA 94306 (415) 856-3344

Steven Farady Stone
Alza Corp.
950 Page Mill Rd.
Palo Alto, CA 94304 (415) 494-5283

John Stoner, Jr.
101 Alma St., No. 608
Palo Alto, CA 94301 (415) 323-2306

Laura M. Terlizzi
Leydig, Voit & Mayer, Ltd.
350 Cambridge Ave., Suite 200
Palo Alto, CA 94306 (415) 324-8999

Charles J. Tonkin
Flehr, Hohbach, Test, Albritton & Herbert
200 Page Mill Rd., Suite 200
Palo Alto, CA 94306 (415) 324-8888

Liza B. K. Toth
Syntex Inc.
3401 Hillview Ave.
P.O. Box 10850
Palo Alto, CA 94303 (415) 855-5986

Eugene H. Valet
Hewlett-Packard Co.
3000 Hanover St.
Palo Alto, CA 94304 (415) 857-1501

Herwig Von Morze#
Syntex Inc.
3401 Hillview Ave.
P.O. Box 10850
Palo Alto, CA 94304 (415) 855-5160

Kenneth L. Warsh
Varian Associates, Inc.
611 Hansen Way
Palo Alto, CA 94303

Thomas M. Webster#
Xerox Corp.
3333 Coyote Hill Rd., M S 20B0
Palo Alto, CA 94304 (415) 494-4266

Douglas L. Weller
Hewlett-Packard Co.
3000 Hanover St.
Palo Alto, CA 94303 (415) 857-3864

James M. Williams
Hewlett-Packard Co.
3000 Hanover
Palo Alto, CA 94303 (415) 857-5949

Gary S. Winer
Alza Corp.
950 Page Mill Rd.
Palo Alto, CA 94306 (415) 494-5223

Ellen J. Wise
Syntex Inc.
3401 Hillview Ave.
P.O. Box 10850
Palo Alto, CA 94303 (415) 855-6593

John C. Yakes
320 Palo Alto Ave., Suite A-1
Palo Alto, CA 94301 (415) 323-2300

C. Michael Zimmerman
Cushman, Darby & Cushman
525 University Ave.
Palo Alto, CA 94301 (408) 326-9040

Terry J. Anderson
1353 Via Zumaya
Palos Verdes Estates, CA 90274
(213) 544-3222

Thomas A. Seeman
33 Malaga Cove Plaza
Palos Verdes Estates, CA 90274
(213) 378-7095

Vernet Charles Kauffman
8431 Montna Dr.
Paradise, CA 95969 (916) 877-0744

Noel B. Hammond#
16442 Parkshire Court
Paramount, CA 90723 (213) 633-3349

Denton L. Anderson
Sheldon & Mak
201 S. Lake Ave., Suite 800
Pasadena, CA 91101 (818) 796-4000

Philip J. Anderson
Christie, Parker & Hale
P.O. Box 7068
Pasadena, CA 91109 (818) 795-5843

Edward O. Ansell
California Institute Of Tech.
1201 E. California Blvd.
Pasadena, CA 91125 (818) 356-4567

Andrew J. Belansky
Christie, Parker & Hale
P.O. Box 7068
Pasadena, CA 91109 (818) 795-5843

Charles Berman
Sheldon & Mak
201 S. Lake Ave., Suite 800
Pasadena, CA 91101 (818) 796-4000

Earl C. Briggs#
606 Michigan Blvd.
Pasadena, CA 91107 (213) 796-4939

Hayden A. Carney
Christie, Parker and Hale
P.O. Box 7068
Pasadena, CA 91109 (213) 795-5843

Nathan Cass
Unisys Corp.
460 Sierra Madre Villa
Pasadena, CA 91109 (818) 351-8784

Norman L. Chalfin#
Calif. Institute Of Tech.
4800 Oak Grove Dr.
Pasadena, CA 91109 (213) 354-6833

Teresa P. Clark
Sheldon & Mak
201 S. Lake Ave., Suite 800
Pasadena, CA 91101 (818) 796-4000

Edwin Roderick Cline
Christie, Parker & Hale
P.O. Box 7068
Pasadena, CA 91109 (818) 795-5843

Thomas J. Daly
Christie, Parker & Hale
P.O. Box 7068
Pasadena, CA 91109 (818) 795-5843

Edward Joseph Darin
301 E. Colorado Blvd., Suite 518
Pasadena, CA 91101 (213) 793-0689

David A. Dillard
Christie, Parker & Hale
P.O. Box 7068
Pasadena, CA 91109 (818) 795-5843

James T. English#
California Institute Of Technology
Jet Propulsion Lab.
4800 Oak Grove Dr.
Pasadena, CA 91103 (213) 354-3318

Michael B. Farber
Sheldon & Mak
201 S. Lake Ave., Suite 800
Pasadena, CA 91101 (818) 796-4000

Felix L. Fischer#
Christie, Parker & Hale
P.O. Box 7068
Pasadena, CA 91109 (818) 795-5843

Lawrence Thomas Fleming#
1830 Reiter Dr.
Pasadena, CA 91106 (213) 796-3288

Shlomo R. Frieman
Sheldon & Mak
201 South Lake Ave., Suite 800
Pasadena, CA 91101 (818) 796-4000

Frederick Gotha
80 South Lake Ave., Suite 823
Pasadena, CA 91101 (818) 796-1849

William Jacquet Gribble
1421 Glengarry Rd.
Pasadena, CA 91105 (213) 254-4142

John Peter Grinnel
Christie, Parker and Hale
P.O. Box 7068
Pasadena, CA 91109 (818) 795-5843

Peter J. Groom
Christie, Parker & Hale
201 S. Lake Ave., Suite 600
Pasadena, CA 91101 (213) 795-5843

William W. Haefliger
201 S. Lake Ave., Suite 512
Pasadena, CA 91101 (213) 684-2707

C. Russell Hale
Christie, Parker & Hale
P.O. Box 7068
Pasadena, CA 91109 (818) 795-5843

Edwin L. Hartz
Christie, Parker & Hale
P.O. Box 7068
Pasadena, CA 91109 (818) 795-5843

Robert S. Jamieson#
California Institute Of Technology
Jet Propulsion Lab.
4800 Oak Grove Dr.
Pasadena, CA 91109 (213) 354-5161

Rowland William Johnston
Christie, Parker & Hale
P.O. Box 7068
Pasadena, CA 91109 (818) 795-5843

J. Leslie Jones, Sr.#
P.O. Box 233
Pasadena, CA 91102 (213) 792-7280

Thomas H. Jones*
NASA
Jet Propulsion Lab.
4800 Oak Grove Dr.
Pasadena, CA 91109 (818) 354-5179

Stephen J. Koundakjian
201 S. Lake Ave.
Union Bk. Bldg., Suite 408
Pasadena, CA 91101 (213) 578-1850

Alfred Waldemar Kozak#
Burroughs Corp.
460 Sierra Madre Villa Ave.
Pasadena, CA 91109 (213) 351-6551

Danton K. Mak
Sheldon & mak
201 S. Lake Ave., Suite 800
Pasadena, CA 91101 (818) 796-4000

Elizabeth Manning
California Institute Of Tech.
1201 E. California Blvd.
Mail Code 307-6
Pasadena, CA 91125 (818) 356-4567

Walter G. Maxwell
Christie, Parker & Hale
P.O. Box 7068
Pasadena, CA 91109 (213) 681-5637

Paul F. McCaul
NASA
NASA Resident Off.-J P L
4800 Oak Grove Dr.
Pasadena, CA 91109 (213) 354-2734

John J. McCormack#
Box 5421
Pasadena, CA 91107 (213) 373-7018

Donald D. Mon
750 E. Green St.
Pasadena, CA 91101 (213) 793-9173

Deidre A. Oppenheimer#
California Inst. Of Tech.
4800 Oak Grove Dr.
Pasadena, CA 91109 (818) 354-1867

Russell R. Palmer, Jr.
Christie, Parker & Hale
P.O. Box 7068
Pasadena, CA 91109 (213) 681-5637

D. Bruce Prout
Christie, Parker & Hale
P.O. Box 7068
Pasadena, CA 91109 (213) 681-5637

L. T. Rahn
Christie, Parker & Hale
P.O. Box 7078
Pasadena, CA 91109 (213) 681-5637

Michael J. Ram
P.O. Box 70517
Pasadena, CA 91107 (818) 796-0281

Howard B. Scheckman
Jet Propulsion Lab/California Institute Of Technology
Pats. & Tech. Utilization
4800 Oak Grove Dr.
Pasadena, CA 91109 (818) 354-7775

Stephen R. Seccombe
Sheldon & Mak
201 S. Lake Ave., Suite 800
Pasadena, CA 91101 (818) 796-4000

Richard Dorland Seibel
Christie, Parker & Hale
P.O. Box 7068
Pasadena, CA 91109 (213) 681-5637

William Douglas Sellers
Sellers and Brace
16 N. Marengo Ave., Suite 510
Pasadena, CA 91101 (213) 681-4514

Jeffrey G. Sheldon
Sheldon & Mak
201 S. Lake Ave., Suite 800
Pasadena, CA 91101 (818) 796-4000

Luther Price Speck, II
California Institute Of Tech.
Jet Propulsion Lab.
4800 Oak Grove Dr.
Pasadena, CA 91103 (213) 354-6060

Milton R. Spielman
Christie, Parker & Hale
P.O. Box 7068
Pasadena, CA 91109 (213) 681-5637

Tobias Leon Stam
1400 Oak Meadow Rd.
Pasadena, CA 91006 (818) 355-9322

Henry P. Stevens#
Calif. Institute Of Technology
Jet Propulsion Lab.
4800 Oak Grove Dr.
Pasadena, CA 91103 (213) 354-3203

Richard Joseph Ward, Jr.
Christie, Parker & Hale
P.O. Box 7068
Pasadena, CA 91109 (818) 795-5843

Leo J. Young
Christie, Parker & Hale
P.O. Box 7068
Pasadena, CA 91109 (818) 795-5843

John H. Kraus
3040 Larkin Rd.
Pebble Beach, CA 93953 (408) 255-4340

Clarence W. Martin#
18499 Fair Oaks Dr.
Penn Valley, CA 95946 (916) 432-4442

Harry Eugene Aine
8601 Signal Ridge Rd.
P.O. Box 304
Philo, CA 95466 (408) 779-0989

Lynn H. Latta
431 Indio Dr.
Pismo Beach, CA 93449 (805) 773-4773

Walter John Adam
7720 Quartz Hill Rd.
Placerville, CA 95667 (916) 626-7073

Paul A. Renick#
8650 Gulana Ave., Apt. L2174
Playa Del Rey, CA 90293 (213) 823-1106

Paul E. Calrow
83 Mozden Lane
Pleasant Hill, CA 94523 (415) 798-5002

Armand G. Guibert#
Scm Corp.
Pat. Dept. Suite 204
171 Mayhew Way
Pleasant Hill, CA 94523 (415) 932-1997

Joseph M. St. Amand, Jr.*
U.S. Navy
Off. Of Naval Research
Pacific Missile Test Center
Point Mugu, CA 93042 (805) 982-3266

Angus C. Fox, III
General Dynamics
Pomona Division
P.O. Box 2507
Pomona, CA 91769 (714) 945-7898

Edward Ronald Grant
1623 Juniper Ridge
Pomona, CA 91766 (714) 622-7008

Edward B. Johnson#
General Dynamics Corp.
Pomona Div.
1675 W. Mission Blvd.
Pomona, CA 91766 (714) 868-1033

Robert Oswald Webster
1255 Westridge Dr.
Portola Valley, CA 94025 (415) 851-0454

Gregory O. Garmong
13126 Silver Saddle Ln.
Poway, CA 92064 (619) 451-0660

Emmette Rudolph Holman#
28401 Ridgethorne Ct.
Rancho Palos Verdes, CA 90274
(213) 541-2141

Arnold W. Lieman
26622 Fond Du Lac Rd.
Rancho Palos Verdes, CA 90274
(213) 378-5086

Irving B. Osofsky
28327 San Nicolas Dr.
Rancho Palos Verdes, CA 90274
(213) 377-5829

Lee W. Tower
19 Saddle Rd.
Rancho Palos Verdes, CA 90274
(213) 548-3709

Alfons Valukonis#
6760 Los Verdes Dr.
Rancho Palos Verdes, CA 90274
(213) 541-3549

Leonard D. Schappert
P.O. Box 3068
Redding, CA 96049 (916) 244-9914

Leo Francis Costello
P.O. Box 7000-343
Redondo Beach, CA 90277 (213) 377-3146

James E. Crawford#
357 Avenue, E.
Redondo Beach, CA 90277 (213) 540-2930

Benjamin De Witt
TRW Inc.
One Space Park
Bldg. E2/Room 7073
Redondo Beach, CA 90278 (213) 535-1243

Thomas N. Giaccherini
TRW Inc.
One Space Park
Bldg. E2/Rm. 7073
Redondo Beach, CA 90278 (213) 535-6999

Sol L. Goldstein
TRW Inc.
One Space Park
Bldg. E2/Rm. 7073
Redondo Beach, CA 90278 (213) 812-1516

John Holtrichter, Jr.
P.O. Box 227
Redondo Beach, CA 90277 (213) 544-3033

Robert W. Keller
TRW, Inc.
One Space Park
Redondo Beach, CA 90278 (213) 536-1295

Monty Koslover#
145 Via Monte Doro
Redondo Beach, CA 90277 (213) 378-7498

William B. Leach
TRW, Inc.
Bldg. E-2, Rm. 7062
1 Space Park
Redondo Beach, CA 90278 (213) 536-1282

Donald Richard Nyhagen
TRW, Inc.
One Space Park
Redondo Beach, CA 90278 (213) 535-1608

Robert J. Stern
TRW Inc.
One Space Park
Redondo Beach, CA 90278 (213) 535-2394

Robert M. Wallace
TRW Inc.
Electronics & Defense Sector
One Space Park
E2/7073
Redondo Beach, CA 90278 (213) 536-4737

Harold H. Wilson#
Patent Prosecution Service
2105 Rockefeller Lane Apt. 7
Redondo Beach, CA 90278 (213) 217-3446

George B. Almeida#
Ampex Corp.
401 Broadway
Redwood City, CA 94063 (415) 367-3331

Richard P. Lange
Ampex Corp.
401 Broadway
Redwood City, CA 94063 (415) 397-3338

Ralph Leonard Mossino
Ampex Corp.
401 Broadway
Redwood City, CA 94063 (415) 367-3333

Elizabeth E. Strnad#
Ampex Corp.
401 Broadway
Redwood City, CA 94063 (415) 367-3658

Joel D. Talcott
Ampex Corp.
401 Broadway
Redwood City, CA 94063 (415) 367-3330

Harry G. Thibault
Ampex Corporation
401 Broadway
Redwood City, CA 94063 (415) 367-3337

Richard Paul Alberi
191 W. Linden St.
Reedley, CA 93654 (209) 638-6059

James H. Griffith
P.O. Box 965
Reseda, CA 91335 (818) 886-2695

Joel G. Ackerman
Stauffer Chemical Co.
1200 S. 47th St.
Richmond, CA 94804 (415) 231-1194

Edwin Hale Baker
Stauffer Chemical Co.
1200 S. 47th St.
Richmond, CA 94804 (415) 231-1193

Michael Joseph Bradley
Stauffer Chemical Co.
1200 S. 47th St.
Richmond, CA 94804 (415) 231-1017

Elliott L. Fineman
Stauffer Chemical Co.
Patent Dept.
1200 S. 47th St.
Richmond, CA 94804 (415) 231-1000

Leona L. Lauder
Stauffer Chemical Co.
1200 S. 47th St.
Richmond, CA 94804 (415) 231-1197

Harry A. Pacini#
Stauffer Chemical Co.
1200 S. 47th St.
Richmond, CA 94804 (415) 231-1202

Gerald Franklin Baker
Baker & Houston
114 S. China Lake Blvd., Suite A
Ridgecrest, CA 93555 (714) 375-1618

Kenneth G. Pritchard
112 C South China Lake Blvd., Suite A
Ridgecrest, CA 93555 (619) 375-4020

William G. Becker
Bourns, Inc.
1200 Columbia Ave.
Riverside, CA 92507 (714) 781-5138

John Hance Crowe#
4333 Orange St., Suite 8
Riverside, CA 92501 (714) 684-5833

Herbert E. Kidder#4
4376 Maplewood Pl.
Riverside, CA 92506 (714) 683-7854

Fritz B. Peterson#
3290 Monroe St.
Riverside, CA 92504 (714) 688-5621

Willard M. Graham#
55 Cypress Way
Rolling Hills Estates, CA 90274
(213) 541-3045

Walter J. Jason
49 Cypress Way
Rolling Hills Estates, CA 90274
(213) 541-7525

John S. Bell
Aerojet-General Corp.
P.O. Box 13618
Sacramento, CA 95853 (916) 355-3788

Joseph E. Gerber
925 Carro Dr., Apt. 3
Sacramento, CA 95825 (916) 485-1730

Mark C. Jacobs
2775 Cottage Way, Suite 5
Sacramento, CA 95825 (916) 485-5588

Raymond Owyang
P.O. Box 19237
Sacramento, CA 95819 (916) 486-2620

Robert G. West
Lothrop & West
555 Capitol Mall, Suite 1525
Sacramento, CA 95814 (916) 444-5412

Robert M. West
Lothrop & West
555 Capitlo Mall, Suite 1525
Sacramento, CA 95814 (916) 444-5412

Richard P. Maloney#
Arundel Associates Inc.
P.O. Box 608
San Anselmo, CA 94960 (415) 331-1502

Freling E. Baker
Baker, Maxham & Jester
110 West C St., Suite 1202
San Diego, CA 92101 (714) 233-9004

Laurence E. Banghart
3864 Mt. Ainsworth Ave.
San Diego, CA 92111 (619) 277-4125

Stanley Alan Becker
Dressler, Goldsmith, Shore, Sutker & Milnamow, Ltd.
11300 Sorrento Valley Rd., Suite 200
San Diego, CA 92121 (619) 546-1555

Douglas A. Bingham
Dressler, Goldsmith, Shore, Sutker & Milnamow, Ltd.
11300 Sorrento Valley Rd., Suite 200
San Diego, CA 92121 (619) 546-1555

Ralph S. Branscomb
Charmasson, Branscomb & Holz
1200 3rd Ave., Suite 1200
San Diego, CA 92101 (714) 236-9500

Carl R. Brown
Brown, Martin, Haller & Meador
110 West C St., 13th Fl.
San Diego, CA 92101 (714) 238-0999

Edward William Callan
3033 Science Park Rd.
San Diego, CA 92121 (619) 457-2340

Donald W. Canady
Dressler, Goldsmith, Shore Suter & Milnamow, Ltd.
11300 Sorrento Valley Rd., Suite 200
San Diego, CA 92121 (619) 546-1555

Henri J. A. Charmasson
Charmasson & Holz
4550 Kearny Villa Rd., Suite 202
San Diego, CA 92123 (619) 569-9515

Morris Cohen
444 West C St., Suite 220
San Diego, CA 92101 (619) 233-3650

John Robert Duncan, Jr.
General Dynamics Corp.
Convair Div.-Mz 103-10
P.O. Box 85357
San Diego, CA 92138 (714) 547-3542

Charles J. Fassbender
Unisys Corp.
10850 Via Frontera
San Diego, CA 92127 (714) 451-4507

Harvey Fendelman*
Dept. Of The Navy
271 Catalina Blvd.
Code 0012
San Diego, CA 92152 (619) 553-3824

William C. Fuess
Dressler, Goldsmith, Shore, Sutker & Milnamow, Ltd.
11300 Sorrento Valley Rd., Suite 200
San Diego, CA 92121 (619) 546-1555

Albert L. Gabriel
Whann & Clevenger, Apc
9939 Hibert St., Suite 206
San Diego, CA 92131 (714) 695-0360

Frank Donald Gilliam
4655 Ruffner St., Suite 150
San Diego, CA 92111 (619) 292-0901

John L. Haller
2560 First Ave., Suite 107
San Diego, CA 92103 (619) 232-6646

Drew S. Hamilton
Knobbe, Martens, Olson & Bear
101 W. Broadway, Suite 1640
San Diego, CA 92101 (619) 235-8550

David J. Harshman
2560 First Ave., Suite 107
San Diego, CA 92103 (619) 232-6646

Andrew B. Hellewell#
Benson Resource Mgt., Inc.
2425 San Diego Ave., Suite 107
San Diego, CA 92110 (619) 260-1494

Andrew F. Hillhouse, Jr.
225 Broadway, Suite 1600
San Diego, CA 92101 (714) 233-0741

Arthur F. Holz
Charmasson & Holz
4550 Kearny Villa Rd., Suite 202
San Diego, CA 92123 (619) 569-9515

William O. Jacobson
Charmasson & Holz
4550 Kearny Villa Rd., Suite 202
San Diego, CA 92123 (619) 569-9515

Michael H. Jester
Baker, Maxham & Jester
110 West C St., Suite 1202
San Diego, CA 92101 (714) 233-9004

Thomas Glenn Keough*
Dept. Of Navy
Naval Ocean Sys. Ctr.
San Diego, CA 92152 (714) 225-6236

Anita M. Kirkpatrick#
Knobbe, Martens, Olson & Bear
Wells Fargo Bank Bldg., Suite 1640
101 West Broadway
San Diego, CA 92101 (619) 235-8550

Bernard L. Kleinke
Laff, Whitesel, Conte & Saret
101 W. Broadway, Suite 1580
San Diego, CA 92101 (619) 232-6060

Henry G. Kohlmann
Clipher Data Products, Inc.
9715 Businesspark Ave.
P.O. Box 85170
San Diego, CA 92138 (619) 693-7252

Peter L. Lagus#
S-Cubed
3398 Carmel Mtn. Rd.
San Diego, CA 92121 (714) 453-0060

John R. Lindsay
Brown & Martin
110 West C St., Suite 1305
San Diego, CA 92101 (714) 238-0999

Roland Augustus Linger
6912 Fisk Ave.
San Diego, CA 92122 (714) 455-9098

Peter Adams Lipovsky*
U.S. Navy
Legal Counsel For Pats.
Code 0012
Naval Ocean Systems Center
San Diego, CA 92152 (619) 553-3821

Neil F. Martin
Brown & Martin
110 West C St., Suite 1305
San Diego, CA 92101 (619) 238-0999

Robert J. Mawhinney
5335 Westknoll Dr.
San Diego, CA 92109 (714) 270-2705

Lawrence A. Maxham
Baker, Maxham, Callan & Jester
110 West C St., Suite 1202
San Diego, CA 92101 (714) 233-9004

James McCafferty
McCafferty, Akers & Williams
225 Broadway, Suite 1500
San Diego, CA 92101 (714) 239-1935

James W. McClain
Brown, Martin, Haller & Meador
110 West C St., 13th Fl.
San Diego, CA 92101 (619) 238-0999

James D. McFarland
Knobbe, Martens, Olson & Bear
101 West Broadway, Suite 1640
San Diego, CA 92101 (619) 235-8550

George W. McLaughlin
4912 Mt. Elbrus Dr.
San Diego, CA 92117 (619) 279-6547

Terrance A. Meador
Brown & Martin
110 West C St.
San Diego, CA 92101 (619) 238-0999

Roy Miller
2801 C Ocean Front Walk
San Diego, CA 92109 (714) 488-6117

Russell Ben Miller
Brown, Martin & Haller
110 West C St., 13th Fl.
San Diego, CA 92101 (619) 238-0999

Hugo F. Mohrlock#
2867 Grandview St.
San Diego, CA 92110 (714) 276-4456

Ralph Gordon Monsees
450 B St., Suite 1490
San Diego, CA 92101 (714) 235-9035

Linda Rae Neyenesch#
3006 Falcon
San Diego, CA 92103 (619) 260-0537

Gregory D. Ogrod
Brown, Martin & Haller
110 West C St., 13th Fl.
San Diego, CA 92101 (619) 238-0999

George Edward Pearson
973 Manor Way
San Diego, CA 92106 (619) 222-7025

Benjamin C. Pollard
Brown & Martin
110 West C St.
San Diego, CA 92101 (619) 238-0999

Katherine Proctor#
Brown, Martin & Haller
110 West C St., 13th Fl.
San Diego, CA 92101 (619) 238-0999

William L. Respess
Gen-Probe Incorporated
9880 Campus Point Dr.
San Diego, CA 92121 (619) 546-8000

J. Ronald Richbourg
Unisys Corp.
10850 Via Frontera
San Diego, CA 92128 (714) 451-4506

Dennis P. Ritz
Charmasson & Holz
4550 Kearny Villa Rd., Suite 202
San Diego, CA 92103 (619) 569-9515

Daniel Robbins
Eastman Kodak Co.
Spin Physics
3099 Science Park Rd.
San Diego, CA 92121 (619) 453-5410

John R. Ross, Jr.
G A Technologies, Inc.
10955 John Jay Hopkins Dr.
San Diego, CA 92014 (619) 455-2008

George J. Rubens
2117 Blackmore Ct.
San Diego, CA 92109 (619) 272-2260

Elliot N. Schubert#
10585 Porto Ct.
San Diego, CA 92124 (619) 576-8222

Herbert R. Schulze
Fulwider, Patton, Rieber, Lee & Utecht
9191 Towne Center Dr., Suite 285
San Diego, CA 92122 (619) 457-1984

Joseph Carl Schwalbach
110 West C St., Suite 1212
San Diego, CA 92101 (714) 234-1002

Ronni L. Sherman
Hybritech Incorporated
11095 Torreyana Road
San Diego, CA 92121 (619) 455-6700

Tom Sherrard
2285 Comstock Rd.
San Diego, CA 92111 (619) 541-7852

Karl H. Sommermeyer
1725 Willow St.
San Diego, CA 92106 (619) 222-8810

John Stan*#
U.S. Dept. Of The Navy
Naval Ocean Systems Center
San Diego, CA 92152 (714) 225-6236

Paul C. Steinhardt
Hybritech Incorporated
11095 Torreyana Rd.
San Diego, CA 92121 (619) 535-8407

Vincent E. Sullivan
110 West C St., Suite 2202
San Diego, CA 92101 (619) 235-0550

Stephen P. Swinton
Luce, Forward, Hamilton & Scripps
110 West A St., Suite 1700
San Diego, CA 92101 (619) 699-2465

Howard C. Tarr
5995 Eldergardens St.
San Diego, CA 92120 (619) 573-2513

Calif Kip Tervo
6387 Caminito Lazaro
San Diego, CA 92111 (619) 234-4034

Thomas J. Tighe
Brown, Martin, Haller & Meador
110 West C St., 13th Fl.
San Diego, CA 92101 (619) 238-0999

Bruno J. Verbeck
Verbeck & Haller
110 West C St., Suite 1212
San Diego, CA 92101 (714) 234-1861

Kenneth J. Woolcott
Hybritech, Inc.
P.O. Box 269006
San Diego, CA 92126 (619) 455-6700

Raul V. Aguilar
Miller & Daar
50 California St., Suite 2910
San Francisco, CA 94111 (415) 392-6280

Gary T. Aka
Townsend & Townsend
One Market Plaza
Steuart St. Tower
San Francisco, CA 94105 (415) 493-2590

Elmer S. Albritton
Flehr, Hohbach, Test, Albritton & Herbert
Four Embarcadero Center, Suite 3400
San Francisco, CA 94111 (415) 781-1989

James W. Ambrosius
Chevron Corp.
Law Dept.
P.O. Box 7141
San Francisco, CA 94120 (415) 894-4712

Ernest M. Anderson
Eckhoff, Hoppe, Slick, Mitchell & Anderson
Four Embarcadero Center, Suite 760
San Francisco, CA 94111 (415) 391-7160

William L. Anthony, Jr.
Townsend & Townsend
Steuart St. Tower
One Market Plaza
San Francisco, CA 94105 (415) 543-9600

Elliot B. Aronson
Townsend & Townsend
One Market Plaza
2000 Steuart Tower
San Francisco, CA 94105 (415) 543-9600

Richard Elliott Backus
Flehr, Hohbach, Test, Albritton & Herbert
Four Embarcadero Center, Suite 3400
San Francisco, CA 94111 (415) 781-1989

Arthur B. Bakalar
2730 Lyon St.
San Francisco, CA 94123 (415) 346-7119

Stephen E. Baldwin
Flehr, Hohbach, Test, Albritton & Herbert
Four Embarcadero Center, Suite 3400
San Francisco, CA 94111 (415) 781-1989

Donald L. Bartels
Fitch, Even, Tabin & Flannery
100 Bush St., 26th Fl.
San Francisco, CA 94104 (415) 391-8950

Thomas Richard Baruch
3954 Clay St.
San Francisco, CA 94118 (415) 386-4668

Robert A. Beck
Fibreboard Corp.
22 Battery St., Suite 404
San Francisco, CA 94111 (415) 362-6900

Robert J. Bennett
Townsend & Townsend
Steuart St. Tower
One Market Plaza
San Francisco, CA 94105 (415) 543-9600

Lowell C. Bergstedt
Technology Licensing International, Inc.
47 Dartmouth St.
San Francisco, CA 94134 (415) 337-5935

Steven D. Beyer
Flehr, Hohbach, Test, Albritton & Herbert
Four Embarcadero Center, Suite 3400
San Francisco, CA 94111 (415) 341-8888

Suzanne L. Biggs
Chevron Research Co.
Pat. Tdmk. & Contracts Div.
555 Market St., Room 411
San Francisco, CA 94105 (415) 894-3171

Michael K. Bosworth
Chevron Corp.
P.O. Box 7141
San Francisco, CA 94120 (415) 894-9575

David Jay Brezner
Flehr, Hohbach, Test, Albritton & Herbert
Four Embarcadero Center, Suite 3400
San Francisco, CA 94111 (415) 781-1989

Theodore G. Brown
Townsend & Townsend
One Market Plaza
Steuart St. Tower, #2000
San Francisco, CA 94105 (415) 543-9600

J. A. Buchanan, Jr.
100 Pine St., Suite 250
San Francisco, CA 94111 (415) 788-7799

Henry C. Bunsow
Townsend & Townsend
One Market Plaza
Steuart St. Tower, 20th Fl.
San Francisco, CA 94105 (415) 543-9600

William H. Callaway
Austgen Biojet
500 Sansome St., Suite 500
San Francisco, CA 94111 (415) 989-8333

Richard K. Cannon
Cartwright, Slobodin, Bokelman, Borowsky, Wartnick, Moore & Harris
101 California St., Suite 2600
San Francisco, CA 94111 (415) 433-0440

Charles M. Carman, Jr.#
1648 Great Highway
San Francisco, CA 94122 (415) 665-8399

Claude J. Caroli
Chevron Corporation
Law Dept. Pat. Tdmk. & Contract Div.
555 Market St.
San Francisco, CA 94105 (415) 894-5863

Anthony J. Castro
2035 9th Ave.
San Francisco, CA 94116 (415) 753-8672

Vincent J. Cavalieri, Jr.
Chevron Corp.
Law Dept., Pat. Div.
P.O. Box 7141
San Francisco, CA 94120 (415) 894-5435

Guy W. Chambers
Townsend & Townsend
Steuart St. Tower, 20th Fl.
One Market Plaza
San Francisco, CA 94105 (415) 543-9600

Robert C. Colwell
Townsend & Townsend
Steuart St. Tower
One Market Plaza
San Francisco, CA 94105 (415) 543-9600

Roger L. Cook
Townsend & Townsend
One Market Plaza
Steuart St. Tower
San Francisco, CA 94105 (415) 543-9600

Luann Cserr
Flehr, Hobbach, Test, Albritton & Herbert
Four Embarcadero Center, Suite 3400
San Francisco, CA 94111 (415) 781-1989

Russel D. Culbertson#
829 14th St.
San Francisco, CA 94114 (415) 863-8564

Debra E. Dahl
Flehr, Hohbach, Test, Albritton & Herbert
Four Embarcadero Center, Suite 3400
San Francisco, CA 94111 (415) 781-1989

Philip A. Dalton, Jr.
Flehr, Hohbach, Test, Albritton & Herbert
Four Embarcadero Center, Suite 3400
San Francisco, CA 94111 (415) 781-1989

Herbert Davis
Chevron Corp.
555 Market St.
San Francisco, CA 94105 (415) 894-2021

Joel J. De Young
Chevron Res. Co.
P.O. Box 7141
San Francisco, CA 94120 (415) 894-5574

Thomas George Dejonghe
Chevron Research Co.
Law Dept.
555 Market St.
San Francisco, CA 94105 (415) 894-3546

James A. Deland#
Townsend & Townsend
One Market Plaza
Steuart St. Tower, 20th Fl.
San Francisco, CA 94105 (415) 543-9600

Samuel D. Delich
Graham & James
One Martime Plaza, Suite 300
San Francisco, CA 94111 (415) 954-0220

Michael E. Dergosits
Limbach, Limbach & Sutton
2001 Ferry Bldg.
San Francisco, CA 94111 (415) 433-4150

Veronica C. Devitt
Limbach, Limbach & Sutton
2001 Ferry Bldg.
San Francisco, CA 94111 (415) 433-4150

Q. Todd Dickinson
Chevron Research Co.
P.O. Box 7141
San Francisco, CA 94120 (415) 894-3867

Anthony Bernard Diepenbrock
Townsend & Townsend
One Market Plaza
Steuart St. Tower
San Francisco, CA 94105 (415) 543-9600

Hubert E. Dubb
Fliesler, Dubb, Meyer & Lovejoy
Four Embarcadero Center, Suite 400
San Francisco, CA 94111 (415) 362-3800

Stephen C. Durant
Fliesler, Dubb, Meyer & Lovejoy
Four Embarcadero Center, Suite 400
San Francisco, CA 94111 (415) 362-3800

Robert Holland Eckhoff
Eckhoff, Hoppe, Slick, Mitchell & Anderson
Four Embarcadero Center, Suite 760
San Francisco, CA 94111 (415) 391-7160

Lawrence Edelman
Hexcel Corp.
650 California St.
San Francisco, CA 94108 (415) 956-3333

William J. Egan, III
Flehr, Hohbach, Test, Albritton & Herbert
Four Embarcadero Center, Suite 3400
San Francisco, CA 94111 (415) 781-1989

Alfred A. Equitz
Limbach, Limbach & Sutton
2001 Ferry Bldg.
San Francisco, CA 94111 (415) 391-7160

Roger W. Erickson
Owen, Wickersham & Erickson, P.C.
433 California St.
San Francisco, CA 94104 (415) 781-6361

Noemi C. Espinosa
Townsend & Townsend
One Market Plaza
Steuart St. Tower, 20th Fl.
San Francisco, CA 94105 (415) 543-9600

Stephen M. Everett
Limbach, Limbach & Sutton
2001 Ferry Bldg.
San Francisco, CA 94111 (415) 433-4150

Jerome N. Field
182 Second St.
San Francisco, CA 94105 (415) 777-5300

Hugh D. Finley
Phillips, Moore, Lempio & Finley
177 Post St., Suite 800
San Francisco, CA 94108 (415) 421-2674

Paul D. Flehr
Flehr, Hohbach, Test, Albrittin & Herbert
Four Embarcadero Center, Suite 3400
San Francisco, CA 94111 (415) 781-7989

Martin C. Fliesler
Fliesler, Dubb, Meyer & Lovejoy
Four Embarcadero Center, Suite 400
San Francisco, CA 94111 (415) 362-3800

Dirks B. Foster
Townsend & Townsend
Stewart St. Tower, 20th Fl.
One Market Plaza
San Francisco, CA 94105 (415) 543-9600

Ralph L. Freeland, Jr.
Burns, Doane, Swecker & Mathis
100 Pine St., Suite 770
San Francisco, CA 94111 (415) 397-8100

Thomas M. Freiburger
Owen, Wickersham & Erickson
433 California St.
San Francisco, CA 94104 (415) 781-6361

Richard C. Gaffney, Sr.
Chevron Corp.
555 Market St.
San Francisco, CA 94105

Elmer W. Galbi
Townsend & Townsend
Steuart St. Tower, 20th Fl.
One Market Plaza
San Francisco, CA 94105 (415) 543-9600

Thomas Allen Gallagher
Majestic, Gallagher, Parsons & Siebert
101 California St., 39th Fl.
San Francisco, CA 94111 (415) 362-5556

Philip A. Girard
Limbach, Limbach & Sutton
2001 Ferry Bldg.
San Francisco, CA 94111 (415) 433-4150

Neil D. Greenstein
Pillsbury, Madison & Sutro
235 Montgomery St.
P.O. Box 7880
San Francisco, CA 94120 (415) 983-6430

James F. Hann
Townsend & Townsend
Steuart St. Tower, 20th Fl.
One Market Plaza
San Francisco, CA 94105 (415) 543-9600

Thomas B. Haverstock
Limbach, Limbach & Sutton
2001 Ferry Bldg.
San Francisco, CA 94111 (415) 433-4150

M. Henry Heines
Townsend & Townsend
Steuart St. Tower, 20th Fl.
One Market Fl.
San Francisco, CA 94105 (415) 543-9600

Alvin E. Hendricson
500 Sutter St., Suite 604
San Francisco, CA 94102 (415) 981-4463

Thomas Oliver Herbert
Flehr, Hohbach, Test, Albritton & Herbert
Four Embarcadero Center, Suite 3400
San Francisco, CA 94111 (415) 781-7989

Walter D. Herrick
512 Wisconsin St.
San Francisco, CA 94107 (415) 282-3393

Robert Charles Hill
235 Montgomery St., Suite 1741
San Francisco, CA 94104 (415) 421-2080

Albert J. Hillman
Townsend & Townsend
Steuart St. Tower
One Market Plaza
San Francisco, CA 94105 (415) 543-9600

Harold C. Hohbach
Flehr, Hohbach, Test, Albritton & Herbert
Four Embarcadero Center, Suite 3400
San Francisco, CA 94111 (415) 781-7989

James S. Hsue
Majestic, Gallagher, Parsons & Siebert
101 California St., 39th Fl.
San Francisco, CA 94111 (415) 362-5556

William Michael Hynes
Townsend & Townsend
Steuart St. Tower
One Market Plaza
San Francisco, CA 94105 (415) 543-9600

James E. Jacobson, Jr.
Fitch, Even, Tabin & Flannery
100 Bush St., 26th Fl.
San Francisco, CA 94104 (415) 391-8950

Donald B. Jarvis
530 Dewey Blvd.
San Francisco, CA 94116 (415) 557-3601

Bruce H. Johnsonbaugh
Eckhoff, Hoppe, Slick, Mitchell & Anderson
Four Embarcadero Center, Suite 760
San Francisco, CA 94111 (415) 391-7160

Frank E. Johnston
Limbach, Limbach & Sutton
2001 Ferry Bldg.
San Francisco, CA 94111 (415) 433-4150

Marion P. Johnston
Howard, Rice, Nemerovski, Canady, Robertson & Falk
Three Embarcadero Center, 7th Fl.
San Francisco, CA 94111 (415) 781-1989

Kenneth M. Kaslow
Limbach, Limbach & Sutton
2001 Ferry Bldg.
San Francisco, CA 94111 (415) 433-4150

Edward J. Keeling
Chevron Research Co.
P.O. Box 7141
San Francisco, CA 94120 (415) 894-2420

Wayne M. Kennard#
Limbach, Limbach & Sutton
2001 Ferry Bldg.
San Francisco, CA 94111 (415) 433-4150

Warren J. Krauss
Sedgwick, Detert, Moran & Arnold
111 Pine St.
San Francisco, CA 94111 (415) 982-0303

Robert E. D. Krebs
Burns, Doane, Swecker & Mathis
100 Pine St., Suite 770
San Francisco, CA 94111 (415) 397-8100

Charles E. Krueger
Townsend & Townsend
Steuart St. Tower, 20th Fl.
One Market Plaza
San Francisco, CA 94105 (415) 543-9600

Warren P. Kujawa
Townsend & Townsend
Steuart St. Tower
One Market Plaza
San Francisco, CA 94105 (415) 543-9600

S. Russell La Paglia
Chevron Research Co.
P.O. Box 7141
San Francisco, CA 94120 (415) 894-2220

Ronald S. Laurie
Townsend & Townsend
Steuart St. Tower, 20th Fl.
One Market Plaza
San Francisco, CA 94105 (415) 543-9600

Paul S. Lempio
Phillips, Moore, Lempio & Finley
177 Post St., Suite 800
San Francisco, CA 94108 (415) 421-2674

George C. Limbach
Limbach, Limbach & Sutton
2001 Ferry Bldg.
San Francisco, CA 94111 (415) 433-4150

Karl A. Limbach
Limbach, Limbach & Sutton
2001 Ferry Bldg.
San Francisco, CA 94111 (415) 433-4150

Katherine Lloyd#
34 Chabot Terrace
San Francisco, CA 94118 (415) 221-8880

David Eugene Lovejoy
Fliesler, Dubb, Meyer & Lovejoy
Four Embarcadero Center, Suite 400
San Francisco, CA 94111 (415) 362-3800

Donald N. Mac Intosh
Flehr, Hohbach, Test, Albritton & Herbert
Four Embarcadero Center, Suite 3400
San Francisco, CA 94111 (415) 781-1989

Martin F. Majestic
Majestic, Gallagher, Parsons & Siebert
101 California St., 39th Fl.
San Francisco, CA 94111 (415) 362-5556

Miriam E. Majofis
Flehr, Hohbach, Test, Albritton & Herbert
Four Embarcadero Center, Suite 3400
San Francisco, CA 94111 (415) 781-1989

C. Woodworth Marsh#
3530 Wawona St.
San Francisco, CA 94116

Arthur L. Martin
Boone, Knudsen & Martin, P.C.
862 Folsom St.
San Francisco, CA 94107 (415) 543-6600

Janet K. Martinez
Foremost-Mc Kesson, Inc.
One Post St.
San Francisco, CA 94104 (415) 983-8348

J. Thomas Mc Carthy
Univ. Of San Francisco
Law School
2130 Fulton St.
San Francisco, CA 94117 (415) 666-6517

James Bruce Mc Cubbrey
Fitch, Even, Tabin & Flannery
100 Bush St., 26th Fl.
San Francisco, CA 94104 (415) 391-8950

Philip L. Mc Garrigle
Chevron Research Co.
P.O. Box 7141
San Francisco, CA 94120 (415) 894-0545

John L. McGannon
Townsend & Townsend
One Market Plaza
Steuart St. Tower, 20th Fl.
San Francisco, CA 94105 (415) 543-9600

Virginia S. Medlen
Limbach, Limbach & Sutton
2001 Ferry Bldg.
San Francisco, CA 94111 (415) 433-4150

Harold Dale Messner
Chevron Research Co.
P.O. Box 7141
San Francisco, CA 94120 (415) 894-4096

Sheldon R. Meyer
Fliesler, Dubb, Meyer & Lovejoy
Four Embarcadero Center, Suite 1740
San Francisco, CA 94111 (415) 362-3800

Virginia H. Meyer
Fitch, Even, Tabin & Flannery
100 Bush St., 26th Fl.
San Francisco, CA 94104 (415) 391-8950

Bernard H. Meyers
First Interstate Bank Of Ca.
405 Montgomery St., Suite 1108
San Francisco, CA 94104 (415) 544-5451

William C. Milks, III
Townsend & Townsend
Steuart St. Tower, 20th Fl.
One Market Plaza
San Francisco, CA 94105 (415) 543-9600

Dennis D. Miller#
1468-A Fifth Ave.
San Francisco, CA 94122 (415) 564-0427

Mark E. Miller
Fliesler, Dubb, Meyer & Lovejoy
Four Embarcadero Center, Suite 1740
San Francisco, CA 94111 (415) 362-3800

Warren E. Miller
1917 Baker St.
San Francisco, CA 94115 (415) 922-7639

Mark Mohler
Townsend & Townsend
Steuart St. Tower, 20th Fl.
One Market Plaza
San Francisco, CA 94105 (415) 543-9600

Carlisle M. Moore
Phillips, Moore, Lempio & Finley
177 Post St., Suite 800
San Francisco, CA 94108 (415) 421-2674

Dix A. Newell
Chevron Research Corp.
P.O. Box 7643
San Francisco, CA 94120 (415) 894-2220

Vernon A. Norviel
Chevron Research Co.
P.O. Box 7141
San Francisco, CA 94120 (415) 894-4849

John F. O'Flaherty
Hexcel Corp.
650 California St.
San Francisco, CA 94108 (415) 956-3333

Melville Owen
Owen, Wickersham & Erickson, P.C.
433 California St.
San Francisco, CA 94104 (415) 781-6361

Gerald P. Parsons
Majestic, Gallagher, Parsons & Siebert
101 California St., 39th Fl.
San Francisco, CA 94111 (415) 362-5556

Howard M. Peters
Phillips, Moore, Lempio & Finley
177 Post St., Suite 800
San Francisco, CA 94108 (415) 421-2674

James W. Peterson
Burns, Doane, Swecker & Mathis
100 Pine St., Suite 770
San Francisco, CA 94111 (415) 397-8100

Clarence R. Pfeiffer
554 Rockdale Dr.
San Francisco, CA 94127 (415) 334-1790

Leonard Phillips
Phillips, Moore, Lempio & Finley
177 Post St., Suite 800
San Francisco, CA 94108 (415) 421-2674

Michael J. Pollock
Limbach, Limbach & Sutton
2001 Ferry Bldg.
San Francisco, CA 94111 (415) 433-4150

David Roy Pressman
1237 Chestnut St.
San Francisco, CA 94109 (415) 776-3960

Alfons Puishes
White & Puishes
1095 Market St., Suite 806
San Francisco, CA 94103 (415) 863-3911

William David Reese
1322 5th Ave.
San Francisco, CA 94122 (415) 661-1836

Baylor G. Riddell
Flehr, Hohbach, Test, Albritton & Herbert
Four Embarcadero Center, Suite 3400
San Francisco, CA 94111 (415) 781-1989

David C. Ripma
627 Mangels Ave.
San Francisco, CA 94127 (415) 587-2811

Gerald B. Rosenberg
Fliesler, Dubb, Meyer & Lovejoy
Four Embarcadero Center, Suite 1740
San Francisco, CA 94111 (415) 362-3800

Steven H. Roth#
Chevron Corp.
555 Market St.
San Francisco, CA 94105 (415) 894-9309

Charles P. Sammut
Visa International
P.O. Box 8999
San Francisco, CA 94128 (415) 570-3166

Ernest Arthur Schaal
Chevron Research Co.
555 Market St.
San Francisco, CA 94120 (415) 894-3695

Milton W. Schlemmer
Flehr, Hohback, Test, Albritton & Herbert
Four Embarcadero Center, Suite 3400
San Francisco, CA 94111 (415) 781-1989

John W. Schlicher
Townsend & Townsend
Steuart St. Tower, 20th Fl.
One Market Plaza
San Francisco, CA 94105 (415) 543-9600

Bruce W. Schwab
Townsend & Townsend
One Market Plaza
Steuart St. Tower
San Francisco, CA 94105 (415) 543-9600

George M. Schwab
Townsend & Townsend
1 Market Plaza
Steuart Tower
San Francisco, CA 94105 (415) 543-9600

Johann G. Seka
Townsend & Townsend
Steuart St. Tower, 20th Fl.
One Market Plaza
San Francisco, CA 94105 (415) 543-9600

Gerald T. Sekimura
Limbach, Limbach & Sutton
2001 Ferry Bldg.
San Francisco, CA 94111 (415) 433-4150

Harold Shain
254 Edgewood Ave.
San Francisco, CA 94117 (415) 863-6723

Philip M. Shaw, Jr.
Limbach, Limbach & Sutton
2001 Ferry Bldg.
San Francisco, CA 94111 (415) 433-4150

Stephen C. Shear
Flehr, Hohbach, Test, Albritton & Herbert
Four Embarcadero Center, Suite 3400
San Francisco, CA 94111 (415) 781-1989

J. Suzanne Siebert
Majestic, Gallagher, Parsons & Siebert
101 California St., 39th Fl.
San Francisco, CA 94111 (415) 362-5556

Jack L. Slobodin
Cartwright, Sucherman & Slobodin, Inc.
160 Sansome St., Suite 900
San Francisco, CA 94104 (415) 433-0440

David N. Slone
Townsend & Townsend
Sterart St. Tower, 20th Fl.
One Market Plaza
San Francisco, CA 94105 (415) 543-9600

Jonathan A. Small
Townsend & Townsend
Steuart St. Tower, 20th Fl.
One Market Plaza
San Francisco, CA 94105 (415) 543-9600

Thomas F. Smegal, Jr.
Townsend & Townsend
Steuart St. Tower
One Market Plaza
San Francisco, CA 94105 (415) 543-9600

Karen S. Smith
1275 Stanyan St.
San Francisco, CA 94117 (415) 566-2037

Neil Arthur Smith
Limbach, Limbach & Sutton
2001 Ferry Bldg.
San Francisco, CA 94111 (415) 433-4150

Ralph Carlisle Smith
838 Arguello Blvd., Apt. 1
San Francisco, CA 94118 (415) 751-6708

Lawrence S. Squires
Chevron Research Co.
P.O. Box 7141
San Francisco, CA 94120 (415) 894-4986

Michael A. Stallman
Limbach, Limbach & Sutton
2001 Ferry Bldg.
San Francisco, CA 94111 (415) 433-4150

Joseph L. Strabala
One Market Plaza
Spear St. Tower, Suite 1900
San Francisco, CA 94105 (415) 981-8083

Walter L. Stumpf, Jr.
Chevron Research Co.
P.O. Box 7141
San Francisco, CA 94120 (415) 894-4410

John P. Sutton
Limbach, Limbach & Sutton
2001 Ferry Bldg.
San Francisco, CA 94111 (415) 433-4150

Reginald J. Suyat
Flehr, Hohbach, Test, Albritton & Herbert
Four Embarcadero Center, Suite 3400
San Francisco, CA 94111 (415) 781-1989

James F. Sweeney
Fitch, Even, Tabin & Flannery
100 Bush St., 26th Fl.
San Francisco, CA 94104 (415) 391-8950

Gerald F. Swiss
Chevron Corp.
P.O. Box 7141
San Francisco, CA 94120 (415) 894-5765

Aldo J. Test
Flehr, Hobbach, Test, Albritton & Herbert
Four Embarcadero Center, Suite 3400
San Francisco, CA 94111 (415) 781-1989

James E. Toomey
Burns, Doane, Swecker & Mathis
233 Sansome St., Suite 1004
San Francisco, CA 94104 (415) 397-8100

Charles E. Townsend, Jr.
Townsend & Townsend
Steuart St. Tower, 20th Fl.
One Market Plaza
San Francisco, CA 94105 (415) 543-9600

Richard F. Trecartin
Flehr, Hohbach, Test, Albritton & Herbert
Four Embarcadero Center, Suite 3400
San Francisco, CA 94111 (415) 781-1989

William K. Turner
Chevron Res. Co.
555 Market St., Rm. 4035
San Francisco, CA 94120 (415) 894-2789

John Klaas Uilkema
Limbach, Limbach & Sutton
2001 Ferry Bldg.
San Francisco, CA 94111 (415) 433-4150

Allen H. Uzzell
Chevron Research Co.
P.O. Box 7141
525 Market St.
San Francisco, CA 94120 (415) 894-4743

Paul W. Vapnek
Townsend & Townsend
Steuart St. Tower, 20th Fl.
One Market Plaza
San Francisco, CA 94105 (415) 543-9600

Paul M. Vuksich
244 Kearny St., 6th Fl.
San Francisco, CA 94108 (415) 421-6438

Carrie L. Walthour
Limbach, Limbach & Sutton
2001 Ferry Bldg.
San Francisco, CA 94111 (415) 433-4150

Calvin B. Ward
Fitch, Even, Tabin & Flannery
100 Bush St., 26th Fl.
San Francisco, CA 94104 (415) 391-8950

Kenneth A. Weber
Townsend & Townsend
Steuart St. Tower, 20th Fl.
One Market Plaza
San Francisco, CA 94105 (415) 543-9600

James C. Weseman#
Limbach, Limbach & Sutton
2001 Ferry Bldg.
San Francisco, CA 94111 (415) 433-4150

Arlington C. White
White & White
969 Mills Bldg.
200 Montgomery St.
San Francisco, CA 94104 (418) 931-1881

Douglas E. White
220 Jackson St.
San Francisco, CA 94111 (415) 421-4402

Robert E. Wickersham
Owen, Wickersham & Erickson, P.C.
433 California St., 11th Fl.
San Francisco, CA 94104 (415) 781-6361

J. William Wigert, Jr.
Limbach, Limbach & Sutton
2001 Ferry Bldg.
San Francisco, CA 94111 (415) 433-4150

Gary S. Williams
Flehr, Hohbach, Test, Albritton & Herbert
Four Embarcadero Center, Suite 3400
San Francisco, CA 94111 (415) 781-1989

Malcolm B. Wittenberg
Limbach, Limbach & Sutton
2001 Ferry Bldg.
San Francisco, CA 94111 (415) 433-4150

Warren S. Wolfeld
Fliesler, Dubb, Meyer & Lovejoy
Four Embarcadero Center, Suite 400
San Francisco, CA 94111 (415) 362-3800

Henry Kissinger Woodward
Flehr, Hohbach, Test, Albritton & Herbert
Four Embarcadero Center, Suite 3400
San Francisco, CA 94111 (415) 781-1989

Edward S. Wright
Flehr, Hohbach, Test, Albritton & Herbert
Four Embarcadero Center, Suite 3400
San Francisco, CA 94111 (415) 362-3800

Jerry G. Wright
Flehr, Hohbach, Test, Albritton & Herbert
Four Embarcadero Center, Suite 3400
San Francisco, CA 94111 (415) 781-1989

Ronald Loh - Hwa Yin
Limbach, Limbach & Sutton
2001 Ferry Bldg.
San Francisco, CA 94111 (415) 433-4150

A. Stephen Zavell
Chevron Research Co.
555 Market St.
P.O. Box 7141
San Francisco, CA 94120 (415) 894-2776

David J. Arthur
IBM Corp.
Intellectual Property Law, 951/029
5600 Cottle Rd.
San Jose, CA 95193 (408) 997-4210

Warren Michael Becker
Fliesler, Dubb, Meyer & Lovejoy
60 S. Market St., Suite 1570
San Jose, CA 95113 (408) 287-8278

Thomas R. Berthold
IBM Corp.
951-029
5600 Cottle Rd.
San Jose, CA 95193 (408) 997-4793

David A. Boone
55 S. Market St., Suite 1000
San Jose, CA 95113 (408) 291-6000

Robert Bruce Brodie
IBM Corp.
5600 Cottle Rd.
San Jose, CA 95193 (408) 997-4223

Kenneth D alessandro
Lyon & Lyon
111 W. St. John St., Suite 400
San Jose, CA 95113 (408) 993-1555

Paul Davis
Spectra-Physics, Inc.
3333 N. First St.
San Jose, CA 95134 (408) 432-3333

Jacques M. Dulin
111 N. Market St., Suite 900
San Jose, CA 95113 (408) 286-0700

Jack Warren Edwards
95 S. Market St., Suite 300
San Jose, CA 95113 (408) 298-8886

Robert J. Grassi
2160 The Alameda
San Jose, CA 95126 (408) 243-7444

Robert O. Guillot
Hoffman, Kafubowski, Guillot & Stafford
777 N. First St., Suite 444
San Jose, CA 95112 (408) 977-1444

Claude A.S. Hamrick
Rosenblum, Parish & Becigalupi
55 Almaden Blvd., 5th Fl.
San Jose, CA 95113 (408) 977-0120

Michael L. Harrison
Harrison & Kaylor
4320 Stevens Creek Blvd., Suite 150
San Jose, CA 95129 (408) 241-2220

Mark A. Haynes
Fliesler, Dubb, Meyer & Lovejoy
60 S. Market St., Suite 1570
San Jose, CA 95113 (408) 287-8278

James E. Hite, III
Rosenblum, Parish & Bacigalupi, P.C.
55 Almaden Blvd., Fifth Fl.
San Jose, CA 95113 (408) 977-0120

David H. Jaffer
Rosenblum, Parish & Bacigalupi
55 Almaden Blvd.
Fifth Floor
San Jose, CA 95113 (408) 977-0120

Ivor J. James, Jr.
2120 Briarwood Dr.
San Jose, CA 95125 (408) 266-7346

Allston L. Jones
Wiseman, Jones & Smith
12 S. First St., Suite 911
San Jose, CA 95113 (408) 294-6824

Yoshio Katayama
P.O. Box 1525
San Jose, CA 95109 (408) 286-0333

Elizabeth A. Lawler#
55 Almaden Blvd., Suite 721
San Jose, CA 95113 (408) 297-9733

Peter R. Leal
IBM Corp.
Pat. Operations, 951/123
5600 Cottle Rd.
San Jose, CA 95193 (408) 256-2332

John James Leavitt#
777 N. First St., Suite 610
San Jose, CA 95112 (408) 286-2262

Simon K. Lee
IBM Corp.
Intellectual Property Law Dept.
951/029
5600 Cottle Rd.
San Jose, CA 95193 (408) 997-4210

I. Robert Mednick
888 N. First St.
San Jose, CA 95112 (408) 287-8278

Armand G. Morin
197 Giddings Ct.
San Jose, CA 95139 (408) 629-7948

Herman H. Murphy#
1231 Rosalia Ave.
San Jose, CA 95117 (408) 248-5724

Henry E. Otto, Jr.
IBM Corp.
5600 Cottle Rd.
951/029
San Jose, CA 95153 (408) 997-4170

Mark J. Protsik#
111 W. St. John St., Suite 620
P.O. Box 2E
San Jose, CA 95109 (408) 297-9733

Louis J. Quick
3560 Parkland Ave.
San Jose, CA 95117 (408) 243-3552

Ronald W. Redo#
1815 Mc Daniel Ave.
San Jose, CA 95126 (408) 292-3407

Kevin G. Rivette
Glaspy, Elliott, Creech, Mc Mahon, Roth & Reed
P.O. Box 5812
San Jose, CA 95150 (408) 371-2332

Jerald E. Rosenblum
Rosenblum, Parish & Bacigalupi, P.C.
55 Almaden Blvd.
San Jose, CA 95113 (408) 977-0120

Paul F. Schenck#
P.O. Box 2-E
San Jose, CA 95109

John F. Schipper
FMC Corp.
San Jose, CA 95109

Otto Schmid, Jr.
IBM Corp.
Intellectual Property Law Dept.
5600 Cottle Rd.
San Jose, CA 95193 (408) 997-4182

Thomas Schneck, Jr.
111 W. St. John St., Suite 620
P.O. Box 2-E
San Jose, CA 95109 (408) 297-9733

Donald E. Schreiber
1338 Ringrose Ct.
San Jose, CA 95121 (408) 225-4244

Melvyn D. Silver
Suden & Silver
111 N. Market St., Suite 715
San Jose, CA 95113 (408) 298-9755

Raymond G. Simkins
General Elec. Co.
175 Curtner Ave.
M/c 822
San Jose, CA 95125 (408) 925-5937

Joseph H. Smith#
Wiseman, Jones & Smith
12 S. First St., Suite 911
San Jose, CA 95113 (408) 294-6824

Robert S. Smith#
1263 Emory St.
San Jose, CA 95126 (408) 287-1894

Edward M. Suden
Suden & Silver
111 N. Market St., Suite 715
San Jose, CA 95113 (408) 298-9755

John P. Taylor
4988 Barron Park Dr.
San Jose, CA 95136 (408) 224-5050

Samuel Edward Turner
1450 Glenwood Ave.
San Jose, CA 95125 (408) 297-4039

William B. Walker, Sr.
Walker & Dulin
1625 The Alameda, Suite 801
San Jose, CA 95126 (408) 288-9245

Joseph Gregory Walsh
IBM Corp.
Intellectual Property Law Dept.
951/029
5600 Cottle Rd.
San Jose, CA 95193 (408) 997-4216

Jack M. Wiseman
Wiseman, Jones & Smith
12 S. First St., Suite 911
San Jose, CA 95113 (408) 294-6824

Egbert Walter Mark#
1516 140th Ave.
San Leandro, CA 94578 (415) 357-8442

Jerry N. Lulejian
Carsel, Craig & Associates
1118 Palm St.
San Luis Obispo, CA 93401 (805) 544-8510

Frank E. Mauritz
1657 Hilliard Dr.
San Marino, CA 91108 (213) 287-7255

John Benefield Young
1419 Vandyke Rd.
San Marino, CA 91108 (818) 287-5988

John A. Bucher
241 N. San Mateo Dr.
San Mateo, CA 94401 (415) 347-8871

James E. Eakin
Harrison, Harrison & Eakin
1700 S. El Camino Real, Suite 405
San Mateo, CA 94402 (415) 571-7500

Donald C. Feix
241 N. San Maeto Dr.
San Mateo, CA 94401 (415) 342-4508

John D. Garvic
520 El Camino Real, Suite 300
San Mateo, CA 94402 (415) 342-0873

David B. Harrison
Harrison, Harrison & Eakin
1700 S. El Camino Real, Suite 405
San Mateo, CA 94402 (415) 571-7500

Frank A. Neal
4237 Bettina
San Mateo, CA 94403 (415) 345-2477

David E. Newhouse
Newhouse & Associates
2855 Campus Dr., 2nd Fl.
San Mateo, CA 94403 (415) 345-4930

Ralph C. Grove#
1149 Crestwood Street
San Pedro, CA 90732 (213) 831-3480

Charles T. Silberberg
1913 W. 9th St.
San Pedro, CA 90732

Arnold Thomas Bertolli
8 Bradcliff Ct.
San Rafael, CA 94901 (415) 453-2763

Larry D. Johnson
185 N. Redwood Dr., Suite 130
San Rafael, CA 94903 (415) 499-8822

Melvin Robert Stidham
1050 Northgate Dr., Suite 100
San Rafael, CA 94903 (415) 472-3164

Draper B. Gregory
111 Deerwood Place, Suite 370
San Ramon, CA 94583 (415) 820-7323

Gary Appel
1170 W. Civic Center Dr.
Sanata Ana, CA 92703 (714) 558-0366

Lawrence D. Sassone#
900 N. Broadway, Suite 725
Sanata Ana, CA 92701 (714) 547-5611

David J. Althoen
18641 Silver Maple Way
Santa Ana, CA 92705 (714) 639-1733

Arthur W. Fuzak
Kimstock, Inc.
2200 S. Yale St.
Santa Ana, CA 92704 (714) 546-6850

Abe Goldstein
Goldstein, Block & Block
11577 Forum Way, Suite A
Santa Ana, CA 92705 (714) 832-2518

Robert H. Himes
1282 Landfair Circle
Santa Ana, CA 92705 (714) 544-3660

Harold C. Horwitz
2127 N. Main St.
Santa Ana, CA 92706 (714) 542-1102

John A. Kane
Kendall Mc Graw Labs., Inc.
P.O. Box 25080
Santa Ana, CA 92799 (714) 660-2083

Donald J. Koprowski
Kawasaki Motors Corp.
2009 E. Edinger Ave.
Santa Ana, CA 92705 (714) 835-7000

William H. May
2401 N. Santiago Ave.
Santa Ana, CA 92706 (714) 835-1949

John George Mesaros
540 N. Golden Circle Dr., Suite 210
Santa Ana, CA 92705 (714) 835-2260

Frank C. Price#
13812 Sandhurst Pl.
Santa Ana, CA 92705 (714) 544-7907

Roland G. Rubalcava
2030 E. Fourth St., Suite 222
Santa Ana, CA 92705 (714) 558-7363

Howard E. Sandler
Wahlco, Inc.
3600 W. Segerstrom Ave.
Santa Ana, CA 92704 (714) 979-7300

Robert L. Sassone
900 N. Broadway, Suite 725
Santa Ana, CA 92701 (714) 542-1111

Robert E. Strauss
Plante, Strauss & Vanderburgh
1020 N. Broadway, Suite 400
Santa Ana, CA 92701 (714) 667-1570

James D. Thackrey#
13852 Dall Ln.
Santa Ana, CA 92705 (714) 731-0705

Catherine J. Bos
1114 State St., Apt. 200
Santa Barbara, CA 93105 (805) 963-1347

Harry W. Brelsford
233 E. Carillo St., Suite C
Santa Barbara, CA 93101 (805) 966-2281

Ronald E. Grubman
Digital Sound Corp.
2030 Alameda Padre Serra
Santa Barbara, CA 93103 (805) 569-0700

Allen H. Sochel
P.O. Box 5548
1485 E. Valley Rd.
Santa Barbara, CA 93108 (805) 969-0354

John Martin Sullivan, Jr.
P.O. Box 658
Santa Barbara, CA 93102 (805) 963-8911

Philip J. Wyatt#
Wyatt Technology Corp.
820 E. Haley St.
P.O. Box 3003
Santa Barbara, CA 93130 (805) 963-5904

Mark A. Aaker#
National Semiconductor Corp.
M/s 16-135
2900 Semiconductor Dr.
Santa Clara, CA 95051 (408) 721-5365

Richard J. Bartlett
444 Saratoga Ave., Apt. 17E
Santa Clara, CA 95050 (408) 247-2980

Eugene T. Battjer
Unisys Corp.
San Tomas At Central Expressway
M/s 12-33
Santa Clara, CA 95052 (408) 987-3000

Charles K. Epps
IBM Corp.
4900 Old Ironside Dr.
Santa Clara, CA 95054 (408) 986-2447

Richard Karl Franklin
Skjerven, Morrill, Macpherson, Franklin & Friel
3600 Pruneridge
Santa Clara, CA 95051 (408) 246-1405

Lloyd B. Guernsey#
FMC Corp.
Mach. Pat., Tdmk & Lic. Dept.
900 Lafayette St., Suite 608
Santa Clara, CA 95050 (408) 241-8320

Jeffrey A. Hall
212 Clinton St.
Santa Clara, CA 95062 (408) 423-1365

David W. Heid
Memorex Corp.
San Tomas At Central Expressway
M/s 1233
Santa Clara, CA 95052 (408) 987-3263

Michael J. Hughes
Hughes, Bedolla & Diener, Inc.
2350 Mission College Blvd., Suite 1150
Santa Clara, CA 95054 (408) 727-9991

Nathan N. Kallman
Skjerven, Morrill, Mac Pherson, Franklin & Friel
3600 Pruneridge Ave., Suite 100
Santa Clara, CA 95051 (408) 246-1405

Michael H. La Cava
Skjerven, Morrill, Mac Pherson, Franklin & Friel
3600 Pruneridge
Santa Clara, CA 95051 (408) 246-1405

Sam E. Laub
Dysan Corp.
5201 Patrick Henry Dr.
Santa Clara, CA 95050 (408) 988-3472

Kenneth E. Leeds
Skjerven, Morrill, Mac Pherson, Franklin & Friel
3600 Pruneridge, Suite 100
Santa Clara, CA 95051 (408) 246-1405

Richard D. Lowe
P.O. Box 296
Santa Clara, CA 95052 (408) 984-1555

Thomas Scott Mac Donald
Skjerven, Morrill, Mac Pherson & Drucker
3600 Pruneridge Ave., Suite 100
Santa Clara, CA 95051 (408) 246-1405

Alan H. Mac Pherson
Skjerven, Morrill, Mac Pherson & Drucker
3600 Pruneridge Ave., Suite 100
Santa Clara, CA 95051 (408) 246-1405

Walter J. Madden, Jr.
Skjerven, Morrill, Mac Pherson, Franklin & Friel
3600 Pruneridge Ave., Suite 100
Santa Clara, CA 95051 (408) 246-1405

Michael J. Mazza#
3211 Scott Blvd., Suite 201
Santa Clara, CA 95051 (408) 727-7077

Douglas R. Millett#
3211 Scott Blvd.
Santa Clara, CA 95054 (408) 727-7077

Alan Johnston Moore
FMC Corp.
1185 Coleman Ave., Box 580
Santa Clara, CA 95052 (408) 289-2477

Michael D. Rostoker
Intel Corp.
3535 Garrett Dr.
Santa Clara, CA 95050 (408) 987-8080

Thomas E. Schatzel
3211 Scott Blvd., Suite 201
Santa Clara, CA 95054 (408) 727-7077

Saul A. Seinberg
Rolm Corp.
Intellectual Property Law Dept.
4900 Old Ironsides Dr.
Santa Clara, CA 95054 (408) 986-2045

Cheryl L. Shavers#
2026 Klamath Ave., Apt. 1
Santa Clara, CA 95051 (408) 246-8310

Carl L. Silverman
Intel Corp.
3065 Bowers Ave.
G R 1-21
Santa Clara, CA 95051 (408) 987-8080

Henry M. Stanley
FMC Corp.
1185 Coleman Ave.
Santa Clara, CA 95050 (408) 289-2133

Lawrence K. Stephens#
Rolm Corp.
M S 107
Intellectual Property Law Dept.
4900 old Ironsides Dr.
Santa Clara, CA 95054 (408) 492-4897

Stanley M. Weir#
P.O. Box 365
Santa Clara, CA 95052 (408) 296-0378

Paul John Winters
National Semiconductor Corp.
2900 M/s 16-180
Santa Clara, CA 95051 (408) 721-6515

Gail W. Woodward#
2166 San Rafael Ave.
Santa Clara, CA 95051 (408) 246-4853

Edel M. Young#
Skjerven, Morrill, Mac Pherson, Franklin & Friel
3600 Pruneridge Ave., Suite 100
Santa Clara, CA 95051 (408) 246-1405

Robert G. Slick
22780 E. Cliff Dr.
Santa Cruz, CA 95062 (408) 475-7160

George H. Nicholson#
1094 Clubhouse Dr.
Santa Maria, CA 93455 (805) 937-0816

Keith D. Beecher
Jessup, Beecher & Slehofer
2001 Wilshire Blvd., Suite 500
Santa Monica, CA 90403 (213) 829-4525

Robert Boyd Block
P.O. Box 1016
Santa Monica, CA 90406 (213) 829-1109

Donald M. Cislo
Cislo & Thomas
233 Wishire Blvd., Suite 900
Santa Monica, CA 90401 (213) 451-0647

John E. Kelly
Pastoriza and Kelly
606 Wilshire Blvd., Suite 512
Santa Monica, CA 90401 (213) 393-0244

William A. Kemmel, Jr.
Monogram Inds., Inc.
1299 Ocean Ave.
Santa Monica, CA 90401 (213) 451-8151

Malcolm J. Romano
Lear Siegler, Inc.
2850 Ocean Park Blvd.
Santa Monica, CA 90406 (213) 452-8886

Seymour M. Rosenberg
2520 La Mesa Way
Santa Monica, CA 90402 (213) 394-0904

Lyle J. Schlyer#
Tosco Corp.
2401 Colorado Ave.
Santa Monica, CA 90406 (213) 207-7012

Kim Effron#
Henkel Research Corp.
2330 Circadian Way
Santa Rosa, CA 95407 (707) 575-7155

Gary F. Grafel
Keegan & Coppin, Inc.
1355 N. Dutlon Ave.
Santa Rosa, CA 94501 (707) 528-1400

Andrew A. Steiner
3000 Cleveland Ave., Suite 101
Santa Rosa, CA 95401 (707) 526-9822

Jane Bieberman De Nuzzo#
4335 Woodstock Rd.
Santa Ynez, CA 93460 (805) 688-3092

Daniel Strugar
9336 Nalini Ct.
Santee, CA 92071 (619) 258-9750

Richard B. Catto#
P.O. Box 243
14711 Bohlman Rd.
Saratoga, CA 95071 (408) 867-4059

Craig W. Hartsell
12291 Kosich Ct.
Saratoga, CA 95070 (408) 252-0780

Robert S. Kelly
19191 Portos Pl.
Saratoga, CA 95070 (408) 867-5648

J. B. Mc Guire
P.O. Box 2488
Saratoga, CA 95070 (408) 741-0311

Wilmur M. Mc Millan#
13748 Saratoga Vista Ave.
Saratoga, CA 95070 (408) 867-3647

Joseph Mednick
13424 Beaumont Ave.
Saratoga, CA 95070 (408) 867-4663

Daniel C. Moyles#
18430 Baylor Ave.
Saratoga, CA 95070 (408) 379-5712

Lester J. Vincent
12311 Saraglen Dr.
Saratoga, CA 95070 (408) 253-4127

Owen Lester Lamb
P.O. Box 66737
Scotts Valley, CA 95066 (408) 438-6657

Edward M. Bayer
16601 Nordhoff St.
Sepulveda, CA 91343 (818) 892-3913

William M. Harris#
15736 Tuba St.
Sepulveda, CA 91343 (213) 892-8706

Harry C. Burgess
5410 Ball Dr.
Sequel, CA 95073 (408) 475-6307

Paul Bernhard Fihe, II#
P.O. Box 126
Soquel, CA 95073 (408) 462-5079

Harold R. Beck#
3970 Cody Rd.
Sherman Oaks, CA 91403 (818) 789-2484

Warner W. Clements#
P.O. Box 5882
Sherman Oaks, CA 91413 (213) 276-7918

Ralph Deutsch#
Deutsch Res. Labs., Ltd.
3647 Scadlock Lane
Sherman Oaks, CA 91403 (213) 789-2779

Roger A. Marrs#
Union Bank Plaza
15233 Ventura Blvd., Suite 806
Sherman Oaks, CA 91403 (213) 788-4115

John M. May
707 Woodland Dr.
Sierra Madre, CA 91024 (213) 355-4617

Benjamin Franklin Spencer
175 N. Mountain Trail Ave.
Sierra Madre, CA 91024 (818) 355-9233

Philip S. Schmidt#
4450 N W Shelley Dr.
Silverdale, CA 98383 (206) 692-5660

Howard R. Boyle#
Kaypro Corp.
533 Stevens Ave.
Solana Beach, CA 92075 (619) 259-4453

Neil K. Nydegger
1332 Sun Valley Rd.
Solana Beach, CA 92075 (619) 481-1979

Melanie A. Calver
630 A Orange Grove Ave.
South Pasadena, CA 91030 (818) 441-2596

Robert M. McManigal
1701 Camden Parkway
South Pasedena, CA 91030 (818) 799-1571

Walter Eugene Buting
Genentech, Inc.
460 Point San Bruno Blvd.
South San Francisco, CA 94080 (415) 266-1715

Thomas E. Byrne
Genentech, Inc.
460 Point San Bruno Blvd.
South San Francisco, CA 94080 (415) 266-1994

Walter H. Dreger#
Genentech, Inc.
460 Point San Bruno Blvd.
South San Francisco, CA 94080 (415) 266-1746

Janet E. Hasak
Genentech, Inc.
460 Point San Bruno Blvd.
South San Francisco, CA 94080 (415) 266-1000

Max D. Hensley
Genentech, Inc.
460 Point San Bruno Blvd.
South San Francisco, CA 94080 (415) 266-1994

Thomas D. Kiley
Genentech, Inc.
460 Point San Bruno Blvd.
South San Francisco, CA 94080

Dennis G. Kleid#
Genentech, Inc.
460 Point San Bruno Blvd.
South San Francisco, CA 94080 (415) 266-1713

James G. Passe
Genencor, Inc.
180 Kimball Way
South San Francisco, CA 94080 (415) 588-3475

Shelley G. Precivale
620 Grand Ave., Apt. 5
South San Francisco, CA 94080 (415) 583-9402

Clarence J. Ott
6403 Embarcadero Dr.
Stockton, CA 95209 (209) 952-3411

Roger B. Webster
Webster & Webster
661 S. Tuxedo Ave.
Stockton, CA 95204 (209) 948-4911

Max Geldin
3386 Canton Way
Studio City, CA 91604 (818) 763-5803

Rodger N. Alleman
Lockheed Missiles & Space Co.
P.O. Box 3504
1111 Lockheed Way
Sunnyvale, CA 94088 (408) 742-0691

Clifton L. Anderson
333 Cobalt Way, Suite 107
Sunnyvale, CA 94086 (408) 245-0820

Richard A. Brown
Jackson, Brown & Efting
465 S. Mathilda Ave., Suite 304
Sunnyvale, CA 94086 (408) 732-3114

Richard Harry Bryer
Lockheed Missiles & Space Co., Inc.
1111 Lockheed Way
Sunnyvale, CA 94088 (408) 742-0691

Russell Adams Cannon#
961 Harney Way
Sunnyvale, CA 94087 (408) 739-6612

Charles D. B. Curry*
Dept. Of The Navy
Office Of Counsel
Strategic Systems Program Off.
1111 Lockheed Way, Bldg. 181-B
Sunnyvale, CA 94089 (408) 756-3662

Brian J. Flynn
Intersil, Inc.
1275 Hammerwood Ave.
Sunnyvale, CA 94086 (408) 743-4300

Yemmanur Jayachandra#
1713 Chitamook Ct.
Sunnyvale, CA 94087 (408) 737-9762

Patrick T. King
Advanced Micro Devices, Inc.
Patent Dept. M/s 68
901 Thompson Place
P.O. Box 3453
Sunnyvale, CA 94088 (408) 732-2400

John J. May#
790 Lucerne Dr.
Sunnyvale, CA 94086 (408) 245-7451

Ronald J. Meetin
Signetics Corporation
Pat. Dept., M/s 54
811 E. Arques Ave.
Sunnyvale, CA 94088 (408) 991-2046

John J. Morrissey, Jr.
Lockheed Missiles & Space Co., Inc.
Dept. 26-02, Bldg. 101
1111 Lockheed Way
Sunnyvale, CA 94086 (415) 742-0691

Brian D. Ogonowsky*
Sunnyvale AFS
AFSCF/DVEC
Sunnyvale, CA 94088 (408) 744-6613

Bruce D. Riter
Schlumberger Computer Aided Systems
1259 Oakmead Parkway
Sunnyvale, CA 94086 (408) 720-7740

J. Vincent Tortolano
Advanced Micro Devices, Inc.
901 Thompson Place
P.O. Box 3453, M S/68
Sunnyvale, CA 94088 (408) 982-6045

Bryant R. Gold
Siemens-Pacesetter, Inc.
12884 Bradley Ave.
Sylmar, CA 91342 (818) 362-6822

Leslie S. Miller
Siemens-Pacesetter, Inc.
12884 Bradley Ave.
Sylmar, CA 91342 (818) 362-6822

Joseph A. Nicassio
13797 De Garmo Ave.
Sylmar, CA 91342 (213) 362-4636

Robert Catlett Smith
Allied Corp.
Law Dept.
15825 Roxford St.
Sylmar, CA 91342 (213) 367-0111

John A. Scillieri
18831 Wells Dr.
Tarzana, CA 91356 (818) 343-6878

Anthony T. Zachary
19601 Greenbriar Dr.
Tarzana, CA 91356 (213) 342-4021

Francis X. Lo Jacono, Sr.#
Box 172B, Rte. 2
Templeton, CA 93465 (805) 466-4314

Louise P. Anderson#
711 Woodbine Ct.
Thousand Oaks, CA 91360 (805) 492-6760

Stanley E. Anderson, Jr.
711 Woodbine Ct.
Thousand Oaks, CA 91360 (805) 492-6760

John J. Deinken
Rockwell Science Ctr.
P.O. Box 1085
Thousand Oaks, CA 91360 (805) 373-4556

Robert E. Geauque
167 Windsong St.
Thousand Oaks, CA 91360 (805) 492-5331

Thomas Edward Kristofferson
161 Verde Vista Dr.
Thousand Oaks, CA 91360 (805) 492-1951

Craig O. Malin
Rockwell Internatl. Corp.
P.O. Box 1085
1049 Camino Dos Rios
Thousand Oaks, CA 91360 (805) 498-4545

Dennis L. Mangrum
325 E. Hillcrest, Bld. C
Thousand Oaks, CA 91360 (805) 495-0113

Steven M. Odre
Amgen
1900 Oak Terrace Ln.
Thousand Oaks, CA 91320 (805) 499-5725

Pamela A. Simonton
Amgen Inc.
1900 Oak Terrace Ln.
Thousand Oaks, CA 91320 (805) 499-5725

Robert D. Weist
AMGEN
1900 Oak Terrace Ln.
Thousand Oaks, CA 91320 (805) 499-5725

Lois C. Babcock#
21535 Hawthorne Blvd., Suite 223
Torrance, CA 90503 (213) 543-2120

G. Joseph Buck
3868 Carson St., Suite 300-15
Torrance, CA 90503 (213) 540-8840

Paul M. Coble
Hughes Aircraft Co.
Bldg. 230, M.S. 2115
P.O. Box 2999
3100 W. Lomita Blvd.
Torrance, CA 90509 (213) 517-5741

Ronald M. Goldman
3004 Oakwood Ln.
Torrance, CA 90505 (213) 326-6177

Terje Gudmestad
Hughes Aircraft Company
3100 West Lomita Blvd.
Torrance, CA 90509 (213) 517-5742

John E. Halamka
Halamka & Halamka
21515 Hawthorne Blvd., Suite 590
Torrance, CA 90503 (213) 316-6100

Noel F. Heal
21535 Hawthorne Blvd., Suite 223
Torrance, CA 90503 (213) 543-2120

Irving Keschner
21535 Hawthorne Blvd., Suite 500
Torrance, CA 90503 (213) 543-5200

Michael M. Schuster#
Hi-Shear Corp.
2600 Skypark Dr.
Torrance, CA 90509 (213) 326-8110

Irwin Shuldiner
2231 W. 235th St.
Torrance, CA 90501 (213) 325-5649

Gerald Singer
Singer & Singer
3142 Pacific Coast Hwy., Suite 208
Torrance, CA 90505 (213) 530-2202

Edward A. Sokolski
3868 Carson St., Apt. 105
Torrance, CA 90503 (213) 540-5631

James M. Steinberger
21535 Hawthorne Blvd., Suite 223
Torrance, CA 90503 (213) 543-2120

Joan S. Trygstad
Rexnord Incorporated
Specialty Fastener Div.
3000 W. Lomita Blvd.
Torrance, CA 90505 (213) 530-2220

Louis R. Price
P.O. Box 727
Tracy, CA 95376 (209) 835-9223

Gilbert E. Moody
Moody & Johnson
250 W. Main St.
Turlock, CA 95380 (209) 632-1086

Gordon K. Anderson#
14632 Pacific St.
Tustin, CA 92680 (714) 730-3460

Harold Leo Jackson
Jackson & Jones, P.C.
17592 Irvine Blvd.
Tustin, CA 92680 (714) 832-2080

Stanley R. Jones
Jackson & Jones
17592 Irvine Blvd.
Tustin, CA 92680 (714) 832-2080

Stuart W. Knight
Central Fed. Savings Bldg.
13522 Newport Ave., Suite 201
Tustin, CA 92680 (714) 730-4808

Forrest E. Logan
275 Centennial Way, Suite 205
Tustin, CA 92680 (714) 730-5553

Steven R. Markl#
14902 Bridgeport
Tustin, CA 92680 (714) 730-8287

Rolf M. Pitts, Jr.
12720 Newport Ave., Apt. 19
Tustin, CA 92680 (714) 544-0174

Robert W. Cramer
Knapp, Petersen & Clarke
70 Universal City Plaza, Suite 400
Universal City, CA 91608 (213) 508-5000

Robert T. Merrick
Baxter Healthcare Corp.
27200 N. Tourney Rd.
P.O. Box 5900
Valencia, CA 91355 (805) 253-7436

Roger A. Williams
American Hospital Supply Corp.
27200 N. Tourney Rd.
Valencia, CA 91355 (805) 253-7579

Bruce D. Jimerson
27375 Coolwater Ranch Rd.
Valley Center, CA 92082 (714) 749-1991

Price Gerald L.
Calvert Eng., Inc.
7051 Hayvenhurst Ave.
Van Nuys, CA 91406 (213) 781-6029

Allan D. Mockabee
14643 Sylvan St.
Van Nuys, CA 91411 (818) 997-0434

Candace P. Olsen#
5836 Bevis Ave.
Van Nuys, CA 91411 (213) 785-7573

William K. Quarles, Jr.
Sunkist Growers, Inc.
P.O. Box 7888
Van Nuys, CA 91409 (213) 986-4800

Allan M. Shapiro
5900 Sepulveda Blvd., Suite 431
Van Nuys, CA 91411 (818) 781-8515

Albert S. Sheppard#
13610 Valerio St.
Van Nuys, CA 91405 (818) 902-1130

Walter Unterberg#
5709 Burnet Ave.
Van Nuys, CA 91411 (818) 780-6333

Lawrence D. Weber
14416 Hamlin St., Suite 209
Van Nuys, CA 91401 (818) 988-3248

Karl M Manheim
2321 Zeno Pl.
Venice, CA 90291 (213) 822-5005

Kenneth Jennings Hovet
1175 New Bedford Ct.
Ventura, CA 93001 (213) 648-2994

Marvin E. Jacobs
Koppel & Jacobs
2151 Alessandro Dr.
Ventura, CA 93001 (805) 648-5194

Donald A. Streck
2319 Alameda Ave., Suite 2F
Ventura, CA 93003 (805) 644-4035

C. Douglas DeFreytas
Thompson & Thompson
14440 Civic Dr.
Victorville, CA 92392 (619) 245-3450

Dennis Blaine Haase
P.O. Box 1587
Visalia, CA 93291 (209) 733-1844

Joseph R. Dwyer
P.O. Box 3183
Vista, CA 92083 (619) 945-0211

Andrew E. Barlay
2033 N. Main St., Suite 750
Walnut Creek, CA 94596 (415) 746-3241

Edward Brosler
3100 Tice Creek Dr., Apt. 2
Walnut Creek, CA 94595 (415) 932-7693

Murray K. Hatch
1511 Treat Blvd., Suite 200
Walnut Creek, CA 94598 (415) 930-0777

Robert T. Kloeppel
2673 Velvet Way
Walnut Creek, CA 94596 (415) 933-3697

William D. Mc Cann
800 S. Broadway, Suite 410
Walnut Creek, CA 94596 (415) 932-7500

Rankin Allen Milliken
1511 Treat Blvd., Suite 200
Walnut Creek, CA 94598 (415) 930-0777

D. Wendell Osborne#
Dow Chemical Co.
2800 Mitchell Dr.
Walnut Creek, CA 94598 (415) 944-2041

John A. Perona#
31 Hanson Lane
Walnut Creek, CA 94598 (415) 935-0758

Thomas H. Smith#
1490 Whitecliff Way
Walnut Creek, CA 94536 (415) 943-6661

Robert R. Stringham#
Dow Chemical Co.
2800 Mitchell Dr.
Walnut Creek, CA 94598 (415) 944-2041

Manfred Maurice Warren
Warren, Chickering & Grunewald
3266 Ptarmigan, #3B
Walnut Creek, CA 94595 (415) 937-6500

Earl Frederick Kotts
914 W. Linden Rd.
Watsonville, CA 95076 (408) 728-9726

Boniard I. Brown
1500 West Covina Parkway, Suite 113
West Covina, CA 91790 (213) 338-0100

Don A. Hart
Honeywell Inc.
T & C S D
1200 E. San Bernardino Rd.
West Covina, CA 91790 (818) 915-9137

D. James Schwedler
1129 E. Walnut Creek Pkwy.
West Covina, CA 91790 (213) 919-3816

Donald C. Glynn
1019 Barrow Ct.
Westlake Village, CA 91361 (805) 497-4488

Warren T. Jessup
Jessup Beecher & Slehofer
875 Westlake Blvd., Suite 205
Westlake Village, CA 91361 (818) 991-7062

Richard S. Koppel
Koppel & Harris
31255 Cedar Valley Dr., Suite 302
Westlake Village, CA 91362 (805) 497-9633

Richard D. Slehofer
Jessup & Beecher
875 Westlake Blvd., Suite 205
Westlake Village, CA 91361 (213) 991-7062

Edgar Waite Averill, Jr.
Averill & Varn
8244 Painter Ave.
Whittier, CA 90602 (213) 698-8039

Al A. Canzoneri#
15912 Silvergrove Dr.
Whittier, CA 90604 (213) 943-4924

Janis E. Kerber#
11917 Rustic Hill Drive
Whittier, CA 90601 (213) 695-0566

John D. Bauersfeld
Kelly, Bauersfeld & Lowry
21031 Ventura Blvd., Suite 919
Woodland Hills, CA 91364 (818) 347-7900

Gerald L. Cline
Litton Industries, Inc.
5500 Canoga Ave., M/s 30
Woodland Hills, CA 91367 (818) 716-3139

Clark E. DeLarvin
22920 Cass Ave.
Woodland Hills, CA 91364

Donald J. Ellingsberg#
23547 Burbank Blvd.
Woodland Hills, CA 91367 (213) 346-9750

Harold E. Gillmann
Litton Industries, Inc.
5500 Canoga Ave., Ms-30
Woodland Hills, CA 91367 (818) 716-3138

Scott W. Kelley
Pastoriza Kelly & Lowry
21031 Ventura Blvd., Suite 919
Woodland Hills, CA 91364 (818) 347-7900

Sebron Koster
23411 Berdon St.
Woodland Hills, CA 91367 (818) 883-5972

Elliott N. Kramsky
5850 Canoga Ave., Suite 400
Woodland Hills, CA 91367 (818) 992-5221

Stuart O. Lowry
Pastoriza, Kelly & Lowry
21031 Ventura Blvd., Suite 919
Woodland Hills, CA 91364 (818) 347-7900

Milton Madoff#
6200 1/2 Nita Ave.
Woodland Hills, CA 91367 (213) 884-0505

Richard J. McMullen#
5155 Llano Dr.
Woodland Hills, CA 91364 (213) 347-5434

John Joseph Posta, Jr.
5850 Canoga Ave., Suite 400
Woodland Hills, CA 91367 (818) 348-1088

L. David Rish
Litton Systems, Inc.
5500 Canoga Ave., M/s 30
Woodland Hills, CA 91367 (818) 716-3138

Edmund W. Rusche, Jr.
Litton Industries, Inc.
Pat. & Licensing Dept.
5500 Canoga Ave., M S-30
Woodland Hills, CA 91367 (818) 716-3138

Robert Malcolm Sperry, Sr.
23390 Ostronic Dr.
Woodland Hills, CA 91367 (818) 887-0836

A. Donald Stolzy
6301 Glade Ave., Apt. K-110
Woodland Hills, CA 91367 (818) 716-0881

Timothy Thut Tyson
4600 Willens Ave.
Woodland Hills, CA 91364 (213) 477-0578

Robert P. Whipple
5212 Lubad Ave.
Woodland Hills, CA 91364 (818) 884-9029

Ralph B. Pastoriza
Pastoriza & Kelly
21031 Ventura Blvd., Suite 919
Woodlands Hills, CA 91364 (818) 347-7900

Douglas A. Chikin
2995 Woodside Rd.
Suite 400-382
Woodside, CA 94062 (415) 851-8118

Robert R. Meades
5472 Club View Dr.
Yorba Linda, CA 92686 (714) 779-1848

John E. Vanderburgh
6002 Ohio St.
Yorba Linda, CA 92686 (714) 777-3419

Lloyd E.K. Pohl
56020 Santa Fe Trail, Suite 0
Yucca Valley, CA 92284 (619) 365-4515

Colorado

John L. Isaac
13725 W. 67th Circle
Arvada, CO 80004

Donald M. Duft
5078 Cottonwood Dr.
Boulder, CO 80301 (303) 530-0456

John R. Flanagan
1919 19th St.
P.O. Box 13129
Boulder, CO 80308 (303) 449-0884

Lorance L. Greenlee
Greenlee & Associates, P.C.
5370 Manhattan Circle, Suite 201
Boulder, CO 80303 (303) 499-8080

Earl Clark Hancock
3445 Penrose Place, Suite 210
Boulder, CO 80301 (303) 447-2060

Robert E. Harris
5305 Spine Rd., Suite B-East
Boulder, CO 80301 (303) 444-5205

Donald W. Margolis
Klaas & Law
1007 Pearl St.
Boulder, CO 80302 (303) 443-1690

Thomas W. O Rourke
O Rourke & Harris
5305 Spine Rd., Suite B-East
Boulder, CO 80301 (303) 444-5205

John V. Pezdek
Micro Motion, Inc.
7070 Winchester Circle
Boulder, CO 80301 (303) 530-8401

Louis G. Puls#
Museion Research Corp.
P.O. Box 3264
Boulder, CO 80307 (303) 499-8065

Charles Edwin Rohrer
6971 Hunter Pl.
Boulder, CO 80301 (303) 447-3721

Charles L. Sharp, Jr.
1919 14th St., Suite 330
Boulder, CO 80302 (303) 444-2456

Francis A. Sirr
360 Seminole Dr.
Boulder, CO 80303 (303) 499-9161

Sally A. Sullivan#
Agrigenetics Research Corp.
3375 Mitchell Lane
Boulder, CO 80301 (303) 443-5900

Ellen P. Winner
Agrigenetics Corp.
3375 Mitchell Lane
Boulder, CO 80301 (303) 443-5900

Carl M. Wright
646 Furman Way
Boulder, CO 80303 (303) 494-8237

Harold K. Johnston
26 Garden Center, Suite 3
Broomfield, CO 80020 (303) 466-1787

Karl E. Bring
Hewlett Packard Co.
Legal Dept.
8245 N. Union Blvd.
Colorado Springs, CO 80918 (303) 590-5400

Richard Dean Dixon
Ford Microelectronics Inc.
10340 Highway 83
Colorado Springs, CO 80908 (303) 528-7612

Lawrence Stephen Galka
108 E. Cheyenne Rd.
Colorado Springs, CO 80906 (303) 633-4444

Linda N.F. Gould
Berniger, Berg, Rioth & Diver
P.O. Box 1716
Colorado Springs, CO 80901 (303) 475-9900

Richard W. Hanes
Holland & Hart
1400 Holly Sugar Bldg.
Colorado Springs, CO 80901 (303) 475-7730

Anthony W. Raskob, Jr.
823 East High St.
Colorado Springs, CO 80903

Floyd Trimble
125 Dolomite Dr.
Colorado Springs, CO 80919 (303) 599-7514

Gregg I. Anderson
Holland & Hart
555 17th St., Suite 2900
Denver, CO 80202 (303) 295-8000

Martin G. Anderson
Martin Marietta Denver Aerospace
P.O. Box 179
M S 1400
Denver, CO 80201 (303) 977-6474

Glenn K. Beaton
Holme, Roberts & Owen
1700 Broadway, Suite 1800
Denver, CO 80290 (303) 861-7000

David E. Boone#
SSheridan, Ross & Mc Intosh
4155 E. Jewell Ave., Suite 714
Denver, CO 80222 (303) 759-9050

Theresa A. Brown#
Sheridan, Ross & Mc Intosh
4155 E. Jewell, Suite 700
Denver, CO 80222 (303) 759-9050

Duane C. Burton
1100 Writer Commons
1720 South Bellaire St.
Denver, CO 80222 (303) 691-9119

William Scott Carson
Dorr, Carson, Sloan & Peterson
3010 E. 6th Ave.
Denver, CO 80206 (303) 333-3010

Curtis H. Castleman, Jr.
Gates Corp.
900 S. Broadway
Denver, CO 80209 (303) 744-4685

Gary J. Connell#
Sheridan, Ross & Mc Intosh
4155 E. Jewell Ave., Suite 714
Denver, CO 80222 (303) 759-9050

Charles C. Corbin
1050 Lafayette, Apt. 304
Denver, CO 80218 (303) 831-7323

Ralph F. Crandell
Holland & Hart
555 17th St., Suite 2900
P.O. Box 8749
Denver, CO 80201 (303) 295-8390

Phillip L. De Arment
Martin Marietta Corp.
Mail No. T-5009
P.O. Box 179
Denver, CO 80201 (303) 971-1190

Robert C. Dorr
Dorr, Carson, SLoan & Peterson
3010 E. 6th Ave.
Denver, CO 80206 (303) 333-3010

Jack E. Ebel
Gates Corp.
900 S. Broadway
P.O. Box 5887
Denver, CO 80217 (303) 744-1911

William Griffith Edwards
1700 Broadway, Suite 1200
Denver, CO 80290 (303) 861-1456

Frank A. Elzi
1638 S. Jasmine St.
Denver, CO 80224 (303) 756-4656

Gregory A. Evearitt
6301 W. Hampden, #6-103
Denver, CO 80227 (303) 989-3675

Gary D. Fields
Fields, Lewis, Pittenger & Rost
1720 Bellaire St., Suite 1100
Denver, CO 80222 (303) 758-8400

Giles Galahad
5320 W. 29th Ave.
Denver, CO 80214 (303) 233-5302

Joseph C. Herring
Herring & Associates
3257 S. Steele St.
Denver, CO 80210 (303) 756-2372

Patrick M. Hogan
Martin Marietta Corp.
Mail D C-1010
P.O. Box 179
Denver, CO 80201 (303) 977-6109

Michael R. Hope
Roath & Brega, P. C.
1700 Writers Center Five
1873 South Bellaire St.
Denver, CO 80222 (303) 691-5400

Conrad B. Houser
Mobil Oil Corp.
Box 17772
Denver, CO 80217 (303) 293-6100

Richard L. Hughes
Sheridan, Ross & Mc Intosh
4155 E. Jewell Ave., Suite 700
Denver, CO 80222 (303) 759-9050

Joseph J. Kelly
Klaas & Law
738 Pearl St.
Denver, CO 80203 (303) 837-1616

Bruce G. Klaas
Klaas & Law
738 Pearl St.
Denver, CO 80203 (303) 837-1616

Robert M. Krone
Manville Service Corp.
Box 5723
Ken-Caryl Ranch
Denver, CO 80217 (303) 978-2101

Ancel W. Lewis, Jr.
Fields, Lewis, Pittenger & Rost
1100 Writer Commons Bldg.
1720 South Bellaire St.
Denver, CO 80222 (303) 482-2841

John R. Ley
Crandell & Polumbus
1451 Larimer St., Suite 300
Denver, CO 80202 (303) 571-1525

John David Lister
Manville Service Corp.
P.O. Box 5723
Denver, CO 80217 (303) 978-2159

Frank C. Lowe
Klaas & Law
738 Pearl St.
Denver, CO 80203 (303) 837-1616

Thomas R. Marsh
Sheridan, Ross & Mc Intosh
4155 E. Jewell Ave., Suite 814
Denver, CO 80222 (303) 759-9050

Timothy J. Martin
Young & Martin
Right Bank Bldg.
1401 Blake St., Suite 201
Denver, CO 80202 (303) 534-1577

C.E. Martine, III
Rothgerber, Appel, Powers & Johnson
One Tabor Center, Suite 2800
1200 Seventeenth St.
Denver, CO 80202 (303) 623-9000

John S. Mc Guire
Roath & Brega, P.C.
1700 Writers Center Five
1873 S. Bellaire St.
Denver, CO 80222 (303) 691-5411

Michael D. Mc Intosh
Sheridan, Ross & Mc Intosh
4155 E. Jewell Ave., Suite 814
Denver, CO 80222 (303) 759-9050

John A. Mc Kinney
Holme, Roberts & Owen
1700 Broadway, Suite 1800
Denver, CO 80290 (303) 861-7000

John H. Miller
Manville Corp.
Ken Caryl Ranch
Denver, CO 80217 (303) 978-3218

Thomas C. Naber
Klaas & Law
738 Pearl St.
Denver, CO 80203 (303) 837-1616

Loren D. Nelson#
Ophir Corp.
240 S. Broadway
Denver, CO 80209 (303) 744-7930

Gregory W. O Connor
Manville Service Corp.
P.O. Box 5723
Denver, CO 80217 (303) 978-2154

William P. O Meara
Klaas & Law
738 Pearl St.
Denver, CO 80203 (303) 837-1616

H.W. Oberg, Jr.
The Gates Corp.
900 S. Broadway
Denver, CO 80217 (303) 744-4743

James P. Oxenham
50 S. Steele St., Suite 580
Denver, CO 80209 (303) 320-4580

Michael J. Pfister
8225 E. Lehigh Ave.
Denver, CO 80237 (303) 779-9018

James E. Pittenger
Field, Lewis, Pittenger & Rost
1100 Writer Commons Bldg.
1720 South Bellaire St.
Denver, CO 80222 (303) 758-8400

Gary M. Polumbus
Crandell & Polumbus
1451 Larimer St., Suite 300
Denver, CO 80202 (303) 571-1525

Cornelius P. Quinn
Johns-Manville Corp.
Box 5723
Ken-Caryl Ranch
Denver, CO 80217 (303) 978-2397

John Edward Reilly
1554 Emerson St.
Denver, CO 80218 (303) 830-2014

Kyle W. Rost
Fields, Lewis, Pittenger & Rost
1100 Writer Commons Bldg.
1720 South Bellaire St.
Denver, CO 80222 (303) 758-8400

Timothy R. Schulte
Manville Corp.
Ken Caryl Ranch
P.O. Box 5723
Denver, CO 80217 (303) 978-2701

Philip H. Sheridan
Sheridan, Ross & Mc Intosh
4155 E. Jewell Ave., Suite 714
Denver, CO 80222 (303) 759-9050

Jack C. Sloan
Dorr, Carson, Sloan & Peterson
3010 E. 6th Ave.
Denver, CO 80206 (303) 333-3010

Brian D. Smith
Fields, Lewis, Pittenger & Rost
1100 Writer Commons
1720 S. Bellaire St.
Denver, CO 80222 (303) 758-8400

Edwin Leroy Spangler, Jr.
1200 United Bank Ctr.
1700 Broadway
Denver, CO 80290 (303) 861-4876

Jon R. Stark
Holme, Roberts & Owen
1700 Broadway
Denver, CO 80290 (303) 861-7000

Ruel C. Terry#
3090 S. High St.
Denver, CO 80210 (303) 759-3826

Roy K. Uenishi#
1255-19th St., Apt. 402
Denver, CO 80202 (303) 294-9455

H. B. Van Valkenburgh, III
Klaas & Law
738 Pearl St.
Denver, CO 80203 (303) 837-1616

Douglass Vincent
Fields, Lewis, Pittenger & Rost
1100 Writer Commons Bldg.
1720 S. Bellaire St.
Denver, CO 80222 (303) 758-8400

Glenn Lowell Webb#
1961 Roslyn St.
Denver, CO 80220 (303) 377-8133

Paul J. White
2212 Osceloa St.
Denver, CO 80212 (303) 433-8771

Ronald C. Williams
2413 Washington St., Suite 250
Denver, CO 80205 (303) 295-0521

Lesley S. Witt
Sheridan, Ross & Mc Intosh
4155 E. Jewell Ave., Suite 714
Denver, CO 80222 (303) 759-9050

Max L. Wymore
3441 S. Ivy Way
Denver, CO 80222 (303) 756-9162

Thomas H. Young
Rothgerber, Appel, Powers & Johnson
One Tabor Center
1200 17th St., Suite 2800
Denver, CO 80202 (303) 623-9000

Lee W. Zieroth
Cohen, Brame & Smith
3500 Amoco Bldg.
1670 Broadway
Denver, CO 80202 (303) 837-8800

David F. Zinger
Sheridan, Ross & Mc Intosh
4155 E. Jewell Ave., Suite 714
Denver, CO 80222 (303) 759-9050

Maxwell C. Freudenberg
P.O. Box 841
Durango, CO 81301 (303) 259-2765

Arthur A. March
Rogers Hoge & Hills
5350 S. D T C Parkway
Englewood, CO 80111 (303) 740-9307

Robert E. Purcell#
5350 S. DTC Parkway
Denver Tech. Center
Englewood, CO 80111 (303) 773-9222

Mary L. Wakimura#
3209 S. Emerson
Englewood, CO 80110 (303) 781-5840

William D. Sabo
MCCA, Inc. 35885
P.O. Box 2870
Estes Park, CO 80517 (303) 586-4120

Carl C. Batz
2207 Charolais Dr.
Fort Collins, CO 80526 (303) 484-9023

Hugh H. Drake
P.O. Box 727
Fort Collins, CO 80522 (303) 493-0123

Dean P. Edmundson
Stuart Professional Park
1136 E. Stuart St., Suite 4201
Fort Collins, CO 80525 (303) 224-9502

Luke Santangelo
Fischer, Brown, Huddleson & Gunn
215 West Oak, Suite 1100
Fort Collins, CO 80521 (303) 482-1056

John P. Bailey
240 Old Y Road
Golden, CO 80401 (303) 526-0099

John Howard Barney#
Tosco Corp.
18200 W. Highway 72
Golden, CO 80403 (303) 425-6021

Thomas S. Birney
Bradley, Campbell & Carney, P.C.
1717 Wash. Ave.
Golden, CO 80401 (303) 278-3300

Terrence L. J. Clausen
Solar Energy Res. Inst.
1617 Cole Blvd.
Golden, CO 80401 (303) 231-7191

Jerald J. Devitt
Devitt & Weiszmann
1301 Arapahoe St., Suite 300
Golden, CO 80401 (303) 279-3344

Stephen A. Gratton
Gratton & Ebel
1250 Orchard Rd.
Golden, CO 80401 (303) 237-5883

Kenneth L. Richardson
Solar Energy Research Inst.
1617 Cole Blvd.
Golden, CO 80401 (303) 231-7724

Ronald Fredric Weiszmann
Devitt & Weiszmann
1301 Arapahoe St., Suite 300
Golden, CO 80401 (303) 279-3344

Wilbur A. E. Mitchell
907 - 10th Ave.
Greeley, CO 80631 (303) 352-3360

Raymond Fink
Bailey, Wilson, Wagenhals & Finegan
Union Tower, Suite 700
164 S. Union Blvd.
Lakewood, CO 80228 (303) 989-5050

Rodney F. Brown
Marathon Oil Co.
P.O. Box 269
Littleton, CO 80160 (303) 794-2601

Monte L. Gleason
P.O. Box 2034
Littleton, CO 80161 (303) 295-5586

Jack L. Hummel
Marathon Oil Co.
7400 S. Broadway
P.O. Box 269
Littleton, CO 80160 (303) 794-2601

Richard A. Kulp#
6861 So. Clayton Way
Littleton, CO 80122 (303) 741-3969

Edna M. O Connor
6390 W. David Dr.
Littleton, CO 80123 (303) 979-7378

Norvell Edward Von Behren
7463 S. Marion St.
Littleton, CO 80122 (303) 794-9516

James E. Nelson
6255 Niwor Rd.
Longmont, CO 80501 (303) 530-7247

James Ralph Young
Premiere Bldg., Suite P-206
700 Florida Ave.
Longmont, CO 80501 (303) 651-1042

Christopher J. Byrne
Hewlett-Packard Co.
815 S.W. 14th St.
Loveland, CO 80537 (303) 667-5000

William W. Cochran, II
Hewlett-Packard Company
815 S.W. 14th St.
Loveland, CO 80537 (303) 667-5000

Jeffery b. Fromm
Hewlett-Packard Co.
Legal Dept.
815 14th St., S.W.
Loveland, CO 80537 (303) 667-5000

William E. Hein
P.O. Box 335
Loveland, CO 80539 (303) 667-6741

Edward L. Miller#
Hewlett-Packard Co.
815 S.W. 14th St.
Loveland, CO 80537 (303) 667-5000

Donald W. Erickson
P.O. Box 482
Road W35
Norwood, CO 81423 (303) 327-4803

James M. Graziano
4662 Weld Co. Rd. 34
Platteville, CO 80651 (303) 535-4743

James R. Young
11078 Cotton Cr. Dr.
Westminster, CO 80030 (303) 469-2680

David S. Woronoff
P.O. Box 1823
Windsor, CO 80550 (303) 686-5458

Connecticut

Robert J. Galiette
10 Wilcox Lane
Avon, CT 06001 (203) 677-4810

Joseph R. Spalla
17 Oxbow Dr.
Avon, CT 06001 (203) 673-9201

Aziz M. Ahsan
Union Carbide Corp.
P.O. Box 875
Bethel, CT 06801 (203) 794-6140

Joesph R. Carvalko, Jr.
93 Redwood Dr.
Bethel, CT 06801 (203) 792-3340

Robert R. Hubbard
13 Fawn Rd.
Bethel, CT 06801 (203) 744-3527

Ronald Erik Brown
229-2 Wintonbury Ave.
Bloomfield, CT 06002 (203) 243-3908

Lawrence A. Cox#
CIGNA Corp.
900 Cottage Grove Rd.
Bloomfield, CT 06002 (203) 726-8930

George B. Yntema
61 Vernon Rd.
Bolton, CT 06040 (203) 643-9358

Charlie Smith
12 Sandra Dr.
Brandford, CT 06405

John R. Doherty
49 Cannon St.
P.O. Box 9296
Bridgeport, CT 06601 (203) 579-8794

James E. Espe
General Elec. Co.
International Pat. Operation
1285 Boston Ave.
Bridgeport, CT 06601 (203) 382-3515

Ernest Gergely
General Electric Corp.
1285 Boston Ave.
Bldg. 23 C W
Bridgeport, CT 06602 (203) 382-4792

Philip G. Luckhardt
General Electric Co.
1285 Boston Ave.
Bridgeport, CT 06602 (203) 382-2625

Charles R. Miranda
Remington Products, Inc.
60 Main St.
Bridgeport, CT 06606 (203) 367-4400

Michael J. Pantuliano
General Electric Co.
1285 Boston Ave.
Bridgeport, CT 06602 (203) 382-2000

Leonard J. Platt
General Electric Co.
Bldg., 21-B W
1285 Boston Ave.
Bridgeport, CT 06602 (203) 382-2685

John R. Rafter, II
General Elec. Co.
1285 Boston Ave., 23 C W
Bridgeport, CT 06601 (203) 382-4706

Bernard E. Shay#
General Electric Co.
1285 Boston Ave.
Bridgeport, CT 06601 (203) 382-2116

Nicholas Skovran, Sr.
Remington Arms Co., Inc.
939 Barnum Ae.
P.O. Box 1939
Bridgeport, CT 06601 (203) 333-3041

Lloyd P. Stauder
Warnaco Inc.
350 Lafayette St.
Bridgeport, CT 06601 (203) 579-8094

Thomas L. Tully
Cifelli, Frederick & Tully
One Lafayette Circle
P.O. Box 1180
Bridgeport, CT 06601 (203) 367-5325

Donald W. Walk
General Electric Co.
Bldg. 23 C W
1285 Boston Ave.
Bridgeport, CT 06601 (203) 382-2000

Edwin V. Ladd, Jr.
184 Maxine Rd.
Bristol, CT 06010

Edward Herbert#
Rte. 44
Canton, CT 06019 (203) 693-2204

Dale Lynn Carlson
Olin Corp.
350 Knotter Dr.
Cheshire, CT 06410 (203) 271-4059

Ralph D Alessandro
Olin Research Ctr.
350 Knotter Dr.
Cheshire, CT 06410 (203) 271-4055

James B. Haglind#
Olin Corporation
350 Knotter Dr.
Cheshire, CT 06410 (203) 271-4057

Thomas P. O Day
Olin Corp.
P.O. Box 586
350 Knotter Dr.
Cheshire, CT 06787 (203) 271-4052

William A. Simons
Olin Corp.
350 Knotter Dr.
P.O. Box 586
Cheshire, CT 06410 (203) 271-4063

Charles G. Nessler
P.O. Box H
Chester, CT 06412 (203) 526-9149

Curtis W. Carlson
277 Cognewaugh Rd.
Cos Cob, CT 06807 (203) 661-2210

Norman L. Balmer
Union Carbide Corp.
39 Old Ridgebury Rd., E-1266
Danbury, CT 06817 (203) 794-6343

Paul L. Bollo
57 North St., Suite 210
Danbury, CT 06810 (203) 798-8360

Robert C. Brown
Union Carbide Corp.
Law Dept. E-3268
39 Old Ridgebury Rd.
Danbury, CT 06817 (203) 794-6207

Shirley L. Church
Union Carbide Corp.
Law Dept. E2-258
39 Old Ridgebury Rd.
Danbury, CT 06817 (203) 794-6130

Gerald L. Coon
Union Carbide Corp.
Law Dept.-Pat. Sect.
39 Old Ridgebury Rd.
Danbury, CT 06817 (203) 794-6221

Aldo John Cozzi
Union Carbide Corp.
Law Dept.-Pat. Sect.
39 Old Ridgebury Rd.
Danbury, CT 06817 (203) 794-6197

Clyde V. Erwin, Jr.
Union Carbide Corp.
Law Dept. E2
39 Old Ridgebury Rd.
Danbury, CT 06817 (203) 794-6350

J. Hart Evans
Union Carbide Corp.
39 Old Ridgebury Rd.
E3-267
Danbury, CT 06817 (203) 794-6201

Robert J. Feltovic
Union Carbide Corp.
Law Dept. - E1 254
39 Old Ridgebury Rd.
Danbury, CT 06817 (203) 794-6242

Reynold Joseph Finnegan
Union Carbide Corp.
Old Ridgebury Rd.
Danbury, CT 06817 (203) 794-6336

Steven H. Flynn
Union Carbide Corp.
Law Dept.
Old Ridgebury Rd.
Danbury, CT 06817 (203) 794-6339

Alvin H. Fritschler
Union Carbide Corp.
Law Dept. Pat. Sect.
39 Old Ridgeway Rd.
Danbury, CT 06817 (203) 794-6116

Henry H. Gibson
Union Carbide Corp.
39 Old Ridgebury Rd.
E1-265
Danbury, CT 06817 (203) 794-6360

Harrie M. Humphreys
Union Carbide Corp.
Law Dept. Pat. Sect.
Old Ridgebury Rd.
Danbury, CT 06817 (203) 794-6109

Haynes N. Johnson
Johnson & Chambliss
158 Deer Hill Ave.
Danbury, CT 06810 (203) 792-9527

Lawrence G. Kastriner
Union Carbide Corp.
Old Ridgebury Rd.
Danbury, CT 06817 (203) 794-2000

Marylin Klosty
Union Carbide Corp.
Law Dept. Pat. Sect.
Old Ridgebury Rd.
Danbury, CT 06817 (203) 794-6237

Stanley Ktorides
Union Carbide Corp.
Old Ridgebury Rd.
Danbury, CT 06817 (203) 794-6122

John Curtis Le Fever
Union Carbide Corp.
Old Ridgebury Rd.
Danbury, CT 06817 (203) 794-6151

Paul W. Leuzzi, III
Union Carbide Corp.
Old Ridgebury Rd.
Danbury, CT 06817 (203) 794-6346

Jean B. Mauro
Union Carbide Corp.
Old Ridgebury Rd.
Danbury, CT 06817 (203) 794-6394

Frederick J. Mc Carthy, Jr.
Union Carbide Corp.
Law Dept. Pat. Sect.
Old Ridgebury Rd.
Danbury, CT 06817 (203) 794-5012

Richard G. Miller
Union Carbide Corp.
Law Dept. Pat. Sect.
Old Ridgebury Rd.
Danbury, CT 06817 (203) 794-6131

Cornelius Francis O Brien
Union Carbide Corp.
Law Dept. E2-260
Old Ridgebury Rd.
Danbury, CT 06817 (203) 794-6160

Gerald R. O Brien, Jr.
Union Carbide Corp.
Law Dept. Pat. Sect.
Old Ridgebury Rd.
Danbury, CT 06817 (203) 794-6245

Thomas I. O Brien
Union Carbide Corp.
Old Ridgebury Rd.
Danbury, CT 06817 (203) 794-6194

Donald M. Papuga
Union Carbide Corp.
Old Ridgebury Rd.
Danbury, CT 06817 (203) 794-6309

Peter W. Peterson
Union Carbide Corp.
39 Old Ridgebury Rd.
Danbury, CT 06817 (203) 794-6171

John S. Piscitello
Union Carbide Corp.
Law Dept. Pat. Sect.
Old Ridgeway Rd.
Danbury, CT 06817 (203) 794-6241

Leo A. Plum
Union Carbide Corp.
Old Ridgebury Rd.
Danbury, CT 06817 (203) 794-6280

Morris N. Reinisch
Union Carbide Corp.
Old Ridgebury Rd.
Danbury, CT 06817 (203) 794-6245

Daniel Reitenbach
Union Carbide Corp.
Law Dept. E-3
Old Ridgebury Rd.
Danbury, CT 06817

George A. Skoler
Union Carbide Corp.
Law Dept. Pat. Sect.
Old Ridgebury Rd.
Danbury, CT 06817 (203) 794-6306

James L. Sonntag
Union Carbide Corp.
Law Dept. Pat. Sect.
Old Ridgebury Rd.
Danbury, CT 06817 (203) 794-6357

Dominic J. Terminello
Union Carbide Corp
Law Dept. Pat. Sect.
39 Old Ridgebury Rd.
E-3271
Danbury, CT 06817 (203) 794-5015

Eugene C. Trautlein
Union Carbide Corp.
39 Old Ridgebury Rd.
Danbury, CT 06817 (203) 794-6303

Clement J. Vicari
Union Carbide Corp.
Old Ridgebury Rd.
Danbury, CT 06817 (203) 794-6224

Gary L. Wamer
First Brands Corp.
Law Dept. J-1
39 Old Ridgebury Rd.
Danbury, CT 06817 (203) 794-4945

S. A. Giarratana
4 Short Lane
Darien, CT 06820 (203) 655-7161

James F. Snowden
241 Hollow Tree Ridge Rd.
Darien, CT 06820 (203) 655-0735

William J. Sapone
United Technologies Corp.
One Riverview Square
East Hartford, CT 06108 (203) 548-2569

Loren P. Stolp
United Technologies Corp.
Pratt & Whitney Aircraft
400 Main St.
East Hartford, CT 06108 (203) 565-6452

Rayumond E. Stone, Jr.
United Technologies Corp.
Pratt & Whitney Mfg. Div.
400 Main St.
East Hartford, CT 06108 (203) 565-2847

John Blevney Willard
363 Ellington Rd.
East Hartford, CT 06108 (203) 289-7471

George C. Butenkoff
169 Wells Rd.
East Windsor, CT 06088 (203) 623-0441

Benton Blair
Rose Lane
East Woodstock, CT 06244 (203) 928-2334

H. Gibner Lehmann#
5 Kent Rd.
Easton, CT 06612 (203) 372-7695

K. Gibner Lehmann#
5 Kent Rd.
Easton, CT 06612 (203) 372-7695

Barry R. Lipsitz
55 Sweetbriar Trail
Easton, CT 06612 (203) 268-4440

Herman A. Michelson
12 Princess Pine Lane
Easton, CT 06612 (203) 374-7346

Jan H. Van Den Beemt
23 Partridge Hill
Essex, CT 06426 (203) 767-8907

Kenneth B. Adolphson
Mattern, Ware, Stoltz & Fressola
34 Sherman Ct.
P.O. Box 783
Fairfield, CT 06430 (203) 255-8881

Alfred A. Fressola
Mattern, Ware, Stoltz & Fressola
34 Sherman Ct.
Fairfield, CT 04643 (203) 255-8881

Robert S. Friedman
General Elec. Co.
3135 Easton Turnpike
Fairfield, CT 06431 (203) 373-3318

Clarence Arthur Green
Perman & Green
425 Post Rd.
Fairfield, CT 06430 (203) 259-1800

Mark F. Harrington
Perman & Green
425 Post Rd.
Fairfield, CT 06430 (203) 259-1800

Maurice M. Klee
1951 Burr St.
Fairfield, CT 06430 (203) 255-1400

David N. Koffsky
Perman & Green
425 Post Rd.
Fairfield, CT 06430 (203) 259-1800

Peter W. Krehbiel
Babcock Industries Inc.
425 Post Rd.
Fairfield, CT 06430 (203) 255-7158

Harry F. Manbeck, Jr.
General Electric Co.
3135 Easton Turnpike
Fairfield, CT 06431 (203) 373-2446

Martey Robert Perman
Perman & Green
425 Post Rd.
Fairfield, CT 06430 (203) 259-1800

David Enoch Pitchenik
113 Wagon Hill Rd.
Fairfield, CT 06430 (203) 255-6410

George Richard Powers
General Electric Co.
3135 Easton Turnpike
Fairfield, CT 06431 (203) 373-2835

Arthur V. Puccini
General Electric Co.
3135 Easton Turnpike - W2E
Fairfield, CT 06431 (203) 373-3374

Harry F. Smith
Perman & Green
425 Post Rd.
Fairfield, CT 06430 (203) 259-1800

Melvin I. Stoltz
Mattern, Ware, Stoltz & Fressola
34 Sherman Ct.
Fairfield, CT 06430 (203) 255-8881

Robert H. Ware
Mattern, Ware, Stoltz & Fressola
34 Sherman Ct.
Fairfield, CT 06430 (203) 255-8881

Peter R. Bahn
744 New Britain Ave.
Farmington, CT 06032 (203) 674-0813

Barry E. Deutsch
Emhart Corp.
426 Colt Highway
Farmington, CT 06032 (203) 678-3322

Charles Morgan Hussey
Emhart Corp.
426 Colt Hwy.
Farmington, CT 06032 (203) 678-3226

H. Samuel Kieser
Emhart Corp.
426 Colt Highway
Farmington, CT 06032 (203) 677-4631

Robert T. Casey
71 Frank St.
Forestville, CT 06010 (203) 583-1515

Jeffrey E. Kushin#
97 Deerfield Dr.
Glastonbury, CT 06033 (203) 633-3066

Edward A. Siegel
16 Byron Dr.
Grandy, CT 06035 (203) 653-4162

Paul R. Audet
American Can Packaging Inc.
American Lane
Greenwich, CT 06830 (203) 552-3260

Ernestine C. Bartlett
American Can Packaging Inc.
American Lane
P.O. Box 2600
Greenwich, CT 06836 (203) 552-3261

Michael A. Ciomek
Amax Inc.
55 Railroad Ave.
Greenwich, CT 06836 (203) 629-6119

David A. Frank
Primerica Corp.
American Lane
Greenwich, CT 06830 (203) 552-4326

James Benton Grant
Cummings & Lockwood
Two Greenwich Plaza
Greenwich, CT 06830 (203) 869-1200

James Joseph Mc Keever#
20 Flower Lane
Greenwich, CT 06830 (203) 622-6550

Alfred E. Miller
150 Mason St.
Greenwich, CT 06830 (203) 869-8663

Gregg C. Benson
Pfizer Incorp.
Eastern Point Rd.
Groton, CT 06340 (203) 441-4901

Robert K. Blackwood#
Pfizer, Inc.
Legal Div.
Eastern Point Rd.
Groton, CT 06340 (203) 441-4905

William Carter Everett
General Dynamics Corp.
Electric Boat Div.
Eastern Point Rd.
Groton, CT 06340 (203) 441-8252

Albert Edward Frost#
Pfizer Inc.
Eastern Point Rd.
Groton, CT 06340 (203) 441-4904

Gregory E. Gardiner
Pfizer Inc.
Eastern Point Rd.
Groton, CT 06340 (203) 441-3176

James M. Mc Manus#
Pfizer Inc.
Eastern Point Rd.
Groton, CT 06340 (203) 445-5611

John S. Oki
Pfizer Incorporated
Eastern Point Rd.
Groton, CT 06340 (203) 441-3500

Paul D. Thomas#
271 Plant St.
Groton, CT 06340

Paul J. Lerner
47 Clapboard Hill Rd.
Guilford, CT 06437 (203) 453-9793

Robert A. Seemann#
89 Earl Ave.
Hamden, CT 06514 (203) 281-6449

James E. Alix
Chilton, Alix & Van Kirk
750 Main St.
Hartford, CT 06103 (203) 527-9211

Dominic J. Chiantera
United Technologies Corp.
Pat. Dept.
Hartford, CT 06101 (203) 548-2526

Ralph H. Chilton
Chilton, Alix & Van Kirk
750 Main St.
Hartford, CT 06103 (203) 527-9211

Alan C. Cohen
United Technologies Corp.
United Technologies Bldg.
Pat. Dept.
Hartford, CT 06101 (203) 548-2505

Peter Louis Costas
Costas, Montgomery & Dorman, P.C.
3 Lewis St.
Hartford, CT 06103 (203) 278-9892

Anthony J. Criso
United Technologies Corp.
Essex Group, Inc.
1 Financial Plaza
Hartford, CT 06101 (219) 461-4495

John David Del Ponti
Emhart Corp.
P.O. Box 2730
Hartford, CT 06101 (203) 678-3440

Lloyd D. Doigan
United Technologies Corp.
United Technologies Bldg.
Hartford, CT 06101 (203) 548-2567

Joseph A. Fischetti#
Mc Cormick, Paulding & Huber
266 Pearl St.
Hartford, CT 06103 (203) 549-5290

Gene D. Fleischhauer
United Technologies Corp.
United Tech. Bldg.
Pat. Dept.
Hartford, CT 06101 (203) 548-2514

Norman Friedland
United Technologies Corp.
One Financial Plaza
Hartford, CT 06101 (203) 548-2534

Robert E. Greenstien
United Technologies Corp.
Hartford, CT 06101 (203) 548-2536

James K. Grogan
Mc Cormick, Paulding & Huber
266 Pearl St.
Hartford, CT 06103 (203) 549-5290

Harry J. Gwinnell
United Technologies Corp.
Patent Dept.
Hartford, CT 06101 (203) 548-2508

Frederick J. Haesche
Mc Cormick, Paulding & Huber
266 Pearl St.
Hartford, CT 06103 (203) 549-5290

Vernon F. Hauschild
Mc Cormick, Paulding & Huber
266 Pearl St.
Hartford, CT 06103 (203) 548-2527

Donald Joseph Hayes
Hayes & Reinsmith
Cityplace
Hartford, CT 06103 (203) 727-9956

Robert P. Hayter
United Technologies Corp.
Patent Dept.
Hartford, CT 06101 (203) 548-2510

Samuel J. Henderson
Schatz, Schatz, Ribicoff & Kotkin
1 Financial Plaza
Hartford, CT 06103 (203) 522-3234

John C. Hilton
Mc Cormick, Paulding & Huber
266 Pearl St.
Hartford, CT 06103 (203) 549-5290

Donald Keith Huber
Mc Cormick, Paulding & Huber
266 Pearl St.
Hartford, CT 06103 (203) 549-5290

Thomas A. Kahrl
Connecticut National Bank
777 Main St.
Hartford, CT 06115 (203) 728-4669

Vernon F. Kalb
Prutzman, Kalb, Chilton & Alix
750 Main St.
Hartford, CT 06103 (203) 527-9211

John C. Linderman
Mc Cormick, Paulding & Huber
266 Pearl St.
Hartford, CT 06103 (203) 549-5290

Russell M. Lipes, Jr.
United Technologies Corp.
Pat. Section
Legal Dept.
Hartford, CT 06101 (203) 548-2504

Robert E. Lucia#
Emhart Corp.
P.O. Box 2730
Hartford, CT 06101 (203) 677-4631

Francis J. Maguire, Jr.
United Technologies Corp.
Pat. Dept.
United Technologies Bldg.
Hartford, CT 06101 (203) 548-2538

Roger B. Mc Cormick
Mc Cormick, Paulding & Huber
266 Pearl St.
Hartford, CT 06103 (203) 549-5290

John H. Midney
Emhart Corp.
P.O. Box 2730
Hartford, CT 06101 (203) 677-4631

Arthur Dean Olson
United Technologies Corp.
Pat. Dept.
Hartford, CT 06101 (203) 548-2517

Jack Michael Pasquale
Mc Cormick, Paulding & Huber
266 Pearl St.
Hartford, CT 06103 (203) 549-5290

Theodore Roy Paulding
Mc Cormick, Paulding & Huber
266 Pearl St.
Hartford, CT 06103 (203) 549-5290

Eric W. Petraske
United Technologies Corp.
United Technologies Bldg.
Hartford, CT 06101 (203) 548-2537

John M. Prutzman
Prutzman, Kalb, Chilton & Alix
750 Main St.
Hartford, CT 06103 (203) 527-9211

James M. Rashid
United Technologies Corp.
Patent Dept.
Hartford, CT 06101 (203) 548-2539

R. William Reinsmith
Hayes & Reinsmith
Cityplace
Hartford, CT 06103 (203) 727-9956

Stephen E. Revis
United Technologies Corp.
Patent Dept.
Hartford, CT 06101 (203) 548-2533

Brian L. Ribando
Litton Industries, Inc.
179 Allyn St., Suite 508
Hartford, CT 06103 (203) 247-1355

Robert P. Sabath
United Technologies Corp.
United Technologies Bldg.
One Financial Plaza
Hartford, CT 06101 (203) 548-2530

Arthur J. Samodovit#
Mc Cormick, Paulding & Huber
266 Pearl St.
Hartford, CT 06103 (203) 549-5290

Stephen A. Schneeberger
United Technologies Corp.
United Technologies Bldg.
Hartford, CT 06101 (203) 548-2509

Emil Richard Skula
United Technologies Corp.
Patent Dept.
United Technologies Bldg.
Hartford, CT 06101 (203) 548-2558

Robert S. Smith
57 Pratt St., Suite 513
Hartford, CT 06103 (203) 249-1857

Spencer T. Smith
Emhart Corp.
Patent Dept.
P.O. Box 2730
Hartford, CT 06101 (203) 678-3426

Troxell K. Snyder
United Technologies Corp.
Patent Dept.
Hartford, CT 06101 (203) 548-2571

Charles E. Sohl
United Technologies Corp.
Pat. Sect.-Legal Dept.
United Technologies Bldg.
Hartford, CT 06101 (203) 548-2506

John M. Swiatocha
United Technologies Corp.
Patent Dept.
Hartford, CT 06101 (203) 548-2513

Thomas S. Szatkowski
Emhart Corp.
Patent Dept.
P.O. Box 2730
Hartford, CT 06101 (203) 678-3443

Roger A. Van Kirk
Prutzman, Kalb, Chilton & Alix
750 Main St.
Hartford, CT 06103 (203) 527-9211

Robert C. Walker
United Technologies Corp.
1 Financial Plaza
Hartford, CT 06101 (203) 548-2503

Melvin Pearson Williams
United Technologies Corp.
Legal Dept.-Pat. Sec.
Hartford, CT 06101 (203) 548-2548

Guy D. Yale
Prutzman, Kalb, Chilton & Alix
750 Main St.
Hartford, CT 06103 (203) 527-9211

Paul D. Greeley
12 Brookfield Dr.
Huntington, CT 06484 (203) 929-6403

Lawrence A. Cavanaugh
91 Randeckers Lane
Kensington, CT 06037 (203) 229-8502

E. Seward Stevens
West St.
Litchfield, CT 06759 (203) 567-0821

Eldon Harmon Luther
Joshuatown Rd.
RFD #2
Lyme, CT 06371

Ronald G. Cummings
308 Race Hill Rd.
Madison, CT 06443 (203) 421-4864

William W. Jones
6 Juniper Lane
Madison, CT 06443 (203) 245-2418

Walter J. Mc Murray
23 Windward Lane
Madison, CT 06443 (203) 245-7576

Donald Francis Bradley
65 Ludlow Rd.
Manchester, CT 06040 (203) 649-7951

Ira S. Dorman
Watkins Centre
935 Main St.
Manchester, CT 06040 (203) 649-1862

William E. Dickheiser
Uniroyal Chemical Co., Inc.
World Headquarters
Benson Rd.
Middlebury, CT 06749 (203) 573-4377

John A. Shedden#
Unionroyal Inc.
World Headquarters
Middlebury, CT 06749 (203) 573-4388

Raymond D. Thompson
Uniroyal Chemical Co., Inc.
World Headquarters
Middlebury, CT 06749 (203) 573-4385

Donald H. Winslow#
Uniroyal, Inc.
World Headquarters
Middlebury, CT 06749 (203) 573-2774

Alan E. Steele
547 Main St.
Middletown, CT 06457 (203) 347-1137

John L. Peterson
18 Alden Place
Milford, CT 06460 (203) 878-3247

John B. Smith, III#
136 Sunnyside Ct.
Milford, CT 06460 (203) 878-7986

Dennis J. Delaney
450 Monroe Turnpike
Monroe, CT 06468 (203) 261-6115

Gerald E. Linden
11 Vincent Dr.
Monroe, CT 06468

John Herbert Mulholland
71 Old Zoar Rd.
Monroe, CT 06468 (203) 358-7665

Ferdinand R. Hirtler#
Uniroyal Chem. Co.
Spencer St.
Naugatuck, CT 06770 (203) 723-3427

Charles Joseph Fickey
7 Siwanoy Lane
New Canaan, CT 06840 (203) 966-8020

William Kaufman
74 Kimberly Place
New Canaan, CT 06840 (203) 966-2113

Norman J. O Malley
163 Fox Run Rd.
New Canaan, CT 06840 (203) 966-5894

Joseph Church Sweet, Jr.
45 Wee Burn Dr.
New Canaan, CT 06840 (203) 966-0909

Alvin Joseph Riddles
Candlewood Isle
Box 34
New Fairfield, CT 06812 (203) 746-3470

Robert H. Bachman
Bachman & La Pointe, P.C.
55 Church St.
New Haven, CT 06510 (203) 777-6628

Raymond Bower
Southern New England Telephone
300 George St.
New Haven, CT 06510 (203) 771-7340

Anthony P. Delio, II
121 Whitney Ave.
New Haven, CT 06510 (203) 787-0595

Frederick F. Fagal#
Southern New England Tele. Co.
227 Church St.
New Haven, CT 06506 (203) 771-2117

Joseph Fleischer
210 Yale Ave.
New Haven, CT 06515 (203) 387-0723

Barry L. Kelmachter
Bachman & Lapointe, P.C.
55 Church St.
New Haven, CT 06510 (203) 777-6628

Gregory P. La Pointe
Bachman & La Pointe
55 Church St.
New Haven, CT 06510 (203) 777-6628

Robert H. Montgomery
Costas, Montgomery & Dorman, P.C.
246 Church St.
New Haven, CT 06510 (203) 787-2708

Donald Russell Motsko
U.S. Repeating Arms Co.
275 Winchester Ave.
New Haven, CT 06511 (203) 789-5873

Edward F. O Connor
121 Whitney Ave.
New Haven, CT 06510 (203) 787-0595

Gregory S. Rosenblatt
Olin Corp.
91 Shelton Ave.
P.O. Box 30-9643
New Haven, CT 06511 (203) 789-5273

W. Saxton Seward
121 Whitney Ave.
New Haven, CT 06510 (203) 787-0595

Walter Spruegel
155 Linden St.
New Haven, CT 06511 (203) 562-1365

Richard S. Strickler
Bachman & La Pointe
55 Church St.
New Haven, CT 06510 (203) 777-6628

Paul Weinstein
Olin Corp.
91 Shelton Ave.
P.O. Box 30-9642
New Haven, CT 06511 (203) 789-5268

Bruce Edward Hosmer
158 1/2 Prospect Hill
New Milford, CT 06776 (203) 355-0869

Murray J. Kessler
Cohen & Kessler
62 Bridge St., Box 427
New Milford, CT 06776 (305) 354-4488

James R. Reck
Loctite Corp.
705 N. Mountain Rd.
Newington, CT 06111 (203) 278-1280

Edward K. Welch, II
Loctite Corp.
705 N. Mountain Rd.
Newington, CT 06111 (203) 278-1280

Eugene Thomas Warzecha#
Timber Lane
Newtown, CT 06470 (203) 426-4358

Albert William Hilburger
3 Laurel Hill Dr. S.
Niantic, CT 06357 (203) 739-0202

Mary M. Krinsky
5 Norway Rd.
North Haven, CT 06473 (203) 288-2360

James S. Rose#
Upjohn Co.
Pat. Law Dept.
D. S. Gilmore Labs.
410 Sackett Pt. Rd.
North Haven, CT 06473 (203) 281-2792

Marie F. Zuckerman#
Dow Chemical U.S.A.
North Haven Laboratories
410 Sackett Point Road
North Haven, CT 06473 (203) 287-3643

Rasma B. Balodis#
R. T. Vanderbilt Co., Inc.
30 Winfield St.
P.O. Box 5150
Norwalk, CT 06856 (203) 853-1400

Thomas R. Bremer
United States Surgical Corp.
150 Glover Ave.
Norwalk, CT 06856 (203) 866-5050

Jay Lionel Chaskin
80 Washington St.
Norwalk, CT 06854 (203) 838-8589

Richard L. Croiter
Continental Can Company, Inc.
800 Connecticut Ave.
Norwalk, CT 06856 (203) 855-5963

Ernest Fanwick
Burndy Corp.
Richards Ave.
Norwalk, CT 06856 (203) 852-8440

Paul A. Fattibene
Perkin-Elmer Corp.
761 Main Ave.
Norwalk, CT 06859 (203) 762-6888

Richard J. Gallagher
James River Corp.
P.O. Box 6000
Norwalk, CT 06856 (203) 854-2511

Edwin T. Grimes
Perkin-Elmer Corp.
Pat. Law Dept.
761 Main Ave.
Norwalk, CT 06856 (203) 762-6803

Burtsell J. Kearns
Burndy Corp.
Richards Ave.
Norwalk, CT 06856 (203) 852-8343

Samuel Kriegel
Purdue Frederick Co.
50 Washington St.
Norwalk, CT 06856 (203) 853-0123

Francis L. Masselle
The Perkin Elmer Corp.
761 Main Ave.
M S 181
Norwalk, CT 06859 (203) 762-4301

Thomas Patrick Murphy
Perkin-Elmer Corp.
Main Ave.
Norwalk, CT 06856 (203) 762-4304

Lionel M. Rodger
23 Bettswood Road
Norwalk, CT 06851 (203) 853-8346

Robert H. Sorensen#
Perkin-Elmer Corp.
Main Ave.
Norwalk, CT 06856 (203) 762-4901

William S. Henry
Electrolux Corp.
51 Forest Ave.
Old Greenwich, CT 06870 (203) 637-1761

Evans Khan
105 Shore Rd.
Old Greenwich, CT 06870 (203) 637-3897

John Lawrence Sullivan
44 Midbrook Lane
Old Greenwich, CT 06870 (203) 637-9539

Victor P. Triolo
1 Nimitz Place
Old Greenwich, CT 06870 (203) 637-0039

John W. Redman#
7 Jean Dr.
Old Lyme, CT 06371

Jerry M. Presson
Harvey Hubbell Inc.
584 Derby Milford Rd.
Orange, CT 06477 (203) 789-1100

Walter C. Bernkopf
General Elec. Co.
41 Woodford Ave.
Plainville, CT 06062 (203) 747-7135

Richard A. Menelly
General Electric Co.
41 Woodford Ave.
Plainville, CT 06062 (203) 747-7153

David G. Coker#
Schlumberger Doll Research
Old Quarry Rd.
Ridgefield, CT 06877 (203) 431-5209

Bernard F. Crowe
17 Langstroth Dr.
Ridgefield, CT 06877 (203) 792-1063

Charles T. Emmons#
327 Old Branchville Rd.
Ridgefield, CT 06877 (203) 544-8523

David E. Frankhouser
Boehringer Ingelheim Corp.
90 East Ridge
P.O. Box 368
Ridgefield, CT 06877 (203) 431-5920

Louis F. Heeb
24 Strawberry Ridge Rd.
Ridgefield, CT 06877 (203) 438-6713

Michael G. Kroposki
Boehringer Ingelheim Ltd.
90 East Ridge
Ridgefield, CT 06877 (203) 438-0311

Clifford L. Tager
Schlumberger-Doll Research
Old Quarry Rd.
Ridgefield, CT 06877 (203) 431-5000

Mary-Ellen M. Timbers
Boehringer Ingelheim Corp.
Legal-Pat. Dept.
90 East Ridge
P.O. Box 368
Ridgefield, CT 06877 (203) 431-5916

Keith G. W. Smith#
Schlumberger-Doll Research
Old Quarry Rd.
Riverfield, CT 06877 (203) 431-5210

Seymour Polansky#
72 Florence Rd.
Riverside, CT 06878 (203) 637-4020

Joanne T. Rauchfuss#
41 Terrace Ave.
Riverside, CT 06878 (203) 637-9391

Willard Weber Roberts
4 Rainbow Dr.
Riverside, CT 06878 (203) 637-0513

Lewis Clifford Brown
P.O. Box 286
Rowayton, CT 06853

John K. Conant
23 Flicker Ln.
Rowayton, CT 06853 (203) 866-1700

Andrew K. Mc Colpin
61 Bluff Ave.
Rowayton, CT 06853 (203) 838-8815

Ronald J. Sasiela#
Coldwater Seafood Corp.
133 Rowayton Ave.
Rowayton, CT 06853 (203) 852-1600

Prescott W. May
12 Bank St.
P.O. Box 903
Seymour, CT 06483 (203) 888-4144

Satya P. Asija
7 Woonsocket Ave.
Shelton, CT 06484 (203) 736-0774

Paul A. Sobel
120 Hunters Creek
Shelton, CT 06484 (203) 336-1811

Charles E. Mc Tiernan
R R #2, Box 553
Sherman, CT 06784 (203) 355-0829

Victor E. Libert
760 Hopmeadow St.
Simsbury, CT 06070 (203) 658-1766

Anthony H. Handal
Handal & Morofsky
80 Washington St.
South Norwalk, CT 06854 (203) 838-8589

James P. Mackinnon
360 Beelzebub Rd.
S. Windsor, CT 06074 (203) 644-0816

George A. Dalin#
197A Heritage Village
Southbury, CT 06488 (203) 264-9418

Edward Harris Eames
89-B Heritage Village
Southbury, CT 06488 (203) 264-6326

Lawrence Hager
Southbury Plaza Prof. Bldg. 201
Southbury, CT 06488 (203) 264-3515

Allen D. Brufsky
181 Old Post Rd.
P.O. Box 59
Southport, CT 06490 (203) 255-8900

Armand Cifelli
Kramer, Brufsky & Cifelli, P.C.
181 Old Post Rd.
P.O. Box 59
Southport, CT 06490 (203) 255-8900

Howard M. Cohn
Kramer, Brufsky & Cifelli
181 Old Post Rd.
P.O. Box 59
Southport, CT 06490 (203) 255-8900

John F. Cullen
86 Fawn Ridge Lane
Southport, CT 06490 (203) 255-3236

Robert J. Eck
245 Daybreak Rd.
Southport, CT 06490 (203) 259-1128

Arthur T. Fattibene
2480 Post Rd.
Southport, CT 06490 (203) 255-4400

Joann L. Villamizar
Kramer, Brufsky & Cifelli, P.C.
181 Old Post Rd.
P.O. Box 59
Southport, CT 06490 (203) 255-8900

Joseph P. Abate
General Signal Corp.
High Ridge Park
Box 10010
Stamford, CT 06904 (203) 357-8800

Gregory J. Battersby
Grimes & Battersby
P.O. Box 1311
Stamford, CT 06904 (203) 324-2828

Howard M. Bollinger
Parmelee, Bollinger & Bramblett
460 Summer St.
Stamford, CT 06901 (203) 327-2650

Garold E. Bramblett, Jr.
Parmelee, Bollinger & Bramblett
460 Summer St.
Stamford, CT 06901 (203) 327-2650

Alice Comins Brennan
American Cyanamid Co.
1937 W. Main St.
Stamford, CT 06904 (203) 348-7331

R. Bruce Brooks#
27 Northill St., Apt. 1F
Stamford, CT 06907 (203) 324-5253

Harvey M. Brownrout
Xerox Corp.
P.O. Box 1600
Stamford, CT 06904 (203) 968-3106

Roland T. Bryan
Two Landmark Square
Stamford, CT 06901 (203) 359-4358

William H. Calnan, IV
American Cyanamid Co.
Pat. Law Dept.
P.O. Box 60
Stamford, CT 06904 (203) 348-7331

Francis Noel Carten
17 Mill Stone Cir.
Stamford, CT 06903 (203) 329-1771

James R. Cartiglia#
St. Onge, Stewart, Johnston & Reens
986 Bedford St.
Stamford, CT 06905 (203) 324-6155

Thaddius J. Carvis, Jr.
St. Onge, Steward, Johnston & Reens
986 Bedford St.
Stamford, CT 06905 (203) 324-6155

John H. Chapman
Chapman & Moran
One Landmark Square
Stamford, CT 06901 (203) 358-9390

Edward A. Conroy, Jr.
American Cyanamid Co.
1937 W. Main St.
Stamford, CT 06904 (203) 348-7331

Eugene Stephen Cooper
CBS, Inc.
227 High Ridge Rd.
Stamford, CT 06905 (203) 327-2000

Charles F. Costello, Jr.
American Cyanamid Corp.
1937 W. Main St.
P.O. Box 60
Stamford, CT 06904 (203) 348-7331

Samuel S. Cross
Kelley, Drye & Warren
Six Stamford Forum
Stamford, CT 06901 (203) 324-1400

F. Eugene Davis, IV
184 Atlantic St.
Stamford, CT 06901 (203) 324-9662

Michael J. De Sha
Pitney Bowes Inc.
Patent Dept.
Walter H. Wheeler, Jr. Drive
Stamford, CT 06926 (203) 351-6234

James Thomas Dunn
56 Big Oak Lane
Stamford, CT 06903 (203) 322-5773

George Vincent Eltgroth
65 Glenbrook Rd.
Stamford, CT 06902 (203) 327-4566

David Fink
6 Todd Lane
Stamford, CT 06905 (203) 322-9130

Denis Arthur Firth#
St. Onge, Steward, Johnston & Reens
986 Bedford St.
Stamford, CT 06905 (203) 324-6155

Eugene L. Flanagan, III
Geo International Corp.
1 Landmark Square
Stamford, CT 06901 (203) 964-1955

Hensley M. Flash
112 Lawn Ave.
Stamford, CT 06902 (203) 348-0289

Joseph M. Fowler
American Actuator Corp.
P.O. Box 384
Stamford, CT 06904 (203) 324-6334

James R. Frederick
Parmelee, Bollinger & Bramblett
460 Summer St.
Stamford, CT 06901 (203) 327-2650

Philip M. French
1035 Washington Blvd.
Stamford, CT 06901 (201) 329-2727

Stewart Jay Fried
Playtex, Inc.
700 Fairfield Ave.
Stamford, CT 06904 (203) 356-8382

Henry Z. Friedlander
85 Riverside Ave.
Stamford, CT 06905 (203) 357-9277

Stephen Gates
Two Landmark Square
Stamford, CT 06901 (203) 359-4358

Douglas M. Gilbert
Get Service Corp.
One Stamford Forum
Stamford, CT 06904 (203) 965-2278

Michael L. Goldman
St. Onge, Steward, Johnston & Reens
986 Bedford St.
Stamford, CT 06905 (203) 324-6155

Alan M. Gordon
American Cyanamid Co.
Pat. Law Dept.
1937 West Main St.
Stamford, CT 06904 (203) 348-7331

David P. Gordon
4 Stanwick Cir.
Stamford, CT 06905 (203) 325-9150

Harold H. Green, Jr.
General Electric Co.
High Ridge Park
P.O. Box 7600
Stamford, CT 06904 (203) 357-4969

Charles W. Grimes
Grimes & Battersby
184 Atlantic St.
Stamford, CT 06901 (203) 324-2828

John Joseph Hagan
American Cyanamid Co.
1937 W. Main St.
Stamford, CT 06904 (203) 348-7331

James K. Hammond
St. Onge, Steward, Johnston & Reens
986 Bedford St.
Stamford, CT 06905 (203) 324-6155

John Edward Hanrahan
American Cyanamid Co.
1937 W. Main St.
Stamford, CT 06904 (203) 348-7331

Gordon L. Hart
American Cyanamid Co.
1937 W. Main St.
Stamford, CT 06904 (203) 348-7331

Robert Hockfield
1 Prospect St.
Stamford, CT 06901 (203) 357-8988

Harold Huberfeld
General Signal Corp.
High Ridge Park
P.O. Box 10010
Stamford, CT 06904 (203) 357-8800

Herbert Girard Jackson
Amer. Cyanamid Co.
Pat. Dept.
1937 W. Main St.
Stamford, CT 06904 (203) 348-7331

Albert C. Johnston
St. Onge, Steward, Johnston, Reens & Noe
5 Landmark Square
Stamford, CT 06901 (212) 324-6155

Joseph J. Kaliko
243 Willowbrook Ave.
Stamford, CT 06902 (203) 348-5171

Michael J. Kelly
American Cyanamid Co.
P.O. Box 60
1937 W. Main St.
Stamford, CT 06904 (203) 356-8562

Barry Jay Kesselman
Xerox Credit Corp.
100 First Stamford Place
Stamford, CT 06904 (203) 325-6634

Milton E. Kleinman
General Signal Corp.
High Ridge Park 2
Stamford, CT 06904 (203) 357-8800

Peter K. Kontler
Kontler & Grimes
One Landmark Square 426
Stamford, CT 06901 (203) 348-6719

Barry Kramer
Kramer & Brufsky
898 Summer St.
Stamford, CT 06905 (203) 348-6231

Warren Kunz
83 Chestnut Hill Lane
Stamford, CT 06903 (203) 329-9351

Joseph Levinson
Parmelee, Bollinger & Bramblett
460 Summer St.
Stamford, CT 06901 (203) 327-2650

Eugene Lieberstein
2151 Long Ridge Rd.
Stamford, CT 06903 (203) 794-6127

Theodore D. Lindgren
GTE Service Corp.
One Stamford Forum
Stamford, CT 06904 (203) 965-3062

Jeffrey S. Mednick
General Signal Corp.
High Ridge Park
Stamford, CT 06904 (203) 357-8800

Robert E. Meyer#
Pitney-Bowes, Inc.
Walter Wheeler Jr. Dr.
Stamford, CT 06926 (203) 356-6138

Eric Y. Munson
460 Summer St.
Stamford, CT 06901 (203) 325-1361

Alphonse R. Noe
American Cyanamid Co.
1937 W. Main St.
P.O. Box 60
Stamford, CT 06904 (203) 348-7331

Spencer Everett Olson
CBS Inc.
227 High Ridge Rd.
Stamford, CT 06905 (203) 327-2000

Robert T. Orner
GTE Service Corp.
One Stamford Forum
Stamford, CT 06904 (203) 965-3045

Charles G. Parks, Jr.
Pitney Bowes Inc.
World Headquarters
Stamford, CT 06926 (203) 351-6236

G. Kendall Parmelee
Parmelee, Bollinger & Bramblett
460 Summer St.
Stamford, CT 06901 (203) 327-2650

David C. Petre
Xerox Corp.
P.O. Box 1600
Stamford, CT 06904 (203) 329-8700

Susan H. Rauch
American Cyanamid Co.
Pat. Law Dept.
1937 W. Main St.
P.O. Box 60
Stamford, CT 06904 (203) 348-7331

George W. Rauchfuss, Jr.
Parmelee, Bollinger, Bramblett & Drumm
460 Summer St.
Stamford, CT 06901 (203) 327-2650

Robert P. Raymond
American Cyanamid Co.
1937 W. Main St.
Stamford, CT 06904 (203) 348-7331

Louis H. Reens
St. Onge, Steward, Johnston & Reens
986 Bedford St.
Stamford, CT 06905 (203) 324-6155

Ronald Reichman
General Signal Corp.
High Ridge Park
Stamford, CT 06904 (203) 357-8800

Jack W. Richards
American Cyanamid Co.
1937 W. Main St.
Stamford, CT 06904 (203) 348-7331

Stuart N. Roth
Olin Corp.
120 Lond Ridge Rd.
Stamford, CT 06904 (203) 356-2480

Charles N.J. Ruggiero
Grimes & Battersby
P.O. Box 1311
184
Stamford, CT 06904 (203) 324-2828

Melvin J. Scolnick
Pitney Bowes, Inc.
Walter H. Wheeler, Jr. Dr.
Stamford, CT 06904 (203) 326-7515

Albert Willis Scribner
Pitney Bowes, Inc.
Walter Wheeler Jr. Dr.
Stamford, CT 06904 (203) 326-7486

Paul Shapiro
Continental Can Company, Inc.
51 Harbor Plaza
Stamford, CT 06904 (203) 964-6175

Lawrence E. Sklar
Pitney Bownes, Inc.
Walter H. Wheeler, Jr. Dr.
Stamford, CT 06904 (203) 326-7493

Harold M. Snyder
Dorr-Oliver Incorporated
77 Havemeyer Lane
Stamford, CT 06904 (203) 358-3492

William D. Soltow, Jr.
Pitney-Bowes, Inc.
Walter J. Wheeler, Jr. Dr.
Stamford, CT 06926 (203) 326-7475

Evelyn M. Sommer
Champion Internatl. Corp.
1 Champion Plaza
Stamford, CT 06921 (203) 358-7680

William J. Speranza#
St. Onge, Steward, Johnston & Reens
986 Bedford St.
Stamford, CT 06905 (203) 324-6155

Ronald J. St. Onge
St. Onge, Steward, Johnston & Reens
986 Bedford St.
Stamford, CT 06905 (203) 324-6155

Ervin B. Steinberg#
55 Bridge St.
Stamford, CT 06905 (203) 324-7900

Mark Paul Stone
184 Atlantic St.
Stamford, CT 06901 (203) 324-9662

Thomas V. Sullivan
41 Malvern Rd.
Stamford, CT 06905 (203) 322-4601

Frank J. Thompson
Thompson & Walsh
111 Prospect St.
Stamford, CT 06901 (203) 348-9987

Estelle J. Tsevdos
American Cyanamid
1937 W. Main St.
P.O. Box 60
Stamford, CT 06904 (203) 348-7331

Frank M. Van Riet
American Cyanamid Co.
1937 W. Main St.
P.O. Box 60
Stamford, CT 06904 (203) 348-7331

Peter Vrahotes
Pitney Bowes Incorp.
Walnut & Pacific St.
Stamford, CT 06904 (203) 351-7566

Donald P. Walker
Pitney Bowes
One Elmcroft
Stamford, CT 06926 (203) 326-7490

Patrick J. Walsh
111 Prospect St.
Stamford, CT 06901 (203) 967-4144

Robert H. Whisker
Pitney Bowes, Inc.
1 Elmcroft
Stamford, CT 06926 (203) 326-7507

Gene S. Winter
St. Onge, Steward, Johnston & Reens
986 Bedford St.
Stamford, CT 06905 (203) 324-6155

Martin David Wittstein
Pitney-Bowes, Inc.
One Elmcroft
Stamford, CT 06926 (203) 326-7478

Henry S. Wyzan#
39 Bradley Place
Stamford, CT 06905 (203) 322-4787

Peter Xiarhos
GTE Service Corp.
One Stamford Forum
Stamford, CT 06904 (203) 965-3053

Peter Kent
Avco Lycoming Div.
550 S. Main St.
Stratford, CT 06497 (203) 385-2352

Robert E. Kline#
Sikorsky Aircraft
North Main St.
Stratford, CT 06497 (203) 386-6046

Roger Thomas Wolfe
60 Fairfax Dr.
Stratford, CT 06497 (212) 907-3060

Milton Zucker
273A Agawam Dr.
Stratford, CT 06497 (203) 377-5835

William C. Nealon
40 Crane Hill Rd.
P.O. Box 65
Suffield, CT 06078 (203) 668-0226

John H. Crozier
1934 Huntington Turnpike
Trumbull, CT 06611 (203) 582-9561

Raghunath V. Date#
45 Gibson Ave.
P.O. Box 194
Trumbull, CT 06611 (203) 268-7188

F. A. Iskander
14 Stirrup Dr.
Trumbull, CT 06611 (203) 268-0256

Ralph L. Manning
Allied Corp.
% Allied Information Systems Co.
Trumbull Industrial Park
Trumbull, CT 06609 (203) 386-2237

Charles Patrick Martin
45 North St.
Trumbull, CT 06611 (203) 378-7504

Eugene F. Miller
19 Colonial Dr.
Trumbull, CT 06611 (203) 261-4178

Peter C. Van Der Sluys
45 Firehouse Road
Trumbull, CT 06611 (203) 261-5396

George M. Yahwak
25 Skytop Dr.
Trumbull, CT 06611 (203) 268-0383

Charles A. Warren
P.O. Box 2416
Vernon, CT 06066 (203) 646-9893

Aldo A. Algieri#
Bristol-Myers Co.
5 Research Parkway
Wallingford, CT 06492 (203) 284-6024

John J. Balser
Bristol-Myers Co.
Pharmaceutical R. & D. Div.
5 Research Parkway
Wallingford, CT 06492 (203) 284-6121

Deborah J. Barnett
Bristol-Myers Co.
5 Research Parkway
P.O. Box 5100
Wallingford, CT 06492 (203) 284-6759

Robert E. Carnahan#
Bristol-Myers Co.
5 Research Parkway
P.O. Box 5100
Wallingford, CT 06492 (203) 284-6053

Lester E. Johnson
Bristol-Myers Co.
P.O. Box 5100
5 Research Parkway
Wallingford, CT 06492 (203) 284-6759

David M. Morse
Bristol-Myers Co.
5 Research Parkway
Wallingford, CT 06492 (203) 284-6997

Richard P. Ryan#
Bristol-Myers Co.
Legal Div., Pat. Dept.
5 Research Parkway
P.O. Box 5100
Wallingford, CT 06492 (203) 284-6075

Neophytos Ganiaris#
Nettleton Hollow Rd.
Washington, CT 06793 (203) 868-2718

William C. Crutcher
Gager, Henry & Narkis
One Exchange Place
P.O. Box 2480
Waterbury, CT 06722 (203) 597-5116

Dallett Hoopes
21 Church St.
Waterbury, CT 06702 (203) 575-1773

Richard A. Dornon
Colt Industries Inc.
Charter Oak Blvd.
West Hartford, CT 06110 (203) 236-0651

Radford W. Luther
Chandler Evans Inc.
Charter Oak Blvd.
West Hartford, CT 06101 (203) 232-0840

Herbert Magil
29 Colony Rd.
West Hartford, CT 06117 (203) 232-4340

Howard Scott Reiter
Colt Industries Inc.
Charter Oak Blvd.
West Hartford, CT 06110 (203) 236-0651

Anne M. Rosenblum
7 Briarwood Rd.
West Hartford, CT 06107 (203) 561-2334

Raymond C. Seligson
65 La Salle Rd.
P.O. Box 1373
West Hartford, CT 06107 (203) 236-5404

William George Rhines, Sr.
48 Blueberry Hill Rd.
Weston, CT 06883 (203) 226-0464

Dieter J. Schaefer#
53 Godfrey Rd.
Weston, CT 06883

Alfred I. Wirtenberg
15 Wilson Rd.
Weston, CT 06883 (203) 544-9270

William A. Aguele
12 Gault Park Dr.
Westport, CT 06880 (203) 226-1563

A. Sidney Alpert
University Patents, Inc.
P.O. Box 901
Westport, CT 06881 (203) 255-6044

James T. Elfstrum
Stauffer Chemical Co.
Patent Dept.
Nyala Farms Rd.
Westport, CT 06880 (203) 222-3235

Francis M. Frazio
26 Pequot Trail
Westport, CT 06880 (203) 227-5059

Walter G. Hensel
8 Woodhill Rd.
Westport, CT 06880 (203) 259-9758

Lloyd L. Mahone
270 Hillspoint Rd.
Westport, CT 06880 (203) 226-4682

Thomas J. Monahan
University Genetics Co.
8 Wright St.
P.O. Box 5117
Westport, CT 06881 (203) 454-8846

Martin M. Novack
1465 Post Rd. East
P.O. Box 901
Westport, CT 06881 (203) 255-4373

Daniel S. Oritz
Stauffer Chemical Co.
Westport, CT 06881 (203) 222-3068

Andrew M. Riddles
26 Oak St.
Westport, CT 06880 (203) 454-0442

Marvin B. Rosenberg
Cambridge Res. & Dev. Gp.
21 Bridge Square
Westport, CT 06880 (206) 226-7400

Barbara A. Shimei
University Patents, Inc.
P.O. Box 901
Westport, CT 06881 (203) 255-6044

Robert Cook Sullivan
Stauffer Chemical Co.
Patent Dept.
Nyala Farm Rd.
Westport, CT 06880 (203) 222-3490

Andrew L. Gaboriault
296 Chestnut Hill Rd.
Wilton, CT 06897 (203) 762-9030

Alfred H. Hemingway, Jr.
221 Linden Tree Rd.
Wilton, CT 06897 (203) 762-3826

Edward R. Hyde
261 Danbury Rd.
P.O. Box 494
Wilton, CT 06897 (203) 762-5444

Richard Kornutik
Nabisco Brands, Inc.
15 River Rd.
Wilton, CT 06897 (203) 762-2500

Mark G. Paulson#
Nabisco Brands Inc.
15 River Rd.
Wilton, CT 06897 (203) 762-2500

Richard H. Berneike
Combustion Engineering, Inc.
1000 Prospect Hill Rd.
Windsor, CT 06095 (203) 285-9106

Michael Alan Cantor
Fishman & Dionne
360 Bloomfield Ave.
Windsor, CT 06095 (203) 688-4470

Arthur F. Dionne
Fishman & Dionne
360 Bloomfield Ave.
Windsor, CT 06095 (203) 688-4470

David S. Fishman
Fishman & Dionne
360 Bloomfield Ave.
Windsor, CT 06095 (203) 688-4470

Arthur Edmond Fournier, Jr.
Combustion Eng., Inc,.
1000 Prospect Hill Rd.
Windsor, CT 06095 (203) 285-9112

William W. Habelt
Combustion Engineering, Inc.
Corporate Pat. Dept.
1000 Prospect Hill Rd.
Windsor, CT 06095 (203) 285-9113

Robert J. Hoffberg
AIW-Alton Iron Works, Inc.
P.O. Box 20
Windsor, CT 06095 (203) 683-0731

Edward L. Kochey, Jr.
Combustion Engineering, Inc.
1000 Prospect Hill Rd.
Windsor, CT 06095 (203) 683-6105

Robert L. Olson
Combustion Engineering, Inc.
1000 Prospect Hill Rd.
Windsor, CT 06095 (203) 683-6104

L. James Ristas
Combustion Engr. Inc.
1000 Prospect Hill Rd.
Windsor, CT 06095 (203) 285-9107

Gurdon R. Abell#
Perrin Rd.
R R 1, Box 202
Woodstock, CT 06281 (203) 974-2442

Delaware

Earl Christensen
3014 Wrangle Hill Rd.
Bear, DE 19701 (302) 834-7341

John C. Andrade
Parkowski, Noble & Guerke
116 West Water St.
P.O. Box 598
Dover, DE 19903 (302) 678-3262

Anthony P. Mentis
44 Candlewicke
Dover, DE 19901 (302) 697-3474

Roger Arnold Hines
3929 Heather Dr.
Greenville, DE 19807 (302) 655-2265

Edward J. Kaliski#
P.O. Box 3661
Greenville, DE 19807

Don A. Erlandson
11 Guenever Dr.
New Castle, DE 19720 (302) 322-3028

Bill N. Baron#
University Of Delaware
Institute Of Energy Conversion
Newark, DE 19716 (302) 451-6229

John S. Campbell#
W.L. Gore & Assoc. Inc.
555 Paper Mill Rd.
P.O. Box 9206
Newark, DE 19714 (302) 738-4880

Dena Meyer
W.L. Gore & Associates, Inc.
551 Paper Mill Rd.
P.O. Box 9206
Newark, DE 19714 (302) 738-4880

Mark G. Mortenson
Lanxide Corp.
Tralee Industrial Park
Newark, DE 19711 (302) 454-0244

John A. Parkins#
14 Vassar Dr.
Newark, DE 19711 (302) 731-5698

William E. Meason#
1315 Marsh Rd.
Northwood, DE 19803 (302) 478-1953

John G. Abramo
Abramo & Abramo
105 W. 9th St.
P.O. Box 668
Wilmington, DE 19899 (302) 652-3504

Samuel V. Abramo
Abramo & Abramo
105 West Ninth St.
P.O. Box 668
Wilmington, DE 19899 (302) 652-3504

William Stanley Alexander
Hercules, Inc.
Hercules Plaza
Wilmington, DE 19894 (302) 594-6942

William K. Baggott
E.I. Du Pont De Nemours & Co.
1007 Market St.
Wilmington, DE 19898 (302) 992-3202

Nathan Bakalar
Connolly & Hutz
1220 Market St.
P.O. Box 2207
Wilmington, DE 19801 (302) 658-9141

Claude L. Beaudoin
508 Whitby Dr.
Wilmington, DE 19803 (302) 478-2449

Richard M. Beck
Connolly & Hutz
1220 Market St.
P.O. Box 2207
Wilmington, DE 19899 (302) 658-9141

Robert W. Black
E.I. Dupont De Nemours & Co.
Du Pont Bldg. Room 8131
1007 Market St.
Wilmington, DE 19898 (302) 992-3216

Samuel S. Blight
E.I. Du Pont De Nemours & Co.
Legal Dept., Pat. Div.
1007 Market St.
Wilmington, DE 19898 (302) 992-4922

Charles L. Board, Sr.
404 Stafford Rd.
Wilmington, DE 19803 (302) 655-6643

Paul U. Bockrath#
1403 Stoneleigh Rd.
Wilmington, DE 19803 (302) 478-7533

Robert R. Bonczek
E.I. Du Pont De Nemours & Co.
1007 Market St., D-1028
Wilmington, DE 19898 (302) 594-3603

Mary Webb Bourke
Connolly, Bove, Lodge & Hutz
P.O. Box 2207
1220 Market St.
Wilmington, DE 19899 (302) 658-9141

Alanson Gray Bowen, Jr.
E.I. Du Pont De Nemours & Co.
Patent Division
Legal Dept.
Wilmington, DE 19898 (302) 992-3227

Albert Frank Bower
Connolly, Bove, Lodge & Hutz
1220 Market St., 10th Fl.
Wilmington, DE 19801 (302) 658-9141

Donald L. Bruton
820 N. French St.
Wilmington, DC 19801 (302) 571-3847

Thomas J. Bucknum
E.I. Du Pont De Nemours & Co.
Legal Dept.
Wilmington, DE 19898 (302) 992-4334

Richard Hurt Burgess
E.I. Du Pont De Nemours & Co., Inc.
1007 Market St.
Wilmington, DE 19898 (302) 774-5325

Lawton Arthur Burrows, Jr.
E.I. Du Pont De Nemours & Co.
1007 Market St.
Wilmington, DE 19898 (302) 774-2729

Frederick Frank Butzi
1102 Dardel Dr.
Wilmington, DE 19803 (302) 478-3357

Sydney R. Chirlin
Young, Conaway, Stargatt & Taylor
P.O. Box 391
Wilmington, DE 19899 (302) 571-6622

Lynne M. Christenbury
E.I. Du Pont De Nemours & Co.
Legal Dept. P-17-1110
1007 Market St.
Wilmington, DE 19898 (302) 992-5481

Robert Collat#
1300 Greenway Rd.
Wilmington, DE 19803 (302) 478-7112

Michael Conner
E.I. Du Pont
Legal Dept.
Barley Mill Plaza
Wilmington, DE 19898 (302) 992-4924

Arthur G. Connolly, Sr.
Connolly, Bove, Lodge & Hutz
1220 Market Bldg.
P.O. Box 2207
Wilmington, DE 19899 (302) 658-9141

James T. Corle
E.I. Du Pont De Nemours & Co.
Legal Dept., B-11208
1007 Market St.
Wilmington, DE 19898 (302) 774-8536

James Anthony Costello
E.I. Du Pont De Nemours & Co., Inc.
1007 Market St.
Wilmington, DE 19898 (302) 992-4926

Richard F. B. Cox#
2009 Woodbrook Dr.
Wilmington, DE 19810 (302) 475-5288

Jeffrey F. Craft
Connolly & Hutz
1220 Market St.
Wilmington, DE 19801 (302) 658-9141

Paul E. Crawford
Connolly & Hutz
P.O. Box 2207
Wilmington, DE 19899 (302) 658-9141

Harry Cress, Jr.#
2516 Foulk Woods Rd.
Wilmington, DE 19810 (302) 475-3157

John E. Crowe
Hercules Inc.
Hercules Plaza
Market St.
Wilmington, DE 19899 (302) 594-6949

Francis J. Crowley
514 Kerfoot Farm Road
Wilmington, DE 19803 (302) 658-2028

Joseph P. Daniszewki#
E.I. Dupont De Nemours & Co.
Central Res. & Dev. Dept.
P14-1126 Barley Mill Plaza
Wilmington, DE 19898 (302) 992-2611

Gerald E. Deitch
E.I. Du Pont De Nemours & Co.
Legal Dept.
10th & Market St.
Wilmington, DE 19898 (301) 774-1000

Louis Del Vechio
E.I. Du Pont De Nemours & Co.
Legal Dept.
1007 Market St.
Wilmington, DE 19898 (302) 992-3204

Hazel L. Deming#
207 Old Mill Lane
Wilmington, DE 19803 (302) 764-5616

Thomas J. Des Rosier
E.I. Du Pont De Nemours & Co.
1007 Market St.
Wilmington, DE 19898

John Edward Dull
E.I. Du Pont De Nemours & Co.
Legal Dept.
1007 Market St.
Wilmington, DE 19898 (302) 774-6183

David Edwards
Hercules Incorporated
Hercules Plaza
1313 N. Market St.
Wilmington, DE 19894 (302) 594-6952

Barry Estrin
E.I. Du Pont De Nemours & Co.
Legal Dept.
Wilmington, DE 19898 (302) 992-3230

John D. Fairchild
Connolly & Hutz
1220 Market Bldg.
P.O. Box 2207
Wilmington, DE 19899 (302) 658-9141

Charles E. Fenny
E.I. Du Pont De Nemours & Co.
Legal Dept
1007 Market St.
Wilmington, DE 19898 (302) 774-9445

Lynn N. Fisher
E.I. Du Pont De Nemours & Co., Inc.
Legal Dept.
1007 Market St.
Wilmington, DE 19898 (302) 992-3221

James J. Flynn
E.I. Du Pont De Nemours & Co.
1007 Market St.
Wilmington, DE 19898 (302) 774-4668

James A. Forstner
E.I. Du Pont De Nemours & Co.
Legal Dept.
Wilmington, DE 19898 (302) 992-3232

Richard T. Foster
Connolly & Hutz
1220 Market Bldg.
P.O. Box 2207
Wilmington, DE 19899 (302) 658-9141

George A. Frank
E.I. Du Pont De Nemours & Co.
1007 Market St.
Wilmington, DE 19898 (302) 992-3218

Hilmar L. Fricke
E.I. Du Pont De Nemours & Co.
Du Pont Bldg., Rm. 7131
1007 Market St.
Wilmington, DE 19898 (302) 774-7892

Richard G. Gantt
E.I. Du Pont De Nemours & Co., Inc.
Experimental Station E357
Wilmington, DE 19898 (302) 695-2870

Andrew G. Golian
E.I. Du Pont De Nemours & Co.
Barley Mill Plaza
Wilmington, DE 19805 (302) 992-3228

David J. Gould
E.I. Du Pont De Nemours & Co.
Legal Dept.
1007 Market St.
Wilmington, DE 19898 (302) 774-2487

Theodore C. Gregory
E.I. Du Pont De Nemours & Co.
Legal Dept.
1007 Market St.
Wilmington, DE 19898 (302) 992-4925

John E. Griffiths
E.I. Du Pont De Nemours & Co.
Legal Dept.
Pat Div.
Wilmington, DE 19898 (302) 992-4941

John Ellsworth Griffiths
2705 Marklyn Dr.
Wilmington, DE 19810 (302) 475-7961

Charles Albert Haase
ICI Americas Inc.
Concord Pike & New Murphy Rd.
Wilmington, DE 19803 (302) 575-3733

William H. Hamby
E.I. Du Pont De Neumours & Co.
1007 Market St.
Legal Dept.
Wilmington, DE 19898 (302) 774-2575

Earl L. Handley
E.I. Du Pont De Nemours & Co.
1007 Market St.
Wilmington, DE 19898 (302) 773-4325

Gerald A. Hapka
E.I. Du Pont De Nemours & Co.
Legal Dept.
1007 Market St.
Wilmington, DE 19898 (302) 774-9466

William R. Hasek#
E.I. Dupont De Nemours & Co., Inc.
1007 Market St.
Wilmington, DE 19898 (302) 774-7354

William P. Hauser#
E.I. Dupont De Nemours & Co.
Electronics Dept.
Barley Mill Plaza 21-2354
Wilmington, DE 19898 (302) 992-3433

Joseph James Heimbach
E.I. Du Pont De Nemours & Co.
700 Farmers Bank Bldg.
Wilmington, DE 19898 (302) 999-4300

John S. Hendrickson
E.I. Du Pont De Nemours & Co.
Legal Dept., Pat. Div.
M-2710
Wilmington, DE 19898

C. Harold Herr
112 Rockingham Dr.
Wilmington, DE 19803 (302) 478-3871

W. Victor Higgs
E.I. Du Pont De Nemours & Co.
1007 Market St.
Wilmington, DE 19898 (302) 774-6278

Frank C. Hilbert, Jr.
E.I. Du Pont De Nemours & Co.
1007 Market St.
Wilmington, DE 19898 (302) 774-2012

Janet K. Hochstetler
E.I. Dupont De Nemours & Co.
1007 Market St.
Wilmington, DE 19898 (302) 774-5874

Donald Allen Hoes
E.I. Du Pont De Nemours & Co. Inc.
Legal Dept.
1007 Market St.
Wilmington, DE 19898 (302) 774-6974

Joseph Lee Hollowell#
P.O. Box 5447
Wilmington, DE 19808

George H. Hopkins
Himont Incorporated
1313 N. Market St.
Wilmington, DE 19894 (302) 594-5596

Joanne L. Horn
Himont Incorporated
1313 N. Market St.
Wilmington, DE 19894 (302) 594-5576

Edward Leigh Hunt, Sr.
1809 Wyckwood Ct.
Wilmington, DE 19803 (302) 656-4189

Frederick D. Hunter
E.I. Du Pont De Nemours & Co., Inc.
1007 Market St.
Wilmington, DE 19898 (302) 774-2383

Donald Wayne Huntley
E.I. Dupont De Nemours & Co.
Legal Dept.
1007 Market St.
Wilmington, DE 19898 (302) 774-3811

Rudolf Edward Hutz
Connolly, Bove, Lodge & Hutz
1220 Market St.
Wilmington, DE 19801 (302) 658-9141

Martha S. Imbalzano#
E.I. Du Pont De Nemours & Co.
Barley Mill Plaza 21
Wilmington, DE 19898 (301) 992-3435

Lawrence Isakoff#
E.I. Dupont De Nemours Co.
Legal Dept.
M-2714
10th & Market Sts.
Wilmington, DE 19898 (302) 774-5302

Roy V. Jackson#
Hercules Incorporated
Hercules Plaza
Wilmington, DE 19894 (302) 594-6958

Robert Jacobs
Jacobs & Crumplar, P.A.
P.O. Box 1271
Wilmington, DE 19899 (302) 656-5445

James L. Jersild
E.I. Du Pont De Nemours & Co.
1007 Market St.
Wilmington, DE 19898 (302) 774-8618

James T. Jones, III
ICI Americas, Inc.
Murphy Road & Concord Pike
Wilmington, DE 19897 (302) 575-3759

John W. Jones
ICI Americas Inc.
Law Dept. - Pat. Section
New Murphy Rd. & Concord Pike
Wilmington, DE 19897 (302) 575-3731

Elliott A. Katz
E.I. Du Pont De Nemours & Co.
1007 Market St.
Wilmington, DE 19898 (302) 774-5330

Michael B. Keehan
Hercules Incorporated
Hercules Tower, Rm 1372
910 Market St.
Wilmington, DE 19899 (302) 575-6954

Don M. Kerr
E.I. Du Pont De Nemours & Co., Inc.
8125 Du Pont Bldg.
1007 Market St.
Wilmington, DE 19898 (302) 774-4706

Joseph P. Klimowicz#
E.I. Du Pont De Nemours Co., Inc.
1007 Market St.
Wilmington, DE 19898 (302) 366-2613

Robert C. Kline
E.I. Du Pont De Nemours & Co.
1007 Market St.
Wilmington, DE 19898 (302) 774-5876

John J. Klocko, III
E.I. Du Pont De Nemours & Co.
Legal Dept.
1007 Market St.
Wilmington, DE 19898 (302) 774-2408

Charles S. Knothe#
106 Fairfax Blvd.
Wilmington, DE 19803 (302) 655-7854

Costas S. Krikelis
P.O. Box 7228
705 Mt. Lebanon Rd.
Wilmington, DE 19803 (302) 478-7895

Charles E. Krukiel
E.I. Du Pont De Nemours & Co.
1007 Market St.
Wilmington, DE 19898 (302) 774-9593

Mark D. Kuller
Hercules Inc.
Hercules Plaza
Wilmington, DE 19894 (302) 594-6923

Nicholas N. Leach#
Morris, Nichols, Arsht & Tunnell
12th & Market Sts.
(p.o. Box 1347)
Wilmington, DE 19899 (302) 658-9200

Dale R. Lovercheck
Hercules Incorporated
Hercules Plaza
Wilmington, DE 19894 (302) 594-6938

James K. Luchs
32 Hayloft Circle
Wilmington, DE 19808 (302) 239-7020

David M. Lukoff
1510 Turkey Run Rd.
Wilmington, DE 19803 (302) 762-6376

Stanley Lukoff#
E.I. Du Pont De Nemours Inc.
Louviers Bldg.
Wilmington, DE 19898 (302) 366-3726

Thomas J. Lundy#
E.I. Dupont De Nemours & Co.
Wilmington, DE 19898 (302) 999-2842

John P. Luther
Hercules Incorporated
Law Dept. - Pat. Sect.
Hercules Plaza
Wilmington, DE 19894 (302) 594-5000

Joshua W. Martin, III
4619 Big Rock Dr.
Wilmington, DE 19802 (302) 764-6353

Harry J. Mc Cauley
505 Country Club Dr.
Woodbrook
Wilmington, DE 19803 (302) 656-8890

J.R. Mc Grath
100 Rue Mandaleine
Wilmington, DE 19807 (302) 652-0161

George M. Medwick
E.I. Dupont De Nemours & Co., Inc.
Legal Dept. Rm. D-7125
1007 Market St.
Wilmington, DE 19898 (302) 774-5873

Pamela Meitner
E.I. Dupont De Nemours & Co.
1007 Market St.
Wilmington, DE 19809 (302) 774-8720

Thomas Michael Meshbesher
Connolly, Bove & Lodge
10th & Market St.
Wilmington, DE 19801 (302) 658-9141

Gregory C. Meyer#
2647 Drayton Dr.
Wilmington, DE 19808 (302) 995-1904

Rosemary M. Miano
ICI Americas Inc.
Concord Pike & New Murphy Rd.
Wilmington, DE 19897 (302) 575-3729

Clinton F. Miller
800 Greenwood Rd.
Wilmington, DE 19807 (302) 652-7435

Suzanne E. Miller
E.I. De Pont De Nemours & Co.
Legal Dept.
1007 Market St.
Wilmington, DE 19898 (302) 992-4949

Bruce W. Morriessey
E.I. Dupont De Nemours & Co.
Legal Dept.
1007 Market St.
Wilmington, DE 19898 (302) 773-3542

James Robert Morrison
E.I. Du Pont De Nemours & Co.
Legal Dept.
1007 Market St.
Wilmington, DE 19898 (302) 774-6882

Carl W. Mortenson
Mortenson & Uebler, P.A.
Lindell Square, Suite 4
1601 Milltown Rd.
Wilmington, DE 19808 (302) 654-2458

James L. Newsom, Jr.#
E.I. Du Pont De Nemours & Co.
Eng. Dev. Lab.
101 Beech St.
Wilmington, DE 19898 (301) 774-1796

Nicholas E. Oglesby, Jr.
Connolly, Bove & Lodge
1800 Farmers Bank Bldg.
Wilmington, DE 19801 (302) 658-9141

Frank R. Ortolani, Sr.
E.I. Du Pont De Nemours & Co.
1007 Market St.
Wilmington, DE 19898 (302) 774-7186

Donald F. Parsons, Jr.
Morris, Nichols, Arsht & Tunnell
Twelfth & Market Sts.
Wilmington, DE 19899 (302) 658-9200

Joanne W. Patterson#
Hercules Incorporated
Patent Section — Law Dept.
Hercules Plaza, 8th Floor
Wilmington, DE 19894 (302) 594-6961

William Leroy Peverill
803 Hopeton Rd.
Wilmington, DE 19807 (302) 656-4747

Harold Pezzner
Connolly & Hutz
Girard Bank Bldg.
Wilmington, DE 19899 (302) 658-9141

Robert Joseph Reichert
E.I. Du Pont De Nemours & Co.
1007 Market St.
Legal Dept.
Wilmington, DE 19803 (302) 774-5127

Norbert Frederick Reinert
E.I. Du Pont De Nemours & Co.
1007 Market St.
Wilmington, DE 19898 (302) 774-6111

Dean R. Rexford#
2323 W. 16th St.
Wilmington, DE 19806 (302) 652-8893

Vernon R. Rice
Du Pont Co.
1007 Market St.
Wilmington, DE 19898 (302) 774-2806

Annette L. Richter
E.I. Du Pont De Nemours & Co.
Legal Dept., P17-1174
Wilmington, DE 19898 (302) 992-3217

Louis H. Rombach
E.I. Du Pont De Nemours & Co.
Legal Dept.
1007 Market St.
Wilmington, DE 19898 (302) 774-5224

Edmund C. Ross, Jr.
Hercules, Inc.
Hercules Plaza
Pat. Div.
Wilmington, DE 19894 (302) 594-6957

Richard A. Rowe#
ICI Americas Inc.
Concord Pike & New Murphy Rd.
Wilmington, DE 19897 (302) 575-3727

Norris E. Ruckman
314 Waycross Rd.
Wilmington, DE 19803 (302) 654-4040

Howard J. Rudge
E.I. Du Pont De Nemours & Co.
Legal Dept.
Wilmington, DE 19898 (302) 773-4187

James H. Ryan
E.I. Du Pont De Nemours & Co.
Du Pont Bldg., Rm. 8134
1007 Market St.
Wilmington, DE 19890 (302) 774-3972

Gary Allyn Samuels
E.I. Dupont De Nemours & Co.
Legal Dept.
1000 Market St.
Wilmington, DE 19898 (302) 774-3225

Jay Willis Sanner
ICI Americas Inc.
Wilmington, DE 19897 (302) 575-3723

George N. Sausen#
E.I. Du Pont De Nemours & Co.
Central Res. & Dev. Dept.
Barley Mill Plaza 14/2108
Wilmington, DE 19898

John F. Schmutz
E.I. Du Pont De Nemours & Co.
10th & Market Sts.
Wilmington, DE 19898 (302) 774-7202

Ralph C. Schreyer#
E.I. Du Pont De Nemours & Co.
Du Pont Exptl. Station
E 336/233
Wilmington, DE 19898 (302) 772-3136

Sol Schwartz
E.I. Du Pont De Nemours & Co.
1007 Market St.
Wilmington, DE 19898 (302) 774-3177

Robert J. Shafer
E.I. Du Pont De Nemours & Co.
Legal Dept.
1007 Market St.
Wilmington, DE 19898 (302) 774-6326

John M. Sheehan
ICI Americas Inc.
Pats. Sect.-Law Dept.
Concord Pike & New Murphy Rd.
Wilmington, DE 19897 (302) 575-3721

Charles J. Shoaf
E.I. Dupont De Nemours & Co.
Legal Dept.
1007 Market St.
Wilmington, DE 19898 (302) 774-6745

David M. Shold#
E.I. Du Pont De Nemours & Co., Inc.
1007 Market St.
Wilmington, DE 19898 (302) 773-3264

Barbara C. Siegell
1220 King St.
Box 7
Wilmington, DE 19899 (302) 571-1700

Faith L.K. Silver#
4607 Little Rock Dr.
Wilmington, DE 19802 (302) 764-5408

Marion Cole Staves
Hercules Incorporated
Law Dept.
Pat. Section
Hercules Plaza
Wilmington, DE 19894 (302) 594-5000

Curtis Wayne Stephens
E.I. Du Pont De Nemours & Co., Inc.
1007 Market Place
Wilmington, DE 19898 (302) 774-5183

S. Grant Stewart
1204 Covington Rd.
Wilmington, DE 19803 (302) 764-5133

Paul R. Steyermark
E.I. Du Pont De Nemours & Co.
Legal Dept.
1007 Market St.
Wilmington, DE 19898 (302) 774-8966

Carol K. Stouffer#
E.I. Dupont De Nemours & Co., Inc.
1007 Market St.
Wilmington, DE 19898 (302) 774-2817

Robert Francis Sullivan#
E.I. Du Pont De Nemours & Co.
Montchanin Bldg., Room 2600
10th & Shipley Sts.
Wilmington, DE 19898 (302) 374-2250

Ivan Gabor Szanto
E.I. Du Pont De Nemours & Co.
1007 Market St.
Wilmington, DE 19898 (302) 774-5202

Wilkin Evans Thomas, Jr.
E.I. Du Pont De Nemours Co., Inc.
Legal Dept.
1007 Market St.
Wilmington, DE 19898 (302) 992-3210

Melford F. Tietze
ICI Americas Inc.
202 & Murphy Rds.
Wilmington, DE 19897 (302) 575-3734

Edwin Tocker
E.I. Dupont De Nemours & Co.
Legal Dept.
10th & Market Sts.
Wilmington, DE 19898 (302) 992-4921

S. Maynard Turk
Hercules Inc.
Hercules Plaza
Wilmington, DE 19894 (302) 594-7000

Ernest A. Uebler
Mortenson & Uebler, P.A.
1601 Milltown Rd.
Wilmington, DE 19808 (302) 998-9400

Frank R. Waite#
16 Sorrel Dr.
Wilmington, DE 19803 (302) 478-4083

George W. Walker
1103 N. Hilton Rd.
Wilmington, DE 19803 (302) 655-4547

Weston B. Wardell, Jr.
E.I. Du Pont De Nemours & Co.
1007 Market St.
Wilmington, DE 19898 (302) 992-3231

Charles A. Weigel
E.I. Dupont De Nemours & Co., Inc.
1007 Market St.
Wilmington, DE 19898 (302) 992-3219

Howard P. West, Jr.
E.I. Du Pont De Nemours & Co.
1007 Market St.
Wilmington, DE 19898 (302) 774-8831

Robert W. Whetzel
Richards, Layton & Finger
One Rodney Square
P.O. Box 551
Wilmington, DE 19899 (302) 658-6541

Douglas E. Whitney
Morris, Nichols, Arsth & Tunnell
P.O. Box 1347
1105 N. Market St.
Wilmington, DE 19899 (302) 658-9200

John Walter Whitson, Jr.#
1300 Grayson Rd.
Wilmington, DE 19803 (302) 762-2524

Don O. Winslow
E.I. Dupont De Nemours & Co.
10th & Market Sts.
Wilmington, DE 19898 (302) 992-3229

Herbert M. Wolfson
E.I. Du Pont De Nemours & Co.
1007 Market St.
Wilmington, DE 19898 (302) 774-5243

District of Columbia

Mark J. Abate#
2515 K St., N.W. Apt. 103
Washington, DC 20037 (202) 338-8075

Frank L. Abbott
3805 Kanawha St.
Washington, DC 20015 (202) 362-0165

David S. Abrams
Roylance, Abrams, Berdo & Goodman
1225 Connecticut Ave., N.W., Suite 204
Washington, DC 20036 (202) 659-9076

Martin Abramson
Pollock, Vande Sande & Priddy
1990 M St., N.W., Suite 800
Washington, DC 20036 (202) 331-7111

Wilsie H. Adams, Jr.
Mc Kenna, Conner & Cuneo
1575 Eye St., N.W., Suite 800
Washington, DC 20005 (202) 789-7652

Reid G. Adler
Finnegan, Henderson, Farabow, Garrett & Dunner
1775 K St., N.W.
Washington, DC 20006 (202) 293-6850

Irwin Morton Aisenberg
Berman, Aisenberg & Platt
1730 R.I. Ave., N.W.
Washington, DC 20036 (202) 293-1404

Richard L. Aitken
Lane & Aitken
Watergate Office Bldg.
2600 Virginia Ave., N.W.
Washington, DC 20037 (202) 337-5556

Jennifer A. Albert
Finnegan, Henderson, Farabow, Garrett & Dunner
1775 K St., N.W.
Washington, DC 20006 (202) 293-6850

Louis Allahut
Naval Air Systems Comm.
A I R - 00C5A
Washington, DC 20361 (202) 692-3418

Douglas J. Alspach
1100 Park St., N.E.
Washington, DC 20002 (202) 544-2620

John C. Altmiller
Kenyon & Kenyon
1025 Connecticut Ave.
Washington, DC 20036 (202) 429-1776

Burton A. Amernick
Pollock, Vande Sande & Priddy
1990 M St., N.W., Suite 800
Washington, DC 20036 (202) 331-7111

Walter D. Ames
Watson, Cole, Grindle & Watson
1400 K St., N.W.
Washington, DC 20005 (202) 628-0088

Larry N. Anagnos#
Antonelli, Terry & Wands
1919 Pennsylvania Ave., N.W., Suite 600
Washington, DC 20006 (202) 828-0300

Roger B Andewelt
U.S. Dept. Of Justice
Antitrust Div.
10th St. & Pa. Ave., N.W.
Washington, DC 20530 (202) 633-2562

Donald R. Antonelli
Antonelli, Terry & Wands
1919 Pennsylvania Ave., N.W., Suite 600
Washington, DC 20006 (202) 828-0300

James E. Armstrong, III
Armstrong, Nikaido, Marmelstein & Kubovcik
1725 K St., N.W., Suite 912
Washington, DC 20006 (202) 659-2930

Ernest-Theodore Arndt
Arndt & Arndt
777 - 14th St., N.W.
Washington, DC 20005 (202) 636-6165

Thomas P. Athridge, Jr.*
U.S. Federal Trade Commission
601 Penna. Ave., N.W.
Suite 2311
Washington, DC 20580 (202) 326-2824

Robert D. Bajefsky
Finnegan, Henderson, Farabow, Garrett & Dunner
1775 K St., N.W.
Washington, DC 20006 (202) 293-6850

Hollie L. Baker
Saidman, Sterne, Kessler & Goldstein
1225 Connecticutt Ave.
Washington, DC 20036 (202) 833-7533

Raymond N. Baker
Shanley & Baker
2233 Wisconsin Ave., N.W.
Washington, DC 20007 (202) 333-5800

Donald W. Banner
Banner, Birch, Mc Kie & Beckett
One Thomas Circle, N.W.
Washington, DC 20005 (202) 296-5500

M. Paul Barker#
Barnes & Thornburg
1815 H St., N.W., Suite 800
Washington, DC 20006 (202) 955-4500

Harry E. Barlow*
U.S. Dept. Of Juistice
Civil Div., Pat. Sect.
550 11th St., N.W.
Washington, DC 20530 (202) 724-7280

John A. Bauer
Roylance, Abrams, Berdo & Goodman
1225 Connecticut Ave., N.W.
Washington, DC 20036 (202) 659-9076

Richard P. Bauer
Cushman, Darby & Cushman
1615 L St., N.W., 11th Fl.
Washington, DC 20036 (202) 861-3642

James Beckers
Staas & Halsey
1825 K St., N.W.
Washington, DC 20006 (202) 872-0123

William Wade Beckett
Banner, Birch, Mc Kie & Beckett
One Thomas Circle, N.W.
Washington, DC 20005 (202) 296-5500

Michael D. Bednarek
Kenyon & Kenyon
1025 Connecticut Ave., N.W.
Washington, DC 20036 (202) 429-1776

John W. Behringer
Sutherland, Asbill & Brennan
1666 K St., N.W.
Washington, DC 20006 (202) 887-3415

Edward L. Bell
Banner, Birch, Mc Kie & Beckett
One Thomas Circle, N.W.
Washington, DC 20005 (201) 296-5500

Townsend Mikell Belser, Jr.
Pollock, Vande Sande & Priddy
1990 M St., N.W., Suite 800
Washington, DC 20036 (202) 331-7111

John M. Belz
Leydig, Voit & Mayer
655 Fifteenth St., N.W., Suite 520
Washington, DC 20005 (202) 737-6770

Karen G. Bender
Morgan, Lewis & Bockus
1800 M St.
Washington, DC 20036 (202) 872-5000

W. Patrick Bengtsson
Cushman, Darby & Cushman
1615 L St., N.W.
Washington, DC 20036 (202) 861-3000

Robert H. Berdo, Sr.
Roylance, Abrams, Berdo & Farley
1225 Conn. Ave., N.W., Suite 315
Washington, DC 20036 (202) 659-9076

Melvin G. Berger
Brand & Leckie
1730 K St., N.W., Suite 100
Washington, DC 20006 (202) 347-7002

Herbert Berl*
U.S. Dept. Of Justice
Comm. Litigation Branch
Civil Div.
Washington, DC 20530 (202) 724-7283

Stanford Warner Berman
Berman, Aisenberg & Platt
1730 R.I. Ave., N.W.
Washington, DC 20036 (202) 293-1404

Eugene L. Bernard
Bernard, Rothwell & Brown, P.C.
1700 K St., N.W.
Washington, DC 20006 (202) 833-5740

Frank L. Bernstein
Sughrue, Mion, Zinn, Mac Peak & Seas
1776 K St., N.W., Suite 500
Washington, DC 20006 (202) 293-7060

Howard L. Bernstein
Sughrue, Mion, Zinn, Mac Peak & Seas
1776 K St., N.W.
Washington, DC 20006 (202) 293-7060

John R. Berres#
Communications Satellite Corp.
950 L. Enfant Plaza, S.W.
Washington, DC 20024 (202) 863-6181

Mark S. Bicks
Roylance, Abrams, Berdo & Goodman
1225 Connecticut Ave., N.W.
Washington, DC 20036 (202) 659-9076

Waddell Alexander Biggart, II
Sughrue, Mion, Zinn, Mac Peak & Seas
1776 K St., N.W.
Washington, DC 20006 (202) 293-7060

Harold J. Birch
Banner, Birch, Mc Kie & Beckett
One Thomas Circle, N.W.
Washington, DC 20005 (202) 296-5500

Donald J. Bird
Cushman, Darby & Cushman
1615 L St., N.W., 11th Fl.
Washington, DC 20036 (202) 861-3000

Benjamin Herman Bochenek*
U.S. Environmental Protection Agency
(le-132G)
401 M St., S.W.
Washington, DC 20460 (202) 382-5460

Mark Boland
Sughrue, Mion, Zinn, Macpeak & Seas
1776 K St., N.W.
Washington, DC 20006 (202) 293-7060

Michael J. Boland#
4424 Lingan Rd., N.W.
Washington, DC 20007 (202) 337-0402

Thomas R. Boland
Vorys, Sater, Seymour & Pease
1828 L St., N.W., Suite 1111
Washington, DC 20036 (202) 296-2929

Donald D. Bosben
Fleit, Jacobson, Cohn & Price
1217 E St., N.W.
Washington, DC 20004 (202) 638-6666

Walter Y. Boyd
Finnegan, Henderson, Farabow, Garrett & Dunner
1775 K St., N.W.
Washington, DC 20006 (202) 293-6850

William D. Breneman
Breneman & Georges
Suite 290 - International Square
1850 K St., N.W.
Washington, DC 20006 (202) 467-5800

Barry E. Bretschneider
Wegner & Bretschneider, P.C.
1233 20th St., N.W.
P.O. Box 18218
Washington, DC 20036 (202) 887-0400

David William Brinkman
Cushman, Darby & Cushman
1615 L St., N.W., 11th
Washington, DC 20036 (202) 861-3033

Alvin Browdy
Browdy & Neimark
417 Seventh St., N.W.
Washington, DC 20004 (202) 628-5197

Roger L. Browdy
Browdy & Neimark
419 Seventh St., N.W.
Washington, DC 20004 (202) 628-5197

Norman V. Brown*
U.S. Navy
Asst. Secretary
S & L - C A G
Washington, DC 20360 (202) 692-5632

Brian Garrett Brunsvold
Finnegan, Henderson, Farabow, Garrett & Dunner
1775 K St., N.W.
Washington, DC 20006 (202) 293-6850

B. Frederick Buchan, Jr.*
U.S. Dept. Of Justice
Civil Division
Commercial Litigation Branch
550 11th St., N.W.
Washington, DC 20530 (202) 724-7276

Mary-Elizabeth Buckles
Memel, Jacobs & Ellsworth
1800 M St., N.W., Suite 1000-N
Washington, DC 20036 (202) 822-3939

George Joseph Budock*
Dept. Of The Air Force
Off. Of The Judge Advocate
1900 Half St., S.W.
Washington, DC 20324 (202) 475-1386

William T. Bullinger
Cushman, Darby & Cushman
1615 L St., N.W., Suite 1100
Washington, DC 20036 (202) 861-3000

John P.D. Bundock, Jr.
Leydig, Voit & Mayer, Ltd.
Metropolitan Square, Suite 520
655 15th St., N.W.
Washington, DC 20005 (202) 737-6770

Kenneth J. Burchfiel
Sughrue, Mion, Zinn, Macpeak & Seas
1776 K St., N.W.
Washington, DC 20006 (202) 293-7060

Daniel P. Burke#
Kenyon & Kenyon
1025 Connecticut Ave., N.W.
Washington, DC 20036 (202) 429-1776

Walter W. Burns, Jr.
Barnes & Thornburg
1815 H St., N.W.
Washington, DC 20006 (202) 955-4500

Roy W. Butrum
Watson, Cole, Grindle & Watson
1400 K St., N.W.
Washington, DC 20005 (202) 628-0088

John J. Byrne
Baker & Mckenzie
815 Connecticut Ave., N.W.
Washington, DC 20006 (202) 298-8290

Thomas Joseph Byrnes*
Dept. Of Justice
Commercial Litigation Branch
Civil Division
Washington, DC 20530 (202) 724-7221

Frank E. Caffoe#
Finnegan, Henderson, Farabow, Garrett & Dunner
1775 K St., N.W., Suite 600
Washington, DC 20006 (202) 293-6850

Charles W. Calkins*
U.S. Army
Corps. Of Engineers
HQUSACE, DAEN-CCP
20 Massachusetts Ave., N.W.
Washington, DC 20314 (202) 272-0039

John T. Callahan#
Ostrolenk, Faber, Gerb & Soffen
1725 K St., Suite 1114
Washington, DC 20006 (202) 457-7785

John J. Camby
4405 Burlington Pl., N.W.
Washington, DC 20016 (202) 362-6463

Alan I. Cantor
Banner, Birch Mc Kie & Beckett
One Thomas Circle, N. W.
Washington, DC 20005 (202) 296-5500

Herbert I. Cantor
Wegner & Bretschneider
P.O. Box 18218
Washington, DC 20036 (202) 887-0400

Jay M. Cantor
Spencer & Frank
1111 Nineteenth St., N.W.
Washington, DC 20036 (202) 828-8000

Dean E. Carlson
Pollock, Vande Sande & Priddy
1990 M St., N.W., Suite 800
Washington, DC 20036 (202) 331-7111

John F. Carney
Lyon & Lyon
1200 17th St., N.W.
Washington, DC 20036 (202) 296-9600

Alex Chartove
Spensley, Horn, Jubas & Lubitz
1050 17th St., N.W., Suite 1212
Washington, DC 20036 (202) 223-5700

Chun-I Chiang
U.S. Dept. Of Justice
550 11th St., N.W.
Washington, DC 20530 (202) 274-7364

Gay Chin
777 Fourteenth St., N.W., Suite 747
Washington, DC 20005 (301) 897-6301

Conrad J. Clark
Berman, Aisenberg & Platt
1730 R.I. Ave., N.W., Suite 809
Washington, DC 20036 (202) 293-1404

John H.O. Clarke
Rivener, Clarke, Scrivener & Johnson
1776 K St., N.W., Suite 610
Washington, DC 20006 (202) 296-2950

Ronald D. Cohn
Fleit, Jacobson, Cohn & Price
1217 E. St., N.W.
Washington, DC 20004 (202) 638-6666

Joseph V. Colaianni
Pennie & Edmonds
1730 Pennsylvania Ave., N.W. Suite 1000
Washington, DC 20006 (202) 393-0177

Lawrence L. Colbert
5139 33rd St., N.W.
Washington, DC 20008 (202) 363-2309

David J. Cole#
Fisher, Christen & Sabol
2000 L St., N.W., Suite 510
Washington, DC 20036 (202) 659-2000

Kendrew H. Colton
Cushman, Darby & Cushman
1615 L St., N.W., Suite 1100
Washington, DC 20036 (202) 861-3000

Chris Comuntzis
Cushman, Darby & Cushman
1615 L St., N.W.
Washington, DC 20036 (202) 861-3623

Gregory A. Conley
Covington & Burling
1201 Pennsylvania Ave., N.W.
Washington, DC 20044 (202) 662-5434

Edward J. Connors, Jr.*
U.S. Department Of Navy
Office Of Counsel
Spawar Code OOC62
Washington, DC 20363 (202) 692-1109

Richard E. Constant, Sr.*
U.S. Dept. Of Energy
1000 Independence Ave., S.W.
Washington, DC 20585 (202) 586-2802

Robert E. Converse, Jr.
Finnegan, Henderson, Farabow, Garrett & Dunner
1775 K St., N.W.
Washington, DC 20006 (202) 293-6850

Francis A. Cooch, IV*
Dept. Of The Army
Walter Reed Army Inst. Of Res.
Walter Reed Army Medical Center
Attn: Pat. Coun.
Washington, DC 20307 (202) 576-4369

Iver P. Cooper
Mackler, Cooper & Gibbs
1220 L. St., N.W., Suite 615
Washington, DC 20005 (202) 842-1690

Cornell Daniel Cornish
1101 New Hampshire Ave., N.W., Suite 301
Washington, DC 20037 (202) 429-9705

David K. Cornwell
Saidman, Sterne, Kessler & Goldstein
1225 Connecticut Ave., Suite 300
Washington, DC 20036 (202) 833-7533

Patrick J. Coyne
Collier, Shannon, Rill & Scott
1055 Thomas Jeff. St., N.W., Suite 308
Washington, DC 20007 (202) 342-8606

Paul M. Craig, Jr.
Barnes & Thornburg
1815 H St., N.W.
Washington, DC 20006 (202) 955-4500

Vincent M. Creedon
Wenderoth, Lind & Ponack
805 15th St., N.W., Suite 700
Washington, DC 20005 (202) 371-8850

Rae E. Cronmiller
NRECA
1800 Mass. Ave., N.W.
Washington, DC 20036 (202) 857-9593

David J. Cushing
Sughrue, Mion, Zinn, Macpeak & Seas
1776 K St., N.W.
Washington, DC 20006 (202) 293-7060

James H. Czerwonky*
National Oceanic & Atmospheric Admin. NMFS
1825 Connecticut Ave., N.W.
Washington, DC 20235 (202) 673-5464

Scott M. Daniels
Armstrong, Nikaido, Marmelstein & Kubovcik
1725 K St., N.W.
Washington, DC 20006 (202) 659-2930

Akin Thornwall Davis
Cushman, Darby & Cushman
1615 L St., N.W.
Washington, DC 20036 (202) 861-3047

Garrett V. Davis#
Breneman & Georges
Suite 290 - International Square
1850 K St., N.W.
Washington, DC 20006 (267) 567-5800

James F. Davis
Howrey & Simon
1730 Pennsylvania Ave., N.W., Suite 900
Washington, DC 20006 (202) 383-6589

Michael Rhodes Davis
Wenderoth, Lind & Ponack
805 Fifteenth St., N.W., Suite 700
Washington, DC 20005 (202) 371-8850

Michael M. De Angeli
Pennie & Edmonds
1730 Pennsylvania Ave., N.W.
Washington, DC 20006 (202) 383-0177

Carl M. De Franco, Jr.
Beveridge, De Grandi & Weilacher
1819 H St., N.W., Suite 1100
Washington, DC 20006 (202) 659-2811

Joseph A. De Grandi
Beveridge, De Grandi & Weilacher
1819 H St., N.W.
Washington, DC 20006 (202) 659-2811

John P. De Luca
Robbins & Laramie
1919 Pennsylvania Ave., N.W.
Washington, DC 20006 (202) 887-5050

Donald B. Deaver
Cushman, Darby & Cushman
1615 L St., N. W., 11th Fl.
Washington, DC 20036 (202) 861-3000

Arthur P. Demers
Pennie & Edmonds
1730 Pennsylvania Ave., N.W.
Washington, DC 20006 (202) 393-0177

Paul Devinsky
Marks, Murase & White
2001 L St., N.W., Suite 750
Washington, DC 20036 (202) 955-4900

Vito J. Di Pietro*
U.S. Dept. Of Justice
Commercial Litigation Branch
Civil Division
550 11th St., N.W.
Washington, DC 20530 (202) 724-7223

Michael H. Dickman
Spencer & Frank
1111 19th St.
Washington, DC 20036 (202) 828-8000

Richard Diefendorf#
Fleit, Jacobson, Cohn & Price
1217 E. St., N.W.
Washington, DC 20004 (202) 638-6666

Clarence M. Ditlow, III
Center For Auto Safety
2001 S. St., N.W., Suite 410
Washington, DC 20009 (202) 328-7700

Kenneth W. Dobyns*
U.S. Navy
Naval Sea Systems Comm.
S E A - OOL5
Washington, DC 20362 (202) 692-7077

Howard D. Doescher
Cushman, Darby & Cushman
1615 L St., N.W.
Washington, DC 20036 (202) 861-3000

John Donofrio#
Finnegan, Henderson, Farabow, Garrett & Dunner
1775 K St., N.W.
Washington, DC 20006 (202) 293-6850

Charles Richard Donohoe
Cushman, Darby & Cushman
1615 L St., N.W.
Washington, DC 20036 (202) 861-3059

James Lawrence Dooley
Cushman, Darby & Cushman
1615 L St., N.W., 11th Fl.
Washington, DC 20036 (202) 861-3000

Daniel K. Dorsey#
Fleit, Jacobson, Cohn & Price
1217 E. St., N.W.
Washington, DC 20004 (202) 638-6666

David E. Dougherty
Willim, Brinks, Olds, Hofer, Gilson & Lione
1730 Pennsylvania Ave., N.W., Suite 430
Washington, DC 20006 (202) 628-4000

James N. Dresser
Beveridge, De Grandi & Weilacher
1819 H St., N.W., Suite 1100
Washington, DC 20006 (202) 659-2811

Patricia M. Drost
Finnegan, Henderson, Farabow, Garrett & Dunner
1775 K St., N.W.
Washington, DC 20006 (202) 293-6850

William Anthony Drucker
1111 19th St., N.W., 3rd Fl.
Washington, DC 20036 (202) 828-8048

Folsom E. Drummond
5415 Conn. Ave., N.W., Apt. 320
Washington, DC 20015 (202) 363-6420

Michael P. Dunnam#
Cushman, Darby & Cushman
1615 L St., N.W., 11th Fl.
Washington, DC 20036 (202) 861-3740

Donald R. Dunner
Finnegan, Henderson, Farabow, Garrett & Dunner
1775 K St., N.W.
Washington, DC 20006 (202) 293-6850

Frank J. Dynda*
U.S. Dept. Of The Navy
Naval Air Systems Command
A I R - 00C5
Washington, DC 20361 (202) 692-7810

Edward E. Dyson
Baker & McKenzie
815 Conn. Ave., N.W.
Washington, DC 20006 (202) 298-8290

Barry A. Edelberg
Browdy & Neimark
419 7th St., N.W., Suite 300
Washington, DC 20004 (202) 628-5197

Gary R. Edwards
Ragan & Mason
900 17th St., N.W.
Washington, DC 20006 (202) 296-4750

Arthur R. Eglington
Shanley & Baker
2233 Wisconsin Ave., N.W.
Washington, DC 20007 (202) 333-5800

Carol P. Einaudi
Finnegan, Henderson, Farabow, Garrett & Dunner
1775 K St., N.W.
Washington, DC 20006 (202) 293-6850

Ronald I. Eisenstein
Sughrue, Mion, Zinn, Macpeak & Seas
1776 K St., N.W.
Washington, DC 20006 (202) 293-7060

Neil Anthony Eisner
1919 Pennsylvania Ave., N.W., Suite 200
Washington, DC 20006 (202) 296-6070

Daniel L. Ellis#
Fidelman & Wolffe
1233 20th St., N.W., Suite 300
Washington, DC 20036 (202) 833-8801

Michael C. Elmer
Finnegan, Henderson, Farabow, Garrett & Dunner
1775 K St., N.W., Suite 600
Washington, DC 20006 (202) 293-6850

Alvin J. Englert#*
U.S. Dept. Of Commerce
Off. Of Gen. Coun.
14th & Constitution Ave., N.W., Rm. 4610
Washington, DC 20230 (201) 377-5394

Reuben Epstein
2800 Quebec St., N.W., Apt. 1202
Washington, DC 20008 (202) 362-7685

Barbara G. Ernst
Bernard, Rothwell & Brown, P.C.
1700 K St., N.W.
Washington, DC 20006 (202) 833-5740

Hubert E. Evans
Armstrong, Nikaido, Marmelstein & Kubovcik
1725 K St., N.W.
Washington, DC 20006 (202) 659-2930

Donald D. Evenson
Barnes & Thornburg
1815 H St., N.W., Suite 800
Washington, DC 20006 (202) 955-4500

John D. Fado*
U.S. Dept. Of Agriculture
Res. & Operations Div., Rm. 23215
Off. Of General Counsel
Washington, DC 20250 (202) 447-2421

Ford F. Farabow, Jr.
Finnegan, Henderson, Farabow, Garrett & Dunner
1775 K St., N.W.
Washington, DC 20006 (202) 293-6850

John J. Fargo
U.S. Dept. Of Justice
Civil Div.
Commercial Litigation Branch
Washington, DC 20530 (202) 724-7415

John Theodore Fedigan
Wenderoth, Lind & Ponack
Southern Bldg., Suite 700
805 Fifteenth St., N.W.
Washington, DC 20005 (202) 371-8850

Morris Fidelman
Fidelman & Wolffe
1233 20th St., N.W., Suite 300
Washington, DC 20036 (202) 833-8801

Edward A. Figg
Bernard, Rothwell & Brown, P.C.
1700 K St., N.W.
Washington, DC 20006 (202) 833-5740

Michael J. Fink*
U.S. Dept. Of Justice
Todd Bldg.
550 11th St., N.W.
Washington, DC 20530 (202) 724-7416

Jay M. Finkelstein
Spensley, Horn, Jubas & Lubitz
1050 17th St., N.W.
Washington, DC 20036 (202) 828-8000

Elton Fisher
Armstrong, Nikaido, Marmelstein & Kubovcik
1725 K St., N.W.
Washington, DC 20006 (202) 659-2930

George M. Fisher
Bernard & Brown, P.C.
1700 K St., N.W.
Washington, DC 20006 (803) 573-1598

William J. Fisher
Banner, Birch, Mc Kie & Beckett
One Thomas Circle, N.W.
Washington, DC 20005 (202) 296-5500

Edward P. Fitts
Casey, Lane & Mittendorf
815 Conn. Ave., N.W.
Washington, DC 20006 (202) 785-4949

Martin Fleit
Fleit, Jacobson, Cohn & Price
1217 E. Street, N.W.
Washington. DC 20004 (202) 638-6666

Karl W. Flocks
3110 Chain Bridge Rd., N.W.
Washington, DC 20016 (202) 363-3862

Howard M. Flournoy
P.O. Box 6113
Washington, DC 20044 (703) 455-7594

John Prince Floyd
Roberts & Floyd
2555 M St., N.W.
Washington, DC 20037 (202) 785-0990

Christopher P. Foley
Finnegan, Henderson, Farabow, Garrett & Dunner
1775 K St., N.W.
Washington, DC 20006 (202) 293-6850

John L. Forrest, Jr.*
U.S. Dept. Of Navy
Naval Air Systems Comm.
A I R - 00C-5-5870
Washington, DC 20361 (202) 692-3455

John D. Foster
Leydig, Voit & Mayer
655 15th St., N.W., Suite 520
Washington, DC 20005

Samuel L. Fox
Saidman, Sterne, Kessler & Goldstein
1225 Connecticut Ave., Suite 300
Washington, DC 20036 (202) 833-7533

Francis B. Francois
2317 Pennsylvania Ave., N.W., Suite 11
Washington, DC 20037 (202) 333-7598

Robert J. Frank
Spencer & Frank
1111 19th St., N.W.
Washington, DC 20036 (202) 828-8000

Martin I. Fuchs#f
Finnegan, Henderson, Farabow, Garrett & Dunner
1775 K St., N.W.
Washington, DC 20006 (202) 293-6850

James R. Gaffey
1000 Connecticut Ave., N.W.
Washington, DC 20036 (202) 337-2295

Arthur Joseph Gajarsa
Joseph, Gajarsa, Mc Dermott & Reiner, P.C.
1300 19th St., N.W., Suite 400
Washington, DC 20036 (202) 331-1955

Donald A. Gardiner, Jr.*
U.S. Dept. Of Interior
Off. Of The Solicitor
18th & C St., N.W.
Washington, DC 20240 (202) 343-4471

Arthur Sellers Garrett
Finnegan, Henderson, Farabow, Garrett & Dunner
1775 K St., N.W., Suite 600
Washington, DC 20006 (202) 293-6850

John C. Garvey
Staas & Halsey
1825 K St., N.W., Suite 816
Washington, DC 20006 (202) 872-0123

Leslie H. Gaston
2500 Que St., N.W.
Washington, DC 20007 (202) 333-1074

Robert James Gaybrick
Finnegan, Henderson, Farabow, Garrett & Dunner
1775 K St., N.W.
Washington, DC 20006 (202) 293-6850

Robert U. Geib, Jr.
535 Washington Bldg.
1435 G. St., N.W.
Washington, DC 20005 (202) 393-7323

Neil D. Gershon
Saidman, Sterne, Kessler & Goldstein
1225 Connecticut Ave., N.W.
Washington, DC 20036 (202) 833-7533

Jim W. Gipple
Gipple & Hale
P.O. Box 40513
Washington, DC 20016 (703) 448-1770

Charles E. B. Glenn*
NASA Headquarters
Off. Of Assoc. Gen. Coun.
Intellectual Property
Mail Code G P
Washington, DC 20546 (202) 453-2421

Raymond C. Glenny#
Saidman, Sterne, Kessler & Goldstein
1225 Connecticut Ave., N.W.
Washington, DC 20036 (202) 833-7533

Robert C. Godbey
Peabody, Lambert & Myers
1150 Connecticut Ave., N.W.
Washington, DC 20036 (202) 457-1033

Alfred N. Goldman
Roylance, Abrams, Berdo & Goodman
1225 Connecticut Ave., N.W., Suite 204
Washington, DC 20036 (202) 659-9076

Ricky S. Goldman
1990 M St., N.W., Suite 540
Washington, DC 20036 (202) 659-2366

Jorge A. Goldstein
Saidman, Sterne, Kessler & Goldsstein
1225 Connecticut Ave., Suite 300
Washington, DC 20036 (202) 833-7533

Richard A. Gollhofer
Staas & Halsey
1825 K St.
Washington, DC 20006 (202) 872-0123

Michael J. Gonet*
U.S. Navy
N A V A I R
A I R - O O C-5-5870
Patents Division
Washington, DC 20361 (202) 692-3456

Thomas W. Gorman
Howrey & Simon
1730 Pennsylvania Ave., N.W.
Washington, DC 20006 (202) 783-0800

Elmer Ellsworth Goshorn
U.S. Dept. Of Navy
Off. Of General Counsel
Washington, DC 20350 (202) 692-7136

Lawrence Jay Gotts
Kirkland & Ellis
655 15th St., N.W.
Washington, DC 20005 (202) 879-5000

Peter W. Gowdey
Cushman, Darby & Cushman
1615 L St., N.W., 11th Fl.
Washington, DC 20036 (202) 861-3078

Barry W. Graham
Finnegan, Henderson, Farabow, Garrett & Dunner
1775 K St., N.W., Suite 600
Washington, DC 20006 (202) 293-6850

Paul Grandinetti
Fisher, Christen & Sabol
2000 L St., N.W., Suite 510
Washington, DC 20036 (202) 659-2000

Stanley Barry Green
Pollock, Vande Sande & Priddy
1990 M St., N.W., Suite 800
Washington, DC 20036 (202) 331-7111

Susan H. Griffen
Finnegan, Henderson, Farabow, Garrett & Dunner
1775 K St., N.W., Suite 600
Washington, DC 20006 (202) 293-6850

Alan M. Grimaldi
Howrey & Simon
1730 Penn. Ave., N.W.
Washington, DC 20006 (202) 383-6989

Robert Groover, III
Saidman, Sterne, Kessler & Goldstein
1225 Connecticut Ave., N.W., Suite 300
Washington, DC 20036 (202) 833-7533

Barry L. Grossman
Banner, Birch, Mc Kie & Beckett
One Thomas Circle
Washington, DC 20005 (202) 296-5500

Francis W. Guay
1713 18th St., N.W.
Washington, DC 20009 (202) 234-3546

Louis Gubinsky
Sughrue, Mion, Zinn, Macpeak & Seas
1776 K St., N.W.
Washington, DC 20006 (202) 293-7060

Charles S. Guenzer
Sughrue, Mion, Zinn, Macpeak & Seas
1776 K St., N.W.
Washington, DC 20006 (202) 293-7060

Thomas S. Hahn*
U.S. House Of Representatives
Comm. On Armed Services
2120 Rayburn House Off. Bldg.
Washington, DC 20515

Charles S. Hall
Finnegan, Henderson, Farabow, Garrett & Dunner
1775 K St., N.W., Suite 600
Washington, DC 20006 (202) 293-6850

James Dillard Halsey, Jr.
Staas & Halsey
1825 K St., N.W.
Washington, DC 20006 (202) 872-0123

Philip G. Hampton, II
Kenyon & Kenyon
1025 Connecticut Ave., N.W.
Washington, DC 20036 (202) 429-1776

John A. Hankins
Kenyon & Kenyon
1025 Connecticut Ave., N.W.
Washington, DC 20036 (202) 429-1776

Donald W. Hanson
Huff & Hanson
1400 K St., N.W., Suite 725
Washington, DC 20005 (202) 466-6386

Larry Harbin
Cushman, Darby & Cushman
1615 L St., N.W., Suite 1100
Washington, DC 20036 (202) 861-3000

Scott C. Harris
Cushman, Darby & Cushman
1615 L St., N.W.
Washington, DC 20036 (202) 641-0727

Fred W. Hathaway
Robbins & Laramie
1919 Pennsylvania Ave., N.W., Suite 506
Washington, DC 20006 (202) 887-5050

Laurence R. Hefter
Finnegan, Henderson, Farabow, Garrett & Dunner
1775 K St., N.W.
Washington, DC 20006 (202) 293-6850

Douglas B. Henderson
Finnegan, Henderson, Farabow, Garrett & Dunner
1775 K St., N.W.
Washington, DC 20006 (202) 293-6850

William R. Henderson
U.S. Dept. Of Navy
Naval Sea Systems Comm.
Off. Of Gen. Counsel
Washington, DC 20362 (202) 692-7077

Bruce J. Hendricks
Bernard, Rothwell & Brown, P.C.
1700 K St., N.W., Suite 800
Washington, DC 20006 (202) 833-5740

Charles E. Hepner
Kenyon & Kenyon
1025 Connecticut Ave., N.W.
Washington, DC 20036 (202) 429-1775

William F. Herbert
Staas & Halsey
1825 K St., N.W.
Washington, DC 20006 (202) 872-0123

David A. Hey#
Antonelli, Terry & Wands
1919 Pennsylvania Ave., N.W., Suite 600
Washington, DC 20006 (202) 828-0300

David W. Highet#
Robbins & Laramie
1919 Pennsylvania Ave., N.W., Suite 506
Washington, DC 20006 (202) 887-5050

Judson R. Hightower*
U.S. Dept. Of Energy
1000 Independence Ave., S.W.
Washington, DC 20585 (202) 586-3499

David W. Hill
Finnegan, Henderson, Farabow, Garrett & Dunner
1775 K St., N.W.
Washington, DC 20006 (202) 293-6850

Joseph A. Hill
Dept. Of Justice
Comm. Litigation Br.
Civil Div.
Washington, DC 20530 (202) 724-7275

David L. Hoffman
Sughrue, Mion, Zinn, Mac Peak & Seas
1776 K St., N.W., Suite 500
Washington, DC 20006 (202) 293-7060

Mark L. Hogge
Fisher, Christen & Sabol
2000 L St., N.W., Suite 510
Washington, DC 20036 (202) 659-2000

Barry I. Hollander
Fisher, Christen & Sabol
2000 L St., Suite 510
Washington, DC 20036 (202) 659-2000

Lawrence M. Holloway
1829 Irving St., N.W.
Washington, DC 20010 (202) 234-6613

John C. Holman
Holman & Stern, Chartered
Meridian Hall
2401 15th St., N.W.
Washington, DC 20009 (202) 483-2234

John E. Holmes
Robbins & Laramie
1919 Pennslyvania Ave., N.W.
Washington, DC 20006 (202) 887-5050

Alvin D. Hooper
AT&T
1120 20th St., N.W.
Washington, DC 20036 (202) 457-2377

Dale H. Hoscheit
Banner, Birch, Mc Kie & Beckett
One Thomas Circle, N.W.
Washington, DC 20005 (202) 296-5500

Frank L. Huband*
Natl. Science Foundation
Off. Of Gen. Coun.
1800 G St., N.W.
Washington, DC 20550 (202) 357-7829

John W. Huckert
7400 Benjamin Franklin Station
Washington, DC 20044 (301) 258-9418

Donald N. Huff
Huff & Hanson
1400 K St., N.W.
Washington, DC 20005 (202) 466-6386

Patrick H. Hume
3830 Macomb St., N.W.
Washington, DC 20016 (202) 244-2772

Nathaniel A. Humphries
Mason, Fenwick & Lawrence
1225 Eye St., N.W., Suite 1000
Washington, DC 20005 (202) 289-1200

Lawrence A. Hymo
Cushman, Darby & Cushman
1615 L St., N.W., 11th Fl.
Washington, DC 20036 (202) 861-3000

Jeffrey L. Ihnen
Robbins & Laramie
1919 Pennsylvania Ave., N.W., Suite 506
Washington, DC 20006 (202) 887-5050

John R. Inge
Sughrue, Mion, Zinn, Macpeak & Seas
1776 K St., N.W., Suite 500
Washington, DC 20006 (202) 293-7060

Edward S. Irons
Irons & Sears, P.C.
1785 Massachusetts Ave., N.W.
Washington, DC 20036 (202) 466-5200

Thomas L. Irving
Finnegan, Henderson, Farabow, Garrett & Dunner
1775 K St., N.W., Suite 600
Washington, DC 20006 (202) 293-6850

Christopher P. Isaac
Finnegan, Henderson, Farabow, Garrett & Dunner
1775 K St., N.W., Suite 600
Washington, DC 20006 (202) 293-6850

Yoji Ito#
Nippondenso Co. Ltd.
% Cushman, Darby & Cushman
1615 L St., N.W.
Washington, DC 20036 (202) 861-3627

Thomas H. Jackson
Banner, Birch, Mc Kie & Beckett
One Thomas Circle, N.W., Sixth Fl.
Washington, DC 20005 (202) 296-5500

Matthew M. Jacob
Wenderoth, Lind & Ponack
Southern Bldg., Suite 700
805 Fifteenth St., N.W.
Washington, DC 20005 (202) 371-8850

Harvey B. Jacobson, Jr.
Fleit, Jacobson, Cohn & Price
1217 E. St., N.W.
Washington, DC 20004 (202) 638-6666

Harvey B. Jacobson, Sr.
Fleit, Jacobson, Cohn & Price
1217 E. St., N.W.
Washington, DC 20004 (202) 393-3380

Kenneth E. Jaconetty#
Wegner & Bretschneider
P.O. Box 18218
Washington, DC 20036 (202) 887-0400

George Jameson*
U.S. Dept. Of The Navy
Patents, Code 1208.2
Naval Research Laboratory
Washington, DC 20375 (202) 767-3428

Harold G. Jarcho
Jarcho & Rusz
1377 K St., N.W., Suite 115
Washington, DC 20005 (301) 434-8357

Edmund Michael Jaskiewicz
1730 M St., N.W., Suite 501
Washington, DC 20036 (202) 296-2295

Thomas H. Jenkins
Finnegan, Henderson, Farabow, Garrett & Dunner
1775 K St., N.W., Suite 600
Washington, DC 20006 (202) 293-6850

Tipton D. Jennings
Finnegan, Henderson, Farabow, Garrett & Dunner
1775 K St., N.W.
Washington, DC 20006 (202) 293-6850

Harold Alden Jewett
5451 - 42nd St., N.W.
Washington, DC 20015 (202) 362-0909

Henry R. Jiles#
Wenderoth, Lind & Ponack
Southern Bldg.
805 15th St., N.W.
Washington, DC 20005 (202) 371-8850

James W. Johnson, Jr.
1631 Jonquil St., N.W.
Washington, DC 20012 (202) 944-6853

William D. Johnston, III
419 Seventh St., N.W., Suite 300
Washington, DC 20004

Raymond N. Jones#
1426 Leegate Rd., N.W.
Washington, DC 20001 (202) 726-0730

Wayne A. Jones
Cushman, Darby & Cushman
1615 L St., N.W., Suite 1100
Washington, DC 20036 (202) 861-3685

Kevin E. Joyce
Cushman, Darby & Cushman
1615 L St., N.W., 11th Fl.
Washington, DC 20036 (202) 861-3000

Chester L. Justus
4208 45th St. N.W.
Washington, DC 20016 (202) 362-0210

Sarah Anne Kagan#
Banner, Birch, Mc Kie & Beckett
1 Thomas Circle, N.W.
Washington, DC 20005 (202) 296-5500

David S. Kalmbaugh
Naval Air Systems Command
A I R - OOC-S-5870
Patent Division
Washington, DC 20361 (202) 692-3456

Ronald P. Kananen
Vorys, Sater, Seymour & Pease
1828 L St., N.W., Suite 1111
Washington, DC 20036 (202) 822-8200

Geoffrey M. Karney
Finnegan, Henderson, Farabow, Garrett & Dunner
1775 K St., N.W.
Washington, DC 20006 (202) 293-6850

Jessie W. Karsted
Clarence A O Brien & Harvey B. Jacobson
1217 E. St., N.W.
Washington, DC 20004 (202) 393-3380

Albert Anthony Kashinski
U.S. Dept. Of Interior
Office Of Solicitor
Washington, DC 20240 (202) 343-5431

Alan J. Kaspar
Communications Satellite Corp.
Comsat Bldg.
950 L Enfant Plaza S., S.W.
Washington, DC 20024 (202) 863-6142

Daniel C. Kaufman
Ely, Ritts, Brickfield & Betts
600 New Hamp. Ave., N.W., Suite 915
Washington, DC 20037 (202) 342-0800

Donald Allen Kaul
Brownstein, Zeidman & Schomer
1025 Conn. Ave., N.W., Suite 900
Washington, DC 20036 (202) 457-6555

Harvey Kaye
Cohen & Burg, P.C.
2555 M St., N.W., Suite 300
Washington, DC 20037 (202) 785-0773

Irving Kayton
2100 Pennsylvania Ave., N.W., Suite 237
Washington, DC 20037 (202) 223-1177

Francis A. Keegan
Ward, Lalos, Leeds, Keegan & Lett
1200 17th St., N.W.
Washington, DC 20036 (202) 466-7660

Herbert B. Keil
Keil & Witherspoon
1101 Conn. Ave., N.W.
Washington, DC 20036 (202) 659-0100

Gabor J. Kelemen
Spencer & Kaye
1111 19th St., N.W.
Washington, DC 20036 (202) 828-8000

Michael L. Keller
Cushman, Darby & Cushman
1615 L St., N.W., 11th Fl.
Washington, DC 20036 (202) 861-3000

Brian C. Kelly
U.S. Dept. Of Navy
Off. Of Gen. Counsel
Naval Research Lab.
Code 1208.2
Washington, DC 20375

Robert F. Kempf*
NASSA
Asst. Gen. Coun.
400 Maryland Ave., S.W.
Washington, DC 20546 (202) 755-3932

Lynn V. Kent#
Banner, Birch, Mc Kie & Beckett
One Thomas Circle, N.W.
Washington, DC 20005 (202) 296-5500

Edward J. Kessler
Saidman, Sterne, Kessler & Goldstein
1225 Connecticut Ave.
Washington, DC 20036 (202) 833-7533

Warren B. Kice
Lane, Aitken, Kice & Kananen
1100 Ring Bldg.
1200 18th St., N.W.
Washington, DC 20036 (202) 466-8090

Bradford E. Kile
Baker & Mc Kenzie
815 Conn. Ave., N.W.
Washington, DC 20006 (202) 298-8290

Robert Kinberg
Spencer & Kaye
1111 19th St., N.W.
Washington, DC 20036 (202) 828-8000

William K. King
Ford Motor Co.
815 Conn. Ave., N.W.
Washington, DC 20006 (202) 785-6032

Allen Kirkpatrick, III
Cushman, Darby & Cushman
1615 L St., N.W., 11th Fl.
Washington, DC 20036 (202) 861-3000

Gordon Kit
Sughrue, Mion, Zinn, Macpeak & Seas
1776 K St., N.W.
Washington, DC 20006 (202) 293-7060

Richard H. Kjeldgaard
Bernard, Rothwell & Brown
1700 K St., N.W.
Washington, DC 20006 (202) 833-5740

Alan P. Klein*
Dept. Of Navy
Code Sea-Oop - Off. Of Pat. Coun.
Naval Sea Systems Command
Washington, DC 20362

Peter L. Klempay#
1003 K Street, N.W., Suite 820
Washington, DC 20001 (202) 737-6610

Richard G. Kline
Beveridge, Degrandi & Kline
1819 H St., N.W., Suite 1100
Washington, DC 20006 (202) 659-2811

G. Lloyd Knight, Jr.
Cushman, Darby & Cushman
1615 L St., N.W., 11th Fl.
Washington, DC 20036 (202) 861-3000

John L. Knoble#
Cushman, Darby & Cushman
1615 L St., N.W.
Washington, DC 20036 (202) 861-3000

Paul N. Kokulis
Cushman, Darby & Cushman
1615 L St., N.W., 11th Fl.
Washington, DC 20036 (202) 861-3000

Epaminondas Philip Koltos*
U.S. Dept. Of The Interior
Office Of The Solicitor
18th & C St., N.W.
Washington, DC 20240

Anne M. Kornbau
Browdy & Neimark
419 Seventh St., N.W., Suite 300
Washington, DC 20042 (202) 628-5197

William Kovensky
Browdy & Neimark
419 7th St., N.W.
Washington, DC 20004 (202) 628-5197

Melvin Kraus
Craig & Antonelli
1919 Pennsylvania Ave., N.W., Suite 600
Washington, DC 20006 (202) 828-0300

Louis F. Kreek
4636 Verplanck Pl. N.W.
Washington, DC 20016 (202) 363-5991

Carla Magda Krivak#
Staas & Halsey
1825 K St., N.W.
Washington, DC 20006 (202) 872-0123

Ronald J. Kubovcik
Armstrong, Nikaido, Marmelstein & Kubovcik
1725 K St., N.W.
Washington, DC 20006 (202) 659-2930

Norman N. Kunitz
Spencer & Kaye
1111 19th St., N.W.
Washington, DC 20036 (202) 828-8000

Joseph Labow*
NASA
Bldg. F B - 6, Rm. 5065
Fourth St. & Maryland Ave.
Washington, DC 20546 (202) 755-1602

David L. Ladd
Wiley & Rein
1776 K St., N.W.
Washington, DC 20006 (202) 429-7000

Jack L. Lahr
Foley, Lardner, Hollabaugh & Jacobs
1775 Pa. Ave., N.W.
Washington, DC 20006 (202) 862-5300

Peter N. Lalos
Lalos, Leeds, Keegan, Lett & Marsh
1200 17th St., N.W.
Washington, DC 20036 (202) 466-7660

Richard A. Lambert*
U.S. Dept. Of Energy
1000 Independence Ave., S.W.
Washington, DC 20585 (202) 252-2806

John Thomas Lanahan
Ward, Lazarus, Grow & Cihlar
1711 N St., N.W.
Washington, DC 20036 (202) 331-8160

Sheldon I. Landsman
Sughrue, Mion, Zinn, Macpeak & Seas
1776 K St., N.W.
Washington, DC 20006 (202) 293-7060

Joseph M. Lane
Lane, Aitken & Kananen
300 Watergate Office Bldg.
2600 Virginia Ave., N.W.
Washington, DC 20037 (202) 337-5556

Thomas Joseph Lannon
105 Whittier St., Northwest
Washington, DC 20012 (202) 726-2882

John C. Laprade
1511 K St., N.W., Suite 831
Washington, DC 20005 (202) 347-3100

James R. Laramie#
Robbins & Laramie
1919 Pennsylvania Ave., N.W.
Washington, DC 20006 (202) 887-5050

Douglas N. Larson
Vorys, Sater, Seymour & Pease
1828 L St., N.W., Suite 1111
Washington, DC 20036 (202) 822-8200

Robert J. Lasker
Watson, Cole, Grindle & Watson
1400 K St., N.W.
Washington, DC 20005 (202) 628-0088

Joseph P. Lavelle
Howrey & Simon
1730 Pennsylvania Ave., N.W.
Washington, DC 20006 (202) 383-6888

Lawrence M. Lavin#
Philips Petroleum Co.
Patent Div.
1825 K St., N.W., Suite 1107
Washington, DC 20006 (202) 785-1252

Irvin A. Lavine
Mason, Fenwick & Lawrence
1225 Eye St., N.W., Suite 1000
Washington, DC 20005 (202) 289-1200

Nina M.S. Lawrence*
NASA
400 Maryland Ave., S.W.
G P - 4
Washington, DC 20546 (202) 755-3930

Dale S. Lazar
Cushman, Darby & Cushman
1615 L St., N.W., 11th Fl.
Washington, DC 20036 (202) 861-3000

Samuel Lebowitz
5736 26th St., N.W.
Washington, DC 20015 (202) 537-0248

Eugene M. Lee#
Fidelman, Wolffe & Waldron
2120 L St., N.W., Suite 300
Washington, DC 20037 (202) 833-8801

Frederick J. Lees*
NASA
Board Of Contract Appeals
400 Maryland Ave., S.W.
Washington, DC 20546 (202) 755-3728

Charles B. Lefkoff#
P.O. Box 28347
Washington, DC 20005 (202) 483-4762

Lewis J. Lenny
1301 20th St., N.W.
Washington, DC 20036 (202) 659-2611

Andrew Matthew Lesniak*
Naval Air Sys. Comm.
Off. Of Pat. Coun.
Washington, DC 20361 (202) 920-2764

Michelle N. Lester#
Cushman, Darby & Cushman
1615 L St., N.W., 11th Fl.
Washington, DC 20036 (202) 861-3000

Jack Q. Lever, Jr.
Lupo, Lipman & Lever, P.C.
2000 K St., N.W., Suite 200
Washington, DC 20006 (202) 429-0625

Alfred Bernard Levine
5225 Wisconsin Ave., N.W., Suite 601
Washington, DC 20015 (202) 293-6390

Charles D. Levine
Banner, Birch, Mc Kie & Beckett
One Thomas Circle, N.W.
Washington, DC 20005 (202) 296-5500

Samuel Levine
6716 Eastern Ave., N.W.
Washington, DC 20012 (202) 723-5491

Sherman Levy
1511 K St., N.W.
Investment Bldg., Suite 808
Washington, DC 20005 (202) 628-7625

Terrell P. Lewis
Browdy & Neimark
419 Seventh St., N.W., Suite 300
Washington, DC 20004 (202) 628-5197

Basil J. Lewris
Finnegan, Henderson, Farabow, Garrett & Dunner
1775 K St., N.W.
Washington, DC 20006 (202) 293-6850

Thomas H. Liddle, III
U.S. Dept. Of Justice
Pennsylvania Ave.
Washington, DC 20530 (202) 724-7969

Donald S. Lilly*
U.S. Postal Service
475 L Enfant Plaza, SW
Room 9134
Washington, DC 20260 (202) 245-4611

Nancy J. Linck
Cushman, Darby & Cushman
1615 L St., N.W., 11th Fl.
Washington, DC 20036 (202) 861-3658

Richard Linn
Marks Murase & White
2001 L St., N.W., Suite 750
Washington, DC 20036 (202) 955-4900

Steven E. Lipman
Lupo, Lipman & Lever, P.C.
2000 K St., N.W., Suite 200
Washington, DC 20006 (202) 429-0625

Raymond F. Lippitt
Cushman, Darby & Cushman
1615 L St., N.W., 11th Fl.
Washington, DC 20036 (202) 861-3000

Charles E. Lipsey
Finnegan, Henderson, Farabow, Garrett & Dunner
1775 K St., N.W.
Washington, DC 20006 (202) 293-6850

Morris Liss
Pollock, Vandesande & Priddy
1990 M St., N.W., Suite 800
Washington, DC 20036 (202) 331-7111

James R. Longacre#
Cushman, Darby & Cushman
1615 L St., N.W., 11th Fl.
Washington, DC 20036 (202) 861-3000

Carl George Love
Cushman, Darby & Cushman
1615 L St., N.W., 11th Fl.
Washington, DC 20036 (202) 861-3000

Ethel G. Love
4513 17th St., N.E.
Washington, DC 20017 (202) 529-8857

Paul T. Lubeck*
U.S. Dept. Of Justice
Antitrust Division
Safeway Bldg. - 704
Washington, DC 20530 (202) 724-7966

Robert Bennett Lubic
2033 M St., N.W.
Washington, DC 20036 (202) 452-8200

Frank A. Lukasik
Federal Bar Bldg. Penthouse
1815 H. St., N.W.
Washington, DC 20006 (202) 785-0491

Raphael V. Lupo
Lupo, Lipman & Lever, P.C.
2000 K St., N.W., Suite 200
Washington, DC 20006 (202) 429-0625

Harry Lupuloff*
National Aeronautics & Space Administration
Off. Of Assoc. Gen. Coun.
Intellectual Property, Code G P
Washington, DC 20546 (202) 453-2421

Harry J. Macey#
Banner, Birch, Mc Kie & Beckett
One Thomas Circle, N.W., Suite 600
Washington, DC 20005 (202) 296-5500

Susan J. Mack
Sughrue, Mion, Zinn, Macpeak & Seas
1776 K St., N.W.
Washington, DC 20006 (202) 293-7060

Thomas J. Macpeak
Sughrue, Mion, Zinn, Macpeak & Seas
1776 K St., N.W.
Washington, DC 20006 (202) 293-7060

Theodore Major*
U.S. Postal Service
475 L Enfant Plaza, S.W., Rm 9226
Washington, DC 20260 (202) 245-4062

Michael A. Makuch
Beveridge, De Grandi & Weilacher
1819 H St., N.W.
Washington, DC 20006 (202) 659-2811

Pamela H. Malech
4115 Davis Pl., N.W. Apt. 105
Washington, DC 20007 (202) 342-8366

John W. Malley
Cushman, Darby & Cushman
1615 L St., N.W., 11th Fl.
Washington, DC 20036 (202) 861-3075

Scott D. Malpede#
Sughrue, Mion, Zinn, Macpeak & Seas
1776 K St., N.W.
Washington, DC 20006 (202) 293-7060

John Raymond Manning*
NASA
Office Of General Counsel
400 Maryland Ave., S.W.
Washington, DC 20546 (202) 453-2416

John G. Mannix*
NASA
Office Of General Counsel
400 Maryland Ave., S.W.
Washington, DC 20546 (202) 755-3954

Robert J. Marchick*
U.S. Dept. Of Energy
1000 Independence Ave., S.W.
Washington, DC 20585 (301) 252-2806

Donald William Marks
Bernard, Rothwell & Brown, P.C.
1700 K St., N.W., Suite 800
Washington, DC 20006 (202) 833-5740

Charles M. Marmelstein
Armstrong, Nikaido, Marmelstein & Kubovcik
1725 K St., N.W., Suite 912
Washington, DC 20006 (202) 659-2930

James H. Marsh, Jr.
Staas & Halsey
1825 K St., N.W.
Washington, DC 20006 (202) 872-0123

Virgil Homer Marsh
Fisher, Christen & Sabol
2000 L St., N.W., Suite 510
Washington, DC 20036 (202) 659-2000

Edgar H. Martin
Cushman, Darby & Cushman
1615 L St., N.W., 11th Fl.
Washington, DC 20036 (202) 861-3000

James Thomas Martin
1700 H St., N.W.
Washington, DC 20006 (202) 296-3650

Larry H. Martin
Leitner, Palan, Martin & Bernstein
2201 Wisconsin Ave., N.W., Suite 300
Washington, DC 20007 (202) 337-5900

Peter B. Martine
Finnegan, Henderson, Farabow, Garrett & Dunner
1775 K St., N.W.
Washington, DC 20006 (202) 293-6850

Raymond N. Matson#
402 Westory Bldg.
605 14th St., N.W.
Washington, DC 20005 (202) 628-8467

Clifton E. Mc Cann
Lane, Aitken & Kananen
300 Watergate Office Bldg.
2600 Virginia Ave., N.W.
Washington, DC 20037 (202) 337-5556

Garland Thomas Mc Coy*
NASA
400 Maryland Ave., S.W.
Washington, DC 20546 (202) 755-3932

Barbara C. Mc Curdy
Finnegan, Henderson, Farabow, Garrett & Dunner
1775 K St., N.W.
Washington, DC 20006 (202) 293-6850

Michael R. Mc Gurk#
Finnegan, Henderson, Farabow, Garrett & Dunner
1775 K St., N.W.
Washington, DC 20006 (202) 293-6850

James F. Mc Keown
Antonelli, Terry & Wands
1919 Pennsylvania Ave., N.W., Suite 600
Washington, DC 20006 (202) 828-0300

Edward F. Mc Kie, Jr.
Banner, Birch, Mc Kie & Beck
One Thomas Circle, N.W.
Washington, DC 20005 (202) 296-5500

Le-Nhung Mc Leland
Armstrong, Nikaido, Marmelstein & Kubovcik
1725 K St., N.W., Suite 912
Washington, DC 20006 (202) 659-2930

Robert G. Mc Morrow
Sughrue, Mion, Zinn, Macpeak & Seas
1776 K St., N.W.
Washington, DC 20006 (202) 293-7060

Paul F. Mc Quade
Finnegan, Henderson, Farabow, Garrett & Dunner
1775 K St., N.W.
Washington, DC 20006 (202) 293-6850

John J. Mc Veigh
Fisher, Wayland, Cooper & Leader
1255 23rd St., N.W.
Washington, DC 20037 (202) 659-3494

Michael J. McGreal
Fisher, Christen & Sabol
2000 L St., N.W.
Washington, DC 20036 (202) 659-2000

Francis G. McKenna
Anderson & Pendleton
1000 Connecticut Ave., N.W.
Washington, DC 20036 (202) 659-2334

Bernard A. Meany, Sr.
Mason, Fenwick & Lawrence
1730 Rhode Island Ave., Suite 310
Washington, DC 20036 (202) 293-2010

Joseph T. Melillo*
U.S. Dept. Of Justice
Antitrust Div.
Intellectual Property Section
Washington, DC 20530 (202) 724-7969

Allen S. Melser
Mason, Fenwick & Lawrence
1225 Eye St., N.W., Suite 1000
Washington, DC 20005 (202) 289-1200

Michael E. Melton
Spensley, Horn, Jubas & Lubitz
1050 17th St., N.W., Suite 1212
Washington, DC 20036 (202) 223-5700

Evelyn K. Merker
Wegner & Bretschneider
1233 20th St., N.W., Suite 300
Washington, DC 20036 (202) 887-0400

A. Donald Messenheimer
Antonelli, Terry & Wands
1919 Pennsylvania Ave., N.W.
Washington, DC 20005 (202) 828-0300

Darryl Mexic
Sughrue, Mion, Zinn, Macpeak & Seas
1776 K St., N.W.
Washington, DC 20006 (202) 293-7060

Richard S. Meyer#
Kenyon & Kenyon
1025 Connecticut Ave., N.W.
Washington, DC 20036 (202) 429-1776

Kenneth John Meyers
Beveridge, De Grandi & Kline
Federal Bar Bldg.
1819 H St., N.W.
Washington, DC 20006 (202) 659-2811

Edward F. Miles*
U.S. Dept. Of The Navy
Code 1208.2
Naval Research Laboratory
Washington, DC 20375

D. Byron Miller, Jr.
Banner, Birch, Mc Kie & Beckett
One Thomas Circle, N.W. Sixth Fl.
Washington, DC 20005 (202) 296-5500

John T. Miller
Wenderoth, Lind & Ponack
805 Fifteenth St., N.W., Suite 700
Washington, DC 20005 (202) 371-8850

Marc A. Miller
Holman & Stern
2401 15th St., N.W.
Washington, DC 20009 (202) 483-2234

Leo Millstein
Comsat
950 L Enfant Plaza, S.W.
Washington, DC 20024 (202) 863-6173

Herbert H. Mintz
Finnegan, Henderson, Farabow, Garrett & Dunner
1775 K St., N.W.
Washington, DC 20006 (202) 293-6850

John H. Mion
Sughrue, Mion, Zinn, Macpeak & Seas
1776 K St., N.W.
Washington, DC 20006 (202) 293-7060

George T. Mobille
Cushman, Darby & Cushman
1615 L St., N.W., 11th Fl.
Washington, DC 20036 (202) 861-3000

Gregory E. Montone
Antonelli, Terry & Wands
1919 Pennsylvania Ave., N.W. Suite 600
Washington, DC 20006 (202) 828-0300

John P. Moran
Staas & Halsey
1825 K St., N.W.
Washington, DC 20006 (202) 872-0123

Thomas J. Morgan
Irons & Sears, P.C.
1785 Massachusetts Ave., N.W.
Washington, DC 20036 (202) 466-5200

Roy L. Morris
MCI Communications Corp.
1133 19th St., N.W.
Washington, DC 20036 (202) 887-2363

Robert H. Morse
Galland, Kharasch, Calkins & Morse, P.C.
1054 31st St., N.W.
Washington, DC 20007 (202) 342-5260

William R. Moser*
U.S. Dept. Of Energy
Off. Of Asst. Gen. Coun.
1000 Independence Ave., S.W.
Washington, DC 20585 (202) 252-2806

Kathleen S. Moss
Wegner & Bretschneider
2023 M St., N.W.
Washington, DC 20036 (202) 887-0400

Gerald J. Mossinghoff
Pharmaceutical Manufacturers Assn.
1100 Fifteenth St., N.W.
Washington, DC 20005 (202) 835-3420

William E. Mouzavires
Lyon & Lyon
1200 17th St., N.W., Suite 405
Washington, DC 20036 (202) 296-6028

Boardman Shaw Mowry*
Naval Air Systems Command
(air O O P)
Washington, DC 20361 (202) 692-2445

Philip J. Moy, Jr.#
Finnegan, Henderson, Farabow, Garrett & Dunner
1775 K St., N.W.
Washington, DC 20006 (202) 293-6850

Douglas P. Mueller#
Wegner & Bretschneider
2030 M St., N.W.
P.O. Box 18218
Washington, DC 20036 (202) 887-0400

E.C. Mulcahy, Jr.
Pharmaceutical Mgfrs. Assn.
1100 15th St., N.W.
Washington, DC 20005 (202) 835-3512

Mary Ann G. Mullen#
Lupo, Lipman & Lever, P.C.
2000 K St., N.W., Suite 200
Washington, DC 20006 (202) 429-0625

Michael M. Murray#
Banner, Birch, Mc Kie & Beckett
One Thomas Circle N.W.
Washington, DC 20005 (202) 296-5500

Robert B. Murray, Jr.
Murray & Whisenhunt
P.O. Box 40574
Washington, DC 20016 (703) 243-0400

Sheridan L. Neimark
Browdy & Neimark
419 Seventh St., N.W., Suite 300
Washington, DC 20004 (202) 628-5197

Frank Louis Neuhauser
Bernard, Rothwell & Brown, P.C.
1700 K St., N.W.
Washington, DC 20006 (202) 833-5740

Richard David Nevius
4000 Cathedral Ave., N.W.
Washington, DC 20016 (202) 333-8652

Mark M. Newman
Craig & Burns
1825 Eye St., N.W.
Washington, DC 20006 (202) 429-0400

Simon M. Newman
1411 Hopkins St., N.W.
Washington, DC 20036 (202) 785-1438

Hugh P. Nicholson*
U.S. Air Force
Legal Off., Pats. Div.
AF/JACP
1900 Half St.
Washington, DC 20324 (202) 693-5710

James A. Niegowski
Banner, Birch, Mc Kie & Beckett
One Thomas Circle, N.W.
Washington, DC 20005 (202) 296-5500

David T. Nikaido
Armstrong, Nikaido, Marmelstein & Kubovcik
1725 K St., N.W.
Washington, DC 20006 (202) 659-2930

Jeffrey Nolton
Wenderoth, Lind & Ponack
805 15th St., N.W.
Washington, DC 20005 (202) 371-8850

Lawrence Geoffrey Norris
Bernard, Rothwell & Brown
1700 K St., N.W.
Washington, DC 20006 (202) 833-5740

James G. O Boyle
Brady, O Boyle & Gates
920 Chevy Chase Bldg.
5530 Wisconsin Ave., N.W.
Washington, DC 20015 (301) 656-3355

Dennis P. O Reilley
Finnegan, Henderson, Farabow, Garrett & Dunner
1775 K St., N.W.
Washington, DC 20006 (202) 293-6850

Charles L. O Rourke
Finnegan, Henderson, Farabow, Garrett & Dunner
1775 K St., N.W., Suite 600
Washington, DC 20006 (202) 293-6850

Walter P. O Rourke, Sr.
2800 Ontario Rd., N.W., Apt. 505
Apt. 505
Washington, DC 20009 (202) 402-7937

Franz O. Ohlson, Jr.
Aerospace Industries Association Of America, Inc.
1725 De Sales St., N.W.
Washington, DC 20036 (202) 429-4625

Robert A. Oleary
3600 Patterson St., N.W.
Washington, DC 20015 (202) 966-3610

Peter D. Olexy
Sughrue, Mion, Zinn, Macpeak & Seas
1776 K St., N.W.
Washington, DC 20006 (202) 293-7060

Warren E. Olsen
Pierson, Semmes, Crolius & Finley
1054 31st St., N.W.
Washington, DC 20007 (202) 333-4000

George E. Oram, Jr.
Armstrong, Nikaido, Marmelstein & Kubovick
1725 K St., N.W., Suite 912
Washington, DC 20006 (202) 659-2930

Karen Lee Orzechowski
Lyon & Lyon
1200 17th St., N.W., Suite 405
Washington, DC 20036 (202) 296-9600

J. Frank Osha
Sughrue, Mion, Zinn, Macpeak & Seas
1776 K St., N.W.
Washington, DC 20006 (202) 293-7060

Perry Palan
Barnes & Thornburg
1815 H St., N.W., Suite 500
Washington, DC 20006 (202) 955-4500

Sherman O. Parrett
Cushman, Darby & Cushman
1615 L St., N.W., 11th Fl.
Washington, DC 20036 (202) 861-3024

Charles C. Parsons
400 First St., N.W.
Washington, DC 20001 (202) 638-3375

Scott F. Partridge
Banner, Birch, Mc Kie & Beckett
One Thomas Circle, N.W.
Washington, DC 20005 (202) 296-5500

Herbert W. Patterson
Finnegan, Henderson, Farabow, Garrett & Dunner
1775 K St., N.W.
Washington, DC 20006 (202) 223-6957

John C. Paul
Finnegan, Henderson, Farabow, Garrett & Dunner
1775 K St., N.W.
Washington, DC 20006 (202) 293-6850

Eugene J. Pawlikowski
U.S. Dept. Of Comm.
Off. Of Gen. Coun.
Hch Bldg., Rm. 5883
Washington, DC 20230 (202) 377-1362

Kenneth E. Payne
Finnegan, Henderson, Farabow, Garrett & Dunner
1775 K St., N.W.
Washington, DC 20006 (202) 293-6850

Donald G. Peck
U.S. Dept. Of Navy
Naval Sea Systems Command
Off. Of Coun., Code OOL5
Washington, DC 20362 (202) 692-7077

Peter Peckarsky
Banner, Birch, Mc Kie & Beckett
One Thomas Circle, N.W.
Washington, DC 20005 (202) 296-5500

Stephen A. Pendorf
Sughrue, Mion, Zinn, Mackpeak & Seas
1776 K St., N.W.
Washington, DC 20006 (202) 293-7060

Glenn J. Perry
Cushman, Darby & Cushman
1615 L St., N.W., 11th Fl.
Washington, DC 20036 (202) 861-3070

Michelle Peters
P.O. Box 25055
Georgetown Station
Washington, DC 20007 (202) 845-8292

Stephen L. Peterson
Finnegan, Henderson, Farabow, Garrett & Dunner
1775 K St., N.W.
Washington, DC 20006 (202) 293-6850

Thomas L. Peterson
Banner, Birch, Mc Kie & Beckett
One Thomas Circle, N.W., Suite 600
Washington, DC 20005 (202) 296-5500

John M. Petruncio*
Department Of Air Force
Office Of Judge Advocate Gen.
1900 Half St.
Washington, DC 20324 (202) 475-1386

George R. Pettit
Pollock, Vandesande & Priddy
1990 M St., N.W.
Washington, DC 20036 (202) 331-7111

Frank V. Pietrantonio
Kenyon & Kenyon
1025 Connecticut Ave. N.W.
Washington, DC 20036 (202) 429-1776

David M. Pitcher
Staas & Halsey
1825 K St., N.W., Suite 816
Washington, DC 20006 (202) 872-0123

Michael Terry Platt
Berman, Aisenberg & Platt
OFC Bldg.
1730 R.I. Ave., N.W.
Washington, DC 20036 (202) 293-1404

Robert C. Platt
Ginsberg, Feldman & Bress, Chartered
1250 Connecticut Ave., N.W.
Washington, DC 20036 (202) 637-9188

William E. Player
Wegner & Bretschneider
P.O. Box 18218
Washington, DC 20036 (202) 887-0400

Robert H. Plotkin
U.S. Dept. Of Justice
Land & Natural Res. Pk.
Washington, DC 20530 (202) 724-8201

Elliott I. Pollock
Pollock, Vande Sande & Priddy
1990 M St., N.W., Suite 800
Washington, DC 20036 (202) 331-7111

James K. Poole
Fisher, Christen & Sabol
2000 L St., N.W., Suite 510
Washington, DC 20036 (202) 659-2015

Martin S. Postman
1377 K St., N.W., Suite 115
Washington, DC 20005 (301) 946-0456

Joseph M. Potenza
Banner, Birch, Mc Kie & Beckett
One Thomas Circle, N.W.
Washington, DC 20005 (202) 296-5505

Theodore Prahinski
Air Force Systems Command
H Q S (j A T)
Andrews Air Force Base
Washington, DC 20334 (301) 981-5372

D. Douglas Price
Fleit, Jacobson, Cohn & Price
1217 E. St., N.W.
Washington, DC 20004 (202) 638-6666

Robert R. Priddy
Pollock, Vande Sande & Priddy
1990 M St., N.W., Suite 800
Washington, DC 20036 (202) 331-7111

Edward M. Prince
Cushman, Darby & Cushman
1615 L St., N.W.
Washington, DC 20036 (202) 861-3044

Sharon E. Pula
Finnegan, Henderson, Farabow, Garrett & Dunner
1775 K St., N.W.
Washington, DC 20006 (202) 293-6850

Melanio R. Quintos
Wegner & Bretschneider
1233 20th St., N.W., 3rd Fl.
P.O. Box 18218
Washington, DC 20036 (202) 887-0400

Steven M. Rabin
Spencer & Frank
1111 19th St., N.W., 12th Fl.
Washington, DC 20036 (202) 828-8000

Richard B. Racine
Finnegan, Henderson, Farabow, Garrett & Dunner
1775 K St., N.W.
Washington, DC 20006 (202) 293-6850

Lawrence R. Radanovic
Watson, Cole, Grindle & Watson
1400 K St., N.W.
Washington, DC 20005 (202) 628-0088

W. Scott Railton
Reed, Smith, Shaw & Mc Clay
1150 Connecticut Ave., N.W., Suite 900
Washington, DC 20036 (202) 457-6100

William W. Randolph*
U.S. Dept. Of Energy
1000 Independence Ave.
Washington, DC 20585 (202) 586-2816

Marvin Reich
1111 Nineteenth St., N.W., 12th Fl.
Washington, DC 20036 (202) 828-8048

George R. Repper
Bernard, Rothwell & Brown, P.C.
1700 K St., N.W.
Washington, DC 20006 (202) 833-5740

Frances Richey#
520 N. St., S.W.
Suite S-1228
Washington, DC 20024 (202) 488-1223

Beatrice N. Robbins#
Robbins & Laramie
1919 Pennsylvania Ave., N.W.
Washington, DC 20006 (202) 887-5050

Frank Edward Robbins
Robbins & Laramie
1919 Pennsylvania Ave., N.W., Suite 506
Washington, DC 20006 (202) 887-5050

John Tyssowski Roberts
1747 Pennsylvania Ave., N.W., Suite 805
Washington, DC 20006 (202) 785-0990

George N. Robillard
Finnegan, Henderson, Farabow, Garret & Dunner
1775 K St., N.W.
Washington, DC 20006 (202) 293-6850

John M. Romary
Finnegan, Henderson, Farabow, Garrett & Dunner
1775 K St., N.W.
Washington, DC 20006 (202) 293-6850

John Marshall Rommel
Beveridge, De Grandi & Kline
1819 H St., N.W.
Washington, DC 20006 (202) 293-6850

Herbert C. Rose#
Leydig, Voit & Mayer, Ltd.
1190 Vermont Ave., N.W., Suite 490
Washington, DC 20005 (202) 842-3771

Stephen J. Rosenman
Finnegan, Henderson, Farabow, Garrett & Dunner
1775 K St., N.W., Suite 600
Washington, DC 20006 (202) 293-6850

Martha L. Ross#
3540 Van Ness St., N.W.
Washington, DC 20008 (202) 362-2158

Gideon Franklin Rothwell, IV
Bernard, Rothwell & Brown, P.C.
1700 K St., N.W.
Washington, DC 20006 (202) 833-5740

Donald Carvar Roylance
Roylance, Abrams, Berdo & Goodman
1225 Conn. Ave., N.W.
Washington, DC 20036 (202) 659-9076

Joseph J. Ruch, Jr.
Sughrue, Mion, Zinn, Macpeak & Seas
1776 K St., N.W.
Washington, DC 20006 (202) 293-7060

Christopher John Rudy
Fisher, Christen & Sabol
2000 L St., N.W., Suite 510
Washington, DC 20036 (202) 659-2000

Eugene Sabol
Fisher, Christen & Sabol
2000 L St., N.W., Suite 510
Washington, DC 20036 (202) 659-2000

Perry J. Saidman
Saidman, Sterne, Kessler & Goldstein
1225 Connecticut Ave.
Washington, DC 20036 (202) 833-7533

Colin G. Sandercock
Roylance, Abrams, Berdo & Goodman
1225 Conn. Ave., N.W., Suite 315
Washington, DC 20036 (202) 659-9076

Albert J. Santorelli
Finnegan, Henderson, Farabow, Garrett & Dunner
1775 K St., N.W., Suite 600
Washington, DC 20006 (202) 293-6850

Harry Morris Saragovitz
700 7th St., S.W.
Washington, DC 20024 (202) 554-3083

Walter A. Scheel
503 H. St., S.W.
Washington, DC 20024 (202) 554-5908

Jonathan L. Scherer
Fleit, Jacobson, Cohn & Price
1217 E. St., N.W.
Washington, DC 20004 (202) 638-6666

Alan E. Schiavelli#
Antonelli, Terry & Wands
1919 Pennsylvania Ave., N.W. Suite 600
Washington, DC 20006 (202) 828-0335

Charles F. Schill
Adduci, Dinan & Mastriani
1140 Connecticut Ave., N.W., Suite 250
Washington, DC 20036 (202) 467-6300

Jerome Schnall#
2801 Quebec St., N.W.
Washington, DC 20008 (202) 362-8872

John W. Schneller
Spencer & Frank
1111 19th St., N.W., 12th Fl.
Washington, DC 20036 (202) 828-8000

William E. Schuyler, Jr.
4801 Massachusetts Ave., N.W., Suite 400
Washington, DC 20016 (202) 966-9086

Nigel L. Scott
Scott & Yallery - Arthur
7603 Georgia Ave., N.W., Suite 200
Washington, DC 20012 (202) 882-5770

Thomas J. Scott, Jr.
Pennie & Edmonds
1730 Pennsylvania Ave., N.W., Suite 1000
Washington, DC 20006 (202) 393-0177

Watson T. Scott
Cushman, Darby & Cushman
1615 L St., N.W., 11th Fl.
Washington, DC 20036 (202) 861-3000

Joseph Scovronek
5930 14th St., N.W., Apt. A1
Washington, DC 20011 (202) 723-0538

Mary Helen Sears
Ginsburg, Feldman & Bress
1250 Connecticut Ave., N.W.
Washington, DC 20036 (202) 637-9000

Robert John Seas, Jr.
Sughrue, Mion, Zinn, Macpeak & Seas
1776 K St., N.W., Suite 500
Washington, DC 20006 (202) 293-7060

David Hopkins Semmes
Pierson, Semmes & Finley
1054 31st St., N.W.
Washington, DC 20007 (202) 965-4570

John Gibson Semmes
3286 M St., N.W.
P.O. Box 3559
Washington, DC 20007 (202) 965-1234

Kurt Shaffert
U.S. Dept. Of Justice
Antitrust Div.
Washington, DC 20530 (202) 376-8603

James J. Shanley
Shanley & Baker
2233 Wisconsin Ave., N.W.
Washington, DC 20007 (202) 333-5800

John P. Shannon, Jr.
Lane & Aitken
Watergate Office Bldg.
2600 Virginia Ave., N.W.
Washington, DC 20037 (202) 337-5556

Linda J. Shapiro
Mason, Fenwick & Lawrence
1225 Eye St., N.W., Suite 1000
Washington, DC 20005 (202) 289-1200

F. Barry Shay
2808 Northampton St., N.W.
Washington, DC 20015 (202) 363-2457

James R. Shay#
Howrey & Simon
1730 Pennsylvania Ave., N.W.
Washington, DC 20006 (202) 783-0800

Peter R. Shearer
Bernard, Rothwell & Brown, p.c.
1700 K St., N.W.
Washington, DC 20006 (202) 833-5740

Sol Sheinbein*
Naval Research Laboratory
Code 1208.2
4555 Overlook Ave., S.W.
Washington, DC 20375 (202) 767-3427

Ronald J. Shore
Antonelli, Terry & Wands
1919 Pennsylvania Ave., N.W., Suite 600
Washington, DC 20006 (202) 828-0300

Darle M. Short
Roylance, Abrams, Berdo & Goodman
1225 Connecticut Ave., N.W.
Washington, DC 20036 (202) 659-9076

Neil B. Siegel
Sughrue, Mion, Zinn, Macpeak & Seas
1776 K St., N.W.
Washington, DC 20006 (202) 293-7060

M. Howard Silverstein
U.S. Dept. Of Agriculture
Off. Of Gen. Coun.
Washington, DC 20250 (202) 447-5474

Jeffrey A. Simenauer#
Finnegan, Henderson, Farabow, Garrett & Dunner
1775 K St., N.W.
Washington, DC 20006 (202) 293-6850

Donald James Singer*
Hq. U.S.A.F.
Jacp
1900 Half St., S.W.
Washington, DC 20324 (202) 475-1386

Leroy G. Sinn
Roberts & Floyd
1747 Pennsylvania Ave., N.W.
Washington, DC 20006 (202) 785-0990

George M. Sirilla
Cushman, Darby & Cushman
1615 L St., N.W., 11th Fl.
Washington, DC 20036 (202) 861-3536

Joseph M. Skerpon
Banner, Birch, Mc Kie & Beckett
One Thomas Circle, N.W.
Washonton, DC 20005 (202) 296-5500

Peter K. Skiff
Wenderoth, Lind & Poncak
805 Fifteenth St., N.W., Suite 700
Washington, DC 20005 (202) 371-8850

Paul J. Skwierawski#
Sughrue, Mion, Zinn, Macpeak & Seas
1776 K St., N.W., Suite 500
Washington, DC 20002 (202) 293-7060

Robert Vincent Sloan
Sughrue, Mion, Zinn, Macpeak & Seas
1776 K St., N.W.
Washington, DC 20006 (202) 293-7060

Michael R. Slobasky
Fleit, Jacobson, Cohn & Price
1217 E. St., N.W.
Washington, DC 20004 (202) 638-6666

Homer Ashby Smith#
Clarence A O Brien & Harvey B. Jacobson
1217 E. St., N.W.
Washington, DC 20004 (202) 393-3380

Randolph A. Smith#
Spencer & Frank
1111 19th St., N.W.
Washington, DC 20036 (202) 828-8000

Richard Harold Smith
Finnegan, Henderson, Farabow, Garrett & Diner
1775 K St., N.W.
Washington, DC 20006 (202) 293-6850

Samuel B. Smith, Jr.
U.S. Air Force
J A C P
1900 Half St.
Washington, DC 20324 (202) 475-1386

Ronald Ralph Snider
Wegner & Bretschneider
Fourth Fl.
2030 M St., N.W.
P.O. Box 18218
Washington, DC 20036 (202) 887-0400

John Pennington Snyder#
Watson, Cole, Grindle & Watson
1400 K St., N.W.
Washington, DC 20005

Allen M. Sokal
Finnegan, Henderson, Farabow, Garrett & Dunner
1775 K St., N.W.
Washington, DC 20006 (202) 293-6850

William I. Solomon
Antonelli, Terry & Wands
1919 Pennsylvania Ave., N.W., Suite 600
Washington, DC 20006 (202) 828-0300

George H. Spencer
Spencer & Frank
1111 19th St., N.W.
Washington, DC 20036 (202) 828-8000

Avrom David Spevack*
U.S. Dept. Of Navy
Assoc. Coun. For Pats.
Code 1208.2
Naval Res. Lab.
Washington, DC 20375 (202) 767-3427

Harry John Staas
Staas & Halsey
1825 K St., N.W., Suite 816
Washington, DC 20006 (202) 872-0123

Alfred A Stadnicki
International Telecommunicatios Satellite Organization
3400 International Dr., N.W.
Washington, DC 20008 (202) 944-6855

A. Fred Starobin
419 7th St., N.W.
Washington, DC 20004 (202) 393-2717

Marvin R. Stern
Holman & Stern
2401 15th St., N.W.
Washington, DC 20009 (202) 483-2234

Richard H. Stern
2101 L St., N.W., Suite 800
Washington, DC 20037 (202) 775-4727

Robert Greene Sterne
Saidman, Sterne, Kessler & Goldstein
1225 Connecticut Ave.
Washington, DC 20036 (202) 833-7533

Gene W. Stockman
Staas & Halsey
1825 K St., N.W.
Washington, DC 20006 (202) 872-0123

Mary B. Stohler#
Bernard, Rothwell & Brown, P.C.
1700 K St., N.W.
Washington, DC 20006 (202) 833-5740

James D. Stokes, Jr.
U.S. Justice Dept.
550 11th St., N.W.
Washington, DC 20004 (202) 724-7279

Leonard Francis Stoll
U.S. Dept. Of Air Force
1900 Half St., S.W.
Washington, DC 20324 (202) 475-1386

Donald R. Stone
Burditt, Bowles & Radzius
1029 Vermont Ave., N.W., Suite 200
Washington, DC 20005 (202) 638-3833

Isaac G. Stone
7928 Orchid St., N.W.
Washington, DC 20012 (202) 882-3176

Michael Stone#
Wenderoth, Lind & Poncak
805 Fifteenth St., N.W., Suite 700
Washington, DC 20005 (202) 371-8850

Donald E. Stout
Antonelli, Terry & Wands
1919 Pennsylvania Ave., N.W.
Washington, DC 20006 (202) 828-0300

Lloyd Joseph Street
Cushman, Darby & Cushman
1615 L St., N.W., 11th Fl.
Washington, DC 20036 (201) 861-3000

Richard L. Stroup
Finnegan, Henderson, Farabow, Garrett & Dunner
1775 K St., N.W.
Washington, DC 20006 (202) 293-6580

Donald R. Studebaker#
Spencer & Frank
1111 19th St., N.W., Suite 1200
Washington, DC 20036 (202) 828-8000

Charles L. Sturtevant
3231 Reservoir Rd., N.W.
Washington, DC 20007 (202) 337-5723

Richard Charles Sughrue
Sughrue, Mion, Zinn, Macpeak & Seas
1776 K St., N.W.
Washington, DC 20006 (202) 293-7060

Stephen T. Sullivan
Finnegan, Henderson, Farabow, Garrett & Dunner
1775 K St., N.W.
Washington, DC 20006 (202) 293-6580

Christian L. Swartz
2000 L St., N.W., Suite 504
Washington, DC 20036 (202) 659-4505

George W. Swenson#
Vorys, Sater, Seymour & Pease
1828 L St., N.W., Suite 1111
Washington, DC 20036 (202) 822-8200

John Randolph Swindler
Banner, Birch, Mc Kie & Beckett
One Thomas Circle, N.W.
Washington, DC 20005 (202) 296-5500

W. Warren Taltavall, III
Cushman, Darby & Cushman
1615 L St., N.W., 11th Fl.
Washington, DC 20036 (202) 861-3533

Bruce A. Tassan
Dickinson, Wright, Moon, Van Dusen & Freeman
1901 L St., N.W., Suite 801
Washington, DC 20036 (202) 457-0160

Rodger L. Tate
Banner, Birch, Mc Kie & Beckett
One Thomas Circle, N.W., Suite 600
Washington, DC 20005 (202) 296-5500

Hosea E. Taylor#
1749 N. Portal Dr., N.W.
Washington, DC 20012 (301) 829-1323

Roger D. Taylor
Finnegan, Henderson, Farabow, Garrett & Dunner
1775 K St., N.W.
Washington, DC 20006 (202) 293-6850

Jennifer A. Tegfeldt
Mason, Fenwick & Lawrence
1225 Eye St., N.W., Suite 1000
Washington, DC 20005 (202) 289-1200

David T. Terry
Antonelli, Terry & Wands
1919 Pennsylvania Ave., N.W., Suite 600
Washington, DC 20006 (202) 828-0300

Melinda B. Thaler
Marks, Murase & White
2001 L St., N.W., Suite 750
Washington, DC 20036 (202) 955-4900

Dirk D. Thomas
Finnegan, Henderson, Farabow, Garrett & Dunner
1775 K St., N.W.
Washington, DC 20006 (202) 293-6850

Francis D. Thomas, Jr.
2317 Pa. Ave., N.W., Suite 11
Washington, DC 20037 (202) 333-7598

Harry B. Thornton
1829 Upshur St., N.W.
Washington, DC 20011 (202) 882-2878

Albert Tockman
Armstrong, Nikaido, Marmelstein & Kubovcik
1725 K St., N.W.
Washington, DC 20006 (202) 659-2930

Allan J. Topol
Covington & Burling
1201 Pennsylvania Ave., N.W.
Washington, DC 20044 (202) 662-5402

Donald E. Townsend*
U.S. Dept. Of Justice
Patent Sect.
550 11th St., N.W.
Washington, DC 20530 (202) 724-7278

William Cecil Townsend*
Dept. Of The Navy
Space & Naval Warfare Systems Command
Code OO C 6
Washington, DC 20363 (202) 692-8458

Francis G. Toye
1817 Sudbury Rd., N.W.
Washington, DC 20012 (202) 829-9007

Wilma F. Triebwasser
Bureau Of National Affairs
Pat. Tdmk. & Copyright Journal
1231 25th St., N.W.
Washington, DC 20037 (202) 452-6356

Peter K. Trzyna
Cadwalader, Wickersham & Taft
1333 New Hampshire Ave., N.W., Suite 700
Washington, DC 20036 (202) 862-2200

Richard C. Turner
Sughrue, Mion, Zinn, Macpeak & Seas
1776 K St., N.W.
Washington, DC 20006 (202) 293-7060

Richard H. Tushin
Watson, Cole, Grindle & Watson
1400 K St., N.W.
Washington, DC 20005 (202) 628-0088

Archie W. Umphlett
Phillip Petroleum Co.
1825 K St., N.W., Suite 1107
Washington, DC 20006 (202) 785-1380

Thomas B. Van Poole, Jr.
Mason, Fenwick & Lawrence
1225 Eye St., N.W., Suite 1000
Washington, DC 20005 (202) 289-1200

George Vande Sande
Pollock, Vande Sande & Priddy
1990 M St., N.W., Suite 800
Washington, DC 20036 (202) 331-7111

J. Derek Vandenburgh#
Wegner & Bretschneider
1233 20th St., N.W.
P.O. Box 18218
Washington, DC 20036 (202) 887-0400

Ralph E. Varndell, Jr.#
Varndell Legal Group
1511 K St., N.W., Suite 431
Washington, DC 20005 (202) 783-3443

David Edwards Varner
Cushman, Darby & Cushman
1615 L St., N.W., 11th Fl.
Washington, DC 20036 (202) 861-3539

Irving Vaughn#
2324 Second St., N.E.
Washington, DC 20002 (202) 832-6051

Jerry Dean Voight
Finnegan, Henderson, Farabow, Garrett & Dunner
1775 K St., N.W.
Washington, DC 20006 (202) 293-6850

Stanley A. Wal#
Antonelli, Terry & Wands
1919 Pennsylvania Ave., N.W., Suite 600
Washington, DC 20006 (202) 828-0326

James S. Waldron#
Fidelman, Wolffe & Waldron
2120 L St., N.W., Suite 300
Washington, DC 20037 (202) 833-8801

James H. Wallace, Jr.
Wiley & Rein
1776 K St., N.W.
Washington, DC 20006 (202) 429-7240

Wallace G. Walter
Cushman, Darby & Cushman
1615 L St., N.W., 11th Fl.
Washington, DC 20036 (202) 861-3000

Mark P. Watson
Lalos, Keegan, Marsh & Kaye
900 17th St., N.W.
Washington, DC 20006 (202) 887-5555

Harold C. Wegner
Wegner & Bretschneider
P.O. Box 18218
Washington, DC 20036 (202) 887-0400

Helmuth A. Wegner
Wegner & Bretschneider
P.O. Box 18218
Washington, DC 20036 (202) 887-0400

Robert G. Weilacher
Beveridge, De Grandi & Weilacher
Federal Bar Bldg. West, Suite 1100
1819 H St., N.W.
Washington, DC 20006 (202) 659-2811

Max M. Weisman
4740 Connecticut Ave., N.W.
Washington, DC 20008 (202) 966-8313

James Price Welch
Armstrong, Nikaido, Marmelstein & Kubovcik
1725 K St., N.W.
Washington, DC 20006 (202) 659-2930

Ashley J. Wells
Spencer & Frank
1111 19th St., N.W.
Washington, DC 20036 (202) 828-8000

William K. Wells, Jr.
Lalos, Keegan, Marsh, Bentzen & Kane
900 Seventeenth St., N.W.
Washington, DC 20006 (202) 887-5555

William K. West, Jr.
Cushman, Darby & Cushman
1615 L St., N.W., 11th Fl.
Washington, DC 20036 (202) 861-3000

William F. Westerman
Armstrong, Nikaido, Marmelstein & Kubovcik
1725 K St., N.W., Suite 912
Washington, DC 20006 (202) 659-2930

John R. Wetherell, Jr.#
Saidman, Sterne, Kessler & Goldstein
1225 Connecticut Ave.
Washington, DC 20036 (202) 833-7533

John T. Whelan
Finnegan, Henderson, Farabow, Garrett & Dunner
1775 K St., N.W., Suite 600
Washington, DC 20006 (202) 293-6850

John M. White#
Cushman, Darby & Cushman
1615 L St., N.W., 11th Fl.
Washington, DC 20036 (202) 861-3708

Paul E. White, Jr.
Cushman, Darby & Cushman
1615 L St., N.W.
Washington, DC 20036 (202) 861-3000

Richard Wiener
Pollock, Vande Sande & Priddy
P.O. Box 19088
1990 M St., N.W.
Washington, DC 20036 (202) 331-7111

Otto M. Wildensteiner
Dept. Of Transportation
C-15
400 7th St., S.W.
Washington, DC 20590 (202) 426-4710

Richard C. Wilder
Finnegan, Henderson, Farabow, Garrett & Dunner
1775 K St., N.W., Suite 600
Washington, DC 20006 (202) 293-6850

Thomas W. Winland
Finnegan, Henderson, Farabow, Garrett & Dunner
1775 K St., N.W., Suite 600
Washington, DC 20006 (202) 293-6850

Morris Wiseman*
Air Force Sys. Comm
Andrew Air Force Base
Washington, DC 20334 (301) 981-5372

John F. Witherspoon
1667 K St., N.W.
Washington, DC 20006 (202) 296-7300

Henry N. Wixon
Saidman, Sterne, Kessler & Goldstein
1225 Connecticut Ave., N.W.
Washington, DC 20036 (202) 833-7533

Franklin David Wolffe
Fidelman & Wolffe
1233 20th St., N.W. 3rd Fl.
P.O. Box 18218
Washington, DC 20036 (202) 833-8801

James A. Wong
307 Park Lane Bldg.
2025 Eye St., N.W.
Washington, DC 20006 (202) 822-8998

Louis Woo
Pollock, Vande Sande & Priddy
P.O. Box 19088
1990 M St., N.W.
Washington, DC 20006 (202) 331-7111

L. Allen Wood, Jr.
Spencer & Frank
1111 19th St., N.W., Suite 1200
Washington, DC 20036 (202) 828-8000

George N. Woodruff#
1000 6th St., S.W., Suite 807
Washington, DC 20024 (202) 554-1228

Laurence Arthur Wright
U.S. Air Force
Pat. Divisions Rd.
1900 Half St., S.W.
Washington, DC 20324 (202) 475-1386

William H. Wright
Henderson & Sturm
1924 N St., N.W.
Washington, DC 20036 (202) 296-3854

Edward R. Yoches
Finnegan, Henderson, Farabow, Garrett & Dunner
1775 K St.
Washington, DC 20006 (202) 293-6850

Karen Stephan Young#
Fidelman, Wolffe & Waldron
2120 L St., N.W., Suite 300
Washington, DC 20037 (202) 833-8801

Richard G. Young
Beveridge, De Grandi & Weilacher
1819 H St., N.W., Suite 1100
Washington, DC 20006 (202) 659-2811

Thomas Zack*
U.S. Dept. Of Commerce
Off. Of Gen. Coun.
14th & Constitution Ave., N.W.
Rm. 4610
Washington, DC 20230

Donald E. Zinn, Sr.
Sughrue, Mion, Zinn, Macpeak & Seas
1776 K St., N.W.
Washington, DC 20006 (202) 293-7060

Joseph J. Zito
Marks, Murase & White
2001 L St., N.W., Suite 750
Washington, DC 20036 (202) 955-4900

Bruce C. Zotter
Finnegan, Henderson Farabow & Garrett
1775 K St., N.W.
Washington, DC 20006 (202) 293-6850

Henry Michael Zykorie
Wenderoth, Lind & Ponack
Southern Bldg.
805 Fifteenth St., N.W.
Washington, DC 20005 (202) 371-8850

Florida

Harry R. Dumont
415 Montgomery Rd., Suite 175
Altamonte Springs, FL 32714
(305) 788-2788

William B. Penn
1080 99th St.
Bay Harbor Islands, FL 33154
(305) 861-6630

Jack Hensel
8319 Roxboro Dr.
Bayonet Point, FL 34667 (813) 868-1955

Herbert K. Anspach
2760 N.W. 29th Dr.
Boca Raton, FL 33434 (305) 483-2403

Harry W. Barron
8221 Glades Rd.
Suite 202
Boca Raton, FL 33434 (305) 488-3000

Burton P. Beatty, Sr.
410 SW 7th Way
Boca Raton, FL 33486 (305) 368-1584

Richard E. Bee
IBM Corp.
Pat. Operations
P.O. Box 1328
Boca Raton, FL 33431 (305) 998-4899

John C. Black
940 S.W. 15th St.
Boca Raton, FL 33432 (305) 391-7715

Charles H. Brown
6300 N.W. 2nd Ave., Apt. 308
Boca Raton, FL 33432 (305) 994-1100

D. Kendall Cooper
IBM Corp.
Intellectual Property Law Dept.
P.O. Box 1328
1000 N.W. 51st St.
Boca Raton, FL 33432 (305) 998-4708

Don Daniel Doty
6065 S. Verde Trail Apt. G-211
Boca Raton, FL 33433 (305) 488-2362

Leonard Forman
6680 Burning Wood Dr., Apt. 264
Boca Raton, FL 33433 (406) 368-8406

Olney M. Gardiner
701 Marine Dr.
Boca Raton, FL 33487 (305) 391-4199

George E. Grosser
IBM Corp.
951 NW 51st St.
Area 4318
Boca Raton, FL 33431 (305) 998-0430

Edward Halle
4001 N. Ocean Blvd., Apt. 603-B
Boca Raton, FL 33431 (305) 394-5384

Anthony G.F. Hawkins#
IBM Corp.
Dept. 91K, Bldg. 234-2
ZIP 4318
P.O. Box 1328
Boca Raton, FL 33432 (305) 998-4395

Rubin Hoffman
7615 Sierra Dr. West
Boca Raton, FL 33433 (305) 392-2784

Paul T. Kashimba
IBM Corp.
Pat. Operations (4318)
P.O. Box 1328
951 N.W. 51st St.
Boca Raton, FL 33432 (305) 998-2000

Robert Lieber
IBM Corp.
Pat. Oper. (91K/219-1)
P.O. Box 1328
Boca Raton, FL 33432 (305) 998-0054

H. Geoffrey Lynfield
7050 N.E. 7th Ave.
Boca Raton, FL 33431 (305) 997-5825

Douglas R. Mc Kechnie
IBM Corp.
Zip 4318 P.O. Box 1328
Boca Raton, FL 33431 (305) 998-4708

Philip T. Mintz
198 NW 67th St., Apt. 508
Boca Raton, FL 33431

Raymond R. Skolnick#
10658 180th Ct. South
Boca Raton, FL 33434 (305) 483-6186

Harold Hugh Sweeney, Jr.
IBM Corp.
Dept. 91K
Box 1328
Boca Raton, FL 33432 (305) 998-2928

Richard A. Tomlin
IBM Corp.
951 NW 51st St.
Boca Raton, FL 33431 (305) 998-9786

Philip A. Wadsworth
IBM Corp.
951 N.W. 51st St., 4301
Boca Raton, FL 33431 (305) 998-0430

William E. Zitelli
IBM Corp.
Intellectual Property Law Dept.
951 NW 51st St.
Internal Zip 4318
Boca Raton, FL 33432 (305) 982-1880

Shaler G. Smith, Jr.#
27339 King s Kew SW
Bonita Springs, FL 33923 (813) 992-3969

Edward Grant Haggett, Jr.
1607 S.W. 14th Ave.
Leisureville
Boynton Beach, FL 33435 (305) 732-4066

James F. Higgins
13 Afton Place
Boynton Beach, FL 33462 (305) 965-3251

Arthur T. Mc Keon#
2086 SW 13th Way
Boynton Beach, FL 33435 (305) 737-6057

Victor F. Volk#
646 Snug Harbor Dr.
Boynton Beach, FL 33435 (305) 732-4934

Clement J. Paznokas
10424 Spoonbill Rd. W.
Bradenton, FL 33529 (813) 792-3796

Frederick Shapoe
6501-17th Ave. West W-318
Bradenton, FL 34209 (813) 792-2555

Richard A. Zambo
205 N. Parsons Ave.
Brandon, FL 33511 (813) 681-3220

Vincent Augustus White
900-5506 U S 41N
Brooksville, FL 33512 (904) 796-5090

Charles E. Vautrain, Jr.#
23465 Harborview Rd., Apt. 724
Charlotte Harbor, FL 33952 (813) 629-6940

Daniel J. Hanlon, Jr.
2351 Irish Lane, Apt. 45
Clearwater, FL 33515 (813) 796-1250

John L. Harris
470 Palm Island, NE
Clearwater, FL 34630 (813) 446-1953

Leon R. Horne
3502 Bimini Lane Apt. J-1
Coconut Creek, FL 33063 (305) 972-3698

James R. Hulen
Concept, Inc.
12707 U.S. Hwy. 19 So.
Clearwater, FL 33516 (813) 536-2791

Herbert William Larson, Sr.
1307 U.S. 19 South, Suite 102
Clearwater, FL 33546 (813) 538-3800

Donald Levy
1849 B Bough Ave.
Clearwater, FL 33520 (813) 535-7241

Charles E. Lykes, Jr.
The Legal Arts Bldg.
501 S. Fort Harrison Ave.
Clearwater, FL 33516 (813) 441-8308

Joseph C. Mason, Jr.
1307 U.S. 19 South, Suite 102
Clearwater, FL 33546 (813) 538-3800

Ronald E. Smith
Arbor Shoreline Center
1307 U.S. 19 South
Suite 102
Clearwater, FL 34624 (813) 538-3800

Miguel A. Valdes
Smith & Valdes
1477 S. Missouri Ave.
Clearwater, FL 33516 (813) 446-4004

Ernest S. Cohen
1103 Bahama Bend G1
Coconut Creek, FL 33066 (305) 975-3465

Allen B. Curtis#
3204 Portofino Point, Apt. H2
Coconut Creek, FL 33066 (305) 979-2098

Jacob H. Steinberg
1502 Cayman Way
Coconut Creek, FL 33066 (305) 977-8073

Edward Stern
2614 Apt. B1
Coconut Creek, FL 33066 (305) 972-5118

Robert J. Van Der Wall
2951 S. Bayshore Dr., Suite 811
Coconut Grove, FL 33133 (305) 445-6500

Meredith P. Sparks
5129 Granada Blvd.
Coral Gables, FL 33146 (305) 661-5756

Leroy Greenspan
10844 NW 7th St.
Coral Springs, FL 33071 (305) 755-3262

Mark P. Kahler
10932 N.W. 13th Ct.
Coral Springs, FL 33065

Charles W. Lanham, Jr.
330 Cornell Dr.
Daytona Beach, FL 32018 (904) 677-0272

Franklin Mohr
145 N. Halifax Ave.
Apt. 711
Daytona Beach, FL 32018 (904) 252-0973

John H. Schneider
105 Regency Dr.
Daytona Beach, FL 32019 (904) 761-8261

James R. Hagen#
143 Sepp Rd.
De Bary, FL 32713 (305) 668-8342

William R. Lawton#
1436 S.E. 12th Ave.
Deerfield Beach, FL 33441 (305) 426-0465

Arthur J. Greif
Brunswick Corp.
2000 Brunswick Lane
Deland, FL 32724 (904) 736-1700

Roger L. Martin
1456 Farmington Ave.
Deltona, FL 32725 (305) 574-7530

Stanley M. Miller
748 Broadway, Suite 201
Dunedin, FL 33528 (813) 733-8825

Raymond G. Brodahl
1029 Grant St.
Englewood, FL 33533 (813) 475-7643

William A. Mikesell, Jr.
1000 Lee St.
Englewood, FL 33533 (813) 475-5228

Harold N. Powell
P.O. Box 64
Englewood, FL 33533 (813) 474-2842

Douglas J. Drummond#
309 N. 6th St.
Flagler Beach, FL 32036 (904) 439-2154

Alvin S. Blum#
2350 Del Mar Place
Fort Lauderdale, FL 33301 (305) 462-5006

Winfield J. Brown, Jr.
Motorola, Inc.
8000 W. Sunrise Blvd.
Fort Lauderdale, FL 33322 (305) 738-2860

Kevin P. Crosby#
Malin, Haley & Mc Hale, P.A.
One East Broward Blvd., Suite 1609
Fort Lauderdale, FL 33301 (305) 763-3303

Dale P. Di Maggio
Malin, Haley & Mc Hale, P.A.
One Financial Plaza, Suite 2110
Fort Lauderdale, FL 33394 (305) 763-3303

Joseph T. Downey
Motorola, Inc.
8000 W. Sunrise Blvd.
Fort Lauderdale, FL 33322 (305) 475-6449

William Joseph Flynn
Oltman & Flynn
915 Middle River, Suite 415
Fort Lauderdale, FL 33304 (305) 563-4814

Herman I. Hersh
3900 Galt Ocean Dr.
Fort Lauderdale, FL 33308

P. Gregory Jones
Malin & Haley, P.A.
One Financial Plaza, Suite 2110
Fort Lauderdale, FL 33394 (305) 763-3303

Adam A. Jorgensen#
Oltman & Flynn
915 Middle River Drive, Suite 415
Fort Lauderdale, FL 33304 (305) 563-4814

Louis V. Lucia
5200 N. Ocean Blvd.
Fort Lauderdale, FL 33308 (305) 782-5634

Eugene F. Malin
Malin & Haley, P.A.
One Financial Plaza, Suite 2110
Fort Lauderdale, FL 33394 (305) 763-3303

Ralph M. Martin
Oltman & Flynn
915 Middle River Dr., Suite 415
Fort Lauderdale, FL 33304 (305) 563-4814

Martin J. Mc Kinley
Motorola, Inc.
8000 W. Sunrise Blvd.
Fort Lauderdale, FL 33322 (305) 475-3861

Daniel K. Nichols
Motorola, Inc.
Pat. Dept.
8000 W. Sunrise Blvd.
Fort Lauderdale, FL 33322 (305) 475-6449

John Harold Oltman
Oltman & Flynn
915 Middle River Dr., Suite 415
Fort Lauderdale, FL 33304 (305) 563-4814

Melvin K. Silverman
407 Third Ave., N.E.
Ft. Lauderdale, FL 33301 (305) 764-5155

Ralph H. Swingle
5200 N. Ocean Blvd., Rm. 1502
Fort Lauderdale, FL 33308 (305) 782-3170

Carl V. Wisner, Jr.
2709 N.E. 26th Terrace
Fort Lauderdale, FL 33306 (305) 564-2137

Erwin A. Yaeger
1800 N.E. 43rd St.
Fort Lauderdale, FL 33308 (305) 771-2643

Joseph Zallen
2455 E. Sunrise Blvd., Suite 1105
Fort Lauderdale, FL 33304 (305) 565-9506

Frank P. Cyr
1004 La Paloma Blvd.
Ft. Myers, FL 33903 (813) 731-1583

Robert H. Heise
994 N. Waterway Dr.
Fort Myers, FL 33919 (813) 482-0613

David G. Thompson#
16956-1 Mc Gregor Blvd.
Fort Myers, FL 33908 (813) 466-1616

Henry Ovington Wright
769 Entrada Drive S.
Fort Myers, FL 33901 (813) 481-1827

M. Katherine Baumeister#
8502 S.W. 52nd Pl.
Gainesville, FL 32608 (904) 377-8503

Charles A. Bevelacqua
1806 N.W. 93rd Dr.
Gainesville, FL 32607 (904) 377-2806

John R. O Malley
University Of Florida
Larsen Hall
Gainesville, FL 32611 (904) 392-4922

David R. Saliwanchik#
529 N.W. 60th St., Suite B
Gainesville, FL 32607 (904) 388-1533

Roman Saliwanchik
529 N.W. 60th St., Suite B
Gainesville, FL 32607 (904) 338-1533

Merrill Wilcox#
2911 Nw 30th Terr.
Gainesville, FL 32605 (904) 376-1174

Kelly O. Corley
P.O. Box 273
Gonzalez, FL 32560 (904) 477-3041

Carl Fissell, Jr.#
Coulter Electronics, Inc.
590 Coulter Way
Hialeah, FL 34983 (305) 995-0131

Gerald R. Hibnick
Coulter Electronics, Inc.
590 W. 20th St.
Hialeah, FL 33010 (305) 885-0131

Philip T. Liggett
3114 S. Ocean Blvd.
Highland Beach, FL 33431 (305) 278-8582

Charles A. Cohen#
1201 So. Ocean Dr.
Hollywood, FL 33019 (305) 922-4138

Laurence A. Greenberg
Lerner & Greenberg, P.A.
P.O. Box 2480
Hollywood, FL 33022 (305) 925-1100

Herbert L. Lerner
Lerner & Greenberg, P.A.
P.O. Box 2480
Hollywood, FL 33022 (305) 925-1100

Peter F. Hilder
624 Hampshire Lane
Holmes Beach, FL 33510 (813) 778-3903

Donna Brooks
P.O. Box 11296
Jacksonville, FL 32211 (904) 721-1986

James D. Dee
Brockway, Inc.
225 Water St., 17th Fl.
Jacksonville, FL 32202 (904) 791-4172

Richard E. Klein
Livermore, Klein & Lott
701 Fisk St., Suite 225
Jacksonville, FL 32204 (904) 359-0500

Nathaniel L. Leek
819 Point La Vista
Jacksonville, FL 32207 (904) 398-4115

Thomas A. Redding
2276 Meadowlark Ct.
Jacksonville, FL 32216 (904) 221-2940

Thomas C. Saitta
Atwater & Fagan
335 East Bay St.
Jacksonville, FL 32202 (904) 358-2011

Earl L. Tyner
Baldwin & Yeager
1305 Barnett Bank Bldg.
Jacksonville, FL 32202 (904) 355-9631

Arthur G. Yeager
1305 Barnett Bank Bldg.
Jacksonville, FL 32202 (904) 355-9631

Alexander Raymond Field
376 River Edge Rd.
Jupiter, FL 33477 (305) 744-9140

James O. Harrell
NASA
Mail Code Pt-Pat
Kennedy Space Center, FL 32899
(305) 867-2544

Leon Robbin
1111 Cradon Blvd. A503
Key Biscayne, FL 33149 (305) 361-5067

Joseph B. Allen, III#
617 Whitehead St.
Key West, FL 33040 (305) 296-5031

S. H. Hartz
Route 2 Box 39A
Keystone Hgts., FL 32656

George R. Jones
638 E. Wood Dr.
Kissimmee, FL 32741 (813) 427-3300

Harris Cade Lockwood
4101-11 Northgate Dr.
Kissimmee, FL 32741 (813) 427-1014

Delavan Palmer Smith
O. Box 1118
La Belle, FL 33935 (813) 675-3168

Alfred E. Wilson
975 Caloosa Estates Dr.
La Belle, FL 33935 (813) 675-1676

Henry M. Fendrich
Stromberg-Carlson Corp.
400 Rinehart Rd.
Lake Mary, FL 32746 (305) 849-3055

Alan H. Norman
243 Broadmoor Rd.
Lake Mary, FL 32746 (305) 322-8423

Michael F. Oglo
Stromberg-Carlson Corp.
400 Rinehart Rd.
Lake Mary, FL 32746 (305) 849-3000

Jerry A. Thiebeau#
810 Summer St.
Lake Worth, FL 33461 (305) 965-9684

John S. Brown#
1510 W. Ariana St.
Box 462
Lakeland, FL 33803 (813) 686-7548

David D. Centola
125 Hypoluxo Rd.
Lantana, FL 33462 (305) 588-8821

Michael J. Colitz, Jr.
217 Harbor View Lane
Largo, FL 33540 (813) 585-4058

Harold D. Shall
1101 Belcher Rd.
South, Suite B
Largo, FL 33641 (813) 536-2711

Samuel W. Kipnis
100 Sands Point Rd., Apt. 105
Longboat Key, FL 33548 (813) 388-1621

Sidney Alfred Ochs
4350 Chatham Dr.
Longboat Key, FL 33548 (813) 383-1428

Joseph E. Kerwin
480 Century Dr.
Marco Island, FL 33937 (813) 394-0432

Raymond F. Kramer
55 Primrose Ct.
Marco Island, FL 33937 (813) 394-1702

Dennis L. Cook*
Harris Corp.
GASD Div. Counsel
P.O. Box 94000
Melbourne, FL 32902 (305) 727-4127

John L. De Angelis, Jr.
Harris Corp.
Government Systems Sector M S 2/138
P.O. Box 37
Melbourne, FL 32902 (305) 729-3353

Harry Martin Fleck, Jr.
Harris Corp.
1025 W. Nasa Blvd.
Melbourne, FL 32919 (305) 727-9155

Leslie J. Hart
Harris Corp.
P.O. Box 883
Melbourne, FL 32901 (305) 724-2580

Charles C. Krawczyk
Harris Corp.
1025 W. Nasa Blvd.
Melbourne, FL 32919 (305) 727-9156

Joel I. Rosenblatt
Harris Corp.
1025 W. NASA Blvd.
Melbourne, FL 32919 (305) 727-9100

Robert I. Smith
1825 Marywood Rd., Apt. 67
Melbourne, FL 32935 (305) 259-5904

William A. Troner
Harris Corp.
P.O. Box 883
Melbourne, FL 32901

Thomas N. Twomey
Harris Corp.
Semiconductor Sector
P.O. Box 883
Melbourne, FL 32901 (305) 729-4508

Bernard H. Breymann
1925 Brickell Ave.
Penthouse 9
Miami, FL 33129 (305) 856-3948

Curtis D. Carlson#
Fowler, White, Burnett, Hurley, Banick & Strickroot, P.A.
25 W. Flagler St.
Miami, FL 33130 (305) 358-6550

Michael C. Cesarano
Steele, Gould & Fried
Southeast Financial
200 So. Biscayne Blvd., Suite 2920
Miami, FL 33131 (305) 358-0007

Henry W. Collins
Cordis Corp.
P.O. Box 025700
Miami, FL 33102 (305) 551-2707

John H. Faro
Amer. Hospital Supply Corp.
1851 Delaware Parkway
P.O. Box 520672
Miami, FL 33152 (305) 633-6461

James A. Gale
Morgan, Lewis & Bockius
5300 Southeast Financial Center
200 S. Biscayne Blvd.
Miami, FL 33131 (305) 579-0414

Allegra A. Genest#
11302 S.W. 133 Place
Miami, FL 33186 (305) 386-4779

Robert C. Kain, Jr.
168 S.E. First St.
Miami, FL 33131 (305) 374-8418

John Cyril Malloy
Consolidated Bank Bldg.
168 S.E. First St., 8th Fl.
Miami, FL 33131 (305) 374-8418

Sybil Meloy
1915 Brickell Ave., Apt. C-1108
Miami, FL 33129 (305) 652-2276

Raymond L. Robinson
801 Brickell Ave., Suite 1200
Miami, FL 33131 (305) 374-4192

Joseph Matthew Roehl
9730 S.W. 148th Ave.
Miami, FL 33196 (305) 382-2184

Jesus Sanchelima
235 S.W. Le Jeune Rd.
Miami, FL 33134 (305) 447-1617

Paul S. Seward
235 S.W. Le Jeune Rd.
Miami, FL 33134 (305) 447-1617

James E. Wetterling, Jr.
Consolidated Bank Bldg.
168 S.E. First St., 8th Fl.
Miami, FL 33131 (305) 374-8418

Erwin Myles Barnett
7960 Hawthorne Ave.
Miami Beach, FL 33141 (305) 865-6996

Milton H. Gross
2862 Fairgreen Dr.
Miami Beach, FL 33140 (305) 532-5617

Abraham J. Nydick
1008 Morton Towers
1500 Bay Rd.
Miami Beach, FL 33139 (305) 673-3074

Phillip H. Pohl
1020 Meridian Ave., Apt. 916
Miami Beach, FL 33139 (305) 531-7102

Jack Edward Dominik
Dominik, Stein, Saccocio & Reese
6175 N.W. 153rd St., Suite 225
Miami Lakes, FL 33014 (305) 556-9889

Kevin P. Fenton#
8306 Dundee Terrace
Miami Lakes, FL 33016 (305) 825-4929

Richard S. Ross
6175 N.W. 153rd St., Apt. 225
Miami Lakes, FL 33014 (305) 556-9889

Richard M. Saccocio
Dominik, Stein, Saccocio & Reese
6175 N.W. 153rd St., Suite 2225
Miami Lakes, FL 33014 (305) 556-9889

Steven R. Scott
5433 State Rd. 218
Middleburg, FL 32068 (904) 282-0163

Gregory A. Nelson#
7849 Alhambra Blvd.
Miramar, FL 33023 (305) 966-4093

James N. Buckner
5409 Foxhound Dr.
Naples, FL 33942 (813) 643-0955

J. Wesley Haubner
1337 Osprey Ave.
Naples, FL 33942 (813) 775-1657

Jack W. Heberling, Jr.#
422 Cypress Way East
Naples, FL 33942 (813) 598-2489

Marshall M. Holcombe
2610 Half Moon Walk
Naples, FL 33940 (813) 262-5009

Merrill Nels Johnson
800 Harbour Dr.
Naples, FL 33940 (813) 262-8502

Ernest H. Schmidt
5740 24th Ave., SW
Naples, FL 33999 (813) 455-3846

John Kenneth Wise
660 Regatta Rd.
Naples, FL 33940 (813) 261-8683

Lloyd F. Seebach#
18 Stymie Lane
New Smyrna Beach, FL 32069 (904) 427-6966

William F. Smith#
353 De Soto Dr.
New Smyrna Beach, FL 32069 (904) 428-0100

Tiobias E. Levow
1700 N.E. 191st St., Apt. 109
North Miami Beach, FL 33179 (305) 949-8690

Harry Levy#
3475 N. Country Club Dr.
North Miami Beach, FL 33180 (305) 931-5802

Robert M. Schwartz
2020 N.E. 163rd St., Suite 300
North Miami Beach, FL 33162 (305) 994-9100

Liber J. Montone#
9242 Vanderbilt Dr.
N. Naples, FL 33940 (813) 597-8781

William Thomas Clarke
2935 S W 32nd Ave.
Ocala, FL 32674 (904) 622-7127

Stanley C. Felton
611 S.E. 9th Ave., Apt. 34
Ocala, FL 32671 (904) 622-7947

Leon Chasan
490 Hickorynut Ave.
Oldsmar, FL 34677 (813) 785-6179

Robert W. Adams*
Dept. Of Navy
Intellectual Property
NTSC Code 004P C
Orlando, FL 32813 (305) 646-4745

Herbert L. Allen, Jr.
Duckworth, Allen & Dyer, P.A.
1 South Orange Ave.
Orlando, FL 32801 (305) 841-2330

James H. Beusse
Duckworth, Allen & Dyer, P.A.
1 South Orange Ave.
Orlando, FL 32801 (305) 841-2330

Robert W. Duckworth
Duckworth, Allen & Dyer, P.A.
One South Orange Ave.
Suite 600
Orlando, FL 32801 (305) 841-2330

Warren L. Franz
Duckworth, Allen & Dyer P.A.
P.O. Box 3791
Orlando, FL 32802 (305) 841-2330

William J. Iseman
Martin Marietta Corp.
P.O. Box 5837
Orlando, FL 32855 (305) 356-2625

Edward M. Livingston
Pan American Bank Bldg., Suite 1000
P.O. Box 1706
Orlando, FL 32802 (305) 422-3912

John F. Miller#
2532 Overlake Ave.
Orlando, FL 32806 (305) 859-0749

David Olsen#
3602 Country Lakes Dr.
Orlando, FL 32812 (305) 859-0630

James L. Simon
Bogin, Munns, Munns & Simon
P.O. Box 2807
Pan American Bank Bldg., Suite 1001
Orlando, FL 32802 (305) 425-1812

Michael L. Slonecker
Martin Marietta Orlando Aerospace
P.O. Box 5837
Mail Point - 186
Orlando, FL 32855 (305) 356-3405

Joseph V. Truhe, Sr.#
8764 Granada Blvd.
Orlando, FL 32819 (305) 876-4509

Macdonald J. Wiggins
Hobby & Wiggins
1327 N. Mills Ave.
Orlando, FL 32803 (305) 896-5995

Lewis J. Lamm
66 Fairway Dr.
Ormond Beach, FL 32074 (704) 672-9559

Carl C. Mueller
2327 Bonnieview Dr.
Ormond Beach, FL 32074 (904) 441-3163

Jerrell P. Hollaway#
Consumer Engineering, Inc.
1330 Meadowbrook Rd., N.E.
Palm Bay, FL 32905 (305) 727-7625

Charles E. Wands
Antonelli, Terry & Wands
1520 Bottlebrush Dr., N.E., Suite 1
Palm Bay, FL 32905 (305) 725-4760

Theodore Bishoff
3546 S. Ocean Blvd., Apt. 826
Palm Beach, FL 33480 (305) 588-8192

Harry S. Colburn, II
Alley, Maass, Rogers, Lindsay & Chauncey
321 Royal Poinciana Plaza, S.
P.O. Box 431
Palm Beach, FL 33480 (305) 659-1770

George A. Teacherson#
P.O. Box 762
Palm Beach, FL 33480 (305) 439-7005

Ronald E. Matteson
P.O. Box 235
Palm City, FL 33490 (305) 287-4384

Kevin P. Redmond
6960 SW Gator Trail
Palm City, FL 34990 (305) 283-1507

Oistein J. Bratlie
P.O. Box 2889
Palm Coast, FL 32037 (904) 445-0910

James D. Haynes
2595 S.R. 584, Suite V
Palm Harbor, FL 33563 (813) 787-3555

Melvin Yedlin
2390 Republic Dr.
Palm Harbor, FL 33563 (813) 796-5434

Harvey A. David*
U.S. Navy
Office Of Patent Counsel
Naval Coastal Sys. Lab.
Panama City, FL 32407 (904) 234-4156

Thomas Y. Awalt, Jr.
Deep Seven Co.
14260 Inncravity Pt. Rd.
Pensacola, FL 32507 (904) 492-0250

Alexander Kozel
5920 San Gabriel Dr.
Pensacola, FL 32504 (904) 477-3461

Thomas N. Wallin
Monsanto Co.
P.O. Box 12830
Pensacola, FL 32575 (904) 968-8266

John William Whisler
Monsanto Co.
P.O. Box 12830
Pensacola, FL 32575 (904) 968-8272

Elmer L. Zwickel
8250 Burgundy Dr.
Pinellas Park, FL 33565 (813) 546-1367

William A. Newton
11601 N.W. 18th St.
Plantation, FL 33323 (305) 475-3197

Roy N. Envall, Sr.
1360 N.E. 27th Terrace, Apt. 6
Pompano Beach, FL 33062 (305) 782-0346

Walter A. Modance
808 Cypress Blvd., Apt. 304
Pompano Beach, FL 33069 (305) 971-2362

William D. O Connor
651 S.W. 6th St., C T - 609
Pompano Beach, FL 33060

Stanley W. Sokolowski
2605 E. Atlantic Blvd.
Pompano Beach, FL 33062 (305) 782-6539

Aaron Mack Scharf
812 Spinnakers Reach Rd.
Ponte Vedra, FL 32082 (904) 285-5110

Harvey W. Rockwell#
127 SW Graham St.
Port Charlotte, FL 33952 (813) 629-1013

Oscar B. Waddell
11050 Elderberry Dr.
Port Richey, FL 33568 (813) 863-0356

D. Chester Wintemute#
1202 Farmington Lane
Port Richey, FL 33568 (813) 849-2970

Thomas H. Buffton#
1838 S.E. Westmoreland Blvd.
Port St. Lucie, FL 33452 (305) 335-3785

William G. H. Finch#
3025 Morningside Blvd.
Port St. Lucie, FL 33452 (305) 335-5147

Cornelius J. Husar
2442 SE Floresta Dr.
Port St. Lucie, FL 34984 (305) 879-4661

Harry T. Berriman#
2730 Luna Ct.
Punta Gorda, FL 33950 (813) 637-1841

Edward J. Brenner
586 Bal Harbor Blvd.
Punta Gorda, FL 33950 (813) 637-1075

Robert J. Norton
Norton & Marryott
First National Bank Bldg., Suite 408
Punta Gorda, FL 33950 (813) 639-0311

Jon L. Liljequist
5770 Pine Tree Drive
Sanibel, FL 33957

John F. Ahern
6342 Midnight Pass Rd. Apt. 467
Sarasota, FL 34242 (813) 346-1635

Marshall J. Breen
3345 Spring Mill Circle
Sarasota, FL 33579 (813) 924-8142

Richard G. Bremer#
5636 Pipers Waite
Sarasota, FL 33580 (813) 377-8595

Ralph Husack#
6714 Roxbury Dr.
Sarasota, FL 33581 (813) 924-4586

William Genther Lambrecht
Williams, Parker, Harrison, Dietz & Getzen
1550 Ringling Blvd.
Sarasota, FL 33578 (813) 366-4800

Carl William Laumann, Jr.
Rte. 3, Box 57
Sarasota, FL 34243 (813) 753-4716

David C. Noller#
1040 Sylvan Dr.
Sarasota, FL 33580 (813) 365-1935

Charles J. Prescott
2033 Wood St., Suite 115
Sarasota, FL 33577 (813) 957-4208

Raymond H. Quist
2314 Tulip St.
Sarasota, FL 34239 (813) 957-4208

William Julius Van Loo, Jr.#
1727 Bahia Vista St.
Sarasota, FL 33579 (913) 955-3887

Jack N. Mc Carthy#
655 Bimini Rd.
Satellite Beach, FL 32937 (305) 773-2081

Ruth M. Rife#
220 Hedgecock Ct.
Satellite Beach, FL 32937 (305) 725-1381

George B. Oujevolk
P.O. Box 273
Sebring, FL 33871 (813) 385-5232

Roy E. Raney
14245-93rd Ave. North
Seminole, FL 33542 (813) 595-1473

Edwin W. Uren#
6224 Kelvin Ct.
Spring Hill, FL 33526 (904) 683-0355

Walter J. Monacelli
720 36th Ave. North
St. Petersburg, FL 33704 (813) 525-5759

Samuel B. Pritchard
1774 63rd Ave., S.
St. Petersburg, FL 33712 (813) 867-8744

Hugh E. Smith
1935 1st Ave. S.
St. Petersburg, FL 33712 (813) 822-7592

Richard A. Wahl
1981 Hawaii Ave., N.W.
St. Petersburg, FL 33703 (813) 527-4100

Wilbur J. Kupfrian
1706 NW Fork Rd.
Stuart, FL 33494 (305) 692-1922

Jane C. Mc Gregor
2600 S.E. St. Lucie Blvd.
Stuart, FL 33494 (305) 283-9543

David M. Schiller
46 Rio Vista Dr.
Stuart, FL 33494 (305) 287-2509

Martin J. Carroll
1206 Beach Blvd.
Sun City Center, FL 33570 (813) 634-7010

Arthur L. Morsell, Jr.
1212 Fordham Dr.
Sun City Center, FL 33570 (813) 634-6062

Leslie G. Noller
623 Allegheny Dr.
Sun City Center, FL 33570 (813) 634-3158

Louis Robertson
1411 Nashua Circle
Sun City Center, FL 33570 (813) 634-6105

Jerry A. Miller#
8571 NW 21st Ct.
Sunrise, FL 33322 (305) 742-0972

Horace Schow, II
2816 Roscommon Dr.
Tallahassee, FL 32308 (904) 893-4028

David Schonberg#
6091 N.W. 61st Ave.
Tamarac, FL 33319 (305) 722-4387

Donald Ruh Bahr
Questor Corp.
5750 A North Hoover Blvd.
Tampa, FL 33630 (813) 887-5274

George T. Breitenstein
Frijouf, Rust & Pyle
201 East Davis Blvd.
Tampa, FL 33606 (813) 254-5100

Ralph W. Burnett
GTE Data Services Inc.
111 Madison St.
Tampa, FL 33601 (813) 224-3228

Arthur W. Fisher, III
6304 Benjamin Rd., Suite 500
Tampa, FL 33614 (813) 885-2006

Robert Frank Frijouf
Frijouf, Rust & Pyle, P.A.
201 East Davis Blvd.
Tampa, FL 33606 (813) 254-5100

James Wm. Grace
Walter Industries, Inc.
P.O. Box 31601
Tampa, FL 33631 (813) 871-4456

Charles A. Mc Clure
P.O. Box 1168
Tampa, FL 33601 (813) 251-2868

C. Douglas Mc Donald
Duckword, Allen, Dyer & Pettis, P.A.
403 E. Kennedy Blvd.
P.O. Box 1528
Tampa, FL 33601 (813) 229-8176

John Joseph Mc Laughlin
Wagner, Cunningham, Vaughan & Mc Laughlin, P.A.
708 Jackson St.
Tampa, FL 33602 (813) 223-7421

David L. Partlow
Frijouf, Rust & Pyle, P.A.
201 East Davis Blvd.
Tampa, FL 33606 (813) 254-5100

David W. Pettis, Jr.
Duckworth, Allen, Dyer & Pettis, P.A.
P.O. Box 1528
Tampa, FL 33601 (813) 229-8176

Ray S. Pyle
Frijouf, Rust & Pyle, P.A.
201 East Davis Blvd.
Tampa, FL 33606 (813) 254-5100

Carolyn B. Ray
Frijouf, Rust & Pyle, P.A.
201 East Davis Blvd.
Tampa, FL 33606 (813) 254-5100

Benjamin P. Reese, II
Stein, Reese & Prescott
312 East Harrison St.
Tampa, FL 33602 (813) 229-2122

Stefan V. Stein
Stein & Reese
800 Freedom Savings Blvd.
220 E. Madson St.
Tampa, FL 33602 (813) 229-2122

Charles W. Vanecek#
Walter Industries, Inc.
P.O. Box 31601
Tampa, FL 33631 (813) 871-4456

Sherman H. Barber
3111 Heron Shores Dr.
Venice, FL 33595 (813) 493-6645

Edward T. Connors
521 W. Venice Ave., Apt. 21
Venice, FL 33595 (813) 488-0391

Henry B. Kellog
511 Verdi St.
Venice, FL 33595 (813) 485-7807

Jerome F. Kramer
3339 Cardinal Dr.
Vero Beach, FL 32960 (305) 231-6516

Morris B. Lore#
851 Seminole Lane
Vero Beach, FL 32960 (305) 231-2866

Carroll F. Palmer
1845 20th St.
Vero Beach, FL 32960 (305) 569-0145

Martin Kalikow
2260 Sunderland Ave.
Wellington, FL 33414 (305) 798-6128

Hal H. Mc Caghren
120 S. Olive Ave., Suite 555
West Palm Beach, FL 33458 (305) 655-1963

J. Rodman Steele, Jr.
Steele, Gould & Fried
700 Tower B. Forum I I I
1665 Palm Beach Lakes Blvd.
West Palm Beach, FL 33401 (305) 471-1449

Bruce K. Thomas
1420 Northampton Terrace
West Palm Beach, FL 33414 (305) 793-1140

Elsie T. Apthorp
1350 Orange Ave. Apt. 217
Winter Park, FL 32789 (305) 628-5751

John William Pease
1730 Winchester Dr.
Winter Park, FL 32789 (305) 647-4825

Julian Carroll Renfro
1350 Orange Ave., Suite 213
Winter Park, FL 32789 (305) 628-3600

Georgia

William R. Alford
265 Cedar Springs Dr.
Athens, GA 30605 (404) 353-7744

Donald R. Andersen
Mozley, Finlayson & Anderson
1800 Coastal States Bldg.
260 Peachtree St.
Atlanta, GA 30303 (404) 522-2010

Albert S. Anderson
Jones & Askew
P.O. Box 56326
Atlanta, GA 30343 (404) 688-7500

Anthony Bartholomew Askew
Jones & Askew
P.O. Box 56326
Atlanta, GA 30343 (404) 688-7500

Jason A. Bernstein
Jones, Askew & Lunsford
230 Peachtree St., Suite 2000
Atlanta, GA 30303 (404) 688-7500

Thomas R. Boston
Coca-Cola Co.
P.O. Drawer 1734
Atlanta, GA 30301 (404) 676-6682

James K. Boudreau
Coca-Cola Co.
P.O. Drawer 1734
Tower 2140
Atlanta, GA 30301 (404) 898-4872

William D. Brooks
Coca-Cola Co.
P.O. Drawer 1734
Atlanta, GA 30301 (404) 676-2103

Eduardo M. Carreras
Coca-Cola Co.
One Coca-Cola Plaza NW
Atlanta, GA 30313 (404) 676-3272

William R. Cohrs
King & Spalding
2500 Trust Co. Tower
25 Park Place, NE
Atlanta, GA 30303 (404) 572-3386

Carl M. Davis, II
Jones, Askew & Lunsford
P.O. Box 56326
Atlanta, GA 30343 (404) 688-7500

Todd Deveau
Newton, Hopkins & Ormsby
1010 The Equitable Bldg.
100 Peachtree St.
Atlanta, GA 30303 (404) 688-1788

Erwin Doerr#
Mead Corp.
Mead Packaging Div.
1040 W. Marietta St., N.W.
Atlanta, GA 30302 (404) 897-6752

Michael V. Drew
P.O. Box 50415
Atlanta, GA 30302 (404) 987-3520

James L. Ewing, IV
Kilpatrick & Cody
3100 Equitable Bldg.
100 Peachtree St.
Atlanta, GA 30043 (404) 572-6500

Roger T. Frost
Jones, Askew & Lunsford
P.O. Box 56326
Atlanta, GA 30343 (404) 688-7500

Dale V. Gaudier
Neptune International
4360 Chamblee Dunwoody Rd., Suite 410
Atlanta, GA 30341 (404) 458-1212

Michael J. Gilroy
Coca-Cola Co.
P.O. Drawer 1734
Atlanta, GA 30301 (404) 676-3207

Joel S. Goldman
Scientific Atlanta, Inc.
One Technology Park
P.O. Box 105600
Atlanta, GA 30348 (404) 441-4872

Jamie L. Graham
Jones, Askew & Lunsford
230 Peachtree St., Suite 2000
Atlanta, GA 30303 (404) 688-7500

Gregory T. Gronholm
Jones & Askew
230 Peachtree St., Suite 2000
Atlanta, GA 30303 (404) 688-7500

John R. Harris
Jones, Askew & Lunsford
230 Peachtree St., N.W., Suite 2000
Atlanta, GA 30303 (404) 688-7500

Patrick F. Henry
2601 1st National Bank Bldg.
2 Peachtree St.
Atlanta, GA 30383 (404) 658-1754

William George Hervey
Needle & Rosenberg, P.C.
133 Carnegie Way, Suite 400
Atlanta, GA 30303 (404) 688-0770

Gwenetta Douglas Hill
986 Oglethorpe Ave., S.W.
Atlanta, GA 30310 (404) 755-3288

Thomas Arthur Hodge
Jones, Askew & Lunsford
P.O. Box 56326
Atlanta, GA 30343 (404) 688-7500

L. Harmon Hook
30 Perimeter Center East
Suite 200
Atlanta, GA 30346 (404) 395-6877

George M. Hopkins
Hurt, Richardson, Garner, Todd & Cadenhead
1400 Peachtree Place Tower
999 Peachtree St., NE
Atlanta, GA 30309 (404) 870-3999

Louis T. Isaf
5770 Powers Ferry Rd., N.W.
Suite 202
Atlanta, GA 30327 (404) 951-2623

John L. James
Jones, Askew & Lunsford
2000 Peachtree Center Tower
230 Peachtree St.
Atlanta, GA 30303 (404) 688-7500

James D. Johnson
Cyt Rx Corp.
150 Technology Parkway
Technology Park Atlanta
Atlanta, GA 30092 (404) 368-9500

Harold D. Jones, Jr.
Jones, Askew & Lunsford
230 Peachtree St., Suite 2000
Atlanta, GA 30303 (404) 255-9629

Brij M. Kapoor
230 Peachtree St., N.W.
Suite 1920
Atlanta, GA 30303 (404) 522-1122

James W. Kayden
Newton, Hopkins & Ormsby
1010 Equitable Bldg.
100 Peachtree St.
Atlanta, GA 30303 (404) 688-1788

David P. Kelley
17 Dove Hill
3873 Roswell Rd., N.E.
Atlanta, GA 30342 (404) 364-0136

Robert Bruce Kennedy
Thomas & Kennedy
100 Galleria Parkway, NW, Suite 590
Atlanta, GA 30339 (404) 951-0931

Steven D. Kerr
Thomas & Kennedy
100 Galleria Parkway, Suite 590
Atlanta, GA 30339 (404) 951-0931

William C. Lee, III
Coca-Cola Co.
P.O. Drawer 1734
Atlanta, GA 30301 (404) 676-2121

Harry I. Leon#
Leon & Leon
924 Bowen St., N.W.
Atlanta, GA 30318 (404) 352-3882

Vivian L. Leon#
Loen & Leon
924 Bowen St., N.W.
Atlanta, GA 30318 (404) 352-3882

Robert A. Lester
Coca-Cola Co.
310 N. Ave.
P.O. Box 1734
Atlanta, GA 30301 (404) 898-2530

Dale Lischer
Jones, Askew & Lunsford
P.O. Box 56326
Atlanta, GA 30343 (404) 688-7500

J. Rodgers Lunsford, III
Jones, Askew & Lunsford
P.O. Box 56326
Atlanta, GA 30343 (404) 688-7500

Frank Madonia
Neptune Internatl. Corp.
30 Perimeter Park
Atlanta, GA 30341 (404) 458-1212

Harold L. Marquis
Mead Packaging
950 W. Marietta St., N.W.
Atlanta, GA 30318 (404) 875-2711

John R. Martin
Coca-Cola Co.
P.O. Drawer 1734
310 N. Ave., NW
Atlanta, GA 30301 (404) 898-2102

William H. Needle
The Carnegie Bldg., Suite 400
133 Carnegie Way NW
Atlanta, GA 30303 (404) 688-0770

Edward Taylor Newton
Newton, Hopkins & Ormsby
100 Peachtree St., Suite 1010
Atlanta, GA 30303 (404) 688-1788

William Joseph Ormsby, Jr.
Newtonl, Hopkins & Ormsby
100 Peachtree St., Suite 1010
Atlanta, GA 30303 (404) 688-1788

Patrea L. Pabst
Kilpatrick & Cody
3100 Equitable Bldg.
100 Peachtree St.
Atlanta, GA 30043 (404) 572-6508

Salvatore P. Pace
Coca-Cola Co.
P.O. Drawer 1734
Atlanta, GA 30301 (404) 676-2121

Frank A. Peacock
Jones & Askew
P.O. Box 56326
Atlanta, GA 30343 (404) 688-7944

Billy J. Powell
1447 Peachtree St., NE
Suite 710
Atlanta, GA 30309 (404) 892-8046

Frederick W. Powers, III
Siemens-Allis, Inc.
223 Perimeter Pkwy.
Atlanta, GA 30338 (404) 393-8200

John S. Pratt
Kilpatrick & Cody
3100 Equitable Bldg.
100 Peachtree St.
Atlanta, GA 30043 (404) 572-6500

Robert E. Richards
Jones & Askew
P.O. Box 56326
Atlanta, GA 30343 (404) 688-7944

Larry A. Roberts
Jones & Askew
230 Peachtree St., Suite 2000
Atlanta, GA 30303 (404) 688-7500

Walter A. Rodgers
Rodgers & Rodgers
1114 Gas Light Tower
235 Peachtree St., NE
Atlanta, GA 30303 (404) 523-6059

Walter M. Rodgers, Jr.
1114 Gas Light Tower
235 Peachtree St., NE
Atlanta, GA 30303 (404) 523-6059

Sumner C. Rosenberg
133 Carnegie Way, NW, Suite 400
Atlanta, GA 30303 (404) 688-0770

Martin C. Ruegsegger
Advanced Mobile Phone Service, Inc.
2030 Powers Ferry Rd.
Atlanta, GA 30067 (404) 951-7300

Stephen M. Schaetzel
Jones & Askew
P.O. Box 56326
230 Peachtree St., Suite 2000
Atlanta, GA 30303 (404) 688-7944

George Marshall Thomas
Thomas & Kennedy
100 Galleria Parkway, Suite 590
Atlanta, GA 30339 (404) 951-0931

John J. Timar
Newton, Hopkins & Ormsby
1010 Equitable Bldg.
100 Peachtree St.
Atlanta, GA 30303 (404) 688-1788

James F. Vaughan
Newton, Hopkins & Ormsby
1010 The Equitable Bldg.
100 Peachtree St.
Atlanta, GA 30303 (404) 688-1788

Charles L. Warner, II
Jones & Askew
P.O. Box 56326
Atlanta, GA 30343 (404) 688-7500

Walter Lewis Williamson
Bell South, Suite 1800
1155 Peachtree St., NE
Atlanta, GA 30367 (404) 249-2603

Jeffrey E. Young
Jones & Askew
P.O. Box 56326
Atlanta, GA 30343 (404) 688-7500

Eugene S. Zimmer
Jones & Askew
P.O. Box 56326
Atlanta, GA 30343 (404) 688-7500

Nathan D. Field#
2905 Arrowhead Dr., Apt. E-6
Augusta, GA 30909 (404) 736-7292

V. Lee Ringler
The 500 Bldg.
501 Greene St., Suite 210
Augusta, GA 30901 (404) 724-4000

James W. Wallis, Jr.
Southwire Co.
P.O. Box 1000
Carrollton, GA 30119 (404) 832-5375

Van C. Wilks
P.O. Box 1900
Carrollton, GA 30117 (404) 832-3838

James B. Middleton
P.O. Box 1968
Decatur, GA 30031 (404) 377-5327

Robert L. Jay#
6934 Clear Lake Ct.
Doraville, GA 30360 (404) 395-1015

Marla J. Church
Elan Pharmaceutical Res. Corp.
1300 Gould Dr.
Gainesville, GA 30501 (404) 534-8239

Michael C. Smith
P.O. Box 2222
La Grange, GA 30241 (404) 882-6664

James Allan Hinkle
Hinkle & Bull
175 Gwinnett Dr.
Suite 300
Lawrenceville, GA 30245 (404) 995-8877

Robert Paul Barton
Lockheed-Georgia Co.
82 South Cobb Dr.
Marietta, GA 30063 (404) 424-3388

Eric R. Katz
Lockheed-Georgia Co.
86 S. Cobb Dr.
Dept. 85-01, Zone 35
Marietta, GA 30063 (404) 424-3143

John J. Sullivan
Lockheed - Georgia Co.
86 S. Cobb Dr.
Marietta, GA 30060 (404) 424-3143

Stanley L. Tate
Lockheed Georgia Co.
86 S. Cobb Dr.
Marietta, GA 30063 (404) 424-3143

Edward Walter Somers
AT&T Technologies, Inc.
2000 Northeast Expressway
Norcross, GA 30071 (404) 447-2040

W. Ferrel Bentley, Jr.#*
Federal Communications Comm.
P.O. Box 65
Powder Springs, GA 30073 (404) 943-6425

Don Porter Bush
395 Saddle Horn Circle
Roswell, GA 30076 (404) 998-5834

Jeremiah J. Duggan
Kimberly-Clark Corporation
1400 Holcomb Bridge Rd.
Roswell, GA 30076 (404) 587-8626

Robert W. Hampton
1015 Windsor Trail
Roswell, GA 30076 (404) 594-7834

William D. Herrick
Kimberly-Clark Corp.
P.O. Box 103002
1400 Holcomb Bridge Rd.
Roswell, GA 30076 (404) 587-8096

Karl V. Sidor
Kimberly-Clark Corp.
1400 Holcomb Bridge Rd.
Bldg. 200/1
Roswell, GA 30076 (404) 587-7253

Patrick C. Wilson
Kimberly-Clark Corp.
1400 Holcomb Bridge Rd.
Roswell, GA 30076 (404) 587-7214

William S. Mc Curry, Jr.
242 Wiley Bottom Rd.
Savannah, GA 31411 (912) 598-0079

Richard G. Stahr
5 Milledge Lane
Savannah, GA 31411 (912) 598-1260

Albert M. Heiter
807-E Mallory St.
St. Simons Island, GA 31522 (912) 638-6193

Virginia M. Mc Guffey
5011 Post Rd. Ct.
Stone Mountain, GA 30088 (404) 498-0325

Earl D. Harris
P.O. Box 498
Watkinsville, GA 30677 (404) 769-7717

Hawaii

William Bruce Walter#
82 Halaulani Pl.
Hilo, HI 96720 (808) 935-4797

Mark S. Holmes
315 Sand Island Rd.
Honolulu, HI 96819 (808) 842-5555

Martin E. Hsia
Cades, Schutte, Fleming & Wright
1000 Bishop St.
Honolulu, HI 96813 (808) 521-0200

Kazuo Kiyonaga#
1840-A Ninth Ave.
Honolulu, HI 96816

George W.T. Loo
755 McNeill St., Suite B202
Honolulu, HI 96817 (808) 847-1056

Seth M. Reiss
P.O. Box 442
Honolulu, HI 96809 (808) 523-9122

James E. Smith
Innovation Associates, Inc.
P.O. Box 25546
Honolulu, HI 96825 (808) 923-5481

Idaho

William Johnson Bethurum, III
Hewlett-Packard Co.
11413 Chinden Blvd.
Boise, ID 83714 (208) 323-3047

Frank J. Dykas
210 W. Mallard Dr., Suite C
Boise, ID 83706 (208) 345-1122

Thomas G. Faull#
Techno Consultants
P.O. Box 1662
Boise, ID 83701 (208) 344-2320

Paul F. Horton
1700 Vista Ave.
P.O. Box 5388
Boise, ID 83705 (208) 345-0241

John D. Merris
2324 Scyene Way
Boise, ID 83712 (208) 343-2311

Leslie G. Murray#
Hewlett Packard
P.O. Box 39
Boise, ID 83707 (208) 323-3597

Stephen R. May
Hopkins, French, Crockett, Springer & Hoopes
P.O. Box 1219
Idaho Falls, ID 83402 (208) 523-4445

Ignacio Resendez
U.S. D O E
Idaho Operations Off.
550 Second St.
Idaho Falls, ID 83401 (208) 526-1633

Warren Charles Porter
19 College Ave.
Rexburg, ID 83440 (208) 356-4616

Nicholas T. Bokides
P.O. Box 28
Weiser, ID 83672 (208) 549-0611

Illinois

Donna I. Bobrowicz
Abbott Laboratories
Dept. 377, Ap6D
Abbott Park, IL 60064

Thomas D. Brainard
Abbott Laboratories
D-377/a P 6D-2
Routes 43 & 137
Abbott Park, IL 60064 (312) 937-6367

Donald L. Corneglio, Jr.
Abbott Laboratories
Dept. 377
A P 6D
Abbott Park, IL 60064 (312) 937-6365

Steven R. Crowley#
Abbott Laboratories
D441/A P 6D
Abbott Park, IL 60064 (312) 937-2335

Robert W. Stevenson
Abbott Laboratories
Patent Dept.
D-377
Abbott Park, IL 60064 (312) 937-6366

James L. Wilcox
Abbott Laboratories
D9R K, A P 6C
Routes 137 & 43
Abbott Park, IL 60064 (312) 937-5437

Charles W.B. Connors
Magneco/Metrel, Inc.
206 Factory Rd.
Addison, IL 60101 (312) 543-6660

Mathew R.P. Perrone, Jr.
204 S. Main St.
Algonquin, IL 60102 (312) 658-5140

Powell L. Sprunger
Box 236-A, Rte. 3
Haegers Bend Rd.
Algonquin, IL 60102 (312) 558-4581

Russel P. Steele
P.O. Box 5
101 First Street
Alpha, IL 61413 (309) 529-4701

Phillip J. Kardis
100 E. Beltline Pkwy.
Alton, IL 62002 (618) 465-6650

Stanley M. Parmerter
C P C Internatl. Inc.
Moffett Tech. Ctr.
Box 345
Argo, IL 60501 (312) 458-2000

John M. Albrecht*
U.S. Dept. Of Energy
Off. Of Pat. Coun.
9800 South Cass Ave.
Argonne, IL 60439 (312) 972-2179

Thomas G. Anderson*
U.S. Dept Of Energy
9800 S. Cass Ave.
Argonne, IL 60439 (312) 972-2164

Arthur Alan Churm*
U.S. Dept. Of Energy
9800 South Cass Ave.
Argonne, IL 60439 (312) 972-2165

Robert J. Fisher*
U.S. Atomic Energy Comm.
Chicago Oper. Off.
9800 S. Cass Ave.
Argonne, IL 60439 (312) 972-2176

Hugh Walker Glenn, Jr.*
U.S. Dept. Of Energy
9800 S. Cass Ave.
Argonne, IL 60439 (312) 972-2178

Paul A. Gottlieb*
U.S. Dept. Of Energy
9800 S. Cass Ave.
Argonne, IL 60439 (312) 972-2169

Michael J. Higgins*
U.S. Dept. Of Energy
9800 S. Cass Ave.
Argonne, IL 60439 (312) 972-2308

William Lohff
Argonne National Laboratory
9700 S. Cass Ave.
Argonne, IL 60439 (312) 972-6408

Walter L. Rees*
U.S. Dept. Of Energy
9800 S. Cass Ave.
Argonne, IL 60439 (312) 972-2163

James W. Weinberger*
U.S. Dept. Of Energy
9800 S. Cass Ave.
Argonne, IL 60439 (312) 972-2173

William C. Clarke
1264 North Race Ave.
Arlington Heights, IL 60004 (312) 577-1678

Michael J. Femal
Square D Co.
812 N. Belmont Ave.
Arlington Heights, IL 60004 (312) 397-2600

Marvin Adolph Henrickson
710 E. Crabtree Drive
Arlington Heights, IL 60004 (312) 255-2435

Ronald J. La Porte
1722 Stratford Rd.
Arlington Heights, IL 60004 (312) 392-1030

Charles F. Lind
2210 E. Sherwood Rd.
Arlington Heights, IL 60004 (312) 392-9324

E. Jerome Maas
1716 E. Hawthorne
Arlington Heights, IL 60004 (312) 392-7469

Jos. F. Shekleton
710 E. Waverly
Arlington Heights, IL 60004 (312) 259-4890

Robert A. Stenzel
American National Can Co.
516 E. Fairview
Arlington Heights, IL 60005 (312) 255-6868

Richard L. Dornfeld#
Walker Process Corp.
840 N. Russell Ave.
Aurora, IL 60506 (312) 892-7921

Jack Larsen
Cunningham & Wood
P.O. Box 1486
(302 Downer Place)
Aurora, IL 60507 (312) 897-3375

Frederick W. Neill
600 Fifth St.
Aurora, IL 60505 (312) 851-2306

Bradford S. Allen
240 Castle Ct.
Barrington, IL 60010 (312) 381-5845

Morris P. Burkwall, Jr.#
Quaker Oats Co.
617 W. Main St.
Barrington, IL 60010 (312) 381-1980

Richard J. Karas#
American Can Co.
433 N. Northwest Hwy.
Barrington, IL 60010 (312) 381-1900

John A. Krieger#
American Can Tech. Center
433 N. Northwest Hwy
Barrington, IL 60020 (312) 381-1900

Kajane Mc Manus#
Vigil, Meroni & White
417 N. Hough St.
Barrington, IL 60010 (312) 382-6500

Jacque Louis Meister
102 N. Cook St.
Barrington, IL 60010 (312) 381-7743

Beverly A. Vandenburgh
102 N. Cook St.
Barrington, IL 60010 (312) 381-1831

Thomas Raymond Vigil
Vigil, Meroni & White
836 South Northwest Hwy.
Barrington, IL 60010 (312) 382-6500

Thomas B. Lindgren
950 Cottonwood Lane
Bartlett, IL 60103 (312) 894-7396

Alan A. Pitas#
409 S. Jefferson St.
Batavia, IL 60510 (312) 840-4657

Stephen F. Skala#
3839 S. Wenonah Ave.
Berwyn, IL 60402 (312) 788-5021

Rolland R. Hackbart#
730 Thompson Ct.
Buffalo, IL 60089 (312) 634-0779

Henry A. Weber
671 Aberdeen Lane
Buffalo Grove, IL 60089 (312) 537-6740

Arne R. Jarnholm
Rostoker, Inc.
3333 E. 143rd Street
Burnham, IL 60633 (312) 730-1392

Raymond R. Kimpel
Summers, Watson & Kimpel
404 W. Church St
Champaign, IL 61820 (217) 352-7629

Delphine Kranz#
University Of Illinois
211 Coble Hall
801 S. Wright Street
Champaign, IL 61820 (217) 333-9389

Donald G. Flaynik, Jr.
223 E. Eames
Channahon, IL 60410 (815) 467-4333

Hugh A. Abrams
Neuman, Williams, Anderson & Olson
77 W. Washington St.
Chicago, IL 60602 (312) 346-1200

Roger Aceto
Viskase Corporation
6855 West 65th Street
Chicago, IL 60638 (312) 496-4732

Paul L. Ahern
Leydig, Voit & Mayer, Ltd.
One I B M Plaza, Suite 400
Chicago, IL 60611 (312) 822-9666

John L. Alex
Lockwood, Alex, Fitz Gibbon & Cummings
Suite 1515
3 First Natl. Plaza
Chicago, IL 60602 (312) 782-4860

Richard Elmont Alexander
Alexander, Unikel, Zalewa & Tenenbaum
55 W. Monroe St.
Chicago, IL 60603 (312) 726-7800

D. Dennis Allegretti
Allegretti, Newitt, Witcoff & Mc Andrew, Ltd.
125 S. Wacker Dr.
Chicago, IL 60606 (312) 372-2160

Caliste Jay Alster
Trexler, Bushnell, Giangiorgi & Blackstone, Ltd.
141 W. Jackson Blvd.
Suite 3440
Chicago, IL 60604 (312) 427-8082

Irwin C. Alter
Alter & Weiss, Ltd.
33 N. La Salle St.
Suite 2215
Chicago, IL 60602 (312) 337-2100

Louis Altman
Laff, Whitesel, Conte & Saret
401 N. Michgan Avenue
Chicago, IL 60611 (312) 661-2100

James M. Amend
Kirkland & Ellis
200 E. Randolph
Chicago, IL 60601 (312) 861-2154

Dalton L. Anderson
P.O. Box 912
Chicago, IL 60690 (312) 856-3611

David Albert Anderson
Willian, Brinks, Olds, Hofer, Gilson & Lione Ltd.
One I B M Plaza, Suite 400
Chicago, IL 60611 (312) 822-9800

Richard H. Anderson
Mason, Kolehmainen, Rathburn & Wyss
20 N. Wacker Dr.
Suite 4200
Chicago, IL 60606 (312) 621-1300

Theodore W. Anderson, Jr.
Neuman, Williams, Anderson & Olson
77 W. Washington St.
Chicago, IL 60602 (312) 346-1200

Stephen R. Arnold
Wallenstein, Wagner, Hattis, Strampel & Aubel
100 S. Wacker Dr.
Suite 2100
Chicago, IL 60606 (312) 641-1570

Leo Jhel Aubel
Wallenstein, Wagner, Hattis, Strampel & Aubel
100 S. Wacker Dr.
Chicago, IL 60606 (312) 641-1570

F. David Aubuchon
Navistar International Corp.
401 N. Michigan Avenue
Chicago, IL 60611 (312) 836-2320

Richard C. Auchterlonie
Arnold, White & Durkee
Quaker Tower, Suite 800
321 N Clark Street
Chicago, IL 60610 (312) 744-0090

Jack Axelrood
Morton Chemical Div. Of Morton-Norwich Prods. Inc.
333 W. Wacker Dr.
Chicago, IL 60606 (312) 807-3189

Karen E. Ayd
The Quaker Oats Co.
345 Merchanise Mart Plaza
Chicago, IL 60654 (312) 222-7803

Y. Judd Azulay
Azulay & Azulay, P.C.
205 W. Wacker Dr.
Suite 1600
Chicago, IL 60606 (312) 236-6965

Michael H. Baniak
Willian, Brinks, Olds, Hofer, Gilson & Lione Ltd.
One I B M Plaza, Suite 400
Chicago, IL 60611 (312) 822-9800

Mark T. Banner
Allegretti, Newitt, Witcoff & Mc Andrews, Ltd.
125 S. Wacker Dr.
Chicago, IL 60606 (312) 372-2160

James P. Barr
900 N. Lake Shore Drive ,Suite 2114
Chicago, IL 60611 (312) 280-1652

Robert M. Barrett#
Hill, Van Santen, Steadman & Simpson
70th Fl.
Sears Tower
Chicago, IL 60606 (312) 876-0200

Alan L. Barry
Wallenstein, Wagner, Hattis, Strampel & Auber, Ltd.
100 S. Wacker Dr.
Suite 2100
Chicago, IL 60606 (312) 641-1570

Lawrence J. Bassuk
Wallenstein, Wagner, Hattis, Stramdel & Aubel, Ltd.
100 S. Wacker Dr.
Chicago, IL 60606 (312) 641-1570

Sarah E. Bates
Kirkland & Ellis
200 E. Randolph Dr.
Chicago, IL 60601 (312) 861-2266

Robert W. Beart
Illinois Tool Works Inc.
Patent Dept.
8501 W. Higgins Rd.
Chicago, IL 60631 (312) 693-3040

Gregory Buckingham Beggs
Neuman, Williams, Anderson & Olson
77 W. Washington St.
Chicago, IL 60602 (312) 346-1200

Robert S. Beiser
Welsh & Katz, Ltd.
135 S. La Salle St.
Suite 1625
Chicago, IL 60603 (312) 781-9470

Stanley Belsky
175 E Delaware Place Apt. 8608
Chicago, IL 60611 (312) 440-1638

Glen P. Belvis
Willian, Brinks, Olds, Hofer, Gibson & Lione Ltd.
One I B M Plaza
Chicago, IL 60611 (312) 822-9800

Marvin N. Benn
Hamman & Benn
25 E. Washington St.
Suite 600
Chicago, IL 60602 (312) 372-2920

Joel W. Benson
Willian, Brinks, Olds, Hofer, Gilson & Lione, Ltd.
One I B M Plaza, Suite 4100
Chicago, IL 60611 (312) 822-9800

Martin J. Benson#
4107 N. Spaulding Ave.
Chicago, IL 60618 (312) 267-8241

Robert H. Benson
Leydig, Voit & Mayer Ltd.
One I B M Plaza, Suite 4600
Chicago, IL 60611 (312) 822-9666

Jack Charles Berenweig
Willian, Brinks, Olds, Hofer, Gilson & Lione Ltd.
One I B M Plaza, Suite 4100
Chicago, IL 60611 (312) 822-9800

Paul H. Berghoff
Allegretti, Newitt, Witcoff & Mc Andrews, Ltd.
125 S. Wacker Dr.
Chicago, IL 60606 (312) 372-2160

Michael G. Berkman
Trexler, Bushnell & Wolters, Ltd.
141 W. Jackson Blvd.
Suite 3440
Chicago, IL 60604 (312) 427-8082

Louis Bernat
135 South La Salle
Suite 1135
Chicago, IL 60604 (312) 346-3798

Denis A. Berntsen
Allegretti, Newitt, Witcoff & Mc Andrews, Ltd.
125 S. Wacker Dr.
Suite 3100
Chicago, IL 60606 (312) 372-2160

Albert W. Bicknell
Marshall, O Toole, Gerstein, Murray & Bicknell
Two First National Plaza
Chicago, IL 60603 (312) 346-5750

William Joseph Birmingham
Neuman, Williams, Anderson & Olson
77 W. Washington St.
Chicago, IL 60602 (312) 346-1200

Raiford A. Blackstone, Jr.
Trexler, Bushnell, Giangiorgi & Blackstone, Ltd.
141 W. Jackson Blvd.
Suite 3440
Chicago, IL 60604 (312) 427-8082

Todd P. Blakely
Neuman, Williams, Anderson & Olson
77 W. Washington
Chicago, IL 60602 (312) 346-1200

James Bernard Blanchard
Willian, Brinks, Olds, Hofer, Gilson & Lione Ltd
One I B M Plaza , Suite 4100
Chicago, IL 60611 (312) 822-9800

Robert Emory Blankenbaker
Amoco Corp.
200 East Randolph Dr.
Chicago, IL 60601 (312) 856-2054

Gunar John Blumberg
Amoco Corp.
200 E. Randolph Dr.
Mail Code 1906
Chicago, IL 60601 (312) 856-5967

Joel H. Bock
Kinzer, Plyer, Dorn, & Mc Eachran
55 East Monroe St.
Chicago, IL 60603 (312) 726-4421

Daniel A. Boehnen
Allegretti, Newitt, Witcoff & Mc Andrews, Ltd.
125 S. Wacker Dr.
Chicago, IL 60606 (312) 372-2160

William J. Bohler
Willian, Brinks, Olds, Hofer, Gilson & Lione Ltd.
One I B M Plaza, Suite 4100
Chicago, IL 60611 (312) 822-9800

Jeffrey S. Boone
Akzo America Inc.
300 S. Riverside Plaza
Chicago, IL 60606 (312) 906-7590

Michael F. Borun
Marshall, O Toole, Gerstein, Murray & Bricknell
Two First National Plaza
Chicago, IL 60603 (312) 346-5750

George S. Bosy
Neuman, Williams, Anderson & Olson
77 W. Washington St.
Chicago, IL 60602 (312) 346-1200

Earhart Bouton
Amsted Industries, Inc.
205 N. Michigan Ave.
Boulevard Towers South-44th Fl.
Chicago, IL 60601 (312) 819-8481

John C. Brezina
Brezina & Buckingham P.C.
111 W. Jackson Blvd.
15th Fl.
Chicago, IL 60604 (312) 427-2922

Henry L. Brinks
Willian, Brinks, Olds, Hofer, Gilson & Lione Ltd.
One I B M Plaza, Suite 4100
Chicago, IL 60611 (312) 822-9800

Donald J. Brott
Marshall, O Toole, Gerstein, Murray & Bicknell
Two First National Plaza
Chicago, IL 60603 (312) 346-5750

Paul L. Brown
Emrich & Dithmar
Suite 3000
150 N. Wacker Dr.
Chicago, IL 60606 (312) 368-8575

Robert E. Browne, Sr.
Wallenstein, Wagner, Hattis, Strampel & Aubel, Ltd.
100 S. Wacker Dr.
Suite 2100
Chicago, IL 60606 (312) 641-1570

Albert J. Brunett
Hill, Van Santen, Steadman & Simpson
70th Fl.
Sears Tower
Chicago, IL 60606 (312) 876-0200

Thomas W. Buckman
Illinois Tool Works, Inc.
8501 W. Higgins Rd.
Chicago, IL 60631 (312) 693-3040

Angelo J. Bufalino
Lockwood, Alex, Fitz Gibbon & Cummings
Suite 1515
Three First National Plaza
Chicago, IL 60602 (312) 782-4860

George Edward Bullwinkel
Burditt, Bowles & Radzius, Ltd.
333 W. Wacker Dr.
Chicago, IL 60606 (312) 781-6667

Mark J. Buonaiuto
Mc Dermott, Will & Emery
111 West Monroe St.
Chicago, IL 60603 (312) 372-2000

Marshall A. Burmeister
Burmeister, York, Palmatier, Hamby & Jones
135 S. Lasalle St.
Chicago, IL 60603 (312) 782-6663

Patrick G. Burns
Welch & Katz, Ltd.
135 S. La Salle St.
Suite 1625
Chicago, IL 60603 (312) 781-9470

Richard Bushnell
Trexler, Bushnell, Grangiorgi & Blackstone, Ltd.
141 W. Jackson Blvd.
Chicago, IL 60604 (312) 427-8082

Joseph Peter Calabrese
Neuman, Williams, Anderson & Olson
77 W. Washington
Chicago, IL 60602 (312) 346-1200

Charles G. Call
Allegretti, Newitt, Witcoff & Mc Andrews, Inc.
125 S. Wacker Dr.
Chicago, IL 60606 (312) 372-2161

James V. Callahan
Allegretti, Newitt, Witcoff & Mc Andrews, Ltd.
125 S. Wacker Dr.
Chicago, IL 60606 (312) 372-2160

Kevin R. Casey
Allegretti, Newitt, Witcoff & Mc Andrews, Ltd.
125 S. Wacker Dr.
Chicago, IL 60606 (312) 372-2160

Myron C. Cass
Silverman, Cass, Singer & Winburn, Ltd.
105 W. Adams St.
27th Fl.
Chicago, IL 60603 (312) 726-6006

John Jerome Cavanaugh
Neuman, Williams, Anderson & Olson
77 W. Washington St.
Suite 2000
Chicago, IL 60602 (312) 346-1200

Richard A. Cederoth
Neuman, Williams, Anderson & Olson
77 W. Washington St.
Chicago, IL 60602 (312) 346-1200

Talivaldis Cepuritis
Dressler, Goldsmith, Shore, Sutker & Milnamow, Ltd.
1800 Prudential Plaza
Chicago, IL 60601 (312) 527-4025

Edward J. Chalfie
Epton, Mullin & Druth, Ltd.
140 South Dearborn St.
Suite 1200
Chicago, IL 60603 (312) 984-1000

Daniel R. Cherry
Welsh & Kratz, Ltd.
135 S. La Salle St.
Suite 1625
Chicago, IL 60603 (312) 781-9470

Ernest Cheslow
Dressler, Goldsmith, Shore, Sutker & Milnamow, Ltd.
1800 Prudential Plaza
Chicago, IL 60601 (312) 527-4025

John William Chestnut
Tilton, Fallon, Lungmus & Chestnut
100 S. Wacker Dr.
Suite 960
Chicago, IL 60606 (312) 263-1841

Davis Chin
105 W. Madison St.
Suite 1707
Chicago, IL 60602 (312) 726-6448

Daniel N. Christus
Wallenstein, Wagner, Hattis, Strampel & Aubel, Ltd.
100 S. Wacker Dr.
Suite 2100
Chicago, IL 60606 (312) 641-1570

John J. Chrystal
Ladas & Parry
104 S. Mich. Ave.
Chicago, IL 60603 (312) 236-9021

Jeffrey L. Clark
Wood, Dalton, Phillips, Mason & Rowe
20 N. Wacker Dr.
Suite 2200
Chicago, IL 60606 (312) 346-1630

James W. Clement
Clement & Ryan
150 N. Michigan Ave., Suite 1250
Chicago, IL 60601 (312) 663-1200

John L. Cline
Willian, Brinks, Olds, Hofer, Gilson & Lione Ltd.
One I B M Plaza , Suite 4100
Chicago, IL 60611 (312) 822-9800

Charles S. Cohen
Gordon & Glickson, P.C.
444 N. Michigan Avenue, Suite 3600
Chicago, IL 60611

Eric C. Cohen
Welsh & Katz, Ltd.
135 S. La Salle St.
Suite 1625
Chicago, IL 60603 (312) 781-9470

Paul A Coletti
Willian, Brinks, Olds, Gilso
Chicago, IL 60611

Paul A. Coletti
Willian, Brinks, Olds, Hofer,Gilson & Lione Ltd.
One I B M Plaza
Chicago, IL 60611 (312) 822-9800

John B. Conklin
Leydig, Voit & Mayer, Ltd.
One I B M Plaza, Suite 4600
Chicago, IL 60611 (312) 822-9666

James J. Conlon
Conlon & Kerstein
Suite 1220
205 W. Randolph St.
Chicago, IL 60606 (312) 726-0545

P. Phillips Connor
Hill, Van Santen, Chiara & Simpson
70th Fl., Sears Tower
233 S. Wacker Dr.
Chicago, IL 60601 (312) 876-0200

Robert F.I. Conte
Laff, Whitesel, Conte & Saret
401 N Michigan Avenue, Suite 2000
Chicago, IL 60611 (312) 661-2100

Granger Cook, Jr.
Cook, Wetzel & Egan, Ltd.
135 S. La Salle St.
Suite 3300
Chicago, IL 60603 (312) 236-8500

Ronald B. Coolley
Arnold, White, & Durkee
Quaker Tower, Suite 800
321 N Clark Street
Chicago, IL 60610 (312) 744-0090

Gordon Russell Coons
Leydig, Voit & Mayer, Ltd.
One I B M Plaza, Suite 4600
Chicago, IL 60611 (312) 822-9666

Terrace J. Coughlin
39 S. La Salle
Room 820
Chicago, IL 60603 (312) 332-7374

Lawrence J. Crain
Welsh & Kratz, Ltd.
135 S. La Salle St.
Suite 1625
Chicago, IL 60603 (312) 781-9470

Philip J. Crihfield
Sidley & Austin
One First National Plaza
Chicago, IL 60603 (312) 853-7650

John A. Crook
Willian, Brinks, Olds, Hofer, Gilson & Lione Ltd.
One I B M Plaza, Suite 4100
Chicago, IL 60611 (312) 822-9600

Wannell M. Crook
Willian, Brinks, Olds, Hofer, Gilson & Lione Ltd
One I B M Plaza, Suite 4100
Chicago, IL 60611 (312) 822-9600

John R. Crossan
Cook, Wetzel & Egan, Ltd.
135 S. La Salle St.
Chicago, IL 60603 (312) 236-8500

Eugene M. Cummings
Lockwood, Alex, Fitz Gibbon & Cummings
Suite 1515
3 First Natl. Plaza
Chicago, IL 60602 (312) 782-4860

Robert P. Cummins
Katten, Muchin & Zavis
525 West Monroe St.
Chicago, IL 60606 (312) 902-5525

Stanley C. Dalton
Wood, Dalton, Phillips, Mason & Rowe
20 N. Wacker Dr.
Suite 2200
Chicago, IL 60606 (312) 346-1630

Andreas M. Danckers#
Laff, Whitesel, conte & Saret
401 N Michigan Ave, Suite 2000
Chicago, IL 60611 (312) 611-2106

Rodney A. Daniel
Willian, Brinks, Olds, Hofer, Gilson & Lione Ltd
One I B M Plaza, Suite 4100
Chicago, IL 60611 (312) 822-9800

Howard Helseth Darbo
Lee, Smith & Zickert
150 S. Wacker Dr.
Suite 950
Chicago, IL 60606 (312) 726-1982

Britton A. Davis
Haight & Hofeldt
55 E. Monroe St.
Suite 3614
Chicago, IL 60603 (312) 263-2353

Basil Emanuel Demeur
Knechtel & Demeur
Suite 1925
20 N. Wacker Dr.
Chicago, IL 60606 (312) 726-5342

John D. Dewey
Lockwood, Alex, Fitz Gibbon & Cummings
3 First National Plaza
Chicago, IL 60602 (312) 782-4860

Richard E. Dick
Dick & Harris
200 W. Madison St.
Suite 3200
Chicago, IL 60602 (312) 726-4000

Robert M. Didrick
U S G Corp.
101 S. Wacker Dr., Dept. 157
Chicago, IL 60606 (312) 321-5805

Melinda Lois Dierstein
1337 W. Fargo Ave., Apt.11B
Chicago, IL 60626 (312) 465-7376

Thomas E. Dorn
Kinzer, Plyer, Dorn & Mc Eachran
55 E. Monroe St.
Chicago, IL 60603 (312) 236-1112

Vasilios D. Dossas
Neuman, Williams, Anderson & Olson
77 W. Washington St.
Chicago, IL 60602 (312) 346-1200

James R. Dowdall
Neuman, Williams, Anderson & Olson
77 W. Washington St.
Chicago, IL 60602 (312) 346-1200

W. Dennis Drehkoff
Tilton, Fallon, Lungmus & Chestnut
100 S. Wacker Dr.
Suite 960
Chicago, IL 60606 (312) 263-1841

Max Dressler
Dressler, Goldsmith, Shore, Sutker & Milnamow, Ltd.
1800 Prudential
Chicago, IL 60601 (312) 527-4025

Grantland G. Druchas
Willian, Brinks, Olds, Hofer, Gilson & Lione Ltd.
One I B M Plaza, Suite 4100
Chicago, IL 60611 (312) 822-9800

Jean M. Dudek
Mason, Kolehmainen, Rathburn & Wyss
20 N. Wacker Dr.
Suite 4200
Chicago, IL 60606 (312) 621-1300

Christine A. Dudzik#
Marshall, O Toole, Gerstein, Murray & Bicknell
20 South Clark
Suite 2100
Chicago, IL 60603 (312) 346-5750

Jeffery M. Duncan
Willian, Brinks, Olds, Hofer, Gilson & Lione Ltd
One I B M Plaza, Suite 4100
Chicago, IL 60611 (312) 822-9800

George F. Dvorak
Balogh, Osann, Kramer, Dvorak, Genov & Traub
53 W. Jackson Blvd.
Chicago, IL 60604 (312) 922-6262

Thomas E. Earle#
Amoco Corp.
Pats. & Licensing Dept.
Mail Code 1907A
200 E. Randolph Dr.
Chicago, IL 60601 (312) 856-5932

Donald E. Egan
Cook, Wetzel & Egan, Ltd.
135 S. La Salle St.
Chicago, IL 60603 (312) 236-8500

C. Lyman Emrich, Jr.
Emrich & Dithmar
150 N. Wacker Dr.
Chicago, IL 60606 (312) 368-8575

Ronald L. Engel#
360 E. Randolph
Chicago, IL 60601 (312) 565-4970

Patrick D. Ertel
Wood, Dalton, Phillips, Mason & Rowe
20 N. Wacker Dr.
Suite 2200
Chicago, IL 60606 (312) 346-1630

Francis A. Even
Fitch, Even, Tabin & Flannery
135 South Lasalle St.
Suite 900
Chicago, IL 60603 (312) 372-7842

Irving Faber
Faber & Cunniff, Ltd.
Suite 900
29 S. La Salle St.
Chicago, IL 60603 (312) 782-5140

Martin Faier
135 S. La Salle St.
Suite 4005
Chicago, IL 60603 (312) 332-2060

Jerome F. Fallon
Tilton, Fallon, Lungmus & Chestnut
209 South Lasalle St.
Chicago, IL 60604 (312) 263-1841

Mark I. Feldman
Rudnick & Wolfe
203 N. La Salle St.
Chicago, IL 60601 (312) 368-7084

Robert W. Fieseler
Neuman, Williams, Anderson & Olson
77 W. Washington St.
Chicago, IL 60602 (312) 346-1200

Thomas J. Filarski
Willian, Brinks, Olds, Hofer, Gilson & Lione Ltd.
One I B M Plaza, Suite 4100
Chicago, IL 60611 (312) 822-9800

Jacques John Filliung#
Kinser, Plyer, Dorn & Mc Eachran
55 E. Monroe St.
Chicago, IL 60603 (312) 726-4421

Morgan L. Fitch
South Chicago Saving Bank
9200 S. Commercial AVenue
Chicago, IL 60617 (312) 768-1400

Morgan Lewis Fitch, Jr.
Fitch, Even, Tabin & Flannery
135 So. Lasalle St.
Chicago, IL 60603 (312) 372-7842

James T. Fitz Gibbon
Lockwood, Alex, Fitz Gibbon & Cummings
3 First Natl. Plaza
Suite 1515
Chicago, IL 60602 (312) 346-6540

John Francis Flannery
Fitch, Even, Tabin & Flannery
135 S. Lasalle St.
Chicago, IL 60603 (312) 372-7842

Clarence J. Fleming
Jones, Day, Reavis & Pogue
225 W. Washington St.
Suite 2600
Chicago, IL 60606 (312) 782-3939

F. Frederick Fondriest
Amoco Corp.
200 E. Randolph Dr.
Chicago, IL 60601 (312) 856-3966

John Severin Fosse
30 N. La Salle St.
Suite 3030
Chicago, IL 60602 (312) 332-0080

William H. Frankel
Neuman, Williams, Anderson & Olson
77 W. Washington St.
Suite 2000
Chicago, IL 60602 (312) 346-1200

Roger J. French
Trexler, Bushnell, Giangiorgi & Blackstone, Ltd.
141 W. Jackson Blvd.
Chicago, IL 60604 (312) 427-8082

Eugene F. Friedman
One North La Salle St.
Suite 2025
Chicago, IL 60602 (312) 782-8882

James A. Gabala
Haight & Hofeldt
224 S. Michigan Ave.
Suite 600
Chicago, IL 60604 (312) 939-7909

Bruce M. Gagala
Leydig, Voit & Mayer
One I B M Plaza, Suite 4600
Chicago, IL 60611 (312) 822-9666

John W. Gaines
2300 Equitable Bldg
401 N. Michigan Avenue
Chicago, IL 60611 (312) 836-2312

Paul H. Gallagher
105 W. Adams St.
Chicago, IL 60603 (312) 332-7335

Priscilla F. Gallagher
Jones, Day, Reavis & Pogue
225 W. Washington
Chicago, IL 60606 (312) 782-3939

Edward P. Gamson
Dressler, Goldsmith, Shore, Sutker & Milnamow, Ltd.
1800 Prudential Plaza
Chicago, IL 60601 (312) 527-4025

John R. Garrett
Hill, Van Santen, Steadman & Simpson, P.C.
70th Fl.
Sears Tower
Chicago, IL 60606 (312) 876-0200

Stephen D. Geimer
Dressler, Goldsmith, Shore, Sutker & Milnamow, Ltd.
1800 Prudential Plaza
Chicago, IL 60601 (312) 527-4025

Henry J. Gens
7040 W School St.
Chicago, IL 60634 (312) 282-2967

Timothy H. Gens
Olson & Hierl
20 N. Wacker Dr.
Suite 3000
Chicago, IL 60606 (312) 580-1180

James Alan Geppert
Borg-Warner Corp.
200 S. Michigan Ave.
Chicago, IL 60604 (312) 322-8597

Gerald S. Geren
Hill, Van Santen, Steadman & Simpson
70th Fl.
Sears Tower
Chicago, IL 60606 (312) 876-0200

Allen H. Gerstein
Marshall, O Toole, Gerstein, Murray & Bicknell
Two First National Plaza
Suite 2100
Chicago, IL 60603 (312) 346-5750

Milton S. Gerstein#
6629 N. Francisco Avenue
Chicago, IL 60645 (312) 372-2926

George Henry Gerstman
Pigott & Gerstman, Ltd.
2 N. La Salle St.
Chicago, IL 60602 (312) 263-4350

Richard Allen Giangiorgi
Trexler, Bushnell, Giangiorgi & Blackstone, Ltd.
141 W. Jackson Blvd.
Suite 3440
Chicago, IL 60604 (312) 427-8082

Edward D. Gilhooly
20 N. Wacker Dr.
Suite 1925
Chicago, IL 60606 (312) 726-5342

Arthur Gwyer Gilkes
Leydig, Voit & Mayer, Ltd.
One I B M Plaza, Sutie 4600
Chicago, IL 60611 (312) 822-9666

Murray A. Gleeson
Mc Caleb, Lucas & Brugman
230 W. Monroe St.
Suite 2040
Chicago, IL 60606 (312) 236-4711

Jerome Goldberg
175 W. Jackson Blvd.
Suite 1629
Chicago, IL 60604 (312) 922-5031

Melvin M. Goldenberg
Jones, Day, Reavis & Pogue
225 W. Washington
Chicago, IL 60606 (312) 269-4104

R. Howard Goldsmith
Dressler, Goldsmith, Shore, Sutker & Milnamow, Ltd.
1800 Prudential Plaza
Chicago, IL 60601 (312) 527-4025

Frank P. Grassler
G.D. Searle & Co.
P.O. Box 5110
Chicago, IL 60680 (312) 470-6077

Richard O. Gray, Jr.
Winburn & Gray, Ltd.
Suite 1353
111 W. Washington St.
Chicago, IL 60602 (312) 346-7998

Raymond William Green
Willian, Brinks, Olds, Hofer, Gilson & Lione, Ltd.
One I B M Plaza, Suite 4100
Chicago, IL 60611 (312) 822-9800

Robert F. Green
Leydig, Voit & Mayer, Ltd.
One I B M Plaza, Suite 4600
Chicago, IL 60611 (312) 822-9666

Martin R. Greenstein
Baker & Mckenzie
2800 Prudential Plaza
Chicago, IL 60601 (312) 861-2770

Roger D. Greer
Welsh & Katz, Ltd.
135 S. La Salle St.
Suite 1625
Chicago, IL 60603 (312) 781-9470

Craig C. Groseth
Willian, Brinks, Olds, Hofer, Gilson & Lione, Ltd.
One I B M Plaza, Suite 4100
Chicago, IL 60611 (312) 822-9800

Dennis A. Gross
Hill, Van Santen, Steadman, & Simpson
233 S. Wacker Dr.
70th Fl
Chicago, IL 60606 (312) 876-0200

J. Arthur Gross
Hill, Van Santen, Steadman, & Simpson
70th Fl.
Sears Tower
Chicago, IL 60606 (312) 876-0200

Lewis S. Gruber
400 E. Randolph, Apt. No. 3911
Chicago, IL 60601 (312) 938-8740

Kevin W. Guynn
Hill, Van Santen, Steadman & Simpson, P.C.
70th Fl.
Sears Tower
233 S. Wacker Dr.
Chicago, IL 60606 (312) 876-0200

Stanton Thomas Hadley
USG Corp.
101 S. Wacker Dr.
Chicago, IL 60606 (312) 606-4000

Alan H. Haggard
Bell & Howell Co.
6800 McCormick Rd.
Chicago, IL 60645 (312) 676-7808

Edward A. Haight
Haight & Hofeldt
55 E. Monroe St.
Suite 3614
Chicago, IL 60603 (312) 263-2353

Timothy J. Haller
Niro, Scavone, Haller & Niro, Ltd.
200 W. Madison Plaza
Suite 3500
Chicago, IL 60606 (312) 236-0733

Jack Roger Halvorsen
Illinois Tool Works, Inc.
Patent Dept.
8501 W. Higgins Rd.
Chicago, IL 60631 (312) 693-3040

James John Hamill
Fitch, Even, Tabin & Flannery
135 S. La Salle St.
Suite 900
Chicago, IL 60603 (312) 372-7842

John W. Harbst#
Dressler, Goldsmith, Shore, Sutker & Milnamow, Ltd.
1800 Prudential Plaza
Chicago, IL 60601 (312) 527-4025

Robert L. Harmon
William, Brinks, Olds, Hofer, Gilson & Lione, Ltd.
One I B M Plaza, Suite 4100
Chicago, IL 60611 (312) 822-9800

James M. Harris
4800 S. Chicago Beach No. 2406N
Chicago, IL 60615 (312) 536-6772

Richard D. Harris
Dick & Harris
200 W. Madison St.
Suite 3200
Chicago, IL 60606 (312) 726-4000

Bruce Hart
401 E. 32nd Street, Apt. 1608
Chicago, IL 60616

Herbert D. Hart, III
Neuman, Williams, Anderson & Olson
77 W. Washington St.
Chicago, IL 60602 (312) 346-1200

H. Michael Hartmann
Leydig, Voit & Mayer, Ltd
One I B M Plaza , Suite 4600
Chicago, IL 60611 (312) 822-9800

Russell E. Hattis
Wallenstein, Wagner, Hattis, Strampel & Aubel
100 S. Wacker Dr.
21st Fl.
Chicago, IL 60606 (312) 641-1570

John J. Held, Jr.
Allegretti, Newitt, Witcoff & Mc Andrews, Ltd.
125 S. Wacker Dr.
Chicago, IL 60606 (312) 372-2160

Stephen B Heller
Leydig, Voit, Osann, Mayer & Holt, Ltd.
One I B M Plaza, Suite 4600
Chicago, IL 60611 (312) 822-9666

James R. Henes
Amoco Corp.
200 E. Randolph Dr.
Chicago, IL 60601 (312) 856-2808

Stephen L. Hensley
Amoco Corp.
200 East Randolph Dr.
Chicago, IL 60601 (312) 856-2764

Brett A. Hesterberg
Leydig, Voit, & Mayer
One I B M PLaza, Suite 4600
Chicago, IL 60611

Michael A. Hierl
Olson & Hierl
20 N. Wacker Dr.
Suite 3000
Chicago, IL 60606 (312) 580-1180

Joseph B. Higgs
Container Corp. Of Amer.
1st Natl. Plaza
Chicago, IL 60603 (312) 580-5413

James Joseph Hill
Emrich & Dithmar
150 N. Wacker Dr.
Suite 3000
Chicago, IL 60606 (312) 368-8575

Thomas E. Hill
Emrich & Dithmar
150 N. Wacker Dr.
Suite 3000
Chicago, IL 60606 (312) 368-8575

James Darwin Hobart
Hill, Van Santen, Steadman, Chiara & Simpson
70 th Fl.
Sears Tower
233 S. Wacker Dr.
Chicago, IL 60606 (312) 876-0200

James M. Hoey
Clausen, Miller, Gorman, Caffrey & Witous, P.C.
10 S. La Salle St.
Chicago, IL 60603 (312) 855-1010

Roy E. Hofer
Willian, Brinks, Olds, Hofer, Gilson & Lione, Ltd.
One I B M Plaza , Suite 4100
Chicago, IL 60611 (312) 822-9800

John R. Hoffman
Hosier, Niro & Daleiden, Ltd.
208 S. La Salle St.
Chicago, IL 60604 (312) 236-0733

Richard B. Hoffman
Tilton, Fallon, Lungmus & Chestnut
100 South Wacker Dr.
Suite 960
Chicago, IL 60606 (312) 263-1841

Thomas J. Hoffmann
Suite 1250
135 S. La Salle St.
Chicago, IL 60603 (312) 726-2800

Kevin D. Hogg
Marshall, O Toole, Gerstein, Murray & Bicknell
Two First National Plaza
Suite 2100
Chicago, IL 60603 (312) 346-5750

Matthew R. Hooper
Amoco Corporation
Pats. & Lic. Dept.
200 E. Randolph Dr.
Mail Code 1907
Chicago, IL 60601 (312) 856-5911

Allen Joseph Hoover
Dressler, Goldsmith, Shore, Sutker & Milnamow, Ltd.
1800 Prudential Plaza
Chicago, IL 60601 (312) 527-4025

Gerald D. Hosier
Hosier & Sufrin
100 S. Wacker Dr.
Suite 224
Chicago, IL 60606 (312) 726-1762

Joseph N. Hosteny, III
Niro, Scavone, Haller & Niro, Ltd.
200 W. Madison Plaza
Suite 3500
Chicago, IL 60606 (312) 236-0733

Clemens Hufmann
Mason, Kolehmainen, Rathburn & Wyss
20 N. Wacker Dr.
Suite 4200
Chicago, IL 60606 (312) 621-1300

Bradley J. Hulbert
Allegretti, Newitt, Witcoff & Mc Andrews, Ltd.
125 S. Wacker Dr.
Chicago, IL 60606 (312) 372-2160

James Pickrell Hume
Willian, Brinks, Olds, Hofer, Gilson,& Lione, Ltd.
One I B M PLaza
Chicago, IL 60611 (312) 822-9800

Charles J. Hunter
Quaker Oats Co.
345 Merchandise Mart Plaza
Chicago, IL 60654 (312) 222-7801

Kareem M. Irfan#
Arnold, White, & Durkee
321 N Clark Street, Suite 800
Chicago, IL 60610 (312) 744-0090

Jerold A. Jacover
Willian, Brink, Olds, Hofer, Gilson & Lione, Ltd.
One I B M Plaza, Suite 4100
Chicago, IL 60611 (312) 822-9800

Melvin F. Jager
Lee, Smith & Jager
Suite 950
150 S. Wacker Dr.
Chicago, IL 60606 (312) 726-1982

Robert V. Jambor
Kinzer, Plyer, Dorn, Mc Eachran & Jambor
55 E. Monroe St.
Suite 3905
Chicago, IL 60603 (312) 726-4421

Anthony J. Janiuk
Amoco Corp.
200 East Randolph Dr.
Mail Code 1907
Chicago, IL 60601 (312) 856-7972

L. Michael Jarvis
Allegretti, Newitt, Witcoff & Mc Andrews, Ltd.
125 S. Wacker Dr.
Suite 3100
Chicago, IL 60606 (312) 372-2160

James Joseph Jennings, Jr.
Borg-Warner Corp.
200 S. Michigan Ave.
Chicago, IL 60604 (312) 322-8575

Everett A. Johnson
333 N. Michigan Ave.
Chicago, IL 60601 (312) 782-4010

Harold V. Johnson
Willian, Brinks, Olds, Hofer, Gilson & Lione, Ltd.
One I B M Plaza, Suite 4100
Chicago, IL 60611 (312) 822-9800

Neal C. Johnson
International Harvester Co.
401 N Michigan Ave
Chicago, IL 60611 (312) 836-2313

Richard Leiter Johnston
135 S. Lasalle St., Rm. 1150
Chicago, IL 60603 (312) 236-2953

Robert Bruce Jones
Fitch, Even, Tabin & Flannery
135 S. Lasalle St.
Chicago, IL 60603 (312) 372-7842

Premkumar K. Joshi
1441 W. Farwell Ave. Apt 3B
Chicago, IL 60626 (312) 761-0720

Paul G. Juettner
Juettner, Pyle, Lloyd & Verbeck
221 North La Salle St., Suite 850
Chicago, IL 60601 (312) 236-8123

Thomas R. Juettner
Juettner, Pyle, Lloyd & Verbeck
221 North La Salle St., Suite 850
Chicago, IL 60601 (312) 236-8123

Konrad H. Kaeding
Willian, Brinks, Olds, Hofer, Gilson & Lowe
One I B M Plaza, Suite 4100
Chicago, IL 60611 (312) 822-9800

Robert L. Kahn
Suite 1200
189 W. Madison St.
Chicago, IL 60602 (312) 372-2552

Leonard J. Kalinowski#
Emrich, Lee, Brown & Hill
Suite 3000
150 N. Wacker Dr.
Chicago, IL 60606 (312) 368-8575

Ronald C. Kamp
Fmc Corp.
200 E. Randolph Dr.
Chicago, IL 60601 (312) 861-6655

Henry S. Kaplan
Dressler, Goldsmith, Shore, Sutker & Milnamow, Ltd.
1800 Prudential Plaza
Chicago, IL 60601 (312) 527-4025

Richard A. Kaplan
Hume, Clement, Brinks, Willian & Olds, Ltd.
One I B M Plaza, Suite 4100
Chicago, IL 60611 (312) 822-9800

Nathan N. Karus
6649 N. Maplewood Ave.
Chicago, IL 60645 (312) 262-5192

A. Sidney Katz
Welsh & Katz
135 S. La Salle St.
Suite 1625
Chicago, IL 60603 (312) 781-9470

Robert D. Katz
Hume, Clement, Brinks, Willian & Olds, Ltd.
One I B M Plaza, Suite 4100
Chicago, IL 60611 (312) 822-9800

David D. Kaufman
1800 Prudential Plaza
Chicago, IL 60601 (312) 527-4025

Edward M. Keating
Kinzer, Plyer, Dorn & Mceachran
55 E. Monroe St.
Mid-Continental Plaza
Chicago, IL 60603 (312) 726-4421

Esther O. Kegan
Kegan & Kegan
79 W. Monroe St.
Suite 1320
Chicago, IL 60603 (312) 782-6495

Dolores T. Kenney
Olson & Hierl
20 N. Wacker Dr.
Suite 3000
Chicago, IL 60606 (312) 580-1180

Evan M. Kent
Dressler, Goldsmith, Shore, Sutker & Milnamow, Ltd.
1800 Prudential Plaza
Chicago, IL 60601 (312) 527-4025

Paul A. Kerstein
Conlon & Kerstein
Suite 1220
205 W. Randolph St.
Chicago, IL 60606 (312) 726-0545

John Kilyk, Jr.#
Leydig, Voit & Mayer, Ltd.
One I B M Plaza, Suite 4100
Chicago, IL 60611 (312) 822-9666

Charles C. Kinne
Allegretti, Newitt, Witcoff & Mc Andrews
125 S. Wacker Dr.
Chicago, IL 60606 (312) 372-2160

Layton F. Kinney#
Sherwin-Williams Res. Ctr.
10909 S. Cottage Grove
Chicago, IL 60628 (312) 821-3559

James B. Kinzer
Kinzer, Plyer, Dorn & Mc Eachran
Suite 3611
55 E. Monroe St.
Chicago, IL 60603 (312) 236-1112

Jerome B. Klose
Merriam, Marshall & Bicknell
Suite 2100
2 1st Natl. Plaza
Chicago, IL 60603 (312) 346-5750

Thomas A. Kmiotek
2243 N. Mango Avenue
Chicago, IL 60639 (312) 622-3074

Robert Edward Knechtel
Knechtel, Valentino, Demeur & Dallas
Suite 4020
20 N. Wacker Dr.
Chicago, IL 60606 (312) 726-5342

Leonard S. Knox
2300 Lincoln Park W., Suite 707
Chicago, IL 60614 (312) 472-3325

Philip M. Kolehmainen
Mason, Kolehmainen, Rathburn & Wyss
20 N. Wacker Dr.
Chicago, IL 60606 (312) 621-1300

Ludwig E. Kolman
Phelan, Pope & John, Ltd.
180 N. Wacker Dr.
Suite 500
Chicago, IL 60606 (312) 621-0700

Fred Paul Kostka
Amsted Industries, Inc.
3700 Prudential Palza
Chicago, IL 60601 (312) 645-1695

Nick C. Kottis
Wood, Dalton, Phillips, Mason & Rowe
2200 Civic Opera Bldg.
20 N. Wacker Dr.
Chicago, IL 60606 (312) 346-1630

John W. Kozak
Leydig, Voit & Mayer, Ltd.
One I B M Plaza, Suite 4600
Chicago, IL 60611 (312) 822-9666

Jill H. Krafte
Leydig, Voit, Osann, Mayer & Holt, Ltd.
One I B M Plaza, Suite 4600
Chicago, IL 60611 (312) 833-9666

Otto Raymond Krause
Hill, Van Santen, Steadman, Chiara & Simpson
7000 Sears Tower
Chicago, IL 60606 (312) 876-0200

Richard A. Kretchmer
Standard Oil Co.
200 E. Randolph Dr.
Chicago, IL 60601 (312) 856-5921

Joseph Krieger
Mason, Kolehmainen, Rathburn & Wyss
Suite 3200
20 N. Wacker Dr.
Chicago, IL 60606 (312) 621-1300

Frederick J. Krubel
International Harvester Co.
401 N Michigan Avenue
Chicago, IL 60611 (312) 836-2313

Robert G. Krupka
Kirkland & Ellis
200 E. Randolph Dr.
Chicago, IL 60601 (312) 861-2156

Linda A. Kuczma
Wallenstein, Wagner, Hattis, Strampel & Aubel
100 S. Wacker Dr.
21st Fl.
Chicago, IL 60606 (312) 641-1570

Samuel Kurlandsky
U.S. Gypsum Co.
101 S. Wacker Dr.
Chicago, IL 60606 (312) 321-5811

Richard M. La Barge
Marshall, O toole, Gerstein, Murray & Bicknell
2 First National Plaza
Chicago, IL 60603 (312) 346-5750

Robert G. Ladd#
Amoco Corp.
200 E. Randolph Dr.
Chicago, IL 60601 (312) 856-5965

Charles A. Laff
Laff, Whitesel, Conte & Saret
401 N. Michigan Avenue, Suite 2000
Chicago, IL 60611 (312) 649-0200

Jay Calvin Langston, Jr.
Keil & Weinkauf
Suite 1216
135 S. La Salle St.
Chicago, IL 60603 (312) 265-5030

James V. Lapacek
S & C Electric Company
6601 N. Ridge Blvd.
Chicago, IL 60626 (312) 338-1000

Ronald E. Larson
Allegretti, Newitt, Witcoff & Mc Andrews, Ltd.
125 S. Wacker Dr.
Chicago, IL 60606 (312) 372-2160

Daniel W. Latham
Quaker Oats Co. 25-8
P.O. Box 9001
Chicago, IL 60604 (312) 222-7554

Jon B. Leaheey
Neuman, Williams, Anderson & Olson
77 W. Washington St.
Chicago, IL 60602 (312) 346-1200

George F. Lee
Emrich, Lee, Brown & Hill
150 N. Wacker Dr.
Chicago, IL 60606 (312) 368-8575

William M. Lee, Jr.
Lee, Smith & Jager
Suite 950
150 S. Wacker Dr.
Chicago, IL 60606 (312) 726-1982

Wm. Marshall Lee
Lee, Smith & Jager
Suite 950
150 S. Wacker Dr.
Chicago, IL 60606 (312) 726-1982

Edward Alan Lehman
Hill, Van Santen, Steadman, Chiara & Simpson
70th Fl.
Sears Tower
233 S. Wacker Dr.
Chicago, IL 60606 (312) 876-0200

Solomon J. Lehrer
22 W. Monroe St.
Chicago, IL 60603 (312) 281-8145

David Lesht
Lockwood, Alex, Fitz Gibbon & Cummings
Suite 1515
Three First National Plaza
Chicago, IL 60602 (312) 782-4860

Norman Lettvin
208 S. La Salle St.
Suite 1670
Chicago, IL 60604 (312) 782-8862

Russell E. Levine
Kirkland & Ellis
200 E. Randolph Dr.
Chicago, IL 60601 (312) 861-2466

Seymour Levine
5515 N. Virginia Ave
Chicago, IL 60625 (312) 878-0595

Timothy E. Levstik
Fitch, Even, Tabin Flannery & Welsh
Suite 900
135 S. La Salle St.
Chicago, IL 60603 (312) 372-7842

Harry M. Levy
Dithmar, Stotland, Stratman & Levy
Suite 1200
189 W. Madison St.
Chicago, IL 60602

C. Frederick Leydig, Jr.
Leydig, Voit & Mayer, Ltd.
One I B M Plaza, Suite 4600
Chicago, IL 60611 (312) 822-9666

Henry Lilienheim
Ladas & Parry
104 S. Michigan Ave.
Chicago, IL 60603 (312) 236-9021

Robert L. Lindgren
Mc Dougall, Hersh & Scott
Suite 1540
135 S. La Salle St.
Chicago, IL 60603 (312) 346-0338

Charles M. Lindrooth
Hill, Van Santen, Steadman, Chiara & Simpson
70th Fl.
Sears Tower
233 S. Wacker Dr.
Chicago, IL 60606 (312) 876-0200

John R. Linton
Natl. Assoc. Of Realtors
430 N Michigan Avenue
Chicago, IL 60611 (312) 329-8369

Richard G. Lione
Hume, Clement, Brinks, Willian & Olds, Ltd.
One I B M Plaza, Suite 4100
Chicago, IL 60611 (312) 822-9800

Steven G. Lisa#
Hosier & Sufrin, Ltd.
100 S. Wacker Dr.
Suite 224
Chicago, IL 60606 (312) 726-1762

Carl S. Lloyd
Kirkland & Ellis
Suite 5600
200 E. Randolph St.
Chicago, IL 60601 (312) 861-2090

Robert A. Lloyd
Juettner, Pyle, Lloyd & Verbeck
221 North La Salle St.
Suite 850
Chicago, IL 60601 (312) 236-8123

Fred Stark Lockwood
Lockwood, Alex, Fitz Gibbon & Cummings
Suite 1515
3 First Natl. Plaza
Chicago, IL 60602 (312) 782-4860

John K. Lucas
Willian, Brinks, Olds, Hofer, Gilson & Lione, Ltd.
One I B M Plaza, Suite 4100
Chicago, IL 60611 (312) 822-9800

William E. Lucas
McCaleb, Lucas & Brugman
230 W. Monroe St.
Chicago, IL 60606 (312) 236-4711

Peter S, Lucyshyn
A.B. Dick Co.
5700 W. Touhy Ave.
Chicago, IL 60648 (312) 763-1900

Van Metre Lund
Neuman, Williams, Anderson & Olson
77 W. Washington St.
Chicago, IL 60602 (312) 346-1200

John B. Lungmus
Tilton, Fallon, Lungmus & Chestnut
209 S. La Salle St.
Chicago, IL 60604 (312) 263-1841

Kathleen A. Lyons
Niro, Scavone, Haller & Niro
200 W. Madison
Suite 3500
Chicago, IL 60606 (312) 236-0733

William Howard Magidson
Standard Oil Co.
200 E. Randolph Drive
Chicago, IL 60680 (312) 856-3967

David W. Maher
Isham, Lincoln & Beale
19 S. La Salle St.
Chicago, IL 60603 (312) 558-5229

Timothy J. Malloy
Allegretti, Newitt, Witcoff & Mc Andrews, Ltd.
125 S. Wacker Dr.
Chicago, IL 60606 (312) 372-2160

Dale A. Malone#
Allegretti, Newitt, Witcoff & Mc Andrews, Ltd.
125 S. Wacker Dr.
Chicago, IL 60606 (312) 372-2160

Stephen J. Manich
Haight & Hofeldt
224 S. Michigan Ave.
Suite 600
Chicago, IL 60604 (312) 939-7909

Basil P. Mann
Merriam, Marshall & Bicknell
Two First National Plaza
Chicago, IL 60603 (312) 346-5750

John Mcgregor Mann, Sr.
Mann, Mc Williams, Zummer & Sweeney
175 W. Jackson Blvd.
Suite 1541
Chicago, IL 60604 (312) 427-1351

Edward Manzo
Cook, Wetzel & Egan, Ltd.
135 S. La Salle St.
Chicago, IL 60603 (312) 236-8500

Joseph Robert Marcus
Welsh & Katz, Ltd.
135 S. La Salle St., Suite 1625
Chicago, IL 60603 (312) 781-9470

Frank H. Marks
4940 E. End Ave.
Chicago, IL 60615 (312) 684-3124

William A. Marshall
Merriam, Marshall & Bicknell
Two First National Plaza
Suite 2100
Chicago, IL 60603 (312) 346-5750

William A. Marvin
Fitch, Even, Tabin & Flannery
135 S. La Salle
Suite 900
Chicago, IL 60603 (312) 372-7842

Lloyd W. Mason
Wegner, Mc Cord, Wood & Dalton
20 N. Wacker Dr.
Chicago, IL 60606 (312) 346-1630

Albert L. Matthews
180 N. La Salle St.
Chicago, IL 60601 (312) 726-9467

Mart C. Matthews
Quaker Oats Co.
Law Dept.
P.O. Box 9001
Suite 25-6
Chicago, IL 60604 (312) 222-7574

Phillip H. Mayer
Leydig, Voit & Mayer, Ltd.
One I B M Plaza, Suite 4600
Chicago, IL 60611 (312) 822-9666

George P. Mc Andrews
Allegretti, Newitt, Witcoff & Mc Andrews, Ltd.
125 S. Wacker Dr.
Suite 3100
Chicago, IL 60606 (312) 372-2160

William Thomas Mc Clain
Standard Oil Co.
200 E. Randolph Dr.
Chicago, IL 60601 (312) 856-5930

John B. Mc Cord
Wegner, Stellman, McCord, Wiles & Wood
20 N. Wacker Dr.
Rm. 2200
Chicago, IL 60606 (312) 346-1630

Michael D. Mc Coy
Neuman, Williams, Anderson & olson
77 W. Washington St.
Chicago, IL 60602 (312) 346-1200

William E. Mc Cracken
Wood, Dalton, Phillips, Mason & Rowe
20 North Wacker Dr.
Suite 2200
Chicago, IL 60606 (312) 346-1630

Dugald Stewart Mc Dougall
Mc Dougall, Hersh & Scott
135 S. La Salle St.
Suite 1540
Chicago, IL 60603 (312) 346-0338

Daniel C. Mc Eachran
Kinzer, Plyer, Dorn, & Mc Eachran
55 E. Monroe St.
Chicago, IL 60603 (312) 726-4421

Gary W. Mc Farron
Cook, Wetzel & Egan, Ltd.
135 S. La Salle St.
Suite 3300
Chicago, IL 60603 (312) 236-8500

Michael R. Mc Kenna
25 E. Washington
Suite 1221
Chicago, IL 60602 (312) 372-7777

Janice M. Mc Lain
Amoco Corp.
Mail Code 1906
200 East Randolph Dr.
P.O. Box 87703
Chicago, IL 60601 (312) 856-4138

F. William Mc Laughlin
Wood, Dalton, Phillips, Mason & Rowe
20 N. Wacker Dr.
Chicago, IL 60606 (312) 346-1630

Terrence W. Mc Millin
Marshall, O Toole, Gerstein, Murray & Bicknell
Two First National Plaza
Suite 2100
Chicago, IL 60603 (312) 346-5750

Warren D. Mc Phee
Mason, Kolehmainen, Rathburn & Wyss
20 N. Wacker Dr.
Chicago, IL 60606 (312) 621-1300

Dennis Michael Mc Williams
Mann, Mc Williams, Zummer & Sweeney
175 W. Jackson Blvd.
Suite 1215
Chicago, IL 60604 (312) 427-1351

Thomas Francis Mc Williams
Mc Williams, Mann & Zummer
53 W. Jackson Blvd.
Chicago, IL 60604 (312) 427-1351

Michael B. McMurray
Jenner & Block
One I B M Plaza
Chicago, IL 60611 (312) 229-350

Richard B. Megley
F M C Corp.
200 E. Randolph Dr.
Chicago, IL 60601 (312) 861-6650

Raymond M. Mehler
Lockwood, Alex, Fitz Gibbon & Cummings
Suite 1515
Three 1st Natl. Plaza
Chicago, IL 60602 (312) 782-4860

Maureen C. Meinert
Peterson, Ross, Schloerb & Seidel
200 E. Randolph Dr.
Suite 7300
Chicago, IL 60601 (312) 861-1400

Irving H. Melnick
Gerlach, O brien & Kleinke
29 South La Salle St.
Chicago, IL 60603 (312) 332-6930

Charles F. Meroni, Jr.
Hill, Van Santen, Steadman, Chiara & Simpson
70th Fl.
Sears Tower
Chicago, IL 60606 (312) 876-0200

Henriette Mertz
5425 E. View Park
Chicago, IL 60615 (312) 493-6078

Gerson E. Meyers
Dressler, Goldsmith, Shore, Sutker &
Milnamow, Ltd.
1800 Prudential Plaza
Chicago, IL 60601 (312) 527-4025

William J. Michals
ITT Midwest Patent Operations
100 S. Wacker Dr.
Suite 1530
Chicago, IL 60606 (312) 236-7373

Reuben Miller
9027 S. Luella Ave
Chicago, IL 60617 (312) 731-5049

John P. Milnamow
Dressler, Goldsmith, Shore, Sutker &
Milnamow, Ltd.
1800 Prudential Plaza
Chicago, IL 60601 (312) 527-4025

Dean A. Monco
Hill, Van Santen, Steadman, Chiara & Simpson,
P.C.
70th Fl.
Sears Tower
Chicago, IL 60606 (312) 876-0200

Donald D. Mondul
Illinois Tool Works, Inc.
8501 West Higins Rd.
Chicago, IL 60631 (312) 693-3040

Marvin Moody
Hill, Van Santen, Steadman, Chiara & Simpson
70th Fl.
Sears Tower
233 S. Wacker Dr.
Chicago, IL 60606 (312) 876-0200

Carl E. Moore, Jr.
Merriam, Marsall & Bicknell
Two First National Plaza
Suite 2100
Chicago, IL 60603 (312) 346-5750

Terry D. Morgan#
Arnold, White & Durkee
321 N. Clark Street, Suite 800
Chicago, IL 60610 (312) 902-3640

Jeffrey M. Morris
Mc Coy & Morris
Suite 1200
27 E. Monroe St.
Chicago, IL 60603 (312) 372-4390

Louis A. Morris
Akzo America Inc.
300 S. Riverside Plaza
Chicago, IL 60606 (312) 906-7589

William Joseph Morris
Morris & Stella
1950 Avondale Centre
20 North Clark St.
Chicago, IL 60602 (312) 782-2345

John S. Mortimer
Wegner, Mc Cord, Wood & Dalton
20 N. Wacker Dr.
Chicago, IL 60606 (312) 346-1630

Charles H. Mottier
Leydig, Voit & Mayer, Ltd.
One I B M Plaza, Suite 4600
Chicago, IL 60611 (312) 822-9666

Russell N. Muehleman#
Hill, Van Santen, Steadman, Chiara & Simpson
70th Fl.
Sears Tower
233 S. Wacker Dr.
Chicago, IL 60606 (312) 876-0200

Kurt Mullerheim#
Kurt & Witherspoon
135 S. La Salle St.
Chicago, IL 60603 (312) 236-5030

Edward W. Murray
Neuman, Williams, Anderson & Olson
77 West Washington St.
Chicago, IL 60602 (312) 346-1200

Owen Joseph Murray
Merriam, Marshall & Bicknell
Suite 2100
2 1st Natl. Plaza
Chicago, IL 60603 (312) 346-5750

William E. Murray
Standard Oil Co.
200 E. Randolph Dr.
Chicago, IL 60601 (312) 856-5990

James B. Muskal
Leydig, Voit & Mayer, Ltd.
One I B M Plaza, Suite 4600
Chicago, IL 60611 (312) 822-9666

Richard John Myers
Alexander & Zalewa, Inc.
900 Xerox Centre
55 W. Monroe St.
Chicago, IL 60603 (312) 726-7800

James J. Myrick
Fitch, Even, Tabin, Flannery & Welsh
135 S. La Salle St.
Chicago, IL 60603 (312) 372-7842

Wayne E. Nacker
Fitch, Even, Tabin, Flannery & Welsch
135 S. La Salle St., Rm. 900
Chicago, IL 60603 (312) 372-7842

John C. Nahrwold
American Bar Association
1155 E. 60th Street
Chicago, IL 60626 (312) 947-4118

James J. Napoli#
Mason, Kolehmainen, Rathburn & Wyss
20 North Wacker Dr.
Suite 4200
Chicago, IL 60606 (312) 621-1300

James P. Naughton
Neuman, Williams, Anderson & Olson
77 W. Washington St.
Chicago, IL 60602 (312) 346-1200

Jon O. Nelson
Allegretti, Newitt, Witcoff & Mc Andrews, Ltd.
125 S. Wacker Dr.
Chicago, IL 60606 (312) 372-2160

Sidney Neuman
Neuman, Williams, Anderson & Olson
Suite 2000
77 W. Washington St.
Chicago, IL 60602 (312) 346-1200

George B. Newitt
Allegretti, Newitt, Witcoff & Mc Andrews, Ltd.
125 S. Wacker Dr.
Chicago, IL 60606 (312) 372-2160

Gerald M. Newman
Schoenberg, Fisher & Newman, Ltd.
Suite 1600
222 S. Riverside Plaza
Chicago, IL 60606 (312) 648-2310

Mary Nicolaides
233 East Erie Street, Suite 1804
Chicago, IL 60611 (312) 337-1835

Keith K. Nicolls
Mc Caleb, Lucas & Brugman
230 W. Monroe St.
Chicago, IL 60606 (312) 236-4711

Raymond N. Nimrod
Neuman, Williams, Anderson & Olson
77 W. Washington
Chicago, IL 60602 (312) 346-1200

Raymond P. Niro
Niro, Scavone, Haller & Niro, Ltd.
200 W. Madison Plaza
Suite 3500
Chicago, IL 60606 (312) 236-0733

Steven H. Noll
Hill, Van Santen, Steadman, Chiara & Simpson, P. C.
70th Fl.
Sears Tower
Chicago, IL 60606 (312) 876-0200

Joan I. Norek
One North La Salle St.
Chicago, IL 60602 (312) 782-6025

Geoffrey M. Novelli
Trexler, Bushnell, Giangiorgi
141 W. Jackson Blvd.
Chicago, IL 60604 (312) 427-8082

John S. O Brien
Gerlach & O brien
29 S. La Salle St.
Suite 635
Chicago, IL 60603 (312) 332-6930

Joseph P. O Halloran
Quaker Oats Co.
Law Dept.-Pat. Sec. 25-8
P.O. Box 9001
Chicago, IL 60604 (312) 222-7111

Edward M. O Toole
Merriam, Marshall & Bicknell
Two First National Plaza
Chicago, IL 60603 (312) 346-5750

John P. O'Brien
Illinois Tool Works, Inc.
8501 W. Higgins Rd.
Chicago, IL 60631 (312) 693-3040

Margaret M. O'Brien
Laff, Whitesel, Conte & Saret
John Hancock Center, Suite 3430
Chicago, IL 60611 (312) 649-0200

Patrick J. O'Shea
5453 W. North Avenue
Chicago, IL 60639 (312) 622-3755

William P. Oberhardt
Neuman, Williams, Anderson & Olson
77 West Washington
Suite 2000
Chicago, IL 60602 (312) 346-1200

Helen A. Odar#
Kirkland & Ellis
200 East Randolph Dr.
Chicago, IL 60601 (312) 861-2000

Paul M. Odell
Dressler, Goldsmith, Shore, Sutker & Milnamow, Ltd.
1800 Prudential Plaza
Chicago, IL 60601 (312) 527-4025

Glenn W. Ohlson
Lee, Smith & Jager
Suite 950
150 S. Wacker Dr.
Chicago, IL 60606 (312) 726-1982

Dietmar H. Olesch
1141 West Morse Avenue
Chicago, IL 60626

Wallace L. Oliver, Jr.
Standard Oil Co.
200 E. Randolph Dr.
Chicago, IL 60601 (312) 856-5543

Arne M. Olson
Olson & Hierl
20 N. Wacker Dr.
Suite 3000
Chicago, IL 60606 (312) 580-1180

Arthur Andrew Olson, Jr.
Neuman, Williams, Anderson & Olson
77 W. Washington St.
Chicago, IL 60602 (312) 346-1200

Edward W. Osaan, Jr.
29 S. La Salle St.
Suite 420
Chicago, IL 60603 (312) 782-5937

Charles S. Oslakovic
Leydig, Voit & mayer, Ltd.
One I B M Plaza, Suite 4600
Chicago, IL 60611 (312) 822-9666

Algird R. Ostis
Container Corp. Of America
One First National Plaza
Chicago, IL 60603 (312) 580-5412

John S. Pacocha
Marvin Glass & Associates
815 N. LaSalle Street
Chicago, IL 60610 (312) 664-8855

Francois Newell Palmatier
Burmeister, York, Palmatier, Hamby & Jones
135 S. Lasalle St., Rm. 1046
Chicago, IL 60603 (312) 782-6663

John S. Paniaguas
Mason, Kolehmainen, Rathburn & Wyss
20 North Wacker Dr.
Suite 4200
Chicago, IL 60606 (312) 621-1300

Joseph H. Paquin, Jr.#
One I B M Plaza, Suite 4100
Chicago, IL 60611 (312) 822-9800

Boris Parad
International Harvester
401 N Michigan Avenue, Suite 2300
Chicago, IL 60611 (312) 836-2314

Todd S. Parkhurst
Schiff, Hardin & Waite
7200 Sears Tower
233 S. Wacker Dr.
Chicago, IL 60606 (312) 876-1000

Raymond E. Parks#
Fmc Corp.
200 E. Randolph Dr.
Chicago, IL 60601 (312) 861-6652

John A. Parrish
Jenner & Block
One I B M Plaza
Chicago, IL 60611 (312) 222-9350

Timothy T. Patula
Epton, Mullin & Druth, Ltd.
140 S. Dearborn St.
Chicago, IL 60603 (312) 984-1000

Thomas D. Paulius
Lockwood, Alex, Fitz Gibbon & Cummings
Three First National Plaza
Suite 1515
Chicago, IL 60602 (312) 782-4860

John J. Pavlak
Willian, Brinks, Olds, Hofer, Gilson & Lione, Ltd.
One I B M Plaza, Suite 4100
Chicago, IL 60611 (312) 822-9800

Louise S. Pearson
Kirkland & Ellis
200 E. Randolph Dr.
Chicago, IL 60601 (312) 861-2484

Joan Pennington
Mason, Kolehmainen, Rathburn & Wyss
20 N. Wacker Dr.
Suite 4200
Chicago, IL 60606 (312) 621-1300

Donald A. Peterson
Neuman, Williams, Anderson & Olson
77 W. Washington St.
Chicago, IL 60602 (312) 346-1200

Philip C. Peterson
Mason, Kolehmainen, Rathburn & Wyss
Rm. 4200
20 N. Wacker Dr.
Chicago, IL 60606 (312) 621-1312

Thomas F. Peterson
Ladas & Parry
104 S. Mich. Ave.
Chicago, IL 60603 (312) 236-9021

Ronald C. Petri
Standard Oil Co.
200 E. Randolph Dr.
Chicago, IL 60601 (312) 856-7747

Philip T. Petti
Neuman, Williams, Anderson & Olson
77 West Washington St.
Chicago, IL 60602 (312) 346-1200

Mark E. Phelps
Leydig, Voit & Mayer, Ltd.
One I B M Plaza, Suite 4600
Chicago, IL 60611 (312) 822-9666

Richard S. Phillips
Wood, Dalton, Phillips, Mason & Rowe
20 N. Wacker Dr.
Chicago, IL 60606 (312) 346-1630

Charles Francis Pigott, Jr.
Pigott, Gerstman & Gilhooly, Ltd.
2 North La Salle St.
Chicago, IL 60602 (312) 263-4350

R. Steven Pinkstaff#
Fitch, Even, Tabin & Flannery
135 S. La Salle St.
Chicago, IL 60603 (312) 372-7842

Daniel C. Pinkus
4752 W. Lake St.
Chicago, IL 60644 (312) 626-0487

Michael Piontek
221 North La Salle St., Suite 850
Chicago, IL 60601 (312) 236-8123

Alfred H. Plyer, Jr.
Kinzer, Plyer, Dorn & Mc Eachran
55 E. Monroe St.
Chicago, IL 60603 (312) 726-4421

Donald J. Pochopien#
Allegretti, Newitt, Witcoff & Mc Andrews, Ltd.
125 S. Wacker Dr.
Suite 3100
Chicago, IL 60606 (312) 372-2160

James W. Potthast
1210 Three Illinois Center
Chicago, IL 60601 (312) 565-1260

Nicholas A. Poulos
Neuman, Williams, Anderson & Olson
77 W. Washington St.
Chicago, IL 60602 (312) 346-1200

Russell Weston Pyle
Juettner, Pyle, Lloyd & Verbeck
221 North La Salle St.
Suite 850
Chicago, IL 60601 (312) 236-8123

Joseph R. Radzius
Burditt & Calkins
135 S. La Salle St.
Chicago, IL 60603 (312) 641-2121

Irving Shale Rappaport
Bally Manufacturing Corp.
President's Plaza Two
8700 W. Bryn Mawr Ave.
Chicago, IL 60631 (312) 399-1300

Ralph Robert Rath
Wallenstein, Wagner, Hattis, Strampel & Aubel, Ltd.
100 S. Wacker Dr.
Suite 2100
Chicago, IL 60606 (312) 641-1570

William E. Recktenwald
Wood, Dalton, Phillips, Mason & Rowe
20 N. Wacker Dr.
Suite 2200
Chicago, IL 60606 (312) 346-1630

Edward W. Remus
Allegretti, Newitt, Witcoff & Mc Andrews, Ltd.
125 S. Wacker Dr.
Suite 3100
Chicago, IL 60606 (312) 372-2160

Robert H. Resis
Allegretti, Newitt, Witcoff & Mc Andrews, Ltd.
125 S. Wacker Dr.
Chicago, IL 60606 (312) 372-2160

Donald P. Reynolds
Welsh & Katz, Ltd.
135 S. La Salle St.
Suite 1625
Chicago, IL 60603 (312) 781-9470

Jerry A. Riedinger
Allegretti, Newitt, Witcoff & Mc Andrews, Ltd.
125 S. Wacker Dr.
Chicago, IL 60606 (312) 372-2160

Daniel M. Riess
Lockwood, Alex, Fitz Gibbon & Cummings
Suite 1515
Three First National Plaza
Chicago, IL 60602 (312) 782-4860

William T. Rifkin
McDougall, Hersh & Scott
135 S. LaSalle St.
Chicago, IL 60645 (312) 346-0338

Reed F. Riley
Standard Oil Co.
200 E. Randolph Dr.
Chicago, IL 60601 (312) 856-5947

Thomas J. Ring
Three Illinois Center-Suite 1210
303 East Wacker Dr.
Chicago, IL 60601 (312) 663-1207

Ralph F. Risse#
Hill, Van Santen, Steadman & Simpson
Sears Tower
70th Fl.
Chicago, IL 60606 (312) 876-0200

W. William Ritt, Jr.#
F M C Corp.
200 E. Randolph Dr.
Chicago, IL 60601 (312) 861-6657

Richard A. Robbins
1482-H W. Summerdale Avenue
Chicago, IL 60644 (312) 878-0605

Kenneth E. Roberts
U.S. Gypsum Co.
Pat. Dept.
101 S. Wacker Dr.
Chicago, IL 60606 (312) 321-5806

Melvin A. Robinson
Hill, Van Santen, Steadman & Simpson
70th Fl.
Sears Tower
Chicago, IL 60606 (312) 876-0200

Robert H. Robinson
U.S. Gypsum Co.
101 S. Wacker Dr.
Chicago, IL 60606 (312) 321-5802

David I. Roche
Illinois Tool Works, Inc.
8501 W. Higgins Road
Chicago, IL 60631 (312) 693-3040

Keith Von Rockey
Mc Dougall, Hersh & Scott
135 S. La Salle St.
Chicago, IL 60603 (312) 346-0338

Howard B. Rockman
Laff, Whitesel, Conte & Saret
401 N. Michigan Avenue, Suite 2000
Chicago, IL 60611 (312) 649-0200

Amy L.H. Rockwell
Wallenstein, Wagner, Hattis, Strampel & Aubel, Ltd.
100 South Wacker Dr.
Suite 2100
Chicago, IL 60606 (312) 641-1570

Robert Lee Rohrback
Mason, Kolehmainen, Rathburn & Wyss
20 N. Wacker Dr.
Chicago, IL 60606 (312) 621-1330

Harry J. Roper
Neuman, Williams, Anderson & Olson
77 W. Washington St.
Chicago, IL 60602 (312) 346-1200

Gary M. Ropski
Willian, Brinks, Olds , Hofer, Gilson & Lione, Ltd.
One I B M PLaza, Suite 4100
Chicago, IL 60611 (312) 822-9800

John E. Rosenquist
Leydig, Voit & Mayer, Ltd.
One I B M Plaza, Suite 4600
Chicago, IL 60611 (312) 822-9666

Thomas I. Ross#
Hill, Van Santen, Steadman & Simpson
70th Fl.
Sears Tower
233 S. Wacker Dr.
Chicago, IL 60606 (312) 876-0200

Seymour Rothstein
Allegretti, Newitt, Witcoff & Mc Andrews, Ltd.
125 S. Wacker Dr.
Chicago, IL 60606 (312) 372-2160

Charles L. Rowe
Wood, Dalton, Phillips, Mason & Rowe
20 N. Wacker Dr.
Chicago, IL 60606 (312) 346-1630

Stephen G. Rudisill
Arnold, White & Durkee
Quaker Tower, Suite 800
321 N. Clark Street
Chicago, IL 60610 (312) 744-0090

Douglas W. Rudy
F M C Corp.
Mach. Pat., Tdmk & Lic. Dept.
200 East Randolph Dr.
Chicago, IL 60601 (312) 861-6654

Charles W. Rummler
Suite 960
100 South Wacker Dr.
Chicago, IL 60606 (312) 236-3418

Donald W. Rupert
Neuman, Williams, Anderson & Olson
77 West Washington St.
Chicago, IL 60602 (312) 346-1200

Charles W. Ryan
Clement & Ryan
Suite 2150
150 N. Michigan Ave.
Chicago, IL 60601 (312) 663-1200

James T. Ryan
Greenberger, Krauss & Jacobs
Suite 2700
180 N. La Salle St.
Chicago, IL 60601 (312) 346-1300

Robert C. Ryan
Allegretti, Newitt, Witcoff & Mc Andrews, Ltd.
125 S. Wacker Dr.
Chicago, IL 60606 (312) 372-2160

William A. Ryan
Continental Materials Corp.
Suite 2940
111 E. Wacker Dr.
Chicago, IL 60601 (312) 938-2222

James D. Ryndak
Jenner & Block
One I B M Plaza
Chicago, IL 60611 (312) 222-9350

James P. Ryther
Mc Dougall, Hersh & Scott
135 S. Lasalle St.
Chicago, IL 60603 (312) 346-0338

Alan B. Samlan
Suite 1925
20 N. Wacker Dr.
Chicago, IL 60606 (312) 726-5342

Ronald A. Sandler
Epton, Mullin & Druth, Ltd.
140 S. Dearborn St.
Chicago, IL 60603 (312) 984-1000

Larry L. Saret
Laff, Whitesel, Conte & Saret
401 N Michigan AVenue, Suite 2000
Chicago, IL 60611 (312) 661-2100

Lester J. Savit
Jones, Day, Reavis & Pogue
Suite 2600
225 W. Washington
Chicago, IL 60606 (312) 782-3939

William J. Scanlon
Fitch, Even, Tabin & Flannery
135 S. La Salle St.
Suite 900
Chicago, IL 60603 (312) 372-7842

Nate Frank Scarpelli
Marshall, O Toole, Gerstein, Murray & Bicknell
Two First Natl. Plaza
Chicago, IL 60603 (312) 346-5750

Thomas G. Scavone
Niro, Scavone, Haller & Niro, Ltd.
200 W. Madison Plaza
Suite 3500
Chicago, IL 60606 (312) 236-0733

Paul E. Schaafsma
Olson & Hierl
20 N. Wacker Dr.
Suite 3000
Chicago, IL 60606 (312) 580-1180

Julian Schachner
Borg-Warner Corp.
200 S. Mich. Ave.
Chicago, IL 60604 (312) 322-8577

John A Schaerli
Union Special Corp.
400 N Franklin St.
Chicago, IL 60610 (312) 266-4030

James A. Scheer
Welsh & Katz
135 S. La Salle
Chicago, IL 60603 (312) 781-9470

Leland P. Schermer
Mc Dermott, Will & Emery
111 W. Monroe St.
Chicago, IL 60603 (312) 372-2000

Dennis R. Schlemmer
Leydig, Voit & Mayer, Ltd.
One I B M Plaza, Suite 4600
Chicago, IL 60611 (312) 822-9666

James S. Schlifke
209 W. Jackson Blvd.
Suite 801
Chicago, IL 60606 (312) 322-1703

Stanley A. Schlitter
Kirkland & Ellis
200 E. Randolph Dr.
59th Fl.
Chicago, IL 60601 (312) 861-2201

Richard J. Schlott
Amoco Corporation
200 E. Randolph Drive, Mail Code 1904
P.O. Box 87703
Chicago, IL 60680 (312) 856-5071

Jerold B. Schnayer
Welsh & Katz, Ltd.
135 S. La Salle St.
Suite 1625
Chicago, IL 60603 (312) 781-9470

Homer J. Schneider
Leydig, Voit & Mayer, Ltd.
One I B M Plaza, Suite 4600
Chicago, IL 60611 (312) 822-9666

Robert J. Schneider
Mc Dermott, Will & Emery
111 W. Monroe St.
Chicago, IL 60603 (312) 372-2000

Richard A. Schnurr
Marshall, O Toole, Gerstein, Murray & Bicknell
Two First National Plaza
Suite 2100
Chicago, IL 60603 (312) 346-5750

Ekkehard Schoettle
Standard Oil Co.
200 East Randolph Dr.
Chicago, IL 60601 (312) 856-5637

Jerry A. Schulman
Epton, Mullin & Druth, Ltd.
140 S. Dearborn
Chicago, IL 60603 (312) 984-1000

Robert K. Schumacher
Fitch, Even, Tabin & Flannery
135 S. Lasalle St.
Suite 900
Chicago, IL 60603 (312) 372-7842

James J. Schumann
Fitch, Even, Tabin & Flannery
135 S. Lasalle St.
Suite 900
Chicago, IL 60603 (312) 372-7842

Gerald S. Schur
Welsh & Katz, Ltd.
135 S. La Salle St.
Suite 1625
Chicago, IL 60603 (312) 781-9470

Richard J. Schwarz
Hill, Van Santen, Steadman & Simpson
233 S. Wacker Dr.
Chicago, IL 60606 (312) 876-0200

Robert J. Schwarz
Lockwood, Alex, Fitz Gibbon & Cummings
Suite 1515
Three First National Plaza
Chicago, IL 60602 (312) 782-4860

Theodore R. Scott
Jones, Day, Reavis & Pogue
225 W. Washington St.
Chicago, IL 60606 (312) 269-4103

Max Shaftal
200 W. Madison
Suite 3200
Chicago, IL 60606 (312) 726-4000

Steven M. Shape
Niro, Scavone, Haller & Niro, Ltd.
200 W. Madison Plaza
Suite 3500
Chicago, IL 60606 (312) 236-0733

Jeffrey S. Sharp
Marshall, O Toole, Gerstein, Murray & Bicknall
Two First National Plaza
Suite 2100
Chicago, IL 60603 (312) 346-5750

Gerald T. Shekleton
Welsh & Katz, Ltd.
135 S. La Salle St.
Suite 1625
Chicago, IL 60603 (312) 781-9470

Berton Scott Sheppard
Leydig, Voit & Mayer, Ltd.
One I B M Plaza, Suite 4600
Chicago, IL 60611 (312) 822-9666

Richard J. Sheridan
Morton Thiokol, Inc.
110 N. Wacker Dr.
Chicago, IL 60606 (312) 621-3480

Edward S. Sherman#
Union Carbide Corporation
Films-Packaging Division
6733 West 65th Street
Chicago, IL 60638 (312) 496-4735

John Howland Sherman
Hill, Van Santen, Steadman & Simpson
70th Fl. Sears Tower
233 S. Wacker Dr.
Chicago, IL 60606 (312) 876-0200

Stephen F. Sherry
Allegretti, Newitt, Witcoff & Mc Andrews, Ltd.
125 S. Wacker Dr.
Chicago, IL 60606 (312) 372-2160

Charles W. Shifley
Allegretti, Newitt, Witcoff & Mc Andrews, Ltd.
125 S. Wacker Dr.
Chicago, IL 60606 (312) 372-2160

Joseph E. Shipley
Fitch, Even, Tabin & Flannery
135 S. La Salle St.
Suite 900
Chicago, IL 60603 (312) 372-7842

Jack Shore
Dressler, Goldsmith, Shore, Sutker & Milnamow
1800 Prudential Plaza
Chicago, IL 60601 (312) 527-4025

Tony T. Shu
Lee & Shu
208 S. La Salle St.
Suite 1400
Chicago, IL 60604 (312) 641-3303

Alvin D. Shulman
Marshall, O Toole, Gerstein, Murray & Bicknell
Two First National Plaza
Chicago, IL 60603 (312) 346-5750

John H. Shurtleff
Suite 411
140 S. Dearborn St.
Chicago, IL 60603 (312) 236-5032

Steven P. Shurtz
Willian, Brinks, Olds, Hofer, Gilson & Lione, Ltd.
One I B M Plaza, Suite 4100
Chicago, IL 60611 (312) 822-9800

Joel E. Siegel
Dressler, Goldsmith, Shore, Sutker & Milnamow, Ltd.
1800 Prudential Plaza
Chicago, IL 60601 (312) 527-4025

Gustavo Siller, Jr.*
William, Brinks, Olds, Hofer, Gilson & Lione, Ltd.
One I B M Plaza, Suite 4100
Chicago, IL 60611 (312) 822-9800

Howard E. Silverman
Dick & Harris
200 W. Madison
Suite 3200
Chicago, IL 60606 (312) 726-4000

James C. Simmons
Morton Thiokol, Inc.
Pats. & Tdmk. Dept.
110 N. Wacker Dr.
Chicago, IL 60606 (312) 807-2191

John D. Simpson
Hill, Van Santen, Steadman, & Simpson
70th Fl.
Sears Tower
233 S. Wacker Dr.
Chicago, IL 60606 (312) 876-0200

Herbert Jay Singer
Silverman, Cass & Singer, Ltd.
105 W. Adams St.
Chicago, IL 60603 (312) 726-6006

David H. Sitrick#
Welsh & Katz, Ltd.
135 S. La Salle St.
Suite 1625
Chicago, IL 60603 (312) 781-9470

Robert W. Slater
Mc Dougall, Hersh & Scott
135 South La Salle St.
Chicago, IL 60603 (312) 346-0338

Robert E. Sloat
Standard Oil Company
200 E. Randolph Drive, M.C. 1906
P.O. Box 5910-A
Chicago, IL 60680 (312) 856-5275

Jamie S. Smith
Allegretti, Newitt, Witcoff & Mc Andrews, Ltd.
125 S. Wacker Dr.
Chicago, IL 60606 (312) 372-2160

Noel Irving Smith
Neuman, Williams, Anderson & Olson
77 W. Washington St.
Chicago, IL 60602 (312) 346-1200

Thomas Eugene Smith
Lee, Smith & Zickert
150 S. Wacker Dr.
Chicago, IL 60606 (312) 726-1982

Marvin Smoolar
Laff, Whitesel, Conte & Saret
401 N. Michigan Ave.
Chicago, IL 60611 (312) 649-0200

William A. Snow
100 South Wacker Dr.
Hartford Plaza
Chicago, IL 60606 (312) 236-3418

James R. Sobieraj
Willian, Brinks, Olds, Hofer, Gilson & Lione, Ltd.
One I B M Plaza, Suite 4100
Chicago, IL 60611 (312) 822-9800

Julius L. Solomon#
2400 Lakeview, Suite 1508
Chicago, IL 60614 (312) 525-0425

Joseph M. Sorrentino
Dressler, Goldsmith, Shore, Sutker & Milnamow, Ltd.
1800 Prudential Plaza
Chicago, IL 60601 (312) 527-4025

Steven J. Soucar
Dressler, Goldsmith, Shore, Sutker & Milnamow, Ltd.
1800 Prudential Plaza
Chicago, IL 60601 (312) 527-4025

Ann W. Speckman
33 North La Salle St.
Chicago, IL 60602 (312) 332-7267

Thomas William Speckman
Speckman & Trittipo
33 N. La Salle St.
Chicago, IL 60602 (312) 332-7266

George S. Spindler
Standard Oil Co.
200 E. Randolph Dr.
M C 1004
Chicago, IL 60601 (312) 856-5420

James Alexander Sprowl
Fitch, Even, Tabin & Flannery
Rm. 900
135 S. La Salle St.
Chicago, IL 60603 (312) 372-7842

Frank J. Sroka
Standard Oil
200 E. Randolph Dr.
Chicago, IL 60601 (312) 856-5939

Thomas Albert Stansbury
Mason, Albright & Stansbury
180 N. La Salle St.
Chicago, IL 60601 (312) 726-4141

James G. Staples
Baker & Mc Kenzie
130 E. Randolph Dr.
Chicago, IL 60601 (312) 861-2766

Lewis T. Steadman
Hill, Van Santen, Steadman & Simpson
70th Fl.
Sears Tower
233 S. Wacker Dr.
Chicago, IL 60606 (312) 876-0200

Edward J. Steeve#
7122 N. Odell
Chicago, IL 60631 (312) 631-1781

Roger H. Stein
Neuman, Williams, Anderson & Olson
77 W. Washington
Suite 2000
Chicago, IL 60602 (312) 346-1200

Martin L. Stern
Laff, Whitesel, Conter & Saret
401 N. Michigan Avenue, Suite 2000
Chicago, IL 60611 (312) 661-2100

Allan J. Sterstein
Willian, Brinks, Olds, Hofer, Gilson & Lione, Ltd.
One I B M Plaza, Suite 4600
Chicago, IL 60611 (312) 822-9800

Thomas K. Stine
Wood, Dalton, Phillips, Mason & Rowe
20 N. Wack Dr.
Suite 2200
Chicago, IL 60606 (312) 346-1630

Harold Victor Stotland
Emrich & Dithmar
Suite 3000
150 N. Wacker Dr.
Chicago, IL 60606 (312) 372-2552

Harry V. Strampel
Wallenstein, Wagner, Hattis, Strampel & Aubel
100 S. Wacker Dr.
21st Fl.
Chicago, IL 60606 (312) 641-1570

J. Terry Stratman
Emrich & Dithmar
Suite 3000
150 N. Wacker Dr.
Chicago, IL 60606 (312) 372-2552

William A. Streff, Jr.
Kirkland & Ellis
200 E. Randolph Dr.
59th Fl.
Chicago, IL 60601 (312) 861-2126

Richard J. Streit
Ladas & Parry
104 S. Mich. Ave.
Chicago, IL 60603 (312) 236-9021

William Creighton Stueber
Hill, Van Santen, Steadman, Chiara & Simpson
70th Fl.
Sears Tower
233 S. Wacker Dr.
Chicago, IL 60606 (312) 876-0200

Barry W. Sufrin
Hosier & Sufrin, Ltd.
100 South Wacker Dr.
Suite 224
Chicago, IL 60606 (312) 726-1762

Dennis K. Sullivan
Internatioal Harvester Co.
401 N Michigan Ave.
Chicago, IL 60611 (312) 836-2311

Marshall W. Sutker
Dressler, Goldsmith, Shore, Sutker & Milnamow, Ltd.
1800 Prudential Plaza
Chicago, IL 60601 (312) 527-4025

Philip C. Swain
Kirkland & Ellis
200 E. Randolph Dr.
Chicago, IL 60601 (312) 861-2000

James R. Sweeney
Mann, Mc Williams, Zummer & Sweeney
175 W. Jackson Blvd.
Suite 1215
Chicago, IL 60604 (312) 427-7900

Slawomir Z. Szczepanski
Willian, Brinks, Olds, Hofer, Gilson & Lione, Ltd.
One I B M Plaza, Suite 4100
Chicago, IL 60611 (312) 822-9800

Julius Tabin
Fitch, Even, Tabin & Flannery
135 S. Lasalle St.
Chicago, IL 60603 (312) 372-7842

Marina A. Tanzer
Ladas & Parry
104 S. Michigan Ave.
Chicago, IL 60603 (312) 236-9021

Jay G. Taylor
Haight & Hofeldt
224 S. Michigan Ave.
Suite 600
Chicago, IL 60604 (312) 939-7909

David R. Terrill
852 W. Buckingham Place
Chicago, IL 60657 (312) 525-7458

Frank R. Thienpont
230 W. Monroe St.
Chicago, IL 60606 (312) 236-4711

Andrew L. Tiajoloff
53 W. Jackson Blvd.
Suite 1616
Chicago, IL 60604 (312) 922-6262

Timothy L. Tilton
Tilton, Fallon, Lungmus & Chestnut
100 S. Wacker Dr.
Suite 960
Chicago, IL 60604 (312) 263-1841

Stanley J. Tomsa
Mason, Kolehmainen, Rathburn & Wyss
20 N. Wacker Dr.
Suite 4200
Chicago, IL 60606 (312) 621-1300

James Earl Tracy
Borg-Warner Corp.
200 S. Mich. Ave.
Chicago, IL 60604 (312) 322-8594

Richard R. Trexler
Trexler, Bushnell & Wolters, Ltd.
141 W. Jackson Blvd.
Chicago, IL 60600 (312) 427-8082

Maria P. Tungol
Amoco Corporation
Pats. & Licensing Dept.
200 E. Randolph Drive
P.O. Box 5910-A
Chicago, IL 60680 (312) 856-5946

Alan L. Unikel
Alexander, Unikel, Bloom, Zalewa & Tenenbaum, Ltd.
55 W. Monroe St.
Suite 900
Chicago, IL 60603 (312) 726-7800

Charles C. Valauskas
Karon, Morrison & Savikas, Ltd.
5720 Sears Tower
233 South Wacker Dr.
Chicago, IL 60606 (312) 876-6000

Brett A. Valiquet
Hill, Van Santen, Steadman, Chiara, Simpson & Simpson
Sears Tower
70th Fl.
Chicago, IL 60606 (312) 876-0200

James Van Santen
Hill, Van Santen, Steadman, Chiara & Simpson
70th Fl. Sears Tower
233 S. Wacker Dr.
Chicago, IL 60606 (312) 876-0200

William A. Van Santen, Jr.
Wood, Dalton, Phillips, Mason & Rowe
20 N. Wacker Dr.
Suite 2200
Chicago, IL 60606 (312) 346-1630

Paul M. Vargo
Dressler, Goldsmith, Shore, Sutker & Milnamow Ltd.
1800 Prudential Plaza
E. Randolph St.
Chicago, IL 60601 (312) 527-4025

Maynard P. Venema
Mid-America Legal Foundation
20 N. Wacker Dr.
Chicago, IL 60606 (312) 263-5163

George E. Verhage
U S G Corp.
101 S. Wacker Dr.
Chicago, IL 60606 (312) 321-5801

Timothy J. Vezeau#
Mason, Kolehmainen, Rathburn & Wyss
20 N. Wacker Dr.
Suite 4200
Chicago, IL 60606 (312) 621-1300

David H. Vickrey
Akzo American Inc.
300 S. Riverside Plaza
Chicago, IL 60606 (312) 786-0400

Robert A. Vitale
Niro, Scavone, Haller & Niro, Ltd.
200 W. Madison St.
Suite 3500
Chicago, IL 60606 (312) 236-0733

Daniel W. Vittum, Jr.
Kirkland & Ellis
200 E. Randolph Dr.
Chicago, IL 60601 (312) 861-2160

Gregory J. Vogler
Allegretti, Newitt, Witcofff & McAndrews, Ltd
10 S. Wacker Dr.
Chicago, IL 60606 (312) 372-2160

Richard L. Voit
Leydig, Voit & Mayer, Ltd.
One I B M Plaza, Suite 4600
Chicago, IL 60611 (312) 822-9666

Charles F. Voytech
Haight, Hofeldt, Davis & Jambor
224 S. Michigan Ave.
Chicago, IL 60604 (312) 263-2353

Robert E. Wagner
Wallenstein, Wagner, Hattis, Strampel & Aubel, Ltd.
100 S. Wacker Dr. Suite 2100
Chicago, IL 60606 (312) 641-1570

Robert J. Wagner
Amoco Corporation
200 E. Randolph Dr.
Chicago, IL 60680 (312) 856-5941

Richard B. Wakely
Fitch, Even, Tabin & Flannery
135 S. La Salle St.
Suite 900
Chicago, IL 60603 (312) 372-7842

Sidney Wallenstein
Wallenstein, Wagner, Hattis, Strample & Aubel, Ltd.
100 Wacker Dr., Suite 2100
Chicago, IL 60606 (312) 641-1570

Gomer Winston Walters
Haight & Hofeldt
224 S. Michigan Ave.
Suite 600
Chicago, IL 60604 (312) 939-7909

Ronald Lee Wanke
Jenner & Block
One I B M Plaza, 44th Floor
Chicago, IL 60611 (312) 222-9350

Robert M. Ward
Cook, Wetzel & Egan, Ltd.
135 S. La Salle St.
Suite 3300
Chicago, IL 60603 (312) 236-8500

Michael O. Warnecke
Neumann, Williams, Anderson & Olson
77 W. Washington St.
Chicago, IL 60602 (312) 346-1200

Phillip H. Watt
Fitch, Even, Tabin & Flannery
135 S. La Salle St.
Chicago, IL 60603 (312) 372-7842

William A. Webb
Willian, Brinks, Olds, Hofer, Gilson & Lione, Ltd.
One I B M Plaza, Suite 4100
Chicago, IL 60611 (312) 822-9800

Ernest A. Wegner
Wood, Dalton, Phillips, Mason & Rowe
20 N. Wacker Dr., Suite 2200
Chicago, IL 60606 (312) 346-1630

Sandra B. Weiss
Jones, Day, Reavis, & Pogue
225 W, Washington St., 26th Fl.
Chicago, IL 60606 (312) 782-3939

Donald L. Welsh
Welsh & Katz, Ltd.
135 S. La Salle St.
Suite 1625
Chicago, IL 60603 (312) 781-9470

William M. Wesley
Neuman, Williams, Anderson & Olson
77 W. Washington St.
20th Fl.
Chicago, IL 60602 (312) 346-1200

James Michael Wetzel
Cook, Wetzel & Egan, Ltd.
135 S. La Salle St.
Suite 3300
Chicago, IL 60603 (312) 236-8500

George F. Wheeler
Morton Thiokol, Inc.
110 N. Wacker Dr.
Chicago, IL 60606 (312) 807-2183

Gerald K. White
Morton Thinkol, Inc.
110 N. Wacker Dr.
Chicago, IL 60606 (312) 807-2186

J. Warren Whitesel
Laff, Whitesel, Conte & Saret
401 N. Michigan Avenue, Suite 2000
Chicago, IL 60611 (312) 649-2100

Christopher L. Wight
Willian, Brinks, Olds, Hofer, Gilson & Lione, Ltd.
One I B M Plaza, Suite 4100
Chicago, IL 60611 (312) 822-9800

Bradford Wiles
Wood, Dalton, Phillips, Mason & Rowe
20 N. Wacker Dr., Suite 2200
Chicago, IL 60606 (312) 346-1630

Keith B. Willhelm
Leydig, Voit & Mayer, Ltd.
One I B M Plaza, Suite 4600
Chicago, IL 60611 (312) 822-9666

James Talbot Williams
Neuman, Williams, Anderson & Olson
77 W. Washington St.
20th St.
Chicago, IL 60602 (312) 346-1200

Dennis J. Williamson#
Mc Dougall, Hersh & Scott
135 S. La Salle St.
Chicago, IL 60603 (312) 346-0338

Clyde F. Willian
Willian, Brinks, Olds, Hofer, Gilson & Lione, Ltd.
One I B M Plaza, Suite 4100
Chicago, IL 60611 (312) 822-9800

John T. Winburn
Silverman, Cass & Singer, Ltd.
105 W. Adams St.
27th Fl.
Chicago, IL 60603 (312) 726-6006

Glen P. Winton
Welsh & Katz, Ltd.
135 S. La Salle St.
Suite 1625
Chicago, IL 60603 (312) 781-9470

David L. Witcoff
Kirkland & Ellis
200 E. Randolph Dr.
Chicago, IL 60601

Sheldon W. Witcoff
Allegretti, Newitt, Witcoff & McAndrews, Ltd.
125 S. Wacker Dr.
Chicago, IL 60606 (312) 372-2160

James C. Wood
Wood, Dalton, Philllips, Mason & Rowe
20 N. Wacker Dr., Suite 2200
Chicago, IL 60606 (312) 346-1630

Richard L. Wood
Welsh & Katz, Ltd.
135 S. La Salle St.
Chicago, !L 60603 (312) 781-9470

A. Lewis Worthem, Jr.
USG Corp.
101 S. Wacker Dr.
Chicago, IL 60606 (312) 606-4000

John S. Wrona
13351 Baltimore Ave.
Chicago, IL 60633 (312) 646-0022

Walther E. Wyss
Mason, Kolehmainen, Rathburn & Wyss
20 N. Wacker Dr.
Chicago, IL 60606 (312) 621-1300

Foster York
Burmeister, York, Palmatier, Hamby & Jones
135 S. La Salle St.
Chicago, IL 60603 (312) 782-6663

Richard W. Young
Kirkland & Ellis
Suite 5900
200 E. Randolph Dr.
Chicago, IL 60601 (312) 861-2290

Richard A. Zachar
Pope, Ballard, Sheppard & Fowle, Ltd.
69 W. Washington St.
Suite 3200
Chicago, IL 60602 (312) 630-4225

Philip J. Zadeik
Baker & Mc Kenzie
2800 Prudential Plaza
Chicago, IL 60601 (312) 861-2852

James D. Zalewa
Alexander, Unikel, Bloom, Zalewa & Tennenbaum, Ltd.
55 W. Monroe St.
Chicago, IL 60602 (312) 726-7800

Mari-Kathleen F. Zaraza
5634 N. Kerbs Ave.
Chicago, IL 60646 (312) 777-5598

James P. Zeller
Marshall, O Toole, Gerstein, Murray & Bicknell
Two First Natl. Plaza
20 South Clark St.
Chicago, IL 60603 (312) 346-5750

Lloyd L. Zickert
Smith, Lee, & Jager
150 S Wacker Drive
Chicago, IL 60606 (312) 726-1982

Robert L. Zieg
Borg-Warner Corp.
200 S. Michigan Ave.
Chicago, IL 60604 (312) 322-8595

Anthony S. Zummer
Mann, Mc Williams, Zummer & Sweeney
175 W. Jackson Blvd.
Suite 1215
Chicago, IL 60604 (312) 427-1351

Don W. Weber
212 E. Main Street
Collinsville, IL 62234 (618) 345-8424

Wayne Morris Russell
344 Division St.
Crete, IL 60417 (312) 672-5568

Floyd Boberg Harman
406 Mchenry Ave.
Crystal Lake, IL 60014 (815) 459-0153

Hideo Tomomatsu#
987 Darlington Lane
Crystal Lake, IL 60014 (815) 459-2162

Russel D. Acton
1219 Robinson St.
Danville, IL 61832 (217) 446-1945

Charles A. Minne#
2223 Denmark Road
Danville, IL 61832 (217) 443-6403

Augustus G. Douvas
1606 Coachmans Rd.
Darien, IL 60559 (312) 985-6122

A. Samuel Oddi
Northern Illinois University
College of Law
De Kalb, IL 60115 (815) 753-1980

Philip L. Bateman
Samuels, Miller, Schroeder, Jackson & Sly
406 Citizens Bldg.
Decatur, IL 62523 (217) 429-4325

Michael F. Campbell#
A.E. Staley Manufacturing Co.
2200 East Eldorado St.
Decatur, IL 62525 (217) 421-2613

John F. Dunn
325 Millikin Court
Decatur, IL 62523 (217) 429-4000

Gary W. Granzow
A. E. Staley Manufacuring Co.
Horizon Chem. Div.
2200 E. Eldorado St.
Decatur, IL 62525 (217) 421-2711

James B. Guffey
A.E. Staley Manuf. Co., Pat. Dept.
2200 E. Eldorado St.
Decatur, IL 62525 (217) 421-2640

Charles J. Meyerson
175 Park Place
Decatur, IL 62522 (217) 432-4097

Lori D. Tolly#
2282 Valley View Place
Decatur, IL 62522 (217) 425-1702

John Daniel Wood
A.E. Staley Manufacturing Co.
2200 East Eldorado Street
Decatur, IL 62521 (217) 432-4411

Robert A. Benziger
Travenol Laboratories, Inc.
One Baxter Parkway
Deerfield, IL 60015 (312) 291-4184

Michael P. Bucklo
Baxter Travenol Laboratories, Inc.
One Baxter Parkway
Deerfield, IL 60015 (312) 948-2422

John A. Caruso
Baster Travenol Labs., Inc.
Law Dept. 3-2 E
One Baxter Pkwy.
Deerfield, IL 60015 (312) 948-4905

Barry L. Clark
1759 We-Go Trail
Deerfield, IL 60015 (312) 945-1932

James F. Coffee
320 Earls Ct.
Deerfield, IL 60015 (312) 945-4479

Susan B. Fentress
Baxter Health Care Corp.
1 Baxter Parkway
Deerfield, IL 60015 (312) 948-3149

Paul C. Flattery
Baxter Healthcare Corp.
One Baxter Pkwy.
Deerfield, IL 60015 (312) 948-4940

William B. Graham
Baxter Healthcare Corp.
One Baxter Pkwy.
Deerfield, IL 60015 (312) 948-3000

Robert E. Hartenberger
Baxter Healthcare Corp.
One Baxter Pkwy.
Deerfield, IL 60015 (312) 948-9779

Marjorie Decou Hunter
Baxter Travenol Labs., Inc.
One Baxter Pkwy.
Deerfield, IL 60015 (312) 948-4934

Mary R. Jankousky
Baxter Healthcare Corp.
One Baxter Pkwy.
Deerfield, IL 60015 (312) 948-4283

John P. Kirby, Jr.
Baxter Travenol Labs., Inc.
Law Dept.-Bldg. 2-2E
One Baxter Pkwy.
Deerfield, IL 60015 (312) 948-4942

Charles R. Mattenson
Baxter Travenol Labs., Inc.
One Baxter Parkway
Deerfield, IL 60015 (312) 948-4928

Bradford R.L. Price
Baxter Travenol Labs., Inc.
Law Dept., Bldg. 2-2E
One Baxter Parkway
Deerfield, IL 60015 (312) 948-4948

Daniel D. Ryan, III
Baxter Travenol Labs., Inc.
One Baxter Parkway
D F 2-2E
Deerfield, IL 60015 (312) 948-4928

Thomas R. Schuman
Baxter Travenol Labs., Inc.
Law Dept.
One Baxter Pkwy.
Deerfield, IL 60015 (312) 948-4946

Maynard L. Youngs
Travenol Labs., Inc.
One Baxter Pkwy.
Deerfield, IL 60015 (312) 948-4916

Harold W. Bergendorf
U O P Inc.
Algonquin & Mt. Prospect Rds.
Box 5017
Des Plaines, IL 60017 (312) 391-2518

George J. Cannon
Xerox Corp.
3000 Des Plaines Ave.
Des Plaines, IL 60018 (312) 635-2240

Richard J. Cordovano
U O P Inc.
Algonquin & Mt. Prospect Rds.
Des Plaines, IL 60017 (312) 391-2024

John Glenn Cutts, Jr.
U O P, Inc.
Box 5017
Des Plaines, IL 60017 (312) 391-2035

J.H. Hall
Allied Signal-Uop, Inc.
25 E. Algonquin Rd.
Des Plaines, IL 60016 (312) 391-2033

Ronald H. Hausch
Uop Incorporated
25 E. Algonquin Rd.
Des Plaines, IL 60017 (312) 391-2516

Thomas F. Lysaught, Jr.
GTE Directories Corp.
1865 Miner St.
Des Plaines, IL 60016 (312) 391-5122

Thomas Kerr Mc Bride
U O P Inc.
10 U O P Plaza
Algonquin & Mt. Prospect Rds.
Des Plaines, IL 60016 (312) 391-2018

Richard R. Morris
U O P Inc.
20 U O P Plaza
Algonquin & Mt. Prospect Rd.
Des Plaines, IL 60016 (312) 391-2030

Raymond Harry Nelson
U O P Inc.
10 U O P Plaza
Algonquin & Mt. Prospect Rd.
Des Plaines, IL 60016 (312) 391-2032

Kevin M. O Brien#
U O P Inc.
10 U O P Plaza
Des Plaines, IL 60016 (312) 391-2021

William H. Page, II
U O P Inc.
10 U O P Plaza
Des Plaines, IL 60016 (312) 391-2012

William E. Parry
U O P Inc.
Box 5017
Des Plaines, IL 60017 (312) 391-2521

Kenneth J. Pedersen
1411 Van Buren Ave.
Des Plaines, IL 60018 (312) 635-8967

Paul F. Pedigo#
U O P Inc.
10 U O P Plaza
Des Plaines, IL 60016 (312) 391-2020

Eugene I. Snyder
Signal U O P Group
10 U O P Plaza
Des Plaines, IL 60016 (312) 391-2061

John F. Spears, Jr.
Signal U O P Group
10 U O P Plaza
Des Plaines, IL 60016 (312) 391-2037

Mark O. Thomas
905 Center St., Suite 405
Des Plaines, IL 60016 (312) 293-1923

Harold N. Wells, Jr.
Uop Inc.
25 E. Algonquin Rd.
Des Plaines, IL 60017 (312) 391-2000

Frederick Morrow Arbuckle#
Arbuckle & Associates, Ltd.
1160 Barneswood Dr.
Downers Grove, IL 60515 (312) 852-9730

Anthony M. Berardi
4140 Main St.
Downers Grove, IL 60515 (312) 964-8820

George R. Clark
1501 Almond Ct.
Downers Grove, IL 60515 (312) 852-7754

Carmen B. Patti
Rockwell International
Legal Dept.
P.O. Box 1494
Downers Grove, IL 60515 (312) 960-8055

Bruce E. Burdick
Olin Corporation, T-189
Shamrock Street
East Alton, IL 62024 (618) 258-2362

William C. Grabarek
443 W. Pierce St.
P.O. Drawer G
Elburn, IL 60119 (312) 365-5404

Douglas J. Scheflow
Scheflow, Rydell & Travis
63 Douglas Ave.
Suite 200
Elgin, IL 60120 (312) 695-2800

Edward C. Vandenburgh, III
6988 S. Pleasant Hill Rd.
Elizabeth, IL 61028 (312) 277-5598

Stephen A. Kozich
Videojet Systems International
2200 Arthur Ave.
Elk Grove Village, IL 60007 (312) 593-8800

Robert F. Van Epps
Caluwaert, Panegasser & Van Epps
579 W. North Ave.
Suite 201
Elmhurst, IL 60126 (312) 279-7300

Donald L. Barbeau#
Biomega
1830 Sherman Ave.
Evanston, IL 60201 (312) 869-6003

David C. Hannum
840 A Forest Ave.
Evanston, IL 60202 (312) 937-4686

Callard Livingston
P.O. Box 591
Evanston, IL 60204 (312) 866-7100

Fred Tuttle Williams
2425 Central St.
Evanston, IL 60201 (312) 491-6131

Frank J. Uxa, Jr.
3019 Mac Heath Crescent
Flossmoor, IL 60426 (312) 799-5965

Willis J. Jensen
Duo-Fast Corp.
3702 N. River Rd.
Franklin Park, IL 60131 (312) 678-0100

Robert D. Teichert
Ekco Housewares Co.
9234 W. Belmont Ave.
Franklin Park, IL 60131 (312) 678-8600

Philip Jerome Zrimsek
Micro Switch
A Div. Of Honeywell Inc.
Freeport, IL 61032 (312) 763-1900

Allen L. Landmeier
Smith & Landmeier, P.C.
15 N. Second St.
P.O. Box 127
Geneva, IL 60134 (312) 232-2880

John L. Schmitt
P.O. Box 656
Geneva, IL 60134 (312) 232-1244

Kenneth Todd Snow, Sr.
East End Dr.
Box 175
Gilberts, IL 60136 (312) 426-3363

J. Leo Astrup#
21 W. 604 Monticello
Glen Ellyn, IL 60137 (312) 469-3055

George H. Lee
572 Hickory Rd.
Glen Ellyn, IL 60137 (312) 469-2044

Roger W. Nolan, Jr.
621 Linden
Glen Ellyn, IL 60137 (312) 469-4492

Norman M. Shapiro
1198 Royal Glen Dr.
Glen Ellyn, IL 60137 (312) 932-0148

Donald W. Carlin
Kraft Inc.
Kraft Ct.
Glenview, IL 60025 (312) 998-2488

Ralph E. Clarke, Jr.#
Zenith Electronics Corp.
1000 N. Milwaukee Ave.
Glenview, IL 60025 (312) 391-8099

John Harding Coult
Zenith Electronics Corp.
1000 Milwaukee Ave.
Glenview, IL 60025 (312) 391-8015

Florian S. Gregorczyk
Illinois Tool Works, Inc.
3650 W. Lake Ave.
Glenview, IL 60025 (312) 657-4848

Fred L. Johnson
Kraft Corp.
Kraft Court
Glenview, IL 60025 (312) 998-2475

Jack Kail
Zenith Radio Corp.
Pat. Dept.
1000 Milwaukee Ave.
Glenview, IL 60025 (312) 391-7000

John J. Kowalik
411 Crabtree Lane
Glenview, IL 60025

Ronald H. Kullick
Kraft, Inc. Corp.
Kraft Ct.
Glenview, IL 60025 (312) 998-2493

Cornelius J. O Connor
Zenith Radio Corp.
1000 Milwaukee Ave.
Glenview, IL 60025 (312) 391-8001

Paul W. O Malley, Jr.
712 Carriage Hill Dr.
Glenview, IL 60025 (312) 724-1550

A. Andrew Olson, III
2105 Linneman St.
Glenview, IL 60025 (312) 729-3614

John J. Pederson
Zenith Radio Corp.
1000 Milwaukee Ave.
Glenview, IL 60025 (312) 391-7995

Irving D. Ross, Jr.
Signode Corp.
3600 W. Lake Ave.
Glenview, IL 60025 (312) 724-6100

Benjamin Schlosser
330 Michael Manor
Glenview, IL 60025 (312) 966-0062

Charles J. Sindelar
Zenith Electronics Corp.
1000 Milwaukee Ave.
Glenview, IL 60025 (312) 671-7550

Edward Leonard Benno
17960 W. Hwy 120
Grays Lake, IL 60030 (312) 223-4906

John G. Heimovics
1958 Mc Craren Rd.
Highland Park, IL 60035 (312) 831-2548

Thomas W. Tolpin
969 Judson Ave.
Highland Park, IL 60035 (312) 433-2556

Stephen Z. Weiss
1354 Sunnyside
Highland Park, IL 60035 (312) 831-2548

James Ramsey Hoatson, Jr.
Six Godair Park
Hinsdale, IL 60521 (312) 323-2624

Edward Ptacek
17 W. 367 S. Frontage Rd.
Hinsdale, IL 60521 (312) 655-0977

Jack L. Uretsky
206 N. Grant
Hinsdale, IL 60521 (312) 323-5990

Arthur H. Bransky
18205 Hart Dr.
Homewood, IL 60430 (312) 799-4028

Edwin C. Lehner
1224 Hillview Rd.
Homewood, IL 60430 (312) 798-5764

L.A. Combs
6508 Blackhawk Trail
Indian Head Park, IL 60545 (312) 246-3293

Ernest S. Kettelson
Suite 212
57 N. Ottawa St.
Joliet, IL 60431 (815) 727-4735

John L. Parker
1900 Douglas St.
Joliet, IL 60435 (815) 725-4725

Richard T. Lauterbach#
632 S. Tenth Ave.
La Grange, IL 60525 (312) 352-5332

Harold R. Schwappach
717 Terry Lane
La Grange, IL 60525 (312) 352-5325

John R. Diver
868 Larchmont Lane
Lake Forest, IL 60045 (312) 234-8314

Dorothy R. Thumler#
551 S. Beverly Pl.
Lake Forest, IL 60045 (312) 295-2551

Philip Hill
3256 Ridge Rd.
P.O. Box 187
Lansing, IL 60438 (312) 895-4404

Joseph W. Holloway
Rt. 3, Box 175
Liberty, IL 62347 (217) 656-4355

Paul David Burgauer#
1110 Woodview Dr.
Libertyville, IL 60048 (312) 362-0034

William Elliott Dominick
1260 Lake St.
Libertyville, IL 60048 (312) 362-8612

Gildo E. Fato
515 Ash St.
Libertyville, IL 60048 (312) 362-0567

Robert E. O Neill*
1641 N. Milwaukee Ave.
Libertyville, IL 60048 (312) 367-5180

Joy Ann G. Serauskas#
One Fairfax Lane
Lincolnshire, IL 60015 (312) 945-2169

Thomas E. Torphy
Tenneco Automotive
100 Tri State International, Suite 300
Lincolnshire, IL 60015 (312) 940-6037

David L. Neer
7329 Keeler
Lincolnwood, IL 60646 (312) 679-5607

John W. Cornell
Molex Incorporated
2222 Wellington Ct.
Lisle, IL 60532 (312) 969-4550

Louis A. Hecht
Molex Inc.
2222 Wellington Ct.
Lisle, IL 60532 (312) 969-4550

John Leary#
Western Electric Co.
2600 Warrenville Rd.
Lisle, IL 60532 (312) 260-4540

Russell L. Mc Ilwain
1830 Middleton Ave.
Lisle, IL 60532 (312) 964-7767

Robert P. Miller
Western Electric Co., Inc.
2600 Warrenville Rd.
Lisle, IL 60532 (312) 260-4524

Edward J. Brosius
13058 Thistle Ct.
Lockport, IL 60441 (312) 460-0032

Morando Berrettini
200 W. 22nd
Lombard, IL 60148 (312) 691-1222

Marvin M. Chaban
1067 Apple Lane
Lombard, IL 60148 (312) 629-8126

Jon Carl Gealow
2903 N. Bay View Lane
Mc Henry, IL 60050 (815) 385-2329

Terry R. Mohr
Mohr, Lewis & Reilly
420 N. Front St.
Mc Henry, IL 60050 (815) 385-1313

Mark A. Appleton
Laff, Whitesel, Conte & Saret
P.O. Box 250
Moline, IL 61265 (309) 762-8568

Joel S. Carter
Deere & Co.
John Deere Road
Moline, IL 61265 (309) 752-6221

Duane A. Coordes
Deere & Co.
John Deere Road
Moline, IL 61265 (309) 752-4383

Charles L. Dennis, II.
Deere & Company
John Deere Road
Moline, IL 61265 (309) 765-5615

John O. Hayes
Deere & Co.
John Deere Road
Moline, IL 61265 (309) 765-4967

Michael S. Hlavaty
Deere & Co.
John Deere Rd.
Moline, IL 61265 (309) 765-4232

Raymond L. Hollister
Deere & Co.
John Deere Road
Moline, IL 61265 (309) 765-4451

Kevin J. Moriarty
Deere & Company
John Deere Road
Moline, IL 61265 (309) 752-4048

William A. Murray
Deere & Co.
John Deere Rd.
Moline, IL 61265 (309) 752-4421

John M. Nolan
Deere & Co.
Administrative Center
John Deere Rd.
Moline, IL 61265 (309) 752-4371

Jimmie Ralph Oaks
Deere And Co.
John Deere Road
Moline, IL 61265 (309) 752-4392

Robert T. Payne
Deere And Company
John Deere Road
Moline, IL 61629

John G. Tolomei
Deere & Company
John Deere Road
Moline, IL 61265 (309) 752-4465

Andrew J. Bootz
504 S. Albert
Mt. Prospect, IL 60056 (312) 255-6280

James P. Gaughan
1068 Mt. Prospect Plaza
Mt. Prospect, IL 60056 (312) 398-7779

Ray Edward Snyder
Tower Center
200 E. Evergreen
Mt. Prospect, IL 60056 (312) 398-1525

Robert William Welch
22 South William
Mt. Prospect, IL 60056 (312) 259-3071

Samuel Shiber#
P.O. Box 371
Mundelein, IL 60060 (312) 949-0424

Rae K. Stuhlmacher
19 Edgemont
Mundelein, IL 60060 (312) 566-7442

Donald J. Breh
6 S 200 Country Dr.
Naperville, IL 60540 (312) 357-3664

Aubrey L. Burgess
1191 Banbury Circle
Naperville, IL 60540 (312) 961-2470

Donald G. Epple
Nalco Chemical Co.
1 Nalco Center
Naperville, IL 60566 (312) 961-9500

Richard J. Godlewski
AT&T Bell Laboratories
Rm 6B-220
Naperville-Wheaton Rd.
Naperville, IL 60566 (312) 979-2004

Thomas J. Goodwin
1428 Briarwood Dr.
Naperville, IL 60540 (312) 357-5184

Frederick S. Jerome
Amoco Corp.
P.O. Box 400
Warrenvile Rd. & Mill St.
Naperville, IL 60566 (312) 420-5456

Mathew L. Kalinowski#
734 S. Sleight St.
Naperville, IL 60540 (312) 355-1504

William H. Kamstra
Bell Telephone Laboratories, Inc.
Naperville-Wheaton Rd.
Naperville, IL 60566 (312) 979-2005

Louis H. Le Mieux
Nalco Chemical Co.
One Nalco Center
Naperville, IL 60566 (312) 961-9500

Bruce R. Mansfield
400 Olesen Dr.
Naperville, IL 60540 (312) 420-1608

Carole A. Mickelson#
Standard Oil Co.
Mail Station H-7
P.O. Box 400
Naperville, IL 60566 (312) 420-4966

Robert A. Miller
Nalco Chemical Co.
One Nalco Center
Naperville, IL 60566 (312) 961-9500

John C. Moran
Bell Telephone Laboratories, Inc.
Naperville-Wheaton Rd.
Naperville, IL 60566 (312) 979-2001

Frederick W. Padden
AT&T Bell Laboratories
Naperville-Wheaton Rd.
Naperville, IL 60566 (312) 979-4637

Kenneth H. Samples
AT&T Bell Laboratories
Naperville-Wheaton Rd.
Naperville, IL 60566 (312) 979-2006

Werner Ulrich
AT&T Bell Laboratories
Naperville-Wheaton Rd.
Naperville, IL 60566 (312) 979-3255

Peter Visserman
AT&T Bell Laboratories
Naperville-Wheaton Rd.
Naperville, IL 60566 (312) 979-5036

David Volejnicek
AT&T Bell Laboratories
Naperville-Wheaton Rd.
Naperville, IL 60566 (312) 979-2155

Ross T. Watland
AT&T Bell Laboratories
Naperville-Wheaton Rd.
Naperville, IL 60566 (313) 979-2003

Herman Wissenberg#
28 W411 87th St.
Naperville, IL 60540 (312) 420-2343

Fred Gordon Thelander
209 April Ln.
North Aurora, IL 60542 (312) 896-8306

Martin Lewis Katz
Abbott Laboratories
14th St. & Sheridan Rd.
North Chicago, IL 60064 (312) 937-6364

Michael J. Roth
Abbott Laboratories
14th St. & Sheridan Rd.
North Chicago, IL 60064 (312) 937-6366

Dennis K. Shelton
Abbott Laboratories
Pat. & Tdmk Dept. D-377
14th St. & Sheridan Rd.
North Chicago, IL 60064 (312) 937-6365

Steven F. Weinstock
Abbott Laboratories
Abbott Park
North Chicago, IL 60064 (312) 937-6367

Peter Andress, Jr.
International Minerals & Chemical Corp.
2315 Sanders Rd.
Northbrook, IL 60062 (312) 564-8600

Robert A. Brown
2530 Shannon Rd.
P.O. Box 2127
Northbrook, IL 60065 (312) 272-3182

Lawrence William Brugman
2625 Techny Rd., Apt. 712
Northbrook, IL 60062

John A. Doninger
Premark International Inc.
2211 Sanders Rd.
Northbrook, IL 60062 (312) 498-8486

Thomas L. Farquer
International Minerals & Chemical Corp.
2315 Sanders Rd.
Northbrook, IL 60062 (312) 205-2268

Sidney Norman Fox
555 Skokie Blvd.
Northbrook, IL 60062 (312) 498-3322

S. David Hoffman
Underwriters Labs., Inc.
333 Pfingsten Rd.
Northbrook, IL 60062 (312) 272-8800

Gregory J. Mancuso
Dart & Kraft, Inc.
Div. Pat. Coun.
2211 Sanders Rd.
Northbrook, IL 60062 (312) 498-8477

Joyce R. Niblack
Niblack & Niblack, P.C.
555 Skokie Blvd., Suite 205
Northbrook, IL 60062 (312) 291-9900

Robert L. Niblack
Niblack & Niblack, P.C.
555 Skokie Blvd., Suite 205
Northbrook, IL 60062 (312) 291-9900

George M. Perry
2530 Crabtree Lane
Northbrook, IL 60062

Howard E. Post
Internatl. Minn. & Chems. Corp.
2315 Sanders Rd.
Northbrook, IL 60062 (312) 564-8600

Leigh Bannister Taylor
Dart & Kraft, Inc.
Gen. Pat. Coun.
2211 Sanders Rd.
Northbrook, IL 60062 (312) 498-8470

James E. Wolber
Internatl. Minerals & Chem. Corp.
2315 Sanders Rd.
Northbrook, IL 60062 (312) 564-8600

William Garrettson Ellis
635 Woodland Lane
Northfield, IL 60093 (312) 446-7234

Norman R. Smith
Stephan Co.
22 Frontage Rd.
Northfield, IL 60093 (312) 446-7500

Andrew Frank Zikas
Stephan Co.
22 W. Frontage Rd.
Northfield, IL 60093 (312) 446-7500

Joseph M. Gartner
5 Baybrook Ct.
Oak Brook, IL 60521 (312) 654-0826

Kenneth Wayne Hadland
Mc Donald S Corp.
Mc Donald S Plaza
Oak Brook, IL 60521 (312) 887-3538

Julius F. Harms
1st. Natl. Bank & Trust Co. Of Oakbrook
One Mc Donald S Plaza
Oak Brook, IL 60521 (312) 654-2030

James F. Lambe
Nalco Chemical Co.
2901 Butterfield Rd.
Oak Brook, IL 60521 (312) 887-7500

Edward T. Mc Cabe
Swift-Eckrich, Inc.
1919 Swift Dr.
Oak Brook, IL 60522 (312) 574-7015

Edward J. Mooney, Jr.
Nalco Chemical Co.
2901 Butterfield Rd.
Oak Brook, IL 60521 (312) 887-7500

William Patrick Porcelli
Interlake, Inc.
2015 Spring Rd.
Oak Brook, IL 60521 (312) 986-6652

John G. Premo
Nalco Chemical Co.
2901 Butterfield Rd.
Oak Brook, IL 60521 (312) 887-7500

Neil M. Rose
Sunbeam Corp.
2001 South York Rd.
Oak Brook, IL 60521 (312) 850-5290

Edmond T. Patnaude
Fidler, Patnaude & Batz
Terrace Executive Center Court C
1 S 376 Summit Ave.
Oak Brook Terrace, IL 60181 (312) 627-4552

James N. Videbeck
Patnaude, Batz & Videbeck
1 S. 376 Summit Ave.
Court C
Oak Brook Terrace, IL 60181 (312) 627-4552

Robert L. Chandler#
9704 S. Kenneth Ave.
Oak Lawn, IL 60453 (312) 423-3386

Joseph P. Krause
5812 W. 101 First St.
Oak Lawn, IL 60453 (312) 423-4578

Charles A. Doktycz#
1023 N. Kenilworth
Oak Park, IL 60302 (312) 383-3384

Philip H. Kier
321 Home Ave.
Oak Park, IL 60302 (312) 972-3989

Robert George Petrinec#
617 S. East Ave.
Oak Park, IL 60304 (312) 848-8955

John Vander Weit, Jr.
Vander Weit & Gabber, Ltd.
20200 Governors Dr.
Suite 112
Olympia Fields, IL 60461 (312) 747-9291

M. Russell Bramwell
800 E. Northwest Hwy.
Suite 326
Palatine, IL 60067 (312) 359-5404

Anthony L. Cupoli
37 Peppertree Dr.
Palatine, IL 60067 (312) 961-9500

Clifford A. Dean
719 Greenwood Dr.
Palatine, IL 60067 (312) 359-1816

Larry I. Golden
Square D Co.
Executive Plaza
Palatine, IL 60067 (312) 397-2600

Richard T. Guttman
Square D Co.
Executive Plaza
1415 S. Roselle Rd.
Palatine, IL 60067 (312) 397-2600

Albert S. Johnston
Square D Co.
Pat. Dept.
Executive Plaza
Palatine, IL 60067 (312) 397-2600

Norton Lesser
Square D Co.
Executive Plaza
1415 S. Roselle Rd.
Palatine, IL 60067 (312) 397-2600

Stephen A. Litchfield
Square D Co.
Executive Plaza
Palatine, IL 60067 (312) 397-2600

Charles W. MacKinnon
1533 California St.
Palatine, IL 60067 (312) 397-9022

Edward C. Threedy
Threedy & Threedy
2277 N. Circle Dr.
Palatine, IL 60067 (312) 303-1130

Ralph C. Medhurst
416 Illinois St.
Park Forest, IL 60466 (312) 856-5954

Herman E. Smith
402 Douglas St.
Park Forest, IL 60466 (312) 748-3963

Robert James Black#
1400 Renaissance Dr., Suite 205
Park Ridge, IL 60068 (312) 635-6371

Richard W. Carpenter
1400 Renaissance Dr., Suite 205
Park Ridge, IL 60068 (312) 635-6357

Robert W. Dudley
634 North Overhill
Park Ridge, IL 60068 (312) 825-4633

Frank B. Hall
855 N. Northwest Hwy.
Park Ridge, IL 60068 (312) 825-2501

Frederick J. Otto
Otto & Snyder
3 South Prospect
Suite 6
Park Ridge, IL 60068 (312) 698-1160

Harold J. Rathburn
1892 De Cook Ave.
Park Ridge, IL 60068 (312) 825-1267

Robert Donald Silver
125 E. Kathleen Dr.
Park Ridge, IL 60068 (312) 825-5262

Sterling R. Booth, Jr.#
Caterpillar Inc.
100 N.E. Adamas St.
Peoria, IL 61629 (309) 675-5136

J.W. Burrows#
Caterpillar Inc.
100 N.E. Adams St.
Peoria, IL 61629 (309) 675-5676

Larry G. Cain#
Caterpillar Inc.
100 N.E. Adams St.
Peoria, IL 61629 (309) 675-4922

Clavin E. Glastetter#
Caterpillar Inc.
100 N.E. Adams Street
Peoria, IL 61629 (309) 675-5124

Eugene C. Gooldale
Caterpillar Tractor Co.
100 N.E. Adams St.
Peoria, IL 61629 (309) 675-5089

John W. Grant#
Caterpillar Tractor Co.
100 N.E. Adams St.
Peoria, IL 61629 (309) 675-5613

Frank L. Hart
Caterpillar Inc.
100 N.E. Adams St.
Peoria, IL 61629 (309) 675-5313

William B. Heming
Caterpillar Inc.
Peoria, IL 61629 (309) 675-5509

Alan J. Hickman#
Caterpillar Inc.
100 N.E. Adams St.
Peoria, IL 61629 (309) 675-4517

Joseph W. Keen
Caterpillar Tractor Co.
100 N.E. Adams St.
Peoria, IL 61629 (309) 675-5753

Federick L. Knop, Jr.#
Caterpillar Tractor Co.
100 N.E. Adams St.
Peoria, IL 61629 (309) 675-4015

Charles E. Lanchantin, Jr.#
Caterpillar Tractor Co.
100 N.E. Adams St.
Peoria, IL 61629 (309) 675-4013

Robert A. McFall#
Caterpillar Tractor Co.
100 N.E. Adams St.
Peoria, IL 61629 (309) 675-4610

Robert E. Muir
Caterpillar Tractor Co.
100 N.E. Adams St.
Peoria, IL 61629 (309) 675-4073

Stephen L. Noe
Caterpillar Tractor Co.
100 N.E. Adams Street
Peoria, IL 61619 (309) 689-5589

Oscar G. Pence
Caterpillar Tractor Co.
100 N.E. Adams St.
Peoria, IL 61629 (309) 675-4460

William C. Perry#
Caterpillar Tractor Co.
100 N.E. Adams St.
Peoria, IL 61629 (309) 675-5083

Kenneth A. Rhoads#
Caterpillar Tractor Co.
100 N.E. Adams St.
Peoria, IL 61629 (309) 675-4015

Curtis P. Ribando*#
1815 N. University Street
USDA-ARS-NCR
Peoria, IL 61604 (309) 685-4011

William Scott Thompson
Caterpillar Tractor Co.
100 N.E. Adams St.
Peoria, IL 61629 (309) 675-4452

Ralph Eugene Walter
301 S.W. Adams
Commercial Natl. Bank Bldg., Suite 700
Peoria, IL 61602 (309) 676-1381

Loyal O. Watts#
Caterpillar Tractor Co.
100 N.E. Adams St.
Peoria, IL 61629 (309) 675-4923

Claude F. White#
Caterpillar Tractor Co.
100 N.E. Adams St.
Peoria, IL 61629 (309) 675-0901

Anthony N. Woloch
Caterpillar Tractor Co.
Pat. Dept. Ab6A
100 N.E. Adams St.
Peoria, IL 61629 (309) 675-5210

Evan D. Roberts
122 N. Second St.
Box 325
Peotone, IL 60468 (312) 258-6318

B. R. Tongren
Clinton, Tongren & Grim
103 E. Main St.
P.O. Box 549
Peotone, IL 60468 (312) 258-6335

Bessie A. Lepper
517 South 43rd Street
Quincy, IL 62301 (217) 222-5372

D. James Bader
3677 Sauk Trail
Richton Park, IL 60471 (312) 481-3100

Nicholas Anthony Camasto
545 Thatcher
River Forest, IL 60305 (312) 366-0604

Francis J. Lidd#
247 Lawton Rd.
Riverside, IL 60546 (312) 447-2476

Glenn C. Sechen
424 Selborne Rd.
Riverside, IL 60546 (312) 447-7271

Thomas E. Currier
Connolly, Oliver, Coplan, Close & Worden
124 N. Walter St., Suite 300
Rockford, IL 61104 (815) 968-7591

Robert M. Hammes, Jr.
Barber-Colman Co.
P.O. Box 7040
Rockford, IL 61125 (815) 397-7400

Ted E. Killingsworth, Jr.
Sundstrand Corporation
4751 Harrison Ave.
Rockford, IL 61101 (815) 226-6307

William D. Lanyi
Sundstrand Corporation
4751 Harrison Avenue
P.O. Box 7003
Rockford, IL 61125 (815) 226-6000

Leroy W. Mitchell
Leydig, Voit & Mayer, Ltd.
815 N. Church Street
Rockford, IL 61103 (815) 963-7661

Edward A, Morsbach
303 N. Prospect Street
Rockford, IL 61107 (815) 226-1351

Michael C. Payden
Leydig, Voit & Mayer, Ltd.
815 North Church Street
Rockford, IL 61103 (815) 963-7661

Vernon J. Pillote
303 N. Main St., Suite 712
Rockford, IL 61101 (815) 964-9312

James A. Wanner
Sundstrand Corporation
4751 Harrison Avenue
P.O. Box 7003
Rockford, IL 61125 (815) 226-7456

Harold A. Williamson
Sundstrand Corp.
4751 Harrison Ave.
Rockford, IL 61101 (815) 526-7407

Michael S. Yatsko
Sundstrand Corporation, Dept. 912-6
4751 Harrison Avenue
Rockford, IL 61125 (815) 226-6000

George P. Edgell
Gould Inc.
10 Gould Center
Rolling Meadows, IL 60008 (312) 640-4697

Robert J. Fox
Gould Inc.
10 Gould Center
Rolling Meadows, IL 60008 (312) 640-4693

Kay H. Pierce
Gould Inc.
Gould Center
Rolling Meadows, IL 60008 (312) 640-4663

Edward E. Sachs
Gould Inc.
10 Gould Center
Rolling Meadows, IL 60008 (312) 460-4543

Lester N. Arnold
1409 Wright Blvd.
Schaumburg, IL 60193 (312) 893-1620

Thomas G. Berry#
Motorola, Inc.
1303 E. Algonquin Rd.
Schaumburg, IL 60196 (312) 576-5066

Douglas A. Boehm#
Motorola, Inc.
1303 E. Algonquin Rd.
Schaumburg, IL 60196 (312) 576-5214

Frank J. Cerny, Jr.#
Motorola Incorporated
1301 E. Algonquin Rd.
Room 2918
Schaumburg, IL 60196 (312) 576-2443

Robert J. Crawford#
Motorola, Inc.
1303 E. Algonquin Rd.
Schaumburg, IL 60196 (312) 576-5212

James W. Gillman
Motorola, Inc.
1303 E. Algonquin Rd.
Schaumburg, IL 60196 (312) 576-5223

Raymond A. Jenski#
Motorola, Inc.
1303 E. Algonquin Rd.
Schaumburg, IL 60196 (312) 576-5223

Jose W. Jimenez
Motorola, Inc.
Corp. Off.-Motorola Center
1303 E. Algonquin Rd.
Schaumburg, IL 60196 (312) 576-4545

Bernard L. Kramer
21 Kristin Dr.
Apt. 414
Schaumburg, IL 60195 (312) 884-8841

Phillip H. Melamed
Motorola, Inc.
1303 E. Algonquin Rd.
Schaumburg, IL 60196 (312) 576-5218

John H. Moore
Motorola, Inc.
1303 E. Algonquin Rd.
Schaumburg, IL 60196 (312) 576-5213

Steven G. Parmelee
Motorola, Inc.
1303 E. Algonquin Rd.
Schaumburg, IL 60196 (312) 576-0860

James S. Pristelski
Motorola, Inc.
1303 E. Algonquin Rd.
Schaumburg, IL 60196 (312) 576-5218

Vincent Joseph Rauner
Motorola, Inc.
1303 E. Algonquin Rd.
Schaumburg, IL 60196 (312) 576-5220

Edward Milton Roney, III
Motorola, Inc.
1303 E. Algonquin Rd.
Schaumburg, IL 60196 (312) 576-5222

Donnie Rudd
Rudd & Associatestd.
1030 W. Higgins Rd.
Schaumburg, IL 60195 (312) 882-8655

Anthony J. Sarli, Jr.
Motorola, Inc.
1303 E. Algonquin Rd.
Schaumburg, IL 60196 (312) 397-5000

Melvin A. Schechtman
Motorola Incorporation
1931 Prairie Square
Suite 327
Schaumburg, IL 60195 (312) 397-2641

Donald B. Southard
Motorola, Inc.
1303 E. Algonquin Rd.
Schaumburg, IL 60172 (312) 576-5214

Raymond J. Suberlak
A D P Dealer Services
920 E. Algonquin Rd.
Schaumburg, IL 60195 (312) 397-1700

Charles L. Warren
Motorola, Inc.
1303 E. Algonquin Rd.
Schaumburg, IL 60196 (312) 576-5222

Douglas B. White
Bianchi & White
1501 Woodfield Rd.
Suite 201 North
Schaumburg, IL 60195 (312) 885-3730

Craig M. Bell
Nutra Sweet Co.
4711 Golf Rd.
P.O. Box 1111
Skokie, IL 60076 (312) 982-7354

Roger M. Fitz-Gerald
Bell & Howell Co.
5215 Old Orchard Rd.
Skokie, IL 60077 (312) 470-7645

Mary J. Kanady
G.D. Searle & Co.
5200 Old Orchard Rd.
Skokie, IL 60077 (312) 470-6501

J. Timothy Keane
G.D. Searle & Co.
5200 Old Orchard Rd.
Skokie, IL 60077 (312) 470-6500

William Gresham Lawler, Jr.
Brunswick Corp.
One Brunswick Plaza
Skokie, IL 60077 (312) 470-4321

Paul D. Matukaitis
G.D. Searle & Co.
Corporate Pat. Dept.
5200 Old Orchard Rd.
Skokie, IL 60077 (312) 470-6300

John J. Mc Donnell
G. D Searle & Co.
Corporate Patent Dept.
4711 Golf Rd.
Skokie, IL 60076 (312) 982-7306

Raymond C. Nordhaus
8301 Karlov
Skokie, IL 60076 (312) 674-5739

Donald Steen Olexa
Brunswick Corp.
One Brunswick Plaza
Skokie, IL 60077 (312) 470-4055

John M. Sanders
Nutra Sweet Co.
4711 Gold Rd.
Skokie, IL 60076 (312) 982-8347

Algis A. Tirva
AT&T Teletype Corp.
5555 Touhy Ave.
Skokie, IL 60077 (312) 982-3680

Wayne Golomb
624 S. Second Street
Springfield, IL 62704 (217) 544-4980

John C. Albrecht
1044 N. Second Ave.
St. Charles, IL 60174 (312) 377-2415

Edward F. Jurow
1044 N. 6th Ave.
St. Charles, IL 60174 (312) 377-0484

Joseph P. Sauber, Jr.#
508 Cedar St.
St. Charles, IL 60174 (312) 584-6536

Mark D. Hilliard
Panduit Corp.
17301 Ridgeland Ave.
Tinley Park, IL 60477 (312) 532-1800

Charles R. Wentzel
Panduit Corp.
17301 Ridgeland Ave.
Tinley Park, IL 60477 (312) 532-1800

Michael T. Murphy
Rust-Oleum Corp.
11 Hawthorn Parkway
Vernon Hills, IL 60061 (312) 367-7700

R. Warren Comstock
Outboard Marine Corp.
100 Sea Horse Dr.
Waukegan, IL 60085 (312) 689-5229

Mark W. Croll
Outboard Marine Corp.
100 Sea-Horse Dr.
Waukegan, IL 60085 (312) 689-6187

Robert Kingsley Gerling
Outboard Marine Corp.
100 Sea Horse Dr.
Waukegan, IL 60085 (312) 689-5247

James P. Hanrath
415 W. Washington St., Suite 216
Waukegan, IL 60085 (312) 249-1420

Dorsey L. Baker
28 W. 340 Indian Knoll Trail
West Chicago, IL 60185 (312) 231-3578

Walter Lothar Schlegel, Jr.
1 So. 311 Edgewood Walk
West Chicago, IL 60185 (312) 231-0821

Francis V. Cunningham#
5316 Central Ave.
Western Springs, IL 60558 (312) 246-5158

William K. Serp
4027 Harvey Ave.
Western Springs, IL 60558 (312) 246-7856

James W. Ove
50 West 57th St.
Westmont, IL 60559 (312) 964-6063

Fred R. Ahlers
1125 Delles Rd.
Wheaton, IL 60187 (312) 668-4378

John L. Hutchinson#
1 S. 505 Bayberry Ln.
Wheaton, IL 60187 (312) 668-6056

Margaret M. Parker#
1302 Scott St.
Wheaton, IL 60187 (312) 668-6550

Cedric M. Richeson
Ashburn & Richeson
Suite 900
2100 Manchester Rd.
Wheaton, IL 60187 (312) 462-4467

Emily A. Richeson#
Ashburn & Richeson
2100 Manchester Rd.
Suite 900
Wheaton, IL 60187 (312) 462-4467

Hugh M. Gilroy
1641 Hunter Dr.
Wheeling, IL 60090 (312) 577-6143

John Max Brown#
811 Lacrosse Ave.
Wilmette, IL 60091 (312) 256-2760

Sheldon Lee Epstein
P.O. Box 400
Wilmette, IL 60091 (312) 948-9292

John Warren Mc Caffrey
1200 Greenwood Ave.
Wilmette, IL 60091

Walter Christoph Ramm
444 Skokie Rd., Suite 302
Wilmette, IL 60091 (312) 256-5425

William Brandt Ross
1334 Isabella
Wilmette, IL 60091 (312) 856-2070

Morris Spector
1500 Sheridan Rd.
Suite 1-H
Wilmette, IL 60091 (312) 251-5061

Raymond A. Andrew
2 Kent Rd.
Winnetka, IL 60093 (312) 446-7868

Robert Gottschalk
P.O. Box 8436
545 Lincoln Ave.
Winnetka, IL 60093 (312) 446-5230

Waino M. Kolehmainen
1218 Oak St.
Winnetka, IL 60093 (312) 446-2729

John C. Shepard
575 Sunset Rd.
Winnetka, IL 60093 (312) 965-8660

Indiana

Ken C. Decker
Allied Corporation
401 N. Bendix Dr.
Allied, IN 46634 (219) 237-2455

F. Kristen Koepcke
Hillenbrand Inds, Inc.
Rt. 46
Batesville, IN 47006 (812) 934-7361

Brian J. Leitten
Hillenbrand Industries
Batesville, IN 47006 (812) 934-7330

Steve M. McLary
Hillenbrand Industries, Inc.
Highway 46
Batesville, IN 47006 (812) 934-7904

Gary M. Gron
Cummins Engine Co., Inc.
Box 3005
Columbus, IN 47202 (812) 377-3554

Robert T. Ruff
Cummins Engine Co., Inc.
Box 3005
Columbus, IN 47202 (812) 372-5936

Stephen H. Friskney
1725 Country Club Road
Connersville, IN 47331 (317) 825-5457

Thomas J. Page, Jr.
P.O. Box 29242
Cumberland, IN 46229

Mary G. Boguslaski
Miles Laboratories., Inc.
1127 Myrtle St.
P.O. Box 40
Elkhart, IN 46515 (219) 264-8384

R. Norman Coe
Miles Laboratories, Inc.
1127 Myrtle St.
Elkhart, IN 46515 (219) 262-7937

Daniel W. Collins
Miles Laboratories, Inc.
P.O. Box 40
Elkhart, IN 46515 (219) 262-6454

Louis E. Davidson
Miles Laboratories, Inc.
P.O. Box 40
1127 Myrtle St.
Elkhart, IN 46515 (219) 264-8393

John H. Engelmann
Miles Inc.
1127 Myrtle
Elkhart, IN 46514 (219) 262-7482

George A. Foster, Jr.#
Miles Incorporated
P.O. Box 40
Elkhart, IN 46515 (219) 293-3815

John J. Gaydos
442 Communicana Bldg.
421 S. Second St.
Elkhart, IN 46516 (219) 294-1516

Charles J. Herron
Miles Laboratories, Inc.
1127 Myrtle St.
Elkhart, IN 46514 (219) 264-8384

Jerome L. Jeffers
Miles Inc.
P.O. Box 40
Elkhart, IN 46515 (219) 264-8394

Andrew L. Klawitter
Miles Laboratories, Inc.
1127 Myrtle St.
Elkhart, IN 46514 (219) 262-7148

David S. Saari#
711 Christian Avenue
Elkhart, IN 46517 (219) 262-6470

Jennifer L. Skord
Miles Laboratories
P.O. Box 40
Elkhart, IN 46515 (219) 262-6453

Harry Thomas Stephenson
3201 Eastlake Drive North
Elkhart, IN 46514 (219) 264-5112

Richard W. Winchell
Miles Labs., Inc.
1127 Myrtle St.
Elkhart, IN 46514 (219) 262-7748

Warren Dale Flackbert
401 N. Weinbach Avenue, Suite D
Evansville, IN 47711 (812) 477-2434

George H. Morgan#
309 Springhaven Drive
Evansville, IN 47710 (812) 428-7102

Robert H. Uloth#
Mead Johnson And Co.
2404 Pennsylvania Ave.
Evansville, IN 47721 (812) 426-6720

David L. Ahlersmeyer
Jeffers, Irish & Hoffman
1500 Anthony Wayne Bank Bldg.
Fort Wayne, IN 46807 (219) 426-1700

Edward P. Armstrong
Central Soya Co., Inc.
1300 Ft. Wayne Natl. Bk. Bldg.
Fort Wayne, IN 46801 (219) 425-5477

Eugene G. Botz#
8234 Ravinia Dr.
Fort Wayne, IN 46825 (219) 489-3055

George R. Caruso
8405 Lima Road
Fort Wayne, IN 46818 (219) 489-6233

Lawrence E. Freiburger#
3220 Walden Run
Fort Wayne, IN 46815 (219) 486-3006

Bobby B. Gillenwater
Barnes & Thornburg
600 One Summit Square
Fort Wayne, IN 46802 (219) 423-9440

George A. Gust
Gust, Ricker & Welch
1416 Anthony Wayne Bank Bldg.
Fort Wayne, IN 46802 (219) 426-3400

John F. Hoffman
Jeffers, Hoffman & Niewyk
1500 Anthony Wayne Bank Bldg.
Fort Wayne, IN 46802 (219) 426-1700

Albert L. Jeffers
Jeffers, Irish & Hoffman
1500 Anthony Wayne Bank Bldg.
Fort Wayne, IN 46802 (219) 426-1700

Douglas E. Johnston
Tourkow, Crell, Rosenblatt & Johnston
814 Anthony Wayne Bldg.
Fort Wayne, IN 46802 (219) 426-0545

Stan C. Kaiman
Tokheim Corp.
1602 Wabash Ave.
P.O. Box 360
Fort Wayne, IN 46801 (219) 423-2552

Ralph E. Krisher, Jr.
3409 Rosewood Dr.
Fort Wayne, IN 46802 (219) 432-4744

David A. Lundy
Lundy & Associates
1020 Anthony Wayne Bk. Bldg.
Fort Wayne, IN 46802 (219) 422-1534

Anthony Niewyk
Jeffers, Irish & Hoffmans
1500 Anthony Wayne Bank Bldg.
Fort Wayne, IN 46802 (219) 426-1700

Joseph Edward Papin
General Elec. Co.
Component Motor Div.
1635 Broadway, P.O. Box 2204
Fort Wayne, IN 46801 (219) 428-3281

George Pappas
Jeffers, Irish & Hoffman
1500 Anthony Wayne Bank Bldg.
Fort Wayne, IN 46802 (219) 426-1700

Charles S. Penfold
3217 Oswego Avenue
Fort Wayne, IN 46805 (219) 426-5958

Roger Mark Rickert
Gust, Rickert & Welch
1416 Anthony Wayne Bank Bldg.
Fort Wayne, IN 46802 (219) 426-3400

Richard L. Robinson
Jeffers, Hoffman & Niewyk
1500 Anthony Wayne Bldg.
Fort Wayne, IN 46802 (219) 426-1700

Richard Thompson Seeger
1067 Delaware Avenue
Fort Wayne, IN 46805 (219) 422-9829

Robert D. Sommer#
1601 Wall Street
P.O. Box 1601
Fort Wayne, IN 46801 (219) 461-4254

Arlyce R. Stearns
1200 Anthony Wayne Bank Bldg.
Fort Wayne, IN 46802 (219) 426-4512

John Matthew Stoudt, III
General Electric Co.
1635 Broadway, P.O. Box 2204
Fort Wayne, IN 46801 (219) 428-3287

Robert L. Walker
Lundy & Walker
1020 Anthony Wayne Bank Bldg.
Fort Wayne, IN 46802 (219) 422-1534

George Hyman, Jr.
R.R. 2
Galveston, IN 46923 (219) 859-4622

Joseph P. Kulik, Jr.
50766 Heather Hill Lane
Granger, IN 46530 (219) 272-1829

James L. Wilson
3534 43rd Street
Highland, IN 46322 (219) 924-4468

Edward P. Archer
University Of Indiana
Indianapolis Law School
735 W. New York St.
Indianapolis, IN 46202 (317) 264-4998

Charles W. Ashbrook
Eli Lilly & Co.
Lilly Corporate Center
Indianapolis, IN 46285 (317) 276-6015

David H. Badger
William, Brinks, Olds Hofer Gilson, & Lione
One Indiana Square, Suite 3160
Indianapolis, IN 46204 (317) 636-0886

William F. Bahret
Woodard, Emhardt, Naughton,Moriaty &
McNett
One Indiana Square, Suite 2000
Indianapolis, IN 46204 (317) 634-3456

Bruce J. Barclay
Eli Lilly & Co.
Lilly Corporate Center
307 E. Mc Carty Street
Indianapolis, IN 46285 (317) 276-3474

Spiro Bereveskos
Woodard, Emhardt, Naughton, Moriarty &
McNett
One Indiana Square, Suite 2000
Indianapolis, IN 46204 (317) 634-3456

Clifford W. Browning
Woodard, Emhardt, Naughton, Moriarty &
McNett
One Indiana Square, Suite 2000
Indianapolis, IN 46204 (317) 634-3456

Mary Spalding Burns
Barnes & Thornburg
1313 Merchants Bank Bldg.
11 S. Meridan Street
Indianapolis, IN 46204 (317) 638-1313

John H. Calhoun, Jr.
6100 N. Keystone Avenue
Suite 333
Indianapolis, IN 46220 (317) 255-3438

Francis E. Cislak#
5331 N. Kenwood Ave.
Indianapolis, IN 46204 (317) 255-7115

Richard R, Clapp#
Barnes & Thornburg
1313 Merchants Bank Bldg.
11 S. Meridan Street
Indianapolis, IN 46204 (317) 231-7461

William R. Coffey
Barnes & Thornburg
11 S. Meridian Street, Suite 1313
Indianapolis, IN 46204 (317) 231-7280

James A. Coles
Barnes & Thornburg
11 S. Meridan Street, Suite 1313
Indianapolis, IN 46204 (317) 638-1313

Paul S. Collignon
6710 N. Riley
Indianapolis, IN 46220 (317) 251-7659

Richard D. Conard
Barnes & Thornburg
11 South Meridan St., Suite 1313
Indianapolis, IN 46204 (317) 231-7285

Robert A. Conrad
Eli Lilly & Co.
Lilly Corporate Center
Indianapolis, IN 46285 (317) 276-6013

Gerald V. Dahling
Eli Lilly & Co.
Corporate Center
Indianapolis, IN 46285 (317) 276-2965

Mark R. Daniel
Eli Lilly & Co.
Lilly Corporate Center
Patent Division
Indianapolis, IN 46285 (317) 276-3589

James M. Dulacher
Woodard, Emhardt, Naughton, Moriarty & McNett
One Indiana Square, Suite 2000
Indianapolis, IN 46204 (317) 634-3456

Richard J. Egan
Dow Consumer Products Inc.
9550 N. Zionsville Rd.
Indianapolis, IN 46268 (317) 873-7286

C. David Emhardt
Woodard, Emhardt, Naughton, Moriarty & McNett
One Indiana Square, Suite 2000
Indianapolis, IN 46204 (317) 634-3456

Carl A. Forest
Emhardt Industries,Inc.
Mallory Components Group
3029 E. Washington St., P.O. Box 706
Indianapolis, IN 46204 (317) 261-1470

R. Randall Frisk
Woodard, Emhardt, Naughton, Moriarty & McNett
One Indiana Square, Suite 2000
Indianapolis, IN 46204 (317) 634-3456

Roland A. Fuller, III
Barnes & Thornburg
11 S. Meridian St., Suite 1313
Indianapolis, IN 46204 (317) 231-7274

Gerald Howard Glanzman
Willian, Brinks, Olds, Hofer, Gilson & Lione
One Indiana Square, Suite 3160
Indianapolis, IN 46204 (317) 636-0886

David W. Gomes
Mallory Components Group
Emhardt Industries, Inc.
3029 E. Washington St.
Indianapolis, IN 46204 (317) 261-1470

Edward P. Gray
Eli Lilly & Co.
Lilly Corporate Center
Pat. Div./e P G
Indianapolis, IN 46285 (317) 276-3785

Ronald S. Hansell
Amax Coal Co. Regulatory Affairs
Beechbank
1205 W. 64th St.
Indianapolis, IN 46260 (317) 253-9624

Nancy J. Harrison
Eli Lilly & Co.
Lilly Corporate Center
Dept. M C 529
Indianapolis, IN 46285 (317) 276-2308

Thomas Q. Henry
Woodard, Emhardt, Naughton, Moriarty & McNett
One Indiana Square, Suite 2000
Indianapolis, IN 46204 (317) 634-3456

Peter Peck-Koh Ho
502 East Ohio Street
Indianapolis, IN 46204 (317) 634-6158

Charles W. Hoffmann
Emhart Industries, Inc.
3029 E. Washington St.
Indianapolis, IN 46204 (317) 261-1417

Jerry E. Hyland
Barnes & Thornburg
11 South Meridian Street
1313 Merchants Bank Bldg.
Indianapolis, IN 46204 (317) 231-7288

Ettore V. Indiano
8445 Keystone Crossing
Suite 102
Indianapolis, IN 46240 (317) 257-6263

Thomas P. Jenkins
Barnes & Thornburg
11 South Meridian Street
Indianapolis, IN 46204 (317) 231-7260

Joseph A. Jones
Eli Lilly & Co.
Patent Division
Lilly Corporate Center
Indianapolis, IN 46285 (317) 276-5183

Scott V. Kissinger
Ice Miller Donadio & Ryan
One American Square, Box 82001
Indianapolis, IN 46282 (317) 236-2466

Dilip A. Kulkarni
R C A Corporation
Consumer Electronics Div.
600 N. Sherman Drive, P.O. Box 1976
Indianapolis, IN 46204 (317) 267-1970

Steven R. Lammert
Barnes & Thornburg
1313 Merchants Bank Bldg.
Indianapolis, IN 46204 (317) 261-9258

Ron K. Levy
Eli Lilly & Co.
Lilly Corporate Center
Indianapolis, IN 46285 (317) 261-5383

Daniel J. Lueders
Woodard, Emhardt, Naughton, Moriarty & McNett
One Indian Square, Suite 2000
Indianapolis, IN 46204 (317) 634-3456

Lawrence M. Lunn
Wick, White & Lunn
6125 U.S. Highway 31 South
Indianapolis, IN 46227 (317) 788-4000

William C. Martens, Jr.
Eli Lilly & Co.
307 E. Mc Carty St.
Indianapolis, IN 46285 (317) 261-2573

John C. McNett
Woodard, Emhardt, Naughton, Moriarty & McNett
One Indiana Square, Suite 2000
Indianapolis, IN 46204 (317) 634-3456

Robert F. Meyer
Mallory Components Group
Emhart Industries, Inc.
3029 E. Washington St.
Indianapolis, IN 46204 (317) 261-1469

John V. Moriarty
Woodard, Emhardt, Naughton, Moriarty & McNett
One Indiana Square, Suite 2000
Indianapolis, IN 46204 (317) 634-3456

Michael A. Morra
Bell Telephone Labs., Inc.
6612 E. 75th Street
P.O. Box 1008
Indianapolis, IN 46206 (317) 845-6012

Dwight Edward Morrison
250 Williams Drive
Indianapolis, IN 46260 (317) 251-0909

Joseph A. Naughton, Jr.
Woodard, Emhardt, Naughton, Moriarty & McNett
One Indiana Square, Suite 2000
Indianapolis, IN 46204 (317) 634-3456

Anthony Nimmo
Barnes & Thornburg
1313 Merchants Bank Bldg.
Indianapolis, IN 46202 (317) 638-1313

Paul Overhauser
810 Fletcher Trust Bldg.
Indianapolis, IN 46204

Kathleen R.S. Page
Eli Lilly & Co.
307 E. McCarty St.
Indianapolis, IN 46239 (317) 467-4518

George W. Pendygraft
Baker & Daniels
810 Fletcher Trust Bldg.
Indianapolis, IN 46204 (317) 636-4535

David B. Quick
Woodard, Emhardt, Naughton, Moriarty & McNett
One Indiana Square, Suite 2000
Indianapolis, IN 46204 (317) 634-3456

Charles R. Reeves
Woodard, Emhardt, Naughton, Moriarty & McNett
One Indiana Square, Suite 2000
Indianapolis, IN 46204 (317) 634-3456

Richard A. Rezek
Barnes & Thornburg
11 South Meridan, Suite 1313
Indianapolis, IN 46204 (317) 261-9283

Andrew J. Richardson
Barnes & Thornburg
1313 Merchants Bank Bldg.
11 S. Meridian
Indianapolis, IN 46204 (317) 261-9290

James L. Rowe
7775 Spring Mill Road
Indianapolis, IN 46260 (317) 251-0070

William B. Scanlon
Eli Lilly & Co.
307 E. McCarty St.
Indianapolis, IN 46285 (317) 261-3159

Jack Schuman
Mantel, Mantel, & Reiswerg
717 E. 86th Street
Indianapolis, IN 46240 (317) 255-5797

Everet F. Smith
Barnes & Thornburg
1313 Merchants Bank Bldg.
11 S. Meridian St.
Indianapolis, IN 46204 (317) 638-1313

Calvin Norris Sparrow
Eli Lilly and Co.
307 E. McCarty St.
Indianapolis, IN 46285 (317) 261-3173

Robert A. Spray
7114 E. 71st Street
Indianapolis, IN 46256 (317) 841-0113

Donald R. Stuart
Eli Lilly & Co.
307 E. McCarty St.
Indianapolis, IN 46285 (317) 261-5183

Houston L. Swenson
Eli Lilly And Co.
307 E. McCarty St.
Indianapolis, IN 46204 (317) 261-2923

Mary A. Tucker
Eli Lilly & Company
Lilly Corporate Center
Indianapolis, IN 46285 (317) 261-3881

Vincent O. Wagner
Woodard, Emhardt, Naughton, Moriaty & McNett
One Indiana Square, Suite 2000
Indianapolis, IN 46204 (317) 634-3456

Charles W. Walton, III
Ransburg Corporation
3939 W. 56th Street
Indianapolis, IN 46254 (317) 298-5191

Richard H. Weber#
3216 W. 46th Street
Indianapolis, IN 46208 (317) 291-0354

Leroy Whitaker
Eli Lilly & Co.
Lilly Corporate Center
Indianapolis, IN 46285 (317) 261-2719

Craig A. Wood
22 E. Washington Street
Suite 316
Indianapolis, IN 46204 (317) 637-5245

Harold Raymond Woodard
Woodard, Emhardt, Naughton, Moriarty & McNett
One Indiana Square, Suite 2000
Indianapolis, IN 46204 (317) 634-3456

Stepen E. Zlatos
Woodard, Emhardt, Naughton, Moriarty & McNett
One Indiana Square, Suite 2000
Indianapolis, IN 46204 (317) 634-3456

Norman L. Roelke
Jeffboat, Incorporated
1030 E. Market Street
Jeffersonville, IN 47130 (812) 288-0294

Joseph J. Phillips#
Cabot Corporation Legal Div.
1020 W. Park Ave.
Kokomo, IN 46901 (317) 456-6112

John Robert Nesbitt
Purdue National Bank Bldg.
Suite 1014
Lafayette, IN 47906 (317) 742-8121

Linda M. Chinn#
3335 Elm Swamp Rd.
Lebanon, IN 46052 (317) 482-7550

Richard G. Kinney
1000 E. 80th Place
Suite 425 South Tower
Merrillville, IN 46410 (219) 736-2110

Walter Leuca
1000 E. 80th Place, Suite 524
Merrillville, IN 46410 (219) 769-3080

Martin B. Barancik
General Elec. Corp.
Highway 69 South
Mt. Vernon, IN 47620 (812) 838-7966

Edwin R. Acheson, Jr.
Borg Warner Automotive, Inc.
5401 Kilgore Avenue
Muncie, IN 47302 (317) 286-6579

Gilbert E. Alberding
Ball Corp.
345 S. High St.
Muncie, IN 47302 (317) 747-6422

Perry G. Cross
Cross, Marshall, Schuck, De Weese, Cross & Feick
200 E. Wash. St.
Muncie, IN 47305 (317) 289-6151

Frank A. Steldt
1346 Pebble Brook Drive
Nobleville, IN 46060 (317) 896-5560

Ronald Perry Shipman
106 W. Benton St.
Oxford, IN 47971 (317) 385-2170

Marmaduke A. Hobbs
P.O. Box 367
105 N. Shelby Street
Salem, IN 47167 (812) 883-6145

William Nicholas Antonis
The Bendix Corp.
401 N. Bendix Dr.
South Bend, IN 46634 (219) 237-2450

Thomas J. Dodd
Oltsch, Knoblock & Hall
625 J M S Bldg.
South Bend, IN 46601 (219) 234-6091

Ryan M. Fountain
Barnes & Thornburg
100 N. Michigan, Suite 600
South Bend, IN 46601 (219) 233-1171

James D. Hall
Oltsch, Knoblock & Hall
625 Jms Bldg.
South Bend, IN 46601 (219) 234-6091

Eugene Conrad Knoblock
Oltsch, Knoblock & Hall
625 Jms Bldg.
South Bend, IN 46601 (219) 234-6091

Leo H. McCormick, Jr.
Bendix Corp.
401 N. Bendix Drive
South Bend, IN 46634 (219) 237-2452

David R. Melton
Barnes & Thornburg
100 N. Michigan
South Bend, IN 46601 (219) 233-1171

Larry J. Palguta
Bendix Corp.
401 N. Bendix Drive
P.O. Box 4001
South Bend, IN 46634 (219) 237-2451

Charles V. Sweeney
Barnes & Thornburg
100 N. Michigan
South Bend, IN 46601 (219) 233-1139

David W. Van Story#
18077 State Rd., 23
South Bend, IN 46637 (219) 272-7496

Glen A. Weirich
2012 Leer
South Bend, IN 46613 (219) 233-6582

Ronald D. Welch
Allied Corporation
401 N. Bendix Drive
South Bend, IN 46620 (219) 237-2453

H. J. Barnett
2901 Ohio Blvd., Room 150
Terre Haute, IN 47803 (812) 232-6362

Robert Hastings Dewey#
405 South 34th Street
Terre Haute, IN 47803 (812) 234-6976

Wendell R. Guffey
International Mineral And Chemicals Corporation
1401 South Third Street
Terre Haute, IN 47802 (812) 232-0121

Kent R. Fase
2756 Hearthstone
Valparaiso, IN 46383 (219) 464-9676

Margaret L. Geringer#
Bristol-Myers
c/o Zimmer, Inc.
P.O. Box 708
Warsaw, IN 46580 (219) 372-4275

Wendell E. Miller#
1907 Crescent Dr.
Warsaw, IN 46580 (219) 267-2729

Paul D. Schoenle
Zimmer, Inc.
P.O. Box 708
Warsaw, IN 46580 (219) 372-4234

Iowa

Spencer Lorraine Blaylock
Research Foundation, Inc.
Iowa State Univ.
315 Beardsher Hall
Ames, IA 50011 (515) 294-4741

Robert O. Richardson
1445 - 14th Street
Bettendorf, IA 52722 (319) 359-0626

William Topping Metz
Jackson & Metz
306 Tama Bldg.
Burlington, IA 52601 (319) 752-2241

Gary L. McMinimee
Wunschel, Eich & McMinimee
805 North Main
Carroll, IA 51401 (712) 792-9241

Allan L. Harms
Wenzel, Piersall, Riccolo & Harms, P.C.
4080 First Avenue, N.E.
Cedar Rapids, IA 52402 (319) 393-8900

John C. McFarren
Rockwell International Corp.
Pat. Dept. M/s 124-214
400 Collins Road, N.E.
Cedar Rapids, IA 52498 (319) 395-8208

Macka L. Murrah
Rockwell International Corp.
400 Collins Road, N.E., Mail Station 124-214
Cedar Rapids, IA 52402 (319) 305-8208

James C. Nemmers
Shuttleworth & Ingersoll
500 M N B Bldg.
P.O. Box 2107
Cedar Rapids, IA 52406 (319) 365-9461

Haven Ely Simmons
Simmons, Perrine, Albright & Ellwood
1200 Merchants National Bank Bldg.
Cedar Rapids, IA 52401 (319) 366-7641

A. James Valliere
Norand Corporation
550-2nd Street, S.E.
Cedar Rapids, IA 52401 (319) 369-3132

Stephen W. Southwick
Iowa Southern Utilities Co.
300 Sheridan Ave.
Centerville, IA 52544 (515) 437-4400

Albert E. Arnold, Jr.
1916 W. 38th St.
Davenport, IA 52806

John E. Cepican
Henderson & Sturm
101 W. 2nd Street, Suite 204
Davenport, IA 52801 (319) 323-9731

H. Vincent Harsha
Henderson & Sturm
101 W. Second Street, Suite 204
Davenport, IA 52801 (319) 323-9731

Harold M. Knoth
Henderson & Sturm
1111 Davenport Bank Bldg.
Davenport, IA 52801 (319) 323-2465

Morton S. Adler
Alder, Brennan, Joyce & Steger
317 Sixth Avenue
Des Moines, IA 50309 (515) 244-1391

James D. Birkenholz
974-73rd Street, Suite 10
Des Moines, IA 50309 (515) 223-1335

Robert Lee Farris
Massey-Ferguson Inc.
P.O. Box 1813
Des Moines, IA 50306 (515) 247-2100

Richard L. Fix
Henderson & Sturm
1213 Midland Financial Bldg.
206 Sixth Avenue
Des Moines, IA 50309 (515) 288-9588

Mark Davig Hansing
Zarley, McKee, Thomte, Voorhees & Sease
2400 Ruan Center
Des Moines, IA 50309 (515) 288-3667

Kirk M. Hartung
Zarley, McKee, Thomte, Voorhees & Sease
2400 Ruan Center
Des Moines, IA 50309 (515) 288-3667

H. Robert Henderson
Henderson & Sturm
1213 Midland Financial Bldg.
206 Sixth Avenue
Des Moines, IA 50309 (515) 288-9589

David J. Henry
Davis, Hockenberg, Wine, Brown & Koehn
2300 Financial Center
666 Walnut Street
Des Moines, IA 50309 (515) 243-2300

Kent A. Herink
Davis, Hockenberg, Wine, Brown, Koehn & Shors
2300 Financial Center
666 Walnut Street
Des Moines, IA 50309 (515) 243-2300

Michael R. Hoffmann
1000 Des Moines Bldg.
Des Moines, IA 50309 (515) 243-4148

Rudolph L. Lowell
2300 Financial Center
666 Walnut Street
Des Moines, IA 50309 (515) 243-2300

Bruce Welcher McKee
Zarley, McKee, Thomte, Voorhees & Sease
2400 Ruan Center
Des Moines, IA 50309 (515) 288-3667

G. Brian Pingel
Davis, Hockenberg, Wine, Brown & Koehn
2300 Financial Center
Des Moines, IA 50309 (515) 243-2300

Edmund John Sease
Zarley, McKee, Thomte, Voorhees & Sease
2400 Ruan Center
Des Moines, IA 50309 (515) 288-3667

Michael O. Sturm
Henderson & Sturm
1213 Midland Financial Bldg.
Des Moines, IA 50309 (515) 288-9589

Michael G. Voorhees
Zarley, McLKee, Thomte, Voorhees & Sease
2400 Ruan Center
Des Moines, IA 50309 (515) 288-3667

Donald H. Zarley
Zarley, McKee, Thomte, Voorhees & Sease
2400 Ruan Center
Des Moines, IA 50309 (515) 288-3667

Thomas E. Frantz#
Sheaffer Eaton Inc.
301 Avenue H
Fort Madison, IA 52627 (319) 372-3300

Lucas J. DeKoster
1106 Main Street
Hull, IA 51239 (712) 439-2511

Dale A. Kubly
Fisher Controls Co., Inc.
205 S. Center St.
Marshalltown, IA 50158 (515) 754-2135

Ray V. Bailey
Millers Bay
R R 2 - Box 190
Milford, IA 51351 (712) 337-3571

Frank B. Hill
Bandag, Inc.
Bandag Center
Muscatine, IA 52761 (319) 262-1373

Allan P. Orsund#
Maytag Co.
403 W. 4th St. N.
Newton, IA 50208 (515) 792-7000

Richard L. Ward#
The Maytag Co.
One Dependability Square
Newton, IA 50208 (515) 791-8851

David C. Larson
Stoller & Larson
P.O. Box 441
Spirit Lake, IA 51360 (712) 336-4210

Gregory G. Williams
P.O. Box 108
Spirit Lake, IA 51360 (712) 336-4870

Kansas

Thomas M. Scofield
4901 College Blvd.
Leawood, KS 66211 (913) 491-6474

Kenneth W. Iles
13229 West 107th Terrace
Lenexa, KS 66210 (913) 469-8629

William R. Price
Suite 1810 United Kentucky Bldg.
One Riverfront Plaza
Louisville, KS 40202 (502) 587-6961

John O. Mingle
2408 Buena Vista
Manhattan, KS 66502 (913) 537-0838

Paul H. Harder#
4 Hickory Court
Newton, KS 67114 (316) 283-8262

Charles L. Johnson, Jr.
201 E. Loula, Suite 203
P.O. Box 545
Olathe, KS 66061 (913) 764-8773

D.A.N. Chase
Linde, Thomson, Fairchild, Langworthy, Kohn & Van Dyke, P.C.
9300 Metcalf, Suite 1000
Overland Park, KS 66212 (913) 649-4900

William B. Day
Linde, Thomson, Fairchild, Langworthy, Kohn & Van Dyke, P.C.
9300 Metcalf, Suite 1000
One Glenwood Place
Overland Park, KS 66212 (913) 649-4900

Douglas J. Edmonds
Brown, Koralchik & Fingersh
P.O. Box 25550
Overland Park, KS 66225 (913) 451-8500

Chung L. Feng#
4005 W. 104th Terrace
Overland Park, KS 66207 (913) 649-9212

Joan O. Herman
Linde, Thompson, Fairchild, Langworthy, Kohn & Van Dyke, P.C.
9300 Metcalf, Suite 1000
One Glenwood Place
Overland Park, KS 66212 (913) 649-4900

Claude W. Lowe
10051 Roe Avenue
Overland Park, KS 66207 (913) 642-6212

Keith Dillon Moore
P.O. Box 3563
Shawnee Mission, KS 66203 (913) 299-8526

Bruce J. Clark
Davis, Wright, Unrein, Hummer & McCallister
3715 S.W. 29th Street
Topeka, KS 66614 (913) 273-4220

Edward Linus Brown, Jr.
200 E. First Street, Suite 303
Wichita, KS 67202 (316) 263-6400

John W. Carpenter
401 Bitting Bldg.
Wichita, KS 67202 (316) 267-8381

Lee W. Huffman
Boeing Military Airplane Company
3801 South Oliver
Wichita, KS 67210 (316) 526-7618

Ronald L. Lyons
2251 N. Bramblewood, Suite 401
Wichita, KS 67226 (316) 687-3370

Harold J. Pfountz
Coleman Co., Inc.
250 N. Saint Francis St.
Wichita, KS 67202 (316) 261-3197

Phillip A. Rein
1005 N. Market
Wichita, KS 67210 (316) 263-8421

John H. Widdowson
401 Bitting Bldg.
Wichita, KS 67202 (316) 267-8381

Kentucky

Theresa F. Camoriano
11508 Arbor Drive East
Anchorage, KY 40223 (502) 244-2705

Michael Ross Dowling
433 16th St. P.O. Box 1689
Ashland, KY 41101 (606) 325-7682

Stanley M. Welsh
Ashland Oil,Inc.
P.O. Box 391, B L-5
Ashland, KY 41101 (606) 329-5931

Richard Coale Willson
Ashland Oil Company
2000 Ashland Drive
Ashland, KY 41114 (606) 329-4153

Philip R. Cloutier
906 Wrenwood
Bowling Green, KY 42101 (502) 782-3560

Bryan W. LeSieur
P.O. Box 57
Brownsville, KY 42210 (502) 597-2132

P. Joseph Clarke, Jr.
Clarke & Clarke
120 North Third Street, Box 297
Danville, KY 40422 (606) 236-2240

Carl E. Knochelmann, Jr.
98 Garvey Avenue
Elsmere, KY 41018 (606) 342-9029

Charles A. McCrae
One Riverde Plaza
P.O. Box 332
Greenup, KY 41144 (606) 473-9855

John Arthur Brady
I B M Corp., Intellectual Prop. Law Dept.
740 New Circle Rd.
Lexington, KY 40511 (606) 232-4785

William J. Dick
I B M Corp.
740 New Circle Rd,, 952/035-2
Lexington, KY 40511 (606) 232-5292

John Warren Girvin, Jr.
I B M Corporation, Dept. 952, Bldg. 035-2
740 New Circle Road, N.W.
Lexington, KY 40311 (606) 232-5292

J. Ralph King
King & Liles, Psc
Southcreek Park, Suite B200
2365 Harrodsburg Road
Lexington, KY 40504 (606) 223-4050

Frank C. Leach, Jr.
P.O. Box 22455
Lexington, KY 40522 (606) 254-1395

Laurence R. Letson
2468 Heather Court
Lexington, KY 40503 (606) 278-1216

John J. McArdle, Jr.
I B M Corporation, Intellectual Property Law (952/035-2)
740 New Circle Road, NW
Lexington, KY 40511 (606) 232-3939

Warren D. Schickli
King, Liles & Schickli
3070 Harrodsburg Road, Suite 210
Lexington, KY 40503 (606) 223-4050

Jack E. Toliver
Clark Material Sys. Tech. Co.
300 Security Bldg., Short & Mill Sts.
Lexington, KY 40507 (606) 252-5686

Francis H. Boos, Jr.
10509 Timberwood Circle, Suite 100
Louisville, KY 40223 (502) 425-8896

Ralph B. Brick
Polster, Polster & Lucchesi
2303 Tuckaho Road
Louisville, KY 40207 (502) 895-4672

Donald L. Cox
Lynch & Cox
1800 Meidinger Tower
Louisville, KY 40202 (502) 589-4215

James P. Dowd#
8820 Tranquil Valley Lane
Louisville, KY 40299

Robert W. Fletcher
Hilliard-Lyons Pat. Management, Inc.
10509 Timberwood Circle
Louisville, KY 40223 (502) 429-0015

Frank P. Giacalone#
4623 Fox Run Road
Louisville, KY 40207 (502) 896-1539

James R. Higgins, Jr.
Middleton & Reutlinger
2500 Brown & Williamson Tower
Louisville, KY 40202 (504) 584-1135

Harold N. Houser
General Electric Co. Appliance Park
A P 2-225
Louisville, KY 40225 (502) 452-4653

Charles G. Lamb
507 Oak Branch Road
Louisville, KY 40223 (502) 774-7755

Martin R. Levy
2114 Dogoon
Louisville, KY 40223 (502) 423-9478

William J. Mason
Brown & Williamson Tobacco Corp.
1600 W. Hill Street
Louisville, KY 40232 (502) 566-1022

Maurice L. Miller, Jr.
Robert, Miller & Thomas
200 Whittington Parkway, Suite 101
Louisville, KY 40222 (502) 425-2802

Harry B. O'Donnell, III
200 W. Broadway, Suite 606 B, Portland Fed. Bldg.
Louisville, KY 40202 (502) 583-7336

Thomas R. Payne
General Electric Appliance Park 2-315
Louisville, KY 40225 (502) 452-4606

Herbert Peter Price
Lynch, Cox, Gilman & Mahan, P.S.C.
1800 Meidinger Tower
Louisville, KY 40202 (502) 589-4994

Radford Monroe Reams, III
General Electric Co. Appliance Park
Ap1-230
Louisville, KY 40225 (502) 452-3331

Edward Miller Steutermann
1332 South 2nd Street
Louisville, KY 40208 (502) 636-0466

Frederick P. Weidner, Jr.
General Electric Co.
A P 2-225
Louisville, KY 40225 (502) 452-5875

Jon C. Winger
O'Donnell, Steutermann & Winger
200 W. Broadway, Suite 612
Louisville, KY 40202 (502) 589-7023

Bruce A. Yungman
Hilliard-Lyons Pat. Mgmt., Inc.
545 South Third Street
Louisville, KY 40202 (502) 588-8452

Nathan J. Cornfeld
2139 Griffith Ave.
Owensboro, KY 42301 (502) 684-2668

Louisiana

Thomas E. Balhoff
Mathews, Atkinson, Guglielmo Marks & Day
P.O. Box 3177
Baton Rouge, LA 70821 (504) 387-6966

Allen D. Darden
P.O. Box 4412
Baton Rouge, LA 70821 (504) 346-0285

Patricia J. Hogan#
Ethyl Corporation
451 Florida Blvd.
Baton Rouge, LA 70801 (504) 388-7023

Donald Lewis Johnson
5120 E. Bluebell Drive
Baton Rouge, LA 70808 (504) 924-0703

William David Kiesel
Roy, Kiesel, Patterson & McKay
666 South Foster Drive
Baton Rouge, LA 70806 (504) 927-9908

Paul H. Leonard, III
10639 Rondo Avenue
Baton Rouge, LA 70815 (504) 927-6991

Robert Allen Linn
Ethyl Corp. Pat. & Tdmk. Div.
451 Florida Blvd.
Baton Rouge, LA 70801 (504) 388-7526

Eugene Donald Mays
Ethyl Corporation
451 Florida Street
Baton Rouge, LA 70801 (504) 388-7635

Shelton B. McAnelly#
1340 Monterrey Blvd.
Baton Rouge, LA 70815 (504) 926-0897

Timothy J. Monahan
Roy, Kiesel, Aaron & West
2355 Drusilla Lane
Baton Rouge, LA 70809 (504) 927-9908

Willard G. Montgomery
Ethyl Corporation Patent & Trademark Div.
451 Florida Blvd., 9th Floor
Baton Rouge, LA 70801 (504) 388-7937

Joseph Daniel Odenweller
Ethyl Corp.
451 Florida Blvd.
Baton Rouge, LA 70801 (504) 388-8188

J. Bradley Overton
Ethyl Corporation
451 Florida Blvd.
Baton Rouge, LA 70801 (504) 388-7599

Joel R. Penton
P.O. Box 16420B
Baton Rouge, LA 70893 (504) 387-4626

Phillip M. Pippenger
Ethyl Corporation
451 Florida Blvd.
Baton Rouge, LA 70801 (504) 388-7096

Llewellyn Allen Proctor
11481 Sheraton Dr.
Baton Rouge, LA 70815 (504) 275-8689

David L. Ray
2051 Silverside Drive, Suite 205
Baton Rouge, LA 70808 (504) 343-8813

Reginald F. Roberts, Jr.#
P.O. Box 515
Baton Rouge, LA 70821 (504) 343-8500

John F. Sieberth
Ethyl Corp., Pat. & Tdmk. Div.
451 Florida Blvd.
Baton Rouge, LA 70801 (504) 388-7925

Edgar E. Spielman, Jr.
Ethyl Corp. Pat. Tdmk. Div.
451 Florida Blvd.
Baton Rouge, LA 70801 (504) 388-7604

Frederick A. Stolzle, Jr.
Holladay & Stolzle
2103 Government Street
P.O. Box 66437
Baton Rouge, LA 70896 (504) 388-0001

Robert C. Tucker
Roy, Kiesel, Aaron & West
2355 Drusilla Lane
Baton Rouge, LA 70895 (504) 927-9908

Joseph L. Lemoine, Jr.
Onebane, Donohoe, Bernard, Torian, Diaz, McNamara & Abell
Southwest National Bank Bldg.
102 Versailles Blvd., Suite 600
Lafayette, LA 70502 (318) 237-2660

Seth M. Nehrbass#
1225 St. John Street
Lafayette, LA 70506 (318) 234-2347

William W. Stagg
Durio, McGoffin & Stagg
P.O. Box 51308
Lafayette, LA 70505 (318) 233-0300

Mason M. Campbell#
3105 Edenborn Avenue, No. 911
Metairie, LA 70002 (504) 456-6424

George A. Bode
2314 Broadway
New Orleans, LA 70125 (504) 861-8288

Michael D. Carbo
Adams & Reese
4500 One Shell Square
New Orleans, LA 70139 (504) 581-3234

Stephen R. Doody
Sessions, Fishman, Rosenson, Boisfontaine, Nathan & Winn
201 St. Clarles Avenue, 36th Floor
New Orleans, LA 70170 (504) 582-1530

Robert John Edward
McDermott International Inc.
1010 Common St.
New Orleans, LA 70112 (504) 587-5722

Raul V. Fonte
Freeport-McRan Inc.
1615 Poydras Street
New Orleans, LA 70112 (504) 582-4234

Charles C. Garvey, Jr.
Pravel, Gambrell, Hewitt, Kimball & Krieger
963 International Trade Mart
2 Canal Street
New Orleans, LA 70130 (504) 524-7207

Michael L. Hoelter
McDermott, Inc.
1010 Common Street, P.O. Box 60035
New Orleans, LA 70160 (504) 587-5709

Thomas St. Paul Keaty
Keaty & Keaty
1100 Poydras Street, Suite 1840, Energy Centre
New Orleans, LA 70163 (504) 585-7601

David M. Kelly
Keaty & Keaty
1818 I T M
New Orleans, LA 70130 (504) 581-1706

Robert C. Mai#
McDermott Incorporated
1010 Common Street
P.O. Box 60035
New Orleans, LA 70114 (504) 587-5721

Edward D. Markle
Adams & Reese
4500 One Shell Square
New Orleans, LA 70139 (504) 581-3234

William R. Mustain, Jr.#
Sessions, Fishman, Rosenson, Boisfontaine & Nathan
601 Poydras, Suite 2500, Pan Amer. Life Center
New Orleans, LA 70130 (504) 581-5055

Alexander H. Plache
Jones, Walker, Waechter, Poitevent, Carrere & Denegre
225 Baronne Street
New Orleans, LA 70112 (504) 581-6641

C. Emmett Pugh
Pugh & Associates
639 Loyola Avenue, Suite 1660
New Orleans, LA 70113 (504) 587-0000

Stanley L. Renneker
McGlinchey, Stafford, Mintz, Cellini & Lang
630 Camp Street
New Orleans, LA 70130 (504) 586-1200

Leonard Scott Sauer#
3201 St. Charles Ave. Apt. 210
New Orleans, LA 70115 (504) 899-4313

Lloyd N. Shields
Simmon,Peragine, Smith & Redfearn
3000 Energy Centre
New Orleans, LA 70163 (504) 569-2030

Gregory C. Smith
Pravel, Gambrell, Hewitt, Kimball & Krieger
1400 Poydras Center
650 Poydras Street
New Orleans, LA 70130 (504) 524-7207

Raymond C. Von Bodungen
1009 Opelousas Avenue
New Orleans, LA 70114 (504) 361-4247

Dan R. Howard
Dow Chemical Co., Pat. Dept., B-2507
P.O. Box 400, Highway 1
Plaquemine, LA 70765 (504) 389-8914

James M. Pelton
Dow Chemical Company, Pat. Dept., Bldg. 2507
P.O. Box 400
Plaquemine, LA 70765 (504) 389-1807

Arthur J. Young
Dow Chemical Company
P.O. Box 400
Plaquemine, LA 70765 (504) 389-1807

Daniel N. LaHaye
10317 Stewart Place
River Ridge, LA 70123

John M. Harrison
1400 Youree Dr.
Shreveport, LA 77101 (318) 222-4553

John D. Jeter#
1403 Teche Drive
St. Martinville, LA 70582 (318) 349-5017

Maine

Charles A. Cutting
Upton Rd.
Andover, ME 04216 (207) 392-3741

David Francis Gould
McCabe Nurseries
2220 Ohio St.
Bangor, ME 04401 (207) 947-7822

Daniel H. Kane, Jr.
Fenton, Griffin, Chopman, Smith & Fenton
109 Main St.
Bar Harbor, ME 04609 (207) 288-3331

Stuart A. White
19 Sherman St., P.O. Box 141
Island Falls, ME 04747 (207) 774-0317

Harris M. Isaacson
Isaacson, Isaacson & Hark
40 Pine St.
Lewiston, ME 04240 (207) 786-4271

William Rowsell Hulbert
Fish & Richardson
P.O. Box 90
Lincolnville, ME 04849 (207) 236-3508

Daniel L. Peabody
Lake Region Professional Center
P.O. Box L, Route 302
Naples, ME 04055 (207) 693-6030

Thomas L. Bohan
371 Fore Street
Portland, ME 04101 (207) 773-3132

Martin J. Robles
Pierce, Atwood, Scribner, Allen, Smith & Lancaster
One Mounment Square
Portland, ME 04101 (207) 773-6411

Laforest S. Saulsbury
519 Congress St.
Portland, ME 04101 (207) 773-8463

Harold W. Lockhart
Star Route #64, Box 122
South Bristol, ME 17302

Wolfgang G. Fasse
Indian Pond Lane
P.O. Box K
St. Albans, ME 04971 (207) 938-4422

James W. Bock
Box 356
Swans Island, ME 04685 (207) 526-4368

Abbott Spear
Main Street
Warren, ME 04864 (207) 273-2768

Francis M. Di Biase#
Scott Paper Co.
89 Cumberland St.
Westbrook, ME 04092 (207) 856-6911

Maryland

Nancy Ann Coleman
420 Hillcrest Dr.
Aberdeen, MD 21001 (301) 272-3175

Todd E. Stevenson*
U.S. Army Test & Evaluation Comm.
Aberdeen Proving Grounng, MD 21005 (301) 278-3805

Richard C. Reed
P.O. Box 233
Accokeek, MD 20607 (301) 292-5618

Saul Elbaum*
U.S. Army
2900 Powder Mill Rd.
Adelphi, MD 20783 (202) 394-3790

Alan J. Kennedy*
U.S. Army
Electronics Res. & Dev. Comm.
2800 Powder Mill Rd.
Adelphi, MD 20783

Paul E. Maslousky#
2512 Hughes Rd.
Adelphi, MD 20783 (301) 439-4987

Thomas E. McDonald*
U.S. Dept. Of The Army, Intellectual Property Div.
2800 Powder Mill Rd.
Adelphi, MD 20783

Guy M. Miller*#
Department Of The Army, Hdqrs. U.S. Army Lab. Comm.
2800 Powder Mill Rd.
Adelphi, MD 20783 (202) 394-3790

Rudolph V. Rolinec#
2208 Lackawanna Street
Adelphi, MD 20783 (301) 434-6634

Joseph Shortill*
Dept. Of The Army, U.S. Army Lab. Command, Attn: SLCIS-CC-
2800 Powder Mill Rd.
Adelphi, MD 20783 (202) 394-1105

Edward L. Stolarun*
U.S. Army Lab. Command
2800 Powder Mill Rd.
Adelphi, MD 20783 (202) 394-1105

Nicholas J. Aquilino
1917 Hidden Point Rd.
Annapolis, MD 21401 (301) 757-6367

Samuel W. Engle
2610 Vantage Cove
Annapolis, MD 21401 (301) 224-3473

Albert Harrison Helvestine
1729 Fairlop Trail
Epping Forest, Route 1
Annapolis, MD 21401 (301) 849-8467

Harry A. Herbert, Jr.
1821 Manor Green Ct.
Annapolis, MD 21401

Fendall Marbury#
9 Neal Street
Annapolis, MD 21401 (301) 266-8254

Richard E. Rice
Niles, Barton & Wilmer
P.O. Box 589, 410 Stevern Ave. Suite 401
Annapolis, MD 21404 (301) 261-1138

Jere W. Sears
310 Melvin Ave.
Annapolis, MD 21401 (301) 268-2563

Charles William Helzer
694 White Swan Dr., P.O. Box 309
Arnold, MD 21012 (301) 261-1242

Robert E. Bushnell
Fay, Sharpe, Beall, Fagan, Minich & McKee
200 N. Rolling Rd.
Baltimore, MD 21228 (301) 684-1120

James B. Eisel
Martin Marietta Lab.
1450 S. Rolling Rd.
Baltimore, MD 21227 (301) 247-0700

Walter G. Finch
Fidelity Bldg., Suite 1501-03
206 N. Charles St.
Baltimore, MD 21201 (301) 539-8170

Leonard J. Kerpelman
2403 W. Rogers Ave.
Baltimore, MD 21409 (301) 367-8855

Robert Thomas Killman
107 Midhurst Rd.
Baltimore, MD 21212 (301) 377-8116

Charles L. Kraft, II#
P.O. Box 18122
Baltimore, MD 21220

Bruce L. Lamb
Allied Corporation Law Dept.
1400 Taylor Ave.
Baltimore, MD 21284 (301) 339-3170

Dinah H. Lewitan
3318 A Clarks Lane
Baltimore, MD 21215 (301) 358-9510

Richard M. McMahon
Scally, Scally & McMahon, P.A.
8901 Harford Rd.
Baltimore, MD 21234 (301) 661-8590

Mark W. Noel#
6110 Bellona Ave.
Baltimore, MD 21212 (301) 433-9481

Charles F. Obrecht, Jr.
Obrecht & Obrecht
7 N. Calvert St., 906 Munsey Bldg.
Baltimore, MD 21202 (301) 685-6938

Bernard Joseph Ohlendorf
2924 Christopher Ave.
Baltimore, MD 21214 (301) 426-0216

James O. Olfson
General Elevator Co., Inc.
P.O. Box 1702
Baltimore, MD 21203 (301) 789-0200

Robin J. Pecora
204 E. Preston St.
Baltimore, MD 21202 (301) 539-1990

leslie Rajkay#
2447 Pickwick Rd.
Baltimore, MD 21207 (301) 448-2961

Bradley S. Thomas#
9332 Ramblebrook Rd.
Baltimore, MD 21236 (301) 529-3881

Robert M. Trepp
Allied Corporation Law Dept.
1400 Taylor Ave.
Baltimore, MD 21284 (301) 339-3170

Robert Halper
3118 Calverton Blvd.
Beltsville, MD 20705 (301) 572-4719

Vance Y. Hum#
Cheung Laboratories, Inc.
5026 Herzel Place
Beltsville, MD 20705 (301) 937-5677

Beverly K. Johnson*
U.S. Dept. Of Agriculture, A R S, Bldg. 5
Room 415
Beltsville, MD 20705 (301) 344-4032

David Robert Sadowski*#
USDA - Ars - Oci
Room 411, Bldg. 005, Barc-W
Beltsville, MD 20705 (301) 344-4302

Viviana Amzel#
5929 Anniston Rd.
Bethesda, MD 20817

William M. Blackstone#
5225 Pooks Hill Rd.
Bethesda, MD 20814 (301) 493-6733

Arthur Edward Dowell, III
5121 Scarsdale Rd.
Bethesda, MD 20815 (301) 229-1815

Thomas G. Ferris*
Dept. Of Health & Human Services
Patent Branch Ogc, Dhhs
5A-03 Westwood Bldg.
Bethesda, MD 20892 (301) 496-7056

Gerald M. Forlenza
6401 Rockhurst Rd.
Bethesda, MD 20817 (301) 530-8518

Bernice W. Freundel#
9212 Bardon Rd.
Bethesda, MD 20014 (301) 897-5380

William T. Fryer, III
7507 Clarendon Rd.
Bethesda, MD 20814 (301) 656-9479

Claude Funkhouser
8808 Ridge Rd.
Bethesda, MD 20817 (301) 365-4137

Edward H. Gerstenfield
Jones & Gerstenfield, P.A.
7316 Wisconsin Ave.
Bethesda, MD 20814 (301) 654-8911

Paul Louis Gomory
5609 Ogden Rd.
Bethesda, MD 20816 (301) 320-4327

Emory Lowell Groff, Jr.
Groff & Groff
4720 Montgomery Lane, Suite 1003, Montgomery Bldg.
Bethesda, MD 20814 (301) 675-2774

Alvin Guttag
6612 Whittier Blvd.
Bethesda, MD 20817 (301) 229-0170

Jackson T. Hawkins*
U.S. Navy, DTNSRDC
Code 1740.2
Bethesda, MD 20084 (202) 227-3767

Glenna M. Hendricks*
Dept. Of Health & Human Service Patent Branch
5333 Westwood Bldg., Room 5A03
Bethesda, MD 20892 (301) 496-7056

Kenneth C. Hutchison*#
Naval School Of Health Sciences, Bldg. 141, B C C, Room B25
Bethesda, MD 20814 (202) 295-6089

Milford A. Juten
4008 61st St. N.W.
Bethesda, MD 20816 (301) 229-2876

Jack H. Linscott
7203 Fairfax Rd.
Bethesda, MD 20814 (301) 652-1620

Alfred C. Marmor#
7604 Quintana Court
Bethesda, MD 20817 (301) 365-2075

Luther A. Marsh*
U.S. Navy, Office Of Naval Research
David Taylor Naval Ship R & D Center.
Bethesda, MD 20084 (202) 227-1834

Wilmer Mechlin
4733 Bethesda Ave., Suite 350
Bethesda, MD 20816 (301) 652-6580

Robert W. Michell
7932 Maryknoll Ave.
Bethesda, MD 20817 (301) 229-4979

Herbert W. Mylius
Martin Marietta Corp.
6801 Rockledge Dr.
Bethesda, MD 20817 (301) 897-6134

Barry J. Nace
Paulson & Nace
4550 Montgomery Ave., Suite 800-N
Bethesda, MD 20814 (301) 986-9100

Lawrence A. Neureither#
5807 Greentree Rd.
Bethesda, MD 20817 (301) 530-6137

Anthony A. O'Brien
Groff & O'Brien
4720 Montgomery Lane, Suite 1003, Montgomery Bldg.
Bethesda, MD 20814 (301) 657-2774

Kevin F. O'Brien#
Groff & O'Brien
4720 Montgomery Lane, Suite 1003, Montgomery Bldg.
Bethesda, MD 20814 (301) 657-2774

Thomas S. O'Dwyer*
Dept. Of The Navy, Office Of Pat. Coun, Code 005P
David Taylor Research Center
Bethesda, MD 20084 (202) 227-1837

Leroy Bruce Randall*
Dept. Of Health & Human Service
Westwood Bldg., Rm, 5A-03
Bethesda, MD 20892 (301) 496-7056

Robert R. Redmon
4701 Sangamore Rd.
Bethesda, MD 20016 (301) 320-5500

Peter E. Rosden
5024 Wissioming Rd.
Bethesda, MD 20816 (301) 229-2288

William H. Schultz#
4920 Redford Rd.
Bethesda, MD 20816 (301) 656-7334

David W. Selesnick#
6516 Elgin Lane
Bethesda, MD 20817 (301) 229-5871

Sheldon J. Singer
7315 Wisconsin Avenue
Bethesda, MD 20814 (301) 654-6505

Kenneth W. Sprague#
6619 Lone Oak Dr.
Bethesda, MD 20817 (301) 365-0121

Bernard Stickney
6021 Berkshire Dr.
Bethesda, MD 20014 (301) 530-1504

Walter Stolwein#
5211 Roosevelt Street
Bethesda, MD 20014 (301) 530-0921

Lazar D. Wechsler#
10105 Ashburton Lane
Bethesda, MD 20817 (301) 530-8697

Lorraine A. Weinberger
5404 Linden Court
Bethesda, MD 20814 (301) 530-0429

Frederic K. Wine
9116 Friars Rd.
Bethesda, MD 20817 (301) 897-8843

Jonathan D. Zischkau
8315 N. Brook Lane Apt. 606W
Bethesda, MD 20814 (301) 657-2310

Leon Zitver
6502 E. Halbert Rd.
Bethesda, MD 20817 (301) 229-6725

Donald J. Arnold#
2811 Sudberry Lane
Bowie, MD 20715 (301) 464-1452

Francis E. Blake
6403 S. Homestake Dr.
Bowie, MD 20715 (301) 262-0549

Morris Kaplan
3014 Tyson Lane
Bowie, MD 20715 (301) 464-8970

F. Richard Malzona
14300 Gallant Fox Lane, Suite 219
Bowie, MD 20715 (301) 262-6349

Alexander Skopetz
11911 Galaxy Lane
Bowie, MD 20715 (301) 262-6012

Mayer Weinblatt#
3326 Memphis Lane
Bowie, MD 20715 (301) 262-3264

John K. Donaghy
6108 McKay Dr.
Brandywine, MD 20613 (301) 372-8685

John Root Hopkins, Jr.
Route 4 Box 289A, Dark Road
Cambridge, MD 21613 (301) 228-6380

Luther Weston Gregory
Downes & Gregory
115 Lawyers Row
Centreville, MD 21617 (301) 758-0680

M. David Kreider
2512 Crest Ave.
Cheverly, MD 20785 (301) 773-1869

Roland Alfred Anderson
3810 Club Dr.
Chevy Chase, MD 20015 (301) 656-4175

Harold S. Block#
6412 Ruffin Rd.
Chevy Chase, MD 20815 (301) 652-8915

George W. Boys
4811 Wellington Dr.
Chevy Chase, MD 20815 (301) 652-5972

Rupert Joseph Brady
Brady, O'Boyle And Gates
920 Chevy Chase Bldg., 5530 Wisconsin Ave., N.W.
Chevy Chase, MD 20815 (301) 656-3326

Peter T. Dracopoulos
7407 Bybrook Lane
Chevy Chase, MD 20815 (301) 652-3330

Lawrence I. Field
3214 Pauline Dr.
Chevy Chase, MD 20815 (301) 656-3903

William L. Gates
Brady, O'Boyle And Gates
920 Chevy Chase Bldg., 5530 Wisconsin Ave., N.W.
Chevy Chase, MD 20815 (301) 656-3326

Harold L. Jenkins
4407 Walsh St.
Chevy Chase, MD 20815 (301) 657-2471

Sam Meerkreebs
5509 Greystone St
Chevy Chase, MD 20815 (301) 652-7960

Andrea G. Nace#
6208 Garnett Dr.
Chevy Chase, MD 20815 (301) 657-9393

Thomas A. Robinston
8101 Connecticut Avenue, S-502
Chevy Chase, MD 20815 (301) 986-8030

Raphael Semmes
31 Quincy Street
Chevy Chase, MD 20815

Natalie Trousof
4701 Willard Avenue Apt. 721
Chevy Chase, MD 20815 (301) 656-1795

Michael W. Werth
14 Grafton Street
Chevy Chase, MD 20815 (301) 986-0793

Frederick L. Matteson, Jr.
5918 Chillum Gate Rd.
Chillum, MD 20782 (301) 559-2412

Charles L. Harness
6705 Whitegate Rd.
Clarksville, MD 21029 (301) 531-6189

Bennett G. Miller, Sr.
P.O. Box 155
Cobb Island, MD 20625 (301) 259-4569

William G. Christoforo
10609 Blue Bell Way
Cockeysville, MD 21030 (301) 666-0824

Boyce C. Dent
Ward Machinery Co.
10615 Beaver Dam Rd.
Cockeysville, MD 21030 (301) 584-7700

Jeffrey R. Melnikoff
10311 Greentop Rd.
Cockeysville, MD 21030 (301) 628-2975

John D. Randolph
7100 Baltimore Avenue
College Park, MD 20740 (301) 927-3035

Edward J. Cabic
W.R. Grace & Co.
7379 Route 32
Columbia, MD 21044 (301) 531-4512

David E. Heiser
W.R. Grace & Company, Washington Research Center
7379 Route 32
Columbia, MD 21044 (301) 531-4519

William W. McDowell, Jr.
W.R. Grace & Co.
7379 Route 32
Columbia, MD 21044 (301) 531-4514

Arthur Paul Savage
W.R. Grace And Co. Research Division
7379 Route 32
Columbia, MD 21044 (301) 531-4511

Steven T. Trinker
W.R. Grace & Co., Pat., Dept.
7379 Route 32
Columbia, MD 21044 (301) 531-4120

Howard J. Troffkin
W.R. Grace & Co., Pat. Div.
7379 Route 32
Columbia, MD 21044 (301) 531-4516

Warren Hall Willner
836 Hazel
Crownsville, MD 21032 (301) 923-3591

Randall G. Erdley
15201 Springfield Rd.
Darnestown, MD 20874 (301) 948-8631

Harold A. Dixon
16204 Deer Lake Rd.
Derwood, MD 20855 (301) 926-1739

J. D. Miller
10031 Kaylorite Street
Dunkirk, MD 20754 (301) 855-8891

J.H. Fielding Jukes
22 Lynnbrook Terr.
Easton, MD 21601 (301) 822-9122

Donald E. Bullock#
1513 Shore Drive
Edgewater, MD 21037 (301) 261-7049

William M. Henry#
3669 First Ave.
Edgewater, MD 21037 (301) 798-0205

Frank W. Lane#
Red Bird Farm
162 Russell Rd.
Elkton, MD 21921 (301) 398-0724

Morton J. Rosenberg
3444 Ellicott Center Dr., Suite 105
Ellicott City, MD 21043 (301) 465-6678

John R. Utermohle#
National Security Agency Attn: R (pA)
Ft. George G., MD 20755 (301) 859-6647

Thomas O. Maser*
National Security Agency Attn. R (pa)
Ft. George G. Meade, MD 20755
(301) 859-6647

Joseph A. Finlayson, Jr.
8410 Indian Head Highway
Fort Washington, MD 20744 (301) 567-7230

Frank W. Miga
8007 Carey Branch Dr.
Fort Washington, MD 20744 (301) 567-6840

Richard Salvatore Sciascia
13218 Park Lane
Fort Washington, MD 20744 (301) 292-2970

Harry E. Thomason
609 Cedar Avenue
Fort Washington, MD 20744 (301) 292-5122

Paul E. ODonnell, Jr.*
U.S. Army
Medical Res. & Dev. Comm.
Fort Detrick
Frederick, MD 21701 (301) 663-2065

Isaac A. Angres#
6 War Admiral Court
Gaithersburg, MD 20878 (301) 926-2742

Mark S. Berninger#
Life Technologies Inc.
8717 Grovement Circle
Gaithersburg, MD 20877 (301) 258-8205

James C. Haight
20413 Cherrystone Court
Gaithersburg, MD 20879 (301) 842-1690

Nathan Kaufman#
427 Christopher Ave., Apt. 22
Gaithersburg, MD 20760 (301) 948-8147

David W. Lotterer
P.O. Box 3175
Gaithersburg, MD 20878 (301) 948-7550

Michael W. York
5508 Griffith Rd.
Gaithersburg, MD 20879 (301) 253-4217

Edward D. C. Bartlett#
14026 Burntwoods Rd.
Glenelg, MD 21737 (301) 442-2203

Robert D. Marchant*
NASA, Goddard Space Flight Center
Off. Of Pat. Coun., Code 204
Greenbelt, MD 20771 (301) 286-9279

Harvey Ostrow*#
NASA, Code 925
Goddard Space Flight Center.
Greenbelt, MD 20771 (301) 344-8107

Ronald F. Sandler*
NASA, Goddard Space Flight Center
Greenbelt, MD 20771 (301) 344-9275

James R. Garrett#
4300 College Hgts., Dr.
Hyattsville, MD 20782 (301) 779-1384

Richard J. Keegan
4312 Hamilton Street
Hyattsville, MD 20781 (301) 779-2016

John H. Mack
7208 Hitching Post Lane
Hyattsville, MD 20783 (301) 422-8898

Edward M. Woodberry#
7401 N. Hampshire Ave., Apt 903
Hyattsville, MD 20783 (301) 434-2609

Jo Anne S. Beery
211 Kearney Dr.
Joppa, MD 21085 (301) 679-2010

Morton J. Frome
Katz, Frome, Slan & Bleecker, P.A.
10605 Concord St., Suite 300
Kensington, MD 20895 (301) 949-5200

Albert Henry Kirchner
P.O. Box 101
Kensington, MD 20895 (301) 493-8322

Harry E. Kitchen#
3514 Plyers Mill Rd., Suite 104
Kensington, MD 20895 (301) 933-5440

Charles B. Parker
9918 Kensington Pkwy.
Kensington, MD 20895 (301) 942-3350

John O. Tresansky
9604 Old Spring Rd.
Kensington, MD 20895 (301) 946-0775

Elizabeth Lassen
American Electronics, Inc.
9332 Annapolis Rd.
Lanham, MD 20788 (301) 459-4343

Robert Edmund Archibald
Johns Hopkins Univ., Applied Physics Lab., Johns Hopkins Rd.
Laurel, MD 20707 (301) 953-5632

Leander F. Aulisio#
13203 Claxton Dr.
Laurel, MD 20708 (301) 776-7905

Mary L. Beall
The Johns Hopkins University, Applied Physics Lab.
Johns Hopkins Road
Laurel, MD 20707 (301) 953-5641

Carl I. Brundidge
Johns Hopkins University Applied Physics Labs.
Johns Hopkins Road
Laurel, MD 20707 (301) 953-5632

Howard W. Califano
Johns Hopkins University Applied Physics Lab.
Johns Hopkins Rd.
Laurel, MD 20707 (301) 953-5641

Gordon C. Fell#
8716 Granite Lane
Laurel, MD 20708 (301) 953-7392

Larry Chauncey Hall
Westvaco Corporation
Johns Hopkins Rd.
Laurel, MD 20707 (301) 792-9100

Thomas A. Lupica#
9444 Canterbury Riding
Laurel, MD 20707 (301) 953-7986

Mishrilal L. Jain
101 W. Ridgely Rd., Suite 1B, Ridgely Prof. Bldg.
Lutherville, MD 21093 (301) 252-7166

Stanley T. Krawczewicz
6103 87th Avenue
New Carrolton, MD 20784 (301) 577-7352

Joseph H. O'Toole
4008 Eland Rd.
Phoenix, MD 21131 (301) 592-2399

Walter O. Ottesen
5 Nantucket Garth
Phoenix, MD 21131 (301) 592-8383

Edwin T. Yates, Jr.#
19 Club View Lane
Phoenix, MD 21131 (301) 667-4977

Rodney D. Bennett, Jr.
10609 Crossing Creek Rd.
Potomac, MD 20854 (301) 299-4204

Maurice U. Cahn
Hall, Myers & Rose
10220 River Rd., Suite 200
Potomac, MD 20854 (301) 365-8000

Alan M. Friedman#
8510 Wilkesboro Lane
Potomac, MD 20854 (301) 299-6631

S. Rolfe Gregory#
11603 Milbern Dr.
Potomac, MD 20854 (301) 299-6368

William D. Hall
Hall, Myers & Rose
10220 River Rd. Suite 200
Potomac, MD 20854 (202) 365-8000

Geoffrey R. Myers
Hall, Myers & Rose
200 Semmes Bldg.
10220 River Rd.
Potomac, MD 20854 (301) 299-2320

Gordon S. Parker
8901 Falls Rd.
Potomac, MD 20854 (301) 299-5659

Leonard Rawicz
11806 Prestwick Rd.
Potomac, MD 20854 (301) 983-1536

Howard L. Rose
Hall, Myers & Rose
200 Semmes Bldg., 10220 River Rd.
Potomac, MD 20854 (301) 365-8000

Joel Stearman
8506 Wild Olive Dr.
Potomac, MD 20854 (301) 424-2877

Tak Ki Sung
11106 Candlelight Lane
Potomac, MD 20854 (301) 983-0842

Frank P. Flury
5811 Baltimore Avenue
Riverdale, MD 20737 (301) 927-3400

Arthur Livingston Branning
11301 Commonwealth Dr. Apt. 302
Rockville, MD 20852 (301) 881-6673

Sidney Cater
Old Georgetown Village
11339 Empire Lane
Rockville, MD 20852 (301) 984-1239

Ira Charles Edell
Epstein & Edell
932 Hungerford Dr., Suite 6A, Jackson Place South
Rockville, MD 20850 (301) 424-3640

Robert Howard Epstein
Epstein & Edell
932 Hungerford Dr., Suite 6A, Jackson Place South
Rockville, MD 20850 (301) 424-3640

Roy D. Frazier#
14109 Bauer Dr.
Rockville, MD 20853 (301) 871-7324

Joseph A. Genovese
Control Data Corp.
6003 Executive Blvd.
Rockville, MD 20852 (301) 468-8547

Bernard Helfin#
13918 Marianna Dr.
Rockville, MD 20853 (301) 871-7157

Margaret R. Howlett
12903 Atlantic Avenue
Rockville, MD 20851 (301) 881-7056

Daniel M. Kennedy#
1140 Rockville Pike, Suite 750
Rockville, MD 20852 (301) 770-1767

Donald A. Kettlestrings
414 Hungerford Dr., Suite 211
Rockville, MD 20850 (301) 279-7577

George H. Krizmanich#
14609 Melinda Lane
Rockville, MD 20853

O.A. Neumann
6821 Old Stage Rd.
Rockville, MD 20852 (301) 881-4053

David B. Newman, Jr.
932 Hungerford Dr., Suite 35-A
Rockville, MD 20850 (301) 294-2310

Rex Logan Sturm
Brown & Strum
260 E. Jefferson St.
Rockville, MD 20850 (301) 762-2555

Marvin S. Towsend
8 Grovepoint Court
Rockville, MD 20854 (301) 279-0660

William Britton Moore
531 Little John
Sherwood Forest, MD 21405 (301) 849-8303

Edward C. Allen
1903 Gatewood Place
Silver Spring, MD 20903 (301) 434-3575

Harold Ansher
11703 Fulham Street
Silver Spring, MD 20902 (301) 649-3752

Norton Ansher
13102 Middlevale Lane
Silver Spring, MD 20906 (301) 946-4642

Jack E. Armore
11817 Mentone Rd.
Silver Spring, MD 20906 (301) 946-1360

Leon J. Bercovitz
10036 Renfrew Rd.
Silver Spring, MD 20901 (301) 593-4675

Hyland Bizot
2702 Harmon Rd.
Silver Spring, MD 20902 (301) 933-2199

Milton Buchler
9505 Ocala St.
Silver Spring, MD 20901 (301) 587-6673

Alice Lee Chen#
3013 Birchtree Lane
Silver Spring, MD 20906 (301) 460-8921

W. B. Childs
17 Piping Rock Rd.
Silver Spring, MD 20904 (301) 384-9343

Justin P. Dunlavey
2027 Forest Hill Dr.
Silver Spring, MD 20903 (301) 439-3354

Nathan Edelberg
11012 Lombard Rd.
Silver Spring, MD 20901 (301) 593-2040

Charles W. Fallow
9822 Capitol View Avenue
Silver Spring, MD 20910 (301) 589-5891

Samuel Feinberg
12915 Goodhill Rd.
Silver Spring, MD 20906 (301) 946-8372

William Feldman
932 Schindler Dr.
Silver Spring, MD 20903 (301) 593-3253

Bernard A. Gelak#
115 Delford Ave.
Silver Spring, MD 20904 (301) 622-2676

Harold A. Gell, Jr.
13720 Lockdale Rd.
Silver Spring, MD 20906 (301) 460-0756

Lawrence Glassman
2203 Quinton Rd.
Silver Spring, MD 20910 (301) 358-9278

Benjamin J. Goldfarb
1001 Playford Lane
Silver Spring, MD 20901 (301) 593-3162

Herman Lewis Gordon
306 Ellsworth Dr.
Silver Spring, MD 20910 (301) 588-8968

Quinton E. Hodges
12508 White Dr.
Silver Spring, MD 20904 (301) 622-0167

Roger D. Johnson*
U.S. Dept. Of Navy Off. Of Pat. Coun. Code C72W
Naval Surface Warfare Center
10901 New Hampshire Avenue
Silver Spring, MD 20903 (202) 394-2174

Murray Katz
11435 Monterrey Dr.
Silver Spring, MD 20902 (301) 949-3878

David Klein
11418 Monterrey Dr.
Silver Spring, MD 20902 (301) 949-3185

Bernard Konick
11617 Fullham Street
Silver Spring, MD 20902 (301) 649-3180

Roy Lake#
1221 Burton St.
Silver Spring, MD 20910 (301) 588-7602

Joseph R. Liberman
609 Gilmoure Dr.
Silver Spring, MD 20901 (301) 593-1062

Eli Lieberman#
1712 Overlook Dr.
Silver Spring, MD 20903 (301) 434-2827

Hung C. Lin#
8 Schindler Court
Silver Spring, MD 20903 (301) 454-6853

Charles R. Malandra, Jr.
Singer Company, Link Simulation System Div.
11800 Tech Rd.
Silver Spring, MD 20904 (301) 622-8586

Raymond E. Martin
423 St. Lawrence Dr.
Silver Spring, MD 20901 (301) 593-0555

Alex Mazel#
15036 Condover Court
Silver Spring, MD 20906 (301) 598-5888

Aldrich F. Medbery#
8121, Suite 802
Silver Spring, MD 20910 (301) 565-2332

William Misiek
405 Royalton Rd.
Silver Spring, MD 20901 (301) 593-3953

Milton Osheroff
1316 Fenwick Lane
Silver Spring, MD 20910 (301) 588-3586

Richard P. Plunkett
615 Lycoming St.
Silver Spring, MD 20901 (301) 585-2891

Joseph S. Reich#
8503 Sundale Dr.
Silver Spring, MD 20910 (301) 585-0901

Elbert L. Roberts#
13103 Brittany Dr.
Silver Spring, MD 20904 (301) 384-3255

Samuel B. Rothberg#
1121 University Blvd., West Apt.811
Silver Spring, MD 20902 (301) 649-5030

Irving J. Rotkin
10202 Lariston Lane
Silver Spring, MD 20903 (301) 434-8882

Gersten Sadowsky
12400 Conn. Ave.
Silver Spring, MD 20906 (301) 949-7095

Paul Joseph Schmitz
1121 W. University
Silver Spring, MD 20902 (301) 649-4654

David S. Scrivener
1714 Overlook Dr.
Silver Spring, MD 20903 (307) 434-7722

Lewis A. Thaxton#
520 Beaumont Rd.
Silver Spring, MD 20904 (301) 384-8224

Thomas C. Tokos
1707 Sanford Road
Silver Spring, MD 20902 (301) 593-8536

Kenneth E. Walden*
U.S. Dept. Of Navy, Naval Surfaces Weapons Center
Code C72W
Silver Spring, MD 20903 (202) 394-2174

Joseph Christian Warfield, Jr.
14607 Edelmar Dr.
Silver Spring, MD 20906 (301) 598-6299

William T. Webb#
14420 Cantrell Rd.
Silver Spring, MD 20904 (301) 384-6398

Fredrick A. Wein*
Dept. Of The Navy, Naval Surface Weapons Center Dept.
10901 New Hampshire Ave.
Silver Spring, MD 20903 (202) 394-2174

Morris O. Wolk
1113 Easecrest Dr.
Silver Spring, MD 20902 (301) 649-1311

Stephen F.K. Yee
Georges Breneman, Hellweg & Yee
8720 Georgia Ave., Suite 704
Silver Spring, MD 20910 (301) 589-1210

Charles Sukalo
4806 Silver Hill Rd.
Suitland, MD 20746 (301) 420-7707

Louis A. Scholz
850 Route 32
Sykesville, MD 21784 (301) 795-9100

George E. Schmitkons#
4511 Simmons Lane
Temple Hill, MD 20748 (301) 894-1288

Anthony V. Ciarlante#
4310 Brinkley Rd.
Temple Hills, MD 20748 (301) 423-4148

Leonard Bloom
401 Washington Ave., Suite 803
Towson, MD 21204 (301) 337-2295

Dennis A. Dearing
Black & Decker (U.S.) Inc.
701 E. Joppa Rd.
Towson, MD 21204 (301) 583-3503

Joseph Bruce Hoofnagle, Jr.
The Black & Decker Corp.
701 E. Joppa Rd.
Towson, MD 21204 (301) 583-2704

Jacob A. Manian
39 Acorn Ct., Apt.202
Towson, MD 21204 (301) 296-9224

John F. McClellan, Sr.
Loyola Federal Bldg.
22 W. Penn. Ave.
Towson, MD 21204 (301) 821-7900

Edward D. Murphy
Balck And Decker Mfg. Co.
701 E. Joppa Rd.
Towson, MD 21204 (301) 583-2867

Kevin B. Nachtrab
401 Washington Ave.
Towson, MD 21204 (301) 337-2295

Reginald F. Pippin, Jr.
7806 Ruxway Rd.
Towson, MD 21204 (301) 666-1400

Harold Weinstein
Black And Decker Mfg. Co.
701 E. Joppa Rd.
Towson, MD 21204 (301) 583-2886

Charles E. Yocum
Black And Decker Mfg, Co.
701 E. Joppa Rd.
Towson, MD 21204 (301) 583-2956

Alfred C. Perham#
8409 Thornberry Dr, West
Upper Marlboro, MD 20772 (301) 627-2234

Jess Joseph Smith, Jr.
Burroughs & Smith, Drawer 519
14718 Main St.
Upper Marlboro, MD 20870 (301) 627-2558

Robert Myers McKinney#
52 Ridge Rd.
Westminster, MD 21157 (301) 848-3325

Eugene F. Osborne, Sr.
P.O. Box 423
Westminster, MD 21157 (301) 848-0861

Robert W. Boyle#
13809 Bethpage Lane
Wheaton, MD 20906 (301) 460-3322

Herbert S. Cockeram#
3804 Delano Street
Wheaton, MD 20902 (301) 933-9319

Isaac J. Gorman#
2714 Henderson Ave.
Wheaton, MD 20902 (301) 942-2497

Massachusetts

Barry Raymond Blaker#
4 Algonquin Rd.
Acton, MA 01720 (617) 263-3440

Arthur W. Fisher
Digital Equipment Corp.
100 Nagog Park
Acton, MA 01702 (617) 264-6805

Dewitt C. Seward, III#
P.O. Box 261
Acton, MA 01720 (617) 263-3871

Richard W. Toelken#
32 Elm Street
Agawam, MA 01001 (413) 730-2046

William J. Driscoll
90 Main Street
Andover, MA 01810 (617) 475-6371

Edward A. Gordon
90 Main St.
Andover, MA 01810 (617) 475-6371

Kenneth A. Green#
19 Burton Farm Dr.
Andover, MA 01810 (617) 475-8423

Lowell H. Mc Carter
Allied Health & Scientific Products Co.
1 Burtt Rd.
Andover, MA 01810 (617) 470-1790

James W. Mitchell
18 Copely Dr.
Andover, MA 01810 (617) 475-3121

Herbert Warren Arnold
151 Mystic St. Apt. M-5
Arlington, MA 02174 (617) 646-7210

Joseph J. Gano#
31 Davis Ave.
Arlington, MA 02174 (607) 643-9319

Kent R. Johnson
2 Old Colony Lane, #11
Arlington, MA 02174 (617) 643-1538

John D. Karagounis
687 Tiffany St.
Attleboro, MA 02703 (617) 222-0290

James Patrick McAndrews
Texas Instruments Inc.
34 Forest St.
Attleboro, MA 02703 (617) 699-3245

Frank M. Ward, III
8 Blackstone Road
Attleboro, MA 02703 (617) 266-1024

Robert T. Dunn
4 Cedar Ridge Dr.
Bedford, MA 01730 (617) 275-6146

Andrew T. Karnakis
Millipore Corp.
80 Ashby Rd.
Bedford, MA 01730 (617) 275-9200

Thomas P. Melia
Raytheon Co., Missile System Div.
Hartwell Rd.
Bedford, MA 01730 (617) 274-7100

William J. O'Brien
49 Glenridge Dr.
Bedford, MA 01730 (617) 275-6289

Stuart B. Zigun
33 Lido Lane
Bedford, MA 01730 (617) 275-6921

William N. Anastos
28 Longmeadow Rd.
Belmont, MA 01930 (617) 281-0440

James L. Diamond
22 Pine St.
Belmont, MA 02178 (617) 684-6148

Herbert Malsky
96 Washington St.
Belmont, MA 02178 (617) 484-1315

Andrew M. McGinnis
489 Common St.
Belmont, MA 02178 (617) 489-2650

Aubrey C. Brine
244 E. Lothrop St.
Beverly, MA 01915 (617) 927-4200

Elton T. Barrett
21 Hattie Lane
Billerica, MA 01821 (617) 667-6064

Lawrence A. Chaletsky
Cabot Corp.
Concord Rd.
Billerica, MA 01821 (617) 663-3455

George M. Medeiros
5 Andover Rd., Suite 23
Billerica, MA 01821 (617) 663-3467

Gerald Altman
Morse, Altman & Dacey
85 E. India Row., Suite 5F
Boston, MA 02110 (617) 523-3515

Robert M. Asher
Dike, Bronstein, Roberts, Cushman & Pfund
130 Water St.
Boston, MA 02109 (617) 542-8492

Michael E. Attaya
Cesari & McKenna
Union Wharf East
Boston, MA 02109 (617) 523-8100

Barbara A. Barakat
Pahl, Lorusso & Loud
60 State Street
Boston, MA 02109 (617) 723-8880

Steven M. Bauer
Weingarten, Schurgin, Gagnebin & Hayes
Ten Post Office Square
Boston, MA 02109 (617) 542-2290

Doris M. Bennett
Stone & Webster
245 Summer St.
Boston, MA 02107 (617) 589-8600

Michael J. Bevilacqua
Hale & Dorr
60 State St.
Boston, MA 02109 (617) 742-9100

Richard J. Birch
Thompson, Birch, Gauthier & Samuels
225 Franklin St., Suite 3300
Boston, MA 02110 (617) 426-8989

Arthur Z. Bookstein
Wolf, Greenfield & Sacks, P.C., Federal Reserve Plaza
600 Atlantic Ave.
Boston, MA 02210 (617) 720-3500

William E. Booth#
Fish & Richardson
One Financial Center, Suite 2500
Boston, MA 02110 (617) 542-5070

Joseph H. Born
Cesari & McKenna
312 Union Wharf E.
Boston, MA 02109 (617) 523-8100

Sewall P. Bronstein
Dike, Bronstein, Roberts, Cushman & Pfund
130 Water St.
Boston, MA 02109 (617) 545-8492

Donald Brown
Dike, Bronstein, Roberts, Cushman & Pfund
130 Water St.
Boston, MA 02109 (617) 542-8492

Linda M. Buckley
Dike, Bronstein, Roberts, Cushman & Pfund
130 Water St.
Boston, MA 02109 (617) 542-8492

William A. Cammett
Cesari & McKenna
312 Union Wharf East
Boston, MA 02109 (617) 523-8100

Adrienne M. Catanese
Gaston Snow & Ely Bartlett
One Federal St.
Boston, MA 02110 (617) 426-4600

Robert A. Cesari
Cesari And McKenna
Union Wharf
Boston, MA 02109 (617) 523-8100

Franklin H. Chasen
44 School St., Suite 710
Boston, MA 02108 (617) 367-9943

Stephen Y. Chow
Cesari & McKenna
312 Union Wharf East
Boston, MA 02108 (617) 227-8100

Francis J. Clark
Kendall Company
One Federal St.
Boston, MA 02110 (617) 574-7915

Paul T. Clark
Fish & Richardson
1 Financial Center, Suite 2500
Boston, MA 02110 (617) 542-5070

James E. Cockfield
Lahive & Cockfield
60 State St.
Boston, MA 02109 (617) 227-7400

Jerry Cohen
Cohen & Burg, P.C.
33 Broad St.
Boston, MA 02109 (617) 742-7840

David C. Conlin
Dike, Bronstein, Roberts, Cushman & Pfund
130 Water St.
Boston, MA 02109 (617) 542-8492

G. Eugene Dacey
Morse, Altman, Dacey & Benson
One Exeter Plaza
Boston, MA 02116 (617) 523-3515

David Markham Driscoll
Wolf, Greenfield & Sacks, P.C., Federal
Reserve Plaza
600 Atlantic Ave.
Boston, MA 02210 (617) 720-3500

Albert P. Durigon
Weingarten, Schurgin, Gagnebin & Hayes
Ten Post Office Square
Boston, MA 02109 (617) 542-2290

James M. Emery#
Fish & Richardson
28 State St.
Boston, MA 02109 (617) 523-7340

Willis Marion Ertman
Fish & Richardson
One Financial Center
Boston, MA 02110 (617) 542-5070

Michael I. Falkoff
Bromberg, Sunstein & McGregor
31 Milk St.
Boston, MA 02109 (617) 426-6464

David L. Feigenbaum
Fish & Richardson
One Financial Center
Boston, MA 02111 (617) 542-5070

James J. Foster
Wolf, Greenfield & Sacks, P.C.
600 Atlantic Ave.
Boston, MA 02210 (617) 720-3500

John W. Freeman
Fish & Richardson
One Financial Center
Boston, MA 02111 (617) 542-5070

Timothy A. French
Fish & Richardson
One Financial Center
Boston, MA 02111 (617) 542-5070

R. W. Furlong
Fish & Richardson
One Financial Center
Boston, MA 02111 (617) 542-5070

Charles L. Gagnebin, III
Weingaten, Schurgin, Gagnebin & Hayes
Ten Post Office Square
Boston, MA 02109 (617) 542-2290

Robert Trafton Gammons
Dike, Bronstein, Roberts, Cushman & Pfund
130 Water St.
Boston, MA 02109 (617) 542-8492

Edward R. Gates
Wolf, Greenfield & Sacks, P.C.
600 Atlantic Ave.
Boston, MA 02210 (617) 720-3500

Herbert L. Gatewood
Lappin, Rosen & Goldberg
One Boston Place, 41st Floor
Boston, MA 02108 (617) 437-1000

Maurice Edward Gauthier
Sammuels, Gauthier, Stevens & Kehoe
225 Franklin St.
Boston, MA 02110 (617) 426-9180

William C. Geary, III
Thompson ,Birch, Gauther & Samuels
225 Franklin St.
Boston, MA 02110 (617) 426-9180

Lawrence Gilbert
Boston University
881 Commonwealth Ave., 5th Floor
Boston, MA 02215 (617) 353-2212

Susan G.L. Glovsky
225 Friend St.
Boston, MA 02114 (617) 227-8539

David S. Goldenberg
Cesari & McKenna
312 Union Wharf East
Boston, MA 02109 (617) 523-8100

Elmer J. Gorn
Russell & Tucker
89 State St.
Boston, MA 02109 (617) 227-3835

William F. Gray
Weingarten, Schurgin, Gagnebin & Hayes
Ten Post Office Square
Boston, MA 02109 (617) 542-2290

Lawrence M. Green
Wolf, Greenfield & Sacks, P.C., Federal
Reserve Plaza
600 Atlantic Ave.
Boston, MA 02210 (617) 720-3500

George L. Greenfield
Wolf, Greenfield & Sacks, P.C., Federal
Reserve Plaza
600 Atlantic Ave.
Boston, MA 02210 (617) 720-3500

Marvin Curtis Guthrie
Massachusetts General Hospital Office Of
Tech. Affairs
75 Blossom Court
Boston, MA 02114 (617) 726-8608

Paul J. Hayes
Weingarten, Schurgin, Gagnebin & Hayes
Ten Post Office Square
Boston, MA 02109 (617) 542-2290

Therese A. Hendricks
Wolf, Greenfield & Sacks, P.C., Federal
Reserve Plaza
600 Atlantic Ave.
Boston, MA 02210 (617) 720-3500

Gilbert H. Hennessey, III
Fish & Richardson
One Financial Center
Boston, MA 02111 (617) 542-5070

Steven J. Henry
Wolf, Greenfield & Sacks, P.C.
600 Atlantic Ave.
Boston, MA 02215 (617) 720-3500

Mark J. Herbert
Fish & Richardson
One Financial Center Suite 2500
Boston, MA 02111 (617) 542-5070

Robert Eliot Hillman
Fish & Richardson
One Financial Center
Boston, MA 02111 (617) 542-5070

Jason M. Honeyman#
Gaston & Snow
One Federal St.
Boston, MA 02110 (617) 426-4600

David E. Hoppe
294 Beacon St.
Boston, MA 02116 (617) 262-4509

Robert J. Horn, Jr.
Kenway & Jenney
60 State St.
Boston, MA 02109 (617) 227-6200

Alvin Isaacs
Kendall Company
One Federal St.
Boston, MA 02110 (617) 574-7920

Alfred Russell Johnson
weingarten, Schurgin, Gagnebin & Hayes
Ten Post Office Square
Boston, MA 02109 (617) 542-2290

Richard A. Jordan
Cesari & McKenna
312 E. Union Wharf
Boston, MA 02109 (617) 523-8100

Herert S. Kassman
Cohen & Burg, P.C.
33 Broad St.
Boston, MA 02109 (617) 742-7840

Andrew F. Kehoe
Samuels, Gauthier, Stevens & Kehoe
225 Franklin St., Suite 3300
Boston, MA 02110 (617) 426-9180

Edger H. Kent
Fish & Richardson
One Financial Center
Boston, MA 02111 (617) 542-5070

Herbert P. Kenway
Kenway & Jenney
60 State St.
Boston, MA 02109 (617) 227-6300

Daniel Kim
Bromberg, Sunstein & McGregor
Ten West St.
Boston, MA 02111 (617) 426-6464

Philip G. Koenig
Dike, Bronstein, Roberts, Cushman & Pfund
130 Water St.
Boston, MA 02109 (617) 542-8492

Ronald Joseph Kransdorf
Wolf, Greenfield & Sacks, P.C.
600 Atlantic Ave.
Boston, MA 02210 (617) 720-3500

Walter Joseph Kreske, Sr.
60 State St., Suite 3330
Boston, MA 02109 (617) LA3-4420

Paul E. Kudirka
Wolf, Greenfield & Sacks, P.C., Federal Reserve Plaza
600 Atlantic Ave.
Boston, MA 02210 (617) 720-3500

John A. Lahvie, Jr.
Lahvie & Cockfield
60 State St., Suite 510
Boston, MA 02109 (617) 227-7400

James B. Lampert
Hale & Dorr
60 State St.
Boston, MA 02109 (617) 742-9100

Mark G. Lappin
Lahive & Cockfield
60 State St., Suite 510
Boston, MA 02109 (617) 227-7400

Victor F. Lebovici
Weingarten, Schurgin, Gagnebin & Hayes
Ten Post Office Square
Boston, MA 20109 (617) 542-2290

G. Roger Lee
Fish & Richardson
One Financial Center
Boston, MA 02111 (617) 542-5070

William H. Lee
Wolf, Greenfield & Sacks, P.C., Federal Reserve Plaza
600 Atlantic Ave.
Boston, MA 02210 (617) 720-3500

W. Hugo Liepmann
Lahive & Cockfield
60 State St., Suite 510
Boston, MA 02109 (617) 227-7400

Ernest V. Linek
Dike, Bronstein, Roberts, Cushman & Pfund
130 Water St.
Boston, MA 02109 (617) 542-8492

Ralph A. Loren
Lahive & Cockfield
60 State St., Suite 510
Boston, MA 02109 (617) 523-8100

Anthony M. Lorusso
Pahl, Lorusso & Loud
60 State St.
Boston, MA 20109 (617) 723-8880

Jeremiah Lynch
Lahive & Cockfield
60 State St.
Boston, MA 02109 (617) 227-7400

Paul F. Lynch#
240 Commercial St.
Boston, MA 02109 (617) 367-3006

Gregory A. Madera
Fish & Richardson
One Financial Center
Boston, MA 02111 (617) 542-5070

Stephen G. Matzuk
Weingarten, Schurgin, Gagnebin & Hayes
Ten Post Office Square
Boston, MA 02109 (617) 542-2290

William R. McClellan
Wolf, Greenfield & Sacks, P.C., Federal Reserve Plaza
600 Atlantic Ave.
Boston, MA 02210 (617) 720-3500

John P. McGonagle
McGonagle & McGrowan
11 Beacon St.
Boston, MA 02108 (617) 227-7755

John F. McKenna
Cesari And McKenna
Union Wharf East
Boston, MA 02109 (617) 523-8100

Donald W. Meeker#
P.O. Box 8
Prudential Center Station
Boston, MA 02199 (617) 267-4332

John B. Miller
Gadsby & Hannah
One Post Office Square
Boston, MA 02109 (617) 357-8700

Anthony J. Mirabito
Cesari & McKenna
312 Union Wharf East
Boston, MA 02109 (617) 523-8100

James E. Mrose
Thomas & Mrose
468 Park Dr.
Boston, MA 02215 (617) 262-6452

Charles Francis Murphy
Massachusetts General Hospital Off. Of Tech. Affairs
Fruit St.
Boston, MA 02114 (617) 726-8608

George W. Neuner
Dike, Bronstein, Roberts, Cushman & Pfund
130 Water St.
Boston, MA 02109 (617) 542-8492

Robert F. O'Connell
Dike, Bronstein, Roberts, Cushman & Pfund
130 Water St.
Boston, MA 02109 (617) 542-8492

Martin J. O'Donnell
Cesari & McKenna
Union Wharf East
Bostor, MA 02109 (617) 523-8100

Thomas C. O'Konski
Cesari & McKenna
Union Wharf East
Boston, MA 02109 (617) 523-8100

Michael L. Oliverio
Wolf, Greenfield & Sacks, Federal Reserve Plaza
600 Atlantic Ave.
Boston, MA 02210 (617) 720-3500

Louis Litman Orenbuch
Wolf, Greenfield & Sacks, P.C.
201 Devonshire St.
Boston, MA 02110 (617) 426-6131

Ronald G. Ort
Cesari & McKenna
Union Wharf East
Boston, MA 02109 (617) 523-8100

Henry D. Pahl, Jr.
Pahl, Lorusso & Loud
60 State St.
Boston, MA 02109 (617) 723-8880

Sam Pasternack
Kenway & Jenney
60 State St.
Boston, MA 02109 (617) 227-0700

Edward F. Perlman
Wolf, Greenfield & Sacks, P.C., Federal Reserve Plaza
600 Atlantic Ave.
Boston, MA 02210 (617) 720-3500

Scott K. Peterson
Cesari & McKenna
Union Wharf East
Boston, MA 02109 (617) 523-8100

Charles E. Pfund
Dike, Bronstein, Roberts, Cushman & Pfund
130 Water St.
Boston, MA 02109 (617) 542-8492

Edmund R. Pitcher
Lahive & Cockfield
60 State St. Suite 510
Boston, MA 02109 (617) 227-7400

Frank P. Porcelli
Fish & Richardson
One Financial Center
Boston, MA 02111 (617) 542-5070

David J. Powsner
Lahive & Cockfield
60 State St.
Boston, MA 02109 (617) 227-7400

Eric L. Prahl
Cesari & McKenna
Union Wharf East
Boston, MA 02109 (617) 523-8100

Terrance J. Radke
Weingarten, Schurgin, Gagnebin & Hayes
Ten Post Office Square
Boston, MA 02109 (617) 542-2290

Alfred H. Rosen
Wolf, Greenfield & Sacks, P.C.
201 Devonshire St.
Boston, MA 02110 (617) 426-6131

Gary E. Ross
Lahive & Cockfield
60 State St.
Boston, MA 02109 (617) 227-7400

Robert Bernard Russell
Russell & Tucker
89 State St.
Boston, MA 02109 (617) 227-3835

William W. Rymer, Jr.
Fish & Richardson
One Financial Center
Boston, MA 02111 (617) 542-5070

Stanley Sacks
Wolf, Greenfield & Sacks, P.C., Federal Reserve Plaza
600 Atlantic Ave.
Boston, MA 02210 (617) 720-3500

I. Stephen Samuels
Samuels, Gauthier, Stevens & Kehoe
225 Franklin St., Suite 3300
Boston, MA 02110 (617) 426-9180

Edward J. Scahill, Jr.
Kendall Co.
One Federal St.
Boston, MA 02101 (617) 423-2000

Stanley M. Schurgin
Weingarten, Schurgin, Gagnebin & Hayes
Ten Post Office Square
Boston, MA 02109 (617) 542-2290

Edward R. Schwartz
Wolf, Greenfield & Sacks, P.C., Federal Reserve Plaza
600 Atlantic Ave.
Boston, MA 02210 (617) 720-3500

Robert A. Shack
Gillette Company
Prudential Tower Bldg.
Boston, MA 02199

Donald J. Shade
18 Pinckney St.
Boston, MA 02114 (617) 742-4527

Patricia A. Sheehan
Cesari & McKenna
Union Wharf East
Boston, MA 02109 (617) 523-8100

David Silverstein
Samuels, Gauthier, Stevens & Kehoe
225 Franklin St., Suite 3300
Boston, MA 02110 (617) 426-9180

John M. Skenyon
Fish & Richardson
One Financial Center
Boston, MA 02111 (617) 542-5070

Mandel E. Slater
Gillette Co.
Prudential Tower Bldg.
Boston, MA 02199 (617) 421-7885

Thomas V. Smurzynski
Lahive & Cockfield
60 State St., Suite 510
Boston, MA 02109 (617) 227-7400

Richard L. Stevens
Samuels, Gauthier, Stevens & Kehor
225 Franklin St., Suite 3300
Boston, MA 02110 (617) 426-9180

Bruce D. Sunstein
Bromberg, Sunstein & McGregor
31 Milk St., Suite 810
Boston, MA 02109 (617) 426-6464

C. Hall Swain
Hale & Dorr
60 State St.
Boston, MA 02109 (617) 742-9100

Robert Kanof Tendler
33 Broad St.
Boston, MA 02109 (617) 723-7268

David J. Thibodeau, Jr.
Cesari & McKenna
Union Wharf East
Boston, MA 02109 (617) 523-8100

Mary C. Thomson
Thomson & Mrose
468 Park Dr.
Boston, MA 02215 (617) 362-6452

David A. Tucker
Russell & Tucker
99 Chauncy St.
Boston, MA 02111 (617) 338-5411

Gary A. Walpert
Lahive & Cockfield
60 State St., Suite 510
Boston, MA 02109 (617) 227-7400

Richard J. Warburg#
Fish & Richardson
One Financial Center
Boston, MA 02111 (617) 542-5070

Steve J. Weissburg
Pahl, Lorusso & Loud
60 State St.
Boston, MA 02109 (617) 723-0880

John L. Welch
Wolf, Greenfield & Sacks, P.C., Federal Reserve Plaza
600 Atlantic Ave.
Boston, MA 02210 (617) 720-3500

Gregory D. Williams
Dike, Bronstein, Roberts, Cushman & Pfund
130 Water St.
Boston, MA 02109 (617) 542-8492

John Noel Williams
Fish & Richardson
One Financial Center
Boston, MA 02111 (617) 542-5070

Stephen P. Williams
Dike, Bronstein, Roberts, Cushman & Pfund
130 Water St.
Boston, MA 20109 (617) 542-8492

Charles C. Winchester, Jr.
Fish & Richardson
One Financial Center
Boston, MA 02111 (617) 542-5070

David Wolf
Wolf, Greenfield & Sacks, P.C., Federal Reserve Plaza
600 Atlantic Ave.
Boston, MA 02210 (617) 720-3500

William Hugh McNeill#
13 Silver Brook Rd.
Boxford, MA 01921 (617) 887-9623

Richard E. Favreau#
11 Carriage Dr.
Brewster, MA 02631 (617) 385-5502

Joseph H. Killion
4 Cypress St.
Brookline, MA 02146 (617) 232-0951

Bernard H. Lemlein
173 Mason Terrace
Brookline, MA 02146 (617) 232-7383

David Prashker
417 Washington St.
Brookline, MA 02146 (617) 232-7509

James R. O'Connor
T R W Assemblies & Fasteners Group
31 Burlington Mall Rd.
Burlington, MA 01803 (617) 273-0770

Wilfred J. Baranick
Badger Co., Inc.
One Bdwy.
Cambridge, MA 02140 (617) 494-7245

Herbert L. Bello
Schiller, Padiscio & Kusmer
125 Cambridge Park Dr.
Cambridge, MA 02140 (617) 499-2770

Philip P. Berestecki
Badger Co., Inc.
One Broadway
Cambridge, MA 02142 (617) 494-7203

Robert L. Berger
Polaroid Corp.
549 Technology Square
Cambridge, MA 02139 (617) 577-2202

David L. Berstein
Genetics Institute, Inc.
87 Cambridge Park Dr.
Cambridge, MA 02140 (617) 876-1170

Sylvia Lyle Boyd
P.O. Box 1050
Cambridge, MA 02238 (617) 924-4380

Sybil A. Campbell#
549 Technology Square
Cambridge, MA 02139 (617) 577-2576

Francis J. Caufield
Polaroid Corp., Pat. Dept.
549 Technology Square
Cambridge, MA 02139 (617) 577-3532

Alfred Ernest Corrigan
Polaroid Corp.
549 Technology Sq.
Cambridge, MA 02139 (617) 577-2974

Bruce M. Eisen
Genetics Institute, Inc.
87 Cambridge Park Dr.
Cambridge, MA 02140 (617) 876-1170

John W. Ericson
Polaroid Corp.
545 Technology Square
Cambridge, MA 02139 (617) 577-2713

Robert M. Ford
Polaroid Corp.
549 Technology Square
Cambridge, MA 02139 (617) 577-2215

Robert W. Hagopian
Orion Research Inc.
840 Memorial Dr.
Cambridge, MA 02139 (617) 864-5400

Beverly E. Hjorth#
4 Chester St. Unit 4B
Cambridge, MA 02140 (617) 547-9155

Karl Hormann
Polaroid Corp.
549 Technology Square
Cambridge, MA 02139 (617) 577-3912

David A. Jacobs
M.I.T., Box 91
Cambridge, MA 02139 (617) 272-7400

Ellen J. Kapinos
Genetics Institute, Inc.
87 Cambridge Park Dr.
Cambridge, MA 02140 (617) 876-1170

John J. Kelleher
Polaroid Corp.
549 Technology Square
Cambridge, MA 02139 (617) 577-3372

Philip George Kiely
Polaroid Corp.
549 Technology Sq.
Cambridge, MA 02139 (617) 577-3691

Toby H. Kusmer
Schiller, Padiscio & Kusmer
125 Cambridge Park Dr.
Cambridge, MA 02140 (617) 499-8770

Gaetano D. Maccarone
Polaroid Corp., Pat. Dept.
549 Tech. Square
Cambridge, MA 02139 (617) 577-4592

James E. Maslow
M I T
77 Mass. Ave., E 19/722
Cambridge, MA 02139 (617) 253-6966

Lawrence H. Meier
Schiller, Pandiscio
125 Cambridge Park Dr.
Cambridge, MA 02140 (617) 499-2770

Stanley H. Mervis
Polaroid Square
549 Technology Sq.
Cambridge, MA 02139 (617) 577-2281

Charles Mikulka
Polaroid Corp.
549 Technology Square
Cambridge, MA 02139 (617) 577-3011

John Joseph Moss
116 Bp. Allen Dr.
Cambridge, MA 02139 (617) 864-3900

Mark J. Pandiscio
Schiller, Pandiscio & Kusmer
125 Cambridge Park Dr.
Cambridge, MA 02140 (617) 499-2770

Nicholas A. Pandiscio
Schiller, Pandiscio & Kusmer
125 Cambridge Park Dr.
Cambridge, MA 02140 (617) 499-2770

Leslie J. Payne#
Polaroid Corp.
549 Technology Square
Cambridge, MA 02139 (617) 577-4714

Robert F. Peck
Polaroid Corp.
549 Technology Square
Cambridge, MA 02139 (617) 577-2905

Frank R. Perillo
59 Dana St.
Cambridge, MA 02138

William D. Roberson
Polaroid Corp. Patent Dept.
549 Technology Sq.
Cambridge, MA 02139 (617) 577-2218

Edward S. Roman
Polaroid Corp.
549 Technology Sq., 6th Floor
Cambridge, MA 02139 (617) 577-2518

Sheldon W. Rothstein
Polaroid Corp.
575 Technology Sq., 3rd Floor
Cambridge, MA 02139 (617) 577-2793

Bradley N. Ruben
Ceramics Process Systems Corp.
840 Memorial Dr.
Cambridge, MA 02139 (617) 354-2020

Robert J. Schiller
Schiller, Pandiscio & Kusmer
125 Cambridge Park Dr.
Cambridge, MA 02140 (617) 499-2770

Arthur A. Smith, Jr.
Mass. Institute Of Tech., Room E 19-722
77 Mass. Ave.
Cambridge, MA 02139 (617) 253-6966

H. Eugene Stubbs#
P.O. Box 1050
Cambridge, MA 02238 (617) 924-4380

David R. Thornton
Polaroid Corp. Pat. Dept.
549 Technology Sq.
Cambridge, MA 02139 (617) 577-2519

John S. Vale
Polaroid Corp. Pat. Dept.
549 Tech. Sq.
Cambridge, MA 02139 (617) 577-3013

Louis George Xiarhos
Polaroid Corp. Pat. Dept.
549 Technology Sq.
Cambridge, MA 02139 (617) 577-4314

Irwin P. Garfinkle
366 River Rd.
Carlisle, MA 01741 (617) 369-9074

Sherman Gilbert Davis#
39 Emerson Way
Centerville, MA 02632 (617) 771-2241

Michael N. Raisbeck
85 High St.
Chelmsford, MA 01824 (617) 250-1236

William Nitkin
850 Boylston St.
Chestnut Hill, MA 02167 (617) 232-1854

Arsen Tashjian
15 Glenland Rd.
Chestnut Hill, MA 02167 (617) 731-4070

E. Mackay Fraser
L F E
55 Green Street
Clinton, MA 01510 (617) 835-1011

Robert E. Ross
Box 76
Cohasset, MA 02025 (617) 383-1340

Ralph L. Cadwallader
30 Raymond Rd.
Concord, MA 01742 (617) 369-2906

Albert D. Ehrenfrid#
Metritape, Inc.
33 Bradford St.
Concord, MA 01742 (617) 369-7500

Kevin S. Lemack
747 Main St., Suite 200
Concord, MA 01742 (617) 369-0230

Henry Cooper Nields
Concord Professional Center
747 Main St., Suite 200
Concord, MA 01742 (617) 369-0230

Carlo Bessone#
G.T.E. Service Corp.
100 Endicott St.
Danvers, MA 01923 (617) 777-1900

Edward Joseph Coleman
GTE Service Corp.
100 Endicott St.
Danvers, MA 01923 (617) 777-1900

Martha A. Finnegan
G T E Products Corporation
100 Endicott Street
Danvers, MA 01923 (617) 750-2311

William E. Meyer
GTE Products Corp.
100 Endicott St.
Danvers, MA 01923 (617) 750-2384

John A. Odozynski#
G T E Service Corp.
100 Endicott St.
Danvers, MA 01923 (617) 777-1900

James Theodosopoulos
G T E Service Corp.
100 Endicott St.
Danvers, MA 01923 (617) 777-1900

Robert E. Walter
G T E Service Corp.
100 Endicott St.
Danvers, MA 01923 (617) 777-1900

Scott R. Foster
P.O. Box 136
Dover, MA 02030 (617) 421-7889

Lewis M. Smith, Jr.
Whitcomb Hill Rd., P.O. Box 77
Drury, MA 01343

Raymond J. DeVellis
900 Mayflower St.
Duxbury, MA 02332 (617) 585-3377

Daniel R. Radin#
233 Hutchville Rd.
E. Falmouth, MA 02536 (617) 563-2655

Peter Lorillard Tailer
P.O. Box 1327
Edgartown, MA 02539 (617) 693-3658

Paul J. Murphy
48 McKinley St.
Everett, MA 02149 (617) 387-3444

Irwin A. Shaw
Quaker Fabric Corp.
941 Grinnel St.
Fall River, MA 02721 (617) 678-1951

Frederick M. Murdock
7 Saconesset Rd.
Falmouth, MA 02540 (617) 584-6070

Terrence Martin
P.O. Box 540
Foxboro, MA 02035 (617) 549-6155

Jack H. Wu
Foxboro Co., Pat. Dept. 187 (52-1J)
38 Neponset Ave.
Foxboro, MA 02035 (617) 549-6295

Mark A. Hofer
Integrated Genetics, Inc.
31 New York Ave.
Framingham, MA 01701 (617) 872-8400

Esther A. H. Hopkins
135 Stony Brook Rd.
Framingham, MA 01701 (617) 872-8148

Barry D. Josephs
Dennison Mfg. Co., Pat. Dept.
300 Howard St.
Framingham, MA 01701 (617) 879-0511

Irving Martin Kriegsman
883 Edgell Rd.
Framingham, MA 01701 (617) 877-8588

John T. Meaney
184 Summer St.
Framingham, MA 01701 (617) 872-7227

Arthur B. Moore
Dennison Mfg. Co., Pat. Dept.
300 Howard St.
Framingham, MA 01701 (617) 879-0511

John E. Toupal
116 Concord St.
Framingham, MA 01701 (617) 872-3781

Leonard J. Janowski
12 Northgate Rd.
Franklin, MA 02038 (617) 528-9967

Julian Lee Siegel*
U.S. Air Force Esd/ Ja
Hanscom AFB, MA 01731 (617) 861-4077

Henry S. Miller, Jr.
Harvard Professional Bldg.
Ayer Rd.
Harvard, MA 01451 (617) 772-0011

Carol A. Karolow#
C A K Inc.
P.O. Box 88
Hathorne, MA 01937 (617) 777-5022

John M. Brandt
60 Thaxter St.
Hingham, MA 02043 (617) 749-2889

Michael J. Sayles
10 Hancock Rd.
Hingham, MA 02043 (617) 740-1568

Rufus M. Franklin
54 Centerwood Dr.
Holden, MA 01520 (617) 829-5766

Robert D. Donley
3 Thayer St.
Hopedale, MA 01747 (617) 473-5313

R. Bruce Balance
Monsanto Company
730 Worcester St.
Indian Orchard, MA 01151 (413) 730-2827

Herbert Edward Farmer
95 Whitcomb Ave.
Jamaica Plain, MA 02130 (617) 522-7464

Tatsuya Ikeda
American Hoechst Corp., Patent Dept.
289 N. Main Street
Leominster, MA 01453 (617) 537-8131

William Lee Baker
W.R. Grace Co.
55 Hayden Ave.
Lexington, MA 02173 (617) 861-6600

David B. Bernstein
Hamilton, Brook, Smith & Reynolds
Two Militia Dr.
Lexington, MA 02173 (617) 861-6240

David Edward Brook
Hamilton, Brook, Smith & Reynolds
Two Militia Dr.
Lexington, MA 02173 (617) 861-6240

Paula A. Campbell
Hamilton, Brook, Smith & Reynolds
Two Militia Dr.
Lexington, MA 02173 (617) 861-6240

Stacey L. Channing
W.R. Grace & Company
55 Hayden Avenue
Lexington, MA 02173 (617) 861-6600

William R. Clark
Raytheon Company
141 Spring St.
Lexington, MA 02173 (617) 862-4845

Walter F. Dawson
Raytheon Co.
141 Spring St.
Lexington, MA 02173 (617) 862-6600

Giulio A. DeConti
Hamilton , Brook, Smith & Reynolds
Two Militia Dr.
Lexington, MA 02173 (617) 861-6240

Peter J. Devlin
Raytheon Company
141 Springs St.
Lexington, MA 02173 (617) 860-2535

Fred Fisher
7 Springdale Rd.
Lexington, MA 02173 (617) 466-3523

Melvin E. Frederick
16 Westwood Rd.
Lexington, MA 02173 (617) 862-1564

Michael P. Gilday#
497 Massachusetts Ave.
Lexington, MA 02173 (617) 862-2501

Patricia Granahan
Hamilton, Brook, Smith & Reynolds
Two Militia Dr.
Lexington, MA 02173 (617) 861-6240

George Grayson
9 Suzanne Rd.
Lexington, MA 02173 (617) 862-2517

Thomas O. Hoover
Hamilton, Brook, Smith & Reynolds
2 Militia Dr.
Lexington, MA 02173 (617) 861-6240

John D. Hubbard
W.R. Grace & Co.
55 Hayden Ave.
Lexington, MA 02173 (617) 861-6600

Christopher L. Maginniss
Raytheon Co.
141 Spring St.
Lexington, MA 02173 (617) 860-4848

Donald E. Mahoney
Raytheon Company
141 Spring St.
Lexington, MA 02173 (617) 862-6600

Philip J. McFarland
Raytheon Co.
141 Spring St.
Lexington, MA 02173 (617) 862-6600

Bart G. Newland
W.R. Grace & Co.
55 Hayden Ave.
Lexington, MA 02173 (617) 861-6600

Leo R. Reynolds
Hamilton, Brook, Smith & Reynolds
Two Militia Dr.
Lexington, MA 02173 (617) 861-6240

Martin Michael Santa
Raytheon Co.
141 Spring St.
Lexington, MA 02173 (617) 860-2695

Richard M. Sharkansy
Raytheon Co.
141 Spring St.
Lexington, MA 02173 (617) 860-2697

Michael L. Sheldon
Hamilton, Brook, Smith & Reynolds
Two Militia Dr.
Lexington, MA 02173 (617) 861-6240

James M. Smith
Hamilton, Brook, Smith & Reynolds
Two Militia Dr.
Lexington, MA 02173 (617) 861-6240

John J. Wasatonic
W.R. Grace & Co.
55 Hayden Ave.
Lexington, MA 02173 (617) 861-6600

Richard A. Wise
Hamilton, Brook, Smith & Reynolds
Two Militia Dr.
Lexington, MA 02173 (617) 861-6240

Charles Edward Parker#
37 Birchwood Lane
Lincoln, MA 01773 (617) 025-0950

Frank Abbott Steinhilper
Stonehedge
Lincoln, MA 01773 (617) 259-0613

Joseph S. Nelson#
1 Druid Circle
Longmeadow, MA 01106 (413) 567-3418

Joseph E. Funk, Sr.
Wang Laboratories, Inc.
One Industrial Ave.
Lowell, MA 01851 (617) 459-5000

Kenneth L. Milik#
Wang Lab., Inc.
One Industrial Ave.
Lowell, MA 01851 (617) 459-5000

Gordon E. Nelson
Wang Lab., Legal Dept.
1 Industrial Dr.
Lowell, MA 01851

John H. Pearson, Jr.
Pearson & Pearson
12 Hurd Street
Lowell, MA 01852 (617) 452-1971

John Howard Pearson
Pearson & Pearson
12 Hurd St.
Lowell, MA 01852 (617) 452-1971

Sally L. Pearson#
Pearson & Pearson
12 Hurd St.
Lowell, MA 01852 (617) 452-1971

Michael H. Shanahan
Wang
One Industrial Ave.
Lowell, MA 01851 (617) 967-6020

Francis L. Conte
General Electric Co.
1000 Western Ave.
Lynn, MA 01910 (617) 594-2701

Derek P. Lawrence
General Electric Co.
1000 Western Ave.
Lynn, MA 01905 (617) 594-4627

Donald R. Castle
Harington Way
Manchester, MA 01944 (617) 526-7329

Paul John Cook
9 Tanglewood Rd.
Manchester, MA 01944 (617) 526-7149

Donald N. Halgren
35 Central St.
Manchester, MA 01944 (617) 526-8000

George A. Herbster
27 Skytop Dr.
Manchester, MA 01944 (617) 526-7033

Kenneth Wright Brown
75 Harbor Ave.
Marblehead, MA 01945 (617) 631-9135

George W. Crowley
74 Atlantic Ave.
Marblehead, MA 01945 (617) 639-1111

Martin Kirkpatrick
P.O. Box 1109
Marblehead, MA 01945 (617) 631-1334

Thomas C. Stover, Jr.
13 Essex St.
Marblehead, MA 01945 (617) 631-1111

John M. Gunter
Digital Equipment Corp.
2 Results Way
Marlboro, MA 01752 (617) 467-7083

Maureen L. Stretch
Digital Equipment Corp.
One Iron Way
Marlboro, MA 01752 (617) 467-6601

Albert P. Cefalo
Digital Equipment Corp., Legal Dept.
111 Powdermill Rd.
Maynard, MA 01754 (617) 493-8571

Ronald James Clak
P.O. Box 190
Maynard, MA 01754 (617) 493-3683

Gary D. Clapp
Digital Equipment Corp.
111 Powdermill Rd., M S O / C5
Maynard, MA 01754 (617) 493-8943

William C. Cray
Digital Equipment Corp., Law Dept.
111 Powdermill Rd.
Maynard, MA 01754 (617) 493-2469

Richard M. Kotulak
Digital Equipment Corp.
111 Powdermill Rd.
Maynard, MA 01754 (617) 493-2093

Gerald E. Lester
Digital Equipment Corp., Legal Dept., Ms/m6
111 Powdermill Rd.
Maynard, MA 01754 (617) 493-6571

Robert C. Mayes
Digital Equipment Corp.
111 Powdermill Rd., MSO/ c5
Maynard, MA 01754 (617) 897-5111

Maura K. Moran
Digital Equipment Corp.
111 Powdermill Rd.
Maynard, MA 01754 (617) 493-3665

Ronald E. Myrick
Digital Equipment Corp.
111 Powdermill Rd.
Maynard, MA 01749

Richard J. Paciulan
Digital Equipment Corp.
111 Powdermill Rd.
Maynard, MA 01754 (617) 493-6501

Vincenzo D. Pitruzzella
Digital Equipment Corp.
111 Powdermill Rd.
Maynard, MA 01754 (617) 493-6604

Timothy C. Pledger
Digital Equipment Corp.
111 Powdermill Rd.
Maynard, MA 01754 (508) 493-6355

David G. Pursel
Digital Equipment Corp.
111 Powdermill Rd.
Maynard, MA 01754 (617) 493-9763

Ronald T. Reilling
Digital Equipment Corp.
111 Powdermill Rd.
Maynard, MA 01754 (617) 493-2991

Richard F. Schuett
Digital Equipment Corp.
111 Powdermill Rd.
Maynard, MA 01754 (617) 493-6502

Thomas C. Siekman
Digital Equipment Corp., Legal Dept.
111 Powdermill Rd.
Maynard, MA 01754 (617) 493-4422

Penelope A. Smith
Digital Equipment Corp.
111 Powdermill Rd.
Maynard, MA 01754 (617) 493-4293

David J. Koris#
Corning Glass Works, Inc.
Medfield Industrial Park
Medfield, MA 02052 (617) 359-7711

Brian D. Voyce
Ciba Corning Diagnostics Corp.
63 North St.
Medfield, MA 02052 (617) 359-7711

John P. Morley
320 Upham St.
Melrose, MA 02176 (617) 665-7755

Edwin H. Paul, Jr.
Waters Associates
34 Maple Street
Milford, MA 01757 (617) 478-2000

William Robert Sherman
15 Gay St.
Nantucket, MA 02554 (617) 228-3880

Richard L. Ballantyne
Prime Computer Inc.
Prime Park
Natick, MA 01760 (617) 655-8000

Anthony N. Fiore
Prime Computer, Inc., Legal Dept.
Prime Park, MS 15-36
Natick, MA 01760 (617) 655-8000

Lawrence E. Labadini*
U.S. Army
R & D Laboratoris, Kansas St.
Natick, MA 01760 (617) 651-4510

Ronald J. Paglierani
Prime Computer, Inc.
Prime Park, M S 15-36
Natick, MA 01760 (617) 655-8000

Ronald S. Cornell
Duracell Inc.
37 A Street
Needham, MA 02194 (617) 449-7600

James B. McVeigh, Jr.#
Duracell Inc.
37 A Street
Needham, MA 02194 (617) 449-7600

Frederick August Goettel, Jr.#
Gould Inc., Circuit Protection Div.
374 Merrimac St.
Newburyport, MA 01950 (617) 462-3131

Richard T. Oakes#
Gould Inc.
374 Merrimac St.
Newburyport, MA 01945 (617) 462-3131

Frederick H. Brustman
48 Lantern Lane
Newton, MA 02159 (617) 527-6146

Robert Louis Goldberg
Dike, Bronstein, Roberts, Cushman & Pfund
2345 Washington St.
Newton, MA 02162 (617) 244-4990

Edward W. Porter
Dragon Systems, Inc., Chapel Bridge Park
55 Chapel St.
Newton, MA 02158 (617) 965-5200

Barbara Z. Terris#
269 Franklin St.
Newton, MA 02158 (617) 965-4940

Walter Juda#
Prototech Co.
70 Jaconnet St.
Newton Highlands, MA 02161
(617) 965-2720

Milton Edwin Gilbert
Barry Wright Corp.
One Newton Executive Pk.
Newton Lower Falls, MA 02162
(617) 965-5800

Vincent Hilary Sweeney#
Sprague Electric Co.
Marshall St.
North Adams, MA 02147 (413) 664-4411

Joseph A. Cameron
A T & T Bell Labs.
1600 Osgood St.
North Andover, MA 01845 (617) 681-6018

John A. Haug
11 Ryder Circle
North Attleboro, MA 02760 (617) 699-3314

Carole M. Calnan
77 Ellis Ave.
Norwood, MA 02062 (617) 769-3240

Richard Paul Crowley
901 Main Street
Osterville, MA 02655 (617) 428-4000

Erwin F. Berrier, Jr.
General Electric Co.
One Plastics Ave.
Pittsfield, MA 01201 (413) 448-4781

John R. Castiglione
201 Wendell Ave.
Pittsfield, MA 01201 (413) 494-4531

Spencer D. Conard
General Electric Plastics
One Plastics Ave.
Pittsfield, MA 01201 (413) 448-7662

Sidney Greenberg
18 Glenn Drive
Pittsfield, MA 01201 (413) 448-8336

John W. Harbour
General Electric Company
One Plastics Ave.
Pittsfield, MA 01201 (413) 494-6378

William F. Mufatti
General Electric Co.
One Plastics Ave.
Pittsfield, MA 01201 (413) 494-4707

Gordon Needleman
1359 Hancock St., Suite 8
Quincy, MA 02169 (617) 471-5632

Owen J. Meegan
65 Dearborn St.
Salem, MA 01970 (617) 745-0219

Donald A. Teare
339 First Parish Rd.
Scituate, MA 02066 (617) 545-6681

Frank J. Fleming#
8 South Pleasant St., Box 152
Sharon, MA 02067 (617) 784-2775

James H. Grover
63 Norwood St., P.O. Box 296
Sharon, MA 02067 (617) 742-9100

Charles E. Cullen, Jr.
27 Main Circle
Shrewsbury, MA 01545 (617) 845-6510

Edmund F. Chojnowski
P.O. Box 1188
S. Lancaster, MA 01561 (617) 368-8826

Nicholas J. Cafarelli
60 Audubon Street
Springfield, MA 01108 (413) 739-5045

John J. Dempsey
Chapin, Neal And Dempsey, P.C.
1331 Main Street
Springfield, MA 01103 (413) 736-5401

William J. Farrington
Monsanto Co.
730 Worcester Street
Springfield, MA 01151 (413) 730-2811

John Edward Flanagan
95 State St., Suite 508
Springfield, MA 01103 (413) 733-2189

Chester Edwin Flavin
Ross, Ross & Flavin
120 Maple Street
Springfield, MA 01103 (413) 733-3194

William Darrell Fosdick#
P.O. Box 80545
Springfield, MA 01138

Donald S. Holland
1391 Main Street
Springfield, MA 01103 (413) 737-5523

Thomas E. Kelley
Monsanto Co.
730 Worcester Street
Springfield, MA 01151 (413) 730-2046

Linda L. Lewis
Monsanto Co.
730 Worcester Street
Springfield, MA 01151 (413) 730-2046

Leonard S. Michelman
Michelman & Feinstein
1333 East Columbus Avenue, P.O. Box 2992
Springfield, MA 01101 (413) 737-1166

Michael John Murphy
Monsanto Co.
730 Worcester Street
Springfield, MA 01151 (413) 730-2091

Kenwood Ross
Ross, Ross & Flavin
120 Maple Street, Room 207
Springfield, MA 01103 (413) 733-3194

Charles R. Fay
194 Justice Hill Rd.
Sterling, MA 01564 (617) 422-7146

Milton D. Bartlett#
566 Boston Post Rd.
Sudbury, MA 01776 (617) 443-2125

Jeffery D. Marshall#
21 Central
Topsfield, MA 01983 (617) 396-8880

William F. White
1 Orchard Lane
Topsfield, MA 01983 (617) 887-2184

Robert S. Sanborn
Box 357
Vineyard Haven, MA 02568 (617) 693-2098

John E. Herlihy
925 Main St., Box 54
Walpole, MA 02081 (617) 668-9360

William G. Auton*
U.S.A.F., A F J A C P B
Bldg. 104
424 Trapelo Rd.
Waltham, MA 02154 (617) 377-4072

Stanton E. Collier*
U.S. Dept. Of The Air Force
A F J A C P B - Bldg. 104
424 Trapelo Rd.
Waltham, MA 02154 (617) 377-4072

Frances P. Craig#
G T E Service Corp.
100 First Ave.
Waltham, MA 02254 (617) 466-3522

Douglas E. Denninger
260 Bear Hill Rd.
Waltham, MA 02154 (617) 890-5678

Richard J. Donahue*
Dept. Of The Air Force
A F J A C P B - Bldg. 104
424 Trapelo Rd.
Waltham, MA 02154 (617) 377-4072

Faith F. Driscoll
Honeywell Information Systems Inc.
333 Wyman St.
Waltham, MA 02154 (617) 895-6165

Ivan L. Ericson
G T E Service Corp.
100 First Ave.
Waltham, MA 02254 (617) 466-3308

Jacob N. Erlich*
U.S. Air Force
424 Trapelo Rd.
Waltham, MA 02154 (617) 377-4075

William R. Griffin
Hmm Associates
255 Bear Hill Rd.
Waltham, MA 02154 (617) 890-6933

Charles Hieken
Fish & Richardson
470 Totten Pond Rd.
Waltham, MA 02154 (617) 890-0110

Joseph S. Iandiorio
60 Hickory Dr.
Waltham, MA 02154 (617) 890-5678

David Malcolm Keay#
G T E Service Corp.
100 First Ave.
Waltham, MA 02254 (617) 890-9200

Frank J. Lamattina*
H Q U.S. Air Force
A F J A C P B, Bldg. 104
424 Trapelo Rd.
Waltham, MA 02154 (617) 861-4072

William A. Linnell
Honeywell Incorporated
333 Wyman St.
Waltham, MA 02154 (617) 895-6115

Willard R. Matthews, Jr.*
U.S.A.F.- A F J A C P B
424 Trapelo Rd.
Waltham, MA 02154 (617) 861-4075

Herbert E. Messenger
Thermo Electron Corp.
101 First Ave., P.O. Box 459
Waltham, MA 02254 (617) 890-8700

Robert Evan Meyer
3314 Stearms Hill Rd.
Waltham, MA 02154

Jules Jay Morris*
Dept. Of The Air Force
A F J A C P B Bldg. 104
424 Trapelo Rd.
Waltham, MA 02154 (617) 377-4072

Robert Lang Nathans*
U.S. Air Force, A F J A C P B - Bldg. 104
424 Trapelo Rd.
Waltham, MA 20154 (617) 377-4072

James Lewis Neal
Thermo Electron Corp.
101 First Ave., P.O. Box 459
Waltham, MA 02254 (617) 890-8700

William E. Noonan
60 Hickory Dr.
Waltham, MA 02154 (617) 890-5678

Nicholas Prasinos
333 Wyman St.
Waltham, MA 01543 (617) 895-6129

Joseph S. Romanow
G T E Service Corp. Legal Dept.
100 First Ave.
Waltham, MA 02254 (617) 466-3371

John S. Solakian
Honeywell Inc.
333 Wyman St.
M A 40-734
Waltham, MA 02154 (617) 895-6140

William Stepanishen*
U.S. Air Force, A F J A C P B
424 Trapelo Rd.
Waltham, MA 02154 (617) 377-4072

Francis I. Sullivan, Jr.
G T E Government Systems Corp.
100 First Ave
Waltham, MA 02254 (617) 466-3311

Robert Eugene Walrath
GTE Service Corp.
100 First Ave.
Waltham, MA 02254 (617) 466-3348

John S. Yeo
GTE Service Corp.
100 First Ave.
Waltham, MA 02254 (617) 466-3599

Jack M. Young
92-B Charles River Rd.
Waltham, MA 02154 (617) 893-3827

Mark Goldberg*
Dept. Of The Army Materials Tech. Labs.
Arsenal St.
Watertown, MA 02172 (617) 923-5276

Norman E. Saliba
Ionics Inc.
65 Grove St.
Watertown, MA 02172 (617) 926-2500

Edward J. Collins
62 Claypit Hill Rd.
Wayland, MA 01778 (617) 358-5645

Gerald J. Cechony
89 Killdeer Island
Webster, MA 01570 (617) 943-0907

Robert P. Cogan
E G & G, Inc.
45 William St.
Wellesley, MA 02181 (617) 237-5100

Leo M. Kelly
E.G. & G., Inc.
45 William St.
Wellesley, MA 02181 (617) 237-1650

Moisey M. Lerner#
Tomin Corp.
P.O. Box 206
Wellesley, MA 02181 (617) 444-6529

James F. Baird
South Main Street
West Brookfield, MA 01585 (617) 867-2441

Thomas W. Underhill
P.O. Box 553, 88 Howes Lane
West Chatham, MA 02669 (617) 945-1427

Simon L. Cohen
Data General Corp.
4400 Computer Dr.
Westboro, MA 01580 (617) 870-7768

Robert L. Dulaney
Data General Corp.
4400 Computer Dr.
Westboro, MA 01580 (617) 366-8911

Jacob Frank
Data General Corp.
4400 Computer Dr.
Westboro, MA 01580 (617) 820-7781

Joel Wall
Data General Corp.
4400 Computer Dr.
Westboro, MA 01581 (617) 870-7777

Philip Colman
384 Glen Rd.
Weston, MA 02193

Richard F. Benway
265 Washington St.
Westwood, MA 02090 (617) 329-6611

William M. Anderson
51 Board Reach Apt. T-102A
Weymouth, MA 02191 (617) 331-5293

Allan R. Redrow
97 Walker Street
Whitinsville, MA 01588 (617) 234-6883

Arthur K. Hooks#
1341 Green River Rd.
Williamstown, MA 01267

I. David Blumenfeld
General Electric Co.
Aircraft Instrument Dept.
50 Fordham Rd.
Wilmington, MA 01887 (617) 937-4727

Abraham Ogman
Avco Corporation
201 Lowell Street
Wilmington, MA 01887 (617) 657-4421

Francis A. DiLuna#
13 Utica St.
Woburn, MA 01801 (617) 935-4339

Gerry A. Blodgett
Blodgett & Blodgett, P.C.
43 Highland St.
Worcester, MA 01608 (617) 753-5533

Norman S. Blodgett
Blodgett & Blodgett, P.C.
43 Highland St.
Worcester, MA 01608 (617) 753-5533

Thomas C. Blodgett#
Blodgett & Blodgett, P.C.
43 Highland St.
Worcester, MA 01609 (617) 753-5533

Frank S. Chow
Norton Co.
1 New Bond St.
Worcester, MA 01606 (617) 795-2087

Wilfred F. Desrosiers, Sr.#
Blodgett & Blodgett, P.C.
43 Highland St.
Worcester, MA 01608 (617) 753-5533

Arthur A. Loiselle, Jr.
Norton Company Pat. Dept.
1 New Bond St.
Worcester, MA 01606 (617) 853-1000

Nicholas I. Slepchuk, Jr.#
Blodgett & Blodgett, P.C.
43 Highland St.
Worcester, MA 01609 (617) 753-5533

Norvell E. Wisdom, Jr.
Norton Company Patent Dept.
1 New Bond Street
Worcester, MA 01606 (617) 795-5000

Michigan

Michael A. Mohr
Amway Corp.
7575 E. Fulton Rd.
Ada, MI 49355 (616) 676-5416

Marion Duane Ford
Wacker Silicones Corp. Patent Dept.
3301 Sutton Rd.
Adrian, MI 49221 (517) 263-5711

Elizabeth M. Anderson#
Warner Lambert Co.
2800 Plymouth Rd.
Ann Arbor, MI 48105 (313) 996-7304

Ronald A. Daignault
Warner-Lambert Co.
2800 Plymouth Rd.
Ann Arbor, MI 48105 (313) 996-7530

James H. Dautremont
2124 Brockman
Ann Arbor, MI 48104 (313) 663-0058

Jemes M. Deimen
325 E. Eisenhower Pkwy. Suite 2
Ann Arbor, MI 48104 (313) 994-5947

Jerry F. Janssen
Warner-Lambert Co.
2800 Plymouth Rd.
Ann Arbor, MI 48105 (313) 996-7073

Daniel H. Sharphorn
1228 Olivia Ave.
Ann Arbor, MI 48104 (313) 764-3423

James E. Stephenson
301 E. Liberty St. Suite 555
Ann Arbor, MI 48104 (313) 662-5653

Joan V. Thierstein
Warner-Lambert Co., Legal Dept.
2800 Plymouth Rd.
Ann Arbor, MI 48105 (313) 996-7190

Dennis H. Rainear
Dow Corning Corp.
2200 W. Salzburg Rd.
Auburn, MI 48611 (517) 496-6306

Burton Howard Baker
Whirlpool Corp. 2000 M63
Benton Harbor, MI 49022 (616) 926-3412

Franklin Clyde Harter
Whirlpool Corp.
2000 M-63
Benton Harbor, MI 49022 (616) 926-5020

Gene A. Heth
Whirlpool Corp.
2000 M-63 North
Benton Harbor, MI 49022 (616) 926-5600

Robert L. Judd
Whirlpool Corp.
2000 M-63
Benton Harbor, MI 49022 (616) 926-3511

Edward A. Ketterer, III
Whirlpool Corp. Res. & Eng. Ctr.
Monte Rd.
Benton Harbor, MI 49022 (616) 926-5015

James Stanley Nettleton
Whirlpool Corp.
Monte Rd.
Benton Harbor, MI 49022 (616) 926-5080

Charles D. Putman
Whirlpool Corp.
Admin. Ctr.
Benton Harbor, MI 49022 (616) 926-3466

Robert O. Rice
Whirlpool Corp.
Monte Rd.
Benton Harbor, MI 49022 (616) 926-5013

Thomas J. Roth
Whirlpool Corp.
2000 M-63
Benton Harbor, MI 49022 (616) 926-5604

Thomas E. Turcotte
Whirlpool Corp.
Administrative Center, Monte Rd.
Benton Harbor, MI 49022 (616) 926-5021

Luis Miguel Acosta
Harness, Dickey & Pierce
1500 N. Woodward Ave.
Birmingham, MI 48011 (313) 642-7000

Thomas E. Anderson
Gifford, Groh, Van Ophem, Sheridan, Sprinkle & Dolgorukov
280 N. Woodward, Suite 210
Birmingham, MI 48011 (313) 647-6000

John A. Artz
Harness, Dickey & Pierce
1500 N. Woodard Ave.
Birmingham, MI 48011 (313) 642-7000

John R. Benefiel
360 Birmingham Place, 401 S. Woodward Ave.
Birmingham, MI 48011 (313) 644-1455

Ernest A. Beutler, Jr.
Harness, Dickey And Pierce
1500 N. Woodward Ave.
Birmingham, MI 48011 (313) 642-7000

Charles H. Blair
Harness, Dickey And Pierce
1500 N. Woodward Ave., Suite 300
Birmingham, MI 48011 (313) 642-7000

John A. Blair
Harness, Dickey And Pierce
1500 N. Woodward Ave.
Birmingham, MI 48011 (617) 642-7000

Daniel H. Bliss
Harness, Dickey And Pierce
1500 N. Woodward Ave.
Birmingham, MI 48011 (313) 642-7000

Robert L. Boynton
Harness, Dickey And Pierce
1500 N. Woodward Ave., Suite 300
Birmingham, MI 48011 (313) 642-7000

Michael P. Brennan
Harness, Dickey & Pierce
1500 N. Woodward Ave.
Birmingham, MI 48011 (313) 642-7000

Christopher M. Brock
Harness, Dickey & Pierce
1500 N. Woodward Ave.
Birmingham, MI 48011 (313) 642-7000

Richard L. Carlson
Harness, Dickey & Pierce
1500 N. Woodward Ave.
Birmingham, MI 48011 (313) 642-7000

Edward R. Casselman
Harness, Dickey & Pierce
1500 N. Woodward Ave.
Birmingham, MI 48011 (313) 642-7000

William J. Coughlin#
Harness, Dickey & Pierce
1500 N. Woodward Ave.
Birmingham, MI 48011 (313) 642-7000

T.I. Davenport
1055 Larchelea
Birmingham, MI 48009 (313) 643-3645

Michael R. Dinnin, Jr.
Harness, Dickey & Pierce
1500 N. Woodward Ave.
Birmingham, MI 48011 (313) 642-7000

D. Edward Dolgorukov
Gifford, Groh, Van Ophem, Sheridan, Sprinkle & Dolgorukov
280 N. Woodward Ave., Suite 210
Birmingham, MI 48011 (313) 647-6000

Robert A. Dunn
Harness, Dickey & Pierce
1500 N. Woodward Ave., Suite 300
Birmingham, MI 48011 (313) 642-7000

Hugh L. Fisher
Fisher, Gerhardt, Crampton & Groh
16231 W. 14 Mile Rd.
Birmingham, MI 48009 (313) 645-2430

Paul Fitzpatrick
851 S. Glenhurst
Birmingham, MI 48009 (313) 644-1022

Stephen J. Foss
Harness, Dickey & Pierce
1500 N. Woodward Ave.
Birmingham, MI 48011 (313) 642-7000

George Edward Frost
291 Hupp Cross Rd.
Birmingham, MI 48010 (313) 647-5508

Ernest I. Gifford
Gifford, Groh, Sheridan, Sprinkle & Dolgorukov, P.C.
280 N. Woodward Ave., Suite 210
Birmingham, MI 48011 (313) 647-6000

Charles T. Graham
Harness, Dickey & Pierce
1500 N. Woodwad Ave.
Birmingham, MI 48011 (313) 642-7000

Irvin L. Groh
Gifford, Groh, Van Ophem, Sheridan, Sprinkle & Dolgorukov
280 N. Woodward, Suite 210
Birmingham, MI 48011 (313) 647-6000

Don K. Harness
Harness, Dickey And Pierce
1500 N. Woodward Ave.
Birmingham, MI 48011 (313) 642-7000

Ronald L. Hofer
Harness, Dickey & Pierce
1500 N. Woodward Ave.
Birmingham, MI 48011 (313) 642-7000

Paul A. Keller
Harness, Dickey, & Pierce
1500 N. Woodward Ave.
Birmingham, MI 48011 (313) 642-7000

Charles M. McCuen
1344 Brookwood Lane
Birmingham, MI 48009 (313) 540-2877

James P. Meloche
401 S. Woodward Ave., Suite 360
Birmingham, MI 48011 (313) 644-2114

H. Keith Miller
Harness, Dickey & Pierce
1500 N. Woodward Ave
Birmingham, MI 48011 (313) 642-7000

Cyrus G. Minkler
Harness, Dickey & Pierce
1500 N. Woodward Ave.,Suite 210
Birmingham, MI 48011 (313) 642-7000

William R. Nolte#
191 Baldwin
Birmingham, MI 48009 (313) 647-7660

Steven L. Oberholtzer
Harness, Dickey & Pierce
1500 N. Woodward, Suite 300
Birmingham, MI 48011 (313) 642-7000

Joseph Richard Papp
Harness, Dickey & Pierce
1500 N. Woodward Ave.
Birmingham, MI 48011 (313) 642-7000

Alfred Lawrence Patmore, Jr.
Gifford, Groh, Van Ophem, Sheridan, Sprinkle & Dolgorukov
280 N. Woodward Ave. Suite 210
Birmingham, MI 48011 (313) 647-6000

Charles R. Penninger
Harness, Dickey And Pierce
1500 N. Woodward Ave.
Birmingham, MI 48011 (313) 642-7000

Jeffrey A. Sadowski
Harness, Dickey & Pierce
1500 N. Woodward Ave.
Birmingham, MI 48011 (313) 642-7000

G. Gregory Schivley
Harness, Dickey & Pierce
1500 N. Woodward Ave.
Birmingham, MI 48011 (313) 642-7000

John D. Scofield
Harness, Dickey & Pierce
1500 N. Woodward Ave.
Birmingham, MI 48011 (313) 642-7000

James V. Sheridan
Gifford, Van Ophen, Sheridan & Springle
280 N. Woodward Ave.
Birmingham, MI 48011 (313) 647-6000

John Lyons Shortley
5890 Snowshoe Circle
Birmingham, MI 48010 (313) 642-2791

John A. Sinclair
Harness, Dickey & Pierce
1500 N. Woodard Ave. Suite 300
Birmingham, MI 48011 (313) 642-7000

Cass L. Singer
Gifford, Van Ophem, Sheridan, Sprinkle & Nabozny, P.C.
280 N. Woodward Suite 210
Birmingham, MI 48011 (313) 647-6000

John V. Sobesky
Harness, Dickey & Pierce
1500 N. Woodward Ave.
Birmingham, MI 48011 (313) 642-7000

Douglas W. Sprinkle#
Gifford, Van Ophem, Sheridan & Sprinkle, P.C.
280 N. Woodward Suite 210
Birmingham, MI 48011 (313) 647-6000

Gregory A. Stobbs
Harness, Dickey & Pierce
1500 N. Woodward Ave.
Birmingham, MI 48011 (313) 642-7000

W.R. Duke Taylor
Harness, Dickey & Pierce
1500 N. Woodward Ave.
Birmingham, MI 48011 (313) 642-7000

Richard P. Vitek
Harness, Dickey & Pierce
1500 N. Woodward Ave.
Birmingham, MI 48011 (313) 642-7000

Richard A. Walker
Kuhlman Corp.
P.O. Box 288
Birmingham, MI 48011 (313) 649-9300

Ronald W. Wangerow
Harness, Dickey & Pierce
1500 N. Woodward Ave.
Birmingham, MI 48011 (313) 642-7000

Donald L. Wenskay#
Harness, Dickey & Pierce
1500 Woodward Ave.
Birmingham, MI 48011 (313) 642-7000

Marjory G. Basile
Miller, Candield, Paddock & Stone
P.O. Box 2014, 1400 N. Woodward, Suite 100
Bloomfield Hills, MI 48303 (313) 645-5000

James H. Bower
2138 Randall Lane
Bloomfield Hills, MI 48013 (313) 642-2171

Raymond J. Eifler, Sr.
Cross & Tucker
505 North Woodward
Bloomfield Hills, MI 48013 (313) 644-4343

James B. Raden
I T T Automotive, Inc.
505 N. Woodward Ave., Suite 2500
Bloomfield Hills, MI 48013 (313) 540-6805

Robert P. Seitter
I T T Automotive, Inc.
505 N. Woodward Ave., Suite 2500
Bloomfield Hills, MI 48011 (313) 540-9666

George N. Shampo
1355 Trowbridge Rd.
Bloomfield Hills, MI 48011 (313) 644-8697

John C. Sterritt
105 Harlan Dr.
Bloomfield Hills, MI 48011 (313) 642-8560

Bertram F. Claeboe
6307 Baldwin Circle
Brighton, MI 48116 (313) 227-2416

William D. Suomi
2485 Hunter Rd.
Brighton, MI 48116 (313) 229-6765

Paul D'Arc Garty#
P.O. Box 121
Brooklyn, MI 49230

John Calvin Wiessler
Clark Equipment Co.
Circle Dr.
Buchanan, MI 49107 (616) 697-8104

Kenneth C. Witt
401 W. Front St.
Buchanan, MI 49107 (616) 695-3178

Lon H. Romanski
Suite 5, Crandell Bldg. 210 1/2 N. Micthell Ave.
Cadillac, MI 49601 (616) 775-0171

Douglas E. Mark
2152 Chevychase Dr.
Davison, MI 48423 (313) 653-6470

Peter Abolins
Ford Motor Co. Parklane Towers E., Suite 911
One Parklane Blvd.
Dearborn, MI 48126 (313) 337-3341

Glenn S. Arendsen
Ford Motor Co. Office Of Gen. Coun.
W H Q 1010
Dearborn, MI 48121 (313) 322-4898

Jerome R. Drouillard
Ford Motor Company
Parklane Towers - 911 East
Dearborn, MI 48126 (313) 845-5101

Herbert Epstein
Ford Motor Co. The American Rd.
P.O. Box 1899, Room 1078
Dearborn, MI 48121 (313) 322-4397

Roger E. Erickson
Ford Motor Co. Off. Of Gen. Coun.
1078 Whq, P.O. Box 1899
Dearborn, MI 48121 (313) 337-5462

Paul K. Godwin, Jr.
Ford Motor Co.
911 E. Parklane Towers
Dearborn, MI 48126 (313) 337-8718

Donald Joseph Harrington
Ford Motor Co. Patent Dept.
Parklane Towers E., Suite 911
One Parklane Blvd.
Dearborn, MI 48126 (313) 323-1908

Olin B. Johnson
Ford Motor Co. Off. Of The Gen. Coun.
Parklane Towers E. Suite 911
One Parklane Blvd.
Dearborn, MI 48126 (313) 323-1903

William Edwin Johnson
Ford Motor Co.
911 Parklane Towers East
Dearborn, MI 48126 (313) 323-2023

Casimir R. Kiczek
27109 Kingswood Dr.
Dearborn, MI 48127 (313) 563-6587

Anthony T. Lesnick
5021 Horger
Dearborn, MI 48126 (313) 584-3166

Dwight A. Lewis
Ford Motor Co.
The American Rd.
Dearborn, MI 48124 (313) 323-2024

Allan J. Lippa
Ford Motor Co.
911 East Parklane Towers
Dearborn, MI 48126 (313) 594-1145

Joseph William Malleck
Ford Motor Co. Parklane Towers E., Suite 911
One Parklane Blvd.
Dearborn, MI 48126 (313) 322-2830

Adolph Gustav Martin
13732 Michigan Ave.
Dearborn, MI 48126 (313) 581-4444

Roger L. May
Ford Motor Co., Off. Of Gen. Coun.
One Parklane Blvd., Suite 911, Parklane Towers East
Dearborn, MI 48126 (313) 676-3894

Peter D. Mc Dermott
Ford Motor Co. One Parklane Blvd., Suite 911 E
Dearborn, MI 48126 (313) 322-4281

Robert Ellsworth McCollum
Ford Motor Co., Parklane Tower E., Suite 911
One Parklane Blvd.
Dearborn, MI 48126 (313) 323-1904

Frank G. McKenzie
Ford Motor Co., Parklane Towers, Suite 911
One Parklane Blvd.
Dearborn, MI 48126 (313) 323-0903

Lorraine S. Melotik#
Ford Motor Co.
One Parklane Blvd.
911, Parklane Towers E.
Dearborn, MI 48126 (313) 337-1069

Richard J. Mossburg
Ford Motor Credit Co. Legal Off.
P.O. Box 1732
Dearborn, MI 48121 (313) 322-5636

Allan J. Murray
3120 Lindenwood Dr.
Dearborn, MI 48120 (313) 271-0084

Clifford Lincoln Sadler
Ford Motor Co.
911 Parklane Towers East
Dearborn, MI 48126 (313) 323-1823

Robert D. Sanborn
Ford Motor Co.
911 Parklane Towers East
Dearborn, MI 48126 (313) 594-1145

Daniel M. Stock
Ford Motor Co., Suite 911- Parklane Tower East
One Parklane Blvd.
Dearborn, MI 48126 (313) 323-1289

Jay C. Taylor
1525 Belmont
Dearborn, MI 48128 (313) 274-3829

Keith L. Zerschling
Ford Motor Co. Suite 911- Parklane Towers E.
One Parklane Blvd.
Dearborn, MI 48126 (313) 322-7725

Martin J. Adelman
Wayne State University
468 West Ferry
Detroit, MI 48202 (313) 577-3943

Thomas W. Baumgarten, Jr.
Unisys Corp.
One Burroughs Place
Detroit, MI 48205 (313) 972-7118

Edward Joseph Biskup
General Motors Corp.
New Center One Bldg., 3031 W. Grand Blvd.
Detroit, MI 48202 (313) 974-1307

Bernard J. Cantor
Culler, Sloman, Cantor, Grauer, Scott & Rutherford, P.C.
2400 Penobscot Bldg.
Detroit, MI 48226 (313) 964-0400

Robert A. Choate
Barnes, Kisselle, Raisch Choate, Whittemore & Hulbert, P.C.
1520 Ford Bldg.
Detroit, MI 48226 (313) 962-4790

Robert C. Collins
Barnes, Kisselle, Raisch Choate, Whittemore & Hulbert, P.C.
1520 Ford Bldg.
Detroit, MI 48226 (313) 962-4790

Howard N. Conkey
General Motors Corp. Pat. Sect. New Center One Bldg.
3031 W. Grand Blvd. P.O. Box 33114
Detroit, MI 48232 (313) 974-1340

Alfonse J. D'Amico
Barnes, Kisselle, Raisch Choate, Whittemore & Hulbert, P.C.
1520 Ford Bldg.
Detroit, MI 48226 (313) 962-4790

Chester L. Davis, Jr.
Barnes, Kisselle, Raisch Choate, Whittemore & Hulbert, P.C.
1520 Ford Bldg.
Detroit, MI 48226 (313) 962-4790

Albert F. Duke
General Motors Corp.
3031 W. Grand Blvd. Suite 450
P.O. Box 33114
Detroit, MI 48202 (313) 974-1358

Dean L. Ellis
General Motors Corps. Pat. Sect.
3031 W. Grand Blvd., P.O. Box 48232
Detroit, MI 48232 (313) 974-1323

B. Lynn Enderby
General Motors Corp.
1344 W. Grand Blvd.
Detroit, MI 48202 (313) 974-1594

Joseph W. Farley
710 Buhl Bldg.
Detroit, MI 48226 (313) 961-5190

Douglas D. Fekete
General Motors Corp. Patent Section
P.O. Box 33114
Detroit, MI 48232 (313) 974-1313

Warren E. Finken
General Motors Corp. Pat. Sect. Suite 450
3031 W. Grand Blvd., P.O. Box 33114
Detroit, MI 48232 (313) 974-1304

Francis James Fodale
General Motors Corp. Pat. Sect. New Center One Bldg.
3031 W. Grand Blvd.
Detroit, MI 48232 (313) 974-1362

Basil C. Foussianes
Barnes, Kisselle, Raisch Choate, Whittemore & Hulbert, P.C.
1520 Ford Bldg.
Detroit, MI 48226 (313) 962-4790

William H. Francis
Barnes, Kisselle, Raisch Choate, Whittemore & Hulbert, P.C.
1520 Ford Bldg.
Detroit, MI 48226 (313) 962-4790

Herbert Furman
General Motors Corp. Pat. Sect. New Center One Bldg.
3031 W. Grand Blvd., P.O. Box 33114
Detroit, MI 48232 (313) 974-1336

Richard D. Grauer
Cullen, Sloman, Cantor, Grauer, Scott & Rutherford, P.C.
2400 Penobscot Bldg.
Detroit, MI 48226 (313) 964-0400

Patrick M. Griffin
General Motors Corp. Pat. Sect.- New Center One Bldg.
3031 W. Grand Blvd., P.O. Box 33114
Detroit, MI 48232 (313) 974-1330

William J. Griffith
Barnes, Kisselle, Raisch Choat, Whittemore & Hulbert, P.C.
1520 Ford Bldg.
Detroit, MI 48226 (313) 962-4790

George Arthur Grove
General Motors Corp. Patent Section
P.O. Box 33114
Detroit, MI 48232 (313) 974-1322

Elizabeth F. Harasek
General Motors Corp. Pat. Sect.-New Center One Bldg.
3031 W. Grand Blvd.
Detroit, MI 48202 (313) 974-1365

George H. Hathaway
Detroit Edison
2000 2nd Ave.
Detroit, MI 48226 (313) 237-8958

Lewis R. Hellman
A N R Pipeline Co.
500 Renaissance Center
Detroit, MI 48243 (313) 496-3773

Ernest E. Helms
General Motors Corp. Patent Section
P.O. Box 33114
Detroit, MI 48232 (313) 974-1346

Robert F. Hess
Federal-Mogul Corp.
P.O. Box 1966
Detroit, MI 48235 (313) 354-9926

William H. Honaker
Cullen, Sloman, Canton, Grauer, Scott & Rutherford, P.C.
2400 Penobscot Bldg.
Detroit, MI 48226 (313) 964-0400

Tim G. Jaeger
General Motors Corp. New Center One Bldg. - Suite 450
3031 West Grand Blvd. Pat. Section
Detroit, MI 48232 (313) 974-1329

Robert L. Kelly
Cullen, Solman, Cantor, Grauer, Scott & Rutherford, P.C.
2400 Penobscot Bldg.
Detroit, MI 48226 (313) 964-0400

Roger R. Kline#
12780 E. Outer Dr.
Detroit, MI 48226 (313) 237-8958

Arthur Nicholas Krein
General Motors Corp. Pat. Sect. New Center One Bldg.
3031 W. Grand Blvd., P.O. Box 33114
Detroit, MI 48232 (313) 974-1339

Bradford Laughlin
15410 Artesian
Detroit, MI 48223 (313) 273-3762

Charles E. Leahy
General Motors Corp. Pat. Section
P.O. Box 33114
Detroit, MI 48232 (313) 974-1369

Robert E. Luetje#
Kolene Corp.
12890 Westwood Ave.
Detroit, MI 48223 (313) 273-9220

Doonan Dwight McGraw
General Motors Corp. Pat. Sect. New Center One Bldg.
3031 W. Grand Blvd., P. O. Box 33114
Detroit, MI 48232 (313) 974-1366

Creighton Roland Meland, Sr.
General Motors Corp. Pat. Section. New Center One Bldg.
3031 W. Grand Blvd., P.O. Box 33114
Detroit, MI 33114 (313) 974-1327

Mark A. Navarre
General Motors Corp. New Center One Bldg.
3031 W. Grand Blvd., P.O. Box 33114
Detroit, MI 48232 (313) 974-1349

Gary L. Newtson
Chrysler Corp.
P.O. Box 1118
Detroit, MI 48288 (313) 956-4293

Robert S. Nolan
Cullen, Sloman, Canton, Grauer, Scott & Rutherford, P.C.
2400 Penobscot Bldg.
Detroit, MI 48226 (313) 964-0400

Robert John Outland
General Motors Corp. Patent Section, New Center 1 Bldg.
3031 W. Grand Blvd., P.O. Box 33114
Detroit, MI 48232 (313) 974-1361

Kevin R. Peterson
Burroughs Corp., Burroughs Place
Detroit, MI 48232 (313) 972-7982

Ronald Lloyd Phillips
General Motors Corp. Pat. Section-New Center One Bldg.
3031 W. Grand Blvd., P.O. Box 33114
Detroit, MI 48232 (313) 974-1355

Lawrence Bruce Plant
General Motors Corp. Pat. Sect. New Center One Bldg.
3031 W. Grand Blvd., P.O. Box 33114
Detroit, MI 48202 (313) 974-1350

Ralph T. Rader
Cullen, Sloman, Cantor, Grauer, Scott & Rutherford, P.C.
2400 Penobscot Bldg.
Detroit, MI 48226 (313) 964-0400

Arthur Raisch
Barnes, Kisselle, Raisch Choate, Whittemore & Hulbert, P.C.
1520 Ford Bldg.
Detroit, MI 48226 (313) 962-4790

Frederick M. Ritchie
General Motors Corp. New Center One Bldg. Pat. Sect.
3031 W. Grand Blvd., P.O. Box 33114
Detroit, MI 48232 (313) 974-1344

Charles R. Rutherford
Cullen, Sloman, Cantor, Grauer, Scott & Rutherford, P.C.
2400 Penobscot Bldg.
Detroit, MI 48226 (313) 964-0400

Donald F. Scherer
General Motors Corp. Pat. Section
3031 W. Grand Blvd., P.O. Box 33114
Detroit, MI 48232 (313) 974-1353

Jerold I. Schneider
Cullen, Sloman, Cantor, Grauer, Scott & Rutherford, P.C.
2400 Penobscot Bldg.
Detroit, MI 48226 (313) 964-0400

William Adolph Schuetz
General Motors Corp. Pat. Sect. New Center One Bldg.
3031 W. Grand Blvd., P.O. Box 33114
Detroit, MI 48232 (313) 974-1306

Lee A. Schutzman
General Motors Corp.
3044 W. Grand Blvd.
Detroit, MI 48202 (313) 556-4417

Saul Schwartz
General Motors Corp. Pat. Sect. New Center One Suite 450
P.O. Box 33114
Detroit, MI 48232 (313) 974-1317

Raymond E. Scott
Cullen, Sloman, Cantor, Grauer, Scott & Rutherford, P.C.
2400 Penobscot Bldg.
Detroit, MI 48226 (313) 964-0400

Lawrence J. Shurupoff
Federal- Mogul Corp.
P.O. Box 1966
Detroit, MI 48235 (313) 354-9439

Robert M. Sigler, Jr.
General Motors Corp. Pat. Sect, New Center One Bldg
3031 W. Grand Blvd., P.O. Box 33114
Detroit, MI 48232 (313) 974-1359

Robert A. Sloman
Cullen, Sloman, Cantor, Grauer, Scott & Rutherford, P.C.
2400 Penobscot Bldg.
Detroit, MI 48226 (313) 964-0400

Dale R. Small
400 Tower, Suite 500
Renaiannce Center
Detroit, MI 48243 (313) 962-6192

Dennis G. Stenstrom
Burroughs Corp.
One Burrough Place
Detroit, MI 48232 (313) 972-7378

Christopher A. Taravella
Chrysler Corp., Litigation & Insurance Dept.
P.O. Box 1919
Detroit, MI 48288 (313) 956-2878

Randy W. Tung
General Motors Corp. Pat. Section
3031 W. Grand Blvd
Detroit, MI 48202 (313) 974-1326

Charles Kenneth Veenstra
General Motors Corp. Pat. Sect. New Center One Bldg.
3031 W. Grand Blvd., P.O. Box 33114
Detroit, MI 48232 (313) 974-1333

Robert James Wallace
General Motors Corp. Pat. Section
P.O. Box 33114
Detroit, MI 48232 (313) 974-1309

William J. Waugaman
Barnes, Kisselle, Raisch Choate, Whittemore & Hulbert, P.C.
1520 Ford Bldg.
Detroit, MI 48226 (313) 962-4790

Charles Richard White
General Motors Corp. New Center One Bldg. Pat. Section
Detroit, MI 48232 (313) 974-1335

Thomas K. Ziegler
Cullen, Sloman, Cantor, Grauer, Scott & Rutherford
2400 Penobscot Bldg.
Detroit, MI 48226 (313) 964-0400

Alan James Steger
529 Greenwood S.E.
E. Grand Rapids, MI 49506 (616) 458-4994

Everett R. Casey
5845 Old Orchard Trail
Orchard Lake, MI 48033 (313) 682-0400

John F. Rohe
226 Park Ave.
Petoskey, MI 49770 (616) 347-7327

Robert H. Elliott, Sr.#
15688 Northville Forest Dr.
Plymouth, MI 48170 (313) 420-2465

John J. Cantarella
1004 Joslyn Ave.
Pontiac, MI 48055 (313) 858-8871

Robert J. Madden
5292 Rosamond Lane
Pontiac, MI 48054 (313) 681-3354

Donald E. Overbeek
210 E. Centre Ave.
Portage, MI 49002 (616) 327-8041

Donald L. Waller
801 Ironwood Dr., Apt. 250
Rochester, MI 48063 (313) 652-9047

Dale A. Winnie
404 N. Main, P.O. Box 426
Romeo, MI 48065 (313) 752-6519

Peter D. Keefe
Box 259
Roseville, MI 48066 (313) 927-1187

Donald P. Bush
1608 Vinsetta Blvd.
Royal Oak, MI 48067 (313) 542-3669

John F. Learman
Learman & McCulloch
5291 Colony Dr. North
Saginaw, MI 48603 (517) 799-5300

John K. McCulloch
Learman & McCulloch
5291 Colony Dr. North
Saginaw, MI 48603 (517) 799-5300

John J. Swartz
Le Fevre & Swartz
908 Court St.
Saginaw, MI 48607 (517) 793-8540

William G. Abbatt
Brooks & Kushman
2000 Town Center, Suite 2000
Southfield, MI 48075 (313) 358-4400

Lynn Lawrence Augspurger
Jenkins, Augspurger, Reebel & Zameck
15999 W. Twelve Mile Rd., P.O. Box 5069
Southfield, MI 48084 (313) 559-2828

William D. Blackman
3000 Towm Center Suite 1145
Southfield, MI 48075 (313) 353-4321

George L. Boller
Rhodes & Boller
24700 Northwestern Hwy. Suite 412
Southfield, MI 48075 (313) 356-3457

Ernie L. Brooks
Brooks & Kushman
2000 Town Center Suite 2000
Southfield, MI 48075 (313) 358-4400

Pamela S. Burt#
3000 Town Center Suite 1145
Southfield, MI 48075 (313) 353-4321

Mark A. Cantor
Brooks & Kushman
2000 Town Center Suite 2000
Southfield, MI 48075 (313) 358-4400

Lynn E. Cargill
Brooks & Kushman
2000 Town Center Suite 2000
Southfield, MI 48075 (313) 358-4400

Joseph P. Carrier#
3000 Town Center, Suite 1145
Southfield, MI 48075 (313) 353-4321

Frank H. Cullen
Master Data Center, Inc.
29100 Northwestern Hwy., Suite 300
Southfield, MI 48034 (313) 352-5810

Paul J. Ethington
Reising, Ethington , Barnard, Perry & Brooks
3000 Town Center, Suite 2121
Southfield, MI 48075 (313) 358-0004

Christopher J. Fildes
Brooks & Kushman
2000 Town Center Suite 2000
Southfield, MI 48075 (313) 358-4400

Robert B. Gerhardt
24700 Nothwestern Hwy. Suite 300
Southfield, MI 48075 (313) 352-5810

Kevin J. Heinl
Brooks & Kushman
2000 Town Center, Suite 2000
Southfield, MI 48075 (313) 358-4400

Myron B. Kapustij
29451 Greenfield, Suite 107
Southfield, MI 48076 (313) 569-1421

James A. Kushman
Brooks & Kushman
2000 Town Center, Suite 2000
Southfield, MI 48075 (313) 358-4400

Earl J. LaFontaine
Brooks & Kushman
2000 Town Center Suite 2000
Southfield, MI 48075 (313) 358-4400

J. Gordon Lewis
Eaton Corp., Corporate Res. & Dev. Detroit Center
26201 Northwestern Hwy., P.O. Box 766
Southfield, MI 48037 (313) 354-5028

Thomas A. Lewry
Brooks & Kushman
2000 Town Center, Suite 2000
Southfield, MI 48075 (313) 358-4400

Harold William Milton, Jr.
Reising, Ethington, Barnard, Perry, Brooks & Milton
3000 Town Center, Suite 2121
Southfield, MI 48075 (313) 358-0004

Ronald M. Nabozny
Brooks & Kushman
2000 Town Center, Suite 2000
Southfield, MI 48075 (313) 385-4400

Vett Parsigian
Bendix Corp., Bendix Center
P.O. Box 5060
Southfield, MI 48037 (313) 827-6058

Floyd K. Reynolds
24915 Thorndyke
Southfield, MI 48034 (313) 352-4460

Alex Rhodes
24700 Northwestern Hwy. Suite 412
Southfield, MI 48075 (313) 356-4949

Paul S. Rulon#
Eaton Corporation, Corp. Res. & Dev. Detroit Center
26201 Northwestern Hwy.
P.O. Box 766
Southfield, MI 48037 (313) 354-5057

Markell Seitzman
Allied Corp.
20650 Civic Center Dr., P.O. Box 5060
Southfield, MI 48086 (313) 827-6280

David R. Syrowik
Brooks & Kushman
2000 Town Center, Suite 2000
Southfield, MI 48075 (313) 358-4400

Robert C.J. Tuttle
Brooks & Kushman
2000 Town Center, Suite 2000
Southfield, MI 48075 (313) 358-4400

Loren H. Uthoff, Jr.#
Eaton Corporation
26201 Northwestern Hwy.
Southfield, MI 48037 (313) 354-2871

Irving M. Weiner
3000 Town Center Suite 1145
Southfield, MI 48075 (313) 353-4321

Russel C. Wells
Allied Corporation
20650 Civic Center Drive, P.O. Box 5060
Southfield, MI 48037 (313) 827-5149

William Houseal
Kinney, Cook, Lindenfeld & Kelley
P.O. Box 24, Law & Title Bldg., 811 Ship St.
St. Joseph, MI 49085 (616) 983-0103

Charles M. Kaplan
724 Nanagosa Trail, Route 2
Suttons Bay, MI 49682 (616) 271-6868

Joseph J. Goluban
20700 Ecorse Rd.
Taylor, MI 48180 (313) 388-5988

E. Dennis O'Connor
Masco Corp. Res & Dev.
26855 Trolley Industrial Dr.
Taylor, MI 48180 (313) 291-3500

Steven L. Permut
Masco Corp.
21001 Van Born Rd.
Taylor, MI 48180 (313) 274-7400

Leon E. Redman
Masco Corporation
21001 Van Born
Taylor, MI 48180 (313) 274-7400

Malcolm L. Sutherland
Masco Corp.
21001 Van Born Rd.
Taylor, MI 48180 (313) 274-7400

Edgar A. Zarins
Masco Corp.
21001 Van Born Rd.
Taylor, MI 48180 (313) 274-7400

Douglas S. Bishop
Elhart, Bishop & Thomas, P.C.
329 S. Union At Lake St.
Traverse City, MI 49685 (616) 946-4100

Andrew Raymond Basile
Basile & Hanlon, P.C.
1650 W. Big Beaver Rd., Suite 210
Troy, MI 48084 (313) 649-0990

John G. Batchelder
Basile & Hanlon, P.C.
1650 W. Big Beaver Rd., Suite 210
Troy, MI 48084 (313) 649-0990

Wilfred S. Bobier#
Basile & Hanlon, P.C.
1650 W. Big Beaver Rd., Suite 210
Troy, MI 48084 (313) 649-0990

E.W. Christen
Reising, Ethington, Barnard, Perry & Milton
3290 W. Big Beaver, Suite 510
Troy, MI 48084 (313) 649-3060

Ronald W. Citkowski
Energy Conversion Devices, Inc.
1675 W. Maple Rd.
Troy, MI 48084 (313) 280-1900

David Bowerman Ehrlinger
Krass & Young
2855 Coolidge, Suite 210
Troy, MI 48084 (313) 649-3323

John Charles Evans
Reising, Ethington, Barnard, Perry & Milton
3290 W. Big Beaver
Troy, MI 48084 (313) 649-3060

William L. Fisher
1051 Naughton St., Room 202
Troy, MI 48084 (313) 689-8550

Lloyd Mason Forster
755 W. Big Beaver Rd., Suite 2218
Troy, MI 48084 (313) 362-1115

Herman Foster
Budd Company
3155 W. Big Beaver Rd.
Troy, MI 48084 (313) 643-3530

Denise M. Glassmeyer
Basile & Hanlon, P.C.
1650 W. Big Beaver Rd., Suite 210
Troy, MI 48084 (313) 649-0990

Richard M. Goldman
Energy Conversion Devices, Inc.
1675 W. Maple Rd.
Troy, MI 48084 (313) 280-1900

William M. Hanlon, Jr.
Basile & Hanlon, P.C.
1650 W. Big Beaver Rd., Suite 210
Troy, MI 48084 (312) 649-0990

Warren D. Hill
Reising, Ethington, Barnard, Perry & Milton
3290 W. Big Beaver Rd. Suite 510
Troy, MI 48084 (313) 693-2345

James R. Ignatowski
Gifford Van Ophem & Sprinkle
755 West Big Beaver, Suite 1313
Troy, MI 48084 (313) 362-1210

Steven G. Jonas
Newcor, Inc.
3270 W. Big Beaver Rd.
Troy, MI 48084 (313) 643-7730

Edward G. Jones
6738 Fredmoor St.
Troy, MI 48098 (313) 828-8032

Kenneth I. Kohn
Reising, Ethington, Barnard, Perry & Brooks
3290 W. Big Beaver, Suite 510
Troy, MI 48084 (313) 649-3060

Allen M. Krass
Krass, Young & Schivley
2855 Coolidge, Suite 210
Troy, MI 48084 (313) 649-3323

Lyman R. Lyon
755 W. Big Beaver Rd., Suite 2219
Troy, MI 48084 (313) 362-2600

Robert D. Marshall, Jr.
Krass & Young
2855 Coolidge, Suite 210
Troy, MI 48084 (313) 649-3323

Gerald Edward McGlynn, Jr.
1650 W. Big Beaver Rd.
Troy, MI 48084 (313) 643-9500

Malcolm Robert McKinnon
McKinnon & McKinnon
755 W. Big Beaver Rd., Suite 2219
Troy, MI 48084 (313) 362-0115

John P. Moran
Ex-Cell-OCorporation
2855 Coolidge
Troy, MI 48084 (313) 649-1000

Donald Alvin Panek
The Valeron Corporation
750 Stephenson Hwy.
Troy, MI 48084 (313) 589-1000

Owen E. Perry
Reising, Ethington, Barnard, Perry & Milton
3290 W. Big Beaver Rd. Suite 510
Troy, MI 48084 (313) 649-3060

Paul J. Reising
Reising, Ethington, Barnard, Perry & Milton
3290 W. Big Beaver Rd. Suite 510
Troy, MI 48084 (313) 649-3060

Judith M. Riley
Basile & Hanlon, P.C.
1650 W. Big Beaver Rd. Suite 210
Troy, MI 48084 (313) 649-0990

Marvin S. Siskind
Energy Conversion Devices, Inc.
1675 W. Maple Rd.
Troy, MI 48084 (313) 280-1900

Robert M. Storwick
Krass & Young
2855 Coolidge Suite 210
Troy, MI 48084 (313) 649-3323

Stanley C. Thorpe
Reising, Ethington, Barnard, Perry & Milton
3290 Big Beaver, Suite 510
Troy, MI 48084 (313) 649-3060

Edward J. Timmer
Reising, Ethington, Barnard, Perry & Milton
3290 W. Big Beaver, Suite 510
Troy, MI 48084 (313) 649-3060

Theodore Van Meter
Vickers Corporation
1401 Crooks Rd.
Troy, MI 48084 (313) 280-3388

Remy J. Van Ophem
Gifford, Van Ophem & Sprinkle
755 W. Big Beaver, Suite 1313
Troy, MI 48084 (313) 362-1210

Neal A. Waldrop
755 W. Big Beaver Rd.
Troy, MI 48084 (313) 362-3620

Arnold S. Weintraub
3001 W. Big Beaver Rd. Suite 504
Troy, MI 48084 (313) 649-3850

Donald L. Wood
Krass & Yong
2855 Collidge, Suite 210
Troy, MI 48084 (313) 649-3323

Thomas N. Young
Krass & Young
2855 Cooldige
Troy, MI 48084 (313) 649-3323

Gregory T. Zalecki
275 E. Big Beaver Suite 203
Troy, MI 48083 (313) 528-9390

Jerry Gunther Beck
5328 Fairway Court
W. Bloomfield, MI 48033 (313) 851-3228

Arthur E. Bahr
Carboloy Inc.
11177 E. Eight Mile Rd.
Walled Lake, MI 48089 (313) 497-5116

Frank D. Risko#
Ex-Cell-O Corp.
850 Ladd Rd.
Walled Lake, MI 48088 (313) 624-7800

David L. Kuhn*
U.S. Army, Tank-Automotive Command
Attn: A M S T A- L P
Warren, MI 48397

Richard Paul Mueller
Occidental Chemical Corp., Patent/Legal Dept.
21441 Hoover Rd.
Warren, MI 48089 (313) 497-6892

Damian Porcari*#
U.S. Army Tank-Automotive Command
Attn: S M D Y S - L P
Warren, MI 48090 (313) 574-8683

Walter Potoroka, Sr.
11955 E. Nine Mile Rd.
Warren, MI 48089 (313) 497-4202

J. Santo
11161 Hanover
Warren, MI 48093 (313) 977-1712

Gail S. Soderling*
U.S. Army Tank-Automotive Command
Attn: A M S T A -L P
Warren, MI 48090 (313) 574-8682

Peter Arthur Taucher*
U.S. Army Tank-Automotive Command
Attn: A M S T A - L P
12 Mile & Van Dyke
Warren, MI 48090 (313) 573-2552

Stanley J. Ference
Ference, Ference & Cicirell
8623 N. Wayne Rd. Suite 255
Westland, MI 48185 (313) 422-4666

Clair D. Goodman, Jr.
22259 Loretta
Woodhaven, MI 48183 (313) 568-4298

John C. Demeter
B A S F Corporation Chem. Divs. Pat. Dept.
1419 Biddle Ave.
Wyandotte, MI 48192 (313) 246-6194

Norbert M. Lisicki
B A S F Wyandotte Corp.
1419 Biddle Ave.
Wyandotte, MI 48192 (313) 246-6191

Dorothy B. McKenzie
B A S F Corporation
1419 Biddle St.
Wyandotte, MI 48192 (313) 246-6194

Joseph Dale Michaels
B A S F Wyandotte Corp.
1419 Biddle Ave.
Wyandotte, MI 48192 (313) 246-6188

Bill C. Panagos
B A S F Corp.
1419 Biddle Ave.
Wyandotte, MI 48192 (313) 246-6496

Bernhard R. Swick
B A S F Wyandotte Corp. Patent Dept.
1419 Biddle Ave.
Wyandotte, MI 48192 (313) 246-6190

Charles S. Saxon
Eastern Michigan University
511 Pray-Harrold
Ypsilanti, MI 48197 (313) 487-2454

Minnesota

Frederick J.B. Wall#
13670 Tomahawk Dr. South
Afton, MN 55001 (612) 436-8668

Henry C. Kovar#
P.O. Box 571
Anoka, MN 55303 (612) 427-4679

Edward P. Heller, III
Control Data Corp.
8100 34th Ave. South
Bloomington, MN 55440 (612) 853-8872

Roger W. Jensen
8127 Pennsylvania Circle
Bloomington, MN 55438 (612) 944-7525

Anthony A. Juettner
8430 Pennsylvania Rd.
Bloomington, MN 55438 (612) 942-8252

Joan S. Keps
8540 Irwin Rd.
Bloomington, MN 55437 (612) 897-1166

Oliver A. Ossanna, Jr.
2100 Overlook Dr.
Bloomington, MN 55420 (612) 884-2848

Edward L. Schwarz
Control Data Corp.
H Q S04H
Box O
Bloomington, MN 55440 (612) 853-3268

Donald A. Jacobson
151 W. 126th St.
Burnsville, MN 55337 (612) 894-1055

James H. Wills
Kalina & Wills
4111 Central Ave, N.E. Suite 102 South
Columbia Heights, MN 55421
(612) 788-1681

Darrell K. Morse
Morse, Clinton & OGorman
7200 - 80th St. South
Cottage Grove, MN 55016 (612) 459-6644

David C. Bohn#
Rosemount Inc.
12001 Technology Dr.
Eden, MN 55344 (612) 828-3823

Michael J. Schneider
Rosemount Inc.
12001 West 78th St.
Eden Prairie, MN 55344 (612) 828-3370

James R. Cwayna
5200 Wilson Rd.
Edina, MN 55424 (612) 927-5533

Vernon Alfred Johnson
Toro Mfg. Corp., One Corp. Center
7401 Metro Blvd.
Edina, MN 55435 (612) 887-8911

James Edward Olds
7401 Metro Blvd., Suite 445
Edina, MN 55435 (612) 831-0793

Joseph E. Ryan
4449 Fondell Dr.
Edina, MN 55435 (612) 920-3426

Jerold M. Forsberg
6 Old Still Rd.
Grand Rapids, MN 55744

William C. Flynn
932 Westbrooke Way, #5
Hopkins, MN 55343 (612) 933-6937

William O. Ney
18 Williams Woods
Mahtomedi, MN 55115 (612) 426-2504

Augustus Jeter Hipp
Gnb Incorporated
1110 Highway 110
Mendota Heights, MN 55118 (612) 681-5417

John W. Adams
520 Norwest Midland Bldg., 401 2nd Ave. South
Minneapolis, MN 55401 (612) 339-4861

Philip G. Alden
Williamson, Bains Moore & Hansen, Suite 668
Minneapolis, MN 55402 (612) 332-2587

Richard D. Allison
Palmatier & Sjoquist
2000 Northwestern Financial Center, 7900 Xerxes Ave. South
Minneapolis, MN 55431 (612) 831-5454

Robert M. Angus
Control Data Corp.
P.O. Box O
Minneapolis, MN 55440 (612) 853-3266

Oliver F. Arrett
Vidas & Arrett, P.A.
2925 Multifoods Tower
33 S. Sixth St.
Minneapolis, MN 55402 (612) 339-8801

Michael B. Atlas
Control Data Corp.
P.O. Box O
Minneapolis, MN 55440 (612) 835-7546

William C. Badcock
Dorsey & Whitney
2200 First Bank Place East
Minneapolis, MN 55402 (612) 340-2635

Herman Hershel Bains
Williamson, Bains, Moore & Hansen
608 Building, Suite 668., 608 Second Ave. South
Minneapolis, MN 55402 (612) 332-2587

Robert C. Baker
Sturm & Baker, Ltd.
940 Northwestern Financial Center, 7900 Xerxes Ave. South
Minneapolis, MN 55431 (612) 831-9510

Richard Otto Bartz
Burd, Bartz & Gutenkauf
1300 Foshay Tower, 821 Marquette Ave.
Minneapolis, MN 55402 (612) 332-6581

Mary P. Bauman
Fredrikson & Byron, P.A.
1100 International Center, 900 Second Ave South
Minneapolis, MN 55402 (612) 347-7108

Robert C. Beck
Medtronic, Inc.
7000 Central Ave.,N.E.
Minneapolis, MN 55432 (612) 574-3337

Jerry F. Best
Onan Corp.
1400 73rd Ave. N.E.
Minneapolis, MN 55432 (612) 574-5802

Clyde C. Blinn
Honeywell Inc. Off. Of General Coun.
Honeywell Plaza
Minneapolis, MN 55408 (612) 870-2886

William A. Braddock
Kinney & Lange, P.A.
625 Fourth Ave. South, Suite 1500
Minneapolis, MN 55415 (612) 339-1863

Joseph F. Breimayer
Medtronic, Inc.
7000 Central Ave.
Minneapolis, MN 55432 (612) 574-3278

Ronald J. Brown
Dorsey & Whitney
2200 First Bank Place E.
Minneapolis, MN 55402 (612) 340-2879

Robert L. Buckley, Jr.#
Toro Co.
8111 Lyndale Ave. South
Minneapolis, MN 55420 (612) 888-8801

John W. Bunch
Merchant, Gould, Smith, Edell, Welter & Schmidt, P.A.
1600 Midwest Plaza Bldg.
Minneapolis, MN 55402 (612) 332-5300

L. Paul Burd
Burd, Bartz & Gutenkauf
1300 Foshay Tower, 821 Marquette Ave.
Minneapolis, MN 55402 (612) 332-6581

Alan G. Carlson
Merchant, Gould, Smith, Edell, Welter & Schmidt
801 Nicollet Mall, 16th Floor
Minneapolis, MN 55402 (612) 332-5300

John A. Clifford
Merchant, Gould, Smith, Edell, Welter & Schmidt, P.A.
1600 Midwest Plaza Bldg.
Minneapolis, MN 55402 (612) 332-5300

Timothy R. Conrad
Merchant & Gould
1600 Midwest Plaza Bldg.
Minneapolis, MN 55402 (612) 332-5300

Omund R. Dahle
Honeywell Inc. Off, Of Gen. Counsel
Honeywell Plaza
Minneapolis, MN 55408 (612) 870-2864

Robert W. Doyle
Doyle International Law Office., Ltd.
4530 I D S Center
Minneapolis, MN 55402 (612) 338-7511

Reed A. Duthler
Medtronic Inc.
7000 Central Ave.
Minneapolis, MN 55432 (612) 574-3351

Wayne B. Easton
510 Plymouth Bldg.
Minneapolis, MN 55402 (612) 333-8723

Robert T. Edell
Merchant, Gould, Smith, Edell, Welter & Schmidt, P.A.
1600 Midwest Plaza Bldg.
Minneapolis, MN 55402 (612) 332-5300

Lewis P. Elbinger
Honeywell Inc. Off. Of Gen. Coun.
Honeywell Plaza
Minneapolis, MN 55408 (612) 870-2030

Michael D. Ellwein
Pillsbury Co., Pillsbury Center - M.S. 3764
200 South Sixth St.
Minneapolis, MN 55402 (612) 330-4679

Robert A. Elwell
Kinney & Lange, P.A.
625 Fourth Ave. South, Suite 1500
Minneapolis, MN 55415 (612) 339-1863

Gene O. Enockson
1600 Quebec Ave., N.
Minneapolis, MN 55427 (612) 545-2963

David R. Fairbairn
Kinney & Lange, P.A.
625 Fourth Ave. South, Suite 1500
Minneapolis, MN 55415 (612) 339-1863

Douglas B. Farrow
Graco Inc.
P.O. Box 1441
Minneapolis, MN 55440 (612) 623-6769

Alfred N. Feldman
Honeywell Inc.
Honeywell Plaza, P.O. Box 524
Minneapolis, MN 55408 (612) 870-2889

Harold D. Field, Jr.
Leonard, Street & Deinard
100 S. Fifth St., Suite 1500
Minneapolis, MN 55402 (612) 337-1533

Robert C. Freed
Merchant, Gould, Smith, Edell, Welter & Schmidt, P.A.
1600 Midland Plaza Bldg.
Minneapolis, MN 55402 (612) 332-5300

Grady J. Frenchick
Medtronic, Inc.
7000 Central Ave., N.E.
Minneapolis, MN 55432 (612) 574-3271

Norman P. Friederichs, Jr.
Merchant, Gould, Smith, Edell, Welter & Schmidt, P.A.
1600 Midwest Plaza Bldg.
Minneapolis, MN 55402 (612) 332-5300

David N. Fronek
Dorsey & Whitney
2200 First Bank Place East
Minneapolis, MN 55402 (612) 340-2629

Keith J. Goar
10800 Lyndale Ave. Apt. 250
Minneapolis, MN 55420 (612) 881-1601

John D. Gould
Merchant, Gould, Smith, Edell, Welter & Schmidt, P.A.
1600 Midwest Plaza Bldg.
Minneapolis, MN 55402 (612) 332-5300

Alan G. Greenberg
1050 Midland Bk. Bldg.
Minneapolis, MN 55401 (612) 333-7191

Leo Gregory
Reif & Gregory
1500 Dain Tower, 527 Marquette Ave.
Minneapolis, MN 55402 (612) 333-7522

Robert W. Gutenkauf
Burd, Bartz, & Gutenkauf
1300 Foshay Tower, 821 Marquette Ave.
Minneapolis, MN 55402 (612) 332-6581

James R. Haller
Fredrikson & Byron, P.C.
1100 International Centre, 900 Second Ave. South
Minneapolis, MN 55402 (612) 347-7017

Curtis B. Hamre
Merchant, Gould, Smith, Edell, Welter & Schmidt, P.A.
1600 Midwest Plaza Bldg.
Minneapolis, MN 55402 (612) 332-5300

Conrad A. Hansen
Moore & Hansen, Northstar East, Suite 668
608 Second Ave. South
Minneapolis, MN 55402 (612) 332-5300

Henry L. Hanson
Honeywell Inc.
Honeywell Plaza
Minneapolis, MN 55408 (612) 870-2215

James V. Harmon
1200 First Bank Place West
120 South Sixth St.
Minneapolis, MN 55402 (612) 339-1400

Orrin M. Haugen
Haugen & Nikolai, P.A.A Minn. Corp.
900 Second Ave. South, Suite 820
Minneapolis, MN 55402 (612) 339-7461

John M. Haurykiewicz
Faegre & Benson
2300 Multifoods Tower, 33 South Sixth St.
Minneapolis, MN 55402 (612) 371-3168

Gerald E. Helget
Palmatier & Sjoquist, P.A.
7900 Xerxes Ave. South, Suite 2000
Minneapolis, MN 55431 (612) 831-5454

Stuart R. Hemphill
Dorsey & Whitney
2200 First Bank Place East
Minneapolis, MN 55402 (612) 340-2734

Randall A. Hillson
Merchant, Gould, Smith, Edell, Welter & Schmidt
1600 Midwest Plaza Bldg.
Minneapolis, MN 55402 (612) 332-5300

Robert J. Jacobson
Palmatier & Sjoquist
2000 Northwestern Financial Ctr.
7900 Xerxes Ave. South
Minneapolis, MN 55431 (612) 831-5454

Hugh D. Jaeger
3209 W. 76th St. Suite 207
Minneapolis, MN 55435 (612) 830-1197

Cecillia M. Jaisle#
2728 Chowen Ave., S.
Minneapolis, MN 55416 (612) 920-7452

Harold D. Jastram
Oppenheimer, Wolff & Donnelly Plaza Vii
45 S. 7th Street, Suite 3400
Minneapolis, MN 55402 (612) 344-9281

Bruce A. Johnson
Medtronic Inc.
7000 Central Ave.
Minneapolis, MN 55432 (612) 574-3275

Clayton Russell Johnson
3121 Dakota Ave.
Minneapolis, MN 55416 (612) 926-6939

Eugene L. Johnson
Dorsey & Whitney
2200 First Bank Place East
Minneapolis, MN 55402 (612) 340-2625

Daniel R. Johson#
Recognition Specialties Of Amer.
2828 Anthony Lane S.
Minneapolis, MN 55418 (612) 788-9681

Trevor B. Joike
Honeywell Co.
Honeywell Plaza
Minneapolis, MN 55408 (612) 870-2877

Steven E. Kahm
3901 S. Cedar Lake Rd.
Minneapolis, MN 55416 (612) 920-1117

Gregory P. Kaihoi
Fredrikson & Byron, P.A.
1100 International Centre, 900 Second Ave.
Minneapolis, MN 55402 (612) 347-7077

Alan D. Kamrath
Wicks & Nemer, P.A.
1407 Soo Line Bldg.
Minneapolis, MN 55402 (612) 339-8501

Harold J. Kinney
Kinney & Lange, P.A.
625 Fourth Ave. South, Suite 1500
Minneapolis, MN 55415 (612) 339-1863

Michael E. Kiteck, Jr.
5050 Excelsior Blvd.
Minneapolis, MN 55416 (612) 922-5223

Robert J. Klepinski
3120 James Ave. South
Minneapolis, MN 55408 (612) 825-7911

Daniel J. Kluth
Merchant, Gould, Smith, Edell, Welter & Schmidt, P.A.
1600 Midwest Plaza Bldg., 801 Nicollet Mall
Minneapolis, MN 55402 (612) 332-5300

Franklin J. Knoll
5316 1st Ave. S.
Minneapolis, MN 55419 (612) 827-4889

Robert R. Kooiman#
Rosemount Inc.
P.O. Box 35129
Minneapolis, MN 55435 (612) 937-3350

Alan W. Kowalchyk
Merchant, Gould, Smith, Edell, Welter & Schmid, P.A.
1600 Midwest Plaza Bldg.
Minneapolis, MN 55402 (612) 332-5300

Marc G. Kurzman
Kurzman Shapiro & Marehan
601 W. Butler Square, 100 N. 6th St.
Minneapolis, MN 55403 (612) 333-4403

Frederick E. Lange
Kinney & Lange, P.A.
625 Fourth Ave. South, Suite 1500
Minneapolis, MN 55415 (612) 339-1863

Michael B. Lasky
Merchant, Gould, Smith, Edell, Welter & Schmidt, P.A.
1600 Midwest Plaza Bldg., 801 Nicollet Mall
Minneapolis, MN 55402 (612) 332-5300

Ronald L. Laumbach
Cargill, Incorporated
P.O. Box 9300
Minneapolis, MN 55440 (612) 475-6366

Donald J. Lenkszus
Honeywell Inc.
Honeywell Plaza, Mn 12-8251
Minneapolis, MN 55408 (612) 870-2877

Thomas Anthony Lennon
7101 York Ave., S.
Minneapolis, MN 55435 (612) 835-3944

George A. Leone, Sr.
Honeywell Inc. Off. Of Gen. Coun.- M N 12-8251
Honeywell Plaza
Minneapolis, MN 55408 (612) 870-5200

Robert J. Lewis
Pillsbury Co.
Pillsbury Center. 3276
200 S. Sixth St.
Minneapolis, MN 55402 (612) 330-8548

L. Meroy Lillehaugen
General Mills, Inc.
9200 Wayzata Blvd.
Minneapolis, MN 55426 (612) 540-2283

Walter C. Linder
Kinney & Lange, P.A.
625 Fourth Ave. South, Suite 1500
Minneapolis, MN 55415 (612) 339-1863

Robert E. Lowe
Litton Industries, Inc.
P.O. Box 9461
Minneapolis, MN 55440 (612) 553-2337

Ronald E. Lund
Pillsbury Co.
Pillsbury Center., 200 S. Sixth St.
Minneapolis, MN 55402 (612) 330-4589

Steven W. Lundberg
Merchant, Gould, Smith, Edell, Welter & Schmidt, P.A.
1600 Midwest Plaza Bldg.
Minneapolis, MN 55402 (612) 332-5300

Laurence Joseph Marhoefer
Honeywell Inc.
Honeywell Plaza
Minneapolis, MN 55408 (612) 870-6641

Joseph P. Martin#
Toro Co.
8111 Lyndale Ave., S.
Minneapolis, MN 55420 (612) 888-8801

Daniel W. Mc Donald
Merchant, Gould, Smith, Edell, Welter & Schmidt, P.A.
1600 Midwest Plaza, 801 Nicollet Mall
Minneapolis, MN 55402 (612) 332-5300

Wendy M. Mc Donald
Merchant, Gould, Smith, Edell, Welter & Schmidt, P.A.
1600 Midwest Plaza Bldg.
Minneapolis, MN 55402 (612) 332-5300

Albin Medved
Honeywell Inc. Office Of Gen. Counsel
Honeywell Plaza
Minneapolis, MN 55408 (612) 870-6459

Charles G. Mersereau
Honeywell Inc.
Honeywell Plaza
Minneapolis, MN 55408 (612) 870-2875

James W. Miller
Toro Co.
8111 Lyndale Ave., South
Minneapolis, MN 55420 (612) 887-8903

Malcolm L. Moore
Williamson, Bains, Moore & Hansen
608 Bldg. Suite 668
Minneapolis, MN 55402 (612) 332-2587

Lawrence M. Nawrocki
3989 Central Ave., N.E., Suite 605, P.O. Box 21369
Minneapolis, MN 55421 (612) 781-3319

Theodore F. Neils
Honeywell Inc.
Honeywell Plaza
Minneapolis, MN 55408 (612) 870-2892

Bruce A. Nemer
Wicks And Nemer, P.A.
1407 Soo Line Bldg., 5th & Marquette
Minneapolis, MN 55402 (612) 339-8501

Frederick W. Niebuhr
Haugen & Nikolai, P.A.
820 International Centre, 900 Second Ave. South
Minneapolis, MN 55402 (612) 339-7461

Thomas J. Nikolai
Haugen & Nikolai, P.A.
820 International Centre, 900 Second Ave. South
Minneapolis, MN 55402 (612) 339-7461

Michael J. O'Loughlin
615 Minnesota Federal Bldg.
Minneapolis, MN 55402 (612) 338-7509

John A. O'Toole
General Mills, Inc.
9200 Wayzata Blvd.
Minneapolis, MN 55426 (612) 540-2422

Kenneth D. Ohm
General Mills, Inc.
9200 Wayzata Blvd.
Minneapolis, MN 55440 (612) 540-2284

Robert A. Pajak#
Honeywell Inc.
Honeywell Plaza
Minneapolis, MN 55408 (612) 870-2723

H. Dale Palmatier
Peterson, Palmatier, Sturm, Sjoquist & Baker, Ltd.
940 Northwestern Financial Center, 7900 Xerxes Ave. South
Minneapolis, MN 55431 (612) 831-7777

James H. Patterson
Dorsey & Whitney
2200 First Bank Place East
Minneapolis, MN 55402 (612) 340-2627

Stuart R. Peterson
Peterson, Wicks, Nemer & Kamrath, P.A.
Suite 1407 Soo Line Bldg.
Minneapolis, MN 55402 (612) 339-8501

Thomas E. Popovich
Dorsey & Whitney
2200 First Bank Place East
Minneapolis, MN 55401 (612) 340-2964

Robert J. Rickett
3304 Skycroft Dr.
Minneapolis, MN 55418 (612) 330-3640

John L. Rooney
Medtronic, Inc.
3055 Old Highway Eight
P.O. Box 1453
Minneapolis, MN 55440 (612) 574-3279

Matthew K. Ryan
Williamson, Bains, Moore & Hansen
608 Bldg, Suite 668
Minneapolis, MN 55402 (612) 332-2587

Zbigniew P. Sawicki
Kinney & Lange, P.A.
625 Fourth Ave. South, Suite 1500
Minneapolis, MN 55415 (612) 339-1863

Janet P. Schafer
3989 Central Ave, N.E. Suite 605
Minneapolis, MN 55421 (612) 781-3319

Brian F. Schroeder#
Schroeder & Siegfried, P.A.
2340 I D S Center
Minneapolis, MN 55402 (612) 339-0120

Everett J. Schroeder
Schroeder & Siegfried, P.A.
2340 IDS Center
Minneapolis, MN 55402 (612) 229-0116

Michael D. Schumann
Merchant, Gould, Smith, Edell, Welter & Schmidt, P.A.
1600 Midwest Plaza
Minneapolis, MN 55402 (612) 323-5300

Ronald J. Schutz
Robins, Zelle, Larson & Kaplan
1800 International Centre, 900 Second Ave. South
Minneapolis, MN 55402

Lew Schwartz
121 Washington Ave. South, Suite 504
Minneapolis, MN 55401 (612) 332-3023

Micheal L. Schwegman
Merchant, Gould, Smith, Edell, Welter & Schmidt, P.A.
801 Nicollet Mall, 1600 Midwest Plaza Bldg.
Minneapolis, MN 55402 (612) 332-5300

Clayton M. Scott
Federal Cartridge Corp.
2700 Foshay Tower
Minneapolis, MN 55402 (612) 333-8255

John G. Shudy, Jr.
Honeywell Inc. Off. Of Gen. Coun.
Honeywell Plaza
Minneapolis, MN 55408 (612) 870-6419

Wayne A. Sivertson
Kinney & Lange, P.A.
625 Fourth Ave. South
Minneapolis, MN 55415 (612) 339-1863

Paul L. Sjoquist
Palmatier, Sturm, Sjoquist, & Baker, Ltd.
940 Northwestern Financial Ctr., 7900 Xerxes Ave. South
Minneapolis, MN 55431 (612) 831-7777

Phillip H. Smith
Merchant, Gould, Smith, Edell, Welter & Schmidt, P.A.
1600 Midwest Plaza Bldg.
Minneapolis, MN 55402 (612) 332-5300

Patrick John Span
Henkel Corp.
7900 W. 78th St.
Minneapolis, MN 55435 (612) 828-8305

Charles E. Steffey
Faegre & Benson
2300 Multifoods Tower
33 South Sixth St.
Minneapolis, MN 55402 (612) 371-5300

Walter J. Steinkraus
Vidas & Arrett, P.A.
2925 Multifoods Tower, 33 S. Sixth St.
Minneapolis, MN 55402 (612) 339-8801

Douglas A. Strawbridge
Merchant, Gould, Smith, Edell, Welter & Schdmit, P.A.
1630 Midwest Plaza Bldg.
Minneapolis, MN 55402 (612) 332-5300

Warren A. Sturm
Palmatier, Sturm, Sjoquist & Baker, Ltd.
7900 Xerxes Ave. South, Suite 940
Minneapolis, MN 55431 (612) 831-7777

John P. Sumner
Merchant, Gould, Smith, Edell, Welter & Schmidt,
1600 Midwest Plaza Bldg., 801 Nicollet Mall
Minneapolis, MN 55402 (612) 332-5300

John S. Sumners
Merchant, Gould, Smith, Edell, Welter & Schmidt, P.A.
1600 Midwest Plaza Bldg.
Minneapolis, MN 55402 (612) 332-5300

Dudley W. Swedberg#
3035 Long Meadow Circle
Minneapolis, MN 55420 (612) 854-4018

David K. Tellekson
Merchant, Gould, Smith, Edell, Welter & Schmidt, P.A.
1600 Midwest Plaza
Minneapolis, MN 55402 (612) 332-5300

Jon F. Tuttle
Dorsey & Whitney
2200 First Bank Place East
Minneapolis, MN 55402 (612) 340-2631

William T. Udseth
Honeywell Inc.
2701 4th Ave. South
Minneapolis, MN 55408 (617) 870-5158

Albert L. Underhill#
Merchant & Gould
1600 Midwest Plaza
Minneapolis, MN 55402 (612) 332-5300

Charles J. Ungemach
Honeywell Inc. Off. Of Gen. Counsel
M N 12-8251, Honeywell Plaza
Minneapolis, MN 55408 (612) 870-6409

Robert O. Vidas
Vidas & Arrett, P.A.
2925 Multifoods Tower, 33 South Sixth St.
Minneapolis, MN 55402 (612) 339-8801

Scott O. Vidas
Vidas & Arrett, P.A.
2925 Multifoods Tower, 33 S. Sixth St.
Minneapolis, MN 55402 (612) 339-8801

Paul A. Welter
Merchant, Gould, Smith, Edell, Welter & Schmidt, P.A.
1600 Midwest Plaza Bldg.
Minneapolis, MN 55402 (612) 332-5300

Nicholas E. Westman
Kinney & Lange, P.A.
625 Fourth Ave. South, Suite 1500
Minneapolis, MN 55415 (612) 339-1863

Jack W. Wicks
Wicks & Nemer, P.A.
1407 Soo Line Bldg.
Minneapolis, MN 55402 (612) 339-8501

Douglas J. Williams
Merchant & Gould
1600 Midwest Plaza Bldg.
Minneapolis, MN 55402 (612) 332-5300

Warren D. Woessner
Merchant, Gould, Smith, Edell, Welter & Schmidt
801 Nicollet Mall
Minneapolis, MN 55402 (612) 332-5300

Mark A. Wurm
Schroeder & Siegfried, P.A.
2340 I D S Center
Minneapolis, MN 55402 (612) 339-0120

James L. Young
Kinney & Lange, P.A.
625 Fourth Ave. South, Suite 1500
Minneapolis, MN 55415 (612) 339-1869

Malcolm D. Reid
15612 Highway 7
Minnetonka, MN 55343 (612) 938-3472

J. Michael Rosso
2599 Mississippi St.
New Brighton, MN 55112 (612) 633-5685

Gerhard A. Ellestad
301 N. Orchard St.
Northfield, MN 55057 (507) 645-5495

Robert B. Moffatt
615 W. Third Ave.
Pine City, MN 55063 (612) 629-6066

Roger L. Schneider
Honeywell Inc.
Solid State Electronics Div.
12001 State Highway 55
Plymouth, MN 55441 (612) 541-2037

James Michael Anglin
I B M Corp., Dept 917
Hwy. 52 & 37th St., N.W.
Rochester, MN 55901 (507) 253-4661

Keith T. Bleuer
1663 Wilshire Dr., N.E.
Rochester, MN 55901 (507) 288-4978

Bradley A. Forrest
I B M Corp. Dept. 917
Highway 52 & 37th St. N.W.
Rochester, MN 55901 (507) 287-7974

Frederick W. Kellog#
Mayo Medical Ventures
200 First St. S.W.
Rochester, MN 55905 (507) 284-4916

Homer L. Knearl
I B M Corp. Dept. 917
Highway 52 & 37th St. N.W.
Rochester, MN 55901 (507) 253-5331

Robert Wyman Lahtinen
I B M Corp.
3605 Hwy. 52 North
Rochester, MN 55901 (507) 253-5331

William J. Ryan
Michaels Seeger Rosenblad & Arnold
550 Norwest Bank Bldg., 21 First St. S.W.
Rochester, MN 55902 (507) 288-7755

Donald F. Voss
2541 12th Ave., N.W.
Rochester, MN 55901 (507) 282-0396

Charles A. Johnson
Unisys Corp.
2276 Highcrest Rd.
Roseville, MN 55113 (612) 635-7702

Alan Maclean Staubly
17081 Sunset Ave.
Shakopee, MN 55379 (612) 445-4900

Charles L. Rubow
3170 N. Victoria St.
Shoreview, MN 55112 (612) 484-0895

Sten E. Hakanson
119 Tenth Ave. North
South St. Paul, MN 55075 (612) 455-8740

Thomas B. Tate
450 Southview Blvd., P.O. Box 41
South St. Paul, MN 55075 (612) 457-6750

Albert W. Watkins#
30 E. Minn St., P.O. Box 593
St. Joseph, MN 56374 (612) 363-4673

Arthur Sulby Caine
4033 Xenwood Ave.
St. Louis Park, MN 55416 (612) 929-3465

Robert L. Kaner
1410 Colorado Ave., South, Apt. 205
St. Louis Park, MN 55416 (612) 546-8861

Cruzan Alexander
P.O. Box 33427
St. Paul, MN 55133 (612) 733-1511

David W. Anderson
3 M Company Off. Of Pat. Coun.
P.O. Box 33427
St. Paul, MN 55133 (612) 733-2221

John C. Barnes
3M Co. Off. Of Pat. Coun.
P.O. Box 33427
St. Paul, MN 55133 (612) 733-1519

William B. Barte
3M Company Off. Of Pat. Coun.
P.O. Box 33427
St. Paul, MN 55133 (612) 733-7519

Carolyn A. Bates
3M Company Off. Of Pat. Coun.
P.O. Box 33427
St. Paul, MN 55133 (612) 733-1535

William D. Bauer
3M Co.
3M Center, P.O. Box 33427
St. Paul, MN 55133 (612) 733-1532

Jennie G. Boeder
3 M Company
P.O. Box 33427
St. Paul, MN 55133 (612) 733-3084

Warren R. Bovee
3M Company Off. Of Pat. Coun.
P.O. Box 33427
St. Paul, MN 55133 (612) 733-1513

Glenn William Bowen, Sr.
Unisys Corp.
P.O. Box 64525
St. Paul, MN 55164 (612) 456-2682

Richard E. Brink
3 M Co. Off. Of Pat. Coun.
P.O. Box 33427
St. Paul, MN 55133 (612) 733-1517

Stephen W. Buckingham
3 M Company Off. Of Pat. Coun.
P.O. Box 33427, 2501 Hudson Rd.
St. Paul, MN 55133 (612) 733-3379

Robert Withy Burns
3M Company Off. Of Pat. Coun.
P.O. Box 33427, 3M Center
St. Paul, MN 55144 (612) 733-1555

Linda M. Byrne
Merchant, Gould, Smith, Edell, Welter & Schmidt, P.A.
1000 Norwest Center
St. Paul, MN 55101 (612) 298-1055

Gerald Frank Chernivec
Minn. Mining & Mfg. Co.
2501 Hudson, P.O. Box 55133
St. Paul, MN 55133 (612) 733-8398

David R. Cleveland
3 M Company Off. Of Patent Coun.
P.O. Box 33427
St. Paul, MN 55133 (612) 733-1539

Stanley G. De La Hunt
3 M Company
3 M Center
St. Paul, MN 55133 (612) 733-1508

Mark J. DiPietro
Merchant, Gould, Smith, Edell, Welter & Schmidt, P.A.
Suite 1000 Norwest Center
St. Paul, MN 55101 (612) 298-1055

Anthony G. Eggink
3100 First Natl. Bank Bldg., 332 Minnesota St.
St. Paul, MN 55101 (612) 298-1171

William G. Ewert
3 M Company Off. Of Pat. Coun.
P.O. Box 33427
St. Paul, MN 55133 (612) 733-1533

Frederick A. Fleming
1860 Highland Pkwy.
St. Paul, MN 55116 (612) 690-4656

Richard Francis
3M Company Off. Of Pat. Coun.
P.O. Box 33427
St. Paul, MN 55133 (612) 733-7519

Mark William Gehan
757 Fairmount Ave.
St. Paul, MN 55105 (612) 733-1509

Donald C. Gipple
3M Company Off. Of Pat. Coun.
P.O. Box 33427
St. Paul, MN 55133 (612) 733-1538

Philip M. Goldman#
3M Company Off. Of Pat. Coun.
P.O. Box 33427
St. Paul, MN 55133 (612) 733-4247

Charles E. Golla
Merchant, Gould, Smith, Edell, Welter & Schmidt, P.A.
1000 Norwest Center
St. Paul, MN 55101 (612) 298-1055

Randall J. Gort
3M Company
P.O. Box 33427
St. Paul, MN 55133 (612) 733-1868

Kenneth Thomas Grace
Sperry Univac
3333 Pilot Knob Rd.
St. Paul, MN 55165 (612) 456-2682

Robert E. Granrud
1809 Colvin Ave.
St. Paul, MN 55116 (612) 698-7621

Robert W. Hoke, II
3 M Co.
3M Center, P.O. Box 33427
St. Paul, MN 55133 (612) 736-9155

Susan M. Howard
3M Co.
3M Center 270-2S-06
St. Paul, MN 55144 (612) 733-6394

William L. Huebsch
3M Company, P.O. Box 33427
2501 Hudson Rd.
St. Paul, MN 55144 (612) 733-2835

Harold Hughesdon#
3M Co.
3M Center Bldg. 220-4W-01
St. Paul, MN 55144 (612) 733-2973

Dale E. Hulse
Minn. Mining & Mfg. Co. Off. Of Pat. Coun.
P.O. Box 33427
St. Paul, MN 55133 (612) 736-9631

Marvin Jacobson
Jacobson & Johnson
200 So. Robert St. Suite 204
St. Paul, MN 55107 (612) 222-3775

Carl Lowell Johnson
Jacobson & Johnson
200 So. Robert
St. Paul, MN 55107 (612) 222-3775

Robert H. Jordan
Minn. Mining & Mfg. Co.
P.O. Box 33427, 2501 Hudson Rd.
St. Paul, MN 55133 (612) 733-6866

Thomas W. Kenyon
1129 Portland Ave.
St. Paul, MN 55104 (612) 291-7200

Walter N. Kirn, Jr.
3M Company Off. Of Pat. Coun.
P.O. Box 33427
St. Paul, MN 55133 (612) 733-1523

Charles Houlton Lauder
Minn. Mining & Mfg. Co.
P.O. Box 33427
St. Paul, MN 55133 (612) 733-1510

James V. Lilly
3M Company Off. Of Pat. Coun.
P.O. Box 33427
St. Paul, MN 55133 (612) 733-1543

Mark A. Litman
3M Co. Off. Of Pat. Coun.
P.O. Box 33427
St. Paul, MN 55133 (612) 733-1515

Douglas B. Little
Minn. Mining Mfg. Co., 3M Center
P.O. Box 33427
St. Paul, MN 55133 (612) 733-1501

Gary F. Lyons
3M Center
P.O. Box 33428
St. Paul, MN 55133 (612) 733-0243

James B. Marshall, Jr.
3M Co.
2501 Hudson Rd.
St. Paul, MN 55133 (612) 733-1505

Michael L. Mau
Merchant, Gould, Smith, Edell, Welter & Schmidt, P.A.
1000 Northwestern Bank Bldg.
St. Paul, MN 55101 (612) 298-1055

Joan M. Mullins#
15 W. Annapolis
St. Paul, MN 55118 (612) 733-5291

Adonis A. Neblett
Minn. Mining & Mfg. Co.
P.O. Box 33427
St. Paul, MN 55133 (612) 736-4790

Edward T. Okubo
3M Co. Off. Of Pat. Coun.
P.O. Box 33427
St. Paul, MN 55133 (612) 733-1534

William R. Power
Burlington Northern Railroad Co.
176 E. 5th St.
St. Paul, MN 55101 (612) 298-2619

Terryl K. Qualey
3M Center, Bldg. 220-12W
P.O. Box 33427
St. Paul, MN 55133 (612) 733-1940

David A. Roden
P.O. Box 33800
St. Paul, MN 55133 (612) 733-4876

Cecil C. Schmidt
Merchant, Gould, Smith, Edell, Welter & Schmidt, P.A.
1000 Norwest Center
St. Paul, MN 55101 (612) 298-1055

Leland D. Schultz
Minn. Mining & Mfg. Co. Off. Of Patent Coun.
3M/3M Center, P.O. Box 33427
St. Paul, MN 55133 (612) 736-9722

Donald M. Sell
3 M Company Off. Of Pat. Coun.
3 M Center 220-12W-01
P.O. Box 33427
St. Paul, MN 55133 (612) 733-1514

Lorrain R. Sherman
3 M Co.
3M Center, P.O. Box 33427
St. Paul, MN 55133 (612) 733-1507

Michael S. Sherrill
Merchant, Gould, Smith, Edell, Welter & Schmidt, P.A.
1000 Norwest Center
St. Paul, MN 55101 (612) 298-1055

James A. Smith
Minn. Mining & Mfg. Co.
3M Center, P.O. Box 33427
St. Paul, MN 55133 (612) 733-1512

Joseph A. Speldrich#
Sperry Corp.
P.O. Box 64525
St. Paul, MN 55101 (612) 456-2936

Robert W. Sprague#
Minn. Mining & Mfg. Co.
P.O. Box 33427
St. Paul, MN 55133 (612) 733-0052

Roger R. Tamte
Minn. Mining & Mfg. Co.
2501 Hudson Rd., P.O. Box 33427
St. Paul, MN 55133 (612) 733-1520

Kathleen R. Terry
University Of Minnesota Pat. & Licensing
570 Administrative Services Center
1919 University Ave.
St. Paul, MN 55104 (612) 373-2609

John F. Thuente
University Of Minnesota
572 Administrative Services Center
1919 University Ave.
St. Paul, MN 55104 (612) 373-2012

Carole Truesdale
3M Co. 3M Center Off. Of Pat. Coun.
P.O. Box 33427
St. Paul, MN 55133 (612) 736-4151

Douglas L. Tschida
2819 Hamline Ave. South, Hamline Center
Suite 113
St. Paul, MN 55113 (612) 636-3727

Janice L. Umbel
Merchant, Gould, Smith, Edell, Welter & Schmidt
Norwest Center, Suite 1000
St. Paul, MN 55101 (612) 298-1055

J. Wade Van Valkenburg, Jr.
494 Curfew St.
St. Paul, MN 55104 (612) 645-5463

Allen W. Wark
3M/ Office Of Pat. Coun.
P.O. Box 33427
St. Paul, MN 55133 (612) 733-0097

David L. Weinstein
Minn. Mining & Mfg. Co. Off. Of Pat. Coun., 3M Center
P.O. Box 33427
St. Paul, MN 55133 (612) 736-2681

Janet R. Westrom
Merchant, Gould, Smith, Edell, Welter & Schmidt, P.A.
1000 Norwest Bank Bldg.
St. Paul, MN 55101 (612) 298-1055

William F. Wittman#
3M Company
201-15-13 3M Center
St. Paul, MN 55144 (612) 733-4698

Michael T. Koller
9225 84th St. North
Stillwater, MN 55082

Louis B. Oberhauser, Jr.
1421 E. Wayzata
Wayzata, MN 55391 (612) 473-2521

Donald R. Sjostrom
807 Twelve Oaks Center
Wayzata, MN 55391 (612) 475-3611

John M. Zangs, Jr.
1156 Allen Ave.
West St. Paul, MN 55118 (612) 457-5059

Gilbert B. Gehrenbeck#
5482 E. Bald Eagle Blvd.
White Bear Lake, MN 55110 (612) 429-9713

Arthur Mendel#
4525 Oak Leaf Dr.
White Bear Lake, MN 55127 (612) 429-1029

Richard J. Renk
768 Terrace Ln.
Winona, MN 55987 (507) 452-2461

Robert Leslie Marben
7101 Windgate Rd.
Woodbury, MN 55125 (612) 735-5481

Mississippi

Whitfield Price#
P.O. Box 25
Clinton, MS 39056 (601) 924-3113

Benjamin E. Long#
United Technologies Automotive Gp., American Bosch Electrical Prdts.
P.O. Box 2228, McCrary Rd.
Columbus, MS 39701 (601) 328-4150

Alexander F. Norcross, Sr.
Hoffman, Wetzel & Ellis
P.O. Drawer I, 1701 24th Ave.
Gulfport, MS 39502 (601) 864-6400

Dewitt L. Fortenberry, Jr.
Edmonson, Biggs, & Jeliffe
P.O. Box 865
Jackson, MS 39205 (601) 960-3600

Richard J. Hammond
Institute For Technology Dev.
700 N. State St., Suite 500
Jackson, MS 39202 (601) 960-3600

G. Dempsey Ladner
1605 Kent Ave.
Jackson, MS 39211 (601) 969-2438

Paul J. Richardson
Mississippi Power & Light Co.
P.O. Box 1640
Jackson, MS 39205 (601) 969-2390

Lewis H. Wilson
803 South Forest Ave.
Long Beach, MS 39560 (601) 863-4470

Thomas M. Phillips*
Dept. Of The Navy
Naval Ocean Res. & Dev. Activity
Nstl Station, MS 39529

Missouri

Jack W. White
254 White Tree Lane
Ballwin, MO 63011 (314) 527-0085

Elmer J. Fischer
56 York Dr.
Brentwood, MO 63144 (314) 997-0251

Carol H. Clayman
Monsanto Co.
700 Chesterfield Village Pkwy.
Chesterfield, MO 63198 (314) 537-6047

Harvey A. Gilbert
1269 Rogue River Court
Chesterfield, MO 63017 (314) 532-5615

Richard E. Haferkamp
Rogers, Eilers & Howell
11 S. Meramec Ave. Suite 610
Clayton, MO 63105 (314) 727-5188

Arthur S. Morgenstern
710 S. Meramec Ave.
Clayton, MO 63105 (314) 727-7169

Peter Nelson Davis
University Of Missouri-Columbia
Columbia, MO 65211 (314) 882-2624

Donald Edward Gillihan
315 N. Washington
Farmington, MO 63640

Charles R. Landholt#
679 Waterfall Dr.
Florissant, MO 63034 (314) 831-7343

William R. O'Meara#
1065 Jefferson St.
Florissant, MO 63031 (314) 837-9050

Robert J. Owens#
1930 Pyrenees Dr.
Florissant, MO 63033 (314) 837-1081

Glen R. Simmons
Route 2
Green City, MO 63545 (816) 874-4332

Kent Barta
801 Monroe
Jefferson City, MO 65101 (314) 635-5315

Charles N. Blitzer
Marion Laboratories
9300 Ward Parkway
Kansas City, MO 64114 (816) 966-4086

Joseph B. Bowman
Kokjer, Kircher, Bradley, Wharton, Bowman & Johnson
2414 Commerce Tower
911 Main St.
Kansas City, MO 64105 (816) 474-5300

Don M. Bradley
Kokjer, Kircher, Bradley, Wharton,
2414 Commerce Tower
911 Main St.
Kansas City, MO 64105

Mark E. Brown
Litman, Day & McMahon
1215 Commerce Bank Bldg., 922 Walnut St.
Kansas City, MO 64106 (816) 842-1587

James D. Christoff
Schmidt, Johnson, Hovey & Williams
1400 Merchantile Bank Tower, 1101 Walnut St.
Kansas City, MO 64106 (816) 474-9050

John M. Collins
Schmidt, Johnson, Hovey & Williams
1400 Merchantile Tower, 1101 Walnut St.
Kansas City, MO 64106 (816) 474-9050

Dennis A. Crawford#
Litman, Day & McMahon
1215 Commerce Trust Bldg., 922 Walnut St.
Kansas City, MO 64106 (816) 842-1587

Steven R. Dickey
Schmidt, Johnson, Hovey & Williams
1400 Merchantile Bank Tower, 1101 Walnut St.
Kansas City, MO 64106 (816) 474-9050

John Andrew Hamilton
437 Law Bldg., 1207 Grand Ave.
Kansas City, MO 64106

John P. Hazzard
Fermenta Animal Health Co.
7410 N.W. Tiffany Springs Pkwy., P.O. Box 901350
Kansas City, MO 64190 (816) 891-5514

Robert D. Hovey
Hovey, Williams, Timmons & Collins
1400 Merchantile Bank Tower
Kansas City, MO 64106 (816) 474-9050

Michael Bryan Hurd
Kokjer, Kircher, Bradley, Wharton, Bowman & Johnson
2414 Commerce Tower, 911 Main St.
Kansas City, MO 64105 (816) 474-5300

Donald E. Johnson
1400 Merchantile Bank Tower, 1101 Walnut St.
Kansas City, MO 64106 (816) 474-9050

William Blaine Kircher
Kokjer, Kircher, Bradley, Wharton, Bowman & Johnson
2414 Commerce Tower, 911 Main St.
Kansas City, MO 64105 (816) 474-5300

Carter H. Kokjer
Kokjer, Kircher, Bradley, Wharton, Bowman & Johnson
2414 Commerce Tower, 911 Main St.
Kansas City, MO 64105 (816) 474-5300

Malcolm A. Litman
Litman, Day & McMahon
922 Walnut, Suite 1215
Kansas City, MO 64106 (816) 842-1587

Robert E. Marsh
Blakwell, Sanders, Matherny, Weary & Lombardi
Two Perhing Square, 2300 Main Street- Suite 1100
Kansas City, MO 64141 (816) 274-6800

John C. McMahon
Litman, Day & McMahon
922 Walnut, 1215 Commerce Tower
Kansas City, MO 64106 (816) 842-1587

Ronald V. Muller
10736 Blue Ridge Blvd.
Kansas City, MO 64134 (816) 765-5858

Robert E. Mulloy, Jr.
H & R Block, Inc.
4410 Main St.
Kansas City, MO 64111 (816) 753-6900

Thomas E. Roszak
Krigel & Krigel
980 City Center Square, 1100 Main
Kansas City, MO 64105 (816) 474-7800

Gordon D. Schmidt
Schmidt, Johnson, Hovey & Williams
1400 Merchantile Bk., Tower, 1101 Walnut St.
Kansas City, MO 64106 (816) 474-9050

Stephen D. Timmons
Schmidt, Johnson, Hovey & Williams
1400 Merchantile Bk. Tower, 1101 Walnut St.
Kansas City, MO 64106 (816) 474-9050

J. David Wharton
Kokjer, Kircher, Bradley, Wharton, Bowman & Johnson
2414 Commerce Tower, 911 Main St.
Kansas City, MO 64105 (816) 474-5300

Warren N. Williams
Schmidt, Johnson, Hovey & Williams
1101 Walnut St. Suite 1400
Kansas City, MO 64106 (816) 474-9050

Neal O. Willmann
Marion Laboratories, Inc.
10236 Marion Park Dr.
Kansas City, MO 64136 (816) 966-5000

Michael Yakimo, Jr.
819 Walnut St. Suite 404
Kansas City, MO 64106 (816) 556-9404

Arvid V. Zuber#
Sherman, Wickens, Lysaught & Speck, P.C.
Top Of City Center Square, 12th & Baltimore, P.O. Box 26530
Kansas City, MO 64196 (816) 471-6900

William Thomas Black
1549 Soutlin Dr.
Kirkwood, MO 63122 (314) 966-8988

F. Travers Burgess
14 Taylor Woods
Kirkwood, MO 63122 (314) 822-0849

Ronald N. Compton
Summers, Compton, Wells & Hamburg, P.C.
8909 Ladue Rd.
Ladue, MO 63124 (314) 991-4999

Norman Grant Steanson, Jr.
514 East 26th Ave.
N. Kansas City, MO 64116 (816) 472-1116

Philip E. Hodur#
607 Seib Dr.
O'Fallon, MO 63366 (314) 272-1881

John R. Miller
Route 1
Owensville, MO 65066 (314) GE7-2776

Chester Leslie Davis, Sr.
Perry, MO 63462 (314) 565-3570

Donald Myers
University Of Missouri-Rolla
301 Harris Hall
Rolla, MO 65401 (314) 341-4568

Richard L. Marsh
Dayco Corp.
Dayco Technical Center, Battlefield Rd. At Scenic Dr.
P.O. Box 3258
Springfield, MO 65808 (417) 881-7440

Henry Wayne Cummings
124 North Main
St. Charles, MO 63301 (314) 946-0076

Albert J. Greene#
8700 Ezra Dr.
St. John, MO 63114 (314) 426-5328

Michael E. Howell
Bauman & Liles
224 N. Seventh St.
St. Joseph, MO 64501 (816) 364-4224

William I. Andress
Monsanto Co.
800 N. Lindbergh Blvd.
St. Louis, MO 63167 (314) 694-3165

Drew C. Baebler
Hullverson, Hullverson & Frank
1010 Market St., Suite 1550
St. Louis, MO 63101 (314) 421-2313

Stephen Walter Bauer
Senniger, Powers, Leavitt & Roedel
611 Oliver St., Suite 2050
St. Louis, MO 63101 (314) 231-0109

Andrew Joseph Beck
Sherwood Medical Co.
1831 Olive St.
St. Louis, MO 63103 (314) 621-7788

George R. Beck
Monsanto Co. Pat. Dept.
800 N. Lindbergh Blvd.
St. Louis, MO 63167 (314) 694-3187

Paul A. Becker, Sr.#
Emerson Electric Co., White- Rodgers Div.
9797 Reavis Rd.
St. Louis, MO 63123 (314) 577-1513

David Bennett
Monsanto Co.
800 N. Lindbergh Blvd.
St. Louis, MO 63167 (314) 694-3412

Jon H. Beusen
Monsanto Co.
800 N. Lindbergh Blvd.
St. Louis, MO 63167 (314) 694-3191

Edward A. Boeschenstein
Gravely, Lieder & Woodruff
705 Oliver St.
St. Louis, MO 63101 (314) 621-1457

James C. Bolding
Monsanto Co.
800 N. Lindberg Blvd.
St. Louis, MO 63166 (314) 694-3484

Wendell W. Brooks
Monsanto Co.
800 N. Lindbergh Blvd.
St. Louis, MO 63167 (314) 694-3181

William George Bruns
Gravely, Lieder & Woodruff
705 Oliver St.
St. Louis, MO 63101 (314) 621-1457

Robert M. Burton#
Burton International Biomed
P.O. Box 13135
St. Louis, MO 63119 (314) 644-7332

Joseph C. Carr, Jr.
Mi Tek Industries, Inc.
11710 Old Ballas Rd.
St. Louis, MO 63141 (314) 567-7127

James H. Casey*
U.S. Army Aviation System Comm.
S A V - J P, Legal Office
4300 Goodfellow Blvd.
St. Louis, MO 63120 (314) 263-3591

John M. Charnecki
6401 West Court
St. Louis, MO 63116 (314) 752-0244

Arnold Harvey Cole
Monsanto Co.
800 N. Lindbergh Blvd.
St. Louis, MO 63167 (314) 694-3131

Henry Croskell
Permea Inc., Monsanto Co.
11444 Lackland Rd.
St. Louis, MO 63145 (314) 694-0212

William B. Cunningham, Jr.
Polster, Polster & Lucchesi
763 S. New Ballas Rd.
St. Louis, MO 63141 (314) 872-8118

Paul Michael Denk
763 S. New Ballas Rd.
St. Louis, MO 63141 (314) 872-8136

Samuel Digirolamo
Harverstock, Garrett & Roberts
611 Olive St.
St. Louis, MO 63101 (314) 241-4427

William Harry Duffey
Monsanto Co., Pat. Dept.
800 N. Lindbergh Blvd.
St. Louis, MO 63167 (314) 694-3128

Rey Eilers
Polster, Polster & Lucchesi
763 S. New Ballas Rd.
St. Louis, MO 63141 (314) 872-8118

Joseph Alan Fenlon, Jr.
8003 Forsyth Blvd., Suite 220
St. Louis, MO 63105 (314) 721-4378

Grace J. Fishel
11933 Westline, Suite 170
St. Louis, MO 63146 (314) 878-0440

Donald J. Fitzpatrick
Interco Incorporated
101 S. Hanley Rd.
St. Louis, MO 63105 (314) 863-1100

Stanley Nelson Garber
Sherwood Medical Co.
1831 Olive St.
St. Louis, MO 63103 (314) 621-7788

Robert M. Garrett
Haverstock, Garrett & Roberts
611 Olive St., Suite 1610
St. Louis, MO 63101 (314) 241-4427

Peter S. Gilster
Kalish & Gilster
1614 Paul Brown Bldg., 818 Olive St.
St. Louis, MO 63101 (314) 436-1331

Michael E. Godar
Senniger, Powers, Leavitt & Roedel
611 Olive St.
St. Louis, MO 63101 (314) 231-0109

Lynden Neal Goodwin
Mallinckrodt, Inc.
675 McDowell Blvd., P.O. Box 5840
St. Louis, MO 63134 (314) 895-2910

Edward P. Grattan
Monsanto Co.
800 N. Lindbergh Blvd.
St. Louis, MO 63167 (314) 694-3337

Robert C. Griesbauer
Monsanto Co.
800 N. Lindbergh Blvd.
St. Louis, MO 63166 (314) OX4-3194

Jerome Arthur Gross
818 Olive St., Suite 1610
St. Louis, MO 63101 (314) 241-7678

Charles Baker Haverstock
Haverstock, Garrett & Roberts
611 Olive St., Suite 1610
St. Louis, MO 63101 (314) 241-4427

Randall M. Heald
2108 Serenidad Lane
St. Louis, MO 63043

Edward J. Hejlek
Senniger, Powers, Leavitt & Roedel
611 Olive St., Suite 2050
St. Louis, MO 63101 (314) 231-0109

William H. Hellwege, Jr.#
9324 White Ave.
St. Louis, MO 63144 (314) 962-0057

Richard G. Heywood
Robbins & Heywood
314 N. Broadway, Suite 1230
St. Louis, MO 63102 (314) 421-6010

Virgil B. Hill
Ralston Purina Co.
Checkerboard Square
St. Louis, MO 63164 (314) 982-2164

Ronald W. Hind
Cohn, Powell & Hind, P.C.
7700 Clayton Rd., Suite 103
St. Louis, MO 63117 (314) 645-2442

Dennis R. Hoerner, Jr.
Monsanto Co.- B B 4F
700 Chesterfield Village Pkwy.
St. Louis, MO 63198 (314) 537-6099

Arthur Eugene Hoffman
Monsanto Co.
800 N. Lindbergh Blvd.
St. Louis, MO 63166 (314) 694-2714

James F. Hollander
Senniger, Powers, Leavitt, & Roedel
611 Olive St., Suite 2050
St. Louis, MO 63101 (314) 231-0109

John M. Howell
Rogers, Eilers & Howell
11 S. Meramec Ave. Suite 610
St. Louis, MO 63105 (314) 727-5188

Lawrence J. Hurst
Ralston Purina Co.
835 S. 8th St.
St. Louis, MO 63188 (314) 982-2307

Frank B. Janoski
Coburn, Croft & Putzell
One Mercantile Center, Suite 2900
St. Louis, MO 63101 (314) 621-8575

Roger Rance Jones
Monsanto Co.
800 N. Lindbergh Blvd.
St. Louis, MO 63166 (314) 694-3138

Ralph W. Kalish
Kalish & Gilster
818 Oliver St., Suite 1614
St. Louis, MO 63101 (314) 436-1331

Neal Kalishman
7777 Bonhomme Ave., Suite 2300
St. Louis, MO 63105 (314) 726-6545

Patrick D. Kelly
P.O. Box 22157
St. Louis, MO 63116 (314) 352-6874

Joseph D. Kennedy
Monsanto Co.
800 N. Lindbergh Blvd.
G4N D
St. Louis, MO 63167 (314) 694-3163

John W. Kepler, III
Cohn, Powell & Hind, P.C.
7700 Clayton Rd.
St. Louis, MO 63117 (314) 645-2442

Roy J. Klostermann
Mallinckrodt, Inc.
675 McDownnell Blvd., P.O. Box 5840
St. Louis, MO 63134 (314) 895-2915

Michael Kovac
7 Williamsburg Estates
St. Louis, MO 63131 (314) 576-4344

Paul C. Krizov
Monsanto Co.
800 N. Lindbergh Blvd.
St. Louis, MO 63167 (314) 694-3185

William E. Lahey
Senniger, Powers, Leavitt & Roedel
611 Olive St., Suite 2050
St. Louis, MO 63101 (314) 231-0109

Donald G. Leavitt
Senniger, Powers, Leavitt & Roedel
611 Olive St.
St. Louis, MO 63101 (314) 231-0109

Thomas B. Leslie
Haverstock, Garrett & Roberts
611 Olive St.
St. Louis, MO 63101 (314) 241-4427

Roy A. Lieder
Gravely, Lieder & Woodruff
705 Olive St., Suite 1302
St. Louis, MO 63101 (314) 621-1457

Lawrence L. Limpus
Monsanto Co.
800 N. Lindbergh Blvd.
St. Louis, MO 63167 (314) 694-3145

James Charles Logomasini
Monsanto Co.
800 N. Lindbergh Blvd.
St. Louis, MO 63167 (314) 694-3327

Raymond C. Loyer, Sr.
Monsanto Co.
800 N. Lindbergh Blvd.
St. Louis, MO 63131 (314) 694-3190

Lionel L. Lucchesi
763 S. New Ballas Rd.
St. Louis, MO 63141 (314) 872-8118

Charles Emery Markham#
Emerson Electric Co. Pat. Dept. Station 2209
8000 W. Florissant., P.O. Box 4100
St. Louis, MO 63136 (314) 553-2209

Robert B. Martin
Monsanto Co.
800 N. Lindbergh Blvd.
St. Louis, MO 63167 (314) 694-3068

John E. Maurer
Senniger, Powers, Leavitt & Roedel
611 Olive St., Suite 2050
St. Louis, MO 63101 (314) 231-0109

James D. McNeil
Monsanto Co.
800 N. Lindbergh Blvd.
St. Louis, MO 63167 (314) 694-2832

Scott J. Meyer
Monsanto Co. Bldg. G4N D
800 N. Lindbergh Blvd.
St. Louis, MO 63167 (314) 694-3117

Mc Pherson D. Moore
Rogers, Eilers & Howell
11 S. Meramec Ave. Suite 610
St. Louis, MO 63105 (314) 727-5188

John J. Muller, II
A C F Industries, Inc.
Carter Automotive Div. 9666 Olive St. Rd.
St. Louis, MO 63132 (314) 997-7400

Terry J. Owens
4012 Washington Blvd.
St. Louis, MO 63108 (314) 533-3179

Paul Leonard Passley
Monsanto Co.
800 N. Lindbergh Blvd.
St. Louis, MO 63167 (314) 694-3192

Harold R. Patton
Monsanto Co.
800 N. Lindbergh Blvd. Bldg. F 2 W F
St. Louis, MO 63167 (314) 694-3166

Veo Peoples, Jr.
1221 Locust St., Suite 1000
St. Louis, MO 63103 (314) 231-9775

J. Philip Polster
Polster, Polster & Lucchesi
763 S. New Ballas Rd.
St. Louis, MO 63141 (314) 872-8118

Philip B. Polster
Polster, Polster & Lucchesi
763 S. New Ballas Rd.
St. Louis, MO 63141 (314) 872-8118

Irving Powers
Senniger, Powers, Leavitt & Roedel
611 Olive St.
St. Louis, MO 63101 (314) 231-0109

Rudyard Kent Rapp
7 Villa Coublay
St. Louis, MO 63131 (314) 567-5703

Edward H. Renner
Cohn, Powell & Hind, P.C.
7700 Clayton Rd. Suite 103
St. Louis, MO 63117 (314) 645-2442

Sidney B. Ring
Petrolite Corp.
369 Marshall Ave.
St. Louis, MO 63119 (314) 968-6206

Glenn K. Robbins
Robbins & Heywood
314 N. Broadway
St. Louis, MO 63102 (314) 421-6010

Herbert Barton Roberts
Haverstock, Garrett & Roberts
611 Olive St.
St. Louis, MO 63101 (314) 241-4427

John K. Roedel, Jr.
Senniger, Powers, Leavitt & Roedel
611 Olive St., Suite 1553
St. Louis, MO 63101 (314) 231-0109

Edmund C. Rogers
Rogers, Eilers & Howell
11 S. Meramec Ave. Suite 610
St. Louis, MO 63105 (314) 727-5188

John T. Rogers
Rogers, Howell, Renner, Moore & Haferkamp
11 S. Meramec, Suite 610 Commerce Bank Bldg.
St. Louis, MO 63105 (314) 727-5188

Richard B. Rothman
Lewis & Rice
611 Olive St., Suite 1400
St. Louis, MO 63101 (314) 231-5833

Stuart N. Senniger
Senniger, Powers, Leavitt & Roedel
611 Olive St., Suite 2050
St. Louis, MO 63101 (314) 231-0109

Richard H. Shear
Monsanto Co.
800 N. Lindbergh Blvd.
St. Louis, MO 63167 (314) 694-3175

Frank D. Shearin
Monsanto Co.
800 N. Lindbergh Blvd., A2SA
St. Louis, MO 63167 (314) 694-5656

Richard J. Sher
302 Pebble Valley Dr.
St. Louis, MO 63141 (314) 434-9024

Gordon F. Sieckman
Monsanto Co.
800 N. Lindberge Blvd.
St. Louis, MO 63167 (314) 694-6067

Montgomery W. Smith
Sherwood Medical Law Dept.
1831 Olive St.
St. Louis, MO 63103 (314) 621-5700

Kenneth Solomon
Rogers, Howell, Moore & Haferkamp
7777 Bonhomme, Suite 1700
St. Louis, MO 63105 (314) 727-5188

Howard Cromwell Stanley
Monsanto Co.
800 N. Lindbergh Blvd.
St. Louis, MO 63167 (314) 694-3291

Larry R. Swaney
Monsanto Co.
800 N. Lindbergh Blvd.
St. Louis, MO 63167 (314) 694-8055

Stanley M. Tarter
Rogers, Howell, Moore & Haferkamp
7777 Bonhomme, Suite 1700
St. Louis, MO 63105

Arthur H. Tischer*
Dept. Of The Army, U.S. Army Aviation Sys. Comm.
Attn. A M S A V - J P
4300 Goodfellow Blvd.
St. Louis, MO 63120 (314) 263-3591

Gregory E. Upchurch
Polster, Polster & Lucchesi
763 S. New Ballas
St. Louis, MO 63141 (314) 872-8118

John D. Upham
7101 Stanford
St. Louis, MO 63130 (314) 727-8784

Mark F. Wachter
800 N. Lindbergh, G4N E
St. Louis, MO 63167 (314) 694-8651

Edward R. Weber
Pope & Weber, P.C.
1221 Locust St., Suite 1030
St. Louis, MO 63103 (314) 241-8465

Robert W. Welsh
Ralston Purina Co.
Checkerboard Square
St. Louis, MO 63188 (314) 982-3065

Robert E. Wexler
Petrolite Corp.
369 Marshall Ave.
St. Louis, MO 63119 (314) 968-6050

Bryan K. Wheelock
Rogers, Howell, Moore & Haferkamp
11 S. Meramec, Suite 610
St. Louis, MO 63105 (314) 727-5188

James W. William, Jr.
Monsanto Co., G4N D
800 N. Lindbergh Blvd.
St. Louis, MO 63167 (314) 694-5402

J. Russell Wilson
11 S. Meramec, Suite 1010
St. Louis, MO 63105 (314) 725-3200

Norman L. Wilson, Jr.*
U.S. Army Aviation Systems Command, A M S A V - J P
4300 Goodfellow Blvd.
St. Louis, MO 63120 (314) 263-3591

Frederick M. Woodruff
Gravely, Lieder & Woodruff
705 Olive St., Suite 712
St. Louis, MO 63101 (314) 671-1457

Terry A. Witthaus
Route 1, Box 245
Washington, MO 63090

Montana

Paul J. Van Tricht
1134 N. 24th St.
Billings, MT 59101 (406) 259-7631

David A. Veeder
Veeder, Broeder & Michelotti, P.C.
First Bank Bldg. Suite 805
Billings, MT 59101 (406) 248-9156

Richard C. Conover
404 First National Bank Bldg. P.O. Box 1329
Bozeman, MT 59715 (406) 587-4240

William D. West
36 North Last Chance Gulch, Suite 500
Helena, MT 59601 (406) 449-8941

Arthur L. Urban
Box 4045
Red Lodge, MT 59068 (406) 446-1585

Nebraska

Gene D. Watson
Nebr. Public Power Dist.
P.O. Box 499, 1414-15th St.
Columbus, NE 68601 (402) 563-5566

Scott K. Reed
Franklin Pierce Law Center
2 White St.
Concord, NE 03301 (603) 228-1541

Vincent Laurence Carney
P.O. Box 80836, 125 S. 52nd St.
Lincoln, NE 68501 (402) 489-0377

Austin P. Dodge
225 N. Brown Ave.
Minden, NE 68959 (308) 832-1121

John A. Beehner
Zarley, Mckee, Thomte, Voorhees & Sease
2120 S. 72nd St., 1111 Commercial Federal Tower
Omaha, NE 68124 (402) 392-2280

Kevin Lynn Copple
1832 Harney St. Suite 203
Omaha, NE 68102 (402) 444-1722

Mark D. Frederiksen
Zarley, McKee, Thomte, Voorhees & Sease
1111 Commercial Federal Tower, 2120 South 72nd St.
Omaha, NE 68124 (402) 392-2280

E. Robert Newman
Hendeson & Sturm
990 Woodmen Tower
OMaha, NE 68102 (402) 342-1797

George R. Nimmer
P.O. Box 674
Omaha, NE 68101 (402) 342-3077

Dennis L. Thomte
Zarley, Mckee, Thomte, Voorhees & Sease
1111., 2120 S. 72nd St.
Omaha, NE 68124 (402) 392-2280

J. Michael Walker
9909 Harney Parkway, South
Omaha, NE 68114 (402) 633-5408

James D. Welch
10328 Pinehurst Ave.
Omaha, NE 68124 (402) 391-4448

Bernard G. Fehringer
R R 2, Box 76
Sidney, NE 69162 (308) 254-2028

Nevada

Archie M. Cooke#
96 Arrowhead Dr.
Carson City, NV 89706

Herbert C. Schulze
P.O. Box 6070
Incline Village, NV 89450 (702) 831-3700

Robert W. Bass#
P.O. Box 85035
Las Vegas, NV 89135 (702) 733-3834

Andrew R. Juhasz
5462 Rondonia Circle
Las Vegas, NV 89120 (702) 454-5606

John Arthur Koch*
U.S. Energy R. & D. Admin.
Nevada Operations Off., P.O. Box 14100
Las Vegas, NV 89109 (702) 734-3584

Edward John Quirk
Seiler, Quirk & Tratos.
550 E. Charleston Blvd. Suite D
Las Vegas, NV 89104 (702) 386-1778

John E. Roethel
550 E. Charleston Blvd. Suite D
Las Vegas, NV 89104 (702) 386-1778

Jerry Roland Seiler
Seiler, Quirk & Tratos
550 E. Charleston Blvd. Suite D
Las Vegas, NV 89104 (702) 386-1778

New Hampshire

Dennis G. Maloney
R F D #1 Box 22
Chester, NH 03036 (603) 887-2569

Homer O. Blair
Franklin Pierce Law Center
2 White St.
Concord, NH 03301 (603) 228-1541

Thomas O. Field, Jr.
Franklin Pierce Law Center
2 White St.
Concord, NH 03301 (603) 228-1541

Susan H. Hage
1 King St.
Concord, NH 03301 (603) 224-4918

William O. Hennessey
Franklin Pierce Law Center
2 White St.
Concord, NH 03301 (603) 228-1541

Joseph T. Majka#
116 Pleasant St.
Concord, NH 03301 (603) 225-6742

William H. Mandir#
Franklin Pierce Law Center
2 White St.
Concord, NH 03301 (603) 228-1541

Paul D. Parnass#
21 Chesterfield Dr.
Concord, NH 03301 (603) 225-3442

Robert Harvey Rines
Rines And Rines
81 North State St.
Concord, NH 03301 (603) 228-0121

Robert Shaw
209 East Side Dr.
Concord, NH 03301 (603) 228-1541

William S. Van Royen
5 Warren St.
Concord, NH 03301 (603) 224-8811

Frederick M. Zullow#
128 Loudon Rd. Apt. 32R.
Concord, NH 03301 (603) 228-0494

Frank C. Henry
Red Hill Pond Rd.
Ctr. Sandwich, NH 03227 (603) 284-7091

Ronald B. Willoughby
Fisher, Moran, Willouthby & Clancy
42 Main St. P.O. Box 70
Dover, NH 03820 (603) 742-6131

H. Art Turner#
368 High St.
Hampton, NH 03842 (603) 926-8047

Charles M. Allen
1 Ripley Rd.
Hanover, NH 03755 (603) 643-5897

Charles E. Snee, III
Amca International Corp.
Hanover, NH 03755 (603) 643-5454

John Adams Thierry
Murray Hill Rd.
Hill, NH 03243 (603) 744-3540

George W. Dishong
Brynt Rd.
Jaffrey, NH 03452 (603) 532-7206

Janine J. Weins#
Weins & Weins
6 Allen St.
Lebanon, NH 03766 (603) 448-1922

Michael J. Weins
Weins & Weins
6 Allen St.
Lebanon, NH 03766 (603) 448-1922

Lee A. Strimbeck
42 Cottage St.
Littleton, NH 03461 (603) 444-2919

Michael J. Bujold
Hayes, Davis & Soloway
175 Canal St.
Manchester, NH 03031 (603) 668-2430

Anthony G.M. Davis
Hayes, Davis & Soloway
175 Canal St.
Manchester, NH 03031 (603) 668-2430

Oliver W. Hayes
Hayes, Davis & Soloway
175 Canal St.
Manchester, NH 03101 (603) 668-2430

Norman Peter Soloway
Hayes, Davis & Soloway,P.A.
175 Canal St.
Manchester, NH 03031 (603) 668-2430

Thomas N. Tarrant
917 Elm St.
Manchester, NH 03101 (603) 669-4504

Edgar O. Rost
Canal St.
Meredith, NH 03253 (603) 279-4970

Louis Etlinger
15 Apache Rd.
Nashua, NH 03063 (603) 883-3884

Albert Gordon
International Shoe Mach. Corp.
Simon & Ledge Sts.
Nashua, NH 03060 (603) 883-5500

Robert P. Outerbridge
Spraco, Inc.
2 E. Split Brook Rd.
Nashua, NH 03060 (603) 888-1050

William F. Poter, Jr.
Sanders Assco. Inc.
N H Q 1-719, Daniel Webster Highway, South
Nashua, NH 03601 (603) 885-5187

Richard I. Seligman
Sanders Assco. Inc.
Daniel Webster Hwy. South
Nashua, NH 03061 (603) 885-5186

Stanton D. Weinstein
Sanders Assco. Inc.
N H Q 1-719, C.S. 868
Daniel Webster Hwy., South
Nashua, NH 03061 (603) 885-2643

Larry B. Dufault#
Dufault & Dufault
Main St., P.O. Box 306
New London, NH 03257 (603) 526-4472

Marc R.K. Bungeroth
Battle & Bungeroth
Main St. Prof. Bldg., P.O. Box 769
North Conway, NH 03860 (603) 356-6966

C. Yardley Chittick
Rd 1, Box 390
Ossipee, NH 03864 (603) 522-3275

Bayard Jones
Durand Rd.
Randolph, NH 03570 (603) 466-5149

Theodore C. Virgil
East Rumney Rd.
Rumney, NH 03266 (603) 786-9401

Robert E. Brunson#
General Electric Co.
130 Main St.
Somersworth, NH 03878 (603) 749-8371

Merrill Franklin Steward
19 - S Province Rd.
Stafford, NH 03884 (603) 664-2709

Vicent W. Youmatz
R R 1
Box 121
Suncook, NH 03275 (603) 485-9010

John Meredith Leach
Box 544
Walpole, NH 03608 (603) 756-3744

New Jersey

Michael D. Loughnane#
54 Grey Ave.
Allendale, NJ 07401 (201) 327-7922

W. Patrick Quast
Quast & Torrente
One De Mercurio Dr.
Allendale, NJ 07401 (201) 327-0006

John J. Torrente
Quast & Torrente
One De Mercurio Dr.
Allendale, NJ 07401 (201) 327-0006

Adel A.A. Ahmed
Cedar Grove Rd. Box. 68
Annandale, NJ 08801 (201) 735-4791

Albert J. Mrozik, Jr.
617 Bond St.
Asbury Park, NJ 07712 (201) 774-7987

Lucian C. Canepa
15 Fieldstone Dr.
Basking Ridge, NJ 07920 (201) 766-1726

John J. Kissane
Amer, Telephone & Telegraph Co.
245 N. Maple Ave.
Basking Ridge, NJ 07920 (201) 221-6559

D. Laurence Padilla
A T & T Communication, Inc.
295 N. Maple Ave.
Basking Ridge, NJ 07920 (212) 393-4207

Donald F. Wohlers
30 Archgate Rd.
Basking Ridge, NJ 07920 (201) 647-4424

Gloria K. Koenig
664 Broadway
Bayonne, NJ 07002 (201) 436-2247

Ellen T. Dec#
Knickerbocker Dr.
Bell Meade, NJ 08502 (201) 359-2457

Michael Y. Epstein
359 Griggstown Rd.
Belle Mead, NJ 08502 (201) 359-8453

Richard S. Roberts
24 Camden Rd.
Belle Mead, NJ 08502 (201) 359-2980

Frank Cozzarelli, Jr.
Cozzarelli, Mautone, Nardachone & McTigue
286 Union Ave.
Belleville, NJ 07109 (201) 751-4100

Charles F. Gunderson#
1622 N. Marconi Rd.
Belmar, NJ 07719 (201) 681-0464

Glen Erin Books#
A T & T
One Oak Way
Berkeley Heights, NJ 07922 (201) 771-2244

Seymour E. Hollander
A T & T
One Oak Way 4E A 129
Berkeley Heights, NJ 07922 (201) 771-2200

Frank T. Johmann
49 Hampton Dr.
Berkeley Heights, NJ 07922 (201) 322-5429

Martin S. Landis
A T & T Technologies, Inc.
One Oak Way
Berkeley Heights, NJ 07928 (201) 771-2222

Hugh Linton Logan
50 Fern Place
Berkeley Heights, NJ 07922 (201) 464-6794

Stanley I. Rosen
One Oak Way
Berkeley Heights, NJ 07922 (201) 771-2182

William Ryan
A T & T Technologies, Inc.
One Oak Way, Room 4W - A115
Berkeley Heights, NJ 07922 (201) 771-2218

Peter J. Tribulski, Jr.
A T & T
One Oak Way
Berkeley Heights, NJ 07922 (201) 771-2134

Thomas Arthur Lennox
Lennox & Robertson
100 State Hwy 73, P.O. Box 127
Berlin, NJ 08009 (609) 767-6767

Paul R. Gauer
56 Watsessing Ave.
Bloomfield, NJ 07003 (201) 743-7050

Sidney Shaievitz
Schievitz & Berowitz
554 Bloomfield Ave.
Bloomfield, NJ 07003 (201) 743-7753

Louis Anthony Vespasiano
622 Bloomfield Ave.
Bloomfield, NJ 07003 (201) 748-7300

David A. Verner#
933 W. Meadow Dr.
Bound Brook, NJ 08805 (201) 469-2961

Clifford G. Frayne
44 Princeton Ave.
Brick Town, NJ 08723 (201) 840-9595

Gordon W. Kerr
321 Sawmill Rd.
Bricktown, NJ 08723 (201) 528-7298

Salvatore J. Abbruzzese
Thomas & Betts Corp.
1001 Frontier Rd.
Bridgewater, NJ 08807 (201) 707-2367

Charles H. Davis, Sr.
936 Brown Rd.
Bridgewater, NJ 08807 (201) 722-7176

Robert M. Rodrick
Thomas & Betts Corp.
1001 Frontier Rd.
Bridgewater, NJ 08807 (201) 707-2364

Leonard S. Selman
167 Mark Dr.
Bridgewater, NJ 08807 (201) 658-3898

Herbert M. Shapiro
92 Chelsea Way
Bridgewater, NJ 08807 (201) 725-2584

Edwin M. Szala#
Natl. Starch & Chem. Corp.
10 Finderne Ave.
Bridgewater, NJ 08807 (201) 685-5129

Louis J. Virelli, Jr.
National Starch & Chem. Corp.
10 Finderne Ave.
Bridgewater, NJ 08807 (201) 685-5197

Henry Joseph Walsh
1001 Brown Rd.
Bridgewater, NJ 08807 (201) 526-0160

Walter Weick
283 Farmer Rd.
Bridgewater, NJ 08807 (201) 526-1329

Nathan Carl Schwartz
3612 Ocean Ave. Apt. 8
Brigantine, NJ 08203 (609) 266-1385

Edger W. Adams, Jr.
7 Woodland Rd.
Brookside, NJ 07926 (201) 543-4606

J. Llewellyn Mathews
Apell & Mathews, P.C.
Lakehurst Rd. P.O. Box 95
Brown Mills, NJ 08015 (609) 893-3122

Joseph F. Flayer
389 Route 46
Budd Lake, NJ 07828 (201) 691-9000

Ralph Thomas Lilore
163 Eileen Dr.
Cedar Grove, NJ 07009 (201) 777-8876

Elizabeth Anne Bellamy#
Berlex Labs. Inc.
110 E. Hanove Ave.
Cedar Knolls, NJ 07927 (201) 540-8700

George W. Johnston, Jr.
100 Poplar Dr.
Cedar Knolls, NJ 07927 (201) 539-7835

John Joseph Archer
57 Elmwood Ave.
Chatham, NJ 07928 (201) 635-8921

Michael Anthony Caputo
Celanese Corp.
One Main St.
Chatham, NJ 07928 (201) 635-4365

Richard A. Craig
28 Mountain View Rd.
Chatham, NJ 07928 (201) 635-7761

David L. Davis
27 Pembrooke Rd.
Chatham, NJ 07928 (201) 635-1490

Fong S. Lin
22 Edgehill Ave.
Chatham, NJ 07928 (201) 635-8142

Andrew D. Maslow
Celanese Corp.
One Main St.
Chatham, NJ 07928 (201) 635-4305

Don Houghton Phillips
Celanese Corp.
One Main St.
Chatham, NJ 07928 (201) 635-4348

Andrew F. Sayko, Jr.
Celanese Corp.
One Main St.
Chatham, NJ 07928 (201) 635-4151

Robert M. Shaw
Celanese Specialty Operation
One Main St.
Chatham, NJ 07928 (201) 635-4181

Arthur Joseph Torsiglieri
2 Linden Lane
Chatham, NJ 07928 (201) 377-4321

George Nesbitt Ziegler
437 Fairmount Ave.
Chatham, NJ 07928 (201) 635-8928

Nicholas J. De Benedictis
1910 Morris Dr.
Cherry Hill, NJ 08003 (609) 795-1436

Charles Fredric Duffield
409 Route 70 East, Suite 218
Cherry Hill, NJ 08034 (609) 428-5338

Arnold Golden
108 Kingsdale Ave.
Cherry Hill, NJ 08003 (609) 424-2016

Max Goldman
134 Mansfield Blvd. South
Cherry Hill, NJ 08034 (609) 428-8853

Harry Warren Hargis, III#
116 Fenwick Rd.
Cherry Hill, NJ 08034 (609) 428-6808

William S. Hill
101 Bentwood Dr.
Cherry Hill, NJ 08034 (609) 428-7968

Samuel Kane
2 Spring Court
Cherry Hill, NJ 08003 (609) 795-7726

Norman E. Lehrer
1205 North Kings Highway
Cherry Hill, NJ 08034 (609) 429-4100

Robert K. Youtie
200 Uxbridge Dr.
Cherry Hill, NJ 08034 (609) 428-0010

Roy M. Porter, Jr.
444 E. Main St. Box 426
Chester, NJ 07930 (201) 584-3763

Gary F. Danis#
69 Acorn Dr.
Clark, NJ 07066 (201) 388-4872

Philip R. Arvidson
B A S F Corp.
1255 Brcad St.
Clifton, NJ 07015 (201) 365-3692

Cynthia Berlow#
414 Dwas Line Rd.
Clifton, NJ 07012 (201) 773-9568

Evelyn Berlow#
414 Dwas Line Rd.
Clifton, NJ 07012 (201) 773-9568

Michael Robert Chipaloski#
B A S F Corp., Inmont Div.
1255 Broad St.
Clifton, NJ 07015 (201) 365-3413

Anthony Frank Cuoco
66 Mount Prospect Ave.
Clifton, NJ 07013 (201) 779-0833

Joseph A. Giampapa
1054 Clifton Ave.
Clifton, NJ 07013 (201) 778-3203

Siegmar Silber
66 Mount Prospect Ave.
Clifton, NJ 07013 (201) 779-2580

Robert F. Tavares
Givaudan Corp.
100 Delawanna Ave.
Clifton, NJ 07014 (201) 363-8281

Linda A. Vag#
Givaudan Corp.
100 Delawanna Ave.
Clifton, NJ 07014 (201) 365-8165

John J. Herguth
Foster Wheeler Corp.
Perryville Corporate Park
Clinton, NJ 08809 (201) 730-4240

Robert J. Seman
Foster Wheeler Corp. Law Dept.
Perryville Corp. Park
Clinton, NJ 08809 (201) 730-4057

Arthur B. Larsen#
104 Heulitt Rd.
Colts Neck, NJ 07722 (201) 431-0558

Michael S. Jarosz
44 Old Glen Rd.
Convent Station, NJ 07961 (201) 267-1676

Rudolph J. Jurick
1 Canfield Terr.
Convent Station, NJ 07961 (201) 538-9248

Thomas R. Farino, Jr.
Corner Applegarth & Prospect Plains Rd.
Cranbury, NJ 08512 (609) 655-2700

Kenneth Watov
10 Fairway Dr.
Cranbury, NJ 08512 (609) 799-3394

Marthe L. Gibbons#
Miller & Gibbons
P.O. Box 93
Cranford, NJ 07016 (201) 232-2557

Homer J. Hall#
Rutgers University
S C I L S
Cranford, NJ 07016 (201) 276-4311

John J. Lipari
Lipari, Mulkeen, Keefe & Champi
6 North Ave.
Cranford, NJ 07011 (201) 276-4766

Benita J. Rohm
512 Springfield Ave.
Cranford, NJ 07016 (201) 276-3344

Boris M. Pismenny#
14 Park Ave.
Cresskill, NJ 07626 (201) 567-7954

Harold H. Card, Jr.*
U.S. Army Armament Comm. Res. & Dev. Comm.
Dover, NJ 07801 (201) 724-6590

Edward F. Costigan*
U.S. Dept. Of Army, H Q, Arnament, Munition & Chem Com.
AMSMC-GCL(d) Bldg. 3
Dover, NJ 07801 (201) 724-6594

A. Victor Erkkila*
U.S. Army, Picatinny Arsenal
Dover, NJ 07801 (201) 328-4586

Michael C. Sachs*
U.S. Army A M C C O M
Attn: A M S M C - G C L (d), Picatinny Arsenal
Dover, NJ 07801 (201) 724-2689

Ronald M. Spann*
Dept. Of Army, A R D C, Picatinny Arsenal, Bldg. 455
Dover, NJ 07801 (201) 724-2872

Aurora A. Legarda#
145 Lenox Ave.
Dumont, NJ 07628 (201) 384-9599

Joel F. Spivak
38 Yorktown Rd.
E. Brunswick, NJ 08816 (201) 257-6635

Robert L. Stone
13 Meadowlark Lane
East Brunswick, NJ 08816 (201) 254-2674

Joseph J. Borovian, Sr.#
Sandoz Corp.
52 Route 10
East Hanover, NJ 07936 (201) 386-8532

Elaine P. Brenner
Nabisco Brands Inc.
100 De Forest Ave
East Hanover, NJ 07936 (201) 503-3510

Thomas C. Doyle#
Sandow, Inc.
59 Route 10
East Hanover, NJ 07936 (201) 386-8177

Thomas O. Mc Govern, Sr.#
Sandoz Inc.
59 Route 10
East Hanover, NJ 07936 (201) 386-8480

Gerald D. Sharkin
Sandoz, Inc.
59 Route 10
East Hanover, NJ 07936 (201) 386-8483

Richard E. Vila
Sandoz Corp.
59 Route 10
East Hanover, NJ 07936 (201) 386-7852

Frederick Howard Weinfeldt
Sandoz Corp.
59 Route 10
East Hanover, NJ 07936 (201) 386-8420

Martha A. Michaels
26 Wiltshire Dr.
East Windsor, NJ 08520 (609) 443-3611

Martin Sachs
Sachs & Sachs, P.A.
614 U.S. # 130, P.O. Box 968
East Windsor, NJ 08520 (609) 448-2700

Samuel L. Sachs
Sachs & Sachs, P.A.
614 U.S. # 130, P.O. Box 968
East Windsor, NJ 08520 (609) 448-2700

Bill C. Giallovrakis
40 South St.
Eatontown, NJ 07724 (201) 542-2700

Michael J. Zelenka*#
U.S. Army Elect. Comm.
Fort Monmouth
Eatontown, NJ 07703 (201) 532-3062

James J. Farrell
Lever Brothers Co., Res. Center
45 River Rd.
Edgewater, NJ 07020 (201) 943-7100

Milton L. Honig
Lever Brothers Co.
45 River Rd.
Edgewater, NJ 07020 (201) 943-7100

Mary S. King#
Lever Brothers Co.
45 River Rd.
Edgewater, NJ 07033 (201) 558-4669

Mathew J. McDonald
Lever Brothers Co., Res. & Dev. Center
45 River Road
Edgewater, NJ 07020 (201) 943-7100

Gernald J. McGowan, Jr.
Lever Brothers Co.
45 River Rd.
Edgewater, NJ 07020 (201) 943-7100

Omri M. Behr
Behr & Adams
325 Pierson Ave.
Edison, NJ 08837 (201) 494-5240

Michael W. Ferrell#
Engelhard Corp.
33 Wood Ave. South
Edison, NJ 08810 (201) 623-6083

Philip M. Geren
Engelhard Corp.
Menlo Park C N 28
Edison, NJ 08817 (201) 321-5387

Robert A. Green
11 Perry Rd.
Edison, NJ 08817 (201) 572-0368

Inez L. Moselle#
Engelhard Corp. Minerals & Chemical Div.
Menlo Park Cn28
Edison, NJ 08818 (201) 321-5120

Arthur L. Plevy
Plevy & Gittesorth
146 Route 1, North
Edison, NJ 08817 (201) 572-5858

Ralph W. Selitto, Jr.
Plevy, Gittes & Selitto
146 Route 1,North
Edison, NJ 08817 (201) 572-5858

James M. Nickels
280 Overlook Ave.
Elberon, NJ 07740 (201) 222-6457

Richard L. Cannaday
47 W. Grand St.
Elizabeth, NJ 07202 (201) 355-0499

Jack Gerber
60 Watson Ave. Apt 1-H
Elizabeth, NJ 07202 (201) 351-9098

Ellen P. Trevors
C P C International Inc.
International Plaza, P.O. Box 8000
Englewood Cliffs, NJ 07632 (201) 894-2716

Norbert Ederer
1-17 35th St.
Fair Lawn, NJ 07411 (201) 797-5815

Robert B. Feuchtbaum
17-17 Broadway, Route 4
Fair Lawn, NJ 07410 (201) 794-3770

Roger L. Fidler
7-31 Henderson Blvd.
Fair Lawn, NJ 07410 (201) 796-9324

Frederick I. Levine
38-27 Wilson St.
Fair Lawn, NJ 07410 (201) 797-8014

Robert D. Polucki
Rapicom, Inc.
7 Kingsbridge Rd.
Fairfield, NJ 07006 (201) 575-6010

Robert Edward Smith
Singer Co.
70 New Dutch Lane
Fairfield, NJ 07006 (201) 882-6184

Nicholas Anthony Gallo, III
Fcs Labs. Rd. 3 Box 95
Darts Mill Vlg.
Flemington, NJ 08822 (201) 782-3353

Merle V. Hoover#
174 Thatcers Hill Rd.
Flemington, NJ 08822 (201) 782-7624

John W. Ditsler
Exxon Res. & Eng. Co.
P.O. Box 390
Florham Park, NJ 07932 (201) 765-1595

Joseph J. Dvorak
Exxon Res. & Eng. Co.
180 Park Ave., P.O. Box 390
Florham Park, NJ 07932 (201) 765-2255

Ronald David Hantman
Exxon Res. & Eng. Co.
P.O. Box 390
Florham Park, NJ 07932 (201) 765-3647

William F. Kelly, Jr.
6 Village Rd.
Florham Park, NJ 07932 (201) 377-2353

Deborah L. Mellott
Exxon Res. & Eng. Co.
P.O. Box 390
Florham Park, NJ 07932 (201) 765-2347

Reuben Miller
Exxon Res & Eng. Co.
P.O. Box 390
Florham Park, NJ 07932 (201) 765-4658

Richard E. Nafeldt
Exxon Res & Eng. Co.
P.O. Box 390
Florham Park, NJ 07932 (201) 765-4668

Henry E. Naylor
Exxon Res. & Eng. Co.
P.O. Box 390
Florham Park, NJ 07932 (201) 765-4693

Robert J. North#
Exxon Res. & Eng. Co.
P.O. Box 390
Florham Park, NJ 07932 (201) 765-4734

Roy John Ott
Exxon Res. & Dev. Co.
P.O. Box 390
Florham Park, NJ 07932 (201) 765-4661

Paul E. Purwin
Exxon Res. & Eng. Co.
P.O. Box 390
Florham Park, NJ 07932 (201) 765-4624

Kenneth Robert Schaefer
Exxon Enterprises
P.O. Box 390
Florham Park, NJ 07932 (201) 765-1566

John J. Schlager
Exxon Res. & Eng. Co.
P.O. Box 390
Florham Park, NJ 07932 (201) 765-3937

Jay Simon
Exxon Res. & Eng. Co.
P.O. Box 390
Florham Park, NJ 07932 (201) 765-1580

Frank A. Sinnock
Exxon Chemical Co.
200 Park Ave.
Florham Park, NJ 07932 (201) 765-5151

James H. Takemoto
Exxon Res. & Eng. Co.
P.O. Box 390
Florham Park, NJ 07932 (201) 765-1581

Arthur F. Whitley
Bro-Whit Assoc. Inc.
24 Puddingstone Way
Florham Park, NJ 07932 (201) 377-7433

Maurice Leander Williams
Exxon Res. & Eng. Co.
P.O. Box 390
Florham Park, NJ 07932 (201) 765-1601

Richard P. Dyer
1530 Palisade Ave. Apt. 23R
Fort Lee, NJ 07024 (201) 592-6019

Lester Horwitz
6 Horizon Rd.
Fort Lee, NJ 07024 (201) 224-0244

Jack Matalon
Sun Chemical Corp.
222 Bridge Plaza South, P.O. Box 1302
Fort Lee, NJ 07024 (201) 224-4600

Chonstantine A. Michalos
330 New York Ave.
Fort Lee, NJ 07024 (201) 944-9330

John Roger Drayer*#
U.S. Army Cecom
Attn: A M S E L - E D - C C
Fort Monmouth, NJ 07703 (201) 532-1424

James J. Drew*
U.S. Army Communications Elect. Command
Attn: AMSEL-LG-LP
Fort Monmouth, NJ 07703 (201) 532-3384

Victor J. Ferlise*
U.S. Army Communi. Electronics Command
Attn: A M S E L L - L G
Fort Monmouth, NJ 07703 (201) 532-3045

Roy E. Gordon*#
U.S. Army Communications Elect. Command.
Attn: Drsel-Lg-Ls
Fort MonMouth, NJ 07703 (201) 532-3187

Edward P. Griffin, Jr.*
U.S. Dept. Of Army, Comm. Group & Hq.
Cecom, Legal Office
Fort Monmouth, NJ 07703 (201) 532-3187

Sheldon Kanars*
U.S. Army Electronics Comm.
Atnn: Drsel-Lg-Ls
Fort Monmouth, NJ 07703 (201) 532-4112

John Kenneth Mullarney*
Dept. Of The Army Legal Dept., Pat. Prosecution Branch
Attn:AMSEL-LG-L
Fort Monmouth, NJ 07703 (201) 532-1459

Kenneth J. Murphy*
U.S. Army Electronic Comm. Legal Off.
D R S E L - L G - L P
Fort Monmouth, NJ 07703 (201) 532-3062

Kalman S. Pollen*
U S A C E O M
Fort Monmouth, NJ 07703 (201) 532-1104

John T. Rehberg*
U.S. Army Communi. Electrics Comm.
A M S E L - L G - L S
Fort Monmouth, NJ 07703 (201) 532-3187

Maurice W. Ryan*
Dept. Of The Army
H Q. C E C O M, A M S E L - L G - L S
Fort Monmouth, NJ 07703 (201) 532-3187

Paul F. Koch, II
48 Wyker Rd.
Franklin, NJ 07416 (201) 827-3020

Mary M. Allen
Becton, Dickinson & Co.
One Becton Dr.
Franklin Lakes, NJ 07417 (201) 848-7094

Richard E. Brown#
Becton Dickinson & Co.
One Becton Dr.
Franklin Lakes, NJ 07417 (201) 848-7110

Robert Joseph Dockery
Becton Dickinson & Co.
One Becton Dr.
Franklin Lakes, NJ 07417 (201) 848-7104

Robert Paul Grindle
Becton Dickinson & Co.
One Becton Dr.
Franklin Lakes, NJ 07417 (201) 848-7115

Robert M. Hallenbeck
Becton, Dickinson & Co.
One Becton Dr.
Franklin Lakes, NJ 07417 (201) 848-7114

Aaron Passman
Becton, Dickinson & Co.
One Becton Dr.
Franklin Lakes, NJ 07417 (201) 848-7096

John L. Voellmicke
700 Calusa Trail
Franklin Lakes, NJ 07417 (201) 891-8096

Ira M. Adler
11 Broadway
Freehold, NJ 07728 (201) 577-9090

Leonard R. Fellen
Craig Rd. Professional Bldg.
Craig Rd.
Freehold, NJ 07728 (201) 431-0473

Patrick J. Pinto#
37 West Main St.
Freehold, NJ 07728 (201) 431-7662

Sylvia Jean Chin
470 Long Hill Rd.
Gillette, NJ 07933 (201) 647-2283

Levonna Herzog#
16 Evergreen Court
Glen Ridge, NJ 07028 (201) 748-4329

Howard E. Thompson, Jr.
51 Ridgewood Ave.
Glen Ridge, NJ 07028 (201) 743-6364

Victor D. Behn
41 Beech Rd.
Glen Rock, NJ 07452 (201) 444-4853

Mitchell D. Bittman
Sequa Corp.
Three University Plaza
Hackensack, NJ 07601 (201) 343-1122

Richard M. Goldberg
Klauber & Jackson
One University Plaza
Hackensack, NJ 07601 (201) 487-5800

David A. Jackson
Klauber & Jackson
One University Plaza
Hackensack, NJ 07601 (201) 487-5800

Arthur Jacob
Samuelson & Jacob
25 E. Salem St., P.O. Box 686
Hackensack, NJ 07602 (201) 488-8700

Stefan J. Klauber
One University Plaza
Hackensack, NJ 07601 (201) 343-3999

Cyrus D. Samuelson
Samuelson & Jacob
235 Moore St., P.O. Box 686
Hackensack, NJ 07602 (201) 488-8700

Rebecca Yablonsky
One University Plaza
Hackensack, NJ 07601 (201) 343-3999

David Teschner
Amerace Corp.
Newburgh Rd.
Hackettstown, NJ 07840 (201) 852-1122

Kenneth J. Bossong
401 Briarwood Ave.
Haddonfield, NJ 08033

Cary A. Levitt
Archer & Greiner, P.C.
One Centennial Square, P.O.Box 3000
Haddonfield, NJ 08033 (609) 795-2121

Robert D. Thompson
Pennington & Thompson
Suite 105, One Centennial Square
Haddonfield, NJ 08033 (609) 795-0882

Stanley Howard Zeyher
660 Clinton Ave.
Haddonfield, NJ 08033 (609) 429-1063

Melvyn M. Kassenoff
Sandoz Inc.
59 Route 10
Hanover, NJ 07936 (201) 386-8477

Timothy G. Rothwell
Sandoz, Inc.
59 Route 10
Hanover, NJ 07936 (201) 386-8762

Sheldon H. Parker
365 St. Nicholas Ave.
Haworth, NJ 07641 (201) 387-1663

Frederick B. Luludis
69 Virginia Ave.
Hazlet, NJ 07730 (201) 264-5183

Edward Perley Barthel
Chrysler Motors Corp.
12000 Chrysler Dr.
Highland Park, NJ 48288 (313) 956-4644

John J. Jones
121 Orchard Ave.
Hightstown, NJ 08520 (609) 448-7061

George J. Seligsohn
7 Sjagbark Lane
Hightstown, NJ 08520 (609) 448-2495

Donald M. Boles
333 Piermont Ave.
Hillsdale, NJ 07642 (201) 358-0324

Robert J. Molnar#
63 Large Ave.
Hillside, NJ 07642 (201) 666-1915

Dennis P. Tramaloni#
440 Ardmore Rd.
Ho-Ho-Kus, NJ 07423 (201) 652-3431

Andrew M. Wilford
1028 Willow Ave.
Hoboken, NJ 07030 (201) 420-0974

Oleg Edward Alber
A T & T Bell Labs.
P.O. Box 679
Holmdel, NJ 07733 (201) 949-3158

Barry H. Freeman
Bell Telephone Labs. Inc.
Crawfords Corner Rd.
Holmdel, NJ 07733 (201) 949-6043

Richard Blake Havill
A T & T Bell Labs.
Crawfords Corner Rd.
Holmdel, NJ 07733 (201) 949-5555

Alfred E. Hirsch, Jr.
A T & T Bell Labs.
P.O. Box 679
Holmdel, NJ 07733 (201) 949-3222

Roy C. Lipton
Bell Telephone Labs. Inc.
Crawfords Corner Rd.
Holmdel, NJ 07733 (201) 949-5010

David R. Padnes
Bell Labs.
Crawfords Corner Rd.
Holmdel, NJ 07733 (201) 949-5857

Erwin W. Pfeifle
Bell Telphone Labs., Inc.
Crawfords Corner Rd.
Holmdel, NJ 07733 (201) 949-6268

Gregory C. Ranieri
A T & T Bell Labs.
Crawfords Corner Rd., Room 3K-230
Holmdel, NJ 07733 (201) 949-6559

Richard J. Roddy
A T & T Bell Laboratories
Crawfords Corner Rd.
Holmdel, NJ 07733 (201) 949-3578

Ronald D. Slusky
A T & T Information Systems Inc.
Crawfords Corner Rd.
Holmdel, NJ 07733 (201) 834-2517

Donnie E. Snedeker
A T & T Bell Labs.
Crawfords Corner Rd.
Holmdel, NJ 07733 (201) 949-6097

Thomas Stafford
Bell Telephone Labs.
Crawfords Corner Rd.
Holmdel, NJ 07733 (201) 949-5780

Samuel R. Williamson
A T & T
Crawfords Corner Rd. 3K-213
Holmdel, NJ 07733 (201) 949-6710

Joseph D. Lazar
395 Province Line Rd.
Hopewell, NJ 08252 (609) 466-3480

Stanley Dubroff
12 Cooper Dr.
Howell, NJ 07731 (201) 363-3991

Jack H. Stanley#
726 Fermere Ave.
Interlaken, NJ 07712 (201) 531-0472

Sidney S. Kanter
1064 Clinton Ave.
Irvington, NJ 07111 (201) 371-3030

George S. Seltzer
63 Kuna Terrace
Irvington, NJ 07111 (201) 373-7235

Bruce F. Jacobs
Engelhard Industries Div,
70 Wood Ave S.
Iselin, NJ 08830 (201) 632-6209

Karl F. Milde, Jr.
Siemens Corp. Res. & Support, Inc.
186 Wood Ave South
Iselin, NJ 08830 (201) 321-3440

Jeffrey P. Morris
Siemens Corporate Res. & Support Inc. Pat. Operations
186 Wood Ave. South
Iselin, NJ 08830 (201) 321-3400

Volker R. Ulbrich
Siemens Corporate Res. & Pat. Operation
186 Wood Ave. South
Iselin, NJ 08830 (201) 321-3963

Hugh S. Wertz
336 A Newport Way
Jamesburg, NJ 08831 (609) 655-3609

Walter Katz
Congoleum Corp.
195 Belgrove Dr.
Kearny, NJ 07032 (201) 991-1000

Vincent H. Gifford
Schering Plough Corp.
2000 Galloping Hill Rd.
Kenilworth, NJ 07033 (201) 558-4661

Carver C. Joyner, Sr.
Schering Plough Corp.
2000 Galloping Hill Rd.
Kenilworth, NJ 07033 (201) 558-4667

Warrick Edward Lee, Jr.
Schering-Plough Corp. Pat. Dept. K5-3
2000 Galloping Hill Rd.
Kenilworth, NJ 07033 (201) 558-4659

David John Mugford
Schering-Plough Corp.
2000 Galloping Hill Rd.
Kenilworth, NJ 07033 (201) 558-4066

Emill Scheller
Schering-Plough Corp.
2000 Galloping Hill Rd.
Kenilworth, NJ 07033 (201) 558-4835

Mary U. O'Brien
7 Oakwood Trail
Kinelon, NJ 07405 (201) 492-0064

Lawrence Paul Benjamin
123 Fairfield Rd.
Kingston, NJ 08528 (609) 924-4292

Alfred Charled Hill#
9 Shirley Terrace
Kinnelon, NJ 07405 (201) 838-2731

N. Paul Klaas
51 Hoot Owl Terr.
Kinnelon, NJ 07405 (201) 838-3724

Royal N. Ronning, Jr.
98 Longview Ave.
Lake Hiawatha, NJ 07034 (201) 263-8604

Neil D. Edwards
7 Turnbridge Row
Lakehurst, NJ 08733 (201) 657-2699

Stephen M. Hoffman
Advance Developments Co.
480 Oberlin Ave., S.
Lakewood, NJ 08701 (201) 364-8855

Harold Christoffersen
53 Merritt Dr.
Lawrenceville, NJ 08648 (609) 882-9123

Morris J. Cohen
1 Ivy Glen Lane
Lawrenceville, NJ 08648 (609) 896-9036

Vicent A. Mallare
Hydrocarbon Research, Inc.
134 Franklin Corner Rd. P.O. Box 6047
Lawrenceville, NJ 08648 (609) 896-2142

Thomas F. Meagher
G-10 Shirley Lane
Lawrenceville, NJ 08648 (609) 896-1972

David A. Saar
8 Bennington Dr.
Lawrenceville, NJ 08648 (609) 896-4831

Donald C. Caulfield
Exxon Chem. Co.
1900 E. Linden Ave.
Linden, NJ 07036 (203) 474-2238

Harvey L. Cohen
Exxon Chem Co.
1900 E. Linden Ave.
Linden, NJ 07036 (201) 474-2330

Davis B. Dwinell#
American Flange & Mfg. Co. Inc.
1100 W. Blancke St.
Linden, NJ 07036 (201) 862-5000

Harold Einhorn
Exxon Chemical Co.
1900 E. Linden Ave.
Linden, NJ 07036 (201) 474-2256

Diane E. Furman
Exxon Chem. Co.
1900 E. Linden Ave., P.O. Box 710
Linden, NJ 07036 (201) 474-2418

Melvin E. Libby
American Flange & Mfg. Co. Inc.
1100 W. Blanke St.
Linden, NJ 07036 (201) 862-5000

Robert A. Maggio
Exxon Chem. Co.
P.O. Box 710
Linden, NJ 07036 (201) 474-2297

John J. Mahon, Jr.
P.O. Box 710
Linden, NJ 07036 (201) 474-2518

William G. Muller
Exxon Chem. Co.
P.O. Box 710
Linden, NJ 07036 (201) 474-2266

Jack B. Murray, Jr.
Exxon Chem. Co.
P.O. Box 710
Linden, NJ 07036 (201) 474-2271

Albert M. Parker
American Flange & Mfg. Co. Inc.
1100 W. Blancke St., P.O. Box 167
Linden, NJ 07036 (201) 862-5000

Susan D. Schneider#
207 Belhaven Ave.
Linwood, NJ 08221 (609) 340-4011

Thomas W. Kennedy
Singer Co.
Kearfott Div.
1150 Mc Bride Ave.
Little Falls, NJ 07424 (201) 256-4000

Donald G. Marion
Singer Co.
Kearfott Division
1150 Mc Bride Ave.
Little Falls, NJ 07424 (201) 785-5307

Herbert A. Stern
Litton Inds., Inc.
275 Patterson Ave., P.O. Box 444
Little Falls, NJ 07424 (201) 256-5550

H. Gordon Dyke
134 Point Rd.
Little Silver, NJ 07739 (201) 842-7156

Robert Lawrence Lehman#
192 Winding Way
Little Silver, NJ 07739 (201) 741-1537

Stephen B. Coan#
72 Sykes Ave.
Livingston, NJ 07039 (201) 992-3133

James Warren Falk
Bell Communication Res. Inc.
290 W. Mount Pleasant Ave.
Livingston, NJ 07039 (201) 740-6100

Edward M. Fink
Bell Communications Research, Inc.
290 W. Mount Pleasant Ave.
Livingston, NJ 07039 (201) 740-6420

Allen N. Friedman
Bell Communications Research
290 W. Mount Pleasant Ave.
Livingston, NJ 07039 (201) 740-6160

Stephen M. Gurey
34 Scarsdale Dr.
Livingston, NJ 07039 (201) 533-0821

John G. Kovalich#
11 Carteret Rd.
Livingston, NJ 07039 (201) 992-5640

Marvin A. Naigur
Foster Wheeler Energy Corp., Legal Dept.
110 S. Orange Ave.
Livingston, NJ 07039 (201) 533-2515

John T. Peoples
Bell Communications Research
290 W. Mount Pleasant Ave.
Livingston, NJ 07039 (201) 740-6155

Howard R. Popper
Bell Communications Research, Inc.
290 W. Mount Pleasant Ave.
Livingston, NJ 07039 (201) 740-6150

Irving Skeist#
Skeist Labs. Inc.
112 Naylon Ave.
Livingston, NJ 07039 (201) 994-1050

John Edward Wilson
Foster Wheeler Corp.
110 S. Orange Ave.
Livingston, NJ 07039 (201) 533-2616

Max Yarmovsky
64 Bryant Dr.
Livingston, NJ 07039 (201) 992-9515

Mark L. Hopkins
17 Cambridge Dr.
Long Valley, NJ 07853 (201) 876-9252

Richard C. Billups
Schering-Plough Corp.
One Giralda Farms
Madison, NJ 07940 (201) 822-7378

John H.C. Blasdale#
Schering-Plough Corp.
One Giralda Farms
Madison, NJ 07940 (201) 822-7398

Eric S. Dicker#
Schering-Plough Corp.
One Giralda Farms
Madison, NJ 07940 (201) 822-7383

Thomas D. Hoffman
Schering-Plough Corp. Pat. Dept.
One Giralda Farms
Madison, NJ 07940 (201) 822-7379

Steinar V. Kanstad#
Schering-Plough Corp.
One Giralda Farms
Madison, NJ 07940 (201) 822-7373

Anita W. Magatti
Shering-Plough Corp. Pat. Dept. M-1-3W
One Giralda Farms P.O. Box 1000
Madison, NJ 07940 (201) 822-7389

John J. Maitner#
Schering-Plough Corp. Pat. Dept. 3-West
One Giralda Farms
Madison, NJ 07940 (201) 822-7358

Edward H. Mazer
Schering Plough Corp.
One Giralda Farms
Madison, NJ 07940 (201) 822-7303

James R. Nelson
Schering - Plough Corp. Pat. Dept. 3-West
One Giralda Farms
Madison, NJ 07940 (201) 822-7376

Robert Ira Pearlman
6 Coursen Way
Madison, NJ 07940 (201) 398-2520

Gerald Stuart Rosen
Schering Plough Corp. Pat. Dept. M-1-3W
One Giralds Farms., P.O. Box 1000
Madison, NJ 07940 (201) 822-7386

James A. Curley
10 Hoffman St.
Maplewood, NJ 07040 (201) 761-5966

Michael H. Wallach
36 New England Rd.
Maplewood, NJ 07040 (201) 762-4316

Charles Ira Brodsky
9 S. Main St.
Marlboro, NJ 07746 (609) 431-1333

Eric B. Janofsky#
23 Marigold Lane
Marlboro, NJ 07746 (201) 577-8205

Ilene Lapidus Janofsy
23 Marigold Lane
Marlboro, NJ 07746 (201) 577-8205

Theodore J. Criares
1997 Wash. Valley Rd.
Martinsville, NJ 08836 (201) 356-1978

Joseph M. Weinberger#
56-B Galewood Dr.
Matawan, NJ 07747 (201) 583-1137

Edward K. Kaprelian#
Kaprelian Res & Dev. Lowery Lane
Mendham, NJ 07945 (201) 543-7011

James G. Morrow
29 Bissett Place
Metuchen, NJ 08840 (201) 321-3966

William E. Ringle
Box 335
Metuchen, NJ 08840

Henry J. Nix
127 Union Ave.
Middlesex, NJ 08846 (201) 469-6677

John Abraham Caccuro
Bell Telephone Labs.
200 Laurel Ave.
Middletown, NJ 07748 (201) 957-3284

Daniel David Dubosky
209 Pelican Rd.
Middletown, NJ 07748 (201) 671-5611

Alan Huang#
4 Burdge Dr.
Middletown, NJ 07748

Carl P. Steinhauser#
564 Godwin Ave.
Midland Park, NJ 07432 (201) 445-0478

Henry Freeman
825 Ridgewood Rd.
Millburn, NJ 07041 (201) 376-0213

Charles S. Phelan
51 Circle Dr.
Millington, NJ 07946 (201) 647-9226

Robert S. Salzman
P.O. Box 314
Millington, NJ 07946 (201) 647-0969

Frank A. Jones#
Chicopee Mfg. Co. Res. Div.
Ford Ave.
Milltown, NJ 08850 (201) 524-7331

Clinton Blake Townsend
71 S. Mountain Ave.
Montclair, NJ 07042 (201) 744-2346

Raymond P. Wallace#
77 Orange Rd. Apt. 85
Montclair, NJ 07042 (201) 744-5405

Anton B. Weber#
247 Midland Ave.
Montclair, NJ 07042 (201) 783-9796

William V. Ebs
9 Westmorland Ave.
Montvale, NJ 07645 (201) 391-9154

William C. Long
Halcon Sd Group, Inc.
One Philips Parkway
Montvale, NJ 07645 (201) 573-8100

Howard J. Newby#
11 Sunnyside Dr.
Montvale, NJ 07645 (201) 391-4686

David Lawrence Rae
The B O C Group Inc.
85 Chestnut Ridge Rd.
Montvale, NJ 07645 (201) 573-0800

Roger M. Rathbun
The B O C Group, Inc.
85 Chestnut Ridge Rd.
Montvale, NJ 07645 (201) 573-0800

J. Bowen Ross, Jr.
Grand Met U S A, Inc.
100 Paragon Dr.
Montvale, NJ 07645 (201) 573-4102

Donald C. Simpson
Seven East Main St.
Moorestown, NJ 08057 (609) 234-9590

Michael B. Einschlag
48 Vista Dr.
Morganville, NJ 07751 (201) 972-7795

Stephen I. Miller
9 Vista Dr.
Moroganville, NJ 07751 (201) 558-4669

Sandra Gusciora Field
Warner-Lambert Co.
201 Tabor Rd.
Morris Plains, NJ 07950 (201) 540-2000

Charles A. Gaglia, Jr.
Warner-Lambert Co.
201 Tabor Rd.
Morris Plains, NJ 07950 (201) 540-4401

Albert H. Graddis
Warner- Lambert Co.
201 Tabor Rd.
Morris Plains, NJ 07950 (201) 540-2576

Henry C. Jeanette
Warner - Lambert Co.
201 Tabor Rd.
Morris Plains, NJ 07950 (201) 540-2599

Marian F. Kadlubowski
Warner - Lambert Co.
201 Tabor Rd.
Morris Plains, NJ 07950 (201) 540-4015

Anne M. Kelly
Warner - Lambert Co.
201 Tabor Rd.
Morris Plains, NJ 07950 (201) 540-2602

Charles E. Lents
Warner - Lambert Co.
201 Tabor Rd.
Morris Plains, NJ 07950 (201) 540-2151

Gary M. Nath
Warner - Lambert Co.
201 Tabor Rd.
Morris Plains, NJ 07950 (201) 540-4422

Howard Olevsky
Warner - Lambert Co.
201 Tabor Rd.
Morris Plains, NJ 07950 (201) 540-6817

Maxwell A. Pollack#
121 Glenbrook Rd.
Morris Plains, NJ 07950 (201) 539-1724

Stephen Raines
Warner - Lambert Co.
201 Tabor Rd.
Morris Plains, NJ 07950 (201) 540-4420

Daniel A. Scola, Jr.
Warner - Lambert Co.
Corporate Pat. Dept., 201 Tabor Rd.
Morris Plains, NJ 07950 (201) 540-5960

Thomas L. Adams
Behr & Adams
10 Park Place, Suite 307
Morristown, NJ 07853 (201) 538-4112

Joseph J. Allocca
116 Hillcrest Ave.
Morristown, NJ 07960 (201) 267-7274

Roger W. Bailey#
Electronic Data Systems Corp.
95 Madison Ave.
Morristown, NJ 07960 (201) 397-4982

Melanie L. Brown
Allied-Signal Inc.
Columbia Rd. & Park Ave., P.O. Box 2245R.
Morristown, NJ 07960 (201) 455-4851

Ernest D. Buff
Allied Corp.
Columbia Rd. & Park Ave.
Morristown, NJ 07960 (201) 455-3445

Bruce M. Collins
Mathews Woodbridge Goebel Pugh & Collins, P.A.
22 Park Place, P.O. Box 112-M
Morristown, NJ 07960 (201) 267-3444

Salvatore R. Conte
6 Quaker Ridge Rd.
Morristown, NJ 07960 (203) 267-9785

David P. Cooke
Allied-Signal Inc.
P.O. Box 2245R
Morristown, NJ 07960 (201) 455-2817

Anibal Jose Cortina
Allied-Signal Inc.
P.O. Box 2245R
Morristown, NJ 07960 (201) 455-3415

Roger H. Criss
Allied-Signal Inc. Law Dept.
P.O. Box 2245R
Morristown, NJ 07960 (201) 455-4796

Julius J. Denzler
Schenck, Price, Smith & King
10 Washington St.
Morristown, NJ 07960 (201) 539-1000

Edward D. Dreyfus
A T & T
100 Southgate Parkway
Morristown, NJ 07960 (201) 898-8547

Jay Philip Friedenson
Allied Signal Inc. Law Dept.
P.O. Box 2245R
Morristown, NJ 07960 (201) 455-2037

Gerhard Helmut Fuchs
Allied Signal Inc.
P.O. Box 2245R
Morristown, NJ 07960 (201) 455-3451

Ronald G. Goebel
Mathews, Woodbridge, Goebel, Push & Collins, P.A.
22 Park Place
Morristown, NJ 07960 (201) 267-3444

Ronald Gould
Shanley & Fisher
131 Madison Ave.
Morristown, NJ 07960 (201) 285-1000

Gus Theodore Hampilos
Allied Signal, Inc.
Columbia Rd. & Park Ave.
Morristown, NJ 07960 (201) 455-3453

Robert A. Harman
Mathews, Woodbridge, Goebel, Laughlin & Reichard, P.A.
22 Park Place, P.O. Box 112-M
Morristown, NJ 07960 (201) 267-3444

Francis Bradford Henry
Mennen Co.
Hanover Ave.
Morristown, NJ 07960 (201) 631-9361

Patrick L. Henry
Allied-Signal Inc.
P.O. Box 2245R
Morristown, NJ 07960 (201) 455-4705

John Dennis Kaufmann
American Telephone & Telegraph
1 Speedwell Ave., 1776 On-The-Green, Room 10B38
Morristown, NJ 07960 (201) 898-6977

Erwin Koppel
Allied Corp. Law Dept.
P.O. Box 2245R
Morristown, NJ 07960 (201) 455-6186

Marianne M. Kriman#
Allied Corp.
Park Ave. & Columbia Rd.
Morristown, NJ 07960 (201) 455-2380

Anthony Lagani, Jr.
78 Springbrook Rd.
Morristown, NJ 07960 (201) 540-0847

Richard T. Laughlin
Laughlin & Markensohn
22 Park Place
Morristown, NJ 07960 (201) 539-0080

Neal T. Levin
Henkel Corp.
350 Mt. Kemble Ave. P.O. Box 1931
Morristown, NJ 07960 (201) 267-1000

Roy H. Massengill
Allied Corp.
P.O. Box 2245R
Morristown, NJ 07960 (201) 455-5127

Howard G. Massung
Allied Corp. Law Dept.
Box 2245R
Morristown, NJ 07960

David Martin McConoughey
Allied Corp.
P.O. Box 2245R
Morristown, NJ 07960 (201) 455-5228

Richard A. Negin
Allied Chem. Corp.
P.O. Box 2245R
Morristown, NJ 07960 (201) 455-3790

Leslie G. Nunn, Jr.
Diamond Shamrock Chemicals Co.
350 Mt. Kemble Ave. C N 1931
Morristown, NJ 07960 (201) 267-1000

Arthur J. Plantamura
Allied Corp.
P.O. Box 2245R
Morristown, NJ 07960 (201) 455-3781

Jonathan Plaut
Allied Corp.
P.O. Box 2332R, Columbia Rd. & Park Ave.
Morristown, NJ 07960 (201) 455-6570

Stanley N. Protigal
Allied Corp. Law Dept.
Box 2245R
Morristown, NJ 07960 (201) 455-3548

Martha Greenewald Pugh
Mathews, Woodbridge, Goebel, Laughlin, Pugh & Collins, P.A.
22 Park Place, P.O. Box 112-M
Morristown, NJ 07960 (201) 267-3444

Gerard P. Rooney
Allied Corp.
P.O. Box 2245R
Morristown, NJ 07960 (201) 455-2502

John L. Stavert
A T & Technologies, Inc.
475 South St.
Morristown, NJ 07960 (201) 631-6838

Anthony J. Stewart
Allied Corp.
P.O. Box 2245R
Morristown, NJ 07960 (201) 455-4033

Richard C. Stewart#
Allied Corporation
Columbia Rd. & Park Ave., P.O. Box 2245R
Morristown, NJ 07960 (201) 455-3766

William F. Thornton
Allied Corp. Law Dept.
P.O. Box 2245R
Morristown, NJ 07960 (201) 455-3174

Peter F. Willig
63 Hill St.
Morristown, NJ 07960 (201) 538-3034

Richard C. Winter
Michaelson, Einschlag, Ostroff & Winter
Post House Rd.
Morristown, NJ 07960 (201) 766-9458

Edith R.T. Grill
7 Raynor Rd.
Morristownship, NJ 07960 (201) 267-9153

Bernard John Murphy#
10 Rockaway Terrace
Mountain Lakes, NJ 07046 (201) 334-9267

William Frank Pinsak
74 Tower Hill Rd.
Mountain Lakes, NJ 07046 (201) 335-2530

Robert F. Rotella
11 Hillcrest Rd. P.O. Box 72
Mountain Lakes, NJ 07046 (201) 334-1189

Herbert Smith Sylvester
16 Hillcrest Rd.
Mountain Lakes, NJ 07046 (201) 334-2606

F.W. Wyman#
34 Pollard Rd.
Mountain Lakes, NJ 07046 (201) 334-6331

Erwin Klingsberg#
1597 Deer Path
Mountainside, NJ 07092 (201) 232-1108

Herbert I. Sherman
1492 Deer Path
Mountainside, NJ 07092 (201) 232-1063

Riggs T. Stewart
1170 Foothill Way
Mountainside, NJ 07097 (201) 654-3067

Henry T. Brendzel
A T & T Bell Lab.
600 Mountain Ave.
Murray Hill, NJ 07974 (201) 582-4110

Peter A. Businger
Bell Labs.
600 Mountain Ave.
Murray Hill, NJ 07974 (201) 582-2908

David I. Caplan
A T & T Bell Labs.
600 Mountain Ave. Room 3B-521
Murray Hill, NJ 07974 (201) 582-4937

Larry R. Cassett
B O C Group, Inc.
100 Mountain Ave.
Murray Hill, NJ 07974 (201) 771-6434

Jack Saul Cubert
A T & T Bell Labs., Intellectual Property Matters Org.
600 Mountain Ave. Room 3B-529
Murray Hill, NJ 07974 (201) 582-3809

David A. Draegert
B O C Group, Inc. Pat. Tdmk & Licensing Dept.
100 Mountain Ave.
Murray Hill, NJ 07974 (201) 771-6402

Samuel H. Dworetsky
A T & T Bell Labs.
600 Mountain Ave.
Murray Hill, NJ 07974 (201) 771-2236

James H. Fox
Bell Labs. Inc.
600 Mountain Ave.
Murray Hill, NJ 07974 (201) 582-2936

Jerry W. Herndon
A T & T Bell Lab.
600 Mountain Ave.
Murray Hill, NJ 07974 (201) 582-4888

George S. Indig
Bell Tele. Labs., Inc.
600 Mountain Ave. Room 3B-325
Murray Hill, NJ 07974 (201) 582-6117

William L. Keefauver
Bell Laboratories Room 3B-309
600 Mountain Ave.
Murray Hill, NJ 07971 (201) 582-2233

Richard D. Laumann
Bell Telephone Labs. 3B-525
600 Mountain Ave.
Murray Hill, NJ 07974 (201) 582-4323

John Patrick Mc Donnell
A T & T Bell Labs. Room 3B-502
600 Mountain Ave.
Murray Hill, NJ 07974 (201) 582-5444

Walter G. Nilsen, Sr.
Bell Telephone Labs.
600 Mountain Ave.
Murray Hill, NJ 07974 (201) 582-3329

Everett Joseph Olinder
Bell Telephone Labs.
600 Mountain Ave.
Murray Hill, NJ 07974 (201) 582-2300

Kurt C. Olsen
Bell Labs. 3C-511
600 Mountain Ave.
Murray Hill, NJ 07974 (201) 582-4110

Irwin Ostroff
3 Lackawanna Blvd.
Murray Hill, NJ 07974 (201) 464-0248

Eugen E. Pacher
Bell Telephone Labs.
600 Mountain Ave.
Murray Hill, NJ 07974 (201) 582-5337

Bruce S. Schneider
A T & T Bell Labs.
600 Mountain Ave.
Murray Hill, NJ 07974 (201) 582-6358

Bernard Tiegerman
A T & T Bell Lab.
600 Mountain Ave.
Murray Hill, NJ 07974 (201) 582-6866

Michael J. Urbano
A T & T Bell Labs.
600 Mountain Ave.
Murray Hill, NJ 07974 (201) 582-4530

Wilford L. Wisner
A T & T Bell Labs.
600 Mountain Ave.
Murray Hill, NJ 07974 (201) 953-7221

Steven Paul Berman
Johnson & Johnson Off. Of Gen. Coun.
One Johnson & Johnson Plaza
New Brunswick, NJ 08933 (201) 524-5596

Nancy A. Bird
Johnson & Johnson
One Johnson & Johnson Plaza
New Brunswick, NJ 08933 (201) 524-5578

Audley A. Ciamporcero, Jr.
Johnson & Johnson
One Johnson & Johnson Plaza
New Brunswick, NJ 08933 (201) 524-5520

Andrea L. Colby
Johnson & Johnson
One Johnson & Johnson Plaza
New Brunswick, NJ 08933

Geoffrey G. Dellenbaugh
Johnson & Johnson
One Johnson & Johnson Plaza
New Brunswick, NJ 08933 (201) 524-5545

Wayne R. Eberhardt
Johnson & Johnson
One Johnson & Johnson Plaza
New Brunswick, NJ 08933 (201) 524-5524

Wayne R. Eberhardt
Johnson & Johnson
One Johnson & Johnson Plaza
New Brunswick, NJ 08933 (201) 524-5524

Richard J. Grochala
Johnson & Johnson
One Johnson & Johnson Plaza
New Brunswick, NJ 08933 (201) 524-5599

Howard E. Heller#
3 Lansing Place
New Brunswick, NJ 08901 (201) 249-6169

Michael A. Kaufman
Johnson & Johnson
One Johnson & Johnson Plaza, Room # W H 3230
New Brunswick, NJ 08933 (201) 524-5559

Leonard Kean
Johnson & Johnson
One Johnson & Johnson Plaza
New Brunswick, NJ 08933 (201) 524-5599

Benjamin Franklin Lambert
Johnson & Johnson
One Johnson & Johnson Plaza
New Brunswick, NJ 08933 (201) 524-5594

David J. Levy
Johnson & Johnson
One Johnson & Johnson Plaza
New Brunswick, NJ 08933 (201) 524-5547

Jason Lipow
Johnson & Johnson
One Johnson & Johnson Plaza, Room # W H 3231
New Brunswick, NJ 08933 (201) 524-5561

Gale F. Matthews
Johnson & Johnson
One Johnson & Johnson Plaza
New Brunswick, NJ 08933 (201) 524-2802

Charles Joseph Metz
Johnson & Johnson
One Johnson & Johnson Plaza
New Brunswick, NJ 08933 (201) 524-5494

Robert L. Minier
Johnson And Johnson
501 George St.
New Brunswick, NJ 08903 (201) 524-9411

Lawrence D. Schuler
Johnson & Johnson
One Johnson & Johnson Plaza
New Brunswick, NJ 08933 (201) 524-5592

Joseph F. Shirtz
Johnson & Johnson
One Johnson & Johnson Plaza
New Brunswick, NJ 08933 (201) 524-5588

Lewis Stein
Johnson & Johnson
One Johnson & Johnson Plaza
New Brunswick, NJ 08903 (201) 524-6465

Michael Q. Tatlow
Johnson & Johnson
One Johnson & Johnson Plaza
New Brunswick, NJ 08933 (201) 524-5565

Donal B. Tobin
Johnson & Johnson
One Johnson & Johnson Plaza
New Brunswick, NJ 08933 (201) 524-5567

W. Brinton Yorks, Jr.
Johnson & Johnson
One Johnson & Johnson Plaza
New Brunswick, NJ 08933 (201) 524-5522

John G. Schwartz#
731 Mabie St.
New Milford, NJ 07646 (201) 261-2624

John A. Casper
120 Passaic St.
New Providence, NJ 07974 (201) 464-2339

Edwin Blauvelt Cave
25 Alden Rd.
New Providence, NJ 07974 (201) 665-0780

Chris P. Konkol#
B O C Group Inc.
100 Mountain Ave.
New Providence, NJ 07974 (201) 771-6446

Sylvan Sherman
280 Woodbine Circle
New Providence, NJ 07974 (201) 665-1815

R. Hain Swope
B O C Group, Inc. Pat. Tdmk & Lic. Dept.
100 Mountain Ave.
Murray Hill
New Providence, NJ 07974 (201) 464-8100

Daniel H. Bobis
Popper Bobis & Jackson
17 Academy St.
Newark, NJ 07102 (201) 623-1000

Silvio J. De Carli#
Carenter, Bennett & Morrissey
Three Gateway Center
Newark, NJ 07102 (201) 622-7711

Steven W. Grill
Immunomedics
5 Bruce St.
Newark, NJ 07103 (201) 456-4779

Ira J. Hammer
Crummy, Del Deo, Dolan, Griffinger & Vecchione
One Gateway Center
Newark, NJ 07102 (201) 622-2235

Karl Huber
215 Central Ave.
Newark, NJ 07103 (201) 642-6670

Jacob Klapper#
N.J. Institute Of Tech.
323 High St.
Newark, NJ 07102 (201) 645-5491

Rene Oliveras
5 Commerce St.
Newark, NJ 07102 (201) 622-1881

Bernhard D. Saxe
Immunomedics, Inc.
100 Bergen St. Bldg. 5
Newark, NJ 07103 (201) 456-4779

Harries A. Mumma, Jr.
24 Spruce Road
No. Caldwell, NJ 07006 (201) 228-1832

Burton E. Levin
7855 Boulevard East
North Bergen, NJ 07047 (201) 854-2458

James K. Mc Neal, III#
915 Shore Rd.
Northfield, NJ 08225 (609) 641-8909

Matthew Boxer
Hoffmann-La Roche Inc.
340 Kingsland St.
Nutley, NJ 07110 (201) 235-5171

Norman C. Dulak
Hoffmann-La Roche Inc., Patent Law Dept.
340 Kingsland St.
Nutley, NJ 07110 (201) 235-2441

William H. Epstein
Hoffmann-La Roche, Inc.
340 Kingsland St.
Nutley, NJ 07110 (201) 235-3723

William M. Farley
Hoffmann- La Roche, Inc.
340 Kingsland St.
Nutley, NJ 07110 (201) 235-4205

Richard A. Gaither
Hoffmann-La Roche Inc.
340 Kingsland St.
Nutley, NJ 07110 (201) 235-2147

George M. Gould
Hoffman- La Roche, Inc.
340 Kingsland St.
Nutley, NJ 07110 (201) 235-3741

William G. Isgro
Hoffman -La Roche Inc.
340 Kingsland St.
Nutley, NJ 07110 (201) 235-4393

Bernard S. Leon
Hoffman - La Roche, Inc.
340 Kingsland St.
Nutley, NJ 07011 (201) 235-4378

Richard J. Mazza
Hoffman- La Roche Inc.
340 Kingsland St.
Nutley, NJ 07110

Julie Mae Prlina
Hoffman - La Roche Inc.
Nutley, NJ 07110 (201) 235-5000

Jon Sheldon Saxe
Hoffman - La Roche, Inc.
340 Kingsland St.
Nutley, NJ 07110 (201) 235-4387

Alan R. Stempel
Hoffmann - La Roche Inc.
340 Kingsland St.
Nutley, NJ 07110 (201) 235-4205

Robert A. Walsh
I T T Defense Tech. Corp.
500 Washington Ave.
Nutley, NJ 07110 (201) 284-5706

Mary C. Werner#
I T T Defense Tech. Corp.
500 Washington Ave.
Nutley, NJ 07110 (201) 284-5711

Albert Lewis Gazzola#
Witco Chemical Corp.
100 Bauer Dr.
Oakland, NJ 07436 (201) 337-5812

David H. Leroy#
70 Yuwpo Ave.
Oakland, NJ 07436 (201) 337-8247

Robert M. Skolnik
45 Burnt Mill Circle
Oceanport, NJ 07757 (201) 542-1252

Bart J. Zoltan#
152 De Wolf Rd.
Old Tappan, NJ 07675 (201) 768-3580

Richard A. Joel
466 Kinderkamack Rd.
Oradell, NJ 07649 (201) 599-0588

Gunter W. Koch#
Becton, Dickinson & Co.
Mack Center II, Mack Center Dr.
Paramus, NJ 07652 (201) 967-3908

Robert G. Pollock
Becton, Dickinson & Co.
Mack Center Dr.
Paramus, NJ 07652 (201) 967-3836

Richard J. Rodrick
Becton, Dickinson & Co.
Mack Center Dr.
Paramus, NJ 07652 (201) 937-3930

Edward R. Weingram
East 210 Route 4
Paramus, NJ 07652 (201) 843-6300

Brain J. Wieghaus
20 Gilbert Ave.
Paramus, NJ 07652 (201) 261-2429

Louis E. Marn
Marn & Jangarathis
400 Lanidex Plaza
Parisppany, NJ 07054 (201) 884-2122

Richard D. Goldstein
Sony Corporation Of America
Sony Dr. M.D. 3-76
Park Ridge, NJ 07656 (201) 930-7315

Stephen W. White#
E.I. Dupont De Nemours Co. Photo Products Dept.
Cheesequake Rd.
Parlin, NJ 08859 (201) 257-4600

Morris Irwin Pollack
Litton Industries Corp.
270 Passaic Ave.
Passaic, NJ 07055 (201) 777-5500

Harold F. Bennett#
R D 2 Box 55W
Pennington, NJ 08534 (609) 737-8789

Jerome Rosenstock
76 West Shore Dr.
Pennington, NJ 08534

Donald N. Timbie, Sr.
35 E. Curlis Ave.
Pennington, NJ 08534 (609) 737-0492

Frank Samuel Troidl
4 Mallard Dr.
Pennington, NJ 08534 (201) 737-9180

Gregory J. Winsky
Franklin Computer Corp.
Rte. 73 & haddonfield Rd.
Pennsauken, NJ 08110 (609) 488-0666

Diamond C. Ascani#
1013 Carroll Ave.
Pennsville, NJ 08070 (609) 678-3446

Ronald Brian Sherer
Ingersoll-Rand Co.
942 Memorial Pkwy.
Phillipsburg, NJ 08865 (201) 589-7664

Betty B. Tibbott#
410 Ohio Ave.
Phillipsburg, NJ 08865 (201) 454-8726

David W. Tibbott
Ingersoll-Rand Co. Pat. Dept.
942 Memorial Pkwy.
Phillipsburg, NJ 08865 (201) 859-7700

Wayne O. Traynham
Ingersoll-Rand Co.
942 Memorial Parkway
Phillipsburg, NJ 08865 (201) 859-7741

Walter C. Vliet
Ingersoll-Rand Co.
942 Memorial Parkway
Phillipsburg, NJ 08865 (201) 859-7728

Edward Goldberg*
U.S. Army Amccom Amsmc- Gcl D, Bldg. 3
Picatinny Arsenal, NJ 07806 (201) 724-6590

Francis H. Deef
38 Madison Ave.
Piscataway, NJ 08854 (201) 572-5853

Max Fogiel#
61 Ethel Rd. West
Piscataway, NJ 08854 (201) 819-8880

Donna R. Fugit#
Colgate-Palmolive Co.
909 River Rd.
Piscataway, NJ 08854 (201) 878-7535

Robert S. Mac Wright#
Rutgers University Off. Of Corp. & Inds. Res Ser.
P.O. Box 1089
Piscataway, NJ 08854 (201) 932-2074

James P. Scullin#
Nuodex Inc.
Turner Place, P.O. Box 365
Piscataway, NJ 08854 (201) 981-5385

Arthur N. Trausch, III
Ingersoll-Rand Co. Pat. Dept.
91 New England Ave.
Piscataway, NJ 08854 (201) 981-0411

Abraham Wilson
1340 Stelton Rd.
Piscataway, NJ 08854 (201) 985-0002

John E. Callaghan
1120 Park Ave.
Plainfield, NJ 07060 (201) 757-0606

George L. Kensinger
29 Farragut Rd. Suite 23B
Plainfield, NJ 07062 (201) 754-7506

Roderick Bruce Anderson
A T & T International
P.O. Box 900
Princeton, NJ 08540 (609) 639-2307

Stanford J. Asman
A T & T Engineering Res. Center
P.O. Box 900
Princeton, NJ 08628 (609) 882-9247

Harley R. Ball#
R C A Corp. David Sarnoff Res Center
201 Washington Rd.
Princeton, NJ 08540 (609) 734-2344

Donald Jay Barrack
Squibb Corp.
P.O. Box 4000
Princeton, NJ 08543 (609) 921-4328

Clement A. Berard, Jr.
General Electric Co. Aerospace Pat. & Licensing
P.O. Box 432
Princeton, NJ 08543 (609) 734-2491

Allen Bloom
Liposom Co. Inc.
One Research Way
Princeton Forrestal Center
Princeton, NJ 08540 (609) 452-7060

Glenn Huber Brestle
RCA Corp.
CN 5312
Princeton, NJ 08543 (609) 734-2356

William J. Burke
David Sarnoff Research Center, Inc.
CN 5300
Princeton, NJ 08543 (609) 734-2560

David N. Caracappa
RCA Corp.
CN 5312
Princeton, NJ 08543 (609) 734-3278

Dean W. Chace
G E & R C Licensing Management Oper.
2 Independence Way, Route 2023
Princeton, NJ 08540 (609) 734-9434

Richard G. Coalter
RCA Corp.
CN 5312
Princeton, NJ 08543 (609) 734-2038

Samuel Cohen
24 Littlebrook Rd., North
Princeton, NJ 08540 (609) 924-4561

Stephen B. Davis
Squibb Corp.
P.O. Box 4000
Princeton, NJ 08540 (609) 921-4338

Lawrence C. Edelman
R C A Corp. C N 5312
Princeton, NJ 08540 (609) 734-3066

Richard E. Elden
Fmc. Corp.
P.O. Box 8
Princeton, NJ 08540 (609) 452-2300

Peter Max Emanuel
Ge/rca Licensing Mang. Operation, Inc.
Two Independence Way, P.O. Box 2023
Princeton, NJ 08540 (609) 734-9586

Nathan Feldstein#
Surface Technology, Inc.
P.O. Box 2027
Princeton, NJ 08540 (609) 452-2929

Anthony J. Franze
387 Gallup Rd.
Princeton, NJ 08540 (609) 683-9733

Theodore R. Furman
Squibb Corp.
P.O. Box 4000
Princeton, NJ 08543 (609) 921-5735

Bernard Gerb
127 Meadowbrook Dr.
Princeton, NJ 08540 (609) 921-9078

Dominic J. Giancola
12 Cameron Court
Princeton, NJ 08540 (609) 924-6686

Lester L. Hallacher
RCA Corp.
2 Independence Way
Princeton, NJ 08640 (609) 734-9629

Carol R. Harney
Mathews, Woodbridge, Goebel, Pugh & Collins
100 Thanet Circle Suite 306
Princeton, NJ 08540 (609) 924-3773

James B. Hayes
G E & R C A Licensing Management
Two Independence Way, P.O. Box 2023
Princeton, NJ 08540 (609) 734-9570

Shabtay S. Henig
GE & RCA Licensing Mang. Operation, Inc.
21W, P.O. Box 2023
Princeton, NJ 08543 (609) 734-9751

Eric P. Herrmann
G E & R C A Licensing Mange. Operations, Inc.
P.O. Box 2023, 2 Independence Way
Princeton, NJ 08540 (609) 734-9754

Frank Ianno
FMC Corp.
P.O. Box 8
Princeton, NJ 08542 (609) 452-2300

Dennis H. Irlbeck, Sr.
G E & R C A Licensing Mang. Operation Inc.
2 Independence Way
Princeton, NJ 08540 (609) 734-9763

Robert D. Jackson#
FMC Corp.
Princeton, NJ 08542 (609) 452-2300

Herbert Leonard Jacobson
General Electric Co.
P.O. Box 2023
Princeton, NJ 08540 (609) 734-9422

George J. Koeser
E.R. Squibb & Sons, Inc.
P.O. Box 4000
Princeton, NJ 08540 (609) 921-4293

Irwin M. Krittman
RCA Corp.
CN 5312
Princeton, NJ 08543 (609) 734-2359

Ronald H. Kurdyla
G E / R C A Licensing Mange.
2 Independence Way
Princeton, NJ 08540 (609) 734-9444

Catherin L. Kurtz#
Liposome Co. Inc.
One Research Way, Princeton Forrestal Center
Princeton, NJ 08540 (609) 452-7060

Joseph J. Laks
RCA Corp.
CN 5312
Princeton, NJ 08543 (609) 734-2191

Robert E. Lee, Jr.
Squibb Corp.
P.O. Box 4000
Princeton, NJ 08540 (609) 921-4691

Thomas F. Lenihan#
RCA Pat. Corp.
CN 5312
Princeton, NJ 08543 (609) 734-3218

Lawrence S. Levinson
E.R. Squibb And Sons, Inc.
P.O. Box 4000
Princeton, NJ 08540 (609) 921-4330

Robert B. Levy
AT & T Engineering Res. Center
P.O. Box 900
Princeton, NJ 08620 (609) 639-2305

Allen L. Limberg
RCA Corp.
CN 5312
Princeton, NJ 08543 (609) 734-2462

Thomas H. Magee
RCA Corp.
CN 5312
Princeton, NJ 08543 (609) 734-2769

Nicholas P. Malatestinic
E.R. Squibb & Sons, Inc.
P.O. Box 4000
Princeton, NJ 08540 (609) 921-4301

C. Lance Marshall, Jr.
R C A Corp.
P.O. Box 2023
Princeton, NJ 08540 (609) 734-9421

William H. Meagher, Jr.
G E & R C A Licensing & Mange. Operations Inc.
Two Independence Way, P.O. Box 2023
Princeton, NJ 08540 (609) 734-9400

William H. Meise
General Electric Co. Pat. Aerospace & Lic.
P.O. Box 432
Princeton, NJ 08543 (609) 734-3016

Carlos Nieves
G E & R C A Licensing Mange. Operations Inc.
Two Independence Way
Princeton, NJ 08540 (609) 734-9626

Kenneth N. Nigon
GE & RCA Licensing Management Operation, Inc.
P.O. Box 2023
Princeton, NJ 08543 (609) 734-9839

Carl V. Olson
23 Broadripple Dr.
Princeton, NJ 08540 (609) 924-3341

Paul Jorgan Rasmussen
G E & R C A Licensing Mange. Operation, Inc.
P.O. Box 2023, Independence Way
Princeton, NJ 08540 (609) 734-9883

Harry S. Reichard
York Technology, Inc.
1101 State Rd. Bldg. Q
Princeton, NJ 08540 (609) 924-7676

Clarene C. Richard
RCA Corp.
P.O. Box 2023, 2 Independence Way
Princeton, NJ 08540 (609) 734-9444

Burton Rodney
Squibb Corp.
P.O. Box 4000
Princeton, NJ 08540 (609) 921-4336

Jerald E. Roehling
RCA Corp.
P.O. Box 2023
Princeton, NJ 08540 (609) 734-2991

Albert Russinoff
119 Heather Lane
Princeton, NJ 08540 (609) 924-3473

Thomas M. Saunders
Liposome Co. Inc.
One Research Way, Princeton Forrestal Center
Princeton, NJ 08540 (609) 452-7060

Henry I. Schanzer
RCA Corp.
CN 5312
Princeton, NJ 08543 (609) 734-3268

Franklin Schoenberg
241 Dodds Lane
Princeton, NJ 08540 (609) 921-2965

Eugene George Seems
FMC Corp.
P.O. Box 8
Princeton, NJ 08540

Edward J. Sites
Union Camp Corp.
3401 Princeton Pk. P.O. Box 3301
Princeton, NJ 08543 (609) 896-1200

Raymond E. Smiley
General Electric Co. Aerospace Pat. & Lic.
P.O. Box 432
Princeton, NJ 08543 (609) 734-3282

William Squire
RCA Corp.
CN 5312
Princeton, NJ 08543 (609) 734-2862

Daniel Ethan Sragow
GE & RCA Licensing Mang. Operation, Inc.
Two Independence Way, P.O. Box 2023
Princeton, NJ 08540 (609) 734-9891

Norman St. Landau, II
Durand Heher Imbraico
105 College Rd. East
Princeton, NJ 08540 (609) 924-0800

Leon Theodore Stark

P.O. Box 901
Princeton, NJ 08540 (609) 896-4014

Henry Irwin Steckler
RCA Corp.
CN 5312
Princeton, NJ 08543 (609) 734-2389

Richard H. Steen
P.O. Box 2178
Princeton, NJ 08540 (609) 871-5800

Scott J. Stevens
RCA Corp.
CN 5312
Princeton, NJ 08543 (317) 267-6631

Stanley Allen Strober
Mobil Oil Corp.
P.O. Box 1031
Princeton, NJ 08540 (609) 737-5484

Annette M. Tobia
56 Battle Rd.
Princeton, NJ 08540 (609) 924-9123

Joseph S. Tripoli
RCA Corp.
CN 5312
Princeton, NJ 08543 (609) 734-2992

Robert Lance Troike
RCA Corp.
CN 5312
Princeton, NJ 08543 (609) 734-2472

James M. Trygg
RCA Corp.
CN 5312
Princeton, NJ 08543 (609) 734-3268

Howard F. Vandenburgh
RCA Corp.
CN 5312
Princeton, NJ 08543 (609) 734-2775

Stephen Venetianer
E.R. Squibb & Son, Inc.
P.O. Box 4000
Princeton, NJ 08540 (609) 921-4299

Eugene M. Whitacre
GE / RCA Licensing Management Operation, Inc.
Two Independence Way, P.O. Box 2023
Princeton, NJ 08540 (609) 734-9566

Richard C. Woodbridge
Mathews, Woodbridge, Goebel, Pugh & Collins
357 Nassau St.
Princeton, NJ 08540 (609) 924-3773

Royal E. Bright
M & T Chem. Inc.
P.O. Box 1104
Rahway, NJ 07065 (201) 499-2153

Charles M. Caruso
Merck & Co. Inc. Pat. Dept.
P.O. Box 2000
Rahway, NJ 07065 (201) 574-4830

Theresa Y. Cheng
Merck & Co. Inc. Pat. Dept.
126 E. Lincoln Ave.
Rahway, NJ 07065 (201) 574-4982

Menotti J. Lombardi, Jr.
2367 Jowett Pl.
Rahway, NJ 07065 (201) 382-0578

Gabriel Lopez
Merck & Co., Inc. Pat. Dept.
P.O. Box 2000
Rahway, NJ 07065 (201) 574-4417

Frank M. Mahon#
Merck And Co. Inc.
P.O. Box 2000
Rahway, NJ 07065 (201) 574-4372

Stanley A. Marcus
M & T Chemical Inc.
P.O. Box 1104
Rahway, NJ 07065 (201) 499-2153

Roy D. Meredith#
Merck & Co. Inc.
P.O. Box 2000
Rahway, NJ 07065 (201) 574-4678

Salvatore C. Mitri
Merck & Co. Inc.
P.O. Box 2000
Rahway, NJ 07065 (201) 574-4454

Mario Anthony Monaco
Merck & Co. Inc
P.O. Box 2000
Rahway, NJ 07065 (201) 574-5295

William H. Nicholson
Merck & Co. Inc.
P.O. Box 2000
Rahway, NJ 07065 (201) 574-5315

Richard B. Olson
Merck & Co. Inc. Pat. Dept.
P.O. Box 2000
Rahway, NJ 07065 (201) 574-4057

Richard S. Parr
Merck & Co. Inc.
P.O. Box 2000
Rahway, NJ 07065

Donald J. Perrella
Merck & Co. Inc. Pat. Dept.
126 E. Lincoln Ave.
Rahway, NJ 07065 (201) 574-5593

Hesna J. Pfeiffer
Merck And Co. Inc.
P.O. Box 2000
Rahway, NJ 07065 (201) 574-4251

Manfred Polk
Merck & Co. Inc.
P.O. Box 2000
Rahway, NJ 07065 (201) 574-4285

Alice Ota Robertson
Merck & Co. Inc.
P.O. Box 2000
Rahway, NJ 07065 (201) 574-4372

David L. Rose
Merck & Co., Inc. Pat. Dept.
126 E. Lincoln Ave.
Rahway, NJ 07065 (201) 574-4777

Roger G. Smith#
Merck & Co. Inc.
P.O. Box 2000, 126 E. Lincoln Ave.
Rahway, NJ 07065 (201) 574-4817

Raymond M. Speer
Merk & Co. Inc.
P.O. Box 2000
Rahway, NJ 07065 (201) 574-4481

Michael C. Sudol, Jr.
Merck & Co. Inc.
126 E. Lincoln Ave.
Rahway, NJ 07065 (201) 574-5158

Daniel T. Szura
Merck & Co., Inc.
126 E. Lincoln Ave. P.O. Box 2000
Rahway, NJ 07065 (201) 574-4678

Richard A. Thompson
Merck & Co. Inc.
126 E. Licoln Ave.
Rahway, NJ 07065 (201) 574-4331

Jack L. Tribble
Merck & Co. Inc.
126 E. Lincoln Ave.
Rahway, NJ 07065 (201) 574-5321

Walter Francis Jewell#
23 Berry Lane
Randolph, NJ 07869 (201) 328-0371

Jesse Woldman
Thomas And Betts Corp.
920 Route 202
Raritan, NJ 08869 (201) 685-1600

Roger A. Clapp
70 Highway 35
Red Bank, NJ 07701 (201) 758-0800

Peter L. Michaelson
Michaelson, Einschlag, Ostroff & Winter
208 Maple Ave., P.O. Box 8489
Red Bank, NJ 07701 (201) 530-6671

Revis Gale Rhodes, Jr.
Evans, Koelzer, Marriott, Osborne, Kreizman & Bassler
One Harding Rd., P.O. Box B B
Red Bank, NJ 07701 (201) 741-9550

Charles L. Thomason
Evans, Koelzer, Osborn & Kreizman
One Harding Rd., P.O. Box B B
Red Bank, NJ 07701 (201) 741-9550

Donald R. Heiner
245 Main St.
Ridgefield, NJ 07660 (201) 440-8040

Herbert Lewis Davis, Jr.
136 Washington Place
Ridgewood, NJ 07450 (201) 444-6365

Andrew J. Nugent
403 Colonal Rd.
Ridgewood, NJ 07450 (201) 445-5543

Douglas J. Kirk
R.D. 3, Box 376
Runyon Mill Rd.
Ringoes, NJ 08551

Tobias Lewenstein#
653 Woodside Ave.
River Vale, NJ 07675 (201) 391-1014

George M. Kachmar#
Weiner, Ostrager, Fieldman & Mantel
114 Essex Suite
Rochelle Park, NJ 07662 (201) 368-1300

John V. Regan
4 Lemore Circle
Rocky Hill, NJ 08553 (609) 924-5924

Chester A. Williams, Jr.
R D 2 Box 472
Rocky Hill, NJ 08559 (609) 397-0117

John G. Gilfillan, III
Carella, Byrne, Bain & Gilfillan
6 Becker Farm Rd.
Roseland, NJ 07068 (201) 994-1700

John J. Halak
3 Becker Farm Rd. P.O. Box 26
Roseland, NJ 07068 (201) 992-9777

Raymond J. Lillie
Carlla, Byrne, Bain & Gilfillan
6 Becker Farm Rd.
Roseland, NJ 07068

Jeremiah Gerard Murray#
Carella, Byrne, Bain & Gilfillan
6 Becker Farm Rd.
Roseland, NJ 07068 (201) 994-1700

Elliot Michael Olstein
Carella, Byrni, Bain & Gilfillan
6 Becker Farm Rd
Roseland, NJ 07068 (201) 994-1700

Richard S. Serbin
61 Monroe Ave.
Roseland, NJ 07068 (201) 865-7500

Raymond W. Barclay
95 Raymond Ave.
Rutherford, NJ 07070 (201) 939-1295

Charles C. Marshall
52 Chestnut St. P.O. Box 306
Rutherford, NJ 07070 (201) 460-9595

Frank A. Santoro
1500 Park Ave.
S. Plainfield, NJ 07080 (201) 561-6868

Anthony D. Cipollone
49 Market St., P.O. Box 542
Saddle Brook, NJ 07662 (201) 845-6626

Edward A. Steen
Inco Patents & Licensing
Park 80 West - Plaza Two
Saddle Brook, NJ 07662 (201) 368-4848

Robert D. Farkas
1776 Martine Ave.
Scotch Plaines, NJ 07076 (201) 889-5200

Alexander T. Kardos
1557 Ashbrook Dr.
Scotch Plains, NJ 07076 (201) 756-9437

Charles W. Seabury#
550 Forest Rd.
Scotch Plains, NJ 07076 (201) 322-2136

S. Michael Bender
Harman Cove Towers, Suite 127
Secaucus, NJ 07094 (201) 867-6807

Thomas Cifelli, Jr.
22 Cayuga Way
Short Hills, NJ 07078 (201) 379-4187

John W. Fisher
Bell Telephone Labs., Inc.
101 J.F. Kennedy Parkway
Short Hills, NJ 07078 (201) 564-2188

Clayton S. Gates
287 Taylor Rd., S.
Short Hills, NJ 07078 (201) 467-9866

Louis S. Gillow
23 Wellington Ave.
Short Hills, NJ 07078

Kenneth B. Hamlin
Bell Telephone Laboratories, Inc.
101 JFK Parkway
Short Hills, NJ 07078 (201) 564-2044

Joseph Patrick Kearns, Jr.
Bell Telephone Labs. Inc.
101 John F.Kennedy Pkwy.
Short Hills, NJ 07078 (201) 564-2533

Leonhard F. Marx
386 Hartshorn Dr.
Short Hills, NJ 07078 (201) 376-7543

Jonathan E. Myers
19 Canterbury Lane
Short Hills, NJ 07078 (201) 379-4354

Daniel E. Nester
Bell Labs.
101 J.F. Kennedy Parkway
Short Hills, NJ 07078 (201) 564-2985

Harry L. Newman
Bell Telephone Labs., Inc.
101 J.F. Kennedy Parkway
Short Hills, NJ 07078 (201) 564-2527

Rosemary A. Ryan
A T & T Bell Labs.
101 J.F. Kennedy Pkwy. Room 424
Short Hills, NJ 07078 (201) 564-2148

Harold F. Wilhelm
41 Cambridge Dr.
Short Hills, NJ 07078 (201) 376-8205

Patrick E. Roberts
42 Dorchester Way
Shrewsbury, NJ 07702

Elliot A. Lackenbach
311 H H Gates Rd.
Somerset, NJ 08873 (201) 873-3231

Peter J. Butch, III
Golden, Lintner, Rothschild, Spagnola & Di Fazio
1011 Route 22 West, P.O. Box 897
Somerville, NJ 08876 (201) 722-6300

Hugh C. Crall
American Hoechst Corp.
Route 202-206 North
Somerville, NJ 08876 (201) 231-2842

Robert J. Ferb
15 W. High St., P.O. Box 8109
Somerville, NJ 08876 (201) 722-4033

Kenneth A. Genoni
Hoechst Celanese Corp.
Route 202-206 North
Somerville, NJ 08876 (201) 231-4413

Kenneth R. Glick
G E Semiconductor
Route 202
Somerville, NJ 08876 (201) 685-6397

Karen E. Klumas
American Hoechst Corp. Pat. Dept.
Route 202-206 North
Somerville, NJ 08876 (201) 231-4268

Emery Marton
American Hoechst Corp.
Rt. 202-206 North
Somerville, NJ 08876 (201) 231-3002

Birgit E. Morris
G E Semiconductor Pat. Coun.
Route 202
Somerville, NJ 08876 (201) 685-6000

Barbara L. Renda
360 Colonial Rd.
Somerville, NJ 08876 (201) 534-4857

Linda L. Setescak#
Hoechst-Roussel Pharmaceutical
Route 202-206 North
Somerville, NJ 08876 (201) 231-3122

Michael J. Tully
American Hoechst Corp.
Route 202-206 North
Somerville, NJ 08809 (201) 231-3232

Raymond R. Wittekind
American Hoechst Corp.
Route 202-206 North
Somerville, NJ 08876 (201) 231-3391

Richard R. Gipson
11-B Parkwood Dr.
South Amboy, NJ 08879 (201) 727-1891

David P. Alan
81 Second St.
South Orange, NJ 07079 (201) 762-6543

Neil O. Eriksen
Cohn & Cohn
14 S. Orange Ave.
South Orange, NJ 07079 (201) 762-6444

Arthur L. Lessler
540 Old Bridge Turnpike
South River, NJ 08882 (201) 254-5155

George Geier
43 Whippoorwill Lane
Sparta, NJ 07871 (201) 729-6332

John J. Hart
214 Remsen Ave.
Spring Lake, NJ 07762 (201) 449-5881

Daniel Dewitt Sharp
319 - 12th Ave.
Spring Lake Hgts, NJ 07762 (201) 449-8167

Gary T. Gann
Association Management Corp.
66 Morris Ave.
Springfield, NJ 07081 (201) 379-1100

Tennes I. Erstad
P.O. Box 100
Stanton, NJ 08885 (201) 735-9456

Herbert Joseph Winegar
281 A R D 1
Stockton, NJ 08559

Charles B. Barris
Cleanese Research Co.
86 Morris Ave.
Summit, NJ 07901 (201) 522-7260

Henry Carpenter Dearborn
89 Canoe Brook Parkway
Summit, NJ 07901 (201) 273-9319

Leonard P. Prusak
P.O. Box 339
Summit, NJ 07901 (201) 233-7092

Arthur I. Degenholtz#
32 Vandelinda Ave.
Teaneck, NJ 07666 (201) 692-1292

Arnold D. Litt
222 Cedar Lane
Teaneck, NJ 07666 (201) 836-9100

Reno A. Del Ben
25 George St.
Tenafly, NJ 07670 (201) 567-9400

Frederick R. Cantor
P.O. Box 88
Titusville, NJ 08560 (609) 737-3302

David W. Pearce-Smith
84 Kilian Place
Totowa, NJ 07512 (201) 595-8540

William R. Rohrbach
57 Waughaw Rd.
Towaco, NJ 07082 (201) 263-9097

John J. Kane
Suite D., One Highgate Dr.
Trenton, NJ 08618 (609) 882-7575

Nathan Levin
416 Highgate Dr.
Trenton, NJ 08618 (609) 883-6033

William L. Muckelroy
Ewing Professional Bldg.
1901 N. Olden Ave. Ext., Suite 3A
Trenton, NJ 08618 (609) 882-2111

Albert Sperry
Sperry, Zoda & Kane
One Highgate Dr. Suite D
Trenton, NJ 08618 (609) 882-7575

Frederick A. Zoda
Sperry, Zoda & Kane
One Highgate Dr. Suite D
Trenton, NJ 08618 (609) 882-7575

Martin M. Glazer
Haidri, Glazer & Kamel
2333 Morris Ave. Suite C-14
Union, NJ 07083 (201) 688-8700

Amirali Y. Haidri
Haidri, Glazer & Kamel
2333 Morris Ave. Suite C-14
Union, NJ 07083 (201) 688-8700

Alan M. Kamel
Haidri, Glazer & Kamel Ideal Professional Park
2333 Morris Ave. Suite A-14
Union, NJ 07083 (201) 688-8700

Ira Meislik
52 Upper Montclair Plaza
Upper Montclair, NJ 07043 (201) 744-0288

Mitchell P. Novick
52 Upper Montclair Plaza, P.O. Box 876
Upper Montclair, NJ 07043 (201) 744-5150

James N. Blauvelt
60 Fells Rd.
Verona, NJ 07044 (201) 239-5839

David H. Klein
21 Valhalla Way
Verona, NJ 07044 (201) 857-0667

Maurice W. Levy
102 Morningside Rd.
Verona, NJ 07044 (201) 239-2206

Walter C. Morrison#
Meadowyck Lane 28, Route 4
Vincetown, NJ 08088 (609) 859-2091

Jane Massey Licata
5 Red Oak Dr.
Voorhees, NJ 08043

Edward A. Petko
617 Palmer Ave.
W. Allenhurst, NJ 07711 (201) 531-4993

Lionel Norman White
35 Crest Drive
W. Millington, NJ 07946 (201) 647-4096

Andrew N. Parfomak#
35 Orchard St.
Wallington, NJ 07057 (201) 779-6246

Robert B. Ardis
A T & T Bell Laboratories
10 Independence Blvd.
Warren, NJ 07060 (201) 580-6054

Michael Bard
American Telephone & Telegraph Company
P.O. Box 4911, Room No. 3A-E14
Warren, NJ 07060 (201) 580-5969

Maurice M. De Picciotto
A T & T
10 Independence Blvd., P.O. Box 4911
Warren, NJ 07060 (201) 580-5966

Joseph F. Di Prima
Merck & Co. Inc. Pat. Dept.
10 Independence Blvd. P.O. Box 4911
Warren, NJ 07060 (201) 580-5966

Geoffrey D. Green
Amer. Telephone & Telegraph Co.
10 Independence Blvd., P.O. Box 4911
Warren, NJ 07060 (201) 580-5958

William C. Hosford#
97 Stirling Rd.
Warren, NJ 07060 (201) 647-5564

George Wesley Houseweart
A T & T
10 Independence Blvd.
Warren, NJ 07060 (201) 580-6420

David L. Hurewitz
A T & T
10 Independence Blvd. Room 3A- U22
Warren, NJ 07060 (201) 580-5967

Horst M. Kasper
13 Forest Dr.
Warren, NJ 07060 (201) 757-2839

Arthur S. Rosen
A T & T Co.
10 Independence Blvd.
Room No. 3A -U22
Warren, NJ 07060 (201) 580-5964

Eugene Zagarella, Jr.
22 Wilderness Trail
Warren, NJ 07060 (201) 647-3557

Keith E. Gilman#
121 Overlook Ave.
Wayne, NJ 07470 (201) 694-9556

Marilyn J. Maue#
G A F Corp.
1361 Alps Rd.
Wayne, NJ 07470 (201) 628-3544

J. Gary Mohr
G A F Corp.
1361 Alps Rd.
Wayne, NJ 07470 (201) 628-3546

Edward P. Schmidt
124 Lake Dr. W.
Wayne, NJ 07470 (201) 694-1118

Joshua J. Ward
G A F Corp.
1361 Alps Rd.
Wayne, NJ 07470 (201) 628-3529

Robert B. Green
5 Bonn Place
Weehawken, NJ 07087 (201) 863-3983

Bernard Olcott
62 Hackensack Plank Rd.
Weehawken, NJ 07087 (201) 863-4200

Arthur I. Spechler
16 Alluvium Lakes Dr.
West Berlin, NJ 08091

Mitchell G. Condos
307 Northfield Ave.
West Orange, NJ 07052 (201) 731-4375

George H. Fritzinger#
Ritzit Corp.
15 Standish Ave.
West Orange, NJ 07052 (201) 736-3311

Elijah H. Gold#
10 Roosevelt Ave.
West Orange, NJ 07052 (201) 736-3021

Theodore O. Groeger#
2 Collamore Circle
West Orange, NJ 07052 (201) 763-0660

Clay Holland, Jr.
51 Brookside Rd.
West Orange, NJ 07043 (201) 736-4689

J. Russell Juten
7 Undercliff Terr.
West Orange, NJ 07052 (201) 731-6474

Burton I. Levine
71 Burnett Terr.
West Orange, NJ 07052 (201) 522-6416

George J. Minish
Minish & Williams
614 Eagle Rock Ave., P.O. Box 236
West Orange, NJ 07052 (201) 736-9622

Patrick J. Osinski
Organon Inc.
375 Mt. Pleasant Ave.
West Orange, NJ 07052 (201) 575-4350

John T. OHalloran
45 Fran Ave.
West Trenton, NJ 08628 (609) 882-8251

Thomas E. Arther#
408 Everson Place
Westfield, NJ 07040 (201) 283-0238

John E. Coakley#
100 Summit Court
Westfield, NJ 07090 (201) 233-1431

Sidney David
Lerner, David, Littenberg, Krumholz & Mentlik
600 S. Ave. West.
Westfield, NJ 07090 (201) 654-5000

Stephen B. Goldman
Lerner, David, Littenberg, Krumholz & Mentlik
600 S. Ave. West
Westfield, NJ 07090 (201) 654-5000

Charles A. Harris
8 Stoneleigh Park
Westfield, NJ 07090 (201) 232-6043

Margaret B. Kelley
1321 E. Broad St.
Westfield, NJ 07090 (201) 232-4491

Paul H. Kochanski
Lerner, David, Littenberg, Krumholz & Mentlik
600 South Ave. West
Westfield, NJ 07090 (201) 654-5000

Arnold H. Krumholz
Lerner, David, Littenberg, Krumholz & Mentlik
Westfield, NJ 07090 (201) 654-5000

Lawrence Irwin Lerner
Lerner, David, Littenberg, Krumholz & Mentlik
600 South Ave. West
Westfield, NJ 07090 (201) 654-5000

Joseph Saul Littenberg
Lerner, David, Littenberg, Krumholz & Mentlik
600 South Ave.
Westfield, NJ 07090 (201) 654-5000

Jerome Edward Luecke
831 Nancy Way
Westfield, NJ 07090 (201) 654-3097

William L. Mentlik
Lerner, David, Littenberg, Krumholz & Mentlik
600 South Ave.
Westfield, NJ 07090 (201) 654-5000

Marcus J. Millet
lerner, David, Littenberg, Krumholz & Mentlik
600 South Ave. West
Westfield, NJ 07090 (201) 654-5000

Theodore Moss
611 Embree Crescent
Westfield, NJ 07090 (201) 654-4318

Richard R. Muccino
Lerner, David, Littenberg, Krumholz & Mentlik
600 South Avenue West
Westfield, NJ 07090 (201) 654-5000

John R. Nelson
Lerner, David, Littenberg, Krumholz & Mentlik
600 South Ave. West
Westfield, NJ 07090 (201) 654-5000

Richard Manford Rabkin
245 Delaware St.
Westfield, NJ 07090 (201) 233-2467

Richard I. Samuel
Patlex Corp.
533 South Ave. West
Westfield, NJ 07090 (201) 654-6620

Elwood Joseph Schaffer
636 prospect St.
Westfield, NJ 07090 (201) 232-1840

Stanley J. Silverberg
320 Orenda Circle
Westfield, NJ 07090 (201) 232-7275

William Hays Smyers
229 Sylvania Pl.
Westfield, NJ 07090 (201) 233-2284

David Harry Tannenbaum
816 Boulevard
Westfield, NJ 07090 (201) 233-2844

William J. Ungvarsky
728 Boulevard
Westfield, NJ 07090 (201) 233-3458

Roy H. Wepner
Lerner, David, Littenberg, Krumholz & Mentlik
600 South Ave. West
Westfield, NJ 07090 (201) 654-5000

Graig H. Evans
56 Addison Ave.
Westmont, NJ 08108 (609) 858-4341

Richard P. Moon
739 Upper Way
Wharton, NJ 07885 (201) 366-3450

Lester H. Birnbaum
A T & T Bell Labs.
1 Whippany Rd.
Whippany, NJ 07981 (201) 386-6377

Charles Edger Graves
A T & T Bell Labs.
Whippany Rd .
Whippany, NJ 07981 (201) 386-4475

Ruloff F. Kip, Jr.
A T & T Bell Labs.
Whippany Rd.
Whippany, NJ 07981 (201) 386-6834

Robert O. Nimtz
Bell Telephone Labs. Inc.
Whippany Rd.
Whippany, NJ 07981 (201) 386-6377

Alfred George Steinmetz
A T & T Bell Labs. Inc.
Whippany Rd.
Whippany, NJ 07981 (201) 386-2718

Kenneth P. Glynn
P.O. Box 343
Rt. 22 At Rt. 523
Whitehouse, NJ 08888 (201) 534-6110

Nicholas L. Pollis, Jr.
Kaser, Pollis & Kaser, P.A.
P.O. Box 405, 354-56 S. Main St.
Williamstown, NJ 08094 (609) 629-5000

Arthur Frederick
Curtis - Wright Corp. 1 Passaic St.
Wood, NJ 07075 (201) 777-2900

Bryan C. Diner#
103 Maple Hill Dr.
Woodbridge, NJ 07095 (201) 636-3234

Ezra Sutton
Weinstein & Sutton Plaza 9
900 Route 9
Woodbridge, NJ 07095 (201) 634-3520

Carl R. Horten
Ingersoll Rand Co.
200 Chestnut Ridge Rd.
Woodcliff Lake, NJ 07675 (201) 573-3117

Alfred E. Riccardo
168 Pascack Rd.
Woodcliff Lake, NJ 07675 (201) 391-0079

New Mexico

Edward L. Amonette#
5715 El Prado N.W.
Albuquerque, NM 87107 (505) 344-4143

James H. Chafin*
U.S. Dept. Of Energy
Texas & K Street
Albuquerque, NM 87115 (505) 844-8231

Anne D. Daniel*
U.S. Dept. Of Energy
P.O. Box 5400
Albuquerque, NM 87115 (505) 846-3123

Robert W. Harris
Poole, Tinnin & Martin, P.C.
P.O. Box 1769
Albuquerque, NM 87103 (505) 842-8155

Milton R. Kestenbaum
11009 Country Club Dr. N.E.
Albuquerque, NM 87111 (505) 844-8015

Dudley W. King
2510 Gen. Arnold N.E.
Albuquerque, NM 87112 (505) 299-0043

Cedric Hudson Kuhn
5605 Equestrain Dr., N.W.
Albuquerque, NM 87120 (505) 898-9404

John R. Landsdowne
300 Central Ave. S.W. Suite 2500 West
P.O. Box 25891
Albuquerque, NM 87125 (505) 247-9532

George H. Libman*
U.S. Dept. Of Energy
Albuquerque Operations Off.
P.O. Box 5400
Albuquerque, NM 87115 (505) 844-8231

Robert Charles Lucke#
611 Lead S W Apt. 311
Albuquerque, NM 87102

Armand McMillan*
Dept. Of Energy Albuq. Operations Off.
Box 5400 Kirkland A F B
Albuquerque, NM 87115 (505) 844-8231

Richard E. Norton
300 Central Ave. S.W. Suite 2500-W
Albuquerque, NM 87102 (505) 247-3911

Karla J. Ojanen
621 Georgia, Southeast
Albuquerque, NM 87108 (505) 266-8880

Deborah A. Peacock
Rodey, Dickason, Solan, Akin & Robb, P.A.
20 First Plaza Suite 700
Albuquerque, NM 87103 (505) 765-5900

Richard S. Robins#
4433 Magnolia Dr. N.E.
Albuquerque, NM 87111 (505) 298-8099

Donald P. Smith
Singer, Smith & Powell, P.A.
P.O. Box 25565
Albuquerque, NM 87125 (505) 247-3911

Albert Sopp*
Univ. Of New Mexico, Room 102 Scholes Hall
Albuquerque, NM 87131 (505) 277-7646

Robert W. Weig
4701 Cutler N E
Albuquerque, NM 87110 (505) 884-3066

Charles C. Wells, Jr.*
Air Force Contract Management Div.
A F C M D / J A T, Kirtland A F B
Albuquerque, NM 87117 (505) 844-9762

John L. Wiegreff
94 Horner St.
Belen, NM 87002 (505) 864-7286

Hiram B. Gilson#
601 W. Orchard Apt. 45
Carlsbad, NM 88220 (505) 885-3296

Dennis H. Shaw*
U.S. Air Force Off. Of Staff Judge Advocate
Air Force Contract Management Division
Kirtland A F B, NM 87117 (505) 844-0631

John A. Darden, III
Darden, Valentine & Driggers
200 W. Las Cruces Ave.
Las Cruces, NM 88001 (505) 526-6655

William A. Eklund
P.O. Box 822
Los Alamos, NM 87544 (505) 667-3766

Samuel M. Freund
11 Timber Ridge Rd.
Los Almos, NM 87544 (505) 667-9701

Paul D. Gaetjens
Los Almos National Lab. Ms-D412
Los Almos, NM 87545 (505) 667-9701

Robert Dean Krohn#
Los Almos Natl. Lab. Ms 674
Box 1663
Los Almos, NM 87545 (505) 667-5011

Edward C. Walterscheid
Univ. Of California, Los Almos Natl. Lab.
P.O. Box 1663
Los Almos, NM 87545 (505) 667-3970

Ray G. Wilson
Los Almos National Lab.
ADLC/ PL, MS M326
P.O. Box 1663
Los Almos, NM 87545

Milton D. Wyrick
Los Almos National Lab. P.O. Box 1663
L C / P L, M S D412
Los Almos, NM 87545 (505) 667-3302

George W. Prince
2243 Calle Cacique
Santa Fe, NM 87501 (505) 982-9167

James Elbert Snead, III
Jones, Gallegos, Snead & Wertheim, P.A.
215 Lincoln Ave. P.O. Box 2228
Sante Fe, NM 87501 (505) 982-2691

Edward C. Jason
P.O. Box 311
Williamsburg, NM 87942 (505) 894-6929

New York

Charles L. Guettel#
420 Sand Creed Rd. Apt.1-229
Albany, NY 12205 (518) 438-8078

Susan F. Gullotti
Hinman Straub Pigors & Manning, P.C.
90 State St.
Albany, NY 12207 (518) 436-0751

Fredric T. Morelle
Schmeiser & Morelle
8 Automation Lane
Albany, NY 12205 (518) 458-1850

Albert L. Schmeiser
Schmeiser & Morelle
8 Automation Lane
Albany, NY 12205 (518) 458-1850

Charles T. Watts
Schmeiser, Morelle & Watts
8 Automation Lane
Albany, NY 12205 (518) 458-1850

Jay R. Yablon
New York State Legislative Commission On
Science & Tech.
99 Washington Ave. Suite 704
Albany, NY 12210 (518) 455-5081

Roger S. Benjamin
25 Concord Rd.
Ardsley, NY 10502 (914) 693-6788

Harry Falber
Ciba-Geigy Corp.
444 Saw Mill River Rd.
Ardsley, NY 10502 (914) 347-4700

Meredith C. Findlay
CIBA - GEIGY Corp., Patent Dept.
444 Saw Mill River Rd.
Ardsley, NY 10502 (914) 347-4700

Irving M. Fishman
Ciba - Geigy Corp.
444 Saw Mill River Rd.
Ardsley, NY 10502 (914) 347-4700

Michael W. Glynn
Ciba - Geigy Corp
444 Saw Mill River Rd.
Ardsley, NY 10502 (914) 478-3131

Norbert Gruenfeld#
Ciba - Geigy Corp
444 Saw Mill River Rd.
Ardsley, NY 10502 (914) 347-4700

Luther A.R. Hall#
Ciba - Geigy Corp.
444 Saw Mill River Rd.
Ardsley, NY 10502 (914) 347-4700

Karl Francis Jorda
Ciba - Geigy Corp.
444 Saw Mill River Rd.
Ardsley, NY 10502 (914) 347-4700

Paul John Juettner
598 Ashford Ave.
Ardsley, NY 10502 (914) 693-2969

Kevin T. Mansfield#
Ciba - Geigy Corp.
444 Saw Mill River Rd.
Ardsley, NY 10502 (914) 478-3131

Frederick H. Rabin
Ciba - Geigy Corp.
444 Saw Mill River Rd.
Ardsley, NY 10502 (914) 478-3131

Edward Mccreery Roberts
CIBA - GEIGY Corp.
444 Saw Mill River Rd.
Ardsley, NY 10502 (914) 478-3131

Leon E. Tenenbaum
67 Prospect Ave.
Ardsley, NY 10502 (914) 693-3115

Ernest F. Weinberger
51 Eastern Dr.
Ardsley, NY 10502 (914) 693-1591

Carl C. Kling
20 Annandale St.
Armonk, NY 10504 (914) 273-9274

Joe L. Koerber
IBM Corp.
Old Orchard Rd.
Armonk, NY 10504 (914) 765-3680

Murray Nanes
IBM Corp.
Old Orchard Rd.
Armonk, NY 10504 (914) 765-3585

John F. Osterndorf
IBM Corp.
Old Orchard Rd.
Armonk, NY 10504 (914) 765-3582

David J. Zobkiw
1621 Hoover Rd.
R.R. 3
Attica, NY 14011 (716) 591-2854

Salvatore A. Alamia
42 Fire Island Ave.
Babylon, NY 11702 (516) 669-1730

William C. Gerstenzang
Professional Bldg.
Baldwin, NY 10505 (914) 628-1008

William J. Fox
Whitlock Corp.
42-40 Bell Blvd.
Bayside, NY 11361 (212) 423-5550

Marvin D. Genzer
New Indian Hill Rd.
Bedford, NY 10506 (914) 234-7741

Ethel L. Morgan#
Appleby Dr.
Bedford, NY 10506 (914) 234-7989

Edward S. Drake#
Box 104
Bedford Hills, NY 10507 (914) 232-3051

Harmon S. Potter
15 Rogers Ave.
Bellport, NY 11713 (516) 286-0096

John P. Kozma
Grumman Aerospace Corp.
M.S. B29-05
Bethpage, NY 11714 (516) 575-3830

Douglas M. Clarkson
48 Helen St.
Binghamton, NY 13905 (607) 729-5709

Barry L. Haley
Singer Co.
Link Flight Simulation Div.
Corporate Dr.
Binghamton, NY 13902 (607) 721-6525

Charles S. Neave#
One Aberystwyth Pl.
Binghamton, NY 13905 (607) 797-8292

Karl Pechmann#
126 Rosedale Dr.
Binghamton, NY 13905 (607) 724-7273

Richard Glenn Stephens
318 Security Mutual Life Bldg.
80 Exchange St.
Binghamton, NY 13901 (607) 723-8295

Andrew Taras#
25 Clifton Blvd.
Binghamton, NY 13903 (607) 722-9315

John W. Young
Young & Slocum
22 Riverside Dr.
Binghamton, NY 13901 (607) 722-3426

Saul R. Bresch
28 Cypress Lane
Briarcliff Manor, NY 10510 (914) 762-4193

George E. Ham#
284 Pine Rd.
Briarcliff Manor, NY 10510 (914) 762-5682

Homer James Bridger#
2410 Barnes Ave.
Bronx, NY 10467 (212) 655-3844

Ferdinard F.E. Kopecky
3356 Bronx Blvd.
Bronx, NY 10467 (212) 231-3659

Joseph Nachay
1975 Bathgate Ave., Apt. 2H
Bronx, NY 10457 (212) 294-8440

Joseph J. Orlando
1263 Havemeyer Ave.
Bronx, NY 10462 (212) 864-4232

Alvin S. Rohssler
1790 Bruckner Blvd.
Bronx, NY 10473 (212) 991-8259

Janet C. Lentz
21 Sycamore St.
Bronxville, NY 10708 (914) 793-6049

Adrian T. Calderon
658 52 St.
Brooklyn, NY 11220 (718) 972-5688

Gary Cohen
141 Joralemon St.
Brooklyn, NY 11201 (212) 624-6669

Stanley G. Harvey
5502 Kings Hwy.
Brooklyn, NY 11203 (212) 451-0756

Mark H. Jay#
P.O. Box 020083
General Post Office
Brooklyn, NY 11202 (718) 625-0399

Steven L. Krantz
1102 71st St.
Brooklyn, NY 11228 (212) 680-8306

Michael Richard Leibowitz
125 E. 18th St.
Brooklyn, NY 11226 (212) 338-6442

Ronald Lianides#
350 65th St.
Brooklyn, NY 11220 (212) 884-6600

James J. Long#
6808 Tenth Ave.
Brooklyn, NY 11219 (212) 745-2185

Donald Malcolm
40 E. 17th St.
Brooklyn, NY 11226

Daniel R. McGlynn
266-76th St.
Brooklyn, NY 11209 (718) 238-6720

Israel Nissenbaum
1038-56th St.
Brooklyn, NY 11219 (212) 633-4044

Nikolay Parada#
1066 E. 13th St.
Brooklyn, NY 11230 (212) 252-4221

Gasper P. Quartararo#
Witco Corp. Argus Chem. Div.
633 Court St.
Brooklyn, NY 11231 (212) 858-5678

Eugene G. Reynolds
2024 Batchelder St.
Brooklyn, NY 11229 (212) 743-9462

Lowell M. Rubin
141 Argyle Rd.
Brooklyn, NY 11218 (212) 282-4377

Edwin D. Schindler
1523 E. 4th St.
Brooklyn, NY 11230 (718) 336-9346

Robert Sherman
415 Beverly Rd.
Brooklyn, NY 11218 (718) 436-0605

Perry Teitelbaum
Goodman & Teitelbaum
26 Court St. Suite 1400
Brooklyn, NY 11242 (212) 643-0400

Richard T. Treacy#
34 Plaza St. Apt. 1004
Brooklyn, NY 11238 (212) 622-0413

William Paul Vafakos
967 E. 17th St.
Brooklyn, NY 11230 (212) 253-5014

Guy J. Agostinelli, III#
Kavinoky & Cook
120 Delaware Ave.
Buffalo, NY 14202 (716) 856-9234

Edwin J. Bean, Jr.
Christel, Bean & Linihan, P.C.
600 Guaranty Bldg.
28 Church St.
Buffalo, NY 14202 (716) 853-7778

John B. Bean
Bean, Kaufman & Bean
1313 Liberty Bk. Bldg.
420 Main St.
Buffalo, NY 14202 (716) 852-7405

Arthur S. Cookfair#
89 Highgate Ave.
Buffalo, NY 14214 (716) 832-4088

Thomas F. Daley
LTV Aerospace & Defense
Sierra Research Div.
247 Cayuga Rd.
Buffalo, NY 14225 (716) 681-6206

Joseph P. Gastel
722 Ellicott Square Bldg.
Buffalo, NY 14203 (716) 854-6285

Robert Hutter#
205B Clinton Hall
State Univ. of NY
Buffalo, NY 14261 (716) 636-4236

R. Craig Kauffman
Bean, Kauffman & Bean
1313 Liberty Bk. Bldg.
420 Main St.
Buffalo, NY 14202 (716) 852-7405

Martin Gates Linihan, Jr.
Christel, Bean & Linihan
600 Guaranty Bldg.
28 Church St.
Buffalo, NY 14202 (716) 853-7778

E. Herbert Liss
Trico Products Corp.
817 Washington St.
Buffalo, NY 14203 (716) 852-5700

Anna E. Mack
Christel, Bean & Linihan
600 Guaranty Bldg.
28 Church St.
Buffalo, NY 14202 (716) 853-7778

Michael E. Mckee
Christel, Bean & Linihan
600 Guaranty Bldg.
28 Church St.
Buffalo, NY 14202 (716) 853-7778

Laird F. Miller
National Gypsum Co.
1650 Military Rd.
Buffalo, NY 14217 (716) 873-9750

James J Ralabate
5792 Main St.
Williamsville
Buffalo, NY 14221 (716) 634-2280

Kenneth Richard Sommer
Sommer & Sommer
1022 Ellicott Square
Buffalo, NY 14203 (716) 853-7761

Peter K. Sommer
Sommer & Sommer
1022 Ellicott Square
Buffalo, NY 14203 (716) 853-7761

Alan H. Spener
Reichert-Jung Inc.
Box 123
Buffalo, NY 14240 (716) 891-3008

John Charles Thompson
Christel, Bean & Linihan
600 Guaranty Bldg.
28 Church St.
Buffalo, NY 14202 (716) 853-7778

Ivan A. Mc Corkendale
190 Parrish St., Ext. 51
Canandaigua, NY 14424 (716) 394-2192

Gennaro L. Pasquale
One Old Country Rd. Suite 385
Carle Place, NY 11514 (516) 741-6704

Alfred B. Engelberg
Sedgewood Club
R.D. 12
Carmel, NY 10512 (914) 225-7099

Robert K. Bair#
3 Brashear Place
Castleton, NY 12033 (518) 732-7132

William S. Wolfe
2893 Bingley Rd.
Cazenovia, NY 13035 (315) 445-4726

David M. Warren
655 Oakland Ave.
Cedarhurst, NY 11516 (516) 295-2054

John Finnie
152 Little Neck Rd.
Centerport, NY 11721 (516) 757-3085

William G. Valance#
331 Oakland Ave.
Central Islip, NY 11722 (516) 234-7698

Anthony Amaral, Jr.
575 Quaker Rd.
Chappaqua, NY 10514 (914) 238-8164

Allen A. Meyer
P.O. Box 134
Chappaqua, NY 10514 (212) 473-7326

Howard G. Rath#
6 Valley View Rd.
Chappaqua, NY 10514 (914) 238-9304

Frank R. Agovino
Hazeltine Corp.
500 Commack Rd.
Commack, NY 11725 (516) 266-5613

Kenneth P. Robinson
Hazeltine Corp.
Commack, NY 11725 (516) 261-5673

Reginald J. Falkowski
Dresser-Rand Co.
Barton Steuben Place
Corning, NY 14830 (607) 937-6444

Alexander R. Herzfeld
Corning Glass Works
Patent Operations
SPFR 212
Corning, NY 14831 (607) 974-3381

Clinton S. Janes, Jr.
Corning Glass Works
SPFR 212
Corning, NY 14831 (607) 974-3323

William E. Maycock
Corning Glass Works
Sullivan Park FR-212
Corning, NY 14830 (607) 974-3386

Alfred L. Michaelsen
Corning Glass Works
Corning, NY 14830 (607) 974-3054

Clarence R. Patty, Jr.
311 Steuben St.
Corning, NY 14830 (607) 962-5663

Milton M. Peterson
Corning Glass Works
Patent Dept.
Sullivan Park FR 2
Corning, NY 14831 (607) 974-3378

William J. Simmons, Jr.
Corning Glass Works
Patent Dept.
SPFR 2
Corning, NY 14831 (607) 974-9000

Kenneth M. Taylor, Jr.
Corning Glass Works
SPFR 212
Corning, NY 14831 (607) 974-3675

Burton Ralph Turner
Corning Glass Works
SPFR 212
Corning, NY 14831 (607) 974-3375

Kees Van Der Sterre
Corning Glass Works
Sullivan Park
SPFR 212
Corning, NY 14831 (607) 974-3294

Richard N. Wardell
Corning Glass Works
SPFR 212
Corning, NY 14831 (607) 974-3321

Walter S. Zebrowski
Corning Glass Works
SPFR 2
Corning, NY 14831 (607) 974-3170

Kenneth William Greb#
Smith Corona Corp.
839 Rte. 13, S.
Cortland, NY 13045 (607) 753-6011

William E. Mear, III#
SCM Corp.
839 Rte. 13, S.
Box 2020
Cortland, NY 13045 (607) 753-6011

Paul S. Hoffman
139 Grand St. P.O. Box 40
Croton-On-Hudson, NY 10520
(914) 271-5191

Jonathan B. Schafrann
17 Arlington Dr.
Croton-On-Hudson, NY 10520
(212) 687-6625

Joseph Martin Weigman
9N Dove Court, P.O. Box 343
Croton-On-Hudson, NY 10520
(914) 271-6346

Michael F. Brown
Georgetown Manor Inc.
150 W. Industry Court
Deer Park, NY 11729 (516) 242-0700

Vincent J. Ranucci
Eaton Corp. A I L Division Pat. Law Dept.
Commack Rd.
Deer Park, NY 11729 (516) 243-7201

Ralph Cohen
58 Murray Ave.
Delmar, NY 12054 (518) 439-4685

Annette M. Sansone
31 Brookedge Rd.
Depew, NY 14043

Harold I. Popp
6666 Old Lakeshore Rd.
Derby, NY 14047 (716) 947-4795

Salvatore J. Levanti#
10 Parkside Dr.
Dix Hills, NY 11746 (516) 499-5376

Richard L. Miller#
12 Parkside Dr.
Dix Hills, NY 11746 (516) 499-4343

Ilya Zborovsky#
6 Schoolhouse Way
Dix Hills, NY 11746 (516) 243-3818

Robert F. Sheyka
Stauffer Chemical Co.
Livingston Ave.
Dobbs Ferry, NY 10522 (914) 693-1200

John Edward Dumaresq
214 Manor Rd.
Douglaston, NY 11363 (718) 229-8373

Eric J. Sheets#
240-01 43rd Ave.
Douglaston, NY 11361 (212) 428-8721

Bernard Stimler
53-16 244 St.
Douglaston, NY 11362 (212) 229-3507

Robert C. Woodbury
Morten & Woodbury
87 E. Fourth St.
P.O. Box 800
Dunkirk, NY 14048 (716) 366-8050

Frederick J. Teeter#
32 Ginger Ct.
East Amherst, NY 14051 (716) 688-5280

Elmer J. Lawson#
Box 137, Best Rd.
East Greenbush, NY 12061 (518) 286-2893

Gilbert P. Weiner
Pall Corp.
2200 Northern Blvd.
East Hills, NY 11548 (516) 484-5400

Harvey Lunenfeld#
8 Patrician Dr.
East Northport, NY 11731 (516) 754-1000

Norman R. Bardales
IBM Corp.
Dept. N50/251-2
1701 North St.
Endicott, NY 13760 (607) 757-6747

Shelley M. Beckstrand
IBM Corp.
1701 North St.
Endicott, NY 13760 (607) 757-6760

John H. Bouchard
IBM Corp.
Dept. N50 Bldg. 251-2
1701 North St.
Endicott, NY 13760 (607) 757-6753

Lawrence R. Fraley
IBM Corp.
Intellectual Property Law Dept.
N50/251-2
1701 North St.
Endicott, NY 13760 (607) 757-6755

John S. Gasper
IBM Corp.
1701 North St.
Endicott, NY 13760 (607) 757-6757

Gerald R. Gugger#
IBM Corp.
P.O. Box 6
Endicott, NY 13760 (607) 755-3665

Max J. Kenemore
IBM Corp.
N50/251-2
1701 North St.
Endicott, NY 13760 (607) 755-7186

Mark E. Levy
IBM Corp.
Endicott Laboratory
P.O. Box 6
Endicott, NY 13760 (607) 752-3475

David L. Adour
417 Echo Lane
Endwell, NY 13760 (607) 748-7207

Paul M. brannen#
609 Lacey Dr.
Endwell, NY 13760 (607) 785-3128

Francis V. B. Giolma
3704 Frazier Rd.
Endwell, NY 13760 (607) 748-9242

George Weldon Killian#
17 Charing Cross
Fairport, NY 14450

J. Addison Mathews
415 Thayer Rd.
Fairport, NY 14450 (716) 223-1738

Stuart L. Melton
33 Skelby Moor Ln.
Fairport, NY 14450

Harry G. Martin, Jr.
123 Woodmancy Ln.
Fayetteville, NY 13066 (315) 637-3428

Herbert W. Taylor, Jr.#
17 Lynacres Blvd.
Fayetteville, NY 13066 (315) 446-0306

John P. Zacharias
145 Main St. P.O. Box H
Fishkill, NY 12524 (914) 896-8643

John Maier, III
Main St. Box C-1
Fleischmanns, NY 12430 (914) 254-5663

Nicholas J. Garogalo
259-09 81st Ave.
Floral Park, NY 11004 (212) 347-1619

Alfred W. Barber
32-44 Francis Lewis Blvd.
Flushing, NY 11358 (212) 463-3306

Seymour Gerald Bekelnitzky
147-09 72nd Dr.
Flushing, NY 11367 (718) 261-6475

Alvin Engelstein
67-20C 193rd Lane
Flushing, NY 11365 (718) 454-6757

George M. Kaplan
58-40 190 St.
Flushing, NY 11365 (212) 357-6230

Barnwell Rhett King
73-40 188th St.
Flushing, NY 11366 (718) 454-1798

Paul S. Martin
189-54 43rd Rd.
Flushing, NY 11358 (212) 358-5465

Nathaniel Altman#
68-37 Yellowstone Blvd.
Forest Hills, NY 11375 (212) 268-4541

J. Philip Anderegg
50 Exeter St.
Forest Hills, NY 11375 (212) 268-0206

Alexander Mencher
69-42 Ingram St.
Forest Hills, NY 11375 (212) 268-3807

Julius Balogh
1046 Lorraine Dr.
Franklin Square, NY 11010 (516) 775-0077

Paul W. Garbo#
48 Lester Ave.
Freeport, NY 11520 (516) 378-0393

Diane R. Bentley#
Scully, Scoot, Murphy & Presser
200 Garden City Plaza
Garden City, NY 11530 (516) 742-4343

Marvin Bressler
Scully, Scott, Murphy & Presser, P.C.
200 Garden City Plaza
Garden City, NY 11530 (516) 742-4343

Ralph E. Bucknam
Bucknam & Archer
600 Old Country Rd.
Garden City, NY 11530 (516) 549-0956

Mark J. Cohen
Scully, Scott, Murphy & Presser
200 Garden City Plaza
Garden City, NY 11530 (516) 742-4343

Frank S. Di Giglio
Scully, Scott, Murphy & Presser, P.C.
200 Garden City Plaza, Suite 212
Garden City, NY 11530 (516) 742-4343

Paul J. Esatto, Jr.
Scully, Scott, Murphy & Presser
200 Garden City Plaza
Garden City, NY 11530 (516) 742-4343

Fernanda Misani Fiordalisi
Bucknam & Archer
600 Old Country Rd.
Garden City, NY 11530 (516) 222-8885

Richard Gail Geib
64 Whitehall Blvd.
Garden City, NY 11530 (516) 248-7443

Oswald Gray Hayes
Scully, Scott, Murphy & Presser
200 Garden City Plaza
Garden City, NY 11530 (516) 742-4343

Norton Steele Johnson
Scully, Scott, Murphy & Presser
200 Garden City Plaza
Garden City, NY 11530 (516) 742-4343

Kenneth Lee King
Scully, Scott, Murphy & Presser
200 Garden City Plaza
Garden City, NY 11530 (516) 742-4343

Lawrence S. Lawrence
Bases, Russo, Lawrence, Ciovacco, Walsh & Feder, P.C.
1415 Kellum Pl.
Garden City, NY 11530 (516) 248-9400

Robert Mauer
1205 Franklin Ave.
Garden City, NY 11530 (516) 248-5803

William E. McNulty
Scully, Scott, Murphy & Presser, P.C.
200 Garden City Plaza
Garden City, NY 11530 (516) 742-4343

Jules M. Mencher
1205 Franklin Ave. Suite 300
Garden City, NY 11530 (516) 294-6090

Stephen D. Murphy
Scully, Scott, Murphy & Presser
200 Garden City Plaza, Suite 212
Garden City, NY 11530 (516) 742-4343

Joseph A. Osofsky
Bucknm & Archer
600 Old Country Rd.
Garden City, NY 11530 (516) 222-8885

Leopold Presser
Scully, Scott, Murphy & Presser, P.C.
200 Garden City Plaza
Garden City, NY 11530 (516) 742-4343

William C. Roch
Scully, Scott, Murphy & Presser
200 Garden City Plaza
Garden City, NY 11530 (516) 742-4343

Anthony C. Scott
Scully, Scott, Murphy & Presser
200 Garden City Plaza
Garden City, NY 11530 (516) 742-4343

John Francis Scully
Scully, Scott, Murphy & Presser
200 Garden City Plaza
Garden City, NY 11530 (516) 742-4343

John S. Sensny
Scully, Scott, Murphy & Presser
200 Garden City Plaza
Garden City, NY 11530 (516) 742-4343

Edgar N. Jay
11 Raynham Rd.
Glen Cove, NY 11542 (516) 571-3593

Stewart F. Moore
Five Oakwood Dr.
Glen Falls, NY 12801 (518) 798-3844

John A. Seifert
116 Chicken Valley Rd.
Glen Head, NY 11545 (516) 676-2937

Charles P. Boberg
165-5 Robert Gardens, N.
Glens Falls, NY 12801 (518) 792-4217

Thomas J. Cione#
Norton & Christensen
60 Erie St. Box 308
Goshen, NY 10924 (914) 294-7949

Alan D. Akers#
Moore Business Forms Inc.
300 Lang Blvd.
Grand Island, NY 14072 (716) 773-0326

Robert F. Hause
3663 West River Pky.
Grand Island, NY 14072 (716) 773-7038

Albert B. Cooper
Unisys Corp.
Lakeville Rd. & Marcus Ave.
Great Neck, NY 11020 (516) 754-3083

Ralph A. Johnston#
Sperry Corp.
Marcus & Lakeville Rd.
Great Neck, NY 11020 (516) 574-2612

Samson B. Leavitt
66 Essex Rd.
Great Neck, NY 11023 (516) 487-8417

Albert Levine
37 Ruxton Rd.
Great Neck, NY 11023 (516) 466-6663

Seymour Levine
Sperry Corp.
Marcus Ave. & Lakeville Rd.
Great Neck, NY 11020 (516) 574-3061

Julius Marx
68 Wooleys Lane
Great Neck, NY 11023 (516) 466-3182

Emanuel R. Posnack
46 Elm St.
Great Neck, NY 11021 (516) 487-6746

Abner Sheffer
7 Piccadilly Rd.
Great Neck, NY 11023 (516) HU2-5423

Howard A. Taishoff
105 Oxford Blvd.
Great Neck, NY 11023 (516) 482-4455

Howard Paul Terry
Sperry Corp.
Lakeville Rd. & Marcus Ave.
Great Neck, NY 11020 (516) 574-2519

Eli Weiss
6 Farmers Rd.
Great Neck, NY 11024 (516) 482-6123

James Anthony Kane, Jr.
34 O Connell Court
Great River, NY 11729 (516) 581-0322

Michael V. Leahy#
Klarman-Leahy Associates
9 Boulevard Ave.
Greenlawn, NY 11740 (516) 757-7115

Edward A. Onders
Hazeltine Corp.
E. Pulaski Rd.
Greenlawn, NY 11740 (516) 261-7000

James F. Young
15 Duncan Lane
Halesite, NY 11743 (516) 421-4130

Alfred E. Page, Jr.
Clune & White
480 Momaroneck Ave.
Harrison, NY 10528 (914) 698-8200

Jack Oisher
22 Kenneth Rd.
Hartsdale, NY 10530 (914) 332-0222

Robert John Patterson
107 Harvard Dr.
Hartsdale, NY 10530 (914) 948-6190

Edward D. Weil#
6 Amherst Dr.
Hastings-On-Hudson, NY 10706 (914) 478-3370

Alfred M. Walker
742 Veterans Memorial Highway
Hauppauge, NY 11788 (516) 361-8737

Andrew J. French
Shea, French & Shea
21 Kane Ave.
Hempstead, NY 11550 (516) 489-2760

William Lars Ericson
933 Westwood Dr.
Herkimer, NY 13350 (315) 866-0831

George I. Spieler
1515 Hewlett Ave.
Hewlett, NY 11557 (516) 374-1963

John Patrick Mc Gann
234 Jerusalem Ave.
Hicksville, NY 11801 (516) 931-0734

Michael B. Zapantis
86-53 Dunton St.
Holliswood, NY 11423 (212) 460-7707

Ira D. Blecker
I B M Corp. Intellectual Property Law Dept.
Dept. 901/Bldg. 300-482, Route 52
Hopewell Junction, NY 12533
(914) 894-2580

Edward William Brown
I B M Corp.
Dept. 901,Bldg. 300-482, Route 52
Hopewell Junction, NY 12533
(914) 894-3665

John D. Crane
I B M Corp.
Bldg. 300-482. Route 52
Hopewell Junction, NY 12533
(914) 894-3667

Wesley De Bruin
I B M Corp.
Route 52 D 901, B/300-482
Hopewell Junction, NY 12533
(914) 894-4713

Anne V. Dougherty
IBM Corp.
Route 52
Bldg. 300-482
Hopewell Junction, NY 12593
(914) 894-6919

William T. Ellis*
I B M Corp. Intellectual Property Law Dept.
Route 52, Dept. 901, B/300-482
Hopewell Junction, NY 12533

Graham S. Jones, II
I B M Corp. Dept. 901, Bldg. 300-482
Route 52
Hopewell Junction, NY 12533
(914) 894-3668

Jeffrey S. La Baw#
I B M Corp.
Route 52, East Fishkill Facility
Hopewell Junction, NY 12533
(914) 894-6667

Douglas A. Lashmit
R.D. 2, Box 107
N. Kensington Dr.
Hopewell Junction, NY 12533
(914) 894-3668

Steven J. Meyers
I B M Corp. Pat. Operation, Route #52
Dept. 901 Bldg. 300-482
Hopewell Junction, NY 12533
(914) 894-2210

Daniel P. Morris
I B M Corp.
Route 52
Hopewell Junction, NY 12533
(914) 894-7239

Henry Powers
IBM Corp.
Route 52
Hopewell Junction, NY 12533
(914) 894-2210

George Oscar Saile, Jr.
IBM Corp.
Dept. 901
Bldg. 300-482
Hopewell Junction, NY 12533
(914) 894-3402

Robert E. Sandt
IBM Corp.
Dept. 901, Bldg. 300-482
Hopewell Junction, NY 12533
(914) 894-6919

John A. Stemwedel
IBM Corp.
Dept. 901, Bldg. 300-482
Hopewell Junction, NY 12533
(914) 894-3664

Wolmar J. Stoffel
IBM Corp.
Dept. 901
Hopewell Junction, NY 12533
(914) 894-3669

George Tacticos
IBM Corp.
Intell. Property Law Dept.
Dept. 901, Bldg. 300-482
Route 52
Hopewell Junction, NY 12533
(914) 894-2581

Allen C. Miller, Jr.
542 Warren St.
Hudson, NY 12534 (518) 828-6805

Ralph Roger Barnard
200 E. Buffalo St.
Ithaca, NY 14850 (607) 273-1711

George M. Dentes
Hines & Dentes
417 N. Cayuga St.
Ithaca, NY 14850 (607) 273-6111

Joseph A. Edminster
College of Engineering
221 Carpenter Hall
Cornell Univ.
Ithaca, NY 14853 (607) 255-8972

Jacob M. Levine#
170-05 Cedarcroft Rd.
Jamaica, NY 11432 (718) 657-2504

Gregory J. Mason
Long Island Rail Rd.
Jamaica Station
Jamaica, NY 11435 (718) 990-8288

Marianne R. Rich
184-52 Radnor Rd.
Jamaica, NY 11432 (212) 380-0576

Ronald J. Baron
Hoffman & Baron
350 Jericho Turnpike
Jericho, NY 11753 (516) 822-3550

Rocco S. Barrese
Hoffman, Dilworth, Barrese & Baron
350 Jericho Turnpike
Jericho, NY 11753 (516) 822-3550

Gerald T. Bodner
Hoffman & Baron
350 Jericho Turnpike
Jericho, NY 11753 (516) 822-3550

Herbert L. Boettcher
Nolte, Nolte & Hunter, P.C.
350 Jericho Turnpike
Jericho, NY 11753 (516) 935-0180

George J. Brandt, Jr.
Hoffman, Dilworth, Barrese & Baron
350 Jericho Turnpike
Jericho, NY 11753 (516) 822-3550

Peter G. Dilworth
Hoffman, Dilworth, Barrese & Baron
350 Jericho Turnpike
Jericho, NY 11753 (516) 822-3550

James Albert Eisenman
Hoffman & Baron
350 Jericho Turnpike
Jericho, NY 11753 (516) 822-3550

Irving N. Feit
Hoffman & Baron
350 Jericho Turnpike
Jericho, NY 11753 (516) 822-3550

Burton S. Heiko
15 Steuben Dr.
Jericho, NY 11753 (516) 938-1929

Charles R. Hoffman
Hoffman & Baron
350 Jericho Turnpike
Jericho, NY 11753 (516) 822-3550

Edward B. Hunter
Nolte, Nolte & Hunter, P.C.
350 Jericho Turnpike
Jericho, NY 11753 (516) 935-0180

James C. Jangarathis
Two Jericho Plaza
Jericho, NY 11753 (516) 939-6161

A. Thomas Kammer#
Hoffman & Baron
350 Jericho Turnpike
Jericho, NY 11753 (516) 822-3550

Frank Makara
29 Orange Dr.
Jericho, NY 11753 (516) 938-7744

Arlene D. Morris
Hoffman & Baron
350 Jericho Turnpike
Jericho, NY 11753 (516) 822-3550

Albert Charles NolteJr
Nolte & Nolte, P.C.
350 Jericho Turnpike
Jericho, NY 11753 (516) 935-0180

Alan M. Sack
Hoffman & Baron
350 Jericho Turnpike
Jericho, NY 11753 (516) 822-3550

Leonard William Suroff
12 Turnpike Ave.
Jericho, NY 11753 (516) 681-3900

Dominick G. Vicari
Hoffman & Baron
350 Jericho Turnpike
Jericho, NY 11753 (516) 822-3550

Bernard Albert Chiama
P.O. Box 477
Keuka Park, NY 14478 (315) 536-8486

Fred I. Nathanson
Apt. 403, 123-33 83rd St.
Kew Gardens, NY 11415 (718) 793-1986

Max Moskowitz
144-15 70 Rd.
Kew Gardens Hills, NY 11367
(212) 793-3726

George Edmund Clark
I B M Corp. Intellectual Property Law
Neighborhood Rd.
Kingston, NY 12401 (914) 385-4833

Joseph J. Connerton
I.B.M. Corp.
Neighborhood Rd.
Kingston, NY 12401 (914) 385-7055

John D. Moore
I B M Corp. Pat. Operation - R54 058
Neighborhood Rd.
Kingston, NY 12401 (914) 385-4834

Frederick D. Poag
I B M Corp. Pat. Operation R 54-058
Neighborhood Rd.
Kingston, NY 12401 (914) 385-7057

Mark S. Walker
I B M Corp.
Intellectual Property Law, R54/058,
Neighborhood Rd.
Kingston, NY 12401 (914) 385-7057

Robert E. Heslin
Heslin & Rothenberg, P.C.
8 Wade Rd.
Latham, NY 12110 (518) 785-5507

James J. Lichiello
Heslin & Rothenberg
8 Wade Rd., P.O. Box 310
Latham, NY 12110 (313) 661-1936

Kevin P. Radigan
Heslin & Rothenberg, P.C.
8 Wade Rd. P.O. Box 310
Latham, NY 12110 (518) 785-5507

Jeffrey Rothenberg
Heslin & Rothenberg, P.C.
8 Wade Rd. P.O. Box 310
Latham, NY 12110 (518) 785-5507

Benjamin Theodore Sporn
49 Lismore Rd.
Lawrence, NY 11559 (516) 239-6932

Karl W. Brownell#
745 Hillview Ct.
Lewiston, NY 14092 (716) 754-7047

Peter F. Casella#
987 Elliott Dr.
Lewiston, NY 14092 (716) 297-3658

Donald C. Studley
686 Scovill Dr.
Lewiston, NY 14092 (716) 754-8383

Mario D Arrigo
209 Second St.
Liverpool, NY 13088 (315) 451-2383

Robert J. Jarvis
20 Collegeview Dr.
Loudonville, NY 12111 (518) 458-2286

Nicholas Noviello, Jr.
Route 6
Mahopac, NY 10541 (914) 628-4400

Robert Black
127 Eagles Crescent
Manhasset, NY 11030 (516) 484-3423

William V. Pesce
428 Broadway
Massapequa Park, NY 11762 (516) 799-4077

Joseph P. Hammond
Kollmorgen Corp.
P C K Technology Div., 322 S. Service Rd.
Melville, NY 11747 (516) 454-4410

Raymond J. Keogh#
P C K Technology Div.
322 S. Service Rd., Kollmorgen Corp.
Melville, NY 11747 (516) 454-4541

Edward Herbert Loveman
150 Broad Hollow Rd. Suite 315
Melville, NY 11747 (516) 421-1122

John F. Mc Cormack#
P C K Technology Div.
322 S. Service Rd.
Melville, NY 11747 (516) 454-4511

Emma Shleifer#
11 Montrose Place
Melville, NY 11747 (516) 491-7139

Jacob Frederick Murbach
18 Alfred Rd. W.
Merrick, NY 11566 (516) 379-2776

Philip Sands
2156 Seneca Dr. South
Merrick, NY 11566 (516) 378-6264

Jack D. Slobod
Markovits, Blustein, Gottlieb & Slobod
One North St. Box 2025
Middletown, NY 10940 (914) 343-0505

Myron Amer
114 Old Country Rd.
Mineola, NY 11501 (516) 742-5290

Jerome Bauer
Bauer & Amer. P.C.
114 Old Country Rd.
Mineola, NY 11501 (516) 746-1291

William D. Denson
190 Willis Ave.
Mineola, NY 11501 (516) 747-0300

Joseph L. Grosso
Grosso, Cordaro & Petrilli
114 Old Country Rd.
Mineola, NY 11501 (516) 248-1150

Betty A. Maier#
153 Andrews Rd.
Mineola, NY 11501 (516) 746-6235

Allen Roosevelt Morganstern
98 Willis Ave.
Mineola, NY 11501 (516) 742-9000

Richard Raab#
198 Roselle St.
Mineola, NY 11501 (516) 248-7095

Murray Schaffer
114 Old Country Rd.
Mineola, NY 11501 (516) 248-1050

Richard P. Fennelly
1386 Quarry Dr.
Mohegan Lake, NY 10547 (914) 962-5026

Louis B. Seidman
One Langeries Dr.
Monsey, NY 10952 (914) 356-0316

Sandra M. Kotin
Ingber & Lagarenne
230 Broadway
P.O. Box 111
Monticello, NY 12701 (914) 794-4400

Nathaniel G. Sims
Tall Timber Rd. R.D. 3
Mt. Kisco, NY 10549 (914) 666-4929

Albert F. Kronman
142 N. Columbus Ave.
Mt. Vernon, NY 10553 (914) 667-6755

Thomas R. Morrison
142 N. Columbus Ave.
Mt. Vernon, NY 10553 (914) 667-6755

Emmanuel J. Lobato#
First National Bank Bldg.
49 Beekman Ave.
N. Tarrytown, NY 10591 (914) 631-7177

Arthur Dresner
16 Sundbury Dr.
New City, NY 10956 (914) 634-6286

Aaron B. Karas
28 Woodside Dr.
New City, NY 10956 (914) 634-8020

Robert Frederick Dropkin
Special Metals Corp.
Middle Settlement Rd.
New Hartford, NY 13413 (315) 798-2023

Ralph M. Savage
42 Clinton Rd.
New Hartford, NY 13413 (315) 797-4777

Robert J. Speidel
Five Laurelwood Rd.
New Hartford, NY 13413 (315) 732-2565

Harvey Citrin
P.O. Box 3485
New Hyde Park, NY 11040 (516) 574-3226

David M. Rosenblum
170 Evans St.
New Hyde Park, NY 11040 (516) 328-7317

Kurt Eugene Richter
Morgan, Finnegan, Pine, Foley & Lee
345 Park Ave.
New Park, NY 10154 (212) 758-4800

Howard I. Podell#
28 Beachfront Lane
New Rochelle, NY 10805 (914) 235-0641

William Harold Saltzman
2 Trenor Dr.
New Rochelle, NY 10804 (914) 965-3110

Samuel B. Abrams
Pennie & Edmonds
1155 Ave. Of The Americas
New York, NY 10036 (212) 790-9090

Peter A. Abruzzese
ITT Corp.
320 Park Ave.
New York, NY 10022 (212) 940-1618

Bruce L. Adamas
Burns, Lobato & Adams. P.C.
140 Cedar St. Suite 1805
New York, NY 10006 (212) 608-0190

Andrew P. Adler
Fish & Neave
875 Third Ave.
New York, NY 10022 (212) 715-0600

David Aker
Gottlieb, Rackman & Reisman
1430 Broadway
New York, NY 10017 (212) 892-2890

Lawrence C. Akers
Pfizer Inc.
235 E. 42nd St.
New York, NY 10017 (212) 573-7743

Lawrence Alaburda
Fitzpatrick, Cella, Harper & Scinto
277 Park Ave. 42nd Floor
New York, NY 10172 (212) 758-2400

Robert F. Alario
Carter-Wallace, Inc.
767 Fifth Ave.
New York, NY 10153 (212) 758-4500

Ronald W. Alice
ITT Corp.
320 Park Ave.
New York, NY 10022 (212) 940-1622

Prabodh I. Almaula
Bristol Myers Co.
345 Park Ave.
New York, NY 10154 (212) 546-3649

Stanley Louis Amberg
Davis, Hoxie, Faithfull & Hapgood
45 Rockefeller Plaza
New York, NY 10111 (212) 757-2200

Philip D. Amins
1776 Broadway
New York, NY 10018 (212) 765-5800

Morton Amster
Amster, Rothstein & Engelberg
90 Park Ave.
New York, NY 10016 (212) 697-5995

Steven M. Amundson
Curtis, Morris & Safford, P.C.
530 Fifth Ave.
New York, NY 10036 (212) 840-3333

Richard J. Ancel
Colgate-Palmolive Co.
300 Park Ave.
New York, NY 10022 (212) 310-2959

John C. Andres
Morgan & Finnegan
345 Park Ave.
New York, NY 10154 (212) 758-4800

Adriane M. Antler#
Pennie & Edmonds
1155 Ave. Of The Americas
New York, NY 10036 (212) 790-9090

Steven Anzalone
Kenyon & Kenyon
One Broadway
New York, NY 10004 (212) 425-7200

Lawrence E. Apolzon
Weiss, David, Fross, Zelnick & Lehrman, P.C.
750 Third Ave., 25th Floor
New York, NY 10017 (212) 953-9090

Henry W. Archer
5 Tudor City Place
New York, NY 10017 (212) 687-8341

Seth J. Atlas
Morgan & Finnegan
345 Park Ave.
New York, NY 10154 (212) 758-4800

Peter D. Aufrichtig
300 E. 42nd St.
New York, NY 10017 (212) 557-5040

M. Arthur Auslander
Auslander & Thomas
505 Eighth Ave.
New York, NY 10018 (212) 594-6900

Jack Babchik
Bergadano, Zichello & Babchik
420 Lexington Ave.
New York, NY 10176 (212) 972-5560

George E. Badenoch
Kenyon & Kenyon
One Broadway
New York, NY 10004 (212) 425-7200

James William Badie
Wyatt, Gerber, Shoup, Scobey & Badie
261 Madison Ave.
New York, NY 10016 (212) 687-0911

Bradford J. Badke
Kenyon & Kenyon
One Broadway
New York, NY 10004 (212) 425-7200

Robert Louis Baechtold
Fitzpatrick, Cella, Harper, & Scinto
277 Park Av.
New York, NY 10172 (212) 758-2400

W. Edward Bailey
Fish & Neave
875 Third Ave.
New York, NY 10022 (212) 715-0600

Charles P. Baker
Fitzpatrick, Cella, Harper & Scinto
277 Park Ave.
New York, NY 10172 (212) 758-2400

Victor N. Balancia
Pennie & Edmonds
1155 Ave. Of The Americas
New York, NY 10036 (212) 790-6541

Geraldine F. Baldwin
Pennie & Edmonds
1155 Ave. Of The Americas
New York, NY 10036 (212) 790-9090

Edmond R. Bannon#
Robin, Blecker & Daley
330 Madison Ave.
New York, NY 10017 (212) 682-9640

Richard M. Barnes
Fish & Neave
875 Third Ave.
New York, NY 10022 (212) 715-0600

Elizabeth Marion Barnhard
Abelman, Frayne, Rezac & Schwab
708 Third Ave.
New York, NY 10017 (212) 949-9022

Steven J. Baron
685 Third Ave.
New York, NY 10017 (212) 878-6231

David K. Barr
Fish & Neave
875 Third Ave.
New York, NY 10022 (212) 715-0600

Richard S. Barth
Sprung Horn Kramer & Woods
600 Third Ave.
New York, NY 10016 (212) 661-0520

Mark T. Basseches
Colvin, Miskin, Basseches & mandelbaum
420 Lexington Ave.
New York, NY 10170 (212) 682-7280

Paul T. Basseches#
Colvin, Miskin, Basseches & Mandelbaum
420 Lexington Ave.
New York, NY 10170 (212) 682-7280

William L. Batjer
Hopgood, Calimafe, Kalil, Balaustein & Judlowe
60 E. 42nd St.
New York, NY 10165 (212) 986-2480

Charles P. Bauer
Kane, Dalsimer, Kane, Sullivan & Kurucz
420 Lexington Ave.
New York, NY 10170 (212) 687-6000

Laura Anne Bauer
Fitzpatrick, Cella, Harper & Scinto
277 Park Ave.
New York, NY 10172 (212) 758-2400

Walter Jacob Baum
I.T.T. Corp.
320 Park Ave.
New York, NY 10022 (212) 940-1636

Charles Elliott Baxley
Hart, Baxley, Daniels & Holtol
84 Williams St.
New York, NY 10038 (212) 688-0380

Steven H. Bazerman
Moore, Berson, Lifflander & Mewhinney
595 Madison Ave.
New York, NY 10022 (212) 838-0600

Thomas Augustus Beck
Felfe & Lynch
805 Third Ave.
New York, NY 10022 (212) 688-9200

Peter H. Behrendt
100 W. 12th St.
New York, NY 10010 (212) 989-0174

Bernard Belkin
180 E. 79th St.
New York, NY 10021 (212) 988-8058

Vito Victor Bellino
American Home Products Corp.
685 Third Ave.
New York, NY 10017 (212) 878-6255

Leora Ben-Ami
Dewey, Ballatine, Bushby, Palmer & Wood
140 Broadway
New York, NY 10005 (212) 820-1186

Donald R. Bentz
Curtis, Morris & Safford, P.C.
530 5th Ave.
New York, NY 10036 (212) 840-3333

Egon E. Berg
American Home Products Co.
685 3rd Ave.
New York, NY 10017 (212) 878-6230

Michael N. Berg
Darby & Darby
405 Lexington Ave.
New York, NY 10174 (212) 697-7660

Michael J. Berger
Amster, Rothstein & Ebenstein
90 Park Ave.
New York, NY 10016 (212) 697-5995

Peter Lewis Berger
Levisohn Lerner & Berger
535 5th Ave.
New York, NY 10017 (212) 697-8520

George B. Berka#
Striker, Striker & Stenby
360 Lexington Ave.
New York, NY 10017 (212) 687-5068

Richard G. Berkley
Brumbaugh, Graves, Donohue & Raymond
30 Rockefeller Plaza
New York, NY 10111 (212) 408-2554

Jerome M. Berliner
Ostrolenk, Faber, Gerb, & Soffen
260 Madison Ave.
New York, NY 10016 (212) 685-8470

Richard Kent Bernstein
555 Madison Ave.
New York, NY 10022 (212) 750-0544

E. Janet Berry
274 Madison Ave. Room 401
New York, NY 10016 (212) 679-0581

William I. Bertsche
11 Broadway
New York, NY 10004 (212) 344-2930

Vincent F. Bick, Jr.
Exxon Corp.
1251 Ave. Of Americas Room 4804
New York, NY 10020 (212) 333-6416

Jordon B. Bierman
Bierman & Muserlian
757 Third Ave.
New York, NY 10017 (212) 752-7550

Linda G. Bierman
Synerflex Group, Inc.
215 E. 59th St.
New York, NY 10022 (212) 486-5300

Patrick J. Birde
Kenyon & Kenyon
One Broadway
New York, NY 10004 (212) 425-7200

Donald T. Black
Revlon, Incorporation
767 Fifth Ave.
New York, NY 10153 (212) 572-8572

Charles A. Blank
Felfe & Lynch
805 Third Ave.
New York, NY 10022 (212) 688-9200

Paul H. Blaustein
Hopgood, Calimafde, Kalil, Blaustein & Judlowe
60 E. 42nd St. 40th Floor
New York, NY 10165 (212) 986-2480

Herbert Blecker
Robin Blecker & Daley
330 Madison Ave.
New York, NY 10017 (212) 682-9640

Ronald A. Bleker
W.R. Grace & Co.
1114 Ave. Of The Americas
New York, NY 10036 (212) 819-5580

Sharon A. Blinkoff
Bristol-Myers Co.
345 Park Ave.
New York, NY 10154 (212) 546-3167

Edward M. Blocker
Blum, Kaplan, Friedman, Silberman & Beran
1120 Ave. Of The Americas
New York, NY 10036 (212) 704-0400

Israel Blum
Morgan & Finnegan
345 Park Ave.
New York, NY 10154 (212) 758-4800

Norman Blumenkopf
33 East 85th St.
New York, NY 10028 (212) 988-0319

William Tilden Boland, Jr.
Kenyon & Kenyon
One Broadway
New York, NY 10004 (212) 425-7200

James M. Bollinger
Hopgood, Calimafde, Kalil & Blaustein
60 E. 42nd St.
New York, NY 10165 (212) 986-2480

Jay A. Bondell
Wolder, Gross & Yavner
41 E. 42nd St.
New York, NY 10017 (212) 687-3233

Stevan J. Bosses
Fitzpatrick, Cella, Harper & Scinto
277 Park Ave.
New York, NY 10172 (212) 758-2400

Alan T. Bowes
Kenyon & Kenyon
One Broadway
New York, NY 10004 (212) 425-7200

Douglas G. Brace
Kenyon & Kenyon
One Broadway
New York, NY 10004 (212) 425-7200

Charles W. Bradley, Jr.
Davis, Hoxie, Faithfull & Hapgood
45 Rockefeller Plaza
New York, NY 10111 (212) 757-2200

James W. Brady, Jr.#
Fish & Neave
875 Third Ave.
New York, NY 10022 (212) 715-0608

Abraham M. Bragin
504 Grand St.
New York, NY 10002 (212) 982-0131

Charles R. Brainard
Kenyon & Kenyon
One Broadway
New York, NY 10004 (212) 425-7200

Bradford S. Breen
Davis, Hoxie, Faithfull & Hapgood
45 Rockefeller Plaza
New York, NY 10111 (212) 757-2200

Albert J. Breneisen
Kenyon & Kenyon
One Broadway
New York, NY 10004 (212) 425-7200

Sindey R. Bresnick
Fitzpatrick, Cella, Harper & Scinto
277 Park Ave.
New York, NY 10172 (212) 758-2400

Jeffrey K. Brinck
Milbank, Tweed, Hadlay & Mc Cloy
1 Chase Manhattan Plaza 46th Floor
New York, NY 10005 (212) 530-5280

Joseph J. Brindisi
Morgan & Finnegan
345 Park Ave.
New York, NY 10154 (212) 758-4800

Richard Harold Brink
Bristol-Myers Co.
345 Park Ave.
New York, NY 10154 (212) 546-3637

Marilyn Brogan
Curtis, Morris & Safford, P.C.
530 Fifth Ave.
New York, NY 10036 (212) 840-3333

Adam L. Brookman#
Curtis, Morris & Safford
530 Fifth Ave.
New York, NY 10036 (212) 840-3333

Lorimer P. Brooks
Brooks, Haidt, Haffner & Delahunty
99 Park Ave.
New York, NY 10016 (212) 697-3355

Jay M. Brown
Davis, Hoxie, Faithfull & Hapgood
45 Rockefeller Plaza
New York, NY 10111 (212) 757-2200

G.M. Brumbaugh, Sr.
Brumbaugh, Graves, Donohue & Raymond
30 Rockefeller Plaza
New York, NY 10111 (212) 757-2200

Granville M. Brumbaugh, Jr.
Brumbaugh, Graves, Donohue & Raymond
30 Rockefeller Plaza
New York, NY 10112 (802) 496-3549

William John Brunet
Fitzpatrick, Cella, Harper & Scinto
277 Park Ave.
New York, NY 10172 (212) 758-2400

Charles E. Bruzga
Ostrolenk, Faber, Gerb, & Goffen
260 Madison Ave.
New York, NY 10016 (212) 685-8470

James F. Bryan
Cluett, Peabody & Co. Inc.
530 Fifth Ave.
New York, NY 10036 (212) 930-3025

Peter H. Bucci
Davis, Hoxie, Faithfull & Hapgood
45 Rockefeller Plaza
New York, NY 10111 (212) 757-2200

Richard S. Bullitt
Fish & Neave
875 Third Ave.
New York, NY 10022 (212) 715-0600

Harvey E. Bumgardner, Jr.
80 Eighth Ave. Suite 303
New York, NY 10011 (212) 243-7763

Henry T. Burke
Wyatt, Gerber, Shoup, Scobey & Badie
261 Madison Ave.
New York, NY 10016 (212) 687-0911

Jacob B. Burke
25 Central Park W. 11F
New York, NY 10023 (212) 245-8657

Robert E. Burns
Burns, Lobato & Adams, P.C.
140 Cedar St.
New York, NY 10006 (212) 608-0190

Daniel N. Calder
Amster, Rothstein & Ebenstein
90 Park Ave.
New York, NY 10016 (212) 697-5995

John Michael Calimafde
Hopgood, Calimafde, Kalil, Blaustein & Lieberman
60 E. 42nd St., Lincoln Bldg. Suite 4004
New York, NY 10165 (212) 986-2480

Edward T. Callahan
225 Central Pk. West, Suite 1010
New York, NY 10023 (212) 595-5315

Joseph A. Calvaruso
Morgan & Finnegan
345 Park Ave.
New York, NY 10154 (212) 758-4800

Francis T. Carr
Kenyon & Kenyon
One Broadway
New York, NY 10004 (212) 425-7200

John J. Carrara
Westvaco Corp.
299 Park Ave.
New York, NY 10171 (212) 688-5000

David Michael Carter
Morgan & Finnegan
345 Park Ave.
New York, NY 10154 (212) 758-4800

Thomas G. Carulli
Cooper, Dunham, Griffin & Moran
30 Rockefeller Plaza
New York, NY 10112 (212) 977-9550

Frederick C. Carver
Brumbaugh, Graves, Donohue & Raymond
30 Rockefeller Plaza
New York, NY 10112 (212) 408-2528

Anthony J. Casella
274 Madison Ave.
New York, NY 10016 (212) 725-2450

Richard L. Catania#
Cooper, Dunham, Griffin & Moran
30 Rockefeller Plaza
New York, NY 10112 (212) 977-9550

Joseph J. Catanzaro
Pennie & Edmonds
1155 Ave. Of The Americas
New York, NY 10036 (212) 790-6352

Clario Ceccon#
McGlew & Tuttle, P.C.
28 W. 44th St.
New York, NY 10036 (212) 840-6350

John Thomas Cella
Fitzpatrick, Cella, Harper & Scinto
277 Park Ave.
New York, NY 10172 (212) 758-2400

Jules Louis Chaboty#
260 Seaman Ave. Apt. 7-4
New York, NY 10033 (212) 567-8594

Michael I. Chakansky
White & Case
1155 Ave. Of The Americas
New York, NY 10036 (212) 819-8290

Christopher E. Chalson
Morgan & Finnegan
345 Park Ave.
New York, NY 10154 (212) 758-4800

Hugh A. Chapin
Kenyon & Kenyon
One Broadway
New York, NY 10004 (212) 425-7200

Marshall J. Chick
Frishauf, Holtz, Goodman & Woodward, P.C.
261 Madison Ave.
New York, NY 10016 (212) 972-1400

Edward G.H. Chin
Hertzog, Calamari & Gleason
100 Park Ave.
New York, NY 10017 (212) 481-9500

Leighton K.M. Chong
186 Hester St. Apt. 3
New York, NY 10013 (212) 226-6560

Joseph H. Church#
552 W. 184th St.
New York, NY 10033 (212) 928-0828

Jay Saul Cinamon
Abelman, Frayne, Rezac & Schawb
708 Third Ave,
New York, NY 10017 (212) 949-9022

Lester W. Clark
Cooper, Dunham, Griffin & Moran
30 Rockefeller Plaza
New York, NY 10112 (212) 977-9550

Richard S. Clark
Brumbaugh, Graves, Donohue & Raymond
30 Rockefeller Plaza
New York, NY 10112 (212) 408-2558

Kevin Barry Clarke
Carter-Wallace, Inc.
767 Fifth Ave.
New York, NY 10153 (212) 758-4500

Ronald A. Clayton
Fitzpatrick, Cella, Harper & Scinto
277 Park Ave.
New York, NY 10172 (212) 758-2400

Peter Timothy Cobrin
Cobrin & Godsberg, P.C.
366 Madison Ave.
New York, NY 10017 (212) 687-6090

Nicholas Lazaros Coch
Anderson, Russell, Kill, & Olick, P.C.
666 Third Ave.
New York, NY 10017 (212) 850-0743

Joseph S. Codispoti
Kenyon & Kenyon
One Broadway
New York, NY 10004 (212) 425-7200

Brian D. Coggio
Pennie & Edmonds
1155 Sixth Ave.
New York, NY 10036 (212) 790-9090

Julian Harris Cohen
Roberts, Spiecens & Cohen
38 E. 29th St. Third Floor
New York, NY 10016 (212) 684-7766

Martin J. Cohen
Cohen & Silverman
500 Fifth Ave. Suite 3000
New York, NY 10110 (212) 944-9770

Myron Cohen
Cohen, Pontani & Lieberman
551 Fifth Ave. Suite 1210
New York, NY 10176 (212) 687-2770

J. Bradley Cohn
314 E. 41 St.
New York, NY 10017 (212) 697-5088

Christine Cole
80 Wall Street Suite 415
New York, NY 10005 (212) 425-3158

Henry D. Coleman
DeRosa, Vandenberg & Glenman
71 Broadway
New York, NY 11201 (212) 269-3020

Arthur B. Colvin
420 Lexington Ave.
New York, NY 10170 (212) 986-4056

Barry Arnold Cooper
Gottlieb, Rackman & Reisman
1430 Broadway
New York, NY 10018 (212) 869-2890

Gordon D. Coplein
Darby & Darby
405 Lexington Ave.
New York, NY 10174 (212) 692-7660

Michael Alexander Corman
Mandeville & Schweitzer
230 Park Ave.
New York, NY 10169 (212) 986-3377

Thomas Johnson Corum
Colgate-Palmolive Co.
300 Park Ave.
New York, NY 10022 (212) 310-3145

Laura A. Coruzzi
Pennie & Edmonds
1155 Ave. Of The Americas
New York, NY 10036 (212) 790-6431

James V. Costigan
Hedman, Gibson, Costigan & Hoare, P.C.
1185 Ave. Of The Americas
New York, NY 10036 (212) 302-8989

Abigail F. Cousins#
Eslinger & Pelton, P.C.
600 Third Ave.
New York, NY 10016 (212) 972-4433

Thomas L. Creel
Kenyon & Kenyon
One Broadway
New York, NY 10004 (212) 425-7200

Richard L. Crisona
Duker & Barrett
90 Broad St.
New York, NY 10004 (212) 809-7700

Robert Gladden Crooks
American Standard, Inc.
40 West 40th St.
New York, NY 10018 (212) 703-5141

Alfred A. D Andrea, Jr.
Levisohn, Lerner & Berger
535 Fifth Ave.
New York, NY 10017 (212) 692-8520

Felix L. D Arienzo, Jr.
Mandeville & Schweitzer
230 Park Ave. Suite 2200
New York, NY 10169 (212) 986-3377

J. Robert Dailey
Morgan & Finnegan
345 Park Ave.
New York, NY 10154 (212) 758-4800

J. David Dainow
Rosen, Dainow & Jacobs
489 5th Ave.
New York, NY 10017 (212) 692-7000

James J. Daley
Robin, Blecker & Daley
330 Madison Ave.
New York, NY 10017 (212) 682-9460

Lynne Darcy
Lever Brothers Co.
390 Park Ave.
New York, NY 10022 (212) 906-4571

Joseph A. De Girolamo
Morgan & Finnegan
345 Park Ave.
New York, NY 10154 (212) 758-4800

Frank J. De Rosa
De Rosa, Vandenberg & Coleman
71 Broadway, Suite 2200
New York, NY 10006 (212) 269-3020

Donald E. Degling
Fish & Neave
875 Third Ave.
New York, NY 10022 (212) 715-0600

G. Thomas Delahunty
Brooks, Haidt, Haffner & Delahunty
99 Park Ave.
New York, NY 10016 (212) 697-3355

Gregg A. Delaporta
Kenyon & Kenyon
One Broadway
New York, NY 10004 (212) 425-7200

Richard L. Delucia
Kenyon & Kenyon
One Broadway
New York, NY 10004 (212) 425-7200

Manette Dennis
Pennie & Edmonds
1155 Ave. Of The Americas
New York, NY 10036 (212) 790-9090

Daniel A. DeVito#
Brumbaugh, Graves, Donohue & Raymond
30 Rockfeller Plaza
New York, NY 10112 (212) 408-2548

Jean K. Dexheimer
Davis, Hoxie, Faithfull & Hapgood
45 Rockefeller Plaza
New York, NY 10111 (212) 757-2200

Leonard P. Diana
Fitzpatrick, Cella, Harper & Scinto
277 Park Ave. 42nd Floor
New York, NY 10172 (212) 758-2400

John Anderw Diaz
Morgan & Finnegan
345 Park Ave.
New York, NY 10154 (212) 758-4800

Gerard F. Diebner
Hopgood, Calimafde, Kalil, Blaustein & Judlowe
60 E. 42nd St.
New York, NY 10165 (212) 986-2480

William H. Dippert
Kane, Dalsimer, Sullivan, Kurucz, Levy, Eisel & Richard
420 Lexington Ave.
New York, NY 10170 (212) 687-6000

Colin J. Dobbyn
Skadder, Arps, Slate, Meagher & Flom
919 Third Ave.
New York, NY 10017

Arnold B. Dompieri
Davis, Hoxie, Faithfull & Hapgood
45 Rockefeller Plaza
New York, NY 10111 (212) 757-2200

Lorraine M. Donaldson
Pfizer, Inc.
235 E. 42nd St.
New York, NY 10017 (212) 573-2858

Frederick J. Dorchak
Fitzpatrick, Cella, Harper & Scinto
277 Park Ave.
New York, NY 10172 (212) 758-2400

Earl M. Douglas
Morgan & Finnegan
345 Park Ave.
New York, NY 10154 (212) 758-4800

Donald Smith Dowden
Eslinger & Pelton, P.C.
600 Third Ave.
New York, NY 10016 (212) 972-4433

Thomas P. Dowling
Morgan & Finnegan
345 Park Ave.
New York, NY 10154 (212) 758-4800

Joel K. Dranove
401 Broadway
New York, NY 10013 (212) 431-7660

Dimitros T. Drivas
Hopgood, Calimafde, Kalil, Blaustein & Judlowe
60 E. 42nd St.
New York, NY 10165 (212) 986-2480

Mark Dryer
Pfizer Inc.
235 E. 42nd St.
New York, NY 10017 (212) 573-7482

Samuel J. DuBoff
Bristol-Myers Co.
345 Park Ave.
New York, NY 10154 (212) 546-3632

William F. Dudine, Jr.
Darby & Darby
405 Lexington Ave.
New York, NY 10174 (212) 697-7660

Louis C. Dujmich
Kenyon & Kenyon
One Broadway
New York, NY 10004 (212) 425-7200

Christopher Cooper Dunham
Cooper, Dunham, Griffin & Moran
30 Rockefeller Plaza
New York, NY 10112 (212) 977-9550

Robert Secrest Dunham
Cooper, Dunham, Clark, Griffin & Moran
30 Rockefeller Plaza
New York, NY 10112 (212) 977-9550

Gerard F. Dunne
Wyatt, Gerber, Shoup, Scobey & Badie
261 Madison Ave.
New York, NY 10016 (212) 687-0911

Kathleen G. Dussault
Kenyon & Kenyon
One Broadway
New York, NY 10004 (212) 425-7200

Leroy Eason
Western Electric Co. Inc.
222 Broadway
New York, NY 10038 (212) 669-4722

Daniel Simon Ebenstein
Amster, Rothstein & Ebenstein
90 Park Ave.
New York, NY 10016 (212) 697-5995

William F. Eberle
Brumbaugh, Graves, Donohue & Raymond
30 Rockefeller Plaza
New York, NY 10112 (212) 408-2358

Michael Ebert
Hopgood, Calimafde, Kalil, Blaustein & Judlowe
60 E. 42nd St.
New York, NY 10165 (212) 986-2480

Joseph T. Eisele
Kane, Dalsimer, Sullivan, Kurucz, Levy, Eisele & Richard
420 Lexington Ave.
New York, NY 10170 (212) 687-6000

Richard A. Elder
American Home Products Corp.
685 Third Ave.
New York, NY 10017 (212) 878-6218

J. David Ellett, Jr.
Pennie & Edmonds
1155 Ave. Of The Americas
New York, NY 10036 (212) 790-9090

Robert L. Epstein
James & Franklin, P.C.
60 E. 42nd St. Suite 1217
New York, NY 10165 (212) 867-7260

Lewis H. Eslinger
Cooper & Dunham
30 Rockefeller Plaza
New York, NY 10112 (212) 977-9550

Barry Leonard Evans
Curtis, Morris & Safford, P.C.
530 5th Ave.
New York, NY 10036 (212) 840-3333

William R. Evans
Ladas & Parry
26 West 61st St.
New York, NY 10023 (212) 708-1945

Alfred P. Ewert
Morgan & Finnegan
345 Park Ave.
New York, NY 10154 (212) 758-4800

Robert Charles Faber
Ostrolenk, Faber, Gerb
260 Madison Ave.
New York, NY 10016 (212) 685-8470

Sidney G. Faber
Ostrolenk, Faber, Gerb & Soffen
260 Madison Ave.
New York, NY 10016 (212) 685-8470

Gary J. Falce
American Standard Inc.
40 West 40th St.
New York, NY 10017

Allan A. Fanucci
Pennie & Edmonds
1155 Ave. Of The Americas
New York, NY 10036 (212) 790-6537

Martin Allen Farber
866 United Nations Plaza
New York, NY 10017 (212) 758-2878

John Farley
Fish & Neave
875 Third Ave.
New York, NY 10022 (212) 715-0600

Henry M. Farnum
225 E. 46th St.
Executive House
New York, NY 10017 (212) 371-8679

Vincent M. Fazzari
Felfe & Lynch
805 Third Ave.
New York, NY 10022 (212) 688-9200

Ronald C. Fedus
Cooper, Dunham, Griffin & Moran
30 Rockefeller Plaza
New York, NY 10112 (212) 977-9550

Henry M. Feiereisen#
275 Madison Ave.
New York, NY 10016 (212) 557-3060

William S. Feiler
Morgan, Finnegan, Pine, Foley & Lee
345 Park Ave.
New York, NY 10154 (212) 758-4800

Aaron M. Feinberg
Enzo Biochem, Inc.
325 Hudson St.
New York, NY 10013 (212) 741-3838

Marvin Feldman
Feldman & Feldman, P.C.
12 E. 41 St.
New York, NY 10017 (212) 532-8585

Stephen Edward Feldman
12 East 41 St.
New York, NY 10017 (212) 532-8585

Peter F. Felfe
Felfe & Lynch
805 Third Ave.
New York, NY 10022 (212) 688-9200

Richard P. Ferrara
Davis, Hoxie, Faithfull & Hapgood
45 Rockefeller Plaza
New York, NY 10111 (212) 757-2200

Albert E. Fey
Fish & Neave
875 Third Ave.
New York, NY 10022 (212) 715-0600

Robert W. Fiddler
Fiddler & Levine
7814 Empire State Bldg.
350 5th Ave.
New York, NY 10118 (212) 279-6088

Paul Fields
McAulay, Fields, Fisher, Goldstein & Nissen
405 Lexington Ave.
New York, NY 10174 (212) 986-4090

Robert D. Fier
Kenyon & Kenyon
One Broadway
New York, NY 10004 (212) 425-7200

Edward V. Filardi
Brumbaugh, Graves, Donohue & Raymond
30 Rockefeller Plaza
New York, NY 10112 (212) 408-2556

James A. Finder
Ostrolenk, Faber, Gerb & Soffen
260 Madison Ave.
New York, NY 10016 (212) 685-8470

George B. Finnegan, Jr.
Morgan, Finnegan, Pine, Foley & Lee
345 Park Ave.
New York, NY 10154 (212) 758-4800

Robert H. Fischer
Fitzpatrick, Cella, Harper & Scinto
277 Park Ave.
New York, NY 10172 (212) 758-2400

Julius Fisher
McAulay, Fields, Fisher, Goldstein & Nissen
The Chrysler Bldg.
405 Lexington Ave.
New York, NY 10174 (212) 986-4090

Barry H. Fishkin
Rathheim, Hoffman, Kassel & Levie
61 Broadway
New York, NY 10006 (212) 943-3100

Joseph M. Fitzpatrick
Fitzpatrick, Cella, Harper & Scinto
277 Park Ave.
New York, NY 10172 (212) 758-2400

Dennis M. Flaherty#
Rosen, Dainow & Jacobs
489 Fifth Ave.
New York, NY 10017 (212) 692-7058

Porter F. Fleming#
Hopgood, Calimafde, Kalil, Blaustein & Judlowe
60 E. 42nd St.
New York, NY 10165 (212) 986-2480

Gerald James Flintoft
Pennie & Edmonds
1155 Ave. Of The Americas
New York, NY 10036 (212) 790-9090

John Aloysius Fogarty, Jr.
Kenyon & Kenyon
One Broadway
New York, NY 10004 (212) 425-7200

John Dennis Foley
Morgan & Finnegan
345 Park Ave.
New York, NY 10154 (212) 758-4800

Frank W. Ford, Jr.
Brumbaugh, Graves, Donohue & Raymond
30 Rockefeller Plaza
New York, NY 10112 (212) 408-2524

Charles D. Forman
General Instrument Corp.
320 W. 57th St.
New York, NY 10018 (212) 974-8772

James D. Fornari
Jarblum, Solomon & Fornari
650 Fifth Ave.
New York, NY 10019 (212) 265-1200

Leo Fornero
The Sanforized Co.
530 5th Ave.
New York, NY 10036 (212) 697-7272

David R. Francescani
Kenny Group, Inc.
65 Broadway
New York, NY 10006 (212) 770-4950

Peter James Franco
28 W. 44th St.
New York, NY 10036 (212) 944-6168

Irene J. Frangos
Fish & Neave
875 Third Ave., 29th Floor
New York, NY 10022 (212) 715-0600

Bertram Frank
521 5th Ave.
New York, NY 10175 (212) 697-7387

Howard M. Frankfort#
Darby & Darby, P.C.
405 Lexington Ave.
New York, NY 10174 (212) 697-7660

Robert M. Freeman
Brooks, Haidt, Haffner & Delahunty
99 Park Ave.
New York, NY 10016 (212) 697-3355

Lawrence G. Fridman#
Darby & Darby
405 Lexington Ave.
New York, NY 10174 (212) 647-7660

Thomas E. Friebel
Pennie & Edmonds
1155 Ave. Of The Americas
New York, NY 10036 (212) 790-9090

Jacob Friedlander
Leboeuf, Lamb, Leiby & Mac Rae
520 Madison Ave.
New York, NY 10022 (212) 715-8000

Abraham Friedman
420 Lexington Ave. Suite 2930
New York, NY 10170 (212) 867-2000

Alex Friedman
Blum, Kaplan, Friedman, Silberman & Beran
1120 Ave. Of The Americas
New York, NY 10036 (212) 704-0400

Morton Friedman
Witco Corp.
520 Madison Ave.
New York, NY 10022 (212) 605-3848

Michael T. Frimer
J.P. Stevens And Co. Inc.
1185 Ave. Of The Americas
New York, NY 10036 (212) 930-2404

Stephen H. Frisauf
Frishauf, Holtz, Goodman & Woodward
261 Madison Ave.
New York, NY 10016 (212) 972-1400

William S. Frommer
Curtis, Morris & Safford, P.C.
530 5th Ave.
New York, NY 10036 (212) 840-3333

Grover F. Fuller, Jr.
Fitzpatrick, Cella, Harper & Scinto
277 Park Ave.
New York, NY 10172 (212) 758-2400

Richard Guerard Fuller, Jr.
Brumbaugh, Graves, Donohue & Raymond
30 Rockefeller Plaza
New York, NY 10112 (212) 408-2520

James W. Galbrith
Kenyon & Kenyon
One Broadway
New York, NY 10004 (212) 425-7200

John James Gallagher, Sr.
La Morte Buyers & Co. Inc.
3147 One World Trade Ctr.
New York, NY 10048 (212) 432-0400

Thomas A. Gallagher
Klein & Vibber, P.C.
274 Madison Ave.
New York, NY 10016 (212) 689-3441

Peter D. Galloway
Ladas & Parry
26 West 61st St.
New York, NY 10023 (212) 708-1905

Melvin C. Garner
Darby & Darby, P.C.
405 Lexington Ave.
New York, NY 10174 (212) 697-7660

Joseph D. Garon
Brumbaugh, Graves, Donohue & Raymond
30 Rockefeller Plaza
New York, NY 10112 (212) 408-2540

Mark E. Garscia
Ostrolenk, Faber, Gerb & Soffen
260 Madison Ave.
New York, NY 10016 (212) 685-8470

Christopher B. Garvey
Fish & Neave
875 Third Ave.
New York, NY 10022 (212) 715-0600

Bradley B. Geist
Brumbaugh, Graves, Donohue & Raymond
30 Rockefeller Plaza
New York, NY 10112 (212) 408-2562

John M. Genova
Hopgood, Calimafde, Kalil, Blaustein & Judlowe
60 E. 42nd St.
New York, NY 10165 (212) 986-2480

Kenneth P. George
Amster, Rothstein & Ebenstein
90 Park Ave.
New York, NY 10016 (212) 697-5995

Eliot S. Gerber
Wyatt, Gerber, Shoup, Scobie & Badie
261 Madison Ave.
New York, NY 10016 (212) 687-0911

Stanley J. Gewirtz#
Solid State Systems Inc.
435 W. 119th St.
New York, NY 10027 (212) 222-2058

Thomas L. Giannetti
Fish & Neave
875 Third Ave.. 29th Floor
New York, NY 10022 (212) 715-0600

Mark D. Giarrantana#
Kenyon & Kenyon
One Broadway
New York, NY 10004 (212) 425-7200

Thomas Martin Gibson
Hedman, Gibson, Costigan & Hoare, P.C.
1185 Ave. Of The Americas
New York, NY 10036 (212) 302-8989

Douglas J. Gilbert
Fish & Neave
875 Third Ave.
New York, NY 10022 (212) 715-0652

Stephen Paul Gilbert
Stiefel, Gross & Kurland, P.C.
551 Fifth Ave.
New York, NY 10176 (212) 687-1360

William J. Gilbert
Fish & Neave
875 Third Ave.
New York, NY 10022 (212) 715-0600

Alan D. Gilliland
Curtis, Morris, & Safford
530 5th Ave.
New York, NY 10036 (212) 840-3333

Theresa M. Gillis
Dewey, Ballantine, Bushby, Palmer & Wood
140 Broadway
New York, NY 10005 (212) 820-1635

Paul H. Ginsburg
Pfizer Inc.
235 E. 42nd St.
New York, NY 10017 (212) 573-2369

Ann L. Gisolfi
Stiefel, Gross & Kurland, P.C.
551 Fifth Ave.
New York, NY 10176 (212) 687-1360

Howard M. Gitten
Blum, Kaplan, Friedman, Silberman & Beran
1120 Ave. Of The Americas
New York, NY 10036 (212) 873-8633

Franklin M. Gittes
Skadden, Arps, Slate, Meagher & Flom
919 Third Ave.
New York, NY 10022 (212) 735-3760

Marvin S. Gittes
366 Madison Ave.
New York, NY 10017 (212) 949-8787

Steven D. Glazer
Davis, Hoxie, Faithfull & Hapgood
45 Rockefeller Plaza
New York, NY 10111 (212) 757-2200

Stephen C. Glazier
Greenberg, Irwin, Weisinger, P.C.
540 Madison Ave. 7th Floor
New York, NY 10022 (212) 838-6670

Adda C. Gogoris
Darby & Darby
405 Lexington Ave.
New York, NY 10174 (212) 697-7660

Jules Edward Goldberg
McAulay, Fields, Fisher, Goldstein & Nissen,
Chrysler Bldg.
405 Lexington Ave.
New York, NY 10174 (212) 986-4090

S. Delvalle Goldsmith
Ladas & Parry
26 West 61st St.
New York, NY 10023 (212) 708-1910

Martin E. Goldstein
McAulay, Fields, Fisher, Goldstein & Nissen
405 Lexington Ave.
New York, NY 10174 (212) 986-4090

Michael A. Gollin
Sive, Paget & Riesel
460 Park Ave. 10th Floor
New York, NY 10022 (212) 421-2150

Donald J. Goodell
Pennie & Edmonds
1155 Ave. Of The Americas
New York, NY 10036 (212) 790-6307

Edward W. Goodman
Toren, McGeady & Associates P.C.
521 5th Ave.
New York, NY 10175 (212) 867-2912

Herbert H. Goodman
Frishauf, Holtz, Goodman & Woodward, P.C.
261 Madison Ave.
New York, NY 10016 (212) 972-1400

Beverly B. Goodwin
Darby & Darby, P.C.
405 Lexington Ave.
New York, NY 10174 (212) 697-7660

Lawrence B. Goodwin
Davis, Hoxie, Faithfull & Hapgood
45 Rockefeller Plaza
New York, NY 10111 (212) 757-2200

Jennifer Gordon
Pennie & Edmonds
1155 Ave. Of The Americas
New York, NY 10036 (212) 790-9090

Marvin Norman Gordon
Hopgood, Calimafde, Kalil, Blaustein & Lowe
60 E. 42nd St.
New York, NY 10165 (212) 986-2480

Philip H. Gottfried
Amster, Rothstein & Ebenstein
90 Park Ave.
New York, NY 10016 (212) 697-5995

George Gottlieb
Gottlieb, Rackman & Reisman
1430 Broadway
New York, NY 10018 (212) 869-2890

James W. Gould
Morgan & Finnegan
345 Park Ave.
New York, NY 10154 (212) 758-4800

Eben M. Graves
Brumbaugh, Graves, Donohue & Raymond
30 Rockefeller Plaza
New York, NY 10112 (212) 408-1500

Arthur D. Gray
Kenyon & Kenyon
One Broadway
New York, NY 10004 (212) 425-7200

William O. Gray, III
Ostrolenk, Faber, Gerb & Soffen
260 Madison Ave.
New York, NY 10016 (212) 685-8470

Edward W. Greason
Kenyon & Kenyon
One Broadway
New York, NY 10004 (212) 425-7200

Orville N. Greene
Greene & Durr Rm. 4410
10 E. 40th St.
New York, NY 10016 (212) 686-9009

Ernest A. Greenside
Greenside & Schaffer, P.C.
875 Ave. Of The Americas
New York, NY 10001 (212) 868-1790

Richard S. Gresalfi
Kenyon & Kenyon
One Broadway
New York, NY 10004 (212) 425-7200

Gerald W. Griffin
Cooper, Dunham, Griffin & Moran
30 Rockefeller Plaza
New York, NY 10112 (212) 977-9550

Murray M. Grill
Colgate-Palmolive Co.
300 Park Ave.
New York, NY 10022 (212) 310-3135

Marc S. Gross
Stiefel, Gross & Kurland, P.C.
551 Fifth Ave.
New York, NY 10176 (212) 687-1360

Meyer A. Gross
41 E. 42 St.
New York, NY 10017 (212) 687-3232

Joseph H. Guth#
Cooper, Dunham, Clark, Griffin & Moran
30 Rockefeller Plaza
New York, NY 10112 (212) 977-9550

Charles Guttman
Guttman & Rubenstein
25 West 43rd St. Suite 1412
New York, NY 10036 (212) 382-1112

Gaylord Paul Haas, Jr.
150 E. 52nd St. 25th Floor
New York, NY 10022 (212) 906-3067

Alfred L. Haffner, Jr.
Brooks, Haidt, Haffner & Delahunty
99 Park Ave.
New York, NY 10016 (212) 697-3355

Theodore Hafner
1501 Broadway
New York, NY 10036 (212) 354-7800

Harold Haidt
Brooks, Haidt, Haffner & Delahunty
99 Park Ave.
New York, NY 10016 (212) 697-3355

James F. Haley, Jr.
Fish & Neave
875 Third Ave.
New York, NY 10022 (212) 715-0600

C. Bruce Hamburg
Jordan & Hamburg
122 E. 42nd St.
New York, NY 10168 (212) 986-2340

Thomas M. Hammond
Morgan & Finnegan
345 Park Ave.
New York, NY 10154 (212) 758-4800

Francis C. Hand
Kenyon & Kenyon
One Broadway
New York, NY 10004 (212) 425-7200

Joseph H. Handelman
Ladas & Parry
26 West 61st St.
New York, NY 10023 (212) 708-1880

Edward J. Handler, III
Kenyon & Kenyon
One Broadway
New York, NY 10004 (212) 425-7200

Walter E. Hanley, Jr.
Kenyon & Kenyon
One Bdwy.
New York, NY 10004 (212) 425-7200

Norman D. Hanson
Felfe & Lynch
805 Third Ave.
New York, NY 10022 (212) 688-9200

Cyrus S. Hapgood
Davis, Hoxie, Faithfull & Hapgood
45 Rockefeller Plaza
New York, NY 10111 (212) 757-2200

Stephen J. Harbulak
Pennie & Edmonds
1155 Ave. Of The Americas
New York, NY 10036 (212) 790-9090

Carroll G. Harper
Fitzpatrick, Cella, Harper & Scinto
277 Park Ave.
New York, NY 10172 (212) PL8-2400

Thomas Emil Harrison, Jr.
Gulf & Western Inds., Inc.
1 Gulf & Western Plaza
New York, NY 10023 (212) 333-4918

Paul C. Hashim#
Davis, Hoxie, Faithfull & Hapgood
45 Rockfeller Plaza, 28th Floor
New York, NY 10111 (212) 757-2200

Constance J. Hassberg#
American District Telegraph Co.
1 World Trade Center, Suite 9200
New York, NY 10048 (212) 558-1229

Rita V. Hauck
Squibb Corp.
40 W. 57th St.
New York, NY 10019 (212) 621-7083

Edger H. Haug
Curtis, Morris & Safford, P.C.
530 Fifth Ave.
New York, NY 10036 (212) 840-3333

Elliott E. Haymovitz
150 E. 58th St. 39th Floor
New York, NY 10154 (212) 891-8197

Edward Alan Hedman
Hedman, Gibson, Costigan & Hoare, P.C.
1185 Ave. Of The Americas
New York, NY 10036 (212) 302-8989

Samson Helfgott
General Electric Co.
570 Lexington Ave.
New York, NY 10022 (212) 750-3569

Paul Harold Heller
Kenyon & Kenyon
One Broadway
New York, NY 10004 (212) 425-7200

Harold C. Herman
Herman, McGinnis & Kass P.C.
149 Madison Ave.
New York, NY 10016 (212) 889-5610

Kenneth B. Herman
Fish & Neave
875 Third Ave.
New York, NY 10022 (212) 715-0600

Gerald E. Hespos
Hedman, Casella, Gibson & Costigan
501 5th Ave.
New York, NY 10017 (212) 687-0340

Robert J. Hess
Striker, Striker & Stenby
360 Lexington Ave.
New York, NY 10017 (212) 687-5068

Thomas V. Heyman
Dewey, Ballantine, Bushby, Palmer & Wood
140 Boradway
New York, NY 10005 (212) 820-1890

Ronald Budd Hildreth
Brumbaugh, Graves, Donohue & Raymond
30 Rockefeller Plaza
New York, NY 10112 (212) 408-2544

Joel Hirschel
253 West 72nd St.
New York, NY 10023 (212) 799-7767

Joseph Hirshfeld
Western Electric Co. Inc.
222 Broadway
New York, NY 10038 (212) 669-2523

George Philip Hoare, Jr.
Hedman, Gibson, Costigan & Hoare, P.C.
1185 Ave. Of The Americas
New York, NY 10036 (212) 302-8989

Frank P. Hoffman
Bristol-Myers Inc.
345 Park Ave.
New York, NY 10154 (212) 546-5698

Lawrence A. Hoffman
Darby & Darby
405 Lexington Ave.
New York, NY 10174 (212) 697-7660

Norman N. Holland
Holland, Armstrong, Wilkie & Previto
Empire State Bldg.
New York, NY 10007 (212) 736-2080

Norbert P. Holler
Gottlieb, Rackman & Reisman, P.C.
1430 Broadway
New York, NY 10018 (212) 869-2890

Gezina Holtrust
Ladas & Parry
10 Columbus Circle
New York, NY 10019 (212) 245-2600

Leonard Holtz
Frishauf, Holtz, Goodman & Woodward, P.C
261 Madison Ave.
New York, NY 10016 (212) 972-1400

Irving Holtzman
Bristol-Myers Co.
345 Park Ave.
New York, NY 10154 (212) 546-3666

Francis J. Hone
Brumbaugh, Graves, Donohue & Raymond
30 Rockefeller Plaza
New York, NY 10112 (212) 408-2534

William J. Hone
Davis, Hoxie, Faithfull & Hapgood
45 Rockefeller Plaza
New York, NY 10111 (212) 757-2200

Roy C. Hopgood, Jr.
Hopgood, Calimafde, Kalil, Blaustein & Judlowe
60 E. 42nd St.
New York, NY 10165 (212) 986-2480

Leonard Horn
Sprung Horn Kramer & Woods
600 3rd Ave.
New York, NY 10016 (212) 661-0520

Steven Horowitz
Bass & Ullman
747 Third Ave.
New York, NY 10017 (212) 751-9494

Ethan Horwitz
Darby & Darby, P.C.
405 Lexington Ave.
New York, NY 10174 (212) 697-7660

William T. Hought
Polachek, Saulsbury & Hough
110 W. 34th St. Suite 601
New York, NY 10001 (212) 563-3088

Donald J. Howard
Mobil Oil Corp.
150 E. 42nd St.
New York, NY 10017 (212) 883-2742

John A. Howson
Fish & Neave
277 Park Ave.
New York, NY 10017 (212) 826-1050

Richard A. Huettner
Kenyon & Kenyon
One Broadway
New York, NY 10004 (212) HA5-7200

Christopher A. Hughes
Morgan & Finnegan
345 Park Ave.
New York, NY 10154 (212) 758-4800

Jeffrey H. Ingerman
Fish & Neave
875 Third Ave.
New York, NY 10022 (212) 715-0600

Richard A. Inz
Fish & Neave
875 Third Ave.
New York, NY 10022 (212) 715-0600

Robert M. Isackson
Fish & Neave
875 Third Ave. 29th Floor
New York, NY 10022 (212) 715-0600

Shahan Islam#
Morgan & Finnegan
345 Park Ave.
New York, NY 10154 (212) 758-4800

Robert E. Isner
Nims, Howes, Collison & Isner
500 Fifth Ave. Suite 3200
New York, NY 10110 (212) 382-1400

Alan Israel
Kirsthstein, Kirschstein, Ottinger & Israel, P.C.
551 5th Ave.
New York, NY 10176 (212) 657-3750

Robert R. Jackson
Fish & Neave
875 Third Ave.
New York, NY 10022 (212) 715-0600

Albert L. Jacobs
Jacobs & Jacobs, P.C.
521 5th Ave.
New York, NY 10175 (212) 687-1636

Albert L. Jacobs, Jr.
Jacobs & Jacobs, P.C.
521 5th Ave.
New York, NY 10175 (212) 687-1636

James David Jacobs
Rosen, Dainow & Jacobs
489 Fifth Ave.
New York, NY 10017 (212) 692-7000

Seth H. Jacobs
Davis, Hoxie, Faithfull & Hapgood
45 Rockefeller Plaza
New York, NY 10111 (212) 757-2200

Allan J. Jacobson
General Instrument Corp.
320 W. 57th St.
New York, NY 10019 (212) 974-8750

Howard R. Jaeger
Pfier Inc.
235 E. 42nd St.
New York, NY 10017 (212) 573-1229

Harold James
James & Franklin
60 E. 42nd St. Suite 1217
New York, NY 10165

Isaac Jarkovsky
Bristol Myers Co.
345 Park Ave.
New York, NY 10154 (212) 546-3653

Marius J. Jason
Felfe & Lynch
805 Third Ave.
New York, NY 10022 (212) 688-9200

Saul Jecies
605 Third Ave., 39th & 40th St.
New York, NY 10158 (212) 972-1100

Jesse J. Jenner
Fish & Neave
875 Third Ave.
New York, NY 10022 (212) 715-0600

Herbert H. Jervis#
Fitzpatrick, Cella, Harper & Scinto
277 Park Ave.
New York, NY 10172 (212) 758-2400

John E. Johnnidis
1010 5th Ave.
New York, NY 10028 (212) 628-0058

Harry Chapman Jones, III
Pennie & Edmonds
1155 Ave. Of The Americas
New York, NY 10036 (212) 790-9090

Frank J. Jordan
Jordan & Hamburg
122 E. 42nd St.
New York, NY 10168 (212) 986-2340

John J. Jordan
Nynex Materiel Enterprs. Co.
441 9th Ave. 7th Floor
New York, NY 10001 (212) 502-7192

Stephen Barry Judlows
Hopgood, Calimafde, Kalil, Blaustein & Judlowe
60 E. 42nd St.
New York, NY 10165 (212) 986-2480

David L. Just
Lucas & Just
205 E. 42nd St.
New York, NY 10017 (212) 682-4980

Jeffrey M. Kaden#
Blum, Kaplan, Friedman, Silberman & Beran
1120 Ave. Of The Americas
New York, NY 10036 (212) 704-0400

Leonard R. Kahn#
137 E. 36th St.
New York, NY 10016 (516) 222-2221

Stephen D. Kahn
Davis, Hoxie, Faithfull & Hapgood
45 Rockefeller Plaza
New York, NY 10111 (212) 757-2200

Eugene J. Kalil
Hopgood, Calimafde, Kalil, Blaustein & Judlowe
Licoln Bldg.
60 E. 42nd St.
New York, NY 10017 (212) YU6-2480

Nicholas N. Kallas
Fitzpatrick, Cella, Harper & Scinto
277 Park Ave.
New York, NY 10172 (212) 758-2400

David A. Kalow
Lieberman Rudolph & Nowak
292 Madison Ave.
New York, NY 10017 (212) 532-4447

Joseph S. Kaming
156 E. 65th St.
New York, NY 10021 (212) 535-0245

Daniel H. Kane
Kane, Dalsimer, Kane, Sullivan & Kurucz
420 Lexington Ave.
New York, NY 10170 (212) 687-6000

David Schilling Kane
Kane, Dalsimer, Kane, Sullivan & Kurucz
420 Lexington Ave.
New York, NY 10170 (212) 687-6000

Manabu Kanesaka#
Jordan & Hamburg
122 E. 42nd St.
New York, NY 10168 (212) 986-2340

Harold I. Kaplan
Blum, Kaplan, Friedman, Silberman & Beran
1120 Ave. Of The Americas
New York, NY 10036 (212) 704-0400

Pamela D. Kasa
7 Charles St.
New York, NY 10014 (212) 546-3040

David S. Kashman
Gottlieb, Rackman & Reisman, P.C.
1430 Broadway
New York, NY 10018 (212) 869-2890

Gabriel P. Katona
Schweitzer & Cornman
230 Park Ave.
New York, NY 10169 (212) 986-3377

Arthur A. Katz
Warshaw, Burstein , Cohen, Schlesinger & Kuh
555 5th Ave.
New York, NY 10017 (212) 972-9100

Otto S. Kauder
201 E. 21st Street
New York, NY 10010 (212) 858-5678

Ivan S. Kavrukov
Cooper, Dunham, Clark, Griffin & Moran
30 Rockefeller Plaza
New York, NY 10112 (212) 977-9550

Walter Carl Kehm
Dweck & Sladkus
666 5th Ave.
New York, NY 10103 (212) 246-6666

Fred A. Keire
Curtis, Morris, & Safford
530 Fifth Ave.
New York, NY 10036 (212) 840-3333

John J. Kelly, Jr.
Kenyon & Kenyon
One Broadway
New York, NY 10004 (212) 425-7200

John T. Kelton
Darby & Darby, Chrysler Bldg.
405 Lexington Ave.
New York, NY 10174 (212) 697-7660

Gail M. Kempler#
Kenyon & Kenyon
One Broadway
New York, NY 10004 (212) 425-7200

William B. Kempler
Ladas & Parry
26 West 61st St.
New York, NY 10023 (212) 708-3465

William A. Kennaman, Jr.
Sherier & O Connor
122 E. 42nd St.
New York, NY 10168 (212) 682-1986

W. Houston Kenyon, Jr.
Kenyon & Kenyon
One Broadway
New York, NY 10004 (212) HA5-7200

John E. Kidd
Pennie & Edmonds
1155 Ave. Of The Americas
New York, NY 10036 (212) 790-9090

Gerald H. Kiel
Toren, Mcgeady & Stanger
521 5th Ave.
New York, NY 10175 (212) 867-2912

William F. Kilgannon
Davis, Hoxie, Faithfull & Hapgood
45 Rockefeller Plaza
New York, NY 10111 (212) 757-2200

Arthur M. King#
Gen. Elec. Co.
570 Lexington Ave.
New York, NY 10022 (212) 750-2240

Allen R. Kipnes
Hedman, Gibson, Costigan & Hoare, P.C.
1185 Ave. Of The Americas
New York, NY 10036 (212) 302-8989

Robert F. Kirchner
Curtis, Morris & Safford, P.C.
530 Fifth Ave.
New York, NY 10036 (212) 840-3333

Jules P. Kirsch
Cooper, Dunham, Clark, Griffin & Moran
30 Rockefeller Plaza
New York, NY 10020 (212) 977-9550

David B. Kirschstein
Kirschstein, Kirschstein, Ottinger & Corbrin, P.C.
666 5th Ave.
New York, NY 10103 (212) 581-8770

Richard B. Klar
Morgan & Finnegan
345 Park Ave.
New York, NY 10154 (212) 758-4800

Felix Klass
Celanese Corp.
1211 Ave. Of The Americas
New York, NY 10036 (212) 764-8525

Arthur O. Klein
Klein & Vibber
274 Madison Ave.
New York, NY 10016 (212) 689-3441

Milton Klein#
Abelman, Frayne, Rezac
708 Third Ave.
New York, NY 10017 (212) 919-9022

Michael Klotz#
151 E. 83rd St.
New York, NY 10028 (201) 861-3971

Charles J. Knuth#
Pfizer Inc.
235 E. 42st St.
New York, NY 10017 (212) LR3-2774

Ronald A. Koatz
301 E. 45th St.
New York, NY 10017

Kenneth A. Koch
Asarco Inc.
120 Broadway
New York, NY 10271 (212) 669-1437

Claire A. Koegler#
Fitzpatrick, Cella, Harper & Scinto
277 Park Ave.
New York, NY 10172 (212) 758-2400

Moonray Kojima
62 West 39th St.
New York, NY 10018 (212) 575-5818

Joseph G. Kolodny
Sprung, Felfe, Horn, Lynch & Kramer
600 3rd Ave.
New York, NY 10016 (212) 687-0911

Richard C. Komson
Morgan, Finnegan, Pine, Foley & Lee
345 Park Ave.
New York, NY 10154 (212) 758-4800

Gary L. Kosdan
Pennie & Edmonds
1155 Ave. Of The Americas
New York, NY 10036 (212) 790-9090

Robert Edward Kosinski
Hopgood, Calimafde, Kalil, Blaustein & Judlowe
60 E. 42nd St.
New York, NY 10165 (212) 986-2480

Theodore J. Koss, Jr.
Striker, Striker & Stenby
360 Lexington Ave.
New York, NY 10017 (212) MU7-5068

Henry W. Koster
Curtis, Morris & Safford, P.C.
530 Fifth Ave.
New York, NY 10036 (212) 840-3333

Howard K. Kothe
One Beekman Place
New York, NY 10022 (212) 755-9696

Beth Kovitz
Blum, Kaplan, Friedman, Silberman & Beran
1120 Ave. Of The Americas
New York, NY 10036 (212) 704-0400

Thomas J. Kowalski,
Brooks, Haidt, Haffner & Delahunty
99 Park Ave.
New York, NY 10016 (212) 697-3355

Ira Jay Krakower
Arthur, Dry & Kalish Rockefeller Center
1230 Ave. Of The Americas
New York, NY 10020 (212) 489-4538

Nathaniel D. Kramer
Sprung, Felfe, Horn, Lynch & Kramer
600 3rd Ave.
New York, NY 10016 (212) 661-0520

John A. Krause
Fitzpatrick, Cella, Harper & Scinto
277 Park Ave.
New York, NY 10172 (212) PL8-2400

Stuart E. Krieger
Bristol-Myers Inc.
345 Park Ave.
New York, NY 10154 (212) 546-3646

Friedrich Kueffner
Toren, McGeady & Stanger
521 5th Ave.
New York, NY 10175 (212) 867-2912

Eve Kunen
Hopgood, Calimafde, Kalil, Blaustein & Judlowe
60 E. 42nd St.
New York, NY 10165 (212) 986-2480

Lawrence Gerald Kurland
Stiefel, Gross, Kurland, P.C.
551 5th Ave.
New York, NY 10176 (212) 687-1360

John Kurucz
Kane, Dalsimer, Kane, Sullivan & Kurucz
420 Lexington Ave.
New York, NY 10170 (212) 687-6000

John L. La Pierre
Pfizer, Inc.
235 E. 42nd St.
New York, NY 10017 (212) 573-1229

Mitchell L. Lampert
Lampert & Lampert
641 Lexington Ave.
New York, NY 10022 (212) 644-5770

Thomas Langer
Frishauf, Holtz, Goodman & Woodward
261 Madison Ave.
New York, NY 10016 (212) 972-1400

Andrew S. Langsam
Mandeville & Schweitzer
230 Park Ave.
New York, NY 10169 (212) 986-3377

Marina T. Larson#
Brumbaugh, Graves, Donohue & Raymond
30 Rockefeller Plaza
New York, NY 10112 (212) 408-2508

Charles J. Laughon, Jr.
Fish & Neave
875 Third Ave.
New York, NY 10022 (212) 715-0600

Peter C. Lauro#
Curtis, Morris & Safford, P.C.
530 Fifth Ave.
New York, NY 10036 (212) 840-3333

John J. Lauter
Pennie & Edmonds
1155 Ave. Of The Americas
New York, NY 10036 (212) 790-9090

Stanton T. Lawrence, Jr.
Pennie & Edmonds
330 Madison Ave.
New York, NY 10017 (212) 986-8686

Stanton T. Lawrence, III
Pennie & Edmonds
1155 Ave. Of The Americas
New York, NY 10036 (212) 790-9090

William F. Lawrence
Curtis, Morris Safford, P.C.
530 5th Ave.
New York, NY 10036 (212) 840-3333

Gary D. Lawson
Esso Middle East
1251 Ave. Of The Americas
New York, NY 10020 (212) 398-3000

Steven R. Lazar
Le Boeuf, Lamb, Leiby & Mac Rae
520 Madison Ave.
New York, NY 10022 (212) 715-8270

Joseph Louis Lazaroff
Amer. Telephone & Telegraph Co.
222 Broadway
New York, NY 10038 (212) 669-2512

Kleon M. Le Fever
N. Amer. Phillips Corp.
100 E. 42nd St.
New York, NY 10017 (212) 850-5114

Eric M. Lee
875 Third Ave.
New York, NY 10022 (212) 715-0600

Jerome G. Lee
Morgan, Finnegan, Pine, Foley & Lee
345 Park Ave.
New York, NY 10154 (212) 758-4800

Steven J. Lee#
Kenyon & Kenyon
One Broadway
New York, NY 10005 (212) 425-7200

Susan Lee
Pennie & Edmonds
1155 Ave. Of The Americas
New York, NY 10036 (212) 790-9090

Paul Lempel
Kenyon & Kenyon
One Broadway
New York, NY 10004 (212) 425-7200

Michael J. Lennon
Kenyon & Kenyon
One Broadway
New York, NY 10004 (212) 425-7200

Joseph B. Lerch
Darby & Darby, P.C.
405 Lexington Ave.
New York, NY 10174 (212) 697-7660

Henry R. Lerner
Levisohn, Niner & Lerner
535 Fifth Ave.
New York, NY 10017 (212) 682-1322

Hallie R. Levie
Felfe & Lynch
805 Third Ave. 25th Floor
New York, NY 10022 (212) 688-9200

Harry H. Levin
345 W. 58th St., 12th Floor
New York, NY 10019 (212) 586-6218

Alan H. Levine
Fiddler & Levine
350 5th Ave.
New York, NY 10118 (212) 239-4162

Martin A. Levitin
Bryan & Levitin
100 Park Ave. 34th Floor
New York, NY 10017 (212) 972-8600

Gerald Levy
Kane, Dalsimer, Kane, Sullivan & Kurucz
420 Lexington Ave.
New York, NY 10170 (212) 687-6000

Bert J. Lewen
Arthur, Dry & Kalish, P.C.
1230 Ave. Of The Amers.
New York, NY 10020 (212) 841-9344

Arthur L. Liberman
International Flavors & Fragrances, Inc.
521 W. 57th St.
New York, NY 10019 (212) 765-5500

Arthur M. Lieberman
Lieberman, Rudolph & Nowak
292 Madison Ave.
New York, NY 10017 (212) 532-4447

Bernard Lieberman
Colgate Palmolive Co.
300 Park Ave.
New York, NY 10022 (212) 310-3120

Lance J. Lieberman
Cohen, Pontani & Lieberman
551 5th Ave. Suite 1210
New York, NY 10176 (212) 687-2770

Stanley H. Lieberstein
Ostrolenk, Faber, Gerb, & Soffen
260 Madison Ave.
New York, NY 10016 (212) 685-8470

Maria C.H. Lin
Celanese Corp.
1211 Ave. Of The Americas
New York, NY 10036 (212) 719-8631

Nels T. Lippert
Fitzpatrick, Cella, Harper & Scinto
277 Park Ave.
New York, NY 10172 (212) 758-2400

Joseph H. Lipschutz
440 E. 79th St.
New York, NY 10021

Randy Lipsitz
Blum, Kaplan, Friedman, Silberman & Beran
1120 Ave. Of The Americas
New York, NY 10036 (212) 704-0400

Gordon K. Lister
Curtis, Morris & Safford, P.C.
530 Fifth Ave.
New York, NY 10036 (212) 840-7171

Nelson Littell, Jr.
Hammond & Littell, Weissenberger & Muserlin
420 Lexington Ave.
New York, NY 10170 (212) 682-1750

Nelson Littell, Sr.
Hammond & Littell, Weissenberger & Muserlian
420 Lexington Ave.
New York, NY 10170 (212) 682-1750

James B. Litton
117 W. 58th St., Apt 9D
New York, NY 2122477823

Nannellyn W. Lloyd#
Pfizer Inc.
235 E. 42nd St.
New York, NY 10017 (212) 573-2100

Anthony F. Lo Cicero
Amster, Rothstein & Engelberg
90 Park Ave.
New York, NY 10016 (212) 697-5995

Emmanuel J. Lobato
Burns. Lobato & Adams, P.C.
140 Cedar St. Suite 1805
New York, NY 10006 (212) 608-0190

Joel Y. Loewenberg#
Bio-Technology Gen. Corp. Pats. Dept.
375 Park Ave. Suite 3303
New York, NY 10152 (212) 319-8944

Denise L. Loring
Fish & Neave
875 Third Ave.
New York, NY 10022 (212) 715-0600

Regina A. Loughran
Fitzpatrick, Cella, Harper & Scinto
277 Park Ave.
New York, NY 10172 (212) 758-2400

Karen A. Lowney#
Pennie & Edmonds
1155 Ave. Of The Americas
New York, NY 10036 (212) 790-9090

Donald C. Lucas
Lucas & Just
205 E. 42nd St
New York, NY 10017 (212) 682-4980

William D. Lucas
Eyre, Mann, Lucas & Just
205 E. 42nd St.
New York, NY 10017 (212) 682-4980

Peter A. Luccarelli, Jr.
Cooper, Dunham, Clark, Griffin & Moran
30 Rockefeller Plaza
New York, NY 10112 (212) 977-9550

S. Peter Ludwig
Darby & Darby, P.C.
405 Lexington Ave.
New York, NY 10174 (212) 697-7660

John T. Lumb#
Pfizer Inc., Legal Dept.
235 East 42nd St.
New York, NY 10017 (212) 573-3878

Joel E. Lutzker
Amster, Rothstein & Engelberg
90 Park Ave.
New York, NY 10016 (212) 697-5995

John E.D. Lynch
Felfe & Lynch
805 Third Ave. 25th Floor
New York, NY 10022 (212) 688-9200

Thomas D. Mac Blain
Brumbaugh, Graves, Donohue & Raymond
30 Rockefeller Plaza
New York, NY 10112 (212) 489-3346

George W. Mac Donald, Jr.
Cole & Deitz
175 Water Street
New York, NY 10038 (212) 269-2500

Terrence Mac Laren
Siemens Corporate Research & Support, Inc.
767 Fifth Ave.
New York, NY 10153 (212) 935-9797

Leonard B. Mackey
I.T.T. Corp.
320 Park Ave.
New York, NY 10022 (212) PL2-6000

Kenneth Edward Madsen
Kenyon & Kenyon
One Broadway
New York, NY 10004 (212) 425-7200

Barry C. Magidoff
Sutton & Magidoff
521 5th Ave.
New York, NY 10175 (212) 490-8533

Jay H. Maioli
Eslinger & Pelton, P.C.
600 Third Ave.
New York, NY 10016 (212) 972-4433

Bernard Malina
Lincoln Bldg. Suite 1814
60 E. 42nd St.
New York, NY 10165 (212) 697-0444

Adley F. Mandel
American Home Products Corp.
685 Third Ave.
New York, NY 10017 (212) 878-6223

Howard F. Mandelbaum
Colvin, Miskin, Basseches & Mandelbaum
420 Lexington Ave.
New York, NY 10170 (212) 697-8873

Hubert Turner Mandeville
Mandeville & Schweitzer
230 Park Ave.
New York, NY 10169 (212) 986-3377

Harry C. Marcus
Morgan, Finnegan, Pine, Foley & Lee
345 Park Ave.
New York, NY 10154 (212) 758-4800

Michael G. Marinangeli#
Penthouse
331 E. 82nd St.
New York, NY 10028 (212) 628-1979

Michael E. Marion
Lieberman, Rudolph & Nowak
292 Madison Ave.
New York, NY 10017 (212) 532-4447

James M. Markarian
322 W. 57th St. Apt. 29L
New York, NY 10019 (212) 315-1361

James G. Markey#
Doyle & Roth Mfg. Co. Inc.
26 Broadway
New York, NY 10004 (212) 269-7840

Ernest Figdor Marmorek
Marmorek & Bierman
420 Lexington Ave., Suite 2006
New York, NY 10170 (212) 867-9680

Walter G. Marple, Jr.
Pennie & Edmond
1155 Ave. Of The Americas
New York, NY 10036 (212) 790-9090

Jonathan A. Marshall
Pennie & Edmonds
1155 Ave. Of The Americas
New York, NY 10036 (212) 790-9090

Thomas M. Marshall
I.T.T. Corp.
320 Park Ave.
New York, NY 10022 (212) 940-1616

Patricia A. Martone
Fish & Neave
875 Third Neave
New York, NY 10022 (212) 715-0600

Dennis A. Mason#
Abelman, Frayne & Rezac
708 Third Ave.
New York, NY 10017 (212) 949-9022

Clifford J. Mass
Roberts, Spiecens & Cohen
38 East 29th St. Third Floor
New York, NY 10016 (212) 684-7766

Marla J. Mathias#
Cooper, Dunham, Griffin & Moran
30 Rockefeller Plaza
New York, NY 10112 (212) 977-9550

James J. Maune#
Brumbaugh, Graves, Donoham & Raymond
30 Rockefeller Plaza
New York, NY 10112 (212) 489-3364

Richard L. Mayer
Kenyon & Kenyon
One Broadway
New York, NY 10004 (212) 425-7200

William J. Mc Nichol, Jr.#
Kenyon & Kenyon
One Broadway
New York, NY 10004 (212) 425-7200

John Q. Mc Quillian
Kenyon & Kenyon
One Broadway
New York, NY 10004 (212) 425-7200

Lloyd McAulay
McAulay, Fields, Fisher, Goldstein & Nissen
405 Lexington Ave.
New York, NY 10174 (212) 986-4090

Philip J. McCabe
Kenyon & Kenyon
One Broadway
New York, NY 10004 (212) 425-7200

Maria E. McCormack
Pennie & Edmonds
1155 Ave. Of The Americas
New York, NY 10036 (212) 790-9090

Bernard X. McGeady
Toren, McGeady & Stanger
521 5th Ave.
New York, NY 10175 (212) 867-2912

Hansel L. McGee
New York City Civil Court
851 Grand Concourse
New York, NY 10451 (212) 590-3598

Hilda McGlew#
McGlew & Tuttle, P.C.
28 W. 44th St.
New York, NY 10036 (212) 840-6350

John J. McGlew
McGlew & Tuttle, P.C.
28 W. 44th St.
New York, NY 10036 (212) 840-6350

Kevin J. McGough
Morgan & Finnegan
345 Park Ave.
New York, NY 10154 (212) 758-4800

Robert McKay
Pennie & Edmonds
1155 Ave. Of The Americas
New York, NY 10036 (212) 790-9090

Charles E. McKenney
Pennie & Edmonds
330 Madison Ave.
New York, NY 10017 (212) 986-8686

Kevin C. McMahon
Davis, Hoxie, Faithfull & Hapgood
45 Rockfeller Plaza
New York, NY 10111 (212) 757-2200

Edward A. Meilman
Ostrolenk, Faber, Gerb, & Soffen
260 Madison Ave.
New York, NY 10016 (212) 685-8470

Michael Nicholas Meller
50 East 42nd St.
New York, NY 10017 (212) 953-3350

George A. Mentis
Bristol-Myers Co.
345 Park Ave.
New York, NY 10154 (212) 546-3656

Kenneth E. Merklen
Gulf & Western Inds. Inc.
1 Gulf & Western Plaza
New York, NY 10023 (212) 333-4920

Lewis Messulam
INCO Limited
One New York Plaza
New York, NY 10004 (212) 612-5880

Clyde Christian Metzger
Pennie & Edmonds
1155 Ave. Of The Americas
New York, NY 10036 (212) 790-9090

Dominic M. Mezzapelle
Bristol-Myers Co.
345 Park Ave.
New York, NY 10154 (212) 546-3651

Martin P. Michael
Rubin, Baum, Levin, Constant & Friedman
645 Fifth Ave.
New York, NY 10022 (212) 759-2700

Peter C. Michalos
Empire State Bldg.
350 Fifth Ave. Suite 6902
New York, NY 10118 (212) 564-0305

Eric Michelson#
615 Fort Washington Ave.
New York, NY 10040 (212) 923-5533

Carl Miller
3612 Woolworth Bldg.
233 Broadway
New York, NY 10007 (212) 267-5252

Charles E. Miller
Pennie & Edmonds
1155 Ave. Of The Americas
New York, NY 10036 (212) 790-9090

David Byron Miller
Arthur, Dry & Kalish, P.C.
1230 Ave. Of The Americas
New York, NY 10020 (212) 841-9364

Joel Miller
Hopgood, Calimafde, Kalil, Blaustein & Judlowe
60 E. 42nd St.
New York, NY 10165 (212) 986-2480

Richard N. Miller
Colgate Palmolive Co.
300 Park Ave.
New York, NY 10022 (212) 310-3124

Harold J. Milstein
Kuhn Muller & Bazerman
1412 Broadway
New York, NY 10018 (212) 221-0864

Douglas A. Miro
Amster, Rothstein & Engelberg
90 Park Ave.
New York, NY 10016 (212) 697-5995

Howard Charles Miskin
Colvin, Miskin, Basseches & Mandelbaum
420 Lexington Ave.
New York, NY 10170 (212) 697-8873

S. Leslie Misrock
Pennie & Edmonds
1155 Ave. Of The Americas
New York, NY 10036 (212) 790-9090

John A. Mitchell
Curtis, Morrie & Safford, P.C.
530 Fifth Ave.
New York, NY 10036 (212) 840-3333

Marvin B. Mitzner
Stein, Davidoff, Malito & Teitler
100 E. 42nd St.
New York, NY 10017 (212) 557-7200

Guido Moeller#
General Electric Co.
International Pat. Operation, 570 Lexington Ave.
New York, NY 10022 (212) 750-2151

Robert A. Molan
Morgan, Finnegan, Pine, Foley & Lee
345 Park Ave.
New York, NY 10154 (212) 758-4800

Gray R. Molnar
American Home Products Corps
685 Third Ave.
New York, NY 10017 (212) 878-5000

Dennis J. Mondolino#
Hopgood, Calimafde, Kalil, Blaustein & Judlowe
60 E. 42nd St.
New York, NY 10168 (212) 682-1986

James H. Monroe#
Pfizer Inc.
235 E. 42nd St.
New York, NY 10017 (212) 573-2369

Raphael Monsanto
Kenyon & Kenyon
One Broadway
New York, NY 10004 (212) 425-7200

Thomas Francis Moran
Cooper, Dunham, Clark, Griffin & Moran
30 Rockefeller Plaza
New York, NY 10112 (212) 977-9550

William R. Moran
333 E. 43rd St. Apt. 909
New York, NY 10017 (212) 986-5801

Robert C. Morgan
Fish & Neave
875 Third Ave.
New York, NY 10022 (212) 715-0600

Eugene Moroz
Morgan, Finnegan, Pine, Foley & Lee
345 Park Ave.
New York, NY 10154 (212) 758-4800

Francis E. Morris
Pennie & Edmonds
1155 Ave. Of The Americas
New York, NY 10036 (212) 790-9090

Robert H. Morse
303 W. 66th St.
New York, NY 10023 (212) 787-8578

Jacob F. Moskowitz#
P.O. Box 245-A Planetarium Station
New York, NY 10023 (212) 787-2365

Daniel J. Muccio
Kane, Dalsimer, Kane, Sullivan & Kurucz
420 Lexington Ave. Room 2710
New York, NY 10170 (212) 867-6000

Charles Gilmore Mueller
Brooks, Haidt, Haffner & Delahunty
99 Park Ave.
New York, NY 10016 (212) 697-3355

Keith E. Mullenger
Pennie & Edmonds
1155 Ave. Of The Americas
New York, NY 10036 (212) 790-9090

Edward F. Mullowney
Fish & Neave
875 Third Ave.
New York, NY 10022 (212) 715-0635

John D. Murnane
Brumbaugh, Graves, Donohue & Raymond
30 Rockefeller Plaza
New York, NY 10112 (212) 489-3358

Francis J. Murphy
Hopgood, Calimafde, Kalil, Blaustein & Judlowe
60 E. 42nd St.
New York, NY 10165 (212) 986-2480

Francis X. Murphy#
Pfizer Inc.
235 E. 42nd St.
New York, NY 10017 (212) LR3-2405

Kittie A. Murray#
Felfe & Lynch
805 Third Ave. 25th Floor
New York, NY 10022 (212) 688-9200

Charles A. Muserlian
Bierman & Muserlian
757 Third Ave.
New York, NY 10017 (212) 752-7550

Alfred Musumeci
Toren, Mcgeady & Stanger P.C.
521 5th Ave.
New York, NY 10175 (212) 867-2912

George Nalaboff
Manufacturers Hanover Trust Co.
55 Water St. 59th Floor
New York, NY 10015 (212) 623-1183

John E. Nathan
Fish & Neave
875 Third Ave.
New York, NY 10022 (212) 750-2107

Howard Natter
Natter & Natter
25 W. 43rd St.
New York, NY 10036 (212) 840-8300

Seth Natter
Natter & Natter
25 W. 43rd St.
New York, NY 10036 (212) 840-8300

Gregor N. Neff
Curtis, Morris & Safford, P.C.
530 Fifth Ave.
New York, NY 10036 (212) 840-3333

Albin J. Nelson
Kenyon & Kenyon
One Broadway
New York, NY 10004 (212) 425-7200

Carol A. Nemetz
Darby & Darby, The Chrysler Bldg.
405 Lexington Ave.
New York, NY 10174 (212) 697-7660

Thomas R. Nesbitt, Jr.
Brumbaugh, Graves, Donohue & Raymond
30 Rockefeller Plaza
New York, NY 10112 (212) 489-3332

Robert Neuner
Brumbaugh, Graves, Donohue & Raymond
30 Rockefeller Plaza
New York, NY 10112 (212) 489-3338

Christopher Nicastri#
Kenyon & Kenyon
One Broadway
New York, NY 10004 (212) 425-7200

Mitchell B. Nisonoff
Mudge Rose Guthrie & Alexander
20 Broad Street
New York, NY 10005 (212) 701-1152

Joseph H. Nissen
McAulay, Fields, Fisher, Goldstein & Nissen
405 Lexington Ave.
New York, NY 10174 (212) 986-4090

Angelo Notaro
350 Fifth Ave. Suite 6902
New York, NY 10118 (212) 564-0200

Keith D. Nowak
Lieberman, Rudolph & Nowak
292 Madison Ave.
New York, NY 10017 (212) 532-4447

Francis M. O Connor
Shenier & OConnor
122 E. 42nd St.
New York, NY 10168 (212) 682-1986

Thomas A. O Rourke
Wyatt, Gerber, Shoup, & Badie
261 Madison Ave.
New York, NY 10016 (212) 687-0911

James L. OBrien
Bendix Corp.
767 Fifth Ave.
New York, NY 10153 (212) 906-3009

John A. OBrien
Fitzpatrick, Cella, Harper & Scinto
277 Park Ave.
New York, NY 10172 (212) 758-2400

Robert D. OBrien
Bristol Myers Co.
345 Park Ave.
New York, NY 10154 (212) 546-3667

Thomas P. OHare
Morgan & Finnegan
345 Park Ave.
New York, NY 10154 (212) 758-4800

John H. Olding
The Bar Bldg.
36 W. 44th St.
New York, NY 10036 (212) 840-1385

Milton J. Oliver
Frishauf, Holtz, Goodman & Woodward
261 Madison Ave.
New York, NY 10016 (212) 972-1400

Paul J. Olivo
300 E. 40th St.
New York, NY 10016 (212) 867-4727

Stephen Martin Olko#
Olko Engineering
15 West 36th St.
New York, NY 10018 (212) 279-2822

Frances L. Olmsted
Ladas & Parry
26 W. 61st St.
New York, NY 10023 (212) 708-1890

John R. Olsen
The Ritz Paris Enterprises
75 Rockefeller Plaza, Suite 1804
New York, NY 10019 (212) 265-5545

Kenneth Olsen
Schlumber Limited
277 Park Ave.
New York, NY 10172 (212) 350-9425

Michael K. ONeill
Fitzpatrick, Cella, Harper & Scinto
277 Park Ave.
New York, NY 10172 (212) 758-2400

Jerry Oppenheim
Oppenheim & Rothstein, P.C.
477 Madison Ave.
New York, NY 10022 (212) 935-8770

Harold W. Ordway#
Pfizer Inc.
235 E. 42nd St.
New York, NY 10017

Glenn F. Ostrager
Barst & Mukamal
2 Park Ave 19th Floor
New York, NY 10016 (212) 686-3838

Bertram Ottinger
Kirschstein, Kirschstein, Ottinger & Cobrin, P.C.
666 5th Ave.
New York, NY 10103 (212) 581-8770

Glenn A. Ousterhurst
Fish & Neave
875 Third Ave.
New York, NY 10022 (212) 715-0600

Bernard Ouziel
Colt Industries Inc.
430 Park Ave.
New York, NY 10022 (212) 940-9622

James N. Palik
Pennie & Edmonds
1155 Ave. Of The Americas
New York, NY 10036 (212) 790-9090

Ralph R. Palo#
Hubbell, Cohen, Stiefel & Gross
551 5th Ave.
New York, NY 10176 (212) 687-1360

Richard K. Parsell
Kenyon & Kenyon
One Broadway
New York, NY 10004 (212) 425-7200

Walter Patton
American Home Products Corp.
685 Third Ave.
New York, NY 10017 (212) 878-5000

Robert E. Paulson
Morgan, Finnegan, Pine. Foley & Lee
345 Park Ave.
New York, NY 10154 (212) 758-4800

Martin B. Pavane
Schechter, Brucker & Pavane, P.C. Empire State Bldg.
350 Fifth Ave. Suite 4510
New York, NY 10118 (212) 224-6600

John B. Pegram
Davis, Hoxie, Faithfull & Hapgood
45 Rockefeller Plaza
New York, NY 10111 (212) 757-2200

Anthony Pellicano
Morgan, Finnegan, Pine, Foley & Lee
345 Park Ave.
New York, NY 10154 (212) 758-4800

Russell G. Pelton
Eslinger & Pelton
522 Fifth Ave.
New York, NY 10036 (212) 221-1500

William E. Pelton
Eslinger & Pelton, P.C.
522 Fifth Ave.
New York, NY 10036 (212) 221-1500

Jonathan B. Penn
Curtis, Morris & Safford
530 Fifth Ave.
New York, NY 10036 (212) 480-3333

Frederick W. Pepper
Morgan & Finnegan
345 Park Ave.
New York, NY 10154 (212) 758-4800

Lawrence S. Perry
Fitzpatrick, Cella, Harper & Scinto
277 Park Ave.
New York, NY 10172 (212) 758-2400

Sandra M. Person
Bristol-Myers Co.
345 Park Ave. 6th Floor, Rm. 58
New York, NY 10154 (212) 546-3655

David H. Pfeffer
Morgan, Finnegan, Pine, Foley & Lee
345 Park Ave.
New York, NY 10154 (212) 758-4800

Martin Pfeffer
Western Electric Co., Inc.
222 Broadway
New York, NY 10038 (212) 669-4763

Helmuth L. Pfluger
15 West 72nd St.Suite 21-K
New York, NY 10023 (212) 877-3897

Peter J. Phillips
Brumbaugh, Graves, Donohue & Raymond
30 Rockefeller Plaza, Suite 4400
New York, NY 10112 (212) 408-2582

Margaret A. Pierri
Fish & Neave
875 Third Ave.
New York, NY 10022 (212) 715-0600

Granville Martin Pine
Morgan, Finnegan, Pine, Foley & Lee
345 Park Ave.
New York, NY 10154 (212) 758-4800

David C. Plache
Debevoise & Plimpton
875 Third Ave.
New York, NY 10022 (212) 909-6911

Sheldon Plamer
777 Third Ave.
New York, NY 10017 (212) 421-4600

David W. Plant
Fish & Neave
875 Third Ave. 29th Floor
New York, NY 10022 (212) 715-0600

Roland Plottel
30 Rockefeller Plaza
New York, NY 10112 (212) 489-7073

Brian M. Poissant
Pennie & Edmonds
1155 Ave. Of The Americas
New York, NY 10036 (212) 790-9090

Steven B. Pokotilow
Blum, Kaplan, Friedman, Silberman & Beran
1120 Ave. Of The Americas
New York, NY 10036 (212) 740-0400

Thomas C. Pontani
Cohen, Pontani & Lieberman
551 Fifth Ave., Suite 1210
New York, NY 10176 (212) 687-2770

Joseph J. Previto
Holland, Armstrong, Wilkie & Previto
Empire State Bldg.
New York, NY 10001 (212) 736-2080

Peter H. Priest
Davis, Hoxie, Faithfull & Hapgood
45 Rockefeller Plaza
New York, NY 10111 (212) 757-2200

David M. Quinlan
Eslinger & Pelton, P.C.
522 Fifth Ave.
New York, NY 10036 (212) 221-1500

James A. Quinton
Stiefel, Gross, Kurland & Pavane, P.C.
551 Fifth Ave.
New York, NY 10176 (212) 687-1360

Samuel S. Rabkin
211 E. 53rd St.
New York, NY 10022 (212) 935-3739

Michael Irwin Rackman
Gottlieb, Rackman & Reisman, P.C.
1430 Broadway
New York, NY 10018 (212) 869-2890

Rory J. Radding
Pennie & Edmonds
1155 Ave. Of The Americas
New York, NY 10036 (212) 790-9090

Arnold I. Rady
Morgan, Finnegan, Pine, Foley & Lee
345 Park Ave., 29th Floor
New York, NY 10154 (212) 758-4800

Joseph J.C. Ranalli
Pennie & Edmonds
330 Madison Ave.
New York, NY 10017 (212) 986-8686

Robert F. Randle
Columbia University
1419 I & B
New York, NY 10027 (212) 280-4614

Richard M. Rasati
Kenyon & Kenyon
One Broadway
New York, NY 10004 (212) 425-7200

Martin G. Raskin
Steinberg & Raskin
60 E. 42nd St.
New York, NY 10165 (212) 682-2324

Dana M. Raymond
Brumbaugh, Graves, Donohue & Raymond
30 Rockefeller Plaza
New York, NY 10112 (212) 408-2518

Pasquale Razzano
Curtis, Morris & Safford, P.C.
530 Fifth Ave.
New York, NY 10036 (212) 840-3333

Daniel J. Reardon
Graham, Geoffrey & Reardon
535 Fifth Ave.
New York, NY 10017 (212) 599-0039

John A. Reilly
Curtis, Morris & Safford, P.C.
530 5th Ave.
New York, NY 10036 (212) 840-3333

Barry D. Rein
Pennie & Edmonds
1155 Ave. Of The Americas
New York, NY 10036 (212) 790-9090

Jesse David Reingold
Jacobs & Jacobs, P.C.
521 5th Ave.
New York, NY 10175 (212) 687-1636

James Reisman
Gottlieb, Rackman & Reisman, P.C.
1430 Broadway
New York, NY 10018 (212) 869-2890

Morris Relson
Darby & Darby, P.C.
405 Lexington Ave.
New York, NY 10174 (212) 697-7660

C. Cornell Remsen, Jr.
Bierman, Bierman & Peroff
437 Madison Ave.
New York, NY 10022 (212) 752-7550

Henry J. Renk
Fitzpatrick, Cella, Harper & Scinto
277 Park Ave.
New York, NY 10172 (212) 758-2400

James Madison Rhodes, Jr.
Hopgood, Calimafde, Kalil, Blaustein & Judlowe
60 E. 42nd St.
New York, NY 10165 (212) 986-2480

John Richards
Ladas & Parry
26 West 61st St.
New York, NY 10023 (212) 708-1915

Peter C. Richardson
Pfizer Inc.
235 E. 42nd St.
New York, NY 10017 (212) 573-7805

Francis Elbert Rinehart
1025 5th Ave.
New York, NY 10028 (212) 861-6419

Deborah S. Rittman
Kenyon & Kenyon
One Broadway
New York, NY 10004 (212) 425-7200

Leonard John Robbins
Ladas & Parry
26 West 61st St.
New York, NY 10023 (212) 708-1800

Alan K. Roberts
Roberts, Spiecens & Cohen
38 E. 29th St. Thrid Floor
New York, NY 10016 (212) 684-7766

Lee C. Robinson, Jr.
Curtis, Morris & Safford, P.C.
530 5th Ave.
New York, NY 10036 (212) 840-3333

Patricia S. Rocha
Morgan, Finnegan, Pine, Foley & Lee
345 Park Ave., 22nd Floor
New York, NY 10154 (212) 758-4800

Stephan A. Roen
420 E. 64th St. Suite E 12K
New York, NY 10021 (212) EL5-1825

Laurence S. Rogers
Fish & Neave
875 Third Ave.
New York, NY 10022 (212) 715-0600

James J. Romano, Jr.
747 3rd Ave.
New York, NY 10017 (212) 838-0119

Joseph E. Root, III
Kenyon & Kenyon
One Broadway
New York, NY 10004 (212) 425-7200

Daniel M. Rosen
Rosen, Dainow & Jacobs
489 5th Ave.
New York, NY 10017 (212) 692-7000

Jeffrey A. Rosen
Sargoy, Stein & Hanft
105 Madison Ave.
New York, NY 10016 (212) 889-1420

Lawrence Rosen
260 Madison Ave.
New York, NY 10016 (212) 684-4727

Neal Lewis Rosenberg
Amster, Rothstein & Ebenstein
90 Park Ave.
New York, NY 10016 (212) 697-5995

Lawrence Rosenthal
Blum, Kaplan, Friedman, Silberman & Beran
1120 Ave. Of The Americas
New York, NY 10036 (212) 704-0400

James E. Rosini
Kenyon & Kenyon
One Broadway
New York, NY 10004 (212) 425-7200

Warren H. Rotert
Morgan, Finnegan, Pine, Foley & Lee
345 Park Ave.
New York, NY 10154 (212) 758-4800

Jesse Aaron Rothstein
Amster, Rothstein & Engelberg
90 Park Ave.
New York, NY 10016 (212) 697-5995

Tyler S. Roundy
5 West 63rd St.
New York, NY 10023 (212) 873-4175

John Webb Routh
Amer. Home Products Corp.
685 3rd Ave.
New York, NY 10017 (212) 878-6227

Philip E. Roux#
Hedman, Gibson, Costigan & Hoare, P.C.
100 Park Ave.
New York, NY 10017 (212) 697-7300

Mark D. Rowland
Fish & Neave
875 Third Ave.
New York, NY 10022 (212) 715-0600

Harry Ernest Rubens
164 E. 91st St.
New York, NY 10028 (212) 876-9669

Allen I. Rubenstein
Gottlieb, Rackman & Reisman, P.C.
1430 Broadway
New York, NY 10018 (212) 869-2890

Kenneth Rubenstein
Marmorek, Guttman & Rubenstein
420 Lexington Ave.
New York, NY 10170 (212) 867-9680

Stephen J. Rudy
Chicago Pneumatic Tool Co.
6 E. 44th St.
New York, NY 10017 (212) 850-6900

Mary A. Ryan
Brumbaugh, Graves, Donohue & Raymond
30 Rockefeller Plaza
New York, NY 10112 (212) 408-2572

James E. Ryder
Kuhn Muller & Bazerman
1412 Broadway
New York, NY 10018 (212) 221-0864

Eugene C. Rzucidlo
Sprung, Horn, Kramer & Woods
600 Third Ave.
New York, NY 10016 (212) 661-0520

A. Thomas S. Safford
Curtis, Morris & Safford, P.C.
530 Fifth Ave.
New York, NY 10036 (212) 840-3333

James J. Salerno
American Standard, Inc.
40 West 40th Street
New York, NY 10018 (212) 840-5136

Martin I. Samuels
172 York Ave. Suite 16G
New York, NY 10028 (212) 427-2924

Paul Sandler
Consolidated Edison Corp. Of N.Y.
4 Irving Place
New York, NY 10003 (212) 460-6614

Anthony M. Santini#
Bierman & Muserlian
757 Third Ave.
New York, NY 10017 (212) 752-7550

Leonard J. Santisi
Curtis, Morris & Sanford, P.C.
530 5th Ave.
New York, NY 10036 (212) 840-3333

Ronald R. Santucci#
Kane, Dalsimer, Kane, Sullivan & Kurucz
420 Lexington Ave.
New York, NY 10170 (212) 687-6000

Maria A. Savio
Darby & Darby, P.C.
405 Lexington Ave.
New York, NY 10174 (212) 697-7660

Peter Saxon
Fitzpatrick, Cella, Harper & Scinto
277 Park Ave.
New York, NY 10172 (212) 758-2400

Ira J. Schaefer
Sprung, Horn, Kramer & Woods
600 Third Ave.
New York, NY 10016 (212) 697-5995

Robert Schaffer
Klein & Vibber
274 Madison Ave. Suite 1801
New York, NY 10016 (212) 689-3441

Frank F. Scheck
Pennie & Edmonds
1155 Ave. Of The Americas
New York, NY 10036 (212) 790-9090

Robert C. Scheinfeld
Brumbaugh, Graves, Donohue & Raymond
30 Rockefeller Plaza 44th Floor
New York, NY 10112 (212) 408-2500

John R. Schiffhauer
Morgan & Finnegan
345 Park Ave.
New York, NY 10154 (212) 758-4800

Martin W. Schiffmiller
Kirschstein, Kirschstein, Ottinger & Israel, P.C.
551 Fifth Ave.
New York, NY 10176 (212) 697-3750

Richard L. Schmalz, Sr.
Westvaco Corp.
299 Park Ave.
New York, NY 10171 (212) 688-5000

Casper Carl Schnider, Jr.
Davis, Hoxie, Faithfull & Hapgood
45 Rockefeller Plaza
New York, NY 10111 (212) 757-2200

Howard I. Schuldenfrei
441 Lexington Ave. Suite 409
New York, NY 10017 (212) 286-9460

Jeffrey A. Schwab
Abelman, Frayne, Rezac & Schwab
708 Third Ave.
New York, NY 10017 (212) 949-9022

Herbert Frederick Schwartz
Fish & Neave
875 Third Ave.
New York, NY 10022 (212) 715-0653

Fritz L. Schweitzer, Jr.
Mandeville & Schweitzer
230 Park Ave.
New York, NY 10169 (212) 689-6967

Howard Myles Schwiger
11 Park Place
New York, NY 10007 (212) 233-8820

Paul C. Scifo
233 Broadway, Suite 4703
New York, NY 10279 (212) 513-1122

Lawrence F. Scinto
Fitzpatrick, Cella, Harper & Scinto
277 Park Ave.
New York, NY 10172 (212) 758-2400

Earl L. Scott
Scott & Farrell
84 William St. Suite 1000
New York, NY 10038 (212) 809-3880

Walter Scott
Kenyon & Kenyon
One Broadway
New York, NY 10004 (212) 425-7200

Thomas L. Secrest
Fish & Neave
875 Third Ave. 29th Floor
New York, NY 10022 (212) 715-0600

Rochelle K. Seide
Brumbaugh, Graves, Donohue & Raymond
30 Rockefeller Plaza
New York, NY 10112 (212) 408-2592

Robert J. Seligman
State Of New York Mortgage Agency
260 Madison Ave.
New York, NY 10016 (212) 340-4200

W. Joseph Shanley, Jr.
General Electric Co.
570 Lexington Ave.
New York, NY 10022 (212) 750-2107

Philip Thomas Shannon
Pennie & Edmonds
1155 Ave. Of The Americas
New York, NY 10036 (212) 790-6301

Stephen B. Shear
Felfe & Lynch
805 Third Ave.
New York, NY 10022 (212) 688-9200

Henry L. Shenier
Shenier & O Connor
122 E. 42nd St.
New York, NY 10168 (212) 682-1986

Richard S. Shenier
Shenier & OConnor
122 E. 42nd St.
New York, NY 10168 (212) 682-1986

Charles I. Sherman
Wickes Companies, Inc.
1285 Ave. Of The Americas
New York, NY 10019 (212) 603-1916

Theodore F. Shiells
Curtis, Morris & Safford, P.C.
530 Fifth Ave.
New York, NY 10036 (212) 840-3333

Guy W. Shoup
Waytt, Gerber, Shoup
216 Madison Ave.
New York, NY 10016 (212) 687-0911

James K. Silberman
Blum, Kaplan, Friedman, Silberman & Beran
1120 Ave. Of The Americas
New York, NY 10036 (212) 704-0400

Morton S. Simon
Bristol Myers Co.
345 Park Ave. Suite 6-54
New York, NY 10154 (212) 546-3645

Philip Y. Simons
Freeman Wasserman & Schneide
90 John St.
New York, NY 10038

Robert B. Simonton
Sterling Drug Inc.
90 Park Ave.
New York, NY 10016 (212) 907-3035

William F. Simpson
P.O. Box 1255
New York, NY 10008 (212) 451-0090

Stuart J. Sinder
Kenyon & Kenyon
One Broadway
New York, NY 10004 (212) 425-7200

Alvin Sinderbrand
Curtis, Morris & Safford, P.C.
530 Fifth Ave.
New York, NY 10036 (212) 840-3333

John Patrick Sinnott
American Standard Inc. Pat. & Tdmk. Dept.
40 West 40th St.
New York, NY 10018 (212) 840-5402

Brandon N. Sklar
Davis, Hoxie, Faithfull & Hapgood
45 Rockefeller Plaza
New York, NY 10111 (212) 757-2200

Elizabeth O. Slade
Bristol-Myers Co.
International Pat. Dept.
345 Park Ave.
New York, NY 10154 (212) 546-3634

Alan D. Smith
Fish & Neave
875 Third Ave.
New York, NY 10022 (212) 715-0600

Arthur V. Smith
Curtis, Morris & Safford, P.C.
530 Fifth Ave.
New York, NY 10036 (212) 840-3333

Charles B. Smith
Fish & Neave
875 Third Ave.
New York, NY 10022 (212) 715-0637

Robert B. Smith
Brumbaugh, Graves, Donohue & Raymond
30 Rockefeller Plaza
New York, NY 10112 (212) 408-2500

Stephen R. Smith
Morgan, Finnegan, Pine, Foley & Lee
345 Park Ave.
New York, NY 10154 (212) 758-4800

Martin Smolowitz
Two Pennsylvania Plaza, Suite 1500
New York, NY 10121 (212) 244-3100

Gerald Sobel
Kaye, Scholer, Fierman, Hays & Handler
425 Park Ave.
New York, NY 10022 (212) 407-8515

Marvin C. Soffen
Ostrolenk, Faber, Gerb, & Soffen
260 Madison Ave.
New York, NY 10016 (212) 685-8470

Stephen A. Soffen
Ostrolenk, Faber, Gerb, & Soffen
New York, NY 10016 (212) 685-8470

L. Teresa Solomon
Fish & Neave
875 Third Ave.
New York, NY 10022 (212) 715-0600

William F. Sonnekalb, Jr.
Davis, Hoxie, Faithfull & Hapgood
45 Rockefeller Plaza
New York, NY 10111 (212) 757-2200

Mark H. Sparrow
Jacobs & Jacobs, P.C.
521 5th Ave.
New York, NY 10175 (212) 687-1636

Thomas E. Spath
Davis, Hoxie, Faithfull & Hapgood
45 Rockefeller Plaza
New York, NY 10111 (212) 757-2200

William J. Spatz
Wickes Companies, Inc.
1285 Ave. Of The Americas
New York, NY 10019 (212) 603-1912

Charles B. Spencer
Kenyon & Kenyon
One Broadway
New York, NY 10004 (212) 425-7200

Reuben Spencer
8 Peter Cooper Rd.
New York, NY 10010 (212) 254-6690

Philip Sperber
Refac Techn. Dev. Corp.
122 E. 42nd St.
New York, NY 10168 (212) 687-4741

Camil Peter Spiecens
Roberts, Spiecens & Cohen
38 East 29th Street
New York, NY 10016 (212) 684-7766

Allen J. Spiegel#
Pfizer Inc.
235 E. 42nd St.
New York, NY 10017 (212) 573-2841

Seymour L. Spira#
1123 Broadway
New York, NY 10010 (212) 255-3346

Milton Springut
Blum, Gersen, Bushkin, Gaims, Gaines, Jonas & Stream
270 Madison Ave.
New York, NY 10016 (212) 683-6383

Arnold Sprung
Sprung, Horn, Kramer, & Woods
600 Third Ave.
New York, NY 10016 (212) 661-0520

Lawrence A. Stahl
Fitzpatrick, Cella, Harper & Scinto
277 Park Ave.
New York, NY 10172 (212) 758-2400

Leo Stanger
Toren, Mcgeady, Stanger, Goldberg & Kiel, P.C.
521 5th Ave.
New York, NY 10175 (212) 867-2912

Michael Stark
Cooper, Dunham, Griffin & Moran
30 Rockefeller Plaza
New York, NY 10112 (212) 977-9550

Jeffrey S. Steen#
Hopgood, Calimafde, Kalil, Blaustein & Judlowe
60 E. 42nd St.
New York, NY 10165 (212) 986-2480

Daniel H. Steidl
Davis, Hoxie, Faithfull & Hapgood
45 Rockefeller Plaza
New York, NY 10111 (212) 757-2200

Mitchell A. Stein
Lieberman, Rudolph & Nowak
292 Madison Ave.
New York, NY 10017 (212) 532-4447

Harold D. Steinberg
Steinberg & Raskin
60 E. 42nd St.
New York, NY 10165 (212) 682-2324

Jules H. Steinberg
W.R. Grace & Co.
1114 Avenue Of The Americas
New York, NY 10036 (212) 819-5520

Kenneth J. Stempler
Stempler & Cobrin, P.C.
501 Fifth Ave. Suite 1900
New York, NY 10017 (212) 687-6090

Gidon D. Stern
Pennie & Edmonds
1155 Ave. Of The Americas
New York, NY 10036 (212) 790-9090

Henry Sternberg
Wofsey, Certilman, Haft, Lebow & Balin
805 Third Ave.
New York, NY 10020 (212) 418-5226

Maurice B. Stiefel
Stiefel, Gross, Kurland & Pavane, P.C.
551 Fifth Ave.
New York, NY 10176 (212) 687-1360

Klaus P. Stoffel
Striker, Striker & Stenby
360 Lexington Ave.
New York, NY 10017 (212) 687-5068

Robert S. Stoll
Stoll, Wilkie, Previto & Hoffman
Empire State Bldg.
New York, NY 10001 (212) 736-0290

Samuel J. Stoll
Stoll, Wilkie, Previto, & Hoffman
Empire State Bldg.
New York, NY 10001 (212) 736-0290

John M. Striker
Striker, Striker & Stenby
360 Lexington Ave.
New York, NY 10017 (212) 687-5068

John M. Striker
Striker, Striker & Stenby
360 Lexington Ave.
New York, NY 10017 (212) 687-5068

Harold L. Stults
Curtis, Morris & Stafford, P.C.
530 Fifth Ave.
New York, NY 10036 (212) 840-3333

R. Neil Sudol
De Rosa, Vandenberg & Coleman
71 Broadway, Suite 2200
New York, NY 10006 (212) 269-3020

Joseph C. Sullivan
Kane, Dalsimer, Kane, Sullivan & Kurucz
420 Lexington Ave.
New York, NY 10170 (212) 687-6000

Robert C. Sullivan, Jr.
Darby & Darby, P.C.
405 Lexington Ave.
New York, NY 10174 (212) 697-7660

Paul J. Sutton
Miskin & Sutton
Graybar Bldg. 420 Lexington Ave.
New York, NY 10170 (212) 697-5090

Michael Jon Sweedler
Darby & Darby, P.C.
405 Lexington Ave.
New York, NY 10174 (212) 697-7660

John F. Sweeney
Morgan, Finnegan, Pine, Foley & Lee
345 Park Ave.
New York, NY 10154 (212) 758-4800

Lawrence J. Swire
I.T.T. Corp.
320 Park Ave.
New York, NY 10022 (212) 940-1632

Judith L. Sykes
Bristol Myers Co.
345 Park Ave.
New York, NY 10154 (212) 546-4702

George B. Synder
Curtis, Morris & Safford, P.C.
530 5th Ave.
New York, NY 10036 (212) 840-3333

Henry Y.S. Tang
Brumbaugh, Graves, Donohue & Raymond
30 Rockefeller Plaza
New York, NY 10112 (212) 408-2586

John Alton Taylor
307 W. 79th St. Room 1043
New York, NY 10023

Charles E. Temko
Temko & Temko
19 W. 44th St.
New York, NY 10036 (212) 840-2178

Arthur S. Tenser
Brumbaugh, Graves, Donohue & Raymond
30 Rockefeller Plaza
New York, NY 10112 (212) 408-2542

Berj A. Terzian
Pennie & Edmonds
1155 Ave. Of The Americas
New York, NY 10036 (212) 790-6505

William J. Thomashower
Kaplan, Thomashower & Landau
747 Third Ave.
New York, NY 10017 (212) 593-1700

Roger S. Thompson
116 Pinehurst Ave. Apt. D-14
New York, NY 10033 (212) 923-5145

Robert T. Tobin
Kenyon & Kenyon
One Broadway
New York, NY 10004 (212) 425-7200

Thomas W. Tobin
Wilson, Elser, Moskowitz, Edelman & Dicker
420 Lexington Ave.
New York, NY 10170 (212) 490-3000

Leonard M. Todd#
424 W. 119th St.
New York, NY 10027 (212) 865-3435

William G. Todd
Hopgood, Calimafde, Kalil, Blaustein & Judlowe
60 E. 42nd St.
New York, NY 10165 (212) 986-2480

John J. Tomaszewski
Asarco Incorporated
180 Maiden Lane
New York, NY 10038 (212) 510-1943

David Toren
Toren, McGeady, Stanger, Goldberg & Kiel P.C.
521 5th Ave.
New York, NY 10175 (212) 867-2912

John O. Tramontine
Fish & Neave
875 Third Ave.
New York, NY 10022 (212) 715-0600

Robert L. Tucker
Amster, Rothstein & Engelberg
90 Park Ave.
New York, NY 10016 (212) 697-5995

Marvin Turken
Jordon & Hamburg
122 E. 42nd St., Suite 3303
New York, NY 10168 (212) 986-2340

Helen Tzagoloff
152 E. 94th Street Apt. 4J
New York, NY 10128 (212) 289-5902

John D. Vandenberg#
De Rosa, Vandenberg & Coleman
71 Broadway, Suite 2200
New York, NY 10006 (212) 269-3020

Edward E. Vassallo
Fitzpatrick, Cella, Harper & Scinto
277 Park Ave.
New York, NY 10172 (212) 758-2400

John Charles Vassil
Morgan, Finnegan, Pine, Foley & Lee
345 Park Ave.
New York, NY 10154 (212) 758-4800

Vincent J. Vasta, Jr.
489 Fifth Ave. 32nd Floor
New York, NY 10017 (212) 661-2430

Anthony P. Venturino
Morgan & Finnegan
345 Park Ave.
New York, NY 10154 (212) 758-4800

Bartholomew Verdirame
Morgan, Finnegan, Pine, Foley & Lee
345 Park Ave.
New York, NY 10154 (212) 758-4800

Thomas J. Vetter
Fish & Neave
875 Third Ave.
New York, NY 10022 (212) 715-0600

Mark E. Waddell
Cohen, Pontani & Lierberman
551 Fifth Ave.
New York, NY 10176 (212) 687-2770

Maxim H. Waldbaum
Darby & Darby, P.C. Chrysler Bldg.
405 Lexington Ave.
New York, NY 10174 (212) 697-7660

George H. Wang
Boyle, Vogeler & Haimes
30 Rockefeller Plaza
New York, NY 10112 (212) 265-5100

Patrick D. Ward
720 West End Ave. Room 408-C
New York, NY 10025 (212) 316-6000

Gene Warzecha
Bristol Myers Co.
345 Park Ave.
New York, NY 10154 (212) 546-3108

Milton J. Wayne
Burgess, Ryan & Wayne
370 Lexington Ave.
New York, NY 10017 (212) 683-8150

Frances H. Weber
Amster, Rothstein & Engelberg
90 Park Ave.
New York, NY 10016 (212) 697-5995

David Weild, III
Pennie & Edmonds
1155 Ave. Of The Americas
New York, NY 10036 (212) 790-9090

Paul J. Weiner
415 Madison Ave. 7th Floor
New York, NY 10017 (212) 486-9080

Samuel Henry Weiner
Ostrolenk, Faber, Gerb & Soffen
260 Madison Ave.
New York, NY 10016 (212) 685-8470

Steven I. Weisburd
Ostrolenk, Faber, Gerb & Soffen
260 Madison Ave.
New York, NY 10016 (212) 685-8470

Allen Gardner Weise
Brumbaugh, Graves, Donohue & Raymond
30 Rockefeller Plaza
New York, NY 10112 (212) 408-2532

Morris L. Weiser
Grundig Electric Corp.
935 Broadway
New York, NY 10010 (212) 254-9851

Walter G. Weissenberger#
Hammond & Littell, Weissenberger & Dippert
489 Fifth Ave.
New York, NY 10017 (212) 682-1750

Tiberiu Weisz
Kenyon & Kenyon
One Broadway
New York, NY 10004 (212) 425-7200

Paul B. West
Ladas & Parry
26 West 61st St.
New York, NY 10023 (212) 708-1980

Thomas H. Whaley
Cooper, Dunham, Clark, Griffin & Moran
30 Rockefeller Plaza
New York, NY 10112 (212) 977-9550

John P. White
Cooper, Dunham, Clark, Griffin & Moran
30 Rockefeller Plaza
New York, NY 10112 (212) 977-9550

Vivienne T. White
382 Central Park West Apt. 11X
New York, NY 10025

Mary A. Whiting
Natoli & Pocchia
325 Broadway Suite 304
New York, NY 10007 (212) 619-8087

Richard Whiting
Davis, Hoxie, Faithfull & Hapgood
45 Rockefeller Plaza
New York, NY 10111 (212) 757-2200

George Ward Whitney
Brumbaugh, Graves, Donohue & Raymond
30 Rockefeller Plaza
New York, NY 10112 (212) 408-2530

Arthur E. Wilfond
Amer. Home Products Corp.
685 3rd Ave.
New York, NY 10017 (212) 878-6218

Alexander C. Wilkie, Jr.
Stoll, Wilkie, Previto & Hoffman
5200 Empire State Bldg.
New York, NY 10001 (212) 736-2080

Charles D. Wingate
Kenyon & Kenyon
One Broadway
New York, NY 10004 (212) 425-7200

Ira B. Winkler#
Hopgood, Calimafde, Kalil, Blaustein & Judlowe
60 E. 42nd St.
New York, NY 10165 (212) 986-2480

Nord F. Winnan
P.O. Box 3279
Church St. Station
New York, NY 10008 (609) 443-0456

Drew M. Wintringham
Kenyon & Kenyon
One Broadway
New York, NY 10004 (212) 425-7200

Scott A. Wisser
Kenyon & Kenyon
One Broadway
New York, NY 10004 (212) 425-2200

Eric C. Woglom
Fish & Neave
875 Third Ave.
New York, NY 10022 (212) 715-0600

Michael I. Wolfson
Blum, Kaplan, Friedman, Silberman & Beran
1120 Ave. Of The Americas
New York, NY 10036 (212) 704-0400

Penina Wollman#
11 Riverside Dr.
New York, NY 10023 (212) 595-9011

Milton M. Wolson
Malina & Wolson
60 E. 42nd St.
New York, NY 10165 (212) 986-7410

William Redin Woodward
Frishauf, Holtz, Goodman, & Woodward
261 Madison Ave.
New York, NY 10016 (212) 972-1400

Douglas William Wyatt
Wyatt, Gerber, Shoup, Scobey & Badie
261 Madison Ave.
New York, NY 10016 (212) 687-0911

Leon R. Yankwich
Fish & Neave
875 Third Ave.
New York, NY 10022 (212) 715-0600

Stanley J. Yavner
Wolder, Gross & Yavner
41 E. 42nd St.
New York, NY 10017 (212) 687-3233

Francis W. Young
Akzo America Inc.
111 West 40th St.
New York, NY 10018 (212) 382-5536

Philip Young
19 West 34th St.
New York, NY 10001 (212) 244-0028

S.C. Yuter
Rosen, Dainow & Jacobs
489 5th Ave.
New York, NY 10017 (212) 692-7000

Henry J. Zafian
Fish & Neave
875 Third Ave.
New York, NY 10022 (212) 715-0600

Michael E. Zall
Ostrolenk, Faber, Gerb & Soffen
260 Madison Ave.
New York, NY 10016 (212) 685-8470

Ira L. Zebrak
Darby & Darby, Chrysler Bldg.
405 Lexington Ave.
New York, NY 10174 (212) 697-7660

Charles J. Zeller
Bristol Myers Co.
345 Park Ave.
New York, NY 10154 (212) 546-3648

Steve T. Zelson
Pennie & Edmonds
1155 Ave. Of The Americas
New York, NY 10036 (212) 790-9090

Neil M. Zipkin
Amster, Rothstein & Engelberg
90 Park Ave.
New York, NY 10016 (212) 697-5995

Terry S. Zisowitz
McAulay, Fields, Fisher, Goldstein & Nissen
405 Lexington Ave.
New York, NY 10174 (212) 986-4090

Norman H. Zivin
Cooper, Dunham, Clark, Griffin & Moran
30 Rockefeller Plaza
New York, NY 10112 (212) 977-9550

Anthony M. Zupcic
Fitzpatrick, Cella, Harper & Scinto
277 Park Ave.
New York, NY 10172 (212) 758-2400

William J. Crossetta, Jr.
Dunn & Associates
P.O. Box 96
Newfane, NY 14108 (716) 433-1661

Michael L. Dunn
Dunn & Associates
P.O. Box 96
Newfane, NY 14108 (716) 433-1661

Howard M. Ellis
Dunn & Associates
P.O. Box 96
Newfane, NY 14108 (716) 433-1661

William Gerald Gosz
Occidental Chemical Corp.
P.O. Box 189
Niagara Falls, NY 14302 (716) 773-8459

Stanley J. Herowski, Jr.#
4037 Lewiston Rd.
Niagara Falls, NY 14305 (716) 284-0361

James F. Mudd#
Occidental Chemical Corp.
P.O. Box 189
Niagara Falls, NY 14302 (716) 773-8432

Wallace F. Neyerlin
724 Division Ave.
Niagara Falls, NY 14305 (716) 284-6181

James F. Tao
Occidental Chemical Corp.
P.O. Box 189
Niagara Falls, NY 14302 (716) 773-8400

Stanley I. Laughlin
P.O. Box 2030
North Babylon, NY 11703 (516) 669-1999

Daniel J. Roock#
216 Patricia Dr.
North Syracuse, NY 13212 (315) 457-3167

Jack Frank Kramer
4 Eden Court
North Woodmere, NY 11581

Arnold L. Albin
11 Robert Lennox Dr.
Northport, NY 11768 (516) 757-1766

Mellor Alfred Gill
1 Sea Cove Rd.
Northport, NY 11768 (516) 261-9028

Randall L. Reed
Six West Park Pl.
Norwich, NY 13815 (607) 366-1800

Richard J. Schulte
Norwich Eaton Pharmaceuticals, Inc.
17 Eaton Ave.
Norwich, NY 13815 (607) 335-2276

Martin C. Parkinson#
6 North Delaware Dr.
Nyack, NY 10960 (914) 358-3123

Edward A. Ruestow
36 Valley Rd.
Old Westbury, NY 11568 (516) 626-3565

Peter C. Schechter#
One Skerratt Lane
Ossining, NY 10562 (914) 762-7243

Frank R. Trifari#
8 Justamere Dr.
Ossining, NY 10562 (914) 941-8347

Cyril A. Krenzer
Moran & Krenzer, P.C.
18 Lake St.
P.O. Box 161
Owego, NY 13827 (607) 687-4548

Kenneth F. Dusyn
Dusyn & Dusyn
97-11 83rd St.
Ozone Park, NY 11416 (212) 843-0461

Richard W. Watson#
Garlock Inc.
1666 Division St.
Palmyra, NY 14522 (315) 597-4811

Vern G. DeVries#
American Cyanamid Co.
Lederle Labs.
Pearl River, NY 10965 (914) 735-5000

Sergei S. Brozski#
219 Walnut St.
Peekskill, NY 10566 (914) 739-9522

Peter R. Ruzek
RFD 2 Sprout Brook Rd. Box 440
Peekskill, NY 10566 (914) 739-4275

Robert Joseph Bird
2070 Five Mile Line Rd.
Penfield, NY 14526 (716) 381-8920

Robert Maurice Phipps
1118 Whalen Rd.
Penfield, NY 14526 (716) 377-7185

Ewan Campbell Mac Queen
886 Piermont Ave.
Piermont, NY 10968 (914) 359-0417

James Roy Frederick
44 Creek Ridge
Pittsford, NY 14534 (716) 381-7699

William Thomas French
26 E. Jefferson Rd.
Pittsford, NY 14534 (716) 586-3649

Joseph C. Mac Kenzie#
241 Butler Rd.
Pittsford, NY 14534 (716) 334-1299

Howard S. Robbins
15 Stonington Dr.
Pittsford, NY 14534 (716) 385-6514

Theodore B. Roessel
2933 Clover St.
Pittsford, NY 14534 (716) 381-4014

Donald D. Schaper
14 Stonegate Ln.
Pittsford, NY 14534 (716) 381-0589

John R. Schovee
10 Burr Oak Dr.
Pittsford, NY 14534 (716) 248-8261

Bernard Snyder
21 Netto Lane
Plainview, NY 11803 (516) 433-9687

Henry G. Mc Comb
Three Champlain Dr.
Plattsburgh, NY 12901 (518) 563-0701

Reginald Vincent Craddocd
50 Bogart Ave.
Port Washington, NY 11050 (516) 944-8408

William J. Navin
81 Ivy Way
Port Washington, NY 11050 (516) 767-1437

William D. Gregory#
Dept. of Physics
Clarkston Univ.
Potsdam, NY 13676 (315) 268-2396

Robert William Berray, Sr.
IBM Corp.
Dept 447, Bldg. 414
P.O. Box 950
Poughkeepsie, NY 12602 (914) 433-1161

Joseph A. Biela
IBM Corp.
Dept. 447, Bldg. 414
P.O. Box 950
Poughkeepsie, NY 12602 (914) 433-1172

Carlton B. Fitchett#
28 S. White St.
Poughkeepsie, NY 12601 (914) 452-7894

Edward S. Gershuny
12 Round Hill Rd.
Poughkeepsie, NY 12603 (914) 462-3609

Bernard M. Goldman
IBM Corp.
Dept. 447, Bldg. 414
Box 390
Poughkeepsie, NY 12602 (914) 433-1162

Floyd A. Gonzalez
IBM Corp.
Intell. Property Law Dept.
Dept. 447, Bldg 414
P.O. Box 950
Poughkeepsie, NY 12602 (914) 433-1156

Robert J. Haase
16 Mark Vincent Dr.
Poughkeepsie, NY 12603 (914) 471-7227

John F. Hanifin
17 Thornwood Dr.
Poughkeepsie, NY 12603 (914) 454-9409

Edwin Lester
IBM Corp.
P.O. Box 950
Poughkeepsie, NY 12602 (914) 433-1176

William J. Mc Ginnis, Jr.
IBM Corp.
Dept. 447, Bldg. 414
P.O. Box 950
Poughkeepsie, NY 12602 (914) 433-1174

James Edward Murray, Jr.
IBM Corp.
P.O. Box 950
Poughkeepsie, NY 12602 (914) 463-9290

William S. Robertson, Jr.
IBM Corp.
Patent Operations
P.O. Box 950
Poughkeepsie, NY 12602 (914) 463-9286

Joseph L. Spiegel
Moran, Spiegel, Pergament & Brown
272 Mill St.
Poughkeepsie, NY 12601 (914) 452-7400

Joseph Bernard Taphorn
Eight Scenic Dr.
Poughkeepsie, NY 12603 (914) 462-3262

Charles A. Hugget
Indian Hill Rd.
Route 3
Pound Ridge, NY 10576 (914) 764-8370

William Nelson Barret
IBM Corp.
2000 Purchase St.
Purchase, NY 10577 (914) 697-7556

Paul Denney Carmichael
IBM Corp.
2000 Purchase St.
Purchase, NY 10577 (914) 697-7233

Mark F. Chadurjian
IBM Corp. Intellectual Property Law
2000 Purchase St.
Purchase, NY 10577 (914) 697-7372

E. Ronald Coffman
IBM Corp.
2000 Purchase St.
Purchase, NY 10577 (914) 697-7252

Ronald L. Drumheller
I B M Corp.
2000 Purchase St.
Purchase, NY 10577 (914) 697-6781

Gunter A. Hauptman
I B M Corp. 1F-21
2000 Purchase St.
Purchase, NY 10577 (914) 697-7395

John W. Henderson, Jr.
I B M Corp.
2000 Purchase St. 1G-14
Purchase, NY 10577 (914) 697-7367

Victor Siber
IBM Corp.
2000 Purchase St.
Purchase, NY 10577 (914) 697-7385

Roger S. Smith
I B M Corp.
2000 Purchase St.
Purchase, NY 10577 (914) 697-7244

Saverio P. Tedesco
IBM Corp.
2000 Purchase St.
Purchase, NY 10577 (914) 697-7375

Bernard N. Wiener
IBM Corp.
2000 Purchase St.
Purchase, NY 10598 (914) 697-7260

Yen Sung Yee
IBM Corp.
2000 Purchase St.
Purchase, NY 10577 (914) 697-7369

Gustave Goldstein#
22 James Dr.
Putnam Valley, NY 10579 (914) 526-3588

Irving Karmin#
32-22 92nd St.
Queens, NY 11369 (212) 779-3576

Paul E. Dupont#
Sterling-Winthrop Res. Inst.
Columbia Turnpike
Rensselaer, NY 12144 (518) 445-8292

Thomas Lynn Johnson#
Sterling-Winthrop Res. Inst.
Columbia Turnpike
Rensselaer, NY 12144 (518) 445-8290

Theodore C. Miller#
Sterling - Winthrop Res. Inst.
Rensselaer, NY 12144 (518) 445-8284

Frederick W. Stonner#
Sterling - Winthrop Res. Inst.
Columbia Turnpike
Rensselaer, NY 12144 (518) 445-8291

William Gatewood Webb#
Sterlin - Winthrop Res. Inst.
Columbia Turnpike
Rensselaer, NY 12144 (518) 445-8294

B. Woodrow Wyatt#
Sterling - Winthrop Res. Inst.
Rensselaer, NY 12144 (518) 445-8282

Robert Ochis
P.O. Box 195
Rexford, NY 12148 (609) 778-8837

Herbert Dubno
5676 Riverdale Ave.
Riverdale, NY 10471 (212) 884-6600

Robert W.J. Usher#
5355 Henry Hudson Pkwy.
Riverdale, NY 10471 (212) 796-3361

Alan Michael Abrams
3156 Elmwood Ave.
Rochester, NY 14618

Dennis R. Arndt
Eastman Kodak Co.
343 State St.
Rochester, NY 14650 (716) 726-3896

Gerald E Battist
111 Farm Brook Dr.
Rochester, NY 14625 (716) 724-4969

John Edward Beck
Xerox Corp.
Patent Dept.
Xerox Square-020
Rochester, NY 14644 (716) 423-3868

Bernard David Bogdon
Bausch & Lomb, Inc.
One Lincoln First Square
P.O. Box 54
Rochester, NY 14601 (716) 338-6610

Jeffrey L. Brandt
Eastman Kodak Co.
Patent Dept.
343 State St.
Rochester, NY 14650 (716) 726-3168

Robert Francis Brothers
Eastman Kodak Co.
Patent Dept.
343 State St.
Rochester, NY 14650 (716) 724-4792

Judith L. Byorick
Xerox Corp.
Xerox Square
Rochester, NY 14644 (716) 423-4564

Henry Merritt Chapin
239 Avalon Dr.
Rochester, NY 14618 (716) 442-1593

Ronald F. Chapuran
Xerox Corp.
Patent Dept.
Xerox Square-020
Rochester, NY 14644 (716) 423-4445

George Herman Childress
Eastman Kodak Co.
343 State St.
Rochester, NY 14650 (716) 722-7256

Robert A. Chittum
Xerox Corp.
Xerox Square-020
Rochester, NY 14644 (716) 423-4636

Thomas H. Close
Eastman Kodak Co.
343 State St.
Rochester, NY 14650 (716) 477-5272

Robert F. Cody
Eastman Kodak Co.
343 State St.
Rochester, NY 14650 (716) 726-3087

Harold E. Cole
Eastman Kodak Co.
343 State St.
Rochester, NY 14650 (716) 722-9225

Mark Costello
Xerox Corp.
Xerox Square-020
Rochester, NY 14644 (716) 423-5006

Torger N. Dahl
Eastman Kodak Co.
343 State St.
Rochester, NY 14650 (716) 724-4899

William J. Davis
Eastman Kodak Co.
Patent Dept., Bldg. 83
343 State St.
Rochester, NY 14650 (716) 477-7419

William F. Delaney, Jr.
Eastman Kodak Co.
343 State St.
Rochester, NY 14650 (716) 724-4960

Dennis M. Deleo
Eastman Kodak Co.
343 State St.
Rochester, NY 14650 (716) 724-7804

William C. Dixon, III
Eastman Kodak Co.
343 State St.
Patent Dept
Rochester, NY 14650 (716) 477-7418

John R. Everett
Eastman Kodak Co.
Bldg. 83
343 State St.
Rochester, NY 14650 (716) 722-2776

Roger A. Fields
Eastman Kodak Co.
343 State St.
Rochester, NY 14650 (716) 726-2995

Philip Karnes Fitzsimmons
Shlesinger, Fitzsimmons & Shlesinger
183 E. Main St., Ste. 1323
Rochester, NY 14604 (716) 325-4618

Henry Fleischer
Xerox Corp.
Xerox Square-020
Rochester, NY 14644 (716) 423-4225

Kay Donald Fosnaught
Eastman Kodak Co.
Patent Dept.
343 State St.
Rochester, NY 14650 (716) 724-3167

Richard Dauster Fuerle
Eastman Kodak Co.
343 State St.
Rochester, NY 14650 (716) 722-9194

Norman H. Geil
Eastman Kodak Co.
Legal Dept.
343 State St.
Rochester, NY 14650 (716) 724-5129

Samuel Richard Genca
2990 Culver Rd.
Rochester, NY 14622 (716) 266-4480

Robert A. Gerlach
Eastman Kodak Co.
Patent Dept.
343 State St.
Rochester, NY 14650 (716) 722-9430

Frank R. Gollon
133 Danbury Circle
Rochester, NY 14618 (716) 244-8814

Howard J. Greenwald
700 Executive Office Bldg.
36 W. Main St.
Rochester, NY 14614 (716) 454-1200

Steve W. Gremban
Eastman Kodak Co.
343 State St.
Rochester, NY 14650 (716) 722-9194

Douglas Ian Hague
Eastman Kodak Co.
343 State St.
Rochester, NY 14650 (716) 724-4181

Ralph E. Harper
Gleason Corp.
30 Corporate Woods
P.O. Box 22856
Rochester, NY 14692 (716) 272-6041

Joseph J. Hawley
Eastman Kodak Co.
343 State St.
Rochester, NY 14650 (716) 722-9271

Stanley Seamans Hazen
Eastman Kodak Co.
Training Dept., Bldg. 2
1669 Lake Ave
Kodak Park
Rochester, NY 14650 (716) 477-7502

William A. Henry, II
Xerox Corp.
Xerox Square - 20A
Rochester, NY 14644 (716) 423-3086

Tom Hiatt
123 Scotch Ln.
Rochester, NY 14617 (716) 467-4237

Ronald P. Hilst
Eastman Kodak Co.
343 State St.
Rochester, NY 14650 (716) 724-3391

Eugene C. Holloway
Harris, Beach, Wilcox, Rubin & Levey
Two State St.
Rochester, NY 14614 (716) 232-4440

Paul R. Holmes
87 Farm Brook Dr.
Rochester, NY 14625 (716) 381-2946

Dwight J. Holter
Eastman Kodak Co.
343 State St.
Rochester, NY 14650 (716) 724-2883

Hugo F. Huedepohl
Eastman Kodak Co.
Kodak Off., B-6
343 State St.
Rochester, NY 14650 (716) 724-2883

John David Husser
Eastman Kodak Co.
343 State St.
Rochester, NY 14650 (716) 477-5256

David F. Janci
Eastman Kodak Co.
Patent Dept.
343 State St.
Rochester, NY 14650 (716) 722-9139

Joe Allen Jones
Eastman Kodak Co.
343 State St.
Rochester, NY 14650 (716) 724-4437

Ronald S. Kareken
Eastman Kodak Co.
343 State St.
Rochester, NY 14622 (716) 724-4669

Lawrence P. Kessler
Eastman Kodak Co.
343 State St.
Rochester, NY 14650 (716) 477-3421

Thomas F. Kirchoff
Eastman Kodak Co.
343 State St.
Rochester, NY 14650 (716) 722-9349

William H. J. Kline
25 Indian Spring Ln.
Rochester, NY 14618 (716) 244-6627

Richard Elliott Knapp
Eastman Kodak Co.
Patent Dept.
343 State St.
Rochester, NY 14650 (716) 722-9424

Peter H. Kondo
Xerox Corp.
Pat. Dept.
Xerox Square-020
Rochester, NY 14644 (716) 423-4308

Warren W. Kurz
Eastman Kodak Co.
343 State St.
Rochester, NY 14650 (716) 722-2396

Craig E. Larson
Bausch & Lomb
One Lincoln First Square
P.O. Box 54
Rochester, NY 14601 (716) 338-6613

Joshua Gerald Levitt
Eastman Kodak Co.
343 State St.
Rochester, NY 14650 (716) 722-9426

James Lord Lewis
Eastman Kodak Co.
343 State St.
Rochester, NY 14650 (716) 477-5158

Alfred Paul Lorenzo
Eastman Kodak Co.
343 State St.
Rochester, NY 14650 (716) 477-3413

Martin Lukacher
2000 Lincoln First Tower
Rochester, NY 14604 (716) 454-2790

Seymour Manello
265 Warren Ave.
Rochester, NY 14602 (716) 244-4738

Owen D. Marjama
Marjama & Pincelli, P.C.
488 White Spruce Blvd.
Rochester, NY 14623 (716) 272-8230

Paul L. Marshall
Eastman Kodak Co.
Patent Dept.
343 State St.
Rochester, NY 14650 (716) 477-2625

Thomas R. Marton#
102 Southland Dr.
Rochester, NY 14623 (716) 424-3815

Norman Dean Mc Claskey
Eastman Kodak Co.
343 State St.
Rochester, NY 14650 (716) 724-2720

Frederick E. Mc Mullen
Xerox Corp.
Pat. Dept.
Xerox Square-020
Rochester, NY 14644 (716) 423-3715

Ruth E. Merling
204 Oakwood Rd.
Rochester, NY 14616

Howard Anthony Miller#
3156 Elmwood Ave.
Rochester, NY 14618

Dennis P. Monteith
Eastman Kodak Co.
343 State St.
Rochester, NY 14650 (716) 726-3537

De Witt M. Morgan
Bausch & Lomb Inc.
42 East Ave.
P.O. Box 743
Rochester, NY 14603 (716) 338-6612

Paul F. Morgan
Xerox Corp.
Xerox Square-020
Rochester, NY 14644 (716) 423-3015

John A. Morrow
Eastman Kodak Co.
343 State St.
Rochester, NY 14650 (716) 726-3533

Samuel Elmore Mott, III
Xerox Corp.
Pat. Dept
Xerox Square-020
Rochester, NY 14644 (716) 423-3980

Jan A. Muddle#
Eastman Kodak Co.
343 State St.
Rochester, NY 14650 (716) 477-5595

Robert Constantine Najjar#
Eastman Kodak Co.
Research Labs, B-83
Kodak Park
Rochester, NY 14650 (716) 722-0267

Irving Newman
Eastman Kodak Co.
343 State St.
Rochester, NY 14650 (716) 722-9343

John S. Norton#
Bausch & Lomb Inc.
42 East Ave.
P.O. Box 743
Rochester, NY 14603 (716) 338-6611

William F. Noval
Eastman Kodak Co.
343 State St.
Rochester, NY 14650 (716) 477-4027

Raymond L. Owens
Eastman Kodak Co.
343 State St.
Rochester, NY 14650 (716) 477-4653

Armin B. Pagel
Eastman Kodak Co.
Patent Dept.
343 State St.
Rochester, NY 14650 (716) 726-3424

Eugene Onotrio Palazzo
Xerox Corp.
Xerox Square-20A
Rochester, NY 14644 (716) 423-4687

Bruce Stanton Peachey#
Eastman Kodak Co.
343 State St.
Rochester, NY 14650 (716) 722-9134

Frank Pincelli
Marjama & Pincelli, P.C.
488 White Spruce Blvd.
Rochester, NY 14623 (716) 272-8230

Morton Arnold Polster
Gleason Corp.
30 Corporate Woods
P.O. Box 22856
Rochester, NY 14692 (716) 272-6040

Robert Lloyd Randall
Eastman Kodak Co.
Patent Dept.
343 State St.
Rochester, NY 14650 (716) 726-2132

Arthur Henry Rosenstein
Eastman Kodak Co.
343 State St.
Rochester, NY 14650 (716) 722-9342

Norman Rushefsky
Eastman Kodak Co.
Patent Dept.
343 State St.
Rochester, NY 14650 (716) 477-3765

Thomas B. Ryan#
Gleason Corp.
30 Corporate Woods
P.O. Box 22856
Rochester, NY 14692 (716) 282-6044

Joseph R. Sakmyster
Xerox Corp.
Pat. Dept.
Xerox Square
Rochester, NY 14644 (716) 423-4705

Stephen B. Salai
Cumpston & Shaw, P.C.
850 Crossroads Office Bldg.
Two State St.
Rochester, NY 14614 (716) 325-5553

Milton Saunders Sales
Eastman Kodak Co.
343 State St.
Rochester, NY 14650 (716) 722-0355

Dana Murray Schmidt
Eastman Kodak Co.
343 State St.
Rochester, NY 14650 (716) 722-9151

George W. Shaw
Cumpston & Shaw, P.C.
850 Crossroads Office Bldg.
Two State St.
Rochester, NY 14614 (716) 325-5553

Charles Shepard
Stonebraker, Shepard & Stephens
500 Allens Creek Rd.
Rochester, NY 14618 (716) 248-3390

Benjamin B. Sklar, Jr.#
Xerox Corp
100 Clinton Ave., S.
Rochester, NY 14644 (716) 423-4554

James Alfred Smith
Eastman Kodak Co.
Patent Dept.
343 State St.
Rochester, NY 14650 (716) 722-1498

Robert H. Sproule
Eastman Kodak Co.
Patent Dept.
343 State St.
Rochester, NY 14650 (716) 726-9416

Eugene S. Stephens
Stonebraker, Shepard & Stephens
75 College Ave.
Rochester, NY 14607 (716) 244-7910

Elliott Stern
Eastman Kodak Co.
343 State St.
Rochester, NY 14650 (716) 724-5107

Hoffman Stone
1600 Midtown Tower
Rochester, NY 14604 (716) 454-2790

Donald W. Strickland
Eastman Kodak Co.
343 State St.
Rochester, NY 14650 (716) 372-1179

Herman J. Strnisha
Eastman Kodak Co.
343 State St.
Rochester, NY 14650 (716) 722-9134

Carl Otis Thomas
Eastman Kodak Co.
343 State St.
Rochester, NY 14650 (716) 722-9127

Leonard W. Treash, Jr.
Eastman Kodak Co.
343 State St.
Rochester, NY 14650 (716) 477-7600

Richard L. Troutman#
Eastman Kodak Co.
Research Labs, Bldg. 82, Rm. 601
1999 Lake Ave.
Rochester, NY 14615 (716) 477-6177

James L. Tucker
Eastman Kodak Co.
343 State St.
Rochester, NY 14560 (716) 722-9332

John B. Turner#
Eastman Kodak Co.
343 State St.
Rochester, NY 14650 (716) 726-2115

S. C. Van Houten
141 Glen View Ln.
Rochester, NY 14609 (716) 482-9518

Ogden H. Webster
Eastman Kodak Co.
343 State St.
Rochester, NY 14650 (716) 724-4437

Bernard Donald Wiese
Eastman Kodak Co.
343 State St.
Rochester, NY 14650 (716) 722-9021

David M. Woods
Eastman Kodak Co.
343 State St.
Rochester, NY 14650 (716) 726-2180

Harold S. Wynn#
3700 East Ave., Ste. 116
Rochester, NY 14618 (716) 381-5374

William P. Keegan
P.O. Box 293
Rockaway Park, NY 11694 (212) 634-3080

James P. Malone
1 Hillside Ave.
Rockville Centre, NY 11570 (516) 766-3810

Jack W. Benjamin#
257-27 149th Rd.
Rosedale, NY 11422 (718) 723-1001

Raymond W. Augustin
Collard, Roe & Galgano, P.C.
1077 Northern Blvd.
Roslyn, NY 11576 (516) 365-9802

Allison Charles Collard
Collard, Roe & Galgano, P.C.
1077 Northern Blvd.
Roslyn, NY 11576 (516) 365-9802

Thomas M. Galgano
Collard, Roe & Galgano, P.C.
1077 Northern Blvd.
Roslyn, NY 11576 (516) 365-9802

Kurt Kleman#
1077 Northern Blvd.
Roslyn, NY 11576 (516) 627-9104

Chou H. Li#
379 Elm Dr.
Roslyn, NY 11576 (516) 484-1719

Erwin S. Teltscher
69 Diana S Trail
Roslyn, NY 11576 (516) 621-8520

Daniel H. Brown
21 Fenway
Roslyn Estates, NY 11576 (516) 621-1452

Jackson B. Browning
51 Island Dr.
Rye, NY 10580

James Martin Heilman
Heilman & Heilman
10 Hix & Oakland Beach Aves.
Rye, NY 10580 (914) 967-2095

John L. Sniado#
2 Eve Lane
Rye, NY 10580 (914) 967-3628

Theodore C. Jay
160 Brush Hollow Crescent
Rye Brook, NY 10573 (914) 939-6697

William L. Luc#
IBM Corp.
900 King St.
Rye Brook, NY 10562 (914) 934-4264

James W. Fitzsimmons
58 Washington St.
P.O. Box 414
Saratoga Springs, NY 12866 (518) 587-9656

Howard N. Aronson
Lackenbach, Siegel, Marzullo & Aronson
One Chase Rd. Penthouse Suite
Scarsdale, NY 10583 (914) 723-4300

Leonard Cooper
Lackenbach, Siegel, Marzullo & Aronson, P.C.
One Chase Rd. Penthouse Suite
Scarsdale, NY 10583 (914) 723-4300

Edward R. Freedman
Lackenbach, Siegel, Marzullo & Aronson
One Chase Rd. Penthouse Suite
Scarsdale, NY 10583 (914) 723-4300

Phyllis C. Klass#
25 Griffen Ave.
Scarsdale, NY 10583 (914) 725-0327

David Barry Koss
164 White Rd.
Scarsdale, NY 10583 (914) 725-0542

Melvin H. Kurtz
93 Walworth Ave.
Scarsdale, NY 10583 (914) 723-7029

Armand E. Lackenbach
Lackenbach, Seigel, Marzullo, Presta & Aronson
One Chase Rd. Penthouse Suite
Scarsdale, NY 10583 (914) 723-4300

Walter Lewis
12 Wakefield Rd.
Scarsdale, NY 10583 (914) 723-3274

Henry Anthony Marzullo, Jr.
Lackenbach, Siegel, Marzullo & Aronson, P.C.
One Chase Rd.
Scarsdale, NY 10583 (914) 723-4300

James J. OConnell
170 Boulevard
Scarsdale, NY 10583 (914) 725-6141

Jack Posin
624 Fort Hill Rd.
Scarsdale, NY 10583 (914) 723-0061

James E. Siegel
Lackenbach, Siegel, Marzullo, Presta & Aronson, P.C.
1 Chase Rd. Penthouse Suite
Scarsdale, NY 10583 (914) 723-4300

Gerhard K. Adam
823 State St.
Schenectady, NY 12307 (518) 346-7085

Jane M. Binkowski
General Electric Co.
P.O. Box 8, Corp. Res. & Dev.
Schenectady, NY 12308 (518) 387-6289

Richard V. Burgujian
General Electric Co. Bldg. K1 Room 3A69
P.O. Box 8
Schenectady, NY 12301 (518) 387-6275

Donald R. Campbell
General Elect. Co. Corp. Res. & Dev.
1 River Rd.
Schenectady, NY 12301 (518) 387-7714

Joseph T. Cohen
1320 Lexington Ave.
Schenectady, NY 12309 (518) 372-9481

Francis T. Coppa
General Electric Co. Corp. Res. & Dev. Center
P.O. Box 8, K1-3A68
Schenectady, NY 12301 (518) 387-6283

James Clark Davis, Jr.
General Electric Co.
P.O. Box 8
Schenectady, NY 12301 (518) 387-6480

Peter D. Johnson#
1100 Merlin Dr.
Schenectady, NY 12309 (518) 785-5035

Geoffrey H. Krauss
General Elec. Co.
P.O. Box 43
Schenectady, NY 12345 (518) 385-3289

Bernard Joseph Lacomis
General Electric Co.
120 Eric Blvd.
Schenectady, NY 12305 (518) 385-5367

Michael A. Lamanna
General Electric Co.
Bldg. 500- Room 218
1 River Rd.
Schenectady, NY 12345 (518) 385-4867

James Magee, Jr.
G E Company
P.O. Box 8
Schenectady, NY 12301 (518) 385-8510

William Howard Pittman
General Electric Co. Corp. Res. & Dev. Center
P.O. Box 8
Schenectady, NY 12301 (518) 387-5258

Paul Edward Rochford
General Electric Co.
P.O. Box 8, Bldg. K1
Schenectady, NY 12301 (518) 385-8168

Marvin Snyder
General Electric Co.
P.O. Box 8, Bldg. K-1, Room 3A58
Schenectady, NY 12301 (518) 385-3826

Jerome Carmen Squillaro
General Electric Co.
1 River Rd., Bldg. 500 - Rm. 218
Schenectady, NY 12345 (518) 385-4650

William A. Teoli
General Electric Co. Corp. Res. & Dev.
P.O. Box 8
Schenectady, NY 12301 (518) 387-5872

Richard J. Traverso
General Electric Co. Corp. Res. & Dev.
P.O. Box 8, Bldg. K-1, 3A68
Schenectady, NY 12301 (518) 387-6276

Paul Richard Webb, II
General Electric Co.
P.O. Box 8
Schenectady, NY 12301 (518) 387-5892

Julius J. Zaskalicky
3-9 Netherlands Village Apts.
Schenectady, NY 12308

Kevin R. Kepner#
92A Broad St.
Schuylerville, NY 12871 (518) 695-6866

Leo I. Malossi
10 Park Lane
Scotia, NY 12302 (518) 399-5569

James W. Underwood#
9 Daphne Dr.
Scotia, NY 12302 (518) 399-4445

Michael J. Doyle
General Elec. Co.
One Noryl Ave.
Selkirk, NY 12158 (518) 475-5204

John C. Fox
Philips Ecg, Inc.
50 Johnston St.
Seneca Falls, NY 13148 (315) 568-5881

Frederick H. Rinn#
39 Maple St.
Seneca Falls, NY 13148 (315) 568-6926

Daniel Monroe Schaeffer
6 Frederick Dr.
Shoreham, NY 11786 (516) 744-2323

John P. Murphy
3237 E. Lake Rd.
Skaneateles, NY 13152 (315) 685-6608

Shirley K. Morse#
Krumkill Rd.
Slingerlands, NY 12159 (518) 482-6712

Leonard Belkin
202 E. Main St.
Smithtown, NY 11787 (516) 360-3235

Philip Furgang
49 S. Main St.
Spring Valley, NY 10977 (914) 352-2244

Louis A. Tirelli
52 S. Main St.
Spring Valley, NY 10977 (914) 352-4247

Carol K. Lackenbach
507 Maguire Ave.
Staten Island, NY 10309 (212) 948-8928

Frank V. Ponterio#
766 Pelton Ave.
Staten Island, NY 10310 (212) 981-7953

Raymond J. Kenny
Inco Res. & Dev. Ctr. Inc.
Sterling Forest
Suffern, NY 10901 (914) 753-2761

Leo C. Kranzinski
P.O. Box 556
Suffern, NY 10901 (212) CO7-4321

Miriam W. Leff
INCO (u.s.)
P.O. Box 200
Suffern, NY 10901 (914) 578-5634

Francis John Mulligan, Jr.
INCO U.S.
Box 200
Suffern, NY 10901 (914) 578-5694

George J. Darsa
15 Deer Path Lane
Syosset, NY 11791 (516) 921-1948

Morris Krapes
8 Jackson Ave.
Syosset, NY 11791 (516) 496-8466

Michael I. Kroll
171 Stillwell Lane
Syosset, NY 11791 (516) 692-2753

Marilyn L. Olshansky
5 Belmont Circle
Syosset, NY 11791 (516) 921-8060

Steven J. Winick
Ademco
165 Eileen Way
Syosset, NY 11791 (516) 921-6704

Carl W. Baker
General Electric Co.
Electronics Park, 6-102
Syracuse, NY 13221 (315) 456-3682

Dana F. Bigelow
Carrier Corp.
6304 Carrier Pkwy.
P.O. Box 4800
Syracuse, NY 13221 (315) 433-4642

Richard V. K. Bruns
Bruns & Wall
217 Montgomery St.
Syracuse, NY 13202 (315) 422-7383

Arthur A. Chalenski, Jr.
Mackenzie, Smith, Lewis, Michell & Hughes
600 Onondaga Savings Bk. Bldg.
Syracuse, NY 13202 (315) 474-7571

Donald Francis Daley
Carrier Corp.
6304 Carrier Pkwy.
P.O. Box 4800
Syracuse, NY 13221 (315) 433-4819

Frank N. Decker, Jr.
1342 Broad St.
Syracuse, NY 13224 (315) 446-2522

Marvin A. Goldenberg
811 State Tower Bldg.
Syracuse, NY 13202 (312) 422-1191

Robert H. Kelley
Carrier Corp.
Patent Dept.
6304 Carrier Pkwy.
P.O. Box 4800
Syracuse, NY 13221 (315) 433-4609

Richard B. Lang
General Electric Co.
Electronics Park
Syracuse, NY 13221 (315) 456-2519

Richard R. Lloyd#
Bristol-Myers Co.
P.O. Box 4755
Syracuse, NY 13221 (315) 432-2399

Charles Stevens Mc Guire
840 James St.
Syracuse, NY 13203 (314) 471-0361

Bernhard P. Molldrem, Jr.
Bruns & Wall
512 Hills Bldg.
Syracuse, NY 13202

Robert James Mooney
General Electric Co.
Electronics Park
Syracuse, NY 13221 (315) 456-2691

August Edward Roehrig, Jr.
Wall & Roehrig
710 Hills Bldg.
217 Montgomery St.
Syracuse, NY 13202 (315) 422-7383

Lawrence P. Trapani
Bruns & Wall
710 Hills Bldg.
217 Montgomery St.
Syracuse, NY 13202 (315) 422-7383

John R. Varney
Hancock & Estabrook
One Mony Plaza
Syracuse, NY 13202 (315) 417-3151

Thomas J. Wall
Bruns & Wall
710 Hills Bldg.
217 Montgomery St.
Syracuse, NY 13202 (315) 422-7383

Anne E. Barschall
U.S. Philips
580 White Plains Rd.
Tarrytown, NY 10591 (914) 332-0222

Steven R. Biren
U.S. Philips Corp.
580 White Plains Rd.
Tarrytown, NY 10591 (914) 332-0222

Thomas Allen Briody
U.S. Philips Corp.
580 White Plains Rd.
Tarrytown, NY 10591 (914) 332-0222

F. Brice Faller
U.S. Philips Corp.
580 White Plains Rd.
Tarrytown, NY 10591 (914) 332-0222

Bernard Franzblau
U.S. Philips Corp.
580 White Plains Rd.
Tarrytown, NY 10591 (914) 332-0222

Gregory P. Gadson
U.S. Philips Corp.
580 White Plains Rd.
Tarrytown, NY 10591 (914) 332-0222

Barbara R. Greenberg#
449 Martling Ave.
Tarrytown, NY 10591 (914) 332-4452

Jeffrey M. Greenman
Technicon Instruments Corp.
511 Benedict Ave.
Tarrytown, NY 10591 (914) 333-6093

Jack E. Haken
U.S. Philips Corp.
580 White Plains Rd.
Tarrytown, NY 10591 (914) 332-0222

Robert J. Kraus
U.S. Philips Corp.
580 White Plains Rd.
Tarrytown, NY 10591 (914) 332-0222

Robert T. Mayer
U.S. Philips Corp.
580 White Plains Rd.
Tarrytown, NY 10591 (914) 332-0222

Paul R. Miller
U.S. Philips Corp.
580 White Plains Rd.
Tarrytown, NY 10591 (914) 332-0222

Stephen E. Rockwell
177 White Plains Rd.
Apt. 63-X
Tarrytown, NY 10591 (914) 631-0820

Norman N. Spain
U.S. Philips Corp.
580 White Plains Rd.
Tarrytown, NY 10591 (914) 332-0222

William J. Streeter
Nor. Amer. Philips Corp.
580 White Plains Rd.
Tarrytown, NY 10591 (914) 332-0222

Algy Tamoshunas
U.S. Philips Corp.
580 White Plains Rd.
Tarrytown, NY 10591 (914) 332-0222

David Robertson Treacy
U.S. Philips Corp.
580 White Plains Rd.
Tarrytown, NY 10591 (914) 332-0222

Walter J. Olszewski#
Union Carbide Corp.
P.O. Box 44
Tonawanda, NY 14151 (716) 879-2722

Walt Thomas Zielinski
International Paper Co. Corp. Res. Center
Long Meadow Rd.
Tuxedo Park, NY 10987 (914) 351-2101

Charles V. Grudzinskas#
501 N. Broadway
Upper Nyack, NY 10960 (914) 358-9232

Margaret C. Bogosian
Brookhaven National Lab. Bldg. 355
Upton, NY 11973 (516) 282-7338

Vale P. Myles
Brookhaven National Labs. Bldg. 355
Upton, NY 11973 (518) 282-3312

Harold D. Berger
30 Eastwood Lane
Valley Stream, NY 11581 (516) 791-7179

Bernard S. Hoffman#
63 South Dr.
Valley Stream, NY 11581 (516) 791-2488

Leonard H. King
P.O. Box 67
Valley Stream, NY 11582 (516) 997-7050

Kenneth P. Johnson
808 Sequoia Ln.
Vestal, NY 13850 (607) 748-8752

John L. Young
General Electric Company
260 Hudson River Rd.
Waterford, NY 12188 (518) 266-2471

Fred L. Denson
14 E. Main St.
P.O. Box 801
Webster, NY 14580 (716) 265-2710

Gary G. Henry#
335 Brooksboro Dr.
Webster, NY 14580

Wayne Henry Lang#
60 Lee Pl.
Wellsville, NY 14895 (716) 593-39C2

Daniel Jay Tick
463 Dunster Court
West Hempstead, NY 11552 (516) 485-0481

William J. Eppig
1175 Montauk Hwy.
West Islip, NY 11795 (516) 587-8778

Thomas T. Kashiwabara#
St. Regis Paper Co.
West Nyack Rd.
West Nyack, NY 10994 (914) 578-7180

Joseph Michael Maguire
St. Regis Paper Co. Tech. Center
West Nyack Rd.
West Nyack, NY 10994 (914) 578-7038

William R. Robinson
14 Wheeler Place
West Nyack, NY 10994 (914) 358-7368

Herbert S. Ingham
Perkin-Elmer Corp.
1101 Prospect Ave.
Westbury, NY 11590 (516) 683-2216

Eugene Sheek Lovette
460 Canterbury St. Suite 11590
Westbury, NY 11590 (516) 334-6870

James N. Hulme
Kelly & Hulme
277 Mill Rd.
Westhampton Beach, NY 11978
(516) 288-2876

Robert Augustin Kelly
86 Main St.
Westhampton Beach, NY 11978
(516) 288-2876

Curtis Ailes
Ailes & Ohlandt
175 Main St.
White Plains, NY 10601 (914) 949-7677

Mitchell E. Alter
General Foods Corp.
250 North St.
White Plains, NY 10625 (914) 335-9207

I. Walton Bader
Bader & Bader
65 Court St.
White Plains, NY 10601 (914) 682-0072

Charles L. Bauer
Texaco Dev. Corp.
2000 Westchester Ave.
White Plains, NY 10625 (914) 253-4041

Albert Brent
Texaco Dev. Corp.
2000 Westchester Ave.
White Plains, NY 10625 (914) 253-4541

Robert B. Burns
Texaco Dev. Corp.
2000 Westchester Ave.
White Plains, NY 10625 (914) 253-4542

Mark A. Campbell#
175 Main St. 8th Floor
White Plains, NY 10601 (914) 683-8223

Barbara T. D'Avanzo
General Foods Corp.
250 North St.
White Plains, NY 10625 (914) 335-9226

Daniel James Donovan
General Foods Co.
250 North St.
White Plains, NY 10625 (914) 335-9228

Robert J. Eichelburg
15th Floor
44 S. Broadway
White Plains, NY 10601 (914) 948-3616

Eugene E. Geoffrey, Jr.
Graham, Geoffrey & Reardon
470 Mamaroneck Ave. Room 402
White Plains, NY 10605 (914) 997-1520

Ronald G. Gillespie
Texaco Development Corp.
2000 Westchester Ave.
White Plains, NY 10650 (914) 253-4537

Mitchael J. Goldblatt
General Foods Corp.
250 N. St.
White Plains, NY 10625 (914) 683-2500

Myron Greenspan
Lilling & Greenspan
123 Main St.
White Plains, NY 10601 (914) 684-0600

Linn I. Grim
General Foods Corp.
250 North St.
White Plains, NY 10625 (914) 335-7806

Joseph T. Harcarik
General Foods Corp.
250 North St.
White Plains, NY 10625 (914) 335-9219

Joyce P. Hill
General Foods Corp.
250 North St.
White Plains, NY 10625 (914) 335-9188

C. Garman Hubbard
25 Hazelton Dr.
White Plains, NY 10605 (914) 949-9488

A. Kate Huffman
Lilling & Greenspan
123 Main St. Suite 936
White Plains, NY 10601 (212) 365-6665

Gerald Edward Jacobs
General Foods Corp.
250 North St.
White Plains, NY 10625 (914) 683-2362

Glenn E. Karta
Vogt & O'Donnell
707 Westchester Ave.
White Plains, NY 10604 (914) 328-0055

Kevin E. Kavanagh
Texaco Dev. Corp.
2000 Westchester Ave.
White Plains, NY 10650 (914) 253-7941

Robert Knox, Jr.
Texaco Development Corp.
2000 Westchester Ave.
White Plains, NY 10605 (914) 253-4043

Robert A. Kulason
Texaco Dev. Corp.
2000 Westchester Ave.
White Plains, NY 10650 (914) 253-4042

Bruce E. Lilling
Lilling & Greenspan
123 Main St. Suite 936
White Plains, NY 10601 (914) 684-0600

Burton Lawrence Lilling
Lilling & Greenspan
123 Main St. Suite 936
White Plains, NY 10601 (914) 684-0600

Kenneth E. Macklin
180 South Bdwy.
White Plains, NY 10605 (914) 949-6550

Thomas R. Madden#
Reichold Chemical Inc.
525 North Broadway
White Plains, NY 10603 (914) 682-5884

Thomas A. Marcoux
General Foods Corp.
250 North St.
White Plains, NY 10625 (914) 335-9220

Basam E. Nabulsi#
General Foods Corporation
250 North St.
White Plains, NY 10625 (914) 335-9225

James J. O'Loughlin
Texaco Dev. Corp.
2000 Westchester Ave.
White Plains, NY 10650 (914) 253-7943

John F. Ohlandt, Jr.
175 Main St.
White Plains, NY 10601 (914) 949-3389

Walter R. Pfluger, Jr.
Amf Incorporated
777 Westchester Ave.
White Plains, NY 10604 (914) 694-2916

David R. Plautz#
Vogt & O'Donnell
707 Westchester Ave.
White Plains, NY 10604 (914) 328-0055

Irwin Pronin
123 Main St. Suite 900
White Plains, NY 10601 (914) 948-1556

Michael J. Quillinan
General Foods Corp.
250 North St.
White Plains, NY 10625 (914) 335-9209

Charles B. Rodman
Rodman & Rodman
7-11 South Broadway
White Plains, NY 10601 (914) 949-7210

Philip L. Rodman
Rodman & Rodman
7-11 South Broadway
White Plains, NY 10601 (914) 949-7210

Thomas Richard Savoie
General Foods Corp.
250 North St.
White Plains, NY 10625 (914) 335-9222

Richard D. Schmidt
General Foods Corp.
250 North St. R A-6N
White Plains, NY 10625 (914) 335-9223

Rolf E. Schneider
125 Lake Street Apt. 7D-S
White Plains, NY 10604 (914) 332-0222

James M. Serafino
General Foods Corp.
250 North St.
White Plains, NY 10625 (914) 335-9198

Carl G. Seutter
15 Idlewood Rd.
White Plains, NY 10605 (914) 253-4567

Louis S. Sorell
Texaco Dev. Corp.
2000 Westchester Ave.
White Plains, NY 10650 (914) 253-7970

Martin J. Spellman, Jr.
Spellman & Joel
44 S. Broadway
White Plains, NY 10601 (914) 997-0200

Bruno P. Struzzi
General Foods Corp.
250 North St.
White Plains, NY 10625 (914) 335-9210

William H. Vogt, III
Vogt & O'Donnell
707 Westchester Ave.
White Plains, NY 10604 (914) 328-0055

Sam D. Walker
General Foods Corp.
250 North St.
White Plains, NY 10625 (914) 335-9221

Leo Zucker
50 Main St. 8th Floor
White Plains, NY 10606 (914) 761-7799

Charles J. Brown, Jr.
Brown, Kelleher, Zwickel & Wilhelm
Main St. P.O. Box 489
Wilhelm, NY 12496 (518) 734-3800

Allen J. Jaffe
340 Sprucewood Terr.
Williamsville, NY 14221 (716) 632-0633

David A. Stein#
91 Exeter Rd.
Williamsville, NY 14221 (716) 634-9836

Robert R. Strack
Eisenman, Allsopp & Strack
100 Crossways Park W.
Woodbury, NY 11797 (516) 364-3190

Dale A. Bauer
10 Mayflower Dr.
Yonkers, NY 10710 (914) 779-7599

Edward F. Levy
136 La Salle Dr.
Yonkers, NY 10710 (212) 940-0770

Louis C. Smith, Jr.
1200 Warburton Ave. Apt. 22
Yonkers, NY 10701 (914) 423-6721

George P. Ziehmer
243 Scarsdale Rd.
Yonkers, NY 10707 (914) 779-6178

Jack M. Arnold
IBM Corp.
P.O. Box 218
Yorktown Heights, NY 10598 (914) 241-4044

Marc A. Block
IBM Corp.
P.O. Box 218
Yorktown Heights, NY 10598 (914) 241-4060

Douglas W. Cameron
IBM Corp.
P.O. Box 218
Yorktown Heights, NY 10598 (914) 241-4288

Frank Chadurjian
I B M Corp.
P.O. Box 218
Yorktown Heights, NY 10598 (914) 241-4042

Thirumal R. Coca
I B M Corp. Intellectual Property Law Dept.
74-D32, P.O. Box 218
Yorktown Heights, NY 10598 (914) 241-4268

Thomas P. Dowd
I B M Corp.
P.O. Box 218 - D/ 48-78
Yorktown Heights, NY 10598 (914) 241-4062

Mark J. Egyed#
I B M Corp.
Thomas J. Watson Res. Center M/s 24-257
Box 218
Yorktown Heights, NY 10598 (914) 945-1543

Philip J. Feig
IBM Corp.
Intellectual Property Law Dept.
T.J. Watson Research Center, P.O. Box 218
Yorktown Heights, NY 10598 (914) 241-4055

John J. Goodwin
I B M Corp.
Intellectual Property Law Dept.
P.O. Box 218
Yorktown Heights, NY 10598 (914) 241-4045

Terry J. Ilardi
I B M Corp.
T.J. Watson Research Center
P.O. Box 218
Yorktown Heights, NY 10598 (914) 241-4093

John A. Jordan
I B M Corp.
Intellectual Property Law Dept.
P.O. Box 218., 74-E50
Yorktown Heights, NY 10598 (914) 241-4058

Thomas J. Kilgannon, Jr.
I B M Corp.
Thomas Watson Research Center
Box 218
Yorktown Heights, NY 10598 (914) 945-3025

Marc D. Schechter
IBM Corp. Intell. Property Law
T.J. Watson Res. Center, P.O. Box 218
Yorktown Heights, NY 10598 (914) 241-4278

Roy Ramon Schlemmer, Jr.
I B M Corp.
T.J. Watson Res. Center, Box 218
Yorktown Heights, NY 10598 (914) 241-4057

J.E. Stanland#
IBM Corp.
Box 218
Yorktown Heights, NY 10598 (914) 241-4059

Alexander Tognino
IBM Corp. Yorktown Pat. Operations
P.O. Box 218
Yorktown Heiths, NY 10598 (914) 241-4278

North Carolina

Robert M. Wolters
8840 Reigate Ln.
Apex, NC 27502 (919) 779-5138

David M. Carter
Patla, Straus, Robinson & Moore, P.A.
29 North Market
P.O. Box 7625
Asheville, NC 28807 (704) 255-7641

Lawrence A. Nielsen#
416 Ridgecrest Dr.
Chapel Hill, NC 27514 (919) 967-3572

W. Thad Adams, III
2180 First Union Plaza
301 S. Tryon St.
Charlotte, NC 28282 (704) 375-9249

Herbert M. Adrian, Jr.
P.O. Box 220214
Charlotte, NC 28222 (704) 536-8651

Blas P. Arroyo#
Bell, Seltzer, Park & Gibson
1211 E. Morehead St.
P.O. Drawer 34009
Charlotte, NC 28234 (704) 377-1561

John J. Barnhardt, III
Bell, Seltzer, Park & Gibson
1211 E. Moorehead St.
P.O. Drawer 34009
Charlotte, NC 28204 (704) 377-1561

Herman O. Bauermeister
7413 Traelight Ch. Rd.
Charlotte, NC 28212 (704) 545-4645

Paul B. Bell
Bell, Seltzer, Park & Gibson
1211 E. Morehead St.
P.O. Drawer 34009
Charlotte, NC 28234 (704) 377-1561

Michell S. Bigel
I B M Corp.
1001 W.T. Harris Blvd., W.
Charlotte, NC 28257 (704) 594-8300

Robert John Blanke
Celanese Corp.
P.O. Box 32414
Charlotte, NC 28232 (704) 554-2686

Julian E. Carnes, Jr.
Bell, Seltzer, Park & Gibson
1211 Morehead St.
P.O. Drawer 34009
Charlotte, NC 28234 (704) 377-1561

Robert M. Chiaviello
I B M Corp.
1001 W.T. Harris Blvd.
Charlotte, NC 28257 (704) 594-8305

Ralph H. Dougherty
1515 Mockingbird Ln., Ste. 410
Charlotte, NC 28209 (704) 527-7734

Charles B. Elderkin
Bell, Seltzer, Park & Gibson, P.A.
1211 E. Morehead St.
P.O. Drawer 34009
Charlotte, NC 28234 (704) 377-1561

Carl Davis Farnsworth#
511 Briarpatch Ln.
Charlotte, NC 28211 (704) 364-8293

Richard Ferguson#
2701 Briarcliff Pl.
Charlotte, NC 28207 (704) 375-4828

Floyd A. Gibson
Bell, Seltzer, Park & Gibson, P.A.
1211 E. Morehead St.
P.O. Drawer 34009
Charlotte, NC 28234 (704) 377-1561

Karl O. Hesse
I B M Corp.
1001 W.T. Harris Blvd.
Charlotte, NC 28257 (704) 594-8302

James B. Hinson
3126 Milton Rd., Ste. 222B
Charlotte, NC 28215 (704) 536-6594

Clifton Tredway Hunt, Jr.
P.O. Box 15039
Charlotte, NC 28211 (704) 365-2844

Samuel Gilliland Layton, Jr.
Bell, Seltzer, Park & Gibson, P.A.
1211 E. Morehead St.
P.O. Drawer 34009
Charlotte, NC 28234 (704) 377-1561

Ronald T. Lindsay
Bell, Seltzer, Park & Gibson
1211 E. Morehead St.
P.O. Drawer 34009
Charlotte, NC 28234 (704) 377-1561

Raymond O. Linker, Jr.
Bell, Seltzer, Park & Gibson, P.A.
1211 E. Morehead St.
P.O. Drawer 34009
Charlotte, NC 28234 (704) 377-1561

Richard A. Lucey
616 Law Bldg.
730 E. Trade St.
Charlotte, NC 28202 (704) 334-4137

Daniel R. Mc Connell
6618 Fairview Rd., Ste. 202
Charlotte, NC 28210 (704) 366-9696

Fritz Y. Mercer, Jr.
Dozier, Miller, Pollard & Murphy
701 E. Trade St.
Charlotte, NC 28202 (704) 372-6373

James D. Myers
Bell, Seltzer, Park & Gibson, P.A.
1211 E. Morehead St.
P.O. Drawer 34009
Charlotte, NC 28234 (704) 377-1561

Charles B. Park, III
Bell, Seltzer, Park & Gibson, P.A.
1211 E. Morehead St.
P.O. Drawer 34009
Charlotte, NC 28204 (704) 377-1561

Francis M. Pinckney
Richards, Shefte & Pinckney
1208 Cameron Brown Bldg.
301 S. McDowell St.
Charlotte, NC 28204 (704) 375-9181

Harris Emerson Potter#
Bell, Seltzer, Park & Gibson, P.A.
1211 E. Morehead St.
P.O. Drawer 34009
Charlotte, NC 28234 (704) 377-1561

Wilton Rankin
Sodyeco, Inc.
P.O. Box 669246
Charlotte, NC 28266 (704) 827-4351

F. Michael Sajovec
Bell, Seltzer, Park & Gibson
1211 E. Morehead St.
P.O. Drawer 34009
Charlotte, NC 28234 (704) 377-1561

Karl S. Sawyer, Jr.
Richards, Shefte & Pinckney
1208 Cameron Brown Bldg.
301 S. Mc Dowell St.
Charlotte, NC 28204 (704) 375-9181

Kenneth A. Seaman
IBM Corp. Dept. 18A/680
Dept 18A/680
1001 W.T. Harris Blvd.
Charlotte, NC 28257 (704) 594-3003

Donald Miller Seltzer
Bell, Seltzer, Park & Gibson
1211 E. Morehead St.
P.O. Drawer 34009
Charlotte, NC 34009 (704) 377-1561

Dalbert Uhrig Shefte
Richards, Shefte & Pinckney
1208 Cameron Brown Bldg.
301 S. Mc Dowell St.
Charlotte, NC 28204 (704) 375-9181

Kenneth D. Sibley
Bell, Seltzer, Park & Gibson
1211 E. Morehead St.
P.O. Drawer 34009
Charlotte, NC 29204 (704) 377-1561

Forrest Dale Stine
Celanese Corp.
P.O. Box 32414
Charlotte, NC 28232 (704) 554-2000

John L. Sullivan, Jr.
Bell, Seltzer, Park & Gibson
1211 E. Morehead St.
P.O. Drawer 34009
Charlotte, NC 28234 (704) 377-1561

Philip Summa
Bell, Seltzer, Park & Gibson
1211 E. Morehead St.
P.O. Drawer 34009
Charlotte, NC 28234 (704) 377-1561

Joell T. Turner
Bell, Seltzer, Park & Gibson, P.A.
1211 E. Morehead St.
P.O. Drawer 34009
Charlotte, NC 28234 (704) 377-1561

Richard K. Warther
Bell, Seltzer, Park & Gibson
1211 E. Morehead St.
P.O. Drawer 34009
Charlotte, NC 28234 (704) 377-1561

David W. Westphal#
Bell, Seltzer, Park & Gibson
1211 E. Morehead St.
P.O. Drawer 34009
Charlotte, NC 28234 (704) 377-1561

Lynn E. Barber
Olive & Olive, P.A.
500 Memorial St.
P.O. Box 2049
Durham, NC 27702 (919) 683-5514

Franklin H. Cocks#
Dept. of Mechanical Engineering
Research Drive
Duke Univ.
Durham, NC 27706 (919) 684-2832

Steven J. Hultquist
Olive & Olive, P.A.
500 Memorial St.
P.O. Box 2049
Durham, NC 27702 (919) 683-5514

Richard E. Jenkins
3710 University Dr., Ste. 200
Durham, NC 27707 (916) 968-8216

B. B. Olive
Olive & Olive
500 Memorial St.
P.O. Box 2049
Durham, NC 27702 (919) 683-5514

Charles E. Smith
School of Law
1801 Fayetteville St.
North Carolina Central Univ.
Durham, NC 27707 (919) 683-6348

A. Michael Tucker
Olive & Olive, P.A.
500 Memorial St.
P.O. Box 2049
Durham, NC 27702 (919) 683-5514

Edward F. Sherer
B A S F Corp.
Fibers Div.
Sand Hill Rd.
Enka, NC 28728 (704) 667-7738

Tommy R. Vestal
BASF Corp.
Fibers Div.
Sand Hill Rd.
Enka, NC 28728 (704) 667-6451

Walter L. Beavers
338 N. Elm St.
Greensboro, NC 27401 (919) 275-7601

William Glenn Dosse
AT&T Technologies Inc.
P.O. Box 25000
Greensboro, NC 27420 (919) 279-4026

Judith E. Garmon#
1000 Wachovia Bldg.
P.O. Box 2974
Greensboro, NC 27402 (919) 273-4422

Sidney Gundersen
Western Electric Co. Inc.
Guilford Center
P.O. Box 25000
Greensboro, NC 27420 (919) 697-5504

Wallace M. Kain
AT&T Co.
Box 25000
Greensboro, NC 27420 (919) 697-5894

Howard A. Maccord, Jr.
Burlington Industries Inc.
Legal Dept.
3330 W. Friendly Ave.
P.O. Box 21207
Greensboro, NC 27420 (919) 379-4517

John B. Maier
Burlington Industries Inc.
Legal Dept.
3330 W. Friendly Ave.
P.O. Box 21207
Greensboro, NC 27420 (919) 379-2134

Robert Yaeger Peters
AT&T Co.
P.O. Box 25000
Greensboro, NC 27420 (919) 697-3254

Charles R. Rhodes
Rhodes, Coats & Bennett
1000 Wachovia Bldg.
201 N. Elm St.
Greensboro, NC 27402 (919) 273-4422

Edward W. Rilee, Jr.
Rhodes, Coats & Bennett
1000 Wachovia Bldg.
201 N. Elm St.
Greensboro, NC 27402 (919) 273-4422

James J. Trainor, Jr.
AT&T Technologies
P.O. Box 25000
Greensboro, NC 27420 (919) 279-4080

Harvey Zeller
AT&T
P.O. Box 25000
Greensboro, NC 27420 (919) 279-4081

Robert Harding, Jr.
Carolina Village
Box 11
Hendersonville, NC 28739 (704) 697-6452

Stephen Hoynak
544 Broadway
Hendersonville, NC 28739 (704) 692-0480

J. David Abernethy
Siecor Corp.
489 Siecor Park
Hickory, NC 28603 (704) 327-5354

Roy Bratton Moffitt
1928 Main Ave., S.E.
Hickory, NC 28601 (704) 328-2171

Hugh C. Bennett, Jr.
907 English Rd.
P.O. Box 660
High Point, NC 27261 (919) 883-2111

Richard W. Lacher
Rt. 1, Box 128
Jackson Springs, NC 27281 (919) 673-8071

John F. Hohmann
Rte. 1, Box 1265
Mother Vineyard Rd.
Manteo, NC 27954 (919) 473-2254

Timothy R. Kroboth
10238 Woodview Circle
Matthews, NC 28105 (704) 847-6904

Manford R. Haxton
2893 W. Pine St.
Mount Airy, NC 27030 (919) 789-5034

William J. Stellman
P.O. Box 786
Pinehurst, NC 28374 (919) 692-7088

John F. Verhoeven
P.O. Box 1882
Pinehurst, NC 28374 (919) 295-3638

Robert W. Habel
P.H. Glafelter Co.
Ecusta Div.
One Ecusta Rd.
P.O. Box 200
Pisgah Forest, NC 28768 (704) 877-2140

Paul M. Enlow
149 Ferrington Post
Pittsboro, NC 27312 (919) 542-3666

Henry S. Huff
402 Oakwood Dr.
Pittsboro, NC 27312 (919) 542-5839

Arthur D. Begun
8124 Brookwood Ct.
Raleigh, NC 27612 (919) 787-2748

David E. Bennett
Mills & Coats, P.A.
909 Glenwood Ave.
P.O. Box 5
Raleigh, NC 27602 (919) 832-3946

Freddie K. Carr#
Univ. of North Carolina
820 Clay St.
Raleigh, NC 27605 (919) 733-4643

Larry L. Coats
Mills & Coats
909 Glenwood Ave.
P.O. Box 505
Raleigh, NC 27602 (919) 832-3946

Richard S. Faust
Bell, Seltzer, Park & Gibson
Branch Bank & Trust Bldg., Suite 1207
P.O. Box 309
Raleigh, NC 27602 (919) 829-0616

John B. Frisone
8828 Woody Hill Rd.
Raleigh, NC 27612 (919) 848-6834

Joseph H. Heart
Bell, Seltzer, Park & Gibson
1030 Washington St.
P.O. Box 10867
Raleigh, NC 27605 (919) 829-0616

James M. Kuszaj
Ogletree, Deakins, Nash, Smoak & Stewart
P.O. Box 31608
Raleigh, NC 27622 (919) 787-9700

William Charles Lawton
P.O. Box 182
333 Fayetteville St. Mall
Raleigh, NC 27602 (919) 832-9661

John G. Mills, III
Mills & Coats, P.A.
909 Glenwood Ave.
P.O. Box 5
Raleigh, NC 27605 (919) 832-3946

Robert G. Rosenthal
Bell, Seltzer, Park & Gibson
1030 Washington St.
P.O. Box 10867
Raleigh, NC 27605 (919) 829-0616

George E. Ziegler#
4709-H Edwards Mill Rd.
Raleigh, NC 27612 (918) 787-8887

Thomas Joseph Bird, Jr.
General Electric Co.
Semiconductor Business
3026 Cornwallis Rd.
Research Triangle Park, NC 27709
(919) 540-3117

Joscelyn G. Cockburn
IBM Corp.
972/002
Cornwallis Rd.
P.O. Box 12195
Research Triangle Park, NC 27709
(919) 543-9036

Stanley C. Corwin
General Electric Co.
3026 Cornwallis Rd.
P.O. Box 13049
Research Triangle Park, NC 27709
(919) 544-8120

Edward H. Duffield
IBM Corp.
Dept 972
P.O. Box 12195
Research Triangle Park, NC 27709
(919) 543-4710

Irving Melvin Freedman
General Electric Co.
One Micron Dr.
Research Triangle Park, NC 27709
(919) 549-3115

Thomas F. Galvin
IBM Corp.
972/002
P.O. Box 12195
Cornwallis Rd.
Research Triangle Park, NC 27709
(919) 543-4184

Charles T. Joyner
Glaxo Inc.
Five Moore Dr.
Research Triangle Park, NC 17709
(919) 248-2192

John E. Leonarz
P.O. Box 12011
Research Triangle Park, NC 27709
(919) 541-6181

Gerald Ray Woods
IBM Corp.
972/002
P.O. Box 12195
Research Triangle Park, NC 27709
(919) 543-7204

Donald L. Weinhold
Weinhold & Mc Canless, P.A.
318 N. Main St.
Salisbury, NC 28144 (704) 637-2235

Frank Coombs Parker
605 Glenwalden Circle
Tryon, NC 28782 (704) 859-5629

Abner Milton Cornwell, Jr.
111 E. Wade St.
Wadesboro, NC 28170 (704) 694-5143

Fletcher C. Eddens#
1730 Fairway Dr.
Wilmington, NC 28403 (919) 762-0600

Herbert J. Bluhm#
4101 Tangle Ln.
Winston-Salem, NC 27106 (919) 723-7412

Stephen M. Bodenheimer, Jr.
R. J. Reynolds Tobacco Co.
1100 Reynolds Blvd.
Winston-Salem, NC 27102 (919) 773-5014

August J. Borschke
R. J. Reynolds Tobacco
R. J. Reynolds Bldg.
Winston-Salem, NC 27102 (919) 773-5491

William S. Burden#
Sara Lee Corp.
470 Hanes Mill Rd.
P.O. Box 2760
Winston-Salem, NC 27102 (919) 744-3612

Eric J. Groen
AMP Inc.
3700 Reidsville Rd.
Winston-Salem, NC 27102 (919) 727-5511

Charles Yount Lackey
3540 Buena Vista Rd.
Winston-Salem, NC 27106 (919) 744-3704

Grover M. Myers
R. J. Reynolds Inds., Inc.
1100 Reynolds Blvd.
Winston-Salem, NC 27102 (919) 773-2694

Robert W. Pitts
3700 Reidsville Rd.
P.O. Box 55
Winston-Salem, NC 27102 (919) 727-5051

North Dakota

Mack L. Thomas
Melroe Co.
112 N. University Dr.
Fargo, ND 58102 (701) 241-8717

David G. Adams#
Devil's, Highway 57
Ft. Totten, ND 58335 (701) 766-4211

Robert Elick Kleve
1103 24th Ave. South
Grand Forks, ND 58201 (701) 772-4311

Ohio

Donald J. Bobak
Renner, Kenner, Greive, Bobak & Taylor
1610 First National Tower
Akron, OH 44308 (216) 376-1242

Robert Walter Brown
Goodyear Tire & Rubber Co.
Patent and Trademark Dept. 823
1144 E. Market St.
Akron, OH 44316 (216) 796-6389

Ford Whitman Brummer, Jr.
270 Stratford Rd.
Akron, OH 44313 (216) 867-5993

Richard Harvey Childress
Goodyear Tire & Rubber Co.
Dir of Patents & Trademarks
1144 E. Market St.
Akron, OH 44316 (216) 796-4786

Gregory N. Clements
B.F. Goodrich Co.
3925 Embassy Pkwy.
Akron, OH 44313 (216) 374-3229

John Younge Clowney
319 Overwood Rd.
Akron, OH 44313 (216) 836-3506

Mack Dickson Cook, II
Cook & Cook
900 First National Tower
Akron, OH 44308 (216) 376-1005

Alan A. Csontos
Uniroyal Goodrich Tire Co.
Patent Law Dept. D/0530, Ugb-6
600 S. Main St.
Akron, OH 44397 (216) 374-2951

Robert T. Cunningham, Jr.
Goodyear Tire & Rubber Co.
1144 E. Market St.
Akron, OH 44316 (216) 796-7880

Theodore Joseph Dettling
574 Castle Blvd.
Akron, OH 44313 (216) 836-4742

Marc R. Dion, Sr.
Goodyear Tire & Rubber Co.
1144 E. Market St.
Akron, OH 44316 (216) 796-8251

Albert C. Doxsey
1006 Bunker Dr. Apt. A-107
Akron, OH 44313 (216) 666-6430

Lonnie R. Drayer
Goodyear Tire & Rubber Co.
1144 E. Market St.
Akron, OH 44316 (216) 796-7037

Thoburn T. Dunlap#
B.F. Goodrich Co.
3925 Embassy Pkwy.
Akron, OH 44313 (216) 374-2555

Lee A. Germain#
Loral Systems Group Div. Of Loral Corp.
1210 Massillon Rd.
Akron, OH 44315 (216) 796-2070

Merle William Goodyear
Goodyear Tire & Rubber Co.
1144 E. Market St.
Akron, OH 44316 (216) 796-2959

Edward G. Greive
Renner, Kenner, Greive, Bobak & Taylor
1610 First National Tower
Akron, OH 44308 (216) 376-1242

Richard H. Haas
Goodyear Tire & Rubber Co.
1144 E. Market St.
Akron, OH 44316 (216) 796-4503

Daniel N. Hall
Firestone Tire & Rubber Co.
Law Dept.
1200 Firestone Pkwy.
Akron, OH 44317 (216) 379-7543

Everett R. Hamilton
Renner, Kenner, Greive, Bobak & Taylor
1610 First National Tower
Akron, OH 44308 (216) 376-1242

John Dabney Haney
B.F. Goodrich Co.
500 S. Main St.
Akron, OH 44318 (216) 379-2366

William R. Holland
807 Citi Center
146 South High St.
Akron, OH 44308 (216) 535-4114

Byron John Hook
1940 Revere Rd.
Akron, OH 44313 (216) 666-9269

Daniel J. Hudak
Tell Bldg. Suite 408
7 W. Bowery St.
Akron, OH 44308 (216) 535-2220

Joseph Januszkiewicz
B.F. Goodrich Co.
3925 Embassy Pkwy.
Akron, OH 44313 (216) 374-3082

George A. Kap
B.F. Goodrich Co.
3925 Embassy Parkway
Akron, OH 44313 (216) 374-3237

Thomas A. Kayuha
Ohio Edison Co.
76 S. Main St.
Akron, OH 44308 (216) 384-5802

Phillip Lee Kenner
Hamilton, Renner, & Kenner
1610 First National Tower
Akron, OH 44308 (216) 376-1242

Louis F. Kreek, Jr.
Oldham, Oldham, Hudak, Weber & Sands Co. L.P.A.
Twin Oaks Estate
1225 W. Market St.
Akron, OH 44313 (216) 864-5550

Frederick K. Lacher
908 Akron Savings Bldg.
Akron, OH 44308 (216) 535-5522

Samuel B. Laferty
382 Rankin St.
Akron, OH 44311 (216) 376-1956

Thomas P. Lewandowski
Goodyear Tire & Rubber Co.
1144 E. Market St.
Akron, OH 44316 (216) 796-4219

James Robert Lindsay, Sr.
B.F. Goodrich Co.
3925 Embassy Pkwy.
Akron, OH 44313 (216) 374-2167

David M. Lowry
Tramonte, Kot, Davis & Lowry
411 Wolf Ledges, # 100
Akron, OH 44311 (216) 434-1112

Frank C. Manak, III
General Tire & Rubber Co.
Patent Dept.St.
One General St.
Akron, OH 44329 (216) 798-5284

Harold S. Meyer
1154 Jefferson Ave.
Akron, OH 44313 (216) 864-9095

Paul E. Milliken
Goodyear Aerospace Corp.
1210 Massillon Rd.
Akron, OH 44315 (216) 796-3894

Donald O. Nickey
Goodyear Tire & Rubber Co.
1144 E. Market St.
Akron, OH 44316 (216) 796-3151

Nils E. Nilsson
77 Fir Hill No. 4-B-12
Akron, OH 44304 (216) 384-5624

Edwin W. Oldham
Oldham, Oldham, Hudak, Weber & Sands Co. L.P.A.
Twin Oaks Estate
1225 W. Market St.
Akron, OH 44313 (216) 864-5550

Robert L. Oldham
1250 Weathervane Lane
Akron, OH 44313 (216) 867-5363

Vern L. Oldham
Oldham, Oldham, Hudak, Weber & Sand, Co. L.P.A.
Twin Oaks Estates
1225 W. Market St.
Akron, OH 44313 (216) 864-5550

Debra L. Pawl
B.F. Goodrich Co.
3925 Embassy Pkwy.
Akron, OH 44313 (216) 374-2339

Harry F. Pepper, Jr.
B.F. Goodrich Co.
500 S. Main St.
Akron, OH 44318 (216) 374-2951

John G. Pere
2143 Pressler Rd.
Akron, OH 44312 (216) 733-6095

Joe A. Powell
B.F. Goodrich Co.
3925 Embassy Pkwy.
Akron, OH 44313 (216) 374-3049

Jack L. Renner
Hamilton, Renner & Kenner
1610 First National Tower
Akron, OH 44308 (216) 376-1242

Alvin T. Rockhill, III
Goodyear Tire & Rubber Co.
Dept. 823
1144 E. Market St.
Akron, OH 44316 (216) 796-8252

David M. Ronyak
B.F. Goodrich Co.
3925 Embassy Pkwy.
Akron, OH 44313 (216) 374-3363

George W. Rooney, Jr.
Roetzel & Andress
75 E. Market St.
Akron, OH 44308 (216) 376-2700

Frank C. Rote, Jr.
General Tire & Rubber Co.
One General St.
Akron, OH 44329 (216) 798-5230

James A. Rozmajzl
Goodyear Tire & Rubber Co.
1144 E. Market St.
Akron, OH 44316 (216) 796-7417

Ernst H. Ruf
Firestone Tire & Rubber Co.
Legal Dept.
1200 Firestone Pkwy.
Akron, OH 44317 (216) 379-6851

Gordon B. Seward
Monsanto Co.
260 Springside Dr.
Akron, OH 44313 (216) 688-8257

Nestor W. Shust
B.F. Goodrich Co.
3925 Embassy Pkwy.
Akron, OH 44313 (216) 374-4014

Raymond J. Slattery, III
Goodyear Tire & Rubber Co.
1144 E. Market St.
Akron, OH 44316 (216) 796-7263

Robert W. Stachowiak
1685 Far View Rd.
Akron, OH 44312 (216) 644-2940

Perry Reese Taylor, Jr.
Renner, Kenner, Grieve, Bobak & Taylor
1610 First National Tower
Akron, OH 44308 (216) 376-1034

David A. Thomas
Firestone Tire & Rubber Co.
1200 Firestone Parkway
Akron, OH 44317 (216) 379-6850

John A. Tomich
Renner, Kenner, Greive, Bobak & Taylor
1610 First National Tower
Akron, OH 44308 (216) 376-1242

Frank J. Troy, Sr.
Firestone Tire & Rubber Co.
1200 Firestone Pkwy.
Akron, OH 44317 (216) 379-6178

Samuel W. Waisbrot#
2036 Thornhill Dr.
Akron, OH 44313 (216) 836-1615

Ray I. Weber
Oldham, Oldham, Hudak, Weber & Sand Co. L.P.A.
1225 W. Market St.
Akron, OH 44313 (216) 864-5550

Louis Jerome Weisz
Oldham, Oldham, Hudak, Weber & Sand Co., L.P.A.
1225 W. Market St.
Akron, OH 44313 (216) 864-5550

Bruce H. Wilson
Society Bldg., Suite 1000
Akron, OH 44308 (216) 253-1900

Ronald P. Yaist
Goodyear Tire & Rubber Co.
1144 E. Market St.
Akron, OH 44316 (216) 796-4409

Henry Charles Young, Jr.
Goodyear Tire & Rubber Co.
1144 E. Market St.
Akron, OH 44316 (216) 796-2956

Edmund J. Wasp
Nordson Corp.
555 Jackson St.
Amherst, OH 44001 (216) 988-9411

Ralph L. Humphrey
1258 Prospect Rd.
Ashtabula, OH 44004 (216) 998-1112

Robert A. Sturges
3497 Nautilus Trail
Aurora, OH 44202 (216) 562-9134

Jeffrey V. Bamber
Rosenhoffer, Nichols & Schwartz
250 E. Main St.
Batavia, OH 45103 (513) 732-0770

Horace N. Harger
2465 Shade Park Dr.
Bath, OH 44210 (216) 666-2722

Olaf Nielsen
1427 Hillandale Dr.
Bath, OH 44210 (216) 666-4338

Serle I. Mosoff
22000 Halburton Rd.
Beachwood, OH 44122 (216) 991-8070

Charles Clyde Allshouse, Jr.
176 Debs Dr.
Beavercreek, OH 45385 (513) 429-1268

Bradly D. Beams
Superx Drugs
334 E. Columbus Ave.
Bellefontaine, OH 43311 (513) 593-1040

Sheldon R. Shulte
823 S. Remington Rd.
Bexley, OH 43209 (614) 237-6723

Stanley M. Clark
8049 Robin Ln.
Brecksville, OH 44141 (216) 526-8809

Daniel S. Kalka
7610 Traymore Ave.
Brooklyn, OH 44144 (216) 351-4719

John B. Frease
Frease & Bishop
519 National City Bldg.
315 Tuscarawas St., W.
Canton, OH 44702 (216) 455-0331

Joseph Frease
Frease & Bishop
519 National City Bldg.
315 W. Tuscarawas St.
Canton, OH 44702 (216) 455-0331

Michael Sand
Sand & Hudak Co. L.P.A.
4450 Belden Village Ave., N.W., Suite 201
Canton, OH 44718 (216) 492-1925

Melvin Wiviott
1599 Ambridge Rd.
Centerville, OH 45459 (513) 434-1379

Lawrence R. Kempton
5224 Maple Spring Dr.
Chagrin Falls, OH 44022 (216) 338-3661

Arthur S. Collins#
12039 Fowlers Mill Rd.
Chardon, OH 44024 (216) 286-3140

Forrest L. Collins
7515 Avon Lane
Chesterland, OH 44026 (216) 729-0277

Emil F. Sos, Jr.
12690 Opalock Dr.
Chesterland, OH 44026 (216) 729-1252

Wilson G. Palmer#
1440 Valley Dr.
Chillicothe, OH 45601 (614) 775-6750

Robert Raymond Yurich
7862 Lake Rd.
Chippewa Lake, OH 44215 (216) 769-3047

George W. Allen
Procter & Gamble Co.
Patent Div.
P.O. Box 39175
Cincinnati, OH 45247 (513) 245-2912

Robert B. Aylor
Procter & Gamble Co.
Ivorydale Technical Center
5299 Spring Grove Ave.
Cincinnati, OH 45217 (513) 627-5144

Mark W. Bailey
Senco Products, Inc.
8485 Broadwell Rd.
Cincinnati, OH 45244 (513) 388-2913

Gerald A. Baracka#
National Distillers & Chem. Corp.
11500 Northlake Dr.
Cincinnati, OH 45249 (513) 530-6561

J. Wesley Blumenshine
Procter & Gamble Co.
6100 Center Hill Rd.
Cincinnati, OH 45224 (513) 659-2964

George P. Brandenburg
Fox & Clear Co. L.P.A.
800 American Bldg.
30 E. Central Pkwy.
Cincinnati, OH 45202 (513) 421-6644

Frederick H. Braun
Proctor & Gamble Co.
6100 Center Hill Rd.
Cincinnati, OH 45224 (513) 659-6332

Herbert C. Brinkman, Jr.
Wood, Herron & Evans
2700 Carew Tower
441 Vine St.
Cincinnati, OH 45202 (513) 241-2324

Thomas J. Burger
Wood, Herron & Evans
2700 Carew Tower
441 Vine St.
Cincinnati, OH 45202 (513) 241-2324

Charles Lyman Burgoyne#
7410 N. Mingo Lane
Cincinnati, OH 45243 (513) 793-5869

David F. Chalmers#
Procter & Gamble Co.
Ivorydale Technical Center
5299 Spring Grove Ave.
Cincinnati, OH 45217 (513) 627-5515

J. Robert Chambers#
Wood, Herron & Evans
2700 Carew Tower
411 Vine St.
Cincinnati, OH 45244 (513) 241-2324

C. Thomas Cross
2959 Madison House
2324 Madison Rd.
Cincinnati, OH 45208 (513) 475-2993

David K. Dabbiere
Procter & Gamble Co.
P.O. Box 39175
Cincinnati, OH 45247 (513) 245-2924

Rose Ann Dabek
Procter & Gamble Co.
Winton Hill Technical Center
6071 Center Hill Rd.
Cincinnati, OH 45224 (513) 659-5593

James P. Davidson
Frost & Jacobs
2500 Central Trust Center
201 E. Fifth St.
Cincinnati, OH 45202 (513) 651-6993

J. Michael Dixon
Merrell Dow Pharmaceuticals Inc. Pat. Dept.
Patent Dept.
2110 E. Galbraith Rd.
P.O. Box 156300
Cincinnati, OH 45215 (513) 948-7960

Donald Dunn#
1845 Greenbriar Place
Cincinnati, OH 45237 (513) 351-5022

Jack J. Earl
7618 Carriage Lane
Cincinnati, OH 45242

C. Richard Eby
Cincinnati Milacron, Inc.
4701 Marburg Ave.
Cincinnati, OH 45209 (513) 841-8431

Barry W. Elledge#
Pharmacology Dept.
231 Bethesda Ave.
Univ. of Cincinnati
Cincinnati, OH 45267 (513) 559-8919

Douglas E. Erickson
GE Aircraft Engines
One Neuman Way, Mail Drop H-17
Cincinnatti, OH 45215 (513) 243-9840

Richard Henry Evans
Wood, Herron & Evans
2700 Carew Tower
441 Vine St.
Cincinnati, OH 45202 (513) 241-2324

Thomas M. Farrell
Cincinnati Milacron, Inc.
4701 Marburg Ave.
Cincinnati, OH 45209 (513) 841-8536

Julius P. Filcik
Procter & Gamble Co.
Winton Hill Technical Center
6090 Center Hill Rd.
Cincinnati, OH 45224 (513) 977-7503

Lynn T. Fletcher#
10561 Margate Terrace
Cincinnati, OH 45241 (513) 733-4894

Stanley H. Foster
Frost & Jacobs
2500 Central Trust Center
201 E. Fifth St.
Cincinnati, OH 45202 (513) 651-6975

Janice L. Fowler
Procter & Gamble Co.
Miami Valley Labs
P.O. Box 39175
Cincinnati, OH 45247 (513) 245-2924

Donald Francis Frei, Sr.
Wood, Herron & Evans
2700 Carew Tower
441 Vine St.
Cincinnati, OH 45202 (513) 241-2324

George Galanes#
Procter & Gamble Co.
Winton Hill Technical Center
6060 Center Hill Rd.
Cincinnati, OH 45224 (513) 659-4173

Edmund Frederick Gebhardt
2959 Annwood St.
Cincinnati, OH 45206 (513) 751-2717

Roger A. Gilcrest
Frost & Jacobs
2500 Central Trust Center
201 E. Fifth St.
Cincinnati, OH 45202 (216) 651-6746

Steven J. Goldstein
Procter & Gamble Co.
Miami Valley Labs
P.O. Box 398707
Cincinnati, OH 45239 (513) 245-2701

John V. Gorman
Procter & Gamble Co.
6090 Center Hill Rd.
Cincinnati, OH 45224 (513) 977-6148

William H. Gould
8601 Long Lane
Cincinnati, OH 45231 (513) 521-2378

Milton B. Graff
Procter & Gamble Co.
Miami Valley Labs
P.O. Box 398707
Cincinnati, OH 45239 (513) 245-2659

Edward A. Grannen, Jr.#
11186 Lincolnshire Dr.
Cincinnati, OH 45240 (513) 851-5938

John W. Gregg
Cincinnati Milacron, Inc.
4701 Marburg Ave.
Cincinnati, OH 45209 (513) 841-8344

Simon Groner
2011 Carew Tower
Cincinnati, OH 45202

Kurt L. Grossman
Wood, Herron & Evans
2700 Carew Tower
441 Vine St.
Cincinnati, OH 45202 (513) 241-2324

Eric W. Guttag
Procter & Gamble Co.
Winton Hill Technical Center
3210 Center Hill Rd.
Cincinnati, OH 45224 (513) 659-2736

Donald E. Hasse
Procter & Gamble Co.
Ivorydale Technical Center
5299 Spring Grove Ave.
Cincinnati, OH 45217 (513) 627-5144

James Harry Hayes
Frost & Jacobs
2500 Central Trust Center
201 E. Fifth St.
Cincinnati, OH 45202 (513) 651-6800

Robert L. Hearn
10567 Gloria Ave.
Cincinnati, OH 45231 (513) 851-9032

William A. Heidrich, III
Quantum Chemical Corp.
Legal Dept.
11500 Northlake Dr.
Cincinnati, OH 45249 (513) 530-6552

Ronald L. Hemingway
Procter & Gamble Co.
6071 Center Hill Rd.
Cincinnati, OH 45224 (513) 659-5593

Nathan D. Herkamp
GE Aircraft Engines
One Neuman Way, Mail Drop H-17
Cincinnati, OH 45215 (513) 243-3701

Bart S. Hersko
Procter & Gamble Co.
Miami Valley Labs
P.O. Box 398707
Cincinnati, OH 45239 (513) 245-2889

Joseph V. Hoffman
Frost & Jacobs
2500 Central Trust Center
201 E. Fifth St.
Cincinnati, OH 45202 (513) 651-6800

Charles Marshall Hogan
8071 Village Dr.
Cincinnati, OH 45142 (513) 489-2377

Maynard R. Johnson
Merrell Dow Pharmaceuticals, Inc.
Patent Law Dept.
2110 E. Galbraith Rd.
Cincinnati, OH 45215 (513) 948-7967

Joseph R. Jordan
2800 Carew Tower
441 Vine St.
Cincinnati, OH 45202 (513) 891-4455

David J. Josephic
Wood, Herron & Evans
2700 Carew Tower
441 Vine St.
Cincinnati, OH 45202 (513) 241-2324

Thomas L. Kautz
Wood, Herron & Evans
2700 Carew Tower
441 Vine St.
Cincinnati, OH 45202 (513) 241-2324

Howard T. Keiser
2525 Ranchvale Dr.
Cincinnati, OH 45230 (513) 232-2717

James M. Kipling#
Kenner Products
1014 Vine St.
Cincinnati, OH 45202 (513) 579-4808

John J. Kolano#
Merrell Dow Pharmaceuticals Inc.
2110 E. Galbraith Rd.
Cincinnati, OH 45215 (513) 948-7964

William G. Konold
Wood, Herron & Evans
2700 Carew Tower
441 Vine St.
Cincinnati, OH 45202 (513) 241-2324

Lewis H. Lanman
Emery Industries, Inc.
1300 Carew Towers
Cincinnati, OH 45202 (513) 762-6264

E.S. Lee, III
French, Marks, Short, Weiner & Valleau
105 E. Fourth St., Suite 700
Cincinnati, OH 45202 (513) 621-2260

Leonard W. Lewis#
Procter & Gamble Co.
Winton Hill Technical Center
6100 Center Hill Rd.
Cincinnati, OH 45224 (513) 659-6033

James D. Liles
Frost & Jacobs
2500 Center Trust Center
201 E. Fifth St.
Cincinnati, OH 45202 (513) 651-6707

Elmer K. Linman
Procter & Gamble Co.
6300 Center Hill Rd., Rm. E2E08
Cincinnati, OH 45224 (513) 977-6327

Jerrold J. Litzinger
Senco Products, Inc.
8485 Broadwell Rd.
Cincinnati, OH 45244 (513) 388-2912

Gregory J. Lunn
2700 Carew Tower
441 Vine St.
Cincinnati, OH 45202 (513) 241-2324

James D. Lykins#
886 Pinewall Dr.
Cincinnati, OH 45230 (513) 474-4729

Peter J. Manso
Wood, Herron & Evans
2700 Carew Tower
441 Vine St.
Cincinnati, OH 45202 (513) 241-2324

Nancy S. Mayer
Procter & Gamble Co.
6071 Center Hill Rd.
Cincinnati, OH 45224 (513) 977-6147

Raymond A. Mc Donald
Merrell Dow Pharmacenticals Inc.
Patent Dept.
2110 E. Galbraith Rd.
Cincinnati, OH 45215 (513) 948-7963

Robert J. Mc Nair, Jr#
2920 Blue Haven Terr.
Cincinnati, OH 45238 (513) 662-1448

John W. Melville
Frost & Jacobs
2500 Central Trust Center
201 E. Fifth St.
Cincinnati, OH 45202 (513) 651-6984

Terrence E. Miesle#
Hilton-Davis Chemical Group
Sterling Drug Inc.
P.O. Box 37869
Cincinnati, OH 45222 (513) 841-8440

Steven Miller
Procter & Gamble Co.
6100 Center Hill Rd.
Cincinnati, OH 45224 (513) 659-4490

Douglas C. Mohl
Procter & Gramble Co.
Sharon Woods Technical Center
11511 Reed Hartman Hwy., H B 3N58
Cincinnati, OH 45241 (513) 530-3991

Walter S. Murray
6050 Stirrup Rd.
Cincinnati, OH 45244 (513) 231-3093

A. Ralph Navaro, Jr.
Wood, Herron & Evans
2700 Carew Tower
441 Vine St.
Cincinnati, OH 45202 (513) 241-2324

Darla P. Neaveill
Procter & Gamble Co.
5299 Spring Grove Ave.
Cincinnati, OH 45217 (513) 627-5946

Stephen L. Nesbitt
Merrell Dow Pharmaceuticals Inc.
Patent Dept.
2110 E. Galbraith Rd.
Cincinnati, OH 45215 (513) 948-7825

Thomas H. O Flaherty
Procter & Gamble Co.
Ivorydale Technical Center
5299 Spring Grove Ave.
Cincinnati, OH 45217 (513) 763-6911

Daniel H. Owings
713 Locust Corner Rd.
Cincinnati, OH 45245 (513) 752-9289

James W. Pearce
Pearce & Schaeperklaus
105 E. Fourth St., Suite 1110
Cincinnati, OH 45202 (513) 241-1021

Walter A. Peterson#
8222 Monte Dr.
Cincinnati, OH 45242 (513) 984-2637

John D. Poffenberger
Wood, Herron & Evans
2700 Carew Tower
441 Vine St.
Cincinnati, OH 45202 (513) 241-2324

John M. Pollaro
Procter & Gamble Co.
Winton Hill Technical Center
6100 Center Hill Rd.
Cincinnati, OH 45224 (513) 659-7324

John David Rice
National Distillers & Chem. Corp.
4900 Este Ave.
Cincinnati, OH 45232 (513) 482-2400

Lynda E. Roesch
Dinsmore & Shohl
2100 Fountain Square Place
511 Walnut St.
Cincinnati, OH 45202 (513) 621-6747

Steven J. Rosen
General Electric Co.
One Neumann Way, Box 156301
Cincinnati, OH 45215 (513) 243-8925

John Jeffrey Ryberg
Procter & Gamble Co.
Winton Hill Technical Center
6300 Center Hill Rd., Rm. E2E40
Cincinnati, OH 45224 (513) 659-4637

Lee H. Sachs
General Electric Co.
Aircraft Engines Business Group
Mail Drop F-17
Cincinnati, OH 45215 (513) 243-4609

Jack D. Schaeffer
Procter & Gamble Co.
Miami Valley Co.
P.O. Box 39175
Cincinnati, OH 45247 (513) 245-2671

Roy F. Schaeperklaus
Pearce & Schaeperklaus
105 E. Fourth St., Suite 1110
Cincinnati, OH 45202 (513) 241-1021

John G. Schenk
Kinney & Schenk
105 E. Fourth St., Suite 1306
Cincinnati, OH 45202 (513) 721-3440

David E. Schmit
Frost & Jacobs
2500 Central Trust Center
201 E. Fifth St.
Cincinnati, OH 45202 (513) 651-6985

Edlyn S. Simmons#
Merrell Dow Pharmaceuticals, Inc.
2110 E. Galbraith Rd.
Cincinnati, OH 45215 (513) 948-7829

Thomas J. Slone, Sr.
Procter & Gamble Co.
Winton Hill Technical Center
6100 Center Hill Rd.
Cincinnati, OH 45224 (513) 659-7317

Ronald J. Snyder
Procter & Gamble Co.
6090 Center Hill Rd.
Cincinnati, OH 45224 (513) 659-6748

David S. Stallard
Wood, Herron & Evans
2700 Carew Tower
441 Vine St.
Cincinnati, OH 45202 (513) 241-2324

William J. Stein
Merrell Dow Pharmaceuticals, Inc.
2110 E. Galbraith Rd.
P.O. Box 156300
Cincinnati, OH 45215 (513) 948-7965

Albert E. Strasser
Frost & Jacobs
2500 Central Trust Center
201 E. Fifth St.
Cincinnati, OH 45202 (513) 651-6977

Gary D. Street
Merrell Dow Pharmaceuticals, Inc.
Patent Dept.
2110 E. Galbraith Rd.
Cincinnati, OH 45215 (513) 948-7695

Stephen S. Strucnk
General Electric Co.
Aircraft Engines Business Group
Box 156301
One Neuman Way, Mail Drop F-17 1
Cincinnati, OH 45215 (513) 243-4903

David L. Suter
Procter & Gamble Co.
11511 Reed Hartman Hwy.
Cincinnati, OH 45230 (513) 530-3993

Gary M. Sutter
Procter & Gamble Co.
Winton Hill Technical Center
6071 Center Hill Rd., Rm. F3A14
Cincinnati, OH 45224 (513) 659-7939

Bruce Tittel
Wood, Herron & Evans
2700 Carew Tower
441 Vine St.
Cincinnati, OH 45202 (513) 241-2324

Kenneth Dale Tremain
National Distillers & Chemical Corp.
1500 Northlake Dr.
Cincinnati, OH 45249 (513) 530-6550

Edward J. Utz
1306 Fourth & Walnut Bldg.
36 E. Fourth St.
Cincinnati, OH 45202 (513) 241-8829

Gregory A. Welte
General Electric Co.
One Neuman Way, Mail Drop F-17
Cincinnati, OH 45215 (513) 243-3701

Leonard Williamson
Procter & Gamble Co.
P.O. Box 41520
Cincinnati, OH 45241 (513) 530-3387

Charles R. Wilson
4729 Cornell Rd.
Cincinnati, OH 45241 (513) 489-7484

Monte D. Witte
Procter & Gamble Co.
Winton Hill Technical Center
6060 Center Hill Rd.
Cincinnati, OH 45224 (513) 659-7807

Richard C. Witte
Procter & Gamble Co.
Ivorydale Technical Center
5299 Spring Grove Ave.
Cincinnati, OH 45217 (513) 627-5666

Daniel P. Worth
7066 Sprucewood Court
Cincinnati, OH 45241 (513) 777-5631

Daniel A. Yaeger
1391 Devils Back Bone Rd.
Cincinnati, OH 45238 (513) 941-2725

Jerry J. Yetter#
Procter & Gamble Co.
Ivorydale Technical Center
5299 Spring Grove Ave.
Cincinnati, OH 45217 (513) 659-5666

Gibson R. Yungblut
Frost & Jacobs
2500 Central Trust Center
201 E. Fifth St.
Cincinnati, OH 45202 (513) 651-6980

Kim William Zerby
Procter & Gamble Co.
Miami Valley Labs
11810 E. Miami River Rd.
P.O. Box 39175
Cincinnati, OH 45247 (513) 245-2858

Kenneth R. Adamo
Jones, Day, Reavis & Pogue
North Point
1901 Lakeside Ave.
Cleveland, OH 44114 (216) 586-7120

James A. Baker
Parker Hannifin Corp.
17325 Euclid Ave.
Cleveland, OH 44112 (216) 531-3000

Joseph B. Balazs
Parker-Hannifin Corp.
17325 Euclid Ave.
Cleveland, OH 44112 (216) 531-3000

Russell E. Baumann
Clevite Industries Inc.
17000 St. Clair Ave.
Cleveland, OH 44110 (216) 481-7221

Gordon P. Becker
Alcan Aluminum Corp.
100 Erieview Plaza
Cleveland, OH 44114 (216) 523-8250

Robert Emil Bielek
3298 Rumson Rd.
Cleveland, OH 44118 (216) 566-2482

Alfred Carpenter Body
Body, Vickers & Daniels
2000 Terminal Tower
Cleveland, OH 44113 (216) 623-0040

Armand Paul Boisselle, Sr.
Renner, Otto, Boisselle & Lyon
One Public Square, 12fth Floor
Cleveland, OH 44113 (216) 621-1113

Don W. Bulson
Renner, Otto, Boisselle & Lyon
One Public Square, 12th Floor
Cleveland, OH 44113 (216) 621-1113

David A. Burge
P.O. Box 22975
Cleveland, OH 44122 (216) 921-8900

James E. Carson
Carson, Smith & Chandler
2020 Superior Bldg.
815 Superior Ave., N.E.
Cleveland, OH 44114 (216) 771-5818

Albert E. Chrow
Eaton Corp.
Eaton Center
1111 Superior Ave.
Cleveland, OH 44114 (216) 523-4131

Kenneth A. Clark
Calfee, Halter & Griswold
1800 Society Bldg.
Cleveland, OH 44114 (216) 781-2166

George J. Coghill
10211 Lakeshore Blvd.
Cleveland, OH 44108 (216) 451-2323

Hal D. Cooper
Jones, Day, Reavis & Pogue
1700 Huntington Bldg.
Cleveland, OH 44115 (216) 348-3939

Edward M. Corcoran
General Electric Co.
Lighting Business Group
Nela Park, Noble Rd.
Cleveland, OH 44112 (216) 266-3640

Joseph J. Corso
Pearne, Gordon, Mc Coy & Granger
1200 Leader Bldg.
Cleveland, OH 44114 (216) 579-1700

Calvin G. Covell
Tarolli, Sundheim & Covell
1111 Leader Bldg.
Cleveland, OH 44114 (216) 621-2234

Joseph G. Curatolo
Standard Oil Co.
200 Public Square, 36-F
Cleveland, OH 44114 (216) 586-8460

William E. Currie
3945 Northampton Rd.
Cleveland, OH 44121 (216) 382-4706

Edward Kent Daniels, Jr.
Body, Vickers & Daniels
2000 Terminal Tower
Cleveland, OH 44113 (216) 623-0040

David B. Deioma
Pearne, Gordon, Mc Coy & Granger
1200 Leader Bldg.
Cleveland, OH 44114 (216) 579-1700

Richard H. Dickinson, Jr.
Pearne, Gordon, McCoy & Granger
Sixth & Superior Ave.
Cleveland, OH 44114 (216) 579-1700

James D. Donohoe
L T V Steel Co. Inc.
P.O. Box 6778
Cleveland, OH 44101 (216) 622-5624

Merton H. Douthitt
Watts, Hoffmann, Fisher & Heinke Co.
100 Erieview Plaza, Suite 2850
Cleveland, OH 44114 (216) 623-0775

Neil A. Du Chez
Renner, Otto, Boiselle & Lyon
One Public Square, 12th Floor
Cleveland, OH 44113 (216) 621-1113

Richard Joseph Egan
Baldwin, Egan & Fetzer
816 Hanna Bldg.
Cleveland, OH 44115 (216) 621-2956

Michael F. Esposito
Standard Oil Co.
200 Public Square, 36-3454-F
Cleveland, OH 44114 (216) 586-8022

Larry W. Evans
B.P. America
200 Public Square, 36-3556F
Cleveland, OH 44114 (216) 586-8450

Christopher Brendan Fagan
Fay & Sharpe
1113 E. Ohio
Cleveland, OH 44114 (216) 861-5582

Regan J. Fay
Jones, Day, Reavis & Pogue
North Point
901 Lakeside Ave.
Cleveland, OH 44144 (216) 586-7327

Robert Jesse Fay
Fay & Sharpe
1113 East Ohio Bldg.
1717 E. Ninth St.
Cleveland, OH 44114 (216) 861-5582

Robert Joseph Fetzer
Baldwin, Egan & Fetzer
816 Hanna Bldg.
Cleveland, OH 44115 (216) 621-2956

Thomas E. Fisher
Watts, Hoffman, Fisher & Heinke Co., L.P.A.
100 Erieview Plaza
Cleveland, OH 44114 (216) 623-0775

Laurence D. Fogel#
B P America R & D
4440 Warrensville Center Rd.
Cleveland, OH 44128 (216) 581-5982

William Albert Gail
Pearne, Gordon, McCoy & Granger
1200 Leader Bldg.
Cleveland, OH 44114 (216) 579-1700

John X. Garred
Fay, Sharpe, Bell, Fagan, Minnich & McKee
400 National City Bank Bldg.
Cleveland, OH 44114 (216) 861-5582

Teresan W. Gilbert#
Standard Oil Co.
200 Public Square
Cleveland, OH 44114 (216) 586-8475

Charles Byron Gordon
Pearne, Gordon, McCoy & Granger
1200 Leader Bldg.
Cleveland, OH 44114 (216) 579-1700

Howard D. Gordon
Eaton Corp.
Eaton Center
1111 Superior Ave.
Cleveland, OH 44114 (216) 523-4134

Charles H. Grace
Eaton Corp.
Eaton Center
1111 Superior Ave.
Cleveland, OH 44114 (216) 523-4127

Louis V. Granger
Pearne, Gordon, Sessions, McCoy & Granger
1200 Leader Bldg.
Cleveland, OH 44114 (216) 579-1700

Thomas J. Gray
4515 St. Clair Ave.
Cleveland, OH 44103 (216) 391-7070

Vincent A. Greene
P.O. Box 14072
Cleveland, OH 44114 (216) 423-3511

Bruce E. Harang
B P America
200 Public Square, 36-F
Cleveland, OH 44114 (216) 586-8565

M. Elaine Harmon
Woodling, Krost & Rust
655 Union Commerce Bldg.
925 Euclid Ave.
Cleveland, OH 44115 (216) 241-4150

Edward A. Hayman
Squire, Sanders & Dempsey
1800 Huntington Bldg.
Cleveland, OH 44115 (216) 687-8500

Richard D. Heberling
Baldwin, Egan, Hudak & Fetzer
816 Hanna Bldg.
Cleveland, OH 44115 (216) 621-2956

Lowell L. Heinke
Watts, Hoffmann, Fisher & Heinke
100 Erieview Plaza., Suite 2850
Cleveland, OH 44114 (216) 623-0775

Kevin J. Heyd
Watts, Hoffmann, Fisher & Heinke Co.
100 Erieview Plaza, Suite 2850
Cleveland, OH 44114 (216) 623-0775

Stephen A. Hill
Pearne, Gordon, Mc Coy & Granger
1200 Leader Bldg.
Cleveland, OH 44114 (216) 579-1700

John R. Hlavka
Watts, Hoffmann, Fisher & Heinke Co., L.P.A.
100 Erieview Plaza, Suite 2850
Cleveland, OH 44114 (216) 623-0775

D. Peter Hochberg
1510 Ohio Savings Plaza
1801 E. Ninth St.
Cleveland, OH 44114 (216) 771-3800

James T. Hoffmann
Watts, Hoffmann, Fisher & Heinke, Co. Ltd.
1805 E. Ohio Bldg.
Cleveland, OH 44114 (216) 623-0775

William Neill Hogg
Building 3
23200 Chagrin Blvd., Suite 605
Cleveland, OH 44122 (216) 765-8890

Roy F. Hollander
Watts, Hoffmann, Fisher & Heinke Co. L.P.A.
100 Erieview Plaza, Suite 2850
Cleveland, OH 44114 (216) 623-0775

James A. Hudak
Hanna Bldg.
1422 Euclid Ave., Suite 818
Cleveland, OH 44115 (216) 566-9700

Robert R. Hussey
1610 Euclid Ave., 2nd Floor
Cleveland, OH 44115 (216) 687-1111

Christopher D. Jackson
Squire, Sanders & Dempsey
1800 Union Commerce Bldg.
Cleveland, OH 44115 (216) 687-8718

Roger A. Johnston
1444 W. Tenth St., Suite 406
Cleveland, OH 44113 (216) 523-4132

Leslie J. Kasper
Eaton Corp.
Eaton Center
1111 Superior Ave.
Cleveland, OH 44114 (216) 523-4299

Raymond F. Keller
Standard Oil Co.
200 Public Square 36-F-3454
Cleveland, OH 44114 (216) 586-8474

Sherman J. Kemmer
Diamond Point International, Inc.
15300 Pearl Rd.
Cleveland, OH 44136 (216) 238-0200

S.I. Khayat
3122 Euclid Ave.
Cleveland, OH 44115 (216) 431-1636

Gordon D. Kinder
Maky, Renner, Otto & Boisselle
One Public Square, 12th Floor
Cleveland, OH 44114 (216) 621-1113

Herbert D. knudsen
Standard Oil Co.
Midland Bldg., 928- Tt
Cleveland, OH 44115 (216) 575-5618

Thomas E. Kocovsky, Jr.
Fay & Sharpe
1113 E. Ohio Bldg.
1717 E. Ninth St.
Cleveland, OH 44114 (216) 861-5582

Bruce B. Krost
Woodling, Krost & Rust
655 Union Commerce Bldg.
925 Euclid Ave.
Cleveland, OH 44115 (216) 241-4150

Mark M. Kusner
Body, Vickers & Daniels
2000 Terminal Tower
Cleveland, OH 44113 (216) 623-0040

James John Lazna
Body, Vickers & Daniels
2000 Terminal Tower
Cleveland, OH 44113 (216) 623-0040

Ernest W. Legree
General Electric Co.
Nela Park, Noble Rd.
Cleveland, OH 44112 (216) 266-2251

Leonard L. Lewis
Pearne, Gordon, Sessions, McCoy, Granger & Tilberry
1200 Leader Bldg.
Cleveland, OH 44114 (216) 579-1700

William S. Lightbody
655 Union Commerce Bldg.
925 Euclid Ave.
Cleveland, OH 44115 (216) 241-4150

Alfred D. Lobo
Lobo & Greene
1260 Leader Bldg.
Cleveland, OH 44114 (216) 566-1661

James Andre Lucas
Harshaw/Filtrol Partnership
30100 Chagrin Blvd.
Cleveland, OH 44124 (216) 202-9229

John F. Luhrs
29925 Fairmount Blvd.
Cleveland, OH 44124 (216) 449-1375

Charles Stafford Lynch
Standard Oil Co.
200 Public Square, 36-F-3454
Cleveland, OH 44114 (216) 586-8469

Charles Bahlmann Lyon
Calfe, Halter & Griswood
1800 Society Bldg.
Cleveland, OH 44114 (216) 781-2166

James A. Mackin*
NASA Lewis Research Center
Office of Patent Coun.
21000 Brookpart Rd.
Cleveland, OH 44135 (216) 433-4000

Walter Maky
Maky, Renner, Otto & Boisselle
One Public Square, 12th Floor
Cleveland, OH 44113 (216) 621-1113

Vytas R. Matas
Mc Dermott Incorporated
26250 Euclid Ave. Suite 927
Cleveland, OH 44132 (216) 261-4044

John P. Maxey
Fay & Sharpe
1113 E. Ohio Bldg.
Cleveland, OH 44115 (216) 861-5582

Robert John Mc Closky
Eaton Corp.
739 E. 140th St.
Cleveland, OH 44114 (216) 523-6778

John F. Mc Devitt
General Electric Co.
Nela Park. Noble Rd.
Cleveland, OH 44112 (216) 266-2595

John P. Mc Mahon
General Electric Co.
Nela Park, Noble Park
Cleveland, OH 44112 (216) 266-8694

William Charles McCoy, Jr.
Pearne, Gordon, Sessions, McCoy & Granger
1200 Leader Bldg.
Cleveland, OH 44114 (216) 579-1700

Robert E. McDonald
Sherwin-Williams Co.
101 Prospect Ave.
Cleveland, OH 44115 (216) 566-2432

James William McKee
Fay & Sharpe
1113 E. Ohio Bldg.
1717 E. Ninth St.
Cleveland, OH 44114 (216) 861-5582

Jean M. Miller
Jones, Day, Reavis & Pogue
1700 Huntington Bldg.
Cleveland, OH 44115 (216) 348-3939

John E. Miller, Jr.
Standard Oil Co.
200 Public Square, 36-F-3454
Cleveland, OH 44114 (216) 586-8467

Stanley Ross Miller
White Consolidated Inds., Inc.
11770 Berea Rd.
Cleveland, OH 44111 (216) 252-3700

Richard J. Minnich
Fay & Sharpe
1113 E. Ohio Bldg.
1717 E. Ninth St.
Cleveland, OH 44114 (216) 861-5582

Michael H. Minns
Cleveland Electric Illuminating
55 Public Square
Cleveland, OH 44101 (216) 622-9800

Jay F. Moldovanyi
Fay & Sharpe
1113 E. Ohio Bldg.
1717 E. Ninth St.
Cleveland, OH 44114 (216) 861-5582

William D. Mooney
Standard Oil Co.
1650 Midland Bldg.
Cleveland, OH 44115 (216) 575-5508

Christopher H. Morgan
Parker - Hannifin Corp.
17325 Euclid Ave.
Cleveland, OH 44112 (216) 531-3000

Norman T. Musial*
National Aeronautics & Space Admin.
21000 Brookpart Rd.
M.S. 500/318
Cleveland, OH 44135 (216) 433-4000

Timothy E. Nauman
Fay, Sharpe, Fagan, Minnich & McKee
400 National City Bank Bldg.
Cleveland, OH 44114 (216) 861-5582

Frank J. Nawalanic
Midland - Ross Corp.
20600 Chagrin Blvd.
Cleveland, OH 44122 (216) 491-8400

Timothy J. O Hearn
Jones, Day, Reavis & Pogue
North Point
901 Lakeside
Cleveland, OH 44114 (216) 586-1080

I. Monica Olszewdki
Watts, Hoffman, Fisher & Heinke Co. L.P.A.
100 Erieview Plaza, Suite 2850
Cleveland, OH 44114 (216) 623-0775

Lawrence R. Oremland
Calfee, Halter & Griswold
1800 Society Bldg.
Cleveland, OH 44114 (216) 781-2166

Fred Ornstein
Isler & Ornstein
933 Leader Bldg.
Cleveland, OH 44114 (216) 621-2054

Stanley E. Ornstein
Isler & Ornstein
457 Leader Bldg.
Cleveland, OH 44114 (216) 621-2054

Donald L. Otto
Maky, Renner, Otto & Boisselle
One Public Square, 12th Floor
Cleveland, OH 44113 (216) 621-1113

David J. Pasz
Kraig, De Wolfe & Pasz
33 Public Square, Suite 1117
Cleveland, OH 44113 (216) 696-4009

Richard S. Pauliukonis#
6660 Greenbriar Dr.
Cleveland, OH 44130 (216) 842-0828

David R. Percio
Cleveland Electric Illuminating Co.
55 Public Square
Cleveland, OH 44113 (216) 622-9800

Sue E. Phillips
Standard Oil Co.
200 Public Square
Cleveland, OH 44114 (216) 586-8614

Gary R. Plotecher
B P American Inc.
200 Public Square, 36 F 3056
Cleveland, OH 44114 (216) 586-8464

Wayne D. Porter, Jr.
Three Commerce Park Square
23200 Chagrin Blvd., Suite 605
Cleveland, OH 44122 (216) 765-1500

Kenneth George Preston, Jr.
T R W, Inc.
23555 Euclid Ave.
Cleveland, OH 44117 (216) 383-2634

Linn J. Raney
Watts, Hoffmann, Fisher & Heinke Co. L.P.A.
100 Erieview Plaza, Suite 2850
Cleveland, OH 44114 (216) 623-0775

Carl A. Rankin
Pearn, Gordon, Sessions, McCoy, Granger & Tilberry
1200 Leader Bldg.
Cleveland, OH 44114 (216) 579-1700

John William Renner
Maky, Renner, Otto & Boisselle
One Public Square, 12th Floor
Cleveland, OH 44113 (216) 621-1113

James A. Rich
Alcan Aluminum Corp.
100 Erieview Plaza
Cleveland, OH 44114 (216) 523-6874

Michael M. Rickin
Reliance Elec.
P.O. Box 22280
Cleveland, OH 44122 (216) 266-7613

Patrick R. Roche
Fay & Sharpe
400 National City Bank Bldg.
Cleveland, OH 44114 (216) 861-5582

Donald A. Rowe
Eaton Corp.
Eaton Center
1111 Superior Ave.
Cleveland, OH 44114 (216) 523-4299

Charles R. Rust
Woodling, Krost & Rust
655 Union Commerce Bldg.
925 Euclid Ave.
Cleveland, OH 44115 (216) 241-4150

Peter D. Sachtjen
545 Hanna Bldg.
Cleveland, OH 44115 (216) 696-6663

Robert L. Sahr
B P American Inc.
200 Public Square, 36 F 3454
Cleveland, OH 44114 (216) 586-8555

Frank M. Sajovec, Jr.
Eaton Corp.
Eaton Center
1111 Superior Ave.
Cleveland, OH 44114 (216) 523-4136

Maurice R. Salada
T R W, Inc.
Automotive Worldwide Sector
1900 Richmond Rd.
Cleveland, OH 44124 (216) 291-7392

Dan J. Sammon
Watts, Hoffmann, Fisher & Heinke Co. L.P.A.
100 Erieview Plaza, Suite 2850
Cleveland, OH 44114 (216) 623-0775

Walter C. Sanison, Jr.#
1510 Ohio Savings Plaza
Cleveland, OH 44114 (216) 771-3800

Thomas Paul Schiller
Pearne, Gordon, Sessions, McCoy & Granger
1200 Leader Bldg.
Cleveland, OH 44114 (216) 579-1700

Philip L. Schlamp
General Electric Co.
Nela Park, Noble Rd.
Cleveland, OH 44112 (216) 266-2585

Robert N. Schmidt
1721 Fulton Rd.
Cleveland, OH 44113 (216) 781-4096

Thomas M. Schmitz
S C M Corporation
900 Huntington Bldg.
Cleveland, OH 44115 (216) 344-8401

Frederic B. Schramm
3570 Warrensville Center Rd.
Cleveland, OH 44122 (216) 283-0075

Stephen J. Schultz
Watts, Hoffmann, Fisher & Heinke Co. L.P.A.
100 Erieview Plaza, Suite 2850
Cleveland, OH 44114 (216) 623-0775

Albert P. Sharpe, III
Fay & Sharpe
400 National City Bank Bldg.
Cleveland, OH 44114 (216) 861-5582

Howard G. Shimola
Pearne, Gordon, Sessions, McCoy, Granger & Tilberry
1200 Leader Bldg.
Cleveland, OH 44114 (216) 579-1700

Gene Edwin Shook, Sr.*
NASA Lewis Research Center
Office of Patent Coun. 500-311
21000 Brookpart Rd.
Cleveland, OH 44135 (216) 433-4000

Laura F. Shunk
Pearne, Gordon, Sessions, McCoy, Granger & Tilberry
1200 Leader Bldg.
Cleveland, OH 44114 (216) 579-1700

Milton Lawrence Simmons
Ferro Corp.
One Erieview Plaza
Cleveland, OH 44114 (216) 641-8580

Warren A. Sklar
Maky, Renner, Otto & Boisselle
One Public Square, 12th Floor
Cleveland, OH 44113 (216) 621-1113

J. Helen Slough
200 Westview Towers
21010 Center Ridge Rd.
Cleveland, OH 44116 (216) 331-4892

Richard G. Smith
Fay, Sharpe, Fagan, Minnich & McKee
400 National City Bank Bldg.
Cleveland, OH 44114 (216) 861-5582

Susan A. Smith
Arter & Hadden
1100 Huntington Bldg.
Cleveland, OH 44115 (216) 696-1100

Jeffrey J. Sopke
Pearne, Gordon, Sessions, McCoy, Granger & Tilberry
1200 Leader Bldg.
Cleveland, OH 44114 (215) 579-1700

Barry L. Springel
Jones, Day, Reavis, & Pogue
North Point
901 Lakeside Ave.
Cleveland, OH 44114 (216) 586-7236

Kent N. Stone*
NASA Lewis Research Center
21000 Brookpark Rd.
Cleveland, OH 44135 (216) 433-2318

Greg Strugalski
Tarolli, Sunfheim & Covell
1111 Leader Bldg.
Cleveland, OH 44114 (216) 621-2234

Robert Brandt Sundheim
Yount & Tarolli
1111 Leader Bldg.
Cleveland, OH 44114 (216) 621-2234

Harold Duane Switzer
Jones, Day, Reavis & Pogue
North Point
901 Lakeside Ave.
Cleveland, OH 44114 (216) 586-7283

Paul E. Szabo
Yount & Tarolli
1111 Leader Bldg.
Cleveland, OH 44114 (216) 621-2234

Steven W. Tan
Sherwin-William Co.
Legal Dept.
101 Prospect Ave., N.W.
Cleveland, OH 44115 (216) 566-2487

Thomas L. Tarolli
Yount & Tarolli
1111 Leader Bldg.
Cleveland, OH 44114 (216) 621-2234

Joseph R. Teagno
Teagno & Hudak
The Standard Bldg., Suite 1312
Cleveland, OH 44113 (216) 566-9700

Richard Hadley Thomas
S C M Corp.
925 Euclid Ave.
Cleveland, OH 44115 (216) 344-8407

Otto Tichy
Judson Manor Rd.
1890 E. 107th St.
Cleveland, OH 44106 (216) 231-5569

John C. Tiernan
Fay, Sharpe, Fagan, Minnich & McKee
400 National City Bank Bldg.
Cleveland, OH 44144 (216) 861-5582

James H. Tillberry
Pearne, Gordon, Sessions, McCoy, Granger & Tilberry
1200 Leader Bldg.
Cleveland, OH 44114 (216) 579-1700

Clayton J. Toddy
Eaton Corp.
Eaton Center
1111 Superior Ave.
Cleveland, OH 44114 (216) 523-4140

Frederick L. Tolhurst
Parker-Hannifin Corp.
17325 Euclid Ave.
Cleveland, OH 44112 (216) 531-3000

Barry L. Tummino
Yount & Tarolli
1111 Leader Bldg.
Cleveland, OH 44114 (216) 621-2234

James Vincent Tura
Sherwin - Williams Co.
101 Prospect Ave., N.W.
Cleveland, OH 44115 (216) 566-2487

Lawrence C. Turnock, Jr.
Amer. Iron Ore Assn.
514 Bulkley Bldg.
Cleveland, OH 44115 (216) 241-8261

Marvin L. Union
Eaton Corp.
Eaton Center
1111 Superior Ave.
Cleveland, OH 44114 (216) 523-4126

David J. Untener
Standard Oil Co.
200 Public Square, 36th Floor
Cleveland, OH 44114 (216) 586-8472

Michael W. Vary
Jones, Day, Reavis & Pogue
North Point
901 Lakeside Ave.
Cleveland, OH 44114 (216) 586-1241

Robert Vernon Vickers
Body, Vickers & Daniels
2000 Terminal Tower
Cleveland, OH 44113 (216) 623-0040

Allan William Vogele
T R W, Inc.
1900 Richmond Rd.
Cleveland, OH 44124 (216) 291-7393

Richard R. Walling
Wooding, Krost & Rust
530 National City Bank Bldg.
Cleveland, OH 44114 (216) 241-4150

James L. Wamsley, III
Jones, Day, Reavis & Pogue
North Point
901 Lakeside Ave.
Cleveland, OH 44114 (216) 586-7251

James G. Watterson
Watts, Hoffmann, Fisher & Heink Co. L.P.A.
100 Erieview Plaza, Suite 2850
Cleveland, OH 44114 (216) 623-0775

Howard Lee Weinshenker
American Greetings Corp.
Trademark & License Coun.
10500 American Rd.
Cleveland, OH 44144 (216) 252-7300

Robert Philip Wright
L T V Steel Co. Inc.
25 W. Prospect Ave.
Cleveland, OH 44115 (216) 622-5604

Jeffrey A. Wyand
B P American Inc.
200 Public Square, 36 F 3454
Cleveland, OH 44114 (216) 586-4141

Thomas E. Young
Body, Vickers & Daniels
2000 Terminal Tower
Cleveland, OH 44113 (216) 623-0040

Vincent Edward Young
Standard Oil Co.
200 Public Square, 36-F
Cleveland, OH 44114 (216) 586-8453

J. Herman Yount & Tarillo, Jr.
Yount & Tarillo
1111 Leader Bldg.
Cleveland, OH 44114 (216) 621-2234

David P. Yusko
B.P. America
200 Public Square
Cleveland, OH 44114 (216) 586-8461

Thaddeus Arthur Zalenski
L T V Steel Co., Inc.
25 W. Prospect Ave.
Cleveland, OH 44115 (216) 622-5626

Robert J. Zellner
Mc Gean-Rohco, Inc.
1250 Terminal Tower
Cleveland, OH 44113 (216) 621-6425

Alexander Zinchuk#
Yount & Tarolli
1111 Leader Bldg.
Cleveland, OH 44114 (216) 621-2234

Thomas S. Baker, Jr.
88 E. Broad St., Suite 1590
Columbus, OH 43215 (614) 221-4137

Edwin Baranowski
Porter, Wright, Morris & Arthur
41 S. High St.
Columbus, OH 43215 (614) 227-2188

Barry S. Bissell
Battelle Development Corp.
505 King Ave.
Columbus, OH 43201 (614) 424-7798

John D. Boos
4642 Burbank Dr.
Columbus, OH 43220 (614) 457-7327

Anthony Dominic Cennamo
P.O. Box 956
Columbus, OH 43216 (614) 221-0888

Richard A. Crane
2850 Columbus Ave.
Columbus, OH 43209 (614) 231-8717

Daniel H. Dunbar
Kremblas, Foster, Millard & Watkins
50 West Broad St.
Columbus, OH 43215 (614) 464-2700

Philip M. Dunson
1446 Friar Lane
Columbus, OH 43221 (614) 457-4314

Edward Paul Forgrave
Biebel, French & Nauman
50 W. Broad St., Suite 620
Columbus, OH 43215 (614) 464-2902

Frank H. Foster, III
Kremblas, Foster, Millard & Watkins
50 W. Broad St., Suite 2700
Columbus, OH 43215 (614) 464-2700

Palmer Fultz
4386 Stinson Dr.
Columbus, OH 43214 (614) 459-7356

John L. Gray
Emens, Hurd, Kegler & Ritter
65 E. State St.
Columbus, OH 43215 (614) 462-5438

Herbert M. Hanegan
Ashland Chemical Co.
P.O. Box 2219
Columbus, OH 43216 (614) 889-3105

David L. Hedden
Ashland Chemical Co.
P.O. Box 2219
Columbus, OH 43216 (614) 889-4265

Francis Thomas Kremblas, Jr.
Kremblas, Foster, Millard & Watkins
50 W. Broad St., Suite 2300
Columbus, OH 43215 (614) 464-2700

Eugene John Mahoney
Porter, Wright, Morris & Arthur
37 W. Broad St., Suite 1100
Columbus, OH 43215 (614) 227-2026

John P. Mann, III
Dresser Industries, Inc.
Jeffery Minn. Mach. Div.
274 E. First Ave.
Columbus, OH 43201 (614) 297-3089

Dwight A. Marshall
Bell Labs, Inc.
6200 E. Broad St.
Columbus, OH 43213 (614) 860-2127

William J. Mase
Battelle - Columbus Labs
505 King Ave.
Columbus, OH 43201 (614) 424-6331

George P. Maskas
Borden, Inc.
180 E. Broad St.
Columbus, OH 43215 (614) 225-4363

Sidney Wayne Millard#
Kremblas, Foster, Millard & Watkins
50 W. Broad St., Suite 2300
Columbus, OH 43215 (614) 464-2700

William Vernon Miller
145 N. High St.
Columbus, OH 43215 (614) 228-4531

Jerry K. Mueller, Jr.
Mueller & Smith, L.P.A.
7700 Rivers Edge Dr.
Columbus, OH 43085 (614) 436-0600

Clarence Henry Peterson#
1491 Kirkley Rd.
Columbus, OH 43221 (614) 451-2243

Patrick P. Phillips
Schottenstein, Zox & Dunn
41 S. High St., Suite 2600
Columbus, OH 43215 (614) 462-2224

Mary E. Picken
Ashland Chemical Co.
P.O. Box 2219
Columbus, OH 43216 (614) 889-4694

Philip J. Pollick
3316 N. High St.
Columbus, OH 43202 (614) 267-7966

William C. Rambo
Porter, Wright, Morris & Arthur
41 S. High St.
Columbus, OH 43215 (614) 227-2215

Priscilla N. Ratliff#
1965 Glenn Ave.
Columbus, OH 43212 (614) 481-0124

Kenneth Earl Shaweker
Battelle Development Corp.
505 King Ave.
Columbus, OH 43201 (614) 424-7453

Gerald Lee Smith
Mueller & Smith, L.P.A.
7700 Rivers Edge Dr.
Columbus, OH 43085 (614) 436-0600

Robert E. Stebens
50 W. Broad St., Suite 1930
Columbus, OH 43215 (614) 228-6359

Thomas L. Sweeney
180 E. Broad St., Suite 1700
Columbus, OH 43215 (614) 461-4300

Charles H. Thieman#
Kremblas, Foster, Millard & Watkins
50 W. Broad St., Suite 3400
Columbus, OH 43215 (614) 464-2700

Kenneth P. Van Wyck
Borden, Inc.
Law Dept.
180 E. Broad St., 27th Floor
Columbus, OH 43215 (614) 225-3369

Vernon F. Venne
Ashland Chemical Co.
P.O. Box 2219
Columbus, OH 43216 (614) 889-3975

Kenneth R. Warburton
Battelle Columbus Labs
Legal & Patent Dept.
505 King Ave.
Columbus, OH 43201 (614) 424-6585

Robert B. Watkins
2941 Kenny Rd., Suite 260
Columbus, OH 43221 (614) 457-5700

Klaus H. Wiemann
Battelle Columbus Labs.
505 King Ave.
Columbus, OH 43201 (614) 424-6589

James B. Wilkens
The Ohio State Univ.
1314 Kinnear Rd.
Columbus, OH 43212 (614) 422-6079

John Frank Jones
2724 Cedar Hill Rd.
Cuyahoga Falls, OH 44223 (216) 923-2952

Denbigh S. Matthews
1240 Lincoln Ave.
Cuyahoga Falls, OH 44223 (216) 923-5620

Robert W. Becker
Becker & Becker Inc.
211 S. Main St.
Dayton, OH 45402 (513) 228-7801

Jerome P. Bloom
5751 N. Webster St.
P.O. Box 14553
Dayton, OH 45414 (513) 890-2079

Thomas A. Boshincki
Bieble, French & Nauman
2500 Kettering Tower
Dayton, OH 45423 (513) 461-4543

James T. Candor
Candor, Candor & Tassone
P.O. Box 2305
Dayton, OH 45429 (513) 298-8606

J.T. Cavender
NCR Corp.
1700 S. Patterson Blvd.
Dayton, OH 45479 (513) 445-2970

Robert L. Clark
NCR Corp.
1700 S. Patterson Blvd.
Dayton, OH 45479 (513) 445-2913

Robert L. Deddens
2621 Far Hills Ave.
Dayton, OH 45419 (513) 293-9696

Dorothy I. Dianis
7810 Capitol Hill Lane
Dayton, OH 45459

John William Donahue
Biebel, French & Nauman
2500 Kettering Tower
Dayton, OH 45423 (513) 461-4543

Edward Dugas
NCR Corp.
Intellectual Property.
1700 S. Patterson Blvd.
Dayton, OH 45479 (513) 445-2988

Robert Kern Duncan*
4390 Baker Rd.
Dayton, OH 45424 (513) 233-3073

H. Talman Dybvig
Dybvig & Dybvig
2600 Far Hills Ave.
Dayton, OH 45419 (513) 299-5529

Roger S. Dybvig
Dybvig & Dybvig
2600 Far Hills Ave.
Dayton, OH 45419 (513) 299-5529

Thomas William Flynn
Bieble, French & Nauman
2500 Kettering Tower
Dayton, OH 45423 (513) 461-4543

Douglas S. Foote
NCR Corp.
Law Dept.
Dayton, OH 45479 (513) 445-2968

Nathaniel R. French
Biebel, French & Nauman
2500 Kettering Tower
Dayton, OH 45423 (513) 461-4543

Donald P. Gillette
55 Westpark Rd.
Dayton, OH 45459 (513) 433-6563

Irvin V. Gleim
Box 482
Dayton, OH 45459 (513) 433-3831

James F. Gottman
Killworth, Gottman, Hagan & Scheff
1400 One First National Plaza
Dayton, OH 45402 (513) 223-2050

Joseph John Grass
Monarch Marking Sys. Inc.
P.O. Box 608
Dayton, OH 45401 (513) 865-2021

Timothy W. Hagan
Killworth, Gottman, Hagan & Schaeff
1400 One First National Plaza
Dayton, OH 45402 (512) 322-3205

Wilbert Hawk, Jr.
NCR Corp.
1700 S. Patterson Blvd.
Dayton, OH 45479 (513) 445-2960

Louis E. Hay
847 Woodhill Rd.
Dayton, OH 45431 (513) 253-1645

Gilbert N. Henderson
Biebel, French & Nauman
2500 Kettering Tower
Dayton, OH 45423 (513) 461-4543

Ernest T. Hix
53 E. Thruston Blvd.
Dayton, OH 45409 (513) 298-1594

Michael A. Jacobs
NCR Corp.
Law Dept. WHQ 5
1700 S. Patterson Blvd.
Dayton, OH 45479 (513) 445-2928

William R. Jacox
Jacox & Meckstroth
2310 Far Hills Bldg.
Dayton, OH 45419 (513) 298-2811

Stephen F. Jewett
NCR Corp.
1700 S. Patterson Blvd.
Dayton, OH 45479 (513) 445-2972

Richard L. Kelly
Beibel, French & Nauman
2500 Kettering Tower
Dayton, OH 45423 (513) 461-4543

Richard A. Killworth
Killworth, Gottman, Hagan & Schaeff
1400 One First National Plaza
Dayton, OH 45402 (513) 223-2050

Richard William Lavin
NCR Corp.
1700 S. Patterson Blvd.
Dayton, OH 45479 (513) 445-2968

Mark P. Levy#
Smith & Schnack
2000 Courthouse Plaza, N.E.
P.O. Box 1817
Dayton, OH 45401 (513) 226-6500

Theodore David Lienesch
Smith & Schnacke
2000 Courthouse Plaza, N.E.
P.O. Box 1817
Dayton, OH 45401 (513) 266-6500

Ralph L. Marzocco
7606 Springboro Pike
Dayton, OH 45449 (513) 435-5460

Alan F. Meckstroth
Jacox & Meckstroth
2310 Far Hill Bldg.
Dayton, OH 45419 (513) 298-2811

George J. Muckenthaler
NCR Corp.
Dayton, OH 45479 (513) 445-2964

Joseph G. Nauman
Biebel, French & Nauman
2500 Kettering Tower
Dayton, OH 45423 (513) 461-4543

Bruce E. Peacock
Biebel, French & Nauman
2500 Kettering Tower
Dayton, OH 45423 (513) 461-4543

Jack R. Penrod
NCR Corp.
Law Dept.
1700 S. Patterson Blvd.
Dayton, OH 45479 (513) 445-6740

Albert H. Reuther
Becker & Becker Inc.
712 Harries Bldg.
137 N. Main St.
Dayton, OH 45402 (513) 228-7801

Casimer K. Salys
NCR Corp.
Intellectual Property
1700 S. Patterson Blvd.
Dayton, OH 45479 (513) 445-2982

Albert L. Sessler, Jr.
NCR Corp.
Law Dept.
1700 S. Patterson Blvd.
Dayton, OH 45479 (513) 445-2965

Charles N. Shane, Jr.
Mead Corp.
Courthouse Plaza, N.E.
Dayton, OH 45462 (513) 222-6323

James S. Shannon#
410 Delaware Ave.
Dayton, OH 45405 (513) 274-9764

Andrew Martin Solomon
Smith & Schnacke
2000 Courthouse Plaza, N.E.
P.O. Box 1817
Dayton, OH 45401 (513) 226-6500

Lloyd B. Stevens, Jr.
1359 Tattersall Rd.
Dayton, OH 45459 (513) 433-0946

Richard C. Stevens
Biebel, French & Nauman
2500 Kettering Tower
Dayton, OH 45423 (513) 461-4543

Carl A. Stickel
216 Marathon Ave.
Dayton, OH 45405 (513) 274-5573

Joseph V. Tassone
Candor, Candor & Tassone
P.O. Box 2305
Dayton, OH 45429 (513) 298-8606

Elmer Wargo
NCR Corp.
1700 S. Patterson Blvd.
Dayton, OH 45479 (513) 445-2969

Gerald T. Welch
Killworth, Gottman, Hagan & Schaeff
1400 One First National Plaza
Dayton, OH 45402 (513) 223-2050

Norman R. Wissinger
3103 Winding Way
Dayton, OH 45419 (513) 299-2003

Reuben Wolk
3849 Seiber Ave.
Dayton, OH 45405 (513) 278-1258

Daniel D. Mast
3520 Cackler Rd.
Delaware, OH 43015

Patricia A. Coburn
Adria Labs
Div. of Erbamont, Inc.
5000 Post Rd.
Dublin, OH 43017 (614) 764-8121

Edward B. Dunning
Sherex Chemical Co., Inc.
P.O. Box 646
Dublin, OH 43017 (614) 764-6601

William Kammerer
8839 Bingham Ct., N.
Dublin, OH 43017

Walter H. Schneider
P.O. Box 917
Dublin, OH 43017 (614) 889-5747

Clyde H. Haynes
1823 West River Rd.,N
Elyria, OH 44035 (216) 324-4706

Richard J. Killoren#
169 Locust Dr.
Fairborn, OH 45324 (513) 878-1272

John Jay Freer
Eltech Systems Corp.
625 East St.
Fairport Harbor, OH 44077 (216) 357-4055

Girard R. Jetton, Jr.
P.O. Box 1003
Findlay, OH 45839 (419) 423-4680

James H. Sutton
Cooper Tire & Rubber Co.
Lima & Western Aves.
Findlay, OH 45840 (419) 424-4322

Kevin M. Foley
Heper Industries, Inc.
160 Industrial Dr.
P.O. Box 338
Franklin, OH 45005 (513) 746-3603

Richard Charles Darr
915 Croghan St.
Fremont, OH 43420 (419) 332-4106

Joseph R. Black, Jr.
4640 Wendler Blvd.
Gahanna, OH 43230 (614) 475-1183

Charles E. Moore#
55 Beechtree Lane
Granville, OH 43023

Jack C. McGowan
300 Hamilton Center Bldg.
222 High St.
Hamilton, OH 45011 (513) 868-2731

John W. Teare
26 Hollytree Ct.
Hamilton, OH 45011 (513) 868-2564

William Preston Hickey
1970 C R 32
Helena, OH 43435 (419) 457-5525

Timothy B. Gurin#
Picker International, Inc.
595 Miner Rd.
Highland Heights, OH 44143 (216) 473-3570

Ronald B. Brietkrenz#
1888 Bellus Rd.
Hinckley, OH 44223 (216) 225-2583

Ronald R. Stanley
2443 Kellogg Rd.
Hinckley, OH 44233 (216) 278-7933

Woodrow W. Ban
7363 Winsted
Hudson, OH 44236 (216) 656-2953

George W. Moxon, II
38 West Case Dr.
Hudson, OH 44236 (216) 653-9815

John M. Romanchik, Jr.
6107 Elmwood Ave.
Independence, OH 44131 (216) 524-8855

Joseph P. Burke
4050 Benfield Dr.
Kettering, OH 45429 (513) 293-4998

Arthur Richard Parker#
900 Renwood Dr.
Kettering, OH 45429 (513) 298-7156

John W. Ball, Jr.
12030 Lake Ave.
Lakewood, OH 44107 (216) 228-5064

Gustalo Nunez
4463 Oberlin Ave.
Lorain, OH 44053 (216) 282-9109

Edward J. Holler, Jr.
R R #2, Hilltop Dr.
Magnolia, OH 44643 (216) 866-2289

Robert R. Teall
Tappan Co.
222 Chambers Rd.
Mansfield, OH 44906 (419) 755-2455

Myton E. Click
7440 Coder Rd.
Maumee, OH 43537 (419) 865-6775

Neil F. Katz
Hill Internatl., Inc.
6809 Mayfield Rd., Suite 1072
Mayfield Heights, OH 44124 (216) 259-3737

Robert B. Henn
Henn & Cain
45 Public Square
Medina, OH 44256 (216) 722-6640

Arthur L. Wolfe#
Intrnational Science Service
7999 Dartmoor Rd.
Mentor, OH 44060 (216) 951-4369

Edward L. Brown, Sr.#
3011 Central Ave.
Middletown, OH 45044 (513) 422-7721

Robert J. Bunyard#
Armco Inc.
703 Curtis St.
Middletown, OH 45043 (513) 425-5973

Larry A. Fillnow#
Armco Inc.
703 Curtis
Middletown, OH 45043 (513) 425-2494

Robert H. Johnson
Armco Inc.
703 Curtis St.
Middletown, OH 45043 (513) 425-2432

Bruce M. Thomas
4201 Fisher Ave.
Middletown, OH 45042 (513) 425-0777

Thomas F. McGann#
30-D South Terrace Ave.
Newark, OH 43055 (614) 522-3958

Francis D. Thomson#
1586 Stonewall Dr.
Newark, OH 43055

Richardson Blackburn Farley
Hoover Corp.
101 E. Maple St.
North Canton, OH 44720 (216) 499-9200

Gerald H. Kreske
Hoover Co.
101 E. Maple St.
North Canton, OH 44720 (216) 499-9200

A. Burgess Lowe
Hoover Co.
101 E. Maple St.
North Canton, OH 44720 (216) 499-9200

James D. Wolfe
1202 Lake Breeze Dr.
North Canton, OH 44720 (216) 896-1881

Roy Davis
5916 Chapel Rd.
North Madison, OH 44057 (216) 428-2628

Ralph E. Jocke
6029 Barton Rd.
North Olmsted, OH 44070 (216) 777-3671

Robert August Wiedemann
401 Citizens National Bank Bldg.
East Main St.
Norwalk, OH 44857 (419) 668-8211

Theodore A. Te Grotenhuis
7315 Columbia Rd.
Olmsted Falls, OH 44138 (216) 235-3528

John L. Shailer#
1720 Hiner Rd.
Orient, OH 43146 (614) 871-1972

Eugene Nebesh
1949 W. Pleasant Valley Rd.
Parma, OH 44134 (216) 886-6112

Henry Kozak
30779 Shaker Blvd.
Pepper Pike, OH 44124 (216) 831-8394

Richard Roy Drown#
1041 Louisiana Ave.
Perrysburg, OH 43551 (419) 874-4418

David H. Wilson, Jr.
424 E. Front St.
Perrysburg, OH 43551 (419) 241-1314

Timothy E. Tinkler
Ricerca, Inc.
7528 Auburn Rd.
P.O Box 1000
Plainsville, OH 44077 (216) 357-3428

Leslie Hamilton Blair
21111 W. Wagar Circle
Rocky River, OH 44116 (216) 331-0298

Albert L. Ely, Jr.
18951 Inglewood Rd.
Rocky River, OH 44116 (216) 333-5467

George V. Woodling
22077 Lake Rd.
Rocky River, OH 44116 (216) 331-0463

William E. Nobbe
375 Colony Rd.
Rossford, OH 43460 (419) 666-2365

Frederic E. Naragon
248 E. State St.
P.O. Box 317
Salem, OH 44460 (216) 337-9578

Daniel Glenn Blackhurst
3353 Lansmere Rd.
Shaker Heights, OH 44122 (216) 283-8129

A. Joseph Gibbons
14429 Drexmore Rd.
Shaker Heights, OH 44120 (216) 921-2912

Sheldon B. Greenbaum#
Goal Oriented Strategies, Inc.
24139 Shelburne Rd.
Shaker Heights, OH 44122 (216) 831-2079

Robert Donovan Hart
3264 Kenmore Rd.
Shaker Heights, OH 44122 (216) 751-3864

William Isler
16100 Van Aken Blvd.
Shaker Heights, OH 44120 (216) 921-7513

Russell L. Root
3518 Stoer Rd.
Shaker Heights, OH 44122 (216) 283-6361

Ramon Doyle Foltz
6812 Solon Blvd.
Solon, OH 44139 (216) 349-2151

Woodrow W. Portz
Portz & Portz
5530 S.O.M. Center Rd.
Solon, OH 44139 (216) 248-4500

Harlan E. Hummer
P.O. Box 21180
South Euclid, OH 44121 (216) 381-8198

P. Adrian Medert
P.O. Box 21310
S. Euclid, OH 44121 (216) 932-1310

Donald A. Bergquist#
Patent Service Co.
17145 Misty Lake Dr.
Strongsville, OH 44136 (216) 238-1210

Gregory Dziegielewski
4449 Falconhurst Court
Sylvania, OH 43560 (419) 885-2711

John B. Molnar
6832 S. Fredericksburg Dr.
Sylvania, OH 43560 (419) 882-5712

John C. Purdue
6600 Sylvania Ave.
Sylvania, OH 43560 (419) 885-3370

Vincent L. Barker, Jr.
Willian, Brinks, Olds, Hofer, Gilson & Lione
930 National Bank Bldg.
Toledo, OH 43604 (419) 244-6578

David Robert Birchall
3700 Heathesdowns Blvd.
Toledo, OH 43614 (419) 385-1627

Howard G. Bruss, Jr.
Owens-Illinois, Inc.
One Sea-Gate
Toledo, OH 43666 (419) 247-2036

Ronald E. Champion
Owens-Corning Fiberglas Corp.
Law Dept.
Fiberglas Tower
Toledo, OH 43659 (419) 248-6898

Kenneth F. Cherry#
1533 East Gate
Toledo, OH 43614 (419) 385-9165

William J. Clemens
Marshall & Melhorn
Four Seagate, 8th Floor
Toledo, OH 43604 (419) 249-7100

Freeman Crampton
Caller # 10003
Toledo, OH 43699 (419) 241-9322

Richard Donovan Emch
Emch, Schaffer, Schaub & Porcello Co. L.P.A.
One Sea Gate, Suite 1980
P.O. Box 916
Toledo, OH 43692 (419) 243-1294

Harry O. Ernsberger, Sr.
5036 Rolandale Ave.
Toledo, OH 43623 (419) 882-3843

Dean T. Fisher
6924 Perivale Park Rd.
Toledo, OH 43617 (419) 841-4773

Donald R. Fraser
Marshall & Melhorn
Four Seagate, 8th Floor
Toledo, OH 43604 (419) 249-7100

Malcolm W. Fraser
4040 W. Bancroft St.
Toledo, OH 43606 (419) 536-3190

R. La Mar Frederick
Owens-Corning Fiberglas Corp.
Fiberglas Tower
Toledo, OH 43659 (419) 248-8650

Ted C. Gillespie
Owens-Corning Fiberglas Corp.
Fiberglas Tower
Toledo, OH 43659 (419) 248-8461

Thomas H. Grafton
3333 Christie Blvd.
Toledo, OH 43606 (419) 531-6614

Allen D. Gutchess, Jr.
1806 Madison, Suite 408
Toledo, OH 43624 (419) 243-4353

David T. Innis
5003 Rudgate Blvd.
Toledo, OH 43623 (419) 885-1867

Hugh Adam Kirk
4120 Tantara Dr.
Toledo, OH 43623 (419) 882-5995

Henry K. Leonard
2132 Burroughs Dr.
Toledo, OH 43614 (419) 385-5548

Robert M. Leonardi
Dana Corp.
4500 Dorr St.
Toledo, OH 43615 (419) 353-4791

Richard S. Mac Millan
Mac Millan, Sobanski & Todd
905 First Federal Plaza
701 Adams St.
Toledo, OH 43624 (419) 255-5900

Catherin B. Martineau
Emch, Schaffer, Schaub & Porcello Co. L.P.A.
One Sea Gate, Suite 1980
P.O. Box 916
Toledo, OH 43962 (419) 243-1294

Frank B. Mc Donald
Dana Corp.
P.O. Box 1000
Toledo, OH 43697 (419) 535-4655

Thomas A. Meehan
William, Brinks, Olds, Hofer, Gilson & Lione
930 National Bank Bldg.
Toledo, OH 43604 (419) 244-6578

David D. Murray
William, Brinks, Olds, Hofer, Gilson & Lione
930 National Bank Bldg.
Toledo, OH 43604 (419) 244-6578

John Roger Nelson
Owens-Illionis, Inc.
One Sea-Gate
Toledo, OH 43666 (419) 247-1081

Phillip S. Oberlin
Libbey-Owens-Foods Co.
811 Madison Ave.
P.O. Box 799
Toledo, OH 43695 (419) 247-3918

Patrick P. Pacella
Owens-Corning Fiberglas Corp.
Fiberglas Tower
Toledo, OH 43659 (419) 248-8230

Robert E. Pollock
Dana Corp.
P.O. Box 1000
Toledo, OH 43697 (419) 535-4653

James F. Porcello, Jr.
Emch, Schaffer, Schaub & Porcello Co. L.P.A.
One Sea Gate, Suite 1980
P.O. Box 916
Toledo, OH 43692 (419) 243-1294

David C. Purdue
Willian, Brinks, Olds, Hofer, Gilson & Lione
930 National Bank Bldg.
Toledo, OH 43604 (419) 244-6578

Philip M. Rice
Emch, Schaffer, Schaub & Porcello Co. L.P.A.
One Sea Gate, Suite 1980
P.O. Box 916
Toledo, OH 43692 (419) 243-1294

Paul Joseph Rose, Jr.
Owens-Corning Fiberglas Corp.
Fiberglas Tower
Toledo, OH 43659 (419) 248-8214

George R. Royer
421 N. Michigan St.
Toledo, OH 43624

Joseph Dennis Ryan, Sr.#
3018 Middlesex Dr.
Toledo, OH 43606 (419) 535-6563

Robert F. Rywalski
Willian, Brinks, Olds, Hofer, Gilson & Lione
930 National Bank Bldg.
Toledo, OH 43604 (419) 244-8862

Charles R. Schaub
Emch, Schaffer, Schaub & Porcello Co. L.P.A.
One Sea Gate, Suite 1980
P.O. Box 916
Toledo, OH 43692 (419) 243-1294

Charles F. Schroeder
2317 Valley Brook Dr.
Toledo, OH 43615 (419) 244-3344

Ralph J. Skinkiss
2954 Shetland Rd.
Toledo, OH 43617 (419) 841-2158

Mark J. Sobanski
Mac Millan, Sobanski & Todd
905 First Federal Plaza
701 Adams St.
Toledo, OH 43624 (419) 255-5900

Lawrence A. Steward
3730 Upton Ave.
Toledo, OH 43613 (419) 475-6107

D. Henry Stoltenberg
4443 Indian Rd.
Toledo, OH 43615 (419) 536-3969

Paul F. Stutz
964 Spitzer Bldg.
Toledo, OH 43604 (419) 241-4211

Charles Williams Swope
600 Toledo Bldg.
316 N. Michigan St.
Toledo, OH 43624 (419) 241-3197

Oliver E. Todd, Jr.
Mac Millian, Sobanski & Todd
905 First Federal Plaza
701 Adams St.
Toledo, OH 43624 (419) 255-5900

Edward A. Vangunten
3619 Brookside
Toledo, OH 43606 (419) 537-9992

Donald K. Wedding
University Of Toledo
2801 W. Bancroft
Toledo, OH 43606 (419) 841-7286

Kenneth H. Wetmore
Owens- Corning Fiberglas Corp.
Fiberglas Tower
Toledo, OH 43659 (419) 248-8788

Robert E. Witt
316 N. Michigan St., Suite 312
Toledo, OH 43624 (419) 241-3251

William Weigl
Hobart Corp.
World Headquaters
Troy, OH 45374 (513) 332-2111

Mark F. Smith
8168 Mellowtone Circle
West Chester, OH 45069 (513) 874-4002

Michael Leo Gill
Nordson Corp.
28601 Clemens Rd.
Westlake, OH 44145 (216) 892-1580

James Walter Adams, Jr.
Lubrizol Corp.
29400 Lakeland Blvd.
Wickliffe, OH 44092 (216) 943-4200

Karl Bozicevic
Lubrizol Corp.
Patent Dept.
29400 Lakeland Blvd.
Wickliffe, OH 44092 (216) 943-4200

James L. Cordek#
Lubrizol Corp.
29400 Lakeland Blvd.
Wickliffe, OH 44092 (216) 943-4200

Charles A. Crehore
Lubrizol Corp.
29400 Lakeland Blvd.
Wickliffe, OH 44092 (216) 943-4200

Joseph P. Fischer#
Lubrizol Corp.
29400 Lakeland Blvd.
Wickliffe, OH 44092 (216) 943-4200

Robert A. Franks
Lubrizol Corp.
29400 Lakeland Blvd.
Wickliffe, OH 44092 (216) 943-4200

Roger Y.K. Hsu
Lubrizol Corp.
29400 Lakeland Blvd.
Wickliffe, OH 44092 (216) 943-4200

Denis A. Polyn
Lubrizol Corp.
29400 Lakeland Blvd.
Wickliffe, OH 44092 (203) 943-4200

William C. Tritt#
Lubrizol Corp.
29400 Lakeland Blvd.
Wickliffe, OH 44092 (216) 943-4200

Gordon L. Vyrostek#
Lubrizol Corp.
29400 Lakeland Blvd.
Wickliffe, OH 44092 (216) 943-4200

John H. Hornickel
Figgie International Inc.
4420 Sherwin Rd.
Willoughby, OH 44094 (216) 946-9000

Frank B. Robb
Robb & Robb
37750 Euclid Ave.
Willoughby, OH 44094 (216) 951-2211

Richard B. O Planick
Rubbermaid Inc.
1147 Akron Rd.
Wooster, OH 44691 (216) 264-6464

Vernon A. Slabey
1551 Saunders Dr.
Wooster, OH 44691 (216) 264-4482

Gerald R. Black
491 Garden Dr.
Worthington, OH 43085 (614) 460-5254

David A. Greenlee
Lexitech Consultants
878 Blind Brook Dr.
Worthington, OH 43085 (614) 885-6135

Gary L. Loser
General Electric Co.
Specialty Materials Dept.
6325 Huntley Rd.
P.O. Box 568
Worthington, OH 43085

Robert R. Schroeder
General Electric Co.
6325 Huntley Rd.
Worthington, OH 43085 (614) 438-2438

George Wolken, Jr.
6602 Hawthorne St.
Worthington, OH 43085 (614) 885-1411

Charles E. Bricker*
U.S. Air Force
AF/JACPD
Area B, Bldg. 11, Room 100
Wright-Patterson A.F.B., OH 45433
(513) 255-5052

Bernard E. Franz*
U.S. Air Force
AF/JACPD
Wright-Patterson A.F.B., OH 45433
(513) 255-2838

Gerald B. Hollins*
U.S. Air Force
AF/JACPD
Wright-Patterson A.F.B., OH 45433
(513) 255-2833

Thomas Louis Kundert*
U.S Air Force
AF/JACPD
Wright-Patterson A.F.B., OH 45433
(513) 255-2838

Edward W. Nypaver*
U.S. Air Force
Patent Infringement Investigiation
Office of the Judge Advocate
Wright-Patterson A.F.B., OH 45433

Bobby D. Scearce*
U.S. Air Force
AF/JACPD
Wright-Patterson A.F.B., OH 45433
(513) 255-2838

Frederic L. Sinder*
U.S. Air Force
AF/JACPD
Wright-Patterson A.F.B., OH 45433
(513) 255-5052

James D. Thesing*
Office of Staff Judge Advocate
AF/JACPD
Wright-Patterson A.F.B., OH 45433
(513) 255-2838

Richard C. Harpman#
Harpman & Harpman
Federal Plaza East
400 Cty Centre One
Youngstown, OH 44503 (216) 747-1484

Webster B. Harpman
Harpman & Harpman
Federal Plaza East
400 City Centre One
Youngstown, OH 44503 (216) 747-1484

Oklahoma

George E. Bogatie#
Phillips Petroleum Co.
287 Patent Library Bldg.
Bartlesville, OK 74004 (918) 661-0560

Howard W. Bost#
1334 Quail Dr.
Bartlesville, OK 74006 (918) 333-0699

Karlheinz K. Brandes#
Phillips Petroleum Co.
263 Patent Library Bldg.
Bartlesville, OK 74004 (918) 661-0563

James D. Brown#
French & Doescher
P.O. Box 2443
Bartlesville, OK 74005 (918) 661-0536

Kenneth A. Cannon#
Phillips Petroleum Co.
204 Patent Library Bldg.
Bartlesville, OK 74004 (918) 661-0649

Lyell Henry Carver
1312 S.E. Evergreen Dr.
Bartlesville, OK 74006 (918) 333-0678

John R. Casperson
Phillips Petroleum Co.
204 Patent Library Bldg.
Bartlesville, OK 74004 (918) 661-0522

Paul S. Chirgott
Phillips Petroleum Co.
204 Patent Library Bldg.
Bartlesville, OK 74006 (918) 661-0526

Charles G. Cruzan#
918 S.E. King Dr.
Bartlesville, OK 74006 (918) 335-0285

Richard C. Doss#
Phillips Petroleum Co.
250 Patent Library Bldg.
Bartlesville, OK 74004 (918) 661-0671

John M. Fish, Jr.
Phillips Petroleum Co.
259 Patent Library Bldg.
Bartlesville, OK 74004 (918) 661-0524

Louis N. French
2136 Starlight Court
Bartlesville, OK 74006 (918) 333-3434

Bernhard H. Geissler
Phillips Petroleum Co.
216 Patent Library Bldg.
Bartlesville, OK 74004 (918) 661-0521

L. Barret Goodson
1601 Cherokee Pl.
Bartlesville, OK 74003 (918) FE6-1859

Mary P. Haddican
Phillips Petroleum Co. Room 208 P.L. Bldg.
Bartlesville, OK 74003 (918) 661-0536

Bion E. Hitchcock
Phillips Petroleum Co.
236 Patent Library Bldg.
Bartlesville, OK 74004 (918) 661-0534

James H. Hughes
Fractionation Research, Inc.
P.O. Drawer F.
Bartlesville, OK 74005

John W. Klooster
Phillips Petroleum Co.
204 Patent Library Bldg.
Bartlesville, OK 74004 (918) 661-6600

Robert C. Lutton#
Phillips Petroleum Co.
204 Patent Library Bldg.
Bartlesville, OK 74004 (918) 661-0675

John W. Miller#
Phillips Petroleum Co.
204 Patent Library Bldg.
Bartlesville, OK 74004 (918) 661-0565

John D. Olivier
Phillips Petroleum Co.
222 Patent Library Bldg.
Bartlesville, OK 74004 (918) 661-0526

Jack Ewart Phillips
Phillips Petroleum Co.
208 Patent Library Bldg.
Bartlesville, OK 74004 (918) 661-0520

Stephen E. Reiter
Phillips Petroleum Co.
210 Patent Library Bldg.
Bartlesville, OK 74004 (918) 661-0518

Allen W. Richmond
Phillips Petroleum Co.
208 Patent Library Bldg.
Bartlesville, OK 74004 (918) 661-0561

Archie Lew Robbins
Phillips Petroleum Co.
208 Patent Library Bldg.
Bartlesville, OK 74004 (918) 661-0546

William G. Roberts#
1912 Crestview
Bartlesville, OK 74003 (918) 661-3921

William R. Sharp
Phillips Petroleum Co.
204 Patent Library Bldg.
Bartlesville, OK 74004 (918) 661-0519

J. Michael Simpson
Phillips Petroleum Co.
204 Patent Library Bldg.
Bartlesville, OK 74004 (918) 661-9593

Charles F. Steininger
Phillips Petroleum Co.
208 Patent Library Bldg.
Bartlesville, OK 74004 (918) 661-0532

Robert Velgos#
1211 S. Osage
Bartlesville, OK 74003 (918) 366-8039

James Williams Williams
Phillips Petroleum Co.
208 Patent Library Bldg.
Bartlesville, OK 74004 (918) 661-0512

Hal B. Woodrow
Williams, Phillips & Umphlett
Phillips Research Center
261 Patent Library Bldg.
Bartlesville, OK 74004 (918) 661-6600

John Arthur Young
1525 Hillcrest Dr.
Bartlesville, OK 74003 (918) 336-9249

James R. Duzan
Halliburton Services
1015 Bois D'Arc
Duncan, OK 73536 (405) 251-3487

Robert A. Kent
Halliburton Service
P.O. Box 1431
Duncan, OK 73536 (405) 251-3760

John Hall Tregoning
2502 Linwood Lane
Duncan, OK 73533 (405) 255-2274

Thomas R. Weaver
1415 N. 12th
Duncan, OK 73533 (405) 252-2710

Paul W. Hemminger
P.O. Box 1466
Eufaula, OK 74432 (918) 689-5383

John Te Selle
College Of Law
P.O. Box 2848
Univ. of Oklahoma
Norman, OK 73070 (405) 360-2365

Edward L. Bowman
Box 284
Nowata, OK 74048 (918) 273-2342

F. Wesley Turner#
P.O. Box 599
Okay, OK 74446 (918) 683-5840

William G. Addison
Kerr-McGee Corp.
135 Robert S. Keer Ave.
P.O. Box 25861
Oklahoma City, OK 73125 (405) 270-2821

Lucian W. Beavers
Laney, Dougherty, Hessin & Beavers, P.C.
101 Park Ave., Suite 900
Oklahoma City, OK 73102 (405) 232-5586

Glen M. Burdick
Cantrell, Mc Carthy, Kice & Moore
6401 N.W. Grand Blvd., Suite 402
Oklahoma City, OK 73116 (405) 360-9133

Charles Alan Codding
Dunlap, Codding & Peterson
9400 N. Broadway, Suite 420
Oklahoma City, OK 73114 (405) 478-5344

Clifford C. Dougherty, III
Laney, Dougherty, Hessin & Beavers, P.C.
101 Park Ave., Suite 900
Oklahoma City, OK 73102 (405) 232-5586

Clifford Clark Dougherty, Jr.
Laney, Dougherty, Hessin & Beavers, P.C.
101 Park Ave., Suite 900
Oklahoma City, OK 73102 (405) 232-5586

Jerry J. Dunlap
Dunlap, Codding & Peterson
9400 N. Broadway, Suite 420
Oklahoma City, OK 73114 (405) 478-5344

Michael C. Felty#
Fenton, Fenton, Smith, Renau & Moon
211 N. Robinson
One Leadership Sqaure, Suite 800
Oklahoma City, OK 73102 (405) 235-4671

E. Harrison Gilbert, III
Laney, Dougherty, Hession & Beavers, P.C.
101 Park Ave., Suite 900
Oklahoma City, OK 73102 (405) 232-5586

Robert Marion Hessin
Laney, Dougherty, Hessin & Beavers, P.C.
101 Park Ave., Suite 900
Oklahoma City, OK 73102 (405) 232-5586

Neal R. Kennedy
Laney, Dougherty, Hessin & Beavers, P.C.
101 Park Ave., Suite 900
Oklahoma City, OK 73102 (405) 232-5586

William R. Laney
Laney, Dougherty, Hessin & Beavers, P.C.
101 Park Ave., Suite 900
Oklahoma City, OK 73102 (405) 232-5586

Mary M. Lee
Dunlap, Codding & Peterson
9400 N. Broadway, Suite 420
Oklahoma City, OK 73114 (405) 478-5344

Billy D. Mc Carthy
Cantrell, McCarthy, Kice & Moore
6401 N.W. Grand Blvd., Suite 402
Oklahoma City, OK 73116 (405) 842-0000

Alan T. Mc Collom
Laney, Dougherty, Hessin & Beavers P.C.
101 Park Ave., Suite 900
Oklahoma City, OK 73102 (405) 232-5586

Gary S. Peterson
Dunlap, Codding & Peterson
9400 N. Broadway, Suite 420
Oklahoma City, OK 73114 (405) 478-5344

Robert K. Rhea#
305 Western Tower
5350 South Western
Oklahoma City, OK 73109 (405) 235-2067

John P. Ward
Kerr-McGee Corp.
135 Robert S. Keer Ave.
P.O. Box 25861
Oklahoma City, OK 73125 (405) 270-2823

Louis W. Watson
6007 Sharon Lane
Oklahoma City, OK 73149 (405) 634-9495

Ronald J. Carlson
Conoco Inc.
Patent & Licensing
1000 S. Pine
P.O. Box 1267
Ponca City, OK 74603 (405) 767-2657

Robert B. Coleman, Jr.
2513 Mockingbird Lane
Ponca City, OK 74604 (405) 762-1371

Richard W. Collins
Conoco, Inc.
1000 S. Pine
P.O. Box 1267
Ponca City, OK 74603 (405) 767-4768

Henry H. Huth
1704 Monument Rd.
Ponca City, OK 74604 (405) 765-5169

Joseph C. Kotarski
Conoco, Inc.
1000 S. Pine
P.O. Box 1267
Ponca City, OK 74603 (405) 767-3153

William James Miller
120 N. Second
P.O. Box 547
Ponca City, OK 74602 (405) 765-6697

Charles E. Quarton
Conoco, Inc.
1000 S. Pine
P.O. Box 1267
Ponca City, OK 74603 (405) 767-3153

A. Joe Reinert
Conoco, Inc.
1000 S. Pine
P.O. Box 1267
Ponca City, OK 74603 (405) 767-4724

Bayless E. Rutherford, Jr.#
906 N. Fourth
Ponca City, OK 74601 (405) 765-3138

Cortlan R. Schupbach, Jr.
Conoco Inc.
1000 S. Pine
P.O. Box 1267
Ponca City, OK 74603 (405) 767-3109

Richard K. Thomson
Conoco Inc.
1000 South Pine
P.O. Box 1267
Ponca City, OK 74603 (405) 767-2715

Walter M. Benjamin
2620 N. Boston Place
P.O. Box 6099
Tulsa, OK 74148 (918) 582-7257

Gary M. Bond
Amoco Corp.
P.O. Box 591
Tulsa, OK 74102 (918) 660-3625

Scott H. Brown
Amoco Corp.
P.O. Box 591
Tulsa, OK 74102 (918) 660-3302

Robert R. Cochran
3532 E. 71st Place
Tulsa, OK 74136 (918) 492-2505

William S. Dorman
Dorman & Kachigian, Inc.
1146 E. 64th St.
Tulsa, OK 74136 (918) 747-1080

Mildred K. Flowers#
3913 E. 32nd Place
Tulsa, OK 74135 (918) 742-4316

John Dean Gassett
228 W. 17th Place
Tulsa, OK 74119 (918) 584-4187

Frederick P. Gilbert
1401 N.B.T. Bldg.
Tulsa, OK 74103 (918) 582-8201

Ray F. Hamilton, III
Sneed, Lang, Adams, Hamilton, Downie & Barnett
114 E. 8th St., 6th Floor
Tulsa, OK 74119 (918) 583-3145

James R. Head
Head, Johnson & Stevenson
228 W. 17th Place
Tulsa, OK 74119 (918) 584-4187

Charles S. Holmes
Doyle & Holmes
1414 S. Galveston
Tulsa, OK 74127 (918) 582-0090

Fred E. Hook
Amoco Corp.
P.O. Box 591
Tulsa, OK 74102 (918) 660-3548

Paul H. Johnson
Head, Johnson & Stevenson
228 W. 17th Place
Tulsa, OK 74119 (918) 584-4187

Stephen A. Littlefield
Dowell Schlumberger Inc.
P.O. Box 2710
Tulsa, OK 74101 (918) 250-4368

Charles L. Lunsford#
8410 S. Quebec
Tulsa, OK 74136 (918) 481-0336

Robert E. Massa
102 Brookcrest Square Office Bldg.
1535 S. Memorial Dr.
Tulsa, OK 74112 (918) 664-2525

Arthur McIlroy
3425 S. Zunis
Tulsa, OK 74105 (918) 747-1909

William H. Montgomery
Lummus Crest Inc.
4343 S. 118th E. Ave.
Tulsa, OK 74146 (918) 665-4236

William F. Norris
P.O. Box 470012
Tulsa, OK 74147 (918) 627-3330

Serge Novovich
Telex Corp.
P.O. Box 1526
Tulsa, OK 74101 (918) 627-2333

George L. Rushton#
Cities Service Co.
Patent Dept.
Box 300
Tulsa, OK 74102 (918) 561-8540

Daniel Silverman#
Head, Johnson & Stevenson
228 W. 17th Place
Tulsa, OK 74119 (918) 584-4187

Timothy D. Stanley
Amoco Corp.
P.O. Box 591
Tulsa, OK 74005 (918) 660-3236

Robert B. Stevenson
5200 S. Yale, Suite 300
Tulsa, OK 74135 (918) 492-8727

Arthur L. Wade
3208 S. Utica
Tulsa, OK 74105 (918) 742-2282

Lawrence R. Watson
Head, Johnson & Stevenson
228 W. 17th Place
Tulsa, OK 74119 (918) 584-4187

L. Wayne White
Amoco Corp.
P.O. Box 591
Tulsa, OK 74102 (818) 660-3548

Oregon

Franis I. Gray
Tektronix, Inc.
4900 S.W. Griffith Dr.
P.O. Box 500, Y3-121
Beaverton, OR 97077 (503) 643-8220

Robert S. Hulse
Tektronix, Inc.
Patent Dept.
4900 S.W. Griffith Dr.
P.O. Box 500
Beaverton, OR 97005 (503) 643-8167

William S. Lovell
Tektronix, Inc.
4900 S.W. Griffith Dr.
P.O, Box 500 Y3-121
Beaverton, OR 97077 (503) 643-8172

Russell D. Mickiewicz#
Tektronix, Inc.
4800 S.W. Griffith Dr.
11-155, Box 500
Beaverton, OR 97077 (503) 627-3455

Thomas J. Spence
Tektronix, Inc.
4900 S.W. Griffith Dr.
P.O. Box 500
Beaverton, OR 97077 (503) 643-8124

John D. Winkelman
13750 S.W. Latigo Circle
Beaverton, OR 97005 (503) 644-8196

Dudley B. Smith
P.O. Box 938
Brokings, OR 97415 (503) 469-4595

John F. Ingman
107 Oakway Mall, Suite A
Eugene, OR 97401 (503) 342-8184

Kenneth M. Durk#
2070 N.E. Laura Ct.
Hillsboro, OR 97123 (503) 648-4322

David G. Alexander#
P.O. Box 1559
Klamath Falls, OR 97601

Charles F. Robert#
Route 4, Box 4086
La Grande, OR 97850 (503) 963-9765

Robert E. Howard
Entek Manufacturing Inc.
250 N. Hansard Ave.
Lebanon, OR 97355 (503) 259-3901

Robert S. Thompson
328 N. Davis St.
P.O. Box 753
Mc Minnville, OR 97128 (503) 472-4721

Neil J. Driscoll
813 Mason Way, Unit 5
Medford, OR 97501 (503) 772-7106

Henry George O Donohoe#
13755 S.E. Maple Lane
Milwaukie, OR 97222 (503) 654-9675

William D. Haffner#
310 Walnut Dr.
Monmouth, OR 97361 (503) 838-5219

Thomas A. Tarr#
227 N.E. San Bay-O Circle
Newport, OR 97365 (503) 265-2081

Paul S. Angello
Stoel, Rives, Boley, Fraser & Wyse
900 S.W. Fifth Ave., Suite 2300
Portland, OR 97204 (503) 294-9314

Mark L. Becker
Klarquist, Sparkman, Campbell, Leigh, Whinston & Dellett
1620 Willamette Center
121 S.W. Salmon St.
Portland, OR 97204 (503) 226-7391

Daniel J. Bedell
Dellett, Smith-Hill & Bedell
1425 One Main Place,
101 S.W. Main
Portland, OR 97204 (503) 224-0115

William A. Birdwell
Spears, Lubersky, Campbell, Bledsoe, Anderson & Young
520 S.W. Yamhill St., Suite 800
Portland, OR 97204 (503) 226-6151

James Campbell, Jr.
Klarquist, Sparkman, Campbell, Leigh, Whinston & Dellett
1620 Willamette Center
121 S.W. Salmon St.
Portland, OR 97204 (503) 226-7391

Daniel P. Chernoff
Chernoff, Vilhauer, McClung & Stenzel
600 Benjamin Franklin Plaza
One S.W. Columbia
Portland, OR 97258 (503) 227-5631

Glen A. Collet#
200 Market Bldg., Suite 963
P.O. Box 1833
Portland, OR 97207 (503) 295-2472

William Y. Conwell
Klarquist, Sparkman, Campbell, Leigh, Whinston & Dellett
1620 Willamette Center
121 S.W. Salmon St.
Portland, OR 97204 (503) 226-7391

Keith A. Cushing
2929 S.E. Clinton St.
Portland, OR 97020 (503) 236-8680

Jack E. Day#
2400 S.W. Fourth Ave., Suite 100
Portland, OR 97201 (503) 222-1321

John Philip Dellitt
Dellett, Smith-Hill & Bedell
1425 One Main Place
101 S.W. Main
Portland, OR 97204 (503) 224-0115

Jon Macleod Dickinson
Kolisch, Hartwell & Dickinson
520 S.W. Yamhill St., Suite 200
Portland, OR 97204 (503) 224-6655

Eugene M. Eckelman
100 Farley Bldg.
2400 S.W. Fourth Ave.
Portland, OR 97201 (503) 222-1321

Eugene D. Farley
100 Farley Bldg.
2400 S.W. Fourth Ave.
Portland, OR 97201 (503) 222-1321

Bruce J. Ffitch#
100 Farley Bldg.
2400 S.W. Fourth Ave.
Portland, OR 97201 (503) 222-1321

William O. Geny
Chernoff, Vilhauer, McClung & Stenzel
600 Benjamin Franklin Plaza
One S.W. Columbia
Portland, OR 97258 (503) 227-5631

James D. Givnan, Jr.#
209 Sylvan Westgate Bldg.
5319 S.W. Westgate Dr.
Portland, OR 97221 (503) 292-5758

Francine H. Gray
Kolisch, Hartwell & Dickinson
520 S.W. Yamhill St., Suite 200
Portland, OR 97204 (503) 224-6655

Robert L. Harrington
1100 S.W. Sixth Ave., Suite 1106
Portland, OR 97204 (503) 248-9149

Mortimer H. Hartwell, Jr.
Kolisch, Hartwell & Dickinson
520 S.W. Yamhill St., Suite 200
Portland, OR 97204 (503) 224-6655

Donald B. Haslett
Chernoff, Vilhauer, McClung & Stenzel
600 Benjamin Franklin Plaza
One S.W. Columbia
Portland, OR 97258 (503) 227-5631

Peter E. Heuser
Kolisch, Hartwell & Dickinson
520 S.W. Yamhill St., Suite 200
Portland, OR 97204 (503) 224-6655

John J. Horn
Stoel, Rives, Boley, Jones & Grey
900 S.W. Fifth Ave., Suite 2300
Portland, OR 97204 (503) 294-9493

Patrick W. Hughey
Stoel, Rives, Boley, Jones & Grey
900 S.W. Fifth Ave., Suite 2300
Portland, OR 97204 (503) 294-9222

S.S. Jacobson
800 Oregon National Bldg.
610 S.W. Alder St.
Portland, OR 97205 (503) 223-1107

Alexander C. Johnson, Jr.
Marger & Johnson, P.C.
621 S.W. Morrison St., Suite 855
Portland, OR 97205 (503) 222-3613

Kenneth S. Klarquist
Klarquist, Sparkman, Campbell, Leigh, Whinston & Dellett
1620 Willamette Center
121 S.W. Salmon St.
Portland, OR 97204 (503) 226-7391

Ramon A. Klitzke, III
Klarquist, Sparkman, Campbell, Leigh, Whinston & Dellett
1620 Willamette Center
121 S.W. Salmon St.
Portland, OR 97204 (503) 226-7391

J. Pierre Kolisch
Kolisch, Hartwell & Dickenson
520 S.W. Yamhill St., Suite 200
Portland, OR 97204 (503) 224-6655

James S. Leigh
Klarquist, Sparkman, Campbell, Leigh. Whinston & Dellett
1620 Willamette Center
121 S.W. Salmon St.
Portland, OR 97204 (503) 226-7391

Timothy A. Long
Hyster Co.
P.O. Box 2902
Portland, OR 97208 (503) 280-7068

Jay K. Malkin
Klarquist, Sparkman, Campbell, Leigh, Whinston & Dellett
1620 Willamette Center
121 S.W. Salmon St.
Portland, OR 97204 (503) 226-7391

Jerome S. Marger
Marger & Johnson, P.C.
621 S.W. Morrison St., Suite 855
Portland, OR 97205 (503) 222-3613

Flory L. Martin
Klarquist, Sparkman, Campbell, Leigh, Whinston & Dellett
1620 Willamette Center
121 S.W. Salmon St.
Portland, OR 97204 (503) 226-7391

John M. Mc Cormack
Kolisch, Hartwell & Dickinson
520 S.W. Yamhill, Suite 200
Portland, OR 97204 (503) 224-6655

Charles D. McClung
Chernoff, Vilhauer, McClung & Stenzel
600 Benjamin Franklin Plaza
One S.W. Columbia
Portland, OR 97258 (503) 227-5631

Mark M. Meininger#
Stoel, Rives, Boley, Fraser & Wyse
900 S.W. Fifth Ave., Suite 2300
Portland, OR 97204 (503) 224-3380

Erich W. Merrill, Jr.
Miller, Nash, Wiener, Hager & Carlsen
111 S.W. Fifth Ave., Suite 3500
Portland, OR 97204 (503) 224-5858

William D. Noonan
Klarquist, Sparkman, Campbell, Leigh, Whinston & Dellett
1620 Willamette Center
121 S.W. Salmon St.
Portland, OR 97204 (503) 226-7391

Mark D. Olson#
100 Farley Bldg.
2400 S.W. Fourth Ave.
Portland, OR 97201 (503) 222-1321

Oliver D. Olson#
100 Farley Bldg.
2400 S.W. Fourth Ave.
Portland, OR 97201 (503) 222-1321

David P. Petersen
Klarquist, Sparkman, Campbell, Leigh, Whinston & Dellett
1620 Willamette Center
121 S.W. Salmon St.
Portland, OR 97204 (503) 226-7391

Donald James Piggott#
1117 S.W. Jefferson
Portland, OR 97201

Richard J. Polley
Klarquist, Sparkman, Campbell, Leight, Whinston & Dellett
1620 Willamette Center
121 S.W. Salmon St.
Portland, OR 97204 (503) 226-7391

Kenneth J. Powell
Chernoff, Vilhauer, McClung & Stenzel
600 Benjamin Franklin Plaza
One S.W. Columbia
Portland, OR 97258 (503) 227-5631

Patrick J. Reynolds#
6739 S.W. Canyon Rd.
Portland, OR 97225 (503) 297-3234

Lee R. Schermerhorn
100 Farley Bldg.
2400 S.W. Fourth Ave.
Portland, OR 97201 (503) 222-1321

Carole Shlaes
Klarquist, Sparkman, Campbell, Leigh, Whinston & Dellett
1620 Willamette Center
121 S.W. Salmon St.
Portland, OR 97204 (503) 226-7391

John Smith - Hill
1425 One Main Place
101 S.W. Main
Portland, OR 97204 (503) 224-0115

Joseph B. Sparkman, Jr.
Klarquist, Sparkman, Campbell, Leigh, Whinston & Dellett
1620 Willamette Center
121 S.W. Salmon St.
Portland, OR 97204 (503) 226-7391

John P. Staples
Chernoff, Vilhauer, McClung & Stenzel
600 Benjamin Franklin Plaza
One S.W. Columbia
Portland, OR 97258 (503) 227-5631

Dennis E. Stenzel
Chernoff, Vilhauer, McClung & Stenzel
600 Benjamin Franklin Plaza
One S.W. Columbia
Portland, OR 97258 (503) 227-5631

James G. Stewart
Kolisch, Hartwell & Dickinson
520 S.W. Yamhill St., Suite 200
Portland, OR 97204 (503) 224-6655

John W. Stuart
Omark Industries, Inc.
5550 S.W. Macadam Ave.
Portland, OR 97201 (503) 796-1444

Robert D. Varitz
Kolisch, Hartwell & Dickinson
520 S.W. Yamhill St., Suite 200
Portland, OR 97204 (503) 224-6655

Jacob Ernest Vilhauer, Jr.
Chernoff, Vilhauer, McClung & Stenzel
600 Benjamin Franklin Plaza
One S.W. Columbia
Portland, OR 97258 (503) 227-5631

Edward B. Watters#
1425 One Main Place
101 S.W. Main St.
Portland, OR 97204 (503) 224-0115

Arthur Lewis Whinston
Klarquist, Sparkman, Campbell, Leigh, Whinston & Dellett
1620 Wilamette Center
121 S.W. Salmon St.
Portland, OR 97204 (503) 226-7391

Edward B. Anderson
P.O. Box 2248
Roseburg, OR 97470 (503) 267-6404

Charles H. Hilke
Friel & Hilke
205 Equitable Center
530 Center St., N.E.
Salem, OR 97301 (503) 362-1322

Pennsylvania

Frank Kahn
1865 Edmund Rd.
Abington, PA 19001 (215) TU7-4568

Arthur H. Swanson
1836 Harding Ave.
Abington, PA 19001 (215) 659-9388

Andrew Alexander
Aluminum Co. of America
Alcoa Center, PA 15069 (412) 337-2771

David W. Brownlee
Aluminum Co. of America
Patent Div.
Alcoa Center, PA 15069 (412) 337-2773

Thomas J. Connelly
Aluminum Co. of America
Legal Patent Div.
Route 780
Alcoa Center, PA 15069 (412) 337-2759

Glenn E. Klepac
Aluminum Co. of America
Patent Div.
Alcoa Center, PA 15069 (412) 337-2770

Elroy Strickland
Alcoa Tech. Corp.
Alcoa Tech. Center
Alcoa Center, PA 15069 (412) 337-2758

Max L. Williamson
Aluminum Co. of America
Alcoa Center, PA 15069 (412) 337-2768

Nicholas F. Coates#
595 Golf Course Rd.
Aliquippa, PA 15001 (412) 375-5881

Russell Lee Brewer
Air Products & Chems. Inc.
P.O. Box 538
Allentown, PA 18105 (215) 481-7289

Geoffery L. Chase
Air Prods. & Chems. Inc.
P.O. Box 538
Allentown, PA 18105 (214) 481-7265

Richard Arthur Dannells, Jr.
Air Products & Chems, Inc.
P.O. Box 538
Allentown, PA 18105 (215) 481-8820

George G. Dower#
3016 Lindberg Ave.
Allentown, PA 18103 (215) 437-4714

E. Eugene Innis
Neuman, Williams, Anderson & Olson
1929 Brookhaven Dr. E.
Allentown, PA 18103 (215) 434-7915

Willard Jones, III
Air Products & Chem. Inc.
Allentown, PA 18195 (215) 481-4587

Wendy W. Koba
A T & T Bell Labs. Room 2A- 217
555 Union Blvd.
Allentown, PA 18103 (215) 439-7837

Lucie H. Laudenslager#
1391 Springhouse Rd.
Allentown, PA 18104 (215) 398-0091

Michael Leach
Air Products & Chemical, Inc.
P.O. Box 538
Allentown, PA 18105 (215) 481-8519

William F. Marsh
Air Products. & Chem, Inc.
P.O. Box 538
Allentown, PA 18105 (215) 481-8660

Scott W. Mc Lellan
A T & T Bell Labs.
1247 S. Cedar Crest Blvd.
Allentown, PA 18103 (215) 770-3942

Veronica O Keefe
624 Ridge Ave.
Allentown, PA 18102 (215) 434-6900

Stanford J. Piltch
3009 Greenleaf St.
Allentown, PA 18104 (215) 434-4288

Mark L. Rodgers
Air Products & Chem. Inc. Pat Dept.
Box 538
Allentown, PA 18105 (215) 481-8817

Thomas G. Ryder
Air Products & Chem, Inc.
P.O. Box 538
Allentown, PA 18105 (215) 481-7851

Seymour Traub
Traub, Butz & Fogerty, P.C.
133 North 5th St.
Allentown, PA 18102 (215) 820-0677

Peter V.D. Wilde
A T & T Bell Labs.
555 Union Blvd.
Allentown, PA 18103 (215) 439-5416

E. Kears Pollock
2447 Trotter Dr.
Allison Park, PA 15011 (412) 655-8555

Real J. Grandmaison
Henkel Corp.
300 Brookside Ave.
Ambler, PA 19002 (215) 628-1139

Henry E. Millson, Jr.
Amchem Products, Inc.
300 Brookside Ave.
Ambler, PA 19002 (215) 628-1108

Ernest G. Szoke
Henkel Corp.
300 Brookside Ave.
Ambler, PA 19002 (215) 628-1323

Percy P. Lantzy
R.D. 3
22 Sandy Dr.
Annville, PA 17003 (717) 838-9357

James H. Rich, Jr.
122 Mill Creek Rd.
Ardmore, PA 19003 (215) 649-5066

John B. Sowell
Sperry Corp.
182 Midfield Rd.
Ardmore, PA 19003 (215) 649-4815

James Irwin
1706 Fifth Ave.
Arnold, PA 15068 (412) 339-2225

Nicholas Montalto
2742 Apple Valley Lane
Audubon, PA 19403 (215) 666-0292

Eugene Chovanes
Jackson & Chovanes
One Bala Plaza, Suite 319
Bala-Cynwyd, PA 19004 (215) 667-4392

John J. McAleese, Jr.
Bray, McAleese, McGoldrick & Susanin
1 Bala Plaza
Bala-Cynwyd, PA 19004 (215) 667-7080

Allen E. Polson#
6 Broadview Dr.
Barrington, PA 02806 (401) 245-2151

Tracey G. Benson
Miller, Kistler & Campbell, Inc.
124 North Allegheny St.
Bellefonte, PA 16823 (814) 355-5474

William M. Epes
401 Margo Lane
Berwyn, PA 19312 (215) 644-4042

Robert Clifford Nicander
575 Bair Rd.
Berwyn, PA 19312 (215) 644-2729

Francis A. Varallo#
241 Country Rd.
Berwyn, PA 19312 (215) 644-6066

William Z. Warren
R.D. 1, Box 1091
Bethel, PA 19507 (717) 933-5343

Donald Stephen Ferito
6065 Murray Ave.
Bethel Park, PA 15102 (412) 835-2805

Nickolas C. Kotow
6390 Churchill Rd.
Bethel Park, PA 15102 (412) 831-7362

James L. Sherman
5860 Dashwood Dr.
Bethel Park, PA 15102 (412) 833-6730

Anson W. Biggs#
329 Carver Dr.
Bethlehem, PA 18017 (215) 868-5920

George A. Heitczman
O Hare & Heitczman
18 E. Market St.
Bethlehem, PA 18018 (215) 691-5500

John I. Iverson
Bethlehem Steel Co.
701 E. 3rd St.
Bethlehem, PA 18016 (215) 694-6401

John F. Lushis, Jr.
Bethehem Steel Corp.
8th & Eaton Ave.
Bethlehem, PA 18016 (215) 694-7990

Richard T. Muller
5 Wall St.
Bethlehem, PA 18018 (215) 865-7872

William B. Noll, Sr.
Bethlehem Steel Corp. Homer Research Labs.
Law Dept.
Bethlehem, PA 18016 (215) 694-6457

Joseph J. O Keefe
O Keefe & Wilkinson
68 E. Broad St., P.O. Box 1426
Bethlehem, PA 18016 (215) 867-9700

Roger W. Robinson
Bethlehem Steel Corp.
Martin Tower
Bethlehem, PA 18016 (215) 694-7423

John J. Selko
Bethlehem Steel Corp. Law Dept.
Homer Labs.
Bethlehem, PA 18016 (215) 694-6429

John S. Simitz#
2446 Greencrest Dr.
Bethlehem, PA 18017 (215) 691-2167

Thomas Frank H. Thomson
Fuller Co.
2040 Ave. C.
Bethlehem, PA 18001 (215) 264-6555

Charles Alexander Wilkinson
O Keefe & Wilkinson
68 E. Broad St. P.O. Box 1426
Bethlehem, PA 18016 (215) 867-9700

Edmund M. Chung#
Unisys Corp.
Township Line & Union Meeting Rd.
Blue Bell, PA 19424 (215) 542-4411

Edward James Dwyer
3 Tally Ho Lane
Blue Bell, PA 19422 (215) 643-7288

Alvin Morton Esterlitz
One Sentry Pkwy. Suite 6000
Blue Bell, PA 19422 (215) 828-4620

Thomas J. Scott
Sperry Corp.
P.O. Box 500
Blue Bell, PA 19424 (215) 542-4116

Mark T. Starr
Unisys Corp.
Township Line & Union Meeting Rds., P.O. Box 500
Blue Bell, PA 19424 (215) 542-4411

John Shaw Stevenson#
879 Crestline Dr.
Blue Bell, PA 19422 (215) 279-6274

Marshall M. Truex
Sperry Corp.
Townshipline & Jolly Roads
Blue Bell, PA 19422 (215) 542-4111

Frank Joseph Vinci, Jr.
Sperry Corp.
P.O. Box 500
Blue Bell, PA 19424 (215) 542-4921

Mervyn L. Young
Unisys Corp.
Township Line & Union Meeting Rd. P.O. Box 500 M S C 1 S W 19
Blue Bell, PA 19424 (215) 542-4111

W. Melville Van Sciver
250 N. Central Blvd.
Broomall, PA 19008 (215) 356-7284

Robert S. Barton
714 Old Lancaster Rd.
Bryn Mawr, PA 19010 (215) 525-3784

George J. Harding, III
921 Mt. Pleasant Rd.
Bryn Mawr, PA 19010 (215) 525-2039

George C. Atwell
Atwell & Morrow
P.O. Box 829, 421 N. Main St.
Butler, PA 16003 (412) 283-9333

William Hintze
228 Allendale Way
Camp Hill, PA 17011 (717) 761-1361

William J. Keating
The Dickson School Of Law
150 South College St.
Carlisle, PA 17013 (717) 243-4611

Arthur A. Murphy
Dickingson School Of Law
S. College St.
Carlisle, PA 17013 (717) 243-4611

Francis K. Richwine
1213 Stratford Dr.
Carlisle, PA 17013 (717) 245-2810

Frederick A. Tecce#
216 West Pomfret St.
Carlisle, PA 17013 (717) 249-5617

Arba G. Williamson, Jr.#
40 Swallow Hill Rd.
Carnegie, PA 15106 (412) 279-5036

H. Barry Moyerman
235 Bridge St.
Catasauqua, PA 18032 (215) 264-9779

Robert De Majistre
Seton Co. Brandwine One Bldg.
Route 1 & 202
Chadds Ford, PA 19317 (215) 358-9010

Edward J. Newitt#
Box 151, Rd 3
Chadds Ford, PA 19317 (215) 388-7190

James E. Shipley
100 Heyburn Rd.
Chadds Ford, PA 19317 (215) 358-1225

Joseph W. Molasky
Chalfont Centre
4 Limekiln Pike
Chalfont, PA 18914 (215) 822-3324

Howard Burger#
255 North Rd., Unit 62
Chelmsford, PA 01824 (617) 250-0245

Rama B. Nath#
Westinghouse Elec. Corp. Pat. Dept, R & D Center
Beulah Rd.
Churchill, PA 15235 (412) 256-7767

Edward Franklin Possessky
Westinghouse Elec. Corp. R & D Center, Pat. Dept.
1300 Beulah Rd.
Churchill, PA 15235 (412) 256-5249

Walter G. Sutcliff
Westinghouse Elec. Corp. Pat. Dept.
R & D Center, 1310 Beulah Rd.
Churchill, PA 15235 (412) 256-5242

James Riesenfeld
Lutron Electronics Co. Inc.
Suter Rd., Box 205
Coopersburg, PA 18036 (215) 282-3800

Robert F. Palermo
301 South Walnut St.
Dallastown, PA 17313 (717) 244-6795

William E. Denk#
81 Steeplechase Rd.
Devon, PA 19333 (215) 688-2976

Edward J. Feeney, Jr.
151 Steeplechase Rd.
Devon, PA 19333 (215) 687-6583

Mary A. Capria
106 Locust Lane
R.D. 5
Dillsburg, PA 17019 (717) 432-3460

Greorgy J. Gore
2 E. Court St.
Doylestown, PA 18901 (215) 348-1442

John T. Marvin
3 Blythewood Rd.
Doylestown, PA 18901 (215) 345-0237

Armand M. Vozzo, Jr.
Monney & Vozzo
22 S. Clinton St.
Doylestown, PA 18901 (215) 345-4656

George E. Bodenstein#
3255 Pebblewood Ln.
Dresher, PA 19025 (215) 576-0189

Frank J. Earnheart
Selas Corp. Of America
2034 Limekiln Pike, P.O. Box 200
Dresher, PA 19025 (215) 283-8368

Charles E. Bartsch#
1001 Foss Ave.
Drexel Hill, PA 19026 (215) 589-5188

Robert J. Mc Donnell
Mc Donnell & Mc Donnell P.A.
4750 Township Line
Drexel Hill, PA 19026 (215) 446-3290

Thomas R. O Malley
4018 Bloomfield Ave.
Drexel Hill, PA 19026 (215) 259-6712

Richard L. Joyce
Box 69, R.D. #2
Duncannon, PA 17020 (717) 834-4563

Aaron Nerenberg
810 Pinewood Rd.
Elkin Park, PA 19117

John M. Fray
603 Crescent Ave.
Ellwood City, PA 16117 (412) 752-3729

Winfrid O.E. Schellin
1154 Little Lehigh Dr.
Emmaus, PA 18049 (215) 967-5293

Edward W. Goebel, Jr.
Mac Donald, Illig, Jones & Britton
600 1st Natl. Bank Bldg.
Erie, PA 16501 (814) 453-7611

Ralph Hammar
103 W. 10th St.
Erie, PA 16501 (814) 452-3494

Charles Lester Lovercheck
Lovercheck & Lovercheck
931 State St.
Erie, PA 16501 (814) 454-5218

Wayne L. Lovercheck
Lovercheck & Lovercheck
931 State St.
Erie, PA 16501 (814) 454-5218

Philip P. Mc Cann
Lord Corporation
2000 W. Grandview Blvd. P.O. Box 10038
Erie, PA 16514 (814) 868-0924

Albert S. Richardson, Jr.
General Electric Co.
2901 E. Lake Rd.
Erie, PA 16531 (814) 875-3366

James W. Wright
Lord Corp.
2000 West Grandview Blvd.
Erie, PA 16514 (814) 868-0924

Heinrich Goretzky
Mystic Hills, R.D. 3
Export, PA 15632 (412) 327-5362

Donald J. Smith
R.D. 3
Export, PA 15632 (412) 327-0077

Paul T. Teacher
P.O. Box 487
Export, PA 15632 (412) 327-3891

Donald D. Joye#
226 Llandovery Dr.
Exton, PA 19341 (215) 524-0296

Paul Lipsitz#
205 Suffolk Rd.
Flourtown, PA 19031 (215) 233-4620

Kenneth P. Lauria#
Mars Electronics
801 Carpenters Crossing
Folcroft, PA 19032 (215) 534-4200

Frank A. Wolfe#
909 Fifth Ave.
Ford City, PA 16226 (412) 763-7125

Imre Balogh
Rorer Group Inc.
500 Virginia Dr.
Fort Washington, PA 19034 (215) 962-3309

Thomas M. Ferrill, Jr.
Ferrill & Logan
550 Pinetown Rd. Suite 430
Fort Washington, PA 19034 (215) 643-5750

Mitchell John Halista
Honeywell Inc. Mail Sta 225 Pat. Dept
1100 Va. Dr.
Fort Washington, PA 19034 (215) 641-3731

David J. Johns
Highland Off. Center Suite 430
550 Pinetown Rd.
Fort Washington, PA 19034 (215) 643-5750

Ernest B. Lipscomb, III
Rorer Group Inc.
500 Virginia Dr.
Fort Washington, PA 19034 (215) 628-6541

John W. Logan, Jr.
Ferrill & Logan
550 Pinetown Rd. Suite 430
Fort Washington, PA 19034 (215) 643-5750

Thomas E. Merchant
William H. Rorer, Inc.
500 Virginia Dr.
Fort Washington, PA 19034 (215) 628-6325

Thomas E. Merchant
William H. Rorer, Inc.
500 Virginia Dr.
Fort Washington, PA 19034 (215) 628-6325

James A. Nicholson#
Wm. H. Rorer, Inc.
500 Virginia Dr.
Fort Washington, PA 19034 (215) 628-6396

Gilbert W. Rudman
Rorer Group Inc.
500 Virginia Dr.
Fort Washington, PA 19034 (215) 628-6023

Charles Frederick Osgood
508 Gurney Rd.
Franklin, PA 16323 (814) 432-2893

Albert F. Maier, Jr.
Maier & Maier
820 Main St.
Freeland, PA 18224 (717) 636-1140

W. Wyclif Walton
710 Dixon Lane
Gladwyn, PA 19035 (215) 642-1482

Rudolph J. Eisinger
P.O. Box 44
Glen Mills, PA 19342 (215) GL9-2435

Harrison H. Young, Jr.
529 Custis Rd.
Glenside, PA 19038 (215) 572-6362

Jon M. Lewis
205 Coulter Bldg.
Greensburg, PA 15601 (412) 836-4730

Nils H. Ljungman, Jr.
229 South Maple Ave. P.O. Box 130
Greensburg, PA 15601 (412) 836-2305

Thomas N. Ljungman#
229 S. Maple Ave. P.O. Box 130
Greensburg, PA 15601 (412) 836-2305

Philip D. Freedman
204 State St.
Harrisburg, PA 17101 (717) 233-1000

Dennis J. Harnish
Environmental Hearing Board, Blackstone Bldg.
112 Market St.
Harrisburg, PA 17120 (717) 787-3483

Thomas Hooker
101 N. Front St.
Harrisburg, PA 17101 (717) 232-8771

Gerald K. Kita
A M P, Inc.
P.O. Box 3608
Harrisburg, PA 17105 (717) 986-5664

Adrian John Larue
A M P Inc. Pat. Div. 140-62
P.O. Box 3608
Harrisburg, PA 17105 (717) 986-5468

Katherine A.O. Nelson
A M P Inc.
P.O. Box 3608
Harrisburg, PA 17105 (717) 986-5470

Anton Ness
A M P Inc. Pat. Div. (140-62)
P.O. Box 3608
Harrisburg, PA 17105 (717) 986-5477

Allan B. Osborne
A M P, Inc.
P.O. Box 3608
Harrisburg, PA 17105 (717) 986-5459

Gerald Post#
4108 Beechwood Lane
Harrisburg, PA 17112 (717) 652-3960

Jay Louis Seitchik
A M P, Inc.
P.O. Box 3608
Harrisburg, PA 17105 (717) 986-5461

David L. Smith
A M P, Inc.
P.O. Box 3608
Harrisburg, PA 17105 (717) 564-0100

Thomas G. Terrell#
A M P, Inc.
P.O. Box 3608
Harrisburg, PA 17105 (717) 986-5472

Gerald D. Ames#
113 Franklin Ave.
Hatboro, PA 19040 (215) 672-4348

Daniel L. De Joseph
Ninth Floor
Northeastern Bldg.
Hazelton, PA 18201 (717) 455-6308

Anthony J. Rossi
Exide Corp.
101 Gibraltar Rd.
Horsham, PA 19044 (215) 441-7705

Alfred J. Snyder, Jr.
Exide Corp.
101 Gibraltar Rd.
Horsham, PA 19044 (215) 441-7302

Bernard J. Burns#
2100 Winthrop Rd.
Huntingdon, PA 19006 (215) 947-1518

Howard I. Forman
Albidale-Windmill Circle
P.O. Box 66
Huntingdon, PA 19006 (215) 947-4154

Harold K. Hauger
Norwin Medical Center
28 Fairwood Dr.
Irwin, PA 15642 (412) 864-6050

Leon Edelson
Foxcroft Square Apt. 813
Jenkintown, PA 19046 (215) 576-5211

William H. Eilberg
820 Homestead Rd.
Jenkintown, PA 19046 (215) 885-4600

Harold L. Greenwald#
506 Rodman Ave.
Jenkintown, PA 19046 (215) 887-6142

Michael I. Ozalas
Ozalas & Mc Kinleyue
41 Broadway
Jim Thorpe, PA 18229 (717) 325-3616

Charles G. Rudershausen#
109 Taylor Lane
Kennett Square, PA 19348 (215) 444-3773

Michael F. Beausang, Jr.
Butera, Beausang, Moyer & Cohen
700 Valley Forge Plaza
King Of Prussia, PA 19406 (215) 265-0800

Stanley Bilker
2000 Valley Forge Circle Apt. 927
King Of Prussia, PA 19406 (215) 783-2678

Francis M. Linguiti
Ratner & Prestia
500 N. Gulph Rd. Suite 412
King Of Prussia, PA 19406 (215) 265-6666

Henry J. Policinski
General Elec. Co. Pat. & Assigned Components
150 South Warner Rd. Suite 300
King Of Prussia, PA 19406 (215) 964-7655

John S. Stephen Bobb#
P Q Corp.
P.O. Box 258
Lafayette, PA 19444 (215) 825-5000

James R. Bell
136 Crosswick Lane
Lancaster, PA 17601 (717) 397-6698

Vincent J. Coughlin, Jr.
379 Buch Ave.
Lancaster, PA 17601 (717) 560-9194

Robin M. Davis
Armstrong Pat. Dept.
150 North Queen St.
Lancaster, PA 17604 (717) 396-4122

Martin Fruitman
311 E. Orange St.
Lancaster, PA 17602 (717) 397-2314

Barry E. Haverstick
2916 Columbia Ave.
Lancaster, PA 17603 (717) 392-0446

George L. Herr
1340 Hunter Dr.
Lancaster, PA 17601 (717) 393-1195

Clifford B. Price, Jr.
43 Tennyson Dr.
Lancaster, PA 17602 (717) 397-0611

Albert H. Sheaffer
208 N. President Ave. Apt. 10
Lancaster, PA 17603 (717) 394-2899

Theodore L. Thomas
Blakenger, Grove & Chillas P.C.
33 E. Orange St.
Lancaster, PA 17602 (717) 291-1200

Sylvia A. Gosztonyi
R.D. 2, Box 48A
Landenberg, PA 19350 (215) 274-8829

Robert G. Danehower
17 Church Rd.
Landsdale, PA 19446 (215) 855-5924

Frank E. Manson#
857 Sunnylea Rd.
Lansdale, PA 19446 (215) 855-1842

Phileman J. Moore#
Valley Stream Apt. Q - 101
Lansdale, PA 19446 (215) 368-3045

Lawrence R. Burns
16 Romar Ave.
Latrobe, PA 15650 (412) 539-2271

John J. Prizzi
Kennametal Incorporation
P.O. Box 231
Latrobe, PA 15650 (412) 539-5331

John S. Friderichs
Heatex Of Philadelphia
100 Stevens Suite 290
Lester, PA 19113 (215) 595-1052

Murray J. Ellman
39 Four Leaf Rd.
Levittown, PA 19056 (215) 946-4083

Florence U. Reynolds#
60 Yellowood Dr.
Levittown, PA 19057 (215) 945-3336

Alan N. Mc Cartney
Consolidation Coal Co.
4000 Brownsville Rd.
Library, PA 15129 (412) 854-6631

James Charles Simmons
7353 Hillcrest Dr.
Macungie, PA 18062 (215) 395-9060

Alfred R. Brady
Ecolair Inc.
Two Country View Rd.
Malvern, PA 19355 (215) 648-8630

George D. Hobbs, II
Centocor
244 Grant Valley Parkway
Malvern, PA 19355 (215) 296-4488

John Steffen Munday
386 Conestoga Rd.
Malvern, PA 19355 (215) 296-9487

Donald L. Rose
434 Lincoln Ave.
Mars, PA 16046 (412) 625-1917

Colleen A. Dettorre#
R D #3, Box 36 A
McDonald, PA 15057

John H. Perkins
207 Lakeside Dr.
McKees Rocks, PA 15136 (412) 787-2150

Nicholas A. Vonneuman
1396 Lindsay Lane
Meadowbrook, PA 19046 (215) 886-6244

Robert B. Famiglio
201 N. Jackson St. P.O. Box 546
Media, PA 19063 (215) 565-4730

Han-Jolyon Lammers
431 Kirk Lane
Media, PA 19063 (302) 575-3725

Robert S. Lipton
201 N. Jackson St.
P.O. Box 546
Media, PA 19063 (215) 565-4730

Delbert E. Mc Caslin#
12 Wyncroft Dr. Middletown Township
Media, PA 19063 (215) 566-1964

Anthony J. Mc Nulty
115 N. Monroe St.
P.O. Box 605
Media, PA 19063 (215) 565-6700

Eugene E. Renz, Jr.
Court House Square East
229 N. Olive St.
Media, PA 19063 (215) 565-6090

Sheldon Kapustin
1011 Prospect Ave.
Melrose Park, PA 19126 (215) ME5-2925

Charles J. Kelley#
P.O. Box 46
Mendenhall, PA 19357 (215) 388-7238

Warren B. Gilbert#
6527 Edwards Dr.
Mercersburg, PA 17236 (717) 328-3270

William R. Glisson
542 Prescott Rd.
Merion Station, PA 19066 (215) 644-4792

Granville Y. Custer, Jr.#
501 Manor Dr.
Middletown, PA 17057 (717) 944-9607

Frederick W. Raring#
A M P, Inc.
2901 Fulling Mill Rd.
Middletown, PA 17057 (717) 986-5460

M. Richard Page
Sharpoint Inc.
P.O. Box 187
Mohnton, PA 19540 (215) 777-7854

Oscar B. Brumback
1340 Towerlawn Dr.
Monroeville, PA 15146 (412) 372-1162

Hymen Diamond
4409 Ruth Dr.
Monroeville, PA 15146 (412) 372-3555

Lucinda A. Fuerle#
1710 Mountain View Dr.
Monroeville, PA 15146 (412) 325-2971

William David Palmer
2317 Haymaker Rd.
Monroeville, PA 15146 (412) 373-0717

Robert J. Zinn
1060 Route 309
Montgomeryville, PA 18936 (215) 646-7800

Philip W. Humer#
8 Glenwood South Gate
Morrisville, PA 19067 (215) 295-0479

John M. O Meara
1019 N. Pennsylvania Ave.
Morrisville, PA 19067 (215) 542-6276

Alphonso Henry Caser
20 S. Maple St.
Mount Carmel, PA 17851 (717) 339-3464

Daniel S. Buleza#
4412 West Run Rd.
Munhall, PA 15120 (412) 462-1452

Elliott V. Nagle#
3404 Oakdale Dr.
Murrysville, PA 15668 (412) 327-1514

Anthony J. Santantonio
5053 Northlawn Dr.
Murrysville, PA 15668 (412) 327-1015

Michael J. Delaney
510 E. Main St.
Nanticoke, PA 18634 (717) 735-3950

Stephen Ross Green
Gamble, Verterano, Mojock, Piccione & Green
First Federal Plaza Suite 500
New Castle, PA 16101 (412) 658-2000

Joseph A. Brown
New Holland Inc.
500 Diller Ave.
New Holland, PA 17557 (717) 354-1439

James J. Kennedy
Sperry Corp.
Sperry New Holland Div.
New Holland, PA 17557 (717) 354-1447

Darrell F. Marquette#
Sperry Corp. Sperry New Holland Div.
500 Diller Ave.
New Holland, PA 17557 (717) 354-1353

Larry W. Miller
Sperry Corp. Sperry New Holland Div.
500 Diller Ave.
New Holland, PA 17557 (717) 354-1353

John B. Mitchell
Sperry New Holland
500 Diller Ave.
New Holland, PA 17557 (717) 354-1447

Frank Allyn Seemar
Sperry New Holland
500 Diller Ave.
New Holland, PA 17557 (717) 354-1341

Adolfo Arturo Mangieri#
1170 Seventh St.
New Kensington, PA 15068 (412) 337-7795

Daniel A. Sullivan, Jr.
150 Chaney Court
New Kensington, PA 15068 (412) 335-8121

Louis B. Applebaum
24 Toppa Blvd.
Newport, PA 02840 (401) 846-1241

Prithvi C. Lall*
U.S. Dept. Of Navy
Naval Underwater Systems Center, Bldg 142
Newport, PA 02840 (401) 841-4736

Arthur A. Mc Gill*
Dept. Of Navy, Pat. Coun.
Naval Underwater Systems Ctr.
Newport, PA 02840 (401) 841-4736

Michael J. Mc Gowan*
Off. Of Naval Research, Naval Underwater Systems Center
Building 142
Newport, PA 02841 (401) 841-4736

Stephen D. Krefman
S P S Technologies, Inc.
Newtown-Yardley Rd.
Newtown, PA 18940 (215) 860-3072

John J. Simkanich
122 Chesapeske Dr.
Newtown, PA 18940

Allen E. Amgott
315 Earls Lane
Newtown Square, PA 19073 (215) 353-5857

Joseph M. Corr
63 Charter Oak Dr.
Newtown Square, PA 19073 (215) 353-9040

M. Delcina M. Esser#
Arco Chemical Co.
3801 W. Chester Pike
Newtown Square, PA 19073 (215) 359-2906

John Bernard Sotak#
12191 Church Dr.
North Huntingdon, PA 15642 (412) 863-5277

Gordon Gale Menzies
210 W. Main St.
North Kingstown, PA 02852 (401) 295-1230

Raymond S. Visk#
1122 Jack S. Run Rd.
North Versailles, PA 15137 (412) 824-4974

Albert C. Martin
Leeds & Northrup Co.
Sumneytown Pike
North Wales, PA 19454 (215) 643-2000

William G. Miller, Jr.#
Leeds & Northrup Co.
Unit Of General Signal, Sumneytown Pike
North Wales, PA 19454 (215) 643-2000

William Henry Deitch
666 10th St.
Oakmont, PA 15139 (412) 828-8547

Edward B. Foote
911 Washington Ave.
Oakmont, PA 15139 (412) 828-4883

J. Spencer Overholser#
P.O. Box 106
Oley, PA 19547 (215) 987-6079

William Earl Cleaver
225 Orchard Rd.
Paoli, PA 19301 (215) 296-3381

Michael P. Abbott
Seidel, Gonda, Goldhammer& Abbott, P.C.
1800 Two Penn Center Plaza
Philadelphia, PA 19102 (215) 568-8383

Marc S. Adler
Rohm & Haas Co.
Independence Mall West
Philadelphia, PA 19105 (215) 592-3416

Robert L. Andersen
F M C Corp.
2000 Market St.
Philadelphia, PA 19103 (215) 299-6967

Stanford M. Back
Sun Refining & Marketing Co. Pat. & Licenses - 27/10P
1801 Market St.
Philadelphia, PA 19103 (215) 977-3072

Alexis Barron
Synnestvedt & Lechner
2600 One Reading Center, 1101 Market Center
Philadelphia, PA 19107 (215) 923-4466

Carl W. Battle
Rohm & Haas Co. Pat. Dept. 7th Floor
Independence Mall West
Philadelphia, PA 19105 (215) 592-3052

Robert C. Beam
Paul & Paul
2900 Two Thousand Market St.
Philadelphia, PA 19103 (215) 568-4900

James B. Bechtel
Paul & Paul
2000 Market St., Suite 2900
Philadelphia, PA 10103 (215) 568-4900

Stuart E. Beck
Trachtman, Jacobs & Beck
1500 Chestnut St. Suite 1908
Philadelphia, PA 19102 (212) 569-9800

Frank Joseph Benasutti
Benasutti & Murray
Suite 2701, the A R A Tower
11th & Market St.
Philadelphia, PA 19107 (215) 923-6100

Philip E. Berens
Suite 1401, Lewis Tower Bldg.
Philadelphia, PA 19102 (215) 735-2425

Alan H. Bernstein
Caesar, Rivise, Bernstein, Cohen & Pokotilow, Ltd.
21 S. 12th St. Suite 800
Philadelphia, PA 19107 (215) 567-2010

John P. Blasko
Benasutti & Murray Suite 2701, A R A Tower
11th & Market St.
Philadelphia, PA 19107 (215) 561-4300

Stephen J. Bor
Master, Donsky, Soffian & Allen
230 S. Broad St.
Philadelphia, PA 19102 (215) 546-9800

Stuart S. Bowie
7th Floor, 1530 Chestnut St.
Philadelphia, PA 19102 (215) 567-1530

Steven D. Boyd
Benasutti & Murray, Suite 2701 One Reading Center
1101 Market St.
Philadelphia, PA 19107 (215) 923-6100

Robert Sherman Bramson
Schnader, Harrison, Segal & Lewis
1600 Market St. Suite 3600
Philadelphia, PA 19103 (215) 751-2066

David M. Bunnell
Pennwalt Corp.
Three Parkway Pat. Dept.
Philadelphia, PA 19102 (215) 587-7696

James R. Burdett
Woodcock, Washburn, Kurtz, Mackiewicz & Norris
One Liberty Place, 46th Floor
Philadelphia, PA 19103 (215) 568-3100

Michael S. Bush#
Woodcock, Washburn, Kurtz, Mackiewicz & Norris
One Liberty Place, 46th Floor
Philadelphia, PA 19103 (215) 568-3100

Jean A. Buttmi
128 E. Chesnut Hill Ave.
Philadelphia, PA 19118 (215) 247-6422

Abraham D. Caesar
Caesar, Rivise, Bernstein, Cohen & Pokotilow, Ltd.
21 S. 12th St. Suite 800
Philadelphia, PA 19107 (215) 567-2010

John W. Caldwell
Woodcock, Washburn, Kurtz, Mackiewicz & Norris
One Liberty Place, 46th Floor
Philadelphia, PA 19103 (215) 568-3100

Carol Grobman Canter
Smith, Kline, Beckman Corp. Corporate Pat. & Tdmk N-160
One Franklin Plaza, P.O. Box 7929
Philadelphia, PA 19101 (215) 751-6148

Charles Mclean Carter
Reading Co.
1101 Market St.
Philadelphia, PA 19107 (215) 922-3303

Paul Checkovich
General Electric Co.
P.O. Box 8555
Philadelphia, PA 19101 (215) 354-5915

John S. Child, Sr.
Synnestvedt & Lechner
2600 One Reading Center, 1101 Market St.
Philadelphia, PA 19107 (215) 923-4466

Scott J. Childress#
2202 Hopkinson House
Philadelphia, PA 19106 (215) 627-3471

Thomas D. Christenbury
Miller & Quinn
1125 Land Title Bldg. Broad & Chesnut St.
Philadelphia, PA 19109 (215) 563-1810

Donald S. Cohen
Woodcock, Washburn, Kurtz, Mackiewicz & Norris
One Liberty Place, 46th Floor
Philadelphia, PA 19103 (215) 568-3100

Gary M. Cohen
Weiser & Stapler
230 S. 15th St. Suite 500
Philadelphia, PA 19102 (215) 875-8383

Lester H. Cohen
613 Avon St.
Philadelphia, PA 19116 (215) 676-5078

Stanley H. Cohen
Caesar, Rivies, Bernstein, Cohen & Pokotilow, Ltd.
21 S. 12th St. Suite 800
Philadelphia, PA 19107 (215) 567-2010

Peter J. Cronk
Woodcock, Washburn, Kurtz, Mackiewicz & Norris
One Liberty Place, 46th Floor
Philadelphia, PA 19103 (215) 568-3100

C. Marshall Dann
Dann, Dorfman, Herrell & Skillman, P.C.
1310 The Fidelity Bldg. 123 S. Broad St.
Philadelphia, PA 19109 (215) 545-1700

John P. Donohue, Jr.
Woodcock, Washburn, Kurtz, Mackiewicz & Norris
One Liberty Place, 46th Floor
Philadelphia, PA 19103 (215) 568-3100

John C. Dorfman
Dann, Dorfman, Herrell & Skillman, P.C.
1310 The Fidelity Bldg. 123 S. Broad St.
Philadelphia, PA 19109 (215) 545-1700

Jordan Joseph Driks
Rohm & Haas Co. Independence Mall West
6th & Market St.
Philadelphia, PA 19105 (215) 592-2478

James Albert Drobile
Schnader, Harrison, Segal & Lewis
1600 Market St. Suite 3600
Philadelphia, PA 19103 (215) 751-2242

Thomas J. Durling
Seidel, Gonda, Goldhammer & Abbott, P.C.
1800 Two Penn Center
Philadelphia, PA 19102 (215) 568-8383

John F.A. Earley
Harding, Earley, Follmer & Frailey
1910 Two Mellon Bank Center
Philadelphia, PA 19102 (215) 568-2606

John F.A. Earley, III
Harding, Earley, Follmer & Frailey
1910 Two Mellon Bank Center, Broad St. & S. Penn Square
Philadelphia, PA 19102 (215) 568-2606

William H. Edgerton#
Smith, Kline Corp.
P.O. Box 7929, 1500 Spring Garden St.
Philadelphia, PA 19101 (215) 854-5180

Christopher Egolf
F M C Corp.
2000 Market St.
Philadelphia, PA 19103 (215) 299-6979

Dianne B. Elderkin
Woodcock, Washburn, Kurtz, Mackiewicz & Norris
One Liberty Place, 46th Floor
Philadelphia, PA 19103 (215) 568-3100

William H. Elliott, Jr.
Synnestvedt & Lechner
2600 One Reading Center, 1101 Market St.
Philadelphia, PA 19107 (215) 823-4466

Gerry J. Elman
Elman Assoc. Bourse Bldg.
Suite 900
Philadelphia, PA 19105 (215) 592-3416

Henry Robinson Ertelt
F M C Corp. Chemical Pats. Licensing
2000 Market St.
Philadelphia, PA 19103 (215) 299-6969

Susan B. Evans
Jackson & Evans
2043 Walnut St.
Philadelphia, PA 19103 (215) 557-8101

Vincent L. Fabiano#
Smith, Kline, Beckman, Corp. Pat. N 160
P.O. Box 7929
Philadelphia, PA 19101 (215) 751-6143

Martin L. Faigus
Caesar, Rivise, Bernstein, Cohen & Pokotilow, Ltd
21 S. 12th St. Suite 800
Philadelphia, PA 19107 (215) 567-2010

Edward M. Farrell
1328 Land Title Bldg.
Philadelphia, PA 19110 (215) 567-2230

Michael B. Fein
Rohm & Haas Co.
Independence Mall West
Philadelphia, PA 19105 (215) 592-3595

Charles C. Fellows
F M C Corp. Chemical Pat. & Licensing Dept.
2000 Market St.
Philadelphia, PA 19103 (215) 299-6970

Richard D. Foggio#
Smith- Kline Corp.
1500 Spring Garden St.
Philadelphia, PA 19101 (215) 854-5184

Frank A. Follmer
Harding, Earley, Follmer & Frailey
1910 Two Girard Plaza
Philadelphia, PA 19102 (215) 568-2606

Albert L. Free
Synnestvedt & Lechner
2600 One Reading Center, 1101 Market St.
Philadelphia, PA 19107 (215) 923-4466

William Freedman
Millman & Jacobs
2940 P S F S Bldg., 12 S. 12th St.
Philadelphia, PA 19107 (215) 592-6565

Allan H. Fried
Pennwalt Corp. Pat. Dept.
Three Parkway
Philadelphia, PA 19102 (215) 587-7573

Harvey D. Fried
Steele, Gould & Fried
1700 Market St. Suite 3232
Philadelphia, PA 19103 (215) 563-8020

Joel S. Goldhammer
Seidel, Gonda, Goldhammer & Abbott
1800 Two Penn Center Plaza
Philadelphia, PA 19102 (215) 568-8383

Stuart M. Goldstein
Clark, Ladner, Fortenbaugh & Young
32nd Floor, 1818 Market St.
Philadelphia, PA 19103 (215) 241-1885

Edward C. Gonda
Seidel, Gonda & Goldhammer, P.C.
Two Penn Center Plaza, Suite 1800
Philadelphia, PA 19102 (215) 568-8383

Lewis F. Gould, Jr.
Steel, Gould & Fried
1700 Market St. Room 3232
Philadelphia, PA 19103 (215) 563-8020

Mark A. Greenfield
128 E. Chesnut Hill Ave.
Philadelphia, PA 19118 (215) 247-6422

Stephen P. Gribok
Steele, Gould & Fried
1700 Market St. Suite 3232
Philadelphia, PA 19103 (215) 563-8020

Patrick J. Hagan
Dann, Dorfman, Herrell & Skillman, P.C.
1310 The Fidelity Bldg.
123 S. Broad St.
Philadelphia, PA 19109 (215) 545-1700

Linda E. Hall
Smith, Kline, Beckman Corp.
One Franklin Plaza, P.O. Box 7929
Philadelphia, PA 19101 (215) 751-5180

Robert H. Hammer, III
Seidel, Gonda, Goldhammer & Abbott, P.C.
Two Penn Center Plaza, Suite 1800
Philadelphia, PA 19102 (215) 568-8383

Carl A. Hechmer, Jr.
Pennwalt Corp. Pat. Dept.
3 Pkwy.
Philadelphia, PA 19102 (215) 587-7700

William E. Hedges
The Warwick
17th & Locust Sts. Suite 408
Philadelphia, PA 19103 (215) 546-4834

Louis M. Heidelberger
Dilworth, Paxson, Kalish & Kauffman
2600 The Fidelity Bldg.
Philadelphia, PA 19109 (215) 875-7230

Dale M. Heist
Woodcock, Washburn, Kurtz, Mackiewicz & Norris
One Liberty Place, 46th Floor
Philadelphia, PA 19103 (215) 568-3100

Roger Wayne Herrell
Dann, Dorfman, Herrell & Skillman
1310 The Fidelity Bldg. 123 S. Broad St.
Philadelphia, PA 19109 (215) 545-1700

James Edward Hess
Sun Refining & Marketing Co. Pat. Lics.
Ten Penn Center, 1801 Market Place
Philadelphia, PA 19103 (215) 977-3075

Marc Hodak#
4120 Walnut St.
Philadelphia, PA 19104 (215) 386-9908

James D. Hodnett
1130 Suburban Sta. Bldg.
1617 J.F.K. Blvd.
Philadelphia, PA 19103 (215) 561-2650

Robert G. Hoffman
Pennwalt Corp.
Three Parkway
Philadelphia, PA 19102 (215) 587-7694

Richard Kay Jackson
Wyeth - Ayerst Labs.
P.O. Box 8299
Philadelphia, PA 19101 (215) 341-2310

Arthur A. Jacobs
Trachtman, Jacobs & Beck
1908 Pennsylvania Bldg. 1500 Chestnut St.
Philadelphia, PA 19102 (215) 569-9800

Morton C. Jacobs
Millman & Jacobs
2940 P S F S Bldg., 12 S. 12th St.
Philadelphia, PA 19107 (215) 592-6565

John Jamieson, Jr.
Panitch, Schwarze, Jacobs & Nadel
2000 Market St. Suite 1400
Philadelphia, PA 19103 (215) 567-2020

Charles H. Johnson
F M C Corp. Pat. & Lic. Dept.
2000 Market St.
Philadelphia, PA 19103 (215) 299-6983

Donald R. Johnson
Sun Refining & Marketing Co. Pat. & Lics.
Ten Penn Center
Philadelphia, PA 19103 (215) 977-3074

Philip S. Johnson
Woodcock, Washburn, Kurtz, Mackiewicz & Norris
One Liberty Place, 46th Floor
Philadelphia, PA 19103 (215) 568-3100

Howard Kaiser
2701 Welsh Rd. Suite 1-D
Philadelphia, PA 19152 (215) 969-3704

John William Kane, Jr.
Scott Paper Co. Scott Plaza II
Philadelphia, PA 19113 (215) 521-5000

Leslie L. Kasten, Jr.
Panitch, Schwarze, Jacobs & Nadel
2000 Market St. Suite 1400
Philadelphia, PA 19103 (215) 567-2020

Howard S. Katzoff
520 C Lombard St.
Philadelphia, PA 19147 (215) 922-6177

Robert W. Kell
The Warwick
17th & Locust, Suite 409
Philadelphia, PA 19103 (215) 546-4834

Robert M. Kennedy
F M C Corp.
2000 Market St.
Philadelphia, PA 19103 (215) 299-6966

Nelson E. Kimmelman
Edelson, Udell, Kimmelman & Farrell
1328 Land Title Bldg.
Philadelphia, PA 19110 (215) 567-2230

William T. King
Synnestvedt & Lechner
2600 One Reading Center, 1101 Market St.
Philadelphia, PA 19107 (215) 923-4466

Bruce H. Kleinstein
Science Information Ser, Inc.
1500 Locust St. Suite 3504
Philadelphia, PA 19102 (215) 732-0426

Louis Frank Kline, Jr.
Rohm & Haas Co.
Independence Mall W.
Philadelphia, PA 19105 (215) 592-2992

C. Frederick Koening, III
Volpe & Koenig, P.C. Benjamin Franklin Bus. Ctr.
Ninth & Chestnut St. Suite 206
Philadelphia, PA 19107 (215) 238-0088

Robert A. Koons, Jr.
Schnader, Harrison, Segal & Lewis
1600 Market St. Suite 3600
Philadelphia, PA 19103 (215) 751-2180

Dennis M. Kozak
Atlantic Richfield Co.
1500 Market St.
Philadelphia, PA 19101 (215) 557-2362

Richard E. Kurtz
Woodcock, Washburn, Kurtz, Mackiewicz & Norris
One Liberty Place, 46th Floor
Philadelphia, PA 19103 (215) 568-3100

Rene A. Kuypers
Benasutti Associates, Ltd.
1020 Suburban Sta. Bldg.
1617 John F. Kennedy Blvd.
Philadelphia, PA 19103 (215) 564-4370

Jerome H. Lacheen
2400 Lewis Tower Bldg. 225 S. 15th St.
Philadelphia, PA 19102 (215) 545-6300

William E. Lambert, III
456 Flamingo St.
Philadelphia, PA 19128 (215) 483-8410

Gregory J. Lavorgna
Seidel, Gonda & Goldhammer, P.C.
Two Penn Center Plaza, Suite 1800
Philadelphia, PA 19102 (215) 568-8383

Peter Y. Lee
Synnestvedt & Lechner
3131 P S F S Bldg. 12 S. 12th St.
Philadelphia, PA 19107 (215) 923-4466

Edward T. Lentz
Smith, Kline, Beckman Corp.
1500 Spring Garden St.
Philadelphia, PA 19101 (215) 751-3120

Gary H. Levin
Woodcock, Washburn, Kurtz, Mackiewicz & Norris
One Liberty Place, 46th Floor
Philadelphia, PA 19103 (215) 568-3100

John Lezdey
Steele, Gould & Fried, Suite 3232
2000 Market St.
Philadelphia, PA 19103 (215) 563-8020

Charles H. Lindrooth
Synnestvedt & Lechner
2600 One Reading Center, 1101 Market St.
Philadelphia, PA 19107 (215) 923-4466

Alan D. Lourie
Smith, Kline, Beckman Corp.
P.O. Box 7929, 1500 Spring Garden St.
Philadelphia, PA 19101 (215) 751-5189

John Jacob Mackiewicz
Woodcock, Washburn, Kurtz, Mackiewicz & Norris
One Liberty Place, 46th Floor
Philadelphia, PA 19103 (215) 568-3100

Paul Maleson
Maleson, Rosenberg & Bilker
1407 Lewis Tower Bldg.
Philadelphia, PA 19102 (215) 735-2678

Joseph A. Marlino, Sr.#
Smith, Kline, Beckman Corp.
P.O. Box. 7929
Philadelphia, PA 19101 (215) 751-5182

Eric S. Marzlug
Caesar, Rivise, Bernstein, Cohen, & Pokotilow, Ltd
Suite 800, Stephen Girard Bldg. 21 S. 12th St.
Philadelphia, PA 19107 (215) 567-2010

Nicholas J. Masington, Jr.
Exide Electronic
2 Penn Center Plaza
Philadelphia, PA 19102 (215) 422-4040

Barbara V. Maurer
Weiser & Stapler
1510 Two Penn Center Plaza
Philadelphia, PA 19102 (215) 563-6600

Ronald J. Mc Caully
Wyeth Labs., Inc.
Box 8299
Philadelphia, PA 19101 (215) 688-4400

Robert G. Mc Morrow
Dann, Dorfman, Herrell & Skillman
1310 The Fidelity Bldg. 123 S. Broad St.
Philadelphia, PA 19109 (215) 545-1700

John F. Mc Nulty
Paul & Paul
Two Thousand Market St., Suite 2900
Philadelphia, PA 19103 (215) 568-4900

Thomas F. Mccaffery
Rohm & Haas Co.
Independence Mall West
Philadelphia, PA 19105 (215) 591-3691

James C. McConnon
Paul & Paul, 2900
Two Thousand Market St.
Philadelphia, PA 19103 (215) 568-4900

Austin R. Miller
Miller, Frailey & Prestia
1125 Land Title Bldg. Broad & Chesnut St.
Philadelphia, PA 19110 (215) 563-1810

Max R. Millman
1919 Chestnut St.
Philadelphia, PA 19103 (215) 563-4857

William D. Mitchell#
Pennwalt Corp. Pat. Dept.
3 Pkwy.
Philadelphia, PA 19102 (215) 587-7445

Daniel A. Monaco
Seidel, Gonda & Goldhammer, P.C.
Two Penn Center Plaza, Suite 1800
Philadelphia, PA 19102 (215) 568-8383

Charles W. Morck, Jr.
3468 St. Vincent St.
Philadelphia, PA 19149 (215) 624-3678

Lisa Mumma Morgan
Morgan, Lewis & Bockius
2000 One Logan Square
Philadelphia, PA 19103 (215) 963-4940

William H. Murray
Benasutti & Murray
Suite 2701, The A R A Tower, 11th & Market St.
Philadelphia, PA 19107 (215) 923-6100

Alan S. Nadel
Seidel, Gonda, Goldhammer & Panitch, P.C.
Three Penn Center Plaza, Suite 600
Philadelphia, PA 19102 (215) 568-8383

Frederick W. Neitzke
Paul & Paul
2000 Market St. Suite 2900
Philadelphia, PA 19103 (215) 568-4900

Wallace D. Newcomb
Panitch, Schwarzel, Jacobs & Nadel
2000 Market St. Suite 1400
Philadelphia, PA 19103 (215) 567-2020

Pauline Newman
F M C Corp.
2000 Market St.
Philadelphia, PA 19103 (215) 299-6973

Andrew Louis Ney
Weiser, Stapler & Spivak
1510 Two Penn Ctr. Plaza
Philadelphia, PA 19102 (215) 563-6600

Norman Leon Norris
Woodcock, Washburn, Kurtz, Mackiewicz & Norris
One Liberty Place, 46th Floor
Philadelphia, PA 19103 (215) 568-3100

Vincent T. Pace
Dann, Dorfman, Herrell & Skillman
1310 Fidelity Bldg. 123 S. Broad St.
Philadelphia, PA 19109 (215) 545-1700

Francis Arthur Paintin
Woodcock, Washburn, Kurtz, Mackiewicz & Norris
One Liberty Place, 46th Floor
Philadelphia, PA 19103 (215) 568-3100

Ronald L. Panitch
Panitch, Schwarze, Jacobs & Nadel
2000 Market St. Suite 1400
Philadelphia, PA 19103 (215) 567-2020

Henrik D. Parker
Woodcock, Washburn, Kurtz, Mackiewicz & Norris
One Liberty Place, 46th Floor
Philadelphia, PA 19103 (215) 568-3100

Peter J. Patane
Denny & Patane
2140 Land Title Bldg.
Philadelphia, PA 19110 (215) 563-3205

Henry N. Paul, Jr.
Paul & Paul
2000 Market St. Suite 2900
Philadelphia, PA 19103 (215) 568-4900

Steven R. Peterson
Dilworth, Paxson, Kalsih & Kauffman
2600 The Fidelity Bldg.
Philadelphia, PA 19109 (215) 875-8540

Donald R. Piper, Jr.
Dann, Dorfman, Herrell & Skillman, P.C.
1310 The Fidelity Bldg. 123 S. Broad St.
Philadelphia, PA 19109 (215) 545-1700

Bernard F. Plantz
Pennwalt Corp. Pennwalt Bldg.
3 Parkway
Philadelphia, PA 19102 (215) 587-7573

Harris A. Platt
Panitch, Schwarze, Jacobs & Nadel
2000 Market St., Suite 1400
Philadelphia, PA 19103 (215) 567-2020

Robert Charles Podwil
Wolf, Block, Schorr & Cohen
15th & Chestnut Sts., 12th Fl. Packard Bldg.
Philadelphia, PA 19102 (215) 977-2198

Manny Pokotilow
Caesar, Rivise, Bernstein, Cohen & Pokotilow, Ltd
21 S. 12th St. Suite 800 - Stephen Girard Bldg.
Philadelphia, PA 19107 (215) 567-2010

Joseph F. Posillico
Woodcock, Washburn, Kurtz, Mackiewicz & Norris
One Liberty Place, 46th Floor
Philadelphia, PA 19103 (215) 568-3100

Albert W. Preston, Jr.
Woodcock, Washburn, Kurtz, Mackiewicz & Norris
One Liberty Place, 46th Floor
Philadelphia, PA 19103 (215) 568-3100

Jack D. Puffer
Boeing Vertol Co.
P.O. Box 16858
Philadelphia, PA 19142 (215) 499-9400

Charles N. Quinn
Miller & Prestia
1125 Land Title Bldg. Broad & Chestnut St.
Philadelphia, PA 19110 (215) 563-1810

Polly E. Ramstad
Rohm & Haas Co.
Independence Mall West
Philadelphia, PA 19105 (215) 592-2423

Coleman Robert Reap
Atlantic Richfield Co.
1500 Market St.
Philadelphia, PA 19101 (215) 557-2360

Steven J. Rocci
Woodcock, Washburn, Kurtz, Mackiewicz & Norris
One Liberty Place, 46th Floor
Philadelphia, PA 19103 (215) 563-3100

Andrew G. Rodau
259 W. Johnson St. Apt. U-4
Philadelphia, PA 19144 (215) 849-7504

Gordon Samuel Rogers
Howson & Howson
13th Floor, Three Parkway
Philadelphia, PA 19102 (215) 963-9200

Evelyn M. Sabino#
3722 W. Country Club Rd.
Philadelphia, PA 19131 (215) 877-1333

Edward A. Sager
Pennwalt Corp., Pennwalt Bldg.
3 Parkway
Philadelphia, PA 19102 (215) 587-7688

Martin F. Savitzky
Synnestvedt & Lechner
2600 One Reading Center, 1101 Market St.
Philadelphia, PA 19107 (215) 923-4466

J. Walter Schilpp
Paul & Paul
2000 Market St. Suite 2900
Philadelphia, PA 19103 (215) 568-4900

William Schmonsees
F M C Corp.
2000 Market St.
Philadelphia, PA 19103 (215) 299-6977

William W. Schwarze
Panitch, Schwarze, Jacob & Nadel
2000 Market St. Suite 1400
Philadelphia, PA 19103 (215) 567-2020

William E. Scott*#
U.S. Dept. Of Agriculture, A.R.S. Res, & Dev,Div.
600 E. Mermaid Lane
Philadelphia, PA 19118 (215) 233-6597

Arthur Harris Seidel
Seidel, Gonda & Goldhammer, P.C.
Two Penn Center Plaza, J F K Blvd. Suite 1800
Philadelphia, PA 19102 (215) 568-8383

Arthur G. Seifert
Wyeth Labs.
P.O. Box 8299
Philadelphia, PA 19101 (215) 341-2314

Henry H. Skillman
Dann, Dorfman, Herrell & Skillman, P.C.
1310 Fidelity Bldg. 123 S. Broad St.
Philadelphia, PA 19109 (215) 545-1700

Alex R. Sluzas
Paul & Paul
2900 Two Thousand Market St.
Philadelphia, PA 19103 (215) 568-4900

William W. Smith#
9200 Stenton Ave.
Philadelphia, PA 19118 (215) 247-3927

James A. Spady
3814 Walnut St.
Philadelphia, PA 19104 (215) 898-4758

Karl L. Spivak
Suite 3232-I V B Bldg.
1700 Market St.
Philadelphia, PA 19103 (215) 568-8020

Alfred Stapler
7334 Rural Lane
Philadelphia, PA 19119 (215) 242-2698

Brain W. Stegman
Rohm & Haas Co. Pat. Dept.
Independence Mall West
Philadelphia, PA 19105 (215) 592-6818

Barry Allen Stein
Caesar, Rivise, Bernstein, Cohen & Pokotilow, Ltd.
21 S. 12th St. Suite 800
Philadelphia, PA 19107 (215) 567-2010

Terence P. Strobaugh
Rohm & Haas W.
Independence Mall West
Philadelphia, PA 19105 (215) 592-3677

Arthur M. Suga
Pennwalt Corp. Patent Dept.
Three Parkway
Philadelphia, PA 19102 (215) 587-7690

Stuart R. Suter
Smith, Kline Corp.
One Franklin Plaza
Philadelphia, PA 19101 (215) 751-5186

John T. Synnestvedt
Synnestvedt & Lechner
2600 One Reading Center, 1101 Market St.
Philadelphia, PA 19107 (215) 923-4466

Kenneth P. Synnestvedt#
Synnestvedt & Lechner
2600 One Reading Center, 1101 Market St.
Philadelphia, PA 19107 (215) 923-4466

George Tarnowski
Wyeth Labs.
P.O. Box 8299
Philadelphia, PA 19121 (215) 341-4245

John E. Taylor, III#
Rohm & Haas Co. Pat. Dept.
Independence Mall W.
Philadelphia, PA 19105 (215) 592-3294

Joseph A. Tessari
Paul & Paul
2900 Two Thousand Market St.
Philadelphia, PA 19103 (215) 568-4900

Blucher Stanley Tharp, Jr.
324 South St.
Philadelphia, PA 19107 (215) 922-5539

E. Arthur Thompson
Paul & Paul
2000 Market St. Suite 2900
Philadelphia, PA 19103 (215) 568-4900

Jacob Trachtman
Trachtman, Jacob & Beck
1500 Chestnut St. Pennsylvania Bldg. Suite 1908
Philadelphia, PA 19102 (215) 569-9800

Walter B. Udell
Maleson, Udell, Rosenberg, Bilker & Farrell
1407 Lewis Tower Bldg.
Philadelphia, PA 19102 (215) 735-2678

R. Duke Vickrey
Scott Paper Co. Scott Plaza
Philadelphia, PA 19113 (215) 521-5809

Franklin J. Visek
Boeing Vertol Co.
P.O. Box 16858
Philadelphia, PA 19142 (215) 522-2663

Anthony S. Volpe
Volpe & Koenig, P.C.
Benjamin Franklin Bus. Center, Ninth & Chestnut St. Suite 206
Philadelphia, PA 19107 (215) 238-0088

P. Michael Walker
Harding, Earley, Follmer & Frailey
1910 Two Mellon Bank Center
Philadelphia, PA 19102 (215) 568-2606

Robert B. Washburn
Woodcock, Washburn, Kurtz, Mackiewicz & Norris
One Liberty Place
Philadelphia, PA 19103 (215) 568-3100

Richard D. Weber
Synnestvedt & Lechner
2600 One Reading Center, 1101 Market St.
Philadelphia, PA 19107 (215) 923-4466

Gerard J. Weiser
Weiser & Stapler
1510 Two Penn Ctr. Plaza
Philadelphia, PA 19102 (215) 563-6600

Bernard M. Weiss#
108-110 Almatt Place
Philadelphia, PA 19115 (215) 676-2280

Louis Weistein
Weinstein & Kimmelman
2410 Two Mellon Bank Center
Philadelphia, PA 19102 (215) 557-9797

John A. Weygandt
Scott Paper Co. Scott Plaza
Philadelphia, PA 19113 (215) 522-5815

Janice E. Williams
Smith, Kline, Beckman Corp.
One Franklin Plaza
Philadelphia, PA 19103 (215) 751-5187

Douglas E. Winters
Rohm & Mall West
Independence Mall West
Philadelphia, PA 19105 (215) 592-6745

Robert Wiser
American Home Products Corp.
Wyeth Labs., Div. P.O. Box 8299
Philadelphia, PA 19101 (215) 688-4400

Zachary T. Wobensmith, III
Dann, Dorfman, Herrell & Skillman
1310 Fidelity Bldg. 123 S. Broad St.
Philadelphia, PA 19109 (215) 545-1700

Joseph H. Yamaoka
1731 Lombard St.
Philadelphia, PA 19146 (215) 735-4381

Stephen A. Young
General Electric Co.
3198 Chestnut St.
Philadelphia, PA 19101 (215) 823-2942

Daniel C. Abeles
Westinghouse Elec. Corp. Law Dept., I P S
1310 Beulah Rd.
Pittsburgh, PA 15235 (412) 256-5227

John M. Adams
Buchanan Ingersoll
57th Floor, 600 Grant St.
Pittsburgh, PA 15219 (412) 562-1067

Godfried R. Akorle#
P P G Industries Inc.
One P P G Plaza
Pittsburgh, PA 15272 (412) 434-2469

Lynn J. Alstadt
Buell, Ziesenheim, Beck & Alstadt
322 Blvd. Of The Allies Suite 500
Pittsburgh, PA 15222 (412) 471-1590

Edward Charles Arenz
Westinghouse Electric Corp.
R & D Center
Bldg. 801
Pittsburgh, PA 15235 (412) 256-5244

Brian W. Ashbaugh
Rose, Schmidt, Chapman, Duff & Hasley
900 Oliver Bldg.
Pittsburgh, PA 15222 (412) 434-8859

Klaus J. Bach#
Westinghouse Electric Corp.
1310 Beulah Rd. R & D Center
Pittsburgh, PA 15235 (412) 256-5257

Fred J. Baehr, Jr.
Westinghouse Elec. Corp. Law Dept. 1 PS
1310 Beulah Rd.
Pittsburgh, PA 15235 (412) 256-5267

Kent E. Baldauf
Parmelee, Miller, Welsh & Kratz, P.C.
Suite 721, 301 Fifth Ave. Bldg.
Pittsburgh, PA 15222 (412) 281-2931

Paul A. Beck
Buell, Ziesenhiem, Beck & Alstadt
322 Blvd. Of The Allies, Suite 500
Pittsburgh, PA 15222 (412) 471-1590

William A. Behare
Baskin, Flaherty, Elliott & Mannino, P.C.
29th Floor, One Mellon Bank Center
Pittsburgh, PA 15219 (412) 562-8600

Richard M. Bies
217 Park Entrance
Pittsburgh, PA 15228 (412) 343-7421

Byron A. Bilicki
Kirkpatrick & Lockhart
1500 Oliver Bldg.
Pittsburgh, PA 15222 (412) 355-8653

Harry Donald Bishop#
8360 Remington Dr.
Pittsburgh, PA 15237 (412) 364-6713

Walter J. Blenko, Jr.
Ecker, Seamans, Cherin & Mellott
600 Grant St. 42nd Floor
Pittsburgh, PA 15219 (412) 566-6189

Paul Bogdon
100 Ross St.
Pittsburgh, PA 15219 (412) 391-7133

George John Bohrer#
188 Oak Park Place
Pittsburgh, PA 15243 (412) 561-6563

Stanley R. Bramham#
Westinghouse Electric Corp. Law Dept.
Intellectual Property Section
1310 Beulah Rd.
Pittsburgh, PA 15235 (412) 256-5260

John S. Brams#
Carother & Carothers
1900 Commonwelth Bldg.
Pittsburgh, PA 15222 (412) 471-3575

Thomas M. Breininger
P P G Industries
One P P G Place
Pittsburgh, PA 15272 (412) 434-3741

Edgar Wallace Breisch, Jr.
524 Olive St.
Pittsburgh, PA 15237 (412) 366-1621

John Reeder Bronaugh
Rockwell Internatl. Corp.
600 Grant St.
Pittsburgh, PA 15219 (412) 565-2902

Francis E. Browder
3931 Greensburg Pike
Pittsburgh, PA 15221 (412) 351-0942

David C. Brueing
Webb, Burden, Robinson & Webb, P.A.
515 Oliver Bldg.
Pittsburgh, PA 15222 (412) 471-8815

Joseph L. Brzuszek
Westinghouse Elec. Corp.
R & D Center, Bldg. 801 Law Dept.
1310 Beulah Rd.
Pittsburgh, PA 15235 (412) 256-5259

Eugene Franklin Buell
Buell, Ziesenheim, Beck & Alstadt
322 Blvd. Of The Allies Suite 500
Pittsburgh, PA 15222 (412) 471-1590

Richard L. Byrne
Webb, Burden, Robinson & Weeb
515 Oliver Bldg.
Pittsburgh, PA 15222 (412) 471-8815

Joseph John Carducci
708 Scurbgrass Rd.
Pittsburgh, PA 15243 (412) 563-6732

Floyd Barber Carothers
Carothers & Carothers
1900 Commonwealth Bldg.
Pittsburgh, PA 15222 (412) 471-3575

Ernest A. Carpenter
Dravo Corp.
One Oliver Plaza
Pittsburgh, PA 15222 (412) 566-3126

Raymond S. Chisholm
276 Trotwood Dr.
Pittsburgh, PA 15241 (412) 835-0418

Daniel P. Cillo
Westinghouse Electric Corp. Law Dept.
Intellectual Property
R & D Center, Churchill Borough
Pittsburgh, PA 15235 (412) 256-5264

Frederick H. Colen
Reed, Smith, Shaw & McClay
435
Pittsburgh, PA 15219 (412) 228-4164

Andrew J. Cornelius
Baskin, Flaherty, Elliott & Mannino, P.C.
500 Grant St. 29th Floor - One Mellon Bank
Pittsburgh, PA 15219 (412) 562-8641

Bruce H. Cottrell
P P G Industries Inc.
One P P G Place
Pittsburgh, PA 15272 (412) 434-2882

John E. Curley
P P G
One P P G Place
Pittsburgh, PA 15272 (412) 434-3794

Louis A. De Paul
Westinghouse Elec. Corp. Law Dept. I P S
1310 Beulah Rd.
Pittsburgh, PA 15235 (412) 256-5228

Zigmund L. Dermer
Westinghouse Electric Corp.
R & D Center, 1310 Beulah Rd.
Pittsburgh, PA 15235 (412) 256-5247

George D. Dickos
Kirkpatrick & Lockhart
1500 Oliver Bldg.
Pittsburgh, PA 15222 (412) 355-6785

Alan M. Doernberg
Fisher Scientific Co. Law Dept.
711 Forbes Ave.
Pittsburgh, PA 15219 (412) 562-8451

Marvin R. Dunlap
Dickie, McCamey & Chilcote, P.C.
Two PPG Place
Suite 400
Pittsburgh, PA 15222 (412) 281-7272

Eugene Francis Dwyer
Seven Creighton Ave.
Pittsburgh, PA 15205 (412) 922-6579

Russell A. Eberly
P P G Industries Inc.
One Gateway Center
Pittsburgh, PA 15222 (412) 434-2934

William A. Elchik
Westinghouse Electric Corp.
1300 Beulah Rd.
Pittsburgh, PA 15235 (412) 256-5215

Daniel W. Ernsberger
Behrend & Ernsberger
2400 Grant Bldg.
Pittsburgh, PA 15219 (412) 391-2515

Christine R. Ethridge
350 Porter Bldg.
Pittsburgh, PA 15219 (412) 596-2405

Julian Falk
2131 5th Ave.
Pittsburgh, PA 15219 (412) 471-0774

D. Leigh Fowler, Jr.
126 Lebanon Hills Dr.
Pittsburgh, PA 15228 (412) 341-1041

Joseph C. Gil
Mobay Chemical Corp.
Mobay Rd.
Pittsburgh, PA 15205 (412) 777-2342

Eugene P. Girman
Girman & Bacharach
513 2nd Ave.
Pittsburgh, PA 15219 (412) 391-8713

Douglas Gene Glantz
Alcoa
138 Alleyne Dr.
Pittsburgh, PA 15215 (412) 784-1216

Lee R. Golden
1000 Allegheny Bldg.
Pittsburgh, PA 15219 (412) 471-4722

David C. Hanson
Webb, Burden, Robinson & Webb, P.A.
515 Oliver Bldg.
Pittsburgh, PA 15222 (412) 471-8815

Gordon R. Harris
Buell, Zesenheim, Beck & Alstadt
322 Blvd. Of The Allies
Pittsburgh, PA 15222 (412) 471-1590

Gene Harsh
Mobay Chemical Corp.
Mobay Rd.
Pittsburgh, PA 15205 (412) 777-2340

George E. Hawranko
Westinghouse Electric Corp. Res. Dev, Center
1310 Beulah Rd.
Pittsburgh, PA 15235 (412) 256-5222

Rea C. Helm
U.S. Steel Corp.
600 Grant St.
Pittsburgh, PA 15230 (412) 433-2986

Richard E.L. Henderson#
Mobay Chemical Corp.
Patent Dept.
Mobay Rd.
Pittsburgh, PA 15205 (412) 777-2000

George J. Hurlston#
1110 Kelton Ave.
Pittsburgh, PA 15216 (412) 561-4637

Lee P. Johns
Westinghouse Elec. Co.
1310 Beulah Rd.
Pittsburgh, PA 15235 (412) 256-5213

Barbara E. Johnson
Webb, Burden, Robinson & Webb
Suite 515, Oliver Bldg.
Pittsburgh, PA 15222 (412) 471-8815

Chester Arthur Johnston, Jr.
P P G Industrise, Inc.
One P P G Place
Pittsburgh, PA 15272 (412) 434-2931

Herbert L. Jones
5264 Beelermont Place
Pittsburgh, PA 15217 (412) 687-4092

John W. Jordan, Iv.
Gigsby, Gaca & Davies, P.C.
One Gateway Center Tenth Floor
Pittsburgh, PA 15222 (412) 281-0737

Alan P. Kass
Baskin, Flaherty, Elliott & Mannino, P.C.
2900 One Mellon Bank Center
Pittsburgh, PA 15219 (412) 562-8776

Koichiro Kato#
Westinghouse Elec. Corp.
1310 Beulah Rd. Churchill
Pittsburgh, PA 15235 (412) 256-3189

Harry B. Keck
Murray, Keck & Zurawsky
1802 Frick Bldg.
Pittsburgh, PA 15219 (412) 765-1580

Deane E. Keith
Gulf Oil Corp.
715 Gulf Bldg., P.O. Box 1166
Pittsburgh, PA 15230 (412) 263-5101

Suzanne Kikel#
Wean United Inc.
948 Ft. Duquesne Blvd.
Pittsburgh, PA 15222 (412) 456-5491

Michael J. Kline#
209 N. Dithridge Ave.
Pittsburgh, PA 15213 (412) 683-1413

William G. Kratz, Jr.
Parmelee, Miller, Welsh & Kratz, P.C.
301 5th Ave. Suite 721
Pittsburgh, PA 15222 (412) 281-2931

William L. Krayer
U.S. Steel Corp. Law Dept., Room 6082
600 Grant St.
Pittsburgh, PA 15230 (412) 433-2984

Ronald L. Kuis
Kikpatrick & Lockhart
1500 Oliver Bldg.
Pittsburgh, PA 15222 (412) 355-8614

James R. Kyper
Kirkpatrick, Lockhart, Johnson & Hutchison
1500 Oliver Bldg.
Pittsburgh, PA 15222 (423) 556-542
/phA

Donald Raymond Lackey
Westinghouse Electric Corp. R & D Center
1300 Beulah Rd.
Pittsburgh, PA 15235 (412) 256-7537

Edward B. Lee, III
Lindsay, Mchinnis, Mccandless & McCabe
Suite 200, Standard Life Bldg.
345 4th Ave.
Pittsburgh, PA 15222 (412) 471-2420

Robert P. Lenart
Westinghouse Elec. Corp.
1310 Beulah Rd.
Pittsburgh, PA 15235 (412) 256-5792

Donald C. Lepiane
P P G Industries, Inc.
One P P G Place
Pittsburgh, PA 15272 (412) 434-2930

Mark Levin
P P G Inds, Inc.
One Gateway Center
Pittsburgh, PA 15222 (412) 434-3792

Edward L. Levine
Joy Mfg. Co.
301 Grant St.
Pittsburgh, PA 15219 (412) 562-4838

John William Linkhauer
1425 Porter Bldg.
Pittsburgh, PA 15219 (412) 232-3500

Carl R. Lippert
513 Guyasuta Rd.
Pittsburgh, PA 15215 (412) 781-5355

William Henry Logsdon
Webb, Burden, Robinson & Webb, P.A.
535 Smithfield St. 515 Oliver Bldg.
Pittsburgh, PA 15222 (412) 471-8815

Ronald S. Lombard
2001 Clark Bldg.
Pittsburgh, PA 15222 (412) 261-3939

Daniel J. Long
Koppers Co. Inc.
1450 Koppers Bldg.
Pittsburgh, PA 15219 (412) 227-2653

Charles M. Lorin
Westinghouse Elec. Corp.
Beulah Rd.
Pittsburgh, PA 15235 (412) 256-7507

Michael Patrick Lynch
Westinghouse Elec. Corp. Res. & Dev. Ctr.
Churchill Borough
Pittsburgh, PA 15235 (412) 256-7521

Debra A. Mangus#
5326 Pocusset St. Apt. 24
Pittsburgh, PA 15217 (412) 422-0188

Georgh E. Manias#
H.H. Robertson Co.
P.O. Box 2793
Pittsburgh, PA 15220 (412) 928-7548

Michael I. Markowitz
253 Curry Hollow Rd.
Pittsburgh, PA 15236 (412) 653-3180

Albert G. Marriott
Rockwell International Corp.
600 Grant St.
Pittsburgh, PA 15219 (412) 565-2972

Edward I. Mates
P P G Industries, Inc.
One P P G Place
Pittsburgh, PA 15272 (412) 434-2936

Douglas K. Mc Claine
Mine Safety Appliances Co.
P.O. Box 426
Pittsburgh, PA 15230 (412) 273-5316

Clement L. Mc Hale
Westinghouse Elec. Corp.
1310 Beulah Rd.
Pittsburgh, PA 15235 (412) 256-7554

Thomas F. McKnight
Neville Chemical Co. Neville Island
Pittsburgh, PA 15225 (412) 331-4200

Charles Leroy Menzemer
Westinghouse Elec. Corp. Bldg. 801 Rm. 4C66
1300 Beulah Rd. R & D Center
Pittsburgh, PA 15235 (412) 256-7566

Alex Mich, Jr.
Westinghouse Elec. Corp. Pat. Dept.
1310 Beulah Rd.
Pittsburgh, PA 15235 (412) 256-7570

Thomas G. Michalek
1079 Greetree Rd. Suite 5
Pittsburgh, PA 15220 (412) 343-8889

Julian Kennedy Miller
4747 Bayard St.
Pittsburgh, PA 15213 (412) 682-2469

Thomas G. Miller
Parmelee, Miller, Welch & Kratz, P.C.
301 5th Ave. Bldg. Suite 721
Pittsburgh, PA 15222 (412) 281-0512

Dennis G. Millman
P P G Industires Inc.
One P P G Place
Pittsburgh, PA 15272 (412) 434-2936

William C. Mitchell
2349 Casswell Dr.
Pittsburgh, PA 15102 (412) 777-8404

Martin J. Moran
Westinghouse Elec. Corp.
1310 Beulah Rd.
Pittsburgh, PA 15235 (412) 256-7693

Dennis M. Morgenstern
Rose, Schmidt, Dixon & Hasley
Suite 900, Oliver Blvd.
Pittsburgh, PA 15222 (412) 434-8870

George Douglas Morris
P P G Inds. Inc.
One Gateway Center
Pittsburgh, PA 15222 (412) 434-3797

Joyce L. Morrison
Eckert, Seamans, Cherin & Mellott
600 Grant St. 42nd Floor
Pittsburgh, PA 15219 (412) 566-6000

Lee D. Moses
Messer, Shilobod & Crenneyo.
One Gateway Center, 12th Fl.
Pittsburgh, PA 15222 (412) 281-7200

Clair Xavier Mullen, Jr.
Crucible, Inc.
Parkway West & Rte. 60, P.O. Box 88
Pittsburgh, PA 15230 (412) 923-2955

Thomas Henry Murray
2230 Koppers Bldg.
Pittsburgh, PA 15219 (412) 765-1580

Ronald F. Naughton
Rockwell Internatl. Corp.
600 Grant St.
Pittsburgh, PA 15219 (412) 565-2977

Meyer Neishloss
1010 Summer Pl.
Pittsburgh, PA 15243 (412) 561-8810

Bidyut K. Niyogi#
3823 Henley Dr.
Pittsburgh, PA 15235 (412) 243-8959

William J. O Rourke, Jr.
Joy Mfg. Co.
301 Grant St.
Pittsburgh, PA 15219 (412) 562-4632

Russell D. Orkin
Webb, Burden, Robinson & Webb, P.A.
515 Oliver Bldg.
Pittsburgh, PA 15222 (412) 471-8815

William E. Otto
Westinghouse Elec. Corp. Pat Dept.
1310 Beulah Rd.
Pittsburgh, PA 15235 (412) 256-7515

Michael G. Panian
Westinghouse Elec. Corp. R & D Center, Bldg. 801
1310 Beulah Rd. Law Dept. Intellectual Prop. Sec
Pittsburgh, PA 15235 (412) 256-5233

Barbara J. Park
P P G Industries, Inc.
One P P G Place
Pittsburgh, PA 15272 (412) 434-3131

William H. Parmelee
Parmelee, Miller, Welsh & Kratz, P.C.
301 Fifth Ave. Bldg. Suite 721
Pittsburgh, PA 15222 (412) 281-2931

Daniel Patch
Wean United, Inc.
948 Ft. Duguesne Blvd.
Pittsburgh, PA 15222 (412) 456-5490

Lee Patch, #
Reed, Smith, Shaw & McClay
P.O. Box 2009
Pittsburgh, PA 15230 (412) 288-4220

John R. Pegan
United States Steel Corp.
600 Grant St. Room 1412
Pittsburgh, PA 15230 (412) 433-2994

Edward L. Pencoske
Kirkpatrick & Lockhart
1500 Oliver Bldg.
Pittsburgh, PA 15222 (412) 355-8645

Joel R. Petrow
Westinghouse Elec. Corp. Research & Dev. Center
Pat. Dept.
Pittsburgh, PA 15235 (412) 256-5212

Linda Pingitore
P P G Industries, Inc.
One P P G Place
Pittsburgh, PA 15272 (412) 434-3704

Clifford A. Poff#
2230 Koppers Bldg.
Pittsburgh, PA 51219 (412) 765-1580

Lawrence S. Pope
Mobay Chemical Corp.
Patent Dept.
Penn Lincoln Pkwy., W.
Pittsburgh, PA 15205 (412) 777-2341

Aron Preis
Mobay Chemical Corp.
Penn. Lincoln Pkwy., W.
Pittsburgh, PA 15205 (415) 777-2343

Stanley J. Price
Price & Adams, Ltd.
4135 Brownsville Rd., P.O. Box 98127
Pittsburgh, PA 15227 (412) 882-7170

James O. Ray, Jr.#
11682 Althea Dr.
Pittsburgh, PA 15235 (412) 243-4513

George Raynovich, Jr.
Wheeling- Pittsburgh Steel Corp.
Four Gateway Center
Pittsburgh, PA 15222 (412) 288-3517

Stanley J. Reisman
Law & Finnance Bldg.
Pittsburgh, PA 15219 (412) 232-0433

Charles Frank Renz
Westinghouse Elec. Corp.
1310 Beulah Rd.
Pittsburgh, PA 15235 (412) 256-5216

William F. Riesmeyer, III
U S X Corp.
600 Grant St. Room 1569
Pittsburgh, PA 15230 (412) 433-2842

Alvin E. Ring
Buell, Ziesenheim, Beck & Alstadt
322 Blvd. of the Allies, Suite 500
Pittsburgh, PA 15222 (412) 471-1590

Rita M. Rooney#
Eckert, Seamans, Cherin & Mellott
600 Grant St.
Pittsburgh, PA 15219 (412) 566-6000

Thomas W. Roy
Mobay Chemical Corp.
Penn Lincoln Pkwy., W.
Pittsburgh, PA 15205 (412) 777-2345

William J. Ruano
402 St. Clair Bldg. 1725 Washington Rd.
Pittsburgh, PA 15241 (412) 835-3111

Donald M. Satina
Westinghouse Elec. Corp.
R & D Center, Pat. Dept. 1310 Beulah Rd.
Pittsburgh, PA 15235 (412) 256-5374

C. William Schildnecht
Dravo Corp.
One Oliver Plaza
Pittsburgh, PA 15222 (415) 566-3225

Dean Schron
Westinghouse Elec. Corp.
R & D Center, Beulah Rd.
Pittsburgh, PA 15235 (412) 256-5237

Ansel M. Schwartz*
Reed, Smith, Shaw & McClay
P.O. Box 2009
Pittsburgh, PA 15230

Frederick L. Segal
Suite 700 Manor Bldg.
564 Forbes Ave.
Pittsburgh, PA 15219 (412) 391-2263

Donnan L. Seidel#
P P G Industries, Inc.
One P P G Place
Pittsburgh, PA 15272 (412) 434-3798

Carl T. Severini
P P G Industries, Inc.
One P P G Place
Pittsburgh, PA 15272 (412) 434-2938

Forest Charles Sexton
U.S. Steel Corp.
600 Grant St. Room 1412
Pittsburgh, PA 15230 (412) 433-2872

Thomas R. Shaffer
Buell, Ziesenheim, Beck & Alstadt
322 Blvd. Of The Allies Suite 500
Pittsburgh, PA 15222 (412) 471-1590

Ronald H. Shakely
Mine Safety Appliance Co.
600 Penn Ctr. Blvd.
Pittsburgh, PA 15235 (412) 273-5215

Michael I. Shamos
605 Devonshire St.
Pittsburgh, PA 15213 (412) 681-8398

Thomas F. Shanahan
225 Foxcroft Rd.
Pittsburgh, PA 15220 (412) 279-3431

Arnold Barry Silverman
Ecker, Seamans, Cherin & Mellott
600 Grant St., 42nd Fl.
Pittsburgh, PA 15219 (412) 566-6000

Andrew C. Siminerio
P P G Industries, Inc.
One P P G Place
Pittsburgh, PA 15272 (412) 434-4645

William R. Sittig, Jr.
Midtick & Giltinan, P.C.
816 Fifth Ave.
Pittsburgh, PA 15219 (412) 391-3334

Richard A. Speer
Rockwell Internatl. Corp.
600 Grant St.
Pittsburgh, PA 15219 (412) 565-7107

Kenneth J. Stachel
P P G Industries, Inc.
One P P G Place
Pittsburgh, PA 15272 (412) 434-3186

Patrica K. Staub
308 Harvester Circle
Pittsburgh, PA 15241 (412) 831-2106

Arland T. Stein
Reed, Smith, Shaw & McClay
747 Two Mellon Bank Center, P.O. Box 2009
Pittsburgh, PA 15219 (412) 288-3100

Irwin M. Stein
P P G Industries, Inc.
One P P G Place
Pittsburgh, PA 15272 (412) 434-3799

Richard A. Stoltz
Westinghouse Elect. Corp.
1310 Beulah Rd.
Pittsburgh, PA 15235 (412) 256-5202

Blair Ross Studebaker
Westinghouse Elec. Corp. R & D Center, Pat. Dept.
1300 Beulah Rd.
Pittsburgh, PA 15235 (412) 256-5211

Michael R. Swartz
205 Royal Oak Ave.
Pittsburgh, PA 15235 (412) 244-9205

Jame J. Tedjeske
Schneider Enterprises Inc.
121 Seventh St.
Pittsburgh, PA 15222 (412) 288-7037

Gordon Howard Telfer
Westinghouse Elec. Corp. R & D Center
1310 Beulah Rd.
Pittsburgh, PA 15235 (412) 256-5208

Gary P. Topolosky
Webb, Burden, Robinson & Webb, P.A,
515 Oliver Bldg. 535 Smithfield St.
Pittsburgh, PA 15222 (412) 471-8815

Thomas R. Trempus
105 Chalet Dr.
Pittsburgh, PA 15221

Frederic C. Trenor, II
Meyer, Darragen
2000 Frick Bldg.
Pittsburgh, PA 15219 (412) 261-6000

James G. Uber
Reed, Smith, Shaw & McClay
P.O. Box 2009
Pittsburgh, PA 15230 (412) 288-4134

David S. Urey
U.S. Steel Corp.
600 Grant St. Room 1412
Pittsburgh, PA 15230 (412) 433-2873

James C. Valentine
130 Morrison Dr.
Pittsburgh, PA 15216 (412) 563-6513

Patrick J. Viccaro
Allegheny Ludlum Steel Corp.
1000 Six P P G Place
Pittsburgh, PA 15222 (412) 394-2839

Robert F. Wagner
Dickie, Mccarney & Chilcote
2 P P G Place
Pittsburgh, PA 15222 (412) 392-5393

David A. Warmbold
Joy Manufacturing Co.
301 Grant St.
Pittsburgh, PA 15219 (412) 562-4662

John M. Webb
Webb, Burden, Robinson & Webb, P.A.
535 Smithfield St., 515 Oliver Bldg.
Pittsburgh, PA 15222 (412) 471-8815

William Hess Webb
Webb, Burden, Robinson And Webb
515 Oliver Bldg., 535 Smithfield St.
Pittsburgh, PA 15222 (412) 471-8815

Edward F. Welsh
Parmelee, Miller, Welsh & Kratz, P.C.
301 Fifth Ave Bldg. Suite 721
Pittsburgh, PA 15222 (412) 281-2931

Richard V. Westerhoff
Eckert, Seamans, Cherin & Mellott
600 Grant St. 42nd Floor
Pittsburgh, PA 15219 (412) 566-6090

Thomas C. Wettach
Reed, Smith, Shaw & McClay
435 Sixth Ave. P.O. Box 2009
Pittsburgh, PA 15219 (412) 288-3102

Lyndanne M. Whalen
Mobay Chemical Corp.
Patent Dept.
Mobay Rd.
Pittsburgh, PA 15205 (412) 777-2347

Edward J. Whitfield
P P G Industries, Inc.
One P P G Place
Pittsburgh, PA 15272 (412) 434-2937

Walter H. Williams#
Neville Chemical Co. Neville Island
Pittsburgh, PA 15225 (412) 331-4200

John K. Williamson
Westinghouse Elec. Corp. Law Dept. I P S
1310 Beulah Rd.
Pittsburgh, PA 15235 (412) 256-5219

Bruce J. Wolstoncroft
523 Siesta Court
Pittsburgh, PA 15205 (412) 787-5073

Robert D. Yeager
Kirkpatrick & Lockhart
1500 Oliver Bldg.
Pittsburgh, PA 15222 (412) 355-8605

Herbert J. Zeh, Jr.
Koppers Co. Inc.
Koppers Bldg.
Pittsburgh, PA 15219 (412) 227-2655

Frederick B. Ziesenheim
Buell, Ziesenheim, Beck & Alstadt
322 Blvd. Of The Allies, Suite 500
Pittsburgh, PA 15222 (412) 471-1590

Lawrence G. Zurawsky
Zurawsky & Keck
Suite 415 Lawyers Bldg. 428 Forbes Ave.
Pittsburgh, PA 15219 (412) 281-7766

Fred C. Battles#
516 Morgantown St.
Point Marion, PA 15474 (412) 725-5414

Jack L. Foltz
Sun Co., Inc.
100 Matsonford Rd.
Radnor, PA 19087 (215) 293-6392

Anthony Potts, Jr.
Sun CO.
100 Matson Ford Rd.
Radnor, PA 19087 (215) 293-6867

Richard O. Church
Quittner & Church
152 N. 6th St. P.O. Box 1459
Reading, PA 19603 (215) 372-4631

Leonard M. Quittner
152 North Sixth St.
Reading, PA 19601 (215) 372-4631

Donald C. Watson
111 Montrose Blvd.
Reading, PA 19607 (215) 777-4660

Everett H. Murray, Jr.
141 Browning Lane
Rosemont, PA 19010 (215) 525-4497

Lucretia R. Quatrini#
409 South Wilbur Ave.
Sayre, PA 18840 (717) 888-7022

Harry D. Anspon#
29 Beaver St.
Sewickley, PA 15143 (412) 741-6029

Thomas W. Brennan#
132 Apache Dr.
Shickshinny, PA 18655 (717) 256-3016

John R. Ewbank
1150 Woods Rd.
Southampton, PA 18966 (215) 357-3977

Robert J. Mooney
736 Second St. Pike
Southampton, PA 18966 (215) 322-1980

Mary E. Bak
Howson & Howson
Spring House Corp. Center, P.O. Box 457
Spring House, PA 19477 (215) 540-9200

Henry Hansen
Howson & Howson
Spring House Corp. Center, P.O. Box 457
Spring House, PA 19477 (215) 540-9200

Leonard R. Hecker#
Mc Neil Pharmaceutical
Spring House, PA 19477 (215) 628-5528

Stanley B. Kita
Howson & Howson
Spring House Corp. Center, P.O. Box 457
Spring House, PA 19477 (215) 540-9200

Wilson Oberdorfer
Howson & Howson
Spring House Corp. Center, Box 457
Spring House, PA 19477 (215) 540-9200

Karl F. Ockert
Rohm & Haas Co. Research Lab.
727 Norristown Rd.
Spring House, PA 19477 (215) 641-7822

George A. Smith, Jr.
Howson & Howson
Spring House Corp. Center, P.O. Box 457
Spring House, PA 19477 (215) 540-9200

Karl E. Geci
Twin Springs
962 Million Dollar Highway
St. Marys, PA 15857 (814) 781-3409

Virgil P. Quirk#
828 Johnsonburg Rd.
St. Marys, PA 15857 (814) 834-3610

Richard L. Hanson
Calder Square
P.O. Box 10361
State College, PA 16805 (814) 863-1160

Thomas E. Sterling
P.O Box 14, 234 E. College Ave.
State College, PA 16801 (814) 238-9455

Arthur W. Collins
514 Scool Ln.
Swarthmore, PA 19081 (215) 543-1620

Alexander D. Ricci
Betz Labs, Inc.
4636 Somerton Rd.
Trevose, PA 19047 (215) 355-3300

Patrick C. Baker, II
Ratner & Prestia, P.C.
P.O. Box 980, 412 Leighton Bldg. 500 N. Gulph Rd.
Valley Forge, PA 19482 (215) 265-6666

Lawrence A. Husick
P.O. Box 980
Valley Forge, PA 19482 (215) 265-6666

Michael F. Petock
46 The Commons At Valley Forge
1220 Valley Forge Rd. P.O. Box 856
Valley Forge, PA 19481 (215) 935-8600

Ernest G. Posner
P Q Corp.
Bldg. 11, Valley Forge Executive Mall
Valley Forge, PA 19482 (215) 293-7354

Paul F. Prestia
Ratner & Prestia
412 Leighton Bldg. 500 N. Gulph Rd. P.O. Box 980
Valley Forge, PA 19482 (215) 265-6666

Allan Ratner
Ratner & Prestia
Box 980
Valley Forge, PA 19482 (215) 265-6666

Donald M. Mac Kay
1098 Maple Ave.
Vernon, PA 15147 (412) 795-1569

Marden S. Gordon#
331 Radnor Chester Rd.
Villanova, PA 19085 (215) 687-5972

George F. Mueller
701 Knox Rd.
Villanova, PA 19085 (215) 688-0640

Marvin C. Gaer
U.S. Navy, Naval Air Dev. Ctr.
Code 4033
Warminster, PA 18970 (215) 441-2031

Walter Scott John, III
130 W. Lancaster Ave.
Wayne, PA 19087 (215) 688-5426

Frederick J. Olsson
Suite 216, Bldg. No. 8
Valley Forge Executive Mall
Wayne, PA 19087 (215) 687-1676

Hugh N. Rocks
216 W. Second St
Waynesboro, PA 17268 (717) 762-4021

Robert B. Frailey
498 Eaton Way
West Chester, PA 19380 (215) 692-9825

Roger R. Horton
1105 Ashford Ln.
West Chester, PA 19382 (215) 399-0172

A. Newton Huff
407 Eaton Way
West Chester, PA 19380 (215) 431-0167

Wayne C. Jaeschke
1104 General Lafayette Dr.
West Chester, PA 19380 (215) 793-3545

Marvin J. Powell
1157 Mallard Rd.
West Chester, PA 19382 (215) 436-5144

Laurence A. Weinberger
1414 Morstein Rd.
West Chester, PA 19380 (215) 431-1703

Lewis J. Young
1202 Clearbrook Rd.
West Chester, PA 19380 (215) 696-1076

Anthony J. Dixon
601 N. Broad St.
West Hazleton, PA 18201 (717) 455-7112

Floyd S. Scheier
391 McKinney Rd.
Wexford, PA 15069 (412) 935-6746

Ruth Moyerman
Steele, Gould & Fried
514 Fullerton Ave.
Whitehall, PA 18052 (215) 432-3000

George P. Baier
American Standard, Inc.
P.O. Box 67
Wilmerding, PA 15148 (412) 825-1020

B. Max Klevit
B 221 Cedarbrook Hill Apts.
Wyncote, PA 19095 (215) TU6-9146

James Fay Hall, Jr.
926 Remington Rd.
Wynnewood, PA 19096 (215) 642-1610

E. Barron Batchelder
120 Valley Greene Circle
Wyomissing, PA 19610 (215) 374-2277

Donald D. Denton
2200 Yardley Rd.
Yardley, PA 19067 (215) 493-1532

Henry H. Gage
3 Sutphin Pines
Yardley, PA 19067 (215) 736-2496

Daniel E. Kramer
2009 Woodland Dr.
Yardley, PA 19067 (215) 493-4280

Paul B. Weisz#
3 Delaware Rim Dr.
Yardley, PA 19067 (215) 493-3551

Edward J. Hanson, Jr.
Dentsply International Inc.
570 W. College Ave.
York, PA 17405 (717) 845-7511

Charles H. Just
1933 Woodstream Dr.
York, PA 17402 (717) 755-5301

Samuel M. Learned, Jr.
149 East Market St.
York, PA 17401 (717) 846-9290

Charles J. Long
Smith Lecates & Campbell
124 E. Market St.
York, PA 17401 (717) 845-9641

Daniel J. O Connor
52 S. Duke St.
York, PA 17401 (717) 845-6593

Sidney N. Rosenfeld
Borg - Warner Corp. Air Conditioning Group
P.O. Box 1592
York, PA 17405 (717) 771-7432

Richard G. Weber
Precision Components Corp.
Box M - 101
500 Lincoln St.
York, PA 17405 (717) 848-1126

David E. Wheeler
Dentsply International Inc.
570 West College Ave.
York, PA 17405 (717) 845-7511

Rhode Island

Daniel A. Curran
P.O. Box 432
Adamsville, RI 02801 (401) 635-8519

Louis S. Coppolino#
21 Houghton St.
Barrighton, RI 02806 (401) 246-1788

Robert J. Doherty
10 George St.
Barrington, RI 02806 (401) 431-1320

Herbert B. Barlow, Jr.
Barlow & Barlow, Ltd.
1150 New London Ave.
Cranston, RI 02920 (401) 463-6830

Kurt R. Benson
Salter & Michaelson
Heritage Bldg., 321 South Main St.
Providence, RI 02903 (401) 421-3141

H. Edward Foerch
I T T Grinnell Corp.
260 W. Exchange St.
Providence, RI 02901 (401) 831-7000

Ralph Douglas Gelling
Textron Inc.
40 Westminster St.
Providence, RI 02903 (401) 457-2320

David D. Mc Kenney
I T T Grinnell Corp.
260 W. Exchange St.
Providence, RI 02901 (401) 831-7000

Leonard Michaelson
Salter & Michaelson
Heritage Bldg. 321 South Main St.
Providence, RI 02903 (401) 421-3141

Elliot A. Salter
Salter & Michaelson
Heritage Bldg. 321 S. Main St.
Providence, RI 02903 (401) 421-3141

Albert P. Davis
2080 Boston Neck Rd.
Saunderstown, RI 02874 (401) 294-3410

Burnett W. Norton
1600 Division Rd.
West Warwick, RI 02893 (401) 884-9920

South Carolina

William A. Callahan#
Route 1 Box A-142
Aiken, SC 29801 (803) 652-3454

Allen F. Westerdahl*
U.S. Energy R. & D. Admin
Savannah River Oper. Off. P.O. Box A
Aiken, SC 29801 (803) 725-2497

Alfons G. Hutter
337 Lake Ridge Lane
Anderson, SC 29624 (803) 225-9828

William S. Rambo
R.R. 4, Providence Villas
Anderson, SC 29624 (803) 224-9667

Warley L. Parrott
P.O. Box 594
Camden, SC 29020

William A. Dallis, Jr.
Dallis & Dreyfoos
124 Meeting St. P.O. Box 1840
Charleston, SC 29402 (803) 577-9425

Billy C. Killough
14 N. Adgers Wharf
Charleston, SC 29401 (803) 577-9890

Larry Harold Kline
18 Broad St. Suite 805
Charleston, SC 29401 (803) 577-7715

James Spool
Siebe North, Inc.
4090 Azalea Dr.
Charleston, SC 29405 (803) 554-0660

John Hamilin Roberts#
4833 Holbird Dr.
Charleston Hgts., SC 29405 (803) 744-7587

Donald H. Feldman#
210-1 Cochran Rd.
Clemson, SC 29631 (803) 654-5483

Davis T. Moorhead#
#4 Holiday East
Clemson, SC 29631 (803) 654-7501

F. Rhett Brockington#
4016 Mac Gregor Dr.
Columbia, SC 29206 (803) 787-7922

Mark C. Dukes#
2730 Kiawah Ave.
Columbia, SC 29205 (803) 256-2029

Timothy D. Harbeson
1205 Pendleton St.
Columbia, SC 29201 (802) 734-0457

Benoni O. Reynolds
P.O. Drawer 6924, 5219 Trenholm Rd. Suite 101
Columbia, SC 29260 (803) 782-9449

Edgar E. Ruff#
2809 Wales Rd.
Columbia, SC 29206 (803) 788-6659

Frank L. Valenta, Jr.
Capitol Complex
P.O. Box 11549, 1000 Assembly St.
Columbia, SC 29211 (803) 758-2697

Benjamin G. Weil
P.O. Box 48
Darlington, SC 29532 (803) 395-1875

William D. Lee, Jr.
W.R. Grace & Co. Cryovac Div.
P.O. Box 464
Duncan, SC 29334 (803) 433-2334

Mark B. Quatt
W.R. Grace & Co. Cryovac Div.
P.O. Box 464
Duncan, SC 29334 (803) 433-2817

John J. Toney
W.R. Grace & Co. Cryovax Div.
Hwy. 290 & I - 85
Duncan, SC 29334 (803) 433-2332

Wesley L. Brown
Saint-Amand, Thompson & Brown
210 S. Limestone St. P.O. Box 936
Gaffney, SC 29340 (803) 489-6052

James M. Bagarazzi
Dority & Manning
700 E. North Street Suite 15
Greensville, SC 29601 (803) 271-1592

Ralph Bailey
Bailey & Hardaway
125 Broadus Ave.
Greenville, SC 29601 (803) 233-1338

Julian W. Dority & Manning
700 E. North St.
Greenville, SC 29601 (803) 271-1592

Thomas W. Epting
10 Whittington Dr.
Greenville, SC 29615 (803) 297-1362

Cort R. Flint, Jr.
P.O. Box 10827, Federal Station, 501 E. Mc Bee Ave.
Greenville, SC 29601 (803) 232-4261

John B. Hardaway, III
Bailey & Hardaway
125 Broadus Ave.
Greenville, SC 29601 (803) 233-1338

Bryan F. Hickey
Haynsworth, Marion, Mckay & Guerard
75 Beattie Place, C & S Tower, 11th Floor
Greenville, SC 29601 (803) 240-3246

Martin K. Lindemann#
102 Independence Dr.
Greenville, SC 29615 (803) 288-4799

Wellington M. Manning, Jr.
Dority & Manning
700 E. North St. Suite 15
Greenville, SC 29601 (803) 271-1592

Ralph M. Mellom
Ogletree, Deakins, Nash, Smoak & Stewart
P.O. Box 2757, 1000 E. North St.
Greenville, SC 29602 (803) 242-1410

Richard M. Moose
Dority & Manning
700 E. North St. Suite 15
Greenville, SC 29601 (803) 271-1592

Roger F. Bley
103 Amhurst Dr.
Greenwood, SC 29646 (803) 223-0565

C. Gordon Mc Bride
P.O. Box 966, 644 S. Fourth St.
Hartsville, SC 29550 (803) 332-0193

James K. Everhart, Jr.
43 N. Port Royal Dr.
Hilton Head, SC 29928 (803) 681-2189

Melvin C. Flint
59 S. Sea Pines Dr.
Hilton Head, SC 29928 (803) 671-3182

Willard M. Hanger
200 Professional Bldg.
Hilton Head Island, SC 29602 (803) 242-1410

Milford W. Mac Donald
28 Angel Wing Dr.
Hilton Head Island, SC 29928 (803) 785-5291

Raymond F. Mac Kay#
82 Crosstree Dr, N.
Windmill Harbour
Hilton Head Island, SC 29928

Mathew P. Mc Dermitt, Sr.
23 Rusty Rail Lane
Hilton Head Island, SC 29928

Wilbur C. Tupman
133 Devil S. Elbow Lane
Hilton Head Island, SC 29928 (803) 757-3036

Francis W. Crotty
2 Pine Point Rd.
Lake Wylie, SC 29710

Carl B. Fox, Jr.
P.O. Box 2907
Myrtle Beach, SC 29578 (803) 293-7777

H. Hume Mathews
303 Club Dr.
Myrtle Beach, SC 29577 (803) 449-7612

William j. Smith#
P.O. Box 814
Newberry, SC 29108 (803) 276-4946

Terry B. Mc Daniel
Westvaco Corp.
P.O. Box 5207, Virginia Ave.
North Charleston, SC 29406 (803) 745-3568

William R. Hovis
Hovis & Duncan
P.O. Box 10970
Rock Hill, SC 29731 (803) 324-1122

Howard G. Garner, Jr.
2019 Bethel Road
Simpsonville, SC 29681 (803) 963-9810

Edward L. Bailey
180 Library St.
Spartanburg, SC 29301 (803) 582-3733

Earle Rollins Marden
Milliken Research Corp.
P.O. Box 1927
Spartanburg, SC 29301 (803) 573-1599

Terry T. Moyer
Milliken Res, Corp.
P.O. Box 1927
Spartanburg, SC 29302 (803) 573-2266

Henry William Petry
Milliken Res. Corp.
P.O. Box 1927, Iron Ore Rd.
Spartanburg, SC 29304 (803) 573-2666

Luke John Wilburn, Jr.
364 S. Pine St. P.O. Box 5445
Spartanburg, SC 29304 (803) 585-6688

South Dakota

Ivar M. Kaardal
101 South Main, Suite 418
Sioux Falls, SD 57102 (605) 334-5898

Tennessee

L. Aubrey Goodson
6005 Old Jonesboro Rd.
Bristol, TN 37620 (615) 878-6356

David J. Hill
6400 Lee Highway
Chattanooga, TN 37422 (615) 892-7120

Douglas T. Johnson
Miller & Martin
Georgia Ave. Suite 1000
Chattanooga, TN 37402 (615) 756-6600

James L. Johnson
543 Pioneer Bank Bldg.
Chattanooga, TN 37402 (615) 266-5531

Alan Ruderman
806 Maclellan Bldg.
Chattanooga, TN 37402 (615) 267-6980

Stephen T. Belsheim
322 Main St. Suite 100
Clarksville, TN 37040 (615) 647-8337

Louis Milton Deckelmann#
303 Woodhaven Lane
Cliton, TN 37716 (615) 457-2491

Abram Wooldridge Hatcher
Nathan Smith Rd.
P.O. Box 112
College Grove, TN 37046 (615) 368-7256

Adrian J. Good
Great Lakes Res. Div.
P.O. Box 1031
Elizabethton, TN 37643 (615) 543-3111

Carl F. Peters
415 E. K Street
Elizabethton, TN 37643 (615) 543-5022

Robert L. Taylor#
P.O. Box 681
Elizabethton, TN 37644 (615) 926-8595

Remega G. Hyder#
Route 1 Box 228
Hampton, TN 37658 (615) 725-3482

David Schwendinger#
Route 2 Box 1620, Pine Ridge Dr.
Heiskel, TN 37754 (615) 457-8642

John P. Dority
104 Cherry Hill Dr.
Hendersonville, TN 37075 (615) 824-8068

William J. Hite, III
4234 Sweden Dr.
Hermitage, TN 37076 (615) 871-0950

Gary C. Bailey
Tennessee Eastman Co.
P.O. Box 511
Kingsport, TN 37662 (615) 229-4941

George P. Chandler
Tennessee Eastman Co.
Eastman Rd. P.O. Box 511
Kingsport, TN 37662 (615) 229-3620

Malcolm Graeme Dunn
Colonial Square
154 Cherokee St., P.O. Box 414
Kingsport, TN 37660 (615) 378-4403

William P. Heath, Jr.
Tennessee Eastman Co.
P.O. Box 511
Kingsport, TN 37662 (615) 229-3634

Charles R. Martin
3701 Hemlock Park
Kingsport, TN 37663 (615) 239-9975

Cecil D. Quillen, Jr.
Tenn. Eastman Kodak Co. Legal Dept.
P.O. Box 511
Kingsport, TN 37662 (615) 229-4305

Daniel B. Reece, III
Tenn. Eastman Co.
P.O. Box 1972
Kingsport, TN 37662 (615) 229-2097

Thomas R. Savitsky
Tenn. Eastman Co. Legal Dept Bldg 75
P.O. Box 511
Kingsport, TN 37662 (615) 229-4305

Martin J. Skinner#
836 Nelson Dr.
Kingsport, TN 37763 (615) 376-6894

Donald W. Spurrell
Tenn. Eastman Co.
P.O. Box 1972
Kingsport, TN 37662 (615) 229-2802

John Frederick Stevens
Tenn. Eastman Co. Div. Of Eastman Kodak Co.
P.O. Box 511
Kingsport, TN 37662 (615) 229-3618

John Frederick Thomsen
Tenn, Eastman Co.
Eastman Rd. P.O. Box 511
Kingsport, TN 37662 (615) 229-2282

Clyde L. Tootle
Tenn. Eastman Co. Div. Of Eastman Kodak Co.
P.O. Box 511
Kingsport, TN 37662 (615) 229-2094

Mark S. Graham
Luedeka, Hodges & Neely, P.C.
830 First American Center, 507 Gay St.
Knoxville, TN 37922 (615) 546-4305

Paul E. Hodges
Luedeka, Hodges & Neely, P.C.
830 First American Center
Knoxville, TN 37902 (615) 546-4305

James C. Kesterson
Pitts & Kesterson
1116 Weisgarber Rd.
Knoxville, TN 37919 (615) 584-0105

Geoffrey D. Kressin
Watson, Kressin & Erickson, P.C.
703 S. Gay St.
Knoxville, TN 37914 (615) 637-1809

Katherine Parks Lovingood
2242 Knollcrest Lane
Knoxville, TN 37920 (615) 573-8982

Edwin M. Luedeka
Luedeka & Neely, P.C.
830 First American Center, 507 Gay St.
Knoxville, TN 37902 (615) 546-4305

Andrew S. Neely
Luedeka & Neely, P.C.
830 First American Center, 507 Gay St.
Knoxville, TN 37902 (615) 546-4305

Robert E. Pitts
Pitts, Ruderman & Kesterson
P.O. Box 51295
Knoxville, TN 37950 (615) 584-0105

Douglas R. Scott
Robertshaw Controls Co. Fulton Sylphon Div.
P.O. Box 400
Knoxville, TN 37901 (615) 546-0550

William T. Dixson, Jr.#
111 Caldwell St.
Mc Minnville, TN 37110

Howard Roy Berkenstock, Jr.
Richards Medical Co.
1450 Brooks Rd.
Memphis, TN 38116 (901) 396-2121

Larry W. Mc Kenzie
Walker & Kenzie, P.C.
6363 Poplar, Suite 434
Memphis, TN 38119 (901) 685-7428

John J. Mulrooney
410 Brinkley Plaza, 80 Monroe Ave.
Memphis, TN 38103 (901) 526-7777

Loyal W. Murphy, III#
5050 Poplar Ave. Suite 1214
Memphis, TN 38157 (901) 767-4701

John Russell Walker, III
Walker & Mc Kenzie
6363 Poplar Ave. Suite 434
Memphis, TN 38119 (901) 685-7428

Jonny O. Younghanse#
4405 Ross Rd.
Memphis, TN 38115 (901) 362-8976

Harrington A. Lackey
4235 Hillsboro Rd. Suite 203
Nashville, TN 37215 (615) 292-5665

Lawrence C. Maxwell
Trabue, Sturdivant & Dewitt
life & Casualty Tower, Fourth & Church, 26th Floor
Nashville TN 37219 (615) 244-9270

Gary V. Pack
Service Merchandisce Co. Inc.
P.O. Box 24600, 1283 Murfresboro Rd.
Nashville, TN 37202 (615) 366-3215

Mark J. Patterson
Manier, White, Herod, Hollabaugh & Smith, P.C.
700 Metropolitan Federal Bldg. 230 Fourth Ave., N.
Nashville, TN 37219 (615) 244-0030

J. Bruce Synnott, Jr.
4960 Edmondson Pike Apt. S-4
Nashville, TN 37211 (615) 833-8558

Ira C. Waddey, Jr.
Waddey & Jennings
500 Church St.
Nashville, TN 37219 (615) 244-7545

Casey F. Wilson#
2203-A 25th Ave. South
Nashville, TN 37212 (615) 298-4038

Irving Barrack*
U.S. Dept. Of Energy Federal Off. Bldg.
P.O. Box E
Oak Ridge, TN 37831 (615) 576-1072

David E. Breeden*#
U.S. Dept. Of Energy Oak Ridge Oper. Off.
Admin. Road. P.O. Box E
Oak Ridge, TN 37831 (615) 576-1082

James D. Griffin#
Martin Marietta Energy Systems, Inc.
P.O. Box Y
Oak Ridge, TN 37849 (615) 574-4178

Stephen David Hamel*
U.S. Dept. Of Energy
P.O. Box E
Oak Ridge, TN 37830 (615) 576-1073

Herman L. Holsopple, Jr.#
102 Potomac Circle
Oak Ridge, TN 37830 (615) 482-1623

Earl L. Larcher*#
U.S. Dept. Of Energy
P.O. Box E.
Oak Ridge, TN 37830 (615) 576-1080

Fred O. Lewis*#
U.S. Dept. Of Energy
P.O. Box 3
Oak Ridge, TN 37830 (615) 576-1081

Robert Maxwell Poteat*
U.S. Dept. Of Energy
P.O. Box E
Oak Ridge, TN 37830 (615) 576-1070

James M. Spicer#
Martin Marietta Energy Systems Inc.
P.O. Box X
Oak Ridge, TN 37831 (615) 574-4180

Carmack Waterhouse
108 Orange Ln.
Oak Ridge, TN 37870 (901) 483-4870

Bruce M. Winchell
Martin Marietta Energy Systems Inc.
P.O. Box Y
Oak Ridge, TN 37831 (615) 576-6885

David S. Zachry, Jr.
874 W. Outer Dr.
Oak Ridge, TN 37830 (615) 483-6725

Vivian G. Hyder#
P.O. Box 147
Rogersville, TN 37857 (615) 345-3950

Texas

F. Vern Lahart
D L M, Inc
P.O. Box 3000
Allen, TX 75002 (214) 248-6300

Michael Anthony Sileo, Jr.
Texet Corp.
301 Texet Dr.
Allen, TX 75002 (214) 727-1111

Russell J. Egan
Intermedics Inc.
4000 Technology Dr.
Angleton, TX 77515 (409) 848-4073

Frederick S. Frei
Snider & Moore
505 Ryan Plaza Dr. Suite 337
Arlington, TX 76011 (817) 861-1122

Leroy Franklin Halley
6216 Lake Ridge Rd.
Arlington, TX 76016 (817) 451-6192

Charles William Mc Hugh
311 W. Abram. Suite 100
Arlington, TX 76010 (817) 461-3113

Peggy L. Smith#
4814 Tamanaco Court
Arlington, TX 76017 (817) 467-9512

Andrea P. Bryant
I B M Corp. Dept. 932
11400 Burnet Rd.
Austin, TX 78758 (512) 838-1003

Robert M. Carwell
I B M Corp.
Intellectual Property Law 932/815
11400 Burnet Rd.
Austin, TX 75758 (512) 823-1017

James L. Clingan, Jr.
Motorola, Inc.
3501 Ed Bluestein Blvd.
Austin, TX 78721 (512) 928-7004

Dudley R. Dobie
Fulbright & Jaworski
600 Congress Ave. Suite 2400
Austin, TX 78701 (512) 474-5210

John A. Fisher
Motorola Inc. Dept F-4
3501 Ed Bluestein Blvd.
Austin, TX 78721 (512) 928-7002

Richard J. Gross
Arnold, White & Durkee
2300 One American Center, 600 Congress Ave.
Austin, TX 78701 (512) 474-2580

Curtis A. Henschen*
Federal Bureau Of Investigation
300 East 8th
Austin, TX 78767 (512) 478-8501

Daniel S. Hodgins
Arnolds, White & Durkee
2300 One American Center
Austin, TX 78701 (512) 474-2583

Charles D. Huston
Arnold, White & Durkee
2300 One American Center, 600 Congress Ave.
Austin, TX 78701 (512) 474-2583

John Lionel Jackson
I B M Corp. Intellectual Property Law Dept. 932
11400 Burnet Rd.
Austin, TX 78758 (512) 823-1000

Jerry M. Keys
Brown, Maroney, Rose, Barber & Dye
1300 One Republic Plaza, 333 Guadalupe
Austin, TX 78701 (512) 472-5456

Robert L. King
Motorola, Inc.
3501 Ed Bluestein Blvd.
Austin, TX 78721 (512) 928-7001

Thomas E. Kirkland
9200 Quail Hill Circle
Austin, TX 78758 (512) 339-7278

Julius B. Kraft
I B M Corp. Austin Pat. Oper.
11400 Burnet Rd.
Austin, TX 78759 (512) 838-3300

Douglas H. Lefeve
I B M Corp. Dept. 932
11400 Burnet Rd.
Austin, TX 78758 (512) 838-9302

Joseph F. Long
8912 Laurel Grove
Austin, TX 78758 (512) 835-0880

Jonathan P. Meyer
Motorola Inc. Pat. Dept.
3501 Ed Bluestein Blvd., Mail Stop F7
Austin, TX 78721 (512) 928-7002

David L. Mossman
Motorola, Inc.
3501 Ed Bluestein Blvd.
Austin, TX 78721 (512) 928-6570

Jeffrey V. Myers
Motorola, Inc.
3501 Ed Bluestein Blvd.
Austin, TX 78721 (512) 928-7003

Floyd R. Nation
Arnold, White & Durkee
1440 Texas Comm. Bank Bldg. 700 Lavaca St.
Austin, TX 78701 (512) 474-2583

Louis T. Pirkey
Arnold, White & Durkee
1440 Texas Comm. Bank Bldg. 700 Lavaca St.
Austin, TX 78701 (512) 474-2583

Kenneth R. Priem
10305 Mourning Dove Circle
Austin, TX 78750 (512) 258-9031

William D. Raman
Arnold, White & Durkee
1440 Texas Comm. Bank Bldg. 700 Lavaca St.
Austin, TX 78701 (512) 474-2583

Ernst William Schultz
8106 Middle Court
Austin, TX 78759 (512) 345-2453

John N. Shaffer, Jr.#
Bannerot, Shaffer & Armstrong
1250 Capital Of Texas Hwy. S., Bldg. Two, Suite 560
Austin, TX 78746 (512) 327-8930

Marilyn D. Smith
I B M Corp. Pat. Oper. 932/815
11400 Burnet Rd.
Austin, TX 78758 (512) 823-1005

Horace St. Julian
I B M Corp.
11400 Burnet Rd.
Austin, TX 78758 (512) 823-1014

Thomas E. Tyson
I B M Corp. Pat. Oper. 932/815
11400 Burnet Rd.
Austin, TX 78758 (512) 823-1004

Joseph F. Villella, Jr.
I B M Corp. Pat. Oper. 932/815
11400 Burnet Rd.
Austin, TX 78758 (512) 823-1014

Ben C. Cadenhead
Exxon Chemical Co.
P.O. Box 5200
Baytown, TX 77522 (713) 425-5401

Wayne Hoover
P.O. Box 1463
Baytown, TX 77522 (713) 420-1358

John F. Hunt
Exxon Chemical Co.
P.O. Box 5200
Baytown, TX 77522 (713) 425-2400

Myron Bernard Kurtzman
Exxon Chem. Co. Pat. Law Dept.
P.O. Box 5200
Baytown, TX 77522 (713) 428-0327

Steven H. Markowitz
Exxon Chem, Co. Baytown Polymers Center
P.O. Box 5200
Baytown, TX 77525 (713) 425-5954

Terry B. Morris
Exxon Chem. Co.
P.O. Box 5200
Baytown, TX 77522 (713) 425-1355

Charles E. Smith
Exxon Chemical Co. Baytown Polymers Center
P.O. Box 5200
Baytown, TX 77522 (713) 425-1354

Eugene T. Wheelock
Exxon Chem. Co. Baytown Polymers Ctr.
P.O. Box 5200
Baytown, TX 77522 (713) 425-5202

William Earl Kinnear
1810 Washington Blvd.
Beaumont, TX 77705 (713) 833-6386

Alan J. Atkinson
4409 Acacia
Bellaire, TX 77401 (713) 661-2177

James L. Bailey
Texaco Dev. Corp.
P.O. Box, 4800 Fournace Place
Bellaire, TX 77401 (713) 432-2629

Bill B. Berryhill
5001 Bissonnet, Suite 100
Bellaire, TX 77401 (713) 661-3305

Harold J. Delhommer
Texaco Dev. Co.
P.O. Box 430
Bellaire, TX 77401 (713) 432-2638

Al Harrison
6565 W. Loop South Suite 710
Bellaire, TX 77401 (713) 664-1778

Kenneth H. Johnson
5555 W. Loop South, Suite 452
Bellaire, TX 77401 (713) 664-6982

Cynthia L. Kendrick
Texaco Dev. Corp.
P.O. Box 430, 4800 Fournace Place
Bellaire, TX 77401 (713) 432-2623

Richard Albert Morgan
Texaco Dev. Corp.
P.O. Box 430, 4800 Fournace Place
Bellaire, TX 77401 (713) 432-2624

Jack H. Park
Texaco Dev. Corp.
P.O. Box 430, 4800 Fournace Place
Bellaire, TX 77401 (713) 432-2620

Roy H. Smith, Jr.
6565 W. Loop South Apt. 710
Bellaire, TX 77401 (713) 664-1778

James J. Smolen
5959 W. Loop South, Suite 500
Bellaire, TX 77401 (713) 432-0681

Delmar L. Sroufe
Sroufe & Payne
4710 Bellaire Blvd. Suite 230
Bellaire, TX 77401 (713) 666-2288

Harold Herbert Flanders
J.M. Huber Corp.
P.O. Box 2831, 1100 Penn
Borger, TX 79007 (806) 274-6331

Alec H. Horn
J.M. Huber Corp.
1100 Penn Ave.
Borger, TX 79007 (806) 274-6331

Glwynn Robinson Baker
5410 C.R. 510, Route 5
Brazoria, TX 77422 (409) 798-4839

Melvin W. Barrow
304 West Texas St.
Brazoria, TX 77422 (409) 758-7580

Carlos A. Torres
P.O. Box 756, 2131 Peachridge Rd.
Brookshire, TX 77423 (713) 391-7300

Robert S. Nisbett
P.O. Box 1131, 478 W. Wellington
Carthage, TX 75633 (214) 693-3747

Sidney T. Walker
35 Robinhood
Clute, TX 77531 (409) 265-2901

Donald Raymond Cassady
Celanese Corp.
P.O. Box 9077
Corpus Christi, TX 78469 (512) 241-2343

G. Turner Moller
Guaranty Bank Plaza
Corpus Christi, TX 78401 (512) 883-7257

Ralph M. Prichett#
Celanese Chem. Co.
P.O. Box 9077
Corpus Christi, TX 78408 (512) 241-2343

Stewart N. Rice
Wood, Boykin, Wolter, Smith & Hatridge
2000 First City Bank Tower, F C B 249
Corpus Christi, TX 78477 (512) 888-9201

John K. Abokhair
Cosden Tech. Inc.
P.O. Box 410
Dallas, TX 75221 (214) 750-2585

Richard E. Aduddell
Kanz, Scherback & Timmons
South Tower- Suite 1030, Plaza Of The Americas
Dallas, TX 75201 (214) 969-7576

Garland Paul Andrews
Richard, Harris, Medlock & Andrews
1201 Elm St. Suite 4500, Renaissance Tower
Dallas, TX 75270 (214) 939-4500

Richard A. Bachand
Texas Instrument, Inc. Pat. Dept. M/s 219
P.O. Box 655474
Dallas, TX 75265 (214) 995-5491

Alva Harlan Bandy
Hubbard, Thurman, Turner & Tucker
2100 One Galleria Tower
Dallas, TX 75240 (214) 233-5712

Robert M. Betz
12660 Hillcrest Rd. Apt. 4101
Dallas, TX 75230 (214) 991-2813

Terry M. Blackwood
Rockwell International Corp.
P.O. Box 10462
Dallas, TX 75207 (214) 996-6494

Arthur C. Boos
Johnson, Bromberg & Leeds
2600 Lincoln Plaza
Dallas, TX 75201 (214) 740-2782

John F. Booth, Sr.
Crutsinger & Booth
1000 Thanksgiving Tower, 1601 Elm St.
Dallas, TX 75201 (214) 741-4484

Robert G. Boydston
1030 Frito-Lay Tower
Dallas, TX 75235 (214) 350-7741

Stanton Connell Braden
Texas Instruments M S 219
P.O. Box 655474
Dallas, TX 75265 (214) 995-5921

James P. Bradley
Richards, Harris, Medlock & Andrews
1201 Elm St. Suite 4500 - Renaissance Tower
Dallas, TX 75270 (214) 939-4500

Randall C. Brown
Hubbard, Thurman, Turner & Tucker
2100 One Galleria Tower
Dallas, TX 75240 (214) 233-5712

Michael J. Caddell
Fina Oil & Chemical Co.
8350 N. Central Expwy
Dallas, TX 75206 (214) 750-2888

Thomas Lee Cantrell
Schley, Catrell, Kice, Garland & Moore
5001 L B J Freeway, Suite 705
Dallas, TX 75244 (214) 387-3804

Gregory W. Carr
Gardere & Wynne
1500 Diamond Shamrock Tower
Dallas, TX 75201 (214) 979-4542

Albert W. Carroll#
3008 Primrose Lane
Dallas, TX 75234 (214) 247-2738

Roger N. Chauza
Baker, Smith & Mills P.C.
2001 Roos Ave, 500 L T V Center
Dallas, TX 75201 (214) 220-8287

M. Norwood Cheairs
Amer. Petrofina Co. Of Tex
8350 North Central Expwy.
Dallas, TX 75221 (214) 750-2532

Robert D. Clamon#
Dresser Inds. Inc.
P.O. Box 718
Dallas, TX 75221 (214) 740-6508

Roger C. Clapp
Richard, Harris, Medlock & Andrews
1201 Elm St. Suite 4500, Renaissance Tower
Dallas, TX 75270 (214) 939-4500

Alan B. Clay#
Richards, Harris, Medlock & Andrews
1201 Elm St. Suite 4500, Renaissance Tower
Dallas, TX 75270 (214) 939-4500

William B. Clemmons, Jr.
Hubbard, Thurman, Turner & Tucker
2100 One Galleria Tower
Dallas, TX 75240 (214) 233-5712

James T. Comfort
Texas Instruments, Inc.
P.O. Box 655,474 M/s 219, 13510 N. Central Expressway
Dallas, TX 75243 (214) 995-4400

John M. Cone
Strasburger & Price
4300 Inter First Plaza, 901 Main St.
Dallas, TX 75202

Charles S. Cotropia
Richards, Harris, Medlock & Andrews
1201 Elm St. Suite 4500, Renaissance Tower
Dallas, TX 75270 (214) 939-4500

Roland O. Cox#
Otis Engineering Corp.
P.O. Box 819052
Dallas, TX 75381 (214) 323-3883

George L. Craig
Texas Instruments Inc.
P.O. Box 655474, MS 219. 13500 N. Central Expwy
Dallas, TX 75265

Thomas Lynn Crisman
Johnson & Swanson
100 Founders Square, 900 Jackson St.
Dallas, TX 75202 (214) 977-9614

Morgan L. Crow#
Dresser Industries Inc.
P.O. Box 24647
Dallas, TX 75224 (214) 333-3211

Albert M. Crowder, Jr.
E-Systems, Inc.
6250 L B J Freeway
Dallas, TX 75240 (214) 661-1000

Gerald G. Crutsinger, Sr.
Crutsinger & Booth
1000 Thanksgiving Tower
Dallas, TX 75201 (214) 741-4484

Jennifer R. Daunis
Hubbard, Thurman, Turner & Tucker
2100 One Galleria Tower
Dallas, TX 75240 (214) 233-5712

Thomas George Devine
Texas Instruments, Inc.
P.O. Box 655474, M/s 219
Dallas, TX 75226 (214) 995-5315

James Owen Dixon
Crutsinger & Booth
1000 Thansgiving Tower
Dallas, TX 75201 (214) 741-4484

Richard L. Donaldson
Texas Instruments Inc.
P.O. Box 225474
Dallas, TX 75265 (214) 995-5921

Terrence Dean Dreyer
Frito- Lay Inc.
P.O. Box 660634
Dallas, TX 75266 (214) 353-3814

Larry B. Dwight
4350 Beltway Dr.
Dallas, TX 75244 (214) 991-2222

Robert H. Falk
Hubbard, Thurman, Turner & Tucker
2100 One Galleria Tower
Dallas, TX 75240 (214) 233-5715

Thomas R. Felger
Otis Eng. Corp.
P.O. Box. 819052
Dallas, TX 75381 (214) 418-3882

Edward Gerald Fiorito
Dresser Industries, Inc.
1600 Pacific Ave.
Dallas, TX 75201 (214) 740-6901

Thomas R. Fitzgerald
Texas Instruments
P.O. Box 655474, M S 219
Dallas, TX 75265 (214) 995-1370

Michael David Folzenlogen
6774 Inverness
Dallas, TX 75214 (214) 823-5599

Gene Wilgus Francis, Jr.
Brice & Mankoff
300 Crescent Court 7th Floor
Dallas, TX 75201 (214) 855-3700

H. Mathews Garland
Johnson & Swanson
100 Founders Square, 900 Jackson St.
Dallas, TX 75202 (214) 740-6508

William E. Geyer
1015 Elm St. Suite 2100
Dallas, TX 75202 (214) 748-8187

Kenneth Roy Glaser
Glaser, Griggs & Schwartz
5430 L B J Freeway, Suite 1540
Dallas, TX 75240 (214) 770-2400

David C. Godbey
Hughes & Luce
2800 Momentum Place, 1717 Main St.
Dallas, TX 75201 (214) 939-5581

Lowell W. Gresham
Baker, Mills & Glast, P.C.
500 L T V Center, 2001 Ross Ave.
Dallas, TX 75201 (214) 220-8503

Dennis T. Griggs
Glaser, Griggs & Schwartz
5430 L B J Freeway, Suite 1540
Dallas, TX 75240 (214) 770-2400

Stephen A. Grimmer
Haynes & Boone
3100 First Republic Bk Plaza, 900 Main
Dallas, TX 75202 (214) 670-0550

Rene E. Grossman
Texas Instruments Inc.
P.O. Box 655474, Ms 219
Dallas, TX 75265 (214) 995-1345

Norman L. Gundel
Crutsinger & Booth
1000 Thanksgiving Tower, 1601 Elm St.
Dallas, TX 75201 (214) 220-0444

William R. Gustavson
Richards, Harris, Medlock & Andrews
4500 Renaissance Tower, 1201 Elm St.
Dallas, TX 75270 (214) 939-4510

Alfred E. Hall
Gardere & Wynne
717 North Harwood, Suite 1500
Dallas, TX 75201 (214) 979-4689

Herbert J. Hammond
Gardere & Wynne
717 N. Harwood Suite 1500
Dallas, TX 75201 (214) 979-4612

Eugenia S. Hansen
Richards, Harris, Medlock & Andrews
4500 Renaissance Tower, 1201 Elm St.
Dallas, TX 75270 (214) 939-4500

Roy W. Hardin
Richards, Harris, Medlock & Andrews
4500 Renaissance Tower, 1201 Elm St.
Dallas, TX 75270 (214) 939-4508

William David Harris, Jr.
Richards, Harris, Medlock & Andrews
4500 Renaissance Tower
1201 Elm St.
Dallas, TX 75270 (214) 939-4500

Andrew M. Hassell
Sigalos & Levine
2700 First Republic Bank Tower
Dallas, TX 75201 (214) 953-1420

John N. Hazelwood
6616 Rolling Vista Dr.
Dallas, TX 75248 (214) 661-2788

Leo N. Heiting#
Texas Instrument Inc.
P.O. Box 655474 M.S. 219
Dallas, TX 75265 (214) 995-5493

Kenneth C. Hill
Texas Instruments Inc.
Box 665474, M S 219
Dallas, TX 75265 (214) 995-5492

William Eugene Hiller
Texas Instrument, Inc. M.S. 219
13500 N. Central Expwy.
Dallas, TX 75231 (214) 995-1364

David L. Hitchcock
Richard, Harris, Medlock & Andrews
4500 Renaissance Tower, 1201 Elm St.
Dallas, TX 75270 (214) 939-4511

Carlton H. Hoel
Texas Instruments Inc.
13500 N. Central Expwy.
Dallas, TX 75265 (214) 995-1349

John M. Holland
Hubbard, Thurman, Turner & Tucker
2100 One Galleria Tower
Dallas, TX 75240 (214) 233-5712

Gary C. Honeycutt
Texas Instruments, Inc.
P.O. Box 655474, M.S. 219
Dallas, TX 75265 (214) 995-1363

E. Mickey Hubbard
Hubbard, Thurman, Turner, Tucker & Glaser
1200 N. Dallas Bk. Tower, Lbj Freeway At Preston Rd.
Dallas, TX 75230 (214) 233-5712

Marc A. Hubbard#
Hubbard, Thurman, Tuner & Tucker
2100 One Galleria Tower
Dallas, TX 75240 (214) 233-5712

Carl D. Hughes, Jr.
900 Jackson, Suite 500
Dallas, TX 75202 (214) 761-9342

William D. Jackson
Richards, Harris, Medlock & Andrews
4500 Renaissance Tower
Dallas, TX 75270 (214) 939-4500

Jonathan E. Jobe, Jr.
Hubbard, Thurman, Turner & Tucker
2100 One Galleria Tower
Dallas, TX 75240 (214) 233-5712

David L. Joers
Crutsinger & Booth
1000 Thanksgiving Tower, 1601 Elm St.
Dallas, TX 75201 (214) 220-0444

David H. Judson
Hughes & Luce
2800 Momentum Place
Dallas, TX 75201 (214) 939-5672

Jack A. Kanz
Kanz & Timmons
South Tower-South 2230 Plaza
Dallas, TX 75201 (214) 742-3022

H. Dennis Kelly
Gardere & Wynne
1500 Diamond Shamrock Tower
Dallas, TX 75201 (214) 979-4858

Martin Korn
Richards, Harris, Medlock & Andrews
Renaissance Tower, 1201 Elm St. Suite 4500
Dallas, TX 75270 (214) 939-4500

Ronald R. Kranzow
Frito-Lay, Inc.
P.O. Box 35034
Dallas, TX 75235 (214) 351-7589

Robert Elsworth Lee, Jr.
Atlantic Richfield Co.
P.O. Box 2819
Dallas, TX 75221 (214) 422-3397

David N. Leonard
Leonard & Lott, Three Lincoln Centre
5430 L B J Freeway. Suite 875
Dallas, TX 75240

Harold Levine
Sigalos & Levine
1300 Republic Bk, Tower
Dallas, TX 75201 (214) 745-1751

Alan W. Lintel
Baker, Smith & Mills
2001 Ross Ave. Suite 500
Dallas, TX 75201 (214) 220-8285

John M. Lorenzen
Dresser Inds. Inc.
P.O. Box 718
Dallas, TX 75221 (214) 746-6967

Robert D. Lott
Leonard & Lott, Three Lincoln Centre
5430 L B J Freeway, Suite 875
Dallas, TX 75240 (214) 960-7447

George R. Love
Cowles, Sorrells, Patterson & Thompson
1800 One Main Place
Dallas, TX 75250 (214) 747-8291

Hulit L. Madinger
4207 Cobblers Lane
Dallas, TX 75252

Raymond T. Majesko
Dresser Inds. Inc.
1505 Elm
Dallas, TX 75201 (214) 746-6916

David L. Mc Combs
Hubbard, Thurman, Turner & Tucker
2100 One Galleria Tower
Dallas, TX 75240 (214) 233-5712

William K. Mc Cord
Hubbard, Thurman, Turner &
1200 N. Dallas Bank Turner &
Dallas, TX 75230 (214) 233-5712

Nina L. Medlock
Baker, Miller, Mills & Murray
1000 Pacific Place, 1910 Pacific Ave.
Dallas, TX 75201 (214) 969-0300

Virgil Bryan Medlock, Jr.
Richards, Harris, Medlock & Andrews
Renaissance Tower, 1201 Elm St. Suite 4500
Dallas, TX 75270 (214) 939-4500

Harold Eugene Meier
Gardere & Wynne
1500 Diamond Shamrock Tower
Dallas, TX 75201 (214) 748-7211

Norman R. Merrett
Texas Instruments Inc.
13500 N. Central Expwy. M/s 219V
Dallas, TX 75265 (214) 995-5491

Albert C. Metrailer
Atlantic Richfield Co. Pat. Dept.
P.O. Box 2819
Dallas, TX 75221 (214) 422-3340

Jerry Woodrow Mills
Baker, Smith & Mills, P.C.
500 L T V Center, 2001 Ross Ave.
Dallas, TX 75201 (214) 220-8200

Samuel M. Mims, Jr.
Texas Instruments Inc.
13500 N. Central Expy.
Dallas, TX 75265 (214) 995-4292

Michael Molins
Richards, Harris, Medlock & Andrews
4500 Renaissance Tower, 1201 Elm St.
Dallas, TX 75270 (214) 939-4518

John W. Montgomery
Sigalos & Levine, P.C.
1300 Republic Bank Tower
Dallas, TX 75201 (214) 745-1751

Mark A. Montgomery#
Cosden Technology, Inc.
P.O. Box 410
Dallas, TX 75221 (214) 750-2971

Stanley R. Moore
Johnson & Swanson
100 Founders Square, 900 Jackson St.
Dallas, TX 75202 (214) 977-9616

Peter J. Murphy
3000 South Center
Dallas, TX 75201 (214) 220-9066

P. Weston Musselman, Jr.
Jenkens & Gilchrist
3200 Allied Bank Tower
Dallas, TX 75202 (214) 855-4764

Dale B. Nixon
Richards, Harris, Medlock & Andrew
Renaissance Tower, 1201 Elm St. Suite 4500
Dallas, TX 75270 (214) 939-4500

Michael A. O Neil
Gardere & Wynne
1500 Diamond Shamrock Tower
Dallas, TX 75201 (214) 748-7211

James D. Olsen
Sun Exploration & Production Co.
P.O. Box 2880
Dallas, TX 75221 (214) 890-6090

John L. Palmer, Sr.
Eastman Kodak Co.
6300 Cedar Spring Rd.
Dallas, TX 75235 (214) 353-4551

William R. People
Dresser, Inds. Inc.
1505 Elm St.
Dallas, TX 75201 (214) 746-6910

Jefferson F. Perkins
Baker, Smith & Mills, P.C.
500 L T V Center, 2001 Ross Ave.
Dallas, TX 75201 (214) 979-9030

Loren W. Peters
Boyd & Du Bose
2001 Bryan Tower, Suite 2700
Dallas, TX 75201 (214) 922-0099

James A. Phillips#
Sunoco Energy Dev. Co.
12700 Park Central Place
Dallas, TX 75251

John P. Pinkerton
Hubbard, Thurman, Turner & Tucker
13355 Noel Rd. 2100 One Galleria Tower
Dallas, TX 75240 (214) 233-5712

Harry C. Post, III
2001 Bryan Tower, Suite 771
Dallas, TX 75201

J. Hughes Powell, Jr.
3315 Hanover St.
Dallas, TX 75225 (214) 361-6242

D. Carl Richards
Richards, Harris, Medlock & Andrews
Renaissance Tower, 1201 Elm St. Suite 4500
Dallas, TX 75270 (214) 939-4500

Jonathan W. Richards
Thompson & Knight
3300 First City Center, 1700 Pacific Ave.
Dallas, TX 75201 (214) 969-1700

Richard K. Robinson
Texas Instruments Inc.
P.O. Box 225474, M S. 219
Dallas, TX 75265 (214) 995-5491

Daniel Rubin
Three Lincoln Centre
5430 L B J Freeway, Suite 1540
Dallas, TX 75240 (214) 770-2400

Stephen S. Sadacca
L T V Corp.
P.O. Box 225090
Dallas, TX 75265 (214) 266-1908

Joseph A. Schaper
Richards, Harris, Medlock & Andrews
Renaissance Tower, 1201 Elm St. Suite 4500
Dallas, TX 75270 (214) 939-4500

William John Scherback
Kanz, Scherback & Timmons
South Tower, Suite 2230 Plaza Of The Americas
Dallas, TX 75201 (214) 696-7576

Larry C. Schroeder
Texas Instruments Inc.
P.O. Box 655474, Ms 219
Dallas, TX 75265 (214) 995-1365

Richard L. Schwartz
Glaser, Griggs & Schwartz, Three Lincoln Centre
5430 L B J Freeway, Suite 1540
Dallas, TX 75244 (214) 770-2400

Jerry T. Selinger
Baker, Smith & Mills
500 L T V Center, 2001 Ross Ave.
Dallas, TX 75201 (214) 979-9030

V. Lawrence Sewell
Rockwell Internat. Corp.
M/s 407-111
P.O. Box 10462
Dallas, TX 75207 (214) 996-2656

Melvin Sharp
Texas Instruments
P.O. Box 225474
Dallas, TX 75265 (214) 995-5865

John Louis Sigalos
Sigalos & Levine
1300 Republic Natl. Bk. Tower
Dallas, TX 75201 (214) 745-1751

Harry E. Simpson#
Dresser Industries Inc.
P.O. Box 24647, Security Div.
Dallas, TX 75224 (214) 333-3211

Arthur M. Sloan
2828 Forest Lane, Suite 1069
Dallas, TX 75234 (214) 241-9343

Douglas A. Sorensen
Texas Instruments Inc.
P.O. Box 225474, M S 219
Dallas, TX 75265 (214) 995-5315

Robert P. Stecher
Richard, Harris, Medlock & Andrew
Renaissance Tower, 1201 Elm St. Suite 4500
Dallas, TX 75270 (214) 939-4500

Mark W. Stockman
Shank, Irwin, Conant, Lipshy & Casterline
2200 Lincoln Plaza, 500 N. Akard
Dallas, TX 75201 (214) 720-9700

Tom Streeter
P.O. Box 50201
Dallas, TX 75250 (214) 699-8102

Frederick J. Telecky, Jr.
Texas Instruments Inc.
P.O. Box 225474
Dallas, TX 75265 (214) 995-5315

Peter J. Thoma
Kanz, Scherback & Timmons, South Tower-Suite 1030
Plaza Of The Americas
Dallas, TX 75201 (214) 969-7576

Daniel V. Thompson
Richard, Harris, Medlock & Andrews
2900 One Main Place
Dallas, TX 75250 (214) 742-8013

Ronald V. Thurman
Hubbard, Thurman, Turner & Tucker
2100 One Galleria Tower
Dallas, TX 75240 (214) 233-5712

W. Thomas Timmons
Kanz, Scherback & Timmons
Plaza Of The Americas, Suite 1030 South
Dallas, TX 75201 (214) 969-7576

Laurey Dan Tucker
Hubbard, Thurman, Turner & Tucker
2100 One Galleria Tower
Dallas, TX 75240 (412) 233-5712

Robert W. Turner
Hubbard, Thurman, Turner & Tucker
2100 One Galleria Tower
Dallas, TX 75240 (214) 233-5712

Roy L. Van Winkle
Kanz, Scherback & Timmons South Tower, Suite 1030
Plaza Of The Americas
Dallas, TX 75201 (214) 969-7576

Harry J. Watson
Hubbard, Thurman, Turner & Tucker
2100 One Galleria Tower
Dallas, TX 75240 (214) 233-5712

Stuart L. Watt
Richards, Harris, Medlock & Andrews
Renaissance Tower, 1201 Elm St. Suite 4500
Dallas, TX 75270 (214) 939-4500

Fred A. Winans
Dresser Inds. Inc.
1600 Pacific Ave.
Dallas, TX 75201 (214) 740-6919

Mack J. Casner
P.O. Box 24073
El Paso, TX 79914 (915) 533-2681

Cecil Freedman
228 Lomont Dr.
El Paso, TX 79912 (915) 581-3650

James A. Arno
Alcon Lab. Inc.
6201 S. Freeway
Fort Worth, TX 76134 (817) 551-8260

James E. Bradley
Felsman, Bradley, Gunter & Kelly
2850 Continenal Plaza, 777 Main St.
Fort Worth, TX 76102 (817) 332-8143

Gregg C. Brown
Alcon, Labs. Inc.
6201 South Freeway
Fort Worth, TX 76134 (817) 551-8663

Andrew J. Dillon
Gearhart Inds, Inc.
P.O. Box 1936
Fort Worth, TX 76101 (817) 293-1300

James C. Fails
Wofford, Fails & Zobal
110 W. 7th St. Suite 500
Fort Worth, TX 76102 (817) 332-1233

Robert A. Felsman
Felsman, Bradley & Gunter
2850 Continental Plaza, 777 Main St.
Fort Worth, TX 76102 (817) 332-8143

George Galerstein
Bell Heliopter Textron, Inc.
P.O. Box 482
Fort Worth, TX 76101 (817) 280-2834

Elton F. Gunn#
5612 Oakmont Lane
Fort Worth, TX 76112 (817) 390-8640

Charles D. Gunter, Jr.
Felsman, Bradley & Gunter
Suite 2850 Continental Plaza, 777 Main St.
Fort Worth, TX 76102 (817) 332-8143

Melvin A. Hunn
Felsman, Bradley & Gunter
777 Main St. Suite 2850
Fort Worth, TX 76102 (817) 332-8143

Geoffrey A. Mantooth
Wofford, Fails & Zobal
110 W. Seventh, Suite 500
Fort Worth, TX 76102 (817) 332-1233

William L. Martin, Jr.
Tandy Corp.
1800 One Tandy Center, P.O. Box 1718
Fort Worth, TX 76102 (817) 390-3700

Robert E. Roehrs#
3729 Hulen Park
Fort Worth, TX 76109 (817) 924-8446

Wm T. Wofford
Wofford, Fails & Zobal
110 W. 7th, Suite 500
Fort Worth, TX 76102 (817) 332-1233

Arthur Fred Zobal
Wofford, Fails & Zobal
110 W. 7th, Suite 500
Fort Worth, TX 76102 (817) 332-1233

Charles E. Schrman
General Dynamics Corp.
P.O. Box 748
Forth Worth, TX 76101 (817) 777-1711

A. Cooper Ancona#
Dow Chem. Co.
B-1210, Pats. Dept.
Freeport, TX 77541 (409) 238-7259

James G. Carter#
Dow Chemical Co.
B-1210 Bldg. Pat. Dept.
Freeport, TX 77541 (713) 238-7250

Carol J. Cavender#
Dow Chemical Co. Pat. Dept.
B-1210
Freeport, TX 77541 (409) 238-3581

Benjamin G. Colley#
Dow Chemical Co.
B-1210
Freeport, TX 77541 (713) 238-7239

James H. Dickerson, Jr.
Dow Chemical Co.
B-1210
Freeport, TX 77541 (409) 238-7084

William Douglas Miller
The Dow Chemical Co. Pat. Dept. B-1210
Freeport, TX 77541 (409) 238-2266

C. Ray Holbrook, Jr.
County Of Galveston
County Courthouse
Galveston, TX 77550 (713) 766-2244

Elizabeth F. Sporar#
4601 Ursuline
Galveston, TX 77550 (409) 762-0239

Thomas D. Copeland, Jr.
1900 Melody Lane
Garland, TX 75042 (214) 278-8012

Kenneth G. Lupo#
4407 Vintage Way
Garland, TX 75042 (214) 272-9186

James M. Cate
L T V Corp. L T V Aerospace & Defense Co.
1902 W. Freeway
Grand Prairie, TX 75051 (214) 266-1908

James Duke Willborn
1517 W. North Ccarrier Pkwy. Suite 140
Grand Prairie, TX 75050 (214) 641-3996

C.M. Kucera
Route 3 Box 157B
Hallettsville, TX 77964 (512) 789-5597

Albert Julius Adamcik
Shell Dev. Co.
P.O. Box 2463
Houston, TX 77001 (713) 241-4729

Margaret E. Anderson
Browning, Bushman, Zamecki & Anderson
5718 Westheimer, Suite 1800
Houston, TX 77057 (713) 266-5593

Rodney M. Anderson
Texas Instruments Inc.
P.O. Box 1443, M S. 676
Houston, TX 77001 (713) 274-3654

Ernest R. Archambeau, Jr.
2210 Lexford Lane
Houston, TX 77080 (713) 973-8741

Andres M. Arismendi, Jr.
Shell Dev. Co.
P.O. Box 2463
Houston, TX 77001 (713) 241-3997

Gordon T. Arnold#
6014 Deerwood
Houston, TX 77057 (713) 789-8250

Tom Arnold
Arnold, White & Durkee
P.O. Box 4433
Houston, TX 77210 (713) 787-1400

Martin S. Baer
5746 Valkeith Dr.
Houston, TX 77096 (714) 729-0100

Douglas Baldwin, Jr.
Shell Dev. Co.
P.O. Box 2463, One Shell Plaza
Houston, TX 77001 (713) 241-3716

Edmund Francis Bard, III
6565 W. Loop South, Suite 710
Houston, TX 77401 (713) 684-1778

James Allen Bargfrede
2323 S. Voss, Suite 123
Houston, TX 77057 (713) 781-0422

Hardie R. Barr*
N A S A
Johnson Space Center, Mail Code A L 3
Houston, TX 77058 (713) 483-4871

Kenneth D. Baugh
1802 Calumet
Houston, TX 77004 (713) 528-0496

Robert E. Bayes
14403 Broadgreen Dr.
Houston, TX 77079 (713) 497-7184

William James Beard
Halliburton Services
P.O. Box 42800, 2135 Hwy. 6 South
Houston, TX 77242 (713) 496-8331

Emil J. Bednar
6131 Paisley
Houston, TX 77096 (713) 771-2941

Keith A. Bell
Exxon Production Res. Co.
P.O. Box 2189
Houston, TX 77001 (713) 965-7994

Christopher R. Benson
Arnold, White & Durkee
750 Bering Dr. P.O. Box 4433
Houston, TX 77210 (713) 789-1400

Peter A. Bielinski
Shell Dev. Co.
P.O. Box 2463
Houston, TX 77001 (713) 241-4991

Johan A. Bjorksten#
9117 Almeda - Genoa Rd.
Houston, TX 77075

Thomas L. Blasdell
Vinson & Elkins
1001 Fannin
Houston, TX 77002 (713) 651-2951

Nelson A. Blish
Cooper Inds. Inc.
1001 Fannin, Suite 4000
Houston, TX 77002 (713) 739-5858

Ronald G. Bliss
Fulbright & Jaworski
1301 Mc Kinney St. 51st Floor
Houston, TX 77010 (713) 651-5151

Mark G. Bocchetti
1800 Augusta, Suite 400
Houston, TX 77057 (713) 266-6000

Kirby L. Boston
Merichem Co.
4800 Texas Comm. Tower
Houston, TX 77002 (713) 224-3030

Margaret Anne Boulware
Vanden, Eickenroht, Thompson & Boulware
One Riverway Suite 1100
Houston, TX 77056 (713) 961-3525

Raymond H. Bradley#
850 Regal
Houston, TX 77034 (713) 944-1973

Patricia N. Brantley
Arnold, White & Durkee
P.O. Box 4433
Houston, TX 77210 (713) 787-1449

Patricia A. Breland
Arnold, White & Durkee
P.O. Box 4433
Houston, TX 77210 (713) 787-1438

Michael P. Breston
2600 S. Gessner Suite 425
Houston, TX 77063 (713) 953-2990

Sylvester W. Brock, Jr.
Exxon Co. U.S.A.
P.O. Box 2180
Houston, TX 77252 (713) 656-4864

Walter R. Brookhart
Browning, Bushman, Zamecki & Anderson
5718 Westheimer, Suite 1800
Houston, TX 77057 (713) 266-5593

Timothy L. Burgess
Three Riverway Suite 1300
Houston, TX 77056 (713) 622-7312

Karen T. Burleson
Arnold, White & Durkee
P.O. Box 4433
Houston, TX 77210 (713) 787-1411

Gary L. Bush
Dodge, Bush & Moseley
950 Echo Lane, Suite 180
Houston, TX 77024 (713) 827-1054

C. James Bushman
Browning, Bushman, Zamecki & Anderson
5718 Westheimer, Suite 1800
Houston, TX 77057 (713) 266-5593

Wendy K.B. Buskop
1330 Augusta Dr. Apt. 26
Houston, TX 77057

Jesus D. Cabello
Compaq Computer Corp.
20555 F.M. 149
Houston, TX 77070 (713) 374-2634

Stephen H. Cagle
Arnold, White & Durkee
400 One Bering Park, 750 Bering Dr.
Houston, TX 77057 (713) 787-1400

Rodney K. Caldwell
Arnold, White & Durkee
P.O. Box 4433
Houston, TX 77210 (713) 787-1441

Norman R. Carlson
13707 Tosca Lane
Houston, TX 77079 (713) 464-0269

Salvatore J. Casamassima
Exxon Co. U.S.A. Pat. Dept.
P.O. Box 2180
Houston, TX 77252 (713) 656-3437

Ronald L. Clendenen
Shell Dev. Co.
P.O. Box 2463
Houston, TX 77001 (713) 241-4738

Ned L. Conley
Butler & Binon
1600 First Interstate Bank Plaza
Houston, TX 77002 (713) 237-3195

Robert T. Cook
Arnold, White & Durkee
P.O. Box 4433
Houston, TX 77210 (713) 787-1452

Harold Wade Coryell#
Shell Dev. Co.
P.O. Box 2463, Oil Shell Plaza
Houston, TX 77001 (713) 241-4708

Charles M. Cox
Pravel, Gambrell, Hewitt, Kimball & Krieger
1177 W. Loop South, 10th Floor
Houston, TX 77027 (713) 850-0909

Hubert E. Cox, Jr.
Exxon Production Res Co.
P.O. Box 2189
Houston, TX 77001 (713) 965-7965

Jack Darrell#
7505 Fannin Suite 214
Houston, TX 77054 (713) 797-1345

Charles H. De La Garza
Arnold, White & Durkee
P.O. Box 4433
Houston, TX 77210 (713) 789-7600

Paul L. De Verter, II
Fulbright & Jaworski
1301 Mc Kinney St. Suite 5100
Houston, TX 77010 (713) 651-5151

Marc L. Delflache
Pravel, Gambrell, Hewitt, Kimball & Krieger
1177 W. Loop South Suite 1010
Houston, TX 77027 (713) 850-0909

Stephen D. Dellett
Arnold, White & Durkee
P.O. Box 4433
Houston, TX 77210 (713) 787-1423

Harold Louis Denkler
823 Patchester
Houston, TX 77079 (713) 462-2996

Robert W.B. Dickerson
Dickerson & Lee
5100 Westheimer, Suite 100
Houston, TX 77056 (713) 627-0252

James George Dieter
Bechtel, Inc.
5400 Westheimer Court
Houston, TX 77056 (713) 235-3405

John H. Dodge, II
Dodge, Bush & Moseley
950 Echo Lane, Suite 180
Houston, TX 77024 (713) 827-1054

Paul I. Douglas
Shell Dev. Co.
One Shell Plaza, Room 1120, P.O. Box 2463
Houston, TX 77001 (713) 241-5342

N. Elton Dry
Pravel, Gambrell, Hewitt, Kimball & Krieger
1177 W. Loop S. Suite 1010
Houston, TX 77027 (713) 850-0909

Arthur M. Dula
Dula, Shields & Egbert
Texas Comm. Tower, 69th Floor
Houston, TX 77002 (713) 227-9000

George H. Dunn, III
Vinson & Elkins
2900 First City Tower, 1001 Fannin
Houston, TX 77002 (713) 750-1073

William D. Durkee
Arnold, White & Durkee
P.O. Box 4433
Houston, TX 77210 (713) 787-1444

Lewis Hamilton Eatherton, III
Exxon Co. U.S.A.
P.O. Box 2180
Houston, TX 77001 (713) 656-4864

John S. Egbert
Dula, Shield & Egbert
6900 Texas Comm. Tower
Houston, TX 77002 (713) 227-9000

Marvin B. Eickenroht
Vanden, Eickenroht, Thompson & Boulware
One Riverway, Suite 1100
Houston, TX 77056 (713) 961-3525

James J. Elacqua
Arnold, White & Durkee
P.O. Box 4433
Houston, TX 77210 (713) 787-1465

Clarence E. Eriksen
Arnold, White & Durkee
P.O. Box 4433
Houston, TX 77210 (713) 787-1400

A.H. Evans
Vinson & Elkins
1001 Fannin, 2927 First City Tower
Houston, TX 77002 (713) 651-2356

Lawrence E. Evans, Jr.
Gunn, Lee & Jackson, Suite 500
6200 Savoy Dr.
Houston, TX 77036 (713) 977-5000

William B. Farney
Arnold, White & Durkee
750 Bering Suite 200
Houston, TX 77057 (713) 787-1400

John B. Farr
1409 Upland Dr.
Houston, TX 77043 (713) 486-4205

Edward K. Fein*
N.A.S.A. Johnson Space Center
Mail Code A L 3
Houston, TX 77058 (713) 483-4871

Donald H. Fidler
Fidler & Assco., P.C.
8955 Katy Freeway, Suite 201
Houston, TX 77024 (713) 468-3997

Albert M.T. Finch
Shell Oil Co. Pats. & Licensing
P.O. Box 2463
Houston, TX 77252 (713) 241-0401

Melvin F. Fincke
2210 Pelham Dr.
Houston, TX 77019 (713) 523-0855

Richard D. Fladung
Pravel, Gambrell, Hewitt, Kimbrell & Krieger
1177 W. Loop South, Suite 800
Houston, TX 77027 (713) 850-0909

Fred Floersheimer
N L Indus. Inc.
3000 North Belt East
Houston, TX 77032 (713) 987-5152

Raymond C. Floyd
Exxon Chemical Co.
13501 Katy Freeway
Houston, TX 77079 (713) 870-6714

Ernest A. Forzano
Shell Co.
One Shell Plaza, P.O. Box 2463
Houston, TX 77001 (504) 588-4753

James B. Gambrell
Pravel, Gambrell, Hewitt, Kimball & Krieger
1177 W. Loop S. Suite 1010
Houston, TX 77027 (713) 850-0909

Henry N. Garrana
Schlumberger Well Services
5000 Gulf Freeway
Houston, TX 77023 (713) 928-4071

M.H. Gay
Vinson & Elkins
First City Tower, 1001 Fannin
Houston, TX 77002 (713) 651-2350

Henry C. Geller#
Shell Oil Company
P.O. Box 2463
Houston, TX 77001 (713) 241-3760

James E. Gilchrist
Exxon Minerals Co.
P.O. Box 4508, 13111 Northwest Fwy.
Houston, TX 77210 (713) 895-1166

Jefferson D. Giller
Fulbright & Jaworski
1301 Mc Kinney St. 51st Floor
Houston, TX 77010 (713) 651-5151

George E. Glober, Jr.
Exxon Production Res. Co.
P.O. Box 2189
Houston, TX 77001 (713) 965-4554

Edward William Goldstein
Arnold, White & Durkee
750 Bering Dr.
Houston, TX 77057 (713) 787-1400

Jack C. Goldstein
Arnold, White & Durkee
P.O. Box 4433
Houston, TX 77210 (713) 787-1400

Kenneth D. Goodman
Arnold, White & Durkee
750 Bering Dr, Suite 400
Houston, TX 77057 (713) 787-1400

Rosanne Goodman#
Fulbright & Jaworski
1301 Mc Kinney St. 51st. Floor
Houston, TX 77010 (713) 651-5151

W. Gary Goodson
Marathon Oil Co.
P.O. Box 3128
Houston, TX 77253 (713) 629-6600

Alan H. Gordon
Arnold, White & Durkee
P.O. Box 4433
Houston, TX 77057 (713) 787-1492

Nancy J. Gracey#
Shell Oil Co.
P.O. Box 2463
Houston, TX 77001 (713) 241-3901

John G. Graham
Texas Instrument Inc.
Box 1443
Houston, TX 77001 (713) 274-3657

Robert L. Graham
2616 South Loop West Suite 640
Houston, TX 77054 (713) 668-3325

Donald R. Greene
16160 Seahorse Dr.
Houston, TX 77062 (713) 480-2575

D. Arlon Groves
333 Clay St., 4800 Three Allen Center
Houston, TX 77002 (713) 654-9728

C. Donald Gunn
6200 Savoy Dr. Suite 500
Houston, TX 77036 (713) 977-5000

Donald F. Haas
Shell Oil Co.
One Shell Plaza, P.O. Box 2463
Houston, TX 77001 (713) 241-3356

Francis J. Hagel#
Schlumberger Well Services
P.O. Box 2175
Houston, TX 77252 (712) 928-4337

Frederick D. Hamilton
Vinson & Elkins
1001 Fannin
Houston, TX 77002 (713) 651-2732

Wayne M. Harding
Arnold, White & Durkee
P.O. Box 4433
Houston, TX 77210 (713) 787-1400

Robin A. Hartman
3212 Smith, Suite 103
Houston, TX 77006 (713) 521-1135

Howard Wayne Haworth
9400 Doliver Apt. 23
Houston, TX 77063 (713) 974-4028

Jack W. Hayden
1003 Wirt Rd. Suite 311
Houston, TX 77055 (713) 465-5015

John T. Headly
Arnold, White & Durkee
750 Bering
Houston, TX 77057 (713) 787-1432

Loren G. Helmreich
Carwell & Helmreich, Eleven Greenway Plaza
Suite 2127, Summit Tower
Houston, TX 77046 (713) 850-0505

Lester L. Hewitt
Pravel, Gambrell, Hewitt, Kimball & Krieger
1177 W. Loop S. Suite 1010
Houston, TX 77027 (713) 850-0909

Alden D. Holford
2450 Fondren Rd. Suite 312
Houston, TX 77063 (713) 266-0050

Henry Welcker Hope
Fulbright & Jaworski
1301 Mc Kinney St. 51st Floor
Houston, TX 77010 (713) 651-5151

Jeffrey M. Hoster
Exxon Production Res. Co.
P.O. Box 2189
Houston, TX 77252 (713) 965-4991

Roy F. House#
5726 Ettrick St.
Houston, TX 77035 (713) 721-2117

Eugene Y. Hsiao
Hsiao & Thurmond
7745 San Felipe Suite 204
Houston, TX 77063 (713) 783-5061

Walter D. Hunter#
12118 Attlee Dr.
Houston, TX 77077 (713) 558-4265

W.F. Hyer
Hyer & Mathews, P.C.
2401 Fountainview, Suite 500
Houston, TX 77057 (713) 266-6000

Rita M. Irani
Pravel, Gambrell, Hewitt, Kimball & Krieger
1177 W. Loop S. Suite 1010
Houston, TX 77027 (713) 850-0909

James Lonnie Jackson, Sr.
Gunn, Lee & Jackson
6200 Savoy Dr. Suite 500
Houston, TX 77036 (713) 977-5000

John A. Jacobi
Tenneco Gas Pipeline Group
P.O. Box 2511
Houston, TX 77001 (713) 757-3670

George Byron Jamison
Jamison & Mc Gregor
4600 Republic Bank Center
Houston, TX 77002 (713) 224-1212

Paul M. Janicke
Arnold, White & Durkee
P.O. Box 4433
Houston, TX 77057 (713) 787-1455

Catherine K. Jen
3303 Louisiana, Suite 216
Houston, TX 77006 (713) 523-8811

Edward L. Jensen
6363 Woodway, Suite 310
Houston, TX 77057 (713) 782-1311

Kenneth C. Johnson
Exxon Co. U.S.A.
P.O. Box 2180
Houston, TX 77252 (713) 656-1719

William Erby Johnson, Jr.
Browning, Bushman, Zamecki & Anderson
5718 Westheimer, Suite 1800
Houston, TX 77057 (713) 266-5593

Larry C. Jones
Fulbright & Jaworski
1301 Mc Kinney St. 51st Floor
Houston, TX 77010 (713) 651-5151

Barry C. Kane#
Western Geophysical Co. Of Amer.
P.O. Box 2469
Houston, TX 77252 (713) 789-9600

Kenneth A. Keeling
2916 W. T.C. Jester Blvd.
Houston, TX 77018 (713) 680-1447

Christopher D. Keirs
Arnold, White & Durkee
P.O. Box 4433
Houston, TX 77210 (713) 787-1400

Albert B. Kimball, Jr.
Pravel, Gambrell, Hewitt, Kimball & Krieger
1177 W. Loop S. Suite 1010
Houston, TX 77027 (713) 850-0909

John R. Kirk, Jr.
Baker & Kirk
1020 Holcombe Suite 444
Houston, TX 77030 (713) 790-9316

William A. Knox#
8310 Ashcroft Dr.
Houston, TX 77096

Paul E. Krieger
Pravel, Gambrell, Hewitt, Kimball & Krieger
1177 W. Loop S. Suite 1010
Houston, TX 77027 (713) 850-0909

Kenneth E. Kuffner
Arnold, White & Durkee
P.O. Box 4433
Houston, TX 77210 (713) 789-7600

William L. Lafuze
Vinson & Elkins.
2900 First City Tower
Houston, TX 77002 (712) 236-2595

John A. Langworthy
3605 Maroneal
Houston, TX 77025 (713) 655-0133

Lee R. Larkin
Andrews & Kurth
Texas Commerce Tower, 600 Travis St.
Houston, TX 77002 (713) 220-4054

Casimir F. Laska
Vanden, Eickeroht, Thompson & Boulware
One Riverway, Suite 1100
Houston, TX 77056 (713) 961-3525

Sydney M. Leach
Arnold, White & Durkee
P.O. Box 4433
Houston, TX 77210 (713) 789-7600

Richard F. Lemuth
Shell Development Co.
One Shell Plaza, P.O. Box 2463
Houston, TX 77001 (713) 241-3554

Bart E. Lerman
Baker & Kirk, P.C.
1020 Holcombe, Suite 444
Houston, TX 77030 (713) 790-9316

Mitchell D. Lukin
Baker & Botts
One Shell Plaza, 900 Louisiana
Houston, TX 77002 (713) 229-1234

Daniel N. Lundeen
Pravel, Gambrell, Hewitt, Kimball & Krieger
1177 W. Loop S. Suite 1010
Houston, TX 77027 (713) 850-0909

Craig M. Lundell
Arnold, White & Durkee
P.O. Box 4433
Houston, TX 77210 (713) 789-7600

Keith E. Lutsch
Pravel, Gambrell, Hewitt, Kimball & Krieger
1177 W. Loop S. Suite 1010
Houston, TX 77027 (713) 850-0909

John F. Lynch
Arnold, White & Durkee
P.O. Box 4433
Houston, TX 77210 (713) 789-7600

Michael L. Lynch
Arnold, White & Durkee
P.O. Box 4433
Houston, TX 77210 (713) 787-1400

Gregory L. Maag
Butler & Binion
1600 Fist Interstate Bank Plaza
Houston, TX 77002 (713) 237-3130

Michael E. Macklin
Arnold, White & Durkee
P.O. Box 4433
Houston, TX 77210 (713) 789-7600

Robert B. Mahley
The M.W. Kellogg Co.
The Greenway Plaza
Houston, TX 77046 (713) 960-2147

Robert J. Marett
Marett & Marrett
1800 W. Loop S. Suite 800
Houston, TX 77027 (713) 622-2800

Rodney L. Marett
Marrett & Marett
1800 W. Loop South, Suite 800
Houston, TX 77027 (713) 622-2800

Fredrik Marlowe
Shell Dev. Co.
900 Louisiana, P.O. Box 2463
Houston, TX 77001 (713) 241-4746

Marvin J. Marnock
13630 Indian Creek Rd.
Houston, TX 77079

Thomas F. Marsteller, Jr.
3000 Post Oak Blvd. Suite 1400
Houston, TX 77056 (713) 963-9168

Julian C. Martin
Vinson & Elkins
First City Tower
Houston, TX 77002 (713) 651-2490

Guy E. Matthews
2401 Fountainview, Suite 500
Houston, TX 77057 (713) 266-6000

Marvin F. Matthews*
N A S A Johnson Space Center
Mail Code A L 3
Houston, TX 77058 (713) 483-4871

Roger L. Maxwell
Cooper Inds. Inc. First City Tower
P.O. Box 4446, Suite 4000
Houston, TX 77210 (713) 739-5785

Douglas H. May, Jr.
N L, Off. Of General Coun.
P.O. Box 60087
Houston, TX 77205 (713) 987-5156

John H. Mc Carthy#
Shell Dev. Co.
P.O. Box 2463
Houston, TX 77001 (713) 241-4798

Carl O. Mc Clenny
6154 Willer s Way
Houston, TX 77057 (713) 782-3620

Guy L. Mc Clung, III
Vanden, Eickerson, Thompson & Boulware
One Riverway, Suite 1100
Houston, TX 77056 (713) 961-3525

Cecil A. Mc Clure
Shell Dev. Co.
P.O. Box 2463
Houston, TX 77001

Pamela J. Mc Collough
Shell Dev. Co. Pats. & Licensing
P.O. Box 2463, 900 Louisiana
Houston, TX 77001 (713) 241-4091

Patrick H. Mc Collum
Dresser Atlas
P.O. Box 1407, 10201 Westheimer
Houston, TX 77001 (713) 972-6024

Martin L. Mc Gregor, Jr.
Jamison & Mc Gregor
700 Louisiana, Suite 3990
Houston, TX 77002 (713) 224-1212

Michael T. Mc Lemore
Arnold, White & Durkee
P.O. Box 4433
Houston, TX 77210 (713) 789-7600

Leonard P. Miller
Shell Dev. Co.
P.O. Box 2463
Houston, TX 77001 (713) 241-2538

Thomas A. Miller
Arnold, White & Durkee
Box 4433
Houston, TX 77210 (713) 789-7600

Peter E. Mims
Vinson & Elkins
1001 Fannin, Suite 2913
Houston, TX 77002 (713) 651-2732

Eric-Paul Mirabel
Butler & Binion
1600 First Interstate Bank Plaza
Houston, TX 77002 (713) 237-3149

Raul R. Montes
Exxon Production Res. Co.
P.O. Box 2189
Houston, TX 77252 (713) 965-4064

Fay E. Morisseau, III
Vinson & Elkins
1001 Fannin
Houston, TX 77002 (713) 651-2740

Paula D. Morris
Arnold, White & Durkee
P.O. Box 4433
Houston, TX 77210 (713) 787-1472

David L. Moseley
Arnold, White & Durkee
750 Bering Dr. Suite 400
Houston, TX 77057 (713) 789-7600

Richard L. Moseley
4394 Varsity Lane
Houston, TX 77004 (713) 747-9780

Neal J. Mosely
2916 W. T.C. Jester Blvd.
Suite 101
Houston, TX 77018 (713) 680-9676

James J. Mullen
Shell Oil Co. Legal Dept.
P.O. Box 2463
Houston, TX 77001 (713) 241-5765

Kurt Sheridan Myers
Butler & Binion
1600 First Interstate Bank Plaza
Houston, TX 77002 (713) 237-3274

Michael A. Nametz
Exxon Production Res Co.
P.O. Box 2189
Houston, TX 77001 (713) 965-4064

Francis D. Neruda
Baeder Neruda Interests Inc.
1300 Post Oak Blvd. Suite 765
Houston, TX 77056 (713) 993-0010

Bruce E. Newell
Enserch Engineers & Constructors
10375 Richmond Ave.
Houston, TX 77042 (713) 954-4612

Nick A. Nicholas, Jr.
Gunn, Lee & Jackson
6200 Savoy Dr. Suite 500
Houston, TX 77036 (713) 977-5000

Theron H. Nichols#
8723 Nairn St.
Houston, TX 77074 (713) 772-2112

John D. Norris
Arnold, White & Durkee
P.O. Box 4433
Houston, TX 77210 (713) 789-7600

William C. Norvell, Jr.
Norvell & Assco.
6363 Woodway
Houston, TX 77057 (713) 266-1914

Herbert E. O Niell
Exxon Production Res. Co.
P.O. Box 2189
Houston, TX 77252 (713) 965-4811

David M. Ostfeld
Chamberlain, Hrdlicka, White, Johnson & Williams
1100 Milam, 28th Floor
Houston, TX 77002 (713) 658-1818

Ronald L. Palmer
Baker & Botts
3000 One Shell Plaza
Houston, TX 77002 (713) 229-1227

David L. Parker
Arnold, White Durkee
750 Bering Dr.
Houston, TX 77057 (713) 789-7600

Michael L. Parks & Moss
2100 Eleven Greenway Plaza
Houston, TX 77046 (713) 626-9870

Joe P. Parris#
Cameron Iron Works, Inc.
P.O. Box 1212
Houston, TX 77251 (713) 939-3021

Alton W. Payne, Jr.
Dodge, Bush & Moseley
950 Echo Lane, Suite 180
Houston, TX 77024 (713) 827-1054

James M. Peppers#
1800 Augusta Suite 400
Houston, TX 77057 (713) 781-9595

Larry B. Phillips, III
Anderws & Kurth
4200 Texas Comm. Tower
Houston, TX 77002 (713) 220-4090

Richard F. Phillips
Exxon Production Res. Co.
P.O. Box 2189, (3120 Buffalo Speedway)
Houston, TX 77252 (713) 940-4938

John A. Poindexter
Thelen, Marrin, Johnson & Bridges
1300 Texas American Bank Bldg.
Houston, TX 77002 (713) 654-8877

Bernarr R. Pravel
Pravel, Gambrell, Hewett, Kimball & Krieger
1177 W. Loop South Suite 1010
Houston, TX 77027 (713) 850-0909

Frank Burruss Pugsley
Baker & Botts
3000 One Shell Plaza
Houston, TX 77002 (713) 229-1577

Richard T. Redano
Arnold, White & Durkee
750 Bering Dr.
Houston, TX 77057 (713) 787-1400

James A. Reilly
12510 Mossycup Dr.
Houston, TX 77024 (713) 465-3249

Bernard A. Reiter
2401 Fountain View Dr. Suite 500
Houston, TX 77056 (713) 266-6000

James W. Repass#
Fulbright & Jaworski
1301 Mc Kinney St. 51st Floor
Houston, TX 77010 (713) 651-5151

Ronald R. Reper
Shell Dev. Co.
P.O. Box 2463
Houston, TX 77001 (713) 241-3247

Glenn W. Rhodes
Arnold, White & Durkee
P.O. Box 4433
Houston, TX 77095 (713) 787-1400

Eugene N. Riddle
Dodge, Bush & Moseley
950 Echo Lane, Suite 180
Houston, TX 77024 (713) 827-1054

Carl G. Ries
8015 Lorrie Dr.
Houston, TX 77025 (713) 661-2149

James H. Riley
Pravel, Gambrell, Hewitt, Kimball & Krieger
1177 W. Loop South Suite 1010
Houston, TX 77027 (713) 850-0909

W. Ronald Robins
Vinson & Elkins
2918 First City Tower
Houston, TX 77002 (713) 651-2452

Murray Robinson
Butler & Binion
1600 First Interstate Bk Plaza
Houston, TX 77002 (713) 237-3198

Kenneth A. Roddy#
2916 W. T.C. Jester Blvd.
Suite 108
Houston, TX 77108 (713) 680-9676

David Alan Rose
Butler & Binion
1600 First Interstate Bank Plaza
Houston, TX 77002 (713) 237-3640

Stave Rosenblatt
24 Greenway Plaza, Suite 1515
Houston, TX 77046 (713) 626-1555

Alan D. Rosenthal
Baker & Botts
3000 One Shell Plaza
Houston, TX 77002 (713) 229-1584

Carl A. Rowold
Reed Rock Bit Co.
6501 Navigation Blvd.
Houston, TX 77011 (713) 924-5256

Robert E. Sandfield
12 Hackberry Lane
Houston, TX 77027 (713) 871-0023

Ezra L. Schacht#
1620 W. Main St.
Houston, TX 77006 (713) 523-0515

John E Schaich
Arnold, White & Durkee
750 Bering Dr. Suite 400
Houston, TX 77057 (713) 787-1400

Wilbur Allison Schaich
Norvell & Assco,
6363 Woodway, Suite 275
Houston, TX 77057 (713) 266-1914

Russell E. Schlorff*
N A S A
Johnson Space Center, Mail Code A L 3
Houston, TX 77058 (713) 483-4871

John S. Schneider
Exxon Co.
P.O. Box 2180
Houston, TX 77001 (713) 656-5004

Michael B. Schroeder
Arnold, White & Durkee
750 Bering Dr.
Houston, TX 77057 (713) 787-1450

Willem G. Schuurman
Arnold, White & Durkee
P.O. Box 4433
Houston, TX 77210 (512) 474-2583

Eddie E. Scott
Cooper Industries, Inc.
1001 Fannin, Suite 4000, P.O. Box 4446
Houston, TX 77210 (713) 739-5534

Mark V. Seeley
Arnold, White & Durkee
P.O. Box 4433, 750 Bering, Suite 400
Houston, TX 77057 (713) 787-1502

Sue Z. Shaper
Vaden, Eickenroht, Thompson & Boulware
One Riverway, Suite 2420
Houston, TX 77056 (713) 961-3525

William E. Shull
Butler & Binion
1600 First Interstate Bk Plaza
Houston, TX 77002 (713) 237-3645

Rand N. Shulman
Shell Oil Co.
900 Louisiana
Houston, TX 77001 (713) 241-3165

Peter J. Shurn, III
Arnold, White & Durkee
750 Bering Dr. Suite 400
Houston, TX 77056 (713) 789-7600

Paul F. Simpson
Butler & Binion
1600 First Interstate Bk Plaza
Houston, TX 77002 (713) 237-2074

Thomas L. Sivak
Shell Oil Co.
One Shell Plaza, P.O. Box 2463
Houston, TX 77000 (713) 241-7452

Henry L. Smith, Jr.
Tenneco Inc.
1010 Milam, P.O. Box 2511
Houston, TX 77001 (713) 757-4166

Mark A. Smith
Shell Dev. Co.
P.O. Box 2463
Houston, TX 77001 (713) 241-2094

Jack R. Springgate
Vinson & Elkins
1001 Fannin, 2924 First City Tower
Houston, TX 77002 (713) 651-2150

Darryl M. Springs
Baker, Kirk & Bissex, P.C.
1020 Holcombe Blvd. Suite 1444
Houston, TX 77030 (713) 790-9316

William Arnold Stout
Fulbright & Jaworski
1301 Mc Kinney St. 51st. Floor
Houston, TX 77010 (713) 651-5151

Michael O. Sutton
Arnold, White & Durkee
P.O. Box 4433
Houston, TX 77210 (713) 789-7600

Keith M. Tackett
Shell Oil Co. One Shell Plaza, 11th Floor
P.O. Box 2463
Houston, TX 77252 (713) 241-2976

Jeffrey W. Tayon
Butler & Binion
1600 First Interstate Bank Plaza
Houston, TX 77002 (713) 237-3238

Alan R. Thiele
Cooper Industries, Inc.
P.O. Box 4446
Houston, TX 77210 (713) 739-5855

E. Eugene Thigpen
1115 Augusta Dr. Apt. 36
Houston, TX 77057 (713) 266-2371

Klaus D. Thoma
Hollrah, Lange & Thoma
1331 Lamar, Suite 1570
Houston, TX 77010 (713) 650-1500

Jennings B. Thompson
Vaden, Eickenroht, Thompson & Boulware
One Riverway, Suite 1100
Houston, TX 77056 (713) 961-3525

Marcus L. Thompson, Jr.
Andrews & Kurth
4200 Texas Comm. Tower
Houston, TX 77002 (713) 220-4180

Ben D. Tobor
Tudzin & Tobor
777 N. Eldridge. Suite 650
Houston, TX 77079 (713) 870-1173

D.C. Toedt, III
Arnold, White & Durkee
P.O. Box 4433
Houston, TX 77210 (713) 787-1400

Anastassios Triantaphyllis
Butler & Binion
1600 First Interstate Bank Plaza
Houston, TX 77002 (713) 237-3285

Timothy N. Trop
Arnold, White & Durkee
750 Bering Dr.
Houston, TX 77057 (713) 787-1409

Yung-Yi G. Tsang#
Shell Dev. Co. Pat. & Licensing Dept.
P.O. Box 2463
Houston, TX 77001 (713) 241-0956

James G. Ulmer
Baker & Botts
3000 One Shell Plaza
Houston, TX 77002 (713) 229-1358

Frank Samuel Vaden, III
Vaden, Eickenroht, Thompson & Boulware
One Riverway, Suite 1100
Houston, TX 77056 (713) 961-3525

Paul Van Slyke
Pravel, Gambrell, Hewitt, Kimball & Krieger
1177 W. Loop South, 10th Floor
Houston, TX 77027 (713) 650-0909

Dean F. Vance
Shell Dev. Co.
One Shell Plaza, P.O. Box 2463
Houston, TX 77001 (713) 241-3356

Craig W. Walford
M.W. Kellogg Co.
Three Gateway Plaza
Houston, TX 77046 (713) 960-2150

Charles F. Walter
9131 Timberside Dr.
Houston, TX 77025

John D. Watts#
10700 N.W. Fwy. Suite 105
Houston, TX 77092 (713) 957-0986

Russell D. Weaver
Hollrah, Lange & Thoma
1331 Lamar Suite 1570
Houston, TX 77010 (713) 650-1500

Wayne E. Webb, Jr.
Pravel, Gambrell, Hewitt, Kimball & Krieger
1177 W. Loop South, Suite 1010
Houston, TX 77027 (713) 850-0909

James F. Weiler
Fulbright & Jaworski
1301 Mc Kinney St. 51st Floor
Houston, TX 77010 (713) 651-5151

Walter T. Weller
Shell Dev. Co.
P.O. Box 2463
Houston, TX 77001 (713) 241-3921

Albert S. Weycer
Weycer, Kaplan, Pulaski & Zuber
1414 Summit Tower, 11 Gateway Plaza
Houston, TX 77046 (713) 961-9045

Travis G. White
Arnold, White & Durkee
750 Bering Dr.
Houston, TX 77057 (713) 787-1551

Robert V. Wilder
Compaq Computer Corp.
20555 F.M. 149
Houston, TX 77070 (713) 374-2124

Danny L. Williams
Arnold, White & Durkee
P.O. Box 4433
Houston, TX 77210 (713) 787-1493

Horace C. Wilson, Jr.
Pravel, Gambrell, Hewitt, Kimball & Krieger
1177 W. Loop South Suite 1010
Houston, TX 77027

David S. Wise
Butler & Binion
1600 First Interstate Bk Plaza
Houston, TX 77002 (713) 237-3169

Donald G. Wolff
Halliburton Service
P.O. Box 721110, 5950 North Course Dr.
Houston, TX 77272 (713) 561-1471

Denise Y. Wolfs
Shell Oil Co. Legal Dept. Pats & Licensing
P.O. Box 2463
Houston, TX 77252 (713) 241-4789

Russell T. Wong
Arnold, White & Durkee
750 Bering Dr. Suite 400
Houston, TX 77057 (713) 787-1458

Stephen R. Wright
15723 Windy Glen Dr.
Houston, TX 77095 (713) 550-4452

William D. Zahrt, II
Shell Dev. Co.
P.O. Box 2463
Houston, TX 77252 (713) 241-1150

E. Richard Zamecki
Browning, Bushman, Zamecki & Anderson
5718 Westheimer
Houston, TX 77057 (713) 266-5593

Ren D. Zentner
University Of Houston Law Dept.
4800 Calhoun Rd.
Houston, TX 77004 (713) 749-2256

Roland A. Dexter
8319 Laurel Leaf
Humble, TX 77346 (713) 852-2677

James Harold Barksdale, Jr.
I B M Corp. Intellectual Property Law Dept.
220 Las Colinas Blvd.
Irving, TX 75039 (214) 556-5202

Henry C. Goldwin
1405 Ben Dr.
Irving, TX 75061 (214) 438-4013

William Lloyd Jones
2413 Crestview Circle
Irving, TX 75062 (214) 255-2361

Thomas V. Malorzo
N C H Corp.
2727 Chemsearch Blvd. P.O. Box 152170
Irving, TX 75015 (214) 438-0211

Marcus S. Rasco#
P.O. Box 153729
Irving, TX 75015 (214) 570-7993

John J. Sheedy
Curtis Mathes Corp.
1411 Greenway Dr.
Irving, TX 75038

Jacob W. Sietseman#
Frito- Lay, Inc.
900 N. Loop 12
Irving, TX 75601 (214) 579-2261

C.W. Crady, Jr.
22334 Wetherburn Lane
Katy, TX 77449 (713) 574-4939

David A. Roth
2803 Kings Forest Dr.
Kingwood, TX 77339 (713) 358-3280

Walter Joe Lee
301 Ligustrum St.
Lake Jackson, TX 77566 (713) 297-6253

David E. Cotey
Texas Eastman Co.
P.O. Box 7444
Longview, TX 75607 (214) 236-5454

Wendell Coffee
P.O. Box 3726
Lubbock, TX 79452 (806) 763-9252

Raywood H. Blanchard
Adobe Wells C. Club
Box 149
Mission, TX 78572 (312) 687-8050

John M. Duncan
3027 La Quinta Dr.
Missouri City, TX 77459 (713) 437-0141

Clarence E. Keys
P.O. Box 1378
Monahans, TX 79756 (915) 943-4531

Tom F. Pruitt
4635 Northeast Stallings Dr.
Suite 104
Nacogdoches, TX 75961 (409) 560-1088

Theodore E. Bieber
1431 Kingstree Lane
Nassau Bay, TX 77058 (713) 333-2673

Evert Allen Autrey
155 N. Mesquite Ave.
New Braunfels, TX 78130 (512) 625-0332

Frank Streightoff
413 Cantebury Dr.
New Braunfels, TX 78130 (512) 629-6320

Marcus L. Bates, Sr.#
111 W. 10th St.
Odessa, TX 79760 (915) 333-2121

Robert Christian Peterson
Texas Comm. Bank Bldg.
P.O. Box 2626, Suite 507
Odessa, TX 79760 (915) 332-0463

Fred Sylvester Valles
El Paso Products Co.
619 N. Grant St.
Odessa, TX 79760 (915) 333-8595

Margareta Le Maire#
El Paso Products Co.
9802 Fairmont Parkway, P.O. Box 2006
Pasadena, TX 77505 (713) 474-3211

John Paul Robinson, Jr.
801 W. Ellaine
Pasadena, TX 77506 (713) 477-0240

Robert Daniel Winn
Medico Drug Co.
1405 W. U.S. Expressway 83
Pharr, TX 78577 (512) 787-5931

Richard M. Byron
Atlantic Richfield Co.
P H O 400, 2300 W. Plano Pkwy
Plano, TX 75075 (214) 754-3386

J. Richard Konneker
6601 Garfield Dr.
Plano, TX 75023

Dennis O. Kraft
2612 Chadbourne Dr.
Plano, TX 75023

Roderick W. Mac Donald
Atlantic Richfield Co.
3000 Plano Parkway
Plano, TX 75075 (214) 422-3337

Michael E. Martin#
Atlantic Richfield Co.
2300 W. Plano Pkwy
Plano, TX 75075 (713) 486-3511

Jasper C. Rowe
2605 Cielo Dr.
Plano, TX 75074 (214) 280-4465

Thomas P. Schur
Frito-Lay, Inc.
7701 Carpenter Rd.
Plano, TX 75024 (214) 353-3822

F. Lindsey Scott
1448 Scarborough
Plano, TX 75075

Erving A. Trunk
P.O. Box 863543
Plano, TX 75086

Calvin A. Rising, Sr.
4121 Everglades
Port Arthur, TX 77642 (409) 982-7159

Richard Warren Anderson#
417 Ridgewood Dr.
Richardson, TX 75080 (214) 231-3951

Robert J. Crawford
Rockwell International Corp.
1200 N. Alma Rd.
Richardson, TX 75081 (214) 966-5492

Drude Faulconer
Aetna Tower
2350 Lakeside Blvd. Suite 850
Richardson, TX 75081 (214) 437-1204

Howard R. Greenberg
Rockwell Internatl. Corp.
1200 N. Alma Rd.
Richardson, TX 75081 (214) 996-5470

Gregory M. Howison
Ross & Howison
740 E. Campbell Rd., 900 Park Pacific One
Richardson, TX 75081 (214) 231-9510

Thomas P. Hubbard, Jr.#
1321 Apache
Richardson, TX 75080 (214) 231-2871

Warren Henry Kintzinger
777 S. Central Expwy. Suite 3E.
Richardson, TX 75080 (214) 234-3914

Bruce C. Lutz
404 Arborcrest Dr.
Richardson, TX 75080 (214) 231-4687

Monty L. Ross
Ross & Howison
900 Park Pacific One, 740 E. Campbell Rd.
Richardson, TX 75081 (214) 231-9510

Ronnie D. Wilson
100 N. Central Expressway, Suite 710
Richardson, TX 75080 (214) 699-0041

Guy W. Caldwell
Cox & Smith Inc.
600 National Bk. Of Comm. Bldg.
San Antonio, TX 78205 (512) 226-7000

Donald R. Comuzzi
Cox & Smith, Inc.
600 National Bk. Of Comm. Bldg.
San Antonio, TX 78205 (512) 226-7000

James N. Ezzell
Groce, Locke & Hebdon
2000 Frost Bk. Tower
San Antonio, TX 78205 (512) 231-6691

E. Manning Giles
7926 Broadway, Apt. 108
San Antonio, TX 78209 (512) 826-7539

Gary W. Hamiton
Gunn, Lee & Jackson
711 Navarro, Suite 720
San Antonio, TX 78205 (512) 222-2336

Charles W. Hanor
Cox & Smith, Inc.
600 N B C Bldg.
San Antonio, TX 78205 (512) 226-7000

Ted D. Lee
Gunn, Lee & Jackson, P.C.
711 Navarro, Suite 720
San Antonio, TX 78205 (512) 222-2336

Ann Livingston
Gunn, Lee & Jackson
711 Navarro, Suite 720
San Antonio, TX 78205 (512) 222-2336

John R. Merkling
5802 N.W. Expressway
San Antonio, TX 78201 (512) 733-6235

Mark H. Miller
Gunn, Lee & Jackson, P.C.
711 Navarro, Suite 720
San Antonio, TX 78205 (512) 222-2336

Martha F. Mims
Cox & Smith
600 N B C Bldg.
San Antonio, TX 78205 (512) 226-7000

Jack V. Musgrove
Gunn, Lee & Jackson
711 Navarro, Suite 720
San Antonio, TX 78205 (512) 222-2336

E. Suzanne Parr#
6842 Cerro Bajo
San Antonio, TX 78239 (512) 655-4280

Gale R. Peterson
Cox & Smith, Inc.
600 National Bank Of Comm.
San Antonio, TX 78205 (512) 226-7000

Leland A. Sebastian
8030 Misty Canyon
San Antonio, TX 78250 (512) 522-0534

Thomas E. Sisson
Sisson & Smith
6243 N.W. I H - 10, Suite 840
San Antonio, TX 78201 (512) 735-2200

Richard J. Smith
Sisson & Smith, P.C.
6243 N.W. I H - 10, Suite 840
San Antonio, TX 78201 (512) 735-2200

John C. Stahl
P.O. Box 13236
San Antonio, TX 78213 (512) 344-7479

Gustav N. Van Steenberg
Southwest Research Institute
8500 Culebra Rd.
San Antonio, TX 78284 (512) 684-5111

Andrew S. Viger
Mc Camish, Ingram, Martin & Brown
1200 Two Republic Bank Plaza
San Antonio, TX 78205 (512) 225-5500

Mark R. Wisner
Cox & Smith, Inc.
600 Natl. Bank Of Comm. Bldg.
San Antonio, TX 78205 (512) 226-7000

Ronald H. Evans
6607 Saffron Hills Dr.
Springs, TX 77379 (713) 251-8101

Stephen L. Borst
Schlumberger Limited
200 Macco Blvd.
Sugar Land, TX 77478 (713) 274-8299

Mary Jane Gaskin
3 Box Turtle Lane
The Woodlands, TX 77380 (713) 367-6621

John W. Adee#
P.O. Box 131505
Tyler, TX 75713 (214) 566-4707

Ronald B. Sefrna
P.O. Box 567, 325 S. Fannin
Tyler, TX 75701 (214) 592-5965

Charles C.M. Woodward
Route 25
Box 692
Tyler, TX 75707 (214) 566-2873

Lawrence E. Johnson#
410 Mbank Tower
Waco, TX 76701 (817) 756-7041

Utah

Allen H. Erickson#
Morton Thiokol Inc. Aerospace Group
P.O. Box 524
Brigham City, UT 84302 (801) 863-2501

Edward E. Mc Cullough#
Thiokol Corp.
P.O. Box 524
Brigham City, UT 84302 (801) 863-2501

James W. Young
Young Concepts
P.O. Box 1088
Centerville, UT 84014 (801) 292-1241

J. David Nelson
Day, Barney & Tycksen
45 East Vine St.
Murray, UT 84107 (801) 262-6800

A. Roy Osburn
Mallinckrodt, Mallinckrodt, Russell & Osburn
914 First Security Bank Bldg.
Ogden, UT 84401 (801) 393-0331

Thompson E. Fehr*
Defense Reutilization & Marketing Region
500 W. 12th St.
Orden, UT 84407 (801) 399-7759

Joseph William Brown#
J.W. Brown & Assoc.
698 E. 2320 North
Provo, UT 84604 (801) 373-1367

Robert B. Crouch
1165 Cedar Ave.
Provo, UT 84604 (801) 373-4758

H. Tracy Hall#
1711 N. Lambert Lane
Provo, UT 84604 (801) 374-0300

George H. Mortimer
3684 N. Little Rock Terr.
Provo, UT 84604 (801) 224-5647

Robert A. Bingham
I R E C O Inc.
Crossroads Tower, 11th Floor
Salt Lake City, UT 84144 (801) 364-4800

Richard F. Bojanowski
Suite 735, Judge Bldg.
8 East Broadway
Salt Lake City, UT 84111 (801) 533-0727

Laurence Blair Bond
Trask, Britt & Rossa
P.O. Box 2250
Salt Lake City, UT 84111 (801) 532-1922

William S. Britt
Trask, Britt & Rossa
525 S. 300 East, P.O. Box 2250
Salt Lake City, UT 84110 (801) 532-1922

Berne S. Broadbent
Workman, Nydegger & Jensen
57 W. 200 South, Third Floor
Salt Lake City, UT 84101 (801) 533-9800

Kent S. Burningham
Workman, Nydegger & Jensen
57 W. 200 South Third Floor
Salt Lake City, UT 84124 (801) 533-9800

Jon C. Christiansen
Van Cott, Bagley, Cornwell & Mc Carthy
50 S. Main St., Suite 1600
Salt Lake City, UT 84144 (801) 532-3333

Grant R. Clayton
Workman, Nydegger & Jensen
57 West 200 South, Third Floor
Salt Lake City, UT 84101 (801) 533-9800

Kay S. Cornaby
Jones, Waldo, Holbrook & Mc Donough
1500 First Interstate Plaza, 170 South Main St.
Salt Lake City, UT 84101 (801) 521-3200

B. Deon Criddle
1399 South 700 East, Suite 10
Salt Lake City, UT 84101 (801) 485-9811

Robert R. Finch
4322 Vallejo Dr.
Salt Lake City, UT 84124 (801) 278-8184

Lynn Grant Foster
602 E. 3rd South
Salt Lake City, UT 84102 (801) 364-5633

Andrew C. Hess
Callister, Duncan & Nebeker
Kennecott Bldg., Suite 800
Salt Lake City, UT 84133 (801) 530-7318

Allen R. Jensen
Workman, Nydegger & Jensen
America Plaza II, 3rd Floor, 57 W. 200 South
Salt Lake City, UT 84101 (801) 533-9800

Mathew D. Madsen#
123 Second Ave., Apt. P-105
Salt Lake City, UT 84103 (801) 521-6769

Craig J. Madson
Workman, Nydegger & Jensen
American Plaza II, Third Floor, 57 W. 200 South
Salt Lake City, UT 84110

Philip A. Mallinckrodt
Mallinckrodt, Mallinckrodt, Russell & Osburn
10 Exchange Place, Suite 1010
Salt Lake City, UT 84111 (801) 328-1624

Robert R. Mallinckrodt
Mallinckrodt, Mallinckrodt, Russell & Osburn
10 Exchange Place, Suite 1010
Salt Lake City, UT 84111 (801) 328-1624

Michael D. McCully
2533 Catalina Dr.
Salt Lake City, UT 84121 (801) 942-1883

Lloyd C. Metcalf
Workman, Nydegger & Jensen
57 W. 200 South
Salt Lake City, UT 84101 (801) 533-9800

Charles J. Moxley
Roe, Fowler & Moxley
340 E. Fourth South
Salt Lake City, UT 84111 (801) 328-9841

Rick D. Nydegger
Fox, Edwards & Gardiner
P.O. Box 3450
Salt Lake City, UT 84110 (801) 521-7751

Charles L. Roberts
Workman, Nydegger & Jensen, American Plaza II, Third Floor
57 West 200 South
Salt Lake City, UT 84101 (801) 533-9800

Thomas J. Rossa
Trask & Britt
P.O. Box 1978
Salt Lake City, UT 84110 (801) 532-1922

Piero G. Ruffinengo
849 18th Ave.
Salt Lake City, UT 84103 (801) 328-1936

M. Reid Russell
559 East South Temple
Salt Lake City, UT 84102 (801) 532-1601

David O. Seeley
Workman, Nydegger & Jensen
American Plaza II, 3rd Fl.
57 West 200 South
Salt Lake City, UT 84106 (801) 533-9800

Marlin Ralph Shaffer, Jr.
Suite 1000 Continental Bank Bldg.
200 S. Main St.
Salt Lake City, UT 84101 (801) 359-7771

Marcus G. Theodore
Suite 701, Valley Tower
50 West Broadway
Salt Lake City, UT 84101 (801) 359-8622

David V. Trask
Trask & Britt
P.O. Box 1978
Salt Lake City, UT 84110 (801) 532-1922

Joseph A. Walkowski, Jr.
Eastman Christensen Co.
1937 South 300 West
Salt Lake City, UT 85115 (801) 487-4545

David A. Westerby
Utah Power & Light Co.
1407 W. North Temple
Salt Lake City, UT 84111 (801) 535-4265

Evan R. Witt
Workman, Nydegger & Jensen
57 West 200 South
Salt Lake City, UT 84101 (801) 533-9800

John L. Wood
Snow, Christensen & Marhieau
10 Exchange Place, 11th Floor
Salt Lake City, UT 84145 (801) 521-9000

H. Ross Workman
Workman, Nydegger & Jensen
American Plaza II, 3rd Fl.
57 West 200 South
Salt Lake City, UT 84110 (801) 533-9800

Terry M. Crellin
Thrope, North & Western
9662 South State St.
Sandy, UT 84070 (801) 566-6633

Vaughn W. North
Thorpe, North & Western
9662 S. State St.
Sandy, UT 84070 (801) 566-6633

Calvin E. Thorpe
Thorpe, North & Western
9662 S. State
Sandy, UT 84070 (801) 566-6633

Marion Wayne Western
Thorpe, North & Western
9662 S. State St.
Sandy, UT 84070 (801) 566-6633

David Ferber#
P.O. Box 99
Springdale, UT 84767 (801) 772-3237

Vermont

Mary Ann Mento#
Rural Route 1
Box 1945
Arlington, VT 05250 (802) 375-6795

John H. Gallagher
R.D. 1
Beebe Lake
Brandon, VT 05733 (802) 273-2797

Bailin L. Kuch
General Elec. Co. Armament Dept.
Lakeside Ave.
Burlington, VT 05401 (802) 657-6592

Stephen John Limanek
75 De Forest Heights
Burlington, VT 05401 (802) 862-1200

Donald T. Steward, Sr.#
P.O. Box 63
Cambridgeport, VT 05141 (802) 869-2754

Francis Joseph Thornton
R D No. 1
Charlotte, VT 05445 (802) 425-2410

Howard J. Walter, Jr.
I B M Corp.
1000 River St.
Essex Junction, VT 05452 (802) 769-9555

Clarence L. Carlson#
R.F.D. 2
Box 301
Lyndonville, VT 05851 (802) 626-8583

James J Cannon, Jr.
Barnumville Rd.
P.O. Box 1508
Manchester Center, VT 05255

Thomas N. Neiman#
R. D. 3
Meadow Ridge Lane
Milton, VT 05468 (802) 893-2342

Theodore E. Galanthay
27 Pheasant Way
S. Burlington, VT 05403 (802) 769-8843

Robert E. Dunn
205 Harbor Rd.
Shelburne, VT 05482 (802) 985-2120

J. Franklin Jones, Jr.
P.O. Box 737
103 Summer St.
Springfield, VT 05156 (802) 885-5127

William O. Moeser
Brownell & Moeser
7 Wall St.
Springfield, VT 05156 (802) 885-4591

Rudolph J. Anderson, Jr.
R R 3
Logging Hill Rd.
Stowe, VT 05672 (802) 253-9827

George E. Kersey
Brook Rd.
P.O. Box 64
Strafford, VT 05072 (802) 763-8502

Lawrence Shaper
Cream St.
Thetford Center, VT 05075 (802) 785-4403

Virgin Islands

Thomas V. Crevling#
P.O. Box 8650
St. Thomas, VI 00801 (809) 774-3272

Virginia

William Earl Fears
Fears & Agor
Box 210
Accomac, VA 23301 (804) 787-1560

Wescott B. Northam
P.O. Box 55
Accomac, VA 23301 (804) 787-1511

Duane L. Antton#
Bacon & Thomas
625 Slaters Ln. Suite 400
Alexandria, VA 22314 (703) 683-0500

David D. Bahler
Parkhurst & Oliff
277 South Wash. St.
Alexandria, VA 22314 (703) 836-6400

Thomas E. Beall, Jr.
Fay, Sharpe, Beall, Fagan, Minnich & Mc Kee
104 E. Hume Ave.
Alexandria, VA 22301 (703) 684-1120

Stephen A. Becker
Lowe, Price, Le Blanc, Becker & Shur
427 N. Lee St.
Alexandria, VA 22320 (703) 684-1111

Stephen A. Bent#
Schwartz, Jeffery, Schwaab, Mark, Blumenthal & Evans
P.O. Box 299
Alexandria, VA 22314 (703) 836-9300

Eugene Berman
5903 Mt. Eagle Dr. Apt. 714
Alexandria, VA 22303 (703) 960-0078

William P. Berridge
Parkhurst & Oliff
277 S. Washington St. # 212, P.O. Box 19928
Alexandria, VA 22320 (703) 836-6400

David A. Blumenthal
Schwartz, Jeffery, Schwaab, Mack, Blumenthal & Evans
1800 Diagonal Rd. Suite 510
Alexandria, VA 22314 (703) 836-9300

Evon C. Blunk, Sr.
8701 Highgate Rd.
Alexandria, VA 22308 (703) 780-8186

Bruce J. Boggs, Jr.#
Burns, Doane, Swecker & Mathis
699 Prince St.
Alexandria, VA 22314 (703) 836-6620

Eugene Michael Bond
Box 1251
Alexandria, VA 22313 (202) 842-0010

William J. Bond#
Burns, Doane, Swecker & Mathis
699 Prince St.
Alexandria, VA 22313 (703) 836-6620

Joseph A. Boska#
7802 Strathdon Court
Alexandria, VA 22310 (703) 971-8366

Alfred W. Breiner
Breiner & Breiner
115 North Henry St. P.O. Box 19290
Alexandria, VA 22314 (703) 684-6885

Theodore A. Breiner
Breiner & Breiner
115 North Henry St.
Alexandria, VA 22314 (703) 684-6885

John C. Brosky
Burns, Doane, Swecker & Mathis
Washington & Prince Sts.
P.O. Box 1404
Alexandria, VA 22313 (703) 836-6620

Kevin C. Brown#
Parkhurst & Oliff
277 S. Washington St.
Alexandria, VA 22314 (703) 836-6400

William J. Bundren
7469 Towchester Court
Alexandria, VA 22310 (703) 971-5585

Frederick F. Calvetti
Stevens, Davis, Miller & Mosher
515 N. Washington St.
Alexandria, VA 22314 (703) 549-7200

Robert Warren Carlson
6406 May Blvd.
Alexandria, VA 22310 (703) 971-4839

John A. Carroll#
7119 Vantage Dr.
Alexandria, VA 22306 (703) 765-7713

Herbert T. Carter, Sr.#
2115 Shiver Dr.
Alexandria, VA 22307 (703) 768-0185

Perry Carvellas
Sherman & Shalloway
413 N. Wash St.
Alexandria, VA 22314 (703) 549-2282

Donald Clarke Casey
Lowe, Prince, Le Blanc, Becker & Shur
427 Lee St., N.
Alexandria, VA 22314 (703) 684-1111

Lance W. Chandler#
Burns, Doane, Swecker & Mathis
Washington & Prince Sts.
P.O. Box 1404
Alexandria, VA 22313 (703) 836-6620

Chung-Chin Chen#
Bacon & Thomas
625 Slaters Lane, 4th Floor
Alexandria, VA 22314 (703) 683-0500

Joseph A. Cooke
5903 Mount Eagle Dr. Suite 518
Alexandria, VA 22303 (703) 960-9327

Clifton B. Cosby#
710 Parkway Terr.
Alexandria, VA 22302 (703) 549-8862

Thomas J. D Amico
Stevens, Davis, Miller & Mosher
515 N. Washington St. Box 1427
Alexandria, VA 22314 (703) 549-7200

Joseph De Benedictis
Bacon & Thomas
625 Slaters Ln., 4th Fl.
Alexandria, VA 22314 (703) 683-0500

Benton S. Duffett, Jr.
Burns, Doane, Swecker & Mathis
699 Prince St.
Alexandria, VA 22313 (703) 836-6620

Joseph D. Evans
Schwartz, Jeffrey, Schwaab, Mack, Blumenthal & Evans, P.C.
1800 Diagonal Rd. Suite 510
Alexandria, VA 22313 (703) 836-9300

John J. Feldhaus#
Schwartz, Jeffery, Schwaab, Mack, Blumenthal & Evans.
1800 Diagonal Rd. Suite 510
Alexandria, VA 22313 (703) 836-9300

Richard E. Fichter
Bacon & Thomas
4th Floor, 625 Slaters Lane
Alexandria, VA 22314 (703) 683-0500

Milton M. Field
108 S. Columbus St.
Alexandria, VA 22314 (703) 683-4700

George Fine
5787 Winston Court
Alexandria, VA 22311 (703) 820-3081

Glen B. Foster#
1719 Preston Rd.
Alexandria, VA 22302 (703) 671-0975

Joel Mark Freed
Burns, Doane, Swecker & Mathis
699 Prince St.
Alexandria, VA 22313 (703) 836-6620

Vincent J. Frilette#
515 N. Washington St.
Alexandria, VA 22314 (703) 549-0264

Erich J. Gess
Burns, Doane, Swecker & Mathis
699 Prince St.
Alexandria, VA 22313 (703) 836-6620

Robert Paul Gibson*
U.S. Army Materiel Commd
5001 Eisenhower Ave
Alexandria, VA 22333 (202) 274-8040

Vincent Gerard Gioia
Schwartz, Jeffrey, Schwaab, Mack, Blumenthal & Evans
P.O. Box 299, Suit 510, 1800 Diagonal Rd.
Alexandria, VA 22313 (703) 836-9300

Philip Goodman#
307 Yoakum Pkwy.
Alexandria, VA 22304 (703) 751-3631

Israel Gopstein
Lowe, Price, Le Blanc, Becker & Shur
427 Lee St., N.
Alexandria, VA 22320 (703) 684-1111

Rosemary M. Graniewski#
7806 Ridgecrest Dr.
Alexandria, VA 22308 (703) 765-5595

William J. Griffin#
Burns, Doane, Swecker & Mathis
P.O. Box 1404
Alexandria, VA 22313 (703) 836-6620

Ronald L. Grudziecki
Burns, Doane, Swecker & Mathis
699 Prince St.
Alexandria, VA 22313 (703) 836-6620

Richard C. Harris
Stevens, Davis, Miller & Mosher
515 N. Washington St.
Alexandria, VA 22314 (703) 549-7200

Benjamin J. Hauptman
Lowe, Price, Le Blanc, Becker & Shur
427 North Lee St.
Alexandria, VA 22320 (703) 684-1111

Patricia W. Heenan#
910 Vicar Lane
Alexandria, VA 22302 (703) 751-1922

Karl F. Hoback#
Sherman & Shalloway
413 N. Washington St.
Alexandria, VA 22314 (703) 549-2282

Herman J. Hohauser
625 Slaters Lane 4th Floor
Alexandria, VA 22314 (703) 683-0500

George A. Hovanec, Jr.
Burns, Doane, Swecker & Mathis
699 Prince St.
Alexandria, VA 22313 (703) 836-6620

Kirk M. Hudson
Parkhurst & Oliff
277 S. Washington St.
Alexandria, VA 22314 (703) 836-6400

Deborah S. Humble
6000 Fort Hill Rd.
Alexandria, VA 22307 (703) 329-1236

Robert Danny Huntington
Burns, Doane, Swecker & Mathis
699 Prince St.
Alexandria, VA 22313 (703) 836-6620

Donald D. Jeffery
Schwartz, Jeffery, Schwaab, Mack, Blumenthal & Evans
Suite 510, 1800 Diagonal Rd. P.O. Box 299
Alexandria, VA 22313 (703) 836-9300

Lance G. Johnson#
211 N. Union St. P.O. Box 1878
Alexandria, VA 22313 (703) 836-6070

J. Ernest Kenney
Bacon & Thomas
625 Slaters Lane, 4th Floor
Alexandria, VA 22314 (703) 683-0500

Jeffrey M. Ketchum#
Beal Law Off.
104 E. Hume Ave.
Alexandria, VA 22301 (703) 684-1120

Joseph M. Kileen
510 King St. Suite 408
Alexandria, VA 22314 (703) 838-0400

Carl E. Kneuertz#
1204 S. Washington St. Suite 114
Alexandria, VA 22314 (703) 549-5518

William A. Knoeller
Stevens, Davis, Miller & Mosher
515 N. Washington St.
Alexandria, VA 22314 (703) 549-7200

Robert J. Koch#
Schwartz, Jeffery, Schwaab, Mack, Blumenthal & Koch, P.C.
111 N. Alfred St. P.O. Box 299
Alexandria, VA 22313 (703) 836-9300

John Kominski
1402 Key Dr.
Alexandria, VA 22302 (703) 751-3026

Alan Edward Kopecki
Burns, Doane, Swecker & Mathis
699 Prince St. P.O. Box 1404
Alexandria, VA 22313 (703) 836-6620

Holly D. Kozlowski#
Lowe, Price, Le Blanc, Becker & Shur
427 N. Lee St.
Alexandria, VA 22320 (703) 684-1111

Charlotte M. Kraebel
340 Commerce St.
Alexandria, VA 22314 (703) 683-6226

Kenneth E. Krosin
Lowe, Price, Le Blanc, Becker & Shur
427 N. Lee St.
Alexandria, VA 22314 (703) 684-1111

Robert I. Lainof
1513 King St.
Alexandria, VA 22314 (703) 548-7777

James A. Le Barre
Burns, Doane, Swecker & Mathis
699 Prince St.
Alexandria, VA 22313 (703) 836-6620

Robert E. Le Blanc
Lowe, Price, Le Blanc, Becker & Shur
427 North Lee St.
Alexandria, VA 22320 (703) 684-1111

Jameson Lee
Burns, Doane, Swecker & Mathis
Suite 100, 699 Prince St.
Alexandria, VA 22314 (703) 836-6620

Joseph R. Lentz#
609 S. St Asaph St.
Alexandria, VA 22314 (703) 549-4453

Allan M. Lowe
Lowe, Price, Le Blanc, Becker & Shur
427 N. Lee St.
Alexandria, VA 22320 (703) 684-1111

James C. Lydon
Parkhurst & Oliff
277 S. Washington St. Suite 212
Alexandria, VA 22314 (703) 836-6400

Jerry C. Lyell
5743 Harwich Court #22
Alexandria, VA 22311 (703) 379-1523

Peter G. Mack
Schwartz, Jeffery, Schwaab, Mack, Blumenthal & Evans.
Suite 510, 1800 Diagonal Rd. P.O. Box 299
Alexandria, VA 22313 (703) 836-9300

Joseph R. Magnone
Bruns, Doane, Swecker & Mathis
699 Prince St.
Alexandria, VA 22313 (703) 836-6620

Platon Nick Mandros
Burns, Doane, Swecker & Mathis
699 Prince St.
P.O. Box 1404
Alexandria, VA 22313 (703) 836-6620

Eugene Mar
Bacon & Thomas
625 Slaters Lane, 4th Floor
Alexandria, VA 22314 (703) 683-0500

William Lowrey Mathis
Burns, Doane, Swecker & Mathis
699 Prince St.
Alexandria, VA 22314 (703) 836-6620

John R. Mattingly#
Fay, Sharpe, Beall, Fagan, Minnich & Mc Kee
103 E. Hume Ave.
Alexandria, VA 22301 (703) 684-1120

Erin Alice Mc Dowell#
Aiken Advanced Sys.
5901 Edsall Rd.
Alexandria, VA 22304 (703) 370-0900

Christopher L. Mc Kee#
Parkhurst & Oliff
277 Washington St., S.
Alexandria, VA 22314 (703) 836-6400

Frederick G. Michaud, Jr.
Burns, Doane, Swecker & Mathis
699 Prince St.
Alexandria, VA 22313 (703) 836-6620

Samuel C. Miller
Burns, Doane, Swecker & Mathis
P.O. Box 1404, Washington & Prince Sts.
Alexandria, VA 22313 (703) 836-6620

Thomas J. Moore
Bacon & Thomas
625 Slaters Lane, 4th Floor
Alexandria, VA 22314 (703) 683-0500

Ellsworth H. Mosher
Stevens, Davis, Miller & Mosher
515 North Washington St.
Alexandria, VA 22314 (703) 549-7200

Robert G. Mukai
Burns, Doane, Swecker & Mathis
P.O. Box 1404, Wash. & Prince Sts.
Alexandria, VA 22313 (703) 836-6620

David R. Murphy
Quaintance, Murphy & Presta
6462 Little River Turnpike, Suite E
Alexandria, VA 22312 (703) 354-4600

Chittaranjan N. Nirmel
Lowe, Price, Le Blanc, Becker & Shur
427 North Lee St.
Alexandria, VA 22320 (703) 684-1111

Joseph M. Noto#
Schwartz, Jeffery, Schwaab, Mack, Blumenthal & Evans P.C.
Suite 510, 1800 Diagonal Rd. P.O. Box 299
Alexandria, VA 22313 (703) 836-9300

James A. Oliff
Parkhurst & Oliff
P.O. Box 19928
Alexandria, VA 22320 (703) 836-6400

Glenn S. Ovrevik#
7912 Telegraph Rd.
Alexandria, VA 22310 (703) 971-1824

Thomas J. Pardini
Parkhurst & Oliff
P.O. Box 19928, 277 S. Wash. St.
Alexandria, VA 22320 (703) 836-6400

Gayle Parker
8346 Orange Ct.
Alexandria, VA 22309 (703) 780-4314

Roger W. Parkhurst
Parkhurst & Oliff
277 S. Washington St.
Alexandria, VA 22314 (703) 836-6400

Thomas P. Pavelko
Stevens, Davis, Miller & Mosher
P.O. Box 1427
Alexandria, VA 22313 (703) 549-7200

Patricia Q. Peake
515 N. Washington St.
Alexandria, VA 22314 (703) 549-7609

James A. Poulos
Stevens, Davis, Miller & Mosher
515 Washington St., N.
Alexandria, VA 22314 (703) 549-7200

Frank Paul Presta
Quaintance, Murphy & Presta
6462 Little River Turnpike, Suite E.
Alexandria, VA 22312 (703) 354-4600

Robert Lee Price
Lowe, Price, Le Blanc, Becker & Shur
427 North Lee St.
Alexandria, VA 22320 (703) 684-1111

Teresa Stanek Rea
Burns, Doane, Swecker & Mathis
699 Prince St.
Alexandria, VA 22313 (703) 836-6620

David D. Reynolds
Burns, Doane, Swecker & Mathis
699 Prince St.
Alexandria, VA 22313 (703) 836-6620

Julie Ilisa Ring#
Burns, Doane, Swecker & Mathis
699 Prince St.
Alexandria, VA 22313 (703) 836-6620

Muzio B. Roberto*
U.S. Army Materiel Dev. And Readiness Comm.
5001 Eisenhower AVe.
Alexandria, VA 22333 (202) 274-8147

L. Lawton Rogers, III
510 King St. Suite 408
Alexandria, VA 22314 (703) 836-0400

Joseph M. Rolnicki#
Bacon & Thomas
625 Slaters Lane
Alexandria, VA 22314 (703) 683-0500

Marc A. Rossi#
Schwartz, Jeffery, Schwaab, Mack, Blumenthal & Evans
King St. Station, Suite 510, 1800 Diagonal Rd.
Alexandria, VA 22313 (703) 836-9300

Peter T. Rutkowski
Burns, Doane, Swecker & Mathis
699 Prince St.
Alexandria, VA 22313 (703) 836-6620

Michael G. Savage#
Burns, Doane, Swecker & Mathis
699 Prince St.
Alexandria, VA 22313 (703) 836-6620

Stanley D. Schlosser
Schwartz, Jeffery, Schwaab, Mack, Blumenthal & Evans
1800 Diagonal Rd. Suite 510
Alexandria, VA 22313 (703) 836-9300

Matthew L. Schneider#
Burns, Doane, Swecker & Mathis
699 Prince St.
Alexandria, VA 22313 (703) 836-6620

Robert M. Schulman
Burns, Doane, Swecker & Mathis
699 Prince St.
Alexandria, VA 22313 (703) 836-6620

Richard L. Schwaab
Schwartz, Jeffery, Schwaab, Mack, Blumenthal & Evans
P.O. Box 299, Suite 510, 1800 Diagonal Rd.
Alexandria, VA 22313 (703) 836-9300

Arthur Schwartz
Schwartz, Jeffery, Schwaab, Mack, Blumenthal & Evans.
P.O. Box 299, Suite 510, 1800 Diagonal Rd.
Alexandria, VA 22313 (703) 836-9300

David J. Serbin
Schwartz, Jeffery, Schwaab, Mack, Blumenthal & Koch
111 N. Alfred St. P.O. Box 299
Alexandria, VA 22313 (703) 836-9300

Edwin A. Shalloway
Sherman & Shalloway
413 N. Wash St.
Alexandria, VA 22314 (703) 549-2282

Anthony W. Shaw
Burns, Doane, Swecker & Mathis
699 Prince St. Suite 100
Alexandria, VA 22313 (703) 836-6620

Fred W. Sherling
1233 N. Picket St.
Alexandria, VA 22304 (703) 370-2445

Leonard W. Sherman
Sherman & Shalloway
413 N. Wash. St.
Alexandria, VA 22314 (703) 549-2282

Chandrakant C. Shroff#
600 N. Picket St.
Alexandria, VA 22304 (703) 823-2024

Henry Shur
Lowe, Price, Le Blanc, Becker & Shur
427 N. Lee St.
Alexandria, VA 22320 (703) 684-1111

Regis E. Slutter
Burns, Doane, Swecker & Mathis
699 Prince St.
Alexandria, VA 22313 (703) 836-6620

Peter H. Smolka
Burns, Doane, Swecker & Mathis
P.O. Box 1404, 699 Prince St.
Alexandria, VA 22313 (703) 836-6620

H. Jay Spiegel
Sherman & Shalloway
413 N. Wash. St.
Alexandria, VA 22314 (703) 549-2282

Stanley C. Spooner
Stevens, Davis, Miller & Mosher
515 North Wash. St. P.O. Box 1427
Alexandria, VA 22313 (703) 549-7200

Mary C.O. Stauss#
7701 Tauxemont Rd.
Alexandria, VA 22308 (703) 765-4859

Richard A. Steinberg#
Sherman & Shalloway
P.O. Box 788, 413 N. Washington St.
Alexandria, VA 22314 (703) 549-2282

Norman H. Stepno
Burns, Doane, Swecker & Mathis
699 Prince St.
Alexandria, VA 22313 (703) 836-6620

William D. Stokes
707 Prince St.
Alexandria, VA 22314 (703) 548-0210

Michael J. Strauss
Lowe, Price, Le Blanc, Becker & Shur
427 North Lee St.
Alexandria, VA 22314 (703) 684-1111

Brereton Sturtevant
1227 Morningside Lane
Alexandria, VA 22308 (703) 765-8598

Robert S. Swecker
Burns, Doane, Swecker & Mathis
P.O. Box 1404
Alexandria, VA 22313 (703) 836-6620

Joerg-Uwe V. Szipl#
Schwartz, Jeffery, Schwaab, Mack, Blumenthal & Evans
111 N. Alfred St. P.O. Box 299
Alexandria, VA 22313 (703) 836-9300

Marian T. Thomson
7126 Devonshire Rd.
Alexandria, VA 22307 (703) 765-1893

Stanley H. Tollberg#
203 Yoakum Parkway Apt. 726
Alexandria, VA 22304 (703) 370-5538

Raymond Irving Tompkins
8314 Ashwood Dr.
Alexandria, VA 22308 (703) 360-3188

Bruce H. Troxell
Bacon & Thomas
625 Slaters Lane, 4th Floor
Alexandria, VA 22314 (703) 683-0500

Mark E. Ungerman#
277 S. Washington St. Suite 202
Alexandria, VA 22313 (703) 836-6070

Peter J. Van Bergen
Schwartz, Jeffery, Schwaab, Mack, Blumenthal & Evans
111 N. Alfred St.
Alexandria, VA 22314 (703) 836-9300

Leonidas Vlachos#
4007 Javins Dr.
Alexandria, VA 22310 (703) 960-2126

Ursula Von Usslar#
Schwartz, Jeffery, Schwaab, Mack, Blumenthal & Evans
Suite 510, 1800 Diagonal Rd. P.O. Box 299
Alexandria, VA 22313 (703) 836-9300

Edward Phillip Walker
Parkhurst & Oliff
277 S. Washington St.
Alexandria, VA 22314 (703) 836-6400

Eric H. Weisblatt#
Burn, Doane, Swecker, & Mathis
699 Prince St.
Alexandria, VA 22313 (703) 836-6620

John Remon Wenzel
Quaintance, Murphy & Presta
6462 Little River Turnpike, Suite E
Alexandria, VA 22312 (703) 354-4600

Farrell Roy Werbow
Stevens, Davis, Miller & Mosher
515 Washington St., N.
Alexandria, VA 22314 (703) 549-7200

John A. Weresh#
Burns, Doane, Swecker & Mathis
699 Prince St.
Alexandria, VA 22313 (703) 836-6620

Gerald H. Werfel*
Defense Logistics Agency
Office of General Council
Cameron Station
Alexandria, VA 22304 (202) 274-6815

Milton S. Winters
Watergate At Landmark
205 Yoakum Pkwy. Apt. 621
Alexandria, VA 22304 (703) 370-1466

Charles R. Wolfe, Jr.
Bacon & Thomas
625 Slaters Lane, 4th Floor
Alexandria, VA 22314 (703) 683-0500

Jim Zegeer
801 N. Pitt St. Suite 108
Alexandria, VA 22314 (703) 684-8333

John R. Spielman
P.O. Box 871
Amherst, VA 24521 (804) 946-5266

Gladden L. Brilhart#
8206 Galahad Court
Annandale, VA 22003 (703) 560-9577

James E. Bryan
8209 Briar Creek Dr.
Annandale, VA 22003

Ernest F. Chapman
4840 Kingston Dr.
Annandale, VA 22003 (703) 256-3808

I.J. Crickenberger
Springdale Professional Center
5027 Backlick Rd.
Annandale, VA 22003 (703) 256-2244

Harold P. Deeley, Jr.
4921 S. Centaurs Court
Annandale, VA 22003 (703) 978-7531

Carl D. Ouarforth#
3902 Oliver Ave.
Annandale, VA 22003 (703) 256-2695

Vincent L. Ramik
Diller, Ramik & Wight, P.C.
Merrion Square, Suite 101
1345 Mc Whorter Pl.
Annandale, VA 22003 (703) 642-5705

Terrance L. Siemens#
4108 Yerkes Place
Annandale, VA 22003 (703) 978-4883

Dinesh Agarwal
Shlesinger, Arkwright & Garvey
3000 South Eads St.
Arlington, VA 22206 (703) 684-5600

Penrose Lucas Albright
Pravel, Gambrell, Hewitt, Kimball & Krieger
P.O. Box 2246
Arlington, VA 22202 (703) 979-3242

George Alfred Arkwright, Jr.
Shlesinger, Arkwright, Garvey & Fado
3000 S. Eads St.
Arlington, VA 22202 (703) 684-5600

John B. Armentrout
1600 S. Eads St. Suite 1007N
Arlington, VA 22202 (703) 521-1143

James S. Bailey
6007 Williamsburg Blvd.
Arlington, VA 22207 (703) 536-8298

Joseph Jay Baker
Murray & Whisenhunt
1925 N. Lynn St. Suite 906
Arlington, VA 22209 (703) 243-0400

James L. Bean
O Neil & Bean
1601 N. Kent St. Suite 910
Arlington, VA 22209 (703) 525-3131

William E. Beaumont
Oblon, Fisher, Spivak, Mc Clelland & Maier
Crystal Square Five - Suite 400, 1755 S. Jeff. Davis Hwy.
Arlington, VA 22202 (703) 521-5940

John E. Benoit
Benoit, Smith & Laughlin
2001 Jefferson Davis Hwy. Suite 503
Arlington, VA 22202 (703) 521-1677

Joseph W. Berenato, III
Shlesinger, Arkwright, Garvey & Fado
3000 S. Eads St.
Arlington, VA 22202 (703) 684-5600

Frederick L. Berget
Nies, Webner, Kurz, & Bergert
1911 Jeff. Davis Hwy. Suite 700
Arlington, VA 22202 (703) 521-6590

Jerry William Berkstresser
Shoemaker & Mattare, Ltd.
2001 Jeff. Davis Hwy. Suite 1203, C P - 1
Arlington, VA 22202 (703) 521-5210

Bruce H. Berstein
Sandler & Greenblum
2920 Glebe Rd., S.
Arlington, VA 22206 (703) 739-0333

Lois P. Besanko#
Dressler, Goldsmith, Shore, Sutker & Milnamow, Ltd.
2001 Jeff. Davis Hwy. Suite 810
Arlington, VA 22202 (730) 521-1880

Richard George Besha
Nixon & Vanderhye, P.C.
2000 N. 15th St. Suite 409
Arlington, VA 22201 (703) 875-0400

George Ladow Black#
I B M Corp.
1755 S. Jeff. Davis Hwy. Suite 605
Arlington, VA 22202 (703) 769-2421

William A. Blake
Jones, Tullar & Cooper, P.C.
P.O. Box 2266, Eads Station
Arlington, VA 22202 (703) 521-5200

Samuel H. Blech
Oblon, Fisher, Spivak, Mc Clelland & Maier
1755 S. Jeff. Davis Hwy.
Arlington, VA 22202 (703) 521-5940

Peter A. Borsari#
2001 Jeff. Davis Hwy. Suite 603
Arlington, VA 22202 (703) 920-2377

Charles Paul Boukus, Jr.
2001 Jeff. Davis Hwy. Suite 202
Arlington, VA 22202 (703) 920-6120

Wade J. Brady, III
Indyk, Pojunas & Brady
2001 Jeff. Davis Hwy. Suite 409
Arlington, VA 22202 (703) 769-2990

Alan Edward Joseph Branigan
Griffin, Branigan & Butler
775 S. 23rd St.
Arlington, VA 22202 (703) 979-5700

Charles A. Brown
727 23rd St. South
Arlington, VA 22202 (703) 521-4536

Charles E. Brown
727 23rd St. South
Arlington, VA 22202 (703) 892-2791

Laurence Ray Brown
2001 Jefferson Davis Hwy. Suite 408
Arlington, VA 22202 (703) 521-7200

Scott L. Brown#
603 South 23rd St.
Arlington, VA 22202 (703) 920-6027

William S. Brown
2001 Columbia Pike Apt. 404
Arlington, VA 22204 (703) 920-5916

Harold W. Burnam, Jr.
Griffin, Branigan & Butler
775 S. 23rd St.
P.O. Box 2326
Arlington, VA 22202 (703) 979-5700

F. Prince Butler
Griffin, Brannigan & Butler
775 S. 23rd St.
Arlington, VA 22202 (703) 979-5700

Robert Allen Cahill
General Elec. Co.
2001 Jeff. Davis Hwy.
Arlington, VA 22202 (202) 637-4525

James A. Cairns
Sandler & Greenblum
2920 S. Glebe Rd.
Arlington, VA 22206 (703) 739-0333

Marvin A. Champion
1600 S. Eads St. Apt. 836-N
Arlington, VA 22202 (703) 521-1947

Kathleen H. Claffy#
2301 Jeff. Davis Hwy. Apt. 716
Arlington, VA 22202 (703) 979-6230

Herbert Cohen
Wigman & Cohen
Crystal Plaza 3, Suite 200, 1735 Jeff. Davis Hwy.
Arlington, VA 22202 (703) 892-4300

Sheldon I. Cohen
2009 N. 14th St. Suite 708
Arlington, VA 22201 (703) 522-1200

Robert R. Cook#
Oblon, Fisher, Spivak, McClelland & Maier
1755 Jeff. Davis Hwy. Suite 400
Arlington, VA 22202 (703) 521-5940

George M. Cooper
Jones, Tullar & Cooper, P.C.
2001 Jeff. Davis Hwy. Suite 1002
Arlington, VA 22202 (703) 521-5200

James F. Cottone#
Crystal Plaza One. Suite 1008
2001 Jeff. Davis Hwy.
Arlington, VA 22202 (703) 920-6772

Melvin L. Crane#
1201 S. Eads St.
Arlington, VA 22202 (703) 892-0300

Arthur R. Crawford
Nixon & Vanderhye P.C.
2000 N. 15th St. Suite 409
Arlington, VA 22201 (703) 875-0406

Robert Thompson Crawford
3233 North Pershing Dr.
Arlington, VA 22201 (703) 524-9781

Charles A. Cross#
865 N. Jefferson St.
Arlington, VA 22204 (703) 243-7394

Robert F. Custard
2001 Jeff. Davis Hwy., Suite 301-Crystal Plaza 1
Arlington, VA 22202 (703) 521-1666

Felix J. D Ambrosio
Jones, Tullar & Cooper, P.C.
P.O. Box 2266, Eads Station, 2001 Jeff. Davis Hwy.
Arlington, VA 22202 (703) 521-5200

William J. Daniel
2009 N. 14th St. Suite 701
Arlington, VA 22201 (703) 527-0068

Kenneth E. Darnell#
1515 Jeff. Davis Hwy. Suite 303
Arlington, VA 22202 (703) 892-0462

Bryan H. Davidson
Nixon & Vanderhye, P.C.
2000 North 15th St. Suite 409
Arlington, VA 22201 (703) 875-0421

Anthony J. De Laurentis
2001 Jeff. Davis Hwy. Suite 603
Arlington, VA 22202 (703) 920-2377

Paul V. Del Giudice#
Shoemaker & Mattare, Ltd.
2001 Jeff. Davis Hwy. Suite 1203
Arlington, VA 22202 (703) 521-5210

Donald L. Dennison
Dennison, Meserole, Pollack & Scheiner
1911 Jefferson Davis Hwy.
Arlington, VA 22202 (703) 521-1155

Robert I. Dennison
Dennison, Meserole, Pollack & Scheiner
1911 Jefferson Davis Hwy.
Arlington, VA 22202 (703) 521-1155

George A. Depaoli
Depaoli & O Brien
1911 Jefferson Davis Hwy.
Arlington, VA 22202 (703) 521-2110

John Bernard Dickman, III#
2001 Jeff. Davis Hwy. Suite 1203
Arlington, VA 22202 (703) 521-1320

A. Yates Dowell, III
Dowell & Dowell
2001 Jeff. Davis Hwy. Suite 705
Arlington, VA 22202 (703) 521-5550

A. Yates Dowell, Jr.
Dowell & Dowell
2001 Jeff. Davis Hwy. Suite 705
Arlington, VA 22202 (705) 521-5550

Ralph A. Dowell
Dowell & Dowell
2001 Jeff. Davis Hwy. Suite 705
Arlington, VA 22202 (703) 521-5550

Frederic C. Dreyer#
2301 Jeff. Davis Hwy. Apt. 1029
Arlington, VA 22202 (703) 920-1295

Kevin J. Dunleavy#
Larson & Taylor
727 S. 23rd St.
Arlington, VA 22202 (703) 920-7200

Stuart D. Dwork
Oblon, Fisher, Spivak, Mc Clelland & Maier, P.C.
1755 S. Jefferson Davis Hwy. Crystal Square Five-Suite 400
Arlington, VA 22202 (703) 521-5940

Ellen Marcie Emas
5057 12th St. South
Arlington, VA 22204 (703) 578-3620

Charles R. Engle
George Mason Univ. Law School
3401 N. Fairfax Dr.
Arlington, VA 22201 (703) 841-2614

Roger J. Erickson*
U.S. Dept. Of Navy Off. Of Naval Res.
800 N. Quincy St.
Arlington, VA 22217 (202) 696-4001

William T. Estabrook
Kemon & Estabrook
2001 Jeff. Davis Hwy.
Arlington, VA 22202 (703) 920-2162

Anna P. Fagelson
2301 S. Jeff. Davis Hwy.
Arlington, VA 22202 (703) 892-4449

Robert W. Faris
Nixon & Vanderhye
2000 N. 15th St. Suite 409
Arlington, VA 22201 (703) 875-0400

Walter C. Farley
1100 Wilson Blvd. Suite 1701
Arlington, VA 22209 (703) 528-5282

Donavon Lee Favre
1600 S. Joyce St. Suite C 503
Arlington, VA 22202 (703) 920-2962

Stanley Paul Fisher
Oblon, Fisher, Spivak, Mc Clelland & Maier, P.C.
Crystal Square, Suite 400, 1755 S. Jeff. Davis Hwy.
Arlington, VA 22202 (703) 521-5940

Rodger H. Flagg#
2101 Crystal Plaza Arcade, Box 135
Arlington, VA 22202 (703) 553-0501

Cynthia L. Foulke#
4326 N. Pershing Dr., Apt. 3
Arlington, VA 22203 (703) 528-3482

William F. Frank#
1911 Jeff. Davis Hwy.
Arlington, VA 22202 (703) 920-7062

Stuart D. Frenkel
Depaoli & Obrien, P.C.
1911 Jeff. Davis Hwy. Suite 1005
Arlington, VA 22202 (703) 521-2110

James D. Frew*
U.S. Dept. of Navy
Office of Naval Research
800 N. Quincy St.
Arlington, VA 22217 (202) 696-4002

Reuben Friedman
Oblon, Fisher, Spivak, Mc Clelland & Maier
1755 Jeff. Davis Hwy. Suite 400
Arlington, VA 22202 (703) 521-5940

Raymond L. Gable
Neuman, Williams, Anderson & Olson
2001 Jeff. Davis Hwy. Suite 903
Arlington, VA 22202 (703) 892-8787

William George Gapcynski
3833 N. Military Rd.
Arlington, VA 22207 (703) 525-6809

William C. Garvert*
Dept. of the Navy
Office of Naval Research
800 N. Quincy St.
Arlington, VA 22217 (202) 696-4000

George A. Garvey
Shlesinger, Arkwright & Garvey
3000 S. Eads St.
Arlington, VA 22202 (703) 684-5600

Peter J. Georges
Russell, Georges & Breneman
745 S. 23rd St. Suite 304
Arlington, VA 22202 (703) 521-1760

Karen M. Gerken#
Hoffman, Wasson & Fallow
522 Hayes Bldg. 2361 Jeff. Davis Hwy.
Arlington, VA 22202 (703) 920-1434

Arthur P. Gershman
Wigman & Cohen
1735 Jeff. Davis Hwy.
Arlington, VA 22202 (703) 892-4300

Charles L. Gholz
Oblon, Fisher, Spivak, Mc Clelland & Maier
1755 S. Jeff. Davis Hwy. Suite 400
Arlington, VA 22202 (703) 521-5940

Walter C. Gillis, Jr.
Larson & Taylor
727 23rd St. South
Arlington, VA 22202 (703) 920-7200

Stewart L. Gitler
Hoffman, Wasson & Fallow
2361 Jeff. Davis Hwy. Suite 522
Arlington, VA 22202 (703) 920-1434

Roger P. Glass
Sandler & Greenblum
2920 S. Glebe Rd.
Arlington, VA 22206 (703) 739-0333

Robert F. Gnuse
Oblon, Fisher, Spivak, Mc Clelland & Maier, P.C.
1755 Jeff. Davis Hwy. Cryatal Square Five, Suite 400
Arlington, VA 22202 (703) 521-5940

Thomas T. Gordon
2001 Jeff. Davis Hwy. Suite 301
Arlington, VA 22202 (703) 521-1666

Neil F. Greenblum
Sandler & Greenblum
2920 S. Glebe Rd.
Arlington, VA 22206 (703) 739-0333

Thomas J. Greer, Jr.
727 23rd St. South
Arlington, VA 22202 (703) 892-2410

Craig M. Gregersen
General Electric Co.
2001 Jeff. Davis Hwy. Suite 505
Arlington, VA 22202 (202) 637-4329

Edwin E. Greigg
727 23rd St. Suite 220
Arlington, VA 22202 (703) 892-0300

Ronald E. Greigg
727 23rd St. South, Suite 220
Arlington, VA 22202 (703) 892-0300

Benjamin F. Griffin, Jr.
Griffin, Branigan & Butler
P.O. Box 2326, 775 S. 23rd St.
Arlington, VA 22202 (703) 979-5700

Sarojini B. Grigsby#
1311 S. Norwood St.
Arlington, VA 22204 (703) 979-6315

Arnold G. Gulko
Dressler, Goldsmith, Shore, Sutker & Milnamow
2001 Jeff. Davis Hwy. Suite 810
Arlington, VA 22202 (703) 521-1880

Michael S. Gzybowski#
Sandler & Greenblum
2920 S. Glebe Rd.
Arlington, VA 22206 (703) 739-0333

H. Walter Haeussler
Jones, Tullar & Cooper
2001 Jeff. Davis Hwy. Suite 1002
Arlington, VA 22202 (703) 521-5200

George D. Hall
612-20th St. South
Arlington, VA 22202 (703) 979-3399

James D. Hamilton
Oblon, Fisher, Spivak, Mc Clelland & Maier
Crystal Square 5-Suite -400, 1755 S. Jeff. Davis Hwy.
Arlington, VA 22202 (703) 521-5940

Douglas R. Hanscom
Jones, Tullar & Cooper, P.C.
P.O. Box 2266, Eads Station
Arlington, VA 22202 (703) 521-5200

Watson D. Harbaugh
Sixbey, Friedman & Leedom, P.C.
2001 Jeff. Davis Hwy.
Arlington, VA 22202 (703) 521-2610

Robert F. Hargest, III
Witherspoon & Hargest
745 S. 23rd St. Suite 204
Arlington, VA 22202 (703) 521-0511

Boris Haskell
Paris & Haskell
2316 S. Eads St.
Arlington, VA 22202 (703) 979-4870

Dos T. Hatfield
4527 N. Rock Spring Rd.
Arlington, VA 22202 (703) 538-4527

Brion P. Heaney#
Millen & White, P.C.
503 Crystall Mall, Bldg. 1, 1911 Jeff. Davis Hwy.
Arlington, VA 22202 (703) 892-2200

James W. Hellwege
Young & Thompson
745 South 23rd St. Suite 200
Arlington, VA 22202

Heinrich Wolfgang Herzfeld#
P.O. Box 2445
Arlington, VA 22202 (703) 230-2836

David Harmon Hill
B-1205 Riverhouse II, 1400 S. Joyce St.
Arlington, VA 22202 (703) 920-7001

William R. Hinds
727 Twenty-Third St. South
Arlington, VA 22202 (703) 920-7500

Martin Paul Hoffman
Hoffman, Wasson & Fallow
522 Hayes Bldg. 2361 Jeff. Davis Hwy.
Arlington, VA 22202 (703) 920-1434

Alan Holler
Oblon, Fisher, Spivak, Mc Clelland & Maier
Crystal Square 5-Suite 400, 1755 Jeff. Davis Hwy.
Arlington, VA 22202 (703) 521-5940

William Harry Holt
727 23rd St. South, Suite 218
Arlington, VA 22202 (703) 553-0030

James T. Hosmer
Nixon & Vanderhye
2000 N. 15th St.
Arlington, VA 22201 (703) 875-0400

Ross Franklin Hunt, Jr.
Larson & Taylor
727 23rd St. South
Arlington, VA 22202 (703) 920-7200

Douglas E. Jackson
Larson & Taylor
727 S. 23rd St.
Arlington, VA 22202 (703) 920-7200

William E. Jackson
Larson & Taylor
727 23rd St. South
Arlington, VA 22202 (703) 920-7200

Fred Jacob
General Electric Co.
2001 Jeff. Davis Hwy.
Arlington, VA 22202 (202) 637-4314

Gary M. Jacobs
Fitzpatrick, Cella, Harper & Scinto
2011 Crystal Dr. Suite 905
Arlington, VA 22202 (703) 769-5040

Kenneth E. Jacobs
1600 S. Eads St. Apt. 311-S
Arlington, VA 22202 (703) 892-5167

Julius Jancin, Jr.
I B M Corp.
1755 S. Jeff. Davis Hwy. Suite 605
Arlington, VA 22202 (703) 769-2401

David Leonard Johnson, Jr.
1600 S. Eads St. 708S
Arlington, VA 22202 (703) 920-3489

James L. Jones, Jr.#
626 South 23rd St.
Arlington, VA 22202 (703) 979-4586

Jerry T. Kearns#
Washington Patent Service Corp.
2101 Crystal Plaza Arcade, Suite 269
Arlington, VA 22202 (703) 892-5713

Michael J. Keenan
Nixon & Vanderhye, P.C.
2000 N. 15th St. Suite 409
Arlington, VA 22201 (703) 875-0400

Steven B. Kelber#
Oblon, Fisher, Spivak, Mc Clelland & Maier
Crystal Square Five - Suite 400, 1755 S. Jeff. Davis Hwy.
Arlington, VA 22202 (703) 521-5940

Richard D. Kelly
Oblon, Fisher, Spivak, Mc Clelland & Maier
1755 S. Jeff. Davis Hwy. Suite 400
Arlington, VA 22202 (703) 521-5940

Solon B. Kemon
Shlesinger, Arkwright & Garvey
3000 S. Eads St.
Arlington, VA 22202 (703) 684-5600

William B. Kerkam, Jr.
Kerkham, Stowell, Kondrack, & Clarke, P.C.
1235 S. Jeff. Davis Hwy.
Arlington, VA 22202 (703) 920-8980

Richard Benedict Kirk#
Kirk & Smith
2001 Jefferson Davis Hwy. Suite 309
Arlington, VA 22202 (703) 521-1820

Werner Warren Kleeman
1735 Jeff. Davis Hwy. Suite 200 - Crystal Square 3
Arlington, VA 22202 (703) 892-4300

Maurice H. Klitzman
I B M Corp.
1755 S. Jeff. Davis Hwy. Suite 605
Arlington, VA 22202 (703) 769-2415

Steven Kreiss
Wigman & Cohen, P.C.
1735 Jeff. Davis Hwy. Suite 200, Crystal Square 3
Arlington, VA 22202 (703) 892-4300

Dennis L. Kreps#
2001 Jeff. Davis Hwy. Suite 602
Arlington, VA 22202 (703) 979-3315

Walter Kruger
Brisebois & Kruger
2361 Jefferson Davis Hwy. Suite 612
Arlington, VA 22202 (703) 521-1550

Eckhard H. Kuesters
Oblon, Fisher, Spivak, Mc Clelland & Maier
1755 Jeff. Davis Hwy. Suite 400
Arlington, VA 22202 (703) 521-5940

James A. Kunkle#
I B M Corp.
1755 S. Jeff. Davis Hwy. Suite 605
Arlington, VA 22202 (703) 769-2403

Philip Elledge Kurz
Nies, Webner, Kurz & Bergert
1911 Jeff. Davis. Hwy. Suite 700
Arlington, VA 22202 (703) 521-6590

Alfons F. Kwitnieski*
Office of Naval Research
800 N. Quincy St.
Arlington, VA 22217 (202) 696-4000

Dennis H. Lambert
Lambert & Skolnik
2001 Jeff. Davis Hwy. Suite 705
Arlington, VA 22202 (703) 451-1227

Munson H. Lane, Jr.#
1515 N. Courthouse Rd. Suite 402
Arlington, VA 22201 (703) 522-6762

Roberts Browing Larson
Larson & Taylor
727 23rd St., S.
Arlington, VA 22202 (703) 920-7200

Lawrence E. Laubscher, Jr.
Laubscher, Philpitt & Laubscher
745 South 23rd St. Suite 300
Arlington, VA 22202 (703) 521-2660

Lawrence Edwin Laubscher
Laubscher, Philpitt & Laubscher
745 South 23rd St. Suite 300
Arlington, VA 22202 (703) 521-2660

James H. Laughlin, Jr.
Benoit, Smith & Laughlin
2001 Jeff. Davis Hwy. Suite 501
Arlington, VA 22202 (703) 521-1677

Jean-Paul P.M. Lavalleye#
Oblon, Fisher, Spivak, Mc Clelland & Maier, P.C.
1755 S. Jeff. Davis Hwy.
Arlington, VA 22202 (703) 521-5940

Henry S. Layton#
2111 Jeff. Davis Hwy. Suite 1114 South
Arlington, VA 22202 (703) 979-5441

James E. Ledbetter
Sandler & Greenblum
701 S. 23rd St.
Arlington, VA 22202 (703) 521-7800

Saul Leitner
Leitner, Greene & Christensen, P.C.
Suite 203 Crystal Square 3, 1735 Jeff. Davis Hwy.
Arlington, VA 22202 (703) 486-8100

Marion P. Lelong
Depaoli & Obrien, P.C.
1911 Jeff. Davis Hwy. Suite 1005
Arlington, VA 22202 (703) 521-2110

Herbert Levine
522 Hayes Bldg. 2361 Jeff. Davis Hwy
Arlington, VA 22202 (703) 920-1434

Richard C. Litman
1725 S. Jeff. Davis Hwy. Suite 801
Arlington, VA 22215 (703) 920-6000

James H. Littlepage
Nies, Webner, Kurz & Bergert
1911 Jeff. Davis Hwy. Suite 700
Arlington, VA 22202 (703) 521-6590

George A. Loud
Pahl, Lorusso & Loud
2001 Jeff. Davis Hwy. Suite 1003
Arlington, VA 22202 (703) 979-1960

Leonard L. Lourie*
Office of Naval Research
800 N. Quincy St., Code 312
Arlington, VA 22217 (202) 692-3730

Warren N. Low
Low & Low
2316 S. Eads St.
Arlington, VA 22202 (703) 979-4870

Gregory J. Maier
Oblon, Fisher, Spivak, Mc Clelland & Maier
1755 S. Jeff. Davis Hwy. Suite 400- Crystal Square Five
Arlington, VA 22202 (703) 521-5940

Alfred J. Mangels
2001 Jeff. Davis Hwy. 408 Crystal Plaza I
Arlington, VA 22202 (703) 521-7200

Gary L. Manuse
P.O. Box 227
Eads St. Station
Arlington, VA 22202 (703) 920-4756

Brian J. Marton#
Shlesinger, Arkwright, Garvey & Fado
3000 S. Eads St.
Arlington, VA 22202 (703) 684-5600

William B. Mason
Mason, Mason & Albright
2306 S. Eads St.
Arlington, VA 22202 (703) 979-3242

Helen M. Mc Carthy
1200 N. Nash St. Apt. 1132
Arlington, VA 22209 (703) 522-1228

William F. Mc Carthy, Jr.*
U.S. Dept. of Navy
Office of Naval Research
800 N. Quincy St.
Arlington, VA 22217 (202) 696-4003

C. Irvin Mc Clelland
Oblon, Fisher, Spivak, Mc Clelland & Maier
1755 S. Jeff. Davis Hwy., Crystal Square 5- Suite 400
Arlington, VA 22202 (703) 521-5940

Thomas E. Mc Donnell*
Dept. of the Navy
Office of Naval Research
Code O O C C I P
800 N. Quincy St.
Arlington, VA 22217 (202) 696-4006

Barbara A. Mc Dowell#
I B M Corp.
1755 S. Jeff. Davis Hwy. Suite 605
Arlington, VA 22202 (703) 769-2407

John J. Mc Glew, Jr.#
Shlesinger, Arkwright, Garvey & Fado
Crystal Plaza 1, Suite 607
Arlington, VA 22202 (703) 521-1500

Angelo J. Mele
4007 N. 27th St.
Arlington, VA 22207 (703) 385-6767

Charles J. Merek#
Research Publication
2221 Jeff. Davis Hwy.
Arlington, VA 22202 (703) 920-5050

James J. Merek#
Shlesinger, Arkwright, Garvey & Fado
3000 S. Eads St.
Arlington, VA 22202 (703) 684-5600

William H. Meserole, Jr.
Dennison, Meserole, Pollack & Scheiner
1911 Jeff. Davis Hwy.
Arlington, VA 22202 (703) 521-1155

Calvin H. Milans
Jones, Tullar & Cooper
2001 Jeff. Davis Hwy.
Arlington, VA 22202 (703) 521-5200

I. William Millen
Millen & White, P.C.
503 Crystal Mall Bldg. 1, 1911 Jefferson Davis Hwy.
Arlington, VA 22202 (703) 892-2200

Davidson Church Miller
2538 23rd Rd., N.
Arlington, VA 22207 (703) 525-1563

Robert C. Miller
Oblon, Fisher, Spivak, Mc Clelland & Maier,
Suite 400- Crystal Square, 1755 S. Jeff. Davis Hwy.
Arlington, VA 22202 (703) 521-5940

George H. Mitchell, Jr.
304 Crystal Plaza One, 2001 Jeff. Davis Hwy.
Arlington, VA 22202 (703) 486-2252

James C. Mitchell#
Jones, Tullar & Cooper, P.C.
2001 Jeff. Davis Hwy. Suite 1002
Arlington, VA 22202 (703) 521-5200

A. Louis Monacell#
6613 N. 29th St.
Arlington, VA 22213 (703) 241-8421

Ruth N. Mordouch
Littlepage & Webner, P.C.
727 23rd St. South, Suite 216
Arlington, VA 22202 (703) 920-2544

John R. Moses
Quaintance & Murphy
2001 Jeff. Davis Hwy.
Arlington, VA 22202 (703) 521-2400

Martin G. Mullen
Sandler & Greenblum
2920 S. Glebe Rd.
Arlington, VA 22206 (703) 739-0333

Kimbley L. Muller
Russell & George
745 South 23rd St. Suite 304
Arlington, VA 22202 (703) 521-1760

Richard D. Multer
Le Blanc, Nolan, Shur & Nies
1911 Jefferson Davis Hwy.
Arlington, VA 22202 (703) 521-6590

Richard Murray#
2301 Jeff. Davis Hwy. Apt. 1029
Arlington, VA 22202 (703) 920-1295

George C. Myers, Jr.
Wigman & Cohen, P.C.
Crystal Plaza 3 - Suite 200, 1735 Jeff. Davis Hwy.
Arlington, VA 22202 (703) 892-4300

Arthur I. Neustadt
Oblon, Fisher, Spivak, Mc Clelland & Maier
Suite 400 - Crystal Square Five, 1755 S. Jeff. Davis Hwy.
Arlington, VA 22202 (703) 521-5940

Frank G. Nieman*
U.S. Navy
Office of Naval Research
800 N. Quincy St.
Arlington, VA 22217 (202) 692-4007

John Dirk Nies
Nies, Webner, Kurz & Bergert
1911 Jefferson Davis Hwy. Suite 700
Arlington, VA 22202 (703) 521-6590

William G. Niessen#
5122 N. 16th St.
Arlington, VA 22205 (703) 525-2053

Joseph P. Nigon
Suite 301-Crystal Plaza 1
2001 Jeff. Davis Hwy.
Arlington, VA 22202 (703) 521-1666

Larry S. Nixon
Nixon & Vanderhye, P.C.
2000 N. 15th St. Suite 409
Arlington, VA 22201 (703) 875-0400

Jerome J. Norris
2001 Jeff. Davis Hwy.
Arlington, VA 22202 (703) 521-1666

Sheri M. Novack
Sandler & Greenblum
2920 S. Glebe Rd.
Arlington, VA 22206 (703) 739-0333

Harold L. Novick
Larson & Taylor
727 S. 23rd St.
Arlington, VA 22202 (703) 920-7200

Mark E. Nusbaum
Nixon & Vanderhye, P.C.
2000 N. 15th St. Suite 409
Arlington, VA 22201 (703) 875-0400

William E. O Brien
Depaoli & O Brien, P.C.
1911 Jeff. Davis Hwy.
Arlington, VA 22202 (703) 521-2110

Paul T. O Neil
O Neil & Bean
1601 N. Kent St., Suite 910
Arlington, VA 22209 (703) 525-3131

Norman F. Oblon
Oblon, Fisher, Spivak, Mc Clelland & Maier, P.C.
1755 S. Jeff. Davis Hwy. Suite 400, Crystal Square Five
Arlington, VA 22202 (703) 521-5940

Philip J. Ostaszewski
P.O. Box 15172
Arlington, VA 22215 (703) 521-6759

Robert A. Ostmann
Bldg. 1, 2001 Jefferson Davis Hwy.
Arlington, VA 22202 (703) 521-1221

Magdalene J. Palumbo#
1204 S. Oakcrest Rd.
Arlington, VA 22202 (703) 521-4089

Lutrelle F. Parker, Sr.
2016 S. Fillmore St.
Arlington, VA 22204 (703) 979-7696

Robert J. Patch
Young & Thompson
745 S. 23rd St.
Arlington, VA 22202 (703) 521-2297

Marvin Petry
Larson & Taylor
727 23rd St. South
Arlington, VA 22202 (703) 920-7200

Fred C. Philpitt
Crystall Plaza I - Suite 702
2001 Jeff. Davis Hwy.
Arlington, VA 22202 (703) 521-1555

Robert G. Pierce#
2910 S. Glebe St. Suite 209
Arlington, VA 22206 (703) 684-7331

James William Pike
Bacon & Thomas
1755 Jeff. Davis Hwy. Suite 300
Arlington, VA 22202 (703) 979-9340

Allan R. Plumley, Sr.
3231 N. Albemarle
Arlington, VA 22207 (703) 536-9610

Leonard W. Pojunas, Jr.
Indyke, Pojunas & Brady
2001 Jeff, Davis Hwy. Suite 409
Arlington, VA 22202 (703) 769-2990

David Pollack
Dennison, Meserole, Pollack & Scheiner
1911 Jeff. Davis Hwy.
Arlington, VA 22202 (703) 521-1155

Robert T. Pous
Oblon, Fisher, Spivak, Mc Clelland & Maier
1755 Jeff. Davis Hwy. Suite 400
Arlington, VA 22202 (703) 521-5940

S. Matthew Prastein
Argonne Natl. Lab.
1611 N. Kent St. Suite 201
Arlington, VA 22209 (703) 235-9215

David G. Rasmussen
Burroughs Corp.
1235 Jeff. Davis Hwy. Suite 1304
Arlington, VA 22202 (703) 521-0178

Aaron B. Retzer
Jones, Tullar & Cooper, P.C.
Box 2266 Eads Station
Arlington, VA 22202 (703) 521-5200

John Summerfield Roberts, Jr.
2001 Jeff. Davis Hwy. Suite 504, Bldg. 1
Arlington, VA 22202 (703) 521-0600

Arne Ros#
% Loichot Army Navy Dr.
Arlington, VA 22202

Curtis G. Rose#
I B M Corp. Pat. Operations
1755 S. Jeff. Davis Hwy. Room 605
Arlington, VA 22202 (703) 769-2417

Leo Ross
6340 12th Place, North
Arlington, VA 22205 (703) 532-7145

James L. Rowland
Sandler & Greenblum
2920 S. Glebe Rd.
Arlington, VA 22206 (703) 739-0333

William C. Rowland
Wigman & Cohen
1735 Jeff. Davis Hwy. Suite 200, Crystal Square 3
Arlington, VA 22202 (703) 892-4300

Stephen M. Roylance#
Millen & White, P.C.
503 Crystal Mall 1, 1911 Jeff. Davis Hwy.
Arlington, VA 22202 (703) 892-2200

Rene S. Rutkowski
Low & Low
2316 S. Eads St.
Arlington, VA 22202 (703) 979-4870

Herman Karl Saalbach
2001 Jeff. Davis Hwy. Suite 301, Crystal Plaza 1
Arlington, VA 22202 (703) 521-1666

Mitchell Saffian#
Dennison, Meserole, Pollack & Scheiner
1000 Crystal Mall, 1911 Jeff. Davis Hwy.
Arlington, VA 22202 (703) 521-1155

Abraham A. Saffitz
P.O. Box 2311
Arlington, VA 22202 (703) 521-8121

Donald M. Sandler
Sandler & Greenblum
2920 South Glebe Rd.
Arlington, VA 22206 (703) 739-0333

Jeffrey D. Sanok#
R C A Patent Operation
2001 Jeff. Davis Hwy.
Arlington, VA 22202 (703) 521-1716

Thomas P. Sarro
Larson & Taylor
727 23rd St., S.
Arlington, VA 22202 (703) 920-7200

Jane Y. Sasai#
Nies, Webner, Kurz & Bergert
1911 Jeff. Davis Hwy. Suite 700
Arlington, VA 22202 (703) 521-6590

Paul E. Sauberer#
2527 N. Lexington St.
Arlington, VA 22202 (703) 538-4449

Joseph Scafetta, Jr.
745 South 23rd St. Suite 304
Arlington, VA 22202 (703) 521-1804

Burton S. Scheiner
Dennison, Meserole, Pollack & Scheiner
1000 Crystal Mall, 1911 Jeff. Davis Hwy.
Arlington, VA 22202 (703) 521-1155

Eric Paul Schellin, Sr.
2001 Jeff. Davis Hwy.
Arlington, VA 22202 (703) 521-1666

Billy A. Schulman
Larson & Taylor
727 S. 23rd St.
Arlington, VA 22202 (703) 920-7200

Ira J. Schultz
Dennison, Meserole, Pollack & Scheiner
1911 Jeff. Davis Hwy. Suite 1000
Arlington, VA 22202 (703) 521-1155

Stanley D. Schwartz
Schwartz & Weinrieb
1109 Crystal Plaza - Bldg. 1, 2001 Jeff. Davis Hwy.
Arlington, VA 22202 (703) 521-5250

Timothy R. Schwartz#
Oblon, Fisher, Spivak, Mcclelland & Maier
1755 S. Jeff. Davis. Hwy. Suite 400
Arlington, VA 22202 (703) 521-5940

Mitchell W. Shapiro
Shapiro & Shapiro
1100 Wilson Blvd. Suite 1701
Arlington, VA 22209 (703) 276-0700

Nelson Hirsh Shapiro
Shapiro & Shapiro
1100 Wilson Blvd. Suite 1701
Arlington, VA 22209 (703) 276-0700

Samuel Shipkovitz
1900 S. Eads St. Apt. 923
Arlington, VA 22202 (703) 521-2345

B. Edward Shlesinger, Jr.
Shlesinger, Arkwright & Garvey
3000 S. Eads St.
Arlington, VA 22202 (703) 684-5600

Harry B. Shubin
Millen & White, P.C.
503 Crystal Mall Bldg. 1, 1911 Jeff. Davis Hwy.
Arlington, VA 22202 (703) 892-2200

Jacob Shuster
Zalkind & Shuster
2001 Jefferson Davis Hwy.
Arlington, VA 22202 (703) 920-2170

Melvin J. Sliwaka, Jr.*
Space & Naval Warfare Sys. Command
2511 Jeff. Davis Hwy.
Arlington, VA 22217 (202) 692-2894

John Coventry Smith, Jr.
Benoit, Smith & Laughlin
2001 Jefferson Davis Hwy. Suite 501
Arlington, VA 22202 (703) 521-8622

John F. Smith
Shoemaker & Mattare Ltd.
2001 Jeff. Davis Hwy. Suite 1203, Bldg. 1
Arlington, VA 22202 (703) 521-5210

Richard Darwin Smith#
Kirk & Smith
2001 Jeff. Davis Hwy. Suite 309
Arlington, VA 22202 (703) 521-1820

Yvonne H. Smith#
1515 Jeff. Davis Hwy. Apt. 414
Arlington, VA 22202 (703) 521-0380

Eric S. Spector
Jones, Tullar & Cooper, P.C.
2001 Jeff. Davis Hwy. Suite 1002
Arlington, VA 22202 (703) 521-5200

Marvin Jay Spivak
Oblon, Fisher, Spivak, Mcclelland & Maier
1755 S. Jeff. Davis Hwy. Crystal Square, Suite 400
Arlington, VA 22202 (703) 521-5940

Edward J. Stachura#
5709 18th Rd. N.
Arlington, VA 22205 (703) 538-6976

Clarence F. Stanback, Jr.
2009 N. 14th St. Suite 307
Arlington, VA 22201 (703) 524-3800

William H. Steinberg
General Elec. Co.
2001 Jeff. Davis Hwy.
Arlington, VA 22202 (202) 637-4312

Milton Sterman
Oblon, Fisher, Spivak, Mc Clelland & Maier
1755 S. Jeff. Davis Hwy.
Arlington, VA 22202 (703) 521-5940

Wayne B. Stone, Jr.
Colton & Stone, Inc.
2111 Jeff. Davis Hwy. Suite 2N
Arlington, VA 22202 (703) 920-2270

Robert C. Sullivan
P.O. Box 2402, Eads Station
Arlington, VA 22202 (703) 521-9361

Vincent J. Sunderdick
Oblon, Fisher, Spivak, Mcclelland & Maier
1755 S. Jeff. Davis Hwy. Suite 400, Crystal Square 5
Arlington, VA 22202 (703) 521-5940

Alfonso T. Suro Pico#
1117 N. Rockingham St.
Arlington, VA 22205 (703) 536-8465

Hoge Tyler Sutherland
E.I Du Pont De Nemours & Co.
2001 Jeff. Davis Hwy. Suite 606
Arlington, VA 22202 (703) 521-1800

Yusuke R. Takeuchi#
2111 Jeff. Davis Hwy. Suite 819-N
Arlington, VA 22202 (703) 920-3375

John Paul Tarlano*
U.S. Dept. Of The Navy, Strategic Sys. Prog Off.
P.O. Box 15187
Arlington, VA 22215 (202) 695-4308

David L. Tarnoff#
Hoffman, Wasson & Fallow
522 Hayes Bldg. 2361 Jeff. Davis Hwy.
Arlington, VA 22202 (703) 920-1434

Andrew E. Taylor
727 23rd St. South
Arlington, VA 22202 (703) 920-7200

A. Robert Theibault, Sr.
408 Crystal Plaza One, 2001 Jeff. Davis Hwy.
Arlington, VA 22202 (703) 521-2100

James R. Thein
2001 Jeff. Davis Hwy. Suite 705
Arlington, VA 22202 (703) 920-9161

Murray Tillman#
Oblon, Fisher, Spivak, Mc Clelland & Maier
1755 S. Jeff. Davis Hwy. Suite 400, Crystal Sq. 5
Arlington, VA 22202 (703) 521-5940

Mike L. Tompkins#
Eastman Kodak Co.
727 Twenty-Third St. South
Arlington, VA 22202 (703) 920-7200

Ralph S. Turoff
2001 Jeff. Davis Hwy. Suite 705
Arlington, VA 22202 (703) 521-5550

William C. Tyrrell#
Griffin, Branigan & Butler
775 S. 23rd St.
Arlington, VA 22202 (703) 979-5700

L.S. Van Landingham, Jr.
2001 Jeff. Davis Hwy. Suite 507
Arlington, VA 22202 (703) 979-4244

Robert A. Vanderhye
Nixon & Vanderhye
2000 N. 15th St. Suite 409
Arlington, VA 22201 (703) 875-0404

Frederick D. Vastine#
Oblon, Fisher, Spivak, Mcclelland & Maier
1755 S. Jeff. Davis Hwy. Suite 400- Crystal Sq. 5
Arlington, VA 22202 (703) 521-5940

William M. Wannisky
Fitzpatrick Cella, Harper & Scinto
2011 Crystal Dr. Suite 905
Arlington, VA 22202 (703) 769-5040

Mitchell B. Wasson
2361 Jeff. Davis Hwy. Suite 522
Arlington, VA 22202 (703) 920-1434

Harold W. Weakley#
515 S. Lexington St.
Arlington, VA 22204 (703) 578-4430

Steven R. Wegman
Griffin, Branigan & Butler
775 S. 23rd St.
Arlington, VA 22202 (703) 979-5700

Steven W. Weinrieb
Schwartz & Weinrieb
2001 Jeff. Davis Hwy.
Arlington, VA 22202 (703) 521-5250

Gilbert Leo Wells
Wells & Wells
P.O. Box 2445
Arlington, VA 22202 (703) 521-2726

Walter F. Wessendorf, II
P.O. Box 15846, Crystal City
Arlington, VA 22215 (703) 892-0110

Fred Smith Whisenhunt, Jr.
Murray, Whisenhunt & Ferguson
906 Waterview Bldg. 1925 N. Lynn St.
Arlington, VA 22202 (703) 243-0400

John L. White
Millen & White
503 Crystal Mall Bldg. 1, 1911 Jeff. Davis Hwy.
Arlington, VA 22202 (703) 892-2200

Victor M. Wigman
Wigman & Cohen, P.C.
1735 Jeff. Davis Hwy. Suite 200, Crystal Sq-3
Arlington, VA 22202 (703) 892-4300

Ann C. Williams*
Off. Of Naval Res, Dept. Of The Navy
800 N. Quincy St.
Arlington, VA 22217 (202) 696-4005

Harry S. Williams#
2001 Jeff. Davis Hwy., Suite 304
Arlington, VA 22202 (703) 685-0582

J. Paul Williamson
Arnold, White & Durkee
2001 Jeff. Davis Hwy. Suite 808
Arlington, VA 22202 (703) 920-8720

Charles L. Willis
Oblon, Fisher, Spivak, Mc Clelland & Maier
Crystal Sqaure-5, Suite 400
1755 Jeff. Davis Hwy.
Arlington, VA 22202 (703) 521-5940

Fred L. Witherspoon, Jr.
Witherspoon & Hargest
745 S. 23rd St.
Arlington, VA 22202 (703) 521-0511

Robert M. Wohlfarth, Sr.*
Dept. Of Navy Stategic Sys, Prog Off.
P.O. Box 15187
Arlington, VA 22217 (202) 695-4308

Mark H. Woolsey
1600 S. Joyce St. B-510
Arlington, VA 22202 (703) 521-5558

Christopher P. Wrist
Fitzpatrick, Cella, Harper & Scinto
2011 Crystal Dr. Suite 905
Arlington, VA 22202 (703) 769-5040

John G. Wynn*
Off. Of Naval Res. Code. 1611
800 N. Quincy St.
Arlington, VA 22217 (202) 696-4004

Michael M. Zadrozny#
Shlesinger, Arkwright, Garvey & Fado
3000 S. Eads St.
Arlington, VA 22202 (703) 684-5600

Albert M. Zalkind
Zalkind & Shuster
2001 Jeff. Davis Hwy. Bldg. 1
Arlington, VA 22202 (703) 920-2170

Anthony J. Zelano
Millen & White
1911 Jeff. Davis Hwy. Suite 503
Arlington, VA 22202 (703) 892-2200

John F. Pitrelli
6133 Covered Bridge Rd.
Burke, VA 22015 (703) 941-3475

Gerald E. Smallwood
P.O. Box 196
Burke, VA 22015 (703) 455-2054

Morris Sussman
P.O. Box 1013
Callao, VA 22435 (804) 529-7155

Ormand R. Austin
G E Fanuc Auto. P.O. Box 8106
Charlottesville, VA 22906 (804) 978-5632

Keith F. Goodenough
P.O. Box 5231
Charlottesville, VA 22905 (804) 971-7100

Fred L. Kelly
11325 Canterbury Rd.
Chester, VA 23831 (804) 748-4054

George M.J. Sarofeen
12501 Brook Lane
Chester, VA 23831 (804) 748-5483

Lawrence W. Langley#
910 Cardinal Dr.
Christiansburg, VA 24073 (703) 382-9322

Robert A. Halvorsen
Box 182
Clarksville, VA 23927 (804) 374-2507

William Allen Marcontell
Westvaco Corp. Res. Center
Covington, VA 24426 (703) 969-5520

Marquerite O. Dineen*
Off. Of Naval Res. Code. C73
Naval Surface Weapons Ctr.
Dalgren, VA 22448 (703) 663-7121

John D. Lee#
15694 Beacon Court
Dumfries, VA 22026

Stanislaus Aksman
Mobil Oil Corp.
3225 Gallows Rd.
Fairfax, VA 22037 (703) 849-4066

Walter R. Baylor*
U.S. Army Lab. Command
7701 Willowbrook Rd.
Fairfax, VA 22039 (703) 250-9284

Bennett A. Brown
Gilliam, Sanders & Brown
10560 Main St.
Fairfax, VA 22030 (703) 591-3500

Walter L. Carlson#
3519 Kirkwood Dr.
Fairfax, VA 22031 (703) 280-2358

Gene A. Church#
Church Associates, Inc.
4037 Autumn Court
Fairfax, VA 22030 (703) 644-5260

Ronald J. Cier
Mobil Oil Corp.
3225 Gallows Rd.
Fairfax, VA 22037 (703) 849-4112

Alan B. Croft
Hazel, Beckhorn & Hanes
4084 University Dr.
Fairfax, VA 22030 (703) 273-6644

Ben W. Delos Reyes
13136 Morning Spring Lane
Fairfax, VA 22033 (703) 968-6706

Henry Louis Ehrlich
Mobil Oil Corp.
3225 Gallows Rd.
Fairfax, VA 22037 (703) 849-4075

Joseph P. Flanagan
Mohasco Corp.
4401 Fair Lakes Court
Fairfax, VA 22033 (703) 968-8041

Michael G. Gilman
Mobil Oil Corp.
3225 Gallows Rd.
Fairfax, VA 22037 (703) 849-4094

Van D. Harrison, Jr.
Mobil Oil Corp.
3225 Gallows Rd.
Fairfax, VA 22037 (703) 849-4110

Laurence P. Hobbes
Mobil Oil Corp.
3225 Gallows Rd.
Fairfax, VA 22037 (703) 849-4125

Malcolm Keen
Mobil Oil Corp.
3225 Gallows Rd.
Fairfax, VA 22037 (703) 849-4060

Edward F. Kenehan
Mobil Oil Corp.
3225 Gallows Rd.
Fairfax, VA 22037 (703) 849-4089

James H. Knebel
4615 Tapestry Dr.
Fairfax, VA 22032 (703) 425-3525

Frank J. Kowalski
Mobil Oil Corp.
3225 Gallows Rd.
Fairfax, VA 22037 (703) 849-4081

James W. Lawrence#
10237 Brigade Dr.
Fairfax, VA 22030 (703) 273-3453

Ronald H. Lazarus
Eskovitz, Lazarus & Pitrelli
7023 Little River Turnpike, Suite 202
Fairfax, VA 22003 (703) 354-0561

Gerald Lee Lett
Sedam & Shearerhneider
Suite 155, Fair Lakes One, 12500 Fair Lakes Circle
Fairfax, VA 22033 (703) 631-5000

Charles A. Malone
Mobil Oil Corp.
3225 Gallows Rd.
Fairfax, VA 22037 (703) 849-4064

Giedre M. Mc Candless#
9031 Pixie Court
Fairfax, VA 22031 (703) 978-2661

Alexander J. Mc Killop
Mobil Oil Corp.
3225 Gallows Rd.
Fairfax, VA 22037 (703) 849-4070

Lawrence O. Miller
Mobil Corp.
3225 Gallows Rd.
Fairfax, VA 22037 (703) 849-4087

James P. O Sullivan, Sr.
Mobil Oil Corp.
3225 Gallows Rd.
Fairfax, VA 22037 (703) 849-4071

Channing L. Pace
2900 Hideaway Rd.
Fairfax, VA 22030 (703) 280-2118

Alverna M. Paulan
Mobil Oil Corp.
3225 Gallows Rd.
Fairfax, VA 22037 (703) 849-4062

James F. Powers, Jr.
Mobil Oil Corp.
3225 Gallows Rd.
Fairfax, VA 22037 (703) 849-4079

Ernest Roy Purser, Sr.#
4915 Gadsden Dr.
Fairfax, VA 22032 (703) 323-5248

Peter W. Roberts#
Mobil Oil Co.
3225 Gallows Rd.
Fairfax, VA 22037 (703) 849-4129

Dennis P. Santini
Mobil Oil Corp.
3225 Gallows Rd.
Fairfax, VA 22037 (703) 849-4074

Marina V. Schneller
Mobil Oil Corp.
3225 Gallows Rd.
Fairfax, VA 22037 (703) 849-4088

Claude E. Setliff
11203 Clara Barton Dr.
Fairfax, VA 22039

Charles J. Speciale
Mobil Oil Corp.
3225 Gallows Rd.
Fairfax, VA 22037 (703) 849-4098

Richard D. Stone
Mobil Oil Corp.
3225 Gallows Rd.
Fairfax, VA 22037 (703) 849-4120

Hastings S. Trigg
3707 John Barnes Lane
Fairfax, VA 22033 (703) 620-8869

Edward John Trojnar
Mobil Oil Corp.
3225 Gallows Rd.
Fairfax, VA 22037 (703) 849-4096

Edward Hatch Valance
Mobil Oil Corp.
3225 Gallows Rd.
Fairfax, VA 22037 (703) 849-4058

Lowell G. Wise
Mobil Oil Corp.
3225 Gallows Rd.
Fairfax, VA 22037 (703) 849-4084

William V. Adams*
U.S. Army Pat. Copyrights & Tdmk. Div.
5611 Columbia Pike
Falls Church, VA 22041 (703) 756-2434

Robert F. Altherr, Jr.*
U.S. Army Off. Of Judge Advocate Gen. Att. J A L S - P C
5600 Columbia Pike
Falls Church, VA 22041 (202) 756-2623

Ralph J. Alvey
3709 S. George Mason Dr. Apt. 708E
Falls Church, VA 22041 (703) 998-5910

James C. Arvantes
6539 Cedarwood Court
Falls Church, VA 22041 (703) 256-0515

John P. Beauchamp, Jr.#
6629 Kirby Court
Falls Church, VA 22043 (703) 533-0987

Werten F.W. Bellamy#*
Dept. Of The Army, Off. Of Judge Adv. Gen.
5611 Columbia Pike
Falls Church, VA 22041 (202) 756-2434

Mark W. Binder#
Sixbey, Friedman & Leedom
7653 Leesburg Pike
Falls Church, VA 22043 (703) 790-9110

Anthony L. Birch
Birch, Stewart, Kolasch & Birch
P.O. Box 209, 301 N. Wash St.
Falls Church, VA 22046 (703) 241-1300

Herbert M. Birch
Birch, Stewart, Kolasch & Birch
301 N. Wash St.
Falls Church, VA 22046 (703) 241-1300

Terrell C. Birch
Birch, Stewart, Kolasch & Birch
301 N. Wash. St., Box 209
Falls Church, VA 22046 (703) 241-1300

Matthews P. Blischak#
Birch, Stewart, Kolsach & Birch
301 N. Washington St.
Falls Church, VA 22046 (703) 241-1300

Robert Brown, Jr.
3329 Wilkins Dr.
Falls Church, VA 22041 (703) 820-8017

William R. Browne#
3245 Rio Dr., Condo 406
Falls Church, VA 22041 (703) 671-3608

Frank Cacciapaglia, Jr.#
P.O. Box 5050
Falls Church, VA 22044 (703) 536-2323

Terry L. Clark
Birch, Stewart, Kolesch & Birch
301 N. Washington St.
Falls Church, VA 22046 (703) 241-1300

Dennis Philip Clarke
Kerkam, Stowell, Kondracki & Clarke, P.C.
6404 R - Seven Corners Place
Falls Church, VA 22044 (703) 534-6600

Thomas W. Cole
Sixbey, Friedman & Leedom
7653 Leeburg Pike
Falls Church, VA 22043 (703) 790-9110

Kenneth L. Crosson
5113 Leesburg Pike, Suite 100
Falls Church, VA 22041 (703) 385-1010

Vincent M. De Luca
Birch, Stewart, Kolasch & Birch
301 N. Wash. St.
Falls Church, VA 22046 (703) 241-1300

Vangelis Economou
Kerkam, Stowell, Kondracki & Clake
6404R Seven Corners Place
Falls Church, VA 22044 (703) 534-6600

C. Joseph Faraci#
Birch, Stewart, Kolesch & Birch
301 N. Washington St.
Falls Church, VA 22046 (703) 241-1300

William L. Feeney
Kerkam, Stowell, Kondracki & Clarke, P.C.
6404R Seven Corners Place
Falls Church, VA 22044 (703) 534-6600

Gerald Joseph Ferguson, Jr.
5205 Leesburg Pike
Falls Church, VA 22041 (703) 820-4500

Barbara A. Fisher#
Birch, Stewart, Kolasch & Birch
301 N. Washington St.
Falls Church, VA 22046 (703) 241-1300

Michael J. Foycik, Jr.
One Skyline Place, Suite 310
5202 Leesburg Pike
Falls Church, VA 22041 (703) 820-4500

Abraham Frankel
3401 Glen Carlyn Dr.
Falls Church, VA 22041 (703) 820-7140

Stuart Jay Friedman
Sixbey, Friedman & Leedom, P.C.
7653 Leeburg Pike
Falls Church, VA 22043 (703) 790-9110

Richard Gerard#
220 Midvale St.
Falls Church, VA 22046 (703) 534-3094

Charles Gorenstein
Birch, Stewart, Kolasch & Birch
301 N. Wash. St. Box 747
Falls Church, VA 22046 (703) 241-1300

Michael P. Hoffman
5205 Leesburg Pike, Suite 310
Falls Church, VA 22041 (703) 820-4500

Edward J. Kelly
2921 Rosemary Ln.
Falls Church, VA 22042 (703) 534-1238

John C. Kerins#
Kerkham, Stowell, Kondracki & Clarke
6404 R Seven Corners Place
Falls Church, VA 22046 (703) 534-6600

William L. Klima#
Birch, Stewart, Kolasch & Birch
301 N. Wash. St. P.O. Box 747
Falls Church, VA 22046 (703) 241-1300

Donald Craig Kolasch
Birch, Stewart, Kolasch & Birch
301 N. Wash. St.
Falls Church, VA 22046 (703) 241-1300

Joseph Arlen Kolasch
Birch, Stewart, Kolasch & Birch
301 N. Wash. St. Box 209
Falls Church, VA 22046 (703) 241-1300

Edward J. Kondracki
Kerkam, Stowell, Kondracki & Clarke, P.C.
6404 R Seven Corners Place
Falls Church, VA 22044 (703) 534-6600

Joan K. Lawrence
Sixbey, Friedman & Leedom, P.C.
7653 Leedburg Pike
Falls Church, VA 22043 (703) 790-9110

Charles M. Leedom, Jr.
Sixbey, Friedman & Leedom, P.C.
7653 Leeburg Pike
Falls Church, VA 22043 (703) 790-9110

Ronni S. Malamud
5205 Leesburg Pike, Suite 310
Falls Church, VA 22041 (703) 820-4500

Fred C. Mattern, Jr.
Birch, Steward, Kolasch & Birch
301 N. Washington St.
Falls Church, VA 22046 (703) 241-1300

Kenneth F. Mc Clure
306 Lawton
Falls Church, VA 22046 (703) 532-3577

Joe M. Muncy, Jr.#
Birch, Stewart, Kolasch & Birch
P.O. Box 747, 301 N. Wash. St.
Falls Church, VA 22046 (703) 241-1300

Gerald M. Murphy, Jr.
Birch, Stewart, Kolasch & Birch
301 N. Wash. St. P.O. Box 747
Falls Church, VA 22046 (703) 241-1300

Michael K. Mutter
Birch, Stewart, Kolasch & Birch
301 W. Wash. St. P.O. Box 209
Falls Church, VA 22046 (703) 241-1300

Meyer Perlin#
2016 Dexter Dr.
Falls Church, VA 22043 (703) 893-9556

David S. Safran
Sixbey, Friedman & Leedom, P.C.
7653 Leesburg Pike
Falls Church, VA 22043 (703) 790-9110

Steven C. Schnedler
Kerkam, Stowell, Kondracki & Clarke
6404 R Seven Corners Place
Falls Church, VA 22044 (703) 534-6600

Daniel W. Sixbey
Sixbey, Friedman & Leedom
7653 Leesburg Pike
Falls Church, VA 22043 (703) 790-9110

James M. Slattery
Birch, Stewart, Kolasch & Birch
301 N. Wash. St. P.O. Box 747
Falls Church, VA 22046 (703) 241-1300

Raymond C. Stewart
Birch, Stewart, Kolasch & Birch
301 N. Wash. St. Box 747
Falls Church, VA 22046 (703) 241-1300

Harold L. Stowell
Kerkham, Stowell, Kondracki & Clarke
6404 R Seven Corners Place
Falls Church, VA 22044 (703) 534-6600

Leonard R. Svensson
Birch, Stewart, Kolasch & Birch
301 N. Wash. St.
Falls Church, VA 22046 (703) 241-1300

Bernard L. Sweeney
Birch, Stewart, Kolasch & Birch
301 N. Wash. St. Box 747
Falls Church, VA 22046 (703) 241-1300

Marc S. Weiner#
Birch, Stewart, Kolasch & Birch
301 N. Wash. St. P.O. Box 747
Falls Church, VA 22046 (703) 241-1300

Harry Wong, Jr.#
Birch, Stewart, Kolasch & Birch
301 N. Wash. St. P.O. Box 747
Falls Church, VA 22046 (703) 241-1300

Barry N. Young
Kerkam, Stowell, Kondracki & Clarke
6404 R Seven Corners Place
Falls Church, VA 22044 (703) 534-6600

John E. Becker*
Dept. Of The Army, U.S. Army Belvoir Res. & Dev. Ctr.
S T R B E - L
Fort Belvoir, VA 22060 (703) 664-5411

Aubrey J. Dunn*#
U.S. Army C E C O M
C N V E O, Attn: A M S E L - L G - P - N V E O
Fort Belvoir, VA 22060 (703) 664-5513

Max L. Harwell*#
U.S. Army Communication Electronics Command
Center for Night Vision & Electronics
Fort Belvoir, VA 22060 (703) 664-2223

Darrell E. Hollis*
U.S. Army Corps Of Eng.
Kingman Bldg. Cehec-Oc
Fort Belvoir, VA 22060 (202) 355-3671

Gary W. Hudiburgh*
U.S. Army Corps. Of Engineers
Kingman Bldg.
Fort Belvoir, VA 22060 (202) 355-2160

Milton Wayne Lee*#
U.S. Army Electronics Res. & Dev. Command
Night Vision & Electro-Optics Lab
Fort Belvoir, VA 22060 (706) 664-2223

Roger F. Phillips*
Dept. Of The Army Off. Of Chief Coun.
U.S. Army Belvoir Res. & Dev. Ctr.
Fort Belvoir, VA 22060 (703) 664-5411

Edwin D. Grant
P.O. Box 225
Fredericksburg, VA 22404

David G. Mc Connell#
5514 Rudy Lane
Fredericksburg, VA 22401

John E. Pruitt, Jr.
Rawlings, Pruitt & Bieber
405 Amelia St.
Fredericksburg, VA 22401 (703) 373-7444

Richard M. Foard
P.O. Box 356
Gloucester, VA 23061 (804) 693-5665

John R. Janes
Goode, VA 24556 (703) 586-4244

Robert Mistrot Meith
605 Clear Spring Rd.
Great Falls, VA 22066 (703) 444-4908

Joseph G. Seeber
P.O. Box 750
Great Falls, VA 22066 (703) 430-1702

Harold W. Adams*
N A S A, Langley Res. Center
278/Off. Chief Coun.
Hampton, VA 23665 (804) 865-3725

George Franke Helfrich*
N A S A Langley Res. Center, Mail Stop 279
Hampton, VA 23665 (804) 865-3725

William H. King*
N.A.S.A. Langley Res. Center
Langley Station, Bldg. 1192, Rm. 106
Hampton, VA 22365 (804) 827-3725

Wallace J. Nelson*
N.A.S.A. Lrc, Ms 279
Hampton, VA 23365 (804) 865-3725

Howard J. Osborn*
N.A.S.A. Langley Res. Ctr. Ms 279
Hampton, VA 23365 (804) 865-3725

Raymond L. Greene
Route 658
James Store, VA 23080 (804) 725-4047

Americus Mitchell
P.O. Box 1335
Kilmarnock, VA 22482 (804) 435-3489

Bonnie L. Deppenbrock#
706 Wage Dr., S.W.
Leesburg, VA 22075 (703) 777-3452

Archie R. Borchelt#
P.O. Box 655
Locust Grove, VA 22508 (703) 339-1204

Robert C. Lampe, Jr.
General Elec. Co. Mobile Commun. Bus. Div.
Mountain View Rd. P.O. Box 4096
Lynchburg, VA 24502 (804) 528-7400

Michael Masnik
General Electric Co.
Mountain View Rd.
Lynchburg, VA 24502 (804) 528-7500

H. Stanley Muir, III#
2217 Cambridge Place
Lynchburg, VA 24503 (804) 384-9376

James J. Williams
4704 Alclif Dr.
Lynchburg, VA 24503 (804) 384-1696

Jesse L. Abzug
I B M Corp.
9500 Goodwin Dr.
Manassas, VA 22110 (703) 367-2121

Harold H. Dutton, Jr.
8711 Plantation Lane, Suite 301
P.O. Box 3110
Manassas, VA 22110 (703) 369-1922

John E. Hoel
I B M Corp.
9500 Goodwin Dr.
Manassas, VA 22110 (703) 367-4308

Shelley Krasnow
P.O. Box 70, 9617 Center St.
Manassas, VA 22110 (703) 368-2101

John H. Merchant
9115 Grant Ave.
Manassas, VA 22110 (703) 368-3519

Joseph C. Redmond, Jr.
I B M Corp.
9500 Godwin Dr.
Manassas, VA 22110 (703) 367-3896

Daniel R. Alexander#
6862 Elm Street, Suite 300
Mc Lean, VA 22101 (703) 556-0122

Gerald H. Bjorge
1300 Alps Dr.
Mc Lean, VA 22102 (703) 893-7890

Francis X. Bradley, Jr.
7909 Falstaff Rd.
Mc Lean, VA 22101 (703) 356-0319

Charles Martin Bredehoft
6107 Woodland Terrace
Mc Lean, VA 22101 (703) 534-2189

James J. Brown
Gipple & Hall
6667-B Old Dominion Dr.
Mc Lean, VA 22101 (703) 448-1770

J. Howard Flint, Jr.#
1114 Dead Run Dr.
Mc Lean, VA 22101 (703) 356-7453

Terry M. Gernstein
1015 Salt Meadow Lane
Mc Lean, VA 22101 (703) 790-5945

John S. Hale
Gipple & Hale
6667-B Old Dominion Dr.
Mc Lean, VA 22101 (703) 448-1770

William F. Hamrock
6862 Elm Street, Suite 300
Mc Lean, VA 22101 (703) 893-6928

Harrison L. Hinson#
935 Douglass Dr.
Mc Lean, VA 22101 (703) 356-3041

H.M. Hougen
Dynacorp.
1313 Dolley Madison Blvd.
Mc Lean, VA 22101 (703) 790-2796

Milton Kaufman
8360 Greenboro Dr., Apt. 424
Mc Lean, VA 22102 (703) 790-0427

James L. Kohnen
6714 Whittier Ave.
Mc Lean, VA 22101 (703) 442-8203

Warren H. Mc Inteer#
Dynamic Systems, Inc.
8200 Greensboro Dr.
Mc Lean, VA 22102 (703) 734-3062

Garo A. Partoyan
Mars, Inc.
6885 Elm St.
Mc Lean, VA 22101 (703) 821-4900

Edward A. Pennington
1493 Chain Bridge Rd. Suite 300
Mc Lean, VA 22101 (703) 442-4800

Alvin B. Peterson
6203 Nelway Dr.
Mc Lean, VA 22101 (703) 356-7259

Joseph Edward Rusz
1102 Roberta Court
Mc Lean, VA 22101 (703) 356-5696

Robert K. Schaefer
7735 Falstaff Rd.
Mc Lean, VA 22102 (703) 790-9486

James H. Tayman, Jr.#
1959 Virginia Ave.
Mc Lean, VA 22101

Glenn E. Wise#
6450 Georgetown Pike
Mc Lean, VA 22101 (703) 521-4879

James Creighton Wray
1493 Chain Bridge Rd., Apt. 300
Mc Lean, VA 22101 (703) 442-4800

Daniel E. Wyman#
8370 Greensboro Dr., Apt. 307
Mc Lean, VA 22102 (703) 442-0039

Robert F. Ziems
6862 Elm St.
P.O. Box 1055
Mc Lean, VA 22101 (703) 556-0122

Lawrence D. Bush, Jr.
1120 Hanover Green Dr. P.O. Box 788
Mechanicsville, VA 23111 (804) 730-2200

Robert L. Spicer, Jr.#
2905 Greenlake Circle
Mechanicsville, VA 23111 (804) 798-4402

William O. Quesenberry*
P.O. Box 117
Merry Point, VA 22513 (804) 462-5738

Robert S. Auten
Route 1 Box 302
Moneta, VA 24121 (703) 297-7562

Nelson W. Edgerton#
Route 4 Box 127
Moneta, VA 24112 (703) 721-3325

Adolph Charles Hugin
7602 Boulder St.
N. Springfield, VA 22151 (703) 569-2233

Clifford N. Rosen#
Newport News Shipbuilding
4101 Washington Ave.
Newport News, VA 23607 (804) 380-2307

Samuel F. Allison
1522 Leaview Ave.
Norfolk, VA 23503 (804) 588-8020

Richard Alan Anderson
Allied Corp.
P.O. Box 31
Petersburg, VA 23803 (804) 520-3651

Hanna S. Burke#
Allied Signal Corp. Fibers Div. Tech. Center
P.O. Box 31
Petersburg, VA 23803 (804) 520-3619

William H. Thrower, Jr.
Allied Corp.
P.O. Box 31
Petersburg, VA 23804 (804) 520-3622

Robert James Buttermark#
6000 Derwent Rd.
Powhatan, VA 23139 (804) 375-9131

Mark J. Regen
Telenet Communications Corp.
12490 Sunrise Valley Dr.
Reston, VA 22096 (703) 689-6905

David R. Schultz
Mobil Land Dev. Corp.
11800 Sunrise Valley Dr., Suite 1400
Reston, VA 22091 (703) 620-4780

Roland H. Shubert
P.O. Box 2339
Reston, VA 22090 (703) 435-4141

Alfred E. Smith#
1933 Red Lion Court
Reston, VA 22091 (703) 860-1709

Victor J. Toth
2719 Soapstone Dr.
Reston, VA 22091 (703) 476-5515

Charles Lamont Whitham
11507 Purple Beech Dr. P.O. Box 3740
Reston, VA 22090 (703) 620-2734

Michael E. Whitham#
11866 D Sunrise Valley Dr.
Reston, VA 22091 (703) 391-2510

Virginia S. Andrews
9302 Belfort Rd.
Richmond, VA 23229 (804) 740-2319

Alan M. Biddison
Reynolds Metals Co. Law Dept.
P.O. Box 27003
Richmond, VA 23261 (804) 281-2410

Ivan Christoffel#
A.H. Robins Co. Inc.
1407 Cummings Dr.
Richmond, VA 23220 (804) 257-2947

William B. Cridlin, Jr.
A.H. Robins Co. Inc.
1407 Cummings Dr.
Richmond, VA 23220 (804) 257-2184

John L. Dewey#
7012 Hunt Club Lane Apt. 1834
Richmond, VA 23228 (804) 269-6869

Walter M. Dotts, Jr.#
2232 Park Ave.
Richmond, VA 23220 (804) 355-6392

John Willard Gibbs, Jr.
204 Dryden Lane
Richmond, VA 23229 (804) 741-5855

Donald Edward Gillespie
A.H. Robins Co. Inc.
1407 Cummings Dr.
Richmond, VA 23220 (804) 257-2193

Arthur Lawrence Girard
8310 Poplar Hollow Trail
Richmond, VA 23235 (804) 320-2964

John F.C. Glenn
8915 Tolman Rd.
Richmond, VA 23229 (804) 740-8409

Edward H. Gorman, Jr.
A.H. Robins Co.
1407 Cummings Dr.
Richmond, VA 23261 (804) 257-2125

Susan Addington Hutcheson#
Philip Morris Inc. Res. & Dev.
P.O. Box 26583
Richmond, VA 23261 (804) 274-2162

George Esler Inskeep#
Phillip Morris Inc.
4201 Commerce Rd. Box 26583
Richmond, VA 23261 (804) 274-2789

Auzville Jackson, Jr.
Staas & Halsey
8652 Rio Grande Rd.
Richmond, VA 23229 (804) 740-6828

George William King
A.H. Robins Co. Inc.
1407 Cummings Dr.
P.O. Box 26609
Richmond, VA 23261 (804) 257-2191

Louis R. Lawson, Jr.#
6416 Roselawn Rd.
Richmond, VA 23226 (804) 288-3991

Robert Chamberlayne Lyne, Jr.
Reynolds Metals Co.
6601 W. Broad St.
Richmond, VA 23261 (804) 281-2837

Alan T. Mc Donald
Reynolds Metals Co.
6601 W. Broad St.
Richmond, VA 23261 (804) 281-3791

Arthur I. Palmer, Jr.
Philip Morris, Inc.
P.O. Box 26583
Richmond, VA 23261 (804) 274-2822

Norman B. Rainer#
2008 Fon Du Lac Rd.
Richmond, VA 23229 (804) 282-7109

James Eric Schardt
Philip Morris Inc.
P.O. Box 26583
Richmond, VA 23261 (804) 274-2822

Willard R. Sprowls
Box 128, Route 1
Richmond, VA 23231 (804) 795-2611

Paul Yale Virkler
5500 Riverside Dr.
Richmond, VA 23225 (804) 232-4778

Casmir A. Nunberg#
R.R. 1, Box 167
Rixeyville, VA 22737 (703) 937-5520

Myles T. Hylton
Gentry, Locke, Rakes & Moore
800 Colonial Plaza, 10 East Franklin Rd.
Roanoke, VA 24001 (703) 982-8000

Thomas Dean Simmons
Gentry, Locke, Rakes & Moore
800 Colonial Plaza, P.O. Box 1018
Roanoke, VA 24005 (703) 982-8000

John Krout
Gipple & Hale
1925 N. Lynn St. Suite 905
Rosslyn, VA 22209 (703) 243-5800

Leonard C. Mitchard
Murray & Whisenhunt
1925 N. Lynn St.
Rosslyn, VA 22209 (703) 243-0400

J. Howard Flint
Round Hill, VA 22141 (703) 554-2817

Arnold E. Renner
General Electric Co. Drive Syst. Oper.
1501 Roanoke Blvd.
Salem, VA 24153 (703) 387-7144

Jesse B. Grove, Jr.
P.O. Box 207
Scottsville, VA 24590 (804) 286-3134

Kathleen E. Crotty
7817 Ravenel Court
Springfield, VA 22151 (703) 321-8051

Thomas De Benedictis, Sr.#
7123 Kerr Dr.
Springfield, VA 22150 (703) 569-1184

Dong Whee Kim#
7415 Jenna Rd.
Springfield, VA 22153 (703) 569-0020

Neil F. Markva
8322-A Traford Lane
Springfield, VA 22152 (703) 644-5000

Richard P. Matthews
8410 Terra Woods Dr.
Springfield, VA 22153 (703) 455-7087

Wallace F. Poore#
Rt. 1 Box 286-A
Troy, VA 22974 (804) 589-3261

Robert M. Clark
S T A C
111 Center Streets S. Suite B
Vienna, VA 22180 (703) 281-6351

Murriel E. Crawford
Dickstein, Shapiro & Morin
8300 Boone Blvd. Suite 800
Vienna, VA 22180 (703) 847-9190

Joseph A. Geiger#
438 Park St., N.E.
Vienna, VA 22180 (703) 281-2639

Michael C. Greenbaum
Dickstein, Shapiro & Morin
8300 Boone Blvd. Suite 800
Vienna, VA 22180 (703) 847-9190

Donald A. Gregory
Dickstein, Shapiro & Morin
8300 Boone Blvd. Suite 800
Vienna, VA 22180 (703) 847-9190

Daniel R. Gropper
2763 Stone Hollow Dr.
Vienna, VA 22180 (703) 560-7157

Gary M. Hoffman
Dickstein, Shapiro & Morin
8300 Boone Blvd.
Vienna, VA 22180 (703) 847-9190

Phillip G. Lookadoo
Wickwire, Gavin & Gibbs, P.C.
8230 Boone Blvd. Suite 400
Vienna, VA 22180 (703) 790-8750

Jon L. Roberts#
Lewis, Mitchell & Moore
1950 Gallows Rd.
Vienna, VA 22180 (703) 790-9200

Roberts G. Sheridan#
8605 Pepperdine Dr.
Vienna, VA 22180 (703) 560-4921

Malcolm R. Uffelman#
1808 Horseback Trail
Vienna, VA 22180 (703) 938-6184

Billy J. Wilhite#
2520 Rocky Branch Rd.
Vienna, VA 22180 (703) 938-1246

Irwin M. Lewis
1800 N. Alanton Dr.
Virginia Beach, VA 23454 (804) 481-0645

George W. Hager, Jr.
397 Woodstone Court
Warrenton, VA 22186 (703) 849-4086

Woolridge Brown Morton, Jr.
Route 1, Box 586
Warsaw, VA 22572 (804) 333-3311

Rupert B. Hurley, Jr.#
Badische Corp.
P.O. Drawer D
Williamsburg, VA 23185 (804) 887-6820

Richard J. Scanlan, Jr.#
12028 Wm. & Mary Circle
Woodbridge, VA 22192 (703) 494-2718

Washington

Keith D. Gehr#
35820 57th Ave. South
Auburn, WA 98002 (206) 939-5997

Salim A. Kassatly
18600 N.E. August Ave.
Battle Ground, WA 98604 (206) 687-1662

George B. Fox
Seed & Berry
1700 Skyline Tower, 18900 N.E. Fourth St.
Bellevue, WA 98004 (206) 622-4900

R. Reams Goodloe, Jr.
Resources Conservation Co.
3101 N.E. Northup Way
Bellevue, WA 98004 (206) 828-2400

Lawrence Adams Jackson
Jackson & Richardson
411 - 108th Ave. N.E.
Bellevue, WA 98004 (206) 455-5575

Robert W. Jenny#
77 Cascade Key
Bellevue, WA 98006 (206) 747-2936

Roy Edwin Mattern, Jr.
Mattern & Kessler
13410 S.E. 32nd
Bellevue, WA 98005 (206) 641-9000

Don R. Mollick*
Deits, Mollick & Moravan
108th Ave. N.E. Suite 601
Bellevue, WA 98004 (206) 545-2700

Gregory W. Moravan
Mollick & Moravan
Suite 302, Westridge Bldg. 11671 S.E. 1st St.
Bellevue, WA 98005 (206) 454-2700

Harry A. Richardson, Jr.
Jackson & Richardson
1750 One Bellevue Center, 411 108th Ave. N.E.
Bellevue, WA 98004 (206) 455-5575

Mikio Ishimaru
John Fluke Mfg. Co. Inc.
P.O. Box C 9090
Everett, WA 98206 (206) 356-5819

George T. Noe
John Fluke Mfg. Co. Inc.
P.O. Box C 9090, M S 203 A
Everett, WA 98206 (206) 356-6172

John D. Mc Auliffe
Abrahams & Mc Auliffe
33430 13th Place S. Suite 208
Federal Way, WA 98003 (206) 952-3756

Mark J. Zovko, Jr.
36504 28th Ave. South
Federal Way, WA 98003 (206) 838-1909

Robert Keith Sharp
906 N. Ledbetter St.
Kennewick, WA 99336 (509) 783-4796

Marshall F. Gehring
25825 104th Ave. S.E. Suite 375
Kent, WA 98301 (206) 746-0463

Robert L. Jepsen#
2216 Beta St.
Lacey, WA 98503 (206) 438-5166

George W.F. Simmons
Route 2 Box 3377
Lopez, WA 98261 (206) 468-2068

J. Dennis Malone
Route 1, Box 1443
Lopez Island, WA 98261 (206) 468-2807

Norris E. Faringer
140 Orchard Dr.
Naches, WA 98937 (509) 965-3094

William I. Beach#
232 W. 10th St.
Port Angeles, WA 98362 (206) 452-8628

Lawrence H. Poeton#
213 S. Oak St.
Port Angeles, WA 98362 (206) 457-1321

Donna J. Thies
Hughes & Cassidy P.S.
15042 N.E. 40th St. Suite 205
Redmond, WA 98052 (206) 453-5701

David L. Tingey
1100 Maple Ave, S.W.
Renton, WA 98055 (206) 271-7690

David J. Brown#
55 Jadwin Ave. Apt. 110
Richland, WA 99352 (509) 946-0541

Joseph James Hauth#
Batelle Memorial Institute
P.O. Box 999
Richland, WA 99352 (509) 375-2981

Dorothy L. Sander
Rockwell International Corp. Rockwell Hanford Oper.
P.O. Box 800
Richland, WA 99352 (509) 373-1272

Robert Southworth, III*
U.S. Dept. Of Energy
825 Jadwin Ave.
Richland, WA 99352 (509) 376-7225

Kent B. Roberts
326 Delafield Ave.
Richmond, WA 99352 (509) 375-2981

Daniel T. Anderson
Boeing Comm. Airplane Co.
Box 3707, Mail Sta. 7E-25
Seattle, WA 98124 (206) 251-0229

Ronald M. Anderson
Christensen, O Connor, Johnson & Kindness
2700 Westin Bldg. 2001 Sixth Ave.
Seattle, WA 98121 (206) 441-8780

William C. Anderson
Boeing Military Airplane Co.
P.O. Box 3707, M/s 32-50
Seattle, WA 98124 (206) 284-5112

David L. Baker
P.O. Box 15507
Seattle, WA 98115 (206) 259-2295

Delbert J. Barnard
Barnard, Pauly & Kaser, P.S.
6000 Southcenter Blvd. Suite 240
Seattle, WA 98188 (206) 246-0568

Robert J. Baynham
Seed & Berry
6300 Columbia Center
Seattle, WA 98104 (206) 622-4900

Robert Willis Beach
Beach & Brown
3107 Eastlake Ave. East
Seattle, WA 98102 (206) 325-6789

Benjamin F. Berry
Seed & Berry
6300 Columbia Center
Seattle, WA 98104 (206) 622-4900

Michael W. Bocianowski
Christensen, O Connor, Johnson & Kindness
2701 Westin Bldg. 2001 Sixth Ave.
Seattle, WA 98121 (206) 441-8780

Thomas F. Broderick
Christensen, O Connor, Johnson & Kindness
2701 Westin Bldg. 2001 Sixth Ave.
Seattle, WA 98121 (206) 441-8780

Ward Brown
Beach & Brown
3107 Eastlake Ave. East
Seattle, WA 98102 (206) 325-6789

Carroll L. Bryan, II#
Bryan, Schniffrin & Mc Monagle
2701 First Ave.
Seattle, WA 98121 (206) 448-8100

Edward W. Bulchis
Seed & Berry
6300 Columbia Center
Seattle, WA 98104 (206) 622-4900

David V. Carlson
Seed & Berry
6300 Columbia Center
Seattle, WA 98104 (206) 622-4900

Morris A. Case
921 S.W. 152nd St.
Seattle, WA 98166 (202) 243-1240

Joseph O. Chalverus
12553 39th N.E.
Seattle, WA 98125 (206) 367-7816

Thomas D. Cohen#
3613 N.E. 43rd
Seattle, WA 98105 (206) 523-9342

George M. Cole
1254 Bank Of Calif. Center
Seattle, WA 98164 (206) 622-3740

Harry Maybury Cross, Jr.
Dowrey & Cross
1254 Bank Of Calif. Ctr.
Seattle, WA 98164 (206) 624-6535

Daniel D. Crouse
Christensen, O Connor, Johnson & Kindness
2701 Westin Bldg. 2001 Sixth Ave.
Seattle, WA 98121 (206) 441-8780

Paul C. Cullom, Jr.
Boeing Comm. Airplane Co.
P.O. Box 3707, Mail Stop. 7E - 25
Seattle, WA 98124 (206) 251-0275

Nicolaas De Vogel#
Boeing Corp.
P.O. Box 3707, M/s 7E-25
Seattle, WA 98124 (206) 251-0398

David H. Deits
Seed & Berry
6300 Columbia Center
Seattle, WA 98104 (206) 622-4900

Bernard A. Donahue
Boeing Comm. Airplane Co.
P.O. Box 3707 M.S. 7E 25
Seattle, WA 98124 (206) 251-0443

Carl Gordon Dowrey, Sr.
Dowrey, Cross & Cole
401 Second Ave. South, Suite 630
Seattle, WA 98104 (206) 624-6535

Christopher Ogden Duffy
3031 The Bank Of Calif. Center
Seattle, WA 98164 (206) 623-8088

Patrick M. Dwyer
Garrison & Stratton, P.S.
2100 Fifth Ave.
Seattle, WA 98121 (206) 441-3440

William O. Ferron, Jr.
Seed & Berry
6300 Columbia Center
Seattle, WA 98104 (206) 622-4900

Michael J. Folise
Seed & Berry
6300 Columbia Center
Seattle, WA 98104 (206) 622-4900

Conard Oliver Gardner
Boeing Comm. Airplane Co.
P.O. Box 3707, M.S. 7E-25
Seattle, WA 98124 (206) 251-0384

Pryor A. Garnett
Graybeal, Jensen & Puntigam
1020 United Airlines Bldg. 2033 Sixth Ave.
Seattle, WA 98121 (206) 448-3200

David Louis Garrison
Garrison & Stratton, P.S.
2100 Fifth Ave.
Seattle, WA 98121 (206) 441-3440

John O. Graybeal
Graybeal, Jensen & Puntigam
1020 United Airlines Bldg. 2033 Sixth Ave.
Seattle, WA 98121 (206) 448-3200

Robert L. Gullette
Boeing Aerospace Co.
P.O. Box 24346, Mail Stop 7A-52
Seattle, WA 98124 (206) 865-5245

Henry William Haigh#
18982 Marine View Dr. S.W.
Seattle, WA 98166 (206) 243-6291

Scott G. Hallquist
Immunex Corp.
51 University St.
Seattle, WA 98101 (206) 462-0362

James P. Hamley
Boeing Comm. Airplane Co.
P.O. Box Licensing Staff, Mail Stop 7E-25
Seattle, WA 98124 (206) 251-0262

John C. Hammer
Boeing Aerospace Co.
P.O. Box 3999 Mail Stop 85-74
Seattle, WA 98124 (206) 773-2572

H. Gus Hartmann#
Boeing Comm. Airplane Co.
P.O. Box 3707, Mail Stop. 7E-25
Seattle, WA 98124 (206) 251-0237

Thomas Waldo Hennen
Boeing Aerospace Co.
P.O. Box 3999, M S 85-74
Seattle, WA 98124 (206) 773-2572

Eugene O. Herberer
Barnard, Pauly & Kaser, P.S.
6000 Southcenter Blvd. Suite 240
Seattle, WA 98188 (206) 246-0568

Lynn H. Hess
Boeing Commercial Airplane Co.
P.O. Box 3707, M.S. 7E-25
Seattle, WA 98124 (206) 251-0342

Edward V. Hiskes
4717 Stone Way N.
Seattle, WA 98103 (206) 634-0226

Ronald D. Hochnadel
234 27th Ave. East
Seattle, WA 98112 (206) 322-8085

Robert A. Jensen
Graybeal, Jensen & Puntigan
2033 6th Ave. Suite 1020
Seattle, WA 98121 (206) 448-3200

Lee E. Johnson
Christensen, O Connor, Johnson & Kindness
2700 Westin Bldg. 2001 Sixth Ave.
Seattle, WA 98121 (206) 441-8780

Bruce A. Kaser
Barnard, Pauly & Kaser, P.S.
6000 Southcenter Blvd. Suite 240
Seattle, WA 98188 (206) 246-0568

Gary Stanley Kindness
Christensen, O Connor, Johnson & Kindness
2701 Westin Bldg. 2001 Sixth Ave.
Seattle, WA 98121 (206) 623-8780

Steven P. Koda
Seed & Berry
6300 Columbia Center, 701 Fifth Ave.
Seattle, WA 98104 (206) 622-4900

Walter Louis Larsen#
P.O. Box 18302
Seattle, WA 98118 (206) 723-0870

Debra K. Leith#
Wegner & Bretschneider
1914 North 34th St., Suite 400
Seattle, WA 98103 (206) 547-7148

Kenneth M. Mac Intosh
1155 Dexter Horton Bldg.
Seattle, WA 98104 (206) 464-7045

David J. Maki
Seed & Berry
6300 Columbia Center
Seattle, WA 98104 (206) 622-4900

John B. Mason
Christensen, O Connor, Johnson & Kindness
2701 Westin Bldg. 2001 Sixth Ave.
Seattle, WA 98121 (206) 623-8780

Clinton L. Mathis
360 Central Bldg.
Seattle, WA 98104 (206) 624-5131

Paul T. Meiklejohn
Seed & Berry
6300 Columbia Center
Seattle, WA 98104 (206) 622-4900

Raymond E. Metter#
Boeing Co.
P.O. Box 3707, M/s 20-75
Seattle, WA 98124 (206) 655-1615

Jeffrey J. Miller
Seed & Berry
6300 Columbia Center
Seattle, WA 98104 (206) 622-4900

Landon C.G. Miller
Touche Ross & Co.
4220 N.E. 203rd Place
Seattle, WA 98155 (206) 292-1800

J. Peter Mohn
Boeing Aerospace Co.
M/s 85-74, P.O. Box 3999
Seattle, WA 98124 (206) 773-2572

Jerald E. Nagae
Christensen, O Connor, Johnson & Kindness
2701 Westin Bldg.
Seattle, WA 98121 (206) 623-8780

John M. Neary
Boeing Aerospace Co.
P.O. Box 3999
Seattle, WA 98124 (206) 773-2384

Harry D. Nelson
Boeing Aerospace Co. Div. Of Boeing Co.
P.O. Box 3999, M.S. 85-74
Seattle, WA 98124

Suzanne M. Niedzwiecki
Seed & Berry
6300 Columbia Center
Seattle, WA 98104 (206) 622-4900

Bruce E. O Connor
Christensen, O Connor, Johnson & Kindness
2701 Westin Bldg. 2001 Sixth Ave.
Seattle, WA 98121 (206) 623-8780

Jeffrey B. Oster
Seed & Berry
6300 Columbia Center
Seattle, WA 98104 (206) 622-4900

Gary E. Parker#
Zymo Genetics, Inc.
4225 Roosevelt Way N.E.
Seattle, WA 98105 (206) 547-8080

Steven W. Parmelee
Genetic Systems Corp.
3005 First Ave.
Seattle, WA 98121 (206) 728-4900

Joan H. Pauly
Barnard, Pauly & Kaser, P.S.
6000 Southcenter Blvd. Suite 240
Seattle, WA 98188 (206) 246-0568

Roberta A. Picard#
Wegner & Bretschneider
1914 North 34th St., Suite 400
Seattle, WA 98103 (206) 547-7148

Clark A. Puntigam
Cole, Jensen & Puntigam, P.S.
2033 6th Ave. Suite 1020
Seattle, WA 98121 (206) 622-3740

Mary Y. Redman
Boeing Aerospace Co.
P.O. Box 3999, M S
Seattle, WA 98124 (206) 773-2572

Glenn P. Rickards
Dowrey & Cross, P.S.
1254 Bank of California Center
Seattle, WA 98164 (206) 624-6535

George C. Rondeau, Jr.
Seed & Berry
6300 Columbia Center
Seattle, WA 98104 (206) 622-4900

Katie E. Sako#
Christensen, O Connor, Johnson & Kindness
2601 6th Ave. Suite 2700
Seattle, WA 98121 (206) 441-8780

Laurence Arthur Savage
Boeing Assco. Products
P.O. Box 3707, Mail Stop 7E-14
Seattle, WA 98124 (206) 822-0227

Joe M. Scott#
5202 37th Ave. S.W.
Seattle, WA 98126 (206) 935-5595

Thomas W. Secrest, III
1023 N.E. 62nd
Seattle, WA 98115 (206) 523-2464

Richard W. Seed
Seed & Berry
6300 Columbia Center
Seattle, WA 98104 (206) 622-4900

Kathryn Ann Seese#
Neo Rx Corp.
410 W. Harrison St.
Seattle, WA 98119 (206) 281-7001

Richard G. Sharkey#
Seed & Berry
6300 Columbia Center
Seattle, WA 98104 (206) 622-4900

Ford E. Smith
Garrison & Assco.
2100 Westin Bldg. 2001 Sixth Ave.
Seattle, WA 98121 (206) 728-5920

Ronald E. Suter
Boeing Comm. Airplane Co.
P.O. Box 3707, M.S. 7E-25
Seattle, WA 98124 (206) 251-0275

Earl R. Tarleton
Christensen, O Connor, Johnson & Kindness
27th Floor, 2001 Sixth Ave.
Seattle, WA 98121 (206) 441-8780

Kenneth William Thomas
Boeing Aerospace Co.
P.O. Box 3999
Seattle, WA 98124 (773) 257-2 /p
hA

Michael G. Toner
Christensen, O Connor, Johnson & Kindness
2701 Westin Bldg. 2001 6th Ave.
Seattle, WA 98121 (206) 441-8780

Albert G. Tramposch#
Seed & Berry
6300 Columbia Center
Seattle, WA 98104 (206) 622-4900

James R. Uhlir
Christensen, O Connor, Johnson & Kindness
Westin Bldg., Suite 2701
2001 6th Ave.
Seattle, WA 98121 (206) 441-8780

Jeannette M. Walder
Boeing Military Airplane Co.
P.O. Box 3707, M S 32-50
Seattle, WA 98188

Daryl B. Winter
Christensen, Oconnor, Johnson & Kindness
2700 Westin Bldg. 2001 6th Ave.
Seattle, WA 98121 (206) 441-8780

Edward M. Yoshida
Imre Corp.
130 Fifth Ave. N.
Seattle, WA 98109 (206) 448-1000

B. Peter Barndt
Paulsen Center, Suite 711
Spokane, WA 99201 (509) 747-8830

Keith S. Bergman
418 Symons Bldg. S. 7 Howard St.
Spokane, WA 99204 (509) 838-2851

Leon Gilden
5210 S. Cree Dr.
Spokane, WA 99206 (509) 924-4075

Michael A. Glenn
Wells, St. John & Roberts
815 Wash. Mutual Bldg.
Spokane, WA 99201 (509) 624-4276

Randy A. Gregory
Wells, St. John & Roberts, P.S.
W. 601 Main Ave. Suite 815
Spokane, WA 99201 (509) 624-4276

Mark W. Hendricksen
Layman, Loft, Arpin & White
820 Lincoln Bldg.
Spokane, WA 99201 (509) 455-8883

Mark S. Matkin
Wells, St. John & Roberts, P.S.
815 Wash. Mutual Bldg.
Spokane, WA 99201 (509) 624-4276

James L. Price#
Wells, St. John & Roberts
815 Wash. Mutual Bldg.
Spokane, WA 99201 (509) 624-4276

David P. Roberts
Wells, St. John & Roberts, P.S.
815 Wash. Bldg.
Spokane, WA 99201 (509) 624-4276

Richard J. St. John
Wells, St. John & Roberts
W. 601 Maine St. Suite 815
Spokane, WA 99201 (509) 624-4276

J. Robert Cassidy
Hughes & Cassidy, P.S.
P.O. Box 439
Sumas, WA 98295 (206) 988-2061

Robert Bruce Hughes
Hughes & Cassidy, P.S.
P.O. Box 439, 617 Cherry St.
Sumas, WA 98295 (206) 988-2061

Patrick Donlan Coogan
Weyerhaeuser Co.
Tacoma, WA 98477 (206) 924-2061

John M. Crawford
Weyerhaeuser Co. Pat. Dept.
Tacoma, WA 98477 (206) 924-5611

Kenneth S. Kessler
Mattern & Kessler
543 Broadway
Tacoma, WA 98402 (206) 383-1751

James F. Leggett
Leggett & Kram
1901 South I St.
Tacoma, WA 98405 (206) 272-7929

Bryan C. Ogden
Weyerhaeuser Co. Pat. Dept.
Tacoma, WA 98477 (206) 924-2062

Susan L. Preston
Weyerhaeuser Co.
Tacoma, WA 98477 (206) 924-3461

George A. Cashman
4407 Carriage Hill Dr.
Yakima, WA 98908

West Virginia

James S. Lovell#
136 Oakwood Rd.
Charleston, WV 25314 (304) 343-0989

John R. Hanway
Box 683
Fairmont, WV 26554 (304) 366-2621

William L. Jarvis#
1717 C Mileground
Morgantown, WV 26505 (304) 296-5209

Michael F. Fronko#
P.O. Box 465
Parkersburg, WV 26102 (614) 678-2868

Thomas Braden Hunter
Borg-Warner Chem. Inc.
International Center
Parkersburg, WV 26102 (304) 424-5564

Allen F. Millikan#
800 W. Virginia Ave.
Parkersburg, WV 26101 (304) 422-7116

George J. Neilan
2324 Woodland Ave.
South Charleston, WV 25303 (304) 744-8702

William K. Cox
Weirton Steel Corp.
400 Three Springs Dr.
Weirton, WV 26062 (304) 797-2387

Charles D. Bell
Bell, Mc Mullen & Cross
67 Seventh St.
Wellsburg, WV 26070 (304) 737-0771

Wisconsin

E. Frank Mc Kinney
Appleton Papers Inc.
P.O. Box 359, 825 E. Wisconsin Ave.
Appleton, WI 54912 (414) 735-8660

Benjamin Mieliulis
Appleton Papers Inc.
P.O. Box 359, 825 E. Wisconsin Ave.
Appleton, WI 54912 (414) 734-9841

Paul S. Phillips, Jr.#
Appleton Papers Inc.
P.O. Box 359,
Appleton, WI 54912 (414) 735-8661

Robert E. Emery
P.O. Box 667, 41 Broad St,
Bayfield, WI 54814 (715) 779-5354

David J. Archer#
Beloit Corp.
1 St. Lawrence Ave.
Beloit, WI 53511 (608) 364-7018

Raymond W. Campbell
Beloit Corp.
1 St. Lawrence Ave.
Beloit, WI 53511 (608) 364-7042

A. Richard Koch#
1660 Morgan Terr.
Beloit, WI 53511 (608) 362-3508

Gerald Albert Mathews#
Beloit Corp.
1 St. Lawrence Ave.
Beloit, WI 53511 (608) 364-7044

Dirk J. Veneman
Beloit Corp.
1 St. Lawrence Ave.
Beloit, WI 53511 (608) 364-7040

Aaron Lee Hardt, Sr.
Rexnord Inc. Legal Dept. (cH)
350 N. Sunny Slope Rd.
Brookfield, WI 53005 (414) 797-5687

James L. Kirschnik
4445 N. 12th St.
Brookfield, WI 53005 (414) 781-2050

James E. Lowe, Jr.
R T E Corp. Brookfield Lakes Corp. Center
175 N. Patrick Blvd.
Brookfield, WI 53005 (414) 792-9300

Richard C. Ruppin, Sr.
Rexnord Inc.
350 N. Sunny Slope
Brookfield, WI 53005 (414) 797-6900

Vance A. Smith
Rexnord Inc.
350 N. Sunny Slope Rd.
Brookfield, WI 53005 (414) 797-5663

Willis B. Swartwout, III
Swartwout & Eichfeld, S.C.
15850 W. Bluemound Rd. Suite 220, P.O. Box 1068
Brookfield, WI 53005 (414) 786-8614

Glenn S. Thompson#
4508 W. Calumet Rd.
Brown Deer, WI 53223 (414) 354-4213

Francis J. Bouda
13319 Centerville Rd.
Cleveland, WI 53015 (414) 693-8202

Arthur M. Streich
1307 Milwaukee St.
Delafield, WI 53018 (414) 646-8182

William Arthur Denny
Denny & Yanisch
13500 Watertown Plank Rd.
Elm Grove, WI 53122 (414) 797-8777

Joseph S. Heino
Denny & Yanisch
13500 Watertown Plank Rd.
Elm Grove, WI 53122 (414) 797-8777

Donald Edwin Porter
15025 Westover Rd.
Elm Grove, WI 53122 (414) 782-3324

Donald Cayen
104 S. Main St.
Fond Du Lac, WI 54935 (414) 921-2288

Robert C. Curfiss
Mercury Marine Div. Of Brunswick Corp.
1939 Pioneer Rd.
Fond Du Lac, WI 54935 (414) 929-5419

Olin T. Sessions
Mercury Marine
1939 Pioneer Rd.
Fond Du Lac, WI 54935 (414) 929-5261

John A Reiter, Jr.#
247 North St. Route 2 Box 9
Fountain City, WI 54629 (608) 687-3401

T. Lloyd Lafave
Weber, Raithel, Malm & La Fave
5900 N. Port Washington Rd. Suite 210
Glendale, WI 53217 (414) 964-5250

Gordon Paul Ralph
6642 N. Atwahl Dr.
Glendale, WI 53209 (414) 352-7475

Peter Paul Kozak
1008 S. Van Buren
Green Bay, WI 54301 (414) 437-9235

Joseph M. Recka
Recka, Joannes & Faller S.C.
211 S. Monroe
Green Bay, WI 54301 (414) 435-8159

Dennis J. Verhaagh
P.O. Box 995
Green Bay, WI 54305 (414) 435-6423

Ronald Clarence Klett#
P.O. Box 232
Greendale, WI 53129

James T. Barr
P.O. Box 153
125 Main St.
Hancock, WI 54943 (715) 249-5182

Lewis L. Lloyd
U.S. Marine Corp. & Bayliner Marine Corp.
105 Marine Dr.
Hartford, WI 53027 (414) 673-2200

M. Paul Hendrickson
403 Main St. P.O. Box 508
Holmen, WI 54636 (608) 526-4422

Gary L. Griswold
Cove Rd.
Route No. 3
Hudson, WI 54016 (612) 733-8904

Howard M. Herriot
Consigny, Andrews, Hemming & Grant, S.C.
303 E. Court St. P.O. Box 1449
Janesville, WI 53547 (608) 755-5050

Edward R. Antaramian
Antaramian, Easton & Antaramian
2221 - 63rd St.
Kenosha, WI 53140 (414) 654-8669

Neil E. Hamilton
8306 42nd Ave.
Kenosha, WI 53142 (414) 694-6283

David J. Richter
Snap-On Tools Corp.
2801 80th St.
Kenosha, WI 53141 (414) 656-5322

Richard T. Naruo
451 Audubon Rd.
Kohler, WI 53004 (414) 452-4215

Mark D. Schuman#
113 Church St.
Kohler, WI 53004 (414) 452-6126

William J. Beres
Trane Co. Pat. Dept.
3600 Pammel Creek Rd.
La Crosse, WI 54601 (608) 787-4177

Peter D. Ferguson
The Trane Co.
3600 Pammell Creek Rd.
La Crosse, WI 54601 (608) 787-3405

Robert J. Harter#
Trane Co. Pat. Dept.
3600 Pammell Creek Rd.
La Crosse, WI 54601 (608) 787-3860

David L. Polsley
Trane Co. Pat. Dept.
3600 Pammel Creek Rd.
La Crosse, WI 54601

Raymond Lynn Balfour
Rayovac Corp.
601 Rayovac Dr.
Madison, WI 53711 (608) 275-4584

Howard W. Bremer
Wisconsin Alumni Res. Foundation
P.O. Box 7365, 614 N. Walnut St.
Madison, WI 53707 (608) 263-2831

Barry U. Buchbinder#
Agrigenetics Advanced Science Co.
5649 E. Buckeye Rd.
Madison, WI 53716 (608) 221-5000

William J. Connors
Nicolet Instrument Corp.
5225 Verona Rd.
Madison, WI 53711 (608) 271-3333

Harry C. Engstrom, Jr.
Isaksen, Lathrop. Esch, Hart & Clark
P.O. Box 1507, 122 W. Wash. Ave.
Madison, WI 53701 (608) 257-7766

David T. Flanagan
Wisconsin Dept. Of Justice, State Capitol
Madison, WI 53704 (608) 266-7971

Carl E. Gulbrandsen
Haight & Hofeldt
3 South Pinckney St. Suite 715
Madison, WI 53703 (608) 258-7888

M. Henry Halle
Isaksen, Lathrop, Esch, Hart & Clark
122 W. Wash. Ave. Suite 1000
Madison, WI 53701 (608) 257-7766

David J. Houser
Wisconsin Alumni Res. Foundation
614 N. Walnut St.
Madison, WI 53705 (608) 263-9395

James A. Kemmeter
Isaksen, Lathrop, Esch, Hart & Clark
P.O. Box 1507, 122 W. Wash. Ave.
Madison, WI 53701 (608) 257-7766

Trayton L. Lathrop
Isaksen, Lathrop, Esch, Hart & Clark
P.O. Box 1507. 122 W. Wash. Ave.
Madison, WI 53701 (608) 257-7766

Theodore J. Long
Isaksin, Lathrop, Esch, Hart & Clark
122 W. Wash. Ave.
P.O. Box 1507
Madison, WI 53701

Costa Perchem
25 W. Main St. Suite 711
Madison, WI 53703 (608) 256-0681

Douglas E. Pfrang#
Nicolet Instru. Corp.
5225 Verona Rd.
Madison, WI 53711 (608) 271-3333

Charles S. Sara
Isaksen, Lathrop, Esch, Hart & Clark
122 W. Wash. Ave. P.O. Box 1507
Madison, WI 53701 (608) 257-7766

H. Keith Schoff
2257 E. Wash. Ave.
Madison, WI 53704 (608) 241-2616

Nicholas J. Seay
Isaksen, Lathrop, Esch, Hart & Clark
122 W. Wash. Ave. P.O. Box 1507
Madison, WI 53702 (608) 257-7766

David E. Stewart
De Witt, Sundby, Huggett, Schumacher & Morgan, S.C.
6515 Grand Teton Plaza, Suite 120
Madison, WI 53719 (608) 255-8891

Joseph T. Hepp#
2325 Riverside Ave.
Marinette, WI 54143 (715) 735-9671

Raymond Joseph Miller#
326 Winnebago Ave.
Menasha, WI 54952 (414) 722-7804

William Arthur Autio
N 85 W 17231 Lee Place
Menomonee Falls, WI 53051 (414) 251-2863

John Kay Crump
945 W. Heritage Court N. 207
Meqoon, WI 53092 (414) 241-9279

Karl William Marquardt
3905 Sumac Circle
Middleton, WI 53562 (608) 836-8526

Robert B. Benson
Allis-Chalmers Mfg. Co.
P.O. Box 512
Millwaukee, WI 53201 (414) 475-4038

Frank S. Andrus
Andrus, Sceales, Starke & Sawall
735 N. Water St.
Milwaukee, WI 53202 (414) 271-7590

Russell J. Barron
Foley & Lardner
777 E. Wisconsin Ave.
Milwaukee, WI 53202 (414) 289-3752

Ronald E. Barry
Frisch, Dudek & Slattery, Ltd.
825 N. Jefferson St.
Milwaukee, WI 53202 (414) 273-4000

Keith M. Baxter
General Electric Co. Med. Sys. Group
P.O. Box 414
Milwaukee, WI 53201

John L. Beard
Michael, Best & Friedrich
250 E. Wisconsin Ave.
Milwaukee, WI 53202 (414) 271-6560

Glenn A. Buse
Michael, Best & Friedrich
250 E. Wisc. Ave. Suite 2000
Milwaukee, WI 53202 (414) 271-6560

Donald G. Casser
Quarles & Brady
411 E. Wisconsin Ave.
Milwaukee, WI 53202 (414) 277-5000

Robert E. Clemency
Michael, Best & Friedrich
250 E. Wisconsin Ave.
Milwaukee, WI 53202 (414) 271-6560

John C. Cooper, III
Whyte & Hirschboeck S.C.
111 East Wisconsin Ave. Suite 2100
Milwaukee, WI 53202 (414) 271-8210

James R. Custin
Nilles, Custin & Kirby, S.C.
777 E. Wisconsin Ave. Suite 3070
Milwaukee, WI 53202 (414) 276-0977

Thomas William Ehrmann
Quarles & Brady
411 E. Wisconsin Ave.
Milwaukee, WI 53202 (414) 277-5000

Arnold J. Ericsen
Whyte & Hirschboeck S.C.
111 East Wisconsin Ave. Suite 2100
Milwaukee, WI 53202 (414) 271-8210

Gary A. Essmann
Andrus, Sceales, Starke & Sawall
735 N. Water St.
Milwaukee, WI 53202 (414) 271-7590

George A. Evans
735 N. Water St.
Milwaukee, WI 53202 (414) 271-7590

William L. Falk#
Foley & Lardner
777 East Wisconsin Ave.
Milwaukee, WI 53202 (414) 289-3682

Daniel Dawson Fetterley
Andrus, Sceales, Starke & Sawall
735 N. Water St.
Milwaukee, WI 53202 (414) 271-7590

Paul W. Fish
Jos. Schlitz Brewing Co.
235 W. Galena St.
Milwaukee, WI 53212 (414) 224-5306

John D. Franzini
Quarles & Brady
411 E. Wisconsin Ave.
Milwaukee, WI 53202 (415) 277-5000

John F. Friedl
Deutz-Allis Corp.
P.O. Box 933
Milwaukee, WI 53201 (414) 475-2146

Raymond E. Fritz, Jr.
Thomas, Mc Kinnon, Securities, Inc.
731 N. Water St.
Milwaukee, WI 53202 (414) 271-5670

Henry C. Fuller, Jr.
Fuller, Puerner & Hohenfeldt, S.C.
633 W. Wisconsin Ave.
Milwaukee, WI 53203 (414) 271-6555

Joseph A. Gemignani
Michael, Best & Friedrich
250 E. Wisconsin Ave.
Milwaukee, WI 53202 (414) 271-6560

Alexander M. Gerasimow
Johnson Controls, Inc.
P.O. Box 591
Milwaukee, WI 53201 (414) 228-3218

Thomas Edward Goss, Sr.
Allis-Chalmers Corp.
P.O. Box 512
Milwaukee, WI 53201 (414) 475-4466

George E. Haas
Quarles & Brady
411 E. Wisconsin Ave.
Milwaukee, WI 53202 (414) 277-5000

Cyril M. Hajewski
Quarles & Brady
411 E. Wisconsin Ave.
Milwaukee, WI 53202 (414) 277-5000

Ralph G. Hohenfeldt
Fuller, Puerner & Hohenfeldt
633 W. Wisconsin Ave.
Milwaukee, WI 53203 (414) 271-6555

Guenther W. Holtz
Andrus, Sceales, Starke & Sawall
735 N. Water St.
Milwaukee, WI 53202 (414) 271-7590

Robert H. Jacob#
733 N. Van Buren St.
Milwaukee, WI 53202 (414) 271-6669

Joseph J. Jochman, Jr.
Andrus, Sceales, Starke & Sawall
735 N. Water St.
Milwaukee, WI 53202 (414) 271-7590

Ira Milton Jones
1037 N. Astor St.
Milwaukee, WI 53202 (414) 276-4210

Lee H. Kaiser
Allis Chalmers Corp.
P.O. Box 512
Milwaukee, WI 53201 (414) 475-4340

Nicholas A. Kees
Fuller, Puerner & Hohenfeldt
633 W. Wisconsin Ave.
Milwaukee, WI 53203 (414) 271-6555

Roman A. Klitzke
Marquette Univ. Law School
1103 W. Wisconsin Ave.
Milwaukee, WI 53233 (414) 224-7094

Thomas O. Kloehn
Quarles & Brady
411 E. Wisconsin Ave.
Milwaukee, WI 53202 (414) 277-5000

Thad F. Kryshak
Quarles & Brady
411 E. Wisconsin Ave.
Milwaukee, WI 53202 (414) 277-5000

Michael A. Lechter
Foley & Lardner
777 E. Wisconsin Ave.
Milwaukee, WI 53202 (414) 271-2400

Allan W. Leiser
Quarles & Brady
411 E. Wisconsin Ave.
Milwaukee, WI 53202 (414) 277-5000

Stephen P. Malak
Spring-Mornne, Inc.
2040 W. Wisconsin
Milwaukee, WI 53233 (414) 342-4809

Philip P. Mann
Michael, Best & Friedrich
250 E. Wisconsin Ave.
Milwaukee, WI 53202 (414) 271-6560

Andrew S. Mc Connell
Andrus, Sceales, Starke & Sawall
735 N. Water St.
Milwaukee, WI 53202 (414) 271-7590

Michael J. Mc Govern
Quarles & Brady
411 E. Wisconsin Ave.
Milwaukee, WI 53202 (414) 277-5000

Edward W. Mentzer
Rexnord Inc.
777 E. Wisconsin Ave.
Milwaukee, WI 53204 (414) 643-2504

Philip G. Meyers
Foley & Lardner, First Wisconsin Ctr.
777 E. Wisconsin Ave.
Milwaukee, WI 53202 (414) 289-3761

Bayard H. Michael
Michael, Best & Friedrich
250 E. Wisconsin Ave.
Milwaukee, WI 53202 (414) 271-6560

Mark L. Mollon
General Electric Co. Med Sys. Group
P.O. Box 414
Milwaukee, WI 53201 (414) 544-3175

William H. Nehrkorn
Heil Co.
3000 W. Montana St.
Milwaukee, WI 53215 (414) 647-3205

Arthur Lowell Nelson
Allis- Chalmers Mfg. Co.
P.O. Box 512
Milwaukee, WI 53201 (414) 475-2118

James E. Nilles
Nilles, Custin & Kirby, S.C.
777 E. Wisconsin Ave. Suite 3070
Milwaukee, WI 53202 (414) 276-0977

James P. O Shaughnessy
Foley & Lardner
777 E. Wisconsin Ave.
Milwaukee, WI 53202 (414) 289-3568

Mark W. Pfeiffer#
Quarles & Brady
411 E. Wisconsin Ave.
Milwaukee, WI 53202 (414) 277-5000

David R. Price
Michael, Best & Friedrich
250 E. Wisconsin Ave.
Milwaukee, WI 53202 (414) 271-6560

Paul R. Puerner
Fuller, Puener & Hohenfeldt, S.C.
633 W. Wisconsin Ave. Suite 1810
Milwaukee, WI 53203 (414) 271-6555

Joseph D. Radtke
740 N. Plankinton
Milwaukee, WI 53203 (414) 271-1231

Hugh R. Rather
5400 N. Diversey Blvd.
Milwaukee, WI 53217 (414) 332-5045

Michael D. Rechtin
Reinhart, Boerner, Van Deuren, Norris & Rieselbach, S.C.
111 E. Wisconsin Ave. Suite 1800
Milwaukee, WI 53202 (414) 271-1190

Andrew O. Riteris
Michael, Best & Friedrich
250 E. Wisconsin Ave.
Milwaukee, WI 53202 (414) 271-6560

John Phillip Ryan
Johnson Controls, Inc.
5757 N. Green Bay Ave.
Milwaukee, WI 53201 (414) 228-2311

Barry E. Sammons
Quarles & Brady
411 E. Wisconsin Ave.
Milwaukee, WI 53202 (414) 277-5000

Eugene R. Sawall
Andrus, Sceales, Starke & Sawall
735 N. Water St.
Milwaukee, WI 53202 (414) 271-7590

Merl E. Sceales
Andrus, Sceales, Starke & Sawall
735 N. Water Street, Suite 1102
Milwaukee, WI 53202 (414) 271-7590

Charles L. Schwab
Deutz-Allis Corp.
Box 933
Milwaukee, WI 53201 (414) 475-3181

Carl R. Schwartz
Quarles & Brady
411 E. Wisconsin Ave.
Milwaukee, WI 53202 (414) 277-5000

Jean Seaburg
8081 N. 105th St.
Milwaukee, WI 53224 (414) 355-1234

Neal Seegert
Foley & Lardner
777 East Wisconsin Ave.
Milwaukee, WI 53202 (414) 289-3552

Larry L. Shupe
Johnson Controls Inc. X-73
5757 N. Green Bay Ave. P.O. Box 591
Milwaukee, WI 53201 (414) 228-2310

James O. Skarsten
Cross & Trecker Corp.
11000 Theodore Trecker Way
Milwaukee, WI 53214 (414) 476-8300

David B. Smith
Michael, Best & Friedrich
250 E. Wisconsin Ave.
Milwaukee, WI 53202 (414) 271-6560

George H. Solveson
Andrus, Sceales, Starke & Sawall
735 N. Water St.
Milwaukee, WI 53202 (414) 271-7590

Warren T. Sommer
3137 N. Cramer St.
Milwaukee, WI 53211 (414) 962-7076

Glen O. Starke
Andrus, Sceales, Starke & Sawall
735 N. Water St. Suite 1102
Milwaukee, WI 53202 (414) 271-7590

Richard C. Steinmetz, Jr.
Allen-Bradley Co.
1201 S. 2nd St.
Milwaukee, WI 53204 (414) 671-2000

Douglas E. Stoner
General Electric Co.
P.O. Box 414
Milwaukee, WI 53201 (414) 544-3439

C. Thomas Sylke
Whyte & Hirschboeck S.C.,
111 E. Wisconsin Ave. Suite 2100
Milwaukee, WI 53202 (414) 271-8210

Michael E. Taken
Andrus, Sceales, Starke & Sawall
735 N. Water St.
Milwaukee, WI 53202 (414) 271-7590

V. Robins Tate
1840 N. Prospect Ave. Apt. 808
Milwaukee, WI 53202

Larry G. Vande Zande#
Eaton Corp.
4201 N. 27th St.
Milwaukee, WI 53216 (414) 449-6264

Allan B. Wheeler
Wheeler, Morsell, House & Wheeler
606 West Wisconsin Ave. Suite 1506
Milwaukee, WI 53203 (414) 278-7733

Fred Wiviott
Whyte & Hirschboeck, S.C.
111 E. Wisconsin Ave. Suite 2100
Milwaukee, WI 53202 (414) 271-8210

Thomas M. Wozny
Andrus, Sceales, Starke & Sawall
735 North Water St.
Milwaukee, WI 53202 (414) 271-7590

Elroy J. Wutschel
Kirst, Surges & Wutschel
522 N. Water St.
Milwaukee, WI 53202 (414) 276-0220

Elwin J. Zarwell
Quarles & Brady
780 N. Water St.
Milwaukee, WI 53202 (414) 277-5000

John L. Chiatalas
Kimberly-Clark Corp.
401 N. Lake St.
Neenah, WI 54956 (414) 721-2985

Gregory E. Croft
Kimberly-Clark Corp.
401 N. Lake St.
Neenah, WI 54956 (414) 721-3616

Paul A. Leipold
Kimberly-Clark Corp.
401 N. Lake St.
Neenah, WI 54956 (414) 721-2358

Douglas L. Miller
Kimberly-Clark Corp.
401 N. Lake St.
Neenah, WI 54956 (414) 721-2000

Eckhard C.A. Schwarz#
Biax-Fiberfilm Corp.
1066 American Dr.
Neenah, WI 54956 (414) 722-3180

Donald L. Traut
Kimberly-Clark Corp.
401 N. Lake St.
Neenah, WI 54956 (414) 721-2433

Thomas D. Wilhelm#
104 E. Wisconsin Ave.
Neenah, WI 54956 (414) 725-1297

Paul Y.P. Yee
Kimberly-Clark Corp.
401 North Lake St.
Neenah, WI 54956 (414) 721-2435

Donald C. Mc Gaughey
5615 S. Frances Ave.
New Berlin, WI 53151 (414) 425-3493

Jack G. Kinn
640 Glenview Ave.
Oconomowoc, WI 53066 (414) 567-6070

Carl M. Lewis
642 Winter St.
Onalaska, WI 54650 (608) 783-3924

James J. Getchius
711 Jackson St.
Oshkosh, WI 54901 (414) 235-0601

R. Jonathan Peters
3169 Ryf Rd.
Oshkosh, WI 54904

Chester J. Giuliani
Box 53
Phelps, WI 54554 (906) 548-2138

Robert T. Johnson#
603 Collins St.
Plymouth, WI 53073 (414) 892-8556

Jerome Donald Drabiak
S.C. Johnson & Son, Inc.
1525 Howe St. M S 077
Racine, WI 53403 (414) 631-2000

J. William Frank, III
S.C. Johnson & Son, Inc. Pat. Sec.
Patent Section
1525 Howe St.
Racine, WI 53403 (414) 631-2673

Robert D. Godard#
J.I. Case Co.
700 State St.
Racine, WI 53404 (414) 636-6873

Arthur John Hansmann
312 7th St.
Racine, WI 53403 (414) 632-2818

Peter N. Jansson
254 Main St. Suite M
Racine, WI 53403 (414) 632-6900

Joseph T. Kivlin, Jr.
S.C. Johnson & Son, Inc.
1525 Howe St.
Racine, WI 53403 (414) 544-2442

Richard E. Rakoczy
S.C. Johnson & Son, Inc.
1525 Howe St. Mail Sta. 077
Racine, WI 53403 (414) 631-2909

Stanley E. Sutherland
S.C. Johnson & Son Inc.
1525 Howe St.
Racine, WI 53403 (414) 631-2995

Tipton L. Randall#
Zimpro/Passavant
301 W. Military Rd.
Rothschild, WI 54474 (715) 359-7211

Fabian A. Brusok
3519 High Cliff Circle
Sheboygan, WI 58083 (414) 458-6833

Ray Gunnar Olander
Bucyrus-Erie Co.
1100 Milwaukee Ave.
South Milwaukee, WI 53172 (414) 768-4099

Robert J. Steininger
531 Linden Circle
South Milwaukee, WI 53172 (414) 272-0530

John Lawrence Beales
Peterson Builders, Inc.
P.O. Box 650
Sturgeon Bay, WI 54235 (414) 743-5577

Arnold J. De Angelis
527 Oakwood Dr.
Thiensville, WI 53092 (414) 242-5893

Harold Warner Grothman
621 Grand Ave.
Thiensville, WI 53092 (414) 242-1574

Ronald W. O Keefe
G E Co. Med. Sys. Div.
3000 N. Grandview Blvd.
Waukesha, WI 53186 (414) 544-3155

James R. Sommers
Hunter & Sommers
259 South St.
Waukesha, WI 53187 (414) 547-7788

John W. Kelley
Kelley, Weber & Piezt, S.C.
530 Jackson St.
Wausau, WI 55401 (715) 845-9211

Russell E. Weinkauf
1403 Steuban St.
Wausau, WI 54401 (715) 842-7066

Robert C. Jones
3860 North 102 St.
Wauwatosa, WI 53222 (414) 461-2366

Thomas F. Kirby
6434 Betsy Ross PL.
Wauwatosa, WI 53213 (414) 257-3746

Russell L. Johnson#
P.O. Box 161
Weyauwega, WI 54983 (414) 867-3482

Wyoming

Phillip D. Koontz
1472 N. 5th St. P.O. Box 820
Laramie, WY 82070 (307) 721-8917

James P. De Clercq
202 Park Ave.
Wheatland, WY 82201 (307) 322-5457

PART FOUR
Appendixes and Master Index

Appendix A

U.S. PATENT CLASSIFICATIONS

The U.S. Patent Classifications are an alphabetical list of the subject headings referring to specific classes and subclasses of the U.S. patent classification system. The Classifications are intended as an initial means of entry into the Patent and Trademark Office's (PTO's) classification system and should be particularly useful to those who lack experience in using the classification system or who are unfamiliar with the technology under consideration.

The Classifications are to searching a patent what the card catalog is to looking for a library book. It is the only way to discover what exists in the field of prior art. The Classifications are a star to steer by, without which no meaningful patent search can be completed.

Although continual changes are made in the classification system, the following pages will serve as an invaluable tool for 1) preparing in advance for a patent search, and 2) obtaining an understanding of how inventions are classified by the PTO.

Before you begin your search, use the classifications that follow to prepare your direction. First, look for the term which you feel best represents your invention. If a match cannot be found, look for terms of approximately the same meaning, for example, the essential function or the effect of use or application of the object of concern.

Once you have recorded the identifying numbers of possibly pertinent classes and subclasses, go to a Patent Depository Library (PDL). Here you may wish to look first at the *Manual of Classification,* a loose-leaf PTO volume listing the numbers and descriptive titles of more than 300 classes and 95,000 subclasses used in the subject classification of patents, with an index to classifications.

The Classifications are arranged with subheadings that can extend to four levels of indentation. A complete reading of a subheading includes the title of the most adjacent higher heading, and so on until there are no higher headings. Some headings will reference other related or preferred entries with a "(see...)" phrase.

New classes and subclasses are continuously based upon breaking developments in science and technology. Old classes and subclasses are rendered obsolete by technological advance. In fact, if you have suggestions for future revisions of the Classifications, find omissions or errors, you are encouraged to alert the PTO. Send your suggestions to Editor, *Index to U.S. Patent Classification,* Office of Documentation, U.S. Patent and Trademark Office, Washington, D.C. 20231.

Before the Index begins, a table of abbreviations and a section on the classes arranged in alphabetical order is included. It can be helpful to look over this section before jumping into the work of locating your field of search. It will help familiarize you with class titles and groupings.

Table of Abbreviations

A . . . Z	An Alpha designation following the official numeric subclass identifies an unofficial subclass, i.e., a grouping of patents selected out from an official subclass by an examiner and then made an indented subclass under the official subclass, usually for purposes of further breaking down a concept. For example, in Class 273 Amusement Devices, Games, Subclass 58A collects balls which are solid and which are covered. Topics not specifically selected out are left in a residual subclass, designated R. For example, 58R is generic to balls not provided for elsewhere. Unofficial subclass groupings are used in the Examiners' Files; they are not available in the Public Search Room.
Ctg.	"Containing" may be abbreviated "ctg" in long chemical phrases.
D	Design classes are preceded by the letter D, as in D2.
DIG.	A digest, indicated by DIG. and a subclass number, is an unofficial collection of patents based on a concept which relates to a class but not to any particular subclass of that class. In this Index 123 DIG. 12 is a collection of Hydrogen fueled engines in Class 123, Internal Combustion Engines. Digests have been created over the years by examiners to facilitate their searches within the arts under their jurisdiction. They are unofficial collections and are not available in the Public Search Room.
PLT	Plant patents are identified as class PLT.
TM	A few trademark names have been included where a particular field of search could best be suggested by such. These are identified by the letters TM.
+	The plus sign following the subclass indicates that the entry includes that subclass and all subclasses indented thereunder.
X-art	Cross reference art collections may be described in the text as "X-art" collections. These are official collections of patents based on a concept which relates to a class but not to any particular subclass of that class. For example, Class 425 contains molding apparatus described by a functional structure and cross-reference art collection 801 collects button molding patents into one group.
*	The asterisk following a subclass designates a cross-reference art collection.

Users are encouraged to make suggestions for future revisions of this Index. Identification of errors and omissions are welcomed. When recommending new entries, please include the classification with the descriptor, if possible. Send suggestions to *Index to U.S. Patent Classification*, Office of Documentation, U.S. Patent and Trademark Office, Washington, D.C. 20231.

The Index is available from the Superintendent of Documents, U.S. Government Printing Office, Washington, D.C. 20402. Each subscription to the *Manual of Classification* includes one copy of the Index and any new edition issuing during the life of the subscription. Responsive to the many requests for multiple copies, the Government Printing Office now sells the Index as a separate stand-alone document.

Classes Arranged in Alphabetical Order

Class	Title of Class
51	Abrading
181	Acoustics
357	Active Solid State Devices, e.g., Transistors, Solid State Diodes
156	Adhesive Bonding and Miscellaneous Chemical Manufacture
226	Advancing Material of Indeterminate-Length
244	Aeronautics
366	Agitating
102	Ammunition and Explosives
86	Ammunition and Explosive-Charge Making
330	Amplifiers
272	Amusement and Exercising Devices
273	Amusement Devices, Games
446	Amusement Devices, Toys
119	Animal Husbandry
223	Apparel Apparatus
2	Apparel
221	Article Dispensing
236	Automatic Temperature and Humidity Regulation
4	Baths, Closets, Sinks and Spittoons
136	Batteries, Thermoelectric and Photoelectric
384	Bearing or Guides
5	Beds
6	Bee Culture
402	Binder Device Releasably Engaging Aperture or Notch of Sheet
8	Bleaching and Dyeing; Fluid Treatment and Chemical Modification of Textiles and Fibers
10	Bolt, Nail, Nut, Rivet and Screw Making
412	Bookbinding: Process and Apparatus
281	Books, Strips and Leaves
36	Boots, Shoes and Leggings
12	Boot and Shoe Making
175	Boring or Penetrating the Earth
215	Bottles and Jars
188	Brakes
14	Bridges
15	Brushing, Scrubbing and General Cleaning
300	Brush, Broom and Mop Making
24	Buckles, Buttons, Clasps, Etc.
441	Buoys, Rafts, and Aquatic Devices
17	Butchering
79	Button Making
40	Card, Picture and Sign Exhibiting
502	Catalyst, Solid Sorbent, or Support Therefor, Product or Process of Making
59	Chain, Staple and Horseshoe Making
297	Chairs and Seats
194	Check-Actuated Control Mechanisms
436	Chemistry: Analytical and Immunological Testing
435	Chemistry: Molecular Biology and Microbiology
530	Chemistry, Peptides or Proteins; Lignins or Reaction Products Thereof
23	Chemistry, Analytical and Physical Processes
260	Chemistry, Carbon Compounds
204	Chemistry, Electrical and Wave Energy
429	Chemistry, Electrical Current Producing Apparatus, Product and Process
71	Chemistry, Fertilizers
518	Chemistry, Processes Which Include a Fischer-Tropsch Reaction; or Purification, Recovery or Conversion of the Products of Such Processes
585	Chemistry, Hydrocarbons
423	Chemistry, Inorganic
279	Chucks or Sockets
209	Classifying, Separating and Assorting Solids
134	Cleaning and Liquid Contact with Solids

Classes Arranged in Alphabetical Order (Continued)

Class	Title of Class
160	Closures, Partitions and Panels, Flexible and Portable
292	Closure Fasteners
192	Clutches and Power-Stop Control
118	Coating Apparatus
401	Coating Implements with Material Supply
427	Coating Processes
372	Coherent Light Generators
133	Coin Handling
431	Combustion
367	Communications, Electrical: Acoustic Wave System and Devices
340	Communications, Electrical
343	Communications, Radio Wave
501	Compositions: Ceramic
252	Compositions
106	Compositions, Coating or Plastic
7	Compound Tools
159	Concentrating Evaporators
193	Conveyers, Chutes, Skids, Guides and Ways
406	Conveyors, Fluid Current
198	Conveyers, Power-Driven
147	Coopering
30	Cutlery
407	Cutters, For Shaping
83	Cutting
408	Cutting by Use of Rotating Axially Moving Tool
329	Demodulators and Detectors
433	Dentistry
232	Deposit and Collection Receptacles
222	Dispensing
202	Distillation: Apparatus
203	Distillation: Processes, Separatory
201	Distillation: Processes, Thermolytic
424	Drug, Bio-Affecting and Body Treating Compositions
514	Drug, Bio-Affecting and Body Treating Compositions
34	Drying and Gas or Vapor Contact with Solids
369	Dynamic Information Storage or Retrieval
360	Dynamic Magnetic Information Storage or Retrieval
172	Earth Working
434	Education, Demonstration, Cryptography
381	Electrical Audio Signal Processing and Systems
364	Electrical Computers and Data Processing Systems
339	Electrical Connectors
310	Electrical Generator or Motor Structure
377	Electrical Pulse Counters, Pulse Dividers or Shift Registers: Circuits and Systems
338	Electrical Resistors
307	Electrical Transmission or Interconnection System
320	Electricity, Battery and Condenser Charging and Discharging
200	Electricity, Circuit Makers and Breakers
174	Electricity, Conductors and Insulators
361	Electricity, Electrical Systems and Devices
337	Electricity, Electrothermally or Thermally Actuated Switches
335	Electricity, Magnetically Operated Switches, Magnets and Electromagnets
324	Electricity, Measuring and Testing
318	Electricity, Motive Power Systems
323	Electricity, Power Supply, or Regulation Systems
322	Electricity, Single Generator Systems
191	Electricity, Transmission to Vehicles
219	Electric Heating
313	Electric Lamp and Discharge Devices
314	Electric Lamp and Discharge Devices, Consumable Electrodes

Classes Arranged in Alphabetical Order (Continued)

Class	Title of Class
315	Electric Lamp and Discharge Devices, Systems
445	Electric Lamp or Space Discharge Component or Device Manufacturing
363	Electric Power Conversion Systems
187	Elevators
227	Elongated-Member-Driving Apparatus
474	Endless Belt Power Transmission Systems and Components
229	Envelopes, Wrappers and Paperboard Boxes
371	Error Detection/Correction and Fault Detection/Recovery
37	Excavating
411	Expanded, Threaded, Headed, and Driven Fasteners, or Locked or Coupled Bolts or Nuts
92	Expansible Chamber Devices
149	Explosive and Thermic Compositions or Charges
168	Farriery
256	Fences
42	Firearms
169	Fire-Extinguishers
182	Fire Escapes, Ladders, Scaffolds
43	Fishing, Trapping and Vermin Destroying
383	Flexible Bags
141	Fluent Material Handling, with Receiver or Receiver Coacting Means
303	Fluid-Pressure Brake and Analogous Systems
137	Fluid Handling
416	Fluid Reaction Surfaces (i.e., Impellers)
239	Fluid Sprinkling, Spraying and Diffusing
99	Foods and Beverages: Apparatus
426	Food or Edible Material: Processes, Compositions and Products
410	Freight Accommodation on Freight Carrier
44	Fuel and Igniting Devices
110	Furnaces

Class	Title of Class
261	Gas and Liquid Contact Apparatus
55	Gas Separation
48	Gas, Heating and Illuminating
409	Gear Cutting, Milling, or Planing
935	Genetic Engineering: Recombinant DNA Technology, Hybrid or Fused Cell Technology and Related Manipulations of Nucleic Acids
33	Geometrical Instruments
65	Glass Manufacturing
294	Handling, Hand and Hoist-Line Implements
54	Harness
56	Harvesters
237	Heating Systems
432	Heating
165	Heat Exchange
108	Horizontally Supported Planar Surfaces
368	Horology: Time Measuring Systems or Devices
405	Hydraulic and Earth Engineering
362	Illumination
382	Image Analysis
494	Imperforate Bowl, Centrifugal Separators
254	Implements or Apparatus for Applying Pushing or Pulling Force
376	Induced Nuclear Reactions, Systems and Elements
336	Inductor Devices
373	Industrial Electric Heating Furnaces
123	Internal-Combustion Engines
63	Jewelry
403	Joints and Connections
277	Joint Packing
289	Knots and Knot Tying
280	Land Vehicles
278	Land Vehicles, Animal Draft Appliances

Classes Arranged in Alphabetical Order (Continued)

Class	Title of Class
296	Land Vehicles, Bodies and Tops
298	Land Vehicles, Dumping
301	Land Vehicles, Wheels and Axles
69	Leather Manufactures
122	Liquid Heaters and Vaporizers
210	Liquid Purification or Separation
70	Locks
184	Lubrication
74	Machine Elements and Mechanisms
308	Machine Elements, Bearings and Guides
282	Manifolding
493	Manufacturing Container or Tube from Paper; or Other Manufacturing from a Sheet or Web
440	Marine Propulsion
414	Material or Article Handling
73	Measuring and Testing
124	Mechanical Guns and Projectors
186	Merchandising
266	Metallurgical Apparatus
75	Metallurgy
72	Metal Deforming
164	Metal Founding
228	Metal Fusion Bonding
76	Metal Tools and Implements, Making
148	Metal Treatment
29	Metal Working
196	Mineral Oils: Apparatus
208	Mineral Oils: Processes and Products
299	Mining or *in situ* Disintegration of Hard Material
328	Miscellaneous Electron Space Discharge Device Systems
16	Miscellaneous Hardware
332	Modulators

Class	Title of Class
91	Motors, Expansible Chamber Type
185	Motors, Spring, Weight and Animal Powered
180	Motor Vehicles
49	Movable or Removable Closures
800	Multicellular Living Organisms and Unmodified Parts Thereof
370	Multiplex Communications
84	Music
163	Needle and Pin Making
420	Nonferrous Alloys or Metallic Compositions
351	Optics, Eye Examining, Vision Testing and Correcting
353	Optics, Image Projectors
356	Optics, Measuring and Testing
352	Optics, Motion Pictures
350	Optics, Systems and Elements
89	Ordnance
532	Organic Compounds—Part of the Class 532-570 Series
534	Organic Compounds—Part of the Class 532-570 Series
536	Organic Compounds—Part of the Class 532-570 Series
540	Organic Compounds—Part of the Class 532-570 Series
544	Organic Compounds—Part of the Class 532-570 Series
546	Organic Compounds—Part of the Class 532-570 Series
548	Organic Compounds—Part of the Class 532-570 Series
549	Organic Compounds—Part of the Class 532-570 Series
556	Organic Compounds—Part of the Class 532-570 Series
558	Organic Compounds—Part of the Class 532-570 Series
560	Organic Compounds—Part of the Class 532-570 Series

Classes Arranged in Alphabetical Order (Continued)

Class	Title of Class
562	Organic Compounds—Part of the Class 532-570 Series
564	Organic Compounds—Part of the Class 532-570 Series
568	Organic Compounds—Part of the Class 532-570 Series
570	Organic Compounds—Part of the Class 532-570 Series
331	Oscillators
224	Package and Article Carriers
53	Package Making
162	Paper Making and Fiber Liberation
355	Photocopying
354	Photography
358	Pictorial Communication; Television
138	Pipes and Tubular Conduits
285	Pipe Joints or Couplings
111	Planting
PLT	Plants
47	Plant Husbandry
264	Plastic and Nonmetallic Article Shaping or Treating: Processes
425	Plastic Article or Earthenware Shaping or Treating: Apparatus
419	Powder Metallurgy—Processes
60	Power Plants
100	Presses
290	Prime-Mover Dynamo Plants
283	Printed Matter
101	Printing
422	Process Disinfecting, Deodorizing, Preserving or Sterilizing, and Chemical Apparatus
623	Prosthesis (i.e., Artificial Body Members) Parts Thereof or Aids and Accessories Therefor
375	Pulse or Digital Communications
417	Pumps

Class	Title of Class
150	Purses, Wallets, and Protective Covers
250	Radiant Energy
430	Radiation Imagery Chemistry—Process, Composition or Product
104	Railways
238	Railways, Surface Track
213	Railway Draft Appliances
258	Railway Mail Delivery
105	Railway Rolling Stock
246	Railway Switches and Signals
295	Railway Wheels and Axles
220	Receptacles
346	Recorders
62	Refrigeration
235	Registers
152	Resilient Tires and Wheels
404	Road Structure, Process and Apparatus
901	Robots
418	Rotary Expansible Chamber Devices
415	Rotary Kinetic Fluid Motors or Pumps
464	Rotary Shafts, Gudgeons, Housings and Flexible Couplings for Rotary Shafts
109	Safes, Bank Protection and Related Devices
234	Selective Cutting (e.g., Punching)
225	Severing by Tearing or Breaking
112	Sewing
270	Sheet-Material Associating
271	Sheet Feeding or Delivering
413	Sheet Metal Container Making
114	Ships
116	Signals and Indicators
241	Solid Material Comminution or Disintegration
206	Special Receptacle or Package

Classes Arranged in Alphabetical Order (Continued)

Class	Title of Class
267	Spring Devices
365	Static Information Storage and Retrieval
249	Static Molds
52	Static Structures, e.g., Buildings
428	Stock Material or Miscellaneous Articles
125	Stone Working
126	Stoves and Furnaces
127	Sugar, Starch and Carbohydrates
248	Supports
312	Supports, Cabinet Structures
211	Supports, Racks
604	Surgery
128	Surgery
520	Synthetic Resins or Natural Rubbers—Part of the Class 520 Series
521	Synthetic Resins or Natural Rubbers—Part of the Class 520 Series
522	Synthetic Resins or Natural Rubbers—Part of the Class 520 Series
523	Synthetic Resins or Natural Rubbers—Part of the Class 520 Series
524	Synthetic Resins or Natural Rubbers—Part of the Class 520 Series
525	Synthetic Resins or Natural Rubbers—Part of the Class 520 Series
526	Synthetic Resins or Natural Rubbers—Part of the Class 520 Series
527	Synthetic Resins or Natural Rubbers—Part of the Class 520 Series
528	Synthetic Resins or Natural Rubbers—Part of the Class 520 Series
455	Telecommunications
178	Telegraphy
179	Telephony
135	Tents, Canopies, Umbrellas and Canes
87	Textiles, Braiding, Netting and Lace Making
26	Textiles, Cloth Finishing
19	Textiles, Fiber Preparation
68	Textiles, Fluid Treating Apparatus
38	Textiles, Ironing or Smoothing
66	Textiles, Knitting
28	Textiles, Manufacturing
57	Textiles, Spinning, Twisting and Twining
139	Textiles, Weaving
374	Thermal Measuring and Testing
130	Threshing
131	Tobacco
132	Toilet
81	Tools
173	Tool Driving or Impacting
291	Track Sanders
212	Traversing Hoists
190	Trunks and Hand Carried Luggage
334	Tuners
82	Turning
400	Typewriting Machines
199	Type Casting
276	Type Setting
27	Undertaking
171	Unearthing Plants or Buried Objects
251	Valves and Valve Actuation
293	Vehicle Fenders
98	Ventilation
333	Wave Transmission Lines and Networks
177	Weighing Scales
166	Wells
157	Wheelwright Machines
305	Wheel Substitutes for Land Vehicles

Classes Arranged in Alphabetical Order (Continued)

Class	Title of Class
231	Whips and Whip Apparatus
242	Winding and Reeling
140	Wireworking
245	Wire Fabrics and Structure
217	Wooden Receptacles
144	Woodworking
142	Wood Turning
269	Work Holders
378	X-Ray or Gamma Ray Systems or Devices

DESIGN CLASSES ARRANGED IN ALPHABETICAL ORDER

Class	Title of Class
D30	Animal Husbandry
D2	Apparel and Haberdashery
D22	Arms, Pyrotechnics, Hunting, Fishing, and Trapping Equipment
D4	Brushware
D25	Building Units and Construction Elements
D29	Devices and Equipment Against Fire Hazards, for Accident Prevention and for Rescue
D7	Equipment for Preparing or Serving Food or Drink not Elsewhere Specified
D13	Equipment for Production, Distribution or Transformation of Electricity
D23	Fluid Distribution Equipment, Sanitary, Heating, Ventilation and Air Conditioning Equipment, Solid Fuel
D1	Edible Products
D6	Furnishings
D21	Games, Toys, and Sports Goods
D11	Jewelry, Symbolic Insignia and Ornaments
D26	Lighting
D15	Machines, Not Elsewhere Specified
D34	Material or Article Handling Equipment
D10	Measuring, Testing or Signalling Instruments
D24	Medical and Laboratory Equipment
D99	Miscellaneous
D17	Musical Instruments
D19	Office Supplies, Artists' and Teachers' Materials
D9	Packages and Containers for Goods
D28	Pharmaceutical, Cosmetic Products, and Toilet Articles
D16	Photography and Optical Equipment
D18	Printing and Office Machinery
D14	Recording, Communication or Information Retrieval Equipment
D20	Sales and Advertising Equipment
D5	Textile, or Paper Yard Goods; Sheet Material
D27	Tobacco and Smokers' Supplies
D8	Tools and Hardware
D12	Transportation
D3	Travel Goods and Personal Belongings
D32	Washing, Cleaning, or Drying Machine

A

	Class	Subclass
A-frame Structure	D25	29
Abacus	434	203+
Design	D18	6+
Abattoir (see Butchering)		
Abdominal		
Supporters	450	94+
Abietic Acid	260	97+
Esters	560	7
Sulfonated	260	503.5
Abietyl Alcohol	568	714
Ablative Fluids	169	DIG. 2
Abort		
Indicator of aborted start	431	15
Motor with abort means	60	379
Abrading(see File or Filing;grinding)	51	
Abradant-filled bags	51	294
Attachments for cutlery	51	241R+
Comminuting solid material by	241	
Apparatus	241	31+
Hard material in situ	299	
Process	241	1+
Cutlery combined	30	138+
Dental apparatus	433	142
Motor driven	433	125
Filter block cleaning by	210	353
Frames and mounts	51	166R+
Fruit peeler and parer	99	588+
Grain hulling		
Apparatus	99	600+
Processes	426	482+
Graters for vegetables	241	273.1+
Hand held power tool	D 8	61+
Horseshoe calk sharpeners	168	46
Kerf cutting in the earth	299	29+
Lubricant by fluid for feeding	184	60
Machinery design	D15	124+
Machines	51	2R+
Manicuring		
Compound tools	132	75.6
Implements	132	76.4
Mask or stencil	51	262R
Utilizing	51	310+
Material	51	293+
Methods	51	281R+
Processes	51	281R+
Razor combined	30	138+
Saw sharpening	76	31+
Severing by	51	98R
Sharpeners	D 8	91+
Test by	73	7
Tool carrier	51	330+
Tool, flexible member	51	394+
Tool, rigid	51	204+
Tool making	51	293+
Traveling band	51	135R+
Tumbling device	51	163.1+
Water purification		
Chemical feeders	241	39
Chemical treatment	210	198.1+
Wearing in	29	89.5
With glass working	65	61
Work feeder	51	215R+
Work holder	51	216R+
Work rests	51	238R+
Work tables	51	240R+
Abrasives	51	293+
Belts	51	357
Detergents having	252	89.1+
Discs	51	358+
Lubricants having	252	11
Metal stock particulate	75	231
Silicon carbide	51	307+
Stock material	428	143+
Metallic	428	932*
Tungsten carbide	51	307+
Absorbents and Absorbers (See Sorbents)		
Absorber shading, manually adjusts	126	DIG. 1
Bandages	128	155+
Design	D24	49
Carbohydrate treatment with	127	55
Chemical purification	127	49
Carbon	502	416
Carbon compound purifying processes	260	708
Catamenial & diaper	604	358+
Designs	D24	50
Cigar & cigarette tips	131	331
Coating	126	901*
Compositions	502	400+
Chemical agents	252	184
Colloid resolving or inhibiting	252	322+
Drying	252	194
Getters for electric lamps etc	252	181.1+
Liquid containing	252	1
Refrigerant	252	69
Converting mineral oil using	208	46+
Dehumidification of gas with refrigeration	62	271
Processes	62	94
Dehydrating mineral oil with	208	188
Distilled vapor treating	203	41
Apparatus	202	183+
Drying solids (See blotter)	34	95+
Sheet form for printed matter	101	419
Web form for printed matter	101	417+
Web or sheet form absorbent	34	94
Electric battery gas vents	429	86
Gas separators	55	
Drier combined with	34	71
Gas storing	206	.6+
Liquefied gas handling	62	48
Producing	62	17+
Liquid separating, by	210	660+
Chromatography	210	665+
Material separation by		
Carbon compound physical treatment	260	708
Chemical sugar refining processes	127	49
Combined drying	34	71
Dehydrating oil	208	188
Distillation process combined	203	41
Distillation system combined	202	183+
Drying	34	95
Drying vapor treatment	34	80+
From gases	55	
Liquid purification	210	660+
Chromatography	210	656+
Mineral oil separation	208	341+
Physical sugar refining processes	127	55
Purification of fats etc	260	427
Smoking device smoke treating	131	203+
Tobacco products smoke treating	131	331
Neutron	376	339
Other drying means combined	34	71
Purification of fats oils waxes	260	427
Receptors	604	358+
Refrigeration producing	62	476+
Automatic	62	141+
Processes	62	101+
Smoking devices used with	131	203+
Surgical pads	604	358+
Vacuumizing a chamber	62	268+
Wipers & brushes	15	
Absorber, Nuclear Energy	250	515.1+
Radar	342	1
Absorption (See Absorbents)		
Type frequency measurements	324	81
Absorptive Concrete Form Lining	249	189
Abuse, Drugs of	436	901*
Abutment for Bridges	14	26
AC DC Converter	363	
Acceleration		
Electric		
Auto stop light	340	22+
Motor system	318	384+
Inertia switch	200	61.45 R
Vehicle light system	315	77+
Accelerator	74	513
Fermentations with	435	1+
Hand control with	74	482
Polymers prepared from	520	1+
Accelerometer		
Electrical	324	
Piezoelectric	310	329
Non-electrical	73	514+
Accident Preventing Devices (See Safety Devices)	D29	
Acclimatization of Ferments	435	245
Accordion		
Musical instrument	84	376R
Design	D17	3+
Music digest	84	DIG. 15
Pleating or folding	270	30+
Coating device for accordiom folding wall	401	153
Straps, tighteners	24	68AS
Accounting		
Binder device releasably engaging aperture or notch of sheet	402	
Book strips & leaves	281	
Calculating	235	
Computer controlled, monitored	364	406
Manifolding	282	
Perforated article assorting	209	613
Printed matter	283	
Punch selector	234	
Typing	400	
Accumulator	425	DIG. 1
Boiler steam storage	122	35
Calculating totalizer	235	
Variator with accumulator	235	61FC
Combustion products power plants	60	731
Cookers fireless	126	273.5
Electric battery	429	149+
Fluid pressure compensator	138	26+
Heat absorber structure	126	400
Heat radiator	237	75
Heat storage	165	18
Regenerator	165	4+
Hill planting	111	51
Hydraulic	138	26+
Power plant	60	413+
Power plant	60	659
Motor steam exhaust	60	659
Pump combined	417	540+
Refrigerant	62	503
Automatic	62	174
Storage battery	429	
Vehicle heater	237	44
Water heater combined	126	375
Acetals	568	590+
Cyclic 5 membered ring	549	430+
Cyclic 6 membered ring	549	369+
Acetanilide	564	218
Acetate	536	69+
Cellulose	536	69+
Esters	560	231+
Lead	556	105
Light metal salts	562	607
Polyvinyl (See also synthetic resin or natural rubber)	526	319
Silk (See artificial silk)		
Acetic Acid	562	607
Dehydration of	203	16
Acetoacetic Acid	562	577
Esters	560	178
Acetone	568	382
Fermentative production	435	150+
Purification or recovery	568	411
Acetonitrile	558	435
Acetophenone	568	335
Acetosalicylic Acid	560	143
Acetyl		
Chloride	260	544R
Acetylation (See Acetate)		
Amines	564	137+
Ketene production	568	301
Phenols	560	130+
Textiles	8	121
Acetylene (See Triple Bond Hydrocarbon)		
Aldehydes from	568	458
Burner		
Lantern	362	160
Cylinder	206	.6+
Electrostatic field or electrical discharge production	204	171
Esters from	560	242
Gas tanks	220	3
Generation from carbide	48	
Package or receptacle	206	.6+
Separating from gases	55	63+
Six membered hetero N compounds from	546	253
Synthesis of	585	534+
Synthetic resins from (See also synthetic resin or natural rubber)	526	285
Vinyl halides from	570	233
Achromatic Image		
From chromatic	430	367
From organic compound	430	402
To organic compound	430	565
With chromatic	430	364
Acid (See Type of Acid)		
Absorbing agents for	252	189
Acidizing or formation destroying in well treating	252	8.553
Activated (See clay)		
Anhydride		
Carbocyclic or acyclic	260	545R+
Cyclic polycarboxylic	549	231+

Entry	Class	Subclass
Electrolytic synthesis of	204	103+
Inorganic, phosphorus compound	423	304+
Inorganic, sulfur compound	423	511+
Catalyst regeneration with	502	27
Organic	502	28
Containers	206	524.1+
Detergents containing	252	142+
Electrolytic synthesis of	204	103+
Organic	260	
Carboxylic halides	260	544R
Halides	260	543R
Higher fatty electric discharge treatment	204	167
Higher fatty electromagnetic wave treatment	204	157.6+
Hydrocarbon purification using	585	866
Phosphate fertilizer	71	33+
Acidophilus Milk, Cream, Buttermilk	426	71
Acorn Type Electronic Tube	313	
Acoustics (See Sound)	181	
Building construction	52	144+
Cells		
Altering optical element	350	358
Isolation of motion picture housing and supports	352	35
Coupler		
Combined telephone-phonograph	379	78
Design	D14	57
Oscillators	331	155
Electrical, wave systems & devices	367	
Airborne shock-wave detection	367	906*
Collision avoidance	367	909*
Coordinate determination	367	907*
Doppler compensation systems	367	904*
Material level detection	367	908*
Noise reduction in nonseismic receiving system	367	901*
Particular well-logging apparatus	367	911*
Portable sonar devices	367	910*
Side lobe reduction or shading	367	905*
Sonar time varied gain control	367	900*
Speed of sound compensation	367	902*
Transmit-receive circuitry	367	903*
Energy measuring	73	861.25+
Fluid pressure leak detector	73	40.5 A
Holography, vibration measuring	73	603
Measuring instruments	73	570+
Stock material	181	284+
Target for golf	273	181C
Transducer making	29	594
Transducer making	29	602A
TV		
Mechanical-optical scanning by acoustic wave	358	201
Picture projection device	358	235
Wall construction	52	144+
Acridine	546	102+
Anthrone or anthraquinone nuclei	546	
Azo compounds	534	753+
Acridone Dye	8	662
Acriflavine	546	104+
Acrilan T M (polyacrylonitrile Copolymers)	526	342
Acrolein		
Polymers (See also synthetic resin or natural rubber)	526	315
Acronycines	546	62+
In drug	514	285
Acrylon T M (polyacrylonitrile Methyl Methacrylate)	526	329.4
Acrylonitrile (See also Synthetic Resin or Natural Rubber)	526	341
Actinic Glass	501	900*
Developing by heat	65	30.1+
Actinide Series Metals (See Radioactive)		
Alloys	420	1+
Compositions	252	625
Electrolysis	204	1.5
Heat treatment	148	132
Metallurgy	75	84.1 R
Nuclear fuels	376	409+
Organic compounds	534	11+
Actinometer	356	213+
Actinomycetales	435	169
X-art collections	435	822*
Actinospectacin	549	361
Activating Carbon	502	180+
Activation		
Lamp filaments	445	6
Of catalyst before use in hydrocarbon synthesis	585	906*
Vitamins in foods	426	72+
Active Solid State Devices (See Transistor)	357	
Actuated and Actuation (See Type of Device Actuated as Door Gate Valve, etc or Means Employed to Actuate as Cam, Gear, Motor Pawl, Speed Responsive Device, Spring Thermostat)		
Acupuncture Test, Electric	128	735
Acupuncture, X-art	128	907*
Acyclic Hydrocarbon (See Diolefin; Olefin; Parafin; Triple Bond)		
Acylation	260	691
Acyloins	568	303+
Adamantane Synthesis	585	352
Adapter		
Electrical connector socket	439	638+
Pipe joint	285	176+
Adding		
Dispenser cycle totalizer	222	36+
Machine	235	
Design	D18	3+
With recorder	235	58R+
Pencil	235	64
Addressing Machine	101	47+
Address printers	D18	13+
Plates	101	369+
Adducts		
Stabilizing an enzyme by	435	188
Urea	260	96.5 R+
Adenine	544	277
Adhesive		
Applying		
Stacking apparatus with means to apply adhesive to articles	414	904*
To pluck fowls	17	11.1 A
With laminating	156	
Bonding (See extensive digest list)	156	
Apparatus	156	346+
Methods	156	60+
Brackets attached by	248	205.3+
Carrier component having adherent surface	224	901*
Compositions		
Alkali metal silicate	106	74+
Biocide with	424	
Carbohydrate gum	106	205+
Cellulose liberation liquor	106	123.1
Core oils	106	38.2+
Protein	106	124+
Rubber ctg (See synthetic resin or natural rubber)	520	1+
Starch	106	210+
Synthetic resin ctg (See synthetic resin or natural rubber)	520	1+
Anerobic	523	176
Dental	523	116
Optical cement	523	168
Surgical	523	118
Vermin catching, trapping	424	77
Digest	24	DIG. 11
Fastener for diaper	604	389+
Insect catching and destroying	424	77
Insect trap	43	114+
Jewelry	63	DIG. 1
Manifolding digest	282	DIG. 2
Metal working involving	29	DIG. 1
Moisteners	118	
Envelope sealing combined	156	441.5+
Process of forming	427	207.1+
Separator for sheet feeding	271	33
Tape	428	261+
Coated on both sides	427	208
Coated on outermost layer	428	343
Coating process	427	207.1
Holder with edge for tearing	225	6+
Laminated	428	343+
Removable layer	428	40+
Rolls	206	411
Testing	73	150A
Toys, detachably adhesive	446	85+
X-art, special receptacle or package	206	813*
Adipic Acid	562	590
Adjuncts		
Static mold	249	205+
Sink head or hot top	249	202
Adjustable (See Device to be Adjusted)		
Adrenalin T M	564	365
Medicine containing	424	113
Adrenochrome	548	512
Adrenocorticotrophin Acth	530	306
Adsorbent (See Sorbent)		
Advancing		
Device for tool	173	141+
Automatic	173	4+
Cyclic self acting	173	19
Impacting device combined	173	112+
Material of indeterminate-length	226	
Advertising		
Aerial	40	212+
Calendars	40	107
Card support	D 6	512
Design	D20	10+
Display cards	40	124.1
Mirror or reflector devices	350	600+
Mounting bracket for sign	D 8	354+
Printed matter	283	56
Show boxes and cards	D20	40
Sign exhibiting	40	
Skywriting	40	213
Telegraphophones & annunciators	369	
Vehicle body	296	21
Adzes	30	308.1
Design	D 8	76
Aeration		
Aerator	D23	213+
Agitation mixers	366	101+
Bait container	43	57
Dispensing with gas agitating	222	195
Distillation separatory		
Distillate	202	203
Vapor	202	201
Faucet attachment for	239	428.5
Digest	261	DIG. 22
Food, cereal puffing	99	323.4
Liquid purification		
Contact surface means	210	150+
Process, biological	210	620+
Oxidation	210	758+
With flotation	210	703+
Submerged fluid inlet	210	220+
Oxygen transfer technique	435	818*
Pumping by aerating liquid column	417	108+
Sewage process	210	620+
Diffusers	261	DIG. 7
Rotating	261	DIG. 71
Aerial (See Antenna)		
Bombs	102	382+
Cableway	104	112+
Camera	354	65+
Films, radiation imagery	430	928*
Fire fighting apparatus	244	136
Photo instruments	33	1A
Sighting	33	229+
Aerogram Design	D19	2
Aerometer	73	30
Aeronautics	244	
Aeroplane (See Aircraft)	244	
Aerosol T M	252	305
Aerostatic Exhibiting Devices	40	212
Skywriting	40	213
Aerosteam Engines	60	200.1+
Aesculin	536	18.1
Affixing Apparatus		
Label pasting and paper hanging	156	
Sheet metal container making	413	
Afghan Blanket	D 6	603
African Violets	PLT	69
Afterburner, Reaction Motor with	60	261
Aftertreatment of Polymers		
Chemically from ethylenic monomers only	525	326.1+
Physically	528	480+
Agar	536	3
Age Hardening		
Ferrous alloys	148	12.3
Ferrous alloys	148	142
Nonferrous alloys	148	12.7 R
Nonferrous alloys	148	158+
Permanent magnet material	148	102
Agglomerating		
Apparatus	425	222
Inorganic compounds	23	313R+
Metal powder or scrap	425	222
Ores	75	3+
Preventing		
Agents	252	381+
Comminution combined	241	16
Aggregate and Pellet	425	DIG. 101
Agitating (See Compacting, Jarring, Jogging, Mixer, Stirrer, Vibrators)	366	

	Class	Subclass
Abrading tumbler	51	163.1^{+}
Processes	51	313^{+}
Sand blast combined	51	422
Agitator discharging	366	184
Receptacle moving	366	187
Agitator feeding	366	150^{+}
Condition responsive control of	366	151
Liquid injector in mixing chamber	366	167
Suction	366	163^{+}
Tangential to mixing chamber	366	165
Agitator mixing chamber movable	366	219^{+}
Oscillating	366	237^{+}
Rotating	366	220
Agitator mixing chamber stationary	366	341
With stirrer	366	241
Agitator plural mixing chamber		
Mortar	366	14
Moving	366	235
Rubber & heavy plastic	366	91
Agitator stirrer	366	343^{+}
Cover combined	366	242^{+}
Bathtubs	4	584^{+}
Bearing lubricant	384	395
Axle bearing	384	384
Churn butterworker combined	99	459
Coating implement combined	401	4
Comminuting combined	241	98
Concentrating evaporator	159	25.1
Distillation appartus combined	202	
Distillation processes combined	201	
Distillation processes combined	203	
Electrode agitator	204	222^{+}
Liquid electrode	204	221
Rotary receptacle	204	213^{+}
Electrolytic cell	204	273
Diaphragm type	204	261
Feeders and conveyors		
Article dispensers	221	200^{+}
Comminutor feeder	241	249
Dispensing	222	
Fluid current conveyor	406	86^{+}
In intake receptacle	406	134^{+}
Gravity flow drier	34	166
Kiln feeder	34	203
Reciprocating conveyors	198	750^{+}
Pushers	198	736^{+}
Sifter feeder	209	245
Stationary receptacle drier	34	179
Fuels and oils		
Cracking assistant	208	146^{+}
Distillation assistant	201	
Gas generator fuel	48	85.1^{+}
Oil distillation process	208	347^{+}
Oil vaporizing apparatus	196	123^{+}
Furnace grate	126	155^{+}
Gas injection, effected by	366	101^{+}
Dispensing	222	195
Gas and liquid contact	261	
Mortar mixers	366	10^{+}
Mortar mixing processes	366	3^{+}
Heating or cooling combined		
Boiling apparatus	99	409
Canning apparatus	99	371
Coffee and tea making apparatus	99	287
Congealed product producers	62	342^{+}
Congealed product producing	62	68^{+}
Cooking apparatus	99	348
Cooled heat transfer liquid	62	435
Cooled withdrawable liquid	62	392
Drying gas agitator	34	241
Gravity flow tube drier	34	166
Heat exchange apparatus	165	109.1
Heating furnace	432	80
Kiln zigzag flow	34	203^{+}
Mixing	366	144^{+}
Mortar mixer heater or cooler	366	22^{+}
Processes	366	7
Rotary drier	34	108^{+}
Stationary receptacle drier	34	179^{+}
Vibrating drier	34	164
Water heater	126	387
Zigzag flow drier	34	171^{+}
Insect catcher		
Oscillating plant leaf agitator	43	143
Rotating plant leaf agitator	43	142
Liquid heater cleaner	122	380
Liquid jet	366	167^{+}
Amalgamator agitator	209	66
Dispensing assistant	222	195
Flotation apparatus	209	170
Fluid current conveyor	406	
Injector for gas and liquid contact	261	76^{+}
Mortar mixing apparatus	366	10
Mortar mixing processes	366	3
Multiple jet gas and liquid contact	261	40^{+}
Multiple screen sifter	209	312
Radiant energy treatment combined	250	428^{+}
Reciprocating sifter	209	321
Sifter cleaner	209	380
Superposed screen sifter	209	318
Separator		
Amalgamators	209	50^{+}
Centrifugal filter	210	360.1^{+}
Electrical liquid purification	204	307
Filter	210	407
Liquid suspension	209	155^{+}
Magnetic	209	225^{+}
Sifter	209	233^{+}
Spittoons	4	278
Textile		
Fluid treating	68	
Fluid treating processes	8	159
Vibratory support or receptacle	366	108^{+}
Dispenser part	222	196^{+}
Drying apparatus	34	164
Filled bag or receptacle	53	525
Gas generator fuel	48	85.1
Heat exchange receptacle	165	86^{+}
Movable dispensing container	222	161
With glass making apparatus	65	178
Agraffe	84	211
Agriculture & Agricultural (See Plants, Animal, or Implement Type)		
Implements		
Design	D 8	1^{+}
Dibbles	111	99
Fruit gatherers	56	328R^{+}
Hand cultivators	172	371^{+}
Hand cutters	56	239^{+}
Hand rakes	56	400.1^{+}
Machinery		
Boring or penetrating the earth	175	
Diggers	171	
Digging & recovering buried objects	171	
Harrows	172	
Harvesters	56	
Hydraulic & earth engineering	405	
Not hand-held	D15	10
Planters	111	
Plows	172	
Wells	166	
Post or column	D25	131
Aibn (azobisisobutyronitrile)	526	219.6
Aileron	244	90R
Aim (See Sight)		
Aiming		
Direct sighting	33	227^{+}
Optical	356	247^{+}
Radar, external device	342	61^{+}
Gun	342	67
Air (See Gas; Pneumatic)		
Air injected liquid heaters & vaporizers	122	5.5 R
Airbag passenger restraints	280	728^{+}
Aircooled engines	123	41.56^{+}
Analysis	73	23
Auxiliary inlet carburetors	261	DIG. 1
Blast or suction		
Dispensed material entraining	222	630^{+}
In conduit or guide	406	
Dispensed material propelling	222	195
Fiber liberation	19	9
Filter cleaning	210	407
Solid material comminution apparatus	241	38^{+}
Solid material comminution process	241	15^{+}
Blast or suction cleaning		
Blast only	15	405^{+}
Both blast and suction	15	345^{+}
Convertible type	15	328^{+}
Ejectors for unpiling articles	414	121
Installed or fixed position devices.	15	301^{+}
Methods	134	
Tube blowers	15	316R^{+}
With dirt filter	15	347^{+}
With mechanical agitator for work	15	363^{+}
Blow molding needle or nozzle	425	535
Circulating shoe	36	3R
Cleaner (See air purification)		
Compressors	415	
Conditioning (See cooling; drying; disinfection; filter; heat; moistener; purifying; sterlizing)	D23	351^{+}
Cooling	62	404^{+}
Air coolers	261	DIG. 43
Aircooling	261	DIG. 3
Automatic	62	186^{+}
Compressing and expanding	62	401^{+}
Apparatus	62	401^{+}
Automatic	62	172
Process	62	86^{+}
Processes	62	89^{+}
Dielectric capacitor	361	326^{+}
Disinfecting or sterilizing	422	4
Apparatus	422	120^{+}
Using heat transfer other than air	422	38
Displacement		
Funnels	141	297^{+}
Portable receptacle filling	141	37
Distributor		
Register floor	98	103^{+}
Register wall	98	108
Wick type fuel burner	431	298^{+}
Drying	55	
Compositions	252	194
Ejection	425	DIG. 102
Feeding (See air mixer)		
Filled nuclear counter	376	108^{+}
System	250	374
Flow meter	73	861^{+}
Foil	244	123^{+}
Heating	126	99R^{+}
Industrial blast stoves, etc	432	214^{+}
Impregnating inhaler	128	203.12
Inductor	336	199^{+}
Instruments	33	DIG. 2
Line lubricators	261	DIG. 35
Manifold	198	955*
Mattresses	5	449^{+}
Mixer		
Gas heating illuminating	48	180.1^{+}
Internal combustion engine	123	527^{+}
Liquid contact	261	
Solid contact	34	
Ventilators	98	38.1
Moistener	261	
Cabinet combined	312	31^{+}
Fireplace attachments	126	134
Furnace hot air	126	113
Heat exchanger combined	165	60
Radiator combined	237	78R
Slow diffuser	239	34^{+}
Stovepipe attachments	126	313
Ventilating floor register	98	105
Ventilating wall register	98	109
Monitor radioactivity	250	374^{+}
Motor expansible chamber	91	
Dental plugger	433	120
Turbine	415	
Wind	416	
Pillows	5	441
Plane (See aircraft)	244	
Plugger dental	433	120
Port (See airport)		
Pressure indicator		
Pressure gage	73	700^{+}
Tire deflation	116	34R
Propeller	416	
Aircraft type	416	
Pump (See pump)		
Purification	55	
Chemical	423	210
Raid shelter	109	
Screw	416	176
Separator	55	
Cleaning	55	282^{+}
Speed measurement	73	181^{+}
Suction cleaning	15	300R^{+}
Supply device for breathing	128	204.18^{+}
Combined with fire fighting nozzles	239	270
Support for shaping-treating item	425	DIG. 2
Syringe thermal medicator	604	113^{+}
Tank		
Filling portable receptacles	141	
Fluid pressure dispenser	222	394^{+}
High pressure	220	3
Ship raising	114	52^{+}
Spring wheels	152	10

	Class	Subclass
Tire inflater	222	3+
Tire inflater wheel carried	152	418+
Testing apparatus	73	37+
Tracer for pattern control	409	126+
Vehicle heating effected by circulating air	237	12.3 A
Washing apparatus	261	75+
Air Brake		
Fluid pressure operator	188	151R+
System control	303	
System testing	73	39
Vehicle fender combined	293	5
Vehicle momentum responsive	188	141
Air Brush	D23	223+
Air Conditioning (See Cooler)		
Air Gun	124	56+
Air Hammer Motor	173	90+
Noise muffler	181	230
Air Lock		
Caissons	405	9
Closures for	49	68
Diving bell combined	405	192
Pneumatic conveyer	406	169
Air Wick T M Composition		
Dispenser	239	44+
Airbag Passenger Restraints	280	728+
Steering wheel mount	280	777
Aircraft (See Airship; Balloon)	244	
Air cushion	180	116+
Air intake guard	55	306
Altitude indicator	33	328+
Angle or direction instrument	D10	67
Artillery	89	
Synchronized with propeller	89	133+
Autogiro T M	244	17.11+
Toys	446	36+
Autorotating wing	244	17.11+
Balloons	244	31+
Gas cell construction	244	128
Boarding ramp design	D12	
Bomb sights	33	229
Buildings	52	86+
Tent	135	87+
Cabin air pressure maintenance	98	1.5
Cable or rail mounted	104	23.1+
Carrier, ship	114	261+
Control automatic	244	175+
Crash signal		
Radio	342	385+
Water dye	116	211
Design	D12	319+
Dirigible	244	24+
Discharging	244	137.1+
Inflatable evacuation slide	244	905
Drop bomb retarders	102	386+
Flares parachute	102	337+
Foil construction	244	123
Gas & liquid contact apparatus	261	DIG. 2
Ground effect	180	116+
Gun sights	33	233+
Gyroplane	244	17.11+
Hang glider	244	900+
Hangar	52	86+
Movable wall	52	64+
Helicopter	244	17.11+
Ice prevention	244	134R
Jet propulsion	60	200.1+
Kites	244	154
Landing gear or strut	244	100R+
Spring	267	2+
Lift modifiers	244	198+
Lights	362	62
Loading	244	137.1+
Locating lights	362	62
Maintenance vehicle	D12	14+
Mines aerial	102	405
Mobile stair for boarding	D34	30
Model airplane systems	244	190
Motion retarding device	244	113
Parachute	244	142+
Motor mount	248	554+
Navigation instruments	73	178R+
Operator training or testing	434	30+
Radio navigation simulation	434	239+
View simulation	434	38+
Ordnance mounts	89	37.16+
Pilot operated	244	220+
Pilotless control	244	190+
Propeller or airplane system	60	904*
Pyrotechnic	102	335+
Radio beacon for	342	385+
Radio controlled radar	342	36+
Radio course or bearing for	342	407+
Radio glide slope	342	410+
Radio receiver	455	345
Rockets	102	384
Simulations		
Inflatable toys	446	225
Velocipede	280	1.21
Velocipede occupant propelled	280	1.12
Stores ejectors	244	137.4
Bomb, flare, or signal	89	1.51+
Streamlining	244	130
Structural element strength measurement	73	802
Take off condition measuring	73	178T+
Take off signal calculation	364	427
Towed target	273	360+
Toys	446	230+
Design	D21	87+
Flying	446	34+
Ultralight	244	900+
Airfield	244	114R+
Airflow Responsive Switch	200	81.9 R
Airfoil		
Construction	244	123+
Lift modifiers	244	198+
Railway traction regulator	105	74
Shape	244	35R+
Airing Device for Bedclothes	5	506
Airport		
Airway signs, ground indicia	40	217
Arrangement landing field	244	114R
Bulletin boards	40	446+
Electric signaling systems	340	945+
Radio		
Beacon for	342	385+
Course or bearing	342	407+
Glide slope	342	410+
Security by metal detector systems	340	568+
Traffic analysis computerized	364	439
Traffic control systems (See airway)		
Airship (See Aircraft)		
Cabin pressure control	98	1.5
Control	244	96
Gas cell construction	244	128
Hull construction	244	125
Load attachment	244	127
Mooring devices	244	115+
Propulsion and sustenance	244	24
Skin construction	244	126
Airway Traffic Control Systems		
Beacons		
Electric signaling	340	947+
Illuminating	362	35
Radar type	342	385+
Rotating	342	398+
Mechanical	116	28R+
Radio		
Beacon	342	385+
Course	342	407+
Glide slope	342	410+
With position indicating	342	456
Aisle		
Inter-aisle article transfer	414	266+
Ajmaline	546	40
Alanine	562	575
Alarm (See Monitor; Signal)		
Boiler	122	504.2
Burglar		
Electric, approach operated	340	541+
Mechanical	116	75+
Clock	368	244
Closure position	49	13+
Coffin	340	573
Design	D10	106
Gas		
Electrical	340	632+
Mechanical	116	67R
Heat exchanger combined	165	11.1
Light responsive	340	600+
Photoelectric system	250	200+
Lock combined	116	8+
Design	D 8	332
Noise exposure	340	540
Over telephone circuit	379	37+
Pipe joint combined	285	93
Register mechanism	235	128
System electrical	340	500+
Valve combined	137	558
Watches	368	244
Alathon T M (polyethylene)	526	352
Album	281	
Book binder sheet	402	79
Design	D19	26+
Display	40	152+
Fasteners	24	2
Leaves display	40	152+
Phonograph record		
Cabinet type	312	9+
Folder carrier indicia	40	359
Package type	206	311
Albumin	530	362+
Coating or plastic compositions containing	106	124+
Foods containing	426	656
Tests for	436	88*
Alcamine (See Alkamine)		
Alcohol	568	700+
Aldehyde	568	496+
Amines	564	503+
Beverages	426	592
Apparatus	99	275+
Breath alcohol	436	900*
Distillation		
Processes	203	
Stills	202	232+
Systems	202	152+
Electrolytic production	204	77
Ether	568	672+
Fermentation	435	161+
From olefines	568	895+
Fuel containing		
Higher	44	77+
Lower	44	53+
Hydrogen purification using	585	864
Hydrometers	73	32R+
Stove	126	43
Treatment		
Aging for beverages	426	
Aldehyde	568	420+
Acyclic	568	448+
Amine	564	502
Amine or ammonia reaction product	564	248
Aromatic	568	425+
Cellulose ether or ester composition containing	106	187
Fatty acid or ester containing	106	184
Fuel containing	44	77+
Nitrogen compound	44	75
Hydrogen purification using	585	864
Polymer (See synthetic resin or natural rubber)		
Sulfonic acid aromatic	260	511
With carboxylic acid esters		
Acyclic	560	174+
Alicyclic	560	126
Aromatic	560	51+
Aldols	568	496+
Aldoximes	564	253+
Ale (See Beer)		
Alfalfa	426	636
Alfin Catalysts	526	175
Algicide	71	67
Algin	536	3
Algoecide Algicide	71	67
Alicyclic Carboxylic Acid Esters	560	1+
Alicyclic Hydrocarbon	585	22+
Purification of	585	800+
Synthesis of	585	350+
Alignment		
Screw drivers	81	451+
Tools for radio	81	
Alizarin	260	383
Alkadiene (See Diolefin)		
Alkali	420	564
Aluminates	423	593+
Amids	423	351
Carbonates	423	421+
Preparation by electrolysis	204	87
Chemical agents for absorbing	252	193
Earth metal		
Alloys	420	415
Lead	420	564
Carbonates	423	430+
Crystallization of compounds	23	304
Electrolytic synthesis	204	69+
Halides	423	
Hydroxides and oxides	423	635
Electrolic synthesis of	204	100
Electrolic treatment of	204	153
Mercury alloys by electrolysis	204	125
Nitrates	149	

	Class	Subclass
Containing	149	61$^+$
Explosive	149	
Organic compounds	260	
Oxides	423	635
Peroxides	423	583
Phosphates	423	305$^+$
Pigments fillers or aggregates	106	306
Pyrometallurgy	75	67R
Refractory compositions ctg	501	123$^+$
Sulphates	423	554$^+$
Fusion of organic compound with	260	393
Hydroxide preparation	423	641
By electrolysis	204	98$^+$
Metal		
Alloys	420	400
Lead	420	564
Carbonates	423	419R$^+$
Chemical agents containing	252	192
Chemical agents for absorbing	252	193
Crystallization of compounds	23	302R$^+$
Electrolytic synthesis	204	68
Electrolytic synthesis of hydroxides and oxides	204	98$^+$
Electrolytic treatment of hydroxides and oxides	204	153
Halides	423	462$^+$
Hydroxides and oxides	423	592
Inorganic compound recovery	423	179
Mercury alloys by electrolysis	204	125
Misc. chemical manufacture	156	DIG. 71
Nitrates	149	
Explosive containing	401	118$^+$
Organic compounds	260	
Oxy-halogen salt	423	472
Explosive	149	77
Phosphates	423	304$^+$
Preparation	75	66
Pyrometallurgy	75	66
Solvents, sodium chloride & potassium chloride	422	902*
Sulphates	423	551
Nitrates	423	
Organic	260	
Oxids		
Packing	206	524.1$^+$
Secondary battery		
Chromium cobalt or cadmium	429	
Iron	429	
Nickel	429	
Alkaline Earth		
Carbonates	423	430$^+$
Metals & cds	156	DIG. 78
Oxides and hyclroxides	423	635
Silicates	423	326
Sulfates	423	554$^+$
Sulfides	423	554
Alkaloids		
Biocides containing	424	
Cinchona	546	134$^+$
Ergot	546	67$^+$
Medicines containing	424	
Opium	546	44$^+$
Solanum	546	124
Tobacco	546	282
Undetermined constitution	424	195.1
Alkamine		
Acyclic	564	503
Alkanoic esters	560	253
Monocyclic carboxylic acid esters	560	110
Monocyclic carboxylic acid esters	560	106$^+$
Nitro carboxylic acid esters	560	20$^+$
Nitro carboxylic acid esters	560	156
Alkane (See Hydrocarbon; Parafin)	585	700$^+$
Alkenes (See Olefines)		
Alkines (See Acetylene; Triple Bond Hydrocarbon)		
Alkyd Resins (See Synthetic Resin or Natural Rubber)		
Alkylate Detergent Synthesis	585	455$^+$
Alkylation		
Acyclic hydrocarbons	585	709$^+$
Alicyclic hydrocarbons	585	375
Aromatic hydrocarbons	585	446$^+$
Mineral oil	208	46$^+$
Phenols	568	780$^+$
Allantoin	548	308
In drug	514	390
Allochiral Arrangements		
Bulb or tube special package	206	422
Closure actuators	49	299
Closure link system	49	111$^+$
Tool with series of teeth	83	852
Tool-face, metal deforming	72	198
Alloy		
Age or precipitation hardened	148	405
Amorphous	148	403
Bearing compositions	252	12$^+$
Compositions	420	
Making when casting	164	55.1$^+$
Continuously casting	164	473
Deformation of	72	700
Dispersion strengthened	75	951*
Electrodes for electrolytic apparatus	204	293
Electrolytic		
Coating from aqueous bath	204	43.1$^+$
Synthesis from aqueous bath	204	123$^+$
Synthesis from fused bath	204	71
Treatment	204	140$^+$
Electron emissive cathode of	420	
Ferrous (See ferrous alloys)	420	8$^+$
Iron	420	8$^+$
Loose metal particles	75	251$^+$
Mechanical memory	148	402
Metal treatment	148	
Metallurgy	420	
Powder metallurgy		
Processes	419	
Products	75	228$^+$
Pyrophorous	420	416
Railway surface track	238	150
Shape memory	148	402
Alloying		
Fusing dopant with substrate	437	134$^+$
Alpha		
Beta gamma survey meter	250	336.1
Scintillation type	250	361R$^+$
Hand counter	250	374
Proportional counter	376	153$^+$
Similar counter tubes	313	93
System	250	374
Scintillation detector		
Crystal mounting	250	361R
Electric	250	361R
Non-electric	250	483.1
With sample holder	250	361R$^+$
Survey meters	250	374
Electric	250	361R
Nonelectric	250	483.1
Scintillation type	250	361R$^+$
Alpha Globulins	530	392
Alphabet		
Block	434	172
Design, type	D18	24$^+$
Letter training	434	159$^+$
Playing cards	273	299
Specialized font	382	
Stencil	101	127$^+$
Alphanumeric		
Display		
By matrix	340	748$^+$
Cathode ray tube	340	756$^+$
Altars	81	
Design	D 6	329$^+$
Alternating Current (See Type of Equipment Using)		
Converters to direct current	363	
Electromagnet with armature	335	243$^+$
Electromagnetic switches	335	99$^+$
Generator	310	159$^+$
Prime mover driven	290	5
Lamp or electronic tube	315	246$^+$
Meter	324	137$^+$
Motor	310	159$^+$
Polyphase to arc lamp	314	31
Polyphase to lamp or electronic tube	315	137$^+$
System		
Telautograph	178	19
Telegraph	178	66.1$^+$
Transmission	307	
Alternating Mechanical Motions (See Gearing)	74	
Alternator Structure	310	
Altimeter (See Altitude Instrument)	D10	70
Electrical echo systems	367	87$^+$
Radio wave systems	342	120
Altitude Instrument		
Barometric	73	384$^+$
Design	D10	70$^+$
Directional radio	342	462
Light ray type	33	282
Aerial bomb sight	33	229
Optical	356	3$^+$
Solar position locator	33	268$^+$
Telescopes	350	537$^+$
Vertical and horizontal angle	33	281
Vertical angle	33	282$^+$
Passive sonic or supersonic object detection type	367	118$^+$
Radar type	51	229$^+$
Radiant energy	342	120$^+$
Radiosonde	340	345$^+$
System	340	870.1
Sonic or supersonic echo system, ie active sonar	367	87
With position indicating	342	462
Alto		
Wind instrument (See instrument)		
Alum	423	544$^+$
Alumina	423	625
Refractory containing	501	127$^+$
Magnesium compound combined	501	118$^+$
Aluminates	423	592$^+$
Aluminosilicate		
Alicyclic compound synthesis	585	475
By isomerization	585	481$^+$
Compounds with plural metals	423	118
Saturated compound synthesis	585	722
Saturated compound synthesis	585	739
Silicon and silicon compounds	423	328$^+$
Unsaturated compound synthesis	585	533
Unsaturated compound synthesis	585	666
Aluminum	75	68R$^+$
Alloys	420	528
Copper	420	529
Compounds		
Aluminates	423	592
Azo compounds containing	534	692$^+$
Azoles containing	548	101$^+$
Carbocyclic or acyclic compounds containing	556	170$^+$
Diarylene ortho diketones containing	260	366
Diazine containing	544	225
Diazo compounds containing	534	562$^+$
Halides	423	462
Higher fatty acids containing	260	414
Inorganic	423	
Misc. chemical manufacture	156	DIG. 61
With 3 &/or 4 compounds	156	DIG. 104
Nitrides	423	409
Organic	556	170$^+$
Chelates	556	175
Clay or zeolite containing	556	173
Salicylates	556	175
Sesquihalides	556	180
Oxazine containing	544	64
Oxides	423	625
Proteins containing	530	400$^+$
Pyridine nucleus compounds containing	546	2$^+$
Sulphates	423	544$^+$
Thiazine containing	544	4
Triarylmethanes containing	260	387
Electrolytic coating of	204	58
Cleaning or etching combined	204	33
Electrolytic synthesis	204	67
Electrothermic processes	75	10.1$^+$
Heat treatment	148	159
Metal working using	29	DIG. 2
Pigments	106	290
Plate	428	650$^+$
Pyrometallurgy	75	68R
Working & heat treatment	148	11.5 A
Including age hardening	148	12.7 A
Amalgam Mixer, eg Dental Filling	366	602*
Amalgamator (See Mercury)		
Coated surface adhesion	209	50$^+$
Mercury suspension	209	174$^+$
Other separators combined	209	12$^+$
Selective differentiation	209	48
Amantadine, in Drug	514	656
Amberlite T M (ion Exchange Resins)	521	25$^+$
Ambulance	296	19$^+$
Ambulatory Pneumatic Splints	128	83.5
Amides	564	
Anthraquinone nuclei	260	377
Higher fatty acid	260	404$^+$
Sulphonated	260	401
Sulfa drug type	260	397.6$^+$
Amidines	564	225$^+$
Aromatic	564	244$^+$
N prime-aryl formamidines	564	245
Polyamidines	564	243

U.S. PATENT CLASSIFICATIONS

	Class	Subclass
Amidino Hydrazines or Hydrazones	564	226+
Amidoximes	564	229
Amine	564	
Acyclic	564	463+
Aldehyde or ketone ctg	564	502
Hydroxy or ether ctg	564	503+
Plural amine ctg	564	511+
Sulfur	564	500+
Unsaturated	564	509
Aldehyde (See azomethine)		
Aromatic	564	305+
Aralkyl	564	374+
Aryloxy alkanol	564	349+
Aryloxy alkyl	564	347+
Benzyl	564	384+
Diaryl amines	564	433+
Diphenyl ether or sulfides	564	430
Monocyclic	564	305
Phenethyl	564	376+
Polycylo ring system	564	426+
Boron ctg	564	8+
Bonded directly to nitrogen	564	9+
Boron-nitrogen rings	564	10+
Carbocyclic	564	305+
Oxides	564	297+
Phenol addition salts	564	280+
Phosphorus bonded		
Directly to nitrogen	564	12+
In same ring	564	13
Indirectly to amino nitrogen	564	15+
Phosphorous in a ring	564	106
Preparation		
Amination	564	395+
Dealkylation of secondary or tertiary amine	564	486
From amides	564	414
From amides	564	488
From esters	564	393
From esters	564	468
From heterocyclic compound	564	413
From organic acid acid halide or salt	564	394
Halogenation or dehalogenation	564	412
Nitration	564	411
Nitrosation	564	410
Purification or recovery	564	497+
Reduction	564	448+
Ring alkylation or deakylation	564	409
Separation of aromatic isomere	564	424+
Stabilized or preserved	564	5+
Aminimines	564	464+
Amino Acids	260	
Acyclic	562	553+
Aromatic	562	433+
Protein derived	562	516
Tests for	436	89*
Amino Carboxylic Acid Esters		
Acyclic	560	155+
Alicyclic	560	125
Aromatic	560	19+
Amino Optical Isomers		
Racemization	564	302
Resolution	564	303
Aminophenols	564	305+
Aminoplast (See Synthetic Resin or Natural Rubber)		
Aminopyrine	548	368
Ammeter	324	76R+
Ammonia		
Chemical agents for binding	252	193
Derivatives inorganic	423	
Developing	430	150
Machines	354	300
Distillation processes	203	
Electrolytic synthesis	204	102
Gas generators	422	148
Medicator	128	203.12
Production	423	352+
Soda apparatus	422	148
Stills	202	232+
Systems for distilling	202	152+
Ammonium		
Crystallization of compounds	23	302A
Electrolytic synthesis of compounds	204	102
Fertilizer containing compounds	71	
Halides	423	470
Nitrates	423	396
Explosive or thermic composition containing	149	46+
Ore treatment with compounds	75	103
Perchlorates	423	476
Explosive or thermic composition containing	149	76
Phosphates	423	305+
Quaternary	564	281+
Sulphates	423	545+
Ammonolysis	564	
Acyclic amine production	564	469+
Anthraquinone amine production	260	382
Aromatic amine production	564	395
Ammunition (See Cap; Magazine; Shell; Ordnance)	102	
Armor piercing	102	517+
Belts	89	35.1+
Blasting	102	301+
Cartridges	102	430+
Combustible	102	431+
X-art collection	102	700*
Holder	224	223
Bandoleer	224	203
Covered	224	239
Packing	86	47+
Charge wave forming	102	305+
X-art collection	102	701*
Compression ignition	102	702*
Coolants	102	704*
Design	D22	112+
Belt or belt attachable holders	D 3	100
Drop bombs	102	382+
Explosive & thermic compositions or charges	149	
Explosive charge making	86	
Explosive devices & ammunition	102	
Feeding devices	86	45+
Firearms	42	
Flechette	102	703*
Fuses, primers & igniting devices	102	200+
Heat shield	244	158R
Loading	86	23+
Making	86	10+
Mines	102	401+
Narcotic containing	102	512
Non-ricocheting	102	398
Ordnance	89	
Mechanical guns & projectors	124	
Powder bags	102	
Powder form	102	282
Projectile	102	501+
Pyrotechnics	102	335+
Receptacle	206	3
Body or belt attached	224	223
Sabots & rifling bands	102	520+
Separated explosive constituents	102	705*
Shells	102	473+
Testing	73	167
Torpedoes, aircraft	244	14
Torpedoes, ships	114	20.1+
Wads	102	448+
Well torpedoes	102	301+
Amorphous Alloy or Metal Stock	148	403
Amoxicillin	540	331
Ampere Meter	324	76R+
Amphetamine	564	381
In drug	514	654
Amphibian Vehicles		
Airborne	244	50
Fluid cushion	180	116+
Design	D12	324
Landing gear	244	101
Ampicillin	540	336
Ampicillin, in Drug	514	198
Amplidyne	322	92
Amplifier	330	
A C raw power or bias supply	330	114+
Adjustable coupling impedance		
Automatic	330	144+
Automatic thermal	330	143
Anode voltage control	330	128
Anode voltage supply	330	202
With input electrode bias	330	203
Artificial line coupling	330	57
Beam power	313	299
Diode	313	298
With beam forming electrodes	313	299+
Bias voltage control	330	129+
In feedback amplifier	330	96
In push pull amplifier	330	182
Bootstrap	330	156
Bridge, wheatstone	330	146
Cascode type	330	72
Impedance network in coupling circuit	330	175
Buffer	330	
Radio transmitter with	455	91+
Bypass of an amplifier stage	330	151
Capacitive amplifier device system	330	7
Input	330	185
Interstage	330	177+
L in anode or grid circuit	330	179
L in series between stages	330	178
Output	330	192
R in anode and grid circuit between stages	330	180
R in series between stages	330	178
Cascaded differently coupled stage	330	152+
Including a cathode follower	330	153
Including an interstage resonant network	330	154
Including interstage transformer coupling	330	154
Cascode	330	70+
Cathode self biasing circuit	330	142
Cathode to cathode coupling of adjacent stages	330	89
Cathode to cathode feedback	330	88+
Class D type	330	251
Combined with diverse type art device	330	1R
Compressor thru bias control	330	129+
Output	330	193+
Phase inverter	330	117
Push pull	330	119
Compressor thru impedance control	330	144+
Interstage	330	158+
Push pull	330	119
Compressor thru thermal impedance control	330	143
Screen grid input	330	161
Screen grid output	330	163
Coupling from cathode		
Interstage	330	172+
Interstage transformer	330	168
Coupling from grid	330	162+
Coupling from screen grid	330	162+
Coupling impedance automatically varied	330	144+
Diode	330	145
Electronic tube	330	145
Thermal	330	143
Coupling network		
H filter	330	176
Lattice filter	330	175
P I filter	330	176
R C	330	180
T filter	330	176
Wheatstone bridge	330	175
Coupling to cathode		
Bootstrap	330	156
Input	330	186+
Coupling to plate or auxiliary electrode	330	160+
Coupling to screen grid	330	160+
Current mirror type	330	257
Current mirror type	330	288
D C coupling		
Cathode input	330	187
Cathode input interstage	330	159
Cathode output	330	194
Cathode output interstage	330	173
Grid output	330	163
Input	330	191
Interstage	330	181+
Output	330	198
Plate or aux electrode input	330	161
Push pull interstage	330	121
Design	D14	96
Detector for AVC	330	140
Biased	330	138
Triode	313	293+
Differential	330	69
Diode as coupling element	330	164
Diode power pentode	313	298
With two cathodes	313	5
Diode type	330	287
With plural power supplies	315	166
Direct coupled power	313	3
Distortion bucking in	330	149
Distributed parameter coupling	330	53+
Diverse type amplifying devices	330	3
Dynamic coupled power	313	3
Dynamo- electric rotary linear	330	58
Electric	330	
Automatically varied impedance in feedback path	330	86

	Class	Subclass
Auxiliary electrode	330	111
Cascade amplifiers	330	98+
Cathode	330	87+
Cathode impedance feedback	330	87+
Cathode-cathode feedback	330	88+
Combined with bias voltage control of FL bl amplifier	330	85
Combined with bias voltage control of signal amplifier	330	96
Current and voltage feedback	330	102
Current feedback	330	105
DC signal feedback path	330	97
In push pull amplifier	330	83
Electromechanical telephone	381	161+
Electronic tube type	330	
Frequency responsive	330	109
Frequency responsive cathode impedance	330	94
Negative feedback	330	75+
Non- linear cathode impedance	330	95
Non- linear impedance	330	110
Phase shift means in loop path	330	107
Plural channel amplifiers	330	84
Positive feedback	330	112
Pulse amplifier electronic tube type	328	53+
Transformer	336	
Transistor making		
Apparatus	29	569.1
Process	437	
Transistor structure	357	
Transistor system (See amplifier, transistor)	330	250+
Transistor system non-linear	307	200R+
Traveling wave tube systems	330	43
Traveling wave type tube structure	315	3.5
Traveling wave type tube structure	315	39.3
Two way telephone repeater	379	338+
Electrocardiograph	128	902*
Electrolytic	330	207R
Electromechanical transducer coupling element	330	174
Positive and negative feedback in push pull amplifiers	330	82
Positive and negative feedback in same path	330	101
Positive feedback included with cathode impedance feedback	330	93
Push pull amplifiers	330	81+
Electron beam tube systems	330	44+
Electronic tube as coupling element	330	164
Feedback	330	75+
Amplifier in feedback path	330	85
Filamentary cathode	330	206
AC supply	330	115
Plural	330	201
With input electrode supply	330	205
Gas tube amplifying device system	330	41
Guitar	84	1.14+
Hum bucking in	330	149
Indicating amplifier condition	330	2
Input	330	185+
Input coupling	330	185+
Push pull	330	122
Integrated circuits	330	307
Interstage coupling	330	157+
Push pull	330	120+
Light controlled or activated means in amplifier system	330	59
Magnetic linear	330	8
Magnetic switching	307	415
Magnetic voltage magnitude control	323	
External or operator controlled	323	329
Input responsive	323	302
Output responsive	323	249
Self regulating	323	310
Magneto-resistive type amplifier device	330	62
Magneto strictive means in amplifier system	330	60
Magnetron systems	330	47+
Masertype	330	5
Measuring amplifier condition	330	2
Microphone	381	111+
Modulator- demodulator systems	330	10
Neutralization by feedback	330	76+
Neutralization by feedback transistor	330	292
Neutron	376	171
Noise bucking in	330	149
Output coupling	330	192+
Push pull	330	122
Parametric type capacitive reactor	330	7
Parametric type distributed parameter solid element	330	5
Parametric type saturable core reactor	330	8
Combined with meter	324	118
Pentode	330	
Pentode	313	300
Phase inverters	330	116
Phase inverters	330	117
Phase inverters transistors	330	301
Piezoelectric crystal coupling element	330	174
Pilot current controlled	330	52
Plural amplifier channel systems	330	124R+
Plural input sources	330	147
Plural output loads	330	148
Power supply	330	199+
AC raw	330	114+
Control or regulation	330	127+
Polyphase	330	113
Push pull	330	123
Push pull control or regulation	330	123
Preselector	330	
Printed circuits in amplifier system	330	66
Protection means	330	298
Push pull systems	330	118+
Radio frequency	330	
Power amplifier	330	
Shielded	330	68
With gain control	330	129+
Resistance coupled	330	180
Resistive type amplifying device systems	330	61R
Saturable core reactor systems		
External or operator controlled	323	329
Input responsive	323	302
Linear	330	8
Non- linear	307	401
Output responsive	323	249
Self regulating	323	310
Voltage magnitude control	323	
Screen grid voltage control	330	128
Secondary emitter vacuum tube systems	330	42
Semiconductor device (See amplifier, transistor)		
Series energized tubes	330	70+
Sound	181	
Diaphragms and mountings	181	148+
Electromechanical audio	381	161+
Megaphones	181	177+
Receiver telegraph	178	100
Recording and reproducing	369	
Speaker	381	111+
Stagger tuned	330	154
Sum & difference amplifiers	330	69
Testing amplifier condition	330	2
Thermal impedance in amplifier signal path	330	143
Diode type	330	287
Electrostatic field control	330	277
Feedback	330	291+
Four electrode type	330	307
Graded conductivity type	330	250+
Input to collector	330	250+
Intrinsic zone	330	250+
Light control	330	308
Negative feedback	330	293
Neutralization by feedback	330	292
Plural channel	330	295
Plural input	330	295
Plural output	330	295
Push- pull	330	262+
Radiant energy bombarding	330	308
Three electrode type	330	309
Unbalanced to balanced	330	301
Volume control	330	278+
Time constant circuit in bias control system	330	141
Tone control automatic		
Diode as variable coupling impedance	330	145
Electronic tube as variable coupling impedance	330	145
Feedback frequency selective network	330	109
Feedback frequency selective network in cathode impedance	330	94
Impedance control in feedback path	330	86
Phase shift means in feedback amplifier	330	107
Variable coupling impedance	330	144+
Variable coupling thermal impedance	330	143
Tone control manual	330	
Input coupling adjustment	330	185+
Interstage coupling adjustment	330	157+
Output coupling adjustment	330	192+
Plural amplifier channels different frequencies	330	126
Potentiometer common to signal and feedback path	330	108
Push pull input coupling adjustment	330	122
Push pull interstage coupling adjustment	330	120+
Push pull output coupling adjustment	330	122
Unicontrol means	330	155
Transductor linear	330	8
Transductor voltage magnitude control general	323	
External or operator controlled	323	329
Input responsive	323	302
Output responsive	323	249
Self regulating	323	310
Transformer coupling		
Additional impedance input	330	189
Additional impedance interstage	330	167
Additional impedance output	330	196
Additional reactive coupling	330	166
Adjustable inductance	330	169
Cathode input	330	186
Cathode input interstage	330	158
Cathode output	330	193
Cathode output interstage	330	168
In cascade differently coupled stages	330	154
Input	330	188+
Interstage	330	165+
Output	330	195+
Push pull input	330	122
Push pull interstage	330	120
Push pull output	330	122
Stagger tuned	330	154
Structure input	330	190
Structure interstage	330	171
Structure output	330	197
Tuned input	330	189
Tuned interstage	330	167
Tuned output	330	196
With shielding	330	170
Transistor, linear	330	250+
Balanced to unbalanced	330	301
Cascaded	330	310+
Combined diverse type combined	330	299+
Combined with semiconductor diode	330	299+
Common collector	330	250+
Complementary types cascaded	330	310+
Complementary types parallel input	330	263+
DC coupled cascaded	330	310+
Transistor system, non-linear	307	200R+
Travelling wave tube systems	330	43
Twin triode power	313	301
With two cathodes	313	6
Unicontrol of coupling	330	155
Vapor tube amplifying device system	330	41
Voltage delay in bias control amplifier	330	132
Volume control automatic	330	
Amplifier in feedback path	330	85
Anode voltage control	330	128
Bias control	330	129+
Diode impedance variable	330	145
Electronic tube impedance variable	330	145
Feedback	330	75+
Power supply regulation	330	127+
Push pull amplifier bias or power supply control	330	123
Variable coupling impedance	330	144+
Variable coupling thermal impedance	330	143
Variable feedback impedance	330	86
Volume control manual		

U.S. PATENT CLASSIFICATIONS

Entry	Class	Subclass
Anode voltage adjustment	330	128
Bias adjustment	330	129
Cathode self bias adjustment	330	142
Input coupling adjustment	330	185$^+$
Input coupling adjustment push pull amplifier	330	122
Interstage coupling adjustment	330	157$^+$
Interstage coupling adjustment push pull amplifier	330	120$^+$
Output coupling adjustment	330	192$^+$
Output coupling adjustment push pull amplifier	330	122
Potentiometer common to signal and feedback paths	330	108
Power or bias supply adjustment	330	127$^+$
Power or bias supply adjustment in push pull amplifier	330	123
Unicontrol means	330	155
With high frequency structure	330	53$^+$
With long line	330	53$^+$
With volume compression thru bias control	330	129$^+$
With volume compression thru impedance control	330	144$^+$
With volume compression thru thermal impedance control	330	143
Amplitude		
Comparison, compressional wave, geophysical	367	30
Limiter		
Amplifier AVC	330	129$^+$
Clipper		
Grid control type	328	54
Telephone system	330	
Amplitude Modulation	332	31R$^+$
Analog carrier wave communications	455	
Peak compressional wave, geophysical	367	29
Pulse carrier wave communications	375	24
Pulse or digital communication	375	
Ampoule (See Container)		
Exhaust	141	
Filling	141	31
Medicament containing	424	451$^+$
Opener	51	204
Structure	215	32$^+$
Surgical	604	200$^+$
Hypodermic	604	415
Amprotropine, in Drug	514	534
Amusement Devices	273	
Design	D21	240$^+$
Elevator	272	6$^+$
Exercising devices	272	
Games	273	
Houses	272	2
Mechanical guns & projectors	124	
Mirrors and reflectors	350	600$^+$
Railway	104	53$^+$
Car in wheel	104	77
Roundabout stationary	272	43$^+$
Roundabout transported	272	29
Trip illusions	272	17$^+$
Toys (See toys)	446	
Water guns	222	78$^+$
Amyl		
Alcohol	568	840
Chloride	570	181
Mercaptan	568	61$^+$
Nitrate	558	480
Amylaceous Material	127	
Fermentative analysis or testing	435	4$^+$
Anabasine	546	193
Anaerobic		
Cultivation	435	801*
Curing adhesive	156	322
Fermentation	435	
Fertilizer	71	10
Sewage treatment	210	601$^+$
And aerobic	210	605
Anaglyphs	350	132
Analgesic, Audio	600	26$^+$
Analogue Computer		
Electrolytic tank for determining equipotential lines	235	61R
Analytical and Analytical Control		
Apparatus		
Chemical	422	50$^+$
Electrolytic	204	400$^+$
Compositions	330	299
Electrical	330	299
Electron microscope	330	299
Human diagnostics	128	630$^+$
Mass spectrometry	250	281$^+$
Method		
Chemical	436	43
Fermentative	435	3
Using invisible light	250	336.1
Using X rays	378	
Analyzers (See Analytical and Sensing)		
Electric waves and harmonics	324	77R
Cathode ray oscilloscopes	324	121R
Mirror galvanometer oscilloscopes	324	97
Oscillographs	346	108$^+$
Sound	73	584$^+$
Speech	381	41$^+$
Refrigerator, sorbent type	62	495$^+$
Sound	181	
Frequency components	324	77R
Wave separation	333	167$^+$
Analyzing (See Sensing)		
Anastigmat Lenses	350	409
Anatase	423	610
Anatomical Gauges	33	511$^+$
Anatomy, Teaching	434	267$^+$
Anchor and Anchorage		
Animal powered apparatus	185	25
Bolt		
Concrete set type	52	698$^+$
Bridges		
Arch	14	26
Suspension	14	21
Building component	52	698$^+$
Cable rail	104	117
Cycle locked to	70	234
Decoy	43	2$^+$
Design	D12	215
Dredges		
Bank spud anchor	37	74
Bottom spud anchor	37	73
Earth control device combined	405	259$^+$
Expansible well type	166	206$^+$
Earth boring combined	175	230
Fishing trolley	43	27.2
Freight on freight carrier	410	101$^+$
Land	52	155$^+$
Land vehicle anchors	188	7
Marine structure to bed	405	224$^+$
Marine type mines	102	411$^+$
Masonry	52	698$^+$
Pile	405	244
Cast in situ enlargement	405	237$^+$
Pole base	52	292$^+$
Railway surface track		
Bolt	238	377
Clamp type rail anchors	238	343$^+$
Concrete for rail seat	238	265
Rail anchors	238	315$^+$
Screw pedestals	238	120
Tie anchors	238	106
Tie plate anchors	238	297$^+$
Ship anchors	114	294$^+$
Stovepipe	126	318
Strand ends	24	122.6
Hoist rope	212	95
Line driven planters	111	49
Suspension bridges	14	21
Textile warp	139	47
Tank-car body	105	362
Wall tile		
Andirons	126	298
And robot	901	1*
Android	901	1*
Design	D23	407
Androsterone	260	397.4
Anemometer	73	861.85
Direction determination combined	73	189
Anerobic Sealant Resin Compositions	523	176
Aneroid Barometers	73	386$^+$
Anesthetic		
Anesthesia gas scavenging system	128	910*
Compositions	424	
Explosive devices containing	102	512
Feeder	128	203.12
Anethol	568	658
Angle Iron	52	720
Angle Polysection Measure	33	1AP
Measuring instrument	D10	65$^+$
Angling (See Fishing; Trapping)		
Anhydrite Anhydrous Gypsum	106	109
Calcining processes	106	100$^+$
Plasters	106	109
Aniline	564	305$^+$
Formaldihide polymers (See synthetic resin or natural rubber)		
Animal		
Artificial	428	16
Having lights	362	124
Attack resistant stock	428	907*
Beds	D30	118
Bee culture	6	
Blocking repelling or chasing	49	58$^+$
Butchering	17	
Depilating compositions	17	1D
Cages	119	17$^+$
Calls and callers	446	397
With mouthpiece	446	207$^+$
Canopy frame supported by	135	90
Care and handling	D30	
Catching and holding device	119	151$^+$
Dips and sprays	424	
Draft appliance	278	
Monorail vehicle	105	143
Droppings handler	294	1.3$^+$
Extracts		
Medicine containing	424	95$^+$
Farriery	168	
Feed devices	119	51R$^+$
Design	D30	121$^+$
Fertilizer	71	15$^+$
Bones	71	19$^+$
Fishing, trapping, & vermin destroying	43	
Food	426	635$^+$
Apparatus	99	
Design	D 1	
Mixer	366	603*
Protein	426	635$^+$
Grooming	D30	158$^+$
Harness	54	
Husbandry	119	
Design	D30	
Marker	40	300$^+$
Design	D30	155
Medicines for treating	424	
Poisons	424	
With food bait	424	410
Pokes	119	136$^+$
Preserving or mounting including teaching taxidermy	434	296$^+$
Releasing apparatus	119	15.5 R
As target	119	15.6
Saddlery	D30	134$^+$
Shears	30	194$^+$
Shelter or pen	D30	108$^+$
Shipping crates	D30	109$^+$
Stop in building	52	101
Stop in sewage system	210	131
Taming or training devices	119	29
Tissue treatment	8	94.1 R$^+$
Toys	446	268$^+$
Traps	43	58$^+$
Design	D22	119$^+$
Watering devices	119	72$^+$
Design	D30	121$^+$
Whips & whip apparatus	231	
Animal Cell Culture	435	240.2$^+$
Animal Cell Growth Media	435	240.3$^+$
Animal Cell Per se	435	240.2$^+$
Animated Cartoons (See Motion Picture)		
Anisidine	564	443
Ankle		
Supports	128	166
Anklets	2	239$^+$
Annealing		
Boxes	266	262$^+$
Casting plus annealing in mold	164	122$^+$
Containers for material heated	432	254.1$^+$
Glass apparatus	65	348$^+$
Cutter associated	65	176
Glass making and	65	
Glass process	65	117$^+$
Heating apparatus	432	120$^+$
Metal	148	13$^+$
Metal casting and working combined	148	2
Metal casting combined	148	3
Metal working combined	148	11.5 R$^+$
Metallurgical apparatus	266	249$^+$
Misc. chemical manufacture	156	DIG. 73
Annular Article, Container for	206	303
Annunciator and Systems	369	
Drops	340	815.29
Electrical	340	815.1$^+$

	Class	Subclass
Systems	340	286R+
Telephone switchboard	379	315+
Mechanical	116	200+
Timer electric	340	309.15+
Train position	246	124
Anode		
Electrolytic	204	280+
Basket type cell	204	259
Support	204	297R
Support combined	204	286+
Gas and vapor lamp	313	326+
Plural consumable electrode eg arc	314	36+
Space discharge devices electronic tubes	313	326+
Anodic Oxidation	204	56.1+
Multiple layers	204	42
Answer (See Question)		
Answer Back		
Annunciator system	340	313+
Telegraph system	178	4.1 B+
Antenna	D14	86+
Active antenna as reflector	343	832
Aircraft, with	343	705
Alarm combined	343	894
Moving means for scanning	343	760
Array	343	700R+
Automobile, with	343	711+
Balanced V-wave	343	735
Biconical horn	343	773
Body attached or carried	343	718
Bracket supported	343	892
Cage	343	896
Clover leaf type	343	741+
Coated	343	873
Cone	343	829+
Connectors for electrical	439	
Counterpoise, with	343	846+
Coupled to plural lead ins	343	858
Coupling for electrical	333	24R+
Plural channel	333	1+
Coupling, with	343	850+
Design	D14	86
Dielectric lens	343	911R
Dielectric type	343	785
Dipole	343	793+
Folded	343	803+
Director	343	912+
Doublet balanced	343	793+
Drooping ground plane	343	829+
Duplex system, in	370	24+
Electromagnetic wave refractor	343	753+
Embedded	343	873
Filter in an active antenna	343	722
Filter in coupling, with	343	850+
Fishbone	343	811
Foundry	164	374+
Ground plane	343	829+
Grounding structure combined	343	846+
Helical	343	895
High frequency loop	343	741+
Horn	343	786
Housing, with	343	872+
Ice clearer or preventer, with	343	704
Impedance matching, with	343	860+
Plural path	343	852
Indicator combined	343	894
Moving means for scanning	343	760
Insulators for	174	137R+
Bushing type	174	152A+
Inverted V-wave	343	736
Knockdown	343	880
Lens	343	909
Lens combined	343	753+
Lightning protector combined	343	904
Lightning protectors for		
Grounding switch	200	153R+
Miscellaneous	361	117+
Spark gap type	313	325
Thermal current	337	28+
Liquid column type	343	700R
Loading	343	749+
Loop	343	866+
Collapsible	343	871
High frequency	343	741+
Rotatable	343	869
Making	29	600
Mast for	52	649
Guy wire	52	148
Pivoted or extensible	52	111+
Mast type	343	874
Mesh type	343	897
Orienting	343	757+
Pillbox	343	780
Planar linear array	343	824
Plural	343	893
Polarization		
Converter combined	343	756
Filter combined	343	756
Polarizer	343	909
Potted	343	873
Rabbit-ears	343	805
Radio		
Cabinet combined	343	702
Receiver, in	455	269+
Transmitter, in	455	129
Reel type	343	877
Reflector	343	912+
Collapsible	343	915
With reflector	343	833+
Resonant V-type	343	809
Retractible into support	343	899
Rhombic	343	733+
Rod type	343	900+
Decorative element	52	301
Extensible	52	632
Reel type extender	242	54A
Support for	248	511+
Telescoping	343	901+
Vehicle attached	52	110
Rotator	D14	90
Scanning	343	757+
Shield, with	343	841+
Signal measuring structure, with	343	703
Skirt	343	829+
Sleeve type	343	790+
Slot type	343	767+
Space discharge device, with	315	34
Space discharge device, with	343	701
Spiniferous	343	899
Spiral	343	895
Scanning	343	759
Steering	342	368+
Wheel	343	829+
Stub	343	831
Submerged	343	719
Support for	248	
Support for	211	
Support, with	343	878+
Electric stress distributing	343	885
Static discharge	343	885
Streamlined support	343	887
Support rotatable	343	882
Sweeping	343	757+
Switch for	200	
Switch, with	343	904
Switching between antenna and line	343	876
Telescoping	343	901+
Tower type	343	874+
Transceiver, in	455	82
Transmission line for	174	68R+
Electrically long	333	236+
Transmission line with	343	905
Traveling wave	343	731+
Tuning	343	745+
Turnstile	343	797+
Underground	343	719
Underwater type	343	719
Vehicle combined	343	711+
Water-borne	343	709+
Wave	343	731+
Guide type	343	772+
Whip	343	900+
Woven type	343	897
Anthanthrones	260	359
Anthracene (See Aromatic Hydrocarbon)	585	400+
Oxidation of	260	385
Purification of	585	800
Anthranilic Acid	562	458
Anthranols	260	351+
Anthrapyridones	546	76
Anthrapyrimidines	544	248
Anthraquinones	260	365+
Acridine nucleus containing	546	102+
Anthraquinonebenzacridones	546	58
Azo compound ctg.	534	654+
Carbazole nucleus containing	548	416+
Dye compositions containing	8	675+
Piperazines containing	544	380+
Pyrazole nucleus containing	548	357
Pyridine nucleus containing	546	
Pyrimidines containing	544	294
Triazines containing	544	187+
Anthrones (See Anthraquinones)	260	351+
Antiabrasion Cable Device	174	136
Antiaircraft (See Gun)		
Antibiotic Compositions	424	
Antibodies	530	387
Anticaking Agents	252	381+
Anticannibalism Devices for Poultry	119	97R
Anticathode	378	143
Anticorrosion		
Compositions	252	387+
Coating or plastic	106	14.5+
Detergents containing	252	146+
Heat exchange fluid containing	252	68
Well treating	252	8.555
Magnetic field apparatus	422	186.2
Metal protection		
Reactive coating	148	6+
Within nuclear reactor	376	305
Mineral oil conversion combined	208	47
Silverware tray incorporated	206	553
Vapor treating apparatus	312	31+
Anticreep		
Rim and wheel	152	384
Spring wheel drive	152	66
Wheel drive	152	68
Tire and tube	152	500
Antielectrolysis		
Electrical distribution system	307	95
Object protection		
Apparatus	204	196+
Processes	204	147+
Antiexplosion Agents	252	386
Explosive combined	149	
Antifading Radio Receiver	455	52
Antifoaming		
Agents	252	358
Devices for boilers	122	459
Antifouling Plastic Compositions	106	15.5+
Antifreeze		
Aircraft deicing	244	134R
Compositions	252	71+
Coating	106	13
Heating system protectors	237	80
Tester	73	32R+
Antifreeze Agent Supply	138	34+
Antifriction		
Bearing	384	445+
Compositions	252	12+
Fluid bearing combined	384	101+
Linear	384	7+
Plain bearing combined	384	126+
Closure guide	49	425+
Compositions	252	9+
Keepers	292	340+
Locks	70	464
Antigen	424	88+
Antigen-antibody Testing	436	500+
In vitro	436	500+
Enzyme labeled	435	4+
In vivo	424	
Antigen, Chemical Modification of Protein	530	403+
Antiglare		
Light projectors	362	351+
Mirrors and reflectors	350	278+
Windows	350	284
Windshields	296	97R
Antihalation Layer	430	510+
Antihemorrhagic Composition	424	
Quinones	260	396R
Vitamin	260	396R
Anti-induction		
Cathode ray tube circuits	315	8
Inductor devices with shields	336	84R+
Resistors	338	61+
Shielding tubes and spark plugs	315	85
Telegraph interference balancing	178	69B
Telephone circuitry	379	416+
Transmission line	174	32+
Anti-ingestible	252	365+
Antijimmy Locks	70	418
Antikink Devices		
Appliance cords		
Extensible cords	174	69
Springs	267	74
Antiknock		
Artificial organic fuel	44	68+
Compositions	252	386
Cyclomatic compounds	556	1+
Fuel combined	44	50+

U.S. PATENT CLASSIFICATIONS

	Class	Subclass
Gaseous compositions	252	373
Engine accessories	123	198A
Exhaust gas treatment	422	168+
Measuring knock	73	35
Tetraalkyl lead	556	95
With preservative or stabilizer	556	3+
Antileak Joint	105	424
Antimagnetic Timepiece	368	293
Antimicrobial Activity		
Fermentative analysis or testing	435	32
Antimony	75	69
Alloys	420	576
Antimonates	423	592+
Carbocyclic or acyclic compounds containing	556	64+
Halides	423	462+
Misc. chemical manufacture	156	DIG. 94
Oxides	568	477
Pigments fillers or aggregates	106	303
Pyrometallurgy	75	69
Sulphides inorganic	423	511
Antinoise		
Bed	5	309
Closure buffer	16	82+
Cranks and wrist pins	74	604
Ear shields	128	152
Flywheels and rotors	74	574
Gears	74	443
Land vehicle		
Animal draft gear	278	61+
Axle connection	280	138
Tongue	280	108
Levers and linkage	74	490
Lock mufflers	70	463
Machinery support		
Bases	248	678+
Bracket	248	674+
Suspension	248	637+
Resilient	248	610+
Motor support vehicle	248	560+
Mufflers	181	212+
Pawl and rachet	74	576
Railway		
Rolling stock	105	452
Track	238	382
Wheels	295	7
Reaction motor nozzle	239	265.13
Rotary motors or pump	415	119
Supports	248	562+
Typewriter		
Muffler	400	689+
Platen	400	661+
Antioronze Agent or Process	430	929*
Antioxidant		
Compositions	252	397+
Edible oils and fats	426	601+
Fats, fatty oils, fatty acids, ester type waxes	260	398.5
Polymers treated (See class 523, 524)		
Solid polymers containing (See synthetic resin compositions)		
Antipanic Door Bolts	292	92+
Antiphone Ear Sound Stoppers	128	152
Antipick Locks	70	419+
Antipyrine	548	368
Antirachitic Vitamin		
D compound	260	397.2
Foods and processes	426	
Medicines	514	167+
Antirattle	16	DIG. 6
Lock	70	463
Keeper	292	341.12+
Window	292	76
Car	49	428+
Antireflection Films	350	164+
Antiricochet Projectiles	102	398
Antiscorbutic Vitamin		
C compound	549	315
Foods and processes	426	72+
Medicines	514	474
Antiseptics (See Disinfection; Preserving)	422	
Compositions	424	
Detergents combined	252	106+
Medicators	604	19+
Water softener combined	252	175+
Antiside Draft Device (See Equalizer)		
Animal draft	278	3+
Harvesters	56	321
Antisiphon for Fluid Handling Systems	137	215+

	Class	Subclass
Antislip or Antiskid		
Compositions, coating or plastic	106	36
Horseshoe calks	168	29+
Ladder terminals and feet	182	107+
Overshoes	36	7.6+
Pavements	404	19+
Block surface	404	19+
Shoe		
From foot	36	58.5
Tread	36	59C
Supports		
Illumination	362	389
Tires	152	208+
Armored	152	167+
Track sanders	291	
Traction mats	238	14
Anti-smoking Product or Device	131	270
Antismut		
Compositions, coating repellent	106	2
Printing combined	101	416R+
Slip sheet		
Apparatus	34	94
Processes	34	6
Antisnoring		
Body restraints	128	135
Chin supporters	128	164
Mouth restraints	128	136
Antispasmodics	424	
Antisplash		
Vehicle mud guards	280	152R+
Antistatic		
Digest, fuel & igniting devices	44	DIG. 2
Electrically conductive coating compositions	252	500+
Polymers treated (See class 523, 524)		
Process for synthetic fibers	57	901*
Antitheft Devices, eg Locks	70	
Antithumb Sucking	128	133+
Antitoxins	424	85+
Syringes	604	187+
Antivermin	43	124+
Animal treatment	119	156+
Perches fowl	119	25
Antivibration (See Antinoise)		
Anvil	D 8	46
Assembling and disassembling	29	283
Blacksmith	72	476+
Bolt and rivet making	10	22
Hammer attached	72	407+
Hammer or press	72	462+
Hollow rivet setting	227	61+
Machine combined	29	560+
Nail extractor	254	19
Pile or post	173	128
Saw setting	76	73
Saw swaging	76	57
Tool driving	173	128
Vises combined	72	457+
Apa-6	540	312
Apartments	52	234+
Rolling partitions	160	351
Hanging or drape type	160	350
Aperture Card	493	944*
Apiaries	6	1+
Apparel (See Type of Article Worn; Clothes)	2	
Adornment, attachable	D11	
Apparatus	223	
Boot and shoe making	12	
Knitting	66	
Patterns	33	2R+
Weaving	139	
Blouse, shirt	D 2	208+
Boots shoes and leggings	36	
Design	D 2	264+
Making	12	
Cape, stole, shawl	D 2	179+
Catamenial and diaper	604	358+
Design	D24	50
Ceremonial robe	D 2	79
Cleaning		
Brushing devices	15	
Ironing or pressing	38	
Methods	8	137+
Washing devices	68	
Coat, jacket, vest	D 2	183+
Collar making	83	901*
Containers for	206	278+
Making	493	938*
Making garment bag	493	935*

	Class	Subclass
Making with support	493	939*
Costume	D 2	79
Design	D 2	1
Hosiery	D 2	329+
Negligees	D 2	12+
Dress or suit	D 2	49+
Female support undergarments	450	1+
Fire fighters	169	48+
Footwear	D 2	264+
Hand coverings	D 2	610+
Headwear	D 2	509+
Knitted	66	169R+
Layout and measurement means	33	403+
Footwear	33	4+
Garments	33	11+
Leather manufactures	69	
Light thermal electrical treatment	128	379+
Lingerie	D 2	1+
Locks for	70	59
Marking & measuring instruments	33	2R+
Neckwear	D 2	600+
Respirators	128	200.24+
Skirt	D 2	223+
Special woven fabrics	139	384R+
Stockings, socks	D 2	329+
Swimming suit	D 2	40+
Travel bags for	206	278+
Undergarments	D 2	1+
Vestment	D 2	79
Appendage	425	DIG. 3
Apple Trees	PLT	34
Appliances		
Cooking	D 7	323+
Food preparation, household	D 7	
Major household (See types)	D15	
Applicator		
Brushing	15	
Material supply combined	401	268+
Coating by absorbent	427	429
Coating by brush	427	429
Coating or cleaning implement	15	104R+
Material supply combined	401	
Daubing	15	
Dispenser combined	141	110+
Electrical	128	783+
Electrolytic	204	224R
Hair dye	D28	7+
Kinesitherapy	128	24R+
Light	128	395+
Loaders	15	257.5+
Material supply combined	401	
Medicators	604	19+
Medicinal	D24	63
Polishing	15	
Material supply combined	401	
Powder compact	132	82R
Pad overlying material including	401	130
Separable from combined supply container	401	118+
Thermal	128	399+
Wiping	15	
Appliers (See Applicator)		
Abrading sheet	51	275
Antiskid device	152	213R+
Boot and shoe cement	118	410+
Boot and shoe lining	12	39
Box bail	493	88
Box end enclosure	493	102+
Cigar and cigarette making	131	88+
Cosmetic	132	88.7
Hose clamp	81	9.3+
Label pasting and paper hanging	156	
Metallic leaf	156	540+
Receptacle closure	53	287+
Sealing wax	401	1+
Dispenser with heating means	222	146.2
Dispenser with illuminator or burner	222	113
Sheet metal container making components	413	
Ship caulking	114	224
Skid chain	81	15.8
Surgical clip	29	243.56
Tools for assembling/disassembling	29	700+
Wire to articles	140	93R+
Winding on	242	
Woodwork glueing and pressing	156	
Applique	428	914*
Design	D 5	63+
Sewn	112	
Apricot Trees	PLT	39

Entry	Class	Subclass
Apron (See Associated Machine)		
— Apparel	2	48+
— — Design	D 2	226+
— — Design, combined with dress	D 2	85+
— Cigar and cigarette machine	131	47+
— Drawbridge	14	47
— Dump car	105	279
— Lathe	82	22+
— Ship leak stopper	114	229
— Vehicle fender	293	41
— Vehicle storm protector	296	78R+
Aqua Regia		
— Detergent containing	252	101
— Gold or silver recovery	75	118R
Aquaplanes (See Ski; Sled; Surfboard; Tobaggon)	441	65+
— Design	D21	224+
Aquarium	119	5
— Design	D30	101+
— Filter	210	416.2
— Immersion water heater	D23	316
Aquasols (See Sol)		
Aquatic Culture	119	2+
Aquatic Devices (See Boats; Buoys; Lifesaving; Swimming)	441	
Aquatic Plant Destruction	71	66
Aqueducts	405	84
Aqueous Fluids		
— Earth boring compounds	252	8.51
Aqueous Polymerization Systems (See Class 523, 524, 526)		
Araldite T M	528	87+
Arbor (See Mandrel; Shaft)		
— Garden	47	44+
— Press	29	251
— Processes of making watch	29	178
— Watch and clock	368	322
— Work	279	2R
Arbutin	536	4.1
Arc		
— Curved line scribers	33	27.1+
— — Straight line	33	26
— Electric		
— — And incandescent lamp	313	8
— — Blowout lightning arrestors	361	117+
— — Blowout magnet	335	210+
— — — Arc system having	314	20
— — — Electronic tube system having	315	344+
— — — Lamp having	313	153+
— — — Switch having	200	147R
— — Chemical apparatus	422	186.21
— — — With magnetic field	422	186.3
— — Chemical processes nitrogen	204	178+
— — Chemical processes with magnetic field	204	156
— — Consumable electrode eg lamp	314	
— — Consumption feed type	314	59+
— — Cored carbon lamp	314	
— — Economizer combined	314	25
— — Extinction in a fusible element actuated switch	337	273+
— — Extinction in lightning arrester	361	117+
— — Extinction or prevention in switch	200	10
— — Extinction or prevention in switch	337	110
— — Extinction or prevention in switch	200	144R+
— — Feeding electrodes to	314	
— — Fume condenser for	314	26+
— — Fume flow director	314	26+
— — Furnaces	373	60
— — — Induction	373	7
— — — Resistance	373	5+
— — — Resistance with induction	373	6
— — Globes for	362	363
— — Lamps	313	
— — — Electrode composition	252	500+
— — — Electrode holder	403	
— — — Electrode structure	313	326+
— — — Hangers	362	416
— — — Manufacture	445	
— — Metal heating	219	121R+
— — Metal heating and casting	164	250.1
— — — In continuous casting apparatus	164	505+
— — Plural diverse heating means	373	1
— — — With induction	373	4
— — — With resistance	373	3
— — Projectors using	362	261+
— — Recorder	346	150
— — Regulating consumable electrode	314	
— — Regulating generally	315	
— — Search light	362	261
— — Shade or bowl supports for	362	433+
— — Space discharge devices, manufacture	445	
— — — Spot knocking	445	5
— — Stage spotlights	362	261+
— — Supply systems	315	
— — Suspended support for	248	317+
— — Suspension supports for	362	391+
— — Telegraph receiver	178	92
— — Tip cleaner for	314	23+
— — Tower or post for supporting	52	40
— — Treating materials with	422	186.21
— — Ventilator for	314	26+
— — Welding	219	136+
— Furnace		
— — Chemical	422	186.21+
— — Welding electrode holder	219	138+
— Gas lamp	313	567+
— — Circuit	315	
— — Structure	313	567+
— Metal heaters	219	121R+
— Saw machines	83	782
— Slicer blade	144	167
— Welding	219	121R+
— — Electrode	219	145.1+
Arch	52	86
— Arch bar bogie	105	204+
— Bridge		
— — Arch	14	24+
— — Truss and arch	14	2
— Doorway	52	85
— Fire tube boilers		
— — Water arch	122	89
— — Water arch subjacent	122	79
— Furnace	110	
— — Arch per se	110	331+
— — Door arch	110	181+
— Garden	47	44
— Masonry	52	89
— Monolithic	52	88
— Support for footwear	36	71
— — Design	D 2	314
— — Shoe with	128	581+
— — Ventilated	36	3R
— — With bandage structure	128	166.5
Archery		
— Arrows and darts	273	416+
— Blow guns	124	62
— — Design	D22	102
— Bow	124	23R+
— — Cross bow	124	25
— Design	D22	107
— Design	D22	107
— Elastic band guns	124	22
Architecture Teaching	434	72+
Architrave	52	211+
Arctics Rubber	36	7.3
Area		
— Calculators	33	121+
— Integrators	33	121+
— — Velocity integrators	73	227
— Monitors		
— — Nuclear energy	250	336.1
— — Scintillation type	250	361R+
Areaway	52	107
Arecoline	546	318
Arena		
— Amusement	272	3+
— Construction	D25	12
— Lighting	362	227+
Areometers	73	441+
Argol	562	585
Argon	423	262
— Lamps	313	567+
Arithmetic Teaching	434	191+
— Calculators	235	61R+
Arm		
— Artificial	623	57+
— Phonograph	369	244+
— — Acoustical	369	158
— Protectors	2	16+
— Robot	901	
— Signal		
— — Railway semaphore	246	482+
— — Vehicle direction	116	52+
Arm Band	D 2	624+
Arm Rest	248	118
— Chair folding	297	35+
— Chair or seat	297	411+
— Chair reclining	297	323
— Crutches	135	73
— Land vehicle	296	71
— Pianist	84	469
— Sofa bed	5	52
— Stringed instruments	84	328
— Writing aid	248	118.1+
Armature		
— Banding	310	271
— — Process	29	598
— — Winding lathes	242	7.4
— Coil	310	195+
— — Ring winding	242	4BE+
— — Taping	242	5+
— Cores	310	216+
— Dynamo	290	
— — Differential wound in arc lamp	314	76
— — Disk	310	268
— — Drum	310	265
— — Multiple	290	6
— — Pole	310	269
— — Ring	310	267
— Electromagnet	335	279+
— — Plunger type	335	255+
— Electromagnetic switch	335	80+
— Electromagnetic switch	335	95+
— Electromagnetically actuated in arc lamp	314	119+
— Fault tester	324	545+
— Growler	324	537+
— Making		
— — Commutator	29	597
— — Electromagnet	29	602R+
— — Rotor	29	598
— Manufacture of	29	
— Massage device vibrator	128	41
— Railway switch motor reciprocating	246	225+
— Railway switch motor rotary	246	240+
— Testers	324	555+
— Wedges	310	261+
— Winding apparatus, drum	242	7.4
— — Transverse slot	242	7.5 R
— Winding methods and apparatus	242	
— — Apparatus, drum winding	242	7.4
— — Apparatus, transverse slot winding	242	7.5 R
— — Assemblying apparatus	29	732+
— — Combined operations	29	605
— — Method, drum or transverse slot winding	242	7.3+
— — Miscellaneous	29	605
— — Ring	242	5
Armchair	D 6	334+
Arming Devices	102	221+
— Bomb, flare, & signal dropping	89	1.55
Armor and Armored		
— Body armor	2	2.5
— Cable	174	102R+
— — Fittings	174	50+
— Conduit	138	123+
— — Electric conductive sheath	174	102C+
— — Liner	138	137+
— — Spirally wound	138	134+
— Lock protection	70	417
— Piercing ammunition	102	517+
— Railroad ties	238	95+
— Shields	89	36.1+
— Shields and protectors	109	49.5
— — Port combined	109	58.5
— Tires resilient	152	167+
— Vehicle	296	188
— — Railway cars	105	394
— — Ships	114	9+
— Wall structure	109	78
Armpit Shield	2	53+
— Fastener	24	151
Arms,	D22	
— Making	29	1.1+
Aromatic	260	350R+
— Hydrocarbon	585	24+
— — Hydrogenation of	585	266+
— — Purification of	585	800+
— — Synthesis of	585	400+
Aromatization		
— Hydrocarbon synthesis	585	407+
— Mineral oil	208	46+
Arranging or Piling Apparatus (See Assembling; Placers)		
— Article	414	28+
— Article unpiling	414	112+
— Articles on conveyors	198	373+
— Bulk material by conveyor rotation	198	631
— Bulk material by conveyors	198	633+
— Cigarette making combined	131	282

Class Subclass

Conveyors from static support 198 508
Dried or gas treated material 34 38
Grain stackers 130 20
Horticultural tools D 8 1+
Piles arrangements holders or spacers 206
Sheet material associating or folding 270
Stackers for fluid current conveyors 406
Tobacco stringers or unstringers 414 26
Type 276 36+
Arrester
Cash register key 235 26
Lightning 361 117+
Combined with fuse 337 31+
Design D10 105
Loom letoff 139 111+
Loom take up 139 314
Metallurgical fume 266 144+
Spark 110 119+
Fixed whirler or rotator deflector separators 55 447+
Illuminating burners having 362 380
Arrhythmia Detector 128 702
Arrow 273 416+
Carriers 224 916*
Design D22 115
Making 29 1.2+
Sight 33 265
Arsanilic Acid 556 74
Arsenic 423
Alloys 420 579
Arsenates 423 602+
Azoles containing 548 102
Biocides containing 424 133+
Organic 514 504
Carbocyclic or acyclic compounds containing 556 64+
Dearsenizing ores 75 6+
Halides 423 462+
Medicines containing 424 133+
Organic compounds of 514 504+
Misc. chemical manufacture 156 DIG. 81
Oxides 423 617
Pyridine nucleus compounds 546 72
Pyrometallurgy 75 84
Sulphides 423 511
Arsenites 423 601
Arseno Carbocyclic or Acyclic Compounds 556 65+
Arsphenamines 556 66+
Art (See Subject Involved)
Collage 428 39
Etching process 156 658
Non-uniform coating 427 256+
Ornamental stitching 112 439
Sectional layer 423 410+
Superposed elements 428 77+
Artery, Artifical 623 1
Article (See Type of Article)
Advertising design D20 10+
Carrier 224
Hunting coats 2 94
Controlled burner 126 52
Cooling
Magazine type cabinet 312 36
Dispensing 221
Fabric delivery roll holder 242 55.2+
Magazine type 312 35
Match and toothpick 206 102+
Packages 206 233
Pellet pill tablet pocket use 206 528+
With heating or cooling 221 150R
With treating means 221 135+
Handling (See process or machine with which associated) 414
Conveyors 198
Hand or hoist line implements 294
Land vehicle compartment 296 37.1+
Stacking
Dispenser 221 175
Article Support Brackets 248 200+
Specially mounted or attached 248 205.1+
Corner 248 220.1
Interlocked bracket and support 248 220.2+
Post or column attached 248 218.4+
Support being pierced or cut 248 216.1+
Articulator 433 54
Artificial (See Imitation, Synthetic)
Arms 623 57+
Artery 623 1
Boards
Laminated 428 105+
Particle 428 403
Body member 623
Electrical actuation 623 24+
Fluid actuation 623 26
Bones 623 16
Butter
Making 426 603+
Renovating or treating
Apparatus 99 452
Process 426 663
Ear 623 10
Eye prosthesis 623 4+
Fauna 428 16
Feet 623 53+
Fiber
Making 264 165+
Plasticizing 8 130.1+
Fingernails 132 73
Fish bait 43 42
Trap hook combined 43 35
Flies 43 42.24+
Flora 428 17+
Flower 428 24+
Fog 252 305
Fuel (See fuel) 44 1R+
Garland 428 10
Grass 428 17
Heart valve 623 2
Horizon
Levels and plumbs 33 365+
Light ray type measurement 33 320+
Intelligence 364 513
Speech signal processing 364 513.5
Larynx 623 9
Leather 428 904*
Legs 623 27+
Musk 568 375
Nose 623 10
Reef 405 21+
Respiration 128 28+
Seaweed 405 24
Silk
Cellulose ester or ether containing composition 106 169+
Cellulose or derivative 536 56+
Chemical modification 8 129+
Cuprammonium compositions 106 167
Dyeing processes & compositions 8
Filament forming apparatus 425 66
Filament forming apparatus 425 67+
Filament forming apparatus 425 76
Filament forming apparatus 425 382.2
Filament forming process 264 165+
Nitrogen containing 536 30+
Regenerated cellulose 106 168
Spinning or twisting process 57 76
Spinning or twisting strand structure 57 243+
Swelling or plasticizing 8 130.1+
Viscose compositions 106 164+
Snow 106
Exhibitor 40
Sod 428 17
Stone
Dyeing 8 523
Firing 264 56+
Inorganic settable ingredients 106 85+
Molding process 264 333
Precious stone 501 86
Static mold 249 140
Water glass plaster of paris 106 77
Water glass portland cement 106 76
Stone decorative face 52 311+
Teeth 433 167+
Molds 425 2
Molds 425 175
Static mold 249 54
Three dimensional product 428 15+
Tobacco 131 353
Trachea 623 9
Tree 428 18+
Turf 428 17
Wreath 428 10
Artillery 89
Calculators ordnance control 235 400
Sights 33 235+
Artist (See Painters and Painting)
Canvas carriers 294 137
Fine arts materials D19 35+
Materials, design D19
Paint boxes 206 1.7+
Palette D19 35+
Artwork
Sheet bearing 428 195
Asbestos
Catalyst carrier 502 439+
Chemical treatment 423
Comminution 241 4
Fiber liberation 162 153+
Non-structural laminates 428 443+
Ascarite 252 408.1
Ascorbic Acid 549 315
Tests for 436 129*
Ascot D 2 605+
Aseptic Cultivation 435 800*
Ash
Ashpan boiler feed heaters 122 413
Cans 110 165R+
Cigars and cigarettes with ash holder 131 349+
Furnaces 110 165R+
Holder combined with smoking device 131 174+
Sifters 209 233+
Design D23 420
Stoves 126 242+
Tray 232 43.1+
Combined with other devices 131 231+
Design D27 8
Gas separation, used in 55 385G
Luminous 250 462.1
Vehicle, built in 296 37.9
Vehicle mounted 131 329+
Asparagus Harvester 56 327A
Asphalt 208 22+
Binders in coating compositions 106 DIG. 7
Cement 106 273R
Compositions 106 273R+
Alkali metal silicate 106 83
Carbohydrate 106 216
Carbohydrate gum 106 207
Casein 106 145
Cellulose ether or ester 106 191
Coating or plastic emulsions 106 277
Emulsions 252 311.5
Fat or fatty acid salt 106 269
Fatty oil 106 246+
Feathers hair or leather 106 156
Gelatine or glue 106 134
Ink 106 31
Mold 106 38.8
Mold coating 106 38.25
Natural resin 106 232+
Natural resin and drying oil 106 225
Natural resin and nondrying oil 106 229
Plaster of paris 106 116
Polish 106 10
Portland cement 106 96
Prolamine 106 152
Protein miscellaneous 106 160
Starch 106 212
Making, treating and recovery 208 39+
Melting furnaces 126 343.5 A
Mixers 366 1+
Design D15 19
Pavements 404 17+
Paving machinery 404 83+
Polymers containing or derived from (See synthetic resin or natural rubber)
Aspirator
Body fluids 604 19+
Flowing liquid aspirates gas 261 DIG. 75
Body inserted 604 35
Jet pump 417 151+
Mixing 137 888+
Static mold 249 102+
Aspirin 560 143
In drug 514 165
Assembling
Apparatus 29 700+
Caps and cartridges 86 12
Chain links 59 25
Chains 59 1+
Chains sprocket 59 7
Cleaning combined 134 59
Electrostatic type 29 900*
Liquid purification 210 232
Matrices type casting 199 9+
Nailing and stapling combined 227 19+
Press type filter 210 230

Entry	Class	Subclass
Supporters garment	223	49
Tire and rim	157	
Welding combined	228	4.1+
Bolts nuts and washers	10	155R+
Boxes		
Closure cap and liner	413	56+
Laminating	156	
Electric discharge devices	445	23+
Hinges	29	11
Lamps and vacuum tubes	445	23
Apparatus	445	60+
Processes	445	1+
Mold parts	164	339+
Processes		
Batteries	29	623.1+
Electrostatic type	29	900*
Receptacle with closure	413	2+
Apparatus	413	26+
Sheet material	270	
Shoe heels	12	50+
Shoe uppers	12	52
Textile fibers	19	144+
Well	166	373+
Assembly Means		
Fluid handling	137	315
Pipe joint	285	18
Associating Sheet Material	270	
Tobacco leaf	131	327
Assorter (See Associated Process, Article, or Machine)		
Brush making	300	18
Cleaning machines combined	15	3+
Coin	453	3+
Livestock	119	155
Printing machines combined	101	
Solid material	209	
Astrolabe	33	1R
Astrology	434	106
Astronomical Instrument		
Clock	368	15+
Globes	33	268+
Optical	356	138+
Solar position locator	33	268+
Telescopes	350	537+
Tellurions	33	268+
Astronomy Teaching	434	284+
Astrophysical Instruments		
Angle measurement	33	281+
Solar	33	268+
Spectroscope	356	300+
Telescope	350	537+
Astygmatism		
Ophtahalmic lenses for correction	351	176
Test chart	351	241
Athletic		
Equipment (See type)		
Mens shoes	D 2	309
Protective hand covering	D29	20+
Shoes, making	12	142P
Shorts	D 2	42
Atmometer	374	39+
Atmosphere		
Special atmosphere	156	DIG. 89
Utilizing special		
Glass making	65	32
Refrigeration	62	78
Atmospheric Electricity	307	149
Chemically inert or reactive	425	815*
Collecting	310	308+
With condenser	320	1
Discharging	361	212
Grounding	174	6+
Lightning rods	174	2+
Atomic Energy	376	
Fission reactions	376	349
Fusion reactions	376	100
Power utilization	376	336
Power utilization	376	402
Power utilization	376	370
Power utilization	376	391
Transmutation reactions	376	156
Well processes	166	247
Atomic Hydrogen Welding	219	75
Atomizer	239	
Container or bottle with	D28	91.1
Electrically heated	219	271
Electrically heated	219	380
Fumigator	422	305+
Gas liquid contact	261	78.1
Non-carboretor	261	78.2
Internal combustion engine supply	123	434+
Liquid fuel burner	239	
Medical	D24	62
Medicator	128	200.14+
Perfume	239	355+
Sprayer	239	337+
Temperature responsive control	236	8
Atropine	546	131
In drug	514	304
Attack Defeating Locks		
Mechanism	70	1.5+
Key lock	70	416+
Attempering		
Canned food processors	99	367+
Conveyor combined	99	361
Attenuator	333	81R
Pads	333	24R+
Resistance structure	338	
Voltage magnitude control resistors	323	369
Attic		
Fans	416	
Inlet	98	39.1
Inlet and outlet	98	33.1
Outlet	98	42.2
Ventilators	98	29+
Attitude or Pitch Indicator	33	328+
Attrition Comminutor		
Apparatus	241	284
Suspended particles	241	39
Processes	241	26
Suspended particles	241	5
Auction		
Livestock tags	40	300+
Audience		
Radio survey	455	2
Reaction apparatus		
Loudness meter	73	646
Voting machines	235	51+
Television survey	358	84
Audio Amplifier	330	
Audio-visual Teaching Machine	D19	60
Audiometer	73	585
Design	D24	21+
Audion T M		
Amplifier, electronic tube type	330	
Cathode ray tube structure	313	364+
Miscellaneous systems	328	
Oscillation generators	331	
Structure	313	
Audiphone	381	68+
Amplifier for	330	
Auditorium	52	6
Acoustics	181	30
Auger (See Bit; Drill)		
Boring machines wood	408	199
Earth boring	175	394+
Land anchors	52	157
Manufacture		
Dies	72	470+
Machines	76	2+
Processes	76	102
Twisting stock	72	299
Mortising machines wood	144	69+
Post base	52	157
Wood	408	199+
Auramine	564	269
Auricle	181	129+
Aurines	260	395
Autoclaves	422	242
Cooker (See steaming food)		
Autofrettage	29	1.11
Autogenous Bonding		
Of particles	264	123+
Of running length fibers	156	180+
Of self-sustaining laminae	156	308.2+
Autogenous Welding	228	196+
Electric	219	78.1+
Autogiro	244	17.11+
Aerial toy	446	36+
Airplane combined	244	8
Autograph Register	D18	21
Autoharp T M	84	8+
Automatic (See Controller & Control)		
Automated machine		
Broaching	409	245+
Gear cutting	409	2+
Milling	409	80+
Planing	409	289+
Camera		
Exposure control	354	400+
Focusing	354	410+
Cutouts for electric circuit	200	
Electromagnetic	335	
For lamp circuit	315	
Dispenser	222	52+
With cutoff operator	222	14+
With recorder register or indicator	222	23+
Drying and gas or vapor contact with solids	34	43+
Earth boring means	175	24+
Fire alarms	116	101+
Electric	340	500+
Flame	340	577+
Heat	340	584+
Smoke	340	628+
With indicator	116	5
Fish butchering	17	54
Flexible partitions and panels eg curtains	160	1+
Food treating machine	99	486+
Glass making apparatus	65	160+
Grain hullers and scourers	99	488
Grapples	294	110.1+
Gun (See gun)		
Heat exchange	165	13
Heating and cooling	165	14
Heating systems	237	2R+
Metal deforming		
Mining control	299	1
Apparatus	72	6+
Musical instruments	84	2+
Comb type	84	94.1+
Electrical tone generation	84	1.3
Stringed eg pianos	84	7+
Two different instruments	84	2+
Wind eg organs	84	83+
Nuclear reactions	376	207
Press control	100	43
Refrigeration	62	132+
Solid material comminution or disintegration	241	33+
Teller machine		
Using cryptographic code	380	24
Temperature or humidity regulation	236	
Cereal puffing apparatus	99	323.4+
Cooking apparatus	99	325+
Distillation separatory processes	203	2
Distillation separatory systems	202	160
Electronic tube or lamp	315	117
Humidor	312	31+
Hygrostats	73	335+
Infuser, eg percolator	99	281+
Liquid heaters and vaporizers	122	446+
Mineral oil distillation condensers	196	141
Mineral oil distillation vaporizors	196	132
Thermostats	60	527+
Water heater vessels	126	374
Tool driving or impacting device	173	2+
Volume control		
Amplifier coupling impedance	330	144+
Amplifier having of bias or power supply	330	127+
Amplifier thermal coupling impedance	330	143
In radio receiver	455	234+
Weigher	177	60+
Cigar and cigarette making	131	280
Automation (See Robot)	901	1*
Automobile (See Land Vehicle; Motor Vehicle)		
Air bag passenger restraints	280	728+
Beds	5	118+
Cribs	5	94
Body crushers	100	901*
Carburetor (See carburetor)	261	
Compound repair tools	7	100+
Cooler	62	243
Door handle	D 8	300+
Electric motors	D13	1
Fender covers	280	762+
Frame straightener	72	705
Headlight	362	61+
Design	D26	28+
Dimming system	315	82+
Electric system for	315	82+
Mutually responsive automatic dimmers	315	82+
Retractable	362	65
Sealed beam	313	113+
Heater	237	12.3 R
Heater and cooler	165	42+
Hood light	362	80

	Class	Subclass
Horn	D10	116+
Insignia	D11	95+
License tags	40	200
Changeable year section	40	207+
Lift or hoist	269	58+
Lights combined with part	362	61+
Ceiling or roof light	362	74
Mirrors	350	600+
Plural	350	612+
Plural reflections	350	618+
Miscellaneous equipment design	D12	155+
Name plate fasteners	40	20R+
Radio	455	345
Portable	455	346
Portable transmitter	455	99
Registration plates	D20	13+
Roll bars	280	756
Safety belt or harness	297	464+
Motor vehicle system responsive	180	268+
Passive	280	802+
Safety promoting means,	180	271+
Seats	297	
Crash	297	216
Design	D 6	334+
With body modification	296	63+
Shipped as freight	410	4+
Group of bodies shipped	410	43
Signals	116	28R+
Electric	340	22+
Lamp supply systems	315	77+
Stop electric	340	71+
Simulators	434	65+
Steering post or wheel		
Fluid handling	137	352
Lock for pilot wheel control or linkage	70	253
Suspension systems	280	688+
Stub axle mounts	280	660+
Tailpipe muffler	181	227
With exhaust pipe	181	228
Tape storage for cars	D 3	35
Toy	446	431+
Sounding	446	409+
Transmission jack	254	DIG. 16
Trunk light	362	61+
Switch	200	61.62+
Turntables	104	44
Ventilation	98	2+
Wheels and axles for land vehicles	301	
Aligning tools, hand	29	273
Lock for spare or mounted wheel	70	259
Window defrosting (See window)	52	171+
Autophon	84	83+
Autosyn (See Selsyn)		
Averaging, Aromatic Hydrocarbon Synthesis	585	474
Aviaries (See Bird)		
Awakener		
Bedstead related	5	131+
Clock	368	12
Alarm	368	244+
Person contacting	116	205
Time operated	368	12
Time alarm except clock	111	91+
Electric	340	309.15+
Awl	D 8	47
Leather	30	366
Needles	223	104
Punch type	30	366
Sewing		
Implement	112	169
Machine	112	48
Shoe making	12	103
Surgery	128	339+
Awning	160	45+
Design, fabric type	D25	57
Head rod bracket	248	273
Non rigid structure	160	45+
Storage or shield combined	160	22
Rigid structure	52	74+
Design	D25	57
Window closing	49	71
Ax	30	308.1+
Design	D 8	76
Diaphragm	92	96+
Holder	224	234
Making		
Blanks	76	103
Dies	72	470+
Machines	76	7+
Processes	76	103
Rack	211	13+
Type cutlery	30	318
Axle		
Box		
Manufacture dies forging	72	356
Manufacture dies forging railway car	72	343+
Railway car	29	168
Railway car mounting	105	218.1+
Wheel combined	301	109+
Gauge	33	193
Housings	74	607
Lubricating jack	254	32
Manufacture		
Assembly or disassembly tools	29	700+
By metal deforming process	72	
Dies forging railway car	72	343+
Dies forging vehicle	72	356+
Lathes turning	82	8
Puller	29	244+
Railway	295	36R+
Rear	180	88
Removing device	29	277
Rethreading dies	408	20+
Rethreading dies	408	22+
Shafts	301	124R
Skein	301	134+
Spindle	301	131+
Vehicle	301	124R+
Design	D12	160
Vehicle wheel combined	301	1+
Axminster		
Carpet	139	399+
Making apparatus	139	2+
Azaporphyrins	540	121+
Azeotropes	203	50+
Azetididiones	540	356
Azetidines	548	950+
Azetidinones	540	200+
Azides	423	
Inorganic metal azides	149	35
Explosive or thermic containing	149	35
Organic radical containing	260	349
Azidocillin	540	331
Azimuth Instrument		
Ammunition	102	372
Horizontal and vertical angle	33	281
Horizontal angle	33	285
Optical	356	138+
Solar locating	33	268+
Telescope	350	537+
Azines	544	1+
Azines	564	249
Azo compounds	534	751+
Heavy metal containing	534	701+
Aziridines	548	954+
Azo Compounds	534	573+
Anthraquinone containing	534	654+
Dye compositions containing	8	662+
Several dyes	8	639+
Textile printing	8	445+
Fiber-reactive	534	617+
Indane containing	534	659
Quaternary ammonium ctg.	534	603+
Resorcinol containing	534	682+
Salicylic acid containing	534	660+
Stilbene containing	534	689+
Azoles	548	100+
Acridine nucleus containing	546	26+
Aluminum containing	548	101+
Arsenic containing	548	102
Azo compounds	534	769+
Heavy metal containing	534	710+
Boron containing	548	110
Heavy metal containing	548	101+
Phosphorus containing	548	111+
Silicon containing	548	110
Azomethine	564	271+
Benzoselenazoles	548	121
Benzothiazoles	548	152
Benzoxazoles	548	219
Heavy metal containing	556	32+
Pyridines	546	268
Quinolines	546	152
Reduction of	260	689
Azoxy Compound	534	566+

B

	Class	Subclass
B Naphthol	568	735+
Monoazo compounds	534	840+
Preparation from aryl halides	568	739
Baby		
Bath	D23	278
Bottle warmer	126	261+
Chemical reaction	126	263
Electric	219	429+
Carriage	280	47.38+
Convertible to cradle or crib	280	31
Folding	280	639+
Mosquito nets and canopies	5	416
Steering	280	47.11
Chairs	297	
Combined with table	297	136+
Convertible high and low	297	345+
Cradle	5	101+
Cribs	5	93R+
Diapers	604	358+
Design	D24	50
Doorway safety guard	49	50+
Removable	49	463+
Fences	256	24+
Flatware for	D 7	141
Food or bottle warmer	D 7	326
Harness (See definition notes)	119	96+
Highchair	D 6	339
Tray	D 6	509
Incubators	128	1B
Jumpers	297	274+
Pacifier	128	359+
Design	D24	45
Pen	256	25
With floor	5	99C
Playpen	D 6	331
Rattles	446	419
Design	D21	65
Rubber pants	2	400+
Absorbent pad holders	604	393+
Design	D 2	10
Safety garments	128	134
Shoes		
Bronzing	204	20
Teething device	D24	45
Training toilet	D23	296+
Walker	272	70+
Vehicle	280	87.2
With seat	297	5+
Bacillus	435	832*
Hydrolase from	435	221+
Bacitracins	530	320
Back		
Pad harness	54	66
Rest	297	452+
Bed	5	70
Boat	114	363
Design	D 6	502
Scratcher	128	62R
Backband		
Harness	54	4
Backfire	261	DIG. 6
Backfire Preventer	48	192
Backgammon	273	248
Backgrounds Photographic	354	291
Backing Dental		
Instrument	433	141+
Making	29	160.6
Metalware shaping	72	54+
Backlash Take up		
Between meshing gears	74	409
Sectional gear	74	440+
Milling work feeds	409	146
Backpack	224	153+
Backrest	D 6	502
Backspace on Typewriter	400	308+
Backstay on Car Tops	296	144
Backstop		
Baseball	273	26A
Projectile	273	410
Target	273	404+
Backup Auto Lights	362	257+
Backwash		
Filter cleaning	210	108
Filter with	210	333.1
Fluid cleaning airpump	210	411
In multi-way valve	210	425
Sand bed with rehabilitation means	210	275+
With additional cleaner	210	393
Bacon	426	645
Packaging	53	DIG. 1
Preservation	426	332
Bacteria	435	
Fertilizer preparation with	71	6+
Liquid purification by	210	601+
Virus culture on	435	235+
Bactericidal		
Compositions	424	

	Class	Subclass
Bacteriophage	435	235
Measuring or testing	435	5
Bacteriostatic		
Compositions	424	
Badge	40	1.5$^+$
Design	D11	95$^+$
Film radioactivity	250	475.2
Film radioactivity	250	472.1
Badminton (See Tennis)	273	411
Rackets	273	67R
Baffle (See Process or Machine with which Associated)	416	
Acoustical muffler structure per se	181	264$^+$
Acoustical muffler with	181	264$^+$
Amalgamator agitator	209	187
Animal muzzle	119	131
Cleaning and liquid contact with solids	134	182
Decanter	210	513$^+$
Diverse separators	210	294$^+$
Drop water	122	188
Fire tube boilers	122	44R$^+$
Fuel	122	503
Furnace	110	322$^+$
Loud speaker	181	175$^+$
Mercury coated for amalgamating	209	54
Particulate material separators	210	285$^+$
Pipe and tube	138	37$^+$
Railway dumping car	105	279
Tank gas and liquid contact	261	123
Tower gas and liquid contact	261	108$^+$
Turbine stator	415	148$^+$
Vane	415	216$^+$
Vehicle body top board	296	33
Water tube boiler	122	235F$^+$
Bag (See Pouch; Receptacle; Sack)		
Abrading material filled	51	294
Airship gas cell construction and arrangement	244	128
Balloons	244	31$^+$
Inflation	244	98
Belt or body attached	224	205
Straps crisscross shoulders	224	209$^+$
Theft or loss resistant	150	102
Blood	604	403
Brief bag or case	190	900*
Closures	383	42$^+$
Clothespin		
Design	D32	36
Package	D 9	305
Colostomy	604	338$^+$
Explosive powder	102	282
Fabric	139	389$^+$
Fastener	24	30.5 R$^+$
Shirring cord	383	72
Traveling bag handle	16	110R
Feed	119	65$^+$
Flexible	383	
Closures	383	42
Fountain syringe with tubing	206	69
Garment	206	278
Gas separator	55	361$^+$
Golf	206	315.3
Holder	248	95$^+$
Sorting racks as for mail bags	211	12
Hot water	383	901*
Ice	383	901*
Inflatable for raising vessels	114	54
Locks for	70	64$^+$
Mouth support	141	391
Paper		
Closures	383	42$^+$
Making of bag	493	186$^+$
Opening	206	601$^+$
Pocketbook	150	100$^+$
Closures	150	118$^+$
Saddlebag	224	191$^+$
Sewing machine	112	10
Filled bags	112	11
Filling and closing	53	139
Sleeping	2	69.5
With bed structure	5	413
Striking or punching	272	77$^+$
Tea or coffee	426	77
Traveling	206	278$^+$
Design	D 3	30.1$^+$
Turner or reverser	223	39$^+$
Wheeled traveling	280	37
Bagasse Furnace	110	235$^+$
Baggage	190	
Cars	105	355$^+$

	Class	Subclass
Design	D 3	30.1$^+$
Identification tags	40	6
Rack on vehicle, illumination	362	73
Sample case with terraced trays	190	16
Supports	190	18R
Trunks	190	19$^+$
Vehicle attached carrier	224	273$^+$
Wheeled	280	29$^+$
With picnic or lunch unit	190	12R
Bail		
Applying to paper bag	493	88
Ear		
Metallic receptacle	220	91$^+$
Wooden bucket	217	126
Handle		
Basket	217	125
Lantern	362	399
Metallic receptacle	220	94R$^+$
Making or forming wire	140	75
Railway coupling link	213	75R$^+$
Bailer		
Cistern	294	68.22$^+$
Hoisting bucket type	294	68.22$^+$
Well	166	162$^+$
Bait		
Animal food and preparation	426	1
Food		
With poison	424	410
Glass drawing	65	352$^+$
Holder	43	55
Poison	424	84
Baked Products	426	549$^+$
Design	D 1	
Bakelite T M	528	129
Baker and Bakery		
Ovens	432	120$^+$
Peels	294	49
Products	D 1	
Baking		
Molds	99	372$^+$
Ovens domestic	126	273R$^+$
Pans	220	
Balanced to Unbalanced		
Coupling network with frequency characteristic	333	25$^+$
In amplifier stage	330	116
Powder	426	562$^+$
With phase inverter	330	116
Balancer	D10	82
Beam scales	177	246$^+$
Equal arm weigher	177	190$^+$
Horology	368	169$^+$
Making	29	178
Sash	16	193$^+$
Balancing Machine Parts (See Counterbalance)		
Comminuting elements combined	241	292
Determining balance by computer	364	463
Flywheels and rotors combined	74	573R
Grindstone weights	51	169
Instrument calibrating	73	1R$^+$
Rotor unbalance testing	73	66
Weights for land wheel	301	5R
Wheels vehicle	301	5R
Balata (See Rubber)		
Balconies	52	73
Window connected	182	53$^+$
Bale		
Band tightener and twister	140	93.2$^+$
Portable & detachable tensioner	254	199$^+$
Tightener and sealer	140	150$^+$
Grappling hooks	294	26
Round hay bale handling	414	24.5
Tie	24	16R$^+$
Design	D 8	394$^+$
Making	140	73
Baling	100	
Ball		
Amusement devices	273	58R$^+$
Design	D21	204$^+$
And socket joints (See joints)		
Bearing for typewriter type bar	400	447
Dental tool support	433	64
Vehicle steering mechanism	180	258
Baseball curvers	273	28
Bearing		
Hinge	16	275$^+$
Inking mechanism	400	270.2
Making	29	148.4 B
Mounting for ball tool	401	212
Radial	384	452$^+$

	Class	Subclass
Retainer design	D15	143
Thrust	384	590$^+$
Billiard ball spotting racks	273	22
Bowling		
Ball gauge	33	509$^+$
Ball return	273	47$^+$
Grip testing	33	510
Calculator	235	68
Carriers	224	919*
By hand	294	137$^+$
Cash register check	235	18
Chaser grinder		
Frictionally driven	241	103$^+$
Loose ball	241	173$^+$
Clock pendulum	368	179
Cocks toilet tank	137	409$^+$
Games	273	118R$^+$
Golf ball making	156	146
Design	D21	204$^+$
Making	29	148.4 B
Abrading processes	51	289S
Glass	65	21.1$^+$
Hollow metal spheres	72	348
Laminating	156	
Spheroid winding	242	3
Winding	242	2
Wooden balls	142	1
Mill	241	170$^+$
Point pen	401	209$^+$
Point stylus	81	9.2
Racks	211	14$^+$
Pool table attached	273	10$^+$
Sport or game balls	273	58R$^+$
Tea	426	77$^+$
Receptacle type	99	323
Tethered bowling ball	273	40
Time balls	116	200
Ball Cock	137	409
Ballast		
Aircraft	244	93$^+$
Cleaning		
Excavating combined	171	16
Loose material separation	209	
Railway grading combined	37	104
Resistor	338	20$^+$
Ship	114	121$^+$
Tamper	104	10$^+$
Ballers	425	332$^+$
Kitchen hand tool	D 7	100
Balloon		
Aircraft	244	31$^+$
Barrage, antiaircraft	89	36.16
Aerial mine carrying	102	409
Montgolfier type	244	31$^+$
Pyrotechnic device combined	102	356
Railway car attached	104	22
Rocket combined	102	347
Support with antenna	343	706
Toy	446	220$^+$
Design	D21	84
Ballot	283	5
Box	232	2$^+$
Registering	235	57
Counting or voting machine	235	51$^+$
Marking, teaching	434	306
Baluster	52	720
Balustrade	256	59$^+$
Bamboo Seat	D 6	369
Banana		
Banana plugs electric	439	825$^+$
Insulated	439	625$^+$
Band		
Endless band cutting	83	925EB*
Instruments	84	1$^+$
Lessons	84	470R$^+$
Saw	83	788$^+$
Saw	30	380
Blades	83	835$^+$
Spreading in radio tuner	334	
Filter	333	167$^+$
Using distributed impedance only	333	219$^+$
Band Width		
Reduction system		
Facsimile	358	260$^+$
Multiplex	370	118
Pulse or digital communications	375	122
Radio	455	72
Recording	369	60
Magnetic	360	8$^+$
Speech	381	29$^+$
Television	358	133$^+$

	Class	Subclass
Regulator		
Amplitude modulation	332	37R
Phase or frequency modulation	332	19
Bandage	D24	49
Body treating	604	304+
Controlled release of medication	604	896+
Electric	128	82.1
Package for coiled type	206	389+
Scissors	30	286
Splints	128	87R
Surgical	128	155+
Application	128	82+
Tire patches	152	371+
Webbing	128	156
Winding	242	60
Banding		
Bale	100	1+
Box covering	493	111+
Brush broom and mop making	300	15
Hat making	223	22
Label pasting and paper hanging	156	
Paper bunch	493	386+
Bandoleer	224	203
Bandsaw	30	380
Banister	256	59+
Banjo	84	269+
Clocks	368	229
Banking, Aircraft Control	244	75R+
Banks & Banking		
Burglar traps	43	59
Checks and deposit slips	283	57+
Coin handling	453	
Deposit apparatus	D99	34+
Depository	D99	43
Depository with verification means	109	24.1
Photographic or microfilming	346	22
Time record-marking	346	22
Indicator for degree of	33	328+
Piggy bank	D99	35+
Protection devices	109	2+
Safes	109	
Toy	446	8+
Banners		
Flag type	116	173+
Design	D11	165+
Rod	D11	181+
Sign type	40	
Baptismal Font	D99	25
Bar (See Rod)		
Boring	408	199+
Counters	312	140.1+
Crow	254	120
Horseshoe blank	59	62+
Lingual	433	190
Manufacture (See particular operation)		
Compound metal bars	228	4.1+
Drawing	72	274+
Extruding metal	72	253.1+
Indirectly	72	273.5
Horseshoe blank rolling	59	63+
Juxtapose and bond metal bars	228	101+
Rolling	72	199+
Masonry reinforcing	52	720+
Metal shape	428	544+
Design	D25	119+
Metallic plural layers	428	615+
Mining bar cutter	299	79+
Needle for sewing machine	112	222
Sash bar	52	777+
Splice rail joint	238	243+
Insulated	238	159+
Towel	211	105.1+
Type printing	101	401.1
Forming machines	101	401.6
Processes of forming	101	401.4
Wrecking	254	120
Barb		
Applying to wire, barbing	140	58+
Cutting fence barbs	29	7.1
Fences	256	2+
Hayfork type of harpoon having	294	127+
Nails having barbs	411	456
Staple barbing	59	73+
Tag fastener	40	22
Wire	140	59
Barbecues	99	419+
Cooker	D 7	332+
Fork	30	322+
Turners	99	419+
Barber		
Cabinets	312	209
Chairs	297	68+
Design	D 6	334
Implement with light	362	115
Poles	40	538
Design	D20	16
Tweezers	128	354
Barbital	544	307
Barbituric Acids	544	299+
In drug	514	270
Barges	114	26+
Cranes	212	190+
Barium		
Alkali earth metal	23	304+
Naphthenate	562	511
Bark		
Debarking	144	340
Extract	560	68+
Descaling agent ctg	252	83+
Osier peelers	144	207
Rossing	144	208R
Stripping machine	144	208R
Branches	144	207
Barkhausen Kurz Oscillator	331	92+
Barn	119	16
Barometer	73	384+
Design	D10	55
Barometric Material Feed		
Animal watering trough	119	77
Dispenser trap chamber	222	457
Gas liquid contact device	261	73
Inkstand or inkwell	222	585+
Liquid level miscellaneous	137	453+
Lubricator	184	84
Supply container and independent applicator	401	120
Barrel (See Cask; Container)		
Bungs	217	98+
Charring	432	224
Cleaning		
Brushing implements	15	164+
Brushing machines	15	57+
Methods	134	22.1+
Washing apparatus	134	43+
Wipers	15	211+
Closures	217	76+
Design, package simulates shape	D 9	325
Design, shipping	D34	39
Drying	34	104+
Methods	34	21
Electrolytic apparatus	204	213+
Filling	141	
Firearm	42	76.1+
Breech hinged	42	8
Forward sliding	42	10+
Moving and recoiling	89	160+
Muzzle loading	42	51
Recoiling with revolving chambers	89	157
Revolver hinged	42	63+
Side swinging	42	12+
Stock fastenings	42	75.1+
Swinging barrel locks	42	44+
Heads	217	76+
Interior illuminators	362	154
Laterally directed		
Drop bomb	102	383
Shell	102	383
Liquid level indicators	73	290R+
Making	147	
Sheet metal container	413	1+
Metal receptacle	220	
Ordnance	89	14.5+
Moving and recoiling	89	160+
Practice	89	29
Recoil checks	89	42.1+
Organs	84	86
Paper receptacle	102	480
Periscope for examining interior	350	540+
Plugs	217	110
Collapsible bulb	604	212+
Removing bungs	137	324+
Syringe pump hand held	604	187+
Tapping	137	317+
With cutter or punch	222	81+
Watch	368	142+
Wooden receptacle	217	72+
Insulated	217	131
Barrette	132	48R
Design	D28	39+
Barrier (See Ceiling; Floor; Gate; Wall)		
Building construction	52	
Static mold or form for making	249	18+
Flight deck	244	110R
Neutron	376	287
Neutron	250	518.1
Neutron	376	347
Floating	405	63+
Impact absorbing closure	49	9
Race track	119	15.5 R
Radiation	250	505.1
Process	55	16
Solid, for gas separation	55	158
Subterranean moisture	405	38
Barrier Layer		
Coating	437	+
By electrolysis	204	14.1+
Dry rectifier	357	
Composition	252	62.3 R
Manufacture	29	569.1+
Electrolytic condenser or rectifier	361	433+
Electrolyte	252	62.2
Manufacture	29	570.1
Material P N type	148	33
Making P N type	437	
Barrow		
Dumping	298	2+
Hand	294	15+
Wheel	280	47.31
Bascule Bridges	14	
Base Exchange		
Compositions	252	193
Water treating	423	328+
Silicates	423	
Water purification	210	660+
Apparatus	210	263+
Baseball	273	25+
Balls	273	60R
Bats	273	72R
Card or title games	273	298
Cover sewing machine	112	121.28
Game board	D21	28
Glove	2	19
Design	D29	21
Projector	124	
Simulated game	273	88+
Bases (See Foundation)	248	346
Apparel apparatus	223	120
Bed plate	52	292+
Compositions	252	193
Curved wall	52	247
Dispenser supports	222	173+
Electric lamp	439	611+
Electric switch	200	293+
Flatiron	248	117.2+
Inorganic	423	
Lanterns and lamps	362	190
Air preheating type	362	172
Chimney	362	314
Design	D26	93+
Shade	362	441+
Machinery	248	678
Organic	260	
Pole	52	292+
Concrete type	52	294+
Prop and brace	248	357+
Robot	901	
Staff	248	519+
Stand	248	188.1+
Swinging for movable receptacle	248	144
Vehicle runner	280	28
Basin (See Bowl)		
Auxiliary in ship lifting locks	405	4+
Bath	4	619+
Design	D23	284+
Bath fittings	4	191+
Cabinet combined	312	228
Canal locks	405	85+
Dispenser drip	222	108+
Drinking fountain catch	239	28+
Model	73	148
Receptacle type	220	
Sewerage catch	210	532.1+
Grated inlet	210	163+
With strainer	210	299+
Surgical receptors	604	317+
Basket (See Container; Creel)		
Closures	217	124
Cloth leather or rubber	150	48+
Design	D11	143+
Creel	D 3	38
Electrolytic cell electrode	204	259
Forming	147	48
Ski pole	280	824

	Class	Subclass
Wire	220	19
Wooden	217	122$^+$
Basketball Devices	273	1.5 R
Balls	273	65R
Teaching	434	248
Bass		
Bars for violin	84	276
Drums	84	411R$^+$
Finders	84	470R$^+$
Violin	84	274$^+$
Viols	84	274$^+$
Bassinet	D 6	390
Bassoons	84	380R$^+$
Basting	99	345$^+$
Brush	D 4	130
Bat (See Batt)		
Forming apparatus for hats	19	148
Game and amusement device	273	67R$^+$
Baseball	273	72R
Design	D21	211$^+$
Harvesting reel	56	219$^+$
Batch Charger		
Furnace type	414	167$^+$
Glass furnace combined	65	335
Batch Mixer		
Bread pastry and confection	366	69$^+$
Batch feeder and furnace	65	335
Mortar	366	1$^+$
With heat exchanger	165	109.1
Bath	4	
Bird	119	1
Design	D30	123
Brush	D 4	132
Cloth	D28	63
Design	D23	277$^+$
Fused salt	266	120
Heater	D23	318$^+$
Material supply coated or impregnated	15	104.94
Mitt	15	227
Design	D28	63
Photographic (See photography)		
Quench	266	130$^+$
Sponges	15	209R$^+$
With soap	401	201
Room accessories	D 6	524
Sponges	15	209R
Design	D28	63
With soap	401	201
Steam or sauna	D24	37
Therapeutic	128	365$^+$
Tub or footbath, theraputic	D24	38
Bathing		
Garments	2	67$^+$
Buoyant	441	88$^+$
Hat	D 2	510$^+$
Suit	D 2	36$^+$
Trunks	D 2	42$^+$
Pools	4	488$^+$
Swimming pool purification	210	169
Shoes	36	8.1
Sandals	36	11.5
Shower	4	596$^+$
Shower nozzles	239	548$^+$
Tubs	4	538
Bathometer	73	300
Bathroom		
Disinfection	4	222$^+$
Fixture liners	4	DIG. 18
Prison bathrooms	4	DIG. 15
Ventilation	4	209R$^+$
Bathtub (See Receptacles)	4	538$^+$
Antislip mat	4	582
Bath and basin fittings	4	191$^+$
Design	D23	277$^+$
Fittings	4	191$^+$
Seat	4	578
Soap dishes with soap handling means	206	77.1
Support	4	571$^+$
Therapeutic	128	369$^+$
Water cutoff device	4	191$^+$
Bating	8	94.17
Fermentative	435	265
Batiste	139	426R
Baton	D21	100
Illuminated	362	102
Music conductors type	84	477B
Batt		
Bonding with	156	62.6$^+$
Forming articles from	264	116

	Class	Subclass
Impregnation of workpiece by	264	136
Mechanical molding of article	264	257
Multilayer metal recepticle with	220	452
Special package for reel or roll	206	417
Battery (See Electret)	136	
Applications	136	291*
Assembling components	29	730$^+$
Methods	29	623.1$^+$
Bonding apparatus	228	58*
Bonding process	228	901*
Cable terminals	429	179$^+$
Carrier	224	902*
By hand	294	149$^+$
By hand	294	903*
Cases	429	164$^+$
Charge indicator	324	427$^+$
Battery attached	429	90$^+$
Charger	320	2$^+$
Design	D13	5$^+$
Material electrolyte	429	189$^+$
Material regenerating	429	17$^+$
Circuit applications	136	293*
Clamps for terminals	439	754$^+$
Deferred action	429	110$^+$
Depolarizer	252	
Design	D13	8$^+$
Electro chemical	429	
Electrode receptacle		
Conducting	429	239$^+$
Nonconducting	429	235$^+$
Electrolytic cell internal battery	204	248$^+$
Electronic tube having	315	55
Filler	429	233
Flash lights using	362	157$^+$
Grid	429	233$^+$
Filling	141	32$^+$
Making	29	2
Pasting	141	32$^+$
Handling hand and hoist line implements	294	
Holddowns, vehicle mounted	248	503$^+$
Holders in automobiles	180	68.5
Hydrometers	429	90$^+$
Internal or electrolytic		
Cell	204	248$^+$
Cleaning process	204	144
Protecting apparatus	204	197
Protecting process	204	148
Water treatment	204	150
Jar	429	163$^+$
Metal platemaking	29	623.1
Motor fed systems	318	139
Nuclear	310	301$^+$
Photoelectric	136	243$^+$
Plate forming	204	6
Plate making	29	2
Plugs	429	89
Prime mover dynamo plant with	290	50
Radioisotope-powered	136	202
Radioisotope-powered	376	317
Receptacle acidproof	206	524.1$^+$
Regenerating material	429	19$^+$
Separator	429	247$^+$
Solar cell	136	243$^+$
Space satellite applications	136	292*
Stoppers	429	89
Storage	429	149$^+$
Surgical application	128	391$^+$
Switches	320	2$^+$
Terminal	429	179$^+$
Applying or removing apparatus	29	246
Protector	429	65
Tester	D10	77
Testing	324	426$^+$
Accessories	429	90$^+$
Calibrating, treating, testing	136	290*
Charging combined	320	47
Hydrometer	73	441$^+$
Integral with battery	429	90$^+$
Thermoelectric	136	200$^+$
Vent caps	429	89$^+$
Batting	19	296
Package for	206	389$^+$
Battle Lanterns		
Flashlight type	362	157$^+$
Bay Window	52	201
Bayonet	30	
Gun combined	42	86
Holder	224	232$^+$
Holder	224	232
Joint (See joints)	362	35

	Class	Subclass
Beacon		
Aircraft	362	262
Radio	342	385$^+$
Directive radio	342	350$^+$
Rotating	342	398$^+$
Floating marine	441	1$^+$
Illuminating lights	362	35
Radar transponder	342	42$^+$
I F F	342	45
Signal lights	340	
Air craft	340	981$^+$
Bead and Beading		
Adsorbent or catalytic	502	527
Bead chain making	59	2
Garment supporting	2	300
Glass compositions	501	33$^+$
Glass manufacturing	65	21.1$^+$
Apparatus	65	142
Lathing corner beads	52	255$^+$
Metal beading	72	67$^+$
Metal molding of	164	179
Curving and	72	298$^+$
Tube forming and	72	149$^+$
Metal tube corrugating	72	367
Necklace	63	2
Design, strung	D11	11
Rosary	235	123
Design	D99	26
Sewing on with sewing machine	112	104$^+$
Shoe		
Heel beading machines	12	49
Upper beading machines	12	57.1
Spreaders for tires	254	50.1$^+$
Spread holders	81	15.3
Tire mounting combined	157	1.17$^+$
Stringing	223	48
Apparatus	29	241
Method	29	433
Tire beads	152	539
Window bead fasteners	16	220
Beakers (See Receptacles)	422	102
Dispensing	222	566$^+$
Molten metal	222	591$^+$
Measuring	73	426$^+$
Beam		
Brake	188	219.1$^+$
Design, architectural	D25	126$^+$
I-shaped beam cutter	83	DIG. 2
Locomotive truck buffer	105	173
Making by rolling	72	365$^+$
Masonry	52	723
Mortar and block	52	433
Plow	172	681$^+$
Power amplifier	313	299
Rectifier	313	298
With beam forming electrodes	313	299$^+$
With two cathodes	313	5
Railroad ties I beam	238	65$^+$
Two part	238	57
Scales	177	246$^+$
Static mold	249	50
Support for brackets	248	228
Pipe or cable type	248	72
Beam Lead Frame or Device Structure	357	69$^+$
Making	29	827
Beam-trammel Distance Device	33	158
Beaming	28	190$^+$
Beanie	D 2	256
Beans	426	629
Bearing (See Journal)	384	
Brasses or linings	384	276$^+$
Clutch throwout	192	110B
Connecting rod adjustable	74	594
Design	D15	143
Hydrostatic bearing with gearing	409	904*
Lathe headstock	82	30
Manufacture		
Assembly apparatus roller and bearing ball	29	724$^+$
Processes	29	149.5 R$^+$
Processes of grinding	51	289R$^+$
Materials	252	12$^+$
Piano pedal	84	228
Pullers	29	244$^+$
Rudder post	114	169
Separator, imperforate bowl centrifugal	494	83
Testing frictional resistance	73	9$^+$
Turntable railway	104	46
Typewriter bar	400	448$^+$
Typewriter carriage	400	354$^+$

Entry	Class	Subclass
Watch and clock	368	324
Wringers	68	269R
Beat Note in Detector		
Autodyne detector	455	321+
Heterodyne detector	455	313+
Heterodyne detector	329	153+
Homodyne	455	324
Beater		
Agitator	366	343+
Cleaning	15	89+
Implements	15	141.1+
Rotary	15	141.2
Cloth finishing	26	25+
Comminuting processes	241	27
Drum and cymbal	84	422R
Fiber liberation and preparing	19	85+
Brakes and beaters	19	30
Decorticating	19	33
Heat exchanging	165	92
Meat tenderer	17	25+
Paper fiber engine	241	97
Pastry and confection	366	69+
Rotary beater mill	241	185R+
Perforate casing	241	86+
Screen	209	299
Clearer	209	383
Shoe welt	12	67.2
Textile braider and	87	36
Wire twister and	140	32
Portable machine	140	42
Beauty Parlor Equipment	D28	9+
Bed (See Associated Machine or Tool)	5	
Activated sludge	210	623+
Animal	119	1
Design	D30	118
Attachments and accessories	5	508+
Backpack convertible to	224	156
Baggage convertible to	190	2
Bedding	D 6	382
Bedstead	5	131+
Camping land vehicle	296	174
Clothes holder	24	72.5
Clothing	5	482+
Bed attached	5	498
Weight supporting	5	505
Combination furniture	5	2R+
Cooling	5	284
Heater combined	165	46
Mattress	5	421+
Refrigerating	62	261
Crib	5	93R+
Design	D 6	382
Doll or toy	446	482
Exercising device	272	144+
Fan combined	416	146R
Filter (See filter)		
Frame component	D 6	503+
Garment		
Gown type	2	114
Pajama type	2	83
Glass molding		
Planar platen	65	256+
Heater	126	205
Cooling combined	165	46
Electric	219	217
Mattress	5	421+
Mattress electric	219	217
Medicator combined	604	113+
Surgical	128	376+
Horticultural	47	18
Hotbed	47	19
Invalids	5	60+
Jacket, apparel	D 2	12+
Light combined	362	130
Light thermal electrical application	128	376+
Machine or tool part		
Addressing machines	101	57+
Apparel plaiting, fluting, shirring	223	31
Bed and platen printing	101	287+
Clothes pressing platen	38	17+
Clothes pressing roll	38	44+
Clothes washing scrubbing	68	63+
Clothes washing squeezing	68	94+
Coopering stave jointing	147	26
Glass molding	65	361
Inkers for printing devices	101	335+
Intaglio printing bed and cylinder	101	158+
Lathe	82	32
Metal bending	72	214+
Multicolor printing bed and cylinder	101	186+
Numbering and printing	101	72+
Planer reciprocating bed	409	321+
Planers woodworking endless	144	128
Planographic copying	101	131+
Planographic printing bed and cylinder	101	146
Printing member and inker	101	104+
Rolling contact printing bed and cylinder	101	214+
Selective or progressive bed and platen printing	101	93+
Solid material stratifier	209	422+
Ticket printing	101	66+
Mattress or cushion	5	448+
Design	D 6	596+
Midrib and center strip	5	192
Modular bedroom lighting	362	801*
Mosquito nets and canopies	5	414+
Pan	4	450+
Design	D24	57
Particulate material separator	210	263+
Pavement or road	404	27+
Person restrainer for	128	134+
Quilts	D 6	603+
Railway car with	105	316+
Ships	114	188+
Sleeping bags	5	413
Sofa	5	12R+
Spread or cover	D 6	596+
Supported horizontal surface	108	49
Tray	D 6	406
Trunks convertible	190	2
Waterbed	5	451
Window ventilator combined	98	89
Wire bottom frame attaching	140	110
Bedbug Trap	43	123
Bedpan	4	450+
Design	D24	57
Rinser in flush toilet	4	300.2
Bedplate (See Bases)		
Bedroom Lighting, eg Modular Combinations	362	801*
Bedspread	D 6	596
Bedsteads		
Design	D 6	382
Folding	5	174+
Bee Culture	6	
Beeswax purification	260	420+
Feeder	6	5
Hive	6	1+
Honeycomb type receptacle	6	10+
Smokers	43	127+
Beef Jack		
Closure operators	49	69
Trolley transfer	104	97
Beer	426	592
Apparatus for making	99	275+
Beeswax Purification	260	420+
Beet		
Harvesting devices	171	
Digging then topping	171	26+
Lifters	171	50+
Topping then digging	171	26+
Toppers	56	
Washers	15	3.1+
Beetling	26	26
Beheader Fowl	17	12
Belaying Pin	114	221R
Bell		
And hopper charger	414	204
Combined with blast furnace	266	184
Animal and sleigh	116	170
Automatic musical instruments	84	103
Bicycle	116	166
Church	116	150
Design	D10	116
Combined with escutcheon	D 8	350+
Electric	340	392+
Electrically actuated signals	340	392+
Electrically simulated church bell	340	398
Gas holder		
Inverted bell and tank	48	176
Moving bell	48	179
Sectional telescoping bell	48	177
Glasses for protection of plants	47	26+
Highway crossing type	246	296
Mechanical	116	148+
Musical bell	84	406
Design	D17	22
Musical bell with striker	84	407
Ornamental	428	11
Railway crossing protection	246	111+
Signals & indicators	116	148+
Sound producing	116	148+
Swinging eg church	116	150
Typewriter margin signal	400	712+
Bellows	D23	384
Accordions	84	376R
Cameras	354	187+
Dispenser	222	206+
Fluid pressure responsive	92	3+
Making	29	454
Metal tube corrugating	72	54+
X-art collection	493	940*
Meter	73	262+
Organs	84	355+
Pump	417	472+
Fumigator combustion air	422	305
Belt (See Machine with which Associated)		
Apparel	2	311+
Compressor type	450	154
Design	D 2	627+
Design, garter or sanitary	D 2	625
Elements of supporters	2	338+
Buckle attached	D 2	627+
Carrier mounted onto	224	163
Cartridge feeding	89	35.1+
Cigar and cigarette type machine	131	55
Comminutor type	241	200
Conveyors (See conveyor, endless)		
For dredges	37	69
Couplers	24	31B+
Hook	24	35
Lacing	24	34
Design	D 2	627+
Drive train design	D15	148
Drives	474	
Endless for vehicle fender	293	20
Fastener	24	31R+
Structure for nuclear reactor moderator	376	302
Gearing	474	
Coating to prevent slippage	106	36
Hanger	D 6	315+
Holder for cartridges, fishing lures	D 3	100
Holder or rack	D 6	315+
Holster combined	D 3	101
Hook inserters	29	243.51
Lifting tool	474	130
Making	156	137+
Of leather	69	
Special sewing machine	112	121.27
Mechanisms		
Belt and sprockets	474	
Guards	474	144
Money	224	229
Paper making		
Endless drying	34	243R
Fourdrinier	162	348
Pulleys	474	
Registers	235	125
Reinforced (See reinforcement, fabric)	474	268+
Safety	182	3+
Aircraft	244	122B
Vehicle seat	297	464+
Hand vehicle	280	801+
Motor vehicle	180	268+
Seat belt safety buckle	D11	200+
Shifter	474	101+
For variable speed drive	474	80+
Starting gas pumps and fans	417	223
Stock material	428	
Conveyor type	198	
Drive type	474	237+
Knitted	66	169R+
Laminated	428	
Woven	139	383R+
Support in leather working machine	69	41
Tightener (See notes under)	474	101+
Abrasive belt	51	148
Band type twisting apparatus	57	105
Belt type drying apparatus	34	118
Railway wheel and axle drive	105	105+
Tool holder carried by	224	904*
Tree trunk guards	47	24
Trusses surgical	128	99.1+
V-type	474	237+
Bench		
Cabinet combined	312	235R
Greenhouse	47	18

	Class	Subclass
Ladder combined	182	33
Shoemakers	12	122
Table	108	
Wash tub or machine	68	236
Woodworking	144	286R+
Clamps	269	
Design	D 6	396+
Dogs	144	306+
Support for	144	286R
Tool chest combined	144	285
Bench Mark	52	103
Bed plate	52	292+
Bending (See Crimping; Folding)		
Brake	72	310
Chain making machines	59	27
Sheet metal	59	15
Hoof and shoe expanders	168	47
Horseshoe making machines	59	36+
Lock washer making machines	10	73
Metal		
Machines and processes for	72	
Machines combining with other metal shaping	29	34R+
Wire	140	
Working, with bending	29	DIG. 3
Nut making machines	10	74
Other metal shaping	72	
Scale removal by flexing	29	81R
Sheet metal seaming by shaping	228	144
Sheet metal seaming machines	72	48+
Sheet metal seaming processes	228	137
Staple making and setting apparatus	227	82+
Staple making and setting	29	400R
Staple making machines	59	71
Strength of materials testing by	73	849+
Sweep arm bender	72	217
Tubes	72	367+
Wood	144	254+
Beneficiating Ores	75	1R+
Apparatus	266	168+
Bentonite	106	DIG. 4
Benz C Fluoran	549	224+
Benzaldehyde	568	425
Benzanthrones	260	352+
Benzene (See Aromatic Hydrocarbon)		
Benzidine	534	822+
Disazo compounds from	534	822+
Disazo pyrazoles from	534	760
Tetrakisazo compounds from	534	808
Trisazo compounds from	534	813+
Benzoates		
Ammonia	562	493
Benzyl	560	106
Bismuth	556	78
Caffeine	544	274
Lithium	562	493
Magnesium	562	493
Mercury	556	132
Benzocaine	560	19
In drug	514	535
Benzodianthrones	260	361
Benzofuranes	549	462+
Benzoic		
Acid	562	493
Anhydride	260	546
Benzoin	568	331
Benzol (See Aromatic Hydrocarbon)		
Motor spirits	585	14
Benzomorphans, in Drug	514	295
Benzophenazines	544	343
Benzophenone	568	332
Benzopyrenequinones	260	362
Benzoquinoline	546	101
Benzoquinolzines	546	95
Benzoquinone	260	396R
Benzoselenazoles	548	121
Benzothiazines	544	49+
Benzothiazoles	548	152+
Benzothiophenes	549	49+
Benzotrichloride	570	185
Preparation	570	191+
Benzoxazoles	548	217+
Benzyl		
Alcohol	568	715
Butyrate	560	254
Disulphide	568	25
Ether	568	659
Salicylate	560	71
Thiocyanate	558	10
Benzylamine	564	391
Bergamot Oils	512	5

	Class	Subclass
Berry		
Clipper		
Catcher combined	56	331
Holder combined	30	124+
Nipper type	30	175+
Crates	217	40
Berth		
Bunk beds	5	9R+
Self leveling	114	192+
Sleeping car	105	316+
Beryllium		
Alloy compositions	420	401
Electrolytic synthesis from fused bath	204	65
Preparation	75	84
Bessemer Converter	266	243+
Bessemerizing		
Copper matte	75	75
Copper smelting combined	75	73
Iron or steel	75	59.12+
Beta		
Crystal mounting	250	361R
Gamma survey meter	250	336.1
With sample holder	250	336.1
Beta Alanine	562	576
Beta Globulins	530	394
Betaines	260	501.11+
Bevel		
Gear	74	640+
Intercontrolled blades	33	455
Protractor	33	455
Square	33	474
Beverages	426	
Apparatus	99	275+
Beer or pop can	206	139
Can cover	220	903
Carbonated	426	590+
Carbonater	261	
Carbonating and flavoring	99	323.1+
Coaster or mat	D 7	45
Cooler	62	389+
Foam forming or inhibiting	137	170.1+
With selection from plural materials	222	144.5
Electrolytic treatment	204	138+
Mixer blender, household	D 7	376+
Mixing		
By mechanical agitating	366	
Milk shake type	366	197+
By nozzle shape	239	549
Preparation or dispensing machines	D 7	300+
Reusable infusion receptable for preparing beverages	99	323
Storage receptacle	D 7	76
Vending machines	D20	5
Bezel		
Design	D26	139
Test instrument	73	431
Watch	368	295
Bias Voltage Control in an Amplifier	330	129+
Bib	2	49R+
Dental patient	433	137
Design	D 2	226+
Bicarbonate	423	419R
Drugs or bio-affecting	424	156
Bichromate Cells	429	
Bicycle	280	200+
Attached carrier	224	30R+
Bell	116	166
Brake	188	24.11+
Coaster	192	6R
Convertible	280	7.1+
Cranks and pedals	74	594.1+
Cyclometers (See odometer)		
Design	D12	111
Dust and mud guards	280	152.1+
Exercising devices	272	73
Generator	310	
System	322	1
Handle bars	74	551.1+
Handle or grip design	D 8	DIG. 8
Horn	116	137R+
Hub ball bearing	384	545
Lights	362	72+
Generator bulb system	315	76
Generator control	322	1
Generator per se	310	
Supports	362	382+
With wheel driven generator	362	193+
Locks	70	233+
Wheel	70	225+

	Class	Subclass
Making		
Frame assembling	29	700+
Methods	29	428+
Pedal crank bearings	384	431
Propelled marine pedomotors	440	30
Racks	211	17+
Rack or holder, design	D12	115
Reflector	350	97+
Seats	297	195+
Simulations	280	1.11 R+
Umbrella for	135	88
Wheel	301	5R+
Guards	280	160.1
Scrapers and cleaners	280	158.1
Bidet	4	443+
Design	D23	295+
Bier	27	27
Wheeled	280	47.34+
Extensible	280	640
Tiltable	280	47.17+
Bifocal		
Contact lens	351	161
Spectacle lens	351	168+
Bile Acids	260	397.1
Bilge Discharge	114	183R+
Bilirubin	436	97*
Billboard Type Signs	D20	39+
Billets	428	577+
Fault removal	29	526.2+
Harness	24	182
Hame strap	54	28
Hame tugs	54	32+
Piercing	72	325+
Billfold	150	132+
Leather, design	D 3	56
Billiards	273	2+
Balls	273	59R
Chalk	273	17
Cues	273	68+
Tips	273	70
Design, cues and accessories	D21	210
Register operating device	235	91B
Stick or rest	273	23
Tables		
Beds	273	6
Covers	273	13
Design	D21	232
Tops	273	6
Timing device	368	3
Billing Machines (See Printers)		
Bills in Knotting	289	11
Billy Club	273	84R
Design	D22	117
Guns	42	1.16
Biltmore Sticks	33	483
Bimetallic Stock (See Thermostats)	428	615+
Bonded layers	228	101+
Casting	164	91+
Juxtapose and bond layers	228	101+
Making by extrusion	72	258
Shape or structure	428	577+
Ingots	428	585
Bin (See Box; Receivers)		
Charging or discharging	414	288+
Deforming	72	700
Discharging (See dispensing)	222	
Charging and	414	288
Granaries		
Hopper with port	52	192+
Sifter associated	209	370+
Ventilated grain bins	98	55
Binary Compounds, Inorganic	423	509+
Uranium containing	423	253+
Binaural	381	1+
Binder and Binding		
Baling press combined	100	8+
Binder device releasably engaging aperture or notch of sheet	402	
Depository	402	73+
Expander	402	80R
Filler	402	80R
Fly-strip fly-leaf	402	80R
Marker position holder	402	80R
Books	412	
Design	D19	26+
Cooking apparatus combined	99	350
Covering with metal	29	33.2
Severing base combined	29	33.5
Edge binding guide for sewing	112	137
Handle or grip design	D 8	DIG. 8
Harvesting	56	432+

U.S. PATENT CLASSIFICATIONS

Entry	Class	Subclass
Cutter combined	56	67+
Hat making	223	22
Post	439	775+
Press	100	
Resilient sectional tire	152	300+
Ski	280	611+
Snow shoe	36	126
Table top edge	52	783+
Laminated top	52	783+
Tobacco product	131	365
Vehicle load	410	96+
Water ski	441	70
Wire or band tensioners		
Combined with sealers	140	93.2
Portable & detachable	254	199+
Binding		
With flexible filament band or strand	100	1+
Binding Assays	436	500+
Binding Proteins	530	387
Bingo Games	273	269
Binoculars	350	145
Carrier	224	909*
Cases	206	316
Design	D16	133+
Microscope	350	514
Telescope	350	545
Bio-affecting Compositions (See Drug, Bio-affecting and Body Treating Compositions)	514	
Biochemical Oxygen Demand Tests, Bod	436	62*
Biocidally Protected Polymers	523	122
Biocides	424	
Coating or plastic composition	106	15.5+
Detergent with	252	106+
Mineral oil derived	208	2
Water softening or purifying agent	252	175+
Biodegradable		
Plant receptacle	47	74+
Biology		
Coating processes	427	4
Preservation	47	
Teaching	434	295+
Biopsy	128	739
Biopsy device	128	751+
Biosynthesis	435	
Biotin	548	303
In drug	514	387
Biphenyl (See Aromatic Hydrocarbon)		
Biphosphates, Inorganic	423	
Bird		
Antiroosting structure	52	101
Artificial	428	16
Feather	428	6
Baths	119	1
Design	D30	123
Cage	119	17+
Design	D30	114+
Perches	119	26
Perches, design	D30	119+
Calls	446	204+
Feeders or waterers	D30	121+
Food	D 1	
Houses	119	23
Design	D30	110+
Perches or cage attachments	D30	119+
Repellents		
Electric fence	256	10
Electric prods	231	2E
Scarecrows	40	
Training	119	
Traps	43	
Birefringent Element		
Color televison	358	61
Polarized light examination	356	365
Birth Control (See Contraceptives)		
Biscuit Package	206	830*
Biscuit Shredded	426	560
Apparatus	425	289+
Bismuth	75	
Alloys containing	420	577
Misc. chemical manufacture	156	DIG. 79
Organic compounds containing	556	64+
Pyrometallurgy	75	70
Bisphthalimides	548	461+
Bisulphates, Inorganic	423	520+
Bisulphides	423	511
Bisulphites	423	520
Reducing composition	252	188.21
Bit (See Cutter; Drill)		
Bench plane	30	493
Bit clamps	30	492
Earth boring	175	327+
Frame	408	53
Gauges	33	201
Harness bridle	54	7+
Design	D30	136
Weaning type	119	134
Ice boring	175	18
Ice cutting in situ	299	24+
Key bitting	70	409+
Mining cutter	299	79
Stock	81	28+
Wood boring double ended bit stock	408	36
Wood boring movable work machines	408	62+
Stationary bitstock	144	98
Wood boring stationary work machines	408	72R+
Inclined bitstock	144	100
Tobacco pipe	131	227+
Wood auger	408	199+
Bitts	114	218
Bituminous		
Abrasives containing	51	305
Compositions	106	273R+
Emulsions	252	311.5
Mineral oil recovery from	208	400+
Polymers containing (See asphalt, polymers containing)		
Roads	404	17+
Bivalve Opener, Marine Animal	17	74+
Black Liquor	106	123.1
Blackboard	248	441.1+
Cabinet combined	312	230
Compasses	33	27.2+
Compositions	106	32.5
Design	D19	52+
Easels	248	441.1+
Erasers	15	208+
Writing surface	106	32.5
Blacking (See Coating)		
Box	15	258
Applicator combined	401	118+
Holder	15	259
Stand	15	265+
Blackjack	273	84R
Carrier	224	914*
Bladder Instrument	128	328
Blade (See Associated Machine)		
Electric switch	200	271+
Hand manipulated cutlery	30	
Holder for blades	30	329+
Making	76	104R
Saws	76	25R
Metalworking cutter	407	
Propeller	416	
Adjustable	416	147+
Stoneworking saw		
Reciprocating	125	18
Rotary	125	15
Blanching		
Fruits and vegetables	426	506+
Grain apparatus	99	600+
Grain process	426	506+
Nuts	99	623+
Blankets		
Animal	D30	145
Bed	5	482+
Electrically heated	219	212
Horse	54	79
Design	D30	145
Household linen	D 6	603
Nuclear fertile material	376	172
Printers	428	909*
Retaining devices	24	72.5
Blanking		
Cathode ray tube circuit	315	384
Metal workpiece for deformation	72	324+
Blanks (See Stock)	428	542.8
Cartridges	102	530+
Cigar	131	364
Horseshoe	59	62+
Metal	428	577+
Double blanks	29	DIG. 2
Holder for during deformation	72	293+
Metallic receptacle	220	62
Nut	411	427
Paper envelope	229	75
Design	D19	3+
Phonograph record	428	64+
Printed	283	
Processes and		
Chain	59	35.1
Garment collar	2	143
Printing members	101	401.1
Sprocket chain	59	8
Staple	59	77
Tools and implements	76	101R+
Weldless chain	59	12
Shelf bracket of single blank sheet material	248	248
Shoe sole prepared for attachment	36	22R
With upper structure	36	12+
Shoe upper	36	47+
Spike	10	62
Blast		
Abrading by sand	51	410+
Boiler cleaning by sand	122	395
Cleaning	15	300R+
Beater and or brush combined	15	363+
Suction combined	15	345+
Decorticating fibrous material	19	9
Filter cleaning	210	407+
Fluid flow dispensing	222	630+
Forges	110	195
Furnace (See shaft, furnace)	266	197+
Furnace reduction of iron & steel	75	40+
Gas agitation	222	195
Insect powder dusters	43	132.1+
Metal treating	266	
Ore beneficiating	75	8
Pipes for locomotives	60	685+
Jet pump type	417	155+
Pneumatic conveyors	406	
Track sanders	291	3+
Blasting	102	301+
Caps	102	275.12
Cartridges	102	314+
Well	102	301+
Blaugas	48	211
Bleachers	52	8+
Design	361	2+
Bleaching	8	101+
Apparatus textiles	68	
Compositions		
Cake, tablet or powder form	8	524+
Detergents combined	252	94+
Oxidative	252	186.1+
Reductive	252	188.1+
Dyeing and	8	931*
Electrolytic	204	133
Foods and beverages	426	253+
Apparatus	99	467+
Flour	426	253+
Paper stock	162	1+
Photographic	430	430+
Resin bleach	8	DIG. 6
Silicones	8	DIG. 1
Sugar	127	64
Textile bleaching	8	101+
Tobacco	131	290
Electrical or radiant energy	131	299
Vinyl sulfones & precursors thereof	8	DIG. 2
Waste paper	162	4+
Bleeder Resistor	338	
Blenders		
Household mixer-blenders	D 7	378
Blending		
Dispensing combined	222	145
Mineral oil	208	
Mixing	366	
Tobacco leaf	131	327
Cigar making combined	131	39
Feeding combined	131	108+
Blends of Polymers (See Synthetic Resin or Natural Rubber)		
Cellular	521	134+
Blind or Blinders (See Shutters, Blind or Sightless)		
Adjusters	160	176R
Animal worn	119	104
Design	D30	144+
Harness bridle attached	54	10+
Cord latch for venetian blind	D 8	394
Eye shield type	2	15
Design	D29	18
Fasteners	292	
Flying (See flying)		
Hunter concealing	43	1
Radar	342	24+
Using light ray	340	555+

	Class	Subclass
Using sound or supersonic	367	87+
Rollers	160	284
Slat holders	292	345
Stitch sewing	112	267.1+
Machines	112	176+
Work guide for	112	140
Tilter unit for venetian blind	D 8	358
Window	160	
Blind Landing Systems		
Aircraft control	244	75R+
Geometrical instrument	33	227
Light	362	145
Flashing beacon	340	947+
Radio	343	
Beacon	342	385+
Beacon fixed course or bearing	342	407+
Localizer	342	413
Direction finding receiver	342	417+
Glide slope	342	410+
Blind or Sightless, Aids for		
Braille typewriter	400	122+
Canes	135	65+
Converting information to sound	434	116
Photocell scanners for reading machines	250	555
Radar type	342	24+
Tactile	434	113+
TV including aid for blind	358	94
Using sound	367	87+
Writing guides	434	117
Blister Package	206	461+
Block		
Butcher	269	289R+
Construction (See blocks; toy; panel)	52	
Break away section	52	98+
Column attached	52	474+
Glass	52	306+
Keys on angled surfaces	52	589+
Keys on single surface	52	593
Layered masonry	52	612
Masonry	52	596+
Mortar bonded masonry	52	415+
Parallel course masonry	52	561+
Paving	404	34+
Reinforced block	52	600+
Settable material on masonry	52	444
Static molds	249	144+
Through passage	52	606+
Tied	52	282
With drain	269	13
With passage	52	606+
Electrical connector		
Ceiling	439	450
Ceiling quick detachable	439	450
For floor, wall, roof	D25	138+
Fuse receptacle	439	830+
Hat forming	223	24+
Design	D15	135+
Keel and bilge	405	7
Log deck	83	708
Log turners combined	83	710+
Molding	425	
Holder	206	39
Planes	30	478+
Pulley	254	401+
Block and tackle, design	D 8	360
Railway rail seats		
Spiking blocks	238	286
Wood cushion blocks	238	285
Sanding	51	204+
Flexible	51	358+
Sectional pipe	138	100+
Tackle pulley	254	401+
Toy and educational		
Building	446	85+
Design	D21	108
Educational display	434	403
Teaching arithmetic	434	208
Teaching music	84	476
Teaching spelling	434	167+
Block Copolymer (See Synthetic Resin or Natural Rubber)		
Blocking Layer (See Barrier Layer)		
Blood		
Albumin from	530	363
Analysis		
Apparatus	422	44
Chemical processes	436	66+
Equipment design	D24	21
Centrifuge for	494	16+
Computer controlled, monitored	364	416
Counter	377	10
Fertilizers containing	71	17
Filtration equipment	D24	21
Flow	128	691
Foods containing	426	647
Gas analysis	436	68*
Hemoglobinometer	356	40+
Occult blood	436	66*
Oxygenating	422	45+
Oxygenators	261	DIG. 28
Plasma	424	101
Pressure	128	672
Light radiation tests	128	667
Measurer	D24	21+
Recorder	128	900*
Ultrasonic radiation tests	128	662
Stock feed	426	635
Storage bags	604	403
Transfusion	604	7
Treatment	604	4
Blood Coagulation Factors	530	381+
Blood Proteins	530	380+
Blotter	34	95+
Design	D19	98
Rulers combined	34	89.2
Blouses	2	106+
Design blouses and shirts	D 2	208+
Combined with skirt	D 2	112+
Evening, combined with skirt	D 2	74
Shirts	2	113+
Blow Film	264	555+
Blow Molding	425	522+
Process	264	500+
Of parison	264	523+
Blower	D23	383
Conveyer type	406	96+
Furnace solid fuel feeder	110	104R+
Gas pumps and fans	415	
Open fireplace	126	135+
Snow	37	244
Testing	73	168
Wind supply for organs	84	355
Blowing Glass	65	
Blowing-agent for Synthetic Resin or Natural Rubber	521	50+
Blowout		
Magnets	335	210+
Arc in switch	200	147R
Arc lamp having	314	20
Electronic tube system having	315	344+
Space discharge device having	313	153+
Preventers		
Actuators for	251	1.1+
Burners	239	553+
Cigarette lighter	431	310
Gate valves	251	1.1+
Well casing heads	166	75.1+
Well coupled rod type	277	5+
Well valve type	251	1.1+
Blowpipe		
Cutting heads	266	904*
Earth boring	175	14
Making	29	157R+
Notch & transverse cutting	266	902*
Structure	239	398+
With guide means	266	58+
Blowtorches	355	78+
Combined fuel tank	431	344
Heated	431	203+
Flame deflected by air	431	252
Heated feed	431	218+
Blueprint		
Lamps electric (See type of lamp)		
Paper, sensitive composition	430	540
Printing machine	355	78+
Processes photographic	430	368+
Bluing		
Compositions	8	648
Packages	206	.5
Boa		
Sewing machine for making	112	121.17
Board		
Base	52	241
Cake	269	289R+
Clapboard	52	518+
Clapboard positioning gages	33	646+
Composite or veneer boards	428	106+
Croquet boards	273	57
Finger board on musical instrument	84	314R
Game boards	273	236+
Design	D21	14+
Teaching	434	128
Ironing board	38	103+
Convertible to ladder	182	28+
Design	D32	66
Mere support surface	108	
Lee boards	114	126
Molding		
Article holding	52	27
Corner	52	288
Panel edging	428	21+
Trim	52	716
With filler strip	52	312
Music educational boards or charts	84	471R+
Pastry and cutting boards		
Cutting blocks and boards	269	289R+
Pastry boards	269	302.1
Plaiting boards for apparel	223	33
Plaster board	52	443+
Reed boards for organs	84	360+
Side boards pivoted for convertible wagon boxes	296	14
Sideboard with refrigerator	62	258
Sink drainboards	4	637
Sounding board		
Frame & string plates combined	84	184
String plates combined	84	187
Structure of	84	192+
Springboard	272	66
Switchboard		
Power type	361	332+
Telephone multiple at exchange	379	310+
Telephone multiple section type	379	313+
Telephone type details	379	319+
Textile weaving thread boards	57	358+
Top and side board fasteners	296	36
Top board details	296	32+
Top boards pivoted for convertible wagon boxes	296	13
Washboards	68	223+
Clothes supporting and spotting boards	68	240
Design	D32	67
Support for washboard	248	27.5
Boat (See Ship)	114	
Airborne	244	
Capstans	254	266+
Chock	D 8	356+
Cleats	114	218
Design	D 8	382+
Cranes	212	190+
Design	D12	300+
Heating	432	262+
Land vehicle for carrying	280	414.1+
Launching	114	344
Life boats	114	348+
Liferafts	441	35+
Lowering apparatus	114	365+
Propulsion	440	
Rope fastening hardware	D 8	356
Ships	114	
Design	D12	300+
Small	114	
Storing mother ships	114	258+
Tillers (See stearing, marine)		
Toy	446	160+
Design	D21	130
Bobbers Fishing	43	43.11
Bobbin	242	118+
Bobbin winding on sewing machine	112	279
Braiding and bobbinet machines	87	55+
Changing		
In loom	139	241+
In spinning or twisting frame	57	266+
Packages	242	159+
Packing in receptacles	53	235+
Sewing machine	242	118.4+
Design	D15	78
Stripping and handling	28	292+
Replenishing in sewing machine	112	186
Shuttle clearing combined	139	262
Supporters and holders	242	130+
Textile twisting	57	129+
Winding	242	18R+
Bobbinet		
Lace	87	4
Machines	87	27
Bobby Pins (See Hairpin)	132	50R
Design	D28	39+
Bobeche	D26	142+
Bobs, Plumb	33	392+
Bod, Biochemical Oxygen Demand Tests	436	62*

Entry	Class	Subclass
Bodine Vibrator	366	600*
Bodkin	223	103
Body		
Artificial body members	623	
Attached antenna	343	718
Attached article carriers	224	148+
Bio-affecting compositions & drugs	424	
Braces orthopedic	128	78
Carrier, hand	294	140
Catcher	182	137+
Chute combined	182	48+
Ladder combined	182	101+
Cooler	62	
Bandaging combined	128	82.1
Display	62	246+
Heat exchange structure	165	46
Ice bag type	62	530
Kinesiatric	128	24.1
Medicator	604	113+
Open surface support	62	458
Orthopedic	128	68.1
Surgical instrument combined	128	303.1+
Therapeutic	128	362+
Cremators	110	194
Fluid collection	604	317+
Garment	2	
Footwear	36	
Footwear design	D 2	264
Life preservers	441	88+
Life preserving design	D21	238
Therapeutic	128	379+
Harness (See notes to def of)	119	96+
Heaters or warmers	604	113+
Apparel	219	211
Boots and shoes	36	2.6
Chemically reactive material for		
Irreversible reaction	44	3.1+
Readily reversible reaction	252	70
Electric	219	527
Heat exchanger	165	46
Receptacle for thermal material	220	3.1
Stoves	126	204+
Therapeutic	128	362+
Treating material	604	291+
Land vehicles	296	
Leveling device vehicle	280	6R+
Life preservers	441	80+
Massagers	128	24R+
Orthopedics	128	78
Protectors	2	2+
Trap		
Bank protection	109	3+
Burglar	43	59
Vehicle fender combined	293	15+
Body Treating Compositions (see Drug, Bio-affecting and Body Treating Compositions)	514	
Treatment and care	128	604
Artificial body members	623	
Dentistry	433	
Methods	604	49+
Surgery	128	
Toilet, ie, grooming articles	132	
Undertaking	27	
Valve	251	366+
Vehicle (See type of vehicle)		
Bogie	105	182.1+
Boiler	122	
Carried by land vehicle	280	5R
With leveling device	280	7
Cleaning	122	379+
Air blast	15	406+
Combined cooking stove	126	5
Water back on fluid fuel cooking stove	126	53+
Electric	219	281+
Electrolytic type	219	284+
Feeding fuel and water	122	446+
Feeding water	122	451R+
Purification	210	311+
Food apparatus	99	403+
Fuel magazine	122	30
Furnace structure for		
Water tube attached baffles	110	322+
Heating radiator combined with	237	16+
With automatic control	237	7
Hot water	126	361
Liquid level sight glasses	73	328+
Guard combined	73	326
Locomotive	105	43+
Articulated	105	40
Making processes	29	157.4
Nuclear reactor type	376	370
Open type	126	344+
Safety devices	122	504+
Scale preventing compositions	252	175+
Supports	122	510
Vehicle with steam heating boiler	237	41+
Vehicle with water heating boiler	237	35+
Wash boiler attachments	68	237+
Water gages	73	328+
Boiling Point Determining	374	16+
Boiling Water Reactor Nuclear	376	370
Bologna	426	646
Packaging	426	392+
Bolometer	374	32
Bolster		
Bearing	384	239
Bed pillow	5	434+
Railway truck	105	226+
Spring	267	6+
Vehicle	280	143+
Bolt	411	378+
Closure fastener	292	2+
Guard	292	346
Safe	109	59R+
Design	D 8	387+
Eye	D 8	367+
Fabric rolls	206	389+
Holder	81	13
Lock	70	229
Locked to nut	411	190+
Bolt compresses inward	411	305+
Bolt includes axial opening		
In free end for expander	411	271
In free end for expander	411	325
Recess in outer surface	411	217
Locked to substructure	411	81+
Stud bolt	411	107
Making	10	11R+
Rearward sliding breech	42	69.2+
Sliding lock	D 8	339+
Toggle	411	340+
Wrenches	81	52+
For bolts having cavity	81	436
Bomb	102	
Aerial sighting	33	229+
Drop	102	382+
Automatic flight control	102	384
Dropping	89	1.51+
Racks	89	1.59
Shelter	109	1S
Underground building	52	169.6+
Sights	33	229
Calculators for	235	400
Optical	356	29
Snuffer	169	
Bonded Layers	428	
Bonded Masonry	52	415
By pressure per se	100	
Bonding and Bonds (See Grounds Electric; Welding)		
Autogenous laminating	156	308.2+
Battery bonding apparatus	228	58*
Battery bonding process	228	901*
Dental	433	226+
Glass working by	65	36+
Apparatus	65	152+
Laminating by	156	
Heat shield	228	59*
Metal to non-metal	228	903*
Preform with fluent hardenable material		
Static mold	249	83+
Rail	238	14.1+
Juxtapose & bonding	228	101+
Making, general	228	4.1+
Thermite molding	164	54
Shaping metal concurrently	228	265
Tube end closing	228	60*
Using flame	228	902*
Wire bonding	228	904*
Bone		
Adsorbents from	502	416+
Artifical	623	16
Conduction earphone	381	151
Conduction hearing aid		
Electric	381	68.3
Through teeth	181	127
Crushers	128	319+
Dyeing	8	523
Fertilizers	71	19+
Glue	106	125+
Growth stimulators	128	419F
Removal apparatus	17	56+
Process	17	46
Saws	128	317
Trephine	128	310
Screws	128	92R
Splints	128	92R
Vises & fracture apparatus	128	83+
Bonnets		
Apparel	2	204
Design	D 2	515
Horse	54	80
Book	281	15R+
Accounting books	283	63A
Animated	446	147+
Books, strips, & leaves	281	
Cabinet	312	
Sectional unit type	312	107+
Showcase type rack inclosing	312	128+
With support for books	312	233
Carriers	294	149+
Case		
Design	D 6	396+
Casing in	412	5
Cover	206	450
Design	D19	26
Handle, combined	294	138
Making	412	3
Design	D19	26+
Disinfecting & preserving	422	243+
Holders	248	441.1+
Cabinet combined	312	233
For book or leaf	281	45+
Furniture design	D 6	419
Telephone type	248	441.1+
With illumination	362	98
Index cutting	83	904*
Looseleaf or ring bound	D19	27
Magazine cover removal	83	925MG*
Making machine	412	9
Leave or signature folding or associating	270	
Sewing	112	21+
Stapling	227	
Manifolding	282	22R+
Books and leaves	282	8R+
Marks	116	234+
Book holder combined	281	42
Design	D19	34
Match	206	104+
Filling and closing	53	394
Methods	53	395
Picture leaf	352	99
Printed matter	283	63R+
Accounting and listing	283	66R
Calendar	283	4
Indexed	283	42+
Transportation	283	33
Rack	211	42+
Shelf	108	59+
Stands	248	441.1+
Stereoscope combined	350	133+
Straps	294	149+
Strips & leaves	281	
Trimmers	83	925A*
Typewriting machines	400	24+
Bookend	D19	34.1
Booking Tobacco Leaves		
Cutting combined	83	510+
Smoothing combined	131	326
Stemming combined	131	316+
Boomerang	273	426
Projecting holder	124	44
Booms	52	
Crane	212	266
Floating	441	47+
Power sprayer	239	159+
Booster		
Booter fluid in conveyor	406	93
Brakes with power assist	303	113
For tilting dump vehicle	298	19B
Boot (See Shoes)	36	
Anti-slipping devices	36	59R+
Design	D 2	271+
Design, protective	D30	146+
Blowout	152	367
Cleaner and cleaning	15	
Demonstration devices	434	397
Foot elevators	36	81
Footwear	36	

	Class	Subclass
Hook	D 2	643
Horse	54	82
Design	D30	146+
Jack	223	114+
Design	D 2	642
Laces	24	143R+
Design	D 2	316
Making	12	
Forms	12	128R+
Heel machines	12	42R+
Jigging	12	DIG. 3
Lasting machines	12	7+
Dash pots	12	DIG. 2
Fluid activated	12	DIG. 1
Processes	12	142R+
Sole machines	12	17R+
Toe and heel stiffening	12	61R+
Tools	12	103+
Upper machines	12	51+
Welt and rand machines	12	67R+
Mold	425	119
Mold	425	393
Polish	106	3+
Protectors	36	72R+
Putting on or removing	223	113+
Design	D 2	641+
Quarter boots	D30	146+
Rack	211	34+
Retaining	36	58.5+
Riding	36	131
Shoe covering	D 2	271
Ski	36	117
Sole gauges	33	3R
Tree holders	12	123.5
Treeing machines	12	53.2
Vehicle body	296	76
Booth	52	261+
Article supporting	52	27+
Collapsible	52	64+
Design	D25	16
Guard	109	9
Illuminated	52	28
Ordnance shield type	89	36.14
Spray coating	118	300+
Booth per se	98	
Spray washing	134	
Bootstrap Amplifier	330	156
Boranes	423	294
Explosive compositions	149	22
Organo boranes	568	1+
Borate Esters	558	286+
Borax	423	277+
Recovery	423	179
Borders		
Coating	427	284
Coping for vertical structures	52	300
Earth supported	52	102
Garden	47	33
Printing elements for	101	400
Borehole Exploration		
Compressional wave type	367	86
Core permeability testing	73	38
Earth boring combined	175	40+
Inclinometer	33	304+
Ore detection electric	568	1+
Physical condition	73	151+
Seismic acoustical	181	
Sounding	33	126+
Test electric	324	323+
Boring (See Auger; Bit)		
Bar	408	199+
Drilling and	408	
Earth	175	
Earth boring compositions	252	8.51
Embroidering machine	112	89
Pattern controlled	112	85
Head	408	199+
Lathe drill holders	408	238+
Metal cutters rotary	407	103+
Wood	408	
Augers	408	199+
Borneol	568	820
Boron and Compounds		
Acid and acid anhydride	423	277
Binary compounds	423	276+
Borates	423	
Carbide	423	291
Carbocyclic or acyclic compounds	568	1+
Detergents containing	252	89.1+
Elemental	423	298
Esters containing	423	277
Explosive or thermic composition containing	149	22
Fuels containing	44	76
Misc. chemical manufacture	156	DIG. 86
Nuclear fuel containing	252	636+
Bosom Shirt	2	118+
Machine	223	2+
Botany	47	
Bottle (See Carboy; Receptacles)	215	
Advertising indicia	40	310+
Design	D20	37
Atomizer	D28	91.1
Attachments	215	100R
Baby bottle warmer	D 7	326
Bail stopper bottle filling apparatus	53	265
Breakers	241	99
Brushes internal	15	164+
Cap seal	425	809*
Capping	53	287+
Caps	215	316+
Case or holder	206	139
Design	D 9	455
Cleaning	15	
Filling combined	141	85+
Liquid contact	134	
Processes	134	
Closures	215	200+
Cap type	215	316+
Childproof	215	201+
Content indicating	215	365+
Cooled receptacle for	62	457
Liquid contacting bottle	62	373+
Cooler of inverted bottle type	165	132
Design	D 9	
Dispensers	221	92+
Dispensing from	222	
Dynamic dispensing means	D 9	300
Filling	141	
Inspection optical	356	240
With rotation	356	428
Making		
Glass	65	
Molding and label applying	425	500+
Plastic	425	522+
Nursing	215	11.1+
Design	D24	47
Nipple appliers	29	235.5
Openers		
Cap combined	215	228
Design	D 8	33+
Hand manipulated types	81	3.7+
Mounts or supports	81	3.25+
Non manual operation	81	3.2
Other tools combined	7	151+
Plural or combined	81	3.9+
Stopper combined	215	228
Stopper removal facilitated	215	295+
Pressurized	D 9	300
Pull stopper bottle filling apparatus	53	264
Rack supports for	211	74+
Shape	215	1R
Nesting	215	10
Siphon bottle filling apparatus	141	14+
Stand, rack or tray for	D 7	70
Sterilizer	219	429+
Stoppers or plugs	215	355+
Supports for nursing	248	102+
Tap		
Cutter and or punch	222	80+
Thermos T M	215	13R
Warmer	126	261+
Chemical reaction	126	263
Electric	219	429+
Wipers internal	15	211+
Wrapper for	229	89+
Bougie Surgical	128	341
Medicator, soluble	604	288
Methods of introducing	604	54+
Bouillon Cube	426	589
Meat containing	426	641+
Bouquet		
Imitation	428	17+
Supports	248	27.8
For personal wear	24	5+
For personal wear design	D 2	624+
With holder	428	23
Bourdon Tube		
Gauge fluid pressure	73	732+
Making	29	157R+
Switch electric	200	81.8
Thermometer	374	203
Boutonniere	D11	117+
Bow	D22	107
Apparel trim	2	244+
Making	223	46
Arrow projector	124	23R+
Compound	124	DIG. 1
Design	D22	107
Carrier for bow or arrow	224	916*
Decorative	428	4+
Dental bite impression device	433	44
Garment support combined	2	300
Hair ribbon fastener	132	47
Land vehicle top	296	98
Necktie	2	144+
Design	D 2	606
Ox	54	77
Piano	84	256+
Ribbon	132	47
Saws	30	507+
Sight	33	265
String on bridge truss	14	9+
Thrusters, ship	114	151
Violin	84	282
Guides	84	283
Bow Tie	D 2	606
Bowden Wire (See Shaft, Flexible)	74	501R+
Link and lever system combined	74	501R+
Mechanical movement combined	74	82
Bowking	8	139
Bowl (See Basin; Pot; Receptacles)		
Arenas	272	3+
Basins	4	619+
Compartmented	D 7	27
Dry closet	4	420+
Fish	D30	101+
Household article	D 7	23+
Kitchen	220	
Lamp	362	154+
Reflector	362	341+
Lavatory	4	650+
Nut bowl	D 7	98
Railway amusement ride	104	68
Roundabouts	272	46+
Separator, imperforate bowl centrifugal	494	43+
Shaving	220	
Smoking pipe	131	226
Cleaners	131	246
Detachable from neck cup	131	222
Detachable from neck cup smoke treating type	131	214
Feeder	131	180
Lined or coated	131	220
Lined or coated material traps	131	204
Reversible	131	221
Spaced inner bowl	131	196
Storage means	131	180
Toilet		
Closures	4	253
Couplings and supports	4	252R+
Covers	4	234+
Disinfection	4	222+
Drip catchers	4	252A
Obstruction removers	4	257
Ventilation	4	216
Water closet		
Bowl	4	420+
Design	D23	295+
Plunger	4	420+
Seat combined	4	234+
Side receptacle	4	341+
Siphon	4	421+
Tank combined	4	300+
Urinal	4	311+
Valved	4	434+
Washout	4	420+
Bowling	273	37+
Ball	273	63R+
Case	D 3	36
Grip test	33	510
Bowling alley equipment	D21	233
Bowling game tables	D21	233
Games	273	37+
Pins	273	82R
Digest	76	DIG. 1
Shoes	36	130
Teaching	434	249
Box (See Bin; Chest; Crate; Receptacle; Safe)		
Annealing	266	262+
Axle process of manufacture of		

	Class	Subclass
Railway car irons	29	168+
Ballot	232	2+
Registering	235	57
Beam	52	731
Blacking box (See blacking box)	15	258
Closure		
Paper	229	124+
Paperboard box	229	124+
Wooden	217	56+
Cooled commodity containing	62	371+
Core	164	228
Couch	297	192+
Covering	493	111+
Deposit and collection	232	
Design	D 9	414+
Egg candling	356	64+
Electric battery	429	96+
Fare	232	7+
File	206	425+
Fire (See firebox)		
Flower attachable to window	47	68+
Folded blank structure	229	16R+
For transport, goods handling	D34	40+
Hat	206	8+
Housings for electrical conductors	174	50+
Vacuum or fluid containing	174	17LF+
Ice	62	459+
Air controller combined	62	420+
Joint clamp	269	111+
Land vehicle axle	301	109+
Letter boxes	232	22
Lubricating railway car journal	384	160+
Making machine		
Assembling and nailing	227	19+
Grooving handhold	144	136D
Nail driving	227	
Paper box	493	52
Sheet metal bending	72	394+
Sheet metal stamping	72	343+
Staple forming and setting	227	82+
Staple setting	227	
Metallic	220	
Outlet or junction type	220	3.2+
Sheet metal container making methods	413	1+
Meter	73	201
Circuit protectors	361	364+
Miter	83	746+
Money	446	8+
Opener		
Cutting	30	2
Prying	254	18+
Paper	229	6R+
Folded blank	229	16R+
Pipe and box joints	285	128+
Presses		
Expressing	100	127
Movable bale	100	221+
Railway axle box mounting	105	218.1+
Registering	235	100
Ballot	235	57
Rheostat plug	338	77
Savings	229	8.5
Sewing	223	107
Special articles and packages	206	
Splints	128	93
Extension appliances	128	86
Switch	200	293+
Ties	229	125.22
Toilet kits	132	79F+
Powder	132	82R
Toy money	446	8+
Design	D99	34+
Traps for animals	43	60+
Wooden	217	5+
Egg cells	217	18+
Ice	217	130
Insulated	217	128+
Boxing Gloves	2	18
Brace (See Suspenders)		
Body		
Bandages	128	155+
Combined with corset	450	96
Fracture apparatus	128	83+
Orthopedic	128	68+
Shoe attached skates	280	11.36
Shoulder and back	2	44+
Trusses	128	95.1+
Walking irons	128	83.5
Boilers	122	493
Closure fasteners	292	338+
Design		
Dispenser spouts	222	573
Handle combined	222	475
Door	254	39
Drill	81	28+
Freight on carrier	410	121+
Bar wall-to-wall	410	143+
Panel wall-to-wall	410	129+
Yieldable	410	117+
Making	29	150
Orthodontic	433	2+
Instruments	433	3+
Pole or wall	52	146+
Wing type	52	153+
Rail		
Tie plates	238	292+
Track fastenings	238	336+
Supports	248	351+
Clothesline	248	353
Vehicle spring	267	66+
Bracelet	63	3+
Design	D11	3+
Identification	40	625+
Bracket	248	200+
Antenna on	343	892
Article carried for storage support	248	360
Article support	248	200+
Specially mounted or attached	248	205.1+
Bicycle carrier	224	39+
Book or music holder support	248	441.1+
Clamped to support	248	225.31+
Dental engine	433	103+
Design	D 8	354+
Dispenser casing	222	180+
Eaves trough to wall	248	48.2
Guides for sliding panels and doors	16	90
Ladder rung	182	220
Light supports	362	432
Design	D26	138+
Machinery supports	248	674+
Making and assembly	29	150
Mine car axle	295	42
Nursing bottle type	248	103+
Orthodontic	433	8+
Pipe or cable	248	65+
Ray generation machine support	250	522.1
Receptacle support	248	311.2
Rotary shaft to wall	384	442+
Staff	248	534+
Stand alternative	248	126
Machinery supports	248	676
Stand combined	248	121+
Static mold form	249	219R
Stove shelf	126	333
Thermometer supports	374	208
Track combined for sliding panels and doors	16	94R+
Vehicle attached	224	42.45 R+
Watch and clock	248	115
Whip socket	280	175
Brad	411	439+
Braiding	87	
Guide for sewing	112	139
Packages	206	389+
Sewing machine	112	23
Shaft packing material	277	227+
Tire carcass material	152	548
Trimming design	D 5	7+
Braille	434	113+
Typewriter	400	122+
Brake (See Machine or Device Combined)	188	
Airplane	244	110A+
Skids	244	108
Back pedaling	192	5
Beam	188	219.1+
Fulcrum	188	231+
Guide	188	233.3
Beams and bars for vehicles	188	219.1+
Bending brake	72	457
Bicycle	188	24.11+
Caliper type	188	24.12
Coaster	192	6R
Disc type	188	26
Block	188	250R+
Centrifugal casting machine combined	164	294
Closure safety	49	322
Clutch combined	192	12R+
Cooling	301	6CS
Deformable impact absorber	188	371+
Device with fluid	92	8+
Closed loop system	188	297+
Combined with spring	267	
Fluid spring	267	64.11+
Mechanical spring	267	217+
Open loop system	92	8
Disc type	188	71.1+
Door checks	16	82+
Electric		
Electrodynamic torque type	310	92+
Locomotive	105	61
Motor braking	318	362+
Motor with brake	310	77
Electrodynamic	310	92+
Elevator	187	73+
Expansible chamber	92	15+
Flexible panel or closure		
Automatic control	160	8
Roll type	160	291+
Fluid		
Internal resistance	188	266+
Internal resistance mechanical combined	188	271
Systems	303	
Testing fluid pressure systems	73	39
Inertial vibration damper	188	378+
Internal resistance motion retarder	188	266+
Master cylinder	60	533+
Motor control combined	192	1+
Power stop mechanisms	192	116.5+
Reel type	242	99
Fishing rod	242	84.5 R+
Ship propeller shaft	440	74
Shoes	188	250R+
Composition ctg synthetic resin or natural rubber	523	149+
Spinning or twisting element	57	113
Strand tension	242	156
System control	291	15
Testing	73	121+
Fluid pressure systems	73	39
Track sander control combined	291	14
Transmission control combined	192	4R+
Vehicle fender combined	293	2+
Wheel lock of brake type	70	228
Wheeled skate	280	11.2
Bran	426	618+
Comminution processes	241	6+
Removal from grain		
Apparatus	99	600+
Processes	426	482+
Separation	209	
Comminution combined	241	7
Brandboards	68	224
Branding		
Electrically heated stamp	219	228
Food cooking apparatus combined	99	430
Slice toaster or broiler type	99	388
Ink	106	20+
Instruments (See pyrographic)		
Iron		
Electric	219	228
External heat	101	31
Fuel burner	126	402+
Stamp	219	228
Brass		
Compositions	420	477
Knuckles	273	84R
Plating	204	44
Brasses		
Bearings with brasses	384	191+
Lubricating means	384	162+
Lubricating reservoir	384	163+
Musical instruments	84	387R+
Brassieres	450	1+
Combined with corset or girdle	450	7+
Design	D 2	24
Design combined with girdle	D 2	3
Brazing (See Soldering)		
Electric heating	219	85R
Metal working with	29	DIG. 4
Methods	228	101+
Bread		
Box	D 7	82
Compositions and processes	99	
Cutters	83	761
Design	D 1	129+
Kneading board, household	D 7	46
Loaf design	D 1	129
Testing	73	169
Toaster	99	385+

	Class	Subclass
Breaker		
Circuit (See circuit)		
Cornstalk harvester	56	52
Fiber preparation	19	35+
Ice breaker ships	114	40+
Metal	225	93+
Sheet, strip, rod, strand	225	93+
Solid disintegrator	241	
Stalk choppers	56	500+
Strips for tires	152	542
Wood chip	144	243+
Disintegrator	241	
Breaking Emulsion		
Agents	252	358
Mineral oil	252	328+
Combined processes	208	187+
Processes	252	319+
Breaking or Tearing	225	
Container opening	206	601+
Device	119	29
Harness	54	71+
Breaking Strength Testing Machines	73	788+
Breakwater	405	21+
Breast		
Artificial body member	623	7+
Board		
Plow moldboard	172	754+
Washboard	68	225
Drills	81	28+
Harness	54	20
Protectors	2	2
Pump	604	73+
Shields	128	132R+
Strap	54	58+
Design	D30	134+
Breasting Shoe Heels	12	47+
Nailing combined	12	43
Breath Alcohol	436	900*
Breathers		
Caps	220	367+
Crank case	92	78+
Internal combustion engine	123	41.86
Plugs	220	367+
Roof		
Tanks	220	85R
Tank	220	367+
Valves, safety	137	455+
Breech	42	16+
Block	89	17+
Breeching	54	5
Holdbacks	278	128+
Breeder Reactor Nuclear	376	171
Briar Pipe	131	230
Brick (See Block; Panel)		
Carrier	294	62+
Checker	165	9.1+
Cleaning	125	26
Coking ovens	202	267R
Composition	501	141+
Cutting after firing	125	23R
Double course wall	52	561+
Veneer	52	434+
Drying apparatus		
Combined with kilns	432	128
Furnace, solid fuel	110	338+
Glazing	427	376.2
By tie	52	582+
On plural face	52	589+
Mold	425	398
Mold	425	412+
Static	249	117+
Molding machines	425	130+
Molding machines	425	218+
Pavements	404	34+
Presses	425	253+
Presses	425	363+
Presses	425	383+
Presses	425	413
Cutting while green	425	289
Cutting while green	83	
Road modules	404	34+
Sanding apparatus	118	308+
Through passage	52	606+
Brickkiln	432	120+
Bricklaying Machine	52	749
Brickmaking	425	
Brickset Cookstove	126	8
Brickwork		
Imitation		
Static mold	249	16
Brides Dress	D 2	50+

	Class	Subclass
Bridge	14	
Arch	14	24+
Cable traction railways at swing bridges	104	188
Closure shiftable to bridge pit	49	33
Covering	14	74
Cranes horizontally swinging	212	205+
Horizontally	212	226
Deck	14	73
Dental	433	167+
Draw bridge	14	31+
Duplex bridge telegraph system	370	28
Electrical		
Amplifier (See amplifier, bridge)		
Bridge networks	323	365
Testing bridge	324	57R+
Eye glass bridges	351	124+
Fire hose bridges for spanning railroads	104	275+
Floating	14	27+
Floor bridges for vaults & safes	109	87
Gangways	14	69.5+
Girder	14	17
Glass furnace barrier	65	342+
Glass furnace in	65	342
Guitar bridges	84	298
Tailpiece combined	84	299
Irons, process of making	72	
Masonry or concrete	52	263
Arch	52	86+
Musical bridge details	84	307+
Piano bridges	84	209+
Piers	14	75+
Portable	14	2.4+
Railway bridge warnings or telltales	246	486
Railway safety bridges between cars	105	458+
Railway track tread bridges	238	218+
Steel or wood	14	
Stringed instruments (See music)	84	307+
Design	D17	21
Guitar	84	298
Guitar tailpiece	84	299
Piano	84	209
Structural		
Steel or wood	14	
Suspension	14	18+
Truss	14	3+
Wheatstone (See electrical)		
Bridle		
Bit		
Design	D30	136+
Brush and broom	15	168+
Harness	54	6R+
Design	D30	134+
Pipe or cable supporting bracket	248	69
Brief Case	190	900*
Design	D 3	76+
Molded integral handle	D 3	73
Zipper	D 3	71
Locks	70	67+
Bright Polishing by Etching	156	664+
X-art	156	903*
Brightening (See Burnishing; Polishing)		
Coatings	427	158
Electrolytic	204	36
Electrolytic	204	140
Radiation imaging, brightener ctg	430	933*
Brim		
Curler	223	14
Trimmer	223	16
Wirer	223	17
Brinell Testers	73	78+
Brines		
Purifying	210	
Briquette (See Block)		
Metal for furnace charging	75	256
Solid fuel	44	14+
Briquetting	425	579
Distillation combined	201	5+
Fuel	44	10R+
Nuclear	264	.5
Iron melting with	75	44R
Meat	17	32
Metal eg powder scrap		
Apparatus	425	78+
Apparatus	100	
Processes	419	61+
Processes sintering	419	
Ore beneficiating	75	3+
Bristle	15	159R+

	Class	Subclass
Brushware, design	D 4	130+
Dressing and assorting	300	18
Fastening in brooms and brushes	300	
Trimming	300	17
Broach or Broaching	409	243+
Bolt heads	10	20
Buttons	79	1
Cutters for	407	13+
Dental instruments for	433	102
Gear cutting machines	409	58+
Nuts	10	81
Broadcasting		
Container & scattering means for non-fluid material	239	650+
Insect powder dusters or sprayers	239	
Planting	111	10
Planting with drilling	111	8+
Radiant energy systems eg radio	455	
Broadcloth	139	426R
Wool	26	19+
Brocade	139	416
Broiler		
Attachments	126	14
Drip collecting	99	444+
Gas stove combined	126	41R
Griddles	99	422+
Grids	99	450
Slice	99	385+
Bromacil	544	313
Bromate	423	475
Bromides Inorganic	423	462+
Bromine	423	500+
Halocarbons	570	101+
Halogenation of organic compounds	260	694
Ore treatment with	75	102
Bromural	564	45
Bronchoscopes	128	4+
Bronze	420	470
Aluminum	420	471
Pigment	106	290
Brooch	D11	40+
Brooder	119	31+
Automatically controlled heater	236	6
Heater, design	D23	334
Heating system	237	14+
Automatic control	237	3+
Incubator combined	119	30
Broom	15	159R+
Cabinet	312	206+
Cleaning material combined	401	268+
Design	D 4	135+
Non-bristle	D32	40+
Holder	15	146
Lawn or grass	56	400.17+
Making	300	
Machines	300	12+
Sewing machines	112	6
Rack	211	65+
Support	248	110+
Whisk broom	D 4	135
Brucine	546	35
Brush or Brushing (See Associated Process, Machine or Device)	15	
Attachments	15	246+
Bottle or jar closure supporting brush	215	208+
Container closure supporting brush	206	15.2
Drip cups and shields	15	248R+
For moving surface cleaning	15	256.5+
Bristle brush with scraper	D 4	118
Broom	D 4	135
Brushes and brooms	15	159R+
Cabinet	312	206+
Cleaner for	15	38
Cleaning implements	15	104R+
Making containers, X-art	493	942*
Cleaning machines	15	21R+
Cleaning or coating	15	
Material supply combined	401	268+
Cleaning processes	134	6+
Clothes and textiles washing machine scrubber	68	
Design, bristled	D 4	130+
Clothes, hat, fingernail	D 4	
Dispenser clearing or striking	222	352
Dispensing, design	D 4	114+
Dustpan combined	15	104.8
Fly fan	416	501
For vacuum or floor polisher	D32	15+
Ginning saw picker	19	60

Entry	Class	Subclass
Hardware	16	2+
Implement	15	104R+
Light thermal and electrical application to body	128	393
Machine	15	3+
Air blast or suction combined	15	363+
Brush cleaner combined	15	48
Fruit, vegetable, meat or egg	15	3.1+
Receptacle cleaner	15	56+
Shoe	15	36+
Shoe blacking and shining	15	30+
Street cleaner	15	78+
Textile cloth finishing	26	29R+
Thread finishing	28	217+
Magnetoelectric	310	248+
Holder	310	239+
Making	300	
Individual wire bristle or tooth setting	227	79
Machines	300	2+
Massage	128	44+
Multiple-tip multiple-discharge	401	28
Non-bristle	D32	40+
Paint	15	159R+
Design	D 4	
Reservoir attached	401	268+
Pastery or basting	D 4	
Plural	D 4	119+
Powered	D 4	100
Rack	211	65+
Receptacle	206	361+
Roughening by wire brush	29	76R
Scraper can attachment	220	90
Sifter cleaner combined	209	385+
Simulative	D 4	124+
Solid material assorting	209	615
Support	248	110+
Tobacco		
Feeding combined	131	109.1
Leaf	131	324+
Leaf stemming with smoothing or cleaning	131	315+
Toilet bowl	D 4	
Toilet kit	132	85+
Material supply combined	401	118+
Shaving brush type including toilet article, eg mirror	132	80R+
Soap and brush combined	401	123+
Tooth brush type including toilet article, eg mirror	132	84R
Tooth	15	167.1+
Tooth paste supply combined	401	268+
With massage tool	15	110
Washboard surface	68	227
Wick trimmers	431	120
Work supports	15	268
Shoe	15	36+
X-art collection	425	805*
Bubble		
Amusement devices	446	15+
Bubble domain caculating	364	714
Magnetic bubbles	365	1+
Solid separation processes	209	164+
Specific gravity test	73	439
Tower (See dephlegmators)		
Porous mass	261	94+
Wet baffle	261	108+
Bucket (See Pail, Tub)	220	
Cloth leather and rubber	150	48+
Conveyer (See type conveyer)		
Design	D32	53
Coal scuttle	D23	410
Minnow	D22	136
Excavating	37	
Clamshell	37	183R+
Orange peel	37	182
Fire extinguisher	169	34
Heater for dinner bucket	126	266+
Hoist line	414	564+
Vertically swinging shovel	414	565
Vertically swinging support	414	680+
Minnow	43	56+
Paperboard	229	910+
Rotary kinetic fluid motor	415	
Sap	47	50+
Staved wooden	217	72+
Tank type meter with rotary bucket	73	217+
Turpentine and rubber collecting	47	11
Type conveyer	198	701+
Dispensing	222	369
Dispensing endless	222	371
Wind motor collapsible	416	142+
Wooden	217	
Bucking Bar	72	457
Buckle	24	163R+
Belt attached buckled, clasp, slide	D 2	627+
Design	D 2	3
Harness	D30	139
Design apparel	D11	200+
Making	29	3
Seat belt	D11	200+
Shoe	D11	200+
Turnbuckle	403	43+
Buckstay Construction	52	86+
Bucky Grid	378	154
Budding and Grafting	47	7
Buddles	209	458+
Buffers (See Springs)		
Amplifier	330	
Radio transmitter with	455	91+
Bedstead	5	309
Bridge	14	48
Loom picker and picker stick	139	167+
Manicuring	132	76.4+
Compound tool	132	75.6
Polishing wheels	51	358+
Railway		
Draft gear	213	220+
Dump car door	105	285
Freight car interior end	105	374
Locomotive buffer beams	105	173
Store service	186	24+
Tire combined	152	158
Track	104	249
Trunk and baggage	190	37
Typewriter	400	686+
Warship	114	13
Water closet combined	4	248
Bug Type Telegraph Key	178	79+
Bugle	84	387R
Buhrstone or Burrstone	241	296+
Mills	241	244+
Builder Mechanism for Bobbins and Cops	242	26.1+
Builders		
Detergents	252	89.1+
Elevators	182	141+
Traveling scaffold	182	12+
Hardware	16	
Design	D 8	
Buildings	52	
Amusement	272	2+
Animal barns and sheds	119	16
Apparatus for moving material to a position for erection or repair	414	10+
Assembly or disassembly	52	127.1+
Auditorium or stadium feature	52	6+
Bay window	52	201
Burial vault	52	128+
Lifting or handling	52	124.1+
Construction elements	D25	
Convertible	52	64+
Cupola	52	200+
Drip deflector	52	97
Foundation	52	292+
Heated and cooled	165	48.1
Ventilated	165	16
Jail type structure	52	106
Lifting or handling of primary component	52	122.1+
Position adjusting	52	126.1+
Molding in situ	264	31+
Multi-room or multi-level	52	234+
Non-rectangular	52	236.1+
Panel or facer	52	474+
Prefabricated	52	
Subenclosure	52	79.1+
Railroad car roof construction	52	45+
Refrigerated	62	259.1+
Shipbuilding	114	65R+
Static mold	249	13+
Stepped or stair	52	182+
Synthetic resin component	52	309.1+
Tents and canopies	135	
Toy	446	476+
Construction	446	108+
Design	D21	114
Ventilation	98	29+
With terranean relationship	52	169.1+
Work holder positions in installed location	269	904*
Bulb		
Discharge tube	220	2.1 R+
Electronic device envelope	220	2.1 R+
Making of glass	65	
Expansible chamber device	92	92
Fluorescent lamp	313	484+
Gas or vapor lamp	220	2.1 R+
Incandescent lamp	220	2.1 R+
Reflector	313	113+
Radio tube	220	2.1 R+
Bulkhead		
Compartment and	114	78
Door and	114	116+
Bulldozer		
Blade	172	701.1+
With vehicle mount	172	811+
Cab	180	89.12
Design	D15	23+
Bullet	D22	116
Bullet (See Ammunition; Projectiles)	102	501+
Making	29	1.22+
Resistant		
Tanks	220	415+
Setting	86	43
Speed electrically measured	324	160
Timing electrically	324	178+
Timing recorder	346	38
Bulletin Board	D19	52+
Frame	D19	54
Bulletproof	428	911*
Apparel, guards and protectors	2	2.5
Plural layers	428	
Walls and panels	109	78+
Bulls Eye Liquid Level Gauge	73	331
Bumper		
Automobile	293	102+
Guards	293	142+
Design	D12	167+
Fluid	293	108
Horizontal bar	293	120+
Lights	362	82
Closure checks	16	82+
Railway car end	213	220+
Railway car stops	104	254+
Release of closure latch	49	364
Vehicle, design	D12	163+
Buna T M (See also Synthetic Resin or Natural Rubber)	526	339
Bunching (See Bundling)	493	386
Vegetable end cutting combined	99	635
Bundle Formers		
Hay	56	401+
Bundling	100	
Compacted trash	428	2
Compression	53	523+
Plug tobacco	131	112+
Raking and	56	341+
Wood splitting and	144	192
Bung		
Extractors	81	3.7+
Plug type closure for dispenser	222	544+
Wooden barrel	217	98+
Bunk		
Bed	5	9R+
Log bunks for logging truck	105	160
Buns	426	556+
Bunsen Burner	431	354
Buoy	441	1+
Design	D10	107
Signal	116	107+
Submarine escape	114	323+
Submarine marker	114	326+
Buoyancy Motors	60	495+
Tide or wave	60	497+
Buoyant Devices		
Convertible, inflatable	441	125+
Life preservers	441	80+
Garment type	441	88+
Seats	441	126+
Bur		
Dental	433	165
Pipe reamer	408	227
Removers		
Horseshoe making	59	59
Textile fiber	19	84
Bureau	312	
Baggage convertible	190	3+
Plural		
Burette, Chemical Apparatus	422	100
Graduated container	222	158
Measuring vessel	73	426+
Burgee	116	173+
Burglar		
Alarm		

	Class	Subclass
Combined with other function	116	6+
Electric	340	541+
Electric approach operated	340	541+
Electric switch for	200	
Indicator combined	116	5
Mechanical	116	75+
Trap	43	59
Burglarproof Structures	109	
Burial		
Body preparation	27	21+
Coffins	27	2+
Casing portable	27	35
Garment	2	64
Shoes	36	8.2
Vault	52	128+
Vault design	D99	1+
Burner (See Furnace; Heat)		
Bunsen	431	351
Cap	431	129+
Cap	431	144+
Carbon black apparatus	422	150+
Control	431	18+
Design	D26	114
Disinfecting material preparation	43	127+
Dispensers having	222	113
Extinguisher	431	129+
Extinguisher	431	144+
Automatic control	431	33+
Flame protector with	362	376
Flashback prevention	431	22
Flashback prevention	431	346
Fluid distributor for	239	
Fluid fuel	431	
Distributor for	239	
Stoves	126	
Insect destroying	43	144
Light distributor with	362	93
Lime light	431	347
Distributor for	239	
Making	29	157R+
Nozzle	239	
Operation process	431	2+
Pilot with	431	278+
Pot type	431	331+
Pyrotechnic single use	102	335+
Rotary	239	214+
Solid fuel	110	
Stoves	126	
Tobacco smoking devices	131	330+
Toilet closets having	4	111.4+
Torch	239	398+
Torch	141	345
Capillary mass	431	327
Welding	239	398+
Welding	228	
Wick	431	298+
Burning		
Electric heater for	219	227+
Insulation from wire	81	9.4
Lime, cement, etc. kilns	432	
Burning in	29	89.5
Burnishing (See Polishing)		
Electrolytic coating combined	204	36
Implement		
Mutilating eraser combined	7	124+
Shoe	12	104+
Machines	29	90R+
Paper tube	493	467
Photograph	354	351+
Shoe	12	70+
Wearing in	29	89.5
Burr (See Bur)		
Bus (See Land Vehicles)		
Baggage rack ladders	182	127
Bar (See conductor, electric)	174	
Attachment means for switches	337	191
Distribution or control panels	361	355
Duct	174	50
Electrical housings for	361	341
Electronic service distribution	361	361
Printed circuit board	361	407
Design	D12	84
Destination signs	40	446+
Motor changed	40	470+
Bushing	16	2+
Abrading processes	51	290+
Applying and removing		
Couplings to conduits	29	237
Grommets	227	55+
Processes	29	428+
Pulling or pushing type	29	244+
Resilient bushings	29	235+
Tube and coextensive core	29	234
Bearing type	384	276+
Bung wooden barrel	217	113
Glass dispensing orifice	65	1+
Glass filament intrusion	65	325+
Insulating	174	152R+
Arcing or stress distributing	174	142+
Fluid or vacuum containing	174	31R
For antenna	174	151+
In line insulator	174	167
Railway wheel	295	35
Wooden barrel	285	130
Business		
Computer controlled, monitored	364	401+
Forms		
Carbon copy sets	282	28R
Carbonless copy sets	503	200+
Printed forms	283	
Teaching	434	107+
Bust Support (See Brassiere)	450	1+
Bustle	2	210
Design	D 2	699
Butadiene (See Diolefin Hydrocarbon)		
Liquid polymer	585	507+
Solid polymer (See synthetic resin or natural rubber)		
Synthesis of	585	601+
By cracking	585	615
Butchering	17	
Fowl	17	11+
Hog scalders	17	15
Hog scrapers	17	16+
Marine animals	17	53+
Meat tenderers	17	25+
Processes	17	45+
Sausage stuffers	17	35+
Supports & shackles	17	44+
Butter		
Apparatus	99	452+
Compositions	426	
Cutters	83	
Design	D 1	
Dish		
Cooled	62	457
Design, general	D 7	84
Design, refrigerating	D15	89
Fruit	426	616
Mold	249	
Cutter combined	425	289+
Packaging	426	392+
Peanut	426	633
Preservation	426	317
Processes manipulative	426	
Worker	99	452+
Butterfly		
Resonator	331	95
Tuner	334	67
Valve, fluid handling, unbalanced	137	484
Valve, rotary	251	305
Button	24	572+
Abrading cleaning shields	51	265
Apparel, design	D11	222+
Art collection	425	801*
Attaching		
Processes	2	265
Sewing machine	112	104+
Bell buttons mechanical	116	172
Design	D11	222+
Lapel	D11	95+
Eyelet and rivet setting	227	55+
Fabric	24	92
Fastener making	29	4
Fasteners	24	90R+
Feeder	227	31
Sewing machine	112	113
Fluorescent and phosphorescent	250	462.1
French cuff button	24	102FC
Hinged leaf	24	97+
Hole (See buttonhole)	24	659+
Cutters	30	118+
Marking guides	33	575+
Hook	D 2	643
Loop	24	660
Making	79	
Making, metal	29	4
Metallic	79	3+
Pearl & composition	79	6+
Setting machine	227	55
Shank buttons	79	2
Strap or cable button	24	114.5
Switch	200	329+
Telegraph key	178	110
Whip	231	6
Work supports	79	18
Buttoner	24	40
Design, shoe	D 2	643
Machine for shoes	12	58.3
Buttonhole	24	659+
Cutting	83	905*
Hand tool	30	118
Marker	33	575+
Sewing machines	112	65+
Processes and seams	112	264.1
Shoe closures	36	52
Stitch machine	112	157
Workholder in making	269	
Buttress, Vertical Wall	405	112
Butyl Alcohol	568	840
Fermentative preparation	435	160+
Preparation by olefine hydration	568	895+
Preparation from alkyl halides	568	891+
Preparation from alkyl sulfates	568	888
Butyn T M	560	49
Buzzers		
Electric	340	392+
Toys		
Noisemaking tops	446	213+
Sounding	446	397+
Spinning and whirling	446	236+
BX TM Cable	174	109
Fitting	174	50+
Staples	411	457+
Driving implement	227	140+
Insulated	174	159+
By Product Recovery (See Source Material or Material Recovered)		
Byte	370	83
Assembly and formatting	370	99
C		
C-clamp	D 8	73
C-vinyl Aromatic Hydrocarbon Synthesis	585	435
Cab		
Closure operators		
Door and window	49	72
Window	49	324+
Locomotive		
Dust guards	98	28
Ventilation	98	3
Window condensation preventers	98	91
Motor vehicle body	180	89.1+
Railway	105	456
Signal or train control	246	167R+
Block signal systems	246	20+
Drawbridge protection	246	119
Grade crossing track protection	246	117
Automatic	246	115
Switch stands	246	394
Train dispatching central signal	246	4
Train dispatching train order	246	6
Squirts	239	174
Cabin and Stateroom		
Pressurized	98	1.5
Aircraft structure combined	244	59
Heated and cooled	165	15
Ship	114	71
Furniture arrangement	114	189
Toy construction logs	446	106
Cabinet	312	
Bath tubs	4	538+
Bed combined	5	2B
Card or sheet magazine	312	50
Card or sheet retainer	312	183+
Cooking stove	126	37R
Dark cabinet	354	307+
Dispensing (See dispensing)		
Escutcheons	70	452
File	312	183+
Instrument sterilizing	422	
Ironing table support	108	33+
Joint of sectional cabinets	312	111
Light combined	362	133
Light etc application to body	128	371+
Locks for	70	78+
Metallic receptacle spaced wall	220	445+
Music	84	DIG. 17
Office machines	D18	
Outside players for music instruments	84	108
Phonograph type	312	8
Console type	D14	18+

Entry	Class	Subclass
Record holders	312	9$^+$
Photography fluid treating	354	297$^+$
Radio type	312	7.1
Design	D14	18$^+$
Receiver	455	347$^+$
Transceiver	455	90
Transmitter	455	128
With antenna	343	702
Safes	109	
Combined	109	23$^+$
Convertible	109	22
Incidentally movable	109	45$^+$
Plural compartment	109	53$^+$
Shields and protectors	109	49.5
Supports and mountings	109	50$^+$
Wall and panel structure	109	58$^+$
Show case	312	114$^+$
Shower bath	4	596$^+$
Design	D23	283
Sign exhibiting reel	40	515
Sound or video recording etc instrument	369	75.1$^+$
Slot for disc insertion	369	77.1$^+$
Structure	312	
Television	358	254
Toy	446	482
Type case accessory	276	44
Wall or ceiling mountable	312	242
Cable and Cable Making	57	
Advancing material of indeterminate length	226	
Block and tackle	254	389$^+$
Bridge suspension	14	22
Burglar alarm cable controlled	116	94
Button		
Strap chain cable attachment	24	114.5
Supports pipe or cable	248	49$^+$
Supports suspended	248	317$^+$
Car (See traction below)		
Clamp		
Bridge suspension	14	22
Cord and rope	24	115R$^+$
Pipe or cable	248	49$^+$
Clip (See cable clamp)	104	200$^+$
Coaxial	333	239$^+$
Composite plastic cable molding	425	500$^+$
Compound die expressed plastic	425	113$^+$
Concentric conductors	174	102R$^+$
Connector	403	
Container for	206	389$^+$
Conveyor		
Hoist linear traversing	212	76$^+$
Ship coaling type	212	72
Material or article handling	414	
Power driven	198	
Endless	198	804$^+$
Endless single cable	198	643
Store service systems	186	14$^+$
Cutting	30	
Fishing line	43	43.12
Ships anchor cable	114	221R
Well cable	166	54.5$^+$
Well torpedo cable	89	1.14
Drum and cable mechanism	74	506
Antifriction	254	901*
Slip clutch	254	903*
Specific material	254	902*
Wave motion responsive actuator	254	900*
Electric	174	
Loaded	178	45$^+$
Telegraphy	178	63R
Thermal responsive circuit breaker incorporated	337	415
Electric conductor making and or joining	156	47$^+$
Electrical assembly	29	755
Emergency elevator	187	71
Fairing	114	243
Gear drum and cable mechanism	74	505
Grip (See clamp above)		
Cable driving stops railway	104	239
Car control vehicle fender	293	3
Motor placement railway rolling stock	105	134
Guides (See guides, cable line)		
Hauling or strand placing	254	134.3 R$^+$
High noise for burglar alarm	174	68R$^+$
Hoisting		
Apparatus for hauling or hoisting load	254	264
Crane overhead travelling cable driven	212	205$^+$
Crane rotary drive	212	246
Elevator car brakes and catches	187	81$^+$
Grapples pivoted jaws separate cable actuator	294	111
Hook tackle	294	82.11$^+$
Inclined elevator brake	187	13$^+$
Ladders extension	182	207$^+$
Ladders track type	182	63$^+$
Lever actuator pushing and pulling	254	127$^+$
Linear traverse	212	76$^+$
Material or article handling		
Combined carriers endless	414	564$^+$
Elevator	414	592$^+$
Traversing type	414	560$^+$
Pushing and pulling implements or apparatus	254	264
Ship coaling type linear traverse	212	72
Testing physical	73	158
Trucks	254	4R
Vehicle body lifter	254	47$^+$
Impregnating		
Apparatus	118	
Indefinite length electrical conductor	156	47$^+$
Metal casting, continuous	164	462$^+$
Joining wire	140	111$^+$
Joints and couplings (See joints, pipes, shafts, rods, plates)		
Laying apparatus	405	154$^+$
Protector from abrasion	174	136
Rail	104	112$^+$
Railway	104	240
Switch cable and pully connection	246	332
Releaser	187	72
Stop		
Drum automatic control slack cable	254	272
Ship	114	199$^+$
Traction railway	104	239
Submarine		
Loaded	178	45
Support for pipe or cable	248	49$^+$
Bracket	248	65$^+$
Suspended	248	58$^+$
Table equalizers	108	87
Tension device	254	199$^+$
Tension equalizer	187	1A
Tension reliever	114	215
Tightener belt	474	101$^+$
Towing		
Land vehicle train	280	480
Marine propulsion	440	34$^+$
Ship	114	242$^+$
Traction		
Closure operator	49	347
Hauling apparatus	254	264$^+$
Lever single throw	254	127$^+$
Linear hoists	212	76$^+$
Linear hoists ship coaling type	212	72
Overhead cranes	212	214$^+$
Railway rails	104	112$^+$
Railway rails combined with rigid	104	87
Railway rolling stock suspended	105	151
Railway traction	104	173.1$^+$
Railway turntable actuator	104	39
Rotary cranes	212	246
Store service systems	186	14$^+$
Tube or rod splicing	425	802*
Twisting wire	140	149
Underground conduits for pipes or cables	138	105
Winding & reeling	242	
Wireworking article making or forming	140	71R$^+$
Cable Tool Rig	173	81$^+$
Cacao	426	631$^+$
Cachets	604	403$^+$
Cacodyl Compounds	556	70$^+$
Cadmium	75	
Alloys	420	525
Batteries	429	
Compounds	23	
Electrolytic coating with	204	50.1
Electrolytic synthesis	204	114$^+$
Fluorescent or phosphorescent compositions containing	252	301.6 R
Misc. chemical manufacture	156	DIG. 92
Organic compound containing	556	118$^+$
Pigments fillers or aggregates	106	301
Pyrometallurgy	75	71
Caesium	75	
Pyrometallurgy	75	84
Cafeteria		
Coffee service unit	D34	14
Tray dispenser unit	D34	14
Caffeine	544	274
Extraction from coffee	426	427$^+$
Cage		
Animal mouth guard	119	133
Bank protection	109	10$^+$
Bearing		
Ball	384	523$^+$
Making	29	148.4 C
Radial	384	512$^+$
Thrust	384	623
Bird and animal confining	119	17$^+$
Design	D30	114$^+$
Fish and animal traps	43	58$^+$
Design	D22	121
Material handling		
Hand type	294	26.5
Hoist line type	294	68.1$^+$
Mills	241	188R
Plant supports	47	45
Queen bee	6	9
Shower bath spray	4	601$^+$
Solid fuel grate	110	294
Water cooled	122	373
Type antenna	343	896
Washing apparatus having	68	
Caisson		
Box or open	405	11$^+$
Diving bell type	405	185$^+$
Ordnance	89	40.7
Pressurized	405	8$^+$
Shaft lining	405	133
Ship raising	114	49
Ship raising salvage	114	53
Ship repair	405	12
Shoring, temporary	405	272
Cake		
Boxes (See paper receptacles)	229	
Compositions	426	552$^+$
Cover	312	283$^+$
Cutter or server	D 7	142
Design	D 1	129$^+$
Dish or cover	D 7	83
Enfolding with enfolding with cloth before	100	101
Inedible	D11	184
Pans	249	DIG. 1
Stripping cloth from cake after pressing	100	298
Trimmers	100	94$^+$
Calcimines (See Kalsomines)		
Calcining		
Furnaces	432	
Portland cement processes	106	100$^+$
Calcium (See Alkali, Earth Metal)	75	
Carbide	423	
Acetylene generator carbide feed.	48	38$^+$
Cartridges	48	59$^+$
Compounds inorganic	423	
Calculator (See Computer)	235	61R$^+$
Design	D18	6$^+$
Instruments simulating	33	1SB
Specific gravity test	73	443
Vehicle controlling	364	424$^+$
Calendar	40	107$^+$
Clocks	368	28$^+$
Coin controlled	40	107$^+$
Design	D19	20$^+$
Desk article combined	40	358
Holder design	D19	20$^+$
Pads	40	121
Printed matter type	283	2$^+$
Watches	368	28
Calender		
Roll	29	110$^+$
Paper manufacture	100	161$^+$
Textile	38	63$^+$
Calendering		
Cloth finishing	38	52
Laundry	38	44$^+$
Paper	100	161$^+$
Plastics	425	363$^+$
Calesthenic Wands	272	93$^+$
Calibrating		
Battery calibration	136	290*
Instrument	73	1R$^+$
Radar systems	342	165
Caliper	33	143R$^+$

Class Subclass

- Assorter combined 209 602+
- Brake 188
  - Bicycle 188 24.11+
  - Disc type 188 71.1+
- Design D10 73
- Gauge 33 179.5 A+

Calk
- Forming on horseshoes 59 65
  - Combined machines 59 39
- Horseshoe calk making 59 66+
- Horseshoe type 168 29+
- Sharpener 51 172
  - Tool 168 46
- Shoe device 36 67R
- Tire antiskid device 152 229+

Calking
- Compositions 106
- Dispenser 222
  - Gun 222 326+
- Edge sealing laminated glass 156 107
- Implements 81 8.1
  - Metal packing 72 462+
  - Pointers for masonry 15 235.3
  - Ship calking 114 224
- Metal working 72 462+
- Methods 114 86
- Packing fibrous material 81 8.1
- Puttying devices 425 87
- Puttying devices 425 458
- Seaming ships combined 114 86

Calling Card Receiver D 7 37+

Calliope 84 330

Callus Culture 435 240.48+

Calomel 423 491
- Medicines containing 424 146

Calorimeter 374 31+

Calutron
- Mass spectrometer 250 281+
- Structure 313 359.1

Cam (See Apparatus Using) 74 567+
- Actuated or operated 74
- In bolt to nut lock 411 272

Camber
- Correction
  - Bending for 72 704
  - Bending three point jacks 72 386+
  - Forging for 72 704
  - Vehicle running gear spring 280 688+
  - Vehicle running gear stub axle 280 661
- Test wheel gauge 33 203

Cambric 139 426R

Camel
- Ship raising 114 49
  - Salvage 114 53

Cameo Molds 249 104

Camera 354 288
- Animated cartoon 352 87
- Carrier or holder 224 908*
  - Hand 294 139
- Cases 150 52J
- Cinematographic D16 3+
- Compound lens system with 350 502
- Copy making, eg mechanical negative 430 951*
- Copying 355 18+
- Design D16 1+
- Electronic still 358 909
  - Scanned semiconductor matrix 358 213.11+
- Exposure control, automatic 354 410+
- Film winding spools 242 71
  - Motion picture 242 210
- Flashlight 354 126+
- Focusing, automatic 354 400+
- Lens mount 354 286
  - Angularly adjustable camera front or lens mount 354 189
- Lucida 350 121
- Microscope 350 511
- Motion picture 352
  - Film winding 242 179+
- Multiple exposure on single plate 354 120+
- Obscura 350 122
- Panoramic 354 94+
- Periscope 350 541
- Photographing internal body organs ultrasonically 128 660+
- Reflex 354 152
- Reproducing 355 18+
- Schmitt
  - Lens and mirror 350 443
  - Motion picture 352 69
  - Television 358 209+
- Shutter 354 226+
  - Automatic control 354 456+
  - Between lens 354 233
  - Mechanical test 73 5
  - Optical test 73 5
  - Timer 354 267.1+
- Special purpose D16 2
- Stereoscopic 354 112+
- Submarine motion picture 352 132
- Supports 354 293+
  - Tripods 248 163.1+
- Telescope 350 557
- Television D14 78
- Time lapse 352 84
- Tripod 354 293+
- Twin lens reflect 354 151
- View finders 354 219+
  - Geometrical 33 290+
  - Motion picture 354 219+
  - Perpendicular to objective lens D16 9

Camisole D 2 23

Camouflage
- Canopies 135 87+
- Coat hunters 2 94
- Coating composition 106
- Coating non-uniform 427 256+
- Illusion 273
- Including open mesh 428 255+
- Ordnance 89 36.1
- Stock material 428 919*
- Tents 135 87+
- Warship 114 15

Camp and Camping
- Beds 5 112+
- Bodies land vehicles 296 156+
- Chairs folding 297 16+
- Ground mat 5 417+
- Kit combined or convertible to table 190 12A
- Lunch kits 206 541+
- Sleeping bag 2 69.5
  - With bed structure 5 413
- Stools folding 297 166
- Stoves cooking 126 25R+
- Stoves heating 126 59
- Tables convertible baggage 190 12R
- Tent 135 87+
  - Heater 126 59
- Trailer design D12 101+
- Vehicle design D12 100

Camphene (See Alicyclic Hydrocarbon) 585 350
- Terpene isomerization 585 355

Camphor 568 339+

Camphoric Acid 562 504

Camping (See Camp)

Camshafts 74 567
- Drive means internal combustion engine 123 90.31
- Overhead location 123 90.27

Can (See Container)
- Beer, pop, or food 206 139
- Carrier by hand 220
- Crushers 100 902*
  - Presses in general 100
- Cutting to scrap 83 923
- Design D 9
- Filling and closing 53 266R+
- Insulating jacket 220 903
- Labeling machines 156
- Leakage tester 73 40+
  - Discarding type 73 45+
- Making
  - Body straightening 72 367+
  - Fusion-bonding apparatus 228
  - Metal cap preparing machine 413 56+
  - Sheet metal container making 413
- Metal and miscellaneous 220
- Milk 220
- Milking machine 119 14.46
- Nuclear fuel containing 376 409
- Openers 30 400+
  - Closure combined 220 260
  - Cutter type combined 7 152+
  - Design D 8 33+
  - Electric 30 401
  - Key holder combined D 3 64
  - Other tool combined 7 151
  - Receptacle puncturing and closing 220 278
  - With dispensing features 222 81+
- Packing in portable receptacles 53 531+
- Processors 99 359+
- Resilient wall 222 206+
- Testers leak detector 73 52
- Vacuumizing & sealing methods 53 403+
- Valves for conveying
  - Bin charge or discharge 414 288+
  - Chute retarders 193 32
  - Chute switches 193 31R
  - Cooker inlet or outlet 99 366
  - Horizontal axis rotary conveyor 198 670
  - Horizontal axis spiral 198 778
  - Rotary conveyor 198 803.1
    - Thrower 198 642
- Washers and cleaners
  - Brushing machine 15 70+
  - Liquid contact 134
- With valved outlet 222

Canal 405 84+
- Amusement park 104 73
- Boats
  - Form 114 60
  - Structure 114 70
- Locks 405 85+

Canceler
- Fare box ticket puncher combined 232 8
- Stamp
  - Barrel bung 217 114
  - Hand 101 371
  - Machine 101 233+

Cancer Treatment
- 5-flurouracil 544 313

Candelabra 431 295
- Electric 362 410+
  - Design D26 24

Candle 431 288
- Candlestick D26 9+
- Compositions for making 44 7.5
- Design D26 6
- Electric lamp 362 157+
- Electric socket simulating 200 51.1
- Extinguishers 431 144+
- Forming 425 803*
- Fumigants 424 40
- Holders 431 289+
  - Christmas tree 362 249+
  - Design D26 9
- Imitation 362 810*
- Lanterns using 362 161
- Lighters (See igniter and ignition)
- Lighting imitation of 362 810*
- Making
  - Casting or molding apparatus 425 117+
  - Casting processes 264 271.1+
  - Dipping processes 427 442+
- Physical and ornamental structure 431 126
- Shade or bowl supports for 362 447
- Simulated 431 125
- Snuffers 431 144+
  - Design D29 2
  - Miners 7 104+

Candling 356 52+

Candling, instrument D10 48

Candy
- Coating apparatus 118 13+
- Compositions 426
- Cutting machine 83
- Design D 1 127+
- Kneading-pulling equipment 366 70
- Making, apparatus 99 450.1
- Making, processes 426
  - Cotton candy 425 9
- Preservation 426 321+
  - Stick, container 493 943*

Cane 135 65+
- Design D 3 5+
- Detonating toy 446 402+
- Fabrics 139 424
- Gun combined 42 52
- Light combined 362 102
- Locks for cane racks 70 59
- Racks 211 62+
- Simulated umbrella case 135 18
- Stand convertible 248 155
- Stand or stool converting to 248 155+
- Stripper 130 31R
- Therapeutic appliers combined 128 394
- Umbrellas combined 135 17
- Whip convertible 231 3

Canister
- Design D 7 79+
- Gas separation 55
- Projectile 102 507+

	Class	Subclass
Canning		
Cooking and subsequent	99	356
Cooking filled receptacles	99	359+
Filling and closing	53	266R+
Food preserving apparatus	99	467+
Food preserving processes	426	392+
Hermetic	426	392+
Cannon	89	
Toy	42	55
Cannula	604	239
Attaching means	604	240+
Medicator combined	604	187+
Structure	604	239
Canoe	114	347+
Chair	114	363
Design	D12	302
Paddles	440	101
Sails	114	103+
Canopy	135	87+
Bed	5	414+
Camp	5	113
Hammock and canopy with common support	5	128
Hammocks	5	121
Body or belt attached	224	186
Building light type	350	258+
Cabinet type	312	3+
Coffin	27	9
Design	D25	56
Flexible panel combined	160	19+
Frames	135	101+
Light shade	362	351+
Light support	362	404+
Padlock shields	70	56
Reflector	362	341+
Rigid type	52	74+
Support type	248	345
Tent-type design	D21	253+
Cant Hook	294	17
Canteen		
Design	D 3	30.1+
Flow controller or closure	222	544+
Jacket or spaced wall	220	415+
Jacketed dispensing type	222	131
Kit with	206	547
Separable cup or funnel	141	379+
Cantilever		
Bridge	14	7+
Building	52	73+
Design	D25	61
Spring	267	41
Canting Element in Bolt to Nut Lock	411	274+
Canvas		
Awning stripe	139	417
Fabric	139	426R
Shoes rubber sole	36	9R
Caoutchouc (See Rubber)		
Cap		
Ammunition	102	204
Making	86	10+
Apparel	2	195+
Bathing	D 2	68
Bathing design	D 2	510+
Design (See headwear)	D 2	509+
Protective against blows	2	410+
With eye shield	2	10
With light support	362	106
Beading and crimping	72	102+
Blasting	102	275.12
Bottle seal	425	809*
Brush and broom type	15	175
Burner		
Extinguishing	431	146
Chimney		
Coping	52	244
Cowls	98	61+
With valve	98	59
Closure		
Auto radiator or gas tank	D12	197+
Bottle	215	200+
Depressible oil cup cap	184	89
Design	D 9	435+
Flexible bag	383	80
Inflation stem air & dust cap	138	89.1+
Inflation stem air cap	138	89.3
Inflation stem dust cap	138	89.4
Inflation stem dust cap and tool combined	152	431
Lining apparatus	156	
Lining combined with metal working	413	
Lining process	156	262
Making from sheet metal	413	8+
Apparatus	413	56+
Removers	81	3.7+
Screw type	220	288
Slip type	220	352
Soldering to can	228	
Coating implement with material supply combined	401	
Column	52	301
Detonating toys	446	398+
Mechanical projector combined	124	2
Pistol simulating	42	54+
Electric fuse terminal	337	248
Fastenings for saw teeth	83	840+
Fountain pen bifurcate nib-type	401	243+
Gas or radiator	D12	197+
Insulator	174	188+
Lock	70	158+
Making		
Ammunition	86	10
Die shaping sheet metal	72	343+
Spirally grooving metal	72	105+
Mast	212	266
Nail screw and nut		
Nut caps	411	429+
Pile driving	173	128+
Pile protector	405	255
Pipe vented	98	122
Radiator		
Ornament	428	31
Ornament bird or insect	40	411+
Removing	29	245
With thermometer	209	684
Spinning machines	57	74
Strand twisting and laying	57	127
Surfacing	52	465+
Tower	52	649
Typewriter key	400	490+
Ventilating	98	122
Wheel hub	301	108R
Whip	231	6
Capacitance Interelectrode Testing Electronic Tube	324	409+
Cathode ray tube	324	404
Capacitor (See Condenser; Electret)		
Air dielectric	361	326+
Design	D13	21
Neutralizing	361	212+
Capacity Measuring Bridge	324	60R+
Cape Body Garment	2	88+
Design	D 2	179+
Combined	D 2	180+
Combined with dress	D 2	89+
Combined with evening dress	D 2	53+
Combined with suit	D 2	87+
Capillary		
Active agents (See wetting)	252	351
Heat exchange pipe	165	104.26
Tube, refrigeration using	62	511
Capotasto	84	318
Capping		
Bottles	53	287+
Nails and screws	10	156
Screw closure applying		
Soldering iron having guide	228	25+
Capric Acid	260	413+
Caprylic Acid	260	413+
Capsaicin, in Drug	514	627
Capstans	254	266+
Capsule (See Container)		
Composite stock material with liquid	428	321.5
Filling and closing	53	266R+
Making by dipping	425	269+
Making, X-art	425	804*
Medicine vehicle	424	451+
Microcapsule (See microcapsule)		
Nonmetallic collapsible wall	222	107
Sheet metal making	72	347+
Stock with liquid	428	321.5
Car (See Vehicle)		
Airship	244	30
Bodies and tops	296	
Control		
Electric railway external	104	295+
Elevator	187	28+
Elevator brakes and catches	187	73+
Elevator fluid governors	187	68+
Railway switches and signals	246	
Vehicle fenders combined	293	2+
Derailer railway	246	163
Car attached	104	261
Elevator or hoist	414	227+
Dumb waiter type	187	5
Superposed	187	16
Fire escape reciprocating moving rope	182	12+
Fire escape reciprocating stationary rope	182	189+
Fire escape vertically movable	182	40
Hand	105	162
Heated and cooled	165	42
Automatic control	165	25
Land vehicle	280	
Loading or unloading	414	373+
Railway unloading using tractor-trailer	414	333
Self	414	467+
Train of unloading vehicles	414	339
Motor	180	
Propulsion (See electric)		
Elevator	187	17+
Railway	105	
Railway cable crossing conduit	104	187
Railway cable curves	104	190
Railway systems	104	287+
Road vehicle	180	
Store service	186	
Pusher		
Implement	254	35+
Railway hand crank	105	90.1
Railway	105	
Amusement	104	53+
Cable mule	104	176
Elevator	104	128
Forging dies	72	343+
Ladder	182	93+
Pilot or signal	246	166
Stop	104	249+
Toy simulation	446	431+
Yard	104	26.1+
Refrigeration	62	239+
Air controller combined	62	407+
Alternately usable energy source	62	236
Separable cooling means	62	237
Variable capacity ice bunker	62	327+
Replacer	104	262+
Retarder railway	188	62+
Stock	119	7+
Store service cable	186	18
Vehicle unloading	414	373+
Dumping tilting track section	414	354+
Caramel	127	29
Flavoring compositions	426	658+
Making	127	34
Carbamic Acid	562	555
Carbamic Acid Esters	560	24+
Acyclic	560	132+
Alicyclic	560	115
Nitrogen ctg	560	157+
Sulfur ctg	560	148
Carbanilide	564	55
Carbazide	564	34
Carbazole	548	440+
Azo compounds	534	790
Carbazones & Semicarbazones	564	36
Carbenicillin	540	338
Carbide (See Calcium, Carbide)	423	
Acetylene from	48	
Actinide	423	249+
Alloys	75	236+
Cartridges	48	59+
Ceramic compositions containing	501	87+
Metal impregnated	428	539.5
Silicon	423	345
Tungsten	423	
Uranium	423	256
Carbitol T M	568	613+
Carbocyanine Dyes	546	152+
Photographic emulsions containing	430	541
Carbocyclic	260	350R+
Carbodiimides	564	252
Polymers from (See synthetic resin or natural rubber)		
Structure in cellular polymer	521	901*
Carbohydrate, Sugar and Starch	127	
Coating or plastic compositions	106	162+
Derivatives	536	1.1+
Nitrogen containing	536	18.7+
Diastolic mashing	435	93+
Fermentative purification	435	262+
Polymers containing (See synthetic		

	Class	Subclass
resin or natural rubber)		
Wave energy preparation	204	157.68
Carbolic Acid (See Phenols)		
Carbon	423	445$^+$
Absorbents	502	416$^+$
Carbonizing processes	502	416$^+$
Regeneration	502	20$^+$
Activated	502	180$^+$
Agglomerating	23	314
Apparatus for making	422	150$^+$
Compositions containing synthetic resin or natural rubber (See classes 523, 524)		
Compounds	260	
Preparation by wave energy	204	157.6
Dioxide	423	415R
Hydrogenation of	518	700$^+$
Refrigerating with	62	384$^+$
Refrigerating with, automatic	62	165$^+$
Solidifiers	62	8$^+$
Synthesis gas containing	252	373
Disulphide	423	443$^+$
Biocides containing	424	161
Viscose preparation	536	60$^+$
Electric furnaces for treatment		
Electric lamp filament	252	500$^+$
Electrode (See electrode)		
Arc lamps consumable	314	60
Furnace	373	88$^+$
Space discharge device	313	326$^+$
Electrolytic electrodes	204	294
Fibre	423	447.1
Processes of manufacture	423	449
Product per se	423	447.2
Filaments	313	341$^+$
Fluorocarbon preparation from	570	150
Layer impregnated or coated	428	408
Leaf and strip	282	
Preparation by coating	427	153$^+$
Typewriter	400	497$^+$
Lubricants containing	252	16$^+$
Microphone, granular	381	180$^+$
Moderator structure	376	350
Moderator structure for nuclear reactors	376	458
Monoxide	423	439
Carboxylic acid production		
Acyclic acid	562	517$^+$
Alicyclic acid	562	497
Aromatic acid	562	406
Utilizing	562	
Generator	48	85
Hydrogen production utilizing	423	648R
Hydrogenation of	518	700$^+$
Neutralizing agents for	252	189$^+$
Organic acid production utilizing	562	408$^+$
Synthesis gas containing	252	373
Paints containing	106	261
Paper	428	488.1$^+$
Business forms	282	28R
Container	206	215
Design	D19	1
Flux	148	23$^+$
Pigments	106	307
Preparation by carbonization	201	
Preparation by electrochemistry	204	173
Prevention in mineral oil conversion	208	48R
Refractory compositions	501	99$^+$
Removal (See decarbonizing)		
Saver arc lamp	313	335
Tetrachloride	570	101$^+$
Carbon Black	423	445$^+$
Carbonate	423	419R$^+$
Electrolytic synthesis	204	87$^+$
Esters of carbonic acid	558	260$^+$
Fertilizers containing	71	
Fire extinguishing compositions containing	252	7
Carbonation		
Alcoholic beverage processes with	426	592
Beverage	261	
With flavoring means	99	323.2
Carbonation equipment	D 7	300$^+$
Carbonators	261	DIG. 7
Colloid systems foams	252	307
Fire extinguishing compositions	252	8
Foam control in handling gas charged liquids	137	170.1$^+$
Gas and liquid contact apparatus	261	
Carbonic or Chlorcarbonic Acid Esters	558	260
Carbonium Ion Production in Hydrocarbon Synthesis	585	942*
Carbonizing (See Carburizing)		
Absorbents	502	416$^+$
Barrel interiors	432	224
Distillation	201	
Apparatus	202	
Fabrics	8	140
Molded article	264	29.1
Fiber formation	264	29.2
Fiber, fabric or textile	423	447.1
Ovens	202	215$^+$
Processes	201	
Mineral oils	208	46$^+$
Textiles	8	140
Apparatus	68	2
Carbonless Transfer Paper	503	200$^+$
Carbonyl Halides	260	544R
Carbonylamides	564	123$^+$
Stabilized or preserved	564	4
Carbonyls	423	416
Carboxamides	564	123$^+$
Acyclic acid amides	564	192
Aldehyde or ketone ctg acid	564	199$^+$
Halogen ctg acid	564	209$^+$
Hydroxy or ether ctg acid	564	201$^+$
Lower fatty acid	564	215$^+$
Nitrogen ctg acid	564	193$^+$
Unsaturated acid	564	204$^+$
Alicyclic acid amides	564	188$^+$
Aminimine ctg	564	147
Aromatic acid amides	564	161$^+$
Aldehyde or ketone ctg acid	564	169
Diphenyl methane ctg acid	564	181
Hydroxy or ether ctg acid	564	170$^+$
Phenoxy alkanoic ctg acid	564	175
Monocyclic ctg acid	564	182$^+$
Nitrogen ctg acid	564	163$^+$
Polycyclo ctg acid	564	180
Sulfur ctg acid	564	162
Hydazine ctg	564	148$^+$
Plural carboxamides	564	152$^+$
Preparation		
By acid hydrolysis only of nitrile	564	129
By catalytic hydration only of nitrile	564	126
By nitration of amides	564	146
From acyclic nitrile	564	130
Unsaturated	564	131
From carbon monoxide or dioxide	564	132
From carboxlic acids of functional derivative	564	133
From acids	564	138$^+$
From anhydrides	564	144
From esters	564	134
From cyano compound	564	124$^+$
From hcn or cyanogen	564	125
From acid halides	564	142$^+$
Carboxylate (See Carboxylic Acid)		
Carboxylic Acid Esters	560	1$^+$
Aromatic	562	405$^+$
Fermentative production of	435	136$^+$
Hydrophenanthrene nucleus ctg	562	403$^+$
With preservative	560	2$^+$
Carboy (See Bottle)	D 9	367$^+$
Dispenser casing	222	183
Jacketed	222	131
Spaced wall or jacket	215	12.1$^+$
Metallic	220	445$^+$
Carburetor		
Carbureting processes heating gas	48	219
Hydrogen	48	199R
Water	48	205
Deicers for	261	DIG. 2
Drawing excess fuel from carbureting passage	261	DIG. 21
Fluid amplifiers in	261	DIG. 69
Gas and liquid contact apparatus	261	
Heating and illuminating gas generator	48	144$^+$
Cupola water	48	79$^+$
Hydrogen	48	116$^+$
Oil and steam	48	96
Water	48	109
Heating and illuminating gas mixers	48	180.1$^+$
Internal combustion engine		
Additional air supply	123	26
Charge forming	123	434$^+$
Spray nozzles	239	
Liquified petroleum gas (lpg)	48	180.1$^+$
Manifolds	261	DIG. 36
Primers	261	DIG. 8
Primers, ticklers	261	DIG. 73
Processes, carburetors	261	DIG. 45
Surge prevention in	261	DIG. 5
Venturi	261	DIG. 12
With supercharging blowers	261	DIG. 51
With valves	261	DIG. 52
With vented bowl	261	DIG. 67
Carburizing		
Apparatus	266	249$^+$
Continuous casting combined	164	473
Metal casting combined	164	55.1$^+$
Metal treatment	148	
Molten iron or iron alloys	75	48
Carbylamines	558	302$^+$
Carcass		
Labels or tags	40	625$^+$
X-art collection	40	908*
Splitting, butchering	17	23
Card	D19	34.1
Addressing plate combined	101	369
Assorting	209	613
Case and holder	150	147$^+$
Case design	D 3	56$^+$
Coin	229	92.9
Pocket type	206	39$^+$
Trays	206	425$^+$
Typewriter paper	400	521$^+$
Clothing		
Fiber preparation	19	114
Grinding	51	242$^+$
Wireworking	140	97$^+$
Cordage	242	50
Design	D19	1$^+$
Dispensing		
Feeding or delivering	271	
Packages	206	233
Pocket case	206	39$^+$
Exhibiting	40	
Design	D20	10$^+$
Magazine	312	50
Packages	206	44.11
Racks	211	50$^+$
Trays	206	425
Fabric winding	242	61
File directory	40	371$^+$
Greeting		
Design	D19	1
Postal	229	92.8
Subject matter	283	1R
Toys	446	147$^+$
Grinders	51	242$^+$
Holder or file	D19	75
Index	40	380
Indexed	283	36$^+$
Label affixing machine	156	
Letter sheets	229	92.1
Match	44	44
Folder	206	104$^+$
Music teaching	84	481
Playing	273	292$^+$
Design	D21	42$^+$
Postal packets	229	92.8
Design	D19	1$^+$
Protrusion to receive identification	40	360
Rack	211	44
Picture or business card display	40	124
Rack stacked article	211	50
Receiver plate	D 7	1$^+$
Record machine control	235	487
Retaining cabinet	312	183$^+$
Selection		
Magnetic code	209	569
Peripheral shape	209	608
Sewing machine	112	4
Stop motions for jacquard looms	139	338
Support	D 6	396
Tables folding	108	115$^+$
Tray-type container	206	557$^+$
Verifying	73	156
Weaving pattern	139	335
Workmens, time	346	134
Card or Sheet Dispenser	221	33$^+$
Cardboard		
Boxes	229	6R$^+$
Making machine	493	52
Cardenolides	540	105
Cardiac (See Heart)		
Cardiovascular diagnosis	128	668
Cardigan	D 2	47$^+$
Carding	19	98$^+$
Card clothing making	140	97$^+$
Safety pins	227	25

	Class	Subclass
Tobacco		
Cigar and cigarette making combined	131	109.1
Leaf disintegrating	131	311+
Stemming	131	321
Twisting combined	57	327
Cardiographs (See Electrocardiographs)		
Carillons	84	406+
Carline	52	46+
Carotene Synthesis	585	351
Carousel	D21	249
Carpet		
Attachment (See fasteners)	16	1R
Anti-slip (See fasteners)	16	1R
Stiffener	16	1R
Axminster	139	399+
Making	139	2+
Cleaning machinery	D32	15+
Coated	428	96
Cushion and padding	D 5	
Design, carpet per se	D 6	582
Design, yard goods for	D 5	
Dyeing	8	929*
Electrically heated	219	528+
Fastener	16	4+
Fringed	428	115
Installation compound tools	7	103+
Knitted	66	194
Laminated	428	96
Pile or nap type	428	85+
Seam bonding	156	304.7
Sewing machines	112	7+
Sheet runners	D 5	
Stretcher	294	8.6
Mechanical advantage utilizing	254	200+
Suction cleaner	15	300R+
Sweeper	15	41R+
Design	D32	50+
Tufting	139	399+
Weight	16	1R
Wilton brussels or velvet	139	391+
Making	139	37+
Woven	139	391+
Yardgoods design	D 5	
Carport or Garage	D25	34
Carriage		
Baby	280	47.38+
Design	D12	128
Powered	180	166
Blanket or robe design	D 6	603+
Feed		
Calculator	235	63R
Lathe	82	21R+
Sheet metal spinning pattern controlled	72	81
Steeping-motor drive for	400	903*
Firearm	42	94
Fluid motor tide or wave	60	497+
Manufacture		
Irons forging dies	72	472+
Ordnance	89	37.1+
Tool or work		
Band saw	83	406.1
Circular saw reciprocating	83	435.1
Circular saw rotary	83	409.1
Circular saw swinging	83	730
Coopering crozing staves		
Endless stave	147	19
Sliding stave	147	20
Swinging stave	147	21
Coopering jointing staves saw		
Sliding	147	30
Tilting	147	31
Diminishing work piece slicer	83	703+
Mechanical movement	74	
Rotary gearing	74	353
Swinging gearing	74	354
Saw making reciprocating tool	76	33
Saw making rotating tool		
Pivoted	76	42
Sliding	76	43
Sawmill	83	713+
Sheet metal container seaming	413	69+
Shingle sawing	83	704+
Textile apparatus doffing or donning	57	268+
Textile apparatus drawing	57	320
Checks	57	320
Traveling cutter for sheets and bars	83	614
Wire fabric making	140	29

	Class	Subclass
Wood turning	142	47
Traversing hoist		
Holders	212	122
Release	212	112
Typewriter	400	352+
Vehicle		
Land	280	
Land bodies and tops	296	
Land dumping	298	
Motor	180	
Carrier (See Bucket; Conveyor; Receptacles)		
Article or package carrier	224	
Baby	D 3	31
Bottle	294	
Multiple type grapple	294	87.1+
Braiding textiles	87	33+
Brakes for single rail conveyor type	188	42
Brick	294	62+
Camera plate	354	178+
Cartridge	224	223+
Bandoleer	224	203
Covered	224	239
Case for sewing machine or typewriter	312	208
Casket	27	27
Casting apparatus	164	323+
Chutes skids guides and ways	193	
Cooking apparatus	99	324+
Corpse	27	28
Dumping scows	114	27+
Earth boring cuttings type	175	
Elevator type material handling	414	592+
Carrier operated feed	414	609+
Inclined	414	595+
Endless		
Article magazine	312	97
From static support	198	506+
Auxiliary item storing section	198	347
Between plural conveyers	198	620+
Bottle brushing machine	15	60
Bucket	198	701+
Can body feeding	413	45+
Check released article delivery	194	
Coin trap closure	232	56
Cutting harvesters	56	181+
Dragline scoop combined	414	569
Drying house	34	207
Fire escape	182	42+
Flexible pocket	198	715
Furnace charging reciprocating	414	156
By driven force	414	198
Furnace charging type	414	147+
Grain separator straw	130	21
Hoist combined	414	564
Hopper bottom forming	198	550.1
Having gravity conveyer section	198	550.1+
Item holder	198	375+
Item holder	198	688.1+
Loading machine type, buckets	198	509
Mold for metal	164	324
Plural mold cigarette making	131	87
Plural rotary conveyer	198	608
Potato digger	171	94
Potato digger	171	111+
Rack	211	121+
Raking and loading harvesters	56	345+
Reciprocating abrading tool	51	61
Rotary abrading tool	51	76R
Sheaf	56	477
Sheet drying apparatus	34	162
Showcase	312	134
Traveling band abrader	51	138
Vehicle loading external	414	398
Vehicle unloading external	414	390+
Wood planer	144	123
Feed tables for shearing machines	414	677
Fire escape chute or tower	182	48+
Fire or heat shield	182	47
Fluid current	406	184+
Furnace charging or discharging	414	147+
Furnace fuel feeders	110	101R+
Glass mold	65	361
Grain separator straw	130	21+
Hand held	294	137+
Harness trace	54	54+
Harvester	56	474+
Corn	56	121
Hose or nozzle support ladder combined	248	78

	Class	Subclass
Mechanical gun projectiles	124	41R+
Occupant propelled vehicle combined	280	202+
Package and article	224	
Person	224	158+
Hand held	294	140+
Photographic fluid treatment	354	297+
Pivoted		
Cutting machine knife	83	597+
Screw threading machine chaser	408	150+
Shoe nailing machine	227	41
Pneumatic dispatch	406	184+
Railway drawbar	213	
Reciprocating		
Abrading band	51	142
Cutting machine knife	83	613+
Envelop machine	493	186
Shoe nailing machine	227	99+
Reel	242	85+
Design	D 8	358+
Design, fishing	D22	137+
Fabric	242	64
Hose	242	86+
Rotating		
Acetylene generator carbide feed	48	48
Article heating	432	138
Bottle brushing machine	15	63+
Can body feeding	413	50
Cotton picker spindle	56	44+
Drying apparatus	34	184+
Furnace charging reciprocating	414	156
Needle making	163	4
Photographic fluid treating	354	313
Plural mold cigarette making	131	87
Selective printing plate	101	62
Shoe nailing machine	227	99+
Slide	353	117
Tobacco plug making	131	115
Wood shaping work	144	154
Selective conveyor delivery	414	227+
Controlled by condition of site	414	232
Sewing machine article attaching	112	104+
Sewing machine work	112	121.12
Sewing machine work	112	121.15
Shuttle thread carrier supports	139	206+
Sleeping compartment combined	224	923*
Spool supports	242	136
Store service	186	
Strap	294	149+
Swinging		
Drying apparatus	34	184+
Telephone	379	64+
Textile fluid treatment	68	
Traversing hoist rope	212	117+
Truss pad	128	106.1
Type casting matrix carriers	199	
Umbrella	224	186+
Folded umbrella	224	915*
Vehicle attached	224	273+
Vehicle loading and unloading	414	340+
Forming a train of vehicles	414	339
Forward movement onloads	414	338
Forward movement unloads	414	337
Partly independent of vehicle	414	334+
Tractor-trailer to unload	414	333
Vertically swinging	414	680+
Tilting	414	743
Weapon (See type of weapon)	224	
Weft replenishing feeler	139	269+
Wheeled drum	242	95
Carrier Current		
Modulator	332	
Radio system	455	
Telegraph system	178	66.1+
Multiplex	370	45
Telephone system	379	64+
Combined with power line	379	66
Multiplex	370	69.1
Carrier Wave Repeater or Relay	455	7+
Frequency division multiplexing	370	75
Pulse or digital	375	3+
Time division multiplexing	370	97
Carroting	8	112
Apparatus	69	28
Carrousel	272	28R+
Organ	84	84+
Cart (See Vehicle)		
Brakes	188	19
Harness saddles	54	39
Shopping	280	47.26
Nesting	280	33.99 R
Carton	229	

	Class	Subclass
Cutter	83	926K*
Filling and sealing	53	
Flap deflector or closer	414	906*
Forming machines	493	52
Openers	53	381R+
Dispensing carton	222	80+
Openers, cutlery	30	2
Tools	30	2
Cartoon Motion Picture Photography	352	50
Cartridges (See Detonating, Signal)		
Ball point	401	209+
Belt with cartridge holder	D 3	100
Belts	89	35.1
Beverage infuser	99	295
Carbide feed acetylene generator	48	47
Carbide for acetylene generator	48	59
Carriers body or belt attached	224	223+
Bandoleer	224	203
Covered flaps	224	239
Case	102	464+
Case necking die shaping	72	352+
Edge trimmer	29	1.32
Extracting tools	81	3.5
Explosive	102	430+
Belts, disintegrating	89	35.2+
Blasting	102	314+
Case making	29	1.3+
Caseless	102	431+
Chutes	193	25AC+
Combustible	102	700*
Feeding ammunition making	86	45+
Feeding gun (See magazine)	89	33.1+
Loading	86	23+
Making	86	10+
Packing	86	47
Priming	86	32+
Pyrotechnic	102	346
Film pack or cartridge	430	471
Film pack with structure	354	275+
Fuse electric	337	228+
Ink rupturable	401	132+
Material containing for dispenser	222	325+
Tiltable holder for	222	165
Motion picture film	242	197+
Picture film	D14	10+
Tape recording or motion picture film	242	197+
Container for	206	387
Carvacrol	568	781
Carving		
Wood processes	144	363+
Cascade		
Connection for alternating current motors	318	49+
Refrigeration	62	335
Automatic	62	175
Case (See Casing; Receptacles)		
Blacking box and brush	15	258
Blacking or brush retained	401	118+
Body and belt attached	224	191+
Body electric applicators	128	801+
Cardcase	150	147+
Cartridge	102	464+
Cigar	206	242+
Combination	206	38
Clock	368	276+
Design	D10	1+
Crank	123	
Fishing rod	43	26
For transport, handling of goods	D34	40+
Hardening	148	
Lighted showcase	362	125+
Paper cell making	493	90
Pass or identification card	40	10D
Design	D19	9+
Piano	84	177+
Automatic	84	80+
Pocket for smoking pipe	206	244
Protective	150	52R+
Reed organ	84	352+
Sample	190	16+
Shift		
Typewriter ribbon mechanism	400	216.6+
Typewriting machines	400	251+
Show	312	114+
Design, store furniture	D 6	
Special	206	
Musical instrument	206	314+
Pocket and personal use	206	37+
Spectacle	206	5+
Tennis racket	273	74
Type	276	44+
Umbrella	135	33C
Imitation cane combined	135	18+
Upending bed	5	159R+
Vehicle foldable to traveling	280	37
Violin	206	14
Watch	368	276+
Design	D10	30+
Watch manufacturing processes	29	179
Wooden	217	5+
Barrel	217	73
Bottle mailing	217	127
Egg	217	18+
Insulated	217	128+
Cased Glass Making	65	60.1
Casein	530	360
Coating or plastic compositions containing	106	138+
Foods	426	656
Casement		
Housed roll	160	26
Metal window fasteners	292	DIG. 33
Cash Register and Recorder	D18	4+
Coin box	235	100
Combined	235	2+
Computer controlled, monitored	364	405
Fare	235	33+
Recorders	235	4+
Registers	235	7R+
Cashew Nut Shell Liquor	562	475
Polymers from (See also synthetic resin or natural rubber)	528	1+
Cashmeres, Design	D 6	603+
Casing (See Case; Envelope; Guard; Housing; Package)		
Abrading apparatus	51	268+
Bearings	384	
Body light	128	396+
Boiler	122	494
Stand type water tube	122	19
Vertical sectional	122	224
Book making	412	1
Brushes and brooms	15	171+
Car stove protective	126	57
Closure fastening bolt	292	337
Clothesline type reel	242	100+
Clutch	192	112
Coffin	27	35
Dispenser	222	173+
Article	221	282+
For single mixed drink	222	129.1+
Door and arch	110	181+
Cooling	110	180
Driven spoke reinforcing	301	83
Plural spoke series	301	77
Electric conductor and insulator	174	50+
Sleeve and end cap type joints	174	93
Vacuum or fluid containing	174	17R+
Expansible chamber device	92	261
Filter	210	541
Flashlight battery type	362	157+
Flat	362	200+
Tubular	362	202+
Food preserving	426	321+
Sausage	426	332+
Food process	17	49
Rotary	415	219R
Heat exchange		
Chamber	165	72+
Horizontal	165	157+
Motor vehicle radiator	165	148+
Trickler	165	115
Vertical	165	157
Hot air furnace	126	114
Lantern	362	362+
Machine elements	74	606R+
Guards	74	608+
Marine screw propeller	440	76
Measuring instrument	73	431
Design	D10	46+
Metal rolling apparatus	72	237+
Metal working with centrifugal casing	29	DIG. 6
Metal working with composite casing	29	DIG. 8
Meter	73	273+
Meter circuit protectors	361	368+
Muffler and sound filter	181	282+
Pipe joint coupling	285	45+
Pneumatic conveyor material intake fan	406	96+
Rotary	415	89
Railway draft friction		
Drawbar	213	65
Spring combined	213	31+
Shoes and wedges	213	24
Railway motor placement	105	140
Axle mounting and stator combined	105	137
Ray generator	250	522.1
Reel carrier, hand or body	242	96
Rotary shafting	464	170+
Flexible shaft combined	464	52+
Sash cord guides	16	211+
Wheel and	16	215
Sewing machine casing	112	231
Shirred sausage casing	206	802*
Sifting	209	370+
Sight glasses for liquid level gauges	73	325+
Sign illuminated	40	541+
Signal for vehicle illuminated	116	49
Stove hood	126	302+
Switch	200	293+
Thermal current	337	186
Switchboard	361	364+
Telephone	379	440
Thermometer indicating tube	374	174+
Tiltable receptacle convertible support	248	135
Tire	152	151+
Armored	152	196+
Armored antiskid	152	169
Cushion and pneumatic combined	152	155+
Cushion enclosed core	152	310+
Pneumatic	152	450+
Vehicle stove air	237	30
Well	166	242
Wheel hub reinforcing	301	106+
Cask (See Barrel)		
Drum or	D34	39
Making machines	147	1
Pitching	118	
Tilting apparatus	248	133+
Casket	27	2+
Carriers	27	27
Design	D99	1+
Grave rest	27	26
Locks	70	63
Lowering devices	27	32+
Portable casings for caskets	27	35
Casserole (See Vessel)	215	
Design	D 7	19+
Cassettes		
Camera	354	275+
Film and plates	206	455
Design	D14	11
Recorder	D14	6+
Rolls	206	387
X ray	378	182
Cast Iron		
Alloys	420	13+
Heat treated stock	148	400
Heat treatment	148	138+
Metallurgy	75	28+
Caster		
Brake	188	1.12
Design	D 8	375
Condiment	D 7	52+
Glass or cup	211	74+
Wheel type	16	18R+
Casting		
Battery electrodes	164	
Coating combined	29	527.3
Fishing	43	18.1+
Forging combined	29	34R+
Methods	29	527.5
Glass	65	
Marine structure in sitiy	405	222+
Metal	164	
Continuous	164	418+
Die	29	DIG. 1
Digest	29	DIG. 5
Founding	164	
Centrifugal apparatus	164	286+
Centrifugal processes	164	114+
Heat treatment combined	148	3
Joining articles together	164	91+
Rheo-casting	164	900*
Working combined	29	527.5+
Pile in situ	405	232+
Plastics, apparatus	425	256
Metal working combined	29	527.1

U.S. PATENT CLASSIFICATIONS

	Class	Subclass
Molds, dynamic	425	
Sound records	425	406+
Plastics, processes	264	299+
Centrifugal	264	310+
Sound records	264	106+
Pottery machines	425	263+
Pottery machines	425	459
Tunnel lining in situ	405	150+
Type	199	
Castoff		
Garment supporter	2	342
Sewing machine thread	112	56
Castor Oil	260	398+
Dehydrating	260	405.5
Castrating Instrument	128	306
Cat Whisker	357	
Catafalque	27	1
Catalin T M (phenoplast)	528	129
Catalysts	502	100+
Combustion products generator with	60	723
Enzyme containing	435	183+
Fermenting processes using	435	41+
For wave energy process	522	6+
Plant	71	65+
Polymerization using transition metal (See also synthetic resin or natural rubber)	526	906*
Comminution of	526	906*
In presence of hydrocarbon	526	905*
Monomer		
Hydrocarbon additive affects polymer properties	526	903*
In bulk	526	902*
In vapor state	526	901*
Part supported on polymer	526	904*
Specified means of reacting components	526	907*
Vapor phase in absence of transition metal	526	913*
Redox catalyst	526	915*
Regeneration	502	20+
Mineral oil conversion combined	208	
Specified particle size	526	908*
Ziegler-natta	526	159+
Catalytic		
Apparatus	422	177+
Igniting device	431	268
Mineral oil conversion with	208	46+
Solids contacting	208	146+
Catamenial Receptor	604	358+
Adhesive fastener for	604	387
Cataplasmic Medicament Depositor	604	304+
Catapult	124	
Aircraft combined	244	63
Catch (See Associated Device)		
Basin		
Dispenser type drip	222	108+
Dispenser type inverted container	141	364
Fountain type	239	28+
Sewage type	210	532.1+
Door and window	292	
Dumping vehicles land type	298	38
Holder for fish	43	55
Catcher (See Drip, Catcher)		
Aerial projectile projector and	273	318+
Animal husbandry	119	151+
Band saw machine	83	814+
Body	182	137+
Dispenser waste	222	108+
Fish automatic	43	15+
Harvester		
Cornstalk chopper	56	61
Cornstalk motor type with chopper	56	16.6
Cornstalk topper	56	56+
Cutter and endless carrier with discharging	56	184
Cutter with detachable	56	5
Cutting combined	56	202+
Cutting reel combined	56	199+
Dump rake and cutter combined	56	166
Fruit	56	329
Shelled grain	56	207
Insect	43	133+
Inverted container drainage	141	364
Railway droppings	104	133
Stove and furnace soot	126	280
Traps	43	58+
Fish	43	100+
Insect adhesive type	43	114+
Vehicle brakes on track type	188	63
Catenary		
Trolley wire supports	191	41
Cataract Correction Lenses	351	167
Caterpillar T.m. Track (See Track)		
Cathanode	313	305
Cathartic Composition	424	
Cathead		
Automatic	254	373
Drum	242	117
Catheter		
Milker	119	14.19+
Surgical	604	96+
Couplings	604	280+
Design	D24	54
Flexible	604	280+
Genital	604	264+
Inflatable	604	96+
Intravenous	604	53
Manipulative	604	53
Cathode		
Arc lamp	313	326+
Electrolytic	204	280+
Electronic tube eg radio tube	313	346R+
Forms for galvanic apparatus	204	280+
Gas or vapor lamp	313	326+
Making for discharge device	445	46+
With assembly	445	35+
Ray tube		
Circuit blanking	315	384
Facsimile systems	358	256+
Focus coil and	313	442
Focus coil per se	335	213+
Image pick-up tube	313	365+
Manufacturing	445	36+
By photography	430	23+
Music application	84	DIG. 6
Oscillographs	346	110R
Oscilloscopes	324	121R
Repair	445	2
Screen	313	461
Used in teaching	434	307
Used in teaching	434	323
Storage	313	391+
Structure	313	364+
Symbol generation on screen	340	720
Systems	315	1+
Television receiver	358	188
Television scanner	358	199+
Television systems	358	83+
Television transmitter	358	186+
Work holder for	269	908*
Sputtering		
Apparatus	204	298
Processes	204	192.1+
Catnip Ball	119	29.5
Catsup	426	615
Cattle Guard		
Fences	256	14+
Track type automatic return gate	49	133+
Caulking Gun	D 8	14.1
Caustic		
Alkalis	423	641
Cauter		
Electric	219	232+
Electric	219	236
Liquid or gaseous fuel	126	405+
Surgical combined	128	303.18
Cave	52	169.1+
Caviar	426	643
Canned	426	131
Cavity Resonator	333	227
In cathode ray tube	315	5+
In electronic tube	315	39+
Magnetron	315	39.51+
Spectrometer electrical	324	316+
Ceiling	52	
Block	439	450
Article supported from	52	39
Quick detachable	439	450
Cargo ship	114	76
Spaced roof	52	22
Suspended	52	484+
Waffle type	52	337
Lights	362	147+
Wood buildings		
Centerpieces	52	311
Celcon (resin from Trioxane and Ethylene Oxide)	528	250
Celery		
Blanching	47	3
Harvesters	171	21+
Planters	111	2+
Salt	426	638
Celeste	84	DIG. 4
Organ stops	84	371
Cell		
Acetylene gas generator	48	
Airship gas	244	128
Box or receptacle		
Making, paper	493	90
Metallic	220	21
Paperboard	229	120.2+
Wooden	217	18+
Cartridge belt block	224	239
Dry	429	163+
Electric	429	
Chemical	429	
Photo	136	243
Railway rolling stock systems	105	51
Thermo	136	200+
Electrolytic	204	194+
Circuit interruption type	361	435
Testing	204	400+
Galanic, gas type	429	12+
Jail	52	106
Leeching or diffusing	127	7
Photo electric	136	243+
Queen bee	6	9
Storage battery	429	149+
Support for removable cell	429	96+
Cellar	52	169.1+
Cello (See Violin)	84	274
Cellophane		
Compositions	106	168
Making article from	493	
Cellosolve T M	568	678
Cellular (See Porous)		
Glass body making	65	22
Glass composition	501	39
Laminate	428	304.4+
Forming by spaced sealing	156	290+
Magazine type		
Cabinet	312	97.1
Dispenser	221	69
Plant or animal material, polymer ctg or derived from (See synthetic resin or natural rubber)		
Polymer (See also synthetic resin or natural rubber)	521	134
Spring mattresses	5	475+
Celluloid T M	106	169+
Cellulose and Derivatives	536	56+
Compositions	106	163.1+
Esters	536	58+
Compositions	106	169+
Ethers	536	84+
Compositions	106	169+
Fermentation of	435	179
Fermentative treatment	435	277+
Hydrolysis to sugars	127	37
Liberation	162	1+
Nitrogen containing	536	30+
Cement (See Adhesive)		
Compositions containing	106	85+
Refractory	501	124
Cycle tire	106	33
Earth treatment with	405	266+
Injector	405	269
Magnesia sorel	106	106+
Manufacturing processes	106	100+
Portland roman	106	100+
Roads	404	17+
Making	404	72+
Sheet structure	404	17+
Shoe sole attached by	36	19.5
Cementing		
Implements	15	235.4+
Lining and coating pipes	118	105
Lining and coating pipes	118	125
Lining and coating pipes	118	404+
Lining and coating pipes	425	110+
Lining and coating pipes	425	460
Shoe making	118	410+
Well cementing processes	166	285+
Center		
Drilling work support	408	72R+
Gauge	33	670+
Lathe centers	82	33R
Making in work for mounting	408	199+
Milling tail stock	409	242
Woodworking lathe	142	53
Center of Gravity Measuring	73	65

	Class	Subclass
Center Punch	30	366
Locating	33	670^{+}
Sheets or bars	30	366
Centerboard	114	127^{+}
Centerers	82	45
Centering Devices		
Aligning tool	29	271^{+}
Circular saw on work holder	76	79
Locating	82	45
Mold for arch or tunnel		
Repositioning means for progressive molding	425	63
Using subterranean feature	425	59
With vehicle to move apparatus	425	62
Slide in cylinder	175	325
Well casing type	166	241
Centerless Grinding	51	103R
Propelling work through	51	74R^{+}
Work rest for	51	238R
Central Heating Systems	237	13
Centrifugal Force		
Assorting	209	642
Brake systems fluid	384	168
Brakes	188	180^{+}
Centrifugally operated devices		
Abrading tools	51	332^{+}
Bearings	384	135^{+}
Bobbin winding	242	46.5
Cable drum drive or brake	254	267
Fuse arming	102	237^{+}
Governor (See governor)		
Light support	362	384
Lubricator force feed	184	43
Lubricator gravity feed	184	77
Pipe and tube cleaners	15	104.14
Switch	200	80R
Classifying by fluid suspension	209	148
Clutches	192	103R^{+}
Comminution apparatus	241	275
Comminution process	241	5
Conveyers and feeders		
Feeder	209	915*
Pneumatic	406	100^{+}
Structure, rotary	198	803.1
Structure, thrower	198	638
Dispensing	222	410^{+}
Drying apparatus	34	58^{+}
Drying process	34	8
Film formers or spreaders		
Coating machine	118	56
Coating process	427	240
Evaporators	159	6.1
Fluid treatment cleaning and washing textile fibers yarns	68	147^{+}
Forming or casting		
Liquid comminuting	425	6^{+}
Liquid comminuting	425	425
Metal apparatus	164	286^{+}
Metal process	164	114^{+}
Molding apparatus	425	425
Process	264	310^{+}
Pulp	162	384^{+}
Sugar crystals	127	19
Grain hullers	99	519
Gun	124	4^{+}
Lubricators	184	70
Meter fluid flow	73	861.29
Mufflers and sound filters	181	274
Nut crackers	99	571
Pumps		
Gas rotary	415	
Gas screw combined	415	143
Liquid rotary	415	203^{+}
Material agitating	366	263^{+}
Scatterers or sprayers		
Fluid sprayer	239	214^{+}
Fuel burners	239	214^{+}
Gas and liquid contact	261	83^{+}
Rotary	239	681^{+}
Spray fluid actuated	239	381
Separating colloids	252	349
Separator (See separators and separating, centrifugal)		
Spinning apparatus	57	76^{+}
Centrifuge	D24	22
Centripetal Apparatus		
Dispenser rotary central discharge	222	411
Extractors with inward flow	210	360.2^{+}
Pumps rotary	415	120^{+}
Cephaeline	546	96
Cephalosporin	540	215^{+}
In drug	514	200^{+}
Cepharanthrine	546	31
Ceramics		
Compositions	501	1^{+}
Fluxes	106	313
Firing apparatus	432	
Firing processes	264	56^{+}
Forming and molding	425	
Uniting combined	264	241
Cerates	424	
Cereals	426	618^{+}
Apparatus	99	
Comminution	241	6^{+}
Design	D 1	
Preservation	426	321^{+}
Puffing, processing	99	323.4^{+}
Ceremonial Robe	D 2	79
Ceresin	208	24^{+}
Cerium	75	
Alloys	420	416
Inorganic compounds	423	
Recovery	423	21.1^{+}
Pyrometallurgy	75	84
Cermet	75	230^{+}
Certificate		
Design	D19	9
Of deposit	D19	9
Cesium		
Compounds	23	
Radioactive compositions	252	625
Cesspool	210	542
Sewage purifying	210	532.1^{+}
Cevadine	546	34
Chafe		
Guard positioned between artice and bearer	224	907*
Chafe Iron Vehicle	280	161^{+}
Chafing Dish	D 7	355
Chain	59	78^{+}
Bracelets	63	4
Cable stoppers	114	200
Chain saw machines	83	788^{+}
Mining	299	29^{+}
Saw element	83	830^{+}
Design	D 8	499^{+}
Ornamental	D11	13
Door fastener	292	264
End holders	24	116R
Endless chain coupling	464	49
Endless chain pumps	415	5
Endless earth boring tool	175	89^{+}
Fabrics	245	4
Flexible shafts	464	
General purpose	D 8	499^{+}
Grates	110	269^{+}
Jewelry attachments	63	21^{+}
Design	D11	13
Key ring	70	457
Link formation by soldering	72	192
Lubricators	184	15.1^{+}
Making	59	1^{+}
Molds for chains	249	57
Ornamental	59	80^{+}
Design	D11	13
Propellers	416	7^{+}
Marine type	440	95^{+}
Reaction, nuclear	376	
Skid chains	152	231^{+}
Applying tool	81	15.8
Design	D12	154
Sprocket	474	152^{+}
Bicycle	D12	123
Design	D15	148^{+}
Guards	474	144^{+}
Stirrer, chain-type	366	607*
Stitch sewing machine	112	197
Tire	152	208^{+}
Design	D12	154
Traction railways	104	172.1^{+}
Warp threads	28	175^{+}
Watch chain snap hooks	24	240
Welding	59	31^{+}
Electric	219	51
Wheel combined for traction	301	42
Wood mortising cutter	144	72^{+}
Chain Saw	30	381^{+}
Chair	297	
Baggage convertible to	190	8
Barbers	297	68^{+}
Design	D 6	334^{+}
Bath	4	578
Cabinet combined	312	235R
Chair lift	D12	52
Chaisse lounge	D 6	360
Design, seat	D 6	334^{+}
Doll or toy	446	482
Dry closet	4	465
Elevator landing	187	75^{+}
Exercising	272	144
Fan combined	416	146R
Folding	297	16^{+}
Folding vehicle occupant seated	280	650
Kinesitherapy	128	33^{+}
Ladders	182	33
Light thermal or electrical application	128	376^{+}
Locks	70	261
Milking stool type	297	175^{+}
Pedestal design	D 6	364
Railway car type making	29	16
Railway surface track	238	187^{+}
Concrete pedestal bar	238	116
Crossings	246	467
Insulated rail joint	238	154^{+}
Reversible rail or english type	238	266
Switches	246	453
Two rail	238	20
Reclining	297	68^{+}
Road	404	136
Round making	144	8
Self leveling	114	194
Theater seat	297	311^{+}
Wheel chair	280	647
Chair Lift Design	D12	52
Chaise Lounge	D 6	360
Chalk		
Chalkboard cleaning, chalk or chalkdust holder	434	417
Composition	106	19
Cord	33	414
Cue chalkers	273	17^{+}
Holder	401	88^{+}
Chamber (See Process, Machine or Device with which Associated)		
Automatic weighers	177	60^{+}
Chemical apparatus	422	
Elements or details	422	310
Cleaning and liquid contact with solids	134	
Closet	4	300^{+}
Disinfecting	4	222^{+}
Concentrating evaporator	159	12
Closed	159	22^{+}
Dissolver-mixer sprayer	239	267^{+}
Distilling	202	
Drying and gas or vapor contact with solids	34	
Internal resistance brake	188	266^{+}
Liquid separation or purification	210	
Motor		
Expansible chamber type	91	
Mufflers and sound filters	181	212^{+}
Musical instrument		
Piano	84	24^{+}
Pipe organ	84	331^{+}
Reed organ	84	351^{+}
Pipe or tube pressure compensator	138	30
Pneumatic conveyor	406	
Pot type	4	479^{+}
Preserving, disinfecting, and sterilizing	422	243^{+}
Pumps	417	
Expansible chamber type	417	437^{+}
Pressure compensator	417	540^{+}
Rotary	418	
Resilient tire	152	
Steaming textiles	68	5C^{+}
Submarine dredger	37	56
Therapeutic treatment with medicaments	128	202.12
Tobacco smoke treating	131	211^{+}
Volume or rate of flow meter	73	861^{+}
Expansible chamber type	73	232^{+}
Chambray	139	416
Cord stripe	139	417
Chamfer		
Barrel chamfering machines	147	13^{+}
Barrel closure	217	80
Bench planes for chamfering wood	30	484^{+}
Coopering tools for chamfering	147	42
Gear chamfering machines	409	8
Chamois	8	94.1 R^{+}
Chance and Game Apparatus		
Baseball	273	93R

U.S. PATENT CLASSIFICATIONS

Entry	Class	Subclass
Board games	273	243+
Chance devices	273	138R+
Check connector reciprocating	194	293+
Check connector turning	194	295+
Check lock releasing	194	247+
Guns mechanical	124	33
Chandelier	362	405+
Design	D26	72
Hanger	362	406+
Change		
Bag for photographers	354	308+
Auxiliary plate magazine	354	282
Coin machines	D20	1+
Gate	232	14
Maker	453	1+
Changeable Exhibitor		
Amusement cards and pictures	446	147+
Card picture and sign	40	446+
Electric annunciators	340	815.1+
Systems	340	286R+
Figure toys, features change	446	321
Illuminated	362	253
Indicator combined	116	307
Information displaying	40	446
With sound	40	455
Channel		
Ditch	405	118
Lining	405	268
Electric signals	340	
High frequency	333	
Radio	455	
Telegraph	178	
Telephone	379	
Fishway	405	81+
For electric conductor	174	68R+
Glass conditioning	65	346
Guide	112	50
Iron structural members	52	720
Navigable	405	84+
Preserving channel for water control	405	21+
Rail and tie		
Flange down railroad ties	238	59+
Flange up railroad ties	238	62+
Rail joints splice bars in	238	161
Railroad rails	238	127
Railroad rails joints	238	176
Two part railroad ties	238	56
Stock	428	122
Metallic	428	595
Type holding printing	276	37
Channeling Machine and Tool		
Ditcher	37	80R+
Mining	299	
Endless flexible cutter	299	82+
Machine	299	29+
Rotary cutter	299	79+
Sewing leather	112	45
Shoe sole	12	27+
Shoe sole combined with trimming	12	85.2
Chaplet	164	398+
Character Recognition		
Computer means	382	
Radiant energy, optical or pre-photocell system	382	58
Characters		
Calendar slide	40	109
Interchangeable sign	40	618
Markers for calendar	40	110
Printing	101	
Sensing	382	
Controlling display	340	806+
Controlling punch	234	59+
Operating typewriter	400	61+
Photoelectric	382	65
Television	358	294+
Telegraphic printing system	178	30
Type casting	199	
Charcoal		
Absorbents containing	502	416+
Briquettes	44	10C+
Broilers	99	385+
Design	D13	3.2
Production by distillation	201	
Stoves	126	25R
Charge		
Dispensing	222	
Explosive		
Ammunition making and	86	
Ordnance pneumatic combined	89	7
Forming		
Fuel injection	123	445+
Internal combustion	123	434+
Internal combustion combined generating	123	3
Charge Transfer Device		
Computers	364	862
Delay line	307	607
Logic circuit	307	444
Memory	365	183
Shift register	377	57
Structure	357	24
Wave transmission lines	333	165
Charger and Charging		
Arc furnace device	373	60
Arc between spaced electrodes	373	62+
Charging or discharging	373	79+
Battery	320	2+
Design	D13	5+
Bin or tank	414	288+
Electric furnace	373	
Glass furnace device	373	33+
Induction furnace device	373	142+
Resistance furnace device	373	115
Filters, liquid	210	456
Precoat adding means	210	193
For capacitor	320	1
For capacitor	307	109+
For nuclear dosimeter	320	1
For nuclear reactor	376	264
For pocket chamber nuclear	320	1
Furnace type	414	147+
Heating	110	101R+
Metallurgical	266	176+
Gas generators	48	86R
Lamp bulb		
Gas	141	66
Gas combined	445	
Liquid purification	210	519+
Means to add treating material	210	198.1+
Tangential	210	512.1
Magazine	42	87+
Metal under pressure	164	303
Molding apparatus for nonmetals	425	542+
Block	425	258+
Glass	65	
Nuclear reactor	414	146
Pneumatic motor railway	104	157+
Receptacles	141	
Portable	141	
Refrigeration apparatus	62	292
Process	62	77
Ship	414	137
Still	202	262+
Closure combined	202	251
Processes	201	40
Charlotte Russe	426	658+
Charm	63	23
Jewelry	D11	79+
Charpy Test	73	844
Charring (See Carbonizing)		
Chart		
Apparel layout	33	12+
Astrological	434	106+
Astronomical	434	284
Color display	434	98+
Educational	434	430
Holder for map or chart	40	904*
Light ray type	33	268
Menstruation	40	109
Music	84	471R+
Recorder	D10	46
Chase		
Gem setting	63	26+
Printing	101	391
Bed and platen shifting	101	63+
Securing or mounting means	101	390
Chasers	408	215+
Type setting	276	38+
Chasing Mill	241	107+
Ball type		
Frictional drive	241	103+
Loose ball	241	173+
Chassis		
Amplifier	330	65+
Electrical devices	361	380+
Radio parts type	361	422
Radio system	455	347+
Television	358	254+
Vehicle	280	781+
Occupant propelled type	280	281R+
Chatelaine		
Bag case or frame carrier	224	268
Safety hooks	24	4
Check		
Bank	283	58
Design	D19	11
Cash registers		
Check employing or controlled	235	17+
Printing checks	235	3
Coin type	40	27.5
Controlled apparatus	194	
Cash registers	235	3
Cereal puffer	99	323.8
Dispensers	222	2
Lock releasing slots	194	247+
Motion picture machine	352	104
Self photograph	354	76+
Telephone systems	379	143+
Voting machines	235	53
Weighers	177	125
Workmens time recorder	346	
Door	16	82+
Closer combined	16	49+
Design	D 8	330+
Gun breach gas	89	26
Gun recoil	89	42.1+
Harness underchecks	54	57
Labels tags and	283	81
Labels, tags, design	D20	22+
Pay check with statement of deductions	282	DIG. 1
Protectors		
Embossing	101	24
Fraud preventing coating	427	7
Writer machines	D18	5
Check-out Counter	186	59+
Checkbook Cover Design	D19	26+
Checker Brick	165	9.1+
Checkers	D21	53
Game board design	D21	24
Checkhooks	54	61+
Checking and Unchecking Device	54	70
Underchecks	54	57
Checking Watermarks	356	432+
Checkout Counter	D 6	396+
Checkreins	54	16+
Cheek Plate Railway Draft	213	54+
Cheese		
Compositions and processes	426	582+
Cutter	83	
Cutter & expresser	425	308
Design	D 1	
Making apparatus	99	452+
Packaging	426	392+
Preservation	426	330.2
Presses	100	
Textile		
Fluid treatment	8	155.2
Packages	242	159+
Treating apparatus	99	452+
Cheesecloth	139	426R
Chelate		
Aluminum containing	556	175
Chelation ion exchange	423	DIG. 14
Detergents having agent	252	DIG. 11
Drug compositions	424	DIG. 6
Food or edible material	426	271
Heavy metals containing	556	1+
Printing dye with metal chelating group	8	452
Tobacco smoke separator or treater	131	334
Chelidamic Acid	546	299
Chemical Modification, Protein	530	402
Chemical Reaction Heat Producers		
Irreversible reaction	44	3.1+
Readily reversible reaction	252	70
Chemiluminescent	252	700
Lighting	362	34
Tests involving	436	172*
Chemistry & Chemical (See Particular Compound, Composition, or Art)	422	
Analytical and analytical control methods	436	43+
Apparatus	422	44+
Carbon compounds	260	
Carbon compounds	560	
Chemical manufacture, misc.	156	
Chemical oxygen demand tests, cod.	436	62*
Chemical symbol character	400	900*
Classifying, separating, & assorting solids	209	
Cleaning & liquid contact with solids	134	1+
Apparatus	134	42+

Entry	Class	Subclass
Coating apparatus	118	
Coating processes	427	
Printing	101	
Cooling by chemical reaction	62	4
Distillation		
Apparatus	202	
Separatory processes	203	
Thermolytic processes	201	
Dyestuffs	8	
Electrical current producing apparatus, product & process	429	
Electrical radiant or wave energy	204	
Fermentation	435	
Fertilizers	71	
Foodstuffs	99	
Inorganic	423	
Measuring instrument	D10	81
Metallurgical	75	
Reactions	376	
Modification of synthetic resins or natural rubbers	525	
Modification of textiles & fibers	8	115.51+
Fluid treating apparatus	68	
Manipulative fluid treatment	8	147+
Photographic chemistry, processes & products	430	
Physical processes	23	293R+
Refrigeration	62	4
Sterilizers, refrigeration combined	62	78
Synthetic resins (See synthetic resin or natural rubber)	520	1+
Teaching	434	298
Chemotherapy Drugs		
5-flurouracil	544	313
Chenille	223	45
Misc. manufacture	28	144
Spinning twisting twining		
Apparatus	57	24
Strand structure	57	203
Tufting sewing machine	112	80.1+
Woven pile fabrics	139	393+
Strands	139	395
Cherry Seeder	30	113.1+
Chess Pieces	D21	52
Chest (See Box)		
Cedar	206	278.1
Medicine	206	828*
Protector	2	92
Tool	206	349+
Wind	84	54
Woodworking tool and workbench combined	144	285
Chest of Drawers	D 6	432
Cheviot	26	19+
Chevron	2	246
Badge	40	1.5+
Chewing Gum	426	3+
Package	206	800*
Tobacco combined	131	347+
Chicken (See Fowl)		
Chiffon, Foamable Edible Material	426	571
Chiffoniers, Trunks	190	19
Childproof Bottle Closures	215	201+
Chilean Type Mill	241	107+
Chill		
Brake shoes	188	260
Fastener bedstead	5	289
Metal founding	164	371+
Chilling iron processes	164	127
Consumable	164	357
Patterns chill supporting	164	230
Sand mold or core having	164	352+
Static mold having	249	111
Chime		
Bells	84	406+
Clocks	368	273
Electric	340	392+
Musical	84	402+
Design	D17	22
Electric tone type	84	1.1+
Signal	116	141
Chimney		
Building flue heaters	98	46
Cap, masonry	52	244
Cap, ventilating	98	67
Cleaning	134	1+
Sweep or brush	15	162+
Lamp chimney lantern	362	180
Chimney element	362	312+
Chimney element design	D26	118
Masonry smoke flue	52	245+
Static mold for making	249	17
Ventilation	98	58+
Design	D23	374
Chin		
Rest		
Firearm stocks	42	71.1+
Violin	84	279
Violin combined with support	84	278
Strap		
Hats	132	58
Surgical	128	164
Support		
Corpse	27	25
Surgical	128	164
China Oil	260	398+
Compositions	106	244+
Synthetic resin or natural rubber ctg (See synthetic resin or natural rubber)		
Chinaware	D 7	1+
Display stand	D 6	512+
Vase	D11	143
Chintz	139	426R
Chip		
Breaker	144	243+
Cutting tool with	407	2+
Cutting	83	906*
Making		
Comminuting type	241	83+
Slicer type	144	162R+
Poker	40	27.5
Design	D21	53
Chiropody		
Bandages	128	157
Kinesitherapy	128	33
Pedicuring instrument	132	73+
Shields	128	153
Chisel		
Cold	30	168
Design	D 8	47
Dovetail	144	88
Mortising	144	67+
Sawtooth cutting machine	76	28
Stoneworking	125	41
Tenoning	144	202
Wood turning	142	42+
Pattern guide combined	142	21
Woodworking implement	30	167+
Chitin	423	413
Chloracetic Acid	562	602
Chloral	568	495+
Hydrate	568	844
Chloramine	423	383
Chloramphenicol	564	213
Chloranil	260	396R
Chlorate	423	475
Electrolytic synthesis	204	95
Per compounds	204	82+
Oxidizing compositions	252	186.1+
Chlorazene Chloramine T	260	543R
Chlorhydrins	568	841+
Hydrolysis of	568	859
Chloride		
Binary halides inorganic	423	462
Biocides containing	424	
Dissolver, sodium chloride & potassium chloride	422	902*
Halocarbons	570	101+
Hydrochloric acid	423	481+
Organic acid halides	260	543R+
Oxidizing compositions	252	186.1
Salts inorganic	423	462+
Sulfur	423	462
Chlorination		
Halogenation	260	694
Hydrocarbon	570	101+
Hydrometallurgy	75	111+
Sulphating combined	75	110
Water purification	210	198.1+
Processes	210	754+
Chlorine	423	500
Biocides containing	424	149
Bleaching with	8	101+
Electrolytic	204	133
Electrolytic synthesis	204	128
Feeder	137	87+
Chlorine Di, -monoxide Oxidizing Composition	252	187.21
Chlorocarbonic Acid Esters	558	280+
Chloroform	570	181
Preparation from acyclic halides	570	101+
Chlorophenols	568	774+
Chlorophyll	540	145
Chloroprene	570	189
Preparation	570	216
Solid polymers (See synthetic resin or natural rubber)		
Chlorpheniramine, in Drug	514	357
Chock		
Boat supports	114	381
Design	D12	217
Railway car stops	104	257+
Brake type	188	36+
Ship spar	114	101
Vehicle fender combined	293	7
Wheel on ground	188	32
Chocolate	426	631+
Confectionery containing	426	631+
Molding	425	
Refiner	241	
Choke		
Coil	336	
Firearm barrel	42	79
Input filter	333	181
Plug	184	52
Tube		
Having plurality of leaves	261	DIG. 58
Interchangeable	261	DIG. 57
Leaves flexible	261	DIG. 59
Longitudunally reciprocating along air passage	261	DIG. 61
Having movable parts	261	DIG. 62
Suction operated	261	DIG. 64
Throttle-operated	261	DIG. 63
Cholanic Acid	260	397.1
Cholecalcipherol, in Drug	514	167
Cholesterol Compounds	260	397.2
Cholesterol, lipids, triglycerides	436	71*
Choline	564	293
Choline Acetate	560	253
Chopper		
Cornstalk type	56	60+
Cotton chopper plow	172	
Chopper heads	172	518+
DC to AC	363	124+
Food	241	
Block or board	269	289R+
Meat tenderer	17	31
Stalk	56	500+
Chopping		
Blocks and boards	269	289R+
Chord		
Played by single key of keyboard	84	443+
Selector for zithers	84	286+
Christmas Tree		
Candle holders	431	296
Container or wrapper for	493	956*
Decorative lights	362	249+
Design	D11	118+
Design	D11	118
Holder	248	511+
Design	D11	130.1
Fluorescent	250	462.1
Imitation	428	7+
Lamps	D26	25+
Light string	D13	25
Ornaments	428	7+
Design	D29	1
Series lamp system	315	185R+
Chromanes	549	398+
Chromate Processes		
Metal coating	148	6.2+
Photographic	430	289
Chromatography	210	656+
Gas separation	55	386
Process	55	67
Testing	73	23.1
Hydrocarbon purification by	585	825
Liquid	210	656+
Gel or liquid-liquid	210	635
Testing	73	61.1 C
Physical recovery methods	435	803*
Chromium	75	
Alloys (non-ferrous)	420	428
Alum	423	544+
Chromates	423	592+
Electrolytic synthesis	204	89
Ferro	420	34+
Organic compounds	556	57+
Oxid pigments	106	302
Oxides and hydroxides	423	607
Electrolytic synthesis	204	97

	Class	Subclass
Plating		
Electrolytic	204	51
Electrolytic plural layers	204	41
Metal stock	428	666
Preparation	75	84
Electrolytic	204	105R
Chromoscopes		
Color picture viewer	350	174
Colorimeter	356	402+
Chromosome Multiplication, Cross-breeding	47	DIG. 1
Chronograph		
Electrical or electronic	368	114
With speed determination	324	178+
Recorder	346	
Watches	368	101
Chronometer		
Structure	368	62+
Testing	73	6
Chronoscope		
Electric	368	107
Chrysanthemic Acid, in Drug	514	572
Annular for wood turning	142	52
Counterpoised	142	51
Electromagnetic holders	335	289+
Hollow mandrel guide	142	54
Key	81	16
Nailing machine	227	149
Chuck (See Socket)	279	
Church Bell	116	148+
Electrically simulated	340	398
Churn	366	
Agitator type	366	
Butterworker combined	99	452+
Design	D 7	371
Chute (See Conveyer)	193	
Amusement railway	104	69
Assorting		
Gaseous suspension	209	149
Gauging apertures	209	682
Household sifting bin	209	375
Liquid suspension reciprocating	209	196
Liquid suspension stationary	209	202
Manual	209	702+
Rolling incline	209	69
Billiard rack, pocket and	273	11R
Cabinet		
Showcase type	312	124
Stock and sample	312	121
Can making machine body feeding	413	70+
Card magazine	312	60
Conveyer having gravity section	198	523+
Deposit and collection receptacle	232	44+
Dumping vehicle	298	7
Fire escape	182	48+
Fruit gatherer	56	328R+
Furnace fuel feeder	110	116+
Fixed detaching member	56	340
Pole type pivoted jaw	56	334
Pole type pivoted knife	56	336
Hopper RR car bridge for side	105	252
Hopper RR car center	105	249
Mortar mixing discharging	366	68
Nail etc feeder inclined	10	165
Planter vibrating	111	76
Plural destination conveyor &	198	359
Showcase	312	124
Stock and sample	312	121
Spindle lathe automatic feed	142	20
Stratifiers liquid treatment	209	458
Reciprocating	209	437+
Reciprocating rotating	209	435+
Rotating	209	444
Spiral	209	434
Stratifiers movable bed	209	479+
Stratifiers pneumatic treatment	209	477
Movable bed	209	471+
Textile liquid treating receptacle conditioning	68	178
Pleating in receptacle	68	177
Typesetting converging	276	20+
Vehicle body endgate combined	296	61
Wooden silo	52	195
Cibanite T M (aminoplast)	528	230
Cigar and Cigarette	131	
Ashtray	D27	8+
Bands for	D20	22+
Case	206	242+
Combination	206	236+
Design	D27	43+
Cutter	30	109+
Appliance combined	131	248+
Cigarette case combined	206	238+
Design	D27	51
End cutter	83	
X-art	83	926C*
Design	D27	1+
Extinguisher	131	256
Filter end	131	331+
Filter for	131	331+
Ash tray combined	131	234
Cigarette case combined	206	236+
Dispensing case combined	206	236+
Electric heating coil	219	260+
End cutter combined	131	249
Magazine cigarette dispenser combined	221	136+
Matches	44	42+
Part of cigar or cigarette	131	351
Smoking device combined	131	185
Holder	206	242
Design	D27	2+
Lighter	D27	36+
Liquid gas cartridge for	D23	202
Machines	D20	1+
Magazine type cabinet	312	72+
Making	131	280+
Turning cigarettes end for end	198	952*
Packages	206	242+
Manufacturers container	493	910*
Matches combined	206	85+
Packaging	53	148+
Papers container	206	237
Pocket lighter		
Butane	431	130+
Butane	431	254
Catalytic	431	147
Frictional, chemical, percussive	431	267+
With cover	431	144+
With cover	431	129+
Protective cover for packs	D27	48+
Smokers tools	D27	51
Snuffer	D27	8
Vending machine	221	
Coin operated	194	
Design	D20	1+
With igniting means	221	136+
Cinchona Alkaloids	546	134+
Cinchonine	546	134
Cinchophen	546	170
Cine Film	430	934*
Cinema (See Motion Picture)		
Cinematography	D16	
Cineol	549	397
Cinerary Urns	27	1
Circuit (See System, Electric)		
Applications, battery	136	293*
Breaker	335	6+
Design	D13	34
System	361	1+
Closer	200	
Annunciator	340	815.28
Arc lamp having	314	131
Arc lamp shunt closing or cut out	314	10+
Changeable exhibiter control	40	463+
Lamp or electronic tube combined with	315	32+
Lamp or electronic tube system	315	362+
Closing apparatus	200	
Controller (See system having)		
Clock electric	368	52
Condenser structure	361	271+
Electronic tube type	328	85+
Flashing light etc	40	902*
Phase regulation	323	212
Regulators	323	
Rheostat or resistance	338	
Switch structure	200	
Transformer	323	355
Trolley conductor section switching	191	16+
Tuners	334	
Interruption		
Electrical systems	307	96+
Electrolytic	361	435
Inverter systems	363	109
Rectifier systems	363	108+
Regulator electrical systems	307	96+
Therapeutic apparatus	128	423R+
Maintenance		
Signal box transmission	340	292
Telegraph lines	178	69R
Maker and breaker	200	
Annunciators	340	815.28
Arc lamp cutout or shunt	314	10+
Arc lamp system	314	131
Check control telephone	379	146+
Electromagnetic switching systems	361	139+
Key telegraph	178	101+
Lamp or electrode tube system	315	362
Motor system	318	
Switchboard details	361	332+
Telegraph manipulating	178	75
Telephone	379	414+
Printed	428	901*
Manufacture including etching	156	901*
On organic base	156	902*
Protector		
Lightning arrester	361	117+
Meter	324	110
Retainer automatic telegraph repeater	178	72
Work holder for electrical circuit	269	903*
Circular		
Knitting machine		
Independent needles	66	8+
United needles	66	79+
Looms	139	13R+
Saw		
Dressing and jointing	76	48
Metal sawing	83	594
Wood sawing	83	469+
Cistern	52	169.1+
Cleaners	15	246.5
Hoisting bucket type	294	68.22+
Molds for	425	59+
Sewage purification	210	532.1+
Citazinic Acid	546	296
Citral	568	448
Citric Acid	562	584
Citrus Fruit		
Cutters	30	123.5+
Machines	99	539+
Plants	PLT	45
Civetone	568	375
Clack Valve	251	298+
Clamp & Clamping Device (See Clasp; Fastener)	24	455+
Adjustable	403	DIG. 1
Apparel apparatus	269	
Applier		
Hose clamp	81	9.3
Surgical clip	29	243.56
Binder device releasably engaging aperture or notch of sheet	402	
Book or leaf holder	281	45+
Bracket clamped to support	248	225.31+
Bracket with article holding clamp	248	316.1+
Building components	52	584
Anchor or tie	52	698+
Clasp, clip, or support clamp	24	455+
Design	D19	65
Clip board	24	67.3+
Closure fastener	292	256+
Rod clamp type	292	305+
Cord rope and strand anchors	24	115R+
Bridge cable	14	22
Sash cord	16	205
Dental implement	433	153+
Dam	433	138+
Design	D 8	72
Hose	24	19
Appliers	81	9.3
With support	248	75+
Jar lid	D 9	435+
Locked	70	19
Metal deforming apparatus	72	293+
Moderator structure for nuclear reactor	376	458
Paper fastener	24	67R+
Portable	269	
Railway track fastening	238	338+
Saw setting	76	78R+
Scaffold binders	24	16R+
Shoemakers	12	103
Static mold	249	219R
Strand tensioning	242	149+
Surgical implement	128	346
Design	D24	27
Testing strength of specimen with	73	856+
Woodworking	269	
Clamped Work (See Holders)		
Abrading machine	51	217R+

Entry	Class	Subclass
Cutting & punching sheets & bars	83	
Cutting tables	83	451+
Drawcut	83	642+
Drawcut	83	644+
Pivoted cutter	83	597+
Punching or pricking	83	
Reciprocating cutter	83	613+
Table with	83	
Daguerreotype plates	354	2
Engraving	81	4
Fish dressing	17	70
Fowl supporting	17	44.3
Sewing machine presser	112	235+
Shoemaking	12	
Specimen testing by stress or strain application	73	856+
Vegetable and meat cutting	269	
Vises	269	
Pipe or cable	269	
Wire fabric making	140	51
Wood bending	144	269
Wood turning	142	
Woodworking	269	
Machine	144	278R
Clamshell Bucket	294	68.23+
Clappers for Bells (See Bells)		
Clarinet	84	382+
Clasp	24	455+
Clothespin (See clothespin)		
Design	D32	61+
Cuff holder	24	43+
Design apparel	D 2	200+
Fastener combined	24	326+
For nuclear reactor moderator structure	376	304
General purpose	D 8	382+
Jewelry	D11	86+
Pen or pocket	24	11R+
Pen or pencil	D19	41+
Project-retract means associated	401	104+
Utilitarian, garment use	D 2	200+
Classifying Solids	209	
Cutting or comminuting combined		
Apparatus	241	68+
Cereal processes	241	9+
Fluid applying apparatus	241	38+
Gas applying processes	241	19
Liquid applying processes	241	20
Processes	241	24
Vegetable and meat apparatus	99	510
Vegetable and meat processes	426	518+
Dorr type	209	462
Clathrates (See Definition Note (3))	428	402.2
Clavier	84	DIG. 25
Clavulanic Acid	540	349
Claw Bar Lever Nail Extractor	254	25+
Clay		
Acid activated	502	81
Burning in kilns	264	56+
Cleaner	209	
Comminuting combined	241	
Compositions	501	141+
Abrasive	51	308
Adsorbents	502	407+
Catalyst	502	80+
Catalyst carrier	502	439+
Porcelain	501	141+
Refractory	501	94+
Molding and casting	425	
Pigeon	273	362+
Projector	124	6+
Pipe or tube	138	175+
Shaping processes	264	
Firing ware combined	264	60+
Treatment	501	145+
Chemical	501	146+
Cleaner and Cleaning (See Clearer)		
Abrading file	51	414
Antismut device combined	101	423+
Applicator	15	104R+
Material supply combined (See implements below)	401	
Ash tray combined with cleaning device	131	232
Attachment	15	246+
Curtain or shade	160	11
Electric insulator	174	211
Electrode tip	314	23+
Hand rake	56	400.8+
Harrow teeth	172	90
Liquid level sight glass	73	324
Printing apparatus	101	425
Razor	30	41
Solid material comminuter member	241	166+
Textile knitting apparatus	66	168
Tobacco smoking device	131	184.1
Type casting apparatus	199	62
Typewriting machine	400	701+
Band saw pulley	83	168
Boiler with means for	122	379+
Boot and shoe scrapers	15	237+
Bottle washing machine attachment	15	104.9
Bread and pastry	15	3.11
Bread and pastry	15	3.16
Brushes for, design	D 4	
Brushing scrubbing wiping general cleaning	15	
Burner	431	121
Burner tip	362	458
Button cleaning	51	171
Garment shields	51	265
Carpet	15	3+
Sweeper type	15	41R+
Cistern	15	246.5
Comb		
Brush machine	15	39
Combined	132	119
Implement	15	104.5+
Combustion means	110	236+
Composition		
Abrasive	51	293+
Detergent	252	89.1+
Flue	44	5
Lubricant combined	252	11
Metal fluxing	148	23+
Soap	252	369+
Soap products	252	368
Conveyor	198	493+
Currycomb	119	88
Design	D30	159
Design	D32	1+
Dispenser	222	148+
Receiver inlet cleaning attachment	141	90
Distillation apparatus carbon removers	202	241
Processes	201	2
Drive belt and chain	474	92
Earth boring means with	175	84
Earth boring tool with	175	313
Electric trolley head	191	62
Electrolysis		
Cleaning or etching	204	129.1+
Coating with cleaning or etching	204	32.1+
Electrostatic	15	1.5 R+
Engine cylinder	15	104.11
Filter	210	407+
Flue		
Brush and broom	15	162+
Flue attached	15	249
Scraper	15	242+
Fluid handling devices	137	237+
Fluid nozzle	239	106+
Foundry apparatus with	164	158
Fruits		
Processes	426	478+
Fur treatment	69	23+
Gas separator	55	282+
By liquid	55	242+
Electrical precipitator	55	108+
Process	55	96
Ginning saw	19	64
Glass mfg apparatus	65	27
Gun bore	15	104.5+
Cartridge contained	102	442
Gun combined	42	95
Harvesting binder needle cleaner	56	448
Hat cleaning	223	23
Heat exchange apparatus	165	95
Hoof	168	48R
Implement	15	104R+
Brush broom or mop with material supply	401	268+
Coated or impregnated with cleaning material	15	104.93+
Container, srapper for	493	942*
Erasing (See eraser)	15	424+
Material supply combined	401	
Padlike, bladelike or apertured with material supply	401	261+
Porous pad with material supply	401	196+
Stationary and with material supply	15	104.92
Stranded fabric mop or mop head	15	229.1+
With fixed handle	15	229.2+
With pivoted handle	15	229.6+
Intestine	17	43
Liquid contact		
Solids apparatus	134	
Textile apparatus	68	
Textile processes	8	137+
Liquid heaters and vaporizers	122	379+
Water tube heater	122	361
Machines	15	3+
Lights on	362	91
Marine hull	114	222
Metal working with (See metal debris)	29	DIG. 7
Cutter engaging cleaner	29	DIG. 97
Rotating, reciprocating, or oscillating cleaner	29	DIG. 98
Nozzle	239	106+
Oil vaporizing apparatus carbon removers	196	122
Processes	208	48R
Paper making felt & wire combined	162	274+
Pen	15	423
Piano action combined	84	64
Pipe and tube (See apparatus or process using)		
Brushing machines	15	88
Implements	15	104.3+
Liquid contacting apparatus	134	166R+
Processes	134	1+
Tobacco smokers	131	243+
Piston ring groove	15	104.12
Railway		
Snowplows	37	214+
Switch point	246	444
Track	104	279+
Track sweepers	15	54+
Receptacle		
Brush and broom implement	15	164+
Brushing machines	15	56+
Liquid contacting	134	
Wiper implement	15	211+
Refrigeration apparatus	62	303
Sand blast nozzle combined	291	12
Scale removing from metal	29	81R
Scouring with abradant	51	16+
Solids cleaning apparatus cleaner combined	34	85
Sound record	15	
Spark plug	15	104.11
Steam boilers	122	379+
Stove flue	126	16
Stratifier	209	487
Street		
Air blast or suction	15	300R+
Flushers	239	159+
Sweepers	15	78+
Teeth	433	216
Textile fiber working combined	19	262
Roll cleaning	15	256.51
Textile weaving weft break sensing fork	139	379
Tobacco smoking device	131	243+
Tobacco treatment		
Fluids or fluent material	131	300+
Leaf	131	325
Stemming and cleaning	131	315+
Tool driving or impacting combined	173	57+
Automatic control	173	3
Track sander delivery pipe	291	42+
Vacuum	15	300R+
Nozzles	15	415R+
Processes	134	21
Tool driving motor fluid caused	173	60
Valve	137	244+
Vegetables		
Processes	426	478+
Washing-drying machines	D32	6+
Welding apparatus combined	228	18
Well	166	170+
Processes	166	311+
Screen with washing point or shoe	166	157+
Wheel scraper vehicle attached	280	158R+
Wood saw	83	169
Clearance Reduction		
Compressor	92	60.5
Regulator	417	274+
Internal combustion engine	123	48R
Clearer (See Cleaner)		
Diggers potato	171	114+

	Class	Subclass
Fork	294	50+
Harvesters	56	314+
Hat making	223	11
Plows	172	606+
Propeller screw	440	73
Railway track	104	279+
Rakes		
— Hand	56	400.8+
— Horse drawn	56	395
Roadway snow excavators	37	196+
Sifters	209	379+
Textile machinery		
— Drawing roll	19	258+
— Loom shuttle	139	261+
— Spinning	57	300+
— Warp beaming	28	173
Cleat		
Insulator electrical	174	157
— Bracket for	248	67.5
Ship	114	218
Shoe sole edge gripping	36	7.7
Shoe spike or	D 2	317
Sliding car door	49	472
Traction for wheels	301	43+
Cleaver	30	308
Clerestory	52	18
Clevis	403	157
Ball socket	403	142
Coupler	403	302
Hoist line		
Insulators	174	207
Links	59	86+
Thill couplings	278	52+
Clew	114	115
Clicker	242	84.51 R
With slip clutch	242	219
Climber		
Plants		
— Rose	PLT	2+
— Vine	PLT	54+
Playground apparatus	D21	242+
Pole post or rope	182	133+
Tire attached surface rail climber	301	90
Toy figure	446	314+
Clincher Rim	152	382+
Clinometer	33	365+
Clip		
Applier	29	243.5+
— Surgical	29	243.56
Bottle attachment	215	101
Design		
— Apparel	D11	200+
— Hair	D28	39+
— Jewelry	D11	40+
— Paper, pen, pencil	D19	56
Leaf spring	267	53
Making and joining		
— Fabric wire	29	243.56
— Spring head	29	172
— Wire	140	82+
Nose animal	119	135
Paper	24	455+
Paper fastener	D19	65
Pedal crank combined with shoe type	74	594.6
Pen and pencil	24	10R+
— Design	D19	56
— Project-retract means associated	401	104+
Rafters and	33	DIG. 16
Rail fastenings	238	378
Structural unit fasteners	403	
Surgical	128	337
Wire fabric	245	3
Clippers		
Clip applying	29	243.5+
Cloth finishing cutters	26	7+
Fur hair removing	69	25
Hair	D28	52+
Hand manipulated cutters	30	
Harvesters	56	
Hat making	223	19
Nails	D28	60+
With hair catching	30	124+
Cloak	2	
Clock	368	62+
Atomic	331	3
Atomic frequency control for	331	3
Cases	224	164+
— Clock movement casing	206	18
— Design	D10	1+
Cordless electric	310	191
Crystal mounting	29	807
Design	D10	1+
Digital exhibitor	D10	15
Electric motor for	310	162+
Floor standing model	D10	16
Illumination combined	362	23+
Keys	81	122+
Power substitution for electric	307	64
Processes for making	29	177+
Radio	455	344
Radio encased	455	231
Radio or phonograph combined	D14	20
Radio station selector	334	
Spring assembly	29	228
Staking	29	231+
Supports	248	114+
Testing	73	6
Time clock	D10	41
Transistor operated	318	127
Winding key	D 8	347+
Work control (See time operated and controlled)	368	1
— Fuses	102	276
— Switch operation	200	35R+
— Time locks	70	272+
Clocking Device (See Timers)		
Clod Crusher	172	
Cultivator and roller	172	170+
Harrow and roller	172	170+
Land roller	404	122+
Plow and roller	172	170+
Soil elevator and treater	172	33
With separating after earth working.	172	32
Clog, Footwear	D 2	292
Platform heel	D 2	324
Platform sole	D 2	322
Close (See Closer; Closing; Closure)		
Closed Circuit Grinding		
Apparatus	241	80
— Fluid applying	241	61
— Gas swept grinder	241	52+
Processes	241	24
— Cereal	241	10
— Gas applying	241	19
— Liquid applying	241	20
Closed-die Metal Shaping	72	343+
Forging	72	352+
Closer (See Circuit; Closure; Fly; Operator)		
Door	16	71+
— Check combined	16	49+
— Design	D 8	330
Fly closer lasting tool	12	113
Paper		
— Bag closing	493	186
— Box closing	493	52
Slot		
— Cable railway	104	194
— Lever	74	566
— Pneumatic tube	104	161
Closet		
Sanitary	4	300+
— Basin combined	4	664+
— Bed combined	5	90
— Bowl couplings supports and drip collectors	4	252R+
— Bowl outlet obstruction remover	4	257
— Bowls	4	420+
— Cleaning devices	4	257
— Closure	4	253
— Covers for seats	4	242+
— Design	D23	295+
— Disinfection	4	222+
— Dry closet and furnace combined	4	111.1+
— Dry type	4	449+
— Dry type with drier or burner	4	111.1+
— Flushing tanks	4	353+
— Head foot and body rests	4	254
— Seats	4	237+
— Static mold	249	58
— Tub and basin combined	4	663
— Ventilation	4	209R+
Warming	126	18
— Stove base combined	126	55
— Stove pipe heated	126	17
Closing		
Ammunition loading and shell closing	86	25+
Apparel pocket attachment	2	252
Bottom opening hopper door equalizer	298	37
Conductor conduit	191	25
— Third rail	191	31
Inkstand (See inkwell)		
— Pen, actuated for opening	15	257.75
Metal tube ends		
— Welding apparatus	228	60
Paper box or tube		
— Folding setting up end closing	493	183+
Receptacle filling and closing	53	266R+
— Cooking and filling with cooked food	99	356
— Multifolded closure	53	378+
Closure (See also Article Having Closure eg Window)	160	
Adaptable for various sizes	220	287
Apparel		
— Buttons and fasteners	24	90R+
— — Design	D 2	222+
— Coat	2	96
— Collar	2	141R+
— Design	D 2	200+
— Flies	2	234
— Lacing devices	24	140
— Pocket	2	252
— Separable fasteners	24	572+
— Shirt	2	128
— Shoe	36	50+
— Skirt placket	2	218
— Skirt waistband	2	219
— Sleeves or legs	2	270
— Slit	24	437+
— Trouser waistband	2	235
Applicator attached	401	126+
Bayonet		
— Metallic receptacle	220	293+
— Wooden barrel bung	217	107
Boiler header	122	360+
— Tube	122	364+
Bottles & jars	215	33+
— Designs	D 9	367+
Box		
— Designs	D 9	414+
Breech	89	17+
Building construction with	52	204+
Bung hole	217	98+
Cabinet		
— Foldaway device support combined	312	22+
— Inner and outer	312	291
— Sectional unit	312	109+
Check	16	82+
— Door closer combined	16	49+
— Photographic plate holder	354	281
Closet bowl	4	253
— Obstruction remover combined	4	255+
Coffin	27	14+
Cooking ovens	126	190+
Disc manufacture of	493	379+
Dispensing	222	544+
Distilling apparatus	202	242+
Fastener	292	
— Barrel	217	89
— Bottle and jar	215	273+
— Breakable neck bottle and jar	215	35+
— Burglar alarm combined	116	12+
— Cork retaining wire	140	85
— Envelope	229	77+
— House letter box	232	23
— Keepers for	292	340+
— Lantern reflector combined	362	374
— Letter sheet	229	92.5+
— Marine hatch	114	203
— Metallic receptacles	220	315
— Paperboard box	229	125.19+
— Pivoted type closure bottle and jar	215	237+
— Railway track joint	238	312
— Safes	109	59R+
— Sectional metallic receptacle	220	7
Fences	256	
Flexible, eg sliding desk top type	206	816*
Flexible and portable	160	
Flexible bag closure, misc.	383	42+
Floating roof type	220	216+
Foot pedal actuated	220	262
Friction held	220	352
Gas retorts	48	124
Grate		
— Fireplace summer front	126	140
— Stove draft control	126	160
Hinged metal receptacle	220	334
House paint can	220	354

	Class	Subclass
Humidifier	312	31.1
Ice bunker	98	7
Lock	70	77+
Magnetic	220	230
Mail box	232	45+
Mailing sheet	229	92.5+
Manhole	404	25+
Mesh sifter element	209	391
Motor vehicle	49	
Having vehicle system responsive to its position	180	286
Related to theft prevention	180	289
Responsive to vehicle movement	180	282+
With vehicle feature	296	146+
Movable or removable	49	
Operator (See closer)	49	324+
Alarm	116	100
Burglar alarm	116	85+
Cabinet	312	319
Car dumping tilting track	414	354+
Curtain shade screen (See appropriate subclasses in)	160	
Elevator door	187	51+
Furnace door	110	176+
Gate	49	324+
Pen-operated inkwell	15	257.75
Power plant X-art	60	903*
Railway dumping car	105	286+
Railway dumping car control	105	311.1+
Railway passenger car	105	341+
Receptacle and stand	248	147
Receptacle pivotally mounted on rack	211	83
Seat and cover	4	251
Sectional cabinet equalizers	312	110
Showcase	312	139
Stand combined	248	147
Stove door	126	192
Tilting receptacle and stand	248	134
Ornamental, metal receptacle	220	376
Partitions & panels, flexible & portable	160	
Permanent	220	66+
Pipe or conduit	138	89+
Railway tie end closure	238	104
Protective grill or safety guard	49	50+
Purses	150	118+
Railway rail joint	238	312
Receptacle		
Bag, flexible	383	42+
Barrels	217	76+
Baskets	217	124
Batteries	429	163+
Bottles and jars	215	200+
Bucket paper	229	910+
Bucket staved wooden	217	76+
Cigar cigarette smoking device holder	131	242
Collapsible material guide combined	222	528
Compartmented letter box	232	25+
Deposit and collection receptacle	232	44+
Dispenser	222	544+
Electrical boxes and housings	174	65R
Electrical outlet and junction box	220	3.8
Envelope	229	76+
Flexible closure	206	816*
Folded blank paper bucket	229	910+
Lantern	362	375
Lubricator valve combined	184	80
Metallic receptacles	220	200+
Milk pail	220	200+
Movable material guide combined	222	531+
Nuclear fuel	376	434
Oil cup	184	88.1+
Paper or web container	493	962*
Paperboard box	229	124+
Purses	150	118+
Rack and closure	211	71+
Sectional for dispenser	222	502
Sifter	209	372
Sliding apertured cap type	222	522+
Surface condenser	165	73+
Tube	229	93
Usable as cup or funnel	141	381
Water heater overflow combined	126	384
Wooden boxes	217	56+
Safes	109	64+
Screw-type	220	288
Showcase	312	138R+
Rack interconnected	312	129+
Tray interconnected	312	127
Soldered or sealed	220	361
Sorter for container closure	209	928*
Static mold	249	204
Stopper type	220	307
Tab tops	220	271
Tie string	493	962*
Time controlled	49	29+
Tire casing	152	515+
Transparent	220	377
Boiler type sight glass	73	330+
Door peep	49	171
Sight glass	73	334
Stove oven door	126	200
Window	49	
Valve	493	962*
Vault	404	25+
Weatherstrip seal	49	475+
Well	166	325+
Zipper, ie receptacle securement	206	810*
Cloth (See Fabric; Textile)		
Abrasive filled	51	294
Bags	383	
Buttons covered	24	92
Making	79	5
Cleaning or polishing	15	208+
Cutting cloth stock	83	925CC*
Finishing	26	
Beating	26	25+
Cutting	26	7+
Expanding device for webs	26	71+
Inspecting	26	70
Napping	26	29R+
Pile fabrics	26	2R+
Rubbing	26	27+
Sheet stretching frame	38	102+
Shrinking	26	18.5+
Singeing	26	51+
Stretching or spreading & working	26	51+
Tenter for webs	26	89+
Web spreader	26	87+
Web stretcher	26	71+
Biaxial	26	72+
Fireproof	427	223+
Packages or reels of	242	222+
Purses	150	
Reeling and unreeling	242	62
Clothes, Clothing (See Apparel)		
Bed clothing	5	482
Boots and shoes	36	
Design	D 2	264+
Brushes, bristle	D 4	130+
Brushes, non-bristle	D28	9+
Card clothing	19	114
Applying	140	97+
Cleaning machines	D32	1+
Devices for putting on & removing	223	111+
Dryers (See driers drying)		
Handling devices		
Laundry sticks	294	23.5
Tongs	294	8.5
Hanger	223	85+
Holder on chairs	297	190
Knitted	66	169R+
Design	D 2	43
Design	D 2	44+
Label	D 5	63+
Patterns	33	2R+
Pounder	294	23.5
Rack	211	
Bed attachment	5	504+
Sewing methods	112	262.1+
Thermal and electrical treatment	128	379+
Clothesline		
Cleaners	15	256.6
Guide pulley	254	389+
Isolated supports	211	119.1+
Props	248	353
Reels	242	100+
Reels	D 8	358+
Support or attachment	D32	60+
Clothespin	24	455+
Bag design	D32	36
Clasps	24	455+
Design	D32	61+
One piece rigid	24	570+
Pivoted	24	489+
Resilient	24	530+
Track or way guided	24	522+
Wire working making	140	83
Woodworking making	144	9
Clothesline		
Cleaners	15	256.6
Guide pulley	254	389+
Isolated supports	211	119.1+
Props	248	353
Reels	242	100+
Reels	D 8	358+
Support or attachment	D32	60+
Clotting (See Coagulating)		
Analysis by coagulometer	422	73
Test with blood clotting factor	435	13
Cloud Chamber	250	335
Clover Huller	130	2
Cloxacillin	540	327
Club		
Carrier for policemans weapon	224	914*
Dumbbell and exerciser	272	122+
Game	273	67R+
Design	D21	210+
Golf	D21	214+
Clutch (See Pawl; Rachet)	192	
Brake and	192	12R+
Coaster brakes	192	6R
Steering by driving	180	6.2+
Crane	212	170+
Eddy current	310	105+
Electric clutch control system	310	92+
Automatic	310	94+
Electromagnetic	310	103+
Freewheeling	192	41R+
Gear combined	192	20+
Lathe carriage feeds	82	21R+
Lathe transmissions	82	29B
Mechanical movements	74	
Metal rolling mill drives	72	249
Transmission and clutch controls	192	3.51+
Impact delivering	173	93.5
Intermittent grip type movement	74	111+
Exhibiting device		
Electrical control	40	463+
Mechanical control	40	446+
Magnetic		
Automatic electric friction	192	40
Operator electric friction	192	84R
Motor combined	192	
Railway wheel and axle drive	105	130
Reversing mechanisms		
Planers	409	336
Tapping	408	123
Side, in bolt to nut lock	411	296+
Take up	192	110R
Exhibiting device		
Electrical control	40	463+
Mechanical control	40	446+
Transmission combined	192	3.51
Crane	212	170+
Coagulating	252	315.1+
Blood		
Composition for	424	
Properties of fluids determining	73	54+
Surgical	128	334R+
Electrical	128	303.17
Synthetic resin or natural rubber	528	480+
Coal		
Bag or carrier	294	149+
Bin		
Building construction	52	192+
Charging or discharging	414	288
Locomotive tender	105	237
Briquette making	44	10C+
Buckets or hods	220	2
Cleaner or separator	209	
Coking	201	
Apparatus	202	
Comminutor	241	
Compositions	44	
Conversion to hydrocarbon	585	943*
Cutter	299	79+
Endless flexible	299	82
Rotary	299	89
Feed		
Boiler control	122	449
Boiler water grate combined	122	376
Furnace structure	110	101R+
Locomotive tender combined	105	232+
Metallurgical furnace combined	266	176+
Stokers	110	267+
Fertilizer containing	71	24
Gas making	48	
Liquefaction	208	400+

	Class	Subclass
Locomotives (See locomotive)		
Mineral oil recovery	208	400+
Mining	299	
Piling	414	28+
Ships	114	26+
Shovels	294	49+
Stove	126	
Coanda and Coanda Effect	239	DIG. 7
Flow affected by	137	803
Coaptation Splints	128	85
Coaptator	128	335+
Coast Artillery (See Gun)		
Coaster		
Brake	192	5+
Drink mat or holder	D 7	45
Glass or tumbler	248	346.1
Wagon	280	87.1+
Coat		
Apparel	2	93+
Boat mast	114	93
Fluid control device using		
Design	D 2	183
Design, dress &	D 2	91+
Design, evening dress or suit	D 2	53+
Design, 3 pc. ensemble	D 2	87+
Forms	223	68+
Hangers	223	92+
Design	D 6	315
Coated Article (See Type of Article)	428	
Electrolytically produced	204	14.1+
Explosive or thermic stock	149	3+
Metal particle	428	570
Metal stock	428	615+
Coating		
Anticorrosive	106	14.5+
Apparatus	118	
Detearing	118	639
Electromagnetic	118	623
Electrostatic	118	621+
Lubrication	184	
Match dipping	144	50+
Mold metal casting	118	
Mold plastics	118	
Paper hanging combined	156	574+
Pipes	118	708+
Pipes	118	105
Pipes	118	125
Pipes	118	404+
Pipes	425	110+
Pipes	425	460
Printing	101	
Typecasting	199	
Typesetting	276	
Typewriting machines	400	
Semiconductor vapor doping	118	900*
Xerographic transfer	118	644+
Applicator or applier	15	
Material supply combined	401	
Separable from supply	401	118+
Classifying separating assorting solids by	209	47+
Coated surface or mass	209	49+
Cleaning by	134	4
Combined with other manufacturing operations (See particular art)		
Composition (See composition coating)		
Window glass to prevent deposit and freezing of moisture	106	13
Electrical barrier layer	437	+
Composition	252	62.3 R
Electro less	427	304
Implement	15	104R+
Material supply combined (See brush, pens, etc.)	401	
Metal founding combined	164	
Mold making with	164	14
Mold surface	164	72+
Workpiece	164	75
Paper making combined		
Apparatus	162	265+
Process	162	135+
Paper undried	162	158+
Plastic	106	
Printing	101	
Processes	427	
Abrasive tool	51	295
Catalysts	502	100+
Cathode sputtering	204	192.1+
Dyeing combined	8	495+
Electrically conductive	427	58+
Electro less	427	304
Electrolytic	204	14.1+
Electron emissive	427	77+
Electrophoretic or electro osmotic	204	180.2+
Electrostatic	427	13
Foods	426	289+
Metals by chemical action	148	6+
Molds	427	133+
Textiles chemical modification combined	8	115.6
Recovery	427	345
Removal		
By electric spark or arc	219	383+
Repellent	106	3
Manufacture	65	24+
Robot	901	43*
Synthetic resin or natural rubber (See class 523, 524)		
Testing	73	150R
Textile operation combined		
Braiding, netting or lace making	87	23
Covering or wrapping	57	7
Spinning twisting or twining	57	295+
Warp preparation	28	178+
With glass mfg	65	60.1
Coaxial Cables	174	102R+
Making	29	828
Switch for	200	
Wave guide type	333	239+
Cobalamin, in Drug	514	52
Cobalt	75	
Alloys	420	435
Electrolytic		
Coating	204	48
Synthesis	204	112+
Metal stock	428	668+
Organic compounds containing	556	138+
Pyrometallurgy	75	82
Radiation imagery	430	936*
Cocaine	546	130
Cochineal Solution	252	408.1
Cock (See Valve)		
Abrading machines	51	28
Barrel bung	217	99+
With cutter or punch	222	81+
Carbide feed generator	48	40
Design	D23	233+
Dispensing	222	544+
Nozzle terminal	239	445
Pinch	251	4+
Sea	114	198
Support	248	67
Test liquid level	73	297
Time actuated	137	624.11+
With time indicator	137	552.7
Cocker	56	401+
Firearms	42	40+
Cockpit	244	119+
Cocktail Shaker	220	1E
Agitating feature	366	
Design	D 7	51
Cocktail Stick	D 7	42
Cocoa Preparation	426	631
Apparatus	99	467
Cocoanut	426	617
Comminuting	241	
Shelling	99	568+
Cod, Chemical Oxygen Demand Tests	436	62*
Cod Liver Oil	260	398+
Extraction	260	412.1
Medicine containing	424	107
Code and Code Signaling		
Card or tape punching	83	
Code conversion	234	69+
Finger print as code for lock	70	277
Punch selection	234	94+
Punch selector	234	94+
Selective	234	
Synthetic speech from	381	51+
Code conversion	234	69
Coded record		
Magnetic record medium	360	131+
Record controlled calculators	235	419+
Records per se	235	487+
Sensors	235	433+
Cryptography	434	56+
Fraud preventing	283	73
Printed matter	283	17
Typewriter	400	89+
Digital data error correction	371	30+
Electric	340	
Boring (See auger; bit)	178	
Error/fault detection techniques	371	
Handling & transmission of coded intelligence	178	2R
Modulation system	375	59+
Modulator	332	11R
Telephone	375	8+
Radio	375	
Transmitter	375	59+
Teaching	434	222
Train control	246	175
Train dispatching	246	7+
Visual or audible	116	18+
Codeine	546	44
In drug	514	282
Coffee	426	594+
Apparatus	99	568+
Balls	426	279
Cafeteria service unit	D34	14
Coating	426	289+
Grinder	D 7	372+
Grinding mill	241	
Maker	D 7	300+
Household type	D 7	300+
Pot	99	279+
Handle, asymmetric	D 8	DIG. 9
Strainer	210	464+
Preservation	426	330.2
Roaster	D 7	323+
Substitutes	426	596+
Cofferdam	405	11
Coffin	27	2+
Alarm systems	340	573
Casings portable	27	35
Design	D99	1+
Hearses	296	18+
Lid fasteners	27	DIG. 1
Lowering devices	27	32
Cable	254	264+
Coherers	329	100
Resistors for	338	1
Coil (See Spring)		
Bimetallic thermometer	374	204
Covers and cases	150	54R
Electrical	336	
Electromagnet	335	299+
Armature combined	335	220+
Electrotherapeutic	128	1.5
In audio reproducer	381	192+
Making	29	605+
Making laminated	156	
Surgical	128	1.3+
Switch	335	250
Winding apparatus	242	7.7+
Winding method	242	7.3
Electromagnetic switch	335	299+
Winding apparatus	242	7.7+
Winding method	242	7.3
Evaporator rotatable heater	159	25.1
Fabric		
Cleaning implement	15	225
Wire	245	6
Wire mesh	140	92.3+
Form electric	336	199+
Holder winding and reeling	242	129
Induction	336	
Body treatment	128	1.3+
Check operated switch combined	194	239+
Coupling networks	333	24R+
Electric systems	323	355
Furnace core-type	373	160
Furnace coreless	373	152+
Heaters	219	6.5+
In telephone terminal	379	391+
Metal heating	219	6.5+
Telephone movably actuated	381	192+
Telephone repeating type	379	344+
Transformer	336	
Tuners radio	334	71+
Liquid cooler	62	399
Liquid heater central standpipe water tube	122	244+
Liquid heater flue type internal water tube	122	169+
Liquid heater water tube	122	247+
Loading communication lines	178	46
Making miscellaneous	29	606+
Pipe making	72	135
Resistance electrical	338	296+
Shield electric	174	35CE
With coil	336	84R+

Entry	Class	Subclass
Springs		
Bed bottom with attached slats	5	245
Bed bottom with connecting fabric	5	235
Spiral bed bottom	5	256
Top connection bed bottom	5	272
Thermostat compound bar	428	616
Transformer	336	
Telegraph system	178	64
Tuners radio	334	59+
Coiling (See Winding and Coiling)		
Coin and Coining (See Check)	72	359
Box		
Closure	232	55+
Fare	232	7+
Savings	232	4R+
Savings paper	229	8.5
Toy	446	8+
Brassiere with coin pocket	450	89
Card and container	D18	3
Cash register, coin refund device	235	7A
Change holding machines	D99	34+
Change makers	453	1+
Check-controlled apparatus	194	
Coin ejection	194	342
Coin return	194	344+
Controlled by coin size or weight	194	334+
Controlled by machine operation	194	200
Empty dispensing magazine	194	200
Coin slot closers	194	351
Different values for different coins	194	227
Multiple coin	194	229+
Multiple coin, electrical	194	230+
Chute	193	DIG. 1
Coin-controlled		
Arcade-type viewer projector	D16	16
Billard table	273	11C
Circuit maker-breaker digest	200	DIG. 3
Corn popper	99	323.6
Lock	70	DIG. 41
Phonograph	D14	15
Jukebox	D14	26
Print machines	101	DIG. 23
Telegraphy system	178	2F
Telephony system	379	143+
Telephone design	D14	55
Vending machines	D20	1+
Coining	72	359
Counting apparatus	453	58+
Deliverer	232	64+
Delivering apparatus	453	18+
Detector	194	302+
Dispensing receptacle	453	18+
Handling	453	
Controlled apparatus	194	
Holder or tray	206	.8
Design	D99	24
Pocketbook-type design	D99	34
Wrapper or holder	493	945*
Mats	206	.8+
Metal work by deflection		
Helix	72	135+
Orbiting tool	72	66
Spiral	72	146+
Meters	194	
Money holder-receptacle	232	1D
Packaging	453	59
Packing apparatus	53	532+
Purse	150	150+
Design	D 3	42+
Keyholder combined	D 3	43
Receptacles and packages	206	.8+
Register, electrically operated	377	7
Sorters	453	3+
Stacking	453	61+
Symbolic token or ornament	D11	95
Tester	73	163
For coin controlled apparatus	194	302+
Scales	177	51
Tokens or coin type checks	40	27.5
Traps	232	55+
Wrappers	229	87.2
Wrapping machines	53	212+
Coke		
Apparatus for making	202	96+
Autothermic	202	91+
Autothermic and thermolytic distillation combined	202	85+
Separatory & thermolytic distillation combined	202	84

Entry	Class	Subclass
Processes of making	201	
Mineral oil feed	208	46+
Refining combined	201	17
Pushing machines	414	160+
Quenchers	239	750+
Colamine	564	503
Colander	210	415
Plunger combined	241	84.2
Colchicine, in Drug	514	629
Cold Cream	514	772+
Cold Frame	47	19
Cold Squirting	72	267
Cold Trap	62	55.5
Cold Welding	228	115+
Collage		
Representative of real object	428	39
Collagen	530	356+
Collapsible		
Bed	5	12R+
Core bar	164	400
Receptacle	222	92+
Sign	40	610
Steering post	280	777
Toy	446	487+
Building	446	478
Tube	138	119
Wall		
Receptacle	222	92
Weigher	177	126
Work holder	269	901*
Collar		
Animal restraining	119	106
Design	D30	152
Apparel	2	129+
Coat	2	98+
Design	D 2	602+
Design, fur	D 2	601
Ironing	223	52.1+
Making	223	2+
Making, X-art	83	901*
Protectors	2	60
Receptacle flexible	150	53
Shirt attached	2	116
Weaving	139	385.5
Buttoners	24	40
Buttons	24	101R+
Horse	54	19R+
Design	D30	137
Hame combined	54	18R
Shaping	69	3
Stuffing	69	4
Lamp shade support	362	443
Puller	29	700+
Rod collar joint	403	230+
Stovepipe	126	315
Supports	2	132
Collators	270	52+
With feeding	271	287
Collecting Mechanism (See Device with which Combined)		
Carbon black	422	150+
Collection Receptacle and Chamber	232	
Abrading device	51	270
Earth boring combined	175	
Railway passenger car dust guard	105	352
Stove ashpan	126	242+
Street sweeping machine	15	83+
Wiping daubing polishing implement	15	221
Collector		
Electric		
In arc lamp	314	129
Ring	310	232
To moving object	191	
Vehicle transmission	191	45R+
Overlapping roof shingles or tiles	126	DIG. 2
Collet		
Chuck	279	46R+
Split socket	279	41R+
Horological hair spring	368	177
Insertion / removal tool	29	213R
Collimator		
Image projector having	353	102
Light beam projector having	362	257+
Faceted reflector	362	297+
Optical system	356	138+
Signal reflector having	350	102+
Testing device having	356	153
Collisions (See Protection)		
Avoidance	367	909*
Aircraft systems	342	29
Ship systems	342	41

Entry	Class	Subclass
Collocating Gauge	33	613+
Collodion	106	169+
Colloids	252	302+
Radioactive	252	634+
Cologne	D28	5
Colophony (See Rosin)		
Color & Coloring (See Dyeing; Pigment)		
Application training	434	81+
Assorting by	209	580
Charts		
Color display	434	98+
Unknown color test	356	421+
Filter	350	311+
Photographic chemically defined	430	6+
Flowers by absorption of dye	427	4
Food	426	540+
Preserving or modifying color	426	262
Fresh flowers	47	58
Glass heat developed	65	111+
Lanterns	362	166+
Signal	362	168
Luminous, inorganic	252	301.6 R+
Luminous, organic	252	301.16
Mineral oil	208	12
Musical instrument accessory	84	464R
Paper	8	
Photography (See photography)	430	357+
Separation record making	430	356
Seams	112	17
Colorimeter	356	402+
Colorimetric Radiation Dosimeter	250	474.1
Colostomy		
Bags	604	338+
Receptacle engaging stoma	604	332+
Colter	172	
Bearings	384	157
Plow combined	172	165+
Rolling	172	518+
Column (See Mast; Pole; Post)	52	720+
Aerated column pump	417	108+
Design	D25	126+
Fractionating	261	75+
Combined	202	158+
Mineral oil	196	139
Still combined	202	153+
Making	29	155C
Static mold	249	48
Comb and Combing	D28	21+
Cleaner		
Brush machine	15	39
Comb combined	132	119
Implement	15	104.5+
Cornstalk huskers	56	114+
Cotton harvester	56	34+
Curry	119	86+
Design	D30	159
Fiber liberating	19	29
Fiber working	19	115R+
Grids combined	15	142
Hair	132	11R+
Design	D28	21
Hat fasteners	132	61
Honey foundation	6	11
Honey frames	6	10
Kits toilet	132	121
Leather manufactures	69	27
Making wire	140	100
Making wood	144	26
Making, X-art	83	908*
Making, X-art	425	805*
Music	84	94.1+
Scissors combined	7	136+
Seed gatherers	56	127+
Surgical	128	393
Textile		
Fiber preparation	19	115R+
Thread finishing	28	222+
Warp preparing	28	212+
Comber Board	139	86
Combinational Code		
Conversion, for punching	234	69+
Punching, selective	234	
Combustion (See Thermal)	431	
Attenuating sound	431	114
Burner control	431	18+
Chamber	431	350+
Exhaust pump	431	157
Fluid fuel burner	431	308+
Fluid fuel burner	431	350+
Outlet forms jet nozzle	431	158

	Class	Subclass
Pre-chamber	123	253$^+$
Control of nitrous oxides	60	900*
Earth boring by	175	12
Earth in situ	166	256
Engine	123	
Exhaust treatment	60	272$^+$
Combustion products with	60	39.5$^+$
Fluid injector	239	584$^+$
Fluid motor combined	60	616$^+$
Free piston device	60	596
Vacuum generated	60	397
Waste heat driven motor	60	597$^+$
Pump drive	417	364
Pump regulator	417	34
Reaction motor	60	200.1$^+$
Reaction motor with aircraft	244	74
Rotary piston	123	200$^+$
Temperature measurement	374	144$^+$
Turbine combined	60	598$^+$
Turbine combined, product	60	39.34$^+$
Failure responsive fuel safety cutoff	137	65$^+$
Fumigant	424	40$^+$
Processes	431	2$^+$
Product		
Exhaust treatment	60	272$^+$
Fire extinguishing agent	169	12
Generator	60	722$^+$
Power plant	60	39.1$^+$
Pulsating	431	1
Pump	417	73$^+$
Tube		
Comforters	5	502$^+$
Bedding	D 6	596
Comminuting	241	
Refrigeration combined	62	320
Refuse treatment means	110	222
Rock in situ	299	
Shredding metal	29	4.5 R$^+$
Shredding metal and sintering	419	33
Strainer combined	210	173$^+$
Combined	201	7$^+$
Tobacco	131	311$^+$
Feeding combined	131	109.1
Vegetables and meat	241	
Comminutor Element	241	293$^+$
With cooperating surface	241	261.1
Commode		
Dry closet	4	449$^+$
Communication Equipment (See Type of Equipment Desired)		
Aids for handicapped	434	112$^+$
Design	D14	52$^+$
Communications		
Demodulators & detectors	329	
Electrical	340	
Alarms	340	500$^+$
Audible, eg simulated noise in trainers	340	384R$^+$
Character recognition systems	382	
Code converters	340	347R$^+$
Code transmitters	340	345$^+$
Compressional wave	367	
Continuously variable indicating, eg telemetering	340	870.1$^+$
Digital comparator systems	340	146.2$^+$
Echo systems	367	87$^+$
Elevator	187	130$^+$
Error checking systems	371	
Image analysis	382	
Selective, eg remote control	340	825$^+$
Signals, visual	340	815.1$^+$
Tactual signals	340	407
Traffic & vehicle	340	22$^+$
Underwater	367	131$^+$
Well bore, compressional wave	367	81$^+$
Well bore, electrical	340	853$^+$
Light wave	455	600$^+$
Modulated carrier wave communications systems	455	
Modulators	332	
Multiplexing	370	
Non-electric miscellaneous	116	
Pulse or digital	375	
Radio	455	
Recorders	346	
Recording registers & calculators	235	
Sound recorders	369	
Telecommunications	455	
Telegraphic carrier wave	375	
Telegraphy	178	
Telephony	379	
Television, facsimile	358	
Commutation Book and Ticket	283	26
Commutator		
Bars	310	233$^+$
Brushes	310	248$^+$
Holders	310	239
Flanging tube to hold segments	29	243.52
Grinders	51	244
Making	29	597
Assembling apparatus	29	732$^+$
Structure	310	233$^+$
Turning portable lathe	82	4R
Type motors		
Alternating current	310	173$^+$
Universal	310	158
Compacting (See Jarring)		
Agitating	366	
Beverage infuser combined	99	287
Burnishing	29	90R$^+$
Cooking apparatus combined	99	349$^+$
Earth boring by	175	19$^+$
Foundry mold sand	164	37$^+$
Glass manufacturing	65	111$^+$
Leather hammering	69	1
Machinery design	D15	20
Metals	419	61
With sintering	419	
Nuclear fuel	264	.5
Pipe making combined	425	110$^+$
Pipe making combined	425	262
Powder metallurgy	419	38
Slip or slurry	419	40
Presses and pressing	100	
Foundry mold material	164	207$^+$
Receptacle filling combined	141	71$^+$
Bag	53	523$^+$
Refuse treatment means	110	223
Textile thread	26	18.6
Trash bundle	428	2
Compactor	D15	20
Compacts	132	82R$^+$
Absent special toilet article	401	126$^+$
Design	D28	78
Illuminated	362	135$^+$
Compander	333	14
Compressor amplifier		
Thru bias control	330	129$^+$
Thru impedance control	330	144$^+$
Thru thermal impedance control	330	143
Expander amplifier		
Thru bias control	330	129$^+$
Thru impedance control	330	144$^+$
Thru thermal impedance control	330	143
In electrical recording or reproducing	369	174
Comparators		
Configuration	356	388$^+$
Microscope subcombination	350	507$^+$
Optical	356	
Projection type	356	391$^+$
Compartment		
Aircraft		
Freight	244	118.1$^+$
Passenger	244	118.5$^+$
Safety lowering device	244	140$^+$
Bed element	5	308
Boiler	122	37$^+$
Building	52	234$^+$
Cigar cigarette smoking device holder	131	242
Land vehicle bodies	296	24R
Luggage	296	37.1$^+$
Marine	114	78
Motormans	105	342
Pillow	5	442
Receptacle		
Apparel pocket	2	253
Bag, flexible	383	38
Bag, multiple pocket	383	38$^+$
Baggage convertible to furniture type bureau	190	3$^+$
Barrel	217	75
Beverage infuser	99	316$^+$
Bottles and jars	215	6
Cabinet	312	
Collapsible	222	94
Cover for barrel	217	82
Dispenser	222	129$^+$
Drying apparatus rotary drum	34	109
Envelope	229	72
Fire extinguishers	169	71$^+$
House type letter box	232	21
Ice in wooden receptacle	217	130
Kilns and driers	34	209$^+$
Letter box	232	24$^+$
Material holder for heating	432	213$^+$
Medication containing	604	403$^+$
Metallic	220	20$^+$
Nested drum driers	34	128
Oil vaporizer	196	111
Paper box	229	120.2$^+$
Paperboard box	229	120.2$^+$
Pitcher	222	129$^+$
Purse	150	112$^+$
Showcase	312	117$^+$
Sifter	209	373
Sliding tray type paper box	229	10
Spaced wall or jacket	220	445$^+$
Special receptacles	206	
Traveling bags	190	109$^+$
Type cases	276	44$^+$
Washing machine tumbling drum	68	139$^+$
Wooden box	217	7$^+$
Safes	109	53$^+$
Movable	109	45$^+$
Secret cabinet with	312	204
Sofa bed element	5	58
Tub	4	514
Outlet fittings	4	208
Supply fittings	4	193
Water closet flush tank	4	363$^+$
Compass		
Course recorder	346	8
Design	D19	38
Divider	33	143R$^+$
Gyromagnetic	33	316$^+$
Gyroscopic	33	324$^+$
Magnetic	33	355R$^+$
Repeater stations combined	340	870.15$^+$
Saws	83	835$^+$
Scriber	33	27.2
Beam type	33	27.3
Scriber compound	33	26
Compensating		
Abrading tool wear	51	155
Alternating current phase	323	212
Backlash	409	146
Gear cutter	409	5
Brake wear	188	214$^+$
Clock and watch balances	368	171
Regulating	368	170
Clock pendulums	368	182
Clutch	192	110R
Compass magnetic	33	356$^+$
Conductor electric	174	12R$^+$
Flow meter	73	254
Fluid pressure	138	26$^+$
Gearing	74	710$^+$
Gearless differential	74	650
Gun sight	33	237
Internal combustion engine	123	192R
Lamp electric	314	13$^+$
Musical instrument winding	84	131
Pump pressure	417	540$^+$
Railway switch interlocking	246	152$^+$
Thermometer	374	197
For thermocouple circuit reference junction	374	181
Transmission electric	333	18
Valve	137	625.34$^+$
Vehicle	180	76
Vehicle steering	180	6.24$^+$
Weighing scale temperature	177	226$^+$
Compensator		
Expansion or contraction		
Static mold	249	82
Compliance Device	33	169R
Robot	901	45*
Composing		
Typesetting	276	
Sticks	276	38$^+$
Composite Article (See Laminate)		
All metal or with adjacent metals	428	615$^+$
Making by powder metallurgy	419	5
Making core and winding	242	7.1
Multipart article shaping	264	45.1$^+$
Static mold for making	249	83$^+$
Composition (See Type)	252	
Abrasive	51	293$^+$
Absorbent	502	400$^+$
Algicides	71	67
Alloy	420	
Antibiotic mixture	424	114

	Class	Subclass
Antifog	106	287.1^{+}
Antihemorrhagic	424	
Aquatic plant destruction	71	66^{+}
Asphalt, tar, pitch	106	
– Mineral oil only	208	22
Biocidal	424	
– Coating or plastic combined	106	15.5^{+}
– Mineral oil only	208	2
Candles	44	7.5
Catalytic	502	100^{+}
Ceramic	501	1^{+}
Chemiluminiscent	252	700
– Combined light source	362	34
Coating	106	
– Match splint	149	3^{+}
– Metal base reactive	148	6^{+}
– Mold	106	38.22^{+}
Colloidal	252	302^{+}
Core		
– Metal casting	106	38.2^{+}
– Plastic molding	106	38.2^{+}
Crayons	106	19
Defoliant	71	69^{+}
Desuckering	71	78
Detergent	252	89.1^{+}
Diagnostic biological	424	2^{+}
Dielectric fluid	252	570^{+}
– Hydrocarbon	585	6.3^{+}
Distilling apparatus	202	267R
Dye	8	
Electric insulation fluid	252	570^{+}
– Hydrocarbon	585	6.3
Electrically conductive	252	500^{+}
Electrode (See electrodes)		
Electrolyte	204	
– Battery	429	188^{+}
– Condenser or rectifier	252	62.2
Electron emissive	252	500^{+}
Embalming	424	75
Emulsifying	252	351^{+}
Explosive	149	
Fertilizer	71	
Fire extinguishing	252	2^{+}
Flux	148	23^{+}
Food and beverage	99	
Friction for matches	149	
Fuel	44	
– Gaseous	48	197R^{+}
– Jet	149	1
– Mineral oil only	208	15^{+}
Growth regulating for plants	71	65^{+}
Herbicides	71	65^{+}
High boiling	252	71
Humidistatic	252	194
Hydrocarbon	585	1^{+}
Inorganic materials only	106	286.1
Insulators		
– Fluent electric	252	570^{+}
Light transmission modifying	252	582
Low pour point	252	71
Lubricant	252	9^{+}
– Mineral oil only	208	18^{+}
Luminous, inorganic	252	301.6 R^{+}
Luminous, organic	252	301.16
Mantle	252	492
Match	149	
Medicinal	424	
Mold forming	106	38.2^{+}
Optical filter	252	582^{+}
Photo conductive	430	56^{+}
Photographic		
– Developing	430	464^{+}
– Sensitizing	430	570^{+}
– Sensitizing and developing	430	486^{+}
Plastic	106	
– Mineral oil only	208	22^{+}
Pyrotechnic	149	37^{+}
Radioactive	252	625
– Medicinal	424	1.1
– Production	376	171
Rubber containing (See synthetic resin or natural rubber)		
Tanning	8	94.19 R^{+}
Textile treating	252	8.6^{+}
Thermoelectric	136	236.1^{+}
Undertaking	424	75
Utilizing wave energy	522	71^{+}
Water softening	252	175^{+}
Wax	106	
– Mineral oil only	208	20^{+}
Welding flux	148	23^{+}

	Class	Subclass
Wick	502	400
X-ray contrast	424	4
Compound (See Type)		
Casting metal	164	91
– Electrolytic production of	204	59R^{+}
Hydrocarbon	585	16^{+}
Lumber	52	782^{+}
– Mail receiving	52	364^{+}
Metallic stock	428	615^{+}
Organic	260	
– Electrolytic production of	204	72^{+}
– Electromagnetic production of	204	157.15^{+}
– Electrostatic or electric discharge production of	204	165^{+}
– Fermentative production of	435	
– Mineral oils	208	
– Sugar and starch	127	
Refrigeration	62	332^{+}
– Automatic	62	175
Steam engines	91	152^{+}
Temperature sensing member	428	616
– In thermometer	374	205^{+}
Tools	7	
Compressed Air (See Pneumatic and Fluid Operated)		
Vehicle brakes	303	
Compressing Apparatus (See Compressor, Molding, Pressing, Shaping)	417	
Barrel	147	4^{+}
Harvesters		
– Binding combined	56	432^{+}
– Shocker combined	56	430
Refrigeration	62	498^{+}
– Automatic	62	190^{+}
Rims	157	2
Seams		
– Die	413	31^{+}
– Folding combined	413	27^{+}
– Roller	413	31^{+}
– Tube side seam	72	48^{+}
Snow	37	225^{+}
Tobacco plug making	131	111^{+}
Compression Member (See Column)		
Spring	267	70^{+}
Compressional Wave		
Generators	116	137R
Submarine signal	367	131^{+}
Compressor		
Body		
– Brassiere		
– Brassiere design	D 2	24
– Brassiere or chest bandage	450	1^{+}
– Corset or girdle	450	94^{+}
– Corset or girdle design	D 2	2^{+}
– For upper leg or thigh	450	101
Design of machine	D15	7^{+}
Gas pumps	417	
Refrigerator compressor muffler	181	403*
Tube pinching valve	137	384.2^{+}
Comptometer	235	82C
Computer		
Applications & systems	364	400^{+}
Automatic control of recorder mechanism	360	69^{+}
Cassette	360	131^{+}
Cassette or cartridge holder	206	387
Character recognition	382	
Coded document sorting	209	547^{+}
Computerized switching in telephone systems	379	284
Controlled display	364	518^{+}
Converting analog signal to digital	360	32
Data processing equipment	D14	100^{+}
Disc	360	135
Drum	360	136
Dynamic magnetic information storage or retrieval	360	
Electrical	364	
– Analog	364	800^{+}
– Applications		
– – Computational systems	364	400^{+}
– – Data processors, general	364	200
– – Data processors, special	364	900
– – Measuring system, electrical	364	481^{+}
– – Measuring systems, physical	364	550^{+}
– Calculators		
– – Analog	364	800^{+}
– – Digital	364	700^{+}
– Control systems	364	130^{+}
– General purpose digital data processor	364	200

	Class	Subclass
– Hybrid type	364	600^{+}
– Programming	364	300
– Special purpose digital data processor	364	900
– Systems controlled by data bearing records	235	375^{+}
Equipment	D14	100^{+}
Error checking system in computer	371	
General processing of digital signal	360	39^{+}
General processing of TV signal	360	33.1^{+}
General recording or reproducing	360	55^{+}
Geophysical system with digital computer	367	60
Head	360	110^{+}
– Mounting	360	104^{+}
– Transport	360	101^{+}
Holder for magnetic disc assembly	206	444
Memory system subcombinations	365	
– Data processing	364	900
Microprocessor	364	200
– Camera photographic operation	354	412
Modulating & demodulating	360	29^{+}
Monitoring or testing progress of recording	360	31
Non-electrical	235	
– Fluidic	235	200R
– Mechanical	235	61R
– – Record controlled	235	419
Ordnance	235	400^{+}
Programmed data processors	364	200
Radar system combined	342	195
Record controlled, electronic	235	375^{+}
Record copying	360	15^{+}
Record editing	360	13^{+}
Record medium	360	131^{+}
Record transport	360	81^{+}
Restrictive access over telephone line	379	95
Signal splitting	360	22
Special purpose television system application	358	93^{+}
Tape	360	134
Teaching computer science	434	118
TV application	358	903*
Computing		
Distance rolling contact	33	142
Nuclear reaction control	376	217
Registers	235	
Scales	177	25^{+}
– Weighing	177	25^{+}
Solar	33	268
Tape length	33	140
Workmens labor and wage recorder	346	
Concealed Data		
Printed matter	283	72^{+}
Concealment (See Camouflage)		
Warship	114	15
Concentrating		
Beverage	426	425^{+}
Cathode ray beams	313	441^{+}
Evaporators	159	
Ionized gas for thermo nuclear reaction	376	100
Liquids in liquids		
– Apparatus	422	256^{+}
– Processes	23	306^{+}
Milk	426	587^{+}
Ores	75	1R
– Rare element, inorganic	423	
– Separating solids	209	
Rubber		
– Natural	528	937*
– Synthetic or natural	523	335
Sugar crystallization combined	127	16
Concentric Tube or Cylinder		
Burner gas	431	195^{+}
Electric cable flexible	174	102R^{+}
– Fluid vacuum or air	174	28^{+}
– With insulation spacer	174	99R^{+}
Expansible chamber device	92	51^{+}
Flue	122	160^{+}
Header water tube	122	362
Heat exchanger	165	154^{+}
Pipe joint	285	133.1^{+}
Pipe joint flexible pipe	285	149
Transmission line	174	103^{+}
– Involving line parameters	333	236^{+}
– With nonsolid insulation	174	28^{+}
Concertina	84	375
Keyboard type	84	376R
Concrete		
Beam or joist	52	723^{+}

	Class	Subclass
Block machine	249	117$^+$
Compositions	106	89$^+$
Filling subterraneah cavity	405	267
Freight car	105	405
Making concrete blocks	264	333
Marine structure cast in situ	405	222$^+$
Mixers	366	1$^+$
Mold	249	
Mold	425	62
Mold	425	63
Mold	425	87
Mold	425	458
Tie wire puller	254	18$^+$
Piles	405	256$^+$
Casting in situ	405	233$^+$
Sheet	405	275
Pipe machine	425	110$^+$
Pipe machine	425	262
Pipes	138	175$^+$
Composite	138	153
Polymer containing (See synthetic resin or natural rubber)		
Prestressed making	264	228
Radiation barrier nuclear	376	293
Radiation barrier nuclear	376	287
Radiation barrier nuclear	250	515.1
Railway surface track	238	5
Center way concrete	238	7
Pedestals	238	115$^+$
Stringers	238	25
Rammers	425	424
Rammers	425	425
Rammers	425	456
Rammers	425	469
Reinforcement	52	720$^+$
Reinforcing	264	271.1
Reinforcing bar	D25	164
Ribbed construction	52	319$^+$
Structures	52	
Arch	52	88
Beam or joist	52	723
Corner	52	250$^+$
Faced	52	378
Hollow	52	380$^+$
Lift slab	52	125.1
Moisture remover	52	310
Position adjusting	52	126.1$^+$
Reinforcement	52	720$^+$
Ribbed	52	319$^+$
Stair	52	189
Void former	52	577
Condensation (See Alkylation; Polymerization)		
Preventer		
Electric heaters	219	203
Windows and windshields	52	171$^+$
Windows by ventilation	98	90$^+$
Condensed		
Milk	426	587
Condensed Billing		
Typewriter		
Line spacing	400	546$^+$
Paper feeding	400	595$^+$
Condenser		
Atmospheric condensate	62	272$^+$
Automatic	62	150$^+$
Process	62	80$^+$
Electric		
Amplifying device system	330	7
Anti induction telephone or system	379	414$^+$
Bushing insulating	174	143
Charging and discharging	320	1
Circuit element	361	271$^+$
Electrolytic type	361	433$^+$
Induction apparatus combined	361	268$^+$
Laminated making	D22	115$^+$
Lamp and electronic tube supply circuit combined with	315	227R$^+$
Making	29	25.41$^+$
Making electrolytic	29	570.1
Motor system	318	781$^+$
Nonelectrolytic type	361	271$^+$
Phase regulation	323	212
Reactance tube system	333	213
Synchronous motors	310	162$^+$
Testing	324	60R
Treatment living body	128	783$^+$
Variable type	361	277$^+$
With variable resistor	323	370
X ray system combined with	378	103

	Class	Subclass
Furnace electric	373	56$^+$
Microscopes	350	414$^+$
Lens construction	350	374$^+$
Mounting, vehicle type	180	68.4$^+$
Press	100	
Refrigeration including	62	
Steam engines	60	643$^+$
Steam heaters	261	DIG. 1
Textile fiber preparation		
Article screen	19	148
Feed with	19	89
Multiple rotor feed with	19	88
Web forming screen	19	304$^+$
Vapor	165	110$^+$
Arc lamp fumes	314	26$^+$
Beverage infuser	99	293$^+$
Cooking apparatus basting	99	347
Drying combined	34	73$^+$
Heat exchanger	165	110$^+$
Jet gas and liquid contact apparatus	261	76$^+$
Jet gas pump combined	417	173$^+$
Jet heat exchanger	165	112$^+$
Jet power plant exhaust	60	264$^+$
Jet power plant exhaust	60	266$^+$
Mercury vapor lamp	313	34
Mineral oil distillation	208	347$^+$
Mineral oil distillation apparatus	196	138$^+$
Mineral oil distillation with vaporizing apparatus	196	98$^+$
Refrigerant	62	506$^+$
Steam	165	110$^+$
Still separatory	202	164
Still separatory preheater	202	180
Still separatory vapor treating	202	185R$^+$
Water heater combined with	126	381$^+$
Water heater with liquid supply	126	380
Zinc pyrometallurgy	75	88
Condiment	426	650$^+$
Holder and dispenser		
Design	D 7	52$^+$
General types	222	
Hand manipulable shaker type	222	142.1$^+$
Conductor		
Electric	174	
Coated	428	375$^+$
Coated	427	58$^+$
Coated or covered	174	110R
Coated welding electrodes	427	59$^+$
Coated with metal	428	615$^+$
Coated with plural coatings	427	118$^+$
Coating	427	117$^+$
Coating by extruding plastic	425	113
Composition alloy	420	
Composition metallic	75	
Composition miscellaneous	252	500$^+$
Composition treated metal	148	400$^+$
Freeing of ice and snow	191	62
High frequency systems	333	
Inductive disturbance reduction (See anti induction)		
Lightning protection	174	2$^+$
Making indefinite length	156	47$^+$
Making or joining running length	156	47$^+$
Piercing detachable tap connector	439	431$^+$
Placing apparatus	254	134.3 R$^+$
Plate vehicle supply type	191	13
Rail joint interposed	238	14.5$^+$
Railways for electric	191	
Sectional vehicle supply type	191	14$^+$
Supports	248	49$^+$
Tubes and conduits	174	
Vehicle supply type	191	22R$^+$
Whipping apparatus	242	7.6
Feeding metal rolling guide	72	250$^+$
Furnace spark and smoke	110	145$^+$
Furnace spark returning	110	120
Gas flow to separator	55	418$^+$
Downstream of separator	55	410$^+$
Lightning	D10	105
Sound muffler fluid guiding	181	212$^+$
Spoons with fluid	30	141
Thermal flow tubular textile treatment	68	193
Conduit (See Hose; Pipe; Tube)	138	
Boiler blowoff	122	383$^+$
Boiler circulation	122	408R$^+$
Boiler feed heater	122	428$^+$
Chain making	59	78.1
Connector electric	439	207$^+$
Coupling plural	285	131$^+$

	Class	Subclass
Pipe and plate	285	130
Culvert	405	124$^+$
Drainage or irrigation	405	43$^+$
Drill oil	408	57$^+$
Earth boring tool	175	393
Electric conductor	174	
Design	D13	13
Electric transmission to vehicles	191	23R$^+$
Collectors	191	48
Fire tube circulation	122	59
Flue spiral	122	183
Fluid current conveyor	406	191$^+$
Design	D23	266
Flume	405	119$^+$
Heat exchange	165	177$^+$
Irrigation	405	43$^+$
Laying	405	154$^+$
Milker hand	119	14.22
Pneumatic conveyor	406	191$^+$
Railway crossing cable	104	185$^+$
Railway slotted	104	140$^+$
Railway switches	246	419
Railway truck in	104	139
Static mold	249	144$^+$
Structure	138	177
Thermometer	374	147$^+$
Track clearers	104	280
Trap thermostatic	236	60
Tubular	138	
System	137	
Valved	251	
Ventilating	98	56$^+$
Cone		
Edible design	D 1	116$^+$
Ice cream	D 1	116
Mills	241	244$^+$
Paper stock refining	241	244$^+$
Seger	374	106
Shell mill	241	244$^+$
Wind	242	27$^+$
Packages	242	175
Winding	242	27$^+$
Confectionery		
Coating apparatus	118	13$^+$
Compositions	426	660$^+$
Cutting machines	83	
Design	D 1	
Hand molds	249	
Packaging	426	392$^+$
Confetti Devices	446	475
Conformator Gauge	33	175$^+$
Congelation Refrigeration (See Gel)		
Confections with molding	425	9
Confections with molding	425	118
Confections with molding	425	332$^+$
Dewaxers	208	37$^+$
Dry ice making	62	10
Drying process step	34	5
Earth control solidification	405	130
Fluid storage cavity walls	405	56
Food preparation combined	426	524$^+$
Gas solidifiers	62	8$^+$
Shaping combined	62	35
Jellies	426	576$^+$
Liquid	62	340$^+$
Automatic	62	135$^+$
Ice cube trays and molds	249	66R$^+$
With release means	62	300$^+$
Process	62	66$^+$
Connecting Rod (See Pitman)	74	579R
Locomotive	105	84
Piston connection	92	187$^+$
Railway track switch	246	452
Connection (See Coupling; Fastener; Hook; Joint; Securing Means)	24	
Aircraft structure	244	131
Beehive	6	8
Bridge truss	14	14
Carbon electrobe	403	DIG. 5
Cell electrode connector	429	161
Clamp, adjustable	403	DIG. 9
Construction toy	446	85$^+$
Cutlery handle and blade	30	340$^+$
Drive belt combined	474	253$^+$
Electrical	439	
Arc lamp electrode to	314	129
Battery	429	121$^+$
Battery terminal assembling	29	246
Box with	174	59$^+$
Brush with connector (See electric connector)	310	249

	Class	Subclass
Insulated pipe joint	285	47$^+$
Insulator with	174	145
Juxtapose & bond rails	228	101$^+$
Lamp with	439	611$^+$
Lamp with	313	331$^+$
Making processes	29	825$^+$
Meter circuit protector	324	110
Overhead cable with	174	44
Rail bond making	228	4.1$^+$
Railway bonds	238	14.1$^+$
Sliding contact	191	1R
Switch	200	
Telephone terminal switch	379	422$^+$
Telephone testing	379	19
Vehicle transmission	191	
Electrical connector	439	
Design	D13	24$^+$
Winding implement	242	7.6
Winding machine	242	7.17$^+$
Winding method	242	7.3$^+$
Frangible		
Brake shoe fastener	188	241
Damper release	126	287.5
Fire extinguisher	169	42
Lock	70	422
Overload release	464	32$^+$
Reactive building component	52	232
Thermal actuated electric	337	401
Thermal electrical current	337	142$^+$
Grating	52	665
Handle to shaft	74	548$^+$
Intercell connector	429	160
Load-responsive release	403	DIG. 3
Magnetic	403	DIG. 1
Metal treatment	403	DIG. 2
Panel joints, readily disengageable	403	DIG. 1
Pipe shaft rod plate etc		
Flexible shaft	464	
Liquid heater tube	122	511
Packing	277	
Pipe	285	
Pipe valved	251	142$^+$
Piston rod	403	230$^+$
Articulated with piston	403	52$^+$
Rail fence	256	65$^+$
Rod	403	
Shaft end attached	403	230$^+$
Spout attaching	222	567$^+$
Tobacco smoking device	131	225
Wall safes	109	79
Wire fence	256	47$^+$
Wooden joist	403	230$^+$
Through intermediate member	403	187$^+$
Plow	172	681$^+$
Releasable	172	261$^+$
Release, load-responsive	403	DIG. 3
Release, quick	403	DIG. 4
Sail	114	113
Selectional pile	405	251$^+$
Spectacle and eye glass	351	140$^+$
Bridge	351	133$^+$
Temple	351	121$^+$
Sprocket chain	474	218
Tool handle	403	
Vehicle		
Animal draft	278	
Articulated land vehicles	280	504$^+$
Railway draft	213	
Wheel		
Land vehicle axle	301	
Railway axle	295	
Resilient tire	152	375$^+$
Consolidation		
Powder metallurgy	419	38
Slip or slurry	419	40
Constant Velocity Joint	464	904$^+$
Constrictor		
Nozzles	239	546
Trap	43	87
Contact		
Actuated		
Numbering members	101	88
Railway derailment signal or control	246	171$^+$
Diaphragm horn rotary striker ball	116	146
Electric		
Body applicator	128	783$^+$
Circuit makers and breakers	200	
Code transmitters rotary	178	83$^+$
Commutators	310	233$^+$
Composition	252	500$^+$
Connectors	439	
Flashlight direct battery	362	203
Lamp electrode holders	403	
Lamp electrodes in	313	235
Making processes	29	874$^+$
Railway automatic block signal	246	20$^+$
Railway derailment signal trip	246	171$^+$
Railway switch and signal actuation	246	218$^+$
Rectifier	357	
Rectifier composition	252	62.3 R
Rectifier manufacture	29	569.1$^+$
Semiconductor making process	437	180$^+$
Process	437	
Switch contact details	200	238$^+$
Switchboard movable	361	332$^+$
Telegraphic keys	178	101$^+$
Telephone electrode	381	205$^+$
Telephone switch rotary	379	362$^+$
Train mechanism signal or control	246	192A$^+$
Train telephone and telegraph continuous	246	9$^+$
Transmission to vehicle collectors	191	45R$^+$
Gas and liquid	261	
Building ventilation	98	30
Building ventilation	98	100
Car ventilation inlet	98	17
Car ventilation inlet & outlet	98	12
Concentrating evaporators	159	
Condensers	165	110$^+$
Defecating apparatus for sugar	127	12
Distilled mineral oil condensing	196	140
Heat exchange	261	127$^+$
Gas or vapor with solids	34	
Agitation	222	195
Disinfecting sterilizing	422	
Pneumatic conveyor material intake	406	108$^+$
Pneumatic conveyor material suspension	406	86$^+$
Liquid and solid		
Beverage infusers	99	279$^+$
Boiler or deep fat fryer	99	403$^+$
Chemical extracting or leaching	422	261$^+$
Cleaning and miscellaneous	134	
Cooker with basting	99	345$^+$
Cooking fluid automatic control	99	330
Corn popping apparatus with flavoring	99	323.8
Drying processes for liquid treated material	34	9$^+$
Filled receptacle cooker	99	359$^+$
Fryer or egg boiler control	99	336
Hydraulic conveyors	406	
Pigment and vehicle in pen	401	40$^+$
Preserving disinfecting and sterilizing	422	
Soap and water in cleaning implement	401	40$^+$
Solid metal treating apparatus	266	114$^+$
Textile treating apparatus	68	
Making clocks	368	164
Period electric switch	200	19R$^+$
Point electric switch	200	275$^+$
Preventing		
Drying apparatus	34	94
Drying processes	34	6
Printing antismut devices	101	420$^+$
Stenciling machine internal inker	101	120
Rails	191	22DM$^+$
Rectifier	357	
Composition	252	62.3 R
Manufacture	29	569.1
Process	437	
Rolls electrical transmission	191	45R$^+$
Sterilization by metal	426	322
Type microphone	381	178$^+$
Wheels electrical transmission	191	45R$^+$
Contact Lens for Eyes	351	160R$^+$
Applicators	294	1.2
Case or cover	206	5$^+$
Fitting instruments	351	247
Synthetic resin compositions	523	106*
Synthetic resin or natural rubber	525	937*
Container (See Ampoule; Barrel; Basket; Can; Capsule; Holder; Receptacle; or Under Item Contained)		
Acetylene generator carbide	48	29
Acid-proof	206	524.5
Adjustable bottom	206	45.16
Metallic	220	93
Wooden	217	64
Baggage	190	
Baked cereal product container type	426	138$^+$
Bottles and jars	215	
Design	D 9	367$^+$
Breaking	206	601$^+$
Cabinet structure	312	
Coupon is detachable	206	831
Dispensing	222	
Drying houses and	34	201$^+$
Edible design	D 1	
Envelopes	229	
Flexible bags	383	
Food sterilizing in	426	521
Freight	220	1.5
Hermetic food packaging	426	392$^+$
Insulating jacket	220	903
Latching-type	206	405
Loose material filter	210	263$^+$
Manufacturing from paper	493	
Decoration article	493	955$^+$
Pallet	493	964$^+$
Pliable	493	916$^+$
Rigid	493	901$^+$
Textile	493	937$^+$
Material	428	35
Medical syringes bulb combined	604	212$^+$
Medicator	604	232$^+$
Metallic	220	
Opening	206	601$^+$
Opening	206	601$^+$
Ornamental	428	35
Paperboard boxes	229	
Plastic metal working	72	272
Poison alarms	116	72
Separation	206	602
Sheet metal container making	413	1$^+$
Shock protecting	206	521
Snap-opening actions	206	404
Special packaging	206	
Of coiled items	206	389$^+$
Specialty carriers & boxes	D 3	30.1$^+$
Open top bins, trays	D34	40$^+$
Sterilizer apparatus	422	
Storage convertible to display	206	44R
Tearing	206	601$^+$
Toilet kits	132	79R$^+$
Absent special toilet article, eg mirror	401	118$^+$
Bristled brush design	D 4	130$^+$
Toothpicks with thread supply	132	92R
Toys	446	71$^+$
Type static mold	249	117
Wooden	217	
Contamination, Reduction of	435	800*
Contour		
Non-condition responsive sorting by	209	940*
Contraceptives		
Condom	604	349
I U D	128	130
Pessaries	128	127$^+$
Spermacides	424	DIG. 14
Control or Controller (See Operator; Pattern Control; Program; Regulator; Time, Operated & Controlled)		
Aircraft	244	75R$^+$
Airship	244	96$^+$
Bulldozer positioning	172	812
Car controlled fenders	293	2$^+$
Circuit		
Clock operator	368	184
Motor system	318	
Railway signal	246	28R$^+$
Sectional conductor vehicle transmission	191	14$^+$
Switch	200	
Clutches and power stop control	192	
Power stop control	192	116.5$^+$
Condition responsive metal deforming machine	72	6$^+$
Driftage fence	256	12.5
Drill press	408	8
Electric		
Battery charging & discharging	320	2$^+$
Car steps railway	105	444
Electric furnace	373	
Electrochemical apparatus	204	193$^+$
Fluid pressure brake	303	20
Fluid pressure brake with fluid control	303	15$^+$

U.S. PATENT CLASSIFICATIONS

	Class	Subclass
Gear cutting	409	290+
Generator	322	
Limit stop	192	142A
Locomotive railway	105	61
Motor	318	
Motor and brake	192	2
Motor and clutch	192	
Music tone generation	84	1.1+
Musical instrument	84	113
Musical instrument	84	171
Phonograph turntable stop mechanism	369	237
Prime mover dynamo plants	290	7+
Radiant energy	250	
Light wave	455	603
Railway block signal systems	246	20
Stop mechanism	192	127
Switch	200	
Telegraphy remote control	178	4.1 R
Telephone systems	379	
Textile making	87	15
Transmission and brake	192	9
Tuner remote control	334	8+
Turntable railway	104	38
Vehicle brake	188	137+
Vehicle brake operator	188	182
Elevator	187	100+
Fluid		
Feed in solid fuel furnace	110	188+
Pressure brake systems	303	
Pressure regulators	137	505+
Valve actuation	251	
Humidity	236	44R
Lock		
Control and machine elements	70	174+
Time	70	267+
Machine elements		
Lever and linkage	74	469+
Trips	74	2+
Motor or pump (See regulator)		
Brake combined	192	1+
Clutch combined	192	
Pump regulators	417	279+
Motor vehicle	180	315+
Mechanism	180	315+
Radiated wave responsive means	180	167+
Velosity responsive means	180	170+
Nuclear reactor	376	207
Nuclear reactor	376	220
Nuclear reactor	376	339
Nuclear, control component	376	327
Parachute	244	152
Pipe		
Fluid pressure brake	303	86
Fluid pressure brake automatic	303	29
Fluid pressure brake automatic and direct	303	25+
Fluid pressure brake charging	303	66+
Fluid pressure brake releasing	303	81+
Power stop	192	116.5+
Press	100	43+
Robot	901	2+
Speed (See speed)		
Temperature (See cooling; heating)		
Automatic	236	
Controlled Release of Medicines	604	890+
Conversion of Power, Electrical	363	
Conversion Reaction, Nuclear	376	156
Conversion Systems, Electrical	363	
Converter		
Bessemer	266	243+
Electric current	363	
Cryogenic	363	14
Frequency	310	160
Apparatus	310	160
System	363	157+
Heptode pentagrid	313	300
Metallurgical	266	243+
Octode	313	300
Pentagrid	313	300
Triode heptode	313	298
With two cathodes	313	6
Triode hexode	313	298
With two cathodes	313	6
Convertible (See Type of Device)		
Armchair to bed	297	115+
Auto bodies	296	
Chair	297	118+
To cane	297	118
To stacked bunks	297	62
To table	297	119+
Cutlery	30	122
Garment supporter	2	301
Horizontally supported		
Planar surfaces	108	11+
Plural seats to single bed	297	63+
Racks	211	2+
Refrigeration related	62	326+
Stand to cane	248	155+
Vehicle to chair or seat	280	30+
Walker to vehicle	280	7.1+
Converting, Mechanical Process of	29	401.1
Conveyor (See Carrier; Chute; Feeder)	193	
Air blast &/or vacumm	29	DIG. 78
Article support for label affixing	156	556+
Bearing support for roller	384	418+
Belt (See endless below)		
Acetylene generator carbide feed	48	39
Automatic gauging apertured	209	681
Automatic gauging pocketed	209	685
Automatic gauging slotted	209	665+
Coated surface separator receiver with	209	63
Magazine dispenser	221	76+
Magazine dispensing slender article	221	76+
Manual assorting	209	705
Manure spreader type	239	650+
Mattress filling	53	255+
Paper pasting strip cutter feed	156	510
Power driven	198	
Sifters	209	307+
Sifting liquid treatment	209	272
Stratifiers liquid treatment	209	428+
Textile smoothing transfer	38	8+
Butchering	17	24
Centrifugal fluid distributor feed	239	215
Chain or belt	29	DIG. 73
Chute	193	2R+
Apertures automatic gauging	209	682
Drying apparatus	34	165+
Gaseous suspension assorting	209	149
Gauging automatic by rolling	209	696
Liquid suspension assorting	209	196
Reciprocating	209	196
Stationary	209	202
Manual assorting	209	703+
Movable bed stratifiers	209	479+
Liquid treatment, spiral	209	434
Plural destination conveyor &	198	359
Pneumatic stratifiers		
Movable bed	209	471+
Stationary bed	209	477
Pockets automatic gauging with	209	684
Power driven conveyor combined	198	523+
Stationary bed liquid treatment		
Stratifier	209	458+
Stratifiers reciprocating	209	437+
Stratifiers rotating	209	444
Reciprocating	209	435
Chutes skids guides ways	193	
Coated surface separator receivers with	209	65+
Cooking apparatus	99	443R
Beverage infuser	99	289R
Boiler or deep fat fryer	99	404+
Corn popping	99	323.9
Filled receptacle cooker	99	360+
Griddle	99	423+
Molding support	99	427
Smoke treating	99	467+
Spit or empaling type	99	420
Toaster or broiler	99	386+
Waffle iron type cooker	99	373
Design	D34	29+
Drying		
Hollow article	34	105
Houses containers with	34	201+
Removable shelf or tray handling	34	194
Stationary receptacle with	34	179+
Treated material handling	34	236
Dumping scow platform	114	34
Egg candling	356	55+
Electrolytic apparatus work	204	198+
Endless		
Adjustable material deflector for	222	290
Bucket ditcher longitudinal	37	89
Bucket ditcher transverse	37	90
Chain	198	643
With pusher	198	725+
Container sterilizing	422	
Cooperating comminuting	241	200
Dispenser	222	415
Dispensing trap chamber on	222	371
Dragline scoop combined	414	570
Drying house plural run	34	207
Drying house single run both active	34	208
Electrolytic movable electrode work	204	202+
Endless digger	37	192R
Flexible sheet drying	34	162
Furnace type charging	414	147+
Furnace type reciprocating	414	586+
Charger combined	414	198
Gangways	14	70
Grain hulling	99	616
Hoist combined	414	564
Material deflector	222	408
Plow ditcher longitudinal	37	101+
Plow ditcher longitudinal and transverse	37	107
Plow road grader conveyor wheel combined	37	112
Plow road grader transverse	37	110
Pneumatic stacker combined	406	77+
Power driven	271	194+
Product remover	83	155.1
Rotary digger	37	190
Self loading excavating vehicle	37	8
Sheet delivering	271	198+
Sheet delivering to curtains	271	67
Sheet delivering to flies	271	66+
Sheet delivering to other conveyors	271	69+
Sheet feeding	271	264+
Sheet feeding and delivering	271	7
Sheet feeding and separating	271	12+
Sheet feeding delivering and separating	271	6
Vehicle unloading external	414	373+
Horizontal movement	414	353
Reorienting support	414	356
Repositioning support	414	351
Wheel excavator longitudinal	37	96
Wheel excavator transverse	37	97
Excavator	37	
Ditchers	37	80R+
Railway snow	37	208
Road grader	37	108R+
Roadway snow	37	237+
Flexible	29	DIG. 99
Fluid current	406	
Boiler or deep fat frying	99	406
Diamagnetic separation of solids	209	212
Drying apparatus treated material	34	57R
Drying process material suspension or	34	10
Electrostatic separation of solids	209	127.1+
Fluid conveyor or applicator	29	DIG. 63
Fluid handling system cleaners	137	238+
Insect dusting machine with	43	141
Insect powder dusters	239	650+
Mortar mixer processes	366	3+
Mortar mixers	366	10
Paramagnetic separation of solids	209	213+
Railway snow excavator	37	202
Rotary meat comminutor combined	241	38+
Sand blast abrading	51	410+
Separating operations combined	209	12+
Separation of solids	209	132+
Separation of solids by adhesion	209	45+
Separation with heat treatment	209	11
Sheet delivering	271	306
Sifting	209	233+
Solids comminuting apparatus with	241	38+
Solids comminuting processes with	241	15+
Strand or web feeding	226	97
Stratifiers	209	422+
Track sander pipe cleaning	291	43
Track sanders	291	3+
Treatment precedent to separation	209	3+
Treatment subsequent to separation	209	10
Fluid pressure testing for leakage	73	41+
Garbage and sewage furnace	110	255+
Gas as a	29	DIG. 81
Gaseous suspension assorting feeding	209	147+

Entry	Class	Subclass
Glass annealing	432	121^{+}
Harvester		
Cutting binding combined	56	131^{+}
Cutting combined	56	153^{+}
Cutting threshing combined	56	122^{+}
Marine	56	9
Motor cutter and detachable	56	3^{+}
Motor cutter combined	56	14.5
Heat exchange	165	120^{+}
Chamber	165	120^{+}
Linear motor	198	619
Liquid suspension assorting feeding	209	184
Liquid treatment stratifiers rotating with	209	452
Liquid treatment stratifiers stationary with	209	461^{+}
Lubricator	184	12
Force feed	184	61^{+}
Systems	184	6^{+}
Material for	198	957*
Mining machine	299	64^{+}
Moving wave	198	630
Potato digger spiral	171	113
Power driven	198	
Press combined	100	144^{+}
Quencher for thermolytic distillation	202	230
Receptacle positioning for filling	141	129
Reciprocating		
Discharge assistants sets of	222	275^{+}
Dispensing	222	409
Dispensing trap chamber	222	361^{+}
Furnace charging	414	194
Furnace fuel feed	110	109
Plural source dispensing unitary	222	137
Power driven	198	750^{+}
Pusher	198	736^{+}
Power driven pusher	198	736^{+}
Series discharge assistants	222	243^{+}
Rotary and	222	233^{+}
Sheet delivering	271	84^{+}
Sheet feeding	271	267^{+}
Sheet separating and feeding	271	14
Thermolytic distillation	202	119
Transverse to material flow	222	200
Vibrating dispenser bottom	222	199
Vibrating trough	198	750
Refrigeration combined		
Cooled article	62	378^{+}
Door mounted	62	377
Process	62	62^{+}
Solid refrigerant	62	379^{+}
Residue or sediment mover	210	523^{+}
Roaster type combined peeler or parer	99	483
Robot cooperating with	901	7*
Roller	193	35R^{+}
Live rolls	198	780^{+}
Rotary	198	624
Rotary		
Acetylene generator carbide feed	48	48^{+}
Discharge assistants sets of	222	271^{+}
Dispensing	222	410^{+}
Dispensing trap chamber	222	367^{+}
Drying apparatus	34	184^{+}
Electrolytic with work	204	199^{+}
Gravity flow dryer with	34	166
Heat exchange cylinder with	165	120^{+}
Heat exchange horizontal	165	88
Heat exchange vertical	165	89^{+}
Mercurial adhesion separating	209	70
Power driven	198	
Rotating drum sifter	209	296
Series dispensing means	222	236^{+}
Series dispensing means reciprocating and	222	233^{+}
Sheet delivering	271	314^{+}
Sheet feeding	271	275^{+}
Scales combined	198	502.1^{+}
Screw	198	657^{+}
Acetylene generation carbide feed	48	54
Cam or screw conveyor	29	DIG. 1
Centrifugal fluid distributor feed	239	218.5
Ditcher	37	82
Earth boring tool	175	394^{+}
Furnace fuel feeder	110	110
Pneumatic conveyor feed	406	56^{+}
Potato digger	171	134
Rotary discharge assistant	222	412^{+}
Series rotary discharge assistants	222	240
Thermolytic distillation	202	118
Self loading vehicle	414	467^{+}
Sheet feeding and delivering	271	
Sifter	209	261^{+}
Discharging	209	257
Discharging and valve	209	256
Feeding	209	247^{+}
Feeding and discharging	209	241
Feeding hopper	209	245
Plane reciprocating with	209	320^{+}
Rotating drum combined	209	293^{+}
Superposed plane with intermediate	209	316
Store service	186	
Thermolytic distillation retort	202	117^{+}
Threshing machine dust	130	29
Thrower	198	638^{+}
Furnace charging	414	193
Scattering unloader	239	672^{+}
Track sander	291	32
Transversing wall aperture	198	950*
Turning cigarettes end for end	198	951*
Typesetting machine	276	16^{+}
Washer type combined peeler or parer	99	623^{+}
Washing apparatus having		
Brushing scrubbing etc	15	3^{+}
Liquid contact with solids	134	
Washing apparatus having endless	137	124^{+}
Bottle brushing	15	60^{+}
Control by work or work holder	134	48
Sequential stations work transfer	134	67^{+}
Work transfer	134	70^{+}
Wood cutting machine dust	144	252R
Wooden flight making machine	144	10
Convolute Wind	242	
Laminated article making	156	
Paper tube machine	493	303^{+}
Cooker & Cooking		
Barbecue unit	D 7	332^{+}
Commercial equipment	D 7	323^{+}
Digester	422	307^{+}
Electric ovens	219	391^{+}
Electric vessels	219	438^{+}
Electric vessels and stands	219	429^{+}
Food	99	324^{+}
Non-cooking heat treatment	99	483
Heat source	D 7	362^{+}
Household appliance	D 7	323^{+}
Household container	D 7	354^{+}
Popcorn	D 7	325
Solar	D 7	324
Solid fuel	D 7	332
Starch	127	28
Stoves	126	
Cookie Cutter	D 7	43
Coolant	102	704*
Cooler & Cooling (See Congelation; Quenchers; Refrigeration)	62	
Abrasive tool or support	51	356
Air	62	404^{+}
Automatic	62	186^{+}
Process	62	89^{+}
Air conditioning equipment	D23	351^{+}
Article in dispenser	221	150R
Automatic cooling and heating	165	14
Bearings	384	313^{+}
Beds	5	284
Metal rolling	72	251
Body	604	113^{+}
Treating material	604	113^{+}
External	604	291^{+}
Brakes	188	264R
Cabinet combined	312	236
Castings	164	122^{+}
Continuous	164	443^{+}
Chemical reaction or solids dissolving	62	4
Cleaning or polishing tool or support	15	230.1
Clutches	192	113R
Coating treatment	427	398.1^{+}
Compositions	252	70
Antifreeze type	252	71
Vaporization or expansion	252	67^{+}
Congealed material drying	34	5
Control	236	
Cooling and heating apparatus	165	58^{+}
Conveyer art	198	952*
Design	D23	
Towers	261	DIG. 11
Discharge nozzle	239	132.3
Reaction motor	239	127.1^{+}
Dispenser	222	146.6
Distillate	203	
Apparatus combined	202	
Drying combined with	34	13
Cold gas contact	34	20
Cooling means	34	62^{+}
Vapor condensers	34	73^{+}
Earth boring combined	175	17
Electric lamp	313	11^{+}
Electric lamp or electronic tube in system	315	112^{+}
Electrical apparatus	174	15R^{+}
Electron tube	313	11^{+}
Emulsion breaking by	252	346^{+}
Evaporation, cooling by		
In hydrocarbon synthesis	585	940*
Fluid handling systems or devices with	417	243^{+}
Discharge nozzles	239	132.3
Foundry mold	164	348
Furnace	432	233^{+}
Arc, electrical	373	74
Sidewall	373	76
Design	D23	329
Doors or casings	110	180
Doors or fronts by water	122	497^{+}
Electrical induction	373	165
Coreless	373	151^{+}
Industrial by water	122	6R^{+}
Metallurgical shaft type	266	190^{+}
Resistance, electrical	373	113^{+}
Fusion-bonding apparatus combined	228	46
Gas and liquid contact	261	127^{+}
Glass manufacturing		
For apparatus	65	355
Melt cooling	65	137
Product cooling	65	348
Grinder	51	266^{+}
Heating combined or by heat exchange	165	
Humidity regulation	236	
Light projectors	362	294
Liquid	62	389^{+}
Automatic	62	177^{+}
Discrete commodity contacting	62	373^{+}
Indirect heat transfer liquid	62	430^{+}
Process	62	98^{+}
Liquid purification	210	774^{+}
Apparatus	210	175^{+}
Precipitation	210	737
Lubricators	184	104.1
Material during filling	53	127
Milk treating apparatus	99	455
Mold stereotype	164	144
Mold type	199	56
Motor or pump		
Combustion products power plant	60	39.83
Compressor intercooler	417	243
Condensers	165	110
Electric locomotive	105	59
Internal combustion	123	41.1^{+}
Land vehicle system	180	68.4^{+}
Marine propulsion	440	88
Power plant combustion exhaust	60	310
Power plant combustion exhaust	60	317
Power plant combustion exhaust	60	320^{+}
Power plant combustion exhaust	60	685
Power plant flue ejected exhaust	60	685^{+}
Radiator	165	121^{+}
Shaft coupling	464	17
Turbine	415	175^{+}
Needle of sewing machine	112	281
Nozzles	239	132.3
Reaction motor discharge	239	127.1^{+}
Pipe joint	285	41
Pipe smoking	131	195
Plastic metal working	72	253.1^{+}
Power plant X-art	60	912*
Radio type tubes		
Integral cooler	313	11^{+}
Rolling mill	72	201
Sewage treatment process	210	612^{+}
Tires	152	153
With comminution	241	65
By fluid or lubricant and other methods	241	17
By fluid or lubricant method	241	15
By gas or vapor method	241	18^{+}
Fluid applying means	241	38^{+}
Material cooling means	241	65^{+}
Material cooling methods	241	23
Work holder electric metal heating	219	158^{+}

	Class	Subclass
X ray generating apparatus with	378	141
X ray tubes	313	11$^+$
Cooling Tower	261	
Ice prevention	261	DIG. 86
Coop	119	19
Coopering (See Barrels)	147	
Coordinate Determination	367	907*
Cop and Cop Winding	242	18R$^+$
Bobbins or spools	242	118.3$^+$
Filling receptacles with	53	235$^+$
Finishing apparatus	28	284$^+$
Sorter	209	927*
Wound packages	242	159
Copal	260	97$^+$
Sand type	164	349$^+$
Making	164	6$^+$
Coping	52	300
Earth supported	52	102
Tubular	52	244
Copper	75	72
Alloys	420	469
Aluminum	420	489
Compound inorganic	423	
Biocides containing	424	140$^+$
Cuprates	423	592$^+$
Halides	423	462$^+$
Hydrocarbon purification by	585	849
Oxide or hydroxide	423	604
Photoelectric cell battery	136	265
Compound organic	556	110$^+$
Medicine	514	184
Electrolytic synthesis	204	106$^+$
Froth flotation	209	901*
Heat treatment	148	160
Special compositions	148	13.2
Hydrometallurgy	75	117
Misc. chemical manufacture	156	DIG. 76
Plate	428	674$^+$
On iron	428	676$^+$
Plating	204	52.1
Pyrometallurgy	75	72$^+$
Copy		
Advancing means	226	168$^+$
Holders (See copyholders)	248	441.1$^+$
Advancing means	226	12
Multiple	282	9R$^+$
Photocopying equipment	D16	27$^+$
Copyholder	248	441.1$^+$
Card holder on typewriter	400	521$^+$
Guide static for moving copy	226	196$^+$
Line guide type	40	352$^+$
Movable copy	40	342$^+$
Sheet or web holder on typewriter	400	578$^+$
Typewriter attachable	248	442.2
Copying (See Pattern Control)		
Camera	355	18$^+$
Coated surface	427	144
Compositions	106	14.5
Machines (See pattern control)		
Pads and forms	118	264$^+$
Paper or carbon paper	282	28R
Coated	427	153$^+$
Photocopying machines	355	
Planographic		
Apparatus	101	131$^+$
Elements	101	473
Processes	101	468$^+$
Presses		
Design	D18	13$^+$
Wood turning	142	7$^+$
Cord (See Strand)		
Electric (See conductor)		
Fabric		
Bed bottom	5	190$^+$
Braided netted or lace	87	
Hammock	5	122
Knitted	66	169R$^+$
Woven	139	383R$^+$
Fastener or tie	24	115R$^+$
Bale or package	24	18
Box closure	229	125.22
Flexible bag	383	71
Garment supporter	2	341$^+$
Hat	132	61
Knotter combined	289	
Necktie	24	58
Design	D 2	607
Sash	16	202$^+$
Separable	24	660
Window sash combined	49	445$^+$
Finishing apparatus	28	219$^+$

	Class	Subclass
Guide	254	389
Knotter combined	289	15
Reeling	242	157R
Sash	16	210$^+$
Hat	2	189
Opener, terr member	206	616$^+$
Operating (See device operated)		
Packaging	28	289$^+$
Pliant, deformable sheet retainer	402	8$^+$
Shortener (See take up)		
Spun twisted or twined	57	
Spun twisted or twined design	D 5	7$^+$
Straightener		
Electric conductor type	174	69
Insulator combined	174	154
Spring device	267	74
Telephone cord type	379	441
Supporting		
Light type	362	407
Package and article carrier	294	149$^+$
Picture frame combined	40	153
Picture type	248	489$^+$
Sash balance	16	193$^+$
Sheet retainer also serves to insert binder strand thru sheet	402	7
Tool cord type	248	51$^+$
Terminal	439	775$^+$
Tire		
Antiskid devices	152	221$^+$
Carcass material	152	548
Carcass material arrangement	152	558
Making	156	117
Tying or knotting	289	
Harvester compressing and binding	56	433$^+$
Harvester raking and bundling	56	343
Winding and reeling	242	47$^+$
Cordeau Detonant	102	275.1$^+$
Composition	149	105$^+$
Corduroy	139	392
Core (See Casting, Plastics; Mandrel)		
Assembling apparatus	29	234
Box	164	228$^+$
Planes	30	479
Sections hinged	164	233
Brush and broom	15	204$^+$
Casting		
Boxes	164	228$^+$
Cement pipe mold	425	59$^+$
Cement pipe mold	425	406$^+$
Chaplets	164	398$^+$
Materials	106	38.2$^+$
Means to expand or contract	249	178$^+$
Metal founding sand	164	369$^+$
Molds metal separable	249	142$^+$
Molds sand combined	164	365$^+$
Of plural sections	249	184$^+$
Of resilient material	249	183
Plastic molds	425	468
Removing	164	132
Removing apparatus	164	345$^+$
Static mold	249	175$^+$
Cutters auger type	408	199$^+$
Cutting		
Annular drills	408	204$^+$
Button making	79	16
Earth	175	403$^+$
Tubular wood saw	408	204$^+$
Wood disk	408	204$^+$
Dam	405	109
Deforming metal		
Destructible core	72	57$^+$
Flexible core	72	466
Distillation retort	202	225
Electromagnet	335	297
Circuit breaker armature	335	281$^+$
Circuit breaker multiple	335	180$^+$
Flexible for tube bending	72	466
Magnetic		
Electromagnet with armature	335	220$^+$
Reactor	336	233$^+$
Transformer	336	233$^+$
Materials	106	38.2$^+$
Molding machines for foundries	164	228$^+$
Nuclear reactor	376	347
Nuclear reactor	376	409
Pile driving type	405	245$^+$
Pillow	5	439
Plates for storage batteries	429	209$^+$
Railway pedestal block	238	117
Reeling fabrics	242	68$^+$

	Class	Subclass
Screw threading taps		
Cam	408	158
Wedge	408	179
Static mold having		
Removable	249	142
Strand structure covered	57	210$^+$
Supports	164	397$^+$
Textile covering processes	57	3$^+$
Textiles covering	57	3$^+$
Tires casing enclosed	152	310$^+$
Corer		
Hand manipulated	30	113.1$^+$
Plural cooperating blades	30	174
Machine	99	547
Doffer parer combined	99	541$^+$
Doffer parer segmenter combined	99	543
Parer combined	99	541
Parer slicer combined	99	543
Parer slicer combined	99	592
Segmenter combined	99	545
Slicer combined	99	552
Cork		
Board making	264	124
Bottle stoppers	215	355$^+$
Coating or plastic compositions containing	106	204.1
Fastener making	140	85
Hoof covering	168	26$^+$
Paving	404	17$^+$
Polymers ctg (See synthetic resin or natural rubber)		
Press	144	284
Pullers	81	3.7$^+$
Removing implement	D 8	42
Wire applying to	140	94$^+$
Corkscrew	81	3.45
Hayfork manipulator	294	121
Making	140	86
Corn		
Cob splitter	225	93$^+$
Ear holder	294	5
Files	132	76.4
Foods	426	627$^+$
Flakes	426	621
Grinding	241	
Harvesting	56	51$^+$
Holder for eating	D 7	42
Husking		
Implements	130	4
Machines	130	5R
Means to butter ear corn	401	12
Medicines for	424	
Pads for	128	153
Planter	111	
Clutches	192	23
Popping apparatus	99	323.5$^+$
Shellers	130	6$^+$
Green corn	99	567
Stringers weaving	139	19
Stripper	30	121.5
Corncob		
Pipe	131	230
Splitter	225	93$^+$
Corner		
Building component	52	272$^+$
Concrete or plaster	52	250$^+$
Embedded protector	52	254$^+$
Panel edge binder	52	264$^+$
Trim or shield	52	288
Window frame post	52	282
Cabinets	312	238
Connector		
Folded wall extension	229	190$^+$
Folded web	229	186$^+$
Structure		
Bedstead fastenings	5	288$^+$
Block	52	284$^+$
Bolted table leg	108	156
Bracket type support	248	220.1
Building component	52	272
Coffins	27	10
Curbs and gutters	404	8
Elastic flat bedbottom	5	205$^+$
Land vehicle body	296	29$^+$
Panel type switchboard	361	362
Razors	30	76
Resilient tires sectional	152	183
Trunk shields and buffers	190	37
Cornet	84	388$^+$
Design	D17	11
Drainage	84	397

Entry	Class	Subclass
Mouthpieces	84	398+
Mutes	84	400
Tremolos	84	401
Cornice		
Curtain shade or screen combined	160	19+
Design, fabric	D 6	575
Design, rigid	D25	55
Design, structural	D25	55
Metal sheathing	52	287
Trim or shield	52	288
Corona		
Discharge process	430	937*
Chemical apparatus	422	907*
Prevention		
Conductors	174	127
Dynamos	310	196
Corpse		
Carriers	27	28
Coolers		
Display type	62	246+
Open support	62	458
Cremating furnace	110	194
Undertaking	27	
Corrosion Preventing		
Agents	252	387+
Coating composition	106	14.5+
Distilling combined	203	6+
In hydrocarbon synthesis	585	950*
In nuclear reactors	376	305
Mineral oil conversion combined	208	47
Processes	422	7
Refrigeration with	62	85
Corrugated Structure		
Building panel	52	630
Metal	428	603+
Pipes		
Flexible	138	121+
Reinforced	138	173
Railway ties	238	82
Receptacle wooden straps for	217	67
Reflector lantern combined	362	297+
Sheet metal	428	603+
Ship	114	80
Stock material	428	179
Making apparatus	156	205+
Making process	264	286+
Wrappers for bottles	229	90
Corrugating (See Crimping)		
Land furrowing implement	172	
Metal	72	190+
Curving and	72	177
Slitting and	29	6.1
Tube	72	102+
Web or sheet	72	196
Paper		
Crepe paper making	162	111+
Processes for	264	286+
Plastic & earthenware shaping	425	396
Ruffler for sewing machine	112	132+
Corsage	D11	117+
Corset	450	94
Combined with brassiere	450	7+
Design	D 2	3
Covers	2	110
Design	D 2	2
Making	223	53
Reinforcing	223	4
Cortisone, in Drug	514	179
Corundum	423	
Abrasive composition	51	309
Refractory composition	501	127+
Magnesium compound containing	501	119
Corynanthine	546	53
Cosmetic Product	D28	4+
Applicator, bristled brush	D 4	
Applicator, general	D28	7+
Box	206	542
Brush	D 4	
Cold cream	514	772+
Compacts	401	126+
Design	D28	78
Facial	514	844*
Including toilet article, eg mirror	132	83R
With light	362	135+
Lipsticks	424	64
Container	206	385
Lipstick in container	D28	76+
Package for	206	803*
Powder applicator	D28	8
Powder in container	D28	76+
Shaped appliers	132	88.7
Design	D28	76+
Cosmic Ray Detection	250	336.1
Cloud chamber	250	
Detector tubes	376	153+
Detector tubes	313	93
Photographic film	250	475.2
Scintillation	250	361R+
Costume	D 2	79
Garment prior to 1900	D 2	26
Cots	5	110+
Canopies	5	414+
Cotter	411	513+
Nut locks		
Design	D 8	382
In bolt to nut lock	411	213
In bolt to nut lock	411	320
Making	29	5
Pullers	29	247+
Railway car wheel	295	50
Shaft joint type	403	378+
Slot punching	10	14
Cotton		
Chopper type plows	172	
Fiber preparation	19	
Harvesting	56	28+
Ripening	47	5
Cotton Cady Maker	425	9
Couch		
Box type	297	192+
Furniture design	D 6	334
Hammock	5	124+
Kinestherapy	128	33
Light thermal and electrical applying	128	376+
Orthopedic	128	70+
Roll	162	314
Tub combined	4	547
Coulombmeter		
Ballistic galvanometer	324	97
Electrolytic	324	94
Coumarins	549	283
Coumarone	549	434
Resins (See also synthetic resin or natural rubber)	526	266
Counter & Counting (See Recorders; Registers; Shelf & Shelving; Tally; Showcase)	D10	97
Alarm for operation	116	73
Applications	377	1
Bank protection trapping mechanism	109	4
Blood	377	10
Cabinet type	312	140.1+
Check-out counter	186	59+
Coin	453	58+
Box	235	100
Decade		
Electronic tube type	377	103
Gas tube	315	
Electromechanical	377	82
Electromotive force generator excitation control	322	86+
Electronic	377	
Faucet operation	235	94R
Geiger muller	250	374+
Indicating means	377	112
Tube per se	313	93
Mechanical	235	
Mechanism for package	235	98R
Mechanism for turnstiles	235	93
Nuclear for fluid	250	336.1
Railway block signal in-and-out	246	77+
Railway car actuated circuit control by	246	247
Recorder combined	346	14R+
Registers	235	
Rotation	235	103+
Scintillation	250	361R+
Sheets	235	89R
Shoe	36	69
Showcase	312	114+
Cabinet type	312	140.1+
Protectors and guards	312	137
Stiffener		
Boot and shoe	36	68
Shoe heel and	36	69
Systems	377	27
Typewriter words	235	102
Counterbalance (See Counterpoise; Counterweight)		
Adjustable bracket with	248	274+
Cable hauling drum with	254	312
Centrifugal metal casting device with	164	287
Closet seat	4	241
Crane	212	195+
Cranks and wristpins	74	603+
Cultivator beam seat	172	431+
Electric lamp etc electrode	314	98
Elevator	187	94
Pedals	74	594.5
Pitman and connecting rod	74	589+
Railway cable car	104	174+
Railway wheels with	295	6
Robot	901	48*
Spring biased tool	172	705+
Stand and adjustable bracket	248	123.1
Standard type support adjustable with	248	162.1
Suspended support adjustable with	248	323+
Upending bed	5	164R+
Valves	251	281+
Weighted tool	172	611
Weights per se	16	1C
Weights support type	248	364
Counterbore		
Boring head	408	223+
Counterpane	5	482
Counterpoise (See Counterbalance)		
Disappearing gun mount	89	39
Oval pattern lathe chucks with	142	51
Radio antenna	343	846+
Countersink		
Augers	408	223+
Drill combined	408	223+
Pipe and plate packed joint	285	211+
Reamers	408	223+
Screw pipe joint external packing and	285	349+
Screws	411	378
Countertop		
Tile	D23	308
Counterweight (See Counterbalance)		
Carcass splitting saw	17	23
Chute	193	24
Chute vertical swing	193	18+
Drilling machine	408	235
Elevator fluid governors	187	70
Grinder mount	51	166R
Light supports with	362	401+
Material handling hoist	414	601
Material handling hoist with	414	673
Railway car stop bumpers with	104	255
Railway switch and signal mechanical locking with	246	153+
Railway switch stands automatic with	246	288
Sash	16	216+
Sash balances cord and	16	194
Suspended support with	248	331
Vertically swinging shovel with	414	719+
Couple		
Delivery twist textile spinning	57	60+
Strand controlled starting and stopping	57	82
Printing	101	183+
Coupler for Organ (See Organ)		
Coupling (See Connection)		
Agricultural implement to tractor	280	504+
Bolt or nut substructure	411	81+
Bolt to nut	411	190+
Boxes to electric fixtures	174	61+
Circuit	333	24R+
Cord and cable		
End holders	24	115R+
Ship tension reliever	114	216+
Detachable from container	206	831
Draft		
Articulated land vehicle	280	504+
Cable	104	193
Horse drawn streetcar	278	2.1
Plow type	172	677+
Railway	213	75R+
Railway tools	294	18
Thill	278	52+
Electric		
Amplifier (See amplifier)		
Brake	310	93
Circuit	333	24R+
Clutch type	310	92+
Network type	333	24R+
Wave guide	333	202+
With antenna	343	850+
Fluid	60	325+
Clutch combined	192	3.21+

Entry	Class	Subclass
Gearing combined	74	730+
Impeller making	29	156.8 FC
Rafting and booming timber type	441	50+
Railway tools	294	18
Railway wheel and axle	295	39+
Rod pipe and shaft		
Closet bowl	4	252R+
Dredger pipe supports	37	72
Electrical pipe type	439	191+
Flexible shaft	464	
Nozzle type dispenser	222	567+
Packing	277	
Pipe type	285	
Railway draft pipe support	368	15+
Rod type	403	
Scaffolding pole	403	
Sprinkler pipe	239	266+
Strainer and pipe	4	288
Valved pipe	60	530+
Coupon		
Envelope	229	70
Tallies	283	51+
Course for Animals	119	15.5 R
Cover (See Closure; Covering; Guard; Lid; and Notes to)	220	200+
Baseball field	273	27
Bath and closet	4	293+
Book	281	29
Design	D19	26+
Indicia carrying	283	64
Multiple book removable	281	17
Removable book	281	19R
Removable books and leaves	281	4
Brush and broom	15	247
Button	24	113R
Cloth	24	92
Cloth leather rubber	150	52R+
Clothesline	211	119.18
Corset	2	110
Design	D 2	122
Dentists engine hand piece	433	116
Furniture		
Armrest	297	227+
Billiard and pool table	273	13
Chair and seat	297	219+
Garment hanger	223	98
Hinge	16	250+
Horizontal building entry	52	19+
Insulating jacket for beverage container	220	903
Ironing board	38	140
Keyhole	70	455
Magazine cover removal	83	925MG*
Magazine cover removal	83	925MG*
Mixing chamber	366	347
Pillow	5	490
Pipe smokers	131	176+
Plant	47	26+
Protective for cigarette pack	D27	48+
Protective for matches	D27	31
Radiator	237	79
Receptacle		
Barrel	217	81+
Cloth leather and rubber	150	52R+
Covered dishes	215	
Covers	215	200+
Electric outlet box	174	66+
Oil cup lubricators	184	90+
Sap bucket	47	54
Sap buckets spout combined	47	51
Spittoon	4	267+
Roof	52	
Seat toilet	4	242+
Sewing machine	D15	75
Ship opening	114	201R+
Stair	52	177+
Tarpaulin type	52	3+
Tent	135	115
Tree	47	20+
Trunk guard	47	23+
Typewriter		
Cabinet combined	312	208
Vault pavement type	404	25+
Design	D25	36
Vehicle		
Dust	296	136
Load	296	100+
Roll up	296	98
Steering wheel dust covers	150	52M+
Steering wheel grip increasing	74	558
Covering (See Apparel; Clothes; Coating; Garment)		
Applying by laminating	156	
Book	412	3
Bridge	14	74
Coffin	27	19
Floor		
Laminated	428	54
Slatted	52	664+
Grave	27	30
Insulating for electric conductors	174	
Making and applying	156	
Button covering	79	4+
By laminating	156	
Extruding metal sheath on wire	72	268
Extruding plastic on core	425	113
Indefinite length electrical conductor	156	47+
Metal covering	29	33.2
Paper	493	
Sheath to tube or rod	228	126+
Shoe heel	12	49.1
Strap	12	59
Textile spinning	57	3+
Wire	29	428+
With sheet metal	29	243.57
Shelf	428	904.4
Spring leaf	267	37.1+
Wall	428	904.4
Covert	139	416
Cowl		
Ventilating	98	61
Design	D23	374+
Intake and outlet	98	35
Crack-off	225	
Glass	225	94+
Cracker		
Firecracker	102	361
Food	99	
Design	D 1	128+
Cracking		
Hydrocarbon synthesis		
Aromatic	585	483+
Diolefin	585	613+
Olefin	585	648+
Paraffin	585	752
Triple bond	585	534+
Oils	208	46+
Electrostatic or electric discharge	204	172
Crackled Coating	427	257
Crackled Glass Making	65	111
Cradle		
Battery	429	96+
Bed	5	101+
Design	D 6	382
Grain	56	324
Land vehicle convertible	280	31
Receptacle stand	248	139
Squeezer type washing machine	68	114
Cramps		
Block tie	52	712+
Crane (See Derrick)		
Automatic control	212	153+
Cut-off	212	149+
Cyclic	212	161
Sway	212	147+
Cab	212	165
Travelling bridge	212	206
Collapsible	212	182+
Cornstalk cutter with discharger	56	72+
Counterweight	212	195+
Removable	212	178+
Davit	114	368
Design	D34	33
Electric control	212	124+
Excavator with cable operated boom	37	116
Floating	212	190+
Invalid lift	5	87
Locomotive with turntable	105	28
Material or article handling	414	561+
Remote or dual control	212	160
Rotary	212	223+
Self-erecting	212	176
Climbing	212	199+
Ship mounted	212	190
Transmission of electricity to	191	
Traveling	212	71+
Bridge	212	205+
Trolley type railway with transfer	104	98
Trucks for overhead	105	163.1+
Vehicle mountable	212	180
Window	49	148
Crank		
Foot operated	74	594.1
Hand lever	74	545+
Mechanical movement including	74	
Operated (See device operated)		
Paddle		
Impeller	416	78
Marine	440	26+
Water motor	416	78
Pedals combined	74	594.1+
Velocipede design	D12	123
Wrist pin combined	74	595+
Crankcase		
Breathers	92	78+
Drainers	184	1.5
Housings	74	606R
Internal combustion engine	123	196CP
Crankshaft		
Bearing		
Antifriction	384	457
Plain	384	429+
Thrust	384	250
Lapping	51	73R
Making	29	6
By turning	82	9
Straightening	72	389+
Twisting	72	299
Methods	72	371
Crash		
Fabric	26	26
Seat	297	216
Signal for airplane		
Radio	342	385+
Water dye	116	211
Crate	D 9	414+
Animal neck stocks combined	119	99
Animal shipping, design	D30	109
Fermentation apparatus	435	287+
Freight car combined	105	373
Shooks	217	43R
Wooden receptacle	217	36+
Cravats	2	144+
Fasteners	24	49R+
Crayon	401	49+
Compositions	106	19
Holder	401	88+
Advancing means combined	401	55+
Modifying thermometers	374	183*
Molding device	425	117
Molding device	425	425
Cream		
Cold	514	772+
Separating and gas	55	189+
Separating decanter	210	514+
Separators	D 7	369+
Treating processes	426	334+
Whipping	261	DIG. 16
Cream of Tartar	562	585
Baking powder	426	562+
Creaming Latex		
Natural rubber	528	937*
Synthetic or natural rubber	523	335
Creaser (See Folder)		
Boot and shoe making		
Loose upper shaping	12	54.1
Box, cut and crease	493	59+
Envelope	493	186+
Horseshoe	59	36+
Plaiting fluting shirring and cross creasing	223	29
Sewing machine	112	131
Leather	112	51
Creche	D11	122+
Credit Card Systems	235	380+
Printers using credit cards with means to check validity of card before printing	101	DIG. 18
Creel (See Basket)		
Bobbin supporter and holder	242	131
Fishing	43	55
Design	D 3	38
Creeper		
Antislip shoe attachment	36	59R+
Baby	297	5+
Electric wire placing	254	134.5+
Horseshoe calk	168	30
Repairmans	280	32.6
Cremation Furnace	110	194
Crematory Urns	27	1
Design	D99	5
Creosote	568	716
Crepe Paper	428	153+
Making	264	282

Entry	Class	Subclass
Creping		
By mechanical manufacture	28	155$^+$
Cellulose fibers	8	117
Differential	8	114.5
Cresidine	564	443
Creslan T M (polyacrylonitrile Copolymers)	526	342
Cresol	568	716$^+$
Cresotinic Acid	562	475
Cresting	52	57
Crib		
Bed	5	93R$^+$
Design	D 6	382
Blanket design	D 6	603$^+$
Chair convertible	297	7$^+$
Infant, powered	180	166
Land vehicle convertible	280	31
Storage, with changing or discharging means	414	288$^+$
Cribbage Board	235	90
Design	D10	46.1
Cribbing	405	273
Shaft	405	272
Crimped		
Filament compound metal	428	603
Filament fiber or strand	428	369$^+$
Staple length fiber	428	362
Metal closure	220	309
Twisted and coated or impregnated	57	250$^+$
Twisted synthetic	57	247
Crimping (See Corrugating)		
Ammunition loading	86	39$^+$
Apparatus textile strands	57	282$^+$
False twist	57	282$^+$
Miscellaneous strands	28	247$^+$
Plastic filaments	425	66
Plastic filaments	425	303
Plastic filaments	425	325$^+$
Twisting and coating or impregnating	57	286
Barbed wire	140	59$^+$
Boot and shoe making		
Loose upper shaping	12	54.1$^+$
With diverse operations	12	52.5
Box edge making	493	109
Cigar or cigarette end forming	131	89$^+$
Closure cap structure	215	324$^+$
Glass reshaping apparatus	65	269$^+$
Hair crimper	132	31R
Design	D28	35
Process	132	7
Metal bending by rolls	72	199$^+$
Paper in wrapping machine	53	370
Paper receptacle end structure	229	5.6
Pie making implement	425	293
Sewing machine running stitch	112	174$^+$
Sleeve or tube connector	29	282
Hand tools	29	282
Textile fabrics	26	21
Textile fibers	19	66.1$^+$
Textile strands		
Plastic filaments	264	168
Twisting synthetic strands	57	282
Wire	140	105$^+$
Cringle	114	114
Crocheting		
Fabric	66	193
Knitting	66	85R
Needles	66	118
Sewing	112	116
Crook	119	152
Crop		
Riding	231	2R$^+$
Thinners	47	1.43
Croquet		
Design	D21	210$^+$
Game	273	56$^+$
Mallet	273	83
Cross		
Bail for harness	24	172$^+$
Bearer for railway car under frame	105	419
Braced pole arm	52	697
Chain antiskid for wheels	152	223$^+$
Decorative item	D11	96
With illumination	362	121
Head		
Bearing slide	384	11
Die cutting machine	83	624$^+$
Mine elevator guide	187	96
Motor valve actuator	91	218$^+$
Shoe heel nailing machine	227	135
Modulation (See intermodulation)		
Rail for metal planer	409	325
Tree ship spar	114	92
Winding	242	43R
Fishing reels	242	84.4
Cross Laid Textile	28	100$^+$
Cross Talk		
Suppression in telephone system	379	417$^+$
Telegraphy	178	69R
Cross Walk Pavement	404	17$^+$
Crossbar		
Closure fastener	292	259R$^+$
Exercising device	272	102
Telephony selective switch	335	112
Crossbow		
Design	D22	107
Mechanical projector	124	25
Crossing		
Continuous rail signals	246	375$^+$
Protection railway	246	111$^+$
Railway		
Cables	104	185$^+$
Conduit	104	141
Drawbridge signal	246	118
Extensible gates	49	124
Gates on opposite approaches	49	93
Train actuated gates	49	263$^+$
Road system	404	1
Signals car displayed	246	208
Signals highway automatic	246	293$^+$
Street road bed railway	238	8
Structure railway	246	454$^+$
Trolley	191	37
Crossover		
Electric insulator midline spacer	174	147
Portable railway track	238	12
Preventer for skis	280	817
Crotch	114	99
Crotonic Acid	562	598
Crowbar		
Design	D 8	88$^+$
Freight car door operator	49	275
Pushing and pulling element	254	133R
Tool combined with others	7	166$^+$
Track tool	104	10
Croweacin Acid	549	436
Crown		
Boiler crown sheet	122	496
Protecting	122	495
Bridle loop	54	13
Gearing interchangebly locked	74	351
Gem setting	63	27
Light shade or bowl support	362	452
Tooth	433	218$^+$
Shaping instruments	433	156$^+$
Watch and clock crown making	29	179
Crowning Device		
Dental instrument	433	141$^+$
Piano sounding board	84	196
Croze and Crozing		
Barrel	147	13$^+$
Closure	217	80
Staves	147	18$^+$
Crt Recorder	346	158
Electrical memories	365	118
Crucible (See Vessel)		
Electric furnace	373	
Arc furnace device	373	72
Electroslag remelting device	373	45
Glass furnace device	373	30$^+$
Induction furnace device	373	138$^+$
Crucible	373	156$^+$
Crucible or hearth	373	163
Resistance furnace device	373	109$^+$
Including crucible or hearth structure details	373	122
Furnaces	432	156$^+$
Heating general application	432	262$^+$
Laboratory	422	102
Metallurgical	266	275$^+$
Special material or treatment	156	DIG. 83
Thermit	266	167
Crucifix Design	D99	27
Crullers	426	556
Crumb Brush	D 4	
Crupper	54	22
Crusher	241	
Elements	241	291$^+$
Grinding wheel dressing	125	11R
Hollow metal body	100	
Nut edible	99	568$^+$
Road in situ	404	90$^+$
Surgical instruments	128	319$^+$
Tobacco stem	131	324
Vegetable masher	241	169.2
Crust		
Pie-crust sheet metal mold	99	432
Pie dough	426	391
Crustacean Culture	119	2
Crutch	135	68$^+$
Design	D 3	8
Stilts	432	258$^+$
Cryogenic Trap	62	55.5
Liquified gas content receptacles	220	901*
Crypt (See Mausoleum)		
Cryptography		
Code receivers	178	89$^+$
Code transmitters	178	79$^+$
Codes	178	113
Fraud preventing	283	73
Printed matter	283	17
Signals	116	18$^+$
Signals railway cab	246	175
Teaching	434	55$^+$
Typewriting	400	89$^+$
Crystal		
Composition, inorganic luminescent	252	6$^+$
Composition, organic luminescent	252	301.16
Detector radio per se	357	
Filter	333	187$^+$
In sample holder	378	79
Liquid crystal composition	252	299.1$^+$
Liquid optical filter	350	330$^+$
Liquid stock material	428	1
Metal working, making	29	DIG. 17
Microphones	381	173
Pickups	369	144
Piezoelectric	310	311$^+$
Scintillation	250	458.1
Sheet, ribbon or tube	156	DIG. 88
Web or cage crystal growth	156	DIG. 84
Crystallization		
Apparatus	422	245$^+$
Distilling processes with	203	48
Growing crystal from fluid-semiconductor device forming	437	81$^+$
Hydrocarbon recovered	585	812$^+$
Inorganic compounds process	23	295R$^+$
Organic compounds process	260	707
Single crystal	156	600$^+$
Sugar		
Apparatus	127	15$^+$
Processes	127	58$^+$
Cubicle	52	329
Cue		
Billiard	D21	210$^+$
Chalkers	273	17$^+$
Racks	211	68
Rests	273	23$^+$
Structure	273	67R$^+$
Trimmers	30	494
Cuff		
Apparell design	D 2	222.1
Handcuff	70	16$^+$
Holder	24	41$^+$
Design	D11	222$^+$
Making	223	2$^+$
Ironing	223	52.1$^+$
Protector	2	60
Receptacles	150	53
Shirt	2	123$^+$
Cuff Link	24	41
Design	D11	222$^+$
Culinary		
Processes and apparatus	99	
Cultivator		
Flame type	47	1.44
Crop thinning	47	1.43
Methods	47	58
Weed killer	126	271.1$^+$
Harrow	172	
Metalworking making	29	14
Plow type	172	
Cultures		
Animal husbandry		
Acquariums	119	5
Aquatic	119	2
Crustation	119	2
Fish	119	3
Mollusk	119	4
Silkworm	119	16
Bacterial culture	426	43

	Class	Subclass
Bee culture	6	1$^+$
Incubators or racks	435	809
Plant husbandry		
Algae culture	47	1.4
Electroculture	47	1.3
Mushroom culture	47	1.1
Water culture	47	59$^+$
Tissue cell culture	435	41
Virus culture apparatus	435	284
Culvert	405	124
Static mold	249	11$^+$
Cup (See Dish; Receptacle)	215	
Cleaning attachments drip	15	248R$^+$
Corrugating wall	72	176
Dispenser	D 6	516
Drinking	215	
Collapsible	220	8
Design	D 7	6$^+$
Dispensers	D 6	516
Stand, rack, or tray for	D 7	70
Strainer detachable	210	464$^+$
Drinking fountain bubble	239	31$^+$
Holders	312	43
Drip (See drip cup)		
Edge		
Making machine	493	109
Trimming	83	178$^+$
Egg	211	14
Design	D 7	7
Footed	D 7	11
Holder	222	
Ice cream	D 1	116$^+$
Inkwell desk-attached	108	26.2
Lacteal		
Extractors	604	74
Pads	604	346
Measuring	73	426
Design	D 7	50
Metal working, cup formed & bottom removed	29	DIG. 9
Oil cup closure	184	88.1$^+$
Pessary	128	131
Sounding combined	33	126
Thermometer combined	374	157
Sewing machine feed	112	18
Shaving	220	
Sounding with sampling	33	126.4 R
Stamping from sheet	72	343$^+$
Suction		
Bracket support	248	205.5$^+$
Light supports	362	397
Surgical vacuum	604	313$^+$
Umbrella drip	135	48
Cupboard		
Latch	D 8	331$^+$
Cupola		
Building	52	208
Design	D25	35$^+$
Gas generators	48	62R$^+$
Metallurgical	266	197$^+$
Discharging	266	195$^+$
Forehearths	266	166
X-art	266	900*
Ore treating	266	219
Cuprammonium Acetate		
Hydrocarbon purification by	585	846
Cuprammonium Cellulose	536	57
Compositions	106	167
Fiber modification	8	123
Cupreine	546	134
In drug	514	305
Curb		
Design	D25	119
Feeler	D10	117
Flower bed edging	D25	38$^+$
Hydrants	137	272$^+$
Pin for watch balance regulating	368	176
Street	404	7$^+$
Catch basin with strainer	210	299$^+$
Electric conductor containing	174	39
Static mold for making	249	8
Curbing Chain Making	59	28
Curd Breaker	99	452$^+$
Curette and Scraper	128	304
Curing		
Hay and fodder	426	636
Tobacco		
Electrical or radiant energy	131	299
Fluids or fluent material	131	300$^+$
Curler		
Apparel ironers	223	52.2$^+$
Curling iron for hair	D28	38
Curling irons self heating by burner	126	408$^+$
Electric heaters	219	222$^+$
Hair	132	31R$^+$
Design	D28	35$^+$
Hat brim shaping apparatus	223	14
Metal edge curling rolls	72	102
Textile fiber working	19	66.1
Yarn twister	57	29
Current Electric		
Alternating in telegraph systems	178	66.1$^+$
Alternating rotary field and armature dynamo	290	5
Belts	128	387
Collectors		
Dynamo	310	219$^+$
Moving vehicle	191	
Rotatable connector	439	13$^+$
Conductive fluids	174	9F
Control		
Arc discharge	315	
Electrical separation of liquids	204	305$^+$
Electrolytic apparatus	204	193$^+$
Electrostatic discharge	422	186.28$^+$
Railway signal systems	246	
Converting	363	
Current conversion	363	13$^+$
Direction indicators polarity testers	324	133
Distribution in vehicle transmission	191	2$^+$
Electric music instrument	84	1.1$^+$
Food preservation	426	244$^+$
Generating stress strain test	73	763$^+$
Generating thermometers	374	179
Measurement of electricity	D10	75$^+$
Meters electric	324	76R$^+$
Modifying thermometers plural zone	374	110$^+$
Averaging	374	115
Distinct temperatures	374	166$^+$
Operated telautograph	178	19$^+$
Operated thermometers	374	163$^+$
Digital output	374	170$^+$
Pulsating in telegraph systems	178	66.1
Railway block signals	246	
Regulation for X rays	378	109
Regulation of alternating current generators	322	27
Regulators	323	
Reversing in telegraphy	178	16
Shifting switch in arc lamps	314	2
Superposed telegraph system	178	49
Supply to electric furnaces	373	
Arc furnace	373	102$^+$
Glass furnace	373	39$^+$
Induction furnace	373	147$^+$
Supply to electric railways	191	
Transformers	336	
Current Fluid		
Brake operators	188	155
Centrifugal fluid distributor operated by	239	214.13
Classifying separating and assorting solids	209	132$^+$
Cleaning implements operated by	15	104.6 R$^+$
Conveyer for railway snow excavator	37	202
Conveyers	406	
Dredges operated	37	68
Scouring	37	79
Evaporators blast or current	159	16.1
Evaporators film type	159	8$^+$
Evaporators spray type	159	4.1
Fumigators	422	305$^+$
Massage	128	66
Measuring		
Flow direction	73	188
Ships logs	73	181
Motor	416	
Dynamo with	290	
Pump with	417	334
Rotary fluid distributor operated by	239	222.17
Steam injection	159	16.3
Submerged combustion	159	16.2
Valve reciprocating disk rotating	137	330$^+$
Currycomb	119	86$^+$
Design	D30	159
Currying		
Compositions	252	8.57
Process	8	94.21$^+$
Cursor, Display Input Control	340	709$^+$
Curtain		
Bedstead	5	163
Cabinet	312	3$^+$
Camera shutter	354	241$^+$
Design, window	D 6	575
Lace or net	87	
Design	D 5	11$^+$
Loop threader for curtain rods	223	105
Material (See textile)		
Net or mesh fabric	D 5	11$^+$
Piano pianissimo device	84	220
Portable flexible panel type	160	
Projection lantern dimmer	362	278
Pulls		
Knob and cord	16	122
Tassels	428	28
Rail or runner	16	87R$^+$
Railway sleeping car	105	324
Releaser for train vestibule	105	23$^+$
Rigid tieback	D 8	368
Ring	16	108$^+$
Design	D 8	367$^+$
Rod	211	86$^+$
Bracket	248	261$^+$
Design	D 8	363$^+$
Design	D 8	380
Decorative element	52	301
Design	D 8	376$^+$
Shade roller bracket combined	248	252$^+$
Single	211	105.1$^+$
Rope for pulling	160	340$^+$
Sheet delivery	271	67
Shower bath	4	608$^+$
Stretcher	38	102.1$^+$
Support or rack with	211	180
Theater curtain derrick hoist	254	266$^+$
Valve head	137	625.28$^+$
Vehicle		
Design	D12	183
Storm front	296	83
Tops	296	138$^+$
Curtain Wall	52	235
Curved wall	52	245$^+$
Picture window	52	766$^+$
Curve		
Analysis, chart &	33	1C
Cable railway	104	189$^+$
French	33	177
Ruler and scriber		
Gauge	33	177
Line	33	27.1$^+$
Compass	33	27.2$^+$
Conic section	33	27.8
Rose engine	33	27.11
Sine curve	33	27.1
Spiral	33	27.9
Line combined	33	26
Pantographic	33	23.1$^+$
Surface	33	21.1$^+$
Track	238	15$^+$
Curved Work-Forming Means		
Gear cutting	409	1$^+$
Glass molding	65	286$^+$
Metal bending	72	380
Milling	409	64$^+$
Pattern curvilinear draw	164	215$^+$
Planing	409	309$^+$
Plate cutting	82	72$^+$
Saw table gauges	83	471$^+$
Stave jointing traveling saw	147	34
Wood shaping guide	144	152$^+$
Cushion (See Mattress)	5	448$^+$
Billiard table	273	9
Bowling alley	273	53
Chrysanthemum	PLT	75
Dash board mounted	280	727$^+$
Design	D 6	596
Elevator well end	187	67
Land vehicle letdown top	296	125$^+$
Pillow	5	434$^+$
Pin cushion	D 3	27
Pump pressure compensator	417	540$^+$
Rail seat	238	283$^+$
Railway wheels	295	11$^+$
Saddle	54	41
Spring device	267	
Reciprocating bed	267	75
Well rig walking beam	74	581$^+$
Cushioned Device		
Axle box	105	224.1
Bedpans	4	456$^+$
Boxes	217	5$^+$
Chair seat	297	455

Entry	Class	Subclass
Connecting rods	74	583
Container	493	904*
For piston of expansible chamber device	92	85R
Gun stock	42	74
Design	D22	111
Heel	36	35A+
Horseshoe	168	12+
Design	D30	147+
Impacting device	173	139
Power hammer and presses	72	431+
Railway draft devices	213	7+
Shell	102	473+
Shoe		
Heel	36	35A+
Heel antislip	36	59R+
Heel ventilated	36	3R+
Insole	36	44
Insole design	D 2	318
Pads	36	71
Sole	36	28
Sole ventilated	36	3B
Sole	36	28+
Tires	152	151+
Tool driving means	173	162R
Typewriter key cap and stem	400	491+
Wheel and axle	301	2
Window sash	49	428
Cuspidor		
Cleaner	15	73
Dental apparatus combined	433	97
Design	D34	2
Stove attachment	126	32
Structure	4	258+
Cut Film Holders	354	275+
Magazine	354	174
Cut Pile Fabric		
Cutting	26	13
Finishing	26	2R+
Woven	139	
Cuticle		
Cutlery	30	
Combined	132	73.5
Cutlery	30	
Blades	30	346+
Cutting tools	30	165+
Design, miscellaneous	D 7	137+
Handles & blade connections	30	340+
Holder for	D 7	73
Holders for detachable blades	30	329+
Making	76	104R+
Dies for	72	470+
Sheathed	30	151+
Silverware sorting	209	926*
Cutoff		
Brake system flow retarder and	303	84R
Expansible chamber motor		
Admission	91	247+
Multiple expansion	91	152+
Fluent material	251	
Automatic weigher combined	177	60+
Fluid distribution valve actuation	251	
Food depositing machine	425	447+
Food depositing machine	222	
Hopper cutoff to sifter	209	246
Lubrication gravity feed	184	66+
Milking machine vacuum line	119	14.8
Nozzle with preset	239	67+
Receptacle filling	141	192+
Sand feeder	291	31
Slit resilient diaphram dispenser	222	490
Water distribution spout and	137	872+
Fluent material dispenser		
Biasing closure	222	511+
Contents pressure operated	222	491
Flow controller	222	544+
Gravity actuated	222	500
Motor operated	222	504
Preset cutoff	222	14
Rate of flow responsive	222	59
Relatively movable actuator for	222	505
Sectional	222	502+
Snap action	222	498+
Fluid pressure regulator combined	137	456+
Fuel feed		
Automatic burner control	431	18+
Boiler	122	447
Burner nozzle	239	569+
Selectively preset	239	67+
Milking machine vacuum line	119	14.8
Steam engines	91	247+
Cutout (See Circuit, Maker & Breaker)		
Arc lamp circuit	314	10+
Electric vehicle supply	191	8
Fluid heater	236	21R
Lamp circuit	315	119+
Muffler		
Acoustic	181	236
Valve structure	137	625.44+
Sign letters	40	602
Switches	200	
Cutter & Cutting (See Abrading; Filing; Sawing; Trimming)	83	
Abrading	51	
Agricultural		
Citrus fruit cutter hand tool	30	113.1+
Colters	172	
Corncob splitting	225	93
Cornstalk	56	53+
Cutter type harvester	56	229+
Design	D 8	1+
Fruit gatherers	56	328R+
Green corn	130	9R
Harrow disk	172	518+
Harvester	56	
Planting machines with seed cutters	111	
Plow moldboard furrow slice	172	758+
Potato cutters for planting	83	
Potato diggers	171	26+
Pruning saws	30	166R
Scythes	30	309
Sickles	30	309
Sod cutters	172	19+
Stalk choppers	56	500+
Threshing band cutters	130	1
Tree hacks for debarking	30	121
Vine cutters	56	229+
Block and board	269	289R
Board, household	D 7	46
Brick	83	
Broaching machine	409	243+
Buttonholes	30	118+
Machines	83	905
Cable	30	
Fishing line	43	43.12
Ships anchor cable	114	221R
Well cable	166	54.5+
Well torpedo cable	89	1.14
Candle	83	
Candy	83	
Cigar tip	30	109+
Design	D27	51
Cleaner for inside of pipe	15	104.13+
Coating processes with	427	289+
Combined with other tools	7	158+
Computer controlled, monitored product manufacturing	364	475
Container opening	206	601+
Cutlery hand manipulated	30	
Eraser combined	15	105.53
Hair planers	30	30+
Hair thinning shears	30	195
Knives	30	272R+
Manicure	30	26+
Nippers	30	175+
Razors	30	32+
Saws	30	166R+
Scrapers	30	169+
Shears	30	194+
Dental instruments	433	144+
Dermatome	128	305.5
Design hand held tools	D 8	14+
Abrading	D 8	90+
Bayonet, dagger, hunting	D22	118
Cleaning	D32	35+
Combined with key holder	D 3	64
Hair	D28	52+
Kitchen	D 7	99+
Knife, fork, spoon	D 7	137+
Letter opener	D 8	102+
Pencil sharpener	D19	73+
Razor	D28	45+
Silverware	D 7	137+
Dispenser with	222	80+
Article dispenser	221	30+
Article impaler	221	213+
Gas or vapor dispensing	222	5
Soap dispenser	241	602+
Earth and stone		
Bush hammers and chisels	125	41
Core type posthole diggers	294	50.5+
Earth boring	175	
Earth boring bits	175	327+
Earth core drills	175	403+
Earth excavating	37	
Earth reamers	175	406+
Earth scoops	37	118R+
Earth working	172	
Grinding wheel dressing	125	11R
Millstone dressing machine	125	27+
Millstone picks	125	42
Mining machines	299	29+
Precious stone working	125	30R
Rock drilling	175	
Rockworking in situ in earth	299	
Shave type mining machine	299	34
Stone chisels	125	6+
Stone dressing	125	2+
Stone planing	125	9
Stone sawing	125	12+
In place in earth	299	
Stone turning	125	10+
Stoneworking	125	
Street surface removal	299	36
Tunneling machines	299	29+
Well drilling	175	
Edge	83	425.2
Eraser combined	15	105.53
Eraser mutilating	30	169+
Burnisher combined	7	124
Feeding and cutting	83	401+
Food		
Butter	83	
Cheese	83	
Citrus fruit	30	113.3
Cleavers	30	308
Cooking molds with trimmer or divider	99	430
Corer	99	547+
Annular cutter hand tool	30	178+
Citrus fruit hand tool	30	113.1+
With segmenters	99	545
With segmenters hand tool	30	302
With two blades	30	174
Fish cutters	54	63
Indiscriminate comminution	241	
Indiscriminate cutting	83	
Meat saws (See wood saws)		
Parers for fruit	99	588+
Pastry making	425	289+
Pie segmenters	30	114
Tub or block with	241	199+
Vegetable end	99	635+
Waffle irons with excess trimmer	99	375
Furbishing	409	63
Gear cutting, milling, planing	409	
Glass (See search notes)	83	879+
Machines with charging molding		
Machines with shaping and	65	174+
Guide combined	30	286+
Hair guide for cutting	132	45R
Hard material in situ	299	
Boring	175	
Cutter	299	79+
Disintegrating machine	299	29+
Ice boring	175	18
Ice working tool	299	24
Heated (See 4 note to class 30)	83	171
Heel and sole edge	12	85+
I-beam cutter	83	DIG. 2
Ice		
Chippers	30	164.5
Harvesting implements	299	24+
Heat cutting	62	320
Heat cutting electrically	30	140
Implements	299	24+
Picks	30	164.5+
Scraper hand held	30	136+
Scrapers for removing ice from sidewalks	37	266+
Ship ice breaker	114	42
Snow plows with ice scrapers	294	54.5
Inflatable tube	83	DIG. 3
Insulation	81	90.1+
With remover	81	9.4+
Leather manufactures	69	
Making cutters	76	
Edge by electric weld deposition	219	77
Manicure compound tool with	132	75.4+
Metalworking	407	
Air current generated by	29	DIG. 82
Ambulatory, with fluent conduit	29	DIG. 64

	Class	Subclass
Auger lip cutting	76	3
Ax making	76	7$^+$
Boring head	408	234
Boring machines	408	
Broaches	407	13$^+$
Cable sheath cutters	30	90.1$^+$
Can openers	30	400
Casting combined	29	527.6
Casting device with	164	262$^+$
Chain making machine with	59	16$^+$
Chasing screw thread	408	215$^+$
Cold chisels	30	168
Contained supply reservoir	29	DIG. 65
Countersinks	408	223$^+$
Cutter making processes	76	101SM$^+$
Cutters for machine mounting	407	
Die shaping sheet metal and	72	324$^+$
Drilling machines	408	
Drills	408	199
Drive	29	DIG. 55
Engaging cleaner	29	DIG. 97
Expanded metal making	29	6.1
Fan coaxial with	29	DIG. 83
Feed	29	DIG. 57
Fence barb cutting	29	7.1
File cutting	76	12$^+$
File resharpening sand blast	51	414
Files and filing	29	76R$^+$
Flame cutting	148	9R
Flame cutting apparatus	266	48$^+$
Fluid channel in	29	DIG. 92
Fluid spreader contacts	29	DIG. 69
Forging machine with cutting machine	29	34R
Forging machines with incidental cutting	72	324$^+$
Gear cutting machines	409	1$^+$
Hand manipulated tools for	30	165$^+$
Heat employed to cut	148	9R
Hood encased	29	DIG. 86
Horseshoe making machine with	59	37$^+$
Immersion complete	29	DIG. 71
Immersion partial	29	DIG. 74
Lathes	82	2R$^+$
Milling machines	409	64$^+$
Nail making machines with	10	28$^+$
Needle making machine with	163	1$^+$
Pin making machine with	163	6
Pipe and rod cutters	30	92$^+$
Planing machines	409	288$^+$
Printers leads	83	
Punching	83	
Reamers	408	227
Rifling	409	306
Rolling and cutting machines	72	203$^+$
Rotating	29	DIG. 67
Saw tooth cutting	76	28$^+$
Sawing	83	
Scrap cutting	83	923
Screw cutting lathe	82	5$^+$
Screw threading dies	408	215$^+$
Screw threading machines	408	72R$^+$
Screw threading taps	408	215$^+$
Screw threading with cut off	408	28$^+$
Shapers	409	288$^+$
Sharpeners abrading	51	246
Sharpeners for cutting tools	76	82$^+$
Sharpeners for saw teeth	76	30$^+$
Sheet and bar cutting	83	
Sheet metal shaping die with	72	324$^+$
Shredding	29	4.5 R$^+$
Staple making machine with	59	71$^+$
Stereotype concave surface finishing	409	309
Tap or die	408	215$^+$
Tire upsetting machine with cutter	29	22
Tube and sheath splitters	30	90.1$^+$
Tube cutters for well tubing	166	55
Turning to taper	82	15$^+$
Turning with profiled cutter	82	13
Turning with radial cutters in hollow head	82	20
Type casting mechanism with	199	54$^+$
Valve refitting portable tool	409	64$^+$
Wire	83	
Wire cutting with tag machine	493	343$^+$
Wire fabric making with slat	140	26
Wire straightening and	140	139$^+$
Wire stretching implements with	140	123.6
Wrench attached cutter	81	181$^+$
Oiler	83	169
Paper	D15	127
Pencil sharpener (See pencil sharpener)	30	451$^+$
Machine	144	28.1$^+$
Pipe hand tools	30	92$^+$
Plastic block apparatus	425	289$^+$
Plastic material implements	30	115
Punching	234	
Receptacle openers		
Implement	30	2$^+$
Metallic receptacle with	220	260
Refrigeration related	62	294
Rotary (See notes to)	30	347
Rotating axially moving tool	408	
Safety devices	83	DIG. 1
Selective	234	
Pattern controlled	234	59
Severing by tearing or breaking	225	
Sewing machines combined	112	128
Thread	112	285$^+$
Shaping devices	407	
Sheet and bar	83	
Cigar receptacle with	206	238$^+$
Cigar wrapper cutter	131	33$^+$
Cigarette mouthpiece making	131	91
Cloth finishing	26	7$^+$
Dispensing package with	206	216
Fabric reels with	83	
Label machine strip severing	156	510$^+$
Measurer fabric and cord with	83	522
Paper aperture making punch for sheet retainer	402	1
Paper bag making with	493	227$^+$
Paper box making with	493	56$^+$
Printing machine with	101	224$^+$
Sheet material associating and folding with	270	21.1
Sheet material folding with	493	340
Stenciling machine	101	117
Tobacco leaf disintegrating	131	311$^+$
Tobacco plug	131	117$^+$
Tobacco stem	131	317$^+$
Shoemaking	12	
Soap	83	
Solid material comminution or disintegration	241	
Sound waves	83	14
Sound waves	83	701
Stone working	125	
Street surface removal		
Ice	299	24
Surgical	128	305$^+$
Design	D24	28
Electrical	128	303.14$^+$
Test		
Cutting edge testing	73	104$^+$
Shear stress, eg Izod, charpy	73	844
Textile and strand	83	913
Beria type cutter	83	403
Bobbin and cop winding with	242	19
Buttonhole cutter	30	118$^+$
On sewing machine	112	68
Cloth finishing	26	7$^+$
Cordage winding with	242	48
Fabric reeling with	242	56R
Knitting strand controlled stop	66	159
Measurer for fabric with	33	127$^+$
Severing cutter for sewing machine	112	130
Sewing machine shuttle with	242	21
Shears	30	194$^+$
Spool or twine holders with	83	649$^+$
Textile knitting	66	125R$^+$
Textile spinning strand control stop	57	86$^+$
Textile weaving	139	43
Textile weaving thread control	139	260$^+$
Thread cutter for sewing machine	112	285$^+$
Trimmer for sewing machine	112	129
Tufting sewing machine	112	80.1$^+$
Well torpedo cable	89	1.14
Threshing	130	
Tobacco and tobacco products	131	
Cigar end cutters	83	
Cigar or tobacco receptacle with	206	238$^+$
Cigar tip cutters hand	30	109
Tools with additional tools	7	158$^+$
Monkey wrench	7	139$^+$
Plier type	7	129$^+$
Plier type	7	132$^+$
Torch	266	48$^+$
Feed dispersing	239	398$^+$
Tracheotome	128	305.3
Trepan	128	305.1
Trepanning	408	703
Turning	82	
Vehicle body	296	2
Woodworking	144	36$^+$
Coopering	147	
Draw knives	30	313
Knives	30	272R$^+$
Pencil sharpener	30	451$^+$
Reciprocable along elongated edge	83	746$^+$
Sawing	83	
Spokeshaves	30	281
Turning	142	
Work attached device	83	743$^+$
Cyanamides	423	368
Catalysts containing	502	200
Electrolytic synthesis		
Aqueous bath	204	91
Fused bath	204	63
Fertilizers containing	71	55$^+$
Organic radical containing	564	103
Cyanates	423	364$^+$
Esters	560	301
Cyanide		
Catalysts containing	502	200
Detergents containing	252	141
Electrolytic synthesis		
Aqueous bath	204	91
Fused bath	204	63
Leaching ores with	75	105$^+$
Electrolysis combined	204	
Organic	558	303$^+$
Preserving, disinfecting, sterilizing with	422	35
Salts inorganic	423	364
Cyanine Dyes		
Cyanoethylation of fibers	8	DIG. 13
Photographic compositions containing	430	581$^+$
Cyano Group, Formation of	558	308$^+$
Cyanoacetic Acid	558	443
Cyanogen		
Ammonia from compounds	423	352
Biocides containing	424	129
Electrolytic synthesis		
Aqueous bath	204	91
Fused bath	204	63
Hydrocyanic acid	423	364
Salts	423	364$^+$
Cyanoguanidines	564	104
Cyanohydrins	558	451
Cyanuric Acid	544	192
Cycle (See Velocipede)		
Bicycle type	280	200$^+$
Design	D12	111
Design	D12	107$^+$
Dicycle type	280	208
Electric bicycle signal	340	134
Electric light	362	72
Locks	70	233
Mud guards	280	152.1$^+$
Parallel connected	280	209
Polycycles	280	282
Rack or holder	D12	115
Saddles	297	195$^+$
Sidecar	D12	116
Single axis or wheel	280	205$^+$
Stirling	62	6
Theft preventing alarms	116	33
Tricycle type	280	200$^+$
Convertible	280	7.15$^+$
Design	D12	112$^+$
Simulations	280	1.11 R$^+$
Vehicle energy actuated signal	116	56$^+$
Cyclic Hydrocarbon (See Alicyclic Hydrocarbon; Aromatic Hydrocarbon)		
Cyclic Peptides	530	317$^+$
Cyclic Polymerization, Hydrocarbon Synthesis	585	415$^+$
Aromatic hydrocarbon synthesis	585	415$^+$
Cyclic Thiocarbonates	549	30$^+$
Dithiocarbonates, monothiocarbonate	549	30
Trithiocarbonates	549	36
Cyclization		
Mineral oil	208	46$^+$
Cyclobutyl Carboxylic Acid Esters	560	123
Cyclohexamide	546	
In drug	514	328

	Class	Subclass
Cyclohexanol	568	835
Cyclohexanone	568	376
Cyclohexylamine	564	462
Cyclometers	235	95R+
Cyclone Separator	406	173
Gas separation	55	337
Liquid seoaratiom	210	512.2
Cycloolefines	585	23
Synthesis	585	350+
By dehydrogenation	585	379+
Cycloparaffins	585	20+
Synthesis	585	350+
Cyclopentadiene Synthesis	585	350+
By alkylation	585	446
Cyclopentanohydrophenanthrene Compounds	260	397+
Heterocyclic	540	2+
Cyclopentyl Carboxylic Acid Esters	560	121+
Cyclopropane	585	350+
Cyclopropyl Carboxylic Acid Esters	560	124
Cycloserine	548	244
Cyclotrons		
Energizing systems	328	234
Nuclear fusion reactions use in	376	100
Structure	313	62
Systems including	328	234
Tubes	313	62
Cylinder		
Airplane with sustaining rotor	244	10
Bracket support	248	230+
Caster	16	22+
Compressor		
Movable cylinder with reciprocating piston	417	460+
Multiple cylinder	417	521+
Rotary cylinder	417	462
Door closing and liquid checking	16	57
Earthworking	172	554
Escapement	368	129
Expansible chamber device	92	169+
Concentric cylinders	92	51+
Lubricating means	92	153+
Movable cylinder	91	196+
Moving cylinder multple	92	117R+
Gauges	33	178R
Glass cylinder making	65	
Glass handling	414	24
Grinding ie internal grinding	51	290
Handles, hollow cylinder	414	910*
Head packing	277	
Heat exchange means	165	181
Holding member	292	261
Internal combustion engine		
Construction	123	193R
Cooling	123	41.72+
Multiple	123	52R+
Oscillating	123	42
Reciprocating	123	50R
Rotary	123	43R+
Joint for rotating cylinder	285	134+
Lagging	165	135+
Lock	70	357+
Housing	70	449
Key	70	406
Lubricator	184	18+
Machine gun	89	13.5+
Material treating		
Abrading floors	51	176
Abrading with rotary flexible tool	51	358+
Abrading with rotary rigid tool	51	206R+
Acetylene generator	48	
Baffle	210	512.1
Brushing	15	3+
Comminuting	241	293+
Corn sheller	130	6
Drying and cooling with rotary cylinder	34	63
Fiber carding	19	112
Fiber combing	19	126+
Filter element	210	497.1
Huller	99	600+
Leather treatment	69	37
Movable filter	210	402+
Paper stuff straining	210	402+
Photographic printing	355	104+
Potato digger with screen	171	111+
Reactor compartment	210	207+
Scouring	51	22+
Textile treatment with fluid	68	5R+
Threshing	130	27R
Warp shedding	139	61
Web forming	162	323+
Weighted land roller	404	111+
Winding	242	7.21
Meter having expansible chamber	73	232+
Music boxes automatic playing	84	95.1+
Comb actuator	84	95.1+
Organ pin cylinder	84	86
Separable from keyboard	84	106
Panel hanger	16	89
Pattern	139	317+
Percussive tool rammer	173	125
Pistol or rifle revolving cylinder	42	59+
Pump (See pump fluid)		
Multiple cylinder	417	521+
Rotary cylinder	417	462
Rotary cylinder with reciprocating piston	417	460+
Rotor sustaining airplane	244	21
Sheet handling		
Associating	270	60
Folding and associating	270	32+
Folding, associating & printing	270	1.1+
Solid separation devices	209	
Steam locomotive	105	42
Syringe	604	218+
Telescoping	92	51+
Toothed cylinder manufacture	29	23
Vehicle support resilient	152	
Wood and metal treating		
Barrel stave cutter	147	35
Corrugating	72	
Cutter	144	221+
Planer with rotary cutter	144	116+
Slicer with rotary cutter	144	172+
Cylindrical		
Photographic printing apparatus		
Cymbal Beater	84	422R
Cymel T M (aminoplast)	528	254
Cymene (See Aromatic Hydrocarbon)	585	350+
Cyproheptadine, in Drug	514	325
Cysteine Ester, in Drug	514	550
Cystine	562	557
Preparation from protein	562	516
Cystoscope	128	7
Cytosine	544	317

D

	Class	Subclass
D D T, in Drug	514	748
D D V P, in Drug	514	136
Dacron Insulated Cable	174	100D
Dacron T M (See also Synthetic Resin or Natural Rubber)	528	308.1
Dado		
Cutter	144	222
Lapped multiplanar surfacing	52	536
Machine	144	133R
Daggers	30	295
Daguerreotypy	354	2+
Analysis and analytical control Apparatus	422	50+
Apparatus design	D16	
Dairy, Food Treatment	99	452+
Dairy-type Bottles	D 9	
Dam	405	107+
Dental	433	136+
Mortar	52	421
Power extraction from water	405	78
Damask	139	416+
Dampening (See Damper; Moistener; Steaming)	68	5R
Consumable electrode electric lamp		
Electromagnet feeding with damping	314	128
Feed with damping	314	99+
Instrument mechanism damping	73	430
Bourdon tube	73	739+
Musical instrument damping	84	
Overhead electric conductor with vibration damping	174	42
Paper making calender	100	73+
Planographic printing	101	147+
Copying apparatus	101	132.5
Pressure compensator		
Pipes and tubular conduits	138	26+
Pumps combined	417	540+
Vibration damping support	248	562+
Nonresilient	248	636
Resilient	248	562+
Suspension	248	610+
Damper		
Ear sound stoppers	128	152
Electric measuring instruments for	324	125
Electric musical instrument	84	1.13
Rate of attack	84	1.26
Volume control	84	1.27
Electric winding for alternating or direct current dynamos	310	183+
Flywheels and rotors having	74	574
Furnace draft	110	163
Heat exchanger	165	69
Impact absorption	49	9
By flexible barrier	49	9
By plastic deformation	188	371+
By resilient deformation	188	268
Inertial vibration damper	188	378+
Microphone or diaphragm	381	158
Piano damper head and stems	84	255
Harmonic	84	234
Held	84	218
Lifted	84	217
Railway truck	105	193
Stove or furnace flue	126	285R+
Open evaporating pans having	159	41
Time controlled	126	285.5
Weighing scale	177	184
Zither	84	287+
Damping Diode	313	317
With emissive cathode	313	310
With thermionic cathode	313	310
With two cathodes and anodes	313	1
Dancing		
Figure toys	446	330+
Shoes	36	8.3
Teaching	434	250
Dandy Roll	162	314
Dark		
Cabinet photography	354	307+
Lantern with shutter or screens	362	167+
Room	354	307+
Illuminators	362	293
Ventilators	98	29+
Darning		
Knitting	66	2
Last	223	100
Sewing machines	112	121
Design	D15	66+
Elements	112	236
Weaving	139	33.5
Dart	273	416+
Design	D22	115
Amusement devices	D21	49
Projector	D22	107
Projector	124	22
Dashboard		
Land vehicle	296	70+
Lighting		
Automobile	362	23+
Trolley car or rail car	362	77
Motor vehicle	180	90
Dasher		
Food	366	69+
Gas and liquid contact	261	32+
Dashpot		
Check valve	137	514+
Closure check	16	82+
Closure check brake type	188	266+
Door check and closer	16	49+
Electrode feed retarder	314	99+
Electromagnetic electrode operator combined with	314	128
Gas & liquid contact apparatus	261	DIG. 18
Spring combined with	267	195+
Switch contact retarder	200	34
Valve retarder	251	48+
Water closet elements	4	248
Water closet tank valve	4	388+
Dasymeters	73	30
Data Comparing		
Calculator	235	431
Cards perforated sorting	209	613
Condition responsive	209	554
Lens	333	248+
Meter fluid flow	73	196+
Punching machine	234	34
Data Conversion		
Analog to digital	340	347M
Card or tape punch	234	69+
Cryptographic	380	
Digital to analog	340	347M
Digital to digital	340	347DD
Radix conversion	235	311
With card or type punch	234	69+
With computation	235	310+

	Class	Subclass
With data processing	364	900
With visual display	340	700+
Data Handling		
Selective punching	234	
Storage	234	55+
Data Processing, Electrical (See Computer, Data Conversion, Error Checking)	364	
Applications	364	400+
Computers, generic control	364	130+
Controlled by record means	235	375+
Electrical computers and	364	
General purpose digital data processors	364	200
Measuring, testing, or monitoring	364	550+
Microprocessors	364	200
Miscellaneous digital systems	364	900
Plural diverse storage systems	364	900
Programmed processing	364	200
Programming	364	300
Selective punching, for	234	55+
Special purpose digital data processors	364	900
Synchronous transmission of digital data	375	106
Video color signal derived by data processor	358	81
Data Storage		
Dynamic information storage or retrieval	369	
Magnetic	360	
Registers electrical	377	
Registers mechanical	235	
Static information storage or retrieval	365	
Dauber	15	209R+
Animal antivermin treatment	119	157
Polisher combined	15	107+
Kit including polish	401	118+
Supply container closure attached	401	126+
Davenports, I.e. , Sofa Beds	5	12R+
Davit	114	368+
Davy Lamp	362	164+
Day Beds	5	12R+
DC to AC Chopper	363	124+
Dead Bolt for Lock	70	113+
Dead Center Overcoming	74	36
Dead End Connector Cord Holder	24	115R+
Dead Man Type Anchor	52	166
Dead Mans Switch	191	1R
Switch structure	200	
Deadeners (See Dampening)		
Gearing having plural sections	74	443
Railway		
Car wheel	295	7
Cushioning of track	238	283+
Rolling stock	105	452
Track	238	382
Dealkylation (See Cracking)		
Aromatic hydrocarbon synthesis	585	483
Foam breaking	252	321
Mineral oil	208	46+
Debarking	144	340
Debris Accumulating Pencil Sharpener	30	453+
Debris Control	29	DIG. 5
Chutes	29	DIG. 102
Receptacle, remover	29	DIG. 79
Remover, catcher, deflector	29	DIG. 94
Remover, plural type	29	DIG. 61
Separators from workpiece	29	DIG. 53
Decade Counter		
Electronic tube type	377	103
Gas tube	315	
Oscillator	377	
Decahydronaphthalene (See Alicyclic Hydrocarbon)		
Naphthalene hydrogenation	585	266+
Synthesis	585	350+
Hydrogenation of unsaturated hydrocarbon	585	266+
Decal (See Decalcomania)		
Decalcomania (See Transfer)	D20	11
Exposable adhesive layer	428	40+
Fraud or tamper detecting	428	915*
Made by coating	427	147+
Process of transfer	156	230+
Stock material	428	914*
Decalin (See Alicyclic Hydrocarbon)		
Decanter		
Coffee	222	475.1
Design	D 9	384+
Distillation systems separatory	202	204
Filter combined	210	299+
Heating means	210	182+
Gas and liquid contact	261	7+
Liquid separation	210	513+
Starch	127	27
Decanting	210	767+
Filtering combined	210	806
Decapitator for Marine Animals	17	63
Decarbonizing		
Absorbents	502	20+
Catalysts	502	20+
Mineral oil vaporizing combined	208	347+
Apparatus	196	122
Conversion combined	208	48R
Processes	134	39
Heating combined	134	20
Stills	201	2
Apparatus	202	241
Decarboxylation Aromatic Polycarboxylic Acid	562	479
Deceleration (See Acceleration)		
Decelerometer	73	514+
Decibel Meter		
Meter movement	324	76R+
Sound loudness measurement	73	646
Decimal Point on Calculators	235	61DP
Binary coded	377	33
Elimination	235	63DE
Locators	235	64.3
Deciphering (See Cryptography)		
Deck		
Bridge	14	6
Railroad		
Freight cars with convertible decks	105	370
Passenger cars double decked	105	340
Stock cars with	119	9
Ship	114	85
Type insulator for antenna	174	158R
Deckling		
Paper making		
Finishing	162	286
Web forming	162	310
Pinking shears	30	229+
Sheet		
Cutting	83	
Tearing	225	
Decladding Nuclear Fuel	252	625
Decoding (See Cryptography)		
Decollator	270	
Web	270	52.5
Decolorizing (See Bleaching)		
Chemical	260	701+
Liquids	210	660+
X-art collection	210	917+
Decomposing Fat	260	415+
Decoration (See Article by Name; Ornament)		
Bow rosette, pompon	428	4+
Collapsible	428	12
Displays and exhibitors	40	406+
With special effects	40	427+
Feathers and plumes	428	6
Festoon	428	10
Collapsible	428	9
Filamentary	428	542.6
Flora, artificial	428	17+
Lamp housing	362	362+
Lights	362	227+
X-art collection	362	806*
Party type	428	7+
Pole, cap	52	301
Roof finial or cresting	52	57
Sheet-form	428	542.6
Trimming for clothing	2	244
Trimming for hat	2	186
Trimstrip for building	52	717.1+
Decorticating	19	5R+
Decoupling Filter	333	181+
Decoy		
Design	D22	125+
Trapping	43	2+
War training	434	11
Deep Drawing Metal	72	347+
Deer Sling	224	101+
Defecating Apparatus	127	11+
Defect Coating	430	938*
Defiberizers	241	
Deflating Tool for Resilient Tires	81	15.4
Deflection		
Cathode ray stream	313	421+
Cathode ray stream	315	399
Gas separation	55	434+
Liquid treatment	55	220+
Processes	55	1
Lamp arc	313	153+
Consumable electrode	314	20
Solid separator		
Gas suspension	209	143+
Liquid suspension	209	210
Testing	73	849+
Wind motor wheel	416	9+
Deflector		
Bearing lubricator	384	322+
Blowpipe nozzle	239	461+
Breath	128	139
Drip	52	97
Fluid distributor	239	461+
Fluid fuel burner nozzle	239	461+
Furnace air	110	310+
Gas separation	55	434+
Processes	55	1
Pipe	138	37+
Plant	172	517
Scouring dredger with current	37	79
Ship armor	114	10
Sprinkler combined	239	461+
Water motor	416	9+
Windmill	416	9+
Window wind	98	44
Deflocculation		
Liquid separation	210	696+
Solid separation	209	5
Defoliants	71	69+
Deforming		
Metal	72	
Casting combined	29	527.5+
Plastics		
Apparatus	425	
Processes	264	
Defribrillators	128	419D
Defroster		
By ventilator	98	90
Electrical resistance heaters	219	203
Heat applying window cleaner	15	250.5
Material applying window cleaner	15	250.1
Refrigerator design	D 8	29.2
Surface heater	126	271.1+
Vehicle power plant heater combined	237	12.3 A
Window	52	171+
Ventilator type	98	2
Defrosting		
Defrosting device	D 8	29.2
Heat pump	165	17
Refrigeration apparatus	62	272+
Automatic	62	151+
Automatic, frost initiated	62	140
Frost indicator	62	128
Process	62	80+
Degasifying Liquids	55	159+
Process	55	36+
Degassers	261	DIG. 19
Degaussing (See Demagnetizing)		
Means in cathode ray tube circuit	315	8
Television	361	150
Torpedo nets	114	240R
Degradability Enhanced Synthetic Resin or Natural Rubber Composition	523	124+
Cellular product	521	916
Degreasing		
Distillation		
Still and extractor	202	168+
Still extractor	202	170
Solids	134	
Textiles	8	139+
Degumming	8	138
Dyeing combined	8	
Dehalogenation		
Preparation of epoxy compounds	549	518+
Preparation of olefines	585	641+
Preparation of unsaturated esters	560	213
Unsaturation by	260	696
Dehorners		
Electric heater type	128	303.1
Hand tool	30	
Dehumidifier	D23	359+
Dehydrating Oil		
Fatty oils	260	405.5
Mineral oil	208	187+
Electrical separation	204	188+
Electrophoretic or electroosmotic	204	181.8+

	Class	Subclass
Dehydration (See Drier)		
Acid anhydrides by	260	545R+
Acyclic ketones from alcohols	568	403+
Aldehydes from alcohols	568	485+
Compositions	252	194
Concentrating evaporating	159	
Distillation	203	12+
Fatty oils or acids	260	405.5
Foods	426	443+
Apparatus	99	467
Gas or vapor contact with solids	34	
Gas separation	55	
Refrigeration combined	62	271
Refrigeration combined process	62	93+
Mineral oils	208	187+
Electrical	204	188+
Electrophoretic or electroosmotic	204	181.8+
Olefine production by	585	638
Refrigerant	62	474+
Automatic	62	195
Process	62	85
Dehydroabietic Acid	562	404
Dehydrogenation	260	696
Acyclic ketones from alcohols	568	403+
Aldehydes from alcohols	568	485+
Aromatic hydrocarbon synthesis	585	440
Diolefin hydrocarbon synthesis	585	616+
Esters from alcohols	560	239
Mineral oil	208	46+
Olefines from hydrocarbons	585	654+
Synthesis aromatic from alicyclic	585	430
Dehydrohalogenation		
Alicyclic hydrocarbon synthesis	585	359
Diolefin hydrocarbon synthesis	585	612
Olefin hydrocarbon synthesis	585	641+
Dehydrothiotoluidine Azo Compounds	534	800
Deicer (See Thawing)		
Aircraft	244	134C
Antenna	343	704
Carburetors	261	DIG. 2
Combustion product as motive fluid	60	39.93
Trolley collector head	191	62
Deinking		
Paper stock	162	4+
Textiles	8	137
Delasting	12	15.1
Delay Networks		
Delay lines including		
A lumped parameter	333	138+
Elastic bulkwave propagation means	333	141+
Elastic surface wave propagation means	333	150+
Long line elements	333	156+
Electronic tube system	328	55
Delinter	19	40+
Carbonizing processes	8	140
Delinting Cotton Seed	47	1R
Processes	47	58
Fluid treatment	8	140
Delivery		
Collection receptacle	232	
Coins	232	64+
Letter box chute	232	53
Conveyer	198	
Carrier controlled selective	198	570+
Dumping vehicle with chute	298	7
Letter box	232	17+
Mail from aircraft	258	1.2+
Planting chute	111	76
Railway mail	258	
Railway selective	104	88+
Record holder	312	15+
Sheet	271	278+
Store service	186	3
Delrin T M (polyoxymethylene)	528	270
Delustering		
Cellulose ether or ester compositions	106	192
Viscose	106	166
Coating or impregnating processes	427	170
Demagnetizing	361	267
Demagnitizer	D13	99
Erase, head	360	118
General recording biasing, erasing	360	66
Sorting treatment	209	8
Demijohns (See Carboy)	D 9	367+
Demodulator	329	
Demonstrating Apparatus or Product	434	365+
Demonstrator Radioactivity (See Radioactivity Demonstrator)		
Demountable Rim	301	10R+

	Class	Subclass
Demulsifying Compositions	252	358
Denatured Compositions	252	365
Alcohol	252	366
Denim	139	416
Densitometer		
Emulsion exposing printer	354	20
Emulsion opaqueness measuring	356	436+
Density		
Analysis gas	73	30
Analysis liquids	73	32R+
Bed art	5	DIG. 2
Dent Remover Sheet Metal	72	457
Dental and Dentistry	433	
Amalgams and alloys	420	526
Amalgam mixer, eg dental filling	366	602*
Apparatus, fixed	D24	4+
Apparatus, portable	D24	10+
Cabinet	312	209
Cassettes	378	168
Chairs	297	68+
Design	D 6	334
Compositions	106	35
Containers for dental use	206	63.5
Dentures	433	167+
Die shaping	72	
Engines	433	103+
Equipment stands	433	25+
Fillings	433	226+
Floss	132	93
Holders	132	91+
Design	D28	64
Impression devices	433	34+
Impression material	106	38.2+
Instruments	433	25+
Medicines	424	
Molding devices	425	2
Molding devices	425	175+
Molding devices	249	54
Flasks	425	175+
Flasks for metal casting	164	376
Molds	425	175+
Molding processes	264	16+
Orthodontic devices	433	2+
Practice	433	
Processes manufacturing	29	160.6
Spittoons	4	263+
Spotlights	362	257+
X-art collection	362	804*
Supply packages	206	63.5
Teaching devices & methods	434	263
Teeth	433	167+
Tool container	206	368+
Waste receptacles	206	63.5
Dentifrices	49	
Dentiphone		
Acoustic hearing aid	181	127
Electrical hearing aid	381	68.3
Dentistry (See Dental)	433	
Design	D24	
Deodorant		
Body	424	65+
Cosmetics containing	424	65+
Fertilizer containing	71	3
Milk	426	488
Non body	424	76.1+
Organic compound treatment	260	708
Preserving, disinfecting, and sterilizing	422	5
Deodorizer or Ozonizer	D23	366
Deoxidant Compositions	252	188.1+
Descaling compositions containing	252	81
Water softening or scale inhibiting compositions ctg	252	178
Dephlegmators		
Mineral oil	196	139
Still combined	202	
Depilating		
Animal carcass	17	47
Process	17	47
Compositions		
Electric needle	128	303.18
Needle supports	128	303.19
Fermentative	435	265
Hides and skins	8	94.16
Mechanical	19	2+
Depolarizer	252	
Depolarizing		
Compositions	252	
Depolymerization to Obtain Hydrocarbon Mixture	585	241
Deposit		
Certificates	283	59

	Class	Subclass
Change gates	232	14
Collection and deposit receptacle	232	
Cabinet combined	312	211+
Depository design	D99	28+
Letter boxes	232	22
Milk bottle collection	232	42
Receptacles for	232	
Registering or receipt printing combined	109	24.1+
Photograph or microfilm record	346	22
Safes or depositories	109	
Design	D99	28+
Receipting means included	109	24.1
Depositor		
Bread etc making depositor	425	447+
Implement	222	
External body surface	604	289+
Medicine	604	47+
Solid materials	604	57+
Planting	111	
Dibbling	111	89+
Hill	111	34+
Toy figure for coins	446	10+
Depressor		
Necktie	24	53+
Railway crossing cable	104	186
Tongue	128	15+
Atomizer combined	128	200.15
Depth Bomb	102	390+
Depth Gauge		
Compressional wave	367	99+
Depth measuring instrument	D10	46+
Fluid level	33	
Geometrical	33	126+
Geophysical exploration	324	323+
Hydrophone	367	141+
Liquid	73	290R+
Radar	342	120+
Other than air, eg, underwater	342	118
Sound	181	
Derailer		
Actuated cab signal or train control	246	170+
Cycle or motorcycle	D12	124
Enclosure or guard	D12	127
Guards		
Railway	104	242+
Store service	186	30
Railway		
Car attached	104	261
Interlocking	246	163
Derailleur (See Derailer)		
Derectifiers Electric	363	
Dereverberator	381	66
Dermatological Device		
Needle surgical	128	303.18
Support	128	303.19
Surgical instrument	128	355
Dermatome	128	305.5
Derrick		
Extensible or movable	52	111+
Pushing and pulling elements		
Cable hoists	254	283+
Screw	254	99
Skeleton tower	52	648+
Desalting Sea Water	210	642
Distillation	203	10
Fractional crystallization	62	532+
Direct contact refrigeration	62	533+
Reverse osmosis	210	652
Descaling Metal	29	81R
Combined with metal deformation	72	40
Deserpidine	546	55
Desiccation (See Dehydration; Drier)		
Design (See Design Classification)		
Advertizing	D20	
Apparel & haberdashery	D 2	
Arms, pyrotechnic articles, articles for hunting, fishing, and pest killing	D22	
Artists materials	D19	35
Brushware	D 4	
Building units & construction elements	D25	
Care & handling of animals	D30	
Cases not specified elsewhere	D 3	
Caskets	D99	1+
Coating nonuniform	427	256+
Containers	D 9	
Devices & equipment against fire hazards, for accident prevention and for rescue	D29	

	Class	Subclass
Electrolytic		
Coating	204	18.1
Lighting	26	
Embroidery and trimmings	D 5	
Equipment for production, distribution or transformation of electricity	D13	
Exercising equipment	D21	191+
Fishing equipment	D22	134+
Fluid distribution equipment, sanitary, heating, ventilation and air conditioning equipment, solid fuel	D13	3.2
Foodstuffs	D 1	
Furnishings	D 6	
Games and gambling articles	D21	1+
Hardware	D 8	
Household articles, not elsewhere specified	D 7	
Hunting articles	D22	
Jewelry, symbolic insignia and ornaments	D11	
Knitting & netting	D 5	
Lighting	D26	
Linoleum	D 5	
Luggage & special containers &	D 3	76+
Machines, not specified elsewhere	D15	
Agricultural	D15	10+
Cleaning	D32	1+
Construction	D15	10+
Drive train	D15	148+
Drying	D32	8
Engine	D15	1+
Food or drink related	D 7	
Heat treatment	D15	144
Lubricator or oil collector	D15	150+
Material working, etc	D15	122+
Packaging or wrapping	D15	145+
Pump or compressor	D15	7+
Refrigeration	D15	79+
Sewing	D15	66+
Vibrator or separator	D15	147
Washing	D32	6+
Measuring, testing or signaling instruments	D10	
Medical & laboratory equipment	D24	
Miscellaneous articles not specified elsewhere	D99	
Musical instruments	D17	
Office machinery	D18	
Office materials	D19	
Packages & containers for the transport or handling of goods	D 9	
Paper manufactures	D 5	
Pharmaceutical & cosmetic products, toilet articles & apparatus	D28	
Photographic, cinematographic and optical equipment	D16	
Printing	D18	13+
Office machinery	D18	
Office photocopiers	D16	27
Recording, communication or information retrieval equipment	D14	
Safes and vavlts	D99	28+
Sewing equipment	D 3	18+
Signaling instruments	D10	
Signs	D20	10+
Smoking articles	D27	
Sports equipment	D21	191+
Stationery, artists & teachers materials, & office equipment not elsewhere specified	D19	
Teachers materials	D19	59
Tents	D21	253+
Testing instruments	D10	
Tobacco & smokers supplies	D27	
Tombstone or monument	D99	17+
Tools & hardware	D 8	
Toys	D21	59+
Transportation, containers for	D 9	
Transporation or hoisting	D12	
Travel cases	D 2	
Umbrellas	D 2	5+
Vending machines	D20	1+
Weaving	D 5	47+
Desilverizing Lead	75	79
Desizing	8	138
Dyeing combined	8	
Desk		
Article including card, picture or sign	40	358
Inkwell combined	108	26.2
Baggage convertible	190	11+
Bed combined	5	2R+
Chair or seat combined	312	235R
Design	D 6	396
Inkwell combined	312	232
Kneehole	312	194+
Letter holder, desk-type	D19	75+
Pad	248	346+
Design	D19	99
Paper rack or file	211	11
Piano case	84	180
Fall board combined	84	178
School desk	D 6	396
School seat combined	297	135+
Table	108	
Typewriter housing or support combined	312	208
Wall desk	D 6	555
Desoldering Apparatus	228	19+
Desoxy Morphine	546	46
Desoxycholic Acid	260	397.1
Desserts	426	573
Design	D 1	
Destructible Feature		
Static mold	249	61+
Desuckering	71	78+
Desulfurizing Ores	75	6+
Desuperheaters	261	DIG. 13
Detacher		
Cable guide block combined	254	410
Comb combined	19	
Fruit gatherer combined	56	339+
Gin combined		
Rotary	19	53
Vibratory	19	52
Harness	54	69
Horse	278	21+
Detearing		
Apparatus	118	639
Detector and Detecting	329	
Amplifier triode	313	293+
Beat note		
Autodyne detector	455	321+
Heterodyne detector	455	313+
Heterodyne detector	329	153+
Homodyne	455	324
Coating or impregnating	427	7
Diode radio system	329	203+
Crystal type	329	205R+
Electronic tube type	329	203+
Fiber manipulation combined	19	
Grid leak system	329	175+
Heads		
Crystal mounting	250	361R
Scintillation	250	361R+
With sample holder	378	38
Infinite impedance system	329	173
Leak	137	455+
Leak by fluid pressure	73	40+
Leak system in situ	137	551+
Lock combined	70	439
Loom		
Feeler	139	269+
Shuttle	139	203+
Stopping	139	336+
Metal bending	72	6+
Metal detector for airport security	340	568+
Metal detector for library books	340	572
Magnetic locators	324	200+
Musical instrument automatic eg player	84	115+
Nuclear reactor condition	376	245
Of submarines		
Sonar, echo type	367	87
Non-echo receiver circuitry	367	135+
Non-echo type	367	118+
Ore	324	323+
Plate system	329	186+
Pneumatic dispatch	406	10+
Radiant energy ray	250	336.1
Radio receiver	329	
Radio tube structure	313	
Regenerative systems		
Pentode tube type	329	170+
Screen grid tube type	329	170+
Triode tube type	329	170+
Thread finishing combined	28	227
Warp preparation combined		
Loom replenishing	28	205
Stopping	28	189
Winding and reeling combined		
Bobbin and cop	242	36+
Cone wind	242	28+
Cordage	242	49
Fabrics	242	57
Detent		
Check controlled device lock releasing	194	247+
Check label and tag holder with slide and	40	17+
Chuck socket type with	279	9R+
Closure fastener tripper with	292	333+
Control lever and linkage system	74	527+
Detention device combined with signal alarm box	340	304+
Electrode feed control for electric lamp and discharge	314	83+
Freight car stake	105	384
Joint with	403	326+
Lock		
Door biased latch bolt with	70	467+
Padlock with	70	20+
Pipe joint		
Auxiliary	285	82+
Essential holding means	285	305+
Swivel socket with	285	276+
Railway switch detention device	246	399
Rod joint or coupling sleeve with	403	300+
Rotating electric snap switch with	200	410+
Tool handle fastenings with		
Hinged	403	52+
Rigid	403	
Tuners with	334	88
Typewriter irregular line spacing	400	563+
Valve holder	251	297+
Wheel attaching device	301	119
Whiffletree with	278	108
Detergent Compositions	252	89.1+
Detergent alkylate synthesis	585	455+
Detinning	75	64
Electrolytic	204	146
Deposition combined	204	120+
Detonators		
Characteristics testing for	73	35
Compositions	149	
Explosive devices	102	
Design	D22	112
Safes and bank protection	109	36+
Sequentially fired	102	217
Signal and indicators		
Burglar alarms door & window operated	116	87+
Burglar alarms door securing	116	15
Burglar alarms locks	116	11
Burglar alarms portable	116	83
Burglar alarms sash fastening	116	17
Burglar drop type	116	78
Lamp ignition combined	116	7
Periodic	116	23
Railway torpedo	246	487
Combined with other signal	246	217
Combined with switch	246	217
Placing mechanism	246	210+
Toys	446	398+
Detoxicant	424	10
Deuterium Heavy Hydrogen	423	648R+
Catalytic recombiner	422	213+
Compounds inorganic	423	644
Electrolytic production	204	101
Thermo nuclear fuel	376	151
Deuterium Oxide Heavy Water	423	580
Catalytic recombiners	422	213+
Electrolytic production	204	101
Nuclear moderator	376	220
Nuclear moderator	376	220
Nuclear moderator	376	350
Developing		
Copy camera	355	27+
Dyes on fiber	8	666+
Photographic apparatus	354	297+
Design	D16	33+
Photographic processes	430	434+
Compositions	430	464+
Sensitizing combined	430	434+
Tank	354	331+
Devitrified Glass-ceramics	501	2
Dew Point (See Hygrometer)		
Dewaxing		
Mineral oil	208	28+
Apparatus	196	14.5
Wax paper	162	4+

	Class	Subclass
Dextran	536	112+
Nitrogen containing	536	51
Dextrin	127	38
Fermentative preparation	435	
Dextromethorphan, in Drug	514	289
Dextrose	127	30
Diagnostic		
Biological test compositions and methods	424	2+
Medical (See medical & surgical equipment)	128	630
Acupuncture	128	907*
Eye	128	645
Test by visual stimulus	128	745
Feedback to patient of biological signal other than brain electric signal	128	905*
Heart	128	695
Instrument design	D24	17
Multiphasic diagnostic clinic	128	906*
Respiratory	128	716
Skin sensitivity to allergens or radiation	128	743
Skin touch or pain response	128	744
Teaching	434	262+
Dial		
Calipers	33	147R+
Clock	368	228+
Calender combined	368	28
Hands gear train	368	221
With hands	368	228
Distance measurement tape	33	139+
Gauge	33	501+
Gauge or meter design	D10	102+
Handwheel	74	553
Illuminator	362	23+
Weighing machine	177	177+
Indicating	116	200+
Indicator hand	D10	127
Indicators	33	172R
Knitting multiple needle machine	66	19+
Knitting pile loop formers	66	92
Lock operator	70	332
Micrometers	33	147R
Mirror combined	350	113
Music charts	84	474
Operator	74	10R+
Optical element combined	350	110
Registers cash	235	19+
Registers fare	235	34
Parallel axis	235	38
Single axis	235	43
Registers parallel axis disk combined	235	111
Registers single axis hand combined	235	112
Single axis	235	120+
Structure	116	334+
Switch multiple	200	11R+
Telephone	340	345+
Design	D14	66
In instrument	379	362+
System	379	258+
Thermometers	374	187
Time stamps	346	81+
Vehicle	D12	192
Watch face	D10	123+
Dialyzer	210	321.2
Process	210	644+
Using liquid membrane	210	643
Sugar		
Apparatus	127	10
Electrolytic	204	138
Processes	127	54
Diameters		
Determining	33	178R+
Diaminotriaryl	260	393
Disazo compounds	534	677+
Diaminotriphenyl Methane	260	393+
Diamond (See Gems Jewel)		
Artifical producing	423	446+
Cutting	125	30R
Earth boring tool	175	329+
Misc. chemical manufacture	156	DIG. 68
Tool	125	39
Digest	76	DIG. 12
Making	76	101R
Dianisidine	564	309
Diaper	604	358+
Bacteria inhibitor	604	360
Deodorant containing	604	359
Design	D24	50

	Class	Subclass
Materials	604	367+
Pail	D32	37
Securing	604	386
Adhesive	604	389
Velcro T M	604	391
Diaphragm		
Acoustic structure	181	157+
Acoustic with mounting	181	148+
Barrel plug	217	111
Battery electric	429	247+
Camera		
Automatic control	354	410+
Dispenser type inkwell discharge assistant	222	581
Electrolytic cells	204	252+
Electrolytic electrodes	204	282+
Electrolytic elements	204	295+
Electrolytic treatment of water sewage	204	151
Electrophoretic	204	301
Fire tube	122	83
Flexible connecting	403	50+
Fluid pressure responsive	92	96+
Material	92	103R
Mounting	92	98R+
Plural	92	48+
Galvanic cells and storage batteries	429	247+
Gauge fluid pressure	73	715+
Horns	116	142R
Vehicle energy actuated	116	59
Iris type valve	251	212
Light valves	350	266+
Meter fluid flow	73	262+
Casing	73	274
Mounting	73	278
Structure	73	279+
Microphone	381	168+
Ribbon	381	176
Structure	181	158+
Telephone transmitter granular	381	180+
Mixer gas	48	184
Operated valve	251	12+
Barrel bung	217	104
Pop	48	53.3
Pop safety	137	469+
Packing rod	74	18+
Packing rotary rod	277	88+
Paper moulding	162	401+
Pen-actuated means in inkwell	15	257.72
Pen, fountain filling means	401	145
Phonograph sound box	181	161
Photographic	354	226+
Piano sounding boards	84	192+
Railway vestibule connection	105	15+
Regulator fluid pressure	137	505.36
Shutters photographic	354	226+
Switch fluid pressure	200	83R
Toys sounding	446	416
Valve	251	331
Biased	137	540+
Collapsible	137	843+
Seal	251	335.1+
Vibrator surgical	128	39
Wringer	68	242
Diapositives, ie Finished Picture	354	354+
Diarylketone	568	332
Amines	564	328
Carboxylic acids	562	460
Diarylmethine Amines	564	315+
Benzhydrols or benzthiols	564	327
Benzophenones or benzothiophenones	564	328
Diamino diphenyl methanes	564	330+
Diastase	435	195
Diathermy		
Bandaging	128	82.1
Kinesitherapy	128	24.1+
Medicators	604	20+
Orthopedics	128	68.1
Surgical	128	362+
Surgical instruments	128	303.1+
Surgical magnetic	128	1.5
Surgical receptors	604	358+
Diatomaceous Earth (See Silica)		
Diazines	544	224+
Azo compounds	534	767
Diazo	534	588+
Dye compositions	8	664+
Reactions	436	903*
Diazoamino	534	550+
Dye compositions	8	665

	Class	Subclass
Diazoles-1, 2	548	356+
Diazoles-1, 2	548	300+
Diazoles-1, 3	548	300+
Diazomethanes	534	558+
Diazonium	534	558+
Diazooxide	534	556+
Diazotate	534	556+
Dibbling	111	89+
Dibbles	111	99
Dibenzanthrone	260	357
Dibenzo¢1, 4!thiazepine	540	550
Dice	273	146
Agitators	273	145R
Dichlorodifluoromethane	570	134
Dickey	2	103
Dicloxacillin	540	327
Dictating Machine (See Sound, Recording and Reproducing)	D14	3
Dictograph T M (See Phonograph)		
Dicyandiamide	564	104
Electrolytic synthesis		
Aqueous bath salt	204	91
Fused bath	204	63
Salt	260	
Dicycle	280	208
Die (See Mold)		
Cutting	83	
Fence barbs	29	7.1
Punching	83	
Saw tooth forming	76	29
Saw tooth sharpening	76	30
Sheets and bars	83	
Tobacco stem	131	323
Tobacco wrapper	83	
Electroforming	204	6
Expressing		
Bread, pastry, confection	425	287+
Bread, pastry, confection	425	381.2
Bread, pastry, confection	425	461+
Cores founding	164	174
Earthenware	425	376R+
Earthenware	425	461+
Metal plastic	72	253.1+
Plastics	425	381.2
Soap	425	289+
Soap	425	340+
Soap	425	383+
Foodstuff cutting		
Reciprocating	83	613
Rotary	83	591
Forming	76	4
Machines	72	
Holder		
Metal shaping	72	462+
Sheet punching	83	
Horseshoe making	59	60
Machinery part	D15	138
Metal		
Bolt, nail, nut, rivet, screw making	10	
Casting	29	DIG. 1
Chain welding	59	33
Coiling spiral	72	141
Deforming	72	462+
Design	D15	136+
Forging anvil adjustable	72	418
Forging presses enclosed	72	343+
Horseshoe	59	60
Injecting	164	303+
Process for making	76	107R
Screw threading platen	72	88
Through-die	72	467+
Wire cutting	83	
Wire joining	140	116
Work-orbiting screw threader	72	191+
Paper box	493	167+
Punches		
Sheets and bars	83	
Wood match making	144	53
Woodworking	144	197
Punching dies	83	651+
Screw threads	72	88+
Shaping		
Closure applying	53	341+
Embossing members	101	16+
Metal can seaming	413	31+
Metal sheet shaping	72	462+
Printing embossing hot	101	8+
Stretch press	72	302
Wire fabric	140	107
Shoe sole	12	38

	Class	Subclass
Textile	57	138
Dielectric		
Capacitors in	361	301+
Fluent compositions	252	570+
Hydrocarbon	585	6.3
Heating apparatus	219	6.5+
Heating methods	219	10.41+
Hygrometer	324	61R
Lens	333	248+
Diels Alder Synthesis (See Diene Synthesis)		
Hydrocarbon synthesis	585	361
Diene Synthesis		
Higher unsaturated fatty acid	260	404.8
Maleic acid	562	498+
Maleic acid anhydride	549	262
Carbocylic compound	549	234+
Natural resins	260	101
Polymerization	585	507+
Differential		
Gearing		
Cash register key operated	235	14+
Change speed and	74	700+
Conveyers having different speed zones	198	792
Plural conveyer sections	198	577
Plural sections with differing speeds	198	579
Elevator rope drive sheave	187	23
Multiple driving or driven elements	74	674+
Planetary	74	710
Register single axis	235	119
Register transfer mechanism	235	136
Sectional rotary bodies	74	444
Speed responsive device	73	507
Gearless	74	650
Material treatment		
Comminution of mixed solids	241	14
Sorting pretreatment	209	4+
Mechanism		
Brake operator movement	188	134
Chemical feed pressure control	137	100+
Fire extinguisher automatic valve	169	22
Flow meter pressure type	73	861.42+
Railway draft springs	213	26+
Sewing machine stitch forming feed	112	312+
Ships logs pressure type	73	182+
Telegraph system duplex	370	27
Telephone transmitter electrodes	381	168+
Motor	91	415+
Motor vehicle with	180	76
Differential Amplifier	330	69
Differentiating Circuit	333	19
Calculators having	235	61R+
Electronic tube type	328	127+
Diffraction	350	162.11+
Grating	350	162.17+
Diffuser		
Slow diffuser holder	239	34+
Sugar making	127	3+
Processes	127	43+
Diffusing (See Extracting)		
Air sterilizing	422	120+
Aircraft structure	244	136
Cabinets	312	31+
Fans	416	
Fumigators	422	305+
Gases for separation	55	158
Process	55	16
Thermal diffusion	55	209
Thermal diffusion process	55	81
Liquids through membranes	210	634+
Process in semiconductor device making	437	141
Sacchariferous material	127	3+
Sprinkling and spraying	239	
Digallic Acid	560	70
Digester & Digestion (See Extracting)		
Apparatus	422	307+
Fermentative	435	
Liquid purification	210	601+
Mineral oils	208	46+
Pulp apparatus	162	233+
Pulp processes	162	1+
Still combined	202	107
Digger	171	
Design	D15	10+
Hand tool	D 8	
Ditcher	37	80R+
Elevator	414	596+
Dredger	37	54+
Endless excavator	37	191R+
Fire arm combined	42	93
Posthole		
Implement	294	50.6+
Machine	175	
Potato	171	
Rotary excavator	37	189+
Tunnel excavating	299	29+
Digital		
Clock	D10	15
Motor art digest	91	DIG. 1
Readouts	73	901*
Digital Data		
Communications	375	
Computer systems, general	364	200
Error correction	371	30+
Processing machimes	364	900
Synchronous transmission	375	106+
Transmission of coded intelligence	178	2B
Digital Memory	365	
Speech reproduction from	381	51+
Digitalis Glycosides	536	6.1
Dihydroabietic Acid	562	404
Dihydronovobiocin	536	13
Dihydrostreptomycin	536	15
Diketo Purines	544	267+
Diketopyrimidines	544	309+
Dilatometry	374	55
Dilator		
Surgical instruments	128	341+
Surgical instruments combined	128	303.11+
Diluents Coating Compositions	106	311
Dimer (See Polymerization)		
Dimmer for Lamps	362	257+
Headlight systems	315	82+
Lamp systems	315	291+
Structure	323	905*
Dimpling Sheet Metal	72	414+
Testing ductility by	73	87
Dinas Brick	501	141
Dining Car	105	327
Dining Room Store Service	186	38+
Diode	313	
Beam power amplifier	313	298
With two cathodes	313	5
Damping	313	317
With emissive cathode	313	310
With thermionic cathode	313	310
With two cathodes and anodes	313	1
Duo pentode	313	298
Triode	313	303
Duplex hi mu triode	313	303
With two cathodes	313	5
Duplex pentode	313	298
With plural cathodes	313	5
Duplex triode	313	303
With plural cathodes	313	5
Duplex twin	313	307
Germanium	357	
Glow discharge	313	567+
High mu triode	313	303
With two cathodes	313	5
Light emitting	362	800*
Pentode	313	298
With two cathodes	313	5
Power amplifier pentode	313	298
With two cathodes	313	5
R F	313	317+
With emissive cathode	313	310
With thermionic cathode	313	310
Sharp cutoff pentode	313	298
With two cathodes	313	5
Triode pentode	313	298
With plural cathodes	313	6
Triple	313	1
Triple high mu triode	313	303
With plural cathodes	313	5
Triple triode	313	303
With plural cathodes	313	5
Twin	313	306
With two cathodes	313	1
Diolefin Hydrocarbon	585	16+
Purification	585	800+
Synthesis of	585	601+
Dioramas	272	9+
Dioxane	549	377
Dioxazine	544	63+
Dioxazoles	548	124
Dioxirane	549	200
Dip Pipe	202	255+
Dip-cup-provided Inkwell	15	257.7+
Dispenser type	222	576+
Dipole Antenna	343	793+
Dipper		
Excavators	414	685+
Dipping (See Coating)		
Animals	119	158
Apparatus for molding	425	269+
Molds or forms	425	275
Channel pumps	415	88+
Hat	223	10
Match making	144	60
Cutting and framing combined	144	52+
Frames	144	62+
Framing combined	144	58
Processes for forming by	264	301+
Trap chamber dispensing	222	356+
Endless belt	222	371
Rotary	222	369
Dips Animal	424	
Direct Current Distribution	307	
Pulsating telegraph systems	178	66.1+
Transmission to vehicles	191	
Direction		
Radio signaling	342	350+
Direction Indicator		
Design	D10	65+
Earth boring combined	175	45
Fluid flow velocity combined	73	188
Geographical	33	320
Navigation	73	178R
Radar	342	147
Register fare	235	48
Ships course	116	19
Traffic		
Barrier	404	6+
Director	404	9+
Vehicle	116	35R+
Design	D26	28+
Electrically operated system	340	73
Director Antenna	343	912+
Dirigible	244	24+
Disappearing Instrument Cabinet	312	21+
Disassembly (See Assembling)		
Apparatus	29	700+
Processes	29	426.1+
Battery	29	
Repairing combined	29	402.3+
Disc Recording or Reproducing		
Television	358	342+
Color	358	322
Disconnecting Devices (See Stop Mechanism)		
Discriminator	329	110+
Dish (See Cup; Tray)	215	
Butter refrigerating	62	457
Design, general	D 7	84
Design, refrigerating	D15	89
Cleaner		
Brushing	15	74
Liquid contacting	134	
Compartmented	D 7	27
Cover	215	200+
Covered	D 7	16
Design	D 7	23+
Design, simulative	D 7	5
Drainer		
Compartmented	220	20+
Rack	211	41
Wire receptacle	220	19
Heater	126	246
Making		
Clay	425	263+
Clay	425	459
Paper and paper board	162	387+
Sheet metal container	413	
Stand	248	128+
Design	D 6	310+
Washer	134	
Design	D32	2+
Disher		
Ice cream cone	425	118
Ice cream scoops	425	221
Ice cream scoops	425	276+
Dishwasher	D32	2+
Dishwashing racks	D32	55
Disinfectants & Disinfecting	422	
Apparatus for treating air	422	120+
Baths closets sinks and spittoons	4	222+
Sewer	4	220

	Class	Subclass
Spittoon	4	261
Strainer or stopper cover	4	294
Urinal	4	309+
Cabinet combined	312	31
Chemical holders for flush toilet	4	228+
Disinfectants	424	
Electrolytic	204	130+
Ozonizer	422	186.7+
Preparation and distribution	43	127
Animal treating	119	156+
Dusters or sprayers	239	
Fumigators vermin destroying	43	125+
Intermittent discharge type	422	123+
Receptacle attachment	220	87+
Telephone attachment	379	452
Thermometer case	206	306
Tobacco	131	290
Electrical or radiant energy	131	299
Vapor and fume generator	422	305+
Disintegrator (See Comminuting; Crusher)	241	
Etching with electric arc	219	68+
Machining with electron beam	219	68
Paper fiber	241	
Peat	44	31
Rock in situ	299	
Sugar treatment combined	127	4
Textile fiber preparation	19	82+
Tobacco leaf	131	311+
Feeding combined	131	109.1
Disk		
Agricultural implement	172	518+
Bearing for	384	460
Bearing support for	384	157
Harvester cutter	56	255+
Harvester cutter with conveyor	56	157+
Hedge trimmer	56	235
Planting by drilling	111	88
Plow moldboard	172	167
Scraper for	172	558+
Sharpener combined	172	437
Amusement roundabout	272	46+
Animal powered motor	185	18
Brake	188	71.1+
For railway vechile	188	58+
For velocipede	188	26
Brush rotary	15	180
Boot clean black and polish	15	34+
Handle mount	15	28+
Street sweeper	15	87
Calendar	40	113+
Carbon black making collector	422	150+
Card exhibitor	40	495+
Cathode ray tube stream concentrator	313	441+
Clock geographical	368	27
Closure seal	292	308+
Comminuting and grinding		
Abrading tool	51	358+
Cherry stoner	99	566
Comminutor	241	
Corn shellers	130	7
Disc grinding	51	
Floor surfacing machine	51	177
Grain huller	99	600+
Scouring machine	51	24+
Counting mechanism	235	98R
Coin	453	58+
Cutting		
Annular drills	408	204
Barrel stave jointing cutter	147	29
Button cutting	79	16
Can opener	30	435
Earth boring tool	175	373
Pipe cutter external	30	101+
Portable auger	408	204
Roller hand tool	30	307
Saw sharpening	76	45
Stone sawing	125	20
Sweep auger	408	199+
Tool sharpener	76	82
Tubular saw wood	408	204+
Wood	408	204
Wood dovetailing	144	89
Wood planer rotary cutter	144	118+
Wood sawing knife disc	83	469+
Wood turning chisel feed pattern	142	44
Educational elements	434	402
Electric conductor insulation	174	111
Electric generator armature	310	268+
Electrode consumable	314	44

	Class	Subclass
Evaporator moving film support	159	9.1+
Feed for wood lathe	142	19
Feed for wood saw mill carriage	83	403.1
Feeder for nails	10	169
Flowmeter variable restriction	73	861.55+
Fluid distributor rotary	239	224
Gas and liquid contact impeller	261	84+
Gear sectional body	74	444+
Gearing frictional wheel and	74	194+
Laminated fabric disk making	156	
Magazines for target	124	46+
Making or working by rolling metal	72	67+
Metal rolling apparatus with disk platen	72	80+
Motion picture	352	102+
Paper closure assembling with container	493	108
Projectile holders	124	42+
Railway wheels	295	1+
Recorder record receiver	346	137
Shaft coupling yielding element	464	98+
Non metallic	464	92+
Sound record	369	272+
Apparatus for molding with label applying	425	500+
Stacking racks for	211	49.1
Vehicle land wheel	301	63R+
Making	29	159.1
Dislodging of Anchors	114	297
Dispatch Pneumatic	406	
Dispatching of Trains	246	2R+
Dispensing (See Notes to Main Class)	222	
Articles	221	
Beverage and food machines	D 7	300+
Cabinet		
Article	221	
Mixed drinks	222	129.1+
Spool holder and thread cutter	83	
Tearing	225	
Twine holder with cutter	83	
Coin handling	453	
Confetti	446	475
Container opening	206	601+
Dental apparatus	433	80+
Deposit & collection receptacles	232	
Device with wheels, skids, etc.	222	608+
Filling portable receptacles with fluent material	141	
Fire extinguishers	169	
Gas dispensing article	222	3+
Inkwell	222	576+
Molten metal	222	591+
Continuous casting	164	437+
Planting apparatus	111	25+
Refrigerated liquids	62	389+
Sprinkling, spraying, & diffusing fluids	239	
Store dispensing service	186	
Tape		
Gummed	83	
Length measuring	33	127+
Length measuring with cutting	83	522
Magazine type dispenser	312	39
Roll or spool	206	389+
With fixed severing edge	225	
Track sanders	291	
Trap chambers	222	424.5+
Movable or conveyer type	222	344+
Unwinding and cutting fabrics	242	55+
Vending machines	D20	1+
Web or strand feeding	226	
Dispersing (See Colloids; Emulsifying)		
Agents	252	351+
Solids combined	252	363.5
Agitating	366	
Colloids	252	302+
Radioactive	252	634+
Compositions including agents for		
Bituminous material	106	278
Carbohydrate gum	106	208
Casein	106	146
Cellulose	106	203
Glue or gelatine	106	135
Natural resin	106	236+
Pigments	106	308R
Prolamine	106	153
Protein	106	161
Starch	106	213
Synthetic resin (See class 523, 524)		
Wax	106	271

	Class	Subclass
Dyes including agents for	8	
Fermentation apparatus	435	287+
Foam control	435	812*
Medium for coating or plastic compositions	106	311
Solid disintegrating	241	
Display		
Animated	340	724+
Bank counters	109	10+
Boxes	217	11
Closures hinged	217	58
Compartmented	217	10
Design	D 9	414+
Folding	217	9
Sliding closure	217	63
Cabinet	312	114+
Frozen food, produce or meat	D 6	432+
With display opening	312	234
Card picture and sign	40	
Design	D20	10+
Design misc	D20	10+
Cathode ray tube	340	720+
Changeable exhibitor	40	446+
Container convertible for storage	206	44R
Device with gas or liquid movement	40	406+
Drying apparatus combined	34	88
Easels	248	441.1+
Educational devices	434	365
Electro-optic	350	330+
Electroluminescent	315	169.3
Envelopes	229	71
Forms	D20	10+
Furniture		
Design	D 6	
Garment forms	223	66+
Gas panel	315	169.4
Heads up	340	705
Integrated with circuit	340	718+
Letters sheets	229	92.3
Light systems electric control	340	286R+
Liquid crystal	350	330+
Matrix	340	752+
Monogram	340	756+
Packages	206	44R+
Printed advertising	283	56
Racks	211	
For picture or business card	40	124
Receptacles	206	44R+
Refrigerators	62	246+
Open access	62	458
Show cases	312	114+
Stands	261	DIG. 14
T V channel	358	192.1
Three dimensional movable figure	40	411+
Timepiece	368	223+
Trays	206	557+
Vehicle body feature	296	21
Wooden containers	217	9+
Wrappers	229	92.3
Disposal		
Closets sinks spittoons	4	
Crematory	110	194
Cutlery combined	30	124+
Incinerator	110	235+
Incinerator garbage	110	235+
Nuclear radioactive waste	252	626+
Razor combined	30	41
Receptacles collection	232	
Sewage treatment	210	
Sewerage	137	
Disproportionation		
Alicyclic hydrocarbon synthesis	585	375+
Aromatic synthesis	585	470+
Hydrogen exchange (See dehydrogenation; hydrogenation)		
Olefin synthesis	585	643+
Paraffin synthesis	585	708
Dissemination (See Scattering)		
Dissolves in Motion Pictures	352	91R
Dissolving (See Solvent)	423	658.5
Actinide series compounds	423	249+
Inorganic compounds	423	210+
Solids to produce cooling	62	4
Distance Measuring Devices	33	125R+
Speed integrator	73	490
Distillation		
Apparatus	202	
Azeotropic	203	50+
Beverages alcoholic	426	
Bromine or iodine	423	500
Chemical apparatus combined	422	189+

	Class	Subclass
Convertive	203	49
Distillate treatment (See vapor)		
Destructive distillation	201	29
Distillation combined	203	
Mineral oils	208	349
Extractive	203	50$^+$
Filming	203	89
Filming	203	72
Flash	203	88
Fluidized bed	201	31
Heating and illuminating gas	48	
Mineral oils	208	347$^+$
Conversion combined	208	46$^+$
Rectification	208	350$^+$
Separatory	203	
Stills	202	81$^+$
Mineral oil	196	104$^+$
Thermolytic	201	
Distillery		
Waste as fertilizer	71	26
Stock feed	426	635
Distortion Bucking in an Amplifier	330	149
Distortion Control		
Amplification	330	
Metal deforming	72	701
Modulation	332	
Telegraphy	178	69A
Telephone circuit	379	414
Wave transmission lines and networks	333	
Distributing Material		
Ammunition filamentary material	102	504$^+$
Bolt nail nut rivet screw making	10	162R$^+$
Coating implement with material supply	401	
Conveyer power driven thrower	198	638$^+$
Conveyer thrower powerdriven	198	638$^+$
Distilland in retort	201	40
Drying		
Agitators	34	241$^+$
Drum rotary	34	130$^+$
Kilns	34	218$^+$
Kilns plural	34	210$^+$
Processes	34	11
Stationary receptacle	34	24
Ensilage	406	164$^+$
Fluid centrifugal	239	214$^+$
Gear friction pressure balancing	74	196
Gear worm	74	427
Gearing pressure	74	410
Hay	414	25
Letter boxes compartment	232	26
Mobile orchard type	239	77$^+$
Pamphlets from aircraft	40	216
Pastry machine combined	425	289$^+$
Railway track layers	104	5$^+$
Road material	404	101$^+$
Scattering unloader	239	650$^+$
Sifters	209	254
Stratifiers	209	498
Type cases	276	44$^+$
Type casting machine	199	33$^+$
Type casting machine combined	199	14$^+$
Type setting machine	276	22$^+$
Type setting machine combined	276	2$^+$
Wire nail making	10	44
Distribution		
Boxes for electric conductors	174	50$^+$
Electric current	307	
Fluid (See ventilation)		
Gas or mist	102	367$^+$
Handling	137	
Motors expansible chamber	91	
Motors valve actuation	91	
Nozzles	239	
Gas in gas mains	48	190$^+$
Distribution Amplifier	330	54
Distributor Electric		
Auto	200	6R$^+$
Ignition circuit	315	210
Ignition system	123	146.5 R$^+$
Conductors		
Combined	174	73R
Overhead	174	43
Underground	174	38
Ignition		
Structure	200	23$^+$
System	123	143R$^+$
System	315	211$^+$
Insulators combined	174	140R
Signalling transmitters	340	364
Switch multiple circuit	200	1R$^+$
Switch periodic	200	19DR$^+$
Systems miscellaneous	307	
Telegraphy multiplex	370	46$^+$
Telephone switchboards	379	319$^+$
Transmission to vehicle	191	2$^+$
X ray tubes potential stress	378	139
Ditch Filler	37	142.5
Ditcher	37	80R$^+$
Elevator	414	596$^+$
Filler	37	142.5
Dithiazoles	548	123
Dithionites		
Reducing composition	252	188.22
Dithiosulfurous Acid Esters	560	310
Diversity Receiver	455	132$^+$
Divider		
Assorter discharging	209	493
Dough, severing apparatus general	225	93
Fleece	19	151
Fluid flow porportional	137	118$^+$
Gauges point markers	33	665
Center	33	670$^+$
Proportional	33	663$^+$
Harvester track clearing	56	314
Opposed contact	33	148R
Road	404	6$^+$
Scribing	33	18.1$^+$
Diving Apparatus	405	185$^+$
Artificial gill	128	200.25
Board	D21	236
Helmets	2	2.1 R
Suits	2	2.1 R
Air or oxygen supply	128	201.29$^+$
Buoyant or swimming feature	441	88$^+$
Submarine working device	405	186$^+$
Divining Rods	324	800*
Dobby Loom	139	66R$^+$
Dock		
Drydocks	405	4$^+$
Floating	114	45$^+$
Doctor Blade (See Device or Apparatus with which Associated)		
Document	283	
Sales	283	60R
Documentation Computer Application	364	419
Doffing		
Bobbin and cop winding combined	242	41
Carding combined	19	106A
Conveyer section	198	622
Ginning combined		
Delinters	19	42$^+$
Saw	19	58$^+$
Magnetic solids separator or classifier combined	209	229$^+$
Spinning, twisting, or twining combined	57	266$^+$
Preparation for	57	276$^+$
Vegetable cutter or comminutor combined		
Parer and corer	99	542
Parer corer and segmenter	99	543
Parer corer and slicer	99	543
Peeler or parer	99	588$^+$
Dog		
Collar	119	106
Design	D30	152
With leash or tether	119	118
Driven ratchet bar combined	74	169
Foodstuff	D 1	
For bolt	70	467$^+$
Furnace fire	126	298
Harness	54	
House	D30	108$^+$
Kennels	119	19
Muzzle	119	30$^+$
Sawmill	83	721
Set works end	83	730
Sewing machine feeding	112	324
Turning lathe	82	41$^+$
Woodworking bench	144	306$^+$
Doily	428	81$^+$
Embroidery, design	D 6	613$^+$
Furniture protector	D 6	613$^+$
Dollies		
Article supporting	280	47.34$^+$
Design	D34	23
Hoisting trucks	254	2R$^+$
Rivet	227	
Washing machine	68	138
Dolls	446	268$^+$
Aquatic	446	156$^+$
Assembling and disassembling	29	805
Design	D21	148$^+$
Eating drinking nursing	446	304$^+$
Sleeping	446	345$^+$
Talking crying	446	297$^+$
Wheeled	446	270$^+$
Voices	446	297$^+$
Pneumatic	446	188$^+$
Wheeled	446	269$^+$
Inflatable	446	226
Dolomite Refractory Compositions	501	113
Dome		
Arcuate design	D25	19
Building structures	52	80$^+$
Fire tube	122	116
Geodesic design	D25	13
Lantern	362	182
Tubular	362	173
Lights vehicle	362	61
Steam	122	508
Separator	122	492
Superheater	122	486
Track sander hopper	291	40
Dominoes	273	292$^+$
Donning		
Empty spinning bobbins	57	266$^+$
Preparatory	57	276$^+$
Door (See Closure)		
Bar for jamming closed	D 8	331$^+$
Bells electric	340	392$^+$
Braces	52	291
Brakes, track or guideway	16	DIG. 2
Cabinet		
Sectional unit	312	109$^+$
Showcase	312	138R$^+$
Channel for sliding door	D25	119$^+$
Checks	16	82$^+$
Checks and closures	16	49$^+$
Chimes	D10	118
Coin controlled	194	
Curved cast archway	52	85
Design	D25	48$^+$
Electric		
Contacts for burglar alarms	200	61.93
Elevator	187	51$^+$
Flexible roll	160	238$^+$
Folding	160	229R$^+$
Sliding together	160	222$^+$
Frame	49	504
In situ structure	52	204$^+$
Guards	49	50$^+$
Handles (See --, knob)	16	110R$^+$
Latch	16	DIG. 32
Opening apparatus	16	DIG. 7
Hanger or track	16	87R$^+$
Hinges	16	221$^+$
Jail	49	15$^+$
Keyhole		
Fluorescent	250	466.1
Illuminator	362	100$^+$
Knob	16	121$^+$
Design	D 8	
With spindle attachment	292	347$^+$
Knockers	116	148
Design	D 8	401$^+$
Land vehicle	296	146$^+$
Latch (See closure; fastener)		
Latch bolts, biased	70	144
Dog for bolt	70	467$^+$
Lock	70	91$^+$
Keepers	292	340$^+$
Railway	105	395
Mat	15	215$^+$
Motor vehicle (See closure)		
Operators	49	324$^+$
Ornamental panel	52	311$^+$
Overhead door	16	DIG. 1
Panel warp correction	52	291
Plate or sign for	D20	43$^+$
Plural panels	52	455$^+$
Pneumatic-type closer	D 8	330
Railway		
Dumping car	105	280$^+$
Emergency	105	348$^+$
Platform trap	105	426$^+$
Vehicle restrained	246	304$^+$
Releasers	292	341.16
Removers	414	684.3
Traversing hoist	212	166
Sectional panel	52	455$^+$

Entry	Class	Subclass
Ships	114	116+
Silencer	D 8	400
Sliding	49	404+
Stop	D 8	402
Stove, furnace or oven		
Casings furnace	110	181
Cooling furnace	110	180
Furnace	110	173R+
Stove	126	190+
Switches	200	61.62+
Lamp or electronic tube system	315	84
Track	D 8	377
Trim	52	211+
Troughs	49	408
Ventilating	98	87
Stove	126	198
Work holder for door and frame	269	905*
Dop Jewel Holder	51	229
Doping		
To form semiconductor P-N type		
junction	437	16+
Alloying	437	134+
Diffusing	437	141+
Fusing dopant with substrate	437	134+
Using energy beam	437	16+
While depositing material	437	81+
Doping Agent Source Material	252	950*
For vapor transport	252	951*
Doppler Compensation Systems	367	904*
Dorr Classifier	209	462
Dosage-related Cabinet	312	234+
Dosing Device		
Dispensing	222	
Indicator for medicine	116	308
Medicators	604	
Dot		
Dot type mosaic screen cathode ray		
tube	313	472
Printer	400	124
Double Antibody Test	436	540
Double Bond Shift (See Isomerization)		
Doubler	202	199
Doublet Antenna	343	793+
Doubling		
Twisting strands		
Covering or wrapping	57	14
Delivery twist	57	59+
Ends or hanks	57	26+
Receiving twist	57	66+
Winding bobbins and cops	242	42
Fault detecting	242	38
Fault detecting load	242	40
Douche	604	36+
Hand held	604	212
Nozzle with separate ingress &		
egress	604	39
Treating material introduced	604	54
Dough		
Compositions	426	549+
Crimping devices	425	293
Cutting machines	83	
Dividing machines	425	289+
Forming, molding and working	425	197+
Forming, molding and working	425	200+
Kneading machines	366	69+
Kneading machines	425	197+
Kneading machines	425	200+
Mixing machines	366	69+
Packaging or wrapping	426	392+
Presses	425	
Raisers	126	281+
Rollers	425	294
Rollers	425	329
Testing	73	169
Doughnut	426	496
Cookers		
Deep fat fryer type	99	403+
Forming or shaping combined	99	354
Opposed heated surface type	99	382
Support combined	99	442
Deep fry process	426	439
Design	D 1	120
Dough fermentation	426	18
Doup Heddle	139	51+
Dovetailing		
Bedstead corner fastening	5	300
Design	D 6	503+
Calks	168	35
Woodworking	144	85+
Dowel		
Jigs	408	72R+
Pins	403	292+
Making	144	12
Module connector	52	585
Road joint	404	47+
Downdraft		
Furnaces	110	315
Hot air furnace	126	103
Liquid heaters and vaporizers		
Fire tube	122	97
Water tube	122	286
Stove	126	76
Downspout	52	16
Draft		
Air current		
Back draft preventer	98	119
Boiler	122	
Control to combustion chamber	431	20
Downdraft furnace	110	315
Forced draft fumigant burner	43	125+
Heating stove downdraft	126	76
Hot air furnace downdraft	126	103
Intermediate in superimposed fire		
box with air or steam	110	267+
Stove damper	126	290
Superimposed fire box and		
intermediate draft	110	317
Appliance		
Animal	278	
Articulated vehicle train	280	400
Cable hoist	212	76+
Cable railway	104	193
Horse drawn sweep	185	23
Monorail animal draft	105	143
Railway	213	
Wheeled horse rake draft		
dumping	56	386+
Equalizer		
Animal draft appliance	278	3+
Horse drawn sweep	185	22
Gauges	73	700+
Regulator		
Boiler	122	38
Damper	126	285R+
Damper automatic	236	45
Furnace	110	147+
Spark arrester combined	110	123
Drafting (See Drawing)		
Board		
Cabinet combined	312	231
Implement	D19	35+
Box for	206	371
Curved ruler type	33	177
Scriber type	33	18.1+
Straightedge type	33	403+
Table	D 6	420
Textile fibers	19	236+
Drag		
Animal restraining	119	107
Classifier	209	462
Conveyer chains	198	717+
Conveyor chains for viscous fluids	198	643
Drilling planting machine	111	84
Earth working	172	
Fishing reel	242	84.5 R
Land vehicle type of	280	19
Line scraper	172	26.5
Metal casting sand	164	349+
Oil distributor	114	234
Reciprocating saw machine	83	746+
Saw guide	83	821
Sea anchor	114	311
Design	D12	215
Ships log	73	184+
Textile spinning and twisting	57	113
Bobbin	57	72
Flier type bobbin	57	70
Drain and Drainage		
Building construction	52	302+
Cabinets with	312	229
Conduit electrical	191	26
Dispenser	141	364
Footwear with	36	3R
Fryer deep fat	99	410+
Machine fluid treating	68	208
Photographic wet plates	354	280
Pipe cleaner	15	104.3 R+
Pump	417	437
Road or pavement	404	2+
Roll type clothes wringer	68	271
Roofing interior	52	553
Gutter	52	11+
Safes or banks	109	28
Sand	405	50
Sewerage	210	163+
Shield	D23	261
Sink	4	650+
Soil	405	36+
Static mold having fluid	249	141
Sugar centrifugals	127	19
Surgical	604	264+
Table surgical	108	24
Washing machines	68	208
Wringer roll	68	271
Drain Board (See Dish, Drainer)		
Sink accessory	D32	56
Tub with	4	637
Attachment	4	656
Design	D23	277+
Draperies	160	330+
Hook	16	93D
Surgical	128	
Tie back, fabric	D 6	578
Tie back, rigid	D 8	368
Draw		
Press	72	343+
Drawbar		
Articulated vehicles combined	280	400+
Bedstead corner fastening	5	298
Elastic extension device inclosing	267	72
Railway	213	62R+
Car forging dies	72	343+
Car making	72	
Drawbridge	14	31+
Electrical transmission to	191	9
Floating	14	29+
Protection switch or signal	246	118+
Drawer		
Bed combined	5	308
Cabinet	312	
Horizontal sliding	312	330R+
Cash register operation	235	22
Element of sofa bed	5	58
Key set cash register operated by	235	10
Knob labels and tags	40	331+
Locks	70	85+
Plate or sign for	D20	43+
Pull labels and tags	40	325
Refrigerated	62	382
Air blocking when open	62	266
Drawing (See Drafting)		
Abrading reel	51	75
Board		
Cabinet combined	312	231
Design	D19	52
Straight edge mounting	33	430+
Die	72	467
Easels	248	441.1+
Etching combined with	156	658
Glass	65	193+
Process	65	66+
Implement	D19	35+
Manufacturing glass		
Product by	65	
Material, design of	D19	35+
Mechanical instruments	33	
Mercury vapor lamp starting by	313	170+
Metal	72	
Cup or shell	72	347+
Method & apparatus	29	DIG. 11
Push drawing	72	343
Rolling combined	72	206
Sheet die shaping	72	380+
Spinning	72	82+
Wire or tube	72	274+
Pen	401	267
Design	D19	41+
Perspective instruments	33	1K
Reverse drawing on metal	72	708*
Textile fiber	19	236
Textile spinning, twisting, twining		
Combined	57	315+
Rollers	57	97
Triangles	33	474
Visual arts	434	85+
Drawknives	30	313
Drawplate		
Diamond setting	76	107R
Drawstring	24	266
Dredge		
Excavating	37	54+
Submerged vessel raising	114	55
Dredge Top Dispenser	222	480

	Class	Subclass
Dress	D 2	49+
Dressmaker	D 2	195
Garment	2	71+
Forms	223	68+
Hangers	223	92+
Protector	2	46+
Shield pressing	223	54
Weights	2	273
Maternity	2	76
Sifting screen	209	392+
Wheel guards	280	160.1
Dresser, Fish Preparation Tool	D 7	106+
Dresser, Furniture	D 6	432+
Dressing		
Brushes and brooms	300	18
Meat and fowl	17	
Saw making		
Jointing and gaging combined	76	46
Jointing combined	76	47R+
Separating screen	209	392+
Separating solids	209	
Stone working	125	2+
Diamond tool for	125	39
Grindstones	125	11R
Millstone	125	27+
Surgical (See bandaging)		
Tobacco		
Fluid or fluent material	131	300+
Smoothing brushing rolling	131	324+
Stemming combined	131	315+
Warps	28	178+
Dressmaker	D 2	195
Dried Fruits and Vegetables	426	640+
Drier and Drying (See Dehydration)	34	
Blotter or towel type	34	95+
Non-drying device combined	34	89.1+
Cigar and cigarette making		
Tip or mouthpiece applying	131	92
Wrapper sealing	131	68
Clothes	34	
Automatic	34	43+
Clothes drying rack	D32	58
Machine design	D32	8
Coating combined with drying	427	372.2+
Collar cuff and bosom making	223	3
Composition	106	310
Drying oils combined	106	264
For coating	106	310
Dress coat or skirt forming and stretching	223	69
Dry closet	4	111.1+
Fishing line	242	104
Fruits and vegetables dehydrating	426	471+
Garbage and sewerage furnace	110	224+
Gas and vapor contact with solids	34	
Hair dryer	D28	12+
Hand dryer	D28	54.1
Methods	34	3+
Milk dehydrating	426	471+
Mortar mixer combined	366	22+
Paper article making		
Box making	493	141+
Envelope making	493	265
Rack	211	
Refuse treatment means	110	224+
Shoe and boot	34	104
Solids apparatus	34	43+
Textiles		
Thread finishing, heating or drying	28	219
Washing machine combined	68	20
Web spreader combined	26	92
Web stretching combined	26	106
Tobacco	131	300+
Wood bending	144	254
Drift		
Indicator (See direction indicator)		
Pin	29	275
Drill (See Bit)		
Bit	D15	138+
Dental	433	165+
Chucks	279	
Cigar tip perforators	131	254
Dentists	433	165
Design	D24	12
Drilling machines	408	
Drills	408	199+
Design	D15	139
Design hand tool	D 8	59
Earth boring	175	327+
Expansible	175	263+
Grain planting	111	14+
Grinding processes	51	288
Making and sharpening machines	76	5R
Making blanks and processes	76	108R
Press	408	72R+
Press type combined metal working machine	29	26R+
Pyramidal end	125	1
Rack for	211	69
Rail portable	408	77
Receptacles special for	206	379
Rock	175	327+
Trepan	128	305.1
Twist for metalworking	408	230
Twisting stock to make	72	299
Vise attached	29	560.1
Well	175	
Woodworking	408	199+
Drilling (See Boring)		
Borehole and drilling study	73	151
Brush making tuft setting and	300	3
Button making	79	11+
Pearl button surfacing and	79	6
Earth	175	
Grain drill chute	193	9
Machines		
Dental	433	103+
Driving or impacting	173	
Radial	408	236+
Rock	175	
Mining implement	299	79+
Muds		
Analysis	73	153
Compositions	252	8.51
Earth boring with	175	65+
Planting	111	14+
Broadcasting and	111	8+
Liquid or gas	111	6+
Plant setting	111	3
Solid material	408	
Electric	310	
Well	175	
Marine platform	405	195+
Drink		
Preparation or dispensing machines	D 7	300+
Registers	235	94R
Drinking Fountain	239	24+
Animal watering devices	119	72+
Design	D30	121+
Design	D 7	304
Tube	239	33
Filter combined	210	251
Drinking Straw	239	33
Drinking Vessel		
Container for collapsible type	206	218+
Drinking Water		
Chemical purification	210	198.1+
Process	210	601+
Filtering	210	348+
Decanting combined	210	294+
Drip		
Building attached deflector	52	97
Catcher (See drip pan)		
Closet bowl	4	252A
Coating implement adjunct	401	15
Cooking apparatus	99	444+
Dispenser	222	108+
Fluid sprayer or sprinkler	239	120+
Gas separation apparatus	55	280
Griddle	99	425
Inverted container support with	141	364
Refrigerator defrost water	62	285+
Refrigerator ice melt	62	459+
Refrigerator ice melt filter	62	318
Toaster	99	400
Waffle type cooker	99	375
Water heating vessels	126	383+
Coffee maker	99	306
Collector	431	119
Cooled		
Ice melt gas contactor	62	312+
Ice melt heat exchanger	62	460+
Wet wall type refrigerator	62	278
Cup		
Cleaning attachment	15	248R+
Umbrella	135	48
Liquid diffusers	239	38
Meters	604	251
Drip sensor	604	253
Filter	604	252
Flow control	604	246
Pan (See drip catcher)		
Liquid fuel cooking stove	126	51
Lubricators	184	106
Vehicle body	296	38
Plate		
Boiler feed heater	122	417
Water heater	126	355
Drive in Theatres	52	6+
Anchor	52	155
Supported building component	52	292+
Earth piercer	52	155+
Terranean relationship	52	169.1+
Drive or Driver		
Barrel hoop	147	7+
Barrel tap	81	27
With dispensing	222	81+
Bolt or nut	81	52+
Centrifugal separators	210	360.1+
Conveyor endless	198	854+
Earth boring tool combined	175	170+
Impact	175	135
Elongated-member driving apparatus	227	
Golf club	273	77R
Design	D21	214+
Hammer	173	90+
Design	D 8	75+
Hill planting machine belt feed	111	19
Locomotive	105	26.5+
Mallets	81	19
Nail		
Implements	227	140+
Machine	227	
Pile	173	90+
Railway car wheel or axle	105	96+
Railway turntable actuator	104	41
Screw driver implement	81	436+
Screw driving machine	81	54
Sewing machine	112	220+
Shoe lasting and nailing	12	13.1
Tap driving ratchet	408	120+
Tool	173	
Track spike	104	17.1
Trackmans car	105	86+
Turning work	82	40R+
Well point	175	
Driven Headed and Screw Threaded Devices	411	
Railway spikes	238	366+
Drop		
Annunciator systems	340	286R+
Annunciators	340	815.29+
Bomb	102	382+
Forging	72	435+
Fluid operated	72	453.1+
Hammer	72	435+
Pick up for overhead railway car	104	122
Store service	186	22+
Telephone switchboard restorers	379	315+
Dropper		
Animal waterer	119	72.5
Cash register indicator tablet	235	25
Dispenser	222	420+
Medicine dropper type	141	24
Medicine	604	295+
Combined with bottle, design	D 9	338
Combined with closure, design	D 9	447
Droppings Catcher		
Dispenser	222	108+
Railway	104	133
Drugs	424	
Bio-affecting & body treating compositions	514	
Acronycines	514	285
Allantoin	514	390
Amantadine	514	656
Amphetamine	514	654
Ampicillin	514	198
Amprotropine	514	534
Aspirin	514	165
Atropine	514	304
Barbituric acid	514	270
Benzocaine	514	535
Benzomorphans	514	295
Biotin	514	387
Capsaicin	514	627
Cephalosporins	514	200+
Chlorpheniramine	514	357
Cholecalciferol	514	167
Chrysanthemic acid	514	572
Cobalamin	514	52
Codeine	514	282

	Class	Subclass
Colchicine	514	629
Cortisone	514	179
Cupreine	514	305
Cycloheximide	514	328
Cyproheptadine	514	325
Cysteine ester	514	550
D D T	514	748
D D V P	514	136
Dextromethorphan	514	289
Dyphylline	514	263
Ephedrine	514	653
Estradiol	514	182
Fluspirilene	514	278
Glaucine	514	284
Glaumine	514	669
Griseofulvin	514	462
Hexachlorophene	514	735
Hydrocortisone	514	179
Isoniazid	514	354
Malathion	514	122
Melatonin	514	415
Meperidine	514	330
Methadone	514	648
Methapyrilene	514	336
Methomyl	514	477
Morphinans	514	289
Morphine	514	282
Nandrolone	514	178+
Niacinamide	514	355
Nicotinamide	514	355
Nicotine acid	514	356
Nortestosterone	514	178
Novocaine	514	535
Oxolinic acid	514	291
Parathion	514	132
Penicillin G	514	199
Perdnisolone	514	179
Perimidines	514	269
Phenylephrine	514	653
Phenyltoloxamine	514	651
Pilocarpine	514	397
Pimozide	514	323
Piromidic acid	514	303
Procaine	514	535
Progesterone	514	177
Psoralen	514	455
Pteridine	514	249
Purines	514	261
Quinicine	514	314
Quinidine	514	305
Quinine	514	305
Quinoxaline	514	249
Riboflavines	514	251
Salinomycin	514	460
Scopolamine	514	291
Tartaric acid	514	574
Tetracycline	514	152
Tetramisole	514	368
Theophilline	514	263
Thiamines	514	276
Tocopherols	514	458
Tripelennamine	514	352
Tryptophan	514	419
Tyrosine	514	567
Uracil	514	274
Vinblastine	514	283
Vincamine	514	283
Viquidil	514	314
Childproof bottle closures	215	201+
Drugs of abuse	436	901*
Radioactive	424	1.1+
Drums (See Cylinders; Rollers)		
Armature (See armature)	310	265
Brake	188	218R
Cask or	D34	39
Centrifugal (See centrifugal force, separators)		
Container type	220	
Earthworking or smoothing (See rolls & rollers, land)		
Evaporator moving film support	159	9.1
Treated material inside drum	159	9.2
Material separating (See separator)		
Musical	84	411R+
Automatic	84	104
Rotary furnace	432	103+
Metallurgical	266	173
Rotary heat exchange	165	89+
Rotary or tilting support	384	549
Stand	248	128+
Tumbling device for abrading	51	164.1
Winding (See coiling winding)		
Drumstick	84	422S
Dry Cell Battery	429	156
Recharging system	320	4+
Thermoelectric & photoelectric	136	
Dry Cleaning	8	142
Cleaner compositions	252	89.1+
Machinery design	D32	10
Dry Closet	4	449+
Furnace combined	4	111.1+
Dry Disk Rectifier	357	
Dry Dock	405	4+
Floating	114	45+
Dry Ice T M Solid Carbon Dioxide		
Making	62	8+
Process	62	10
Shaping combined	62	35
Refrigeration by	62	384+
Automatic	62	165+
Dryers (See Drier & Drying)		
Drying Japans	106	310
Drying Oil	260	398+
Composition containing	106	252+
Natural resin containing	106	222+
Fatty oil with preservative	106	263
Hydrocarbon	585	945*
Mineral oil	208	1
Polymers ctg (See synthetic resin or natural rubber)		
Drywall Construction	52	344+
Strip	52	459+
Duck		
Calls	446	207+
Cotton textile	139	426R
Decoys	43	3
Duct (See Conduit)		
Humidifier	261	DIG. 15
Ductility Testing	73	87
Dulcimer	84	284
Dumb Waiter	187	3+
Dumbbell	272	122+
Dumdum Bullet	102	507+
Dummy		
Ammunition	102	444+
Bomb	102	395
Shell	102	498
Clothes and other display	40	538
Figure dispenser	222	78
Figure toys	446	268+
Aquatic	446	156+
Inflatable	446	226
Wheeled	446	269+
Pyrotechnics	102	355
Tackling	273	55R
Dumping Mechanism		
Cable hoist	212	79+
Clothes washing machines	68	210
Egg candling trays	356	65
Furnace ash pan	110	167+
Hand shovel	414	722+
Mortar mixer	366	45+
Nuclear reactor	376	261
Portable receptacle	414	403+
Safes	109	46
Shelf to shelf flow drying apparatus	34	172
Stove grates	126	162
Rocking bar	126	177
Shaking combined	126	158
Vertical axis oscillatory	126	171
Vehicle	298	
Horse rake	56	386+
Railway	105	239+
Moving car	105	241.1
Rake and tedder	56	368
Roadway with external	414	354+
Scoop	37	140
Scow	114	27+
Wheeled toy	446	428
Water closet	4	365+
Dune Buggy Design	D12	87
Dunnage		
Element	267	136+
For container	220	429
For freight	410	121
Edge-around	410	155
Honeycomb	410	154
Duplex Diplex System	370	36
Duplex Telegraph System	370	24+
Duplicate Whist and Other Card Hand Holding Apparatus	273	151
Duplicating (See Manifolding)		
Cutting machine	409	93
Keys	409	81+
Machines	D18	13+
Pattern controlled		
Milling machine	409	79+
Sheet or card punching	234	59+
Plural ribbon typewriter	400	204+
Printing machine	101	113
Selecting or progressive	101	90
Ticket printing machine	101	67+
Durez T M (phenoplast)	528	129
Dust		
Carburetor	406	
Collectors		
Abrading machines	51	273
Cabinet with	312	229
Dental engine	433	116
Design	D32	15+
Earth boring cuttings	175	207+
Gas separator with	55	429
Textile spinning apparatus	57	300+
With air blast or suction cleaner	15	347+
Woodworking machines	144	252R
Conveyor		
Threshing machine	130	29
Woodworking	144	252R
Cover		
Billiard and pool table	273	13
Inflation stem type	138	89.4
Inflation stem type air and	138	89.1+
Vehicle top	296	136
Wheel and valve stem with	152	428
Fixation	404	76
Guard		
Car ventilation	98	25+
Locomotive cab	98	28
Motor vehicle	180	84
Passenger railway car body	105	350+
Typewriter bar	400	467+
Vehicle	280	159
Vehicle mud and	280	152R+
Watch and clock keys with	81	123
Laying compositions	252	88
Pan	15	257.1+
Design	D32	74
Separators		
Cleaning	15	
Gas and	55	
With air blast or suction cleaner	15	347+
Dust Core	148	300+
Making	148	104
Structure	336	233+
Dust Pan	D32	74
Duster (See Dispensing; Distributing; Sprayer; Spreader)		
Animal antivermin treatment	119	159
Coating apparatus	118	308+
Feather	15	234
Design	D32	40+
Insect powder	239	
Design	D 8	2
Mobile orchard type	239	77+
Dwelling		
Cleaning		
Installed air blast or suction systems	15	301+
Dye		
Coating or plastic compositions containing	106	
Compositions & processes	8	
Cake, tablet or powder form	8	524+
Compounds anthraquinone	260	365
Anthrone	260	351+
Azo	534	653
Triphenylmethane	260	386+
Envelope sealing flap combined	229	83
Food	426	540+
Hair	8	
Mordant	430	941*
Oxidizable dyes	436	904*
Sensitizer in photoconductive composition	430	60+
Synthetic resin or natural rubber (See class 523, 524)		
Dyeing	8	
Electrolytic	204	134
Fluid treating apparatus	68	
Hair	8	
Mordanting	8	
Sewing combined	112	17
Textile operation combined	28	169
Warp preparation	28	178

	Class	Subclass
Dynamics, Teaching	434	302
Dynamite	149	
Making	149	
Dynamo	310	
Brush holders	310	239+
Brushes	310	248+
Commutators	310	233+
Connecting in parallel	307	43+
Electric machines	310	10+
Electric prime mover plant	290	
Internal combustion engine igniters	123	149R
Nonelectric prime mover plant	290	
Railway locomotive	105	35+
Telegraph systems	178	65
Dynamoelectric		
Educational demonstration	434	380
Electromagnet like linear motor or generator	335	209+
Geophysical transducer	367	140+
Linear	310	12+
Linear	310	15+
Oscillating	310	36+
Mechanical motion converter	310	37
Reciprocating	310	15+
Rotary		
Amplifier system linear	330	58
Bearing adjustable	310	90
Coated	310	45
Combined with other device	310	66+
Connector on	310	71+
Cooling	310	52
Cooling circulating type	310	58
Cooling filtered coolant	310	56
Cooling liquid	310	54
Cooling nonatmospheric gas	310	55
Cooling sealed type	310	57
Cooling self forced	310	60R+
Cooling treated coolant	310	56
Cooling with control	310	53
Cooling with purifying means	310	56
Cover combined	310	89
Dirt proofed	310	88
Explosion proofed	310	88
Flywheel combined	310	74
Housing combined	310	89
Housing for	310	85
Electrical apparatus	361	331
Illumination	310	73
Impregnated	310	45
Inbuilt or incorporated unit	310	67R
Inertia device for	310	74
Lamp as part	310	73
Lead in for	310	71
Lubrication	310	90
Manufacture	310	42
Manufacture	29	596+
Mechanical shield with	310	85
Moisture proofed	310	88
Molded plastic part	310	43
Powdered metal part	310	44
Shaft and armature timing	310	79
Shaft driven switch	310	69+
Shield between armature and field	310	86
Sound proofed	310	51
Submerged	310	87
Support combined	310	91
Terminal on	310	71
Underwater type	310	87
Vibration damped	310	51
With impedance , capacitor or resistor	310	72
With plastic part	310	43
With sintered part	310	44
Dynamometer	73	862
Dynamotor, electric brake	310	93
Electric meter	324	144+
Strength of material test	73	836
Dynatron, Oscillator	331	133
Demodulator or detector	329	149
Dynel T M (polyacrylonitrile Vinyl Chloride)	526	342
Dyphylline, in Drug	514	263

E

	Class	Subclass
E-layers in Thermo Nuclear Reactions	376	126
Ear		
Coupling detachable janney type	213	157
Guards	2	2+
Design	D29	19
Surgical	128	151
Hair cutters	30	29.5
Pieces		
Eyeglass with protective	351	122
Speaking tube combined	181	20
Telephone	381	187
Design	D14	67
Plug	D24	67
Receptacle		
Bail	220	91+
Wooden bucket	217	126
Surgical treatment electrical	128	789
Syringes	604	187+
Trolley support	191	43+
Insulator	191	42
Trumpets	181	129+
Earmuff	2	209
Design	D29	19
Earplug	D24	67
Earring	63	12+
Design	D11	40+
Earth (See Type of Earth Working Equipment)		
Boring	175	
Compositions	252	8.51
Carrier wave communication system	455	40
Combustion in situ	166	256
Electric current through	166	248
Electrolyte or conductor in batteries, having earth feature	429	47
Engineering	405	
Caissons	405	8+
Diving	405	194
Dry docks	405	4+
Earth control	405	258+
Marine ways	405	2
Pipe and cable laying	405	154+
Ship caissons	405	12+
Stable structures in shifting media	405	229
Marine structures	405	195
Tunnel	405	132+
Underground fluid storage	405	53+
Water control	405	52+
Excavating	37	
Getter for lamps etc	252	181.7
Grounds	174	6+
Induction compass	33	361+
Metals preparation	75	84
Electrolytic	204	59R+
Mining of in situ disintegration of hard material	299	
Planting	111	
Pulse or digital communications	375	6
Rare earths	156	DIG. 63
Roadway stabilization	404	76
Sciences computerized, monitored	364	420
Stabilizing for mining	299	11
Transmission teledynamic systems	455	68
Unearthing plants or buried objects	171	
Wells	166	
Working	172	
Earthenware		
Apparatus for shaping	425	
Decorating	427	256+
Final coating nouniform	427	269
Nonuniform coating	427	287
Plural nonuniform coating	427	266
Glazing apparatus	118	
Earthing (See Grounds Electric)	174	6+
Earthquake Proof Construction	52	167
Earthquake Recorders	346	
Seismographs		
Electrical	367	115
Mechanical	181	
Earthworking	172	
Excavating	37	
Earthworm Agitator		
Electric		
Earth grounds	174	6+
Electrocuting	43	98+
Prods	47	1.3
Trapping	43	58+
Easel	248	441.1+
Body supported	248	444
Collapsible	248	460+
Copyholder	248	441.1+
Advanceable copytype	40	342+
Line guide type	40	352+
Painters	248	444+
Photographic enlarging		
For original	355	75+
For photosensitive paper	355	72+
Seat with	D 6	335+
Wall mounted	248	447.1+
Watch and clock	248	442
Easy Out Extractor	81	436+
Eaves Trough	52	11+
Supports	248	48.1+
Ebonite	525	332.6
Ebullioscope	73	53
Eburnamonine	546	50
Eccentric	74	570+
Adjustment of frictional gearing	74	211
Belt tightener	474	112
Chucks and sockets	279	6
Drive mechanical movement	74	116+
Drive shaft for gearing	74	390
Gearing	74	393
Planetary	74	804
Instruments	33	DIG. 8
Pivot cutter	30	104
Weight wheeled toy	446	458
Wheel mounts vehicle	280	229
Ecgonine	546	131
Echo Sounding	367	87
Economics Teaching	434	107+
Economizers		
Feed water	122	412+
Ecraseur	128	320
Edge (See Border; Cutter)		
Beading metal sheet	72	102+
Beveling pencil sharpener	144	28.2
Binding and covering	29	243.58
Binding or protector		
Skirt garment	2	222
Coating method	427	284+
Curling by bending rolls	72	102+
Cutter	83	425.2+
Fabric bed bottom	5	192
Flanging metal	72	352+
Grinding lenses (See templet)		
Guarded cutting		
Cutlery	30	286+
Manicure cutlery	30	27
Razor	30	51+
Hand gardening tool	172	371
Design	D 8	1+
Indenting shoe sole	12	32
Knitted garment antiravel portion	66	172R+
Mattress	5	474
Panel e. g. table top	52	782+
Resilient strip	52	397+
Retainer	52	764+
Reinforcing		
Laminated strand	156	166
Metallic container	220	73+
Runner mill comminutor	241	107+
Serrating		
Paper box making	493	56+
Tool sharpener	76	89.1
Stays pivoted	24	267
Tool clamp for abrading	51	218R+
Trimming stock for apparel	428	122+
Upholstery trim	428	123
Winding flat wire	72	135+
Wire fabric	245	10
Edgewise Sheet-metal Bender	72	411
Edible Foodstuffs	D 1	
Container holder, edible	D 1	
Education and Teaching	434	
Appliance design	D19	59+
Blocks & cards	434	403+
Cryptography	434	55+
Display panel, chart, or graph	434	430+
Erasable surfaces	434	408+
Keyboard operation	434	227+
Effervescent Tablets		
Medicine	424	43+
Egg		
Beater	366	343+
Breaker or cracker	99	568+
Candling	356	52+
Testing instrument	D10	48
Carton		
Design	D 9	341
Making	493	52+
Paper, compartmented box	206	521.1+
X-art collection	493	913*
Paperboard	206	521.1+
Wooden	217	18+
Cleaning	134	
Brushing or scrubbing	15	3.1+
By abrading	51	16+
Compositions and processes	426	614+

Entry	Class	Subclass
Cooker		
Automatic control	99	336
Boiler	99	403+
Supports combined	99	440
Cup	211	14
Cutting and separating		
Apparatus	99	498
Processes	426	299+
Dyeing	426	252
Incubator	119	35+
Brooder combined	119	30
Nest medicated	119	46
Packaging wrapping and casing	426	298
Powdered	426	614
Preservation	426	298+
Apparatus	99	467+
Coating	426	289+
Dehydration	426	471+
Hermetic sealing	426	298+
Packing in liquids and powders	426	392+
Refrigeration	426	524
Racks	211	14+
Separators	99	497
Shaker-mixer	D 7	51
Sorting	209	510+
Tester	73	432.1
Candling	356	52+
Electrical	324	
Specific gravity	73	32R
Treatment	426	614
Turner	99	644
Implement	294	7+
Incubator tray type	119	44
Ejecting or Ejector		
Article dispensing magazine with	221	208+
Stacked card or sheet	312	50+
Bale	100	218
Bobbin and cop winding with	242	41
Boiler blowoff	122	384
Car ventilation with	98	9
Outlet	98	20
Conveyor bucket discharging	198	703+
Cowl ventilator outlet	98	78+
Windvane	98	70
Cutlery combined with material	30	128
Embalming	27	24R
Envelope machine with	493	268
Firearms	42	25
Upward tilting breech	42	47+
Fluid sprayer with collapsible holder	239	328
Furnace smoke box spark	110	127
Grain binder with	56	449
Ingot strippers	164	405
Jet pump	417	151+
Mechanism for key	70	388
With key operating mechanism	70	414
Medicator gas	604	23+
Metal castings	164	344+
Molds with fluent hardening		
material forming	249	66R+
Plastic forming	425	422
Plastic forming	425	436R+
Type casting	199	80+
Movable dispensing pockets with	222	216+
Parachute flare carrying	102	340
Phonograph record holder	312	15+
Plastic die expressing with	425	381.2
Pottery machines with	425	263+
Pottery machines with	425	459+
Power plant exhaust treatment	60	685+
Punch implement	30	368
Punch press	83	123+
Pyrotechnic device with article	102	357
Rocket	102	351
Receptacles		
Dispensing type	222	
Non refillable bottles	215	16
Pocket safes or ticket cases	221	208+
Receptacle filling charge	141	249
Saliva	433	91
Self unloading vehicles with	414	506
Sewing machine bobbin	112	180
Sheet cutting dies with blank	83	123+
Sheet cutting dies with blank	83	652+
Sheet metalware die shaping with	72	344+
Ship bilge discharge	114	184+
Special tools	29	700+
Stereotype casting device with	164	145
Tobacco smoking device with	131	182+
Type casting mechanisms integral line with	199	60+

Entry	Class	Subclass
Type setting machine with	276	12+
Elastic (See Spring)		
Band guns and projectors	124	17+
Electrical cord	174	69
Extension devices	267	69+
Fabric (See fabric)	428	230+
Bed bottom	5	186R+
Fastener, elastometric	269	907*
Shoe upper	36	51
Stocking	2	239+
Knitted	66	178A+
Surgical	128	165+
Strand	57	200+
Core covering processes	57	3+
Covered core	57	225
Making by spinning twisting	57	3+
Elbow		
Bending	72	369
Conduit electric	174	81
Coupling pipe	285	179+
Electret	307	400
Gas separation	55	DIG. 39
Manufacture	29	592E
Static information storage	365	146
Telephone transmitter	381	173
Electric & Electricity		
Actuation of artificial body member	623	24+
Amplifiers	330	
Apparatus, demonstrating	434	379
Arc (See arc, electric)		
Bandaging	128	82.1
Batteries		
Battery & condenser charging & discharging	320	
Boxes for	429	96+
Electro chemical	429	
Thermo electric or photoelectric	136	
Bearings, magnetic supports	310	90.5
Bells	340	392+
Binding posts	439	
Terminal boards	439	709+
Blankets and pads	219	212
Block signaling systems for railways	246	20+
Board games	273	237+
Boiler	219	281+
Boiler	219	284+
Brake		
Electrically operated	188	158+
Frictionless	310	92+
Bridge networks	323	365
Wave filter	333	170
Cabinet combined with electrical features	312	223
Cable (See cable, electric)	174	
Composite cable molding	425	505
Connections or joints	174	
Forming tool	7	107+
Support	248	49+
Candlelabrum	D26	24+
Capacitor devices	361	271+
Car propulsion systems	104	
Cell (See cell, electric)		
Chair, eg electrocution	128	377+
Cigar or cigarette lighter	219	260+
Circuit maker or breaker (See circuit; switch)	200	
Clock	368	76+
Clutches		
Electric motor and clutch control	192	
Electric operator	192	84PM
In automatic type	192	40
Torque transmitting frictionless	310	92+
Coating processes	427	58
Coaxial cable		
Electronic tube combined with	315	39
Transmission lines	333	243+
With fluid or vacuum	174	28+
Coil winding	242	7.7+
Method	242	7.3
Compositions		
Conductive or emissive	252	500+
Computer controlled, monitored item	364	480+
Computers & data processing systems	364	
Condensers (See condenser)	361	271+
Making	29	25.41+
Conductor (See conductor)		
Conduits and tubes	174	
Insulators	174	
Making or joining	156	47+
Running lengths	156	47+

Entry	Class	Subclass
Structure	174	
Transmission to moving object	191	1R
Transmission to vehicles	191	
Conduits & housing (See conduits)		
Housed switchboards	361	332+
Meter circuit protectors	361	364+
Outlet or junction box type	220	3.2+
Pipe joints	285	
Pipes	138	
Structure	174	
Connectors (See connection)	439	
Design	D13	24+
Winding implement	242	7.6
Winding machine	242	7.17+
Winding method	242	7.3
Wire nuts	174	87
Contact (See contact, electrical)		
Contact brush		
Making by plastic operation	264	104+
Making methods	29	826+
Motor or generator	310	248+
Moving arc lamp electrode	314	129
To moving object or vehicle	191	
Contact metal	75	
Composite	428	929*
Container for components in roll form	206	328+
Control of dispensers	222	52+
Interlocked with discharge guide or support	222	75
Controller (See control or controller, electric)		
Conversion systems	363	
Current conversion	363	13+
Frequency conversion	363	157+
Phase conversion	363	148+
Cooking by current through food	99	358
Cooling produced by	62	3
Cord		
Storing	191	12R+
Course control	244	175+
Current distribution	307	
Current producing apparatus, product, process	429	
Demagnetizers	361	267
Depilatories	128	303.18
Design	D13	
Air conditioning	D23	351+
Clocks	D10	1+
Fan	D23	370+
Heating	D23	314+
Lamps	D26	
Measuring testers	D10	75+
Medical xray	D24	2
Soldering iron	D 8	30
Device making processes	29	592R+
Photo chemical etching	430	313+
Diagnostics medical	128	630
Distribution	D13	
Distributor (See distributor, electric)		
Door or gate	49	59
Drill	30	500*
Electrical power conversion systems	363	
Electrical systems & devices	361	
Electrocardiographs	128	697
Electrolytic device	361	433+
Batteries	136	
Chemistry, electrical & wave energy	204	
Electromagnetic operation	335	
Electronic tube (See electric space discharge devices)		
Electrothermally or thermally actuated switches	337	
Electrothermic metallurgical processes	75	10.1+
Elevators, electric control	187	100+
Engine starting motor systems	290	
Motor per se	310	
Equipment, design	D13	
Eye (See photoelectric)		
Fan	310	40.5
Guards	416	247R
With body motion	310	40.5
Fault isolating safety systems	361	1+
Fault location	324	512+
Fault testing	324	500+
Fence	256	10
Insulator	174	137R+
Filter (See filter, electric)		
Fire starting device	D 7	416+

	Class	Subclass
Flashlight	362	208
Fluid heater	219	280+
Fluid motor system incorporating electrical system	60	911*
Furnace	373	
Fuses	337	142+
For explosives	102	
In arc lamp	314	10+
In lamp or electronic tube	315	74+
Gas & liquid contact apparatus	261	DIG. 8
Geiger muller counter	250	374+
Geiger muller counter	313	93
Generation	310	
In lamp or electronic tube	315	55
System	322	
Generator (See generators)	310	
Bicycle driven with lamp	362	193
Nonelectric prime mover driven	290	
System	322	
Generator control	322	44+
Polarity control	322	5+
Power transmitting mechanism control	322	40+
Generator systems	322	
Automatic control	322	11
Automatic control	322	17+
Excitation control	322	59+
Piezoelectric	322	2R
Gyroscopic compass	33	324+
Heating	219	
Distillation processes		
Paratory	203	
Thermolytic	201	19
Electrothermic metallurgy	75	10.1+
Ferrous metal local hardening	148	150
Ferrous metal treatment	148	154
Gas generator retort	48	103
Incubator type heater	236	3
Knife	30	140
Metal working with	29	DIG. 13
Static mold	249	78
Thawing pipes and freeze protection	138	33
Track sanders	291	20
High frequency medical	128	804+
Igniter with burner (See igniter)	431	258+
Ignition, internal combustion engine	123	143R+
Impulse generator		
Condenser discharge	320	1
Dynamoelectric	310	10+
Electromagnetic periodic switch	335	87+
Electronic tube system	328	59+
Gas tube relaxation generator	331	129
Nondynamoelectric	310	300
Periodic switch	200	19R+
Radar system	342	175+
Relaxation oscillation generator	331	143
Signal transmitters	340	345+
Supplied to lamp or electronic tube	315	289
Telegraph system	178	
Telephone call transmitters	379	
Transistor relaxation generator	331	111
Inductor devices	336	
Insulation, electrical	493	949*
Insulator	D13	17+
Interrupter (See interrupter)		
Ionization chamber	313	93
For nuclear fusion reaction	376	100
System	250	335
Junction box	D13	24+
Key (See key, electric)		
Lamp (See lamps, electric)	313	
Consumable electrode arc lamp	314	
Display systems, matrix	340	780
Display systems, monogram	340	760+
Flash lights	362	208+
Heaters using	219	552+
Lamp systems	315	
Making	445	
Making glassworking machine	65	152+
Making glassworking processes	65	36+
Making miscellaneous	445	
Repair apparatus	445	61
Repair methods	445	2
Signs, multiple function	40	553+
Lead frame stock	428	572
Line (See line, electric)		
Liquid purifier insulating or discharging	210	243
Magnetic surgery	128	1.3+
Magnetically operated switches, magnets, & electromagnets	335	
Massage	128	24.1+
Mattress, flexible heater element	219	549
Measuring & testing	324	
Instruments, design	D10	46+
Medical and surgical treatment	128	362+
Medicators	604	20+
Melting furnace	373	
Glass	373	27
Meter (See meter, electric)	324	76R+
Coin controlled	194	
Recording	346	
Miscellaneous	361	
Miscellaneous electron space discharge device systems	328	
Miscellaneous non-linear reactor systems	307	401+
Miscellaneous non-linear solid state device circuits	307	200R+
Motive power (See motor, electric)	310	
Aircraft control actuator	244	175+
Arc electrode actuator	314	69+
Arc electrode actuator	314	94+
Automatic miscellaneous	318	445+
Generating electric locomotives	105	35+
Induction motors	310	166+
Induction motors systems	318	727+
Interchangeably locked gearing operator	74	365
Ironing or smoothing machine	38	38+
Lock operating mechanism	70	277+
Locomotives	105	49+
Motor acceleration	318	384+
Motor braking	318	362+
Motor making processes	29	596+
Motor reversing	318	280+
Motor vehicle	180	65.1+
Piezoelectric	310	311+
Plural motor control	318	255+
Plural motors	318	34+
Potentiometer controller	D13	21
Reciprocating motor systems	318	119+
Speed control	318	301+
Steering gear operator	180	79.1
Synchronous motor systems	318	700+
Systems	318	
Textile twisting apparatus	57	100
Time controlled locks	70	271
Toy machinery	446	
Toys	446	484
Vehicle	446	457
Vehicle	D13	
Motor generators	310	113
System	322	
Motors alternating current	310	159+
Music instrument	84	1.1+
Nail hammers	227	131
Needle	128	303.18
Oil well tubing dewaxer	219	277+
Organ (See organ, electric)	84	1.1+
Orthopedics	128	68.1
Oscillators	331	
Outlet box support	248	DIG. 6
Oven	219	391+
Packaging for components in roll form	206	328+
Pencil sharpener	144	28.1+
Phase control systems	323	212
Photography, electric	430	31+
Plant culture	47	1.3
Plasma control	376	143
Plug (See plug, electric)		
Power conversion systems	363	
Power supply or regulation systems	323	
Inrush current limiters	323	908*
Lamp dimmer structure	323	905*
Medical electronics	323	911*
Optical coupling to semiconductor	323	902*
Precipitators	323	903*
Remote sensing	323	909*
Solar cell systems	323	906*
Starting circuits	323	901*
Temperature compensation of semiconductor	323	907*
Touch systems	323	904*
Two of three phases regulated	323	910*
Precipitator	55	101+
Distillation processes &	201	19
Distillation processes &	201	40
Processes	55	2
Preserving, disinfecting, sterilizing methods	422	22+
Food or beverage	426	237
Apparatus	99	451
Fumigator with electric fan	422	124
Prime-mover dynamo plants	290	
Printed circuits	361	380+
Prods	231	2E
Production	D13	
Protection (See protection, electric)		
Public utility meter	D10	100
Radio tube (See electric space discharge devices)		
Radioactive surgery	128	1.1+
Rail bonds	238	14.1+
Recorder	346	
Electrochemical	346	165
Lock indicator	70	434
Photographic oscillograph	346	109
Spark perforating	346	163
Rectifier		
Electrolytic	361	436+
Electronic tube type structure	313	
Supplying electronic tube or lamp	315	200R+
Supplying radio	328	262+
Systems	363	13+
Relay (See relay, electric)		
Resistor devices (See resistance)	338	
Rivet heaters	219	157
Safety devices (See safety devices)		
Score boards	235	1B
Timing type	340	309.15
Shock apparatus (See shocker)		
Cattle goads	231	2E
Electric chair	128	377+
Fences	256	10
Hazard prevention systems	307	326+
Initiating	272	27R
Patient protection from shock	128	908*
Preventing shock	174	5R
Preventing shock	361	212+
Surgical	128	362+
Therapeutic application	128	783+
Signaling, electric (See signal)	340	
Noise suppression in signal	128	901*
Radio	455	
Railway	246	
Telegraph	178	
Telephone	379	
Television transmission	358	141+
Socket (See socket, electric)		
Soldering	219	129
Iron	219	245+
Iron design	D 8	30
With pressure	219	85R
Solid state barrier layer device structure	357	
Space discharge device (See radio; space discharge device; television)		
Spark perforating	219	384
Recorder	346	163
Steering motor system	318	
Sterilizing and pasteurizing		
Apparatus	99	451
Methods	426	237+
Stove	219	392+
Super conductor metal	75	
Composite	428	930*
Surgical instruments	128	303.1+
Surgical receptors	604	20+
Switches (See switch, electric)	200	
Magnetically operated	335	
Mechanically actuated, general	200	
Pencil sharpener pencil actuated	144	28.5
Thermally actuated	337	
Synthetic resin or natural rubber formation	521	915
Teaching	434	301
Terminals (See terminals, electric)		
Tester	324	
Testing devices (See testing)	324	
Nonelectric test	73	
Radiant energy using	250	
X ray using	378	
Time internal measurements	368	114+
Electrical	324	160+
Toothbrush	D 4	101
Toys	446	484+
Telephones	446	142
Train dispatching systems	246	2R+

	Class	Subclass
Transformation	D13	
Transformers	336	
Transistors	357	
Transmission or interconnection systems (See transmission)	307	
Transmission to vehicles or moving object	191	
Elevators	187	100+
Moving arc lamp	314	129
Tube (See tube, electric)		
Tuners	334	
Valve (See electric space discharge devices)		
Valve actuation	261	DIG. 74
Vehicle		
Locomotive	105	49+
Locomotive generating electric	105	35+
Motor vehicle	180	65.1+
Vibrator		
Electric switch	335	87+
Massage	128	32+
Rectifying system	363	110+
Voltage magnitude control systems	323	
Current regulation	323	
Electronic tube with	323	291
Impedance systems	323	293
Saturable reactor	323	249
Tap changing	323	255
Transformer systems	323	247
Wheatstone bridge	323	293
Wall plug, outlet	439	
Design	D13	30+
Watches	D10	30+
Wave analysis	324	77R
Wave guide	333	239+
In cathode ray tube	315	3+
In electronic tube	315	39+
Wave resonator	333	219+
In cathode ray tube	315	4+
In electronic tube	315	39.51+
Wave transmission lines & networks	333	
Weighing scales	177	210R+
Welding apparatus	D15	144
Whistles	340	406
Wire (See wire, electric)		
Wire nuts	174	87
Wireless power transmission to a radio receiver	455	343
Motors having	318	16
Vehicles having	191	10
Wiring systems	307	
Electric Control Systems		
Alarm signal	340	500+
Annunciators	340	815.1
Antihunting		
Generator systems	322	19
Motor systems	318	448
Motor systems synchronous	318	702
Battery condition responsive	320	43+
Coded record sensors	235	433+
Computer controlled, monitored	364	480+
Current responsive		
Battery charging systems	320	39+
Electronic tube voltage magnitude control systems	323	291+
Generator systems	322	23+
Generator systems	322	27
Impedance voltage magnitude control systems	323	293
Motor systems	318	474+
Motor systems	318	478
Motor systems acceleration	318	390
Motor systems acceleration	318	394+
Motor systems braking	318	367
Motor systems speed control	318	332
Motor systems speed control	318	333
Motor systems synchronous	318	714
Field winding circuits	318	717
Motor systems ward leonard	318	144
Voltage magnitude control systems	323	
Direction responsive	340	
Motor systems	318	489
Display systems	340	789+
Electric sign, matrix	340	780
Electric sign, monogram	340	760+
Error detector/corrector	371	
Fluid level responsive		
Alarm signal	340	618+
Motor systems	318	481
Fluid pressure responsive		
Alarm signal	340	626+
Battery charging systems	320	46
Generator systems	322	35
Motor systems	318	335
Motor systems	318	481
Moisture responsive		
Motor systems	318	483
Motion picture change over	352	135
Multiplexing	370	
Musical instrument	84	
Ordance control data	235	400+
Position or movement responsive		
Motor systems	318	265+
Motor systems	318	466+
Motor systems braking	318	369
Motor systems follow up	318	560+
Motor systems reciprocating	318	127
Motor systems reversing	318	282
Motor systems reversing	318	286
Motor systems synchronous	318	715
Armature winding circuits	318	721
Field winding circuits	318	719
Power factor or phase responsive		
Generator systems	322	20+
Motor systems	318	437+
Motor systems	318	438+
Motor systems induction	318	729
Motor systems synchronous	318	700+
Motor systems synchronous	318	700+
Phase control systems	323	205
Power or load responsive		
Battery charging systems	320	31+
Generator systems	322	24+
Motor systems	318	474+
Motor systems acceleration	318	388
Motor systems acceleration	318	393
Motor systems braking	318	366
Motor systems constant load	318	432+
Motor systems during starting	318	430+
Motor systems load limitation	318	434
Motor systems plural motors	318	98+
Motor systems speed control	318	332
Motor systems synchronous	318	700+
Motor systems ward leonard	318	141+
Public address	381	82+
Radiant energy responsive		
Electronic voltage magnitude control systems	323	
For nuclear reactor	376	215
Generator systems	322	26
Heat responsive test	374	32
Motor systems	318	480
Photoelectric circuits	250	200+
Special ray circuits	250	336.1
Rate of change responsive		
Electronic tube voltage magnitude control systems	323	
Generator systems	322	19+
Motor systems	318	456+
Signal	340	
Alarm	340	500+
Thermal control	340	584+
Sound or mechanical vibration responsive motor systems	318	128
Speed responsive		
Battery charging systems	320	41
Generator systems	322	29+
Motor systems	318	461+
Motor systems braking	318	369
Motor systems dynamic braking	318	379
Motor systems induction primary circuit	318	799+
Motor systems plural motors	318	59+
Motor systems plural motors	318	66+
Motor systems speed control	318	257+
Motor systems speed control	318	268+
Motor systems speed control	318	301+
Motor systems synchronous	318	715
Armature winding circuits	318	721
Field winding circuits	318	719
Motor systems ward leonard	318	146
Motor systems ward leonard	318	147
Synchronization	375	107+
Telegraph systems	178	
Telephone	379	
Television	358	93+
Thermally responsive		
Alarm	340	584+
Battery charging systems	320	35+
Electric heater systems	219	482+
Generator systems	322	33+
Impedance voltage magnitude control systems	323	294
Motor systems	318	471+
Motor systems acceleration	318	399
Motor systems speed control	318	334
Signal system	340	584+
Time or time delay responsive		
Battery charging systems	320	31+
Battery charging systems	320	37+
Generator systems	322	18
Motor systems	318	364
Motor systems	318	447
Motor systems	318	452+
Motor systems	318	484+
Motor systems acceleration	318	391
Motor systems acceleration	318	392
Motor systems acceleration	318	400+
Motor systems braking	318	306+
Motor systems induction	318	727+
Starting	318	778+
Motor systems reversing	318	283+
Motor systems synchronous	318	700+
Motor systems ward leonard	318	141+
Voltage responsive		
Battery charging systems	320	32+
Battery charging systems	320	39+
Generator systems	322	24+
Generator systems	322	28
Motor systems	318	479
Motor systems acceleration	318	388+
Motor systems acceleration	318	394+
Motor systems braking	318	368
Motor systems induction, primary circuit	318	805
Motor systems speed control	318	331
Motor systems speed control	318	333
Motor systems synchronous	318	712+
Armature winding circuits	318	720
Field winding induced voltage	318	718
Motor systems terminal voltage	318	459
Motor systems ward leonard	318	145
Phase control systems	323	212
Voltage magnitude control	323	234
Electric Toothbrush	D 4	101
Electricians Tools	D 8	
Electrifying Fluid	361	225+
Electrocardiograph	128	697
Electrochemistry	204	
Electrocoating	204	14.1+
By cathode sputtering	204	192.1+
By electrophoreses or electro osmosis	204	180.2+
Electroculture	47	1.3
Electrocution		
Chair	128	377+
Trap	43	98+
Insect	43	112
Electrode		
Anode making	445	46+
With assembly	445	35+
Arc furnace	373	88+
Arc lamp	314	
Composition	252	500+
Structure	313	357
Battery, electro chemical	429	209+
Battery, thermo or photoelectric	136	
Brush and holder	310	239+
Brushes	310	248+
Carbon		
Composition for	252	502+
Cored	313	354
Joint and connection digest	403	DIG. 5
Structure	313	326+
Carbon saver	313	335
Cathode making	445	46+
Wire shaping	140	71.5+
With assembly	455	35
Coating	427	58+
Collector rings dynamoelectric	310	232
Collector transmission to vehicle	191	45R+
Commutators dynamoelectric	310	233+
Condenser	361	271+
Cooking apparatus	99	358
Electrical connector for	439	
Chuck or socket	279	
Clasp	403	344
Carbon electrode	403	DIG. 5
Interfitted members	403	345
Mechanical rod joint	403	
Electrical precipitator	55	101+
Processes	55	2+

	Class	Subclass
Electroanalysis	128	635
Electrolytic chemical apparatus	204	193^{+}
Composition	204	291^{+}
Liquid electrode	204	219^{+}
Structure	204	280^{+}
Supports	204	$297R^{+}$
Electrolytic condenser	361	433^{+}
Electrolytic rectifier	361	436^{+}
Electronic tube	313	326^{+}
Electronic tube element	313	326
Electrophoretic or electro osmotic	204	$299R^{+}$
Electrothermic processes	75	10.1^{+}
Consumable metal electrode	75	10.23^{+}
Filament making (See cathode)		
Film	361	433
Food treating apparatus	99	451
Formed as inductive impedance electronic tube	315	40
Grid making	445	35^{+}
Wireworking	140	71.5
Heating fluid in circuit	219	284^{+}
Holders	219	138^{+}
Arc furnace	373	94^{+}
Arc lamp	403	
Arc lamp	314	
Igniter internal combustion engine	123	146.5 R
Igniters	361	247^{+}
With burner	431	264
Immersible electrode in liquid level gauge	73	304R
Ion sensitive	204	416^{+}
Lamp	313	326^{+}
Arc composition	252	500^{+}
Arc consumption feed type	314	60
Arc structure	313	357
Filament composition	252	500^{+}
Filament form	313	341^{+}
Gas or vapor	313	326^{+}
Making	445	
Lightning arresters	361	117^{+}
Magnetic chemical apparatus	422	186.1^{+}
Making	29	592R
Battery grid filling	141	32^{+}
Battery grid of metal	29	2
Coating electron emissive	427	77^{+}
Condenser electrode	29	25.41^{+}
Extruding plastic coating apparatus	425	113^{+}
Filament wireworking	140	71.5
Plastic operation filament		
Plastic operation method	264	104^{+}
Medical diagnostic	128	639^{+}
Mercury cathode arc lamp	313	328
Mercury cathode tubes	313	328^{+}
Mercury vapor lamps	313	326^{+}
Metal for arc lamps	313	328
Metal heating	219	119^{+}
Arc	219	145.1^{+}
Work holders	219	138^{+}
Microphone	381	168^{+}
Nernst glower	313	326^{+}
Composition	252	500^{+}
Ozonizers	422	186.7^{+}
Perforating recording spark	346	163
Pressure or vacuum furnace	373	60^{+}
Arc furnace with internal atmosphere control	373	68
Arc furnace with internal atmosphere control	373	77
Radio tube element	313	326^{+}
Selectively permeable membrane	204	415
Self baked furnace	373	60^{+}
Self-baking	373	89
Soderberg electrode	373	97
Separation or purifying of liquid	204	302^{+}
Electrolytic apparatus	204	225^{+}
Spark plug	123	169EL
Sterilizer or pasteurizer	426	237^{+}
Storage batteries	429	209^{+}
Making	29	623.1^{+}
Paste mixtures	429	149^{+}
Surgical	128	
High frequency therapy	128	804^{+}
High frequency therapy	128	1.5
Instruments	128	303.1^{+}
Tip cleaner for	314	23
Welding	219	145.1^{+}
Electrodeposition	204	
Laminating combined	156	150^{+}
Electrodynamometers	324	144^{+}
Electromagnetic brake	310	92^{+}
Electroencephalograph	128	731
Electroetching	204	129.1^{+}
Electroforming		
Molds or strip plates	204	281
Processes	204	3^{+}
Electroliers	362	405^{+}
Electroluminescent		
Stock material	428	917*
Type face design	D18	26
Electrolysis	204	$1R^{+}$
Electrolytes	204	
Battery	429	188^{+}
Circulation	429	51
Condenser or rectifier	252	62.2
Conductivity tester	324	438^{+}
Drain	429	95
Feeding control	429	63
Fuel cell		
Solid	429	30^{+}
Moveable means	429	70
Non movable means	429	81
Two-fluid	429	101
Electrolytic		
Analogue computer	235	61R
Apparatus and method	204	
Bleaching	204	133
Circuit controller	361	435^{+}
Cleaning processes	204	141.5^{+}
Condenser or rectifier	361	433^{+}
Electrolyte	252	62.2
Manufacture	29	570.1
Detector radio receiver	329	196
Electrodeposition	204	14.1^{+}
Alloys	204	43.1^{+}
Multiple layers	204	40^{+}
Meters	324	94
Methods of producing specific chemical compounds	204	$59R^{+}$
Recorder	346	165
Synthesis	204	$59R^{+}$
Tank for determining equipotential lines	235	61R
Electromagnet (See Magnet)	335	209^{+}
Armature and	335	220^{+}
Brake	310	93
Brake operator and brake	188	161^{+}
Chemical apparatus	422	186.1^{+}
Object protection	422	186.2
Clutch	310	92^{+}
Control for consumable electrode	314	
Deflecting arc in lamp	313	153^{+}
Electric switch	335	2^{+}
Handling implements	294	65.5
Liquid separation or purification	210	222^{+}
Process	210	695^{+}
Making methods	29	602R
Moving consumable electrode	314	
Pickup for musical instrument	84	1.15
Relay systems	361	160^{+}
Relay systems switching	361	160^{+}
Surgical	128	1.3
Switching systems using	361	160^{+}
Systems of supply to	361	160^{+}
Torque device	310	92^{+}
Tuner operated by	334	17^{+}
Electromagnetic Energy		
Used in reaction to prepare inorganic or organic material	204	157.15^{+}
Used in reaction to prepare or treat synthetic resin or natural rubber	522	1^{+}
Electrometallurgy	75	10.1^{+}
Electrometers	324	$76R^{+}$
Electrostatic	324	109
With amplifier	324	123R
Electron (See Electric; Space Discharge Device)		
Apparatus for treating with	250	492.1
Beam	430	942*
Beam heat	156	DIG. 102
Beam tube amplifier systems	330	44^{+}
Counter	250	472.1
Electron spin resonance testing	436	173*
Emissive coating	427	77^{+}
Microscope	250	311
Multipliers		
Amplifier systems	330	42
Photomultipliers	313	523^{+}
Systems	328	243
Tube structure	313	$103R^{+}$
Optics		
Tube structure	313	441^{+}
Ray tube	313	107.5
Tube	D13	22
Twin indicator	313	107.5
With current connections	315	382.1
Space discharge device systems, misc	328	
Electronic & Electronics (See Electric, Radio, Television, Wave Energy)		
Compositing system	358	22
Computer controlled, monitored item	364	480^{+}
Cooker	99	358
Heating	219	10.55 R
Inductive apparatus	219	6.5^{+}
Inductive methods	219	10.41^{+}
Music instrument	84	1.1^{+}
Organs	84	1.1^{+}
Piano tuning	84	454
Receiving tubes	313	
Switch	328	76^{+}
Voltage regulation	323	
Electronic Still Camera	358	909
Scanned semiconductor matrix	358	213.11^{+}
Electronic Tube (See Electric Space Discharge Device)		
Acorn type	313	
Battery type	313	
Bias supply	328	226
Gas tube system	315	
Power packs	307	150
Rectifier system	363	117^{+}
Voltage divider	307	15
Voltage regulation	363	117^{+}
With voltage regulation	363	111^{+}
Interelectrode capacitance testing	324	409^{+}
Metal type	313	248
Metal envelope as electrode	313	246^{+}
Miniature	313	
Mixer	313	300
Pencil type	313	
Ruggedized	313	
Subminiature type	313	
Electrooptical		
Batteries	136	243^{+}
Cathode ray tube system	315	1^{+}
Cathode ray tubes	313	441^{+}
Diagnostic lamps	128	23
Display signs and systems	340	810
Illuminated signs	40	541^{+}
Lamps	313	
Light sensitive circuits	250	206^{+}
Electric lamp control	315	149^{+}
Musical instrument	84	1.18
Photocell		
Battery type	136	243^{+}
Electronic tube type	313	523^{+}
Phosphor screen	313	525^{+}
Resistance type	338	15^{+}
System discharge device control	315	149^{+}
Systems	250	200^{+}
Photographic telegraph recorder	178	15
Code recorder receiver	178	90
Photometers	356	218^{+}
Recorders		
Electrochemical	346	165
Photographic	346	$107R^{+}$
Radiant energy	365	106^{+}
Recording or reproducing sound	369	100^{+}
Signalling	455	600^{+}
Sound recording and reproducing	369	100^{+}
Surgical and medical	128	362^{+}
Television	358	209^{+}
Ultraviolet application	250	365^{+}
X ray tubes	378	121
Electroosmosis or Electrophoresis		
Apparatus	204	$299R^{+}$
Immunoelectrophoresis	436	516*
Processes	204	180.1^{+}
Electrophotography Apparatus	355	$3R^{+}$
Chemical processes and materials	430	31^{+}
Electroplating		
Apparatus	204	194^{+}
Metal working with	29	DIG. 12
Processes	204	14.1^{+}
Electropneumatic Action in Organs		
Automatic	84	88
Pipe organs	84	338
Swells	84	347
Electropolishing	204	129.1^{+}
Electroprinting	204	2
Electroscopes	324	109
Radiation measuring type	250	390^{+}

	Class	Subclass
Electrostatic (See Separator, Solids)		
Assembly apparatus or process	29	900*
Coating	118	620+
Control	118	671
Gas separator	55	101+
Processes	55	2+
Induction generators	310	309+
Measuring and testing	324	327
Printing using	101	DIG. 13
Spraying	239	690+
Electrosynthesis	204	59R+
Electrotherapeutic		
Apparatus	128	362
Electrothermic Processes, Metallurgy	75	10.1+
Consumable metal electrode	75	10.23
Electrical induction	75	10.14
Gaseous treating agent	75	10.39
Plasma	75	10.19
Solid treating agent	75	10.39
Volatilization of metal	75	10.29
Electrotype		
Manufacture of	204	6
Elevator (See Lifters)	187	
Aircraft control	244	75R+
Amusement	272	6+
Automobile lift or hoist	269	58+
Bucket	198	701+
Butchering	17	24
Caster leg	16	32+
Catch devices	187	73+
Design	D34	28+
Door	187	51+
Control	187	108+
Drying apparatus	34	189+
Sheet	34	149
Electric control	187	100+
Electric motive power	187	112+
Extensible fire escape type	182	63
Fish	405	82
Floating dry dock	114	48
Grain storage type	52	245+
With material port	52	192+
Guiding device	187	95
Hanging door wheel mounts	16	99
Hatch mechanism	187	62+
Lighting on steps	362	76
Material handling	414	592+
Mortar mixer combined	366	26
Platform type scaffold	182	141+
Power driven conveyor type	198	
Railway	104	127+
Refrigerated article	62	378
Process	62	63
Refrigerator ice	62	379
Screw	198	657+
Self loading vehicle	414	486+
Shaft closure operators	49	68
Signal or call unit	D10	108
Skeleton towers metallic	52	111+
With platforms	182	141+
Spiral	198	778+
Store service type	186	22+
Ellagic Acid	549	278
Ellipsograph	33	30.1+
Flexible cord type	33	30.5
Pattern follower type	33	30.3
Pivoted circular pattern type	33	30.2
Elongated-member-driving Apparatus	227	
Embalming	27	21+
Compositions	424	75
Embalming equipment or apparatus	D24	6
Emblem or Badge (See Medal)	D11	95+
Embossing (See Impressing; Indenting)	101	3R+
Embossing implement	D18	15+
Laminating fabrics with	156	219+
Plastics	264	293
Tobacco plugs	131	111+
Typewriting machines	400	127+
Woodworking methods	144	358
Embrittlement Test	73	86
Embroidering		
Sewing machine	112	78+
Presser device feature	112	236
Embroidery	112	439
Design	D 5	
Hoops	38	102.2
Machinery	D15	66+
Packages	206	49
Processes	112	266.1
Trimmer cutter	83	910*
Embryotome	128	307
Emergency		
Closure release	49	141
Draft coupling janney type	213	111+
Elevator cables	187	71
Exit door securer with lock	70	92
Gear for aircraft water landing	244	107
Kits, eg flare, first aid, survival packs	206	803*
Lock release	70	465
Operating means for closure bolt	292	92+
Multiple bolt	292	21
Railway car exits	105	348+
Release for clockwork lock operating mechanism	70	274
Resilient tires	152	152
Signal stations railway train dispatching	246	13
Uncoupling device for janney type coupling	213	160
Emergency Power Sources, Electric		
AC DC converter	328	258+
Battery		
Dry	429	163+
Storage	429	149+
Dynamotor	310	113
Electric motor	318	17
Generator	310	
Generator system, portable	322	1
With prime mover	290	
Lighting systems	315	86
Flashlights	362	208
Generator motor	363	
Portable	315	33
Self-powered lamps	362	157
Emery	51	309
Wheels	51	206R+
Disc	51	209R+
Emetine	546	96
Emissivity		
Compensated radiation thermometer	374	128
Measurement	374	9
Empennage	244	87+
Emptying (See Discharging)		
Emulsifying	252	302+
Agitation	366	
Emulsion	252	302+
Biocidal	514	937*
Bituminous	252	311.5
Bituminous coating	106	277
Breaking		
Agents for	252	358
Mineral oil chemical treatment	208	188
Combined	208	188
Processes	252	319+
Cellulose ether or ester	106	170
Fuel	44	51
Photographic	430	495+
Synthetic resin or natural rubber (See class 523, 524)		
Enamel	501	14+
Opacifiers for	106	312
Enameling	427	
Metal working with	29	DIG. 14
Encapsulation of Material		
Gaseous core	521	76
Liquid core	264	4
Using emulsion or dispersion process	264	4.1
Solid core	427	213.3
Encoder Communicating Switch-type	318	602
End		
Gate	160	
Ship dry dock	114	47
Vehicle body	296	50+
Matching woodworking	144	91
Sleeves for electric conductors	174	93
End Effector	901	30+
Endless Belt Conveyor (See Conveyor, Endless)		
Endodontic Apparatus	433	81
Method or material for testing treating, restoring, removing natural teeth	433	224
Endoradiosonde	128	631
Endoscope		
Diagnostic specula	128	4
Surgical cutter	128	303.15
Endosmosis	210	649+
Gas separating process	55	2+
Enema, Receptacle for Syringe	206	364
Energy Accumulating Means, Motor Vehicle	180	165
Engine (See Motor)		
Air steam	60	39.181+
Ammonia	60	509+
Externally applied heat with	60	644.1
Boring cylinder	408	709
Carbonic acid	60	509+
Compound		
Fluid pressure and turbine	60	698+
Internal combustion engine and	60	597+
Combustion products	60	39.1+
Fluid pressure	60	597+
Turbine	60	598+
Motive fluid reheater between	60	679+
Motors in series	60	715+
Multiple expansion chambers	91	152+
Plural pneumatic	60	407+
Compressed air	91	
Demonstrating and teaching	434	389
Dental	433	103+
Design	D15	1+
Ether	60	509+
Expansible chamber	91	
One shot explosion actuated	60	632+
External combustion	60	39.6+
Fire	169	24
Fire escape	182	63+
Gas air	60	39.19+
Gunpowder	123	24R
Hand crank starter	74	550+
Heat exchanging	165	51+
Hot air	60	508+
Expansion of medium	60	516+
Externally applied heat	60	643+
Internal combustion engine	123	
Cooling	123	41.1
Design	D15	1+
Excess air to assist exhaust	60	900*
Exhaust treatment of rotaries	60	901*
Lubricators	123	196R
Nozzle	239	584+
Oil by-pass heat exchange	165	35
Pump drive	417	364
Refrigeration utilizing	62	323.1
Rotary piston	123	200+
Rotary reactor, separator or treater of exhaust	60	902*
Speed responsive throttle control	60	906*
Temperature measurement	374	144+
Locomotive	105	26.5+
Microwave	60	203.1
Muffler for hobby craft engine	181	404*
Plural motors system	60	698+
Power plant	60	
Prime mover dynamo plant	290	
Pulp refining	241	
Reaction motor	60	200.1+
Signal systems ships	116	21
Solar	60	641.8+
Starter gearing	74	6+
Steam	91	
Design	D15	2
Steam traction	180	36+
Strap starter	74	139+
Testing	73	116+
Toy	446	
Valve gear (See valve, actuation)		
Washing machine system	60	908*
Engineering Hydraulic and Earth	405	
Engraving		
Equipment design	D18	13
Photographic process	430	269+
Plate cutting	409	79+
Printing	101	150+
Stippling	72	76
Tools	30	164.9
Enlargers Photographic	355	18+
Ensilage Compaction	100	65+
Enteric Coating of Medicinal Tablets	424	475+
Entrenching Earthworking Apparatus	105	161
Envelope (See Bulb; Casing)		
Battery plate separator	429	136+
Closures	229	76+
Cutting blank form	83	911*
Feeding or delivering	271	2
Making	493	186+
Sealing and stamping combined	156	442
Tear strip opener	493	923*
Window	493	919*
X-art collection	493	917*
Manifolding		
Leaf combined	282	25

Entry	Class	Subclass
Strip connected	282	11.5 R
Moistening and sealing	156	441.5+
Opener	83	912*
Design	D 8	102+
Paper	229	.68R+
Design	D19	3+
Environmental Impact or Decreasing Pollution	422	900*
Survival computerized, monitored	364	418
Enzyme	435	183
Enzyme or microbe electrode	435	817*
Fruit juice treatment	426	51
Separation or purification	435	814*
By sorption	435	815*
Ignition, internal combustion	435	816*
Tobacco treatment	131	308
Epaulet	2	246
Badge type	40	1.5+
Ephedrine	564	364
In drug	514	653
Epicycle		
Land vehicle occupant propelled	280	207
Railway		
Drive wheel	105	100
Wheel	295	3
Epicyclic Gearing	74	640+
Episcope	353	65
Episcotister		
Light valve	350	484+
Photometer	356	217
Epon T M (See Synthetic Resin or Natural Rubber)	528	87+
Epdm (See Synthetic Resin or Natural Rubber)		
Epoxide Resin (See Synthetic Resin or Natural Rubber)		
Epoxy Compounds	549	512+
Equalized and Equalizing		
Animal draft	278	3+
Horses abreast	278	5+
Animal powered sweep	185	22
Bed bottom	5	278
Brake position adjusters	188	204R
Dumping vehicle hopper door closing	298	37
Extension table	108	87
Hot air furnace	126	105R
Locomotive	105	82
Railway brake series	188	46
Railway motor torque	105	135
Railway truck	105	209
Bogie	105	194
Sectional cabinet closure	312	110
Serial valves dependent motion	137	613+
Direct response	137	512+
Stop	137	629+
Trolleys for single rail suspended car	105	152
Vehicle fender car and truck	293	14
Vehicle running gear frame	280	104
Equatorial Telescope	350	568
Eraser	15	424+
Abradant type mutilating	51	
Blackboard	15	208+
Blade type	30	
Design	D19	53+
Disintegrable type	15	424+
Brush combined	15	105.52
Cutter e. g. , pencil sharpener combined	15	105.53
Mechanical	15	3.53
Ruler combined	15	105.51
Sharpener pencil combined	15	105.53
Mutilating type and burnisher	7	124+
Scraper type mutilating	30	169+
Shield	51	262.1
Typewriter attachment	400	695+
Ergolines	546	67+
Ergometer with Feedback	272	DIG. 6
Ergosterol	260	397.2+
Ergot Alkaloids	546	67+
Erosion Preventing	405	15+
Mineral oil conversion combined	208	47
Error		
Checking, electrical	371	
Arithmetic operations	364	737+
Communications synchronization	371	47
Computers	371	16+
Data processing systems, general	364	200
Memory devices	371	21
Program debugging	371	19
Correcting storage on typewriter	400	6
Detecting	371	
Positional servo systems	318	638
Prevention means on communication keyboards	340	365E
Erysodine	546	72
Erysonine	546	72
Erysothiopine	546	72
Erysothiovine	546	72
Erysotrine	546	72
Erysovine	546	72
Erythraline	546	48
Erythramine	546	48
Erythratine	546	48
Erythritol	568	853+
Erythroidine	546	66
Erythromycin	536	7.2+
Escalator	198	321+
Design	D34	30
Escape		
Acetylene generator automatic safety	48	56
Fire	182	
Endless conveyor	182	42+
Single strand stile or pole	182	189+
Tower or chute	182	48+
Gas holder high pressure safety	48	175
Submarine	114	323+
Escapement		
Clock and watch structure	368	124+
Clock electric driving type	368	131
Electrode feed	314	82
Gong striker	116	161+
Meter electric	324	139
Motor	74	1.5
Motor composite with	185	5
Motor spring with	185	38
Motor weight with	185	31
Music leaf turner	84	498+
Type casting machine	199	23
Typewriter feed mechanism	400	319+
Typewriter line locks	400	672+
Escutcheon	D 8	350+
Bushings or lining thimbles	16	2+
Covers or face plates	220	241
Design	D 8	350+
Electrical apparatus	174	66+
Fluorescent or phosphorescent apparatus	250	466.1
Design	D 8	
Door knob handle or rose plate	292	357
Fluid handling	137	377+
Key operated lock	70	452
Pipe joint	285	46
Radio plate for	D14	76
Television, plate for	D14	84
Wall outlet with	D13	30+
Espalier	D25	100
E S R, Electron Spin Resonance	436	173*
Essential Oil	512	5
Biocides containing	424	195.1+
Esters	260	
Carboxylic acid	560	1+
Hydrophenanthrene nucleus containing	560	5+
Cellulose	536	58+
Nitrogen containing	536	30+
Higher fatty acid	260	410+
Resin acid	260	103+
Free acid	560	5+
Starch	536	107+
Nitrogen containing	536	48+
Sulfate	558	20+
Thiocyanic acid	558	10+
Estradiol, in Drug	514	182
Estrone	260	397.4
Estrous Cycle Monitoring	128	738
Etching	156	625+
Apparatus	134	
Bright polishing	156	903*
Composition	252	79.1+
Differential apparatus	156	345
Direction indicator combined	33	305+
Electrolytic	204	129.1+
Electrolytic coating combined	204	32.1+
Glass	156	663
Glass with glass working	65	31
Metal working with	29	DIG. 16
Misc. chemical manufacture	156	DIG. 111
Photo	430	299+
Multicolor printing surface	430	301
Printing plate making	156	905*
Resist compositions	156	904*
Semiconductor making	156	625+
Semiconductor making	156	901*
Photoetching	430	313
Ethanol (See Alcohol)	568	840+
Beverages containing	426	592+
Dehydrating	203	19
Ethers	568	
Cellulose	536	84+
Nitrogen containing	536	43+
Cyclic hetero-o-type	549	200+
Oxy	568	579+
Phenol	568	630+
Thio	568	38+
Vinyl	568	687+
Ethyl		
Acetate	560	265
Cellulose not plastic	536	100+
Formate	560	265
Nitrite nitrous ether	558	488
Ethylene (See Olefin Hydrocarbon)		
Ethylene diamine tetraacetic acid	562	566
Glycol	568	852
Polymer	526	352
Synthesis	585	500+
By cracking	585	648+
Urea	548	300+
Ethylenic Monomers		
Solid polymer solely therefrom	526	72+
Eucalyptol	549	397
Eudiometers	73	23+
Eugenol	568	652
Euphoniums	84	402
Eurhodines	544	347+
Eutectic, Refrigeration Utilizing	62	430+
Automatic	62	139
Process	62	59
Evacuated Chamber		
Means to introduce material into	414	217+
Means to seal chamber	414	292
Evacuating Receptacles	141	65
Air displacement in filling receptacles	141	59
Successive receptacles	141	65
Apparatus	141	65
Electric lamp etc manufacture	445	38+
Apparatus	445	73
Combined operation methods	445	56+
Methods	445	53+
Food preserving	426	404+
Storage receptacle	99	472
Gas filling combined	141	66
Hermetic closing of filled receptacle, method	53	403+
Radiant energy tubes	313	553+
Having heating means	313	549
Evaginated Shoe	12	142A
Sole attaching means	36	17A
Evaporator		
Boiler	122	
Cabinet combined	312	31.1+
Concentrating	159	
Spray type	159	3
Gaseous current	159	4.1+
Gas and liquid contact apparatus		
Porous mass	261	94+
Porous sheet	261	100+
Oil combustion engine	123	522
Refrigerator	62	515+
Automatic	62	216+
Defrost water by heater	62	275+
Defrost water by refrigerator heat	62	279
Making	29	157R+
Sugar manufacture	127	16
Ventilator		
Building	98	30
Car	98	12
Car inlet	98	17
Evaporometer	374	39
Eveners		
Animal draft vehicle equalizer	278	13
Horse detacher	278	32
Grain harvesters, head or butt	56	468
Textile loom	139	74
Excavation (See Digger)	37	
Caisson	405	272
Pressurized	405	8+
Clamshell buckets	37	183R+
Ditch filler	37	142.5+
Ditchers	37	80R+

	Class	Subclass
Dredgers	37	54+
Mine ventilation	98	50
Mining or quarrying with separating	299	7
Orange-peel buckets	37	182+
Railway graders	37	104+
Railway snow excavators	37	198+
Road grader type	37	108R+
Roadway snow excavators	37	196+
Self-loading vehicles	37	4+
Shoring or bracing	405	272+
Snow excavators & melters	37	227+
Stump and removers	37	2R+
Tunnel excavating	299	
Tunnel ventilation	98	49
Excavator		
Dipper	414	690+
Force pump	37	75+
Pipe and cable laying combined	405	154+
Screw	175	394+
Suction pump	37	58+
Excelsior		
Making	144	185+
Package filling	493	967*
Exchange		
Heat	165	
Mortar mixers having	366	22+
Processes	366	4
Temperature regulation	236	
Tobacco smoking device	131	194+
Telephone	379	
Execution Devices		
Electric chairs	128	377+
Medicators	604	19+
Poisons	424	
Surgical instruments	128	303R+
Trap animal		
Choking or squeezing	43	85+
Electrocuting	43	98+
Everset	43	64+
Explosive	43	84
Impaling or smiting	43	77+
Self reset	43	73+
Trap insect		
Adhesive	43	114+
Electrocuting	43	112
Exercising		
Arm wrestling	272	901*
Devices	272	93+
Breathing improvement	272	99
Field sports	272	100
Isometric	272	125
Weights	272	117+
Kinesitherapy	128	25R+
Light thermal and electrical application combined	128	363+
Music	84	465+
Opponent-supplied resistance	272	902*
Poultry feeding combined	119	70
Exfoliation		
Cereals	426	449+
Compositions	252	378R
Decorticating	19	5R+
Seed or seed parts	241	6+
Stone	125	23R+
Exhaust (See Evacuating Receptacles)		
Mufflers	181	212+
Design	D12	194
Operated		
Brakes	188	154
Horns and whistles	116	138
Recirculation	123	568
Silencers	181	212+
For internal combustion engines	60	272+
Steam engine	60	681+
Treatment		
Combustion device feed in	60	683
Combustion engine	60	272+
Excess air to internal combustion engine	60	900+
Serially connected motors	60	679+
Furnace	432	67+
Furnace, metallurgical	266	144+
Furnace, solid fuel	110	203
Motors	60	685+
Rotary engines	60	901*
Exhauster (See Evacuating Receptacles)		
Filled receptacle cooking apparatus	99	359+
Heating etc gas making apparatus	48	173
Water closet tank outlet siphon	4	368+
Exhibiting		
Cards pictures and signs	40	
Design	D20	10+
Design display card	D20	40
Check controlled	194	
Dispensing combined	222	23+
Gem setting feature	63	30
Indicator combined	116	307
Exhibitor		
Cabinet with	312	234
Exit		
Bee hive	6	4R
Door lock	70	92
Fire escape chute	182	48+
Insect	160	12+
Railway car		
Control	105	341+
Emergency	105	348+
Stratifier discharge	209	494+
Expanded Metal	52	670+
Making	29	6.1
Expander and Expanding		
Anchor	411	15+
Expanding or piercing earth	52	155+
Boiler formers	72	393
Boring head cutter	408	199+
Building component	52	67
Cereals	426	449+
Apparatus	99	323.4+
Collapsible taps	408	147+
Exfoliated composition	252	378R
Face plate railway car	105	10+
Felly	157	9+
Gas liquefying and separating	62	36+
Processes	62	9+
Gate valve faces	251	193+
Hoof and shoe	168	47
Joints (See device incorporating)		
Liquefied gas fuel as refrigerant	62	7
Mandrels	269	48.1
Chuck	279	2R
Lathe work mount	82	44
Winding reel	242	63
Winding reel core	242	110
Metal making	29	6.1
Article	428	596+
Orthodontic	433	7
Pipe expanders	72	393
Methods of expanding	29	157.4
Piston ring applier	29	222+
Reamer	408	227+
Earth boring type	175	263+
Refrigeration by	62	
Rod or piston packings	277	138
Sheet metal shaping die	72	393
Tube cutter	30	103+
Wheel felly	157	9
Expansible Cabinets	312	205
Expansible Chamber Devices	92	
Expansion (See Expander)		
Engine, refrigeration producer	62	402+
Automatic	62	172
Liquefied gas producing	62	38+
Process	62	86
Joints		
Between articulated members	403	52+
Bridge	14	16.1
Building	52	573
Distillation apparatus	202	268
Flexible diaphragm or bellows	403	50+
Fluid distributor nozzle	239	397.5
Heat exchanger	165	81+
Pipe	285	298+
Resilient building	52	313
Static mold forming	249	9
Plug		
Metallic receptacle	220	233+
Pipe	138	89+
Reeds warp preparing	28	213
Valve, refrigeration utilizing	62	527+
Automatic	62	222+
Exploring		
Borehole and drilling study	73	151+
Depth sounding	73	290R+
Electro-acoustic	367	99+
Sound	181	
Diagnostics medical	128	630+
Earth boring	175	50
Electrical	324	
Electrical ore testing	324	376
Meteorology	73	170R+
Radar	342	191+
Radiosonde	340	345+
System	340	870.1+
Sound	181	
Explosion Control		
Acetylene generator	48	56
Blast or cupola furnace	266	174
Blowout plugs, frangible or fusible	220	89R
Containers of inflammable substances	220	88R
Davy lamp or wire gauze protected		
Dry cleaning and laundry apparatus	68	209
Drying apparatus	34	51
Gas cooking stoves	126	42
Gas distribution	48	192
Gas holder high pressure safety escape	48	175
Internal combustion engines	123	24R
Milling or grinding apparatus	241	31
Mine safety systems	299	12
Safes and bank protection combined	109	26+
Turbines	60	39.2+
Water back	126	35
Water elevators	417	73+
Explosive (See Ammunition; Bomb; Engine; Gun)	102	
Anchors	114	295
Article molding or shaping	264	3.1
Extruding	264	3.3
Forming or treating particulate material	264	3.4+
Rolling to form sheet or rod	264	3.2
Bag to contain explosive powder	102	282
Cap mechanical gun combined	124	2
Carbon preparation process	423	449
Charge making	86	
Comminution or puffing		
Apparatus	241	301
Cereals apparatus	99	323.4+
Cereals processes and products	426	625+
Processes	241	1
Vegetable processes	426	445
Compositions	149	
Devices	102	
Narcotic containing	102	512
Safes & bank protection combined	109	36+
Earth boring	175	2+
Fuses primers and igniting devices	102	200+
Impact driven fastener, eg nail	411	440+
Making apparatus	422	163
Material incinerator	110	237
Material treating	264	
Metal deforming X-art	72	706*
Metal forming	72	56
Mining	299	13
Operated assembling apparatus	29	254+
Operated one shot motor	60	632+
Operator		
Assembling device	29	254
Metal deforming device	72	430
Well tube perforator	175	2+
Power plant X-art	60	914*
Testing	73	35
Thermic compositions or charges	149	
Well		
Apparatus	166	63
Processes using	166	299
Torpedoes	102	301+
Tubing perforator	175	2+
Exposure		
Camera		
Automatic control system	354	410+
Meter	356	213+
Expressing		
Butter and cheese cutter combined	425	308+
Butter worker combined	425	200+
Die		
Block and earthenware	425	308+
Block and earthenware	425	325+
Metal including powdered	72	253.1+
Mold core making	164	174
Particulate metal	419	67
Pastry and confections	99	
Pastry and confections	425	
Plastic molding	425	381.2
Soap molding	425	308+
Soap molding	425	325+
Drying combined	34	70
Presses	100	104+
Expression Mechanism Musical		
Applying expression lines to record	84	463
Automatic pneumatic expression mechanism for automatic pianos	84	33+

U.S. PATENT CLASSIFICATIONS

Entry	Class	Subclass
Automatic stringed instrument pneumatic action	84	33+
Electric organs	84	1.24+
Piano	84	216+
Tracker bar modified for expression	84	157
Extension		
Cords	439	502+
Tables, co planar surfaces	108	65+
Extensometer		
Dilatometer	374	55
Distance measuring	33	125B+
Dynamometer	73	862.39+
Electrical	324	
Gauges	33	501+
Optical	356	372+
Strain resistance	338	2
Stress strain test	73	760+
Extinguisher		
Boiler safety	122	504.3
Burner cap or cover	431	144+
Cigar and cigarette	131	256
Electric arc prevention	200	144R+
Electric contact arc prevention	361	2
Extinguishing composition	252	2+
Fire	169	
Design	D29	2+
Fluid type for burner	431	145
Furnace fire	169	54+
Ash pan	110	171
Inflammable fluid containers	169	66+
Signalling system electric	340	289
Tobacco ash receiver combined	131	235.1+
Tobacco product combined	131	347
Extracting and Extracts (See Material Recovered)		
Apparatus	422	261+
Beverage	99	279+
Mineral oil	196	14.52
Sugar	127	3+
Textile treating fluid combined with treating	68	19+
Biocidal	424	
Distillation combined		
Retort	202	107
Separatory processes	203	
Still	202	170
Still with	202	168+
Thermolytic processes	201	
Dye compositions	8	438
Dyes natural	8	646
Food	426	655
Chocolate	426	631+
Coffee	426	594
Coffee substitute	426	596
Coffee substitute from fruits or vegetables	426	594+
Condiments and flavors	426	650
Hop	426	600
Malt	426	64
Meat	426	656
Pectin	426	655
Sea food	426	655
Tea	426	435
Inorganic materials	423	210+
Actinide series compounds	423	249+
Crystallization combined	23	299
Crystallization selective combined	23	297
Medicinal animal	424	95+
Mineral oils	208	311+
Conversion combined	208	46+
Ores and metals	75	101R+
Organic materials	260	705
Algins	536	3
Alkaloids cinchona	546	134+
Alkaloids opium	546	44+
Alkaloids tobacco	546	282
Alkaloids undetermined constitution	546	1
Esters carboxylic acid	560	1+
Fats fatty oils ester type waxes higher fatty acids	260	412+
Glycosides	536	18.5+
N-glycosides	536	2+
Pectins	536	2
Synthetic resins from plant extract	528	1+
Tannins	560	68+
Sugar	127	43+
Tanning	8	94.32
Aluminum compound combined	8	94.3
Tobacco	131	297
Extractor (See Wringers)		
Bolt or screw	81	436+
Centrifugal	210	360.1+
Closure remover	81	3.7+
Can opener having	7	152+
Compound tool having	7	151+
Pliers having	7	126+
Disassembling apparatus	29	700+
Easy out	81	436+
Electrolyte treatment	204	233+
Nail	254	18+
Shoe lasting	12	16
Pen	29	808
Press	100	104+
Shell from firearm	42	25
Revolver	42	68
Tilting breech	42	46
Shell projectile or wad	81	3.5
Stump and poles	254	199+
Tooth	433	141+
Wheel spoke	157	12
Extruding (See Expressing Die)		
Hydrostatic extrusion	72	711*
Eye		
Artificial	623	4+
Contact lenses	351	160R+
Cup, protector	D24	66
Deformity correction	128	76.5
Doll	446	392
Movable	446	343+
Dropper	604	295+
Design combined with bottle	D 9	338
Design combined with closure	D 9	447
Dispenser combined	222	420+
Syringe	604	295+
Examining & testing instruments	351	200+
Design	D10	46+
Objective	351	205+
Subjective	351	222+
Test charts and targets	351	239+
Exercising	128	25A
Optometers	351	205+
With testing	351	203
Forming		
Chain making	59	21
Wireworking	140	104
Heddles weaving	139	94
Hook and eye fastener	24	698+
Design apparel	D11	208+
Design, hardware	D 8	367+
Relatively movable hook portions	24	598+
Sewing machine attaching	112	105
Interlocking wire machine	140	114
Neck yoke	278	125
Nose grips	351	132
Integral with bridge	351	131
Pince-nez	351	71+
Photography	354	62
Testing instruments	351	206
Pieces for optical instruments	350	579
Protecting masks	2	10+
Design	D29	18+
Protectors for animals	54	80
Design	D30	144+
Shields		
Electric welder S hood combined	219	147
Thill coupling	278	77+
Toy figure	446	392
Treating		
Douches	604	295+
Electrically	128	793
Exercising	128	76.5
Medicine	424	
Vision testing & correcting	351	
Eye Bolt, Making	29	7
Eye Glasses	351	41
Anti-glare or shading	351	44
Bridge	351	133
Cases	206	5+
Design	D 3	34
Cleansing compositions	252	89.1+
Contact lens	351	160R+
Contact lens fitting	351	247
Correctors	351	140+
Design	D16	102+
Eye shields attached	2	13
Fitting apparatus	351	200+
Gages for	33	200
Hinge	16	228
Holder	248	DIG. 2
Design	D16	129
Rigid container	493	914*
Metal frame making	29	20
Ornamentation for	351	51+
Special tool for	81	3.5+
Protective goggles	2	426+
Temple	351	121
Eye Goggles	2	426+
Anti-glare or shading	351	44
Design	D16	102+
Eyebrow Brush	D 4	
Eyelash Curler	D28	36
Eyelet	24	141+
Apparel fastener	D11	208+
Closure fastener sealing	292	314
Making by die shaping sheet metal	72	343+
Setting	227	51+
Eyeleting		
Machine	227	51+
Paper tag making	493	375+
Sewing machine	112	66
Eyeshade (See Eye Glasses)	D29	18+
F		
Fabric (See Cloth; Metal; Paper; Textile; Web)		
Apparel per se	2	
Bed bottoms	5	186R+
Bolts or rolls for	206	389+
Boots and shoes	36	9R
Braided netted or lace	87	
Burnishing or lustering	26	27+
Cashmeres	D 6	603+
Central or S american patterns	D 5	20+
Chemically modified of	8	115.51+
Coated	428	245+
Coating and printing	427	
Electrolytic	204	21
Coating and uniting	156	
Coating and waterproofing compositions	106	
Cribs	5	98R
Crocheted	66	193
Damask	139	416+
Design	D 5	
Distance measuring	33	127+
Fastener made of	269	906*
Feeding	226	
Felt	428	280
Fence	256	45
Barbed	256	5
Fire retardant composition for	252	608
Chemically modified fiber	8	115.51+
Folding	493	405+
By reciprocating piler	270	30+
Hammocks	5	122+
Insect traps	43	118
Insulated electrical conductors	174	124R
Plural layers	174	122R
Plural layers impregnated	174	121R
Knitted	66	169R+
Design	D 5	47+
Laminated	428	224+
Making	28	100+
Braiding, netting, lace making	87	
Electrolytic	204	
Felting	28	116+
Felting methods	28	116+
Finishing operations	26	
Knitting	66	
Needling methods	28	107+
Fluid needling	28	104+
Sewing	112	
Weaving	139	
Wireworking	140	
Medicated	424	443+
Needled	428	300+
Nonwoven	428	
Ornamenting mechanically	26	69R
Parchmentizing and uniting	156	76
Preserving, disinfecting, and sterilizing		
Biocidal compositions	424	
Fireproofing compositions	252	608
Pressing	38	
Apparel	223	52+
Hats	223	21
Ratine like ornamenting by rubbing	26	27
Ravelers	28	171+
Reinforced (See reinforcement belts and fabric)		
Sewing methods	112	262.1+
Sifting elements	209	401+

	Class	Subclass
Softeners	252	8.6+
Spotting eg by chenille	223	45
Stitched seams	112	262.1+
Stretching frames	38	102.1+
Tensioning, sewing machine	112	121.26
Testing	73	159
Tire	152	DIG. 14
Carcass material	152	563+
Inner tube reinforcing	152	512
Interliners	152	204
Off-drum manufacture	156	906+
Toys covered with	446	269+
Web making	19	296+
Weftless	428	292+
Laminating	156	
Winding	242	55+
Wire	245	2+
Design	D 5	47+
Design for floor or wall	D 5	47+
Making	140	3R+
Working	140	107+
Woven	139	383R+
Fabric Softeners	252	8.6+
Face		
Bandage	128	163+
Bows dental	433	73
Orthodontic	433	5
Coverings	2	206
With therapeutic means	128	380
False	2	206
Hat combined	2	173
Guards (See masks)	2	9
Animal restraining type	119	96+
Lifter orthopedic	128	76B
Mill cutter		
Gear generating	409	10+
Rotary	409	64+
Plates		
Electrical box	174	55+
Electrical box	174	66+
Locks	70	450
Luminous	250	466.1
Receptacle closure	220	241
Train vestibule	105	21+
Train vestibule expanders for	105	10+
Protector	D29	17+
Radial sealing	277	81R+
Facer Construction (See Wall)	52	
Concrete layer	52	378
Disparate layers	52	506+
Facer to shaft	52	474+
Plaster type	52	344+
Holding panel	52	443+
Shingle type	52	518+
Facings		
Corrugated board	156	199+
Frictional gear	74	215
Shirt	2	122
Facsimile (See Copying; Pattern Control)	358	256+
Electric engraving	358	299
Electric lights form mosaic		
Matrix	340	780
Monogram	340	760+
Switching of lights by tape	200	46
Jacquard warp shedding	139	59+
Power to top harnessing	139	85+
Vibrating griff dobby	139	68
Photographs sent by electricity	358	294+
Factice	260	399
Fade Resistance Testing		
By exposure to light	378	38
Fading		
Bleaching processes for photographic purposes	430	461+
Fagots Metal	428	588+
Fail-safe		
Apparatus fault recovery with	371	14
Digital logic	307	442
Electrical communications condition responsive indicating sysem	340	507
Engine fuel injection systems	123	479
Engine fuel pump cut off	123	359
Positional servo systems with	318	563
Fair		
Leaders		
Cable guides	254	389+
Spar or ship timber attached	114	101
Stitch machines leather sewing	112	31
Fall		
Boards for piano cases	84	179
With desk	84	178
Tube fluid or vacuum pumps	417	150
False		
Face	2	206
Hemstitch forming	112	179
Teeth	433	167+
Twisters	57	332+
With drawing means	57	328+
Fancy Roll		
Fiber preparation	19	98+
Fancy Yarn Making	57	317
Fans	417	
Design, hand fan	D 3	1+
Design, ventilation	D23	370+
Fumigating evaporator	422	124
Furnace exhaust	110	162
Heat exchanger	165	87
Heater	219	369+
Hot air furnace with	126	110R
Temperature regulating	236	10
Impeller making	29	156.8 CF
Refrigeration combined	62	404+
Automatic	62	186+
Process	62	89+
Ventilation	98	
Faraday Shield	174	35R
For conductor	174	32+
For radio apparatus	455	300+
Transformer with	336	69+
Fare		
Box	232	7+
Recorder	346	33R
Register combined	346	15+
Register	235	33+
Recorder combined	346	15+
Farinaceous Ceral Dough	426	496
Deep frying doughnuts	426	439
Fermentation	426	18
Farm Machinery Equipment Tools (See Type)		
Cabs for self-propelling machinery	180	89.12
Farriery (See Horseshoes)	168	
Calks	168	29+
Nonmetallic digest	168	DIG. 1
Overshoes	168	1+
Shoeing stands	168	44
Sole pads	168	26+
Tools	168	45
Traction shape digest	168	DIG. 2
Fastener (See Connection; Joint; Key; Lock; Securing Means; Spline; Tie)	292	
Aircraft skin	244	132
Apparel	24	
Button and eyelet attaching	227	
Coat collar	2	100
Design	D11	200+
Fastener attaching methods	2	265+
Hat	132	57+
Metal stiffener with garment fastener	2	263
Necktie	24	49R+
Pocket	2	251
Shoe antislipping device combined	36	62
Shoe sole	36	23+
Stiffener with attaching means	2	257
Beds		
Bedclothes to mattress	5	498
Bedstead corner	5	288+
Sheet having	5	496
Mattress to bed bottom	5	411
Bolts	411	378+
Book binding	281	25R
Bottle or jar guard	215	29
Brake shoe	188	234+
Brush and broom tuft	15	190
Buckles buttons clasps etc	24	
Building		
Sash burglar alarm	116	16
Window bead	16	220
Window pane wood	52	598+
Camera plate holder	354	278+
Check label or tag	40	20R+
Clasp	24	455+
Clip	24	455+
Closure	292	
Bank protection	109	59R+
Bank protection plural closure	109	7
Bottle	215	273+
Bottle pivoted	215	237+
Car journal box latch	384	190.7
Envelope	229	77+
Lantern projector	362	374
Letter sheet wrapper	229	92.5
Lock type	70	77+
Paper box	229	125.19+
Receptacle	220	315
Ship	114	203
Wood barrel	217	89
Composition		
Elastometric	269	907*
Fabric	269	906*
Leather	269	906*
Non-metallic	269	904*
Paper	269	905*
Several materials	269	900*
Thermo-responsive memory	269	909*
Wood	269	905*
Design	D 8	
Harness	D30	134+
Driven headed and screw threaded	411	
Expanding anchor	411	15+
Explosive driving	411	440+
Folding sectional receptacle	220	7
Gun stock and barrel	42	75.1+
Hair	132	46R+
Harness		
Collar	54	21
Design	D30	134+
Hame	54	26+
Pad	54	68
Trace and whiffletree	278	117
Key in lock retaining	70	429+
Making		
Button fasteners	29	4
Lacing studs	29	12
Paper fasteners	29	13
Wire clips	140	82
Musical instrument, stringed	84	297R+
Nail, spike, tack	411	439+
Nuts	411	427+
Paper integral making	493	390+
Plastic	411	904*
Railway car journal brass	384	191.4
Railway tire	295	15+
Railway track	238	310+
Splice bar	238	260+
Rake teeth	56	400
Rivets	411	500+
Sail	114	108
Sash cord	16	202+
Saw teeth	83	838+
Scaffolding pole	403	
Screws	411	378+
Separable-fastener, two part	24	572+
With protruding filaments	24	442+
Sheet fabric or web	24	
Abrading tool cylinder to	51	358+
Abrading tool disk to	51	358+
Abrading tool handle to	51	392+
Bandage	128	171
Bed bottom	5	194+
Billiard or pool table	273	7
Carpet	16	4+
Conveyor gripper	198	803.3+
Copy holders	248	451+
Ironing table cover	38	140
Printing sheet	101	415.1
Reeling & unreeling apparatus	242	74
Shoe animal	168	17+
Shoe last instep block	12	137+
Sign letter fastener	40	620+
Skate	280	11.3+
Ski	280	611
Slide	24	381+
Applying to material edge	29	243.58
Cutting of	83	921*
Making metal	29	408+
Making molded plastic	264	252
Apparatus	425	500+
Sorter	209	929*
Staples	411	457+
Support clamp	24	455+
Table leg	248	188
Tacks	411	439+
Tent cover	135	119
Tenter hooks or clamps	26	93
Web condition responsive control	26	93+
Tool handle	403	
Umbrella	135	36R
Vehicle wheel		
Detachable section	301	9R+
Nonresilient tire	301	87+

U.S. PATENT CLASSIFICATIONS

Entry	Class	Subclass
Velcro-type T M	428	100
Washers	411	531+
With tension indicator	411	8+
With torque responsive driving connection	411	1+
Work holders	269	
Wound strand	242	125+
Wound web	242	74
Zipper	24	381+
Slide	24	415+
Fatigue Testing of Materials	73	808+
By vibration	73	577+
Of structure	73	787
Fats	260	398+
Electrical discharge treatment	204	167
Electromagnetic wave treatment	204	157.6+
Fats & fatty oils with preservative	260	398.5
Fatty drying oil with preservative	106	263
Fermentative treatment	435	271
Foods containing	426	601+
Apparatus for	99	452+
Preservation	426	321+
Liquoring	8	94.23
Ointments	424	
Splitting	260	415+
Fatty Acids (See Soap, Making)	562	606+
Higher	260	413+
Faucets (See Valves)		
Aerators	261	DIG. 22
Barrel tap	137	317+
Brackets	248	212+
Drinking fountain combined	239	25+
Filter combined	210	418+
Fixture	D23	238+
Foam		
Carbonated fluids	137	896+
Carbonated fluids by aspiration	137	888+
Jet nozzle type	239	343
Handle design	D23	230+
Spigot joint means	285	8
Water aerator for	239	428.5
Fault Light Indicator	361	58
Feather		
Apparel trimming making and applying	223	47
Coating	427	4
Coating or plastic compositions containing	106	155+
Dyeing	8	404+
Fertilizers from	71	18
Fiber preparation from	19	4
Fowl plucking machine	17	11.1 R
Imitation	428	6
Implements for general cleaning	15	234
Mattress filling	53	524
Removing, butchering process	17	47
Synthetic resin or natural rubber containing	524	10+
Treating or drying	34	2
Treatment chemical	8	94.1 R+
Working	223	47
Feathering		
Impeller	416	98+
Paddle wheel sustaining aircraft	244	20
Wind motor tail vane	416	9+
Feedback Amplifiers	330	75+
Feeder (See Conveyor; Filler)		
Abradant supplying	51	263+
Abrader coolant	51	266+
Abrading work	51	215R
Abrasive supplying process	51	292
Agitating	366	150+
Interrelated discharging combined	366	131+
Mortar mixers	366	30+
Mortar mixers discharging and	366	27+
Air or steam		
Boilers	122	
Fluid pressure brakes etc	303	
Furnace	110	
Metallurgical furnaces	266	186+
Stoves	126	
Anesthetic	128	203.12
Animal food	119	51R
Bee	6	5
Cage	119	18
Design	D30	121+
Stock car type	119	10
Timer controlled	119	51.11
Bar or tube	226	
Button fastening magazine	227	31+
Sewing machine mounted	112	113
Capillary (See wick)		
Acetylene generator water	48	25+
Coating machine of immersion type	118	401
Coating machine of solid applicator type	118	264+
Decanter and filter	210	294+
Dispensing	222	187
Fluid distributor	239	145
Force feed lubricators	184	64
Gas and liquid contact	261	75+
Water tube boiler	122	242+
Car unloading tilting track with	414	362+
Actuated by moving onto track	414	357
Rollable along underlying support	414	358
Rotates about fixed axis	414	359
Cartridge body or belt attached	224	239
Check controlled delivery	194	
Check row planter	111	34+
Coating implement combined	401	
Fountain pen bifurcate nib type	401	222+
Consumable electrode for lamps etc	314	
Control means	226	10+
Conveyers and feeders		
Accumulator receiving separated item	209	933*
Ambulant	209	935*
Centrifugal feeder	209	915*
Diverse sequential feed steps	209	914*
Driven or fluid conveyer moves item from separating station	209	925*
Endless belt pusher feed item	209	917*
Endless conveyer & feed through	209	923*
Endless feed conveyer, means for holding each item individually	209	912*
Fluid applied to items	209	932*
Free fall item feeding	209	908*
Gravity conveyer moving item from separating station	209	924*
Gravity influencing item moving	209	911*
Having opposed grippers	209	903*
Holding item by magnetic attraction	209	904*
Holding item by suction	209	905*
Hopper	209	910*
Inspection		
Illuminating means for visual	209	938*
Video scanning	209	939*
Item carrying bridge raisable to expose discharge opening	209	941*
Item holding feed magazine insertable in sorting apparatus	209	909*
Magnetic feeder	209	907*
Materials of construction	209	931*
Miscellaneous feed conveyers	209	922*
Moving items to sorting means in spaced relation lengthwise of feed path	209	934*
Operator selects destination of item	209	942*
Oscillating or reciprocating	209	921*
Reciprocating pusher feeds item	209	916*
Rotary feed conveyer	209	919*
Screw feed conveyer	209	913*
Swinging or rotating pusher feeds item	209	918*
Testing plural items as group	209	936*
Trough with at least one endless conveyer	209	923*
Vibratory feed conveyer	209	920*
Cutter or comminutor having		
Sheets and bars	83	
Solid material	241	
Dispensing	222	
Automatic weighers	177	116+
Check controlled	194	
Coating implement combined	401	137+
Corn popper with	99	323.9+
Inkwell combined	222	576+
Receptacle filling and closing	53	266R+
Distillation apparatus	202	262
Dough cutter	425	289+
Drilling machine feed mechanisms	408	129+
Earth boring type	175	162+
Electric conductors etc with fluid	174	15C+
Electrolytic apparatus		
Cells with	204	242+
Electrode	204	225+
Feed water per se or combined		
Boiler	122	451R+
Boiler purification	210	311+
Condenser feed heater type	165	112+
Distillation still heater	202	167
Furnace	110	297+
Gas pressure discharge of traps	137	156+
Heater for boiler	122	412+
Power plant furnace regulation	60	645+
Power plant furnance regulation	60	660+
Festooner	226	104+
Fluid current conveyer	406	108+
Diverse conveyer feeder	406	51+
Single for plural conveyors	406	123
Fuel		
Aircraft structure	244	135R
Coal tender	105	232
Furnace combined	110	101R+
Furnace type	414	147+
Nuclear reactor use	414	146
Internal combustion engine	123	495+
Progressive for furnace	110	267+
Reciprocating piston expansible chamber pump	417	437+
Special machine for sewing	112	121.17
Straw burning furnace	110	196
Trash burning cookstove	126	223
Valve retaining	261	DIG. 24
Valves	261	DIG. 23
Wet fuel furnace underfeed	110	255+
Gas manufacture		
Acetylene generator carbide	48	38
Acetylene generator water	48	4+
Heating gas oil retort	48	106
Glass melt	65	325+
Grain sheller and husker	130	33
Harvester grain adjusters	56	471
Heat exchange	165	120+
Invalid	414	9
Article manipulator analogous with hand, finger movement	414	1+
Liquid	137	
Acetylene generator	48	4+
Ambulant supply sprayers	239	146
Barometric feed to cooler	62	397
Battery electrolyte	429	72+
Bearing and guide material for oil	384	181
Centrifugal distributor	239	215+
Concentrating evaporators	159	43.1+
With alternating operating means	159	43.2
Diffusers slow	239	34+
Electrolytic apparatus	204	193+
Fountain pen	401	222
Fountain pen force feed or filler	401	143+
Fountains	239	16+
Gas and liquid contact apparatus	261	
Liquid contact with solids	134	
Liquid suspension separating	209	155+
Locomotive cab squirts	239	174
Metallurgical furnace	266	186+
Mineral oil distillation	196	135
Purification, material feed controlled	210	101
Purification, treating material addition	210	199+
Self proportioning systems	137	87+
Sprinklers, sprayers, diffusers	239	
Stratifier	209	500+
Water purification chemical	210	198.1+
Lubrication	184	
Main conveyor and collecting wing	198	601
Manicuring device with fluent material	132	74.5
Manufacturing device combined (See device per se)		
Ammunition and explosive charge making	86	45+
Ammunition loading	86	23+
Can body	413	70+
Can head	413	45+
Can head seamer with	413	27+
Cigar and cigarette making	131	38+
Laminated fabric hose making	156	
Leather skiving and splitting fixed knife	69	11+
Metal burnishing continuous	29	90.5
Metal founding molds	249	106
Metal heating and feeding		
Sewing machine for articles attached by sewing	112	104+
Sewing machine work	112	303+
Solder or flux	228	33+
Stock pushing or pulling	414	222+

	Class	Subclass
Textile embroidery hemstitching	112	82
Textile felting	28	127+
Textile fiber working	19	204+
Textile knitted fabric manipulation	66	147+
Textile knitting	66	125R+
Textile strand	226	
Tool driving or impacting device	173	141+
Turning lathe carriage	82	21R+
Web or strand	226	
Wire	226	
Wood shaving fixed knife	144	155+
Wood turning	142	
Woodworking	144	
Woodworking match making	144	50+
Manure spreader load	239	650+
Material handling elevator	414	592+
Mechanical gun magazine	124	45+
Member driving machines magazine	227	82+
Motion picture strip intermittent	352	166+
Music instrument automatic selectors	84	115+
Nail driving machine	227	107
Ordnance cartridge	89	33.1+
Paper tube making axial	493	302
Paper web for recorder	346	136
Paper web forming regulator for	162	259
Pavement marking material	404	94+
Pencil with	401	55
Pin feeders	226	52+
Power driven conveyor and chute	198	523+
Printing inkers	101	335+
Multicolor	101	202+
Railway car stops releasable	104	252+
Raking and loading machine	56	344+
Rivet setting magazine	227	51+
Screw conveyor opposed direction	198	657+
Screw conveyor plural direction	198	608
Separating with	209	
Gas suspension	209	146+
Liquid suspension gravity	209	172+
Paramagnetic	209	231+
Sheet or strip material		
Delivering or	271	
Electrolytic continuous electrode	204	206+
Label pasting machine	156	569+
Magnetic recording, moving head	360	81+
Magnetic recording, stationary head	360	88
Measuring combined	33	127+
Mechanical sound reproducing	369	258+
Endless	369	259
Motion picture	352	166+
Nail and peg driving implement	227	96
Platen combined	281	6+
Shoe nailing and stapling machines	227	
Tack driving implement	227	136
Telegraphic recorder	178	42
Telegraphophones photographic	369	125
Tobacco treatment	131	290+
Typewriter paper	400	578+
Typewriter ribbon	400	223+
U strip staple forming and setting machine	227	82+
Shockers automatic	56	402+
Sifting	209	243+
Discharging and	209	240+
Horizontal drum	209	284+
Sorting automatic with	209	509+
Strand	226	
Stratifiers with	209	497+
Discharging and	209	488+
Suctuon operated feed valve	261	DIG. 49
Threshing band cutter and	130	1
Tool driving or impacting device	173	141+
Track sanders	291	25+
Type case	276	45
Valve		
Acetylene generator gas & water	48	58
Dispenser transfixing	222	510
Dispenser with	222	544+
Dispenser with retarded	222	477
Web	226	
Feeler Mechanism		
Lock condition indicating	70	438
Looms		
Multicolor	139	233
Pattern sensing	139	331
Replenishing	139	269+
Thread cutter combined	139	264
Record sensing	235	475

	Class	Subclass
Type distributors	276	23+
Setting combined	276	5
Feet		
Artificial	623	53+
Furniture	248	188.8
Ladder	182	108+
Surface compensating	182	200+
Stand	248	188.1+
Table		
Leg	248	188.1+
Pedestal supported	108	150
Pedestal supported extension	108	88
Wheel rim carried	305	4+
Fellies		
Expanding tire tightening	157	9+
Land wheel	301	95+
Making	144	
Felling Trees	144	34R
Boots and shoes	36	9R
Paper making	162	
Cleaners for porous material	162	274+
Pickers instrument tuning	84	460
Felt		
Fabric	428	280+
Laminated	428	297+
Paper making		
Endless drying	34	243R
Felting	28	116+
Chemical treatment precedent	8	112
Glass or mineral wool	65	4.1+
Apparatus	65	9
Hat making		
Design	D15	135+
Piano hammer	144	29
Fence	256	
Barbs		
Attaching to wire	140	58+
Making	29	7.1
Top guard	256	11
Compound repair tools	7	117+
Design	D25	38+
Electric	256	10
Insulator	174	137R+
Hareware for	D 8	
Jet engine blast deflector	244	114B
Lighting	362	152
Lock	70	321+
Molds	425	84+
Noise baffle between airport and highway	181	210
Post design	D25	126+
Static mold	249	143+
Wire making	140	3R+
Wrapper for wire fencing	206	389+
Fencing Apparatus	272	98
Fender (See Mudguard)		
Automobile light combined	362	83
Boat	D12	168
Bridge pier combined	405	76
Carpet sweepers	15	45
Covers		
Automobile	280	762+
Design, fireplace	D23	403+
Horse rake	56	399
Land vehicle including railway car	293	
Marine structure	405	212
Plant protecting	172	509+
Railway car combined with draft	213	
Roof	52	24+
Scuff plate	280	153R
Ship	114	219+
Stove	126	201+
Vehicle	293	
Design	D12	186
Magnetic fender securing device	293	DIG. 6
Pusher attachment for tractors	293	DIG. 1
Ferment		
Medicine containing	424	
Fermenting	435	
Animal foods	426	53
Apparatus	435	287+
Beverages	426	7+
Apparatus	99	275+
Fertilizers	71	8+
Tobacco treatment	131	308
Ferments Acclimatization	435	245
Ferris Wheel	272	28R+
Ferrocenes	556	143+
Ferrocyanogen Compounds	423	364+
Ferroelectric		
Battery	136	254

	Class	Subclass
Information storage and retrieval		
Interconnection	365	65
Radiant energy	365	117
Photoconductivity and	365	109
System with particular element	365	145
Ferrotype Plates	354	351
Ferrous Alloys		
Composition	420	8+
Electrolysis		
Aqueous bath coating	204	43.1+
Aqueous bath synthesis	204	123+
Fused bath synthesis	204	71
Treatment	204	140+
Electrothermic processes	75	10.1+
Heat treatment	148	134+
Special composition with	148	14
Working combined	148	12R+
Pyrometallurgy	75	45+
Ferrule	16	108+
Assembly and dissassembly	29	282
Hand tools	29	280+
Percussive or explosive operator	29	255
Design	D 4	199
Digest, brushes & cleaning tools	15	DIG. 4
Making		
Metalworking methods	29	428+
Wireworking	140	76
Ring ferrules	16	109
Ferry		
Boats	114	70
Slips	114	231
Suspended car or load carrier	104	90
Fertile Nuclear Conversion Material (See Nuclear Reactions, Reactors)	376	172
Fertility, Pregnancy	436	510*
Fertility tests	435	806*
Fertilizers	71	
Adjuvant	71	DIG. 1
Apparatus	422	193+
Chelating agent	71	DIG. 2
Defluorination	71	DIG. 3
Distributing machines	222	
Application to plant	47	48.5
Scattering unloaders	239	650+
Distributing machines below ground	111	
In irrigating systems	47	1R
Festoon	428	10
Collapsible	428	9
Paper, making	493	957*
Festooning (See Loopers)		
Fetters		
Animal husbandry	119	127
Butchering	17	44+
Design	D30	151+
Portable locks	70	15+
Fez	D 2	256+
Fiber (See Filament)		
Bicomponent	428	373
Blend	162	181.1
Blend	162	141
Bonding apparatus	156	433
Bonding methods	156	166+
Bristles	15	159A+
Brushes design	D 4	
Chemical modification of	8	115.51
Cleaning or laundering	8	137+
Coating	427	212
Coating with	427	180+
Composite	428	373+
Crimped	428	370
Composite metal	428	607
Composition (See type)		
Crimped	428	369+
Staple length	428	362
Dyeing & bleaching	8	
Fiber optics	358	901+
Fluid treatment of	8	147+
Glass compositions	501	35+
Glass manufacturing	65	1+
High modulus	428	902*
Insulated electric conductor	174	124R
Plural layers	174	122R
Plural layers impregnated	174	121R
Isocyanate & carbonate fiber modification	8	DIG. 11
Layers needled	428	300
Liberation & paper making	162	
Making		
Powder metallurgy	419	4
Micro fiber	428	903*
Optics	350	96.1+

U.S. PATENT CLASSIFICATIONS

	Class	Subclass
Sign	40	547
Polyvinyl halide esters or alcohol fiber modification	8	DIG. 1
Preparation (See textile prep)	19	
Refractory compositions	501	95
Retting		
Chemical	162	1^+
Fermentation	435	279^+
Rods	350	96.1^+
Fiber optics	358	901*
Photocopy device use	355	1
Spinning mechanical	57	315^+
Staple length	428	359^+
Strand structure	57	200^+
Swelling & stretching	8	DIG. 3
Testing	73	159^+
Textile (See type)		
Coated or impregnated	428	245^+
Treatment		
Bleaching	8	101^+
Carroting	8	112
Chemical modification	8	115.51^+
Cleaning or laundering	8	137^+
Coating	427	412
Coating, general processes	427	
Comminution	241	
Dyeing	8	
Electrolytic treatment	204	132^+
Fluid apparatus	68	
Fluid processes manipulative	8	147^+
Liberation chemical	162	1^+
Liberation physical	19	1^+
Swelling or plasticizing	8	130.1^+
With structure or coating	428	364^+
Fiber Board Making	156	62.2^+
Fiber Board Making	264	109^+
Fiber Optic Lamp	D26	27
Fiber-reactive Azo Dyes	534	617^+
Fiberglass		
Coating	427	389.8
Fishing rods	43	18.5
Metallic receptacles digest	220	DIG. 23
Fibrillation		
Cellular materials	264	DIG. 8
Diagnostic testing for	128	705
Textiles fibrillated	57	907*
Fibroin	106	161
Fids	114	221R
Field		
Artillery	89	
Mounts	89	40.1^+
Glasses	350	545^+
Design	D16	133^+
Holders	350	547^+
Optics	350	481
Magnet for electric machine	310	180^+
Field Sport Devices	272	100^+
Fifth Wheel		
Land vehicle		
Articulated	280	433^+
Running gear swinging axle	280	$80R^+$
Figure		
Decorative with illumination	362	808*
Toys	446	268^+
Air actuated	446	199^+
Aquatic	446	156^+
Design	D21	148^+
Knockdown	446	97^+
Spinning or rotating	446	236^+
Toy money box with coin deposit figure release	446	10^+
Velocipede type	280	1.13^+
Wheeled	446	269^+
Filament (See Strand; Wire)		
Bonding apparatus	156	441
Bonding methods	156	60^+
Filament formation	156	167
Filaments only	156	180^+
Filaments winding	156	169^+
Stressed filaments	156	161
Cereal	426	560
Container for non-spooled	206	388
Container or wrapper for coiled	206	389^+
Cutting filament-to-staple fiber	83	913*
Distributing explosive device	102	504
Electric lamp	313	341^+
Compositions	252	500^+
Mounting methods	445	29^+
Electronic tube	313	341^+
Replaceable filament	313	237
In electronic tube	315	46^+

	Class	Subclass
In electronic tube	315	49
Laminating	156	166
Lantern projector combined	362	211^+
Making		
Electric conductor	156	47^+
Electroforming	204	12^+
Electrolytic apparatus	204	206^+
Electrophoretic	204	180.4
Glass apparatus	65	1^+
Glass processes	65	2^+
Liquid comminuting and solidifying apparatus	425	6^+
Liquid comminuting and solidifying processes	264	5^+
Metal casting	164	423^+
Metal drawing	72	274^+
Metal working plastic	72	253.1^+
Molding apparatus	425	66
Molding apparatus	425	67
Molding apparatus	425	76
Molding apparatus	425	382.2
Molding processes	264	165^+
Powder metallurgy	419	4^+
Spinning twisting or twining	57	
Textile misc processes	28	299
Metal	428	606
Coating of	427	216^+
Compound	428	607
Multiple filament lamp and electronic tube	315	64^+
Spun twisted or twined	57	200^+
Design yarn cord	D 5	7^+
Stock material	428	364^+
Supply circuit for electronic tube	315	94^+
Testing	73	160
Tires combined	152	168
Transformer	336	
With anode supply winding	336	170^+
With anode supply	315	94^+
With grid bias	328	226
With grid bias power packs	307	150
File or Filing		
Binder device releasably engaging aperture or notch of sheet	402	
Cabinet, design	D 6	432^+
Cabinet with material retainer	312	183^+
Card spacer	283	36^+
Dental	433	141^+
Directory	40	371^+
Folder	493	947*
Handles	29	80
Hoof trimming	168	48R
Leather	69	1
Implement	69	20
Manicuring	132	76.4^+
Combined in compound tool	132	75.6
Metal and metal working files	29	$76R^+$
Cleaning and resharpening by electrolysis	204	141.5
Cleaning and resharpening by sand blast	51	414
Making	76	12^+
Processes of abrading	51	313^+
Pencil sharpening (See pencil sharpener)	29	78
Rasps and	29	78^+
Receptacle file partition	220	22^+
Riffle file	40	372
Rotary or flip card	D19	75
Sharpening tools by	76	82^+
Saw sharpening	76	31^+
Sorting type desk	211	11
Fillers		
Barometric control of dispenser	222	479
Bars for beds	5	283
Cables conduits with fluid or vacuum	174	8^+
Cables conduits with insulation	174	$68R^+$
Composition	106	$288R^+$
Shoe	106	38
Curtain shade or screen seals	160	40^+
Dispenser and receiver combined	141	
Portable receptacle	141	173^+
Dispenser with refilling of supply	141	18^+
Ditch	37	142.5
Floor mats pivoted link	15	240
Fountain pen combined	401	118^+
Force means in filler	141	20.5
Gas generator	48	74
Insulating heat or sound	252	62
Mattress	5	448^+
Foam materials	5	481

	Class	Subclass
Pastry	99	
Pastry	425	230^+
Printing member	101	402^+
Rail joint containing	238	163
Safe wall structure combined	109	84
Sheet metal cup die shaping	72	54^+
Sound mufflers having	181	282
Synthetic resin or natural rubber (See class 523, 524)		
Wheel web combined	295	23
Filling		
Abrasive bag making	51	294
Ammunition cartridge packing	86	47^+
Ammunition loading	86	23^+
Attachments for metallic receptacles	220	86R
Automatic weigher	177	116^+
Removable weigh chamber	252	62^+
Bags	141	313^+
Bail stopper type	53	265
Barometric control of dispensing	222	479
Battery grid processes	141	
Bottles	141	
Brush making machine	300	9
Capsules	141	237^+
Cigarette making by tube	131	70^+
Confectionery machines	99	
Confectionery machines	425	130^+
Container refilling by dispensing means	141	21^+
Dental	433	226^+
Carrier	433	163
Compositions	106	35
Dispenser and receiver combined	141	
Distillation retorts	201	40
Golf ball making	156	146
Match boxes	53	236
Match making combined with	144	61
Dipping frames	144	63^+
Milk receptacle with funnel	232	43
Paper receptacle closing by plural steps with	53	$266R^+$
Paper receptacle filling &&closing	53	467^+
Plastic molding blank covering &	425	110^+
Portable receptacles	141	
Cooking combined	99	356
Shoe filling appliers	12	18.1^+
Processes	12	148
Storage batteries	429	72^+
Subterranean cavity with cement	405	267
Weaving (See weft)	139	
Film		
Accumulating dispensers	222	423
Discharge assistant type	222	403
Removable from supply container	141	112^+
Badges		
Radioactivity	250	472.1
Burster sounding toy	446	181^+
Processes	159	49
Cassette	D14	11
Cine film	430	934*
Container or wrapper for	206	389^+
Distillation separatory	203	72
Distillation separatory	203	89
Mineral oil	208	360
Editor viewer	352	129
Electrode	361	433^+
Electroforming of	204	12^+
Electroplating of	204	27^+
Nonconducting	204	22
Flotation separation	209	207
Forming apparatus	425	223
Forming apparatus	425	224
Forming process	264	165^+
Heaters and vaporizers	122	39
Holders	354	276^+
Motion picture	352	232^+
Rewinder	242	179
Storage cans	352	78R
Winding and reeling	242	179^+
Photographic	430	495
Acetate recovery from	536	78
Badges	250	475.2
Cabinet	312	20
Cartridge or pack	430	471
With structure	354	275
Cleaner	15	100
Coating apparatus	118	
Coating processes	430	935*
Color	430	541^+
Container	206	316
Developing machines	354	297

Entry	Class	Subclass
— Digest, drying & gas or vapor contact with solids	34	
— Fluorescent screen	250	475.2
— Gold & silver from photographic materials	75	118P
— Guides developing apparatus	354	339
— Holders	354	276+
— Nitrate recovery from	536	40
— Processing machines	354	297+
— Processing, manipulative	354	297+
— Radioactive sensitive	430	942*
Projector	353	
— Motion picture	352	232+
Reel	D14	10
Reflection reducing on optical surfaces	350	164+
— Motion picture	352	233
Splicer	350	164+
— Design	D16	41+
— Winder	242	179
— X-art	83	926J*
Spool	242	71+
Stills separatory	202	236
Stripping sensitized film	430	264*
Type of concentrator and evaporator	159	5+
Vaporizer for mineral oil	196	128
Winding apparatus, motion picture	242	179+
X ray apparatus using	378	
Filter (See Filtering)		
Aid	210	500.1+
— Applying means	210	193
— Applying process	210	777+
Choke input	333	181+
Cigarette or cigar	131	331+
Color filter	350	311+
— Photographic chemically defined	430	7+
Container or wrapper for	493	941*
Crystal	333	187+
Decoupling	333	181+
Diaphragm type cell having	204	264
Dispenser combined	222	189
Electric current frequency	333	167+
Electrical		
— Balancing in radio receiver	455	339
— Blocking type	333	167+
— Digital	364	724
— Frequency modulation for receiver	455	213
— Noise control	455	307
— — Antenna coupling	455	286
Electrolyte feeding combined	204	276+
Electrolyte treatment	204	240
— Electrolyte recirculation	204	238
— Leacher dissolver or extractor	204	235
Furnace door combined	110	179
Gas	55	
— Frozen constituent removal	62	15
— Refrigeration combined	62	317
Glare screen on vehicle body	296	97R
Heat exchanger combined	165	119
Heat ray	350	311+
Heating boiler combined	122	431
Internal battery cell having	204	249
Light ray	350	311+
— Color photographic elements	430	495+
— Color photography	430	357+
— Hand lamp	362	166+
— Infrared	350	1.1+
— Lenticular	354	101
— Projection dimmer	362	257+
— Signal lamp	362	166+
— Ultraviolet	350	1.1+
— X ray	378	156
Liquid		
— Congealing liquid process	62	67
— Design treatment	D23	209
— Frozen constituent removal	62	123+
— Purification or separation	210	500.1+
— Purifying	210	348+
— Refrigeration combined	62	318+
— Sediment testing	73	61R
Oil, fuel entering nozzles or float chamber	261	DIG. 4
Optical	350	311+
— Compositions	252	582+
— Making coating	427	162+
Press	210	224+
Ray generation combined	250	503.1
Refrigerant	62	474+
— Automatic	62	195
— Process	62	85
Smoking device	131	202+
Smoothing	333	181+
Sound	181	175+
— Using electronic tubes	328	138
Still vapor treating	202	200
Tuning fork	333	186+
Filtering		
Electrolytic water or sewage treating combined	204	152
Gas purifying	55	
— Expansible chamber device with	92	78
Liquid purifying	210	
— Expansible chamber device with	92	78
— Mode	333	248+
— Polarization	333	248+
Oil purification	208	177+
— Dehydration	208	187
— Vapor treatment	208	347+
Radiant energy in space	343	909+
Sugar treatment	127	55
Treating drying or solid contacting gas or vapor by	34	82
Finance, Computer Applications	364	408
Finder		
Camera view	354	219+
Distance and vertical angle	33	284
Electric trolley	191	72+
Exposure lens	354	152+
Optical acessories for photography	354	219+
— Ground glass	354	152+
Switches for telephone systems	178	16+
Finger		
Article manipulation by finger like movement	414	1+
Attachable coating implement with material supply	401	6+
Attachable or conforming penholder	15	443
Attachments handling implements	294	25
Automatic musical instrument selector	84	117
Blotter attachment	34	95.2
Boards for stringed instruments	84	314R
Cots or protectors	2	21
Cutting tool, finger supported	30	298
— Blade guard or guide	30	291
— Shears	30	232
Exercisers	272	67+
— Kinesitherapy	128	26
— Music	84	465+
Glove	2	163
Guard for door	49	383
Harvester cutter guard	56	310+
— Bars and	56	307+
— Sickles and	56	298
— Supplemental lifting	56	312+
Interlocking bed corners	5	295
Letter box guard	232	54
Print		
— Fingerprinting outfits	118	31.5
— Identification type printed matter	283	69
— Methods	427	1
— Printed matter alteration preventing	283	69
Release for control lever pawl	74	537+
Ring		
— Blotter supporting	34	95.2
— Jewelry	63	15+
— Jewelry design	D11	26+
— Making and sizing	29	8
— Metal founding molds	249	57
— Penholder supporting	15	443+
— Supports for tobacco users	131	258
Stringed instrument fingering devices	84	315+
Stringed instrument picks	84	322
Supported cutting tool	30	298
Typewriter paper feeder	400	645+
— With feed roller	400	638+
Finger Cot	D 4	103
Fingernail		
Brush	D28	57
Polish	424	61
— Removers	252	162+
Fingerprinting	427	1
Finial (See Product)		
Drapery rod	D 8	378+
Pendant or finial type article	428	28
Finings	426	654
Apparatus	99	277.1+
Finishing		
Apparatus loom mounted	139	291R+
Block concrete or ceramic	264	
Bolt heads	10	19+
Fabric	26	
— Seams stitching combined	112	217
Nails horseshoe	10	38+
— Cutting combined	10	37
Paper	162	204+
Pavement	404	83+
Stereotype plate	29	21
— Planing backs	409	309+
Straps leather	69	17
Thread	28	217+
Type casting combined	199	81+
Fins		
Heat exchanger	165	185
Nuclear fuel having	376	435
Stabilizing ships	114	126
Steering ships	114	152
Submarine diving	114	331+
Vertical on aircraft	244	91
Fire (See Burner; Fireproof)		
Alarms		
— Electric detectors	340	577+
— Electric system automatic	340	577+
— Electric systems	340	287+
— Smoke bells	362	379
— Thermal control	116	101+
— Thermal electric switch	337	298
— Thermometer systems	374	141+
— — Electrical	374	166+
— With indicator	116	5
Brick	75	95
— Design	D23	422
Cigarette lighters	D27	36+
Detector (See alarms)		
Doors furnace	110	173R+
Electric starter device	D 7	416+
Engine	169	24
— Boiler feed heater	122	418+
— Heating system combined	237	12.2
— Hose and ladder type	280	4
— Ladder or escape	182	63+
Escape	182	
— Rope and brake	188	65.1+
Extinguisher	169	
— Boiler fires	122	504+
— Compositions	252	2+
— — Fire retarding	169	45
— Design	D29	2+
— Gas agent	169	11+
— Toy vehicle	446	432
Extinguishing	169	
— Aerial fire apparatus	244	136
— Methods of spraying fluids	239	1+
— Nozzles automatic	169	37+
— Processes chemical	169	43+
— Shipboard	169	5
Fighting	D29	
— Hose	D23	266
— Hose nozzles	239	
— — Design	D23	213+
— Hydrant	D23	234
— Ladder	D25	62+
— Mask	D29	7+
— Vehicle design	D12	13
Kindling	44	34+
— Design solid fuel	D13	3.2
Matches design	D27	30+
Plate	126	148
Pots and linings	126	144+
— Thermostats	236	104
Preventing		
— Devices, design	D29	
— Receptacle attachments	220	88R
— Textile apparatus	68	209
Retarding composition	252	601+
— Biocidal composition with	424	
— Containing gas	252	605
— Dispersion or collodian systems	252	611
— Fiber chemically modified	8	115.51+
— For living matter	252	603
— For synthetic polymer	252	609
— Intumescent	252	606
— Paper-making or fiber liberated	162	159
— Stock material mechanically interengaged strands	428	276
— — Fire or flame proof features	428	921*
— Synthetic polymer use	252	609
— Trees, grass use	252	603
— Woven material use	252	608
Screen	126	202
Starting devices	D 7	416+
Stop	52	317

Entry	Class	Subclass
Telegraphs	340	287+
Tong	294	11
Design	D 8	52+
Tube	122	
Fire-polish Glass	65	
Firearm (See Gun)	42	
Automatic	89	125+
Barrels	42	76.1
Bayonets	42	86
Breech loading	42	2+
Cane guns	42	52
Converted to spring gun	124	28
Design	D22	100+
Firing mechanism	42	69.1+
Imitation, light flashing	362	112
Light combined	362	110+
Magazine chargers	42	87+
Making	29	1.1
Muzzle loaders	42	51
Pistol swords	42	53
Plural triggers	42	42.2
Revolvers	42	59+
Sights	33	233+
Silencers or mufflers	181	223+
Simulating		
Dispenser	222	79
Light	362	157+
Stocks	42	71.1+
Toy	42	54+
Design	D21	145+
Fireback	126	130+
Asbestos	126	128+
Firebox (See Furnace)		
Feeding steam air or water	110	297+
Feeding water to	110	297+
Fire tube	122	44R+
Flue	122	135R+
Fuel disperser installed in	431	159
Liquid heater or boiler combined	122	
Locomotive		
Structure combined	105	37+
Type combustion features	110	
Type water heating	122	
Smoke and gas return to	110	203+
Solid fuel	110	317+
Straw and wet fuel burning	110	196
Water		
Tube	122	235MF+
Wall	122	294
Firebrick	75	95
Carbide containing	501	87+
Design	D23	422
Refractory	501	94+
Firecracker	102	361
Firedamp		
Analyzing and testing apparatus	422	50+
Indicators	73	23+
Firedogs	126	298
Design	D23	407
Firefighting (See Fire)		
Fireless Cookers		
Heat accumulators	126	400
Ovens	126	273.5
Vessels	126	375
Fireplace	126	120
Arches	52	219
Dampers	126	288
Design	D23	343+
Fenders	126	201+
Design	D23	403+
Grates	126	164+
Design	D23	398+
Heating systems having	237	51
Logs		
Electric	272	8R
Gas	431	125+
Gas design	D23	409
Mantel design	D23	404
Screen	126	202
Digest	16	DIG. 11
Shelves	126	336
Fireproof (See Fire, Retarding; Insulating, Heat)		
Miscellaneous compositions	252	601+
Plastic or coating compositions	106	15.5+
Reactive building component	52	232
Stock material	428	920*
Synthetic resin or natural rubber (See synthetic resin or natural rubber)		
Cellular product	521	50+
Fireworks	102	335+
Loading	86	20.1+
Firing		
Clayware processes	264	56+
Devices or mechanisms		
Ammunition & explosive devices	102	200+
Automatic firearms and guns	89	132+
Blasting	102	200+
Drop bomb	102	396
Firearm nonautomatic	42	69.1+
Gun cartridge	102	470+
Mechanical guns	124	31+
Mine	102	424+
Mines marine	102	416+
Ordnance nonautomatic	89	27.11+
Electric	89	28.5+
First Aid		
Cabinets	312	209
Kits	206	570+
X-art collection	206	803*
Fischer-tropsch Crude Processing	208	950*
Fischer-tropsch Synthesis	518	700+
Fish (See Fishing)		
Butchering apparatus	17	53+
Fish bowl design	D30	101+
Hooks	43	43.16
Knife for kitchen use	D 7	106+
Liver extraction		
Fats etc	260	412.1
Medicine containing	424	107
Vitamins	260	
Net	43	7+
Design	D22	135
Fabric and processes for making	87	12
Making apparatus	87	53
Plates	238	243
Special, bolt or nut to substructure lock	411	94
Special culture	119	3
Fisherman's Tool	D22	149+
Fishing	43	4+
Amusement fish ponds	273	140
Automatic hookers	43	15+
Bait, tackle, and catch container	43	54.1+
Bait or lure carrier	224	920*
Minnow buckets	43	56
Design	D22	136
Baskets and creels	43	55
Belt with lure holder	D 3	100+
Containers & carrying cases	D 3	38
Design	D22	134+
Equipment	D22	134+
Files, work holder for	269	907*
Fish traps	43	100+
Hooks	43	43.16
Design	D22	144
Making	29	9
Line	43	44.98
Reel	242	84.1 R
Design	D22	137+
Modifiable drive	242	211+
Rod	43	18.1+
Cover or container for disjointed	43	26
Fly tieing	242	7.19
Holder	224	922*
Method	242	7.2
Snelling	242	7.19
Stringers	224	103
Tackle and accessories	43	4+
Tool		
Compound	7	106+
Electricians	254	134.3 R
Traps	43	100+
Design	D22	121
Trotlines	43	27.4
Holder	43	57.3
Vessel	114	255+
Weighing of fish	D10	87+
Fishing Rod	D22	137+
Fishing Rod Holder	D22	147
Fission Counter Scintillations	250	361R+
Fission Sustaining Nuclear Reactors	376	
Fitting Boots and Shoes	36	8.4
Fittings		
Bath basin and receptacle	4	191+
Design	D23	271+
Electrical		
Boxes housings	174	50+
Conductor insulator combined	174	169+
Making		
Gas and water	29	157R+
Pipe	285	120+
Fluid distribution design	D23	200
Gaslight design	D26	113+
Scaffolding design	D25	66
Suspended support	248	342+
Fixatives		
Artists	106	
Perfume	512	2
Photographic	430	455+
Flag	116	173+
Badge	40	1.5
Design	D11	165+
Design, insignia or jewelry	D11	165+
Pole or post structure	52	720+
Socket	52	298
Staff	116	173+
Decorative element	52	301
Staff type support	248	511+
Flagpole	52	720
Design	D11	181
Holder	D11	182+
Socket	52	298
Flail		
Cotton harvester	56	29
Threshing machine	130	28
Flails Flexible	51	334
Flake Ice Making	62	320+
Flakes		
Coating	427	212+
Electroforming	204	10
Flame (See Fire; Fireproof)		
Arc lamp	313	231.1+
Movable electrode	314	
Arrester	48	192
Dispenser having	222	189
Receptacle having	220	88R
Respirator	128	202.24
Bonding by	228	902*
Deflector		
Deflector in furnace	431	171+
Resistance, coated	427	393.3
Retardant, textile collection	57	904*
Thrower	431	91
Flanger		
Pipe or tube	72	316+
Sheet	72	101
Flannel and Flannelette Napping	26	29R+
Flap		
Bag closures	383	84
Envelope moistening & sealing	156	441.5
Head coverings	2	172
Paper receptacles		
Closure for paperboard box	229	126+
Letter sheet closure	229	92.7
Pneumatic tire	152	514
Shoe sole channel machines		
Layers	12	29
Turners	12	30
Flare		
Burners		
Magnesium strip	431	99
Solid fuel	102	336+
Dropping	89	1.51+
Guns	42	1.15
Package for	206	803*
Pyrotechnic	102	335+
Composition for	149	
Design	D22	112
Parachute	102	337+
Torches	102	336+
Flaring Tool (See Pipe Making)		
Flash		
Camera attachment	D16	42
Electric weld	219	97+
Flash bulb mount	D16	42
Flash removal	425	806*
Spotlight	D26	24+
Trimmers	83	914*
Flash Bulb		
Combustion type	431	358+
Mount design	D16	42
Flashing	52	408+
Exterior type, eg roof	52	58+
Glass	65	60.1
Lamps		
Sign	40	541+
Sign systems	340	791+
Switch in lamp	315	72
Systems	315	
Light as signal	40	902*
Pipe and plate joint roof flashing type	285	42+

	Class	Subclass
Signals	340	331
Flashlight		
Combined with structure of another device	362	157$^+$
Canes	362	102
Pen or pencil	362	118
Electric	362	208
Bicycle operated generator	362	193
Design	D26	37$^+$
Hand generator per se	310	50
Hand operated generator	362	192
Electric generator combined	362	192$^+$
Bicycle	362	193
Flash powder and photoflash	431	357$^+$
Flask		
Adjustable	164	377
Pattern with	164	237$^+$
Plastic molding	425	175$^+$
Flat Tire Alarm		
Electric	340	58
Switch	200	61.22$^+$
Nonelectric	116	34R
Flatiron	38	74$^+$
Design	D32	68$^+$
Heating		
Automatic temperature regulators	236	7
Electric	219	245$^+$
Liquid and gaseous fuel	126	230
Self heating liquid and gaseous fuel	126	411$^+$
Stove and furnace	126	227$^+$
Stands combined	38	142
Steam	38	77.1$^+$
Support	248	117.1$^+$
Ironing table combined	38	107
Flats		
Bed bottoms	5	186R$^+$
Carding	19	113
Grinding	51	243
Stationary	19	104
Stripping	19	110$^+$
Traveling	19	102
Flattening (See Crusher; Straightening Devices)		
Glass	65	67
Sheet material rolling or wrapping combined	53	116$^+$
Textile ironing and smoothing	38	
Tobacco leaf	131	324$^+$
Stemming combined	131	315
Flatware-tableware	D 7	137$^+$
Disposable	D 7	138
Flavanthrone	546	31
Flavones	549	403
Flavors	426	534$^+$
Artificial	426	533$^+$
Flavylium, Basic Dye	8	657$^+$
Animal-derived natural fiber	8	426
Printing permanently	8	454
Flaw Detection		
Magnetic	324	238$^+$
Induced voltage-type sensor	324	240$^+$
Oscillator type	324	237
Utilizing radiant energy	250	
Flax (See Fiber)		
Retting	435	279$^+$
Thresher	130	13
Flechettes	244	3.24$^+$
Ammunition	102	703*
Fleece		
Dividers in fiber preparation	19	151
Knitting		
Fabric or articles	66	194
Needle cooperating wheels	66	94
Unknitted materials	66	191
Fletching Jig	269	38
Flexible		
Concentric electric cable	174	102R$^+$
Fluid vacuum or air	174	28$^+$
With insulation spacer	174	99R$^+$
Flail tools	51	334
Shafts	464	
Tubing	138	118$^+$
Flicker Beam Photometer	356	217
Flier		
Spinning machine	57	67$^+$
Flier elements	57	115$^+$
Flies		
Design fishing bait	D22	125$^+$
Sheet delivering	271	83
Trouser	2	234

	Class	Subclass
Flight Training Simulator	D25	1
Flint		
Composition	420	416
Firearm firing device	42	69.1$^+$
Lighter	431	273$^+$
Special packaging for	206	528$^+$
Flip Flop Oscillator	331	144
Driven type	328	193$^+$
Gas tube type	331	113R
Flippers for Swimming	D21	239
Floating Compasses	33	364
Floats	169	DIG. 1
Ballasting	114	123
Marine pipe or cable laying	405	158$^+$
Submergible floating matter barrier	405	64
Submergible marine structure	405	205
Submergible wave or flow dissipator	405	23
Buoy	441	1$^+$
Closure	220	216$^+$
Controlled		
Alarm	116	110$^+$
Alarm electric	340	623$^+$
Boiler safety device	122	505
Bottle nonrefillable	215	20
Burner liquid fuel	431	64
Closure	49	21$^+$
Dispenser fluid pressure discharge	222	62
Dispenser indicator	222	51
Dispenser material level	222	67$^+$
Flood gate	49	11
Funnel valve	141	199$^+$
Funnel valve air displacement	141	199$^+$
Gas liquid contact	261	70
Liquid feeder traps	137	165$^+$
Liquid level indicator	73	305$^+$
Liquid purifier	210	121$^+$
Liquid purifier level indicator	210	86
Pump electric	417	36$^+$
Pump jet	417	182.5
Pump pneumatic displacement	417	126$^+$
Pump starter	417	36$^+$
Pump throttle	417	36$^+$
Trap thermal	236	53
Valve	137	409$^+$
Valve steam trap	137	192$^+$
Water closet outlet	4	395$^+$
Water closet plural flusher	4	324$^+$
Water closet valved bowl	4	437$^+$
Watering trough	119	78$^+$
Decoy ducks	43	3
Fishing	43	43.1$^+$
Design	D22	146
Fluid pressure gauge	73	713$^+$
Impeller		
Endless chain	416	7
Paddle wheel	416	84
Masonry implement	15	235.4$^+$
Mortar-joint finisher	15	235.3
Oil distributor	114	234
Operated		
Acetylene generator	48	53.4
Beverage infuser	99	320
Brake	188	179
Dispenser indicator	222	51
Heater fluid	236	31
Liquid level gauge	73	305$^+$
Motor fluid	60	495$^+$
Pump tide motor	417	331$^+$
Specific gravity indicator	73	451$^+$
Switch electric	200	84R
Thermal air relief valve	236	62$^+$
Thermal regulator	236	52
Thermometer combined	374	156
Watergate or weir	405	96$^+$
Plural connected, bath	4	DIG. 1
Regulator		
Fluid pressure	137	505.19
Pump starter	417	36$^+$
Pump throttle	417	36$^+$
Skimmer glass furnace	65	206
Structure	73	322.5
Specific gravity tester	73	448$^+$
Supported liquid separator	210	242.1
Traversing hoist	212	190$^+$
Valve	137	409$^+$
Vessel raising and docking	114	44$^+$
Warship battery	114	4
Water closet design	D23	303
Water motor	60	495$^+$

	Class	Subclass
Flocculating and Deflocculating	209	5
Prevention of	210	696$^+$
With precipitation	210	702$^+$
Flocculation		
Liquid separation	210	702$^+$
Solid separation	209	5
Flocking		
Apparatus	118	308$^+$
Coating combined	427	180$^+$
Laminating combined	156	279
Floodgate	49	10$^+$
Floodlights	362	
Design	D26	61$^+$
Dirigible support	362	418$^+$
Floor or Flooring	52	
Animal stall	119	28
Bridge	14	73
Bridges for safes	109	87
Building		
Friction or wear surface	52	177
Joists to wall	52	289
Masonry	53	391$^+$
Mat	52	660$^+$
Nailing beam	52	364$^+$
Parquet	52	390$^+$
Rotatable	52	65
Spacing sleeper	52	480
Static mold for making	249	18$^+$
Supporting wall	52	264
Vertically adjustable	52	126.1$^+$
Yielding sub-structure	52	403
Carpet	428	85$^+$
Design	D 6	582$^+$
Cleaners	D32	15$^+$
Cleaning		
Compositions	252	88
Machinery design	D32	15$^+$
Machines and devices	15	
Disintegrating machine	299	36$^+$
Drying apparatus	34	237$^+$
Endless excavating vehicle	37	7
Heat exchanger	165	56
Heated vehicle	237	43
Jack	254	11$^+$
Lamp	D26	93$^+$
Lighting	362	153
Mat, portable	15	215$^+$
Mop, sponge, non-bristle broom	D32	40$^+$
Outlet or junction box	220	3.3$^+$
Packaging for coiled floor covering	206	389$^+$
Parquet	428	50
Pavement	404	17$^+$
Polishing		
Compositions	106	3$^+$
Railway car	105	422
Auxiliary	105	375
Registers ventilation	98	102$^+$
Scrubbing or brushing machine	D32	15$^+$
Sectional	428	44$^+$
Ship	114	76
Slatted covering	52	660$^+$
Surfacing machines	51	174$^+$
Vehicle floor mat	D12	203
Flora	428	17$^+$
Framework for	211	
Floral		
Holders	47	41R$^+$
Supports	248	27.8
Design, apparel attached	D11	1$^+$
Floss, Dental		
Holder	D28	64
Toilet kit with	132	79E
Flotation		
Beneficiating ores	75	2
Liquid purifying	210	703$^+$
Gas-liquid surface contactor	210	150
Liquid surface outlet	210	221.1
Physical separation agents	252	60$^+$
Separating solids	209	162$^+$
Sugar manufacturing	127	57
Flouncing Lace Goods	87	10
Flour	426	622$^+$
Bleaching	426	253$^+$
Manufacture by comminution	241	6$^+$
Testing	73	169
Flouring or Flour Dusting	425	101
Flow (See Proportional, Feeding)		
Control (See valve, actuation)		
Dryer	34	54
Electrical purification of liquids	204	306
Electrolytic apparatus	204	229$^+$

U.S. PATENT CLASSIFICATIONS

	Class	Subclass
Gas filtration particulate material	55	474
Gas filtration reticulated		
Gas liquid contact porous sheet	261	100+
Gas liquid contact porous mass	261	94+
Gas liquid contact wet baffle	261	108+
Gas separator combined	55	
Heat exchange	165	32+
Pump reversible	417	315
Radiator heat exchange	165	96+
Counter nuclear		
Determine characteristic of fluid	250	336.1
Through counter tube	250	374+
Indicators		
Dispenser combined	222	40
Electric	340	606+
Fluid	116	273+
Refrigerant	62	125
Lines making, metal working	29	DIG. 17
Meters		
Electric current	324	76R+
Fluid	73	861+
Fluid direction combined	73	189
Proportional	137	98+
Recorders	346	72
Regulator	138	37+
Earth boring tool combined	175	232+
General utility type	138	37+
Impact activated valves	251	76
Responsive to element deformation	137	67
Sprinkling and spraying	239	533.1+
Valve combined	251	118+
Responsive		
Alarm	116	112
Alarm electric	340	606+
Dispenser cutoff	222	59+
Dispenser preset cutoff	222	14
Heater closed fluid	236	25R
Pump throttle valve	417	43
Switch electric	200	81.9 R
Valve	137	455+
Retarder		
Brake fluid pressure	303	84R
Fluid in conduit	138	37+
Railway switch indication	246	109
Flower	PLT	
Curbing for flower bed	D25	38+
Holder	47	41R+
Apparel attached	D11	1+
Corsage boutonniere personal wear	24	5+
For plant shrub vine etc	47	44+
Furniture type	D 6	403+
Vases	D11	143+
Wall mounted	D 6	556+
Wreaths and sprays	248	27.8
Imitation	428	17+
Plants	PLT	
Pots	47	66+
Design	D11	143+
Support, depended	D 6	513+
Support, shelf	D 6	556+
Preserving		
Coating or impregnating	427	4
Plant catalyst or stimulant	71	68
Flowmeter for Heat	374	29+
Floxacillin	540	327
Flue		
Boiler	122	
Building attached	52	219
Cleaner		
Brush and broom	15	162+
Brush and broom design	D 4	
Scraper	15	242+
Collar adjustable	126	315
Connection making	29	157.5
Cover		
Design		
Expanders (See expander)		
Gas heating	432	219+
Masonry	52	245+
Mineral oil still	196	116+
Regenerative	431	170
Checker bricks	165	9.1+
Stopper	126	319
Thermolytic retorts having	202	124+
Horizontal	202	138+
Sprinkling and spraying	239	533.1+
Fluent Material Handling	141	
Fluent Tool Metal Shaping	72	54+
Fluid (See Gas; Liquid)		
Amplifier	137	803+
Bearing	384	100+
Breakwater	405	22
Calculator	235	200R+
Cleansing tool drive combined	173	57+
Clutch	192	58R
Collection from body	604	317+
Conveyer (See conveyer)		
Coupling	60	325+
Gearing combined	74	730+
Shaft connection	464	24
Delivery		
Check controlled device	194	
Sprinkling, spraying, diffusing	239	
Track sander	291	3+
Distribution	137	
Conduit and electrical conductor combined	174	47
Cooling and or insulating electrical conductors	174	8+
Expansible chamber motor	91	
Gas separator combined	55	
Outlet nozzles	239	
Reaction motor discharge	239	265.11+
Reaction motor discharge cooling	239	127.1+
Sound muffler	181	212
Drive	60	325+
Earth boring with	175	65+
Electrifying	361	225
Floating matter barrier	405	62
Flow direction measurement	73	188+
Flow indicator	116	273+
Fuel		
Efficiency test	73	113+
Gas analysis	73	23+
Illuminating testing	73	36
Liquid	44	50+
Liquid analysis	73	53+
Fuel burner		
Combustion feature	431	
Dispersing feature	239	
Fuel for heating		
Material or bodies	432	
Structures	237	
Fuel in nuclear reactors	376	151
Fuel in nuclear reactors	376	172
Fuel in nuclear reactors	376	356
Fuel in nuclear reactors	376	356
Governor	73	521+
Elevator	187	68+
Handling equipment	137	
Mounted on vehicular support	137	899+
Systems	137	561R+
Branched passage flow control	137	861+
Heat exchange	165	
Packaged	D 9	
Heater (See heat heater and heating liquid)		
Electric	219	280+
Infrared treatment of	250	432R
Jet texturizer	28	271+
Level gauges	73	290R+
Mattress	5	449+
Measuring vessels	73	426+
Mixture sensing	236	12.1+
Motor (See propeller)		
Brake	303	
Expansible chamber	91	
Manual displacement energized	49	32
Manufacture	29	156.4 R+
Meter	73	861+
Power plant	60	
Rotary kinetic	415	
Dental hardpiece	433	132
Turbine	415	
Operated		
Aircraft control	244	78
Alarm	116	109+
Apparel or garment turning	223	43
Belt and pulley control	474	
Body vibrators or massagers	128	24R+
Boiler fire extinguisher	122	506
Brake	188	151R+
Camera shutter	354	257
Changeable exhibitor	40	477+
Chuck	279	4
Closure by ambient fluid	49	21+
Clutch operator	192	82R+
Depth gauge	73	299+
Die shaping	72	54+
Dumping vehicles	298	22R
Earthenware apparatus	425	
Explosive arming devices	102	222
Fluid scatterer or sprayer	239	214+
Governor	73	521+
Governor modifier	73	521+
Hat shaping	223	13
Impacting device	173	90+
Indicator	116	264+
Linear hoist	212	138
Lubricators	184	14+
Machine elements	74	
Milling machine work feed	409	170
Mortar mixers	366	10+
Process	366	3+
Motors temperature controlled	236	79+
Overhead hoist	212	213
Photographic printing	355	91+
Pipe and tube cleaner	15	104.3
Pipe cleaners	131	244
Piston packing	277	
Pneumatic tire alarm	116	34R
Power hammer and presses	72	407
Press	100	269R+
Pressure gauge	73	700+
Puller or pusher assembler or disassembler	29	252
Pushing and pulling implements	254	93R
Railway car stop	104	256
Railway switches and signals	246	
Railway turntable	104	37
Rate of climb meter	73	179+
Reciprocating cutter	83	639
Rod or piston packing	277	
Rotary hoist	212	250
Sausage stuffer	17	39
Sawmill set works	83	726
Seal with bearing	384	131
Ship steering	114	150+
Ships log	73	181+
Sound record molding device	425	385
Sound record molding device	425	406+
Sprayer	239	237+
Steering gear	180	79+
Switch	200	81R+
Syringe	604	131+
Testing device	73	37+
Thermometer	374	201+
Thermostat	60	530+
Tool turret	29	42
Toy	446	176+
Track sander dispenser	291	24
Oscillator	239	589.1
Trolley retriever	191	85+
Trolley stand	191	67
Valve responsive to flow	137	455+
Valve responsive to pressure	362	384
Vehicle head lamp	362	37+
Well tubing cutter	166	55+
Wire placing apparatus	254	134.4
Work holder for grinder	51	233
Pillow	5	441
Pressure gauges	73	700+
Pressure regulator (See pressure)	137	505+
Proportioning	236	12.1+
Pumps	417	
Ratio maintenance	236	12.1+
Reaction surfaces, ie impellers	416	
Regulator and control		
Volume or rate of flow meters combined	73	199
Samplers	73	863+
Screen	110	179
Shark screen	405	22
Sprinkling, spraying, diffusing	239	
Storage in earth cavity	405	53+
Transmission (See transmission)		
Treatment		
Bleaching and dyeing	8	
Coating implement with material supply	401	
Drying and gas contact with solid	34	
Food apparatus	99	516+
Food, butchering process	17	51
Gas and liquid contact	261	
Gas separation	55	
Metal	148	
Shoe soles	12	41.3
Textile fabric apparatus	68	
Textile fibers	19	66R+
Thread finishing with	28	217+
Ultraviolet treatment of	250	432R
Vortex amplifier	137	812+

	Class	Subclass
Vortex generator	137	808+
Fluke		
Anchor	114	301+
Anchor trippers	114	210
Flume	405	119+
Gates	405	87+
Screen	210	154+
Comminuting	210	173+
Fluoranthene (See Aromatic Hydrocarbon)	585	400+
Fluorene (See Aromatic Hydrocarbon)	585	400+
Fluorenone	568	326
Fluorescein	549	223
Fluorescent		
Cathode ray screen	313	467
Cathode ray tube	313	467+
Coating	427	157
Composition		
Inorganic	252	301.6 R
Mineral oil	208	12
Organic	252	301.16
Radioactive	252	625+
Devices and applications	250	462.1
Electric light	313	484+
Ballast	336	155+
System	315	
Electric lights and electronic tubes	313	495+
Illuminating devices with	362	217+
Screens	250	483.1
Signs	40	542+
Fluorides Inorganic	423	462+
Binary	423	489
Fluorine	423	500
Acids	423	481
Halocarbon containing	570	123+
Organic compound treatment	260	694
Fluorocarbons, Eg, Teflon, Kel-f		
And memory plastics, pipe joints	285	909
Insulated cable and conductors	174	110FC
Plastic article forming, resin	264	127
Fluoroscope and Fluorescent	250	458.1
Article inspection	250	458.1
Fluorescent applications	250	458.1
Fluoroscopic equipment	D24	2
Tables	378	38
Fluosilicates	423	342+
Fluosilicic Acid	423	341
Flush Tank		
Liquid level control	137	409+
Water closet	4	353+
Bowl combined	4	300+
Urinal	4	353+
Flusher		
Nozzle cleaner	239	106+
Sewerage	137	240
Siphon including automatic	137	132+
Street	239	146+
Valve	251	15+
Water closet	4	300+
Design	D23	295+
Fluspirilene, in Drug	514	278
Flute	84	384
Fluter	223	28+
Flux		
Bituminous material	106	279+
Ceramic	106	313
Earth boring combined	175	13
Gum fuel containing	44	58
Metallurgical	75	257
Natural resin	106	240
Protein containing	106	161
Casein	106	147
Gelatine or glue	106	136
Prolamine	106	153
Soldering feeders for	228	33+
Solid metal treating	148	23+
Synthetic resin or natural rubber (See class 523, 524)		
Fluxmeter		
Magnetic flux	324	244+
Fly		
Brush	416	501
Closer lasting tool	12	113
Fishing	43	42.24+
Design	D22	125+
Holders for fishhooks, flies	43	57.1+
Net	54	81
Paper box making	493	52+
Swatter	43	137
Design	D22	124
Textile spinning	57	96

	Class	Subclass
Trap	43	122
Fly Ash	106	DIG. 1
Processes of using	264	DIG. 49
Stock material	428	406
Flying (See Aircraft)		
Blind		
Aircraft controls	244	75R+
Traffic control systems	340	947+
Machine	244	
Design	D12	319+
Design toy	273	362+
Propeller toys	446	36+
Target	273	362+
Design	D22	114+
Drone	D12	16.1
Trainers	434	30+
Flying Tool		
Cutting	83	284+
Wire tool	83	307.1+
Metal deforming	72	184+
Flypaper	43	114+
Catching implements	43	136
Packages	206	447
Flytraps	43	107+
Design	D22	122
Flywheels		
Fan combined	416	60
Fluid	60	330+
Impellers combined	416	60
Machine elements	74	572+
Occupant propelled vehicle power storing	280	217
Sewing machine abraiding attachment	51	257
Foam	252	307
Breaking		
Distillation apparatus	202	264
Distillation process combined	203	20
Processes	252	321
Coating implement with material supply	401	44+
Control	435	812*
Fire extinguishing	252	3
Stabilizing agent containing	252	8.5
Froth flotation agents	252	61
Gas & liquid contact apparatus	261	DIG. 26
Glass making	65	22
Laminate	428	304.4+
In situ foaming	156	77+
Metallic receptacles collection	220	902*
Plastic molding	264	41+
Producing compositions	252	350
Protective coating or zone	252	382
Separating gases from	55	178
Processes	55	87
Fob	D11	79+
Focusing		
Camera	354	195.1+
Automatic	354	400+
Hood	354	287
Flashlight	362	187
Focus coil and cathode ray tube	313	442
Focus coil per se	335	213+
Projector	362	317+
Adjustable light	362	285+
X ray tube combined	378	138
Fodder	426	636
Fog	252	305
Combustion	126	59.5
Dispelling appliances	98	1
By spraying	239	2.1
Horns	116	137R+
Preventing compositions	106	13
Foil (See Sheet)		
Dental filling	433	227+
Fencing	272	98
Laminate		
Manufacture	29	17R+
Metal	428	606+
Folder (See Creaser)		
Apparel apparatus	223	37+
Collar cuff bosom making	223	3
Collar, cuff or neckband ironers	223	52.1+
Plaiting etc with fold guides	223	34
Plaiting fluting shirring with cross creasing	223	29
Boot and shoe making		
Lasting	12	12+
Uppers	12	55
Cigar or cigarette tip or mouthpiece forming	131	89+

	Class	Subclass
Folder carried indicia	40	359
Paper article making		
Bags, envelopes	493	243+
Boxes	493	162+
Sheet material associating or folding		
Associating and folding	270	32+
Associating folding and rotary printing combined	270	4+
Folding and rotary printing	493	321+
Wrapping folding and or printing combined	53	116+
Sheet metal can seam	413	69+
Sheet metal seaming	29	243.5+
Folding	493	405+
Boats	114	353
Chairs	297	16+
Tables	108	115+
Foliage		
Artificial	428	17+
Design	D11	117+
Folic Acid	544	261
Aircraft control	244	76R+
Controlled gearing	74	388R+
Fluid pressure brake valve	303	54+
Gearing	74	388R+
Recorder drive	346	31+
Servomotor fluid	91	358R+
Ship steering	114	144R+
Follower		
Ammunition dispenser (See magazine)	42	49.1+
Cabinet		
Phonograph record holder	312	9+
Cam	74	569
Cooking apparatus combined	99	349+
Dispenser combined		
Container	222	386+
Insertable cartridge	222	326+
Fountain pen	401	141+
Lubricator pump	184	37
Mattress filling apparatus	53	255+
Metallic receptacle	220	93
Followers and partitions	220	22+
Partition follower	220	22+
Pipe joint piston type	285	302
Rack card or sheet	211	51+
Self loading material handler	414	506
Sheet retainer with releasable keeper	402	60+
Smoking device	131	181
Trunk	190	36
Wooden		
Barrel	217	86
Box	217	64
Font		
Sorting	199	40+
Typewriter embossing	400	127+
Typewriter, penetrating	400	135+
Food (See Type)	426	
Animal foodstuffs	D 1	
Animal mastication aids	D30	199
Apparatus	99	
Canning	426	392+
Filling receptacles	141	
Cereal pupping	99	323.4
Conductivity control	99	DIG. 11
Cooking apparatus	99	324+
Dairy food treatment	99	452
Dehydration	426	443+
Design	D 1	
Edible containers	D 1	
Edible laminated product making	99	450.1
Enclosed modified atmosphere treating	99	467
Food mixer-grinder-blenders	D 7	368+
Fruit and vegetable peelers	D 7	147
Hand tools for food prerapation	D 7	99+
Induction heating	99	DIG. 14
Inedible stick or holder &	D 1	105
Live stock	426	635+
Means to treat food	99	485+
Non-cooking heat treatment	99	483
Non-protein nitrogen	426	69
Preparation or dispensing machines	D 7	
Preparing and treating (See type)		
Apparatus	99	
Preserving	426	
Processes, compositions & products	426	
Equipment design	D 7	368+
Storage receptacle, household	D 7	76+
Strip or stick form	D 1	

	Class	Subclass
Temperature responsive	99	DIG. 1
Vinegar making	99	323.12
Foot		
Bandages	128	165+
Bath	D24	38
Closure by foot pedal	220	262
Controlled dibbling apparatus	111	98
Electric applicator	128	795
Flipper	441	64
Guard railway track	238	379+
Levers pedals	74	594.4
Measuring devices	33	3R
Monitor radioactivity	378	38
Miscellaneous	378	38
Scintillation type	250	458.1
Operated		
Anvil vises forging	D21	1+
Closure bolt releasers	292	255
Control lever and linkage systems	74	469+
Cutting tool	30	275
Cutting tool plural blade	30	297
Dispenser	222	179
Earthworking tool lift	172	495+
Gang punching machine	83	620+
Runner vehicle	280	12.1+
Sheet punching machine	83	613+
Ship steering apparatus	114	153
Vehicle steering apparatus	280	87.1
Wheeled vehicle	280	200+
Orthopedics	128	80R+
Piece		
Dustpan	15	257.5
Fork and shovel hand	294	60
Posts for skates	280	11.17
Resilient machinery support	248	615+
Rest (See --, support)	297	423+
Chair reversible back	297	73+
Chair with folding	297	30
Closet	4	254
Control lever system type	74	564
Design	D 6	501
Land vehicle body	296	75
Land vehicle combined	280	291
Piano	84	232
Rocker with	297	271
Substitute	297	428
Support		
Boat	114	363
Life preserver with	441	80+
Skate joined runner & support	280	11.15
Skate resiliently mounted	280	11.14
Surgical table	269	
Supporter	36	71
Warmer		
Heat radiator	237	77
Stove	126	204
Surgical body wear	128	382+
Football	273	65R
Electric scoreboard	340	323R
Game	273	55R
Game board	D21	29
Scoreboard	116	222+
Shoes	36	128
Simulated game	273	94
Tackling dummy	273	55R
Teaching	434	251
Footboard		
Bed	5	53R+
Foot or leg support	5	443+
Motor vehicle body	180	90.6
Railway rolling stock	105	460
Footholds		
Building with	52	184+
Friction surface	52	177+
Cane	135	70
Crutch	135	70
Ladder terminals	182	108+
Pavement	404	19+
Stair covers	428	85+
Walking aids with tips	135	77+
Footing (See Bases; Foundation)		
Footlights		
Cleaning type	15	238+
Footwear	36	
Design, general	D 2	264+
Mens	D 2	308+
Stockings & socks	D 2	329+
Womens	D 2	282+
Foraminous		
Cabinet wall structure	312	213
Force Measuring	73	

	Class	Subclass
Coin controlled devices	194	
Electric	324	
Force Pumps	4	255+
Forceps		
Dental	433	159
Railway traction cable	104	217
Surgical	128	321+
Design	D24	27
Forcer		
Closure applying die	53	341+
Sheet metal die shaping	72	54+
Forehearth	266	166
Forestry	47	
Hand tools design	D 8	1+
Forge	110	195
Tuyeres	110	182.6
Water cooled	122	6.7
Forging	72	
Machines with other tools	29	34R
Metal working with	29	DIG. 18
Nuts and washers	10	76R+
Piston rings	29	156.61
Processes	72	372+
Wrought nails	10	55+
Fork (See Shovel)		
Bicycle type	280	276+
Compound tool with culinary fork	7	112
Compound tool with pitch fork	7	115
Cutlery	30	322+
Design	D 7	137+
Guard	30	323
Hand	294	55.5
Handles	D 8	DIG. 4
Material handling		
Blanks and processes for making	76	111
Cooking spit or impaler	99	419
Excavating scoops	37	120+
Hand	294	55.5+
Hayfork	294	120+
Hayfork type grapple	294	86.4+
Inclined elevator or hoist	414	595+
Tilting	414	598+
Making dies	72	470+
Railway mail delivery	258	21+
Railway mail delivery support	258	11+
Single throw lever fulcrumed	254	131.5
Tedder	56	374
Vertically swinging	414	685+
Table (See cutlery)	30	322+
Blanks & processes for making	76	105
Combined tool	7	112+
Combined with knife	30	148
Dies for making	72	470+
With ejector	30	129
With material holder	30	137
Tuning or musical		
Music instrument	84	409
Tuning	84	457
Wheel occupant propelled vehicle		
Front	280	279+
Rear	280	288
Rear yielding	280	284+
Forklift Truck	414	529
Ramp design	D12	
Formaldehyde	568	448
Dehydrating	203	17
Synthetic resin (See synthetic resin or natural rubber resin or natural rubber)		
Formazans	534	652
Formamide	564	215
Formates		
Alkali forming metal	562	609
Cellulose	536	67
Esters	560	231+
Former (See Die; Molding; Shaping)		
Formic Acid	562	609
Forming		
Metal by plastic shaping	72	
Fluent tool or energy field	72	54+
Forms		
Display	223	66+
Design	D20	10+
Dress	223	68
In situ construction engineering or building type	249	1+
Forte Devices (See Piano)		
Fortune Telling Devices	273	161
Coin controlled	194	
Playing cards	D21	42+
Toys	D21	240

	Class	Subclass
Foundation (See Bases)		
Bee comb	6	11
Building	52	292+
Hair	132	54+
Marine	405	195+
Ballasted	405	207+
Piles	405	231+
Road bed	404	27+
Founding	164	
Foundry Apparatus	164	
Foundry Mold or Mold Core Binder	523	139+
Fountain		
Beverage infuser	99	313+
Gravity feed	99	310+
Comb	132	112+
Design	D23	201
Drinking	D 7	304
Stock watering	D30	121+
Drinking	239	24+
Design	D 7	304
Inkwell (See inkwell)	222	577
Machine		
Bed and platen oscillating printing	101	315
Bed and platen reciprocating printing	101	321
Dampener planographic	101	148
Inker	101	364+
Inker multicolor	101	210
Inker roller	101	363
Printing members and inker	101	330+
Pen (See fountain pen)		
Sprinkling	239	17+
Stencil addressing	101	48+
Stock watering	D30	121+
Stock watering	119	74+
Syringe	604	131+
Gravity feed from plural reservoirs	604	80
Fountain Pen (See Pens)	401	
Ball point	401	209+
Follower fluid	401	142
Bifurcate nib type	401	221+
Ink retainer attachment at tool	401	252+
Filling	141	21+
Filling	141	18+
Projectable and retractable tool	401	99+
Rupturable cartridge	401	132+
Sign or picture carrying	40	334+
Stylographic	401	258+
Fourdrinier Machines	162	348+
Fowl	17	11+
Brooders	119	31+
Butchering	17	11+
Care and handling design	D30	
Confining and housing devices	119	21+
Design	D30	108+
Feeding devices	119	51R+
Foods	426	635
Medicines	424	
Supports and shackles	17	44.1
Watering devices	119	72+
Design	D30	121+
Fowler Flap for Aircraft	244	216
Fracture Apparatus Surgical	128	83+
Frame	52	
Abrading machine	51	166R+
Agricultural implement	172	776
Aircraft		
Airfoil	244	123+
Airship hull	244	125
Fuselage	244	119+
Apparel		
Armpit shield	2	57
Hat	2	180
Apparel apparatus		
Dress coat or skirt stretching	223	69
Hat frame making	223	8
Tie	223	65
Trousers stretching	223	63
Bag holder	248	99+
Bed	5	
Bee		
Comb	6	10
Hive	6	2R+
Camera	354	161
Canopy	135	101+
Car roof	52	45+
Caster	16	31R+
Clock	368	88
Closures partitions panels	160	371+
Awning	160	76

	Class	Subclass
Mount or support	160	369
Nonrigid	160	354
Plural	160	353
Roll type	160	239$^+$
Door	49	504$^+$
Metal	52	782$^+$
Drill press	408	234$^+$
Drying		
Curtain	38	102.1$^+$
Electric lamp	314	130$^+$
Arc	314	130$^+$
Fabric cleaner	15	233
Fabric stretching	38	102.1$^+$
Fastening savings box	232	6
Fireplace	126	139
For building	52	648
For expansible chamber device	92	161
Hair foundation	132	54
Hand and hoistline implements	294	67.1$^+$
Brick carriers	294	63.1$^+$
Grapple	294	113$^+$
Vacuum	294	65
Harvester	56	
Headwear receptacle	206	9
Internal combustion engine	123	195R
Land vehicle		
Body structure	296	187$^+$
Body supporting	280	781$^+$
Design	D12	159
Drop frame convertible to	280	7.11
Dropped body	280	2
Endless track carried by	305	16$^+$
Equalized	280	104$^+$
Roller bearing	280	105
Velocipede	280	281R$^+$
Velocipede steered	280	274$^+$
Leather making work holder	69	19
Light shade or bowl	362	433$^+$
Luggage	190	121$^+$
Metallic receptacle wall	220	84
Mirror or reflector	52	
Mounted motor	180	58$^+$
Music		
Hammer action	84	250
Key	84	432
Piano	84	184
Open work	52	633$^+$
Panel	52	474$^+$
Edging	428	21$^+$
Sectional	52	455$^+$
Pavement light or lens	404	22$^+$
Picture	40	152$^+$
Picture frame molding	D25	119$^+$
Planting drill	111	52$^+$
Portal with building structure	52	204$^+$
Printing	355	122$^+$
Quilting	112	119
Radiator	165	149
Railway		
Bogie side	105	206.1$^+$
Car	105	396$^+$
Locomotive	105	172$^+$
Scroll saw	83	776$^+$
Sewing	112	258$^+$
Sheet drying apparatus	34	163
Sifting machine	209	404$^+$
Sign or bulletin board	D19	52
Slate	434	422
Spreader	294	81.1
Table	248	163.1$^+$
Tent	135	101$^+$
Textile		
Pile tufting tubes	139	10
Warp hand loom	139	34
Textile thread	28	149$^+$
Trolley pole	191	65
Trunk	190	24
Truss surgical	128	99.1$^+$
Typewriter	400	691$^+$
Ventilating register	98	114
Wall decoration	D 6	300$^+$
Window	49	504$^+$
Wireworking	140	92.1
Woodworking		
Bit	408	53
Clamps	269	
Handsaw	30	166R$^+$
Match dipping	144	65
Frangible		
Attachment for metallic receptacle	220	89A
Barrel bung	217	111
Bottle or jar cap	215	250$^+$
Building structures	52	98$^+$
Cartridge fountain pen	401	132$^+$
Container opening	206	601$^+$
Dispenser outlet	222	542
Electric signal box element	340	303
Element in pipe joint	285	3$^+$
Feature		
Static mold	249	61
Fluid handling means	137	797
Locks		
Key bit	70	410
Key operating mechanism	70	422
Material used in glass process	65	23
Metallic receptacle closure	220	265$^+$
Multiple shoe brake connection	188	241
Railway switch element	246	289
Railway train pipe closure	246	199
Derailment operated type	246	172
Valve controlling element	137	68.1$^+$
Frankfurter (See Sausage)		
Packaging	53	DIG. 1
Fraternal Insignia	D11	95$^+$
Fraud Preventor or Detector		
Check controlled apparatus	194	302$^+$
Combined	194	202$^+$
Coating or impregnating	427	7$^+$
Printed matter	283	72$^+$
Stock material	428	916*
Transfer or decalcomania	428	915*
Weaving combined	139	315
Free Fall Item Feeding	209	908*
Free Radical, Hydrocarbon Synthesis Involving	585	942*
Free Wheeling Clutches	192	41R$^+$
Freeze-out Trap	62	55.5
Freezer and Freezing (See Congelation)		
Flowable material	62	340$^+$
Automatic	62	135$^+$
Process	62	66$^+$
Ice cream machine design	D15	82
Pressure compensators	138	27$^+$
For internal combustion engines	123	41.5
Refrigerator-freezer design	D15	81$^+$
Safety systems for fluid handling	137	59$^+$
Work holder using	269	7
Freight Accomodation on Freight Carrier	410	
Lashing	410	96$^+$
Anchor	410	101$^+$
Load brace	410	121$^+$
Bar wall-to-wall	410	143$^+$
Panel wall-to-lading	410	127$^+$
Panel wall-to-wall	410	129$^+$
Retainer	410	77$^+$
Yieldable	410	117$^+$
Particular article	410	2$^+$
Automobile shipment	410	4$^+$
Grouped	410	31$^+$
Massive	410	44
Freight Carrier		
Freight accomodating construction	410	
Land vehicle	280	
Body	296	
Rail car	105	355$^+$
Framing	105	404$^+$
French Curves	33	177
French Dressing	426	615$^+$
French Fryers	99	403$^+$
Freon T M	570	134
Frequency		
Amplitude study mechanical	73	579$^+$
Analyzer	324	77R
Automatic control		
Gain, level, volume	455	234
Plural receivers	455	136
System radio	455	71
Television receiver	358	194.1
Transceiver	455	119
Transmitter	455	313$^+$
Changers	310	127$^+$
Conversion systems	363	157$^+$
Converters	310	
Systems	363	157$^+$
Demodulators	329	110$^+$
Discriminators	329	110$^+$
Divider	363	157$^+$
Electronic tube type	328	15
Electronic tube type	328	25
Dividers oscillator	331	51
Group frequency radio signaling	455	
Measuring	324	78R$^+$
Absorption type	324	81
Heterodyne	324	79R
Wave meter	250	250
Modulators	332	16R$^+$
Multipliers oscillator	331	53
Radio signaling	455	
Response modification	358	904*
Speech range compression or expansion in telephone system	381	29$^+$
Synthesis	328	14
Fret Saws	30	513$^+$
Frets for Fingerboards of Stringed Instruments	84	314R
Control	84	DIG. 3
Friction		
Clutch	192	
Brake position adjuster combined	188	199
Gearing	74	190$^+$
Single speed	74	206
Generators		
Boiler heating	122	26
Electric	310	310
Heater	126	247
Producing compositions	106	36
Removable closure retained by	220	352
Tape	428	343$^+$
Coated on both sides	427	208
Package	206	411
Test		
Between solid bodies	73	9
Liquid viscosity	73	55
Torque	73	862.12
Welding apparatus	228	2
Nonmetallic	156	580.1
Methods	156	73.1$^+$
Sonic or ultrasonic	156	580.1
Welding process, nonmetallic	156	73.5
Sonic or ultrasonic	156	73.1
Fringe		
Apparel trimming	2	244
Making on looms	139	118$^+$
Misc. manufacture	28	145$^+$
Of sewn strands	112	409
Design	D 5	7$^+$
On web or sheet	428	115$^+$
Sewing machine forming	112	64
Trimming design	D 5	
Turning guide for sewing machines	112	149
Woven fabric	139	385
Frisbee T M	446	46$^+$
Frisket	101	421
Frits, Glass Compositions	501	14$^+$
Frog		
Flower stem holder	47	41.13
Design	D11	147
Railway	246	454$^+$
Continuous rail	246	275$^+$
Frog obviating switches	246	417$^+$
Switch connected	246	382$^+$
Switch connected vehicle actuated	246	274
Trolley transfer	104	105
Textile weaving	139	346
Front		
Boards	5	54
Boiler	122	497$^+$
Fireplace	126	138$^+$
Design	D23	343$^+$
Furnace	110	172
Summer for stoves & fireplaces	126	140
Upending bedstead case		
Panel	5	160
Wing	5	161
Winter or storm		
Land vehicle	296	78R$^+$
Railway car	105	353
Frost Indicator for Refrigerator	62	128
Frost Preventing		
Composition	252	70
Coating	106	13
Plants and fruit	47	2
Refrigerator	62	272$^+$
Automatic	62	150$^+$
Process	62	80$^+$
Frosting Lamp Bulbs	156	625$^+$
Froth	252	307
Flotation	209	163$^+$
Agents	252	61
Copper	209	901*
Phosphate	209	902*
Preventing	252	321

U.S. PATENT CLASSIFICATIONS

	Class	Subclass
Fruit	426	616+
Apparatus for treating	99	467
Cleaning		
Brushing or wiping apparatus	15	3.1+
Fluid treatment apparatus	134	
Processes	426	478+
Clipper	30	123.5+
Coffee substitutes from	426	594+
Corer	99	547+
Design	D 7	106
Gatherer	56	328R+
Design	D 8	13
Halving cutters	83	
Juice extractor		
Comminuting citrus fruits	99	495
Design	D 7	48+
Interfitting cup type press	100	213
Presses	100	104+
Peelers or parers	99	588+
Design	D 7	147
Plants	PLT	33+
Preservation	426	321+
Canning	426	392
Coating	426	289+
Dehydration	426	471+
Refrigeration	426	524
Seeding and stoning	99	547+
Sorting machines	209	509+
Stemmer	99	635+
Frying Pan		
Deep fat	99	403+
Heat transfer feature	126	373+
Structure	220	
Surface griddles	99	422+
Ftorafur	544	313
Fuchsine	260	391+
Fuel	44	
Artificial	44	1R+
Boxes	126	283
Can	D23	202+
Cell	429	12
Active material electrode	429	27+
Automatic control means	429	22+
Catalytic electrode structure	429	40+
Cell electrical generator	429	12+
Chemically specified electrode	429	46
Design solid	D13	3.2
Diesel		
Ignition promoter containing	44	57
Mineral oil	208	15
Feed and supply systems	137	
Engine combined generally	60	
Engine internal combustion combined	123	495+
Fuel injection	123	445
Rotary piston	123	200+
Liquid fuel nozzle generally	239	
Solid fuel burners combined	110	101R+
With combustion feature	431	
Gas		
Distribution	48	190+
Generator	48	61+
Gaseous	48	
Heat exchange feature	429	26+
High energy compounds	149	120+
Igniting devices & fuel	44	
Injector combined with rotary piston engine	123	200+
Liquid	44	50+
Explosive or thermic containing	149	1
Material, actinide or radioactive inorganic compound	423	249+
Solidifying	44	7.1+
Mineral oil	208	15+
Nuclear	376	
Alloys for actinide groups	420	1+
Blanket arrangements	376	172
Bundle or pack	376	434
By-product treatment	376	189
Canned	376	412
Carrier, insertable	376	342
Cladding	376	457
Coated or impregnated	376	414
Complementary segments	376	429
Compositions	252	626+
Concentric layers	376	431
Coolant included	376	424
Core arrangements	376	348
Electrolytic production	204	1.5
In fission plenum	376	412
Particles	376	411
Reprocessing	423	249+
Handling	376	260
Heat treatment	75	31
Layered	376	416
Leak detecting	376	250
Warning structure	376	450
Loading and unloading	376	264
Manufacturing		
Chemical milling	156	625+
Chemical, misc	156	67
Cladding with metal bonding	29	400N+
Coating	427	6+
Etching	156	625+
Irradiation or conversion	376	156
Laminating	156	67
Metal fabrication	29	400N+
Metal stock heat treated	148	401+
Metallurgical production	75	84.1 R
Metallurgical production	420	1+
Powder metallurgy	419	
Material, actenide or radioactive		
X-ref patents	376	901*
Alloy	376	422
Organic compounds	534	10+
Mechanical spacing means or securing	376	438
Molten	376	359
Projections, prongs fins	376	454
Metal alloys	420	1+
Seal for	376	451
Segmented	376	426
Non-metal	264	21+
X-art	376	903
Sintered process	419	
Sintered stock	75	228+
Slurry, associated components	376	356
Slurry, conversion process	376	171
Spacing means or securing	376	438
Waste disposal	252	626+
Plural cells	429	18+
Process of operating	429	13+
Pumps	415	
Pump	417	
Service station pump	D15	9.1+
Refrigerant utilized as	62	7
Service station pump	D15	9.1+
Solid and liquid	44	
Solid electrolyte	429	30+
Spread out into a film	261	DIG. 25
Testing consumption	73	113+
Treatment, solid fuel furnace	110	218+
Fulcrum		
Brake beam type	188	231+
Full Stroke Mechanism	74	17.5
Dispensers		
Pump type	222	375
Full Wave Gas Rectifier	313	581
High vacuum rectifier	313	306
Fuller's Earth		
Adsorbent	502	232+
Carbon combined	502	180
Fulling	26	19+
Hide apparatus	69	33+
Inorganic explosive or	149	33+
Thermic composition containing	149	33+
Fulminating Igniter	431	269+
Fumaric Acid	562	595
Fume Arrester		
Consumable electrode discharge devices	314	26+
Gas separation	55	
Hoods	98	115.1
Design	D23	371+
Metallurgical apparatus	266	144+
Textile fluid treating	68	209
Fume Generator	422	305+
Fumigants	424	40+
Fumigation	422	
Animal	119	160
Food	426	312+
Apparatus	99	467
Soils	47	58
Vermin destroying	43	125+
Fungicides		
Animal dips and sprays	424	
Cyanogen compounds	424	129
Fertilizers containing	71	3+
Fumigants	424	40+
Inorganic	424	127+
Oil emulsions	514	938*
Organic	514	1+
Suspending agents & spreaders	514	769+
Fungistatic		
Compositions	424	
Funnel	141	331+
Chemistry lab apparatus	422	99
Dispenser with funnel type outlet	222	460+
Dispenser with separable	141	343
Household article design	D 7	53
Liquid level or depth gage having	73	294
Milk receptacles with funnel filler	232	43
Supports for	248	94
Fur (See Pelts)		
Artificial	428	85+
Cape design	D 2	182
Coat design	D 2	198+
Coat design hip length or above	D 2	188
Collar	D 2	602+
Cutting	83	915*
Fluid treatment	8	94.14
Dyeing	8	404+
Garments	2	65+
Design	D 2	
Leather manufactures derived from furs and related processes	69	23+
Neckwear design	D 2	601+
Receptacles for	206	11
Scarf or muff hangers	223	97
Sewing machines	112	16+
Treating apparatus	69	23+
Furan	549	505
Furfural	549	483
Furfuramide	266	144
Furling	114	104+
Furnace (See Heat)	110	
Absorber with manually adjustable shading means	126	DIG. 1
Arc	373	60+
Ash receiving and handling devices	110	165R+
Automatic temperature control	236	
Bessemer converter	266	243+
Boiler controlled		
Air or steam feed	110	188
Fuel feeder	110	103
Stoker	110	186+
Car type	432	241
Charge conveyor within	414	150+
Charging and discharging	414	147+
Distillation apparatus	202	
Metal charge	75	256
Shaped	428	576
Checker brick	165	9.1+
Chemical reaction	422	198+
Coke	202	215+
Collector includes overlapping roof shingles or tiles	126	DIG. 2
Cremating	110	194
Cupola	432	95+
Metallurgical	266	197+
Metallurgical, X-art	266	900*
Design	D23	329
Distilling and combustible gas generating		
Burning combined	110	229+
Gas generators	48	61+
Mineral oil vaporizing	208	347+
Stills	202	81+
Doors	110	173R+
Draft regulators	110	147+
Electric	373	
Arc furnance	373	60+
Electron beam	373	10+
Zone melting	373	17
Glass furnaces	373	27+
Glow discharge	373	26
Induction furnaces	373	138+
Channel	373	162
Core-type	373	159
Coreless	373	151
Lining	373	164+
Making of lining	264	30+
Zone melting	373	139
Pressure and vacuum furnace	373	60+
Arc with internal atmosphere control	373	68
Furnace body detail	373	71
Resistance furnace	373	109+
Wall cooled furnace		
Arc furnace device	373	75+
Induction core-type	373	165
Induction coreless	373	158
Induction furnace device	373	138+

	Class	Subclass
Resistance furnace device	373	113$^+$
With condonsers	373	56$^+$
Zone melting	373	17$^+$
Electron beam	373	10$^+$
Zone melting	373	17
Fluid fuel burner in	431	159$^+$
Solid fuel combined	110	260$^+$
Fronts	110	172
Fuel feeders	110	101R$^+$
Gas scrubbers	261	DIG. 9
Gases		
Combustion air heating	110	303
Heating boiler feed water	122	420$^+$
Power plant water heating	60	669
Glass		
Annealing	65	349$^+$
Electric	373	27$^+$
Melting	432	195$^+$
Melting and refining	65	347
Pot type	432	156
Incinerator	110	235$^+$
Liquid heating or evaporating		
Air heater combined	126	101
Concentrators	159	
Cooking stove combined	126	
Fireplace combined	126	132$^+$
Open type	126	344$^+$
Pressure generating boilers	122	
Melting	266	200$^+$
Metallurgical		
Apparatus	266	
Processes	75	
Miscellaneous heating	432	
Nuclear	376	366
Open hearth	432	159$^+$
Pots	432	262$^+$
Powdered fuel burning		
Fluid fuel combined	110	260$^+$
Fuel feeders	110	104R$^+$
Furnace structure	110	263$^+$
Puddling	266	203
Recuperator	432	179
Regenerative	432	180
Sawdust refuse and wet fuel	110	235$^+$
Feeders	110	102
Stoves	126	222$^+$
Shaft	432	95$^+$
Metallurgy	266	197$^+$
Smokestacks	110	184$^+$
Solid fuel burning	110	
Stoves	126	
Spark and smoke conductors	110	145$^+$
Spark arresters	110	119$^+$
Stokers	241	600*
Stoves & furnaces	126	
Tuyeres	110	182.5$^+$
Wall manufacture and repair	264	30
Furniture		
Beds	5	
Cabinets	312	
Chairs and seats	297	
Design	D 6	
Doll or toy	446	479$^+$
Design	D21	121
Dry closet combined	4	465$^+$
Education features combined	434	432
Fire escape combined	182	35$^+$
Illuminator combined	362	127$^+$
Insect trap combined	43	109
Music cabinet	84	DIG. 17
Polish	106	3$^+$
Protector or cover	D 6	610$^+$
Racks and stands for	211	27
Ship furniture as life raft	441	125$^+$
Ships	114	188$^+$
Store design	D 6	
Tables	108	
Toy	D21	121
Furoic Acid	549	484$^+$
Furring (See Studding)	52	474$^+$
Settable material backer	52	344$^+$
Furrower	172	
Furrow followers	280	776
Irrigation dam former	172	233$^+$
Planting drill	111	87$^+$
Roller combined	172	176
Furyl Compounds	549	429$^+$
Azo compounds	534	751$^+$
Fuse		
Ammunition and explosive device	102	200$^+$
Blasting	102	200$^+$
Cordeau detonant	102	275.1$^+$
Mine	102	424$^+$
Mine marine type	102	416$^+$
Condition, elecric signal	340	638$^+$
Cutter blasting	86	22
Electric	D13	35
Box	D13	35
Connector combined	337	187$^+$
Thermal	337	401$^+$
Thermal current	337	142
Fusible elements	169	42
Building structures combined	52	232
Making	493	948*
Measuring and testing timing apparatus	73	5
Operated		
Alarm	116	106
Safety devices for fire extinguishers	122	504.3
Safety devices for liquid heaters and vaporizers	122	504.1
Stove and furnace dampers	126	287.5
Proximity	102	211$^+$
Puller tool	81	3.8
Setters, ordnance	89	6$^+$
Fused Cell Formation	435	172.2
X-art collection	935	93*
Fusee	102	358
Fuselage		
Aircraft construction	244	119$^+$
Lifting type	244	36
Fusible		
Boiler safety control	122	504.1
Fire extinguisher	122	504.3
Bolt or nut, in bolt or nut to substructure lock	411	171
Connections	169	42
Metal, in bolt to nut lock	411	257
Plugs for automatic fire sprinkler heads	169	37$^+$
Receptacle attachment	220	89B
Sensor signal system	340	590$^+$
Stove damper release	126	287.5
Switch	337	401$^+$
Thermal current element	200	
Thermal current heater	337	182$^+$
Thermometer	374	106
Valve operator	137	72$^+$
Valves and controllers	222	54
Fire extinguisher	169	19$^+$
Ventilator cowl release	98	86
Fusion (See Melting)		
Bonding apparatus	228	
Coating combined	427	375
Iron and steel making	75	40$^+$
Combined	75	38
Separation combined	422	285$^+$
Sulphur	423	567R
Fusion Bonding		
Glass preforms of	65	36$^+$
Apparatus	65	152$^+$
Fusion Nuclear Reactors	376	100
G		
G Acid	260	512R
Gable, Roof	52	90$^+$
Roof end	52	94$^+$
Gaff		
Fishing	43	5
Gamecock	30	297
Grappling	294	19.3
Ship spars	114	97$^+$
Gag		
Fishing tackle	43	53.5
Runner for bridles	54	14
Side cutting harvester lever	56	271$^+$
Specula	128	12$^+$
Gage (See Gauge)		
Gagger	164	411
Flask with	164	382
Gain Control, Automatic	73	900*
Gaining	144	133R
Gaiters	36	2R
Design	D 2	267
Fastenings	24	
Making	12	142W
Galena Crystal Radio	329	205R
Galley		
Ships stove	126	24$^+$
Type setting	276	40$^+$
Justifying machine	276	33
Gallic Acid	562	476
Gallocyanines	544	99$^+$
Galloons		
Lace	87	10
Gallows (See Execution Devices)		
Galvanized Sheets	428	659
Galvanizing		
Metals	427	433$^+$
Galvanometers	324	76R$^+$
Galvanoplastic Processes	204	6
Gambrels	294	79$^+$
Butchers	17	44$^+$
Roof design	D25	24
Game		
Amusement railways	104	53$^+$
Annunciator systems	340	323R
Board games	273	236
Carrier	224	103
Other than stringer	224	921*
Check or coin control	194	
Computer controlled, monitored	364	410$^+$
Design	D21	1$^+$
Game apparatus	273	
Peg tally board	235	90
Pieces	273	288$^+$
Design, misc	D21	51$^+$
Score boards	116	222$^+$
Spinners & indicators for	D21	22
Tables	D21	121$^+$
Tally sheet	283	49$^+$
Tv, input control	273	
Input control per se	340	709$^+$
Gamecock		
Boxing muff	119	143
Gaff	30	297
Gamma		
Acid	260	509
Geiger survey meter	250	374$^+$
Radiation reflector	376	220
Radiation reflector	376	350
Radiation reflector	376	458
Radiation shield	376	287
Radiation shield	376	347
Radiation survey meter	250	336.1
Scintillation type	250	361R$^+$
Ray detector	250	336.1
Gang		
Button, eyelet, rivet setting machine	227	51$^+$
Circular saw roller feed	83	425.2$^+$
Earthworking disk	172	599$^+$
Plural	172	579$^+$
Scrapers	172	558$^+$
Harvesters	56	6$^+$
Metal working rotary cutter	407	31
Nail feeding delivery	227	
Plows		
Wheeled	172	314
Punching machine	83	620$^+$
Woodworking		
Rotary disc cutter	144	237
Slivering saw	144	189
Gang Bar Type Locking or Latching Means		
Cabinet	312	216$^+$
Sectional unit type	312	107.5
Ganged Radio Tuner	334	
Operator only for tuner	74	10.45
Gangway		
Animal traffic	119	82
Car loading side	105	436
Conveyer skids	193	41
Conveyer skidways	193	38$^+$
Endless conveyer type	14	70
One end attached	14	71.1
Ship attached one end floating	114	258
Unattached	14	69.5
Gantry	212	218$^+$
Gap Jumper Amusement Railway	104	54
Garage	52	
Vehicle handler	414	227$^+$
With ramp	52	174
Garbage		
Boiler plants for destroying	122	2
Carbonization of	201	21
Dump & trash truck design	D12	15
Fat etc recovery from		
Rendering	260	412.6$^+$
Solvent extraction	260	412.8
Fertilizer from	71	14
Grinding and disposal		
Comminuting solids	241	
Design of machine	D 7	375

	Class	Subclass
Grinder	4	DIG. 4
Sinks combined	4	629
Incinerator	110	235^{+}
Insect trap for can	43	120
Receptacle		
Household design	D34	1^{+}
Stand with closure operator	248	147
Truck, compactor type	100	100
Utilization for food	426	655
Wet fuel furnace	110	238
Garden (See Plant Husbandry)		
Garland	428	10
Design	D11	119
Paper, making	493	958*
Garment		
Absorbent pad holding	604	393^{+}
Apparel	2	
Apparatus	223	
Design	D 2	26^{+}
Design costumes	D 2	79
Design costumes bifurcated	D 2	30^{+}
Design lingerie or undergarments	D 2	1^{+}
Design surgical	D 2	
Bag, protector	493	935*
Bandaging type	128	82^{+}
Respirator mask	128	206.12^{+}
Suspensory	128	159
Body discharge shield	604	356^{+}
Buttons buckles clasps etc	24	
Article holders	24	3R^{+}
Shielded buckles	24	184
Cleaning		
Fluid treatment	8	
Fluid treatment apparatus	68	
Demonstration	434	395^{+}
Fastener bonding	156	66
Fastener making	140	81
Forms	223	66^{+}
Foundation type	450	7^{+}
Hangers	223	85^{+}
Making	140	81.5
Knitted	66	171^{+}
Design	D 2	44^{+}
Design dresses, suits, skirts	D 2	43
Life preservers	441	88^{+}
Light etc application to body	128	379^{+}
Parachutes attached to	244	143
Protectors boot and shoe	36	70R
Receptacles for	206	278
Hat and headwear	206	8^{+}
Seat attached holders for	297	190
Stay making	140	91
Stiffener making	223	28^{+}
Supporters		
Blouses	2	107
Brassiere with	450	28^{+}
Combined	2	300
Connected spaced for plural	2	304^{+}
Convertible or reversible	2	301
Corset or girdle with	450	110^{+}
Elements	2	336^{+}
Foundation garment with	450	26
Mens outer shirts	2	117
Partially encircling	2	309
Plural encircling type	2	308
Rigid vertical type	2	302^{+}
Shoulder suspension and encircling	2	310
Strip connected spaced	2	323^{+}
Torso or limb encircling	2	311^{+}
Underwear	2	112
Garnet Paper	51	394^{+}
Garnetting Machines	19	98^{+}
Garters		
Apparel design	D 2	
Corset or girdle with	D 2	2^{+}
Fasteners	D 2	625^{+}
Brassiere with	450	28^{+}
Combined	2	300
Connected spaced for plural	2	304^{+}
Corset or girdle with	450	110^{+}
Elements	2	336^{+}
Encircling	2	311^{+}
Foundation garment with	450	26
Partially encircling	2	309
Plural encircling type	2	308
Rigid vertical type	2	302^{+}
Shoulder suspension and encircling	2	310
Strip connected spaced	2	323^{+}
Gas (See Fluid)		
Ammunition and explosive devices	102	367^{+}
Drop bomb	102	369
Grenade	102	368
Gun cartridge	102	370
Shell	102	367
Analysis		
Chemical apparatus	422	50^{+}
Chemical process	436	153^{+}
Electrical	324	
Physical	73	23^{+}
Spectrometry	250	281^{+}
Apparatus, gas & liquid contact	261	
Batteries	429	12^{+}
Blood gas analysis	436	68*
Burners	431	
Burners	239	
Nozzles per se	239	
Cap for automobile	D12	197^{+}
Cigarette lighter	431	150
Igniter correlated with feed	431	130^{+}
Igniter correlated with feed	431	254
Circulated in circuit	261	DIG. 27
Coal	48	210
Coating contacting	427	248.1^{+}
Compositions	252	372^{+}
Concentrating evaporators	159	
Cooler	62	404^{+}
Automatic	62	186^{+}
Liquid contact with gas	62	304^{+}
Process	62	89^{+}
Deflector for separation	55	434^{+}
Detectors		
Alarms	116	67R
Analysis chemical apparatus	422	50^{+}
Analysis chemical methods	436	153^{+}
Analysis miscellaneous	436	153
Apparatus	435	807*
Electric alarms	340	632^{+}
Nonchemical gas analysis	73	23^{+}
Discharge display panels	313	582^{+}
Discharging nozzles	239	
Dispensing	222	3^{+}
Packaging combined, method	53	403^{+}
Distribution	137	
Airship control	244	96^{+}
Cooling and or insulating electrical conductors	174	16R
Explosion prevention	48	192
Heating and illuminating	48	190^{+}
Leak detectors	137	455^{+}
Leakage detection & prevention	48	193^{+}
Musical instruments	84	330^{+}
Outlet nozzles	239	
Electrochemical processes	204	124^{+}
Engines	123	
Extinguishers	169	11
Foam forming	169	14^{+}
Filling lamp or radio tube	141	66
Filters	55	
Fittings, design	D26	113^{+}
Furnace		
Coking	110	230^{+}
Domestic hot air	126	99R^{+}
Industrial	432	
Power plant combined	60	669
Producer	110	229^{+}
Solid fuel furnace combined	110	260^{+}
Getter (See getters)		
Guns	42	
Bank protection	109	20
Bank protection combined	109	29^{+}
Heaters and stoves	126	
Boilers	122	
Carburetor	261	127^{+}
Cooking stoves	126	39R^{+}
Design	D 7	339^{+}
Design	D23	314^{+}
Heating stoves	126	58^{+}
Hot air furnace	126	116R
Metallurgical apparatus combined	266	138^{+}
Stove burners	431	
Nozzle	239	398^{+}
Stove structure	126	
Heating and illuminating	48	
Acetylene	48	1^{+}
Carburetors	48	144^{+}
Distribution	48	190^{+}
Generators	48	61^{+}
Mixers	48	180.1^{+}
Natural	48	196R^{+}
Processes	48	197R^{+}
Purifiers	48	128^{+}
Retorts	48	119^{+}
Holders and packages	206	.6^{+}
Acetylene generator combined	48	
Heating or illuminating	48	174^{+}
High pressure	220	3
Inhaler combined	128	203.12
Medicator	604	403^{+}
Structures	220	
Telescopic	48	174
Illuminating	48	
Burner nozzles	239	
Inert	55	66
Separating	55	66
Injection		
Bathtub combined	4	542^{+}
Convective distillation	203	49
Iron steel (molten)	75	59.1^{+}
Liquid contact apparatus	261	
Medicators	604	19^{+}
Mortar mixer	366	10^{+}
Process	366	3^{+}
Ionization in thermo nuclear reaction	376	100
Thermolytic distillation	201	31
Thermolytic distillation	201	36^{+}
Lamp brackets, fittings	D26	113^{+}
Light incandescent burners	431	100^{+}
Design	D26	1^{+}
Liquefaction	62	36^{+}
Processes	62	9^{+}
Liquefied handling apparatus	62	45^{+}
Receptacles	220	901*
Liquefiers and solidifiers		
Acetylene generator combined	48	1
Refrigeration	62	
Separating gases	55	209
Separating gases processes	55	81
Liquid contact with	261	
Automatic refrigeration	62	171
Concentrating evaporators	159	
Distillation	203	49
Gas separation combined	55	
Heat exchanger	261	127^{+}
Mineral oil	208	
Percolation control	261	DIG. 81
Refrigerating process	62	121
Refrigeration	62	304^{+}
Sugar treating	127	12
Log heater-fireplace	D23	409
Making		
Coking furnace	110	230^{+}
Compositions	252	4^{+}
Furnace	110	229^{+}
Furnace and boiler combined	122	5
Furnace using liquid fuel	431	161^{+}
Generators chemical	422	164
Heating and illuminating	48	
Life preservers	441	98^{+}
Oil	48	211^{+}
Mantles	431	100
Compositions	252	492
Mask	128	206.12^{+}
Canisters for	55	DIG. 33
Fillers	502	400^{+}
Processes of making corrugated tubes for	264	DIG. 52
Separating gases	55	
X-art collection	425	815*
Meters	73	861^{+}
Coin controlled	194	
Mixers		
Acetylene generator combined	48	3R
Fuel burner	431	354
Heating and illuminating	48	180.1^{+}
Internal combustion engine charge forming device	123	434^{+}
Monitor	73	23^{+}
Photoelectric	250	573
Radioactivity	250	336.1
Scintillation type	250	336.1
Nuclear fusion fuel	376	151
Oil	48	211
Pump	D15	9.1
Water	48	214R^{+}
Operated		
Firearms and guns	89	125^{+}
Rotary kinetic motors	415	
Rotary kinetic pumps	415	
Permeable coated fabrics	427	245
Poison	424	
Pressure storage	137	206^{+}
Pressurized gas treatment of textiles	8	DIG. 15

Entry	Class	Subclass
— Producers		
— — Chemical	422	126
— — Heating and illuminating	48	
— Protecting joint during bonding	228	214
— Public utility meter	D10	99+
— Pumps	415	
— — Impellers	416	
— — Oil or gasoline pump	D15	9.1
— Purifying	55	
— — Chemical	423	210+
— — Compositions, absorbent	502	400
— — Compositions, chemical agents	252	182+
— — Heating and illuminating	48	128
— Ranges	126	1R+
— Rapid producers of	149	
— Rare	423	262
— — Electric furnace atmosphere	373	
— — — Arc furnace	373	60
— — — Resistance furnace	373	110+
— — Electric lamps	313	567+
— — Electronic tube	313	567+
— — Food preservation with	426	512+
— — Lantern	362	263+
— — Liquefying and separating	62	22
— — Purification chemical	423	210+
— — Separating	55	66
— Rectifier full wave	313	581
— Rectifier half wave	313	567
— — With hot cathode	313	629
— Refrigeration produced by	62	401+
— — Automatic	62	172
— — Heat generation type	62	6
— — Process	62	86+
— Samplers	73	863+
— Separation	55	
— — Chemical	423	210+
— — Electrostatic precipitators	55	101+
— — Flowmeter combined	73	200
— — Flue	110	148+
— — Liquid contacting means	55	220+
— — Liquid from	55	159+
— — Non-liquid cleaning means	55	282+
— — Recycle means	55	338+
— — Refrigeration apparatus	62	36+
— — Refrigeration process	62	11+
— Solid contact with	34	
— — Comminuting	241	
— — Distillation	201	
— — Gas separation	55	
— Textile treating with	68	
— Thyratron	313	591+
— Triode	313	567+
— Tube amplifier system	330	41
— Ultraviolet treating		
— — Apparatus	250	432R+
— — Chemical change in	204	124+
— — Sterilizing	422	121+
— Water heaters	126	344+
Gaseous Fuel Burners	431	
— Discharge nozzle	239	
— Igniters	431	
— — Discharge nozzle	239	
— — Heating attachments	126	249+
— — Tool heating attachments	126	231+
— Synthetic	585	14
Gasket	277	
— Applying		
— — Apparatus laminating	156	
— — To metal cap or closure	413	8+
— — — Apparatus	413	58+
— Barrel bung	217	109
— Design	D23	269
— Making by laminating	156	
— Pipe joint combined	285	335+
Gasoline	208	
— Carrier truck	D12	95
— Explosion preventing appliances in containers	220	88R
— Fuel composition containing	44	50+
— Pick up truck mounted	D12	156
— Pump	222	
— — Design	D15	9.1+
— Storage tank systems	220	85S
Gasometer (See Gas Holders)		
Gastroscopes	128	8
Gate	49	
— Design	D25	50+
— — Hinges	D 8	323+
— Drawbridge	14	50+
— End for vehicle bodies	296	50+
— — Standing top	296	106
— Hanger	16	86.1+
— Hinge	D 8	323+
— Latch	D 8	331+
— Motion picture	352	221+
— Operators	49	324+
— Railway highway crossing		
— — Automatic	246	292+
— — Automatic electric	246	125+
— — Fluid motor	246	261
— — Vehicle actuated	246	272
— Register operated by	235	93
— Sorting animal	119	155
— Umbrella joint	135	32
— Valve (See valve)		
— Water gates	405	87+
— — With flow meter	73	215
Gated Beam Discriminator	313	300
Gatherer		
— Brush and broom making		
— — Tuft gathering	300	7
— — Tuft gathering and setting	300	5+
— Conveyor type loading machine	198	506+
— Fish with conveyor to boat	43	6.5
— Glass		
— — Ladling and gathering	65	125
— — Tank furnace gathering pool	65	336+
— Harvesters	56	
— — Cornstalk type	56	119
— — Cornstalk type cutter combined	56	51+
— — Cornstalk type picker or husker combined	56	103+
— — Design	D15	10+
— — Design hand tool	D 8	13
— — Fruit	56	328R+
— — Seed	56	126+
— — Seed motor driven	56	12.9+
— — Standing grain	56	219+
— — Vegetable	56	327R
— Sewing machine attachment	112	132+
— Signature	270	54+
— Stones and rocks	171	63+
— Vertically swinging shovel or fork	414	686+
Gating Circuits	358	153+
— Multiplex		
— — Telegraph	370	46+
— — Telephone	370	112
— Radar	342	94+
— — Automatic range gate	342	95
— Television	178	69R
— Transmit receive switches	333	13
Gauge (See Guide)	33	501+
— Abrading		
— — Process of gage grinding	51	286
— — Work holder	51	220+
— — Work rest	51	239
— Altimeter	73	384+
— Auger attachments	408	72R
— Awl or prick punch combined	30	368
— Bore	33	178F
— Chisel attachments	30	167.2
— Circular size	33	178R+
— Collocating	33	613+
— Combined with diverse tool	7	164
— — With cutter	7	163
— — With handsaw	7	150
— — With hatchet, hammer	7	122
— Contact type	33	501+
— Control for abrading wheel	51	165R
— Control for rolling mill		
— — Elongation	72	250
— — Metal thickness	72	16
— — Strip tension between stands	72	10+
— Cooking apparatus	99	343
— Dental	433	72+
— Depth		
— — Endodontic	433	75
— — Inkstand	15	257.73
— — Liquid	73	290R+
— — Ore electric	324	323+
— — Sounding	33	126+
— — Specific gravity test combined	73	447
— — Submarine, electro-acoustic	367	99+
— — System electric	340	618+
— Design	D10	61+
— Die rolling	72	250+
— Distance single contact	33	169R+
— Feeler type	33	168R
— Fixed distance	33	168R
— Fluid flow	73	861+
— — Direction combined	73	189
— Harvesting cutter	56	121.4+
— Ionization	250	374+
— Marking	33	18.1+
— Nailing combined		
— Nippers combined	30	179
— Pressure		
— — Calibrating	73	4R
— — Fluid	73	700+
— Punch point combined	30	368
— Railway		
— — Tie plate	238	288
— — Track clamp	238	339
— — Truck changeable gauge	105	178
— — Wheel gauge changer	104	33
— Saw making	76	46
— Screw threads	33	199R
— Sewing machine		
— — Buttonhole	112	75
— — Hat	112	15
— — Shoe sole	112	39
— Shears with	30	233+
— Sheet cutting machine	83	268+
— Sheet cutting machine	83	391+
— Sheet cutting machine	83	467R+
— Sheet feeding alignment	271	226+
— Static mold having	249	53R
— Structure	33	501+
— Tool	33	201+
— Track leveling light ray type	33	287
— Tympan	101	413+
— Typewriter		
— — Paper feed edge	400	630+
— Wheel	33	203+
— With point markers	33	666+
— Wood hoop bending	144	258
— Wood saw table	83	522
— Woodworking bench plane	30	481
— Work supported work holder	269	
Gauging	73	
— Automatic sorting by	209	659+
Gauze		
— Fabrics	139	419
— Looms	139	50+
Gear and Gearing	74	20+
— Actuated (See device actuated)		
— Backlash take up	74	409
— — Gear body	74	440+
— Bearing, hydrostatic	409	904*
— Belt and sprocket or pulley	474	
— — Guards for	474	144+
— Bevel		
— — Gearing	74	423+
— — Gearing intersected axis	74	417
— — Gears	74	459.5
— — Pivotally supported gearing	74	385+
— Bolts for closure fasteners	292	279+
— — Multiple	292	22
— Broaching	409	243+
— Casings	74	606R+
— — Design for velocipedes	D12	124
— Chain	474	
— — Design	D 8	499
— Change speed	74	325+
— — Friction	74	190+
— — Planetary	74	750R+
— Crimping	28	279
— Cutting	409	1+
— Differential		
— — Disc type gears	74	444
— — Multiple input or output	74	674+
— — Planetary	74	710+
— Electric motor and gearing	310	80+
— — Control gearing	74	640+
— — Motor control system	318	9+
— Epicyclic	74	640+
— Friction	74	190+
— — Single speed	74	206
— Gauge	33	179.5 R
— Gearing	74	640+
— Geneva		
— — Gear	74	436
— — Gearing	74	84R
— Hinge	16	354
— Holder, work	409	903*
— Indexing	409	221+
— Interchangeably locked	74	325+
— Intermittent motion		
— — Geneva	74	436
— — Irregular teeth	74	437
— — Worm	74	426
— Landing (See landing devices)	244	100R+
— Locks for	70	190+

Entry	Class	Subclass
Making		
Cutter	407	20+
Cutting	409	1+
Forging	72	376+
Grinders	51	
Grinding processes	51	287
Laminating processes	156	
Processes	29	159.2
Rolling	72	67+
Metal rolling type	72	249
Meter	73	261
Milling	409	50+
Motor		
Expansible chamber	418	191+
Internal combustion	123	
Mutilated (See gear and gearing geneva)	74	435
Patterns, tracers and tools	409	2+
Planer type	409	1
Planetary	74	750R+
Speed varying	74	394
Planing	409	42+
Power take off	74	11+
Printing plate stereotype	409	901*
Pump	418	191+
Lubricating	184	31
Used as brake	188	292
Used as clutch	192	61
Rack and pinion	74	422
Pushing and pulling implements	254	95+
Ratio changing	74	393+
Belt and sprocket	474	69+
Expansible pulley utilizing	474	47+
Expansible cone pulley	474	8+
Friction	74	190+
Interchangeably locked	74	325+
Planetary	74	750R+
Reversing	74	640+
Robot	901	25+
Roll with driving	29	115
Running gear land vehicles	280	80R+
Screw, helical or worm	74	640+
And nut	74	424.8 R
Screw geared jacks	254	103
Sectional		
Gears	74	439+
Watch winding and setting	368	197+
Shift transmission control	74	473R
Brake combined	192	4C+
Clutch combined	192	3.51+
Motor control combined	74	843+
Shiftable gear clutch or key (See gear interchangeably locked)		
Alternating rotary clutchable gear	74	322+
Alternating rotary shiftable gear	74	321
Clutch and gear	192	20+
Speed controlled (See type of gearing)		
Speedometer drive type	235	95R+
Spline and	33	DIG. 14
Sprocket	474	
Crank and pedal combined	74	594.2
Spur	74	460+
Steering (See steering)		
Stock material, metal	29	
Motor vehicle	180	79+
Teeth and bodies	74	431+
Bodies	74	434+
Teeth	74	457+
Teeth lubricated	74	468
Testing	73	162
Tire mold	409	902*
Valve		
Expansible chamber motor type	91	
Internal combustion engine	123	90.1+
Work tables	409	92+
Yieldable		
Gearing	74	411
Gears	74	461
Geat, Gate, Git	135	32
Geiger Counter	250	374+
Design	D10	47
In nuclear reactions	376	254
Geiger Muller Tube	313	93
Survey meter	250	374+
With circuit	250	374+
Geissler Tubes	313	567+
Illuminating fixture with	362	217+
Gel		
Colloidal	252	315.1+
Misc. chemical manufacture	156	DIG. 108

Entry	Class	Subclass
Radioactive	252	634
Silica	423	334+
Gel Route to Glass Making	501	12
Gelatine	530	354
Jellies	426	576
Gems	63	32
Artificial	501	86
Changing color by radiation	378	38
Design	D11	89+
Setting methods and apparatus	29	10
Settings	63	26+
Genealogical Charts		
Teaching	434	154+
Generator Motor	310	113
Generators (See Steam, Generators & Boiler)	D13	1+
Acetylene	48	1
Acetylene	48	2
Acetylene	48	3R
Acetylene	48	4+
Chemical	422	198+
Chemical pressure in fire extinguisher	169	6+
Combustible mixture for internal combustion engine	123	3
Electric	310	
Battery charging	320	61+
Bicycle	362	193
Chair operated therapeutic	128	378
Combined in electronic tube	315	55
Compressional wave actuated	367	140+
Correlated with burner feed	431	255
Driven by compressional wave	367	178
Electric meter having	324	76R+
Electrically driven	322	
Electrostatic	310	309
For supplying engine ignition	123	149R
Infrared heat	219	335+
Locomotive	105	35+
Motor generator	310	113
Musical	84	1.1+
Nonelectric prime mover plant	290	
Nonmagnetic	310	300+
Nuclear	310	301+
Reactor	376	100
Reactor	376	347
Oscillator	331	
Piezoelectric	310	311+
Piezoelectric systems	322	2R
Plural generator in system	307	43+
Rotary	310	40R+
Sound wave actuated	367	140+
Superimposed different currents	307	1+
Supplying different circuits	307	18+
Synchronous inductive	310	171
System of control	322	
Systems	322	
Temperature measuring plural	374	110+
Thermoelectric	136	200+
Voltage controlled	322	
Fumigators sterilizing	422	305+
Gas	48	61+
Power calculation for	364	492
Refrigerator	62	
Sinusoidal		
Electronic tube type oscillator	331	
Generator structure	310	
Generator system	322	
Steam	122	
Steam combined with stove	126	5
Vapor pressure thermostat	374	201+
Genetic		
Expression	935	33*
Transfection	935	52*
Transformation	935	52*
Vectors	935	72*
Genetic Engineering	435	172.1+
Apparatus	935	85*
Assay	935	76*
Recombination	435	172.3
X-art collection	935	*
Genetically Engineered Cells	435	240.1
Genetically Engineered Cells	435	243+
Utilization	435	
X-art collection	935	59*
X-art collection	935	66*
Geneva		
Drive	414	905*
Gear	74	436
Gearing	74	84R
Genital		
Body wear bands combining	128	386

Entry	Class	Subclass
Catheter	604	264+
Curette and scraper	128	304
Cutter		
Castrating	128	306
Embryotome	128	307
Urethrotome	128	311
Dilator with light and electrical application	128	303.11+
Kinesitherapic appliance	128	31
Male electric applicator	128	794
Obstetric instruments	128	361
Forceps	128	323+
Orthopedic appliance	128	79
Pessary and inserter	128	127+
Receptor		
Body protruding member	604	346+
Catamenial and diaper	604	358+
Obstetric	604	317+
Urethrorrhoeal	604	327+
Sexual restrainer	128	138R
Stallion shield	119	145
Suspensory bandage	128	158+
Tampon		
Inserters	604	11+
Structure	604	358+
X-art collection	604	904*
Vaginal		
Douche (See douche)	604	36+
Electric applicator	128	788
Geodisic Dome	52	81
Design	D25	13
Geographical		
Clock	368	21+
Teaching	434	130+
Geology, Teaching	434	299
Geometrical Instruments (See Type of Instrument)	33	
Blocks educational design	D21	108
Design	D10	61+
Teaching geometry	434	211
Geophysical Prospecting (See Prospecting)		
Acoustic means	181	101
Digital computer with system	367	60+
Exploration in situ	324	323+
Geochemical exploration	436	25
Nuclear	250	253+
Underwater seismometer spreads	367	15+
Vibration transducers	367	140+
Georgette	139	426R
Geothermal Power Plant	60	641.2
Geraniol	568	875
Masonry	52	723
Germanium, Misc, Chemical Manufacture	156	DIG. 67
Alloys	420	556
Organic compounds	556	81+
Germicides	424	
Germicidal lamp (See ultraviolet; X ray)		
Water softening compositions containing	252	175+
Germination		
Retarding & accelerating compositions	71	65+
Seed germinator	47	61
Germine	546	28
Gettering	437	10+
Getters (See Absorbent)		
Compositions	252	181.1+
Gettering	156	DIG. 66
Processes of use	445	55
Pumps	417	48+
Gib (See Key)		
Slide bearings	384	39
Gill Net	43	10
Gilsonite (See Asphalt; Pitch)	208	22+
Gimbal	248	182
Gimlets	408	230
Gin		
Beverage	426	
Cotton	19	39+
Saw sharpening and gumming	76	32+
Ginger Ale	426	590+
Gingham	139	417
Girder (See Beam)	52	721+
Bridge	14	17
Composite	52	730+
Dissimilar material	52	722+
Design	D25	126+
I-beam	52	729

	Class	Subclass
Making	29	155R+
Mortar bonded	52	433+
Static mold	249	50
Girdles	450	94+
Combined with bra	450	7+
Design	D 2	2
Design bifurcated	D 2	4
Design combined with bra	D 2	3+
For nuclear reactor	376	302
Moderator structure	376	304
Girths		
Harness	54	23
Packing	277	
Gland		
Extracts		
Medicines containing	424	95+
Glare		
Reduction	350	276R+
Automobile screen	296	97R
Filters	350	311+
Glass antireflection coating	350	164+
Polarizers	350	399
Rearview mirror	350	277+
Glass	65	
Actinic composition	501	900*
Developing by heat	65	111+
Bead		
Reflector	350	104+
Reflector making	427	163+
Block construction	52	596+
Translucent component	52	306
Blowing		
Charging glass melt furnace	414	165+
Compositions	501	11+
Cutters (See search notes)	83	879+
Drinking glass	D 7	6+
Electric furnaces	373	27+
Etching	156	663
Fiber, coated	428	357+
Glass-to-metal seal	428	432
Metallic	428	636
Gratings optical	350	162.11+
Grinding processes	51	283R+
Handling cylinders of	414	24
Household glassware	D 7	1+
Laminated and safety	428	426+
Wire glass	428	38+
Manufacturing	65	
Filament or fiber making	65	1+
Process	65	17+
Processes of uniting glass	65	36+
With electric lamp making	445	22
With metal founding	164	
With metal working	29	
Ornamenting		
By abrasive blasting method	51	317+
By grinding	51	
Photochromic	501	13
Plate		
Grinding and polishing	51	
Pot furnaces	432	156+
Pots	432	262+
Product manufacturing computerized	364	473
Scribers	33	18.1+
Stand, rack or tray for	D 7	70
Stemware	D 7	11+
Structural	428	268
Wire mesh	428	38
Treating hard glass	65	
Apparatus	65	348+
Process	65	111+
Window	428	38
Glasses (See Eye Glasses)		
Eye and spectacles	351	41+
Design	D16	102+
Metal frame making	29	20
Optics	350	481
Ground for cameras	354	161
Sight for liquid level gauge	73	323+
Glassware	D 7	1+
Glaucine, in Drug	514	284
Glaze		
Compositions	501	14+
Glaziers Points	411	477+
Setters	227	
Glazing		
Earthenware	427	376.2
Firing	264	60+
Fruits and vegetables	426	289+
Glider		
Aerial toy	446	61+

	Class	Subclass
Aircraft	244	16
Furniture design	D 6	344+
Porch swing	297	282
Hammock couch type	5	124+
Glides	16	42R+
Furniture design	D 8	374
Globe		
Clock	368	23
Design	D10	10
Lamp	362	809*
Design	D26	118+
Lantern	362	363
Design	D26	37
Operators for	362	174+
Manipulating pole	81	53.11+
Operator tube lantern	362	174+
Railway amusement	104	68
Teaching	434	131+
Terrestrial globe	362	809*
Globin	530	385
Globulins	530	386
Glove	2	159+
Baseball	2	19
Boxing	2	18
Buttoners	24	40
Design	D 2	643
Design	D 2	617+
Fingerless	D 2	610+
Forms	223	78+
Knitted	66	174
Circular machine	66	45
Straight machine	66	65
Sewing machine	112	16+
Glow		
Discharge, chemical apparatus	422	907*
Discharge diode	313	567+
Transfer counter	377	103+
Glower (See Electric; Electrode; Filament; Lamp)		
Glucamines	564	507
In drug	514	669
Gluconic Acid	562	587
Glucosamines	536	55.2
Glucose	127	30
Making	127	36+
Tests for	436	95*
Glucosides	536	4.1+
Glue (See Adhesive)		
Compositions	106	125+
Containing	106	125+
Gluing	156	
Applicator	D19	70+
Clamps	269	
Glue heating pots	126	284
Press	100	
Using particular adhesive	156	325+
Glutamates	562	573
Foods ctg	426	656
Preparation by protein hydrolysis	562	516
Glutamic Acid	562	573
Glutamine	562	563
Glutathione	530	332
Glyceric Acid	562	587
Glycerides (See Fats)	260	398+
Rosin	260	104
Glycerine	568	852+
Fermentative production	435	159+
Glycerol (See Glycerine)		
Glycidol	549	554+
Glycine	562	575
Glycols	568	852+
Ethers	568	672+
Oleate	260	410.6
Silicate	556	482+
Stearate	260	410.6
Glycopeptides	530	322
Glycoproteins	530	395
Glycosides	536	4.1+
Cardiac	536	5+
Glyoxal	568	494
Treatment of textiles, polyaldehyde combined	8	DIG. 17
Glyoxaline	548	300+
Glyptal T M (See Alkyd Resins)		
Goad	231	2R
Design animal	D30	156+
Goal, Game Element	D21	48+
Goblet	D 7	11
Goddard Patent Collection	60	915*
Godet		
Strand feeding wheel	226	168+

	Class	Subclass
Goggles	2	426+
Design	D16	102+
Combined with head covering	D 2	509+
Gold		
Alloys	420	507
Beating	72	420+
Carbon compounds	556	110+
Electrolysis		
Coating	204	47.5+
Synthesis	204	109+
Foil	428	606
Hydrometallurgy	75	118R
Laminate	428	457+
Metal-to-metal	428	672
With glass	428	433+
Misc. chemical manufacture	156	DIG. 101
Plate	428	672
Pyrometallurgy	75	83
Golf		
Accessory	D21	234
Bag	206	315.3
Bag cart	D34	15
Holder	248	96
On wheeled carrier	280	47.26
Digest	280	DIG. 6
Ball	273	62
Design	D21	204+
Making	156	146
Means to print on balls	101	DIG. 17
Cart	224	274
Cart motorized	280	DIG. 5
Club	273	77R+
Design	D21	214
Rack or support for	211	70.2
Woodworking machines for making	144	2XA
Courses	273	32R
Miniature	273	176R
Covers & cases	150	52G
Game board	273	245
Pocket or capture feature	D21	18
Simulation	D21	27
Surface projectile game	D21	11
Gauge	33	508
Glove	2	161R
Design	D29	22
Handles or grips	273	81R
Design	D 8	DIG. 6
Holes	273	32R
Putting	273	34R
Powered passenger cart	D12	16
Practice devices	273	35R+
Score registers	273	87R
Shoes	36	127
Simulated game	273	87R
Targets	273	181R
Miniature golf	273	176R
Putting	273	34R
With ball return	273	182R
Teaching devices	434	252
Tees	273	33+
Carriers	224	191+
Design	D21	234
Device for setting	273	32.5
Gondola		
Railway		
Drop bottom	105	244+
Freight	105	406.1+
Inclined bottom	105	256
Side door	105	258+
Gong (See Bell)		
Goniometer		
Crystal testing	356	31
Light ray type		
Horizontal angle	33	285
Vertical and horizontal angle	33	281
Vertical angle	33	282+
Radio	342	428+
Goniometer device	342	441
Straight edge type	33	403+
Governor (See Brake; Control; Regulator)		
Elevator	187	68+
Fire escape	182	191+
Gearing control	74	336.5
Reversing means	74	404.5
Internal combustion engine	123	319+
Pump liquid	417	279+
Speed recorder	346	73
Speed responsive	73	488+
Switch electric	200	80R

	Class	Subclass
Valve	137	47+
Vehicle motion and direction indication	116	38
Wind motor		
Impeller deflection control	416	9
Impeller speed control	416	44
Rotary wheel deflection	416	9+
Gr-n Rubber (See Synthetic Resin or Natural Rubber)		
Gr-s Rubber (See also Synthetic Resin or Natural Rubber)	526	340
Grab (See Grapple)		
Cable traversing hoist		
Load dumping draft rope	212	81
Load suspension	212	84
Clam shell buckets	37	183R+
Dredges	37	55
Furnace charging	414	186
Hoistline grab hooks	294	82.1+
Irons railway rolling stock	105	461
Magnetic	294	65.5
Material handling elevator	414	618+
Magnet and	414	606
Orange peel buckets	37	182
Rotary crane	212	243
Self propelled traversing hoist electric	212	127
Guide bar	212	129
Vertically swinging load support	414	680+
Graders and Grading (See Separator)	209	
Design	D15	23+
Grinding abradant supplier	51	264
Railway	37	104+
Road	37	108R+
Scraper type	172	
Gradometers	33	365+
Graduates	73	426+
Graft Copolymer (See Synthetic Resin or Natural Rubber)	525	50+
Grafting	47	6+
Textile fiber grafting	8	DIG. 18
Grain		
Adjusters	56	466+
Car		
Doors	49	404+
Temporary closures	160	368G
Casse-grain type antenna	343	781CA
Comminuting processes	241	6+
Cradles	56	324+
Drill		
Chutes	193	9
Hoppers	222	
Planters	111	
Drying & gas or vapor contact with solids	34	
Feeder or dispenser	222	
Automatic control of flow	99	488
Food		
Apparatus	99	467
Design	D 1	
Preservation	426	321+
Processes and products	426	549+
Geometrical shapes of grain	51	DIG. 3
Harvesters	56	
Hulling	99	600
By applying fluid	99	518
Land vehicle tank	296	15
Paper grain, digest	229	DIG. 5
Pattern and size		
Grain boundaries in metal stock	148	33.2
Metallurgical linings having specific grain size	266	284
Optical grain sensitizing in photography	430	571
Paper grain digest	229	DIG. 5
Stock, angular grain	428	105
Stock, artifical wood or leather surface	428	151
Stock, parallel grain	428	114
Superimposed grain as color photographic technique	430	365
Wood grain	428	106
Scourers		
Apparatus	99	600+
Processes	426	483+
Scouring by brushing	15	3.1+
Shelled-grain catchers	56	207
Standing-grain gatherers	56	219+
Storage units	D25	14
Bin heaters	D23	334
Ventilation	98	52+
Threshing	130	
Flying grain arrestors	130	27E
Recleaners	130	DIG. 4
Savers	130	DIG. 6
Separtors	130	20+
Tungsten carbide grains in an abrasive compostion	51	DIG. 26
Wheels and casters	56	322
Wood grain, stock material	428	106
Artificial woodsurface	428	151
Pattern arrangement in buildings	52	313
Working roller parallel to grain	144	362
Graining (See Granulating)		
Boxes	51	6
Paint brush texturing implement	D 8	16
Tools	15	210.5
Gramicidins	530	310
Gramophones (See Phonograph)		
Granaries	52	192+
Grandfather Clock	D10	16
Granular		
Detectors for radio	329	100
Detectors for radio	338	1
Microphones	381	180+
Resistance elements		
Brakes	188	268
Electrical	338	223+
Voice transmitter (eg microphone)	381	180+
Granulating (See Crystallizing)		
Common salt	23	303
Evaporation combined	159	45
Inorganic materials	23	313R+
Sugar	127	21
Tobacco	131	360
Grapefruit (See Citrus Fruit)		
Graphic Recorder	D10	61+
Graphite	423	448
Fishing rod	43	18.5
Lubricants containing	252	9+
Moderator structures	376	350
For nuclear reactors	376	458
Graphophone (See Phonograph)		
Grapnels (See Anchor; Grab)		
Grapple (See Grab)	294	86.4+
Fish	43	5
Fork or shovel combined	294	50.5
Fork or shovel type	294	50.6+
Gambrel	294	80
Hand rakes combined	56	400.12
Hooks	294	82.1+
Hand	294	26
Overhead beam	294	85
Submarine	294	66.1+
Tilting handling vehicle	414	444+
Well	294	86.1+
Lifting means combined	166	98
Grappling		
Well, methods	166	
Grasper (See Grapple; Holder)		
Chucks and sockets	279	
Hand and hoist line	294	
Potato digger	171	50+
Grass		
Covering or wrapping	57	4
Spinning twisting twining	57	28+
Grate	126	152R+
Bars	126	167+
Cage	110	294
Design	D23	397+
Fireplace	D23	398+
Hollow air feeding	110	298+
Ore roasting	266	279
Ore roasting, traveling	266	171+
Roller for ore and fuel dressing	209	393
Rotary	126	181+
Stokers	110	268+
Water cooled	122	371+
Grater		
Comminuting surface having opening		
Reciprocating	241	94
Rotating	241	91+
Stationary	241	95
Design	D 7	47
Gratings (See Grille; Register)	52	660+
Diffraction	350	162.11+
Light	350	1.1+
Pavement	404	17+
Sewer inlet	210	163+
Ventilating	98	114
Window	49	50+
Grave		
Casket		
Design	D99	1
Lowering devices	27	32
Rests	27	26
Coverings	27	30
Life signals	27	31
Linings	27	29
Marker	40	124.5
Design	D99	17+
Monuments markers guards	52	103+
Coping	52	101
Vaults	52	128+
Gravelling		
Roads	404	101+
Wells		
Apparatus	166	51
Processes	166	278
Gravitometer	73	382R+
Gravity		
Bucket		
Endless chain water motor	415	5
Self-proportioning systems	137	101.25+
Changing		
Aerated column pump	417	108+
Mineral oil increasing	208	
Conveyers	193	
Moving item from separating station	209	924*
Determination	73	382R+
Feed		
Beverage infuser	99	304+
Boiler water	122	456+
Boiler water heater combined	122	458
Dispensing nongravity combined	222	205
Dispensing nozzles, spouts, etc	222	566+
Molten metal	222	591+
Drier	34	165+
Drier combined type	34	64
Fluid sprinklers	239	376+
Furnace charging type material	414	208
Discharging	414	216+
Liquid cooler	62	397
Lubricators	184	65+
Sprayer pump with	239	329+
Supporting or guiding item moving under influence of	209	911*
Track sander fluid or	291	7+
Trap chamber dispensing	222	424.5+
Levitation & reduced gravity	156	DIG. 62
Operating		
Agitating	366	336+
Aircraft control	244	80
Aircraft stabilizing weights	244	93+
Bolt or nut to substructure lock, mass offset	411	169
Check controlled connectors	194	301
Check controlled lock releasing	194	260+
Clock escapements pendulum type	368	136
Dispenser outlet element	222	500
Fence wire stretcher	256	38
Folding table paired legs	108	131+
Fountain syringes	604	80+
Heating gas carburetors	48	160
Heating gas mixers	48	185
Hinges	16	309+
Hooked end closure fasteners	292	130+
Janney type couplings knuckle openers	213	117+
Janney type couplings locks	213	139
Latch released closure	49	379
Linear traversing hoist	212	92+
Mechanical gun magazine	124	49+
Momentum type brake operators	188	139
Mortar mixers	366	9
Music leaf turners	84	488
Nonrefillable bottles	215	21+
Open link couplings pin support	213	194+
Power plant combined types	60	698+
Railways cable rails	104	113
Sliding bolt closure fasteners	292	183+
Sliding counter protecting shields	109	18
Spittoon	4	284
Spring hinges	16	303
Spring locomotives	105	72
Stamp mills	241	273
Swinging bolt closure fasteners	292	230+
Telephone switch hook	379	424+
Vehicle actuated turntable	104	42
Specific determining	73	32R+
Systems		
Fluid fire extinguishing	169	10
Store service	186	2+

	Class	Subclass
Store service single impulse	186	12
Vessel		
Pneumatic displacement pump valve actuation	417	118+
Grays Telegraphs	178	2R+
Grease Guns		
Dispenser type		
Discharge assistant	222	251+
Follower combined with pump	222	256+
Material operated differential piston	222	253
Movable nozzle connected assistant	222	320+
Receiver discharge coupling	141	383
Replaceable cartridge	222	325+
Lubricators	184	14
Hand operated followers	184	38.1
Hand operated pumps	184	28
Pumps	184	26
Refilling devices	184	105.1
Grease Traps	210	538+
Greenhouses	47	17+
Building per se	52	
Design	D25	15
Greeting Card	283	1R
Graphic and or movable	446	147+
Display	40	124.1
Folding display	40	539
Sounding		
Mechanical tape	369	68
Phonograph records	369	273
Phonographic repetition	369	63+
Sounding toy	446	397+
Grenade		
Explosive	102	482+
Thrower		
Firearm attached	42	105
Mechanical	124	4+
Type fire extinguisher	169	36
Grid		
Battery grid filling	141	32+
Battery grid making	29	2
Static mold	249	60
Bucky	378	154
Cleaning or combing	15	142
Deflectors in gas separation		
Electrical resistance elements	338	283+
Electronic tube	313	348+
Heat exchange device	165	168+
Ice cube tray type	249	203
Leaks		
Circuit	329	175+
Resistor	338	
Making of wire	140	71.5+
Nuclear fuel plates	376	432
Plates for storage batteries	429	233+
Grid Glow (See Electric Space Discharge Device)		
Griddles	99	422+
Household cooker	D 7	363
Gridirons		
Drip collecting	99	444+
Perforated food support	99	450
Slice toasting	99	385+
Griff	139	67+
Grill Cooking		
Slice toasting	99	385+
Solid heated surface	99	422+
Opposed surfaces	99	372+
Grille (See Gratings; Register)	52	660+
Connected panels	52	581
Curtain shade or screen combined	160	104
Design heating, ventilating	D23	
Lattice	52	660+
Louvre	52	471
Ventilation	98	121.1
Making	29	160
Rod for	52	720
Tile holder	52	384+
Grinding (See Abrading)	51	
Attachments	51	241R+
Cutlery	30	138+
Lathe	29	28
Leather working tool sharpener	69	38
Valve	137	243+
Dental	433	25+
Design	D24	12
Design vegetable and meat	D 7	372+
Hand-held tools	D 8	90+
Household mixer-grinders	D 7	372+
Lens	D15	124+

	Class	Subclass
Machinery design	D15	124+
Mill	241	
Alarms	116	71
Paper making combined	162	261
Optical lenses and mirrors	51	284R
Physical recovery methods	435	803*
Superfinish	51	57+
Tool	D 8	62
Grindstone	51	
Truing or turning	125	11R
Grip (See Machine or Device Having Grip; Handle)		
Assembling hand tool	29	278+
Chucks and sockets	279	
Cigar and cigarette holders	131	187+
Clasps	24	
Dashboard	296	71
For invalid in bed	5	445
Golf club	273	81R
Handle attachment	74	551.9
Handling implement	294	
Handwheel attachment	74	558
Metal working with	29	DIG. 19
Nose grips for spectacles	351	132
Integral with bridge	351	131
Pince-nez	351	71+
Paper article making	493	405+
Pulley		
Belt traction	474	9+
Strand catch	254	391
Robot	901	31+
Rope or cable		
Anchor	24	115R+
Grip pulley	254	391
Railway cable	104	202+
Store service	186	15
Strand brake	188	65.1+
Vehicle fender cable grip combined	293	3
Screw driver attachment	81	451+
Switch	200	157
Vehicle seat back	297	183
Gripping (See Device or Machine Utilizing)		
Climbers	182	133+
Clothesline combined	211	119.1+
Conveyor		
Article dispenser	221	210+
Endless	198	626+
Rotary	198	624+
Horizontal rack bar combined	211	124
Receptacle closure removers	81	3.4+
Shears	30	134+
Griseofulvin	549	345
In drug	514	462
Grist Mill Alarm	116	71
Grocery Store		
Cart	280	DIG. 4
Display case design	D 6	432+
Groins and Breakwaters	405	21+
Grommets (See Bushing; Eyelet)	D 8	356+
Eyelet lacing device	24	141+
Lining thimbles	16	2+
Sail cringle	114	114
Tire bead	245	1.5
Grooming Devices	119	83+
Design animal care	D30	158+
Design personal care	D28	
Bristled articles	D 4	130+
Snow or ice surface	37	219+
Grooving		
Gear cutting	409	1+
Ice shave implement	299	24+
Metal planing	409	304
Road or pavement joint	404	87+
Shoe sole slitting and	12	40
Type finishing and	29	24
Wood	144	136R
Slicing	144	173
Turning spiral	142	23+
Ground		
Detectors	340	650+
Ground effect machine, vehicle	180	116+
Vessels	114	67R
Glasses	354	161
Heating	405	131
Surface	126	271.1
Thermal mining	299	3+
Mats	5	417+
Speed calculators	235	61NV
Speed indicators	33	231

	Class	Subclass
Optical	356	27+
Grounds Electric		
Antielectrolysis	307	95
Box or housing	174	51
Conduit or cable	174	78
Earth	174	6+
Earth transmission radio systems	455	40
Pulse or digital communications	375	6
Electrolytic object protection	204	196+
Gas separator having	55	360
Indicator	340	650+
Lightning rod type	174	2+
Static electricity discharge	361	212+
With antenna	343	846+
Group Frequency Radio	455	
Grouper	101	58
Grousers	305	54+
Grouting		
Process	52	744
Growth and Displacement		
Polymerization	585	328
Logarithmic growth phase	435	802*
Guaiacol	568	653
Guanamine	544	205
Guanidines	564	230+
Biguanides	564	233+
Polyguanidines	564	236
Guano	71	21
Guanyl Hydrazines or Hydrazones	564	227+
Guards (See Safety Devices)		
Animal mouth	119	129+
Animal poke face	119	142
Automobile bumper	293	102+
Bed bottom	5	193
Bed clothing dirt	5	501
Bed side	5	425+
Beer faucet hole	137	380
Bookbinding	281	24
Bottle nonrefillable	215	17+
Building and structure parts		
Bank guard booth	109	9
Barbed fence	256	3+
Door and window gratings	49	50+
Fence top	256	11+
Fire screen	126	202+
Insect blocking	52	101
Insect or animal blocker	52	101
Ladder attachment	182	106
Ladder convertible to window	182	21
Pier fenders	14	76
Rail car seat guard control	104	241
Railway car side	105	439+
Road vault cover	404	25+
Burner nozzle	239	288+
Cable bracket for road	248	66
Chair or seat	297	464+
Closure knob attaching screw	292	351
Clothing		
Added to garment supporters	2	300
Bed	5	501
Body protecting	2	2+
Hat cord	2	189
Hatpin point	132	72
Pin fastener	24	156+
Trouser	24	72
Crib movable	5	100
Derailment		
Railway	104	242+
Store service wheeled carrier	186	30
Dust and mud		
Journal box lubricating	384	159
Journal box packed joint	277	130+
Land vehicle	280	152R+
Locomotive cab	98	28
Motor vehicle with	180	84
Railway car door or window	105	350+
Railway car window shield	98	25
Resilient tire splash	152	154
Roller bearing seal	384	477+
Typewriter bar	400	467+
Electric		
Collector trolley	191	72+
Conductor	191	35
Signal box	340	303
Eye glass nose	351	136+
Pince-nez	351	78+
Fastener bolt	292	346
Finger ring	63	15
Firearm	42	83
Fishing tackle		
Bait	43	42.4+

	Class	Subclass
Bait casting guard	43	41.2
Hook	43	43.2+
Harness buckle tongue	24	175
Heat exchanger	165	134.1
Highway fence	256	13.1
Hook	43	43.2
Keyhole	70	423+
Letter box chute	232	54
Light	362	376+
Liquid level sight glass	73	325+
Lock antipick	70	420+
Machine and tools		
Cable guide block	254	403
Cleaning nozzle	15	422
Control lever	74	566
Cutlery with	30	284
Dental	433	116
Fabric reeling	242	76
Fan	416	247R
Fork	30	323
Grinding wheel	51	268+
Hand rake	56	400.11+
Harvester cord knotter	56	448
Harvester cutter	56	307+
Machine casing	74	608+
Manicure edge	30	27
Piano pedal mouse	84	233
Power stop safety device	192	129R+
Razor edge	30	51+
Sewing machine	112	261
Shear plural blade	30	233
Shoe sole sewing machine	112	40
Shovel edge	294	56
Sickle	56	298
Slide	384	15+
Spinning machine stand	57	352
Spinning ring with	57	121
Spool holder thread	242	140
Textile wash board splash	68	225
Textile wringer roller	68	264+
Way	384	15+
Weaving shuttle	139	193
Wood saw	83	440.2
Woodworking cutter	144	251R
Zither chord selector	84	289
Nozzle	239	288+
Plant covers	47	26+
Platform associated	182	113
Pocket match safe with wind	206	97+
Railway coupling	213	154
Railway track	238	17+
Crossing tread	246	456
Fence	256	14+
Foot guard	238	379+
Rail detachable	238	140
Switch nonderailing	246	422
Switch point tread	246	441
Rein	54	73
Design	D30	141
Sheet retainer pin	402	71
Ship		
Fenders	114	219+
Mooring line	114	221R
Torpedo	114	240D+
Smoking tobacco ash	131	174+
Splash for cabinet	312	229+
Tree covers	47	20+
Tree trunk	47	23+
Vehicle rein supports and	280	181
Watch	63	24+
Wheel		
Brake	188	233.3
Brake shoe fastener with	188	237
Land vehicle	280	160+
Land vehicle dust guard with	280	152.2
Sawmill carriage	37	194
Tires resilient splash	152	154
Track sander hopper	291	39
Window or door	D25	53
Guayule Rubber	528	930
Gudgeons		
Partition roll type with roller	160	326
Rotary shafting	464	184
Guide (See Associated Device)		
Agricultural implement part		
Cornstalk type harvester	56	119
Cotton picker spindle	56	41+
Reciprocating cutter	56	305
Apparel folding	223	37+
Bearings	384	
Belt and sprocket	474	

	Class	Subclass
Brush striping type	15	166
Cable line or strand	254	389+
Bobbins and spools combined	242	125+
Cable railway gripper combined	104	202+
Carrier reels combined	242	85+
Cloth stretching clamp chain	26	71+
Fishing rod line guide	43	24
Knitting feed	66	125R+
Knotter	289	15
Sash cord	16	210+
Spool holders combined	242	140
Textile spinning	57	352+
Textile spinning centrifugal pot	57	76+
Textile spinning rail or support	57	136+
Textile spinning receiving twist flier	57	71
Textile strand feeding	226	196
Textile warp	28	172+
Thread stripping guide	28	222+
Winding and reeling	242	157R+
Camera		
Focusing	354	191
Magazine type quarter turned plate	354	183
Sliding plate magazine type	354	180
Closure		
Automobile glass	49	404+
Door and window frames	49	404+
Hanging or drape type	160	330+
Parallelogram type	160	159
Plural panel type	160	181+
Roll panel type	160	238+
Sliding car door	49	404+
Sliding door or panel	16	87R+
Venetian blind	160	172
Compartmented letter box	232	26
Conveyer belt	198	837+
Automatic shew	198	837
Load retainer	198	836
Conveyers	193	
Coopers stave jointing saw	147	34
Copyholder with static	226	196+
Cutting tool	83	821+
Deflector for fluid motor	415	216
Derailment preventing	104	245+
Dispenser with material guide	222	
Movable guide type	222	526+
Drawer	384	20+
Ball bearing	384	18
Cabinet combined	312	350
Combined with drawer	312	330R+
Roller bearing	384	19
Earth boring tool combined	175	408
Electric arc furnace electrode	373	
Arc furnace devices	373	60+
Electrode support	373	94+
With electrode holder or guide guide	373	69
Elevator	187	95+
Fluid or fluent material (See distribution, fluid; feeder; nozzle; pipe)		
Acoustical mufflers	181	233+
Chutes skids guides ways	193	
Conveyer fluid type	406	191+
Metal founding furnace or crucible	266	275+
Nozzle, spout, pouring devices	222	566+
Molten metal	222	591+
Receptacle filling	141	365+
Receptacle filling	141	391
Water control	405	80+
Forging hammer or press head	72	428
Foundry flask	164	385+
Gate hanger		
Bracket	16	90
Roller	16	91
Wheel mounts	16	106
Hair cutting and dressing	132	45R
Hand (See hand grips guards guides rests and straps)		
Hat fastener	132	65+
Line on fishing rod	43	24
Liquid level or depth gauge	73	320
Lock or latch part		
Bolt casing	292	337
Combination setting	70	331
Internal key guides	70	453
Key insertion	70	454
Machine or instrument		
Bottle cleaner	15	66

	Class	Subclass
Cultivator lister furrow	172	26
Cultivator tree or stake guided	172	24
Drill	408	72R+
Drill planter check correcting	111	16+
Peripheral face abrading machine	51	102
Radial face abrading machine	51	128
Radial face opposed abrading machine	51	116
Rotary cylinder abrading tool	51	208
Rotary disk abrading tool	51	210
Shoe blacking polisher	15	266
Sole and heel edge trimmer	12	90
Stationary abrading tool	51	214
Manual cutting tool (See machine or instrument guide)	30	286+
Coring or gage paring knives	30	282
Ice pick	30	164.7
Miter box	83	761+
Nippers	30	179
Razor and sharpener	30	37
Shears	30	233
Tool sharpener	76	88+
Metal can head applying	413	26+
Metal coiling	72	142+
Metal rolling	72	250+
Metal turning template	82	14R
Miter box	83	761+
Monorail system	104	118+
Rolling stock	105	141+
Paper calender doctor combined	100	173+
Pattern		
Gear cutting	409	2
Wood sawing	83	747
Wood shaping	144	137+
Wood turning	142	
Pencil sharpening implement	30	451+
Pencil sharpening machine	144	28.1+
Photography film developing	354	339
Piano key	84	436
Piano key fulcrum combined	84	434
Plain bearings	384	26+
Rail clamp lateral adjusting	238	347
Rail splice bar	238	246+
Railway automatic switch	246	348
Railway rolling stock truck element		
Axle box	105	218.1+
Bogie bolster	105	207
Swinging bolster column mounted	105	191
Saw making		
Band saw type	76	74
Filing	76	36
Screw driver combined	81	451+
Sewing machine		
Glove	112	20
Hat	112	15
Leather channel	112	50
Leather edge and crease	112	51
Leather welt	112	52
Needle	112	227
Work	112	136+
Shaft bearing	384	29
Stage scenery	272	23
Tap metal screw thread type	408	72R+
Tool sharpener	76	88+
Traffic (See traffic lights and guides)		
Type casting	199	3+
Typewriter part		
Keyboard hand guide	400	715+
Ribbon	400	248+
Type bar	400	456+
Upending bedstead shifting fulcrum	5	137
Violin bow	84	283
Washing machine wringer	68	264+
Wave electromagnetic	333	239+
Weaving pattern	139	332
Web or strand	226	
Weigher loading	177	161+
Wheelwright spoke setter type	157	4
Winding and reeling fabrics	242	76
Wood saw	83	821+
Woodworking		
Beveling slicer	144	169
Curved work shaping	144	152
Tool	408	72R+
Work	144	253R
Wrench element		
Pivoted outer jaw traveling fulcrum	81	109
Writing or scribing (See masking; stencils and stenciling)		
Educational	434	164

	Class	Subclass
Scribers	33	18.1^{+}
Straightedge	33	403^{+}
Guided Missile		
Control	244	75R^{+}
By radio (See type of servomotor system)		
Design	D12	16.1
Manned	244	75R^{+}
Radar not within missile	342	62
Space ship	244	158R^{+}
Telemetering	340	870.1^{+}
Torpedo type	114	20.1^{+}
Tracking by antenna	342	75^{+}
Aiming a gun	342	67
Unmanned	244	3.1^{+}
Guillotine (See Execution Devices)		
Shears	83	613^{+}
Guitar	84	267
Bridges	84	298
With tailpiece	84	299
Electric amplifier	84	1.16^{+}
Gum (See Synthetic Resin or Natural Rubber)	520	1^{+}
Chewing gum package	206	800*
Gum ball vending machines	D20	7
Gums and derivatives	536	114
Nitrogen containing	536	52
Resins (See resins)		
Rubber	520	
Tablet form		
Design	D 1	
Gumming		
Envelope manufacture	493	220^{+}
Saw aperture	83	850
Saw making	76	30^{+}
Gun (See Firearm)	42	
Airgun	124	56^{+}
Antiaircraft	89	
Automatic	89	125^{+}
Blowback	89	194^{+}
Firing devices	89	132^{+}
Gas piston type	89	191.1^{+}
Movable barrel	89	160^{+}
Movable chamber	89	155^{+}
Band type clasps	24	2.5
Barrel		
Firearms	42	76.1^{+}
Materials or coatings	42	76.2
Ordanance	89	14.5^{+}
Bayonet combined	42	86
Billy club	42	1.16
Blowgun	124	62
Bore inspection	356	241
Breakdown type	42	40
Cane gun	42	52
Cattle slaughter type	42	1.12
Control calculators	235	400^{+}
Gun training mechanism	89	41.1^{+}
Motor operated	89	41.2^{+}
Cotton	536	35^{+}
Composition containing	149	94^{+}
Over 10%	149	96^{+}
Design	D22	100^{+}
Grease	D 8	14.1
Pistol	D22	104^{+}
Racks	D 6	552^{+}
Sights	D22	109
Toy	D21	146^{+}
Dummy	42	106
Ejectors	42	25
Electrically operated		
Firearms	42	84
Lighting devices	362	110^{+}
Ordnance	89	135
Electron	313	441^{+}
Extractors	42	16^{+}
Firing mechanisms	42	69.1
Revolver	42	65
Upward tilting breech	42	41^{+}
Flare	42	1.15
Fluid pressure adapter	124	58
Foob, ie fire out of battery	89	42.3
Gatling type	89	12
Grenade launchers	42	105
Gun engaging means	102	483^{+}
Handles	89	1.42
Heaters	89	1:12
Howitzer	89	
Implement combined	42	90^{+}
Indicators	42	1.1
K gun	89	1.1

	Class	Subclass
Knife combined	42	53
Loading	89	45
Lubricating or caulking type	D 8	14.1
Machine gun	89	9^{+}
Toy	D21	146^{+}
Magazine	42	87^{+}
Making	29	1.1^{+}
Mechanical	124	
Mount	89	37.1^{+}
Training mechanism	89	41.1^{+}
Mounted	89	
Movable chambers		
Firearms	42	39.5
Ordnance	89	155
Multiple barrel	89	1.41
Nonrecoil	89	1.7
Pen and knife	42	1.9
Port ship	114	173^{+}
Stopper	114	175
Portable	42	
Powder		
Ammunition loading with	86	
Bags	102	282
Engine starters	123	183
Engines	123	24R
Forms	102	283^{+}
Racks	211	64
Rapid fire	124	72^{+}
Recoil operated	89	162
Recoilless	89	1.7
Rests	42	94
Revolver	42	59^{+}
Safety mechanism	42	70.1^{+}
Automatic guns	89	137^{+}
Revolvers	42	66
Semiautomatic	89	4.5^{+}
Shields	89	36.1^{+}
Deflected ray tube	356	253^{+}
Shotguns	42	
Sidearms	D22	
Sights	33	233^{+}
Design	D22	109
Design, telescopic	D16	132
Optical system	356	247^{+}
Stocks	42	71.1^{+}
Teargas	42	1.8
Telescopic gunsight	D16	132
Toy simulating	42	54^{+}
Ammunition	102	281
Machine gun or projector	124	29
Non-detonating	446	473
With sound	446	405^{+}
Training in gunnery	434	16^{+}
Trigger protectors	42	70.7
Underwater	42	1.14
Walking cane combined	4	515^{+}
Water gun	124	56^{+}
Toy	D21	146^{+}
Water pistol	222	79
Well tubing perforator	175	2
Y gun	89	1.1
Gussets		
Garment	2	275
Gut or Gut Treatment	8	94.11
Splitter	83	926A*
Guttapercha	525	331.9^{+}
Gutter	405	119^{+}
Eaves trough	52	11^{+}
Electric conductor underground structure	174	39
Road and pavement	404	2^{+}
Support design	D 8	363
Guy	52	146^{+}
Bed spring and frame	52	272
Gymnastic Devices	272	109^{+}
Coin controlled apparatus	194	
Gypsum	423	554
Calcining	106	100^{+}
Coating or plastic compositions containing	106	109^{+}
Alkali metal silicate	106	77
Gyrating		
Reciprocating sifter		
Actuating means	209	366^{+}
Horizontal and vertical shake	209	326
Horizontal shake	209	332
Gyratory Crusher		
Jaw crushers rotary component	241	207^{+}
Parallel flow through plural zones	241	140
Series flow through plural zones	241	156
Gyro Stabilized		
Article support	248	183

	Class	Subclass
Furniture for ships	114	119
Gyroplane (See Aircraft)	244	17.11^{+}
Gyroscope	74	5R^{+}
Aerial camera combined	354	70
Aircraft control	244	79
Direction indicator	33	318^{+}
Gimbals	248	182
Gun sight combined	33	236
Gyroscopic compass	33	324^{+}
Telemetric system combined	340	870.7^{+}
Monorail rolling stock	105	141^{+}
Suspended	105	150^{+}
Rotors	74	572
Rotors and flywheels	74	
Ship antiroll	114	112
Ship stabilizer	114	122
Ship steering	114	144R
Speed responsive devices	73	504
Torpedo	114	24
Torpedo steering	114	24
Toy	446	233^{+}
Transmission	74	64

H

	Class	Subclass
H Acid	260	509
Habitat, Submarine	114	314
Hack Saw	30	507^{+}
Combined	7	149^{+}
Design	D 8	96
Hanging	83	783^{+}
Hackling		
Combing	19	115R^{+}
Decorticating	19	5R^{+}
Hacks Tree	30	121
Haemocytometer	356	39
Testing lenses	356	124^{+}
Hair		
Artificial furs	428	85^{+}
Artificial structure	132	5
Beauty parlor equipment	D28	10^{+}
Brush	15	160^{+}
Carried hat fasteners	132	60
Clippers	D28	52^{+}
Coating compositions	106	155^{+}
Curlers		
Curling iron	D28	38
Electrically heated	219	222^{+}
Fluid fuel heated	126	408^{+}
Cutters	30	
Design	D 8	57
Design clippers	D28	52^{+}
For inside ear or nose	30	29.5
Hair planers	30	30^{+}
Drying on head		
Apparatus	34	96^{+}
Processes	34	3
Supports for	34	101
Dye applicator	D28	7^{+}
Dyeing and dyes	8	405^{+}
Fasteners	132	46R^{+}
Design	D28	39^{+}
Fertilizer from	71	18
Hairpiece	D28	92^{+}
Inserters	128	330
Jeweled fastener	63	2
Net	132	49
Pins (See hairpins)	132	50R^{+}
Planers	30	30^{+}
Removing (See notes)	30	32
Burial preparation	27	21^{+}
Butchering	17	1D
Coarse or water hair from fur	69	24^{+}
Cutters for inside ear or nose	30	29.5
Depilating untanned skins	8	94.16
Depilatories	8	94.16^{+}
Electric needle	128	303.18
Electric needle supports	128	303.19
Fiber liberating	19	2
Fur treatment	69	24^{+}
Process	17	47
Razors	30	32^{+}
Surgical instruments	128	355
Tweezers	128	354
Shampooing apparatus	4	515^{+}
Shearing, fur finishing	26	15R^{+}
Thinning shears	30	195
Springs	368	175^{+}
Strand making		
Covering by spinning etc	57	4
Spinning etc	57	28^{+}
Synthetic resin or natural rubber containing compositions	524	12

	Class	Subclass
Textile spinning etc	57	29
Thinners	30	195
Design	D28	52$^+$
Toilet preparations	424	70$^+$
Treating process	132	7
Tufting hair in doll or wig	112	80.2
Waving	132	7
Hairpin	132	50R$^+$
Design	D28	39$^+$
Dispenser	132	1A
Hat fastener cord or loop and	132	61
Making	140	87
Packaging	227	25
Half Belts	2	309
Half Wave		
Gas rectifier	313	567$^+$
With hot cathode	313	629
High voltage rectifier	313	317$^+$
With emissive cathode	313	310
With thermionic cathode	313	310
Rectifier system	363	13$^+$
Circuit interrupter for	200	
Dynamoelectric machine	310	10$^+$
Electronic tube for	313	317$^+$
Gas tube type	363	114$^+$
Power packs	307	150
Unidirectional impedance for	357	
Vacuum tube type	363	114$^+$
With filter	363	39$^+$
With voltage regulator	363	84$^+$
Halftone		
Blanks and processes printing	101	401.1
Etching	156	654$^+$
X-art	156	905*
Photographic process	430	396
Photographic screens	350	322$^+$
Chemically defined	430	6$^+$
Printing plates	101	395
Halides (See Material Halogenated)		
Hydrocarbon	570	101$^+$
As azeotropes	203	67
Electromagnetic wave synthesis	204	157.15$^+$
Electrostatic field or electrical discharge synthesis	204	169
Metal	423	462$^+$
Electrolytic synthesis	204	94
Nitroaromatic	568	927$^+$
Nonmetal inorganic	423	462
Organic acid	260	543R$^+$
Rubber hydrohalide	525	332.3
Hall Effect Means in an Amplifier	330	6
Haloamines	564	114$^+$
Acyclic	564	118$^+$
Hydroxy or ether containing	564	119
Plural difluoramime groups	564	121$^+$
Unsaturated	564	120
Alicyclic	564	117
Amidines	564	116
Halogen Compounds (See Material Halogenated)		
Halogenated Carboxylic Acid Esters	560	1$^+$
Acyclic acid esters	560	226$^+$
Of phenals	560	145
Acyclic amino acid esters	560	172
Acyclic carbamic acid esters	560	161
Acyclic oxy acid esters	560	184
Acyclic polycarboxylic acid esters	560	192
Acyclic unsaturated acid esters	560	219
Alicyclic	560	125
Aromatic amino acid esters	560	47
Aromatic carbamic acid esters	560	30
Aromatic polycarboxylic acid esters	560	23
Oxybenzoic acid esters	560	65
Phenoxyacetic acid esters	560	62$^+$
Halogenation (See Halides)	260	694
Halohydrin	568	841$^+$
Halothiocarbonate Esters	558	249
Halowax	570	181
Halter		
Brassiere type garment	450	1$^+$
Feed bags supported on	119	66
Harness	54	24$^+$
Design	D30	134$^+$
Poke with bar and	119	141
Snap releasers	119	114
Hamburger		
Cookers	99	422
Grinders	241	
Molding and shaping (See briquetting meat)		
Hames	54	25$^+$
Collar combined	54	18R
Design	D30	137
Traces and connectors	54	30$^+$
Tugs	54	32$^+$
Hammer	81	20$^+$
Automobile fender straightening	72	705
Burglar alarm	116	88$^+$
Claw	254	26R
Combined with additional tools	7	143$^+$
Design	D 8	75
Drop forging	72	435$^+$
Earth boring tool combined	175	135
Firearm	42	
Forging	72	476$^+$
Heads for piano actions	84	254
Impact clutch type	173	93.5
Implement combined	81	463$^+$
Awl or prick punch	30	358$^+$
Internal combustion charge igniter rocking electrode	123	157
Leather compacting	69	1
Magazine	227	133
Making		
Forging dies for	72	470$^+$
Processes of	76	103
Metal bending	72	462$^+$
Mills	241	185R$^+$
Parallel material flow	241	138
Perforated discharge	241	86$^+$
Process	241	27
Series material flow	241	154
Musical instruments		
Piano	84	236$^+$
Stringed instrument	84	323$^+$
Tuning	84	459
Nut cracker	30	120.1
Pile driver	173	90$^+$
Punching machine	83	
Riveting	72	476$^+$
Road rammer	404	133
Rock drilling	175	135$^+$
Rod encircling type	81	27
Saw stretching machine	76	26
Scale removing	29	81D
Shoe lasting stretcher and	12	109
Stoneworking combined	125	6$^+$
Impact tools	125	40
Tool driving	173	90$^+$
Tube cleaner inside	15	104.7
Tuning for pianos	84	459
Typewriter		
Bar	400	388$^+$
Key wheel	400	154$^+$
Welding	228	24
Woodworking	81	20$^+$
Hammock	5	120$^+$
Berth	105	320
Design	D 6	386$^+$
Swing	297	277
Hamper or Basket	D32	37
Laundry	D32	37
Hand		
Article manipulated by mechanical hand-like movement	414	1$^+$
Artificial	623	57$^+$
Baskets	217	122$^+$
Checker radioactivity	250	336.1
Miscellaneous	250	336.1
Scintillation type	250	361R$^+$
Clock and watch	368	228
Design	D10	127
Coverings	D 2	610$^+$
Doll	446	327$^+$
Dryers electric	219	369$^+$
Design	D28	54.1
Electric applicators	128	796
Exercising appliances	128	26
Gloves mittens and wristlets	2	158$^+$
Design	D 2	617$^+$
Grips (See handle)		
Blotter	34	95.2
Bowling ball	273	64
Coating implement with material supply	401	6$^+$
Crutch hand hold	135	72
Golf club	273	81R$^+$
Hand wheel rim	74	558
Handle bar	74	551.9
Penholder	15	443
Rein holds	54	74
Ski pole	280	821$^+$
Tennis racket	273	75
Wrench	81	300$^+$
Guards and protectors	2	16
Cutting hand tool	30	295
Design	D29	20$^+$
Fork	30	323
Machine safety stop	192	130$^+$
Guides rests and straps		
Coffin lowering strap	27	33$^+$
Railway car strap	105	354
Stringed instrument rest	84	328
Typewriter keyboard guide	400	715$^+$
Levers	74	523$^+$
Looms	139	29$^+$
Manipulated implements (See hand tools)		
Apparel fluting iron	223	36
Atomizer	239	337$^+$
Bars for carrying	294	15$^+$
Bells	116	171
Blowtorch	431	344
Cutlery	30	
Dental	433	141$^+$
Electric flashlight	362	208
Flatiron	38	95
Lantern	362	257$^+$
Material handling	294	
Pick for stringed musical instrument	84	322
Sorting	209	614
Telephone dialing tool	379	456
Toaster or broiler	99	394
Muffs design	D 2	611$^+$
Operated devices (See type of device operated)		
Operated devices agricultural and earth working		
Corn huskers	130	4
Cultivating tools	172	351$^+$
Forks and shovels	294	49$^+$
Harvesting cutter	56	239$^+$
Plant irrigators	47	48.5
Planting dibble	111	99
Planting drill	111	82
Rakes	56	400.1$^+$
Raking and bundling	56	342
Scoop excavator wheeled	37	130
Snow excavator	37	196$^+$
Snow excavator and melter	37	230
Subsoil irrigators	111	7.1$^+$
Organs automatic	84	84$^+$
Setting mechanism for electric clocks	368	60$^+$
Stamps	101	405
Design	D18	14$^+$
Surgical thermal wear	128	381
Tools (See type of tool)	81	
Assembling or disassembling	29	270$^+$
Barrel croze	147	24
Barrel head scriber	147	41
Boot and shoe making	12	103$^+$
Butchering intestine cleaners	17	43
Cherry seeder	30	113.1
Cleaning	15	
Cutlery	30	
Dental	433	25
Fish dressing	17	66$^+$
Forks and shovels	294	49$^+$
Hog scrapers	17	19
Meat and vegetable	30	
Meat tenderers	17	29$^+$
Oyster openers	17	74$^+$
Peach stoner	30	113.1
Pick, miners	125	43
Plural diverse tools	7	
Railway mail delivery	258	3
Raisin seeder	30	113.1
Rake	56	400.1$^+$
Tamper for track laying	104	13$^+$
Tufting eg chenille applying	112	80.3$^+$
Tying cords or strands	289	17
With light	362	119$^+$
Vehicles		
Land occupant propelled by hand	280	242R$^+$
Railway hand car	105	86$^+$
Scoop excavator	37	130
Self loading type	414	444$^+$
Truck ladder	182	127
Truck ladder with erection means	182	63$^+$
Trucks and barrows	280	47.17$^+$
Trucks and barrows dumping type	298	2$^+$
Wheels	74	552$^+$

	Class	Subclass
Hand Cart Design	D34	12$^+$
Hand Mirror	350	640
Plural mirror	350	640$^+$
Handbags	150	100$^+$
Design	D 3	42$^+$
Frame	D 3	54
Latch	D 8	331$^+$
Handcuffs	70	16$^+$
Carrier	224	914*
Handgun (See Pistol; Revolver)		
Handicapped Person Handling	414	921*
Handkerchief	139	383R
Design	D 2	500$^+$
Garment attaching means	24	3R$^+$
Garment worn	2	279
Lace	87	10
Medicated	604	358$^+$
Silk	139	428
Woven fabric	139	420R$^+$
Handle (See Grip)	16	110R
Article carrier	294	
Asymmetric-type		
Coffee, teapot and pitcher	D 8	DIG. 9
Lever type	D 8	DIG. 1
Bar	74	551.1$^+$
Design	D12	178
Handgrip design	D 8	303
Battery handle	16	DIG. 15
Bicycle grip	D 8	DIG. 8
Vehicle equipment design	D12	178
Brushware	D 4	
Cabinet with	312	244
Cane type	135	65$^+$
Cast or molded handles	16	DIG. 19
Cleaning equipment, household	D 7	
Cleaning equipment, machinery	D15	
Composition handles	16	DIG. 18
Cooker	D 7	
Dashboard	296	71
Elongated, for tools	D 8	DIG. 7
Extensible	16	115
Fastenings	16	DIG. 24
Brushing machines	15	22R$^+$
Clamp band	16	DIG. 25
Cutlery blade and handle	30	340$^+$
Cutlery detachable blade	30	329$^+$
Golf club handle	273	80R
Luggage	190	115
Tool	403	
Faucet and valve wheels	D 8	DIG. 3
Fluorescent and phosphorescent	250	465.1
Fly swatter	D22	124
Forming, paper	493	909*
Golf and ski grips	D 8	DIG. 6
Gun	89	1.42
Hand carrying	294	137$^+$
Hand grips, preformed, semipermanent	16	DIG. 12
Handwheel	74	552
Handles for handwheels	74	557
Insulated	16	117
Pipe joint	285	38
Insulated	16	116R$^+$
Coating implement	401	3
Electric	174	46
Kitchen utensil	D 7	
Knife	30	162
Knife and fork	D 8	DIG. 4
Knob-type	D 8	DIG. 2
Locked	70	207$^+$
Luggage or handbag	D 8	300$^+$
Making		
Hollow cutlery type	76	106
Wood	144	11
Mattress	5	466$^+$
Digest, hardware	16	DIG. 28
Movable closure	49	460$^+$
Nozzle	239	525$^+$
Package and receptacle	294	137$^+$
Bail type	220	94R$^+$
Basket	217	125
Closure fastener combined	220	315
Dispenser	222	465.1$^+$
Dispenser discharge assistant combined	222	323$^+$
Dispenser trap chamber combined	222	441
Flexible bag	383	6
Luggage	190	115$^+$
Luggage fastener combined	190	118
Paper box	229	52R
Resilient wall dispenser	222	210
Trunk	190	39
Pan	D 7	
Pull-type	D 8	DIG. 1
Pyrotechnic or explosive articles	102	358$^+$
Shell thrower	102	486
Receptacle type handle (See magazine)	16	110.5
Cane	135	65$^+$
Coating implement	401	
Cutlery	30	125
Firearm magazine	42	7
Insulated electrically	174	46
Manicuring tool	132	75
Razor	30	86
Retractable	16	115
Sporting goods, misc	D21	
Switch	200	329$^+$
Sword	D 8	DIG. 5
Tool and implement	81	489$^+$
Cleaning tool	15	143R$^+$
Curry comb	119	94
Dispenser combined	222	191
Drive control	173	170
Drive fluid passage	173	168$^+$
Flexible abradant	51	392$^+$
Forks and shovels	294	57$^+$
Golf club	273	81R$^+$
Light support	362	399
Manipulating for drive or hammer	173	
Multiple tool	7	167$^+$
Printing stamp	101	405$^+$
Razor	30	85$^+$
Rigid abradant	51	205R
Sadiron	38	90$^+$
Saw	30	166R$^+$
Shears detachable blade and handle	30	260
Wrench	81	177.1$^+$
Valve		
Detachable actuator	251	291$^+$
Disabling means	251	89$^+$
Stops	251	284$^+$
Vehicle seat back grips	297	183
With indicia	40	10R
Handling (See Conveyor)		
Castings with casting device	164	269
Completed work, of sewing machine	112	121.29
Earth boring drilling fluid or cuttings	175	207$^+$
Fluid handling		
Baths, closets, sinks, & spittoons	4	
Pipes or tubular conduits	138	
Valves & valve actuation	251	
Ventilation	98	
Handicapped person	414	921*
Handler-type toys	414	915*
Handlers using parallel links	414	917*
Handlers with spring devices	414	913*
Implements		
Bakers peel	294	49
Combined or convertible	414	912*
Hand and hoist line implements	294	
Hand tools	81	
Sorting manually	209	614
Mechanism		
Coin	453	
Dryer for hollow articles with	34	105$^+$
Dryer rotary drum type with	34	108$^+$
Dryer rotary drum with	34	108$^+$
Dryer shelf or tray	34	194
Dryer with for dried material	34	236
Fish net	43	8
Furnace charging	414	147$^+$
Target	273	406
Trap shooting targets	273	406
Metal forging work	72	419$^+$
Nuclear reactor components	376	146
Nuclear reactor components	376	260
Nuclear reactor components	376	260
Nuclear reactor components	376	308
Charging or discharging	414	146
Photographic fluid treating	354	340
Pipe or cable	414	745$^+$
Submerged	405	158$^+$
Rolling mill work shifter	72	200
Shiftable handler, transmission line for	414	918*
Solid material handling		
Chutes, skids, guides, & ways	193	
Delivery to or from moving vehicle	258	
Elevators	187	
Fluid current conveyers	406	
Food	99	
Furnace ash	110	165R$^+$
Hand & hoist-line implements	294	
Hoists	414	592$^+$
Marine loading or unloading	414	137
Material or article	414	
Package & article carriers	224	
Piling in abutment or row	414	28$^+$
Pneumatic dispatch	406	
Power-driven conveyers	198	
Store service	186	
Traversing hoists	212	
Unpiling	414	112
Toy, material handling	446	424$^+$
Traversing hoists	212	
Vehicles with overhead guard for operator	414	914*
Handpiece (See Hand, Manipulated)		
Dental	433	114$^+$
Handrail	D25	119$^+$
Support	D 8	363
Handsaw	30	166R$^+$
Combined with diverse tool	7	148$^+$
Handwheel	74	552$^+$
Hanger (See Suspension Devices)	224	232$^+$
Bed bottom		
Frame	5	207$^+$
Slat	5	238
Broom type	248	110$^+$
Butchering	17	44.2$^+$
Eaves trough	248	48.1
Garment	223	85$^+$
Attached to garment	2	271
Design	D 6	315$^+$
Making	140	81.5
Gate	16	86.1$^+$
Illuminating light support	362	432$^+$
Movable component	52	64$^+$
Panel movable	16	87R$^+$
Pipe	248	58$^+$
Rack	211	
Railway		
Cable rail	104	115$^+$
Traction cable	104	182
Trolley	104	111
Rod for sheet or leaf handling	294	5.5
Scabbard	224	232
Shaft	384	215$^+$
Sifter support	209	415
Slender article type, including broom	211	60.1
Suspended roof	52	83
Tent	135	119
Train vestibule face plate rod	105	22
Hanging		
Cable traversing hoist	212	133
Canopies	135	905*
Grindstone	51	168$^+$
Paper	156	574$^+$
Design	D 5	
Plant receptacles	47	67$^+$
Ships rudder	114	165
Sign exhibiting	40	617
Wood saw	83	469$^+$
Hanks		
Finishing	28	287$^+$
Sails and rigging	114	114
Spinning twisting and twining	57	25$^+$
Winding	242	53
Harbor		
Channels	405	84$^+$
Floating	114	258
Wharves	405	284$^+$
Hard-hat	2	411$^+$
Hardening		
Aging	148	158$^+$
Iron	148	142
Iron with working	148	12.3
Cast iron	148	141
Electric hardening		
Apparatus	219	50$^+$
Inductive apparatus	219	6.5$^+$
Processes	148	13$^+$
Zone hardening of ferrous metal	148	150
Fat	260	409
Glass	65	111$^+$
Quenching iron	148	143$^+$
Localized heat treated stock	148	902$^+$
Localized treatment	148	145$^+$
Surface treatment	148	152
Working combined	148	12.4

	Class	Subclass
Steel	148	
Hardness Test	73	78+
Electric space discharge devices vacuum	324	413
Hardware Miscellaneous	16	
Brushings or lining thimbles	16	2+
Carpet fasteners	16	4+
Casters	16	18R+
Checks & closers	16	49+
Design of	D 8	
Ferrules, rings & thimbles	16	108+
Gate hangers	16	86.1+
Handles	16	110R+
Hinges	16	221+
Nesting hinge leaves	16	DIG. 29
Sash balances	16	193+
Sash weights	16	216+
Sash-cord fasteners	16	202+
Sash-cord guides	16	210+
Tracks, travelers, panel hangers	16	87R+
Window-bead fasteners	16	220+
Harmidine	546	86
Harmonic		
Electric		
Filtering	333	175+
Generation by oscillators	331	53
Harmonic or reed telegraphs	178	47+
Intensifier for loudspeaker or earphone	381	161
Party line ringing	379	179
Semiautomatic telephone ringing	379	252
Telephone ringing	379	372+
Wave analysis	324	77R
Music		
Tunings or arrangements of string	84	312R
Musical		
Electric organ selective control of tone partials	84	1.19+
Electric organ with tuned generator control	84	1.11+
Piano harmonic dampers	84	234
Relays	335	94
Circuits	361	182+
Selective system	340	825.39
Relays	335	101
Structure	335	243+
Harmonicas	84	377+
Design	D17	12
Harmonica digest	84	DIG. 14
Harmoniums (See Organ, Reed)	84	351+
Harness (See Body, Harness)	54	
Animal stocks	119	101
Body		
Fire escape	182	3+
Land vehicle occupant	280	290
Parachute	244	151R
Bridles	54	6R+
Buckles	24	164+
Design	D30	139
Checking devices	54	70
Collars	54	19R+
Design	D30	134+
Feed bags supported by	119	67
Hames	54	25+
Hand package carrier	294	157+
Loom	139	82+
Drawing in warp	28	203+
Hand pushed	139	30+
Shedding hand pushed	139	57+
Stopping actuated by	139	337
Vibrating griff levelers	139	75
Motions for dobby looms	139	66R+
Pads	54	65+
Saddles	54	37+
Stirrups	54	47+
Trunk	190	27
Protecting	190	26
Harp		
Aeolian ie wind driven	84	330
Autoharp or zither	84	286+
Jews harp	84	375
Toy	446	397+
Musical stringed instrument	84	264+
Design	D17	16
Piano	84	258
Support for lamp	362	417
Harpoons		
Bomb lances	102	371
Fishing	43	6
Gun	89	1.7
Mechanical	124	
Handling implement	294	125+
Line carrying shells	102	504
Harrow	172	
Cultivator combined	172	133+
Design	D15	10+
Disc sharpener	76	85
Abrading attachment	51	246+
Plow combined	172	133+
Roller combined	172	173
Scraper or drag combined	172	197
Harvester		
Agriculture	56	
Bundle discharging carrier	56	474+
Carrier	56	473.5
Compressing and driving	56	432+
Guards	56	DIG. 24
Motorized	56	10.1+
Rake and rakers		
Bundling and	56	341+
Hand rakes	56	400.1+
Horse rake	56	375+
Loading and	56	344+
Tedders and	56	365+
Shocker	56	401+
Stalk choppers	56	500+
Catching and	56	194+
Conveying and	56	153+
Conveying binding, and	56	131+
Raking and	56	193+
Windrowing	56	192+
Design	D15	26+
Knife holder	51	222+
Knife sharpener	51	36
Marine	56	8+
Mining	299	
Potato or beet digger	171	
Tree felling	144	34R
Hasp		
Closure fastener	292	281+
Lock	70	2+
Hassocks	297	462
Design	D 6	349+
Hat	2	175+
Band	2	179
Box	206	8+
Brush	D 4	
Design	D 2	509+
Eye shields attached to	2	10
Fasteners	132	57+
Forms	223	24+
Labels and tags	40	329
Making apparatus	223	7+
Making apparatus	D15	122+
Ornaments		
Design	D 2	263
Pins	132	68+
Design	D11	209
Stickpin	D11	47
Protective packing	206	9+
Racks	211	30+
Design	D 6	315+
Receptacles	206	8+
Safety helmet	D29	12+
Sewing	112	12+
Methods and seams	112	263.1
Shower cap	D 2	510+
Try on linings	2	63
Wires making	140	77
Hatch		
Building	52	20
Closure operators for	49	324+
Elevator	187	62+
Fasteners	292	256.5
Freight car	105	377
Ship	114	201R+
Hatchets	30	308.1+
Carrier	224	234
Design	D 8	76
Making		
Dies	72	470+
Processes and blanks	76	103
Hatpin	132	68
Non ornamental	D11	207
Protective packing	206	9+
Stickpin	D11	47
Hauling (See Pull)		
Cables drum type	254	266+
Hawks, Masonry and Concrete	294	3.5
Hawse Pipes	114	179+
Hawsers (See Rope)		
Hay		
Handling		
Bale accumulator, vehicle carried	414	111
Curing and preserving	426	321+
Distributors	414	25
Forks	294	107+
Grapples expanding pivoted	294	98
Grapples fixed and movable jaw	294	105
Grapples pivoted jaws	294	107+
Harpoon	294	126+
Hay retainer for vertically swinging fork	414	721
Hoist line fork	294	120+
Load binders	280	180
Pitch forks	294	55.5+
Round bale	414	24.5
Stack shapers	414	132
Hayracks		
Brakes for	188	14
Self-loading movable car	414	522
Vehicle body	296	6+
Convertible box and	296	11+
Pelleters	100	
Hcg	436	818*
Head		
Brake		
Beam adjustable head	188	219.6+
Beam combined	188	219.1+
Beam fixed head	188	222.1+
Shoe fastener	188	236
Shoe fastener interlocking	188	242+
Coupling		
Animal draft D type	278	72+
Animal draft L type	278	67
Animal draft T type	278	68+
Animal draft whiffle tree	278	105+
Railway draft cushioned	213	18+
Railway draft link or bar	213	182+
Rod to base plate or head	403	187+
Through intermediate member	403	187+
Tool handle fastening plate	403	
Tool handle fastening split	403	302
Covering with light	362	105+
Drum musical	84	414
Gear telephone	379	430
Implement and tool		
Brush	15	171+
Hand fork and shovel pivoted and adjustable	294	53.5
Mop	15	228+
Plow rotary chopper	172	518+
Razor handle	30	89
Internal combustion engine cooling	123	41.72+
Multiple for bolt		
Hooked end	292	116+
Sliding	292	156+
Sliding and rotary	292	59
Sliding and swinging	292	68
Spring arm	292	91
Swinging	292	213+
Swinging and hooked end	292	56
Nail, spike or tack	411	439+
Design	D 8	388+
Nozzle		
Sprinkler, fire	169	37+
Prop	248	357
Puller carcass	17	1R
Rail		
Detachable	238	143+
Joint	238	221+
Reversible	238	133
Rudder	114	169
Screw machines for working on	10	5+
Sewing machine	112	259
Shampooing apparatus	4	515+
Spar ship	114	94
Stock lathe	82	28R+
Toy	446	397+
Tripod	248	177+
Trolley electric	191	59+
Valve		
Cooperating seat	251	333+
Cooperating stem	251	84+
Rotary plug	251	309+
Structure	251	356+
Swiveled	251	88
Well casing	166	75.1+
Headboards and Sections		
Bedstead extension	5	183
Bedstead folding	5	178+
Bedstead upending	5	148
Invalid bed	5	67+
Sofa	5	53R+

	Class	Subclass
Headed Fastenings	411	500+
Making	10	
Header		
Boiler	122	
Harvester	56	
Manufacture	29	157.4
Radiator	165	153
Heading (See Forging)		
Barrel	147	6
Bars by upsetting	72	470+
Bolts	10	11R+
Cartridge cases	29	1.31
Filling combined	141	73+
Filling combined	141	115+
Nails	10	47+
Cut	10	28+
Wire	10	43+
Pins	163	6
Glass headed	65	142+
Rivets	10	11R+
Screws making	10	2+
Sheet metal cans	413	26+
Apparatus for making	413	69+
Soldering apparatus	228	
Headlight		
Dirigible	362	62+
Sealed beam	313	113+
System electric	315	82+
Testing	356	121+
Vehicle	362	61+
Headphones		
Earpieces	381	187
Electrical	381	183+
Supports	381	188+
In or under pillow	381	158
Mechanical	181	
Pillow type	381	158+
Radio	455	344
Support for phone	381	205+
With radio	455	149
Headrest		
Bath tubs	4	578
Chair	297	391+
Chair reclining	297	61
Coffin	27	13
Design	D 6	501
(See also)	D 6	601
Shampooing apparatus	4	523+
Sofa	5	52
Water closet	4	254
Headset	D14	36
Headwear		
Decorative	2	171+
Design	D 2	509+
Guard and protector	2	410+
Hat fastener	132	57+
Helmet	2	410+
Protective	2	410+
Receptacle	206	8+
Sports	2	425+
Structure	2	171+
Surgical	128	380
Health		
Computer controlled, monitored	364	413+
Lamps	128	395
Heap Still	202	210
Hearing Aid	D24	35
Design	D24	35
Ear trumpets	181	129+
Electrical	381	68
Non electrical	181	129+
Dentiphone	181	127
Ear trumpet	181	129+
Stethoscope	181	126+
Stethoscopes	181	126+
Hearing Therapy		
Voice reflector	D24	35
Hearses	296	16+
Design	D12	82+
Heart		
Artifical	623	3
Artery	623	1
Valve	623	2
Energized magnet actuator	251	65
Motor operated by motivating mass	60	516
Pivoted, line condition change responsive	137	527
Cardiac assist device	128	1D
Expansible chamber pump	417	472
Heart cam, register	235	144HC
Heart-lung digest	128	DIG. 3
Pacemaker	128	419P+
Shaped packaging	D 9	315
Hearth		
Cooking	126	12+
Electric furnace		
Arc	373	60
Electroslag	373	42
Glass	373	27
Induction	373	138
Resistance	373	109
Fireplace	126	143
Fenders	126	203
Material heating	432	
Thermolytic still	202	102+
Heat, Heater & Heating (See Furnace; Radiator)	432	
Absorption meters	374	39
Accumulators	165	
By change of state	165	104.15
Ovens having	126	273.5
Regenerators	165	4
Vessels having	126	375
Air		
Automatic temperature regulation	236	10+
Cookstove combined	126	6
Distributing system combined	237	
Fireplace	126	120+
Furnaces	126	99R+
Gas separation combined	55	
Heat exchange	165	
Heated ventilating flue	98	45+
Illuminating burner attachments	126	248+
Nuclear reactors	376	383
Regenerative	432	214+
Stoves	126	58+
Animal or plant husbandry		
Brooders and incubators	119	30+
Electrically heated branding irons	219	245+
Fuel heated branding irons	126	402+
Greenhouses	47	17+
Hot beds	47	19
Orchard heaters	126	59.5
Preventing plant frosting	47	2
Seed testers	47	16
Sterilizing gathered soil	47	1.44
Tree covers	47	22
Waterers	119	73
Apparel forms	223	
Automatic temperature regulation	236	
Heating and cooling combined	165	14+
Heating systems	237	2A+
Automobile	237	28+
Design	D23	324+
Bathtub	4	545
Bedsteads	5	284
Bending metal by heating	72	54
Body	604	113+
Treating material	604	291+
Brooder and incubator		
Automatic temperature control	236	2+
Automatically controlled heating systems	237	3+
Brooder	119	31+
Combined	119	30
Heating systems	237	14+
Incubator	119	35+
Buildings	237	
Burner feed	431	
Cabinet combined	312	236
Carburetors	261	127+
Internal combustion engine	123	543
Chemical	126	263
Closets		
Dry closets	4	111.1+
Collectors, solar	4	461+
Combustion	431	
Computerized or monitored in product manufacturing	364	477
Cutting (See 4 note to class 30)	30	
Distillation and cracking	201	
Apparatus	202	
Mineral oils	208	46+
Solid carbonaceous material	201	
Drier	34	
Textile web spreader combined	26	92
Textile web stretcher combined	26	106
Earth	405	131
Earth boring by	175	11+
Earth boring combined	175	17
Electric fuses having	337	182+
Electric railway conduit having	191	27
Electrical	219	
Earth boring by	175	16
Element	338	
Furnaces	373	
Heating gas generator	48	103
Metallic element in woven fabric	139	425R
Rivet heaters	219	157
Embossing hot die	101	27
Exchange (See radiator)	165	
Agitating with	366	144+
Mortar mixer	366	22+
Cabinet combined	312	236
Combustion products generator	60	730
Distillation apparatus	202	
Distillation processes	201	
Distillation processes	203	
Gas separators having	55	
Gas washers having	261	127+
Gun barrel cooling	89	14.1
Internal combustion engine	123	41.1+
Mineral oil process	208	347+
Mineral oil stills having	196	134
Nuclear reactor with	376	369
Nuclear reactor with	376	378
Nuclear reactor with	376	391
Nuclear reactor with	376	404
Nuclear reactor with	376	405
Particulate	165	920
Static mold having	249	79+
Temperature control of	236	17+
Tobacco smoking device	131	194+
Fans		
Electric	219	369+
Fire screen or guard		
Fluid screen	110	179
Stove and fireplace	126	202
Fluid fuel burner	431	
Nozzle per se	239	
Fluid handling systems or devices with	137	334+
For glass manufacturing		
Apparatus	65	355
Discharge nozzle	239	128+
Freight car heaters	105	451
Fume		
Generators	422	305+
Furnace, solid or combined fuels	110	
Garments & bodywear (See body, heaters and warmers)		
Boot or shoe	36	2.6
Electrically heated	219	211
Heat exchanger	165	46
Therapeutic body member inclosing	128	402
Therapeutic body wear	128	379+
Gas & liquid contact apparatus	261	DIG. 31
Heaters, spray	261	DIG. 33
Heaters and condensers	261	DIG. 32
Gas manufacture		
Chemical purification	423	210+
Heating and illuminating	48	
Gun	89	1.12
Hair curling straightening or treating		
Drying apparatus	34	96+
Drying processes	34	3
Heated brushes	15	160
Heated combs	132	118
Process	132	7
Heat or cooling, conveyer act	198	952*
Heat pump, reversible	62	324.1+
Automatic	62	160
With supplemental heat source	165	29
Heat storage control	165	18
Heater submerged in liquid	261	DIG. 29
Heating equipment	D23	330+
Heating systems	237	
Humidity regulation	236	
Immersion type	126	367+
Fluid fuel	126	360R
Insulating and screening	52	
Alkali metal silicate compositions	106	75
Bifurcated garments	2	81
Bottles	215	13R
Cement compositions	106	86+
Ceramic compositions	501	94+
Compositions miscellaneous	252	62
Flatiron	38	89
Head protectors	2	7+
Laminated fabric making	156	
Miscellaneous plastic or coating compositions	106	122

U.S. PATENT CLASSIFICATIONS

Entry	Class	Subclass
Safe doors	109	65
Safe walls	109	78+
Stock material	428	920*
Toaster or broiler	99	401
Internal combustion engine charge	123	543
Leather sewing machine	112	41
Liquid	122	
Automatic temperature control	236	20R+
Automatically controlled heating systems	237	2R+
Concentrators	159	
Cookstove combined	126	38+
Discharge nozzle	239	128+
Distillation	203	
Distributing system combined	237	
Electric	219	281+
Fireplace	126	132+
Heat exchange	165	
Heated street oilers	239	128+
Heating system combined	237	
Hot air furnace combined	126	101
Liquid separator heated	210	175+
Open type	126	344+
Pressure generators	122	
Radiator and boiler combined	237	16+
Range fluid fuel water back	126	53+
Range water back	126	34+
Solar	126	417+
Textile washing apparatus combined	68	15+
Liquid separators or purifiers	210	175+
Dispensers with heater	222	146.2+
Dispensers with heating jacket	222	131
Dispensers with illuminator or burner	222	113
Distillation apparatus containing	202	
Mortar mixers	366	22+
Portable receptacle fillers	53	127
Solid material comminution combined	241	
Lubricators having	184	104.1
Material heating	432	
Material treating (See material treated)		
Coating, baking or drying	427	372.2+
Driers	34	
Electrolyte	204	
Electrolytic coating	204	37.1+
Fertilizer	71	44+
Foods	99	
Ozonizers	422	186.8+
Peat fuel	44	33
Road or pavement material	404	93+
Solids separators	209	
Tobacco	131	290+
Water purifiers	210	177+
Mattresses	5	421+
Measuring and responsive instruments		
Calorimeters	374	31+
Electrothermal switches	337	14+
Heat responsive metal stock	428	616
Hygrostats	73	335+
Thermal batteries	136	200+
Thermometers	374	147
Thermostatic switches	337	298+
Metal		
Agglomerating processes	75	3+
Casting apparatus with	164	338.1
Deforming with heating or cooling	72	
Electric heating	219	50+
Electric heating working and welding	219	50+
Electrolytic coating heating	204	37.1+
Founding	164	
Heat treatment	148	
Metallurgical apparatus	266	
Miscellaneous heating	432	
Pyrometallurgy processes	75	20R+
Roasting ores processes	75	6+
Rolling and heating	72	200
Rolling and heating	72	252
Soldering apparatus	228	
Working with	29	DIG. 21
Working with electric heating	29	DIG. 13
Nozzle combined	239	128+
Pile installation combined	405	234
Pipe	165	104.15
Pipe joint	285	41
Power plant combined	60	
Engine heated vehicles	237	12.1+
Processes	432	1+
Producing compositions	252	70
Fuel	44	
Pump	62	324.1+
Register or diffuser	D23	308+
Resolving colloids by	252	346+
Sealing wax applier	401	1+
Searing Leather	69	7.5
Seat combined	297	180
Separator, imperforate bowl centrifugal	494	13+
Shields for aircraft	244	117R+
Shields, general	244	158R
Shoes having heater	36	2.6
Single room heater	D23	330+
Sink	165	
Design	D13	23
Ventilation	174	16HS
Snow excavator and melter	37	227+
Solid fuel burners	110	
Special ray generator combined	250	495.1
Stabilized (See synthetic resin or natural rubber, class 523, 524)		
Sterilizing	422	
Fume generators	422	305+
Stoves	126	
Design	D23	342+
Surgical	604	113+
Systems	237	
Electric	219	482+
Temperature & humidity regulation	236	
Tents having	135	92
Textiles or strands		
Heating or drying threads	28	217+
Singeing textiles	26	3+
Web spreader combined	26	92
Web stretching combined	26	92
Tools and instruments		
Apparel forms	223	
Branding iron	126	402+
Brushes	15	160
Coating implement combined	401	1+
Cutlery	30	140
Dress or coat forms	223	70
Electrically	219	221+
Flatirons	38	82+
Glove forms	223	79
Hair combs	132	118
Hat forms	223	26
Ice cutter	62	320+
Rolls	126	410
Shoe formers	12	129.4
Soldering irons	228	51+
Sterilizer for	422	
Stocking forms	223	76
Tool heating stoves	126	226+
Tools self heated by fluid fuel	126	401
Tools self heated electrically	219	221+
Trouser or sleeve forms	223	73
Windshield cleaners	15	250.5
Track sanders having	291	19+
Transfer mediums	252	70
Expanding or vaporizing type	252	67+
Low freezing or high boiling point	252	71+
Tray heaters	261	DIG. 3
Treating by induction	219	10.41+
Apparatus	219	6.5+
Treatment of metals	148	
Vaporizers	122	
Vehicle systems	237	28+
Power plant combined	237	12.1+
Vulcanizing combined	425	28R+
Vulcanizing combined	425	340+
Vulcanizing combined	425	363
Welding with heat	228	
With heat and pressure	228	228
Wells		
Electric heaters	219	277+
Having heaters	166	57+
Processes using heat	166	302+
Solid material recovering	299	
Heat Producer		
Flameless composition	44	3.1+
Hecogenin	540	19
Hectograph	101	131+
Compositions	106	14.5
Flies, work holder for	101	463.1
Coating processes	427	144
Transfer sheet	428	914*
Paper	428	488.1
Processes	101	463.1
Heddle		
Making		
Cord	29	4.6
Wire	140	72
Needle winding	242	51
Structure	139	93+
Doup	139	52
Hedge		
Fence	256	20
Training	47	4
Trimmer	56	233+
Heel		
Clamping work supports	12	125
Cushioned horseshoes	168	15
Cutting machines		
Die cutting	83	
Heel lifts for shoes	D 2	323
Interfitted sole and	36	24.5
Lasts		
Plates and sockets	12	139
Separate heel block	12	135R+
Punching machines	83	
Shoe	36	34R+
Design	D 2	323+
Heel engaging shoe retainers	36	58.6
Protectors	36	73+
Supporters	36	69
Shoemaking machines	12	42R+
Burnishing	12	70+
Edge trimmers	12	85+
Heel seat forming	12	31.5
Loading	227	125
Nailing	227	140+
Plate attaching	227	
Stiffener	12	61R+
Shoemaking processes	12	147R
Skate attaching clamps	280	11.31+
Spring with soles	36	27
Straps for horseshoes	168	22
Heelless		
Overshoes	36	7.2
Rubber overshoes	36	7.4
Sandals	36	7.7
Helical (See Gearing)		
Bed spring connectors	5	269
Brass music instrument valve movement	84	391
Course		
Amusement ride	104	56+
Filled receptacle cooker	99	365
Textile liquid treatment	68	176
Nuclear fuel structures	376	362
Pump interengaging rotary impellers	418	201+
Rotary pump dipping channel	415	88
Springs	267	166+
Torsion	267	155+
Track sander feeder vertical plunger guide	291	37
Traction railway	104	167
Helicopter	244	17.11+
Airplane combined or convertible	244	6+
Airship combined	244	26
Rotor	416	
Blade positioning means	416	147+
Blade structure	416	223R+
Motor combined	416	20R+
Toy	446	230+
Flying	446	37+
Heliographic Code Signaling	116	20
Heliostats	353	3
Heliotropine	549	436
Helium (See Gas, Rare)		
Isolation by physical processes	55	66
Apparatus for	55	
Liquefaction	62	11+
Refrigeration apparatus for	62	36+
Speech	381	54
Helix (See Screw; Thread)		
Electrical resistance	338	296+
Rheostat	338	143+
Type product		
Static mold	249	59
Helmets	2	410+
Aviators	2	6
Design	D29	12+
Diving with air or oxygen	128	201.27+
Firemans	2	5
Hood type	2	205
Illumination combined	362	105
Miners hat	362	164

	Class	Subclass
Military	2	6
Space suit	2	2.1 A
Sports headgear	2	425+
Helmitol	544	185
Hemacytometers		
Colorimeter type	356	39
Microscope type	356	39+
Hematin	540	145
Hemmers Sewing Machine	112	141+
Hemocytometer (See Haemocytometer)		
Hemoglobin	540	145
Tests for	436	66*
Hemoglobinometer	356	40+
Hemostatic Devices	128	325+
Hemstitch		
Machine	112	81+
False	112	179
Heparin	536	21
Medicines containing	514	56
Heptode		
Mixer	313	300
Pentagrid converter	313	300
Triode	313	298
With two cathodes	313	6
Herbicides	71	65+
Hermetic Sealing		
Closing portable receptacles	53	285+
Gas filling of portable receptacle, method	53	403+
Glass uniting processes	65	36+
Metallic receptacle closures	220	200+
Soldering apparatus	228	
Hernia Pads	128	112.1
Heroin	546	44
Hertzian Wave (See Radio)		
Railway signal systems	246	30
Transmission line systems	333	
Hetaamoxicillin	540	325
Hetacillin	540	325
Heterocyclic Compounds	540	1+
Azo compounds	534	751+
Heavy metal ctg	534	701+
Silver sensitized formation in color photography development	430	376+
Bio-affecting & body treating composition	424	
Wave energy preparation	204	157.69+
Heterodyne Frequency Measurement	324	79R
Heterodyne Receiver	455	130+
Hevea (See Synthetic Resin or Natural Rubber)		
Hexachlorocyclohexane	570	212
Hexachlorophene, in Drug	514	735
Hexamethylene Tetramines	544	185+
Hexestrol	568	729
Hexode Triode	313	298
Hexylresorcinol	568	766
Hides Skins and Leather		
Biocidal saturant for	424	
Chemical or fluid treatment	8	94.1 R+
Manipulative	8	150.5
Coating	427	
Compositions for treating	252	8.57
Deliming	8	94.17
Dyeing	8	436+
Electrolytic treatment	204	135
Manufacture	69	
Removing fat	8	139+
Tanning	8	94.19 R+
Treatment of	8	94.1 R+
High Altitude Compartments		
Aircraft cabin	98	1.5
Aircraft power plant	244	59
High Energy Metal Forming	72	56
High Frequency		
Coaxial cable	333	239+
Triode	313	293+
High Frequency Heater (See Induction, Heating)		
High mu Triode	313	293+
Highchair		
Childs furniture	D 6	329+
Tray	D 6	396+
Highway	404	
Crossing railway signal automatic	246	293+
Electric	246	125+
Guard		
Fence	256	13.1
Pipe or cable bracket	248	66
Guide or barrier	404	6+
Track combined	238	8
Track for nonflanged wheel	238	3+
Traffic director	404	9+
Hilling and Hill Planting	111	
Ridging and covering plows	172	642
Hinge or Hinged Joint (See Device Hinged)	16	221+
Adjustable	16	235+
Belt fastener	24	33R+
Breakaway	16	222
Check and closer combined	16	50
For nuclear reactor moderator structure	376	285
Liquid check	16	54+
Pneumatic check	16	68
Design	D 8	323+
Flaccid closure mount	49	147
Flask sections	164	391+
Headlight hinge	16	DIG. 26
Magnetic	16	DIG. 14
Making assembly or mounting		
Metal	29	11
Wood seat cutting	144	27
Nesting hinge leaves	16	DIG. 29
Pipe joints	285	273
Detachable	285	283
Pivotal sheet retainer	402	26+
Plastic	16	DIG. 13
Rod	403	119+
Adjustable angle	403	83+
Rubber sleeve bearings & hinges	16	DIG. 33
Separable	16	254+
Snap-hinge	16	227
Spring	16	277+
Stepladder	182	174+
Stop	16	374+
Stove door	126	194
Toilet seat	4	240
Cover combined	4	236
Tool handle fastenings		
Adjustable angle	403	84+
Freely swinging	403	52+
Watchcase	368	313
Windshield	296	92
Hinged		
Closures	49	381+
Barrel	217	83
Bottom opening hopper vehicle	298	29+
Building shutters	49	381+
Dispenser	222	556+
Drawbridge gates	14	50+
Fence panels	256	26
Flexible and portable	160	
Gates	49	
Lockets	63	19
Metallic receptacle	220	334
Oil cup	184	90+
Operators for	49	324+
Paperboard box	229	125.8+
Purse	150	119
Railway car dumping	105	280+
Ship port	114	178
Spectacle case	206	6
Vault road or pavement	404	25
Vehicle windows and doors	296	146+
Vehicle windshields	296	86+
Vehicle windshields hinges per se	296	92
Windows sliding and swinging sash	49	208+
Windows swinging sash	49	381+
Wooden box	217	57+
Devices		
Base for folding camera	354	192+
Bed combined	5	2R+
Bellows for roll type camera	354	194
Binder device releasably engaging aperture or notch of sheet	402	
Bracelet multiple	63	9
Bracelet single	63	7+
Buttons and fasteners hinged leaf	24	97+
Calendars hinged leaves	40	119
Card holding clamp	40	13
Changing exhibitor	40	446
Display items	40	530
Endless type signs	40	524
Cultivator center	172	617
Cutter frame for combine	56	125
Demountable wheel rims sectional	301	32
Drawbridge bascule	14	36+
Extension ladder	182	163+
Extension table	108	65+
Fluid fuel water heater	126	357
Folding bedsteads	5	174+
Folding table	108	115+
Freight car sectional deck	105	372
Garment stays	2	262
Horseshoes sectional	168	7+
Invalid beds	5	60+
Ironing table with supports	108	115+
Lamp shade etc drop support	362	451
Lister type cultivator laterally hinged	172	640
Lockets	63	19
Movable on horizontal axis	211	169.1
Plow moldboard	172	736
Pneumatic tire rims sectional	152	414
Propped ladder eg stepladder	182	165+
Revolver barrel	42	63+
Scoop type vehicle fender	293	44+
Screw threading die stocks	408	180+
Sectional board ironing	38	139
Sectional mattress	5	465
Sectional scows	114	29
Shoe lasts	12	136R
Sleeping car berths	105	321
Slice toaster or broiler grids	99	402
Sofa beds	5	32R+
Surgical splints	128	88
Upending bedsteads	5	133+
Waffle iron type cooker	99	380+
Wall supported table	108	134+
Weather strips	49	475+
Hip artifical	623	22+
Hippuric Acid	562	450
Histidine	548	344
History, Teaching	434	154
Hitching		
Animal restraining devices	119	109+
Posts design	D30	154
Harness type straps	54	34
Holders	54	64
Hitching post	D30	154
Vehicle attachments	280	186+
Hive		
Bee	6	1+
Connectors	6	8
Hmx	540	475
Hobbing	409	11
Hobs	407	23+
Hobby Horses	272	52+
Hockey Stick	273	67R
Hod		
Household bin sifter	209	376
Masons	224	44.5
Endless elevator for	198	803.3+
Receptacle with pouring lip	222	572
Hoe	172	371+
Design	D 8	11
Making		
Dies	72	358+
Processes and blanks	76	109
Sausage stuffers	17	35+
Scraper	30	171
Weeder combined	172	375
Hog		
Cholera antigens and sera	424	88+
Oiler	119	157
Scalders	17	15
Scraper combined	17	13+
Scrapers	17	16+
Singers	17	20
Hoist & Hoisting (See Elevator; Lift)		
Alarms for	116	68
Arc lamps	248	317+
Automobile lift	269	58+
Building with	52	122.1+
Cable drum type	254	266+
Derrick	254	283+
Design	D34	33+
Endless or rotary carrier combined	414	564
Fire escape	182	141+
Lateral traverse combined	182	12+
Forklift truck ramp design	D12	
Hoist line implements	294	
Inclined for blast furnaces	414	208
Receptacle moved back & forth along incline	414	168
Invalid lift	5	83+
Beds with	5	61
Load handling	414	592+
Load handling type traversing	414	560+
Measuring and testing cables and ropes	73	158

U.S. PATENT CLASSIFICATIONS

	Class	Subclass
Nuclear reactor control components	376	219
Fuel charging & discharging	376	268
Ordnance loading	89	46
Railway car type	414	391+
Transports from one vehicle to another	414	342
Ramp design	D12	
Skidway combined	414	571+
Sleeping berth vertically movable	105	318
Slings	294	74+
Submerged vessel	114	51
Tilting railway car body	105	273
Traversing	212	
Truck ladder	182	63+
Trucks	254	2R+
Vehicle hoist design	D12	
Vertically adjustable vehicle	280	43+
Vertically swinging load support combined	414	569
Weighing	177	147
Wheel brake one way for	188	82.1+
With weighing scale	177	147
Hoist Line Implements	294	
Hold		
Backs		
Animal draft	278	126+
Breeching	54	5
Hill hold transmission control	192	4R+
Hill hold vehicle	188	82.1+
Hill hold wheel	188	30
Hold-downs	248	500+
Food cooking	99	349+
Food cooking filled receptacle	99	369
Freight retaining on freight carrier	410	
Container twist-lock	410	82+
Punching machines	83	374+
Punching machines	83	438+
Punching machines	83	451+
Supports for machines combined	248	680
Holder (see Container; Receptacles)		
Animal catching and holding	119	151+
Arc furnace electrode	373	94+
Arc lamps electrode	403	
Article locking racks	211	4
Assembly jig	269	37
Bag	248	95+
Golf	248	96
Bait fishing	43	54.1+
Live bait inclosing	43	41
Design	D22	136
Bicycle article carrier spring	224	37+
Billiard cue chalkers	273	18+
Blind slat	292	345
Body and belt attached cartridge	224	239+
Body or clothing attached article	24	3R+
Book and sheet		
Book leaf	281	42
Book or sheet	281	45+
Camera plate	354	276+
Copy (See copyholder)		
Depository	402	73+
Guides	402	24
Label pasting etc	156	
Leaf	281	45+
Manifolding leaf	282	29R
Music and	248	441.1+
Printing flexible sheet	101	415.1
Sheet feeding	271	8.1+
Stencil	101	127.1+
Typewriter card	400	521+
Boot tree	12	123.5
Brackets	248	200+
Article	248	309.1+
Shelf type	248	250
Brush for dynamo	310	239+
Camera roll	354	212+
Candle	431	289+
Cane	211	62
Card	40	10A+
Chain	24	116R+
Cigar and cigarette	131	187+
Design	D27	2+
Wind guard or ash receiver combined	131	175
Cigar tip cutter with	131	250+
Clasp, clip or support clamp	24	455+
Shirt collar	24	900
Cleaner cover	15	247
Closure fastener portable securing bar	292	289+
Coin	206	.8+
Copy (See copyholder)	248	441.1+
Movable copy or line guide	40	341+
Cord rope thread and wire	24	115R+
Fabric delivery roll	242	55.2+
Fishing floats with	43	44.92+
Harness hitching	54	64
Harness reins	54	74
Insulator and conductor embracing	174	154+
Insulator with conductor	174	168+
Measuring and dispensing	33	127+
Piano bridge	84	214
Piano bridge on sounding board with	84	213
Sash cord slack	16	208
Spooled strand material	242	134+
Textile spinning apparatus	57	353
Textile spinning receiver with	57	131
Twine unwinding	242	141+
Tying cords or strands	289	
Cutlery and combined material	30	124+
Dental dam	433	138+
Dental floss	132	91+
Design garment or belt attached	D 3	100+
Hunting or fishing	D 2	14
Dispenser handle and spout	222	475
Ear corn	294	5
Electric lamp etc consumable electrode	314	130+
Fluid diffusers	239	34+
Fluid supply and combined hose	137	355.16+
With nozzle	239	195+
Flypaper	43	115
Garment and fabric		
Article supporting	24	3R+
Bed attached bedclothes	5	498
Bed attached pillow sham	5	498
Bedclothes	24	72.5
Chair attachments	297	190
Closure panel elongated element	160	383+
Closure panel frame type	160	371+
Clothespins (See clothespin)		
Connected substantially spaced plural garment	2	304+
Cuff	24	41+
Drying continuous strips edge	34	158
Drying sheets etc movable	34	163
Garment type combined	2	300
Garment type convertible or reversible	2	301
Garment type elements	2	336+
Garment type partially encircling limb or torso	2	309
Garment type plural encircling	2	308
Garment type strip connected spaced	2	323+
Garment type suspension and encircling	2	310
Glove and fur sewing work	112	18+
Looped fabric sewing work	112	27
Napkin	138	177
Necktie	24	49R+
Rack for paper or textile sheets	211	45+
Rigid vertical type	2	302+
Skirt lifter and	2	217
Sorting rack bag	211	12
Stacked article type rack	211	50+
Torso or limb encircling	2	311+
Towel service cabinet	312	37
Vehicle lap robe	296	77
Work manipulating sewing machine	112	148
Gear cutting, milling, planing	409	903*
Hair		
Combs	132	11R+
Mustache	132	10
Handling type weigher	177	148+
Hatpin point guards or	132	72
Guides combined	132	69
Hollow wound package liquid treating	68	198
Hot dish lifting	294	27.1+
Illumination lens	362	455+
Implement and machine parts (See mandrel)		
Apparel making	223	106+
Axial metalworking tools and movable slide rest	29	52+
Bobbin winding or unwinding	242	130
Bolt	81	13
Brush and broom	15	146
Brush bristle	15	206
Can head seaming roller	413	31+
Chucks or sockets	279	
Cigar making machine with wrapper feed or holder	131	36
Cigar wrapper	131	105
Coil winding or unwinding	242	129
Compound tools	7	167+
Cutlery blade detachable	30	329+
Drill	408	238+
Electric generator and motor brush	310	239+
Engine starter handle	74	551
Fabric reeling or unreeling	242	68+
Illuminating lens	362	455+
Machine tool cutter mount	407	66+
Machine type wrench with bolt	81	55+
Main tapping multiple tool	408	31+
Manicuring file etc with separable	132	76.5
Mechanical gun projectile	124	41R+
Metal deforming die	72	462+
Metal file and rasp	29	80
Metalworking turret with rotary	29	40
Metalworking turret with sliding	29	41
Mirror	52	
Mop	15	147R+
Multiple metalworking tools oscillating	29	56
Pivoted clasp handle engaging	24	489+
Razor with sharpener	30	37+
Reciprocating abrading machine	51	68+
Reciprocating wood saw	83	783+
Rectilinearly movable multiple metalworking tools	29	54+
Rotary reciprocating abrading machine	51	47
Saw making file	76	36
Screw driver attachments	81	451+
Sewing machine attachment	112	257
Sewing machine bobbin	112	231
Sewing machine shuttle	112	196
Sheet punching machine die	83	651
Skein winding or unwinding	242	127
Socket wrench with nut	81	125
Sound box stylus	369	160
Sponge	15	244.1+
Having pivoted handle	15	244.2
Spool winding or unwinding	242	134+
Spring drum type article	242	107+
Stationary metalworking tools and movable work	29	51
Tool handle fastenings	403	
Type casting mechanism matrix	199	49
Wiper dauber and polisher sheet	15	231+
Insect poison	43	131
Key	70	456R+
Music keyboard	84	441
Light guard and	362	376
Live bait inclosing	43	41
Design	D22	136
Lock tumbler	70	328
Permutation compound tumbler	70	317
Nuclear fuel	376	434
Packet	24	17R+
Pencil	401	88+
Penholders	15	435+
Phonograph record	312	9+
Photographic plate	354	276+
Roll film	354	212+
Picture	40	152+
Support type	248	488+
Support type	248	490+
Piles arrangements spacers for	206	
Pocket and personal use	206	37+
Ticket	206	39
Racks	211	
Bridles, saddles, whips, design	D30	143
Radioactive body treatment	128	1.2
Receptacle		
Acetylene carbide feed	48	38+
Acetylene generator and	48	2
Acetylene water feed	48	4+
Animal cage food and water	119	18
Article carrying tray	294	172
Bait or catch	43	55
Blacking box	15	259
Blood	604	403+
Boiler or deep fat fryer	99	403+
Brush and blacking box	15	258
Camera with storage for plate	354	174+
Closet bowl disinfectant	4	222+
Closure remover with receptacle	81	3.8

	Class	Subclass
Collapsible wall dispenser with	222	105
Cover containing a	220	23
Dispenser with cutter and	222	81^{+}
Dispenser with support or	222	173^{+}
Extracting or leaching	422	261^{+}
Fishing hook and tackle	43	54.1^{+}
Flower	47	41R^{+}
Handle	220	96
Handle containing a	16	110.5
Heating and illuminating gas	48	174^{+}
High pressure gas	220	3
Inkstand	248	127^{+}
Dispenser type	222	577
Insect poison	43	131
Inverted container drainage	141	364
Match and toothpick	206	102^{+}
Matchbox	206	121^{+}
Mattress filling bat holder	53	255^{+}
Medicating inhaler gas	128	203.12^{+}
Medicator	604	403
Motor vehicle battery	180	68.5
Plural source dispenser stacking	222	143
Racks for	211	71^{+}
Rotatable assembly of dispensers	222	144
Spittoon flexible material and	4	285
Spray fluid	239	302^{+}
Rein	54	74
Design	D30	141
Whip socket combined	280	177
Scale load	177	253^{+}
Handling type weigher	177	148^{+}
Sleeping car article	105	325
Stringed music instrument with article	84	329
Tail	54	78
Cow	119	105
Telephone	379	453^{+}
Tobacco ash receiver with		
Cigar cigarette or smoking device	131	240.1^{+}
Match	131	239
Tooth	433	141^{+}
Tops with spinning devices	446	262^{+}
Train dispatching safety system signal	246	16
Vacuum-operated load engagers	294	64.1^{+}
Vacuum-type holding means	269	21
Valve head	251	89^{+}
Friction detent	251	297
Vehicle	280	182^{+}
Hand lantern	362	61^{+}
Rein	280	182
Vehicle attachment	280	182^{+}
Washboard soap	68	224
Washing machine transferable clothes	68	10
Welding electrode	219	145.1^{+}
Wheel	157	14
Work (See mandrel)	269	
Abrading	51	216R^{+}
Abrading work rests	51	238^{+}
Abrading work tables	51	240R
Ammunition shell loading	86	44
Apparel apparatus	269	
Book typewriter	400	24^{+}
Boring machines	408	
Brush boring	408	89^{+}
Brushing scrubbing cleaning	15	268
Brushmaking	300	10^{+}
Butchering	17	44^{+}
Can opener	30	400^{+}
Capping nail and screw making	10	158
Chain welding	59	34^{+}
Circular saw	83	409^{+}
Cleaning and liquid contact with solids	134	137^{+}
Collapsible structure	269	901*
Door	269	133
Drilling machines	408	
Electric heating	219	158^{+}
Electrolytic apparatus	204	297R
Feeding stock combined	29	56.6
Forging machines	269	
Gear cutting milling and planing	269	
Glove and fur sewing	112	18^{+}
Hair dryers	34	101
Hair shampooing apparatus	4	515^{+}
Hat sewing machines	112	13^{+}
Horseshoe nail finishing	10	39
Immersion scouring rotary	51	19
Ironing	38	64
Ironing table	38	103^{+}

	Class	Subclass
Leather manufacturing	69	19
Looped fabric sewing	112	27
Magnetic	335	285^{+}
Manual sewing	269	
Meat blocks	269	289R^{+}
Metal working	269	
Metalworking multiple tool laterally moving stock	29	51
Mortising	144	84
Ordnance cartridge feeding	89	34
Photographic washing	354	297^{+}
Printing	101	407R^{+}
Reciprocating abrading machine	51	64^{+}
Rotary	51	31
Rivets for riveting	72	457^{+}
Rotary abrading one way work traverse	51	74R^{+}
Rotary abrading opposed tool	51	80R^{+}
Rotary tool abrading	51	109R
Radial face	51	109R
Reciprocating holder	51	91R
Stationary holder	51	98R
Sand blast abrading moving	51	417^{+}
Sawmill	83	720^{+}
Screw head nicking	10	6
Screw threading machine	408	103^{+}
Sewing machine	112	260
Sheet metal ware making	269	
Shoe machines nailing	227	154^{+}
Shoe machines sole sewing	112	62
Shoe machines soleing	12	36^{+}
Shoemaking	12	123
Stationary axis milling cutter rotating holder	409	165^{+}
Stationary tool abrading	51	150^{+}
Stationary vegetable grater	241	273.1^{+}
Grater reciprocating holder	241	84.4
Stenciling	101	126
Stoneworking	125	35
Vices	269	
Wheel spoke setters	157	4
Wire nail making	10	52
Wood planing beveling	144	114R^{+}
Wood sawing	83	409^{+}
Wood shaping rotary holder	144	153
Wood turning	142	57
Wood turning	142	48
Wood turning polygonal section	142	2^{+}
Wood turning steady rests	142	50
Wood turning tool and	142	48
Woodworking bench dogs	144	306
Woodworking benches	144	286R
Woodworking clamps	269	
Woodworking lath	144	288.5^{+}
Woodworking machine clamps	144	278R
Woodworking stands	269	289R^{+}
Work rotating abrading machine	51	103R^{+}
Work specific	269	909*
Wrought horseshoe nail making	10	69
Wrought spike making	10	60
Holdover, Refrigeration	62	430^{+}
Accumulation in situ	62	59
Automatic	62	139^{+}
Hole		
Puncher, paper	D19	72^{+}
Punches	83	
Holiday Season Decoration	D11	117^{+}
Hollow Article		
Block wall	52	561^{+}
Blocks	52	576
Forming passage	52	503^{+}
Laterally related	52	561^{+}
Void former	52	577
With passage	52	606^{+}
Cast barrier	52	380^{+}
Static mold	249	22
Drying		
Apparatus	34	104^{+}
Billet piercing	72	327
Coating	427	230^{+}
Coating by electrolysis	204	25^{+}
Cooking apparatus	99	
Die expressing plastics	425	378R
Electroforming	204	9
Handles for cutlery	76	106
Process	34	21
Insulated wall	52	404^{+}
Projectile making	29	1.21
Rope machine	57	33
Sheet metal container making	413	
Making by extrusion metal	72	264^{+}

	Class	Subclass
E.g. , can	100	
Making by powder metallurgy process	419	5
Stud wall	52	479^{+}
Holograms & Holographic Systems	350	3.6
Memories, electric	365	125
Radiation imagery process	430	1^{+}
Holsters		
Design	D 3	101^{+}
Handgun	224	193
For either side of torso	224	192
Leather or flexible fabric	224	911*
Metal or rigid material	224	912*
Rotatably or swingably attached	224	198
With closure flap	224	238
With gun positioner	224	243
Knife, bayonet, sword, ice pick	224	232^{+}
Shoulder holster	224	206
Homatropine	546	131
Hominy Mills		
Grain hullers	99	600^{+}
Solids comminution	241	
Homogenizing		
Agitator	366	
Colloid making	252	302^{+}
Comminuting solids	241	
Equipment design	D 7	377
Foods	99	
Restrictor for	138	40^{+}
Homokinetic Coupling	464	904*
Homopolar Dynamoelectric Machines	310	178
Hones	51	211H^{+}
Honey		
Bee culture	6	
Centrifugal separator	210	361^{+}
Modified or artificial	426	658^{+}
Honeycomb (See Article by Name)		
Dunnage	410	121
Foundations	6	11
Frames	6	10
Stock material	428	116^{+}
Metal	428	593
Structure for containers	493	966*
Hood		
Actuated by tool or work approach	29	DIG. 59
Automobile	180	69.2^{+}
Combustion air vents	180	68.1^{+}
Design	D12	173
Hinge	16	355^{+}
Lock	70	240^{+}
Conveyor discharge	198	637
Curtain shade or screen combined	160	19^{+}
Design (See headwear)	D 2	256
Encased cutter	29	DIG. 86
Eye shields	2	10^{+}
Fireplace	126	142
Flash powder light	431	364
Fluorescent and phosphorescent devices	250	492.1
Focusing camera	354	287
Garment combined	2	84
Head covering	2	202^{+}
Metal working with shield or hood movements	29	DIG. 56
Progressive displacement of	29	DIG. 6
Stove	126	299R^{+}
Design	D23	371^{+}
Telephone	379	453
Ventilating	98	115.1
Woodworking cutter	144	252R
Hoof		
Calks	168	29^{+}
Sharpening tools	168	46
Cleaners and trimmers	168	48R
Expander	168	47
Farriery	168	
Gauge	33	195
Hoof crack remedy	128	336
Pads	168	26^{+}
Design	D30	149^{+}
Shoes	168	4^{+}
Design	D30	147^{+}
Hook (See also Connection)	24	230.5 R
Attaching		
Belt hook inserting	29	243.51
Lacing	227	51^{+}
Sewing machine	112	105^{+}
Bicycle carrier	224	42
Bolts hooked end	292	95^{+}
Cant hook hand bar	294	17
Carpet fastener	16	6

U.S. PATENT CLASSIFICATIONS

	Class	Subclass
Carrier by hand	294	137+
Chatelaine safety	24	4
Clasp combined	24	343+
Cuff holder	24	45
Curtain hook	D 8	367+
Cutter combined	7	161+
Design	D 8	367+
Apparel	D11	208+
Curtain	D 8	367+
Draft couplings		
Animal	278	
Railway	213	78+
Eye and	24	598+
Design	D11	208+
Relatively movable hook portions	24	698+
Fish	43	43.16+
Design	D22	144
Gaff combined with tackle	43	5
Making	29	9
Trap hook	43	34+
With artificial bait	43	42+
Hand	294	26
Hardware, eye &	D 8	367+
Harness		
Checkhooks	54	61+
Checkrein loop	54	17
Design	D30	138+
Traces	54	55
Hat fastener rotary	132	62
Hoist line	294	82.1+
Knitting	66	3+
Lacing fastening	24	146
Lacing stud	24	146+
Ladder extensible stop	182	209+
Ladder suspension	182	206
Log turner	83	710+
Chain combined	83	710+
Making		
Fish	29	9
Metal	29	7
Wire	140	80+
Napkin	24	8
Necktie	24	65
Obstetric tractor	128	353
Pelican	294	82.24+
Picture or mirror support	D 8	373+
Hook or eye	D 8	367
Pin combined	24	351+
Pipe joint		
Duplicate end	285	70+
Pipe to plate	285	191
Rail fastening	238	355+
Rod joint		
Socket combined	403	301
Sash cord	16	206
Separable hinge	16	260
Snap	24	231+
Tag label check fastener	40	26
Submarine	294	66.1+
Supports		
Bag mouth holder	248	100
Bracket having	248	290
Bracket having	248	294
Bracket having	248	301
Bracket having	248	303
Bracket having	248	304+
Clamped to	248	227
Faucet engaging	248	213
Horizontal rod engaging	248	215
Ladder engaging	248	211
Overhead beam	294	85
Penetrating	248	216.1+
Rod to radiator	248	234
Suspended	248	322
Suspended	248	339+
Tackle	294	82.11+
Telephone receiver switch	379	426+
Hookahs	131	173
Hoop		
Barrel	217	91+
Coopering		
Drivers	147	7+
Machines	147	43+
Truss	147	49
Embroidery	38	102.2
Making		
Metal	29	7
Wood by bending	144	258
Wood by sawing	83	425+
Sail	114	113
Skirts	2	216
Design	D 2	223+
Design frame	D 2	699
Making	223	6
Toy	446	236+
Rolling	446	450+
Hopper (See Magazine)		
Discharge		
Bottom forming conveyor	198	550.1
Discharge to gravity section	198	550.1+
Gravity discharge holder	198	540+
Dispenser	222	
Dumping land vehicles	298	24+
Dumping railway car	105	247
Flat convertible to	105	243
Feeding and feeder		
Animal	119	52R+
Animal, design	D30	121+
Article	221	
Bag filling	53	570+
Cigarette tube	131	74
Filler tobacco or bunch	131	44
Furnace charging bell and	414	204
Furnace fuel	110	108+
Furnace fuel blower	110	105
Match dipping frames	144	64
Rotary drying drum external type	34	112
Sifter	209	244+
Static mold having	249	108
Tobacco for cigar etc making	131	108+
Track sander	291	38+
Track sander with jet in	291	18
Track sander with oscillating	291	30
Wood slicing	144	180
Planting combined		
Auxiliary frame hill type	111	24
Dibble with revolving	111	90+
Floating auxiliary frame	111	63+
Revolving	111	74
Vibrating	111	75
Power driven conveyor combined		
Bottom forming	198	550.1
Chute and	198	568+
Discharge for	198	523+
Feeder	198	523+
Plural conveyors	198	570+
Hopples	119	126+
Slings combined	119	100
Hops and Hopping	426	11+
Apparatus	99	278+
Plants	PLT	89
Stirrer	366	343+
Vine stripper	130	30D
Horizon Artificial	356	248+
Hormones	260	397+
Animal extracts		
Medicines containing	424	95+
Heterocyclic	540	2+
Sex	260	397+
Hormones, Protein	530	399
Horn	D10	120
Antenna	343	786
Ear trumpet	181	129+
Electric	340	388+
Fertilizer	71	18
Fog	116	137R+
Lamp combined	116	3
Loud speaker	381	156
Making by deforming sheet metal	72	
Megaphone	181	177+
Musical instrument	84	387R+
Design	D17	11
Periodic sounding	116	24
Phonograph	369	163
Shield	119	144
Shoe	223	118+
Design	D 2	642
Shoe inseam trimmer	12	4.5
Stringed instrument resonator	84	294
Structure	116	137R+
Toy	446	209
Vehicle	116	59
Design	D10	120+
Horology	368	
Antimagnetic devices	368	293
Balances	368	169+
Clocks	368	62+
Design	D10	30+
Dial trains	368	220+
Dials	368	232+
Dials hands	368	238+
Escapements	368	124+
Invertible, hour glass type	368	93
Parking meter type	368	90+
Pendulums	368	134+
Safety wheels	368	181+
Watches	368	62+
Horse		
Blankets	54	79
Design	D30	145
Boots	54	82
Design	D30	146+
Collars	54	19R
Design	D30	137
Hames combined	54	18R
Shaping	69	3
Stuffing	69	4
Confining and housing devices	119	15+
Detachers	278	21+
Four legged support	182	181+
Unitary foldable	182	153+
Grooming and currying devices	119	83+
With air draft	15	363+
Harness	54	
Powered motor	185	15+
Composite or multiple	185	3
Rake horse drawn	56	375+
Training apparatus	54	71
Horseshoe	D30	147+
Calk	D30	150
Farriery	168	
Making	59	36+
Nail making		
Cut type	10	36+
Wrought type	10	64+
Nails	411	439+
Horticulture (See Plant Husbandry)		
Horticultural tools	D 8	1+
Illumination	362	805*
Hose (See Stocking)		
Alarm system via fluid hose	340	320
Applying and removing	29	235+
Bridge	104	275
Brush or mop combined	401	289
Carrier reel	242	86+
Clamp applier	81	9.3
Clamps	24	19
Cutoff	251	4+
Connections	285	238+
Dispensed material guide	222	527+
Dispensing guide interlock	222	74+
Fire fighting vehicle	280	4
Holder and fluid supply	137	355.16+
With nozzle	239	195+
Liquid level gauge combined	73	294
Making	156	143+
Nozzle	239	589+
Pantyhose	2	409
Protectors	138	110
Structure	138	177
Design	D23	266
Support	248	75+
Supporters		
Fluid conduit or nozzle	248	75+
Plural garment	2	306
Strip connected spaced holder	2	335
Torso or limb encircling	2	311+
Water hose	D23	266
Wheeled hose reels	242	86.2+
Hospital Furniture and Equipment (See Type)		
Hot Air Furnace	126	99R
Hot Bearing Indicator	116	214
Train hotbox	246	169A
Hot Bed	47	19
Hot Blast Stoves	432	214+
Hot Dog		
Cooker	D 7	323+
Packaging	53	DIG. 1
Hot Houses	47	17+
Engine cooling vents	180	68.3
Hot Pad or Mitt	D29	20+
Hot Plate	219	443+
Design	D 7	362+
Hot Topping	164	122+
Hot Tops	249	106
Hot Water Bottle	383	901*
Design	D24	43
Hot Water Furnace	122	
Hot Wire Meters	324	106
Hounds		
Vehicle	280	139
Draft pole combined	278	43

	Class	Subclass
Hour Glass	368	93
House	52	
Amusement	272	2
Animal	119	15+
Annunciators	340	815.1+
Compound maintenance tools	7	105+
Drying ie kilns	34	201+
Heating and cooling system	165	48.1
Heating system	237	
House trailer	D12	103
Moving	52	143
Portable track	238	13
Numbers		
Fluorescent	250	462.1
Luminous	40	542
Plate	D20	10+
Plumbing (See plumbing)	4	211+
Toy	446	476+
Construction	446	108+
Furnishings	446	479+
Ventilation	98	
Toilet	4	209R+
Houseboat		
Design	D12	315
Housecoat or Robe	D 2	12+
Household Articles	D 7	
Linen	D 6	595+
Measuring implement	D 7	50
Housing		
Abrading apparatus	51	268+
Animal confining and	119	15+
Brush and broom	15	184+
Closure seals	292	317
Combination lock	70	443
Fence supported by	70	322
Electricity conductors	174	50
Fluid or vacuum type	174	17R+
Wall mounted	174	48+
For an electrical device	174	50+
For different electrical devices	361	331+
For specific electrical device (See the type of device)		
Key operated locks	70	448+
Lock on rotary shaft	70	184+
Lock on valve	70	179
Material treating drum with gas flow	34	139
Heating	34	133
Metal rolling apparatus	72	237+
Padlock	70	52
Nonshackle type	70	33
Radio receiver	455	347+
Vehicle	455	343+
Rotary shafting	464	170+
Shaft combined	464	52+
Swinging door locks		
Interfitting keeper and	70	102+
Swinging deadbolt type	70	136
Switchboard	361	331+
Television	358	254
Thermometer	374	208+
Transceiver	455	90
Transmitter	455	128
Tuner with	334	85
Hovering Vehicles	180	116+
Howitzers (See Gun)		
Hub		
Automobile light combined	362	78
Bearing		
Ball	384	544
Roller	384	589
Brake		
Operator	188	114
Transmission	192	6R
Vehicle	188	17+
Velocipede	188	26
Cap	301	108R
Lock	70	259+
Groove railway	295	45+
Impeller	416	204R+
Land vehicle	301	105R+
Compression plural spoke series	301	74+
Compression single spoke series	301	80+
Tension	301	59+
Wrench	301	116
Manufacture and assembly	29	159.3
Boring	408	72R
Special woodworking machines	144	16
Runner vehicle	280	14
Wrench	81	75+
Hub Odometers	235	95B+
Hull		
Airship	244	125
Cleaning implements	114	222
Rotary for propulsion	440	99
Ship	114	56+
Small boat	114	355+
Huller (See Husker)		
Boll	19	35+
Gin saw combined	19	56
Clover	130	2
Grain		
Abrasive or impact	99	600+
Apparatus combined	241	101R
By solid agent	426	483
Other comminution combined	241	7
Nut	99	568+
Skinner or blancher	99	623+
Hulling		
Grain	426	481+
Apparatus	99	600+
Hum Bucking in an Amplifier	330	149
Human Chorionic Gonadotropin, Hcg, Testing	436	818*
Humanities, Computer Applications	364	419
Humidifiers Gas	261	
Air heating furnace combined	126	113
Auto humidifier	261	DIG. 4
Cabinet combined	312	31+
Design	D23	359
Drier combined	34	
Duct	261	DIG. 15
Gas separation combined	55	
Pad retaining means in	261	DIG. 41
Radiator combined	237	78R
Residue prevention in	261	DIG. 46
Slow diffusers	239	34+
Ventilator combined	98	
Humidity		
Meter (See hygrometer)		
Regulation (See air moistener)	236	
Automatic	261	DIG. 34
Drying apparatus combined	34	50
Heat exchange process	165	3
Heat exchanger combined	165	19+
Tobacco treatment	131	303
Humidor	312	31+
Design	D27	45
Humming (See Vibration, Dampers)		
Humus	71	24
Hunting	43	
Belt with cartridge holder	D 3	100
Boats	114	351
Coats	2	94
Decoys	43	2+
Design, general	D22	
Equipment and game carriers	224	
Knife	D22	118
License holder	40	10R+
Badge	40	1.5+
Telescopic gun sight	D16	132
Traps	43	58+
Design	D22	119+
Hurdles	272	101+
Layouts	272	4
Hurdy Gurdy (See Orchestrion)		
Husbandry		
Animal	119	
Design	D30	
Plant	47	
Design	D 8	1
Husker (See Huller)		
Corn	130	4+
Ear separator	99	585
Feeders for	130	33
Grain		
Abrasive or impact	99	600+
By solid agent	426	483
Picker combined	56	103+
Picker cornstalk type combined	56	64+
Power operated	56	14.1+
Process	426	482+
Hybrid Cell Formation	435	172.2
X-art collection	935	89*
Hydantoin	548	308
Hydnocarpic Acid	562	413
Hydrant	D23	234
Charged sprinkler	239	148
Structure	137	272+
Syringe	604	150
Hydrastinine	546	90
Hydrate of Hydrocarbon	585	15
Hydrators Lime	422	162
Hydraulic		
Accumulator	138	26+
Apparatus		
Aircraft control	244	226+
Bar and tube drawing	72	287+
Bulldozer blade control	172	812
Conveyer	406	
Pump	415	
Punching machine	83	639
Pushing and pulling implements	254	93R
Snowplow blade control	37	234
Testing	73	168
Bark rossing	144	208D
Brake	188	151R+
Electric fluid alarm	340	22+
Fluid	252	71+
Fluid alarm	116	28R+
Line valve	137	598
Pulsator	60	533+
Cement		
Composition	106	85+
Glass combined	501	32
Manufacture	106	100+
Clutch coupling or transmission	74	730
Accumulator type	138	26+
Vehicle spring combined	267	2+
Dynamometers	73	862+
Tensile strain testing	73	836
Endless flexible track adjusting	305	10
Engineering	405	52
Caissons	405	8+
Diving	405	185+
Drydocks	405	4+
Marine structures in shifting media	405	195+
Marine ways	405	2
Pipe & cable laying	405	154+
Ship caissons	405	12
Tunnel	405	132+
Underground fluid storage	405	53+
Water control	405	52+
Jack	254	93R
Motor and pump	60	325+
Mains		
Flow control	137	
Structure	138	177
Metal injector	164	314+
Motor regulators	137	505+
Speed responsive	415	30+
Torque responsive	415	30+
Motors (See motor)	91	
Pump combined	60	325+
Power plants	60	325+
Press		
Die shaping	72	
Forging	72	453.1+
Motor pump operator	60	325+
Multiple motors	60	698+
Pressure fluid source & motor	60	325+
Sand mold compactor	164	212
Propulsion		
Motor	60	200.1+
Railway car	104	154
Ship	440	5+
Pump	417	
Ram		
Compressor	417	104
Pump	417	226+
Weighing scale	177	208+
Hydrazide (See Amide)		
Hydrazine	423	407+
Explosive or thermic composition containing	149	36
Organic	564	310+
Reducing compositions	252	188.1
Sulfate	423	388
Hydrazoic Acid	423	351
Hydrazones	564	250+
Hydrides	423	644
Reducing compositions	252	188.25+
Hydrindone	568	327
Hydrins	558	303+
Chlor	568	841+
Hydrolysis of	568	859+
Cyan	558	451+
Hydriodic Acid	423	481
Hydrocarbons	585	
Apparatus of recited composition	585	920*
Apparatus structure recited	585	921*
Reactor fluid manipulating device	585	922*

	Class	Subclass
Reactor shape or disposition	585	924*
Azeotropic distillation with	203	68^{+}
Blended or composition	585	1^{+}
Compound	585	16^{+}
Drying oil hydrocarbon	585	945*
Fischer tropsch synthesis	518	700^{+}
Gas separation	55	
Hydrate of	585	15
Inorganic material conversion	585	943*
Low molecular weight polymer compositions	106	901*
Mineral oils	208	
Opening of hydrocarbon ring	585	940*
Polymerized unsaturates only (See also synthetic resin or natural rubber)	526	335^{+}
Purification	585	800^{+}
Radiant resistant composition	585	944*
Carbonium ion production	585	942*
Recovering hydrocarbon		
Solvent extraction	585	833^{+}
Werner complex formation	585	850
Recycling		
Diluent or mass-action agent	585	901*
Hydrogen to control reaction	585	903*
Solvent and catalyst	585	902*
Synthesis		
Alicyclic	585	350^{+}
Aromatic	585	400^{+}
By-product conversion to feed	585	905*
Condition-responsive control in alicyclic synthesis	585	956*
Cooling by evaporation	585	914*
Corrosion prevention	585	950*
Cyclic polymerization	585	415*
Diels-alder reaction	585	361
Isotope exchange	585	941*
Mass-action phenomenon	585	954*
Pulsed, sonic, plasma processes	585	953*
Rehabilitation of hydrogen acceptor	585	900*
Saturated	585	700^{+}
Specified mixing procedure	585	955*
Start-up procedure	585	951*
Stopping or retarding procedure	585	952*
Unsaturated	585	500^{+}
Waxy hydrocarbon polymer production	585	946*
Hydrocellulose	536	56^{+}
Nitrogen containing	536	30^{+}
Hydrocephalus Pumps	604	8
Hydrochloric Acid	423	481^{+}
Hydrocinnamic Acid	562	496
Hydrocortisone, in Drug	514	179
Hydrocracking		
Mineral oil	208	46^{+}
Paraffin synthesis	585	752
Hydrocyanic Acid	423	364
Electrolytic production of	204	63
Hydrofluoric Acid	423	481
Hydrofluosilic Acid	423	341
Hydrofoils	114	274
Hydrogen	423	648R^{+}
Alloys for storage	420	900*
Carbureting	48	199R
Apparatus	48	116^{+}
Compositions containing	252	372^{+}
Electrolytic synthesis	204	129
Exchange disproportionation		
Hydrogenation by	585	257
Olefin synthesis by	585	656
Generative composition	252	188.1^{+}
Ion		
Testing electrolytic apparatus	204	433
Testing electrolytic processes	204	1R
Testing systems	324	438
Peroxide	423	584^{+}
Electrolytic synthesis	204	84
Electrostatic field or electrical discharge synthesis	204	175
Treatment	430	943*
Sulphide		
Synthetic resin or natural rubber vulcanization with	525	343^{+}
Hydrogenation		
Co	518	700^{+}
Coal	208	400^{+}
Electrolytic	204	73R^{+}
Fatty oils	260	409
Hydrocarbons	585	250^{+}
Aromatic	585	266^{+}
Synthesis	585	250^{+}
Mineral oils	208	
Natural resins	260	100
No2 compound reduction	564	416^{+}
Polymers	525	338^{+}
Hydrolysis		
Alkyl halides to alcohols	568	891^{+}
Alkyl sulfates to alcohols	568	886^{+}
Polyhydric	568	858
Apparatus for carbohydrate	127	1
Carbohydrate	127	36^{+}
Fats oils waxes	260	415^{+}
Fermentative	435	
Halides to phenols	568	796
Hydrometallurgy	75	97R^{+}
Hydrometer	73	441^{+}
Design	D10	84
Design	D10	84
Integral with battery	429	90^{+}
Hydrophenanthrene Carboxylic Acid Esters	560	5^{+}
Hydrophilic Cellular Synthetic Resin or Natural Rubber	521	905*
Hydrophobic Cellular Synthetic Resin or Natural Rubber	521	905*
Hydrophone	367	140^{+}
Hydrophthalic Acid	562	509
Hydroplane Boats	114	271^{+}
Hydroponics	47	59^{+}
Hydroquinone	568	763^{+}
Hydrosols (See Colloids)		
Hydrostatic Extrusion	72	711*
Hydrotherapy Units	D24	38
Hydroxamate Esters	560	312^{+}
Hydroxide (See Hydroxy Compounds)		
Alkali metal electrolytic treatment	204	153
Ammonium electrolytic production	204	102
Bases	423	592
Bindant or reactant	252	193
Electrolytic production of	204	96^{+}
Hydroxylamine Organic	564	300^{+}
O-esters, O-ethers	564	300
Hygiene, Feminine		
Tampons	604	358^{+}
X-art collection	604	904*
Hygrometer	73	335^{+}
Dielectric	324	61R
Electric		
Capacitance change	324	61R
Resistance change	324	65R
Hygroscope	73	335^{+}
Cooler combined	62	130
Automatic	62	126
Hygroscopic Compositions	252	194
Refrigerated gas drying	62	271
Process	62	94
Refrigeration utilizing	62	476^{+}
Automatic	62	141^{+}
Process	62	101^{+}
Vacuumized chamber	62	269
Hygrostat	73	335^{+}
Humidity control	236	44R
Refrigeration control	62	176.1
Hypnotic Composition	424	
Hypo	423	511^{+}
Hypochlorite	423	473
Biocides containing	424	149
Bleaching composition	252	187.1^{+}
Detergent compositions	252	95^{+}
Hypodermic		
Ammunition	102	512
Surgical	604	187^{+}
Design	D24	24
Syringe package	206	364^{+}
Hypohalite Esters	560	300
Hypolon T M (See Synthetic Resin or Natural Rubber)		
Hyponitrite Esters	560	316
Hypophosphites Inorganic	423	
Hyposulphites	423	511
I		
I-beam	52	729
Cutter	83	DIG. 2
Ice	62	
Bag	383	901*
Carrier	294	
Design	D24	43
Boats	114	43
Boring	175	18
Boxes	62	459^{+}
Air controller combined	62	420^{+}
Portable, commodity containing	62	371^{+}
Portable receptacle	62	457
Breakers	114	40^{+}
Bucket	D 7	78
Control on waterways	405	61
Crushers	241	
Hand tool	D 7	101
Household appliance	D 7	374
Crushing apparatus	241	
Cutting (See notes to)	30	164.5
Electrically heated cutter	30	140
Hand held	30	140
Heated cutter	62	320^{+}
Ice cream freezer with	62	321
Machine	83	171
Machines cutting, X-art	83	915.3*
Sawing	83	915.3
Food	426	101^{+}
Gas solidifiers	62	8^{+}
Grooving	83	875^{+}
Harvesting	299	24^{+}
Houses and cellars	62	459^{+}
Air controlling combined	62	420
Structural installation	62	259.1^{+}
Machine for producing	62	340^{+}
Automatic	62	135^{+}
Making (See congelation)	D15	80
Picks and shavers		
Chipper or pick	30	164.5
General	241	
Harvesting cutters and picks	299	24^{+}
Holder for	224	232^{+}
Ice pick or chipper	30	164.5^{+}
Knives and scrapers with material holder or disposal	30	136^{+}
Prevention on aircraft	244	134R
Prevention on cooling tower	261	DIG. 86
Preventive coating	106	13
Processes of making	62	66^{+}
Protecting marine structure from	405	211
Removal from streets	37	197
Thawing		
Compositions for	252	70
From streets	37	227^{+}
Pipes	138	32^{+}
River and harbor	405	61
Tray design	D15	90
Ice Cream	426	101
Agitator for freezer	366	144^{+}
Cabinets for	220	DIG. 8
Cans for refrigeration	62	433
Cone	426	139
Carriers for cones	229	DIG. 7
Design	D 1	116^{+}
Cup	D 1	116
Cutting	83	
Design	D 1	
Disher, cone shaping	425	118
Freezer	62	340^{+}
Automatic	62	135^{+}
Tray	366	144
With ice breaker	62	321
Machine design	D15	82
Molding	426	515^{+}
Package vending cabinets	221	
Process of freezing	62	66^{+}
Product and making	426	565^{+}
Scoop	425	221
Scoop	425	276^{+}
Scoop or plunger	425	187
Serving tool	D 7	100
Stick, on a	D 1	
Ice Skates	280	11.12^{+}
Attachments for	280	809^{+}
Design	D21	225
Scabbards	280	825
Ichthyol	260	504R
Icing	426	659
Edible coating applying machines	118	13^{+}
Processes for applying	99	
Iconometer	33	227^{+}
Camera viewfinder	354	219^{+}
Range & remote distance finding optics	356	3
Iconoscope	313	367^{+}
Television system using	358	217^{+}
Identification		
Agents, chemical or biological analysis	436	
Agents for liquid fuels	44	59
Arm band	40	1.5

	Class	Subclass
Badges	40	1.5+
Design	D20	22+
Card	428	13
Envelope for	206	
Paper envelope for	229	
With printing	283	74+
Checks labels and tags	283	81
Dyeing	8	
Electrical conductors	174	112
Fingerprint, comparisons	356	71
Fraud preventing coating	427	7
Key locks	70	460
Key	40	330
Of medicine or poison	424	2+
Plug tobacco	131	368
Printed matter	283	74+
Signals and indicators	116	
Tag	283	74+
Writing on photographic film	354	105+
During X-ray exposure	378	165
Idlers (See Machine on which used)		
Bearing	384	262
Belt	474	166+
Metal rolling		
Mill	72	199+
Roll	72	199
Pulleys	474	166+
Belt tightener	474	101+
Igepon T M	260	400+
Igniter and Ignition (See Lighter)		
Alarms operated by	116	104+
Electrical	340	385
Ammunition and explosive devices		
Cartridges	102	470+
Compositions for	149	108.6
Land mines	102	424+
Marine mines	102	416+
Shells	102	200+
Burner	431	
Automatic control of	431	18
Electrical	431	254+
Electrical	431	258+
Mechanical	431	129+
Mechanical	431	254+
Catalytic	431	268
Pocket type	431	147
Chemical	431	267+
Cigarette, pocket type	431	129+
Cigarette, pocket type	431	144+
By electrical means	361	235+
Design	D27	36+
Electrical resistance	219	260+
Tobacco appliance with	131	234
Tobacco with	131	351
Combustion engine combined with rotary piston	123	210+
Electric (See engine)		
Blasting cartridge	102	
Cartridges	102	202.2+
Explosive land mines	102	424+
Flash lights solid fuel	431	357+
Flash lights solid fuel	431	357+
Fluid fuel burners	431	254+
Fluid fuel burners	431	256+
Interference suppressor resistor	338	66+
Resistance type	219	260+
Shells	102	206+
Spark type	361	253+
Time controlled	431	73
Electric space discharge devices		
Consumable electrode type	314	
Energizing systems	315	
Engine	123	594+
Dynamo combined	290	
Spark timing control	123	406+
Starting	123	186+
Fire kindling apparatus and shaped compositions	44	34+
Flashlight combined	431	357+
Electric primer ignites	431	362
Fluid-flint lighter	431	144+
Frictional	431	267+
Fuel & igniting devices	44	
Fuse	493	948*
Igniting devices & fuel	44	
Lamp system combined with spark	315	178+
Make and break for internal combustion engines	123	153
Matches	44	42+
Boxes and packages	206	96+
Design	D 9	414+
Matchbook design	D 9	304
Compositions	149	
Scratchers	44	48
Motor vehicle, internal combustion engine equipped	123	143R+
Safety belt responsive to ignition circuit	180	270
Percussive	431	267+
Promoter for motor fuel	44	57
Receptacle combined		
Cigarette or cigar dispenser	221	136+
Match boxes and packages	206	96+
Pocket cigarette or cigar containers	206	85+
Reigniting hot spot	431	347
Speeder for ignition wire manifold	D 8	357
Timing	44	36
Tobacco or tobacco users appliances combined		
Ash trays	131	234
Ash trays design	D27	8+
Cigar end cutter	131	249
Cigars or cigarettes	131	185
Pipes or cigarette holders	131	185
Ignitron (See Electric Space Discharge Device)		
Illuminated		
Changeable exhibitor	40	452
Combined unit	362	253
Clock	368	82
Compact	132	82A
Design	D28	64.1
Electric	362	135
Dispenser	222	113
Fare box	232	13
Fountain	239	18+
House number	40	584+
Self-luminous	250	462.1
Knobs	362	100
Liquid level gauge	73	293
Picture frame	40	152.2
Radio dial	116	241+
Refrigerators	62	264
Semaphore arm	246	483
Sign	40	541+
Combined sources of illumination	362	253
Fiber optic, ie light pipe	40	547
Trap	43	113
Vehicle direction indicator		
Pivoted	116	54
Rotatable casing	116	49
Rotatable pointer	116	48
Weighing scale	177	177+
Illuminating		
Gas (See gas, heating & illuminating)	48	
Liquefying apparatus	62	36+
Liquefying process	62	11+
Purifying	423	210
Illumination	362	
Ammunition and explosive devices with tracer	102	513
Bedroom lighting	362	801*
Buoys	441	13+
Cabinets with	312	223
Dental or surgical spotlight	362	804*
Diagnostic type reflector	128	22
Direction indicator	33	348
Dispenser	222	113
Drying apparatus	34	88
Egg candling	356	52+
Electric lamp supply systems	315	
Endoscopes	128	6+
Fans with	416	5
Fountains	362	101
Gags surgical	128	13
Gas generation	48	
Horticultural	362	805*
Light emitting diode	362	800*
Liquid level or depth gage	73	293
Optical testing	356	
Ornamental or decorative lighting	362	806*
Candle imitation	362	810*
Figure	362	808*
Globe	362	809*
Psychedelic	362	811*
Star	362	807*
Parachute flares	102	337+
Pharyngoscopes	128	11
Photo safe lamp	362	803*
Pivoted type specula	128	18
Pyrotechnics rockets	102	347+
Radio dial	116	241+
Shells with	102	513
Signs	362	812*
Stoves heating and illuminating	126	97
Stoves illuminating lid or top	126	213
Switch position or condition responsive	362	802*
Testing illuminating fluids	73	36
Tongue depressors	128	16
Torches	102	336+
Toys with electric	446	485+
Vehicle simulation	446	438+
Illuminators (See Illumination)		
Illusion		
Amusement	272	8R+
Railway amusement vehicle	104	83
Optical	104	84
Physical	104	85
Image Analysis	382	
Image Dissector Television	358	209
Cathode ray tube system	358	217+
Scanning	358	199+
Tube structure	313	373+
Image, Latent	428	29+
Formed or developed	427	145
Hidden answer	434	328
Hidden answer	434	348
Illuminated signs	40	541+
Imbibition Printing	101	464
Imidazobenzodiazepine	540	562
Imidazobenzodiazepinone	540	498
Imidazoles	548	335+
Fused polycarboxylic imide	546	
Imidazolidines	548	300+
Imidazolidones	548	300+
Imidazolines	548	300+
Imides (See Amides)		
Indole nucleus ctg	548	473+
Naphthostyrils	548	437
Polycarboxylic	546	
Sulfa drug type	260	397.6+
Imidic Acid Esters	558	6+
Hydroximidic acid esters	558	7
Pseudo ureas	558	8
Imidic Acid Halides	260	543.1
Imidocarbonic acid halides	260	543.2
Isocyanide dihalides	260	543.2
Imitations (See Artificial)		
Candle fixture or support	362	392
Cane umbrella case	135	18
Firearm, flashes light	362	112
Marble in plastic	264	73+
Plural modules on wall		
Static mold for making	249	16
Stage	272	8R+
Props	272	21+
Static mold shaped to simulate product of nature	249	55
Stitch on shoe	12	32
Water pistol type dispensers	222	78+
Immersing and Immersion		
Abrading	51	7
Scouring	51	17+
Boiler or deep fat fryer type	99	403+
Clothes washing	68	
Coating	118	400+
Method	427	430.1+
Dish and solids washing	134	43+
Electric heater	219	316+
Electric heater	219	523
Filling portable receptacles	141	110+
Sterilizing	422	301+
Tumbler cleaning with brushing	15	76
Water heater	126	348
Welding apparatus	427	430.1+
Immunizing		
Antigens and sera	424	88+
Cellulose fibers	8	120+
Ferments	435	245
Immunoassay	436	818*
Immunoglobulins	530	387
Immunological Testing	436	500+
Impact		
Absorbtion by flexible barrier		
By flexible barrier	49	9
By plastic deformation	188	371+
By resilient deformation	188	268
Comminutor		
By fluid stream	241	39+
By materials impact	241	5

	Class	Subclass
By moving surfaces	241	27
Hammer mills	241	185R+
Drive device for tool	173	90
Measuring	73	492
Metal extruding by	72	267
Mining machine	299	69
Operated or operation		
Assembling or disassembling	29	275
Bump release of closure	49	364
Clutch	173	93.5+
Cutter	30	277
Mail catcher load release	258	25
Mail catcher load release combined	258	11+
Solid material separation	209	
Valve	251	76
Plates	198	956*
Recorders	346	7
Scale removal by	29	81R
Support movable or disengageable on impact or overload	248	900*
Testing by	73	12+
Hardness	73	82
Shear strength by	73	844
Tools		
Assembling or diassembling	29	275+
Awl or prick punch	30	367
Cutter blade mechanism	30	277
Drilling bits	175	414+
Electric motor	310	15+
Electric motor systems	318	114
Holders	81	487
Internal impact element	81	463+
Jack hammer type	175	189
Mining or disintegrating in situ	299	94
Punch	30	367
Pusher or puller eg explosive operated	29	254+
Stonemasons	125	40+
Tunneling	299	94
Welding apparatus	228	24
Wrenches	81	463
Impaling (See Point; Prong)		
Closure remover combined	81	3.47
Impaling only eg cork screw	81	3.48+
Cooker electrode	99	358
Cooker spit	99	419+
Dispenser	221	213+
Endless conveyor	198	692+
Rake endless chain	56	400.3
Support spiked	211	125
Support stacked article	211	54.1+
Traps	43	77+
Impedance (See Condenser; Inductance; Resistance)		
Alternating current regulators	323	
Anti inductive telephone system	379	416+
Arc lamp frame with	314	132
Branched	333	124+
Coupler	333	24R+
High frequency type	333	
In amplifier coupling (See amplifier)		
Inductances	336	
Loaded signaling circuit	178	45+
Long signaling lines	333	236+
Matching	333	32+
Testing of	324	57R+
Mechanical vibration	73	574+
Transformers	336	
Variable inductance	336	
Variable transformer	336	
With antenna	343	860+
Plural path	343	852
Impellers (See Propellers)	416	
Compressor		
Expansible chamber	418	
Jet pump combined	417	76+
Screw	415	71+
Fluid reaction surfaces	416	
Gas and liquid contact apparatus	261	84+
Making		
Machines	29	23.5
Process	29	156.8 R+
Pump		
Centrifugal	415	213R+
Expansible chamber	418	
Screw	415	71+
Textile impulsing	68	134
Implant		
Controlled release of medicines	604	891
Dental	433	173

	Class	Subclass
Reservoirs	604	93
Securing conduit to body	604	175
Synthetic resin compositions for	523	105+
Synthetic resin used for	525	937*
Implements (See Instrument; Tool)		
Coating or cleaning	15	104R+
Material supply combined	401	
Cutter	30	
Earthworking	172	
Bulldozer	172	811+
Embroidering	112	80.3+
Flatiron	38	74+
Handling	294	
Igniting	431	267+
Making	76	
Pushing	254	
Sampling	73	864+
Snow excavating	37	196+
Splicing	57	23
Impregnated		
Coating implement	15	104.93+
Product of nature	428	15+
Stone	428	540
Textile	428	224+
Wood	428	541
Impregnating		
Abrasive tools	51	295
Apparatus	118	
Beer with carbonic acid	261	
Cigars and cigarettes	131	362
Compositions	106	
Conductors insulated	174	120R+
Fluid for vacuum containing	174	25R+
Covering apparatus combined	57	7
Covering process combined	57	7
Match splints	149	3+
Packing with rubber	277	227+
Plants living	47	57.5
Processes	427	
Spinning or twisting apparatus combined	57	295
Paper or cellulose tape	57	32
Textile process combined	28	167+
Covered strands	57	217
Twisted strands	57	250
Tobacco	131	300+
Twisting process combined	57	295+
Warp preparation combined	28	178+
Impressed		
Book	282	23R
Leaf	282	27R
Impressing		
Dental device	433	68+
Mold	433	34+
Leather	69	2
Tobacco labeling	131	113
Improver		
Bread or flour	426	653
Impulse (See Electric Impulse Generator)		
Incandescent (See Electrode; Illumination; Light)		
Gaseous fuel burners	431	100+
Grates for gas burners	126	92R
Igniters for engines	123	145R+
Incandescent lamp	D26	2+
Lamps	313	315+
Design	D26	
Heaters	219	552+
Making, glass working		
Apparatus	65	152+
Methods	65	36+
With arc	313	8
Mantle	431	100+
Mantle packages	206	68
Incendiary		
Ammunition and explosive devices	102	364+
Compositions for	149	
Incense Burners		
Design	D11	131.1
Design, simulative forms	D11	131.1
Incinerators	D34	1.1
Cooking stove	126	222+
Dry matter	110	235+
Garbage	122	2
Wet matter	110	238
Incline (See Hoists)		
Elevator	187	12
Handling type	414	595+
Portable	187	10
Gravity way	193	

	Class	Subclass
Inclinometer	33	365+
Incubators	119	35+
Automatic control	236	2+
Baby	128	1B
Controlled heating systems	237	3+
Culture plates or container	435	809*
Design	D24	9
Heating systems	237	14+
Switches	200	61
Indamines	260	396R
Indane	585	27
Azo compound ctg.	534	659
Indanones	568	327
Indanthrene	544	339+
Indene Polymers	526	280+
Coumarone indene resins	526	266
Indenting or Indented		
Code recorders	178	92
Laminated fabric process	156	196+
Nails	411	439+
Rod joint	403	274+
Shoe sole sewing machine	112	29
Stippling press	72	76
Stock material	428	156+
Wire substitute for nail	411	439+
Index Card	40	380
Indexes		
Book cutting	83	904*
Calendar	283	3
Map	283	35
Printed matter	283	36+
Protrusion to receive identification	40	360
Tabs	D19	99
Telephone selector list finder	40	371+
Indexing (See Turret)		
Brush making		
Work holder	300	11
Chuck	279	5
Drilling work support	408	71
Graduation scriber	33	19.1+
Milling work holder	409	221+
India Ink	106	20+
Indication Train Position	246	122R+
Indicators (See Measuring; Signal; Testing)		
Angler	43	17
Annunciators	340	815.1+
Arm or pointer structure	116	327+
Article dispensing	221	2+
Automobile turn or stop		
Electrical	340	22+
Mechanical	116	35R+
Switches	200	61.27+
Boards for house signaling systems	340	815.23
Boiler safety	122	504.2
Bottle or jar content	215	365+
Poison	215	367
Brake actuation	116	200+
Electrical	340	69
Electrical deceleration	340	669
Brake condition	188	1.11
Building indicia	52	105
Cabinet with	312	234
Channel depth	116	113
Depth sounding	33	126
Chemical	252	408.1
Heat exchange composition combined	252	68
Chuck jaws combined	279	111
Closure condition	49	13+
Coating implement combined	401	194
Coffee makers	99	285
Content alteration	215	365+
Conveyer combined	198	502.1+
Means to facilitate inspection	198	339.1
Cooking apparatus	99	342+
Counting	235	
Damper position	126	295
Direction	33	300+
Drying apparatus combined	34	89
Earth boring combined	175	40+
Electrical		
Annunciators	340	815.1+
Cable combined	174	11R
Current direction	324	133
Fuse	337	241+
Ground	340	649
Line-signal indicator testing	379	20
Phase	324	83R+
Polarity	324	133
Switch operating	200	56R

	Class	Subclass
Switch position	200	308+
Telephone call	379	27+
Expansible chamber motor combined	91	1
Fire escape combined	182	18
Fisherman	43	17
Fluid handling system or devices with	137	551+
Fluorescent and phosphorescent	250	483.1
Foul line bowling	273	50
Fountain pen combined	401	194
Funnel combined	141	95
Fuse	337	265
Headlight combined	362	61+
Heat exchanger combined	165	11.1
Illumination combined source	362	253
Leakage	137	312+
Liquid heater	126	388
Lock		
Combination	70	330+
Condition	70	432+
Machine tool		
Cutting machine	83	522
Selective punching	234	131
Mail box combined	232	37
Measuring combined	73	
Spacial relations	33	
Mechanical	116	
Metal casting combined	164	150+
Metal deforming apparatus combined	72	31+
Micrometer	33	166
Mountings	248	DIG. 4
Music		
Instruments automatic	84	169
Teaching	84	477R+
Nautical	116	26+
Buoy	441	1+
Life preserver combined	441	89
Nuclear fuel rupture warning	376	251
Nuclear fuel rupture warning	376	450
Water level in boat	116	227+
Nuclear reactor condition	376	245
Photographic film holders	354	276+
Pipes	138	104
Wear	138	36
Pneumatic dispatch	406	34+
Poison	215	367
Refrigeration	62	125+
Resonance	334	30+
Safes	109	38+
Fluid releasing combined	109	31
Separator, imperforate bowl centrifugal	494	10
Speaking tube combined	181	19
Steam engine	346	3+
Steering towed ship	114	246
Strand tension	242	148
Tags stock talley	283	79
Taillight combined	362	61+
Target combined electrical	273	371+
Target combined mechanical	273	378+
Telemetric	340	870.4+
Tester with	73	
Thermostat combined	236	94
Time	368	223+
Train carried station indicator	246	174+
Train defect	246	169R
Hot box	116	214
Tuner	334	86+
Tuning radio	455	154
Type justifying	276	11
Typewriter	400	703+
Line spacing combined	400	548+
Valve	137	551+
Winding		
Spring motor	185	44
Spring motor plural	185	14
Watch	368	210
Weight motor	185	36
Wreck buoys	441	6+
Indicia (See Indicators)		
Mileage transportation	283	24
Indigo	548	457+
Indigoid		
Azo compounds	534	789
Dye compositions containing	8	653
Indium, Metal Working with	29	DIG. 22
Indolines	548	490
Indophenols	260	396R
Carbazole	548	442
Inductance	336	
Capacitor with	361	270+
Lamp or electronic tube system having	315	289+
Resistor with reduced	338	61+
Resonator having only	333	219+
Systems	323	355
Transformer having modifying means for	336	69
Transmission		
Line components with	333	245+
Line with	333	236+
Tuner having lumped impedance	334	71+
Variable	336	
Induction (See Anti-induction)		
Cab signal	246	194
Signal block	246	63R
Coil		
Check operated	194	239+
Communication circuit loading	178	45+
Electric apparatus	336	
Telephone	379	172+
Current transmission to vehicles	191	10
Disturbances in electric conductors and cables reducing	174	32+
Telegraph	178	69R
Drive for electric railways	104	288+
Electrostatic generator	310	309+
Electrothermic processes, metallurgy	75	10.1+
For electrotherapy	128	783+
Furnace	373	138+
Furnace with arc	373	4
Furnace with resistance	373	6
Generators	310	168+
Heaters	219	10.75
Laminating apparatus	156	379.6+
Laminating methods using	156	272.2+
Heating		
Apparatus	219	6.5+
Metal casting device combined	164	250.1
Continuous casting apparatus	164	507
Methods	219	10.41+
Igniting device	361	247+
Inductor device	336	
Motor	310	166+
System	318	727+
Reactor variable	336	
Regulator	336	
Telegraph system	178	43
Telephone system	379	55
Train dispatching	246	8
Transformer structure	336	
Transformer system for metal heating	219	116
Induction Motor	310	166+
Inductor (See Inductance)		
Induline	544	348
Indurating		
Catalysts	502	100+
Compositions	106	12
Metal casting using	164	66.1+
Continuous casting	164	475
Inert Gases		
Separating	55	66
Inertia Controlled		
Aircraft control gravity	244	80
Aircraft control gyroscopic	244	79
Brake	188	266
Brake	188	275
Centrifugal electric switch	200	80R
Closure checks	16	83
Compasses gyroscopic	33	324+
Derailment trip	246	173
Dispenser outlet element	222	500
Drop bomb control	102	382+
Fly wheels	74	572+
Fuses	102	247+
Hydraulic brake	303	24R
Monorail cars	105	141
Operated switches	200	61.45 R
Pendulum operated exhibitor	40	485
Speed responsive devices	73	514+
Torpedo steering	114	24
Toy vehicle drive	446	462+
Valve actuators	137	38+
Vibrating exhibitor	40	613+
Vibration damping	188	378+
Vibration sensing	73	652+
Infant Crawling, Teaching	434	255
Infinite Impedance Detector	329	173
Inflammability Testers	73	36
Inflation, Inflated & Inflatable		
Abrading tool	51	372+
Airship	244	98
Article filling	137	223+
Catheters	604	96+
Clothes form	223	67
Compositions, rapid	149	
Compositions, slow	252	188.1+
Dilator	128	344
Element		
Static mold having	249	65
Freight bracing bag	410	119
Panel attached and urging	410	129
Squeeze including	410	125
Joint packing	277	
Life preservers	441	90+
Life rafts	441	40+
Mattresses	5	449+
Parachute stiffener or opener	244	146
Pessaries	128	129
Pipe seal	138	93
Ship raising device	114	54
Shoe forms	12	114.4
Syringe nozzle	604	275
Tire filling devices	222	3+
Portable receptacle type	141	38
Vehicle attached	417	231+
With nongaseous material	141	38
Tire per se	152	
Tire signal	116	34R
Toys	446	220+
Design	D21	59+
Truss	128	118.1
Valves	137	223+
Information Search & Retrieval		
Dynamic systems	369	
Magnetic	360	
Registers system	377	
Static systems	365	
Infrared (See Heat)		
Detection of	250	338
Drying process with	34	39+
Generator	219	354
Generator	219	553
Generator ultraviolet and	250	495.1
Heater	219	354
Heater	219	553
Lamp	313	110
Photography	430	348+
Radiation imaging	430	944*
Sensitive composition	430	616
Telescope	250	338
Testing with	250	338
Treatment of fluids	250	432R
Infusing (See Extracting and Extracts)		
Apparatus and appliances		
Beverage	99	279+
Beverage, reusable	99	323
Closet or sink combined	4	222+
Medicators	604	131+
Infuser, teabag	D 1	199
Slow diffuser	239	34+
Special receptacle and package	206	.5
Ingot		
Casting	164	
Feeding & orienting	198	373
Handling	193	
Molds	249	174
Composite	249	84
Powder metallurgy	419	
Stripper	164	405+
Structure	428	585+
Design	D15	144
Turning devices for rolling mills	72	227+
Inhalers		
Artificial respirator	128	28+
Inhalants	424	
Medicated	128	203.12+
Inhibitors	252	380+
Initiating Devices	272	27R
Injection		
Condensers	261	76
Food pickling needle	99	532+
Molding	425	542+
Needleless hypodermic	604	68+
Plant injection	47	57.5
Pumps	417	151+
Surgical instruments	604	187+
Injectors		
Bathtub combined	4	542+
Boiler circulation	122	407
Boiler cleaners	122	404
Car ventilator	98	9

U.S. PATENT CLASSIFICATIONS

	Class	Subclass
Embalmer	27	24R
Food pickling needle	254	2R$^+$
Fuel		
Combustion products generator	60	740$^+$
Rotary piston engine combined	123	205$^+$
Furnace draft regulator	110	150
Gas and liquid contact devices	261	76$^+$
Hardenable material into earth formation	405	269
Hypodermic	604	187$^+$
Lubricators	184	39.1
Medicator gas	604	23$^+$
Mineral oil distillation		
Vaporizers	196	127
Vaporizing circulation	196	126
Molding		
Metal apparatus	164	303$^+$
Metal method	164	113$^+$
Nozzle	239	
Plant husbandry	47	57.5
Pneumatic conveyors	406	194
Pumps	417	151$^+$
Resilient tire cement	81	15.5
Textile fluid treating apparatus	68	201
Track bed tampers	104	11
Ink (See Printing)		
Bottle	D 9	
Cartridge rupturable	401	132$^+$
Compositions	106	20$^+$
Electroless coating	106	1.5$^+$
Eradicator	252	89.1$^+$
Guards penholder combined	15	443
Making inkwell	15	257.7
Repellent		
Antismut devices	101	422
Compositions	106	2
Stands (See inkwell)	248	127$^+$
Calendar combined	40	358
Design	D19	93$^+$
Dispenser type	222	577
Pen engaged closure	15	257.75
Pen rack combined	211	69.2$^+$
Testing of	73	150R
Ink Jet Printing	346	75
Carbon compound digest on ink	260	DIG. 38
Markers or driving means thereof	346	140R
Inkers		
Bed and platen printing machine	101	287$^+$
Selective or progressive	101	93$^+$
Boot and shoe sole edge	118	200$^+$
Intaglio printing	101	150$^+$
Multicolor printing	101	171$^+$
Numbering machine bed and platen type	101	78$^+$
Pads and forms	118	264$^+$
Perforator combined e. g. , tattooer	81	9.22
Planographic printing	101	134
Printing	101	335$^+$
Printing members combined	101	327$^+$
Selective or progressive	101	103$^+$
Recorder stylus	346	140R
Rolling contact printing	101	212$^+$
Stencil	101	114$^+$
Stencil type addressing machine	101	48$^+$
Straight line scriber	33	34$^+$
Dotted line	33	39.1$^+$
Hand pen	33	39.2
Telegraph code recorder	178	96
Typewriter	400	191$^+$
Bar	400	470$^+$
Ribbon	400	197$^+$
Inkwell	15	257.7$^+$
Cabinet with	312	232
Desk attached	108	26.2
Dispenser type	222	576$^+$
Inkstand	222	577
Inkstand	248	127$^+$
Movable inkwell	248	128$^+$
Stationary inkwell	248	146$^+$
Pen rack combined	211	69.2$^+$
Inlaid		
Linoleum	428	57$^+$
Laminating	156	
Molding	425	
Ornamentation	428	77$^+$
Parquet floors (See parquetry)	52	315
Treads tires having	152	188
Inner Tubes (See Tire Pneumatic)		
Tire making	156	118$^+$
Innersoles, for Shoes	D 2	318
Inoculation		
Antigens and sera	424	88$^+$
Fermentative	435	30
Inorganic Compounds	23	
Conversion to hydrocarbon	585	943*
Wave energy preparation	204	157.4$^+$
Insect		
Building guard	52	101
Catching nets	43	143
Destroying	43	132.1$^+$
Electrocuting	43	112
Exits in closures and screens	160	12$^+$
Fumigators	43	125$^+$
Netting for beds	5	414$^+$
Preserving, teaching taxidermy	434	296$^+$
Traps	43	107$^+$
Design	D22	122$^+$
Insecticides	424	
Coated or impregnated vehicle	424	411$^+$
Coating or plastic compositions containing	106	15.5$^+$
Container	206	524.1
Detergents containing	252	106$^+$
Fertilizer containing	71	3$^+$
Garment hangers combined	223	86
Insecticide sprayer	D23	213$^+$
Apparatus	239	
X-art collection	43	900*
Method	43	132.1
Water softening or purifying compositions containing	252	175
Inserters (See Nailing; Stapling)		
Assembling apparatus	29	700$^+$
Belt hook	29	243.51
Box lining	493	93$^+$
Sausage stuffers	17	35$^+$
Saw tooth	76	80
Surgical implement	128	330$^+$
Needle	128	340
Tape inserting apparel	223	50
Tire patch inserters	81	15.5$^+$
Wick	431	120
Inserts		
Bearing surface	384	282$^+$
Bedstead	5	280
Brake shoes cast	188	256
Composite metal casting	164	91$^+$
Cooking molds	249	112$^+$
Earth boring bit insert	175	410
Garment	2	400
Gearing	74	451
Holding		
Core	164	231
Mold	164	332$^+$
Pattern	164	236
Nonmetallic composite casting	249	83$^+$
Insignia		
Apparel	2	246
Badges	40	1.5$^+$
Design	D11	95$^+$
Emblem per se	D11	99
Insoles		
Making method	12	146B
Reinforcing machine	12	20
Shoe (See soles)	36	43$^+$
Laminated	36	44
Prepared for securing	36	22R
Ventilated	36	3R
With body treating means	128	383
With local padding	36	71
With shank stiffener	36	76R
Therapeutic	128	383
Inspection & Inspecting (See Specific Subject; Testing)		
Cloth	26	70
Dispenser with means for	222	154$^+$
Drying apparatus with means for	34	88
Gauges	33	501$^+$
Lamp for	362	138
Measuring and testing	73	
Metal founding device with means for	164	150$^+$
Robot	901	44*
Instrument (See Type of Instrument)		
Board	73	866.3
Illumination, automobile	362	61$^+$
Switch boards	361	331$^+$
Telephone switchboard	379	319$^+$
Calibrating	73	1R$^+$
Electrical instruments	324	74$^+$
Casings	73	431
Dental	433	25$^+$
Electric switch combined with	200	56R
Fluorescent or phosphorescent indicating	250	483.1
Flying of aircraft	342	407$^+$
Apparel	33	2R$^+$
Area integrators	33	121$^+$
Direction or force indicator	33	300$^+$
Distance	33	125R$^+$
Landing	342	410$^+$
Scribers	33	18.1$^+$
Straight edge type	33	431$^+$
Navigational	33	457
Straight line light ray type	33	227$^+$
Household measuring	D 7	50
Materials	33	DIG. 11
Measuring	D10	
Mechanical expedients	33	DIG. 12
Mechanism damping	73	430
Medical	D24	8$^+$
Musical	84	1$^+$
Design	D17	
Electrical	84	1.1$^+$
Stringed	84	173$^+$
Wind	84	330$^+$
Panel in vehicle	D12	192
Panel mounted	248	27.1$^+$
Signalling	D10	
Sterilizing, disinfecting apparatus	422	
Support heads for	248	177$^+$
Testing	D10	
Transformers	336	
Insufflators	604	58
Insulation and Insulating		
Building wall	52	404$^+$
By lamina	52	408$^+$
Yieldable	52	393$^+$
Electric	174	
Battery separator	429	247$^+$
Bridged rail joint	246	48
Bushings	174	152R$^+$
Ceramic composition	501	1$^+$
Conductor cover burnishing	29	90.5
Fluent composition	252	570$^+$
Handle	16	116R$^+$
Indefinite length conductor covering	156	47$^+$
Lamp sockets	439	
Magnet or coil tape	242	7.8
Making metal parts for	29	631
Meters and testers	324	555$^+$
Pipe joints or couplings	285	47$^+$
Rail joints	238	152$^+$
Railway rolling stock parts	105	60
Railway tie tie plate and rail	238	107$^+$
Ring winding tape	242	6
Shaft coupling	464	900*
Trolley conductor section	191	39
Trolley conductor support	191	42
X-art	493	949*
Heat		
Alkali metal silicate containing composition	106	75
Boiler outlet flue	122	164
Bottle or jar	215	13R
Building	52	408$^+$
Cabinets	312	214
Ceramic porous composition	501	80$^+$
Compositions	252	62
Container	206	542
X-art	493	903*
Covering laminated type making	156	
Double-wall receptacle, wooden	217	128$^+$
Flatiron body	38	89
Flatiron stand or base	248	117.3$^+$
Fluid sprinkler, sprayer, diffuser	239	397.5
Gas separator	55	267
Handle	16	116R$^+$
Inorganic ingredient containing composition	106	86$^+$
Metal receptacle spaced wall	220	445$^+$
Muffler or sound filter	181	282
Oven	126	273.5
Pore forming composition	106	122
Stock material	428	920*
Tobacco smoking devices	131	194$^+$
Water heater	126	375
Sound (See muffler)		
Aircraft body structure	244	119
Alkali metal silicate containing composition	106	75
Auditorium	181	30
Ceramic porous composition	501	80$^+$

	Class	Subclass
Compositions	252	62
Gas separator	55	276
Gearing	74	443
Inorganic ingredient containing composition	106	86+
Lock silencer or muffler	70	463
Motion picture camera	352	35
Muffler or sound filter	181	207+
Piano resonance device	84	191
Pore forming composition	106	122
Structure	181	
Tape	428	343+
Coating	427	207.1+
Coating opposite sides	427	208
Packaged roll	206	411
Insulation Strippers		
Tube or sheath splitter	30	90.1+
Wire	81	9.4+
Insulin, T.m	530	303
Medicine containing	424	110
Insurance Documents	283	54
Intaglio		
Electrolytic reproduction	204	4+
Ornamentation	428	156+
Photographic production of	430	307+
Printing	101	150+
Combined with another type	101	DIG. 22
Static mold for forming product having	249	104
Intake		
Channel	405	127
Filter for pipe	210	460+
Fluid current conveyor	406	108+
Adjustable	406	113+
Plural	406	117+
Pneumatic straw stacker	406	139+
Gravity positioned sprayer	239	334
Pump rotary center	415	206
Integrated Circuit (See Transistor)		
Integrated Optical Circuit	350	96.11+
Integrators		
Area	33	121+
Calculators	235	61C
Distance	73	490
Ships log combined	73	183
Velocity	73	503
Volume or rate of flow meters		
Area velocity	73	227
Pressure differential type	73	861.43+
Weigher combined	177	16
Intensifiers (See Amplifiers)		
Diaphragms acoustic	181	157+
Photographic	430	487
Telephone	379	156+
Intercalates (See Definition Note (3))	428	402.2
Intercommunication System		
Electric	379	167+
Multiplex	370	
Paging	340	311.1
Selective	340	825.44+
Intercom	D14	92
Telephone-type	D14	58
Intercomparison		
Article or material	434	367
Printed matter	281	1
Interconnection Systems	307	
Interelectrode Capacitance Testing Electronic Tube	324	409+
Interference Film Antireflection	350	162.19
Interference Suppression (See Transmission, Electric)		
Interferometer	356	345+
Digest	33	DIG. 4
Holography	356	347
Interferon	435	811*
Spectroscopy	356	346
Interferon	530	351
Interleukin	530	351
Interliners Tires	152	203+
Interlock and Interlocking		
Clutches and power stop control	192	
Electric switches	200	50R
Machine tool functions		
Cutting machine	83	399+
Selective cutting	234	52
Multiple controlling elements	74	483R+
Nuclear fuel segments	376	429
Railway switches and signals	246	131+
Refrigeration related	62	267
Seams joints and connections		
Block, angle surfaces	52	589+

	Class	Subclass
Block, single surface	52	593+
Building facings or shingles	52	519+
Chain loop	59	3
Flat rod joints	403	
Flexible rod joints	403	206+
Metal container	220	76+
Pavement block	404	47+
Pipe joints	285	330
Road module	404	50+
Sheet metal container making	413	
Tile wall facing	52	589+
Wire fabric interlocking loop	245	5
Intermediate Frequency Amplifier	330	
Transformers for	336	
Tuned with capacitor	334	78+
Tuned with slug	336	130+
Intermittent		
Applied wheel brake	188	85
Clutches	192	
Coloring of thread	8	149
Combustion bursts or flare-ups	431	1
Electric signal clock controlled	368	98
Engine valve	91	
Feed and or discharge		
Disinfecting	422	123+
Dispenser type	222	
Electrode	314	
Fiber picking beaters	19	92
Harvester conveyor belt	56	346+
Planter	111	37+
Rotary web or strand feeder	226	152+
Sprayer discharge	239	99+
Sprayer supply	239	331+
Strip film	352	166+
Web or strand	226	120+
Wire	72	419+
Gear	74	435
Grip (See clutch; panl; ratchet)		
Electrode feeder	314	
Engaging land vehicle drive	280	241
Engaging locomotive drive	105	32
Machine element	74	111+
Railway signal mechanism	246	154
Strip film feed	352	191
Harvester conveyor belt	56	183
Impulse speed responsive	73	506
Land vehicle propeller	280	
Liquid applicator for textiles	68	203
Machine element	74	
Textile spinning operations cyclic	57	319+
Intermodulation		
Distortion test	324	475
Telephone	379	6
Sound recording distortion test	369	58+
Internal Combustion Engines (See Carburetor; Cylinder;decarbonizing; Exhaust, Treatment; Feeder, Fuel; Fuel; Lubrication; Muffler; Piston; Radiator; Starter; Supercharger)	123	
Backfire preventer	48	192
Dynamo combined	290	7
Exhaust treatment	60	272+
Combustion products with	60	39.5
Fluid pressure motor combined	60	598+
Freg piston drive	60	596
Vacuum generated	60	397
Waste heatdriven motor	60	597
Locomotive drive	105	62.1+
Pump drive	417	364
Refrigeration utilizing	62	323.1+
Turbine combined	60	598+
Interpoles	310	186
Interpolymers (See also Synthetic Resin or Natural Rubber)	520	1+
Interrupter		
Electric	200	
Body application system	128	423R+
Electrolytic	361	435
Generator control	322	69+
Sheet feeding	271	263
Systems	307	96+
Transformer combined with	361	433+
Printing		
Bed and platen	101	287+
Inker	101	335+
Multicolor	101	171+
Planographic multicolor	101	135+
Planographic rotary	101	141+
Rolling contact bed and cylinder	101	282+
Rolling contact rotary	101	216+
Sheet feeding	271	256+

	Class	Subclass
Signature gatherer	270	56
Interval Generator Electrical	328	129.1+
Intervalometers	200	35R+
Intervalometers systems	307	141.4
Intestines		
Cleaning	17	43
Removing	17	58+
Intravenous Devices	D24	52
Intrenching	42	93
Tools	294	49+
Cutter combined	7	116+
Invalid		
Beds	5	60+
Chair design	D12	128+
Feeding devices	414	9
Litters and stretchers	5	81R+
Mattresses	5	431+
Inventories	283	55
Computer controlled, monitored	364	403
Invert Sugar	127	41
Inverter (See Derectifier)		
Iodic Acid	423	481
Iodides	423	462
Iodine	423	500+
Medicines containing inorganic	424	150
Medicines containing organic	514	1+
Iodoform	570	181
Ion		
Exchange, liquid purification	210	660+
For actinide series compounds	423	249+
For thermonuclear reactors	376	146
Exchange synthetic polymers	521	25
Treating polymers with	528	482
Generating chamber		
Gas tube	313	567+
Mass spectrometer type (See ion source)		
Vacuum tube having	313	230
Implantation apparatus	315	111.81
Liquid purification	210	660+
Source for mass spectrometry	313	564
Electrical system for	315	111.81
Having heating means	313	550
With spectrometer	250	281+
Trap	335	210+
With cathode ray tube	313	424+
With TV picture tube	313	424+
Ionization		
Chamber radioactivity	250	335+
Gauge	250	374+
Geiger muller tube	313	93
In thermonuclear reactions	376	100
Tissue equivalent	250	472.1
With tissue equivalent gas	250	336.1
Ionomer Resin Systems (See Synthetic Resin or Natural Rubber)		
Ionones	568	378
Iraser		
Amplifiers	330	4.3
Oscillators		
Iridium		
Alloys	420	461
Filaments	75	
Inorganic compound recovery	423	22
Preparation	75	84
Electrolytic	204	109+
Hydrometallurgy	75	121
Iron (See Ferrous Alloys; Ironing Devices; Metal)		
Bridge girder	14	17
Curling hair	D28	38
Dies for making		
Plow and cultivator	72	472
Railway car	72	343+
Railway track	72	343+
Vehicle land type	72	470+
Electrolysis	204	112+
Gas injection	75	59.1
Founding	164	
Inorganic compound	423	
Coating or plastic composition	106	304
Oxide or hydroxide	423	632
Tanning	8	94.28
Making		
Columns and girders	29	155R+
Flanged bar rolling	72	190+
Plow and cultivator	29	14
Misc. chemical manufacture	156	DIG. 74
Ore beneficiating	75	1R+
Ore briquetting and binding	75	3+
Dephosphorizing desulfurizing roasting	75	6+

Entry	Class	Subclass
Electrometallurgical treatment	75	10.1+
Organic compound		
Azo	534	692+
Carbocyclic or acyclic	556	138+
Ferrocenes	556	143+
Protein or reaction product	530	400
Pigments	106	304
Plate	428	681+
Preparation	75	
Electrolytic	204	112+
Inorganic medicinal	424	147
Organic medicinal	514	184
Pyrometallurgy	75	28+
Railway		
Grab	105	461
Jacking poling roping	105	462
Ship	114	79R+
Mast	114	94
Spar	114	100
Soldering	D10	46+
Steam	38	77.1+
Structural elements	52	
Tools		
Branding electrically heated	219	227+
With indicia	101	31
Branding with heating burner	126	402+
Curling electric	219	222+
Curling with heating burner	126	408+
Flat	38	74+
Glass blow	65	300+
Sadiron electric	219	245+
Sadirons with heating burner	126	411
Soldering	228	51+
Soldering electric	219	229+
Soldering with heating burner	126	413+
Vehicle land type		
Body stay	296	42
Chafe	280	161+
Hay rack body	296	9
Running gear	280	149
Top	296	119
Walking	128	83.5
Working (See metal, treatment; metal, working)		
Wrought producing	75	47
Irone	568	378
Ironing Devices		
Attachment		
Cord plug	439	
Cord support	248	51
Board	38	103+
Convertible to ladder	182	28+
Design	D32	66
Collar cuff neckband	223	52.1+
Electrically heated	219	245
Hat type	223	21
Heating stoves		
Liquid or gaseous fuel	126	230
Solid fuel	126	227+
Iron with light	362	117
Ladder convertible to ironing board	182	29+
Laundry iron or press	D32	9
Plaiting fluting shirring type	223	36
Self heating burner	126	411+
Tables for textiles	38	103+
Textile smoothing or pressing	38	
Implements	38	69+
Machines	38	1R+
Thermostats	236	7
Tobacco	131	324+
Stemming combined	131	315+
Irradiation (See Radiant Energy)		
Coating	427	35+
Nuclear conversion (See also nuclear reaction or reactor)	376	156
Irrigation		
Channel	405	118+
Furrow plow and roller	172	176
Mobile, ambulant pipeline	239	726+
Nozzles	239	542
Pipe coupling	285	5+
Plant husbandry		
Germinator for seeds	47	61+
Plant cover shade and screen	47	27
Plant receptacle	47	79+
Subsoil	47	48.5
Subsoil	111	6+
Syringe	604	187+
Isatin	548	485
Isethionic	260	513R
Isinglass (See Glue)		
Isoborneol	568	820
Isobutylene Polymers (See also Synthetic Resin or Natural Rubber)	526	348.7
Isochondodendrine	546	34
Isocyanates		
Esters	560	330+
Fiber modification, carbonate &	8	DIG. 11
Inorganic	423	364
Polymer product derived from or containing (See synthetic resin or natural rubber)		
Cellular porous product	521	50+
Isocyanurate Cellular Product	521	902*
Isodibenzanthrone	260	358
Isoeugenol	568	652
Isoindoles	548	470
Isoindolines	548	482
Isomerization		
Fatty oils	260	405.6
Mineral oils	208	46+
Of hydrocarbons	585	
Alicyclic synthesis		
By double-bond shift	585	377+
By double-bond shift in side chain	585	363
By ring opening or shift	585	353
Aromatic synthesis	585	477+
In addition to alkylation	585	332
In addition to hydrogenation	585	253
Olefin synthesis, double-bond shift	585	664+
Paraffin synthesis	585	734
Isoniazio, in Drug	514	354
Isonitriles	558	302
Isophorone	568	377
Isoprene (See Diolefin Hydrocarbon)	585	601+
Polymerized solid (See also synthetic resin or natural rubber)	526	335
Isopropylphenol	568	781
Isoquinolines	546	139+
Isostatic Pressing		
Powder metallurgy	419	49
Isothiazoles	548	206+
Isothiocyanate Esters	558	17+
Isotope		
Actinide group metals	75	84.1 R
Comparator	250	281+
Detection	250	281+
Exchange, in hydrocarbon synthesis	585	941*
Nuclear conversion reactions	376	156
Nuclear fuel material	376	421
Nuclear fuel material	376	356
Nuclear fuel material	376	171
Radioactive composition	252	625
Radioactive metal composition	420	1+
Separatory distillation of	203	5
Using wave energy to enrich or separate	204	157.2+
Isourea	558	8
Isoxazoles	548	240+
Itaconic Acid	562	595
I U D		
Pessaries	128	127+
X-art collection	525	937*
Izod Test	73	844

J

Entry	Class	Subclass
J Acid	260	509
Jack Shield	174	35R
With jack	439	
Anti inductive shield	439	607+
Metallic cover	439	892+
Mounted on jack support	439	527+
Jack-in-the-box	446	310
Jack-o-lantern	362	257+
Jackets	D 2	183
Boiler	122	
Bottle or jar	215	12.1+
Bullet making	29	1.23
Container, metal	220	415+
Cupola water	48	67+
Dispenser part	222	
For expansible chamber device	92	144
Garment	2	
Design	D 2	
Heat exchange	165	154+
Insulating, for beverage container	220	903
Internal combustion engine	123	41.72+
Mineral oil boiler	196	119
Mold	164	394+
Nuclear fuel	376	457
Safe reinforcement	109	24
Jackknife	30	155+
Jacks		
Bending 3 point	72	389
Boot	223	114+
Floor	254	11+
Hydraulic	254	93R
Crane hoist cable	212	250
Crane mast	212	203
Motor and pump	60	325+
Jack or jack stand	D34	31
Knitting needle	66	123+
Double end machine	66	63
Jacquard machine	66	215+
Pipe or rod	254	29R+
Pump	74	41
Rack and pinion	254	95+
Roller for conveyors	193	42
Screw	254	98+
Shoe last support	12	126+
Smoke for sheds	104	52
Step by step traveling bar	254	105+
Telephone switchboard spring type	379	316
Thill coupling	254	28.5
Traversing	254	84+
Vehicle attached	254	418+
Weighing	177	146
Jacquard		
Card sewing	112	4
Lamp circuit regulation	315	292
Lamp circuit switching	315	316
Looms	139	59+
Harnessing	139	85+
Switch controlled by	200	46
Jails	52	106
Doors	49	15
Jambs for Doors	52	204+
Safe	109	77
Jams	426	573+
Jellies	426	576+
Jardiniere	47	72+
Jarring (See Agitating; Compacting; Impact; Jiggers)		
Cleaning heat exchanger	165	84
Cleaning screen by	55	300
Cleaning sifters	209	381+
Drive device for tool	173	90+
Packing mold	164	39
Receptacle unloading	414	415
Scale removal by	29	81R
Vehicle unloading	414	363
Agitation to loosen load	414	375
Jars (See Pot)		
Acid proof	206	524.1+
Battery	429	164+
Design	D 9	
Drilling rig part	175	293+
Fruit jar vises	81	3.31+
Leyden	361	301+
Receptacle	215	
Support	211	74+
Support associated with funnel	141	328
Wide mouth	D 9	367+
Jaw (See Device Having)		
Braces	433	2
Jeans	2	227+
Jellies	426	576+
Jelutong	525	331.9+
Chewing gum	426	3
Jersey		
Knit cloth	66	169R
Jervine	546	15
Jet		
Agitation in leaching or extracting apparatus	422	224+
Air heat exchanger combined	165	123
Article supporting means combined with	126	249+
Bidet	4	420.4
Boiler cleaner	122	390+
Chimney draft regulator	110	160+
Combustion chamber with jet outlet	431	158
Condenser	261	76
Liquid spray	261	115+
Conveyers	406	144
Material in suspension	406	86+
Dredge combined	37	54+
Fluid nozzle	239	290+
Fuels	149	1
Furnace fuel delivery	110	104R
Liquid heating by	126	379
Marine structure installing	405	226
Mining	299	17

	Class	Subclass
Mixer for tool heater	126	233
Motors	60	200.1+
Aircraft combined	244	4R+
Aircraft propulsion	244	73R+
Aircraft steering	244	52
Discharge nozzle	239	265.11+
Discharge nozzle cooling	239	127.1+
Impeller mounted	416	20R+
Processes of operating	60	204+
Projectile	102	374
Pyrotechnics	102	335+
Reaction motor drive	415	80+
Ship propelling	440	38+
Ship steering	114	151
Sprinkler ambulant	239	747
Sprinkler stationary	239	251+
Toy aircraft	446	211+
Toy boat	446	163
Toy spinning device	446	211
Noise muffling at highways		
Blast deflector fences	244	114B
Screen or fence absorbing sound	181	210
Pile installing combined	405	248+
Pipe or cable laying	405	163
Pulverizer		
Apparatus	241	39+
Process	241	5
Pump	417	151
Pneumatic displacement combined	417	86
Sand hopper combined	291	18
Telephone transmitter fluid type	381	165
Jetties	405	34+
Jewel		
Earring pendent	63	13
Gem	63	32
Setting	29	10
Hand tools for	81	7
Working	125	30R
Holders for while grinding	51	229
Jewelry	63	
Adhesive jewelry	63	DIG. 1
Box, design	D 9	414+
Dresser type	D 3	75
Bracelets	63	3+
Chain	D11	13
Clip, clamp, clasp	D11	86+
Design	D11	1+
Earrings	63	12+
Finger rings	63	15+
Functional fasteners	D11	200+
Garment attached vase	D11	143+
Gem setting	63	26+
Illumination combined	362	104
Lockets	63	18+
Making	29	160.6
Finger-ring forming and sizing	29	8
Gem and jewel setting in metal	29	10
Ornamental pins	63	20
Perfumed jewelry	63	DIG. 2
Plastic jewelry	63	DIG. 3
Setting	D11	91+
Tie clasp or tack	D11	202+
Trays	206	566
Watch and chain attachments	63	21+
Jews Harp	84	375
Toy	446	397
Jibs (See Sails)		
Hoist	212	
Jig		
Drilling	408	72R+
Welding	269	
Woodworking	269	37+
Jiggers (See Agitating; Impact; Jarring)		
Fluid treating of textiles	68	180
Pottery machines	425	263
Pottery machines	425	459
Jigsaw Puzzle	428	33
Jinrikishas	280	63+
Jock Straps	128	158
Jodhpurs	2	227+
Jogger (See Agitating; Impact; Jarring)	271	221+
Joining		
Adhesive for	428	343+
Box corner staying	493	89
Labels	156	
Stamps	156	
Wallpaper	156	574+
Removable layer	428	40
Wire	140	111+
Fabric stay applying clip	140	10

	Class	Subclass
Sleeve deforming machines	29	282
Sleeve deforming tools	29	282
Tire applying with	157	1.14
Joint Packing	277	
Jointing		
Indefinite length conductors	156	49
Saws	76	47B+
Gaging combined	76	46
Staves	147	25+
Joints (See Bond; Connection; Hinge; Seam)	403	
Artifical body member	623	18+
Ball and socket		
Lockable at fixed position	403	90
Pipe	285	261+
Pipe insulated	285	51
Pipe multiple	285	135
Plural	403	56
Road joint, per se	404	47+
Rod to base plate head	403	68
Surgical truss	128	122.1
Thermal expansion sprayer combined	239	397.5
Tripod head	248	181+
Universal	403	122+
Vehicle hitch	280	511+
Vibrating mounts for rod packing	277	100
With static joint	403	76+
Barrel	217	96
Bayonet (See joints closure)		
Pipe joint	285	396
To gun	42	86
Bridge truss	14	14
Building block	52	
Building facings or shingles	52	519
Closure		
Coffin lid	27	17
Safe interfitting	109	74
Safe seals	109	75
Weather strips	49	475+
Electrical	439	
Conduit or cable	174	21R+
Conduits cables and conductors	174	68C
Joining indefinite lengths	156	49
Length	156	47+
Rail bonds	238	14.1+
Rail bonds block signal system	246	48
Rail bonds block signal system	246	57
Rail bonds block signal system	246	75
Switch with connector	200	51R+
Fabric fastening (See notes to)	160	382+
Furnace hot air	126	119
Geometrical instruments		
Pivot joints	33	495+
Squares, pivoted	33	465+
Squares, separable	33	478
Making, general (See welding)	228	4.1+
Caulking ships	114	86
Juxtapose and bonding	228	101+
Metal to nonmetal	228	903
Mortar	52	415+
Riveted	227	51+
Sheet metal container making	413	
Static mold	249	98+
Metal		
Receptacle	220	75
To nonmetal bonded	403	265+
Using thermal characteristic	403	28
Metal casting	164	108+
Mining tooth to head	299	91+
Packing ends	277	220+
Pavement	404	47+
Static mold for making	249	9
Pipes shafts rods and plates (See joints ball and socket)		
Bedstead	5	282R+
Fence rail	256	65
Flexible shaft	464	
Packings	277	
Pipe	285	
Pipe dredging	37	72
Piston rod	403	230+
Articulated with piston	403	52+
Railway rail	238	151+
Reinforcing bar	52	719
Rod	403	
Smoking device	131	225
Thill	278	49
Tool to handle	403	
Toy figure	446	376+
Umbrella frame	135	29+

	Class	Subclass
Valve combined	251	142+
Whip	231	6
Railway track	238	151+
Riveted by hollow rivet	227	51+
Riveted by solid rivet	227	51+
Robot	901	28*
Safe wall	109	79
Sectional cabinets joining means	312	111
Showcase	312	140
Snap fitting	29	453+
Stirrup	403	232.1
Stove	126	98
Toy blocks bars sheets	446	85+
Universal (See joints ball and)	464	106+
Vehicle		
Airplane structure	244	131
Animal draft	278	
Articulated land	280	400+
Body land vehicle	296	29+
Body welt land vehicle	280	153.5
Bumper end connections	293	154
Bumper joints	293	155
Felly and rim	301	99+
Freight car granular material seal	105	424
Ship	114	88
Tops land vehicle	296	121+
Wooden box	217	65
Joist		
Bridging floors	52	695
Connection to wall	52	289
Hanger	52	702
Static mold	249	18+
Journal (See Bearing)	384	
Box		
Dust guards	277	130+
Jacks	254	33+
Design	D15	143
Machine elements	384	
Joystick (Electrical Input Device)		
Code signal generating	340	345
Computer	364	709
Display	340	709
Game device	273	148R
Resistive	338	68+
Writing	178	18
Juice		
Comminution and extraction	99	510+
Dehydration	426	471+
Juice extractor	D 7	48+
Preservation	426	330.5
Presses	100	104+
Sugar crystallizing	127	58
Evaporating and concentrating	127	61
Extracting	127	43+
Purifying	127	46.1+
Separating	127	53+
Treatment	426	305+
Juke Box	369	30+
Check controlled	194	
Design	D14	15
Selective or remote electric control	369	24+
Jumpers		
Baby chair	297	274+
Railway amusement park		
Gap	104	54
Passenger	104	82
Jumping Jack	446	322+
Jumping Shoes	36	7.8
Junction		
Boxes		
Electrical	174	50+
Metallic receptacle	220	3.2+
Pipe combined	285	128
Switch	200	297
Semiconductor P-N type process of forming	437	15+
Semiconductor process of forming	437	15+
Thermoelectric	136	200+
Thermometer	374	179+
Thermometer zone	374	179+
Tube pneumatic piano	84	57
Junk Yard Crushers & Presses	100	1+
Junket	426	578
Jury Rudders	114	168
Justification		
Counter or indicator for	234	4+
Type casting	199	51
Typesetting	276	28+
Circulating	276	8+
Typewriter letter space	400	1+

K

	Class	Subclass
K Acid	260	509

	Class	Subclass
K Guns	89	1.1
Kaleidoscopes	350	4.1$^+$
Design	D21	60
Kalsomines	106	
Kanamycin	536	13.7$^+$
Kaolin (See Clay)		
Karaya Gum	536	114
Kauri Gum	260	97$^+$
Kayaks	114	347
Design	D12	302
Keels		
Blocks for ships in dry dock	405	7
Rudders mounted on or below keel	114	149
Ship	114	140$^+$
Keenes Cement	106	109
Keepers		
Lock or latch	292	340$^+$
Casing interfitting for swinging door	70	102
Extensible for sliding door	70	96$^+$
Movable checkhook	54	62
Kefir	426	583
Keg (See Barrel)		
Kelly	464	163
Kelp		
Alkali metal inorganic recovery from	423	179$^+$
Iodine recovery from	423	500
Kennels	119	19
Keratins	530	357
Fermentative treatment	435	272$^+$
Kerf		
Cutting		
Mining or disintegrating in situ	299	
Woodworking	144	136R
Guide	83	102.1
Spreading	83	102.1
Ketchup (See Catsup)		
Ketenes	568	301$^+$
Ketimines	564	271$^+$
Acyclic	564	276$^+$
Aromatic	564	271
Ketol	568	414$^+$
Ketole (See Indole)		
Ketones	568	303$^+$
Carboxylic acid esters		
Acyclic	560	174$^+$
Alicyclic	560	126
Aromatic	560	51$^+$
Ketone bodies	436	128*
Ketoximes	564	253$^+$
Kettle (See Pot)	220	
Drums	84	419
Furnace	126	345$^+$
Liquid heating	126	373$^+$
Key		
Board		
Code generator	340	365R$^+$
Display control	340	711$^+$
Electric signal systems	340	365R$^+$
Keyboard actvated display	340	711$^+$
Musical instrument transposer	84	445
Musical instruments	84	423R$^+$
Design	D17	1$^+$
Operation training	434	227$^+$
Punch machine, selective	234	123$^+$
Registers	235	145R$^+$
Registers key locks	235	27
Typewriter	400	472$^+$
Typewriter key levers	400	496$^+$
Typewriter key locks	400	663$^+$
Can opening	220	274$^+$
Clock winding	D 8	347$^+$
Code instruction	434	222$^+$
Controlled closure	220	210
Controlled musical instruments	84	1$^+$
Automatic players	84	2$^+$
Cotter type	411	513$^+$
Nail lock	411	356$^+$
Electric		
Signal box transmission	340	287$^+$
Switch	200	
Telegraph	178	101$^+$
Telephone	379	251$^+$
Holders	70	456R$^+$
Case	206	37.1$^+$
Design, cases or rings	D 3	61$^+$
With light	362	116
Hole		
Alarm	116	80
Cover	70	455
Escutcheons	70	452
Guards	70	423$^+$
Illuminator	362	100
Horological apparatus	81	122$^+$
Making	76	114
Winding attached	368	214
Identification tags	283	74$^+$
Lock illuminator	362	100
Lock type (See key hole)	70	393$^+$
Alarm	116	80
Cutting by milling	409	89$^+$
Design	D 8	347$^+$
Holders	70	456R$^+$
Identification	70	460
Insertion guide	70	454
Processes and blanks	76	110
Locking bolt or nut to substructure		
Laterally introduced key	411	137
Longitudinal key	411	110
Locking bolt to nut		
Elongated key, disposed axially	411	216
Elongated key, disposed axially	411	321
Laterally movable key	411	199
Laterally movable key	411	209
Laterally movable key	411	315
Received in each of plural nuts	411	224
Swingable key	411	205
Used with thread lock		
Longitudinal key	411	292
Radial key	411	294
Tangential key	411	300
Mechanism		
Bowed stringed instruments	84	315$^+$
Plucked stringed instruments	84	320$^+$
Morse	178	101$^+$
Operated (See device operated)		
Rail joints utilizing	238	151$^+$
Railway wheel tire fastening	295	19
Ring or fob	D 3	61$^+$
Seater	409	190
Telegraph	178	101$^+$
Torque transmitting	403	355$^+$
Pullers	29	250
Slidable for gearing	74	356$^+$
Keyed Block (See Block)		
Keyer		
In transmitter	375	68$^+$
Alternating current telegraphy	178	66.2
Vacuum tube	328	247
Keystone Blocks	52	575
Kick Plate	16	1R
Kick guard	16	DIG. 2
Kicker	83	707$^+$
Kickers and Blowing Agent Moderators (See Synthetic Resin or Natural Rubber)		
Kier Treatment of Fibers	8	157
Kiln	432	120$^+$
Carbonizing	202	
Drying	34	201$^+$
Kiln	D15	144.1
Kimonos	2	74
Kindlers	44	34$^+$
Kinematographic Apparatus (See Motion Picture)		
Kinescope		
Cathode ray tube	313	364$^+$
Motion picture	352	129
Television system using	358	242$^+$
Kinesitherapy	128	24R$^+$
Design	D24	36
King Pin	403	
Kingbolts	72	472
Kitchen		
Cabinets	312	
With sifter	312	210.5
Utensils	D 7	99$^+$
Kites	244	153R$^+$
Design	D21	88
Kits		
Camp	206	541$^+$
Convertible to table	190	12A
Diagnostic	206	569
Lunch	206	541$^+$
Mess	206	541$^+$
Photographic	354	278$^+$
Pocket stove	126	38
Signal for trainmen	246	488
Toilet kit	132	79R
Absent special toilet articles eg, mirror	401	118$^+$
Klystron	315	5$^+$
Amplifier	330	45
Oscillator	331	83
Reflex	315	5.18$^+$
Knapsack	224	209$^+$
Backpack convertible to other devise	224	153
With belt	224	215
Kneader (See Mixer)		
Bread board	269	302.1
Bread, pastry or confection	366	69$^+$
Implements	416	70R
Surgical	128	60
Vulcanizable gums		
Apparatus	425	197$^+$
Apparatus	425	200$^+$
Knee		
Action		
Running gear springs	280	688$^+$
Shaft transmission mechanism combined	180	73.1
Stub axle mounts	280	96.1$^+$
Vehicle spring devices	267	228$^+$
Artificial	623	39$^+$
Knee bone	623	20
Garment protectors	2	62
Length trousers or overalls	2	228
Lock for composing stick	276	39
Operated control lever or linkage	74	515R
Pads or rests	2	24
Ship	114	87
Support surgical table combined	269	
Vehicle runner	280	27
Kneehole Desk	312	194$^+$
Knife		
Box or holder	D 7	73
Changing	83	926G*
Cutlery	30	138$^+$
Cutting	76	82$^+$
Design	D 7	137$^+$
Electric or battery driven	D 7	139
Hand manipulable cutting	30	
Additional tool combined	7	151$^+$
Handles	D 8	DIG. 4
Harvester	51	36
Clamp	51	222
Holder, body or belt attached	224	232
Knife edge texturing	28	260
Machette or hunting knife	D22	118
Machine knife in situ	51	246$^+$
Processes	51	285
Rest	248	37.3$^+$
Sharpener attachment for food mixer	D 8	35
Sharpener, powered	D 8	63
Solid material comminution	241	291$^+$
Swaging	76	89.2
Switch blades	200	162$^+$
Knit Fabrics and Articles	66	169R$^+$
Caps	2	201
Design	D 5	47$^+$
Knit-deknit texturing	28	218
Neckties	2	147
Knitting		
Darning	66	2
Design	D 5	11$^+$
Fabric manipulation	66	147$^+$
Fabric or article	66	169R$^+$
Feeding	66	125R$^+$
Looper pin & hook	66	3$^+$
Machines		
Elements	66	90$^+$
Needles (See needle & needle)	66	116$^+$
Independent needle machines	66	7$^+$
Needle beds	66	114$^+$
Needle cooperating elements	66	90$^+$
United needle machines	66	78$^+$
Pattern mechanism	66	231$^+$
Reknitting	66	1.5
Sheaths	30	151$^+$
Stopping	66	157$^+$
Additional tool combined	7	158$^+$
Knobs	16	121$^+$
Bed post ornament	5	281
Combination lock operator	70	445
Door	292	347
Attaching devices	292	348$^+$
Bearings	292	356
Design	D 8	300$^+$
Rose plates	292	357
Insulated	16	118
Label combined	40	331$^+$
Lock combined	70	207$^+$

	Class	Subclass
Machine element	74	553
Making		
Deforming sheet metal	72	
Metal	29	161
Pull or	D 8	300+
Signal combined (See knob)	116	
Knockdown		
Beds and cots	5	
Box wooden	217	12R+
Bureaus	312	266+
Cabinets	312	257R+
Cooking rack	99	449
Crate wooden	217	43R+
Desks kneehole type	312	195
Foundry core bar	164	400
Freight car	105	363
Hand rake	56	400.19+
Hold knife or razor	30	332
Ladder	182	194
Pedestal table	108	150
Picture frames	40	155
Razor	30	47
Reel	242	115
Sifter support	209	414
Stand	248	529
Plural leg	248	165
Receptacle	248	150
Stoves	126	9R
Toys	446	85+
Velocipede	280	278
Frames	280	287
Weighing scale	177	126+
Knockers		
Boll hulling combined	19	37
Breaker combined	19	36
Cable hoist	212	116
Door	D 8	401
Sifting element	209	382
Knockouts	220	265+
Knot and Knotting (See Tying)	289	
Book sewing machine	112	22
Detector	28	227
Harvester		
Compressing and binding	56	433+
Cutting conveying and binding	56	131+
Raking and bundling	56	343
Mesh fabric	87	53
Making	87	12
Necktie	2	148+
Oriental	139	400
Making	139	4+
Sash cord	16	207
Sewing machine stitch former	112	156
Strand joint	403	
Wireworking	140	101
Knuckle	213	100R+
Knurling Metal	72	703
By rolling	72	191+
Methods & apparatus	29	DIG. 23
Kraton T M (See Synthetic Resin or Natural Rubber)		
Kraut	426	
Canned	426	131

L

	Class	Subclass
Label		
Clothing label	D 5	63+
Design	D20	22+
Exhibiting	283	81+
Gumming apparatus	118	
And applying	156	
Dynamic information storage and retrieral	360	
Identifying plug tobacco	131	368
Label or tape embossing tool	D18	19+
Labels	283	81+
Magnetic bubbles	365	1+
Read only memories	365	97+
Static information storage and retrieral	365	
Obliterating stamp scarifier	81	9.21
X-art	493	961*
Labeling (See Digests 1-51)	156	
Cigarette packaging machine	131	281+
Fabrics or deformable surfaces	156	DIG. 19
Feeding	156	DIG. 28
Flat rigid surfaces	156	DIG. 1
Magazines	156	DIG. 29
Manual dispenser	156	DIG. 48
Non-flat surfaces	156	DIG. 5
Plug tobacco	131	113+
Tobacco products	131	106
Laboratory Apparatus	422	99+
Burettes	422	100
Cabinet	312	209
Elements	422	99+
Laboratory equipment	D24	
Monitor nuclear	250	472.1
Scintillation type	250	483.1
Labyrinth	272	19
Memory systems		
Dynamic electric information storage and retrival	369	
Dynamic magnetic information storage and retrival	360	
Packing	277	53+
Lace		
Digest art	D 5	47+
Table cloth	D 6	617+
Tipping	425	807*
Laces and Lacings		
Belt end connectors	24	34
Belt operated corset closure	450	119
Design, narrow	D 5	11+
Garment supports	2	326+
Knot or end holders	24	117+
Lacing and unlacing machines	12	58.5
Lacing device	24	140+
Making	87	
Fiber destruction by chemicals	8	114.6
Packages for	206	49
Sewing machine for lacing or whipping	112	121.2
Shoe lacing machine	12	58.5
Shoelace fasteners	24	140+
For ends	24	117+
Shoelaces	24	143R+
Design	D 2	316
Stud making	29	12
Stud setting machine	227	51+
Tipping	72	282
Lacquer (See Varnish)		
Cellulose ether or ester (See also synthetic resin or natural rubber compositions)	106	169
Synthetic resin or natural rubber (See class 523, 524)		
Lacrosse	273	326
Lactams		
Caprolactam	540	485+
Delta lactams eg alpha piperidone	546	243
Gamma lactams, eg alpha pyrrolidine	548	543+
Isatins	548	485
Naphthostyrils	548	437
Lactic Acid	562	589
Preparation from carbohydrates	562	515
Fermentation	435	139
Lactide	549	274
Lactometer	73	32R+
Lactones	549	263+
Poly hetero O		
5 membered ring	549	296
6 membered ring	549	274
7 membered ring	549	267
Lactose	127	31
Ladders	182	194
Attached supports	248	210
Bracket attachments	248	210+
Scaffold or shelf type	248	238
Chute or escape combined	182	49
Collapsible	182	156+
Convertible	182	21+
To chair	182	33+
To scaffold	182	27
Design	D25	62+
Door combined	182	77+
Elevating platform	182	101+
Extension	182	207+
Fire trucks	280	4
Hose or nozzle support	248	76
Attached	248	77+
Land vehicle	182	127
Platform detachable	182	120+
Propped	182	165+
Safety devices for	182	107+
Scaffold support	182	116+
Self sustaining	182	180
Sleeping car	105	326
Wall or floor attached	182	93+
Ladles	141	110
Glass handling	65	324+
Hoist traversing	212	130
Kitchen	30	324+
Design	D 7	104
Metal dispensing type	222	
Molding device with	164	335+
Lagging		
Reels	206	398+
Lake		
Dye fixing with	8	
Mordanting metal	8	
Tannate combined	8	596
Pigments containing	106	289
Laminate (See Class 428 Glossary)	428	
Abrasive tool		
Flexible cylindrical	51	358+
Making	51	297
Rigid cylindrical	51	207
Belt transmission	474	237+
Brushes rotary	15	181
Building constructions	52	408+
Curvilinear	52	267
Building panel	52	662
Conductors insulated	174	120C+
Electrode for electrolysis	204	290R
Glass	428	426+
Making	156	99+
Hinges	16	372
Insoles	36	44
Making	156	
Coating and stripping	264	
Coating apparatus	118	
Coating processes	427	
Metallic	428	615+
Nuclear fuel structures	376	416
Panel fences	256	24+
Pipe	138	140+
Flexible	138	137+
Pulleys	474	166+
Rail bonds	238	14.1+
Shoe parts	12	
Sole leveling	12	33+
Soles	36	30R
Stencils	101	128.21
Wipers, eg blackboard erasers	15	223+
Wood	144	
Lampblack	423	445+
Agglomerating	23	314
Apparatus	422	150+
Electrostatic field or electrical discharge preparation	204	173
Lamps (See also Electrode)		
Arc	313	
Barrel illumination	362	154
Bicycle	362	72
Burner		
Camera attached light	D16	42
Candle in lantern	362	161
Candle simulating	362	157+
Carburetors for	48	144+
Chimney cleaners	15	211+
Chimneys	362	312+
Consumable electrode type	314	
Daylight illuminating fixture	362	1
Daylight lamp structure	313	110+
Fluorescent material type	313	486+
Design	D26	
Diagnostic	128	23
Dimmer structure	323	905*
Discharge device, energizing	315	
Disinfecting, ultraviolet	250	504R+
Dough raiser	126	282
Egg candling		
Box with lamp control	356	67
Lamp attachment	356	68
Electric	313	
Arc lamp	313	41
Arc lamp electrode feeding	314	
Arc lamp gas or vapor	313	567+
Arc lamp mercury vapor	313	163+
Automatic substitution of filament	315	65
Circuit element in multiple filament	315	64+
Connectors for clusters of	439	638+
Consumable electrode	314	
Exhausting	141	65
Filament compositions	252	500+
Filament for gas or vapor lamps	313	341+
Filament shapes	313	341+
Gas filling	141	66
Gas or vapor	313	567+
Glassworking apparatus for making	65	152+

U.S. PATENT CLASSIFICATIONS

	Class	Subclass
Glassworking processes for making	65	36+
Glow discharge	313	567+
Incandescent bulb manufacturing	445	27
Mercury vapor	313	567+
Multiple filament	313	316
Nernst	313	14
Portable self contained	362	157+
Reflector built in	313	113+
Repair apparatus	445	61
Repair processes	445	2
Sockets	439	
Sockets rheostat combined	338	70+
Space discharge device manufacturing	445	
Systems of supply for	315	
Extinguishing devices	431	144+
Flashlight	362	208+
Camera shutter synchronized	354	126
Photographic type	431	358+
Globe or bowl	D26	118+
Heater attachment	126	255+
Tool	126	235
Heating with electric	219	552+
Illuminated		
Sign attachment	40	553+
Sign lamp box	40	564+
Incandescent	D26	2+
Inspection	362	138
Magnesium	431	99
Miners	362	164+
Nonwired supply		
High frequency field	315	248
Induction type	315	248
Radio	315	149+
Transformer in lamp	315	57
Ornamental bulb	220	2.1 R+
Fluorescent	313	493
Gas or vapor	313	567+
Reflector	313	113+
Sign	40	541+
Photographic lighting	362	3+
Photographic safe lamp	362	803*
Pole		
Decorative element	52	301
Post combined with	D26	67+
Shades	362	351+
Design	D26	118+
Making	493	950*
Rotating	362	35
Signal lamp, horn combined	116	3
Signaling devices having	340	
Design	D10	111+
Socket	D13	25
Stands	362	382+
Switchboard	379	315+
Testing of	379	20
Therapeutic	128	396+
Tool heater combined	126	241
Vehicle	362	61+
Water heater type	126	358
Wick trimmers	431	120
Wicks	431	325
Lampworking		
Glass manufacturing	65	
Lances		
Bomb	102	371
Fishing	43	6
Metallurgical	266	225+
Lancet	128	314+
Land		
Anchors	52	155+
Markers for plows	172	126+
Rollers	404	122+
Land Vehicle (See Specific Types)	280	
Air bags	280	728+
Amphibious boats		
Airborne	244	50
Vertically adjustable wheels	280	43+
Animal draft appliances	278	
Attachments	280	727+
Auxiliary article compartments		
Add-on debris carrier	224	273+
Built-in general	296	37.1+
Bodies and tops	296	
On & off freight container	410	52+
Dumping	298	
Making	29	400R+
Motor driven	180	
Passenger safety guards	280	748+
Seat belts	280	801+
Space tire compartment	296	37.2
Tires and wheels resilient	152	
Toilets	4	458+
Vertically adjustable wheels	280	43+
Wheel substitutes	305	
Wheels and axles	301	
Landfill	405	129
Landing Devices		
Aircraft		
Carriers	114	261
Fields	244	114R
Landing gear	244	100R+
Retarding and restraining	244	110R+
Elevator stop	187	32+
Chair	187	75+
Field or platform, floating		
Illumination	362	145+
Mats	404	35+
Mooring	244	114R+
Searchlight	362	145+
Moving train	104	19
Language Teaching	434	156
Forms	283	46
Writing	283	45
Laboratory	434	157
Laboratory	434	319+
Lanolin	260	397+
Purification	260	420+
Recovery from wash liquors	260	397+
Lantern		
Design	D26	37+
Electric	362	261
Flashlight	362	208
Hand lantern support on vehicle	362	61+
Heated dinner bucket and	126	267
Heater and	126	209
Lantern slide binders	40	152
Ornamental	362	806*
Slides	40	152+
Storm	362	159+
Lap Robe	2	46+
Holder vehicle body combined	280	289R
Lapidary Work	125	30R
Lapped Surfacing (See Facer Construction)		
Lappet Looms	139	49
Lapping Abrading		
Compositions	51	293+
Rigid tool	51	204+
Without tool	51	26+
Lapping, Chemical	156	625+
Wheeled vehicle combined	280	289R
Lard		
Rendering	260	412.6+
Substitutes	426	601+
Larding Pins	17	42.1
Lariats	119	153
Amusement type	446	247
Larner Johnson Type Internal Hydraulic Valves	137	219+
Laryngoscopes	128	10
Larynx, Artificial	623	9
Larynx, Electric	381	70
Laser	372	
Amplifiers	330	4.3
Beam	430	945*
Applications, oscillators		
Communication system	455	609+
Instruments with	33	DIG. 21
Laser induced diffusion of impurities into semiconductor substrates		
Device	357	7+
Method	437	16+
Misc. chemical manufacture	156	DIG. 8
Modulator	332	7.51
Oscillators		
Printer		
Beam recorder	346	108
Electric photography	355	3R+
Photocomposition	354	5+
Used in reaction to prepare inorganic compound	204	157.41
Used in reaction to prepare organic compound	204	157.61
Used in reaction to prepare or treat synthetic resin or natural rubber	522	2
Lashes	231	4
Lashings		
Boat	114	381
Cargo on freight carrier	410	96
Freight car	105	375
Lasso	119	153
Lasting		
Continental type machine	12	10.7
Lasts		
Darning	223	100
Delasting machines	12	15.1
Design	D15	137
Inserting machines	12	15
Machines for making	12	7+
Dash pots	12	DIG. 2
Fluid actuated	12	DIG. 1
Nailing combined	12	13.1+
Sewing combined	12	13.2+
Processes for making	12	145+
Removing machines	12	15.1
Shoe	12	133R+
Tools	12	107R+
Vulcanizing boots and shoes	425	383
Wire	12	7.9
Latch (See Lock)		
Actuators for knitting machines	66	111+
Bobbin	112	188
Closure fastener	292	
Closure operator combined	49	279+
Stove door	126	197
Design	D 8	331+
Gate	D 8	331+
Hinge combined	16	231
Locks & other hardware	D 8	330+
Operated clutch	192	22+
Project retract pen	401	99+
Releaser closure fastener	292	254+
Switch combined	200	318
Trip	74	2+
Latex		
Brassiere	450	36
Girdle or corset	450	97
Synthetic rubber or natural rubber containing (class 523, 524)		
Natural rubber latex		
Coagulating	528	936*
Creaming	528	937*
Physical treating	528	934*
Preserving	528	935*
Lath	52	
Holders	144	288.5
Imperforate	52	443+
Slat type	52	342+
Lather Forming		
Foam making	252	307
Gas and liquid contact	261	
Lather maker	366	604*
Spray type	239	343
Lathes	82	2R+
Abrading attachments for	51	259+
Boring horizontal	408	
Centers	82	33R
Chucks	279	
Cutoff	82	46+
Cutters and cutter holders for	407	66+
Design	D15	130
Drill holder for	408	238
Milling attachments for	409	165
Planing attachment for	409	328
Punching attachment for	83	
Spinning and fusing sheet metal	228	
Stoneworking	368	10
Straightening shafts in	72	380+
Turret or capstan	29	33J+
Winding and reeling attachments for	242	
Wood saw feed	83	704+
Wood slicer strip cutting feed	144	165+
Wood turning	142	
Wood veneer lathes	144	209R+
Latrines	4	460+
Latrobes	126	123
Lattice	52	660+
Frame for panels	52	455+
Nuclear fuel cores	376	346
Nuclear fuel cores	376	267
Type product		
Static mold	249	60
Wire		
Machines for making	140	25+
Launches	114	
Launching		
Aircraft	244	63
Boat	114	365
Life boat	114	365+
Marine vessels	405	1
Torpedo	114	238+

	Class	Subclass
Toy glider	446	63$^+$
Toy rocket	446	212
Toy vehicle	446	429$^+$
Laundry		
Apparatus	68	
Design	D32	6$^+$
Bag	D32	36
Net or mesh type for use in washing	383	117
Bluing compositions	8	648
Combined washer-drier machine	D32	5
Dry cleaning machine	D32	10$^+$
Drying	34	
Combined washing	68	19$^+$
Household equipment	D32	35$^+$
Ironing	38	
Machinery	D32	6$^+$
Methods	8	137$^+$
Manipulative	8	147$^+$
Sorting	209	937*
Sticks	294	23.5
Tongs	294	8.5
Lauric Acid	260	413$^+$
Laurolactam	540	451$^+$
Lauryl Alcohol	568	840
Production from coconut acids	568	876$^+$
Lavatory	4	650$^+$
Basin supports	4	643$^+$
Building wall attached	52	34$^+$
Lavatory	D23	284$^+$
Soap dish with soap handling means	206	77.1
Law Enforcement Computer Applications	364	409
Lawn		
Aerator	172	21$^+$
Grass	PLT	88
Mower (See harvester)	56	229$^+$
Bar handle hardware	16	111A
Circumferentially spaced blade	172	552
Design	D15	14$^+$
Disc type	56	255$^+$
Motor operated	56	16.7$^+$
Mower-type tongue & crossbar	16	DIG. 38
Reel type	56	249$^+$
Sharpener	76	82.1
Sharpener abrading in situ	51	246$^+$
Small engine starters	123	179SE
Snowplow combined	37	243
Tool driven, laterally extended bar or blade	172	121
Rakes	56	400.1$^+$
Roller	404	122$^+$
Sprinkler	D23	214$^+$
Sweeper	D15	27
Swings	297	245$^+$
Trimmers	172	13$^+$
Laxative	424	
Layers (See Composition; Sheet)		
Bonded	428	
Electroplating	204	40$^+$
Explosive or thermic composition containing	149	14$^+$
Filling of articles	53	531$^+$
Filter	210	290
Filter screen gas separation	55	486$^+$
Layering	47	5.5
Nuclear fuel structures	376	432
Railway track	104	2$^+$
Shoe sole machine	12	29
Signs multiple	40	615
Static mold having super-imposed metal	249	116
Structurally defined	428	98$^+$
Metallic	428	615$^+$
Winding of bobbins	242	26
Laying		
Building facings or shingles	52	747$^+$
Mines	102	411$^+$
Paving blocks	404	99
Pipe or cable	405	154$^+$
Shoe machine for	12	33$^+$
Layout Device		
Clothes	33	11$^+$
Shoes	33	4$^+$
Lays Loom	139	188R$^+$
Lazy Susan	D 7	2
Lazy Tongs		
Bracket	248	277
Collapsible supports	211	202
Drawbridge	14	45
Elevator motor	187	18
Extension ladder	182	157$^+$
Fire escape	182	40
Linkage system	74	521
Mast or tower	52	109
Pushing and pulling elements	254	122
Vehicle fender	293	27
Window operators	49	363
Leaching (See Diffusing; Extracting)		
Apparatus	422	261$^+$
Electrolytic cell combined	204	233$^+$
Metallurgical	266	101
Sugar starch carbohydrates	127	3$^+$
Processes		
Adsorbent reactivation	502	20$^+$
Adsorbent reactivation water only	502	22
Hydrometallurgy chemical	75	101R$^+$
Inorganic compounds	423	210$^+$
Lead (See Pencil)	75	77
Acetate	556	105
Alloys	420	563
Copper	420	491
Copper tin	420	474
Battery	429	149$^+$
Chamber processes	423	524
Compound inorganic	423	
Carbonate	423	433$^+$
Chromate	423	592
Electrolytic synthesis of carbonate	204	88
Halide	423	462$^+$
Oxide	423	619
Pigment filler or aggregate	106	297$^+$
Compound organic	556	81$^+$
Stabilization	556	3$^+$
Electrolytic synthesis		
Aqueous bath	204	114$^+$
Fused bath	204	66
Fuses	337	290$^+$
Heat treatment	148	
Hydrometallurgy	75	120
Misc. chemical manufacture	156	DIG. 85
Paste for batteries	429	225$^+$
Process of mixing	252	182.1
Process of putting on grid	141	1.1
Plate	428	645
Electrolytic	204	53
Pyrometallurgy	75	77$^+$
Saw adjustments	83	816$^+$
Shields for radioactivity	250	515.1
Shot making	29	1.22
Apparatus for comminuting liquid	425	6$^+$
Processes for comminuting liquid	264	5$^+$
Sounding	33	126.5$^+$
Styphnates	556	109
Tetraethyl	556	95
Fuel containing	44	69
Leaded Glass	428	38
Leaders		
Cattle	119	151$^+$
Halters	54	24$^+$
Fair	114	101
Fishing	43	44.98
Boxes	43	54.1$^+$
Magnetic record tape or wire	360	131$^+$
Leaf (See Spring)		
Book binder sheet	402	79
Burner	D34	1.1
Extension table		
Center	108	83$^+$
Circular segment	108	66
Drop leaf type	108	77$^+$
End	108	69$^+$
Hinges	16	387$^+$
Manifolding	282	26$^+$
Books combined	282	8R$^+$
Envelopes	282	25
Strips and books combined	282	2
Means to select	40	532
Metallic type ornamentation	428	17$^+$
Applying apparatus	156	540$^+$
Applying by transfer	156	230$^+$
Packages	206	71
Transfers	428	914*
Music turners	84	486
Paper	402	79
Book	281	38$^+$
Strips combined	281	2
Printed matter	283	61
Sliding leaf	283	65
Resilient wheel	152	1$^+$
Rests	248	441.1$^+$
Rotatable exhibitors	40	493$^+$
Sheet manipulative devices		
Flip, insert, lift, remove	402	80R
Transfer fork	402	80R
Transfer post	402	47
Supports		
Binder device releasably engaging aperture or notch of sheet	402	
Hinge mounted displays	40	530$^+$
Holders combined	248	441.1$^+$
Newspaper type	281	45$^+$
Paper file	211	169.1
Tobacco treating	131	290$^+$
Leak		
Catcher	137	312$^+$
Dispenser combined	222	108$^+$
Heat exchanger combined	165	70
Detectors		
Composition	252	68
Fluid pressure apparatus	73	40$^+$
Heat exchanger combined	165	11.1
Nuclear reactor control	376	250
Nuclear reactor control	376	450
Pipe joint	285	13
System in situ	137	455$^+$
Floatable matter, eg oil, confining or collecting	405	60
Preventers		
Gas distribution illuminating	48	193$^+$
Pipe joint	285	171$^+$
Stoppers		
Composition	252	72
Composition coating or plastic	106	33
Ship	114	227$^+$
Leakage		
Disposal		
Pipe joint	285	13
Electric current testing	324	500$^+$
Learners Instruments		
Golf	273	35R$^+$
Telegraphy	178	115
Leash		
Animal	119	96
Design, dog	D30	153
Collars	119	106
Design	D30	152
Design, dog	D30	152
Leasing		
Looms	139	98
Stopping	139	350
Warp beam	28	198$^+$
Leather	69	
Artificial	428	904*
Bag, flexible	383	
Bating	8	94.17
Fermentive	435	265
Stretching	69	1.5
Beveling machines	69	9$^+$
Bobbins or spools	242	118.32
Cop type	242	118.7
Coating of processes	427	389
Coated article	428	473
Plural coating	427	412
Post-treatment combined	427	389
Striping bordering or edging	427	284$^+$
Compositions containing (See synthetic resin or natural rubber and class 524-11)	106	155
Degreasing	8	139$^+$
Digest, drying & gas or vapor contact with solids	34	
Dyeing	8	436$^+$
Electrolytic treating of	204	135
Embossing	69	2
Embossing	101	3R$^+$
Extracting oil from	260	412
Fastener made of	269	906*
Fertilizers from	71	18
Fluid treatment, eg tanning	8	94.19 R$^+$
Manipulative	8	150.5
Fulling apparatus	69	33$^+$
Hammering apparatus	69	1$^+$
Handgun holder made of	224	911*
Impregnating compositions for biocidal	424	
Mailbag	383	
Manufacturers, machines	69	1$^+$
Processes	69	21$^+$
Tools	69	20
Work stretcher or holder	69	19$^+$
Pocketbook and portfolio design	D 3	42$^+$
Preserving	8	94.1 R$^+$

	Class	Subclass
Punching	83	
Roughening	69	1
Sewing machines	112	28^{+}
Shoes	36	
Making	12	
Skiving and splitting	69	9^{+}
Softening	69	
Stock material	428	
Laminate	428	473
Stretching frames	69	19.1^{+}
Stuffing with oil	8	94.21^{+}
Tanning	8	94.19 R^{+}
Drums	69	29^{+}
Treating compositions	252	8.57
Workers irons, forging dies	72	343^{+}
Working	69	
Compound tools	7	102^{+}
Leavening Compositions	426	561^{+}
Yeast	435	255^{+}
Leaves (See Leaf)		
Lecher Wires	333	220^{+}
Oscillator combined	331	99
Radio receiver combined	455	130^{+}
Radio transmitter combined	455	91^{+}
Vacuum tube combined	315	39
Lecithin	260	403
Foods	426	662
Lectern	D 6	419
Leer (See Lehr)		
Leeway Navigation Side Slip	73	180
Leg (See Support)		
Artificial	623	27^{+}
Furniture		
Bathtub	4	594
Bedstead	5	310^{+}
Folding bedstead	5	179^{+}
Folding chair	297	16^{+}
Folding stand leg	248	188.6
Folding stool	248	166^{+}
Folding table	108	115^{+}
Plural folding stand	248	166^{+}
Stand	248	188.1^{+}
Stand detachable	248	151
Stove	126	304R^{+}
Table	248	188.1^{+}
Guards and protectors	2	22
Rest	D 6	501
Bed	5	443^{+}
Chair reclining	297	68^{+}
Vehicle occupant propelled	280	291
Surgical appliance		
Bandage	128	165
Electric applicator	128	783
Orthopedic	128	80R^{+}
Leggings	36	2R
Design	D 2	267
Footless stocking	2	242
Design	D 2	330
Legumes	426	629^{+}
Lehr	432	120^{+}
With cooling means	65	349^{+}
Lemon Squeezers	100	104^{+}
Comminutor combined	99	495^{+}
Lemonade	426	590^{+}
Concentrate	426	599
Effervescent	426	591
Fermented	426	7^{+}
Leno	139	50^{+}
Lens	350	409^{+}
Axis finder	356	127
Blocking or working holding devices	51	216LP^{+}
Catadioptric lens	350	442
Camera structure combines	354	
Cathode-ray tube combined	358	
Cathode ray tube		
Structure	313	317^{+}
System	315	14^{+}
Cleaner		
Brushing	15	
Composition	252	89.1^{+}
Scrubbing	134	
Coated	350	165
Coating method	350	320
Coating per se	350	165
Combined with electric lamp	313	110^{+}
Contact for eye	351	160R^{+}
Contact lens fitting	351	247
Design	D16	134^{+}
Echelon lens	350	451
Electromagnetic wave	343	909
Dielectric type	343	911R
With antenna	343	753^{+}
Zoned type	343	910
Electron microscope	250	311
Eye glasses	351	159^{+}
Finder	354	152^{+}
Fog lamp	362	317^{+}
Fresnel lens	350	452
Galilean telescope	350	453
Gauge	33	507
Grinding machines	51	284R
Design	D15	124^{+}
Hoods	350	580^{+}
Intraocular	623	4^{+}
Lenticular elements	350	167
Light fixture	D26	120^{+}
Luneberg type	343	911L
Making	65	37^{+}
Grinding machines	51	
Grinding processes	51	284R
Shaping glass surfaces	65	61
Uniting glass	65	37^{+}
Marker gage	33	200
Mold	425	808*
Moulding glass while hot	65	66^{+}
Moulding plastic	425	
Mounts and holders		
Eye glasses and spectacles	351	41^{+}
Light source combined	362	455^{+}
Sidewalk light transmitting	350	409^{+}
Optical	350	409^{+}
Panoramic lens	350	441
Periscope	350	540^{+}
Projector combined		
Antiglare	362	351^{+}
Reflector combined	362	341^{+}
Reflector signal combined	350	104^{+}
Refracting combined	362	326^{+}
Schmidt lens	350	443
Shades	350	580^{+}
Sidewalk light transmitting	404	22^{+}
Spectacles	351	159^{+}
Lens lining or rims	351	154
Spectacle mountings	351	83^{+}
Telephoto	350	454^{+}
Testers	356	124^{+}
Thermometer tube combined	374	194
Toroidal lens	350	434
Fuse or primer combines	102	200^{+}
Variable focal length	350	423^{+}
Zoom lens	350	423^{+}
Lenticular	430	946*
Lepidine	546	181
Letoff (See Clutch; Ratchet)		
Thread	139	100^{+}
Letter Box	232	17^{+}
Closures and chutes	232	45^{+}
Design	D99	29
Electric switch combined	200	61.63
Letter Opener	D 8	102
Lettering Guides, Stencils	D19	40
Letters Sign Interchangeable	40	618^{+}
Levees	405	107^{+}
Leveling		
Apparel	33	7
Clocks	248	188.4
Furniture	248	188.2^{+}
Caster adjustment	16	19
Illuminated	33	348.2
Load levelers	267	64.16^{+}
Metal straightening	72	160^{+}
Wire fabric	140	107^{+}
Railway trucks	105	164
Self leveling supports	114	191^{+}
Shaft	33	412
Sole machines		
Fastener inserting and laying combined	12	33.1
Laying combined	12	33^{+}
Straightedge geometric instrument	33	451
Wall guide & plumb	33	404^{+}
Suspended railway car	105	149
Track	33	287
Tripod or stand head	248	180
Self leveling	248	182
Vehicle body land type	280	6R^{+}
Body relative to wheels	180	41
Steam boiler body	180	39
Levels (See Leveling)	33	365^{+}
Illuminated	33	348.2
Light ray type	33	290
Liquid		
Alarm electric	340	618^{+}
Alarm mechanical	116	109^{+}
Dispenser material level control	222	64^{+}
Gauge	73	290R^{+}
Indicator	116	227^{+}
Responsive or maintaining systems	137	386^{+}
Material level detection	367	908*
Plumb	33	392^{+}
Spirit level	D10	69
Lever	74	519^{+}
Design	D 8	88
Flexible material tensioning	254	243
Carpet	254	209
Floor jack	254	17
Handle, asymmetric	D 8	DIG. 1
Link systems	74	469^{+}
Locks	70	192^{+}
Switch combined	70	254
Switch combined	200	61.85^{+}
Nail extractor	254	21^{+}
Pushing or pulling implement	254	120^{+}
Rail or tie shifters	254	44
Shaft coupling	403	
Treadle	74	561
Vehicle spring	267	228^{+}
Levigation		
With electric heating	219	7.5
Levitation & Reduced Gravity	156	DIG. 62
Magnetic railway	104	281^{+}
Levulinic Acid	562	577
Esters	560	174
Lewises	294	89
Lewisite	556	70
Lexan T M (See Synthetic Resin or Natural Rubber)		
Leyden Jars	361	301^{+}
Liberating		
Fermentative	435	262^{+}
Fibers chemical	162	1^{+}
Fibers mechanical	19	1^{+}
Library Security by Metal Detection	340	568^{+}
License Plate		
Design	D20	13^{+}
Frames	40	209
Frames	40	152^{+}
Holder with light	D26	31
Jewel reflectors	350	97^{+}
Licorice	426	638
Lid (See Closure; Cover)		
Applying	53	287^{+}
With compacting of contents	100	54^{+}
Burner	431	144^{+}
Car journal	384	189^{+}
Lifter	294	12^{+}
Lock	70	158^{+}
Manhole	404	25^{+}
Purse	150	119
Railway ties tubular type	238	70^{+}
Retort	48	124
Stove	126	211^{+}
Lie Detectors		
Electrocardiograph	128	697
Life Boats	114	348^{+}
Design	D12	316^{+}
Life craft handling devices	114	365^{+}
Life Preservers	441	88^{+}
Aquatic lifesaving devices	441	80^{+}
Design	D21	238
Life vests	441	106^{+}
Racks	114	190
Life Rafts	441	35^{+}
Life Sciences, Computer Applications	364	413^{+}
Life Signals		
Computer controlled, monitored	364	413^{+}
Signalling from burial grave	27	31
Undertaking equipment	27	31
Lifesaving Apparatus		
Aquatic	441	80
Artificial respiration	128	28^{+}
Buoyant devices	99	311
Drowning prevention	441	80
Fenders	293	
Fire escapes and body catchers	182	137^{+}
Parachutes	244	142^{+}
Submarine emergency equipment	114	323^{+}
Supporting and launching boats		
Handling equipment	114	365^{+}
Watercraft lowering drum	254	266^{+}
Lift Modifiers for Aircraft	244	198^{+}
By flap or spoiler	244	213^{+}

	Class	Subclass
Lift Trucks		
Elevator type	187	9R
Handling assistant	414	444
Jack type	254	2R+
Running gear lifts	280	43+
Lift-slab Construction	52	125.1
Lifts, Lifters, Lifting (See Conveyor; Dispensing; Elevator)		
Aircraft	244	34R
Apparel skirt	2	217
Beet	171	50+
Beet	171	103
Belt shifter	474	101+
Bridge lifts	14	42
Canal lock raising	405	86
Car coupling link	213	192+
Car journal box	254	33+
Cartridge	42	17
Clutch for plow	192	62
Copyholder platen	40	349+
Damper musical instrument	84	217
Dobby	139	71+
Door	414	684.3
Traversing hoist	212	166
Drawbar locomotive	213	5
Drum sifter	209	294+
Foot sewing machine	112	237+
Harvester cutter	56	273
Harvester cutter bar	56	283+
Harvester cutter bar finger	56	312+
Harvester cutter bar with rocking	56	276+
Harvester reels	56	364
Hinge combined	16	233
Hoist	254	266
Horse drawn rake	56	397+
Incline plane	254	88
Invalid beds	5	60+
Invalid from bed	5	81R+
Jacquard	139	65
Lantern globe	362	174+
Magnets		
Electric	335	291+
Electro combined	294	65.5
Multiple	254	89R+
Pallets fork lift	108	51.1+
Pawl driving	74	149+
Pawl holding	74	155
Pipe or rod	254	30+
Planer cutter	409	347
Planter tool bar	111	67+
Plate and receptacle	294	27.1+
Plows	172	
Clutch	192	62
Control	172	452+
Plant	172	517
Power take-off	172	492
Servomotor actuator	172	491
Pot lid	294	12
Rail or tie	254	121
Shelf loading type	414	246+
Horizontal & vertical movement	414	260
Ships and boats	405	3+
Spool in winder	242	46
Thread boards	57	360
Traction cable	104	199
Valve lift regulation	123	345+
Vehicle body	254	45+
Ligators	128	326+
Ligatures		
Packaged	206	63.3
Structure	128	335.5
Light (See Electric; Electrooptical; Illumination)		
Arc lamp	314	
Automobile	362	61+
Battery light plants	320	2+
Beam splitting	350	169+
Bicycle	362	72
Building	350	258+
Bulbs	313	315+
Design	D26	2
Buoys floating bodies	441	16+
Burglar alarms light producing	116	7
Christmas	D13	25
Communication	455	600+
Multiplexing	370	1+
Conducting rods	250	227
Body treatment type	128	398
Connecting roos	350	96.1+
Controls	315	
Time controlled	315	360
Decorative	362	806*
Direction instruments	33	1DD
Dispersion	350	168
Dividers	350	169+
Electric lamp	313	
Electric lamp supply systems	315	
Emitting diode	357	
Display matrix	340	782
Filter	350	311+
Compositions	252	582+
Infra-red	350	1.1+
Ultra-violet	350	1.1+
Fixture	D26	24+
Fluorescent screen	250	458.1
Game target and light ray projector	273	310+
Globes	362	363
Grating	350	162.17+
Illumination		
Advertizing signs	40	541+
Extinguishing gas		
Intensifier	250	213R+
Interference	350	163
Invisible light	250	493.1
Lens or glass for fixture	D26	118
Light pipe sign	40	547
Measuring testing meters	356	213+
Electrooptical batteries	136	243+
Invisible light and X rays	250	336.1
Photocell controlled circuit	250	200+
Photocell electronic type	313	523+
Photometers	356	213+
Photosensitive resistors	338	15+
Pyrometers	356	43+
Meters	356	213+
Modulation	350	353+
Mountings	362	382+
Musical instrument producing light effects	84	464R
Parachute flares	102	337+
Pen display control	340	707+
Periscopes with	350	540+
Ports ships	114	177+
Position or condition responsive switch	362	802*
Post combined with	D26	67+
Projectors (See projector)		
Ray type geometrical instruments	33	227+
Reflectors	D26	118+
Responsive explosive igniters	102	201
Marine mines	102	418
Mines	102	427
Rod		
Decorative element	52	301
Seals photographic plate holder	354	277
Sensitive (See photoelectric cells)		
Color film photographic	430	541+
Compositions (See fluorescent, compositions)	252	501.1
Film photographic	430	495+
Film sensitizing	430	570+
Film sensitizing and developing	430	486+
Resistances	252	501.1
Shade holders	362	433+
Ships running light	362	61+
Side or cross walk light transmitting	404	22+
Signs	362	253
Sockets	439	660+
Stage illusion	272	10
Stage lighting systems	315	291+
Stage lighting systems	315	313+
Sterilizing and pasteurizing ultra violet	426	248
Straightline light ray instruments	33	227+
Supports	362	382+
Surgical applications	128	
Switch for portable	200	60
Table	362	97
Tents having light means	135	91+
Tool heater combined	126	232
Torches pyrotechnics	102	336+
Track-type light	D26	61+
Transmission modifying compositions	252	582
Displays color change	252	586
Infrared	252	587
Modification caused by energy other than light	252	583
Produces polarized light	252	585
Ultraviolet	252	588+
Used in reaction to prepare inorganic or organic material	204	157.15+
Used in reaction to prepare or treat synthetic resin or natural rubber	522	1+
Valves	350	353+
Illuminating devices having	362	394+
Motion picture apparatus using	352	204+
Optical testing using	356	
Photo electric devices having	250	229+
Sound recording and reproducing using	369	100+
Waves (See search notes to)	250	
Signaling	455	600+
Lighters (See Igniters)		
Barges	114	26+
Cigar (See igniters)		
Design	D27	36+
Electric	219	260+
Arc or spark	361	247+
Flint packaging	206	528+
Lighting Devices (See Igniter; Illumination; Lamp; Light)		
Lightning		
Arresters	361	117+
Design	D10	105
Gas tube per se	313	567+
Gas tube type	361	120
Manufacture	29	
Resistance	338	21
Space discharge type	313	
Space discharge type combined with impedance	315	58+
With thermal cutout	337	31
Atmospheric electricity	307	149
Collecting with condenser	320	1
Conductor	D10	105
Grounding	174	6+
Rods	174	2+
Static discharging	361	212+
Lignins	530	500+
Compositions containing (See synthetic resin or natural rubber, compositions)	106	123.1
Fiber liberation	162	1+
Lignite		
Mineral oil from	208	400+
Limber	89	40.7
Limbs Artificial	623	27+
Lime	423	635
Burning processes	432	1+
Compositions	106	118+
Adhesive for catching insects	424	77
Alkali metal silicate containing	106	78
Sulfur combined biocide	424	160
Kilns	432	
Light	431	347
Limit Stop	192	138+
Elevator control	187	34+
Phonograph turntable	369	238
Switch	200	47
Limiter		
Grid controlled electronic tube	328	225
Radio amplifier	330	129+
Linchpin		
Land vehicle combined	301	113
Railway axle combined	295	50
Lincomycins	536	16.2+
Line		
Blind		
Flow restrictor combined	138	40+
Fluid conduit	138	89+
Carrying		
Ammunition	102	504
Fire escape	182	10+
Shell	102	504
Chalk	33	414
Clothesline support	211	119.1
Drying reel	242	104
Electric		
Block signal current supply	246	87+
Polystation telephone	379	177+
Automatic	379	182+
Revertive call	379	178
Signal control telephone	379	179+
Telegraph key	178	103
Telegraph line clearing	178	69R
Telegraphic printing	178	31+
Telephone party register	379	117+
Testing long line telephone	379	22+
Finder telephone switch	379	293+
Fishing	43	44.98
Attached bodies	43	43.1
Float holding	43	43.1+
Reel	242	84.4
Rod guide	43	24

	Class	Subclass
Harness	54	36
Lead sounding	33	126.5$^+$
Markers	33	32.1$^+$
Moving markers	33	32.3
With moving sheet	33	32.4
Stationary marker	33	32.1$^+$
With moving sheet	33	32.6
With moving support	33	32.5
Musical expression	84	463
Printing		
Bars	101	396$^+$
Spacer	101	403
Telegraphy	178	31$^+$
Rat guards for ships	114	221R
Scriber	33	18.1$^+$
Parallel	33	41.1$^+$
Separator type	276	35
Shorting		
To protect amplifier	330	51
Spacer typewriter	400	545$^+$
Wire driven planter	111	43$^+$
Linear		
Brake shoe arrangement	188	240$^+$
Electric motor	310	12$^+$
Railway	104	290$^+$
Plant support	47	46
Traversing hoist	212	71$^+$
Linemans		
Protective device	174	5SB
Spurs	182	221
Linen		
Household	D 6	595$^+$
Liner & Lining (See Bushing)		
Aligning		
Type casting mold	199	58
Modified	199	57
Typewriting	400	16$^+$
Ball or rod grinding mill	241	182$^+$
Bearing	384	276$^+$
Brass railway car journal	384	191.2
Roller bearing	384	569$^+$
Boat	114	69
Box	493	93
Temperature barrier	217	128$^+$
Brake and clutch		
Brake	188	251R$^+$
Clutch	192	107R
Making by laminating	156	
Stock material	428	
Casting mold for	249	87$^+$
Coating	427	230$^+$
Electrolytic	204	26
Rotary	415	170R$^+$
Tooth	433	217.1
Denture	433	168.1
Composition	523	120
Earthenware pipe	118	708$^+$
Earthenware pipe	118	105
Earthenware pipe	118	125
Earthenware pipe	118	404
Earthenware pipe	425	110$^+$
Earthenware pipe	425	262
Electric conduit bushing	174	83
Electric furnace	373	
Arc	373	71
Induction	373	138
Resistance	373	137
Garment	2	272
Coat	2	97
Glove	2	164
Hat	2	190
Hat try on	2	63
Grave	27	29
Hat	223	22
Lined		
Paper box	220	403$^+$
Pipe	138	140$^+$
Flexible	138	137$^+$
Pipe coupling	285	55
Railway car journal brasses	384	191.2
Smoking devices	131	219$^+$
Bowl or stem	131	204
Walled in space	52	267$^+$
Curvelinear	52	249
With floor	52	265$^+$
Metal bottle cap making	413	8$^+$
Apparatus	413	56$^+$
Metal mold	118	
Metallurgical receptacle	266	280$^+$
Mold		
Compositions	106	38.22$^+$
Earthenware	425	412$^+$
Machine type-bearing or friction surface	249	87
Rotary	415	196$^+$
Railway car interior	105	423
Reactive furnace liner	75	95
Receptacle		
Closure, laminated	156	
Liquid proofed paper	229	3.1
Metal closure making	413	8$^+$
Apparatus	413	56$^+$
Metallic	220	415$^+$
Wooden	217	3R
Temperature barrier	217	131
Shaft	405	133
Shell	86	17
Shoe	36	55
Applier	12	39
Stove	126	144$^+$
Tunnel	405	150
Umbrella	135	33R$^+$
Wagon	296	39R
Lingerie	2	
Closed lingerie clasps	D11	200$^+$
Design	D 2	1$^+$
Lingoes	139	90
Lingual Bars Dentistry	433	190
Linguistic Computer Applications	364	419
Liniments	424	
Link		
Assembling		
Machines	59	25
Sprocket chain	59	7
Belt transmission	474	240$^+$
Sprocket chain	474	202
Button	24	102R$^+$
Chain	59	84$^+$
Antiskid cross	152	243$^+$
Ornamental	59	82
Extension for closures	292	262$^+$
Lock type	70	93
Fabrics		
Bed bottoms	5	188
Flexible panel structure	160	220
Mats floor covering	52	663
Mats scraping	15	239$^+$
Penetration resistant	2	2.5
Wire	245	9
Insulators	174	208
Lever systems	74	469$^+$
Motions in steam engines	91	218$^+$
Railway drawbar coupling	213	208
Cushion connection	213	66
Separation	59	11
Linkers (See Link, Assembling)		
Sausage	17	34
Stuffer combined	17	33
Linking Warp Chains	28	177
Linoleic Acid	260	413$^+$
Linoleum	428	456
Making	156	
Pattern and texture design	D 5	
Rug	D 6	582$^+$
Sheet	D 5	
Linotype TM Machines	199	1$^+$
Laminating	156	
Molding	425	
Linoxyn	260	406
Compositions	106	244$^+$
Linseed Oil	260	398$^+$
Compositions containing	106	252$^+$
Alkali metal silicate	106	82$^+$
Biocidal or fireproofing	106	18
Bituminous material	106	246$^+$
Carbohydrate	106	215
Carbohydrate gum	106	206
Cellulose	106	199
Cellulose ether or ester	106	171$^+$
Inks	106	28$^+$
Natural resin	106	222$^+$
Other fatty oil	106	250$^+$
Polishes	106	9
Portland cement	106	95
Protein	106	124$^+$
Starch	106	211
Sulfurized or sulfonated	106	249
Synthetic resin (See fatty acid triglycerides)		
Wax	106	245
Lint Remover Brush	D 4	
Lintels	52	118$^+$
Beam per se	52	720
Linter		
Ginning type delinter	19	40$^+$
Lip Seal	277	212R
Spring biased	277	152$^+$
Lipids	436	71*
Lipins (See Fats)		
Fermentative liberation or purification	435	282
Lippmann	430	948*
Lipstick		
Brush	D 4	
Compact	401	49$^+$
With light	362	135$^+$
With toilet article, eg mirror	132	82R
Compositions	424	64
Design	D28	4
Container	206	385
Holder	401	88$^+$
Design	D28	85$^+$
Shaped applier	132	88.7
Liquefaction and Liquefiers		
Gas	62	36$^+$
Handling	62	45$^+$
Process of	62	11$^+$
Receptacles	220	901*
Refrigerant	62	
Generation of gas and	48	1
Lubricant	184	98$^+$
Vapor condensers	165	110$^+$
Liqueurs	426	592$^+$
Liquid		
Aerating	261	
Spraying, sprinkling deffusing combined	239	428.5
Arc preventing	200	153R
Circuit		
Electrolyte	204	237$^+$
Electrolyte combined	204	234$^+$
Gas and liquid contact apparatus	261	29
Gas and liquid contact with refrigeration	62	310
Gas cooling	62	406
Gas cooling process	62	96
Gas separation	55	229
Heat exchange	165	
Heat exchange tank	165	108
With refrigeration	62	430$^+$
Liquid congealed	62	348
Liquid cooling by refrigerated circuit	62	393
Liquid electrode	204	220
Nuclear reactor combined	376	370
Clarifying	210	
Collection, diagnostic	128	760
Comminuting and solidifying	425	6$^+$
Processes	264	5$^+$
Contact		
Article cooling process	62	64
Discrete commodity cooler	62	373$^+$
Electrode cells	204	250$^+$
Electrode moving	204	219$^+$
Switch	200	182$^+$
Switch electromagnetic	335	47$^+$
Switch expansible liquid	337	331
Switch periodic	200	196$^+$
Switches periodic	200	200$^+$
Switches periodic	200	205$^+$
Switches periodic	200	208
Contact with gas	261	
Carbohydrate defecating	127	12
Gas separation combined	55	220$^+$
Gas separation processes	55	84$^+$
Contact with solids	134	
Apparatus for comminution	241	38$^+$
Lubrication for comminution	241	15$^+$
Seeds during comminution	241	12
Seeds prior to comminution	241	8
Coolers	62	389$^+$
Process	62	98$^+$
Decolorizing		
Filter materials	210	500.1$^+$
Particulate material type separator	210	263$^+$
Processes	210	660$^+$
X-art collection	210	917*
Dispensers	222	
Extraction by squeezing	100	104$^+$
Feeding nozzles	261	DIG. 39
Fuel	44	50$^+$
Cookers	126	38$^+$

	Class	Subclass
Heating stoves	126	84+
Nuclear reactions utilizing	376	356
Nuclear reactions utilizing	376	356
Solidifying	44	7.1
Fuel burners (See retort)		
Gas contact	261	
Gas separation combined	55	
Processes	55	84+
Heaters	126	344+
Pressure generators	122	
Level control		
Dispensing controlled by	222	64+
Float valves	137	409+
Tanks	137	386+
Level or depth gauges	73	290R+
Electric	340	618+
X-art collection	367	908*
Telemetric system	340	870.38+
Meters	73	861+
Monitor radioactivity	250	336.1
Proofed	229	3.1
Purification or separation	210	
Refrigerant separation	62	474+
Automatic	62	195
Oil	62	470+
Oil, automatic	62	192+
Oil, process	62	84
Process	62	85
Refrigerated liquid separation	62	319
Process	62	85
Seals		
Bellows or	156	DIG. 93
Dispensing apparatus	222	188
Gas separating apparatus	55	355
Pipe joint	285	10+
Sewer trap	137	247.11+
Water closets	4	300+
Separation	210	
Concentrating evaporators	159	31
Contact and	261	2+
Deaerating	55	159+
Deaerating processes	55	36+
Distillation	203	
Electrical	204	186+
Electrical apparatus for	204	302+
Emulsion breaking	252	319+
Gas separation combined	55	421
Gas separation process combined	55	45
Ice melt and impurities	62	318
Liquid meters	73	200
Liquid-solid	4	DIG. 19
Oil dehydration	208	187+
Oil electrophoresis	204	181.8+
Oil paraffin separation	208	28+
Oil residual from paraffin	208	30+
Selective freezing	62	123+
Selective freezing process	62	67
Sprinkling, spraying, diffusing	239	
Testing	73	53+
Treatment		
Amylaceous liquids	127	69+
Coating implement with material supply	401	
Combs employing	132	11R+
Drying	34	
Electrolytic foods and beverages	426	
Electrolytic water and sewage	204	149+
Fermentation	435	
Food preserving by	426	330+
Food sterilizing and pasteurizing	99	467+
Grain huller apparatus	99	600+
Grain huller process	426	482+
Lighted tobacco extinguisher	131	236
Metal heat treating combined	266	121+
Metal quenching apparatus	266	114+
Continuous strip	266	111+
Metal quenching processes	148	13+
Sound muffling by	181	220
Disparate fluid mingling	181	259+
Sterilization with except foods	422	
Textile apparatus for	68	
Textile combined	28	167+
Textile fiber decorticating	19	7+
Textile ironing	38	77.1+
Textile processes for	8	
Thread finishing	28	217+
Vaporizing	122	
Warp preparing	28	178+
Liquid Crystals		
Article or stock	428	1
Cholesteremic	260	397+

	Class	Subclass
Composition	252	299.1
Display device	350	330+
Light control	350	
Display of message	340	784
Hand and dial	368	84
Monogram	340	765
Time indication	368	242
Optical filter composition	252	582
Liquors	426	592
Malt	426	64
Listers	172	642
Drilling	111	83
Walking cultivator or plow	172	351+
Listing Forms	283	66R
Lite Pipe Sign	40	547
Litharge	423	619
Lithium (See Alkali Metal)	423	179.5
Lithography	101	130+
Lithographic emulsion	430	949*
Lithopone	106	294+
Drying oil containing	106	255
Lithotrites	128	328
Litter	5	81R+
Litter bag	D34	1+
Portable	5	82R+
Wheeled	296	20
Livestock		
Care	119	
Design	D30	
Cars	119	7
Living Product	800	1
Lixiviating (See Leaching)		
Load		
Braced on freight carrier	410	121+
Bridging vehicle	414	460+
Cell	177	
Hauling or hoisting	254	266
Lashed on freight carrier	410	96+
Levelers	267	64.16+
Starters		
Animal draft appliances combined	278	2
Wheeled vehicle attachments	280	151
Loaders & Loading (See Gun; Magazine)		
Aircraft	244	127+
Ammunition	86	23+
Antenna	343	749+
Cartridge belts	86	48
Conveyor chutes	193	3
Dump truck	298	
End gate loading chute combined	296	61
Excavator	37	4+
Fireworks	86	20.1+
Glider pickups	244	2+
Harvester combined with raker	56	344+
Loaded		
Conductors for signaling	178	45+
Line	178	45
Feeding to amplifier	178	45
Phantom circuit having	178	45
With attenuator	178	45
Marine	414	137+
Nuclear reactors	376	260
Nuclear reactors	376	268
Railway car bodies dumping	105	239+
Railway mail delivery	258	
Self loading vehicles	414	467+
Sheaf	56	120
Shoe nailer	221	
Textile fluid treatment	68	210
Trains in motion	104	18+
Vertically swinging shovels	414	680+
Weighing scales	177	145+
Loafer, Footwear	D 2	269
Lobster Claw Clamp	17	1R
Localizer	342	413
Locating (See Detection; Testing)		
Foreign bodies	128	737+
Location determining devices		
Acoustic	181	125
Radar	342	126+
Railway track defect	73	146+
Train protecting combined	246	120+
Sound		
Mechanically	181	125
Submarine	367	87+
Lock (See Definition Notes, Associated Devices; Fastener; Detent; Latch; Seal; Securing Means)	70	
Air		
Caissons	405	8

	Class	Subclass
Diving bell combined	405	192
Dryers	34	92
Dryers vacuum	34	242
Furnace charging	414	162+
Material moved between unlike pressure areas	414	217+
Pneumatic conveyor discharging	406	169
Submarine escape	114	334+
Article dispenser	221	154
Automobile	70	237+
Bicycle	70	233+
Book binder sheet retainers	402	2
Box and chest	70	63
Cabinet with		
Selector operator	312	215
Canal	405	85+
Cane, umbrella, apparel	70	59
Closure	70	77+
Barrel bung	217	106+
Bottles and jars	215	207
Burglar alarm combined	116	8+
Camera plate holders	354	281
Car dump	105	308.1+
Check controlled	194	
Drawers	312	333
Elevator door	187	61
Fasteners	292	
Freight car automatic lock	105	395
Railway car platform trap door	105	435
Safe alarm combined	109	44
Safe bolt work	109	59R+
Safe door poison gas releaser combined	109	30
Special receptacle	206	1.5
Till drawer alarm combined	116	76
Vault covers	404	25+
Combination	70	286+
Design	D 8	331+
Dispensers with	222	153
Dog for bolt	70	467+
Door	70	91+
Door latch bolts, biased	70	144
Electric lamp	439	133+
Expansible chamber device with	92	15+
Fluid	92	8+
Firearm safety mechanism	42	70.1+
Fluid handling systems or devices with	137	383+
Gang bar type for cabinet	312	216+
Sectional unit type cabinet	312	107.5
Hasp	70	2+
Key removal preventing	70	389
Keyholders	70	456R+
Lap robe in car body	296	77
Machine elements valves etc	70	175+
Mechanical gun	124	40
Mounting devices	70	466
Nail	411	439+
Railway spike	238	375
Nut and bolt		
Bolt pivoted end	411	340+
Bolt threadless key type	411	340+
Bolt to nut	411	190+
Railway splice bar	238	262
Openers electric or magnetic	70	277+
Operating mechanism	70	266+
Pad	70	20+
Permutation	70	315+
Pipe joint	285	80
Portable	70	14+
Puzzle	70	289+
Rack combined	211	4+
Railway switches and signals	246	
Register	235	130R
Key locks	235	27
Safety chains	292	262+
Seal tamper detecting	70	440
Latch	292	307R+
Padlock	70	50
Side lock	411	191
Slotted bolt, railway splice bar type	238	252
Special application	70	57+
Stitch making	112	
Switch		
Lock combined	200	61.64+
Operated	200	43.1+
Systems	70	262+
Telephone	379	445
Operated over telephone line	379	103
Time	70	267+
Trunk	70	69+

U.S. PATENT CLASSIFICATIONS

Entry	Class	Subclass
Typewriter	400	663+
Valve	70	175+
Automobile fuel	70	242+
Gravity feed lubricators	184	86
Lock operated	137	384.2+
Ward guard	70	420
Weigher	177	124
Wheel	70	225
Brake	188	69
Caster	16	35R+
Yale	70	357+
Lockets	63	18+
Design	D11	80
Lockout		
Key locks	70	337
Telephone	379	194+
Intercommunicating	379	168+
Polystation	379	184+
Locomotive		
Boilers	122	
Chain grate progressive feed	110	198+
Design	D12	38
Feeding air		
Firebox	110	198+
Feeding air and steam		
Firebox	110	199+
Heater	110	201
Undergrate	110	202
Firebox	110	198+
Fuel feeding stoker	110	105.5
Headlight and signal illuminators	362	61+
Journal	384	187
Light supports automatic	362	66+
Railway	105	26.5+
Safety bridges	105	459
Truck frames	105	172+
Shed fixtures	104	51+
Smoke and gas return	110	198+
Stoker type comminutor	241	276
Tender scoops	417	232
Toy	446	467
Illuminated	446	438+
Smoking	446	25
Sounding	446	410
Valves, gears	91	218+
Ventilation of cab	98	3
Log		
Borehole testing	73	152
Building	52	233
Toy	446	106
Gas burner	431	125
Gas or electric, simulative	D23	409
Loading vehicle	414	460
Ship		
Course recorder	346	8
Navigation	73	181+
Splitter	144	193R+
Wood sawing		
Band saw	83	788+
Log deck block	83	708+
Log transfer	83	707+
Log turner	83	708+
Log turner and deck block	83	708+
Woodworking		
Stay log slicer	144	177+
Stay log veneer lathe	144	214
Logging		
Cable type traversing hoists	212	76+
Pole or tree handlers	414	23
Vehicle	414	460
Wood sawing	83	703+
Logic Gate Active Element Oscillator	331	DIG. 3
Logotype	400	95+
Looking Glasses (See Mirrors)	139	
Looms	139	
Fabric take up	139	304+
Lace	87	27
Loose reed motions	139	191
Moistening weaving materials	139	36
Special-type	139	11+
Traveling wave shed	139	436
Vertical shed	139	18
Loop		
Amusement railway with vertical	104	55
Animal catching lasso	119	153
Antenna	343	866+
High frequency type	343	741+
Belt		
Machine for sewing	112	121.27
Cartridge belt	224	223+
Clothes line endless	211	119.2+
Fastening device		
Billet loops	24	182
Button engaging	24	660+
Button engaging garter type	24	464+
Engaging with camming lever	24	69R+
Hat combined with comb or hairpin	132	61
Lace stud	24	145
Pipe couplings	285	71
Tag attaching pins	40	24
Tie to collar button	24	58
Garment	2	271
Garment supporter	2	321+
Crossed loop	2	327+
Plural loops	2	331
Handle	16	125+
Insulated	16	119
Harness		
Bridle crown	54	13
Checkrein hook	54	17
Design	D30	140+
Strap connector	54	87
Lock		
Element on trunk	70	36
Rail holddown	238	43
Sewing machine	112	32
Railroad splice bar	238	256
Textile pile	26	2R+
Cutting	26	7+
Trap	43	86
Wire fabric interlocking	245	5
Loopers Festooner		
Carpet rag	28	148
Drier combined	34	157
Knitting	66	
Packet	227	77
Sewing	112	
Web or strand feeding	226	104
Wireworking	140	102+
Loose Leaf		
Binder device releasably engaging aperture or notch of sheet	402	
Opening reinforcement	402	79
Lorgnettes	351	56
Design	D16	105
Supports	351	56
Lorol (See Lauryl Alcohol)	568	840
Lotions	424	
Loud Speaker (See Telephone)	381	150+
Baffle	181	155
Design	D14	53+
Public address system	381	82+
Loudness Meter	73	646
Lounges	52	473
Louvers		
Movable	49	74+
Shutters		
Unit design	D25	152+
Ventilating	98	121.1
Low mu Triode	313	293+
Lowering		
Beds	5	65
Coffins	27	32+
Cable	254	266+
Hoist hook	212	89
Life craft	114	365+
Load with automatic return	414	594
Plow	172	464
Support releasable	248	320+
Vehicle tops	296	112+
Lozenges		
Design	D 1	127
Medicinal	D28	1+
Presses	99	353+
Lubrication and Lubricators (See Oiler and Oiling)	184	
Air line lubricators	261	DIG. 35
Ammunition	102	511
Axle jacks	254	32
Compositions (See lubricants)		
Earth boring drilling fluid	175	65+
Engines, internal combustion	123	196R
Poppet valve operating mechanism	123	188A+
Fitting	285	169+
Making	29	157B
Valved	251	142+
Fluid handling	137	246+
Gun	D 8	14.1
Lubricants (See oil)	252	9+
Fuel containing	44	58
Mineral oil only containing	208	18+
Purifying used	208	179
Testing	73	64
Frictional characteristics	73	10
Lubricating apparatus	184	
Gun	D 8	14.1
Oil dispensers	222	
Machine parts		
Band saw pulley	83	169
Bearings	384	
Belt and sprocket	474	91
Brakes	188	264B
Casters	16	36
Clutches	192	113B
Connecting rods	74	587
Crank and wrist pins	74	605
Earth boring tool combined	175	227
Endless flexible track pivot	305	14
Gears	74	467+
Glass making apparatus	65	170
Hinges	16	274
Metal founding apparatus	164	149
Piston rings	277	
Pistons	92	153+
Planes wood	30	483
Sewing machines	112	256
Sewing machines leather	112	43
Shaft coupling	464	7+
Spinning rings and travelers	57	120
Spinning whirls	57	133
Springs leaf	81	3.7
Springs leaf	267	50
Springs leaf covered	267	37.4+
Sprocket chain pivot	474	91
Textile drawing saddles	19	284
Valve actuator	251	355
Valve interface	137	246+
Vehicle wheels spring type	152	2+
Material working		
Ammunition making	86	19
Cutters	407	11
Die expressing	425	197+
Die expressing	425	200+
Dies wire drawing	72	41+
Screw threading	408	
Pipe joint	285	94
Refrigeration apparatus	62	468+
Automatic	62	192+
Process	62	84
Reservoir for packing assembly	277	17+
Separator, imperforate bowl centrifugal	494	15
Textile compositions	252	8.6+
Textile processes	28	169+
Braid	87	1
Coating or impregnating	427	
Strands	57	295+
Strands covered	57	7
Lucite T M (See Synthetic Resin or Natural Rubber)		
Luggage (See Baggage)	190	
Design	D 3	30.1+
Latch	D 8	331+
Travel bag	206	278+
Lumber		
Building component	52	
Compound	52	782+
Denailing	254	18
Impregnated	428	541
Laminated	428	18
Luminal	544	305
Luminaries	362	296+
Luminous (See Fluorescent; Illuminated; Light)	362	
Cathode ray screens	313	461+
Photo process of making	430	23+
Compositions fluorescent or phosphorescent	252	301.16
Inorganic	252	301.6 P+
Organic	252	301.16
Compositions radioactive	252	625+
Fluorescent and phosphorescent	250	458.1
Heads for figure toys	446	391+
House number	250	462.1
Sign	40	541+
Illuminating devices with material	362	84
Lamps	313	
Push button	250	465.1
Toys	446	485
Chemiluminescent	446	219
Lunch		
Heaters	126	261+

	Class	Subclass
Kit or box	206	541+
Design	D 7	76+
With heater	126	266
Wagons	296	22
Luneberg Lens Type	343	911L
Lungs		
Heat-lung digest	128	DIG. 3
Lungmotor respirators	128	28+
Lures		
Fishing	43	4+
Artificial	43	42+
Design	D22	126+
Food bait	426	1
Lustering		
Fabrics	26	27+
Lutidine	546	348
Lymphokines	530	351
Lysergic Acid	546	69
Lysergic Acid Diethylamide Lsd	546	69
Lysine	562	562
Lythrine	546	43

M

	Class	Subclass
Macaroni	426	557
Apparatus		
Die expressing	425	288
Die expressing	425	289+
Die expressing	425	331
Die expressing	425	381.2
Design	D 1	126+
Drying digest	34	
Machete	30	
Design	D22	118
Machine Elements	74	
Automatic operation or control	74	2+
Bearings	384	
Design	D15	143
Belt & pulley drive system	474	
Brakes	188	
Fluid pressure brake & analogous systems	303	
Clutches & power-stop control	192	
Control lever & linkage systems	74	469+
Cylinders and pistons	92	
Design, general	D15	
Engine starters	74	6+
Gearing	74	640+
Gyroscopes	74	5R+
Intermittent grip type movements	74	111+
Lighting	362	89+
Mechanical movements	74	20+
Power take-off	74	11+
Pushing & pulling implements, force multipliers	254	1+
Rotary member or shaft indexing	74	813R+
Shaft operators	74	10R+
Shafting and flexible shaft couplings	464	
Lost motion couplings	464	160+
Rotary shafting	464	179+
Tool driving or impacting	173	
Work holders	269	
Machine Gun	89	9+
Explosive propellant type	89	9+
Charger	89	1.4
Self loading explosion operated	89	125+
Mechanical	124	
Centrifugal	124	4+
Machine Supports	248	637+
Machinery, Toy	446	
Mackinaws	2	93+
Macrame	289	16.5
Magazine (See Handle Receptacle Type)		
Ammunition and explosive		
Body or belt attached	224	223+
Bandoleer	224	203
Covered flap	224	239
Chargers	42	87+
Gun combined	42	2+
Loading	86	47
Mechanical gun or projector	124	45+
Ordnance	89	33.1+
Railway torpedo	246	212+
Revolver combined	42	60
Structure	42	49.1+
Article dispenser (See class defs)	221	
Boiler		
Liquid heater or vaporizer	122	
Camera	354	174+
Coin return, empty dispensing	194	200
Dry closets	4	464+
Electrode	314	5
Fire pots combined with magazine section	D23	386+
Holder as furniture	D 6	396+
Implements and machines with		
Assembling apparatus	29	809+
Check controlled machine	194	
Detonating cane	446	403
Label pasting machine	156	
Digest, apparatus	156	DIG. 29
Nail drivers	227	
Pencil	401	89+
Pencil selectively replenishing feed guide	401	56
Pencil sequentially replenishing feed guide	401	57
Razor	30	40
Riveter	227	
Riveter eyelet	227	51+
Sound recording tapes	D14	11
Stapling	227	
Type cases	276	44+
Type casting machine	199	1+
Type setter	276	15
Type setter and distributor	276	3+
Typewriter paper feed	400	624+
Wire and slat	140	27
Wire and slat portable	140	38
Wireworking	140	53+
Labeling	156	DIG. 29
Lock magazine cylinder & pin tumblers	70	DIG. 5
Motion picture film	352	78R
Paper cover removal	83	925MG*
Photographic	354	174+
Slide projector, design	D16	21
Razor blade holder	206	352+
With means of feeding blades into razor	30	40
Stove and furnace		
Furnace structure	110	293+
Stove	126	
Toaster	99	387
With article removing facilities	312	35
Magenta (See Fuchsine)		
Magic		
Amusement devices	272	8R+
Clock	368	81
Slides	40	152+
Magnesia	423	635
Cement	106	121
Magnesium	75	
Alloys	420	402
Aluminum	420	542
Aluminum copper	420	533
Electrolytic synthesis	204	70
Compound inorganic	423	
Carbonates	423	430+
Inorganic	423	
Oxide and hydroxide	423	635
Electrothermic processes	75	10.1+
Volatilization of magnesium	75	10.33
Heat treatment	148	161
Metal working with	29	DIG. 27
Oxychloride cement	106	106+
Peroxide	423	583
Plate	428	649+
Preparation		
Electrolytic	204	70
Pyrometallurgy	75	67R
Strip type solid fuel	431	99
Magnet and Magnetic (See Type of Magnet)	D13	99
Actuated motor for railway switches	246	227+
Actuated motor for railway switches	246	231+
Actuated switch	335	2+
Alloys	420	
Analog storage systems	365	48
Antimagnetic devices for watches and clocks	368	293
Arc or ray deflecting or focusing		
Arc lamp	313	153+
Arc lamp with discharge deflecting	314	20
Cathode ray tube	313	421+
Cathode ray tube circuits with plural	315	14+
Electron microscope	250	311
Electronic tube controlled by	315	267
Electronic tube controlled by	315	344+
X ray tube	378	137
Board games	273	239
Bracket	248	206.5
Holding article by	248	309.4
Bubbles, magnetic	365	1+
Chargers	335	284
Chuck	279	1R
Holding magnet	335	285+
Circuit and switch control		
Arc blowout	315	344
Arc blowout in switch	200	147R
Electromagnet circuits	361	139+
Electromagnetic switch	335	2+
Electromagnetic switching	361	139+
Operator for brake	188	161+
Railway cab signal	246	178+
Railway cab signal	246	193+
Railway cab signal	246	197
Railway cab signal	246	202
Railway point thrower	246	265
Railway switch actuation	246	225+
Railway train trip	246	360
Transmission to vehicles	191	16
Transmission to vehicles	191	17+
Circuit controller for railway section	191	17+
Arc lamp operator	314	104
Closures	220	230
Clutches	310	92+
Compasses	33	355R+
Compositions	252	62.51
Consumable electrode feed or control	314	
Control for frictional gearing	74	210
Conveyers and feeders		
Holding item by magnetic attraction	209	904*
Magnetic feeder	209	907*
Cooling by	62	3
Coupling	464	29
Detector for radio receiver	329	200
Digital signal processing	360	39+
Disk	360	135
Drum	360	136
Transport systems	360	81+
Dynamic magnetic info storage and retrieval	360	
Automatic control system	360	69+
General recording or reproducing	360	55+
Head	360	110+
Physical elements	360	102+
Record medium	360	131+
Signal processing	360	18+
Special purpose devices	360	1+
Earth boring tool	175	328
Electro	335	209
Lifting	335	291
Removing splinters from the eye	128	1.4
Electrochemistry apparatus	422	186.1+
Explosive igniters		
Fuses	102	206+
Marine mines	102	417
Feeler mechanism loom	139	274
Fender securing device	293	DIG. 6
Flip flops	307	406
Hoist line type	294	65.5
Induction current transmission to vehicles	191	10
Instruments	33	DIG. 1
Joints and connections digest	403	DIG. 1
Key lock	70	413
Levitation railway	104	281+
Lifting handling or holding	335	285
Bracket supports	248	206.5
Conveyors	198	690.1
Enhanced by magnetic hold	198	690.1+
Detachable electric connector	439	38+
Grapples	294	88
Hand and hoist line implements	294	65.5
Light supports	362	398
Material handling	414	606
Metal founding apparatus	164	146
Railway mail delivery apparatus	258	4
Lightning arrester	361	133+
Locomotive traction regulator	105	77+
Loom shuttle actuator	139	134
Magnetic bubbles	365	1
Magnetic memory	365	
Associative memories	365	50
Dynamic information storage &	360	
Formation by shaping	264	104
Laminated type	428	900*
Magnetic bubbles	365	1+
Read-only memories	365	97+

	Class	Subclass
Static information storage & retrieval	365	
With coated metal base	437	+
Magnetic recording digest	29	DIG. 28
Making	29	603+
Material treatment	148	100+
In magnetic field	148	108
Metal casting with	164	498+
Continuous casting metal with	164	466+
Nonmetallic	252	62.51
Particulate material	148	105
Permanent	148	101+
Permanent in magnetic field	148	103
Permanent with age hardening	148	102
Silicon steel	148	110+
Metal working digest	29	DIG. 95
Microphone	381	177+
Mines	102	401+
Motor art digest	91	DIG. 4
Motor operated lock	70	276
Motors with armatures	335	220
Motors with armatures	310	10+
Music leaf turner	84	521
Music tone generator	84	1.15
Package or receptacle	206	818*
Permanent and electromagnet structure	335	209+
Making	29	607+
Press digest	100	917*
Pulse generator	307	419
Read only memories	365	97+
Record and recording	360	1+
Coating	427	128+
Copying	360	15+
Editing	360	13+
General reproducing	360	55+
Medium	360	131+
Monitoring	360	31
Transport systems	360	81+
Reluctance		
Adjustable dynamo structure	310	191
Generator control	322	49+
Seal	277	80
Separation (See separator)		
Auto crankcase drain plug	335	305
Diamagnetic	209	212
Gas	55	100
Gas processes	55	3
Iron particles from solids	209	213+
Liquid	210	222+
Paramagnetic	209	213+
Sorting by	209	636
Edge coding	209	609
Sound recording and reproducing		
General recording or reproducing	360	55+
Phonograph	360	86
Phonograph record	360	135
Static information storage & retrieval	365	
Associative memories	365	49+
Magnetic bubble memories	365	1+
Read-only memories	365	94+
Stock	148	300+
Composite	428	928*
Structure	335	209+
Tape for recording	360	134
Cassette	D14	11
Reel	D14	10
Teaching magnetism	434	301
Television signal recording	360	33.1+
Tester	324	205+
Coin fraud preventing	194	320+
Therapeutic magnetic vibrator	128	424
Therapeutic or probe type	128	1.3+
Thermonuclear reactions use in	376	121
Mirror	376	140
Thin film memories	365	171+
Tool holder-flexible Abrasive	51	362
Toys	446	129+
Trigger devices	307	414
Vertically swinging material handling	414	737
Winding (See making)	242	7.7+
Method	242	7.3
Wire recorders	360	89
Work holders	269	8
Magnetizing		
Circuit or apparatus for	361	143+
Magnets or electromagnets for	335	284
Magnetoelectric Generators	310	69+
Permanent magnet type	310	152+

	Class	Subclass
With ignition system	123	143R+
Magnetometers	324	244
Magnetostrictive		
Electroacoustic transducers	73	570+
Microphone	381	168+
Seismograph type	367	140
Telephone receiver	381	190
Underwater type	367	168
Plural transducer array	367	156
Vibration testing	73	570
Magnets and electromagnets	335	215
Printing using	101	DIG. 5
Magnetron		
Amplifier	330	47+
Modulator	332	5
Oscillators	331	157
Radio transmitter with	455	91+
Structure of tube	315	39
Supersonic wave generator Oscillator	331	154
System in tube	315	32
System miscellaneous	328	230
Tube structure	315	39.51
Ultrsonic wave generator agitator	366	108+
With pulsing system	375	
Magnifying (See Telescopes Microscopes)		
Glasses	350	245+
Design	D16	135
With mount	350	252+
Lenses	350	414+
Weighing scale	350	114+
Liquid depth gauges	73	327
Pocket	350	250
Thermometer tube	374	193
Mail		
Containers		
Bags	383	
Bottle cases	217	127
Coin cards	229	92.9
Envelopes	229	68R+
Letter box closures or chutes	232	45+
Letter boxes	232	17+
Mailbox design	D99	29+
Mailing wrappers	229	92
Tubes	229	93
Delivery railway	258	
Handling		
Aircraft	258	1.2+
Cable car	104	177
Cable car single cable	104	181
Canceler and container combined	101	71
Cancelling stamps	101	371
Chutes	193	8
Elevator	187	4
Mail onto & off vehicle	414	265
Railway	258	
Sorting	209	900*
Sorting flat-type mail	209	900*
Letter sheets	229	92.1+
With return card	229	92.8
Postal cards	229	92.8
Railway delivery	258	
Aircraft in flight	258	1.2+
Signaling feature	258	2
Stamping machines		
Posting	156	
Printing	101	95
Mains and Pipes	138	
Tapping	137	317+
Mainspring		
Appliers and removers	29	228
Fastenings	368	144
Winder, watchmakers	81	7.5
Make-ready	101	401.3
Make-up		
Box	132	79R+
Brush	D 4	
Cape	D28	99
Malachite Green	260	393+
Malathion	558	180
Compositions bio affecting	514	122+
Maleic Acid	562	595
Anhydride	549	262
Carbocyclic compound	549	234+
Maleinides	548	548+
Malleableizing Cast Iron	148	138
Mallet	81	19
Amusement device	273	67R+
Design	D 8	77
Malonic Acid	562	590

	Class	Subclass
Malononitrile	558	453
Malt and Malting	435	185
Beverages	426	16+
Unfermented	426	590+
Diastatic mashing	435	93+
Extracts	426	44
Milk	426	580+
Manacles	70	16+
Mandelic Acid	562	470
Mandolin	84	268
Tailpieces	84	301
Mandrel (See Core)		
Circular saw arbor	144	155
Saw sharpening	76	79
Wood saw	144	155
Dental tool	433	134+
Expanding	279	2R
Winding & reeling (See notes to)	242	72R
Glass reshaping neck forming	65	286+
Hair curler	132	33R+
Metalworking		
Billet piercing	72	358
Chuck	279	2R
Coiling on travelling	72	142
Conical spring forming	72	139
Drawing metal tubes on	72	283
Extruding plastic metal by plunger and	72	253.1+
Indirectly	72	273.5
For tube rolling	72	113
Forging and welding	72	481
Nut and washer forging	10	80
Nut facing arbor	408	22
Plastic metalworking	72	476+
Soldering clamp	269	
Tube pressure welding	228	3.1
Tube rolling with	72	208+
Tube threading	408	104+
Turning work driver	82	43+
Welding drawing die	228	16
Welding tubes by drawing on	228	17.5+
Wire article forms	140	92.1+
Paper bag making	493	250+
Paper bag making	493	175+
Paper tube making	493	269+
Sheet and bar cutting	83	924
Sound record cylinder	369	260
Textile braiding apparatus	87	34
Textile spinning apparatus	57	138
Wood turning hollow	142	54
Manganese	75	
Copper alloys	434	
Glycerophosphate	556	24
Heat treating steel containing	148	137
Manganates	423	592
Oxide and hydroxide	423	606
Plate	428	660
Pyrometallurgy	75	80
Thio carbamates	556	38+
Mangle		
Machine element	74	71+
Oscillating rack	74	80
Textile ironing	38	44+
Manhole, Masonry	52	19+
Cover	404	25+
Electrical distribution	174	37+
Manicuring Devices	132	73+
Cutlery	30	26+
Manicure implements	D28	56+
Manifolding	282	
Mark forming cooperating leaves	503	200+
Stylus	81	9.2
Typewriter	400	497+
Manikin		
Anatomical study	434	267+
Apparel display	223	66+
Doll	446	268+
Taxidermic study	434	296+
Manipulators		
Buggy whip	280	178
Clothes wringer pressure	68	256+
Electric furnace	373	
Arc	373	85+
Inductiom	373	138+
Resistance	373	116
Kinesitherapy	128	61
Light globes	81	53.11+
Metal parts assembling	29	270+
Metal sheets	29	19
Programmable	364	
Programmed	901	

	Class	Subclass
Robot	901	
Targets	273	406
Telegraphic switchboard	178	75
Thread and fabric		
Fluid treatment	8	147+
Forming warp into chains	28	175+
Heddle holder	28	205+
Sewing	112	122+
Textile weaving	139	
Type matrix	199	4+
Wheel rim	301	33+
Work heated	432	122+
Manometer	73	747+
Mansard Roof	D25	25
Mantels	52	36
Design	D23	404
Mantles		
Incandescent	431	100
Compositions	252	492
Making	431	100
Making by coating	427	159
Packages	206	68
Support	431	100+
Manufacturing & Assembling		
Books, making	412	1
Brush, broom, & mop making	300	
Button making	79	
Chain, staple, & horseshoe making	59	
Coopering	147	
Electric lamp & discharge devices, manufacturing & repair	445	
Elongated-member driving apparatus	227	
Metal founding	164	
Metal tools & implements, making	76	
Metal working	29	
Assembling apparatus	29	700+
Methods	29	400R+
Needle & pin making	163	
Package making	53	
Paper article making	493	
Sheet material associating or folding	270	
Tobacco	131	
Wheelwright machines	157	
Manure	71	21
Forks	294	55.5+
Loaders	414	680+
Pouches	119	95
Map		
Educational	434	130+
Astronomical	434	284+
Relief	434	152
Holder for map or chart	40	904*
Illuminated route	340	286R
Jigsaw	273	157R
Road	40	
Structure	283	34+
Terrestrial map design	D19	61
Mapping Radioactive Area	250	336.1
Scintillation type	250	
Maraschino Cherries	426	639
Marble and Marbling (See Marbleizing)		
Coating machines	118	402+
Imitating processes	264	73+
Playing marble machines	425	332+
Runways	446	168+
Working	125	
Abrading	51	
Marbleizing	427	256+
Coating processes	427	256+
Dyeing effect coloring	8	478+
Margarine	426	603+
Apparatus	99	452+
Design	D 1	
Preservation	426	330.6
Margin		
Regulator typewriter	400	342+
Marimba	84	403
Design	D17	23
Marine		
Animal butchering	17	53+
Cargo handling	414	137
Drydocks	405	4+
Earth boring	175	5+
Fertilizer derived from marine animals	71	16
Harvesters	56	8+
Merry go rounds	272	32
Mines	102	406+
Paints	106	15.5+
Projectiles	102	399

	Class	Subclass
Propulsion	440	
Chain propeller	440	95+
Crank paddle	440	26+
Manual power	440	21+
Oscillating propeller	440	13+
Paddle wheel	440	90+
Pedomotors	440	30+
Portable propeller	440	53+
Reciprocating propeller	440	13+
Screw propeller	440	49+
Towing	440	33+
Wave propulsion	440	9+
Wind motors	440	8
Railways extensible cable type	104	114
Searchlights	362	35
Signaling (See signal)		
Vessels and appurtenances	114	
Turrets	114	5+
Marionettes	446	363
Marker		
Animal	40	300+
Animal	128	316
Designs	D30	155
Apparel	33	2R+
Book and paper		
Bookleaf holder	281	42
Bookmark	116	234+
Indexed	283	36+
Music expression and tempo	84	164
Music voice	84	165
Water marking	162	110
Calendar	40	110
Car and train	116	30
Clapboard	33	411
Coating implement with material supply	401	
Direction indicator	33	332
Grave	52	103+
Hem	33	9R+
Land and pavement	52	103+
Planting	111	33
Plow combined	172	126+
Roller	172	539
Tennis court	273	31
Tool	404	93+
Traffic guide	404	12+
Monument	52	103
Point guide	33	574+
Stylus	346	139R+
Point, with gauge	33	666+
Recorder stylus	346	139R+
Scriber	33	18.1+
Line, ink type	33	34+
Parallel line	33	41.1+
Dotted	33	39.1+
Sewing	112	131
Stitch, machine for	112	121.25
Tag or label, X-art	493	961*
Marking		
Composition	106	19
Safe combined	109	25
Scale		
Ruler or square	33	483+
Scribers geometrical	33	18.1+
Marlinspike	57	23
Marmalade (See Jams)		
Marquetry (See Parquetry)		
Marquisette	139	419
Making	139	50+
Martingale	54	35
Mascara		
Cosmetic	D28	4+
Maser Type Amplifier	330	4+
Design	D13	19
Maser Type Oscillator	331	94.1+
Masher	241	169.2
Mashing	435	93
Beverage preparation	426	11
Kitchen hand tool	D 7	101
Mask (See Face Guards)		
Baseball	2	9
False faces	2	206
Fire fighting	D29	7+
Gas mask	425	815*
Head covering with	2	173
Photographic	430	5
Respirators	128	206.12+
Surgical or oxygen	D29	7+
Swimming	D29	8+
Masking		
Abrading	51	310+

	Class	Subclass
Coating	427	259
Dyeing	8	445+
Electrolysis combined	204	15+
Etching combined	156	654+
Tape	428	343+
Making by coating	427	207.1+
Templet	132	88.5
To form nonuniform coating	427	259
Masonry (See Furnace)	52	
Design	D25	
Furnaces	432	247+
Pipes and tubular conduits	138	175+
Mass (See Composition)		
Mass-action phenomenon in hydrocarbon synthesis	585	954*
Metal fibers	428	605
Structurally defined	428	357+
Mass Spectrometry		
Ion source for	313	564
Electrical system for	315	111.81
Having heating means	313	550
With spectrometer	250	281+
Massage		
Kinesiatric	128	24R+
Massage brush	D 4	
Massager	D24	36
Tool brush combined	15	110
Mast (See Pole; Post; Tower)	52	
Article supporting	52	40
Crane	212	266+
Collapsible	212	183
Rotary crane mast jack	212	199+
Extensible	52	111+
Horizontally	182	62.5
Openwork	52	648+
Platform	182	222+
Ship	114	90+
Traveler sail	114	112
Transportable	52	111+
Type antenna	343	874+
Mat (See Pad)		
Airfield landing type	404	35+
Armored tire type detachable	152	171+
Casing construction	152	201
Bathtub	4	581
Bed ground	5	417+
Blasting	102	303
Carpet or floor rug type	D 6	582+
Coin pickup type	206	.8+
Door	15	215+
Scrapers	15	238+
Electric floor	200	86R
Burglar alarm	340	565
Expressing press	100	297
Floor	52	660+
Design	D 6	582+
Marine	114	229
Picture frame	40	158R+
Placemat	D 6	613+
Railway track	238	14
Sink lining mat	D32	57
Stove	126	221
Table or counter	248	346.1
Vehicle floor mat	D12	203
Wiping or cleaning type	15	215+
Boot or shoe scraper	15	238+
Design, vehicle	D12	203
Match	44	42+
Book	206	104+
Design	D 9	304
Making	53	394+
Methods	53	395
Box holders	206	127+
Matchbox design	D 9	414+
X-art	493	915*
Cigar or cigarette combined	131	351
Cigarette package combined	206	85+
Compositions	149	
Design	D27	30+
Holder		
Ash receiver combined	131	239
Cigar cutter combined	131	251
Wall mounted	206	127
Making	144	50+
Protective cover design	D27	31
Receptacle filling with	53	236
Books	53	142+
Safes	206	96+
Design	D27	32+
Scratchers	44	48
Ash receiver	131	234

	Class	Subclass
Cigar cutter	131	249
Lamp combined	431	250
With match mover	431	269
Smoking device combined	131	185
Tobacco receptacle combined	206	85+
Matching Machine Woodworking	144	90R+
Planing combined	144	36+
Material (See Fabric)		
Cloth sheet	D 5	
Material Control		
Conveyor drive	198	854+
Difference detectors	73	
Electrical	324	
Ultraviolet	250	373+
X ray	378	
Keeping in suspension		
Pneumatic conveyor	406	86+
Solid classifying and separating	209	
Stop mechanism	192	125R+
Material Dispensing Brush	D 4	114+
Mates	246	454+
Mathematics		
Forms	283	44
Teaching	434	188+
Matrix		
Brake shoe	188	255+
Dental filling	433	39
Display	340	752+
Embossing material	101	12+
Gas panel	340	752+
Energizing circuit	315	169.4
Memory	365	
Paper pulp dip moulding	162	411+
Plastic mold	425	298+
Printers	428	908*
Printing and process of making	101	401.2+
Forming machine	101	401.5+
Selective	340	825.79+
Static mold	249	
Toe and heel stiffener	12	66
Type casting	199	
Matte	75	75
Mattock		
Design	D 8	11
Hoes	172	375
Making	76	109
Root cutters and grubbers	172	
Mattress		
Air or gas filled	5	449+
Bed type	5	448+
Covers for	5	499
Design	D 6	596+
Design	D 6	596+
Earth control revetment	405	19
Fasteners floppers turners	5	411
Filling and closing	53	524
Fluid filled	5	449+
Invalid	5	431+
Life raft	441	135
Protectors	5	499
Rack	211	28
Sewing machines	112	2.1
Springs	5	478+
Waterbed	5	451
Mausoleums		
Undertaking	27	1
Vaults, masonry or concrete	52	134
Mayonnaise	426	605
Maytansinol	540	462
Maze	272	19
Meal	426	622
Measuring (See Definition Notes)	73	
Acceleration	73	488+
Acetylene generator charge	48	46+
Apparel	33	2R+
Capacity measuring bridge	324	60R+
Winding coil apparatus in	242	7.12
Winding coil method in	242	7.3
Change of orientation	73	505
Coin controlled mechanism	194	
Coin size	194	339+
Conveyor combined	198	502.1+
Direction	33	300+
Dispensing combined	222	
Distance	33	125R+
Distance by electromagnetic radiation	356	4+
Earth boring combined	175	40+
Electric	324	
Amplifier condition	330	2
Amplifier with measuring means	324	326+
Capacitance	324	60R
Voltage or current	324	76R+
Filling with stationary measure	141	
Filling with travelling measure	141	135+
Footwear	33	3R+
Fuel burner charge	431	287
Galvanometer	324	76R+
Geometric	33	
Lens testers	D10	61+
Household implement	D 7	50
Instruments for	D10	
Lens testers	356	124+
Liquefied gas handling combined	62	49
Lubricator	184	83
Medical	D24	8+
Meter electric	324	76R+
Nuclear reactor condition	376	245
Control of reactor, combined	376	207
Object by X-ray energy	378	
Optical		
Optical test	356	
Pyrometer	356	43+
Phase electric	324	83R+
Pressure	73	700+
Printing combined	101	73+
Radar, general	342	
Altitude	342	120
Distance determining	342	120+
Radiant energy	250	
Operational nuclear reactor	376	245
Radio		
Receiver	455	226
Transmitter	455	95+
Range by electromagnetic radiation	356	4+
Refrigeration combined	62	125+
Signal energy in antenna	343	703
Speed	73	488+
Spoon, household	D 7	50
Tapes, design	D10	71
Testing electric properties	324	
Testing physical properties	73	
Using ultrasonic	73	570+
Vessels	73	426+
Volume or rate of flow	73	861+
Wave meter electric	250	250
Weighing scales	177	
Meat	426	641+
Blocks	269	289R+
Resurfacing wood	144	2D
Briquetters	17	32
Butchering	17	
Choppers hand	30	
Cleaning	17	
Brushing or wiping machines	15	3.1+
Liquid contact apparatus	134	
Tenderer	17	25+
Comminutor	241	
Cutting boards	269	289R+
Cutting machine	83	
Foodstuffs design	D 1	
Hanging	17	44+
Preservation	426	321+
Canning	426	392+
Coating	426	289+
Dehydration	426	471+
Packaging	426	392+
Press	100	910*
Refrigeration	426	524+
Products, edible	D 1	
Slicers	D 7	381+
Tenderers	17	25
Pounders	17	30
Mechanical Memory Alloy	148	402
Mechanical Movements	74	
Belt & pulley drive system	474	
Mechanical Removal Assistant	29	DIG. 96
Mechanics Teaching	434	302+
Medal		
Badge	40	1.5
Jewelry	63	20
Ornamental design	D11	95+
Medical & Surgical Equipment		
Acupuncture	128	907*
Applicator	128	362+
Aspirators	604	35
Bandaging	128	82.1
Biological signal amplifier	128	902*
Blood pressure recorder	128	900*
Cabinets	312	209
Centrifuge (e. g. , for blood)	464	16+
Colostomy bags	604	338+
Electric shock protection for patient	128	908*
Electrocardiograph	128	697
Electronics	323	911*
Feedback to patient of biological signal other than brain electric signal	128	905*
Fixed apparatus	D24	1.1+
Hypodermics	604	187+
Piston structure	604	218+
Injection devices	604	187+
Instrument	128	303.1+
Kinesiatric	128	24.1+
Magnetic	128	1.3+
Measuring instruments	D24	21+
Medicinal	604	19+
Multiphasic diagnostic clinic	128	906*
Noise suppression in electric signal	128	901*
Nursing	D24	34+
Orthopedic	128	68.1
Portable apparatus	D24	8+
Powder dispensers	604	58
Pressure infusion	604	131+
Radio telemetry	128	903*
Snake or insect venom extractors	604	314
Syringes (See syringe)		
Telephone telemetry	128	904*
Vacuum removal of material	604	131
From external body surface	604	313+
Venom extractors	604	314
Medicated Devices		
Animal		
Antivermin	119	156+
Overshoes	168	2
Sole pads	168	27
Coated or impregnated base	424	443+
Medicinal applicator	D24	63
Nest eggs	119	46
Papers and fabrics	424	443+
Truss pads	128	114.1
Medicated Food	424	439+
Medicated Matrix	424	484+
Medicator	604	19+
Capsule	604	403+
Controlled release	604	890+
Medicine	424	
Anesthetic	424	
Cabinets	312	209
Bathroom	D 6	559+
Capsules	424	451+
Chest	206	438+
Coated or impregnated base	424	443+
Compositions	424	
Computer controlled, monitored	364	413+
Controlled release means	604	890+
Design, pharmaceutical product	D28	1+
Dose		
Administering devices	604	187+
Containers	206	528
Dispenser with removable doser	141	22+
Pump doser	141	23+
Indicator	116	308
Package	206	828*
Structured dosage unit	424	400+
Tear-off packaging	206	532
Special package	206	828*
Medium mu Twin Triode	313	301
With two cathodes	313	6
Megaphone	181	177+
Electric	381	75
Megohmmeters	324	62+
Melamine (See Synthetic Resin or Natural Rubber)	544	200
Aldehyde resins (See synthetic resin or natural rubber)		
Melatonin, in Drug	514	415
Melodeon	84	351+
Melting		
Cleaning by	134	5
Electric furnace	373	
Glass and slag	373	27+
Fat rendering	260	412.6+
Furnaces	432	161
Heaters	126	343.5 R
Iron processes	75	43+
Purifying combined	75	39
Metal casting combined	164	47+
Metals furnaces	266	200+
Metals furnaces, X-art	266	900*
Nonferrous metals	75	65R
Partial melting	156	DIG. 105
Pots for glass	432	262+

Entry	Class	Subclass
Product manufacturing computerized	364	477
Separating combined	422	285+
Processes	23	308R+
Snow	37	227+
Spraying combined	239	79+
Sugar starch and carbohydrates	127	22
Sulphur	423	567R
Melton Fabric	26	19+
Membrane		
Dental dam	433	136+
Diaphragm	92	196+
Acoustical	181	157+
Collapsible	137	843+
Valve	251	331
Valve biased	137	540+
Pipe organ valve	84	336
Vibrant surgical	128	64
Memorial Tablets	40	124.5
Memory Systems		
Computer storage unit design	D14	107
Dynamic electric information storage & retrieval	369	
Dynamic magnetic information storage & retrieval	360	
Malfunction detection, correction	371	
Plural diverse storage	364	900
Static information storage and retrieval	365	
Analog memories	365	45
Associative memories	365	49+
Magnetic bubbles	365	1+
Read only memory	365	94+
Timeplace	368	111
Menstruation Chart	40	109
Menthols	568	829
Menu (See Printed Matter)		
Holders	40	
Meperidine, in Drug	514	330
Mercaptans	568	31+
Mercaptides	568	61+
Mercaptothiazoles	548	182+
Synthetic resin or natural rubber accelerator	525	349
Mercaptovaline	562	558
Mercerizing	8	125+
Mercurials		
Carbocyclic or acyclic	556	118+
Fluoresceins	549	209
Mercurochrome T M	549	209
Mercury	75	
Alloys	420	526
Electrolytic synthesis	204	124+
Amalgamator	209	
Electrolytic	204	124+
Carbocyclic or acyclic compounds	556	118+
Electric discharge devices		
Arc	313	163+
Rectifier systems	363	114+
Supply circuits	315	
Tube structure	313	567+
Fluoresceins	549	209
Frozen foundry pattern	164	8
Fulminate containing explosives	149	33+
Inorganic compounds	23	
Medicine or poison containing		
Inorganic	424	146
Organic	514	190+
Misc. chemical manufacture	156	DIG. 82
Pump piston	417	99
Pyrometallurgy	75	81
Switches	200	182+
Electromagnetic	335	47+
Periodic	200	196+
Periodic	200	200+
Timeplace	368	251
Periodic	200	205+
Periodic	200	208
Thermal	337	331
Vapor		
Electronic device starting	315	327+
Electronic device system	315	
Lamp	313	567+
Rectifier control only	315	246+
Rectifier system	363	114+
Merocyanines	546	174+
Merrifield Synthesis	530	334
Merry-go-round (See Carrousel)	272	28R+
Mesh (See Fabric)		
Mess Kit	206	541+
Messenger Cable		
Electric cable combined	174	41

Entry	Class	Subclass
Supports	248	61
Metabolism Measured by Breath Test	128	718
Metacentric		
Calculator	235	61E
Navigation	73	178R
Testing	73	65
Metal, Metallic, & Metallurgy	75	
Agglomerations	75	228+
Alkali and alkaline earth preparation electrolytic	204	68+
Alkali preparation	75	66
Alkaline earth preparation	75	67R
Alloys	420	
Amids	423	413
Annealing	148	13+
Apparatus, metallurgical	266	
Babbit (See alloy)		
Beneficiating ores	75	1R+
Blowpipe cutting heads	266	904*
Blowpiper to transversely cut	266	902*
Bonding to non-metal	228	903*
Breakers	225	93+
By crushing	241	
Casting	164	47+
Rheo-casting	164	900*
Chill, heat treatment	148	125
Cleaning		
Brushing and scrubbing	15	
By grinding	51	16+
Liquid contact	134	
Coating	427	
Production by oxidizing	148	6+
Cutting and punching	83	
Deforming	72	
After coating	72	46+
Before coating	29	527.4
By application of heat	72	54
Cleaning, descaling, lubricating	72	39+
Cutting combined	72	324+
Differential temperature	72	342+
Explosive	72	706*
Gas treatment	72	38
Hydrostatic extrusion	72	711*
Magnetism	72	707*
Processes	72	362+
Reverse drawing	72	708*
Superplastic material	72	709*
Vibrating	72	710*
Desurfacing	148	9.5
Detector for library book security	340	568+
Detinning electrolytically	204	146
Electro-thermic processes	75	10.1+
Electrodes (See electrodes)		
Emissive cathode	313	346R
Envelope as electrode	313	246+
Etching by chemicals	156	664+
Electrolytic	204	129.1+
Fastening	24	
Book binding	281	25R
Closure, flexible bags	383	42
Closure paper boxes	229	125.18+
Closure paper envelopes	229	78R
Paper clip digest	24	DIG. 8
Sheet material	24	DIG. 9
Wire	24	DIG. 1
Fences	256	21+
Filaments (See electrodes)		
Foam	428	613
Founding	164	
Fusion bonding	228	
Getter for lamps etc	252	181.1+
Handgun holster made of	224	912*
Hardening	148	
Heating	432	
Electric	219	50+
Hydrometallurgy	75	97R+
In form of wire producing by electroforming	204	12+
Leaf		
Applying apparatus	156	540+
Using hot dies, printing	101	DIG. 4
Applying by transfer	156	230
Making	29	17R+
Packages	206	71
Locators		
Compressional wave, echo type	367	87+
Non-echo type	367	118+
Geophysical	324	323+
Magnetic	324	200+
Security means for airport	340	568+
Melting furnaces	266	900*

Entry	Class	Subclass
Metallurgical apparatus	266	
Molten		
Bleaching & dyeing digest	8	DIG. 19
Gas injection	75	59.1+
Heated at nozzle	222	591+
Submersible dispensing dipper	222	629
Nuclear fuel material	376	422
Pickling	134	
Electrolytic	204	141.5+
Powder		
Coated	428	570
Compacting	419	66+
Composite containing	428	546+
Heat treatment	419	31
Loose	75	251+
Lubricant containing	75	231
Making	75	.5 R
Pigment	106	290
Preparation by chemical means	75	
Fiber utilization in powder metallurgy processes	419	24
Sintered stock material	75	228+
Product manufacturing computerized	364	472
Pulverizing	241	
Pyrometallurgy	75	20R+
Pyrophoric	420	416
Radioactive alloys & metals	75	84.1 R
Radioactive alloys & metals	420	1+
Railroad car roof	52	45+
Receptacles	220	
Coffin	27	5
Coffin wood combined	27	6
Foam	220	902*
Liquified gas content	220	901*
Rupture proof	220	900*
Reels	242	77.3
Refining & recovery	75	
Refractory metal extracting	266	905*
Reinforced		
Composite panel	52	782+
Conduits	138	172+
Integral panel	52	630
Masonry	52	596+
Railway track	238	85
Vehicle bodies	296	30
Sacrificial composite	428	933*
Safety shields	266	903*
Saws and sawing machines	83	
Scrap compacting	100	35+
Scrap preheating or melting	266	901*
Shaping		
Auto radiator making	29	157.3 R+
Bonding concurrently	228	265
Chain link formation	59	35.1
Combined with electrical heating	219	7.5+
Sheet metal container making	413	1+
Sheet pilings	405	276+
Sponge	428	613
Stock		
Physical or chemical properties	148	400+
Structure or configuration	428	544+
Teaching metallurgy	434	298
Tool & implement making	76	
Blanks & processes	76	101R
Button making	79	3+
File cutting	76	12+
Machines	76	2+
Saw-making	76	25R+
Sharpeners	76	81+
Treatment	148	
Apparatus	266	
Coating nonreactive	427	
Coating reactive	148	6+
Electrolytic	204	140+
Joints and connections digest	403	DIG. 2
Preserving	422	7+
Thermit type	75	27+
Vacuum during casting	164	61+
Vibration during casting	164	71.1
Continuous casting	164	478
Treatment protective paint	148	22
Scoring	29	DIG. 33
Type electronic tube	313	248
Walls & panels for bank protection	109	80+
Wool	428	605
Making	29	4.5 A
Working	29	
Adhesives	29	DIG. 1
Adjustable liquid level	29	DIG. 75
Aluminum	29	DIG. 2
Assembling apparatus	29	700+

U.S. PATENT CLASSIFICATIONS

Entry	Class	Subclass
Bending	72	
Digest	29	DIG. 3
Brazing	29	DIG. 4
Button making	79	3^+
By projecting, squirting or dripping	29	DIG. 87
Carbonyls	423	417
Casing		
Centrifugal	29	DIG. 6
Composite	29	DIG. 8
Casting	29	DIG. 5
Combined	29	527.5^+
Die	29	DIG. 1
Founding combined	164	
Heat treatment combined	148	2
Chainmaking	59	6^+
Chip breakers	29	DIG. 52
Cleaner, rotating, oscillating, or reciprocating	29	DIG. 98
Cleaning	29	DIG. 7
Coating combined	29	527.2^+
Conveyor		
Air blast &/or vacuum	29	DIG. 78
Chain or belt	29	DIG. 73
Flexible	29	DIG. 99
Fluid conveyor or applicator	29	DIG. 63
Gas as a	29	DIG. 81
Screw or cam conveyor	29	DIG. 1
Cup formed & bottom removed	29	DIG. 9
Cutter		
Air current generated by	29	DIG. 82
Ambulatory with fluent conduit	29	DIG. 64
Complete immersion	29	DIG. 71
Contained supply reservoir	29	DIG. 65
Drive	29	DIG. 55
Engaging cleaner	29	DIG. 97
Fan coaxial with	29	DIG. 83
Feed	29	DIG. 57
Fluid channel in	29	DIG. 92
Fluid spreader contacts	29	DIG. 69
Hood encased	29	DIG. 86
Partial immersion	29	DIG. 74
Rotating	29	DIG. 67
Debris		
Chutes	29	DIG. 102
Control	29	DIG. 5
Receptacle, remover	29	DIG. 79
Remover, catcher, deflector	29	DIG. 94
Removers, plural type	29	DIG. 61
Separators from workpiece	29	DIG. 53
Design	D15	122^+
Digest	29	DIG. 11
Dip or splash supply	29	DIG. 72
Double blank	29	DIG. 2
Drawing	72	274^+
Electric heating	29	DIG. 13
Electroplating	29	DIG. 12
Electrostatic	29	DIG. 95
Elevated tank supply	29	DIG. 9
Enameling	29	DIG. 14
Etching	29	DIG. 16
Extruding	29	DIG. 47
Fan coaxial with cutter	29	DIG. 83
Filter or separator	29	DIG. 77
Flexible conveyor	29	DIG. 99
Flow line & crystal making	29	DIG. 17
Fluid applicator, rotary or oscillating	29	DIG. 7
Cutting fluid application & debris control	29	DIG. 5
Fluid channel in cutter	29	DIG. 92
Fluid control interrelated with machine tool mechanism	29	DIG. 54
Fluid control valve	29	DIG. 85
Fluid conveyor or applicator	29	DIG. 63
Fluid flow to channeled cutter	29	DIG. 66
Fluid paths, multiple	29	DIG. 87
Fluid spreader contacts cutter	29	DIG. 69
Forging	29	DIG. 18
Forging and welding	228	
Forming article on end of long stock then cutting off	29	DIG. 15
Fusion-bonding apparatus	228	
Gas entrained liquid supply	29	DIG. 89
Gear cutting, milling, planing	409	1^+
Grinding	29	DIG. 19
Heat treatment combined	148	11.5 R^+
Heating	29	DIG. 21
Combined	219	149^+
Local heating	29	DIG. 24
Hollow spindle contains	29	DIG. 68
Hood		
Actuated by tool or work approach	29	DIG. 59
Encased cutter	29	DIG. 86
Progressive displacement of	29	DIG. 6
Shield or hood movement	29	DIG. 56
Indium	29	DIG. 22
Knurling	29	DIG. 23
Machining	29	DIG. 26
Magnesium	29	DIG. 27
Magnetic	29	DIG. 95
Magnetic recording digest	29	DIG. 28
Mechanical manufacture	29	400R
Mechanical removal assistant	29	DIG. 96
Molding combined	29	527.1
Molding with other steps	29	DIG. 29
Movable liquid-carrying trough	29	DIG. 76
Ornamenting	29	DIG. 3
Pans, beds & tables	29	DIG. 101
Particular materials	29	DIG. 25
Plastic state	72	
Plastically shaping	72	
Plural discharge loci	29	DIG. 93
Plural discharge openings	29	DIG. 91
Power stop control for movable element	29	DIG. 51
Pressing powder	29	DIG. 31
Reels	242	78^+
Rolling	72	199^+
Rubbing transfer of solid coating to rotating element	29	DIG. 62
Digest	29	DIG. 32
Scaling	29	DIG. 34
Sheet and foil making	29	$17R^+$
Sheet metal container making	413	
Shot blasting	29	DIG. 36
Shrink fitting	29	DIG. 35
Soldering	219	85R
Spot welding	29	DIG. 38
Spraying	29	DIG. 39
Stamping	29	DIG. 37
Staple making	59	72
Stock & blanks	428	544^+
Strips of repeated articles cut up later	29	DIG. 4
Taper tube making	29	DIG. 41
Tension applied during working	29	DIG. 42
Titanium	29	DIG. 45
Tools and implements making	76	
Assembly, magnetic or indicia tools	29	DIG. 105
Tool or work attached	29	DIG. 104
Upsetting & flanging tube end	29	DIG. 43
Vacuum	29	DIG. 44
Vacuum exhaust type	29	DIG. 84
Vibration	29	DIG. 46
Welding	29	DIG. 48
Wire	140	
Work hardening	29	DIG. 49
Workpiece & debris separators	29	DIG. 53
Metaldehyde	549	353
Metalizing	427	
Electrolytic	204	14.1^+
Metallic		
Metallurgy (See Metal & Metallurgy)	75	
Metampicillin	540	331
Metanilic Acid	260	508
Metaphosphates Inorganic	423	
Metaphosphoric Acid	423	307
Metasilicates	423	
Metatarsal Pads	128	69
Meteorology	73	$170R^+$
Teaching	434	217
Meter and Metering	73	
Design	D10	46^+
Dispenser combined	222	23^+
Dispenser cutoff	222	14^+
Automatic	222	59^+
Dispenser volume or rate of flow	222	71^+
Electric	324	76R
Amplifier combined with meter	324	$123R^+$
Amplifier condition measuring system	330	2
Amplitude modulaton	332	39
Circuit protection	324	110
Coin controlled	194	
Electrolytic	324	94
Frequency	324	$78R^+$
Galvanometer	324	$76R^+$
Leakage	324	555^+
Phase	324	$83R^+$
Phase or frequency modulation	332	20
Polyphase current	324	107^+
Radiant energy	250	
Radio frequency	455	155
Recording	346	
Resistance	324	$57R^+$
Slip	324	
Switchboard	361	364^+
Time interval of energization	324	139
Time lapse	368	114^+
Triphase current	324	107^+
Volt	324	$76R^+$
Wave	250	250
Exposure photographic	356	213^+
Flowmeters	73	861^+
Gas separator combined	55	270
Hardness for roentgen tubes	324	413^+
Illumination	356	213^+
Noise exposure	73	572^+
Noise exposure alarms	340	540
Parking		
Alarm electric	340	51
Coin controlled	194	
Time controlled	368	90^+
Photo	356	443^+
Photography		
Measuring shutter speed	73	5
Volume or rate of flow	73	861^+
Dual pipe connections	285	30
Methacrylate Resins (See Synthetic Resin or Natural Rubber)		
Azeotropic distillation with	203	66
Methadone, in Drug	514	648
Methanol (See Alcohol)		
Distillation destructive	201	
Fischer-tropsch synthesis	518	700
Low freezing point compositions	252	73^+
Purification or recovery	568	913^+
Dehydration	203	18
Methapyrilene, in Drug	514	336
Methicillin	540	339
Methionine	562	559
Methomyl	514	477
Methyl		
Methacrylate	560	205
Synthetic resins (See synthetic resin or natural rubber)		
Orange	534	845
Salicylate	560	71
Violet	260	391
Methylenecitric Acid	549	296
Metol T M	564	443
Metronome	84	484
Metronome design	D10	43
Mica	106	DIG. 3
Chemical treatment of	423	
Comminution	241	4
Exfoliation of	252	378R
Flake	428	363
Laminate	428	454^+
Laminate non-structural	428	446^+
Layer	428	363
Mass, structurally defined	428	363
Pigment containing	106	291
Splitting	125	24
Micellar Solution	252	302^+
Michlers Ketone	564	328
Microbiology (See Molecular Biology & Microbiology)	435	
Microcapsules	428	402.2^+
Gaseous core	521	76
Liquid core	428	402.2
Solid core	428	402.24
Microcard		
Photocopying	355	54
Selective data retrieval	353	27R
Microemulsions (See Microcapsule)		
Lubricants	252	49.5
Microfiber	428	903*
Microfiche		
Camera for producing	355	54
Card rack with pockets	40	124.2
Cartridge	353	27A
Slide transfer mechanism	353	103^+
Film mounts	40	158B
Photocopying	355	78^+
Plural image recording on single plate	354	120
Projectors, picture carrier	353	120
Readers-printer design	D16	28
Reading machines	350	241

Entry	Class	Subclass
Transparent film viewers	40	361
Microfilm (See Microcard, Microfiche)		
Frame mats & mounts	40	159
Image transferred from document	355	64
Means to attach film to spool	242	74$^+$
Photocopying	355	5
Vieners	40	361$^+$
Audio-visual equipment	D16	11$^+$
Projector	353	74$^+$
Reader-printer design	D16	28
Selective data retrieval	353	26R
With detailed optics	350	239$^+$
With image projection on screen	353	
Micrometer	33	164R$^+$
Design	D10	73
Magnifiers	350	114$^+$
Microorganism	435	820$^+$
Microphone	D14	12$^+$
Dynamic	381	177$^+$
Granular	381	180$^+$
Mechanical or acoustical	181	
Muffler at mouthpiece	181	242
Musical instrument attached	84	1.1$^+$
Ribbon	381	176$^+$
Moving coil type	381	177
Telephone	381	168$^+$
Transformer	336	
Microphotography	355	18$^+$
With fiber optics	355	1
Microphotometer	356	213$^+$
Microporous (See Pore Forming)		
Microprocessor	364	200
Camera photographic operation control	354	412
Microprojector	353	39
Microscope	350	507$^+$
Container for slides	206	456
Design	D16	131
Electron	250	311
Generally	350	507$^+$
Illuminator	350	253
Interference	350	509$^+$
Objectives	350	414$^+$
Phase contrast	350	509$^+$
Slide	350	529$^+$
Stage	350	529
Box	206	456
Microtome	83	915.5*
Microwave		
Coating by direct application of microwave	427	45.1
Concentrating evaporators electric field digest	159	DIG. 26
Containers, special for cooking	D 7	354$^+$
Cooking food	426	241
Drying & gas or vapor contact	34	4
Food specially packaged or wrapped	426	107
Induction heating digest	99	DIG. 14
Material subjected to electrical energy	34	1
Oven	219	10.55 R
Design	D 7	351
Thermometer	374	149
Piercing probe	374	155
Radio &microwave absorption wavemeters	250	250
Thermal analysis	374	21
Thermometer	374	122
Used in reaction to prepare inorganic or organic material	204	157.15$^+$
Used in reaction to prepare or treat synthetic resin or natural rubber	522	1$^+$
Mildew Proofing	428	907*
Process	424	405$^+$
Proofed stock	428	907*
Mileage Transportation Indicia	283	24
Military		
Computer controlled, monitored	364	423
Insignia	D11	95$^+$
Milk		
Artificial	426	580$^+$
Bottle		
Box for door or window	232	42$^+$
Brushes	15	164$^+$
Caps	215	200$^+$
Easy removal disk caps	215	298
Bottles baby feeding	215	11.1$^+$
Breast pumps	604	36$^+$
Cans	220	
Design	D 9	367$^+$
With light	362	154
Cartons ,paper	206	621.1$^+$
Holder or rack for	D 7	70$^+$
Liquid proofed	229	3.1
Chocolate	426	584$^+$
Condensing	159	
Dairy food treatment	99	452$^+$
Evaporating	159	
Filter	210	348$^+$
Milking devices	119	14.1$^+$
Claw	119	14.54$^+$
Releasers	119	14.5$^+$
Stools	297	175$^+$
With cooling	119	14.9
Milking machines	119	14.1$^+$
Methods of milking	119	14.2$^+$
Releasers	119	14.5$^+$
Modification	426	317$^+$
Preservation	426	330.2
Canning	426	392$^+$
Dehydration	426	471$^+$
Sterilization & pasturization	426	521$^+$
Protein foods from	426	656$^+$
Receptacles deposit and collection	232	41R$^+$
Receptacles for dairy use	220	
Combined with stool	297	175$^+$
Leg or lap supported	220	17.1$^+$
With strainer	210	464$^+$
Shaped bottle	D 9	367$^+$
Testing	422	74
Processes	436	20
Sediment	73	61R
Miller Hook	213	105$^+$
Millinery		
Hat linings	2	63
Hats caps etc	2	171$^+$
Head protectors	2	2.5$^+$
Manufacturing apparatus	223	7$^+$
Milling (See Comminuting)		
Chemical etching	156	625$^+$
Cutting machine	409	64$^+$
Chemical	156	345
Cutters for	407	30$^+$
Electro chemical		
Nut	10	82
Saw shaping	76	44
Gear cutting	409	1$^+$
Test by	73	7$^+$
Mills		
Coffee mill or grinder	D 7	372$^+$
Condiment grinder, household	D 7	53
Gristmill alarms	116	71
Metal rolling	72	199$^+$
Solid material comminutors	241	
Vegetable and meat comminutors	241	
Millstone		
Dress	241	296
Dressing	125	27$^+$
Gauges	33	196
Picks	125	42
Mimeograph		
Plates or surfaces	101	463.1
Processes	101	463.1
Rotary	101	116$^+$
Mincing		
Meat	241	
Hand cutters	30	
Mine & Mining		
Car		
Axles	295	41$^+$
Bodies	105	364
Brakes four wheel spreading	188	55
Lubrication	184	2
Trucks	105	161
Door electric signal railway	246	125$^+$
Elevator crosshead	187	96
Explosive	102	401$^+$
Aerial	102	405
Airplane sustained	244	14
Laying submarines	114	316$^+$
Laying vessels	114	18
Ship protection	114	240R$^+$
Sweeping implements	114	221R
Towed sweeping devices	114	242$^+$
Machine trucks for	105	161
Miners lamp	D26	37$^+$
Combined with tool	7	104$^+$
Safety	362	164$^+$
Testing of	73	164
Picks	125	43
Carrier	224	
Hand	294	
Prop	405	288$^+$
Ventilation	98	50
Mineral		
Acids	423	
Boron compound	423	277$^+$
Carbon	423	414$^+$
Halogen	423	462$^+$
Nitrogen	423	385$^+$
Phosphorus compound	423	304$^+$
Silicon	423	324$^+$
Sulfur	423	511$^+$
Detectors		
Electric prospecting	324	323$^+$
Electric testing	324	
Glass		
Ceramic compositions	501	11$^+$
Coating	427	216$^+$
Laminating	156	62.2$^+$
Layer or laminated	428	426
Making apparatus	425	6$^+$
Making from slag	65	1$^+$
Oil emulsion		
Biocidal	514	938*
Oils	208	
Apparatus	196	
Compositions lubricating and miscellaneous	252	9$^+$
Electrical separation or purification	204	188$^+$
Electrolytic treatment	204	136
Electrophoretic or electro osmotic treatment	204	181.8$^+$
Electrostatic field or electrical discharge treatment	204	168$^+$
Extracting	208	311$^+$
Fischer-tropsch crude	208	950*
Process & products	208	
Recovering or disintegrating in situ	299	
Substances chemical purification	23	
Fermentative, preexisting cd	435	262$^+$
Fermentative, synthesizing cd	435	168
Mineral Block	D 1	100
Mineral Wool Compositions	501	36
Miniature Electronic Tubes	313	
Mining (See Mine & Mining)	299	
Minnow		
Buckets	43	56$^+$
Design	D22	136
Nets	43	11$^+$
Design	D22	135
Mirrors and Reflectors	350	600$^+$
Beds having	5	280
Bracket support for	248	475.1$^+$
Cabinet combined with	312	224$^+$
Cavity inspecting hand	350	640
Changing effects in mirror	40	900*
Comparing or demonstrating	434	371
Correct answer revealed	434	331
Dental	433	3$^+$
Design	D24	13
Design	D 6	300$^+$
Diagnostic	128	3$^+$
Fireplace heat reflectors	126	141
Fluorescent or phosphorescent devices with reflector	250	467.1
Hand-held	D28	64.1
Hats having	2	185R$^+$
Hook or support for	D 8	354$^+$
Support hardware	D 8	373$^+$
Illuminating devices and electric lamps having	362	296
Lamp structure combined	313	113$^+$
Lights combined with mirrors	362	135
Furniture combined	362	128
Inspection facilitation	362	138
Liquid level gage glass having	73	327
Magnetic, use in thermo nuclear reactions	376	140
Coating apparatus	118	
Coating methods	427	162
Electroforming	204	7
Electroplating	204	19
Molding or blowing glass	65	
Making	350	320
Maze or labyrinth illusion	272	19
Observation	350	600$^+$
Examining inner surface of hollow body	350	640
Examining inner walls of cars	350	640
Firearm sighting	33	233$^+$
Reflex camera	354	152$^+$

	Class	Subclass
One way transparent	350	259
Railway track reflectors	246	474
Rear view	D12	187
Signs having	362	253
Stage having	272	13
Support	248	466+
Frames or holders combined	52	
Surface configuration	428	409
Electroformed	204	7
Electroplated	204	19
Metal	428	687+
Surgeons specula	128	3+
Surgical	128	21+
Light combined	362	138
Oral	128	10+
Telescope attachments	350	574+
Television scanners	358	206+
Toilet kits having	132	79G+
Brush type	132	102+
Powder box and applicator	132	83R
Triple	350	102+
Typewriters having	400	716+
X ray reflectors	378	70
Missiles	D12	16.1
Aerial projectile amusement devices	273	317+
Bow	124	23R+
Control	244	3.1+
Explosive gun ejected	102	501+
Guided	244	3.1+
Mortar	102	372+
Projectile making	29	1.2+
Projectiles, mechanical	124	
Rocket projected	102	374+
Simulated projectile game	273	313+
Slingshot	124	17+
Stabilized	244	3.1+
Mist Treating		
Coating	427	428+
Textile machines	68	5R+
Miter		
Box	83	746
Wood sawing apparatus	83	761+
Mitering Devices		
Metal sawing	269	
Printers leads cutting	83	559
Wood sawing		
Boxes	83	746+
Machines	83	471.2+
Woodworking		
Clamps	269	
Cutters	144	216+
Mitomycins	548	422
Mitt		
Apparel	2	158+
Bath mitt or cloth	D28	63
Fingerless glove	D 2	610
Kitchen hot pad	D29	20
Mitten	D 2	622+
Apparel, knitted hand covers	66	174+
Fabric wipers and polishers	15	227
Knitting machine circular	66	45
Knitting machine straight	66	65
Mixer (See Agitating; Air, Mixer; Kneader)		
Agitating	366	69+
Animal food mixer	366	603*
Auxiliary air inlet for combustible gas mixture	48	189.3
Board for dentists	433	49
Bread manufacturing implement	416	70R
Bread manufacturing machine	366	69+
Carbohydrate dryer mixing tank	127	14
Carbohydrate granulation and mixing	127	21
Combustible mixture supply		
Engine charge forming	123	434+
For starting	123	180R
Gas and liquid	261	
Gas generator	48	180.1+
Acetylene combined	48	3R
Oil retort combined	48	104
Liquid or gaseous fuel burner	431	
Concrete	D15	19
Truck mounted	D12	14
Dental amalgam mixer	366	602*
Design	D 7	376+
Dough machines	366	69+
Electronic tube	313	300
Electronic tube type	328	158+
First detector type, heterodine	455	313+
Faucet aerator	239	428.5
Fluid or vapor		

	Class	Subclass
Agitating device	366	101+
Aspiration	137	888
Beverage mixing nozzles	239	549
Carbohydrate granulation and mixing tank	127	14
Chemical feeders, water purification	210	198.1+
Dispenser with plural sources	222	129
Drying apparatus	34	
Faucets gas charged liquid	137	896+
Fire extinguisher chemical tank		
Automatic receptacles	169	27+
Gas pressure type	169	71+
Portable	169	78+
Fire extinguisher fluid system	169	14+
Fluid distributer	137	597+
Inhaler	128	203.12+
Inhaler vaporizer	128	203.26+
Oronasal	128	203.29+
Plural diverse fluids	137	896+
Sprayers dissolvers and mixers	239	310+
Temperature control mixing valves	236	12.1+
Valve multiway unit	128	203.29
Sequential	137	625.12+
Valves plural	137	602+
Sequential	137	628+
For radio signals	455	313+
Glass manufacture	65	178+
Heptode	313	300
Household mixer-grinders	D 7	376+
In line between carburetor and internal combustions engine	48	189.4+
Mixing slab		
Dental	269	302.1
Pill	269	302.1
Mortar	366	10+
Process	366	3+
Tractor mounted mixing chamber	366	606*
Paint	366	605*
Mobile Equipment		
Medical and laboratory	D24	1.1
Radio	455	154+
Repeater or replay	455	11+
Vechile and tuning	455	152
With vehicle	455	297
Robot	901	1*
Transceiver	455	89
Transmitter	455	95+
Mobile Home Design	D12	101+
Moccasin	36	11
Design	D 2	268+
Making	12	142MC
Model (See Toys)		
Airplanes	244	190
Demonstrating	434	365+
Dental	433	213
Photographic	354	292
Modular Construction	52	
Sub-enclosure section	52	79.1+
Modulator	332	
Bedroom combinations with lighting	362	801*
Carrier type electric wave	332	
Amplitude modulators	332	31R+
Combined with radio transmitter	455	91+
Frequency modulators	332	16R+
Keyers	375	68+
Pulse modulators	332	9R+
Cathode ray intensity	313	441
Demodulator amplifier per se	330	10+
With meter	324	118+
Modulation meter		
Amplitude	332	39
Frequency	332	20
Phase	332	20
Pressure modulating relays or followers	137	82+
Radar type pulse modulators	342	194+
General	342	+
With pulse frequency	342	201+
With phase	342	200
Module (See Block; Brick; Panel; Slab; Tile)		
Road or pavement, per se	404	34+
Static mold forming		
Barrier having plural simulated	249	16
Having hollow portions through	249	144
Intaglio or cameo areas on	249	140
Joined by tie	249	85
Modulus		
Fiber or filament, high	428	902*

	Class	Subclass
Moistener and Moistening		
Abrading combined	51	213
Air	239	34+
For hot air heating systems	237	78R+
Generally	261	
Barrel	217	97
Dispenser with	222	190
Envelope sealer combined	156	441.5+
Fireplace combined	126	134
Furnace hot air combined	126	113
Gas and liquid contact	261	
Heat exchanger combined	165	110+
Implement with material combined	401	
Label pasting or paper hanging machine combined	156	
Radiator combined	237	78R
Sewing machine combined	112	43
Stovepipe combined	126	313
Surgical electric applicator	128	803
Textile smoothing combined	38	3
Flatiron	38	77.1+
Thread finishing	28	217+
Tobacco	131	300+
Ventilator combined		
Floor	98	105
Wall	98	109
Moisture		
Curtains shades and screens operated	160	5
Indicator		
Content in gas	73	29
Electric alarms	340	604
Meteorologlcal	340	602
Electric testing capacitative	324	61R
Electric testing resistive	324	65R
Humidity controller	236	44R
Hygrometer	73	335+
Meteorology	73	170R+
Rain alarm	116	69
Rain gauges	73	171
Testing content of	73	73+
Molasses	127	
Electrolytic treatment	204	138
Foods	426	658+
Animal	426	635
Waste as fertilizer	71	26
Moldboard	172	754+
Bulldozer	172	701.1+
Adjustable	172	811+
Supplemental or attachment type	172	817
Shiftable for hillside	172	204+
Supplemental or attachment type	172	754+
Molding (See Casting; Molds)		
Apparatus and appliances		
Blow molding	425	522+
Bread, pastry, confection	99	426+
Butterworker combined	425	200+
Casting mold making	164	159+
Ceramic materials	425	
Concrete	425	
Dairy cutters combined	425	289+
Dental impression	433	34+
Design	D15	135
Foam	425	4R+
Glass	65	
Heel and toe stiffener	12	64+
Heels	12	48
Injection molding	425	542+
Isostatic molding	425	405R+
Metal founding	164	139+
Plastics	425	
Press	425	406+
Soap	425	289+
Soap	425	290+
Sole blank	12	21+
Sugar draining combined	127	18+
Apparel	223	52+
Building element	52	716+
Article supporting	52	27
Baseboard	52	242
Corner attached	52	282
Corner type	52	288
Design	D25	136+
Embedded corner	52	164+
Filler strip	52	312
Veneer applying press	156	349
Carpet fasteners	16	7
Composition, mold forming	106	38.2
Electrical device in insulation	264	272.11
Metal working with other steps	29	DIG. 29
Picture frame	D25	119+

	Class	Subclass
Plastic		
Apparatus	425	
Composition	523	1+
Processes		
Battery electrodes	264	
Battery grids	164	
Bread, pastry, confections	426	660+
Casting mold making	164	6+
Clay, concrete block, earthen and mineral	264	
Computer controlled, monitored	364	476
Glass	65	66+
Glass blowing combined	65	78+
Laminating combined	156	
Metal founding	164	1+
Metal working combined	29	527.1
Particulate material	264	109
Plastic	264	
Elastic memory	264	230
Sugar starch carbohydrates	127	59
Pulp	162	382+
Pulp	162	218+
Apparatus	162	382+
Tire mold	409	902*
Tobacco	131	
Molds		
Edibles, plastic deformation		
Bread, pastry, confections	425	98+
Bread, pastry, confections	425	238+
Cooking	99	
Electroforming	204	281
Fermentation	435	
Kitchen utensil	D 7	43
Metal casting	164	
Dynamic	164	139+
Sand	164	349+
Static	249	
Type casting	199	
Nonmetal molding		
Bread, pastry, confections	425	98+
Bread, pastry, confections	425	238+
Building structure void former	52	577
Cooking	249	
Dentistry	433	34+
Earthen and ceramic	249	
Earthen and ceramic	425	
Glass	65	
Ice making	249	69+
Plastic	425	
Shaping with inflatable mold	264	45.2
Static	249	
Adjuncts	249	205+
Adjustable	249	155+
Destructible feature	249	61+
Element	249	187R
Having chill	249	111
Having coating	249	114R
Having ejector	249	66R+
Having heating means	249	79+
Having removable liner	249	112
Having static filling means	249	105+
Hydraulic and earth control type product	249	10+
In situ construction building or engineering type	249	1+
Building structure	249	13+
Post	249	51
Product having joint or coupling	249	98+
Providing substitution of alternatively used parts	249	102
Road, sidewalk, curb	249	2+
Tobacco cigar	131	70+
Tobacco plug	131	119
Uniting preform with molding material	249	83+
Mole Traps	43	80
Molecular Beam	156	DIG. 103
Molecular Biology & Microbiology (See Cross Art Collections)	435	
Aeration or oxygen transfer technique	435	818
Anerobic cultivation	435	801*
Aseptic cultivation	435	800*
Chromatography	435	803*
Contamination reduction	435	800*
Continuous fermentation	435	813*
Elimination or reduction of contamination by undesired ferments	435	800*
Enzyme or microbe electrode	435	817*
Enzyme separation or purification	435	814*
By solubility	435	816*
By sorption	435	815*
Fermentation, continuous	435	813*
Fermentation vessels in series	435	819*
Fertility tests	435	806*
Foam control	435	812*
Gas detection apparatus	435	807*
Grinding	435	803*
Growth phase logarithmic	435	802*
Incubators, racks or holders for culture plates or containers	435	809*
Interferon	435	811*
Microorganism (See detailed breakdown after 820)	435	822+
Optical sensing apparatus	435	808*
Papers, test	435	805*
Physical recovery methods	435	803*
Protein, single cell	435	804*
Subcellular parts of microorganisms	435	820*
Test papers	435	805*
Tests, fertility	435	806*
Molecular Sieve (See Zeolites)		
Compositions	502	60
Gas separation	55	389
Process	55	75
Molecularly Oriented Stock	428	910*
Molybdenum	75	84+
Alloys	420	429
Composite metal stock	428	663
Heat treatment	148	133
Lubricating compounds	252	12+
Lubricating compounds	252	25+
Misc. chemical manufacture	156	DIG. 107
Organic compounds	556	57+
Momentum Liquid Piston Pump	417	240
Combustion combined	417	75
Money (See Coin)		
Belt	224	229
Box		
Coins	206	.8+
Fare type	232	7+
Register	235	100
Savings or bank type	232	4R+
Toy	446	8+
Toy design	D99	34+
Paper money sorting	209	534
Monitor		
Body function medical unit	D24	17
Monitoring and control	340	825.6+
Monitor Nuclear Energy	250	336.1
Clothing	378	38
Counters	250	390+
Fluorescent	250	361R+
Hand or foot	378	38
Nuclear reactor control combined	376	207
Fusion reactions	376	143
Thermo nuclear reactions	376	143
Photographic fogging	250	475.2
Photographic picture	250	475.2
Scintillation	250	361R+
With recorder	250	336.1
Monkey, ie Guided Hammer	173	90+
Monkey Wrench		
Compound tool	7	139+
Monochromators		
Gratings	350	162.17+
Prisms	350	286
Monocles	351	41
Design	D16	100+
Lens lining or rim	351	154
Monoclonal Antibody as Test Component	436	548
Monoculars	350	507+
Monocycles	280	78
Monolithic Structure (See Concrete)		
Monoplane (See Aircraft)		
Monorail		
Design	D12	37
Rolling stock	105	141+
Track structure	104	118+
Montan Wax	260	420+
Monument	52	103
Coping	52	102
Design	D99	17+
Mooring Devices		
Aircraft	244	115+
Marine	114	230+
Mop and Mop Making		
Brush combined	15	115
Wringer combined	15	116.1+
Cabinet	312	206+
Coating material combined	401	268+
Fabric mop or mop head	15	229+
Having strands	15	229.1+
With fixed handle	15	229.2+
With pivoted handle	15	229.6+
Fabric type	15	228+
Holder	15	147R+
Wrapper or container	493	942*
Household cleaning tool	D32	35+
Machines	300	16
Wringer accessory	15	260+
Wringer combined	15	119R+
Moped	180	205+
Mordants	8	
Dyeing, textile printing	8	452
Dyeing, textile printing	8	460
Morphinan	546	74
In drug	514	289
Morphine	546	44
In drug	514	282
Morpholines	544	106
Morse		
Keys	178	101+
Signal translating to typeprinting	178	26A
Mortar		
Bonded masonry	52	415+
Gun mounts	89	37.5
Introducing into ground	405	266+
Mixer	366	1+
Mixtures	106	85+
Muzzle loaded tubes	89	1.3
Pestle and	241	199
Tractor mounted mixing chamber	366	606*
Trench	89	1.35
Mortise		
Gauges	33	197
Making	144	67+
Woodworking	144	67+
Mosaic	428	49+
Architectural	52	311
Cathode ray tube	313	329
Electron emissive coating	427	75
Photosensitive tube	313	531
Wood grain pattern	52	313
Mosquito Net and Canopy		
Bed	5	414+
Canopies	135	87+
Mother Ships	114	258
Mothproofing		
Bags or holders	206	524.1
Compositions	424	
Processes	428	907
Moting	19	58+
Motion Picture Apparatus (See Camera)	352	
Animated cartoon camera	352	87
Cabinet	312	60
Change overs for pictures	352	135
With sound	352	1+
Combined with sound	352	1+
Design	D16	1+
Dissolves	352	91R
Editor viewer	352	129
Effects	352	85+
Films	352	232+
Winding and reeling	242	179+
Fire prevention	352	143
Multicolor	352	66
Nonintermittant projectors	352	105+
Optical compensation for motion of picture	352	105+
Optical printer	353	18+
Protection against fire	352	143+
Reels and reel winding	242	179+
Screen	350	117+
Shooting gallery	352	95
Stereoscopic	352	57+
Submarine camera	352	132
Threading devices	352	157+
Three dimensional	352	57+
Titler	352	90
View finder on cameras	354	219+
Winding and reeling means	242	179+
Nonwindable reel	242	55.17+
Reel drive plural	242	205
Motive Fluid Special	60	39.461+
External heat, not steam	60	671
External heat, nuclear energy	60	644.1
Heated in expansible chamber	60	509
Internal generation combined	60	39.15
Pressure fluid source & motor	60	326

	Class	Subclass
Motive Fluid Work Cleansing		
Motive Power System, Electrical	318	
Motor (See Device Associated with; Fluid, Motor)	60	
Actuated devices (See device operated)		
Advancing		
Earth boring	175	92+
Animal powered	185	
Brake combined	192	1+
Closure, motor driven	220	211
Combined types	60	698+
Control, agitators	366	601*
Dental handpiece	433	131+
Design	D13	1
Electric (See associated device; electric, motive power)	310	
Awning operator	160	310
Driving air pump	4	213
Driving compressional wave generator	340	384R+
Driving locomotive	105	49+
Driving sound wave generator	367	140+
Electric meter having	324	76R+
Electric system of supply	318	
Fan driven by	416	170R
Fluid type	310	11
Linear	310	12+
Railway	104	290+
Magnetic conveyor	198	690.1
Motor and gearing	310	80
Motor control gearing	74	640+
Motor control system	318	9+
Nonmagnetic	310	300+
Oscillating	310	36+
Prime mover dynamo plants	290	
Pump driven by	417	410+
Pyromagnetic	310	306+
Railway mail delivery having	258	4
Expansible chamber (See device associated therewith)	91	
Devices	92	
Digital	91	DIG. 1
Electrical control	91	459+
Electrical control	91	361+
Exhaust throttled motor control	91	DIG. 2
Large area valve	91	DIG. 3
Magnets	91	DIG. 4
One shot explosion actuated	60	632+
Plural expansible chamber	91	165+
Plural expansible chamber	91	508+
Fluid (See associated device; fluid, motor)	415	
Internal combustion engine	60	597+
Free piston device	60	596
Vacuum generated	60	397
Waste heat driven motor	60	597
Making	29	156.4 R+
Fluid reaction surfaces	416	
Fuels (See fuels; fluid, fuel)		
Liquid	44	50+
Generators	310	113
Heat		
Valve actuator	251	11
Hydraulic	60	325+
Hydro-pneumatic	91	4R
Impellers	416	
Induction systems	318	727+
Braking	318	757+
With reversing	318	741+
Starting	318	778+
Internal combustion	123	
Fluid motor combined	60	597+
Free piston device	60	596
Vacuum generated	60	397
Waste heat driven motor	60	597
Induction type governor valve	137	479+
Power plant	60	632+
Turbine combined	60	598+
Ion	60	202
Marine	440	84+
Mechanical motors	185	
Meter lights	362	25
Meters	324	76R+
Meters	324	139
Multiple	60	698+
One shot explosion actuated	60	632+
Outboard	440	53+
Container or package for	206	319
X-art collection	440	900*
Polyphase current	310	159+
Power plants	405	
Pumps	417	
Reaction (See jet, motor)	60	200.1+
Rotary expansible chamber devices	418	
Rotary kinetic fluid motor or pump	415	
Single phase current	310	159+
Spring	185	37+
Steering actuating	244	175+
Support	248	637+
System of diverse motors	60	698+
Punch card or tape control	60	703
Testing	73	116+
Tide or wave (See associated device)	60	497+
Pulsating flow drive	416	6
Utilizing natural heat	60	641.1+
Vehicle (See motor vehicle)	180	
Ward leonard systems	318	140+
Weight	185	27+
Motor Home		
Trailer design	D12	103
Vehicle design	D12	100
Motor Vehicle (See Vehicle)	180	
Airstream reactive	180	903*
Attracted to surface by fluid force	180	164
Baby seats	297	250+
Body	180	89.1+
Boiler plants	122	3
With vehicle structure	180	310
Brake	188	2R+
Carriage or crib, infant	180	166
Collision, protecting occupant or motor from	180	232
Control means		
Radiated wave responsive	180	167+
Velocity reponsive	180	170+
Electrical servo mechanism	180	178+
Fluid servo mechanism	180	175+
Control mechanism	180	314+
Crank and gear case lubrication	184	1.5
Demonstrating	434	373+
Differential mechanism	180	76
Door	49	
Energy accumulation by fluid or mechanical menas	180	165
Four wheels driven	180	233+
And steerable	180	234+
Differentially	180	248+
Including electric or fluid motor	180	242+
Frame	180	311+
Ground effect	180	116+
Harvester	56	10.1+
Heat and power	237	12.3 R
Land	280	
Laterally driven	180	199+
Leveling device	180	41
Locomotive	105	26.5+
Moped	180	205+
Operation instruction	434	62+
Parking, driven laterally	180	199+
Parking, programmable, steerable wheels	180	204
Plow	172	
Power, eg, traction motor	180	54.1+
Battery mount	180	68.5
Electric motor	180	65.1+
Four wheel driven	180	242+
Three-wheeled, coaxial wheels	180	214
Three-wheeled, steerable	180	216
Two-wheeled	180	220
Exhaust handling	180	309
Hood	180	69.2+
Motor driven by expansible gas	180	302+
Steam	180	303+
Motor driven by noncompressible fluid	180	305+
Motor powered from external source	180	2.1+
Force of nature, eg, sun, wind	180	2.2
Motor to frame relationship	180	291+
Radiator mounting	180	68.4+
Turbine, combustion product driven	180	301
Propelled by other than wheel	180	7.1+
Jet	180	7.3
Propeller	180	7.4
Screw	180	7.2
Stepper	180	8.1+
Track, portable	180	9+
Vehicle mounted winch	180	7.5
Radiator	165	41
Railway car propulsion	104	287+
Restraining belt or harness, system related	180	268+
Ignition or starter circuits	180	270
Transmission	180	269
Runner supported, eg, snowmobile	180	182+
Substitutable for wheel	180	183+
Endless track	180	184+
Surface engaging propulsion element	180	186+
Running grar	180	85+
Safety promoting means	180	271+
Absent, inattentive or unqualified operator	180	272
Acceleration, deceleration or tilt responsive	180	282+
Perimeter engagement responsive	180	274+
Unauthorized use prevention	180	287+
Vehicle load weight responsive	180	290
Seats with body modification	296	63+
Ship	114	
Ship propulsive systems	440	
Ski or skate type	180	180+
Snow excavator	37	231+
Steam powered tractive vehicle	180	36+
Steerable wheel, driven	180	252+
Five or more wheels	180	23+
Four wheels driven, all steerable	180	234+
Motor-carrying attachment	180	12+
Steam tractive vehicle	180	37+
Three-wheeled vehicle	180	211+
Two-wheeled vehicle	180	223+
Rotary-element driven wheel	180	222
Steering	180	79+
By differential traction	180	6.2
Endless track	180	6.7
By walking attendant	180	19.1+
Electrical power	180	79.1
Fluid power	180	132+
Four wheels driven, articulated frame	180	235
Gear	180	79+
Mechanical power	180	79.3
Shaft	180	78
Utilizing terrestrial guide	180	131
Suction effect	180	164
Supported by, & driving, other vehicle	180	198
Surface effect	180	116+
Traction motor powers external device	180	53.1+
Trailer combined	280	400+
Trains of vehicles	180	14.1+
Transmission mechanism	180	70.1+
Vertical surface transversing device	180	901
Wheel base, special	180	21+
Changeable location or number of wheels	180	209
Collapsible or knockdown	180	208
Five or more wheels	180	22+
Three-wheeled	180	210+
Two-wheeled	180	218+
Tandem	180	219+
Wheel chair, motorized	180	907
Wheel slip during acceleration, detecting	180	197
Motorcycle	180	
Design	D12	110
Frames	280	281R+
Lights	362	72
Props	280	293+
Safety devices for motor vehicles	180	271+
Seat support	267	132
Steering	280	263+
Tandem design	180	219+
Design	D12	109
Three-wheeled evhicle	180	210+
Tracked	180	9.25
Two-wheeled vehicle	180	218+
Mount (See Supports)		
Acoustic diaphragm	181	198+
Aircraft		
Airfoils resilient	244	38
Power plant	244	54
Battery	180	68.5
Bearing		
Ball	384	537+
Roller	384	584+
Bracket special	248	205.1+
Electricity		
Conduit or housing	174	48+

	Class	Subclass
Electric device	174	52R$^+$
Insulator	174	158R$^+$
Eye glass lens	351	41$^+$
Brace arm	351	103
Continuous rim	351	83
Pince-nez type	351	65
Rimless	351	110
Split rim	351	90
Fishing reel	43	22
Gun	89	37.1$^+$
Spring	124	29
Outlet or junction box	220	3.3$^+$
Photograph	40	152$^+$
Picture frame	40	158R$^+$
Design	D 6	300$^+$
Printing stencil	101	127.1$^+$
Safes bank protection devices etc	109	50$^+$
Sound box	369	157
Tool and machine part		
Brush bristle	15	186$^+$
Brushing machine handle	15	22R$^+$
Comminutor	241	285R$^+$
Earthworking	172	
Bulldozer	172	811$^+$
Frame mount abrading	51	166R$^+$
Gripper lever	74	156$^+$
Gripper slide	74	160$^+$
Hand control lever	74	511R
Printing plate	101	382R$^+$
Snowplow	37	231$^+$
Spring for spring motor	185	45
Squeezing machine textile	68	125
Work abrading	51	277
Wringer roller textile	68	272
Tripod	354	293
Camera support	354	293
Ladder prop	182	169
Pivoted ladders	182	104
Induction furnace	373	140$^+$
Support stand	248	163.1$^+$
Trolley wheel	191	63.3
Vehicle for conveyer	198	300$^+$
Wheel and axle	301	
Cultivator, adjustable	172	395$^+$
Land marker, slidable	172	126$^+$
Land vehicle	301	
Panel hangers travelers	16	97$^+$
Plow, adjustable	172	395$^+$
Railway	295	
Railway motor axle mounting	105	136$^+$
Railway rolling stock combined	105	
Railway truck axle box mounting	105	218.1$^+$
Skate wheel	280	11.27$^+$
Vehicle combined	280	29$^+$
Mounting		
Tires on wheels	157	
Mouse		
Guard		
Piano pedal openings	84	233
Traps	43	58$^+$
Mouth		
Bag holders	248	99$^+$
Boiler	122	499$^+$
Doll part	446	395
Electric applicator	128	787
Electrical input device (see joystick)		
Guard animal type	119	129$^+$
Guard human type	128	136$^+$
Harmonica	84	377$^+$
Harp	84	375
Organ	84	93
Organs	84	377$^+$
Pieces		
Acoustical muffler	181	242
Animal bits	54	8$^+$
Applying or forming cigar or cigarette	131	88$^+$
Automatic wind instrument	84	93
Brass musical instrument	84	398$^+$
Cigar or cigarette	131	361$^+$
Clarinet	84	383R
Ear piece combined speaking tube type	181	20
Flexible bag closure	383	80$^+$
For wind instruments	84	398$^+$
Speaking tube	181	21
Telephone	381	157$^+$
Antiseptic	379	452
Woodwind instrument	84	383R
Wash	424	
Movement Mechanical	74	

	Class	Subclass
Belt & pulley drive system	474	
Step by step		
Annunciator systems	340	316$^+$
Annunciators	340	815.8
Boring	408	70$^+$
Copy holders movable copy	40	342$^+$
Copy holders movable marker	40	352$^+$
Electric motor operated temperature controls	236	76
Electromagnetic switch	335	138$^+$
Endless conveyors	198	750$^+$
Pusher	198	736
Floor jacks	254	14
Garment supporters and mortising cutters	144	71
Pencil lead advancing	401	65$^+$
Separable fasteners	24	580$^+$
Shingle sawing machines	83	704$^+$
Strap tighteners	24	70R$^+$
Telephone call systems	379	299$^+$
Watch	368	76
Receptacles	206	18
Ring	368	299
Mover		
Bed	5	510$^+$
Trunk tray	190	29$^+$
Movie Apparatus	D16	1$^+$
Moving Picture (See Motion Picture)	D16	1$^+$
Mower		
Harvesting	56	
Sharpener	76	82.1
Mucic Acid	562	582
Mucilage Holder	15	
Design	D 9	
Muck and Scrap Bar	428	576$^+$
Mud		
Bath	128	365
Drum cleaning	122	393
Guard	280	152R$^+$
Design	D12	184$^+$
Gun	266	273
Lifters for wells	166	162$^+$
Ring	122	369
Cleaning	122	385
Traps		
Washers	134	109
Muff		
Apparel design, earmuff	D29	19
Apparel design, handmuff	D 2	611$^+$
Ear and throat	2	208$^+$
Fur	2	66
Hanger	223	97
Muffle & Muffler (See Insulating; Noise, Muffler)	181	212
Acoustical	181	175$^+$
And heat exchanger combined	165	135
Making	29	157R
Apparel	2	91
Design	D29	19
Dry closet with	4	481$^+$
Exhaust treatment	60	272$^+$
For distilling zinc	266	148$^+$
Hobby craft engine	181	404*
Hood combined	2	203
Lock with silencer	70	463
Mouthpiece, ie telephone	181	242
Refrigerator compressor	181	403*
Screen or fence absorbing jet noise	181	210
Silencer for firearms	181	223
Sound absorbing material combined	181	256$^+$
Jet engine muffler devices	181	222
Straight through continous passage	181	248
Through passage	181	252
Mulching	47	9
Mule	57	320$^+$
Womans footwear	D 2	283
Muller		
Comminutors	241	
Mullion	52	777$^+$
Sash component	52	821$^+$
Multicellular Living Organisms and Unmodified Parts Thereof	800	
Multicolor		
Ink block	101	134
Pen or pencil	401	34$^+$
Box	206	371
Photography processes	430	357$^+$
Printing	101	171$^+$
Intalgio	101	151$^+$
Planographic	101	135$^+$

	Class	Subclass
Stenciling	101	115
Weaving	139	232R$^+$
Multilevel Buildings	52	234$^+$
Multiple		
Effect evaporators	159	17.1$^+$
Expansion engine	91	152$^+$
Semi-compound	91	8$^+$
Filament lamps	313	316
Plant receptacles	47	85$^+$
Telephone switchboard	379	310$^+$
Tip multiple discharge coating implement	401	28
Valves	137	
Actuation	137	
Gas and liquid contact	261	42$^+$
Wire systems for continuous current distribution	307	
Multiplex		
Diplexing	370	37
Duplexing	370	24
Frequency division	370	69.1
General purpose programmable computer system	364	200
Teletypewtiter	370	41
Time division	370	77$^+$
General purpose system	370	
Motion picture	352	81
Nippers with plural cutting blades	30	176
Optical	370	1$^+$
Radio	370	
Shears	30	226
Multiplex shearing position	30	197
Special purpose digital data processors	364	900
Switching	370	53
Telegraph	370	41$^+$
Telemetering	340	870.11
Seismic prospecting	367	78$^+$
Telephone	370	
Multiplier		
Electronic tube	313	103R$^+$
In amplifier circuit	330	42$^+$
Frequency	363	157$^+$
Electronic tube type	328	15$^+$
Electronic tube type	328	20$^+$
Multipliers Pattern Mechanisms	139	326
Multiplying		
Machines	235	61F$^+$
Multipolar Instruments	33	1MP
Multiroom Builders	52	234$^+$
Muntin (See Mullion)	52	766$^+$
Murexide	544	296
Muscular Strength Measuring	73	379$^+$
Coin controlled	194	
Mushroom		
Anchors	114	300
Culture	47	1.1
Fertilizing	71	5
Plant	PLT	89
Music (See Musical Instrument)	84	
Accessories	84	453$^+$
Boxes such as swiss music boxes :..	84	94.1$^+$
Design	D17	24
Cabinets	84	DIG. 17
Cases for music roll	206	308
Cases for musical instruments	206	314$^+$
Cathode ray tube	84	DIG. 6
Chiff	84	DIG. 5
Chime	D17	22
Chimes for clocks	368	273
Chorus, ensemble, celeste	84	DIG. 4
Combined in another device		
Toys figure	446	297$^+$
Toys figure wheeled	446	270$^+$
Toys wheeled	446	409$^+$
Crescendo	84	DIG. 16
Electric key switch feature	84	DIG. 7
Electronic gates for tones	84	DIG. 23
Exercising devices	84	465$^+$
Feedback	84	DIG. 1
Filtering	84	DIG. 9
Foldable, detachable, collapsible	84	DIG. 3
Frequency dividers	84	DIG. 11
Gas discharge tube	84	DIG. 13
Holder		
Design	D 6	417
Music stand	248	441.1$^+$
Piano music desks or racks	84	180
Piano music receptacles	84	181
Instruments	84	
Keyed oscillators	84	DIG. 8$^+$

	Class	Subclass
Leaf turners	40	530
Leaf turners for music	84	486+
Light sensitive resistor	84	DIG. 19
Note sheets for automatic players	84	161+
Page turners	84	486+
Piezoelectrical transducers	84	DIG. 24
Pitch (See tuning)		
Players and talking machines combined	84	4+
Plural speakers	84	DIG. 1
Preference networks	84	DIG. 2
Printed matter	283	47
Reverberation	84	DIG. 26
Service manuals	84	DIG. 28
Sheets	283	47
For mechanical players	84	161+
Side rhythm and percussion devices	84	DIG. 12
Stereo	84	DIG. 27
Striking trains for clocks	368	272
Tape	84	DIG. 29
Teaching devices	84	470R+
Toys	446	397+
Tuning	84	DIG. 18
Tuning devices	84	454+
Typewriting machines	400	117+
Musical Instruments	84	1+
Accessories	84	453+
Automatic or self playing	84	2+
Bars	84	102
Bells	84	103
Combined	84	3+
Combs	84	94.1
Cylinder	84	106
Drums	84	104
Organ	84	84+
Piano	84	13+
Rolls for automatic players	84	161+
Selectors	84	115+
Stringed instrument	84	7+
Wind instrument	84	83+
Bowed		
Automatic device	84	10
Bowed piano	84	256+
Violin, cello, and bass	84	274+
Violin bow guides	84	283
Brasses	84	387R+
Trombone	84	395+
Valve type	84	388+
Cases for	206	314+
Hand carriers	224	910*
Violin case	206	14
Design	D17	
Accordians	D17	3
Instrument cases	D 3	30.1+
Organ cabinet	D17	5+
Percussion instruments	D17	22+
Pianos	D17	7
String instruments	D17	14+
Wind instruments	D17	10+
Drums	84	411R
Electric	84	1.1+
Fret control	84	DIG. 3
Keyboards	84	423R+
Keyed oscillators	84	DIG. 8
Manufacture of musical instrument	29	169.5
Mechanical resonator	84	DIG. 21
Monophonic	84	DIG. 2
Picking or plucking devices	84	320
Recording movement of keys	84	461+
Punched record	234	49+
Punched record	234	109+
Rigid vibrators	84	402+
Bells	84	406+
Service manuals	84	DIG. 28
Side rhythm and percussion	84	DIG. 12
Stringed	84	173+
Banjo	84	269+
Dulcimer	84	284
Guitars	84	267
Harp	84	264+
Mandolin	84	268
Pedal clavier	84	DIG. 25
Piano	84	173+
Violin	84	274+
Zither	84	285
Teaching devices	84	470R+
Top	446	258
Transposers	84	445+
Tuning devices	84	454+
Wind	84	330
Accordian	84	376R+
Accordian digest	84	DIG. 15+
Chord organ	84	DIG. 22
Clarinet	84	382
Flute	84	384
Harmonica	84	377+
Harmonica	84	DIG. 14
Pipe organ	84	331+
Actions and valves	84	332+
Valve tremolos	84	348
Reed organ	84	351+
Actions	84	365+
Stop actions and valves	84	369+
Saxophone	84	385R
Woodwinds	84	380R+
Musk Artificial	568	932+
Muskets	42	
Mustache Holder	132	10
Mustard	426	638
Gas	568	56
Mutation	435	172.1
Mutes		
Banjo	84	273
Brass wind instruments	84	400
Violin	84	310+
Mutoscopes	352	99
Muzzle	119	130+
Dog	D30	152
Mylar T M (See also Synthetic Resin or Natural Rubber)	528	308.1
X-art collection	310	800*

N

	Class	Subclass
N-cl Compound	8	DIG. 2
Oxidizing compositions	252	186.35+
Nacelles		
Engine	123	41.7
Aircraft combined	244	53R+
Nafcillin	540	339
Nail	411	439+
Brush	15	167.3
Buffer	132	76.4+
Compound manicure tool with	132	75.6
Capping nail	10	156+
Design	D 8	391+
Digest	10	DIG. 4+
Extractor	254	18+
Feeding and distributing	227	107
Sorting	10	162A
Files	132	76.4
Design	D28	59
Fluorescent or phosphorescent	250	462.1
Hammer	227	133
Hole plug	52	514
Horseshoe	168	23
Calk	168	38
Insulated, electrically	174	159
Making	10	28+
And driving	227	82+
Placer hammer	81	23+
Polish applicator	D 4	
Sets	81	44
Stripper	193	38
Nailing	227	
Broom making	300	13
Receiving strip	52	364
Reel	242	103
Shoe heel working and	12	43+
Shoe lasting and	12	13.1
Tool	12	108
Nainsook	139	426R
Name Plate	232	46
Design	D20	15
Holders	40	10R+
Name Tag	D20	22+
Nandrolone, in Drug	514	178+
Naphthacetols	564	222
Naphthalene	585	400+
Hydrogenation of	585	266+
Sulfonic acids	260	505R
Naphthalic Acid	562	488
Decarboxylation	562	479
Imides	546	
Naphthazarine	260	396R
Naphthenic Acid	562	511
Naphthionic Acid	260	508
Naphthisatins	548	451
Naphthols	568	735+
Acylamino	564	123+
Acylaminosulfo	260	507R
Amino	564	428+
Aminosulfo	260	509
Azo compounds ctg	534	573+
Carboxylic acids	562	466+
Amides	564	173
Esters	560	56
Nitro	568	707
Sulfo acids	260	512R
Naphthothiophenes	549	43+
Naphthylamine	564	428+
Naphthyridine	546	122
Napkin	D 6	595
Design	D 6	595+
Design, paper	D 6	595
Holder		
Body supported	24	7+
Design	D 7	72
Stands	248	127+
Sanitary	604	358+
Design	D24	49
Materials for	604	367+
Napping, Cloth Finishing	26	29R+
Hat making	223	18
Narceine	549	444
Natural Rubber (See Synthetic Resin or Natural Rubber)		
Naval (See Marine)		
Navigation		
Calculators	235	61NV+
Celestial navigation training	434	111
Computer controlled, monitored	364	443+
Course recorders	346	8
Decca	342	397
Instruments (See notes thereto)	73	178R+
Geometric	33	
Loran	342	388+
Omega	342	396
Radionavigation training	434	239
Tacan	342	399
Underwater signal for position finding	367	137
Nebulizers		
Sprayer structure	239	338
Surgical	128	200.14
Neck		
Appliance for thermal surgical treatment	128	380
Bottle	215	31+
Nonrefillable separate	215	30
Cleaning brush for bottle	15	67
Glass reshaping to form	65	108+
Labelling machine for bottle body &	156	476+
Musical stringed instrument	84	293
Pad for harness	54	67
Stock for animal restraining	119	98
Support bracket engaging receptacle	248	312
Supported shower bath	4	618
Tobacco pipe bowl detachable from	131	222
Trap	131	214
Yoke for animal	D30	137
Draft appliance	278	119+
Restraining poke	119	136+
Necklace	D11	3+
Neckwear		
Collars	2	129
Design	D 2	600+
Ironer engaging neckband for clamping	38	13
Ironers for collars and neckbands	223	52.1+
Ironing table with neckband shaper	38	109+
Napkin holder inclosing	24	9
Shaping forms	223	81+
Shirt neckbands	2	127
Ties	2	144+
Design	D 2	605+
Fasteners	24	49R+
Design	D11	202+
Hanger	D 6	315+
Machines for sewing	112	121.22
Rack or holder	D 6	315+
Needle	369	173
Blasting fuse setting	86	21
Compass	33	355R+
Cord knotter	289	16
Harvester needle cleaner or guard	56	448
Crochet	66	118
For dobby looms	139	68
Gun	42	16+
Hypodermic	604	187+
Pocket carrier for	206	365+
Injecting for pickling	99	532+
Injector liquid applying to textile	68	201
Knitting	66	116+

	Class	Subclass
Design	D 3	28
Machine		
Buttonhole sewing	112	65+
Embroidering	112	78+
Knitting textiles	66	
Independent	66	7+
United	66	78+
Leather sewing	112	28+
Sewing	112	
Stitch forming	112	154+
Textile needling	28	107+
Fluid needles	28	104+
Weaving	139	440+
Making	163	
Abrading process	51	281R+
Crochet	163	2
Eye polishers	51	152
Straightening or bending	72	319
Packages or cases	206	380+
Design	D 9	303
Drill holder type	206	379
Phonograph	369	173
Design	D14	29
Radioactive seed holder	128	1.2
Seine making	87	60
Sewing	223	102+
Holder for	223	109R
Machine	112	222
Machine design	D15	72
Surgical		
Electric	128	303.18
Intravenous	604	30+
Suture	128	339+
Threaders		
Hand needle	223	99
Sewing machine	112	222+
Sewing machine design	D15	72
Winding cordage on seine	242	51
Needling		
Processes	28	107+
Fluid needling	28	104+
Negative		
Developing apparatus	354	297+
Developing methods	430	434+
Photographic article	430	
Making camera copy	430	951*
Printing positives from negatives	355	
Sensitizing	430	425+
Developing	430	401+
Washing or fixing apparatus	354	297+
Negative Feedback Amplifier	330	75+
Negligee	D 2	12+
Neomycins	536	13.2
Neon (See Gas, Rare)	423	262
Lamp	313	567+
Manufacture	445	26
Repair	445	
Apparatus	445	61
Process	445	2
Supply circuit	315	
Tube per se	313	567+
Liquefaction and separation	62	36+
Processes	62	22
Sign	40	545
Tube support	248	50
Tube support	362	217+
Neoprene (See Chloroprene Under Synthetic Resin or Natural Rubber)		
Neosalvarsan	556	67
Nephelometer	356	339
Nernst Lamps	313	14
Supply circuits for	315	
With heater for	315	112+
Nest		
Eggs		
Medicated	119	46
Poultry and fowl	119	45R+
Design	D30	108+
Trap	119	47+
Nested and Nesting		
Beds	5	8
Bottles and jars	215	10
Drying drums or receptacles	34	128
Lock tumbler	70	324
Metallic receptacles	220	23.6
Multi article package	206	499
Multi part key	70	400
Stove kit	126	38
Vehicles	280	33.99 R
Net		
Coated	428	255
Fabric	D 5	11+
Fire escape	182	138+
Fishing	43	7+
Fly	54	81
Game element	D21	48
Hair	132	49
Insect or bird catching	43	134
Making	87	
Making and handling		
Hauling fish	43	8+
Hoist line slings	294	77
Textile	87	
Textile design	D 5	
Vehicle fender barrier type combined	293	40
Vehicle fender scoop type	293	46
Wire	140	3R+
Mosquito for beds	5	414+
Torpedo	114	241
Tying machines	87	53
Wire	245	
Nether Garment		
Apparel	2	210+
Design	D 2	1+
Apparel knitted	66	175+
Netting	D 5	11+
Neutralization of Amplifiers by Feedback	330	76+
Transistor type	330	292
Neutralize Inductive Effect	307	89+
Telegraphy	178	69R
Telephony	379	415+
Neutralizing		
Capacitor	361	271+
Compositions	252	182+
Electric charge	361	212+
Neutron		
Absorbing, reactor control	376	216
Control component structure	376	327
Reflector or shield	376	458
Amplifier	376	171
Apparatus using	250	390+
Beam	430	953*
Chemical apparatus	204	193
Control	376	327
Fuel component structure	376	419
Including testing, etc	376	215
Reflector or shield	376	220
Detection or counting	250	390+
Nuclear reactor combined	376	245
Scintillation	250	458.1
Tubes	376	153+
Fast, reactor	376	171
Fast, reactor	376	348
Flux, distribution control	376	207
Generation	376	108+
Nuclear reactions	376	
Intermediate	376	346
Moderated	376	347
Reflector	376	173
Reflector	376	220
Reflector	376	350
Reflector	376	458
Fluid	376	221
Shield	250	518.1
Nuclear reactor, use in	376	287
Survey meters	250	390+
In fluid	250	336.1
Tubes for	376	153+
Thermal	376	347
Unmoderated	376	171
Unmoderated	376	346
Utilization	376	158
Newel Post	52	301
Newspaper		
Delivery box	D99	29
Vending machine	D20	6
Niacinimide, in Drug	514	355
Nibbling Sheet Material	83	916*
Nickel		
Alloys	420	441
Copper	420	457
Iron chromium	420	34+
Super alloys	420	445
Cadmium batteries	429	
Composite metal stock	428	680
Electrodeposition	204	49
Electrolytic synthesis	204	112+
Heat treatment	148	162
Hydrometallurgy	75	119
Inorganic compounds	23	
Misc. chemical manufacture	156	DIG. 1
Organic compounds	556	138+
Pyrometallurgy	75	82
Nickelodeon	D14	15
Nicking		
Screw heads	10	5
Screw sheet metal cap	10	159
Nicotinamide, in Drug	514	355
Nicotine	546	282
Biocides containing	514	343
Removal from tobacco	131	297
Nicotinic Acid	546	318
In drug	514	356
Night Soil		
Fertilizers	71	12+
Purification	210	600+
Nightgowns	2	69.5+
Design	D 2	17+
Nightshirts	2	114
Nightwear	2	
Nigrosines	544	348
Dyes	8	649
Niobate Ceramic Compositions	501	134
Niobium		
Alloys	420	425
Chemical manufacturing	156	DIG. 87
Nipper		
Cutlery	30	175+
Handles	30	341
Manicure	30	28
Shears combined	30	145
Label pasting	156	569+
Manacle and cuff	70	17
Nail plate	10	178
Pliers combined	7	125+
Pliers combined	7	133+
Shoe lasting	12	14.5
Nipper stretcher	12	9+
Nipper wiper	12	8.5
Textile spinning etc	57	114
Nipple		
Applying tool	29	235.5
Chafe prevention	128	150
Dispenser	222	490
Firearm	42	83
Nursing	215	11.1+
Pediatric, oral stimulator	D24	45
Pipe coupling	285	239+
Punching device	83	
Teething	128	360
Valve	137	846+
Nitramines	564	107+
Acyclic	564	109+
Nitroguanidines	564	108
Nitranilines	564	441
Nitrates		
Esters acyclic or carbocyclic	558	480+
Esters carbohydrate	536	18.7+
Cellulose	536	35+
Explosive containing		
Fertilizers containing	71	58+
Phosphates combined	71	50
Inorganic	423	395
Metallo organic	556	1+
Composition containing	149	23+
Explosive or thermic	149	23+
Nitrogen oxides from	423	400
Organic explosive containing	149	88+
Processes organic compounds	260	688
Nitration	260	688
Cellulose	536	35+
Hydrocarbons	568	924+
Organic nitrate preparation	558	480+
Nitric Acid	423	391+
Concentration	159	
Distillative	203	13
Detergents containing	252	101
Esters (See nitrates)		
Phosphate treatment with	71	39
Nitrides	423	409
Ceramic compositions	501	96
Misc. chemical manufacture	156	DIG. 99
Nitriding		
Iron	148	14
Gaseous medium	148	16.6
Nitrile Imines	558	299
Nitrile Oxides	558	299
Nitriles	558	303+
Acyclic	558	435+
Alicyclic	558	428+
Alpha-imino	558	300+

U.S. PATENT CLASSIFICATIONS

	Class	Subclass
Aromatic	558	388+
By ammoxidation	558	319+
Dimerization of	558	360+
Disproportionation of	558	377
Nitrilotriacetic Acid	562	572
Nitrites		
Esters	558	488
Inorganic	423	385
Nitro Compounds		
Organic	568	924+
Phenols	568	706+
Production general	260	688
Reduction general	260	689
Reduction to primary amines	564	416
Nitrocellulose (See Cellulose, Ether; Ester)	536	35+
Nitrogen	423	351
Acids & acid anhydrides, inorganic	423	385+
Explosive or thermic composition containing	149	74
Compositions containing	252	374+
Electrolytic synthesis	204	128
Electrolytic synthesis of compounds		
Aqueous bath	204	91
Fused bath	204	63
Electrostatic field or electrical discharge compound preparation	204	177+
Fertilizer containing compounds	71	
Fixation		
Ammonia production	423	352+
Arc or spark	422	186.24
Fermentative	435	128
Fertilizer preparation	71	7
Inorganic cyanide production	423	364
Wave energy or corpuscular radiation	422	904*
Fuel containing compounds	44	
Obtaining from air by liquefaction	62	8+
Oxids	423	400+
Explosive or thermic containing	149	74
Rings	260	349
Nitroglycerin	558	486
Explosives	149	101+
Nitronic Acids	568	926+
Nitroparaffins	568	943+
Explosives containing	149	89+
Nitrophenols	568	706+
Explosives containing	149	24+
Explosives containing	149	55+
Explosives containing	149	68
Explosives containing	149	80
Explosives containing	149	105+
Nitrosamines	564	112+
Acyclic	564	113+
Dyes containing	8	665
Nitroso Compounds	568	949
Dyes containing	8	665
Reduction of	260	689
Nitrosoamines	564	112+
Nitrostarch	536	48+
Explosives containing	149	58+
Explosives containing	149	108
Nitroxides	564	300
Nobel Metals		
Organic compounds	556	136+
Phosphorus containing	556	13+
Noils	19	115R
Noise Bucking in Amplifiers	330	149
Noise Deadening (See Insulating, Sound)		
Ear stoppers	128	152
Generator		
Electronic tube type	331	78
Jet engine noise muffling		
Blast deflector fences	244	114B
Highway screen or fence	181	210
Limiters		
Audio type	381	158
In radio receiver	455	213
In radio receiver	455	308+
Noise or unwanted signal reduction in nonseismic receiving system	367	901*
Pulse or digital	375	58
Squelch	455	222
Mathematical, reduction	364	574
Muffler	181	175+
Combustion products generator	60	723
Gas separator combined	55	276
Machine support	248	637+
Railway rolling stock	105	452
Railway surface track	238	382
Railway wheel and axle	295	7
Vehicle antirattler	280	138
Non (See Anti)		
Nonrecoil Guns	89	1.7
Nonrefillable		
Bottle or jar	215	14+
Dispenser	222	147
Light bulbs	313	314
Noodles	426	557
Cutting machines	83	
Noose Traps	43	87
Nortestosterone, in Drug	514	178
Noscapine	546	90
Nose		
Bridges	351	124+
Guard	2	9
Hair cutters	30	29.5
Pads or cushions for spectacles	351	136+
Pince-nez type	351	78+
Integral with bridge	351	131+
On bridge	351	132+
Pieces nail plate	10	178
Ring and clip	119	135
Nosecones	244	158R
Nosing-stair	52	179
Notching (See Cutters; Edge; Pinking)		
Selective	234	47+
Shears	30	229+
Sheet material	83	917*
Wood	144	196
Note Sheet		
Instrument control	84	115+
Musical	84	161+
Musical comb	84	101
Printed	283	47
Swiss or regina music box	84	101
Nottingham Lace	87	4
Novobiocin	536	13
Novocain T M	560	49
In drug	514	535
Nozzle	D23	213+
Agricultural, horticultural, forestry sprayers	D 8	2
Air blast	239	589+
Air moistening	239	543+
Boiler cleaning	122	405
Cotton picker	56	32
Design	D23	213+
Digest, drying & gas or vapor contact with solids	34	
Dispensing (See definition notes)	222	566+
Cutter or punch with	222	80+
Functioning as handle	222	475+
Molten metal	222	591+
Movable (See definition notes)	222	526+
Valve or closure with (See notes to definitions of)	222	544+
Earth boring nozzle	175	424
Extruding		
Die expressing earthenware	425	461+
Die expressing plastic metal	72	367+
Die expressing plastics	425	461+
Filament forming	425	66
Filament forming	425	67+
Filament forming	425	76
Filament forming	425	382.2
Sausage stuffer	17	41+
Fire engine	239	436+
Firemans nozzle operated closure	49	21+
Flow meter		
Pressure differential type	73	861.61+
Weir type	73	216
Flow regulating or modifying	239	533.1+
Fluid handling	137	801
Combined with valve	251	155+
Fluid feeding nozzles digest	261	DIG. 39
Swinging	137	615+
Injection mold nozzle vale	425	DIG. 224
Actuated by contact with mold	425	DIG. 226
Automatic control	425	DIG. 225
Injection nozzle, positioned flush with mold or cavity	425	DIG. 227
Jet motor	60	273
Discharge	239	265.11+
Jet pump		
Regulator pump	417	182+
Structure pump	417	151+
Liquid level gauge	73	294
Making	29	157C
Materials for	239	DIG. 19
Oil filter for fuel entering nozzle or float chamber	261	DIG. 4
Pile installing	405	248
Pneumatic conveyor		
Suction	406	152
Reaction jet motor drive	415	80+
Reaction motor discharge	239	265.11+
Resilient	239	602
Safety air nozzles	239	DIG. 22
Sandblast	51	439
Soda water (See valve, soda water; faucet)		
Spray	239	
Suction cleaner	15	415R+
Supports	248	75+
Syringe	604	275
Textile machine	68	59
Track sander nozzle	291	46+
Cleaner	291	12
Vacuum	15	415R+
Nuclear Energy (See Atomic Energy)	376	
Alloys	75	84.1 R
Alloys	420	1+
Compositions	252	625+
Organic	534	11+
Nuclear energy simulation	434	218
Nuclear Reactions & Reactors	376	
Absorbing material for neutrons	376	339
Control component	376	327
Monitoring with control	376	207
Thermonuclear reaction	376	143
Accelerated particle reactions	376	190
Actinide metal	75	84.1 R
Conversion nuclear	376	156
Adhesively bonded fuel layers	376	416
Alloys, actinide	420	1+
Conversion nuclear	376	156
Altering fuel	376	358
For control of reactor	376	212
Amplifier neutron	376	171
Associated reactor components	376	347
Boiling water reaction	376	370
Bonded fuel layers	376	416
Breeder reactor	376	171
Bundle, fuel component as	376	434
By-products	376	189
Handling reaction products	376	308
Treating reaction products	376	201
Impurities removed	376	146
Treatment	376	201
Carrier		
Control or absorbing material	376	239
Conversion, fuel material or sample	376	261
Insertable carrier	376	342
Fluid	376	354
Molten vehicle	376	359
Ceramic fuels	376	409
Compositions	252	625+
Chemical reactions combined	376	323
Chemically bonded layers or parts	376	416
Circulating fluid	376	361
Flow control	376	370
Fuel charge, discharge	376	264
Plural fluids chemically different	376	366
Tanks for	376	403
Coated fuel layers	376	416
Condition determining	376	245
Detecting fuel leak with warning	376	251
Testing, automatic	376	245
Testing, with reaction control	376	207
Contact, fluid-fluid	376	366
Containment, plasma	376	150
Control	52	223R+
Absorbing material combined	376	215
Absorbing material combined	376	327
Adjusting, moving, or conveying	376	219
Automatic control	376	217
Component mechanism	74	
Compositions	420	
Control rod driving	376	219
Handling	376	260+
Location with respect to reactor	376	353
Material for	376	339
Plasma containment	376	100
Plural components	376	213
Sensing & automatic correlation	376	217
Structure	376	327
Conversion nuclear reaction	376	156
Coolant (See moderator)	376	350
Circulating	376	361

	Class	Subclass
Fuel contact	376	395
Materials	376	904
Core	376	
Arrangements	376	267
Associated with other components	376	348
Charging, discharging	376	264
Fluid flow pattern	376	395
Geometry	376	901
Peripheral clamping	376	302
Structures	376	421
X-art	376	903
Correlating reactor conditions	376	216
Critical reactions	376	347
Area	376	212
Sub critical	376	158
Damage control	376	277
Definition by reactor geometry	376	908
Detecting, reactor condition	376	245
Control of reactor combined	376	207
In motion	376	215
Leak in fuel structure	376	251
Warning material included	376	450
Deuterium heavy hydrogen	423	648R^{+}
Compounds inorganic	423	644
Electrolytic production	204	101
Thermo nuclear fuel	376	151
Deuterium oxide heavy water	423	580^{+}
Catalytic recombiner	422	177^{+}
Electrolytic production	204	101
Moderator, handling	376	220
Moderator, materials conversion	376	179
Moderator, plural components	376	358
Discharging, fuel	376	264
Grapple for	294	86.4^{+}
Grapple for	294	906*
Machine for	414	146
Diverse reactor components or materials	376	347
Control rod structure	376	327
Fuel structures	376	409
Handled with fuel	376	262
Moderator channels containing	376	354
E-layer in plasma containment	376	126
Effective density		
Fuel	376	212
Moderator	376	350
Electrical field plasma control	376	144
Electrodes-concentric	376	145
Electrically charged particles accelerated	376	190
In plasma control	376	105
Encased fuel structure	376	412
Energy-power or heat utilization	376	391
Electrical dynamo plants	290	
Electrical generator motor	310	
Single systems	322	
Electrical, radiant	250	493.1
Heat	237	
Motive fuel heating	60	644.1
Propulsion air	244	53B^{+}
Reaction motors	60	203.1
Thermionic	310	306^{+}
Thermoelectric	136	202
Fertile material or blanket	376	172
Fluid or fluent	376	172
Handling	376	178
Location in reactor	376	172
Fission	376	347
Fission	376	409
Fluid, fluent	376	361
Colloid or slurrie	376	356
Fluidized bed reactor	376	355
Moderator	376	350
Coolant	376	401
Molten	376	359
Plural	376	366
Tank, for	376	403
Fuel (See fuel, nuclear)	376	409
X-art collection	376	901
Gas vapor	376	366
Plural fluids	376	369
Single fluid	376	383
Used exterior of reactor	376	378
Geometry reactor	376	908
Girdle, confining moderator structure	376	302
Handling	376	260
Absorbing material	376	260
Control component	376	260
Control rod motivating	376	219
Fertile material or fuel	376	171
Fuel component element	376	212
Refuelling grapples	294	86.4
Refuelling grapples	294	906*
Refuelling machines	414	146
Refuelling schemes	376	267
Shield or reflector	376	220
Heat use of nuclear energy for	237	
Heatpipes	376	367
Heavy water (See deuterium oxide)	376	221
Hollow layer fuel, concentric helical	376	431
Hydrogen	423	648R^{+}
Compound (See water)	423	580
Electrolytic production	204	101
Thermonuclear fuel	376	152
Impregnated fuel layer	376	414
Impurity (See by-product)	376	310
Ceramic	252	625
Metal	420	1^{+}
Inorganic fuel compounds	423	249^{+}
Ionized gas fuel	376	156
Irradiation conversion	376	156
Isotope conversion production	376	156
Jacket materials	376	457
Jacketed fuel component	376	414
Layered fuel component	376	414
Coated or impregnated	376	416
Concentric helical, hollow	376	431
Linear fusion reactor	376	139
Magnetic field plasma control	376	121
Compressive, mirror, pinch effect	376	140
Linear reaction zone	376	139
Material	376	900
Conversion or irradiation	376	156
Handling	376	260
Measuring reactor condition	376	245
With control of reactor	376	207
Automatic	376	217
Mechanical spacer	376	462
Supports for fuel	376	362
Metal	75	
Nuclear fuel material	376	409
Radioactive alloys	75	84.1 R
And metals	420	1^{+}
Mirro effect plasma control	376	140
Mobile nuclear reactor	376	909
Moderator	376	458
Associated with other component	376	350
Circulating fluid	376	350
Confining means for	376	304
Density varying of	376	350
Flow effect	376	352
Fluid type	376	356
Fuel, encased with	376	423
Dispersed in	376	356
Gas	376	383
Geometry	376	904
Girdle for	376	304
Materials for	376	904
Metal	376	906
Organic	376	905
Organic compounds as	260	
Remaking	376	358
Molten fuel or carrier	376	359
Monitoring nuclear reactions	376	245
With reactor control	376	207
Neutron (See neutron)	376	
Burst or pulse production	376	208
Nonaqueous vapor treatment	376	380
Organic compounds	534	10^{+}
Peripheral support for moderation	376	304
Personnel protection feature	376	277
Plasma (See plasma)	376	100
E-layer containment	376	126
Electric field confinement	376	144
Fuel injection	376	127
Impurity removal	376	146
Linear movement	376	139
Magnetic field confinement	376	121
Plutonium (See fuel nuclear)	75	84.1 R
Alloys	420	1^{+}
Electrolytic production	204	1.5
Inorganic compounds	423	251
Metallurgy	75	84.1 R
Nuclear fuel	376	
Organic compound	534	11^{+}
Pool tank	376	403
Power plant		
Combined with prime mover	60	203.1
Combined with prime mover	60	643^{+}
Pressure vessel	376	294
Concrete	376	295
Pressurized	376	230
Fluid	376	366
Product reactor		
Handling during conversion	376	171
Handling product	376	308
Recovery	376	195
Removal	376	146
Propulsion, using energy for		
Air	244	53R^{+}
Marine	114	
Thrust	376	318
Protecting against radiation damage	376	277
Pulse production, nuclear burst	376	208
Pumps	417	
Gas	415	
Plasma	376	100
Radioactive	376	
Alloys as fuel	376	422
Associated plural reactor components	376	347
Compositions	252	625^{+}
Handling	376	260
Irradiation or conversion of materials	376	156
Waste, disposal	252	626^{+}
Recombiner included catalytic	376	301
Reflector	376	458
Handling	376	220
Reservoir tank, pool	376	403
Rod for carrying	376	239
Control rod element	376	327
Automatic	376	217
Safety devices	376	277
Control rod, reflector, shield	376	220
Monitoring	376	245
Neutron absorbers	376	327
Safety fuse	376	336
Shielding	376	288
Sample specimen		
Irradiation of	376	340
Location in reactor	376	342
Treatment with accelerated particle	376	189
Seal for fuel component	376	451
Leak warning feature	376	450
Sensing of reactor condition	376	245
Control combined	376	207
Separating (See --, by-product)		
Chemically different fluids	376	368
Chemically similar fluids	376	370
Servo systems	376	259
Fuel handling	376	261
Shapes		
Fuel	376	409
X-art	376	903
Shield		
Biological or thermol	376	289
Handling	376	260
Shielding for nuclear reactor	376	287
Structures	250	518.1
Shock absorbers		
Absorbing unit materials	376	339
Control component combined	376	234
Control rod	376	234
Sintered fuel		
Non-metal	264	21
Stock metal	75	228^{+}
Source for radioactive energy	376	
Particles	376	190
Structure	250	493.1
Neutrons	376	108^{+}
Stacked fuel components	376	433
Subcritical reactor	376	158
Support reactor	376	362
Fuel core	376	362
Moderator	376	304
Swimming pool reactor		
Fuel component	376	409
Materials conversion	376	176
Plural reactor components	376	361
Tank, as reactor vessel	376	403
Thermol	376	
Insulation or shield	376	289
Thermonuclear reaction	376	100
Thorium (See fuel, nuclear)	75	84.1 R
Alloys	420	1^{+}
Compounds	423	252
Electrolytic production	204	1.5
Organic compounds	534	11^{+}
Transformation (See conversion)	376	156
Underground nuclear explosion	376	273
Uranium	75	84.1 R

	Class	Subclass
Alloys	420	1$^+$
Breeder material	376	172
Compositions radioactive	252	625$^+$
Organic	534	11$^+$
Electrolytic production	204	1.5
Fertile material	376	172
Inorganic compounds	423	253$^+$
Valves, joints	251	
Systems	137	
Waste disposal, radioactive	252	626$^+$
Weld closure for fuel component	376	451
Well processes	166	247
Nucleic Acids	536	22$^+$
Nucleosides	536	22$^+$
Nucleotides	536	22$^+$
Number Plate Illuminators	362	253
License plate	362	253
License plate	40	300$^+$
License plate	40	204$^+$
Locomotive headlight	362	61$^+$
Sign	40	541$^+$
Telephone dial (See telephone dial)	362	24
Numbering		
Photographic negatives	354	105
Printing machine	101	72$^+$
Ticket type	101	70$^+$
Serial number punching	234	61
Nurling (See Knurling)	72	191$^+$
Nursing		
Apparel		
Blouses	2	104
Brassieres	450	36
Bottles	215	11.1$^+$
Design	D24	47
Nipple applying tool	29	235.5
Nipples for	215	11.1$^+$
Supports for	248	102$^+$
Figure toys	446	304
Nut		
Cracker	30	120.2$^+$
Design	D 7	98
Fastening	411	427$^+$
Food preparation	426	632$^+$
Bleaching	426	253$^+$
Cracker	99	568
Grinding	241	
Skinner or blancher	99	623
Hardware	D 8	397
Lock		
Hand tool for applying	81	10
Washer making	10	73
Locked to bolt	411	190$^+$
Jam nut, ie, plural nuts	411	222$^+$
Used with thread lock	411	259$^+$
Castellated nut	411	280
Distorted nut	411	276
Segmented nut	411	266
Locked to substructure	411	81$^+$
Deformable or deforming nut	411	177$^+$
Cylindrical, inner & outer thread	411	178
Folded nut	411	170
Pierce nut	411	179
Weld nut	411	171
Making	10	72R$^+$
Assembling with bolt	10	155R$^+$
Punching blank	83	
Plants	PLT	30$^+$
Roaster	D 7	323$^+$
Shell liquor synthetic resin	528	1$^+$
Spring nut remover	29	270$^+$
Nutcracker	30	120.2$^+$
Nylon (See Synthetic Resin or Natural Rubber)		
Bleaching & dyeing digest	8	DIG. 21
Chemical modification of	525	420$^+$
Dyeing	8	
Fiber	428	475.5
Chemical modification of	8	115.51
O		
Oar	440	101$^+$
Design	D12	215
Lock	440	106$^+$
Oast	432	120$^+$
Oatmeal Making	241	
Objectives		
Photographic	350	409$^+$
Telescope and microscope	350	414$^+$
Oboe	84	380R$^+$
Observation Boat	114	66
Obstacle Controlled or Operated		
Tire pump element	152	423$^+$
Vehicle step, pivoted	182	91
Spring biased	182	89
Obstetric Appliances		
Embryotome	128	307
Exercising	128	31
Forceps	128	323$^+$
Other instruments	128	361
Receptor	604	317$^+$
Tractors	128	352$^+$
Ocarina	84	380R
Occult Blood	436	66*
Occupational Training	434	219$^+$
Ocean Thermal Energy Conversion, Otec	60	641.7
Octane (See Parafin Hydrocarbon)		
Production by alkylation	585	709$^+$
Octave		
Couplers for pianos	84	235
Couplers for pipe organs	84	340$^+$
Couplers for reed organs	84	373
Octode Converter	313	300
Odometer	235	95R$^+$
Combined with electrical speed indicator	324	166
Design	D10	70
Speedometer	116	62.1
Odorizer	48	195
Gas & liquid contact apparatus	261	DIG. 17
Odor release material	428	905*
Office Equipment		
Desk type name plate	D20	15
Machines	D18	
Machine cabinets	D18	
Miscellaneous, design	D19	65$^+$
Photocopy machines	D16	27$^+$
Offset		
Metal bending	72	386$^+$
Printing	101	130$^+$
Offshore		
Platform	405	195$^+$
Floating	114	264$^+$
Well	166	335$^+$
Offtake		
Distillation	202	254$^+$
Flue boiler feed heater	122	421
Furnace shaft	266	159
Ventilating hood	98	115.1
Ohmmeter	324	62$^+$
Oil (See Particular Type)		
Burners (See notes to)	239	
Nozzles (See notes to)	239	
Retort	431	206$^+$
Burnishing	29	89.5
Can	220	
Cutting spouts	222	81$^+$
Design	D23	211
Dispensers	222	
Resilient wall	222	206$^+$
Cutting	252	9$^+$
Distributor		
Ammunition and explosive device	102	366
Buoys	441	34
Fluid breakwater	405	22
Harbor installations	405	22
Wave or ships hulls	114	232$^+$
Edible	426	601$^+$
Preservation	426	330.6$^+$
Emulsions	252	312
Essential	512	5
Filters	210	348$^+$
Fuel entering nozzles or float chamber	261	DIG. 4
In closed circulating system	210	167
Fuel	44	
Burner nozzle (See notes to)		
Burner solid fuel combined	110	260$^+$
Mineral oil only	208	15$^+$
Gages on lamp stands	431	13
Gas making	48	61$^+$
Glyceride	260	398$^+$
Lamps	431	298$^+$
Leak collection	405	60
Lubricating	D 9	301$^+$
Mineral oil only	208	18$^+$
Refrigerant or energy exchange combined	252	68
Mineral	208	
Oil water skimmers	210	242.3
Paints	106	253$^+$
Spill containment	405	60$^+$
Stone	51	213
Synthetic resin ctg fatty acid triglycerides (See synthetic resin or natural rubber)		
Testers	73	53$^+$
Toilet	4	DIG. 11
Transformer	252	570$^+$
All-hydrocarbon	585	6.3$^+$
Vulcanized factice	106	249
Water gas	48	84
Well	166	
Drilling	175	
Heaters electric	219	277$^+$
Strainer making	29	163.5 R
Oil Shale		
Distillation	201	
Mineral oil from	208	400$^+$
Oiler and Oiling (See Lubrication)		
Abrading	51	213
Conduit in drill bit	408	57
Cutter	83	169
Die screw	408	56
Dispenser	222	
Flow meter	73	280
Heat exchange control	165	13
Oil cup closure	184	88.1$^+$
Saw	83	169
Saw guide	83	169
Street	239	146$^+$
Textile treating composition	252	8.6$^+$
Textile weaving	139	45
Waste	384	181
Wrench combined	81	2
Ointments	424	
Oldham Coupling	464	102$^+$
Oleandomycin	536	7.2$^+$
Olefins	585	16$^+$
Alcohols by hydration of	568	895$^+$
Dismutation	585	364
Esters from	560	243$^+$
Hydrohalogenation of	570	246$^+$
Mineral oils	208	
Polymerized solid (See synthetic resin or natural rubber)		
Purification	585	800$^+$
Synthesis	585	500$^+$
Oleic Acid	260	413$^+$
Oleomargarine (See Margarine)		
Oleoresin	260	97$^+$
Oligomers, Hydrocarbon	585	17$^+$
Synthesis	585	502$^+$
Oliver Hammer for Forging	72	445
Omnibuses (See Vehicle)		
One Shot Motor	60	632$^+$
One-arm Bandit	D21	38
Onium Compounds	260	
Antimonium	556	64$^+$
Arsonium	556	64$^+$
Dyes containing	8	654
Organic acid amine addition	260	501.1$^+$
Oxonium	568	557
Phenol amine addition	564	260
Primary aromatic amines	564	305$^+$
Pyridinium	546	347
Quaternary ammonium	564	281$^+$
Quinolinium	546	182
Sulfate esters	558	20$^+$
Sulfonium	568	18$^+$
Wetting agents	252	351$^+$
Onyx	63	32
Pendants	63	13
Opacifiers	106	312
Open Work (See Grille)		
Building component	52	633$^+$
Decoration	428	542.6
Opener and Opening Device		
Bag opening	493	309$^+$
Closing combined	493	255$^+$
Cutting closing combined	493	199$^+$
Refolding combined	493	244$^+$
Web forming cutting closing	493	199
X-art	493	963*
Bottle or jar		
Cap	215	295$^+$
Stopper in combination	215	228
Stopper removal facilitated	215	295$^+$
Can cutter	30	400$^+$
Closure opener combined	7	152$^+$
Combined	7	152$^+$
Can or bottle	D 8	33
Closure metallic receptacle	220	260
Closure remover	81	3.7$^+$

Entry	Class	Subclass
Combined	7	151$^+$
Pliers combined	7	126$^+$
Envelope	83	912
Cutters	30	233
Inkwell pen actuated	15	257.75
Oyster	17	74$^+$
Paper		
Bag or envelope with tear strip	206	610
Carton or box	30	2
Tear strip or breaking means	206	601$^+$
Parachute	244	149
Railway coupling knuckle	213	115$^+$
Textile fiber	19	80R$^+$
Wood box	408	53
Opening of Hydrocarbon Ring	585	940*
Opera Glass		
Holders	350	547$^+$
Optics	350	481
Operations Research Computerized	364	402
Operator		
Aircraft control	244	75R$^+$
Automatic	74	2$^+$
Cabinet component	312	319$^+$
Fluid pressure	312	223
Chair		
Rocking	297	260
Swinging	272	85$^+$
Clutch	192	82R$^+$
Coin operated toy money box	446	8$^+$
Control lever		
Hand operated	74	491$^+$
Knee operated	74	515R$^+$
Controlled guard mechanism	74	616
Dial	74	10R$^+$
Dumping vehicle		
Sliding and tilting	298	14$^+$
Tilting	298	19R$^+$
Elevator		
Car brake	187	92$^+$
Control mechanism	187	42$^+$
Extension ladder	182	208
Lazy tong	182	158
Figure toy		
Figure support operated	446	360
Link operated	446	359$^+$
Platform operated	446	352
Illumination		
Dirigible light supports	362	37$^+$
Lantern globe	362	174$^+$
Switches and valves	362	394
Insect traps	43	110
Multiple brake	188	105$^+$
Pump operating means tires	152	419$^+$
Railway		
Draft coupling	213	211$^+$
Turntable	104	36$^+$
Register	235	91R
Fare	235	44$^+$
Platform operated	235	99R
Stove or furnace	126	169
Switch multiple circuit	200	17R$^+$
Ophthalmic Lenses or Blanks	351	159$^+$
Light filtering	351	163$^+$
Multifocal	351	168$^+$
Ophthalmometers	351	200$^+$
Objective testing	351	205$^+$
Subjective testing	351	222$^+$
Ophthalmoscope	351	205$^+$
Opium Alkaloids	546	44$^+$
Optics & Optical	350	
Alignment testing	356	138$^+$
Article molding	264	1.1$^+$
Connecting means	350	96.2
Coupling means	350	96.15$^+$
Design	D16	100$^+$
Device		
Photo process making	430	321
Electro-optic display	350	330$^+$
Equipment design	D16	
Eye examining	351	
Fiber bundle	350	96.24
Fiber optics	358	901*
Fiber rods	350	96.1
Filter		
Compositions	252	582
Laminated	350	316
Frequency converter, quasi-optical	307	425
Illusion	272	8R$^+$
Amusement railway	104	84
Thaumatrope	446	244
Zoetrope	352	101
Image projectors	353	
Instruments (See type of instrument)		
Design	D16	130$^+$
Distance	33	125A
Light rods used in systems	250	227
Measuring & testing	356	
Measuring or testing instrument	D10	46$^+$
Medical instrument, fixed	D24	1.1$^+$
Medical instrument, portable	D24	8$^+$
Motion pictures	352	
Sensing apparatus	435	808*
Surveyors instrument	D10	61$^+$
Systems & elements	350	
Teaching	434	303
Thin film optical waveguide	350	96.12$^+$
Transmission cable	350	96.23
Vision testing & correction	351	
Optometer	351	200$^+$
Objective testing	351	205$^+$
Subjective testing	351	222$^+$
Orange Peel Bucket	37	182
Orchard		
Heating stoves	126	59.5
Mobile sprayer or duster	239	77$^+$
Plant husbandry	47	
Design	D 8	1
Orchestrion		
Combined with other instrument	84	170
Rotating pin cylinder type	84	95.1
Order Apparatus		
Facsimile	358	256$^+$
Planographic scriber	33	23.1$^+$
Ordnance (See Ammunition; Firearm; Gun)	89	
Automatic	89	125
Barrels	89	14.5$^+$
Bomb, flare & signal dropping	89	1.51$^+$
Breech closures	89	17$^+$
Cartridge feeding	89	33.1$^+$
Control data	235	375
Explosion opened breech	89	4.5$^+$
Firing devices	89	27.11$^+$
Electric	89	28.5$^+$
Fuse setters	89	6$^+$
Loading	89	45$^+$
Machine guns	89	9
Making	29	1.11
Measuring and testing	73	167
Mounts	89	37.1$^+$
Pneumatic	89	7$^+$
Recoilless gun	89	1.7$^+$
Rocket launching	89	1.8$^+$
Shields	89	36.1$^+$
Sights	33	235$^+$
Submarine	89	5
Ore Treatment	75	
Actinide series compound	423	249$^+$
Apparatus	266	168$^+$
Beneficiating ores	75	1R$^+$
Briquets	75	3$^+$
Comminutor	241	
Detector (See prospecting)		
Dressing classifying	209	
Organ and Organette		
Automatic wind organettes	84	92
Automatic wind organs	84	84$^+$
Barrel	84	86
Chord organs	84	DIG. 22
Coupler		
Automatic organ	84	168
Pipe organ	84	340$^+$
Reed organ	84	373
Design	D17	5
Electric		
Action of wind pipe organ	84	337
Couplers of wind pipe organ actions	84	341
Electropneumatic action of automatic organ	84	88
Electropneumatic action of wind pipe organs	84	338$^+$
Electropneumatic swells of wind pipe organs	84	347
Generation of electric tones	84	1.1$^+$
Piano combined with organ	84	172
Pipe organ nonautomatic	84	331$^+$
Pipes	84	349$^+$
Reed organ nonautomatic	84	351$^+$
Reeds	84	363$^+$
Trackers and sheets for automatic organs	84	148
Organic Compounds (See Carbon)	260	
Misc. chemical manufacture	156	DIG. 113
Nuclear fuel, use as	376	156
Nuclear fuel, use as	376	212
Nuclear fuel, use as	376	247
Nuclear fuel, use as	376	409
Nuclear moderator, use as	376	174
Nuclear moderator, use as	376	346
Nuclear moderator, use as	376	458
Organic compounds	532	
Organic compounds	536	
Organic compounds	544	
Organic compounds	548	
Preparation by wave energy	204	157.15$^+$
Radio active compositions	252	625$^+$
Medical	424	1.1
Organo Sol (See Sol)		
Organometallic Compounds	556	1$^+$
Acetylacetonates	556	40$^+$
Anthraquinones	260	366
Azo compounds	534	692$^+$
Azoles	548	101$^+$
Azomethines	556	32$^+$
Diazines	544	225$^+$
Diazo compounds	534	562
Explosives containing	149	23$^+$
Higher fatty acid salts	260	414
Oxazines	544	64
Phosphorus containing	556	13$^+$
Protein containing	530	400
Pyridine nucleus containing	546	2$^+$
Radioactive	534	10$^+$
Stabilization	556	2$^+$
Thiazines	544	4
Thiocarbamates	556	38$^+$
Triphenylmethanes	260	387
Oriental Knot Fabrics	139	400
Looms	139	4$^+$
Orienter and Orienting (See Turner)		
Antenna	343	757$^+$
Article dispenser	221	156$^+$
Article reorienting device	414	754
Cleaning apparatus	134	43
Conveyor article on	198	373$^+$
Conveyor guide	193	44
Cutter with	99	549
Earth boring tool parts	175	85
Furnace charger bucket	414	153
Tool driving or impacting device	173	39
Vehicle load body	414	383$^+$
Well drills		
Bore deflection	175	61
Formation logging	73	151$^+$
Inclinometer	33	365$^+$
Whipstocks	166	117.5$^+$
Orifice in Conduit	138	40$^+$
Cabinets with	312	204
Orlon T M (polyacrylonitrile)	526	341
Ornaments and Ornamental Devices		
Bedpost	5	281
Block molds	425	412$^+$
Building construction	52	311$^+$
Candles	431	125
Chain	59	80$^+$
Design	D11	13
Coating non-uniform	427	256$^+$
Electroforming	204	8
Electroplating	204	18.1
Elements superposed	428	77$^+$
Fountains	239	17$^+$
Harness	54	76
Heating stove	126	219
Illuminated, X-art	362	806*
Figure	362	808*
Imitation candle	362	810*
Psychedelic light	362	811*
Star	362	807*
Terrestrial globe	362	809*
Jewelry	63	20
Design	D11	1$^+$
Layer sectional	428	44$^+$
Layer varying thickness	428	156$^+$
Lighting & illumination	362	806*
Metal ornamenting	29	DIG. 3
Paper box	229	8
Photographic process	430	269$^+$
Plant beds	47	33
Seasonal or religious	D11	121$^+$
Shoe	2	245
Special occasion	428	7$^+$
Stitched	112	439

	Class	Subclass
Textile cloth	26	69R
Napping	26	30
Shearing	26	16
Vehicle body	428	31
Wall decoration	D11	131+
Orthicon Cathode Ray Tube	313	373+
Television system using	358	217+
Orthodontic Devices	433	2+
Implements	433	3+
Orthopedics		
Ankle supports and braces	128	166
Skates with	280	11.36
Appliance	D24	64
Arch supports	36	71
Boot or shoe	128	581+
Pressure distribution testing	73	172
Shoe making	12	142N
Miscellaneous parts	12	146M
Shoe pads	36	71
Surgical	128	68+
Oscillation Circuit		
Generator	331	
Therapeutic treatment type	128	422
Oscillator	331	
Audio	331	
Coil electric	336	
Electronic	331	
Hartley type	331	170
Transistor	331	108R
Multi phase	331	45
Tuned circuit for	334	
Feed conveyer, oscillating-type	209	921*
Laser beam applications		
Logic gate active element	331	DIG. 3
Phase locked loop having locd indicating or detecting means	331	DIG. 2
Tetrode	331	
Oscillographs (See Oscilloscope)		
Cathode ray tube circuits	315	1+
Cathode ray tube structure	313	364+
Design	D10	76+
Electrical quantity measurement	324	97
Cathode ray type	324	121R
Facsmile systems	358	296+
Photographic recorder	346	108+
Plural function recorder	346	45+
Steam engine indicator	346	3+
Oscilloscope (See Oscillograph)	D10	76
Cathode ray tube system	315	1+
Expanded scale	324	131
Cathode ray	324	121R
Meter	324	97
Cathode ray type	324	121R
Television	358	242+
Tube structure	313	364+
Osmium	75	
Alloys	420	461
Extracting	75	83
Hydrometallurgy	75	121
Osmosis (See Desalting Seawater)		
Osones or Osozones	536	1.1+
Osteal Adjuster	128	69+
Otec, Ocean Thermal Energy Conversion	60	641.7
Otoscope	128	9
Ottoman	297	462
Outboard Motor	440	53+
Container or package for	206	319
X-art collection	440	900*
Outdrives	440	900*
Outer Garment	2	115+
Outlet		
Bath basin fitting	4	195+
Compartment tub	4	208
Box		
Digest	33	DIG. 1
Electric conduit	174	50+
Electric connector	439	892+
Electric junction box	220	3.2+
Support	248	DIG. 6
Conduit wiring	439	207+
Support	248	27.1+
Cover or escutcheon plate	D 8	350+
Decanter	210	109+
Distillation vapor	196	136
Drinking fountain		
Catch basin	239	28+
Faucet combined	239	25+
Muffler	181	238
Nozzle	239	
Outlet plug	D13	28+
Pneumatic stacker tube	406	154+
Pressure control		
Fluid pressure brake	303	59+
Fluid pressure regulator	137	505+
Smoking device stem	131	229
Sprinkler pipe	239	266+
Stratifier	209	494
Ventilation		
Building	98	42.1+
Car	98	19+
Cowl	98	66.1+
Wall mounted electric conduit	174	49
Water closet valve		
Siphon	4	368+
Valve	4	378+
Water closet ventilation	4	218
Outriggers		
Awning type		
Multipart	160	78+
Rigid	160	83R
Swinging	160	81+
Vehicle	280	763.1+
Crane	212	189
Outsoles	432	120+
Ovens and Accessories (See Heat)	432	120+
Coke	202	81+
Cooking	126	
Design, household	D 7	348+
Design, special use	D15	144.1
Distilling	202	
Drying	34	
Electric	219	391+
Lights	362	92
Overalls		
Nether type	2	227+
Union type	2	79
Overcoat	2	85+
Overcoming Dead Center	74	36
Overflow		
Bath and basin fitting	4	198+
Conneyer art	198	954+
Decanter and filter	210	299+
Filling portable receptacle	141	115
Gas and liquid contact baffle	261	114.1
Textile machine	68	208
Trough	261	DIG. 44
Water closet		
Outlet valve	4	390+
Outlet valve float	4	395+
Siphon bowl	4	427+
Water control gate	405	101
Water heater	126	359
Receptor	126	383+
Overlaps	2	121
Overload		
Automatic weighers	177	82
Fluid receptacle filling devices	141	115+
Relay		
Electric trips	335	21+
Thermal current trips	337	70
Thermal trips	337	356+
Release		
Comminution apparatus	241	32
Earthworking implements	172	261+
Bulldozer	172	816
Hydraulic control	172	260.5
Elastic couplings ships	114	217
Flexible shaft couplings	464	30+
Motor vehicle trains	180	14.5
Pitmans and connecting rods	74	584+
Plows	172	261+
Snow plows	37	232+
Tools	81	467+
Torque responsive clutches	192	56R
Trailer couplings	280	449+
Stop	192	150
Spring wheels	152	18
Support movable or disengageable on impact or overload	248	900*
Overshoes		
Horse	168	1+
Design	D30	146+
Design	D30	146+
Mens	36	7.1 R+
Design	D 2	271+
Overthrow Preventer	235	131R
Overwinding Preventer		
Spring motor combined	185	43
Plural motors	185	13
Watch or clock key combined	81	467
Watches combined	368	209
Weight motor combined	185	35
Ox		
Shoe	168	5
Making	59	70
Yoke	54	77
Oxacillin	540	327
Oxadiazoles, 1,2, 3-	548	125
Oxadiazoles, 1,2, 4-	548	131+
Oxadiazoles, 1,2, 5-	548	125
Oxadiazoles, 1,3, 4-	548	143+
Oxalic Acid	562	597
Oxazines	544	63+
Bio-affecting & body treating composition	514	227+
Oxazoles	548	215+
Oxazoles, 1, 2-	548	240+
Oxazoles, 1, 3-	548	215+
Oxazolidones	548	225+
Oxidation (See Substance Produced)		
Absorbent regeneration	502	20+
Alcohols		
Acyclic ketone production	568	402
Aldehyde production	568	471+
Carboxylic acid production	562	418+
Aldehydes to produce acids	562	418+
Aldehydes to produce acids	562	531+
Bleaching (See bleaching)		
Carbon monoxide	423	415R
Compositions	252	186.1+
Detergents	252	95+
Compounds for	149	119
Dyeing	8	649
Oxidizable dyes	436	904*
Electrolytic	204	
Fermentative	435	
Glyceride oils	260	406
Purification by	260	423
Hydrocarbons		
Acyclic acid production	562	542+
Anthracene	260	385
Aromatic acid production	562	408+
Aromatic to acid anhydrides	549	231+
Electrolytic	204	80+
Electromagnetic	204	157.15+
Electrostatic	204	169
Mineral oil	208	3+
Nonaromatic mixtures	568	950+
Preparing acyclic aldehydes	568	469.9
Preparing acyclic carboxylic acids esters	560	241.1
Preparing acyclic carboxylic acids	562	512.2+
Preparing acyclic hydroxy ctg compounds	568	910.5
Preparing acyclic ketones	568	398.8
Preparing fats, fatty oils, ester-type waxes or higher fatty acids	260	398.6
Olefines to epoxy compounds	549	523+
Liquid purification	210	758+
Sewage treatment	210	601+
Metallurgy	75	
Metals	23	
Surface	148	6.3+
Nonmetals	23	
Organic compounds general	260	687R
Purification by	260	702
Resins natural	260	99
Oxides		
Heat producing compositions	149	37+
Iron	423	632
Magnesium	423	635
Metal	423	579
Alkali	423	592
Alkali earth	423	583
Nitrogen	423	400+
Combustion control	60	900*
Electrostatic	204	179
Phosphorus	423	304
Sulfur	423	512R+
Titanium	423	608
Electrosynthesis	204	96+
Oxidized Hydrocarbons of Undetermined Structure	568	959
Oximes	564	253+
Acyclic	564	268
Alpha oxy or oxo	564	258
Aromatic	564	265+
Cyclohexanone	564	267
O-carbamoyl	564	255
O-esters	564	254+

	Class	Subclass
O-ethers	564	256+
Preparation	564	259+
Oximeters	356	41
Ear attached	128	633
Oxindole	548	484+
Oxirane	8	DIG. 8+
Oxolinic Acid, in Drug	514	291
Oxonium	568	557
Oxy Containing Carboxylic Acid Esters		
Acyclic	560	179+
Alicyclic	560	126
Aromatic	560	55+
Oxyacetylene Gas Torches	239	398+
Oxychlorid	423	462
Cement	106	105+
Oxygen	423	579+
Blood oxygenators	261	DIG. 28
Electrolytic synthesis	204	129
Generative composition	252	186.1+
Liquefaction	62	8+
Mask	128	205.25
Rings	549	200+
Azo compound containing	534	751+
Oxazine	544	63+
Oxazole	548	215+
Sulfurized	549	200+
Tent	128	205.26
Tests for determining		
Biochemical oxygen demand, bod	436	62*
Chemical oxygen demand, cod	436	62*
Total oxygen demand, tod	436	62*
Oxyhalogenation of Hydrocarbons	570	224
Oxysulfate	106	105
Oxytocin	530	315
Oyster		
Culture	119	4
Dredger	37	55
Opener	17	74+
Opening process	17	48
Scoop	37	119
Ozone	423	581
Electrostatic field or electrical		
discharge synthesis	204	176
Generators	422	186.7+
Food treatment	426	236
Liquid purification	210	760
Preliminary manufacture	210	192
Ozonides	549	430+
Ozonizer		
Electromagnetic wave energy or	422	186.7+
Gas & liquid contact apparatus	261	DIG. 42
Inhaler	128	202.25
Process	62	78
Refrigeration combined	62	264
P		
Pacemaker	128	419P+
Pacifiers	128	359+
Design	D24	45
Pack		
Advancing and feeding sheets from	271	8.1+
Pack holders	271	145+
Animal		
Load attaching means	224	905*
Camera	354	174+
Firearm	42	88+
Forming by receiving sheets	271	278+
Jogger	271	221+
Metal	75	251+
Metal foil formation	29	19
Nuclear fuel components	376	434
Packet		
Clasping or holding devices	24	17R+
Inserting binding loops	227	77
Making and wrapping	53	206
Postal card type	229	92.8
Parachute	244	148
Saddle	54	37
Package	206	
Adhesive	206	813*
Automatic filling and weighing	177	52+
Blister	206	461+
Button eyelet or rivet	206	348
Chewing gum	206	800*
Coins	206	.8+
Card type	229	92.9
Paper wrapper	229	87.2
Cosmetic article	206	823*
Design	D 9	
Detergent containing	252	90+
Emergency packs	206	803*
Fastenings or ties	24	16R+
Filler for space	206	814*
Finger opening	206	815*
Fire kindling	44	40
First aid packs	206	803*
Flares	206	803*
Flexible closure, eg sliding desk top		
type	206	816*
Follower	206	817*
Land vehicle compartment	296	37.1+
Living fungi ferment containing	435	
Lubricant containing	252	10
Magnet	206	818*
Mail box clips or pockets	232	33
Making	53	
Material	206	819*
Medicinal content	206	828*
Nail and staple	206	338
Package and article carriers	224	
Sewing machine case	312	208
Plant	47	84+
Powder puff	206	823*
Recoil-type retainer	206	825*
Registers operated by	235	98R
Rubber band	206	805*
Rupture means	206	601+
Sausage shirred casing	206	802*
Separable, strip-like plural items	206	820*
Sheet material folding	270	39+
Special shape	206	822*
Stacking member	206	821*
Suction cup adjunct	206	829*
Suspension	206	806*
Tamper-proof	206	807*
Tear apertures for roll, strip, or		
sheet	206	824*
Textile yarn or cloth		
Laundering or cleaning methods	8	154
Liquid treating for cleaning	68	
Thread finishing	28	217+
Ties	24	16R+
Toilet article	206	823*
Towel, packaged	206	812*
Vacuum	206	829*
Water treating agent containing	252	176
Waterproof	206	811*
Wound strand	242	159+
Wrapper	229	87R
Wrapper, packaging bag	D 9	305
X-type	206	809*
Zipper, ie receptacle securement	206	810*
Packaging Methods		
Automatic or triggered control	53	52+
Bags, wrappers, tubes	D 9	305
Baling	100	1+
Binding	100	1+
Cigar and cigarette making	131	283+
Closing packages	53	285+
Collar, cuff and bosom making		
combined	53	117
Containers	D 9	414+
Contents material treating	53	111R+
Depositing articles into receptacles	53	235+
Filling and closing portable		
receptacles	53	266R+
Foods	426	392+
In liquids	426	392+
Group forming into unit	53	147+
Needles and pins	227	25
Plug tobacco making combined	131	112+
Receptacle filling & closing	53	467+
Receptacle opening	53	381R+
Textile strands	28	289+
Wrapping machines	53	203+
Packer		
Well	166	179+
Closing port	166	141
Handling	166	373+
Preset	166	114
Preset conduit carried	166	116
With expanding anchor	166	118+
With outward pressure means	166	101
With passage control	166	142+
With pump or plunger	166	106
Packing Method (See Stack)		
Bags & wrappers	D 9	305
Cartridge	86	47
Cases and materials		
Design	D 9	
For cartridges	206	3
For chemicals	206	524.1+
For cigars and cigarettes	206	242+
For cycles	217	37+
For metal leaf	206	71
For photographic papers	206	455
Cigar bundling	131	107
Dispensing devices	222	143
Felting roll pack	28	125+
Closure with gasket or packing	220	378+
Foods	426	392+
Joints	277	
Bottle cap	215	341+
Closure	220	200+
Cup or flange type	277	212R
Dispenser outlet	222	512
Dispenser outlet	222	542
Pipe joint	285	335+
Prefabricated unit	277	35
Receptacle seam	220	81R
Valve	251	356+
Materials		
For sensitive plates	206	454+
Molded separator sheets	D 9	456
Piston	277	
Ring making	29	156.6
Portable receptacles		
Bags and sacks	53	570+
Cigarette	53	148+
Uniform density	141	71+
Receptacle filling & closing	53	467+
Sand founding molds	164	37+
Tobacco plug	131	112+
Tools	81	8.1
Weather strips	49	475+
Wrappers	229	87R+
Packs		
Change for plates and films	354	276+
Photographic plate	354	276+
Pad		
Abrasive sheet holding	51	358+
Handled	51	392+
Absorbent	604	358
Animal hampering pads	119	108
Body	2	267+
Button covering	24	114
Cleaning material supply combined	401	261+
Closet seat	4	244+
Desk or typewriter	D19	99
Diaper	604	358
Electrically heated	219	528+
Foot	36	71
Furniture	248	188.8
Ground mat	5	420
Hand	2	20
Harness	54	65+
Design	D30	134+
Horseshoe	168	26+
Design	D30	147+
Modified shoe	168	12+
Hot pot holders	2	20
Hand manipulable	294	25
Handle type	16	116R+
Knitted base	66	190+
Lifters	294	27.1+
Of textile material	428	224+
Inking and copying	118	264+
Printing machine having	101	
Knee	2	24
Label pasting	156	
Ladder foot	182	108+
Lock		
Shackles	70	53
Mattress	5	500
Pedal	74	563
Receptor		
Catamenial and diaper	604	358+
Shoe	36	71
Heel cushioning	36	37
Skirt combined	2	215
Sleeping bag with pad	5	413
Stair	428	
Conforming to tread nose	52	179
Table	428	53+
Table, cover or casing for	428	68+
Table, scuff plate	248	346.1
Table, sectional layer for	428	44+
Truss	128	112.1+
Padded		
Bottle wrapper	229	91
Container with shock protection	206	521+
Crate	217	53
Paddle and Paddle Wheel	416	78+
Aircraft propulsion	244	70+

Entry	Class	Subclass
Aircraft sustension		
Heavier than air	244	4R+
Lighter than air	244	27+
Boat propulsion manual	440	21
Design	D12	306
Game apparatus	D21	48
Pumps	417	436
Ship and boat propulsion	440	90+
Retractable blades	440	92+
Ship steering	114	147+
Ships	114	58
Water motors	416	78
Padlock	70	20+
Condition indicator combined		
Recorder	70	435
Register	70	437
Design	D 8	334+
Latch combined	292	
Design	D 8	334+
Pail (See Bucket; Receptacle)		
Milk	220	86R
Making	413	69+
With strainer	210	464+
Scrub	15	264
Winding material applying	242	7.22
Method	242	7.2
Paint	106	
Anticorrosive	106	14.5+
Antifouling	106	15.5+
Artists box	206	1.7
Brushes	15	159R+
Design	D 4	
Material supply combined	401	268+
Drier	106	310
Luminous		
Inorganic	252	301.6 R
Organic	252	301.16
Radioactive	252	625+
Signs	40	542
Mill	241	
Mixer	366	605*
Painters equipment		
Artists boxes	206	1.7+
Brushes	15	159R+
Design	D 4	
Graining tool	D 8	
Roller type	D 4	122+
Easels	248	441.1+
Fabric stretching frames	160	374.1
Stencils (See stencil)		
Teaching	434	84
Removing	134	38
Removing-burning tool	D 8	29.1
Roller tray	D32	53.1
Rollers & trays for	D 4	122+
Scraper	D32	46+
Pajamas	2	83
Design	D 2	8+
Palladium (See Platinum)		
Alloys	420	463
Pallet (See Pawl)		
Bed clothing	5	482
Container with	206	386
Dental mixing slab	433	49
Hawk	294	3.5
Horological drive	368	131
Mold	425	253+
Concrete or ceramic type	425	253
Organ valve	84	368
Synthetic plastic industrial platform	108	901*
Made with deformable integral fastening elements	108	902*
Type for lift trucks	108	51.1+
Palmitic Acid	260	413+
Pamphlet		
Covering	412	4+
Distribution from aircraft	40	216
Pan (See Pot)		
Ash		
Boiler feed preheater combined	122	413
Furnace	110	166+
Stove	126	243+
Bed	4	450+
Boiler	122	394
In steam space	122	435+
With blowoff	122	386
Carbonless	503	200+
Closet bowl combined	4	429+
Closet bowl valve	4	442+
Cooking container	D 7	354+
Drip (See drip pan)		
Drying	34	237+
Housed or supported	34	192+
Dust	15	257.1+
Evaporating and concentrating		
Closed	159	22+
Combined open and closed	159	21
Open	159	32+
Land vehicle drip	296	38
Liquid fuel burner	431	331+
Lubricator drip catching	184	106
Mill	241	107+
Motor vehicle splash	180	69.1
Pan shaped metal bending dies	72	394+
Railway tender coal feed	105	235
Scale	177	262+
Stove drip	126	51
Street sweeper elevator	15	85
Water heating	126	373
Pancake		
Fryer	99	422+
Turner	294	7+
Pancreas Extracts	530	845
Medicine containing	514	3+
Pane, Window (See Panel)		
Fasteners	52	764+
Hermetically sealed	428	34
Laminated glass	428	426+
Parallel panes	52	304
Drier	52	171+
Resiliently spaced	52	398
Vented	52	304
Ventilated	98	96
Panel		
Bed	5	280
Board switchboards	361	331+
Telephone	335	108+
Building components (See block)	52	
Adhered tile	52	390+
Attached handle	52	125.2+
Corrugated component	52	795+
Dimpled	52	792
Edged laminated	52	783+
Edged laminated & separate	52	821+
Edged resilient	52	393
Edgewise connected	52	578+
Embossed	52	792
Hollow	52	785
Integral reinforced	52	630
Internal fastener received	52	787
Multicellular core	52	806+
Ornamental	52	311+
Partition	52	238.1+
Perforate sheet	52	794
Perforate sheet, edgeing separate	52	815
Sandwich type	52	785+
Shaft attached	52	474+
Shield	52	202+
Stonelike	52	596+
Sub-structure attached	52	506+
Casket sliding panel	27	15
Closure fastener operator	292	93
Decorative sectional layer	428	44+
Decorative sheet cover or casing	428	68+
Disc	428	64+
Edge lighted	362	31
Fasteners	24	
Fences	256	24+
Panel element	256	73
Flexible	160	
For floor, wall, roof	D25	138+
Hangers	16	87R+
Heat exchanger	165	168+
Heating stove	126	64
Door transparent	126	200
Joints, readily disengageable	403	DIG. 1
Mirror with backing	52	786
Mounting for instrument	248	27.1+
Movable or removable closures	49	
Partitions &, flexible & portable	160	
Phonograph blanks	428	64+
Portable	160	351
Radiant heating wall	165	49
Railway passenger car		
Convertible open to closed	105	332+
End panel open car	105	347
Static mold form	249	189
Travelers	16	87R+
Wall panel, molding continuous	264	46.4+
Windshield	296	90
Sash convertible to screen	160	128
Panic Bar Latches	292	92+
With lock	70	92
Panification		
Processes	426	549+
Panoramic Cameras	354	94+
Pantographic Scribers	33	23.1+
Curved surface	33	22
Cutting machine for shaping	409	86+
Multiplane	33	24.1
Pantolactone	549	319
Pantothenic Acid	562	569
Pants	2	227+
Design	D 2	28+
Design, lingerie or undergarments		
Pantyhose	2	409+
Papaverine	546	149
Paper		
Article design	D 5	
Article making	493	
Computer controlled, monitored	364	471
Bag making	493	186+
Bag opening	493	309+
Bags	383	
Barrels	229	67
Binder device releasably engaging aperture or notch of sheet	402	
Blueprint	430	540
Bobbins and spools	242	118+
Box	229	
Box making	493	52+
Carbon	428	488.1+
Compositions	106	20+
Paper, design	D19	1+
Cartridge shells	86	11+
Clip or clamp fasteners	D19	65
Clips	24	455+
Coating	427	411+
Stripping or edging	427	284+
Collars	2	136
Combined into book	D19	26+
Container formed of folder paperboard	224	906*
Cutter	D18	34
Cutter, machine-type	D15	127
Disc closure making	493	962*
Disinfecting or sterilizing	422	243+
Driers	34	148+
Processes	34	23+
Dyeing	8	
Electroprinting	204	2
Envelope making	493	186+
Envelopes	229	
Fasteners	24	67R+
Making	29	13
Pin type	24	153+
Festoon making	493	957*
Flypaper holder	43	115
Flypaper package	206	447
Folding or plaiting	493	68+
Heat seal	493	394+
Scraps	100	914*
Tag applying	493	375+
Tube making	493	269+
Garland making	493	958*
Hanging	156	574
Design	D 5	
Interlocked sheets	493	390
Lined metal receptacle	220	460+
Making & fiber liberation	162	
Matchmaking	144	51
Medicated	424	414+
Moistener	134	122R+
Money sorting	209	534
Opening reinforcements	402	79
Other than in book	D19	33
Parchmentized or linen finish	8	119
Photographic	430	270+
Photographic paper packages	206	455
Photoprinting	430	270+
Punch	D19	72
Punching	83	
Roll type dispenser packages	206	389+
With severing edge	225	39+
Self toning	430	559
Sheet dispensing packages	206	449+
With tearing edge	225	27+
Sheet manipulating devices		
Flipper, inserter, lifter,	402	80R
Remover, transfer fork	402	80R
Transfer post	402	47
Sheet stock	428	
Shelf paper	D 5	

	Class	Subclass
Shifting plate	29	270
Stationery	D19	
Strand twisting or twining	57	31+
Stripping photographic	430	259+
Tablet	D19	26
Tearing straight edge	225	91+
Gauge	225	91+
Test papers	435	805*
Testing		
Bursting or rupture	73	838+
Carbon paper	73	14
Flexure	73	849+
Pulp freeness	73	63
Sheet material	73	159
Towel dispensing cabinets		
Sheet type	312	50+
Strip type	312	39+
With strip- tearing edge	225	
Towels	162	100+
Twine and strands	57	260+
Twisting and twining apparatus	57	31+
Methods	57	31+
Typewriter feeding	400	578+
Web winding machines and reels	242	55+
Design	D19	96+
Weights	16	1R+
Working and feeding		
Article manufacturing from sheet or web	493	
Associating or folding sheets	270	
Feeding or delivering sheets	271	
Sheet cutting and punching	83	
Wrappers	229	87R
Wrapping	D 5	
Writing	D19	1+
Paperweight Design	D19	96+
Parachutes	244	142+
Amusement drop devices	272	6+
Balloon	244	32+
Drop bomb	102	387
Pyrotechnic device attached	102	354
Flare	102	337+
Rocket	102	348
Toy	446	49+
Paraffin		
Hydrocarbon	585	16+
Purification	585	800+
Synthesis	585	700+
Wax	208	20+
Treatment or recovery	208	24+
Paraldehyde (See Metaldehyde)		
Parallel Operation of Electric Motors	318	34+
Parametric & Parametrons	307	402
Amplifier, capacitor	330	7
Amplifier, saturable inductive reactor	330	8
Amplifier, solid element wave guide as reactor means	330	5
Frequency converter	307	424
Parasiticides	424	
Detergents containing	252	106+
Fertilizers containing	71	3
Paints containing	106	15.5+
Water purifying agents containing	252	175
Parasol (See Umbrella)	135	20R+
Parathion	558	193
Composition bio affecting	514	132
Paravane		
Torpedo and mine protection	114	240A
Towed	114	242+
Parchmentizing	8	118+
Laminating combined	156	76
Paregoric		
Medicine containing	514	282
Parer (See Peeler)		
Vegetable and fruit	99	588+
Design	D 7	147
Parimutuel Systems	235	
Parison	428	542.8
Parking		
Garages	414	227+
Land vehicle	280	761
Lights	362	61+
Meters	194	
Design	D10	42
Design, casing	D10	42
Timer elements	368	90+
With barrier	194	900
Motor vehicle		
Driven laterally	180	199+
Programmable, steerable wheels	180	204

	Class	Subclass
Theaters	52	6+
Parliamentary Procedure, Teaching	434	234
Parlor Organs		
Combined with pianos	84	172
Parquetry	52	315
Keyed blocks	52	589
Sectional layer	428	44+
Wood	428	50
Tile type	52	384
Adhesively held	52	390
Wood grain pattern	52	313
Parrels	114	96
Particles	428	402+
Metal	75	251+
Micro capsules	428	402.2
Parting Compound	106	38.22+
Static mold coated with	249	115
Utilization glass making	65	24+
Partition		
Bathtub	4	514
Bed	5	513
Building or car		
Freight carrier load brace	410	129+
Movable type	52	243.1+
Safety transaction and display	109	10
Sleeping car	105	323
Stock car	119	11+
Trapping bank type	109	5
Fences	256	
Flexible	160	
Luggage	190	109+
Mattress	5	480
Movable or removable closures	49	
Panels &, flexible & portable	160	
Portable	160	351
Purse	150	112
Rack	211	184
Receptacle	220	22
Clerical implements with dividers	206	371
Receptacle file	220	22+
Party Favor	D11	131
Party Line	379	167+
Call system	379	179+
Revertive	379	179
Pass or Identification Card Case	40	10R
Passe-partout Picture Frame	40	152
Passenger		
Aeronautics		
Accommodations	244	118.5
Loading or discharging	244	137.2
Safety lowering compartment	244	140+
Railway		
Car	105	329.1+
Car framing and structure	105	397+
Jumper amusement park	104	82
Station signals	246	209
Terminals and stations	104	28
Transfer moving train	104	20+
Recorder	346	
Seat depression	346	40
Pasta	D 1	126
Paste	429	209
Mixtures and pasted plates for storage batteries	429	209+
Supporting grids for storage batteries	429	233+
Tube		
Collapsible wall	222	92+
Dispenser-holders	248	108
Supports for	248	108+
Pasteboard	D19	1+
Pastery		
Brushes	D 4	
Inedible decoration	D11	184
Pasteurizing	426	521+
Apparatus	99	467
Filled receptacle type	99	359+
Heating and cooling	165	66
Electrolytic apparatus	204	194+
Fermentation combined	435	
Pasting		
Coating machines	118	
Collar cuff and bosom making	223	3
Envelope making	493	220+
Paper hanging and label pasting	156	
Pastry and Confection Making	426	
Brush	D 4	
Cake and pastry making	426	
Coating apparatus	118	13+
Confection making	426	660+
Packaging for pastery wedge	206	551

	Class	Subclass
Patch and Patching (See Repairing)		
Bullet	86	42
Mechanical repairing	29	402.1+
Pipe repairing	138	98+
Shirt	2	122
Tire	152	367+
Valve stem	152	430
Tool for tire repairing	81	15.5+
Wire screen	245	2
Patent Leather	428	473
Plural coating	427	412
Shoe upper	36	45
Patient Carriers		
Ambulances	296	19
Wheeled stretchers	296	20
Body supported	224	158+
Invalid lift and transfer	5	81R+
Field stretcher	5	82R+
Turners	5	61
Patient Monitoring Computerized	364	415+
Pattern	33	12+
Brassiere	450	92
Corset or girdle	450	156
Fluid flow within nuclear reactor	376	373
Footwear	33	5
Gloves	2	169
Metal founding	164	235+
Metal rolling	72	
Plates for metal founding	164	241+
Sprinkler	239	97+
Textile warp	28	184
Pattern Control (See Copying; Facsimile; Templet)		
Braiding machine	87	14+
Calculator	235	419+
Electric lamp system		
Current or voltage	315	292
Plural load systems	315	316
Electric switch	200	46
Embroidering machine	112	84+
Expansible chamber motor	91	37
Facsimile signal genrated by stylus	358	295
Knitting machine	66	231+
Loom	139	
Metalworking machine		
Can making	413	27+
Gear making	409	1+
Milling	409	79+
Patterns	409	125+
Planing	409	289+
Punch selector	234	59+
Rolling	72	6
Spinning	72	81
Tracers	409	126
Turning	82	
Type casting	199	
Musical instrument	84	1.28
Printing machine	101	93+
Punch machine		
Gang punch	83	71
Punch selector	234	59+
Special sewing machine	112	121.11+
Spray with movable deflector	239	232
Sprayer with cam track	239	236+
Typesetting machine	276	13
Typewriter	400	61+
Valve actuator	251	3
Welding	228	7
Woodworking machine		
Coopering	147	27
Cutting	144	236
Sawing	83	789
Shaping	144	137+
Turning	142	
Patty Molds, Briquetting Meat	17	32
Pavement (See Roads)	404	17+
Apparatus & processes for making	404	83+
Static mold	249	2+
Machinery for making	404	83+
Materials for forming		
Bituminous	106	273R+
Concrete	106	85+
Process for making	404	72+
Street crossing	238	9
Pawl (See Ratchet)	74	575+
Actuated (See device actuated)		
Bolt or nut to substructure lock		
Locking pawl, bolt or nut carried	411	141
Movbble pawl	411	145+
Bolt or nut carried	411	114+
Used with side lock	411	128

	Class	Subclass
Ratchet combined	411	89
Pawl carried by washer	411	138
Pawl carried movably	411	145+
Resilient pawl with side lock	411	125+
Bolt to nut lock and ratchet	411	326
Used with jam nut	411	227+
Paying Devices	114	224
Pay check or envelope with statement of deductions	282	DIG. 1
Peach Stoner	30	113.1+
Peanut	426	632+
Butter	426	633
Oil	260	398+
Pearl Button Making	79	6+
Pearl Essence	544	276
Peat	44	27+
Carbonizing	201	25
Excavator	37	3
Fertilizers from	71	24
Mineral oil from	208	400+
Peavy	294	17
Pebble Mill	241	170+
Pectin	536	2
Fermentative production	435	275
Jellies containing	426	577+
Pedal	74	560+
Clavier	84	DIG. 25
Closure actuator	49	263+
Crank	74	594.4+
Bearing	384	429+
Bearing antifriction	384	458
Combined	74	594.1+
Design	D12	125
Hand control combined	74	481
Link and lever system control	74	512+
Music keyboard		
Auxilliary	84	426
Chord player	84	444
Piano	84	229+
Automatic	84	72+
Reed organ	84	353
Key action	84	366
Wind control	84	357+
Valve actuator	251	295
Pedestal		
Bearing support	384	440
Spring type	248	162.1
Vertically adjustable	248	161+
Horizontal planar surface	108	88
Supporting	108	150
Railway surface track	238	109+
Stand supporting	248	158+
Extension	108	88
Pedicure Implements	D28	56
Pedomotors (See Velocipedes)	235	105
Boat	440	21+
Peeler or Parer (See Cutter; Stripper)		
Corer combined	99	542+
Corer segmenter combined	99	545
Food	99	588
Fruit & vegetable	99	588+
Design	D 7	147
Osier	144	207
Rossing bark	144	208R
Slicer combined	99	592
Corer combined	99	543
Wire stripper	81	9.4+
Peeling		
Bark	144	208R
Bark branch	144	207
Fruit and vegetable	99	584+
Peen and Peening	72	76
Shot blasting	72	53
Thread texturizer	28	277
Peepholes	220	82R+
Door mounted	49	171
Stove	126	200
Drying apparatus (See notes to)	34	88
Measuring sight glass plate type	73	334
Shield mounted	109	58.5
Peg (See Pin)		
Cutters	12	47.1
Implements	12	104.5
Cutting and driving	227	93+
Indicators	116	325+
Receivers lasts	12	140
Sole fastening means	36	24
String fastener musical	84	304+
Tally boards	235	90
Tent	135	118
Violin	84	304+
Peg-board Type Article Supports	248	216.1+
Pelican Hook	294	82.24+
Pellet	D22	116
Pellet (See Pills)		
Depositors	604	57+
Design	D 1	
Medicinal	D28	1+
Hay pellet makers in general	100	
Hypodermic injectors	604	187+
Machine	425	222
Process for food	426	512+
Receptacles for pocket	206	528+
Pellicles for Optical Instruments	368	303+
Peltier Effect (See Thermocouple)		
Pelts	69	22
Carroting	8	112
Chemical or fluid treatment	8	94.14
Cleaning elements made of	15	235
Cutting	83	
Dyeing	8	404+
Fiber liberation from	19	2+
Shearing machines	26	15R+
Stretching frames	38	19+
Treatment	69	23+
Pen		
Advertizing on	40	905*
Animal	119	20
Design	D30	108+
Box for	206	371
Caps	401	243+
Cleaners	15	423
Desig	D19	98
Extractors	29	808
Facsimile	358	296+
Facsimile design	D19	41+
Fountain (See fountain pen)	401	
Filling devices	141	21+
Tattooing	81	9.22
Holder (See penholders)	15	435+
Design	D19	81+
Inkstand combined	211	69.2+
Pill carrier	206	537
Play	256	25
With bottom	5	99C
Points (See penpoints)	15	435+
Rack	211	69.1+
Design	D19	81+
Ruling	401	221
Adjustable gap	401	256+
Adjustable gap fountain	401	233
Type flashlight	362	118
Penamecillin	540	342
Pencil	401	49+
Advertizing on	40	905*
Book cover combined	281	30
Box or case	D 3	30.1+
Box sectioned	206	371
Calculator combined	235	64
Calendar combined	40	335
Clip	D19	56
Compositions	106	19
Cosmetic eg lipstick or eyebrow	401	49+
Shaped	132	88.7
Indicia carrying	40	334+
Calendar	40	335
Mechanical	401	49+
Package	206	214
Pocket fastener	24	10R+
Design	D19	56+
Rack	211	69.1+
Sharpeners (See pencil sharpener)	30	451+
Electric	144	28.1+
Pencil combined	401	50+
Type electronic tubes	313	
Wood making	144	28
Pencil Sharpener	D19	73+
Implement	30	451+
Blade	30	165+
Debris receiver combined	30	453+
Eraser combined	15	105.53
Pencil combined	401	50+
Tool combined	7	160+
Machine	144	
Edge bevelling	144	28.2
Hand manipulable	144	28.11
Pendant		
Earrings	63	13
Jewelry	D11	79+
Watch pendant gearing	368	190+
Watchcase	368	301+
Watchcase making	29	179
Pendulum		
Clock	368	134+
Design	D10	130
Electric	368	165+
Escapement	368	134+
Facsimile synchronization using	358	270
Level and plumb	33	353+
Bubble combined	33	353
Lubricator		
Force feed follower	184	44
Force feed pump	184	30
Gravity feed reciprocating valve	184	78
Gravity feed rotatable valve	184	73
Mill	241	129+
Operated changeable exhibitor	40	485
Scale	177	216+
Testing		
Dynamometer	73	862.38
Tensile stress	73	836
Tide or wave motor	60	497+
Penetrating (See Punch; Puncturing)		
Nailing and stapling	227	
Nozzles	239	271
Perforating drilling	83	
Pole climbers	182	221
Printing by	101	3R+
Sorting	209	688+
Testing by	73	81+
Typewriter	400	135+
Penetration Resistant Stock	428	911*
Plural objective lens	D16	7
Penetrometers	73	59
Penholders	15	435+
Desk type for drafting	206	371
Indicia carrying	40	334+
Calendar	40	335
Picture combined	40	334+
Penicillamine	562	558
Penicillin F	540	314
Penicillin G	540	342
Penicillin G, in Drug	514	199
Penicillin K	540	314
Penicillin N	540	335
Penicillin O	540	314
Penicillin S	540	314
Penicillin V	540	341
Penicillin X	540	342
Penicillins	540	304+
By fermentation	435	43
Medicines containing	514	192+
Penmanship Instruction	434	162+
Pennant	116	173+
Penpoints	15	435+
Cleaner	15	423
Extractors or inserters	29	808
Ink retainer attached	401	252+
Ink retainer integral	401	221
Ruling	401	256+
Reservoir	401	233
Stylus ink retaining	401	292
Wiper	15	423
Pentaerythritol	568	853
Nitrated compositions	149	93
Pentaerythritol Tetranitrate	558	485
Explosives containing	149	93
Pentagrid Heptode Converter	313	300
Pentanediol	568	853
Pentode		
Amplifier	313	300
Cutoff	313	300+
Diode	313	298
With two cathodes	313	5
Diode triode	313	298
Tube structure	313	297+
Video	313	300
Pepper		
Design, shaker or mill	D 7	52+
Hand mill	241	168+
Shaker	222	142.1+
Peptide Synthesis	530	333
Peptides	530	300
Production by fermentation	435	68+
Per Acids	260	502R
Electrolytic production	204	82+
Perambulator (See Baby Carriage)	280	47.38+
Convertible to walker	280	7.1+
Springs	267	5
Perborate Esters	560	305
Perborates	423	
Electrolytic production	204	85
Perbunan T M (See Synthetic Resin or Natural Rubber)		
Percale	139	426R

	Class	Subclass
Percarbonates	423	419R
Percarboxylate Esters	560	302
Perch for Fowl	119	24+
Design	D30	119+
Perchlorates	423	476
Compositions		
Chemical agent type	252	186.1+
Detergent type	252	95+
Explosive	149	42
Explosive	149	75+
Percolator Beverage	99	307+
Electrically heated	219	438+
Stand combined	219	429+
Pitcher type	222	
Pumps	417	208+
Strainer	210	473+
Top transparent	220	82.5
Percussion		
Electric weld	219	95+
Musical instruments	D17	22+
Percussive		
Appliance kinestherapy	128	54+
Fuses	102	272+
Signs	40	579
Perdnisolone, in Drug	514	179
Perforating (See Penetrating; Punch)		
By chemical dissolution	156	644+
Devices		
Cigars	131	253+
Design	D19	72+
Earthenware multitubular pipe	425	290+
Electric heater	219	229+
Fountain pen combined	81	9.22
Glass	65	166
Inker combined, eg tattooer	81	9.22
Plastic block expressing and	425	290+
Sheets or bars	83	
Well tubing in situ	166	55
With glass mfg means	65	166
Electric discharge	219	384
Recorder	346	153.1
Handling perforated article	414	908*
Metal	83	
Billet piercing	72	345
Expanded	29	6.1
Multiperforated making	29	163.5 R
Perforated devices		
Control members (See type)		
Cards	235	487
Musical instrument	84	161+
Tape	178	112
Typewriter tape	400	73+
Drive device for tool	173	90+
Recording printing or typing		
Printing	101	3R+
Recorder	346	78+
Telegraph code recorder	178	92
Typewriter	400	80+
Telegraph code tape		
By telegraph signals	178	70R
Selective, by keyboard	234	
Tobacco products	131	253+
With electric spark or arc	219	384
Perfume Compositions	512	
Design	D28	5+
Encapsulated	512	4
Essential oils	512	5
Fixative for	512	2
Garment hanger combined	223	86
Head coverings with therapeutic device, medicament or perfume	2	171.2
Perfumed jewelry	63	DIG. 2
Preserver, stabilizer for	512	2
Solids, gels and shaped bodies	512	4
Perhalate Esters	560	306
Perhydroxamate Esters	560	311
Perhypohalite Esters	560	300
Perimidine	544	249
In drug	514	269
Periodic Switch		
Electric circuit makers & breakers	200	19R+
Electromagnetic	335	87+
Electric discharge device supply control	315	209R+
Periscopes	350	540+
Compound lens system	350	541+
Submarine combined	114	340
View simulation	434	26
Perlite	501	28
Digest	106	DIG. 2
Permanent Magnet		
Alloy	420	

	Class	Subclass
Permanent Press		
Chemically modified fiber	8	115.51
Coated	427	393.2+
Composotions	252	8.6+
Permanent Wave Machine for Hair	D28	11
Permeability Tuned Transformer	336	130
Permutation		
Locks		
Combination	70	315+
Key	70	382+
Key bitting	70	411
Switch combined	200	42.1+
Switch	200	42.1+
Typewriter	400	100+
Valves	137	269
Permutit T M	423	328
Peroxides (See Oxides)		
Alkali metal	423	582
Bleaching with	8	101+
Electrolytic coating with	204	57
Electrolytic synthesis	204	83+
Electrostatic field or electrical discharge	204	175
Heat producing compositions	149	37+
Hydrogen	423	584+
Organic	568	558+
Acids or salts	260	502R
Oxidizing	252	186.22+
Peroxynitrate Esters	560	304
Perspective Plotters	33	432+
Persulfonate Esters	560	318
Persulphides (See Polysulfide)		
Perthioborate Esters	560	305
Perthiocarboxylate Esters	560	302
Pessaries	128	127+
Pesticides (See Vermin Destroying)	D22	120
Pestle	241	291
Grain hulling	99	600+
Mortar & pestle	241	199+
Petrolatum	208	20+
Treatment or recovery	208	24+
Petroleum	208	
Emulsion breaking	252	319+
Fuels	44	
Mineral oil only	208	15+
Fuels containing	44	
Lubricants	252	9+
Mineral oil only	208	18+
Production		
From tunnels	299	2
Thermal or solvent eg secondary recovery	166	
Wells and related processes	166	
Receptacles and storage with explosion preventing devices (See explosion control)		
Sulfonates	260	504R
Petrothene T M (polyethylene)	526	352
Pew	D 6	334+
PH Measurements Electrolytic	324	438
Phantom Circuit		
With loading coil	178	45
Pharmaceutical Product	D28	1+
Pharyngoscopes	128	10+
Phase		
Changer		
Same current in input & output	323	212
Using controlled electronic tubes	328	24
Control electrical	323	212
Conversion, electrical	363	148+
Indication electrical	324	83R+
Measuring and indicating	324	83R+
Meters	324	83R+
Phase inverter amplifiers		
Balanced to unbalanced coupling	330	116
Transistor amplifiers	330	301
Unbalanced to balanced coupling	330	117
Phase locked loop having lock indicating or detecting means	331	DIG. 2
Phase locked loop in detector circuit	329	122
Regulators for alternating current generators	322	20+
Shift keying	375	52
Shifter		
Electronic tube type	328	155
Shifting and power factor control	323	205
Electronic tube circuit having factor control	315	247
Electronic tube circuit having shifting	315	194+
Meter adjustment	324	138

	Class	Subclass
Wave transmission type	333	138+
Phenacetin	564	223
Phenanthraquinone	260	396R+
Phenanthrene Purification (See Aromatic Hydrocarbon)	585	800+
Phenanthridine	546	108
Phenanthroline	546	88+
Phenates (See Phenols)	568	716+
Phenazines	544	347+
Phenethicillin	540	341
Phenetidine	564	443
Phenobarbital	544	305
Phenolphthalein	549	308
Phenols (See Naphthols)	568	716+
Alkylation	568	780+
Amino	564	443+
Addition salts	564	280
Sulfo	260	509
Carboxylic acid esters	560	67+
Monoazo compounds	534	660+
Halo	568	774+
Mineral oil containing	208	2
Nitro	568	706+
Explosive compositions	149	55+
Explosive compositions	149	68
Explosive compositions	149	80
Explosive compositions	149	99
Explosive compositions	149	103
Explosive compositions	149	105+
Polymers (See synthetic resin or natural rubber)		
Sulfo acids	260	512R
Phenones	568	335
Phenoplast (See Synthetic Resin or Natural Rubber)		
Phenothiazines	544	35+
Phenoxazines	544	102+
Phenoxy Resin (See Synthetic Resin or Natural Rubber)		
Phenylarsonic Acid	556	73
Phenylazonaphthyl	534	839+
Heavy metal containing	534	713+
Phenylazonaphthylazonaphthyl	534	832+
Phenylazophenyl	534	843+
Phenylazophenylazonaphthyl	534	832+
Phenyldimethylpyrazolone	544	356+
Phenylene		
Diamin	564	305
Phenylephrine, in Drug	514	653
Phenylglycine	562	443
Indole preparation from	548	489
Phenylhydrazine	564	310
Phenylquinolin	546	173
Carboxylic acid	546	170
Phenyltoloxamine, in Drug	514	651
Phenylurethane	560	75
Phloroglucin	568	763+
Phones		
Sound power	381	177+
Phonograph	369	
Automatic disk changer	369	178+
Cabinets	312	8+
Check operated switch	194	239
Coin controlled	194	
Design	D14	15
Design	D14	14+
Electrical reproducing or recording	369	128+
Figure toy combined	446	297+
Illuminator	362	87
Indexing devices	369	30
Magnetic	360	86
Motion picture apparatus combined	352	32+
Musical instrument combined		
Automatic musical instrument	84	4+
Comb organ	84	100
Electric tone generation eg organ	84	1.2
Needle	369	173
Photographic	369	125
Radio combined	369	6
Reciprocating tone arm motions	74	28
Record and composition		
Cabinets for	312	8+
Changer	369	178
Composition for	106	37
Cylinder holder or case	206	
Disc blank	428	64+
Holder or case	206	309+
Design holder	D 6	407
Envelope	229	68R+
Jacket	493	946*
Magnetic	360	131+

	Class	Subclass
Molding apparatus	425	406+
Molding processes	264	106+
Packages for	206	307+
Photographic	369	284+
Portable cases	206	314+
Racks for	211	40
Sound on film	369	284+
Structure	428	64+
Synthetic resin or natural rubber containing	523	174
Turntable	369	264+
With picture	352	26
With sound groove	369	272+
Sound on film	369	125+
Television combined	358	189
Turntable type	369	264+
Stop mechanism	369	268+
Phosgene	260	544R
Phosphatase	435	21*
Phosphates		
Detergents containing	252	89.1+
Organo-phosphates	252	174.16
Electrolytic synthesis	204	90
Esters	558	70+
Cellulose	536	62
Phosphorus ester group, formation of	558	89+
Fertilizers containing	71	33+
Froth flotation	209	902*
Inorganic	423	
Phosphatides	260	403
Phosphinic Acids	260	502.4 R+
Phosphonic Acids	260	502.4 R+
Phosphonium	568	9
Phosphorescent (See Fluorescent)		
Applications and devices	250	458.1
Coating processes	427	157+
Compositions		
Inorganic	252	301.6 R
Organic	252	301.16
Radioactive	252	625
Phosphoric Acid (See Phosphates)	423	316
Phosphorus	423	322
Compounds inorganic	23	
Wave energy preparation	204	157.45
Compounds organic	568	8
Heavy metal	556	13+
Wave energy preparation	204	157.73
Explosive or thermic	149	29+
Composition containing	149	29+
Removing from ore	75	6+
X-art collection	436	103*
Photocell		
Angle measuring or axial alignment	356	152
Burner	431	79
Digest, automatic temperature & humidity regulation	236	DIG. 15
Electric lamp	315	149+
Electronic tube	250	200+
Gas tube	315	149+
Gearing	74	
Headlights	315	83
Mixing of fluids	137	93+
Photographic fluid treating apparatus control	354	298
Projection printing & copying	355	41
Illumination system or detail	355	68
Radiant energy applications	250	200+
Range or height finding with light detector	356	4+
Relay	361	160+
Repeater, telegraphy	178	70P
Signal	340	600+
Stop mechanism	192	116.5+
Limit stop	192	142A
Telemetering with photocell sensor	340	870.29
Tracer for	409	125+
TV picture signal generator	358	212
Velocity	356	28+
Photochromic Devices	358	902*
Means to alter light transmission	350	353
Photochromic Glass Compositions	501	13
Photocomposition	354	
Display control	340	711+
Photoconductive	355	
Compositions	430	56+
Resistors	338	15+
Photocopying	355	
Equipment design	D16	27+
Fiber rods used in	355	1
Multiple color electric	355	4

	Class	Subclass
Photoduplication	355	
Photoelectric Cell	D13	99
Applications	136	291*
Camera		
Exposure control system	354	410+
Focusing system	354	402+
Cathode ray tube		
Circuit controlled by	315	10+
With photo sensitive electrode	313	365+
Circuits	136	293*
Circuits miscellaneous	250	206+
Control of electron tube circuit	315	149+
Controlled enlarger	355	41
For color film	355	38
Light control	355	68
Controlled switch	361	173+
Copper oxide	136	265+
Analogous rectifiers	357	30
Electronic type	313	523+
Generators, solar cells	136	243+
Light sensitive devices	250	211R+
Light sensitive semiconductor devices	357	30
Mechanical processes of making Semiconductor type	437	
Music tone generator using	84	1.28
Resistive type	338	15+
Signal in electron tube circuit controlled by	315	134
Signal isolator	250	551
Space-satellite applications	136	292*
Stereoplotters	250	558+
Telegraphophone eg phonographs using	369	100+
Television system using	358	209
Testing, calibrating, treating	136	290*
Tubes, electronic type	313	523+
Photoelectric Panels	136	243
Photoelectric System	250	200+
Counting articles	377	6
Drinking fountain controlled	239	24
Electric motor controlled	318	480
Fault detector	356	237+
Foul line detector	340	556+
Garage door controlled	49	197+
Instruments	33	DIG. 3
Safety stop mechanism	192	129A+
Street lighting controlled	315	149+
Photoengraving		
Etching machine	134	
Printing plates and processes		
Planographic	101	453+
Relief and intaglio	101	401.1+
Stencil type	101	128.21
Processes	430	269+
Chemical etching	156	654+
Electrolytic	204	3
Photoenlargers	355	18+
Photoetching Processes		
Multicolor printing surface making	430	301
Relief or intaglio	430	306+
Photoflash		
Combustion type	431	357+
Bulb	431	358+
Synchronization with shutter	354	129+
Testing	73	5
Photoflood Lamp	313	315+
Photogrammetry		
Aerial cameras	354	65+
Drawing maps from photographs	33	20.4+
Photography & Photographic (See Radiation Imagery)	354	
Accessories	D16	38+
Apparatus	D16	1+
Background	354	291
Calculators	235	64.7
Camera	354	
Automatic exposure control	354	410+
Automatic focusing	354	400+
Camera case	D 3	33
Photographic recorder	346	107R+
Seismic cameras	346	107SC
Supports	354	292
Carriers for photographic goods	224	908*
Cartridge or film pack	430	471+
Cassettes, X-ray	378	182+
Change bags	354	308+
Auxiliary plate magazine	354	282
With developing means	354	297+
Chemistry, processes and materials	430	
Screens	430	6+

	Class	Subclass
Cinematography	D16	
Color	430	
Cameras	354	100+
Facsimile previewers	358	76
Films	430	541+
Filter	350	311
Chemically defined	430	4
Combined with camera	354	100+
Motion picture	352	66
Processes	430	357+
Screens	350	322
Chemically defined	430	7
Shade measuring & testing	356	404
Unpatterned filter	350	311+
Contour photography, chemical	156	58
Copying and photostating	355	
Filters	350	311+
Projectors	355	18+
Daguerreotype	354	2+
Depth illusion		
Motion picture appartus	352	86
Motion picture process	352	38
Photographic process	352	86
Design	D16	
Developing and fixing apparatus	354	297+
Equipment design	D16	33+
Tanks	354	331
Developing and fixing composition	430	468+
Developer	430	464+
Having preservative	430	490
Fixing bath	430	455
Direction of force indicator	33	314
Electric	354	3
Apparatus	355	3R+
Developing apparatus	118	644+
Developing compositions	430	105+
Developing methods	430	31+
Methods	430	31+
Plate charging	361	225+
Sensitizing compositions	430	60+
Sensitizing elements	430	31
Equipment case	D 3	33
Exposure identification	354	105+
Eye	354	62
Testing instruments	351	206
Films or plates	430	490+
Acetate recovery from	536	78
Cartridge	430	471
Coating		
Apparatus	118	
Processes	430	935*
With structure	354	275+
Container for undeveloped film	206	455+
Digest, drying & gas or vapor contact	34	
Holders	354	276+
Nitrate recovery from	536	40
Splicer	D16	41
Transparency	356	443+
Wipers	15	100
X ray holders	378	167
Finished photograph	428	195
Flash		
Bulb	431	357+
Flash lamp	D16	42
Flash synchronizers	354	129+
Gold & silver from photographic materials	75	118P
Lanterns eg safelights	362	803*
Lens	D16	134+
Light meter	D16	39
Lights	362	3+
In enlarger	355	67+
Lamp supply circuit	315	
Shutter controlled	354	129+
Shutter operator controlled	354	129+
Magazine camera	354	174+
Manipulative processing	354	
Models	354	292
Motion picture	352	
Optics	354	
Packaging optical or photographic objects	206	316
Assemblage or kit	206	578
Photochemical etching	430	299+
Photoengraving plates		
Chemical etching	156	654+
Electromechanical etching	358	299
Photoflash	354	129+
Photograph		
Corner mount	493	952*

Entry	Class	Subclass
Mapping & aerial rectifying	353	5+
Operating device	235	91C
Relief modeling	156	59
Viewer	D16	11+
Photoplate carriers	294	137+
Picture frame	40	152+
Design	D 6	300+
Playing cards	273	292+
Design	D21	42+
Print		
Dryers	34	143+
Washer	354	297+
Printer		
Enlarging apparatus	355	18+
Projection	355	18+
Printing	355	
Processing, manipulative	354	
Projector	D16	11+
Range finder	D16	38
Recorder	346	107R+
Recording bank deposits	346	22
Reproduction equipment	D16	27+
Retouching	354	348+
Safe lamp or light	362	803*
Screen	D16	43
Sequential stations for cleaning & liquid contact	134	122P
With work transfer or fluid applying devices	134	64P
Sound recording or reproducing	369	125+
Sound recording record	369	284+
Spectroscopy	356	302
Stands		
Retouching	354	349
Stereoscopic camera	354	112+
Viewer	350	133
Studios	354	290
Synchronizing flash with shutter	354	129
Telegraph recorder	178	15
Code	178	90
Telescope, microscope or periscope with photographic means	350	502
Television	358	345+
Receiver	358	244
Color	358	332+
Transparency viewer	40	361+
Trimmer cutters	83	
Typesetting substitute	354	5+
Viewer	D16	11+
Vignetting		
Contact printing	355	125+
Enlarging	355	40+
X ray	378	
Deflected X or gramma ray	378	70
Photogravure	430	307+
Photolithographic Printing Equipment	D16	27+
Photometers	356	213+
Combined with copying camera	355	68
For color pictures	355	38
Flicker beam	356	217
Microphotometers	356	213+
Photosensitive		
Coating applying	430	
Phototypograph	354	5+
Photovoltaic Devices (See Photoelectric)		
Phthalazines	544	237
Phthaleins	549	308+
Phthalic Acid	562	480+
Anhydride	549	247
Oxidation of aromatic compounds	549	248+
Phthalide	549	307
Phthalimides	548	473+
Phthalocyanines	540	122+
Phthalonitrile	558	421
Physical Training or Testing	434	247+
Physics Teaching	434	300+
Piano	84	
Actions	84	236+
And forte devices for pianos	84	216+
Automatic player piano	84	13+
Bench with music holder	312	35
Cases	84	177+
Combined with organ	84	172
Design	D17	7
Duet piano stools	297	232+
Electric tone generation	84	1.4+
Exercising device for pianists	84	467
Hammer action	84	235+
Harp	84	258
Pedal mechanism	84	225+
Pedestal seat	D 6	334+
Pianissimo or forte devices	84	219
Piano hammer felting	144	29
Resonance apparatus	84	189+
Stools		
Adjustable concurrent vertical and horizontal movements	108	138+
Duet	297	232+
Vertically adjustable	248	161+
Stringing	84	197+
Tuning	84	454
Electrical	84	458
Violin	84	256+
Piccolo	84	384
Pick		
Ice	30	164.5+
Carrier	224	232*
Harvesting	299	24+
Making		
Blanks and processes	76	109
Dies		
Millstone	125	42
Miners	125	43
Carrier	224	
Hand carrier	294	137+
Mining	299	79+
Stringed musical instrument	84	320
Tooth	132	89+
Picker and Picking		
Article dispensing	221	210+
Cigar and cigarette making	131	109.1
Cornstalk type	56	103+
Motorized	56	13.5+
With cutter	56	64+
Cotton	56	36+
Motorized	56	12.5
Motorized	56	13.5+
Fly shuttle	139	159+
Check	139	161R+
Intermediate or plural shed type	139	143
Spindle	139	158
Stick	139	157
Keylock antipick	70	419+
Keys	70	394
Trapping	70	390
Label	156	569+
Pads for envelope gummer	118	236
Piano felt	84	460
Stringed musical instrument	84	320+
Automatic	84	8+
Textile fiber preparing	19	80R+
Tufting sewing machine	112	80.1+
Picket		
Animal restraint	119	121
Barbing	227	
Fence		
Like, type of magnetic field in gas plasma control	376	121
Metallic	256	22
Wire	256	34
Pickling		
Apparatus food	99	516+
Cleaning	134	
Compositions		
Food	426	654+
Metal	252	142+
Electrolytic of metal	204	141.5+
Etching	156	625+
Fruits and vegetables	426	626
Meat	426	331
Pickup		
Balers or bundlers	56	341+
Locomotive driver	105	33
Railway car drop type	104	122
Railway mail	258	
Scales counter weight	177	248
Vacuum cleaner nozzle attached	15	415R+
Pickup Electric	310	15+
Generator	310	15
Electromagnetic	310	15+
Piezoelectric	310	311+
Phonograph type	369	135+
Seismic type	367	178+
Sonar submarine type	367	141+
Vibration testing apparatus	73	649+
Picnic		
Box or kit	206	541+
Luggage for carrying	190	12R
Picoline	546	348
Picramic Acid	564	441
Picric Acid	568	710
Picture		
Bracket support for	248	475.1+
Exhibitors	D 6	300+
Frame	40	152
Design	D 6	300+
Molding	D25	119+
Hook or support for	D 8	367+
Support hardware	D 8	373+
Illusion	40	541
Light pipe, fiber optic	40	547+
Machine cabinet	312	20
Motion	352	240
Photograph (See photography)		
Design	D16	1
Projecting	353	
Puzzle	273	157R
Recording or reproduction	D14	
Stands	248	441.1+
Stereoscope	350	133
Target	273	358
Teaching drawing	434	85+
Toys	446	147+
Transfer (See decalcomania)		
Transmission	358	
Transmitting to a distance (See television)		
Tubes TV	313	364+
Type support	248	466+
Stage	272	10
Travel	272	18
Viewing apparatus		
For photographic transparency or X-ray	40	361+
Including illumination	350	235+
With detailed optics	350	239+
Picture Window (See Curtain, Wall)	52	777+
Pie or Pie Making	426	
Apparatus	99	450.1
Cooking	99	432+
Overflow director or receptor	126	383+
Presses	425	233+
Cooler & stacked article		
Carrier	294	161
Tray	294	161
Display container	206	45.32
Pans	99	DIG. 15
Pie design	D 1	
Shaped packaging	206	551
Slicing guide	D 7	43
Pier		
Bridge	14	75+
Foundation	405	231+
Marine	405	195+
Track pedestal	238	119
Wharves	405	284
Piercing (See Perforating, Devices)		
Piezoelectric	310	311+
Acoustic wave with mechanical energy coupling	310	334
Acoustic wave with resonator	310	322
Agitation apparatus having	366	127
Casings for	174	50+
Casings for	220	2.1 R+
Charging piezoelectric material	361	225+
Cleaning processes using vibrations	134	1
Cleaning processes using vibrations	134	17
Compositions	252	62.9
Crystal		
Making	29	25.35+
Rochelle salt	562	585
Electrical devices, underwater type	367	157+
Electrical systems	310	314
Electrical testing of	324	56
Fluid pressure testing having	73	754+
Galvanometers	324	109
Generator systems	322	2R
Light valve	350	269
Manufacture	29	25.35
Microphones	381	173
Modulator having	332	26
Motor systems	318	116
Phonograph pickup	369	144
Polymers	310	800*
Relay circuit	361	207
Sonic testing having	73	570
Speakers	381	190
Speed responsive apparatus	73	517R
Telemetering using	340	870.3
Transducers	84	DIG. 24
Under water vibration transducers	367	157
Piggy Bank	D99	35+

(Class, Subclass)

- **Pigment** 106 288R+
  - Coating implement containing 401 40+
  - Compositions containing 106
    - Synthetic resin or natural rubber (See class 523, 524)
  - Inkwell containing 15 257.7+
  - Printing 101
  - Solid polymer ctg (See synthetic resin or natural rubber)
- **Pigs** 428 576+
  - Casting apparatus 164
    - Molds 249 174
  - Pipeline 15 104.6 R
- **Pike Poles** 294 19.1+
- **Pile**
  - Arrangement holder or spacer 206
  - Battery
    - Primary 429 152+
  - Building 405 231+
    - Cutter circular saw type 83 743+
  - Drivers 173 90+
  - Fabric 428 85+
    - Axminster type loom 28 214+
    - Double pile cutting 26 13+
    - Knitted 66 194
    - Knitting loop forming 66 91+
    - Laminated 156
    - Loop cutting 26 8R+
    - Loop cutting on loom 139 43
    - Multiple shed weaving 139 21
    - Terry 139 25+
    - Tube loom for axminster type 28 216
    - Tufting 139 2+
    - Warp manipulation 139 35+
    - Weft manipulation 139 116+
    - Woven 139 391+
  - Installing 405 232+
    - In marine environment 405 224+
  - Marine foundation 405 195+
  - Nuclear reactor core arrangements 376 171
  - Nuclear reactor core arrangements 376 347
  - Protection 405 211+
  - Resistor 338 204+
  - Rheostat 338 115
    - Compressible 338 101+
  - Sheet 405 274+
- **Piles and Fagots** 428 588+
- **Piling**
  - Apparatus 414 28+
  - Bleaching and dyeing running lengths 8 152
  - Conveyer arranging material on 198 434+
  - Conveyor establishing groups 198 418+
  - Conveyor for piling bulk material 198 508
  - Conveyor thrower 198 638+
  - Material in drying process 34 38
  - Web in accordion folds 270 30+
- **Pill (See Pellet)**
  - Box or container 206 528+
    - Dispensing single dose 206 528+
    - Medicator capsule 604 403+
    - Pocket pen type 206 537
    - Structured dosage unit 424 464+
    - Tear-off dosages 206 532
  - Coating machine 118 13+
  - Medicinal design D28 1+
  - Presses 425 233+
  - Presses 425 406+
- **Pillow** 5 434+
  - Combined with ground mat 5 419
  - Cover 5 490
  - Design D 6 601
  - Dressing means 5 489
  - Sham 5 491
- **Pilocarpine** 544 336
  - In drug 514 397
- **Pilot (See Igniters)**
  - Aircraft control 244 220+
  - Clutch control 192 35+
  - Earth boring tool type 175 385+
  - Flames and formation of same 431 278+
    - Thermocouple controlling 136 217+
  - Train control 246 166
  - Valve or trap
    - Fluid motor distributing 91 304+
    - Servo-motor 91 461
    - Thermal trap 236 54
    - Thermal trap float controlled 236 55
  - Vehicle fender 293 48+
  - Wheel lock 70 253
- **Pimaric Acid** 260 97+
- **Pimozide, in Drug** 514 323
- **Pin (See Hinge; Joints; Peg; Pintle)**
  - Article holding bracket 248 309.2
  - Bedstead element 5 303
  - Cigar tip perforating 131 254
  - Curb for watch balance 368 176
  - Cushion 223 109R+
  - Draft couplings
    - Animal draft 278
    - Railway vehicles 213 78+
  - Fastener 24 150R+
    - Article holder 24 13
    - Binder device releasably sheet 402
    - Button 24 103
    - Button loop interchangeable 24 89
    - Cotter 411 513+
    - Cuff holder 24 48
    - Flower holder 24 6
    - Hair 132 50R+
    - Hair curler 132 31R+
    - Hat 132 68+
    - Holders 223 109R
    - Necktie holder 24 49CP
    - Ornamental 63 20
    - Ornamental design D11 87
    - Packages 206 380+
    - Packaging 227 25
    - Paper 24 153+
    - Pencil holder 24 12
    - Pins combined 24 356+
    - Tag 40 24
    - Watchcase 368 313
  - Flush pin gauges 33 DIG. 18
  - Game 273 67R+
    - Bowling 273 82R
    - Pinball machine 273 118R+
  - Larding 17 42.1
  - Linchpin
    - Land vehicle 301 113
    - Railway wheel 295 50
  - Looper for knitters 66 3+
  - Making 163 6+
    - Clothespins wire 140 83
    - Clothespins wood 144 9
    - Cotter pins 29 5
    - Glass head pins 65 142+
    - Hairpin 140 87
    - Wood pin points
    - Wood pins 144 12
  - Ornamental pin, jewelry 63 20
  - Rolling 29 110.5
  - Setter
    - Bowling pin 273 42R+
    - Bowling pin spotter 273 46
    - Watchmakers 81 8
    - Wood pin 227
  - Shear 464 32+
    - Plow releasing 172 271
  - Ships implement 114 221R+
  - Skewer 17 1S
  - Sprocket chain pivot 474 206+
  - String tuning musical 84 201
  - Tumblers lock type 70 378
  - Wheel 446 217+
  - Wrist 74 595+
- **Pinacol** 568 852+
- **Pinball** D21 7
- **Pince-nez** D16 107+
- **Pincers**
  - Lasting implements 12 110+
- **Pinch**
  - Bar 254 36+
  - Cocks 251 5
  - Effect, in thermo nuclear reactions 376 121
  - Thread, bolt to nut lock, different or varying pitch 411 307
- **Pinching Kinesiatrically** 128 59
- **Pincushion** 223 109R+
- **Pinene (See Alicyclic Hydrocarbon)** 585 350+
- **Pinhole**
  - Camera 354 202+
- **Pinion (See Rack)** 74 434+
  - Making 29 178
  - Typewriter feed 400 330+
  - Watch 368 322
    - Cannon 368 323
- **Pinking Cutters** 83 918*
  - Machines
    - Reciprocating cutter 83 918*
    - Rotary cutter 83 333+
  - Shears 30 230
- **Pinners Hat Machines** 223 11
- **Pintle**
  - Bedstead element 5 303
  - Caster 16 37
    - Retainers 16 38+
  - Hinge 16 386
    - Coil spring 16 304+
    - Multiple pintle 16 366+
    - Retainer 16 263+
  - Railway car journal lid 384 190.6
- **Pintsch Gas** 48 211+
- **Pinwheels** 446 217+
  - Pyrotechnic 102 359
- **Pipe (See Tube; Tobacco)** 138
  - Bending (See bending)
  - Briar 131 230
  - Caps 138 89+
    - Tobacco 131 176+
    - Ventilating 98 122
  - Chute 193 29+
  - Cleaner
    - Boiler combined 122 379+
    - Brushing machine 15 88
    - Cookstove flue 126 16
    - Design D32 14
    - Fluid handling apparatus 137 238+
    - Implement 15 104.3+
      - Operated by fluid current 15 104.61+
      - Sewer grapple snake 15 104.33
    - Jet pump 15 408
    - Process 134 22.1+
    - Tobacco ash receiver combined 131 232
    - Tobacco smoking 131 243+
    - Tobacco smoking combined 131 184.1
  - Closures or plugs 138 89+
  - Connections
    - Closet bowls 4 252R+
  - Corncob 131 230
  - Couplers & connectors (See joints, pipes, shafts, rods, plates)
    - Electrical 439 207+
    - Electrical box and housing 174 65R
    - Electrical quick detachable 439 195
    - Packing 277
    - Valved 251 142+
  - Coupling 285
    - Valved 251 142+
  - Cutters for pipes and rods 30 92+
    - Design D 8 60
    - Well tubing 166 55
  - Design D27 2+
  - Disinfection of flush 4 225
    - Chemical holder 4 226
  - Dry fire extinguisher 169 17
  - Elimination of in ingot 29 526.2+
  - Fittings 285 120+
    - Making 29 157T
  - Flanging machines 72 115+
  - Flaring 72 317
    - Rotating tool 72 115+
  - Flexible
    - Coupling 285 238+
    - Steam locomotive 105 47
    - Structure 138 118+
    - Valved coupling 251 150
  - Fluid distribution type D23 266
  - Fluid pressure brake control 303
  - Hangers and supports 248 49+
    - Dredgers 37 72
    - For slender article 211 60.1
  - Hawse 114 179+
  - Heat exchanger 165 177+
  - Hickeys (See bending)
  - Holder D27 2
  - Hydraulic ram 417 226+
  - Irrigation, vehicle supported 239 723+
  - Jack for pipes or rods 254 29R+
  - Joints 285
    - Making 29 428+
    - Making 29 157A
    - Valved 251 142+
  - Laying pipe and cable 405 154+
    - Submarine 405 158+
  - Making
    - Bending 72 367+
    - Clay smoking 425 233+
    - Earthenware machine 425 376R+
    - Earthenware mold 249 136+
    - Encircling with wire on tape 242 7.22+
    - Expander 72 317
    - Fittings 29 157R

	Class	Subclass
Flanging	72	317
Foundry mold making	164	176$^+$
Glass	65	
Laminated	156	
Filament winding by	169	
Metal casting	164	
Apparatus	164	421$^+$
Metal fusion-bonding apparatus	228	
Metal joint mold	249	89
Reamer	408	227$^+$
Riveting	227	52
Soldering clamp	269	
Swaging	72	399$^+$
Trimming	408	28$^+$
Organ	84	331$^+$
Automatic	84	91
Pattern sprayer	239	548$^+$
Pocket case for	206	244
Rack	211	70.3$^+$
Smoking device	131	260
Rotary shaft	464	183
Joints	464	18$^+$
Sand delivery	291	41$^+$
Shearing machines	30	92$^+$
Smokers pipe literature	131	DIG. 1
Soap bubble	446	19$^+$
Sprinkler	239	266$^+$
Stand (See tank)		
Flue boiler	122	184$^+$
Superheater	122	475
Water tube boiler	122	321$^+$
Water tube boiler coil or loop	122	244$^+$
Water tube boiler loop	122	281
Water tube boiler spur	122	307$^+$
Stems	131	227$^+$
Stove	126	307R$^+$
Damper	126	292$^+$
Oven	126	17
Water heater	126	364$^+$
Strainers	4	DIG. 13
Structure	138	177
End combined	138	109
Supports	248	49$^+$
Tamper tobacco smokers	131	247$^+$
Thread	285	390$^+$
Protectors	138	96R
Threading		
Cutting	408	
Milling	409	65$^+$
Rolling	72	126
Tobacco	131	330$^+$
Cartridge	131	348
Toy	446	24
Train	239	723$^+$
Waste	4	191$^+$
Well		
Cleaner combined	166	173$^+$
Cutter or perforator	166	55
Destroying	166	373$^+$
Expansible	166	207
Joining	166	373$^+$
Plural concentric	285	138$^+$
Screen combined	166	227$^+$
Structure	166	242
Weakening	166	55
Wrenches	81	52$^+$
Pipeline		
Cleaner	15	104.3
Filters	210	435$^+$
Intake	210	460$^+$
Piperazines	544	358$^+$
Piperidine	546	184$^+$
Piperonal	549	436
Pipette	73	864.1$^+$
Pipette	422	100
Absorption for gas analysis	422	83$^+$
Body contacting, eg, hypodermic needle, syringe	128	
Design	D24	55
Design, chicker baster	D24	60$^+$
Chemical apparatus	422	
Chemical methods	23	
Container cap combined	141	29
Container or container closure	141	22$^+$
Bulb type, eg eye dropper	141	24
Dispensing	222	
Drop formers	222	420
Medicinal	604	295$^+$
Design	D24	55
Successive receivers filled from continuously flowing source	141	130
Pipings	36	57
Pipings, Upholstery Trim	428	123
Pique	139	426R
Piromidic Acid, in Drug	514	303
Pistol		
Grip	42	71.2$^+$
Magazine	42	7
Holsters	224	193
Design	D 3	101
For either side of torso	224	192
Leather or flexible fabric	224	911*
Metal or rigid material	224	912*
Rotatably or swingably attached	224	198*
Shoulder	224	206*
With closure flap	224	238*
With gun positioner	224	243*
Revolvers	42	59$^+$
Design	D22	104$^+$
Swords	42	53
Design	D22	118
Water	222	79
Piston (See Associated Device)		
Cleaner for plugs and pistons	15	104.11$^+$
Door check	16	49$^+$
Expansible chamber device	92	
Annular	92	107$^+$
Multiple	92	61$^+$
Oscillating	92	120$^+$
Filler for fountain pens	401	171$^+$
Fire extinguisher	169	33
Free piston	60	910*
Hammer of tool drive	173	134$^+$
Hypodermic	604	218$^+$
Internal combustion engine	123	193R
Manufacture	29	156.5 R$^+$
Grinding	51	289R
Meter	73	232$^+$
Packing	277	
Digest	33	DIG. 15
Pressure gauge	73	744$^+$
Ring	277	
Assembling and disassembling on piston	29	222$^+$
Gauge, ring & bearing race	33	DIG. 17
Group contracting for working	29	269
Lubrication combined with		
Making	29	156.6
Testing	73	120
Testing for leakage	73	47
Rod (See rod connecting)	74	579R$^+$
Lubricators	184	24$^+$
Valved piston compressor or pump combined	417	545$^+$
Rotary engine	123	200$^+$
Rotary for motors	418	68
Sausage stuffer	17	36$^+$
Syringe	604	218$^+$
Valve	251	324$^+$
Check type	137	538
Valved		
Pump liquid	417	545$^+$
Pit Remover	30	113.1$^+$
Pitch (See Tar)		
Changing		
Conveyor screw	198	659
Impeller	416	25$^+$
Musical instrument	84	
Pump impeller	415	129$^+$
Ship propeller	440	50
Fuel containing	44	
Ionized gas		
Fluent material supplied space discharge device	313	231.1$^+$
Making, treating and recovery	208	39$^+$
Mineral oil only	208	22$^+$
Music device for giving pitch	84	454$^+$
Plastic compositions containing	106	
Production by distillation	201	
Pitcher	222	
Dispensing type with handle	222	465.1$^+$
Handle, asymmetric	D 8	DIG. 9
With flow controller or closure	222	556$^+$
Pitchfork	294	55.5$^+$
Cutter combined	7	115$^+$
Making		
Blanks and processes	76	111
Dies	72	470$^+$
Pitching Casks	118	
Pitchometers	33	530
Pith Removers	130	31R
Pitman (See Connecting Rod)	74	579R$^+$
Bearing	384	266$^+$
Making	29	156.5 R
Piton	248	216.1$^+$
Pitot Meter	73	861.65$^+$
Pitter	99	547$^+$
Pivampicillin	540	336
Pivot (See Hinge; Joints)		
Bearing	384	192$^+$
Knife-edge fulcrum	384	4$^+$
Rotary crane	212	253
Teeth	433	220$^+$
Turntable	104	46
Pku	436	108*
Placard Frames	40	152$^+$
Placemat		
Furniture protector	D 6	613$^+$
Paper	D 6	613$^+$
Placers (See Arranging; Piling)		
Electrical wiring	254	134.3 R
Hammer with nail placer	81	23$^+$
Track ties	104	6
Track torpedo	246	216
Placket	2	218
Plaid	139	417
Plait and Plaiting		
Apparel making	223	28$^+$
Box making	493	162$^+$
Knitting	66	14
Laminated pile fabric making	156	
Tucking	112	145
Planes		
Woodworking	30	478$^+$
Planetarium	434	286
Planetary (See Gear and Gearing)		
Planimeter	33	122
Planing Devices		
Bench planes for wood	30	478$^+$
Hair planer	30	30$^+$
Planer type combined machine	29	30
Planing machine	409	288$^+$
Stave jointing	147	33
Stoneworking	125	9
Wood saw teeth	83	835$^+$
Wood slivering scoring	144	186$^+$
Woodworking	144	114R$^+$
Woodworking combined machine	144	36$^+$
Plank and Planking (See Panel)		
Clamp portable for	269	91$^+$
Pavement	404	17$^+$
Railway truck plank	105	208
Scaffold	182	222$^+$
Ship	114	84
Planographic		
Printing	101	130$^+$
Electrolytic coating	204	17
Multicolor surface making photographic	430	301
Surface making photographic	430	300$^+$
Plans		
Building	52	234$^+$
Store	52	33
Traffic guide	52	174
Plant Cell Culture	435	240.4$^+$
Plant Cell Growth Media	435	240.54$^+$
Plant Cell Per se	435	240.4$^+$
Planters and Planting		
Agricultural	111	
Broadcasting	111	10$^+$
Dibbling	111	89$^+$
Drilling	111	14$^+$
Liquid or gas	111	6$^+$
Plant setting	111	2$^+$
Corn planter clutches	192	23
Plants		
Attack resistant stock	428	907*
Catalysts and stimulants	71	65$^+$
Coating of	427	4
Cutting and comminuting		
Discriminate	99	
Indiscriminate comminuting	241	
Indiscriminate cutting	83	
Extracts (See extracts)		
Food	99	
Gas making	48	
Heat or power		
Aircraft	244	53R$^+$
Boiler	122	1R$^+$
Dynamoelectric	290	
Heat and power combined	237	12.1$^+$
Heating systems	237	13
Power	60	

U.S. PATENT CLASSIFICATIONS

	Class	Subclass
Husbandry (See plant class)	47	
Compound tools	7	114$^+$
Covers, shades and screens	47	26$^+$
Diggers	171	
Fertilizers	71	
Fertilizers	47	48.5
Fruit cutters and corers	30	113.3
Garden hose	138	118$^+$
Garden rake	56	400.1
Greenhouses	47	17$^+$
Harrows	172	
Harvesters	56	
Hotbeds and coldframes	47	19
Nutrient supply	47	62$^+$
Ornamental beds	47	33
Planting	111	
Plows and cultivators	172	
Pollination	47	1.41
Pots and receptacles	47	66$^+$
Powder dusters or sprayers	239	
Presses for balers	100	
Sprayers mixing and dissolving	239	310$^+$
Terrarium	47	69$^+$
Threshing	130	
Tree fumigators	43	126
Tree hacks	30	121
Water culture	47	59$^+$
Lights	362	805*
Living	800	1
Metallurgical	266	142$^+$
Molding	425	
Brickmaking	425	88
Nameplates or cards for	40	10C$^+$
Nuclear	376	
Patented varieties of	PLT	
Planting	111	
Dispensing seed	222	
Grain drill chutes	193	9
Preserving	47	58
Coating	427	4
Cut flower holder	47	41R$^+$
Stalks retting	162	1$^+$
Retting bacteriological	435	279$^+$
Stands	47	39$^+$
Thinners	172	534$^+$
Earth working function	172	900*
Plant sensor activates power	47	1.43
Plaque	428	913.3
Holder	248	488$^+$
Holder	248	490$^+$
Layer non-planar	428	174$^+$
Layer varying thickness	428	156$^+$
Panel type, sectional layer	428	44$^+$
Wall decoration	D11	132$^+$
Plasma	424	101
Blood component	424	101
Drying and freezing	34	5
Gas		
Chemical process employing	204	164$^+$
Injecting fuel into body of plasma	376	127
Ion producer	361	230
Magnetohydrodynamic generator	310	11
Moving or shifting plasma	376	107
Nuclear fusion reactions	376	100
Containment by magnetic field	376	121
Systems	315	111.21$^+$
Plasmid	935	27*
Plaster of Paris (see Plaster)	106	109$^+$
Plaster or Plastered (See Plastering; Plastic)		
Board	52	443$^+$
Attached to furring	52	344$^+$
Edge joint	52	448$^+$
Laminate	428	411.1$^+$
Making	156	39
Corner	52	250$^+$
Cutters		
Hand manipulable	30	165$^+$
Machines	83	
Pointers	15	235.3
Medicated composition	424	
On brick	52	444
Panel strip	52	459
Plaster of paris	106	109$^+$
Alkali metal silicate composition	106	77$^+$
Compositions	106	110$^+$
Processes of burning	106	100$^+$
Splint	128	91R$^+$
Screed	52	364$^+$
Slatted lath	52	341$^+$
Surgical body wear	128	390
Trimmers	433	144
Plastering		
Applicator-material supplied	15	104.93
Fountain-trowel	15	235.4
Machines	425	62
Machines	425	87
Mortar-joint finisher	15	235.3
Combined	15	105.5
Trowel	15	235.4$^+$
Plastic (See Plaster)		
Block and earthenware apparatus	425	
Chain, staple, horseshoe making	59	
Coating	106	
Compositions	106	
Cutters	30	115
Deformation, pressing, or treating		
Metal deformation	72	
Metal fusion bonding	228	
Presses	100	
Shaping or treating apparatus	425	
Sheet metal container making	413	
Static molds	249	
Wireworking	140	
Dyeing of	8	
Jewelry	63	DIG. 3
Metalworking	72	253.1$^+$
Superplastic material	72	709*
Plastics	425	
Screws, bolts, nails etc	411	904*
Shaping or treating articles		
Apparatus	425	
Computer controlled, monitored	364	473
Processes	264	
Structures		
Brush or broom tuft socket	15	193
Coffins	27	7
Fence connections rail	256	66
Fence connections wire	256	50$^+$
Fences	256	19
Lacing eyelet	24	142
Lacing studs	24	148
Railway ties	238	84$^+$
Seat covers	297	219$^+$
Settable material masonry	52	
Splints	128	90$^+$
Synthetic resin building component	52	309.1$^+$
Testing of	73	150R
Plasticizing		
Casein ctg compositions	106	147
Gelatine ctg compositions	106	136
Polymer treating (See synthetic resin or natural rubber)		
Textiles	8	130.1$^+$
Plate		
Antiskid for tires	152	225R$^+$
Baking	99	372
Cleaners		
Brushing	15	77
Wiping	15	102
Compound metal	428	615$^+$
Dental	433	167$^+$
Design, household	D 7	23$^+$
Detector system	329	186$^+$
Distribution for electric wires	174	40R$^+$
Electric		
Conductor	191	13
Consumable electrode	314	44
Face	220	241
Face switch	174	66$^+$
Fluorescent	250	466.1
Switch	200	271
Switchboard	361	346$^+$
Escutcheons	70	452
Door knob	292	357
Face	220	241
Electric housing	174	66
Grinding and polishing	51	
Lock	70	450
Making	65	
Hinge	16	251
Illumination lens perforated	362	456
Joints and couplings (See joints)		
Pipe and plate	285	189$^+$
Pipe and plural plate	285	128$^+$
Pipe and plural plate spaced	285	19$^+$
Rod and plate	403	230$^+$
Through plate insulator	174	151$^+$
With intermediate member	403	187$^+$
Kick	16	1R
Kick guard	16	DIG. 2
Knob rose	292	357
License	40	200$^+$
Lifters	294	27.1$^+$
Lifting clamp	294	901*
Liquid heater or vaporizer	122	208
Nail plates	10	33
Distributors	10	171$^+$
Horseshoe nails	10	41
Name	40	
Houses	40	
Letter box	232	46
Nuclear fuel	376	432
Packages	206	449$^+$
Pattern foundry type	164	241
Photography		
Camera	354	
Coating apparatus	118	
Plate holders	354	276$^+$
Plates	430	495$^+$
X ray plate changers	378	172
X ray plate holders	378	167
Portable closure fastener	292	289$^+$
Printing	101	395
Addressing	101	369$^+$
Casting	164	2$^+$
Coating electrolytically	204	17
Making electrolytically	204	6
Making stereotype	29	21
Mountings	101	382R$^+$
Prism for overhead lighting	362	326$^+$
Push	16	1R
Racks for	211	41
Railway		
Draft cheek	213	54$^+$
Frog	246	462
Rail base plate	238	209$^+$
Switch	246	453
Tie	238	287$^+$
Rectifiers, dry	357	
Sounding for pianos	84	184$^+$
Stencil	101	127$^+$
Plate	101	112
Stock metal	428	615$^+$
Storage battery	429	209$^+$
Stove	126	81
Stovepipe thimble	126	317
Table article	220	
Design	D 7	23$^+$
Design, simulative	D 7	5
Transformer	336	
Tumblers lock type	70	377
Turners	294	6
Watch	368	318
Design	D10	123$^+$
Platen		
Copyholding (See copyholder)		
Heat exchanger	165	168$^+$
Liquid extracting press	100	295$^+$
Memorandum pad type	281	6$^+$
Metal rolling	72	88$^+$
Roll and platen	72	88$^+$
Screw thread	72	88$^+$
Printing press bed and platen type	101	287$^+$
Addressing machine	101	57$^+$
Intaglio	101	163$^+$
Multicolor	101	193$^+$
Numbering	101	78$^+$
Roller and plate inker	101	338$^+$
Roller type inker	101	359$^+$
Selective or progressive	101	93$^+$
Special article	101	41$^+$
Ticket machine	101	66$^+$
Reciprocating press	100	214$^+$
Smoothing and pressing textiles	38	17$^+$
Conveyor belt combined	38	10
Implements	38	71$^+$
Multiple cooperative	38	5
Structure of	38	66
With fluid flow	38	15$^+$
Stamping recorder	346	78
Textile felting	28	139$^+$
Typewriter	400	648$^+$
Card holder combined	400	525$^+$
Case vertical shift	400	264$^+$
Divided	400	585$^+$
Flat	400	23$^+$
Index wheel	400	158$^+$
Key wheel	400	158$^+$
Rocker platen	400	649$^+$
Shift lock	400	274$^+$
Wear distributors	400	554$^+$
Washing textiles		

	Class	Subclass
Roll and platen scrubber type	68	86+
Scrubbing by carrier roll and platen	68	42
Platers Railway Tie	104	16
Platform		
Actuator		
Closure fastener	292	94
Door	49	263+
Register	235	99R
Dumping scow	114	31+
Floating landing	114	263
Gyrostabilized	248	183
Harvester platform adjustments	56	208+
Illuminators vehicle	362	76
Industrial	108	51.1+
Synthetic plastic	108	901*
Deformable integral fastening elements	108	902*
Knock-down	108	51.3
Ladder	182	116+
Land vehicle		
Swinging axle	280	114+
Connections	280	130+
Rocking connections	280	127+
Load-transfer	108	52.1
Marine drilling	405	195+
Pole or tree	182	187+
Portable lift truck	108	51.1+
Railroad car	105	425+
End buffer	213	222
Truck platform	105	211+
Railroad station	104	30+
Scaffold	182	222+
Elevating	182	141+
Tunnel construction	405	148
Scale	177	253+
Shoes	36	19.5
Stove and furnace	126	277+
Synthetic plastic industrial platform	108	901+
Made with deformable integral fastening elements	108	902*
Take-it-or-leave-it	108	52.1
Vehicle illuminators	362	76
Window	182	53+
Plating (See Coating; Electrolytic Devices)		
Electrolytic coating processes	204	14.1+
Piano keys having	84	437
Warship armor	114	11+
Platinum		
Alloys	420	466
Compounds	23	
Organic	556	136+
Electrolysis	204	
Igniting devices	431	268
Misc. chemical manufacture	156	DIG. 101
Pyrometallurgy	75	83
Player		
Automatic music instrument	84	2+
Automatic musical player attachable to keyboard	84	105+
Automatic player piano	84	13+
Musical cord played by single key	84	443+
Page turner operated by player	84	502+
Playground Climber	272	113
Design	D21	242+
Playing Cards	273	292+
Design	D21	42+
Playpen	256	25
Childs furniture		
Design	D 6	331
With floor	5	99C
Pleat		
Apparatus	223	28
Method	223	28
Pleated		
Web or sheet	428	181
Plectrum		
Piano	84	258
Picking device	84	320
Plenum		
Fluid mixer thermal	236	13
Surgical receptor	604	313+
Plexiglass T M (See Synthetic Resin or Natural Rubber)		
Pliabilizing		
Shoe soles	12	41.2+
In shoe	12	41.1
Pliers	81	300+
Assembling tool	29	268
Bending	72	409+
Clip applying	72	409+
Nose ring	72	409+
Compound tool	7	125+
Tong type	294	3
Cotter removers	29	248
Dental	433	159+
Orthodontic	433	4
Design	D 8	52+
Ear tag	72	409+
Eyelet setting	227	144
Hollow rivet setting	227	144
Implements	12	110+
Lever pulling or pushing	29	268
Making		
Dies	72	470+
Pivoted	76	64
Sliding	76	69
Piston ring tools	29	223+
Punches	30	363
Spectacle	81	3.6
Staking	29	232
Tongs	294	
Valve tools engine	29	221
Wire twisters	140	121
Plinth (See Base)		
Pliophen T M (phenoplast)	528	129
Plotters Perspective	33	432+
Plow	172	
Cable hauled	104	169
Conduit or trolley		
Electric current collector	191	48
Railway block signal actuation	246	84
Railway switch actuation	246	252
Trap	104	145
Design	D15	11+
Disk bearing	384	157
Radial antifriction	384	460
Ditcher type	37	98+
Forging dies	72	474+
Gondola car unloading	105	259
Hillside	172	204+
Cable hauled	104	169
Iron making	29	14
Lifting clutch	192	62
Making	29	14
Mole	37	193
Cable laying combined	405	184
Pipe or cable laying	405	180+
Submerged	405	164+
Potato digger	171	71+
Potato digger	171	111+
Road grader type	37	110+
Snow		
Excavator and melter type	37	227+
Railway type	37	214+
Roadway type	37	266+
Subsoil	172	699+
Plucks		
Fowl plucking apparatus	17	11.1 R
Fur plucking apparatus	69	26
Musical instrument (See music)		
Plug		
Barrel bung	217	110+
Blasting	102	333
Blasting cartridge combined with	102	304
Bottle and jar closures	215	200+
Dispensing flow control		
Axial discharge stationary	222	521
Cap carried axial	222	546
Closure	222	563
Rotary	222	554
Rotary screw	222	552
Electric connector or switch		
Boxes and housings plug receptacle type	174	53
Cleaner ie spark plug	15	104.11+
Conductor reel combined with	191	12.4
Electric fuse	337	254+
Electric fuse and socket	337	213+
Electric switch combined with	200	51R
Electric switch type	200	162+
Interposed two part	439	625+
Outlet box and wall plug	439	638+
Quick detachable	439	180
Resistance plug boxes	338	77
Screw	439	662+
Telephone multiple switchboard plug terminal	379	312+
Telephone multiple switchboard test	379	20+
Telephone switchboard	379	319+
Flow meter with orifice and tapered plug	73	861.58
Incandescent lamp seal	220	2.1 R+
Lock cylinder type		
Cylinder and plug assembly	70	367+
Element	70	375
Rotary	70	362+
Sliding	70	361
Sliding and rotary	70	360
Lubricator choke	184	52
Metal receptacle closure expansible type	220	233+
Metallurgical furnace tap hole	266	271
Nail hole	52	514
Nuclear fuel closure	376	451
Pipe and conduit	138	89+
Pipe coupling solid type	285	242
Plugging instruments dental	433	25+
Rail spike receiving	238	370
Spark (See spark plugs)		
Swivel pipe joint with	285	121+
Telephone jack and plug	439	676+
In central office switchboard	379	319+
Tire patch combined with	152	370
Tire plug inserting tool	81	15.5+
Tobacco		
Attaching cigar and cigarette	131	94
Combined	131	366+
Compositions	131	352
Design	D27	1
Inserting cigar and cigarette	131	72
Making	131	111+
Shapes	131	111+
Wrappers or binders	131	365
Valve		
Bath or basin outlet type	4	204
Float operated	4	205+
With inside operator	4	204+
Combined with strainer	4	287
Head	251	356+
Hydrant type	137	272+
Liquid filter with screened type	210	429+
Movable liquid filter with screened type	210	390
Multiple outlet nozzle rotary	239	392+
Nonstop reciprocating	137	528+
Plug cock and seat abrading machine	51	28
Reciprocating	251	318+
Rotary	251	309+
Ship bottom type	114	197
Stopper type	4	295
Well		
Cementing or plugging	166	285+
Closing port	166	141
Handling	166	373+
Preset	166	114
Preset conduit carried	166	116
With expanding anchor	166	118+
With outward pressure means	166	101
With pump or plunger	166	106
Plug in Units		
Contact and socket structure	439	626+
Structural combinations	361	392+
Plumb Bob	33	392+
Design	D10	61+
Plumbago (See Graphite)		
Plumbanes	556	95
Hexaorganodiplumbanes	556	82
Plumbates	423	592+
Plumbers Friend	4	255+
Plumbing		
Baths and closets	4	
Design	D23	200
House fittings	4	DIG. 7
Sewerage	137	356+
Snake	15	104.3 SN
Water flow control	137	
Water pressure regulator	137	505+
Plumbs		
Structure	33	391+
Wall construction guides	33	404+
Wireworking	140	41
Plumes	428	6
Plume abatement	261	DIG. 77
Plunger		
Air pressure guns having	124	65+
Conveyer elements	198	717+
Die or cutter		
Box making	493	167+
Pastry cutter	83	

	Class	Subclass
Plastic metal shaping	72	273
Sheet or bar die cutter	83	
Stationary vegetable or meat cutter	83	401+
Wood slivering	144	190
Drilling or dibbling planters having	111	
Expansion chambers having	138	31
Feed (See type of machine)		
Coal spreader for furnace	110	114
Locomotive tender	105	233
Manual dibblers	111	96
Railway sand feeders	291	36+
Rotary peripheral face abraders	51	82R
Gongs having	116	152+
Mechanical gun plunger release mechanism	124	33+
Press		
Block forming plungers	425	398+
Block forming plungers	425	344+
Block forming plungers	425	412+
Block forming presses		
Portable	425	
Reciprocating mold	425	352+
Reciprocating mold	425	406+
Stationary mold	425	352+
Stationary mold	425	406+
Die expressing metal	72	273
Earthenware die expressers	425	406
Glass mold plungers	65	362
Metal mold sand packers	164	207+
Miscellaneous die expressers presses	425	406+
Pump having	417	437+
Textile fluid treating apparatus having	68	
Vegetable comminutors having	241	
Water closet having	4	300+
Plural Amplifier Channel Systems	330	124R+
Transistor type	330	295
Plush	139	391+
Plutonium Compounds		
Inorganic	423	251
Organic	534	11+
Ply		
Delaminating	156	344
Laminate	428	
Wood	428	106
Wood	428	533+
Laminated	428	411.1+
Sewing machine trimmer combined	112	127
Tires resilient	152	151+
Woven fabrics	139	409+
Plywood (See Veneering)	52	
Making		
Glue appliers and presses	156	
Presses	156	580+
Processes	156	
Processes combined	144	346+
Pneumatic (See Hydraulic)		
Apparel turning	223	43
Burner extinguisher	431	145
Carding stripping	19	109
Chuck	279	4
Cleaners	15	300R+
Clock	368	54
Combined automatically played musical instruments	84	6
Conveyers and feeders		
Agitation and	222	195
Article dispensing	221	278
Conveyer	406	
Pneumatic or liquid stream feeding item	209	906*
Sheet or web feeding or delivery	271	
Stacker	406	
Cotton harvester	56	30+
Motor driven	56	13.1+
Dispatch	406	
Tube or system	D18	35
Endless flexible track for land vehicle	305	34
Face plate expander	105	11
Folder	493	450
Gas separator cleaning	55	282+
Grooming device	15	401+
Gun	124	56+
Harvester	56	12.9+
Jet for marine propulsion	440	38
Mattresses	5	449+
Metal casting injector	164	306+
Musical instrument player attachable to keyboard	84	107
Musical instrument tracker	84	151+
Musical instrument tracker combined with sheet	84	146+
Nail driving device	227	9
Piano automatically played	84	24+
Pipe organ	84	331+
Power plant	60	407+
Pump	417	118+
Air jet combined	417	86
Jet type pump	417	151+
Railway cable gripper	104	203
Railway propulsion system	104	155
Locomotives	105	63+
Selector for automatically played musical instrument	84	115+
Sheet turner for sheet music	84	519
Shoe forms	12	114.4
Shoe sole cushion	36	29
Tamper for track laying	104	10+
Tire (See tire, pneumatic)	152	151+
Toy	446	176+
Typewriter	400	176+
Carriage return	400	182+
Velocipede seat	297	199+
Wheel	152	1+
Wind instrument automatically played	84	83+
Pneumothorax Apparatus	604	26
Pocket		
Alarm watch or clock	368	244
Apparel element	2	247+
In garment supporter	2	300
Shoe with	36	1
Book covers	281	31
Calendars combining card pockets	40	122
Carried		
Brush with recess for comb	132	121
Lighter	431	129+
Lighter	431	146+
Match safe	206	96+
Special receptacle carried in pocket	206	37+
Spittoon carried in pocket	4	259
Ticket case	206	39
Cartridge belt with pocket holder	224	224+
Cartridge belt with tilting	224	224
Compound tool	7	118+
Pocketknife	30	155+
Dispenser with		
Barometric or angle of repose trap	222	457
Movable trap chamber	222	344+
Tiltable container trap only	222	454+
Trap chamber cutoff	222	425+
Game table with		
Ball game	273	123R+
Moving surface type	273	113+
Pool	273	12
Pool with chutes connected	273	11R
Lamps	362	157+
With writing pads	362	99
Letter box with	232	33
Letter box with pivoted trap	232	47+
Machine		
Article dispenser ejector	221	208+
Bevel machine table	144	125
Cutting window stile pockets	144	19
Dryer rotary drums	34	109
Dryer shelves	34	172
Endless conveyor with flexible pocket	198	715
Felting machine roller	28	126+
Locomotive tender draft device	213	6
Loom pushed shed	139	28
Pneumatic conveyer feeder	406	63+
Manifolding strip form pockets for sheets	282	11.5 R
Match safe	206	96+
Design	D27	32+
Paper wrappers with	229	87.5
Picture mounts pocketed	40	159
Railway car post holder	105	408
Railway flat car stake holder	105	390+
Releasable	105	383
Sliding window sash	49	372+
Solid separator employing	209	684+
Stove grates with	126	159
Stoves	126	38
Support with	211	55+
Telescopes	350	546
Watches	368	62+
Wooden receptacle with	217	28
Woven	139	389+
Pocketbook	150	100+
Closure	150	118+
Design	D 3	42+
Pocketknife	30	155+
Compound tool	7	118+
Pod		
Augers		
Wood	408	199+
Looms cloth feeding	139	295
Pogo Stick	272	114
Design	D21	67
Walking	272	70+
Point & Points (See Impaling; Prong; Punch)		
Bands wheel hub	301	107
Glaziers	411	477+
Guards or holders		
Hatpin	132	72
Hatpin guide combined	132	69
Marker guide	33	574+
With gauge	33	666+
Nails spikes tacks	411	493+
Needles (See notes to)	223	102
Parallelizing divider	33	151
Pen (See penpoints)	15	445+
Plow	172	681+
Printers for the blind	400	122+
Protectors pencil sharpener	30	460
Railway switch	246	435R+
Thrower	246	264+
Setter glazier	227	
Spreaders nail spike tack	411	359+
Switch		
Foot guard	238	380
Structure	246	435R+
Thrower railway	246	264+
Trolley transfer	104	104+
Ties and braces mattresses	5	458
Trolley transfer	104	104
Vaccine supply	206	367
Well casing	175	19+
Pointer		
Chance device	273	141R
Indicator type	116	
Arm structure	116	327+
Masonry implement	15	235.3
Pointing (See Calking)		
Barrel hoop lapping and	147	45
Bolts	10	21
Nail cutting and	10	31
Pencil (See pencil sharpeners)	30	451+
Screw threading and	10	4
Screws	10	9
Spike	10	59
Staple cutting bending and	59	74
Wooden pin	144	30
Poise		
Scale		
Automatic	177	235+
Beam	177	246+
Beam automatic	177	212+
Computing	177	35
Poison	424	
Holder or container		
Alarm combined	116	72
Indicating contents	215	367
Vermin destroying	43	131
Poker	294	14
Poker Chips	D21	53
Rack	211	49.1
Pokes	119	136+
Polarimeters	356	367+
Polariscopes	356	366
Polarity Testers	324	133
Polarized Devices and Polarizers		
Electromagnet with armature	335	230+
Electromagnetic wave	343	909
In wave guide	333	21R
With antenna	343	756
Examination devices generally	356	356
Gem and crystal examining	356	30+
Glass	350	370+
Headlight	362	19
Illuminated signs	40	548
Image projectors	353	20
Light polarizer	350	370+
Combined with illuminator	362	19
Light valve	350	374+
Making		

Entry	Class	Subclass
Structural orientation	264	1.3
Radio wave polarizations	343	909+
Spectro-photometers	356	326
Stage illusions using	273	9+
Stereoscope, stereo systems	350	132
Storage and retrieval of information	365	121
Switch electromagnetic	335	78+
Telegraph code recorder	178	93
With antenna	343	756
Testing with polarizers	356	364+
Vision testing means	351	49
Polarographic Chemical Analysis	204	400+
Pole (See Mast; Post)		
Barber	40	607
Boat	440	36
Changer	361	245+
Changing in alternating current motors		
Induction	318	773
Synchronous	318	704
Climbers	182	133+
Leg or shoe attachable	182	221
Collectors trolley	191	64
Couplings	191	60.5
Curtain design	D 8	376+
Electrical conductor	174	45R
For electric wire	52	4
Fire escape	182	189+
Wall attached	182	100
Flagstaffs	116	173+
Hammock ridgepole	5	130
Handlers	414	23
Land vehicle		
Supports	278	86+
Thills	278	33+
Tow	280	493+
Light support	362	431
Adjustable light	362	418+
Magnetoelectric armatures	310	264
Manipulating	294	19.1+
Plate or piece		
Switch	200	271
Telephone magnet	381	199+
Rack type	211	105.1+
Rossing bark	144	208R
Scaffolds for	182	187+
Ski	280	819+
Spray	239	532
Switch electric	200	49
Tent	135	114
Window operating	49	460+
Poler		
Bearing plate for poling railway cars	105	462
Marine propulsion device	440	36
Side track push rods	213	224
Police or Fire Alarms and Signals		
Signal systems	340	287+
Telephone circuit combined	379	37+
Telegraphs	340	287+
Whistle	116	137R+
Policemens Clubs or Batons	273	84R
Carriers for	224	914*
Polishing and Polishing Appliances (See Burnishing)		
Abrading	51	
Boot and shoe machines		
Cleaning and blacking combined	15	30+
Cleaning combined	15	36+
Component parts in making	12	79.5
Coating treating process	427	355
Compositions	106	3+
Dental grinders combined	433	142
Motor driven	433	125
Fabric type	15	209R+
Hat pouncing combined	223	20
Machinery design	D15	124+
Metal burnishing	29	90R+
Needle eye working	51	152
Paper calendering	29	90.1
Scouring machines combined	51	4
Planing combined	144	38
Turning combined	144	47
Shoe stands with polisher guide	15	266
Shoebrush and dauber	401	39
Woodworking machines		
Pollinating	47	1.41
Pollution		
Decreasing, & environmental impact	422	900*
Polo		
Ball	273	58R
Mallets	273	67R
Pony boots	36	111
Polyacrylyl (See Synthetic Resin or Natural Rubber)		
Fiber dyeing	8	
Safety glass employing	428	442
Polyamides (See Synthetic Resin or Natural Rubber)		
Fiber dyeing	8	
Processes of molding	264	DIG. 61
Polyarylmethyl Compounds	260	386+
Amines	564	315+
Ketones	568	332
Carboxylic acids	562	460
Polybenzanthronyl	260	363
Polycarbodiimide (See Synthetic Resin or Natural Rubber)		
Polycarboxylic (See Type Compound)	260	
Acyclic	562	590+
Acyclic esters	560	190+
Alicyclic esters	560	127
Anhydrides	549	231+
Aromatic esters	560	76+
Polymers derived from (See synthetic resin or natural rubber)		
Polycycles	280	282
Polycyclic Hydrocarbons		
Alicyclic synthesis	585	360
Aromatic synthesis	585	400
Polyelectrolytes	252	62.2
Polyelectrolytes	429	188+
Electrolytic chemical processes	204	
Polyepoxide		
Polymers from (See synthetic resin or natural rubber)		
Polyester		
Fiber dyeing	8	
Fibers	8	DIG. 4
Polymers from (See synthetic resin or natural rubber)		
Polygon	33	482
Polygonal Turning and Cutting		
Broaching	409	243+
Drilling and boring	82	1.2+
Wood square hole	408	30
Milling	409	79+
Planing	409	289+
Turning	82	18
Wood	142	2+
Wood		
Polygonal cutters	144	224+
Shaping	144	138
Polyhydric Alcohol	568	852+
Carbohydrates	536	1.1+
Ethers	568	672+
Fermentative production	435	158+
Polyimide		
Polymers from (See synthetic resin or natural rubber)		
Polymer		
Dissolver	422	901*
Liquid hydrocarbon	585	17+
Synthesis	585	502+
Low moleuclar weight hydrocarbon polymer containing composition	106	901*
Replacing hard animal tissue	528	950*
Solid synthetic resin or natural rubber (See synthetic resin or natural rubber)		
Polymerization (See Compound Produced)	260	695
Addition (See also synthetic resin or natural rubber)	526	59+
Apparatus for	526	920*
Fatty oils	260	407
Hydrocarbon synthesis, cyclic		
Alicyclic	585	360+
Aromatic	585	422+
Aromatic cyclic	585	415+
Monomer polymerized in presence of transition metal containing catalyst (See catalyst)		
Olefin, fiber dyeing	8	
Photo image	430	269+
Utilizing electric, radiant		
Within nuclear reactor	376	
Utilizing wave energy	522	1+
Polymixins	530	319
Polyolefin	8	DIG. 9
Polyolefin-type foam		
Injection molding	264	DIG. 83
Polyphase		
Current meters	324	107+
Polyphenylene Oxide (See Synthetic Resin or Natural Rubber)		
Polypod		
Breakwater	405	29
Polypropylene		
Molding foamed articles	264	DIG. 16
Polysaccharides	536	1.1+
Heteropolysaccharides	536	123
Polysulfide	423	511+
Fatty oils	260	399
Organic	568	21+
Resins (See synthetic resin or natural rubber)		
Polysulfone (See Synthetic Resin or Natural Rubber)		
Polyurethane		
Cellular product	521	159+
Filters in gas separation	55	DIG. 13
High resiliency or cold cure	521	904*
Nonstructural laminate	428	423.1
Nontextile dyeing	8	515
Radiation sensitive binder	430	906*
Stock material	428	160
Synthetic fiber dyeing	8	926*
Synthetic resins with solvent		
Alcohols	524	379+
Hydrocarbons	524	474+
Phenols	524	323+
Terminal ethylenic unsaturation	525	920*
Tire or tube tread or carcass	156	128.1
Wave energy preparation or treatment	522	1+
Polyvinyl Acetals	525	61
Safety glass employing	428	437
Pom-pom Gun	89	126
Pompom (See Tassel)	428	4+
Pongee	139	428
Pontoons		
Buoys	441	3
Landing platforms	114	258
Salvage	114	44+
Pool		
Billiard game		
Ball	273	59R
Cues	273	68+
Tables and accessories	273	2+
Glass heating furnace	65	336+
Swimming, purification		
Process	210	749+
Purification	210	169
Solar heated	126	415+
Swimming-type nuclear reactor	376	174
Swimming-type nuclear reactor	376	361
Swimming-type nuclear reactor	376	403
Popcorn		
Machine	D 7	325
Poppers	99	323.5
Presses	425	406+
Sheller	130	
Popgun	124	63+
Design	D21	146+
Poplin	139	426R
Popsicle T M Forming Machine	425	DIG. 219
Porcelain		
Compositions	501	141+
Electric insulating	501	141
Shaping processes	264	
Pore Forming or Porous		
Compositions	106	122
Abrasive	51	296
Alkali metal silicate ctg	106	75
Cement hydraulic setting	106	86+
Ceramic	501	80+
Glass	501	39
Metal casting	164	79
Molding	264	41+
Mortar mixer combined	366	10+
Processes	366	2+
Synthetic resin ctg (See cellular under synthetic resin or natural rubber)		
Wells		
Gravel placing means	166	51
Gravelling	166	278
Porous cementitious filter	166	276
Porous material screens	166	228
Pork	D 1	
Porosity Test	73	38
Porous Devices (See Pore Forming & Type of Device)		
Stock material, cellular layer layer	428	304.4

Entry	Class	Subclass
Porous Forming (See Cellular Under Synthetic Resin or Natural Rubber)		
Porphyrins	540	145
Port		
Boiler	122	6.5
Lifting or handling	52	122.1+
Shield mounted	109	58.5
Ship	114	173+
Valve seat	251	359+
Internal combustion engine	123	188S+
Rotary	251	314+
Portable Circular Saw	30	360+
Portable Houses	52	143+
Portfolio	190	900*
Design	D 3	30.1+
Locks	70	67+
Portland Cement		
Compositions containing	106	89+
Alkali metal silicate containing	106	76
Refractory	501	124
Making processes	106	100+
Position Detectors Indicators and Markers	116	209+
Elevator	116	226
Electric	187	130+
Indexed printed matter	283	37+
Liquid level gauge	73	314
Railway switch stand	246	401+
Steering wheel	116	31
Train	246	122R+
Weaving shuttle	139	341+
Positive Feedback Amplifier (See Regenerative, Amplifier)		
Post (See Column; Mast; Pole)	52	720+
Architectural, design	D25	126+
Bearing	114	169
Binding electrical	439	775+
Drilling machine	408	236
Drivers	175	
Continuous push	254	29R+
Impact	173	90+
Electrical conductor	174	45R
Hitching	119	122+
Design	D30	154
Jacks	254	30+
Kiln furniture	432	253+
Lamp or light combined	D26	67+
Light support	362	431
Railway	104	125
Rotatable machine element	74	504+
Scaffold	248	351+
Sign attachments	40	607
Static mold	249	51
Steering	74	492+
Tool	82	37
Top	52	301
Bed	5	281
Tower	52	648
Transmission tower	D25	126+
Velocipede	280	281R+
Violin sound	84	277
Window corner	52	282
Postal Card or Packet	229	92.8
Design	D19	1+
Posthole Digger		
Implement	294	50.6+
Machine	175	
Pot (See Bowl; Jar; Kettle; Pan)		
Burner	431	331+
Chamber	4	479+
Coffee	99	279+
Coffee or teapot strainers	210	473+
Receptacle spout attached	210	466+
Cooking container	D 7	354+
Dash (See dashpot)		
Fire pot thermostats	236	104
Flower	47	66+
Design	D11	143+
Furnace and metallurgical		
Fire pot	126	144+
Fire pot thermostats	236	104
Glue pot heater	126	284
Material heating furnaces	432	156+
Material heating holders	432	252+
Metal melting and casting	164	335+
Solder pot heater	126	240
Glue heating	126	284
Holder	16	116R+
Liquid heating	126	373+
Railway rail supports	238	110+
Railway signals	246	475
Vulcanizing	425	283
Potassium (See Alkali Metal)		
Chloride dissolver	422	902*
Potato		
Chip machine	99	403+
Slicing combined	99	353
Chips	426	637
Digger	171	
Masher	241	
Implements	241	169.2
Picker and hopper	221	213+
Planters	111	
Preservation	426	321+
Canning	426	321+
Coating	426	289+
Compositions	426	637+
Dehydration	426	471+
Refrigeration	426	524
Washing machines	15	3.1+
Potential		
Regulators for dynamos	322	28
Potentiometer	338	68
Assembling apparatus for making	29	729
Current generating thermometer with	374	179
Galvanometer with	324	98+
Structure	338	68+
Systems	323	364
Testing	324	
Pothead	174	74R+
Fluid or vacuum type	174	19+
Potted Antenna	343	873
Potted Electrical Device	361	331
Pottery		
Extruders	425	289+
Molds	425	406+
Potters wheels	425	263+
Potters wheels	425	459
Potting Electrical Device	264	272.11
Pouch (See Bag; Satchel)		
Animal attachment for manure	119	95
Colostomy	604	332+
Diaper having	604	348
Key containing	206	37.1+
Tobacco	D27	44
Poultry		
Butchering	17	11+
Husbandry	119	
Design	D30	
Litter	119	1
Pouncing Hat Machines	223	20
Pounder		
Meat	17	30+
Textiles fluid treatment	68	215+
Pouring		
Articles with pouring spout	D 7	312+
Automatic weigher discharge	177	115
Dispenser	222	
Receptacle filler	141	
Powder (See Coating; Dispensing; Distributing)		
Ammunition powder bags	102	282
Ammunition powder forms	102	283+
Baking	425	526+
Cans	220	
Dispensing	222	
Coating composition ctg synthetic resin or natural rubber (See also 524-904)	525	934*
Comminuted metal	75	.5 R
Cosmetic	424	69
Compact	D28	78
Design	D28	8
Dispenser	604	58
Duster aircraft borne	244	136
Duster insect killing	239	
Dye	8	524+
Engine driven by	123	24R
Engine starting with	123	183
Explosive compositions	149	
Explosive thawer	126	269
Fluent solid applier pocketed, eg powder puff	401	200+
Formed by electrolysis	204	10
Fusing and spraying	239	79+
Insect	424	
Marking ground with	111	26
Medicator for applying	604	58
Metallic powder containing trunks	106	290
Metallurgy		
Briquetting apparatus	425	78+
Comminuted metal	419	
In blank	428	546+
In laminated blank	428	548+
Heat treating, other than sintering	419	31
Making powders	75	.5 R
Pressing powder with other steps	29	DIG. 31
Sintered stock materials	75	228+
Sintering	419	
Sintering then working	419	28
Paper folder for containing	493	948*
Photographic developer	430	499
Puff		
Box, cosmetic	206	823*
Design	D28	91
Machine for stitching	112	121.24
Sintering		
Metals	419	
Non-metals	264	56+
Non-metals	264	319+
With electric heat	219	149
Toilet kit, powder & puff	401	126+
Toilet kit, toilet article eg mirror included	132	82R+
Train		
Compositions	149	14+
Firecrackers	102	361
Fuses	102	275.1+
Fuses blasting	102	275.1+
Pyrotechnics	102	360
Powdered Fuel Burning		
Fluid fuel combined	110	261+
Furnace structure	110	263+
Metal containing composition	149	37+
Metal cutting processes	148	9R+
Apparatus	266	48
Powdered fuel feeders	110	104R+
Shavings and sawdust	110	102
Water tube boiler combined	122	235R
Power (See Engine; Motor; Device Operated)		
Alternate manual operator	74	625+
Amplifier pentode	313	300
Delivery interrelated control	192	
Factor control	323	205
In lamp or electronic tube circuit	315	247
Measurement	324	83R+
Hammer	173	90+
Forging	72	
Measurement	73	862.3+
Pack		
Conversion system	363	
Radio receiver combined	455	343
Plants (See engine; motor)	60	
Aircraft	244	53R+
Dynamoelectric	290	
Heating system combined	237	12.1+
Internal combustion engine combined	123	3
Testing	73	112+
Stop control	192	116.5
Storing dynamos for equalizing loads	322	4
Supply		
Combined with amplifier	330	199+
Combined with push pull amplifier	330	123
Control with amplifier	330	127+
Control with push pull amplifier	330	123
Or regulation systems	323	
Polyphase with amplifier	330	113+
Supply vibrator	307	150
Invertor system	363	110
Tables and stands	74	16
Take off	74	11+
Crane	212	172
Vehicle (See motor vehicle; land vehicle)	180	
Locomotive	105	26.5+
Marine propulsion	440	
Occupant propelled land vehicle	280	200+
Railway propulsion system	104	287+
Ship	114	
Soil working implements combined	172	
Trackmans car	105	86+
Traction systems	104	165+
Truck ladders	182	127
With erection means	182	63+
Power Transmissions		
Electricity	307	
Miscellaneous space	307	149
To a radio receiver	455	343
Practice Devices		
Barrels for ordnance	89	29

Entry	Class	Subclass
Baseball	273	26R
Cartridges	102	444+
Drop bombs	102	395
Golf	273	35R+
Mines	102	401
Projectile	102	529
Shell	102	498
Telegraphy	178	115
Practicing		
For music	84	470R+
Exercising devices	84	465+
Praxinoscope	352	102
With sound accompaniment	352	2
Precious Stones		
Changing colors by beams of light	250	
Producing synthetic	501	86
Compositions	501	86
Working	125	30R
Precipitation (See Decanter)		
Actinide series compound	423	249+
Electrical		
Distillation processes including	201	19
Distillation processes including	203	40
Electrophoresis	204	180.1+
Gas separation	55	101+
Separation or purification	204	186+
Hydrometallurgy	75	97R+
Leaching metals combined apparatus	266	101
Liquid purification or separation	210	703+
Metallurgical apparatus	266	170
Sacchariferous solution		
Impurities	127	48+
Sucrose	127	47
Suspended matter in marine environment	405	74
Precoating Filter for Liquids	210	777
Apparatus	210	193
Precompressing (See Prestress)		
Prefabricated Housing	52	
Design	D25	35+
Subenclosure	52	79.1+
Preform	428	542.8
Pregnancy, Fertility	436	510*
Pregnandiol	260	397.5
Preheaters		
For boiler feed water	122	412+
Preselector Amplifier	330	
Preservers Life	441	80+
Aquatic lifesaving devices	441	80+
Buoyant devices	441	88+
Vests	441	106+
Design	D21	238
Racks for	114	190
Preserves	426	639
Butters	426	616
Preserving & Preservation (See Material Preserved)	422	
Agents	252	380+
Alcohols	568	701+
Antiseptic composition	424	
Canning	426	392+
Closing	53	285+
Cooking after filling	99	359+
Filling	141	
Filling and closing	53	266R+
Precooking and canning	99	356
Channels for water control	405	15+
Corpse	27	11
Egg preserving apparatus	99	485+
Ethers	568	580+
Fats and oils	260	398.5
Fermentation yeasts	435	
Foods	426	321+
Agent other than organophosphatide	426	654
Color treating	426	262
Fat or oil product	426	610
Product having food preserving agent	426	133
Surface coating of solid food with a liquid	426	310
Halocarbons	570	102+
Hydrocarbons	585	1+
With non-hydrocarbon additive	585	2+
Hydrogen peroxide	423	584
Latex		
Synthetic resin or natural rubber (See class 523, 524)		
Natural rubber latex	528	935*
Preserved artifact	428	542.4
Rubber	524	
Textile treatment	8	490
Press	100	
Apparel	223	52+
Arbor	29	251
Auto body crushers	100	901*
Automatic guards	192	134
Billiard cue	273	71
Binding	100	1+
Box making	493	52
Can crusher	100	902*
Compacting ensilage with silo	100	65+
Computer controlled, monitored	364	476
Concurrent pressing & conveying	100	144+
Conveyer combined	100	144+
Cooker combined	99	349+
Design	D15	123
Die cutting and punching		
Die processes and machines	83	
Sheet and bar punching	83	
Die expressing	425	381.2
Drain means for expressed liquid	100	104+
Dryer	34	143+
For arc light electrodes	425	113+
For nailing and stapling	227	
General purpose	100	
Glass	65	305+
Laundry iron	D32	9
Leather	69	48
Crease impressing	69	2
Forming combined	69	8
Seam	69	7
Lid applying	53	287+
With compacting of contents	100	54+
Macaroni	425	289+
Magnetic	100	917*
Material triggered control	100	43+
Materials treatment combined	100	70R+
Meat		
Briquetting	17	32
Tenderers	17	25+
Metalworking	72	
Assembling and disassembling	29	700+
Die cutting	83	
Draw	72	347+
Drill	408	
Drill press combined	29	26R+
Extruding	72	253.1+
Indirectly	72	253.1
File cutting	76	19
Methods	100	35+
Miscellaneous structure	100	214+
Punch	83	651+
Molding		
Bread, pastry and confection	426	496+
Bread, pastry and confection	425	98+
Bread, pastry and confection	425	238+
Computer controlled, monitored	364	476
Concrete, etc. , blocks	425	253+
Concrete, etc. , blocks	425	356
Cork or bung tapering	144	284
Foundry molds	164	37+
Glass	65	305+
Glass blowing and	65	229+
Metal forging	173	
Plastic block and earthenware	425	
Shoe heel	12	48
Shoe sole blank	12	21
Shoe sole laying	12	33+
Paper pulp	162	396+
Paper web forming	162	358+
Particulate material	100	
Peat	44	28+
Pelleter	100	903*
Platens or pressure surfaces	100	295+
Plural diverse presses	100	137+
Plural presses	100	193+
Portable receptacle lid applying	100	54+
Printing	101	
Design	D16	27+
Reciprocating press construction	100	214+
Scrap paper	100	914*
Sectional chamber filter	210	224+
With means to add treating material	210	204
Sewing machines		
Elements	112	235+
Leather type	112	60+
Tennis racket	273	74
Textile ironing or smoothing	38	
Truss presses	100	913*
Woodworking		
Bending	144	256.1+
Cork and bung	144	284
Cork or bung tapering	144	284
Feed and presser devices	144	252R+
Glue applying combined	156	
Methods	144	381
Veneer	156	580+
Pressure		
Aircraft pressure maintenance	98	1.5+
Balancing within packing assembly	277	3
Bonding		
Cold weld	228	115+
Electric weld	219	78.1+
Compensating, support digest	248	DIG. 1
Compensator and conduit combined	138	26+
Cooker	220	
Automatic electric control	99	330
Design	D 7	358
Electric	219	440
Distillation under	203	73+
Distillation under	203	91+
Equalizers (See pulsation devices)	138	26+
Plural valves	137	629+
Pump	417	540+
Feed		
Burner	431	
Conveyers	406	
Dispensing	222	394+
Lubricators	184	57+
Metal casting machine	164	303+
Pneumatic dispatch	406	148+
Sprinkler or sprayer	239	
Filter	210	348+
Closed circulating system	210	167+
Sectional chamber pressure type	210	224+
With pressure responsive by-pass	210	130+
With pump	210	416.1
Fluid actuator or operator (See device operated, fluid operated, motor)		
Fluid pressure regulator	D23	235
Gauge	73	700+
Calibrating	73	4R
Design	D10	85+
Gun	124	56+
Heat exchanger control	165	13
Mineral oils	208	357+
Mineral oils	208	366+
Regulator and control (See fluid pressure regulator and device associated therewith)		
Aircraft cabin	98	1.5
Boiler feed	122	452
Brake	188	105+
Brake system	303	
Concentrating evaporator	159	2.1
Dispensers	222	
Distillation column	202	160
Electrical separation of liquids	204	306
Electrolytic cell combined	204	229+
Fermentation processes	435	3
Fire extinguisher	169	5+
Fluid heater	236	24.5
Gas distributing system	48	191
Inhaler	128	204.18+
Organ	84	356
Piano	84	39+
Pneumatic conveyer	406	12
Pump	417	279+
Resilient tire	152	415+
Testing	73	4R
Thermostat combined	73	199
Volume or rate of flow meters combined	73	199
Water heating system	237	65
Water purification by chemical feeders	137	98+
X ray electrical system	378	108+
Relieving devices		
Degasifying apparatus	55	189+
Degasifying process	55	55
Fluid	137	455+
Gas separator combined	55	309+
Marine mine	102	412+
Shaft packing	277	29
Testing and calibrating	73	4R
Responsive loading means for joint packing	277	27
Tanks or chamber, high pressure (See gas, holder)	220	3
Fire extinguisher	169	71+

	Class	Subclass
Gas	48	174^{+}
Inhalers	128	203.12^{+}
Thermolytic	201	35
Valve, pressure relief	137	455^{+}
Reducing, regulating pressure	137	505^{+}
Prestressed Structures	52	223R^{+}
Facing sheet	52	222
Pretzel	D 1	120
Slip	425	323
Prevention (See Anti)		
Bank robbery	109	2^{+}
Body catching rebound	182	137^{+}
Button loss	24	90.5
Cable sag, endless	254	388
Condensation ice and frost		
Aircraft	244	134R
Coating compositions	106	13
Compositions	252	70
Plants	47	2
Trolley	191	62
Window	52	171^{+}
Window by ventilation	98	2
Corona	174	127
Fire prevention	169	
Receptacle attachment	220	88R
Textile treating apparatus	68	209
Fraud or alteration		
Check controlled apparatus	194	302^{+}
Coating	427	7
Locks	70	416^{+}
Leakage from gas pipes	48	194
Lost motion		
Gearing	74	409
Piano motion	84	222^{+}
Magnetic compass deviation	33	356^{+}
Nuclear reactor damage	376	277
Offset in printing	101	416R^{+}
Overwinding		
Clock and watch key	81	467
Clocks and watches	368	209
Composite motor	185	13
Spring motor	185	43
Weight motor	185	35
Railroad coupling vertical disengagement	213	153
Register overthrow	235	131R
Safe breaking	109	26
Scale	29	81R
Secondary radiant emission	313	106^{+}
Skate retrogression	280	11.21
Switches		
Arc	200	144R^{+}
Overfusing	337	225^{+}
Unauthorized use	200	43.1^{+}
Locked actuator	200	43.11^{+}
Removable actuator	200	43.4^{+}
Type casting squirt		
Integral line	199	53
Separate type	199	90
Weaving weft vibration	139	373
Prexinoscope	352	102
With sound accompaniment	352	2
Prick Punch	30	366
Sheets or bars	30	366
Pricker for Fruit	83	
Pricking Shoe Soles	83	
Primary Battery	429	122^{+}
Prime Mover Dynamo Plants	290	
Primer and Priming		
Explosive	102	200^{+}
Ammunition making with	86	32
Cartridge combined	102	470^{+}
Layers or zones	149	14
Internal combustion engine	123	187.5 R
Pump	417	435
Rotary	415	11
Primuline Azo Compounds	534	800
Printed Electric Circuits	361	397^{+}
Making		
Combined in an amplifier	330	66
Mechanical manufacture	29	829^{+}
Photo chemical etching	430	311^{+}
Preformed panels	174	68.5
Stock material	428	901*
Printed Matter (See Books, Printers & Printing)	283	
Books, strips, & leaves	281	
Circuit stock	428	601
Printers and Printing	101	
Annular typewriter	400	112^{+}
Antismut devices	101	416R^{+}

	Class	Subclass
Bag printing and making	493	187
Bed & platen machines	101	287
Braille	101	3R^{+}
Typewriter	400	122^{+}
Bread pastries lozenge cutting combined	425	289^{+}
Calculator combined	235	58P^{+}
Record controlled	235	432
Check printing cash register	235	3
Cigar or cigarette making	131	284
Cleaning attachments	101	425
Coating combined		
Nonuniform coating processes	427	256^{+}
Coating implements & material supply	401	
Coin-controlled print machines	101	DIG. 23
Color forming phenomenom	101	DIG. 1
Combined with laminating apparatus	156	384^{+}
Methods	156	277
Contact photographic	355	78^{+}
Control by code on plate	101	DIG. 25
Convertible printing press, eg, lithographic to letter press	101	DIG. 28
Credit cards, meant to check vadility before printing	101	DIG. 18
Cylinder	101	
Braking	101	DIG. 2
Inkers	101	353^{+}
Photo process making	430	300^{+}
Revolving, print on	101	DIG. 16
Work support grippers	101	409^{+}
Data processing equipment	D14	111
Devices hand	101	368^{+}
Drying flexible sheets and webs	34	151^{+}
Duplicating machines	101	113
Electrically operated elements	101	DIG. 15
Electrostatic force	101	DIG. 13
Endless printing belt	101	DIG. 27
Envelope printing and making	493	187
Equipment, design	D18	13^{+}
Paper cutters	D18	34
Sheet or special article, cut and print	101	DIG. 19
Stapling machines	D18	34
Facsimile apparatus	101	DIG. 2
Golf ball, print on	101	DIG. 17
Ink	106	20^{+}
Agitated in reservoir	101	DIG. 8
Axially reciprocating rollers	101	DIG. 14
Maintaining condition of rolls	101	DIG. 9
Thickness or density sensor	101	DIG. 24
Transfer using one or more belt	101	DIG. 7
Inkers	101	335
Intaglio	101	150
And other printing means	101	DIG. 22
Key or wheel		
Time recorder	346	80^{+}
Typewriter	400	165.1^{+}
Leads		
Cutting and punching	83	
Fillers	101	402
Handlers	276	34
Lineup table	33	614^{+}
Lithographic	101	130^{+}
Magnetostrictive forces	101	DIG. 5
Member	101	368^{+}
Collocating or aligning	33	614^{+}
Electrodeposition	204	6
Localized area electrolytic etching	204	129.1^{+}
Localized area plating	204	17
Numbering machine	101	84^{+}
Selective or progressive	101	109^{+}
Selective or progressive with inker	101	103^{+}
Type setting and distributing machine	276	4
With inker	101	327^{+}
Metal leaf applied	101	DIG. 4
Metering blade position, automatic or remote control of	101	DIG. 26
Multicolor	101	171
Offset		
Bed and cylinder	101	251^{+}
Intaglio	101	154
Multicolor	101	177
Planographic multicolor	101	137
Rotary	101	217
Rotary and bed cylinder	101	215
Rotary planographic	101	142^{+}
Smut prevention	101	416R^{+}
On cloth by photography	430	

	Class	Subclass
Overlays and underlays (See make-ready)		
Paper disc closure making	493	962*
Photographic	355	
Planographic	101	130
Plate alignment on press	101	DIG. 12
Plate driving and impression	101	DIG. 11
Plate etching	156	625
Powderless	156	640
Plate making or etching	156	905*
Press attachment cutting		
Punching	83	
Rotary cutter	83	331^{+}
Projection photographic	355	18^{+}
Quoins	254	40^{+}
Reflex photographic type	430	395
Rollers and track	101	DIG. 1
Rolling contact machines	101	212
Rule cutting	83	927*
Rules	83	
Sheet folding, associating and	270	1.1^{+}
Silk screen	101	114^{+}
Process	101	129
Sorting &	101	2
Special article machines	101	35^{+}
Article controlled operation	101	DIG. 3
Stenciling	101	114^{+}
Stereotype printing plate	409	901*
Stick	276	38^{+}
Collocating	33	614^{+}
Surface	101	453^{+}
Electrodeposition	204	6
Localized area electrolytic etching	204	129.1^{+}
Localized area electroplating	204	17
Photographic preparation	430	300^{+}
Telegraph	178	23R^{+}
Automatic	178	4
Textile process and compositions	8	445^{+}
Typecasting	199	
Typesetting	276	
Typewriter printing point indicator	400	709
Typewriting machines	400	
Web tensioning	101	DIG. 21
Weigher sets type	177	13
Prism		
Building lights	350	258^{+}
Making by moulding	65	
Optical	350	286^{+}
Telescope	350	569
Prison Bathrooms	4	DIG. 15
Privy, Building Unit	D25	16
Probang	128	356
Probes Magnetic	128	1.4
Probes Radioactivity	250	472.1
For use with fluids	250	336.1
Geiger muller tube	313	93
With circuit	250	374^{+}
Miscellaneous	250	336.1
Neutron detector type	376	153^{+}
With circuit	250	390^{+}
Scintillation	250	361R^{+}
Procaine	560	49
In drug	514	535
Process Motion Picture Photography	352	38^{+}
Prod		
Animal	D30	156^{+}
Electric	231	2E
Producer Gas	48	
Boiler combined	122	5
Combustion furnace combined	110	229^{+}
Profile (See Pattern; Pattern Control)		
Cutter for turning	407	
Lathe employing	82	13
Measuring	356	376^{+}
Optical tester or comparer	356	391^{+}
Proflavine	546	104^{+}
Progesterone	260	397.3
In drug	514	177
Program (See Pattern Control)		
Burner	431	18^{+}
Clock alarm	368	246
Electric circuit maker & breaker	200	19A^{+}
Retarded	200	33R^{+}
Feeding devices for animals	119	51.13
Fluid handling	137	624.11^{+}
Furnace, solid fuel	110	191^{+}
Heat exchange	165	12
Heating and cooling	165	12
Liquid purification or separation	210	141^{+}
Machine mechanisms trips	74	3.52
Memorandum holders	368	42

Entry	Class	Subclass
Nuclear reaction combined	376	217
Refrigeration	62	157+
Sewing machine	112	121.11+
Sighting	33	229+
Signaling or indicating	116	22R
Code	116	18
Electric lights, matrix	340	780
Electric lights, monogram	340	760+
Electrical	340	309.4+
Electrical code	340	345+
Electrical traffic control	340	907+
Sprayer	239	69
Valve	137	624.11+
Programmable		
Aircraft control systems	244	76R+
Digital computer system, general	364	200
Digital data processing systems special	364	900
Methods or procedures	364	300
Programmed Control		
Assembling machine	29	701
Boring or drilling machine	408	3
Conveying apparatus	198	349+
Cutting machine	83	71
Furnace, solid fuel	110	191
Heating apparatus	432	51
Metal deforming machine	72	7
Metal-fusion bonding	228	7
Metal-fusion bonding	228	102
Metallurgical apparatus	266	59
Metallurgical apparatus	266	96
Selective punch machine	234	13+
Electrical code	340	345+
Toy vehicle	446	436
Typewriter		
Carriage	400	279+
Case-shift	400	252+
Format	400	76+
Format and type selection	400	61+
Liner	400	17+
Sheet or web feed	400	582+
Type-face selection	400	70+
Web feeder	226	9
Projectiles		
Amusement devices, aerial	273	317+
Design	D22	115+
Explosive containing	102	473
Explosively propelled	102	501+
Fire escape	182	50
Flechettes	244	3.24+
Games		
Aerial	273	317+
Simulated missles	273	313+
Surface	273	108+
Launched from toy vehicle	446	435
Lubricating	102	511
Making	29	1.2
Measuring and testing	73	167
Mechanical devices	124	
Shell or cartridge loading	86	
Stabilization	244	3.1+
Thrown detonating toy	446	400
Projection or Projector (See Thrower)		
Aerial projectile		
Catcher combined	273	318+
Target combined	273	348.1+
Comminution		
Centrifugal projection	241	275
Fluid	241	39+
Electric lamp	313	
Film	353	
Motion picture	352	233
Slide	40	513
Fire escape	182	50
Grenade		
Firearm combined	42	105
Hand	102	483+
Ordnance combined	89	1.1
Gymnastic	272	65+
Illumination		
Electric bulbs	313	
Lantern	362	257+
Protector combined	362	317+
Reflectors	362	341+
Reflectors and refractors combined	362	327
Refractors	362	326+
Lantern slides	40	152+
Mechanical	124	
Microfilm reader	D16	11+
Motion picture	352	

Entry	Class	Subclass
Film winding	242	179+
Screen	350	117
Winding & reeling means	242	179+
Optics of image projection	353	
Photographic printing	355	18+
Photographic viewer	D16	11+
Railway mail delivery	258	5+
Screen	350	117+
Surface projectile	273	129R
Ball game combined	273	119R+
Vehicle fender		
Forward	293	24+
Side	293	21+
Weight reading	177	178
Prolamines	530	372
Compositions containing	106	149+
Zein	530	373
Prong and Pronged Devices (See Impaling; Point)		
Butchering hangers	17	44.3
Electrical connectors	439	
Electric plug	439	
Electronic tube base	439	611+
Lamp base	439	611+
Nails		
Heads	411	439+
Points plural	411	457+
Prontosil	534	676
Prontyllin	260	397.7 R
Proofreading Devices	281	1
Prop (See Brace)	248	351+
Awning outriggers	160	45+
Bicycle	280	293+
Clothesline	248	353
Draft pole	278	87
Flower	47	44+
Ladders	182	165+
Collapsible	182	156+
Hinged extension	182	163
Separable extension	182	22+
Wheeled	182	16
Scaffolds building supported	182	87
Shocker	56	431
Thill support	278	85
Tree	47	43
Trunk lid	190	34
Propeller (See Impeller)	416	
Aircraft combined	244	62+
Stabilizing	244	92
Design	D12	214
Locomotive combined	105	66
Making, by milling	409	120
Power plant X-art	60	904*
Screws	416	176
Ship combined	440	
Buoyant	440	98+
Chain	440	95+
Oscillating	440	13+
Portable	440	53+
Reciprocating	440	13+
Screw	440	49+
Testing	73	147
Toy	446	36+
Airplane combined	446	57+
Propicillin	540	341
Propiolic Acid	562	598
Propionic Acid	562	606
Proportional		
Counter	250	374+
Geiger muller tube	313	93+
Radioactivity	250	336.1
Scintillation	250	361R+
Tube structure	376	153+
Feeding		
Automatic	431	90
Dispenser automatic control	222	57
Dispensing from nonserial traps	222	426+
Dispensing from plural containers	222	134
Dispensing from plural outlets	222	482+
Fluid distribution	137	87+
Gas & air mixer for burner	431	354
Automatic	401	90
Mortar mixer	366	16+
Mortar mixing process	366	8
Water purification	210	101
Measuring		
Flow meter	73	202+
Gauge point markers	33	663+
Proportional dividers	33	150
Proportioner, Rectangular	33	DIG. 9
Propulsion Devices and Systems		
Aircraft	244	62+

Entry	Class	Subclass
Ammunition and explosive devices		
Design	D22	112
Land mines	102	362
Propelling charge combined	102	374+
Pyrotechnic rocket	102	347+
Electric of railway cars	104	288+
Land vehicles		
Convertible	280	7.15+
Runner	280	12.1+
Simulations	280	1.11 R+
Skates	280	11.115
Wheel propelled	280	3
Wheeled	280	200+
Marine	440	
Manual power	440	21+
Motor vehicles	180	
Propellers	416	
Railway		
Cable	104	173.1+
Car	104	287+
Locomotives	105	26.5+
Switches and signals	246	
Reaction motors	60	200.1+
Store service		
Cable propulsion systems	186	14+
Self propelled car systems		
Single impulse systems	186	8+
Prospecting (See Measuring; Testing)	73	151+
Acoustic vibratory or seismic wave (See seismographs and seismometers)		
Chemical anaylsis with		
Apparatus	422	50+
Methods	436	25
Earth boring combined	175	58+
Electrical	324	323+
Radar	342	191+
Neutron using	250	253+
Radiant energy	250	253+
Radioactive using	250	253+
Sound using	181	
X ray using	250	253
Prostatic Resector (See Resector)		
Prosthetic Article	D24	33
Prostoglandin Acids	562	503
Esters	560	121
Protection (See Cover; Fender; Prevention)		
Against earth currents for electric railways	204	
Against radiation damage	376	277
Antiabrasion or protective layer	430	961*
Arm guard	D29	20
Baking food	126	22
Barrel bung	217	114
Bonnets for animals	54	80
Book	281	20
Building	52	
Animal blocker	52	101
Coating	52	515+
Earth quake	52	167
Earth supported coping	52	102
Exterior flashing	52	58+
Marker or monument	52	103
Metal monument markers and guards	52	102+
Pole shell	52	727+
Safes and banks	109	
Snow stop	52	24+
Tubular edger	52	244
Window	52	171
Burner heat protector	431	350
Automatic	431	23
Casings		
For cigars	206	242+
For watch movements	206	18
Clock key with dust guard	81	123
Closure edge	49	462
Collapsible steering post	280	777
Drill rod		
Applying well protector to	29	236
Guide or slide for	175	325
Electric devices		
Anti-inductive structures	174	32+
Arc lamps liquid electrode	313	164
Circuit breaker arc	200	144R+
Conductors and insulators	174	
Connector cord terminal	439	135+
Demagnetizing	361	267+
Discharging static charges	361	212+
Implosion of tubes	445	8

	Class	Subclass
Lightning arrester	361	117$^+$
Lightning arrester and thermal current switch	337	28$^+$
Lightning conductors	174	2$^+$
Lightning rods	174	3
Meter circuit	324	110
Plural terminal insulator rod type	174	179
Railway right of way	246	121
Ray energy shields	250	515.1
Safety systems	361	1$^+$
Shock hazard	174	5R
Signal box key	340	302
Signal system circuit maintenance	340	292
Telegraph system circuit maintenance	178	69R
Telephone	379	412$^+$
Anti-inductive devices and systems	379	416$^+$
Thermal current switch arc	337	273$^+$
Third rail	191	30$^+$
Track connections	238	14.5$^+$
Trolley rope	191	71
Voltage distribution regulator	191	2$^+$
Erosion	405	15$^+$
Eye and face guard	D29	17$^+$
Fiber protecting during fluid treatment	8	133
Color	8	685
Fire escapes with fire protection	182	47
Fluid	182	51
Fly nets for animals	54	81
Garment and body part		
Apparel type	2	2$^+$
Aprons bibs shields etc	2	46$^+$
Design	D 2	225$^+$
Arm guard	D29	20
Back and chest safety	2	92
Design	D29	10$^+$
Baseball gloves	2	19
Design	D29	21
Boxing gloves	2	18
Buttonhole	24	659$^+$
Eye and face	D29	17$^+$
Garment protectors	D 2	225
Goggles	2	426$^+$
Design	D16	102$^+$
Hand	D29	20
Head protectors for permanent waving	132	36.1 R
Headwear	D 2	509$^+$
Helmets	2	6
Design	D29	12$^+$
Leg guard	D29	10$^+$
Leggings and gaiters	36	2R
Design	D 2	267
Leggings stocking type	2	242
Overshoes	36	7.1 A$^+$
Design	D 2	271
Shirt bosom	2	120
Shoe, boot, and legging	36	72R$^+$
Skirt edge	2	222
Slippers inside shoe	36	10
Spats	36	2R
Stocking heel from shoe	36	55
Trouser	2	231$^+$
Trouser guards and straps	24	72
Wringer release	68	256$^+$
Hand guard or glove	D29	20$^+$
Headware & helmets	D29	12$^+$
Heat exchanger	165	134.1
Heating system parts freezing	237	80
High voltage for electric apparatus and conductors	361	1$^+$
Horse		
Boots	54	82
Shoes	168	
Hose against abrasion or bending	138	110
Jacket engine freezing	123	41.1$^+$
Leg guard	D29	10
Light modifiers and combined	362	317$^+$
Lightning	361	117$^+$
Marine structure	405	211$^+$
Metals	422	7
Casting device safety	164	152
Molten metal oxidation prevention	75	96
Object against corrosion by electrical neutralization	204	147$^+$
Electric current apparatus	204	196$^+$
Magnetic apparatus	422	186.2
Oxen		
Shoes	168	5

	Class	Subclass
Padlock	70	54$^+$
Pencil sharpener and point		
Piano case	84	183
Piles	405	211$^+$
End caps	405	255
Pipe ends	138	96R
Pipe pressure compensator with freeze protector	138	27
Pipe thawing and freeze protector	138	32$^+$
Protective coating compositions	252	381$^+$
Removable	427	154$^+$
Protectors	30	460
Railway car heating stove	126	57
Roadway		
Curbs or corners	404	7$^+$
Railway automatic switch traffic protected	246	330
Railway crossings signals and gates	246	293$^+$
Railway drawbridge	246	118$^+$
Railway grade crossing	246	111$^+$
Railway right of way defect	246	120$^+$
Railway switch snow protected	246	428
Railway track fastener boxes etc	238	312
Third rail	191	30$^+$
Traffic barrier	404	9$^+$
Traffic guides	404	6$^+$
Vault cover	404	25$^+$
Safe	109	
Safety power stops	192	129R$^+$
Spinner traveler	57	121
Systems for electric lamps	315	119$^+$
Trolley operating rope	191	71
Trunk harness	190	26
Umbrellas	135	36R
Vehicle and parts		
Aircraft fuselage shields etc	244	121
Bumpers	293	102$^+$
Car stove	126	57
Fenders	293	
Impact of collision	180	232
Mudguards, eg fenders	280	152R$^+$
Railway car	105	394
Tire chain link with	152	245
Vehicle wheel	301	37R
Windshield auxiliary	296	95R$^+$
Windshield deflected air currents	296	91
Watchcase	368	286
Antimagnetic	368	293
Weaving machine stop	139	345$^+$
Wood against decomposition	252	380$^+$
Proteins	530	350$^+$
Albumin, protein	436	88*
Amino acid from	562	516
Fermentative purification of	435	68$^+$
Foods containing	426	72$^+$
Plastic and coating compositions	106	
Polymers containing (See synthetic resin or natural rubber)		
Single cell	435	804*
Proteins, Source Materials	530	820$^+$
Prothrombin Time, pt	436	69*
Proton Detector		
Detector tubes	376	153$^+$
Electric signalling	250	336.1
Non-electric signalling	250	472.1
Scintillation	250	483.1
Protoveratrine	546	34
Protractor	33	1R
Design	D10	65
With straight edge	33	424$^+$
Provision Safes		
Evaporative cooling	312	31.1$^+$
Ventilated	98	51
Pruning Implements	30	
Design	D 8	5
Prussian Blue	423	364$^+$
Prussic Acid	423	364
Pseudocumene (See Aromatic Hydrocarbon)		
Psevdojervine	536	18.1
Psoralen, in Drug	514	455
Psychedelic Lighting	362	811*
Psychology	434	236$^+$
Psychrometer (See Hygrometer)		
Pteridine, in Drug	514	249
Pterins	544	257
Public Address Systems	381	82$^+$
Puddings	426	579
Puddlers Balls Working	29	15
Puff		
Hair puff or pompadour comb	132	106

	Class	Subclass
Iron	68	5A$^+$
Powder applying pocketed	401	200
Design	D28	91
Puffing		
Cereals	426	449$^+$
Apparatus	99	323.4
Metals	75	20R
Minerals	252	378R
Pull (See Push)		
Boot and shoe	36	56
Button or knob	16	121$^+$
Bell	116	172
Drawer pull tag combined	40	325
Switch	200	329$^+$
Chain or rod		
Closure operator	292	
Dynamometer	73	862.38$^+$
Exerciser	272	116$^+$
Pulling devices	254	
Assembly apparatus	29	244
Bootjacks	223	114$^+$
Disassembly apparatus	29	244$^+$
Fuse	81	3.8
Hoist line implements	294	
Tack	12	16
Switches electric	200	161
Tab, container opening	206	633
Pull Offs		
Sewing machine thread	112	242$^+$
Leather sewing	112	58
Pulley (See Wheels)		
Band saw		
Cleaners and lubricators	83	169
Block and tackle	254	390$^+$
Design	D 8	360
Endless belt power transmission	474	
Grip	254	391$^+$
Idle pulley bearing	384	417
Lathe combined machine for	29	29
Lathe for	82	2R
Processes	29	159R$^+$
Shaft joint	403	
Sheave, reeling storage	242	47
Tension	242	155R
Pulp		
Bleaching of	162	1$^+$
Bobbins and spools	242	118.32
Cop type	242	118.7
Dyeing of	8	
Fluid treatment of manipulative	8	156
Making	162	1$^+$
Making comminutor	241	280$^+$
Processes	241	28
Processes application of fluid combined	241	21
Stones	51	206R$^+$
Molding of	162	
Testing freeness of	73	63
Pulpit Lectern	D 6	419
Pulpstones	51	206R$^+$
Pulse		
Amplifier	328	53
Communications	375	
Detection	128	687
Generator	307	106$^+$
By periodic interruption	307	132R
By regulatory interruption	307	96$^+$
Slave oscillator	331	47
Light wave	455	608
Modulation	332	9R
Radio system	375	
Multiplexing	370	
Repeater or relay	370	97
Oscillator	331	
Pulsating current systems		
Supply for electron tube circuits	315	246
Telegraph	178	66.1
Therapeutic	128	421$^+$
X ray systems	378	106
Pulsation devices	138	26$^+$
Brake system	303	87
Filter cleaning by fluid	210	412
Liquid purification process	210	637
As treatment of liquid	210	748
Filter cleaning	210	785
Pump liquid	417	540$^+$
Pulsator		
Milker	137	103$^+$
Combined	119	14.1$^+$
Motor		
Hydraulic, pneumatic	60	533$^+$

	Class	Subclass
Pump drive	417	383+
Radio system	375	
Telegraph, telephone system	375	38+
Repeater or relay systems	375	3
Pulverizer	241	
Furnace, solid fuel combined	110	232
Furnace feeder having	110	106
Pump	417	
Agitating	366	262
In vacuum	366	139
Bath closets & sinks combined	4	
Bearing lubrication	384	322+
Boiler cleaning	122	379
Brake system	303	10
Breast	604	73+
Casting apparatus combined	164	303
Centrifugal	415	120+
Chain	198	643
Cistern cleaner	210	207
Cleaning combined	15	3
Clutch	192	58R
Condenser combined	165	112
Container combined	222	372
Convertible to motor	417	237
Diverse energy input	60	712
Engine starting motor	60	629
Pneumatic transmission	60	407+
Convertible to other type pump		
Cooling of internal combustion engine	123	41.63
Design	D15	7+
Diaphragm, modern pumps	261	DIG. 37
Diffusion	417	152+
Dispensing	222	
Double acting internal combustion engine combined	123	62
Dredgers combined	37	58
Drinking fountain combined	239	24
Elevator combined	187	69
Filtration	210	416.1
Fire extinguishing apparatus combined	169	
Fluid pressure dispensing	222	401
By collapsible wall pump	222	209
Fluid transmission combined		
Internal combustion engine	60	597+
Turbine, with supercharging	60	598+
Pneumatic	60	329
Pressure fluid source & motor	60	325+
Fountain or trough	119	76
Fountain pen	401	143
Fumigator	43	125+
Gas and liquid		
Contact apparatus	261	
Gas dispensing	222	3
Gas separator combined	55	467
Getter	417	48+
Getter-ion	417	49
Grease gun	222	256
Heat exchange fluid circulating	165	121
Circulating for refrigeration	62	
Heat exchanger combined	165	120
Heating and illuminating gas mixer delivery	48	186
Hydrocephalus	604	8+
Infusion of body treating materials	604	151+
Intensifier	417	225+
Internal combustion engine combined	123	
Air or gas	415	
Internal combustion engine and fluid pressure motor	60	597+
Jack	74	41
Suction	417	
Jet type	417	157+
Jet type boiler tube blower	15	316R+
Jet type in vacuum cleaner	15	408
Lubricating system	184	26
Magneto-hydrodynamic	417	49
Making	29	156.4 R
Manual	D23	231
Medicator combined	604	151+
Milker	119	14.27
Minnow bucket aerator	43	57
Motor and	60	325
Punch card or computer control	60	368
Peristalic	417	474+
Plastic filament forming	425	66
Plastic filament forming	425	67+
Plastic filament forming	425	76+
Plastic filament forming	425	382.2
Plural pump dispenser	222	255
Pneumatic conveyers	406	96+
Pneumatic locomotive	105	63
Pneumatic piano	84	24
Pneumatic stackers	406	
Pressure fluid source & motor	60	325+
Puff, machine for stitching	112	121.24
Purification units	210	258
Regeneration	210	191
Receptacle filling	141	
Reed organ	84	355
Refrigeration with gas & liquid contact	62	314
Rotary expansible chambner	418	
Rotary kinetic fluid	415	
Service station fuel	D15	9.1+
Sonic	417	240+
Sprayer combined	239	302
Sprinkler system combined	239	
Sputter-ion	417	49
Steam heating system, air venting	237	68
Suction	415	
Tank for heat transfer with testing	73	168
Textile fluid treating apparatus	68	3R
Tire inflation pump	152	415
Two cycle internal combustion engine combined	280	201
Vacuum, rotary expansible chamber devices	418	
Vehicle carried sprinkler or sprayer	239	146
Vehicle combined	280	201
Ventilating in general	98	
Ventilation for baths closets or sinks	4	409
Liquid brake	188	266+
Water heating circulation	237	63
Well structure combined	166	105
Wrench combined	81	2
Punch and Punching	83	
Bag	272	76+
Board	273	139
Filling	53	235+
Making	493	374
Bolt or rivet heading	10	26
Book binder sheets	402	1
Opening reinforcement	402	79
Buttonhole	30	119
Can opener type	30	443
Closure type punch	220	278
Card controlled calculator	235	434+
Cards	235	487
In store service	186	56
Center or prick	30	366
Sheets or bars	30	366
Cutting combined	83	
Design of machines for	D15	128
Eyelet setting with	227	51+
Forging and punching machines	29	34R
Fruit seed removing	99	559+
Hole design	D 8	51
Horseshoe making combined with	59	36+
Implements	30	358
Machines	83	
Gauge work locator	83	268+
Gauge work locator	83	391+
Gauge work locator	83	467R+
Metal dimpling and	72	412+
Paper punch	D19	72
Pattern controlled	83	71
Gang	234	64
Selective	234	59+
Product manufacturing computerized	364	475
Punch making	76	107R
Rasp punching	178	66.1
Refrigerant container opener	62	294
Safety guard machine stop	192	134
Screw head nicking	10	7
Selective	234	
Interposer	234	97+
Sheet metal pressing and	72	324+
Stoneworking	125	23R+
Surgical cutting, tracheotome	128	305.3
Tire punching	29	22
Tufting sewing machine	112	80.1+
Wood cutters	144	196+
Wood disc cutting	408	204
Wood match making	144	50+
Workmens time keeper clock	346	49+
Puncture Closing or Healing		
Compositions for	106	33
Container	220	278
Heat exchange compositions combined	252	72
Repairing pipes	138	97+
Stock material	428	912*
Tires pneumatic	152	502+
Patches	152	367+
Processes of making	156	95+
Puncturing Devices (See Punch)		
Can opener type	30	443
Dispenser combined	222	80+
Fountain pen	401	132+
Penetrating, perforating	83	
Perforating & inking eg tattooer	81	9.2
Receptacle closing and	220	278
Sheet retainer type	402	25
Surgical instrument	128	329R
Punts	114	343+
Pupin Coil	336	
With condenser	333	
With long line	178	45+
Puppets	446	363
Hand	446	327+
Purger, Refrigerant	62	474+
Automatic	62	195
Burner	431	29+
Burner	431	121+
Process	62	85
Purging Sugar Crystals	127	18+
Apparatus	127	20
Purification, Proteins	530	412+
Purification of Synthetic Resin or Natural Rubber	528	480+
Purifiers & Purifying (See Bleaching; Cleaner; Dehydration; Disinfection; Distillation; Evaporator; Extracting; Fermenting; Recovery; Sterilizing)		
Beneficiating ores	75	1R+
Apparatus	266	168+
Gas	55	
Hydrocarbon purification	585	800+
Liquid	210	
Used oil	208	179+
Purines	544	264+
In drug	514	261
Purl Knit Fabrics	66	197
Purling Sewing Machine Stitch	112	161
Purpurpin	260	383
Purses	150	100+
Coin	150	150+
Design	D 3	42
French purse	D 3	42+
Latch	D 8	331+
Push (See Pull)		
Button		
Electric switch	200	341+
Fluorescent	250	465.1
Mechanical bell	116	172
Carts	280	47.26+
Cutter	30	
Plate	16	1R
Pushing devices	254	
Rods or bars		
Closure operator	292	
Side track car	213	224
Push Pull Amplifiers	330	118+
Cascade type	330	71+
With distributed parameter coupling	330	55
With feedback	330	81+
Pusher (See Conveyor, Reciprocating)		
Blocks for saw logs	83	710+
Car pushing implement	254	35+
Check controlled article delivery		
Check connector	194	293+
Gravity forwardly turning part	194	262+
Gravity reciprocating part	194	260+
Thrust type	194	249+
Feed		
Check controlled article delivery	194	
Circular wood saw	83	707+
Combined with other types of conveyer sections	198	570+
Conveyer	198	717+
Drying kiln material	34	206
Furnace fuel	110	109
Label pasting strip server and cutter	156	510
Saw log	83	710+
Load discharging		
Fork tine or shovel clearers	294	50+
Furnace	414	214
Self loading vehicle	414	511
Vehicle, external unloader	414	395+
Locomotive track pusher	105	31

	Class	Subclass
Plastic block press portable mold	425	453+
Railway car propulsion system	104	162+
Roadway snow plow V type	37	272+
Sled	280	23
Stock to be machined	414	14+
Truckmans car drive	105	90.1+
Putlogs or Spacer	182	229
Ladder combine	182	214
Scaffold wall embedded	182	87
Putting on		
Garments	223	111+
Coat accessory	2	101
Garment supporters	2	300
Putty		
Devices	425	458
Including putty supply	425	87
Removing		
Attachment, moving	15	256.5
Implement scraper	15	236R+
Puzzle	273	153R+
Design	D21	104
Jigsaw type	428	33
Locks	70	289
Pvc-isocyanate Resins		
Molding cellular	264	DIG. 3
Pvdf	310	800*
Pycnometers	73	32R
Pyramidon	548	366
Pyrans	549	356+
Pyranthrones	260	360
Pyrazines	544	336+
Pyrazoles	548	373+
Acridine nucleus containing	546	26
Azo compounds	534	751+
Heavy metal containing	534	710+
Barbiturate addition compounds	544	300
Pyrazolone (See Pyrazoles)		
Pyrene (See Aromatic Hydrocarbon)	585	400+
Pyrethrum	560	124
Insecticides containing	514	65+
Pyridine	546	348+
Azo compounds	534	770+
Heavy metal containing	546	2+
Pyridium	534	605+
Pyridoanthraquinones	546	78
Pyridone	546	290
Pyridoxine	546	301
Pyrimidines	544	242+
Pyrocatechin	568	763+
Pyroelectric Lamps		
Structure	313	14
System of supply	315	359
Electric heater for	315	115+
Plural loads	315	114
Pyrogallol	568	763+
Pyrographic		
Burning tools	126	401
Electric	219	227+
Fuel	368	248+
Electric tools	219	227+
Hand tool spark	219	384
Machine spark	219	68+
Pens	219	229
Recorders	346	76R
Telegraph receiver	178	94
Textile finishing	26	3+
Pyroligineous Acid	562	607
Production of	201	
Pyrolysis (See Cracking)		
Acetylene production by	585	534+
Aromatic hydrocarbons by	585	476
Carbon preparation by	423	449+
Natural resins	260	106
Pyromagnetoelectric Devices	310	306+
Pyrometallurgy	75	20R+
Pyrometers	374	100+
Optical	356	43+
Radiation	374	121+
Thermoelectric	136	213+
Relay temperature regulation	236	69+
Pyromucic Acid	549	484
Pyrones	549	416+
Pyronine	549	388+
Pyrophoric Igniters	431	273+
Alloys	420	416
Correlated with burner feed	431	254
Tobacco product combined	131	351
With burner cover	431	146+
Correlated actuation	431	129+
Pyrophosphate Esters	558	152
Formation of	558	127+

	Class	Subclass
Pyrotechnic		
Compositions	149	37+
Devices	102	335+
Flare	D22	112
Illusion amusement devices	272	20
Signals	102	335+
Detonating burglar alarms	116	83
Door fastener	116	11
Railroad torpedoes	246	486
Pyroxylin	536	35+
Apparatus	425	90+
Compositions	106	169+
Pyrrocolines	546	73
Pyrrolidine	548	579
Pyrroline	548	565
Pyrrols	548	400+

Q

	Class	Subclass
Quadrant		
Angle measurement		
Horizontal and vertical angles	33	281
Horizontal angle	33	285
Vertical angle	33	282+
Spinning mule	57	322+
Quadrasonic	381	19+
Quadruplex Telegraphy	370	34+
Quarrying	299	15
Quarter Wave Transmission Line	333	27
Quarters Shoe Blank	36	48+
Quartz (See Piezocrystals)		
Glass manufacturing	65	
Compositions	501	27+
Lamps	313	112
Mercury vapor	313	112
Quaternary Ammonium	564	281
Acyclic	564	291+
Aromatic	564	282+
Azo compound ctg.	534	603+
Quebracho Extract	560	68+
Descaling agent containing	252	83+
Tanning with	8	94.32
Aluminum compound containing	8	94.3
Queen Bee Cells and Cages	6	9
Quenchers and Quenching		
Coating cooling	427	398.3
Coke quenchers	239	750+
Distillation residues	201	39
Autothermic systems	202	95
Thermolytic apparatus	202	227+
Metal heat treatment proceses	148	143+
Metal heating and quenching	266	114+
Apparatus	266	121+
Metal quenching apparatus	266	114+
Still bottom closures	202	253
Question and Answer	434	322+
Correct answer indicated	434	327+
Electrical recording means	434	362
Examination card or sheet	434	363
Grading of response form	434	353+
Responses electrically communicate to monitor or recorder	434	350+
Quick Acting		
Fluid pressure brake systems		
Automatic synchronizing	303	37+
Motor releasing	303	69
Guns	89	125+
Mechanical circuit makers and breakers	200	154
Portable wood working clamps		
Parallel screw	269	140+
U beam	269	
Quick setting portable rail drills	408	72R+
Quick traverse bobbin and cop	242	31
Cone	242	31
Winding	242	43R
Quick wheel release trolley heads	191	63.1
Quick Detachable		
Electric connectors	439	180
Plow elements	172	749+
Shaft coupling	464	901*
Quicksilver (See Mercury)		
Quillstock Spinning	57	4
Quilting	112	117+
Frames with cloth winding means	242	67.1 R+
Quilts	5	482+
Design	D 6	603+
Bedding per se	D 6	595
Quinaldine	546	181
Quinazolines	544	283+
Quinhydrones	260	396R
Quinic Acid	562	508
Quinicine, in Drug	514	314

	Class	Subclass
Quinidine, in Drug	514	305
Quinine	546	134
In drug	514	305
Quinizarin	260	383
Quinoline	546	152+
Azo compounds	534	770+
Quinolinium Compounds	546	182
Quinolizines	546	138
Quinones	260	396R
Anthra	260	365+
Quinonimines	260	396R
Quinophthalone	546	173
Quinoxalines	544	353+
In drug	514	249
Quinuclidines	546	133
Quiver	D 3	36+
Quoins	254	40+
Quonset Hut	52	86

R

	Class	Subclass
R Acid	260	512R
Race		
Bearing		
Assembly apparatus	29	724+
Ball	384	513+
Divided	384	199+
Fastening means	384	541
Manufacturing process	29	148.4 R+
Roller	384	569+
Take up	384	500
Thrust	384	622
Starting barriers	119	15.5 A
Track	272	4+
Water wheels	405	118+
Raceway		
Bearing raceway grinding	51	291
Grinding bearings in	51	130
Sewing machine shuttle	112	196
Racing		
Amusement railway	104	60
Arena	272	4+
Simulated games	273	86R
Rack (See Ratchet; Support)	211	
Adjustable	211	175
Animal husbandry	119	20
Feeding	119	60
Article (See also supports for specific articles)	211	
Ball billiard table attached	273	10+
Battery	429	96+
Card	40	124
Check label and tag	40	19.5
Clothes bed attached	5	504
Gun	211	64
Hat	211	30+
Hose or nozzle	248	89+
Ironing table combined		
Life preserver	114	190
Music piano attached	84	180
Pen or pencil	211	69.1+
Portable	294	143
Radiator	165	67+
Cabinet combined	312	
Canning		
Food cooking	99	359+
Receptacle	211	71+
Receptacle stationary	248	146+
Collapsible	211	195+
Disinfecting, preserving	422	297
Knockdown framework	211	189+
Mechanism	74	422
Oven	126	337R+
Pinion and (See device in which used)	74	422
Mechanical movements	74	20+
Intermittent grip type	74	111+
Pushing and pulling mechanism	254	95+
Rail	238	123
Elevator	187	19
Locomotives	105	29.1+
Switches	104	132
Wheel	295	4
Releasable keeper for sheet retainer	402	60+
Showcase combined	312	128+
Sorting-type	211	10
Spool	211	59.1
Stackable	211	194
Terraced trays	206	211
Vehicle body	296	3+
Racket or Racquet	273	67R+
Croquet	273	83
Design	D21	212

	Class	Subclass
Restringing tension meter	73	862.43
Tennis	273	73R
Racking Hoops	147	44
Radar	342	147+
Design	D10	104
Digital computer combined system	342	195
Light ray radar	356	3+
Simulators	434	2+
Sound or supersonic	367	87+
Radiant Energy (See Heat; Illuminated; Music; Optics; Photography; Radio; Sound; Space Induction; Solar; Television)	250	
Barrier	250	505.1
Detection (See radio)		
Computer controlled, monitored	364	414
Infrared	250	338
Invisible light	250	336.1
Neutrons	250	390+
Nuclear reactions control	376	207
Photoelectric	250	200+
X ray	250	336.1
Devices, railway switches	246	DIG. 1
Electric heater	219	339+
Imagery chemistry (See X-art list)	430	
Binder ctg	430	905*
Binder-free emulsion	430	932*
Initiator ctg	430	913*
Radiation-activated cross linking agent ctg	430	927*
Radiation-chromic compound	430	962*
Spectral sensitizer ctg	430	926*
Material treatment		
Article or object apparatus	250	492.1
Chemical and other apparatus	422	186+
Chemical apparatus	204	193
Coating	427	12+
Coating apparatus	118	620+
Coating with fibers or	427	58+
Particulate material	427	12+
Drying processes etc radiant heating	34	39+
Drying processes etc selected range	34	4
Dyeing apparatus	8	103
Dyeing processes	8	444
Electric discharge apparatus	422	186.21+
Electrical energy drying and gas or vapor contact with solids	34	1
Electrons	250	306+
Fluent material apparatus	250	428+
Food and beverage apparatus	99	451
Refrigeration combined	62	264
Solar drying process and gas or vapor contact with solids	34	93
Sterilizing and pasteurizing processes	426	521+
Sterilizing and pasteurizing processes	426	237+
Tobacco treatment	131	299
Vitamin activation processes	426	72
Medical and surgical applications	128	362+
Application to body	604	20
Bandaging	128	82.1
Computer controlled, monitored	364	414
Injection means	604	21
Instruments	128	303.1+
Kinesitherapy eg massage	128	24.1+
Magnetic	128	1.3+
Medicators	604	20+
Nuclear reaction	376	100
Nuclear reaction	376	156
Orthopedics	128	68.1
Processes	376	156
Nuclear reaction produced energy	376	
Radiation measurer	D10	47
Computer controlled, monitored	364	414
Radiation resistant hydrocarbon composition	585	944*
Responsive		
Battery	136	243
Cathode ray tube	313	365+
Cathode ray tube circuit	315	10+
Circuits miscellaneous	250	200+
Closure operator	49	25
Explosive igniters fuses	102	213
Explosive igniters marine mines	102	416+
Explosive igniters mines	102	427
Photocell electronic tube type	313	523+
Resistor	338	15+
To invisible light X ray etc	250	336.1

	Class	Subclass
Secondary emission		
Cathode ray tube circuit	315	11
Cathode ray tube plural cathodes	313	375
Cathode ray tube with photo sensitive electrode	313	387
Electronic tube	313	103R+
Secondary radiation minimizing		
X ray screen	378	154
X ray tube	378	140
Shielding		
Against invisible rays	250	515.1
Nuclear reactor combined	376	287
Nuclear reactor combined	376	347
Simulation of radiation	434	218
Thermometers	374	121
Toaster	D 7	324+
Radiation Imagery--chemical	430	
Achromatic image from chromatic	430	356
By sound or nondigital compressive forces	430	3
Color imaging process	430	351+
Control feature responsive to test or measurement	430	30
Diazo reproduction	430	141+
Effecting frontal radiation during exposure, eg screening etc	430	396
Electric or magnetic imagery	430	31+
Exposure step or specified pre-exposure step perfecting exposure	430	494
Imaged product chemically defined	430	9+
Imaging affecting physical property of radiation sensitive material, or producing non-planar or printing surface	430	269
Liquid crystal	430	20
Microcapsule	430	138
Micrography	430	8
Non-radiation sensitive image	430	449+
Plural exposure steps	430	394
Post imaging process	430	404
Producing cathode ray tube element	430	23
Product having sound record or process of making	430	140
Radiation sensitive product and composition	430	495+
Regenerating image processing composition	430	398
Registration or layout process	430	22
Silver halide colloid tanning	430	264+
Stripping process or element	430	256+
Thermographic process	430	348+
Transfer procedures between image layer & image receiving layer	430	199
Using reflected radiation	430	395
Visible imaging	430	198
Firing or sintering step	430	198
Radiation only, other than heating by surface contact or convection	430	346
Radiator (See Heat)		
Automobile radiator and light	362	79
Automobile radiator leak testing	73	49.7
Automobile type	165	148+
Design	D12	163+
Brushes	15	159R+
Cap ornaments	428	31
Thermometer type	374	146
Caps	220	200+
Covers		
Automobile	165	98
Incased	165	129
Shields	237	79
Design	D23	330+
Electrical	219	339+
Radiating panel	219	345
Radiating plate	219	345
Fluid heated radiators	237	70+
Gas burning	126	91R
Incandescent fire grate	126	92R
Heat exchangers	165	148+
Heating systems having	237	
Making	29	157.3 R+
Assembling	29	726+
Headers and tubes	29	157.4
Work stands	269	
Relief valves and vents	137	455+
Temperature controlled	236	61+
Sectional	165	130
Supported foot rests	248	218.4+
Supports		

	Class	Subclass
Bases	248	348
Brackets	248	232+
Window mounting brackets	248	209
Work holders	269	
Temperature controlled	236	22
Cooling radiator	236	34+
Heating radiator	236	36+
Vehicle, design	D12	166+
Vehicle mountings	180	68.4+
Aircraft	244	57
Viscosity controlling	165	35+
Wall panel heating and cooling	165	49
Radio		
Aircraft control	244	75R
Aircraft control automatic	244	175+
Amplifiers	330	
Automatic volume control	455	234+
Amplifier stage only with bias control	330	129+
Amplifier stage only with impedance control automatic	330	144+
Amplifier stage only with thermal impedance control automatic	330	143
Band changing	334	47+
Block signal systems hertzian wave	246	30
Block signal systems no line wire	246	38
Block signal systems no line wire	246	44+
Cabinets	312	7.1
Design	D14	18+
With antenna	343	702
Chassis	455	347+
Housed	174	50+
Prewired sub assembly	361	422
Clock combined	D14	20
Condenser	361	271+
Adjustable	361	277+
Electrolytic	361	433+
Console with phonograph	D14	14+
Control of devices		
Aircraft	244	175+
Camera shutter	354	131
Frequency responsive	340	825.72
Motors	318	16
Pulse responsive	340	825.69
Transmitter for	340	696
Detectors	329	
Direction		
Finder receiver	342	417+
Directive	342	350+
Reflected wave systems, ie radar	342	
Electronic tube structure	313	238+
Headphones	455	350+
Indicators	116	241+
Magnetic induction transmission to vehicle	191	10
Miscellaneous	455	
Oscillators	331	
Phonograph combined	369	6+
Pocket	455	351
Transmitter	455	100
Power transmission		
Electric motor	318	16
Miscellaneous	307	149
Vehicle	191	10
Program indicator	40	901*
Pulse or digital communications	375	
Receiver	375	75+
Transmitter	375	59+
Radio and phonograph	369	6+
Receivers	455	130+
Crystal detector	329	205R
Design	D14	68+
Pulse or digital	375	
Regenerative	455	336
Stereo	381	2+
Superheterodyne	455	313+
Superregenerative	455	336
Testing of	455	226
Tuned radio frequency	455	150+
Remote control by radio waves	340	825.72+
Pulse responsive	340	825.69
Remote control of receiver	455	352+
Tuning	455	151+
Remote control of transmitter	455	92
Remote signal course control	244	189
Responsive explosive igniters		
Ammunition & explosive devices	102	214
Mines	102	427
Mines marine type	102	418
Second detector	329	
Shielded	330	68

	Class	Subclass
Static eliminator	455	296+
Supersonic remote control for	455	352
Teledynamic		
Motor control	318	16
Telemetric	340	870.28+
Telegraph space induction	178	43
Onto signal line	178	49
Telemetry	128	903*
Telephone superaudible composite system	379	64+
Torpedo control	114	21.1
Train dispatching inductive telegraphy or telephony	246	8
Transmitter	455	91+
Tubes	313	
Envelope	313	248
Manufacture	445	
Manufacture and repair	445	
Part of tube	313	313
Shield	174	35TS
Testing	324	405+
Tuners	334	
Band spreading	334	
Capacitors variable	361	277+
Dial operator	74	10R+
Electric motor operated	334	20+
Electric wave filters	333	167+
Ganged	334	
Limit stop mechanism	192	138+
Operator only for tuner	74	10.45
Push button	74	10.1+
Using distributed impedance only	333	219+
Wave hf transmission	333	
Wave meter	250	250
With gain control by		
Automatic impedance control	330	144+
Automatic thermal impedance control	330	143
Bias control	330	129+
Radioactive & Radioactivity	250	493.1+
Actinide series metals		
Alloys	420	1+
Compositions	252	636+
Electrolysis	204	1.5
Heat treatment	148	132
Inorganic compounds	423	249+
Metallurgy	75	84.1 R
Organic compounds	534	11+
Alloys	420	1+
Battery	310	301+
Carbocyclic or acyclic compounds	534	10+
Nitro	568	924+
Compositions	252	625+
Molding	264	21
Contamination detector	250	336.1
Demonstrator	250	336.1
Cloud chamber	250	
Electrically	250	336.1
Fluorescent	250	458.1
Drug, bio-affecting & body treating compositions	424	1.1+
Electrical & wave energy, chemistry	204	1.5
Energy measurement	250	336.1
Material	52	
Apparatus	422	903*
Compositions	252	625
Containing electrically conductive or emissive compounds	252	517
For influencing fluent	250	493.1
Material incinerator	110	237
Production	376	156
Treating	264	
Geiger type system	250	374+
Measurement of	436	178+
Medicine	424	1.1
Metallurgy alloys cerium and	420	1+
Nuclear reactions & systems	376	
Neutron type system	250	390+
Photography	250	475.2
Producing in material	376	156
Scintillation	250	361R+
Plastic shaping of radioactive material	264	.5
Powder metallurgy	419	
Pyrometallurgy rare and refractory metals	75	84
Radioactive isotope of another element	423	249+
Rare or radioactive metallurgy articles	75	84+
Technetium radioactive metal	432	2+
Organic compounds	534	14
Test method containing	424	1.1
Antigen-antibody	436	500+
Therapeutic application	128	11+
Treating mixture to obtain	423	2+
Waste disposal storing	252	626+
X-ray or neutralized shield	252	478
Radiochemical (See Radiant Energy)		
Apparatus chemical change	204	193
Compositions radioactive	252	625+
Food treatment		
Preservation	426	237+
Vitamin activation	426	72+
Medicine radioactive	424	1.1
Nuclear reactions	376	
Sound recording and reproduction	369	100+
Tobacco treatment	131	299
Radiography	250	
Radioimmunoassay	436	545*
Radioisotope Powered Generator	310	301+
Radioisotopic Iodine for Testing	436	804*
Radiometers		
Light meters	356	213+
Pyrometers	356	43+
Radio wave meter	250	250
Rotating vane type	356	216
Thermometers	374	121
Ultraviolet	250	372+
Radiosonde	340	345+
System	340	870.1+
Radiotelegraphy	375	
Earth transmission systems	375	6
Radiotelephony	455	
Earth transmission systems	455	40
Radium (See Radioactive)		
Radius & Spiral Measurement	33	1SP
Radius Rod		
Brake	188	191
Spring braces	267	66+
Rafter	52	92+
Clips and	33	DIG. 16
Rafts and Rafting		
Life rafts	441	35+
Rafting and booming	441	48+
Rail (See Railway)		
Airplane	104	23.1+
Amusement	104	53+
Anchors	238	315+
Balloon	104	22
Bed	5	
Camp	5	117
Elements and details	5	286+
Extension	5	184
Folding	5	177
Rotating	5	302
Benders	72	210
Portable	72	210
Billiard table	273	8+
Bonds for electric railways	238	14.5+
By electric weld	219	53+
Manufacture and installation	228	4.1+
Manufacture and installation	228	101+
Cable rails	104	112+
Car propulsion systems	104	287+
Car replacer	104	262+
Car stops	104	249
Car yards	104	26.1+
Chalk and eraser	434	417
Circuit		
Power	191	
Signal	246	
Cleaner	15	54+
Clearer	104	279+
Snow	37	198+
Clip	238	378
Derail guards	104	242+
Drills, portable	408	
Elevators	104	127+
Fasteners	238	310+
Fence	256	59+
Fissure detector	73	146
By abrasion	73	8
Magnetic	324	217+
Grinders	51	178
Guard		
Railway surface track	238	17+
Track	256	14+
Hose bridges	104	275+
Joints	238	151+
Land vehicle		
Dashboard	296	71
Top	296	123
Manufacture	72	
Curving	72	210
Punching	83	
Monorail	104	118+
Moving train	104	18+
Razor		
Form	30	83
Guard	30	80
Seat-guard control	104	241
Seats	238	264+
Shapes	238	122+
Shifters	254	43+
Ship belaying pin	114	218
Sledways	104	134+
Slotted conduit	104	140+
Suspended	104	89+
Swimming pool grab rail	D25	41
Switches	104	130+
Terminals & stations	104	27+
Textile making apparatus	57	136+
Toy railroad	104	DIG. 1
Track clearers	104	279+
Track layers	104	2+
Trackmans-car drive	105	86+
Traction	104	165+
Trains	105	1.4+
Trolley rails	104	106+
Trolley transfer	104	96+
Truck changers	104	32.1+
Trucks	105	157.1+
Tunnel ventilation	98	49
Turn tables	104	35+
Railways (See Rail)	104	
Antileak joints	105	424
Brakes	188	33+
Cabs	105	456
Car design	D12	39+
Car framing and structure	105	396+
Car locks for elevators	187	97
Chair making	29	16
Draft appliances	213	
Electric vehicle, streetcar type	D12	36+
Appliance	D13	14
Floors	105	422
Geometrical instruments	33	1Q
Graders	37	104+
Linings	105	423
Locomotives	105	26.5+
Cab ventillation	98	3
Car ventilation	98	4+
Dust guards	98	28
Mail delivery	258	
Marine	405	2
Monorail	105	141+
Motor placement	105	133+
Pit scale	177	134
Platforms	105	425+
Rail or tie shifters	254	43
Rolling stock	105	
Design	D12	36+
Sounding toy	446	410
Safety bridges	105	458+
Ship	405	2
Side guards	105	439+
Signals & switches	246	
Block system	246	20+
Snow excavators	37	198+
Sound deadeners	105	452+
Special car bodies	105	238.1+
Steps	105	443+
Steps with gate	105	437+
Surface track	238	
Electrical connections	238	14.1+
Fence or guard rails	238	17+
Juxtapose and bonding rails	228	101+
Leveling and spacing gauge	33	338
Rail bond making	228	4.1+
Spike pullers	254	18
Suspended	105	148+
Switches and shifters	246	
Tie clamp	D12	51
Toy railroad	104	DIG. 1
Toy simulation	446	444+
Design	D21	129
Electric	104	288+
Wheels and axles	295	
Axle making	72	
Wheel making	29	168
Wheel or axle drive	105	96+
Rain		
Gauges	73	171

Entry	Class	Subclass
Producers	239	14.1
Raincoats	2	87
Raisers (See Hoist)		
Corpse in coffin	27	12
Dough		
Baking powder	426	562+
Heater	126	281+
Jacks	254	
Track	104	7.1+
Wick	431	304+
Wick	431	315+
Raising Ships	114	44+
Rakes		
Agricultural implements	56	
Blanks and processes for making	76	111
Bundling combined	56	341+
Cutter and ground rake	56	193
Cutter with conveying rake	56	158+
Cutter with detachable rake	56	4
Design	D 8	13
Dies for making	72	470+
Grain separator	130	
Hand	56	400.1+
Horse drawn	56	375+
Loading combined	56	344+
Potato digger conveyor	171	13
Side delivery	56	376
Flume screens	210	154+
Furnace grate	110	285
Hand fork combined	294	52
Making	83	908*
Pan type evaporator	159	33
Scoop type	37	120+
Stove grate	126	173
Vegetable or meat comminutor feeder	241	
Vehicle fender	293	53
Rammers (See Tamper)		
Ammunition loading	86	29+
Earth compacting	404	133
Road construction type	404	133
Foundry mold press type compactor	164	207+
Plural rammers	164	172+
Machine for concrete	425	425
Machine for concrete	425	456
Ordnance loading	89	47
Road and pavement	404	133
Ramp, Loading (See Gangway)		
Ramrod		
Firearm	42	90
Rams		
Hydraulic	417	226+
Ice breaker	114	41
Warship	114	2+
Rand for Shoe	36	78
Making machine	12	67R+
Random Control		
Cutting machine	83	58+
Selective	234	22+
Range		
Shooting	273	348.1+
Range Finders		
By straight line light ray	33	284+
Camera focusing, automatic	354	400+
Radar	342	147+
Distance determining	342	118
Sound wave type	181	
Submarine type	367	99+
Using wave radiation	356	4+
Ranges	126	1R+
Electric		
Heating and cooking	219	392
Ranque Tube	62	5
Rare Earths	156	DIG. 63
Organic compounds	534	15+
Rare Gases (See Gas, Rare)		
Raschig Rings	261	94
Surface contact	261	95
Rasp	29	78+
Making	76	13
Wood	29	78+
Rat		
Guards for mooring lines	114	221R
Hair grooming device	132	55
Trap	43	58+
Ratchet		
Drills	408	120+
Combined	408	120+
Pawl and	74	575+
Bit stock	81	28+
Intermittent grip movements	74	111+
Wrenches	81	60+
Rate of Climb		
Gyro controlled	73	504
Pressure type	73	179
Ratine		
Twisted strands	57	210+
Yarn making	57	3+
Apparatus	57	3+
Rattan Seat	D 6	369
Rattles		
Toy	446	419
Design	D21	65
Vehicle shakers	73	669+
Raunitidine	546	48
Ravelers	28	171
Ray (See Radiant Energy)		
Applications and devices	250	
Cathode ray		
Circuit for tube	315	1+
Structure of tube	313	364+
Television receiver	358	242+
Television scanning	358	199+
Television system	358	83+
Television transmitter	358	217+
Light ray geometrical instrument	33	227+
Light ray projector type game	273	310+
Photocell special ray sensitive	313	523+
Television optical ray	358	209+
Rayon (See Artificial, Silk)		
Razor	30	32+
Blade	30	346.5+
Abrading machines for	51	34BR
Rotary tool, one way traverse	51	80B
Blade design	D28	45+
Manufacturing	76	DIG. 8
Package	206	352+
Sharpeners	76	DIG. 9
Cleaner	15	218
Combined with razor	30	41
Design	D28	45+
Dry shaver	30	43
Electric	30	45
Holder	D 6	526
Hones for	51	211H
Illuminated	362	115
Package or container for	206	208
Design	D 9	342
Kit with razor	206	228
Special package for powered razor	206	351
Safety	30	51+
Sharpeners and straps	30	35+
Design	D 8	92
Sharpening	51	
Processes	51	285
Razor combined	30	35+
Strops	76	81+
Strop dressings	51	DIG. 25
Reaches		
Land vehicle	280	140+
Crossed	280	102
Pivoted and sliding	280	101
Reaction Motor (See Jet, Motor)	60	200.1+
Discharge nozzle	239	265.11+
Reactivation (See Revivification)		
Reactive Power Control	323	205
Reactors	336	
Lamp or electronic tube system having	315	289+
Nuclear	376	
Reader		
Cabinet structures	312	
Coded record & reader, invisible radient energy type	250	556
Document verification or graph reader	250	271
Reader-printer Machines	D16	28
Data processing equipment	D14	111
Reading		
Aids		
Blind or sightless user	434	113+
Bookholder	248	441.1+
Bookmark	116	234+
Copyholder (See copyholder)		
Line indicator	116	240
Magnifier	350	245+
Forms	283	
Teaching	434	178+
Rapid reading	434	179+
Reamers	408	199+
Cleaning implement	15	104.2
Cutlery blade type	30	352
Design	D15	138
Earth boring	175	406+
Expansible	175	263+
Grinding	51	288
Rotary fruit juicer	99	495+
Screw threading, reaming & tapping	408	118
Smoking pipe bowl	131	246
Wood	408	227+
Reapers	56	
Rearrangement (See Isomerization)		
Reatomizers	261	DIG. 55
Recapping	156	95+
Receipts		
Transportation cash fare	283	30+
Severable part	283	100+
Receivers (See Bin; Receptacles)		
Bicycle stand type	224	32R+
Closure removing tool with	81	3.8
Conveyor and chute with hopper	198	540+
Discharge to chute	198	550.1+
Dispenser combined	141	
Dispenser control by weight in	222	56
Dry closet		
Movable magazine	4	467+
Urinal separate	4	463
Ejected bobbin	139	255
Facsimile		
Natural color	358	75+
Optical	358	296+
Fare box	232	12
Firearm	42	16+
For drippings from inverted container	141	364
Hand rake material	56	400.11+
Heterodyne, superheterodyne	455	313+
Light wave	455	600+
Pulse or digital	375	
Pulse shift keying	375	52
Railway mail delivery	258	20+
Support combined	258	11+
Recorder record	346	22
Refrigerant	62	509
Sheet pack	271	207+
Smoking device with ash receiver	131	174+
Stereo	381	2+
Surgical receptors	604	358+
Systems for wireless	455	130+
Telecommunications	455	130+
Telegraph	178	118+
Telephone		
Handset	379	433
Switch in terminal	379	422+
Television		
With sound or message	358	142+
Tobacco users appliance with ash receiver	131	231+
Type casting matrix receiver	199	29+
Vehicle fender	293	39
Wind motor	415	9+
Receptacle (See Bag; Bottle; Bowl; Box; Case; Container; Cup; Dish; Holder; Pail; Tray; Vessel)		
Acids	206	524.1+
Agitating type	366	341
Baggage	190	
Ball mill	241	179+
Bath strainer type	4	290+
Bathtub (See bathtub)	4	538+
Battery	429	163+
Cell assembly support feature	429	186
Container only	429	176
Cop container	429	164+
Cover only	429	175
Handle or lifting device	429	187
Plural covers	429	177
Reactive metallic can	429	166
Seal feature	429	172+
Seal feature	429	174
Terminal feature	429	178+
Tubular electrode	429	164+
Beaker (See beaker)	422	104
Bed, for use with	5	503+
Bed with drawers etc.	5	308
Sofa bed	5	58
Bottles or jars	215	
Cabinet structures	312	
Closet bowl attachment	4	341+
Coin handling, eg change makers	453	
Collection	232	
Concentrating evaporators	159	

	Class	Subclass
Cutlery handle type	30	125
Razor	30	86
Deposit and collection	232	
Cabinet combined	312	211+
Discharging hand type	294	26.5
Dispenser	222	
Dry closet	4	479+
Inclosed	4	476+
Drying chamber type	34	179+
Rotary	34	108+
Earth boring tool combined	175	308
Electric battery	429	163+
Electric plug box or housing	174	50+
Electric wiring	439	501+
Emptying devices	414	403+
Evaporative cooled, porous cover	62	316
Evaporative cooled, porous wall	62	315
Porous wall	312	31+
Filling aide	141	391
Fire escape storing	182	93+
Fire extinguisher	169	
Portable	169	30+
Flexible ladder storing	182	70+
Food and beverage	99	467
Beverage infuser	99	323
Cooking filled receptacles	99	359+
Molding type eg pie & cake pans	99	426
Storage	99	516
Forehearth	266	166
Heating crucibles	432	262+
For metalliferous material	266	275+
Metal casting device combined	164	335+
With pouring modification	222	
Hoisting bucket type	294	68.1+
Hot water bottle or ice bag type	128	403
Ice cooled, portable	62	457
Package type	62	371+
Inkwell (See inkwell)		
Instrument sterilizer	422	
Laminated making	156	
Lid applying	53	287+
With compacting of contents	100	54+
Lining as by laminating	156	
Lock attachment	70	63+
Luggage or baggage	190	
Medicaments	604	403+
Metal working reel	242	83
Metallic	220	
Milking machine can	119	14.46
Nut cracker combined	99	568+
Opening	206	601+
Orifice grommet or bushing	16	3
Paper	229	
Opening	206	601+
Paper type filling and closing	53	467+
Piano case music storing	84	181
Plant or flower	47	66+
Bio degradable	47	74+
Irrigator	47	79+
Terrarium	47	69+
Window box	47	68+
Plumbing fittings	4	191+
Portable with burglar alarm	116	99
Racks	211	126+
Wall or window mounted	211	88
Relocatable with charging & discharging	414	332
Safe and vault type	109	
Sheet metal making	413	
Sifter type	209	233+
Special or packages	206	
Spittoons	4	258+
Spool holder	242	137+
Static mold	249	117
Stratifiers	209	422+
Supports	248	128+
Bag holders	248	95
Bracket type	248	311.2+
Desk and inkwell combined	108	26.2
Inkstand dispensing	222	577
Paste tube type	248	108
Stand eg inkstand		
Movable receptacle	248	128
Stationary inkwell	248	146+
Suspended type	248	318
Tearing	206	601+
Testing	73	52
Textile washing machine	68	
Deformable for squeezer type	68	96
Tumbler type	68	139+
Twine holder	242	146
Wash tub	68	232+
With scrubbing surface	68	233
Water heating type	126	373+
Well	166	162+
Hoist bucket type	294	68.1+
Wheeled	280	
Barrows or trucks	280	47.17+
Wooden	217	
Reclaiming (See Recovery)		
Celluloid	536	38+
Cellulose ester film scrap	536	76+
Cotton waste	8	141
Apparatus	68	1
Fats and oils	260	412.5
Lumber denailing	254	18
Paper	162	4+
Rubber	521	41+
Scrap metal salvaging	29	403.1+
Synthetic resin or natural rubber	521	40+
Tin scrap	75	64
Tobacco	131	96
Recoil		
Checks (See shock)	188	266+
Ordnance	89	42.1+
Operated		
Firearms and guns	89	125+
Package retainer, recoil-type	206	825*
Pads for shoulder arms	42	74
Design	D22	111
Recombination (See Genetic Engineering)	435	172.3
X-art collection	935	
Record		
Changer	369	178+
Controlled calculator	235	419+
Disk for phonograph	369	272+
Disk holder	206	309+
Hand carrier	294	158
Forming sound grooves in	264	106+
Shaving	82	1.1
X-art collection	425	810*
Recording and Recorders	346	
Aerial projectile target	273	348.1+
Borehole indicator	33	304+
Burning	346	76R
Calculator	235	58R+
Charts	346	134+
Coating process	427	146
Control unit	D14	2+
Crt	346	158
Electrical memories	365	118
Direction indicator	33	331
Dispenser combined	222	30
Dynamometer	73	862.27
Electric spark	346	162
Electrolytic	346	165
Equipment design	D14	
Facsimile		
Optical	358	296+
Receiver	358	296+
Float gauge	73	312
Fluid pressure gauge	73	713+
Holograms	350	3.6+
Electrical circuits	365	125
Liquid purification or separation	210	85+
Lock position	70	433+
Magnetic bubbles	365	1+
Mail letter box	232	18
Music	84	461+
Pens	346	140R
Printing	101	
Pyrographic	346	76R
Radioactive	346	152
Railway		
Block signal	246	107
Cab signal or train control	246	185
Train position	246	123
Register combined	235	2+
Reproducer combined		
Electrical wave for measuring	324	111+
For electrical transient	324	102
Scale	177	2+
Sound	369	
Acoustic image	367	7
Cabinet	312	9+
Composition	106	37
Electric music instrument	84	1.28
Electroforming process	204	5
Indexing devices	369	30
Magnetic tape or wire	360	88+
Manufacture	29	169.5
Molding device	425	406+
Molding process	264	106+
Phonograph electric	369	128+
Rack	211	40
Seismic mechanical	346	7
Talking picture apparatus	352	1+
Tape winding means	242	179+
Strip sprocket hole testing	73	157
Tape winding means	242	179+
Unidirectional	242	55.17+
Telegraphy		
Chemical & electrolytic	178	62
Page printing	178	28+
Photographic	178	15
Printing selector	178	34
Receiver	178	89+
Reed printer	178	48
Siphon	178	91
Transmitter	178	17R+
Type wheel printer	178	35
Telephone		
Calling number	379	142
Conversation time	379	114+
Telegraphophone	369	132+
Television	358	335+
Camera with recorder	358	906*
Color	358	310+
Magnetic	360	33.1+
Pause control	358	908*
Track skipper	358	907*
Television system combined	358	296+
Thermoplastic, storage retrieval of information	365	126
Thermoplastic, visual record	346	151
Television	358	344
Color	358	310
Facsimile	358	300
Time		
Plural recordings	346	45+
Printing or punching	346	80+
Transparency	365	127
Verifying	73	156
Voting machine	235	50R
Wire	346	150
Sound type	360	89
X ray photography	378	
Recovery (See Material Recovered)		
Coating excess	427	345
Dyes	8	440
Fats and oils	260	412+
Inorganic actinide compounds	423	249+
Nuclear materials	376	189
Nuclear materials	376	201
Nuclear materials	376	308
Scrap metal salvaging	29	403.1+
Separation of solids	209	
Solvent apparatus	68	18R
Rectangular Proportioner	33	DIG. 9
Rectifiers and Rectifying	D13	4+
Beam power amplifier	313	298
With two cathodes	313	5
Distillation	203	
Apparatus	196	
Apparatus	202	
Liquids	203	
Mineral oil	208	350+
Doubler high vacuum	313	306+
With two cathodes	313	1
Electrical	363	13+
Barrier layer coating	437	+
Barrier layer compositions	252	62.3 R
Electrolytes	252	62.2
Electrolytic devices	361	436
Full wave high vacuum	313	306
Lamp or electronic tube supply	315	200R+
Manufacture	29	569.1+
Process	437	
X ray circuit supply	378	101
Gas and liquid contact apparatus	261	
Gas liquefaction	62	42+
Process	62	32+
Process, plural separation	62	24+
Gas separation processes	55	36+
Half wave system	363	13+
Circuit interrupter for	200	
Dynamoelectric machine	310	10+
Electronic tube for	313	317+
Gas tube type	363	114+
Power packs	307	150
Unidirectional impedance for	357	
Vacuum tube type	363	114+

	Class	Subclass
With filter	363	39$^+$
With voltage regulator	363	84$^+$
Mercury vapor system	363	114$^+$
Rectifier control only	315	246$^+$
Mineral oils	208	350$^+$
Vacuum or gas tube	313	
Recuperator		
Distillation retort combined	202	111
Gas bench	202	148
Horizontal	202	140
Inclined	202	130
Vertical	202	122
Furnace		
Fuel burned in permeable mass	431	170
Gas heating furnaces	432	179
Red Lead	423	619
Redox Catalyst	526	915*
Reducer		
Commutation sparking	310	220$^+$
Machines		
Cabinets	128	371$^+$
Exercising	272	93$^+$
Massage	128	24R$^+$
Pipe joint	285	177
Screw	285	392
Pressure (See pressure, regulator)		
Socket wrench	81	185
Reduction (See Hydrogenation)		
Amines primary production by	564	415$^+$
Apparatus for ores	266	168$^+$
Catalysts treatment	502	100$^+$
Chemical agents	252	188.1
Complex ores	75	21
Detergent	252	105
Hydrogen production by	423	648R
Inorganic sulfide production by	423	
Iron and alloys electrothermic	75	10.1$^+$
Iron and steel	75	29$^+$
Metals separating combined	75	89
Metals without fusion	75	90R
Nitro nitroso or azomethine groups	260	689
Ores with gases	75	91
Separation distillation	203	32
Sulphur production by	423	567R
Thermit type	75	27
Reed		
Musical		
Accordian	84	376R
Harmonica	84	377$^+$
Nonorgan reed instruments	84	375$^+$
Organ reed pipes	84	350
Reed organs	84	351$^+$
Toy	446	207$^+$
Tuning reed for furnishing pitch	84	456
Wood winds	84	380R$^+$
Station selective signalling	340	825.39
Telephone call type	379	360$^+$
Telegraph	178	47$^+$
Telephone call transmitter	379	360$^+$
Textile weaving loom reed	139	192
For irregular warp feed	139	27
Movement thereof separate from raceway motion	139	191
Removable to free trapped shuttle	139	189
Valve	137	855$^+$
Vibration sensing	73	651
Reeding	28	201$^+$
Reef, Artificial	405	21$^+$
Reefing Sails	114	104$^+$
Reel		
Agricultural implements		
Cutting and gathering	56	
Line wire driven planter	111	44
Potato digger	171	111$^+$
Thresher straw carrier	130	23
Antenna	343	877
Antismut printing device	101	418
Baking oven combined	432	141
Body supported	224	162
Cable drum type, driven	254	266$^+$
Carriers	242	85$^+$
Changeable exhibitor	40	514$^+$
Calendar type	40	116$^+$
Electric cord	191	12.2 R$^+$
Suspended support	248	328$^+$
Fishing rod combined	43	20$^+$
Design	D22	137$^+$
Reel seat	43	22
Fishing rod type	242	84.1 R$^+$
Modifiable drive	242	211$^+$
Flexible ladder	182	73$^+$
Holder	D 3	30.1$^+$
Hose type with spray device	239	195$^+$
Portable	239	722$^+$
Measuring tape	33	138$^+$
Midline tightener	24	71.2
Motion picture film winding	242	179$^+$
With projector	352	124
Packages	206	389
Plumb bob	33	393$^+$
Resistors	338	79
With motion picture projector	352	124
Rewind motion picture film	242	210
Sheet web or strand drying apparatus	34	153
Sound recording or reproducing tape drive	242	179$^+$
Sounding	33	126.5$^+$
Strand type fire escapes	182	236$^+$
Strop combined	76	81.4
Tape measure	D10	71
Winding and reeling type	242	77$^+$
Automatically contracting	242	63
Belt type	242	67
Revoluble carrier	242	64
Trundle type	242	94
Winding drum or sand reel type	242	117
Reeling	242	
Information-bearing web machine convertible	242	179$^+$
Textile warp preparing	28	190$^+$
Winding &	60	905*
Reeves Drive	474	8$^+$
Refilling (See Fillers; Filling)		
By dispensing means	141	21
Fountain pen filling mechanisms	401	143$^+$
Prevention of		
Bottles	215	14$^+$
Dispensers	222	147$^+$
Specially related dispenser and receiver	141	348$^+$
Refining (See Purifiers)		
Alcoholic	99	277.1$^+$
Apparatus		
Mineral oil	196	
Carbohydrate manufacture	127	34$^+$
Starch manufacture	127	65$^+$
Sugar manufacture	127	42$^+$
Electrolysis	204	59R$^+$
Fats oils and waxes	260	420$^+$
Fatty oils	260	420$^+$
Recovering	260	412$^+$
Glass making in	65	
Lead	75	78$^+$
Metal casting apparatus combined	164	266
Mineral oils	208	177$^+$
Conversion combined	208	46$^+$
Deasphalting	208	309
Gum or gum former removal	208	255$^+$
Metal contaminant removal	208	251R$^+$
Nitrogen removal	208	254R
Organic acid removal	208	263
Phenol removal	208	263
Sulfur removal	208	208R$^+$
Sweetening	208	189$^+$
Paper making	241	
Smelting combined	266	161$^+$
Sugar	127	
Reflectors (See Mirrors & Reflectors)		
Antenna	343	912$^+$
Collapsible	343	915
Design	D26	118$^+$
Glass bead type	350	104$^+$
Nuclear reactor including for neutrons	376	220
Nuclear reactor including	376	347
For neutrons	376	346
Radar	342	5
Reflection reducing films on optical surfaces	350	164$^+$
Reflectometers	356	445$^+$
Solar heat collecting	126	438$^+$
Warning reflector triangle for highway use	40	903*
Design	D10	114
Reflex Klystron	315	5.18$^+$
Reflux, Vapor Barrier	156	DIG. 9
Refractometer	356	128$^+$
Refractors		
Electric lamp having	313	110$^+$
Fluorescent device having	250	485.1
Illuminating device type	362	326$^+$
Illuminating reflector combined	362	327
Lenses	350	409$^+$
Prisms	350	286$^+$
Refractory Material		
Building block	52	596
Checker brick furnace structure	165	9.1$^+$
Compositions	501	94$^+$
Carbide containing	501	87$^+$
Gas generator including	48	74
Metal heat treatment	148	133
Metallurgy	75	84$^+$
Molding processes	264	56$^+$
Refrigerators (See Cooler and Cooling)		
Cabinet structure combined	312	236
Showcase type	312	116
Car	62	239$^+$
Body structure	105	355$^+$
Compositions to produce		
Chemical	62	532$^+$
Chemical reaction, solids dissolving	62	4
Fuel	62	7
Processes combined	62	114
Processes combined, sorption type	62	112
Refrigerants brines	252	71$^+$
Refrigerants cryogenic and other	252	70
Refrigerants evaporative	252	67$^+$
Compressor muffler	181	403*
Design of machine	D15	81$^+$
Food, refrigeration of	426	524$^+$
Food safes with open evaporative cooling	312	31$^+$
Lights for	362	92$^+$
Occupant type vehicle	62	244
Product produced by	62	1
Showcase type	62	259.3
Space suit	62	259.1
Weighing scale combined	177	143
Refuse (See Trash; Waste)		
Fuel burning	110	235$^+$
Stoves and furnaces	126	222$^+$
Hydrocarbon mixture from	585	240
Incineration	110	235$^+$
Combined with stove	126	222$^+$
Receptacles		
Cleaning	15	56$^+$
Stock material from	428	903.3
Regenerated Cellulose	536	57$^+$
Compositions	106	168
Dyeing with azo	8	662$^+$
Filament forming	264	187$^+$
Formation impurities removal	8	137.5
Swelling or plasticizing of	8	130.1
Viscose production of	536	60$^+$
Regeneration Media	435	240.54$^+$
Regeneration (See Revivification)		
Heat of gas refrigeration cycle	62	6
Ion exchange liquid purifier	210	269$^+$
External regenerant supply	210	190$^+$
Motive fluid	60	526$^+$
Process	210	670$^+$
Tone musical instrument having	84	1.5
Regenerative		
Amplifier	330	112
Between adjacent stages	330	89
Cathode impedance feedback	330	93
In push pull type	330	82
In some path at different frequencies	330	101
Including negative feedback	330	104
Detector systems		
Pentode tube type	329	170$^+$
Screen grid tube type	329	170$^+$
Triode tube type	329	170$^+$
Furnace		
Alkyne synthesis using	585	535
Diolefin synthesis using	585	602
Olefin synthesis using	585	634
Radio receiver	455	336$^+$
Regenerator		
Distillation		
Thermolytic retort system	202	105$^+$
Evaporator	159	24.1
Furnace		
Heated air fed to burner	431	167
Heat exchange	165	4
Vapor inhalers	128	202.26
Reggio Patent Collection	60	913*
Registering		
Apparatus		

Entry	Class	Subclass
Telephone call counting	379	111+
Telephone switching control	379	280+
Registers	235	
Autographic		
Cash recorder	235	4+
Strips	281	5+
Strips and leaves	281	2
Strips and leaves manifolding	282	3R+
Strips manifolding	282	12R+
Closure	49	38+
Dispenser combined	222	23+
Display control	377	112
Dynamometer	73	862.27
Fare or ticket	235	33+
Float gauge	73	310
Heating and ventilating	98	101+
Design	D23	
In telephone system	379	111+
Liquid purification or separation	210	85+
Lock condition	70	436+
Mail letter box	232	18
Position		
Musical instrument	84	136+
Pneumatic conveyor	406	62+
Printing machine bed and platen	101	286
Printing machine rotary	101	248
Sheet material associating	270	51
Typesetting	276	15
Typesetting and distributing	276	3+
Web or strand feeding	226	10+
Recorder combined	346	14R+
Scale	177	15+
Senders telephone	379	288
Operator controlled	379	261
Shift	377	64
Telephone connection control	379	280+
Ventilation	98	101+
Registration Plate		
Automotive design	D20	13+
Regulating		
Apparatus		
For electric motors	318	
For electric railway cars	105	49+
Transformers	336	
Electricity		
Arc lamp	314	
Generator system	322	
Lamp generally	315	
Motor system	318	
Regulation		
Continuous current motors	318	
Of alternating current repulsion motors	318	725
Of electric power distribution	307	
Speed in induction motors	318	799+
Plural speed motors	318	772+
Rotor speed secondary circuit impedence control	318	823+
With starting control	318	779
Regulator		
Band saw tension	83	816+
Beverage infuser	99	305
Brake	188	266+
Carbon chemistry		
Process	260	698+
Depth		
Marine mines	102	412+
Torpedo	114	25
Draft		
Boiler	122	38
Furnace	110	147+
Furnace spark arrester	110	123
Electricity (See controller, circuit; controller, electric)		
Arc furnace	373	60+
Automatic system	307	
Battery charging circuit	320	29
Battery charging circuit	320	49+
Battery charging generator	320	27
Field winding	320	64+
Battery charging generator structure	320	72
Condenser	361	271+
Current	323	
Distribution	307	15
Electric generation	322	
Electrode current or voltage	314	135
Electrode feed electromagnetic	314	115+
Electrode feed rotary motor	314	69+
Electrode shunt	314	10+
Elevator operator actuated	187	43
Elevator speed	187	39
Gas and vapor lamp	313	567+
Heater electric	219	482+
Induction furnace	373	138+
Input responsive	323	299
Lamp	315	291+
Locomotive	105	61
Music instrument	84	1.1+
Phase control system	323	
Potential for dynamos	322	28
Pump	417	44+
Railway block signal	246	34R+
Railway car propulsion	104	288+
Resistance	338	
Resistance furnace	373	135+
Rheostat	338	68+
Speed	361	236+
Surgical	128	387
Transformer	336	
Voltage system	323	
Wave transmission	333	
X ray	378	108
Feed		
Boiler fuel and water	122	446+
Dibbling planter	111	95+
Electrode brake	314	82
Electrode consumable plural	314	37
Electrode consumption type	314	59+
Electrode magnetic operator	314	113+
Electrode separating mechanism	314	65+
Electrode shunt circuit	314	11+
Hill planter	111	17
Paper web forming	162	259
Pneumatic conveyer	406	127+
Automatic	406	29+
Solid material comminution	241	34+
Sprinkler pipe	239	562+
Fermentation process	435	289+
Fluid flow baffle	138	37+
Liquid purification or separation	210	143+
Locomotive traction	105	73+
Musical keyboard touch	84	439+
Nuclear reaction control	376	207
Pendulum electric clock	368	166
Power plant, condition responsive	60	660+
Power plant, system operation	60	645+
Pressure (See pressure)		
Calender roll	100	168+
Fermentation process	435	289+
Fluid by valve control	137	505+
Fluid flow meter combined	73	199
Illuminating gas distribution	48	191
Water heating system	237	65
Pump	417	279+
Fans	416	170R
Gas	417	279+
Gas jet flue	417	157
Gas rotary	415	
Jet	417	182+
Railway cable rail tension	104	117
Shear cut	30	200+
Speed (See speed responsive or controlled)		
Temperature and humidity	236	
Animal watering device	119	73
Arc lamp	313	11+
Automatic heat exchanging	165	13+
Automatic refrigeration	62	132+
Fermentation process	435	289+
Internal combustion engine	123	41.2+
Superheater	122	479R
Tobacco treatment	131	303
Timepeice		
Balance	368	170+
Compensating means	368	202
Electrical frequency	368	200
Electrical rate	368	186
Pendulum	368	181+
Typewriter carriage	400	337+
Reinforcement		
Blades cutting	30	348
Bushings	16	2+
Checks labels tags	40	27
Clothing		
Pocket	2	248
Reinforcing collar cuff bosom	223	4
Stockings	2	241
Stockings knitted	66	182
Under wear	2	400
Comb back	132	155
Fabric	2	261
Making laminated	156	
Seams	112	402+
Woven	139	425R
Ferrules	16	108+
Ladder	182	217+
Lining thimbles	16	2+
Masonry and settable material		
Arches	52	88
Beam or girder	52	723
Blocks modules	52	600+
Column pole, post	52	722+
Mold material	164	411
Openwork	52	633+
Pavement block	404	45
Pavement sheet	404	70
Pipes	138	175+
Post fence	256	51
Precast module	52	600+
Prestressed	52	223R+
Processes	264	271.1
Radiating from column	52	260
Railway tie	238	85+
Ribbed	52	319+
Ribbed slab	52	602
Roadway	404	134+
Rods	52	720+
Safe or vault wall	109	83
Packing joint	277	
Pipe conduit	138	172+
Receptacle		
Flexible bag	383	119
Metallic	220	71+
Metallic edge	220	73+
Safe	109	83
Trunk	190	25
Rings	16	108+
Shoe insole	12	20
Spoons	30	328
Switch point railway	246	437
Textile, reinforcing or tire cords	57	902*
Thimbles	16	108
Tires		
Inner tube	152	511
Tire	152	393+
Tire and valve stem	152	430
Valve head	251	358
Vehicle frame	296	30
Wheels		
Hubs	301	106+
Spokes	301	83
Spokes plural series	301	77
Reins	54	36
Check	54	16+
Guards	54	73
Design	D30	141
Holders		
Design	D30	141
Vehicle attachment	280	182+
Whip socket combined	280	177
Holds	54	74
Supports and guards vehicle type	280	181
Reject Catcher or Deflector		
Glass making apparatus	65	165
Refrigeration producer		
Impurity removal	62	195
Reknitting	66	1.5
Relaxation Oscillator	331	
Driven type	328	193+
Relay		
Auto		
Headlight systems	315	77+
Horn systems	340	22+
Electric		
Amplifier	330	
Casing	200	302.1
Directional safety system	361	1+
Electric thermal responsive system	236	78R
Electromagnetic switch	335	2+
Electromagnetic switching systems	361	139+
Frequency division repeaters	370	75
High speed	335	2+
Light wave repeaters	455	601
Motor operated	335	68+
Pulse or digital repeaters	375	3
Railway signalling system	246	
Systems	361	139+
Systems safety	361	1+
Telegraph repeaters	178	70A+
Telephone central line-signal control	379	317+

	Class	Subclass
Thermal current	337	14$^+$
Thermal responsive system	236	68R
Time division repeaters	370	97
Pneumatic dispatch	406	181$^+$
Pressure modulating relays or followers	137	82$^+$
Structure electromagnetic		
Electromagnetic switch	335	
Electrothermal switch	337	14$^+$
Temperature and humidity control by	236	
Time lag	200	33R$^+$
Release & Releasers (See Uncoupling)		
Article	221	289$^+$
Brakes and clutches		
Brakes	188	
Check controlled lock release	194	
Clutch and brake	192	12R
Clutch operators	192	82R$^+$
Clutches	192	30R$^+$
Drive release and brake	192	144$^+$
Fluid pressure brake motor releasing	303	68$^+$
Fluid pressure system synchronizing	303	36
Overload release	192	150
Phonograph stop mechanism	369	268
Speed responsive clutch operator	192	103R$^+$
Transmission control and automatic brake	192	7$^+$
Typewriter feed mechansim	400	334$^+$
Coating		
Compositions	106	2
Methods	427	
Load-responsive release	403	DIG. 3
Overload (See clutch; crank; latch; lever; overload; trip)		
Automatic trip mechanisms	74	2
Elevator cable	187	72
Hand wheels	74	556
Harpoon type hay fork	294	126$^+$
Hoist line sling	294	75$^+$
Intermittent grip mechanical movements	74	111$^+$
Key holding catch for locks	70	459
Lever carried pawl	74	535$^+$
Lever detent and release	74	529
Lock emergency release	70	465
Locks	70	
Mechanical gun triggers	124	31$^+$
Music leaf turners	84	486$^+$
Pitman automatic release	74	584
Railway switches and signals	246	
Rotatable platform trap	43	72
Store service cable propulsion	186	16
Tiltable platform trap	43	70
Tilting hay fork	294	124
Time lock emergency release	70	274
Time switch latch trip	200	40$^+$
Toy money box figure or mechanism	446	9
Traction railway cable grippers	104	207
Typewriter line spacing	400	565$^+$
Valve	251	66$^+$
Well torpedo lowering	102	313
Quick release	403	DIG. 4
Relief		
Electrochemical process	204	4$^+$
Maps	434	152
Milling	409	69
Non-planar layer ornamented	428	187
Photographic process	430	322$^+$
Picture	40	160
Relievers		
Electric cord strain	439	449$^+$
Internal combustion engine pressure	123	182
Metalworking roll adjustment	72	237$^+$
Photographs in relief making	430	269$^+$
Radiator air valves	236	61$^+$
Receptacle filling displaced air pressure	141	
Safe explosive pressure	109	27
Scale platform gear	177	151$^+$
Ship tension	114	213$^+$
Religion Teaching	434	245$^+$
Religious		
Altar	D 6	396$^+$
Apparel	2	
Article not elsewhere specified	D99	25$^+$
Artifact cruciform type	428	3
Baptismal font	D99	25
Cabinet	312	
Creche	D11	122$^+$
Educational device	434	245$^+$
Incense burner	D11	131.1
Jewelry	63	
Medal	D11	95$^+$
Monument	52	103$^+$
Receptacle	206	19
Tabernacle	D99	25
Relishes	426	615$^+$
Relishing	144	6
Remote Control (See Type of Control or Device Controlled)		
By light waves	455	600
By radio	340	825.72
Pulse responsive	340	825.69
By transmitter	340	696
Handlers	414	909*
Mining machine	299	30
Remote Cutoff		
RF pentode	313	300
Tetrode	313	297$^+$
Remover (See Cleaner; Pull; Push; Scraper)		
Antiskid means on wheels	152	213R$^+$
Bur removing		
Chains	59	59
Pipe	408	199$^+$
Closure	81	3.7$^+$
Can opener cutter combined	7	152$^+$
Other tool combined	7	151$^+$
Plier tool combined	7	126$^+$
Dental crown	433	141$^+$
Disassembly	29	700$^+$
Garment	223	111
Heel shoe	12	50.5
Ingot strippers	164	405
Insulation from wire	81	9.4$^+$
Obstruction conduit	4	255$^+$
Pith	130	31R
Residue from filter	210	407$^+$
Rind	99	584$^+$
Scale (See cleaner and cleaning)	29	81R
Scum or sediment	210	523$^+$
Distinct separators	210	299
Shell nut and vegetable	99	568$^+$
Skin fruit and vegetable	99	584$^+$
Sludge		
Acetylene generator	48	57
Sewage	210	523$^+$
Stem	99	635$^+$
Stump excavators	37	2R
Threshing	130	
Tire	157	11
Rubber tire	157	1.1
Welding flash planers	409	288$^+$
Removing		
Caffein or tannin	426	427$^+$
Cast article from mold	164	131$^+$
Demountable rim	301	14
Disassembling processes	29	426.1$^+$
Tools and appliances	29	700$^+$
Drying	34	
Electric insulator moisture or dirt	174	211
Emulsifying agent from mineral oil	208	177$^+$
Excess coating material	427	235
Fermenting processes		
Fermentate removal	435	262$^+$
Fowl sinew	17	11.3
Garments	223	111$^+$
Growth in water courses	56	8$^+$
Rendering	260	412.6$^+$
Liquid separation combined	210	767$^+$
Apparatus	210	175$^+$
Melting separators for chemicals	422	285$^+$
Press	100	104$^+$
Rennet or Rennin		
Cheese preparation with	426	582
Renovating		
Butter	426	663
Apparatus	99	452$^+$
Coatings	427	140$^+$
Dress		
Dry cleaning	8	142
Dyeing combined	8	441
Feathers	34	2
Repairing (See Darning; Patch)	29	402.1$^+$
Access to fluid handling systems	137	583$^+$
Adhesive bonding	156	94$^+$
Building structure	52	514
Assembly or disassembly	52	127.1$^+$
Coating processes	427	140$^+$
Electrolytic	204	16
Disassembly steps included	29	402.3$^+$
Electric		
Conductor	156	49
Discharge device	445	
Apparatus	445	61
Process	445	2
Lamp	445	
Apparatus	445	61
Process	445	2
Fluid handling systems or devices	137	315$^+$
Furnace wall	264	30
Glass making apparatus	65	27
Liquid purification or separation apparatus	210	232$^+$
Sectional chamber press type	210	230
Metal casting	164	92.1
Patching	29	402.9$^+$
Pile	405	250
Pipe joint	285	15$^+$
Pipes and conduits	138	97
Refrigeration apparatus	62	298$^+$
Process	62	77
Reknitting	66	1.5
Seal	29	402.3
Shaping non-metallic material	264	36
Shoes	12	142Q
Textile twisting	57	261$^+$
Tools for tires	81	15.2$^+$
Welding	228	119
Well	166	277
Repeater	340	425
Clock		
Alarm	368	248
Striker	368	267$^+$
Fire signal system	340	291
Firearms and ordnance	42	2$^+$
Automatic	89	125$^+$
Revolver type	42	59$^+$
Frequency division multiplexing	370	75
Light wave	455	601
Lock operator	70	270
Mechanical guns	124	19
Music leaf turner	84	513$^+$
Pulse or digital	375	3
Radiotelegraphy	375	3$^+$
Telegraph	178	70R$^+$
Telegraph quadruplex	370	35
Telephone	379	338$^+$
Relay	330	
Testing	379	4
Time division multiplexing	370	97
Watch	368	267$^+$
Repellent		
Composition	106	2
Ink	101	422
Invertebrate	424	
Microbial pathogen	424	
Utilizing, glass making	65	24$^+$
Vertebrate	424	
Replacer		
Railway		
Car	104	262$^+$
Tie	104	9
Vehicle battery	104	34
Reproducers and Reproducing		
Control unit	D14	
Cutting machine	409	85$^+$
Duplicating	409	93$^+$
Electroforming processes	204	4$^+$
Milling pattern controlled	409	85$^+$
Photographic	355	
Sound	369	
Diaphragm and mountings	181	148$^+$
Earphone	381	187$^+$
Electric music instrument	84	1.28
Electric phonograph type	369	127$^+$
Electrical	369	135
Electrooptical musical instrument	84	1
Motion picture combined	352	1$^+$
Phonograph	369	
Speakers	381	192$^+$
Speakers, methods of manufacture	29	594
Telephone mechanical	181	138
Television	358	335$^+$
Color	358	310$^+$
Magnetic	360	33.1$^+$
Repulsion Motors	318	725$^+$
Resacetophenone	568	337
Resawing Machines	83	425$^+$

Entry	Class	Subclass
Rescinnamine	546	54
Rescue Devices (See Lifesaving)	D29	
Resector	128	305$^+$
Electric	128	303.14$^+$
Coagulator	128	303.17
Endoscopic	128	303.15
Urethrotome	128	311
Reserpic Acid	546	54
Reserpic acid lactone	546	40
Reserpiline	546	48
Reserpine	546	55
Reservations, Computerized	364	407
Reservoir		
Brush, broom, mop attachment	401	140
Bulb type syringe combined	604	212$^+$
Coating implement hand manipulable	401	
Fluid pressure brake systems	303	
Supplementary reservoir	303	85
Fluid sprayer combined	239	302$^+$
Fountain pen filling mechanisms	401	143$^+$
Liquid fuel burner	431	320$^+$
Liquid fuel burner	431	344
Liquid fuel cooking stove	126	49
Liquid purification or separation	210	
Nuclear reactor combined	376	403
Pneumatic dispatch system	406	
Shaft lubricating	384	401$^+$
Soap bubble device combined	446	15$^+$
Writing pen attachment	401	252$^+$
Reset		
Attack defeating locks with	70	1.7
Check controlled lock	194	247$^+$
Fare registers	235	47$^+$
Manual type for weaving		
Fabric take up	139	316
Warp letoff	139	113
Railway switches and signals		
Automatic switch	246	315
Automatic switch stands	246	280$^+$
Automatic switch wheel reset	246	316
Switch stands	246	395$^+$
Vermin trap		
Automatic	43	73$^+$
Victim reset	43	76
Weaving pattern mechanism	139	328
Reshaping		
Apparatus for glass	65	286$^+$
Reheating combined	65	269$^+$
Processes for glass	65	102$^+$
Resharpening (See Pencil Sharpener)		
Abrading processes	51	281R$^+$
Files by sandblast	51	414
Residue Prevention in Humidifiers Air Conditioners	261	DIG. 46
Resilient		
Comminuting elements	241	102
Diaphragm dispensers	222	490
Material valves	137	843$^+$
Railway tire cushions	295	11$^+$
Supports	248	560$^+$
Tires and wheels	152	
Wall dispensers	222	206$^+$
Resinox T M (phenoplast)	528	129
Resins		
Dyeing	8	506$^+$
Mineral oil	208	22$^+$
Making, treating and recovery	208	39$^+$
Natural		
Abrasives containing	51	300
Compositions containing	106	
Purification, preservation or recovery	260	107
Rubber containing	524	274
Synthetic resins with (See synthetic resin or natural rubber)		
Resin bleach	8	DIG. 6
Synthetic (See also synthetic resin or natural rubber)		
Abrasives containing	51	298
Polymer (See 525 X-art collections)		
Reaction induced by electrical	376	190
Or within a nuclear reactor	376	156
Wave energy processes	522	100$^+$
Resist		
Dyeing, cellulose ether or ester	8	446$^+$
Etching	156	659.1$^+$
Etching	156	904$^+$
Resistance		
Aerodynamic resistance reducing	244	130
Brakes internal resistance	188	266$^+$
Fluid	188	266$^+$
Granular material	188	268
Operators regulator	188	270
Compositions electrical	252	500$^+$
Light sensitive	252	501.1
Electrical	338	
Coatings	427	101$^+$
Compositions	252	500$^+$
Furnace	373	109$^+$
Furnace arc	373	3
Furnace inductuon	373	6
Inductive	336	
Manufacturing	29	610R
Measuring	324	62$^+$
Systems	323	364
Telephone resistance elements	373	109
Heaters	219	200$^+$
Electric furnaces	373	109$^+$
Heating for electric stoves	219	391$^+$
Incandescent lamps	315	71
Meters electric	324	62$^+$
Music keyboard touch regulators	84	440
Of ships apparatus for measuring	73	148
Railway draft appliances springs varying	213	41
Regulators electric	338	
Speed responsive device	73	524
Wind motors steering and resistance devices	416	9$^+$
Resistive Hygrometer	324	65R
Resistor (See Resistance)		
Resonance		
Audio receiver having	381	159
Electric filters	333	175$^+$
Electric frequency meters having oscillating reeds	324	80
Frequency division telegraphy	370	45
Lock tampering defeating device	70	334
Music		
Banjo with resonance device	84	270$^+$
Electric music device with tuned reproducer	84	1.6
Harmonica with resonance chambers	84	378
Piano with resonance device	84	189$^+$
Reed board with resonance device	84	362
Rigid vibrators with resonance device	84	410
Stringed instrument with resonance device	84	294$^+$
Tuned sound generator with electric pickup	84	1.4$^+$
Relays	335	94
Selective system	340	825.39
Relays	335	101
Test by mechanical vibration	73	579$^+$
Resonant Transmission Line	333	236$^+$
Resonator		
Amplifier with	330	
Cathode ray tube including	315	3$^+$
Cavity type		
Amplifier system with	330	56
Cathode ray tube combined	315	4
Magnetron combined	315	40
With electron beam amplifier	330	45
With magnetron amplifier	330	47$^+$
With push pull amplifier	330	55
Distributed parameter type	333	219$^+$
Electronic tube with	315	40
Magnetrons with	315	39.51
Modulator having	332	7
Oscillators with	331	
Radio receiver with	455	130$^+$
Tank circuit	334	
Tuned tank circuit	334	
Resorcinol	568	763$^+$
Azo compound ctg.	534	682$^+$
Respirator	128	200.24$^+$
Rest		
Armrest support	248	118
Writers	248	118.1$^+$
Ash tray with cigar or pipe rest	131	240.1
Bed attaching body rest	5	431$^+$
Billiard cue rest	273	23$^+$
Book combined with cabinet	312	233
Book copy and music leaf	248	441.1$^+$
Copyholder, movable copy	40	341$^+$
Casket grave rest	27	26
Chair or seat armrest	297	411$^+$
Chair or seat footrest	297	423$^+$
Chair or seat headrest	297	391$^+$
Cigar or tobacco pipe support	131	257$^+$
Closet head foot and body rests	4	254
Coffin headrest	27	13
Control lever and linkage foot rest	74	564
Crutch armrest	135	73
Gun rest	42	94
Hand and arm rests	248	118$^+$
Writers aid	248	118.1$^+$
Knee pad apparel	2	24
Land vehicle body footrest	296	75
Pianists exercising armrest rail	84	469
Piano pedal footrest	84	232
Sofa arm	5	52
Stringed instrument arm or hand rest	84	328
Tool		
Lathe	82	35$^+$
Turning axial pattern pivoted	82	12
Wood turning	142	48$^+$
Typewriting bar machine	400	454$^+$
Velocipede footrest	280	291
Violin chin rest	84	278$^+$
Work		
Lathe	82	35
Reaction induced by electrical	51	238R$^+$
Wood turning	142	50
Woodworking workbench	144	287
Restrainers		
Aircraft	244	110R$^+$
Animal	119	96$^+$
Breaking and training harness	54	71$^+$
Bed clothing	5	498
Bedspring	5	261
Combination lock tumblers	70	327
Human	128	133$^+$
Seat occupant	297	464$^+$
Restrictor	138	40$^+$
Valve combined	251	118$^+$
Retainers (See Closure; Holders)		
Animal draft parts	278	
Apparel	2	300$^+$
Bait fishing	43	44.2$^+$
Inclosed baits	43	41
Bandage	128	171
Barrel hoop	217	93
Broken parts		
Check valve	137	540.11
Pipe joint	285	117
Casting sand	164	379$^+$
Circuit	178	72
Dams	405	107$^+$
Freight load on freight carrier	410	
Load bearer, eg container	410	77$^+$
Hair		
Comb	132	144$^+$
Curler or crimper	132	38R$^+$
Pin	132	51
Heat		
Accumulator type stove	126	400
Fireless cooker	126	273.5
Furnace	110	322$^+$
Water accumulator type	126	375
Key in lock	70	429
Levees	405	107$^+$
Load, land vehicle	280	107
Necktie knot	2	152R$^+$
Orthodontic	433	6
Packing	277	189
Packing attached	277	181$^+$
Picture frame cover	40	157
Picture in frame	40	156
Pin	132	52
Railway car stop	104	251
Receptacle closure	220	315
Bottle and jar	215	306
Tethered	220	375
Tethered dispenser type	222	543
Shaft or spindle		
Pintle caster	16	38$^+$
Pintle hinge	16	263
Thill coupling	278	60$^+$
Trace eye end	278	105$^+$
Shoe and boot	36	58.5$^+$
Shoe on last	12	141
Strand	24	115R$^+$
Conductor on insulator	174	168$^+$
Picture suspended	248	489$^+$
Umbrella	135	37$^+$
Water		

	Class	Subclass
Earth control	405	107$^+$
Wharves retaining wall type	405	284
Retarders (See Brakes)		
Airplane	244	110R$^+$
Camera shutter	354	256$^+$
Chute	193	32
Clutch	192	109R
Compositions		
Mortar	106	315
Door	16	82$^+$
Closer combined	16	49$^+$
Drying processes	34	7
Electric switches and circuits		
Electromagnetic latch trip switch with	335	172$^+$
Electromagnetic switch with	335	59$^+$
Electromagnets with	335	239$^+$
Railway car actuated switch	246	248
Tensioner	226	195
Endless web feeding	226	195
Fluent material movement	138	40$^+$
Brake systems	303	84R
Making	29	157R
Sheet delivering	271	182$^+$
Sheet feeding	271	229$^+$
Ships speed in steering	114	145R
Skidway	193	40
Spring	267	195$^+$
Switch with	200	33R$^+$
Ventilator rotary	98	120
Retene (See Aromatic Hydrocarbon)	585	400$^+$
Reticulated (See Type of Article or Device)		
Reticules	33	297
Retinoscopes	351	205$^+$
Retort		
Closed garbage incinerating	110	242
Distilling	202	
Heating and illuminating gas generating	48	
Liquid fuel burners	431	207$^+$
Furnaces	431	161$^+$
Metallurgical	266	153
Mineral oil vaporizing	196	104$^+$
Multiple chamber	432	207$^+$
Rotary	432	103$^+$
Rotary garbage incinerating	110	246
Single chamber	432	212$^+$
Retouching	354	348
Stands	354	349
Retracting (See Retractors)		
Comb	132	129$^+$
With brush	132	121
Handles key controlled	70	208
Knife	30	162
Landing gear airplane	244	102R
Manicuring tool	132	76.2
Money slot toy bank	446	13
Pen tool	401	99$^+$
Wheel or runner land vehicle	280	9$^+$
Retractors	254	
Hose	242	
Dental	433	78
Surgical	128	20
Retreading	264	326
Apparatus	425	201$^+$
Retrievers		
Ball	294	19.2
Trolley	191	85$^+$
Well	294	86.1$^+$
Combined type	166	99
Enclosed receptacle	166	162$^+$
Hoist bucket type	294	68.1$^+$
Retting	162	1$^+$
Bacteriological	435	278
Reveal Cover	52	212
Reverberator	381	63$^+$
Reversible and Reversing		
Belt & pulley drive	474	1$^+$
Bowl for smoking devices	131	221
Camera	354	288
Chisel	144	77
Closure bolt	292	244$^+$
Clutch		
Gearing combined	192	21
Multiple	192	51
One way drive	192	43$^+$
Electric current		
Brush shifters	310	241
External electric car propulsion systems	104	302
Telegraph current reversing	178	16
Telegraph key reversing	178	105
Fluid current		
Filter cleaning, automatic	210	108
Filter cleaning, process	210	767$^+$
Filter cleaning, valved	210	427
Ion exchange or sorption process	210	660$^+$
Pump supply	417	315
Sanders	291	5$^+$
Furnace draft	110	
Garment	2	301
Gearing	74	640$^+$
Vehicle combined	280	236$^+$
Hinge	16	265$^+$
Jaws for wrenches	81	178
Lock	70	462
Mechanical movement	74	144$^+$
Motors		
Expansible chamber	91	
Internal combustion engine	123	41R
Turbine	415	
Planer	409	336
Rail	238	132$^+$
Rolling mill	72	229
Sanders	291	26
Saw teeth	83	836
Scoop	37	139
Seat back, eg walk over	297	94$^+$
Seat bottom and back	297	92$^+$
Ship propulsion	440	
Spittoon	4	279
Stove	126	23
Tie plate	238	301
Tools and implements		
Cutting	56	266
Drill	408	124$^+$
Toy	446	442
Turntables	104	35
Automobile	104	44
Typewriter ribbon	400	218$^+$
Valves	137	309$^+$
Multiway	137	270.5
Wood shaper	144	148
Revertible and Reverting		
Sofa bed back	5	37R$^+$
Stove draft base heating	126	75
Hot air stove	126	69
Magazine type	126	74
Revetments	405	16$^+$
Revivification		
Adsorbent	502	20$^+$
Batteries	429	49$^+$
Individual parts	429	49
Catalyst	502	20$^+$
Revolvers		
Design	D22	104$^+$
Firearms	42	59$^+$
Cylinder loaders	42	89
Firing mechanisms	42	65$^+$
Toy	42	58
Revolving Doors	49	42$^+$
Reweaving	139	1F
Darning looms	139	33.5
Patches	2	243R$^+$
Rewinding (See Winding)		
R F Diode	313	317$^+$
With emissive cathode	313	310
With thermionic cathode	313	310
R F Power Amplifier	330	
Rheo-casting	164	900*
Rheostat	338	68$^+$
For lights	362	295
Rhodamines	549	227
Rhodanate (See Thiocyanate)		
Rhodium (See Platinum)		
Rhodols	549	225
Rhumbatron	333	227$^+$
Cathode ray tube	315	5$^+$
Rib		
Concrete construction	52	319$^+$
For piano sounding boards	84	195
Saw gin	19	62R$^+$
Umbrella	135	31
Ribbon		
Feed		
Cap pistol	42	57
Cigar or cigarette making	131	84.1
Tobacco bunch precompressor	131	42
Typewriter	400	223$^+$
Hair	132	47
Inkers		
Printing (See type machine with which associated)	101	336
Typewriter	400	197$^+$
Instruments, wires &	33	DIG. 7
Joints or couplings	403	
Looms	139	22
Packages	206	389$^+$
Resistance element	338	279$^+$
Shuttles	139	199$^+$
Spool holder	242	71.8$^+$
Typewriter	400	237$^+$
Weaving	139	135$^+$
Winding spools	242	71.8
Riboflavin	544	251
BiO-affecting compositions	514	251
Ribose	536	1.1$^+$
Rice		
Bran	426	618$^+$
Cleaning and polishing	426	481$^+$
Apparatus	99	600$^+$
Ricinoleic Acid	260	413$^+$
Ricks		
Hay straw	414	132
Rickshaw	224	159
Riddles	209	331$^+$
Ridge		
Bed pole	5	130
Gondola car drop bottom		
Central	105	245
Rising	105	246
Ridging		
Plow	172	701
Riding		
Boots	D 2	271$^+$
Equipment	D30	134$^+$
Rifamycins	540	458$^+$
Riffles		
Liquid suspension separator having	209	206
Stratifier having	209	506$^+$
Rifles (See Gun)		
Automatic	89	125$^+$
Carrier	224	913*
Receiver holding butt end	224	149
Sling carrier	224	150
Design	D22	103$^+$
Firearm	42	
Ranges	273	348.1$^+$
Recoilless	89	1.7$^+$
Targets	273	348.1$^+$
Rifling		
Cutting	409	306
Firearm barrel	42	78
Rigging		
Brake	188	2R$^+$
Sails and rigging	114	102$^+$
Ships implements	114	223
Rigs		
Earth boring	175	
Rim		
Breaker	157	1.35
Compressor	157	2
Demountable	301	10R$^+$
Gear		
Segmental	74	448
Separate	74	446$^+$
Handwheel grip or cover	74	558
Integral for tires	152	378R
Locks for	70	260
Making	29	159.1
Pulley	474	166$^+$
Expansible	474	47$^+$
Cone pulley	474	8$^+$
Separable for tires	152	396$^+$
Tire rim bands	152	513
Wheel	301	95
Ring (See Band; Collar)	16	108$^+$
Container for annular article	206	303
Fasteners connectors or supports		
Article holding bracket	248	315
Bolts closure fastener	292	299
Bottle and jar closure	215	274$^+$
Demountable rim	301	26$^+$
Draft pole coupling	278	125
Filter medium	210	495
Gimbal shaft coupling	464	125$^+$
Handled filter	210	471
Insulator cap type	174	192
Insulator pin type	174	198
Metal receptacle closure	292	256.6$^+$
Panel to track attaching	16	87.2$^+$
Pipe conduit bracket	248	69

U.S. PATENT CLASSIFICATIONS

Entry	Class	Subclass
Pipe conduit bracket	248	74.1
Pipe conduit suspended	248	62
Shaft joint	403	
Spring lamp shade support	362	440
Tire to rim	152	409+
Tobacco users finger	131	258
Handle	16	127
Jewelry		
Earrings	63	12+
Design	D11	40+
Finger	63	15+
Design	D11	26+
Trays for	206	566
Key	70	456R+
Design	D 3	61+
Lock cylinder	70	381
Locking, bolt or nut to substructure lock	411	109
Magnetoelectric apparatus	228	179
Armature retaining	310	271
Electric heating	219	51+
Electroforming process	204	9
Grinding process	51	290+
Jewelry forming and sizing	29	8
Packings	29	156.6
Rolling	72	67+
Slip	310	232
Spinning, twisting, twining		
Apparatus	57	21
Process	57	21
Welding process	228	179+
Winding ring wound armature	242	4R+
Wire	140	88
Making by deforming metal	72	
Gem and jewel setting	29	10
Napkin holder	D 7	72
Nose animal	119	135
Piston	277	
Slip for electric current machine	310	232
Snap ring	24	16R
Spinning frames	57	75
Textile spinning element	57	119+
Tire bead grommets	245	1.5
Trapeze	272	61
Trim	301	37R
Watchcase movement	368	299
Welding backup	228	50
Joint integratable	285	21
Ringers and Ringing		
Telephone		
Automatic	379	232+
Party line	379	179+
Rinks, Ice	62	235
Grooming apparatus	37	219+
Rinsing		
Solid work pieces	134	
Textiles and fibers		
Machines	68	
Methods	8	147+
Rip Strips		
For metal containers	220	266
For paper boxes	206	601+
Riprap	405	32+
Risers		
Bed bottom		
Corner riser	5	205+
Frame shiftable on riser	5	209
Railway car folding step	105	448
Railway switch	246	432
Nonderailing	246	421
Sand mold	164	359+
Stair riser carpet fastenings	16	11
Rivet	411	500+
Design	D 8	386
Making	10	11R+
Seal type	292	314
Riveting		
Blind	227	51+
Button riveting machine	227	51+
Design	D15	122+
Driving and heading	227	51+
Electric heating and	219	150R
Hollow rivets	227	51+
Solid rivets	29	243.53+
Riving	144	182+
Shaving combined	144	40
Roads	404	
Cement structure	404	71
Chair	404	136
Curb	404	7
Expansion joints	404	47+
Finishing by compacting	404	113+
Grader	37	108R+
Gravelling means	404	101+
Gutter or drain	404	2+
Land roller	404	122+
Manhole cover	404	25
Pavement	404	17+
Apparatus for making	404	83+
Crushing in situ	404	90
Marking	404	93+
Paving machinery	404	83+
Process for making	404	72+
Railroad	238	2+
Railway graders	37	104+
Reinforcement	404	134
Roadbed	404	27+
Screed or drag	404	118+
Street curb sewage inlet	404	4
Surface marker, signaller	D10	113
Tamper	404	133
Testing	73	146
By abrasion etc	73	8
Traffic barrier	404	6+
Traffic directors in pavement	404	9+
Mirrors and reflectors	404	14
Roaster		
Coffee	99	286
Domestic ovens	126	273R+
Gas ovens and broilers	126	39R+
Metallurgical	266	171+
Miscellaneous materials	432	
Rotary driers	34	108+
Spits	99	419+
Stove broiling attachments	126	14
Vegetable and fruit peelers	99	483
Processes	426	479+
Roasting		
Beneficiating ores	75	7+
Coffee	426	523
Metallurgical apparatus	266	171+
Smelting combined	266	175
Robes		
Apparel	2	69
Design, ceremonial	D 2	79
Design, housecoat	D 2	12+
Locks for	70	60
Vehicle lap robe holders	296	77+
Robot	901	
Arm movement	901	14+
Coating	901	43*
Compliance device	901	45*
Counterbalance	901	48*
Drive systems	901	19+
Gearing	901	25*
Grip	901	31+
Industrial		
Grappling functions	294	86.4+
Magnet or piston controlled	294	88
Pivoted jaws	294	106
Gripper functions, body	623	64
Fluid actuated	623	57+
Material or article handling	414	4+
Horizontal linear motion	414	753
Horizontal swing motion	414	744R
Vertical swing motion	414	730
Motion and control, electrical	318	568
Spray painting	118	323
Structure, control lever and linkage systems	74	469+
Welding functions	219	125.1
Inspection	901	44+
Mechanically actuated	901	11+
Mobile	901	1*
Programmable		
Article handling	364	478
Artificial intelligence	364	513
Speech signal processing	364	513.5
Self organizing controls	364	513
Safety device	901	49*
Sensor	901	46+
Spray painting	901	43*
System	901	6+
Teaching system	901	3+
Track-guided	901	14+
Welding	901	42*
Wrist	901	29*
Rochelle Salt	562	585
Rock		
Bits	175	
Blasting	102	301+
Methods	102	301+
Drilling machines	175	
Rocker		
Arms machine elements		
Poppet valve operating, internal combustion engine	123	90.39+
Bearings		
Pivot	384	154+
Shaft	384	154+
Spinning spindle	384	238
Bogie truck	105	187
Chair	297	258+
Design	D 6	344+
Platform	297	261+
Cradle	5	105+
Fly shuttle	139	150
Platen for typewriter	400	649+
Sled runner	280	26
Rocket		
Aeronautics		
Airplane sustained & propelled	244	12.1+
Airship sustentation and propulsion	244	29
Chemical reaction propulsion methods	60	205+
Design	D12	16.1
Propulsion	244	74
Propulsion airplane sustained	244	15
Steering	244	52
Sustentation	244	23R
Design (pyrotechnic)	D22	112
Launchers	89	1.7
Launchers	89	1.8+
Target design	D22	113
Motors	60	200.1+
Propellants	149	
Solid suspensions	149	17
Propulsion		
Aircraft (See rocket aeronautics)	244	74
Composition for	44	50+
Self sustaining	149	
Marine	440	38+
Missile	102	374
Pyrotechnic	102	347+
Steered ships	114	151
Stick or support combined	102	349
Rocking		
Bar grate	126	176R+
Progressive feed	110	278+
Horse	272	52+
Rockwool Compositions	501	36
Rodenticides	424	
Rods		
Antenna	52	632
Mechanism extensible	52	111+
Vehicle shell attached	52	110
Antenna	343	900+
Base for signal rods	248	347
Brake	188	67
Carpet fastening stair	16	12+
Connecting	74	579R+
Locomotive	105	84
Railway switch	246	452
Construction toy element	446	85+
Control		
Nuclear reaction	376	207
Nuclear reaction	376	219
Nuclear reaction	376	327
Couplings and joints (See joints pipes, shafts, rods & plates)		
Curtain rod ornaments	52	301
Curtain rod threaders	223	105
Threading device	223	105
Cutters	30	92+
Earth boring combined	175	320+
Fishing	43	18.1+
Fluorescent	250	462.1
Lathe feed	82	27
Light transmitting	350	96.1
Illumination	362	32+
Lightning rod	174	3
Lock for control	70	181
Transmission shift	70	247+
Making		
Forming tobacco	131	84.1
Process	72	365+
Rolling screw threads on	72	95+
Wrapping tobacco	131	58+
Metal	428	544+
Mill	241	170+
Nuclear fuel	376	261
Nuclear fuel	376	347

Entry	Class	Subclass
Nuclear fuel	376	409
Piston hollow pump	417	547
Push or pull		
Bell	116	165
Janney coupling	213	162
Poppet valve operating, internal combustion engine	123	90.61
Side track	213	224
Train brake	188	148
Radiator supported	248	234
Reinforcing	52	720+
Road	404	134+
Supporting brackets	248	251+
Supports horizontal	211	123+
Driers	34	240
Locked	211	7
Surgical transparent light applicator	128	398
Surveyors	33	293
Tie railroad	238	50+
Truss or brace (See support)		
Props	248	351+
Radius rod	188	191
Rail joint	238	168
Ullage	33	126.7 R
Welding electric	219	56+
Welding, holders	228	57
Roentgen Ray (See X Ray)		
Roll Bar, Automotive	280	756
Roll-top Desk	D 6	433
Rolling (See Coiling)		
Bread pastry and confection molding	425	
Forging and rolling machines	72	206
Kinesitherapy appliance	128	57+
Metal	72	199+
Pin, combined or convertible	7	111+
Reworking worn metal by rolling		
Apparatus	72	221+
Method	72	365+
Sheet material associating or folding	270	32+
Stock for railways	105	
Rolling Pin	29	110.5
Combined or convertible	7	111
Rolls & Rollers (See Associated Function or Device, Notes to 29-110)	29	110+
Automatic sorting by	209	618+
By sizes	209	673
Bakery item	D 1	129
Bearing type	384	565+
Bearings		
Machine elements	384	
Making	29	148.4 A
Typewriter bar	400	448+
Typewriter carriage	400	354+
Vehicle frame	280	105
Weavers shuttles	139	202
Bed bolster	5	491
Cameras	354	202+
Carding cylinders	19	112
Clasps	24	488
Coating by electrolysis	204	25+
Collocating gauges for	33	657
Combining cylinders	19	128
Combs with tuned teeth	84	96
Comminuting	241	293+
Roll mills	241	221+
Corrugated surface	172	537
Couplings	403	
Currycombs, rotary	119	91
Curtain, shutters, shades & awnings		
Curtain rod support combined	248	252
Flexible & portable panels etc	160	238+
Panel hangers and travelers	16	87R+
Protective shields for counters	109	14
Reeling and unreeling fabrics	242	54R+
Supports	248	266
Vehicle	296	138+
Vehicle covers	296	98
Dispensers having	222	
Distance measuring by rolling		
contact	33	141R+
Fabric	33	129+
Drying apparatus	34	240
Educational devices	434	402
Electroforming reproductions	204	9
Expressing	100	121
Clothes wringers	68	244+
Fluid treated tobacco	131	307
Fabric manufacturing		
Cotton boll hulling	19	35+
Felting	28	130+
Fiber carding	19	98+
Fiber drawing	19	236+
Ginning	19	39+
Napping	26	29R+
Sliver forming	19	152
Spinning etc	57	
Spreader	26	99+
Web condition responsive nip control	26	78
Web edge position correcting	26	77
Weaving	139	
Feeding and conveying		
Cigar and cigarette making	131	
Circular wood saw	83	436
Continuous casting with mold	164	448
Continuous casting without mold	164	442
Copyholders	40	344+
Cutting and punching sheets and bars	83	
Label pasting strip severing	156	510
Leather skiving etc	69	13+
Live roll	198	780
Metal stock and blank	29	724
Mounted on endless conveyor	198	779
Panel hangers and travelers	16	87R+
Raisin seeders	99	558
Reciprocating wood saw	83	425+
Roller jacks	193	42
Rollerways	193	35R+
Self loading vehicle with	414	467+
Shingle sawing machine	83	706
Store service carriers	186	31
Troughing conveyor belt	198	818+
Typewriter	400	636+
Vegetable etc cutter	83	
Wood shaving	144	158
Wood slicers	144	181
Wood splitting	30	166R+
Work for abrading devices	51	
Work for woodworking machine	144	246R+
Film on spool	206	389
Filter residue contacting	210	386
Food	426	549+
Design	D 1	129
Ginning	19	54
Grinders for	51	49
Attachments	51	251+
Frames and mounts	51	179
Grinding processes	51	289S
Hair dressing foundations	132	55
Handlers for rolls	414	911*
Hawse pipe friction	114	181
Heaters fluid fuel	126	410
Heating or cooling	165	89
Irregular surface type	29	121.1+
Checkerboard type	29	121.2
Dandy roll	29	121.3
Metal deforming roller	72	191+
Sandblast type	29	121.8
Kinesitherapy type applicators	128	57
Land		
Corrugated clod crushers	172	537
Cotton chopping plow	172	518+
Cultivator combined	172	133+
Earth working tool with	172	133+
Harrow combined	172	133+
Harrow type	172	518+
Motor vehicle with	180	20
Plow	172	518+
Plow colter	172	518+
Plow combined	172	166
Potato diggers toothed	171	98
Revolving moldboard plow	172	167
Road and pavement	404	122+
Toothed clod crusher	172	540+
Leather working	69	
Machine or tool part		
Cigar making machine	131	27.1+
Clothes wringers	68	244+
Cutting tool blade	30	319
Earth boring drills	175	331+
Gear teeth	74	465
Harvesting pickers or huskers	56	104+
Ironer rolls and covers	29	110+
Key operated locks	70	386
Meat tenderers	17	25+
Mop wringers	15	262
Plural blade cutting tool	30	306+
Shoe sole laying and leveling	12	34
Shoe toe etc stiffener molding	12	65
Textile fluid treating	68	
Textile ironing or smoothing	38	
Textile roll for ironer	38	62
Thermometer combined	374	153
Work and tool forming part of circuit	374	180
Tinners	72	101
Tobacco fluid treatment	131	307
Tobacco treatment	131	324+
Tool clutch	81	183
Warp feeding whip	139	114+
Washboards	68	231
Wringer rolls	29	110+
Making bearing type	29	148.4 A
Material applying		
Belt etc lubricator	184	17
Elevator guide lubricator	184	23
Inkers for printing devices	101	335+
Lubricator	184	101
Magazine label affixing	156	
Painting combined with supply means	401	208+
Printing	101	
Railway car journal box lubrication	384	177+
Rotary shaft lubrication	384	407
Metalworking	72	199+
Beading and crimping	72	102+
Bearing type making	29	148.4 D
Bending	72	
Chain making	59	4
Conical wire spring making	72	138+
Continuous metal casting	164	448+
Corrugating	72	180
Cutting and punching sheets	83	
Forging and	72	206
Grooving	409	304
Horseshoe blanks and bars	59	63+
Horseshoe nail forging	10	66+
Ornamental chain multistrand	59	81
Rolling with other steps	29	DIG. 32
Saw making stretching	76	27
Screw threads	411	411+
Sheet metal container making	413	
Walking-form	72	220
Welding apparatus	228	
Wire feeding	72	419
Wire straightening	72	226+
Wire tension device	72	183
Wrought spike making	10	58
Motor vehicles with	180	20
Painting implement	D 4	
Paper article making		
Calenders	100	155R+
Web forming	162	357
Picking	19	97
Plastic material		
Block etc apparatus	425	363+
Block press with pressure	425	362
Bread, etc. , molding	425	363+
Bread mixers, kneaders, etc	366	69+
Casting rollers	425	223+
Casting rollers	425	363+
Glass molding	65	253+
Rolling pins	29	110.5
Combined other tools	7	111+
Pressure		
Embossing	101	3R+
Leather manufactures	69	
Metal bending	72	
Mill type	241	293+
Mills	241	221+
Plastic block mold with	425	362
Plastic block mold with	425	363
Wood veneer lathes	144	213
Wood veneer presses	156	349
Woodworking blank feeders	144	246R+
Railway		
Cable grippers with guide	104	202+
Car replacers	104	266
Derailment guards	104	245
Live roll traction	104	168
Sled track	104	135
Sounding toy rolling stock	446	410
Suspended	104	89+
Track curves	238	16
Railway car freight contacting	105	392
Railway wheel tread with multiple	295	32
Rolled material		
Bandage rolling	242	60
Bandages	128	170
Cloth bolts boards etc	242	222+

	Class	Subclass
Cores and holders	242	68$^+$
Covers for toilet seats	4	247
Fabric delivery roll holder	242	55.2$^+$
Fabric measuring	33	127$^+$
Paper dispensing packages	206	389$^+$
Reeling and unreeling fabrics	242	55$^+$
Ribbon etc packages	206	389$^+$
Selectors for automatic musical instruments	84	115$^+$
Rollerway		
Bearings for	384	418$^+$
Conveyor	193	35R$^+$
Live roll conveyor	198	780$^+$
Railway live roll traction	104	168
Railway track curves	238	16
Roller jacks	193	42$^+$
Shelf type rack mount	211	151$^+$
Sled track	104	135
Track antifriction	384	549
Seed testers	47	15
Ship fenders	114	220
Shutters photographic	354	226$^+$
Skates	D21	226$^+$
Toothed for earth working	172	122
Toothed for unearthing	171	13$^+$
Vehicle		
Chafe iron	280	161
Fenders diverging	293	18
Pilot type fender	293	49
Vehicle trolley	191	56
Vehicle type	180	20
Skates	280	11.19$^+$
Woodworking		
Saw table gauges	83	471$^+$
Sawing	83	
Spool making	144	14
Turning	142	
Turning and boring	408	
Turning and polishing	144	47
Turning and sawing	144	48
Roman		
Candles	102	357
Cement	106	100$^+$
Rompers	2	400$^+$
Design	D 2	33$^+$
Rongeur	128	312
Roofs and Roofing (See Cover)	52	
A-frame	D25	29
Car	52	45$^+$
Clerestory	52	18
Conical	52	82
Cupola	52	200
Curved or vaulted	D25	18$^+$
Design	D25	56$^+$
Dome	52	80$^+$
Floating for liquid storage tanks	220	216$^+$
Gable end	52	94$^+$
Gambrel	D25	24
Inclined	52	90$^+$
Land vehicle	296	98$^+$
Lapped surfacing or shingles	52	518$^+$
Making		
Laminated fabrics	156	
Roofing tile	425	134
Roofing tile	425	218$^+$
Roofing tile	425	289$^+$
Mansard	D25	25
Movable	52	66
Pipe joint flashed	285	42
Rack for vehicle	D12	157
Running board	52	41$^+$
Sawtooth	52	18
Scaffolds	182	45
Brackets	248	237
Sewage ventilator	4	218
Sheet	428	147$^+$
Suspended	52	83
Tile	428	147$^+$
Truss, work holder for	269	910*
Root		
Crops		
Diggers and harvesters	171	
Flax and stalk puller	171	55$^+$
Feeder and irrigator	47	48.5
Puller		
Grapple type	294	86.4$^+$
Hand fork or shovel type	294	49$^+$
Lever type	254	132
Wheel mounted for stalk pulling	171	55$^+$
Rope	57	200$^+$
Block and tackle	254	389$^+$
Braided	87	8$^+$
Clamps	24	135R$^+$
Design	D 8	72$^+$
Connectors	24	115R$^+$
Covering machines	57	3$^+$
Drum hoist	254	266$^+$
Jumping	272	75
Ladders	182	196$^+$
Making wire	57	3$^+$
Shifters	474	101$^+$
Testing	73	158
Thimble	294	74
Towing	280	480
Untwisted strands	428	364$^+$
Rosanilines	260	391$^+$
Rosary	235	123
Design	D99	26
Teaching	434	246
Roses		
Door knob plate	292	357
Plants	PLT	1$^+$
Spray rose or pipe	239	548$^+$
Gravity sprinkler type	239	376$^+$
Rosette	428	4$^+$
Rosin (See Resins)	260	97$^+$
Detergents including	252	107
Distillation processes	203	
Plant source extraction	47	10$^+$
Soaps	260	105
Mixtures containing	252	367$^+$
Rosindulins	544	343
Rossing Machines, Bark	144	208R
Rotary		
Advancing shaft driving table	464	163$^+$
Earth boring combined	175	19
Beater	241	185R$^+$
Agitator	366	
Cutter (See notes to)	30	347
Cutters for tiller blade	D15	28
Drum for abrading	51	164.1$^+$
Engine	415	
Exhaust reactor separator	60	902*
Exhaust treatment	60	901*
Plural rigidly related blade sets	415	198.1$^+$
Expansible chamber devices	418	
Heat exchanger	165	86$^+$
Kinetic fluid motor or pump	415	
Liquid filter	210	359$^+$
Perforate cage	241	86$^+$
Piston engine	123	200
Plural machines	241	138
Series flow	241	154
Processes	241	27
Rotary saw	30	388$^+$
Sprayers	239	225.1$^+$
Centrifugal force	239	214$^+$
Switches electric	200	
Rotator		
Poppet valve	123	90.1$^+$
Internal combustion engine	123	90.28$^+$
Rotenones	549	328
Rotisserie	99	419$^+$
Rotogravure	101	212$^+$
Printing plates	101	375$^+$
Photographic preparation	430	300$^+$
Rotor		
Electrical	310	261
Meter	324	154R$^+$
Machine element or flywheel	74	572$^+$
Vibration damping means	74	574
With balancing means	74	573R
Motor	60	
Internal combustion	123	
Rotary expansible chamber	418	
Propulsion or sustentation	416	
Aircraft combined	244	
Railway combined	105	66
Ship combined	440	
Pump		
Air or gas	415	
Liquid	415	
Revolution counter	235	103$^+$
Turbine	415	
Unbalance measuring	73	66
Valve (See valves)		
Winding electric armature	242	
Cylinder winding	242	7.4
Method cylinder or drum	242	7.3
Ring winding	242	4R$^+$
Transversely slotted drum	242	7.5 R
Windmill	416	
Rouge	51	307
Cosmetics	424	63$^+$
Polishing	51	309
Rougheners		
Leather	69	6.5
Roulettes	273	142R
Roundabouts	272	28R$^+$
Rounders	5	172
Rounding		
Book back	412	30
Leather machine	69	6
Shoe sole	12	85$^+$
Roving	19	236
Frames	57	315$^+$
Rowing		
Exercising machines for simulating	272	72
Land vehicles	280	224
Land vehicles	280	225
Oars	440	101$^+$
Rowlocks	440	106$^+$
R T, ie Receive Transmit or Transmit Receive	333	13
Rubber (See also Synthetic Resin or Natural Rubber)	520	1$^+$
Bale cutter	83	
Band	206	805*
Cutting	83	
Dyeing		
Preformed	8	513
Electrophoretic processes	204	180.3$^+$
Eraser	15	424$^+$
Natural (See synthetic resin or natural rubber)		
Plant source extraction	47	10$^+$
Product manufacturing, computerized	364	473
Punching	83	
Reclaiming		
Cutting from tire	82	
Roughening file	29	78
Shoes	36	4
Heels	12	147A
Making	12	142E
Overshoes	36	7.3
Making	12	142EV
Soles	12	146BR
Attaching process	12	146B
Stamps	101	368$^+$
Endless band	101	111
Factice	260	399
Handles	101	405$^+$
Reaction induced or within nuclear reactor	376	156
Vulcanization		
Molding processes	264	
Working apparatus	425	200
Rubbing (See Scrubbing)		
Amalgamators	209	67
With fluid suspension	209	191
Animal antivermin treatment	119	157
Cloth finishing	26	27$^+$
Contact circuit maker or breaker	200	164R
Multiple circuit	200	11R$^+$
Contact coating (See pencil)	401	49$^+$
Test	73	7
Thread finishing	28	219$^+$
Washboard surfaces	68	226$^+$
Rubidium		
Pyrometallurgy	75	66
Rubijervine	546	38
Ruching		
Design	D 5	7$^+$
Making	2	244
Packages	206	49
Trimming stock	112	427
Rudder		
Aircraft	244	87$^+$
Bars and pedals aircraft	244	235
Bomb	102	384
Position indicator	116	303
Ship or boat	114	162$^+$
Ruffles	112	132$^+$
Sewing	112	9
Trimming design	D 5	
Rug (See Carpet)		
Braided	428	37
Vehicle floor mat	D12	203
Ruggedized Electronic Tubes	313	
Ruhmkorff Coils	361	270$^+$
Rule and Ruler		
Blotter combined	34	89.2

Entry	Class	Subclass
Design	D10	71+
Geometrical instrument	33	483+
Folding	33	458
Printing	101	400
Punching machine combined	83	
Slide		
Bar	235	70R
Cylinder	235	79.5
Disk	235	84
Ruling Machines	33	19.1+
Ruling Pens	401	256+
Fountain	401	233
Rumble (See Tumbling)		
Coating apparatus	118	417+
Pills and confections	118	19
Coating processes	427	242
Run Preventing		
Swelling or plastizing	8	131
Rungs		
Ladder	182	194+
Collapsible with slidable rungs	182	195
Independent	182	228
Wall attached	182	90
Runner		
Earth working tool with	172	
Gear		
Land vehicle	280	80R+
Motor vehicle	180	85+
Occupant propelled vehicle	280	281R+
Hand rake ground support	56	400.15
Harness bridle gag	54	14
Land vehicle	280	12R+
Combined wheeled	280	8+
Convertible wheeled	280	7.12+
Rigging sails	114	111
Rope		
Cable hoist	212	99
Elevator control	187	45
Railway traction	104	180+
Scoops	37	118R+
Shoes	36	129
Skates	280	11.12+
Spring wire and runner mop holder	15	154
Stitch	112	173+
Track cable hoist	212	90
Umbrella	135	28
Retainer	135	38+
Walking plow	172	360
Land vehicle	280	163+
Lights	362	81
Railway car	105	457
Running Board Lights	362	81
Runway (See Conveyor)		
Engaging fasteners	24	580+
Gangway	14	70+
Marble toy	446	168
Rupturing Devices		
Rupture proofmetal receptacles	220	900*
Seal	292	243
Hasp combined	292	287
Shackle combined	292	331
Testing	73	838+
Rush, Furniture	D 6	369
Rust Inhibiting (See Anticorrosion)		
Removing	29	81R
Cleaning	134	
Electrolytic	204	141.5+
Rut Cutter	37	220
Ruthenium (See Platinum)		
Rutile	423	610
Rutin	536	8
S		
Saber	30	165+
Sabots	102	517+
Saccharic Acid	562	582
Preparation from carbohydrates	562	515
Saccharides	536	1.1+
Sacchariferous Material	127	
Condiments and flavors	426	548+
Fermentation	435	
Saccharimeter		
Chemical analysis	422	68+
Fermentative test	435	4+
Hydrometer	73	32R+
Optical analysis	356	366
Saccharin	548	211
Sachet	D28	5
Slow diffuser	239	53+
Sack (See Bag)		
Cleaning with air	15	304
Closing	53	285+
Sewing filled sacks	112	11
Closures for flexible bags	383	42+
Filling and packing machines	53	570+
Holders	248	95+
Sewing	112	
Suspensory bandage	128	162
Separable	128	161
Turners	414	754+
Weighing support	177	160
Sacramental Receptacle	206	19
Saddlebags	224	191+
Saddles and Appurtenances		
Bed bottom top saddles	5	258
Boot and shoe last	12	127
Building, roof running board or saddle	52	41+
Cable railway	104	115+
Clamp		
Joint & connection	403	234
Joint & connection	403	237
Pipe coupling, end to side or plate	285	197
Pipe elbow, cooperating clamp	285	180
Clutch shipper	192	98
Design	D30	135+
Fiber preparation, rolling fibers through stirrup & saddle	19	273
Garment support cord retainer	2	341+
Locomotive boiler	105	44
Pack	54	37
Rail joint	238	224
Rail seat	238	280
Railway		
Cablerail hanger & saddle	104	115+
Car replacer, one-way saddle	104	270
Freight tank saddl & anchor	105	362
Locomotive steam boiler saddle	105	44
Riding	54	37+
Design	D30	135
Rack, stand, holder design	D30	143
Saddlebag	224	191+
Velocipede	297	195+
Velocipede design	D 6	354
Seat		
Design, post or support	D12	119
Detachable window seat as fire escape	182	59
Saddle actuated simulation of land vehicle	280	1.183
Straddle seat	297	195+
Support, special mounted post attached	248	219.4
Shoemakers work support	12	127
Tank car	105	362
Textile fiber drawing	19	284+
Sadiron (See Flatiron)		
Design	D32	68+
Safelight	362	3+
Photo safe lamp	362	803*
Safes and Safety Deposit (See Box)	109	
Buoyant	441	32
Deposit and collection	232	
Depository with verication means	109	24.1
Photograph as record	346	22
Design	D99	28+
Incidentally movable safe	109	45+
Match	206	96+
Design	D27	32+
Plural compartment	109	53+
Provision, ventilation of	98	51
Ventilation of vaults	109	1V
Wall and panel structures	109	58+
Safety Devices (See Casing; Guards)		
Air bag passenger restraints	280	728+
Ammunition arming	102	221+
Automobile (See motor vehicle, safety promoting means)		
Bank counter partition	109	10+
Bed	5	424+
Belt	182	3+
Aircraft	244	122R
Land vehicle	280	801+
Inflatable	280	733
Motor vehicle related	180	268+
Passive	280	802+
Seat	297	464+
Blasting cartridges	102	325+
Boiler sight glass with valves	73	333
Bolt and rivet making	10	23
Closure obstruction responsive	49	26+
Cooking apparatus	99	337+
Cutting art	83	DIG. 1
Digest art	83	DIG. 1
Electric (See protection, electric)	361	1+
Alarms	340	635+
Arc lamp	314	10+
Distribution cutout	191	8
Electric lamp	315	119+
Electric switches for	200	
Elevator	187	100+
Lantern	362	157+
Lightning arrester	361	117+
Lightning protection	174	2+
Motor regulating temperature	236	89
Shock protective	174	5R
Systems	361	1+
Telegraph system	178	69G
Telephone interface	379	412+
Explosion preventing	241	31
Firearm lock	42	70.5
Revolver	42	66
Fluid handling		
Acetylene generator escape	48	56
Boiler devices	122	504+
Boiler feed heater valve	122	437
Burner controls	431	14
Burner controls	431	18+
Gas cookstove	126	42
Gas distribution	48	192
Gas holder high pressure escape	48	175
Heater cutout	236	21R
In fluid handling line	137	455+
Internal combustion charge forming	123	529
Stove water back	126	35
Valve	137	455+
Furnace, solid fuel	110	193
Fuses electric	337	142+
Making	29	623
Garments		
Arm guard	D29	20
Eye and face protectors	D29	17+
Hand protectors	D29	20+
Headware	D29	12+
Leg guard	D29	10
Glass manufacturing machine	65	159
Hat	D29	12+
Lantern for miners	362	164+
Lowering devices aeronautic	244	138R+
Machine power stop	192	129R+
Drilling or screw-threading machine	408	710
Press combined	100	53
Medicinal composition with	424	2+
Mine safety systems	299	12
Motor regulating temperature	236	89
Nuclear	376	207
Control rod, reflector, shield	376	221
Monitoring	376	245
Neutron absorbers	376	327
Safety fuse	376	336
Shielding	376	287
Paper		
Design	D19	5
Fraud preventing coating	427	7
Printed matter fraud preventing	283	74+
Pins and fastenings		
Carding	227	25
Chatelaine hooks	24	4
Guards finger ring	63	15.8
Guards for pins	24	156+
Guards for watch	63	24+
Pins, design	D11	200+
Purse	150	102
Safety pin making	163	7
Pipe joint release	285	1+
Pressure gauges	73	738+
Razor with sharpener	30	39
Receptacle		
Apparel pocket strap	2	254
Barrel bung automatic valve	217	101
Fare box	232	15+
Letter box	232	31+
Reflective triangle for highway use	40	903*
Robot	901	49*
Roll bars passenger restraints	280	756
Saddle stirrups	54	49
Screen for receptacle filling	141	97
Seat belt passenger restraints	280	801+
Separator, imperforate bowl centrifugal	494	12
Shields	266	903*

Class Subclass

- Steam boiler ... 122 504+
- Theft ... 109
- Trigger for mechanical gun ... 124 40
- Valve on muffler ... 181 237
- Valves
  - High temperature ... 137 75+
    - Line responsive ... 137 457
  - Low temperature drain ... 137 62
  - Pressure relief ... 137 455+
  - Steam boiler combined ... 122 504+
- Vehicle and railroad
  - Brake beam ... 188 210
  - Motor vehicle ... 180 271+
  - Moving train lander ... 104 19
  - Railway block systems ... 246 15+
  - Railway car bridges ... 105 458+
  - Railway car truck support ... 105 217
  - Railway sidings ... 246 374
  - Ship coupler tension release ... 114 217
  - Stock car safeguards ... 119 13
  - Wagon neck yoke coupling ... 278 124
  - Wagon thill coupling ... 278 54
  - Wagon tongue neck yoke retainers ... 278 51
- Watch and clock wheels ... 368 141+

Safety Pin ... D11 206
Safranine ... 544 348
Safrole ... 549 434
Sag Preventers ... 254 388
- Gate or door ... 49 396

Saggers ... 432 258+
Sailboat Design ... D12 303+
Sails and Sailing Devices
- Gate mounted wind balance ... 49 135
- Land vehicles ... 280 213
- Railway vehicles ... 104 24
- Ships and boats ... 114 102+
  - Design ... D12 303+

Salad Dressing ... 426 605
Sales Books ... 282
Salicin ... 536 4.1+
Salicylate ... 436 129*
Salicylic Acid ... 562 477
- Acyclic acyl esters of ... 560 143
- Aroyl esters of ... 560 8+
- Azo compound ctg. ... 534 660+

Salinomycin, in Drug ... 514 460
Saliva Ejectors ... 433 91
Salol ... 560 71
Salt (See Type of Compound)
- Condiments and flavors ... 426 650
- Crusher ... 241
- Crystallizing ... 23 295R
- Evaporating ... 159
- Fused bath ... 266 120
- Inorganic ... 23
- Ion exchange regeneration ... 210 190+
  - Process ... 210 670+
- Licks for cattle and game ... 119 51R
  - Design ... D 1 100
- Mining
  - Extraction and crystallization ... 23 298
  - Thermal or solvent ... 299 5
- Organic ... 260
- Shaker
  - Comminutor combined ... 241 168+
  - Design, mill or ... D 7 52+
  - Dredge top dispenser ... 222 480
  - Hand manipulable type dispenser ... 222 142.1+
  - Sifter type dispenser ... 222 565

Salting
- Apparatus ... 99 494
  - Bread, pastry and confections ... 99 362+
- Nuts ... 426 632+

Salvage (See Recovery; Reclaiming)
- Bell ... 405 185+
- Scrap metal ... 29, 403.1+
- Submarine vessels ... 114 44+
  - Combined with device ... 114 336

Salvarsan ... 556 66
Salves ... 424
Sample Chargers
- Analyzer supply ... 73 864.81+
- Chemical analysis ... 422 63+
- Electron microscope ... 250 440.1
- Nuclear energy detector ... 250 328+

Sample Devices
- Baggage or case ... 190 16+
- Card pasting machines ... 156 559+
- Display package ... 206 44R+
- Trays ... 206 557+

Sampler (See Pipette, Swab) ... 73 863+
- Agitator combined ... 366 140
- Body contacting ... 128
- Capture device ... 73 864+
- Chemical reaction, apparatus ... 422
- Chemical reaction, method ... 23
- Closed evaporating chamber ... 159 30
- Container combined ... 141 22+
- Depth sounding ... 33 126.4 R
- Diagnostic, liquid ... 128 760+
- Diagnostic, non-liquid ... 128 749+
- Dispensing ... 222
  - Dispenser combined ... 141 110+
  - Drop former ... 222 420
    - Movable trap chamber ... 222 356+
- Distance sounding combined ... 33 126
- Earth boring combined ... 175 308+
  - Core ... 175 244+
- Electrical signal ... 307 231+
- Evaporating chamber combined ... 159 30
- Gas separator combined ... 55 270
- Gravitational separator combined ... 210 513+
- Mass spectrometer combined ... 250 288
- Milking machine combined ... 119 14.1+
- Optical sample holder ... 356 244+
- Sample taking, cutting ... 83 919*
- Textile treatment combined ... 68 185+
- Thermometer combined ... 374 157
- Well
  - Packer and valve type ... 166 142+
  - Processes ... 166 264
  - Receptacle type ... 166 162+

Sampling, Nuclear Reactor Condition ... 376 245
Sand
- Blast abrading ... 51 410+
- Box playground ... 272 1R
  - Design ... D21 252
- Excluding bands for wheels ... 301 123
- Foundry mixing ... 366 1+
- Lime brickmaking ... 106 120+
- Liquid filter ... 210 263+
  - Rehabilitation process ... 210 791+
- Mold ... 164 349+
- Molding
  - Metal founding ... 164
  - Plastic block etc apparatus ... 118
- Paper ... 51 394+
  - Abrasive composition for ... 51 293+
  - Applying devices ... 51 275
  - Making and compositions ... 51 293+
- Reel or roll
  - Winding and reeling ... 242 117
  - Woven fabric take up ... 139 307+
- Scattering devices ... 291
- Screener or riddle ... 209 233+
- Traps and separators
  - Paper making ... 210
  - Track sanders ... 291 11R
- Washing ... 209 155+

Sandal ... 36 11.5
- Making ... 12 142S
- Overshoes ... 36 7.5+

Sanders and Sand Feed or Blast
- Boiler cleaner ... 122 395
- Brick and mold ... 118
- Coating appliers ... 118 308+
  - With electric orienting ... 118 640
- Conveyers fluid current ... 406
- Design ... D 8 62
- Laminated fabric sanding ... 156
- Sandblast abrading apparatus ... 51 410+
  - Method ... 51 319+
- Scattering unloaders ... 239 650+
- Track ... 291

Sandglass Design ... D10 44
Sanding
- Belts ... 51 357+
- Blocks ... 51 204
- Discs ... 51 358+
- Drums ... 51 358+

Sandwich Making
- Apparatus ... 99 450.1
- Process ... 426 502+

Sanitary
- Equipment & handling ... D23
- Napkins ... 604 358+
  - Belt ... D 2 625
  - Design ... D24 49
- Shield ... 604 393+

Santalol (See Alicyclic Hydrocarbon) ... 585 350+
Santonin ... 549 299

Sap Buckets ... 47 50+
- Turpentine and rubber ... 47 11

Sapogenins ... 540 17+
Saponifying
- Cellulose ether or ester fibers ... 8 130
- Dyeing textiles combined ... 8
- Fats or oils ... 260 417+
  - Mixtures ... 252 369+
- Natural resins ... 260 105

Saponin ... 536 4.1+
Saran T M ... 526 343
Sarcophagi Molds ... 425
Sarcosine ... 562 575
Sash
- Accessories
  - Balances ... 16 193+
  - Burglar alarms and fasteners ... 116 16+
  - Cord fasteners ... 16 202+
  - Cord guides ... 16 210+
  - Electric switches ... 200 61.62+
  - Weights ... 16 216+
- Garment supporter ... 2 300
- Making ... 144 5+
- Panel ... 52 782+
  - Sectioned ... 52 455+
- Window
  - Locks for ... 70 89+
  - Ventilating ... 98 97

Satchel (See Pouch)
Sateen ... 139 426R
Satellite (See Space Ship) ... 244 158R+
- Aplications, photoelectric cell ... 136 292*

Satin ... 139 426R
- Voltage magnitude control ... 323 253

Saturable Core Reactor
- External or operator controlled ... 323 329
- Input responsive ... 323 302
- Linear amplifier systems ... 330 8
- Non-linear reactor systems ... 307 401
- Output responsive ... 323 249
- Self regulating ... 323 310
- Transformer systems ... 323 250

Saturating
- Fireproofing compositions ... 252 601
- Preserving ... 422
- Transformer ... 336 155
  - External or operator controlled ... 323 329
  - Input responsive ... 323 302
  - Output responsive ... 323 249
  - Regulating system ... 323
  - Self regulating ... 323 310

Sauces ... 426 589+
Sauerkraut (See Kraut)
Sauna
- Bath ... D24 37
- Heater ... D23 315

Sausage ... 426 129
- Casing ... 426 140+
  - Inedible ... 428 36
  - Shirred casing ... 206 802*
  - Wall structure ... 138
- Cookers
  - Electrode type ... 99 358
  - Opposed heated surface type ... 99 380+
  - Spit or impaling type ... 99 419+
  - Support combined ... 99 441
- Grinders ... 241
- Linkers ... 17 34
  - Stuffers combined ... 17 33
- Stuffers ... 17 35+

Savings Boxes ... 232 4R+
- Paper type ... 229 8.5
- Toy ... 446 8+
- With register ... 235 100

Saw ... 30 166R
- Band ... 30 380
- Band saw ... 83 788+
- Barrel head making ... 147 38
- Bone ... 128 317
  - Trephine ... 128 310
- Bow ... 30 507+
- Butchering ... 17 23
- Button blank sawing ... 79 15+
- Chain ... 30 381+
- Chain saw ... 83 788+
- Cleaners and oilers ... 83 169
- Dado ... 144 222
- Drag saw ... 83 771+
- Frame ... 83 859
  - Hand ... 30 166R+
- Fret ... 30 513

	Class	Subclass
Gauge	33	202
Gin	19	55R+
Gummers	76	25R+
Wheels	76	45
Handsaws	30	166R
Combined with other cutter	30	144
Combined with other tools	7	148+
Hole saw	408	204
Making	76	25R+
Methods	76	112
Metal sawing machines	83	
Saws	83	835+
Rotary	30	388+
Scroll	83	783+
Hand	30	513+
Sets	76	58+
Stave jointing	147	28+
Stone sawing	125	12
Surgical	128	317
Table saw	83	477+
Tables	269	289R
Gauges	83	522
Modified for blade	269	83+
Teeth	83	835+
Fastenings	83	840+
Vegetable and meat combined with cutter	83	
Wood sawing devices	83	
Design hand saw	D 8	95+
Design, power saws	D 8	64+
Design, sawing machinery	D15	133
Guards	83	440.2
Guides	83	821+
Wood slivering gang	144	189
Woodworking rotary cutter	144	222+
Sawbucks and Sawhorses		
Buck, ie holding or clamping	269	296+
Horse, ie scaffold or plank supporting	182	181+
Sawdust		
Briquettes	44	10B+
Burners (See furnaces solid fuel burning sawdust refuse and wet fuel)		
Fur cleaning by	69	23+
Making processes	241	28
Sawhorse	D25	67
Sawing (See Saw)		
Button blanks	79	15+
Dog	269	
Feed	83	401+
Metal	83	
Sawmill carriage wheel guards	37	194
Sawmill dogs with carriage	83	721
Stone	125	12+
Wood	83	
Saw table gauges	83	522
Sawtooth Oscillator		
Driven	328	193+
Free running	331	
Saxophones	84	385R
Scabbards (See Sheath)	224	232+
Design	D 3	102
Ice skate	280	825
Sword	224	232+
Design	D 3	102
Scaffolds	182	179
Bracket supports	248	235+
Combined	182	129
Convertible	182	20
Ladder	182	27
Design	D25	62+
Design, element or coupling	D25	68+
Elevating means	182	141+
Foldable	182	152+
Platform	182	222+
Shaft supported	182	128
External	182	187
Scalder		
Cleaning	134	
Fowl	17	11.2
Hog	17	13+
Peeler or parer combined	99	483
Scale (See Weighers)		
Armored tire	152	206
Bathroom	D10	87
Lamellate composition	106	291
Metal scaling	29	DIG. 34
Music		
Automatic note selector	84	152+
Enharmonic	84	451
Keyboard touch plate	84	428
Optical element combined	350	110+
Optical pointers	350	110+
Pans	177	262+
Price	177	25+
Register	235	124
Remover and preventer		
Composition	252	80+
Dentistry	433	143
Electrolytic apparatus	204	196+
Electrolytic process	204	147+
Fish	17	64
Hammer	29	81D
Magnetic apparatus	422	186.2+
Metalworking	29	81R
Water softener	252	175+
Rule or	D10	71+
Ruler	33	494
Square	33	474
Typewriter	400	705+
Line spacing	400	548+
Weighing instrument	177	
Design	D10	87
Scalpels	128	305+
Scandium	156	DIG. 114
Scanning		
Antenna	343	757+
Periodically moving optical element	350	6.1+
Periscope	350	540+
Photoelectric	250	200+
Television	358	199+
Telescope	350	537+
Scarecrow	40	
Noise maker		
Bells	116	148+
Horns	116	137R+
Scarf		
Apparel	2	207
Design fur	D 2	601
Design haberdashery item	D 2	500+
Design shaped	D 2	500+
Design shaped ascot	D 2	605+
Design with evening dress	D 2	53+
Hanger for fur	223	97
Joint	238	230+
Slide holder	D 2	202+
Scarfing (See Skiving)		
Apparatus for metal	266	51+
Blowpipes	239	398+
Processes for metal	148	9R
Scarifier		
Earthworking	172	
With scraper	37	108R+
Stamp	81	9.21
Surgical		
Lancet	128	315
Other instruments	128	333
Receptor combined	604	22
Scattering (See Sprayer; Sprinkling)		
Drying apparatus centrifugal	34	59
From aircraft	244	136
Insect powder dusters	239	
Non-fluid material scattering	239	650+
Planting broadcasting	111	8+
Pneumatic conveyor discharge	406	157+
Road treating material	404	101+
Track sanders	291	
Scavenger & Scavenging		
Anesthesia gas scavenging system	128	910*
Oxygen scavinging	252	186.1+
Pyrometallurgy	75	58
School		
Desks and seats	297	135+
Design	D 6	335+
Design, with drawer	D 6	335+
Schreinerized Fabrics	26	69R
Science Teaching	434	276+
Scillin	536	4.1+
Scissors (See Nipper; Shearing)	30	194+
Buttonhole	30	120
Can opening	30	428
Design	D 8	57
Knife combined	30	146
Manicuring	30	29
Compound tool	132	75.4+
Material holder combined	30	131+
Nipper combined	30	145
Plier combined	7	135+
Sharpening		
Abrading design	D 8	93
Apparatus	76	82.2
Combined with can opener	D 8	35
Power sharpener	D 8	63
Process	51	285
Scleroscope	73	79
Scoop	D 7	104
Bag filling	141	108+
Ditchers	37	103
Dredgers	37	71
Excavating	37	118R+
Ice cream	425	118
Ice cream	425	221
Ice cream	425	276+
Design	D 7	100
Kitchen	294	55
Pump	417	232
Railway graders	37	106+
Rotary dryer	34	188
Shovel	294	55
Sifter support	209	418
Snow excavator	37	264+
Vehicle fender	293	42+
Scooter Vehicle	280	87.1+
Toy	D21	81
Scopolamine	546	91
In drug	514	291
Score Boards		
Bowling	273	37
Changeable	116	222+
Electric sign, matrix	340	780
Electric sign, monogram	340	760+
Computer controlled, monitored	364	411
Electric annunciator	340	323R
Electrical registering type	377	5
For games	116	222
Electrical	340	323R+
Games	235	90
Image projection	353	45
Printed matter	283	48R+
Target combined electrical	273	371+
Target combined mechanical	273	378+
Scorers for Games	D10	46.1
Scoring (See Notes to)	33	18.1+
Implements	30	164.9
Indenting stylus	81	9.2
Machines	83	861+
Metal working	29	DIG. 33
Of paper	493	396
Paper bag making	493	228+
Paper box making	493	59+
Plane	144	186+
Registers	235	
Score line, support	248	DIG. 9
Static mold for making scored item	249	52
Wood slicing combined	144	42
Scotch Block	104	258
Scotch Yoke	74	50
Scourer and Scouring		
Abrading	51	16+
Polishing combined	51	4
Tools	51	391+
Tumbling	51	163.1+
Concentration of natural current	405	73
Dredger	37	75+
Grain	99	600+
Processes	426	483+
Pad, household use	D32	40+
Textile cleaning	8	139+
Scows (See Barges)	114	26+
Scram Rod (See Nuclear)	376	207
Component structure	376	327
Element	376	219
Scrap (See Trash)		
Binder device releasably engaging aperture or notch of sheet	402	
Bindings	281	22
Metal		
Box	428	576
Cutting	83	
Electrolytic stripping apparatus	204	193+
Electrolytic stripping processes	204	146
Pile or fagot	428	588+
Preheating or melting	266	901*
Reclaiming tin by pyrometallurgy	75	64
Tin reclaiming by hydrometallurgy	75	98
Scraper (See Rake; Removers)		
Boiler cleaning	122	387
Bristle brush combined	D 4	116+
Bulldozer	172	811+
Carbon black making	422	150+
Caster wheel attachment	16	41
Cleaning device	15	236R+

	Class	Subclass
Brush combined	15	111+
Brush paint can attachment	220	90
Design, food	D32	40+
Design, food preparation	D 7	99+
Inside pipe or tube	15	104.16+
Machine	15	93R+
Vacuum cleaner nozzle combined	15	401
Windshield wiper combined	15	256.5+
Conveying from heat		
Exchange surface	165	95
Conveyor belt unloader	198	637
Cooking or baking molds combined	249	77
Cutting tools	30	169+
With material holder or disposal	30	136
Drying rack	34	185
Drying rake attached	56	400.5+
Earthworking	172	
Railway snow excavator	37	214+
Transverse endless road grader	37	109
Eraser	30	169+
Burnisher combined	7	124+
Filter cleaning	210	407+
Furnace ashpan	110	170
Grinding or milling member cleaner	241	166+
Heat exchange cylinder	165	97+
Hog scraper	17	16+
Scalder combined	17	13+
Household cleaning tool	D32	35+
Intaglio printing wiper	101	169
Rotary with transfer	101	154
Wiper for rotary	101	157
Land vehicle wheel attached	280	158R+
Leather manufactures	69	
Metallic receptacle combined	220	90
Pipe bowl	131	246
Plow or cultivator	172	610
Porous block filter cleaner	210	116+
Sealed bearing	384	137
Shoe	15	237+
Design, boot or	D32	47
Scraper for vehicle	D12	203
Snowplow	37	266+
Stamp scarifier	81	9.21
Surgical	128	304
Design	D24	28
Wick trimmer	431	120
Woodworking	144	115
Scratcher		
Ash receiver combined	131	234
Back	128	62R
Cigar cutter combined	131	249
Cigar or cigarette combined	131	351
Match	44	48
Match combined	44	45
Safe combined	206	137+
Smoking device combined	131	185
Well	166	170+
Screed		
Plastering	52	364+
Corner	52	255+
Road or pavement	404	118+
Screen (See Shield; Sifter)		
Bed	5	512
Combined	5	163
Cathode ray tube	313	461+
Color lantern	362	166+
Decorative panels	D 6	332
Elevator door	187	55
Fireplace	126	202
Fluid current	210	348+
Dredgers with	37	57
Faucet combined	210	449
Flume	210	154+
Gas filtration	55	
Restrictor with	138	41
Starch making	127	25
Track sander	291	48
Track sander fluid delivery	291	13
Fluorescent	250	483.1
Furnace door fluid screen	110	179
Furniture	160	135
Gas filtering	55	
Glare auto	296	97R
Grid	313	297+
Structure	313	348+
Halftone	350	322+
Chemically defined	430	6+
Insect trap with	43	119
Luminescent		
Metal treatment	148	149
Nuclear reactions	376	153

	Class	Subclass
Nuclear reactions	376	156
Nuclear reactions	376	463
Process of making	430	23+
Optical filter	350	311+
Color photography	354	100+
Infra-red	350	1.1+
Lenticular	354	101
Photographic chemically defined	430	4
Ultra-violet	350	1.1+
Phosphorescent	250	483.1
Photographic	D16	43
Photographic printing	350	322+
Chemically defined	430	6+
Plant	47	26+
Frost preventing	47	2
Tree covers	47	20
Potato digger	171	71+
Potato digger	171	111+
Plow combined	171	71+
Plow combined	171	111+
Projection, ie movie	350	117+
Projector light	362	257+
Racks with	211	180
Safety for filling devices	141	97
Seat combined	297	184
Sifters	209	233
Special ray, eg X-ray	250	483.1
Stuff working		
Forming fiber articles	19	148
Forming fiber webs	19	304+
Paper pulp	210	348+
Track sander	291	48
Typewriter keyboard	400	714+
Urinal	4	292+
Warship armor	114	14
Well screen	166	227+
Earth boring means combined	175	314
Making	29	163.5 R
Window	160	
Design	D25	53
Wire fabric	245	
Making	140	3R+
X ray		
Bucky grid	378	154
Film	378	185
Fluorescent or phosphorescent	250	483.1
Tube with secondary ray	378	140
Screw (See Threaded)	411	
Adjusted or actuated (See --, valve)		
Chair pedestal design	D 6	364+
Dump door railway car	105	307
Hoop driver	147	11
Mechanical pencil lead feed	401	68+
Puller	29	256+
Punch	83	631
Railway track pedestal	238	120
Reciprocating bender	72	454
Stand	248	405+
Stool	108	138
Switch	200	158
Tube and bar drawing	72	287
Valve applier	29	217+
Wrench	81	155+
Apparel fastener	D11	200+
Capping nails & screws	10	156+
Clamp grip and vise		
Belt fastener clamp	24	37
Boot antislip anchor	36	65
Cable grippers railway	104	210
Cable grippers railway	104	223
Clamps	269	86+
Fence rail cross connections	256	69
Fence wire cross connections	256	55
Mop holder	15	153
Rope holders nut clamp	24	135R+
Skate toe or heel	280	11.32
Vises	269	
Closure		
Breech ordnance	89	19+
Dispenser	222	549+
Jar or bottle cap	215	329+
Jar or bottle stopper	215	356+
Metallic receptacle	220	288
Pipe and tubular conduits	138	89+
Watchcase	368	310
Wooden barrel bungs	217	107
Closure remover	81	3.45
Connectors electric	439	
Fixture	439	641+
Outlet box	439	
Switch with	200	51.14+

	Class	Subclass
Conveyor & feeder (See conveyor, screw; feeder)	209	913*
Acetylene generator carbide feed	48	54
Drill feed	408	129+
Elevator drive	187	24+
Furnace type	414	197
Hollow heat exchanging	165	87
Lathe feed	82	27
Locomotive tender feed	105	234
Mill feed	241	246+
Sand feed track sander	291	33
Sausage stuffer	17	40
Screw or cam conveyor	29	DIG. 1
Static receptacle	414	
Stoker feed	110	110
Typewriter feed	400	328+
Vegetable and meat cutting and comminuting feeds	241	
Wood slicer feed	144	179
Design	D 8	387
Driver (See screwdriver)	81	436+
Constructed from specific material	81	900*
Electrical connection formine	7	108
With screw feed	81	57.37
Earth boring bit	175	327+
Excavators	37	81+
Extractor	81	436+
Fastenings		
Bottle or jar closure clamp	215	281+
Bottle or jar closure clamp	215	283
Bottle or jar closure pivoted fastener	215	243
Fluorescent or phosphorescent	250	462.1
Horseshoe	168	23
Latch bolt	292	251
Latch bolt operator	292	
Lock bolt	70	140
Machine i. e. bolt	411	378+
Rail joint chair	238	202
Rail joint slide fastener	238	332
Rail joint tie plate	238	303
Separable button	24	105
Setscrew	403	362
Setscrew knob	292	350+
Spikes railway	238	372+
Tool handle	403	
Watchcase bow pendant	368	304
Wood	411	378+
Gearing	74	424.7
Clutch operator	192	94
Intermittent mechanical movements	74	111+
Mechanical movements	74	20+
Nut with	74	424.8 R
Power limit stop	192	141
Ship steering	114	155+
Steering control	74	499
Jack lifter stretcher & tightener	254	98+
Barrel hoop	217	95
Combined	254	7R
Electrostatic	254	92
Nail extractor	254	7R
Printers quoins	254	41
Sail	114	109+
Stretcher	254	231+
Carpet	254	200+
Traversing jacks	254	85
Making	10	2+
Abrading	51	288
Capping	10	156+
Lathe cutting	82	5+
Machine	29	33R+
Mold	164	216
Molding glass	65	309+
Rolling	72	88+
Rolling process	72	365+
Threading	408	
Wire and inserting	29	564.6
Wood	10	2+
Pipe and rod joints and connectors		
Chucks	279	
Electrical connectors	439	191+
Rod joints	403	343
Sockets	279	9R+
Ties railway screw rods	238	53
Press		
Die cutting	83	631
Dispenser follower	222	390
Forging	72	454
Propeller		
Air	416	176

	Class	Subclass
Aircraft with	244	65$^+$
Ship	440	49$^+$
Ship manual drive	440	21
Pump	415	71$^+$
Agitating	366	266
Setscrew guard	74	617
Setscrew guard knob	292	351
Thread gauge	33	199R
Threading	10	87$^+$
Valve	251	216
Bottle	215	314
Dispenser	222	549$^+$
Screwdriver	81	436$^+$
Combined with diverse tools	7	165$^+$
Combined with diverse wrench	81	437
Constructed from specific material	81	900*
Design	D 8	82
Floor jack	254	13
Eye screw, adapted for	81	901*
Machine	81	54
Scriber (See Scoring)	33	18.1$^+$
Barrel head making	147	37
Hand device	147	41
Scroll		
Actuated chuck	279	114$^+$
Display control	340	724$^+$
Saw	83	783$^+$
Hand	30	513$^+$
Handle	30	525
Type comminutor	241	276
Scrubbing and Scrubber		
Bathing apparatus combined	4	606
Brushes	15	159R$^+$
Holders	15	146
Material supply combined	401	268$^+$
Design, equipment	D32	35$^+$
Floor and wall machines	15	50R$^+$
Gas (See gas, liquid contact with)	261	
Implement, general cleaning	15	104R$^+$
Implements for textile washing	68	220
Kneeling stools	182	230
With tool holder	182	129
Machines for cleaning	15	3
Mops	15	208$^+$
Strand material	15	226
Pails	15	264
Textile washing machines	68	
Washing or cleaning methods	134	6$^+$
Washtub surface	68	233
Sculpting		
Ariticles for	D19	35
Chisel	D 8	47
Hammer	D 8	75$^+$
Knife or spatula	D 8	98$^+$
Sculpture (See Product)		
Design	D11	131$^+$
Fauna	428	16
Layer varying thickness	428	156$^+$
Manipulable	446	268$^+$
Milling machine	409	79$^+$
Photographic process	430	323$^+$
Teaching	434	82$^+$
Scupper	114	182
Scutching	19	5R$^+$
Beaters	19	85$^+$
Scuttle		
Operators	49	324$^+$
Scythe	30	309
Design	D 8	12
Handle fastening	403	
Nib joints	403	
Sea Calming	114	232$^+$
Seafood	426	643$^+$
Preservation	426	321$^+$
Antiseptic and pickling	426	654
Canning	426	392$^+$
Dehydration	426	471$^+$
Refrigeration	426	524
Weed compositions	536	3
Seal, Sealer, & Sealing		
Apparatus for making plastic seals	425	
Molding with reinforcements	425	500$^+$
Band or wire applier	140	150$^+$
Battery	429	172$^+$
Bearing		
Ball combined	384	477$^+$
Ball combined with lubrication	384	462$^+$
Plain	384	130$^+$
Books having	281	18
Bottle cap	425	809*
Caulking ships	114	86

	Class	Subclass
Caulking tools	81	8.1
Metalworking	72	463
Cigar or cigarette wrapper	131	67$^+$
Closure	49	475$^+$
Actuated	49	303$^+$
Operator	49	316$^+$
Conduit cable or conductor ends	174	77R
Container opening	206	601$^+$
Cylindrical cell	429	171$^+$
Electron tube sealing	53	DIG. 3
Envelope sealing & stamping	156	442
Fuel cell	429	35$^+$
Glass		
Compositions	501	15
Electronic tube having	174	50.5$^+$
Gas or vapor lamp having	313	623$^+$
Incandescent lamp having	174	50.5$^+$
Incandescent lamp having	220	2.1 R$^+$
Incandescent lamp having	174	17.5$^+$
Lamp or discharge device making		
Apparatus	65	152$^+$
Processes	65	36$^+$
Hermetic (See air lock)		
Drying or gas contact		
Chamber having	34	242
Processes using	34	15
Inflatable toys having	446	222
Preserving food by	426	392$^+$
Still	202	269
Valve actuator	251	335.1$^+$
High frequency electric sealing	53	DIG. 2
Inflatable closure	49	477
Light		
Plate holders having	354	277
Liquid (See trap)		
Barrel bung	217	105
Bottle and jar closure	215	234
Dispenser type	222	
Flow meter having	73	234$^+$
Gas making cupola having	48	69
Gas separator combined	55	355
Hardenable liquid jar closure	215	233
Liquid seal only	429	78
Machine, ie bolt	285	10$^+$
Metallic receptacle closure	220	216$^+$
Packed rod joint	277	135
Plastic seal jar closure	215	233
Rotary expansible chamber device	418	104$^+$
Sewerage having	137	247.11$^+$
Valve	137	251.1$^+$
Valve interface	137	246$^+$
Liquid barrier	277	135
Centrifugal	277	13$^+$
Lock, closure fastener	292	307R$^+$
Barrel bung	217	114
Bottle cap with separate fastener.	215	273$^+$
Bottle or jar closure	215	200$^+$
Closure fastener bolt	292	327
Closure fasteners with	292	
Electric connector	439	271$^+$
Envelope	229	80$^+$
Envelope fastener	229	79
Fountain pen cap combined	401	245
Lock combined	70	440
Metallic closure	220	200$^+$
Padlock combined	70	50
Safe door jamb	109	64$^+$
Weather strips	49	475$^+$
Zipper	24	418$^+$
Magnetic closure	49	478
Making sheet metal container seams or solder seals	413	
Moistener combined	156	441.5$^+$
Movable shafts		
Packing material	277	
Stuffing boxes	277	
Notary	101	3R$^+$
Nuclear fuel components	376	451
Sealing wax applier	401	1$^+$
Separator, imperforate bowl centrifugal	494	38$^+$
Stamp or	D19	12
Tamper indicating		
Fluid handling	137	384
Pipe joint	285	80
Terminal	429	181$^+$
Wax		
Appliers	401	1$^+$
Compositions	106	218$^+$
Zipper surface sealed	24	389
Sealed Beam Lights	313	113$^+$

	Class	Subclass
Seam		
Adherent making		
Apparatus	156	
Process, butt ending	156	304.1
Bag bottom seamless	383	122
Binding devices for sewing machine	112	137$^+$
Box paper	229	48R
Calking on ships	114	86
Cloth rubber and leather	2	275$^+$
Pressing	69	7
Sewed seams	112	262.1$^+$
Sewing	112	
Shoe upper	36	57
Hosiery	66	179
Simulated	66	181
Metal		
Can side making	413	69$^+$
Container spout	222	574
Fusion-bonding apparatus	228	
Opening device	220	260$^+$
Receptacle	220	75$^+$
Receptacle end structure	220	67
Sheet edge interfolding	29	243.5$^+$
Sheet methods	413	1$^+$
Tube of sheet material	72	48$^+$
Wire fabric having	245	10
Pipe having longitudinal	138	156$^+$
Distinct layers	138	151$^+$
Flexible	138	128
Pipe having spiral	138	154
Distinct layers	138	150
Flexible	138	129$^+$
Receptacle paper crimped end	229	5.6
Searchlight	362	35
Boat type	362	61$^+$
Bulbs	313	
Searing		
Leather	69	7.5
Seasonings	426	650$^+$
Seat	297	
Actuated		
Abrading work holder	51	226
Brake	188	109
Dry closet	4	473
Flush valve	4	250
Seat and cover operator	4	251
Seat subclosure	4	238
Trackmans car	105	94
Water closet	4	408$^+$
Agricultural implement	172	431$^+$
Aircraft	244	122R
Safety lowering devices	244	141
Backless	D 6	349$^+$
Backpack convertible to	224	155
Bath and closet	4	237$^+$
Bathtub	4	578$^+$
Chemical holder in seat	4	229
Covers	4	242$^+$
Covers combined	4	234$^+$
Disinfecting	4	233
Dry closet subclosure	4	485$^+$
Seat and cover operator	4	251
Ventilated	4	217
Bed	5	432$^+$
Belt passenger restraints	280	801$^+$
Safety buckle	D11	200$^+$
Boat	114	363
Cabinet with	312	235R
Convertible to railway berth	105	322
Design	D 6	334$^+$
Grinders, milling	409	64$^+$
Guards and control for amusement railways etc	104	241
Harvester combined	56	323
Heated	4	DIG. 6
Land vehicle	296	63$^+$
Rear seat windshields	296	85
Light combined	362	131
Vehicle	362	75
Light thermal and electrical application	128	376$^+$
Locks	70	261
Mechanical parts		
Bottle or jar closure cap	215	316$^+$
Pivoted seat motor	185	15.5
Rail seats	238	264$^+$
Railway rail joint base plate	238	215
Tool handle fastening		
Adjustable	403	83$^+$
Tool handle fastening rigid	403	
Valve (See valve and valved		

U.S. PATENT CLASSIFICATIONS

	Class	Subclass
devices)		
Occupancy indicator	116	200^{+}
Electric	340	667
Occupant carried	297	4
Occupant guards or restraints	297	464^{+}
Passenger recording by depression	346	40
Railway passenger car		
Convertible open and closed	105	336
Placement	105	345^{+}
Seat back side guard control	105	441
Rubber	4	DIG. 8
Saddle tree and seat connection	54	42
Sofa bed	5	12R^{+}
Toilet	4	237^{+}
Trackmans car	105	94
Vacant seat signalling system	340	667^{+}
Vehicle		
Auxiliary	297	232^{+}
Velocipede	297	195^{+}
With body modification	296	63^{+}
Walk over	297	94^{+}
Seat Belt Restraints	297	464^{+}
Safety buckle	D11	200^{+}
Seaweed, Artificial	405	24
Sebacic Acid	562	590
Second Detector Radio	329	
Secondary		
Battery	429	149^{+}
Charging & discharging systems	320	2^{+}
Emitters	313	103R^{+}
Amplifying system	330	42
Cathode ray tube	313	399
Cathode ray tube	313	387
Cathode ray tube system	315	11
Cathode ray tube system	315	12.1
Prevention	313	106^{+}
Secret		
Communication		
Electronic, funds transfer	380	24
Facsimile	380	18
Radio	380	31^{+}
Radio, pulse or digital	380	37
Telegraph	380	9^{+}
Telephone	380	41
Television	380	10^{+}
Access control	358	349
Compartment type cabinet	312	204
Cryptography photo copying	355	40^{+}
Including fiber optics	355	1
Sectional		
Boilers	122	209R
Radiators	165	130
Sectional Unit Cabinet (See Knockdown)	312	107^{+}
Cooking apparatus	99	449
Horizontally attached receptacle sets	220	23.4
Securing Means (See Connection; Lock; Retainers)		
Closure	292	
Bendable	292	253
Locks	70	
Portable	292	289^{+}
Grille	52	507
Rims land vehicles	152	
Demountable	301	11R^{+}
Sheathing	52	506^{+}
Shingle type	52	443^{+}
Sheet members		
Billiard cloth	273	7
Fabric (See notes to)	160	382^{+}
Sheet members on drums	346	138
Spokes	157	1.55
Tires land vehicles	152	
Tires railway	295	15^{+}
Vehicle parts		
Bodies	296	35.1^{+}
Bodies on & off freight containing	410	77
Mud guards	280	154
Top and side boards	296	36
Wheels railway	295	43^{+}
Securities, Commodities		
Computer controlled, monitored	364	408
Security Instruments	D10	104^{+}
Metal detector for airports	340	568^{+}
Metal detector for library books	340	572
Sediment		
Feed heater valve	122	429
Filter combined	210	299^{+}
Liquid purification or separation		
Automatic discharge	210	112^{+}
Gravitational apparatus	210	513^{+}

	Class	Subclass
Gravitational process	210	800^{+}
Precipitation process	210	702^{+}
Seed		
Coated or impregnated	47	57.6
Comminuting	241	6^{+}
Cooker	422	307^{+}
Disinfecting compositions	424	
Extraction, fatty oil from seed	260	412.2
Germinator	47	61
Apparatus fermentative	435	287
Fermentation combined	435	
Proteins	530	370
Coating or plastic compositions containing	106	154.1
Removing or separating		
Decorticating combined	19	6
Fruit apparatus	99	547
Fruits and vegetables processes	426	484^{+}
Ginning	19	39^{+}
Rod holders	156	DIG. 98
Seedling containers	493	951*
Seedling transplanters	47	77^{+}
Stripping or gathering	56	126^{+}
Motorized	56	12.9^{+}
Threshing type	130	30R
Tapes	47	56
Making, methods of	53	467^{+}
Planting	111	14^{+}
Testers	47	14^{+}
Untreated	800	1
Seed Corn Stringers	139	19
Seed Pulling, Crystalization	156	617R^{+}
Seeding	111	
Crystal growing or forming	23	301
Sugar making or refining	127	60
Seersucker	139	384R
Making	139	24
Seesaw		
Amusement device	272	54^{+}
Combined with roundabout	272	30
Design	D21	251
Amusement railway		
Car drive	104	80
Swinging car	104	81
Track	104	79
Figure toy	446	322
Seger Cone	374	106
Segment		
Closure fastener keeper	292	341
Harvester platform adjuster	56	215^{+}
Nuclear fuel	376	429
Pinion	74	32
Rim gear	74	448
Ship steering mechanism	114	159
Segmenter	30	114
Corer combined	99	545^{+}
Parer combined	99	543
Seine Needles	87	60
Seines	43	14
Seismology		
Computer controlled, monitored	364	421
Seismic ware prospecting (See prospecting)		
Computer combined apparatus	367	60
Earth boring combined	175	1
Instruments	33	1HH
Seismographs or seismometers	181	
Electro-acoustic	367	14^{+}
Mechanical	73	649^{+}
Recording	346	7
Submarine signaling	367	15^{+}
Vibration testing	73	649^{+}
Selective Delivery		
Charging discrete loads to plural static structures	414	266
Conveyor power driven	198	348^{+}
Material treatment		
Differentially comminuted mixed	241	14
Preceding sorting	209	4^{+}
Pneumatic dispatch	406	1^{+}
Railway	104	88
Signal and switch		
Electric annunciator	340	500^{+}
Electric signal	340	825^{+}
Railway automatic switch	246	323
Railway facing switch	246	424^{+}
Railway train dispatching	246	5
Store service	186	20^{+}
Telephone hook switch	379	424^{+}
Telephony automatic switches	335	108^{+}
Time controlled alarm	340	309.15^{+}

	Class	Subclass
Store service		
Cable propulsion	186	15
Gravity system	186	3
Switches	186	20^{+}
Suspended road vehicle unloading	414	377^{+}
Selective Systems	340	825
Selector		
Dispensing magazine		
Stacked article ejector	221	92^{+}
Electric annunciator station	340	286R^{+}
Electric organ tone partials	84	1.19^{+}
Gear shifting	74	342^{+}
Label pasting machine	156	
Lock or latch operator		
For cabinets	312	215
Lock tumbler type	70	305
Musical automatic instrument	84	115^{+}
Musical instrument key	84	442
Phonograph cabinet disk record	312	15^{+}
Pile fabric selected warp	139	404^{+}
Planetary gearing	74	754
Printing telegraph	178	33R^{+}
Printing telegraph page	178	27
Punch	234	
Coded	234	94^{+}
Railway dump car door	105	287
Tabulator column	400	284^{+}
Telephone automatic testing	379	17
Textile warp preparing	28	202
Typewriter case	400	251^{+}
Weaving separate weft	139	453
Zither chord	84	286^{+}
Selenazine	544	1
Selenazole	548	100
Selenium	423	508
Cell		
Generator type	136	243
Resistive type	338	15^{+}
Coating	437	
Electrolytic	204	
Compounds		
Binary	423	509
Organic	260	550
Ternary oxy	423	508
Misc. chemical manufacture	156	DIG. 72
Rectifier	357	
Making	29	569.1^{+}
Process	437	
Self Aligning Bearings		
Ball	384	495^{+}
Plain radial	384	192^{+}
Plain thrust	384	193^{+}
Railway car brasses	384	191.1
Shaft hangers	384	205
Self Baked Electrodes Furnace		
Arc	373	60^{+}
Electrolytic apparatus	204	225^{+}
Self Binding		
Cornstalk cutters	56	67^{+}
Shockers	56	429
Shockers automatic feed	56	403^{+}
Self Healing		
Stock material	428	912*
Tires	152	502^{+}
Self Heating		
Article, by electricity	219	201^{+}
Self Leveling		
Furniture	114	191^{+}
Support	248	182
Tank or boiler	280	7
Vehicle body	280	6R^{+}
Self Loading Vehicle		
Excavator	37	4^{+}
Road vehicle	414	467^{+}
Self Service		
Plural related surfaces horizontally adjustable	108	102^{+}
Stores	52	33
Tables, power driven surface	108	20^{+}
Tables, terraced rotatable	108	94^{+}
Self Starter for Engines		
Dynamoelectric	290	
Mechanical	123	179R
Self Winding		
Clocks	368	148
Spring & weight motors combined	185	7
Spring motor	185	40R^{+}
Plural	185	11^{+}
Toys spinning	446	249
Watches	368	207
Weight motor	185	33

	Class	Subclass
Selsyn	318	692+
Communication systems	340	315
Telautograph	178	19+
Telemetering	340	870.34
Selvage		
Edge coating	427	285
Knitting		
Welt fabric	66	172R+
Welt machines	66	95+
Trimming	26	11+
On loom	139	302+
Weaving		
Needle layed weft	139	430+
Warp traversing	139	54
Weft loop retainers	139	195
Semaphore		
Flags	116	173+
Hand set indicator	116	319
Spectacle	246	484
Structure	246	479+
Switch stand	246	409
Train dispatching	246	18
Semi-compound Steam Engines	91	8+
Semicarbazide	564	34
Semicarbazones	564	36
Semiconductive Alloys	420	903*
Semiconductor Device Amplifier (See Amplifier; Transistor)		
Laser induced diffusion of impurities into semiconductor substrates		
Device	357	7+
Method	437	16+
Making	29	569.1+
Process	437	
Coating by	437	+
Etching by	156	625+
Mesa formation	156	649+
Photo etching by	430	311+
Semiconductor Device Making (see Semiconductor Junction Forming)		
Process of manufacture	437	
Applying corpuscular or electromagnetic energy	437	173
Coating and etching	437	228
Coating or material removal	437	225+
Dicing	437	226+
Directly applying electric current	437	170+
Electroless coating	437	230
Forming schottky contact	437	175+
Including shaping	437	249
Making or attaching electrode	437	180+
Securing completed device to mounting or housing	437	180+
Temperature modifying property of device	437	247+
Self-aligning device	437	984*
Semiconductor Junction Forming		
Diffusing dopant	437	142+
From melt	437	159
From solid source in contact with substrate	437	160+
From vapor phase	437	165+
Shallow	437	950*
Doping with material deposition	437	81+
Fusing dopant with substrate	437	134+
Heteroepitaxy	437	126+
Isolation	437	61+
Liquid and vapor phase epitaxy in sequence	437	117+
Plural device formation	437	51+
Shadow masking	437	80
Single layer multi-steps	437	108+
Sliding liquid phase epitaxy	437	119+
Tipping liquid phase epitaxy	437	125+
Using electric current	437	113
Using energy beam	437	16+
Coherent light beam	437	19
Ion beam	437	20+
Ionized molecules	437	18
Neutron, gamma ray, electron beam	437	17
Using seed liquid phase epitaxy	437	114+
Vacuum growing material using molecular beam	437	105+
Semitrailer	280	423R+
Sensing (See Note 4 to Definition of)	73	
Nuclear reaction condition	376	245
Control combined	376	207
Sensitizing Photographic	430	495+
Coating processes	430	
Developing combined	430	401+
Sensitized elements	430	495+
Separation, Proteins	530	412+
Separator and Separating (See Purifiers; Screen; Sifter)	427	247
Battery	429	247+
Centrifugal		
Adhesion	209	60
Assembling apparatus	29	279
Assorting	209	642
Fermentation apparatus	435	287
Filters	210	360.1
Fluid suspension	209	199
Foraminous basket	68	23R
Imperforate bowl	494	
Auxiliary fluid supplied	494	23+
Bearing	494	83
Bowl	494	43+
Condition responsive	494	1+
Countercurrent flow	494	22
Cyclic or timed	494	11
Filtering	494	36
Heating or cooling	494	13+
Indicating, inspecting, etc.	494	10
Involving gases	494	900*
Involving mercury	494	902*
Involving oil	494	901*
Lubrication	494	15
Process	494	37
Recirculation	494	35
Safety	494	12
Seals and sealing	494	38+
Vibration damping	494	82
Oil deyhydration	208	187
Sugar apparatus	127	19
Sugar manufacture	127	56
Vortical whirl gas	55	447
Coating or impregnating	427	
Cream	220	
Diamagnetic	209	212
Drying	34	
Exhaust vapors	34	79+
Egg	99	497
Electrolytic	204	295+
Fiber thread and fabric		
Loom replenishing	28	202
Pneumatic cleaning	19	205+
Sheet conveying and	271	4+
Sheet feeding and	271	10+
Shoe sole sewing	112	29
Textile fiber liberating	19	1+
Tobacco leaf disintegrating with	131	312
Gas	55	
Beer dispensing	137	170.1+
Boiler	122	488+
Furnace draft regulators	110	148
Heat exchange surface type condenser	165	111+
Liquefaction processes with	62	11+
Liquefiers and	62	36+
Liquid separation		
Combined	55	421
Process combined	55	85
Meters with liquid and	73	200
Pneumatic fiber cleaning	19	205
Separate fluid heated boiler with	122	34
Well combined	166	265+
Glass mold	65	357+
Grain scourers with	99	600+
Grain threshing	130	21+
Liquid	210	
Beverage infusing apparatus	99	279+
Beverage processes milk fat	426	491
Carbon compounds from sulfuric acid	260	703
Chemical extracting or leaching	422	261+
Concentrating evaporator liquid	159	31
Crystal producing and	159	45
Electrical apparatus for	204	302+
Electrical processes for	204	186+
Electrolytic treatment	204	130+
Electrophoresis or electro osmotic	204	181.8+
From solid separators	4	DIG. 19
Gas contact apparatus with	261	2+
Gas operated motor condensate	415	168
Gas separation combined	55	421
Process	55	85
Oil residues from paraffin	208	30+
Paper making	210	
Paraffin from mineral oil	208	28+
Sugar starch etc making	127	
Yeasts ferments etc	435	261
Magnetic for separating iron particles from other substances	209	
Metal founding mold	164	358
Sprue	164	262+
Ore metal and coal	422	
Potato digger bearing wheel	171	112
Refrigeration	62	
Process	62	85
Solids	209	
Adhesion	209	45+
Adhesion combined	209	12+
Comminution combined	241	
Electrostatic	209	127.1+
Flotation	209	162+
Fluid handling combined	137	544+
Fluid suspension	209	132
Combined	209	13
Gaseous	209	133+
Grain scourers	99	518+
Liquid	209	155+
With cereal comminuting	241	9+
With cominuting process	241	19
With cominution	241	38+
With refrigeration	62	317
Fusible from nonfusible by heat	422	307+
Gaseous suspension	209	133
Grain scourers	99	518+
Sifting etc	209	12+
With cereal comminuting	241	9+
With comminuting process	241	19
With comminution	241	38+
With gas separation	55	
Ice cube tray	249	69+
Magnetic	209	212+
Combined	209	12+
Pretreatment	209	8
Sifter actuating elements	209	368
Stratifiers	209	478
Mercurial adhesion	209	45+
Mercurial adhesion or mercurial electrolytic synthesis of metallic oxides	204	99
Mercurial in metallurgy	75	
Mercury alloy electrolytic synthesis	204	124
Mercury suspension	209	174
Metallurgical	75	
Mining and quarrying with	299	7+
Molten metal	266	227+
Preservative compositions	252	381+
Scrap metal salvaging	29	403.3+
Sifting	209	233+
Combined	209	12+
Combined with grain hulling	99	601+
Gas borne comminuted material	241	49+
Screen with comminutor	241	69+
Sifting through comminuting	241	83
Surface	241	83+
Stratifying	209	422+
Combines	209	12+
Electrostatic	209	131
Suspension combined	209	12+
Strand spinning etc	57	354
Structurally defined	429	129+
Tobacco feeding		
Cigar making	131	110
Cigar making bunch	131	40+
Leaf disassociating	131	327
Tobacco stemming	131	313+
Tooth	433	148
Type matrix separator	199	42
Distributing to magazines	199	41
Typesetting line	276	35
Vegetable disintegrator and	241	68+
Weldless chain making link	59	11
Yeasts etc from liquids	435	261
Septic Field	405	36
Septic Tanks	210	513+
Construction	52	
Purification	210	601+
Gravitational separator combined	210	513+
Sera	424	
Serge	139	426R
Serrating Edges for Tool Sharpening	76	89.1
Serrefines	128	337+
Serums	424	
Server and Service	186	
Cart	D34	14+
Strip server for coating machine	118	40+
Table self service		
Plural related surface horizontally adjustable	108	102+

U.S. PATENT CLASSIFICATIONS

Entry	Class	Subclass
Power driven surface	108	20$^+$
Terraced rotatable	108	94$^+$
Timing mechanism for telephone		
Train control service stop	246	184
Type case feeding	276	45
Service Station		
Building type	137	234.6
Canopy combined with pump	D15	9.1$^+$
Dispensers	D15	9.1$^+$
Gasoline pump	D15	9.1$^+$
Lubricator or oil collector	D15	151$^+$
Mobile unit	D12	14$^+$
Servomotor		
Electric system	318	560$^+$
Hydraulic	91	
Sesamin	549	435
Set, Display Pyrotechnic	102	355
Sets		
Nail sets	81	44
Radio telegraphy	375	
Telegraph portable key and sounder combined	178	78
Telephone portable sets for testing lines	379	21
Telephone wall sets	379	435
Setscrew		
Guard	74	617
Shaft	403	362
Socket holder	279	83
Settable Material		
Bolt or nut to substructure lock	411	82
Bolt to nut lock	411	258
Settees	297	232$^+$
Design	D 6	334$^+$
Swinging	297	282
Setters Bowling Pin		
Mechanical	273	42R$^+$
Spots	273	52
Spotters	273	46
Setting		
Ammunition loading bullet	86	43
Broom and mop handle	300	20
Brush tuft	300	2$^+$
Button eyelet and rivet	227	51$^+$
Electric clock hand	368	60$^+$
Eyelets	227	51$^+$
Gem		
Jewelry	63	26$^+$
Jewelry design	D11	91$^+$
Setting method and apparatus	29	10
Watchmakers tools	81	6$^+$
Glaziers points	227	
Leather	69	
Plants	111	2$^+$
Covers	47	28R$^+$
Manual dibbling tools	111	92$^+$
Register zero	235	144R
Rivets	227	51$^+$
Hollow rivets	227	51$^+$
Saw	76	58$^+$
Stage	272	11$^+$
Staple	227	
Switch	246	
Type	276	
Watch stem winding and	368	190$^+$
Watchmakers jewel setting tools	81	7
Wheel spokes	157	3$^+$
Wheel tires	157	5$^+$
Wire spring	140	89
Wood pin	227	
Settling Tanks and Chambers		
Automatic fluid distributing	137	255$^+$
Beer foam separators	137	170.1$^+$
Carbon black making	422	150$^+$
Liquid separation and purification		
Decanter and filter	210	295$^+$
Decanter or settler	210	513$^+$
Liquid suspension type separation of solids	209	155$^+$
Solids separation combined	209	12$^+$
Starch making	127	27
Sugar making	127	57
Seven Grid Limiter Discriminator	313	300
Severing (See Cutting; Sawing)		
Abrading for	51	98R
Breaking	225	1$^+$
Glass (See search notes)	83	879$^+$
With glass manufacturing	65	174
Punching	83	
Stone splitting	125	43
Tearing	225	
Well tubing in situ	166	55
Sewage		
Aerators, diffusers	261	DIG. 7
Aerators, rotating	261	DIG. 71
Electrolytic treatment	204	149$^+$
Fertilizer making	422	193$^+$
Processes and products	71	12$^+$
Furnaces burning	110	238
Purification (See settling tanks)	210	
Sewerage	137	
Manhole	52	19$^+$
Pipe cleaners	15	104.3 R$^+$
Design	D32	14
Sewage purification	210	
Sewer reamer, design	D32	14
Sewer ventilation	112	
With cover	52	20
Sewing and Sewing Machine (See Stitching)	112	
Abrading attachments for	51	256$^+$
Air work handling	112	DIG. 2$^+$
Apparel seams	2	275$^+$
Bobbin winding	112	279
Book bindings	281	27
Cabinet and parts	D15	66$^+$
Carrying case	312	208
Cooling elements	112	280
Covers for	312	208
Design	D15	75
Cutting attachments for	112	130
Fans combined	416	60
Filled sacks	112	11
Filling combined	53	139
Hat sweatbands	2	184
Implements design	D 3	18
Kit	D 9	303
Lights for	362	90
Machine, design	D15	69$^+$
Sleeve-type surface	D15	68
Treadle	D15	67
Machines with tables or stands	112	217.1$^+$
Pneumatic, digest	112	DIG. 3$^+$
Sheet retainer inserts binder strand through sheet	402	7
Shoe lasting and	12	13.2
Shuttle winding	242	20$^+$
Special machines	112	2$^+$
Starting and stopping	112	271$^+$
Stitch forming	112	154$^+$
Suction thread cutting	112	DIG. 1$^+$
Tables with work catcher	112	217.1$^+$
Tension devices	242	147R$^+$
Thread cutting, severing, or breaking	112	285$^+$
Threads	57	903*
Treadle machine conversion	112	217.4
Work manipulating	112	122$^+$
Zigzag	112	157$^+$
Backstitch	112	451
Buttonhole	112	446$^+$
Cam actuation	112	459$^+$
Cam activation	112	459$^+$
Electronic memory	112	453$^+$
Sextant		
Artificial horizon	356	143
Artificial horizon	356	148$^+$
Optical	356	138$^+$
Straight ray	33	268$^+$
Shackle		
Animal restraining	119	128
Design	D30	151$^+$
Butchering	17	44
Chain link	59	86
Closure fastening	292	328$^+$
Dies for making	72	472
Foot and hand for prisoners	70	18
For padlocks	70	53
Lock type	70	18
Vehicle spring end	267	271$^+$
Shades	160	
Eye	2	12$^+$
Design	D29	18$^+$
Hat with	2	177
Lamp	362	351$^+$
Design	D26	118$^+$
Inverted bowl devise	362	453$^+$
Pull or tassel	D 6	581
Support	362	433$^+$
Mounting hardware	D 8	354$^+$
Optical instrument	350	580$^+$
Plant	47	26
Pull of cord or fabric	D 6	581
Window	D 6	581
Shading Devices	118	79
Pens	15	435$^+$
Shaft	464	179
Aligning and leveling	33	412
Arrows	273	416$^+$
Axle land vehicle	301	126
Bearings	384	
Clutch	192	110R
Couplings	403	
Flexible	464	
Crank		
Bearing antifriction	384	457$^+$
Bearing plain	384	250
Lathes	82	9$^+$
Making	29	6
Earth boring means combined	175	320$^+$
Earth, eg mine	405	133$^+$
Flexible	464	
Portable drill	408	127
Furnace	432	95$^+$
Fume arrester with	266	144$^+$
Metallurgical	266	197$^+$
Hanger	384	442$^+$
Housing	464	
Flexible shaft combined	464	52$^+$
Hub connections	403	
Joint packings	277	
Joints (See joints)		
Locks	70	182$^+$
Metallurgical furnace	266	197$^+$
Mucking machines	414	916*
Process abrading	51	289R
Speed recorders	346	
Wagon	278	33$^+$
Wall retainer	405	272$^+$
Shaker		
Salt and pepper shakers	D 7	52$^+$
Shaking		
Apparatus		
Grates	126	152R$^+$
Automobile	73	669$^+$
Devices		
Cleaning machine	15	94
Filled cooking receptacle in oven	99	371
Gas filter cleaning	55	304$^+$
Heat exchanger	165	84
Portable receptacle unloading	414	415
Separator receiver	209	64
Shotting combined for cleaning	15	96
Sifter horizontal or inclined	209	309$^+$
Sifters	209	233$^+$
Sprinkler or sprayer	239	374
Vehicle unloader	414	375
Vehicle unloader tilting track	414	357
Pivotable movement	414	362
Receptacle	366	
Clothes washing machine	68	171$^+$
Table		
Grain separator straw carrier	130	24
Separator receiver	209	64
Sifter	209	309$^+$
Shale		
Distillation	201	
Mineral oil from	208	400$^+$
Shale Oil		
Recovery	208	400$^+$
Sham Holders Pillow	5	492
Shampoo	252	89.1$^+$
Shank		
Buttons	24	92
Cloth	24	92
Making	79	2
Sewing machine attaching	112	108$^+$
Key	70	402
Extensible	70	397
Foldable	70	396
Plural	70	401
Manufacture of metal knobs	29	161
Padlock nonshackle type	70	34
Railway spike shank brace	238	309
Railway track clamp anchor	238	345
Shoe	36	76R
Loose upper shaping	12	54.1$^+$
Machines	12	40.5
Tool and machine parts		
Handle fastenings	403	
Slidably adjustable wrenches	81	129$^+$
Wrench attachments	81	184
Wrench handles and	81	177.1$^+$

	Class	Subclass
Shape Memory Alloy	148	402
Shaping (See Product Itself; Molding)	29	
Abrading or grinding	51	
Boring	408	
Comminution combined	241	3
Cone wound bobbin or cop shaper	242	34
Dental instruments for	433	25$^+$
Die (See die)		
Directly applied fluid pressure differential	264	500$^+$
Drilling	408	
Food	99	
Bread pastry and confection making, food apparatus	99	
Bread pastry and confection making, shaping apparatus	425	
Cooking and	99	324$^+$
Dairy molds	249	
Meat briquetting apparatus	17	32
Hay stack shapers	414	132
Metal		
Bending to angle	72	176$^+$
Casting combined	29	527.5$^+$
Coating combined	29	527.2$^+$
Corrugating	72	180
Deforming	72	
Drawing sheet metal	72	347$^+$
Drawing through die	72	274$^+$
Electric heating and	219	149$^+$
Extruding	72	253.1$^+$
Indirectly	72	273.5
Molten metal	164	
Plastic metal	72	
Powder apparatus	425	78$^+$
Powder processes	419	61$^+$
Pressure or impact forging	72	
Rolling	72	199$^+$
Sheet metal by dies	72	
Spinning	72	82$^+$
Wire fabric	140	107
Wire into articles and fabrics	140	
Milling	409	64$^+$
Nonmetallic		
Briquetting fuel	44	10R$^+$
Coopering	147	
Glass	65	
Ironing and smoothing textiles	38	
Leather	69	
Paper	493	395$^+$
Plastic block and earthenware	425	
Plastics	425	
Sand molds for metal	164	
Stone	125	
Tobacco plug	131	111$^+$
Wood sawing	83	
Wood turning	142	
Woodworking	144	
Pencil sharpeners (See pencil sharpener)		
Planing	409	288$^+$
Screw threading	408	
Stack or pile shaping	414	35
Turning	82	
Shares for Plows	172	681$^+$
Forging dies for making	72	425
Making machines and processes	29	14
Sharpener	83	174$^+$
Abrading or grinding	51	
Design, knife or scissors	D 8	93
Button surfacing with tool	79	8
Combined with cutting machine tool	83	174$^+$
Cutlery with	30	138$^+$
Disk plow	172	439
Drill forming and	76	5R
Harvester cutting reel type	56	250
Horseshoe calk	168	46
Making and sharpening	59	65
Knife or scissors		
Combined with can opener	D 8	35
Powered tool	D 8	63
Metal etching	204	141.5
Pencil (See pencil sharpener)	30	451$^+$
Design of	D19	73
Machine	144	28.1$^+$
Razor with	30	35$^+$
Saw	76	30$^+$
Tool	76	82$^+$
Shaving and Shaver		
Brushes	D 4	135
Design	D 4	114
With drip cup or shield	15	248R$^+$

	Class	Subclass
With soap supply	401	268$^+$
Comminutor type		
Ice	30	136$^+$
Mill	241	83$^+$
Vegetable and meat	241	
Cream	424	
Cups	220	
Electric	30	45
Gear	409	37
Hair planers	30	30$^+$
Design	D28	44
Hat	223	19
Kits	132	80R$^+$
Brush or soap retained	401	118$^+$
Pencil (See pencil sharpener)	30	451$^+$
Razor	30	32$^+$
Blades	30	346$^+$
Dry shaver	30	43
Screw head	10	8
Spoke	30	281
Woodworking	144	155$^+$
Shavings Feeding to Furnace	110	102
Shawls	2	91
Design	D 2	179$^+$
Sheaf Handling		
Binding	56	432$^+$
Carriers	56	474$^+$
Composite construction	52	334
Cornstalk loader	56	120
Carrier	56	121
Shocker	56	401$^+$
Threshing	130	1
Shear Pin	464	32$^+$
Shearing & Shears (See Cutter)		
Bolt and rivet making	10	25
Buttonhole	30	120
Can opener	30	428
Cutlery combined		
Knife	30	146
Nipper	30	145
Design	D 8	93
Gauges for shears	33	631
Ground supported hand operated harvester	56	241
Handle and blade connections	30	341
Holder or disposal combined	30	131$^+$
Manicure	30	29
Plier combined	7	125$^+$
Modified handle	7	131$^+$
Receptacle combined		
Cigar receptacle	206	238$^+$
Dispensing sheet receptacle	206	216
Ticket holder receptacle	30	124$^+$
Shears hand manipulable	30	194$^+$
Stone shearing	125	23R$^+$
Surgical	128	318
Textile		
Cloth finishing	26	15R$^+$
Thread finishing	28	226$^+$
Sheath (See Scabbard; Sheathing)	224	191$^+$
Awl or prick punch	30	368
Cutlery	30	151$^+$
Two or more cutters	30	143
Design	D 3	102
Hair comb	132	143
With brush	132	121
Manicuring tool	132	76.2
Padlock	70	55
Splitters	30	90.1$^+$
Surface supporting	108	150$^+$
Wire strippers	81	9.4$^+$
Sheathing (See Facer Sheathing (See Facer Construction;)		
Earth control	405	272$^+$
Tunnel or shaft lining	405	132$^+$
Electrical conductor	174	68R$^+$
Making		
Indefinite length electrical conductor	156	47$^+$
Metal extrusion	72	268
Packing	277	229$^+$
Railway car		
Freight	105	409$^+$
Passenger	105	401
Ship	114	84
Sheave (See Pulley)		
Reeling storage	242	47.5
Shed		
Animal husbandry	119	16
Buildings	52	
Closing devices for dobby looms	139	69$^+$

	Class	Subclass
For jacquard looms	139	64
Locomotive	104	51$^+$
Shedding		
Loom	139	55.1
Special	139	11$^+$
Sheet and Sheet Material (See Composition)		
Accounting and listing form	283	66R
Advancing material of indeterminate length	226	
Apertured	428	131$^+$
Associating and folding	270	
Spacer for drying	34	6
Bedding	5	482
Bedding material	5	DIG. 1
Contour bedding	5	497
Restraining	5	498
Binder device releasably engaging aperture or notch of sheet	402	
Boiler		
Crown	122	4[illegible]6
Tube	122	512
Bracket shelf of	248	247$^+$
Cloth	D 5	
Coated	428	411.1$^+$
Sheets	446	108$^+$
Composite	428	411.1$^+$
Composite with cellular layer	428	304.4$^+$
Composite with foamed layer	428	304.4
Construction toys of	446	108$^+$
Container for coiled form	206	389$^+$
Control (See pattern control)		
Electric circuit maker	200	46
Stop mechanism	192	126$^+$
Cookie	99	422$^+$
Corner structure	428	81$^+$
Counting	235	89R
Cover with	428	68$^+$
Creped	428	152$^+$
Discontinous coating with	428	195$^+$
Easel of	248	441.1$^+$
Edge feature	428	192$^+$
Edge fold	428	121$^+$
Edge spliced	428	57$^+$
Expanded metal making	29	6.1
Feeding and delivery	271	
Dispensing cabinet magazine	312	37
Magnetic oferations	271	901$^+$
Manifolding	282	
Reverse direction of sheet movement	271	902
Separators, pneumatic	271	90$^+$
Separators, rotary	271	109$^+$
Strippers	271	900
Support rack for paper	211	45$^+$
Support rack stacked	211	50$^+$
Towel service locking rack	211	6
Towel service rack	211	16
Typewriter paper on edge	271	903
Typewriter paper	400	624$^+$
Winding and reeling	242	
Figure toys of	446	387$^+$
Filter	210	348$^+$
Fire escape body catcher	182	138$^+$
Flanged pipe joint packing of	285	363$^+$
Garment hanger	223	87
Gas and liquid contact porous	261	100$^+$
Grooved	428	161$^+$
Holders and securing devices		
Billiard table cloth	273	7
Cleaning fabric	15	231$^+$
Cylindrical for sandpaper	51	358$^+$
Endless conveyor with gripper	198	688.1$^+$
Handled for sandpaper	51	392$^+$
Manifolding pockets	282	11.5 R
Music stands	248	441.1$^+$
Pack holders	271	145$^+$
Printing	101	415.1
Rack	211	45$^+$
Sandpaper disc	51	358$^+$
Shoe last with retainers	12	141
Stack rack	211	50$^+$
Tray	206	557$^+$
Wall mounted tray	211	88
Joined	428	57$^+$
Laminate	428	411.1$^+$
Making		
Coating and uniting	156	
Drying	34	
Electroforming	204	12$^+$
Electrolytic coating metal	204	27$^+$

	Class	Subclass
Glass	65	
Metal and foil	29	17R$^+$
Metal by rolling	72	199$^+$
Pastry rolling	425	363$^+$
Pastry rolling	425	335$^+$
Plastic rolling or pressing	425	
Pulp molding compressor	162	396$^+$
Pulp molding winder	162	283$^+$
Metal		
Can making machines	413	26$^+$
Cap preparing	413	8$^+$
Apparatus	413	56$^+$
Pattern and texture design	D 5	1$^+$
Metal laminate	428	615$^+$
Metal shaping (See particular operation)		
Chain making	59	13$^+$
Container making	413	
Sprocket chain making	59	6
Staple making	59	72
Wire barbing	140	66
Metal structures and articles	405	276$^+$
Bale tie band	24	23R$^+$
Cabinets	312	
Can closures	220	200$^+$
Chain link	59	91
Closure seal	292	310$^+$
Connecting rods	74	588
Containers	220	
Display packages	206	44B$^+$
Drive chain	474	202$^+$
Electroplating	204	27$^+$
Flumes	405	119$^+$
Gearing sectional	74	449
Heat radiator	165	170
Jar closures	215	200$^+$
Joint packing reinforcement	277	
Nails spikes tacks	411	439$^+$
Piling	405	274$^+$
Pipe joint	285	424
Pulley	474	166$^+$
Rack support for receptacle	211	72$^+$
Railway ties	238	63$^+$
Railway wheel web	295	22$^+$
Resilient clasps	24	530$^+$
Sash cord guide casing	16	212
Separable-fasteners	24	572$^+$
Shaft hangers	384	443
Shelf type with support	108	
Diverse support for articles	108	28
Spools	242	118.8
Stove for heating	126	65$^+$
Stove legs	126	306
Nonrectangular	428	80$^+$
Non-structural	428	411.1$^+$
Note musical		
Music teaching	84	483R
Musical automatic instrument	84	146$^+$
Opening cutter	402	1
Pack holders	271	145$^+$
Package of coiled sheets	206	449$^+$
Package of flat sheets	206	449$^+$
Paper dispensing package	206	449$^+$
Paper wrapping	229	87R$^+$
Mailing	229	92
Pavement	404	17$^+$
Combined block and	404	18
Perimeter structure	428	81$^+$
Picture support of	248	441.1$^+$
Piling	405	274$^+$
Printing smut preventing	101	419
Railway tie		
Folded sheet	238	77
Reinforcing	238	93
Receptacle stand of	248	152
Receptacle support of	211	72$^+$
Ribbed	428	161$^+$
Roofing	427	147$^+$
Sorting type rack	211	10$^+$
Spacer	402	80R
Spliced	428	57$^+$
Spot bonded	428	198$^+$
Stand of	248	174
Stencil	101	127
Structurally defined	428	98$^+$
Structurally defined element therein	428	221$^+$
Tearing or breaking	225	
Testing sheet material	73	159$^+$
Bursting strength	73	838$^+$
Flatness, optically	356	371
Transparency, optically	356	432$^+$
Thread for pin type insulator	174	204
Treating		
Abrasive applying	51	275
Brushing machines	15	77
Chemically cleaning	134	
Coating	118	
Coating electrolytically	204	22$^+$
Cutting	83	
Drying apparatus	34	148$^+$
Drying method	34	18
Gas contacting	34	23$^+$
Heat	108	28
Ironing shaking out	38	143
Radiant heating	34	41
Stencil making	101	128.4
Wiping machines	15	102
Winding & reeling	242	
Wrapper for coiled form	206	389$^+$
Wrinkled	428	152$^+$
Shelf and Shelving	108	
Attached laterally of support	108	152
Baggage convertible	190	9$^+$
Bookcase shelving		
Design	D 6	567$^+$
Enclosed	312	
Open	108	59$^+$
Cabinet with racks or shelves	312	
Showcase type	312	128$^+$
Covering	428	68
Apertured	428	131$^+$
Edge fold	428	121$^+$
Design	D 6	567$^+$
Display receptacle	206	45
Drying apparatus		
Removable	34	192$^+$
Shelf to shelf material flow	34	165$^+$
Endless conveyer with coacting support	198	860.1
Plural conveying sections	198	600
Heat exchange	165	168$^+$
Ironing board with folding support	108	115$^+$
Loading or unloading	414	266$^+$
Loading or unloading	414	227$^+$
Main or coin chute	232	63
Paper design	D 5	
Rack, plural shelf	108	59$^+$
Shoe or boot rack	211	36
Stove	126	332$^+$
Supports	248	235$^+$
Terraced shelf rack convertible to table	108	17
With structural installation	108	42$^+$
With upward extending support	108	149
Shell		
Ammunition	102	
Cartridge	102	464$^+$
Design	D22	116$^+$
Explosive	102	473$^+$
Extracting implements	81	3.5
Feeding devices (See gun loading)	86	46
Loading	86	23$^+$
Making	86	10$^+$
Making metal working	29	1.3$^+$
Brake shoe	188	254
Edible		
Cereal	426	138$^+$
Design	D 1	
Electrical connector socket	439	753$^+$
Foundry molds	164	361
Shell openers	30	120.1
Shellac (See Resins)	260	97$^+$
Sheller	30	120.1
Corn	130	6$^+$
Feeders for	130	33
Green	99	567
Cutlery	30	120.1$^+$
Nutcracker	30	120.2
Pea	130	30G
Vegetable cutter combined	99	
Shellfish		
Apparatus	17	53
Apparatus	17	71$^+$
Openers	17	71$^+$
Openers	17	74$^+$
Culture	119	2
Mollusk	119	4
Dredgers	37	55
Opening process	17	48
Scoops	37	119
Shelves (See Shelf & Shelving)		
Sherbets	426	101$^+$
Design	D 1	
Shield (See Guard; Protection; Screen; Stencil)		
Abrading	51	271
Button cleaning	51	265
Anti-magnetic, anti-inductive	174	35R
Apparel	2	2$^+$
Armpit	2	53$^+$
Brassiere	450	37
Corset or girdle	450	153
Design, garment	D 2	225$^+$
Dress shield making	223	54
Armour	109	78$^+$
Ship	114	14
Brush and broom		
Drip	15	248R$^+$
Head	15	175
Coffin	27	20
Corner		
Baggage	109	37
Building	52	288
Counter or display partition	109	11$^+$
Curtain shade or screen	160	19$^+$
Distilling retort	202	224
Electric	174	35R$^+$
Conductor	174	102R$^+$
Ignition	315	85
Insulator	174	139$^+$
Lamp & electronic tube systems	315	85
Radio	455	300$^+$
Resistor	338	64$^+$
Shock preventive	174	5R
Erasing	51	262.1
Faraday	174	35R
Transformer with	336	69$^+$
Fire shield, portable	169	48$^+$
Fume flow in arc lamp	314	26$^+$
Hand		
Handling implements	294	131
Sadiron	38	95
Harness	54	80
Breast strap	54	59
Hitching	119	143$^+$
Loop shields	24	183
Heat	228	59*
Heat for ammunition	244	158R
Aircraft use	244	117R$^+$
Infra-red	350	1.1$^+$
Metal treatment	148	149
Nuclear reactions	376	153
Nuclear reactions	376	156
Nuclear reactions	376	913
Ultra-violet	350	1.1$^+$
Light filters	350	311$^+$
Locked	70	158$^+$
Mast	114	93
Ordnance	89	36.1$^+$
Design	D22	112
Piston splash	92	141
Propeller marine	440	71
Radiator	237	79
Automobile	293	115
Radioactivity	250	515.1
Geiger muller tube having	313	93
Nuclear reactions	376	153
Nuclear reactions	376	156
Nuclear reactions	376	913
Rigid shield or pad positioned between article & bearer	224	907*
Solder splash	228	21
Surgical	128	132D$^+$
Headed securing rod	403	360
Tunnel	405	141
Windshield		
Aircraft	244	121
Harvester	56	190
Vehicle storm fronts	296	78R$^+$
Ventilating cowl	98	71
Windshield, vehicle	296	84R$^+$
X ray	250	515.1
Bucky grid	378	154
Shutter or diaphragm type	378	145
Tube structure	378	140
Variable aperture	378	150
Shifter		
Belt	474	101$^+$
Railway cable	104	198
Brush		
Dynamoelectric	310	241
Generator system	322	54$^+$
Motor system	318	541$^+$

Entry	Class	Subclass
Changeable exhibitor	40	446+
Copyholder copy or line guide (See copyholder)	40	341+
Feeding table for metal rolling	72	222
Railway cable	104	198
Track or tie	254	43
Laying	104	4
Typewriter case	400	251+
Shifting Field Motor	310	172
Shim Rod (See Nuclear, Control)	376	207
Shim rod	376	327
Shim rod	376	219
Shim rod	376	347
Shims		
Bearing adjusting	384	626
Brake slack adjusting	188	201
Shingle	52	518+
Container	206	323+
Gauges	33	648+
Combined with hatchet or hammer	7	122+
Making	83	920*
Making machines woodworking	144	13
Sawing	83	704+
Roofing coating	427	186
Ship	114	
Anchor	114	294+
Automatic controlled steering	114	144R+
Ballasting	114	121+
Bilge discharge	114	183R+
Building	114	65R+
Bulkheads and doors	114	116+
Cable stoppers	114	199+
Caissons	405	12
Canals	405	84+
Centerboards	114	127+
Clamps for ship planking	269	91+
Coaling and loading	414	137
Traversing hoists	212	72
Condensers	165	41+
Course recorder	346	8
Design	D12	300+
Drydocks for	405	4+
Engine room telegraph	116	21
Electrical	340	286R+
Furniture	114	188+
Hatches and covers	114	201R+
Ice boats	114	43+
Ice breakers	114	40+
Indicating and signaling		
Course recorder	346	8
Course signaling	116	19
Navigation instruments	73	178R+
Rudder position	116	303
Running lights	362	61+
Shoal water alarms	116	113
Telegraphs	116	21
Keels	114	140+
Ladders	182	82+
Steps level	182	1
Launching	405	1
Life craft handling aboard	114	365+
Lifting mechanism	114	50+
Liquid purifier type	210	242.1
Mother	114	258
Nautical signalling	116	26+
Ports	114	173+
Propellers	416	
Propulsion	440	
Raising and docking	114	44+
Refrigeration	62	240
Repairing	114	44+
Running lights	362	61+
Sailboats	114	39.1+
Sails and rigging	114	102+
Scows	114	26+
Sounding and depth measuring (See sounding)		
Spars	114	89+
Speed control automatic	114	144R+
Steering mechanism	114	144R+
Stoves	126	24
Submarines	114	6+
Torpedo boats	114	18+
Torpedo guards	114	240R+
Torpedoes	114	20.1+
Towing or pushing	114	242+
Ventilation	114	211
Ventilation of grain cargo	98	53
Warships	114	1+
Shipper		
Clutch	192	98
Marine vessel rudder	114	165
Side cut mower automatic clutch	56	274
Transmission control and brake	192	10
Shirring	223	28+
Cords for flexible bags	383	72
Garment trimming	2	244
Sewing machines	112	132+
Shirt		
Coat combined	2	95
Collar attaching to	2	139+
Design	D 2	208
Mens outer	2	115+
Pajama	2	114
Trousers supported by	2	229
Under	2	113+
Union type bifurcated	2	77
Vest combined	2	103
Shock Absorber (See Dashpot)		
Dial indicator type	33	DIG. 6
Fluid	188	266+
Thrust member, variable volume chamber	188	297+
Harvester reciprocating cutters	56	306
Nuclear reactor control	376	234
Component combined	376	285
Nuclear reactor control	376	327
Recoil ordinance	89	42.1+
Spring type	267	
Bed	5	253
Support type, nonresilient	248	636+
Support type, resilient	248	562+
Suspended machine	248	610+
Testing	73	11
Textile feeding knitting	66	130
Vehicle		
Body suspension frame	280	784
Fender	293	132+
Fluid absorbs	293	107+
Railway	105	392.5
Railway coupling	213	7+
Running gear	280	688+
Seat	297	216
Weighing scales	177	184+
Shocker and Shock Delivery		
Electric		
Amusement devices	272	27R
Animal restraining devices	119	145
Bandaging	128	82.1
Check operated	194	
Electric chairs	128	377+
Electric fences	256	10
Electrocuting animal	43	98
Electrocuting trap insect	43	112
Fishing	43	4
Food preservation	426	237+
Food preservation	426	237+
Food preservation apparatus	99	451
Kinesitherapy eg massage	128	24.1+
Kinesitherapy eg massage	128	68.1
Medicators	604	20+
Miscellaneous	361	225+
Receptors	604	20+
Safes	109	35
Sterilizing method	422	22
Surgical applications	128	362+
Surgical instruments	128	303.1+
Therapy, medical	128	419S
Harvester	56	401+
Shoe		
Animal wear	36	111
Design	D30	146+
Farriery	168	
Making	59	36+
Oxshoe making	59	70
Apparel	36	
Accessory	36	136+
Anti-slipping devices	36	59R+
Athletic	36	114+
Bathing	36	8.1
Bowling	36	130
Box	206	278
Buttoners	24	40
Cement	36	DIG. 1
Cleaning type	15	227
Cleat	36	134
Design	D 2	264+
Foot elevators	36	81
Football	36	128
Heels	36	34R+
Insoles	36	43+
Jumping	36	7.8
Lacing devices	24	140
Light thermal and electrical	128	382+
Making (See boot making)	12	
Non-leather	36	84+
Occupation specialized	36	113
Ornaments	2	245
Patterns	33	3R+
Plastic	36	DIG. 2
Pricking soles	83	866+
Putting on and removing aids	223	113+
Rack	211	34+
Riding	36	131
Rubber	36	4
Running	36	129
Sandal	36	11.5
Sewing machines	112	28+
Shoehorns	223	118+
Shoelace design	D 2	316
Shoelace racks	211	61
Shoemakers bench	12	122
Skate attaching	280	11.3+
Ski	36	117+
Slippers for inside	36	10
Sole attaching means	36	12+
Soles	36	25R+
Strap tighteners	24	68SK+
Support feature	36	88+
Track	36	129
Uppers	36	45+
Water proofed finish	36	98
Brake	188	250R+
Beam combined	188	219.1+
Fasteners	188	234+
Momentum operated wedging	188	136
Railway chock roller type	188	37
Railway side	188	59
Railway top	188	57
Railway track rotary	188	39
Vehicle fender combined	293	8
Vehicle type top	188	29
Wagon type retreating	188	15
Wear compensation	188	215
Wheel rotary	188	80
Bridge truss	14	15
Cleaning (See polishing & cleaning)		
Cleat or spike	D 2	317
Demonstration	434	397
Filling compositions	106	38
Flatiron attachments	38	97
Harvester cutter member	56	303
Track clearers with socket	56	318
Horn	223	118+
Design	D 2	642
Lace	D 2	316
Lace tip applier	29	243.57
Panel hangers or travelers	16	93R+
Pedals with clips for	74	594.6
Plow	172	764
Body shoe and guide	172	764
Polishing and cleaning		
Applicator carried by closure	401	126+
Brushing machine	15	36+
Brushing machine with polish supply	15	30+
Composition	106	3+
Implement with material supply	401	
Mitts cots and shoes	15	227
Shoe blacking stands	15	265+
Shoe impregnated or coated with cleaning material	15	104.94
Rack	211	34+
Railway switch point throwers	246	269
Safety	36	77R
Shine kit	D 3	30.1+
Shine machine	D32	14.1
Shine stand with seat	D 6	334+
Shoehorn	D 2	642
Snow	36	122+
Third rail collector	191	49
Trees	12	114.2+
Design	D 2	314.1
Vehicle (See brake)		
Earth working tool with	172	
Fender with car control	293	8
Railway draft central friction post	213	36
Railway draft friction casing	213	32R+
Railway draft spring and friction	213	24
Two wheeled scoop with	37	131
Water	441	76+
Well casing	166	242
With cutting or scraping means	175	402

U.S. PATENT CLASSIFICATIONS

	Class	Subclass
With screen	166	157
Shooting Gallery Game	D21	2$^+$
Shooting Lines		
Fishing apparatus	43	6
Line carrying ammunition	102	504
Bomb lances	102	371
Line projector or projectile	182	50
Explosive	89	1.34
Shooting Targets	D21	5
Shopping Bag	383	6
Handle attaching	493	226
Shorings		
Earth control	405	272
Mine roof props	405	288$^+$
Trench walls	405	282$^+$
Short Circuit		
For arc lamps	314	10$^+$
For incandescent lamps	315	119$^+$
Safety device	174	5R
Safety system	361	1$^+$
Shortening	426	601$^+$
Shorthand		
Code	283	46
Writing form	283	45
Teaching	434	158$^+$
Typewriter	400	91$^+$
Shot		
Blasting	72	53
Metal working digest	29	DIG. 36
Cartridges	102	448$^+$
Earth boring by	175	54
Making	29	1.22$^+$
Metal making		
Apparatus	425	6$^+$
Processes	264	5$^+$
Shotgun	D22	103
Shotguns (See Rifle)	42	
Design	D22	103
Shotting		
Cleaning machine	15	95$^+$
Methods	134	7
Shoulder		
Braces pads and supports	2	45
Back and	2	44
Body protecting	2	2
Brassiere combined	450	2
Corset or girdle combined	450	96
Garment shaping	2	268
Garment supporting	2	305
Rests for violinists	84	280
Combined with chin rest	84	278
Tang holding chuck	279	89$^+$
Shovel		
Crane	212	243
Excavating	37	
Design	D 8	10
Hand manipulable	294	49$^+$
Design	D 8	10
Making		
Blanks and processes	76	113
Dies	72	470$^+$
Plow or cultivator type	172	768
Railway car loading	414	340$^+$
Repositioning support to load or unload	414	349$^+$
Snow	294	54.5
Train of vehicles	414	339
Showcase		
Cabinet	312	114$^+$
Design	D 6	432$^+$
Outdoor type	312	102
Illumination	362	125
Joint between glass walls	312	140
Show or picture window	52	777$^+$
Show or picture window	52	764$^+$
Shower (See Sprinkling)		
Bath	4	615$^+$
Hot air combined	4	525$^+$
Surgical	128	366
Design	D23	283
Design, cap	D 2	510$^+$
Heads	239	548$^+$
Design	D23	213$^+$
Temperature control of water	236	12.1$^+$
Shrapnel	102	491$^+$
Shredded Cereal Biscuit	426	560
Apparatus	425	289$^+$
Design	D 1	
Shredders and Shredding	241	
Metal	29	4.5 R$^+$
Shrinking		
Metal bands by hammering or pressing	72	399$^+$
Shrink fitting	29	447
Metal working digest	29	DIG. 35
Testing for	73	159$^+$
Textiles	26	18.5
Thread compacting	26	18.6
Tire	72	292
Shrinkproofing Fabrics	26	18.5$^+$
Shuffleboard	273	126R
Design	D21	7$^+$
Shunt		
For electromagnet	335	236$^+$
Resistor	338	49
With meter	324	126
Shutter		
Architectural design	D25	47$^+$
Camera	354	226$^+$
Automatic control	354	410$^+$
Copying camera	355	71
Curtain	354	241$^+$
Delayed releaser	354	237$^+$
Diaphragm	354	270$^+$
Focal plane	354	241$^+$
Image projector	353	88$^+$
Iris diaphragm	354	270$^+$
Iris type	354	241$^+$
Mechanical testing	73	5
Motion picture	352	204$^+$
Optical testing	73	5
Pivoted leaves	354	241$^+$
Sliding	354	245$^+$
Check controlled	194	
Curtain shade or screen combined	160	104
Fastener or hook for	D 8	349$^+$
Ladder convertible	182	21
Operator	49	324$^+$
Projector illuminating	362	277$^+$
X ray	378	160
Shuttle		
Boxes	139	183
Braiding tatting and netting	87	54$^+$
Changing in looms	139	234$^+$
Gripper	139	196.2$^+$
Motions		
Sewing machine	112	185$^+$
Weaving	139	133$^+$
Sewing machine	112	232$^+$
Design	D15	72
Threaders	139	381$^+$
Weaving	139	196.1$^+$
Winding		
Sewing machine	242	20$^+$
Tatting	242	52
Shuttlecock	273	417
Design	D21	207
Sialon Refractory Compositions	501	98
Sickle		
Hand manipulable	30	309
Machine using	56	229$^+$
Side Lights	362	61$^+$
Side Line Eliminators	139	170.3
Side Lock (See Lock)		
Sidearms	D22	100$^+$
Sideboard		
Cabinet	312	
Design	D 6	432$^+$
Fully enclosed storage	D 6	432$^+$
Dumping vehicle	298	
Refrigerator combined	62	258
Sidehill Implement		
Harvester platform adjustment	56	209
Land vehicle body levelling	280	6R$^+$
Plows	172	204$^+$
Steam traction vehicle boiler levelling	180	39
Tank or boiler	280	7
Sideslip	73	180
Siding	52	474$^+$
Shingle type	52	518$^+$
Sieve		
Liquid material	210	348$^+$
Solid material	209	233$^+$
Sifter and Sifting (See Screen)		
Cabinet with	312	210.5
Kitchen utensil	D 7	47
Pneumatic conveyor intake	406	121
Potato diggers	171	71$^+$
Potato diggers	171	111$^+$
Separating solids		
Combined operations	209	12$^+$
Comminuting surface apertured for	241	83$^+$
Comminution and gas borne material	241	49$^+$
Comminution combined	241	68$^+$
Stove ashpan	126	244
Sight		
Bow, arrow	33	265
Dispenser inspection	222	154$^+$
Drift	33	320
Feed lubricator	184	96$^+$
Glass liquid level gauge	73	323$^+$
Gun	33	233$^+$
Design	D22	109
Design, telescopic	D16	132
Fiducial telesopes	356	247$^+$
Illuminated sights	362	110
Letter box signal	232	34$^+$
Line controlled scriber	33	20.1$^+$
Opening in umbrella	135	35S
Projector	362	383
Telescopes	350	567$^+$
Sign and Sign Exhibiting	40	584$^+$
Circuit control flashing light etc.	40	902*
Clerical desk article combined	40	358
Directional design	D20	29$^+$
Electric signal display system	340	700$^+$
Fiber optics, eg light pipe	40	547
Frame	D19	52
Illuminated	40	541$^+$
Illumination itself	362	812*
Mirror having changing effect	40	900*
Mounting bracket	D 8	354$^+$
Out-to-lunch type signs	40	907*
Pen and pencil X-art collection	40	905*
Reflecting triangle for highway use	40	903*
Roadway warning	D20	10$^+$
Sound X-art collection	40	906*
Street	40	612
Supports	248	466$^+$
Towed through sky	40	215
TV or radio program indicator	40	901*
Vehicle motion and direction	116	42$^+$
Warning reflector or flasher	D10	114$^+$
Window shade	160	10
Sign, Illuminated	362	812*
Signal	116	
Acoustic	116	67R$^+$
Bells	116	148$^+$
Horns	116	137R$^+$
Sirens	116	147$^+$
Whistles	116	137R$^+$
Alarms	116	67R$^+$
Amplifier condition	330	2
Annunciators	340	815.1$^+$
Arc lamp	314	9
Automobile horns	116	137R$^+$
Awakening device (See awakener)		
Balloons	116	DIG. 9
Bank protection	109	21
Beverage infuser	99	285
Bicycle bells	116	166
Boiler combined	122	504.2
Brake wear	188	1.11
Buoy	116	26$^+$
Burglar electric	340	541$^+$
Burglar mechanical	116	6$^+$
Cabinet with	312	234
Calculator or register combined	235	128
Check controlled apparatus		
Check in circuit	194	304$^+$
Check operated switch	194	239$^+$
Clock alarm	368	244$^+$
Electrically operated	368	250$^+$
Closure position	49	13
Code signalling	116	18$^+$
Coin operated	194	
Consumable electrode	314	9
Conveyer power driven	198	502.1$^+$
Cooking apparatus	99	342$^+$
Design, instruments	D10	104$^+$
Detonating (See detonating signal)		
Dispenser combined	222	23$^+$
Distance measurement combined	33	136
Cutting combined	83	522
Tearing combined	225	18
Drawbridge	14	49
Dropping, ordnance	89	1.51$^+$
Earth boring combined	175	40$^+$
Electric	340	
Alarms	340	500$^+$
Arc lamp	314	9
Extinguisher combined	340	289

Entry	Class	Subclass
Float gauge combined	73	308
Lamp system	315	129+
Noise or unwanted signal reduction	367	901*
Radio analog	455	
Radio directive	343	
Radio pulse or digital	375	
Telegraphy	178	
Telephony	379	
Elevator		
Electric	187	100+
Electric annunciators	187	121+
Mechanical	116	64
Mechanical alarm	116	68
Expansible chamber motor combined	91	1
Fire alarm		
Extinguisher combined	340	289
Flame	340	577+
Smoke	340	628+
Systems	340	287+
Thermal control	340	584+
Fire escape	182	18
Fire extinguisher combined	169	23
Fishing	43	16
Automatic hooker	43	17
Flags & flagstaffs	116	173+
Flexible panel combined	160	10
Float gauge combined	73	307+
Foghorn	116	137R
Electric actuation	340	384R+
Folder carried indicia	40	359
Foot operated linkage	74	514
Gas separator combined	55	274+
Illuminated by locomotive headlight	362	61+
Index card	283	36+
Lamp for vehicle	D26	28
Lamp or reflector	D10	111+
Lantern	362	166+
Letter box	232	34+
Life in undertaking	27	31
Liquid purification	210	85+
Lock condition	70	441
Masts and flagstaffs	116	173+
Mechanical	116	
Mirror combined	350	97
Nautical	116	26+
Ordnance	89	1.51+
Periodic	116	22R+
Pneumatic dispatch	406	34+
Press combined	100	99
Pyrotechnic	102	335+
Cartridge	102	346
Design	D22	112+
Parachute flares	102	337+
Railway torpedos	246	487
Skyrocket	102	347+
Torches	102	336+
Railway	246	
Audible and other signal	246	217
Audible for switch	246	217
Block signal systems	246	20+
Bridge warning	246	486
Cab signal	246	167R+
Crossing signal on train	246	208
Dispatching combined	246	2R+
Drawbridge protection	246	118+
Electric actuation	246	218+
Fluid motor actuation	246	257+
Grade crossing	246	111+
Highway crossing	246	292+
Highway crossing electric	246	125+
Interlocking	246	131
Mail delivery	258	2
Mechanical motor actuation	246	262+
Passenger at station	246	209
Pilot car	246	166
Roadway defect	246	120+
Signal car	246	166
Torpedo	246	487
Torpedo placing	246	210+
Traction cable	104	179
Train dispatching	246	2R
Train position	246	122R+
Tunnel warning	246	486
Vehicle energy actuation	246	270R+
Whistle	116	137R+
Recorder	346	
Recorder combined	346	17+
Reeling tension device	242	128
Rudder position	116	303
Safes combined	109	38+
Security instrument	D10	104+
Sheet carried indicia	40	360
Ships telegraph	116	21
Speaking tubes	181	19
Street traffic	116	63R+
Support structure	248	347
Telephony		
Central system	379	242+
Divided central	379	310+
Thermostatic regulator combined	236	94
Time	368	243+
Tire deflation	116	34R
Electric	340	58
Torpedo for railway	246	210+
Placing mechanism	246	487
Track sander combined	291	17
Traffic	116	63R
Electric	340	22+
Typewriter ribbon	400	249+
Underwater sound waves	367	131+
Vehicle	116	28R+
Watches alarm	368	244
Water heater	126	388
Water heater vessel combined	126	388
Signaling Teaching	434	223
Signature Gatherers and Conveyors		
Conveyor power driven	198	644
Machine for writing several	33	23.6
Sheet associating	270	54+
Signing machine	D18	21
Silastic T M (See Synthetic Resin or Natural Rubber)		
Silazanes	568	410+
Silencer (See Insulating, Sound; Muffle; Noise Deadening)		
Firearm and gun	89	14.4
Muffler for firearm	181	223
Silent Butler	D32	74
Silica	423	335
Abrasive compositions containing	51	308
Absorbents containing	502	407+
Aerogel	106	287.34
Catalysts containing	502	232+
Cement containing	106	98
Gel	423	338
Dispersing or stabilizing agent combined	252	315.1
Glass compositions containing	501	53+
Glass manufacturing	65	
Lime compositions containing	106	120
Lubricants containing	252	9+
Silicane	556	465+
Silicates & Silicon Compounds, Inorganic	423	324+
Abrasive compositions containing	51	308
Absorbents containing	502	407+
Catalysts containing	502	232+
Coating or plastic compositions containing		
Alkali metal	106	74+
Carbide	501	87+
Glass	501	11+
Lime combined	106	120
Opacifiers for enamels	106	312
Pigments fillers aggregates	106	291
Portland cement type	106	89+
Portland cement type making	106	100+
Refractories	501	94+
Slag	106	117
Detergents containing	252	89.1+
Dyeing	8	523+
Fertilizers containing	71	
Hydrofluosilicic acid	423	341
Lubricants containing	252	9+
Organic compounds	556	400+
Esters	556	482+
Silica	423	335
Silicofluorides	423	324
Silicic Acid	423	325
Silicofluoric Acid	423	341
Silicofluorides	423	324+
Silicol	556	463
Silicon	423	348
Alloys		
Aluminum	420	548
Aluminum copper	420	537
Copper	420	490
Metallic	420	578
Carbide	423	345
Ferro producing	420	578
Misc. chemical manufacture	156	DIG. 64
Organic compounds in compositions	106	287.1+
Polymers (See synthetic resin or natural rubber)		
Room-temperature curable	528	901*
Silicon Steel	148	110+
Silicone	556	465+
Compositions	106	287.13
Sililation of Polymers	525	100+
Silk		
Artificial making		
Apparatus	425	66
Apparatus	425	67+
Apparatus	425	76
Apparatus	425	382.2
Processes	264	165+
Chemical treatment	8	128R+
Degumming	8	138
Fabrics woven	139	428
Fermentative treatment	435	268
Liberating	19	3
Screen printing	101	114+
Process	101	129
Thread finishing	28	217+
Silkworm Culture	119	6
Sill	52	204
Railway car	105	396+
Weatherstrips	49	303+
Sillimanite		
Ceramic containing	501	141+
Porcelain containing	501	141+
Refractory containing	501	128+
Silos		
Circular	52	245+
Compacting ensilage	100	65+
Portable, & means for erecting	414	919*
Prestressed	52	224
With port	52	192+
Siloxanes	556	450+
Resinous solids (See also synthetic resin or natural rubber)	528	10+
Silver	75	
Alloys	420	501
Chest or box for	D 3	75
Composite metal stock	428	673
Electrodeposition	204	46.1
Electrolytic synthesis	204	109+
Mercury alloy	204	126+
Hydrometallurgy	75	118R
Misc. chemical manufacture	156	DIG. 101
Organic compounds	556	110+
Proteins	530	400
Pyrometallurgy	75	83
Silverware		
Chest or box for	D 3	75
Sorter	209	926*
Tray or container	206	557
Simulations		
Article dispensing	221	24
Cabinets with	312	204
Display	40	538+
Material dispensing combined	222	80+
Pyrotechnics	102	335
Supports digest	248	DIG. 8
Toys	446	
Velocipedes	280	1.1 R+
Sine Bar	33	536+
Sinew Removing	17	11.3
Singeing		
Cloth finishing	26	3+
Electric burner	219	223+
Hair combs for	132	118
Hogs	17	20
Thread finishing	28	239
Warps	28	174
Single Grid Twin Plate Control Tube	313	303
Single Roll Crusher	241	221+
Singletree		
Horse detacher		
From thill or evener	278	32
Traces from singletree	278	24+
Sink (See Receptacles)	4	
Cabinet combined	312	228
Design	D23	284+
Disappearing bowl	4	DIG. 2
Drain board or tray	D32	56
Lining mat	D32	57
Strainers and stoppers	4	286+
Structure	4	619+
Traps	4	191+
Tub combined	4	553+
Sinker		
Fishing type	43	43.1+

U.S. PATENT CLASSIFICATIONS

	Class	Subclass
Design	D22	145
Knitting machine	66	104+
Traps	43	68
Sinkhead	249	106
Sintering		
Apparatus, ore	266	176
Stationary grate	266	185
Beneficiating ores by	75	5
Cellular synthetic resin or natural rubber product	521	919*
Ceramic composition	501	1
Clay	264	56+
Glass granules	65	18.1
Apparatus	65	144
Portland type cement making	106	100+
Powder metallurgy		
Processes	419	
Liquid phase	419	47
Products	75	228+
Processes	432	1+
Lime gypsum cement	106	100+
Sinusoidal Generator		
Electronic tube type oscillator	331	
Generator structure	310	
Generator system	322	
Siphon		
Acetylene generator	48	17
Boiler	122	68
Bottle	215	4+
Design	D 7	51
Filling	141	14+
Bowl	4	421+
Code recorder	178	91
Dispensers	222	416
Discharge assistant combined	222	204
Fluid flow meter	73	222
Fluid handling	137	123+
Automatic	137	132+
Lubricator	184	85
Plumbing	137	247.11+
Syringes	604	131+
Water closet		
Siphon bowl	4	421+
Tank	4	368+
Siren	116	147
Design	D10	120
Electric	340	405
Toy	446	205
Sirup (See Syrups)		
Sizing (See Assorter)		
Bread, pastry and confections	99	352+
Chain making	59	29
Coating compositions		
Natural resin containing	106	238
Protein containing	106	124+
Starch containing	106	210+
Coating processes	427	
Dyeing processes combined	8	495+
Finger ring forming and	29	8
Hat making	223	10
Paper tube making	493	269
Textiles chemical modification of	8	115.6
Warp threads	28	178
Skate and Skaters Appliances	280	11.1 R
Clamp for sharpening by abrading	51	228
Design	D21	224+
Ice, powered	180	180+
Roller, powered	180	180+
Sharpening by cutting	76	83
Skaters supports	272	70+
Snow	280	600
Design	D21	224+
Strap	280	11.3
Water	441	65+
Wheels for	301	5.3+
Skeeball Game	273	352
Skein		
Axles	301	134+
Fluid treating package	8	155.2
Reeling	441	26+
Holders	242	127
Skeining	28	291
Skeleton		
Figure toys	446	373+
Keys	70	394
Plug valve	251	310+
Towers	52	648+
Skelping Metal	72	176+
Skewer Pins	17	1S
Ski (See Aquaplane; Sled)	280	601+
Apparel design	D 2	31
Bindings	280	611+
Clamp type having plural diverse axes clampsps	280	DIG. 12
Controlled by skier postion or by remote control	280	DIG. 13
Brake	280	604+
Carriers	224	917*
Carriers, hand	294	147
Case	206	315.1
Clamp	280	814
Design	D21	229+
Handle or grip design	D 8	DIG. 6
Poles and sticks	280	819+
Presses	280	815
Snow, ie, non-water, powered	180	180+
Shoe fasteners	280	611+
Trail groomer	37	219+
Teaching skiing	434	253
Water skis	441	68+
Tow line	114	253+
Work holder for	269	906*
Skid	193	41
Aircraft landing gear	244	108
Anti		
Armored resilient tire	152	167+
Boots and shoes	36	59R+
Devices for resilient tires	152	208+
Chain applier	81	15.8
Inverted vehicle	280	32
Platform	108	51.1+
Railway		
Car stop	104	260
Truck	105	216
Skidding		
Traversing hoist	212	167
Skidway	193	38+
Hoisting rope combined	414	571
Railway	104	134+
Vehicle loading conveyor	414	537+
Skillet	126	373+
Skimmer		
Concentrating evaporator	159	42
Fermentation apparatus combined	435	287+
Floating separator	210	242.1
Froth flotation	210	221.1
Gravitational separator	210	513+
Stratifiers	209	493
Skimobile	180	182
Skin Diving (See Diving Apparatus)		
Skinner		
Animal	17	21
Fish	17	62
Nut	99	584+
Process	17	50
Skins and Leather	69	
Carroting of fur	8	112
Cutting for grafting	128	305.5
Dyeing	8	436+
Electrolytic treatment processes	204	135
Fluid or chemical treatment	8	94.1 R+
Manipulative	8	150.5
Forming integral skin, plastic and non-metallic	264	45.5+
Skip		
Chute for rotary drum	193	10
Furnace charging	414	178
Accumulating means before release	414	168
Loading or unloading	414	639+
Mortar mixer feeding	366	39
Liquid and	366	36
Vertically swinging load handler	414	680+
Skipping		
Exercising devices	272	74+
Figure toys	446	311+
Rope skipping	446	307
Skirt		
Christmas tree base	D11	130
Electrical insulators	174	80
Garment	2	211+
Apron	2	48+
Apron design	D 2	226+
Design	D 2	223+
Forms	223	68+
Hangers	223	95+
Hem etc markers	33	9R+
Hoop skirt making	223	6
Protectors	2	47
Protectors design	D 2	225
Undergarment, design	D 2	19+
Union type	2	71+
Machine pistons	92	208+
Skiving		
Leather	69	9+
Shoe		
Toe and heel stiffeners	12	62+
Upper machines	12	58
Skullcaps	D 2	256
Skylight	52	200
Clerestory	52	18
Dome	52	80
Skyrockets	102	347+
Skywriting	40	213
Slabs (See Panel)		
Mixing	433	49
Reinforced concrete	52	600+
Slack Adjuster	74	522
Brakes	188	196R
Slag		
Coating or plastic compositions	501	28
Inorganic setting material containing	106	117
Electric welding with	219	73
Formation and reduction of ores combined	75	30
Glass electric melting of	373	27+
Portland type cement making from	106	103
Recovering metal values from	75	24
Separating, metallurgical	266	227+
Utilization other than wool		
Apparatus	65	141+
Process	65	19+
Wool		
Compositions	501	36
Making apparatus	425	6+
Manufacturing	65	1+
Slaking		
Lime	423	640
Apparatus	422	162
Slant-board	272	144+
Slashing Strand Treatment	28	178+
Slat and Slatted		
Bed	5	236R+
Blind slat holder	292	345
Curtain shades and screens	160	130+
Slats per se	160	236
Tenoning	144	199+
Floor covering	52	660+
Shutter slats		
Vehicle body top	296	118+
Venetian blind	160	166R+
Washboards	68	230
Wire and slat fabric	245	11
Crate	217	51
Making	140	25+
Slate		
Separating from coal	209	
Writing surface	434	423+
Slaughtering	17	
Sled	280	12R
Brakes	188	8
Automatic operator	188	128
Design	D12	6+
Dumping vehicles	298	4
Earth working tool with	172	387+
Excavating		
Scoops	37	122
Occupant-steered	280	16
Powered, eg, snowmobile	180	182+
Shuffler, eg stepper, type	180	187
Toboggan	280	18
Water	441	65+
Wheels combined	280	8+
Sledways	104	134+
Sleep Sofas	5	12R+
Sleeping		
Awakeners (See awakeners)		
Bag	2	69.5
With bed or tent structure	5	413
Car	105	314+
Sleeve		
Apparel		
Adjustable length	2	269+
Mens shirts	2	125+
Protectors	2	59
Sleeve holders	24	15
Sleeve holders design	D 2	625
Appliers and removers	29	282
Force multiplying operator	29	244+
Hand grippers	29	280+
Manufacturing type	29	234
Manufacturing type resilient	29	235+

	Class	Subclass
Bearings	384	276+
Roller	384	569
Rotary thrust collar and	384	275
Chuck pipe and rod joints	403	300+
Auxiliary detent operated by	285	82+
Chucks or sockets	279	
Duplicate ends intermediate	285	66
Electric conductor	174	84R+
Flexible	403	223
Flexible	285	235
Hinged	403	309+
Lugged pipe locking	285	82
Molded	285	292+
Rotary locking	285	377
Swivel lateral port to	285	190
Electric conductor joint	174	93
Grinding process	51	290+
Lost motion socket type chuck	279	19.6+
Pendant watchcase bow fastener	368	305
Railway axle	295	37
Railway track		
Fastenings	238	377
Rail joint	238	248
Tie plate	238	300
Strand knotter	289	12
Tobacco pipe stem with	131	228
Vehicle wheel stop engaging	301	122
Wrench traveling jaw rocking	81	105
Sleigh	280	12R+
Bell	116	170
Slicer & Slicing (See Cutter)	83	707
Diminishing workpiece	83	703+
Vegetable meat etc	83	
Corer combined	99	545
Parer corer doffer combined	99	543
Wood	144	162R
Scoring combined	144	42+
Turning	142	
Slide		
Amusement	272	56.5 R
Design	D21	242+
Bearings	384	7+
Boxes	229	9+
Folded blank	229	19+
Buckles	24	190
Clasps	24	522+
Educational display	434	405
Fastener (See also zipper)	24	381+
Holder	D16	26
Machine tools		
Lathe	82	21R+
Milling machine	409	145+
Turret	29	36+
Way guard	384	15+
Projector	353	
Slide holder	206	456
Slide only	40	152+
Rest	82	24R+
Surface	108	65+
Table combined	384	7+
Viewer, photographic	D16	17
Slide Rule	D18	9+
Bar type	235	70R
Circular	235	78R+
Cylinder type	235	79.5
Disk type	235	84
Teaching device for music	84	471SR
Sliding		
Pole	182	100
Sling		
Animal restraining	119	100
Design	D30	151
Hopple combined	119	102
With alarm	340	573
Cartridge	224	203+
Centrifugal projector	124	4+
For firearms	42	85
Hand package	294	152+
Hoist line	294	74+
Invalid lift	5	89
Knapsack	224	205
With belt	224	215+
Liquid	239	214+
Surgical	128	94
Vehicle towing	414	563
Slinger		
Centrifugal liquid projector	239	214+
Gas and liquid contact	261	84+
Refrigerator defrost liquid	62	280
Rod packing	277	133
Sand mold compactor	164	198+
Slingshot	124	17+
Design	D22	106
Slip (See Apparel)		
Covers for chairs and seats	297	219+
Drill pipe or rod	175	423
Ferry slip	114	231
Key slip for musical instrument	84	80
Rings electric	310	232
Ship building and repairing	405	1
Undergarment	D 2	19+
Well	166	206+
Packer or plug combined	166	118+
Slipper (See Shoe)		
Fleeced soles	36	26
Inside	36	10
Slitting		
Dental crown	433	145
Shoe machines		
Folding uppers combined	12	55.1
Grooving soles combined	12	40
Welt beveling combined	112	44
Tobacco		
Ash receiver combined	131	233
End cutter combined	131	252
End slitter	131	253+
Molding perforating cutting combined	131	83.1
Webs	83	425+
Winding combined rotary cutter	242	56.2
Sliver and Sliver Making		
Assembling	368	102+
Cans	19	159A
Forming	19	150+
Stopping apparatus control	19	
Wood	144	185+
Slot and Slotted		
Airfoils	244	198+
Antenna	343	767+
Bedstead corner fastening	5	299
Bedstead pivoted leg shoulder and	5	314R+
Forming (See specific tools; grooving)		
Hinge	16	357+
Machine slot closers	74	566
Gauge combined	83	522+
Making		
Milling	409	64+
Planing	409	304+
Punching	83	
Wood cutting	144	136R
Punching bolt cotter slots	10	14
Railway		
Couplings	213	75R+
Pneumatic car propulsion	104	155
Slot closers air sealing	104	161
Slot closers cable traction	104	194
Slot switches cable traction	104	195
Slotted pavement conduit	104	140+
Track slot brake	188	40
Slot machines	194	
Trolley rail two tread slotted tube	104	108
Slub Detecting	28	227+
Slubbing	19	237
Sludge		
Activated sludge		
Collector and remover	210	523+
Gas & liquid contact devices	261	75+
Processes	210	601+
Submerged fluid treating	210	220+
Surface contact aeration apparatus	210	150+
Collector and remover		
Acetylene generator	48	57
Earth boring combined	175	206+
Treatment		
Fat etc recovery from	260	412.5
Mineral oils	208	13
Sluer and Sluing		
Rotary crane	212	223+
Cyclic	212	161
Mechanism	212	245+
Track layer	104	8
Slurry (See Sludge)		
Nuclear fuel	376	356+
Small Arms (See Gun)		
Smelting Metal		
Apparatus	266	
Electrothermic	75	10.1+
Furnaces	432	120+
Pyrometallurgy	75	20R+
Smocking	112	133
Smoke		
Bells	362	379
Box		
Boiler	122	
Draft regulators	110	157+
Spark arresters	110	122+
Coating by	427	248.1+
Conductors for locomotives	110	145+
Curing and smoke houses	99	467+
Detectors		
Design	D29	6
Ionization	250	381
With alarm	340	629
Measuring device	356	438
Optical system	250	574
Photoelectric	250	573+
With alarm	340	630
Disposition	55	
Arc lamps	314	26+
Electrical precipitator	55	101+
Processes	55	2+
Elimination	55	
By return to fire box	110	203+
By separating flue gases	110	148
Heating cars by smoke	237	12.5
Indicators	356	438
Jacks	104	52
Miscellaneous stacks	110	184
Monitors	356	438
Ovens smoke pipe heated	126	17
Preventing		
By materials sprinkled on fuel	44	6
Producers and producing	44	
Buoys	441	19
Cartridges	102	334
Drop bombs	102	334
Food smoking	99	467+
Fumigators	422	305+
Grenades	102	334
Plant frost preventing	47	2
Processes	252	305
Projectiles	102	334
Shells	102	334
Signals	116	214
Smudge pots	126	59.5
Toys	446	24+
Vermin and bee smokers	43	127+
Warfare gases	424	40
Warship smoke screens	114	15
Protection from		
Masks	128	206.12+
Oxygen inhalers	128	205.27+
Purifying and washing	55	282+
Process	55	83+
Recorders	356	438
Reduced smoke or gas generation in cellular polymer	521	903*
Return to fire	110	203+
Screen compositions	252	305+
Separating from gases	55	
Stacks	110	184
Chimneys	98	58+
Masonry and concrete	52	245+
Ships	114	187
Smoke box combined	110	121
Smoke jacks	104	52
Stove pipes	126	307R+
Sticks	17	44.4
Tobacco smoking devices	131	330+
Toy smoking devices	446	24+
Track sanding by smoke and cinders	291	4
Smoking Items (See Cigarette; Tobacco)	D27	51
Anti smoking devices	131	270
Ash receiver and cleaner combined	131	232
Cleaners	131	243+
Design	D27	
Holder for diverse items, eg matchbox attached to ashtray	224	278
Smoothing (See Ironing Devices; Press)	38	
Cigar or cigarette tip or mouthpiece applying	131	89+
Combined with making	131	30
Filter	333	181+
Leather	69	
Metal burnishing	29	90R
Thread finishing	28	219+
Tobacco treatment	131	324+
Leaf stemming combined	131	315+
Smudge Pot	D23	334

	Class	Subclass
Smut Removal from Grain		
Abrasion or impact grain scouring	426	481+
Apparatus	99	516+
Apparatus	99	600+
Brushing etc machines	15	3.1+
Cereal comminuting processes	241	6+
Cereal food	426	481+
Preservation	426	321+
Liquid cleaning of bulk material	134	25.1
Tumbling abrading	51	163.1+
Processes	51	313+
Snaffle		
Bit for animals	54	8
Snakes		
Electricians	254	134.3 R
Plumbers	15	104.3 SN
Snapper	56	113+
Snaps		
Acting regulators	236	48R
Actions	74	20+
Temperature and humidity regulators	236	48R
Valve	251	75
Clothing fasteners	D11	220+
Dispenser outlet element	222	498+
Glass work holder	269	
Halter	119	114
Hook	24	231+
Buckle combined	24	165
Separable fastener type	24	572+
Snap ring	24	16R
Switch	200	402+
Whiplash combined	231	5
Snapshot		
Binder device releasably engaging aperture or notch of sheet	402	
Display	40	152+
Folder	40	152+
Snare Drum	84	415+
Snarling Irons	72	399+
Sneakers, Footwear	D 2	310
Snelling	242	7.19
Method	242	7.2
Snip (See Shears)		
Sniperscope (See Snooperscope)		
Snooperscope		
Infrared converter tube	313	525+
With electric circuit	315	10+
Snore Inhibiting Devices		
Sleep inhibting alarm	340	575
Surgical electrical system	128	419R
Snorkel	128	201.11
Snow		
Apparatus for making	239	14.1
Artificial	106	
Artificial	272	15
Blower	37	244+
Compressor	37	225+
Exhibitor	40	
Fences	256	12.5
Heaters		
Compressor combined	37	226
Earth thawing	405	131
Excavator combined	37	227+
Melting furnaces	126	343.5 R
Railway switch	246	428
River surface	405	61
Surface heaters	126	271.1+
Making		
By spraying water	62	121
Congealing sprayed water	62	74
Melting	37	227+
By electricity	219	213+
Compositions	252	70
Plow (See road and track clearers)	37	196+
Design	D15	11+
Scraper	37	266+
Production	239	2.2
By comminuting ice	241	
Road and track clearers	37	196+
Excavator and melter	37	227+
Railway excavator	37	198+
Railway track clearers	104	279+
Sledded scoops	37	264+
Sledded scrapers	37	270+
Snow compressor	37	225+
Roof stop	52	24+
Shoes	36	122+
Design	D21	228
Shovel	294	54.5
Skates	280	600
Design	D21	224+
Suit	D 2	31
Vehicle trailer	D12	6+
Snubber (See Dashpot; Shock Absorber)		
Crane boom	212	220
Snuffer, Flame	431	129+
Snuffer, Flame	431	144+
Candle	431	35
Candle	431	144
Soap		
Bar	252	174
Designs	D28	8.1+
Holder, design	D 6	532+
Sponge pocketed for	401	201
Bubble devices	446	15+
Detergent containing	252	108+
Biocide combined	252	107
Oxidant combined	252	96+
Packages	252	90+
Dispenser	241	602*
Dye	8	525
Fatty acid	260	413+
From mixture of acids	252	367+
Heat exchange compositions containing	252	79
Holding and feeding		
Applicator combined	401	143+
Bracket-type holder	248	200+
Clothes washer combined	68	17R
Devices	206	77.1
Dispenser type	222	
Suspended-type holder	248	317+
Washboard combined	68	224
Liquid soap dispenser	D 6	542+
Lubricants containing	252	32+
Solids containing	252	17+
Making	252	367+
Apparatus	422	129.1
Cutters for	83	
Molding devices	425	289+
Natural resin acid	260	105
Shaving	252	108+
Kit, brush or soap retained	401	118
Kit, with toilet articles, eg mirror.	132	81
Soccer	273	411
Social Sciences		
Computer controlled, monitored	364	419+
Socket (See Joints, Ball & Socket)	279	
Artificial body members		
Leg sockets	623	33+
Barrel hoop	217	90
Bracket article holder	248	314
Brush and broom support	248	111+
Car body stake	296	43
Castor	16	43
Electric connections	439	
Adaptor	439	
Adaptor socket	439	638+
With resistor	338	220
Fastenings and connections	279	
Cord and rope	24	136L+
Railway screw spike	238	373
Railway spike	238	371
Separable head and	24	572+
Framing	403	187+
Light	439	660+
Machine and tool part		
Brush tuft	15	191R+
Chucks	279	
Tool handle fastenings	403	
Work holder wood	142	57
Masonry anchor	52	704+
Pipe rod and shaft joints	279	
Flexible shaft	464	
Packed rod or shaft	277	
Rod	403	361
Rotatable rod in	403	165
Whip	280	170+
Wrenches	81	121.1+
Socks	D 2	329+
Sod Cutters	172	19+
Soda and Potash	423	
Carbon compound	423	421
Oxygen compound	423	641
Recovery	423	179
Soda Fountain		
Cabinet structure, ie syrup compartments	312	
With dispensing	222	173+
Soda water apparatus (See carbonation; dispensing; foam)		
Design	D 7	51+
Soderberg Electrode Arc Furnace	373	60
Electrolytic apparatus	204	225+
Sodium (See Alkali Metal)	75	66
Amalgam	420	526
Electrolytic synthesis	204	125
Chloride dissolver	422	902*
Organic compounds	260	665R
Valve cooled by	123	41.16
Making	29	156.7 C
Vapor lamps	313	567+
Sofa		
Bed	5	12R+
Construction elements	5	52
Chairs and seats	297	232+
Design	D 6	334
Swing type	297	282
Exercise	272	144+
Table bed type	5	7
Soft Drinks	426	590+
Carbonated	426	590+
Softening		
Leather	69	
Water	210	
Ion exchange process	210	660+
Software, Computer Programs	364	300
Protection		
By data encryption	380	4
Soil		
Handlers or conveyors	172	33
Mixer, machinery design	D15	19
Stabilizing compositions	106	900*
Treatment machinery, design	D15	21
Sol (See Colloids; Dispersing; Suspensoids)	252	302
Radioactive	252	634+
Solanum Alkaloids	546	124
Solar		
Bleaching	8	103
Celestial observation instruments	33	268
Cells	136	243
Systems	323	906*
Concentrating evaporators	159	903+
Cooker or warmer	D 7	324
Crystallization process	23	295S
Distillation of water	203	10+
Driers	34	93
Food treatment	99	
Heat accumulator structures	126	400
Heat collectors	126	417+
Heat exchange		
Automatic control of heat	165	18
Heating and cooling	165	48.1
Heating systems	237	1R
Kinestherapy	D24	39
Living body treatment	128	372+
Panel	136	243
Photovoltaic or photoelectric generator, ie solar cell	136	243
Power plants	60	641.8+
Energizing & storage	60	659
Reflectors, curved surface	350	629+
Refrigeration utilizing	62	235.1
Spacecraft altitude control by solar pressure	244	168
Sterilizing	422	22+
Apparatus	250	428+
Stills	202	234
Distillation digest	203	DIG. 1
Stoves and furnaces	126	417+
Thermoelectric generator	136	206
Solder & Soldering (See Alloy)	228	56.3
Apparatus	228	
Design	D 8	30
Blow torches	431	344
Can body forming and side seam	228	17.5+
Can head or body seam	228	
Electric	219	129
Fuse self soldering	337	152
Heat	219	129
Metal heating	219	85R
Soldering iron	219	227+
With pressure	219	85R
Iron	228	51+
Design	D10	46+
Electrically heated	219	227+
Heater portable furnace	126	236+
Self heating with burner	126	413+
Permanent closure	220	66+
Stick	428	558

	Class	Subclass
Temporary closure	220	200
Solderless		
Terminals	439	
Flexible wire joint	403	206+
Wire joining machines	140	113+
Sole		
Forming and attaching		
Attaching means	36	12+
Machines	12	17R+
Sewing machines	112	29+
Straighteners	12	120.5
Trimmers	12	85+
Pad		
Cleaning devices	15	227
Cushioned horseshoe	168	14
Foot supporters and	36	71
Horseshoe	168	26+
Horseshoe, design	D30	149+
Shoe	36	25R+
Antislipping devices	36	59R+
Design	D 2	317
Attaching means	36	12+
Design	D 2	319+
Design, insoles	D 2	318
Design, protectors	D 2	271
Filling compositions	106	38
Insoles	36	43+
Interfitted heel and	36	24.5
Light etc application to body	128	382+
Outsoles	36	25R+
Protectors	36	73+
Solenoid (See Magnet)		
Solid State Circuits and Devices (See Transistor)		
Solidifying (See Crystallization)		
Borehole inclinometer		
Earth control by	405	263+
Gas	62	8
Liquid by cooling	62	340+
Automatic	62	135+
Process	62	66+
Liquid comminuting combined		
Apparatus	425	222+
Processes	264	5+
Liquid fuel	44	7.1
Sugar processes	127	58+
Solids Dissolving for Cooling	62	4
Solids Separating and Stratifying	209	
Solvent (See Extracting)		
Coating or plastic composition for	106	311
Compositions	252	364
Detergent having	252	89.1+
Extraction to recover hydrocarbon	585	833+
Gas package or receptacle having	206	.7
Polymer dissolver	422	901*
Recovery by drying or distillation		
Concentrating evaporators	159	
Distillation	203	
Distillation apparatus	202	
Drying combined	34	72+
Mineral oils	208	179+
Refrigerant combined	252	69
Sodium chloride & potassium chloride dissolver	422	902*
Treatment of coating	427	335
Well		
Cleaning with	166	304
Solid material recovering	299	5
Sonar and Sonics		
Flow	261	DIG. 78
Gas separation	55	277
Process	55	15
Portable sonar devices	367	910*
Simulators, sonar	434	6+
Sonar design	D10	46+
Sonar time varied gain control systems	367	900*
Sonic tests		
Compressional wave in fluid	367	87+
Gas analysis	73	24
Seismic	367	14+
Vibration	73	570+
Vibrators	261	DIG. 48
Wave bonding, nonmetallic		
Apparatus	156	580.1+
Process	156	73.1+
Wave measurers	33	1P
Sonic		
Used in reaction to prepare inorganic or organic material	204	157.15+
Used in reaction to prepare or treat synthetic resin or natural rubber	522	1+

	Class	Subclass
Soot		
Blower		
Boiler installed	122	390+
Heater installed	165	95
Catcher	126	280
Sorbent (See Absorbents)		
Air contacting with refrigeration	62	271
Process	62	94
Gas adsorption thermometer	374	159
Gas separation	55	387+
Processes	55	74+
Liquefied gas		
Producing	62	17+
Storing	62	48
Refrigeration utilizing	62	476+
Automatic	62	141+
Process	62	101+
Sorbic Acid	562	601
Sorbitol	568	852+
Sorel Cement	106	105+
Sorghum	426	658+
Sorption to Recover Hydrocarbon	585	820+
Sorters and Sorting		
Automatic machines	209	509+
Belts sifters	209	307+
Coin	453	3+
Data processing equipment	D14	110
Electrostatic	209	127.1+
Fluid suspension	209	132
For paper sheets	271	287
Grates or sifters	209	233+
Machines for statistical records		
Coded edges	209	608
Coded holes	209	613
Magnetic	209	212+
Racks	211	10+
Refuse treatment means	110	220
Sheet feeding or delivery	271	
By items following different trajectories	209	638
Thrower	209	642
Sorting & classifying solid items	209	
Bottles	209	522+
By sensing radiant energy	209	576+
Bottles, jars, etc	209	524+
Cigarettes, etc	209	536
Laser	209	579
X-ray	209	589
By size	209	659+
By susceptibility to deform		
Condition responsive means	209	599
Separating means	209	699
Cigarettes	209	535+
Eggs	209	510+
Electrostatic	209	127.1+
Fluid suspension	209	132+
Lumber	209	517+
Magnetic	209	212+
Mail	209	900*
Paper money	209	534
Preparing items for sorting	209	3.1+
Sifting	209	233+
Special items	209	509+
Stratifiers	209	422+
Stratifiers	209	422
Weight operated	209	645
Condition responsive controls	209	592+
Sound (See Insulating; Recording; Reproducers; Acoustics)		
Bell	116	148+
Electric	340	392+
Boxes for phonographs	369	160+
Building structure		
Deadening in ventilators	181	224
Deadening in walls and floors	181	290+
Camera focusing, automatic	354	401
Controls direction of travel		
Airplane	244	76R+
Damping		
Refrigeration related	62	296
Vibrations in overhead electric conductors	174	42
Deadening	181	284+
Direction indicators	181	125
Ear stoppers	128	152
Equipment illuminators	362	86
Facsmile recorders	358	296+
Gas separator combined	55	276
Forming sound grooves in record	264	106
Horn or whistle	116	137R+
Electric	340	384R+

	Class	Subclass
Image converter	367	7+
Imitation	272	14
Intensity measuring	73	646+
Locating devices	181	125
Making devices, manufacture of	29	169.5
Meter	73	
Similar to pressure guage	73	432.1
Motion picture apparatus	352	1+
Music	84	
On film	369	125+
Film strip	369	16+
Mechanical	369	111
Phonograph sound box	369	160+
Post for violins	84	277
Recording and reproducing (See phonographs)	369	
Assembling process	29	169.5
Coating composition	106	37
Design	D14	
Electric	369	128+
Speech from static memory	381	51+
Electroforming	204	5
Molding apparatus	425	406+
Molding process	264	106+
Projection screen	350	117+
Talking picture	352	1+
Tape winding means	242	179+
Records	369	272+
On magnetized wire	360	89
Responsive		
Electric	381	168+
Electrical selector	367	197+
Printing	178	31
Radio controlled	340	825.72+
Pulse responsive	340	825.69
Signaling	181	
Toys	446	175
Responsive explosive igniters	102	211
Mines	102	427
Mines marine	102	416+
Signs X-reference collection	40	906*
Siren	116	147
Speed of sound compensation	367	902*
Telegraph sounder	178	98+
Tone arms	369	244
Electrical pickup	369	135
Mechanical pickup	369	158+
Toy	446	397+
Aerial top	446	213+
Figure	446	297+
Figure wheeled	446	270+
Musician	446	297+
Top	446	258
Air actuated	446	213+
Wheeled	446	409+
Whirling	446	215
Transmission		
Electrical telephone	379	
Mechanical telephone	181	138
Wave analysis	73	645+
Wave exhibitor	73	659
Electric wave analyzer	324	77R
Sounders		
Electromechanical bells and gongs	340	392+
Telegraphic	178	98+
Sounding		
Board for pianos	84	192+
Depth	33	126+
Compressional wave underwater	367	99+
Fathometer T M electrical	367	99+
Rods	33	126.7 R
Soups	426	589
Souring	8	137
Agents	252	193
Soya Bean Products		
Beverage	426	598
Coating or plastic compositions containing	106	154.1
Extraction of oil	260	412.2+
Phosphatides	260	403
Proteins	530	378
Sauce	426	629+
Space Discharge Devices, Electric		
Cathode sputtering coating or forming apparatus	204	298
Cathode sputtering coating or forming process	204	192.1+
Chemical apparatus	204	
Chemical electric discharge	204	164+
Apparatus	422	186.21+
Circuits for gas filled tubes	315	

	Class	Subclass
Circuits for vacuum tubes	328	
Drying and gas or vapor contact with solids	34	1
Electric switch, eg lamp starter switch	337	22+
Furnace pressure or vacuum	373	10
Heaters	219	200+
Heating furnace	373	
Igniters	361	247+
Lamps	313	
Lightning arresters	361	117+
Making	445	
Making gettering	445	
Making miscellaneous	29	592R
Medical and surgical application	128	
Metal heating arc	219	121R+
Nuclear reactions	376	
Ozonizers	422	186.7+
Precipitators	55	101+
Preservation of foods and beverages	426	237+
Rectifier structure	313	
Repair apparatus	445	61
Repair methods	445	2
Spark plugs	123	169R
Structure	313	
Structure of arc lamps and similar devices	314	
Structure of arc lamps feeding electrode	313	357
Structure of electronic tubes	313	
Structure of gas & vapor lamps	313	567+
Structure of lightning arresters	361	117+
Structure of mercury vapor lamps	313	163+
Structure of spark gaps and arc gaps	313	325
Structure of spark plugs	123	169R
Systems	328	
Systems amplifiers	373	10
Systems arc feeding electrode	314	
Systems cathode ray tube	315	1+
Systems combined tube and circuit element	315	32+
Systems gas or vapor tube	315	
Systems miscellaneous	250	
Systems television	358	256+
Testing, vacuum tube	324	405+
Cathode ray tube	324	404
Treating foods and beverages	99	451
Vacuum and pressure furnace	373	
Welding arc	219	136+
Arc	373	77
Induction	373	110+
Resistance	373	140+
Space Induction Systems		
Telegraph	178	43
Telephone	379	55
Train inductive telegraphy or telephony	246	8
Transmission of electricity		
To electric motor	318	16
To vehicle	191	10
Vehicle electrically powered by magnetic induction	191	10
Space Ship	244	158R+
Control systems	244	176
Design	D12	320
With booster rocket	102	374
Spaced Wall Containers		
Bottle or jar	215	12.1+
Cabinets	312	214
Metallic receptacles	220	415+
Pitchers	222	131
Spacer and Spacing		
Article arranging on conveyors	198	434+
Article oplacing on conveyors	198	434+
Concrete reinforcement	52	677
Container wall	220	445
Dental separators	433	148
Hoist bar	294	81.1+
Nuclear fuel structure	376	438
Orthodontic	433	2+
Piling material	206	
Planting implement spacing in hill	111	94
Printing fillers	101	403
Railway		
Guard rail	238	22+
Rail joint	238	227+
Tie	238	105
Sewing machine tucking	112	146
Sheet for drying	34	6
Typewriter		
Back	400	308+
Line	400	545+
Variable	400	303+
Vehicle body top board	296	34
Wire fabric making	140	50
Spacesuit	2	2.1 A
Helmet	2	2.1 A
Refrigeration for	62	259.3
Spades (See Shovel)		
Gun mount	89	40.9
Intrenching	42	93
Spaghetti		
Cutter	83	926E*
Design	D 1	126+
Molding apparatus	425	376R+
Pasta processing	426	451
Shaping apparatus, general	425	376R
Perforated plate	425	331+
Spanners	81	176.1+
Spar	114	89+
Torpedo boats	114	19
Spark		
Arresters	110	119+
Abrader spark guard	51	272
Fixed whirler or rotator deflector separators	55	447+
Flue jet pumps	417	156
Illuminating burners	362	380
Coils	361	268+
Gap	313	325
Manufacturing	445	7
Plugs	313	118+
Circuit element combined with	315	32+
Design	D13	16
Engine	123	169R
Making	445	7+
Tester	D10	77
Prevention dynamoelectric	310	220
In electric switch	200	144R+
Relays preventing	361	160+
Switches arc preventing	361	2+
Sparker (See Igniters)	123	146.5 R+
Internal combustion engine	123	146.5 R+
Toy	446	22+
Sparklers	102	358+
Spats	36	2R
Design	D 2	267
Spatulas	294	49+
Speaker	381	150+
With amplifier	381	120+
Speaking Aid		
Megaphone	181	177+
Public address system	381	82+
Amplifier	330	
Speaking Tube	181	18+
Design	D10	116
Spear	30	
Aerial projectile	273	416+
Design	D22	117
Handling	294	61
Harpoons	43	6
Speargun	D22	102
Special Receptacles	206	
Specific Gravity		
Testing	73	32R+
Speciofoline	546	15
Spectacles (See Eye Glasses)		
Metal frame making	29	20
Temples	351	111+
Spectrographic Analysis	356	300+
Spectrographs	356	302+
Spectroheliographs	356	302+
Spectrometer	356	326
Cavity		
Electron resonance system	324	316+
For color film	355	38
Nuclear resonance system	324	307+
Spectrometer components	324	318+
Radioactivity	250	472.1
Ultraviolet	250	372+
X ray	378	
Spectrometry	356	300+
Mass	250	281+
Spectrophotometer	356	319
Ultraviolet	250	428+
X ray	378	
Spectroscope	356	300
Spectrum (See Spectrometer)		
Specula	128	3+
Design	D24	18
Speech (See Sound)		
Amplifier	330	
Controlled	178	31
In typewriter	178	31
Exhibiting by visual means	73	659
Measuring by electrical means	73	570+
Wave shape comparing	324	77R
Making artificially	381	51+
Teaching	434	185
Voice reflector for therapy	D24	35
Waves		
Analysis	324	77R
Speed		
Measuring device, calibrating	73	5
Motor (See type of motor controlled)		
Recording	346	
Centrifugal governor	346	73
Speed indicator or alarm combined	346	18
Regulator and controller		
A C motor	318	301+
Automatic musical instruments	84	128+
Clutches	192	146
D C motor	318	301+
Electric motor	318	301+
Electric railway vehicles	104	300
Fabric manipulation weaving	139	309+
Induction motor	318	727+
Internal combustion engines	123	319+
Lathe headstock gearing	82	29R
Load hoist	254	267
Machine elements and mechanisms	74	
Belt & pulley drive system	474	
Metalworking machine	29	64
Motor vehicles	180	170+
Pumps	417	1+
Railway trains	246	182R
Spinning twisting twining textile apparatus	57	93+
Treating processes material speed control	34	25
Speed control	34	52
Warp feeding wearing	139	105+
Warp tension control textiles	28	185+
Watches	368	186
Responsive or controlled		
Automatic control of a part	340	53
Automobile mounted systems	307	10R
Brakes	188	180+
Centrifugal switches	200	80R
Devices	73	488+
Drive release and brake	192	147
Drop of elevator sans manual control	186	23
Electrical signal processing systems	361	236+
Electrical systems	361	236+
Electrical systems alarms	340	670+
Elevator brakes and catches	187	89+
Elevator operation	187	38+
Engine to throttle control	60	906*
Flexible shaft couplings	464	1+
Fluid distribution	303	91+
Gearing	74	336R+
Limit stops	192	140
Machine elements trips	74	3
Measuring & testing speed	73	488+
Mechanical movements	74	114
Motor controlling arc lamp	314	62
Motor vehicle	180	170+
Planetary gearing	74	752R
Plural automatic testing	324	73R
Predetermined speed indicator	116	57
Railway speed control systems	246	182R
Railway train brakes	188	126
Servomotor	91	458
Servomotor follow-up devices	91	366
Shutters motion picture apparatus	352	209
Valve actuation	137	47+
Vehicle signals	116	37+
Vehicle speed limit alarms	116	74
Vehicle speed limit indicator	116	62.3
Vehicle speed limit signals	116	57
Wave systems to determine velocity	342	104
Synchronizing (See synchronizing)		
Testing electric	324	160+
Timing electrically	324	178+
Timing electrically	324	179
Timing electrically	324	180
Speed Indication		
Electrical indicators	324	160+

	Class	Subclass
Alarms	340	670$^+$
Limit indicators	116	62.3
Alarms	116	74
Vehicle operated	116	57
Speed of sound compensation	367	902*
Speedometer	73	488$^+$
Aerial optical	356	27$^+$
Calibrating	73	2
Design	D10	98
Odometers	235	95R$^+$
Power take off	74	12
Time interval measurement	368	89$^+$
Electrical means	324	78R$^+$
Electrical means	324	179
Electrical means	324	180
Speed checker	368	2
Speedometer (See Speed Indication)		
Design	D10	98
Spelling, Teaching	434	167$^+$
Spelter (See Zinc)		
Spermaceti Purification	260	420$^+$
Spermacides & Contraceptives	424	DIG. 14
Sphygmoscope, Sphygmograph and Sphygmomanometer	128	677$^+$
Spices	426	638$^+$
Condiment dispenser	D 7	52$^+$
Plural cell	D 7	57
Shaker	D 7	54$^+$
Spigot (See Valve)		
Spike	411	439$^+$
Design	D 8	391
Making	10	56$^+$
Pullers	254	18$^+$
Railroad tie		
Drivers	104	17.1
Railway	238	366$^+$
Ship implement	114	221R
Shoe cleat or	D 2	317
Spillway	405	108
Spindle		
Article holding bracket	248	309.2
Axle	301	131
Rotary	301	126
Bearings	384	
Lathe headstock	82	30
Combination lock	70	329
Cotton picker	56	50
Fiber and thread		
Bobbin cone wind	242	35
Weaving picker spindles	139	158
Weaving shuttle spindles	139	208$^+$
Latch	292	358
Catches	292	359
Lathe headstock	82	30
Saw handle	30	521$^+$
Ship	114	59
Toy top	446	264
Spinning		
Can head seaming	413	27$^+$
Filament forming (See wire)		
Apparatus	425	66
Apparatus	425	67$^+$
Apparatus	425	76$^+$
Apparatus	425	382.2
Processes	264	165$^+$
Solutions	106	165$^+$
Metal	72	82$^+$
Paper	493	395
Textile	57	
Machine forging dies	72	343$^+$
Toys	446	236$^+$
Tube end	72	80$^+$
Closed	228	2
Spinning Reel	D22	141
Spinthariscope	250	458.1
Spiral		
Abrading scrolls	51	209S
Cigar wrapping	131	59
Conveyers and gatherers		
Chute	193	12$^+$
Chute stratifier	209	434
Cooking filled receptacle rollway..	99	365
Cornstalk harvester	56	95
Potato diggers	171	57
Power driven conveyor	198	778
Electrical conductor armor	174	108$^+$
Fasteners and chucks		
Chuck scroll actuated	279	114$^+$
Hat fastener	132	62
Spiral fastener	24	112
Feed for sewing machine	112	121.18
Fixed cutter machine	83	425
Gearing	74	424.5$^+$
Harrow	172	532$^+$
Measurement, radius &	33	1SP
Metal coiling by spiral die	72	141
Muffler baffle	181	279$^+$
Nail shank	411	487$^+$
Nail shank making	10	46
Pipe or tube article		
Flexible spirally wound	138	129$^+$
Spiral seam	138	154
Rotary plow	172	532
Sifters	209	362
Signal bell spring	116	168
Spacer for coaxial cable	174	29
Strip cutting		
Die cutting	83	
Toy propellers	446	43
Toy top spinner	446	241
Tube making and welding	228	145
Apparatus	228	17.7
Paper tube winding	493	299
Sheet metal spiral seam	72	49
Wire spiral stud making	140	92
Wood turning spiral grooving	142	23$^+$
Spirit Level	33	377$^+$
Design	D10	69
Illuminator	33	348.2
Spirometer	128	725
Spits Cooking	99	419$^+$
Spittoon	4	258$^+$
Cookstove attachment	126	32
Design	D34	2
Flushing	4	262$^+$
Dental	4	263$^+$
Pocket	4	259
Splash Board		
Cabinet with	312	229
Splash Control Element Mold	249	206
Splice (See Joints)		
Adhering seam making		
Apparatus	156	502$^+$
Process, without overlap	156	304.1
Bars rail type	238	243$^+$
Insulated	238	159$^+$
Belt	24	38
Drive belt combined	474	253$^+$
Electric conductor	156	49
Electric trolley	191	44.1
Film splicer	83	926J*
Design	D16	41
Support combined	191	44
Keys and locks rail type	238	250
Angle chair	238	198$^+$
Rail section	238	226
Web key	238	181
Sprocket chains	474	218
Twisted textile strand	57	202
Twisting and spinning		
Implements	57	23
Machines	57	28
Processes	57	22
Web	242	58.1$^+$
Spline	403	359
Gear and	33	DIG. 14
Match	44	46
Coating composition	149	3$^+$
Making	144	66
Panel connector	52	586
Surgical	128	87R$^+$
Splint and Splint Making		
Split Preventer	411	912*
Building component	52	514
Split Second Watches	368	102$^+$
Stop watches	368	113$^+$
Splitting		
Atomic fission reactions	376	
Carcass	17	23
Corncob	225	
Leather	69	9$^+$
Stone	125	23R$^+$
Tool blank	76	11
Tube and sheath	30	90.1$^+$
Well tubing	166	55.2
Wood	144	193R$^+$
Bundling combined	144	192
Explosive combined	89	1.14
Spoilers for Aircraft	244	213$^+$
Spoke		
Land vehicle wheel	301	5R$^+$
Making and assembling		
Land wheel	29	159.2
Metal wheel assembling	157	1.5
Securing by forging	157	1.55
Tenon turning	144	206
Wood spoke assembling	157	
Wood tenoning and sawing	144	18
Pulley	474	195$^+$
Railway wheel	295	25$^+$
Spokeshaves	30	281
Sponge and Sponging		
Container	206	209
Cups	118	270
Holder	15	244.1$^+$
Household cleaning tool	D32	35
Implement coating material combined	401	196$^+$
Mop	D32	40$^+$
Pocketed to contain soap	401	200
Resilient tire filler	152	313
Rubber	521	50$^+$
With fabric cover or backing	15	244.3
Spontaneous Combustion (See Explosion Control; Fire, Extinguishing; Ventilating)		
Spool	242	118
Cabinet		
Dispensing	221	
Stacked article	312	45
Container for flanged spool	206	398
Cutlery with	30	124
Holders	242	134$^+$
Tape spool	D19	65$^+$
Holders for sewing	223	106$^+$
Labelling	156	556$^+$
Lifters	242	46
Musical stringed instrument	84	303
Packages of spools	206	389$^+$
Rack	242	134$^+$
Ribbon and film	242	71.8$^+$
Spool winders	242	16$^+$
Stand for sewing item	D 3	23$^+$
Typewriter ribbon	400	242$^+$
Valve	137	625.34$^+$
Wood making machine	144	14
Wound with strand material	242	159$^+$
X-art collection	493	954*
Spoolers	242	16$^+$
Spoon	30	324$^+$
Agitating or beating	366	343
Cutlery combined		
Forks	30	150
Knives	30	149
Knives and forks	30	147
Design	D 7	137$^+$
Fluid conductor combined	30	141
Making		
Blanks and processes	76	105
Dies	72	470$^+$
Measuring	73	426$^+$
Design	D 7	50
Pierced bowls	D 7	140
Sport Equipment (See Type)		
Board games	273	259
Chance controlled	273	244$^+$
Card or title games	273	298
Protective hand covering	D29	20$^+$
Shoe, making	12	142P
Sports, Teaching	434	247$^+$
Spot		
Bowling	273	52
Inserting in bottle cap		
Metal cap making combined	413	8$^+$
Apparatus	413	56$^+$
Welding	219	86.1$^+$
Spotlight	362	257$^+$
Bulb	313	
Flashlight hand	362	157$^+$
Photographic design	D26	61$^+$
Supports	362	382$^+$
Surgical or dental	362	804*
Spotter		
Bowling	273	46
Spotting		
Apparel trimming, e. g. chenille	223	45
Board	68	240
Spout		
Antidrip	222	571
Bucket		
Sap	47	51$^+$
Turpentine and rubber	47	11
Dispensing	222	566$^+$
Molten metal	222	591$^+$

	Class	Subclass
Food or beverage dispenser with	D 7	312+
Track sander cutoff	291	31
Water distributing	137	561R+
Sprag	188	6
Spray		
Coolers for liquids and gases	261	
Gun	D23	213+
Tubes		
For spray coolers	261	115+
Sprayer and Spraying	239	
Animal husbandry		
Antivermin treatment	119	159
Restraining device	119	112
Boiler feed heater	122	436
To steam space	122	438
Cement	366	10+
Process	366	3+
Chute combined	193	11
Coating machine	118	300+
Coating process	427	421+
Concentrating evaporator		
Apparatus	159	3+
Process	159	48.1
Design, spray gun	D23	213+
Distillation		
Apparatus	202	236
Mineral oil process	208	359
Process	203	90
Fuel dispersed in furnace	431	159
Gas and liquid contact device	261	115+
Porous mass	261	98
Refrigeration combined	62	304+
Automatic	62	171
Refrigeration combined process	62	121
Rotating impeller	261	89+
Rotating impeller hollow shaft	261	88
Wet baffle	261	111
Liquid freezer	62	347
Process	62	74
Machinery	D15	13
Material diffusing from aircraft	244	136
Metal spraying	29	DIG. 39
Milk dehydration	426	471+
Mineral oil distillation	196	128
Mobile orchard type	239	77+
Nozzles	239	589+
Flow regulating or modifying	239	533.1+
Powder	239	
Sandblast	51	428+
Shot blast	72	53
Shower cage	4	601+
Steam iron combined	38	77.5
Spreader and Spreading		
Animal training leg	54	72
Coating material combined	401	
Cultivator beam	172	645+
Fabric		
Bed bottom reinforcers	5	228
Buttonhole sewing machine	112	74
Frame for sheets	38	102+
Hammock	5	123
Making of wire strand twisting	140	33
Sewing machine loop	112	187
Sheet ironing	38	143
Web running lengths	26	87+
Biaxial	26	72+
Hand-held agricultural type	D 8	2
Horseshoe hoof	168	6+
Ladder propped	182	176
Leaf spring tool	81	3.7
Leather	69	
Material (See broadcasting; scattering)	239	650+
Bag filling hopper	53	570
Charge transported to, from, or within furnace	414	193
Discharging	414	209+
Plural successive driven devices	414	174+
Furnace fuel feeder	110	111+
Glass tube reshaping	65	283
Hand rake with	56	400.5+
Insecticide composition having	424	
Peach stoner tray	99	547
Plastic film molding device	425	461
Plastic film molding method	264	212+
Sifting feeding	209	254
Mobile orchard type	239	77+
Nail point locking	411	359
Piston ring	29	222+
Road material distributing	404	101
Slit opening and filling	29	270+
Store service	186	13
Tire		
Casing spreaders	254	50.1+
Holder	81	15.3
Rim with	152	400+
Wheel assembling	157	
Spring	267	
Actuated or controlled		
Bottle tube type valve	215	315
Car platform trapdoors	105	434
Check valve	137	511+
Clothes wringer pressure mechanisms	68	256+
Cooker with pressure means	99	351
Counterbalanced pitmans	74	592
Cutting tool joints	30	268+
Dispenser agitator and biasing means	222	230
Dispenser discharge assistant or casing	222	336+
Dispenser outlet element	222	511+
Dispenser resilient wall deflector	222	214
Dispenser trap chamber cutoff	222	449
Fence wire stretcher	256	39
Gong striker	116	164
Guns and projectors	124	16+
Holder for detachable sliding blade	30	336
Interlocked blade and holder	30	338
Invalid bed head	5	78
Locking latch bolts	70	144+
Lubricator with fluid pressure follower	184	41+
Lubricator with follower	184	45.1+
Mechanical gun centrifugal type	124	7+
Mechanical gun fluid pressure	124	66+
Mechanical gun magazine	124	52+
Nonrefillable bottle valve	215	19
Oil cup cover	184	88.1+
Packings	277	138+
Pivoted handles shears	30	212
Pivoted knife blade	30	159
Project-retract fountain pen mechanisms	401	109+
Railway mail projectors	258	7
Receptacle single plane swing closure	220	334+
Regrinding reciprocating disk valve	137	243.2+
Rotary disk valve	251	180+
Rotary plug valve	251	181+
Shears with holder	30	135
Shears with intermediate pivot	30	261+
Single impulse store service	186	10+
Siphon bottle	215	5
Sofa bed assisting	5	56
Spiral spring type bell	116	168
Spring joint shears	30	234+
Stop and nonstop valve combined	251	83
Stylographic fountain pen	401	259+
Swinging door	49	386
Trolley retriever	191	91+
Trolley retriever erector release	191	88+
Trolley stands	191	68+
Upending bed counterbalance	5	164R+
Valve parts	251	337
Yieldable pitmans	74	582
Applier and remover	29	225
Engine valve	29	215+
Bed and chair		
Bed	5	
Chair back supports	297	285+
Chair bottoms and backs	297	455+
Mattress	5	475
Pillow having	5	439
Buffer and check		
Bed bottoms with	5	253
Closure checks	16	85+
Liquid and combined concentric	16	52+
Liquid and combined side	16	59+
Ordnance mounts	89	44.1+
Pneumatic and spring	16	66+
Railway car end	213	221+
Railway draft appliances	213	7+
Textile weaving picker	139	167+
Vehicle bumper	293	102+
Vehicle fender	293	25+
Vehicle seat	297	216
Vehicle springs with	267	195+
Vertically swinging shovel	414	720
Button clamp and fastener		
Bottle and jar closures	215	287+
Card exhibitor clip	40	11R
Closure fastener bolts	292	2+
Closure fasteners	292	
Closure knob attaching	292	353
Closure seals	292	322
Folding paired table leg	108	131+
Folding stand leg	248	188.6
Hook tag fasteners	40	26
Mop holder	15	154
Paper	24	67.3+
Pivoted bottle closure	215	244+
Pivoted clasp	24	489+
Pivoted snap hook	24	233+
Rack series spring grip type	211	120
Railway track	238	349
Railway traction cable	104	209
Retainer	402	19+
Saw teeth latches	83	840+
Separable button	24	106+
Shuttle thread delivery	139	216+
Sliding snap hook	24	239
Stopper for breakable bottle necks	215	36
Tool handle hinged with detent	403	84+
Umbrella runner	135	40+
Vehicle rein holders	280	184+
Vehicle whip sockets	280	172
Coil	267	166+
Animal draft pole supports	278	89
Bed bottoms	5	186R+
Bells	116	168
Bobbin winding drive connection	242	46.8
Bolt to nut lock, lareral key	411	318
Bolt to nut lock, thread lock	411	262
Bolt to nut lock, thread lock and distorted nut	411	289
Bolt to nut lock to grip thread	411	251+
Cushion tires	152	246+
Flat	267	156
Hinges multiple	16	382+
Hinges pintle	16	305
Locomotive	105	67+
Loop forming	140	103
Making	72	138+
Pivoted snap hooks	24	234+
Railway car planks and transoms	105	208+
Railway car spring bolster	105	197.5+
Spring wheels	152	1+
Torsional closers	16	75+
Coil spring loop forming	140	103
Coil spring untangling apparatus	198	953+
Compressor	254	10.5
Hand tool	81	486
Couplings	403	229
Shaft	464	51+
Cushion shoe heels	36	38
Cushion shoe sole	36	28
Insole	36	44
Ventilated	36	3B
Cushioned railway draft devices	213	7+
Devices	267	
Bag closures	383	43
Casters	16	44
Closers	16	71+
Coat collars	2	99
Compressible rheostat	338	99+
Deformable discharge assistants	222	407
Ejector for conveyor trap dispenser	222	225
Extensible	351	126
Eyeglass bridge	351	65
Flexible	351	68
Garment supporter belt etc	2	339
Garment supporters	2	324
Handlers with spring devices	414	913*
Harrow teeth	172	705+
Hoof spreading horseshoe	168	8
Material pressure actuated dispenser outlet	222	491+
Mop holders	15	154
Music instrument spring shank hammer	84	324
Pivoted tooth bar harrows	172	634+
Plural bridges	351	67
Rail seats	238	284
Railway car actuated switches	246	270R+
Railway guard rails	238	18
Railway rail splice bars	238	259
Railway switches automatic	246	314

	Class	Subclass
Railway tie plates	238	302
Railway track ties	238	48
Resilient wall dispensers	222	206$^+$
Sandals	36	7.8
Sash balances	16	197$^+$
Sheet retainer with releasable keeper	402	60$^+$
Shoe heel	36	38
Shoe heel making	12	31.6
Shoe soles with spring heel	36	27
Slitted diaphragm dispenser	222	490
Spring back paper binders	24	67.3
Spring heel horseshoe	168	15
Truss pads	128	119.1
Watch spring packages	206	70
Watchcase	368	311$^+$
Wheel harrows	172	613$^+$
Wheel substitute harrows	172	389$^+$
Wooden crates cushioned	217	54$^+$
Exercising devices	272	135$^+$
Springboards	272	66
Fire escape	182	190$^+$
Fluid	267	113$^+$
Vehicle	267	64.11$^+$
Guns and projectors	124	16$^+$
Hairsprings	368	175
Hammer drive	173	119
Hinges	16	277$^+$
Bracelets	63	8
Horseshoe	168	8
Piano pedal	84	227
Telegraph keys	178	107
Weather strips	49	480$^+$
Jacks	379	
Joints couplings and chucks		
Flexible shaft	464	51
Packed shaft longitudinal	277	102$^+$
Packed shaft sectional	277	138$^+$
Packed shaft socket with follower	277	102$^+$
Pipe coupling detent	285	305$^+$
Railway draft	213	200$^+$
Railway draft appliances cushioned	213	7$^+$
Self grasping socket type chuck	278	22$^+$
Split resilient socket	279	41R$^+$
Spring detent socket	279	79$^+$
Spring saw socket	279	46R$^+$
Swivel pipe in socket	285	276$^+$
Thill	278	52$^+$
Trace and whiffletree	278	102$^+$
Leaf spring spreaders	81	3.7
Machine or tool		
Bobbin winding drive connection	242	46.8
Brake operators	188	166$^+$
Clutch operators	192	89R$^+$
Embroidering machine fabric shifting	112	92
Expansible chamber motors		
Insect catcher	43	135
Internal resistance brakes	188	268
Music leaf turners	84	486$^+$
Pliers crossed handle	81	417
Sawmill set works	83	724
Screw driver	81	451$^+$
Scroll saw spring return	83	752
Typewriter bar	400	436$^+$
Typewriter carriage	400	336$^+$
Woodworking feed rolls	144	247
Wrenches	81	52$^+$
Making		
Applying or removing	29	225$^+$
Conical wire	72	138$^+$
Forging dies	72	343$^+$
Loop forming	140	103
Metal bending	72	135$^+$
Processes	29	173
Processes spring head clip	29	172
Spring set saw making	76	71
Wire spring setting	140	89
Motor or drive		
Bobbin winding drive connection	242	46.8
Breech loading gun	42	5
Clock winding	368	147
Composite type motor	185	9$^+$
Dental pluggers	433	157
Drilling machine	408	125
Expansible chamber motor rotary		
Flying propeller toys	446	39$^+$
Hairsprings	368	175$^+$
Hammer	173	119
Handle bars	74	551.2
Locomotive	105	67$^+$
Motor	185	37$^+$
Pump	417	328
Railway motor support	105	139
Reels spring drum type	242	107$^+$
Shiftable motor and gear	74	209
Spring drum article holders	242	107$^+$
Spring drum clothesline reel	242	107$^+$
Spring mounting for	185	45$^+$
Steam traction engine mount	180	40
Store service	186	10$^+$
Strap actuator with return	74	136$^+$
Tape measure drum	242	107$^+$
Tops	446	260
Toy vehicle	446	464
Truck ladder hoisting	182	63$^+$
Typewriter carriage	400	336$^+$
Nut remover	29	270$^+$
Panels for chairs and seats	267	80$^+$
Railway		
Rail seats	238	284
Tire cushions	295	14
Setting	140	89
Support	267	
Casters	16	44
Centrifugal speed responsive device	73	546$^+$
Lamp shade on bulb	362	444
Lamp shade on chimney	362	445
Lamp shade or bowl support	362	440
Machine foot	248	615$^+$
Machinery base	248	560$^+$
Machinery bracket	248	674$^+$
Machinery suspension	248	637$^+$
Pin and open link or draft coupling	213	200$^+$
Pipe or cable	248	560$^+$
Sash balances	16	197$^+$
Telescoping tube suspended	248	610$^+$
Thill	278	81$^+$
Vehicle motor	248	560$^+$
Vehicle pole	278	88$^+$
Vertically adjustable suspended	248	610
Vibration dampening	248	562$^+$
Testing	73	161
Train vestibule face plate expanders	105	14
Valve		
Head	251	356
Parts	251	337
Vehicle parts	267	
Absorber combined	267	64.15$^+$
Article holders for bicycles	224	37$^+$
Axle skein mounting	301	136
Axle spindle mountings	301	133
Bicycle front fork etc	280	275$^+$
Bicycle rear fork etc	280	283$^+$
Body frame suspension	280	788
Bumper combined	293	116
Design	D12	159
Fluid	267	64.11$^+$
Letdown tops	296	107$^+$
Metallic spring cushion tires	152	247$^+$
Pole supports	278	88$^+$
Railway axle box mounting	105	218.1$^+$
Railway car body	105	453
Railway car truck platform	105	212$^+$
Railway spring bolster	105	197.5$^+$
Railway spring planks	105	208
Rocking running gear	280	112R
Roller skate trucks	280	11.28
Runner type	280	25
Running gear	280	688$^+$
Skates with foot supports	280	11.14
Spring wheels	152	1$^+$
Steam traction engine mounting	180	40
Stub axle mounts	280	660$^+$
Swinging and rocking axle	280	110
Swinging axle	280	113$^+$
Swinging axle connections	280	125$^+$
Thill supports	278	81$^+$
Tires spring and pneumatic	152	156
Traction increasing lugs for wheels	301	51
Trailer draft device	280	483$^+$
Two wheel vehicle	280	65$^+$
Velocipede seat	297	195$^+$
Wheeled vehicle	33	418
Watch etc hairsprings	368	175
Sprinkler	D23	214$^+$
Sprinkling	239	
Carts	239	146$^+$
Embedded	239	201$^+$
Flow regulator or modifier	239	533.1$^+$
Head fire extinguishing	169	37$^+$
Rack combined	211	127
Showcase combined	312	115
Textile treating	68	205R$^+$
Sprocket	474	152$^+$
Chain	474	202$^+$
Making	29	159.2
Chain	59	5$^+$
Railway cable	104	236
Chain cable	104	237
Spud Anchor	37	73$^+$
Spur		
Attached shoe protectors	36	74
Gearing (See gearing spur)		
Horseshoe spur fastening	168	21
Riding	54	83R
Design	D30	157
Stirrups combined with	54	49.5
Tag fasteners	40	25R
Tube		
Boiler	122	
Heat exchangers	165	42
Mineral oil distillation	196	129
Wheel traction increasing	301	43$^+$
Sput	285	201$^+$
Sputtering (See Cathode Sputtering)		
Sputum Cup	4	259
Spyglasses	350	537$^+$
Square	33	474$^+$
Combined		
Fixed straightedge	33	429
Hypotenuse	33	420
Nonpivoted straightedge sliding	33	427
Pivoted straightedge sliding	33	425
Leveling feature	33	451
Squeegee (See Wiper and Wiping)		
Cleaning device	15	245
Brush combined	15	117
Design	D32	41
Material supply combined	401	261$^+$
Wiper combined	15	121
Squeezer		
Animal traps	43	85$^+$
Butter worker	99	452$^+$
Citrus fruit	99	495$^+$
Cotton gin	19	49
Expressing	100	104$^+$
Paste tube	222	95$^+$
Raisin seeders	99	547$^+$
Textile fluid treating apparatus	68	94$^+$
Squeezing		
Liquids from solids	100	35$^+$
Apparatus for, with drain	100	104$^+$
Squirt		
Locomotive cab	239	174
Preventer		
Type casting integral line	199	53
Type casting separate type	199	90
Stabilizing (See Preserving)		
Aircraft		
Propellers	244	92
Weights	244	93$^+$
Colloids	252	351$^+$
Gyroscope	74	5.22
Mineral oil	208	351
Missiles, manned	244	75R$^+$
Missiles, unmanned	244	3.1$^+$
Projectiles	244	3.1$^+$
Ships	114	121$^+$
Space ships	244	158R$^+$
Synthetic resin or natural rubber (See class 523, 524)		
Latex	528	935*
Stable		
Animal	119	15$^+$
Stalls	119	27$^+$
Stalls, design	D30	108$^+$
Stanchions	119	147R$^+$
Stack and Stacker (See Packing)		
Beds stacked	5	8
Bottle and jar	215	10
Calender rolls	100	162R
Dispensing	221	
Cabinet	312	42$^+$
Rack	211	59.2$^+$
Electrical insulator	174	137R$^+$
Furnace and flue		
Draft regulator	110	159$^+$
Feed water heater	122	439$^+$

	Class	Subclass
Marine	114	187
Masonry or concrete	52	245
Smoke stacks	110	184
Spark arrester	110	121
Hay stack shaping	414	132
Horizontally attached receptacle set	220	23
Nuclear fuel components	376	433
Pneumatic	406	
Receptacles	206	501+
Special package, stacked item	206	821*
Rotary rolling contact printer	101	232+
Shaper	414	35
Coal storage type	414	133
Sheet		
Cutting combined	83	86+
Delivering	271	278+
Feeding	271	8.1+
Material folder	83	83+
Oscillating blade	271	264+
Stacked article support	211	49.1+
Adjusts for different sizes	414	900*
Card or sheet	211	50+
With impaling means, follower	211	54.1
With impaling means, terraced	211	57.1
With article counter	414	901*
With impaling means	211	59.1
With load-supporting fluid cushion	414	903*
With means to apply adhesive to articles	414	904*
With operational sequence control for stack pattern forming	414	902*
Stacking for dispensing	221	175+
Threshing	130	20
Stadium (See Theatre)	52	6
Construction	D25	12
Staff		
Flag	116	173+
Horological	368	324
Planting	111	97
Railway signal system	246	19
Support	248	511+
Tobacco ash receiver support	131	259
Stage (See Theater)	52	6+
Appliances	272	21+
Scenery shifting	272	22+
Illusions	272	9+
Lighting	315	312+
Motion picture set	352	88+
Set design	D25	58+
Stagger Tuned Amplifiers	330	154
Stained		
Glass mosaics	428	38+
Stainless Steel	420	8+
Austenitic	420	43
Heat treated stock	148	325+
Heat treating	148	135+
Metallurgy	75	28+
Stains Paint	106	34
Dyeing only	8	402
Stair (See Ladder)	52	182+
Amusement	272	2
Design	D25	62+
Escalators	198	326+
Lighting	362	146
Mobile for boarding aircraft	D34	30
Pads and covers		
Conforming to tread nose	52	179
Rug fasteners	16	10+
Slatted	52	180
Scribers for stair curves	33	29
Static mold for making	249	14
Tread	D25	69
Stake		
Animal hitching	119	121
Posts, design	D30	154
Guided cultivator	172	24
Land vehicle body	296	43
Railway freight car	105	380+
Tent	135	118
Staking		
Hide	69	34
Watches and clocks	29	231+
Stalk and Stalk Handling		
Choppers	56	500+
Fiber liberating	19	10
Harvester		
Cornstalk type	56	51+
Weaving	139	424
Stalls		
Animal	119	27+
Thermolytic distillation	202	97+
Stamp (See Branding)		
Affixers	156	
Alteration preventing or detecting	283	71
Brand burning	219	228
Brand burning	101	27
Design	D19	9+
Dispensers	83	242+
Coin controlled	194	
Hand stamp collocator	33	614+
Handles	101	405+
Holder, household type	D19	67
Imprinting or embossing stamp	D18	15+
Key, workmans time recorder	346	53
Mail cancelling	101	371
Mill	241	270+
Pad	118	264+
Plural parallel blade cutlery	30	305
Pocket receptacle	206	39+
With tearing edge	225	39+
Postage roll dispenser	D19	67+
Printed labels	283	81
Purse with holder	150	106
Racks for hand stamps	211	39
Registers	235	101
Rubber	101	368+
Endless band	101	111
Scarifiers	81	9.21
Time	346	80+
Time with automatic recording mechanism	346	80+
Vending machine for	D20	2
Stamping and Stamped Devices		
Bed bottom risers	5	206
Cigar	131	106
Cigarette package	131	283
Envelope		
Printing and stamping	493	320+
Sealing and stamping	156	
Hand	D18	14+
Metal working	29	DIG. 37
Mill	241	270+
Picture frames	40	154
Plug tobacco	131	368
Sheet metal chain	59	15
Stanchions	119	147R+
Stand (See Holder; Support)	248	127
Apparel apparatus	223	120
Bag holder	248	97+
Beehive	6	6
Bicycle		
Kick stands	280	293+
Locked	70	235
Racks	211	17+
Book and music	248	441.1+
Bracket		
Alternative	248	126
Combined	248	121+
Cabinet structures	312	
Calendar	40	120
Christmas tree	248	519+
Copyholder	248	441.1+
Dental equipment	433	25+
Dish	248	128+
Flatiron	248	117.2+
Accessory combined	38	142
Hat	211	33
Horseshoeing	168	44
Hose	248	80+
Ink (See inkwell)	248	127+
Lamp stove	126	259R+
Light	362	410+
Design	D26	142+
Lock type article supports	70	62
Machinery stands	248	676
Bracket convertible	248	645
Mirror type		
Handle	248	471
Plural mirror	350	631+
Music	248	441.1+
Nozzle	248	80+
Nursing bottle	248	105+
Paste tube	248	109
Photography		
Retouching	354	349+
Picture	248	469+
Plant	47	39+
Plural leg	248	163.1+
Power	74	16
Rack type	248	127+
Railway switch	246	393+
Interlocking	246	144+
Signal	246	476
Vehicle actuating	246	278+
Shoe	211	37
Shoe blacking	15	265+
Spool holders	242	139
Staff type	248	519+
Stand boilers	122	13R+
Water heater type	126	361+
Stand pipe (See tank)	137	593
Spur central	122	307+
Superheater spur tube	122	475
Water tube	122	321+
Water tube coil	122	244+
Water tube flue	122	184+
Water tube loop	122	281
Stove gas burner	126	40
Stove shelf	126	333
Telephone	379	454+
Tire	211	24
Tray or receptacle	211	133
Trolley	191	66+
Type case accessory	276	44
Watch and clock	248	116
Woodworking	269	289R+
Modified for saw blade	269	83
Work (See holders, work)	269	
Dental equipment	433	49+
Remover	D 8	48
X ray type	378	193
Standard		
Batteries	429	125
Bolsters combined	280	143+
Dental engine	433	103+
Feed bag support	119	69
Piano hammer action	84	242
Plow	172	681+
Beam and tool standard connection	403	
Shovel connection	403	
Scaffold	248	127+
Sifter support	209	415
Support	248	158+
Stannate Ceramic Compositions	501	134+
Stapilizing	19	
Artificial fibers	83	913
Spinning combined	57	2
Stapling & Staple Setting		
Applying apparatus	227	
Broom making machines	300	13
Handles on shopping bags	493	926*
Labelling fabrics	156	DIG. 22
Reels	242	103
Sheet material associating combined	270	53
Folding combined	270	37+
Staple	411	457+
Insulated	174	159
Making	59	71+
Making and setting	227	82+
Puller	254	28
Stapler design	D 8	49+
Surgical stapler	227	DIG. 1
Tobacco plug tagging combined	131	113+
Star	428	11
Illuminated	362	121
Ornamental lighting	362	807*
Starch and Derivatives	536	102+
Coating or plastic composition	106	210+
Alkali metal silicate	106	80
Casein	106	139
Glue or gelatine	106	130
Inks	106	25
Mold	106	38.51
Mold coating	106	38.23
Plaster of paris	106	114
Polishes	106	5
Portland cement	106	92
Prolamine	106	150
Protein miscellaneous	106	157
Confection molding apparatus	425	
Esters	536	107+
Manufacture	127	
Nitrogen containing	536	45+
Starter and Starting (See Device Associated Therewith)		
Arc	315	
Arc lamp	314	
Mercury vapor lamp	313	170+
Burner		
Buttonhole sewing machine	112	67
Door	49	276+
Engine or motor		

	Class	Subclass
Electric motor	318	384^{+}
Electric motor	318	430^{+}
Electric motor	318	136^{+}
Electric motor type starter	290	
Expansible chamber motor type starter	91	53
Internal combustion engine	123	$179R^{+}$
Starter gearing	74	6^{+}
Strap starter	74	139^{+}
Fan regulator	416	170R
Fluid flow	4	368^{+}
Gate		
Race track	119	15.5 A
Hydrocarbon synthesis start-up procedure	585	951*
Loaded land vehicle	278	2
Vehicle operator actuated	280	151
Motor vehicle, safty belt responsive to starter circuit	180	270
Phonograph	369	243
Pump regulator	417	1^{+}
Retort burner	431	218
Tool driving or impacting device		
Handle relative movement	173	18
Tool relative movement	173	13^{+}
Trap door	105	435
Valve in expansible chamber motor	91	286
Stateroom	114	189
Static Electricity	307	149
Anti inductive transmission line	174	32
Charges discharging	361	212
Circuit maintenance signaling	340	292
Circuit maintenance telegraphy	178	69R
Collecting with condenser	320	1
Corona prevention	174	127
Discharging	361	212
Elimination	361	212
Generators	310	309^{+}
Grounding	174	6^{+}
Ignition circuits	315	85
Lamp circuits	315	85
Lightning arrestors	361	117^{+}
Lightning rods	174	2^{+}
Metallic bleed off	428	922*
Radio	455	296^{+}
Squelch	455	222^{+}
Shielded cable	174	$102R^{+}$
System safety	361	1^{+}
Telephone circuits anti inductive	379	414^{+}
Static Information Storage & Retrieval	365	
Static Structure eg Buildings	52	
Station		
Match packaging fill and cover	53	236
Power driven conveyor work	198	339.1^{+}
Railway	104	27^{+}
Receptacle fill and close separate	53	276^{+}
Receptacle fill and close single	53	268^{+}
Signal or indicator		
Automatic to cab from highway or	246	174^{+}
Electric station selective systems	340	825.38^{+}
Passengers	246	209
Train dispatching	246	14^{+}
Train dispatching emergency	246	13
Vehicle station indicators	116	29
Train dispatching emergency	246	13
Stationery	D19	
Container	206	215
Design	D19	1
Envelope	D19	3^{+}
Stator Winding of a C Machine	310	195^{+}
Statue	428	15
Design	D11	143^{+}
Fauna	428	16^{+}
Stave		
Barrel	217	88
Making shaping jointing	147	
Stay		
Apparel	2	255^{+}
Glove	2	166
Shoe	36	57
Shoe closure	36	53
Trousers	2	231^{+}
Wire	140	91
Boiler		
Braces combined	122	493
Crown sheet combined	122	496
Bolts for steam boilers	411	379
Box	217	69^{+}
Paper	229	49^{+}
Pivoted edge	24	267
Screw impeller	416	194
Stay bolts	411	379
Vehicle body irons	296	42
Curtain backstays	296	144
Wire	140	90^{+}
Woodworking		
Stay logs	144	177^{+}
Stay logs veneer lathes	144	214
Staying		
Box making	493	89
Marine vessels	114	83
Wire fabric	140	10^{+}
Portable machines	140	18^{+}
Steady Rests		
Abrading	51	238R
Turning	82	39
Steam		
Attachments to ashpans	110	171
Bath	D24	37
Boiler (See boiler)		
Design	D23	318^{+}
Nuclear reactor type	376	370
Cleaning of particulate bed filter	210	274^{+}
Combustion		
Engines utilizing	123	25P
Fluid fuel burners utilizing	431	163
Fluid fuel burners utilizing	431	210^{+}
Solid fuel burners utilizing	110	
Condensers	165	110
Distilling with	203	
Dome		
Boilers having	122	
Steam separators	122	492
Structure	122	508
Superheater	122	486
Engine and motor (See motor)		
Design	D15	2
Expansible chamber type	91	
Locomotive structure combined	105	37^{+}
Motor vehicle		
Steam tractive type	180	36^{+}
Traction motor	180	303^{+}
Recording indicator	346	3^{+}
Safety valves	137	455^{+}
Second type combined	60	698^{+}
Turbines	415	
Vehicle heating combined	237	12.1^{+}
Fumigators	43	130
Gas & liquid contact apparatus	261	DIG. 76
Degasifying liquid	55	198
Degasifying liquid process	55	54
Denuding solids	55	59^{+}
Gas separation	55	263
Generator and boiler	122	
Autothermic kiln combined	202	94
Cookstove combined	126	5
Hot air furnace combined	126	101
Incinerators having	110	234
Still and feed water heater	202	167
Still and stand boiler	202	166
Thermolytic retort combined	202	106
Heating by		
Boilers	122	$31R^{+}$
Cooking ovens	126	20
Direct application to food	126	369
Driers	34	
Feed water	122	412^{+}
Flatirons	38	85
Heaters condensers	261	DIG. 1
Heating systems	237	
Open pan concentrators	159	36^{+}
Power plant and water heater	60	642^{+}
Power plant and water heater	60	669
Tables	126	33
Textile ironers	38	
Track sanders	291	21
Vessels	126	377^{+}
Injector		
Boiler cleaner	122	379^{+}
Furnace draft regulator	110	147^{+}
Lubricators	184	55.1^{+}
Iron	38	77.1^{+}
Pumps	417	151^{+}
Liquid purification	210	198.1^{+}
Rollers	180	20
Sterilizing methods	422	26^{+}
Superheater	122	$479R^{+}$
Traps	137	171^{+}
Treatment	122	459^{+}
Power plant exhaust	60	685^{+}
Combustion device feed in	60	683
Serially connected motors	60	679^{+}
Power plant superheating	60	662
Turbine (See turbine)		
Vacuum pump design	D15	7^{+}
Valve design	D23	233^{+}
Whistles	116	$137R^{+}$
Steamer Bag or Trunk	D 3	70
Steaming and Steamer		
Apparel	223	51
Dress coat or skirt	223	70
Stockings	223	76
Trousers or sleeve	223	73
Food	426	506^{+}
Boilers	99	410^{+}
Infusers	99	293^{+}
Miscellaneous food apparatus	126	369^{+}
Sterilizing apparatus	99	467^{+}
Grain	426	507^{+}
Apparatus	99	467^{+}
Hair, beauty parlor equipment	D28	12^{+}
Solid material comminutors	241	1
Apparatus	241	38^{+}
Cereal processes	241	12
Miscellaneous processes	241	15^{+}
Whole seed processes	241	8
Sterilizing and preserving	422	27^{+}
Wood treatment	118	50
Textile apparatus	68	$5R^{+}$
Implements	68	222
Textile processes	8	149.3
Tobacco	131	300^{+}
Wood bending	144	271
Stearic Acid	260	413^{+}
Steel	420	8^{+}
Austenitic	420	8^{+}
Butchers	76	84
Composite casting	164	91^{+}
Heat treatment	148	134^{+}
Making	266	200^{+}
Metallurgy	75	28^{+}
Pyrometallurgy	75	28^{+}
Stainless	420	8^{+}
Heat treated stock	148	325^{+}
Heat treating	148	135^{+}
Wool making	29	4.5 A
Steeping Starch	127	68
Steeple	52	82
Design	D25	35^{+}
Steering		
Aircraft	244	50
Controls	244	$75R^{+}$
Propulsion	244	51^{+}
Automatically controlled		
Aeronautic	244	175^{+}
Drop bombs	102	384
Ship	114	$144R^{+}$
Torpedo	114	23^{+}
By articulated movement	280	442^{+}
Semitrailer	280	426
Collapsible steering post	280	777
Control lever and linkage systems		
Control and steering combined	74	$484R^{+}$
Handle bars	74	551.1^{+}
Handwheel	74	552^{+}
Steering posts	74	492^{+}
Electric	318	489
Five or more wheels	180	23^{+}
Illuminator control combined	362	37^{+}
Lights for steering switches	362	36
Land vehicle		
Hand	280	47.11
Occupant propelled with steering	280	200^{+}
Occupant steered	280	$80R^{+}$
Lock and switch combined	70	252
Marine		
Centerboards	114	128
Chain propeller	440	95
Crank paddle	440	26
Jet	440	38
Oscillating propeller	440	13
Paddle wheel	440	90
Portable propeller	440	53
Reciprocating propeller	440	13
Screw propeller	440	53
Steering mechanism	114	$144R^{+}$
Torpedoes	114	23^{+}
Towing	114	242^{+}
Motor actuated	244	175^{+}
Motor vehicle	180	79^{+}
By differential traction	180	6.2
Endless track	180	6.7

	Class	Subclass
Caterpillar type	180	9.44+
Five or more wheels	180	22+
Fluid power	180	132
Four wheels driven, articualted frame	180	235
Gear	180	79+
Shaft	180	78
Steerable, driven wheel	180	252
Five or more wheels	180	23
Four wheels driven, all steerable	180	234+
Motor-carrying attachment	180	12+
Steam tractive vehicle	180	37+
Three wheels	180	211+
Two wheels	180	223+
Two wheels, rotary element driven wheel	180	222
Using terrestrial guide	180	131
Walking attendant	180	19.1+
Plows and cultivators		
Automatic	172	2+
By walking attendant	172	669+
Lister furrow guided	172	26
Tree or stake guided	172	24
Remote control		
By radio	340	825.72+
Pulse responsive	340	825.69
Electric steering motor	318	
Torpedo	114	21.3
Self steering wheeled toys	446	441
Sled	280	16
Flexible runner	280	21R+
Wheel	74	552+
Cover or grip	74	558
Dust cover	150	52M+
Lock	70	211
Position indicator	116	31
Windmill	416	9
Steins	215	235+
Metallic	220	334+
Stem		
Flower		
False stem	47	55
Key	70	402+
Extensible	70	397
Foldable	70	393
Plural	70	401
Medicators	604	309
Smoking pipe	131	227+
Absorber type	131	204
Cap and tool	152	431
Cleaners	131	245
Valve		
Coaxial	137	637.2
Detachable actuator	251	128
Inflation type	137	223+
Stop combined	251	284+
Tire combined	152	429+
Watch		
Fasteners	368	307
Winding	368	216
Winding and setting	368	190+
Stemming		
Fruit	99	637+
Tobacco	131	313+
Stemware Container	206	426
Stencil and Stenciling	33	563+
Abrading resist	51	262R
Utilization	51	310+
Addressing machine	101	48+
Blank making by coating	427	143
Brush	D 4	135
Design	D19	40
Lettering	D19	40
Nonuniform coating by stenciling	427	256+
Plate	101	112
Printing devices	101	114+
Photo process making	430	308
Toilet or make up	132	88.5
X-art collection	425	811*
X-art collection	493	953*
Stenographic		
Copyholder (See copyholder)		
Stenograph	D18	20
Telegraphic system	178	21
Typewriter	400	91+
Stenter (See Tenter)		
Step		
Brake vehicle step operator	188	108
Cabinet with	312	235R
Climbing vehicles		

	Class	Subclass
Wheeled	280	29
Escalator	198	326+
Ladder	182	228
Detachable	182	120+
Land vehicle	280	163+
Design	D12	203
Endgate convertible	296	62
Mast	114	93
Passenger recorder	346	40
Rail detachable	238	140
Railway car	105	443+
Gates combined	105	437+
Trap door combined	105	430+
Rotary crane	212	253
Step plate for vehicle	D12	203
Stilt	272	70.2+
Tread (See --; ladder; stair)		
Lighting vehicle step	362	76
Rail	238	136
Sleeping car	105	326
Wall attached	182	90+
Window scaffold combined	182	54
Step-by-step Propulsion	280	1.181
Stepladder		
Convertible to chair or seat	182	33+
Propped	182	165+
Pivoted platform	182	124+
Stepped Building	52	182+
Stepped Shelves	206	45
Stepper		
Land vehicle	305	1+
Motor vehicle	180	8.1+
Runner supported, shuffling	180	187
Simulating vehicle	280	1.181+
Stereocomparagraphs (See Photogrammetry)		
Stereograph (See Stereoscopy)	350	130
Stereophonic	379	101+
Sound transmission	379	90+
Stereopticon	84	24+
Lantern slide mount	40	158B
Stereoscope	D16	12
Stereoscopic photography	354	112+
Viewer	350	133
Stereoscopy		
Anaglyphs	350	130+
Camera still	354	112+
Motion pictures	352	57+
Projector	353	7
Stereograph	350	130+
Television	358	88+
Vectograph T M	350	396
Viewers	350	133+
X ray	378	41
Stereotype		
Casting	164	2+
Finishing	29	21
Mold and trimmer	164	140+
Planing	409	309
Plates	101	368+
Sterile Body Treatment Tool Package	206	210
Sterilizer	D24	9
Telephone	379	452
Sterilizing (See Heat)	422	
Apparatus for treating air	422	120+
Electrical	204	186+
Apparatus	204	302+
Electrolytic	204	130+
Foods	426	521+
Apparatus	99	467+
Gas separator combined	55	279
Heating gathered soil	47	1.42
Ray energy	250	455.1
Fluent material	250	428+
Xray	378	64+
Tobacco	131	290
Electrical or radiant energy	131	299
Water	210	
Steroids	260	397+
Glucosidal	536	5+
Heterocyclic nucleus ctg	540	2+
Sterols	260	397.2
Stethoscope		
Acoustic type	181	126+
Design	D24	20
Electric type	381	67
Stibonium	556	64+
Stick		
Candy design	D 1	127
Candy stick per se	426	294
Cane or umbrella type	135	65+

	Class	Subclass
Design	D 3	5+
Confections	426	134+
Manufacturing	425	110+
Edible stick or strip form	D 1	
Handler	294	87.11
Inedible holder for food	D 1	105
Inserter	227	
Laminating	156	
Laundry	294	23.5
Loom picker	139	157
Match	44	46
Meat supporting	17	44.4
Mountain climbing	272	70
Skaters	280	826
Skein or hank supporting for washing	68	147+
Trickler and sprinkler	68	206
Tobacco supporting	294	5.5
Typesetting	276	38+
Stickpin	D11	47
Hat pin	D11	207
Stiffened Fabrics		
Coated or impregnated	428	245+
Stiffener		
Apparel	2	255+
Collar etc making	223	4
Making	223	27
Shoe counter	36	68
Shoe shanks	36	76R
Shoe toe and heel type machines	12	61R+
Shoe upper metallic seam	36	58
Railway rail joint		
Base plate	238	213
Chair base	238	189
Stills (See Distillation)	202	
Chemical apparatus combined	422	189+
Mineral oils	196	98+
Refrigeration system combined	62	
Stilts	272	70.1+
Body attached	623	28+
Ceramic	432	258+
Design	D21	72
Stippling		
Metal forging	72	76
Processes	427	256+
Roll	29	110+
Stirling Cycle	60	517+
Refrigeration	62	6
Stirrer (See Agitator)	366	343
Chain-type	366	607*
Dispensing		
Discharge assistant combined	222	226+
Discharge assistant type	222	251+
Jarring or vibrating	222	196+
Stationary	222	459
Trap chamber type discharge assistant	222	216+
Drier stationary receptacle	34	179+
Gas agitation	222	195
Heating etc gas generator fuel	48	85.2
Heating furnaces with	432	139+
Hops	366	343
Liquid purification		
Centrifugal filter rotary	210	380.1+
Loose material filter cleaning	210	269+
Mineral oil vaporizing process	208	146
Smoking devices with	131	184.1
Tedder type	56	370+
Stirrup		
Connection or joint	403	232.1
Harness	54	47+
Design	D30	142
Railway coupling	213	199
Saddle	54	47+
Textile drawing pressure device	19	266
Stitching (See Sewing)		
Blind stitch guides	112	140
Blind stitch methods and seams	112	267.1+
Brushes wirebound and	15	189
Embroidering machines	112	78+
Hemstitching	112	81+
Horizontal needle stitch forming	112	93+
Knitting textiles	66	
Leather sewing	112	28+
Sewing machine mechanism	112	154+
Sheet material associating and	270	53
Folding combined	270	37+
Shoe sole		
Hemp sole	112	121.21
Imitation stitch	12	32
Leather sewing	112	28+

	Class	Subclass
Removers	12	39.3
Stitch separating and indenting	12	32+
Stitched article	36	18+
Stapling	227	
Stitch control	112	121.13
Stitch marking	112	121.25
Textile processes with	28	140+
Trimmers combined	112	122.1
Stock (See Blanks)		
Anchors	114	301
Animal husbandry	119	
Animal restraining	119	98
Die	408	120+
Firearm	42	71.1+
Lathe headstocks	82	28R+
Lathe tailstocks	82	31
Live care and handling	119	
Railroad sprinkler stock alarm	239	173
Time controlled feeder	119	51.11
Vehicle body	296	12+
Magnetic	148	300+
Material	428	
Label and tag indicators	283	81
Radiation barrier	250	505.1
Treated metal	148	400+
Metal-to-metal laminate	428	615+
Milling machine tailstocks	409	242
Quotation		
Changeable exhibitor	40	446+
Electric	340	825.26+
Indicator type	116	200+
Projector type	353	46+
Telegraph bulletin	178	24
Telegraph printing	178	23R+
Telegraph ticker tape	178	34+
Railway rolling	105	
Toy	446	467
Sample and		
Cabinet compartment	312	234
Showcase compartments	312	118+
Tray	206	557+
Treatment of paper	162	1+
Woodworking bit	81	28+
Stocking (See Hose)	2	239+
Bandage type	128	165
Cap	D 2	249
Darning	66	2
Design appliances	D 3	26
Forms	223	75+
Putting on or removing	223	112
Hose	D 2	329
Knitting	66	178R+
Design	D 2	329+
Pantyhose	2	409
Pressing or molding	223	60
Protectors (See protection)	2	61
Gaiters and leggings	36	2R
Gaiters and leggings design	D 2	267
Sewing		
Cup type feeder	112	18
Supporters	2	300+
Design	D 2	625+
Wear sole combined	36	9R
Stogies	131	360+
Stoker (See Feeder, Fuel)		
Design	D23	396
Furnace	241	600*
Furnace charger	414	147+
Locomotive blower type	110	105.5+
Locomotive coal crushers	241	276
Progressive feed through combustion chamber	110	267+
Stole, Apparel Design	D 2	179+
Stoma Receptacle	604	332+
Stone		
Artificial	106	74+
Inorganic settable ingredient	106	85+
Compositions impregnating	106	12
Fences	256	19
Gatherer	171	63+
Grinding processes	51	283R
Jewelry use	D11	89+
Precious changing color by radiation	378	38+
Printing surfaces	101	453+
Remover	37	2R
Working	125	
Design	D99	17+
In situ	299	
Stoners for Fruits etc	99	547+
Stool	108	
Contoured	297	

	Class	Subclass
Design	D 6	349+
Folding	108	115+
Kneeling or scrubbing	182	230
Milking	297	175+
Padded	297	461
Piano duet	297	232+
Planar surface	108	
Stop		
Accordian	84	376R
Aircraft		
Barrier type	244	110R
Animal	210	131
Backstop bowling	273	53
Car	104	249+
Door	16	49+
Elevator landing	187	32+
Hinge	16	374+
Ladder	182	209+
Lever	74	526
Light	340	907+
Brake operated switch	200	61.67
Pressure operated switch	200	81R+
Wiring system	315	80
Organ	84	85
Shear	30	271
Target projectile stop combined	273	404
Valve	251	284+
Watches	368	101+
Wick	431	301
Wind instrument	84	369+
Zipper end	24	436+
Stop Mechanism		
Fabric and cord		
Bobbin cone wind	242	28+
Bobbin winding	242	36+
Cloth finishing cutting	26	10R+
Cloth fulling stopping	26	24
Cordage winding	242	49
Fabric measuring	33	136
Fabric reeling	242	57
Nuclear reaction control	376	207
Nuclear reaction control	376	227
Printing bed and cylinder	101	275+
Printing rotary press stack	101	232+
Sewing machine	112	271+
Bobbin winding	242	22
Buttonhole	112	67
Embroidering	112	87
Sheet associating and folding	270	46+
Textile braiding netting and lace making	87	18+
Textile covering	57	19
Textile fiber preparation	19	
Textile knitting	66	157+
Textile spinning and twisting	57	78+
Textile twisting by couple	57	61
Textile weaving	139	336
Thread stretching with stopping	28	242
Warp preparing stopping	28	186+
Machine		
Article elevator	414	674
Brakes	188	
Cable drum control	254	267+
Drill feed	408	5
Drying automatic control	34	55
Elevator	187	28+
Fan regulator	416	170R
Hoist traversing cable	212	86+
Lathe carriage	82	21R+
Metal working	72	1+
Motor vehicle safety	180	271+
Planer tool feed	409	326+
Power driven conveyors	198	854+
Power stop control	192	116.5+
Pump regulator	417	1+
Punching machine shaft drive	83	58+
Punching machine shaft drive	83	69
Punching machine shaft drive	83	203+
Register control	235	132R
Rotary crane	212	149+
Telegraph code recorder	178	95
Tool turret carriage	29	65
Type casting	199	52+
Type distributing	199	43
Typewriter carriage	400	298+
Typewriter case shifts	400	277+
Typewriter line spacing	400	547+
Typewriter tabulator	400	294+
Windmill	416	14
Wire feeding for working	72	461
Stop Watches	368	101

	Class	Subclass
Stopcock Boxes	220	3.2+
Stopper		
Bath and sink	4	286
Bottle and jar	215	355+
Breakable neck	215	33+
Design	D 9	439
Closure for metal receptacle	220	307
Flexible bag	383	96
Flue	126	319
Ingot or pig molds	249	204
Ordnance	89	30
Ship or vessel		
Cable	114	199+
Hawse pipe	114	180
Leak	114	227+
Port	114	174+
With measuring devices	222	424.5+
Tiltable	222	454+
Stopwatch	D10	30
Storage		
Battery	429	
Charging and discharging	320	2+
Connections	439	754+
Electrolyte	429	189+
Half lead	429	
Lead	429	
Plates and grids	429	209+
Building & devices (See garage;silo)	52	
Ballast storage for aircraft	244	94
Bed with stowage	5	503
Bin with power driven conveyor	198	550.1
Cordage	242	47.5
Crib with stowage	5	96
Fire escapes	182	82+
Food storage receptacle	99	467+
Magazine for label pasting	156	
Material handling	414	589
Positioning material for building erection or repair	414	10+
Parachute storage	244	147+
Photographic plate holder	354	174+
Portable storage & means for erecting	414	919*
Ramp	52	175
Storage provision on passenger car	105	334+
Table leaf	108	86+
Table leaf terraced	108	93
Vehicle top	296	140+
Ventilation of storage structure	98	52+
X ray film holders	378	167
Floating vessel	114	256
Hydrogen, alloys for	420	900*
Radio active waste material	252	633
Submerged vessel	114	257
Tank	D23	202
Underground fluid	405	53+
Store Service	186	
Trolley type ladders	182	36+
Using punch cards	186	56
Storefront Design	D25	59
Stores	52	28+
Bay windows	52	201
Self service	52	33
Show window panel	52	777+
Show window panel	52	764+
Toy	446	476
Storm Fronts		
Railway passenger cars	105	353
Vehicle body	296	78R+
Window	52	202
Screen combination	160	90+
Storm Windows	49	61+
Stove	126	
Absorber coating	126	901*
Air preheating	266	138+
Dampers	126	285R+
Automatic	236	45
Combined with furnace features	110	163
Design	D 7	339+
Handle or knob	D 8	300+
Heating		
Electric	219	359+
Illumination	362	92+
Implements	294	9+
Compound	7	109+
Polish	106	3+
Still combined	202	165
Ventilators	D23	371+
Working fluid compositions	126	900*
Stovepipe	126	307R+

	Class	Subclass
Design	D23	394
Elbow bending	72	307
Hood combined	126	301+
Thimbles	126	314+
Water heaters	126	364+
Straight Jackets	128	134+
Straight Line Motions	74	103
Straightedge	33	403+
Design	D10	61+
Leveling	33	341+
Rod joint	403	
Tearing or breaking	225	91
Straightening Devices		
Decurling coated flap of envelope	493	395+
Grain harvester	56	473
Metal		
Automobile frame	72	705
Leveler	72	160+
Stretcher	72	302
Paper web	162	270+
Shoe sole	12	120.5
Tobacco treatment	131	324+
Stemming	131	315+
Trouser leg	2	233
Wire	140	147
Cutting combined	140	139+
Strain		
Connector electrical	439	296+
Gauge	33	DIG. 13
Relief		
Pipe joint	285	114+
Reliever electrical cord terminal	439	449+
Strainer (See Filter; Screen; Sifter)		
Baths closets etc	4	286+
Sinks	4	652
Concentrating evaporator	159	42
Dispenser combined	222	189
Dispensing type sprinkler etc	222	565
Eaves, troughs	52	11
Kitchen utensil	D 7	47
Liquid separation	210	348+
Detachable	210	473+
Faucet attached	210	449
Mold for metal having	164	358
Paper stuff	210	
Press	100	130
Separating yeasts ferments etc	435	287+
Separatory stills	202	178
Sewer inlet	210	163+
Sewerage trap	210	435+
Support	248	94
Textile fluid treating apparatus	68	
Vegetable etc comminutor type	241	
Well casing	166	227+
Strand (See Filament; Wire)		
Advancing material of indeterminate length	226	
Brakes	188	65.1+
Drying apparatus	34	148+
Drying processes		
Diverse	34	18
Gas contact	34	23+
Heat	34	41
Vacuum or pressure	34	16
Feeder	226	
Finishing apparatus	28	217+
Fire escape	182	189+
Floor or wall attached	182	100
Joining (See joints)		
Electric conductor	156	49
Rope splices	57	202
Rope splicing devices	57	22+
Tying	289	
Joint packing	277	227+
Laminating	156	166
Making	28	217+
Braiding	87	
Casting	164	418+
Knitting	66	3+
Sliver assembling	19	157+
Spinning or twisting	57	
Waxed end	57	20
Weaving	139	13R+
Package	206	389+
Wound	242	159+
Packaging	28	289+
Placing or cable hauling	254	134.3 R+
Structure		
Barbed wire	256	6+
Conductors	174	126R+
Conductors insulated	174	110R+
Conductors sheathed	174	102R+
Fence wire	256	46
Textile	428	364+
Textile braided	87	8+
Textile twisted	57	200+
Tearing or breaking	225	
Tire bead grommets	245	1.5
Winding & reeling	242	
Strangler		
Bridle	54	15
Strap		
Animal shoe fastening	168	22
Antiskid for wheels	152	221+
Apparel supporting	2	338+
Spaced holders	2	323+
Auto folding top	296	130
Boot	36	56
Button loops	24	660
Closure holding		
Wooden boxes	217	66+
Clutch	192	80+
End attaching devices	24	265R+
Hand		
Casket lowering	27	33+
Package and article carriers	294	149+
Railway car	105	354
Harness		
Breast	54	58+
Breech	278	128
Hame	54	28
Hitching	54	34
Loops	24	182
Hat	132	58
Machine element	74	570+
Making		
Box straps	140	74
Strap finishing	69	17
Nuclear reactor moderator support	376	304
Safety	182	3+
Tighteners	24	68R+
Bale and package ties	24	19
Portable, detachable tensioners	254	199
Trouser	24	72
Weaving machine		
Harnessing connecting	139	88
Shuttle fly lug connecting	139	153
Wheel brake	188	77R
Wrist watch	224	164
End attaching devices	24	265R+
Stratifying (See Separator, Solids)	209	
Straw		
Drinking	239	33
Drinking device	D 7	42
Holder	D 7	75
Mattresses	5	448+
Stacks forming	414	132
Frame	130	20
Streamlined		
Aircraft	244	117R+
Railway trains	105	1.4
Street (See Road)	404	
Cleaners and sweepers		
Brushing	15	78+
Machinery	D32	15+
With air blast or suction	15	300R+
Flushers	239	146+
Lantern	362	257+
Oilers	239	130
Pavements	404	17+
Signal		
Marker	404	9+
Mirrors and reflectors	404	14
Traffic	116	63R
Signs	40	612
Street light	D26	67+
Design	D20	17
Underground electric conductors	174	39
Streetcar	D12	36+
Appliance	D13	14
Strength Measuring		
By acoustic emission	73	801
By impact or shock	73	12+
By testing model of structure	73	804
Hardness	73	78+
Muscular	73	379
Coin controlled	194	
Optically sensed	73	800
Periodic	73	808+
Alternating tension & compression	73	797
Stress strain	73	788+
To cracking of specimen	73	799
Streptomycin	536	14
Stress Distributing (See Equalizing)		
Conduit electric	174	73R
Insulator electric	174	140R+
Testing	73	760+
X ray tube	378	139
Stretchers & Stretching (See Definition Notes; Tension; Tighteners)	254	199+
Boot and shoemaking		
Loose upper shaping	12	54.1+
With diverse operations	12	52.5
Coating process	427	171+
Coating process combined	427	401
Design	D12	128+
Fabric		
Apparel	223	52+
Carpet	254	200+
Carpet	294	8.6
Carpet fastener combined	16	5
Fastener applier combined	227	12
Frame, eg curtain stretcher	38	102.1+
Framed panel	160	378
Hat shaping	223	15
Hung or draped	160	328+
Ironing table combined	38	108
Smoothing	38	102
Smoothing implements combined	38	70
Smoothing machine combined	38	12+
Wire	140	108+
Working combined	26	51+
Fence incorporated	256	37+
Frames		
Tack puller for	254	18
Textile sheets	38	102+
Horseshoe	168	9
Invalid lift		
Bottoms and slings	5	89
Field type	5	82R+
Wheeled	296	20
Leather		
Belt	69	1.5
Frame	69	19.1+
Lasting machine	12	7+
Lasting tool	12	108+
Shoe spot stretchers	12	115.2+
Translating tool	69	46
Work holder	69	19
Metal curving or straightening	72	302
Metal slitting and	29	6.1
Necktie	223	65
Pelt	69	19.1+
Pushing and pulling implements or apparatus	254	199
Resilient article assembling	29	235
Saw making	76	26+
Screws for ropes and cables	28	240+
Twisting combined	57	310
Ships sail	114	109+
Shoe former	12	128V+
Skein washing combined	68	159
Spinning and twisting combined	57	310
Strap	24	68D
Stretch forming	72	295+
Textile web running lengths	26	71+
Biaxial	26	72+
Heater-dryer and spreader	26	92
Heater-dryer combined	26	106
Roller spreader	26	99+
Spreader	26	87+
Tubular fabric spreader	26	80+
Web condition responsive control	26	74+
Thread finishing	28	240+
With liquid treating	28	246
With stopping	28	242
Umbrella tent	135	20R
Wire (See notes to)	140	123.5
Striated Effects Surface Type	428	30
Striker and Striking		
Ammunition fuse etc	102	272+
Animal trap	43	77+
Clocks	368	243+
Electric	368	250+
Comminuting		
Hammer mills	241	86+
Hammer mills	241	185R+
Hammer mills plural parallel zones	241	138
Hammer mills plural series zones	241	154
Impact processes	241	27
Diaphragm horn with rotary	116	143+
Fixed bell with pivoted	116	167

	Class	Subclass
Gong with pivoted	116	155$^+$
Lock	292	340$^+$
Mechanism for music players	84	323$^+$
Piano	84	236$^+$
Movable figure toys	446	336
Music instruments		
Automatic outside players	84	111
Piano auxiliary	84	224
Piano hammer actions	84	236$^+$
Pneumatic pianos	84	54
Rigid vibrators with	84	404
Stringed with	84	12
Punching bags	272	76$^+$
Railway draft block	213	58$^+$
Surfaces for matches	44	48
Compositions	149	
Typewriter		
Index wheel	400	165.1$^+$
Key wheel	400	174$^+$
String (See Cord; Strand)	57	
Catgut	84	297S
Covering with copper or silver wire	57	3
Draw	24	266
Gut	8	94.11
Insulators	174	150
Stress distributing combined	174	141R
Musical instrument	84	297S$^+$
Covering with wire	57	3
Design	D17	14$^+$
Piano	84	199
Tuning devices	84	455
Nail manufacture	10	45
Nailers	227	95$^+$
Plate piano	84	188
Frame and sounding board combined	84	184
Frame combined	84	185
Sounding board combined	84	187
Shoe	24	143R$^+$
Tag	40	20R
Band type	40	21R
Tear	206	616$^+$
Stringed Instruments	84	
Automatic player pianos	84	7$^+$
Electric pickup	84	1.16
Stringer (See Girder)		
Railway	238	24$^+$
Tie ends	238	42
Seed corn	139	19
Stringing		
Assembling by (See notes to)	29	241
Beads	29	700$^+$
Apparel trimming	223	48
Sewing machine for	112	121.19
Tennis racquet	273	73R
Tobacco	414	26
Striping		
Brush	15	166
With material supply	401	193
Coating		
Implement	401	48
Machine	118	
Processes	427	286$^+$
Fabric weaving	139	417
Stripper and Stripping		
Berry	56	330
Bobbin	28	292$^+$
Cane	130	31R
Coating	118	100$^+$
Coil	242	81
Color stripping	8	102
Cotton	56	33$^+$
Cutlery combined	30	128$^+$
Electrolytic	204	146
Apparatus combined	204	193$^+$
Expressing press cake	100	218
Cloth removing	100	298
Fiber and thread		
Carding	19	108$^+$
Guides	28	222$^+$
File blank	29	77
Filling tube	141	265
Harvesting	56	126$^+$
Insulation from wire	81	9.4$^+$
Molds and castings	164	
Ingot strippers	164	405$^+$
Nail strippers	193	38
Photographic paper and films	430	256$^+$
Prick punch	30	368
Punch press	83	111

	Class	Subclass
Tube or sheath splitter	30	90.1$^+$
Vine and seed	130	30R
Strips (See Tape)		
Adhesively affixing	156	
With cutting	156	510$^+$
Apparel	2	338$^+$
Breaker strip	152	526
Building joint covering	52	459$^+$
Buttonhole	2	266
Carpet fastening	16	16
Continuous binding	281	23
Dispensing	221	33$^+$
Continuous strip	312	39
Towel	312	37
Feeding and delivering	226	
Form pockets	282	11.5 R
Guide sewing machine	112	152
Handling (See feeder; sheet)		
Manifolding	282	12R$^+$
Metallic	428	544$^+$
Compound	428	615$^+$
Strips of repeated articles, cut up later	29	DIG. 4
Paper	281	5$^+$
Leaves books combined	282	2$^+$
Leaves combined	281	2
Printed	283	62
Pasters	156	
Perforated telegraph tape	178	112
Recorder record receiver	346	136
Recording paper	346	136
Servers (See notes & definitions)	226	
Spiral cutting	83	680
Supports	211	45$^+$
Stacked	211	50$^+$
Switchboard	361	331
Telephone	379	325
Tack	411	439$^+$
Tape appliers shoe making	12	59.5
Tearing or breaking	225	
Container opening	206	605$^+$
Telegraph tape	178	111$^+$
Towel supports	211	16
Locked	211	6
Vehicle body wear	296	41
Weather	49	475$^+$
Windshield	296	93
Winding and reeling	242	
Woodcutters for strips		
Bench planes	30	478$^+$
Slicers	144	164$^+$
Veneer lathe	144	215
Stroboscopes		
Electric		
Frequency	324	78R$^+$
General	324	
Phase	324	83R$^+$
Electric lamp systems	315	
Condenser discharge	315	227R$^+$
Optical	356	23$^+$
Stroboscopic Light Synchonizer	356	23$^+$
Condenser supplied lamp	315	227R$^+$
Using electronic tube for trigger	315	208
Using photocell	315	149$^+$
Stroke Regulation		
Dispenser volume varying		
Piston or follower	222	309
Relatively movable actuator	222	287
Expansible chamber device	92	13
Expansible chamber motors	91	277$^+$
Pump type dispensers	222	375
Mechanical movement	74	20$^+$
Piano actions	84	221$^+$
Pumps	417	212$^+$
Lubricating	184	26$^+$
Punches	83	248
Punches	83	527$^+$
Rotary	91	472$^+$
Stroller (See Baby, Carriage)	280	47.38$^+$
Collapsible	280	639$^+$
Seated occupant	280	650
Design	D12	129
Strontium (See Alkali Earth Metal)	75	
Strophanthin	536	6.1
Stropping Devices		
Cutlery	30	138$^+$
Razor	30	35$^+$
Reciprocating tool	51	63
Reciprocating work holder	51	153
Structure	76	81$^+$
Surface renewing	51	276

	Class	Subclass
Work rotating	51	155
Structural Elements (See Reinforcement)		
Brake beams	188	219.1$^+$
Bridge details	14	14$^+$
Fence rail	256	60$^+$
Laminated and analogus structure		
Flexible fabrics	428	224$^+$
Glass	156	99$^+$
Making		
Columns and girders	29	155R$^+$
Rails	238	122$^+$
Rolled flanged bars	72	221$^+$
Sheet piling	405	276$^+$
Stock shapes	428	544$^+$
Molded in situ	264	31$^+$
Pipe	138	177
Railway ties	238	54$^+$
Strands	57	902*
Structured Biocides	424	405$^+$
Structured Insecticides	424	405$^+$
Structured Repellants	424	405$^+$
Strut		
Aircraft landing	244	100R$^+$
Spring	267	2$^+$
Aircraft lifting	244	37
Folding bed	5	139$^+$
Railway dumping car door actuator	105	304
Scaffold	248	351$^+$
Building propped	182	87
Window type	182	62
Sprinkler head collapsing	169	38$^+$
Strychnine	546	35
Stub		
Attached tab	402	500
Axle		
Control dirigible light supports	362	55$^+$
Motor vehicle, driven & steerable wheel	180	253$^+$
Occupant propelled vehicle	280	269
Occupant steered vehicle	280	93$^+$
Railway truck	105	180
Railway truck radial	105	169
Short turn vehicle	280	103
Trailer steering devices	280	443$^+$
Switch railway	246	445$^+$
Stud		
Balance	368	178
Fixture wiring with box	174	62$^+$
Hame bolt	54	26$^+$
Lacing	24	144$^+$
Making metal studs	29	12
Making		
Spiral	140	92
Removers	81	53.2
Studding	52	720$^+$
Composite, stone like component	52	367$^+$
Facer attached	52	474$^+$
Ribbed concrete	52	327$^+$
Settable material backer attached	52	344$^+$
Studio Photographic	354	290
Stuff Working	241	
Stuffed and Stuffing		
Container filling	493	967*
Horse collar	69	4
Mattresses	5	448$^+$
Stuffing devices	53	524
Multiweft fabric stuffer warp	139	415
Packed shaft or rod joint	277	
Pile fabric stuffer warp	139	406
Sausage	17	35$^+$
With linker	17	33
Truss pads	128	120.1
Stuffer Box	28	263
Stump		
Burner	110	239
Design	D23	334
Remover	254	
Excavator	37	2R
Hand and hoist line implements	294	
Traversing hoist	212	
Splitter in situ	144	193E
Stylographic Pens	401	258$^+$
Stylus		
Code transmitter	178	87
Facsimile	358	295
Indenting	81	9.2
Manifolding	81	9.2
Pen material retaining	401	292
Phonograph or sound recorder and reproducing		

	Class	Subclass
Changers	369	171+
For sound box	369	161
Holders	369	170
Stylus feed	369	172
Styptic Pencil	604	309
Composition	424	
Styrene	585	
Copolymers (See also synthetic resin or natural rubber)	526	347
With diolefines	526	340
With cyclodiolefines	526	280
Polymerized liquid	585	428
Polymerized solid (See also synthetic resin or natural rubber)	526	347.2
Synthesis	585	435+
Styrofoam T M	521	79
Mold filling	264	DIG. 1
Pre-expansion	264	DIG. 9
Styrols (See Styrenes)		
Sub-critical Reactor Nuclear	376	158
Subduing Color	8	102
Subirrigation	405	36
Sublimation		
Apparatus combined	422	189+
Furnaces metallurgical	266	144+
Hydrocarbon purification by	585	801
Processes		
Inorganic materials	23	294R+
Organic compounds	260	706
Refrigeration by	62	384+
Automatic	62	165+
Sulfur	423	578R
Submarine		
Amusement railway	104	71
Cable	174	
Loaded	178	45
Telegraph system	178	63R
Dredger chamber	37	56
Earth boring	175	5+
Hooks and grapples	294	66.1+
Ordnance	89	5
Pipe and cable laying	405	158+
Rescue bell	405	189
Signal and light		
Electric signaling	340	850+
Indicator or signal	116	27
Lanterns	362	257+
Toy boat	446	161+
Tunneling	405	136+
Vessels	114	312+
Subscript Character	400	904*
Subsoil		
Drainage	405	36+
Irrigation channels	405	36
Irrigators	111	6+
Plant	47	48.5
Plow	172	699+
Weeding	172	720
Substitutes (See Artificial)		
Foods and beverages		
Butter and lard	426	601+
Butter design	D 1	
Coffee	426	569+
Egg	426	531+
Leather	428	904*
Tobacco	131	359
Vehicle wheel	305	
Subterranean Heater	165	45
Subways (See Tunnel)		
Tunnels	405	132
Ventilation	98	49
Succinic Acid	562	590
Esters	560	190+
Succinic Anhydride	549	233
Succinimides	548	545+
Suckling Appliances	119	71
Sucrose	127	
Inversion	127	41
Precipitation	127	47
Suction (See Pneumatic; Vacuum)		
Abradant remover	51	273
Anchors	114	296
Boxes paper making	162	363+
Cup		
Bracket	248	205.5+
Light support	362	397
Support	248	363
Surgical	604	314+
Dental apparatus	433	91+
Denture	433	184+
Draft appliances for furnaces	110	162

	Class	Subclass
Feeder conveyer holding item by	209	905*
Machines		
Cleaning	15	300R+
Dispensing articles	204	124
Dredgers	37	58+
Insect catchers	43	139+
Label picker	156	569+
Lubricator	184	58+
Milker pulsators	137	103+
Milkers	119	14.27+
Shears combined	30	133
Tobacco feeding	131	110
Operated motor horn	116	138
Pneumatic conveyer	406	151+
Nozzle	406	152
Pump	415	
Pump	417	
Tool holder-flexible abrasive	51	362
Sugar	127	
Cane plant	PLT	89
Crystals washing out	127	63+
Cutting and shaping	425	289+
Electrolytic treatment	204	138
Fermentative liberation	435	94+
Foods containing	99	
Design of foodstuffs	D 1	
Grape	127	36+
Invert	127	41
Purification	435	276
Syrups	426	658+
Suint	260	398+
Removal from wool	8	139.1
Suit	D 2	49+
Space suit	128	402+
Wet suit	2	2.1 R
Suitcase		
Locks	70	69+
Traveling bag	206	278+
Sulfa Drugs	260	397.6
Medicines ctg	514	155+
Sulfadiazine	544	297
Sulfamides	564	79
Sulfanic Acid	260	513.6
Amine salt	260	501.12
Sulfanilamide	260	397.7 R
Sulfanilic Acid	260	508
Sulfapyridine	546	312
Sulfate		
Electrolytic synthesis	204	93
Esters	558	20+
Fertilizers containing	71	
Inorganic	423	544+
Ore treatment with	75	115+
Chloridizing combined	75	110+
Thio	423	514
Sulfation	260	686
Carbocyclic or acyclic cpds	558	20+
Fatty oils or acids	260	400+
Sulfenamides	564	102
Sulfenate Esters	558	62
Sulfides		
Binary inorganic	423	511
Inorganic	423	511+
Containing	424	162
Medicine or poison containing	424	161
Organic	568	38+
Amine	564	100+
Containing	514	706
Sulfinamides	564	101
Sulfinate Esters	558	61
Sulfinic Acid	260	513.7
Sulfinylamines	560	317
Sulfite		
Inorganic	423	519
Liquor		
Carboxylic acids from	562	513
Coating or plastic compositions containing	106	123.1
Compositions	162	83+
Fermentation by yeasts	435	251
Lignins from	530	500+
Paper stock treatment	162	83+
Synthetic resins containing	527	400+
Tanning agents containing	8	94.31
Organic	260	
Reducing composition	252	188.21
Sulfite Esters	558	59
Sulfo Acids	260	503+
Sulfoacetic Acid	260	513R
Sulfohydroxamate Esters	560	303
Sulfonamides	564	80+

	Class	Subclass
Acyclic	564	95+
Aromatic	564	84+
Hydrazine containing	564	81
Plural	564	82+
Sulfa type	260	397.7 R
Sulfonate Esters	558	44+
Sulfonation (See Sulfation)	260	686
Fatty oils	260	400
Natural resins	260	98
Sulfo acid preparation	260	503+
Sulfones	568	28
Sulfa drug type	260	397.6
Sulfonic Acids	260	503+
Anthraquinone	260	370+
Aromatic preservation	260	505P
Fat cleavage with	260	416
Hydrophenanthrene nucleus containing	260	503.5
Mineral oil sludge source	208	13
Triphenylmethane	260	389+
Wetting agent containing	252	353+
Sulfonium	568	18
Sulfophthalic Acid	260	507R
Sulfoxides	568	27
Sulfur		
Acids and acid anhydrides	423	512R+
Electrolytic synthesis	204	104
Binding agents	252	189+
Burner	422	160+
Compounds organic		
Carboxylic acids		
Acyclic	562	512+
Acyclic acid moiety	560	147+
Alicyclic esters	560	125
Aromatic	560	9+
Fats fatty oils ester type waxes higher fatty acids	260	399+
Heterocyclic	549	1+
Dye compositions	8	
Food fumigation with	426	319
Medicine or poison containing	424	
Mining	299	3+
Misc. chemical manufacture	156	DIG. 72
Preparation inorganic	423	
Electrolytic	204	128
Rendering harmless in fuel	208	177+
Salts inorganic	423	
Electrolytic synthesis	204	92+
Sulfureted hydrogen	423	563
Sulfuric acid	423	522+
Apparatus		
Chambers	422	
Concentrating	422	160+
Trains	422	160
Electrolytic synthesis	204	104
Esters	558	20+
Mineral oil sludge source	208	13
Sulfurization	260	685
Fatty oils	260	399
Rubber molding combined	264	
Synthetic resin or natural rubber	525	343+
Synthetic resin from	528	389
Synthetic resin or natural rubber compositions containing	524	1+
Treating agent containing	526	917*
Sulfurous Acid	423	521
Apparatus trains	422	160
Paper stock treatment	162	83+
Sulfuryl Chlorid	423	462
Sulky Vehicle	280	63+
Go cart	280	47.24
Sultones	549	10+
Naphthosultone, seven membered hetero ring	549	10
Five members	549	33
Four members	549	89
Six members	549	15
Sun (See Solar)		
Sun Baths	128	372
Sun Roof, Automobile	296	
Sundial	33	270
Design	D10	45
Sunglasses		
Clip on type	D16	102
Supercharger		
Centrifugal pump	415	
Internal combustion engine	123	434+
Combined	123	559
Fluid pressure motor combined	60	598+
Turbine combined	60	598+
Turbine combined	60	597+

	Class	Subclass
Superconductive Alloys	420	901*
Superfinish Grinder	51	57+
Superheater		
Concentrating evaporator	366	127
Cupola combined	48	64
Steam	122	459+
Locomotive combined	105	45
Power plant combined	60	643+
Superheterodyne Receiver	455	130+
Superphosphates	423	
Fertilizers	71	33+
Superplastic	420	902*
Superregenerative Radio	455	336+
Superscript Character	400	904*
Supersonic Wave	367	
Agitation	366	
Bonding non-metallic		
Apparatus	156	580.1+
Process	156	73.1+
Chemical apparatus	204	193
Chemical change causing		
Apparatus	204	193
Code signaling	116	18+
Controls direction of travel	318	460
Airplane	244	76R+
Electric motor actuated	318	460
Torpedo	114	23
Gas separation	55	277
Process	55	15
Generator	116	137A
Piezoelectric, resonator combined.	310	322
Piezoelectric with mechanical energy coupling means	310	334
Laminating processes	156	73.1+
Measuring using	73	570+
Metal casting using	164	501
Nautical signaling	116	26+
Seismic surveying	181	
Submarine signalling	367	131+
Testing miscellaneous	73	570+
Welding apparatus using	228	1.1
Supervisory Control	340	825.6+
Support (See Article or Device Supported)	248	
Airplane motor	248	554+
Antenna combined	343	878+
Apparel	2	300+
Apparatus	223	120
Design	D 2	624+
Arch	128	166.5
Armrest	248	118
Article by building	52	27+
Articles		
Or object being X rayed	378	38
Plural	211	
Plural locked	70	62
Single	248	
Ash tray smokers	131	240.1+
Athletic	128	157+
Bag	248	95+
Baggage	190	18R
Ball rack	211	14+
Basin	4	643+
Bathtub	4	592+
Bather	4	571+
Bearing	384	428+
Bed linens	24	72.5
Bed attached	5	504+
Beds	5	
Blotter, ie desk pad	248	346+
Boat		
Foot supports	114	363
Seat	114	363
Bobbin	242	130+
Body catchers	182	137+
Body harness	182	3+
Baby jumper	297	275
Body in bed	5	431+
Boiler	122	510
Bookshelf combined	108	59+
Boot rack	211	34+
Brackets (See bracket)	248	200+
Design	D 8	354+
Brake	188	205R+
Brush and broom	248	110+
Plural	211	65+
Cabinet	312	
Cabinet housing	312	253
Selective diverse	312	243
Camera (See --, tripod)	354	293
Cane rack	211	62+
Casket	27	26
Covers	27	18
Catamenial	604	393+
Ceramic stilt	432	258+
Chairs & seats	297	
Chin	128	164
Corpse	27	25
Christmas tree	248	511+
Cigar and cigarette	131	257+
Clock	248	114
Case	368	316
Clothes spotting boards	68	240
Clothesline	211	119.1+
Clutch	192	115
Coating implement fountain combined	401	48
Non-use support	401	131
Cue	211	68
Curtain	160	
Cutlery	30	296R+
Base supported	30	273+
Forks	30	323
Guarded	30	290+
Shears	30	231+
Spoons	30	327
Device with	92	161
Diaper	604	393+
Disinfecting apparatus	422	297
Dispenser	222	173+
Movable supply container	222	160+
Distill and	202	266
Drill rack	211	69
Drum musical	84	421
Drying apparatus	34	239+
Earthworking tool	172	
Bulldozer	172	811+
Easels	248	441.1+
Eaves trough	248	48.1
Egg rack	211	14+
Electric devices (See X-ray)		
Battery cradles	429	96+
Battery mounts automobile	180	68.5
Battery racks	429	96+
Cable	248	49+
Cell assembly support feature	429	186
Conductors in duct	174	99R+
Electrode	314	130+
Electrode	429	208
Fixture rack	211	26
Holder per se	429	100
Insulator	174	158R+
Lamp filament	313	271+
Outlet or junction box	220	3.9+
Support	248	DIG. 6
Overhead conductors	174	40R+
Plating electrode	204	286
Removable cell	429	96+
Secondary battery	429	208+
Telephone	379	454+
Third rail	191	32
Trolley	191	40+
Wire supports	52	40
Eye glass	351	155+
Holder	248	DIG. 2
Fan	416	146R
Field glasses	350	547+
Flashlight	362	157+
Flexible abrasive tool	51	358+
Floral	248	27.8
Fluid handling apparatus	137	343+
Food	99	324+
For receptacle	248	311.2
Founding		
Core	164	397+
Insert in mold	164	332+
Funnel	248	94
Furnace (See --, stove)		
Furniture	211	27
Gaff	114	99
Gate front end	49	396
Glass mold	65	361
Golf bag	248	96
Gun rack	211	64
Hair foundations	132	53+
Hammock	5	127+
Hand stamp rack	211	39
Harness	54	84
Body	119	96+
Reins	280	181+
Trace end	54	56
Hatrack	211	30+
Head	297	391+
Heat exchanger	165	67+
Corpse	27	13
Hinged leaf		
Binder device releasably engaging aperture sheet	402	
Book or leaf	281	45
Display items	40	530+
Movable on horizontal axis	211	169.1
Horizontally supported planar surfaces	108	
Hose conduit	248	75+
Indicator mountings	248	DIG. 4
Inkstand, dispenser type	222	577
Inkstand, movable inkwell	248	128+
Inkstand, stationary inkwell	248	146+
Instrument in panel	248	27.1+
Jock strap	128	158+
Junction box	220	3.9+
Kinesitherapy vibrator	128	33
Light	362	382+
Bowl	362	433+
Chimney	362	314
Inverted bowl	362	453
Lens	362	455
Shade	362	433+
Shade and bowl	362	351+
Light filter	350	318
Machine	248	637+
Mattress rack	211	28
Mirror	248	466+
Frames or holders combined	52	
Nozzle	248	75+
Sprinkler head	169	41
Nuclear reactor	376	461
Fuel core	376	362
Nursing bottle	248	102+
Opera glasses	350	547+
Optical instruments	248	
Orthopedic	128	68+
Outboard motor	248	640+
Outlet box	220	3.9+
Panel	52	474+
Face-to-face panels	52	479+
Retaining means	52	764+
Paste tube	248	108+
Pen or pencil rack	211	69.1+
Phonograph record rack	211	40
Picture	248	466+
Frame combined	40	152.1+
Pipe		
Conduit	248	49+
Coupling	285	61+
Dredger	37	72
External structure combined	138	106+
Joint	285	61+
Smokers	131	257+
Plant	47	44+
Plaster	52	443+
Furring joist or stud	52	344+
Plumbs	33	370+
Pressure compensating	248	DIG. 1
Projection machine	248	637+
Punching bag	272	78
Racks (See rack)	211	
Radiator	165	67+
Railway couplings		
Link supports	213	205+
Pin supports	213	192+
Railway mail delivery	258	23+
Receptacle	248	
Letter box	232	39+
Milk pail lap supported	220	17.1+
Racks	211	71+
Safe	109	50
Stands	248	128+
Tray type	211	126+
Tray type stand	248	127+
Wall or ground supported	220	18
Receptacle covers		
Casket	27	18
Removers	81	3.8+
Wooden	217	60R+
Refrigerating apparatus, resilient	62	295
Refrigerator housing	62	297
Refrigerator, internal	62	465+
Door mounted	62	377
Ice	62	459+
Rod	248	251+
Roller		
End	160	323R

	Class	Subclass
Intermediate	242	73.5
Score line	248	DIG. 9
Screen	160	
Shade	160	
Shipping	410	46
Shoe	211	34
Counter and heel	36	69
Pad and foot	36	71
Shoelace rack	211	61
Sifting	209	409+
Simulation	248	DIG. 8
Skaters	272	70+
Smoking appliance	131	257+
Spectacle	351	155+
Stands (See also stand)	248	127+
Static mold having	249	139
Stove or furnace		
Base	126	305
Cooking stove fluid fuel	126	50
Field stove	126	30
Stand boiler	126	363
Strainer	248	94
Stratifier	209	508
Stringed instruments	84	327
Surgical needle	128	303.19
Tank, water closet	4	419+
Temperature compensating	248	DIG. 1
Thermometer, indicating tube	374	194+
Tire rack	211	23+
Tissue dispenser holder	248	DIG. 5
Tobacco users appliance	131	257+
Toilet bowl	4	252R+
Tool driving or impacting device combined	173	31
Toy	446	227
Figure rotary	446	236
Figure wheeled	446	269
Trash can support	248	DIG. 7
Tree	47	42+
Tripod	248	163.1+
Truss	128	99.1+
Umbrella rack	211	62+
Valve	251	143+
Vehicle outrigger	280	763.1+
Crane	212	189
Vehicle part		
Harvester tongue	56	218
Motor	248	637+
Truck railway safety	105	217
Velocipede	211	17+
Vessel bilge or keel	405	7
Violin	84	280
Walkers	272	70+
Washboard	248	27.5
Watch	248	114
Weighing scales	177	244
Weighted base	248	DIG. 1
Wheel rack	211	23+
Whip	211	67
Design	D30	156+
Window pane	52	777+
Window pane	52	764+
Work (See holders, work)		
Work supported machines		
Lathe	82	4R
Riveting machine	29	243.53+
Wringer	68	239
Wash bench combined	68	236
X ray film or screen	378	167
Dental cassetts	378	168
Table	378	209
X ray tube	378	193
Supporting		
Frame for skater	272	70+
Suppository	424	DIG. 15
Composition	514	966*
Soluble in body	604	288
Surgical		
Composition	424	
Packages	206	529
Suprarenal Extracts		
Medicine containing	424	113
Surface Active Agents	252	351+
Surfaces		
Active agent	252	351+
Aircraft sustentation	244	200
Bearing		
Inserts	384	276+
Treating	384	625
Bed bottom	5	186R+
Carbonizing of iron	148	16.5
Chair or seat		
Changeable supporting surface	297	283
Headrest with changeable surface	297	220+
Spring supported upholstery	267	80+
Cooking mold	99	372+
Sheet or griddle	99	422+
Demonstration	434	365+
Electric insulator	174	212
Electrically charged for safes	109	35
Erasable, eg chalkboard	434	408+
Leaves	281	39
Evaporators sugar making	127	16
Fluid reaction ie impellers	416	
Friction gearing	74	214+
Gauge plates	33	567
Heater	126	271.1+
Water vessel	126	390+
Horizontally supported planar	108	
Laminated type photographic	101	178
Stencil	101	128.21
Pavement	404	17+
Antislip	404	19+
Sheet type	404	18
Surface treating	404	83+
Planographic printing	101	453+
Platen or pressure	100	295
Printing member	101	401
Railway rail tread surface	238	148
Tie plate top surface	238	304
Receptacle stand for sloping surface	248	148
Roll or rollers	29	121.1+
Roughness testing	73	105
Smoothing or pressing textiles	38	
Street mirror or reflector T	404	14
Surgical electric applicator	128	783+
Textile drawing roll	29	110+
Washboard	68	226+
Washtub with scrub surface	68	233
Yielding surface printing member	101	379+
Rolling contact	101	376
Surfacing		
Button	79	7+
Drilling combined	79	6
Grinding	51	
Roads	404	17+
Stone	125	
Surfboard	441	74+
Design	D21	228+
Surgery and Surgical (See Classes 128 and 604)		
Acupuncture	128	907*
Adhesive fastener for absorbent pad	604	389+
Article package	206	438+
Basin	D24	56+
Biological signal amplifier	128	902*
Feedback to patient	128	905*
Blood coagulation	128	DIG. 22
Breathing apparatus without patient cross-contamination	128	909*
Cannula supporters	128	DIG. 26
Catheters (See catheter)		
Cervical collars	128	DIG. 23
Clip appliers	29	243.56
Collagen	128	DIG. 8
Computer controlled, monitored	364	413+
Cyrogenic	128	DIG. 27
Design	D24	
Electric shock, patient protection	128	908*
Fluid amplifiers	128	DIG. 1
Heart-lung	128	DIG. 3
Heat shrinkable film	128	DIG. 18
Infusion monitoring	128	DIG. 13
Intravenous injection support	128	DIG. 6
Invalid beds	5	60+
Mask	D29	7
Medical-surgical bags	128	DIG. 24
Medicament receptacle	604	403+
Mixing receptacles	604	416
Multiphasic diagnostic clinic	128	906*
Pressure infusion	128	DIG. 12
Radio telemetry	128	903*
Servo-systems	128	DIG. 7
Silicone	128	DIG. 21
Sphineters artificial	128	DIG. 25
Splint, clavicle	128	DIG. 19
Splint, inflatable	128	DIG. 2
Spotlights	362	804*
Stapler	227	DIG. 1
Supplies packaged	206	438+
Suppression of noise in electric signal	128	901*
Syringes (See syringe)		
Table	269	
Teaching	434	262
Teflon T M	128	DIG. 14
Tree	47	8
Velcro T M	128	DIG. 15
Surlyn D T M (polyethylene Methacrylic Acid)	526	317.1
Survey Meter Nuclear Energy	250	472.1+
Scintillation type	250	361R
Surveying Instruments	33	290+
Bent line of sight type	356	253+
Angle measuring	356	138+
Range or height finders	356	3+
Optical features	356	247+
Design	D10	61+
Survival in Environment		
Computer controlled, monitored	364	418
Suspended		
Roof	52	83
Swings	272	85+
Suspender		
Artificial leg	623	32
Fasteners design	D 2	640+
Garment		
Design	D 2	626
End elements	2	340
Fasteners for	D11	200+
Plural	2	304+
Saddle type cord retainers	2	341+
Strip connected spaced	2	323+
Trouser attached	2	230
Suspension Devices (See Hangers)		
Battery active material, faure electrodes	429	208
Clothesline single run	211	119.12
Drying		
Hollow articles	34	106
Sheets webs or strands	34	151+
Filter household	210	105
Fluid (See suspensoids)		
Current drying method	34	10
Separating solids	209	132+
Folding bed fulcrum	5	141
Insulator support	174	160
Knob flexible	16	122
Light bowl	362	453+
Picture support	248	489+
Pins for paper	24	154
Pipe or cable	248	58+
Rack for broom and brush	211	66
Radiator bracket	248	233
Roundabout amusement vehicle	272	40+
Special receptacles or packages	206	806*
Stage appliances	272	24
Structures & vehicles (See hanger)		
Boiler regulation	122	450
Bridge cantilever truss	14	8
Bridges	14	18+
Dumping railway car	105	242
Fences	256	23
Railway car body	105	453
Railway car suspended	105	148+
Railways	104	89+
Scaffolds	182	142+
Track	104	123
Unloader for suspended vehicle	414	377+
Vehicle body spring	280	788
Suspended supports	248	317+
Swinging seat	272	85+
Tea ball type beverage infuser	99	322
Telephone supports	379	454+
Tool and machine		
Brush flue cleaner	15	163
Cable drum	254	380+
Car journal box lifting	254	34
Conveyor, load suspending	198	678+
Conveyor loop	198	678+
Cultivator wheeled	172	
Grapple pivoted saw operator	294	112
Harvester knife grinding clamp	51	223
Machinery support	248	637+
Resilient	248	610
Plow wheeled	172	
Scraper flue cleaner	15	243
Sheet delivering conveyors	271	198+
Strop	76	81+
Textile weaving weft control	139	226+
Vehicle springs	267	
Stub axle mounts	280	660+
Vehicle structure combined	280	788

Entry	Class	Subclass
Vehicle systems	280	688+
Suspensoids		
Colloidal	252	302+
Comminuting or grinding	241	
Electrical purification	204	186+
Electrolytic		
Forming processes	204	10
Treatment processes	204	130+
Electrophoretic or electro osmosis using	204	180.2+
Suspensories	128	158
Sustained or Differential Release		
Capsules	424	457+
Implants	424	422+
Tablets	424	468+
Sustained Release Implants	424	422+
Sutures		
Appliers surgical	29	243.56
Surgical	128	334R+
Textile knitting	66	43
Hosiery seam	66	179
Swab	15	208+
Container for	206	209
Culture medium combined	435	292+
Diagnostic	128	759
Lubricator	184	102
Making	19	145+
Medicated	604	1+
Sampling	73	854
Valved pump piston	417	545+
Swaging		
Bolt side	10	18
Chain making shaping and	59	30
File cutting by	76	14
Horseshoe nail side	10	68
Horseshoe shaping and	59	58
Metal forging	72	
Nut forging side	10	79+
Pipe joint	285	382
Pipe to plate joint	285	222
Saw making	76	51+
Saw setting	76	72
Screw threading by	10	152R+
Tool sharpening by	76	89.2
Swarm Appliances	6	7
Swash Plates	74	60
Expansible chamber device	92	71
Flow meter	73	244
Internal combustion engine		
Multiple parallel cylinder	123	58R
Rotating cylinder	123	43R
Motor	91	472+
Pump	417	269+
Swathing Attachments	56	189
Swatters	43	137
Sweater		
Animal	D30	145
Apparel	2	90
Design	D 2	44+
Knit fabric	66	176
Sweats or Sweat Band	2	181+
Hat making	223	22
Sweep		
Animal powered motor	185	19+
Cable drum actuation	254	381+
Circuit		
Cathode ray tube having	315	378
Electronic tube type oscillator	331	
Triggered	328	193+
Cultivator shovels	172	681+
Cutter	30	310+
Can opener	30	435+
Plural blade	30	310+
Foundry mold making device	164	161+
Pivoted arm bending machine	72	217
Plow trailing		
Separately attached wings	172	722+
Winged	172	730
Sheet cutting machine	82	
Wood auger	408	199+
Wood disk cutting machine	408	72R+
Sweeper		
Brush broom vacuum carpet street	15	
Rake eg lawn broom	56	400.17+
Sweeping Compound	252	88
Sweet Potato Musical Instrument	84	380R
Sweetening		
Mineral oil	208	189+
Swelling		
Artificial fibers	8	130.1+
Dyeing combined	8	
Fiber swelling & stretching	8	DIG. 3
Swells		
Electric tone generation with expression device	84	1.27
Piano	84	182
Pipe organ	84	346+
Reed organ	84	372
Swimming		
Aids	441	55+
Buoyant	D21	237+
Foot attached	441	61+
Foot flipper	441	64
Exercising devices for simulating	272	71
Mask	D29	8+
Pool		
Design	D25	2
Grab rail	D25	41
Inground	52	169.7+
Nuclear reactor	376	174
Nuclear reactor	376	361
Nuclear reactor	376	403
Portable design	D21	252
Sea baths	4	487+
Tank baths	4	488+
Water purification	210	169+
Shoes	36	8.1
With fins	441	76+
Structure of pool	52	261
Concrete	52	250
Curved	52	245
Suit	D 2	40+
Teaching	434	254
Swinging Door	49	381+
Swings		
Amusement	D21	246
Chairs and seats	272	85+
With fan	416	146R
Roundabout	272	28R
Switch		
Apparatus for elevators	187	
Automatic		
Electromagnetic	335	6+
Thermal	337	298+
Automatic for acetylene generator	48	7
Battery	320	
Box	220	3.2+
Burglar alarm	200	61.58 R+
System	340	541+
Charging for storage batteries	320	
Combination	200	1R+
Conveyer type		
Chute	193	31R
Rollerway	193	36
Skidway	193	39
Spiral chute	193	13
Zigzag tube	193	28
Door actuated	200	61.62+
Electric (See controller)	200	
Arc lamp system having	314	131
Arc preventing or extinguishing	200	144R+
Automotive vehicle combined with lock	70	237+
Box or housing	174	53+
Check operated	194	211+
Coin	194	239+
Combined with distribution circuit	307	
Design	D13	32+
Display system control	340	806+
Door or window	200	61.62+
Electric illuminating fixture	362	394
Electric lamp having	315	32
Electromagnetic switching systems	361	139+
Electronic tube having	315	56
Electrothermally actuated	337	
Fire actuated	337	298+
Lamp or electronic tube system having	315	362
Light projector combined with	362	253
Magnetically operated	335	
Making	29	622
Mechanically actuated, general	200	
Multiple filament lamp having	315	64
Plate or socket lighting	362	95
Position or condition responsive	362	802*
Post supported electric light combined with	362	431
Supported electric light combined with	362	382+
Telegraph circuit manipulator	178	75
Telephone	379	422+
Telephone check operated	379	146+
Telephone switchboard	379	319+
Telephone system automatic selective	335	108+
Testing of automatic telephone type	379	17
Thermal control of alarm	340	593+
Thermally actuated	335	
Trolley conductor type	191	38
Tuner with	334	47+
Electronic	328	76
Heat controlled	337	298+
Inductor combined with switch	336	105
To neutralize capacitance	307	134
Interference from elimination	307	134+
Knife	200	
Luminous		
Faceplate	250	465.1
Handle or key	250	465.1
Push button	250	465.1
Movable carrier of textile braider with pattern mechanism	87	16
Movable for carrier of textile braiding apparatus	87	37+
Photoelectrically controlled	361	173+
Plates (See face, plates)		
Pneumatic dispatch system	406	181+
Railway	246	
Cable	104	184
Cable slot type	104	195
Foot guard	238	380
Rail joint	238	165
Track	104	130+
Trolley conductor	191	38
Stands	246	393+
Automatic face setting	246	278+
Manual operation with signal actuation	246	144+
Signal	246	476
Stoplight		
Auto	200	61.59
Hydraulic	200	81R+
Store service system	186	19+
Thermal responsive	337	298+
Heated by electric current	337	14+
Tilting track section type	414	354+
Vehicle unloading	414	357
Travelling crane system	414	562
Vertical axis for cutting and conveying harvester	56	170+
Switchboard		
Electric	361	331+
Telephone	379	319+
Divided central	379	310+
Through ringing	379	256
Switches		
Membrane type	200	512
Rolamite	200	503
Switching Systems	307	
Lamp systems	315	313+
Relay systems	361	160+
Signalling	340	286R+
Selective	340	825+
Transmitters	340	345+
Telegraph	178	
Telephone	379	242
Automatic	379	258+
Calling	379	350+
Central	379	242+
Intercommunicating	379	167+
Party line	379	177+
Plural exchanges	379	219+
Switchboard	379	310+
Swivels	403	164+
Bracelets	63	10
Chain	59	95
Making	59	9
Chute horizontal swinging	193	22
Endless chain pumps	198	643
Fish bait artificial	43	42+
Fishhooks	43	44.83+
Panel hanger wheel mount	16	104
Pipe joint	285	272+
Multiple	285	168
Punching bag	272	77+
Railway car suspending	105	156
Rod joint	403	164+
Sleds multiple	280	17
Socket combined	24	136L+
Tool handle	403	52+
Trolley head electric collector	191	60+
Turnbuckle	403	60

U.S. PATENT CLASSIFICATIONS

Entry	Class	Subclass
Valve disc	251	264+
Valve gate	251	193+
Watch chain snap hook	24	240
Weaving machine shuttle	139	119+
Wheeled vehicle running gear	280	86
Wood saw	83	747
Swizzle Stick	D 7	42
Swords	30	165+
Cane	135	66
Carrier	224	232+
Design	D22	118+
Handles	D 8	DIG. 5
Pistol	42	53
Sylphon T M (See Bellows)		
Flowmeter	73	262
Fluid pressure guage	73	729+
Fluid pressure regulator	137	505.36
Making	29	454
Corrugating	72	54+
Pump	417	472+
Synchronizing		
Camera and flash powder	354	129+
Camera and photoflash testing	356	72
Camera shutter release and light	354	129+
Clock	368	52
Clutches	192	52+
Dynamoelectric machine interconnection	307	43+
Electric motor	318	41+
Electric motor	318	85
Fluid pressure brake systems	303	35+
Gearing	74	395
Governor	73	507
Impeller combined	416	34
Motion picture and sound	352	12+
Multiple motor	60	700+
Multiple motor	60	706+
Musical instrument	84	119
Plural radio stations	235	58M
Telegraphy	375	106+
Telegraphy multiplex	370	47+
Television	358	148+
Recording or reproducing color signal	358	319
Synchronous Motor	310	162+
Systems	318	700+
Synchroscope	324	91
Synthesis		
Electro	204	59R+
Inorganic compounds	23	
Organic compounds	260	
Polymers (See synthetic resin or natural rubber)		
Synthetic		
Alumina	423	625
Diamond and making	423	446+
Gem	501	86
Resin (See synthetic resin or natural rubber)		
Rubber (See rubber; synthetic resin or natural rubber)		
Synthetic Resin or Natural Rubber		
Abs (See also prepared from, below)	526	338
Addition polymerization	526	72+
Manipulative process features	526	59+
Adhesive material	520	1+
Compositions (See class 523, 524)		
Admixtures of resins or rubbers	525	50+
Aftertreatments		
Chemical reaction involved	525	50+
To produce ion exchange resin	521	25+
Utilizing wave energy	522	1+
Vulcanization of rubber (products)	525	331.9+
Vulcanization of rubber with sulfur or sulfur compound	525	343+
With asphalt or bituminous material	525	54.5+
With carbohydrate	525	54.2+
With fatty acid or fatty acid glycerol ester	525	50+
With natural resin	525	54.4+
With protein or biologically active polypeptide	525	54.1+
Physical treatment	528	480+
Compositions	523	300+
Compositions: creaming, concentrating or agglomerating latex	523	335
Natural rubber e. g. extraction of rubber from plant source	528	931*
Natural rubber latex	528	934*
Natural rubber latex coagulating	528	936*
Natural rubber latex concentrating	528	937*
Natural rubber latex preserving	528	935*
Alkyd resin (solid polymer)	528	272+
Compositions (See class 523, 524)		
Containing chemically combined fatty acid or fatty acid glyceride	528	295.5
Containing chemically combined natural resin	527	604
Aminoplast resin (See also prepared from aldehydes, below)	528	266
Bakelite T M type resin	528	129+
Balata (See natural rubber, below)		
Blends of solid polymers	525	50+
Block and block type copolymers	524	88+
Buna T M type rubber	526	339
Buna-n (butadiene nitrile rubber)	526	338
Buna T M type rubber	526	339
Catalysts		
Aldehyde polymerization	528	232+
With phenol	528	137+
Chalcogen ring opening	528	406+
Epoxy resin	528	88+
Ethylenic polymerization	526	89+
Graft polymerization	525	244+
Redox	526	915*
Transition metal	526	90+
Polyester	528	274+
With ethylenic monomer	525	11+
Silicone	528	12+
Urethane	528	48+
Cellular or porous	521	50+
Blends of synthetic resin or natural rubber	521	134+
Blowing agent i. e. foaming agent	521	82+
Resin formation in the presence of	521	99+
Carbodiimide polymer	521	901*
Degradability enhanced	521	916*
Dye or pigment containing	521	920*
Electrical or wave energy used during cell forming	521	50.5*
Ethylene polymer	521	143+
Fireproofed	521	82+
Non-urethane resin	521	907*
Resin formation in presence of fireproofing agent	521	99+
Urethane resin	521	906*
Hydraulic cement ctg (See water settable inorganic composition ctg, below)		
Integral skin foam	521	51
Isocyanurate polymer	521	902*
Kicker or blowing agent moderator	521	909*
Latex rubber	521	65+
Natural resin containing	521	84.1
Synthetic resin formation in the presence of	521	109.1
Natural rubber containing	521	150
Latex	521	65+
Nucleating agents for	521	908*
Phenolic resin	521	180+
Pigment containing	521	920*
Polyester resin	521	182+
With unsaturated monomer	521	128
Protein containing	521	84.1
Resin formation in the presence of	521	109.1
Reduced smoke formation	521	903*
Reticulated foam	521	52
Separated reactive materials e. g. :two package system:	521	912*
Sintering process for formation of porus product	521	919*
Styrene polymer	521	146+
Foamed beads	521	56+
Syntactic foam	521	53+
Non resinous hollow particles	523	218+
Urethane polymer	521	155+
Vinyl chloride polymer	521	145
Compositions (See class 523, 524)		
Adhesive (See class 523, 524)		
Anerobic	523	176
Glass substrate	523	168
Antioxidant containing (See class 523, 524)		
Antisatic	524	910*
Asphalt containing	524	59+
Resin produced in presence of	524	705
With epoxy resin	523	450
With polyester-unsaturated monomer	523	518
Biocidally protected	523	122
Bitumen ctg (See asphalt ctg, above)		
Body contact or implantable (intended use)	523	105+
Brake shoe (intended use)	523	149+
Carbohydrate containing	524	27+
Resin produced in presence of	524	732+
With epoxy resin	523	447+
With polyester-unsaturated monomer	523	509
Carbon black containing (See class 523, 524)		
Resin produced in presence of	524	847
Specified dimension	524	495+
With epoxy resin	523	440
With polyester-unsaturated monomer	523	512
Castor oil containing	524	310
Resin produced in presence of	524	769
Cellular plant or animal material containing	524	9+
Resin produced in presence of	524	702
With epoxy resin	523	446
With polyester-unsaturated monomer	523	500+
Cellulose ether or ester ctg (See carbohydrate ctg, above)		
With polyester-unsaturated monomer	523	509
Chinawood oil containing (See fatty acid triglyceride, below)		
Coating (See class 523, 524)		
Antifogging	523	169
Glass enamel	523	170
Ink for glass substrate	523	160
Concrete ctg (See inorganic water-settable, below)		
Contact lens, used for (See also class 527-937)	523	106+
Cork ctg (See cellular plant or animal material, above)		
Degradability enhanced i. e. pro-degradant containing	523	124+
Dispersions (See emulsions, below)		
Drying oil ctg (See fatty acid triglyceride, below)		
Dye ctg (See class 523, 524)		
Emulsions (See class 523, 524)		
Aqueous (See also 524-501)	524	800+
Epoxide resin containing	523	400+
Fatty acid triglyceride ctg	524	313
Resin produced in presence of	524	773+
With epoxy resin	523	455
With polyester-unsaturated monomer	523	511
Feather ctg (See cellular plant or animal material, above)		
Fireproofed (See class 523, 524)		
Flux ctg (See class 523, 524)		
Foundry mold or mold core binder	523	139+
Hair ctg (See cellular plant or animal material, above)		
Heat stabilized (See class 523, 524)		
Inorganic water settable material ctg e. g. hydraulic cement	524	2+
Resin produced in presence of	524	650+
With epoxy resin	523	402+
With polyester-unsaturated monomer	523	501+
Lacquer (See class 523, 524)		
Latex (See class 523, 524)		
Coagulating	528	936*
Concentrating	528	937*
Creaming, agglomerating or coalescing	523	335
Physical treatment of natural rubber latex	528	934*
Preserving or stabilizing	528	935*

Entry	Class	Subclass
Resin produced in presence of water	524	800
Leather ctg (See cellular plant or animal material, above)		
Lignin or tannin containing	524	76
Resin produced in presence of	524	735
With epoxy resin	523	400+
With polyester-unsaturated monomer	523	500+
Linseed oil ctg (See fatty acid triglyceride, above)		
Molding (See class 523, 524)		
Natural resin ctg (See also rosin, below)	524	77
Synthetic resin produced in the presence of	524	764
With epoxy resin	523	400+
With polyester-unsaturated monomer	523	500+
Natural resin ctg (see class 523, 524 treated as polyisoprene)		
Guayule	528	930
Phonograph record (intended use)	523	174
Pigment ctg (See class 523, 524)		
Plasticizer ctg (See class 523, 524)		
Protein or biologically active polypeptide ctg	524	17+
Resin produced in presence of	524	704
With epoxy resin	523	449
With polyester-ethylenic monomer	523	500+
Rosin containing	524	270+
Synthetic resin produced in the presence of	524	764
Tannin ctg (See lignin, above)		
Tar ctg (See asphalt, above)		
Varnish (See class 523, 524)		
Wax containing	524	487+
Oxygen containing wax	524	275+
Oxygen ctg wax with epoxy resin	523	455+
Oxygen ctg wax with polyester unsaturated monomer	523	511
Resin produced in presence of oxygen ctg wax	524	763+
Condensation polymerization	528	
Manipulative process features	526	59+
Coumarone resin	526	266+
Dacron T M type resin	528	308.1+
Ebonite T M type resin	525	332.6+
Elastomer (See rubber, below)		
E P (ethylene-propylene)	526	348+
E P D M (ethylene-propylene-diene monomer)	526	
Cured or vulcanized	525	331.7+
Epoxide resin (See also classes 525, 528)	528	87+
Foams	521	50+
G R - N (acrylonitrile-butadiene rubber)	526	338
G R - S (styrene-butadiene rubber)	526	340
Graft or graft type	525	50+
Guayule (See also natural rubber)	528	930*
Hevea (See natural rubber)		
Hypalon T M type resin	525	333.9
Ion exchange resin	521	25+
Ionomer resin systems	525	919*
Carboxyl bearing	526	317.1+
Kraton T M type resin (styrene-butadiene rubber)	526	340
Lexan T M type resin	528	196+
Lucite T M type resin	526	329.7
Melamine resin	528	254+
Natural rubber (note: for purposes of the class 520 series natural rubber is considered to be polyisoprene)		
Neoprene	526	295
Vulcanized	525	330.9+
Nylon	528	310+
Perbunan T M type rubber (copolymer butadiene-acrylonitrile)	526	338
Phenolic resin	528	129+
Phenoplast	528	129
Phenoxy resin	528	86+
Plexiglas T M type resin (polymethyl methacrylate)	526	329.7
Polyamide resin (See also 528-332+)	528	310+
Polycarbodiimide	528	44+
Polyesters	528	272+
Polyimide	528	310+
Polyphenylene oxide	525	905*
Polysulfide	528	373+
Prepared from		
Acetylene	526	285
Acrolein	526	315
Acrylates	526	328+
Acrylic acid	526	317.1+
Acrylonitrile	526	341+
Aldehyde e. g. homopolymer of saturated etc.	528	230
Ethylenically unsaturated	526	315
Hydrocarbon	528	247
Melamine	528	254+
Urea	528	259+
With amine e. g. aniline etc.	528	266+
With ketone	528	227
With phenol	528	129+
Asphalt or bituminous material	527	500
Unsaturated	526	290
Butadiene	526	335+
Carbohydrates e. g. sugar, starch etc.	527	300+
Unsaturated carbohydrate derivative	526	238.2+
Carboxylic acid or derivative	528	271+
Ethylenically unsaturated	526	317.1+
Cashew nut liquor	528	2+
Cellular plant or animal material	527	100+
Chloroprene	526	295
Coal derived material (See asphalt or bituminous material, above)		
Cork	527	100+
Coumarone	526	266+
Diolefin hydrocarbon	526	335+
Cycloaliphatic	526	308
Fused or bridged ring	526	280+
Epoxide compound	528	403+
Diglycidyl ether of bisphenol compound (See also 525-523+; 525-107+;525-403+)	528	87+
Ethylenically unsaturated	526	273+
Ethylene (homopolymer)	526	352+
Copolymer	526	72+
Ethylenic monomer (solely therefrom)	526	72+
With non-ethylenic monomer (See also class 527)	528	
Formaldehyde (See aldehyde, above)		
Furfural (See also aldehyde, above)	526	270
With non-ethylenic monomer	528	230+
Hydrocarbons	528	396+
Ethylenically unsaturated (See also 526-335;526-346)	526	348+
Indene	526	280+
With coumarone	526	266+
Isobutylene	526	348.7
With butydiene	526	339
Isocyanate compound	528	44+
Isoprene (note: for purposes of the class 520 series natural rubber is considered to be polyisoprene)	526	340.2
Lignin	527	400+
Melamine (See aldehyde, above)		
Metal compounds	528	395
Ethylenically unsaturated monomer	526	240
Organometallic	528	9
Methacrylate	526	319+
Methacrylic acid	526	317.1+
Methylmethacrylate	526	319+
Monoepoxides (See epoxide, above)		
Olefin	526	72+
Oxirane compounds (See epoxide, above)		
Phosphorous compounds	528	398+
Ethylenically unsaturated	526	274+
Plant extract of undetermined constitution	528	1+
Polyepoxide compound (See epoxide, above)		
Propylene (homopolymer)	526	351
Copolymer	526	72+
Protein or biologically active polypeptide	527	200+
Ethylenically unsaturated derivative	526	238.1
Silicon compound	528	
Ethylenically unsaturated	526	279
Si- H or si- C	528	10+
Tannin	527	400+
Ethylenically unsaturated monomer	526	238.2+
Tar or pitch	527	500+
Ethylenically unsaturated derivative	526	290
Tetra fluoroethylene	526	242+
Urea (See aldehyde, above)		
Vinyl chloride	526	344+
Vinyl compounds	526	72+
Vinylidene compounds	526	72+
Purification of	528	480+
Composition	524	310
Reclaiming	521	40+
Rubber	520	1+
Diene rubber	526	72+
Spandex T M type	528	44+
Thiokol T M type	528	374+
Vulcanized	525	332.5+
E P D M	525	331.8
Rubber hydrocholride	525	332.3
Silastic (organosilicon polymer)	528	10+
Silicone resin	528	10+
Compositions	524	858+
Room temperature vulcanizing (rtv rubber)	528	901*
Synthetic rubber (for purposes of class 520, is considered to be a synthetic resin)		
Teflon T M (polytetrafluoro-ethylene)	526	255
Wave energy utilized	522	1+
Synthol	518	700+
Syringe		
Dental	D24	14
Dispenser	222	206+
With follower	222	386+
Hand held	604	187+
Pump	417	437+
Specific gravity tester	73	441+
Sprayer		
Liquid pump feed	239	329+
Piston feed	239	320+
Surgical	604	187+
Air	604	26
Design	D24	60+
Ear	604	187+
Eye	604	295+
Fluid operated	604	131+
Fountain	604	131+
Hand held	604	187+
Hydrant	604	150
Indication of defect or use	604	111
Irrigation	604	187+
Motorized	128	DIG. 1
Nozzle	604	275
Piston	604	218+
Pocket container for	206	364+
Preventing reuse	604	110
Projectile	604	130
Receptacle for	206	364+
Siphon	604	131+
Thermal medicator	604	113+
With light and electrical application	604	20+
Syrups		
Condiments and flavors	426	658+
Electrolytic treatment	204	138
Malt	426	44
Sugars only	127	30+
Aircraft control	244	175+
Stoves and furnaces	126	

T

Entry	Class	Subclass
T Squares	33	474+
Design	D10	65+
Tab		
Bolt or nut, side lock, tab deformable in situ	411	122+
Exhibitor type	40	20R+
Protrusion to receive identification	40	359+
Suspender ends	2	340
Tabernacle	D99	25
Tables	108	17
Animal restraining	119	103

U.S. PATENT CLASSIFICATIONS

	Class	Subclass
Bed	5	3$^+$
Bed combined	5	507
Billiard and pool	273	3R$^+$
Design	D21	232
Cabinet convertible to	312	241
Cabinet with separable table	312	277
Chair combined	297	135$^+$
Chair convertible	297	119$^+$
Cooling	62	458
Cover or cloth	D 6	617
Croquet	273	57
Disappearing instrument or device	312	21$^+$
Examining or surgical	D24	3
Furniture	108	
Design	D 6	396$^+$
Game	273	309
Design	D21	1
Game boards	273	287
Collapsible	273	285$^+$
Interchangeable pattern	273	284
Rotatably mounted	273	280
With separable pattern	273	283
Gauging eg grizzlies	209	680
Gravity assorting in liquid suspension	209	172$^+$
Invalid back rest combined	5	70
Ironing	38	103$^+$
Combined with sewing machines	112	217.1$^+$
Plural surface	38	135$^+$
Single surface	38	137$^+$
Light table, illumination for	362	97
Luggage convertible	190	11$^+$
Machine part		
Box nailing machine	227	154
Cigar and cigarette making	131	41$^+$
Cutting machine	83	
Grain separator straw carrier	130	24
Metal can making and soldering	413	
Metal flanging	72	102
Plastic material cutting	83	
Plastic material cutting	425	289$^+$
Powered table for operating attachments and auxiliaries	74	16
Shingle machine	83	704$^+$
Slicer woodworking	144	171
Typewriter paper feeding	400	647$^+$
Woodworking saw table	269	289R
Modified for blade	269	83
Manual assorting	209	703$^+$
Pads	248	346.1
Phonograph or sound recording and reproducing		
Flat tablet traveling table	369	213
Mandrels	369	260
Tablet	369	267
Rack convertible to table	108	16
Refrigerator combined	62	258
Self levelling	114	195
Steam cooking	126	33
Stratifier	209	422$^+$
Starch making	127	26
Surgical	269	323$^+$
Telegraph instrument	178	114
Tennis	273	30
Toy or doll	446	482
Transfer	104	48$^+$
Electricity transmission to vehicle	191	9
Trunk	190	11$^+$
Turntables	104	35
Automobile	104	44
Plural related surfaces	108	103$^+$
Self service eg lazy susan	108	94$^+$
Self service power driven	108	20$^+$
Single surface	108	139$^+$
Wall	108	152
With structural installation	108	72$^+$
Work (See also table machine part)		
Abrading	51	240R
Butter working	99	452$^+$
Cloth inspecting	26	70
Illuminated operating table	362	33
Ironing textiles	38	103$^+$
Ironing with work catcher	38	111
Leather working	69	19
Milling planing or gear cutting	409	219$^+$
X ray	378	209
Tablets		
Calendar tear off type	40	121
Coffee	426	594
Dye compositions	8	526
Food design	D 1	
Medicinal	424	464$^+$
Design	D28	1$^+$
Memorial	40	124.5
Molding presses for	425	406$^+$
Phonograph	369	272$^+$
Making by turning	82	1.1
Railway signalling system	246	19
Tabulators		
Calculator	235	85R$^+$
Teletypewriter	178	25$^+$
Typewriter	400	284$^+$
Indicating	400	705.4$^+$
Tachometer	73	488$^+$
Design	D10	98
Electric	324	160$^+$
Rotation counter	235	104
Rotation counter, with electrical speed indicator	324	166$^+$
Tack	411	439$^+$
Pullers	254	18$^+$
Shoe lasting machine	12	16
Shoe sole sewing machine	112	30
Strips		
Making	10	29
Nailing implement	227	136
Tackle		
Cable and hook	294	82.11$^+$
Fishing	43	4$^+$
Design	D22	134$^+$
Pushing and pulling implements		
Block and	254	390$^+$
Single throw lever	254	128
Tensioners	254	242
Traversing hoist cable type	212	76$^+$
Taffeta	139	426R
Tag		
Affixing tags	156	DIG. 23
Collision released identification	116	32
Envelopes	229	74
Exhibitor type	283	74$^+$
Fastenings	40	20R$^+$
Inserters	128	330$^+$
Machines		
Cigar	131	106
Cigar etc packaging combined	131	283
Label pasting	156	
Paper tag applying	493	375$^+$
X-art collection	493	961*
Paper wrapping machine	53	128$^+$
Plug tobacco	131	113$^+$
Printed matter	283	74$^+$
Taillights	362	61$^+$
Tailpieces	84	299$^+$
Stringed instrument	84	297R$^+$
Tailpipe		
With muffler	181	227$^+$
With exhaust pipe	181	228
Tailstocks		
Lathe	82	31
Milling machine	409	242
Take up		
Bed bottom cord tightener	5	212
Braiding netting and lace making	87	31
Brake position adjusters	188	196R$^+$
Brake wear	188	79.5 R
Electric cable extensible	174	69
Electricity flexible extension to vehicle	191	12R$^+$
Embroidering machine thread	112	96
Harvesting binder cord knotter	56	450
Hitching post	119	124
Leather sewing machine thread	112	57
Railway freight car shrinkage	105	412
Railway traction cable	104	196
Sewing machine	112	241$^+$
Textile knitting fabric	66	149R$^+$
Trolley rope slack	191	93
Twine holder	242	129.1$^+$
Weaving fabric	139	304$^+$
Weaving stopping	139	339
Weaving warp feeding	139	99
Winding and reeling	242	129.1$^+$
Recording tape or motion picture film	242	179$^+$
Talcum Powders	424	69
Talking Machine (See Phonograph)		
Book	434	317
Tall Oil	260	97.5$^+$
Pulp preparation combined	162	14$^+$
Tally (See Counter)		
Game	116	222$^+$
Printed forms	283	48R$^+$
Register	235	
Strip feeding	400	586$^+$
Tam-o-shanter	D 2	255
Tambourines	84	418
Tamper (See Rammer)		
Blasting charge	86	20.11$^+$
Earthenware pipe	425	456
Female	425	425$^+$
Male	425	469
Railway track	104	10$^+$
Road or earth	404	133
Tobacco		
Pipe	131	248
Pipe combined	131	184.1
Pipe cover combined	131	177
Well torpedo combined	102	304
Tamper-proof Packages	206	807*
Fasteners with anti-tamper means	269	910*
Tampions (See Tompions)		
Tampon	604	358$^+$
Depositor	604	11$^+$
Making	28	118$^+$
Sanitary	604	358$^+$
Materials	604	367$^+$
X-art collection	604	904*
Tandem		
Cycles	280	200$^+$
Design	D12	109
Motor	180	219$^+$
Three tandem wheels	280	239
Cylinder internal combustion engine	123	57R
Horse hitch	278	3$^+$
Power plants	60	698$^+$
Two wheeled skates	280	11.23
Tang		
Drill	408	213
Socket		
Barbed or pronged tang	279	104
Molded or cast in tang	279	105
Offset tang	279	93$^+$
Shouldered tang	279	89$^+$
Tool handle fastenings	403	
Tank (See Vessel)	220	
Cars, trucks, ships		
Ballasting	114	125
Grain tank body	296	15
Marine tanker	114	74R
Railway car	105	358$^+$
Railway tender	105	236
Self levelling vehicle	280	7
Tank truck	280	5R
Vessel raising or docking air tank	114	52$^+$
Water wagon	280	5R
Chemical reaction	422	
Cooled heat transfer, liquid containing	62	430$^+$
Commodity contacting	62	373$^+$
Liquid congealing	62	356
Depth gauges	73	290R$^+$
Filling or emptying with fluids	137	561R$^+$
Fire extinguishers	169	
Fluid distribution	137	561R$^+$
Fowl and hog scalding		
Fowl	17	11.2
Hog	17	15
Hog scalder with scraper	17	13$^+$
Fuel with burner	431	344
Heated	431	203$^+$
Gas and liquid contact		
Contacting	261	119.1$^+$
Liquid supply	261	72.1$^+$
Gas holding	220	3
Heating and illuminating gas	48	174$^+$
Package	206	.6$^+$
With dispensing	222	3$^+$
Gasoline holding for motor vehicle	280	5R
Glass melting	432	195
Heat exchange		
Trickler	165	118
With agitator or conveyer	165	109.1
Heating system expansion	237	66
Ice mold	249	117$^+$
Lined metallic	220	415$^+$
Liquefied gas storing	62	45$^+$
Liquid heating	122	
Cooking vessels	126	344$^+$
Lock valve combined	137	384.2$^+$
Making		
Outlet nozzle attaching	29	157C
Sheet metal	413	1$^+$

	Class	Subclass
Masonry and concrete	52	264
Curved	52	245
Material or article charging or discharging	414	288
Material separating		
Liquid separation	210	
Solids	209	
Military	89	36.8
Endless track drive	180	9.1+
Five or more wheels	180	22+
Gun mounts	89	40.3
Steering by driving	180	6.7
Toy	446	433
Model basin or testing tank	73	148
Mold	249	144+
Nuclear reactor combined	376	403
Oil tank with fire extinguisher	169	66+
Photographic developing	354	331+
Rack for	211	71+
Sand blast equipment	51	436
Self healing	428	912*
Settling (See settling tanks and chambers)		
Shower bath	4	602+
Spring wheel air tank	152	10
Stage or theatrical	272	26
Stock material	428	35
Sugar starch and carbohydrate treating	127	
Supply for ambulant discharge sprayer	239	302+
Swimming pool	4	488+
Water purification	210	169
Tank-type carrier truck	D12	95
Tower	52	40
Underground	405	53+
Volume or rate of flow meter	73	217+
Washing		
Brushing scrubbing general cleaning	15	
Clothes washing machines & tubs	68	
Dish and solids washing machine	134	
Water closet and urinal flushing	4	353+
Disinfection	4	222+
Wheelwright tire setting	157	7
Wooden	217	4
Tanker	114	74R
Tannic Acid	560	68+
Tanning	8	94.19 R+
Apparatus	69	29+
Descaling agent ctg	252	83+
Electrolytic	204	135
Fish skins	8	94.12
Hides	8	94.19 R+
Processes manipulative	8	150.5
Tannins	560	68
Glucosidal	536	4.1+
Tanning with	8	94.19 R+
Tantalate Ceramic Compositions	501	134
Tantalum		
Alloys	420	427
Compound stock	428	662
Electrolysis	204	
Inorganic compound recovery	423	62
Misc. chemical manufacture	156	DIG. 87
Tap		
Barrel	137	317+
Bung	217	98+
Changer transformers	336	137
Transformer systems	323	
Connector electric	439	638+
Dancing shoe	36	8.3
Gas taps illumination	362	381
Piercing electrical connector	439	387+
Pipe	408	215
Screw threading	408	215+
Machine	408	72R+
Wrenches ratchet	408	120+
Wheel bolt	301	117
Tape (See Adhesive; Strips; Taping)		
Adhesive	428	343+
Processes of coating	427	207.1+
Box making	493	116+
Cutting to sever	83	
Inserting in apparel	223	50
Needles	223	103
Record	360	134
Magnetic		
Coated	360	134+
Recorded, moving head	360	83
Recorded, stationary head	360	90
Measuring	33	137R+
Design	D10	71+
Reel	242	84.8+
Package	206	389+
Tape cartridge package	206	387
Record medium	360	134
Recorders sound transport system	360	90+
Recording head	360	110+
Seed	47	56
Machine	53	545+
Planting	111	14+
Sound reproducing, reel drive means	242	179+
Spinning twisting apparatus	57	31+
Spinning twisting process	57	31
Coating or impregnating	57	295+
Spool holders	D19	67+
Strand structure spun or twisted	57	260
Coated or impregnated	57	259
Covered or wrapped	57	235
Tape deck	D14	2+
Tearing or breaking	225	
Container opening	206	605+
Telegraphic	178	111+
Ticker	178	34
Taper Forming (See Pointing)		
Gauges	33	531+
Needle and pin making	163	
Paper tube making	493	296
Pencil (See pencil sharpener)	30	451+
Machine	144	28.1+
Taper tube making	29	DIG. 41
Tapering resistor	338	217+
Rheostat	338	138+
Thread milling	409	65+
Turning	82	15+
Wood	142	
Tapestries	139	408
Taping	242	7.1+
Hollow object	242	7.23
Pail or pipe	242	7.22
Laminating applying	156	
Object making	242	7.2+
Ring winding combined	242	6
Shoe upper making	12	59.5
Strip pasting and applying	156	
Tappers Electric Line	439	387+
Tappet		
Plow	171	71
Poppet valve operating internal combustion engine	123	90.48+
Tapping		
Container with cutter or punch	222	81+
Drilling and	408	
Fluid handling system	137	317+
Mains under pressure	137	317
Metallurgical plugging and	266	271+
Power plant exhaust re expansion	60	677+
Screw threading internally	408	222
Tar (See Pitch)		
Compounds containing		
Abrasive	51	305
Biocidal	424	
Coating or plastic	106	
Fuel liquid	44	61
Polymer ctg (See synthetic resin or natural rubber)		
Distillation	208	42+
Destructive	202	
Production by	201	
Mineral oil	208	22+
Making, treating and recovery	208	39+
Tar Sand		
Mineral oil from	208	400+
Treatment with liquid	208	390+
Target		
Aerial projectile type	273	348.1+
Projector combined	273	317+
Tethered projectile combined	273	331+
Thermal	273	348.1
Animal holder and releaser	119	15.6
Bowling	273	41
Cathode ray	313	461+
Design, hunting or shooting	D22	113+
Disk		
Holder and releaser	124	43
Magazine	124	46+
For nuclear energy treatment	376	156
Golf (See also golf)	273	181R
Moving	273	359+
Surface	273	127R
Surveying	33	293
X ray	378	143
Tartaric Acid	562	585
In drug	514	574
Tassel (See Product)	428	28
Garment trimming	D 5	7+
Making and attaching	223	46
Making	28	147
Ornamented handle	16	121+
Shoe lace tip	24	143R+
Tatting		
Apparatus	87	52
Shuttles	87	58+
Shuttle winding	242	52
Tattooing Pen	81	9.22
Taurine	260	513R
Tawing	8	94.29+
Taxidermy, Teaching	434	296
Exhibition panel	40	160
Taximeter		
Computer controlled, monitored	364	467
Fare recorder	346	
Register combined	346	15+
Fare register	235	33+
Tea	426	592+
Apparatus general	99	485+
Bag	426	77
Design	D 1	199
Balls	426	77+
Extracts	426	597
Infusers	99	279+
Ddesign	D 1	199
Teaching Devices	434	
Audio recording & visual means	434	308+
Audio recording combined	434	319+
Audio visual machines	D19	60
Blocks & cards	434	403
Dental	434	263+
Design	D19	
Graphic materials	434	314+
Design	D19	62+
Image projector combined	434	324+
Keyboard operation	434	227+
Music	84	470R+
Panel, chart or graph	434	330+
Science demonstrator	D19	62+
Telegraphy	178	115
Toy	D21	59+
Vehicle dual control		
Aircraft	244	229
Automobile	180	322
Teapot		
Handle, asymmetric	D 8	DIG. 9
Tear Gas		
Gun	42	1.8
Tearing		
Breaking or	225	
Container opening	206	601+
Container seam	206	631
Label pasting paper hanging	156	510
Machines	225	93+
Paper gauge	225	91+
Paper straight edge	225	91
Sheets, strands or webs	225	
Tear strip		
Cigarette package opening	206	264
Envelope	493	923*
Pliable container	493	930*
Roll, strip, or sheet having	206	824*
Tensile test	73	835
Tobacco stemming	131	319+
Weakened line	206	620+
Teat Cups	119	14.47+
Pulsator combined	119	14.38
Technetium		
Organic compounds	534	14
Tedders	56	370+
Rake combined	56	365+
Teeming (See Casting)		
Teepee	135	100
Tees, Golf	273	33
Device for setting	273	32.5
Teeth (See Tooth)		
Dentistry	433	
Selection guide	433	26
Display package with	206	83
Gearing	74	457+
Lubrication	74	468
Razor guard	30	82
Static mold	249	54
Teething Devices	128	359+
Design	D24	45

	Class	Subclass
Teflon T M	526	255$^+$
Electrical cable insulation	174	110FC
Pipe & tubular conduit digest	138	DIG. 3
Surgery digest	128	DIG. 14
Telautograph	178	18$^+$
Telegraphophones	369	
Telegraphs	178	
Combined with telephone		
Alternative	379	108$^+$
Simultaneous	370	125
Earth transmission	375	6
Optical light transmission	455	600$^+$
Photo telegraphs	358	256$^+$
Radio transmission	375	
Ships telegraphs	116	21
Systems	178	2R$^+$
Pulse	375	
Train telegraphs	246	7$^+$
Writing on telautograph	178	18$^+$
Telemetering	340	870.1$^+$
Radio	128	903*
Telephone	128	904*
Telephone	379	
Answering device		
Answering-recording system	D14	4
Calling number recorder	379	142
Remote inquiry	379	76
Sound recorder or reproducer	379	70$^+$
Attachment	379	441$^+$
Base pad	248	
Index	40	336$^+$
Pad	281	15R
Roll type pad	281	11
Automatic systems	379	258$^+$
Call	379	350$^+$
With recorded message	379	69
Card	40	336$^+$
Common control	379	268
Party line	379	182$^+$
Booths	52	27$^+$
Design	D25	16
Movable wall	52	71
Calling	379	352$^+$
Card attached to telephone	40	336$^+$
Coin collectors for pay stations	194	
Coin operated	379	143$^+$
Computerized switching	379	284
Cordless	379	61
Design	D14	53$^+$
Dial		
Dial structure	379	362$^+$
Illuminated	362	24
Locking	379	445
Pulse transmitter	340	345$^+$
Self luminous	40	337
Telephone system	379	258
Directory	40	371
Earth transmission	455	40
Handset	379	433
Headgear support	379	430
Key systems	379	156$^+$
Light wave telephony	455	600$^+$
Lights for telephones	362	88
Mechanical telephones	181	138
Message counter	379	139
Muffler for mouthpiece	181	242
Over composite line used for other services	379	90$^+$
Pad or book holder combined	248	441.1$^+$
Party line	379	177$^+$
Plural phone systems	D14	58
Push button call transmitters	379	368$^+$
Radio transmission	455	
Receiver and transmitter combined	379	433$^+$
Repeaters	379	338$^+$
Dial pulse	379	341$^+$
Conversion	379	339
Voice frequency	379	338$^+$
Repertory dialers	379	355$^+$
Sets	379	419$^+$
Sterilizer	379	439
Supports	379	454$^+$
Switchboard	379	319$^+$
Switches	379	422$^+$
Telemetry	128	904*
Toy telephones	446	141
Train telephony	246	7$^+$
Telephony Ultrasonic	367	
Telephotography (See Television)		
Telescopes	350	537$^+$
Design	D16	132
Fiducial for sighting	356	247$^+$
Lens	350	409$^+$
Telescopic and Telescoping Devices		
Antenna	343	901$^+$
Awning	160	62
Frames adjustable size	160	372$^+$
Multipart outrigger	160	80
Roll type multipart outrigger	160	71
Bag or garment reversing inside out	223	40
Chute	193	30
Expansible chamber device	92	51$^+$
Freight car stakes	105	389
Garment or clothes hanger	223	89$^+$
Coat or dress	223	94
Gas holders	48	176$^+$
Gun sight	33	245
Design	D16	132
Ladder	182	195
Luggage	190	104$^+$
Metallic receptacles	220	8
Motor	91	169
Pipe joint	285	298$^+$
Ball and socket	285	165
Pipes and tubular conduits	138	120
Racks	211	195$^+$
Railway car vestibule connections	105	17
Railway tie	238	73
Screw type jacks etc	254	102
Spotlight control or mounting	362	418$^+$
Track sander	291	45
Trunk	190	22
Upending bedstead	5	157
Wardrobe	190	15R
Teletype T M	178	
Television	358	
Accounting for use	358	84
Band with reduction	358	133$^+$
Cabinet	312	7.2
Camera		
Color	358	41$^+$
Submarine hook & grapple for	294	66.2
With recorder	358	906*
Channel selector	116	DIG. 31
Color	358	1$^+$
Holographic	358	2
Recording or reproducing	358	310$^+$
Drop out correction	358	314
Editing	358	311
Photographic	358	332$^+$
Recorder fault correction	358	327$^+$
Slow motion	358	312
Time control	358	320$^+$
Signal processing	358	21R$^+$
Computer application of special purpose television systems	358	93
Computer TV applications	358	903*
Crt circuits, focusing the ray	315	382.1
Current supply, diverse type	315	169.1
Degaussing	361	150
Design	D14	77
With communications equipment	D14	54
With phonograph or radio	D14	21
Editing color record	358	311
Editing dynamic magnetic record	360	14.1$^+$
Fiber optics	358	901*
Freight car identifier	358	93$^+$
Frequency response modification	358	904*
Ghost elimination	358	905*
Monochrome to color	246	200R
Magnetic recording of signal	360	33.1$^+$
Misc. electron space discharge device system with TV signal output	328	187
Panoramic	358	87
Photochromic	358	902*
Photographic recorder	358	345$^+$
Color	358	332$^+$
Picture tube	313	364$^+$
Focus coil	313	442
Focus coil per se	335	213$^+$
Photographic screen making	430	23$^+$
Etching	156	644$^+$
With filter on face	313	465$^+$
Program indicator	40	901*
Pseudo color	358	81$^+$
Radar combined object detection	342	55
Receivers	358	188$^+$
Remote control	358	194.1
Stereo control	358	188
Record editing	360	14.1$^+$
Recording or reproducing	358	335$^+$
Camera with recorder	358	906*
Color	358	310$^+$
Noise or dropout correction	358	336
Noise reduction	358	340
Pause control	358	908*
Photographic	358	345$^+$
Photographing displayed image	358	244$^+$
Signal content changing	360	9.1$^+$
Thermoplastic	358	344
Track skippers	358	907*
Signal processing	360	33.1$^+$
Slow motion recording	360	10.1$^+$
Color	358	312
Special systems	358	93$^+$
Stereo color	358	3
Stereoscopic	358	88$^+$
Support stand	D34	12$^+$
Synchronizing	358	148$^+$
Training mechanism for ordnance	89	41.5
Transmitter	358	186$^+$
Tube	220	2.1 A
Tuners	116	DIG. 29
Remote tuner	116	DIG. 3
Two-way over telephone	358	85$^+$
Underwater transmitter	405	185
Video signal processing	358	160$^+$
Wired broadcast systems	358	86
With sound	358	143$^+$
X-ray television	378	99
Telltale (See Indicator; Sampler; Testing)		
Toller	73	863$^+$
Tellurions	434	284
Tellurium (See Selenium and Compounds)		
Misc. chemical manufacture	156	DIG. 72
Telomer (See Alkylation)		
Temperature Regulation or Modifying (See Cooler & Cooling, Furnace, Heat, Stove)	236	
Animal watering devices	119	73
Arc lamp	314	26$^+$
Beverage apparatus infuser	99	281$^+$
Carbon compound chemistry processes	260	700
Compensating, supports	248	DIG. 1
Cooking apparatus	99	331$^+$
Cooking apparatus and material control	99	326$^+$
Dispensing automatic control temperature responsive	222	54
Distillation apparatus condensing	196	141
Distillation apparatus vaporizing	196	132
Drying and gas or vapor contact		
With solid apparatus	34	43$^+$
With solid process	34	26$^+$
With solid process	34	30$^+$
Electronic tube	313	11$^+$
Electronic tube cooling combined circuit element	315	32$^+$
Fastener made of thermo-responsive memory material	269	909*
Fermentation processes	435	3
Heat exchange	165	13
Internal combustion engines	123	41.1$^+$
Lamp or electronic tube system	315	112$^+$
Magnets and electromagnets	335	217
Mercury vapor lamp	313	11$^+$
Pressurized & high temperature liquid treatment of textiles	8	DIG. 16
Sulfur dioxide	423	539
Sulfur trioxide	423	523
Solid material comminution or disintegration	241	17
Solid material comminution or disintegration	241	23
Solid material comminution or disintegration	241	65$^+$
Stabilized oscillator	331	176
Superheater	122	479R
Tobacco treatment	131	303
Water heating vessels	126	374
X ray tube	378	127
Tempering (See Annealing)		
Appliance design	D15	144
Hard glass	65	114
Apparatus	65	348$^+$
Template		
Design	D10	64
Design, artistic	D19	39
Temples		
Eye glass or spectacles	351	111$^+$

	Class	Subclass
Fabric manipulation	139	292$^+$
Weft manipulation on	139	266
Templet (See Pattern)		
Abrading machine		
Peripheral face	51	100R$^+$
Radial face	51	127
Cutting torch	266	58$^+$
Drilling	408	3
Gauge	33	562$^+$
Gear cutting, milling, planing	409	125$^+$
Toilet	132	88.5
Turning	82	14R
Taper	82	17
Woodworking	144	144R$^+$
Tempo Marks on Player Rolls for Automatic Musical Instruments	84	164
Tenderizer		
Kitchen hand tool	D 7	101
Meat tenderers	17	25$^+$
Tenders		
Mother ships etc	114	258
Railway	105	231$^+$
Boiler feed heater	122	434
Draft connections	213	2$^+$
Safety bridge cab and	105	459
Vestibule connection cab and	105	9
Traction engine	280	400$^+$
Tennis		
Ball	273	61R$^+$
Nets courts etc	273	29B$^+$
Racquet	273	73R$^+$
Design	D21	212$^+$
Restringing tension meter	73	862.43
Shoes	36	114$^+$
Simulated game	273	85R
Tenoning		
Chair round with sawing	144	8
Wheel spoke with sawing	144	18
Wheel with boring	408	31$^+$
Wood boring and	408	31$^+$
Woodworking	144	198R$^+$
Cutter tool	408	199$^+$
Turning	144	205$^+$
Tens Carry Mechanism for Registers	235	133R$^+$
Tensile Testing	73	826$^+$
Tensiometer	73	862.39$^+$
Tension (See Stretching)		
Cable, cord, band, rope, sheet, or strand		
Band saw	83	814$^+$
Bed bottom	5	165
Bed bottom margin	5	233
Bobbin and cop winding	242	45
Cable railway	104	117
Cutting combined	83	
Drier heating control	34	49
Electricity thermal fuse	337	190
Elevator rope drive	187	22
Fabric holder for cleaning wiper	15	233
Fabric reeling and unreeling	242	75
Flexible material	254	199
Hack saw	144	156
Harvesting compressing and binding	56	450
Harvesting reel for standing grain	56	224
Laying	405	177
Underwater	405	165
Sewing machine	112	
Ships cable	114	213$^+$
Ships sail rigging	267	69$^+$
Spring devices	267	69$^+$
Strand brakes	188	65.1
Strap	24	68R$^+$
Strop	76	81$^+$
Textile beaming	28	194$^+$
Textile braiding carrier	87	56$^+$
Textile braiding netting	87	61
Textile knitting	66	146
Textile knitting relief	66	160$^+$
Textile spinning driving band	57	105
Textile weaving	139	
Traveling band abrading machine	51	148
Winding and reeling	242	147R$^+$
Wireworking	72	183$^+$
Clothesline tighteners	211	119.1$^+$
Comb cleaner strands	132	119
Dams and levees with stays	405	113
Fabric, sewing machine	112	121.26
Gates	49	34
Harvester endless cutter	56	292
Land vehicle wheel	301	55$^+$
Compression combined	301	54
Metal working, tension applied	29	DIG. 42
Plural blade cutting tool	30	266$^+$
Scissors or shears	30	208$^+$
Railway draft appliance cushion	213	42
Railway dumping car door actuator	105	302
Razor edge	30	52
Ship tension relievers	114	213$^+$
Travelers	114	205
Snap hook tension operated lock	24	242
Spoked wheel assembling	29	159.2
Spring	267	69$^+$
Spring wheel	152	21$^+$
Trolley collector stand	191	70
Umbrella rib	135	23
Vehicle fender	293	40
Vehicle leaf spring	267	48
Tent	135	87$^+$
Cape convertible to	2	89
Convertible to diverse article	135	95
Design	D21	253$^+$
Knapsack convertible to	224	154
Oxygen	128	205.26
Tenter or Tentering		
Frame for textile sheets	38	102$^+$
Textile web running length	26	89$^+$
Biaxial	26	73
Heater or dryer combined	26	92
Web condition responsive control	26	76
Terminals		
Actuating nut adjustable wrench	81	168$^+$
Adapters	429	121
Air		
Electrical	174	4R
Lightning protection type	174	2$^+$
Attachments or feet for ladders	182	201$^+$
Safety devices	182	108$^+$
Cock for fluid nozzle	239	392
Connections	429	
Seal feature	429	185
Data processing equipment	D14	100
Electric (See connection, --; socket)		
Battery	429	179
Conductor end structure	174	74R$^+$
Conductor fluid insulated type combined housing	174	18
Conductor fluid insulated type end structure	174	19
Conductor holding on insulator	174	169$^+$
Cord conductor	439	874$^+$
Cord conductor	439	877$^+$
Electric	439	884$^+$
Fuse cartridge	337	248
Gas and vapor lamp	439	611$^+$
Housing conductor	174	60
Incandescent lamp	439	611$^+$
Insulator combined	174	176$^+$
Meter enclosure for	361	364$^+$
Meter type	324	149
Sheet metal manufacture	29	874$^+$
Telephone multiple switchboard	379	310$^+$
Electrical manufacture of	29	874$^+$
Gate valve pivoted	251	300
Gate valve reciprocating	251	147
Hook for railway dumping car door actuator	105	303
Lacing cord and rope holder	24	117$^+$
Pneumatic dispatch system	406	
Railway	104	27$^+$
Railway revolving passenger transfer	104	21
Switchboard	361	331$^+$
Termite Shield	52	101
Terneplate	428	648
Terpene (See Alicyclic Hydrocarbon)	585	350$^+$
Alcohols	568	875
Esters of lower fatty acids	560	249
Keto	568	338$^+$
Natural resin combined	260	101
Terpin Hydrate	568	832
Terra Cotta	501	141$^+$
Tile, hollow	52	606$^+$
Terrain Model, Educational	434	130
Terrarium	47	69
Design	D11	145
Terrazzo	52	315
Pavement	404	47$^+$
Revealed strips	52	318
Strips	404	47$^+$
Terrets	54	63
Terry Cloth	139	396
Looms	139	25$^+$
Test Tube Tray or Rack	D24	32
Tester		
Electrolyte conductivity	324	439$^+$
Tooth pulp	433	32
Testing (See Definition Notes)	73	
Amplifier condition	330	2
Check controlled apparatus		
Fraud prevention	194	302$^+$
Chemical		
Apparatus	422	50$^+$
Composition	436	
Electrolytic apparatus	204	400$^+$
Electrolytic process	204	1T
Fermentative analysis	435	4$^+$
Process	436	
Cooker combined	99	342$^+$
Coordination	434	258$^+$
Diagnostic		
Composition for	424	2$^+$
Dryer	34	89
Earth boring combined	175	40$^+$
Electric (See also manufacturing of electric devices infra)	324	
Battery	429	90$^+$
Electrolytic	204	400$^+$
Lamp and electronic tubes	324	403$^+$
Meters	324	74$^+$
Meters with amplifier condition testing	330	2
Meters with amplifiers	324	123R$^+$
Ore detection	324	323$^+$
Radiant energy	250	472.1
Telephones and telephone circuits.	379	1$^+$
Fertility tests	435	806*
For high frequency systems	324	58R$^+$
For wire telegraphy	178	69M
Gas separator combined	55	270
Hardness	73	78$^+$
Instruments for	D10	46$^+$
Insulation	324	500$^+$
Manufacture of electric devices	29	593
Electric lamp	445	
Apparatus	445	63$^+$
Process	445	3
Space discharge devices	445	
Apparatus	445	63$^+$
Process	445	3
Material by sifting	209	237
Milk	422	74
Nuclear reactor condition	376	245
Optical	356	
Lens	356	124$^+$
Papers for	435	805*
Plugs for pipes	138	90
Pulse or digital communications	375	10
Radar systems	342	165
Radio receiver	455	226
Radio transmitter	455	115
Radio tube	324	405$^+$
Refrigeration	62	125$^+$
Seed	47	14
Stress strain of materials	73	760$^+$
Thermal	374	
Flammability	374	8
Tetherball Game	273	414
Tethers		
Pipe joint parts	285	119
Tool	279	1R
Tetracyclines	260	351.1
Adducts or complexes	260	351.2
In drug	514	152
Modified A-ring other than 4-dedimethylamino	260	351.4
Modified D-ring other than 7-halo	260	351.3
Tetrafluoroethylene	570	136
Preparation	570	153$^+$
Solid polymers (See also synthetic resin or natural rubber)	526	255
Tetrakisazo	534	806$^+$
Tetralin (See Aromatic Hydrocarbon)	585	400$^+$
Tetramer (See Polymer)		
Tetramisole	548	155
In drug	514	368
Tetrazine	544	179
Tetrazoles	548	250$^+$
Tetrode	313	297$^+$
Gas	313	581$^+$
Gas	313	599
Hot cathode type	313	593
Oscillator	331	

Entry	Class	Subclass
Remote cutoff	313	297+
Tetryl	564	441
Textiles (See Cloth; Fabric)		
Antistatic	57	901*
Bicomponent material	57	905*
Bleaching & dyeing	8	
Braiding netting and lace making	87	
Central or south american patterns	D 5	20+
Chemical modification of	8	115.51+
Cleaning or laundering	8	137+
Cloth finishing (See cloth finishing)	26	
Coating	427	
Apparatus	118	
Electrolytic	204	21
Textile operation combined	28	169+
Warp preparation combined	28	178+
Container for	206	389
Counters of threads in fabric	356	242
Cyanoethylation of fibers	8	DIG. 13
Dry mixtures of textile reagents	8	DIG. 14
Dyeing & bleaching	8	
Fiber preparation	19	
Assembling	19	144
Bleaching	8	101+
Carding	19	98
Chemical modification	8	115.51+
Combing	19	115R
Drafting	19	236
Drying	19	.27
Fluid or special treatment	19	66R
Fluid treating apparatus	68	
Ginning	19	39+
Liberating	19	1+
Manipulative fluid treatment	8	147+
Picking	19	80R
Stapilizing	19	.3+
Stopping	19	.2+
Working	19	65R+
Finishing	26	
Sheet stretching	38	102
Tenter, for webs	26	89+
Web spreader	26	87+
Web stretcher	26	71+
Flame retardant	57	904*
Fluid treating and washing	8	147+
Apparatus	68	
Carbonizing	68	2
Machines	68	3R+
Water reclaiming	68	1
Foamed or fibrillated	57	907*
Glyoxal & polyaldehyde treatment of textiles	8	DIG. 17
Grafting textile fibers	8	DIG. 18
Improving felting properties	8	112
Ironing and smoothing	38	
Ironing tables	38	103+
Pressing or smoothing processes	38	144
Smoothing implements	38	69+
Smoothing machines	38	1R+
Isocyanate & carbonate fiber modification	8	DIG. 11
Jet interlaced or intermingled	57	908*
Kier treatment	8	157
Knitting	66	
Darning	66	2
Fabric manipulation	66	147+
Fabrics or articles	66	169R+
Feeding	66	125R+
Looper pin and hook	66	3+
Needles	66	116+
Beds	66	114+
Independent needle machines	66	7+
Needle cooperating elements	66	90+
United needle machines	66	78+
Pattern mechanism	66	231+
Reknitting	66	1.5
Stopping	66	157+
Knots & knot tying	289	
Line appliances	57	906*
Lubrication	427	
Apparatus	118	
Chemical modification combined	8	115.6
Compositions	252	8.6+
Dyeing combined	8	495+
Textile operation combined	28	167+
Manufacturing	28	100+
Computer controlled, monitored	364	470+
Molten metal, bleaching digest	8	DIG. 19
Mordanting textiles	8	
Nylon	8	DIG. 21
Organic titanium compounds on	8	DIG. 5
Polyester fibers	8	DIG. 4
Polyolefin	8	DIG. 9
Polyvinyl halide esters or alchol fiber modification	8	DIG. 1
Pressurized & high temperature liquid treatment	8	DIG. 16
Pressurized gas treatment	8	DIG. 15
Printing		
Dyeing type	8	445
Printing type	101	
Reinforcing or tire cords	57	902*
Resin bleach	8	DIG. 6
Sewing threads	57	903*
Silicones	8	DIG. 1
Spinning twisting twining	57	
Apparatus	57	1R+
Processes	57	362
Strand structure	57	200+
Stock material	428	224+
Coated or impregnated	428	245+
Swelling & stretching of fibers	8	DIG. 3
Treating compositions	252	8.6+
Biocide containing	424	
Vinyl sulfones & precursors thereof	8	DIG. 2
Wave energy treatment	8	DIG. 12
Weaving	139	
Weighting or mordanting	8	443
Wire fabrics & structure	245	
Texturing Thread	28	247+
Thatch Roof	52	750
Thaumatrope	446	244
Thawing		
Antenna	343	704
Compositions	252	70
Defrosters	62	272+
Automatic	62	151
By ventilation	98	2
Frost indicator combined	62	128
Frosted window panes	52	171+
Process	62	80+
Earth	405	131
With excavating	299	3+
Engines	123	41.1+
Ice tank	62	349+
Process	62	73
Pipes	138	32+
Radiators by own steam	165	73
Theater		
Arena buildings auditoriums	52	6+
Toy	446	82+
Car drive in	52	6+
Chairs indicating or illuminating		
Electric indicator	340	667
Furniture with light	362	131
Indicators	116	200+
Curtains	160	
Derrick hoist	254	266+
Illusion devices	272	8R+
Lights	362	227+
Control systems for	315	
Dimming systems	315	291+
Switching systems	315	313+
Seat	D 6	334+
Stage appliances	272	21+
Design	D25	58
Stages	52	6+
Movable	52	29+
Tickets	283	53
Theelin	260	397.4
Theft		
Alarms		
Electrical	340	568+
Mechanical	116	75
Prevention, motor vehicle	180	287+
Theine	544	274
Theobromine	544	274
Theodolites	33	290+
Design	D10	66
Horizontal and vertical angle	33	281
Horizontal angle	33	285
Optical	356	138+
Optical readout	33	1T
Theophyllines	544	267+
In drug	514	263
Therapeutic		
Lamps	250	504R+
Vitamins	424	
Thermal		
Activating adhesion in labeling	156	DIG. 21
Activating closure	220	201
Activator for bank protector	109	33
Battery and cells	136	200+
Compositions or charges	149	
Controlled		
Boiler feeders	122	451.1+
Closure	49	1+
Dampers	126	287.5
Electrical alarms	340	584+
Electrical signals	340	584+
Flexible panels	160	1+
Freeze responsive drain valves	137	59+
Mechanical alarms	116	101+
Mining	299	3+
Non-fusible closure release	49	2
Refrigeration, heat transmission	62	383
Temperature regulation	236	
Timers	60	516+
Vertically sliding sash		
Weighing scale compensator	177	226
Electric motors	310	306+
Electrical devices		
Batteries	136	200+
Electromagnet and thermal current switch	335	141+
Lightning arresters	361	117+
Meters	324	105+
Multiple circuit switch	200	3
Pyromagneto electric	310	306+
Radio detectors	329	202+
Thermal current switch	337	14+
Thermal switch	337	298+
Expansive instruments	33	DIG. 19
Gas separation	55	209
Processes	55	81
Heat exchange devices	165	
I-aging composition	430	964*
Impedance in amplifier system path	330	143
Insulated rigid containers	493	903*
Liquid level or depth gages	73	295
Motor	60	516+
Neutrons	376	158
Neutrons	376	347
Reactive building component	52	232
Recording	346	76R
Surgical applications	128	362+
Inhalers	128	204.17
Instruments	128	303.1+
Kinesitherapy	128	24.1+
Medicators	604	20+
Orthopedics	128	68.1
Switches	337	298+
Electrically heated	337	14+
Tests	374	
Borehole formation logging	73	154
Calorimeters	374	31+
Gas analysis	73	25+
Volume or rate of flow fluid meters	73	204
Thermionic Devices	313	
Circuit element combined structurally	315	32+
Electric lamps	313	
Thermistor	338	22R
Systems	323	369
Thermit T M		
Casting	164	53
Compositions	149	37
Explosive or thermic	149	37+
Composition containing	149	37+
Track joint welder	104	15
Type reduction	75	27
Thermocouple		
Body treating electric elements		
Body wear	128	391+
Electrical applicators	128	783+
Electric motor	310	306+
Motive power system	318	117
Electric relay circuit	361	160+
Electricity to heat to electricity	322	2R
Generator systems	322	2R
Making (process)	437	
Radio receiver detector	329	201
Refrigerator	62	3
Peltier effect battery	136	203+
Structure	136	200+
Temperature control system	236	69+
Thermometer	374	179+
Combined	374	141
Thermoelectric		
Battery	136	200+
Controls burner	431	80
Motors	310	306+
Expansion and contraction type	60	516+
Systems	318	117

	Class	Subclass
Thermoforming		
— Directly applied fluid pressure differential	264	544
Thermolysis (See Cracking)		
Thermometers		
— Bimetallic stock	428	616^+
— Case for	206	306
— Design	D10	57^+
— Electrically heated	60	516^+
— Expanding fluid	374	201^+
— Expanding solid	374	187^+
— — Bi-metallic	374	204^+
— Heated	60	516^+
— Hydrometer combined	73	442^+
— — With float	73	449
— Liquid gauge combined	73	292
— Making by glass working	65	62
— — And calibrating	374	1^+
— Structure	374	100^+
— Testing of	374	1^+
— Thermostat	D10	50
Thermomigration	437	14
Thermonuclear Reaction Fusions	376	100
Thermophore		
— Compresses	128	399^+
Thermophoric Compositions	252	70
Thermopiles	136	224^+
Thermoregulators	236	
Thermos T M Bottle	215	12.1^+
— Ceramic	215	12.1^+
— Design	D 7	77^+
— Metallic	220	420^+
— — Vacuumizing	220	1R
— Prefilled	220	411^+
Thermostat (See Thermometers)		
Thermostatic		
— Control and regulation		
— — Arc lamps	314	89^+
— — Burner	431	18^+
— — Refrigeration	62	132^+
— — Temperature and humidity	236	
— — Valve cutoff	137	457
— Materials		
— — Composition	252	70
— — Compound metal	428	616^+
— Switches	337	298^+
— — Electrically heated	337	14^+
— — Electromagnet	335	141^+
Thetine	549	89
Thiadiazines	544	8^+
Thiadiazoles, 1,2, 3-	548	127
Thiadiazoles, 1,2, 4-	548	128^+
Thiadiazoles, 1,2, 5-	548	134^+
Thiadiazoles, 1,3, 4-	548	136^+
Thialdehydes	568	20
Thiamines	544	327
— In drug	514	276
Thiazines	544	3^+
Thiazoles	548	146^+
— Azo compounds	534	751^+
— — Heavy metal ctg	534	701^+
— Synthetic rubber or natural rubber accelerator	525	349
— Thiazoles, 1, 2-	548	206^+
— Thiazoles, 1, 2, 3-	548	255^+
— Thiazoles, 1, 2, 4-	548	262^+
— Thiazoles, 1, 3-	548	146^+
Thiazolidine	548	146^+
Thiazoline	548	146^+
Thief Devices (See Sampler)		
Thienamycin	540	351
Thills		
— Coupling jacks	254	28.5
— Land vehicle animal draft	278	33^+
— — Couplings	278	52^+
— — Supports	278	81^+
— Tugs	54	50^+
Thimbles	D 3	29
— Bushing	16	2^+
— Clews	114	115
— Combined with cutting devices	7	121^+
— Conduit end lining	174	83
— Cutter tool combined with	7	121^+
— Hardware	16	108^+
— Insulator pin type socket	174	200
— Making	72	356
— Packed shaft joint	277	112
— Sewing apparatus	223	101
— — Design	D 3	29
— Stovepipe	126	314^+
— Ventilation valved	98	118

	Class	Subclass
Thinner		
— Composition coating or plastic	106	311
— Hair	30	195
— Plant	172	534
— — Cross-art collection	172	900*
Thioacetic Acid	260	502.6
— Esters	558	250
Thioacetol	568	63
Thioamides	564	74^+
Thiobarituric Acids	544	299^+
Thiobiurets	564	22
Thioborate Esters	558	285
Thiocarbamate Esters	558	232^+
Thiocarbamic	260	513.5
Thiocarbanilide	564	26^+
Thiocarbazides or Semicarbazides	564	18^+
Thiocarbazones or Semicarbazones	564	19^+
Thiocarbonic Acid	260	502.6
— Esters	558	243^+
— — Cellulose	536	60^+
Thiocarboxamides	564	74
Thiocarboxylic Acid	260	502.6
— Esters	558	230^+
— — Cellulose	536	60^+
Thiocyanate	423	366
— Esters	558	10^+
Thioethers	568	38^+
Thioglycollic Acid	562	512
— Esters	560	147
Thiohydroxamate Esters	560	312^+
Thioimidic Acid Esters	558	1^+
— Imidothiocarbonates	558	2
— Pseudothioureas	558	4
— Thiohydroximidic acid esters	558	3
Thioindigo	548	457^+
— Bicyclo ring system	549	52
— Indole nucleus ctg	548	464
Thioketones	568	20
Thiokol T M (polyalkylenepolysulfide)	528	388
Thiolsulfinate Esters	560	310
Thiolsulfonate Esters	560	307
Thiolsulfonic Acids	568	21
Thiomorpholines	544	59^+
Thionitrate Esters	558	480^+
Thionitrite Esters	558	488
Thionyl Chlorid	423	462
Thiooxamides	564	77
Thiophene	549	29^+
— Indole nucleus combined	548	464
— Phosphorus attached	549	6
Thiophenols	568	67
— Addition salts	564	280
Thiosulfate Esters	560	308^+
Thiosulfates, Inorganic	423	514^+
Thiosulfenamides	564	100
Thioureas	564	17
— Stabilized or preserved	564	3
— Synthetic rubber or natural rubber accelerator	525	352
Thioxanthenes		
— Chalcogen or nitrogen attached	549	27
Third Rail		
— Conductor	191	$29R^+$
— Shoe	191	49
— Snow excavator	37	204
— Switch	246	419
Thiuram Disulphide	564	76
Thorium	75	
— Alloys	420	1^+
— Composition radio active	252	625^+
— Compound recovery	423	
— Electrolytic production	204	1.5
— Nuclear fuel material	376	181
— Nuclear fuel material	376	412
— Nuclear fuel material	376	901
— Organic compounds	534	11^+
— Pyrometallurgy	75	84.1 R
Thread		
— Catchers		
— — Spinning and twisting frames	57	353
— Counters of threads in fabrics	356	242
— Cutters on sewing machines	112	285^+
— Dental floss holders	132	91^+
— Envelope affixing	493	376
— Frames	28	149^+
— Generating helix or thread	409	65^+
— Guides		
— — Knitting machines	66	$125R^+$
— — Spinning and twisting frames	57	352^+
— In bolt to nut lock		
— — Gripper	411	246^+

	Class	Subclass
— — Jam nut oppositely threaded nut	411	244
— — — Intersecting thread, on bolt	411	245
— — Jam nut reversed threads on one	411	243
— — Thread lock by pitch difference	411	307
— — Thread lock differential thread	411	263
— — Thread lock dissimilar threads	411	308^+
— — — One deforms other	411	309
— — — One deforms other near crest	411	311
— — Thread lock interrupted threads	411	304
— — Thread lock non-circular thread	411	282
— — Thread lock plural threads, oppositely biased	411	312
— — Thread lock rocking thread section, on bolt	411	264
— Lock, bolt or nut to substructure	411	106
— Lock, bolt to nut	411	259^+
— — Including elastic gripping	411	301^+
— Packages	206	389^+
— Protectors for spinning and twisting frames	57	352^+
— Sensing elements on sewing machine	112	278
— Sewing	57	903*
— Sewing machine handling of	112	
— Spooling machines	242	16^+
— Textile manufacture		
— — Coating or dyeing combined	28	217^+
— — Finishing	28	217^+
— — Loom shuttle implements	139	381^+
— — Spinning	57	
— Winding and reeling	242	
Threaded Device (See Bolt, Nut, Screw)		
— Bottle or jar cap	215	329^+
— Bottle or jar stopper	215	356
— — Fastening	215	283^+
— — Interrupted thread fastening	215	281
— Chuck or socket adjusting	279	
— Fastenings	411	
— — Bolt or nut substructure lock	411	81
— — Bolt to nut lock	411	190^+
— — Chuck grip	279	7
— — Element, deformed	411	333^+
— — Making	10	2^+
— — Nuts	411	427^+
— — Wire nail	411	439^+
— Making		
— — Die cutting	408	215^+
— — Drilling and tapping tool	408	215
— — Glass molding presses	65	309^+
— — Grinding apparatus	51	95R
— — Grinding methods	51	288
— — Metal molding apparatus	164	216
— — Milling	409	65^+
— — Platen rolling	72	88^+
— — Rolling	72	104
— — Spirally corrugating tubes	72	103^+
— — Static mold	249	59
— — Swaging	10	$152R^+$
— — Tapping	408	215^+
— — Turning	82	5
— Pipe joint	285	390^+
— — Thimble	285	386^+
— Protectors		
— — Disassembly impact receiving	29	277
— — Pipe	138	96R
— Safe closure	109	72
— Tool adjusting	81	
Threader and Threading		
— Buttons and staples	227	32
— — Setting combined	227	34
— Curtain rod	223	105
— Loom shuttle	139	221^+
— — Implements	139	381^+
— — Replenishing loom	139	259
— Needle		
— — Hand	223	99
— — Sewing machine	112	224
— Sewing machine needle setting combined	112	225
— Stuffer box crimpers	28	268^+
— Textile twisting	57	279^+
Three Dimensional Object		
— Imitation or treated natural	428	15
Three Dimensional Photography (See Stereoscopy)		
Three Phase		
— Current meters	324	107^+
— Motors	310	159^+
Threonine	562	570
Threshers	130	
— Combines	56	122^+

	Class	Subclass
Motorized	56	14.6
Design	D15	10$^+$
Thresholds for Buildings	49	467$^+$
Throstle Frames	57	
Throttle (See Valve)		
Engine speed responsive	60	906*
Governors for internal combustion combustion engines	123	319$^+$
Nozzle	239	546
Pneumatic piano	84	41
Pump regulator	417	1$^+$
Train control	246	186$^+$
Thrower (See Projector)		
Belt	474	101
Conveyor	198	638$^+$
Fluid sprinkling, spraying and diffusing	239	
Or splasher	239	214$^+$
Gaseous suspension classifier	209	153
Grenade		
Firearm attachment	42	105
Hand	102	486
Ordnance	89	1.1
Mechanical	124	
Pneumatic conveyor	406	71
Point switch railways	246	264$^+$
Scattering non-fluid material	239	650$^+$
Water guns	222	79
Throwoff		
Cradle rocker	5	106
Printing machine		
Bed and cylinder	101	283$^+$
Multicolor bed and cylinder	101	191$^+$
Thrum Mechanism	112	69
Thrust		
Antifriction bearing	384	590$^+$
Radial thrust bearing combined	384	452$^+$
Balanced pistons	92	126
Bar		
Typewriter	400	390$^+$
Bearings plain	384	420$^+$
Railway car journal	384	188
Compositions for developing	149	
Thumbtack	411	439$^+$
Applier	81	44
Remover	254	18$^+$
Thymol	568	781
Thyratron (See Space Discharge Device Space Discharge Device)		
Thyroid Extracts		
Medicine containing	424	111
Thyroxin	562	447
Ticket	283	
Booth	D25	16
Cases for pocket	206	39
For card	150	147$^+$
With cutter	30	124
Dispensing package	206	39.4$^+$
Fare boxes	232	7$^+$
Holders	206	39$^+$
Railway cars	40	10R$^+$
Letter box deposit or collection record	232	18
Machine	221	
Magazine	312	50
Making		
Pin ticket tag applying	493	375$^+$
Printing	101	66$^+$
Registering device	235	32
Theatre	283	53
Ticklers, Carburetor Primers	261	DIG. 73
Ticktack	446	417
Tiddlywink Game	273	353
Tide or Wave Motors	60	497$^+$
Channel	405	76$^+$
Combined with other	60	495$^+$
Gravity	60	639$^+$
Weight	185	30
Fluid current motors	416	
Gas pump actuator	417	330$^+$
Liquid piston type	417	100
Generator drive	290	53
Electric control combined	290	42
Liquid pump drive	417	330$^+$
Marine propulsion	440	9$^+$
Tiered Dish	D 7	4
Ties		
Air mattress parallel strip	5	457
Animal hitching	119	118$^+$
Apparel clasp or tack	D11	202$^+$
Apron	2	52
Bale and package	24	16R$^+$
Making of wire	140	73
Between building component	52	582$^+$
Building type	52	698$^+$
Collar combined	2	130
Design	D 2	602
Dowels		
Fence		
Panel	256	28$^+$
Rail	256	62$^+$
Form panels having	249	40$^+$
Forms	223	82
Neck	2	144$^+$
Design	D 2	605$^+$
Design, ascot	D 2	605$^+$
Fastener	24	49R$^+$
Fastener design	D11	202$^+$
Propped ladders	182	175$^+$
Railway surface track	238	29$^+$
Cast rail seats	238	270
Integral rail clamps	238	350
Tie plates	238	287$^+$
Wrought rail seats	238	273
Rod	249	213$^+$
Rod control of lights	362	57$^+$
Ski	280	814$^+$
Stretching	223	65
Vehicle body	296	40
Wire fabric	245	3
Tighteners	254	
Belt, band, chain, cord, hoop, slat, strap, (See pulls & pulling)		
Article carrier strap	294	150
Baling press bale band	100	29$^+$
Barrel hoop	217	94$^+$
Bed bottom elastic slat	5	211$^+$
Belt	24	32
Clothesline	211	119$^+$
Drive belt	192	11
Flexible material	254	199$^+$
Gearing belt	474	101$^+$
Locomotive drive belt	105	105$^+$
Musical drum cord	84	413
Package strap	24	19
Resilient wheel antiskid	152	217$^+$
Strap	24	68R$^+$
Belt conveyor	198	813$^+$
Pipe coupling	285	309$^+$
Saw handle	30	517$^+$
Wheel tire machine	157	8$^+$
Wheel with compression spoke	301	75$^+$
Tile		
Adhesively secured	52	390$^+$
Asphalt	106	273R
Countertop	D23	308
Cutting	125	
Design	D25	138$^+$
Digest	33	DIG. 2
Facing (See facer construction)	52	384
Filler block between sustainers	52	320$^+$
Floor covering	D 5	
Game	273	292$^+$
Hollow block	52	606$^+$
Mah jong	273	292$^+$
Molding machine	425	134$^+$
Molding machine	425	218$^+$
Mosaic design	428	44$^+$
Roof	52	518$^+$
Sheet attached	52	384$^+$
Till Locks	70	85$^+$
Burglar alarm	116	76
Tiller (See Steering)		
Design	D15	11$^+$
For rudders	114	144R$^+$
Tilting Devices		
Dispensers	222	164$^+$
Dispensing stands	248	128$^+$
Dumping receptacles	414	425
Receptacle supports	248	133$^+$
Tables	108	1$^+$
Vehicle body	414	381
Self loading	414	469$^+$
Timbale Iron	99	431
Time (See Horology; Timers)		
Balls	116	200$^+$
Compression or expansion		
Sound	381	34$^+$
Time division multiplex	370	109
Constant circuit		
Controlling electronic tube	328	129.1$^+$
In oscillator	331	
Measuring and indication	368	
Design	D10	40$^+$
Electric phase indication	368	
Electrical systems	324	160$^+$
Radiant energy reflected	342	176$^+$
Railroad train passage	246	108$^+$
Railroad train passing remote point	246	123
Reaction of persons, examiner controlled	273	1R
Rotation counter	235	104
Solar observation instrument	33	268
Teaching telling time	434	304
Operated and controlled		
Aerosol dispenser	222	649$^+$
Aerosol sprayer	239	99
Battery charging or discharging	320	2$^+$
Bedstead occupant awakener	5	131$^+$
Beverage infusor	99	280$^+$
Burner	431	86
Cable winding drum	254	266
Camera	354	238.1$^+$
Cereal puffing apparatus	99	323.7
Chemical reaction apparatus	422	50$^+$
Combustion starting	431	87
Contact printer	355	115
Dispensing articles	221	15$^+$
Doors	49	29$^+$
Drying	34	53
Electric circuit breaker	200	33R$^+$
Electric demand meters	324	103R$^+$
Electric lamp and electronic tube systems	315	360
Electric relays	361	160$^+$
Electric space discharge system	315	360
Electricity motive power system	318	445$^+$
Electronic tube systems	328	72$^+$
Explosive devices	102	276$^+$
Feeding devices for animals	119	51.11$^+$
Feeding devices for fish	119	3
Fire kindler	44	36$^+$
Fluid handling	137	624.11$^+$
Fluid handling with indicator	137	552.7
Fluid sprinkling, spraying, or diffusing	239	70
Furnace damper	126	285.5
Furnace feeding	110	192
Gas contact with solids	34	53
Gas separator	55	271$^+$
Heating and cooling device	165	12
Internal combustion engine	123	198R
Limit stop	192	138$^+$
Liquid purifier or separator	210	138$^+$
Locks	70	267$^+$
Machine mechanisms trip	74	3.5$^+$
Magnets or electromagnets	335	210$^+$
Material treating apparatus	34	53
Memorandum holder	368	42
Mortar mixers	366	60$^+$
Periodic electric relays	361	160$^+$
Projection printer	355	67$^+$
Railway switches and signals interlocked	246	161
Recording devices	346	
Refrigeration	62	231$^+$
Refrigeration, automatic	62	157$^+$
Signal electrical traffic control	340	907$^+$
Signal or indicator	116	
Signal or indicator electrical	340	309.15$^+$
Solar observation instruments	33	268
Spring motor	185	27$^+$
Stove damper	126	285.5
Switch with electromagnetic operator	335	59$^+$
Switch with thermal operator	337	301$^+$
Switching systems	307	141$^+$
Telephone coin collector	379	123
Telephone toll register	379	124$^+$
Time stamps	346	80$^+$
Typewriting machine lock	400	673$^+$
Urinal flushing	4	302$^+$
Valve	137	624.11$^+$
Valve with fluid timer	251	48$^+$
Valve with indicator	137	552.7
Weight motor	185	37$^+$
Well apparatus	166	64
Work holder for manual tearing	225	8
X ray apparatus	378	96
Punch or time clock	D10	41
Switch	200	33R$^+$
Periodic type	200	19R$^+$

	Class	Subclass
Timepieces	368	
Assembly and manufacture	29	
Watchmakers tools	81	6$^+$
Calibrating and testing	73	5$^+$
Holder, not carried on wrist	224	903*
Timer (See Horlolgy; Time, Operated and Controlled)		
Camera shutter	354	226$^+$
Delayed releaser	354	256$^+$
Disparate device controlled	368	1$^+$
Dispensing	222	638$^+$
Expansible chamber motor	91	35
Food cookers	99	327$^+$
Food cookers	99	332$^+$
Food cookers	99	335$^+$
Gas separator combined	55	271$^+$
Desorbing solids	55	162
Metal founding apparatus control	164	157
Parachutes	244	150
Pneumatic dispatch systems	406	18
Rotation counters	235	104
Sandglass design	D10	44
Telephone coin collector	379	123
Telephone toll register	379	124$^+$
Temperature or humidity control	236	46R
Time measuring instrument	D10	40$^+$
Tin	75	85
Alloys	420	557
Aluminum copper	420	530
Copper	420	470
Cans	220	
Compounds		
Inorganic	423	462$^+$
Organic	556	81$^+$
Oxides and hydroxides	423	618
Stannates	423	592$^+$
Electrolysis		
Coating	204	54.1
Synthesis aqueous bath	204	120$^+$
Synthesis fused bath	204	64R
Misc. chemical manufacture	156	DIG. 11
Plate	428	646$^+$
Pyrometallurgy	75	85
Recovery from scrap		
Hydrometallurgy	75	98
Pyrometallurgy	75	64
Tine		
Hand fork tine clearers	294	50$^+$
Hand rakes	56	400.21
Harvester rake teeth	56	400
Harvesting sheaf carriers	56	478
Power driven conveyors		
Endless	198	692$^+$
Reciprocating	198	772
Tinsel	428	7
Apparatus		
Chenille type	57	24
Covering or wrapping	57	5
Chenille type strands	57	203
Christmas type decoration	D11	121$^+$
Making article of tinsel	493	955*
Tip		
Billiard cues	273	70$^+$
Canes	135	70
Cigar	131	361$^+$
Perforators or slitters	131	253$^+$
Cigar cutters	30	109$^+$
Ash receiver combined	131	233
Combined	131	248$^+$
Pocket receptacle combined	206	238$^+$
Cleaners for light burners	362	458
Crutches	135	70
Fishing rods	43	24
Hat attachments	2	190
Lacing	24	143R$^+$
Pile driving	405	253$^+$
Shoe toe	36	77R
Tobacco pipe	131	227$^+$
Tops	446	264
Umbrella	135	44
Rib	135	36R
Vehicle pole and thill	278	50$^+$
Walking aids, general	135	77$^+$
Tipcat Game	273	341
Tipping		
Can cap soldering	228	
Cigar and cigarette making	131	29$^+$
Wrapping	131	61.1$^+$
Dispensing container	222	164$^+$
Handled	222	466$^+$
Lace	29	243.57
Vehicles	414	381$^+$
Self loading	414	469$^+$
Tipple Mine Car	414	381$^+$
Tipup for Fishing	43	17
Tire		
Air impermeable liner	152	DIG. 16
Anti-skid design	D12	154
Bead spreader		
Holders for spread casings	81	15.3
Rubber tire setters	157	1.1$^+$
Tire casing spreaders	254	50.1$^+$
Bead to rim seal	152	DIG. 9
Building	156	110.1$^+$
Drum	156	414$^+$
Carcass materials	152	548$^+$
Arrangement	152	558$^+$
Carrier vehicle attached	224	42.12$^+$
Tire tool combined	157	1.11
Chain	152	208$^+$
Design	D12	154
Clamps	152	DIG. 8
Cord, reinforcing textile strand	57	902*
Cordless	152	452
Covers	150	54R
Crack resistant	152	DIG. 4
Design	D12	134$^+$
Fabrics	152	DIG. 14
Gauge & balancer	D10	86
Grooved rim	152	DIG. 17
Hub tires	152	DIG. 18
Inflation		
Gas dispensing for	222	3$^+$
Gauge	D10	86
Hose holder and fluid supply	137	355.17
Portable receptacle filling type	141	38
Valve stem closures and plugs	138	89.1$^+$
Valved stem coupling	137	231
Valves stem type	137	223$^+$
Vehicle combined devices	152	415$^+$
Vehicle combined gas pumps	417	231$^+$
Winding and reeling hose carriers	242	86$^+$
With nongaseous materials	141	38
Land vehicle wheels	301	5R$^+$
Antiskid devices	152	208$^+$
Resilient wheels and tires	152	
Leak indicator	116	34R
Electric	340	58
Locks for automobile	70	259$^+$
Making and repairing		
Boring and turning	29	29
Forging	72	305
Laminated type	156	
Apparatus for	156	394.1$^+$
Including adhesive bonding	156	110.1$^+$
Metal processes	29	159.1
Metal rolling	72	67$^+$
Metal working	29	22
Mold cores	425	26$^+$
Mold cores	425	49$^+$
Molding with heating and vulcanizing	425	28R$^+$
Molding with heating and vulcanization	425	340$^+$
Molding with heating and vulcanization	425	363
Patches	152	367$^+$
Presses for repairing tires	425	20$^+$
Roughening by filing	29	76R$^+$
Tools for repairing	81	15.2$^+$
Trimming cutter	82	
Vulcanizable gum apparatus	425	11
Vulcanizable gum apparatus	425	28R$^+$
Mold	409	902*
Mounting and demounting apparatus	157	1.1$^+$
Mounting machinery	D15	199
Overlap	152	DIG. 15
Patching	152	367$^+$
Processes	156	95$^+$
Pebble ejectors	152	DIG. 1
Peg leg	152	DIG. 6
Pneumatic	152	450
Inner tube	152	511$^+$
Process for laminating	156	110.1$^+$
Puncture filling materials for	106	33
Repairing press	425	11$^+$
Processes of making metal	29	159.1
Processes of repairing	264	DIG. 74
Racks	211	23$^+$
Velocipede type	211	20$^+$
Velocipede type wall mounted	211	19
Railway wheels		
Loose tires	295	10
Sheet metal integral web and	295	24
Tire cushions	295	11$^+$
Tire fasteners	295	15$^+$
Repair tool	D 8	31
Removers	157	
Resilient tires	152	151$^+$
Anti-skid armored	152	167$^+$
Anti-skid devices	152	208$^+$
Cushion	152	246$^+$
Cushion & pneumatic combined	152	155$^+$
Grooving	157	13
Inflating devices	152	415$^+$
Patches	152	367$^+$
Pneumatic	152	450$^+$
Repair tools for	81	15.2$^+$
Wheel securing means	152	375
Retreading	425	17$^+$
Rims	301	95$^+$
Rubber valves	152	DIG. 7
Sandwich breakers	152	DIG. 19
Slits in treads	152	DIG. 3
Spare tire compartment	296	37.3
Fender wells for	280	152.5
Protective cover for	150	54R
Split rim seal	152	DIG. 1
Static discharge	152	DIG. 2
Testing	73	146
Abrasion	73	8
Pressure gauges	73	700$^+$
Valve leak	73	48
Tire bead grommets	245	1.5
Tread pattern design	D12	136$^+$
Tubeless valves	152	DIG. 11
Valves stem guards	152	DIG. 13
Vulcanizing machinery	D15	199
Water pilled	152	DIG. 5
Wheelwright machines	157	
White sidewalls	152	DIG. 12
Tissue Equivalent Ionization Chamber	250	472.1
With tissue equivalent gas	250	336.1
Tissue or Sheet Dispenser	206	233
Dispenser, interfolded sheets	D 6	518$^+$
Holder	248	DIG. 5
Titanate Ceramic Compositions	501	134$^+$
Titanium	75	
Alloys	420	417
Composite metal stock	428	660
Metal working digest	29	DIG. 45
Compounds inorganic	23	
Compounds organic	556	51$^+$
Copper alloys	420	492
Light sensitive titanium compound	430	947*
Misc. chemical manufacture	156	DIG. 75
Organic titanium compounds on textiles	8	DIG. 5
Oxide and hydroxide	423	608
Pigments fillers and aggregates	106	299$^+$
Preparation	75	84
Electrolytic	204	64T
Pyrometallurgy	75	84
Titlers		
Movie	352	90
Still copy device	355	39
Titrating Devices	422	92$^+$
Toaster		
Bread	99	385$^+$
Automatic	99	325$^+$
Household unit	D 7	328$^+$
Tobacco	131	
Alkaloids	546	282
Barns	432	500
Cigarette making	131	280$^+$
Tobacco feeding	131	108$^+$
Commercial processing machinery	D15	10$^+$
Condition sensing		
Continuous tobacco rod	131	906*
Detecting machine malfunction	131	910*
In feed hopper	131	909*
Pneumatic means	131	904*
Radiation source	131	905*
Rejection by	131	907*
Unique characteristic	131	908*
Curer	D23	334
Derrick hoist for hangers	254	266
Design, tobacco & smoker supplies	D27	
Die cutting	83	510
Drying		
Apparatus	34	
Fluid treatment combined	131	300$^+$
Processes	131	300$^+$

U.S. PATENT CLASSIFICATIONS

Entry	Class	Subclass
Extract for use in drugs	424	197.1
Hangers	294	5.5
Harvesters	56	27.5
Insecticide from	424	197.1
Jar or pouch design	D27	43^+
Packages	206	242^+
Paper wrappers for cigar, cigarette	229	87C
Pipe literature digest	131	DIG. 1
Pipe making apparatus	425	
Plants	PLT	89
Plug or compressed shape making	131	111^+
Plug tobacco label for fastener	40	25A
Pocket receptacle for, convertible or combined	206	236
Puffing process	131	291
Fixing after puffing	131	903*
Inorganic chemical agents used	131	902*
Liquified gas used	131	900*
Organic liquid used	131	901*
Smokers appliances	131	329^+
Spinning	57	28^+
Stalk stripping	130	30R
Storage or curing building	D25	1^+
Stringers or unstringers	414	26
Supplies	D27	
Treatment	131	290^+
Vehicle rack	296	5
Tobias Acid	260	508
Toboggan (See Ski; Sled)	280	18
Design	D12	11
Water	441	65^+
Tocopherol	549	408^+
In drug	514	458
Tod, Total Oxygen Demand Testing	436	62*
Toe		
Cap protectors	36	72R
Caps	36	77R
Toggle (See Particular Device)		
Bolt	411	340^+
Forging press	72	451
Machine element	74	520^+
Toilet	132	
Aerobic decomposing	4	DIG. 12
Animal grooming	119	83^+
Design	D30	158^+
Article	D28	9^+
Container or wrapper	493	960*
Holder	D28	73
Mirror	D28	64.1
Special package	206	823*
Tooth brush	D 4	104^+
Vanity case for	206	235
Bath, closet, sink, spattoon	4	
Beauty aid demonstration	434	377
Teaching hair or wig styling	434	94
Bed	5	90
Receptacle in mattress	5	463
Bowl	4	420^+
Brushes	15	164^+
Disinfecting chemical holders	4	228^+
Brush holder	D 6	551
Brushes, bristled	D 4	130^+
Brushes, non-bristled	D28	9^+
Case design	D 3	39
Case illuminator	362	135
Childs training seat	D23	296
Deformable traps	4	DIG. 16
Disinfection	4	222^+
Electric flushing	4	DIG. 3
Electrolytic	4	DIG. 17
Grooming articles	132	
Having deformable trap	4	DIG. 13
Head covering	2	174
Heated seats	4	DIG. 6
Kit absent special toilet article, eg mirror	401	118^+
Kits	132	$79R^+$
Manicure tool	132	73^+
Oil toilet	4	DIG. 11
Paper		
Dispensing holder	206	389^+
Interleaved folds	221	48^+
Roll holders	242	55.2^+
Design	D 6	523
Prison bathrooms	4	DIG. 15
Seat	4	234^+
Pads and covers	4	242^+
Seat or cover	D23	311^+
Water additive or substitute	4	DIG. 1
Toilet Water, Perfume	D28	5
Tokens		
Checks labels and tags	283	74^+
Coin type	40	27.5
Coin	D11	95^+
Tolidine (See Benzidine)		
Tollers	73	863^+
Toluene (See Aromatic Hydrocarbon)	585	400^+
Oxidation of	562	409^+
Toluidin	564	305
Tombstones (See Monument)	52	103^+
Design	D99	17^+
Name plate for	40	124.5
Picture attachment for	40	124.5
Tompions	89	31
Tone		
Arms of phonographs		
Electrical	369	135^+
Mechanical	369	158
Control means	333	$24R^+$
Equalizer network	333	28R
With amplifier	330	
Automatic impedance control	330	144^+
Automatic thermal impedance control	330	143^+
Cathode feedback	330	94^+
Feedback	330	109^+
Input coupling impedance	330	157^+
Interstage coupling impedance	330	185^+
Output coupling impedance	330	195^+
Musical pitch (See music)		
Toner	430	105^+
Concentration control	118	689^+
Radiation imaging	430	965*
Tong (See Lazy Tongs)		
Agriculture or forestry	D 8	4
Animal catching	119	154
Extension ladder	182	157^+
Fire escapes	182	69
Fireplace equipment	D 8	52^+
Fireplace tong stand	D 6	416
Food preparation & serving	D 7	105
Handling implement	294	3
Clothes	294	8.5
Design	D 7	105
Fire	294	11
Grapple	294	119
Harvester self raking	56	165
Making	76	
Metal drawing appliance	72	290
Roofers sheet metal seaming	29	243.5^+
Tools	81	300^+
Tongue		
Depressors	128	15^+
Atomizers with	128	200.15
Design	D24	19
Envelope closure	229	84
Sealing	229	82
Harvester type	56	218
Tilting platform central cutter	56	188
Knitting needle making	163	3
Land vehicle		
Animal draft	278	33^+
Antivibrators	280	108
Attachments	280	155
Extension actuators	280	46
Trailer, adjustable	280	462
Truck	280	82^+
Lock type shoe sole sewing machine	112	33
Punching tie band	29	21.1
Railway switch point	246	442^+
Rigid vibrator music instruments	84	408
Shoe	36	54
Single throw lever actuated by	414	436
Straddle row cultivator without	172	338^+
Vehicle brake operator		
Auxiliary	188	115^+
Movable	188	119^+
Tonometers		
Hardness testing	128	645^+
Medical diagnostics	128	645^+
Musical pitch	84	454
Sphygmomanometer	128	677^+
Tonsillotome	128	309
Other treatment combined	128	303.16
Tools (See Implements; Instruments)	81	
Abrading and polishing	51	
Advancing means	173	141^+
Assembling	29	700^+
Belt lifters	474	130
Boxes	206	349^+
Chest	312	DIG. 33
Cleaning	15	$104R^+$
Cleansing means	173	57^+
Fish	17	66^+
Hog carcass scrapers	17	19
Ship hull	114	222
Closure applying and removing		
Ammunition capping	86	37^+
Ammunition shell closing	86	28
Bag with opening device	206	610
Bottle and jar closing	53	287^+
Bottle cap with opener	215	228^+
Box openers	254	18^+
Can head with opener	220	260^+
Can openers	30	400^+
Cap removal facilitated	215	295^+
Closure removing	81	3.7^+
Complex for deforming metal	72	394^+
Compound	7	
Magnetic feature, collection	7	900*
Resilient pivot, collection	7	901*
Cooler	62	293
Cutlery	30	
Dental	433	25^+
Orthodontic	433	3^+
Design of	D 8	
Detachable, portable, power-driven	30	500*
Diamond	125	39
Drill having diamond edge	408	145
Holders for abrading and polishing	51	229
Making	76	101R
Making dies	76	107R
Making drills	76	108R
Precious stone working	125	30R
Driving means	173	
Metal deforming machine	72	429^+
Earthworking	172	
Boring or drilling	175	
Hand forks and shovels	294	49^+
Hand manipulated cultivating	172	371^+
Hand rakes	56	400.1^+
Handling hand and hoist line	294	2
Mining and disintegrating in situ	299	79^+
Spade or shovel with cutter	7	116^+
Subsoil irrigators and weed killers	111	7^+
Electrically heated	219	221^+
Elongated handles	D 8	DIG. 7^+
Farriery	168	45
Feed	173	141^+
Earth boring type	175	162
Food handling	D 7	99
Cherry seeding	30	113.1^+
Corn husking	130	4
Corn shellers	130	10
Cutlery	30	
Meat and vegetable preparing	30	
Oyster openers	17	74^+
Peach stoner	30	113.1^+
Raisin seeding	30	113.1^+
Scoring	30	164.9
Hair dressing or manicuring	D28	9
Hand	81	
Assembly or disassembly	29	270^+
Cooler	62	293
Handle	81	489^+
Handles	7	167^+
Handling	294	
Holders		
Carried by belt	224	904*
For dental machines	433	77^+
For metal lathes	82	$36R^+$
Impacting device	173	90^+
Leather	69	20
Boot and shoe making	12	103
Making	76	
By die rolling	72	184^+
Cutting spirally grooved	409	65^+
Grinding to shape or sharpen	51	
Sharpening electrolytically	204	129.1^+
Manicuring	132	73^+
Medical or surgical	D24	
Metal working		
Assembly, magnetic, or indicia tools	29	DIG. 105
Bolt nail nut rivet and screw making	10	
Cutting & punching sheets & bars	83	
Deforming	72	462^+
Design	D15	122^+
Horseshoe making	59	60
Metal tool and implement making	76	
Soldering	228	51^+
Tool or work attached	29	DIG. 104

	Class	Subclass
Wire working	140,	
Motors		
Multipurpose	7	
Nut lock tool	81	10
Pliers and tongs	81	300+
Pushing and pulling	254	
Rack-type holder	D 8	71
Rack-type holder	D 8	349+
Self heating with burner	126	401+
Sharpener	83	174
Shaving or hair clipping	D28	44+
Shoe making	12	103
Stone working	125	36+
Telephone dialing	379	456
Tire		
Bead spreading	254	50.1+
Breakers for split rims	157	1.35+
Repairing	81	15.2+
Repairing with vulcanizing	425	12
Setting and removing	157	1.1+
Valve stem cap with tool	152	431
Top and dies	408	215+
Valve removal tools	29	213E
Vehicle-mounted	83	928*
Vises	269	
Barrel chamfering	147	42
Design, hardware	D 8	
Design, machinery	D15	122+
Turning	142	56
Wrenches	81	52+
Impact	81	463+
Tooth (See Teeth)		
Agricultural and earthworking implements		
Cultivator tooth structure	172	713
Diggers	171	
Excavating tooth	37	142R
Hand cultivating tool	172	378+
Hand rakes	56	400.21
Harrows	172	
Horse drawn rakes	56	400
Planting drill	111	86+
Raking machine	56	344+
Brushes	15	167.1
Design	D 4	104+
Dispensing	132	84R+
Holder	D 6	524+
Kit	132	84R
Material supply combined	401	268+
Receptacle for	206	361+
With massage tool	15	110
Comb teeth cutting	144	26
Combs	132	152+
Dentifrices	424	49+
Dentistry	433	167+
Design	D24	10+
Pulp tester	433	32
Gear	74	457+
Lubrication	74	468
Testing	73	162
Inserted tooth metalworking cutter	407	33+
Kit		
Brush or paste retained	401	118+
Making		
Forging	72	376
Gear cutting	409	1+
Saw tooth	76	25R+
Toothed article	83	908*
Toothed cylinder	29	23
Paste	424	49+
Holder-squeezer	D 6	541
Tube	222	92+
Picks	132	89+
Design	D28	64
Making	144	185+
Packaging	53	236
Receptacle	206	380+
Stone sawing	125	22
Tooth article making	83	908*
Vehicle fender	293	53
Wood sawing	83	835+
Fastenings	83	840+
Toothbrush (See also Tooth)	D 4	104+
Toothpick		
Cocktail-type pick	D 7	42+
Holder, household	D 7	75
Holder, personal	D28	64
Top Boards		
Vehicle body	296	32+
Convertible box	296	13
Securing devices	296	36

	Class	Subclass
Topical Remedies	424	
Veterinary	424	
Toppers		
Beet	99	635+
Digging followed by topping	171	26+
Harvesting grain or corn crops		
Cutter and catcher combined	56	56+
Cutter combined	56	63
Tops		
Bed		
Bedstead post top knob	5	281
Slat frame top	5	244
Chance device	273	147
Counter	312	140.3
Land vehicle	296	98+
Design	D12	156
Metallurgical furnace top	266	199
Still top closure	202	250+
Stove	126	211+
Table panel	52	783+
Toy	446	256+
Design	D21	95+
Motorized	446	259
Torches (See Flare; Illuminating)		
Design	D26	8
Liquid fuel burner	431	327
With common support for igniter	431	127+
Pyrotechnic	102	336+
Torpedo		
Aerial	244	14
Automatic controlled steering	114	23+
Guards	114	240R+
Net type	114	241
Marine	114	20.1+
Boats	114	18+
Guards	114	240A+
Launching	114	238+
Submarine boats	114	316+
Railway signal	246	487
Placing mechanism	246	210+
Remote control	114	21.3
Toy	102	353
Well	102	301+
Torpedo Boats	114	18
Submarine	114	316+
Torque		
Brake testing lever measuring	73	131
Clutch control by	192	54+
Controls hoist	254	275
Dynamometers		
Absorption	73	862.8+
Transmission	73	862.19+
Wrench	73	862.21
Gearing control by	74	337
Railway rolling stock		
Axle arm	105	132.1
Opposing equalizer	105	135
Tubes	74	607
Wrench overload yielding	81	467+
Torsiometers	73	847+
Torsion		
Balances	73	383
Clock pendulums	368	180
Rod for dumping horse rake	56	392
Spring closers	16	75+
Spring locomotive	105	70
Spring trolley erector	191	69
Testing		
Stress strain	73	847+
Vibration	73	650
Vehicle spring	267	273+
Leaf and	267	25+
Total Oxygen Demand, Tod	436	62*
Totalizers	235	
Weigher combined	177	15+
Touch Systems, Electric Power	323	904*
Toupees	132	53+
Tourniquets	128	327
Tow Spreading	28	282
Tow Truck		
Design	D12	14+
Towels	D 6	608
Dispensing	312	37+
Reels	242	55.2+
Sheets	D 6	522
Packaged towel	206	812*
Racks	211	16
Design	D 6	512+
Lock combined	211	6
Stacked article type	211	50+
Tower (See Mast)		
Bisulphite process	423	520

	Class	Subclass
Chemical apparatus	422	
Distillation	202	
Oil fractionating	196	139
Fire escape	182	48+
Gas and liquid contact	261	
Overhead electrical conductor	174	45R
Fire extinguishing	169	25
Railway	104	125
Suspension bridge	14	21
Type antenna	343	874+
Wheeled water	169	25
Towing (See Traction)		
Aerial signs	40	215
Aerial targets	273	360+
Attachments for wheeled vehicles	280	292
Crane, self propelled	212	141
Marine propulsion	440	33+
Ships	114	242+
Vehicle tow chains and ropes	280	480
Tow-to-Top Stapilizing	19	.3+
Toxiferine	540	472
Toys	446	
Aircraft	446	34
Aquatic	446	153+
Buildings	446	476+
Construction toys	446	85+
Container or wrapper	493	959*
Design	D21	59+
Drums	84	420
Figure toys	446	268+
Figure wheeled toys	446	269+
Firearms	42	54+
Flying	446	34+
Handler-type toys	414	915*
Inflatable	446	220+
Design	D21	59+
Magnetic	446	129+
Model airplane controls	244	190
Money boxes	446	8+
Music boxes	84	94.2+
Railway track	238	10E
Remote control of	446	454+
Electric toy trains	104	295+
Scale model vehicle design	D21	128+
Soap bubble devices	446	15+
Sounding toys	446	397+
Sounding wheeled toys	446	409+
Spinning and whirling	446	236+
Torpedoes	102	353
Watches	368	45
Wheeled toys	446	431+
Tr, I.e. Transmit Receive or Transmit	333	13
Tr box in radio set	455	76+
Transmit-receive circuitry	367	903*
Trace		
Harness	102	353+
Carriers	54	54+
Design	D30	134+
Detachers from singletrees	278	24+
Hame and trace connectors	54	30+
Thill or pole connections	278	118
Whiffletree connections	278	102+
Tracer	409	126
Ammunition	102	513
Explosive shell	102	513
Radio active composition as	252	625+
Tracheotomy	128	305.3
Tracing Board	D19	52
Track (See Rail)		
Clearers		
Harvester	56	314+
Railway	104	279+
Snow excavating	37	198+
Wheel path of earthworking tractor	172	833
Endless flexible	305	
Land vehicle combined	280	28.5
Motor vehicle combined	180	9.1+
Shoe making	29	148.3
Step ascending vehicle combined	280	5.22
Endless track	305	
Layers railway	104	2+
Leveling gauge	33	287
Light, track-type	D26	61+
Marine towing	440	35
Material and article handling		
Pivoted track elevator or hoist	414	597
Tilting	414	357
Tilting track elevator or hoist	414	599
Vertically movable	414	354+
Motor vehicle	180	9+

Entry	Class	Subclass
Operated sawmill carriage	83	716
Panel	16	96R+
Railway	104	
Grade crossing protection	246	111+
Guard fence	256	14+
Guard gate	49	133+
Iron forging dies	72	343+
Mechanism automatic signals and gates	246	297+
Mechanism automatic switches	246	322+
Pushers on locomotives	105	31
Surface	238	
Testing	73	8
Trips cab signal or train control	246	201+
Trips electric	246	76
Sanders	291	
Store service	186	34
Wheel substitute	305	
Tracker	84	151+
Box location	84	116
Double tracker for pneumatic player pianos	84	31+
Plural sheets for single tracker	84	120
Tracker combined with sheet	84	146+
Trackway		
Cabinet combined	312	
Disappearing instrument type cabinet	312	21+
External	312	349
Internal	312	350
Showcase type cabinet	312	114+
Rack combined	211	162
Paper or textile sheet type rack	211	46
Shelf type rack	108	143
Wall or window type rack	211	94
Railway	104	
Surface track	238	
Trunk tray	190	29+
Traction		
Engines		
Railway	105	26.5+
Steam, motor vehicle	180	36+
Floor and wall cleaners		
Pump	15	341+
Motor vehicle		
By other than wheel	180	7.1+
Endless flexible track	180	90
Stepper	180	8.1+
Runner supported	180	85+
Endless track	180	190
Endless track substitutable for wheel	180	184
Traction wheel	180	184
Suction effect	180	164
Vertical surface traversing device	180	901*
Wheel, additional traction	180	15
Wheel, traction, attachments	180	16
Operated		
Harvesters	56	211
Railways	104	165+
Stoneworking tools	125	37
Vacuum cleaner	15	341+
Prime mover dynamo plant	290	45
Electric control	290	9+
Railway		
Mats	238	14
Maximum traction type brakes	188	48
Razors		
Blade moving means	30	46
Regulators for locomotives	105	73+
Runner vehicles (See motor vehicle above)		
Traction wheel, occupant propelled vehicle	280	12.13+
Wheel drive	280	12.13+
Surface coating or plastic	106	36
Wheel		
Land vehicle	301	41R+
Railway turntable actuators	104	43
Tractors (See Motor Vehicles)		
Cab	180	89.12
Design	D15	23+
Handling device on tractor device	414	920*
Obstetric	128	352+
Pusher attachment for	293	DIG. 1
Toy simulation	446	434
Tractor mounted mortar mixing chamber	366	606*
Traffic Control		
Accident display	434	305
Air	342	36
Airway	340	52R+
Computer controlled, monitored	364	439+
Barrier	404	6+
Building structure	52	174+
Computer controlled, monitored	364	436+
Director	404	9+
Lights	340	22+
Design	D10	115
Railway	246	
Signals and guides	116	63R
Design	D10	109+
Electric	340	22+
In pavement	404	9+
Reflective triangle for highway use	40	903*
Stationary signs	40	612+
Tragacanth	536	114
Trailer		
Boat carrying	280	414.1
Coupling	280	504+
Design	D12	101+
Hitch	D12	162
Land vehicle	280	400+
Mobile home design	D12	101+
Semitrailer	280	423R+
Two wheeled	280	400+
Dumping	298	5+
Train		
Aircraft	244	3
Automatic stop	246	167R+
Brake		
Fluid automatic	303	47
Fluid multiple motors	303	7+
Fluid multiple motors valves	303	53
Railway	188	34
Railway automatic	188	124+
Railway automatic momentum	188	147+
Vehicle	188	3R
Vehicle automatic	188	112R
Chemical apparatus	422	189+
Clock electric switch	200	35R+
Control automatic	246	167R+
Coupling type pipe or cable support	248	53
Dispatching	246	2R+
Gear		
Yieldable element	74	411
Heating		
Steam car boiler plus train steam	237	42
Water heated by train steam	237	34+
Horology		
Clock	368	62+
Dial	368	220
Watch	368	268
Land vehicle	280	400+
Occupant propelled tractor	280	204
Lighting	362	61+
Motor vehicle	180	14.1+
Railway	105	1.4+
Audible indicators	246	217
Block signal systems	246	20+
Cab signal or train control	246	167R+
Dispatching	246	2R+
Drawbridge protection	246	118+
Electric switch	246	220
Endless	104	25
Fluid motor actuated	246	260
Grade crossing protection	246	111+
Moving	104	18+
Passengers station signals	246	209
Position indication	246	122R
Railway signal structure	246	473R+
Roadway defect protection	246	120+
Switch stands	246	394
Torpedo mechanism	246	210+
Train carried station indicator	246	174+
Vehicle prime mover dynamo plants	290	3
Training	434	
Aircraft	434	30+
Animal	119	29
Harness	54	71+
Hedge	47	4
Ordnance	89	41.1+
Physical on rails	104	62
Vehicle	434	29+
War	434	11
Trampoline	272	65
Design	D21	235
Life saving nets	182	138+
Transactinide Chemistry	423	249+
Transactinide Chemistry	534	11+
Transalkylation (See Disproportionation)		
Transceiver	455	73+
Transducers		
Angular position measurement	33	1PT
Piezoelectrical	84	DIG. 24
Underwater vibration	367	141+
Transductor (See Amplifier Saturable Core Reactor Systems)		
Transfer		
Applying processes	156	230+
Bed, invalid lifter	5	81R+
Coated article	427	146+
Decalcomania (See decalcomania)	428	914*
Designs for hot iron	8	467+
Devices for knitting machines	66	148
Direct contact transfer	156	
Apparatus	156	539+
Methods	156	230+
Dyeing	8	467+
Egg candling	356	55+
Electrical transmission		
Table	191	9
Heat	165	
Identifying or fraud preventing	428	915*
Lamina processes	156	230+
Laminate	428	
Liquid		
Reciprocating valves	137	251.1+
Regulators	137	505.16
Making by coating	427	146+
Metallurgy		
Pyrometallurgy	75	46
Paper, carbonless	503	200+
Paper files and binders		
Picture (See decalcomania)		
Pipes with heat transfer	138	38
Printing		
Bed and cylinder	101	251+
Intaglio	101	154
Multicolor	101	177
Planographic	101	142+
Preparatory designs	101	33+
Rotary	101	217+
Railway		
Passenger	104	20+
Tables	104	48+
Trolley	104	96+
Register	235	133R+
Sheet		
Inserter, remover, fork	402	80R
Transfer post	402	47
Stripping of backing	156	247+
Textile smoothing machine		
Article	38	7+
Textiles weaving		
Bobbin changing	139	242+
Bobbin changing tips	139	244
Thread control cutter	139	267
Tobacco	131	82
Molding or forming with bunch transfer	131	82
Wrapper	131	105
Transportation	283	27+
With fraud prevention or detection	283	72+
Type casting		
Assembling	199	32
Assembling collector & transfer	199	26
Distributing	199	36
Wood sawing		
Log	83	708+
Transfer or Decal	D 5	63+
Transformation, Nuclear	376	100
Transformation, Nuclear	376	156
Transformers	336	
Arc lamp supplied by	314	32
Differential transformer	33	DIG. 5
Electric	336	
For arc lamp systems	314	32
For incandescent lamps systems	315	276+
In amplifier		
Induction heating system	219	6.5+
Induction heating type	219	10.75
Input coupling	330	188+
Interstage coupling	330	165+
Lamp or electronic tube supplied by	315	276+
Metal heating system	219	50+
Movable core type	336	130+
Output coupling	330	195+
Plate	336	
Power	D13	4+
Regulating	336	
Interconnection systems	307	

	Class	Subclass
Systems	323	
Saturating transformer	336	155
External or operator controlled	323	329
Input responsive	323	302
Output responsive	323	249
Regulating system	323	
Structure	336	
Self regulating	323	310
Telegraph system	178	64
Tuned high frequency	333	177+
Transistor (See Amplifier, Electric)	357	
Amplifiers	330	250+
Composition	252	62.3 R
Concentraction gradient	357	90*
Field effect, making (process)	437	
Integrated injection logic	357	92*
Ion implant	357	91*
Making	29	569.1+
Process	437	
Particular internal structure of region	357	88*
Peltier effect	357	87*
Photo-responsive, making	437	
Shorted junctions	357	86*
Step junction	357	89*
Structure	357	
Systems (See system)		
Miscellaneous	307	200R+
Tester	D10	77
Transits	33	290
Optical	356	138+
Translating		
Agitator mixing chamber stationary	366	332+
Agitator stirrer	366	343+
Cover combined	366	255+
Gate	49	404+
Operator	49	324+
Music		
Electric	84	1.1+
Transposing	84	470R+
Plow	172	483+
Steering post	74	496+
Telegraphic printer	178	26R
Transmission (See Gearing)		
Aircraft power plant	244	60
Automotive lock	70	245+
Centrifugal bowl separator with	223	23+
Clutch control with	192	3.51+
Control and brake	192	4R+
Design	D15	149
Electric		
Cranes transmission to	191	
Distribution	307	
Facsimile	358	256+
Filter network	333	167+
Light wave	455	600+
Multiplexing	370	
Power transmission	307	
Pulse or digital communications	375	
Quarter wave transmision line	333	27
Radio	455	91+
Safety system for transmission lines	361	1
Signal box	340	287+
Signal transmitters	340	345+
Telecommunications	455	
Telegraph	178	
Television	358	83+
Transmission to vehicles	191	
Transmit-receive circuitry	367	903*
Fluid		
Lamp electrode	314	61
Power plant	60	
Pump with motor	417	379+
Railway car axle drive	105	96.2
Scales	177	208+
Guide for a shiftable handler	414	918*
Heat		
Heat exchangers	165	185
Liquid heaters and vaporizers	122	367R
Internal combustion engine	123	197R
Lamp electrode power control	314	79+
Latch operated clutch	192	24+
Line	174	103+
Concentric	174	103+
Involving line parameters	333	236+
Resonant	333	236+
With nonsolid insulation	174	28+
Machine element	74	
Belt & pulley drive system	474	
Motor vehicle	180	70.1+
Having two wheels in tandem	180	230
Safety belt responsive to transmission thereof	180	269
Railway car axle box mounted	105	96.1
Rotary crane	212	247+
Tower or post	D25	126+
Transmutation of Elements Fission	376	156
By fusion	376	100
Transom (See Window)		
Operators	49	356
Railway truck	105	208+
Bolster combined	105	202+
Transparency		
Product with abnormal	428	918*
Testing and inspecting sheets	356	432+
Viewer	40	361+
Walled receptacle	220	82R+
Transplanters	111	2+
Big ball tree	47	76+
Biodegradable receptacle	47	74+
Dibbling implements	111	92+
Receptacles	47	73+
Seedling	47	77+
Tree handlers	414	23
Transponders		
Radar	342	42
Transportation (See Vehicle Handling)		
Advancing material of indeterminate length	226	
Printed matter	283	23+
Vehicle design	D12	
Transporting Attachments Harvesters	56	228
Transposers		
Electric conductors	174	33+
Music		
Indicators with transposing dial	84	480
Keyboard transposers	84	445+
Teaching device with transposing dial	84	474
Teaching device with transposing slide	84	473
Tracker and music sheet shifted to change key	84	145
Transuranium Compounds		
Inorganic	423	250+
Organic	534	11+
Trap & Trapping		
Animal etc	43	58+
Bank protection devices	109	3+
Burglar	43	59
Chamber dispensing	222	
Agitator combined with movable	222	226+
Agitator or ejector for movable	222	216+
Barometric type stationary	222	457
Fluid flow discharge from movable	222	636
Into conduit	406	63+
Inkwell dip well	222	576+
Jarring movable	222	197
Movable or conveyor type	222	251+
Removable from container for discharge	141	110+
Rotatable container with	222	170
Stationary	222	424.5+
Stationary with cutoffs	222	425+
Supply movable relatively to	222	162
Tiltable assembly	222	454+
Comminutor projected material	241	82
Currycomb combined	119	87
Deposit receptacle coin	232	55+
Design	D22	119+
Doors		
Ladder combined	182	77+
Manhole cover	404	25+
Railway car platforms	105	426+
Vertically swinging outside		
False or picking key	70	390
Fish, game, vermin traps	D22	119+
Fish traps	43	100+
Design	D22	119+
Fish locating instrument	D10	46+
Hooks	43	34+
Design	D22	144
Lures	D22	125+
Nets, landing spears, etc	D22	135+
Gas & liquid (See seal, liquid)		
Automatic heating radiator with	236	41
Automatic temperature etc regulation	236	53+
Bath etc	4	191+
Beverage infusers	99	301
Boiler	122	
Cigars etc	131	331
Concentrating evaporators	159	31
Decanters with	210	513+
Dispenser inlet or outlet	222	188
Filters with sediment	210	299+
Fluid current conveyor	406	62+
Fluid delivery track sanders	291	11R
Fluid distribution	137	171+
Gas and liquid contact apparatus	261	86
Gas separator combined	55	355
Manufacturing processes	29	158
Nonrefillable bottle	215	15
Sewerage	137	247.11+
Smoking pipes etc	131	201+
Spittoon	4	283
Steam	137	171+
Water closet	4	300+
Ion	335	210+
With cathode ray tube	313	424+
With TV picture tube	313	424+
Nests	119	47+
Railway plow	104	145
Safety pockets for garments	2	254
Steam	137	171+
Thermostatic	236	53+
Target		
Centrifugal projectors	124	4+
Disk holders and carriers	124	42+
Disk magazines	124	46+
Projectors	124	
Trap shooting target handling	273	406
Vehicle fender	293	15+
Trapezes	272	61
Trapping	43	
Trash (See Refuse; Waste)		
Bundle	428	2
Metal	428	576
Burners	126	222+
Furnace structure for garbage and sewage	110	235+
Furnace structure for refuse	110	235+
Can liners	220	404+
Can support	248	DIG. 7
Clearers for plows	172	606+
Travelers		
Animal hitching	119	120
Ship	114	204+
Ship mast	114	112
Textile spinning rings and	57	119+
Track hardware	16	87R+
Traveling		
Bags	206	278
Design	D 3	30.1+
Grates	110	267+
Special sewing machine	112	121.14
Wave loom	139	436
Traverse Rod	D 8	377
Trawl	43	9
Lines	43	27.4
Tray (See Dish; Receptacle)		
Animal stock	119	103
Automobile supported table	108	44+
Bank protection deal	109	19
Cabinet showcase type	312	126+
Cafeteria dispenser unit	D34	14
Circular household article	D 7	23+
Coin pickup or delivering		
Coin deliverers	232	64+
Special receptacles for coins	206	.8+
Compartmented	D 7	38
Dental equipment	433	77+
Designed for glasses	D 7	70
Drying etc	34	237+
Removable tray type	34	192+
Egg candling	356	61
Gas filter tray	55	494
Handle supported	294	172
Special article	206	557+
Heaters	261	DIG. 3
Ice cube	249	117+
Incubator	119	43+
Making wood	144	33
Paper receptacle	229	190+
Photographic fluid treating	354	331+
Rack	211	126+
Wall or window mounted	211	88
Sample case	190	17
Slide projector	40	361+
Special receptacle	206	557+
Tobacco ash receiving	131	231+

	Class	Subclass
Trunk	190	35
Warming	D 7	363
Tread		
Burglar alarm operated by	116	98
Electric circuit controller	200	86R
Burglar alarm	340	565
Railway car wheel	295	31R+
Railway rail with	238	122+
Joint with bridge	238	218+
Surface per se	238	148
Resilient tire		
Antiskid	152	208+
Armored	152	167+
Shoe antislip	36	59C+
Testing	73	146
Abrasion or rubbing	73	8
Vehicle step	280	169
Wheel guard	293	58
Treadle		
Closure actuator	49	263+
Grinding wheel mount	51	167
Levers	74	512+
Machine element	74	561
Sewing machine conversion	112	217.4
Trackmans car drive	105	93
Treadmill		
Closure operating	49	262
Exercising devices	272	69
Motors	185	
Occupant propelled vehicle	280	228
Treasure Locator (See Prospecting)		
Tree		
Covers	47	20+
Felling	144	34R
Portable circular saw	30	360+
Hacks	30	121
Handlers for trees & poles	414	23
Husbandry	47	
Imitation	428	18+
Tree- like ornament lighted	362	123
Impregnation	47	57.5
Plants		
Broadleaf	PLT	51+
Conifer	PLT	50
Fruit	PLT	33+
Nut	PLT	30+
Ship masts		
Cross and trestle	114	92
Shoe	12	114.2+
Stand for supporting	248	519+
Supports and props	47	42+
Surgery	47	8
Trunk		
Guards	47	23+
Insect traps	43	108
Trellis	47	44+
Design	D25	100
Tremolos	381	62
Brass wind instrument tremolo	84	401
Electric oscillator having	331	182
Electric tone generator tremolo	84	1.25
Fan tremolo for reed organs	84	374
Stringed instrument	84	313
Valve tremolo for pipe organs	84	348
Trepan	128	305.1
Trephine	128	310
Trestle		
Scaffold or saw horse	182	181+
Unitary foldable	182	153+
Tire setter	157	7
Trees ship mast	114	92
Type propped extension ladder	182	105
Triangle		
Geometrical instrument	33	474
Musical instrument	84	402+
Reflective triangle for highway use	40	903*
Triarylmethyl	260	386+
Triazines	544	3+
Azo compounds	534	751+
Stilbene ctg	544	193.1
Triazolobenzodiazepinone	540	499
Tricarballylic Acid	562	590
Esters	560	190+
Trichloracetic Acid	562	602
Esters	560	226+
Trickler		
Coffee pots	99	306
Heat exchanger	165	115+
Ice melt cooler	62	312+
Process	62	64
Textile treating	68	205R+
Trickling Filter	261	94+
Liquid purification	210	150+
Processes	210	601+
Tricot Knitting Machine	66	87
Tricycle (See Velocipede)		
Design	D12	112+
Triethanolamine	564	506
Triethylene Diamines	544	351
Trigger (See Trip and Trigger)		
Circuits		
Electronic tube type	328	191+
Gas tube type	315	
Firearms	42	
Triglycerides	436	71*
Trigonometry, Teaching	434	211
Trimelletic Anhydride	549	245
Trimmers (See Cutters)		
Billiard cue	30	494
Brick	425	289+
Brick	83	
Burner wick	431	120
Capacitor	361	271+
Cigar and cigarette	131	
Condenser	361	271+
With inductor	334	78+
Cooking mold	99	430
Embroidery	83	910
Farriery hoof	168	48R
Hat	223	24
Hedgerow	56	233+
Lawn	172	13+
Photograph	83	
Pie	425	293
Plastic block press	425	289+
Sewing machine	112	122+
Attachment	112	130
Sheet cutting	83	
Scrap cutting	83	923
Shoe inseam machine	12	4.3
Shoe sole and heel edge	12	85+
Textile selvage	139	302+
Waffle iron drip	99	375
Trimming		
Apparel	223	44+
Design	D 5	7+
Battery grid	83	903*
Brush and broom	300	17
Cartridge case	29	1.32
Chain sizing and	59	29
Dental plaster	433	144
Design	D 5	7+
Garment	2	244+
Making and attaching	223	44+
Gun	42	85
Harness	54	75+
Design	D30	134+
Hat	2	186
Banding	223	22
Brim	223	16
Horseshoe	59	59
Leather skiving and splitting	69	9.3
Packages	206	389+
Pearl button surfacing	79	10
Screw threading dies	408	215+
Shoe sole and heel machines	12	
Design	D15	122+
Stock material	428	
Woodworking box shaping	144	135
Trinitrotoluene	568	935
Containing	149	
Explosives containing	149	69
Explosives containing	149	107
Triode	313	293+
Amplifier	313	293+
Detector amplifier	313	293+
Diode high mu	313	303
With two cathodes	313	5
Diode pentode	313	298
With plural cathodes	313	6
Direct coupled twin	313	3
Double	313	301
With two cathodes	313	6
Dual grid	313	297
Duo diode	313	303
Duplex diode	313	303
With plural cathodes	313	5
Duplex diode high mu	313	303
With two cathodes	313	5
Gas	313	581+
Hot cathode type	313	592
Germanium	357	
Heptode converter	313	298
With two cathodes	313	6
Hexode converter	313	298
With two cathodes	313	6
High frequency	313	293+
High frequency twin	313	301
With two cathodes	313	6
High mu	313	293+
High mu twin	313	301
With two cathodes	313	6
Low mu	313	293+
Low mu uhf	313	293+
Medium mu	313	293+
Medium mu twin	313	301
With two cathodes	313	6
Power amplifier	313	293+
Power amplifier twin	313	301
With two cathodes	313	6
Triple diode	313	303
With plural cathodes	313	5
Triple diode high mu	313	303
With plural cathodes	313	5
Twin	313	301
With two cathodes	313	6
Twin power amplifier	313	301
With two cathodes	313	6
Two grid	313	297
U H F	313	293+
Trioxane	549	367
Trip and Trigger		
Animal trap	43	58+
Belt conveyor unloader	198	633+
Cable hoist	212	116
Closure fastener	292	332+
Electric switch latch trip		
Clock train	200	39R+
Double snap oscillating	200	70+
Electromagnetic	335	174+
Reciprocating	200	78
Single snap oscillating	200	74
Thermal current	337	70+
Expansible chamber motor		
Trip gear	91	338
Firearm	42	70.1+
Guns and ordnance	89	27.11+
Automatic gun	89	132+
Electric	89	28.5+
Harvester		
Shocker	56	401+
Mechanical gun	124	31+
Odometer	235	97
Operated file cutter	76	15+
Planting		
Depositing mechanisms	111	34+
Railway		
Cable gripper	104	205+
Railway switches and signals		
Derailment contact trips	246	171+
Track trip automatic electric	246	76
Track trips	246	201+
Train trips	246	359+
Released valve	251	66+
Rotary printing member	101	234+
Saw setting		
Pivoted set	76	62
Sliding set	76	67
Ship anchor	114	210
Swinging door lock	70	157
Trip mechanisms	169	DIG. 3
Trips	74	2+
Compressing and binding	56	436+
Vehicle brake operator	188	111
Vehicle fenders		
Dash and wheel	293	33
Drop fender	293	34+
Tripelennamine, in Drug	514	352
Triphenylmethanes	260	389+
Triple		
Bond hydrocarbon		
Purification	585	800+
Synthesis	585	534+
Diode	313	1
Mirrors	350	102+
Tripod	248	163.1+
Camera supports	354	293
Design	D16	45+
Ladder prop	182	169
Support stand	248	163.1+
Heads	248	177+
Trirams	114	61
Trituration	241	

	Class	Subclass
Trocars	604	264$^+$
Trolley		
Crane	212	142.1
Swing boom carries	212	250$^+$
Vertically	212	257
Excavating		
Cable operated trolley supported	37	117
Ladders	182	36$^+$
Linear traversing hoists		
Rope carrier	212	119
Railway		
Collectors	191	50$^+$
Conductors	191	33R$^+$
Nonretracting trolley stops	191	95
Rails	104	106$^+$
Retrievers	191	85$^+$
Suspended single rail	105	148$^+$
Transfer	104	96$^+$
Railway switches	246	419
Trolley actuated controller automatic	246	84
Trolley actuated controllers	246	252
Trolley completed circuits	246	254
Wire support	191	40$^+$
Trombones	84	395$^+$
Electric transmission line	333	219$^+$
Mutes	84	400
Section in high frequency transmission lines	333	33$^+$
Tropanes	546	124$^+$
Trophy	428	542.4
Design	D11	131
Tropine	546	127
Trotlines	43	27.4
Trough		
Animal feeding	119	61$^+$
Design	D30	121$^+$
Belt conveyor	198	818$^+$
Burners	431	52$^+$
Chute unloading	193	6
Closure	49	408
Dryer gravity flow	34	166
Flume	405	119$^+$
Holder for soldering	269	
Railway track rail	238	129
Roof eaves	52	11$^+$
Support	248	48.1
Sprinkler supply	239	724$^+$
Vibrating conveyor	198	771
Wood sawing log transfer	83	707$^+$
Trousers	2	227$^+$
Design	D 2	202$^+$
Undergarment	D 2	6$^+$
Forms	223	72$^+$
Guards and straps	24	72
Hangers	223	95$^+$
Knee protectors	2	62
Leg guards attached to	2	23
Stretching	223	63
Union type	2	79
Trowels	15	235.4
Design	D 8	10
Handle fastening	403	
Truck		
Antifriction thrust bearing		
Truck center	384	594
Truck side	384	595
Brakes	188	21$^+$
Railway	188	47
Design of	D12	82$^+$
Fork lift	414	529$^+$
Garbage compacting	100	100
Hand	280	47.17$^+$
Dumping	298	2$^+$
Self loading	414	467
Hoisting	254	2R$^+$
Ladder	182	63$^+$
Hand truck	182	16
Land vehicle		
Body	296	
Freight accomodating	410	
Hand	280	47.17$^+$
Hand dumping	298	2$^+$
Multiple	280	81R$^+$
Swivel	280	86
Tongue	280	82$^+$
Wheeled skate	280	11.27$^+$
Motor vehicle		
Steered bogie	180	79.5
Plain thrust bearing		
Truck center	384	422
Truck side	384	423
Portable platform for	108	51.1$^+$
Railway	105	157.1$^+$
Brakes	188	47
Changers	104	32.1$^+$
In conduit	104	139
Two car	105	4.1$^+$
Semitrailer	280	423R$^+$
Accomodated on freight carrier	410	56$^+$
Design of	D12	97$^+$
Step climbing	280	5.2
Wheeled vehicle	280	29
Vehicle carrier	410	3$^+$
Design	D12	94$^+$
Trumpet		
Ear trumpet	181	129$^+$
Music	84	387R$^+$
Design	D17	11
Mutes	84	400
Textile		
Formed sliver trumpet	19	157
Sliver forming trumpet	19	150
Stop for sliver controlling trumpet	19	
Thickness working trumpet	19	
Trundles	446	450$^+$
Reel carrying	242	94
Sounding device combined	446	411$^+$
Trunks		
Baggage	190	19$^+$
Design	D 3	70
Lighting in	362	156
Bed convertible to	190	2
Garment type	2	238
Knitted	66	176
Knitted union type	66	175
Insect traps for tree	43	108
Locks for	70	69$^+$
Pneumatic dispatch systems	406	
Telephony	379	
Trays	190	35
Wardrobes	190	13R
Trusses	52	690$^+$
Brake beams	188	219.1$^+$
Bridge	14	3$^+$
Coopering hoops	147	49
Curvilinear or peaked	52	639$^+$
Frame for wood handsaw	30	506
Pole brace	52	146$^+$
Rail joint rod	238	168
Railway car framing		
Freight	105	407
Passenger	105	399
Railway track tie	238	40$^+$
Railway truck bolster	105	229
Ship yardarms	114	96
Surgical bandaging	128	95.1$^+$
Vehicle axle	301	129
Whiffle tree	278	93$^+$
Wire fence brace	256	36
Tryptamine	548	504
Tryptophan	548	496$^+$
In drug	514	419
Tub (See Bucket; Vessel)		
Basin combined	4	553$^+$
Bath	4	538$^+$
Cabinet combined	312	228
Combined	4	546$^+$
Masonry	52	264$^+$
Metallic	220	
Railway amusement	104	59
Staved wooden	217	72$^+$
Surgical	128	369$^+$
Textile fluid treating	68	232$^+$
Vegetable and meat cutter	241	199$^+$
Wooden	217	
Tube (See Conduit; Pipe)	138	
Abrading	51	290$^+$
Axle casing	74	607
Bell	116	169
Bobbin winding	242	46.3
Boiler		
Cleaning	122	379
Fire	122	44R$^+$
Fire and water	122	42$^+$
Flue	122	135R$^+$
Sheets	122	512
Stand	122	13R$^+$
Structure and connection	122	511
Water	122	235R$^+$
Bourdon		
Gauge	73	732$^+$
Switch	200	81.8
Braiding	87	9
With core	87	6
Burner gas and air mixers	431	354$^+$
Capillary, refrigeration utilizing	62	511
Car propulsion pneumatic	104	156
Motor charging	104	159$^+$
Cleaners for steam boiler tubes condensers and coolers	15	317$^+$
Cleaning implements	15	104.3$^+$
Closing end	228	60*
Collapsible tube	138	119
Collapsible tube squeezer holder	D 6	541
Collapsible wall type dispenser	222	92$^+$
Combustion	431	350$^+$
Compressor valve	251	4$^+$
Connections for cycleframes	280	274$^+$
Container or wrapper for coiled form	206	389$^+$
Corrugating	72	176
Cutter	401	152$^+$
Dispenser collapsible	222	92$^+$
Applicator combined with means for collapsing	401	152$^+$
Electric (See space discharge device)		
Assembling	445	
Apparatus	445	66$^+$
Process	445	23$^+$
Circuit systems	315	
Discharge type	313	
With gas or vapor envelope	313	634$^+$
With particular envelope	313	317$^+$
Electric lamp envelopes	220	2.1 R$^+$
Illuminating lamp structure	313	
Lamp support	248	49$^+$
Lantern	362	263$^+$
Manufacture	445	
Evaporating	159	26.1$^+$
In flue	159	39
Expander	72	393
Fire extinguishing	169	35
Flanger	72	317
Geissler	313	567$^+$
Glass molding		
Curved	65	
Heat exchanger	165	177$^+$
Inflatable	83	DIG. 3
Lantern	362	178
Electric	362	263$^+$
Liquid cooler	62	389$^+$
Liquid depth exploring	73	298
Making		
Bending	72	369
Boiler heater	29	157.4
Cigar wrapping	131	58$^+$
Compound, i. e. plural layers	228	126$^+$
Corrugating	72	176
Drawing	72	283
Drawing and welding apparatus	228	17.5$^+$
Expanding	72	393
Flanging	72	317
Laminated	156	
Paper tube making	493	269$^+$
Plastic molding	425	72R$^+$
Rolling apparatus	72	95$^+$
Rolling process	72	365$^+$
Sewing	112	63
Sheet metal	72	49
Upsetting	72	354
Welding apparatus	228	
Welding electric	219	59.1$^+$
Welding process	228	101$^+$
Winding strand on cylinder	242	7.21
Material	428	36
Musical		
Brass	84	393
Organ reed	84	361
Piano pneumatic	84	57
Nasal medicator	128	203.12$^+$
Paint	222	92$^+$
Making of sheet metal	29	17R$^+$
Paper article making		
Cup	493	52$^+$
Folding box	493	162$^+$
Paper receptacle	229	93
Cylindrical	229	93
Paste tube	222	92$^+$
Holder-dispenser	248	108
Squeezer-holder	D 6	541
Plastic molding reinforced	425	500$^+$
Pneumatic dispatch tube or system	D18	35

U.S. PATENT CLASSIFICATIONS

Class Subclass

- Pneumatic stackers 406 154+
- Portable drinking 239 33
  - Filter combined 210 251
- Pump
  - Jet regulator 417 182+
  - Rotary 415 88
  - Vibrating 417 241
- Railway rail 238 134+
- Railway tie 238 70+
  - Plastic reinforced 238 87
  - Plastic reinforced armored 238 97
- Railway track pedestal 238 114
- Ranque 62 5
- Roentgen ray 378 121
- Rolling 72 95+
- Rubber bag or package 206 69
- Sampling 73 864+
- Shield for radio 174 35TS
  - Envelope 313 248
  - Part of tube 313 313
- Siphon 137 123+
- Smoking device 131 191
- Speaking 181 18+
- Splitters
  - Implements 30 90.1+
  - Well tubing 166 55.2
- Support 248 49+
  - Paste type 248 108+
- Switch electric 337 228+
- Testers electric 324 405+
- Thermometer 374 139+
- Tire inner
  - Anticreep 152 500
  - Armored 152 195
  - Casing combined 152 166
  - Leakage test 73 49+
  - Structure 152 511+
- Trolley rail 104 108
- Vacuum (See space discharge device)
- Vortex, cooling 62 5
- Washboard 68 230
- Workpiece 428 586

Tuberculosis Antigens and Sera 424 88
- Bath 4 538+

Tubular
- Light bulbs 313

Tucker
- Cigar tip etc 131 89+
- Grain harvester band 56 458+
- Sewing machine work guide 112 144+

Tuft
- Brushing
  - Detachable 15 194
  - Fasteners 15 190
  - Socket 15 191R+

Tufting
- Brush making machines 300 2+
- Mattress 5 472
- Textile
  - Fiber preparation 19 215+
  - Pile loom replenishing 28 216
  - Pile pattern setting 28 214+
  - Pile weaving 139 2+
  - Sewing machine for 112 80.1+
  - Sewing machine for 112 121.16
  - Tuft making 28 147

Tugs
- Harness
  - Design D30 137
  - Hame 54 32+
  - Saddle tug bearers 54 43
  - Thill 54 50
- Ships 114
  - Propulsion 440 33+

Tulle (See Net)

Tumbler
- Brushing machines 15 75+
- Gear 74 341
  - Cone combined 74 348+
- Kitchen or table 215
- Locks
  - Antipick 70 421
  - Compound 70 316+
  - Key 70 392
  - Key adjustable 70 384
  - Single key tumbler and ward 70 347
  - Single key tumbler type 70 348+
  - Structure 70 323+
  - Tumbler type 70 291+
- Printing
  - Bed 101 334
  - Bed selective or progressive 101 104+
  - Gripper attached 101 411

Tumbling (See Rumble)
- Abrading
  - Machine 51 163.1+
  - Process 51 313+
  - Sand blast machine 51 422
- Heat exchanging 165 88
- Leather making
  - Fluid treating apparatus 69 30
- Skin removal 99 629+
- Solid material comminution
  - Apparatus 241 170+
  - Processes 241 26
- Textiles fluid treatment
  - Fluid extractor combined 68 24
  - Liquid flowing and 68 58+
  - Machines 68 139+
  - Scrubbing and 68 60
  - Scrubbing squeezing and 68 30
  - Squeezing and 68 61
  - Tank heater combined 68 16
- Toy 446 168+
  - Figure 446 324

Tun Dish 266 275

Tuned Radio Frequency Radio Receiver 455 150+

Tuners (See Tuning)
- In radio set 455 150+
- Musical instrument pitch pipe 84 454+
- Radio 334
- With distributed capacitance and inductance 333 219+

Tungsten 75
- Alloys, ferrous 420 8+
- Alloys, non-ferrous 420 430
- Carbide 423
- Composite metal stock 428 665
- Filament compositions 75
- Fluorescent or phosphorescent compositions ctg 252 301.5
- Misc. chemical manufacture 156 DIG. 69
- Organic compounds 556 57+
- Pyrometallurgy 75 84

Tuning
- Forks 84 409
  - Electric driving means 310 15+
  - Electric filter element 333 186+
  - Electric generator 310 15+
  - Electric tone synthesis 84 1.4
  - Oscillator 331 156
  - Tuning device 84 457
- Indicator in radio 455 154
- Music
  - Accessories for giving pitch 84 454+
  - Chord group as in zithers 84 312R
  - Piano tuning device 84 200+
  - Reed organ 84 364
  - Tools 84 454+
  - Trombone tuners 84 396
  - Tuning pegs for strings 84 304+
  - Valve brass tuners 84 394
- Radio 334
  - Antenna 343 745+
  - Dial 116 241+
  - Electric motor operator 334 20+
  - Electrical indicator 455 154
  - Mechanical 74 10R
  - Mechanical limit stop type 192 138+

Tunnel 405 132+
- Apparatus for making 405 138+
- Excavating 299
- Head 202 244
- Railway telltale signals 246 486
- Recovery of fluid 299 2
- Ventilation 98 49

Turban D 2 256

Turbine 415
- Blade making 29 156.8 R
  - Rotor making combined 29 23.5
- Casing making and assembling 29 156.4 R
- Combustion 60 39.1+
- Control and governor 415 13+
- Cooling 415 175+
- Earth boring combined 175 107
- Exhaust treatment 60 685+
- Flow meter 73 861.79+
- Fluid pressure motor combined 60 698+
- Fluid supply 60 643+
- Generator drive 290 52
- Hydraulic 60 330+
- Internal combustion engine combined 60 624
- Internal combustion engine combined 60 597+
- Locomotive drive 105 26.5
- Mixed fluid 60 649+
- Multiple 60 698+
- Nuclear reactor combined 376 378
- Nuclear reactor combined 376 391
- Nuclear reactor combined 376 402
- Nuclear reactor combined 376 179
- Pneumatic 60 643+
- Pump drive 417 405+
- Submarine drive 114 337
- Torpedo drive 114 20.1+

Turfing, eg Chenille 112 80.1+

Turkey Red Oil 260 400

Turn Indicators
- Geographical 33 328+
- Vehicle 116 28R+
  - Electric 340 22+

Turnbuckles 403 43+
- Making 29 175R
  - Forging dies for 72 343+

Turners (See Orienters)
- Beds
  - Mattress and bed 5 412
  - Patient 5 61
- Boot and shoe making
  - Channel flap 12 30
- Cheese 99 644
- Earthenware pipe 414 754+
- Foods and beverages
  - Egg 99 644
  - Griddle 99 424
  - Toaster carrier or grid 99 395
- Incubator egg 119 44
- Locomotive drive wheel 105 85
- Music leaf 84 486+
- Pancake type 294 8
- Plate 294 6
- Weed 172 514+
- Wood sawing
  - Log 83 708+

Turning 82
- Inside out garments bags and tubes 223 39+
  - Bag sewing and 112 10
- Screw threading and 408 20+
- Screw threading and 408 22
- Sewing machine guide for fringe 112 149
- Shoe upper machine 12 57
- Stone 125 10+
- Wood 142
  - Boring and 408 22+
  - Polishing and 144 47
  - Tenon 144 205+

Turnouts
- Rail 238 124
- Railway switch
  - Automatic 246 277
  - Stationary 246 427

Turnstile
- Antenna 343 797+
- Coin trap 232 61
- Design D25 51
- Gate 49 46+
- Letter box trap 232 52
- Operating register 235 93
- Railway car entrance 105 341.5

Turntable
- Automobile 104 44
- Freight carrier installed 410 1
- Gearing windmill 74 381
- Ladder truck 182 65+
- Plural related surfaces 108 103+
  - Terraced 108 94+
- Railway 104 35+
  - Locomotive 105 28
  - Trolley transfer 104 99
- Rotatable base support 248 349
- Single surface 108 139
- Table self serving power driven 108 20+
- Vehicle attached jack 254 421

Turpentine (See Terpene) 585 350+
- Plant source extraction 47 10+

Turret
- Abrading work holders 51
- Aircraft type 89 37.21
- Drilling multiple spindle 408 35
- Metal working plural tools 29 33J+
  - Turrets 29 36+
- Punching pliers 83 364
- Screw threading multiple die 408 35

	Class	Subclass
Warships	114	5+
Tuyeres		
Furnace	110	182.5+
Liquid heaters	122	6.6+
Metallurgical apparatus	266	265+
TV (See Television)		
Tweezers	294	99.2
Design	D28	55
Surgical	128	354
Twill	139	426R
Twine		
Holders	242	141+
Design	D 8	360.1
Making	57	
Polishing	28	219+
Twining	57	
Twist		
Augers	408	230
Drills	408	230
Twisting and Twisters		
Conveyer belt unloading	198	831
Hold-down means	198	837+
Load held between belts	198	620+
Dough	425	319+
Dough	425	323
False	57	284+
Metal	72	299
Bending combined	72	64+
Mop wringing	15	263
Mop combined	15	120R+
Running length	72	64
Stocking support	2	337
Textile	57	
Lace fabric making	87	24+
Slat and wire fabric making	140	30+
Portable	140	39+
Twister heads	140	36
Warp thread uniting	28	210
Wire	140	149
Joining	140	118+
Portable slat and wire making devices	140	39+
Sheaf binding combined	56	132
Shocker combined	56	451+
Slat and wire fabric making	140	30+
Twister heads for slat and wire fabric making devices	140	36
Wrapping machine	53	370
Wringing	68	243
Twitchells Reagent	260	402
Fat cleavage with	260	416
Two Way Amplifier System	370	26+
Anti singing	379	345+
Telephone	379	338+
Echo suppressing	379	345+
Telephone	379	338+
Tying (See Baling, Banding, Bundling, Fastener, Knot & Knotting)		
Baling	100	
Building components	52	698+
Cords and strands	289	
Bags	53	134+
Bags combined with filling	53	134+
Dies for wire	140	116
Sewing machine stitch forming	112	156
Warp replenishing for looms	28	211
Tympani	84	419
Type		
Casting	199	
Metalworking		
Finishing and grooving	29	24
Printing	101	398+
Blanks and processes	101	401.2+
Border or ornamental design	D18	32+
Continuously rotating type head	400	901*
Design	D18	24+
Foreign alphabet type face	D18	25
Groove	101	381
Script fonts	D18	28
Setting	276	
Lead handlers	276	34
Telegraphy		
Wheel recorder	178	38+
Wheel recorder combined	178	35
Typewriter	400	
Cabinet type housing or support	312	208
Carriage feed, stepping motor for	400	903*
Carrying case	312	208
Chemical symbol character	400	900*
Collocating gauge	33	658
Controlled by speech	178	31
Cover	312	208
Cryptographic	400	89+
Design	D18	1+
Manifolding devices	400	497+
Pad	D19	99
Printer combined	101	46
Ribbon	400	237+
Testing	73	14
Subscript or superscript character	400	904*
Table	312	21+
Telegraphic transmitter combined	178	81
Typed-head continuously rotating	400	901*
Typing instruction	434	227+
Web feed, stepping motor for	400	902*
Word counter	235	102
Tyrosine	562	445
In drug	514	567
U		
Udder Shields	119	146
Uhf Transmitter	455	91+
Uhf Triode	313	293+
Ukulele (See Guitar)		
Ullage Rods	33	126.7 R
Ultramarine	106	305
Ultrasonic (See Supersonic)		
Bonding, non-metallic	156	73.1+
Dental device	433	119
Dispensing	433	86
Used in reaction to prepare inorganic or organic material	204	157.15+
Used in reaction to prepare or treat synthetic resin or natural rubber	522	100+
Vibration measuring & testing	73	570
Welding	228	1.1
Ultraviolet (See wave energy)	228	110+
Ultrasonic Repeller	D22	120
Ultraviolet (See Wave Energy)		
Anti-ultraviolet fading	430	931*
Apparatus and applications	250	493.1
Arc lamp generator	313	112
Arc lamp generator	314	
Bleaching processes	8	102
Chemical reactor apparatus	422	186.3
Drying processes	34	4
Dyeing processes	8	444
Filters	350	1.1+
Generators	250	504R+
Lamp supply circuits	315	
Lamps structure	313	112
Photographic applications	354	
Preserving sterilizing and disinfecting methods	422	24
Ray treatment	250	428+
Food preserving	426	
Sterilizing	422	21+
Signaling with	455	600+
Solar drying apparatus	34	93
Sterilizing food & beverages	426	248
Surgical	128	362+
Treatment of fluids	250	428+
Used in reaction to prepare inorganic or organic material	204	157.15+
Used in reaction to prepare or treat synthetic resin or natural rubber	522	1+
Umber	423	
Umbrella	135	20R+
Awnings	160	53
Carrier body and belt attached	224	186+
Folded umbrella	224	915*
Combined with other devices	135	16+
Design	D 3	5+
Frame making	29	25
Handles	D 3	12+
Locks	70	59
Rack	211	62+
Sewing tips to cover	112	104+
Stand	D 6	416
Sticks	135	65+
With electric light	362	102
Unbalanced to Balanced		
In an amplifier stage	330	117
With phase inverter	330	117
Uncoupling		
Railway coupling	213	211+
Cushioned hand controlling	213	17
Janney type	213	159+
Underchecks	54	57
Undercutting		
Block molding machines	425	289+
Patterns	164	245
Underframes		
Land vehicle	280	781+
Dropped frames	280	2
Railway car	105	413+
Tables	248	163.1+
Surface separable from support	108	157+
Undergarments	D 2	1+
Underground		
Antennas	343	719
Conduits	138	105
Electric cable	174	37+
Electric conductor	174	37+
Shelter	52	169.6+
Water locator	181	
Underpinning	405	230
Undertaking	27	
Body preparation	27	21+
Coffins	27	2+
Compositions	424	75
Underwater		
Boat design	D12	308
Camera	D16	2
Earth boring	175	5
Explosive spear	102	371
Guns	42	1.14+
Wells, submerged	166	335
Underwear		
Drawers	2	400+
Knitted	66	170+
Particular to female	2	406+
Particular to male	2	403+
Union type		
Bifurcated	2	78R+
Skirted	2	73
Unisex	2	400+
Waists	2	109+
Unearthing Plants or Buried Objects	171	
Unicyclic Non-repeat		
Cutting machine actuation	83	203+
Selective cutting or punching	234	51
Uniforms	2	
Design bifurcated	D 2	30
Design dress with apron	D 2	85+
Union Suits	2	70+
Unipolar		
Generator system	322	48
Machine structure	310	178
Unitary Construction, Power Plant	60	916*
Uniting (See Assembling)		
Electrolysis	204	16
Nailing and stapling	227	
Glass molds	65	156
Glass processes	65	47+
Plastic molding	264	241+
Preform with molding material	249	83+
Warp threads	28	209+
Universal Joints	403	57+
Ball and socket	403	122+
Making	29	434+
Shafting	464	106+
Unloading and Unloaders		
Aircraft in flight	258	1.2+
Bin unloading by conveyer	198	550.1+
Chutes	193	4+
Conveyor belt	198	637+
Conveyor belt, impinging fluid	198	493+
Dump scows	114	27+
Marine type	414	137+
Moving train	104	18+
Nuclear reactor fuel removal	376	261+
Pump (See regulation)	417	279+
Railway dump cars	105	239+
Railway mail delivery	258	
Scattering non-fluid material	239	650+
Self unloading vehicles	414	467
Vehicle with external means	414	373+
Washing machines	68	210
Unmoderated Nuclear Reactions Fast	376	171
Unmoderated Nuclear Reactions Fast	376	348
Unpiling		
Articles	414	112
Conveyors		
Bucket	198	701+
Bulk material	198	506+
Collecting from static support	198	519+
Establishing & moving groupings	198	418+
Rearranging streams of articles	198	434+
Unreeling	242	54R+
Round hay bale handling	414	24.6
Unsaturation		
Acrylic esters preparation	560	205+

	Class	Subclass
Carboxylic acid esters		
Acyclic	560	205
Alicyclic	560	128
Aromatic	560	8
Monocyclic	560	104
Diolefin preparation by	585	616
Fatty oils or acids	260	405.5
Olifin preparation	585	654^{+}
Untwisting (See Twisting)	57	1UN
Unwinding	242	128
Preventers	206	53^{+}
Unwrapping		
Apparatus	57	1UN
Warp preparation including	28	176
Upholstery		
Bed bottom & pad	5	401^{+}
Chair bottoms and backs	297	455^{+}
Making	29	91^{+}
Machines for sewing	112	121.23
Mattress or cushion	5	471^{+}
Separable cover	5	470
Stuffing	53	524
Pin fasteners	24	152
Trim	428	123
Uppers		
Lacing	12	58.5
Loose shaping	12	54.1
Shoe	36	45^{+}
Design	D 2	264^{+}
Making machine	12	51
Sole attaching outturned	36	16
Upsetting		
Forging press for	72	
Bolt heading	10	11R^{+}
Dies	72	395
Tire	72	305
Tire with cutting punching etc	29	22
Uracil, in Drug	514	274
Uranium	75	
Alloys	420	1^{+}
Compositions, radio active	252	625^{+}
Compound recovery, inorganic	423	253^{+}
Electrolytic production	204	1.5
Misc. chemical manufacture	156	DIG. 109
Nuclear fuel material	376	180
Nuclear fuel material	376	409
Nuclear fuel material	376	901
Organic compounds	534	11^{+}
Pyrometallurgy	75	84.1 R
Ureas	564	32^{+}
Acyclic	564	58^{+}
Adducts	260	96.5 R
Alicyclic	564	57
Aromatic	564	47^{+}
Disazo compounds	534	817^{+}
N-nitro or N-nitroso	564	33
Resins from aldehydes with	528	259^{+}
Stabilized or preserved	564	3
Sulfo acids	260	506
Synthetic rubber or natural rubber accelerator	525	346^{+}
Testing for	436	108*
Tetrakisazo compounds	534	807
Thio	564	17^{+}
Urea per se or salt, preparation	564	63
From ammonia & carbon dioxide	564	67^{+}
From ammonia & carbon monoxide or carbon oxysulfide	564	65
From ammonium carbamate	564	166
From cyano compound	564	64
Purification or recovery	564	73
Ureides (See Ureas)		
Urethane	560	157
Urethrorrhoeal Collectors of Body Fluids	604	317
Urethroscopes	128	7
Urethrotome	128	311
Uric Acid	544	266
Tests for	436	99*
Urinals		
Design	D23	302
Dry closet	4	144.1^{+}
Guards	4	DIG. 5
Water closet	4	301^{+}
Urinalysis Apparatus	422	68^{+}
Test kit	D24	17
Urine collector	D24	54
Urns		
Burial	27	1
Design	D99	5
Top for heating stove	126	219
Urological (See Surgery)		
Urotropin T M	544	185
Usnic Acid	549	461
Utility		
Meter	D10	99^{+}
V		
Vaccination		
Applier	604	46
Packaged	206	367
Shield	128	154
Vaccine	424	88^{+}
Vaccine	424	
Vacuum (See Evacuating Receptacles; Suction)		
Agitating in	366	134
Bottle	215	13R
Design	D 7	77^{+}
Bracket attaching by	248	205.5^{+}
Bracket holding article by	248	309.3
Brakes	188	
Distribution system	303	
Breaking for backflow prevention	137	215^{+}
Bulbs or tubes		
Apparatus for glass bulb manufacture	65	270
Arc lamp enclosed	313	
Electric discharge lamp	220	2.1 R
Electron tube manufacture	445	
Electron tube structure	313	
Evacuating	141	65
Incandescent lamp	313	315^{+}
Manufacture miscellaneous	445	
Process for glass bulb manufacture	65	34^{+}
Chuck	279	3
Cleaners (See air, blast or suction cleaning)	15	300R^{+}
Cleaner attachment	D32	31^{+}
Convertible	15	328^{+}
Design of machine	D32	21^{+}
Nozzles	15	415R^{+}
Pipe joint	285	7
Street sweeper type	15	340
With dirt filter	15	347^{+}
With mechanical agitator for work	15	363^{+}
Cleaning methods	134	21
Coating combined	427	294
Distillation	203	73^{+}
Distillation	203	91^{+}
Mineral oil	208	366
Mineral oil rectification	208	357
Thermolytic	201	35
Evaporators concentrating	159	22^{+}
Food preparation apparatus	99	472
Gauges	73	700^{+}
Liquid level testing	73	303
Hoist	294	64.1^{+}
Holddown	248	362
Insulated conductors	174	8^{+}
Metal treatment		
Air blast &/or vacuum conveyor	29	DIG. 78
Casting	164	61^{+}
Metal working digest	29	DIG. 44
Purifying	75	49
Vaccuum exhaust type	29	DIG. 84
Motor	91	
Systems	60	407^{+}
Pumps	417	
Design	D15	7^{+}
Dredger combined	37	59
Receptacle	220	420^{+}
X-art collection	206	829*
Receptor for body discharges	604	218^{+}
Rectifier full wave high	313	306
Refrigeration	62	
Vapor discharged	62	268^{+}
Vapor discharged, process	62	100
Resuscitation devices	128	30
Skinner for fruit	99	584
Support	248	363
Abrasive	51	362
Tube (See space discharge device)		
Keyer	328	193^{+}
Keyer in transmitter	375	68^{+}
Tester	D10	77
Wall for railway freight car	105	357
Work holders	269	21
Valances		
Bed	5	493
Curtain	D 8	373^{+}
Land vehicle top	296	135
Trunk	190	28
Valises	206	278^{+}
Valley Roof Gutter	52	13^{+}
Valve (See Faucet)	251	
Acetylene generator	48	58
Carbide feed	48	38^{+}
Actuation	251	
Electric	251	129.1^{+}
Digest	261	DIG. 74
Floor controlled spittoon	4	280
Gas & liquid contact apparatus	261	DIG. 53
Gas heated tool controlled	126	234
Gravity controlled spittoon	4	281
Internal combustion engine	123	188R
Poppet	123	90.1
Manual	251	213^{+}
Pneumatic displacement pump	417	118^{+}
Pump controlled	417	510^{+}
Time controlled	137	624.11^{+}
Time variation, internal combustion engine poppet	123	188A^{+}
Wind operated ventilator	98	117
Wind operated ventilator cowl	98	74
With time indicator	137	552.7
Applying and removing apparatus	29	213R^{+}
Automatic		
Drain	137	177^{+}
Fire extinguisher	169	19^{+}
Fluid brake equalizing	303	33
Hose holder and fluid supply	239	196
Sewerage catch basin	210	109^{+}
Sewerage trap	137	247.15^{+}
Trap	137	171^{+}
Window ventilator	98	95
Balanced		
Reciprocating	251	282
Rotary	251	283
Butterfly	251	305^{+}
Check		
Ball	137	329.3
Flow meter	73	276
Pivoted	137	527^{+}
Pump	417	559^{+}
Radiator vent	137	511^{+}
Reciprocating	137	528^{+}
Siphon bowl	4	426^{+}
Wells, in	166	325^{+}
Combined with support	251	143
Design	D23	233^{+}
Diaphragm	350	266^{+}
Dispensing	222	544^{+}
Pressure operated	222	491^{+}
Resilient type	222	490
Electrically actuated	251	129.1^{+}
Circuit makers and breakers	200	61.86
Engine distributing	91	218^{+}
Liquid pump	417	505
Lubricator	137	630.16
Railway switch	246	140^{+}
Electro optical	350	355^{+}
Polorized light	350	374^{+}
Engine igniter combined	123	151
Float	137	409^{+}
Animal water supply	119	78^{+}
Automatic dispenser	222	67^{+}
Bath or basin fitting	4	205
Closed fluid heater	236	31
Flow meter plural chamber	73	221
Flow meter single chamber	73	224
Funnel	141	199^{+}
Funnel air displacement	141	199^{+}
Pneumatic displacement pump	417	126^{+}
Pump regulator	417	40^{+}
Radiator relief	236	65
Steam trap	137	192^{+}
Temperature regulating	236	52
Thermal trap	236	53
Water closet	4	395^{+}
Water purifier	210	97^{+}
Fluid brake system		
Equalizing	303	33^{+}
Motormans	303	50^{+}
Pipeless	303	90
Fluid motor distributing	91	218^{+}
Fluid pressure actuated	251	12^{+}
Boiler feeder	122	451.2
Engine distributing	91	281^{+}
Fluid flow meter diaphragm	73	271
Fluid flow meter piston	73	249
Fluid flow meter plural	73	220

	Class	Subclass
Liquid pump	417	507
Lubricator reciprocating	184	76
Lubricator rotatable	184	72
Radiator vent thermal	236	61
Receptacle filling	141	
Steam trap	137	171$^+$
Trap thermal	236	54
Fluid pressure regulating	137	505$^+$
Fluid pressure relief	137	455$^+$
Receptacle filler	141	
Fluid servomotor	91	418$^+$
Flush		
Actuation	251	15$^+$
Actuation closet combined	4	249$^+$
Siphon bowl	4	422$^+$
Spittoon	4	266
Water closet bowl	4	435$^+$
Gas motor	415	148$^+$
Gate		
Pastry nozzle	425	381.2
Pastry nozzle	425	461$^+$
Pivoted	251	298$^+$
Pivoted check	137	527$^+$
Reciprocating	251	326$^+$
Grinder	51	27$^+$
Gun muzzle	89	31
Inflating		
Chuck or gauge combined	137	223$^+$
Pressure regulation	137	224$^+$
Inlet or supply		
Gas separation	55	418$^+$
Internal combustion engine	123	188M$^+$
Sand blast machine	51	436
Sewerage trap	137	247.13$^+$
Spittoon	4	282
Street curb for drainage	404	4$^+$
Internal combustion engine	123	188R$^+$
Large area valve, motor art	91	DIG. 3
Larner johnson type internal hydraulic	137	219$^+$
Light	350	353$^+$
Light	350	359$^+$
Liquid	137	251.1$^+$
Thermal	236	50
Locks		
Automotive fuel	70	242$^+$
Combined	70	175$^+$
Combined operation	137	384.2$^+$
Lubricator	184	86
Making repairing and assembling		
Applying and removing apparatus	29	213R$^+$
Fluid motor and pump process	29	156.7 R
Process	29	157.1 R
Refitting by milling	409	64$^+$
Measuring	222	424.5$^+$
Mixing		
Plural valves	137	602$^+$
Motormans air brake	303	50$^+$
Multiple	137	
Gas & liquid contact apparatus	261	42$^+$
Multiway unit	137	625$^+$
Bypass	137	599$^+$
Gate	137	625$^+$
Plural valves sequential opening	137	630.16
Plural valves single outlet	137	602$^+$
Unit	137	625$^+$
Music (See music)		
Needle valves	261	DIG. 38
Optical element displaced or rotated	350	484
Oscillating		
Fluid flow meter combined	73	268
Outlet exhaust and discharge		
Bath or basin	4	203$^+$
Bath or basin fitting	4	194
Bath or basin overflow	4	198$^+$
Bath or basin trap	4	197
Filter	210	418$^+$
Filter, faucet attached	210	449
Four cycle engine	123	84$^+$
Furnace ashpan	110	169
Furnace spark arrester	110	129
Gas separation	55	417
Grain huller	99	611
Internal combustion engine	123	188R$^+$
Sewerage catch basin	210	109$^+$
Sewerage trap	137	247.13$^+$
Spittoon	4	266
Spittoon floor orifice	4	276
Water closet	4	378$^+$
Pipe	251	
System	137	
Piston for pumps	417	481$^+$
Pivoted	251	298$^+$
Check	137	527$^+$
Closet bowl	4	421$^+$
Cowls	98	74
Register	98	102
Position indicator	116	277
Pressure reducing and regulating	137	505$^+$
Pressure relief	137	455$^+$
Puppet or poppet	251	82
Abrading machine	51	29
Acetylene generator	48	53$^+$
Coolant containing	123	41.16
Four cycle engine	123	79R
Internal combustion engine	123	188A
Making	29	156.7 R
Radiator vent	137	455$^+$
Thermostatic	236	61$^+$
Reciprocating	251	318$^+$
Animal watering	119	80
Biased check	137	528$^+$
Engine nozzle	239	584$^+$
Furnace ashpan	110	169
Gate	251	326$^+$
Lubricator horizontal	184	79
Lubricator vertical	184	74$^+$
Reversing	137	309$^+$
Refrigeration expansion	62	527$^+$
Automatic	62	222$^+$
Regulator		
Fluid pressure	137	505$^+$
Pump	417	279$^+$
Removal tool	29	213R
Reversing	137	309$^+$
Reciprocating regenerating furnace type	137	309$^+$
Rotary regenerating furnace type	137	311
Rocking and oscillating		
Animal watering	119	79
Four cycle engine	123	81R
Lubricator	184	34
Ventilator cowl	98	74
Wall register	98	106
Rotary or oscillating		
Disk	251	304$^+$
Floor register	98	102
Fluid flow meter	73	265$^+$
Internal combustion engine	123	190R
Lubricator force feed	184	35
Lubricator gravity feed	184	71$^+$
Plug	251	309$^+$
Safety		
Acetylene generator	48	56
Boiler	122	507
Boiler feed heater	122	437
Fluid pressure relief	137	455$^+$
High temperature cutoff	137	457
Muffler	181	237
Pipe break cutoff	137	456$^+$
Seat grinders		
Abrading	51	27$^+$
Milling	409	64$^+$
Seat making	29	157.1 R
Fluid motor	29	156.7 A
Slide		
Abrading machine	51	30
Acetylene generator	48	55
Soda water type		
Combined with plural sources	222	129$^+$
Multiple valve	137	170.1$^+$
Nozzles	239	445$^+$
Valve	251	
With agitating	366	195$^+$
With deflector nozzles	239	499$^+$
With fizzing stream	239	445
With multiple inlet nozzles	239	407
With variable discharge	239	435
Stem cap	138	89$^+$
Combined tool	152	431
Packing	277	
Switch combined	200	61.86
Thermostat and temperature	236	
Burner combined	431	83$^+$
Cutoff	137	457
Freeze responsive drain	137	59$^+$
Throttle and engineers		
Fluid brake control	303	50$^+$
Train control	246	189$^+$
Tire	137	223$^+$
Caps and cores	137	232$^+$
Combined gauge	137	227$^+$
Combined pressure regulator	152	431
Train control pipe	246	190
Well closures	166	316$^+$
Valved Device		
Acetylene generator		
Flap	48	43
Bag closure	383	44
Filling	141	68
Filling	141	315
Barrel bung	217	99$^+$
Bath basin	4	191$^+$
Battery cell, weight actuated	429	85
Battery vent	429	53$^+$
Blow-out type	429	56
Bottle		
Closure	215	311$^+$
Nonrefillable	215	17$^+$
Chute	193	21
Counterweight	193	20
Closet vent	4	216
Coating implement	401	
Cooking stove	126	52
Dispenser	222	
Drinking fountain	239	29
Bubble cup	239	31
Dry closet	4	466
Bowl	4	471$^+$
Elastic	429	54
Excavating scoop	37	136
Filter	210	418$^+$
Float	429	76
Fluid flow meter		
Diaphragm	73	265
Piston	73	248
Proportional	73	203
Fluid flow path	251	
Interlocking	48	83
Plate	48	52
Pop	48	53
Slide	48	58
System	137	
Fountain pen	401	
Funnel	141	344
Air displacement	141	297$^+$
Gas and liquid contact apparatus	261	38$^+$
Gas separator inflow control	55	418$^+$
Gas separator outflow control	55	417
Heating gas purifier	48	170
Hoist bucket	414	657$^+$
Illuminating fixture	362	394$^+$
Inhaler	128	203.12$^+$
Nasal	128	203.12$^+$
Internal combustion engine		
Cooling	123	41.1$^+$
Igniter	123	146
Liquid level gauge	73	332$^+$
Lubricator		
Force feed pump	184	26$^+$
Gravity feed	184	65$^+$
Music brasses	84	388$^+$
Pastry depositor	222	
Pastry depositor	425	381.2
Pastry depositor	425	461$^+$
Piano pneumatic	84	60$^+$
Bleeds	84	59
Pipe	251	
System	137	
Pipe organ		
Action	84	336
Key	84	342
Tremolo	84	348
Pipe taps	137	317
Piston		
Flow meter	73	248
Fluid motor	91	422
Internal combustion engine	123	47R
Pump	417	545$^+$
Pump oscillating	417	484
Pump reciprocating	417	545$^+$
Pneumatic tire		
Tire and valve stem	152	429$^+$
Tool and valve stem cap	152	431
Wheel and valve stem	152	427$^+$
Preserving apparatus	422	
Pump	417	559
Diaphragm	417	480
Regulator	417	279$^+$
Receptacle filler	141	
Refrigeration producer	62	
Refrigerator, air controlling	62	408$^+$
Automatic	62	186$^+$

	Class	Subclass
Blocking air at access opening	62	265+
Sand blast machine	51	436
Ship		
Plug	114	197
Sea cock	114	198
Ventilator	114	212
Sifter		
Discharging	209	258
Feeding hopper	209	246
Sink strainer	4	287
Smoking device	131	223
Sound muffler	181	254
Spray nozzle	239	569+
Deflector	239	506
Multiple outlet	239	562+
Whirler	239	476+
Strainer	210	156
In pipeline	210	418+
Multiple in pipeline	210	294+
Textile pounder	68	218
With plunger	68	217
Track sander	291	6
Traps		
Sewerage	137	247.13+
Steam	137	171+
Ventilation		
Air pump	98	116
Building inlet	98	41.1
Chimney cap	98	59
Thimbles	98	118
Water closet		
Bowl	4	434+
Siphon bowl	4	421+
Tank	4	378+
Vamp and Quarters	36	48+
Trimming machines	12	57.6
Vanadium		
Alloys, ferrous	420	127
Alloys, non-ferrous	420	424
Carbon compounds	556	42+
Compound stock	428	662
Electrolytic preparation	204	64R
Inorganic compound recovery	423	62
Misc. chemical manufacture	156	DIG. 106
Pyrometallurgy	75	84
Vanes		
Aircraft control	244	82
Ammunition and explosive device combined	244	3.24+
Drop bomb combined	102	384
Fuse arming	102	225+
Parachute flare	102	339
Rocket combined	102	348
Fluid flow direction measuring	73	188+
Forming working member for expansible chamber device	92	121+
Gate mounted wind balance	49	135
Meter	73	861.74+
Expansible chamber	73	232+
Pump	418	
Ships log	73	186+
Signaling (See semaphores)		
Turbine guide vane	415	148+
Ventilating cowl	98	61+
Windmill guide vane	416	12+
Vanillin	568	442
Vanity		
Cases or compacts	132	79R
Absent non-coating article eg, mirror	401	118+
Illuminated mirror	362	136
Vapor (See Gas)		
Baths	4	524+
Coating	427	248.1+
Condenser	165	110+
Contact with solids	34	
Cookstove	126	44
Design	D 7	339+
Deposition	118	715+
Process to form semiconductor P-N type junction	437	
Doping, semiconductor	118	900*
Process to form junction	437	
Electric lamp	313	567+
Arc	313	567+
Lanterns	362	263+
Electric space discharge tubes	313	567+
Generator for disinfecting	422	305+
Heating burners	431	161+
Heating burners	431	207+
Heating stoves	126	95
Liquefied or solidified	62	8
Medicators	604	23+
Testing for explosive		
Catalytic	73	25
Miners lamp	362	164+
Therapeutic baths	128	367+
Thermometer	374	201+
Treating	34	
Cabinets	312	31+
Coating processes	427	335+
Distilled vapors treated	203	
Distilled vapors treated	201	29
Gases with separation	55	
Mineral oil vapors treated	208	3
Textile apparatus	68	5R+
Textile processes	8	149.2+
Tube amplifying system	330	41
Vaporizing and Vaporization	122	
Concentrators	159	
Design	D23	360+
Distilling	201	
Distilling	203	
Drying	34	
Electrical heating for	219	271+
Fumigators	43	129+
Heat motors	60	531
Heat or energy exchange agents	252	67+
Internal combustion engines	123	250+
External vaporizing	123	522+
Jet gas pumps	417	152+
Liquids	122	
Medical inhalers	128	203.12+
Mineral oils	208	
Nuclear reactors involved in	376	370
Nuclear reactors involved in	376	371
Nuclear reactors involved in	376	378
Pumps	417	208+
Recovery of actinide compounds by	423	249
Refrigeration	62	
Resolving colloids by	252	346+
Vaporizers	261	DIG. 65
Variable Capacitor	361	277
With variable resistor	323	354
Variable Denier		
Extruding apparatus	425	76
Stretching	28	243
Variometer	336	115+
Radio tuning system	334	62
Varnish		
Asphalt	106	273R+
Fatty oil	106	246+
Natural resin		
Fatty oil	106	220+
Polymer ctg (See synthetic resin or natural rubber)		
Solvent	106	236
Removing	134	38
Materials	252	89.1+
Varnishing	427	
Shells	86	17
Vase	D11	143+
Vasodilators	424	
Vasopressin	530	315
Vats (See Vessel)		
Cheese	99	452+
Metallic and miscellaneous	220	
Textile fluid treating apparatus		
Movable carrier in vat	68	157
Separate centrifuge and vat	68	26
Vat or sulfur dyeing	8	650+
Several dyes including	8	642
Textile printing	8	453
Textile printing	8	461
Wooden	217	4
Vaulted Roof Design	D25	18+
Vaults		
Building lights	350	258+
Burial	52	128+
Lifting or handling	52	124.1+
Design	D99	1+
Dry closet	4	474+
Road or pavement vault cover	404	25+
Design	D25	36
Vcr (See Recording and Recorders)	360	33.1
Vegetable		
Cleaning		
Apparatus brushing or wiping	15	3.1+
Apparatus liquid treatment	134	
Processes	426	478+
Coffee substitutes from	426	594
Comminuting machines	241	
Cutting, comminuting, peeling, paring	99	588+
Fluid or chemical	426	287+
Fertilizers from	71	23+
Fibers		
Liberation chemical	162	1+
Liberation mechanical	19	
Gathering	56	327R
Digger type	171	
Grating	241	95+
Jellies	426	573+
Juices	426	599+
Peeling	99	584+
Pit stem or core removal		
Corers	99	544+
Processes	426	484+
Seeders and stoners	99	547+
Stemmers	99	635+
Preservation	426	321+
Canning	426	392+
Coating	426	289+
Dehydration	426	471+
Refrigeration	426	524
Protein foods	426	656
Stalks		
Retting	162	1+
Retting bacteriological	435	277+
Synthetic resins from extracts	528	1+
Vegetation		
Hydrocarbon recovery from	585	240+
Imitation	428	17+
Treated	428	17+
Vehicle (See Specific Type)		
Air	244	
Air bag passenger restraints	280	728+
Air conditioner	D23	351+
Air cushion	180	116
Cushion design	D12	5
Animal draft appliances	278	
Attachments		
Abrading	51	258
Antennas	343	700R+
Automobile carried	40	591
Body lifters	254	48
Cutting tool	83	928
Electric lamps	315	77+
Fluid sprinkling etc	239	172+
Jack	254	418+
Land	280	727+
Land wheel ornament	301	37R
Load accomodating on freight carrier	410	
Package and article carrier	224	273+
Receptacles	224	273+
Sifter support	209	421
Steps	280	163+
Weigher	177	136+
Automobile (See automobile)		
Bevel gear	74	424
Bicycle	280	200+
Bodies and tops	296	
On & off freight load bearer	410	52+
Body ornament	428	31
Brake operator		
Automatic	188	110+
Fluid pressure	188	152+
Multiple	188	106R+
Speed responsive	188	181R
Spring	188	167+
Vehicle step	188	108
Weight	188	176+
Brake position adjusters		
Supports	188	208
Vehicle body movement	188	190+
Brakes	188	2R+
Clutch combined	192	13R
Cab or body		
On & off freight bearer	410	52
Car (See car)		
Carriers, attachable	224	273+
Carriers, attachable	294	904*
Cleaning	15	53R+
Apparatus	134	123
Railway car brushing machines	15	54
Collision avoidance	367	909*
Compositions coating or plastic	106	311
Computer controlled, monitored	364	424+
Control calculator	364	424+
Convertible	280	30+
Cycle	280	200+
Dials	D12	192

	Class	Subclass
Draft devices	280	400^{+}
Animal	278	
Draft appliances	213	
Dumping	298	
External cooperating means	414	373^{+}
Railway	105	239^{+}
Electric power		
Locomotive	104	35^{+}
Locomotive	104	49^{+}
Motor vehicle, traction motor	180	65.1^{+}
Motors	D13	21
Electricity transmission to	191	
Excavators		
Self loading	37	4^{+}
Vehicle actuated scoops	37	132
Fender	293	
Design	D12	186
Floor mat design	D12	203
Fluid cushion	180	116^{+}
Fluid handling device supported by	137	899^{+}
Freight accomodating	410	
On & off freight container	410	52^{+}
Gas pumps and fans		
Combined	417	331^{+}
Fans combined	416	146R
Operating devices vehicle actuated	417	231^{+}
Ground effect	180	116^{+}
Headlights and spotlights	362	61^{+}
Design	D26	28^{+}
Headrest	D 6	501
Heat and power plant	237	12.3 R
Motor vehicle boiler plants	122	3
Heating	237	28^{+}
Heating and cooling	165	41
Inflatable passenger restraint	280	728^{+}
Instrument panel	D12	192
Land (See land vehicles)	296	
Bodies & tops	296	
Dumping	298	
Freight accomodating	410	
Motor	180	
License plates	40	200^{+}
Lighting	362	61^{+}
Circuits	315	77^{+}
Design	D26	
Loading and unloading		
Chutes	193	5
Dumping	298	
External cooperating means	414	373^{+}
Self loading or unloading vehicles	414	467^{+}
Locks	70	237^{+}
Medicine		
Inorganic	514	769^{+}
Organic	514	772^{+}
X-art collections	514	800^{+}
Mirrors	350	600^{+}
Mirror frames	52	
Plural	350	612^{+}
Plural combined	350	612^{+}
Signal	350	97
Moped	180	205^{+}
Motion picture	352	132
Motor	180	
Boiler plants	122	3
Supports	248	637^{+}
Motorcycle	180	210^{+}
Two-wheeled	180	218^{+}
Tandem	180	219^{+}
Movement display	434	305
Operation teaching	434	30^{+}
Periscopes	350	540^{+}
Railways	104	
Freight accomodating	410	
Rolling stock	105	
Making	29	168^{+}
Refrigerated	62	239^{+}
Automatic	62	133^{+}
Process	62	61
Roof rack	D12	157
Running gear		
Land vehicle	280	80R^{+}
Motor vehicle	180	85^{+}
Railway trucks	105	157.1^{+}
Suspension	280	688^{+}
Stub axle mounts	280	660^{+}
Safety belt or harness	280	801^{+}
Passive	280	802^{+}
System responsive	180	268^{+}
With seat structure	297	468^{+}
Safety guard	280	748^{+}

	Class	Subclass
Safety promoting means	180	271^{+}
Seat belt passenger restraints	280	801^{+}
Seats	297	195^{+}
Shelter support	135	88^{+}
Simulators	434	29^{+}
Snowmobile	180	182^{+}
Special purpose	D12	1^{+}
Springs	267	2^{+}
Steering mechanism moves pivoted lamp	362	37
Steering switches light system	362	36
Suction effect	180	164
Surface effect	180	116^{+}
Suspension	280	688^{+}
Springs	267	2^{+}
Tires, resilient	152	
Tool driving or impacting device combined	173	22^{+}
Tool, mounted	83	928*
Toy	446	431^{+}
Design	D21	128^{+}
Trains		
Draft appliances	213	
Electricity transmission to	191	11
Land	280	400^{+}
Motor	180	142
Braking	188	112R^{+}
Railway	105	1.4^{+}
Braking	188	124
Tricycle	280	200^{+}
Washer	134	123^{+}
Head lamp	239	284.2
Windshield	239	284.1
Water, ie ships	114	
Buoys, rafts, aquatic devices	441	
Marine propulsion	440	
Weigher combined	177	136^{+}
Wheel propelled	280	3
Wheel substitutes	305	
Wheelchair (See wheelchair)	280	242WC
Wheels & axles (See axle; wheel)		
Axle making	72	
Land	301	
Railway	295	
Resilient wheels	152	
Wheel making	29	159R^{+}
Veils	2	207
Design	D 5	47^{+}
Veining		
Textile dyeing	8	478^{+}
Velcro T M	24	306
Belt fasterners	24	31V
Brackets attached by	248	205.2
Buildings, static structures digest	52	DIG. 13
Chairs & seats digest	297	DIG. 6
Separable fastener	428	100
Surgery digest	128	DIG. 15^{+}
Velocipede		
Brake	188	24.11^{+}
Canopies	135	88
Design	D12	107^{+}
Figure toys	446	440
Land vehicle		
Coasters	280	87.1^{+}
Convertible	280	7.1^{+}
Dust and mud guards	280	152.1^{+}
Occupant propelled	280	200^{+}
Occupant steered	280	263^{+}
Simulations	280	1.1 R^{+}
Wheel guards	280	160.1
Windshields	296	78.1
Racks	211	17^{+}
Locking	211	5
Seats for	297	195^{+}
Walkers nonsteered	280	87.2
Velvet	139	391^{+}
Velveteen	139	391^{+}
Vending Machine	D20	1^{+}
Veneer and Veneering		
Laminate	428	411.1^{+}
Metal-to-metal	428	615^{+}
Static mold on product or existing building structure	249	15
Tile adhered	52	390
Tile backer	52	384
Woodworking		
Lathe	144	209R^{+}
Press	156	349^{+}
Venetian Blind (See Blind)	160	166R^{+}
Cleaning machines	15	77
Wiping	15	102

	Class	Subclass
Design	D 6	577
Making	160	405
Assembly	29	24.5^{+}
Method	160	405
Tape cutting	26	11.4
Woven float cutting	26	7
Wood slat slotting	144	136R
Venom Extractors	604	314
Ventilators & Ventilated Items	98	
Abrasive tool or support	51	356
Aircraft cabin having	98	1.5
Automobile having	98	2^{+}
Barrel	217	74
Bath closets sinks spittoons	4	209R^{+}
Bed mattresses	5	468^{+}
Fluid filled	5	453
Forced air heating, cooling	5	423
Blowers	415	
Boot or shoe	36	3R
Building	98	29^{+}
Cabinets having	312	213
Car	98	4^{+}
Chimneys	98	58^{+}
Cleaning or polishing tool or support	15	230.1
Consumable electrode arc lamp having	314	26^{+}
Cooking oven	126	21R
Cowls	98	61^{+}
Crankcase	92	78^{+}
Crates	217	42
Door	98	87
Dry closet		
Inclosed receptacle	4	477
Receptacle	4	482
Valved bowl	4	472
Vault	4	475
Electric batteries	362	74
Electric batteries	429	82^{+}
Cap type	429	89
Gang type	429	87
Non-spill type	429	84
Plug type	429	89
Reactive, absorbable type	429	86
Separate inlet and exhaust	429	83
Stopper	429	89
Weight actuated valve type	429	83
Electric cables conduits boxes	174	16R
Electric insulators	174	187
Equipment	D23	370^{+}
Fans	416	
Fire escapes	182	47
Hats	2	171.3
Sweats	2	181.6
Heating and cooling systems	165	59
Heating systems	237	46^{+}
Hoods and offtakes	98	115.1^{+}
Light projectors	362	294
Locomotive cab	98	3
Dust guards	98	28
Louvers	98	121.1^{+}
Mask	128	205.25^{+}
Mine	98	50
Oven doors	126	198
Pavement vault cover	404	25^{+}
Photographic dark cabinet	354	309
Pressure maintenance	98	1.5
Provision safes	98	51
Railway cars	98	4^{+}
Railway tunnels and subways	98	49
Registers	98	101^{+}
Roof eaves	98	DIG. 6
Ships	114	211^{+}
Shoes, making	12	142V
Solar power	98	900*
Storage	98	52^{+}
Stove attachments	126	80
Fluid fuel stoves	126	84
Platforms	126	279
Stovepipe combined	126	312
Damper combined	126	293
Thimble combined	126	316
Temperature regulated	236	49
Tents having	135	93
Umbrella cover	135	35V
Water closet		
Siphon bowl	4	351^{+}
Urinal	4	306^{+}
Valved bowl	4	348^{+}
Washout bowl	4	348^{+}
Window	98	88.1^{+}
Vents		
Automobile no-draft	49	390^{+}

	Class	Subclass
Ball point pen	401	217
Building with	52	198$^+$
Fountain pen	401	242
Molding devices		
Chills	164	372
Metal founding	164	410
Metal founding core	164	234
Metal injection die having	164	305
Static molds	249	117$^+$
Ordnance	89	30
Pump priming and venting	417	435
Rotary	415	11
Receptacle		
Batteries	429	82$^+$
Bottle closure	215	307$^+$
Check valves for radiator	137	511$^+$
Closures	220	367$^+$
Cutter for vent and dispensing openings	222	85$^+$
Dispenser	222	478$^+$
Dispenser handle with	222	468
Dispenser movable trap chamber with	222	332
Dispenser trap chamber with	222	442
Dispenser with fluid trap seal for	222	188
Dispenser with follower and	222	387
Float closure for dispenser vent	222	69
Fluid operated lubricator	184	59
Fluid pressure brake systems	303	70
Tube bottle closure	215	307$^+$
Water heating	126	389
Wick burner fuel supply	431	321
Rubber or heavy plastic mixer	366	75
Sifting bins with	209	378
Valve		
Closet etc ventilation	4	216
Independent multiple air	137	583$^+$
Multiway rotary plug	137	625.22$^+$
Serial dependent drainage	137	512$^+$
Direct response	137	512$^+$
Selective motion actuation	137	636$^+$
Separate actuators	137	637$^+$
Sequential	137	628$^+$
Venting	425	812*
Wall with	52	302$^+$
Venturi		
Carburetor	261	DIG. 12
Flow meter	73	861.63$^+$
Restrictor type	138	44
Scrubbers	261	DIG. 54
Variable venturi	261	DIG. 56
Veratramine	546	195
Veratridine	546	34
Veratrol	568	648
Verel T M (polyacrylonitrile Vinylidine Chloride)	526	342
Verifier		
Card or tape punch	234	34
Verifying Statistical Cards	73	156
Vermicelli	426	557
Vermiculite	423	
Compositions including	106	122
Alkali metal silicate	106	75
Ceramic	501	141$^+$
Pigment	106	291
Settable cement	106	86$^+$
Expanding	252	378R
Vermilion	423	511
Vermin Destroying	43	124$^+$
Antivermin treatment of animals	119	156$^+$
Traps	43	58$^+$
Currycomb combined	119	87
Design	D22	119$^+$
Vernier (See Machine with which Combined)	33	1D
Ruler type	33	494
Tabular calculator type	235	85R$^+$
Veronal T M	544	307
Vessel (See Crucible; Tank; Tub; Receptacles; Vats)		
Drinking	215	
Design	D 7	6$^+$
Metallic	220	
Heaters		
Electric	219	438$^+$
Water	126	373$^+$
Receptacles		
Dry closets	4	479$^+$
Measuring	73	426$^+$
Metallic and miscellaneous	220	
Portable fire extinguishers	169	30$^+$
Suspended household filter	210	473$^+$
Testing sealed	73	52
Water heating	126	373$^+$
Ships	114	
Insubmergible	114	68$^+$
Raising and docking	114	44$^+$
Submarine	114	321
Staved wooden	217	72$^+$
Insulated	217	131
Vestibule		
Railway train connection	105	8.1$^+$
Stock cars	119	14
Vestment or Ceremonial Robe	D 2	79$^+$
Vests	2	102$^+$
Coat combined with	2	95
Design	D 2	184
Design	D 2	190
Design vestees	D 2	183
Veterinary		
Animal care and handling	D30	
Animal dips and sprays	424	
Animal poisons	424	
Dental instrument	433	1
Diagnostic		
Gag	128	14
Pivoted specula	128	19
Equipment	D24	
Medicines	424	
Obstetric forceps	128	324
Oral dosing device	604	77
Vetograph	350	396
Vibrating Cup	277	97$^+$
Vibrating Motor	310	15$^+$
Piezoelectric	310	311$^+$
Systems	318	114
Vibration (See Jarring; Vibrator)		
Ammunition igniters		
Marine mine	102	416
Mine	102	424
Audio intensifiers	381	161
Dampers		
Centrifugal extractor with	68	23.1$^+$
Centrifugal extractor with	210	144
Centrifugal extractor with	210	363
Electric brake	310	93
Electricity conductors etc overhead	174	42
Flywheels with	74	574
Hammer	173	139
Heat exchanger element	165	69
Inductor	336	100
Interia	188	378$^+$
Internal-resistance motion retarding	188	266$^+$
Machinery bases	248	560$^+$
Machinery brackets	248	560$^+$
Machinery supports	248	638
Machinery suspension	248	610$^+$
Plastic deformation	188	371$^+$
Resilient deformation	188	268
Separator, imperforate bowl centrifugal	494	82
Shaft structure	464	180
Dispenser feeding		
Articles	221	200$^+$
Discharge assistant	222	196$^+$
Grain hulling apparatus	99	609$^+$
Nail etc making	10	172$^+$
Planter with chute	111	76
Planter with hopper	111	75
Supply container	222	161
Earth boring by	175	56
Electrical applications to body	128	424
Electrical liquid separation with	204	307
Grain separator straw carriers	130	25
Gravity feed lubricator cutoff	184	69
Heat exchanger surface	165	84
Kinesitherapy	128	32$^+$
Vibrant membrane	128	64
Liquid purification		
As treatment of liquid	210	748
Filter	210	388$^+$
Flocculation or precipitation	210	738
For cleaning filter	210	785
Membrane process	210	637
Measuring and testing	73	507$^+$
Recording combined	346	7
Metal treatment	148	12.9
During casting	164	71.1
Vibrating X-art	72	710
Continuous casting	164	478
Welding apparatus	228	1.1
Mining by	299	14
Motor or generator	310	51
Pipe joint	285	48$^+$
Rotating shaft for	74	574
Supports, nonresilient	248	636
Supports, resilient	248	562$^+$
Tool driving device	173	162R
Transformer	336	100
Misc. chemical manufacture	156	DIG. 91
Musical instrument	84	
Rigid	84	402$^+$
Tuned sound wave generator or resonator	84	1.4$^+$
Razor	30	44$^+$
Road and earth compacting	404	113$^+$
Road finishing by	404	113$^+$
Signs	40	613$^+$
Textile weaving	139	
Toy	446	3
Air actuated reed	446	207$^+$
Typewriter ribbon mechanism	400	215$^+$
Vegetable etc cutting processes	426	518$^+$
Warp stopping	28	188$^+$
Well processes	166	249
Vibrato	381	62
Vibrator (See Vibration)		
Agitating	366	127
Bodine vibrator	366	600*
Electric	310	15$^+$
Medical	128	32$^+$
Motor systems	318	119$^+$
Motor systems	318	114
Feed conveyer, vibratory type	209	920*
Geophysical	367	189
Mechanical sensor	73	662$^+$
Metal casting device having	164	260$^+$
Power supply	307	150
Inverter system	363	110
Sifting	209	381
Sonar or submarine type	367	141$^+$
Sound mold compactor	164	203$^+$
Ultrasonic measuring & testing	73	570$^+$
Victrola T M (See Phonograph)		
Video Disc Player	369	
Capacitive	369	126
Disc structure	369	276
Disc player mechanism	369	176$^+$
Changer	369	178$^+$
Magnetic	360	33.1
Mechanical	369	127$^+$
Optical	369	100$^+$
Disc structure	369	275
Pause control	358	908*
Picture signal processing	358	342$^+$
Color	358	322
Stored modulated carrier signal	369	61
Video Pentode	313	300
Video Tape Recorder	360	33.1$^+$
Commercial killers	358	908*
Track skippers	358	907*
View Finder		
Camera	354	219$^+$
Viewing Devices	40	361$^+$
Stereoscopic	350	133$^+$
Vignetting		
Photographic	354	296
Contact printing	355	125
Projection printing	355	40$^+$
Vigroles	238	125$^+$
Vinaigrette	D11	184
Vinblastine	540	478
In drug	514	283
Vincamine	546	51
In drug	514	283
Vincristine	540	478
Vinegar	426	650
Apparatus	99	323.12
Vinyl Compounds		
Cyanide	558	462
Esters from acetylene	560	242
Ether	568	687$^+$
Solid polymers	526	59$^+$
Vinylidene Compounds	526	343
Processes of molding articles of vinylidene chloride	264	DIG. 79
Solid polymers	526	59$^+$
Viola (See Violin)		
Violin	84	274$^+$
Bows	84	282
Bridges	84	309

	Class	Subclass
Mutes	84	310^+
Cases	206	14
Chin rests	84	279
Design	D17	17^+
Pianos with bowed strings	84	256^+
Strings and fastenings	84	$297R^+$
Supports and shoulder rests	84	280
Tailpieces	84	302
Violin Cello (See Violin)		
Viquidil, in Drug	514	314
Virus Culture	435	235
Viscose	536	60^+
Compositions	106	164^+
Alkali metal silicate	106	81
Casein	106	141
Gelatine or glue	106	128
Ink	106	26
Magnesium oxy chloride	106	108
Plaster of paris	106	115
Polishes	106	5
Portland cement	106	93
Prolamine	106	151
Protein miscellaneous	106	158
Filament making processes	264	188^+
Ripening	536	61^+
Viscosity		
Control	137	92
Dividing lamina	52	408^+
Measuring	73	54^+
Vises	269	
Anvil	72	457^+
Anvil attached	72	457^+
Daguerreotypy	354	2
Design	D 8	74
Drilling machine	408	$72R^+$
Fruit jar	81	3.31^+
Milling machine	269	
Saw clamping	76	78R
Tool attached	29	560.1
Type casting	199	50
Vision (See Eye; Optics)		
Testing & correction	351	
Visor, Sun	350	
With article carrier	224	312
Visors for Caps	2	195
Hat with	D 2	244^+
Visual Display (See Display)		
Communication	340	700^+
Vitamin		
A	568	824
B1	544	327^+
B2 or riboflavin	544	251
B6	546	301
C	549	315
Composition	424	
D	260	397.2
Design	D28	1^+
E	549	408
Electrostatic treatment	204	166
Foods containing	426	72^+
G (See vitamin b2)		
K	260	396R
Preparation by wave energy	204	157.67
Vitreous (See Glass)		
Vitrification or Devitrification		
Color processes photo	430	198
Glass composition	501	2^+
Glass preform	65	33
Voice		
Artificial electric	381	
Doll	446	297^+
Air operated	446	188^+
Operated circuits relay	361	160^+
Reflector for therapy	D24	35
Void Former	52	577
Within module	52	576
Voile	139	426R
Volleyball	273	411
Voltage		
Differentiator electric	333	19
Electronic tube type	328	127^+
Plural different voltages	307	15
Integrator electric	333	19
Electronic tube type	328	127^+
Magnitude & phase control systems	323	
Regulation electronic	323	
Regulator (See regulator; electricity)		
Stabilization (See regulator)		
Voltmeter (See Galvanometer)		
Amplifier tube for	313	293^+
Volume Control Automatic		
Amplifier type with		
Bias control	330	129^+
Impedance control	330	144^+
Thermal impedance control	330	143
Public address type	381	
Radio receiver type	455	234^+
Input coupling impedance adjustment	330	185^+
Interstage coupling impedance adjustment	330	157^+
Output coupling impedance adjustment	330	192^+
Volume Control Manual		
Amplifier type by		
Bias adjustment	330	129
Cathode self bias impedance adjustment	330	142
Volume Meters		
Calibration of	73	3
Dispenser combined	222	71^+
Check controlled	194	
Preset cutoff	222	14^+
Fluid	73	861^+
Liquid level	73	$290R^+$
Measurers	33	1V
Measuring vessel	73	426^+
Proportional flow control	137	87^+
Receptacle filling		
Stationary measure	141	
Traveling measure	141	135^+
Timing fluid flow	137	552.7
Volumetric content measuring	73	149
Vortex Tube for Cooling	62	5
Vortical Separator	209	
Gas separation	55	447^+
Liquid contact combined	55	235^+
Voting		
Booths	52	36
Flexible wall	52	63
Computer controlled, monitored	364	409
Machines	235	51^+
Teaching	434	306
Vulcanizable Gums		
Apparatus for	425	$28R^+$
Apparatus for	425	113
Apparatus for	425	381.2
Dyeing of	8	
Electrophoresis involving	204	180.3^+
Synthetic resin (See synthetic resin or natural rubber)		
Vulcanizing		
Machinery design	D15	199
Molding apparatus for composite articles	425	500^+
Molding devices	425	363^+
Molding devices	425	383^+
Presses	425	335^+
Processes		
Molding combined	264	
Synthetic resin or natural rubber	525	50^+

W

	Class	Subclass
Wabble Mounted Circular Saw	144	238
Wabbler Plate	74	60
Wadding or Filling for Containers	493	967*
Wads	102	532
Columns	102	532
With cartridge	102	448^+
Waffle Irons	99	372^+
Batter feeding combined	99	373
Design	D 7	359
Waffle Type Ceiling	52	337
Wagering, Computer Controlled	364	412
Waging War	89	1.11
Wagons	280	
Bodies	296	
Design	D12	17^+
Dumping	298	
Ice	62	239^+
Lunch		
Heater	126	268
Ovens	126	276
Bakers	126	276
Toy	D21	134^+
Wainscoting	52	506^+
Waistbands		
Skirts	2	220^+
Closures	2	219
Special sewing machine	112	121.27
Trousers	2	236^+
Union type skirted	2	76
Waists		
Body garments	2	104^+
Wakers (See Alarms)		
Walkers		
Baby	272	70^+
Design	D12	130
Water	441	76^+
Walkie Talkie Radio	455	73^+
Separate transmitter and receiver	455	39^+
Walking		
Aids	135	68^+
Seats	297	5^+
Cane or stick	135	65^+
Irons	128	83.5
Mechanism for dolls	446	377^+
Motor driven	446	355
Walkways	404	17^+
Wall		
Absorbent filtering gas	34	81
Aperture thermometer support	374	208
Board cutting	83	
Bracket attached to wall anchor	248	231.91
Building (See facer construction)	52	
Article support	52	27
Coated	52	515^+
Heat exchanging	165	47^+
Hollow block	52	503^+
Hollow block static mold	249	144^+
Hollow ventilating	98	31
Lateral block courses	52	561^+
Plasterboard	52	344^+
Static mold for making	249	33
Surfacing abrading machine	51	180
Ventilating register	98	106^+
Cabinet	312	245
In wall or panel recess	312	242
Chute	193	33^+
Cleaning machines	15	$49R^+$
Decoration or plaque	D11	132^+
Electric furnace		
Resistance	373	109^+
Resistance in	373	
Water cooled		
Arc	373	75
Induction	373	138
Resistance	373	113
Exterior elevator	187	6^+
Freight car vacuum	105	357
Heating and cooling panel	165	49
Insulator through wall	174	151^+
Ironing board attached laterally of support	108	152
Folding	108	134^+
Ladder with platforms	182	83^+
Mirror plural	350	612^+
Mounted conduits	174	48^+
Outlets	174	48^+
Wiring	174	50^+
Panels, continuous molding of	264	46.4^+
Plug electrical	439	638^+
Protector for furniture	248	346.1
Racks	211	
Receptacle		
Bottle or jar spaced	215	12.1^+
Electrical	174	53^+
Insulated, wooden	217	131
Metallic	220	83^+
Metallic spaced	220	415^+
Metallic transparent	220	82R
Outlet wall mounted	220	3.3^+
Paper	229	4.5
Pitcher spaced	222	131
Purse	150	127^+
Trunk	190	23
Wooden box	217	17
Insulated	217	128^+
Retaining	405	284^+
Cribbing	405	273
Sheet piling	405	274^+
Wave or flow dissapating	405	30^+
Retort	202	223
Safe and bank	109	78^+
Table attached laterally of support	108	152
Folding	108	134^+
Water for liquid heater	122	106
Bridge	122	192
Wallets	150	132^+
Wallpaper	428	346^+
Cleaning device	15	219
Design	D 5	
Steamer	126	271.1
Walnuts		
Novelties, jewelry motif	D11	117

U.S. PATENT CLASSIFICATIONS

	Class	Subclass
Plants	PLT	32
Ward		
Distributing	276	26
Setting combined	276	6
Wardrobe	312	
Trunk type	190	13R+
Warfare		
Electronic	342	13
Gases	424	
Teaching	434	11
Warmers		
Bed mattress	5	421+
Bedstead	5	284
Body	126	204+
Heat exchanger	165	46
Heated shoe	36	2.6
Bottle	99	359
Food	126	
Household appliances	D 7	362+
Foot warming radiators	237	77
Receptacles	220	3.1
Solar	D 7	324
Surgical devices	128	362+
Bandaging	128	82.1
Instruments	128	303.1+
Kinesitherapy	128	24.1+
Medicators	604	113+
Orthopedic	128	68.1
Receptors	604	358+
Warnings (See Indicators; Signal)		
Warning or flashing signs	D10	114+
Warp		
Frames	139	34
Irregular feed	139	24+
Knitting		
Circular independent needles	66	10
Circular united needles	66	81
Fabrics	66	192
Fabrics having unknit materials	66	195
Straight united needles	66	203+
Straight united needles unknit material incorporating	66	84R+
Knotting machines	28	209+
Panel straightener	52	291
Pile weaving	139	37+
Preparation	28	172+
Shedding mechanism	139	55.1
Stop control	139	349+
Winding	242	26.2+
Warships	114	1+
Wash Basin	D23	284+
Sink, beauty parlor	D28	20
Wash Tub (See Receptacles)	68	232+
Attachments	68	237
Wringer support	68	238
With scrub surfaces	68	233+
Washboards	68	223+
Design	D32	67
Supports	248	27.5
Washboiler		
Attachments	68	237
Wringer support	68	238
With thermal flow means	68	191+
Washers Hardware	411	531+
Bolt or nut to substructure lock	411	136+
Apertured plate type	411	155+
Closed loop, thickness varies	411	160+
Plural washers	411	137
One yieldable	411	150
Split ring type	411	152+
Washer in counterbore	411	148
Bolt to nut lock		
Deformable washer	411	313+
Including thread lock	411	260+
Pawl carrying washer	411	330+
Design	D 8	399
Design, fluid handling	D23	269
Hollow rivet setting	227	51+
Making	10	72R+
Punching	83	
Tag attaching	493	375+
Thrust bearing	384	420+
Washing Compositions		
Detergents	252	89.1+
Soap containing	252	108+
Washing Devices (See Scrubbing)		
Clay	209	
Clothes	68	
Coal ore and sand	209	
Combined washer-drier	D32	5
Dishwasher	134	
Earth bore device combined	175	208
Filter	210	409+
Fish	17	65
Fluid treatment	134	
Drying combined	68	19+
Textile	68	
Textile processes	8	137+
Grain	134	
Head	4	515+
Implement with water supply	401	
Machine systems	60	908*
Machinery	D32	1+
Mineral oil	196	46
Photographic	354	297+
Scrubbing combined	15	
Smoke tobacco	131	173
Starch	127	25
Textiles	68	
Vehicle		
Overhead supply and spray	239	209
Stationary or track type	166	157
Well screen with washing point or shoe	166	157+
Washstands	312	228
Cabinet with basin or tub	312	228
Design	D23	284+
Plumbing with	4	625+
With drain	312	229
Waste (See Reclaiming; Recovery)		
Acid from	562	513
Cutting waste product	83	923*
Disposal in soil	405	128+
Electrolytic treatment	204	149+
Fertilizer from	71	25+
Heat utilizing steam generators	122	7R
Lignins from sulfite liquor	530	500+
Matter utilizing in manufacture of fats	260	412+
Nuclear material	252	636+
Phenols from	568	749+
Recovering from cards	19	107
Sorting municipal solid waste	209	930*
Sulfite liquor (See sulfite liquor)		
Tall oil	260	97.7
Watch	368	62+
Assembling		
Of crystal	29	807
Process	29	177+
Staking	29	231+
Attachments	63	21+
Band or strap	224	164+
Bezel with crystal	368	294+
Bracelet	63	3+
Carrier	224	180
Convertible to different device	224	152
Casing for watch movements	206	18
Chains	59	80+
Watch and chain attachments	63	21+
Cleaning	15	
Crystal	D10	132
Design	D10	30+
Electric	D10	30+
Fob or chain	D11	79+
Gauges	33	659
Holder for timepiece not carried on wrist	224	903*
Illuminated	362	23
Movement construction	368	76
Strap	D11	3+
Strap in combination with watch	D10	32
Structure	368	88+
Support	248	114+
Timing testers	73	6
Tools	81	6+
Compound	7	101+
Wrenches	81	122+
Wheel gauges	33	203.1
Watchmens Signal System	340	287+
Time clock recorders	346	
Water		
Additive or substitute	4	DIG. 1
Aeration	261	
Flotation, processes	210	703+
Processes	210	749+
Separation apparatus	210	198.1+
Waste water processes	210	601+
Backs		
Cookstove	126	34+
Cookstove liquid or gaseous fuel	126	53+
Fireplace	126	132+
Bag		
Thermal material	383	901*
Thermal material metallic	220	3.1
Thermal medicator	604	113+
Thermal surgical	128	403
Bath heating apparatus	126	344+
Bed	5	451+
Blowdown of boilers	122	381+
Boiler purifying	210	311+
Bottle	425	813*
Can	D23	212
Carbonated making	261	
Catalytic recombiners	422	177+
Closets	4	300+
Closet design	D23	295+
Pipe joint	285	56
Conditioner or softener	D23	207
Control, channel preservation	405	52+
Control hydraulic engineering	405	52+
By valve in closed conduit	137	219+
Cooler, machine design	D 7	304
Craft		
Aircraft landing gear	244	105
Aircraft water propulsion steering	244	50
Amphibian	244	101
Brake	244	112
Buoys	441	
Marine propulsion	440	
Ships	114	
Toy	446	153+
Depth measuring		
By radiant energy	342	123
Gage	73	290R+
Sounding line	33	126+
Distillation	203	10+
Distilling	203	10+
Distribution (See distributor)		
Design	D23	200
Systems	137	561R+
Electrolytic production	204	101
Faucet attachment	239	428.5
Filter	D23	209+
Gage posts	33	126.7 R
Gas manufacture	48	204+
Cupola	48	78+
Retort	48	108+
Heaters	126	344
Design, heater or boiler	D23	318+
Electric	219	281+
Liquid heaters and vaporizers	122	
Heating systems for buildings	237	56
Compounds	423	580
Electrolytic synthesis	204	101
Engaging aperture or notch of	237	63
Hydraulic engineering (See hydraulic)	405	
Marks in paper making	162	110
Observing	356	23+
Measuring depth of	73	290R+
Meter	73	861+
Nuclear coolant or moderator	376	174
Nuclear coolant or moderator	376	221
Nuclear coolant or moderator	376	350
Pipe		
Carrying RF telephone signal	379	90
Pistol	222	79
Public utility meter	D10	99+
Purification (See purifiers)	210	
Chemical compositions for	252	186.1+
Distilling for	203	10+
Roundabout	272	32
Skimmer and walker	441	65
Softeners	252	175+
Sprinklers	239	
Embedded	239	201+
Pipes	239	550+
Sprinkling cans	239	375+
Substitute or additive	4	DIG. 1
Towers	169	25
Tube boilers	122	235R+
Underground detectors		
Acoustic type	181	
Borehole and drilling study	73	151+
Electric	324	323+
Water culture	47	59+
Wheels		
Current operated	416	
Water Glass		
Compositions		
Bleach	252	99
Detergent	252	135
Molding or coating	106	74

	Class	Subclass
Soap	252	109
Sorbent	502	407
Compounds	423	326
Waterbed	5	451+
Watering Devices for Animals	119	72+
Feeding and	119	51.5
Design	D30	121+
Stock car feeding and	119	10
Waterproof		
Building structure	52	
Coating	52	515+
Drain for window	52	209
Drain or vent	52	302+
Exterior flashing	52	58+
Guttering	52	11+
Liquid deflector	52	97
Resilient seal	52	393+
Collars	2	135
Electroplating nonconducting base	204	31
Fabrics	428	244+
Overcoats	2	87
Processes of waterproofing	264	DIG. 8
Rubber footwear	36	4
Overshoes	36	7.3
Seatcover	297	184
Special receptacle or package	206	811*
Wattmeter		
Dynamometer	324	144+
Induction disk	324	137+
Radiant energy	250	472.1
Photocell type	324	96
Radio type	455	115
Wave		
Acoustic systems & devices	367	
Analysis	324	77R
Computer controlled, monitored	364	417
Buoys	441	1+
Calming oil distributors	114	232+
Buoys	441	34
Energy & wave transmission (See acoustics; electricity; heat; prospecting; radiant energy; radio; sound; tide)	455	
Along wave guide	333	239+
Block signal systems hertzian waves	246	30
Compressional wave signaling	367	
Electric musical instrument	84	1.4
Electric wave transmission	333	
Explosive mine firing	102	427
Fuses, primers, and igniting devices	102	211
Guides electrical	333	
Interference type electric musical instrument	84	1.2
Marine mine firing	102	416+
Preserving disinfecting and sterilizing processes	422	20+
Microwave	422	21
Radio	455	
Pulse or digital	375	37+
Shaping	328	34+
Sound and supersonic generators	116	137A
Surgical or medical applications (See radiant energy)		
Textile treatment	8	DIG. 12
Tuners	334	
Use in nuclear plasma reactions	376	100
Used in reaction to prepare inorganic or organic material	204	157.15+
Used in reaction to prepare or treat synthetic resin or natural rubber	522	1+
Explosive devices	102	366
Fluid breakwaters	405	22
Form analysis		
Electrical	324	77R
Generator		
Earth boring combined	175	56
Electric oscillations	174	68R
Water	405	79
Guide electric	333	239+
Antenna	343	772+
Conduit only	138	118+
Electrical feature	174	68R+
In cathode ray tube	315	4+
In electronic tube	315	39
Making	29	600
Standing wave indicator	324	58R
With frequency characteristic	333	239+
With insulation	138	137+
Meters	250	250
Motors	60	495+
Resonator	333	219+
In cathode ray tube	315	4+
In electronic tube	315	32+
Shaping		
Electronic tube type system	328	34+
Transmission, wireless	455	
Pulse or digital	375	37+
Waving Hair		
Methods	132	7
Wax		
Centrifugal honey extractor filter	210	361+
Coating with	427	
Compositions		
Abrasive	51	305
Biocidal	424	
Coating or plastic	106	
Detergent	252	89.1+
Ester type	260	398+
Inks	106	31
Mold	106	38.8
Mold coating	106	38.25
Polishes	106	10
Synthetic resin or natural rubber ctg (See synthetic resin or natural rubber)		
Dyeing	8	521
Electromagnetic wave treatment	204	157.15+
Electrostatic field or electrical discharge treatment	204	167
Ester type	260	398+
Fermentative treatment	435	271
Matches	44	42
Melting apparatus for honey separation	210	175+
Modeling dental	433	213
Paper	428	486
Paraffin	208	20+
Waxing		
Bobbin winding combined	242	18R+
Coating pads and forms	118	264+
Leather sewing and	112	42
Shoe making	12	79.5
Waxy hydrocarbon polymer production	585	946*
Ways		
Conveyor	193	
Marine	405	1
Railway		
Freight cars for tubular	105	365
Suspended	104	123
Tubular	104	138.1+
Weaning Devices	119	134
Weapons (See Ammunition; Fireamrs; Ordnance)		
Wear		
Building surface	52	177+
Compensating brake beam	188	214+
Compensating radial bearings	384	247+
Compensating wheel bearings	384	261+
Compensators for clutches	192	110R
Distributor for typewriter platen	400	554+
Earth boring bit indicating	175	39
Element removable from pipe fitting	285	16+
Indicator for pipe	138	36
Pavement	404	17+
Plates for railway truck	105	225
Plates hub thrust	384	247+
Resistant surface	428	908.8
Strips vehicle body	296	41
Surface rod joint packing	277	
Take up for wheel brake	188	79.5 R
Wearing in	29	89.5
Weather (See Humidity; Moisture)		
Barometers	73	384+
Computer controlled, monitored	364	420
Control of	239	2.1
Operated curtain shade screen	160	5
Proofing buildings (See water proof)		
Radar	342	26
Rain actuated window closers	49	21+
Strips	49	475+
Land vehicle windshield	296	93
Vanes	73	188+
Design	D10	59
Weatherstrip, metal	428	595
Weavers Implements and Irons	139	380+
Design	D 8	
Dies for forging	72	462+
Slat and wire fabric	140	35
Portable machines	140	48
Warp manipulation	139	35
Web (See Sheet, Fabric)		
Article part		
Coating or forming by electrophoresis etc	204	180.4
Electroforming	204	12+
Electrolytic coating	204	27+
Formed web assembling	19	161.1+
Paper bag making	493	186+
Paper type forming apparatus	162	289+
Paper type forming processes	162	202+
Railway rail joints	238	151+
Railway rails	238	122+
Railway wheels	295	21+
Railway wheels integral hollow	295	28
Textile fiber forming	19	296+
Feeding	226	
Stepping-motor drive for web	400	902*
Machine part		
Calendars double reel and	40	117
Calendars single reel and	40	116
Changeable exhibitors	40	446+
Drier drum external	34	123
Drier plural drums external	34	116+
Drier rotary drum external	34	111
Drier spacers	34	94
Printing	101	
Sheet feeding or delivery	271	
Supplied machines		
Automatic drier with control by	34	49
Box making, folding	493	162+
Drier for	34	148+
Drier spacer web	34	94
Drying processes diverse types	34	18
Drying processes radiant etc	34	41
Drying processes with vacuum	34	16
Drying processes with vapor etc	34	23+
Drying processes with web spacer	34	6
Envelope making	493	186+
Fabric coating and uniting	156	
Fabric reeling web fasteners	242	74
Fabric reeling web roll supply	242	58
Folding or associating	270	
Printing	101	
Severing	83	
Textile fluid treating	68	177+
Typewriter paper feed	400	611+
Tacky web cutting	83	922*
Winding & reeling	242	
Webbing		
Elastic	139	421
Wedge		
Bolt to nut lock		
Thread gripper with tapered section	411	253
Thread lock with tapered surface	411	265+
Coupling and securing		
Bale tie	24	25
Buckle	24	194+
Buckle design	D11	200+
Clasp	24	526
Also pivots	24	491
Clasp design	D11	87
Closure fasteners	292	342+
Cord holder	24	136R+
Harness clamp	24	171
Rail anchor	238	324
Rail clamp	238	353+
Rod joint sleeve	403	274+
Tool socket fastening	403	367+
Wire fence clamp	256	56
Design	D 8	47
Printers quoins	254	42
Pushing and pulling	254	104
Rock splitting in situ	299	20+
Wood splitting	144	192+
Weed Killers	71	65+
Weeders	30	
Electric	47	1.3
Flame or burner type	126	271.1+
Plant cultivators	47	1.44
Fluid introduction into soil	111	6+
Grappling fork or shovel	294	50.6+
Harrows	172	705+
Process of introducing fluid into soil	47	58
Root puller	254	132
Subsurface blade	172	720
Turners	172	514+
Weft		
Inactive		

U.S. PATENT CLASSIFICATIONS

Entry	Class	Subclass
Restrainers for	139	170.3+
Manipulation	139	116+
Pile fabrics	139	392+
Stop controlling	139	370.1+
Stop motions	139	370.1+
Thread cutting devices	139	256R+
Thread replenishing	139	224R+
Thread tensioning devices	139	194
Winding	242	26.4+
Weighers (See Scales)		
Automatic supply control	177	60+
Cigar and cigarette making	131	280
Coin or check controlled	194	
Conveyer pneumatic combined	406	10+
Conveyer power driven combined	198	502.1+
Design	D10	87+
Dispensing	222	77
Gas separation combined	55	270
Illuminated	177	177+
Liquid level gauge	73	296
Material and handling	414	21
Packaging machine combined	141	129+
Receptacle filling combined	177	59
Recording	177	2+
Printing or perforating	346	9+
Scales	177	
With diverse exhibitors	40	458
Sifting	209	239
Sorting	209	645
Condition responsive controls	209	592+
Specific gravity	73	433+
Traversing hoist	212	158
Weighing Machines	D10	87+
Weight (See Notes to Def of)	16	DIG. 8
Aircraft stabilizing	244	93+
Animal restraining	119	107
Bed counterbalance	5	166R+
Bottle valve	215	22+
Brake	188	174+
Cigar making	131	347
Closure fastener	292	344
Clutch	192	89W+
Coin tester	194	339+
Computer controlled, monitored	364	466
Controlled operated and regulated		
Outlet valve	4	386+
Outlet valve latched open	4	386+
Water closet tank	4	300+
Counterbalance		
Chute	193	19
Pitman	74	590+
Rotor combined	74	573R
Support	248	364
Support suspended	248	325
Wheel	301	5R
Dispenser control		
By discharging receiver	222	56
By second material	222	57
By supply container	222	58
Door closure	16	81
Drape	160	349R
Dress	2	273
Drilling machine	408	235
Electric trolley retriever	191	94
Elevator	187	15
Exercising	272	117+
Flatiron accessory	38	98
Frictional gearing	74	209
Hitching	280	187
Hoop driver	147	12
Horseshoe	168	25
Invalid bed	5	78
Land roller	404	130
Lubricator	184	46
Marine propulsion	440	10
Measuring	177	
Design	D10	83+
Motor	185	27+
Composite	185	4+
Motor vehicle load, responsive to	180	290
Paper	16	1R
Design	D19	96+
Phonograph tone arm counterbalancing	369	254
Pickup scale	177	237
Plumb bob	33	392+
Pump	417	328+
Pump tide motor	417	330+
Reducer aircraft	244	5
Sash	16	216+
Weight and wheel	16	219
Sash cord and weight fastener	16	209
Ship ballast	114	124
Sound box	369	168
Surgical vibrator	128	42
Switch	200	85R+
Latch trip	200	41
Toy eccentric	446	207
Traction cable gripper	104	208
Unbalanced		
Rotary gyratory	74	87
Rotary reciprocating movement	74	61
Sifter	209	167
Sifter gyratory	209	366.5
Surgical vibrator	128	34+
Valve		
Pivoted	137	527.8
Reciprocating	137	532+
Weighers (See scales)	177	
Weighted base	248	DIG. 1
Wheel balancing	301	5R
Woodworking presser roll	144	248
Weighting Textiles	8	443
Weir	405	87+
Chemical feeder control	137	101.27
Fish traps	43	101
Flow meter	73	215+
Overflow discharge	239	193+
Welders Helmet	2	8
Combined with welding	219	147
Design	D29	7+
Shields	2	8
With air supply	128	201.22+
Welding		
Adhesively seaming		
Apparatus	156	
Processes	156	
Autogenous	228	196+
Backup rings	228	50
Pipe integratable	285	21
Chain link welding	59	31+
Cold processes of	228	115+
Compositions	148	23+
Critical metal or filler	228	263.11+
Diffusion	228	193+
Electric welding	219	50+
Arc	219	121R
Inductive	219	6.5+
Methods	219	148
Tongs or clamps	219	158+
Wire joining by	140	112
Electrode with metal particle	428	553
Electrolytic	204	
Flux applying	228	214+
Friction, apparatus for	228	2
Friction treatment	156	73.5+
Gas welding		
Apparatus	228	
Methods	228	101+
Torches	239	398+
Metal heat treating and welding	148	127
Metal working and welding	228	141.1+
Digest	29	DIG. 48
Sheet metal containers	413	
Non-metal or non-metal with metal film	228	263.12
Pressure welding		
Apparatus	228	
Processes of cold welding	228	115+
Processes of making tubes by drawing through dies	228	144+
Tubes by drawing through die	228	16
Railway track welders	104	15
Robot	901	42*
Rod	219	145.1+
Alloy composition	420	
Fluxing composition	148	23+
Metal particle containing	428	553+
Solder	228	56.3
Scale removing	29	81R
Sintering	419	
Sonic or ultrasonic, non- metallic	156	580.1
Apparatus	156	580.1+
Process	156	73.1+
Spot welding	219	86.1+
Tongs or clamps	269	
Ultrasonic	228	1.1
Non-metallic	156	73.1
Apparatus	156	580.1+
Non-metallic processes	156	73.1+
Process	228	110+
Weld strength testing		
Bending	73	850
Shear	73	842
Tensile	73	827
Welded metal stock	428	544+
Wells	166	
Battery receptacle	429	47
Blowout preventer valve	251	1.1+
Borehole study	73	151+
Electrical	324	323+
Borehole telemetering in	340	853+
Cable seal	277	
Centering guide	175	325
Cleaning	166	170+
Processes	166	311+
Concentric conduit joint	285	138+
Coupled rod seal	277	5+
Coupling valve	137	515+
Drilling platform	405	195+
Floatable	405	203+
Earth boring compositions	252	8.51
Earth boring for	175	
Fire extinguishing	169	69+
Gas lift valve for	137	155
Gas separating	55	
Grapples	294	86.1+
Grappling methods	166	301
Heater electric	219	277+
Hoist bucket bailer	294	68.22+
Logging	367	25
Computer controlled, monitored	364	422
Particular apparatus	367	911*
Particular transducer	367	912*
Packer sealing portion	277	
Pump		
Aerated column	417	108+
Expansible chamber rotary	418	
Jet	417	151+
Motor combined	417	321+
Pneumatic displacement	417	118+
Reciprocating piston	417	437+
Reciprocating valved piston	417	545+
Rotary centrifugal	415	206
Rotary screw	415	71+
Well structure combined	166	105+
Rotatable packing rod seal	277	31
Screen making	29	163.5 R
Solid material recovered	299	
Torpedoes	102	301+
Treating compositions	252	8.551
Tubing cutter	166	55
Tunnel fluid recovery	299	2
Valves and closures	166	316+
Wiper	15	210R+
Welting		
Device for knitting machines	66	41
Welts		
Glove	2	165
Shoe	36	78
Handling and sewing	112	46
Machines	12	67R+
Sewing guides	112	52
Slitting beveling and sewing	112	44
Sole attaching	36	17R
Support for in leveling	12	33
Vehicle body joint	280	153.5
Werner Complex Formation to Recover Hydrocarbon	585	850
Wet Suit	2	2.1 R
Wetting (See Moistening)		
Compositions	252	351+
Textile spinning with	57	295
Whale Bone	2	255+
Whale Oil	260	398+
Wharves	405	284+
Wheel (See Pulley; Wheeled)		
Aircraft		
Landing gear	244	103R+
Paddle wheel propelled	244	70
Paddle wheel sustained	244	19+
Plane & paddle wheel sustained	244	9
Heavier than air	244	9
Lighter than air	244	27
Prerotation	244	103S
Aligning tools	29	273
Ambulant spraying device	239	155+
Balance		
Testing	73	66
Weight removing tools	254	
Weights	301	5R
Base, special, motor vehicle	180	21+
Brake	188	218R

	Class	Subclass
Caster	16	45+
Clutch		
Hand	192	95
Velocipede free	192	64
Covers	301	37R
Drive railway	105	96+
Elevator		
Street sweeping machine	15	86
Excavating plows		
Conveyor	37	111+
Extinguishers, movably mounted	169	52+
Fish	43	13
Fluid current motor		
Paddle	416	178
Flywheels	74	572+
Gearing		
Bevel	74	386
Laterally spaced wheels	74	432
Radially spaced wheels	74	433
Hand control	74	552+
Harvesters		
Grain wheels and casters	56	322
Horn		
Toothed wheel	116	144+
Impact receiving tools		
Thread protectors	29	277
Inclinometer	33	335+
Insulated hand	16	117
Joints in general	403	
Making	29	159R
Lathes	82	8
Lock (See lock wheel)		
Lock	70	225
Brake	188	69
Caster	16	35R+
Lock for spare	70	259+
Making	29	159R+
Railway	29	168
Wheelwright machines	157	
Marine propulsion		
Hydraulic jet	440	38
Manual paddle	440	21
Paddle	440	90+
Towing ground wheels	440	36
Metal forging		
Dies	72	343+
Metal founding		
Composite castings & joints	249	56
Metal molds	249	56
Panel hanger	16	107
Mounts	16	97+
Plow		
Guide	172	286+
Potato diggers		
Sifter	171	71+
Sifter	171	111+
Potters	425	263+
Potters	425	459+
Power take off	74	13+
Pullers	29	244+
Pulleys		
Friction drive	474	166+
Positive drive	474	152+
Pumps		
Water motor	417	334+
Rack	211	23+
Velocipede type	211	20+
Railway	366	
Railway cable gripper		
Sprocket	104	236
Whelp	104	235
Resilient	152	
Pneumatic	152	53+
Pneumatic and coil	152	35
Pneumatic and leaf	152	34
Pneumatic and rubber	152	30
Rolling	72	67+
Rotary crane		
Bull wheel	212	246
Rotary fender		
Horizontal	293	19
Sash cord guides		
Casing combined	16	215
Multiple wheel	16	213
Sash weight attached	16	219
Ship steering		
Paddle wheel	114	147
Wheel and drum	114	160
Skates	301	5.3
Slip		
Acceleration induced, detecting in motor vehicle	180	197
During braking, regulating	188	2A
Spring	152	1+
Substitutes	305	
Brakes	305	9
Traction (See traction)		
Training for cycle	D12	122+
Trolley		
Guard or finder	191	76
Head	191	63+
Typewriter		
Index	400	165.1+
Key	400	174+
Vehicle		
Bearings	384	416
Land vehicle	301	5R+
Land vehicle design	D12	204+
Locks	70	259+
Marine towing	440	36
Pivoted carrier	280	38+
Railway	295	
Railway trucks	105	157.1+
Resilient	152	
Runner attachment	280	13
Runner combined	280	8+
Vertically adjustable	280	43+
Wheelbarrow	280	47.31
Water motor		
Wheeled bag	248	98
Wheeled hose reels	242	86.2
Wheelwright machines	157	
Wheel holding stands	157	14
Wind motor		
Woodworking		
Hub boring	144	97+
Special work machines	144	15+
Spoke turning	144	206
Workholding stand for	157	14
Wheelbarrow	280	47.31
Design	D34	16
Wheelchair	280	242WC+
Attachments	280	289WC
Design	D12	131
Folding	280	647+
Motorized	180	907
Seating aspect	297	DIG. 4
Wheeled (See Wheel)		
Aircraft landing gear	244	103R+
Ambulance stretchers	296	20
Beds		
Cable drum carriers	254	279
Coasters steered	280	87.1+
Earth working tools	172	
Fire engines	169	24+
Harrows	172	
Harvester	56	322+
Horse rakes	56	396+
Rear delivery	56	384+
Rear delivery revolving	56	380+
Ladders	182	12+
Plows	172	669+
Frames	172	669+
Land markers	172	126+
Pneumatic dispatch		
Carriers	406	185
Receptacles or trucks	280	
Roofing scaffolds	182	36+
Scoops	37	124+
Fork or rake	37	121
Skates	280	11.19+
Stands		
Bag holder	248	98
Movable receptacle	248	129
Store service		
Carriers	186	27+
Toys		
Aircraft	446	230+
Figure	446	269+
Knockdown	446	93+
Miscellaneous	446	431+
Rotating	446	237
Sounding	446	409+
Vehicles	280	
Vehicle	280	
Convertible to chair or seat	280	30+
Convertible to cradle or crib	280	31
Extensible	280	638+
Bier or casket type	280	640
Folding	280	639+
Folding carriage or stroller	280	642
Folding wheel chair	280	647
Fenders	293	43
Hand	280	47.17+
Motor	180	
Pivoted wheel carrier	280	38+
Railway rolling stock	105	
Shipment on freight carrier	410	3+
Shipping receptacle for	206	335
Walkers	280	87.2
Convertible to steered vehicles	280	7.1+
Steered	280	87.1+
Winding and reeling		
Hose reel carriers	242	85+
Reel carriers	242	86.5 R+
Wheelwright Machines (See Tire)	157	
Whetstone		
Compositions	51	293+
Structures	51	211R+
Whey	426	583
Whiffletree	278	90+
Harness connectors	54	53
Whip	231	2R+
Animal prod	D30	156+
Cotton harvesters	56	29
Design	D30	156
Kitchen beater	D 7	103
Machines	231	1
Rack	211	67
Design	D30	143
Sifter clearers	209	383
Sockets	280	170+
Treadmill etc combined	185	15+
Whipping Devices		
Agitating or mixing	366	
Animal powered motor combined	185	24
Sweep	185	20
Goads	231	2R
Design	D30	156+
Initiating devices	272	27R
Sewing machine	112	121.2
Vehicle combined	280	178
Whipstock	166	117.5+
Earth boring combined	175	79+
Whirlers and Whirling		
Gas separation by deflection	55	447+
Nozzles	239	463+
Pneumatic conveyors	406	92
Rockets		
Rotating	102	350
Whirlers with	102	359
Toys	446	236+
Whirligigs (See Whirlers)		
Noise making toys	446	215
Spinning toys	446	236+
Flying propellers	446	36+
Whisk Brooms (See Brooms)	D 4	130+
Whisker Growth	156	DIG. 112
Whiskey	426	
Whistles		
Buoy	116	108
Electric	340	406
Liquid level	137	213+
Musical instrument	84	330+
Periodic	116	24
Security or signal instrument	D10	119
Signal	116	137R+
Steam	116	137R+
Structure	116	137R+
Toy	446	204+
Design	D21	64
White Lead	423	592+
Whiting	423	592
Wick		
Atmosphere	239	34+
Boiler	122	366
Burners	431	298+
Blast lamps having	431	252
Blast lamps having	431	300
Cookstoves having	126	45+
Heating stoves having	431	102+
Mantle or incandescent element combined	431	102
Perforated combustion tube type	431	195+
Porous block type	431	326
Coated	431	325
Combined dispenser	222	187
Compositions	502	400
Heat exchange, capillary heat pipe	165	104.26
Liquid purification or separation, capillary	210	294
Stops	431	301
Structure	431	325
Trimmer or handler	431	120

U.S. PATENT CLASSIFICATIONS

	Class	Subclass
Wickets Doors	49	169+
Wideband Cable	333	239+
Wiener Roaster	99	324+
Electric	99	358
Elongated type	99	441
Wigner Effect	376	350
Wigner Effect	376	351
Wigner Effect	376	358
Wigner Effect	376	458
Wigs	132	53+
Accessory	D28	93
Design	D28	92+
Rooting hair by tufting	112	80.2
Toy	446	394
Willowers		
Textile fiber	19	85+
Wilton		
Carpet	139	391+
Making apparatus	139	37+
Winches (See Windlass)		
Wind		
Guards and shields		
Couch hammock combined	5	125
Harvester combined	56	190
Land vehicle combined	296	78R+
Match receptacles combined	206	102+
Smoking devices combined	131	174+
Instruments, musical	84	330
Design	D17	10+
Motors (See current fluid; motor; windmills)	416	
Control for windmill driven generator	290	44
Generator drive system	290	55
Marine propulsion device	440	8
Ventilator drive device	98	
Musical instruments	84	330+
Automatic	84	83+
Automatic piano	84	24+
Pressure meters	73	861+
Propelled revolving sign	40	479
Illuminated	40	480
Screens for vehicles	296	84R+
Socks and vanes	73	188+
Tunnels	73	147
Winding & Coiling (See Coil; Spring)	242	
Abrading work handling	51	75
Bobbin winding on sewing machine	112	279
Capstans	254	266+
Drop hammer winding lift	72	437+
Drying winding elongated fabrics	34	153
Elevator	187	27
Portable	187	11
Fence wire	256	40+
Invalid transfer hoist	5	88
Metal bending	72	135+
Metal rolling and coiling	72	127+
Midline tighteners	24	71.2
Motion picture film	352	124+
Overwinding prevention		
Clock & watch key	81	467
Clocks & watches	368	209
Composite motor	185	13
Spring motor	185	43
Weight motor	185	35
Railway brake operator	188	145+
Reciprocating window operator	49	445+
Reeling and	60	905*
Sash balance spring drum	16	198
Spring or weight motor rewinding		
Alarm clocks	368	258
Clocks	368	147+
Electric clocks	368	207+
Mainspring tool	81	7.5
Spring motor	185	39+
Stem winding watches	368	216
Watches	368	147+
Weight motor	185	32+
Tool driving or impacting device operation	173	81+
Traversing hoists	212	142
Vehicle body lifter	254	48
Vertically adjustable lights	362	402
Wire drawing	72	289
Wound article making	242	7.1+
Ball making	156	170
Broom making	300	14
Coiling by rolling	72	127+
Crimp seam tube	72	48+
Hair curlers	132	33R+
Inductance making	29	605
Laminated	156	
Magnet electric making	29	605
Match making	144	59
Method	242	7.2+
Nut making	10	74
Paper manufacture apparatus	162	283+
Paper manufacture processes	162	118+
Paper tube	493	269+
Riveted tube	227	52
Slitting web with	242	56.2
Textile	19	149
Textile delivery twist	57	62
Textile receiving twist	57	66+
Textile silvers	19	159R
Textile warp preparing	28	172+
Textile webs	28	103+
Tobacco products	131	59
Transformer making	29	605
Washer making	10	74
Welded tube method	228	145
Well screens	29	163.5 CW
Wire	72	135+
Wireworking combined	140	
Wood bending	144	268
Windlace	49	500
Windlass (See Winding)	242	85+
Elevator	187	27
Ladder extension	182	208
Ladder wheeled	182	63+
Pulling cables	254	266+
Sand reel	242	117
Ship steering	114	160
Wood bending	144	263
Windmills (See Wind, Motors)	416	
Control for windmill driven generator	290	44
Design	D25	1+
Lubrication	184	4
Pump	417	334+
Toy	446	217+
Design	D21	119
Turntable	74	381
Windows (See Closure; Windshield)		
Bay window	52	201
Bead fastener hardware	16	220
Blinds	160	
Design	D 6	577
Length cutter	83	
Bracket support attaching to	248	208+
Cabinets	312	
Casement window operator	D 8	400
Cathode ray tube	313	420
Channel for sliding window	D25	119+
Cleaners belts	52	37
Cleaning		
Sheet holder wiper	15	232
Wiping device	15	220R+
Closure operator for	49	324+
Condensation preventing	52	171+
Bay	52	201
By ventilation	98	2
Frame in situ construction	52	204+
Picture or show	52	777+
Picture or show (See curtain wall)	52	764+
Storm	52	202
Well	52	107
Container having, rigid	493	905*
Curtain or shade	D 6	575+
Curtain supports	248	251+
Defrosting		
Electric heater with fan	219	203
Heating system	237	
Ventilation type	98	90+
Design	D25	52
Double glazed	52	788
Double glazed	52	479
Double hung	49	404+
Electric switch	200	61.62+
Envelope having	493	919*
Flower box	D11	156
Guards	49	50+
Land vehicle	296	146+
Curtains having	296	145
Design	D12	183
Windshields	296	78R+
Letter box	232	22
Locks	70	89+
Milk receptacles	232	42
Mirrors and reflectors	350	600+
Plant receptacles	47	68+
Racks	211	
Scaffolds	182	53+
Screen or guard, design	D25	53
Screens	160	
Shelf supported on	248	236
Lighting	362	125
Stile pocket cutting machine	144	19
Storm and screen	49	61+
Stove	126	190+
Trim	52	211+
Troughs	49	408
Ventilation	98	88.1+
Windrowing	56	192
Windshield		
Automobile	296	84R+
Design	D12	182
Cleaner	15	
Coating compositions	106	13
Defrosting (See condensation, preventer)	52	171
Vacuum motor	91	218+
Wines	426	592
Cooler design	D 7	304
Presses for making	100	104+
Wing		
Aircraft	244	35R+
Autorotating wing	416	131+
Beating for propulsion	244	72
Beating with ship	244	22
Foldable fenders	293	50
Helicopter	416	131+
Lighter than air with	244	25
Bed front	5	161
Drop bombs with	102	384
Hinge leaf multiple	16	389+
Nut	411	435
Pyrotechnics with	102	335+
Rockets with	102	348
Testing wind tunnels	73	147
Toy gliders with	446	61+
Wipe-forming Metal	72	149+
Wiper and Wiping	15	
Coating material supply means combined	401	261+
Dispenser with	222	148+
Discharge assistant carried	222	342
Inlet cleaner	141	90
Movable or conveyor trap	222	345+
Lubrication system with	184	10
Mold coating combined	118	
Printing intaglio	101	150+
Process	134	6+
Shade with	160	11
Sifting with	209	385+
Solder	228	22+
Wire (See Filament; Strand)		
Articles		
Binder devices releasably releasably engaging aperture or notch of sheet	402	
Book or copy holder	248	465.1
Brush and broom supports	248	112
By electrolysis	204	27+
By encasing in tube	228	126+
Chains	59	83
Cleaning implements	15	210B
Clothespin (See clothespin)		
Coating	427	
Curry combs	119	89
Door mats	15	241
Easels	248	465.1+
Electricity conductors	174	
Fences	256	32+
Garment stays	2	264
Handles	16	120
Hinges	16	373
Hook type brackets support penetrating	248	218.1+
Lathing	52	660+
Metallic receptacles	220	19
Miscellaneous strand structure	57	200+
Music score holder	248	465.1
Nursing bottle supports	248	107
Picture easels	248	465.1+
Picture frames	40	153
Pin fasteners	24	159
Plastic material cutters	30	116+
Racks	211	181
Receptacle bail ears	220	92
Receptacle handles	220	95
Receptacle stands	248	153

	Class	Subclass
Reinforced plastic railway ties	238	94
Rock cutting machine	299	35
Rod joint packing with	277	227+
Saw	83	651.1
Shelf brackets	248	249
Specially mounted racks	211	106
Stands	248	175
Stovepipe mounted racks	211	112
Supporting brackets	248	302+
Suspended racks	211	119
Tension cutter	83	200.1
Wireworking	140	
Wooden crate wired slats	217	51
Bonding	228	904*
Clamp	24	115R+
Coated miscellaneous	428	364+
Making	427	
Process	427	117+
With metal particles	428	553+
Working combined	29	527.2+
Container or wrapper for	206	389+
Couplings, fasteners, holders & ties		
Bale and package ties	24	27+
Bale bands with wire connections	24	26
Barrel covers	217	85
Belt fasteners	24	39
Check row planter wire end anchor	111	49
Cord and rope holders	24	131R
Crossed rod or wire joints	403	400
Interfitting	403	346
Folded tuft brushes retaining wire	15	199
Hooks and eyes	24	699
Relatively movable hook portions	24	600+
Insulator embracing conductor ties	174	173
Internal bottle closures extension	215	264+
Lacing terminals	24	122
Mop holders	15	154
Resilient clasps	24	530+
Resilient head snap fastener	24	626
Resilient socket snap fastener	24	678
Well screen spiral wire wound	166	231+
Wire bound and stitched brushes	15	189
Wire bound brushes and brooms	15	207
Wooden box stays	217	71
Wooden box straps	217	68
Cutter	30	
Cutting coiled wire	83	907*
Cutting wires	83	926B*
Dental arch	433	20
Electric		
Conductors	174	
Magnetic body treating coils	128	1.5
Meter hot wire measuring	324	106
Telephony	379	
Tower support	52	40
Electroplating	204	27+
Fabric	245	
Bed bottoms	5	186R+
Coating with rubber etc	204	180.4
Design	D 5	47+
Design for floor or wall	D 5	47+
Electroforming	204	11
Electrolytic coating	204	24
Heddles	139	95
Pile textile weaving	139	37+
Wireworking	140	
Gauge, strain &	33	DIG. 13
Glass, safety glass	428	255+
Manufacturing	65	
Heat treated	148	400+
Coated	428	
Handling running length work	148	155+
Making	148	4+
Instruments, ribbon &	33	DIG. 7
Insulation cutter	30	90.1+
And remover	81	9.4+
Making		
Casting	164	423+
Coating with rubber etc	204	180.4
Covered or wrapped	57	3+
Drawing	72	274+
Drawing dies	72	467
Electroforming	204	12+
Electrolytic coating apparatus	204	206+
Electrolytic coating processes	204	27+
Metal rolling processes	72	365+
Screw thread rolling	72	95+
Wire glass	65	
Metal miscellaneous		
By electrolysis	204	12+
From powders	419	4+
Making miscellaneous	72	
Packaging for	206	389+
Photo systems (See facsimile)		
Recorders sound	360	89
Recording machines	360	89
Wire per se	360	131
Rope making	57	
Sheet attached lathing	52	454
Splicing		
Bale or package tie	24	16R+
Electric conductor	156	49
Joint	403	206+
Textile analogue	57	22+
Trolley wire	191	44.1
Stitching stapling	227	
Stretchers	254	199+
Strippers	81	9.4+
Working and using machines		
Bobbin and cop winding	242	25R
Broom banding	300	15
Coiling	72	135+
Electric wire placing apparatus	254	134.3 R+
Etching to section	134	
Grain etc compressing and binding	56	451+
Harvester binder type	56	132
Hat frame	223	8
Hatbrim wire shaping	223	17
Metal forging and welding feeder	72	419+
Nail making	10	43+
Nailing and stapling	227	
Reels	242	78
Stapling	227	
Textile spinning twisting etc	57	
Working	140	
Wired Music		
Distributing system	379	101
With phonograph	379	87
Electric organs	84	1.1+
Electric phonograph	379	87
Wireless (See Radio)		
Wood (See Plywood)		
Alcohol (See methanol)		
Ball making	142	1
Bending	144	254+
Boring	408	
Buildings	52	
Carbonizing	201	
Carrier	294	149+
Coating	427	
By immersion	427	440
Comminution	241	28
Distillation	201	
Drying and vapor treatment	34	
Dyeing	8	402
Fastener made of	269	905*
Fences	256	19
Gas generation	48	209
Apparatus	48	111+
Hydrocarbon recovery from	585	242
Impregnating		
Processes for	427	
Insulated compartment for box	217	131
Moldings	52	716+
Nails	411	439+
Ornamenting		
Panel track combined	16	95W
Flexible panel attachment	16	87.4 W+
Pulp making		
Comminution processes	241	21
Pulp mill	241	280+
Receptacles	217	
Saturating or indurating		
Compositions	106	12
Compositions mineral oil	208	2
Fireproofing	252	607
Screws metal	411	378+
Making	10	2+
Shavings making	241	
Ships	114	82
Shoe soles	36	33
Attaching means	36	13
Slicing	144	162R+
Splitting	144	193R+
Staining	8	402
Structural		
Composite columns	52	720+
Compound lumber	52	782+
Joist connections	403	230+
Through intermediate member	403	187+
Nailing beam	52	376
Panels	52	782+
Rail pedestal	238	118
Rail seat cushions	238	285
Splices and joints	403	
Wheel making	144	15
Sugar production	127	37
Apparatus	127	1
Turning lathes	142	
Wind musical instruments	84	380R+
Working	144	
Design	D15	122+
Sawing	83	
Tools (See tools)		
Turning	142	
Woodbury Type	101	
With photographic step	430	300+
Woodwinds	84	380R+
Wool		
Chemical modification	8	128R
Cleaning or laundering	8	137+
Devices and machines	68	
Waste cleaning	68	1
Dyeing	8	
Mixed textiles	8	529+
Vat or sulphur dyes	8	650+
Fertilizers containing	71	18
Liberation mechanical	19	2+
Mercerizing	8	128A
Removing grease from	8	139+
Steel wool making	29	4.5 A
Waste		
Cleaning and reconditioning	68	1
Work		
Carrier and support for sewing machine work	112	121.15
Holder	269	
Dental	433	49+
Electric welding	219	158+
For testing by stress or strain application to specimen	73	856+
Measuring		
Dynamometers	73	862+
Power unit efficiency	73	112+
Work table lighting system	362	33
Worm (See Gearing)		
Culture	119	6
Electrical expeller	47	1.3
Gearing	74	425+
Worship		
Cabinets	312	33
Teaching	434	245+
Wort		
Making	426	16
Apparatus	99	278
Wrap-forming Metal	72	296+
Wrapper (See Package)		
Bicycle article carrier	224	34
Coin type	229	87.2
Label pasting	156	
Newspaper	229	92
Paper	229	87R+
Paperboard boxes	229	40
Tobacco	131	365
Cigar etc tube feeding	131	74+
Holders and carriers	131	105
Wrapping and Wrapped		
Chain links	59	92
Chain making	59	19
Cigar and cigarette making	131	27.1+
Packaging combined	131	283
Wrapping materials and devices	131	58+
Containers	D 9	305
Food preservation	426	392+
Machines	53	203+
Receptacle forming & filling	53	558+
Paper	D 5	
Sheet material		
Folding	53	116+
Printing and folding	53	116+
Printing and rolling	53	116+
Rolling	270	32+
Spinning etc apparatus	57	3+
Spinning etc processes	57	3+
Strand structure	57	210+
Textile smoothing implements	38	73
Warp preparing	28	176
Wire fabric stay applying	140	10+
Portable machines	140	16+
Wreath		
Artificial	428	10

U.S. PATENT CLASSIFICATIONS

	Class	Subclass
Design	D11	120
Illuminated	362	122
Natural flora sustaining	47	41R
Wrenches	81	52$^+$
Adapted to turn eye screw	81	901*
Compound	7	138$^+$
Constructed from specific material	81	900*
Design	D 8	21$^+$
Impact	81	463$^+$
Impact delivery clutch	173	93.5$^+$
Making		
Forging dies for	72	470$^+$
Machines	76	10
Methods	76	114
Monkey, combined with other tools	7	139$^+$
Pump and oiler combined with	81	2
Tap, ratchet	408	120$^+$
Wrigglers	239	229
Wringers	68	241$^+$
Bevel gearing for	74	387
Drier combined	34	70
Mechanisms for washer and	74	17
Mop	15	260$^+$
Brush combined	15	116.1$^+$
Combined	15	119R$^+$
Supports	68	239
Wash bench combined	68	236
Wash tub etc attachments	68	238
Wrinkling		
Coatings	427	257
Corrugating metal	72	191$^+$
Paper apparatus	162	280$^+$
Paper processes	162	111$^+$
Wrist Pins	74	595$^+$
Bearing	384	266$^+$
For piston rod connection	384	268$^+$
Wrist Watch (See Watch)		
Design	D10	30$^+$
Fastener	63	3
Wristlets	2	170
Writing		
Forms	283	45
Guides	434	164
For the blind	434	117
Implement (See pen; pencil)	D19	35$^+$
Multi signature	33	23.6
Papers	D19	1
Surfaces	D19	52
Surfaces erasable	434	408$^+$
Cabinet with	312	230
Teaching	434	162$^+$
Telegraphic	178	18$^+$
Wrought Iron		
Producing	75	32
Reactive lining processes	75	95
Treating molten iron	75	47
X		
X Ray	378	
Absorption	378	51
Circuits	378	91
Computerized tomography	378	4
Contrast compositions	424	4$^+$
Diffraction	378	70
Equipment	D24	2
Exposure process	430	967*
Film	430	966*
With cassette	378	182
Filters	378	156
Fluorescence	378	44
Fluoroscope	378	190
Lithography	378	34
Photographic	378	167
Scattering	378	86
Screen		
Bucky grid	378	154
Fluorescent or phosphorescent	250	483.1
Tube with secondary ray	378	140
Shield	252	478
Shields and shielding	250	515.1
Compositions	252	
Shoe fitting machine	378	192
Therapy	378	65
Timer	378	96
Treating	378	64
Tube	378	121
Tubes	378	121
Used in reaction to prepare		
inorganic or organic material	204	157.15$^+$
Used in reaction to prepare or treat		
synthetic resin or	522	1$^+$
Manufacture	445	28
X-type Package	206	809*
Xanthate		
Cellulose fiber treatment	8	122
Esters	558	245$^+$
Cellulose	536	60$^+$
Froth flotation	209	166$^+$
Compositions	252	61
Xanthenes	549	388$^+$
Xanthines	544	267$^+$
Xanthogenation (See Xanthate)		
Xanthones	446	372
Xenol	568	747
Xenon	423	262
Xerography (See Photography, Electric)	430	31$^+$
Dispensing digest	222	DIG. 1
Xeroradiography	378	28
Xylene		
Alkylation to produce	585	446$^+$
Dealkylation	585	483$^+$
Xylenol	568	716$^+$
Xylidine	564	305$^+$
Xylophone	84	404
Design	D17	23
Y		
Y Guns	89	1.1
Yacht (See Ship)		
Yale TM Lock	70	364R
Yard		
Railway car	104	26.1$^+$
Ship spar	114	95$^+$
Stick	33	483$^+$
Yarn (See Thread)		
Doll	446	372$^+$
Fluid treatment of package	8	155$^+$
Printing on	101	172
Structure	57	
Testing	73	160
Yeast	435	
Cell mutation, genetic engineering	435	172.1
Compositions	426	62$^+$
Bread making with	426	18$^+$
Fermenting carbohydrates by	435	
Products, microorganism per se	435	243$^+$
Enzyme, processes for	435	183
Nonferment material combined	435	174$^+$
For stabilizing	435	188
Propagation	435	255$^+$
Apparatus	435	287$^+$
Nonferment material combined	435	288
Yoghurt	426	583
Yohimbine	546	53
Yoke		
Animal draft appliance		
Neck yoke	278	119$^+$
Thill	278	51
Apparel	2	122
Axle lubricator	384	374
Bottle closure	215	237$^+$
Chuck	279	19.1$^+$
Draft	D30	137$^+$
Grapple	294	91
Harness	54	3
Ox	54	77
Motor fluid current	416	244R
Pipe joint	285	274
Railway		
Conduit	104	142$^+$
Draft appliance cushioned	213	47$^+$
Draw bar	213	67R$^+$
Track fastener	238	314
Truck platform	105	214
Watch		
Lever set	368	199
Pendant set	368	197$^+$
Wrench shank grip	81	152
Ytterium Inorganic Recovery	423	21.1$^+$
Z		
Zearalanone	549	270
Zefran T M (polyacrylonitrile Pyrrolidone)	526	264
Zein (See Protein)	530	373
Compositions containing	106	149$^+$
Zeolites (See Molecular Sieve)	423	328
Compositions containing	502	60
Liquid purification	210	263$^+$
External regenerant supply	210	190$^+$
Process	210	660$^+$
Zero Insertion or Suppression		
Selective punching	234	24
Zero Setting Devices		
Dispenser with register	222	32$^+$
Selectively preset cutoff	222	16
Register	235	144R
Zinc	75	
Alloys	420	513
Aluminum	420	540
Aluminum copper	420	431
Copper	420	477
Copper tin	420	476
Copper tin lead	420	475
Batteries	429	
Compound inorganic	423	
Fluorescent compositions containing	252	301.6 R
Halides	423	462
Inorganic	423	
Organic	556	118$^+$
Dithiophosphates	556	25
Thiocarbamates	556	38$^+$
Oxides and hydroxides	423	622
Paint containing	106	254$^+$
Photo conductive	430	87
Pigment filler or aggregate	106	292$^+$
Zincates	423	592$^+$
Distilling processes	75	88
Retorts	266	153
Electrodeposition	204	55.1
Electrolytic synthesis		
Aqueous bath	204	114$^+$
Fused bath	204	66
Electrothermic processes	75	10.1$^+$
Volatilization of zinc	75	10.3$^+$
Hydrometallurgy	75	120
Misc. chemical manufacture	156	DIG. 77
Plating	427	433
Electrolytic	204	55.1
Pyrometallurgy	75	86$^+$
Zipper	24	381$^+$
Fastening to tape	29	766$^+$
Luggage	190	903
Making	29	408$^+$
Plastic casting on tape	264	252
X-art, package closures	206	810*
X-art, plastic articles	425	814*
Zirconate Ceramic Compostions	501	134$^+$
Zirconium		
Alloys	420	422
Compounds		
Organic	556	51$^+$
Recovery of inorganic	423	69
Refractory composition	501	94
Electrolytic production	204	64T$^+$
Electrolytic production	204	105R
Misc. chemical manufacture	156	DIG. 95
Pyrometallurgy	75	84$^+$
Zithers	84	285$^+$
Zone Melting		
Crystalization	156	604$^+$
Electric furnace	373	17
Induction furnace	373	139
Zootechny		
Breaking and training	119	29
Harness	54	
Breaking and training	54	71$^+$
Propagation and care of animals	119	
Traps	43	58$^+$
Zwieback	426	618$^+$
Mortared	52	415$^+$

APPENDIX B

PATENT AND TRADEMARK OFFICE TELEPHONE DIRECTORY

The Patent and Trademark Office (PTO) is a massive bureaucracy, which can be as crepuscular and difficult to deal with as any of Uncle Sam's federal agencies or departments. It employs some 3,235 people who live in a world in which they are trained to approach issues as being either black or white. There is no gray. Inventors are seen as long numbers on applications. Personalities do not exist.

I have found that the best way to avoid making mistakes in PTO applications, procedures, and fees and thereby save myself time, money, and mental torment, is to pick up the phone and, as AT&T suggests, reach out and touch someone. It so happens that as cold, non-responsive, and devitalizing as the federal bureaucracy can be, its official representatives, when approached one-on-one, can be warm, helpful, and sympathetic to your plight.

To assist you in locating the appropriate officials, who are scattered throughout some 14 buildings in Crystal City, Virginia, the complete organizational telephone directory for the PTO is included here. While people may change positions through career moves, the job slots remain constant, and so do the office telephone numbers.

This notwithstanding, once you start playing telephone tag, call the PTO's Office of General Services at (703) 557-0813. General Services is charged with the updating and publishing of the PTO's telephone directory and the furnishing of accurate personnel locator information to the public. As a last resort, do not hesitate to call the Commissioner's Office at (703) 557-3071. An administrative secretary or special assistant will be glad to help you.

OFFICE OF THE ASSISTANT SECRETARY AND COMMISSIONER OF PATENTS AND TRADEMARKS

Assistant Secretary and Commissioner Donald J Quigg rm906 PK2 557-3071
Administrative Secretary Norma M Rose rm906 PK2 557-3071
Executive Assistant to the Commissioner Edward R Kazenske rm906 PK2 557-3071
Program Analyst Ruth A Nyblod rm906 PK2 557-3071
Deputy Assistant Secretary and Deputy Commissioner (Vacant) rm904 PK2 557-3961
Secretary (Vacant) rm904 PK2 557-3961
Special Advisor to the Deputy Assistant Secretary and Deputy Commissioner
Joseph F Nakamura rm902 PK2 557-4047
Assistant Commissioner for Patents Rene D Tegtmeyer rm919 PK2 557-3811
Secretary Sherry D Brinkley rm919 PK2 557-3811
Assistant Commissioner for Trademarks Jeffrey M Samuels rm910 PK2 557-3061
Secretary Sheila G Pellman rm910 PK2 557-3061
Assistant Commissioner for Administration Theresa A Brelsford rm908 PK2 557-2290
Secretary Karon Morris rm908 PK2 557-2290
Assistant Commissioner for Finance & Planning Bradford R Huther rm904 PK2 557-1572
Secretary Vickie T Bryant rm904 PK2 557-1572
Assistant Commissioner for External Affairs Michael K Kirk rm902 PK2 557-3065
Secretary Johnell M Bersano rm902 PK2 557-3065
Assistant Commissioner for Information Systems
Thomas P Giammo rm1004 PK2 557-9093
Secretary (Vacant) rm1004 PK2 557-9093

OFFICE OF THE SOLICITOR

Solicitor Fred E McKelvey rm5C15 CP2 557-4048
Secretary Olga M Suarez rm5C15 CP2 557-4048
Deputy Solicitor Charles E Van Horn rm5C15 CP2 557-2317
Secretary Theresa B Edwards rm5C15 CP2 557-2317
Associate and Assistant Solicitors
Lee E Barrett rm5C15 CP2 557-4035
John W Dewhirst rm5C15 CP2 557-4035
Albin F Drost rm5C15 CP2 557-4035
Robert D Edmonds rm5C15 CP2 557-4035
John C Martin rm5C15 CP2 557-4035
Harris A Pitlick rm5C15 CP2 557-4035
John H Raubitschek rm5C15 CP2 557-4035
Richard E Schafer rm5C15 CP2 557-4035
Nancy C Slutter rm5C15 CP2 557-4035
Charles A Wendel rm5C15 CP2 557-4035
Paralegal Specialists
Teresa N Byerley rm5C15 CP2 557-4046
Patricia D McDermott rm5C15 CP2 557-4031
Maryann B Volkmar rm5C15 CP2 557-4022
Solicitor's Library
Theresa Trierweiler-Cappo rm5C15 CP2 557-4052

OFFICE OF ENROLLMENT AND DISCIPLINE

Director Cameron Weiffenbach rm810 PK1 557-2012
Secretary Betty Kaminsky rm810 PK1 557-2012
Harry I Moatz rm810 PK1 557-2012
Marian E Ford rm810 PK1 557-2012
Patricia M Jordan rm810 PK1 557-2012
Roster Information rm810 PK1 557-1728

BOARD OF PATENT APPEALS AND INTERFERENCES

Chairman Saul I Serota rm12C12 CG2 557-4072
Secretary (Vacant) rm12C12 CG2 557-4072
Vice Chairman Ian A Calvert rm10D10 CG2 557-4000
Secretary Wanda G Banks rm10D10 CG2 557-4000

- General Information
  - Ex parte Appeals rm12C08 CG2 557-4101
  - Interferences rm10C01 CG2 557-4007
- Examiners-in-Chief
  - Neal E Abrams rm12D04 CG2 557-4057
  - James R Boler rm10C12 CG2 557-4009
  - Raymond F Cardillo Jr rm10AD4 CG2 557-7524
  - Marc L Caroff rm10C04 CG2 557-4009
  - Irwin C Cohen rm12B18 CG2 557-4703
  - Jerry D Craig rm10D02 CG2 557-4058
  - Mary F Downey rm10B14 CG2 557-4065
  - Stephen J Emery rm12D12 CG2 557-4023
  - Donald D Forrer rm10C16 CG2 557-4061
  - Charles E Frankfort rm12D10 CG2 557-4059
  - Bradley R Garris rm12D06 CG2 557-7148
  - Melvin Goldstein rm10D08 CG2 557-4068
  - John T Goolkasian rm10A10 CG2 557-4003
  - Kenneth W Hairston rm10A02 CG2 557-4112
  - Paul J Henon Jr rm10C22 CG2 557-4058
  - Thomas J Holko rm10B08 CG2 557-4063
  - Edward C Kimlin rm10C10 CG2 557-4003
  - Errol A Krass rm10B12 CG2 557-7516
  - William F Lindquist rm10C18 CG2 557-4061
  - Charles N Lovell rm10D06 CG2 557-4070
  - William E Lyddane rm12C04 CG2 557-4073
  - Thomas E Lynch rm10B04 CG2 557-7517
  - Harrison E McCandlish rm12B14 CG2 557-4703
  - John P McQuade rm10A20 CG2 557-4088
  - James M Meister rm10B06 CG2 557-4063
  - Edward J Meros rm10C20 CG2 557-4003
  - Andrew H Metz rm12B10 CG2 557-4326
  - Marion Parsons Jr rm12C02 CG2 557-4393
  - William F Pate III rm10B10 CG2 557-7653
  - Irving R Pellman rm10A16 CG2 557-4064
  - Verlin R Pendegrass rm10A08 CG2 557-4067
  - Alton D Rollins rm12B04 CG2 557-4023
  - Gene Z Rubinson rm10A12 CG2 557-4066
  - James A Seidleck rm10A22 CG2 557-4070
  - William A Skinner rm12D08 CG2 557-7147
  - John D Smith rm12B16 CG2 557-4326
  - Ronald H Smith rm12D02 CG2 557-4057
  - William F Smith rm10A06 CG2 557-4066
  - Michael Sofocleous rm10C06 CG2 557-4009
  - Lawrence J Stabb rm10A18 CG2 557-4087
  - Robert F Stahl rm12B02 CG2 557-4393
  - Arthur J Steiner rm10A14 CG2 557-4062
  - Bruce H Stoner Jr rm12B12 CG2 557-4025
  - Henry W Tarring II rm10B02 CG2 557-4001
  - James D Thomas rm10C02 CG2 557-4061
  - Norman G Torchin rm10B16 CG2 557-4009
  - Stanley M Urynowicz Jr rm10C14 CG2 557-4009
  - Sherman D Winters rm10D04 CG2 557-4001
- Programs and Resources Administrator
  - Craig R Feinberg rm12C10 CG 2 557-7169
- Service Branch
  - Chief Clerk of Board T Maxine Duvall rm12C06 CG2 557-4101
    - Deputy Clerk Nannie B Henry rm10C01A CG2 557-4007
    - Deputy Clerk Shirley A Jefferys rm12C08 CG2 557-4101
    - Deputy Clerk Eunice I Price rm10C07 CG2 557-4101
    - Ex parte Legal Clerk Groups 110 and 150

Eleanor R Green rm10C09 CG2 557-4107
Ex parte Legal Clerk Group 130-180
Karen Sweeney rm12C08 CG2 557-3100
Ex parte Legal Clerk Group 120
Donald T Harris rm12C08 CG2 557-4108
Ex parte Legal Clerk Groups 210-220-230-240-250-260-290
Fernando Burgess rm10C09 CG2 557-4109
Ex parte Legal Clerk Groups 310-320-330-340-350
Mabel A Neal rm12C08 CG2 557-4106
Inter partes Legal Clerk
Olivia M Duvall rm10C01 CG2 557-4006
Inter partes Legal Clerk
Carrie Evans rm10C01 CG2 557-4004

OFFICE OF QUALITY REVIEW

Director James D Trammell rm1100 CP6 557-3564
Secretary Carolyn D Ballard rm1100 CP6 557-3564

OFFICE OF ASSISTANT COMMISSIONER FOR PATENTS

Assistant Commissioner Rene D Tegtmeyer rm919 PK2 557-3811
Secretary Sherry D Brinkley rm919 PK2 557-3811
Special Assistant R Franklin Burnett rm919 PK2 557-3054
Secretary Donna E Ellis rm919 PK2 557-3054
Paralegal Specialist (Vacant) rm919 PK2 557-3054
Manual of Patent Examining Procedure Editor Louis O Maassel rm919 PK2 557-3070
Deputy Assistant Commissioner for Patents James E Denny rm917 PK2 557-4279
Secretary Patricia R Appelle rm917 PK2 557-4279
Patent Programs Administrator Michael J Lynch rm917 PK2 557-4279
Patent Policy and Resources Director (Vacant) rm917 PK2 557-4279
Supervisory Petitions Examiner Jeffrey V Nase rm913 PK2 557-4282
Petitions Information rm913 PK2 557-4282

Office of Patent Program and Documentation Control

Director Richard H Rouck rm925 PK2 557-4222
Secretary Carolyn Evans rm925 PK2 557-9182
Program Analyst Carolyn Arrington rm925 PK2 557-9184
Patent Academy Richard McGarr rm502 PK1 557-2086
Paper Correlating Office JoAnn Harris rm925 PK2 557-5148
Paper Correlating Office Margaret Seward rm925 PK2 557-5149
Palm Coordinator Rolf G Hille rm925 PK2 557-9175
Special Program Examination Unit
Supervisory Special Program Examiner Manual A Antonakas rm923 PK2 557-8384

PATENT DOCUMENTATION ORGANIZATION

Administrator for Documentation William S Lawson rm300 CM2 557-0400
Secretary Jim Doyle rm300 CM2 557-0400
Data Base Administrator Philip K Olson rm300 CM2 557-0400
Deputy Administrator for Documentation Edward J Earls rm300 CM2 557-0400
Classification Support Staff Director Sally Middleton rm300 CM2 557-0400
Contract Monitoring and Reclassification Division Chiquita Clark rm967 CM2 557-8877
Projects Monitoring Unit Inez Roberts rm967 CM2 557-5164
Processing Unit I - Janice Burse rm969 CM2 557-3396
Processing Unit II - Cornell Boney rm968 CM2 557-7467
Processing Unit III - Pat Walker rm965 CM2 557-7458
Special Projects Unit Daisy Turner rm964 CM2 557-2590
Data Control Technician Division Sadie Scott rm326 CM2 557-5910
Editorial Division Vernella Crowley rm300C CM2 557-5103
New Document Processing Division Marcia Smith Lobby CP6 557-5110
Preprocessing Branch Jerry Redmond Lobby CP6 557-5116
Final Processing Branch Gail Lewis Lobby CP6 557-5114
Special Processing Natalie Jackson Lobby CP6 557-5111
Weekly Issue Mary Johnson rm1240 CP6 557-7906

Office of Documentation Planning and Support
Director George Chadwick rm300 CM2 557-0400
Office of Documentation Information
Director Jane Myers rm304 CM2 557-0400
Technology Assessment and Forecast (Vacant) rm304 CM2 557-0400
Information Services Evelyn Freeman (Acting) rm322 CM2 557-5666
Information Resources (Vacant) rm322 CM2 557-5666
Patent Index rm322 CM2 557-3951
Patent Depository Library Program Manager
Carole A Shores rm306 CM2 557-9686
Secretary Dot Jenkins rm306 CM2 557-9686
Chemical-Electrical Classification Group Director Donald J Hoffman rm901 CM2 557-2825
Secretary Sandra P Crawford rm903 CM2 557-2820
Unit I Diane B Russell rm971 CM2 557-2753
Unit II Eugene B Woodruff rm935 CM2 557-2826
Unit III Gary Solyst rm923 CM2 557-3505
Unit IV Earl Folsom rm912 CM2 557-0151
Mechanical-General Classification Group Director John W Will rm310 CM2 557-0107
Secretary Christina Boska rm310 CM2 557-0107
Unit I Donald P Rooney rm310 CM2 557-0138
Unit II Robert Craig rm310 CM2 557-0136
Unit III Harold P Smith rm310 CM2 557-2446
Special Projects Unit Director Leslie Wolf rm982 CM2 557-2781
Secretary Melvina Jarrett rm980 CM2 557-0173
Office of International Patent Documentation Director Thomas Lomont rm300 CM2 557-0667
Secretary Glenda Calhoun rm300 CM2 557-0400
Scientific Library Program Manager Henry Rosicky rm2C08 CP34 557-2955
Secretary Gail Owens rm2C08 CP34 557-2955
Administrative Librarian Irene Heisig rm2C08 CP34 557-2955
Foreign Patents Division Barry Balthrop rm2C01 CP34 557-2970
Bindery Unit Ronald Knickerbocker FERN 557-1530
Document Retrieval and Copy Branch Lendoria Roberson rm2C01 CP34 557-3545
Receipts and Records Branch Beverly Brooks Corridor Level, Suite 1821D CM2 557-0186
Reference Service Bernard Hamilton rm2C01 CP34 557-3545
Scientific Literature Division Kay Melvin rm2C06C CP3 557-2957
Technical Services Branch Jesse Gibson rm2C06A CP3 557-2961
Collection Development (Vacant) rm2C01 CP34 557-3092
User Services Branch Dora Weinstein rm2C04 CP34 557-2957
Circulation rm2C01 CP34 557-2957
Computer Searching rm2C01 CP34 557-2957
Interlibrary Loans rm2C01 CP34 557-2957
Reference Service rm2C01 CP34 557-2957
Translations Division Dean Thorne rm2C15 CP34 557-3193
Receptionist Carol Releford rm2C15 CP34 557-3193

CHEMICAL EXAMINING GROUPS

110 General, Metallurgical, Inorganic, Petroleum and Electrical Chemistry, and Engineering rm9C17 CP3 557-2517
Director Dennis E Talbert rm9D17 CP3 557-9600
Secretary Constance L Morgan rm9D19 CP3 557-9600
General Information/Receptionist rm9C17 CP3 557-2517
SAC Dorothy Dawkins rm9C17 CP3 557-3598
111 Metallurgical Methods and Apparatus, Alloys and Metal Stock
L Dewayne Rutledge rm10E02 CP3 557-6722
112 Electro-Chemistry, Process and Apparatus John F Niebling rm10B04 CP3 557-8788
113 Inorganic compounds and non-metallic elements (except radioactive); chemical gas purification processes;
beneficiating ores; hydrometullargy; magnetic and piezoelectric compositions
plaster; single crystals and crystallization John Doll rm9A15 CP3 557-2517
114 Methods for semiconductor treating and manufacturing; batteries;
photovoltaic cells and their methods of operation Brian E Hearn rm10D35 CP3 557-6728
115 Chemical compositions, dying and pigments Paul Lieberman rm10A15 CP3 557-8779

116 Mineral oil processes and products, catalytic compositions, chemistry of hydrocarbons,
fuel and igniting devices, sugar, starch and carbohydrates, cleaning and contact with solids and
carbon compounds Helen Sneed rm9D35 CP3 557-3029
118 Liquid and solid fuels, chemical and biological fertilizers, refractory
glass and cement compositions, lubricating compositions William R Dixon Jr rm10B02 CP3 557-8787
120 Organic Chemistry and Biotechnology rm8C13 CP2 557-3920
Director Samih N Zaharna rm8A07 CP2 557-0661
Secretary Anne Willey rm8A07 CP2 557-0661
General Information/Receptionist rm8C13 CP2 557-3920
SAC Helen Childs rm8C13 CP2 557-3920
121 Heterocyclic organic chemistry, nitriles and azo chemistry Mary C Lee rm9D01 CP2 557-3920
122 Nitrogen containing heterocyclic compounds, mercaptans and
phosphorus esters Donald G Daus rm8B32 CP2 557-3920
125 Medicines, poisons, cosmetics and testing compositions Albert T Meyers rm8B02 CP2 557-3920
126 Organic carboxylic acid and ester compounds, oxy, aldehyde and
keto compounds, phosphorus acid compounds Donald Moyer rm9B32 CP2 557-3920
129 Herbicides, heterocyclic and nitrogen chemistry Glennon H Hollrah rm9A01 CP2 557-3920
130 Specialized Chemical Industries and Chemical Engineering rm8C17 CP3 557-2475
Director Robert F White rm8D19 CP3 557-3804
Secretary Vickie Enos rm8D19 CP3 557-3804
General Information/Receptionist rm8C17 CP3 557-2475
SAC Ruth Lyles rm8C17 CP3 557-9854
131 Adhesive bonding and miscellaneous chemical manufacture Michael W Ball rm9E02 CP3 557-2475
132 Food or edible material, processes, compositions and products Donald E Czaja rm8D01 CP3 557-2475
133 Paper making and fiber liberation; glass manufacture; concentrating evaporators, separatory and
thermolytic distillation processes and apparatus; separating and assorting solids-froth floatation;
and compositions, method and apparatus for etching in chemical manufacture gas, heating and
illumination apparatus and process and mineral oils apparatus David L Lacey rm8B36 CP3 557-2475
135 General molding or treating apparatus; static molds; gas separation;
gas and liquid contact Jay H Woo rm8D35 CP3 557-2475
136 Liquid purification or separation Richard V Fisher rm7A09 CP3 557-2475
137 Processes of plastic and nonmetallic article shaping or treating Jan H Silbaugh rm8A01 CP3 557-2475
139 Coating processes and coating apparatus Norman Morgenstern rm9B02 CP3 557-2475
150 High Polymer Chemistry, Plastics, Coating, Photography, Stock Materials and Compositions rm11C19 CP2 557-6525
Director James O Thomas Jr rm11A04 CP2 557-6533
Secretary Sharon C Graham rm11A04 CP2 557-6533
General Information/Receptionist rm11C19 CP2 557-6525
SAC Ellen Scott rm11C19 CP2 557-6525
151 Mixed synthetic resin compositions, block and graft copolymers and
irradiation of polymers John Bleutge rm11C19 CP2 557-6525
153 Foams, condensation polymers of cellulose, phenols, isocyanates,
polyesters, polyepoxides, stabilization of polymers John Kight III rm11C19 CP2 557-6525
154 Stock materials or miscellaneous articles of manufacture George Lesmes rm11C19 CP2 557-6525
155 Polymer compositions, addition polymers, and carbohydrates Joseph Schofer rm11C19 CP2 557-6525
156 Radiation imagery chemistry - process, composition or product, and
stock materials or miscellaneous articles of manufacture Paul Michl rm11C19 CP2 557-6525
158 Coating apparatus, record receivers, stock materials or miscellaneous
articles, and selected imaging processes and products Ellis Robinson rm11C19 CP2 557-6525
180 Biotechnology rm9C13 CP2 557-0664
Director John E Kittle rm9A09 CP2 557-3637
Secretary Cheryl P Gibson rm9A09 CP2 557-3637
General Information/Receptionist rm9C13 CP2 557-0664
SAC Kathryn Perry rm9C13 CP2 557-6941
181 Chemical apparatus such as analyzers, reactors and sterilizers; processes of chemical and clinical analysis,
sterilizing and preserving; immunology and liquid purifications or separation by living
organisms Barry S Richman rm9B02 CP3 557-6629
182 Clinical chemistry, microbiology, immunology and enzymology,
purification and chemical engineering Robert J Warden rm9B32 CP2 557-7369

183 Peptide and carbohydrate chemistry, and drug, bioaffecting and body treating compositions containing peptides, carbohydrates, antibody, antigen, enzyme, or animal or plant extracts of undetermined constitution Johnnie R Brown rm8D31 CP2 557-3776
184 Multicellular organisms (animal/plant), molecular genetics, cell culture, nucleic acid assays, immunology, hybridoma, molecular biology, microbiology, fermentation and chemical engineering Charles F Warren rm10D32 CP2 557-7387
185 Molecular genetics, catalysis, enzymology, microbiology and chemical engineering Thomas G Wiseman rm9D31 CP2 557-3567

ELECTRICAL EXAMINING GROUPS

210 Industrial Electronics, Physics and Related Elements rm9C17 CP4 557-5080
Director Gerald Goldberg rm9D19 CP4 557-2488
Secretary Teresa E Dugan rm9D19 CP4 557-2488
General Information/Receptionist rm9C17 CP4 557-5080
SAC Charles B Blake rm9C17 CP4 557-7405
211 Acoustics, Electric Charge Devices and Systems, Electromagnetic Control Systems, Music, Optics, Photography, Photocopying, Motion Pictures, Electric and Magnetographic Recorders, and Mechanical Registers L Thomas Hix rm9B40 CP4 557-7674
212 Electrical Motor-generator Structure, Piezoelectric Elements and Devices, Generator Systems, Battery and Condenser Charging and Discharging, Power Supply Regulation, and Conversion Systems Patrick R Salce rm9A01 CP4 557-9695
214 Electrical Switches and Arc Suppression, Heating by Induction, Plasma and Electric Discharge Machining, Industrial Electric Heating Furnaces, Electrical Component Housing and Mounting and Protection of Electrical Systems and Devices A David Pellinen rm8B34 CP4 557-7215
215 Conductors, insulators, inductors, electric photocopying and electrical devices Arthur T Grimley rm9E16 CP4 557-7671
216 Recorders, Scales, Magnets, Magnetic and Thermal Switches, Electric Heating, Electric Welding and Resistors Elliott A Goldberg rm8B02 CP4 557-2323
217 Motor Control, Electrical Transmission Systems, Prime-mover Dynamo Plants, Electrical Elevator Controls, Code Conversion and Horology William M Shoop Jr rm9D01 CP4 557-3231
220/290 Utility and Design Applications rm10C17 CP4 557-2895
Director Kenneth L Cage rm10D17 CP4 557-2877
Secretary Arnette S McGill rm10D19 CP4 557-2478
General Information/Receptionist Group 220 rm10C19 CP4 557-2895
General Information/Receptionist Group 290 rm1106 CP6 557-2864
SAC Joanne Hodge rm10C17 CP4 557-9151
Licensing and Review Hilda Grimes rm10C34 CP4 557-4948
Mildred Lawrence rm10C34 CP4 557-4949
221 Weapons (firearms, ordnance, ammunition, explosive devices), general lubrication, nuclear reactor systems, illumination, aeronautics and ships as well as all classified mechanical applications Deborah L Kyle rm10D01 CP4 557-3253
222 Radio, optic, acoustic, wave communications systems and all classified electrical applications Thomas H Tarcza rm10E16 CP4 557-4922
223 Chemical and including radioactive materials, powder metallurgy, rocket fuels, explosives and thermal and photoelectric batteries as well as all classified chemical applications John Terapane rm10E02 CP4 .. 557-4934
291 Ornamental designs in the area of industrial arts Wallace R Burke rm11E02 CP6 557-4979
292 Ornamental designs for fine arts Bernard Ansher rm11A02 CP6 557-4965
230 Information Processing, Storage and Retrieval rm11C17 CP4 557-2878
Director Earl Levy rm11D37 CP4 557-5088
Secretary Laura Dorsey rm11D37 CP4 557-5088
General Information/Receptionist rm11C17 CP4 557-2878
SAC Katherine A Nelson rm11C17 CP4 557-4174
231 General & Special Purpose Digital Data Processing, Digital Arithmetic, Speech Analysis & Synthesis & Data Presentation Systems Gary V Harkcom rm10B02 CP4 557-7128
232 General & Special Purpose Digital Data Processing Systems Archie E Williams rm11B02 CP4 557-2119
233 Static Information Storage & Retrieval & Elements of Dynamic Magnetic Information Storage & Retrieval Systems Stuart N Hecker rm11D01 CP4 557-0326
234 Ordnance or Weapon System Computers & Special Applications of Computers Including Vehicle Control, Navigation, Measuring, Testing & Monitoring Parshotam S Lall rm11D17 CP4 557-4316
235 Electrical Dynamic Information Storage & Retrieval & Record Controlled Systems (Vacant) rm11D17 CP4 557-2878

236 Computer Control Systems, Miscellaneous Applications of Computers, Computer Aided Product Manufacturing, Artificial Intelligence, Analog & Hybrid Computers & Error Correction & Detection Systems Jerry Smith rm11E10 CP4 557-8041
237 General & Special Purpose Digital Data Processing Systems Gareth D Shaw rm11B40 CP4 557-8047
240 Packages, Cleaning, Textiles and Geometrical Instruments rm6C17 CP4 557-2900
Director Trygve M Blix rm6D37 CP4 557-2906
Secretary Donna Purdham rm6D37 CP4 557-2906
General Information/Receptionist rm6C17 CP4 557-2900
SAC Doretha A Bailey rm6D15 CP4 557-2900
241 Packaging art including glass, fabric, metal, wood, paper and plastic receptacles plus closures Stephen Marcus rm6B02 CP4 557-4719
242 Fluid treating, presses, food apparatus, cleaning, agitating, centrifuges, and web feeding Harvey C Hornsby rm6E02 CP4 557-6116
243 Conduits, bathroom facilities, cleaning apparatus, filling apparatus, switches, and article carriers Henry Recla rm6E10 CP4 557-9891
245 Textiles, winding and reeling, pushing and pulling, bearings, and flexible torque transmitters Stuart S Levy rm5B24 CP4 557-6855
246 Measuring and testing, dynamic information storage or retrieval, optical image projectors and joint packing William Cuchlinski rm6E14 CP4 557-9894
247 Textile and leather manufacture, apparel, and textiles Werner Schroeder rm6D01 CP4 557-3302
250 Electronic and Optical Systems and Devices rm8D17 CP4 557-3311
Director Edward Kubasiewicz rm8D19 CP4 557-2084
Secretary Deborah P Leeper rm8D19 CP4 557-2084
General Information/Receptionist rm8C17 CP4 557-3311
SAC JoAnn Davis rm8C19 CP4 557-4784
251 Lasers, fiber optic devices and antennas William L Sikes rm8E02 CP4 557-2733
252 Electronic modulators, demodulators, oscillators, amplifiers, tuners and wave transmission lines and networks Eugene R Laroche rm7A15 CP4 557-4317
253 Semiconductor devices Andrew J James rm7B02 CP4 557-4835
254 Semiconductor and vacuum tube circuits and systems and electronic and electromechanical counting circuits and systems Stanley D Miller rm7E02 CP4 557-4753
255 Optical measuring and testing systems and photocell circuits Davis L Willis rm8D13 CP4 557-4339
256 Radiant energy systems Craig E Church rm7E16 CP4 557-3453
257 Optical systems and elements and vision testing and correcting John K Corbin rm8E16 CP4 557-2884
260 Communications, Measuring, Testing and Lamp/Discharge Group rm5D19 CP4 557-3321
Director Stephen G Kunin rm5D19 CP4 557-7075
Secretary Iyone L Miles rm5D19 CP4 557-7075
General Information/Receptionist rm5C14 CP4 557-3321
SAC Vivian C Harris rm5D21 CP4 557-2067
261 Telegraphy, telephony and audio systems Jin F Ng rm4D01 CP4 557-7739
262 Television and television facsimile James J Groody rm4B02 CP4 557-7309
263 Multiplex communications, digital communications and telecommunications Robert L Griffin rm5E02 CP4 557-1139
264 Electrical communications and acoustic wave systems John W Caldwell rm5E16 CP4 557-3356
265 Measuring and testing of non-electrical phenomenon Stewart J Levy rm4E02 CP4 557-7603
266 Lamp and discharge devices/systems and image analysis David K Moore rm5B02 CP4 557-6868
267 Measuring and testing of electrical phenomenon Reinhard J Eisenzopf rm5D01 CP4 557-6878
268 Condition responsive communications, measuring and testing Joseph A Orsino rm5D01 CP4 557-7956

MECHANICAL EXAMINING GROUPS

310 Handling and Transporting Media rm5D19 CP3 557-3618
Director Bobby R Gray rm5D19 CP3 557-3677
Secretary LaJuene Desmukes rm5D19 CP3 557-3677
General Information/Receptionist rm5C17 CP3 557-3618
SAC Margaret Stevens rm5D17 CP3 557-3618
311 Dispensing, article dispensing, coin handling, check-controlled apparatus, elevators, and sheet feeding or delivering devices Joseph J Rolla Jr rm6D01 CP3 557-6491
312 Railways and railway equipment, motor vehicle wheels and bodies and article assorting and handling implements Robert B Reeves rm5A01 CP3 557-6765

314 Brakes, fluid pressure brake systems, spring devices and spraying devices Andres Kashnikow rm5D01 CP3 .. 557-3680
315 Aeronautics and marine arts and fluid conveying and fire extinguishers
Joseph F Peters Jr rm6A01 CP3 ... 557-1471
316 Land and motor vehicles Charles Marmor rm5E02 CP3 ... 557-6738
317 Article handling and conveyors and dumping vehicles Robert J Spar rm5D35 CP3 ... 557-6732
320 Material Shaping, Article Manufacturing, Tools rm6C17 CP3 ... 557-3694
Director Donald G Kelly rm6D19 CP3 ... 557-3547
Secretary Danita L Ingram rm6D19 CP3 ... 557-3547
General Information/Receptionist rm6C17 CP3 ... 557-3694
SAC Vera Thomas rm6C17 CP3 ... 557-3694
321 Metal deforming, packaging machinery, butchering and woodworking Robert Spruill rm7E02 CP3 ... 557-9856
322 Electrical connectors, gear cutting, milling and chucks Gil Weidenfeld rm6D35 CP3 ... 557-3698
323 Abrading, workholders, tools and paper manufactures Frederick Schmidt rm7D35 CP3 ... 557-6506
324 Cutting, cutlery, tools, bookmaking, and printed matter Frank T Yost rm6E02 CP3 ... 557-6039
325 Metal founding, metal turning, miscellaneous hardware, fishing, vermin
trapping and welding Nicholas Godici rm7D01 CP3 ... 557-6513
326 Metal working, comminution and wire working Howard Goldberg rm7A01 CP3 ... 557-9871
330 Surgery, Plant and Animal Husbandry, Amusement and Exercise Devices
and Printing rm4C17 CP4 ... 557-3125
Director John J Love rm4D19 CP4 ... 557-3164
Secretary Norma L Watson rm4D19 CP4 ... 557-3164
General Information/Receptionist rm4C17 CP4 ... 557-3125
SAC Clara S Desmukes rm4D21 CP4 ... 557-3125
331 Planting, Surgical Instruments, toys, earth working Robert Hafer rm3A01 CP3 ... 557-3125
332 Exercising and Therapy Devices, Tobacco, Artificial Body Parts,
Bandages and Hand Applicators Richard Apley rm4E16 CP4 ... 557-3125
333 Dentistry, animal husbandry, harvesting and sign exhibiting Gene Mancene rm3E16 CP3 ... 557-3125
334 Amusement games or educational devices Richard Pinkham rm3E02 CP4 ... 557-3125
335 Surgery diagnostics Kyle Howell rm3E02 CP3 ... 557-3125
336 Surgery instruments, medicators or receptors C Fred Rosenbaum rm3B11 CP3 ... 557-3125
337 Printing, typewriting or excavating Edgar Burr rm3D01 CP3 ... 557-3125
340 Solar, Heat, Power and Fluid Engineering Devices rm3C17 CP4 ... 557-3128
Director Donley J Stocking rm3D19 CP4 ... 557-3340
Secretary Carol A Jones rm3D19 CP4 ... 557-3340
General Information/Receptionist rm3C17 CP4 ... 557-3128
SAC Verlene Day rm3C17 CP4 ... 557-2122
341 Expansible chamber motors and fluid power systems Robert Garrett rm3B22 CP4 ... 557-3464
342 Internal combustion engines including charge forming and ignition
systems Charles Myhre rm2D01 CP4 ... 557-5690
343 Combustion power plants, reaction meters, pumps, rotary expansible
chamber devices Carlton R Croyle rm3B02 CP4 ... 557-0398
344 Environmental control, including heating, air conditioning,
refrigeration and ventilation Albert Makay rm3D01 CP4 ... 557-0397
345 Combustion of fuels and substances, heating the environment and
extracting heat from sun Samuel Scott rm3D41 CP4 ... 557-5999
346 Devices and methods for transferring heat from one material to another, turbochargers, superchargers,
poppet valve operation and exhaust gas treatment for internal combustion engines,
rotary internal combustion engines and powerplant of the type using natural heat
Ira Lazarus rm2B02 CP4 ... 557-5679
347 Fluid handling which includes valves, pressure regulators and flow
controllers for liquids and gases Martin Schwadron rm2B42 CP4 ... 557-6777
350 General Construction, Petroleum and Mining Engineering rm3C17 CP3 ... 557-6200
Director Al Lawrence Smith rm3D19 CP3 ... 557-3414
Secretary Carol M Sinclair rm3D19 CP3 ... 557-3414
General Information/Receptionist rm3C17 CP3 ... 557-6200
SAC Joyce G Hill rm3D13 CP3 ... 557-6200
351 Joints and connections, pipe couplings, fences, earth & hydraulic
engineering Randolph A Reese rm4D01 CP3 ... 557-6200
352 Gearing, power transmissions, clutches, machine elements Leslie A Braun rm2B36 CP3 ... 557-6200
354 Building structures and components David A Scherbel rm2D35 CP3 ... 557-6200

355 Supports, racks, fire escapes, ladders, scaffolds, flexible partitions
Ramon S Britts rm4E02 CP3 557-6200
356 Petroleum, mining, highway and bridge engineering, well drilling, and
endless belts Jerome W Massie rm4D35 CP3 557-6200
357 Tables, chairs, cabinets, windows, doors, buckles, buttons, clasps
Kenneth J Dorner rm4A01 CP3 557-6200
358 Fasteners, safes, locks, closure fasteners, beds, control levers and
linkages Gary L Smith rm4D17 CP3 557-6200

OFFICE OF THE ASSISTANT COMMISSIONER FOR EXTERNAL AFFAIRS

Assistant Commissioner Michael K Kirk rm902 PK2 557-3065
Secretary Johnell M Bersano rm902 PK2 557-3065
Director of Congressional Affairs Arthur E White rm902 PK2 557-1310
Congressional Liaison Janie F Cooksey rm902 PK2 557-1310
Office of Public Affairs
Director William O Craig rm1B01 CP3 557-3341
Public Information Specialist Oscar G Mastin rm1B01 CP3 557-3341
Office of Legislation and International Affairs
Director (Vacant) rm902 PK2 557-3065
Legislative and International Intellectual
Property Specialists
Judy W Goans rm902 PK2 557-3065
H Dieter Hoinkes rm902 PK2 557-3065
Lee J Schroeder rm902 PK2 557-3065
G Lee Skillington rm902 PK2 557-3065
Attorney Advisors
Rosemarie G Bowie rm902 PK2 557-3065
Michael S Keplinger rm902 PK2 557-3065
Richard C Owens rm902 PK2 557-3065

OFFICE OF THE ASSISTANT COMMISSIONER FOR TRADEMARKS

Assistant Commissioner Jeffrey M Samuels rm910 PK2 557-3061
Secretary Sheila G Pellman rm910 PK2 557-3061
Deputy Assistant Commissioner for Trademarks Robert M Anderson rm910 PK2 557-3916
Secretary Charlene A Rucker rm910 PK2 557-3061
Trademark Legal Administrator Carlisle E Walters rm910 PK2 557-7464
Secretary Carol P Smith rm910 PK2 557-3061
Staff Assistant (Vacant) rm910 PK2 557-2222
Trademark Program Analyst (Vacant) rm910 PK2 557-2221

Office of Trademark Program Analysis

Director Kimberly Krehely rm3C06 CP2 557-3268
Program Analyst Betty Andrews rm3C06 CP2 557-3268

TRADEMARK EXAMINING OPERATION

Director David E Bucher rm3C06 CP2 557-3268
Secretary Tawana A Hawkins rm3C06 CP2 557-3268
Administrator for Trademark Operations Patricia M Davis rm3C06 CP2 557-3268
Secretary (Vacant) rm3C06 CP2 557-3268
Administrator for Trademark Policy and Procedures Charles J Condro rm3C06 CP2 557-3268
Secretary Lisa Y Harrell rm3C06 CP2 557-3268
Petitions and Classification Attorney
Jessie N Marshall rm3C06 CP2 557-3268
Petitions Assistant Annette L Pray rm3C06 CP2 557-3268
Trademark Procedures and Special Projects Attorney James T Walsh rm3C06 CP2 557-3268
Secretary Nina Bailey rm3C06 CP2 557-3268
Trademark Program Assistant (Vacant) rm3C06 CP2 557-3268
Paralegal Assistant Blake Pearl rm3C06 CP2 557-3268
Paralegal Assistant Doris Kahn rm3C06 CP2 557-3268
Trademark Law Offices
Managing Attorney Law Office I (Vacant) rm3C28 CP2 557-3273
Senior Lead Attorney Deborah S Cohn

Support Staff Manager Carolyn Spriggs
Managing Attorney Law Office II (Vacant) rm2C24 CP2 557-3277
Senior Lead Attorney Donald Fingeret
Support Staff Manager Doshie Day
Managing Attorney Law Office III Myra Kurzbard rm2C22 CP2 557-9560
Senior Lead Attorney Robert Feeley
Support Staff Manager Gwen Stanmore
Managing Attorney Law Office IV Thomas Lamone rm3C13 CP2 557-9550
Senior Lead Attorney David Soroka
Support Staff Manager Elsie Bradley
Managing Attorney Law Office V Paul Fahrenkoph rm2C11 CP2 557-5380
Senior Lead Attorney Christopher Sidoti
Support Staff Manager Thurmond Streater
Managing Attorney Law Office VI Ronald E Wolfington rm3C27 CP2 557-2937
Senior Lead Attorney Mary Sparrow
Support Manager Pearl Clements
Managing Attorney Law Office VII Lynne Beresford rm4C13 CP2 557-5237
Senior Lead Atttorney David Shallant
Support Staff Manager Karen McCray
Managing Attorney Law Office VIII Sidney Moskowitz rm4C11 CP2 557-5242
Senior Lead Attorney Michael Bodson
Support Staff Manager Ada Rollins
Trademark Services Division Director Doreane Poteat rm4C25 CP2 557-5249
Supervisory Trademark Services Assistant Seth M Cheatham rm4C25 CP2 557-5249
Secretary (Vacant) rm4D29 CP2 557-5249
Quality Review Clerk Deborah Ahmed rm4C25 CP2 557-5249
Application & Classification Section Leon Jackson rm4B30 CP2 557-5255
Publication & Issue Supervisor Tony Milligan rm4C23 CP2 557-5247
Post Registration Sec Supervisor Portia Taylor rm4C24 CP2 557-1986
Affidavit Examiners rm4C24 CP2 557-1988
Renewal Examiners rm4C24 CP2 557-1988
Mail Reader/Messenger Lilly Mott rm4C26 CP2 557-5257
Microfilm Section Della Williams rm4D27 CP2 557-5255

TRADEMARK TRIAL AND APPEAL BOARD

Members of the Board
Chairman J David Sams rm1008 CS5 557-3551
Ellen Seeherman rm1008 CS5 557-3551
Robert F Cissel rm1008 CS5 557-3551
Louise E Rooney rm1008 CS5 557-3551
Gary D Krugman rm1008 CS5 557-3551
Janet E Rice rm1008 CS5 557-3551
Rany L Simms rm1008 CS5 557-3551
Elmer W Hanak III rm1008 CS5 557-3551
Attorney-Examiners
Paula T Hairston rm1008 CS5 557-3551
Beth A Chapman rm1008 CS5 557-3551
G Douglas Hohein rm1008 CS5 557-3551
T Jeffrey Quinn rm1008 CS5 557-3551
Carla C Calcagno rm1008 CS5 557-7049
Paralegal Specialist Gladys R Springer rm1008 CS5 557-3551
Clerk of the Board Evelyn R Lopez rm1008 CS5 557-3551
Administrator Erma S Brown rm1008 CS5 557-3551
Legal Technician Sheila H Veney rm1008 CS5 557-3551

OFFICE OF THE ASSISTANT COMMISSIONER FOR ADMINISTRATION

Assistant Commissioner Theresa A Brelsford rm908 PK2 557-2290
Secretary Karon Morris rm908 PK2 557-2290
Deputy Assistant Commissioner for Administration Wesley H Gewehr rm908 PK2 557-3055
Secretary Diane Rich rm908 PK2 557-3055
Program Analyst Joan S Griffey rm908 PK2 557-2290

Office of General Services

Director John D Hassett rm803 PK1 557-0183
Secretary Peggy Fewell rm803 PK1 557-0183
Deputy G William Richardson rm803 PK1 557-0183
Security/Safety Officer (Vacant) rm803 PK1 557-0183
Correspondence and Mail Division Sallye Rayford rm1A03 CP2 557-1689
Deputy (Vacant) rm1A03 CP2 557-2932
Incoming-Outgoing Mail Branch Margaret LaSalle rm1A01 CP2 557-3233
Initial Review & Serializing Branch Shirley Steele rm1B03 CP2 557-3478
Correspondence Branch (Vacant) rm1A03 CP2 557-3226
Facilities Management Division Robert Randolph rm802C PK1 557-7042
Records and Property Management Branch Flo Stanmore rm802C PK1 557-0410
Space and Telecommunications Branch William Morris rm802C PK1 557-7331
Office Services Division Constant G Fearing rm803 PK1 557-0183
Travel Arrangements rm803 PK1 557-0183
Support Services Branch Joe Ragland FERN 557-3560
Transportation Unit John Holmes FERN 557-1531
File Information Unit William Satterwhite rm1D01 CP3 557-6944
Official Search Unit Daisy Johnson FERN 557-9690

Office of Patent and Trademark Services

Director (Vacant) rm7D25 CP2 557-3236
Secretary Teresa Knight rm7D25 CP2 557-3236
Deputy Mary E Turowski rm7D25 CP2 557-3236
Public Search Services Division Catherine Kern (Acting) rm1A01 CP3 557-2276
Secretary (Vacant) rm1A01 CP3 557-2276
Patent Search Branch Bernard Thomas rm1A03 CP3 557-2219
Secretary Barbara Evans rm1A03 CP3 557-2277
Patent Search Room rm1A03 CP3 557-2276
Trademark Search Branch Linda Lynch rm2C08 CP2 557-3281
Assignment Search Branch Diane Russele rm5C22 CP2 557-3826
Program Control Division Mary Brown (Acting) Lobby CP1 557-7261
Secretary Louwilda Turner Lobby CP1 557-7261
Public Service Window Lois Stevenson Lobby CP3 557-2833
Public Service Center Mary Reed rm1A02 CP4 557-5168
Secretary (Vacant) rm1A02 CP4 557-5168
Assignment/Certification Services Division (Vacant)
Assignment Branch Annie Harrell rm7D19 CP2 557-3266
Examination Section Virginia Clark rm5C16 CP2 557-3247
Digest and Recording Lucy Stevenson rm7D19 CP2 557-3259
Quality Control and Status (Vacant) 557-8691
Certification Branch Lannie Anderson rm800 PK1 557-1552
Input and Records Control Mary Gartrell rm800 PK1 557-1587
Microfiche & Printing Frances Morris rm800 PK1 557-1603
Certification Section Dorothea Saunders rm800 PK1 557-1564
Court and Documentation (Vacant) rm800 PK1 557-1564
PCT International Services Division Jane Corrigan rm1248 CP6 557-2003
Receiving Office Branch Terry Johnson rm1248 CP6 557-2003
Searching/Preliminary Examining Authority; and Designated/
Elected Office Branch Lauretta Zirk rm1248 CP6 557-2003
Application Processing Division Willie Bowman rm7D19 CP2 557-3256
Classification & Routing Branch Norma White rm7C20 CP2 557-3260
Administrative Examination Branch Mose Montgomery rm7C10 CP2 557-3254
Special Processing Branch Otis Quick rm7C18 CP2 557-3831

Data Entry Branch Everette Oliver rm7C24 CP2 557-5662
Quality Review and Assembly Branch Delora Dillard rm7C24 CP2 557-3763
File Maintenance & Correspondence Branch Jeanette Gatling rm7C12 CP2 557-1561
Micrographics Division Michael Johnson (Acting) rm7D25 CP2 557-3236
Application Filming Branch (Crystal) Calvin Pullen rm6C22 CP2 557-3079
Patent & Trademark Filming Branch (HCHB) Al Mundy 377-4968
Quality Review Branch (HCHB) Mary Smith 377-5501

Office of Publications

Director Stanley J Bania rm6C07 CP2 557-3794
Secretary Marjorie V Turner rm6C07 CP2 557-0698
Publishing Division Ruth C Mason rm6C17 CP2 557-3283
Deputy Manager Sylvia F Martin rm6C17 CP2 557-6388
Allowed Files and Assembly Branch Yvette E Simms rm6C30 CP2 557-6395
Production Control Branch Willard D Ireland rm6C10 CP2 557-6412
Editorial Branch Marthina Thompson rm6C06 CP2 557-6393
Data Base Query Section (Vacant) rm6C06 CP2 557-6392
Patent Copy Inspection Section Jim Alexander rm6C06 CP2 557-6393
Drafting Branch (Vacant) rm6C14 CP4 557-6404
Statistical Analysis Division Michael Stellabotte rm6C07 CP2 557-1963
Data Base Inspection Branch Melvinia Gary rm6C30A CP2 557-6414
Certificates of Corrections Branch Mary H Allen rm809 PK1 557-0709
Technical Development Division Edwin P Hall (Acting) rm6C07 CP2 557-1992
Patent Maintenance Division CH Griffen rm811 PK1 557-6945

Office of Equal Employment Programs

Director R Jacqueline Dees rm600A PK1 557-1692
Secretary (Vacant) rm600A PK1 557-1692
Secretary Assistant Shawneequa Graham rm600A PK1 557-1692
Supervisory Equal Opportunity Specialist Godfrey Beckett rm600A PK1 557-1692
Equal Opportunity Specialists
Sharon L Carver rm600A PK1 557-1692
Myra P Young rm600A PK1 557-1692
Helen D Mitchell rm600A PK1 557-1692

Office of Management and Organization

Director Sara E Bjorge rm600B PK1 557-5825
Secretary Karen Fuller rm600B PK1 557-5825
Project Managers:
Alvin Dorsey rm600B PK1 557-5825
Jean E Buckhout rm600B PK1 557-5825
Greg P Mullen rm600B PK1 557-5825

Office of Procurement

Director William J Eldridge rm806 PK1 557-0014
Secretary Cristina M Moran rm806 PK1 557-0014
Contract Division Chief Page A Etzel rm806 PK1 557-0014
Small Purchases Division Chief Jean A McLeod rm806 PK1 557-0014

OFFICE OF THE ASSISTANT COMMISSIONER FOR FINANCE AND PLANNING

Assistant Commissioner Bradford R Huther rm904 PK2 557-1572
Secretary Vickie T Bryant rm904 PK2 557-1572

Office of Budget

Director James R Lynch rm805 PK1 557-3875
Deputy Director Miguel B Perez rm805 PK1 557-3875
Secretary Marna Engram rm805 PK1 557-3875

Office of Finance

Director Leonard L Nahme rm802A PK1 557-3051
Secretary Virginia R Clark rm802A PK1 557-3051
Appropriation Accounting Division John L Oliff rm802B PK1 557-2983
Fee Accounting Division Frank S Lane Sr rm1B01 CP2 557-2983
Deposit Account Branch Delores H Riley rm1B01 CP2 557-3227
Financial Management Division Robert M Kopson rm802A PK1 557-3051

Office of Long-Range Planning and Evaluation

Director Frances Michalkewicz rm904 PK2 557-1610
Program Assistant Jeanette Hawthorne rm904 PK2 557-1610
PROJECT XL rm904 PK2 557-1610

Office of Personnel

Personnel Officer Carolyn P Acree rm700 PK1 557-2662
Assistant to the Personnel Officer Nancy C Swanberg rm700 PK1 557-2662
Secretary Mildred Jeter rm700 PK1 557-2662
Classification and Employment Division Cynthia Nelson rm700 PK1 557-3631
Employee and Labor Relations Division (Vacant) rm700 PK1 557-3643
EAP Counselor Lisa Ray rm808C PK1 557-6327
Workforce Planning, Development and Systems Division Thomas H Neuhauser rm601 PK1 557-3431
Personnel Processing Division Beverly Boykin rm700 PK1 557-1208
Office of Labor Law Counsel Lynn Sylvester rm700 PK1 557-9684

OFFICE OF THE ASSISTANT COMMISSIONER FOR INFORMATION SYSTEMS

Assistant Commissioner for Information Systems
Thomas P Giammo rm1004 PK2 557-9093
Secretary (Vacant) rm1004 PK2 557-9093
Deputy Assistant Commissioner for Information Systems Boyd Alexander rm1002 PK2 557-6000
Secretary Carla Bowman rm1002 PK2 557-6000
Technical Coordinator Linda Budney rm1004 PK2 557-9093
APS Project Manager Don LeCrone rm1002 PK2 557-6000
Secretary Paulette Whiteside rm1002 PK2 557-6000
Program Management Support Services Office L Liddle rm1002 PK2 557-6000
Secretary Michele Helms rm1002 PK2 557-6000

Office of System Test and Evaluation

Director Bruce A Reynolds rm1002 PK2 557-4114
Secretary Linda Bilbo rm1002 PK2 557-4114

Office of Automated Patent Systems

Director Jeff Cochran rm1000 PK2 557-6156
Secretary Audrey Jackson rm1000 PK2 557-6156

Office of Automated Trademark Systems

Director Raymond Rahn rm911 PK2 557-3544
Secretary Felicia Palmer rm911 PK2 557-3544

Office of Systems Engineering and Data Communications

Director Stephen Jacobson rm1004 PK2 557-7862
Secretary Eliza Davis rm1004 PK2 557-7862

Office of Electronic Data Conversion and Dissemination

Director David Grooms (Acting) rm914 PK2 557-6154
Secretary (Vacant) rm914 PK2 557-6154

Center for Automated Patent Systems
Center for Trademark and General Systems

Director William Maykrantz (Acting) rm1001 PK2 557-3646
Program Assistant Sylvia Huffman rm1001 PK2 557-3646
System Support and Operations Division John Fancovic rm1001 PK2 557-3646
Applications Systems Management Division Thomas Woomer rm1001 PK2 557-6330
Secretary Lisa O'Donnell rm1001 PK2 557-6330
Archives and Contingency Management Division J Kent Hughes rm914 PK2 557-6132
Site Manager Mel Pears Boyers PA (412) 794-3636

APS Contract Office

Director Barry Brown rm509 CP1 557-5800

Secretary Arnett Wright rm509 CP1 557-5800

Master Index

Master Index

This index is arranged alphabetically under the name that is listed in the appropriate directory chapter. The name may also be listed under a specific keyword within the context of the name. Titles of publications are listed in italic type. Also listed in this index are acronyms followed by the full name in parentheses. Names mentioned within the text of a main entry are designated by a star (★) before the number. All citations refer to the relevant entry number.

A

A. Olivetti Associates, Inc. 137
A. T. Capital Corporation 1720
AAED (American Association of Entrepreneurial Dentists) 3
AAI (American Association of Inventors) 4
Abacus Ventures 1397
ABC Capital Corporation 1274
ABCs of Books and Thinking Skills K-8 2284
ABCs of Reading, Thinking and Literacy 7-12 2285
Aberdeen Proving Ground Installation Support Activity – U.S. Department of the Army 1911
Academic Research and Public Service Centers in California: A Guide 1974
ACC (American Copyright Council) 5
Accel Partners (Princeton) 1574
Accel Partners (San Francisco) 1275
Accent On Living 1975
Access Management Corporation 138
ACE (Association of Collegiate Entrepreneurs) 11
Acorn Ventures, Inc. 1782
Acquisition Newsletter; Business & 2001
ACS (American Copyright Society) 6
ACT (Acting, Creating and Thinking) 2286
Acting, Creating and Thinking 2286
Adler and Company (New York City) 1604
Adler and Company (Sunnyvale) 1276
Administrative Sciences Research Laboratory – Laval University 450
Ads Capital Corporation 1590
Advanced Ceramics and Composites Partnership .... ★479
Advanced Manufacturing Technology 1976
Advanced Robotics Research Institute 1077
Advanced Science & Technology Institute ... 344
Advanced Technologies Center – Lorain County Community College 456
Advanced Technology Development Center – Georgia Institute of Technology 408
Advanced Technology Ventures 1491
Advent Atlantic Capital Company LP 1492
Advent Industrial Capital Company LP 1493
Advent International Corporation 1494
Advent IV Capital Company 1495
Advent V Capital Company 1496
Advertising Age 1977
The Advocap Small Business Center 1259
Adweek/East 1978
AEA (American Entrepreneurs Association) 7
Aeneas Venture Corporation 1497
Aeromedical Research Laboratory .... ★1929
Aeronautical Systems Division (Wright-Patterson Air Force Base); Air Force Systems Command, – U.S. Department of the Air Force 1895
Aerospace Medical Research Laboratory; Naval .... ★1951
Affiliated Inventors Foundation, Inc. 1
AFHRL Newsletter 1979
Agana District Office – U.S. Small Business Administration 777
Agency Sales Magazine 1980
Agents Registered to Practice before the United States Patent and Trademark Office; Roster of Attorneys and 2204
AgPat 2414
Agricultural Research Service – U.S. Department of Agriculture 1871
Ahoskie Incubator Facility 1211
AIF (Affiliated Inventors Foundation, Inc.) 1
Air Development Center; Naval – U.S. Department of the Navy 1943
Air Development Center; Rome – U.S. Department of the Air Force 1907
Air Engineering Center; Naval – U.S. Department of the Navy 1944
Air Force Human Resources Laboratory .... ★1979
Air Force Systems Command – U.S. Department of the Air Force 1893
Air Force Systems Command, Aeronautical Systems Division (Eglin Air Force Base) – U.S. Department of the Air Force 1894
Air Force Systems Command, Aeronautical Systems Division (Wright-Patterson Air Force Base) – U.S. Department of the Air Force 1895
Air Force Systems Command, Space Division – U.S. Department of the Air Force 1896
Air Market News 1981
Air Propulsion Center; Naval – U.S. Department of the Navy 1945
Air Systems Command; Naval – U.S. Department of the Navy 1946
Akron Industrial Incubator 1213
AKZO Chemicals Inc., Research Library 2528
Alabama Capital Corporation 1261
Alabama High Technology Assistance Center – University of Alabama in Huntsville 580
Alabama Industrial Development .... ★681
Alabama International Trade Center .... ★676
Alabama Small Business Development Consortium 676
Alabama State Department of Economic and Community Affairs – Planning and Economic Development 677
Alabama State Department of Economic and Community Affairs – Science Technology and Energy Division 678
Alabama State Department of Environmental Management 679
Alabama State Department of Finance – Division of Purchasing 680
Alabama State Development Office – Industrial Finance Division 681
Alabama State House of Representatives – Small Business Committee 682
Alan Patricof Associates, Inc. 1605
Alaska Economic Development Center – University of Alaska 581
Alaska Industrial Development Authority 685
Alaska Small Business Development Center 345
Alaska State Department of Administration – Purchasing Office 686
Alaska State Department of Commerce and Economic Development 687
Albany District Office – U.S. Small Business Administration 987
Albert L. Emmons 139
Albright Venture Capital, Inc. 1484
Albuquerque District Office – U.S. Small Business Administration 977
Albuquerque Invention Club 2
Albuquerque SBIC 1591
Allergy and Infectious Diseases; National Institute of – U.S. Department of Health and Human Services 1887
Alliance Business Investment Company (Houston) 1783
Alliance Business Investment Company (Tulsa) 1740
Alliance Enterprise Corporation 1749
Allied Bancshares Capital Corporation 1784
Allied Business Investors, Inc. 1277
Allied Corporation, Buffalo Research Center Library 2529
Allied Investment Corporation (Fort Lauderdale) 1422
Allied Investment Corporation (Washington, DC) 1413
Allied-Signal Engineered Materials Research Center, Technical Information Center 2530
Allied Signal, Inc., Syracuse Research Laboratory, Library 2531
Allied Venture Partnership 1414
Allstate Venture Capital 1443
Ally Finance Corporation 1278
Alpha Capital Venture Partners LP 1444
Altoona Business Incubator 1225
American Association for the Advancement of Science .... ★2186
American Association of Entrepreneurial Dentists 3
American Association of Inventors 4
American Association of Small Research Companies—Members Directory 1982
American Bulletin of International Technology Transfer 1983

American Business Leader ★666
American Copyright Council 5
American Copyright Society 6
American Entrepreneurs Association 7
American European Consulting Company, Inc. 140
American Indian Economic Development Fund 141
American Industry 1984
American Institute of Physics ★2200
American International Toy Fair 2490
American Inventor 1985
American Investors Market Finance Directory 1986
American Journal of Small Business ★2055
American Manufacturing Research Consortium ★549
American Research and Development 1498
American Security Capital Corporation, Inc. 1415
American Society of Inventors 8
Americap Corporation 1785
Ameritech Development Corporation 1445
Ames Research Center – National Aeronautics and Space Administration 1853
AMEV Capital Corporation 1606
AMEV Venture Management, Inc. 1607
AMF Financial, Inc. 1279
AML Information Services 142
Amoco Venture Capital Company 1446
Ampersand Ventures 1499
Analog Devices Enterprises 1500
Analytical Reading and Reasoning 2287
Analyze 2288
Anchorage District Office – U.S. Small Business Administration 688
Anchorage Municipal Libraries, Z.J. Loussac Public Library 2532
Anderson-Bache Financial Group 143
Ann Arbor Technology Park 346
Annual Report of the Financial Institutions Bureau 1987
API International Business Development Group 144
APIPAT 2415
Apple Shines 2289
Appliance New Product Digest 1988
Applied Information Resources, Inc. 145
Applied Information Technologies Research Center 347
Applied Physics Laboratory – Johns Hopkins University 443
ARA Corporation 146
Argonne National Laboratory Technology Transfer Center 348
Argonne/University of Chicago Development Corporation ★348
Arizona Enterprise Development Corporation 689
Arizona; Guide to Establishing a Business in ★691
Arizona Small Business Development Center 349
Arizona State Department of Administration – Purchasing Office 690
Arizona State Department of Commerce – Revolving Loans Program 692
Arizona State Department of Commerce – Urban Development Action Grants 693
Arizona State Department of Commerce 691
Arizona State University – Engineering Excellence Program 350
Arizona State University, Daniel E. Noble Science and Engineering Library 2533
Arizona State University Research Park 351
Arkansas Capital Corporation/Arkansas Capital Development Corporation 696
Arkansas Center for Technology Transfer – University of Arkansas 583
Arkansas Development Finance Authority 697
Arkansas Industrial Development Commission 698
Arkansas Inventors Congress, Inc. 9
Arkansas Science and Technology Authority 699
Arkansas Small Business Development Center 352
Arkansas State Department of Finance and Administration – Purchasing Office 700
Arkansas State Library 2534
Armament Center; Close Combat ★1917
Armament Research, Development, and Engineering Center – U.S. Department of the Army 1912
Armbruster Associates – Chemical Consulting Services 147
Arnold Engineering Development Center – U.S. Department of the Air Force 1897
Arnold, White & Durkee, Library 2535
Arrowhead Research Park – New Mexico State University 497
Art and Perception - The Flexible Line 2290
The Art of Perceiving Problems 2291
The Art of Resolving Problems 2292
Arthritis and Musculoskeletal and Skin Diseases; National Institute of – U.S. Department of Health and Human Services 1888
Arthur Fakes 148
ASBDC (Association of Small Business Development Centers) 12
ASBDC News ★12
ASEA—Harvest Partners II 1608
Ashland Chemical Company, Library and Information Services 2536
ASI (American Society of Inventors) 8
Asia; I.C. 2083
Asian American Capital Corporation 1280
Asija Associates, Library 2537
Asset Management Company 1281
Asset Timing Corporation 149
Assistant Secretary for Conservation and Renewable Energy – U.S. Department of Energy 1878
Assistant Secretary for Defense Programs – U.S. Department of Energy 1879
Assistant Secretary for Nuclear Energy – U.S. Department of Energy 1880
Associated Production Music 150
Associated Southwest Investors, Inc. 1592
Associates Venture Capital Corporation 1282
Association for Science, Technology and Innovation 10
Association for Supervision and Curriculum Development ★2342
Association of Collegiate Entrepreneurs 11
Association of Small Business Development Centers 12
Association of University Related Research Parks 353
Association of University Related Research Parks, Reference Library 2538
Association of Venture Founders 13
Associations: National Organizations of the U.S.; Encyclopedia of 2051
ASTI (Association for Science, Technology and Innovation) 10
ASTI Newsletter 1989
Athens Innovation Center 1214
Atlanta; City of – Business Licenses 763
Atlanta District Office – U.S. Small Business Administration 774
Atlantic Energy Capital Corporation 1501
Atlantic Venture Partners (Alexandria) 1827
Atlantic Venture Partners (Richmond) 1828
Atlas II Capital Corporation 1502
Atoka Industrial Incubator 1220
Attorneys and Agents Registered to Practice before the United States Patent and Trademark Office; Roster of 2204
Auburn Technical Assistance Center – Auburn University 354
Auburn University – Auburn Technical Assistance Center 354
Auburn University Libraries, Science and Technology Department 2539
Augusta District Office – U.S. Small Business Administration 876
Austin District Office – U.S. Small Business Administration 1081
Austin Ventures LP 1786
Author Consultation Services 151
The Authority Report 1990
Avdon Capital Corporation 1609
Avenue Area, Inc. 834
Avery Business Development Services 152
AVF (Association of Venture Founders) 13
Aviation Administration; Federal – U.S. Department of Transportation 1968
Aviation Administration Technical Center; Federal ★1968
Aviation Systems Command – U.S. Department of the Army 1913
Avionics Center; Naval – U.S. Department of the Navy 1947
Award; Inventor of the Year ★36

B

B.F. Goodrich Chemical Company, Avon Lake Technical Center, Information Center 2540
B.F. Goodrich Company, Research and Development Center, Brecksville Information Center 2541
Babson College – Board of Research 355
Baby Expo 2491
Bain Capital Fund 1503
Baker & Botts, Law Library 2542
Baker & Mc Kenzie, Library 2543
Baldhard G. Falk & Associates 153
Ball State University – Center for Entrepreneurial Resources 356
Ball State University – SBIR Proposal Assistance 357
Ballistic Missile Defense Systems Command – U.S. Department of the Army 1914
Ballistic Missile Office – U.S. Department of the Air Force 1898
Ballistic Research Laboratory – U.S. Department of the Army 1915
Baltimore District Office – U.S. Small Business Administration 883
Baltimore Medical Incubator 1172
Banc One Capital Corporation 1721
BancBoston Ventures, Inc. 1504
Bancorp Hawaii SBIC, Inc. 1441
Bando McGlocklin Capital Corporaton 1841
Bankamerica Ventures, Inc. 1283
Bankers Capital Corporation 1561
Barberton Incubator 1215
Barclays Techmart-New Technology Transfer Exhibition 2492
Barrington BioIndustries, Inc. 154

Basic Industry Research Laboratory ....★510
Basic Investment Corporation 1829
Basic Thinking Skills..................................2293
Battelle Memorial Institute 358
Battista Research Institute; O.A. 300
Baxter Healthcare Corporation, Information Resource Center..................................2544
Bay Partners 1284
Bay Venture Group......................................1285
Baylor University – Center for Private Enterprise 359
Bayview Research Campus – Johns Hopkins University 444
BC-R&D ....★393
BC Technology Directory ....★393
Be an Inventor...2294
Beard, Inc.; Walter C. 337
Beckwith Associates; Walter C.....................338
Behind Small Business..............................1991
Beiser Inc.; Leo 279
Belden Associates..155
Belgium Ministere des Affaires Economiques......................................★2419
Belvior Research, Development, and Engineering Center – U.S. Department of the Army 1916
Ben Franklin Advanced Technology Center...1226
Ben Franklin Partnership – Pennsylvania State Department of Commerce 1037
Beneficial Capital Corporation....................1610
Benet Weapons Laboratory – U.S. Department of the Army.........................1917
Berry; E. Janet..215
Bessemer Venture Partners........................1611
Bever Capital Corporation..........................1505
Big Lakes Certified Development Company 835
Bio Engineering News................................1992
Bio-Metric Systems, Inc................................156
Biochemical & Biophysical Research Communications..................................1993
Biographical Dictionary of Scientists: Engineers and Inventors........................1994
BioIndustries, Inc.; Barrington 154
BioINVENTION ....★302
BioLabs, Inc. 157
Biolicensing & Patent Report TM; World 2259
Biomedical Business International, Inc....★2076
Biomedical Research and Development Laboratory ....★1929
Biomedical Research Zone 360
BIONICS 158
BioPatents...2416
Bioprocessing Technology.........................1995
BioScience 1996
Biotechnology and Diagnostic Product Packaging...★154
Biotechnology Center – Carnegie Mellon University..367
Biotechnology; Center for Advanced Research in...★533
Biotechnology Law Report 1997
Biotechnology PatentWatch; McGraw-Hill's 2144
Biovest Partners...1286
Birmingham District Office – U.S. Small Business Administration 684
Birmingham Public and Jefferson County Free Library, Government Documents Department...2545
Black Woman Entrepreneur of the Year Award ....★94
Blalack-Loop, Incorporated 159
BNA's Patent, Trademark & Copyright Journal...1998
BNA's Patent, Trademark & Copyright Journal (Online) 2417
BNP Venture Capital Corporation...............1287
Board of Research – Babson College...........355
Boardroom Reports—Breakthrough...........1999
Bob West Associates, Incorporated 160
Bobbe Siegel, Rights Representative...........161
Bohlen Capital Corporation 1612
Boice Dunham Group..................................162
Boise District Office – U.S. Small Business Administration..790
Boise State University – Idaho Business and Economic Development Center 361
The Book for Women Who Invent or Want To 2295
Book Marketing Ltd.; International...............255
Boothe & Associates, Inc.; Byron 170
Boston Capital Ventures LP 1506
Boston District Office – U.S. Small Business Administration..901
Boston Public Library, Government Documents, Microtext, Newspapers 2546
Boston Public Library, Science Reference Department...2547
Bradford Associates 1575
Bradley Agri-Consulting Inc..........................163
Bradley Industrial Incubator Program.........1156
Bradley University – Technology Commercialization Center 362
Brain Muscle Builders - Games to Increase Your Natural Intelligence 2296
Brain Scratchers..2297
Brand Name and Trademark Guide; Jewelers' Circular/Keystone 2118
Brand Names: Who Owns What................2000
BRANDY 2418
Brentwood Associates...............................1288
BREV 2419
Bridge Capital Advisors, Inc. 1576
Bridge Capital Investors 1577
Bridgeport Innovation Center 1151
British Columbia Research Corporation 363
British Toy and Hobby Fair 2493
Brittany Capital Company..........................1787
Broadcast Capital, Inc. 1416
Brooklyn Public Library, Science and Industry Division 2548
Broome County Industrial Development Agency...1202
Broventure Capital Management................1485
Broward County Main Library, Government Documents Department........................2549
Brown and Williamson Tobacco Corporation, Research Library..............2550
Brown Maroney Rose Barber & Dye, Law Library...2551
Brownstein, Zeidman & Schomer, Law Library...2552
Bruce W. McGee and Associates................164
BT Capital Corporation..............................1613
Buagh Center for Entrepreneurship; The John F...★2055
Buckman Laboratories International, Technical Information Center................2553
Buffalo & Erie County Public Library, Science and Technology Department....2554
Buffalo District Office – U.S. Small Business Administration..988
Build Illinois Small Business Development Program – Equity Investment Fund..........791
Bulletin Officiel de la Propriete Industrielle ....★2443
Bulloff; Jack J. 264
Burdini Technological Innovation Center; D.J. ....★403
Bureau of Business Assistance – Florida State Department of Commerce 753
Bureau of Economic and Business Research – University of Illinois 594
Bureau of Industrial Development – Michigan Technological University 477
Bureau of Mines – U.S. Department of the Interior 1939
Bureau of Professional and Occupational Affairs – Pennsylvania State Department...1036
Bureau of Purchases – Connecticut State Department of Administrative Services....726
Bureau of Purchasing – Maine State Department of Administration 872
Burr, Egan, Deleage, and Company (Boston) 1507
Burr, Egan, Deleage, and Company (San Francisco)..1289
Business Achievement Corporation 1508
Business Advocacy and Licensing Assistance – Montana State Department of Commerce 941
Business & Acquisition Newsletter.............2001
Business and Economic Research; Center for – University of Southern Maine 634
Business and Economic Research; Center for – Western Illinois University 666
Business and Professions Division – Washington State Department of Licensing 1117
Business and Technology Center (Columbus) 1216
Business and Technology Park; Cornell – Cornell University....................................388
Business Assistance Programs – University of Rhode Island...629
Business Capital Sources 2002
Business Center for Innovation and Development (Hayden Lake) 1154
Business Center for New Technology (Rockford)..1157
Business Climate Tax Index......................★389
Business Council of Georgia 762
Business Counselor..................................★762
Business Development Associates, Inc........165
Business Development Associates 166
Business Development Division – Oklahoma State Department of Commerce 1020
Business Development Division – Oregon State Economic Development Department..1031
Business Development Service – University of Illinois..595
Business Development Services; Avery.......152
Business Development Specialist – Montana State Department of Commerce 942
Business Entrepreneur; Minority................2151
Business Equity and Development Corporation..1290
Business Ideas & Shortcuts.......................2003
Business Ideas Newsletter.........................2004
Business Incubator (Brooklyn) 1203
Business Indicators; Maine ....★634
Business Information Celaringhouse – Kentucky State Commerce Cabinet.........859
Business Information Referral Center – North Carolina State Department of Commerce 997
Business Infrastructure Development Program – Pennsylvania State Department of Commerce 1038
Business Innovation Fund – Illinois State Department of Commerce and Community Affairs 794

Business Leader; American ★666
Business Leader; Illinois ★666
Business Licenses – City of Atlanta 763
The Business Loan Center 1614
Business Magazine; Hispanic 2080
Business Matters 167
Business; Minorities and Women in 2150
Business News; Home 2081
Business Opportunities; International ★1014
Business Organizations, Agencies, and Publications Directory 2005
Business Plan for Start-ups ★371
Business Planning Consultants 168
Business Planning Guide ★401
Business Plans; Workshops on ★597
Business Research; Center for – University of Puerto Rico 627
Business Review 2006
Business Review; Illinois ★594
Business Sciences Corporation 169
Business Service Center (Albion) 1178
The Business Side of Patents ★299
Business Ventures, Inc. 1447
Business Venturing; Journal of 2120
Business Week—R&D Scoreboard Issue 2007
Businessowners; Newsletter for Independent 2168
Byron Boothe & Associates, Inc. 170
BYTE 2420
BZ/Rights & Permissions, Inc. 171

C

C-I-L Inc., Research Centre Library 2555
C. Scanlon & Company 172
Cable and Howse Ventures (Bellevue) 1837
Cable and Howse Ventures (Palo Alto) 1291
Cable and Howse Ventures (Portland) 1744
California, Berkeley, Chemistry Library; University of 2672
California Capital Investors 1292
California Engineering Foundation 14
California Industrial Development Financing Advisory Commission 702
California Inventors Council 15
California, Law Library; University of Southern 2683
California, Los Angeles, Chemistry Library; University of 2673
California Partners Draper Association CGP 1293
California State and Consumer Services Agency – Department of Consumer Affairs 703
California State and Consumer Services Agency – Office of Procurement 704
California State Assembly Select Committee on Small Business 705
California State Department of Commerce – Office of Small Business 706
California State Library, Government Publications Section 2556
California State Senate Select Committee on Small Business Enterprise 707
California University of Pennsylvania – Mon Valley Renaissance 364
Calista Business Investment Corporation 1265
Call One Program – Iowa State Department of Economic Development 819
Calmac Manufacturing Corporation 173
Camden District Office – U.S. Small Business Administration 971
CAMS, Ltd. 174
Canaan Venture Partners (Menlo Park) 1294
Canaan Venture Partners (Rowayton) 1398
Canada Centre for Mineral and Energy Technology, Office of Technology Transfer 365
Canada, Inc., Patents Library; Du Pont 2578
Canada, Patent Library; Consumer and Corporate Affairs 2572
Canadian Entrepreneurs; Small Business: The Magazine for 2212
Canadian Industrial Innovation Centre/Waterloo 366
Canadian Industrial Innovation Centre/ Waterloo, Resource Centre 2557
The Canadian Inventors Newsletter ★366
Canadian Patent Reporter 2421
Canadian Research 2008
Capital Corporation of America 1750
Capital Corporation of Wyoming, Inc. 1852
Capital Dimensions Ventures Fund, Inc. 1547
Capital Equity Corporation 1476
Capital for Business, Inc. (Kansas City) 1562
Capital for Business, Inc. (St. Louis) 1563
Capital for Terrebonne, Inc. 1477
Capital Funds Corporation 1722
Capital Impact Corporation 1399
Capital Investments, Inc. 1842
Capital Investors and Management Corporation 1615
Capital Management Services, Inc. 1270
Capital Marketing Corporation 1788
Capital Resource Company of Connecticut, LP 1400
Capital Resource Corporation 1838
Capital Sources; Business 2002
Capital Southwest Corporation 1789
The Capital Strategy Group, Inc. 1448
Capitol Services 175
CAPRI (Computerized Administration of Patent Documents Reclassified According to the IPS) 2434
Cardinal Development Capital Fund I 1723
Caribank Capital Corporation 1423
Carlton Group 176
Carnegie Library of Pittsburgh, Science and Technology Department 2558
Carnegie Mellon University – Biotechnology Center 367
Cascade Business Center Corporation 1224
Casper District Office – U.S. Small Business Administration 1139
Catalog of Government Inventions Available for Licensing to U.S. Businesses 2009
A Catalog of Programs for Teaching Thinking 2298
Catalog of Publications and Services 2299
Catalytica, Inc., Information Center 2559
Category Reports 2010
Caterpillar Venture Capital, Inc. 1449
CBS Inc., Law Library 2560
CCC (Copyright Clearance Center) 21
CDA Reporter 2011
Cedar Rapids District Office – U.S. Small Business Administration 832
CEDCO 177
Celanese Chemical Company, Inc., Technical Center Library 2561
Cellulose & Specialties Division Technical Information Services; Procter & Gamble Company, 2647
CEM (Center for Entrepreneurial Management) 16
Centennial Business Development Fund Ltd. 1390
Centennial Engineering & Research, Inc. 178
Centennial Fund II Ltd. 1391
Centennial Fund III LP 1392
Centennial Fund Ltd. 1393
Center City MESBIC, Inc. 1724
The Center (Everett) 1257
Center for Advanced Research in Biotechnology ★533
Center for Applied Science and Technology – University of Pittsburgh 624
Center for Applied Technology – East Carolina University 395
Center for Business and Economic Research – University of Southern Maine 634
Center for Business and Economic Research – Western Illinois University 666
Center for Business and Economic Services – Troy State University 577
Center for Business Development 1179
Center for Business Innovation, Inc. 1195
Center for Business Research – University of Puerto Rico 627
Center for Defense Information ★2029
Center for Economic Development & Business Research – Jacksonville State University 442
Center for Education and Research with Industry 368
Center for Entrepreneurial Development 369
Center for Entrepreneurial Management 16
Center for Entrepreneurial Resources – Ball State University 356
Center for Entrepreneurial Studies – New York University 499
Center for Entrepreneurial Studies and Development, Inc. 370
Center for Entrepreneurship – Eastern Michigan University 396
Center for Entrepreneurship – Wichita State University 669
Center for Entrepreneurship and Economic Development – Pan American University 522
Center for Entrepreneurship and Technology – University of Louisville 602
Center for Industrial Research and Service – Iowa State University 440
Center for Information Age Technology – New Jersey Institute of Technology 494
Center for Innovation & Business Development 371
Center for Innovation and Development – University of Wisconsin—Stout 648
Center for Innovation Management Studies – Lehigh University 453
Center for Innovative Technology 372
Center for Innovative Technology Transfer – State University of New York at Oswego 560
Center for International Science and Technology Policy – George Washington University 407
Center for Law and Economic Studies – Columbia University 382
Center for Private Enterprise – Baylor University 359
Center for Productivity Enhancement – Wichita State University 670
Center for Studies in Innovation and Entrepreneurship – Columbia University 383
Center for Technology Development and Transfer – University of Texas 638
Center for Technology Transfer – Pittsburg State University 525
Center for Technology Transfer and Economic Development – University of Missouri—Rolla 613

Center for the Development of Technological Leadership – University of Minnesota 608
Center for the New West 373
Center for the Productive Use of Technology – George Mason University 404
Center for the Redevelopment of Industrialized States – Michigan State University 476
Center for the Study of Entrepreneurship – Marquette University 461
Center for the Utilization of Federal Technology ....★2009
Center of Research in Administrative Sciences – University of Moncton 614
Centers for Advanced Technology Program – New York State Science and Technology Foundation 983
Central Florida Inventors Club 17
Central Florida Inventors Council 18
Central Florida Research Park 374
Central Industrial Applications Center – Southeastern Oklahoma State University 552
Central New York SBIC 1616
Centre de Recherche Industrielle du Quebec, Industrie Information 2562
Centre for Advanced Technology 375
Ceramics and Composites Partnership; Advanced ....★479
Ceramics Information Center; Metals and ....★2145
Cerberus Group 179
CERES 180
CFB Venture Capital Corporation (Los Angeles) 1295
CFB Venture Capital Corporation (San Francisco) 1296
Challenge 2300
Challenge Boxes 2301
The Challenge of the Unknown 2302
Challenger Industries, Ltd. 181
Challenges for Children 2303
Chamber Innovation Center 1180
Chamber of Commerce of Hawaii – Small Business Center 778
Chandler & Associates Ltd. 182
Charles River Ventures 1509
Charles Scott Pugh, Jr. 183
Charles Shano 184
Charles W. O'Conor and Associates – Wright Wyman, Inc. 185
Charleston Capital Corporation 1770
Charleston District Office – U.S. Small Business Administration 1119
Charleston District Office Library; U.S. International Trade Administration, U.S. and Foreign Commercial Service, ...2693
Charlotte District Office – U.S. Small Business Administration 1004
Charter Venture Group, Inc. 1790
Charterway Investment Corporation 1297
Chase Manhattan Capital Corporation 1617
Chemical Company, Inc., Technical Center Library; Celanese 2561
Chemical Company, Library and Information Services; Ashland 2536
Chemical Consulting Services – Armbruster Associates 147
Chemical Marketing Reporter 2012
Chemical Propulsion Information Agency ....★2024
Chemical Research, Development, and Engineering Center – U.S. Department of the Army 1918
Chemical Venture Capital Associates LP 1618
Chemicals Inc., Research Library; AKZO 2528
Chemicals, Inc., Technical & Business Information Center; M and T 2611
Chemicals Library; Henkel Corporation, Process 2590
Chemistry Library; Louisiana State University, ...2609
Chemistry Library; University of California, Berkeley, ...2672
Chemistry Library; University of California, Los Angeles, ...2673
Chen's Financial Group, Inc. 1791
Cherry Tree Ventures 1548
Chestnut Capital International II LP 1510
Chevron Corporation, Corporate Law Department, Library 2563
Chevron Research Company, Technical Information Center 2564
CHI Research/Computer Horizons, Inc. 186
Chicago Community Ventures, Inc. 1450
Chicago Craft and Creative Industries Show and Convention; International Craft Exposition/ 2499
Chicago District Office – U.S. Small Business Administration 805
Chicago Model Hobby Show 2494
Chicago Public Library Central Library, Business/Science/Technology Division 2565
Chicago Technology Park 376
Chickasaw Capital Corporation 1774
China Exchange Newsletter 2013
China Trade Center 1175
Chinese Patent Abstracts in English Data Base 2422
Chris Enterprises 187
Chrysalis - Nurturing Creative and Independent Thought in Children 2304
Churchill International 1298
Ciba Corning Diagnostics Corporation, Steinberg Information Center 2566
CIBERPAT 2423
Cin Industrial Investments Ltd. 1299
Cincinnati and Hamilton County, Science and Technology Department; Public Library of 2649
Cincinnati District Office – U.S. Small Business Administration 1015
CIRAS (Iowa State University – Center for Industrial Research and Service) 440
Circle Ventures, Inc. 1465
Circles of Creativity 2305
Circuit News 2014
Citicorp Venture Capital Ltd. (Dallas) 1792
Citicorp Venture Capital Ltd. (New York City) 1619
Citicorp Venture Capital Ltd. (Palo Alto) 1300
Citizenship Decision-Making 2306
City of Atlanta – Business Licenses 763
Citywide Development Corporation of Kansas City, Kansas, Inc. 836
Civil Engineering Laboratory; Naval – U.S. Department of the Navy 1948
Claflin Capital Management, Inc. 1511
CLAIMS/CITATION 2424
CLAIMS/CLASS 2425
CLAIMS/Comprehensive Data Base 2426
CLAIMS/Reassignment & Reexamination 2427
CLAIMS/UNITERM 2428
CLAIMS/U.S. Patent Abstracts 2429
Clarion Capital Corporation 1725
Clark Engineering 188
Clarksburg District Office – U.S. Small Business Administration 1120
Clayton County Permits and Licenses 764
Clemson University – Emerging Technology Development and Marketing Center 377
Cleveland Area Development Corporation 189
Cleveland District Office – U.S. Small Business Administration 1016
Cleveland Public Library, Documents Collection 2567
CLINPAT 2430
Clinton Capital Corporation 1620
Close Combat Armament Center ....★1917
CMNY Capital Company, Inc. 1621
Coast Guard Office of Engineering and Development – U.S. Department of Transportation 1967
Coastal Engineering Research Center ....★1924
Coastal Systems Center; Naval – U.S. Department of the Navy 1949
Cobb County Business Licenses 765
The Coca-Cola Company, Law Library 2568
Cognetics 2307
Cold Regions Research and Engineering Laboratory – U.S. Department of the Army 1919
College of DuPage – Technology Commercialization Center 378
College of St. Thomas – Entrepreneurial Enterprise Center 379
Colorado Growth Capital, Inc. 1394
Colorado Housing Finance Authority 714
Colorado School of Mines – Table Mountain Research Center 380
Colorado Small Business Development Center 381
Colorado State Department of Administration – Division of Purchasing 715
Colorado State Department of Economic Development – Small Business Office 716
Colorado State Department of Regulatory Agencies 717
Colorado State Division of Commerce and Development 718
Columbia Capital Corporation 1622
Columbia District Office – U.S. Small Business Administration 1067
Columbia University – Center for Law and Economic Studies 382
Columbia University – Center for Studies in Innovation and Entrepreneurship 383
Columbia University – Office of Science and Technology Development 384
Columbus District Office – U.S. Small Business Administration 1017
Columbus Inventors Association 19
Combined Agencies Corporation 190
Comdisco Venture Lease, Inc. (Rosemont) 1451
Comdisco Venture Lease, Inc. (San Francisco) 1301
Cominco Ltd., Central Technical Library 2569
Commerce Productivity Center; United States Department of Commerce, Office of Productivity, Technology and Innovation, ...2669
Commercial Associates Incorporated 191
Commercial Development Association ....★2011
Commercial/Industrial Liaison – Indiana University-Purdue University Indianapolis 427
Commercializing University Technology ....★299
Commonwealth Technology Information Service ....★405
Communication Certification Laboratory 192
Communications-Electronics Command – U.S. Department of the Army 1920
The Communications Industries Report 2015

Master Index

Communicative Disorders and Stroke; National Institute of Neurological and – U.S. Department of Health and Human Services 1891
Community Development Block Grant Program – Iowa State Department of Economic Development 820
Community Development Corporation Program – Minnesota State Department of Energy and Economic Development 916
Community Economic Betterment Account – Iowa State Department of Economic Development 821
Community Equity Corporation of Nebraska 1569
Compass Materials Group 193
Competitive Enterprise Institute 385
Complete Guide to Making Money With Your Inventions ....★75
Composites Partnership; Advanced Ceramics and ....★479
Compressors; IMPACT 2085
Compu-Mark Rechtsstandlexicon 2431
Compu-Mark U.K. On-Line 2432
Compu-Mark U.S. On-Line 2433
Computer Age Associates Inc. 194
Computer and Telecommunication Laboratory; National ....★1874
Computer Horizons, Inc.; CHI Research/ 186
Computer Horizons, Inc., Library 2570
Computer Research Opportunities; Microelectronics Innovation and – University of California, Los Angeles 588
Computer Software Law ....★148
Computer Systems News 2016
Computerized Administration of Patent Documents Reclassified According to the IPS 2434
COMPUTERPAT 2435
Computers-Applications in the Laboratory; Instruments & 2098
Comquest, Inc. 195
Concord District Office – U.S. Small Business Administration 964
Concord Partners (New York City) 1623
Concord Partners (Palo Alto) 1302
Confederacy of Mississippi Inventors 20
Conference on Innovation and Creativity 2495
Connecticut Business Development Corporation 723
Connecticut Development Authority 724
Connecticut Invention Convention 2264
Connecticut; MacRae's Industrial Directory 2131
Connecticut Product Development Corporation 725
Connecticut Small Business Development Center 386
Connecticut State Department of Administrative Services – Bureau of Purchases 726
Connecticut State Department of Consumer Protection – Division of Licensing and Administration 727
Connecticut State Department of Economic Development – Corner Loan Program 728
Connecticut State Department of Economic Development – Naugatuck Valley Revolving Loan Fund 729
Connecticut State Department of Economic Development – Sales Contact Centers 730
Connecticut State Department of Economic Development – Set-Aside Program 731
Connecticut State Department of Economic Development – Small Business Services 732
Connecticut State Department of Economic Development – Small Manufacturers Loan Program 733
Connecticut State Department of Economic Development – Urban Enterprise Zone Program 734
Connecticut State Department of Economic Development – Urban Jobs Program 735
Connecticut Venture Management Corporation 196
Connections ....★553
Conoco, Inc., Research and Development Department, Technical Information Services 2571
Conservation and Renewable Energy; Assistant Secretary for – U.S. Department of Energy 1878
Consortium House, Ltd. 197
Consortium of Social Science Associations ....★2022
Construction Engineering Research Laboratory – U.S. Department of the Army 1921
Consultant Affiliates, Inc. 198
Consulting Industries International 199
Consumer and Corporate Affairs Canada, Patent Library 2572
Consumer Electronics Show 2496
Consumer Growth Capital, Inc. 1549
Consumer Market Studies ....★303
Consumer Protection Division – Georgia State Department of Agriculture 770
Consumers United Capital Corporation 1417
Contact - Europe 200
Contacto 2017
Continental Illinois Venture Corporation (CIVC) 1452
Continental Investors, Inc. 1303
Contract Management Division – U.S. Department of the Air Force 1899
Control Data Business and Technology Center (Baltimore) 1173
Control Data Business and Technology Center (Bemidji) 1188
Control Data Business and Technology Center (Champaign) 1158
Control Data Business and Technology Center (Charleston) 1246
Control Data Business and Technology Center (Omaha) 1197
Control Data Business and Technology Center (Pueblo) 1150
Control Data Business and Technology Center (St. Paul) 1189
Control Data Business and Technology Center (San Antonio) 1253
Control Data Business and Technology Center (South Bend) 1168
Control Data Business and Technology Center (Toledo) 1217
Control Data Business and Technology Center (West Philadelphia) 1227
Copyright Administration; Office of Patent and – University of Southern California 632
Copyright and Trademark Library; Walter J. Derenberg ....★22
Copyright Clearance Center 21
Copyright Council; American 5
Copyright Industries in the United States ....★5
Copyright Information Center; International 40
Copyright Information Services 201
Copyright Journal; BNA's Patent, Trademark & 1998
Copyright Journal (Online); BNA's Patent, Trademark & 2417
Copyright Law; Educators' Ad Hoc Committee on 23
Copyright Law for Software Development, Publishers, and Distributors ....★148
Copyright Law Reports 2018
Copyright Library; LEXIS Federal Patent, Trademark, & 2454
Copyright Management 2019
Copyright Office; Patent, Trademark and – University of California, Berkeley 587
Copyright, Patent and Trademark Library; WESTLAW 2486
Copyright Policy Development: A Resource Book for Educators ....★201
Copyright Public Information Office; Library of Congress, ...2605
Copyright Research Foundation, Library; Patent, Trademark and 2639
Copyright Society; American 6
Copyright Society; Los Angeles 81
Copyright Society of the U.S.A. 22
Copyright Society of the U.S.A.; Journal of the 2122
Copyrights, and Licensing; Office for Patents, – Harvard University 417
Coral Gables District Office – U.S. Small Business Administration 758
Coramar Inc. 202
Core States Enterprise Fund 1751
Cornell Business and Technology Park – Cornell University 388
Cornell Industry Research Park 1204
Cornell Research Foundation – Office of Patents and Licensing 387
Cornell University – Cornell Business and Technology Park 388
Corner Loan Program – Connecticut State Department of Economic Development 728
Corporate Finance Sourcebook 2020
Corporate Management, Inc. 203
Corporate Venturing News 2021
Corporation Division – Oregon State Department of Commerce 1028
Corporation for Business Assistance in New Jersey 965
Corporation for Enterprise Development 389
Corporation for Innovation Development – New York State Science and Technology Foundation 984
Corporation for Innovation Development 808
Corporation of Inventive Minded People ....★50
Corporations Bureau – Massachusetts State Department of Revenue 893
Corpus Christi District Office – U.S. Small Business Administration 1082
The Cort Thinking Program 2308
COSSA Washington Update 2022
Coup 2023
CPIA Bulletin 2024
CPR (Canadian Patent Reporter) 2421
CR Chambers & Associates 204
Creating and Thinking; Acting, ...2286
Creative and Independent Thought in Children; Chrysalis - Nurturing 2304
Creative Behavior; The Journal of 2350
Creative Education Foundation, Inc. ....★2350
Creative Encounters with Creative People 2309
Creative Genius; Wake Up Your 2409
Creative Homework Assignments; Mind Movers - 2367
Creative Imagination; Expanding 2326
Creative Institute; Spirit of the Future 124
Creative Kids 2310
Creative Problem Solving: A Guide for Trainers and Management 2311

Creative Problem Solving Techniques 2312
Creative Thinking; Do-It-Yourself Critical and 2321
Creative Thinking Skills; OM-AHA! Problems to Develop 2372
Creative Thinking; Springboards to 2390
Creative Thinking to Problem Solving; The Thoughtwave Curriculum - Applying 2407
Creative Writing in Action 2313
Creativity and Innovation 2265
Creativity and Licensing; MICEL-International Market for 2508
Creativity; Circles of 2305
Creativity; Conference on Innovation and 2495
Creativity; Managing Group 2359
Creativity; Odyssey of the Mind: Problems to Develop 2371
Creativity 1,2,3 2314
Creativity; Problems! Problems! Problems! Discussions and Activities Designed to Enhance 2378
Creativity; Warm-up to 2410
Credit; Special Sources of ....★401
Cresheim Company, Inc. 205
Crestar Capital 1830
CRINC (University of Kansas Center for Research, Inc.) 600
Criterion Investments 1793
Critical Thinking and Thinking Skills: State of the Art Definitions and Practice in Public Schools 2315
Critical Thinking: How to Evaluate Information and Draw Conclusions 2316
Croft & Associates International Consultants; J.C. 262
Cross Gates Consultants 206
Crosspoint Investment Corporation 1304
Crosspoint Venture Partners 1305
Croyden Capital Corporation 1624
CRS Venture Directory—Florida 2025
Crystal Resources, Inc. 207
CSUSA (Copyright Society of the U.S.A.) 22
C3I News 2026
CU Business Advancement Centers 719
Cuisine Crafts, Inc. 208
Cummings Research Park 390
Current Controversy 2027
Current Energy Patents 2028
Curriculum Development; Association for Supervision and ....★2342
CVC Capital Corporation 1625
CW Group, Inc. 1626

D

D.C./Delaware; MacRae's Industrial Directory Maryland/ 2133
D.J. Burdini Technological Innovation Center ....★403
Dallas District Office – U.S. Small Business Administration 1083
Dallas Public Library, J. Erik Jonsson Central Library, Government Publications Division 2573
Dandini Research Park 391
Darden Research Corporation 209
Daughters of Invention: An Invention Workshop for Girls 2317
DAVCO Solar Corporation 210
David Taylor Research Center – U.S. Department of the Navy 1941
DC Bancorp Venture Capital Company 1418
Dearborn Capital Corporation 1539
The Decatur Industrial Incubator 1159
Decision-Making; Citizenship 2306
Defense Advanced Research Projects Agency – U.S. Department of the Air Force 1900
Defense Information; Center for ....★2029
The Defense Monitor 2029
Defense Nuclear Agency – U.S. Department of Defense 1877
Defense Programs; Assistant Secretary for – U.S. Department of Energy 1879
Defense R&D 2030
DeKalb County Business Licenses 766
Delaware: A Guide to Procurement Opportunities; Selling to the State of .....★742
Delaware Development Office 737
Delaware Economic Development Authority 738
Delaware Library, Reference Department; University of 2674
Delaware; MacRae's Industrial Directory Maryland/D.C./ 2133
Delaware Small Business Development Center 392
Delaware State Chamber of Commerce 739
Delaware State Department of Administrative Services – Division of Professional Regulations 740
Delaware State Department of Administrative Services – Division of Purchasing 741
Delaware State Development Office 742
Delaware State Division of Revenue (Wilmington) 743
Delta Capital, Inc. 1713
Delta Market Street Incubator 1181
Delta Monroe Street Incubator 1182
Delta Science Modules 2318
Demery Seed Capital Fund 1540
Dental Lab Products 2031
Dental Research Institute; Naval ....★1951
Dental Research; Institute of ....★1929
Dental Research; National Institute of – U.S. Department of Health and Human Services 1889
Denver District Office – U.S. Small Business Administration 721
Denver Public Library, Business, Science & Government Publications Department ...2574
Department of Consumer Affairs – California State and Consumer Services Agency 703
Department of Economic Development – Vermont State Agency of Development and Community Affairs 1102
Department of Transportation, Small Business Innovation Research Program ....★1973
Derenberg Copyright and Trademark Library; Walter J. ....★22
Derwent, Inc. 211
Des Moines District Office – U.S. Small Business Administration 833
Des Plaines River Valley Enterprise Zone Incubator Project 1160
Design and Research Exhibition (DARE) ...2266
Design Engineering 2032
Design Ideas, Publication, and Production Rights; LIZ '89 Licensing Fair for 2507
Designers; Intermountain Society of Inventors and 38
Designfax 2033
Designing Plastics Products ....★323
Detecting and Deducing - Preparing for Logical Thinking 2319
Detroit District Office – U.S. Small Business Administration 911
Detroit Public Library, Technology and Science Department 2575
Deutsche Patent Datenbank 2436
Developers Equity Capital Corporation 1306
Development Directory 2034
DGC Capital Company 1550
Diamond Award; Sydney ....★42
Dickinson School of Law, Sheely-Lee Law Library 2576
Dime Investment Corporation 1307
Direct Loan Program – Illinois Development Finance Authority 792
Directorate of Research and Development Procurement – U.S. Department of the Air Force 1901
Directory of American Research and Technology 2035
Directory of Financing Sources for Mergers, Buyouts & Acquisitions 2036
Directory of Intellectual Property Lawyers and Patent Agents 2037
Directory of Nebraska Manufacturers ....★951
Directory of Operating Small Business Investment Companies 2038
Directory of Oregon Manufacturers ....★1031
Directory of Public High Technology Corporations 2039
Directory of Scientific Resources in Georgia 2040
Directory of Venture Capital Clubs 2041
Disadvantaged and Women Business Enterprises – Pennsylvania State Department of Transportation 1046
The Discovery 2320
Discovery Foundation 393
Discovery Innovation Office ....★393
Discovery Parks Incorporated 394
District of Columbia Department of Administrative Services – Office of Material Management Administration 749
District of Columbia Department of Consumer and Regulatory Affairs 750
District of Columbia Office of Business and Economic Development – Revolving Loan Fund 751
Diversification Report ....★569
Diving and Salvage Training Center; Naval ....★1949
Division for Small Business – New York State Department of Commerce 979
Division of Business Assistance and Development – South Carolina State Development Board 1065
Division of Capital Planning and Operations – Massachusetts State Office of Facilities Management 898
Division of Community Affairs and Development – Oklahoma State Department of Commerce 1021
Division of Economic and Industrial Development – Utah State Department of Community and Economic Development 1096
Division of Licensing and Administration – Connecticut State Department of Consumer Protection 727
Division of Licensing and Enforcement – Maine State Department of Professional and Financial Regulation 873
Division of Licensing and Enforcement – Texas State Department of Labor and Standards 1078
Division of Licensing and Registration – Vermont Secretary of State 1100

Division of Licensing and Testing – Illinois State Department of Registration and Education 803
Division of Occupational and Professional Licensing – Maryland State Department of Licensing and Regulations 880
Division of Occupational and Professional Licensing – Utah State Department of Business Regulation 1095
Division of Plant and Property Management – New Hampshire State Department of Administration Services 962
Division of Policy Research and Analysis – National Science Foundation 1867
Division of Procurement – Indiana State Department of Administration 812
Division of Procurement – Minnesota State Department of Administration 915
Division of Professional Licensing and Regulation – Iowa State Department of Commerce 818
Division of Professional Regulation – Rhode Island State Department of Labor 1061
Division of Professional Regulations – Delaware State Department of Administrative Services 740
Division of Purchases – Kentucky State Finance and Administration Cabinet 861
Division of Purchases – Kentucky State Transportation Cabinet 863
Division of Purchases and Supplies – Virginia State Office of Administration 1111
Division of Purchasing – Alabama State Department of Finance 680
Division of Purchasing – Colorado State Department of Administration 715
Division of Purchasing – Delaware State Department of Administrative Services 741
Division of Purchasing – Florida State Department of General Services 755
Division of Purchasing – Georgia State Department of Administrative Services 769
Division of Purchasing – Idaho State Department of Administration 787
Division of Purchasing – Illinois Secretary of State 793
Division of Purchasing – Michigan State Department of Management and Budget 908
Division of Purchasing – Mississippi State General Services Office 926
Division of Purchasing – Missouri State Office of Administration 934
Division of Purchasing – Nevada State Department of General Services 956
Division of Purchasing – New York State Department of General Services 980
Division of Purchasing – North Dakota State Department of Management and Budget 1005
Division of Purchasing – Oklahoma State Department of Public Affairs 1025
Division of Purchasing – Oregon State Department of General Services 1030
Division of Purchasing – Tennessee State Department of General Services 1074
Division of Purchasing – Texas State Purchasing and General Services Commission 1080
Division of Purchasing – Utah State Department of Administrative Services 1094
Division of Purchasing – Washington State Department of General Administration 1116
Division of Purchasing – West Virginia State Department of Finance and Administration 1122
Division of Purchasing Agents – Massachusetts State Executive Office of Administration and Finance 895
Division of Purchasing and Materials Management – Iowa State Department of General Services 830
Division of Purchasing and Property Control – Wyoming State Department of Administration and Fiscal Control 1143
Division of Purchasing and Public Records – Vermont State Administration Agency 1101
Division of Purchasing and Supply – Hawaii State Department of Accounting and General Services 779
Division of Purchasing, Services, and Supply – Puerto Rico General Services Administration 1054
Division of Registration – Massachusetts State Executive Office of Consumer Affairs and Business Regulation 896
Dixie Business Investment Company 1478
Do-It-Yourself Critical and Creative Thinking 2321
Domestic & International Technology 212
Domestic Capital Corporation 1765
Dougery, Jones and Wilder (Dallas) 1794
Dougery, Jones & Wilder (Mountain View) 1308
Dr. Dvorkovitz and Associates 213
Dravo Engineers Inc., Library 2577
Dryden Flight Research Facility – National Aeronautics and Space Administration 1854
DSV Partners 1578
Du Pont Canada, Inc., Patents Library 2578
Dugway Proving Ground – U.S. Department of the Army 1922
Durant Industrial Incubator 1221
Dvorkovitz and Associates; Dr. 213
DYNIS 2437

E

E.J. Kuuttila & Associates 214
E. Janet Berry 215
EAHCCL (Educators' Ad Hoc Committee on Copyright Law) 23
Earth Resources Laboratory .... ★1863
East Arkansas Business Incubator System, Inc 1145
East Campus Research Park Project – University of Colorado 589
East Carolina University – Center for Applied Technology 395
East Creek Center 1254
East Liberty Incubator 1228
East/West Technology Digest 2042
Eastech Management Company 1512
Eastern Michigan University – Center for Entrepreneurship 396
Eastern Space and Missile Center – U.S. Department of the Air Force 1902
Eau Claire District Office – U.S. Small Business Administration 1129
ECLATX 2438
Economic Action Council – Idaho First National Bank 785
Economic and Business Research; Bureau of – University of Illinois 594
Economic Business Review .... ★546
Economic Development Administration University Center – Indiana University-Purdue University Indianapolis 428
Economic Development & Business Research; Center for – Jacksonville State University 442
Economic Development and Technical Assistance Center – State University of New York at Plattsburgh 561
Economic Development Bond Program – Kansas State Department of Commerce 840
Economic Development Center; Florida – Florida State University 401
Economic Development; Center for Technology Transfer and – University of Missouri—Rolla 613
Economic Development Financing Division – Ohio State Department of Development 1012
Economic Development; Institute for – Pittsburg State University 526
Economic Development Laboratory – Georgia Institute of Technology 409
Economic Focus Newsletter .... ★533
Economic Outlook; Illinois .... ★594
Edelson Technology Partners 1579
Edison Entrepreneur 2043
Edison National Historic Site, Archives; U.S. National Park Service, ... 2694
Edison Venture Fund 1580
Edmonton Research Park 397
EDOC 2439
Education and Learning to Think 2322
Educators' Ad Hoc Committee on Copyright Law 23
Edwards Capital Company 1627
EE: Electonic/Electrical Product News 2044
Effective Questions to Strengthen Thinking 2323
Egyed; Mark J. 286
El Dorado Technology Partners 1309
El Dorado Ventures 1310
El Paso District Office – U.S. Small Business Administration 1084
Electonic/Electrical Product News; EE: 2044
Electric Light & Power 2045
Electronic Decisions, Inc. 1161
Electronic Developments; World 2260
Electronic Engineering Times 2046
Electronic Systems Division – U.S. Department of the Air Force 1903
Electronics 2047
Electronics and Technology Today 2048
Electronics of America 2049
Electronics Show; Consumer 2496
Elementary Science Study 2324
Elk Associates Funding Corporation 1628
Elmira District Office – U.S. Small Business Administration 989
Elron Technologies, Inc. 1629
Emerging High Tech Ventures 2050
Emerging Technology Development and Marketing Center – Clemson University 377
Emmons; Albert L. 139
Employer's Desk Manual .... ★762
Encyclopedia of Associations: National Organizations of the U.S. 2051
Energy Assets, Inc. 1795
Energy Capital Corporation 1796
Energy Conserver; Industrial .... ★409
Energy Corporation; Trans 328
Energy Development Loan Program – Minnesota State Department of Energy and Economic Development 917

Energy-Environmental Research and Development Company 216
Energy Loan Insurance Program – Minnesota State Department of Energy and Economic Development 918
Energy Patents; Current 2028
Energy Related Inventions; National Institute of Standards and Technology, Office of – U.S. Department of Commerce 1875
Energy-Related Inventions Program; Status Report of the 2218
Energy Research; Institute for – Syracuse University 563
Energy Research; Office of – U.S. Department of Energy 1882
Energy Technology Center Library; U.S. Department of Energy, Morgantown 2691
Energy Technology Center; Morgantown – U.S. Department of Energy 1881
Energy Technology Center; Pittsburgh – U.S. Department of Energy 1883
Engelhard Corporation, Technical Information Center 2579
Engineer Topographic Laboratories – U.S. Department of the Army 1923
Engineer Waterways Experiment Station – U.S. Department of the Army 1924
Engineering and Management; Test 2231
Engineering & Physical Sciences Library; University of Maryland, College Park Libraries, ...2676
Engineering & Research, Inc.; Centennial 178
Engineering; Clark 188
Engineering Company; Johnson 268
Engineering Corporation; Integrated Circuit 248
Engineering; Design 2032
Engineering Excellence Program – Arizona State University 350
Engineering Extension – Kansas State University 446
Engineering Issues ....★571
Engineering Journal 2052
Engineering Laboratory; National ....★1874
Engineering Library and Information Services; University of Washington, ...2686
Engineering Library; Platt Saco Lowell Corporation, ...2646
Engineering Library; University of Nebraska, Lincoln, ...2679
Engineering Library; University of New Mexico, Centennial Science and 2682
Engineering Library; University of Texas at Austin, McKinney 2684
Engineering Research and Industrial Development Office – University of North Carolina at Charlotte 619
Engineering Research Center – University of Maryland 604
Engineering Reviews ....★409
Engineering Transportation Library; University of Michigan, ...2678
Engineers and Inventors; Biographical Dictionary of Scientists: 1994
Engineers; Scott-Brown 320
Enhancing Options in Patent Dispute Resolutions ....★234
Enterprise Capital Corporation 1797
Enterprise Center; Entrepreneurial – College of St. Thomas 379
Enterprise; Center for Private – Baylor University 359
Enterprise Center; New ....★548
Enterprise Development Center; Entrepreneurship and – Sangamon State University 546
Enterprise Development; Corporation for 389
Enterprise Development Implementation Packets; State ....★389
Enterprise Development; Institute for New 434
Enterprise Finance Capital Development Corporation 1571
Enterprise Institute; Competitive 385
Enterprise Institute; The New – University of Southern Maine 635
Enterprise Partners 1311
Enterprise Partnership; Rural ....★479
Enterprise Venture Capital Corporation of Pennsylvania 1752
Enterprises; New Market ....★2128
Entrepreneur Center; George Mason – George Mason University 405
Entrepreneur; Edison 2043
Entrepreneur in Residence – Indiana University 426
Entrepreneur Kit ....★371
Entrepreneur Magazine 2053
Entrepreneur; Mind Your Own Business: A Guide for the Information ....★246
Entrepreneur; Minority Business 2151
Entrepreneur Newsletter; HomeBased 2082
Entrepreneur of the Year Award; Black Woman ....★94
Entrepreneur Tax Letter; The Woman 2257
Entrepreneurial Assistance – North Dakota State Economic Development Commission 1006
Entrepreneurial Center; Sol C. Snider – University of Pennsylvania 623
Entrepreneurial Dentists; American Association of 3
Entrepreneurial Development; Center for 369
The Entrepreneurial Economy ....★389
Entrepreneurial Enterprise Center – College of St. Thomas 379
Entrepreneurial Management; Center for 16
Entrepreneurial Manager ....★16
Entrepreneurial Manager's Newsletter 2054
Entrepreneurial Network; North Florida ....★402
Entrepreneurial Network; South Florida ....★402
Entrepreneurial News ....★3
Entrepreneurial Resources; Center for – Ball State University 356
Entrepreneurial Service Center – University of Arkansas 584
Entrepreneurial Services and Seminars – Hub Financial Network 237
Entrepreneurial Studies and Development, Inc.; Center for 370
Entrepreneurial Studies; Center for – New York University 499
Entrepreneurs Association; American 7
Entrepreneurs Association of Austin; Inventors & 46
Entrepreneurs; Association of Collegiate 11
Entrepreneurs (Balch Springs); National Association of Minority 95
Entrepreneurs Club, Inc.; Silicon Valley 117
Entrepreneurs Club; Minnesota 86
Entrepreneurs Council; High Technology 27
Entrepreneurs; National Association of Black Women 94
Entrepreneurs (New York); National Association of Minority 96
Entrepreneurs Organization; Young 136
Entrepreneurs; Small Business: The Magazine for Canadian 2212
Entrepreneurs; Society for Inventors and 118
Entrepreneurs Society of Indiana, Inc.; Inventors & 47
Entrepreneurs; Women 134
Entrepreneurship and Economic Development; Center for – Pan American University 522
Entrepreneurship and Enterprise Development Center – Sangamon State University 546
Entrepreneurship and Technology; Center for – University of Louisville 602
Entrepreneurship Center – Montana State University 485
Entrepreneurship; Center for – Eastern Michigan University 396
Entrepreneurship; Center for – Wichita State University 669
Entrepreneurship; Center for Studies in Innovation and – Columbia University 383
Entrepreneurship; Center for the Study of – Marquette University 461
Entrepreneurship for Students Magazine; Technology, Innovation & 2397
Entrepreneurship; High Tech ....★154
Entrepreneurship; Initiatives for Not-for-Profit – New York University 500
The Entrepreneurship Institute 24
Entrepreneurship Institute; Innovation and – University of Miami 606
Entrepreneurship; International Center for – Georgia State University 414
Entrepreneurship Program – Florida State Department of Commerce 754
Entrepreneurship Program – Vermont State Department of Economic Development 1103
Entrepreneurship Research Group; Small Business and – Laval University 452
Entrepreneurship; The John F. Buagh Center for ....★2055
Entrepreneurship: Theory and Practice 2055
Environmental Medicine; Research Institute of ....★1929
EOS Technologies, Inc. 217
EPAT 2440
Equal Opportunity Finance, Inc. 1473
Equico Capital Corporation 1630
Equitable Capital Corporation 1312
Equity Capital Corporation 1593
Equity Investment Fund – Build Illinois Small Business Development Program 791
Equity Resource Company, Inc. 1466
Erie SBIC 1753
Eslo Capital Corporation 1581
ESS (Elementary Science Study) 2324
Estee Lauder Inc., Information Center 2580
Euclid Partners Corporation 1631
Eureka! ....★366
Europacific Inc. 218
Europe; New From 2158
European Development Corporation LP 1632
European Patent Bulletin ....★2440
Evaluate 2325
Evaluating Patents ....★299
Eveready Battery Company, Inc., Technical Information Center 2581
Evergreen Capital Company, Inc. 1798
Everlast Capital Corporation 1633
Examining Boards Division – Georgia Secretary of State 768
Executive Office Network, Inc. (Malvern) 1229
Executive Report; U.S. 2243
Exim Capital Corporation 1634
Expanding Creative Imagination 2326
Exploring the Lives of Gifted People in the Arts 2327

Exploring the Lives of Gifted People in the Sciences 2328
Explosive Ordnance Disposal Technology Center; Naval – U.S. Department of the Navy 1950
Explosives & Pyrotechnics 2056
Export/Exportador 2057
Export Finance Program – Iowa State Department of Economic Development 822
Extended Care Product News 2058
Eye Cue Puzzles Sets 2329
Eye Institute; National – U.S. Department of Health and Human Services 1886

F

F&S/Japanese Information Technology Industry Letter 2059
F&S/Japanese Venture Capital & OTC Opportunities 2060
Fact Fantasy and Folklore 2330
Fair Capital Corporation 1635
Fairfield Equity Corporation 1636
Fairfield Venture Partners 1401
Fakes; Arthur 148
Falcon Capital Corporation 1714
Falk & Associates; Baldhard G. 153
Fargo District Office – U.S. Small Business Administration 1009
FATD (Federal Applied Technology Database) 2441
FBS Venture Capital Company 1551
FCA Investment Company 1799
Federal Applied Technology Database 2441
Federal Aviation Administration – U.S. Department of Transportation 1968
Federal Aviation Administration Technical Center ....★1968
Federal Highway Administration – U.S. Department of Transportation 1969
Federal Laboratory Consortium Technology Transfer Network ....★601
Federal Procurement – North Dakota State Economic Development Commission 1007
Federal Railroad Administration – U.S. Department of Transportation 1970
Federal Research in Progress 2061
Federal Research Report 2062
Federated Capital Corporation 1541
FEDRIP (Federal Research in Progress) 2061
Fellows Complex 1255
Ferranti High Technology, Inc. 1637
Fiber Optic News; Military 2147
FIM (Foundation for Innovation in Medicine) 25
Finance and Capital Sources; Handbook of Business 2075
Finance Authority of Maine 871
Finance Division – Mississippi State Department of Economic Development 925
Finance Your Business; How to ....★176
Financial Assistance – North Dakota State Economic Development Commission 1008
Financial Group; Anderson-Bache 143
Financial Institutions Bureau; Annual Report of the 1987
Financial Opportunities, Inc. 1474
Financial Programs – Maine State Development Office 874
Financial Resource Guide ....★120
Financial Resources, Inc. 1775
Financial Services Division – Oregon State Economic Development Department 1032
Financial Times Business Reports 2442
Financing Alternatives for Small Business ....★176
Financing Programs – Wisconsin State Department of Development 1134
Financing Sources for Mergers, Buyouts & Acquisitions; Directory of 2036
Financing Your Business ....★455
Finding and Licensing New Products and Technology From the U.S.A. ....★326
Firestone Tire and Rubber Company, Central Research Library 2582
First American Capital Funding, Inc. 1313
First Boston Corporation 1638
First Capital Corporation of Chicago 1453
First Century Partners (San Francisco) 1314
First Century Partnership (New York City) 1639
First Chicago Venture Capital (Boston) 1513
First Chicago Venture Capital (Chicago) 1454
First Coast Technology Park 398
First Connecticut SBIC 1402
First Interstate Capital, Inc. 1315
First Maryland Capital, Inc. 1486
First Midwest Ventures, Inc. 1552
First New York SBIC 1640
First North Florida SBIC 1424
First Ohio Capital Corporation 1726
First Princeton Capital Corporation 1582
First SBIC of Alabama 1262
First SBIC of California (Costa Mesa) 1316
First SBIC of California (Palo Alto) 1317
First SBIC of California (Pasadena) 1318
First SBIC of California (Washington) 1754
First Source Capital Corporation 1467
First Southern Capital Corporation 1479
First United SBIC, Inc. 1514
First Valley Capital Corporation 1755
Fish and Neave, Library 2583
Flame 2063
The FLC News 2064
Fleet Venture Partners 1766
Fleet Venture Partners II 1515
Flight Research Facility; Dryden – National Aeronautics and Space Administration 1854
Flight Test Center – U.S. Department of the Air Force 1904
Flint Industrial Village for Enterprise 1183
Floco Investment Company, Inc. 1771
Florence Incubator 1247
Florida Atlantic Research Park 399
Florida; CRS Venture Directory— 2025
Florida Economic Development Center – Florida State University 401
Florida Entrepreneurial Network – Florida State University 402
Florida Entrepreneurial Network; North ....★402
Florida Entrepreneurial Network; South ....★402
Florida Small Business Development Center 400
Florida State Department of Commerce – Bureau of Business Assistance 753
Florida State Department of Commerce – Entrepreneurship Program 754
Florida State Department of General Services – Division of Purchasing 755
Florida State Department of Professional Regulation 756
Florida State University – Florida Economic Development Center 401
Florida State University – Florida Entrepreneurial Network 402
Florida Venture Capital Handbook ....★401
Fluid Capital Corporation 1594
Fluid Financial Corporation 1595
Focus on Thinking 2331
For Your Eyes Only 2065
Foreign Technology: An Abstract Newsletter 2066
Foreign Trade Fairs New Products Newsletter 2067
Formosa Capital Corporation 1583
Fort Worth District Office – U.S. Small Business Administration 1085
Fortieth Street Venture Partners (Cincinnati) 1727
Fortieth Street Venture Partners (New York City) 1641
Fortunes From Ideas ....★199
Foster Management Company 1642
Foundation for Innovation in Medicine 25
Four Rivers Development, Inc. 837
Fox Valley Technical College – Technical Research Innovation Park 403
Foxwell; W. John 336
FPAT 2443
Franklin Advanced Technology Center; Ben 1226
Franklin Corporation 1643
Franklin Partnership; Ben – Pennsylvania State Department of Commerce 1037
Fredericks Michael and Company 1644
Free Library of Philadelphia, Government Publications Department 2584
French Advances in Science & Technology 2068
Fresh Start Venture Capital Corporation 1645
Fresno District Office – U.S. Small Business Administration 708
Friday Memo 2444
Friedberg & Associates; Henry R. 229
Frontenac Capital Corporation 1455
Frost & Jacobs, Library 2585
Fry Consultants, Inc. 219
Fulcrum Venture Capital Corporation 1419
Fulton-Carroll Center for Industry 1162
Fulton County Business Licenses 767
Fundex Capital Corporation 1646
Future Options - Unlimited 2332
Future Value Ventures, Inc. 1843

G

G C and H Partners 1319
GAF Corporation, Technical Information Services 2586
Galesburg Business and Technology Center 1163
Game Puzzles for the Joy of Thinking 2333
Gardiner Institute for Small Business Studies; P.J. – Memorial University of Newfoundland 471
Gas & Petrochem Equipment; Oil, ...2172
Gas Research Institute ....★2074
Gaylord Research Institute 220
Gellman Research Associates 221
General Electric Company, Technology Center, Library 2587
Geographic Sciences Laboratory ....★1923
Geologic Division; U.S. Geological Survey, – U.S. Department of the Interior 1940
George C. Marshall Space Flight Center – National Aeronautics and Space Administration 1855
George M. Low Center for Industrial Innovation – Rensselaer Polytechnic Institute 534
George Mason Entrepreneur Center – George Mason University 405
George Mason University – Center for the Productive Use of Technology 404

George Mason University – George Mason Entrepreneur Center 405
George Mason University – Institute for Advanced Study in the Integrative Sciences 406
George W.K. King Associates Consulting Engineers 222
George Washington University – Center for International Science and Technology Policy 407
Georgia; Business Council of 762
Georgia; Directory of Scientific Resources in 2040
Georgia Institute of Technology – Advanced Technology Development Center 408
Georgia Institute of Technology – Economic Development Laboratory 409
Georgia Institute of Technology – Georgia Productivity Center 410
Georgia Institute of Technology – Patent Assistance Program 411
Georgia Institute of Technology – Technology Policy and Assessment Center 412
Georgia Institute of Technology, Price Gilbert Memorial Library 2588
Georgia Productivity Center – Georgia Institute of Technology 410
Georgia Secretary of State – Examining Boards Division 768
Georgia Small Business Development Center 413
Georgia State Department of Administrative Services – Division of Purchasing 769
Georgia State Department of Agriculture – Consumer Protection Division 770
Georgia State Department of Community Affairs – Small Business Revitalization Program 771
Georgia State House of Representatives – Industry Committee 772
Georgia State University – International Center for Entrepreneurship 414
Get Rich News 2069
GHW Capital Corporation 1647
GIDEP (Government-Industry Data Exchange Program) 2445
The Gifted Child Today 2334
Gifted Children Monthly 2335
Gigliotti and Associates Incorporated; Michael 290
Glenwood Management 1320
Goddard Space Flight Center – National Aeronautics and Space Administration 1856
Gold Coast Capital Corporation 1425
Golder, Thoma and Cressey 1456
Goodrich Chemical Company, Avon Lake Technical Center, Information Center; B.F. 2540
Goodrich Company, Research and Development Center, Brecksville Information Center; B.F. 2541
Gorham International Incorporated 223
Gorman's New Product News 2070
Government Data Systems 2071
Government-Industry Data Exchange Program 2445
Government Inventions for Licensing: An Abstract Newsletter 2072
Government Research Directory 2073
GRADTECH 224
Grand Rapids Terminal Ward 1184
Granite State Capital, Inc. 1572
Grants and Program Systems; Office of – U.S. Department of Agriculture 1872
Great American Family Expo 2497
The Great Bridge Lowering 2336
Great Plains Development, Inc. 838
Greater Philadelphia Venture Capital Corporation, Inc. 1756
Greater Syracuse Business Incubator Center 1205
Greater Washington Investors, Inc. 1487
Greenville Incubator 1230
Greyhound Corporation, Patent Library 2589
Greylock Management Corporation (Boston) 1516
GRID 2074
Gries Investment Corporation 1728
Grocers Small Business Investment Corporation 1800
Guide for the Submission of Unsolicited Research and Development ....★1939
Guide to Available Technologies ....★324
Guide to Establishing a Business in Arizona ....★691
Guide to Starting a Business in Minnesota ....★920
Gulf Coast Breeder 26
Gulf South Research Institute, Technology Management Division 415
Gulfport District Office – U.S. Small Business Administration 928
Gwinnett County Business Licenses 773

H

H.B. Hindin Associates, Inc. 225
H.J. Barrington Nevitt 226
Hambrecht & Quist Venture Partners 1321
Hambro International Venture Fund (Boston) 1517
Hambro International Venture Fund (New York) 1648
Hamco Capital Corporation 1322
Handbook of Business Finance and Capital Sources 2075
Hanover Capital Corporation 1649
Hardy; William B. 340
Harlingen District Office – U.S. Small Business Administration 1086
Harrisburg District Office – U.S. Small Business Administration 1047
Harrold Organization 227
Harry Prebluda 228
Hartford District Office – U.S. Small Business Administration 736
Harvard Medical School – Office for Technology, Licensing, and Industry-Sponsored Research 416
Harvard University – Office for Patents, Copyrights, and Licensing 417
Harvest Ventures, Inc. (Cupertino) 1323
Harvest Ventures, Inc. (New York City) 1650
Hato Rey District Office – U.S. Small Business Administration 1056
Hawaii; Chamber of Commerce of – Small Business Center 778
Hawaii High Tech Journal ....★418
Hawaii Ocean Science and Technology Park 418
Hawaii; Starting a Business in ....★781
Hawaii State Department of Accounting and General Services – Division of Purchasing and Supply 779
Hawaii State Department of Commerce and Consumer Affairs – Professional and Vocational Licensing Division 780
Hawaii State Department of Planning and Economic Development – Small Business Information Service 781
Hawaii State Department of Taxation 782
Hawaii's Business Regulations ....★781
Haywood County Incubator 1212
Headquarters Contracts and Grants Division – National Aeronautics and Space Administration 1857
Health Research Center; Naval ....★1951
Healthcare Corporation, Information Resource Center; Baxter 2544
Healthcare Technology & Business Opportunities 2076
Heart, Lung, and Blood Institute; National – National Institutes of Health 1866
Helena District Office – U.S. Small Business Administration 946
Helio Capital, Inc. 1324
Henkel Corporation, Process Chemicals Library 2590
Henry R. Friedberg & Associates 229
Herbert E. Mecke Associates, Inc. 230
Hereld Organization 231
Heritage Capital Corporation 1715
Heritage Venture Group, Inc. 1468
Hi-TECH ....★393
Hi-Tech Alert 2077
Hi-Tech Venture Consultants, Inc. 232
Hickory Venture Capital Corporation 1263
High Tech Entrepreneurship ....★154
High Tech International 2446
High Tech Journal; Hawaii ....★418
High-Tech Materials Alert 2078
High Tech Tomorrow 2079
High Technology Associates 233
High Technology Entrepreneurs Council 27
High Technology Program – Missouri State Department of Economic Development 930
Highway Administration; Federal – U.S. Department of Transportation 1969
Highway Traffic Safety Adminstration; National – U.S. Department of Transportation 1971
Hillard-Lyons Patent Management, Inc. 234
Hillman Ventures, Inc. 1325
Hindin Associates, Inc.; H.B. 225
Hispanic Business Magazine 2080
Hobby Industry Convention and Trade Show 2498
Hoechst Celanese Corporation, R.L. Mitchell Technical Information Center 2591
Hoeft Associates 235
Holding Capital Management Corporation 1651
Holland W. Phillips & Associates, Ltd. 236
Home Business News 2081
HomeBased Entrepreneur Newsletter 2082
Honolulu District Office – U.S. Small Business Administration 783
Horatio Alger Association of Distinguished Americans ....★2063
Horn and Hardart Capital Corporation 1426
HOST (Hawaii Ocean Science and Technology Park) 418
Houston District Office – U.S. Small Business Administration 1087
Houston Inventors Association 28
How Can We Teach Intelligence? 2337
How do You Figure That? Ways of Problem Solving 2338
How to Evaluate Information and Draw Conclusions; Critical Thinking: 2316
How to Evaluate Your Technical Idea ....★164
How to Finance Your Business ....★176
How to Invent 2339

How to Protect an Idea Before Patent .... ★60
How to Shape an Idea Into an Invention ...... ★60
Howard University, University Libraries 2592
Howrey & Simon, Library 2593
Hub Financial Network – Entrepreneurial Services and Seminars 237
Hughes & Company; J.P. 263
Hugo Industrial Incubator 1222
Human Development Services; Office of – U.S. Department of Health and Human Services 1892
Human Resources Laboratory; Air Force .... ★1979
Hunting Park West 1231
Hyland/Associates 238
Hynes & Associates; Richard W. 314

I

I.C. Asia 2083
I.K. Capital Loans Ltd. 1326
IAA (Inventors Association of America) 50
IAI (Inventors Association of Indiana) 53
Ibero American Investors Corporation 1652
ICA (Inventors Club of America) 60
ICGC (Inventors Club of Greater Cincinnati) 61
ICI Americas Inc., Atlas Library 2594
ICI (Inventors Council of Illinois) 68
ICS Group International (US) 239
ICSB Bulletin 2084
IC2 Institute – University of Texas at Austin 640
Idaho Business and Economic Development Center – Boise State University 361
The Idaho Company 784
Idaho Economic Development Center .... ★361
Idaho First National Bank – Economic Action Council 785
Idaho Innovation Center 1155
Idaho Library; University of 2675
Idaho Small Business Development Center 419
Idaho State Board of Occupational Licenses 786
Idaho State Department of Administration – Division of Purchasing 787
Idaho State Department of Commerce 788
Idaho State Department of Revenue and Taxation 789
Idaho State University Research Park 420
Idanta Partners 1801
IDB (INPADOC Data Base) 2447
IDEA .... ★529
The Idea Generator 2340
Ideal Financial Corporation 1427
Ideas for Industry; Science of Creating .... ★75
IEG Venture Management, Inc. 1457
IF (Inventrepreneurs' Forum) 77
IFI/Plenum Data Company, Library 2595
II (Innovators International) 34
ILIMA (International Licensing Industry and Merchandisers' Association) 41
Illinois Business Leader .... ★666
Illinois Business Review .... ★594
Illinois Development Finance Authority – Direct Loan Program 792
Illinois Economic Outlook .... ★594
Illinois Institute of Technology – Manufacturing Productivity Center 421
Illinois Institute of Technology – Technology Commercialization Center 422
Illinois Secretary of State – Division of Purchasing 793
Illinois Small Business Development Center 423
Illinois Small Business Development Program; Build – Equity Investment Fund 791
Illinois State Department of Commerce and Community Affairs – Business Innovation Fund 794
Illinois State Department of Commerce and Community Affairs – Small Business Advocacy 795
Illinois State Department of Commerce and Community Affairs – Small Business Assistance Bureau 796
Illinois State Department of Commerce and Community Affairs – Small Business Development Program 797
Illinois State Department of Commerce and Community Affairs – Small Business Energy Management/Loan Program 798
Illinois State Department of Commerce and Community Affairs – Small Business Financing Program 799
Illinois State Department of Commerce and Community Affairs – Small Business Fixed-Rate Financing Fund 800
Illinois State Department of Commerce and Community Affairs – Small Business Micro Loan Program 801
Illinois State Department of Commerce and Community Development – Procurement Assistance Program 802
Illinois State Department of Registration and Education – Division of Licensing and Testing 803
Illinois State Library 2596
Illinois State University – Technology Commercialization Center 424
Illinois Statistical Abstract .... ★594
Illinois Venture Capital Fund 804
ILLITECH – University of Illinois 596
Imagination Celebration/Invention Convention 2267
Imagine That 2341
IMI (Invention Marketing Institute) 44
IMPACT Compressors 2085
IMPACT Pumps 2086
IMPACT Valves 2087
Imperial Ventures, Inc. 1327
The Importance of Trademark to Small Businesses .... ★148
Improving the Quality of Student Thinking 2342
In Business 2088
INCINC (International Copyright Information Center) 40
INCOM (Inventors Council of Michigan) 69
Incubation Association; National Business 98
Incubator Facilities Program – North Carolina Technological Development Authority 1002
Incubator Industries Building (Buffalo) 1206
Incubator Times 2089
Incubators; National Directory of .... ★98
Indian Affairs Council – Indian Business Loan Program 913
Indian Business Loan Program – Indian Affairs Council 913
Indiana Corporation for Science and Technology 809
Indiana Enterprise Center 1169
Indiana Institute for New Business Ventures 810
Indiana Institute of Technology – McMillen Productivity and Design Center 425
Indiana Seed Capital Network 811
Indiana State Department of Administration – Division of Procurement 812
Indiana State Department of Commerce 813
Indiana State Professional Licensing Agency 814
Indiana University – Entrepreneur in Residence 426
Indiana University-Purdue University Indianapolis – Commercial/Industrial Liaison 427
Indiana University-Purdue University Indianapolis – Economic Development Administration University Center 428
Indiana University-Purdue University Indianapolis – Tech Net 429
Indianapolis Center for Advanced Research; National Aeronautics and Space Administration/ 487
Indianapolis District Office – U.S. Small Business Administration 815
Indianapolis-Marion County Public Library, Business, Science and Technology Division 2597
Industrial Applications Center (Boise); National Aeronautics and Space Administration .... ★361
Industrial Applications Center; National Aeronautics and Space Administration – University of Pittsburgh 625
Industrial Applications Center; University of New Mexico National Aeronautics and Space Administration .... ★617
Industrial Contracts and Licensing – Washington University 662
Industrial Development; Alabama .... ★681
Industrial Development; Bureau of – Michigan Technological University 477
Industrial Development Corporation of Lea County 1596
Industrial Development Office – New Hampshire State Department of Resources and Economic Development 963
Industrial Energy Conserver .... ★409
Industrial Equipment News 2090
Industrial Finance Division – Alabama State Development Office 681
Industrial Financing Section – North Carolina State Department of Commerce 998
Industrial Innovation Extension Service – New York State Science and Technology Foundation 985
Industrial Innovation; National Council for 100
Industrial Marketing Arm 240
Industrial Marketing Projects 241
Industrial Organization Research Group – Laval University 451
Industrial Park; Oxmoor West 683
Industrial Park; Pueblo Memorial Airport 720
Industrial Product Bulletin 2091
Industrial Product Ideas 2092
Industrial Property; International Association for the Protection of .... ★42
Industrial Research and Consulting Center – University of New Hampshire 616
Industrial Research and Development Magazine 2093
Industrial Research and Service; Center for – Iowa State University 440
Industrial Research and Technology Transfer; Office of – University of Wisconsin—Milwaukee 647
Industrial Research Assistance Program .... ★489
The Industrial Research Institute, Inc. .... ★2194

Industrial Research; Office of – University of Manitoba 603
Industrial Technology Institute 430
Industry Committee – Georgia State House of Representatives 772
Industry Research Program; University- – University of Wisconsin—Madison 645
Industry/University Cooperative Research Centers – National Science Foundation 1868
Infectious Diseases; Medical Research Institute of ....★1929
Infectious Diseases; National Institute of Allergy and – U.S. Department of Health and Human Services 1887
Influentia 242
Info/Consult 243
Information Age Technology; Center for – New Jersey Institute of Technology 494
Information and Research Division – Oklahoma State Department of Commerce 1022
Information Brokers of Colorado 244
Information Consulting, Incorporated 245
Information Guild 246
Information Policy Program; Technology and – Syracuse University 564
Information Resources, Inc.; Applied 145
Information Services; AML 142
Information Technologies Research Center; Applied 347
Ingenuity 2094
Initiatives for Not-for-Profit Entrepreneurship – New York University 500
InKnowVation Newsletter 2095
Inno-Media 29
Innovation & Business Development; Center for 371
Innovation and Creativity; Conference on 2495
Innovation and Development; Center for – University of Wisconsin—Stout 648
Innovation and Entrepreneurship; Center for Studies in – Columbia University 383
Innovation & Entrepreneurship for Students Magazine; Technology, ...2397
Innovation and Entrepreneurship Institute – University of Miami 606
Innovation and Productivity Strategies Research Program 431
Innovation & Technology Council; Nevada 107
Innovation and Technology; Sessions on ....★289
Innovation Assessment Center – Washington State University 660
Innovation; Association for Science, Technology and 10
Innovation Center – Massachusetts Institute of Technology 464
Innovation Center and Research Park – Ohio University 515
Innovation Center; D.J. Burdini Technological ....★403
Innovation Center; Rhode Island ....★629
Innovation Center; Utah 1093
Innovation Centre/Waterloo; Canadian Industrial 366
Innovation Centre/Waterloo, Resource Centre; Canadian Industrial 2557
Innovation; Creativity and 2265
Innovation Development; Corporation for 808
Innovation Development Institute 30
Innovation; George M. Low Center for Industrial – Rensselaer Polytechnic Institute 534
Innovation Group 31
Innovation in Medical Devices ....★154
Innovation in Medicine; Foundation for 25
Innovation, Inc.; Minnesota Project 89
Innovation Institute 32
Innovation Invention Network 33
Innovation Management Studies; Center for – Lehigh University 453
Innovation Network Foundation 1126
Innovation Network; Technology Information Exchange- 2471
Innovation News 2096
Innovation Office; Discovery ....★393
Innovation Park 432
Innovation Park; Technical Research – Fox Valley Technical College 403
Innovation Place 433
Innovation Research Fund – North Carolina Technological Development Authority 1003
Innovation Research Promotion Program; New York Small Business 978
Innovation Showcase ....★366
Innovation Team Contest; Mini-Invention 2276
Innovation Workshops; National ....★99
Innovations & Inventions Newsletter ....★216
Innovative Technology; Center for 372
Innovators & Associates, Inc. 247
Innovators Forum – Rose-Hulman Institute of Technology 544
Innovators International 34
InnoVen Group 1584
Innovex Inc. 35
INPADOC Data Base 2447
INPAMAR 2448
INPI-MARQUES 2449
Inside R&D 2097
Institut National de la Propriete Industrielle (INPI) ....★2438
Institute for Advanced Study in the Integrative Sciences – George Mason University 406
Institute for Economic Development – Pittsburg State University 526
Institute for Energy Research – Syracuse University 563
Institute for New Enterprise Development 434
Institute for Research on Public Policy 435
Institute for Scientific Information ....★2027
Institute for Technological Studies – Western Michigan University 668
Institute for Technology Development 436
Institute for Technology Development and Assessment – University of Texas Health Science Center at Houston 641
Institute of Advanced Manufacturing Sciences 437
Institute of Dental Research ....★1929
Institute of Surgical Research ....★1929
Institute of Textile Technology, Textile Information Services, Roger Milliken Textile Library 2598
Institutional Venture Partners III 1328
Instruments & Computers-Applications in the Laboratory 2098
Instruments Corp.; Pixel 306
Integrated Circuit Engineering Corporation 248
Integrated Manufacturing Technology 2099
Intellectual Property; International Bureau for the Protection of ....★40
Intellectual Property Lawyers and Patent Agents; Directory of 2037
Intellectual Property Organization; World ....★2477
Intellectual Property Owners, Inc. 36
Intelligence 2100
Intelligent Systems Technology, Inc. 249
Interaction Business Services 250
Intercapco, Inc. 1564
Intercapco West, Inc. 1565
Intercon Research Associates, Ltd. 251
Intercontinental Capital Funding Corporation 1653
Intercontinental Enterprises Ltd. 252
Interdevelopment, Inc. 253
Intergroup Venture Capital Corporation 1654
Intermarketing Group – Management Consultants 254
Intermountain Society of Inventors 37
Intermountain Society of Inventors and Designers 38
Intermountain Ventures Ltd. 1395
International Association for the Protection of Industrial Property ....★42
International Association of Professional Inventors 39
International Book Marketing Ltd. 255
International Bureau for the Protection of Intellectual Property ....★40
International Business Consulting Group 256
International Business Development Group; API 144
International Business Opportunities ....★1014
International Business Resources, Inc. 257
International Center for Entrepreneurship – Georgia State University 414
International Copyright Information Center 40
International Council for Small Business ....★2084
International Craft Exposition/Chicago Craft and Creative Industries Show and Convention 2499
International Creative 258
International Defense Consultant Services, Inc. 259
International Exhibition of Inventions and New Techniques of Geneva 2500
International Gift Fair; New York 2517
International; High Tech 2446
International High Technology Report 2101
International Inc.; Royalco 317
International Invention Register 2102
The International Lawyer 2103
The International Licensing and Merchandising Conference and Expo/Licensing 2501
International Licensing Industry and Merchandisers' Association 41
International New Product Newsletter 2104
International New Products Newsletter 2105
International Paper Capital Formation, Inc. 1776
International Patent and Trademark Association 42
International Product Alert 2106
International Research Centers Directory 2107
International Science and Technology Institute, Inc. 260
International Technology Review 2108
International Technology Transfer; American Bulletin of 1983
International Technology Transfer Directory ....★79
International Toy Fair; SITOY-Seoul 2520
International Toy Fair; Taipei 2524
International Trade Names Dictionary: Company Index 2109
International Transactions in High-Tech Products and Services ....★148
International Venture Capital Institute, Inc. ....★2041

International Venture Capital Institute—Directory of Venture Capital Clubs 2110
International Wealth Success 2111
Interscope Investments, Inc. 1329
Interstate Capital Company, Inc. 1655
Interven II 1330
InterVen II LP 1745
Interwest Partners 1331
Invent! 2112
INVENT AMERICA! 2268
Invent America! Creative Resource Guide ....★132
Invent America Program ....★132
Invent; How to 2339
Invent or Want To; The Book for Women Who 2295
Invention and Investment Institute ....★573
Invention Book; The Unconventional 2408
Invention Club; Albuquerque 2
Invention Contest; Weekly Reader National 2281
Invention Convention; Connecticut 2264
Invention Convention; Imagination Celebration/ 2267
Invention Convention (Los Angeles) 2502
The Invention Convention (Morristown) 2269
Invention Convention (Plattsburgh) 2270
Invention Convention Procedural Manual 2343
Invention Convention (Richardson) 2271
Invention Convention; Toledo Public Schools 2279
Invention Development Society 43
Invention; How to Shape an Idea Into an ....★60
Invention Marketing Institute 44
Invention Network; Innovation 33
Invention News ....★1
Invention; Patenting Your Own ....★164
Invention Program; Midland Public Schools 2275
Invention Program; San Diego 2278
Invention Program; Tualatin 2280
Invention Program; Western New York 2282
Invention Register; International 2102
Invention; Research & 2193
Invention; Society for the Encouragement of Research and 119
Invention Submission Corp. 261
Invention Workshop for Girls; Daughters of Invention: An 2317
Inventioneering 2344
Inventions and New Techniques of Geneva; International Exhibition of 2500
Inventions and Novel Features; INVEX-International Exhibition of 2505
Inventions Available for Licensing to U.S. Businesses; Catalog of Government 2009
Inventions; Complete Guide to Making Money With Your ....★75
Inventions for Licensing: An Abstract Newsletter; Government 2072
Inventions Newsletter; Innovations & ....★216
Inventions Program; Status Report of the Energy-Related 2218
Inventions; Protecting & Selling Ideas & ....★199
Inventions That Made Big Money; Little ....★75
The Inventive Child 2345
The Inventive Imagination to Illumination 2346
The Inventive Innovation to Ingenuity 2347
Inventive Minded People; Corporation of ....★50
Inventive Thinking: A Teacher-Student Handbook 2348
Inventor; American 1985
Inventor Associates of Georgia 45
Inventor; Be an 2294
Inventor of the Month Award ....★120
Inventor of the Year Award ....★36
Inventor USA ....★120
Inventors; American Association of 4
Inventors; American Society of 8
Inventors and Designers; Intermountain Society of 38
Inventors & Entrepreneurs Association of Austin 46
Inventors and Entrepreneurs; Society for 118
Inventors & Entrepreneurs Society of Indiana, Inc. 47
Inventors & Technology Transfer Corporation 48
Inventors Assistance League 49
Inventors Association; Columbus 19
Inventors Association; Houston 28
Inventors Association; Kearney 78
Inventors Association; Lincoln 80
Inventors Association of America 50
Inventors Association of Connecticut 51
Inventors Association of Georgia 52
Inventors Association of Indiana 53
Inventors Association of Metro Detroit 54
Inventors Association of New England 55
Inventors Association of New England, Connecticut Chapter 56
Inventors Association of New England Youth Education Program 2272
Inventors Association of St. Louis 57
Inventors Association of St. Louis Youth Programs 2273
Inventors Association of Washington 58
Inventors Association; Ohio 109
Inventors Association; Tennessee 127
Inventors Association; Texas 128
Inventors; Biographical Dictionary of Scientists: Engineers and 1994
The Inventors Club 59
Inventors Club; Central Florida 17
Inventors Club of America 60
Inventors Club of Greater Cincinnati 61
Inventors Club of Minnesota 62
Inventors Club; Omaha 111
Inventors Clubs of America, Library 2599
Inventors; Confederacy of Mississippi 20
Inventor's Conference; National Fall ....★8
Inventors Congress, Inc.; Arkansas 9
Inventors Congress; Minnesota 87
Inventors Congress; Minnesota Student 2277
Inventors Congress; Oklahoma 110
Inventors Congress; Rocky Mountain 116
Inventors Congress; South Dakota 123
Inventors Connection of Greater Cleveland 63
Inventors Cooperative Association; National 102
Inventors' Council 64
Inventors Council; California 15
Inventors Council; Central Florida 18
Inventors Council of Dayton 65
Inventors Council of Greater Lorain County 66
Inventors Council of Hawaii 67
Inventors Council of Illinois 68
Inventors Council of Michigan 69
Inventors Council of Ohio—Columbus 70
Inventor's Council; Tampa Bay 125
Inventors' Digest 2113
Inventors Education Network 71
Inventors Expo (Camarillo) 2503
Inventors Expo; National 2512
Inventors Expo (Tarzana) 2504
Inventors Foundation, Inc.; Affiliated 1
Inventors Foundation; National 103
Inventor's Gazette ....★50
Inventors Greek-English Guide ....★113
Inventors Group; Midwest 85
The Inventor's Guide 2349
Inventor's Guidebook ....★75
Inventors Hall of Fame Foundation; National ★101
Inventors; Intermountain Society of 37
Inventors; International Association of Professional 39
Inventor's Journal ....★75
Inventors League 72
Inventors League; Nilson's 298
Inventors Licensing and Marketing Agency ....★75
Inventors; Mississippi Society of Scientists and 93
Inventors; National Society of 105
Inventors Network of Columbus 73
Inventors; New York Society of Professional 108
Inventors News 2114
Inventors Newsletter; The Canadian ....★366
Inventors of California 74
Inventors of Greece in U.S.A.; Pan Hellenic Society 113
Inventors Organizations; National Congress of 99
Inventors; Palm Beach Society of American 112
Inventors Program; Young 2283
Inventors Project; Women ....★2295
Inventors Resource Center; Minnesota Inventors Congress, ...88
Inventors; Society of American 120
Inventors; Society of Minnesota 121
Inventors USA; Worcester Area 135
Inventors Workshop ....★2503
Inventors Workshop International Education Foundation (Camarillo) 438
Inventors Workshop International Education Foundation (Tarzana) 75
Inventors Workshop International (Newbury Park) 76
Inventors Workshop; Mississippi 90
Inventrepreneurs' Forum 77
InVenture 2274
InvestAmerica Venture Group, Inc. (Cedar Rapids) 1471
InvestAmerica Venture Group, Inc. (Kansas City) 1566
InvestAmerica Venture Group, Inc. (Milwaukee) 1844
Investing In ....★389
Investing Licensing & Trading Conditions Abroad 2115
Investment Companies—Membership Directory; National Association of 2153
Investments Orange Nassau, Inc. 1332
Investor's Equity, Inc. 1438
Investors Market Finance Directory; American 1986
INVEX-International Exhibition of Inventions and Novel Features 2505
Iowa Business Development Credit Corporation 816
Iowa Business Growth Company 817
Iowa Procurement Outreach Center – Iowa State Department of Economic Development 823
Iowa Product Development Corportation – Iowa State Department of Economic Development 824
Iowa Small Business Advisory Council – Iowa State Department of Economic Development 825
Iowa Small Business Development Center 439
Iowa Small Business Loan Program – Iowa State Department of Economic Development 826

Iowa Small Business Vendor Application Program – Iowa State Department of Economic Development....827
Iowa State Department of Commerce – Division of Professional Licensing and Regulation....818
Iowa State Department of Economic Development – Call One Program....819
Iowa State Department of Economic Development – Community Development Block Grant Program....820
Iowa State Department of Economic Development – Community Economic Betterment Account....821
Iowa State Department of Economic Development – Export Finance Program....822
Iowa State Department of Economic Development – Iowa Procurement Outreach Center....823
Iowa State Department of Economic Development – Iowa Product Development Corportation....824
Iowa State Department of Economic Development – Iowa Small Business Advisory Council....825
Iowa State Department of Economic Development – Iowa Small Business Loan Program....826
Iowa State Department of Economic Development – Iowa Small Business Vendor Application Program....827
Iowa State Department of Economic Development – Self-Employment Loan Program....828
Iowa State Department of Economic Development – Targeted Small Business Loan Guarantee Program....829
Iowa State Department of General Services – Division of Purchasing and Materials Management....830
Iowa State House Committee on Small Business and Commerce....831
Iowa State University – Center for Industrial Research and Service....440
Iowa State University Research Park....441
IPO News....★36
IPS Industrial Products & Services....2116
IPTA (International Patent and Trademark Association)....42
Irving Capital Corporation....1656
ITASCA Growth Fund, Inc.....1553
ITC Capital Corporation....1657
Ivanhoe Venture Capital Ltd.....1333
IVCI Venture Capital Digest....2117
IWIEF (Inventors Workshop International Education Foundation (Tarzana))....75

J

J and D Capital Corporation....1428
J.B. Blood Building....1176
J.C. Croft & Associates International Consultants....262
J.P. Hughes & Company....263
Jack J. Bulloff....264
Jackson District Office – U.S. Small Business Administration....929
Jacksonville District Office – U.S. Small Business Administration....759
Jacksonville State University – Center for Economic Development & Business Research....442
James B. Kobak and Company....1403
James River Capital Associates....1831
James River Corporation, Neenah Technical Information Center....2600
Japan High Tech Review....2450
Japan Industrial Journal....★2518
Japan; New From....2159
Japan/Pacific Associates....265
Japan Patent Information Organization...★2451
Japanese American Capital Corporation....1658
Japanese Information Technology Industry Letter; F&S/....2059
Japanese Venture Capital & OTC Opportunities; F&S/....2060
Japio....2451
Jefferson Proving Ground – U.S. Department of the Army....1925
Jeffrey D. Marshall....266
Jet Propulsion Laboratory – National Aeronautics and Space Administration....1858
Jewelers' Circular/Keystone Brand Name and Trademark Guide....2118
JoAnn Johnson....267
Jobless Newsletter....2119
The John F. Buagh Center for Entrepreneurship....★2055
John F. Kennedy Space Center – National Aeronautics and Space Administration....1859
John Hancock Venture Capital Management, Inc.....1518
John Marshall Law School, Library....2601
Johns Hopkins University – Applied Physics Laboratory....443
Johns Hopkins University – Bayview Research Campus....444
Johnson Engineering Company....268
Johnson; JoAnn....267
Johnson Space Center; Lyndon B. – National Aeronautics and Space Administration....1862
Johnston Associates, Inc.....1585
Joint Venture Expo; Techno Tokyo - International Licensing and....2526
Josef K. Murek Company....269
Joseph W. Prane....270
Josephberg, Grosz and Company, Inc.....1659
Journal of Behavioral Economics....★666
Journal of Business Venturing....2120
The Journal of Creative Behavior....2350
Journal of Small Business Management....2121
Journal of the Copyright Society of the U.S.A.....2122
Jupiter Partners....1334
JURINPI....2452
Just Think Program Series and Stretch Think Program Series....2351
Juvenile Products Manufacturers Association....2506

K

Kansas City District Office – U.S. Small Business Administration....937
Kansas City, Kansas, Inc.; Citywide Development Corporation of....836
Kansas Enterprise Zone Act – Kansas State Department of Commerce....841
Kansas Municipalities; League of....847
Kansas Small Business Development Center....445
Kansas State Department of Administration – Small Business Procurement Program....839
Kansas State Department of Commerce – Economic Development Bond Program...840
Kansas State Department of Commerce – Kansas Enterprise Zone Act....841
Kansas State Department of Commerce – One-Stop Permitting....842
Kansas State Department of Commerce – Small Cities Community Development Block Grant Program....843
Kansas State Department of Commerce – Tax Increment Financing Program....844
Kansas State Procurement Technical Assistance Program....845
Kansas State University – Engineering Extension....446
Kansas State University Foundation....447
Kansas Technology Enterprise Corporation....846
Kansas Venture Capital, Inc.....1472
Kappa Associates....271
Kar-Mal Venture Capital, Inc.....1271
Kaye, Scholer, Fierman, Hays & Handler, Law Library....2602
Kearney Inventors Association....78
Keep Them Thinking....2352
Keller Research Services....272
Ken-Quest Limited....273
Kendall Square Associates....274
Kennedy Space Center; John F. – National Aeronautics and Space Administration....1859
Kentucky Development Finance Authority....858
Kentucky Small Business Development Center....448
Kentucky State Commerce Cabinet – Business Information Celaringhouse....859
Kentucky State Commerce Cabinet – Small Business Division....860
Kentucky State Finance and Administration Cabinet – Division of Purchases....861
Kentucky State Finance and Administration Cabinet – Occupations and Professions Division....862
Kentucky State Transportation Cabinet – Division of Purchases....863
Kevin C. McGuire....275
Keystone Venture Capital Management Company....1757
King Associates Consulting Engineers; George W.K.....222
Kitty Hawk Capital Ltd.....1716
Kleiman and Associates; M.....281
Kleiner, Perkins, Caufield, and Byers....1335
Knowledge Transfer Institute....449
Knox County Business Incubator....1251
Korea Strategy Associates, Inc.....276
Korey International Ltd.....277
Kraft, Inc., Technology Center Library....2603
Kuuttila & Associates; E.J.....214
Kwiat Capital Corporation....1660

L

Laboratory Command – U.S. Department of the Army....1926
Laboratory; Instruments & Computers-Applications in the....2098
Laboratory Times....2123
LACS (Los Angeles Copyright Society)....81
Lailai Capital Corporation....1336
Lambda Funds....1661
Lancaster Economic Development Corporation....1146
Langley Research Center – National Aeronautics and Space Administration....1860
Lansdale Business Center....1232

Las Vegas District Office – U.S. Small Business Administration 960
Lasers/Electro-Optics Patents Newsletter....2124
The Lateral Thinking Machine....2353
LATIPAT....2453
Laval University – Administrative Sciences Research Laboratory....450
Laval University – Industrial Organization Research Group....451
Laval University – Small Business and Entrepreneurship Research Group....452
Law and Economic Studies; Center for – Columbia University....382
Law Department, Library; Chevron Corporation, Corporate....2563
Law Department Library; Union Carbide Corporation,....2666
Law Directory; Martindale-Hubbell....2143
Law Library; Baker & Botts,....2542
Law Library; Brown Maroney Rose Barber & Dye,....2551
Law Library; Brownstein, Zeidman & Schomer,....2552
Law Library; CBS Inc.,....2560
Law Library; Dickinson School of Law, Sheely-Lee....2576
Law Library; Kaye, Scholer, Fierman, Hays & Handler,....2602
Law Library; Pennie & Edmonds,....2640
Law Library; The Coca-Cola Company,....2568
Law Library; Townley & Updike,....2665
Law Library; UNISYS Corporation,....2667
Law Library; United States Trademark Association,....2671
Law Library; University of Southern California,....2683
Law Reports; Copyright....2018
Law School, Library; John Marshall....2601
Lawrence M. Liggett....278
Lawrence, Tyrrell, Ortale, and Smith....1662
Lawrence Venture Associates....1777
Lawyer; The International....2103
League of Kansas Municipalities....847
Lehigh University – Center for Innovation Management Studies....453
Lehigh University – Office of Research and Sponsored Programs....454
Lehigh University – Small Business Development Center....455
Lehigh University Small Business Reporter....★455
Leisy; Robert B.....315
Lenexa Development Company, Inc.....848
Leo Beiser Inc.....279
LES (Licensing Executives Society)....79
Les Nouvelles....2125
Lessons in Logic: Unravelling Common Complexities....2354
Levi Strauss & Company, Corporate Law Library....2604
Lewis B. Weisfeld....280
Lewis Research Center – National Aeronautics and Space Administration....1861
LEXIS Federal Patent, Trademark, & Copyright Library....2454
LEXPAT....2455
Liberty Street Market Place....1233
Library of Congress, Copyright Public Information Office....2605
Licenses and Permits; Minnesota Directory of....★920
Licensing: An Abstract Newsletter; Government Inventions for....2072
Licensing, and Industry-Sponsored Research; Office for Technology, – Harvard Medical School....416
Licensing and Joint Venture Expo; Techno Tokyo - International....2526
Licensing and Marketing Agency; Inventors....★75
Licensing and Merchandising Conference and Expo/Licensing; The International...2501
Licensing and Regulation Office – Louisiana State Department of Health and Human Resources....867
Licensing & Trading Conditions Abroad; Investing....2115
Licensing as a Business Strategy for Small Companies....★299
Licensing Book....2126
Licensing Center; Television....126
Licensing Directory....★41
Licensing Executives Society....79
Licensing Fair for Design Ideas, Publication, and Production Rights; LIZ '89....2507
Licensing; Industrial Contracts and – Washington University....662
Licensing Industry and Merchandisers' Association; International....41
Licensing Law and Business Report....2127
The Licensing Letter....2128
Licensing; MICEL-International Market for Creativity and....2508
Licensing New Products and Technology From the U.S.A.; Finding and....★326
Licensing; Office for Patents, Copyrights, and – Harvard University....417
Licensing; Office of Patents and – Cornell Research Foundation....387
Licensing; Office of Patents and – University of Colorado Foundation Inc.....590
Licensing; Office of Patents and – University of Minnesota....609
Licensing; Office of Technology – Stanford University....557
Licensing Office; Technology – Massachusetts Institute of Technology....466
Licensing to U.S. Businesses; Catalog of Government Inventions Available for....2009
Licensing Today....2129
Life After School....2355
LIFT (Literature is for Thinking)....2356
Liggett; Lawrence M.....278
Lightbulb....★75
Lincoln Inventors Association....80
Linda Hall Library....2606
Linked-Deposit Loan Program – Wisconsin Housing and Economic Development Authority....1132
LitAlert....2456
Literature is for Thinking....2356
Little Inventions That Made Big Money....★75
Little Rock District Office – U.S. Small Business Administration....701
Livingston Capital Ltd.....1802
LIZ '89 Licensing Fair for Design Ideas, Publication, and Production Rights....2507
Logic Number Problems....2357
Looking Glass Logic: Problems and Solutions....2358
Lookout....2130
Lorain County Community College – Advanced Technologies Center....456
Los Alamos Economic Development Corporation....1199
Los Angeles Copyright Society....81
Los Angeles District Office – U.S. Small Business Administration....709
Los Angeles Public Library, Science & Technology Department....2607
Louisiana Equity Capital Corporation....1480
Louisiana Small Business Development Center – Northeast Louisiana University....503
Louisiana Small Business Equity Corporation....865
Louisiana State Department of Commerce and Industry – Small Business Development Corporation....866
Louisiana State Department of Health and Human Resources – Licensing and Regulation Office....867
Louisiana State Office of the Governor – State Purchasing Office....868
Louisiana State University – Office of Technology Transfer....457
Louisiana State University, Business Administration/Government Documents Department....2608
Louisiana State University, Chemistry Library....2609
Louisville District Office – U.S. Small Business Administration....864
Louisville Free Public Library, Reference and Adult Services....2610
Low Center for Industrial Innovation; George M. – Rensselaer Polytechnic Institute....534
Lowcountry Investment Corporation....1772
Lowering Patent Costs....★266
Lubar and Company, Inc.....1845
Lubbock District Office – U.S. Small Business Administration....1088
Lubrizol Enterprises, Inc.....1729
Lyndon B. Johnson Space Center – National Aeronautics and Space Administration....1862

M

M & I Ventures Corporation....1846
M and T Capital Corporation....1663
M and T Chemicals, Inc., Technical & Business Information Center....2611
M. Kleiman and Associates....281
M.W. Probol & Associates....282
Macomb Business and Technology Center....1164
MacRae's Industrial Directory Connecticut....2131
MacRae's Industrial Directory Maine/New Hampshire/Vermont....2132
MacRae's Industrial Directory Maryland/D.C./Delaware....2133
MacRae's Industrial Directory Massachusetts/Rhode Island....2134
MacRae's Industrial Directory Minnesota...2135
MacRae's Industrial Directory New Jersey....2136
MacRae's Industrial Directory New York State....2137
MacRae's Industrial Directory North Carolina/South Carolina/Virginia....2138
MacRae's Industrial Directory Oregon....2139
MacRae's Industrial Directory Pennsylvania....2140
Madison Business Incubator....1260
Madison District Office – U.S. Small Business Administration....1130
Maehara & Associates....283
Magna Pacific Investments....1337
Maine Business Indicators....★634
Maine Capital Corporation....1483
Maine; Finance Authority of....871

Maine Growth Program – Maine State Development Office 875
Maine/New Hampshire/Vermont; MacRae's Industrial Directory 2132
Maine Small Business Development Center 458
Maine State Department of Administration – Bureau of Purchasing 872
Maine State Department of Professional and Financial Regulation – Division of Licensing and Enforcement 873
Maine State Development Office – Financial Programs 874
Maine State Development Office – Maine Growth Program 875
Maine Technology Park 459
Management Advisory Institute 460
Management Consultants – Intermarketing Group 254
Management Corporation; Access 138
Management of Advanced Automation Technology Center – Worcester Polytechnic Institute 673
Management of Technology; MIT Research Program on the – Massachusetts Institute of Technology 465
Management of Technology; Research Institute for the 537
Management Research and Development; National Centre for 488
Managing Group Creativity 2359
Manitoba Research Council, Industrial Technology Centre Library 2612
Mankind Research Unlimited, Inc. 284
Manual on SBIR Proposal Preparation ....★266
Manufacturers Hanover Venture Capital Corporation 1664
Manufacturing Productivity Center – Illinois Institute of Technology 421
Manufacturing Research Consortium; American ....★549
Manufacturing Resource and Productivity Center (Big Rapids) 1185
Manufacturing Sciences; Institute of Advanced 437
Manufacturing Technology; Advanced 1976
Manufacturing Technology: An Abstract Newsletter 2141
Manufacturing Technology; Integrated 2099
Maple City Business and Technical Center 1165
Mapleleaf Capital Corporation 1803
Marcon Capital Corporation 1404
Margiloff & Associates 285
Marine Corps Headquarters – U.S. Department of the Navy 1942
Marine Venture Capital, Inc. 1847
Mark J. Egyed 286
Market Capital Corporation 1429
Marketcorp Venture Associates 1405
Marketeer 2142
Marketing Agency; Inventors Licensing and ....★75
Marketing Center; Emerging Technology Development and – Clemson University 377
Marketing Plan for Start-ups ....★371
Marketing Technological Products to Industry ....★164
Marquette District Office – U.S. Small Business Administration 912
Marquette University – Center for the Study of Entrepreneurship 461
Marshall District Office – U.S. Small Business Administration 1089
Marshall; Jeffrey D. 266
Marshall Space Flight Center; George C. – National Aeronautics and Space Administration 1855
Martindale-Hubbell Law Directory 2143
Marwit Capital Corporation 1338
Maryland Business Assistance Center – Maryland State Department of Economic and Community Development 878
Maryland, College Park Libraries, Engineering & Physical Sciences Library; University of 2676
Maryland/D.C./Delaware; MacRae's Industrial Directory 2133
Maryland Small Business Development Center 462
Maryland Small Business Development Financing Authority 877
Maryland State Department of Economic and Community Development – Maryland Business Assistance Center 878
Maryland State Department of General Services – Purchasing Bureau 879
Maryland State Department of Licensing and Regulations – Division of Occupational and Professional Licensing 880
Maryland State License Bureau 881
Maryland State Treasury Department – State License Bureau 882
Massachusetts Biotechnology Research Park 463
Massachusetts Business Development Corporation 884
Massachusetts Capital Resource Company 885
Massachusetts Community Development Finance Corporation 886
Massachusetts Financial Resources Directory ....★891
The Massachusetts Government Land Bank 887
Massachusetts Industrial Finance Agency 888
Massachusetts Institute of Technology – Innovation Center 464
Massachusetts Institute of Technology – MIT Research Program on the Management of Technology 465
Massachusetts Institute of Technology – Technology Licensing Office 466
Massachusetts Institute of Technology – University Park 467
Massachusetts Microelectronics Center ....★900
Massachusetts Physical Sciences Library; University of 2677
Massachusetts Product Development Corporation 889
Massachusetts/Rhode Island; MacRae's Industrial Directory 2134
Massachusetts Small Business Development Center 468
Massachusetts State Department of Commerce – Massachusetts Suppliers and Manufacturers Matching Service 890
Massachusetts State Department of Commerce – Office of Financial Development 891
Massachusetts State Department of Commerce – Small Business Assistance Division 892
Massachusetts State Department of Revenue – Corporations Bureau 893
Massachusetts State Department of Revenue – Taxpayer Assistance Bureau, Corporation Section 894
Massachusetts State Executive Office of Administration and Finance – Division of Purchasing Agents 895
Massachusetts State Executive Office of Consumer Affairs and Business Regulation – Division of Registration 896
Massachusetts State Executive Office of Consumer Affairs and Business Regulation – Thrift Institution Fund for Economic Development 897
Massachusetts State Office of Facilities Management – Division of Capital Planning and Operations 898
Massachusetts Suppliers and Manufacturers Matching Service – Massachusetts State Department of Commerce 890
Massachusetts Technology Development Corporation 899
Massachusetts Technology Park Corporation 900
Massachusetts Venture Capital Directory ....★891
Massachusetts Venture Capital Fair ....★891
Massachusetts; Venture Economics in ....★2036
Massey Burch Investment Group 1778
Master Search TM 2457
Mastering Reading Through Reading 2360
Materials Alert; High-Tech 2078
Materials Research, Inc. 287
Materials Research Center, Technical Information Center; Allied-Signal Engineered 2530
Materials Technology Laboratory – U.S. Department of the Army 1927
Materiel Command – U.S. Department of the Army 1928
Mathematics Pentathlon 2361
Matrix Partners (Boston) 1519
Matrix Partners (Menlo Park) 1339
Matryxx Corporation 288
Matternville Business and Technology Center 1234
Maui Research and Technology Park 469
Max Think 2362
May Financial Corporation 1804
Mayfield Fund 1340
MBI Venture Capital Investors, Inc. 1567
MBW Management, Inc. 1542
Mc Carthy and Mc Carthy, Library 2613
Mc Gean-Rohco, Inc., Research Library 2614
McAlester Industrial Incubator 1223
McDowell Industrial Business Center 1147
McGee and Associates; Bruce W. 164
McGowan, Leckinger, Berg 1520
McGraw-Hill Publications Online 2458
McGraw-Hill's Biotechnology PatentWatch 2144
McGuire; Kevin C. 275
MCIC Current Awareness Bulletin 2145
McMillen Productivity and Design Center – Indiana Institute of Technology 425
McPherson County Small Business Development Association 849
MCW Research Foundation – Medical College of Wisconsin 470
Meadow Research Venture Capital Group 1597
Meadville Industrial Incubator 1235
Measurement Laboratory; National ....★1874
Mecke Associates, Inc.; Herbert E. 230
Medallion Funding Corporation 1665
Medical College of Wisconsin – MCW Research Foundation 470

Medical Research and Development Command – U.S. Department of the Army 1929
Medical Research and Development Command; Naval – U.S. Department of the Navy 1951
Medical Research Institute; Naval .... ★1951
Medical Research Institute of Chemical Defense .... ★1929
Medical Research Institute of Infectious Diseases .... ★1929
Medical Sciences; National Institute of General – U.S. Department of Health and Human Services 1890
Medical University of South Carolina Library 2615
Melville District Office – U.S. Small Business Administration 990
Memorial University of Newfoundland – P.J. Gardiner Institute for Small Business Studies 471
Memphis and Shelby County Public Library and Information Center, Business/ Science Department 2616
Menlo Ventures 1341
Mental Menus 2363
Merchandise Show; Mid-Year Variety 2509
Merchandise Show; National 2513
Merchandise Show; National Back-to-School 2511
Merchandising Conference and Expo/ Licensing; The International Licensing and 2501
Meridian Capital Corporation 1758
Merrill, Pickard, Anderson, and Eyre (MPAE) 1342
MESBIC Financial Corporation of Dallas 1805
MESBIC Financial Corporation of Houston 1806
MESBIC of San Antonio, Inc. 1807
Mesirow Capital Partners SBIC Ltd. 1458
Metals and Ceramics Information Center .... ★2145
Metalworking Production and Purchasing 2146
Metro-Detroit Investment Company 1543
Metropolitan Capital Corporation 1832
Metropolitan Center for High Technology 1186
Metropolitan Toronto Reference Library, Science & Technology Department 2617
Metropolitan Venture Company, Inc. 1343
MetroTech 472
MG Consulting 289
Miami-Dade Public Library, Business, Science and Technology Department 2618
Miami Valley Capital, Inc. 1730
Miami Valley Research Institute 473
Miami Valley Research Park 474
MICEL-International Market for Creativity and Licensing 2508
Michael Gigliotti and Associates Incorporated 290
Michigan Biotechnology Institute 82
Michigan Energy and Resource Research Association 83
Michigan, Engineering Transportation Library; University of 2678
Michigan Investment Fund 903
Michigan Patent Law Association 84
Michigan Small Business Development Center 475
Michigan State Business Ombudsman 904
Michigan State Department of Commerce – Procurement Assistance 905
Michigan State Department of Commerce – Technology Transfer Network 906
Michigan State Department of Licensing and Regulation 907
Michigan State Department of Management and Budget – Division of Purchasing 908
Michigan State Department of Transportation – Small Business Liaison 909
Michigan State University – Center for the Redevelopment of Industrialized States 476
Michigan Strategic Fund – Seed Capital Companies 910
Michigan Technological University – Bureau of Industrial Development 477
Microelectronics Center of North Carolina 478
Microelectronics Innovation and Computer Research Opportunities – University of California, Los Angeles 588
Mid-America, Inc. 850
Mid-Year Variety Merchandise Show 2509
Midland Public Schools Invention Program 2275
Midwest Inventors Group 85
Midwest Technology Development Institute 479
MIG (Midwest Inventors Group) 85
Military Fiber Optic News 2147
Military Research Letter 2148
Military Robotics 2149
Miller Brewing Company, Scientific and Technical Information Facility 2619
The Million Dollar Idea 2364
Milwaukee District Office – U.S. Small Business Administration 1131
Milwaukee Public Library, Science & Business Division 2620
Mind Games: Puzzles and Logic 2365
Mind Joggers 2366
Mind Movers - Creative Homework Assignments 2367
Mind Your Own Business: A Guide for the Information Entrepreneur .... ★246
Minerva Consulting Group, Inc. 291
Mines; Bureau of – U.S. Department of the Interior 1939
Mini-Invention Innovation Team Contest 2276
Miniatures Industrial Association of America 2510
Minneapolis Business and Technology Center Limited Partnership 1190
Minneapolis District Office – U.S. Small Business Administration 924
Minneapolis Public Library & Information Center, Technology and Science Department 2621
Minnesota Directory of Licenses and Permits .... ★920
Minnesota Entrepreneurs Club 86
Minnesota Fund – Minnesota State Department of Energy and Economic Development 919
Minnesota; Guide to Starting a Business in .... ★920
Minnesota Incorporated; Opportunities 923
Minnesota Inventors Congress 87
Minnesota Inventors Congress, Inventors Resource Center 88
Minnesota; MacRae's Industrial Directory 2135
Minnesota Project Innovation, Inc. 89
Minnesota Science and Technologies Office 914
Minnesota; Selling Your Product to the State of .... ★915
Minnesota Small Business Development Center 480
Minnesota State Department of Administration – Division of Procurement 915
Minnesota State Department of Energy and Economic Development – Community Development Corporation Program 916
Minnesota State Department of Energy and Economic Development – Energy Development Loan Program 917
Minnesota State Department of Energy and Economic Development – Energy Loan Insurance Program 918
Minnesota State Department of Energy and Economic Development – Minnesota Fund 919
Minnesota State Department of Energy and Economic Development – Minnesota State Small Business Assistance Office 920
Minnesota State Department of Energy and Economic Development – Office of Project Management 921
Minnesota State Department of Energy and Economic Development – Small Business Development Loan Program 922
Minnesota State Small Business Assistance Office – Minnesota State Department of Energy and Economic Development 920
Minnesota Student Inventors Congress 2277
Minnesota Technology Corridor 481
Minorities and Women in Business 2150
Minority Broadcast Investment Corporation 1420
Minority Business Enterprise Program – Ohio State Department of Administrative Services 1010
Minority Business Entrepreneur 2151
Minority Development Financing Commission – Ohio State Department of Development 1013
Minority Enterprise Funding, Inc. 1808
Minority Equity Capital Company, Inc. 1666
MIP Equity Advisors, Inc. (Boston) 1521
MIP Equity Advisors, Inc. (Menlo Park) 1344
Missile Command – U.S. Department of the Army 1930
Missile Test Center; Pacific – U.S. Department of the Navy 1965
Mississippi Inventors Workshop 90
Mississippi Research and Development Center 91
Mississippi Research and Technology Park 482
Mississippi Small Business Development Center 483
Mississippi Society of Scientists 92
Mississippi Society of Scientists and Inventors 93
Mississippi State Department of Economic Development – Finance Division 925
Mississippi State General Services Office – Division of Purchasing 926
Mississippi State Research and Development Center 927
Missouri Corporation for Science and Technology – Missouri State Department of Economic Development 931
Missouri Incutech Foundation 1196
Missouri Research Park – University of Missouri 612
Missouri Small Business Development Center 484
Missouri State Department of Economic Development – High Technology Program 930

Missouri State Department of Economic Development – Missouri Corporation for Science and Technology 931
Missouri State Department of Economic Development – Small and Existing Business Development 932
Missouri State Department of Professional Registration 933
Missouri State Office of Administration – Division of Purchasing 934
MIT Research Program on the Management of Technology – Massachusetts Institute of Technology 465
MO-KAN Development, Inc. 935
Mobil Research & Development Corporation, Dallas Research Laboratory, Library.................................2622
Mobion International....................................292
Model Works Industrial Commons.............1236
Mon Valley Renaissance – California University of Pennsylvania.......................364
Moneta Capital Corporation 1767
Monmouth Capital Corporation 1586
Monsanto Company, Patent Department Library..2623
Monsey Capital Corporation.......................1667
Montana College of Mineral Science and Technology, Library 2624
Montana Economic Development Board – Montana State Department of Commerce 943
Montana Science and Technology Alliance – Montana State Department of Commerce 944
Montana State Department of Administration – Purchasing Bureau 940
Montana State Department of Commerce – Business Advocacy and Licensing Assistance 941
Montana State Department of Commerce – Business Development Specialist............942
Montana State Department of Commerce – Montana Economic Development Board 943
Montana State Department of Commerce – Montana Science and Technology Alliance 944
Montana State Department of Natural Resources – Renewable Energy and Conservation Program............................945
Montana State University – Entrepreneurship Center 485
Montgomery Securities 1345
Montpelier District Office – U.S. Small Business Administration 1098
Mora San Miguel Guadalupe Development Corporation...1598
Morgan & Finnegan, Library.......................2625
Morgan, Holland Ventures Corporation......1522
Morgantown Energy Technology Center – U.S. Department of Energy....................1881
Morgantown Energy Technology Center Library; U.S. Department of Energy, ...2691
Morgantown Industrial/Research Park 486
Morgenthaler Ventures..............................1731
Morris J. Root Associates, Limited 293
Motor Enterprises, Inc. 1544
Motor Vehicle Manufacturers Association, Patent Research Library 2626
Motor Vehicle Manufacturers Association Trademark Data Base.........................★2626
Mount Vernon Venture Capital Company...1469
Mountain Ventures, Inc. 1475
MPLA (Michigan Patent Law Association)......84
Multi-Purpose Capital Corporation 1668
Murek Company; Josef K. 269
Murray Associates, Incorporated..................294
Murray Jelling...295
Music; Associated Production 150
Mutual Investment Company, Inc. 1545
Mventure Corp 1809
MVRI (Miami Valley Research Park) 474
Myriad Capital, Inc......................................1346

N

NABWE (National Association of Black Women Entrepreneurs) 94
NAME (National Association of Minority Entrepreneurs (Balch Springs)) 95
NAME (National Association of Minority Entrepreneurs (New York)) 96
NAPPO (National Association of Plant Patent Owners)..97
Narragansett Venture Corporation 1768
NASA Tech Briefs 2152
NASA Technical Reports.........................★1853
Nashville District Office – U.S. Small Business Administration 1076
NATCO (Northern Advanced Technologies Corporation)...505
Natick Research, Development, and Engineering Center – U.S. Department of the Army 1931
National Aeronautics and Space Administration – Ames Research Center 1853
National Aeronautics and Space Administration – Dryden Flight Research Facility 1854
National Aeronautics and Space Administration – George C. Marshall Space Flight Center 1855
National Aeronautics and Space Administration – Goddard Space Flight Center 1856
National Aeronautics and Space Administration – Headquarters Contracts and Grants Division 1857
National Aeronautics and Space Administration – Jet Propulsion Laboratory 1858
National Aeronautics and Space Administration – John F. Kennedy Space Center 1859
National Aeronautics and Space Administration – Langley Research Center 1860
National Aeronautics and Space Administration – Lewis Research Center 1861
National Aeronautics and Space Administration – Lyndon B. Johnson Space Center..1862
National Aeronautics and Space Administration – National Space Technology Laboratories 1863
National Aeronautics and Space Administration – Office of Space Science and Applications 1864
National Aeronautics and Space Administration – Small Business Innovation Research Program 1865
National Aeronautics and Space Administration/Indianapolis Center for Advanced Research 487
National Aeronautics and Space Administration Industrial Applications Center – University of Pittsburgh.............625
National Aeronautics and Space Administration Industrial Applications Center (Boise).......................................★361
National Aeronautics and Space Administration Industrial Applications Center; University of New Mexico ....★617
National Aeronautics and Space Administration Industrial Applications Center; University of North Dakota ....★371
National Aeronautics and Space Administration/Southern Technology Application Center ....★399
National Aeronautics and Space Administration/University of Kentucky Technology Applications Program – University of Kentucky 601
National Association of Black Women Entrepreneurs...94
National Association of Investment Companies—Membership Directory......2153
National Association of Minority Entrepreneurs (Balch Springs) 95
National Association of Minority Entrepreneurs (New York) 96
National Association of Plant Patent Owners 97
National Association of Watch and Clock Collectors, Inc., Watch & Clock Museum Library..2627
National Back-to-School Merchandise Show..2511
National Business Incubation Association......98
National Cancer Institute – U.S. Department of Health and Human Services 1885
National Center for Research in Vocational Education – Ohio State University............513
National Central New Mexico Economic Development District.............................1599
National Centre for Management Research and Development....................................488
National City Capital Corporation 1732
National Coalition for Science and Technology ....★2157
National Computer and Telecommunication Laboratory ....★1874
National Congress of Inventors Organizations 99
National Council for Industrial Innovation 100
National Council of Patent Law Associations 101
National Directory of Incubators..................★98
National Engineering Laboratory............★1874
National Eye Institute – U.S. Department of Health and Human Services 1886
National Fall Inventor's Conference ....★8
National Heart, Lung, and Blood Institute – National Institutes of Health 1866
National Highway Traffic Safety Adminstration – U.S. Department of Transportation 1971
National Innovation Workshops..................★99
National Institute of Allergy and Infectious Diseases – U.S. Department of Health and Human Services 1887
National Institute of Arthritis and Musculoskeletal and Skin Diseases – U.S. Department of Health and Human Services..1888
National Institute of Dental Research – U.S. Department of Health and Human Services..1889
National Institute of General Medical Sciences – U.S. Department of Health and Human Services 1890

National Institute of Neurological and Communicative Disorders and Stroke – U.S. Department of Health and Human Services 1891
National Institute of Standards and Technology – U.S. Department of Commerce 1874
National Institute of Standards and Technology, Office of Energy Related Inventions – U.S. Department of Commerce 1875
National Institutes of Health – National Heart, Lung, and Blood Institute 1866
National Inventors Cooperative Association 102
National Inventors Expo 2512
National Inventors Foundation 103
National Inventors Hall of Fame Foundation .... ★101
National Measurement Laboratory .... ★1874
National Merchandise Show 2513
National Oceanic and Atmospheric Administration – U.S. Department of Commerce 1876
National Patent Council 104
National Premium Incentive Show 2514
National Research Council of Canada 489
National Science Foundation – Division of Policy Research and Analysis 1867
National Science Foundation – Industry/University Cooperative Research Centers 1868
National Science Foundation – Office of Small Business Research and Development 1869
National Science Foundation – Small Business Innovation Research Program 1870
National Society of Inventors 105
National Space Technology Laboratories – National Aeronautics and Space Administration 1863
National Stationery Show 2515
National Venture Capital Association 106
National Venture Capital Association—Membership Directory 2154
National Video Clearinghouse, Inc., NVC Library and Information Service 2628
Naugatuck Valley Revolving Loan Fund – Connecticut State Department of Economic Development 729
Navajo Small Business Development Corporation 1266
Naval Aerospace Medical Research Laboratory .... ★1951
Naval Air Development Center – U.S. Department of the Navy 1943
Naval Air Engineering Center – U.S. Department of the Navy 1944
Naval Air Propulsion Center – U.S. Department of the Navy 1945
Naval Air Systems Command – U.S. Department of the Navy 1946
Naval Avionics Center – U.S. Department of the Navy 1947
Naval Civil Engineering Laboratory – U.S. Department of the Navy 1948
Naval Coastal Systems Center – U.S. Department of the Navy 1949
Naval Dental Research Institute .... ★1951
Naval Diving and Salvage Training Center .... ★1949
Naval Explosive Ordnance Disposal Technology Center – U.S. Department of the Navy 1950
Naval Health Research Center .... ★1951
Naval Medical Research and Development Command – U.S. Department of the Navy 1951
Naval Medical Research Institute .... ★1951
Naval Ocean Systems Center – U.S. Department of the Navy 1952
Naval Oceanographic Office – U.S. Department of the Navy 1953
Naval Ordnance Missile Test Station – U.S. Department of the Navy 1954
Naval Ordnance Station – U.S. Department of the Navy 1955
Naval Research Laboratory – U.S. Department of the Navy 1956
Naval Sea Systems Command – U.S. Department of the Navy 1957
Naval Submarine Medical Research Laboratory .... ★1951
Naval Surface Warfare Center – U.S. Department of the Navy 1958
Naval Training Systems Center – U.S. Department of the Navy 1959
Naval Underwater Systems Center – U.S. Department of the Navy 1960
Naval Weapons Station – U.S. Department of the Navy 1961
Naval Weapons Support Center – U.S. Department of the Navy 1962
Naval Weapons Systems Center – U.S. Department of the Navy 1963
Nazem and Company 1669
NBIA (National Business Incubation Association) 98
NBIA Review 2155
NCII (National Council for Industrial Innovation) 100
NCIO (National Congress of Inventors Organizations) 99
NCNB SBIC Corporation 1717
NCNB Venture Company LP 1718
NCPLA (National Council of Patent Law Associations) 101
NCPLA Newsletter 2156
NCST Quarterly Briefing 2157
Nebraska Energy Fund – Nebraska State Energy Office 952
Nebraska International Trade Directory .... ★951
Nebraska Investment Finance Authority 947
Nebraska, Lincoln, Engineering Library; University of 2679
Nebraska Manufacturers; Directory of .... ★951
Nebraska Secretary of State 948
Nebraska Small Business Development Center 490
Nebraska State Administrative Service Department – State Purchasing Section 949
Nebraska State Department of Economic Development – Research and Development Authority 950
Nebraska State Department of Economic Development – Small Business Division 951
Nebraska State Energy Office – Nebraska Energy Fund 952
Nebraska State Ethanol Authority Development Board 953
Nebraska Statistical Handbook .... ★951
Nebraska Technical Assistance Center – University of Nebraska, Lincoln 615
NECHIA Trade Show/North East Craft and Hobby 2516
Neighborhood Fund, Inc. 1459
Nelson Capital Corporation 1670
Neosho Basin Development Company 851
NERAC, Inc. 491
Neurological and Communicative Disorders and Stroke; National Institute of – U.S. Department of Health and Human Services 1891
Neurological Research & Development 296
Nevada Innovation & Technology Council 107
Nevada, Reno, Government Publications Department; University of 2680
Nevada Small Business Development Center 492
Nevada State Commission on Economic Development 955
Nevada State Department of General Services – Division of Purchasing 956
Nevada State Legislative Counsel Bureau – Research Division 957
Nevada State Office of Community Services – Procurement Outreach Program 958
Nevada State Office of Community Services – Small Business Revitalization Program 959
Nevitt; H.J. Barrington 226
New Building Workspace for Women 1194
New Castle County Economic Development Corporation 744
New England Capital Corporation 1523
New England MESBIC, Inc. 1524
New Enterprise Associates (Baltimore) 1488
New Enterprise Associates (Menlo Park) 1347
New Enterprise Associates (New York City) 1671
New Enterprise Associates (San Francisco) 1348
New Enterprise Associates (Westlake Village) 1349
New Enterprise Center .... ★548
The New Enterprise Institute – University of Southern Maine 635
New From Europe 2158
New From Japan 2159
New From U.S. 2160
New Frontiers/Nuevas Fronteras 2368
New Hampshire Small Business Development Center 493
New Hampshire State Department of Administration Services – Division of Plant and Property Management 962
New Hampshire State Department of Resources and Economic Development – Industrial Development Office 963
New Hampshire, University Library, Patent Collection; University of 2681
New Hampshire/Vermont; MacRae's Industrial Directory Maine/ 2132
New Jersey Commission on Science and Technology 966
New Jersey; Corporation for Business Assistance in 965
New Jersey Economic Development Authority 967
New Jersey Institute of Technology – Center for Information Age Technology 494
New Jersey; MacRae's Industrial Directory 2136
New Jersey Set-Aside Program – New Jersey State Department of Commerce and Economic Development 968
New Jersey Small Business Development Center 495
New Jersey State Department of Commerce and Economic Development – New Jersey Set-Aside Program 968
New Jersey State Department of Commerce and Economic Development – Office of Small Business Assistance 969

New Jersey State Department of the Treasury – Purchase Bureau 970
New Kukje Investment Company 1350
New Market Enterprises ....★2128
New Mexico Business Development Corporation 1600
New Mexico Business Innovation Center, Inc. 1200
New Mexico, Centennial Science and Engineering Library; University of 2682
New Mexico R&D Forum 2161
New Mexico Research and Development Institute 973
New Mexico Research Park 496
New Mexico State Department of Economic Development and Tourism.......................974
New Mexico State Department of Regulation and Licensing 975
New Mexico State Purchasing Division 976
New Mexico State University – Arrowhead Research Park 497
New Orleans District Office – U.S. Small Business Administration 869
New-Penn-Del Regional Minority Purchasing Council, Inc. 1035
New Product Development.........................2162
New Product Digest; Appliance..................1988
New Product Monthly Reports....................2163
New Product News; Gorman's 2070
New Products and Processes 2164
New Products and Technology From the U.S.A.; Finding and Licensing...............★326
New Products Bulletin 2165
New Products News 2166
New Products Newsletter; Foreign Trade Fairs..2067
New Technology Week 2167
A New Way to Use Your Bean - Developing Thinking Skills in Children......................2369
New West Partners (Newport Beach).........1351
New West Partners (San Diego) 1352
New York City District Office – U.S. Small Business Administration 991
New York International Gift Fair 2517
New York Invention Program; Western 2282
New York Public Library, Annex Section, Patents Collection.................................2629
New York Small Business Innovation Research Promotion Program 978
New York Society of Professional Inventors..108
New York State Department of Commerce – Division for Small Business......................979
New York State Department of General Services – Division of Purchasing 980
New York State Job Development Authority 981
New York State Library, Sciences/Health Sciences/Technology Reference Services..2630
New York State; MacRae's Industrial Directory 2137
New York State Office of Permits and Regulatory Assistance 982
New York State Science and Technology Foundation – Centers for Advanced Technology Program 983
New York State Science and Technology Foundation – Corporation for Innovation Development 984
New York State Science and Technology Foundation – Industrial Innovation Extension Service 985
New York State Science and Technology Foundation – Regional Technology Development Corporation.......................986
New York State Small Business Development Center...............................498
New York University – Center for Entrepreneurial Studies 499
New York University – Initiatives for Not-for-Profit Entrepreneurship...........................500
Newark District Office – U.S. Small Business Administration 972
Newark Public Library, Sciences Division 2631
Newsletter for Independent Businessowners 2168
Nicoletti Productions, Inc. 297
NIF (National Inventors Foundation) 103
Nilson's Inventors League 298
NL Industries, Inc., Spencer Kellogg Products, Research Center Library........2632
NL Industries, Inc., Technology Systems, Technical Information Center................2633
Noble Science and Engineering Library; Arizona State University, Daniel E. 2533
Nondestructive Testing Information Analysis Center...★2169
Noone Associates 299
Noro Capital Ltd. 1672
Noro-Moseley Partners 1439
Norstar Venture Capital..............................1673
North America Investment Corporation 1764
North American Funding Corporation.........1674
North Augusta Incubator 1248
North Bennington Business Incubator........1256
North Carolina Biotechnology Center...........995
North Carolina; Microelectronics Center of...478
North Carolina Small Business Development Center...................................501
North Carolina/South Carolina/Virginia; MacRae's Industrial Directory...............2138
North Carolina State Department of Administration – Purchase and Contract Division 996
North Carolina State Department of Commerce – Business Information Referral Center 997
North Carolina State Department of Commerce – Industrial Financing Section...998
North Carolina State Department of Commerce – Small Business Development Division.............................999
North Carolina State House Committee on Small Business 1000
North Carolina State Senate Committee on Small Business 1001
North Carolina State University, D.H. Hill Library, Documents Department............2634
North Carolina Technological Development Authority – Incubator Facilities Program 1002
North Carolina Technological Development Authority – Innovation Research Fund ...1003
North Dakota Small Business Development Center...502
North Dakota State Department of Management and Budget – Division of Purchasing..1005
North Dakota State Economic Development Commission – Entrepreneurial Assistance 1006
North Dakota State Economic Development Commission – Federal Procurement......1007
North Dakota State Economic Development Commission – Financial Assistance.......1008
North East Craft and Hobby Industry Association ....★2516
North East Tier Advanced Technology Center...1237
North Florida Entrepreneurial Network......★402
North Riverside Capital Corporation...........1440
North Star Ventures, Inc. 1554
North Street Capital Corporation 1675
Northeast Louisiana Incubator Center........1171
Northeast Louisiana University – Louisiana Small Business Development Center 503
Northeast Small Business Investment Corporation..1525
Northeastern Capital Corporation...............1406
Northeastern Texas Small Business Development Center...............................504
Northern Advanced Technologies Corporation..505
Northern Illinois University – Technology Commercialization Center 506
Northern Kentucky University Foundation Research/Technology Park 507
Northern Pacific Capital Corporation 1746
Northland Capital Corporation...................1555
Northstar Community Development Corporation..1191
Northwest Business Investment Corporation..1353
Northwest Wisconsin Business Development Corporation – Northwest Wisconsin Business Development Fund 1127
Northwest Wisconsin Business Development Fund – Northwest Wisconsin Business Development Corporation..1127
Northwestern Texas Small Business Development Center...............................508
Northwestern University – Technology Innovation Center 509
Northwestern University/Evanston Research Park 510
Norwest Venture Capital Management, Inc. (Minneapolis) 1556
Norwest Venture Capital Management, Inc. (Portland) 1747
Norwest Venture Capital Management, Inc. (Scottsdale) 1267
Nova Scotia Research Foundation Corporation, Library..............................2635
NPA/Plus (PTS New Product Announcements/Plus) 2465
NPC (National Patent Council).....................104
NRC Directory of Research Activities........★489
NRC (National Research Council of Canada)..489
NTIAC Newsletter 2169
NUANS...2459
Nuclear Agency; Defense – U.S. Department of Defense..........................1877
Nuclear Energy; Assistant Secretary for – U.S. Department of Energy....................1880
NVCA (National Venture Capital Association) 106
NYBDC Capital Corporation.......................1676

O

O.A. Battista Research Institute 300
O A Laboratories and Research, Inc. 301
Oak Investment Partners (Menlo Park) 1354
Oakland Technology Park 511
Occupations and Professions Division – Kentucky State Finance and Administration Cabinet............................862
Ocean Science and Technology Park; Hawaii 418
Ocean State Business Development Authority 1057

Ocean Systems Center; Naval - U.S. Department of the Navy 1952
Oceanographic Office; Naval - U.S. Department of the Navy 1953
Oceanography..2370
O'Conor and Associates; Charles W. - Wright Wyman, Inc. 185
Odyssey of the Mind: Problems to Develop Creativity..2371
Office for Patents, Copyrights, and Licensing - Harvard University 417
Office for Technology, Licensing, and Industry-Sponsored Research - Harvard Medical School 416
Office of Cooperative Research - Yale University..675
Office of Corporate Programs and Technology - University of Pennsylvania 621
Office of Energy Related Inventions; National Institute of Standards and Technology, - U.S. Department of Commerce 1875
Office of Energy Research - U.S. Department of Energy...........................1882
Office of Financial Development - Massachusetts State Department of Commerce 891
Office of Grants and Program Systems - U.S. Department of Agriculture 1872
Office of Human Development Services - U.S. Department of Health and Human Services..1892
Office of Industrial Research - University of Manitoba...603
Office of Industrial Research and Technology Transfer - University of Wisconsin—Milwaukee...........................647
Office of Material Management Administration - District of Columbia Department of Administrative Services....749
Office of Minority Business Enterprise - Pennsylvania State Department of Commerce 1039
Office of Naval Research - U.S. Department of the Navy..1964
Office of Patent and Copyright Administration - University of Southern California 632
Office of Patents and Licensing - Cornell Research Foundation 387
Office of Patents and Licensing - University of Colorado Foundation Inc......................590
Office of Patents and Licensing - University of Minnesota 609
Office of Procurement - California State and Consumer Services Agency 704
Office of Project Management - Minnesota State Department of Energy and Economic Development..........................921
Office of Research and Sponsored Programs - Lehigh University.................454
Office of Research and Technology Transfer Administration - University of Minnesota 610
Office of Research Services - University of Alberta 582
Office of Science and Technology Development - Columbia University 384
Office of Scientific Research - U.S. Department of the Air Force..................1905
Office of Small and Minority Business Assistance - State of South Carolina.....1066
Office of Small Business - California State Department of Commerce 706
Office of Small Business - Tennessee State Department of Economic and Community Development 1073
Office of Small Business and Financial Services - Virginia State Department of Economic Development........................1110
Office of Small Business Assistance - New Jersey State Department of Commerce and Economic Development...................969
Office of Small Business Research and Development - National Science Foundation...1869
Office of Space Science and Applications - National Aeronautics and Space Administration.......................................1864
Office of Sponsored Program Development - State University of New York at Binghamton 559
Office of Sponsored Programs - University of Pennsylvania 622
Office of Sponsored Research - University of Southern Maine 636
Office of State Purchasing - Ohio State Department of Administrative Services...1011
Office of Technology Licensing - Stanford University..557
Office of Technology Transfer - Louisiana State University.......................................457
Office of Technology Transfer - University of Calgary 586
Office of Technology Transfer - University of Washington..643
Office of the Vice Chancellor for Research - University of Illinois at Urbana-Champaign 598
Office of Vice President for Research, Graduate Studies, and International Programs - Oregon State University........520
Official Gazette of the United States Patent and Trademark Office: Patents.............2170
Official Gazette of the United States Patent and Trademark Office: Trademarks.......2171
Ohio Inventors Association..........................109
Ohio Small Business Development Center...512
Ohio State Department of Administrative Services - Minority Business Enterprise Program...1010
Ohio State Department of Administrative Services - Office of State Purchasing....1011
Ohio State Department of Development - Economic Development Financing Division 1012
Ohio State Department of Development - Minority Development Financing Commission..1013
Ohio State Department of Development - Small and Developing Business Division 1014
Ohio State University - National Center for Research in Vocational Education 513
Ohio State University Libraries, Information Services Department 2636
Ohio State University Research Park 514
Ohio University - Innovation Center and Research Park.......................................515
OIA (Ohio Inventors Association) 109
Oil, Gas & Petrochem Equipment..............2172
Oklahoma City District Office - U.S. Small Business Administration 1026
Oklahoma Industrial Finance Authority.......1019
Oklahoma Inventors Congress....................110
Oklahoma Small Business Development Center...516
Oklahoma State Department of Commerce - Business Development Division..........1020
Oklahoma State Department of Commerce - Division of Community Affairs and Development 1021
Oklahoma State Department of Commerce - Information and Research Division 1022
Oklahoma State Department of Commerce - Procurement Section 1023
Oklahoma State Department of Commerce - Special Services Division 1024
Oklahoma State Department of Public Affairs - Division of Purchasing.............1025
Oklahoma State University - Technology Transfer Center.......................................517
Oklahoma State University, Documents Department...2637
Old Stone Capital Corporation....................1769
Olivetti Associates, Inc.; A. 137
OM-AHA! Problems to Develop Creative Thinking Skills..2372
Omaha District Office - U.S. Small Business Administration...954
Omaha Inventors Club 111
OMEC International, Inc. 302
Omego Capital Corporation........................1810
190 Willow Avenue Industrial Incubator 1207
One-Stop Permitting - Kansas State Department of Commerce 842
Operational Test and Evaluation Center - U.S. Department of the Air Force...........1906
Opportunities Minnesota Incorporated 923
Opportunity Capital Corporation.................1355
Orange Nassau Capital Corporation 1526
ORBPAT 2460
Ordnance Missile Test Station; Naval - U.S. Department of the Navy 1954
Ordnance Station; Naval - U.S. Department of the Navy...1955
Oregon Business Development Fund - Oregon State Economic Development Department..1033
Oregon Graduate Center Science Park........518
Oregon International Trade Directory......★1031
Oregon; MacRae's Industrial Directory.......2139
Oregon Manufacturers; Directory of........★1031
Oregon Productivity and Technology Center - Oregon State University 521
Oregon Resource and Technology Development Corporation......................1027
Oregon Small Business Development Center..519
Oregon; Starting a Business in...............★1031
Oregon State Department of Commerce - Corporation Division 1028
Oregon State Department of Economic Development - Regulation Assistance Service..1029
Oregon State Department of General Services - Division of Purchasing..........1030
Oregon State Economic Development Department - Business Development Division 1031
Oregon State Economic Development Department - Financial Services Division 1032
Oregon State Economic Development Department - Oregon Business Development Fund 1033
Oregon State Library 2638
Oregon State University - Office of Vice President for Research, Graduate Studies, and International Programs........520
Oregon State University - Oregon Productivity and Technology Center 521

Organizational Efficiency Research Group – University of Quebec at Hull....................628
OTTO News..2173
Oxford Partners (Santa Monica)..................1356
Oxford Partners (Stamford)........................1407
Oxmoor West Industrial Park........................683

P

P.J. Gardiner Institute for Small Business Studies – Memorial University of Newfoundland..471
Pacific Business Center Program – University of Hawaii 593
Pacific Capital Fund, Inc.............................1357
Pacific Missile Test Center – U.S. Department of the Navy 1965
Pacific Venture Capital Ltd. 1442
Packaged Facts 303
Palm Beach Society of American Inventors..112
Palmer Service Corporation 1527
Palms and Company, Inc. 1839
Pan American University – Center for Entrepreneurship and Economic Development 522
Pan-Atlantic Consultants.............................304
Pan Hellenic Society Inventors of Greece in U.S.A. 113
Pan Pac Capital Corporation.......................1677
Paoli Technology Enterprise Center...........1238
PAPERCHEM 2461
Partners...2174
PATCOP (LEXIS Federal Patent, Trademark, & Copyright Library)............2454
PATDATA...2462
PATDPA (Deutsche Patent Datenbank).....2436
Patent Abstracts; CLAIMS/U.S...................2429
Patent Abstracts in English Data Base; Chinese 2422
Patent Administrators; Society of University..122
Patent Agents; Directory of Intellectual Property Lawyers and...........................2037
Patent Alert ....★142
Patent and Copyright Administration; Office of – University of Southern California.......632
Patent and Technical Communications Services; 3M,...2663
Patent and Trademark Association; International..42
Patent and Trademark Library; WESTLAW Copyright, ...2486
Patent and Trademark Office: Patents; Official Gazette of the United States......2170
Patent and Trademark Office; Roster of Attorneys and Agents Registered to Practice before the United States..........2204
Patent & Trademark Office, Scientific Library; United States 2670
Patent and Trademark Office Society...........114
Patent and Trademark Office: Trademarks; Official Gazette of the United States......2171
Patent Assistance Program – Georgia Institute of Technology............................411
Patent Attorneys...2175
Patent Bulletin; European........................★2440
Patent Classification System; U.S. 2483
Patent Collection; University of New Hampshire, University Library,...............2681
Patent Costs; Lowering ....★266
Patent Council; National.............................104
Patent Datenbank; Deutsche 2436
Patent Department Library; Monsanto Company, ...2623
Patent Dispute Resolutions; Enhancing Options in..★234
Patent Documents Reclassified According to the IPS; Computerized Administration of..2434
Patent; How to Protect an Idea Before........★60
Patent Information Clearinghouse; Sunnyvale...2658
Patent Information Fair..............................2518
Patent Information Organization; Japan...★2451
Patent Information; World..........................2261
Patent Law Association; Michigan.................84
Patent Law Associations; National Council of..101
Patent Library; Consumer and Corporate Affairs Canada, ...2572
Patent Library; Greyhound Corporation,.....2589
Patent Licensing Gazette; World Technology/...2262
Patent Management, Inc.; Hillard-Lyons 234
Patent Model Foundation; U.S. 132
Patent News...★266
The Patent Office (London) ....★2482
Patent Office Professional Association.........115
Patent Owners; National Association of Plant 97
Patent Pending..2373
Patent Report TM; World Biolicensing & 2259
Patent Reporter; Canadian.........................2421
Patent Research Library; Motor Vehicle Manufacturers Association, ...2626
Patent, Trademark & Copyright Journal; BNA's..1998
Patent, Trademark & Copyright Journal (Online); BNA's 2417
Patent, Trademark, & Copyright Library; LEXIS Federal 2454
Patent, Trademark and Copyright Office – University of California, Berkeley 587
Patent, Trademark and Copyright Research Foundation, Library...............................2639
The Patent Trader......................................2176
Patent Trends..★104
Patenting Your Own Invention..................★164
Patentrol Corporation 305
Patents and Licensing; Office of – Cornell Research Foundation 387
Patents and Licensing; Office of – University of Colorado Foundation Inc......................590
Patents and Licensing; Office of – University of Minnesota 609
Patents and Trademarks, Library; Upjohn Company, Corporate 2688
Patents Collection; New York Public Library, Annex Section,..2629
Patents, Copyrights, and Licensing; Office for – Harvard University 417
Patents; Current Energy.............................2028
Patents; Evaluating ....★299
Patents Files; U.S.......................................2484
Patents Index; World 2488
Patents Library; Du Pont Canada, Inc., ...2578
Patents Newsletter; Lasers/Electro-Optics............................2124
Patents Newsletter; Robotics......................2202
Patents Newsletter; Semiconductors/ICs...2208
Patents Quarterly; United States.................2242
Patents; Stack Gas Control.........................2216
Patents; The Business Side of ....★299
PatentWatch; McGraw-Hill's Biotechnology...2144
Pathfinder Venture Capital Funds 1557
PBC Venture Capital, Inc............................1358
PCF Venture Capital Corporation...............1359
Pennie & Edmonds, Law Library 2640
Pennsylvania Capital Loan Fund – Pennsylvania State Department of Commerce 1040
Pennsylvania Energy Development Authority – Pennsylvania State Department of Commerce 1041
Pennsylvania Industrial Development Authority – Pennsylvania State Department of Commerce 1042
Pennsylvania; MacRae's Industrial Directory...2140
Pennsylvania Small Business Development Center...523
Pennsylvania; Starting a Small Business in..★1044
Pennsylvania State Department – Bureau of Professional and Occupational Affairs...1036
Pennsylvania State Department of Commerce – Ben Franklin Partnership 1037
Pennsylvania State Department of Commerce – Business Infrastructure Development Program.............................1038
Pennsylvania State Department of Commerce – Office of Minority Business Enterprise 1039
Pennsylvania State Department of Commerce – Pennsylvania Capital Loan Fund 1040
Pennsylvania State Department of Commerce – Pennsylvania Energy Development Authority 1041
Pennsylvania State Department of Commerce – Pennsylvania Industrial Development Authority 1042
Pennsylvania State Department of Commerce – Revenue Bond and Mortgage Program.................................1043
Pennsylvania State Department of Commerce – Small Business Action Center...1044
Pennsylvania State Department of General Services – Vendor Information and Support Division..................................1045
Pennsylvania State Department of Transportation – Disadvantaged and Women Business Enterprises...............1046
Pennsylvania State University – Technology Transfer Office 524
Pennsylvania State University Libraries, C207 Pattee Library, Documents Section...2641
Perception, Inc. 2374
Permissions, Inc.; BZ/Rights & 171
Peter J. Schmitt SBIC, Inc. 1759
Peterson Finance and Investment Company 1460
Petro-Canada Products Inc., Technical Library...2642
Petrochem Equipment; Oil, Gas &.............2172
Petrolite Corporation, Information Center...2643
PharmIndex..2177
PharmPat 2463
Philadelphia District Office – U.S. Small Business Administration 1048
Philadelphia Ventures 1760
Phillips & Associates, Ltd.; Holland W. 236
Phillips Petroleum Company R&D Library...2644
Phoenix District Office – U.S. Small Business Administration 694
Photocopy Authorization Report................2178
Photocopy Fee Catalog; Publishers' ....★21
PHSIG (Pan Hellenic Society Inventors of Greece in U.S.A.)...................................113

Physics Laboratory; Applied – Johns Hopkins University 443
Pierre Funding Corporation 1678
Pillsbury Company, Technical Information Center 2645
Pioneer Associates 1679
Pioneer Capital Corporation 1680
Pioneer Country Development, Inc. 852
Pittsburg State University – Center for Technology Transfer 525
Pittsburg State University – Institute for Economic Development 526
Pittsburgh District Office – U.S. Small Business Administration 1049
Pittsburgh Energy Technology Center – U.S. Department of Energy 1883
Pixel Instruments Corp. 306
Planning and Economic Development – Alabama State Department of Economic and Community Affairs 677
Planning for Thinking 2375
Plant Patent Owners; National Association of 97
Plastics Products; Designing ....★323
Platt Saco Lowell Corporation, Engineering Library 2646
Playthings 2179
PNC Venture Capital Group 1761
Policy Research and Analysis; Division of – National Science Foundation 1867
Polymer Laboratories, Inc.; Princeton 307
POPA (Patent Office Professional Association) 115
Portland District Office – U.S. Small Business Administration 1034
Positive Enterprises, Inc. 1360
Power Ventures, Inc. 1272
Prane; Joseph W. 270
Pratt's Guide to Venture Capital Sources 2180
Prebluda; Harry 228
Premium Incentive Show 2519
Premium Incentive Show; National 2514
Primus Capital Fund 1733
Princeton Capital Corporation 1198
Princeton Forrestal Center – Princeton University 527
Princeton Polymer Laboratories, Inc. 307
Princeton University – Princeton Forrestal Center 527
Private Placements – State of Wisconsin Investment Board 1128
Problem Solving and Comprehension 2376
Problem Solving in Science 2377
Problems! Problems! Problems! Discussions and Activities Designed to Enhance Creativity 2378
Probol & Associates; M.W. 282
Process Development; Waterloo Centre for 663
Processes; New Products and 2164
Procter & Gamble Company, Cellulose & Specialties Division Technical Information Services 2647
Procurement Assistance – Michigan State Department of Commerce 905
Procurement Assistance Program – Illinois State Department of Commerce and Community Development 802
Procurement Bureau – Wisconsin State Department of Administration 1133
Procurement Office – South Carolina State Budget and Control Board 1063
Procurement Operations Division – U.S. Department of Transportation 1972
Procurement Outreach Program – Nevada State Office of Community Services 958
Procurement Section – Oklahoma State Department of Commerce 1023
Product Alert 2181
Product Alert; International 2106
Product Announcements/Plus; PTS New 2465
Product Bulletin; Industrial 2091
Product Design and Development 2182
Product Development; New 2162
Product Digest; Appliance New 1988
Product Engineering 2183
Product Ideas; Industrial 2092
Product Improvement Checklist 2379
Product Innovation Center 757
Product Monthly Reports; New 2163
Product News; EE: Electonic/Electrical 2044
Product News; Extended Care 2058
Product News; Gorman's New 2070
Product News; Venture/ 2252
Product Newsletter; International New 2104
Production Rights; LIZ '89 Licensing Fair for Design Ideas, Publication, and 2507
Productivity and Design Center; McMillen – Indiana Institute of Technology 425
Products and Processes; New 2164
Products & Services; IPS Industrial 2116
Products and Technology From the U.S.A.; Finding and Licensing New ....★326
Products Bulletin; New 2165
Products Database; Thomas New Industrial 2474
Products; Dental Lab 2031
Products News; New 2166
Products Newsletter; Foreign Trade Fairs New 2067
Products Newsletter; International New 2105
Professional and Occupational License Division – South Dakota State Department of Commerce and Regulation 1069
Professional and Vocational Licensing Division – Hawaii State Department of Commerce and Consumer Affairs 780
Progress Center: University of Florida Research and Technology Park 528
Project: Problem Solving 2380
Project Success Enrichment 2381
Project Summaries 2184
Prolific Thinkers Guide 2382
Propiedad Industrial (RPI); Registro de la ....★2423
Prospect Group, Inc. 1681
Prospectus 2185
Protecting & Selling Ideas & Inventions ....★199
Providence District Office – U.S. Small Business Administration 1062
Providence Public Library, Knight Memorial Branch 2648
Prudential Venture Capital 1682
PTC Research Foundation 529
PTD-BASEN 2464
PTOS (Patent and Trademark Office Society) 114
PTS New Product Announcements/Plus 2465
PTS PROMT 2466
Public Information Contact Directory 2186
Public Library of Cincinnati and Hamilton County, Science and Technology Department 2649
Publishers' Photocopy Fee Catalog ....★21
Pueblo Memorial Airport Industrial Park 720
Puerto Rico Department of Commerce 1052
Puerto Rico Economic Development Administration 1053
Puerto Rico General Services Administration – Division of Purchasing, Services, and Supply 1054
Puerto Rico Licensing Administration 1055
Puerto Rico Small Business Development Center 530
Pugh, Jr.; Charles Scott 183
Pump News 2187
Pumps; IMPACT 2086
Purchase and Contract Division – North Carolina State Department of Administration 996
Purchase Bureau – New Jersey State Department of the Treasury 970
Purchases Office – Rhode Island State Department of Administration 1059
Purchasing and Printing Office – South Dakota State Department of Administration 1068
Purchasing Bureau – Maryland State Department of General Services 879
Purchasing Bureau – Montana State Department of Administration 940
Purchasing Office – Alaska State Department of Administration 686
Purchasing Office – Arizona State Department of Administration 690
Purchasing Office – Arkansas State Department of Finance and Administration 700
Purdue Industrial Research Park 531
Purdue University – Technical Information Service 532

Q

Quaker Chemical Corporation, Information Resources Center 2650
Quarterly Counselor 2188
Queens Borough Public Library, Science & Technology Division 2651
Questech Capital Corporation 1683
Quizzles - Logic Problem Puzzles 2383

R

R and R Financial Corporation 1684
R&D Alert; Space 2214
R&D Firms Directory; Washington D.C. Area 2253
R&D Funding Corporation 1361
R&D Funding Opportunities; Small Business Guide to Federal 2211
R&D; Inside 2097
R&D Management Digest 2189
R&D Scoreboard Issue; Business Week— 2007
R&D Village 533
R.J. Reynolds Tobacco Company, R&D Scientific Information Services Library 2652
Railroad Administration; Federal – U.S. Department of Transportation 1970
Rain Hill Group, Inc. 308
Rand SBIC, Inc. 1685
RAPRA Tradenames 2467
RAPTN (RAPRA Tradenames) 2467
RCS Associates 309
Realty Growth Capital Corporation 1686
Red River Ventures, Inc. 1811
Reed Institute of Research; Walter ....★1929
Reedy River Ventures, Inc. 1773
Reference Desk 310
Regional Financial Enterprises (Ann Arbor) 1546
Regional Financial Enterprises (New Canaan) 1408

Regional Technology Development Corporation – New York State Science and Technology Foundation 986
Register of Regulations – Virginia State Code Commission 1107
Registro de la Propiedad Industrial (RPI) ....★2423
Regulation Assistance Service – Oregon State Department of Economic Development 1029
Regulatory Boards – Tennessee State Department of Commerce and Insurance...1072
Renewable Energy and Conservation Program – Montana State Department of Natural Resources..................................945
Reno District Office – U.S. Small Business Administration..961
Rensselaer Polytechnic Institute – George M. Low Center for Industrial Innovation ...534
Rensselaer Polytechnic Institute – Rensselaer Technology Park..................535
Rensselaer Technology Park – Rensselaer Polytechnic Institute...............................535
Republic Venture Group, Inc. 1812
Rescorp International...................................311
Research and Business Park; Rochester Institute of Technology.............................543
Research and Development.......................2190
Research and Development Authority – Nebraska State Department of Economic Development 950
Research & Development Authority; Tampa Bay Area 565
Research and Development Center; Mississippi 91
Research and Development Directorate – U.S. Department of the Army.................1932
Research & Development Directory...........2191
Research and Development Magazine; Industrial...2093
Research and Development Park; University Center 579
Research & Development Telephone Directory 2192
Research and Development; Tennessee Center for..566
Research & Invention................................2193
Research and Sponsored Programs; Office of – Lehigh University...............................454
Research and Technology; Directory of American 2035
Research and Technology Foundation; Texas..572
Research & Technology Management.......2194
Research and Technology Park – Washington State University...................661
Research and Technology Park; Maui 469
Research and Technology Park; Mississippi 482
Research and Technology Park; Progress Center: University of Florida 528
Research and Technology Park; Utah State University – Utah State University............654
Research and Technology Transfer Administration; Office of – University of Minnesota 610
Research Assistance Program; Industrial ....★489
Research Associates; Gellman 221
Research Authority; South Carolina 549
Research Center; Table Mountain – Colorado School of Mines........................380
Research Centers Directory.......................2195
Research Centers Directory; International..2107
Research Companies—Members Directory; American Association of Small 1982
Research Connection...............................★661
Research Corporation; British Columbia 363
Research Corporation Technologies............536
Research Corporation; University of Tennessee 637
Research Corridor; Rio Grande....................540
Research Corridor; Southern Willamette......555
Research Council of Canada; National 489
Research Division – Nevada State Legislative Counsel Bureau 957
Research '88; SUNY..................................2219
Research Foundation, Inc.; University of Georgia...592
Research Foundation; MCW – Medical College of Wisconsin 470
Research Foundation; PTC.........................529
Research Foundation; University of Delaware 591
Research Horizons....................................2196
Research in Progress; Federal...................2061
Research Institute for the Management of Technology 537
Research Institute; Miami Valley 473
Research Institute of Environmental Medicine ....★1929
Research Laboratory; Basic Industry ....★510
Research Money...2197
Research Park; Arizona State University 351
Research Park; Arrowhead – New Mexico State University......................................497
Research Park; Central Florida 374
Research Park (Charlotte); University..........650
Research Park; Cummings 390
Research Park; Dandini 391
Research Park; Edmonton 397
Research Park; Florida Atlantic...................399
The Research Park Forum........................★353
Research Park; Idaho State University.........420
Research Park; Innovation Center and – Ohio University 515
Research Park; Iowa State University..........441
Research Park (Madison); University...........651
Research Park; Massachusetts Biotechnology...463
Research Park; Miami Valley 474
Research Park; Missouri – University of Missouri 612
Research Park; Morgantown Industrial/ 486
Research Park; New Mexico 496
Research Park; Northwestern University/Evanston................................510
Research Park Project; East Campus – University of Colorado 589
Research Park; Purdue Industrial.................531
Research Park; Riverfront 542
Research Park; Stanford – Stanford University...558
Research Park; Sunset 562
Research Park; Swearingen – University of Oklahoma 620
Research Park; Texas..................................573
Research Park; Texas A&M University – Texas A&M University 570
Research Park; University of Utah – University of Utah 642
Research Park; Virginia Tech Corporate – Virginia Polytechnic Institute and State University...657
Research Parks; Association of University Related 353
Research Parks, Reference Library; Association of University Related 2538
Research Services & Development – University of Waterloo.............................644
Research Services Directory......................2198
Research-Technology Management..........2199
Research/Technology Park; Northern Kentucky University Foundation 507
Research Triangle Park...............................538
Research Unlimited, Inc.; Mankind..............284
Research with Industry; Center for Education and..368
Researched Products..................................312
Resource Directory for Small Business...★1044
Retzloff Capital Corporation 1813
Revenue Bond and Mortgage Program – Pennsylvania State Department of Commerce 1043
Revere AE Capital Fund, Inc. 1687
The Review of Scientific Instruments 2200
Revolving Loan Fund – District of Columbia Office of Business and Economic Development 751
Revolving Loans Program – Arizona State Department of Commerce 692
Rhode Island Financial Assistance Programs – Rhode Island State Department of Economic Development 1060
Rhode Island Innovation Center................★629
Rhode Island; MacRae's Industrial Directory Massachusetts/.......................................2134
Rhode Island Partnership for Science and Technology 1058
Rhode Island Small Business Development Center...539
Rhode Island State Department of Administration – Purchases Office 1059
Rhode Island State Department of Economic Development – Rhode Island Financial Assistance Programs 1060
Rhode Island State Department of Labor – Division of Professional Regulation 1061
Rice University, Division of Government Publications & Special Resources 2653
Richard E. Wolf & Associates.......................313
Richard W. Hynes & Associates...................314
Richmond District Office – U.S. Small Business Administration 1106
Ridgeway Manufacturing Incubator............1239
Rights Alert..2201
Rights & Permissions, Inc.; BZ/ 171
Rights Representative; Bobbe Siegel,..........161
RIMTech (Research Institute for the Management of Technology) 537
Rio Grande Research Corridor.....................540
Risk Taking 2384
Ritter Partners 1362
River Bridge Industrial Center 1240
River Capital Corporation (Alexandria).......1833
River Capital Corporation (Cleveland)........1734
River East...1218
RiverBend 541
Riverfront Research Park 542
Robert B. Leisy..315
Robert Talmage 316
Robertson, Colman & Stephens.................1363
Robotics; Military.......................................2149
Robotics Patents Newsletter......................2202
Robotics Research Institute; Advanced......1077
Rochester District Office – U.S. Small Business Administration 992
Rochester Institute of Technology Research and Business Park 543
Rochester Public Library, Science and Technology Division..............................2654
Rock Hill Incubator 1249
Rocky Mountain Equity Corporation...........1268
Rocky Mountain High Technology Directory...2203

Rocky Mountain Inventors Congress............116
Rome Air Development Center – U.S. Department of the Air Force..................1907
Root Associates, Limited; Morris J. 293
Rose-Hulman Institute of Technology – Innovators Forum...544
Roster of Attorneys and Agents Registered to Practice before the United States Patent and Trademark Office.................2204
Rothschild Ventures, Inc. 1688
Round Table Capital Corporation...............1364
Royalco International Inc.............................317
RSC Financial Corporation.........................1365
Rubber City Capital Corporation.................1735
Rubottom, Dudash and Associates, Inc. 1741
Rural Enterprise Partnership......................★479
Rutgers Minority Investment Company 1587
Rutgers University – Technical Assistance Program..545

S

S and S Venture Associates Ltd. 1689
Sacramento District Office – U.S. Small Business Administration 710
Safeco Capital, Inc. 1430
SAI (Society of American Inventors).............120
St. Louis District Office – U.S. Small Business Administration 938
St. Louis Public Library..............................2655
St. Louis Technology Center 936
St. Paul Small Business Incubator 1192
St. Thomas District Office – U.S. Small Business Administration 1104
Sales Contact Centers – Connecticut State Department of Economic Development ...730
Sales Magazine; Agency............................1980
Salt Lake City District Office – U.S. Small Business Administration 1092
Salvage Training Center; Naval Diving and...★1949
Salween Financial Services, Inc.................1762
San Antonio District Office – U.S. Small Business Administration 1090
San Antonio Venture Group, Inc.................1814
San Diego District Office – U.S. Small Business Administration 711
San Diego Invention Program 2278
San Diego Public Library, Science & Industry Section....................................2656
San Francisco District Office – U.S. Small Business Administration 712
San Joaquin Capital Corporation................1366
San Jose SBIC..1367
Sangamon State University – Entrepreneurship and Enterprise Development Center...............................546
Saugatuck Capital Company......................1409
Saul Soloway 318
SBA 503 Development Company...............1601
SBAC of Panama City, Florida 1431
SBI Capital Corporation..............................1815
SBIC Directory and Handbook of Small Business Finance 2205
The SBIC of Connecticut, Inc. 1410
SBIR Proposal Assistance – Ball State University...357
SBIR Proposal Preparation; Manual on.....★266
Scanlon & Company; C. 172
SCDF Investment Corporation 1481
Schaeffer Business Center.........................1177
Science and Engineering Library; Arizona State University, Daniel E. Noble...........2533
Science and Technology; Center for Applied – University of Pittsburgh 624
Science and Technology Center; University of Maryland 605
Science and Technology Department; Buffalo & Erie County Public Library, ...2554
Science & Technology; French Advances in...2068
Science and Technology; National Coalition for ....★2157
Science and Technology Policy; Center for International – George Washington University 407
Science and Technology Research Center...547
Science & Technology Resource Center – Southwest State University 556
Science Curriculum Improvement Study....2385
Science Modules; Delta..............................2318
Science of Creating Ideas for Industry.........★75
Science Park 548
Science Park Development Corporation 1152
Science Park Library 2657
Science Park; Oregon Graduate Center.......518
Science Study; Elementary........................2324
Science Technology and Energy Division – Alabama State Department of Economic and Community Affairs 678
Science Trends..2206
Scientific and Technical Organizations and Agencies Directory..................................2207
Scientific Conversion, Inc. 319
Scientific Information; Institute for ....★2027
Scientific Instruments; The Review of 2200
Scott-Brown Engineers 320
Sea Systems Command; Naval – U.S. Department of the Navy 1957
SeaGate Venture Management, Inc. 1736
Seaport Ventures, Inc.................................1368
SEARCH Corporation 321
SEARCHLINE 322
Seattle District Office – U.S. Small Business Administration...1113
Secrets and Surprises...............................2386
Security Financial and Investment Corporation..1489
Security Pacific Capital Corporation 1369
Seed Capital Companies – Michigan Strategic Fund 910
Seed One 1737
Seidman Jackson Fisher and Company.....1461
Self-Employment Loan Program – Iowa State Department of Economic Development 828
Selling to the State of Delaware: A Guide to Procurement Opportunities....................★742
Selling Your Product to the State of Minnesota...★915
Semiconductors/ICs Patents Newsletter....2208
Sequoia Capital..1370
SERI (Society for the Encouragement of Research and Invention)...........................119
Sessions on Innovation and Technology...★289
Set-Aside Program – Connecticut State Department of Economic Development ...731
767 Limited Partnership 1690
SFCI (Spirit of the Future Creative Institute)...124
Shano; Charles 184
Shetland Properties of Illinois.....................1166
Shreveport District Office – U.S. Small Business Administration 870
Siegel, Rights Representative; Bobbe..........161
Sierra Ventures 1371
Signal Capital Corporaton 1742
Silicon Mountain Report..............................2209
Silicon Valley Entrepreneurs Club, Inc. 117
The Silver Prescription................................2210
Simplicity: The Key to Success ....★75
Simulations...2387
Sioux Falls District Office – U.S. Small Business Administration 1071
SITADEX..2468
SITOY-Seoul International Toy Fair 2520
Situation Ventures Corporation 1691
Small and Developing Business Division – Ohio State Department of Development 1014
Small and Existing Business Development – Missouri State Department of Economic Development 932
Small Business Action Center – Pennsylvania State Department of Commerce 1044
Small Business Advocacy – Illinois State Department of Commerce and Community Affairs 795
Small Business; American Journal of......★2055
Small Business and Entrepreneurship Research Group – Laval University..........452
Small Business Assistance Act – Wyoming Office of the State Treasurer.................1141
Small Business Assistance Bureau – Illinois State Department of Commerce and Community Affairs 796
Small Business Assistance Division – Massachusetts State Department of Commerce 892
Small Business; Behind.............................1991
Small Business Center – Chamber of Commerce of Hawaii 778
Small Business Committee – Alabama State House of Representatives 682
Small Business Development Agency; Virgin Islands 1105
Small Business Development Center – Lehigh University 455
Small Business Development Center – University of South Florida.......................631
Small Business Development Center; Alaska..345
Small Business Development Center; Arizona 349
Small Business Development Center; Arkansas...352
Small Business Development Center; Colorado 381
Small Business Development Center; Connecticut...386
Small Business Development Center; Delaware 392
Small Business Development Center; Florida..400
Small Business Development Center; Georgia...413
Small Business Development Center (Houston); Texas 574
Small Business Development Center; Idaho...419
Small Business Development Center; Illinois...423
Small Business Development Center; Iowa 439
Small Business Development Center; Kansas...445
Small Business Development Center; Kentucky 448
Small Business Development Center (Lubbock); Texas 575
Small Business Development Center; Maine..458
Small Business Development Center; Maryland..462

Small Business Development Center; Massachusetts 468
Small Business Development Center; Michigan 475
Small Business Development Center; Minnesota 480
Small Business Development Center; Mississippi 483
Small Business Development Center; Missouri 484
Small Business Development Center; Nebraska 490
Small Business Development Center; Nevada 492
Small Business Development Center; New Hampshire 493
Small Business Development Center; New Jersey 495
Small Business Development Center; New York State 498
Small Business Development Center; North Carolina 501
Small Business Development Center; North Dakota 502
Small Business Development Center; Northeastern Texas 504
Small Business Development Center; Northwestern Texas 508
Small Business Development Center; Ohio 512
Small Business Development Center; Oklahoma 516
Small Business Development Center; Oregon 519
Small Business Development Center; Pennsylvania 523
Small Business Development Center; Puerto Rico 530
Small Business Development Center; Rhode Island 539
Small Business Development Center (San Antonio); Texas 576
Small Business Development Center; South Carolina 550
Small Business Development Center; South Dakota 551
Small Business Development Center; Tennessee 567
Small Business Development Center; Utah 653
Small Business Development Center; Vermont 655
Small Business Development Center; Virgin Islands 656
Small Business Development Center; Washington 659
Small Business Development Center; Washington, D.C., ...658
Small Business Development Center; West Virginia 665
Small Business Development Center; Wisconsin 672
Small Business Development Center; Wyoming 674
Small Business Development Centers; Association of 12
Small Business Development Consortium; Alabama 676
Small Business Development Corporation – Louisiana State Department of Commerce and Industry 866
Small Business Development Corporation; Navajo 1266
Small Business Development Division – North Carolina State Department of Commerce 999
Small Business Development Financing Authority; Maryland 877
Small Business Development Loan Program – Minnesota State Department of Energy and Economic Development 922
Small Business Development Program – Illinois State Department of Commerce and Community Affairs 797
Small Business Division – Kentucky State Commerce Cabinet 860
Small Business Division – Nebraska State Department of Economic Development 951
Small Business Energy Management/Loan Program – Illinois State Department of Commerce and Community Affairs 798
Small Business Finance; SBIC Directory and Handbook of 2205
Small Business; Financing Alternatives for ....★176
Small Business Financing Program – Illinois State Department of Commerce and Community Affairs 799
Small Business Fixed-Rate Financing Fund – Illinois State Department of Commerce and Community Affairs 800
Small Business Guide to Federal R&D Funding Opportunities 2211
Small Business in Pennsylvania; Starting a ....★1044
Small Business Incubator Project – Victor Valley College 1149
Small Business Information Service – Hawaii State Department of Planning and Economic Development 781
Small Business Innovation Research Program – National Aeronautics and Space Administration 1865
Small Business Innovation Research Program – National Science Foundation 1870
Small Business Innovation Research Program – U.S. Department of Agriculture 1873
Small Business Innovation Research Program – U.S. Department of Energy 1884
Small Business Innovation Research Program – U.S. Department of the Army 1933
Small Business Innovation Research Program; Department of Transportation, ...★1973
Small Business Innovation Research Promotion Program; New York 978
Small Business; International Council for ....★2084
Small Business Investment Capital, Inc. 1273
Small Business Investment Companies; Directory of Operating 2038
Small Business Liaison – Michigan State Department of Transportation 909
Small Business Management; Journal of 2121
Small Business Micro Loan Program – Illinois State Department of Commerce and Community Affairs 801
Small Business Office – Colorado State Department of Economic Development 716
Small Business Ombudsman – Wisconsin State Department of Development 1135
Small Business Planning ....★1044
Small Business Procurement Program – Kansas State Department of Administration 839
Small Business Reporter; Lehigh University ....★455
Small Business Research and Development; Office of – National Science Foundation 1869
Small Business; Resource Directory for ....★1044
Small Business Revitalization Program – Georgia State Department of Community Affairs 771
Small Business Revitalization Program – Nevada State Office of Community Services 959
Small Business Services – Connecticut State Department of Economic Development 732
Small Business; Speaking of ....★631
Small Business Start-up Guide ....★742
Small Business Studies; P.J. Gardiner Institute for – Memorial University of Newfoundland 471
Small Business: The Magazine for Canadian Entrepreneurs 2212
Small Cities Community Development Block Grant Program – Kansas State Department of Commerce 843
Small Manufacturers Loan Program – Connecticut State Department of Economic Development 733
SME - Society of Manufacturing Engineers 2521
Snider Entrepreneurial Center; Sol C. – University of Pennsylvania 623
Social Science Associations; Consortium of ....★2022
Society for Inventors and Entrepreneurs 118
Society for the Encouragement of Research and Invention 119
Society of American Inventors 120
Society of Manufacturing Engineers ....★2521
Society of Minnesota Inventors 121
Society of University Patent Administrators 122
Software Protection 2213
Sol C. Snider Entrepreneurial Center – University of Pennsylvania 623
Solar Corporation; DAVCO 210
Soloway; Saul 318
Solve - Action Problem Solving 2388
Sound Ideas 2389
South Atlantic Capital Corporation 1432
South Carolina Research Authority 549
South Carolina Small Business Development Center 550
South Carolina State Budget and Control Board – Procurement Office 1063
South Carolina State Development Board – Division of Business Assistance and Development 1065
South Carolina State Development Board 1064
South Carolina/Virginia; MacRae's Industrial Directory North Carolina/ 2138
South Central Kansas Economic Development District 853
South Dakota Inventors Congress 123
South Dakota Small Business Development Center 551
South Dakota State Department of Administration – Purchasing and Printing Office 1068
South Dakota State Department of Commerce and Regulation – Professional and Occupational License Division 1069
South Dakota State Governor's Office of Economic Development 1070
South Florida Entrepreneurial Network ....★402

South Texas SBIC 1816
Southeast Craft and Hobby Show 2522
Southeast Venture Capital Limited 1 1433
Southeastern Oklahoma State University – Central Industrial Applications Center 552
Southern California Innovation Center 1148
Southern Illinois University at Carbondale – Technology Commercialization Center 553
Southern Illinois University at Edwardsville – Technology Commercialization Center 554
Southern Orient Capital Corporation 1817
Southern Willamette Research Corridor 555
Southgate Venture Partners (Charlotte) 1719
Southwest Capital Investments, Inc. 1602
Southwest Enterprise Associates 1818
Southwest Pennsylvania Business Development Center 1241
Southwest State University – Science & Technology Resource Center 556
Southwest Venture Partnerships 1819
Southwestern Craft and Hobby Christmas in July Show 2523
Southwestern Venture Capital of Texas, Inc. 1820
Sovran Funding Corporation 1834
Space and Missile Center; Eastern – U.S. Department of the Air Force 1902
Space and Naval Warfare Systems Command – U.S. Department of the Navy 1966
Space Center; John F. Kennedy – National Aeronautics and Space Administration 1859
Space Center; Lyndon B. Johnson – National Aeronautics and Space Administration 1862
Space Division; Air Force Systems Command, – U.S. Department of the Air Force 1896
Space Flight Center; George C. Marshall – National Aeronautics and Space Administration 1855
Space Flight Center; Goddard – National Aeronautics and Space Administration 1856
Space R&D Alert 2214
Space Science and Applications; Office of – National Aeronautics and Space Administration 1864
Space Technology Center – University of Kansas 599
Space Technology Center – U.S. Department of the Air Force 1908
Space Technology Laboratories; National – National Aeronautics and Space Administration 1863
Spacenews Capsules 2215
Spartanburg Incubator 1250
Speaking of Small Business ....★631
Special Services Division – Oklahoma State Department of Commerce 1024
Special Sources of Credit ....★401
Spier Corp. 323
Spirit of the Future Creative Institute 124
Spokane Business and Technology Center 1258
Spokane District Office – U.S. Small Business Administration 1114
Springboards to Creative Thinking 2390
Springfield District Office – U.S. Small Business Administration 806
Springfield Incubator 1219
Sprout Group (Boston) 1528
Sprout Group (Menlo Park) 1372
Sprout Group (New York City) 1692
SRB Partners Fund Ltd. 1821
Stack Gas Control Patents 2216
Standards and Technology; National Institute of – U.S. Department of Commerce 1874
Stanford Research Park – Stanford University 558
Stanford University – Office of Technology Licensing 557
Stanford University – Stanford Research Park 558
Start Them Thinking 2391
Starting a Business in Hawaii ....★781
Starting a Business in Minnesota; Guide to ....★920
Starting a Business in Oregon ....★1031
Starting a Small Business in Pennsylvania ....★1044
State & Regional Directory 2217
State Enterprise Development Implementation Packets ....★389
State License Bureau – Maryland State Treasury Department 882
State Link Deposit Plan – Wyoming Office of the State Treasurer 1142
State of South Carolina – Office of Small and Minority Business Assistance 1066
State of Wisconsin Investment Board – Private Placements 1128
State Purchasing Office – Louisiana State Office of the Governor 868
State Purchasing Section – Nebraska State Administrative Service Department 949
State University of New York at Binghamton – Office of Sponsored Program Development 559
State University of New York at Oswego – Center for Innovative Technology Transfer 560
State University of New York at Plattsburgh – Economic Development and Technical Assistance Center 561
Statesboro District Office – U.S. Small Business Administration 775
Statistical Abstract; Illinois ....★594
Status Report of the Energy-Related Inventions Program 2218
Sterling Small Business and Technology Center 1167
Stevens Capital Corporation 1529
Stories to Strech Minds 2392
Strategic Defense Initiative Organization – U.S. Department of the Air Force 1909
Strategic Reasoning 2393
Stroke; National Institute of Neurological and Communicative Disorders and – U.S. Department of Health and Human Services 1891
Submarine Medical Research Laboratory; Naval ....★1951
The Success System ....★4
Summit Ventures 1530
Sun-Delta Capital Access Center, Inc. 1559
Sunnyvale Patent Information Clearinghouse 2658
Sunset Research Park 562
Sunwestern Capital Corporation 1822
SUNY Research '88 2219
SUPA (Society of University Patent Administrators) 122
Super Market Investors, Inc. 1848
Superconductivity Research, Development and Commercialization Report 2220
Supergrowth Technology USA 2221
Surgical Research; Institute of ....★1929
Sussex County Department of Economic Development 745
Sutter Hill Ventures 1373
Swearingen Research Park – University of Oklahoma 620
Sydney Diamond Award ....★42
Syncom Capital Corporation 1421
Syracuse District Office – U.S. Small Business Administration 993
Syracuse Incubator 1208
Syracuse University – Institute for Energy Research 563
Syracuse University – Technology and Information Policy Program 564
Systems Technology, Inc.; Intelligent 249

T

T. Rowe Price 1490
TA Associates (Boston) 1531
TA Associates (New York City) 1693
TA Associates (Palo Alto) 1374
Table Mountain Research Center – Colorado School of Mines 380
Taipei International Toy Fair 2524
Talmage; Robert 316
Tampa Bay Area Research & Development Authority 565
Tampa Bay Inventor's Council 125
Tampa District Office – U.S. Small Business Administration 760
Tank-Automotive Command – U.S. Department of the Army 1934
Targeted Small Business Loan Guarantee Program – Iowa State Department of Economic Development 829
Taroco Capital Corporation 1694
Tax Increment Financing Program – Kansas State Department of Commerce 844
Tax Index; Business Climate ....★389
Taxpayer Assistance Bureau, Corporation Section – Massachusetts State Department of Revenue 894
Taylor Research Center; David – U.S. Department of the Navy 1941
TCW Special Placements Fund I 1695
Teachers Teaching Thinking 2394
Teaching and Learning Mathematical Problem Solving: Multiple Research Perspectives 2395
Tech Net – Indiana University-Purdue University Indianapolis 429
Tech Notes 2222
Tech Review; Japan High 2450
TECHEX Americas 2525
Techni Research Associates, Inc. 324
Technical Assistance Program – Rutgers University 545
Technical Idea; How to Evaluate Your ....★164
Technical Information Center; Buckman Laboratories International, ...2553
Technical Information Service – Purdue University 532
Technical Research Innovation Park – Fox Valley Technical College 403
Techniques of Structured Problem Solving 2396
TECHNO-SEARCH 2469
Techno Tokyo - International Licensing and Joint Venture Expo 2526
Technological Leadership; Center for the Development of – University of Minnesota 608
Technological Studies; Institute for – Western Michigan University 668
Technologies Corporation; Northern Advanced 505

Technologies; Guide to Available..............★324
Technologies, Inc.; EOS..............................217
Technology Advance...............................★479
Technology Advancement Program...........1174
Technology and Information Policy Program – Syracuse University..............................564
Technology Application Center – University of New Mexico...617
Technology Application Center; National Aeronautics and Space Administration/Southern.......................★399
Technology Applications Program; National Aeronautics and Space Administration/ University of Kentucky – University of Kentucky..601
Technology Assessment and Forecast Reports Data Base................................2470
Technology Assistance Center; Alabama High – University of Alabama in Huntsville..580
Technology Associates; High.......................233
Technology Business Development – Texas A&M University..569
Technology Center – University of Arkansas at Little Rock...585
Technology Center – University of Scranton..630
Technology; Center for the Productive Use of – George Mason University..................404
Technology; Center for the Utilization of Federal ....★2009
Technology Center; Management of Advanced Automation – Worcester Polytechnic Institute...............................673
Technology Center; Oregon Productivity and – Oregon State University 521
Technology Center; St. Louis 936
Technology Centers International (Montgomeryville)..................................1242
Technology; Centre for Advanced................375
Technology Centre Library; Manitoba Research Council, Industrial..................2612
Technology Commercialization Center – Bradley University....................................362
Technology Commercialization Center – College of DuPage..................................378
Technology Commercialization Center – Illinois Institute of Technology..................422
Technology Commercialization Center – Illinois State University.............................424
Technology Commercialization Center – Northern Illinois University.......................506
Technology Commercialization Center – Southern Illinois University at Carbondale...553
Technology Commercialization Center – Southern Illinois University at Edwardsville..554
Technology Commercialization Center – Western Illinois University........................667
Technology Commercialization Program – University of Illinois at Chicago 597
Technology Corporations; Directory of Public High..2039
Technology Corridor; Minnesota 481
Technology Corridor; Tennessee.................568
Technology Development and Assessment; Institute for – University of Texas Health Science Center at Houston......................641
Technology Development and Marketing Center; Emerging – Clemson University...377
Technology Development and Transfer; Center for – University of Texas...............638
Technology Development Center; Advanced – Georgia Institute of Technology 408
Technology Development; Institute for.........436
Technology Development Institute; Midwest 479
Technology Development; Office of Science and – Columbia University 384
Technology Digest; East/West...................2042
Technology Directory; BC.........................★393
Technology Directory; Rocky Mountain High..2203
Technology Enterprise Development Center – University of Texas at Arlington.............639
Technology Forecasts and Technology Surveys..2223
Technology Foundation; Tennessee..........1075
Technology Foundation; Texas Research and...572
Technology From the U.S.A.; Finding and Licensing New Products and...............★326
Technology Funding, Inc. 1375
Technology Group Ltd...............................1738
Technology Incorporated; University...........652
Technology Industry Letter; F&S/Japanese Information...2059
Technology Information Exchange-Innovation Network................................2471
Technology, Innovation & Entrepreneurship for Students Magazine...........................2397
Technology Innovation Center – Northwestern University.............................509
Technology Innovation Center – University of New Mexico 618
Technology Institute; Advanced Science & 344
Technology Institute, Inc.; International Science and 260
Technology; Integrated Manufacturing.......2099
Technology, Library; Wilfrid Laurier University, Research Centre for Management of New.............................2699
Technology, Licensing, and Industry-Sponsored Research; Office for – Harvard Medical School..........................416
Technology Licensing Office – Massachusetts Institute of Technology....466
Technology Licensing; Office of – Stanford University...557
Technology Management Division; Gulf South Research Institute, ...415
Technology Management News................2224
Technology Management Studies Institute – University of Pittsburgh...........................626
Technology Mart.......................................2225
Technology Newsletter..............................2226
Technology NY..2227
Technology; Office of Corporate Programs and – University of Pennsylvania 621
Technology Park; Ann Arbor 346
Technology Park; Chicago 376
Technology Park; Cornell Business and – Cornell University.....................................388
Technology Park Corporation; Massachusetts...900
Technology Park; First Coast 398
Technology Park; Hawaii Ocean Science and...418
Technology Park; Maine..............................459
Technology Park; Maui Research and 469
Technology Park; Mississippi Research and...482
Technology Park; Northern Kentucky University Foundation Research/.............507
Technology Park; Oakland 511
Technology Park; Progress Center: University of Florida Research and..........528
Technology Park; Rensselaer – Rensselaer Polytechnic Institute...............................535
Technology Park; Research and – Washington State University...................661
Technology Park; Utah State University Research and – Utah State University 654
Technology Policy and Assessment Center – Georgia Institute of Technology 412
Technology Policy; Center for International Science and – George Washington University..407
Technology Report; International High.......2101
Technology Research Corporation...............325
Technology Review; International.............2108
Technology Search International, Inc. 326
Technology Stock Monitor.........................2228
Technology Today; Electronics and...........2048
Technology Transfer Administration; Office of Research and – University of Minnesota...610
Technology Transfer; American Bulletin of International..1983
Technology Transfer and Economic Development; Center for – University of Missouri—Rolla 613
Technology Transfer; Arkansas Center for – University of Arkansas 583
Technology Transfer; Canada Centre for Mineral and Energy Technology, Office of...365
Technology Transfer Center – Oklahoma State University.......................................517
Technology Transfer Center – Wayne State University..664
Technology Transfer Center; Argonne National Laboratory 348
Technology Transfer; Center for – Pittsburg State University.......................................525
Technology Transfer Center, Special Projects Division – University of Michigan 607
Technology Transfer Corporation; Inventors & 48
Technology Transfer Databank..................2472
Technology Transfer Directory; International...★79
Technology Transfer Directory of People ...2473
Technology Transfer Exhibition; Barclays Techmart-New 2492
Technology Transfer; Federal Laboratory Consortium for ....★2064
Technology Transfer Network – Michigan State Department of Commerce 906
Technology Transfer Network; Federal Laboratory Consortium ....★601
Technology Transfer Office – Pennsylvania State University.......................................524
Technology Transfer; Office of – Louisiana State University.......................................457
Technology Transfer; Office of – University of Calgary 586
Technology Transfer; Office of – University of Washington..643
Technology Transfer; Office of Industrial Research and – University of Wisconsin—Milwaukee...........................647
Technology Transfer Society, Library.........2659
Technology Transfer Society—Newsletter.............................2229
Technology Transfer; Workshops on.........★597
Technology Update...................................2230
Technology Update; USSR........................2244
Technology USA; Supergrowth.................2221
Technology Venture Investors....................1376
Technology Week; New.............................2167
Technology; Who's Who in.........................2255
Technology; World Bank of 2487

TECTRA (Technology Transfer Databank) 2472
TEI (The Entrepreneurship Institute) 24
Telecommunication Laboratory; National Computer and ....★1874
Telesciences Capital Corporation 1696
Television Licensing Center 126
Tenneco Ventures, Inc. 1823
Tennessee Center for Research and Development 566
Tennessee Innovation Center 1252
Tennessee Inventors Association 127
Tennessee Small Business Development Center 567
Tennessee State Department of Commerce and Insurance – Regulatory Boards 1072
Tennessee State Department of Economic and Community Development – Office of Small Business 1073
Tennessee State Department of General Services – Division of Purchasing 1074
Tennessee Technology Corridor 568
Tennessee Technology Foundation 1075
Tennessee Venture Capital Corporation 1779
Terrain Analysis Center ....★1923
Tessler and Cloherty, Inc. 1697
Test and Evaluation Command – U.S. Department of the Army 1935
Test Engineering and Management 2231
Testing Laboratories; Weintritt 339
Texas A&M University – Technology Business Development 569
Texas A&M University – Texas A&M University Research Park 570
Texas A&M University – Texas Engineering Experiment Station 571
Texas A&M University, Evans Library—Documents Division 2660
Texas A&M University Research Park – Texas A&M University 570
Texas at Austin, McKinney Engineering Library; University of 2684
Texas Commerce Investment Company 1824
Texas Engineering Experiment Station – Texas A&M University 571
Texas Inventors Association 128
Texas Research and Technology Foundation 572
Texas Research Park 573
Texas Small Business Development Center (Houston) 574
Texas Small Business Development Center (Lubbock) 575
Texas Small Business Development Center (San Antonio) 576
Texas State Department of Labor and Standards – Division of Licensing and Enforcement 1078
Texas State Economic Development Commission 1079
Texas State Purchasing and General Services Commission – Division of Purchasing 1080
Textile Library; Institute of Textile Technology, Textile Information Services, Roger Milliken 2598
Think and Reason 2398
A Thinker's Log 2399
The Thinker's Toolbox 2400
The Thinking Log 2401
Thinking Posters: Keys to Critical Thinking 2402
Thinking Skills: Meanings, Models, and Materials 2403
Thinking Skills Set 2404
Thinking to Write - A Work Journal Program 2405
Thinking Visually 2406
Thomas J. Lipton, Inc., Library/Information Services 2661
Thomas New Industrial Products Database 2474
Thomas Register Online 2475
Thomson & Thomson 327
Thomson, Rogers, Barristers & Solicitors, Library 2662
The Thoughtwave Curriculum - Applying Creative Thinking to Problem Solving 2407
3i Capital 1532
3i Ventures (Boston) 1533
3i Ventures (Newport Beach) 1377
3M, Patent and Technical Communications Services 2663
Threshold Ventures, Inc. 1558
Thrift Institution Fund for Economic Development – Massachusetts State Executive Office of Consumer Affairs and Business Regulation 897
Tidewater Industrial Capital Corporation 1835
Tidewater Small Business Investment Corporation (TBBIC) 1836
TIE-IN (Technology Information Exchange-Innovation Network) 2471
TIES Magazine (Technology, Innovation & Entrepreneurship for Students Magazine) 2397
TLC Funding Corporation 1698
TLC (Television Licensing Center) 126
TM; Master Search 2457
TMA Trademark Report 2476
TMINT 2477
TMRK (Trade Marks) 2478
TMS (Trademark Society) 129
Tobacco Corporation, Research Library; Brown and Williamson 2550
Tokyo - International Licensing and Joint Venture Expo; Techno 2526
Toledo-Lucas County Public Library, Science and Technology Department 2664
Toledo Public Schools Invention Convention 2279
Tomlinson Capital Corporation 1739
Topeka/Shawnee County Development, Inc. 854
Topographic Developments Laboratory ....★1923
Topographic Laboratories; Engineer – U.S. Department of the Army 1923
Toronto Reference Library, Science & Technology Department; Metropolitan 2617
Tower Ventures, Inc. 1462
TOWERS Club, U.S.A.—Newsletter 2232
Townley & Updike, Law Library 2665
Toy and Hobby Fair; British 2493
Toy & Hobby World ....★2129
Toy Fair; American International 2490
Toy Fair; SITOY-Seoul International 2520
Toy Fair; Taipei International 2524
Trade Fairs New Products Newsletter; Foreign 2067
Trade Marks 2478
Trade Name Directory 2233
Trade Names Dictionary: Company Index 2234
Trade Names Dictionary: Company Index; International 2109
Trademark Alert ....★327
Trademark & Copyright Journal; BNA's Patent, ...1998
Trademark & Copyright Journal (Online); BNA's Patent, ...2417
Trademark, & Copyright Library; LEXIS Federal Patent, ...2454
Trademark and Copyright Office; Patent, – University of California, Berkeley 587
Trademark and Copyright Research Foundation, Library; Patent, ...2639
Trademark Association; International Patent and 42
Trademark Association, Law Library; United States 2671
Trademark Association; United States 131
Trademark Data Base; Motor Vehicle Manufacturers Association ....★2626
Trademark Design Register 2235
Trademark Directory 2236
Trademark Guide; Jewelers' Circular/Keystone Brand Name and 2118
Trademark Library; Walter J. Derenberg Copyright and ....★22
Trademark Library; WESTLAW Copyright, Patent and 2486
Trademark Office, Scientific Library; United States Patent & 2670
Trademark Office Society; Patent and 114
Trademark Register ....★2235
Trademark Report; TMA 2476
Trademark Reporter 2237
Trademark Society 129
Trademark Stylesheets ....★131
Trademark to Small Businesses; The Importance of ....★148
Trademarks, Library; Upjohn Company, Corporate Patents and 2688
Trademarks; Official Gazette of the United States Patent and Trademark Office: 2171
TRADEMARKSCAN—FEDERAL 2238
Trademarkscan-Federal 2479
TRADEMARKSCAN—STATE 2239
Trademarkscan-State 2480
Tradenames; RAPRA 2467
Training Systems Center; Naval – U.S. Department of the Navy 1959
Trans Energy Corporation 328
Transcience Associates, Inc. 329
TRANSIN 2481
Transportation Capital Corporation (Boston) 1534
Transportation Capital Corporation (New York City) 1699
Transportation Systems Center – U.S. Department of Transportation 1973
TransTech Services USA 330
Transworld Housewares and Variety Exhibit 2527
Trendwest Capital Corporation 1748
Triad Capital Corporation of New York 1700
Trilon Discovery Center 130
Trinity Ventures Ltd. 1378
Trivest Venture Fund 1763
Troy Incubator Program 1209
Troy State University – Center for Business and Economic Services 577
Tualatin Invention Program 2280
Tucson District Office – U.S. Small Business Administration 695
Tulane University – U.S.-Japan Biomedical Research Laboratories 578
Tuskegee Capital Corporation 1264
21ST Century Research 331
Twenty-first Century Venture Partners 1379
20/20 2240
Twin Ports Capital Company 1849
Two's News 2241

U

U.K. On-Line; Compu-Mark....2432
UK Trade Marks....2482
UKTM (UK Trade Marks)....2482
Umbrella Industrial Development Bond Program – Virginia State Small Business Financing Authority....1112
The Unconventional Invention Book....2408
Underwater Systems Center; Naval – U.S. Department of the Navy....1960
Unicorn Ventures II LP....1588
Unicorn Ventures Ltd....1589
Union Carbide Corporation, Law Department Library....2666
UNISYS Corporation, Law Library....2667
United Catalysts, Inc., Technical Library....2668
United Financial Resources Corporation....1570
United Mercantile Capital Corporation....1603
United Missouri Capital Corporation....1568
United Oriental Capital Corporation....1825
United States Department of Commerce, Office of Productivity, Technology and Innovation, Commerce Productivity Center....2669
United States Patent & Trademark Office, Scientific Library....2670
United States Patents Quarterly....2242
United States Trademark Association....131
United States Trademark Association, Law Library....2671
Universal Financial Services, Inc....1434
University Center Research and Development Park....579
University City Science Center....1243
University-Industry Research Program – University of Wisconsin—Madison....645
University of Alabama in Huntsville – Alabama High Technology Assistance Center....580
University of Alaska – Alaska Economic Development Center....581
University of Alberta – Office of Research Services....582
University of Arkansas – Arkansas Center for Technology Transfer....583
University of Arkansas – Entrepreneurial Service Center....584
University of Arkansas at Little Rock – Technology Center....585
University of Calgary – Office of Technology Transfer....586
University of California, Berkeley – Patent, Trademark and Copyright Office....587
University of California, Berkeley, Chemistry Library....2672
University of California, Los Angeles – Microelectronics Innovation and Computer Research Opportunities....588
University of California, Los Angeles, Chemistry Library....2673
University of Colorado – East Campus Research Park Project....589
University of Colorado Foundation Inc. – Office of Patents and Licensing....590
University of Delaware Library, Reference Department....2674
University of Delaware Research Foundation....591
University of Georgia Research Foundation, Inc....592
University of Hawaii – Pacific Business Center Program....593
University of Idaho Library....2675
University of Illinois – Bureau of Economic and Business Research....594
University of Illinois – Business Development Service....595
University of Illinois – ILLITECH....596
University of Illinois at Chicago – Technology Commercialization Program....597
University of Illinois at Urbana-Champaign – Office of the Vice Chancellor for Research....598
University of Iowa Technology Innovation Center....1170
University of Kansas – Space Technology Center....599
University of Kansas Center for Research, Inc....600
University of Kentucky – National Aeronautics and Space Administration/ University of Kentucky Technology Applications Program....601
University of Louisville – Center for Entrepreneurship and Technology....602
University of Manitoba – Office of Industrial Research....603
University of Maryland – Engineering Research Center....604
University of Maryland, College Park Libraries, Engineering & Physical Sciences Library....2676
University of Maryland Science and Technology Center....605
University of Massachusetts Physical Sciences Library....2677
University of Miami – Innovation and Entrepreneurship Institute....606
University of Michigan – Technology Transfer Center, Special Projects Division....607
University of Michigan, Engineering Transportation Library....2678
University of Minnesota – Center for the Development of Technological Leadership....608
University of Minnesota – Office of Patents and Licensing....609
University of Minnesota – Office of Research and Technology Transfer Administration....610
University of Minnesota – University Research Consortium....611
University of Missouri – Missouri Research Park....612
University of Missouri—Rolla – Center for Technology Transfer and Economic Development....613
University of Moncton – Center of Research in Administrative Sciences....614
University of Nebraska, Lincoln – Nebraska Technical Assistance Center....615
University of Nebraska, Lincoln, Engineering Library....2679
University of Nevada, Reno, Government Publications Department....2680
University of New Hampshire – Industrial Research and Consulting Center....616
University of New Hampshire, University Library, Patent Collection....2681
University of New Mexico – Technology Application Center....617
University of New Mexico – Technology Innovation Center....618
University of New Mexico, Centennial Science and Engineering Library....2682
University of New Mexico National Aeronautics and Space Administration Industrial Applications Center....★617
University of North Carolina at Charlotte – Engineering Research and Industrial Development Office....619
University of North Dakota National Aeronautics and Space Administration Industrial Applications Center....★371
University of Oklahoma – Swearingen Research Park....620
University of Pennsylvania – Office of Corporate Programs and Technology....621
University of Pennsylvania – Office of Sponsored Programs....622
University of Pennsylvania – Sol C. Snider Entrepreneurial Center....623
University of Pittsburgh – Center for Applied Science and Technology....624
University of Pittsburgh – National Aeronautics and Space Administration Industrial Applications Center....625
University of Pittsburgh – Technology Management Studies Institute....626
University of Puerto Rico – Center for Business Research....627
University of Quebec at Hull – Organizational Efficiency Research Group....628
University of Rhode Island – Business Assistance Programs....629
University of Scranton – Technology Center....630
University of South Florida – Small Business Development Center....631
University of Southern California – Office of Patent and Copyright Administration....632
University of Southern California – Urban University Center, Western Research Applications Center....633
University of Southern California, Law Library....2683
University of Southern Maine – Center for Business and Economic Research....634
University of Southern Maine – Office of Sponsored Research....636
University of Southern Maine – The New Enterprise Institute....635
University of Tennessee Research Corporation....637
University of Texas – Center for Technology Development and Transfer....638
University of Texas at Arlington – Technology Enterprise Development Center....639
University of Texas at Austin – IC2 Institute....640
University of Texas at Austin, McKinney Engineering Library....2684
University of Texas Health Science Center at Houston – Institute for Technology Development and Assessment....641
University of Utah – University of Utah Research Park....642
University of Utah, Documents Division....2685
University of Utah Research Park – University of Utah....642
University of Washington – Office of Technology Transfer....643
University of Washington, Engineering Library and Information Services....2686
University of Waterloo – Research Services & Development....644
University of Wisconsin—Madison – The Wisconsin Alumni Research Foundation....646
University of Wisconsin—Madison – University-Industry Research Program....645

University of Wisconsin—Madison, Kurt F. Wendt Library 2687
University of Wisconsin—Milwaukee – Office of Industrial Research and Technology Transfer 647
University of Wisconsin—Stout – Center for Innovation and Development 648
University of Wisconsin—Whitewater – Wisconsin Innovation Service Center 649
University Park – Massachusetts Institute of Technology 467
University Research Consortium – University of Minnesota 611
University Research Park (Charlotte) 650
University Research Park (Madison) 651
University Technology Center (Minneapolis) 1193
University Technology Development Center I (Pittsburgh) 1244
University Technology Development Center II (Pittsburgh) 1245
University Technology Incorporated 652
Upjohn Company, Corporate Patents and Trademarks, Library 2688
Urban Development Action Grants – Arizona State Department of Commerce 693
Urban Enterprise Zone Program – Connecticut State Department of Economic Development 734
Urban Jobs Program – Connecticut State Department of Economic Development 735
Urban University Center, Western Research Applications Center – University of Southern California 633
U.S. Court of Appeals for the Federal Circuit, National Courts' Library 2689
U.S. Department of Agriculture – Agricultural Research Service 1871
U.S. Department of Agriculture – Office of Grants and Program Systems 1872
U.S. Department of Agriculture – Small Business Innovation Research Program 1873
U.S. Department of Agriculture, Southern Regional Research Center 2690
U.S. Department of Commerce – National Institute of Standards and Technology 1874
U.S. Department of Commerce – National Institute of Standards and Technology, Office of Energy Related Inventions 1875
U.S. Department of Commerce – National Oceanic and Atmospheric Administration 1876
U.S. Department of Defense – Defense Nuclear Agency 1877
U.S. Department of Energy – Assistant Secretary for Conservation and Renewable Energy 1878
U.S. Department of Energy – Assistant Secretary for Defense Programs 1879
U.S. Department of Energy – Assistant Secretary for Nuclear Energy 1880
U.S. Department of Energy – Morgantown Energy Technology Center 1881
U.S. Department of Energy – Office of Energy Research 1882
U.S. Department of Energy – Pittsburgh Energy Technology Center 1883
U.S. Department of Energy – Small Business Innovation Research Program 1884
U.S. Department of Energy, Morgantown Energy Technology Center Library 2691
U.S. Department of Health and Human Services – National Cancer Institute 1885
U.S. Department of Health and Human Services – National Eye Institute 1886
U.S. Department of Health and Human Services – National Institute of Allergy and Infectious Diseases 1887
U.S. Department of Health and Human Services – National Institute of Arthritis and Musculoskeletal and Skin Diseases 1888
U.S. Department of Health and Human Services – National Institute of Dental Research 1889
U.S. Department of Health and Human Services – National Institute of General Medical Sciences 1890
U.S. Department of Health and Human Services – National Institute of Neurological and Communicative Disorders and Stroke 1891
U.S. Department of Health and Human Services – Office of Human Development Services 1892
U.S. Department of Justice, Civil Branch Library 2692
U.S. Department of the Air Force – Air Force Systems Command 1893
U.S. Department of the Air Force – Air Force Systems Command, Aeronautical Systems Division (Eglin Air Force Base) 1894
U.S. Department of the Air Force – Air Force Systems Command, Aeronautical Systems Division (Wright-Patterson Air Force Base) 1895
U.S. Department of the Air Force – Air Force Systems Command, Space Division 1896
U.S. Department of the Air Force – Arnold Engineering Development Center 1897
U.S. Department of the Air Force – Ballistic Missile Office 1898
U.S. Department of the Air Force – Contract Management Division 1899
U.S. Department of the Air Force – Defense Advanced Research Projects Agency 1900
U.S. Department of the Air Force – Directorate of Research and Development Procurement 1901
U.S. Department of the Air Force – Eastern Space and Missile Center 1902
U.S. Department of the Air Force – Electronic Systems Division 1903
U.S. Department of the Air Force – Flight Test Center 1904
U.S. Department of the Air Force – Office of Scientific Research 1905
U.S. Department of the Air Force – Operational Test and Evaluation Center 1906
U.S. Department of the Air Force – Rome Air Development Center 1907
U.S. Department of the Air Force – Space Technology Center 1908
U.S. Department of the Air Force – Strategic Defense Initiative Organization 1909
U.S. Department of the Air Force – Weapons Laboaratory 1910
U.S. Department of the Army – Aberdeen Proving Ground Installation Support Activity 1911
U.S. Department of the Army – Armament Research, Development, and Engineering Center 1912
U.S. Department of the Army – Aviation Systems Command 1913
U.S. Department of the Army – Ballistic Missile Defense Systems Command 1914
U.S. Department of the Army – Ballistic Research Laboratory 1915
U.S. Department of the Army – Belvior Research, Development, and Engineering Center 1916
U.S. Department of the Army – Benet Weapons Laboratory 1917
U.S. Department of the Army – Chemical Research, Development, and Engineering Center 1918
U.S. Department of the Army – Cold Regions Research and Engineering Laboratory 1919
U.S. Department of the Army – Communications-Electronics Command 1920
U.S. Department of the Army – Construction Engineering Research Laboratory 1921
U.S. Department of the Army – Dugway Proving Ground 1922
U.S. Department of the Army – Engineer Topographic Laboratories 1923
U.S. Department of the Army – Engineer Waterways Experiment Station 1924
U.S. Department of the Army – Jefferson Proving Ground 1925
U.S. Department of the Army – Laboratory Command 1926
U.S. Department of the Army – Materials Technology Laboratory 1927
U.S. Department of the Army – Materiel Command 1928
U.S. Department of the Army – Medical Research and Development Command 1929
U.S. Department of the Army – Missile Command 1930
U.S. Department of the Army – Natick Research, Development, and Engineering Center 1931
U.S. Department of the Army – Research and Development Directorate 1932
U.S. Department of the Army – Small Business Innovation Research Program 1933
U.S. Department of the Army – Tank-Automotive Command 1934
U.S. Department of the Army – Test and Evaluation Command 1935
U.S. Department of the Army – Water Resources Support Center 1936
U.S. Department of the Army – White Sands Missile Range 1937
U.S. Department of the Army – Yuma Proving Ground 1938
U.S. Department of the Interior – Bureau of Mines 1939
U.S. Department of the Interior – U.S. Geological Survey, Geologic Division 1940
U.S. Department of the Navy – David Taylor Research Center 1941
U.S. Department of the Navy – Marine Corps Headquarters 1942
U.S. Department of the Navy – Naval Air Development Center 1943
U.S. Department of the Navy – Naval Air Engineering Center 1944
U.S. Department of the Navy – Naval Air Propulsion Center 1945
U.S. Department of the Navy – Naval Air Systems Command 1946
U.S. Department of the Navy – Naval Avionics Center 1947
U.S. Department of the Navy – Naval Civil Engineering Laboratory 1948

U.S. Department of the Navy – Naval Coastal Systems Center 1949
U.S. Department of the Navy – Naval Explosive Ordnance Disposal Technology Center 1950
U.S. Department of the Navy – Naval Medical Research and Development Command 1951
U.S. Department of the Navy – Naval Ocean Systems Center 1952
U.S. Department of the Navy – Naval Oceanographic Office 1953
U.S. Department of the Navy – Naval Ordnance Missile Test Station 1954
U.S. Department of the Navy – Naval Ordnance Station 1955
U.S. Department of the Navy – Naval Research Laboratory 1956
U.S. Department of the Navy – Naval Sea Systems Command 1957
U.S. Department of the Navy – Naval Surface Warfare Center 1958
U.S. Department of the Navy – Naval Training Systems Center 1959
U.S. Department of the Navy – Naval Underwater Systems Center 1960
U.S. Department of the Navy – Naval Weapons Station 1961
U.S. Department of the Navy – Naval Weapons Support Center 1962
U.S. Department of the Navy – Naval Weapons Systems Center 1963
U.S. Department of the Navy – Office of Naval Research 1964
U.S. Department of the Navy – Pacific Missile Test Center 1965
U.S. Department of the Navy – Space and Naval Warfare Systems Command 1966
U.S. Department of Transportation – Coast Guard Office of Engineering and Development 1967
U.S. Department of Transportation – Federal Aviation Administration 1968
U.S. Department of Transportation – Federal Highway Administration 1969
U.S. Department of Transportation – Federal Railroad Administration 1970
U.S. Department of Transportation – National Highway Traffic Safety Adminstration 1971
U.S. Department of Transportation – Procurement Operations Division 1972
U.S. Department of Transportation – Transportation Systems Center 1973
U.S. Executive Report 2243
U.S. Geological Survey, Geologic Division – U.S. Department of the Interior 1940
U.S. International Trade Administration, U.S. and Foreign Commercial Service, Charleston District Office Library 2693
U.S.-Japan Biomedical Research Laboratories – Tulane University 578
U.S. National Park Service, Edison National Historic Site, Archives 2694
U.S. Patent Classification System 2483
U.S. Patent Model Foundation 132
U.S. Patents Files 2484
U.S. Small Business Administration – Agana District Office 777
U.S. Small Business Administration – Albany District Office 987
U.S. Small Business Administration – Albuquerque District Office 977
U.S. Small Business Administration – Anchorage District Office 688
U.S. Small Business Administration – Atlanta District Office 774
U.S. Small Business Administration – Augusta District Office 876
U.S. Small Business Administration – Austin District Office 1081
U.S. Small Business Administration – Baltimore District Office 883
U.S. Small Business Administration – Birmingham District Office 684
U.S. Small Business Administration – Boise District Office 790
U.S. Small Business Administration – Boston District Office 901
U.S. Small Business Administration – Buffalo District Office 988
U.S. Small Business Administration – Camden District Office 971
U.S. Small Business Administration – Casper District Office 1139
U.S. Small Business Administration – Cedar Rapids District Office 832
U.S. Small Business Administration – Charleston District Office 1119
U.S. Small Business Administration – Charlotte District Office 1004
U.S. Small Business Administration – Chicago District Office 805
U.S. Small Business Administration – Cincinnati District Office 1015
U.S. Small Business Administration – Clarksburg District Office 1120
U.S. Small Business Administration – Cleveland District Office 1016
U.S. Small Business Administration – Columbia District Office 1067
U.S. Small Business Administration – Columbus District Office 1017
U.S. Small Business Administration – Concord District Office 964
U.S. Small Business Administration – Coral Gables District Office 758
U.S. Small Business Administration – Corpus Christi District Office 1082
U.S. Small Business Administration – Dallas District Office 1083
U.S. Small Business Administration – Denver District Office 721
U.S. Small Business Administration – Des Moines District Office 833
U.S. Small Business Administration – Detroit District Office 911
U.S. Small Business Administration – Eau Claire District Office 1129
U.S. Small Business Administration – El Paso District Office 1084
U.S. Small Business Administration – Elmira District Office 989
U.S. Small Business Administration – Fargo District Office 1009
U.S. Small Business Administration – Fort Worth District Office 1085
U.S. Small Business Administration – Fresno District Office 708
U.S. Small Business Administration – Gulfport District Office 928
U.S. Small Business Administration – Harlingen District Office 1086
U.S. Small Business Administration – Harrisburg District Office 1047
U.S. Small Business Administration – Hartford District Office 736
U.S. Small Business Administration – Hato Rey District Office 1056
U.S. Small Business Administration – Helena District Office 946
U.S. Small Business Administration – Honolulu District Office 783
U.S. Small Business Administration – Houston District Office 1087
U.S. Small Business Administration – Indianapolis District Office 815
U.S. Small Business Administration – Jackson District Office 929
U.S. Small Business Administration – Jacksonville District Office 759
U.S. Small Business Administration – Kansas City District Office 937
U.S. Small Business Administration – Las Vegas District Office 960
U.S. Small Business Administration – Little Rock District Office 701
U.S. Small Business Administration – Los Angeles District Office 709
U.S. Small Business Administration – Louisville District Office 864
U.S. Small Business Administration – Lubbock District Office 1088
U.S. Small Business Administration – Madison District Office 1130
U.S. Small Business Administration – Marquette District Office 912
U.S. Small Business Administration – Marshall District Office 1089
U.S. Small Business Administration – Melville District Office 990
U.S. Small Business Administration – Milwaukee District Office 1131
U.S. Small Business Administration – Minneapolis District Office 924
U.S. Small Business Administration – Montpelier District Office 1098
U.S. Small Business Administration – Nashville District Office 1076
U.S. Small Business Administration – New Orleans District Office 869
U.S. Small Business Administration – New York City District Office 991
U.S. Small Business Administration – Newark District Office 972
U.S. Small Business Administration – Oklahoma City District Office 1026
U.S. Small Business Administration – Omaha District Office 954
U.S. Small Business Administration – Philadelphia District Office 1048
U.S. Small Business Administration – Phoenix District Office 694
U.S. Small Business Administration – Pittsburgh District Office 1049
U.S. Small Business Administration – Portland District Office 1034
U.S. Small Business Administration – Providence District Office 1062
U.S. Small Business Administration – Reno District Office 961
U.S. Small Business Administration – Richmond District Office 1106
U.S. Small Business Administration – Rochester District Office 992
U.S. Small Business Administration – Sacramento District Office 710
U.S. Small Business Administration – St. Louis District Office 938
U.S. Small Business Administration – St. Thomas District Office 1104
U.S. Small Business Administration – Salt Lake City District Office 1092
U.S. Small Business Administration – San Antonio District Office 1090
U.S. Small Business Administration – San Diego District Office 711

U.S. Small Business Administration – San Francisco District Office 712
U.S. Small Business Administration – Seattle District Office 1113
U.S. Small Business Administration – Shreveport District Office 870
U.S. Small Business Administration – Sioux Falls District Office 1071
U.S. Small Business Administration – Spokane District Office 1114
U.S. Small Business Administration – Springfield District Office 806
U.S. Small Business Administration – Statesboro District Office 775
U.S. Small Business Administration – Syracuse District Office 993
U.S. Small Business Administration – Tampa District Office 760
U.S. Small Business Administration – Tucson District Office 695
U.S. Small Business Administration – Washington, D.C., District Office 752
U.S. Small Business Administration – West Palm Beach District Office 761
U.S. Small Business Administration – Wichita District Office 855
U.S. Small Business Administration – Wilkes-Barre District Office 1050
U.S. Small Business Administration – Wilmington District Office 746
U.S. Small Business Administration (Region Eight) 722
U.S. Small Business Administration (Region Five) 807
U.S. Small Business Administration (Region Four) 776
U.S. Small Business Administration (Region Nine) 713
U.S. Small Business Administration (Region One) 902
U.S. Small Business Administration (Region Seven) 939
U.S. Small Business Administration (Region Six) 1091
U.S. Small Business Administration (Region Ten) 1115
U.S. Small Business Administration (Region Three) 1051
U.S. Small Business Administration (Region Two) 994
U.S. Venture Partners 1380
USCLASS 2485
USPMF (U.S. Patent Model Foundation) 132
USSR Technology Update 2244
USTA (United States Trademark Association) 131
USVP-Schlein Marketing Fund 1381
USX Corporation, USS Division, Information Resource Center 2695
Utah, Documents Division; University of 2685
Utah Innovation Center 1093
Utah Small Business Development Center 653
Utah State Department of Administrative Services – Division of Purchasing 1094
Utah State Department of Business Regulation – Division of Occupational and Professional Licensing 1095
Utah State Department of Community and Economic Development – Division of Economic and Industrial Development 1096
Utah State University – Utah State University Research and Technology Park 654
Utah State University Research and Technology Park – Utah State University 654
Utah Technology Finance Corporation 1097

V

Vadus Capital Corporation 1535
Valley Capital Corporation 1780
Valley National Investors, Inc. 1269
Valve News 2245
Valves; IMPACT 2087
Vanderbilt University, Jean and Alexander Heard Library, Science Library 2696
Vega Capital Corporation 1701
VenCap 1573
Vencon Management, Incorporated 332
Vendor Information and Support Division – Pennsylvania State Department of General Services 1045
Venrock Associates (New York City) 1702
Venrock Associates (Palo Alto) 1382
Venture 2246
Venture Capital 2247
Venture Capital & OTC Opportunities; F&S/Japanese 2060
Venture Capital Association—Membership Directory; National 2154
Venture Capital Association; National 106
Venture Capital Clubs; Directory of 2041
Venture Capital Conference ....★289
Venture Capital Digest; IVCI 2117
Venture Capital Directory; Massachusetts ....★891
Venture Capital Fair; Massachusetts ....★891
Venture Capital Fund of New England II 1536
Venture Capital Handbook; Florida ....★401
Venture Capital Institute—Directory of Venture Capital Clubs; International 2110
Venture Capital Institute, Inc.; International ....★2041
Venture Capital Journal 2248
Venture Capital Network, Inc. 133
Venture Capital Resource Directory 2249
Venture Capital Sons of America ....★2247
Venture Capital Sources for Book Publishers 2250
Venture Capital Sources; Pratt's Guide to 2180
Venture Capital: Where to Find It 2251
Venture Capital; Who's Who in 2256
Venture Capitalists—Directory of Members; Western Association of 2254
The Venture Center 1187
Venture Consultants, Inc.; Hi-Tech 232
Venture Development Corporation 333
Venture Directory—Florida; CRS 2025
Venture Economics 334
Venture Economics in Massachusetts ....★2036
Venture Founders; Association of 13
Venture Founders Corporation 1537
Venture Group, Inc. 1435
Venture Investors of Wisconsin, Inc. 1850
Venture Management Associates 335
Venture Opportunities Corporation 1703
Venture/Product News 2252
Venture SBIC, Inc. 1704
Venturecasts ....★333
Ventures; Emerging High Tech 2050
Ventures in Progress 1201
Venturing; Journal of Business 2120
Venturing News; Corporate 2021
Verde Capital Corporation 1436
Vermont Industrial Development Authority 1099
Vermont; MacRae's Industrial Directory Maine/New Hampshire/ 2132
Vermont Secretary of State – Division of Licensing and Registration 1100
Vermont Small Business Development Center 655
Vermont State Administration Agency – Division of Purchasing and Public Records 1101
Vermont State Agency of Development and Community Affairs – Department of Economic Development 1102
Vermont State Department of Economic Development – Entrepreneurship Program 1103
Vicksburg SBIC 1560
Victor Valley College – Small Business Incubator Project 1149
Vimac Corporation 1538
Virgin Islands Small Business Development Agency 1105
Virgin Islands Small Business Development Center 656
Virginia Commonwealth University, University Library Services, Documents and Interlibrary Loan 2697
Virginia; MacRae's Industrial Directory North Carolina/South Carolina/ 2138
Virginia Polytechnic Institute and State University – Virginia Tech Corporate Research Park 657
Virginia State Code Commission – Register of Regulations 1107
Virginia State Corporation Commission 1108
Virginia State Department of Commerce 1109
Virginia State Department of Economic Development – Office of Small Business and Financial Services 1110
Virginia State Office of Administration – Division of Purchases and Supplies 1111
Virginia State Small Business Financing Authority – Umbrella Industrial Development Bond Program 1112
Virginia Tech Corporate Research Park – Virginia Polytechnic Institute and State University 657
Vista Capital Corporation 1383
The Vista Group (New Canaan) 1411
The Vista Group (Newport Beach) 1384
Vocational Education; National Center for Research in – Ohio State University 513

W

W. John Foxwell 336
Wakarusa Valley Development, Inc. 856
Wake Up Your Creative Genius 2409
Walden Capital Partners 1385
Walnut Capital Corporation 1463
Walnut Street Capital Company 1482
Walter C. Beard, Inc. 337
Walter C. Beckwith Associates 338
Walter J. Derenberg Copyright and Trademark Library ....★22
Walter Reed Institute of Research ....★1929
Warburg Pincus Ventures, Inc. 1705
Warm-up to Creativity 2410
Washington Briefs ....★36
Washington D.C. Area R&D Firms Directory 2253
Washington, D.C., District Office – U.S. Small Business Administration 752
Washington, D.C., Small Business Development Center 658

Washington, Engineering Library and Information Services; University of 2686
Washington Small Business Development Center 659
Washington State Department of General Administration – Division of Purchasing 1116
Washington State Department of Licensing – Business and Professions Division 1117
Washington State Department of Trade and Economic Development 1118
Washington State University – Innovation Assessment Center 660
Washington State University – Research and Technology Park 661
Washington Trust Equity Corporation 1840
Washington University – Industrial Contracts and Licensing 662
Watch & Clock Museum Library; National Association of Watch and Clock Collectors, Inc., ...2627
Watchung Capital Corporation 1706
Water Resources Support Center – U.S. Department of the Army 1936
Waterborne Commerce Statistics Center ....★1936
Waterbury Industrial Commons Project 1153
Waterloo Centre for Process Development 663
Waterways Experiment Station; Engineer – U.S. Department of the Army 1924
Wayne State University – Technology Transfer Center 664
WE (Women Entrepreneurs) 134
Weapons Laboaratory – U.S. Department of the Air Force 1910
Weapons Review; World 2263
Weapons Station; Naval – U.S. Department of the Navy 1961
Weapons Support Center; Naval – U.S. Department of the Navy 1962
Weapons Systems Center; Naval – U.S. Department of the Navy 1963
Weekly Reader National Invention Contest 2281
Weintritt Testing Laboratories 339
Weisfeld; Lewis B. 280
Welsh, Carson, Anderson, and Stowe 1707
Wesbanc Ventures Ltd. 1826
West Associates, Incorporated; Bob 160
West Palm Beach District Office – U.S. Small Business Administration 761
West Tennessee Venture Capital Corporation 1781
West Virginia Certified Development Corporation ....★1124
West Virginia Industrial Trade Jobs Development Corporation 1121
West Virginia Small Business Development Center 665
West Virginia State Department of Finance and Administration – Division of Purchasing 1122
West Virginia State Economic Development Authority 1123
West Virginia State Office of Community and Industrial Development 1124
West Virginia State Small Business Office 1125
Westamco Investment Company 1386
Western Association of Venture Capitalists—Directory of Members 2254
Western Financial Capital Corporation 1437
Western Illinois University – Center for Business and Economic Research 666
Western Illinois University – Technology Commercialization Center 667
Western Michigan University – Institute for Technological Studies 668
Western New York Invention Program 2282
Western New York Technology Development Center 1210
Western Research Applications Center; Urban University Center, – University of Southern California 633
Western Venture Capital Corporation 1743
WESTLAW Copyright, Patent and Trademark Library 2486
WFG-Harvest Partners Ltd. 1708
What to Do? 2411
What's Next 2412
White River Capital Corporation 1470
White Sands Missile Range – U.S. Department of the Army 1937
Who's Who in Technology 2255
Who's Who in Venture Capital 2256
Wichita Area Development, Inc. 857
Wichita District Office – U.S. Small Business Administration 855
Wichita Public Library, Business and Technology Division 2698
Wichita State University – Center for Entrepreneurship 669
Wichita State University – Center for Productivity Enhancement 670
Wilfrid Laurier University, Research Centre for Management of New Technology, Library 2699
Wilkes-Barre District Office – U.S. Small Business Administration 1050
William B. Hardy 340
William Blair and Company (Chicago) 1464
William Blair and Company (Denver) 1396
Wilmington Department of Commerce 747
Wilmington District Office – U.S. Small Business Administration 746
Wilmington Economic Development Corporation 748
Wilshire Capital, Inc. 1387
Winfield Capital Corporation 1709
The Wisconsin Alumni Research Foundation – University of Wisconsin—Madison 646
Wisconsin Community Capital, Inc. 1851
Wisconsin Development Fund-Economic Development Program ....★1134
Wisconsin for Research, Inc. 671
Wisconsin Housing and Economic Development Authority – Linked-Deposit Loan Program 1132
Wisconsin Innovation Service Center – University of Wisconsin—Whitewater 649
Wisconsin—Madison, Kurt F. Wendt Library; University of 2687
Wisconsin Small Business Development Center 672
Wisconsin State Department of Administration – Procurement Bureau ...1133
Wisconsin State Department of Development – Financing Programs 1134
Wisconsin State Department of Development – Small Business Ombudsman 1135
Wisconsin State Department of Regulation and Licensing 1136
Wisconsin State Housing and Economic Development Authority 1137
Wisconsin State Rural Development Loan Fund 1138
Wise Owl 2413
Witco Corporation, Technical Information Center 2700
The Withrow Plan of Linked Deposits 1018
Wolf & Associates; Richard E. 313
Wolff Consultants 341
The Woman Entrepreneur Tax Letter 2257
Women & Co. 2258
Women Entrepreneurs 134
Women in Business; Minorities and 2150
Women Inventors Project ....★2295
Women Who Invent or Want To; The Book for 2295
Wood River Capital Corporation (New York City) 1710
Wood River Capital (Menlo Park) 1388
Worcester Area Inventors USA 135
Worcester Polytechnic Institute – Management of Advanced Automation Technology Center 673
Workshops on Business Plans ....★597
Workshops on Technology Transfer ....★597
World Bank of Technology 2487
World Biolicensing & Patent Report TM 2259
World Electronic Developments 2260
World Intellectual Property Organization ....★2477
World Patent Information 2261
World Patents Index 2488
World Technology/Patent Licensing Gazette 2262
World Weapons Review 2263
WPI (World Patents Index) 2488
Wright Wyman, Inc. – Charles W. O'Conor and Associates 185
Writers Publishing Service Co. 342
Wyoming Community Development Authority 1140
Wyoming Economic Development Block Grant Program ....★1144
Wyoming Office of the State Treasurer – Small Business Assistance Act 1141
Wyoming Office of the State Treasurer – State Link Deposit Plan 1142
Wyoming Small Business Development Center 674
Wyoming State Department of Administration and Fiscal Control – Division of Purchasing and Property Control 1143
Wyoming State Economic Development and Stabilization Board 1144

X

Xerox Venture Capital (Los Angeles) 1389
Xerox Venture Capital (Stamford) 1412

Y

Yale University – Office of Cooperative Research 675
Yang Capital Corporation 1711
Ychem International Corporation 343
YEO (Young Entrepreneurs Organization) ...136
Yonkers Public Library, Information Services, Technical & Business Division 2701
Young Entrepreneurs Organization 136
Young Inventors Program 2283
Yuma Proving Ground – U.S. Department of the Army 1938
Yusa Capital Corporation 1712

Z

Use these cards to...

- order extra copies
- share your comments or suggestions with the editor

...or call toll-free
1-800-877-GALE

Gale Research Inc.

Please send me _____copy(s) of **Inventing & Patenting Sourcebook.**
To save 5% on future editions, I've checked the "standing order" space.

Copies			Enter as Standing Order (5% discount)
_____	Inventing & Patenting Sourcebook	$75.00	_____

Institution ____________________

Address ____________________

City, State & Zip ____________________

Attention ____________________
Name Title/Department

Phone (   ) ____________________

Gale Research Inc.

Please send me _____copy(s) of **Inventing & Patenting Sourcebook.**
To save 5% on future editions, I've checked the "standing order" space.

Copies			Enter as Standing Order (5% discount)
_____	Inventing & Patenting Sourcebook	$75.00	_____

Institution ____________________

Address ____________________

City, State & Zip ____________________

Attention ____________________
Name Title/Department

Phone (   ) ____________________

COMMENT CARD

Please use this postage paid card to make suggestions regarding the content, arrangement, indexing, or other features of **Inventing & Patenting Sourcebook.**

Name/Title ____________________

Institution ____________________

Address ____________________

City, State, & Zip ____________________

Phone (   ) ____________________

No postage necessary if mailed in the United States

BUSINESS REPLY MAIL

FIRST CLASS PERMIT NO. 17022 DETROIT, MI 48226

POSTAGE WILL BE PAID BY ADDRESSEE

Order Department
Gale Research Inc.
P.O. Box 441914
Detroit, MI 48244-9980

No postage necessary if mailed in the United States

BUSINESS REPLY MAIL

FIRST CLASS PERMIT NO. 17022 DETROIT, MI 48226

POSTAGE WILL BE PAID BY ADDRESSEE

Order Department
Gale Research Inc.
P.O. Box 441914
Detroit, MI 48244-9980

No postage necessary if mailed in the United States

BUSINESS REPLY MAIL

FIRST CLASS PERMIT NO. 17022 DETROIT, MI 48226

POSTAGE WILL BE PAID BY ADDRESSEE

Inventing & Patenting Sourcebook
Gale Research Inc.
835 Penobscot Building
Detroit, MI 48226-9980

These cards are for your convenience.

Use this card to order your own copy of Inventing & Patenting Sourcebook.

Make it a STANDING ORDER, and you will be sure of receiving new editions promptly... and at a 5% discount. (All editions come to you on a 30-day approval, and you may cancel your standing order at any time.)

...or call toll-free 1-800-877-GALE